T0418592

Encyclopedia of Applied Electrochemistry

Gerhard Kreysa • Ken-ichiro Ota
Robert F. Savinell
Editors

Encyclopedia of Applied Electrochemistry

Volume 1

A–E

With 1250 Figures and 122 Tables

Editors
Gerhard Kreysa
Eppstein, Germany

Ken-ichiro Ota
Yokohama National University, Fac.
Engineering
Yokohama, Japan

Robert F. Savinell
Case Western Reserve University
Cleveland, OH, USA

ISBN 978-1-4419-6995-8 978-1-4419-6996-5 (eBook)
ISBN Bundle 978-1-4419-6997-2 (print and electronic bundle)
DOI 10.1007/978-1-4419-6996-5
Springer New York Heidelberg Dordrecht London

Library of Congress Control Number: 2014934571

Printed on acid-free paper

Springer is part of Springer Science+Business Media (www.springer.com)

Preface

Electrochemistry provides the opportunity to run chemical redox reactions directly with electrons as reaction partners and against the free energy gradient via external electrical energy input into the reaction system. It also serves as a way to efficiently generate energy from stored energy in chemical bonds. Applied electrochemistry has been the basis for many industrial processes ranging from metals recovery and purification to chemical synthesis and separations since the latter half of the 1800s when large-scale electricity generation became possible. Electrochemical processes have a large impact on energy as it has been estimated that these processes consume about 6–10 % of the world's electricity generation capacity. Applied electrochemistry is now impacting industry and society more and more with technologies for waste water treatment, efficient chemical separation, and environmental sensing and remediation. Electrochemistry is the foundation for electrochemical energy storage by batteries and electrochemical capacitors and energy conversion by fuel cells and solar cells. In fact, applied electrochemistry will play a major role in the world's ability to harness and use renewable energy sources. Electrochemistry is also fundamental to biological cell transport and many aspects of living systems and their activities. It is exploited for use in medical diagnostics to detect abnormalities and in biomedical engineering to relieve pain and deliver function.

The application of electrochemistry involves not just a fundamental understanding of the sciences, but also applying engineering principles to device and technology design by considering mass and energy balances, transport processes in the electrolyte and at electrode interfaces, and multi-scale modeling and simulation for predicting and optimizing performance. The interaction of the interfacial reactions, the transport driving forces, and the electric field defines the field of electrochemical engineering. The understanding of the scientific and engineering principles of electrochemical systems has driven advances in the application of electrochemistry especially during the last half century.

The purpose of this collection is to summarize the A–Z of the application of electrochemistry and electrochemical engineering for use by electrochemists and electrochemical engineers as well as nonspecialists such as engineers and scientists of all disciplines, economists, students, and even politicians. Electrochemical fundamentals, electrochemical processes and technologies, and electrochemical techniques are described by many experts in their fields from around the world, many from industry. Each entry is meant to be an introduction and also gives references for further study. With this collection, we hope that current technology and operating practices can be made available for future generations to learn from. We hope that this encyclopedia will stimulate understanding of the current state of the art and lead to advances in new and more efficient technologies with breakthroughs from new theory and materials. We hope you find it of value to your work.

Gerhard Kreysa
Ken-ichiro Ota
Robert F. Savinell
Editors-in-Chief

Acknowledgments

The Editors-in-Chief would like to acknowledge the backing of their institutions in supporting this work. Specifically we want to thank the Technical University of Dortmund (GK), Yokohama National University (KO), and Case Western Reserve University (RFS). We also want to thank the topical editors and authors; it is because of their dedication to their fields and hard work that this collection was possible. We thank our families for their understanding of the importance of this project to us and our profession. Finally, we want to thank the editorial staff of Springer, especially Barbara Wolf who worked long and hard on this project and Kenneth Howell specifically for encouraging us to embark on this project and nudging us along.

About the Editors

Prof. Dr.rer.nat. Dr.-Ing. E.h. Dr.tekn.h.c. Gerhard Kreysa Retired Chief Executive of DECHEMA e.V.

Gerhard Kreysa was born in 1945 in Dresden. He studied chemistry at the University of Dresden and received his Ph.D. in 1970. In 1973, he joined the Karl Winnacker Institute of DECHEMA in Frankfurt am Main. He developed new concepts for the utilization of three-dimensional electrodes, which became prominent for electrochemical waste water treatment in the process industry. He also played a leading role in the clarification of the "cold fusion" affaire in 1989. In 1985, he was appointed as professor in the Chemical Engineering Department at the University of Dortmund. In 1993, he was appointed as honorary professor at the University of Regensburg. From 1985 to 1995, he served as executive editorial board member of the *Journal of Applied Electrochemistry*. He was a recipient of the Chemviron Award in 1980, the Max-Buchner-Research-Award of DECHEMA and the Castner Medal of the Society of Chemical Industry in 1994, and the Wilhelm Ostwald Medal of the Saxon Academy of Sciences in Leipzig in 2006.

From 1992 to 2009, Dr. Kreysa served as chief executive of DECHEMA Society of Chemical Engineering and Biotechnology in Frankfurt am Main, Germany. During this time he also served as general secretary of the European Federation of Chemical Engineering and the European Federation of Biotechnology. He obtained many distinctions: Honorary doctor degrees of Technical University of Clausthal and of the Royal Institute of Technology in Stockholm, Foreign Member of the Royal Swedish Academy of Engineering Sciences, elected honorary fellow of the Institution of Chemical Engineers, and honorary member of the Czech Society of Chemical Engineering. In 2007 he was awarded with the Order of Merit of the Free State of Saxony, and in 2008 he became a member of the German Academy of Technical Sciences (acetech). He has 196 scientific publications and books and has given 312 scientific and public lectures. Despite the numerous duties and responsibilities of his former senior management position, he continues to have a lively interest in the further development of science and engineering and is highly regarded as an advisor on national and international issues.

Prof. Dr. G. Kreysa
Weingasse 22
D-65817 Eppstein

Ken-ichiro Ota

Ken-ichiro Ota is a professor at and the chairman of Green Hydrogen Research Center at Graduate School of Engineering, Yokohama National University, Japan. He received his B.S.E. in applied chemistry in 1968 and Ph.D. in engineering in 1973, both from the University of Tokyo. After graduation, he became a research associate at the university until 1979. In the same year, he became an associate professor at the Yokohama National University, and a professor in 1995. He has worked on hydrogen energy and fuel cell since 1974, focusing on materials science for fuel cells and hydrogen energy system including water electrolysis. In the fuel cell field he has worked on direct methanol fuel cell, molten carbonate fuel cell, and polymer electrolyte fuel cell. Recently, he is developing transition metal oxide–based cathode for polymer electrolyte fuel cell. He is also working on storage and transport of renewable energies by hydrogen technology. He has published more than 190 original papers, 80 review papers, and 50 scientific books. He received the Molten Salt Award in 1998 and the Industrial Electrolysis Award in 2002 from the Electrochemical Society of Japan. He received the Canadian Hydrogen Society Award in 2004 and the Society Award of the Electrochemical

Society of Japan in 2011. He is now the chairman of the National Committee for Standardization of the Stationary Fuel Cells. He was the president of the Hydrogen Energy Systems Society of Japan from 2000 to 2008 and also the president of the Electrochemical Society of Japan from 2008 to 2009. He is now the chairman of the Fuel Cell Development Information Center of Japan.

Dr. Robert F. Savinell Distinguish University Professor, George S. Dively Professor of Engineering, Department of Chemical Engineering, Case Western Reserve University, Cleveland, OH, USA

Dr. Robert F. Savinell received his B.Che. from Cleveland State University in 1973, and his M.S. (1974) and Ph.D. (1977), both in chemical engineering from University of Pittsburgh. He worked as a research engineer for Diamond Shamrock Corporation, then as a faculty member at the University of Akron before joining the faculty at Case Western Reserve University (CWRU) in 1986. Professor Savinell was the Director of the Ernest B. Yeager Center for Electrochemical Sciences at CWRU for ten years and served as Dean of Engineering at CWRU for seven years. Professor Savinell has been engaged in electrochemical engineering research and development for 40 years. Savinell's research is directed at fundamental science and mechanistic issues of electrochemical processes; and at electrochemical technology systems and device design, development, modeling and optimization. His research has addressed applications for energy conversion, energy storage, sensing, and electrochemical materials extraction and synthesis. Savinell has over 120 peer-reviewed and over 168 other publications, eight patents, and has been an invited and keynote speaker at hundreds of national and international conferences and workshops in the electrochemical field. He has supervised over 50 Ph.D./M.S. student projects.

Professor Savinell is the former North American editor of the *Journal of Applied Electrochemistry* and currently is the editor of the *Journal of the Electrochemical Society*. He is a Fellow of the Electrochemical Society, Fellow of the American Institute of Chemical Engineers, and Fellow of the International Society of Electrochemistry.

Section Editors

Electrocatalysis

Radoslav R. Adzic Chemistry Department, Brookhaven National Laboratory, Upton, NY, USA

Primary Batteries

George Blomgren Imara Corporation, Menlo Park, CA, USA

Environmental Electrochemistry

Christos Comninellis EPFL, Lausanne, Switzerland

Organic Electrochemistry

Toshio Fuchigami Department of Electrochemistry, Tokyo Institute of Technology, Midori-ku, Yokohama, Japan

Solid State Electrochemistry

Ulrich Guth Kurt-Schwabe-Institut für Mess- und Sensortechnik e.V. Meinsberg, Ziegra-Knobelsdorf, Germany

High-Temperature Molten Salts

Rika Hagiwara Graduate School of Energy Science, Kyoto University, Sakyo-ku, Kyoto, Japan

Inorganic Electrochemical Synthesis (Including Chlorine/Caustic, Chlorates, Hypochlorite)

Kenneth L. Hardee Research & Development Division, De Nora, Fairport Harbor, OH, USA

Bioelectrochemistry

Dirk Holtmann DECHEMA e.V., Frankfurt am Main, Germany

Electrochemical Instrumentation and Laboratory Techniques

Rudolf Holze Institut für Chemie, Technische Universität Chemnitz, Chemnitz, Germany

Fuel Cells

Minoru Inaba Department of Molecular Chemistry and Biochemistry, Doshisha University, Kyotanabe, Kyoto, Japan

Supercapacitors

Hiroshi Inoue Department of Applied Chemistry, Graduate School of Engineering, Osaka Prefecture University, Sakai, Osaka, Japan

High-Temperature Electrochemistry

Tatsumi Ishihara Department of Applied Chemistry, Kyushu University, Nishi-ku, Fukuoka, Japan

Secondary Batteries

Kiyoshi Kanamura Applied Chemistry Graduate School of Engineering, Tokyo Metropolitan University, Hachioji, Tokyo, Japan

Semiconductor Synthesis and Electrochemistry

Paul A. Kohl Georgia Institute of Technology, School of Chemical and Biomolecular Engineering, Atlanta, GA, USA

Electrolytes

Werner Kunz Institute of Physical and Theoretical Chemistry, Regensburg University, Regensburg, Germany

Electrodeposition (Electrochemical Metal Deposition and Plating)

Uziel Landau Chemical Engineering Department, Case Western Reserve University, Cleveland, OH, USA

Electrochemical Analysis and Sensors

Chung-Chiun Liu Case Western Reserve University, Cleveland, OH, USA

Electrocatalysis

Nebojsa Marinkovic Synchrotron Catalysis Consortium, University of Delaware, Newark, DE, USA

Environmental Electrochemistry

Yunny Meas Vong CIDETEQ, Parque Tecnológico Querétaro, Sanfandila, Pedro Escobedo, CP, México

Photoelectrochemistry

Tsutomu Miyasaka Graduate School of Engineering, Toin University of Yokohama, Kanagawa, Japan

Electrochemical Engineering

Trung Van Nguyen Department of Chemical and Petroleum Engineering, The University of Kansas, Lawrence, KS, USA

Fuel Cells

Thomas Schmidt Electrochemistry Laboratory, Paul Scherrer Institut, Villigen, Switzerland

Advisory Board

Contributors

Abd El Aziz Abd-El-Latif Institute of Physical and Theoretical Chemistry, University of Bonn, Bonn, Germany

Luisa M. Abrantes Departamento de Química e Bioquímica, FCUL, Lisbon, Portugal

Radoslav R. Adzic Chemistry Department, Brookhaven National Laboratory, Upton, NY, USA

Sheikh A. Akbar The Ohio State University, Columbus, OH, USA

Francisco Alcaide Energy Department, Fundación CIDETEC, San Sebastián, Spain

Antonio Aldaz Instituto Universitario de Electroquímica, University of Alicante, Alicante, Spain

Leonardo S. Andrade Universidade Federal de Goiás, Catalão, Brazil

Juan Manuel Artés Institució Catalana de Recerca i Estudis Avançats (ICREA), Institute for Bioengineering of Catalonia (IBEC), Barcelona, Spain

Electrical and Computer Engineering, University of California Davis, Davis, CA, USA

Mahito Atobe Graduate School of Environment and Information Sciences, Yokohama National University, Yokohama, Japan

Nicola Aust BASF SE, Ludwigshafen, Germany

Arseto Bagastyo Advanced Water Management Centre (AWMC), The University of Queensland, Brisbane, QLD, Australia

Helmut Baltruschat Institute of Physical and Theoretical Chemistry, University of Bonn, Bonn, Germany

Cesar Alfredo Barbero Department of Chemistry, National University of Rio Cuarto, Rio Cuarto, Cordoba, Argentina

Scott Barnett Department of Materials Science and Engineering, Northwestern University, Evanston, IL, USA

Romas Baronas Faculty of Mathematics and Informatics, Vilnius University, Vilnius, Lithuania

Damien Batstone Advanced Water Management Centre (AWMC), The University of Queensland, Brisbane, QLD, Australia

Pierre Bauduin Institut de Chimie Separative de Marcoule, UMR 5257 – ICSM Site de Marcoule, CEA/CNRS/UM2/ENSCM, Bagnols sur Ceze, France

Dorin Bejan Department of Chemistry, Electrochemical Technology Centre, University of Guelph, Guelph, ON, Canada

Luc Belloni CEA Saclay, Gif-sur-Yvette, France

Henry Bergman Anhalt University, Anhalt, Germany

Sonia R. Biaggio Universidade Federal de Goiás, Catalão, Brazil

Salma Bilal National Centre of Excellence in Physical Chemistry, University of Peshawar, Peshawar, Pakistan

José M. Bisang Programa de Electroquímica Aplicada e Ingeniería Electroquímica (PRELINE), Facultad de Ingeniería Química, Universidad Nacional del Litoral, Santa Fe, Santa Fe, Argentina

George Blomgren Blomgren Consulting Services, Lakewood, OH, USA

Nerilso Bocchi Universidade Federal de Goiás, Catalão, Brazil

Pierre Boillat Paul Scherrer Institut, Villigen PSI, Switzerland

Nikolaos Bonanos Department of Energy Conversion and Storage, Technical University of Denmark, Roskilde, DK

Antoine Bonnefont Institut de Chimie de Strasbourg, CNRS-Université de Strasbourg, Strasbourg, France

Oleg Borodin Electrochemistry Branch, Sensor and Electron Devices Directorate, U.S. Army Research Laboratory, Adelphi, MD, USA

Stanko Brankovic Electrical and Computer Engineering Department, Chemical and Bimolecular Engineering Department, and Chemistry Department, University of Houston, Houston, TX, USA

Enric Brillas Laboratory of Electrochemistry of Materials and Environment, Department of Physical Chemistry, Faculty of Chemistry, University of Barcelona, Barcelona, Spain

Ralph Brodd Broddarp of Nevada, Inc., Henderson, NV, USA

Michael Bron Institut für Chemie, Technische Chemie, Martin-Luther-Universität Halle-Wittenberg, Halle, Germany

Nigel W. Brown Daresbury Innovation Centre, Arvia Technology Ltd., Daresbury, UK

Felix N. Büchi Paul Scherrer Institut, Villigen PSI, Switzerland

Richard Buchner Institute of Physical and Theoretical Chemistry, University of Regensburg, Regensburg, Germany

Ratnakumar V. Bugga Jet Propulsion Laboratory, Pasdena, CA, USA

Nigel J. Bunce Department of Chemistry, Electrochemical Technology Centre, University of Guelph, Guelph, ON, Canada

Andreas Bund FG Elektrochemie und Galvanotechnik, Institut für Werkstofftechnik, Technische Universitüt Ilmenau, Ilmenau, Germany

Erika Bustos Centro de Investigacióny Desarrollo Tecnológico en Electroquímica, S. C., Sanfandila, Pedro Escobedo, Querétaro, México

Julea Butt School of Chemistry, University of East Anglia, Norwich, UK

Yun Cai Material Science, Joint Center for Artificial Photosynthesis, Lawrence Berkeley National Laboratory, Berkeley, CA, USA

Claudio Cameselle Department of Chemical Engineering, University of Vigo, Vigo, Spain

Maja Cemazar Department of Experimental Oncology, Institute of Oncology Ljubljana, Ljubljana, Slovenia

Vidhya Chakrapani Department of Chemical and Biological Engineering, Rensselaer Polytechnic Institute, Troy, NY, USA

François Chellé Universite Catholique de Louvain, Louvain-la-Neuve, Belgium

Kaimin Chen MECC, Medtronic Inc., Minneapolis, MN, USA

Po-Yu Chen Department of Medicinal and Applied Chemistry, Kaohsiung Medical University, Kaohsiung, Taiwan

Kazuhiro Chiba Tokyo University of Agriculture and Technology, Fuchu, Tokyo, Japan

Masanobu Chiku Department of Applied Chemistry, Graduate School of Engineering, Osaka Prefecture University, Osaka, Japan

YongMan Choi Chemistry Department, Brookhaven National Laboratory, Upton, NY, USA

David E. Cliffel Department of Chemistry, Vanderbilt University, Nashville, TN, USA

Christos Comninellis Institute of Chemical Sciences and Engineering, Ecole Polytechnique Fédérale de Lausanne (EPFL), Lausanne, Switzerland

Ann Cornell School of Chemical Science and Engineering, Applied Electrochemistry, KTH Royal Institute of Technology, Stockholm, Sweden

Serge Cosnier Department of Molecular Chemistry, CNRS UMR 5250 CNRS-University of Grenoble, Grenoble, France

Vincent S. Craig Department of Applied Mathematics, Research School of Physical Sciences and Engineering, Australian National University, Canberra, ACT, Australia

Hideo Daimon Advanced Research and Education, Doshisha University, Kyotanabe, Kyoto, Japan

Manfred Decker Kurt-Schwabe-Institut fuer Mess- und Sensortechnik e.V. Meinsberg, Waldheim, Germany

Dario Dekel CellEra Inc., Caesarea, Israel

I. M. Dharmadasa Electronic Materials and Sensors Group, Materials and Engineering Research Institute, Sheffield Hallam University, Sheffield, UK

Petros Dimitriou-Christidis Environmental Chemistry Modeling Laboratory, Ecole Polytechnique Fédérale de Lausanne (EPFL), Lausanne, Switzerland

Pablo Docampo Clarendon Laboratory, Department of Physics, Oxford University, Oxford, UK

Deepak Dubal AG Elektrochemie, Institut für Chemie, Technische Universität Chemnitz, Chemnitz, Germany

Laurie Dudik Case Western Reserve University, Cleveland, OH, USA

Jean François Dufreche Institut de Chimie Séparative de Marcoule and Université Montpellier, Marcoule, France

Christian Durante Department of Chemical Sciences, University of Padova, Padova, Italy

Prabir K. Dutta The Ohio State University, Columbus, OH, USA

Ulrich Eberle Government Programs and Research Strategy, GM Alternative Propulsion Center, Adam Opel AG, Rüsselsheim, Germany

Obi Kingsley Echendu Electronic Materials and Sensors Group, Materials and Engineering Research Institute, Sheffield Hallam University, Sheffield, UK

Minato Egashira College of Bioresource Sciences, Nihon University, Fujisawa, Kanagawa, Japan

Takashi Eguro Frukawa Battery, Iwaki, Fukushima, Japan

Martin Eichler AG Elektrochemie, Institut für Chemie, Technische Universität Chemnitz, Chemnitz, Germany

Robert Eisenberg Department of Molecular Biophysics and Physiology, Rush University Medical Center, Chicago, IL, USA

Bernd Elsler Johannes Gutenberg-University Mainz, Mainz, Germany

Eduardo Expósito Instituto Universitario de Electroquímica, University of Alicante, Alicante, Spain

Emiliana Fabbri Electrochemistry Laboratory, Paul Scherrer Institute, Villigen, Switzerland

Yujie Feng Harbin Institute of Technology, Harbin, China

Rui Ferreira Instituto de Tecnologia Química e Biológica, Universidade Nova de Lisboa, Oeiras, Portugal

Sergio Ferro Department of Chemical and Pharmaceutical Sciences, University of Ferrara, Ferrara, Italy

Stéphane Fierro Institute of Chemical Sciences and Engineering, Ecole Polytechnique Fédérale de Lausanne (EPFL), Lausanne, Switzerland

Michael A. Filler School of Chemical and Biomolecular Engineering, Georgia Institute of Technology, Atlanta, GA, USA

Alanah Fitch Department of Chemistry, Loyola University, Chicago, IL, USA

Robert Forster School of Chemical Sciences National Center for Sensor Research, Dublin City University, Dublin, Ireland

György Fóti Institute of Chemical Sciences and Engineering, Ecole Polytechnique Fédérale de Lausanne (EPFL), Lausanne, Switzerland

Alejandro A. Franco Laboratoire de Réactivité et de Chimie des Solides (LRCS) - UMR 7314, Université de Picardie Jules Verne, CNRS and Réseau sur le Stockage Electrochimique de l'Energie (RS2E), Amiens, France

Matthias Franzreb Institute of Functional Interfaces, Karlsruhe Institute of Technology, Eggenstein-Leopoldshafen, Germany

Stefano Freguia Advanced Water Management Centre (AWMC), The University of Queensland, Brisbane, QLD, Australia

Bernardo A. Frontana-Uribe Centro Conjunto de Investigación en Química Sustentable, UAEMéx–UNAM, Toluca, Estado de México, Mexico

Instituto de Química UNAM, Mexico, Mexico

Albert J. Fry Weslayan University, Middletown, CT, USA

Toshio Fuchigami Department of Electrochemistry, Tokyo Institute of Technology, Midori-ku, Yokohama, Japan

Akira Fujishima Kanagawa Academy of Science and Technology, Takatsu–ku, Kawasaki, Kanagawa, Japan

Photocatalysis International Research Center, Tokyo University of Science, Noda, Chiba, Japan

Klaus Funke Institute of Physical Chemistry, University of Muenster, Muenster, Germany

Ping Gao AG Elektrochemie, Institut für Chemie, Technische Universität Chemnitz, Chemnitz, Germany

Vicente García-García Instituto Universitario de Electroquímica, University of Alicante, Alicante, Spain

Helga Garcia Instituto de Tecnologia Química e Biológica, Universidade Nova de Lisboa, Oeiras, Portugal

Darlene G. Garey CIDETEQ, Centro de Investigación y Desarrollo Tecnológico en Electroquímica Parque Tecnológico Querétaro, Pedro Escobedo, Edo. Querétaro, México

Armando Gennaro Department of Chemical Sciences, University of Padova, Padova, Italy

Abhijit Ghosh Advanced Ceramics Section Glass and Advanced Materials Division, Bhabha Atomic Research Centre, Mumbai, India

M. Mar Gil-Diaz IMIDRA, Alcalá de Henares, Madrid, Spain

Luc Girard Institut de Chimie Separative de Marcoule, UMR 5257 – ICSM Site de Marcoule, CEA/CNRS/UM2/ENSCM, Bagnols sur Ceze, France

Jean Gobet Adamant-Technologies, La Chaux-de-Fonds, Switzerland

Luis Godinez Centro de Investigación y Desarrollo Tecnológico en Electroquímica S.C., Querétaro, Mexico

Alan Le Goff Department of Molecular Chemistry, CNRS UMR 5250, CNRS-University of Grenoble, Grenoble, France

Muriel Golzio CNRS; IPBS (Institut de Pharmacologie et de Biologie Structurale), Toulouse, France

Ignacio Gonzalez Department of Chemistry, Universidad Autónoma Metropolitana-Iztapalapa, México, Mexico

Heiner Jakob Gores Institute of Physical Chemistry, Münster Electrochemical Energy Technology (MEET), Westfälische Wilhelms-Universität Münster (WWU), Münster, Germany

Pau Gorostiza Institució Catalana de Recerca i Estudis Avançats (ICREA), Barcelona, Spain

Lars Gundlach Department of Chemistry and Biochemistry and Department of Physics and Astronomy, University of Delaware, Newark, DE, USA

Ulrich Guth Kurt-Schwabe-Institut für Mess- und Sensortechnik e.V. Meinsberg, Waldheim, Germany

FB Chemie und Lebensmittelchemie, Technische Universität Dresden, Dresden, Germany

Geir Martin Haarberg Department of Materials Science and Engineering, Norwegian University of Science and Technology (NTNU), Trondheim, Norway

Jonathan E. Halls Department of Chemistry, The University of Bath, Bath, UK

Ahmad Hammad Research and Development Center, Saudi Aramco, Dhahran, Saudi Arabia

Achim Hannappel DECHEMA Research Institute of Biochemical Engineering, Frankfurt am Main, Germany

Falk Harnisch Institute of Environmental and Sustainable Chemistry, Technical University Braunschweig, Braunschweig, Germany

Akitoshi Hayashi Department of Applied Chemistry, Osaka Prefecture University, Sakai, Osaka, Japan

Christoph Held Department of Biochemical and Chemical Engineering, Technische Universität Dortmund, Dortmund, Germany

Wesley A. Henderson Department of Chemical and Biomolecular Engineering, North Carolina State University, Raleigh, NC, USA

Peter J. Hesketh School of Mechanical Engineering, Georgia Institute of Technology, Atlanta, GA, USA

Michael Heyrovsky J. Heyrovsky Institute of Physical Chemistry of the ASCR, Prague, Czech Republic

Takashi Hibino Graduate School of Environmental Studies, Nagoya University, Nagoya, Japan

Yoshio Hisaeda Department of Chemistry and Biochemistry, Kyushu University, Graduate School of Engineering, Fukuoka, Japan

Tuan Hoang University of Southern California, Los Angeles, CA, USA

Dirk Holtmann DECHEMA Research Institute of Biochemical Engineering, Frankfurt am Main, Germany

Rudolf Holze AG Elektrochemie, Institut für Chemie, Technische Universität Chemnitz, Chemnitz, Germany

Michael Holzinger Department of Molecular Chemistry, CNRS UMR 5250, CNRS-University of Grenoble, Grenoble, France

Dominik Horinek Institute of Physical and Theoretical Chemistry, University of Regensburg, Regensburg, Germany

Teruhisa Horita Fuel Cell Materials Group, Energy Technology Research Institute, National Institute of Advanced Industrial Science and Technology (AIST), Tsukuba, Ibaraki, Japan

Barbara Hribar-Lee Faculty of Chemistry and Chemical Technology, University of Ljubljana, Ljubljana, Slovenia

Chang-Jung Hsueh Electronics Design Center, and Chemical Engineering Department, Case Western Reserve University, Cleveland, OH, USA

Chi-Chang Hu Chemical Engineering Department, National Tsing Hua University, Hsinchu, Taiwan

Gary W. Hunter NASA Glenn Research Center, Cleveland, OH, USA

Jorge G. Ibanez Department of Chemical Engineering and Sciences, Universidad Iberoamericana, México, Mexico

Munehisa Ikoma Panasonic, Moriguchi, Japan

Nobuhito Imanaka Department of Applied Chemistry, Faculty of Engineering, Osaka University, Osaka, Japan

Shinsuke Inagi Tokyo Institute of Technology, Midori-ku, Yokohama, Japan

Hiroshi Inoue Osaka Prefecture University, Sakai, Osaka, Japan

György Inzelt Department of Physical Chemistry, Eötvös Loránd University, Budapest, Hungary

Tsutomu Ioroi AIST, Ikeda, Japan

Hiroshi Irie Yamanashi University, Yamanashi Prefecture, Japan

John Thomas Sirr Irvine School of Chemistry, University of St Andrews, St Andrews, UK

Manabu Ishifune Kinki University, Higashi-Osaka, Osaka, Japan

Akimitsu Ishihara Yokohama National University, Hodogaya-ku, Yokohama, Japan

Tatsumi Ishihara Department of Applied Chemistry, Faculty of Engineering, International Institute for Carbon Neutral Energy Research (WPI-I2CNER), Kyushu University, Nishi ku, Fukuoka, Japan

Masashi Ishikawa Kansai University, Suita, Osaka, Japan

Adriana Ispas FG Elektrochemie und Galvanotechnik, Institut für Werkstofftechnik, Technische Universitüt Ilmenau, Ilmenau, Germany

Gaurav Jain MECC, Medtronic Inc., Minneapolis, MN, USA

Metini Janyasupab Electronics Design Center, and Chemical Engineering Department, Case Western Reserve University, Cleveland, OH, USA

Fengjing Jiang Institute of Fuel Cells, School of Mechanical Engineering, Shanghai Jiao Tong University, Shanghai, People's Republic of China

Maria Jitaru Research Institute for Organic Auxiliary Products (ICPAO), Medias, Romania

Jakob Jörissen Chair of Technical Chemistry, Technical University of Dortmund, Germany

Pavel Jungwirth Institute of Organic Chemistry and Biochemistry, Academy of Sciences of the Czech Republic, Prague, Czech Republic

Yoshifumi Kado Asahi Kasei Chemicals Corporation, Tokyo, Japan

Heike Kahlert Institut für Biochemie, Universität Greifswald, Greifswald, Germany

Yijin Kang University of Pennsylvania, Philadelphia, PA, USA

Agnieszka Kapałka Institute of Chemical Sciences and Engineering, Ecole Polytechnique Fédérale de Lausanne (EPFL), Lausanne, Switzerland

Shigenori Kashimura Kinki University, Higashi-Osaka, Japan

Alexandros Katsaounis Department of Chemical Engineering, University of Patras, Patras, Greece

Jurg Keller Advanced Water Management Centre (AWMC), The University of Queensland, Brisbane, QLD, Australia

Geoffrey H. Kelsall Department of Chemical Engineering, Imperial College London, London, UK

Sangtae Kim Department of Chemical Engineering and Materials Science, University of California, Davis, CA, USA

Woong-Ki Kim Faculty of Electrical Engineering and Computer Science, Ingolstadt University of Applied Sciences, Ingolstadt, Germany

Axel Kirste BASF SE, Ludwigshafen, Germany

Naoki Kise Department of Chemistry and Biotechnology, Graduate School of Engineering, Tottori University, Tottori, Japan

Norihisa Kobayashi Chiba University, Chiba, Japan

Svenja Kochius DECHEMA Research Institute, Frankfurt am Main, Germany

Paul A. Kohl Georgia Institute of Technology, School of Chemical and Biomolecular Engineering, Atlanta, GA, USA

Ulrike I. Kramm Technical University Cottbus, Cottbus, Germany

Mario Kročka AG Elektrochemie, Institut für Chemie, Technische Universität Chemnitz, Chemnitz, Germany

Nedeljko Krstajic Faculty of Technology and Metallurgy, University of Belgrade, Belgrade, Serbia

Akihiko Kudo Tokyo University of Science, Tokyo, Japan

Andrzej Kuklinski Fakultät Chemie, Biofilm Centre/Aquatische Biotechnologie, Universität Duisburg-Essen, Essen, Germany

Juozas Kulys Department of Chemistry and Bioengineering, Vilnius Gediminas Technical University, Vilnius, Lithuania

Werner Kunz Institut für Biophysik, Fachbereich Physik, Johann Wolfgang Goethe-Universität Frankfurt am Main, Frankfurt am Main, Germany

Manabu Kuroboshi Okayama University, Okayama, Japan

Jan Labuda Institute of Analytical Chemistry, Faculty of Chemical and Food Technology, Slovak University of Technology, Bratislava, Slovakia

Claude Lamy Institut Européen des Membranes, Université Montpellier 2, UMR CNRS n° 5635, Montpellier, France

Ying-Hui Lee Chemical Engineering Department, National Tsing Hua University, Hsinchu, Taiwan

Carlos A. Ponce de Leon Electrochemical Engineering Laboratory, University of Southampton, Faculty of Engineering and the Environment, Southampton, Hampshire, UK

Jean Lessard Universite de Sherbrooke, Quebec, Canada

Hans J. Lewerenz Joint Center for Artificial Photosynthesis, California Institute of Technology, Pasadena, CA, USA

Claudia Ley DECHEMA Research Institute, Frankfurt am Main, Germany

Meng Li Chemistry Department, Brookhaven National Laboratory, Upton, NY, USA

R. Daniel Little University of California, Santa Barbara, CA, USA

Chen-Wei Liu Institute for Material Science and Engineering, National Central University, Jhongli City, Taoyuan County, Taiwan

Chung-Chiun Liu Electronics Design Center, and Chemical Engineering Department, Case Western Reserve University, Cleveland, OH, USA

Ping Liu Brookhaven National Laboratory, Upton, NY, USA

Yoav D. Livney Faculty of Biotechnology and Food Engineering, The Technion, Israel Institute of Technology, Haifa, Israel

Leonardo Lizarraga Université Lyon 1, CNRS, UMR 5256, IRCELYON, Institut de recherches sur la catalyse et l'environnement de Lyon, Villeurbanne, France

M. Carmen Lobo IMIDRA, Alcalá de Henares, Madrid, Spain

Svenja Lohner Stanford University, Stanford, CA, USA

Manuel Lohrengel University of Düsseldorf, Düsseldorf, Germany

Reiner Lomoth Department of Chemistry - Ångström Laboratory, Uppsala University, Uppsala, Sweden

Daniel Lowy FlexEl, LLC, College Park, MD, USA

Roland Ludwig Department of Food Science and Technology, Vienna Institute of Biotechnology BOKU-University of Natural Resources and Life Sciences, Vienna, Austria

Dirk Lützenkirchen-Hecht Fachbereich C- Abteilung Physik, Wuppertal, Germany

Vadim F. Lvovich NASA Glenn Research Center, Electrochemistry Branch, Power and In-Space Propulsion Division, Cleveland, OH, USA

Johannes Lyklema Department of Physical Chemistry and Colloid Science, Wageningen University, Wageningen, The Netherlands

Hirofumi Maekawa Department of Materials Science and Technology, Nagaoka University of Technology, Nagaoka, Japan

Anders O. Magnusson DECHEMA Research Institute, Frankfurt am Main, Germany

J. Maier Max Planck Institute for Solid State Research, Stuttgart, Germany

Frédéric Maillard Laboratoire d'Electrochimie et de Physico-chimie des Matériaux et des Interfaces, Saint Martin d'Héres, France

Daniel Mandler Institute of Chemistry, The Hebrew University, Jerusalem, Israel

Klaus-Michael Mangold DECHEMA-Forschungsinstitut, Frankfurt am Main, Germany

Yizhak Marcus Department of Inorganic and Analytical Chemistry, The Hebrew University, Jerusalem, Israel

Nebojsa Marinkovic Synchrotron Catalysis Consortium, University of Delaware, Newark, DE, USA

Frank Marken University of Bath, Bath, UK

István Markó Universite Catholique de Louvain, Louvain-la-Neuve, Belgium

Marko S. Markov Research International, Williamsville, NY, USA

Jack Marple Research and Development, Energizer Battery Manufacturing Inc, Westlake, OH, USA

Virginie Marry Laboratoire Physicochimie des Electrolytes, Colloïdes et Sciences Analytiques, CNRS, ESPCI, Université Pierre et Marie Curie, Paris, France

Guillermo Marshall Laboratorio de Sistemas Complejos, Departamento de Ciencias de la Computación, Facultad de Ciencias Exactas y Naturales, Universidad de Buenos Aires, Buenos Aires, Argentina

Carlos Alberto Martinez-Huitle Institute of Chemistry, Federal University of Rio Grande do Norte, Lagoa Nova, Natal, RN - CEP, Brazil

Marco Mascini Dipartimento di Chimica "Ugo Schiff", Università degli Studi di Firenze, Sesto Fiorentino, Firenze, Italy

Rudy Matousek Severn Trent Services, Sugar Land, TX, USA

Hiroshige Matsumoto Kyushu University, Fukuoka, Japan

Kouichi Matsumoto Faculty of Science and Engineering, Kinki University, Higashi-osaka, Japan

Werner Mäntele Johann Wolfgang Goethe-Universität Frankfurt am Main, Institut für Biophysik, Fachbereich Physik, Frankfurt am Main, Germany

Steven McIntosh Department of Chemical Engineering, Lehigh University, Bethlehem, PA, USA

Jennifer R. McKenzie Department of Chemistry, Vanderbilt University, Nashville, TN, USA

Ellis Meng University of Southern California, Los Angeles, CA, USA

Pierre-Alain Michaud Institute of Chemical Sciences and Engineering, Ecole Polytechnique Fédérale de Lausanne (EPFL), Lausanne, Switzerland

Richard L. Middaugh Independent Consultant, formerly with Energizer Battery Co, Rocky River, OH, USA

Alessandro Minguzzi Dipartimento di Chimica, Università degli Studi di Milano, Milan, Italy

Shigenori Mitsushima Yokohama National University, Yokohama, Kanagawa Prefecture, Japan

Tsutomu Miyasaka Graduate School of Engineering, Toin University of Yokohama, Yokohama, Kanagawa, Japan

Kenji Miyatake University of Yamanashi, Kofu, Yamanashi, Japan

Junichiro Mizusaki Tohoku University, Funabashi, Chiba, Japan

Mogens Bjerg Mogensen Department of Energy Conversion and Storage, Technical University of Denmark, Roskilde, Denmark

Charles W. Monroe Department of Chemical Engineering, University of Michigan, Ann Arbor, MI, USA

Vicente Montiel Instituto Universitario de Electroquímica, University of Alicante, Alicante, Spain

Somayeh Moradi AG Elektrochemie, Institut für Chemie, Technische Universität Chemnitz, Chemnitz, Germany

J. Thomas Mortimer Case Western Reserve University, Cleveland, OH, USA

Hubert Motschmann Institute of Physical and Theoretical Chemistry, University of Regensburg, Regensburg, Germany

Christopher B. Murray University of Pennsylvania, Philadelphia, PA, USA

Katsuhiko Naoi Institute of Symbiotic Science and Technology, Tokyo University of Agriculture and Technology, Koganei, Tokyo, Japan

Hiroki Nara Faculty of Science and Engineering, Waseda University, Okubo, Shinjuku-ku, Tokyo, Japan

George Neophytides Electrochemistry Laboratory, Paul Scherrer Institut, Villigen, Switzerland

Roland Neueder Institute of Physical and Theoretical Chemistry, University of Regensburg, Regensburg, Germany

Jinren Ni Peking University, Beijing, China

Ernst Niebur Mind/Brain Institute, Johns Hopkins University, Baltimore, MD, USA

Branislav Ž. Nikolić Department of Physical Chemistry and Electrochemistry, University of Belgrade, Faculty of Technology and Metallurgy, Belgrade, Serbia

Yoshinori Nishiki Development Department, Permelec Electrode Ltd, Kanagawa, Japan

Shigeru Nishiyama Department of Chemistry, Keio University, Hiyoshi, Yokohama, Japan

Toshiyuki Nohira Graduate School of Energy Science, Kyoto University, Kyoto, Japan

Atusko Nosaka Department Materials Science and Technology, Nagaoka University of Technology, Nagaoka, Niigata, Japan

Naoyoshi Nunotani Department of Applied Chemistry, Faculty of Engineering, Osaka University, Osaka, Japan

Tsuyoshi Ochiai Kanagawa Academy of Science and Technology, Takatsu–ku, Kawasaki, Kanagawa, Japan

Photocatalysis International Research Center, Tokyo University of Science, Noda, Chiba, Japan

Wolfram Oelßner Kurt-Schwabe-Institut für Mess- und Sensortechnik e.V. Meinsberg, Kurt-Schwabe-Straße, Waldheim, Germany

Andreas Offenhäusser Institute of Complex Systems, Peter Grünberg Institute: Bioelectronics, Jülich, Germany

Ulker Bakir Ogutveren Anadolu University, Eskişehir, Turkey

Bunsho Ohtani Catalysis Research Center, Hokkaido University, Sapporo, Hokkaido, Japan

Yohei Okada Tokyo University of Agriculture and Technology, Fuchu, Tokyo, Japan

Osamu Onomura Nagasaki University, Nagasaki, Japan

Immaculada Ortiz Department of Chemical Engineering and Inorganic Chemistry, University of Cantabria, Santander, Cantabria, Spain

Juan Manuel Ortiz Instituto Universitario de Electroquímica, University of Alicante, Alicante, Spain

Tetsuya Osaka Faculty of Science and Engineering, Waseda University, Okubo, Shinjuku-ku, Tokyo, Japan

Ken-ichiro Ota Yokohama National University, Fac. Engineering, Yokohama, Japan

Lisbeth M. Ottosen Department of Civil Engineering, Technical University of Denmark, Lyngby, Denmark

Ilaria Palchetti Dipartimento di Chimica "Ugo Schiff", Università degli Studi di Firenze, Sesto Fiorentino, Firenze, Italy

Vladimir Panić University of Belgrade, Institute of Chemistry, Technology and Metallurgy, Belgrade, Serbia

Marco Panizza University of Genoa, Genoa, Italy

Juan Manuel Peralta-Hernández Centro de Innovación Aplicada en Tecnologías Competitivas, Guanajuato, Mexico

Cristina Pereira Instituto de Tecnologia Química e Biológica, Universidade Nova de Lisboa, Oeiras, Portugal

Araceli Pérez-Sanz IMIDRA, Alcalá de Henares, Madrid, Spain

Laurence (Laurie) Peter Department of Chemistry, University of Bath, Bath, UK

Marija Petkovic Instituto de Tecnologia Química e Biológica, Universidade Nova de Lisboa, Oeiras, Portugal

Ilje Pikaar Advanced Water Management Centre (AWMC), The University of Queensland, Brisbane, QLD, Australia

Antonio Plaza IMIDRA, Alcalá de Henares, Madrid, Spain

Dmitry E. Polyansky Chemistry Department, Brookhaven National Laboratory, Upton, NY, USA

Jinyi Qin Department of Environmental Microbiology, Helmholtz Centre for Environmental Research - UFZ, Leipzig, Germany

Jelena Radjenovic Advanced Water Management Centre (AWMC), The University of Queensland, Brisbane, QLD, Australia

Krishnan Rajeshwar The University of Texas at Arlington, Arlington, TX, USA

Nayif A. Rasheedi Research and Development Center, Saudi Aramco, Dhahran, Saudi Arabia

David Rauh EIC Laboratories, Inc, Norwood, MA, USA

Thomas B. Reddy Department of Materials Science and Engineering, Rutgers, The State University of New Jersey, Piscataway, NJ, USA

David Reyter INRS Energie, Matériaux et Télécommunications, Varennes, Quebec, Canada

Marcel Risch MIT, Cambridge, MA, USA

Vivian Robinson ETP Semra Pty Ltd, Canterbury, NSW, Australia

Romeu C. Rocha-Filho Universidade Federal de Goiás, Catalão, Brazil

Manuel A. Rodrigo Department of Chemical Engineering, Faculty of Chemical Sciences and Technology, Universided de Castille la Mandne, Ciudad Real, Spain

Paramaconi Rodriguez School of Chemistry, The University of Birmingham, Birmingham, UK

Alberto Rojas-Hernández Department of Chemistry, Universidad Autónoma Metropolitana-Iztapalapa, México, Mexico

Sandra Rondinini Dipartimento di Chimica, Università degli Studi di Milano, Milan, Italy

Benjamin Rotenberg Laboratoire Physicochimie des Electrolytes, Colloïdes et Sciences Analytiques, CNRS, ESPCI, Université Pierre et Marie Curie, Paris, France

Anna Joëlle Ruff Lehrstuhl für Biotechnologie, RWTH Aachen University, Aachen, Germany

Luís Augusto M. Ruotolo Department of Chemical Engineering, Federal University of São Carlos, São Carlos, SP, Brazil

Jennifer L. M. Rupp Electrochemical Materials, ETH Zurich, Zurich, Switzerland

Yoshihiko Sadaoka Ehime University, Matsuyama, Japan

Gabriele Sadowski Department of Biochemical and Chemical Engineering, Technische Universität Dortmund, Dortmund, Germany

Hikari Sakaebe Research Institute for Ubiquitous Energy Devices, National Institute of Advanced Industrial Science and Technology (AIST), Ikeda, Osaka, Japan

Hikari Sakaebe National Institute of Advanced Industrial Science and Technology (AIST), Ikeda, Osaka, Japan

Mathieu Salanne Laboratoire PECSA, UMR 7195, Université Pierre et Marie Curie, Paris, France

Wolfgang Sand Fakultät Chemie, Biofilm Centre/Aquatische Biotechnologie, Universität Duisburg-Essen, Essen, Germany

Shriram Santhanagopalan National Renewable Energy Laboratory, Golden, CO, USA

Hamidreza Sardary AG Elektrochemie, Institut für Chemie, Technische Universität Chemnitz, Chemnitz, Germany

Kotaro Sasaki Chemistry Department, Brookhaven National Laboratory, Upton, NY, USA

Richard Sass DECHEMA e.V., Informationssysteme und Datenbanken, Frankfurt, Germany

Shunsuke Sato Toyota Central Research and Development Laboratories, Inc., Nagakute, Aichi, Japan

André Savall Laboratoire de Génie Chimique, CNRS, Université Paul Sabatier, Toulouse, France

Elena Savinova Institut de Chimie et Procédés pour l'Energie, l'Environnement et la Santé, UMR 7515 CNRS, Université de Strasbourg-ECPM, Strasbourg, France

Natascha Schelero Stranski-Laboratorium, Institut für Chemie, Fakultät II, Technical University Berlin, Berlin, Germany

Günther G. Scherer Electrochemistry Laboratory, Paul Scherrer Institute, Villigen, Switzerland

Thomas J. Schmidt Electrochemistry Laboratory, Paul Scherrer Institut, Villigen, Switzerland

Jens Schrader DECHEMA Research Institute of Biochemical Engineering, Frankfurt am Main, Germany

Uwe Schroeder Institute of Environmental and Sustainable Chemistry, Technical University Braunschweig, Braunschweig, Germany

Brooke Schumm Eagle Cliffs, Inc., Bay Village, OH, USA

Ulrich Schwaneberg Lehrstuhl für Biotechnologie, RWTH Aachen University, Aachen, Germany

Hans-Georg Schweiger Faculty of Electrical Engineering and Computer Science, Ingolstadt University of Applied Sciences, Ingolstadt, Germany

Hisanori Senboku Hokkaido University, Sapporo, Hokkaido, Japan

Daniel Seo Department of Chemical and Biomolecular Engineering, North Carolina State University, Raleigh, NC, USA

Karine Groenen Serrano Laboratory of Chemical Engineering, University of Paul Sabatier, Toulouse, France

Anwar-ul-Haq Ali Shah Institute of Chemical Sciences, University of Peshawar, Peshawar, Pakistan

Yang Shao-Horn MIT, Cambridge, MA, USA

Hisashi Shimakoshi Department of Chemistry and Biochemistry, Kyushu University, Graduate School of Engineering, Fukuoka, Japan

Yasuhiro Shimizu Graduate School of Engineering, Nagasaki University, Nagasaki, Japan

Youichi Shimizu Department of Applied Chemistry, Graduate School of Engineering, Kyushu Institute of Technology, Kitakyushu, Fukuoka, Japan

Komaba Shinichi Department of Applied Chemistry, Tokyo University of Science, Shinjuku, Tokyo, Japan

Soshi Shiraishi Division of Molecular Science, Faculty of Science and Technology, Gunma University, Kiryu, Gunma, Japan

Pavel Shuk Rosemount Analytical Inc. Emerson Process Management, Solon, OH, USA

Jean-Pierre Simonin Laboratoire PECSA, UMR CNRS 7195, Université Paris, Paris, France

Subhash C. Singhal Pacific Northwest National Laboratory, Richland, WA, USA

Ignasi Sirés Laboratory of Electrochemistry of Materials and Environment, Department of Physical Chemistry, Faculty of Chemistry, University of Barcelona, Barcelona, Spain

Stephen Skinner Department of Materials, Imperial College London, London, UK

Marshall C. Smart Jet Propulsion Laboratory, Pasdena, CA, USA

Henry Snaith Clarendon Laboratory, Department of Physics, Oxford University, Oxford, UK

Stamatios Souentie Department of Chemical Engineering, University of Patras, Patras, Greece

Bernd Speiser Institut für Organische Chemie, Universität Tübingen, Tübingen, Germany

Jacob Spendelow Los Alamos National Laboratory, Los Alamos, NM, USA

Daniel Steingart Department of Mechanical and Aerospace Engineering, Andlinger Center for Energy, the Environment Princeton University, Princeton, NJ, USA

John Stickney University of Georgia, Athens, GA, USA

Margarita Stoytcheva Instituto de Ingenieria, Universidad Autonoma de Baja California, Mexicali, Baja California, Mexico

Svetlana B. Strbac ICTM-Institute of Electrochemistry, University of Belgrade, Belgrade, Serbia

Eric M. Stuve Department of Chemical Engineering, University of Washington, Seattle, WA, USA

Stenbjörn Styring Department of Chemistry - Ångström Laboratory, Uppsala University, Uppsala, Sweden

Seiji Suga Okayama University, Okayama, Japan

Wataru Sugimoto Faculty of Textile Science and Technology, Shinshu University, Nagano, Japan

I-Wen Sun Department of Chemistry, National Cheng kung University, Tainan, Taiwan

Jin Suntivich Cornell University, Ithaca, NY, USA

Hitoshi Takamura Department of Materials Science, Graduate School of Engineering, Tohoku University, Sendai, Japan

Prabhakar A. Tamirisa MECC, Medtronic Inc., Minneapolis, MN, USA

Shinji Tamura Department of Applied Chemistry, Faculty of Engineering, Osaka University, Osaka, Japan

Hideo Tanaka Okayama University, Okayama, Japan

Tadaaki Tani Society of Photography and Imaging of Japan, Tokyo, Japan

Akimasa Tasaka Doshisha University, Kyotanabe, Kyoto, Japan

Tetsu Tatsuma Institute of Industrial Science, University of Tokyo, Tokyo, Japan

Masahiro Tatsumisago Department of Applied Chemistry, Osaka Prefecture University, Sakai, Osaka, Japan

Pierre Taxil Laboratoire de Génie Chimique, Université de Toulouse, Toulouse, France

Justin Teissie CNRS; IPBS (Institut de Pharmacologie et de Biologie Structurale), Toulouse, France

Ingrid Tessmer Rudolf-Virchow-Zentrum, Experimentelle Biomedizin, University of Würzburg, Würzburg, Germany

Anders Thapper Department of Chemistry - Ångström Laboratory, Uppsala University, Uppsala, Sweden

Masataka Tomita Department of Applied Chemistry, Tokyo University of Science, Shinjuku, Tokyo, Japan

Marc Tornow Institut für Halbleitertechnik, Technische Universität München, München, Germany

Taro Toyoda Department of Engineering Science, The University of Electro-Communications, Chofu, Tokyo, Japan

Dimitrios Tsiplakides Department of Chemistry, Aristotle University of Thessaloniki, Thessaloniki, Greece

Pierre Turq Laboratoire Physicochimie des Electrolytes, Colloïdes et Sciences Analytiques, CNRS, ESPCI, Université Pierre et Marie Curie, Paris, France

Makoto Uchida Fuel Cell Nanomaterials Center, Yamanashi University, 6-43 Miyamaecho, Kofu, Yamanashi, Japan

Kai M. Udert Eawag, Dübendorf, Switzerland

Helmut Ullmann FB Chemie und Lebensmittelchemie, Technische Universität Dresden, Dresden, Germany

Soichiro Uno Nissan Research Center/EV System Laboratory, NISSAN MOTOR CO., LTD., Yokosuka-shi, Kanagawa-ken, Japan

Kohei Uosaki International Center for Materials Nanoarchitectonics (WPI-MANA), National Institute for Materials Science (NIMS), Tsukuba, Japan

Ane Urtiaga Department of Chemical Engineering and Inorganic Chemistry, University of Cantabria, Santander, Cantabria, Spain

Francisco J. Rodriguez Valadez Centro de Investigación y Desarrollo Tecnológico en Electroquímica S.C., Querétaro, Mexico

S. C. Sanfandila, Research Branch, Center for Research and Technological Development in Electrochemistry, Querétaro, Mexico

Vladimir Vashook Kurt-Schwabe-Institut für Mess- und Sensortechnik e.V. Meinsberg, Waldheim, Germany

FB Chemie und Lebensmittelchemie, Technische Universität Dresden, Dresden, Germany

Ruben Vasquez-Medrano Department of Chemical Engineering and Sciences, Universidad Iberoamericana, México, Mexico

Nicolaos Vatistas DICI, Università di Pisa, Pisa, Italy

Constantinos G. Vayenas Department of Chemical Engineering, University of Patras, Patras, Achaia, Greece

Danae Venieri Department of Environmental Engineering, Technical University of Crete, Chania, Greece

Philippe Vernoux Université Lyon 1, CNRS, UMR 5256, IRCELYON, Institut de recherches sur la catalyse et l'environnement de Lyon, Villeurbanne, France

Alberto Vertova Dipartimento di Chimica, Università degli Studi di Milano, Milan, Italy

Bernardino Virdis Advanced Water Management Centre (AWMC), The University of Queensland, Brisbane, QLD, Australia

Centre for Microbial Electrosynthesis (CEMES), The University of Queensland, Brisbane, QLD, Australia

Vojko Vlachy Faculty of Chemistry and Chemical Technology, University of Ljubljana, Ljubljana, Slovenia

Rittmar von Helmolt Government Programs and Research Strategy, GM Alternative Propulsion Center, Adam Opel AG, Rüsselsheim, Germany

Regine von Klitzing Stranski-Laboratorium, Institut für Chemie, Fakultät II, Technical University Berlin, Berlin, Germany

Winfried Vonau Kurt-Schwabe-Institut fuer Mess- und Sensortechnik e.V. Meinsberg, Waldheim, Germany

Yunny Meas Vong CIDETEQ, Parque Tecnológico Querétaro, México, Estado de Querétaro, México

Lj Vracar Faculty of Technology and Metallurgy University of Belgrade, Belgrade, Serbia

Miomir B. Vukmirovic Chemistry Department, Brookhaven National Laboratory, Upton, NY, USA

Vlastimil Vyskocil UNESCO Laboratory of Environmental Electrochemistry, Faculty of Science, Department of Analytical Chemistry, Charles University in Prague, Prague, Czech Republic

Jay D. Wadhawan Department of Chemistry, The University of Hull, Hull, UK

Siegfried R. Waldvogel Johannes Gutenberg-University Mainz, Mainz, Germany

Frank C. Walsh Electrochemical Engineering Laboratory, University of Southampton, Faculty of Engineering and the Environment, Southampton, Hampshire, UK

Jia X. Wang Brookhaven National Laboratory, Upton, NY, USA

Masahiro Watanabe University of Yamanashi, Kofu, Yamanashi, Japan

Takao Watanabe Central Research Institute of Electric Power Industry, Yokosuka, Kanagawa, Japan

Andrew Webber Technology, Energizer Battery Manufacturing Inc., Westlake, OH, USA

Adam Z. Weber Lawrence Berkeley National Laboratory, Berkeley, CA, USA

Hermann Weingärtner Lehrstuhl für Physikalische Chemie II, Ruhr-Universität Bochum, Bochum, Germany

Nina Welschoff Johannes Gutenberg-University Mainz, Mainz, Germany

Alan West Department of Chemical Engineering Columbia University, New York, NY, USA

Reiner Westermeier SERVA Electrophoresis GmbH, Heidelberg, Germany

Lukas Y. Wick Department of Environmental Microbiology, Helmholtz Centre for Environmental Research - UFZ, Leipzig, Germany

Andrzej Wieckowski University of Illinois, Urbana, IL, USA

Alexander Wiek AG Elektrochemie, Institut für Chemie, Technische Universität Chemnitz, Chemnitz, Germany

John P. Wikswo Department of Physics and Astronomy, Vanderbilt University, Nashville, TN, USA

Frank Willig Fritz-Haber-Institut der Max-Planck-Gesellschaft, Berlin, Germany

Yuping Wu Department of Chemistry, Fudan University, Shanghai, China

Naoaki Yabuuchi Department of Applied Chemistry, Tokyo University of Science, Shinjuku, Tokyo, Japan

Kohta Yamada Asahi Glass Co. Ltd., Research Center, Kanagawa-ku, Yokohama, Japan

Yoshiaki Yamaguchi Technical Development Division, Global Technical Headquarters, GS Yuasa International Ltd., Kyoto, Japan

Ichiro Yamanaka Department of Applied Chemistry, Tokyo Institute of Technology, Tokyo, Japan

Shigeaki Yamazaki Kansai University, Suita, Osaka, Japan

Harumi Yokokawa The University of Tokyo, Institute of Industrial Science, Tokyo, Japan

H.-I. Yoo Department of Materials Science and Engineering, Seoul National University, Seoul, Korea

Nobuko Yoshimoto Graduate School of Science and Engineering, Yamaguchi University, Yamaguchi, Japan

Akira Yoshino Yoshino Laboratory, Asahi Kasei Corporation, Fuji, Shizuoka, Japan

Jun-ichi Yosida Kyoto University, Kyoto, Japan

Zaki Yusuf Research and Development Center, Saudi Aramco, Dhahran, Saudi Arabia

Junliang Zhang Institute of Fuel Cells, School of Mechanical Engineering, Shanghai Jiao Tong University, Shanghai, People's Republic of China

Fengjuan Zhu Institute of Fuel Cells, School of Mechanical Engineering, Shanghai Jiao Tong University, Shanghai, People's Republic of China

Hongmin Zhu School of Metallurgical and Ecological Engineering, University of Science and Technology Beijing, Beijing, China

Roumen Zlatev Instituto de Ingenieria, Universidad Autonoma de Baja California, Mexicali, Baja California, Mexico

Jens Zosel Kurt-Schwabe-Institut für Mess- und Sensortechnik e.V. Meinsberg, Waldheim, Germany

Sandra Zugmann EVA Fahrzeugtechnik, Munich, Germany

Institució Catalana de Recerca i Estudis Avançats (ICREA), Institute for Bioengineering of Catalonia (IBEC), Barcelona, Spain

Electrical and Computer Engineering, University of California Davis, Davis, CA, USA

Andreas Züttel Empa Materials Science and Technology, Hydrogen and Energy, Dübendorf, Switzerland

Faculty of Applied Science, DelftChemTech, Delft, The Netherlands

A

Activated Carbons

Soshi Shiraishi
Division of Molecular Science, Faculty of Science and Technology, Gunma University, Kiryu, Gunma, Japan

Introduction

Activated carbon is a porous carbon material with developed nano-sized pores and a high specific surface area (>1,000 m^2g^{-1}). Nowadays, the technical term "nanoporous carbon" is very often used to mean "activated carbon," but the activated carbon should be strictly defined as a porous carbon prepared by an activation process consisting of a gasification reaction to form the developed nano-sized pore structure in the carbon matrix. The activated carbons have been widely utilized as industrial materials, for example, an adsorbent, decolorizing agent, deodorant, and catalyst. The details of the preparation method, pore structure, and applications of activated carbons have been described [1]. The most significant electrochemical application of activated carbons is use as an electrode active material for an electric double-layer capacitor (EDLC). This article reviews the use of activated carbons in an EDLC.

Pore Structure and Preparation

A wide variety of activated carbons with different specific surface areas (1,000 ~ 3,000 m^2g^{-1}) can be manufactured by controlling the activation conditions and selecting the proper carbon precursor. The high specific surface area of the activated carbon is derived from the developed pore structure. According to the IUPAC definition, the type of pores can be classified into three groups such as micropores (pore size: <2 nm), mesopores (2~50 nm), and macropores (>50 nm). The many micropores contained in the activated carbons provide the high specific surface area. The term "nanopore" has not yet been rigorously defined, but it usually means pores with less than around a 10 nm dimension. The details of the pore structure in the activated carbons have not been yet clarified, but some models have been proposed. Figure 1 shows one of the proposed models of the pore structure of the activated carbons and a transmission electron microscopic (TEM) image. The pore morphology is very complicated, never simply cylindrical. It is suggested that the slit-like space between the curved hexagonal-carbon plane units corresponds to the micropores. Some researchers have made efforts to conduct a TEM image analysis of the micropores in the activated carbons to clarify the pore structure [2]. However, it is not usually easy to precisely express the pore structure from TEM images; therefore, the surface area or pore size distribution (PSD) is generally determined by a gas-adsorption technique (e.g., nitrogen adsorption at 77 K). An example of the nitrogen adsorption/desorption isotherm at 77 K and the PSD calculated from the isotherm of a typical activated carbon for EDLC use

G. Kreysa et al. (eds.), *Encyclopedia of Applied Electrochemistry*, DOI 10.1007/978-1-4419-6996-5,

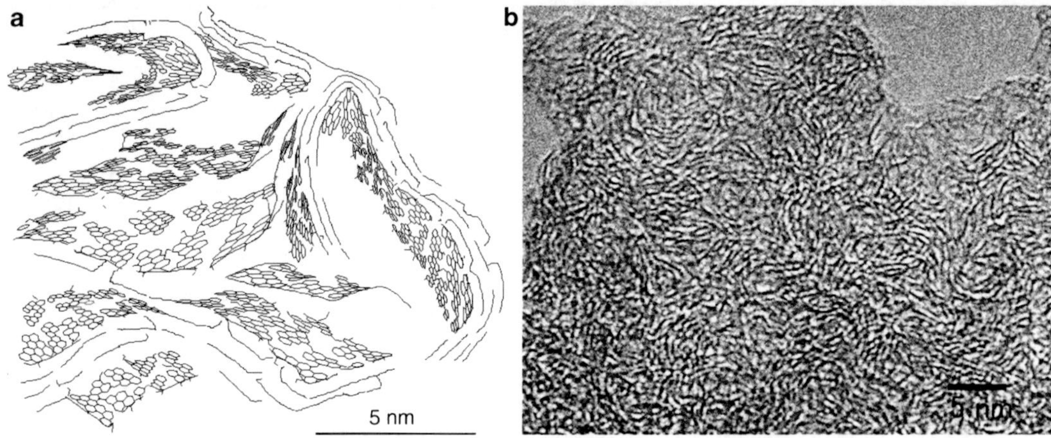

Activated Carbons, Fig. 1 (**a**) Schematic illustration of structural model of activated carbon (presented by courtesy of Professor T. Kyotani, Tohoku University, Japan), (**b**) TEM image of activated carbon (presented by courtesy of Dr. N. Yoshizawa, AIST, Japan)

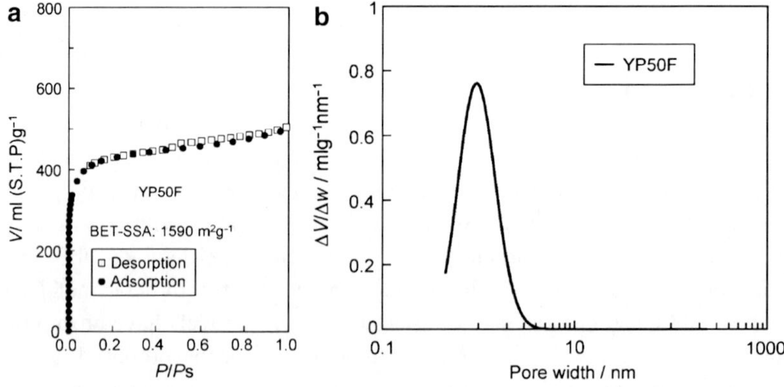

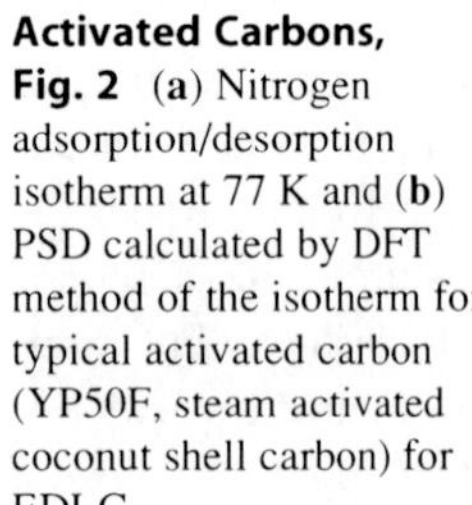

Activated Carbons, Fig. 2 (**a**) Nitrogen adsorption/desorption isotherm at 77 K and (**b**) PSD calculated by DFT method of the isotherm for typical activated carbon (YP50F, steam activated coconut shell carbon) for EDLC

is shown in Fig. 2. The isotherm is the type I representing a microporous structure and the PSD reveals a sharp distribution peak in the micropore region.

The representative carbon precursor is the phenol formaldehyde resin, polyacrylonitrile, pitch, coke, coconut shell, wood, etc. In most cases, these precursors are activated after carbonizing (pyrolysis in inert gas). The activation process for preparing the activated carbons is classified as gas activation (physical activation) and chemical activation. These activation characteristics are summarized in Table 1. In the former, the starting carbon materials are gasified using steam or carbon dioxide gas at 800∼900 °C. The gasified carbon part becomes pores and the resultant carbon residue works as the pore wall. The latter activation process is conducted by the heat treatment of the starting carbon materials or carbon precursor in the inert gas atmosphere together with an activation reagent such as potassium hydroxide (KOH), zinc chloride ($ZnCl_2$), and phosphorous acid (H_3PO_4). The suitable heat treatment temperature depends on the type of activation reagent; for example, for the KOH-activation, it is around 700∼900 °C. Both the gas activation and chemical activation mainly produce micropores, and the surface area or the PSD can be controlled by optimizing the activation conditions.

Activated Carbons, Table 1 Activation types and their characteristics

Gas (physical) activation (heat treatment of carbon in gas)	Chemical activation (heat treatment of carbon with activation reagent)
Steam-activation: Reaction scheme $C + H_2O \rightarrow CO + H_2$ $C + 2H_2O \rightarrow CO_2 + 2H_2$ Commercial activated carbons are usually produced by steam-activation	KOH-activation: • Pore formation reaction is gasification by the pyrolytic product (H_2O, K_2O) of KOH and intercalation of metallic potassium as by-product of the reaction of K_2O with carbon • High surface area is easily obtained with high production yield. However, a washing process using a lot of water is necessary and metallic potassium as by-product is dangerous
CO_2-activation: Reaction scheme $C + CO_2 \rightarrow 2CO$ It is easy to handle in the laboratory. The reaction is relatively slower than steam-activation	Others: $ZnCl_2$-activation, H_3PO_4-activation • These activations are often used for a wood-based precursor • They also require a washing process

Electric Double-Layer Capacitor

The EDLC is an electric energy storage device based on the dielectric property of the electric double layer at the interface between the electrolyte and a nanoporous carbon electrode such as activated carbon. The charge or discharge process is the adsorption/desorption of the electrolyte ion on the carbon surface as shown in Fig. 3. Conway's book states the details of electrochemical capacitors including the EDLC [3].

The electric double-layer capacitance is almost linear to the accessible surface area of the electrolyte ions. Additionally, the chemically/electrochemically stability, electric conductivity, and adequate commercial price are necessary for the EDLC electrode, so the activated carbons are suitable as practical electrode materials. The EDLC has been commercialized as a memory back-up device since the 1970s because of its high cycle efficiency and its long cycle life. Recently, the EDLC is also being considered to be one of the promising systems for electric energy storage because of its excellent power density and cycle life, but the energy density of the EDLC is lower than that of rechargeable batteries such as the lithium ion batteries therefore, it should be further improved for energy saving applications. The energy density of an EDLC can be expressed as follows:

$$E = \frac{CV^2}{2} \tag{1}$$

where E is the energy density of the capacitor, C is the specific double-layer capacitance, and V is the maximum applied voltage. The maximum applied voltage is limited by the interfacial electrochemical activity for the electrochemical decomposition of the electrolyte or carbon electrode itself. The EDLC with a nonaqueous electrolyte (e.g., R_4NBF_4/propylene carbonate or acetonitrile) has an essentially higher energy density than the aqueous type EDLC using sulfuric acid since nonaqueous electrolytes have a wider electrochemical window (≈3 V) compared to the aqueous one (≈1 V). The charging above 3 V of the EDLC causes a serious micropore blocking by the decomposition products [4, 5]. The electrochemical window of an electrolyte on carbon is a key factor that determines the energy storage of the EDLC.

Correlation of Capacitance with Pore Structure

The layer capacitance is strongly influenced by the pore structure of the electrode; therefore, the pore structure and surface chemistry of the activated carbons are very important. Many papers and reviews on these subjects have been published [6–8]. The high capacitance (100 ~ 200 Fg^{-1} for single electrode in organic electrolyte) is due to the high specific surface area (>1,000 m^2g^{-1}) provided by the many micropores. In general, it is believed that there is a proportional correlation between the specific surface area and the electric double-layer capacitance of the activated carbons based on the following equation.

$$C = \int \frac{\varepsilon_0 \varepsilon_r}{\delta} dS \tag{2}$$

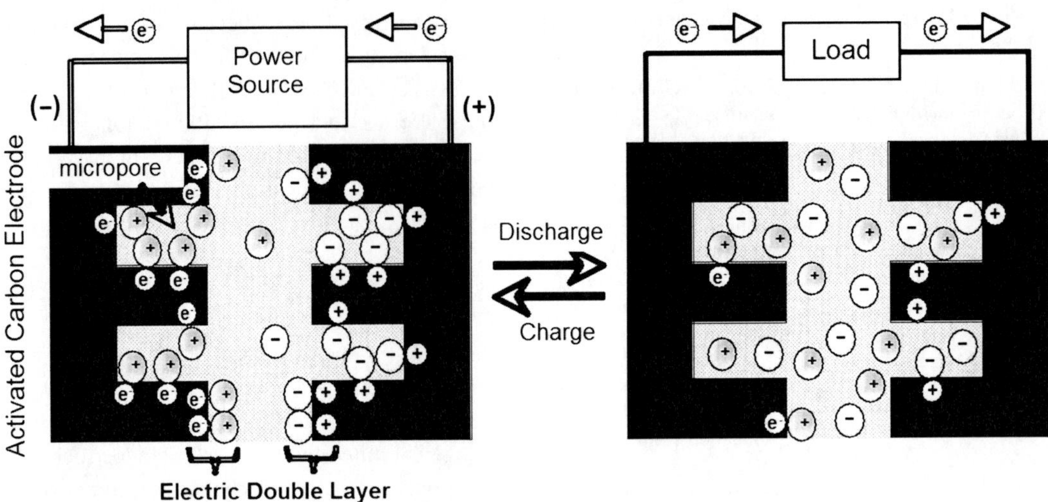

Activated Carbons, Fig. 3 Schematic illustration of charge/discharge state of EDLC

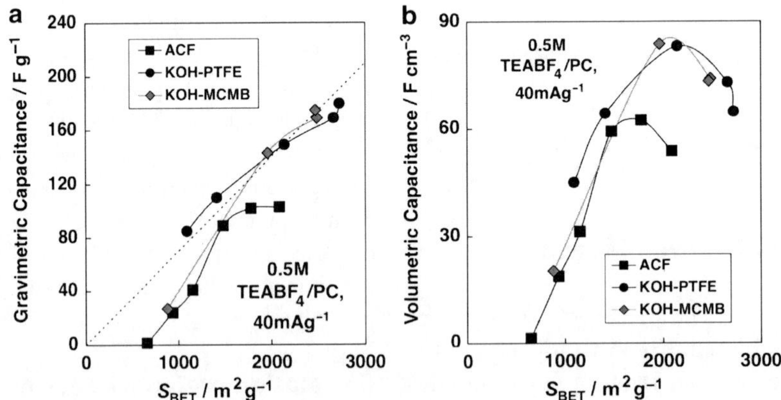

Activated Carbons, Fig. 4 Correlation between (**a**) gravimetric capacitance or (**b**) volumetric capacitance and BET-specific surface area (S_{BET}) of various activated carbons in 0.5 M $(C_2H_5)_4NBF_4$/propylene carbonate solution [21]. The capacitance was measured by the galvanostatic method (40 mAg^{-1}, 1.7 – 3.7V vs. Li/Li^+) using three-electrode cell. ACF steam-activated carbon fibers, KOH-PTFE KOH-activated PTFE-based carbons, KOH-MCMB KOH-activated mesocarbon microbeads, *Dotted line* in Fig. 4a corresponds to the specific capacitance per surface area of 70 mFm^{-2} (7 μFcm^{-2})

where C is the specific capacitance, ε_0 is the permittivity in a vacuum, ε_r is the relative permittivity of the double layer, d is the thickness of the double layer, and S is the specific surface area. Equation 2 is derived from an ideal capacitor consisting of a solid dielectric layer between two parallel plate electrodes such as the practical ceramic or film capacitors.

Figure 4a is the correlation data between the gravimetric capacitance (normalized by electrode weight) and the BET-specific surface area for the activated carbon electrode. This indicates that the gravimetric capacitance is almost linear to the BET-specific surface area regardless of some deviations. The deviation in the region of less than around 1,000 m^2g^{-1} of the specific surface area can be explained by the ion sieving effect. For the activated carbon having a low specific surface area, there are many narrower micropores than the electrolyte ions. Thus, the micropores with a narrow pore size sieve the ions and prevent adsorption. It was found that the presence of

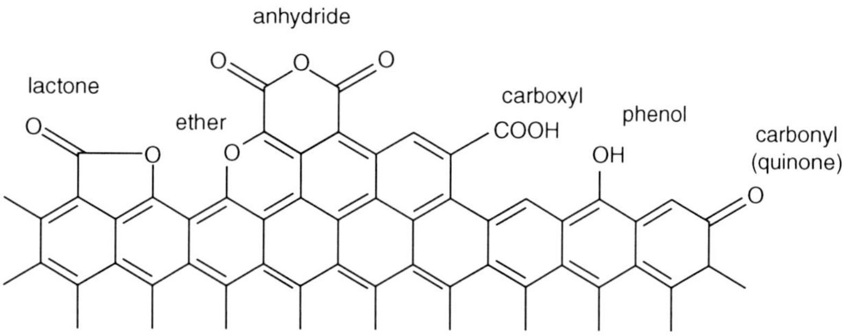

Activated Carbons, Fig. 5 Typical oxygen-containing surface functional groups of activated carbon

mesopores can relax the ion sieving effect and enhance the capacitance and improve the rate performance [9, 10]. However, the mesoporous activated carbon has a low bulk density, which causes a disadvantage of low volumetric capacitance (= capacitance normalized by electrode volume). The volumetric capacitance is more important than the gravimetric capacitance from the viewpoint of energy storage applications. Figure 4a suggests that a higher gravimetric capacitance can be obtained by a carbon electrode with a higher surface area. On the other hand, the correlations for the volumetric capacitance (Fig. 4b) show maximum curves depending on the type of activated carbon. This comes from the trade-off of the surface area and the electrode bulk density. The high surface area and large microporosity are realized by the sufficient activation, but this leads to a lower bulk density. Thus, no longer can only simple activations provide meaningful results for the capacitance or the energy density. Nanoporous carbons (zeolite-templated carbon (ZTC) [11], carbide-derived carbon (CDC) [12], carbon nanotube (CNT) [13, 14]) other than the activated carbons have been developed for EDLCs and have an excellent performance.

Surface Chemistry

In general, the activated carbons contain a few mmol of elemental oxygen in 1 g of the carbon as surface functionalities. Oxygen-containing surface functionalities are also a significant factor that dominates the capacitance of the activated carbon electrode [15]. Typical oxygen-containing surface functional groups are summarized in Fig. 5. The oxygen-containing surface functionalities are considered to increase the capacitance based on the following two reasons. The first one is to improve the wettability of the pore wall by the electrolyte. The second one is the contribution of the pseudo-capacitance derived from the redox reaction of the surface functionalities. The correlation of the oxygen-containing surface functionalities with the capacitance strongly depends on the type of electrolyte. Additionally, it is known that the presence of many oxygen-containing surface functionalities shortens the capacitor's life during aggressive operation such as high-voltage (>3 V) charging at a higher temperature (~70 °C) [16], so the surface modification by oxygen functionalities for improving the capacitance performance should be carefully examined. Nitrogen-containing surface functionalities, such as pyridine, pyrrole-like, or quaternary-like nitrogen, are also focused on because they enhance the capacitance [17–19]. The origin of the high capacitance of the nitrogen-enriched nanoporous carbon is analogous to that of the oxygen-containing surface functionalities. Especially, for the nitrogen-enriched carbon, the electronic effect of heteroatom-doping on the semimetallic carbon matrix can be expected. The electronic effect on the capacitance is also discussed for the boron-doped multi-walled carbon nanotubes [20]. In recent years, it has also been revealed that the nitrogen-containing surface functionalities improve the durability against aggressive operation during high-voltage charging at higher temperatures as shown in Fig. 6. It is suggested that the electrochemical stability of nitrogen-containing surface

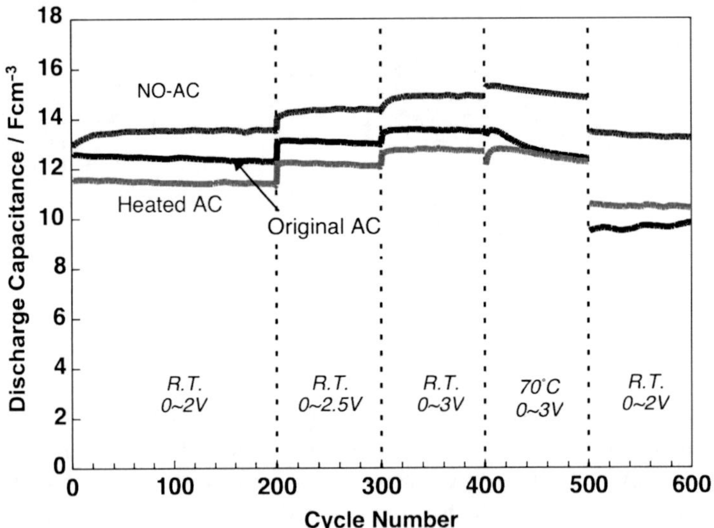

Activated Carbons, Fig. 6 Capacitance-dependence (two-electrode cell, galvanostatic, 80 mAg^{-1}) on charge/discharge cycle for original activated carbon (*AC*), heated AC at 800 °C in N_2 for 2 h, NO-AC (nitrogen-enriched AC) in 0.5 M $TEABF_4$/PC [21]. The charging voltage and temperature: 0 ~ 2 V at room temperature (R.T.) (1 ~ 200 cycle No.), 0 ~ 2.5 V at R.T. (201 ~ 300 cycle No.), 0 ~ 3 V at R.T. (301 ~ 400 cycle No.), 0 ~ 3 V at 70 °C (401 ~ 500 cycle No.), and 0 ~ 2 V at R.T. (501 ~ 600 cycle No.)

functionalities suppresses the electrochemical decomposition during higher voltage charging [21].

Consequently, it can be said that the surface modification using the heteroatom-doping technique of the activated carbon electrode is promising for improving the capacitance and the maximum voltage in order to realize a higher energy density for EDLCs.

References

1. Marsh H, Rodríguez Reinoso F (2006) Activated carbon. Elsevier, Oxford
2. Endo M, Kim YJ, Takeda T, Maeda T, Hayashi T, Koshiba K, Hara H, Dresselhaus MS (2001) Poly (vinylidene chloride)-based carbon as an electrode material for high power capacitors with an aqueous electrolyte. J Electrochem Soc 148:A1135–1140
3. Conway BE (1999) Electrochemical supercapacitors—scientific fundamentals and technological applications. Kluwer Academic/Plenum, New York
4. Ishimoto S, Asakawa Y, Shinya M, Naoi K (2009) Degradation responses of activated-carbon-based EDLCs for higher voltage operation and their factors. J Electrochem Soc 156:A563–A571
5. Ruch PW, Cericola D, Foelske A, Kötz R, Wokaun A (2010) A comparison of the aging of electrochemical double layer capacitors with acetonitrile and propylene carbonate-based electrolytes at elevated voltages. Electrochim Acta 55:2352–2357
6. Frackowiak E (2007) Carbon materials for supercapacitor application. Phys Chem Chem Phys 9:1774–1785
7. Simon P, Burke A (2008) Nanostructured carbons: double-layer capacitance and more. Interface 17:38–43
8. Inagaki M, Konno H, Tanaike O (2010) Carbon materials for electrochemical capacitors. J Power Sources 195:7880–7903
9. Morita M, Watanabe S, Ishikawa M, Tamai H, Yasuda H (2001) Utilization of activated carbon fiber with mesopore structure for electric double layer capacitors. Electrochemistry 69:462–466
10. Shiraishi S, Kurihara H, Shi L, Nakayama T, Oya A (2002) Electric double-layer capacitance of meso/macroporous activated carbon fibers prepared by the blending method. J Electrochem Soc 149:A855–A861
11. Itoi H, Nishihara H, Kogure T, Kyotani T (2011) Three-dimensionally arrayed and mutually connected 1.2-NmNanopores for high-performance electric double layer capacitor. J Am Chem Soc 133:1165–1167
12. Chmiola J, Yushin G, Gogotsi Y, Portet C, Simon P, Taberna PL (2006) Anomalous increase in carbon capacitance at pore sizes less than 1 nanometer. Science 313:1760–1763
13. Futaba DN, Hata K, Yamada T, Hiraoka T, Hayamizu Y, Kakudake Y, Tanaike O, Hatori H, Yumura M, Iijima S (2006) Shape-engineerable and highly

densely packed single-walled carbon nanotubes and their application as super-capacitor electrodes. Nature Mat 5:987–994

14. Honda Y, Haramoto T, Takeshige M, Shiozaki H, Kitamura T, Yoshikawa K, Ishikawa M (2008) Performance of electric double-layer capacitor with vertically aligned MWCNT sheet electrodes prepared by transfer methodology. J Electrochem Soc 155: A930–A935
15. Bleda-Martínez MJ, Maciá-Agulló JA, Lozano-Castelló D, Morallón E, Cazorla-Amorós D, Linares-Solano A (2005) Role of surface chemistry on electric double layer capacitance of carbon materials. Carbon 43:2677–2684
16. Cazorla-Amorós D, Lozano-Castelló D, Morallón E, Linares-Solano A, Shiraishi S (2010) Measuring cycle efficiency and capacitance of chemically activated carbons in propylene carbonate. Carbon 48:1451–1456
17. Frackowiak E, Lota G, Machnikowski J, Vix-Guterl C, Béguin F (2006) Optimisation of supercapacitors using carbons with controlled nanotexture and nitrogen content. Electrochim Acta 51:2209–2214
18. Kodama M, Yamashita J, Soneda Y, Hatori H, Kamegawa K, Moriguchi I (2006) Structure and electrochemical capacitance of nitrogen-enriched mesoporous carbon. Chem Lett 35:680–681
19. Hulicova-Jurcakova D, Kodama M, Shiraishi S, Hatori H, Zhu ZH, Lu GQ (2009) Nitrogen-enriched nonporous carbon electrodes with extraordinary supercapacitance. Adv Funct Mater 219:1800–1809
20. Shiraishi S, Kibe M, Yokoyama T, Kurihara H, Patel N, Oya A, Kaburagi Y, Hishiyama Y (2006) Electric double layer capacitance of multi-walled carbon nanotubes and B-doping effect. Appl Phys A 82:585–591
21. Shiraishi S (2012) Heat-treatment and nitrogen-doping of activated carbons for high voltage operation of electric double layer capacitor. Key Eng Mat 497:80–86

Activity Coefficients

Werner Kunz
Institut für Biophysik, Fachbereich Physik, Johann Wolfgang Goethe-Universität Frankfurt am Main, Frankfurt am Main, Germany

Basic Definitions

At constant temperature and pressure, the chemical potential of a solute component k in a solution can be written as

$$\mu_k = \mu_k^{\infty} + RT\ln a_k = \mu_k^{\infty(m)} + RT\ln m_k\gamma_k$$

where μ_k^{∞} is its chemical potential at infinite dilution, m is molality (moles of solute per kg of solvent), R is the gas constant, T the absolute temperature, and γ_k the activity coefficient in the molality scale at a given molality. Note that the reference state is not the pure solution (most of the pure electrolytes are solid and hence in a different aggregation state) but the infinitely dilute solution:

$$\mu_k^{\infty(m)} = \lim_{x\ (\mathrm{solvent})->1}(\mu_k - RT\ln m_k\gamma_k);$$
$$\lim_{x\ (\mathrm{solvent})->1}\gamma_k = 1$$

where x is the mole fraction of the solvent.

The molality scale is very convenient for experiments, because it is easy to weigh a certain amount of salt and of solvent. However, in the case of electrolytes, a slightly different definition of salt activity is used:

$$\mu_k = \mu_k^{\infty(m)} + \nu RT\ln m_{\pm}\gamma_{\pm}$$

with the stoichiometric coefficient of the salt being

$$\nu = \nu_+ + \nu_- \text{ and } m_{\pm}^{\nu} = m_+^{\nu+}m_-^{\nu-}$$

where m_+ and m_- are the molalities of the cations and anions, respectively. For example, for a 1:1 salt such as NaCl, $\nu = 2$ and $m_{\pm} = m$. Here, $\gamma_{\pm}$ is the *mean* activity coefficient of the salt. Note that usually values of $\gamma_{\pm}$ are compiled in tables, not values of γ.

In contrast to this experimental activity coefficient, theoretical modeling usually gives activity coefficients in the molarity (concentration) scale with

$$\mu^{\mathrm{salt}}(p,T) = \mu^{\mathrm{salt}\ \infty(c)}(p,T) + RT\ \ln c_{\mathrm{salt}}y_{\mathrm{salt}}$$
$$= \mu^{\mathrm{salt}\ \infty(c)}(p,T) + \nu RT\ \ln c_{\pm}y_{\pm}$$

with $c_{\pm}^{\nu} = c_+^{\nu+}c_-^{\nu-}$ and $y_{\pm}^{\nu-} = y_+^{\nu+}y_-^{\nu-}$

For the conversion of different concentration scales and their corresponding activity coefficients, see, e.g., [1, 2]. Usually, the conversion from the molarity to the molality scale is required. The respective mean activity coefficients are related as follows:

$$\gamma_{\pm} = y_{\pm} c \ / \ (m \ d_o)$$

with d_o being the density of the pure solvent.

Typical Examples

In contrast to most activity coefficients of uncharged species, salt activity coefficients can significantly differ from unity. As a result, they must be always carefully considered in order to make proper calculations, whenever the chemical potential of an electrolyte in solution is involved.

Some examples are given in Fig. 2 of the entry "▶ Specific Ion Effects, Evidences." Similar curves were obtained for many other systems (chlorides, iodides, nitrates, chlorates, perchlorates, and many more), c.f. the critical compilation by Hamer and Wu [3]. In Fig. 1 further examples of mean activity coefficients are shown over very wide electrolyte concentration ranges. The picture is adapted from "Thermodynamik der Elektrolytlösungen" [4], an excellent and modern introduction to the field. Unfortunately, it is only available in German. As an English textbook, the recent monograph by L. L. Lee [5] is highly recommended.

In the case of a binary system, the mean salt activity coefficients can be directly related to the molar osmotic coefficient φ of the solution via the Gibbs-Duhem equation:

$$\ln \gamma_{\pm} = (\varphi - 1) - 2 \int (1 - \varphi)/m^{1/2} \ d\, m^{1/2}$$

where m is the molality (moles per kilogram of solvent).

The water activity $a_{water} = x_{water} f_{water}$ (with x_{water} being the mole fraction of water and f_{water} the corresponding activity coefficient) is related to φ through

$$\varphi = -\ln a_{water}/(\nu m \ M_{water})$$

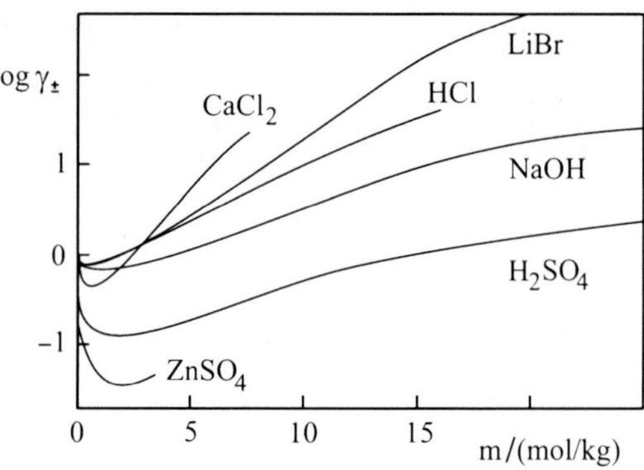

Activity Coefficients, Fig. 1 Concentration dependence of the mean activity coefficient of electrolytes in aqueous solutions at 25 °C (the values are given in the molality scale) (inspired by [4], p. 135)

where M_{water} is the molar mass of water. The osmotic coefficient φ is linked to the osmotic pressure:

$$\Pi = m \ \varphi \nu R \ T \ M_{water}/V_{water}$$

where V_{water} is the molar volume of water. It is difficult to measure directly the osmotic pressure, because of the lack of appropriate membranes due to the small size of the ions. Therefore, the activity of water or of the salt is usually derived from vapor pressure measurements according to the following relation:

$$\ln a_{water} = \ln(p/p*) + (B_{water} - V_{water}) \\ (p - p*)/(RT)$$

where p and p* are the vapor pressures of water in the salt solution and of pure water, respectively, and B_{water} is the second virial coefficient of water vapor.

It should be noted that various other methods are known and applied to determine the activity coefficients of salt solutions. Here, the famous book by Robinson and Stokes [2], initially published in 1959, is still a good reference. Other methods can be found in [1]. Both monographs contain also excellent and extensive discussions of activity coefficients in various experimental, theoretical, and practical aspects.

Where to Find Data

Compilations of data can be found, e.g., in the classical papers by Pitzer and co-workers [1, 6, 7]. A very useful tool is provided by the Joint Expert Speciation System (JESS), a joint project of two groups in Australia and South Africa [8, 9]. Also interesting is the data bank from CERE of the Technical University of Denmark for electrolyte solutions [10] and the ELDAR data bank integrated in the Dechema's DETHERM data bank [11] containing values from which activity coefficients can be calculated.

In the case of *nonaqueous electrolyte solutions*, of course, much less data are known. This is because it is not easy to determine absolute values in solvents different from water, since many types of experimental setups require a standard reference, which is not common for nonaqueous systems. Second, the solubility of salts is low in many nonaqueous systems thus rendering any experimental determinations of activity coefficients difficult. Apart from work done by Russian scientists in the 1950s–70s, e.g., [12], mainly the Regensburg group of Barthel and Neueder published relevant data during the last decades, e.g., [13–16]. A Pitzer parameter compilation for electrolytes in methanol, ethanol, 2-propanol, acetonitrile, acetone, dimethoxyethane, and dimethyl carbonate is given in the appendix of [17].

Modeling

At low concentrations (below c = 0.1 M), all mean activity coefficients sharply decrease very similarly, because of strong cation-anion attraction. This dilute regime can be reasonably described with the help of the Debye-Hückel theory. At higher concentrations, the finite size of the ions lead to significant ion-ion repulsions at short distances (the "excluded volume term"), and the activity coefficient goes up again or, at least, do no longer decrease. In this intermediate regime, ion specificity clearly appears. It is still extremely difficult to describe theoretically this concentration regime. In most cases only a data description with some more or less meaningful parameters is given. The most prominent of this data fitting models is the Pitzer approach, in which a linear combination of exponential functions is very helpful to describe the curves with only few parameters [1, 6, 7].

As a rule of thumb, the following factors influence the curves shown in Fig. 1 and in Fig. 2 of the entry "▸ Specific Ion Effects, Evidences":

- The larger the ion radius, the higher the activity coefficient, because of the excluded volume.
- The stronger the ion solvation, the higher the activity coefficient, because of stronger ion-solvent interactions.
- The stronger the ion pairing, the lower the activity coefficient, because of weaker ion-solvent interactions.

A more quantitative prediction of activity coefficients can be done for the simplest cases [18]. However, for most electrolytes, beyond salt concentrations of 0.1 M, predictions are a tedious task and often still impossible, although numerous attempts have been made over the past decades [19–21]. This is true all the more when more than one salt is involved, as it is usually the case for practical applications. Ternary salt systems or even multicomponent systems with several salts, other solutes, and solvents are still out of the scope of present theory, at least, when true predictions without adjusted parameters are required. Only data fittings are possible with plausible models and with a certain number of adjustable parameters that do not always have a real physical sense [1, 5, 22–27]. It is also very difficult to calculate the activity coefficients of an electrolyte in the presence of other electrolytes and solutes. Even the definition is difficult, because electrolyte usually dissociate, so that extrathermodynamical ion activity coefficients must be defined. The problem is even more complex when salts are only partially dissociated or when complex equilibriums of gases, solutes, and salts are involved, for example, in the case of CO_2 with acids and bases [28, 29].

Future Directions

Today, an impressive number of activity coefficients of electrolytes are known. Not much real innovation is to be expected in the field of experiments, except in the context of special systems such as biological or other confined ones. The general behavior of ions and their consequences on activity coefficients is mainly understood. However, the one problem will remain: the underlying interactions and their balance are very subtle and consequently it will always be difficult to predict quantitatively the values of activity coefficients in liquid solutions.

Cross-References

- ► Conductivity of Electrolytes
- ► Data Banks of Electrolytes
- ► Electrolytes, History
- ► Electrolytes, Thermodynamics
- ► Specific Ion Effects, Evidences
- ► Thermodynamic Properties of Ionic Solutions - MSA and NRTL Models

References

1. Pitzer KS (ed) (1991) Activity coefficients in electrolyte solutions, 2nd edn. CRC Press, Boca Raton
2. Robinson RA, Stokes RH (2003) Electrolyte solutions, 2nd rev edn. Dover, New York
3. Hamer WJ, Wu YC (1972) Osmotic coefficients and mean activity coefficients of Uni-univalent electrolytes in water at 25 °C. J Phys Chem Rev Data 1:1047–1099
4. Luckas M, Krissmann J (2001) Thermodynamik der Elektrolytlösungen. Springer, Berlin
5. Lee LL (2008) Molecular thermodynamics of electrolyte solutions. World Scientific Publishing, Singapore
6. Pitzer KS, Kim JJ (1974) Thermodynamics of electrolytes. IV. Activity and osmotic coefficients for mixed electrolytes. J Am Chem Soc 96:5701–5707
7. Clegg SL, Pitzer KS (1992) Thermodynamics of multicomponent, miscible, ionic solutions: generalized equations for symmetrical electrolytes. J Phys Chem 96:3513–3520
8. http://jess.murdoch.edu.au/jess_home.htm
9. May PM, Rowland D, Königsberger E, Hefter, G (2010) JESS, a joint expert speciation system – IV: a large database of aqueous solution physicochemical properties with an automatic means of achieving thermodynamic consistency. Talanta 81:142–148
10. http://www.cere.dtu.dk/Expertise/Data_Bank.aspx
11. http://www.dechema.de/en/detherm.html
12. Ivanova EF, Aleksandrov VV (1964) Thermodynamic properties of electrolytes in nonaqueous solutions. XV. Solutions of cesium iodide in methanol and cadmium chloride in 1-butanol. Zhurnal Fizicheskoi Khimii 38:878–84
13. Barthel J, Lauermann G, Neueder R (1986) Vapor pressure measurements on non-aqueous electrolyte solutions. Part 2. Tetraalkylammonium salts in methanol. Activity coefficients of various 1–1 electrolytes at high concentrations. J Solution Chem 10:851–867
14. Barthel J, Neueder R, Poepke H, Wittmann H (1999) Osmotic coefficients and activity coefficients of nonaqueous electrolyte solutions. Part 2. Lithium perchlorate in the aprotic solvents acetone, acetonitrile, dimethoxyethane, and dimethylcarbonate. J Solution Chem 28:489–503
15. Nasirzadeh K, Neueder R, Kunz W (2005) Vapor pressures, osmotic and activity coefficients of electrolytes in protic solvents at different temperatures. 3. Lithium bromide in 2-propanol. J Solution Chem 34:9–24
16. Tsurko EN, Neueder R, Kunz W (2007) Water activity and osmotic coefficients in solutions of glycine, glutamic acid, histidine and their salts at 298.15 K and 310.15 K. J Solution Chem 36:651–672
17. Barthel J, Krienke H, Kunz W (1998) Physical chemistry of electrolyte solutions. Modern aspects. Springer, New York
18. Vrbka L, Lund M, Kalcher I, Dzubiella J, Netz RR, Kunz W (2009) Ion-specific thermodynamics of multicomponent electrolytes: a hybrid HNC/MD approach. J Chem Phys 131:154109–1–12
19. Outhwaite CW, Bhuiyan LB, Vlachy V, Hribar-Lee B (2010) Activity coefficients of an electrolyte in a mixture with a high density neutral component. J Chem Eng Data 55:4248–4254
20. Kalyuzhnyi YV, Vlachy V, Dill KA (2010) Aqueous alkali halide solutions: can osmotic coefficients be explained on the basis of the ionic sizes alone? Phys Chem Chem Phys 12:6260–6266
21. Lu X, Zhang L, Wang Y, Shi J, Maurer G (1996) Prediction of activity coefficients of electrolytes in aqueous solutions at high temperatures. Ind Eng Chem Res 35:1777–1784
22. Chen CC, Britt HI, Boston JF, Evans LB (1982) Local composition model for excess Gibbs energy of electrolyte systems. Part I: single solvent, single completely dissociated electrolyte systems. AIChE J 28:588–596
23. Simonin JP, Bernard O, Blum L (1999) Ionic solutions in the binding mean spherical approximation: thermodynamic properties of mixtures of associating electrolytes. J Phys Chem B 103:699–704

24. Papaiconomou N, Simonin JP, Bernard O, Kunz W (2002) MSA-NRTL model for the description of the thermodynamic properties of electrolyte solutions. Phys Chem Chem Phys 4:4435–4443
25. Gering KL, Lee LL, Landis LH, Savidge JL (1989) A molecular approach to electrolyte solutions: phase behavior and activity coefficients for mixed-salt and multisolvent systems. Fluid Phase Equilib 48:111–139
26. Held C, Cameretti LF, Sadowski G (2008) Modeling aqueous electrolyte solutions. Fluid Phase Equilib 270:87–96
27. Held C, Sadowski G (2009) Modeling aqueous electrolyte solutions. Part 2. Weak electrolytes. Fluid Phase Equilib 279:141–148
28. Rumpf B, Xia J, Maurer G (1998) Solubility of carbon dioxide in aqueous solutions containing acetic acid or sodium hydroxide in the temperature range from 313 to 433 K and at total pressures up to 10 MPa. Ind Eng Chem Res 37:2012–2019
29. Papaiconomou N, Simonin JP, Bernard O, Kunz W (2003) Description of vapor–liquid equilibria for CO_2 in electrolyte solutions using the mean spherical approximation. J Phys Chem B 107:5948–5957

Aerospace Applications for Primary Batteries

Ratnakumar V. Bugga and Marshall C. Smart
Jet Propulsion Laboratory, Pasdena, CA, USA

Introduction

Primary batteries are electrochemical devices that convert chemical energy into electrical energy and, unlike rechargeable batteries, are intended for single-use or "one shot" applications. They have a "niche" role in aerospace missions, especially where recharging is either not feasible or operationally difficult and/or where high specific energy density is desired. Planetary exploration missions are broadly categorized as orbiters, landers/rovers, probes, penetrators, and sample return capsules. While rechargeable batteries are exclusively the option for orbiters and small rovers; Probes, penetrators and sample return capsules are typically powered by primary batteries [1]. They are also used in launch vehicles to power pyro devices and on-board electronics guidance and control systems and in several miscellaneous applications, such as launch abort batteries and astronauts' electronic devices, gadgets, and tools.

Primary batteries used in early spacecraft were largely of the aqueous (alkaline) type, e.g., Ag–Zn, which typically exhibit high specific power, relatively low cell voltages, limited life, moderate energy densities, and limited operating temperature range. In the recent missions, however, these aqueous batteries have been replaced by more energetic lithium-based systems, especially with liquid cathodes, such as Li–SO_2 and Li–$SOCl_2$, which yield much higher voltages, energy densities, and longer shelf life and have wider operating temperature range (Table 1). Furthermore, the Li primary cells can operate under microgravity environments and are leak-proof and hermetically sealed, as required in space missions. In contrast to the many performance benefits offered by Li primary batteries, they exhibit relatively low power densities and also display a "voltage delay," i.e., a delayed voltage response to discharge manifested by a slow breakdown of the surface passive film on Li anode [2], which needs to be mitigated by predischarge. Among the various primary battery chemistries available, only a few systems, e.g., Ag–Zn, Li–SO_2, Li–$SOCl_2$, and Li–MnO_2, have been utilized in space missions, and these systems are briefly described below.

Silver–Zinc

The Ag–Zn cell utilizes a powdered zinc anode, a silver oxide cathode, and an aqueous alkaline electrolyte comprised of potassium hydroxide (KOH) (40–45 %) with dissolved zincate (i.e., a salt containing $Zn(OH)_4^{2-}$). Aerospace silver–zinc cells are available in prismatic parallel-plate configuration in sizes ranging up to hundreds of Ah, from different manufactures, including Yardney Technical Products, Eagle Picher Industries, and BST Systems, Inc. [3]. The attractive features of this system are its high-energy density, being probably the highest of the aqueous systems, combined with extremely high power

Aerospace Applications for Primary Batteries, Table 1 Performance characteristics of aerospace primary battery systems

Performance characteristics	Battery systems					
	Ag–Zn	Li–MnO_2	Li–SO_2	Li–$SOCl_2$	Li–BCX	Li–CFx
Cell voltage, V	1.6 V	3.05	2.9	3.7	3.8	3
Cathode Theoretical energy Wh/kg	370	1,005	1,130	1,470	1,405	2,180
Cell-specific energy, Wh/kg	130	238	250	390	414	600 (@ < C/100)
Battery sp. energy, Wh/kg	80	130	130	250	250	400
Battery energy Density, Wh/l	200	180	180	500	500	500
Power density, W/kg	150	150	680	140	140	>100 (SOA < 15)
Shelf life, years	2	5	10	5-10	5	10
Temp. range, °C	−20 to 55	−40 to 60	−40 to 60	−40 to 60	−40 to 60	−40 to 60
Missions/applications	Launch vehicles	Astronaut equipment (camera)	Planetary probes	Planetary probes, Rovers	Astronaut head lamp	EVA, tools, lunar probes

density. The drawbacks, on the other hand, include poor calendar life, electrolyte leakage, orientation sensitivity, poor low-temperature performance, all of which make this system unattractive for planetary missions. They are thus typically used in ground support equipment, such as Launch Abortion Systems (LAS). However, a rechargeable version of this battery (30 V, 50 Ah) was successfully used in the Mars Pathfinder mission in 1997 to power the Mars Lander for ~6 months of surface operations.

Lithium–Sulfur Dioxide

Li–SO_2 is by far the currently most popular primary battery system in planetary exploration missions. It belongs to the class of "liquid cathode" systems, with sulfur dioxide dissolved in a 1 M solution of lithium bromide in acetonitrile serving as the catholyte (cathode as well as electrolyte) and lithium metal as the anode. Sulfur dioxide is reduced over a carbon cathode current collector, and different carbons are typically chosen depending on the desired temperature range of operation, to form insoluble lithium dithionate as the discharge product [4]. Spirally wound designs are generally adopted for realizing high specific energies, especially at high discharge rates. Though other cell sizes may be available, the cell design that has been widely adopted in aerospace batteries of different capacities is the "D-size" cell that delivers ~7.5 Ah. This system offers moderately high-energy densities (>225 Wh/kg and 375 Wh/l) and good low-temperature performance, down to −40 °C. It affords high pulse power capability, as needed in the Entry, Descent, and Landing (EDL) of the planetary surface missions, and an impressively long shelf life of >10 years, as desired for the missions to the outer planets and sample return missions. Also the voltage delay, which is associated with the delayed breakdown of the surface passive film [5] and is often an issue with other "liquid cathode" systems, is significantly less with Li–SO_2 batteries and can easily be mitigated by prior discharge and often achieved by utilizing a depassivation circuit in the power system design. While Honeywell made the batteries for the early missions, SAFT America Inc., fabricated batteries for the more recent missions and is currently the only US manufacturer producing Li–SO_2 cells for space and defense applications.

NASA has utilized Li–SO_2 batteries for a number of planetary probes (Galileo, Cassini's Huygen probe), sample return capsules (Genesis, Stardust), and for the Lander of the Mars Exploration Rover (MER). The Galileo mission to Jupiter consisted of an orbiter and an atmospheric entry probe, the latter to provide an understanding of the Jovian atmosphere. The power source for the Galileo probe was comprised of three Li–SO_2 battery modules, containing 13 D-size cell strings per module, which are required to retain capacity for 8 years. The ground simulation tests confirmed the excellent shelf-life characteristics of these batteries [5]. The Cassini mission also had a similar planetary atmospheric entry probe, named the "Huygen Probe," to perform in-situ observations of Saturn's largest moon, Titan. The Huygen probe also contained batteries made of Li–SO_2 D-size cells of the same pedigree. The five Li–SO_2 batteries on the probe had 1,600 Wh of energy and successfully provided 250 W of power for the planned 3 hours of probe operation. In both cases, a depassivation circuit, i.e., pre-discharge through a load, successfully alleviated the voltage delay.

The sample return capsules for the Stardust and Genesis missions were also successfully powered by Li–SO_2 batteries of similar type and manufactured by Saft; however, they possessed minor design changes. Specifically, the Stardust mission, launched in 1999 to bring back extraterrestrial dust from the Wild-2 Comet, had two Li–SO_2 batteries, i.e., two parallel eight-cell modules with Saft LO26SHX cells. The batteries performed well and supported the Sample Return Capsule (SRC) after 7 years in space. The Genesis SRC (launched in 2001) retrieved samples from the solar wind and was supported by similar Li–SO_2 battery with two parallel eight-cell modules containing LO26SHX cells. Recently, NASA successfully used Li-SO_2 batteries for a different application in the twin Mars Exploration Rovers (2003), i.e., to support power the pyro events during the Entry, Descent, and Landing (EDL) of the spacecraft and the egress of the Rover and the Lander, lasting over 1.5 h. Each spacecraft had five parallel modules of 12 cells (LO26SX) each, to provide an overall energy of ~1,200 Wh [6]. In the absence of a depassivation circuit, these batteries were pre-discharged potentiostatically through a shunt-regulated bus, to mitigate the voltage delay [7].

Lithium–Thionyl Chloride

Li-$SOCl_2$ is a high-energy Li primary system, often used in space applications. This system has lithium as the anode and liquid thionyl chloride ($SOCl_2$) containing lithium tetrachloroaluminate ($LiAlCl_4$) salt as the cathode. Once again, spirally wound designs are generally adopted for realizing high specific energies, especially at moderate to high discharge rates. This system offers considerably higher specific energy (390–410 Wh/kg) and energy density (875–925 Wh/l) than Li–SO_2 cells, but the power densities (typically <100 W/kg) are relatively poor, which also manifests in limited performance capability at low temperatures (-40 °C). Further, the voltage delay tends to be significant due to excessive Li electrode passivation, especially after storage at ambient and warm temperatures.

NASA has utilized this high-energy battery system in several space missions, including the Mars Pathfinder Rover–Sojourner, New Millennium Deep Space-2, and in the Deep Impact mission. Sojourner is a small robotic rover that landed on Mars aboard the Mars Pathfinder lander in 1997, performed science and technology experiments, and transmitted images and data back to the Lander spacecraft. It was powered by solar panels as well as by a Li–$SOCl_2$ battery, fabricated by SAFT, with consisting of nine 12 Ah cylindrical cells in a 3S3P configuration (three parallel strings, each string being three cells in series) for night-time experimentation. The primary batteries supported 4 days of operation, including mobility and science and communication on Mars, before the solar panels provided all power for the Rover and extended the operation for 60 days.

A modified version the Li–$SOCl_2$ system was used for Mars Microprobes as part of the Mars Surveyor Lander Mission (1998). Due to high impact the probes were to encounter during landing and penetration and the ultra-low temperature the aft-bodies were to experience on the Martian surface, the batteries needed to be tolerant to high impacts of ~80,000 g and operate at temperatures of -80 °C. The low-temperature performance was improved with the use of lithium terachloro gallate salt and with a reduction in the salt concentration from 1.0 to 0.5 M, while the impact tolerance was achieved with a pancake electrode design in a cylindrical cell [8]. Though the mission wasn't successful, the ground tests revealed that the batteries had the desired impact tolerance and low-temperature performance.

A more recent mission, Deep Impact (2005), was designed to understand the Comet Tempel 1, by impacting it with a 370-kg mass Impactor and collect/analyze the Comet dust. The Impactor, also equipped with its own scientific equipment, was successfully powered by Li–$SOCl_2$ batteries for one day, for navigation and science and communication, prior to the impact. The batteries were made by SAFT with 216 D-cells (9S24P) housed within a single mechanically and thermally coupled aluminum structure.

Miscellaneous Systems

Another variant of the Li–$SOCl_2$ chemistry has BrCl added to the conventional thionyl chloride electrolyte to form the thionyl chloride–bromine complex (Li–BCX) cathode, which may contribute to improved safety and also provides marginal improvement in cell voltage and specific energy [9]. Wilson Greatbatch Ltd. is the only source for the Li–BCX chemistry, which has been primarily used in some astronauts' electronic devices. This system has similar issues as the Li–$SOCl_2$ system relative to voltage delay and low-temperature performance, with an added uncertainty with regard to its shelf life.

Another Li primary system displaying possibly the highest specific energy is Li–CF_x, which utilizes fluorinated carbon as the solid cathode against Li anode in an electrolyte containing 1 M $LiBF_4$ dissolved in either a propylene carbonate (PC)/dimethoxy ethane (DME) blend or in γ-butyrolactone (GBL). The cells are made in a jelly roll cylindrical configuration, though pouch cells are also currently being developed for enhanced specific energy. Li–CF_x cells and batteries are available mainly from Eagle Picher Industries, Quallion, LCC, and Saft America, Inc. This system has limited use, thus far, in aerospace applications, due to its low power densities (i.e., can only be used at rates less than C/10) and safety concerns prevalent during high rate discharge. Recent efforts have focused on improving the rate capability with the use of sub-fluorinated cathodes [10], suitable electrolyte mixtures, and with fluoride–ion receptor additives for improved low-temperature performance [11].

Future Directions

Among the various primary battery systems available today, both Li–$SOCl_2$ and Li-CF_x are promising and can benefit from further advancements. The Li–$SOCl_2$ system can be improved in terms of rate capability, low-temperature performance, and voltage delay with judicious selection of electrolytes and cathode designs. Their operation can also be extended to ultra-low temperature, for example, from -100 °C to -120 °C, with the addition of low-freezing point oxyhalide blends [12]. The Li–CF_x chemistry displays particularly great promise to exhibit high specific energy and future missions will benefit from the technology it can be made functional over wide range of operating temperatures and at high discharge rates. The use of carbon fluoride with enhanced electronic conductivity, novel electrolyte formulations, thin and improved electrode designs, and modified cell designs already seem promising.

Cross-References

▸ Lithium Primary Cells, Liquid Cathodes
▸ Lithium Solid Cathode Batteries

References

1. Ratnakumar BV, Smart MC (2007) Aerospace applications – planetary exploration missions (orbiters, Landers, rovers and probes). In: Pistoia G, Broussely M (eds) Industrial applications of batteries. From electric vehicles to energy storage and toll collection. Elsevier, Amsterdam, pp 327–387
2. Dey AN (1977) Lithium anode film and organic and inorganic electrolyte batteries. Thin Solid Film 43:131

3. Serenyi R, Skelton J (2004) Proceedings of the 41st power sources conference, # 28.1, Philadelphia, 14 June 2004
4. Linden D, McDonald B (1980) The lithium—sulfur dioxide primary battery — its characteristics, performance and applications. J Power Sources 5:35, Elsevier Sequoia, Lausanne
5. Dagarin BP, Taenaka RK, Stofel EJ (1996) NASA battery workshop, p. 133
6. Ratnakumar BV, Smart MC, Kindler A, Frank H (2003) Potentiostatic depassivation of lithium-sulfur dioxide batteries on mars exploration rovers. J Power Sources 119:906
7. Ratnakumar BV, Smart MC, Ewell RC, Whitcanack LD, Kindler A, Narayanan SR, Surampudi S (2007) Potentiostatic depassivation of lithium-sulfur dioxide batteries on mars exploration rovers. J Electrochem Soc 154:A715–A724
8. Frank H, Deligiannis F, Davies E, Ratnakumar BV, Surampudi S, Russel PG, Reddy TB (1998) NASA battery workshop, Huntsville, 27 Oct 1998
9. Liang CC, Krehl PW, Danner DA, Appl J (1981) Bromine chloride as a cathode component in lithium inorganic cells. Electrochem 11:563–571
10. Whitacre J, Yazami R, Hamwic A, Smart MC, Bennett W, Surya Prakash GK, Miller T, Ratnakumar BV (2006) Low operational temperature Li–CFx batteries using cathodes containing sub-fluorinated graphitic materials. J Power Sources 160(1):577–584
11. Whitacre JF, West WC, Smart MC, Yazami R, Surya Prakash GK, Hamwi A, Ratnakumar BV (2007) Enhanced low-temperature performance of Li–CFx batteries. Electrochem Solid State Lett 10(7): A166–A170
12. West WC, Shevade A, Soler J, Kulleck J, Smart MC, Ratnakumar BV, Moran M, Haiges R, Christe KO, Surya Prakash GK (2010) Sulfuryl and thionyl halide-based ultralow temperature primary batteries. J Electrochem Soc 157:A571

AFM Studies of Biomolecules

Ingrid Tessmer
Rudolf-Virchow-Zentrum, Experimentelle Biomedizin, University of Würzburg, Würzburg, Germany

Introduction

Atomic force microscopy (AFM) provides a prime approach to study biomolecules. Because the technique does not require modification of the samples, biological particles can be studied under near-physiological conditions. Structural information is obtained either by imaging or by mechanical manipulation of intramolecular bonding. Furthermore, AFM can be used to study intermolecular interactions, for instance of receptor-ligand pairs. The forces accessible to AFM cover the broad range from low picoNewton to microNewton, allowing to resolve even the breaking of single bonds. Resolution capabilities of AFM imaging are in the low nanometer range, sufficient to provide information at the level of the individual molecules. I will introduce the principle of AFM measurements, provide a brief overview of different types of applications on biomolecules including example experiments, and finally give a short outlook on recent developments in the field.

Principle of AFM

Atomic force microscopy (AFM) belongs to the family of scanning probe microscopies (SPMs), which use a mechanical probe to investigate the surface of a sample. For investigation by the probe, the sample has to be deposited onto a substrate surface. The first ever SPM was the scanning tunneling microscope (STM), developed by Gerd Binnig and Heinrich Rohrer in 1981 (Nobel Prize in 1986). STM measures a tunneling current between the probe and a conducting sample surface. The STM detection mechanism translates increases and decreases in the current into probe-surface distances and hence height changes to reconstitute a topography image of the sample at atomic resolution. Analogous to STM, the AFM – invented in 1986 by Gerd Binnig, Calvin Quate, and Christoph Gerber – detects direct interactions (forces) between the probe, called the AFM tip, and the molecules and atoms on the substrate surface to build up a topography image, as reflected by its name. These molecular interactions are combinations from the spectrum of non-covalent forces, long-range electrostatic interactions, short-range attractive van der Waals forces, and the very short-range Pauli

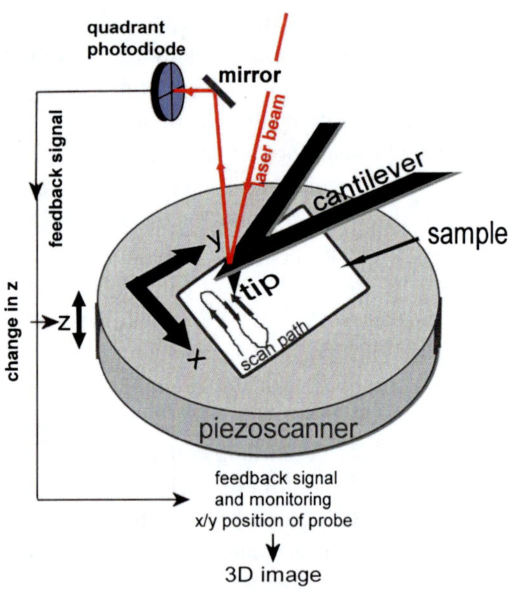

AFM Studies of Biomolecules, Fig. 1 AFM schematic (Adapted from [14])

repulsion when the tip is pressed down into the surface. Since these forces do not require conductance of the sample surface, a major advantage of AFM is its applicability, in particular, to biological samples. Interaction forces are recorded with the help of a cantilever arm, at the bottom end of which the AFM tip is located (Fig. 1). Detection is mostly based on an optical system with a laser beam being reflected from the back of the cantilever onto a position-sensitive photodetector. Attractive or repulsive forces between the AFM tip and the surface bend the cantilever towards the surface or away from the surface, respectively. The resulting cantilever deflection is proportional to and can hence be measured from the detector's electronic difference signal, due to displacement of the laser position on its quadrant photodiode array. There are two major types of AFM experiments, AFM force spectroscopy and AFM imaging, which differently process the cantilever deflection signals to provide information on interaction forces or height of sample features, as outlined in the following two sections.

AFM Force Spectroscopy

AFM can be used to measure the response of a molecular system to an applied force. In these experiments, a molecule is attached at one end to the substrate surface and at its other end to the AFM tip surface (Fig. 2a left). Established surface attachment methods are based, for instance, on bifunctional cross-linkers or biotin-streptavidin-biotin sandwiches. Intramolecular interactions and elastic properties during stretching and relaxation of a biopolymer, such as a protein or DNA molecule, tethered between tip and surface can be recorded upon moving the tip away from or towards the surface. Furthermore, AFM force spectroscopy can be applied to measure intermolecular interactions (Fig. 2a right) between molecules immobilized on the substrate surface (e.g., receptor molecules) and molecules attached to the AFM tip (the corresponding ligand). When the AFM tip is brought down into contact with the sample, the molecule pair can interact, while rupture of the formed bonds can be measured upon retracting the tip, from the resulting force curves (see Fig. 2b).

Detection of the intra- and intermolecular forces in a biological test system is based on the resulting deflection of the cantilever arm towards the surface on attractive interaction forces and away from the surface on repulsive forces. We can translate the cantilever deflection x into an interaction force F using Hooke's law: $F = k\,x$, where k is the cantilever's spring constant. Typical values of AFM force spectroscopy cantilever spring constants range from 10 to several 100 mN/m. These tips and cantilevers for force spectroscopy are usually made from silicon nitride (Si_3N_4), to provide such elastic properties suitable for a sensitive force sensor. Importantly, the spring constant of the applied cantilever has to be determined for each experiment to be able to interpret the measured cantilever deflections. Modern commercial systems offer cantilever spring constant calibration as an integral part of the AFM software, based on the thermal noise spectrum of the cantilever.

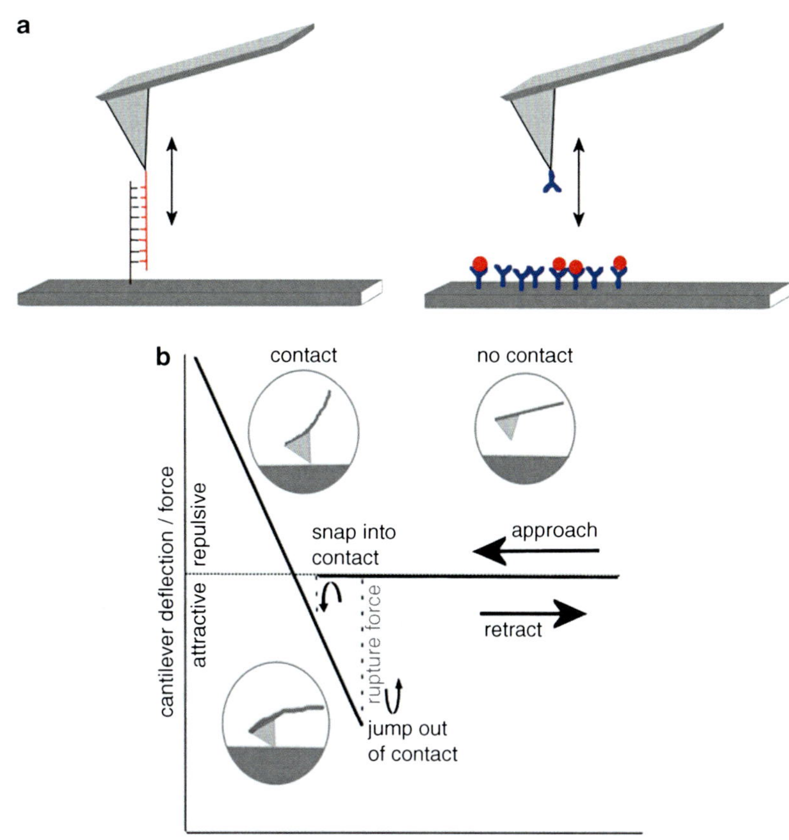

AFM Studies of Biomolecules, Fig. 2 **AFM force spectroscopy** (Adapted from [15]). (**a**) Experimental set-ups of AFM force spectroscopy. (**b**) AFM force-distance curve

To illustrate its strengths and possible applications, I will give two brief representative examples for structural and energetic properties of biomolecular systems obtained from AFM force spectroscopy.

Unfolding of Titin Immunoglobulin Domains

The application of AFM to the unfolding of repeated connected domains of the large muscle protein titin has become maybe *the* classical AFM force spectroscopy experiment (Fig. 3a). In the force curve data, a typical sawtooth pattern can be observed, reflecting the serial complete and abrupt unfolding of the individual immunoglobulin domains [1]. The rupture force required for domain unfolding translates into the parallel breaking of the multiple hydrogen bonds, which hold the beta sheets of these domains together. This type of experiment has provided important insight into structural and mechanical properties of proteins as well as other essential biomolecules such as DNA [2, 3].

Streptavidin-Biotin Receptor-Ligand Interactions

Streptavidin (or avidin)-biotin is probably the most thoroughly investigated receptor-ligand pair. Its strong interactions on the order of ~200 pN together with multiple ligand binding sites of (strept)avidin qualify this system as an important attachment aid for many biological in vitro experiments. In early groundbreaking AFM force spectroscopy experiments, this technique has been applied to the study of protein-protein interactions in this model receptor-ligand system (Fig. 3b) [4]. AFM force spectroscopy has since been applied to a large number of receptor-ligand systems in vitro and on living cells (reviewed, e.g., in [5, 6]). From these data, information on the energy profiles of

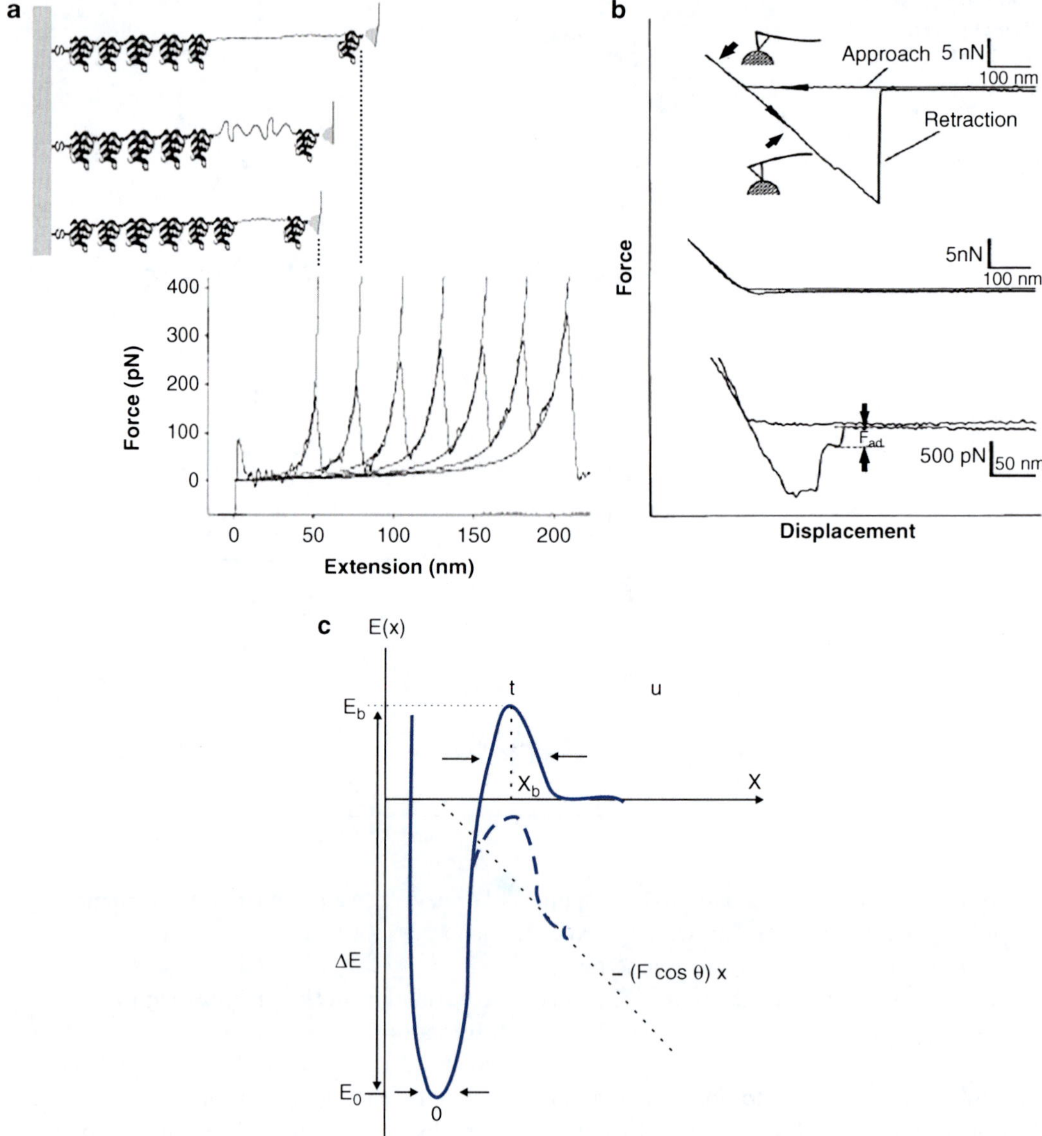

AFM Studies of Biomolecules, Fig. 3 AFM force spectroscopy example applications. (**a**) Classical AFM force spectroscopy experiment of protein unfolding on the model system of repeated immunoglobulin domains of the giant muscle protein titin (Modified from [1]; with permission, copyright @AAAS). (**b**) Force curves from the classical AFM force spectroscopy receptor-ligand system of avidin-biotin, before (*top*) and after flooding with avidin to achieve blocking of ~95 % of all interactions (for example curve shown in the *middle*) and to measure true single molecule interactions in the small percentage of curves showing interactions (*bottom*) (from [4] with permission, copyright @AAAS). (**c**) Energy profile for protein unbinding (bound state *0*, unbound state *u*, transition state *t*). Pulling the AFM tip away from the surface results in a mechanical force *F*, lowering the energy barrier E_b for unbinding along the pulling coordinate *x* by $(F\cos\theta)x$ until bond rupture occurs when $F \geq (k_BT/x_b)$, the so-called force scale (x_b is distance of E_b from the bound state *0* along *x*, k_B Boltzman constant, *T* absolute temperature). From the dependence of bond strength (rupture force) on the applied loading rate (pulling speed), Kramers-Smoluchowski theory provides the height and position of the energy barrier along *x*, mapping the energy landscape *E(x)* of the interaction

the underlying molecular interactions can be obtained (Fig. 3c) [7, 8].

AFM Imaging

For imaging, the AFM tip is scanned across the sample surface, by translating either the AFM tip or the sample stage using a piezo-scanner system. At each pixel position, the degree of deflection of the cantilever arm is detected and translated into height information, as outlined below, so that pixel by pixel and scan line by scan line a topography image of the sample is built up. High features in the sample cause strong bending of the cantilever away from the surface. Since the involved forces can cause damage to tip and sample, modern AFM systems employ a feedback mechanism to continuously reregulate the distance between tip and sample as required to maintain a constant cantilever deflection (Fig. 1).

There are three basic imaging modes, with detection based on slightly different principles, contact, intermittent contact, and noncontact mode (Fig. 4). The most intuitive approach is contact mode imaging, in which the AFM tip directly follows the sample features, often compared to a record player needle following the grooves encoded in a record. In this mode, the deflection of the cantilever directly reflects sample height, as higher features will bend the cantilever away from the surface while holes or lower features in the substrate will result in decreased bending of the cantilever or deflection towards the surface. For many imaging applications, a disadvantage of the contact mode is the lateral force exerted during scanning, so that particles that are not attached strongly enough to the surface can be pushed around by the AFM tip.

A popularly employed imaging mode for biological samples is the intermittent contact mode (also oscillating or tapping mode). In this mode, lateral force components are eliminated by oscillating the AFM tip above the sample surface, so that the tip only directly touches the surface vertically and at the minimum of its oscillation amplitude. Oscillation is mediated via piezo contacts and is applied close to the resonance frequency of the cantilever to optimize the cantilever response to tip-sample interactions. This strategy hugely minimizes the time of contact between tip and sample and the total force exerted on tip and sample. Sample topography is detected as the level of amplitude reduction (or cutting) due to the presence of surface features. This imaging mode also offers access to the additional parameter of phase changes due to tip-sample interactions, indicative of material properties (adhesion, friction).

The third mode, noncontact imaging, also oscillates the AFM tip above the sample, albeit with very small amplitude, to completely avoid direct tip-sample contact. Detection in this mode is based on an increase in oscillation amplitude due to attractive tip-sample interactions. While non-contact imaging is the most gentle imaging mode, it often offers poorer image resolution because of the weakness of the responsible attractive forces compared to the repulsive forces in intermittent contact mode imaging.

Like AFM force spectroscopy, contact mode imaging requires the high spring constants, i.e. high flexibility of Si_3N_4 cantilevers. In contrast, AFM tips and cantilevers for non-contact and intermittent contact imaging applications are typically made from silicon, with higher material stiffness allowing for faster scanning.

In all three imaging modes, image resolution is dictated by the size of the AFM tip, with typical radii of curvature (or sharpness) of between 1 and 20 nm. Combined with its applicability in liquid environment, its high resolution renders AFM imaging ideally suited for studies of diverse biological structures, from whole cells to individual biomolecules. Typical diameters of protein molecules, for instance, are between 5 and 50 nm, allowing for the study of proteins and their interactions at the single-molecule level. The second factor to limit lateral resolution is the size of the image pixels. Image size is chosen as required by the sample, for instance, a typical image size for a protein sample will be $2 \times 2\ \mu m^2$ with typically $1{,}024 \times 1{,}024$ pixels^2, corresponding to a pixel resolution of approximately 2 nm/pixel. Again, the following short examples of applications of topographical AFM imaging may serve to demonstrate its versatility.

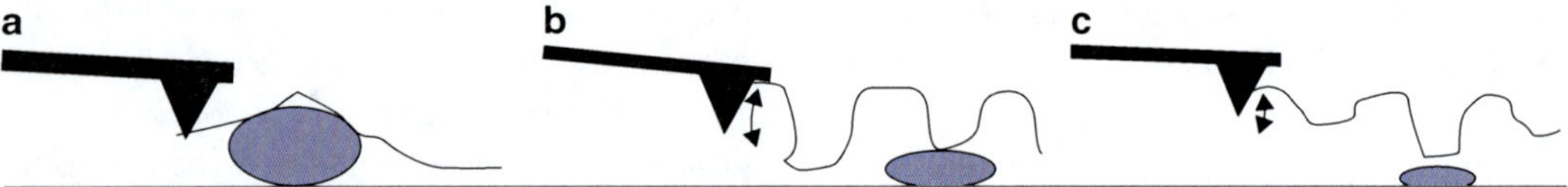

AFM Studies of Biomolecules, Fig. 4 **AFM imaging modes.** (**a**) contact, (**b**) intermittent contact, and (**c**) non-contact mode scanning

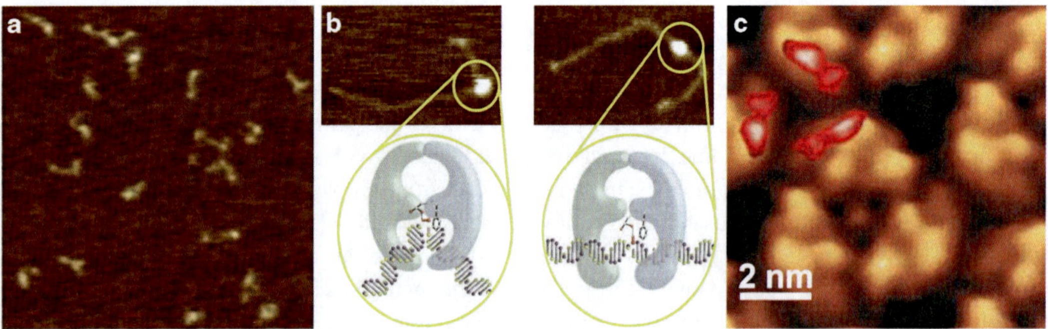

AFM Studies of Biomolecules, Fig. 5 **Examples of AFM imaging applications.** (**a**) Direct visualization of protein molecules by AFM provides insight into their conformational properties; these enzymes, for instance, are in an open, substrate accessible conformation under the applied conditions and show a large level of heterogeneity indicating high flexibility in the molecules [9]. Image size is 250 × 250 nm. (**b**) Quantitative analysis of protein-DNA complexes reveal important structural information that helps interpreting their intricate functions and mechanisms, for example shown here the target site search mechanism of the DNA mismatch repair protein MutS [11]. DNA fragments in the images are approximately 270 nm long. (**c**) A priori symmetry information and averaging applied to high resolution AFM data of bacteriorhodopsin membrane proteins (with permission from [12], copyright© Cell Press, Elsevier)

AFM Topography of Individual Proteins

AFM images provide structural information on proteins and their interactions with other proteins or other molecules such asDNA. From visual inspection of the data, conformational properties of proteins can often be directly derived, as shown for example in Fig. 5a [9]. Quantitative statistical analyses allow for the extraction of further information on protein states and interactions. From the measured volume of particles in the images, their molecular mass can be calculated [10]. This calculation is based on an empirically derived linear relationship obtained with calibration proteins with a range of known molecular weights. From the molecular mass of the protein molecules and complexes, information can be derived on protein oligomeric states, protein-protein interactions, and complex stoichiometries. Single molecule resolution coupled with the analysis of statistical numbers of molecules or molecular assemblies supplies distributions of different states in the investigated system as well as the differential population of these states. Subtle variations in the system, for example changes in incubation conditions, the introduction of a specific point mutation in a protein or modification of a target site in a DNA substrate recognized by a particular protein, allow the extraction of functional mechanisms from these structural data. For instance, AFM imaging has revealed a bending-unbending mechanism in the target site search by the DNA mismatch repair protein MutS (Fig. 5b) [11].

High Resolution Imaging of Ordered Membrane Proteins

Ordered 2-dimensional arrangements of proteins can be imaged by AFM in liquid (buffer) at very low force (<1nN). Power spectra calculated from AFM images of these regular arrays of molecules indicate subnanometer resolution. Such regular structures of proteins are formed, for instance, by bacteriorhodopsin in archaeal purple membrane patches (Fig. 5c) [12]. Averaging of multiple unit cells further enhances resolution and provides error signals showing areas of flexibility in the proteins (red-white in Fig. 5c).

Future Perspectives

The high topographical and force resolution of AFM provides insight into structure and interactions of biological systems at the single molecule level. The resulting statistical analyses reflect energetic and conformational equilibria in the investigated systems and allow the extraction of information, for instance on reaction pathways or molecular flexibility. In recent years, there has also been much effort to develop combinations of AFM with alternative detection methods, providing an additional dimension of information on the sample. For example, combination of AFM topography and recognition imaging [5] or AFM and fluorescence microscopy [13] allow the specific identification of molecules in the images, and the combination of AFM with Raman spectroscopy provides insight into the chemical composition of particular topographical features. Integrative setups are already commercially available for Raman-AFM and fluorescence-AFM combinations from different AFM companies. For the future, exciting hybrid applications of AFM both in the field of force measurements and imaging can be expected.

References

1. Rief M, Gautel M, Oesterhelt F, Fernandez JM, Gaub HE (1997) Reversible unfolding of individual titin immunoglobulin domains by AFM. Science 276:1109–1112
2. Forman JR, Clarke J (2007) Mechanical unfolding of proteins: insights into biology, structure and folding. Curr Opin Struct Biol 17:58–66
3. Rief M, Clausen-Schaumann H, Gaub HE (1999) Sequence-dependent mechanics of single DNA molecules. Nat Struct Biol 6(4):346–349
4. Florin E-L, Moy VT, Gaub HE (1994) Adhesion forces between individual ligand-receptor pairs. Science 264:415–417
5. Hinterdorfer P, Dufrêne YF (2006) Detection and localization of single molecular recognition events using atomic force microscopy. Nat Methods 3(5):347–355
6. Müller DJ, Helenius J, Alsteens D, Dufrêne YF (2009) Force probing surfaces of living cells to molecular resolution. Nat Chem Biol 5(6):383–390
7. Evans E (2001) Probing the relation between force – lifetime – and chemistry in single molecular bonds. Annu Rev Biophys Biomol Struct 30:105–128
8. Dudko OK, Hummer G, Szabo A (2008) Theory, analysis, and interpretation of single-molecule force spectroscopy experiments. PNAS 105(41):15755–15760
9. Lemaire PA, Tessmer I, Craig C, Erie DA, Cole J (2006) Unactivated PKR exists in an open conformation capable of binding nucleotides. Biochemistry 45(30):9074–9084
10. Ratcliff GC, Erie DA (2001) A novel single-molecule study to determine protein-protein association constants. J Am Chem Soc 123:5632–5635
11. Tessmer I, Yang Y, Du C, Zhai J, Hsieh P, Hingorani MM, Erie DA (2008) Mechanism of MutS searching for DNA mismatches and signaling repair. J Biol Chem 283(52):36646–36654
12. Müller DJ, Sass H-J, Müller SA, Büldt G, Engel A (1999) Surface structures of native bacteriorhodopsin depend on the molecular packing arrangement in the membrane. J Mol Biol 285:1903–1909
13. Fronzcek DN, Quammen C, Wang H, Superfine R, Taylor R, Erie DA, Tessmer I (2011) High accuracy FIONA-AFM hybrid imaging. Ultramicroscopy 111:350–355
14. Allen S, Davies MC, Roberts CJ, Tendler SJB, Williams PM (1997) Atomic force microscopy in analytical biotechnology. Trends Biotechnol 15(3):101–105
15. Costa LT, Thalhammer S, Heckl WM (2004) Atomic force microscopy as a tool in nanobiology – part II: force spectroscopy in genomics and proteomics. Cancer Genomics Proteomics 1:71–76

Alkali and Alkaline Earth Metal Production by Molten Salt Electrolysis

Geir Martin Haarberg
Department of Materials Science and Engineering, Norwegian University of Science and Technology (NTNU), Trondheim, Norway

Introduction

Alkali and alkaline earth metals are reactive metals, and their compounds are very stable.

The production of pure alkali metals is difficult due to their extreme reactivity with commonly used substances, such as water. The alkali metals are so reactive that they cannot be displaced by other elements, and molten salt electrolysis is therefore an option for producing many of these metals. Sodium and lithium are the most important alkali metals produced by electrolysis, while magnesium and calcium are the most important alkaline earth metals produced by electrolysis. For calcium and magnesium, thermal processes are currently more important than electrowinning. Magnesium, being the only of these metals that can be used for structural purposes, is a major industrial product. Therefore, a separate chapter is devoted to the electrolysis process for magnesium.

The annual production of sodium is around 100,000 t. The world capacity for lithium production is ~25,000 t per year, while ~20,000 t of calcium is produced annually.

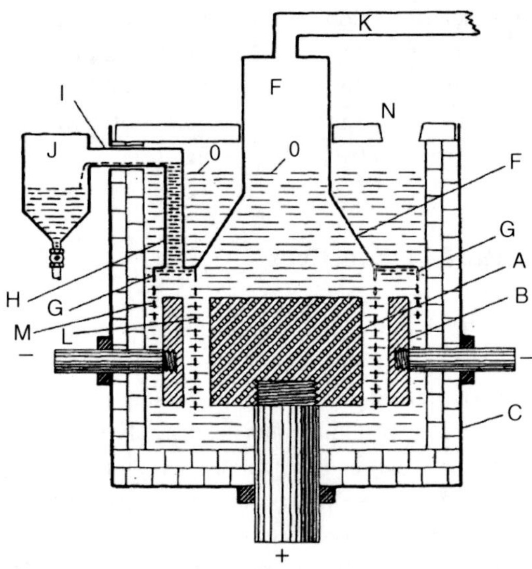

Alkali and Alkaline Earth Metal Production by Molten Salt Electrolysis, Fig. 1 The Downs cell for sodium production by electrolysis [2]. *A* anode, *B* cathode, *G* diaphragm, *H*, *J* sodium collector, *K* chlorine outlet

Technology

In 1888, the Castner process was industrialized for the production of sodium by electrolysis in molten NaOH at ~330 °C [1]. The Downs cell for the electrolytic production of sodium was patented in 1922 [2], and soon after, it replaced the Castner process for Na production, and a modified Downs cell is still being used industrially. The original Downs cell is shown in Fig. 1. A modified version of the Downs cell is shown in Fig. 2. The modified Downs cell is equipped with four anodes and four cathodes. The steel gauze diaphragm prevents direct contact between the products; chlorine bubbles evolved on graphite anodes and liquid sodium deposited on steel cathodes.

The modified Downs cell is equipped with four anodes and four cathodes, and one version is shown in Fig. 2. The steel gauze diaphragm prevents direct contact between chlorine bubbles and liquid sodium.

The traditional Downs cell for sodium electrolysis was operated at ~50 kA in molten $CaCl_2$–NaCl (60–40 wt%) at ~580 °C. Modern cells are using a modified molten chloride electrolyte (28 wt% NaCl – 25 wt% $CaCl_2$ – 47 wt% $BaCl_2$) at ~600 °C [3]. The main challenge for successfully performing the electrolysis is the fact that both products are lighter than the electrolyte, and any contact between them will lead to an exothermic reaction producing NaCl. The cathodic current density is ~1 A/cm^2 while the current load may be ~45 kA. The cell voltage is ~7 V and the specific energy consumption is typically ~10 kWh/kg Na. The current efficiency may be as high as 90 %.

Some calcium will codeposit at the cathode. Deposited and unalloyed Ca will mix with the electrolyte because of the higher density than Na. The following reaction occurs to the right hand:

$$Ca + 2NaCl = CaCl_2 + 2Na \quad (1)$$

The purity of produced Na is high, and most of the calcium is removed by filtration following electrolysis.

Electrolysis is not the preferred process for producing potassium, mainly because of the

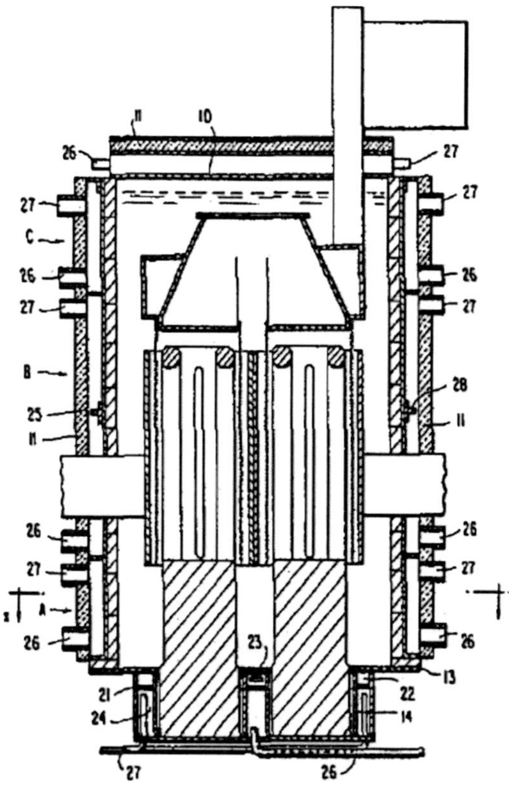

Alkali and Alkaline Earth Metal Production by Molten Salt Electrolysis, Fig. 2 A modified version of the Downs cell for sodium electrowinning [3]

high solubility and electronic conductivity in molten chloride electrolytes. Potassium may be produced through electrolysis of potassium hydroxide, much like the original process developed by Humphrey Davy [4].

Lithium metal is produced electrolytically in a Downs cell from a mixture of molten lithium chloride and potassium chloride at ~450 °C.

Calcium is produced by electrolysis in molten hydroxide electrolytes or by a thermal process. In the case of electrolysis in molten hydroxide electrolyte, hydrogen evolution is likely to be an important cathodic side reaction.

Current Efficiency

Metals are soluble in molten salts. Alkali halides and alkali metals as well as alkaline earth halides and alkaline earth metals constitute real binary systems as shown in Fig. 3 [5]. There are mutual solubilities at temperatures above the critical temperature for miscibility. However, in some cases such as the cesium systems, there is no miscibility gap. In some cases, dissolution of metals gives rise to electronic conduction, and the alkali systems are typical of this class of mixtures of molten salts and metals. Figure 4 shows specific electrical conductivity measured in mixtures of molten binary sodium and potassium systems in the salt-rich region. The electronic conductivity is dominating at a few mol% of dissolved metal. Therefore, the current efficiency may be very low. Newer experimental results have been published more recently for many of the binary alkali halide alkali metal systems [6].

The current efficiency for metal deposition may be around 90 %. The main reason for loss in current efficiency is the back reaction between dissolved metals and chlorine produced at the anode. Also the presence of electronic conduction will cause current inefficiency. The dissolution of metals in itself will also contribute to additional loss in current efficiency. Some metal may be lost due to the reactivity of the alkali and alkaline earth elements. Some evaporation losses of metals may also take place due to the relatively high process temperatures.

The so-called metal fog may also be important for the loss in current efficiency. Metal fog consists of small metal droplets which are formed by homogeneous nucleation from a supersaturated solution of dissolved metal, and these droplets are easily consumed by chlorine from the anode.

The back reaction can be written as follows

$$\mathrm{Na\ (diss)} + 1/2\ \mathrm{Cl_2\ (diss)} = \mathrm{NaCl\ (liq)} \quad (2)$$

The reaction is likely to occur outside the diffusion layer near the cathode, so that the diffusion of dissolved Na is the rate-determining step. Hence, the amount of dissolved Na in the bulk of the electrolyte is reduced very much, which means that the contribution from electronic conduction is less important for the loss in current efficiency. A relatively low content of

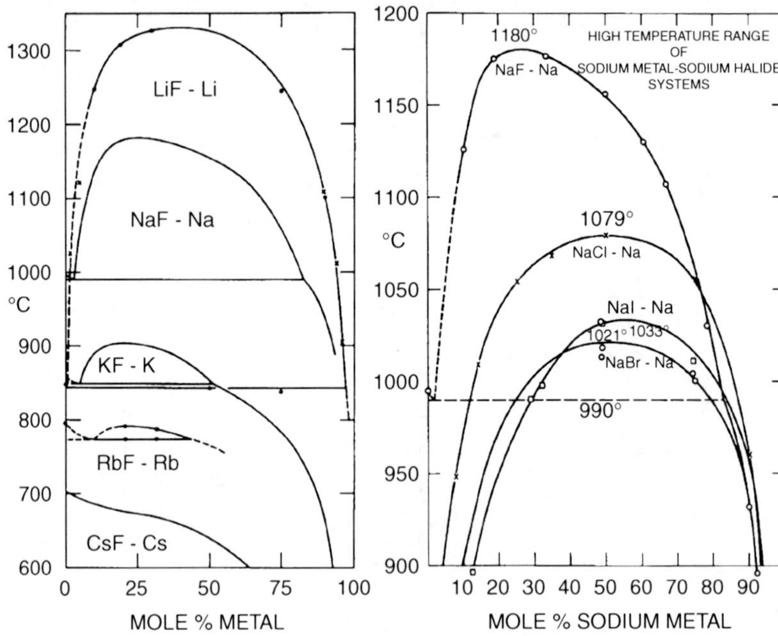

Alkali and Alkaline Earth Metal Production by Molten Salt Electrolysis, Fig. 3 Binary phase diagrams of alkali fluorides and alkali metals and sodium halides and sodium from Bredig [5]

NaCl is desired in order to reduce the solubility of Na. However, a too low content of NaCl will cause codeposition of Ca.

Electrode Reactions

Chlorine gas is evolved on graphite anodes. Normally, a relatively high overvoltage of several hundred millivolts may be expected. Any moisture present in the electrolyte will give rise to the formation of dissolved oxygen containing species which will be oxidized at the anode. Hence, the graphite anode is slowly consumed and must be replaced eventually. Water may hydrolyze according to

$$H_2O + 2NaCl = 2HCl + Na_2O \quad (3)$$

Dissolved Na_2O may be oxidized at the carbon anode:

$$Na_2O + 1/2\ C = 1/2\ CO_2 + 2\ Na^+ + 2\ e^- \quad (4)$$

Dissolved HCl may be reduced to hydrogen at the iron cathode.

Anode effect is a phenomenon that may occur at high current density. During anode effect, the anode is covered by a gas film due to deterioration of the wetting of the anode by the electrolyte, and this will cause an increased anode potential.

The cathode process for sodium deposition is very fast, and very low overvoltage is expected at normal current densities. The kinetics of the process was determined by AC impedance measurements at high frequencies combined with chronopotentiometry at short times [7].

Future Directions

It is likely that more efficient electrolysis processes will be developed.

Alkali metals are excellent battery materials, and batteries are likely to become more important. Therefore, the market for lithium and sodium should see an increase.

Alkali and Alkaline Earth Metal Production by Molten Salt Electrolysis, Fig. 4 Specific electrical conductivity of salt rich mixtures of NaCl-Na and KCl-K from Bredig [5]

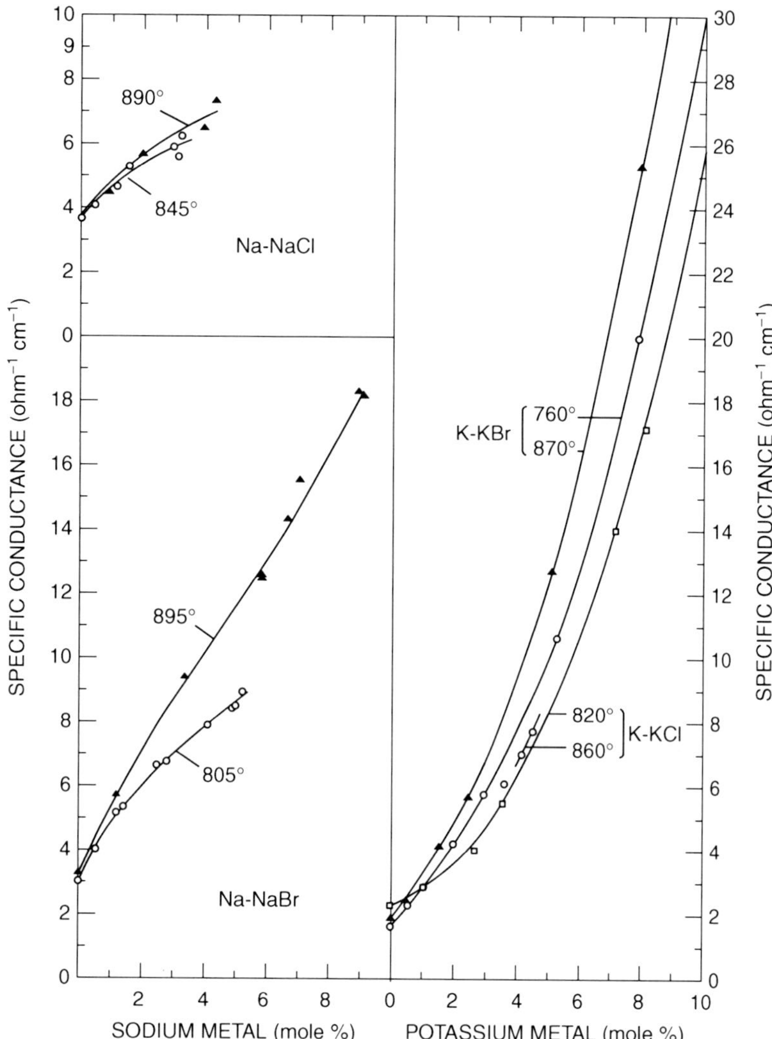

Cross-References

▶ Refractory Metal Production by Molten Salt Electrolysis

References

1. Castner HY (1891) US Patent 452030
2. Downs JC (1924) Electrolytic process and cell. US Patent 1501756
3. Hinrichs W, Lange L, Hammer F, Ludwig H (1986) Device and process for the fused-salt electrolysis of alkali metal halides. US Patent 4584068
4. Klemm A, Hartmann G, Lange L (2012) Sodium and sodium alloys. In: Ullmann's encyclopedia of industrial chemistry, vol 33, p 275. http://en.wikipedia.org/wiki/Humphry_Davy
5. Bredig MA (1964) In: Blander M (ed) Molten salt chemistry. Interscience, New York, p 367
6. Haarberg GM, Egan JJ (1996) Electron mobilities in solutions of alkali metals in molten alkali halides. In: Proceeding of the Tenth International Symposium on Molten Salts, vol. 96-7. Los Angeles, The Electrochemical Society, Pennington, p 468
7. Kisza A, Kazmierczak J, Børresen B, Haarberg GM, Tunold R (1995) The kinetics of the sodium electrode reaction in molten sodium chloride. J Electrochem Soc 142(4):1035

Alkaline Membrane Fuel Cells

Dario Dekel
CellEra Inc., Caesarea, Israel

This chapter reviews a new type of solid electrolyte low-temperature fuel cell, the alkaline membrane fuel cell. The principles and main components of this fuel cell technology are described, with a major focus on the electrocatalysts for both electrodes. Finally, the latest published results on operation of the first developed alkaline membrane fuel cells are reviewed.

Introduction

The alkaline membrane fuel cell (AMFC) technology, also referred to as anion exchange membrane fuel cell (AEMFC), is in principle similar to the proton exchange membrane fuel cell (PEMFC), with the main difference that the solid membrane is an anion exchange membrane (AEM) instead of a proton exchange membrane (PEM). With an anion exchange membrane in the alkaline membrane fuel cell, the OH^- is being transported from the cathode to the anode, opposite to the H^+ conduction direction in the proton exchange membrane fuel cell. The schematic diagram below shows this main difference between the proton exchange membrane fuel cell and the anion exchange membrane fuel cell.

As it can be seen in Fig. 1, in the case of a proton exchange membrane fuel cell (left scheme), the H^+ conducts through a solid cation conducting membrane (called proton exchange membrane) from the anode to the cathode, while in the case of an alkaline membrane fuel cell (right scheme) the OH^- is transported through a solid anion conducting membrane (called anion exchange membrane) from the cathode to the anode.

Fuel cell reactions for both types of fuel cells are compared and described below:

PEMFC	AMFC	
(1) $H_2 \rightarrow 2H^+ + 2e^-$	$H_2 + 2OH^- \rightarrow 2H_2O + 2e^-$	anode
(2) $\frac{1}{2} O_2 + 2e^- + 2H^+ \rightarrow H_2O$	$\frac{1}{2} O_2 + H_2O + 2e^- \rightarrow 2OH^-$	cathode
(3) $H_2 + \frac{1}{2} O_2 \rightarrow H_2O$	$H_2 + \frac{1}{2} O_2 \rightarrow H_2O$	overall reaction

Although the overall reaction (3) is the same for both types of fuel cells, some differences can be observed in the fuel cell electrode reactions (1) and (2):

(a) The transported ion through the membrane is a proton in the case of PEMFC, while it is an OH^- in the case of AMFC;
(b) While in PEMFC water is generated on the cathode side, in AMFC water is generated in the fuel side (1);
(c) While in PEMFC there is no need for a water reactant to fulfill fuel cell reactions, in AMFC water is a reactant (2).

In principle, the advantages of alkaline membrane fuel cells over proton exchange membranes are related to the slightly alkaline pH cell environment of the AMFC and mainly include (a) enhanced electrokinetics of the oxygen reduction reaction, allowing for the use of cheaper, non-platinum metal catalysts like manganese, iron or silver; (b) extended range of available cell and stack materials; and (c) wider choice of fuels, in addition to hydrogen, such as borohydride, hydrazine, and ammonia.

However, although the alkaline membrane fuel cell combines all the advantages of the mature liquid alkaline fuel cell (AFC) and proton exchange membrane fuel cell technologies, some problems remain and there are still concerns regarding the performance that can be achieved by alkaline membrane fuel cells. These concerns seem to be mostly related to the low OH^- conductivities exhibited in the anion exchange membranes developed during the initial years of the alkaline membrane fuel cell research [1]. Nevertheless, while these issues do exist, high conductive anion exchange membranes have been developed in recent years [2], and the alkaline membrane fuel cell technology is now becoming a major focus of research and development [3].

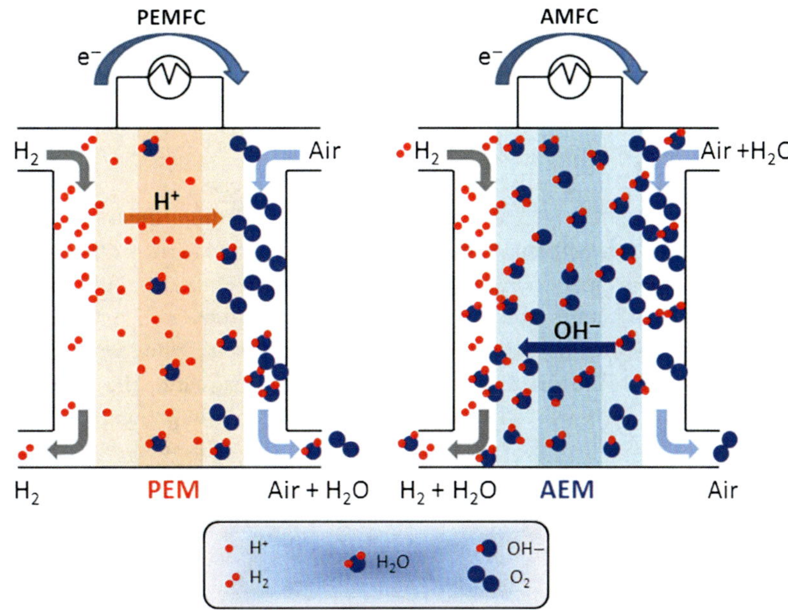

Alkaline Membrane Fuel Cells, Fig. 1 Schematic diagrams of a proton exchange membrane fuel cell, PEMFC (*left*), and an alkaline membrane fuel cell, AMFC (*right*)

In the following sections, a general review of the state of the art in alkaline membrane fuel cell technology is presented, and anion exchange membranes and electrocatalysts for this new technology are described. In addition, a short review of today's alkaline membrane fuel cell performance is also presented.

Anion Exchange Membranes

Due to the ample potential of fuel cell commercialization using non-platinum alkaline membrane fuel cells, studies on anion exchange membranes for this technology are now an emerging field both in terms of research and development.

As it has been pointed out in past years [1], the most critical concerns of the alkaline membrane fuel cell technology were the low conductivity and the poor stability of the early anion exchange membranes. In past years, significant advances were achieved [3, 4], promoting the development of the alkaline membrane fuel cell technology.

A thorough description of the anion exchange membranes is discussed in detail in a chapter within this encyclopedia [2].

Electrocatalysts for Hydrogen-Air Alkaline Membrane Fuel Cells

With these latest advances in anion exchange membranes, the need for suitable catalysts for alkaline membrane fuel cells gains salience. Keeping the advantages of the solid electrolyte polymer membrane-based fuel cells, alkaline membrane fuel cell technology opens the door for the use of non-precious metal catalysts, having the potential to overcome the high fuel cell barrier costs. However, the field of electrocatalysts for both cathode and anode in alkaline membrane fuel cell technology is only now being explored [5]. In the following sections, catalyst developments for oxygen reduction reaction and hydrogen oxidation reaction in alkaline membrane fuel cell are briefly reviewed.

Oxygen Reduction Reaction (ORR) Catalysts

The oxygen reduction reaction overpotential loss in alkaline membrane fuel cells is very similar to that in proton exchange membrane fuel cells, i.e., the cathode overpotential loss remains an important factor limiting the efficiency and performance of alkaline membrane fuel cells [6, 7]. However, switching to an alkaline medium as in alkaline membrane fuel cells allows for the use of a longer list of non-Pt and non-PGM catalysts (non-platinum group metal) with an oxygen reduction reaction activity similar to that of Pt.

Many publications can be found in the literature on the preparation and testing of non-Pt and non-PGM catalysts [8–10]; however, most of the research has been focused on acidic medium for proton exchange membrane fuel cells. There are limited data on catalyst evaluation in alkaline medium [11–14] and even less on catalyst evaluation in real alkaline membrane fuel cells.

Jiang et al. [15] reported that the ORR activities in alkaline medium of a Pd-coated Ag/C were three times higher than the corresponding activities on the Pt/C, as measured in their rotating disk electrode tests. Piana et al. [16] reported that the specific current of their K18 non-PGM catalyst is about three times higher than Pt/C and also its Tafel slope is lower, as it is observed for other non-Pt catalysts [17]. He et al. [18] reported that the kinetic current density of their non-noble metal catalyst based CuFe-Nx/C material was comparable or even higher than a commercial Pt/C catalyst.

Mamlouk et al. [19] tested their Co-based catalysts for oxygen reduction reaction in both liquid KOH rotating disk electrode and alkaline membrane fuel cell tests. In rotating disk electrode in alkaline solution medium, the authors reported that the catalysts exhibited relatively good performance with onset potentials 120 mV more negative than platinum. In alkaline membrane fuel cell tests, they reported a promising oxygen reduction reaction activity, at 50 mA/cm^2 differences in cell potential between Pt- and Co-based cathodes. However, the overall performance of the Pt-based cell used as reference was significantly lower than the one reported elsewhere based on Pt catalysts (for instance, [20]).

Sleightholme et al. [21] studied the oxygen reduction reaction of silver in alkaline solution as well as in anion exchange membranes. By using cyclic voltammetry, they reported that oxygen reduction currents were lower in the Ag/anion exchange membrane system than in the Ag/NaOH system, attributing that to the lower oxygen permeability of the anion exchange membrane. Unfortunately, the authors did not report data on oxygen reduction reaction activity and power performance of the silver in a working alkaline membrane fuel cell.

In summary, little work has been done in past years to better understand electrocatalysts in alkaline medium and to develop new effective catalysts for this new and promising alkaline membrane fuel cell technology. Moreover, almost no work has been done in real alkaline membrane fuel cells with catalysts other than platinum. The development of non-Pt and non-Pt group metal (non-PGM) catalysts toward oxygen reduction reaction for alkaline membrane fuel cells now requires further research in order to make this a real affordable technology.

Hydrogen Oxidation Reaction (HOR) Catalysts

Whereas research on oxygen reduction reaction catalysts in the alkaline medium has only now begun, studies on hydrogen oxidation reaction catalysts for alkaline membrane fuel cells constitute an entirely unexplored field.

The kinetics of the HOR on Pt catalysts in a PEMFC are so fast that the cell voltage losses at the anode are negligible [22]. This seems not to be the case in an alkaline membrane fuel cell [3].

In one of the very few studies on hydrogen oxidation reaction activities of platinum in both acidic and alkaline media, Sheng et al. [6] found that in alkaline electrolyte, the hydrogen oxidation reaction kinetics are several orders of magnitude slower than in acid electrolyte. Recently, this finding has been confirmed and quantified by Rheinländer et al. [23] who reported that ultra low loadings Pt in alkaline medium may exhibit a prohibited potential loss of about 150 mV. Moreover, when looking for non-Pt catalyst candidates toward HOR, it was found that Pd-based catalysts exhibit 5–10 times lower activity than Pt in alkaline medium [23]. Those fundamental studies of hydrogen oxidation reaction for alkaline membrane fuel cells highly emphasize the need for alternative, inexpensive HOR catalysts for the successful development of the alkaline membrane fuel cell technology.

One of the very few studies investigating non-Pt group metal (non-PGM) catalysts for hydrogen oxidation reaction in H_2/air alkaline membrane fuel cells was carried out by Lu et al. [24]. In a full study that involved the development of anion exchange membranes and ionomers, the author reported that by decorating Ni nanoparticles with Cr, they succeeded to tune the electronic surface of Ni, making it possible to operate in anodes of alkaline membrane fuel cells. Although the authors did not report any ex situ activity measurements, they reported a single preliminary test in real alkaline membrane fuel cells, showing a maximum peak power of 50 mW/cm^2 obtained at a cell temperature of 60 °C [25]. Although the power density is very low, this is one of the first published data using non-PGM catalysts for the anodic hydrogen oxidation reaction, demonstrating the potential of the alkaline membrane fuel cell technology.

Alkaline Membrane Fuel Cell Performance

Hydrogen-Based Alkaline Membrane Fuel Cells

Recent developments in highly conductive anion exchange membranes have contributed to the growing interest in alkaline membrane fuel cell technology. Alkaline membrane fuel cell first results have been published in recent years. Some of them show a performance high enough to be seriously considered as an indicator of the potential of alkaline membrane fuel cell technology for practical applications. Table 1 summarizes the alkaline membrane fuel cell results reported in the literature on H_2/O_2 and H_2/air.

As can be seen in Table 1, maximum power densities of up to 200 mW/cm^2 have been obtained. The increasing interest in alkaline membrane fuel cell technology is now encouraged by these promising results.

One of the best indicators of the potential of this flourishing technology comes from the industrial sector. Yanagi and Fukuta [20] showed a maximum peak power of 450 mW/cm^2 and 340 mW/cm^2 for H_2/O_2 and H_2/CO_2-free air, respectively, obtained at 50 °C. Even though those results were obtained with Pt/C catalysts (0.5mg_{Pt}/cm^2), they show a power high enough for practical applications, such as backup power for stationary applications.

High peak power density results were also shown by Gottesfeld [33], who reported results obtained at CellEra [34], which pioneers work toward the development and commercialization of alkaline membrane fuel cell technology. The author reported a peak power density of 750 mW/cm^2 (at 1.7A/cm^2) and 550 mW/cm^2 (at 1.4A/cm^2) with 2 bar of H_2/O_2 and H_2/air, respectively, at 80 °C.

At the same time, Kim [35] reported interesting results with his polyphenylene-based membranes. With 3 mg/cm^2 Pt black catalyst in both anode and cathode, the author reached a maximum power density of 577 mW/cm^2 (at 1A/cm^2) and 450 mW/cm^2 (at ~0.8A/cm^2) at 80 °C under H_2/O_2 and H_2/air conditions, respectively. This power density is very similar to the one achieved and reported by Yanagi et al. [20, 36].

Although these Pt-based alkaline membrane fuel cell results already show good performances enough for practical fuel cell applications [36], the need for alternative, inexpensive catalysts for the development of the alkaline membrane fuel cell technology should be remarked. One of the very few results presented on hydrogen-based alkaline membrane fuel cell based on non-Pt catalysts was recently obtained at CellEra [34]. Dekel [3, 5] reported a peak power density of 700 mW/cm^2 (at 1.5 A/cm^2) with 3barg/1barg of H_2/air at 80 °C for an alkaline membrane fuel cell based on Pt catalyst at anode, and a peak power density of 500 mW/cm^2 (at 1.6 A/cm^2) with 3barg/1barg of H_2/air at 80 °C for an entirely non-Pt alkaline membrane fuel cell.

In summary, extremely rapid and promising achievements in alkaline membrane fuel cell technology have been shown. Altogether, based on the recent results obtained with first prototypes, it seems that alkaline membrane fuel cell technology is not just a future promise but a present reality.

Alkaline Membrane Fuel Cells, Table 1 Peak power densities measured in lab-scale hydrogen-based alkaline membrane fuel cells

Author(s) (Reference)	Max. power density (mW/cm^2)	Current density[a] (mA/cm^2)	Gases	Cell temp. (°C)	Catalyst loading-type (mg/cm^2) Anode		Cathode	
Jung et al. [26]	77	158	H_2/air	70	0.5	Pt	0.5	Pt
Tang et al. [25]	70	170	H_2/O_2	70	0.4	Pt/C	0.4	Pt/C
Luo et al. [27]	180	400	H_2/O_2 1 bar	70	0.4	Pt/C	0.4	Pt/C
Switzer et al. [28]	~90	~170	H_2/O_2	80	2	Pt	2	Pt
	~40	~60	H_2/air 2 bar					
Gu et al. [29]	138	~280	H_2/O_2	50	0.2	Pt	0.2	Pt
	196	~380	2.5 bar	80	0.5	Pt	0.5	Pt
Park et al. [30]	28	60	H_2/air	60	0.5	Pt/C	0.5	Pt/C
Park et al. [31]	30	~65	H_2/air	60	0.5	Pt/C	2	Ag/C
Varcoe et al. [32]	94	~280	H_2/O_2	50	4	PtRu	4	Pt
Lu et al. [24]	50	100	H_2/O_2 1.3 bar	60	5	Ni-Cr	1	Ag

[a]Measured at max. peak power

Alkaline Membrane Fuel Cells Based on Other Fuels

As it was already mentioned, one of the advantages of alkaline membrane fuel cell technology is that different fuels other than hydrogen can be used. The alkaline character of the alkaline membrane fuel cell allows working with fuels such as borohydride and ammonia, which are prohibitive for the acidic alkaline membrane fuel cells (see for instance, [37]). An additional positive characteristic of alkaline membrane fuel cells is that the direction of OH^- transport across the anion exchange membrane opposes the fuel flux, thereby reducing fuel crossover. Furthermore, as water is being generated at the anode side in alkaline membrane fuel cells, system complexity is reduced as water is removed by the aqueous fuel stream.

Since a discussion of alkaline membrane fuel cells based on fuels other than hydrogen is beyond the scope of this work, only a short list of references is provided as a suggestion for further readings. Matsuoka et al., among others, showed interesting first results for alcohol-based AMFCs [32, 38–43]. Some initial work has also been done with alkaline membrane fuel cells based on glycerin [44], hydrazine [45, 46] and even on ammonia [47, 48]. Finally, an investigation of alkaline membrane fuel cells based on the popular borohydride was recently presented by Qu et al. [49].

Future Directions

Alkaline membrane fuel cells have been rapidly developed within the past 5 years. This technology promises to solve the cost barriers of the proton exchange membrane fuel cell, which is the main pain point of fuel cell technology.

Latest improvements done in alkaline membrane fuel cell technology on the whole are more than impressive. While hydrogen-air alkaline membrane fuel cells achieved a practical performance for few applications, there are yet development challenges which must be addressed prior to their large-scale introduction. These development challenges mainly include (a) anion exchange polymers (membrane and ionomers) with improved stability at higher temperatures and (b) platinum-free highly efficient catalysts toward hydrogen oxidation reaction.

Advances in developments in those two fields will assure rapid entrance to existing market opportunities. The first alkaline membrane fuel cell commercial developer already exists. It will be interesting to see which companies will ultimately decide to switch to this emerging and promising alkaline membrane fuel cell technology and join current efforts to make fuel cells a real affordable technology.

Cross-References

▶ Alkaline Membrane Fuel Cells, Membranes
▶ Anion-Exchange Membrane Fuel Cells, Oxide-Based Catalysts

References

1. Pivovar BS (2006) Alkaline membrane fuel cell workshop final report. http://www1.eere.energy.gov/hydrogenandfuelcells/pdfs/amfc_dec2006_workshop_report.pdf. Accessed 6 Nov 2012
2. Dekel D (2012) Alkaline membrane fuel cells: Membranes. In: Savinell R, Ota K, Kreysa G (ed) Encyclopedia of applied electrochemistry. Springer, Berlin/Heidelberg. www.springerreference.com, doi:10.1007/SpringerReference_349797. Accessed 15 Dec 2012
3. Dekel DR (2012) Latest advances in Alkaline Membrane Fuel Cell (AMFC) technology. In: Carisma 2012: 3rd Carisma international conference, September 3rd 2012, Copenhagen. http://www.hotmea.kemi.dtu.dk/upload/institutter/ki/hotmea/carisma%202012/abstracts/dekel%20carisma%202012.pdf. Accessed 21 Dec 2012
4. Pivovar BS (2011) Alkaline membrane fuel cell workshop final report. http://www.nrel.gov/docs/fy12osti/54297.pdf. Accessed 6 Nov 2012
5. Dekel D (2012) Alkaline Membrane Fuel Cell (AMFC) materials and system improvement – state-of-the-art. ECS Trans 50:2051–2052
6. Sheng W, Gasteiger HA, Shao-Horn Y (2010) Hydrogen oxidation and evolution reaction kinetics on platinum: acid vs alkaline electrolytes. J Electrochem Soc 157(11):B1529–B1536. doi:10.1149/1.3483106
7. Strmcnik D, Kodama K, van der Vliet D, Greeley J, Stamenkovic VR, Marković NM (2009) The role of non-covalent interactions in electrocatalytic fuel-cell reactions on platinum. Nat Chem 1(6):466–472
8. Zelenay P, Brosha E, Choi JH, Davey J, Garzon F, Hamon C, Piela B, Ramsey J, Uribe F (2005) Non-precious metal catalysts. DOE hydrogen program, FY 2005 progress report. http://www.hydrogen.energy.gov/pdfs/progress05/vii_c_7_zelenay.pdf. Accessed 13 Nov 2012
9. Wu G, More KL, Johnston CM, Zelenay P (2011) High-performance electrocatalysts for oxygen reduction derived from polyaniline, iron, and cobalt. Science 332:443–447. doi:10.1126/science.1200832
10. Suo Y, Zhuang L, Lu J (2007) First-principles considerations in the design of Pd-alloy catalysts for oxygen reduction. Angew Chem Int Ed 46:2862–2864. doi:10.1002/anie.200604332
11. Jiang L, Hsu A, Chu D, Chen R (2009) Oxygen reduction reaction on carbon supported Pt and Pd in alkaline solutions. J Electrochem Soc 156(3): B370–B376. doi:10.1149/1.3058586
12. Blizanac BB, Ross PN, Markovic NM (2006) Oxygen reduction on silver low-index single-crystal surfaces in alkaline solution: rotating ring disk$_{Ag(hkl)}$ studies. J Phys Chem B 110:4735–4741. doi:10.1021/jp056050d
13. Chatenet M, Genies-Bultel L, Aurousseau M, Durand R, Andolfatto F (2002) Oxygen reduction on silver catalysts in solutions containing various concentrations of sodium hydroxide: comparison with platinum. J Appl Electrochem 32:1131–1140
14. Shimizu Y (2012) Anion-exchange membrane fuel cells: oxide-based catalysts. In: Savinell R, Ota K, Kreysa G (ed) Encyclopedia of applied electrochemistry. Springer, Berlin/Heidelberg. www.springerreference.com, doi: 10.1007/SpringerReference_303656. Accessed 17 Jan 2012
15. Jiang L, Hsu A, Chu D, Chen R (2010) A highly active Pd coated Ag electrocatalyst for oxygen reduction reactions in alkaline media. Electrochim Acta 55:4506–4511. doi:10.1016/j.electacta.2010.02.094
16. Piana M, Catanorchi S, Gasteiger HA (2008) Kinetics of non-platinum group metal catalysts for the oxygen reduction reaction in alkaline medium. ECS Trans 16(2):2045–2055
17. Meng H, Jaouen F, Proietti E, Lefèvre M, Dodelet JP (2009) pH-effect on oxygen reduction activity of Fe-based electro-catalysts. Electrochem Commun 11(10):1986–1989. doi:10.1016/j.elecom.2009.08.035
18. He Q, Yang X, He R, Bueno-López A, Miller H, Ren X, Yang W, Koel BE (2012) Electrochemical and spectroscopic study of novel Cu and Fe-based catalysts for oxygen reduction in alkaline media. J Power Sources 213:169–179. doi:10.1016/j.jpowsour.2012.04.029
19. Mamlouk M, Kumar SMS, Gouerec P, Scott K (2011) Electrochemical and fuel cell evaluation of Co based catalyst for oxygen reduction in anion exchange polymer membrane fuel cells. J Power Sources 196(18):7594–7600. doi:10.1016/j.jpowsour.2011.04.045
20. Yanagi H, Fukuta K (2008) Anion exchange membrane and ionomer for alkaline membrane fuel cells (AMFCs). ECS Trans 16(2):257–262. doi:10.1149/1.2981860
21. Sleightholme AES, Varcoe JR, Kucernak AR (2008) Oxygen reduction at the silver/hydroxide-exchange membrane interface. Electrochem Commun 10:151–155. doi:10.1016/j.elecom.2007.11.008
22. Neyerlin KC, Gu W, Jorne J, Gasteiger HA (2007) A study of the exchange current density for the hydrogen oxidation and evolution reactions.

J Electrochem Soc 154:B631–B635. doi:10.1149/1.2733987
23. Rheinländer P, Henning S, Herranz J, Gasteiger HA (2012) Comparing hydrogen oxidation and evolution reaction kinetics on polycrystalline platinum in 0.1 M and 1 M KOH. ECS Trans 50(2):2163–2174
24. Lu S, Pan J, Huang A, Zhuang L, Lu J (2008) Alkaline polymer electrolyte fuel cells completely free from noble metal catalysts. Proc Natl Acad Sci 105(52):20611–20614. doi:10.1073/pnas.0810041106
25. Tang DP, Pan J, Lu SF, Zhuang L, Lu JT (2010) Alkaline polymer electrolyte fuel cells: principle, challenges, and recent progress. Sci China 53(2):357–364. doi:10.1007/s11426-010-0080-5
26. Jung MJ, Arges CG, Ramani V (2011) A perfluorinated anion exchange membrane with a 1,4-dimethylpiperazinium cation. J Mater Chem 21:6158–6160. doi:10.1039/c1jm10320b
27. Luo Y, Guo J, Wang C, Chu D (2011) Tunable high-molecular-weight anion-exchange membranes for alkaline fuel cells. Macromol Chem Phys 212:2094–2102. doi:10.1002/macp.201100218
28. Switzer EE, Olson TS, Datye AK, Atanassov P, Hibbs MR, Fujimoto C, Cornelius CJ (2010) Novel KOH-free anion-exchange membrane fuel cell: performance comparison of alternative anion-exchange ionomers in catalyst ink. Electrochim Acta 55:3404–3408. doi:10.1016/j.electacta.2009.12.073
29. Gu S, Cai R, Luo T, Chen Z, Sun M, Liu Y, He G, Yan Y (2009) A soluble and highly conductive ionomer for high-performance hydroxide exchange membrane fuel cells. Angew Chem 121:6621–6624. doi:10.1002/ange.200806299
30. Park JS, Park GG, Park SH, Yoon YG, Kim GS, Lee WY (2007) Development of solid-state alkaline electrolytes for solid alkaline fuel cells. Macromol Symp 249–250:174–182. doi:10.1002/masy.200750329
31. Park JS, Park SH, Yim SD, Yoon YG, Lee WY, Kim CS (2008) Performance of solid alkaline fuel cells employing anion-exchange membranes. J Power Sources 178:620–626. doi:10.1016/j.jpowsour.2007.08.043
32. Varcoe JR, Slade RCT, Yee ELH, Poynton SD, Driscoll DJ, Apperley DC (2007) Poly(ethylene-co-tetrafluoroethylene)-derived radiation-grafted anion-exchange membrane with properties specifically tailored for application in metal-cation-free alkaline polymer electrolyte fuel cells. Chem Mater 19(10):2686–2693. doi:10.1021/cm062407u
33. Gottesfeld S (2011) Breaking the fuel cell cost barrier. DOE AMFC workshop. http://www1.eere.energy.gov/hydrogenandfuelcells/pdfs/amfc_050811_gottesfeld_cellera.pdf. Accessed13 Nov 2012
34. www.cellera-inc.com. Accessed 22 Dec 2012
35. Kim YS (2011) Resonance-stabilized anion exchange polymer electrolytes. US DOE hydrogen and fuel cells program and vehicle technologies program annual merit review. http://www.hydrogen.energy.gov/pdfs/review11/fc043_kim_2011_o.pdf. Accessed 15 Nov 2012
36. Fukuta K (2011) Electrolyte materials for AMFCs electrolyte materials for AMFCs and AMFC performance. DOE AMFC workshop. http://www1.eere.energy.gov/hydrogenandfuelcells/pdfs/amfc_050811_fukuta.pdf. Accessed 15 Nov 2012
37. Uribe FA, Gottesfeld S, Zawodzinski TA (2002) Effect of ammonia as potential fuel impurity on proton exchange membrane fuel cell performance. J Electrochem Soc 149(3):A293–A296. doi:10.1149/1.1447221
38. Matsuoka k, Iriyama y, Abe T, Matsuoka M, Ogumi Z (2005) Alkaline direct alcohol fuel cells using an anion exchange membrane. J Power Sources 150:27–31. doi:10.1016/j.jpowsour.2005.02.020
39. Yang CC, Chiu SJ, Lin CT (2008) Electrochemical performance of an air-breathing direct methanol fuel cell using poly(vinyl alcohol)/hydroxyapatite composite polymer membrane. J Power Sources 177:40–49. doi:10.1016/j.jpowsour.2007.11.010
40. Yu EH, Scott K (2004) Direct methanol alkaline fuel cell with catalysed metal mesh anodes. Electrochem Commun 6:361–365. doi:10.1016/j.elecom.2004.02.002
41. Yang CC (2007) Synthesis and characterization of the cross-linked PVA/TiO2 composite polymer membrane for alkaline DMFC. J Membr Sci 288:51–60. doi:10.1016/j.memsci.2006.10.048
42. Huang A, Xia C, Xiao C, Zhuang L (2006) Composite anion exchange membrane for alkaline direct methanol fuel cell: structural and electrochemical characterization. J Appl Polym Sci 100:2248–2251. doi:10.1002/app.23579
43. Zhao TS, Li YS, Shen SY (2010) Anion-exchange membrane direct ethanol fuel cells: status and perspective. Front Energy Power Eng China 4(4):443–458. doi:10.1007/s11708-010-0127-5
44. Ragsdale SR, Ashfield CB (2008) Direct-glycerin fuel cell for mobile applications. ECS Trans 16(2):1847–1854. doi:10.1149/1.2982025
45. Zhang F, Zhang H, Ren J, Qu C (2010) PTFE based composite anion exchange membranes: thermally induced in situ polymerization and direct hydrazine hydrate fuel cell application. J Mater Chem 20:8139–8146. doi:10.1039/c0jm01311k
46. Tanaka M, Fukasawa K, Nishino E, Yamaguchi S, Yamada K, Tanaka H, Bae B, Miyatake K, Watanabe M (2011) Anion conductive block poly(arylene ether)s: synthesis, properties, and application in alkaline fuel cells. J Am Chem Soc 133:10646–10654. doi:10.1021/ja204166e
47. Lan R, Tao S (2010) Direct ammonia alkaline anion-exchange membrane fuel cells. Electrochem Solid State Lett 13(8):B83–B86. doi:10.1149/1.3428469
48. Suzuki S, Muroyama H, Matsui T, Eguchi K (2012) Fundamental studies on direct ammonia fuel cell employing anion exchange membrane.

J Power Sources 208:257–262. doi:10.1016/j.jpowsour.2012.02.043

49. Qu C, Zhang H, Zhang F, Liu B (2012) A high-performance anion exchange membrane based on biguanidinium bridged polysilsesquioxane for alkaline fuel cell application. J Mater Chem 22:8203–8207. doi:10.1039/c2jm16211c

Alkaline Membrane Fuel Cells, Membranes

Dario Dekel
CellEra Inc., Caesarea, Israel

In this chapter a new type of solid electrolyte membrane for low-temperature fuel cell application, the anion exchange membrane (AEM), is reviewed. The properties, advantages, and challenges of the anion exchange membranes are discussed.

Introduction

Anion exchange membranes (AEMs) are used as solid polymer electrolyte in the alkaline membrane fuel cell (AMFC) technology, also referred to as anion exchange membrane fuel cell (AEMFC). In this relatively novel fuel cell technology, the OH^- is being transported from the cathode to the anode through the anion exchange membrane. In other words, the membrane acts as an anion-conducting membrane between both electrodes of the fuel cell.

The fuel cell reactions for an alkaline membrane fuel cell are described below:

$$H_2 + 2OH^- \Rightarrow 2H_2O + 2e^- \quad \text{anode} \quad (1)$$

$$\frac{1}{2}O_2 + H_2O + 2e^- \Rightarrow 2OH^- \quad \text{cathode} \quad (2)$$

$$H_2 + \frac{1}{2}O_2 \Rightarrow H_2O \quad \text{overall reaction} \quad (3)$$

As it can be seen in these reactions, in anion exchange membrane fuel cells, two molecules of water are generated in the anode (see reaction "1"), while a molecule of water is consumed in the cathode (see reaction "2"). This needs to be taken into account when developing anion exchange membranes for alkaline membrane fuel cells, since the water distribution in the membranes is different as compared to proton conducting fuel cells. This different water distribution significantly influences the water management in the alkaline membrane fuel cell, affecting in turn the properties of the required polymer needed for the development of the anion exchange membrane.

In the following sections, a general review of the state of the art in anion exchange membranes is presented, focusing on polymer preparation and properties, as developed in recent years.

Anion Exchange Membranes: Background

Due to the ample potential of fuel cell commercialization using non-Pt alkaline membrane fuel cells, studies on anion exchange membranes for this technology are now an emerging field both in terms of research and development. Figure 1 shows the main research locations (both universities and companies) in which alkaline membrane fuel cell-related fields were or still are being developed [1]. More than 90 % of the indicated specific locations (denoted by yellow dots) focus on the development of anion exchange membranes.

In fuel cells, ionic membranes are called proton exchange membranes (PEMs) or anion exchange membranes (AEMs), depending on the type of ionic groups attached to the membrane matrix. Proton exchange membranes, also called cation exchange membranes (CEM), contain negatively charged groups, such as $-SO_3^-$, $-COO^-$, and $-PO_3^{2-}$, fixed to the polymer backbone and allow the transport of H^+. In contrast, anion exchange membranes contain positively charged groups, such as $-NR_3^+$, $-PR_3^+$, and $-SR_2^+$, fixed to the polymer backbone and allow the transport of OH^-, as it is shown in Fig. 2.

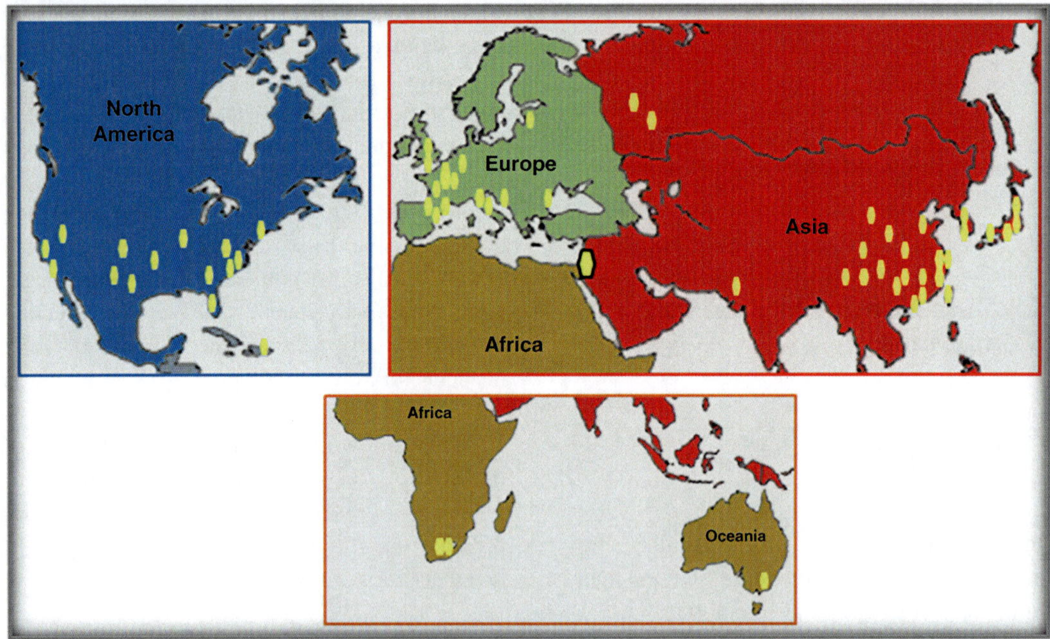

Alkaline Membrane Fuel Cells, Membranes, Fig. 1 AMFC-related research and development locations worldwide [1]

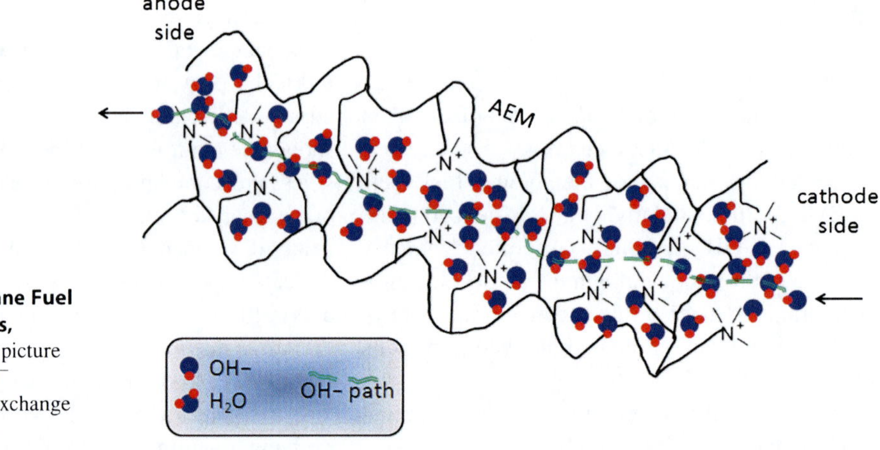

Alkaline Membrane Fuel Cells, Membranes, Fig. 2 Schematic picture of transport of OH^- through an anion exchange membrane (AEM)

As it has been pointed out in past years [2], the most critical concerns of the alkaline membrane fuel cell technology are the low conductivity and the relatively poor stability of the anion exchange membranes that exist in the first years of the alkaline membrane fuel cell development. Before discussing these main concerns, a short review of anion exchange membrane preparation methods is presented.

Preparation of Anion Exchange Membranes for Fuel Cells

The main routes that have been used to prepare anion exchange membranes can be grouped into three different approaches, each using a different starting point to eventually form the membrane [3]: (a) from a monomer containing a moiety that can be copolymerized with nonfunctionalized monomer; (b) from a polymer film, which can be modified by introducing positively charged groups; and finally (c) from polymer or polymer blends, by introducing positively charged groups, followed by the dissolving of a polymer and casting it into a film.

If the anion exchange membrane is prepared from a monomer (approach "a"), polysulfone, polyethylene, or styrene and divinylbenzene are most commonly used for traditional hydrocarbon type membranes, from which the membrane is usually prepared by chloromethylation followed by quaternary amination. Numerous references exist in the alkaline membrane fuel cell literature for homogeneous anion exchange membrane preparation using this method (see, for instance, [4, 5]).

Some efforts have also been directed towards the preparation of anion exchange membranes from polymer films (approach "b"). Generally, these polymers, e.g., polymer films of hydrocarbon polyethylene and polypropylene or fluorocarbon origin (ETFE, PVDF), are insoluble in solvents. The method most commonly used consists in grafting vinyl monomers, such as styrene, onto polymer films followed by subsequent chemical modifications, such as chloromethylation-amination instead of sulfonation [6–8].

Relatively little work has been done with anion exchange membranes prepared by casting a polymer already containing functional groups (approach "c"). Also in this approach polystyrene and polyethylene polymers are usually used; however other less common backbones such as polyvinyl acetate and chitosan have also been used to prepare anion exchange membranes [9–11].

It should be noted that although anion exchange membranes were initially developed for electrodialysis, it was not until last decade that initial work on anion exchange membranes for alkaline membrane fuel cells begun. In the last 4 years, research on anion exchange membranes is proliferating, mainly after first polymers with high anion conductivity were presented.

Conductivity of Anion Exchange Membranes

Ion Exchange Capacity and Intrinsic Conductivity

A few types of anion exchange membranes for alkaline membrane fuel cells were initially developed in the early 2000s, and most of them showed very low OH^- conductivities. Agel et al. [9] developed their epichlorohydrin-based anion exchange membrane, quaternized with 1,4-diazabicyclo[2, 2, 2]octane (DABCO) and trimethylamine (TMA). The ion exchange capacity (IEC) of such membranes was reported to be around 0.5 meq/g, with OH^- conductivities lower than 1 mS/cm. Another example came from Li and Wang [12], who reported the preparation of a polyethersulfone-cardo-based AEM (see Fig. 3) by chloromethylation and posterior reaction with trimethylamine and ion exchange with sodium hydroxide to achieve the final anion exchange membrane. The ion exchange capacity of the membranes was reported to be 1.25 meq/g.

Recently, the number of studies devoted to the development of anion exchange membranes has significantly increased. There are numerous examples of relatively high-ion exchange capacity membranes, with respectable OH^- conductivity that exceeds the minimum conductivity requirements, for instance, for fuel cell stationary applications.

Park et al. [13] synthesized their chloromethylated polysulfones (see Fig. 3), cross-linked with amine/diamine mixtures to control water swelling and conductivity. The corresponding anion exchange membranes exhibited OH^- conductivities of up to 17 mS/cm. Tanaka et al. [14] reported the development of new anion exchange membranes synthesized from aromatic multiblock copolymers, poly (arylene ether)s containing quaternized ammonio-substituted fluorene groups. The quaternized multiblock copolymers produced ductile, transparent anion exchange membranes with an ion exchange capacity of up to 2 meq/g and an OH^- conductivity of around 35 mS/cm, measured in liquid water at 80 °C. Another

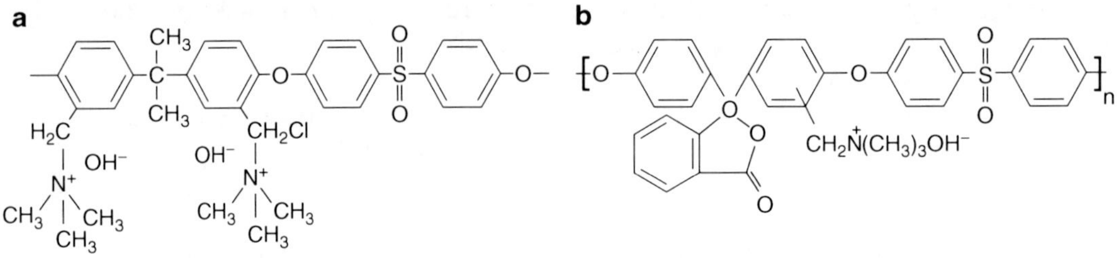

Alkaline Membrane Fuel Cells, Membranes, Fig. 3 Chemical structure of (**a**) quaternary ammonium polysulfone-based anion exchange membrane and (**b**) polyethersulfone-cardo-based anion exchange membrane

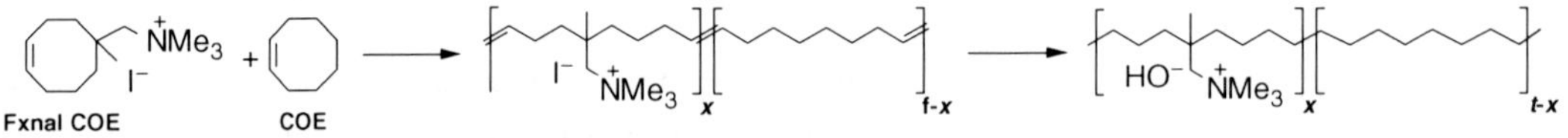

Alkaline Membrane Fuel Cells, Membranes, Fig. 4 Scheme of a simplified synthesis of a polyethylene-based anion exchange polymer with quaternary ammonium functional groups

example of anion exchange membrane with very high-ion exchange capacity and OH^- conductivities was recently reported by Robertson et al. [4], who synthesized an ammonium-functionalized polyethylene-based anion exchange membrane (see Fig. 4), showing ion exchange capacity as high as 2.3 mmol/g, with OH^- conductivity of 65 mS/cm, measured after cross-linking the polyethylene-based copolymer.

Effective Anion Conductivity

While increasing intrinsic conductivity in anion exchange membranes is still focus of research, a major factor influencing effective conductivity is carbon dioxide (CO_2). In contrast to alkaline fuel cells (AFCs) with liquid KOH electrolyte, in which carbonate precipitation occurs, in alkaline membrane fuel cells there is no place for salt precipitation, since in solid electrolytes there are no free cation species available for precipitation. However, CO_2 still affects the alkaline membrane fuel cell general performance by decreasing the effective conductivity of the anion exchange membranes [15–18], as it is indicated below:

$$\sigma^{eff} = \sigma/\alpha \tag{4}$$

where σ^{eff} is the effective anion conductivity through the anion exchange membrane, σ is the intrinsic conductivity of the anion exchange membrane, and α is a conductivity factor given by:

$$\alpha = f\left\{C\left(CO_3^{\ 2-}\right), C\left(HCO_3^{\ -}\right), F_M, \lambda, T\right\} \tag{5}$$

As it is shown in Eq. 5, the conductivity factor α is function of the local concentration of $CO_3^{\ 2-}$ and $HCO_3^{\ -}$ species in the membrane, $C(CO_3^{\ 2-})$ and $C(HCO_3^{\ -})$, respectively; the chemistry of the functional group of the AEM, F_M; the hydration level of the membrane, λ; and the temperature, T. $C(CO_3^{\ 2-})$ and $C(HCO_3^{\ -})$ are in turn related to the concentration of CO_2 in air, $C(CO_2)$, according to electrolyte equilibrium equations described elsewhere [19]. In general, the conductivity factor α increases with increasing $C(CO_3^{\ 2-})$ and $C(HCO_3^{\ -})$ (means with increasing C_{CO2}) [16] and decreases with increasing temperature T.

Attempts to quantify the overall effect of CO_2 on the effective anion conductivity in an anion exchange membrane have been done. For instance, Kim [20] and Lin et al. [21] reported effective anion conductivity reduction of 10–50 %, depending on temperature. Figure 5 shows a comparison of anion exchange membrane conductivity measured in full OH^- form

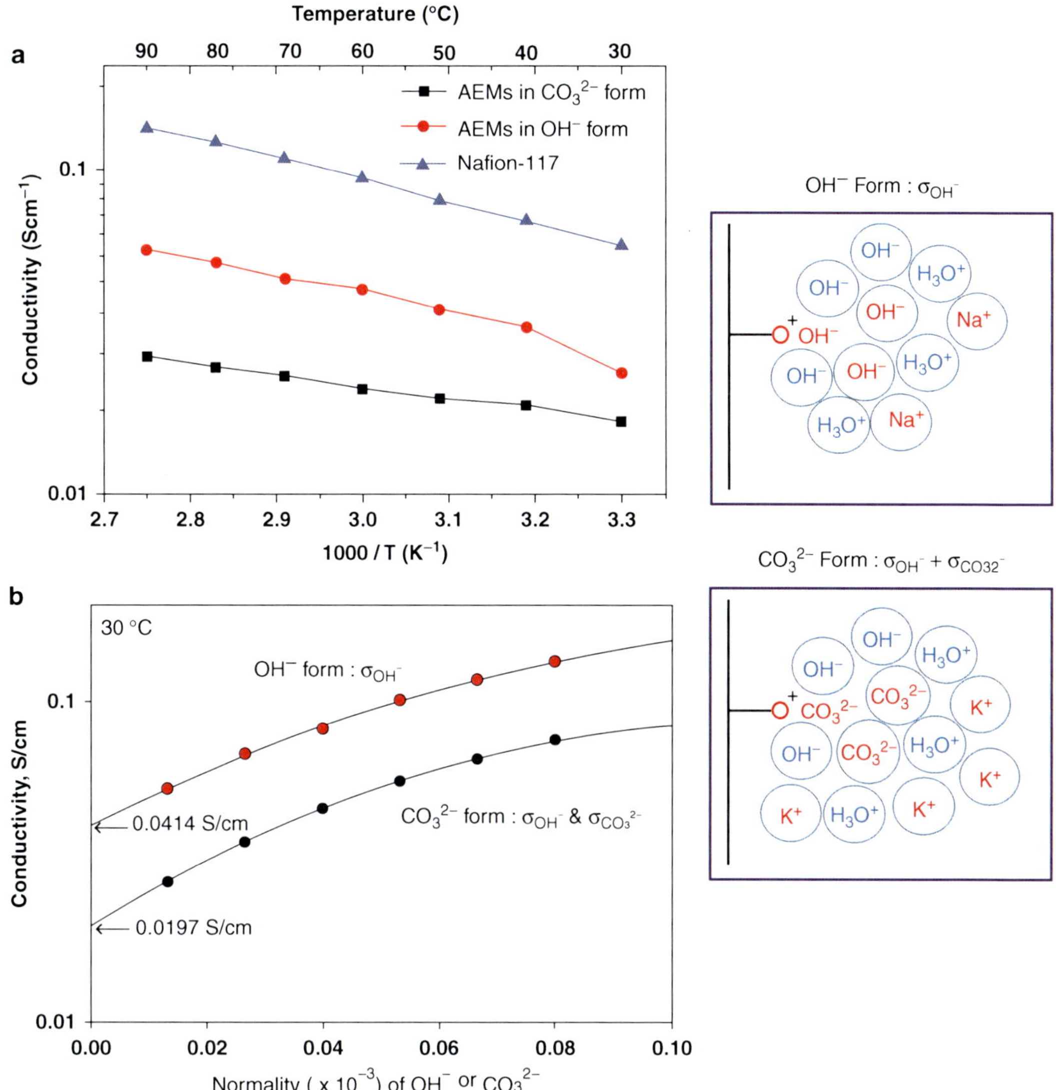

Alkaline Membrane Fuel Cells, Membranes, Fig. 5 Anion conductivity of an anion exchange membrane in OH^- form and in CO_3^{2-} form (measured in CO_2-free water at 30 °C) [20]

and in CO_3^{2-} form (fully equilibrated with ambient air) [20].

Stability of Anion Exchange Membranes

Stability challenges of anion exchange membranes have been known for quite a long time [22, 23]. In spite of the high conductivities achieved in recently developed anion exchange membranes, their stability is still a concern. Numerous research groups have now concentrated on stabilizing the functional groups and backbones of anion exchange membranes in alkaline conditions.

Several degradation mechanisms of anion exchange membranes were proposed in the literature. Varcoe [24, 25] summarized those mechanisms in the following schematic diagram:

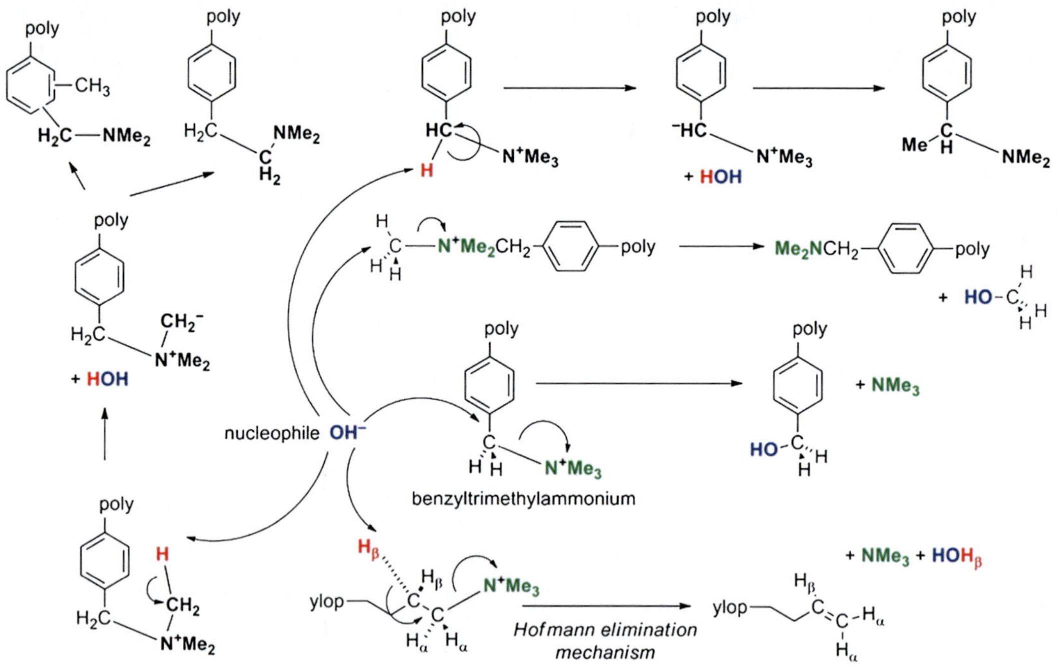

Alkaline Membrane Fuel Cells, Membranes, Fig. 6 Mechanisms of degradation of a quaternary ammonium-functionalized anion exchange membrane [24]

The main reason for the instability of the anion exchange polymer is the attack of the OH^-, which is generated and transported during the operation of the alkaline membrane fuel cell, on the functional group itself and on its bond to the backbone. Due to this attack, the functional group is destroyed or detached from the alkaline exchange polymer, as it is shown in Fig. 6. This in turn diminishes the number of functional groups that can effectively transport OH^- for cell operation, resulting in the eventual loss of ion exchange capacity and conductivity of the anion exchange membrane.

A few studies can be found in the literature testing polymer stability, by following the change in ion exchange capacity and membrane conductivity through time [26–28]. In those works, the anion exchange membranes are immersed in hot alkaline aqueous solutions to simulate the environment of an alkaline membrane fuel cell. Figure 7 shows some examples of stability measurements performed with different anion exchange membranes.

As it can be seen in Fig. 7, in some cases stabilities of more than 2,000 h were achieved, demonstrated by slight or no changes in the membrane ion exchange capacity or conductivity during time. This represents a significant advance in anion exchange membrane stability, as compared to the stability previously reported [2]. The stability improvement reported in past years was mainly accomplished by the use of higher stable polymers as backbones in the anion exchange membranes.

From the different possible degradation mechanisms described by Varcoe [24, 25] and others [20, 29–31], the general consensus in the literature is that significant further advances in alkaline membrane fuel cell durability can be achieved with an improvement in the chemical stability, both of the functional group itself and of its bond to the anion exchange membrane backbone [32, 33].

Several different functional groups have been already used to functionalize AEMs in past years. The following table (see Table 1)

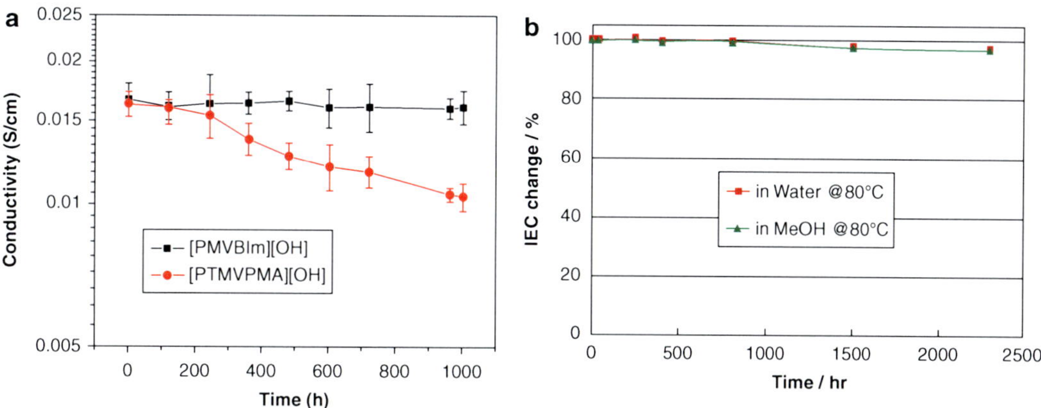

Alkaline Membrane Fuel Cells, Membranes, Fig. 7 Anion exchange membrane ex situ stability measurements. (**a**) Conductivity versus time [26]; (**b**) ion exchange capacity (*IEC*) versus time [27]

summarizes some of the functional groups used in AEMs.

In spite of the large diversity of cationic functional groups used in the anion exchange membrane literature, just a few authors have investigated and tested their stability. Pivovar [53, 54] and Einsla et al. [29] studied different functional groups and reported several accelerated stability results using nuclear magnetic resonance measurements. Some results of those measurements are shown in Fig. 8.

The authors concluded that the ether-linked quaternary ammonium (indicated in Fig. 8 as cation #2) has shown the highest stability as compared to other tested functional groups. This conclusion seems to be consistent with similar findings shown in the literature [36].

While important advances in anion exchange membrane stability have been already achieved, showing stability acceptable for practical applications, there is still a need to further increase the operation temperature of the alkaline membrane fuel cell to significantly improve cell power densities. Due to the high dependence of degradation rates on temperature, further research on anion exchange membrane functional groups with higher stabilities is needed.

Most recent studies focus on functional groups with high stable resonance structures, such as guanidinium [47] and imidazolium [26]. Figure 9 shows functional group candidates with resonance structure stabilizing the positive charge:

Future work based on these, or other functional groups, may increase the stability of anion exchange membranes to more than 5,000 h, as required for practical applications of the alkaline membrane fuel cells.

Future Directions

Anion exchange membranes have been rapidly developed within the past 5 years. These membranes are used in alkaline membrane fuel cell, which promises to solve the cost barriers of the fuel cell technology.

Latest improvements done in anion exchange membranes are more than impressive. While conductivity of anion exchange membranes has already achieved, or even exceeded, the minimum conductivity required for practical fuel cell applications, the stability of the functional groups of the membrane in real alkaline membrane fuel cells still requires further work [58]. The main development challenges which must be addressed prior to their large-scale introduction mainly include development of anion exchange membranes and anion exchange ionomers with improved stability at higher temperatures. Advances in developments in those two fields will assure rapid entrance to existing market opportunities.

Alkaline Membrane Fuel Cells, Membranes, Table 1 Functional groups used to prepare anion exchange membranes for alkaline membrane fuel cells

Functional group	General chemical structure	References as example
Ammonium-based groups		[33, 34]
	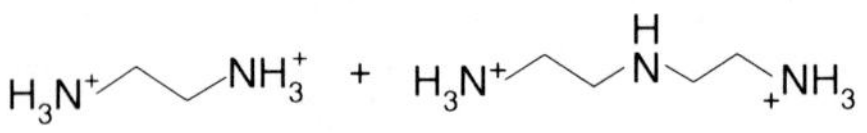	[35]
	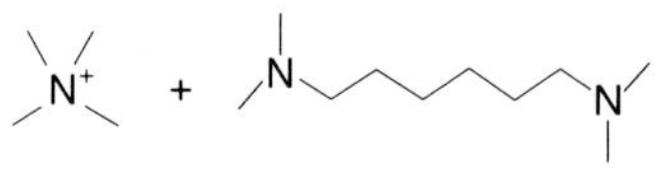	[13]
	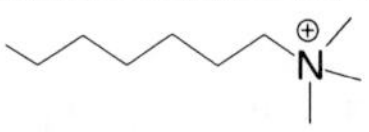	[36]
	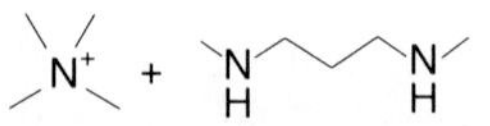	[37]
	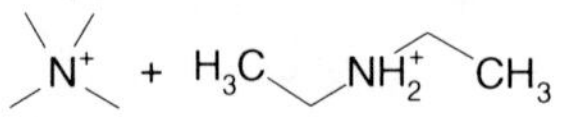	[38]
	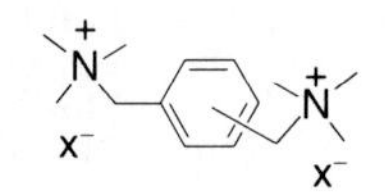	[4]
	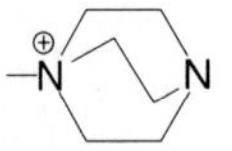	[8, 22, 39, 40]
	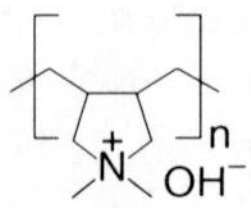	[41, 42]
	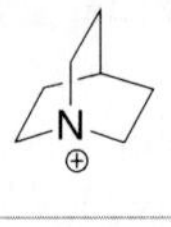	[40]
	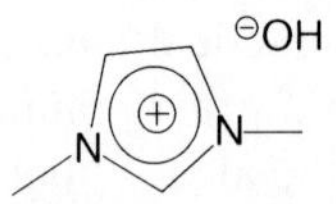	[21, 26, 43, 44]
	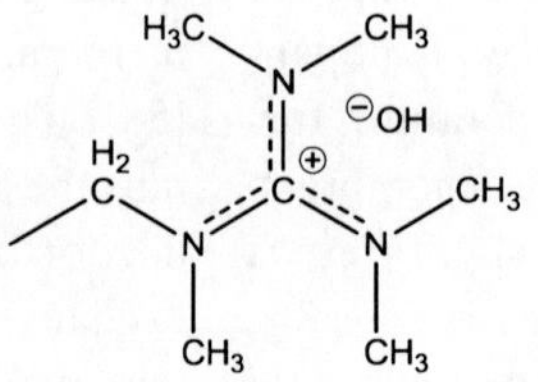	[45, 46]

(*continued*)

Alkaline Membrane Fuel Cells, Membranes, Table 1 (continued)

Functional group	General chemical structure	References as example
	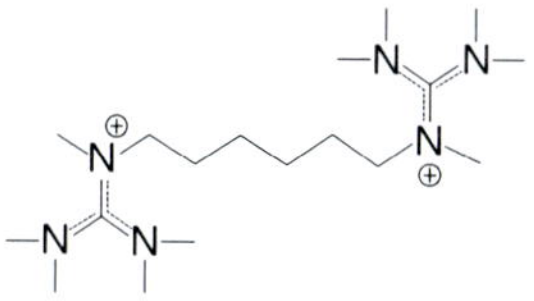	[47]
Phosphonium-based groups	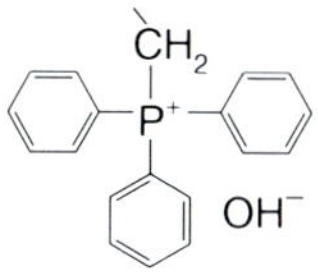	[48]
	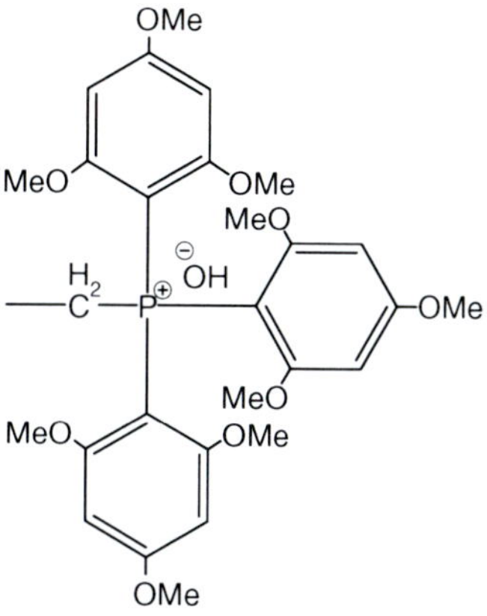	[49]
	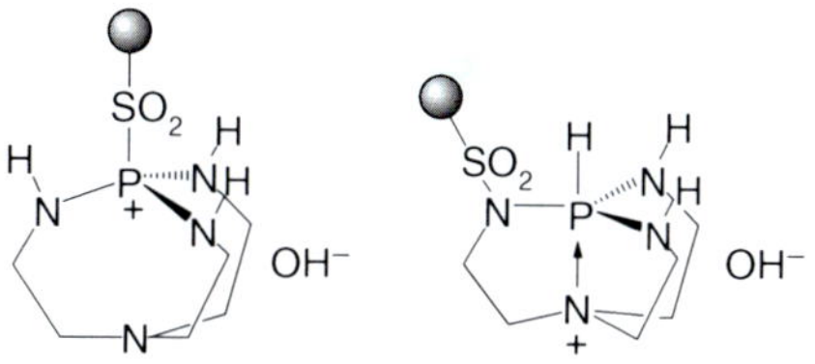	[50]
Sulfonium-based groups	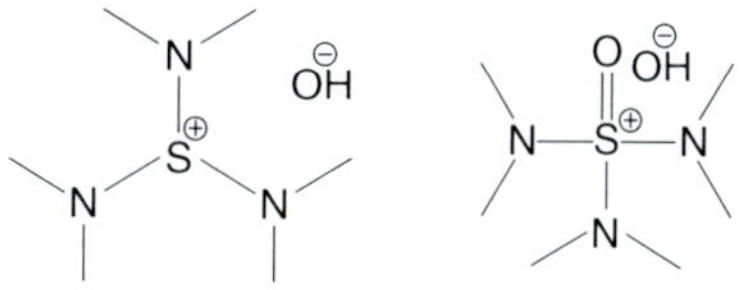	[51]
	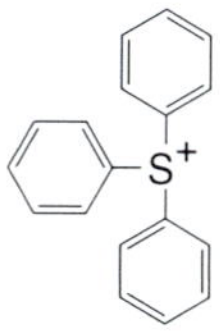	[52]

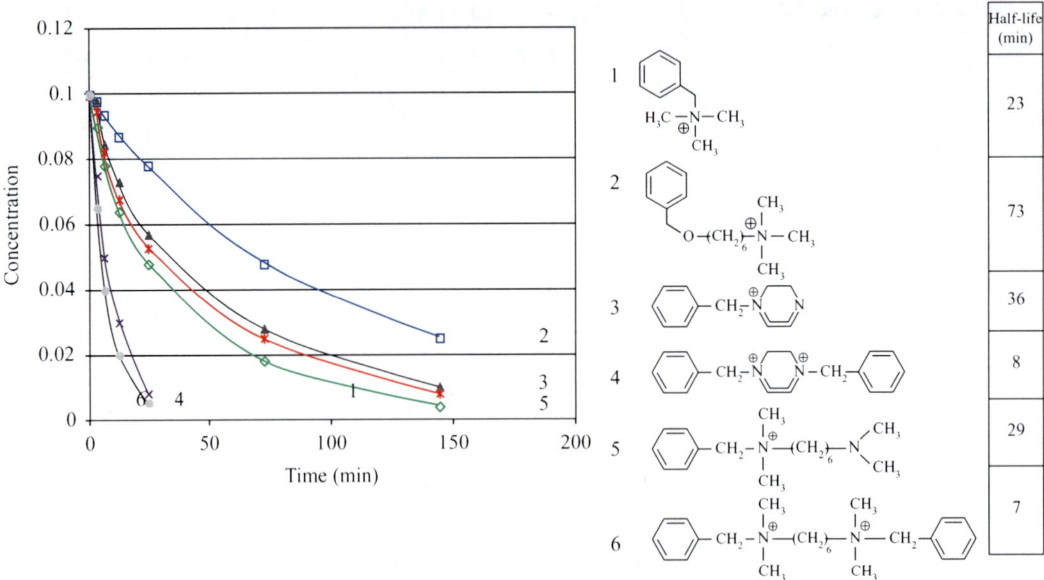

Alkaline Membrane Fuel Cells, Membranes, Fig. 8 Nuclear magnetic resonance measurements of degradation of cations used as functional groups in anion exchange membranes (0.1M cation concentration, in 2M NaOH at 160 °C) [55]

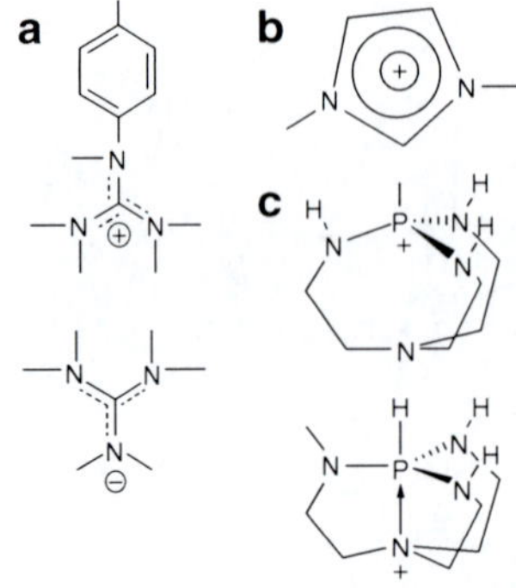

Alkaline Membrane Fuel Cells, Membranes, Fig. 9 Functional group candidates with resonance structure stabilizing the positive charge: (**a**) imidazolium [26], (**b**) guanidinium [45, 56 (phenyl guanidinium), 57], and (**c**) phosphatranium [50]

Cross-References

▸ Alkaline Membrane Fuel Cells
▸ Anion-Exchange Membrane Fuel Cells, Oxide-Based Catalysts

References

1. Dekel DR (2012) Latest advances in alkaline membrane fuel cell (AMFC) technology. In: 3rd Carisma international conference, 3 Sept 2012, Copenhagen. http://www.hotmea.kemi.dtu.dk/upload/institutter/ki/hotmea/carisma%202012/abstracts/dekel%20carisma%202012.pdf. Accessed 16 Dec 2012
2. Pivovar BS (2006) Alkaline Membrane Fuel Cell Workshop Final Report. http://www1.eere.energy.gov/hydrogenandfuelcells/pdfs/amfc_dec2006_workshop_report.pdf. Accessed 6 Nov 2012
3. Strathmann H (2004) Ion-exchange membrane separation processes, vol 9, Membrane science and technology series. Elsevier, Amsterdam
4. Robertson NJ, Kostalik Iv HA, Clark TJ, Mutolo PF, Abruña HD, Coates GW (2010) Tunable high performance cross-linked alkaline anion exchange membranes for fuel cell applications. J Am Chem Soc 132:3400–3404. doi:10.1021/ja908638d
5. Yan J, Hickner MA (2010) Anion exchange membranes by bromination of benzylmethyl containing poly(sulfone)s. Macromolecules 43:2349–2356. doi:10.1021/ma902430y
6. Varcoe JR, Slade RCT (2005) Prospects for alkaline anion-exchange membranes in low temperature fuel cells. Fuel Cells 5:187–200. doi:10.1002/fuce.200400045

7. Yan J, Hickner MA (2009) Efficient synthesis and properties of anion exchange membranes. Polym Preprints 50:272–273
8. Fang FJ, Yang Y, Lu X, Ye M, Li W, Zhang Y (2012) Cross-linked, ETFE-derived and radiation grafted membranes for anion exchange membrane fuel cell applications. Int J Hydrogen Energy 37:594–602. doi:10.1016/j.ijhydene.2011.09.112
9. Agel E, Bouet J, Fauvarque JF (2001) Characterization and use of anionic membranes for alkaline fuel cells. J Power Sources 101:267–274
10. Wan Y, Peppley B, Creber KAM, Bui VT, Halliop E (2008) Quaternized-chitosan membranes for possible applications in alkaline fuel cells. J Power Sources 185:183–187
11. Xiong Y, Fang J, Zeng QH, Liu QL (2008) Preparation and characterization of cross-linked quaternized poly (vinyl alcohol) membranes for anion exchange membrane fuel cells. J Membrane Sci 311:319–325. doi:10.1016/j.memsci.2007.12.029
12. Li L, Wang Y (2005) Quaternized polyethersulfone Cardo anion exchange membranes for direct methanol alkaline fuel cells. J Membrane Sci 262:1–4. doi:10.1016/j.memsci.2005.07.009
13. Park JS, Park GG, Park SH, Yoon YG, Kim GS, Lee WY (2007) Development of solid-state alkaline electrolytes for solid alkaline fuel cells. Macromol Symp 249–250:174–182. doi:10.1002/masy.200750329
14. Tanaka M, Fukasawa K, Nishino E, Yamaguchi S, Yamada K, Tanaka H, Bae B, Miyatake K, Watanabe M (2011) Anion conductive block poly(arylene ether) s: synthesis, properties, and application in alkaline fuel cells. J Am Chem Soc 133:10646–10654. doi:10.1021/ja204166e
15. Adams LA, Poynton SD, Tamain C, Slade RCT, Varcoe JR (2008) A carbon dioxide tolerant aqueous-electrolyte-free anion-exchange membrane alkaline fuel cell. ChemSusChem 1:79–81. doi:10.1002/cssc.200700013
16. Inaba M, Matsui Y, Saito M, Tasaka A, Fukuta K, Watanabe S, Yanagi H (2011) Effects of carbon dioxide on the performance of anion-exchange membrane fuel cells. Electrochemistry 79:322–325
17. Unlu M, Zhou J, Kohl PA (2009) Anion exchange membrane fuel cells: experimental comparison of hydroxide and carbonate conductive ions. Electrochem Solid State Lett 12:B27–B30. doi:10.1149/1.3058999
18. Wang Y, Li L, Hu L, Zhuang L, Lu J, Xu B (2003) A feasibility analysis for alkaline membrane direct methanol fuel cell: thermodynamic disadvantages versus kinetic advantages. Electrochem Commun 5:662–666. doi:10.1016/S1388-2481(03)00148-6
19. Siroma Z, Watanabe S, Yasuda K, Fukuta K, Yanagi H (2011) Mathematical modeling of the concentration profile of carbonate ions in an anion exchange membrane fuel cell. J Electrochem Soc 158:B682–B689. doi:10.1149/1.3576120
20. Kim YS (2010) Resonance-stabilized anion exchange polymer electrolytes. In: Annual merit review and peer evaluation meeting, DOE hydrogen program and vehicle technologies program. http://www.hydrogen.energy.gov/pdfs/review10/fc043_kim_2010_o_web.pdf. Accessed 7 Nov 2012
21. Lin B, Qiu L, Qiu B, Peng Y, Yan F (2011) A soluble and conductive polyfluorene ionomer with pendant imidazolium groups for alkaline fuel cell applications. Macromolecules 44:9642–9649. doi:10.1021/ma202159d
22. Bauer B, Strathmann H, Effenberger F (1990) Anion-exchange membranes with improved alkaline stability. Desalination 79:125–144
23. Sata T, Tsujimoto M, Yamaguchi T, Matsusaki K (1996) Change of anion exchange membranes in an aqueous sodium hydroxide solution at high temperature. J Memb Sci 112:161–170
24. Varcoe JR, Slade RCT, Yee ELH, Poynton SD, Driscoll DJ, Apperley DC (2007) Poly(ethylene-co-tetrafluoroethylene)-derived radiation-grafted anion-exchange membrane with properties specifically tailored for application in metal-cation-free alkaline polymer electrolyte fuel cells. Chem Mater 19:2686–2693. doi:10.1021/cm062407u
25. Poynton SD, Zeng R, Kizewski J, Ong AL, Varcoe JR (2012) Development of alkaline exchange ionomers for use in alkaline polymer electrolyte fuel cells. ECS Trans 50(2):2067–2073
26. Qiu B, Lin B, Qiu L, Yan F (2012) Alkaline imidazolium- and quaternary ammonium-functionalized anion exchange membranes for alkaline fuel cell applications. J Mater Chem 22:1040–1045. doi:10.1039/c1jm14331j
27. Yanagi H, Fukuta K (2008) Anion exchange membrane and ionomer for alkaline membrane fuel cells (AMFCs). ECS Trans 16(2):257–262. doi:10.1149/1.2981860
28. Luo Y, Guo J, Wang C, Chu D (2011) Tunable high-molecular-weight anion-exchange membranes for alkaline fuel cells. Macromol Chem Phys 212:2094–2102. doi:10.1002/macp.201100218
29. Einsla BR, Chempath S, Pratt LR, Boncella JM, Rau J, Macomber C, Pivovar BS (2007) Stability of cations for anion exchange membrane fuel cells. ECS Trans 11(1):1173–1180
30. Chempath S, Einsla BR, Pratt LR, Macomber CS, Boncella JM, Rau JA, Pivovar BS (2008) Mechanism of tetraalkylammonium headgroup degradation in alkaline fuel cell membranes. J Phys Chem C 112(9):3179–3182
31. Schwesinger R, Link R, Wenzl P, Kossek S, Keller M (2006) Extremely base-resistant organic phosphazenium cations. Chem Eur J 12(2):429–437. doi:10.1002/chem.200500837
32. Dekel D (2012) Latest advances in alkaline membrane fuel cell (AMFC) technology. In: Carisma 2012 – 3rd Carisma international conference, Copenhagen, 3 Sept 2012. http://www.hotmea.kemi.dtu.dk/upload/institutter/ki/hotmea/carisma%202012/abstracts/dekel%20carisma%202012.pdf. Accessed 17 Dec 2012

33. Pivovar BS (2011) Alkaline membrane fuel cell workshop final report. http://www.nrel.gov/docs/fy12osti/54297.pdf. Accessed 6 Nov 2012
34. Fujimoto CH, Hickner MA, Cornelius CJ, Loy DA (2005) Ionomeric poly(phenylene) prepared by Diels–Alder polymerization: synthesis and physical properties of a novel polyelectrolyte. Macromolecules 38(12):5010–5016. doi:10.1021/ma0482720
35. Miyazaki K, Sugimura N, Kawakita KI, Abe T, Nishio K, Nakanishi H, Matsuoka M, Ogumi Z (2010) Aminated perfluorosulfonic acid ionomers to improve the triple phase boundary region in anion-exchange membrane fuel cells. J Electrochem Soc 157(11):A1153–A1157. doi:10.1149/1.3483105
36. Tomoi M, Yamaguchi K, Ando R, Kantake Y, Aosaki Y, Kubota H (1997) Synthesis and thermal stability of novel anion exchange resins with spacer chains. J Appl Polym Sci 64:1161–1167
37. Hao JH, Chen C, Li L, Yu L, Jiang W (2000) Preparation of solvent-resistant anion-exchange membranes. Desalination 129:15–22
38. Pan J, Li Y, Zhuang L, Lu J (2010) Self-crosslinked alkaline polymer electrolyte exceptionally stable at 90 °C. Chem Commun 46:8597–8599. doi:10.1039/C0CC03618H
39. Faraj M, Elia E, Boccia M, Filpi A, Pucci A, Ciardelli F (2011) New anion conducting membranes based on functionalized styrene–butadiene–styrene triblock copolymer for fuel cells applications. J Polym Sci A Polym Chem 49:3437–3447. doi:10.1002/pola.24781
40. Stoica D, Ogier L, Akrour L, Alloin F, Fauvarque JF (2007) Anionic membrane based on polyepichlorhydrin matrix for alkaline fuel cell: synthesis, physical and electrochemical properties. Electrochim Acta 53:1596–1603. doi:10.1016/j.electacta.2007.03.034
41. Daikoku Y, Isomura T, Fukuta K, Yanagi H, Yamaguchi M (2011) Anion-exchange membrane and method for producing the same. US Patent Appl, US 2011/0281197 A1. http://appft1.uspto.gov/netacgi/nph-Parser?Sect1=PTO1&Sect2=HITOFF&d=PG01&p=1&u=/netahtml/PTO/srchnum.html&r=1&f=G&l=50&s1=20110281197.PGNR. Accessed 12 Nov 2012
42. Yao W, Tsai T, Chang YM, Chen M (2001) Polymer-based hydroxide conducting membranes. US Patent 6,183,914
43. Chen D, Hickner MA (2012) Degradation of imidazolium- and quaternary ammonium-functionalized poly(fluorenyl ether ketone sulfone) anion exchange membranes. ACS Appl Mater Interfaces. doi:10.1021/am301557w
44. Deavin OI, Murphy S, Ong AL, Poynton SD, Zeng R, Hermanac H, Varcoe JR (2012) Anion-exchange membranes for alkaline polymer electrolyte fuel cells: comparison of pendent benzyltrimethylammonium- and benzylmethylimidazolium-head-groups. Energy Environ Sci 5:8584–8597. doi:10.1039/c2ee22466f
45. Kim DS, Labouriau A, Guiver MD, Kim YS (2011) Guanidinium-functionalized anion exchange polymer electrolytes via activated fluorophenyl-amine reaction. Chem Mater 23:3795–3797. doi:10.1021/cm2016164
46. Wang J, Li S, Zhang S (2010) Novel hydroxide-conducting polyelectrolyte composed of an poly(arylene ether sulfone) containing pendant quaternary guanidinium groups for alkaline fuel cell applications. Macromolecules 43:3890–3896. doi:10.1021/ma100260a
47. Qu C, Zhang H, Zhang F, Liu B (2012) A high-performance anion exchange membrane based on bi-guanidinium bridged polysilsesquioxane for alkaline fuel cell application. J Mater Chem 22:8203–8207. doi:10.1039/c2jm16211c
48. Gu S, Cai R, Luo T, Yan Y (2008) Synthesis and characterizations of quaternary phosphonium polysulfone anion exchange membrane for alkaline fuel cell. In: 214th ECS Meeting, Abstract #1107. http://ma.ecsdl.org/content/MA2008-02/11/1107.full.pdf+html. Accessed 20 Dec 2012
49. Gu S, Cai R, Luo T, Chen Z, Sun M, Liu Y, He G, Yan Y (2009) A soluble and highly conductive ionomer for high-performance hydroxide exchange membrane fuel cells. Angew Chem Int Ed Engl 121:6621–6624. doi:10.1002/ange.200806299
50. Kong X, Wadhwa K, Verkade JG, Schmidt-Rohr K (2009) Determination of the structure of a novel anion exchange fuel cell membrane by solid-state nuclear magnetic resonance spectroscopy. Macromolecules 42(5):1659–1664. doi:10.1021/ma802613k
51. Pivovar BS, Thorn DL (2009) Anion-conducting polymer, composition, and membrane. US Patent 7,582,683
52. Zhang B, Gu S, Wang J, Liu Y, Herring AM, Yan Y (2012) Tertiary sulfonium as a cationic functional group for hydroxide exchange membranes. RSC Adv 2:12683–12685. doi:10.1039/C2RA21402D
53. Pivovar BS (2010) Fundamentals of hydroxide conducting systems for fuel cells and electrolyzers. In: 2010 annual merit review meeting, DOE hydrogen program. http://www.hydrogen.energy.gov/pdfs/review10/bes016_pivovar_2010_o_web.pdf. Accessed 12 Nov 2012
54. Long H, Kim K, Pivovar BS (2012) Hydroxide degradation pathways for substituted trimethylammonium cations: a DFT study. J Phys Chem C 116(17):9419–9426
55. Herring AM, Pivovar BS (2011) Anion exchange membranes for fuel cells. http://www1.eere.energy.gov/hydrogenandfuelcells/pdfs/amfc_110811_herring.pdf. Accessed 21 Dec 2012

56. Zhang F, Zhang H, Ren J, Qu C (2010) PTFE based composite anion exchange membranes: thermally induced in situ polymerization and direct hydrazine hydrate fuel cell application. J Mater Chem 20:8139–8146. doi:10.1039/c0jm01311k
57. Kim YS (2012) Resonance-stabilized anion exchange polymer electrolytes. US DOE Hydrogen and fuel cells program and vehicle technologies program annual merit review. http://www.hydrogen.energy.gov/pdfs/review12/fc043_kim_2012_p.pdf. Accessed 13 Nov 2012
58. Dekel D. Alkaline membrane fuel cells. In: Savinell R, Ota K, Kreysa G (eds) Encyclopedia of applied electrochemistry: springer reference (www.springerreference.com). Springer, Berlin/Heidelberg, 0. doi: 10.1007/SpringerReference_303632 2012-12-04 06:42:39 UTC

Alkaline Primary Cells

Ralph Brodd
Broddarp of Nevada, Inc., Henderson, NV, USA

Introduction

The three main primary cylindrical battery systems are based on zinc metal anode and manganese dioxide cathode but are classified by the electrolyte composition. The Leclanché cell is based on an ammonium chloride–zinc chloride (NH_4Cl–$ZnCl_2$) electrolyte. The zinc chloride cell is based on a zinc chloride ($ZnCl_2$) electrolyte. Both are classified as generic carbon–zinc cells and differ from the alkaline cell in that the alkaline cell is based on KOH electrolyte. The common features in each are the use of zinc as the negative or anode and manganese dioxide as the positive or cathode electrodes. The two main classifications of primary alkaline batteries are (1) cylindrical and (2) coin or button cells. Coin or button cells will be discussed later.

The alkaline zinc–manganese dioxide cell was introduced in 1959 as a high-performance primary cell to replace the Leclanché (carbon–zinc) cell that was developed by Georges Leclanché in 1860 and is still the battery of choice in the developing countries because of its low cost. The zinc chloride cell was introduced with improved performance over the Leclanché cell but at a slightly higher price. Alkaline cells have the lowest internal resistance and faster electrode reaction kinetics and contain no mercury. This translates into higher voltage at high-rate discharge, longer service life, and a more environmentally acceptable chemistry than the competing carbon–zinc cell systems. Alkaline Zn–MnO_2 cell has become the primary battery of choice to power portable devices and equipment.

The alkaline Zn–MnO_2 cell has higher capacity compared to the cells with Leclanché and zinc chloride electrolyte. In the D-size cell configuration, it can deliver about 15 Ah compared to about 7 Ah for the zinc chloride cell and 5 Ah for the Leclanché or carbon–zinc version. In addition, the zinc chloride and Leclanché versions exhibit a significant fall off in capacity on higher-rate discharges.

The alkaline cell will deliver essentially its full capacity under similar conditions. The main competition for the alkaline cell is the lithium–iron sulfide (Li–FeS_2) cell. It has an open-circuit voltage of 1.5 V, better high-rate capability, and better shelf life. However it is available only in the AA- and AAA-size cells and considerably more expensive than the alkaline cell. To a first approximation, it is estimated that 12–15 billion alkaline cells are sold each year, mainly in developed countries. By comparison, the sales of Leclanché cells are estimated to be 30–35 billion cells worldwide but mainly in developing countries.

Alkaline Zn–MnO_2 Cell Construction and Performance

Alkaline cells are available worldwide in a variety of sizes and shapes. The term "alkaline cell" is used to describe the specific cell containing zinc alloy powder anode, nanostructured electrolytic manganese dioxide cathode, and concentrated potassium hydroxide (KOH) electrolyte. The construction of the D-size cell is given in Fig. 1 and is typical of the constructions of all alkaline cylindrical cells from the

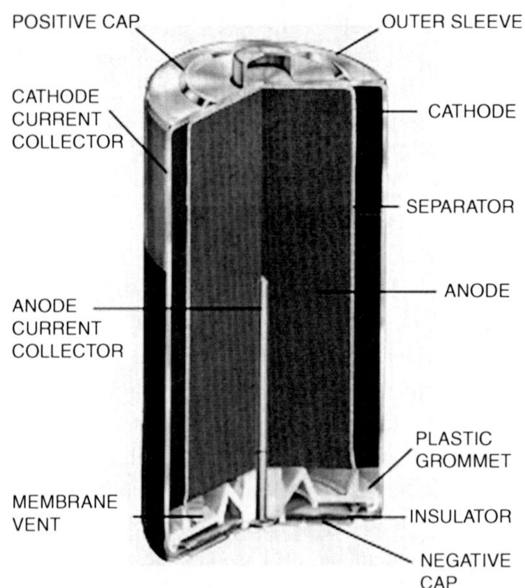

Alkaline Primary Cells, Fig. 1 Cutaway view of the alkaline D-size Zn–MnO_2 cell. Other size cylindrical cells, e.g., C-, AA-, AAA-, AAAA-, and N-size cells, have identical internal structure modified for the smaller size (Courtesy of Duracell)

smallest cell, N-size, to the largest, D-size. The anode, or negative electrode, is a zinc metal powder in a polymer gel such as carboxymethylcellulose, and the cathode, or positive electrode, is an electrolytic/synthetic manganese dioxide with graphite and acetylene black and is bonded with Portland cement or other binder. Good high-rate performance and low cost make the alkaline cell attractive to power portable devices of all types. The AA- and D-size alkaline cells have become the general purpose cells of choice except in developing countries where the Leclanché or carbon–zinc is still the battery of choice because of its lower cost.

The zinc and manganese dioxide electrode reaction kinetics in alkaline electrolyte are faster than the same reactions in the Leclanché and zinc chloride electrolyte. These differences in reaction rate determine the differences in cell performance. The alkaline cell delivers superior capacity, higher-rate discharge capability, lower internal resistance, lower leakage, and longer shelf life. Alkaline cells have excellent high-rate performance and deliver nearly full capacity, even at continuous high-rate discharges.

The advantages of the alkaline cell over the carbon–zinc cells are summarized below (Table 1). The main disadvantage of the alkaline cell is its higher price over the carbon–zinc battery systems.

Advantages of alkaline Zn–MnO_2 battery system compared to carbon–zinc batteries	
Longer shelf life	Lower internal resistance
Up to 10× longer service at high drains	Good low-temperature performance
Rugged construction	Excellent leak resistance
Superior high-rate discharge	Higher conductivity electrolyte
Higher energy density	Lower leakage rate

The alkaline cell has an open-circuit voltage of 1.5 V that can deliver 150 Wh/kg and 460 Wh/l. The reactions have fast kinetics and can deliver full capacity, even at high-rate discharges. Since its introduction in 1959, there has been a steady increase in performance of the alkaline cell as new materials and cell components were incorporated into the structure. The present alkaline cell designs are based on the use of nanostructured electrolytic manganese dioxide, a thinner polymer gasket seal with sealant to increase internal volume and improve shelf life. Mercury has been eliminated by using new zinc alloy compositions. These improvements have resulted in about a 40 % improvement in performance over the same-size cells produced in 1959.

Alkaline Cell Chemistry

$$\text{Negative: } Zn + 2\ OH^- - 2e = Zn(OH)_2 \quad (1)$$

$$\text{Positive: } MnO_2 + H_2O + e = MnOOH + OH^- \quad (2)$$

$$3\ MnOOH + e = Mn_3O_4 + OH^- + H_2O \quad (3)$$

Alkaline Primary Cells, Table 1 List of common cell sizes and ANSI battery system designations

IEC cell	Common name	System	Unit cell dimensions (millimeters)
N		Alkaline	12.0 × 30.2
AAAA	6R61	9 V carbon–zinc	8.3 × 42.5
	6LF22	9 V alkaline	
AAA	R03	Carbon–zinc	10.5 × 44.5
	LR03	Alkaline	
AA	R6	Carbon–zinc	14.5 × 50.5
	LR6	Alkaline	
C	R14	$Zn–MnO_2$	26.2 × 50
	LR14	Alkaline	
D	R20	Leclanché	34.2 × 61.5
	LR20	Alkaline	
L	Coin	Zinc–manganese dioxide	Variable
S	Coin	Zinc–silver oxide	Variable
Z	Coin/button	Zinc–air	Variable

$$\text{Overall}: Zn + 2\ MnO_2 + 2\ H_2O = Zn(OH)_2 + 2\ Mn_3O_4 \quad (4)$$

E = 1.5 V, specific energy = 150 Wh/kg, energy density = 460 Wh/l

Equations 1, 2, 3, and 4 describe the basic chemistry of the alkaline $Zn–MnO_2$ battery system. Depending on cell balance, ratio of anode to cathode reaction capacity in the cell, further reaction is possible below the 0.9 V endpoint.

Zinc Anode (Zn)

The zinc anode in the alkaline cell consists of powdered zinc metal alloyed with other metals such as indium, lead, and bismuth to lower the corrosion rate and lengthen the shelf life of the cell. The modern alkaline cell does not contain mercury as an inhibitor since the use of mercury has been banned worldwide by government edict. The zinc powder has controlled particle size distribution to maximize performance. The electrolyte is concentrated KOH with gelling agents such as CMC (carboxymethylcellulose) to stabilize the anode structure. An organic zinc corrosion inhibitor may be added to the electrolyte to prolong shelf life. The zinc anode gel mix supplies the electrolyte for cell operation. The contact to the zinc anode gel is generally a high-purity brass or silicon bronze nail or strip. The collector may be plated to prevent gassing from the corrosion reaction of the anode current collector with the electrolyte. A porous cellulosic or nonwoven vinyl polyolefin separator prevents direct electrical contact between the anode and cathode.

Manganese Cathode (MnO_2)

The cathode consists of high-purity electrolytic manganese dioxide (EMD) with a graphite–carbon conductive matrix held together with a binder, usually Portland cement or similar material. The EMD is produced by electrolysis from an acid bath made by dissolving calcined natural manganese dioxide in sulfuric acid followed by the removal of heavy metal impurities such as iron, copper, cobalt, nickel, chromium, and molybdenum. The electrolysis is carried out using titanium sheet electrodes at near-boiling acidified aqueous $MnSO_4$. This electrolytic process produces a high-purity, nanostructured MnO_2 material which is the key for the excellent high current performance and long shelf life exhibited by the alkaline cell. The EMD is a hard porous material, ground to a specified particle size to optimize cell performance. It is essential that EMD has a low level of heavy metal impurities that otherwise would dissolve and deposit on the anode and shorten shelf life of the cells. There are several different structures for MnO_2, but the γ-MnO_2 produced by the electrolytic process (EMD) is preferred for its stability and nanostructured characteristic and good high-rate, high-efficiency discharge. The EMD, graphite, and acetylene black are mixed and are molded into a cylinder under high pressure to provide a bond between the steel can and each individual particle of active mass.

The same active materials, powdered zinc metal, electrolytic manganese dioxide, and KOH electrolyte are used in both the standard cylindrical alkaline and the miniature coin cells discussed later. In contrast, the lower-performance, low-cost natural MnO_2 in the Leclanché cells is used as it is mined, after the removal of dirt, rocks, and other debris that result from the mining process. It has lower performance but is also at lower cost than EMD.

In 2008, Panasonic introduced a new cell construction, the Oxyride™ alkaline cell, with several improvements. The process for making EMD was modified to produce a highly purified electrolytic MnO_2 containing titanium dioxide (TiO_2) leading to a new high-purity manganese dioxide cathode active mass with superior performance on high current drains. In addition, the Oxyride™ cathode active material incorporated nickel oxyhydroxide in the cathode to improve high current performance. The Oxyride™ cell construction was modified with a thinner can wall thickness, thinner seal, and a longer current collector, to improve current distribution, and increased cathode active materials contained in the cell. These improvements responded to the demand for increased high-rate discharge capability imposed by new electronic devices and the competition from the rechargeable Ni–MH and Ni–Cd along with the 1.5 V lithium cells with higher current capability and higher capacity for longer life.

Cell Construction and Manufacturing Processes

The alkaline cell assembly operations consist of high-speed precision processing of the various cell components into the final product that has long storage life, high reliability, and low cost. The cell assembly starts with mixing and molding the cathode consisting of a manganese dioxide (EMD), graphite for conductivity, and carbon black to hold the electrolyte and a binding agent, such as Portland cement for physical integrity and long life. In a separate operation the zinc powders are mixed with a gelling agent such as carboxymethylcellulose with the KOH electrolyte in separate vessels. The polymer gasket, with sealant, incorporates a safety venting mechanism to release excess internal pressure generated by abuse conditions such as short-circuiting, incorrect insertion into devices, or disposal in fire. The internal constructions for the smaller N-, AAAA-, AAA-, AA-, and C-size cell constructions are essentially identical to that for D-size cell, adjusted for the outer cell dimensions.

The alkaline cell incorporates a high-performance electrolytic manganese dioxide–carbon–graphite mixture cathode, potassium hydroxide electrolyte, and a high-surface-area zinc powder anode with a porous cellulosic separator. The active materials are held in a nickel-plated steel can and hermetically sealed using a polymer compression seal. A thin coating of asphalt, or similar material, may be used in the seal area to smooth out defects to ensure a good seal. The compression should not exceed the elasticity of the polymer grommet. The cells have a nominal 1.5 V open-circuit voltage and a low internal resistance.

A generalized view of the overall assembly process for cylindrical cells essentially follows the sequence given in Fig. 2. The details of the alkaline cell manufacturing processes are closely held as proprietary and vary in detail from manufacturer to manufacturer but produce essentially the same-size cell with approximately the same ampere-hour capacity. After assembly, the cells are aged/stored for a set period of time, typically 3 or 4 weeks, under controlled conditions. At the end of the storage period, their voltage is measured. If their voltage is outside a predetermined range, it is indicative that an internal defect exists and the cell is rejected and discarded. This same general manufacturing process is used to produce the D, C, AA, and AAA cells as well as the AAAA-size cells used in the 9 V alkaline battery and N-size used for laser pointers.

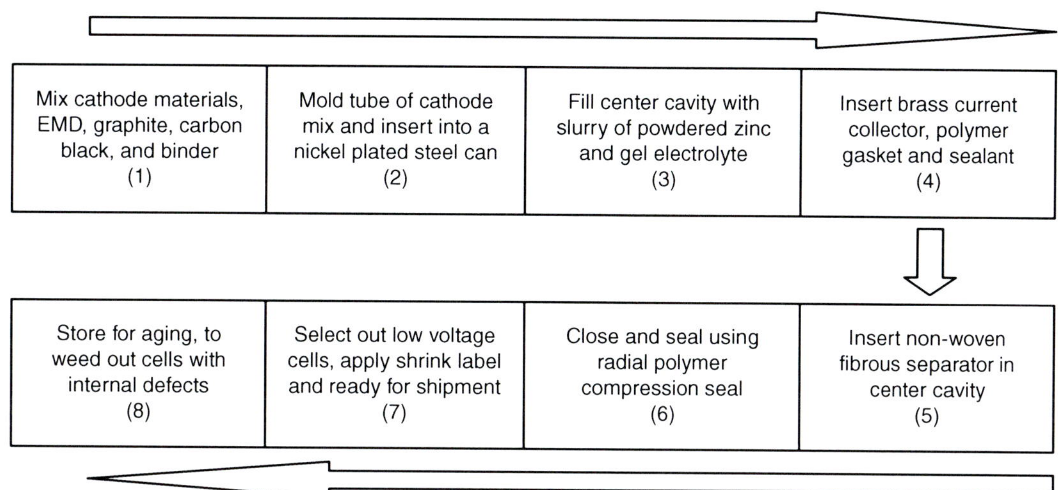

Alkaline Primary Cells, Fig. 2 Generalized alkaline cell assembly process

Performance

Alkaline cells are hermetically sealed in contrast to the Leclanché and zinc chloride electrolyte, termed carbon–zinc cells, that are vented to allow hydrogen gas, generated by the corrosion of zinc with the electrolyte, to escape. The open vent results in a shorter shelf life for the carbon–zinc cells. The superior performance of the alkaline cell is directly related to the superior characteristics of the nanostructured electrolytic manganese dioxide materials and their electrode reactions in alkaline KOH vs. natural MnO_2 and acid $ZnCl_2$–NH_4Cl electrolyte of the carbon–zinc cells. Electrolytic MnO_2 materials give significantly higher capacity and voltage maintenance over the course of discharge, compared the carbon–zinc cells that use natural MnO_2 as is after mining and separation of the MnO_2 materials from the rocks, debris, etc. The main competition for the alkaline cell is the higher-performance, higher-cost, organic electrolyte-based, lithium–iron sulfide (Li–FeS_2) primary cell and the lower-cost, lower-performance, Leclanché or carbon–zinc electrolyte and zinc chloride electrolyte Zn–MnO_2 cell.

Typical discharge curve of the AA-size alkaline cell is given in Fig. 3 at three discharge rates. The manganese dioxide chemistry determines the character of the discharge curve and the capacity of the cell. The general shape of the discharge curve is the same for all cell sizes, except that the smaller volume cells deliver less energy and have shorter useful life. The cells are termed primary cells as opposed to secondary cells because they can be discharged only once and cannot be recharged for reuse. During discharge, the basic structure of the manganese dioxide changes irreversibly as it goes from a 4-valent structure in MnO_2 to a 2.75-valent structure Mn_3O_4. It cannot be returned to its initial 4-valent structure and composition by reversing the current flow. The performance of the common AAAA-, AAA-, AA-, C-, and D-size alkaline cells is summarized in Fig. 4 as the time the cell can deliver useful capacity based on the 0.9 V cutoff voltage, and Fig. 5 details the performance the D-size cell at various temperatures from −10 °C (14 °F) to 45 °C (113 °F).

The alkaline cell competes against the lower-cost carbon–zinc cells. Figure 6 compares the performance of the D-size and AA-size Leclanché (carbon–zinc) cell with the same-size alkaline cell on various discharge rates at 21 °C (70 °F). The superior performance of the alkaline cell at all discharge rates is clear. Since both systems use zinc and manganese dioxide chemistry, the shape of the discharges are similar.

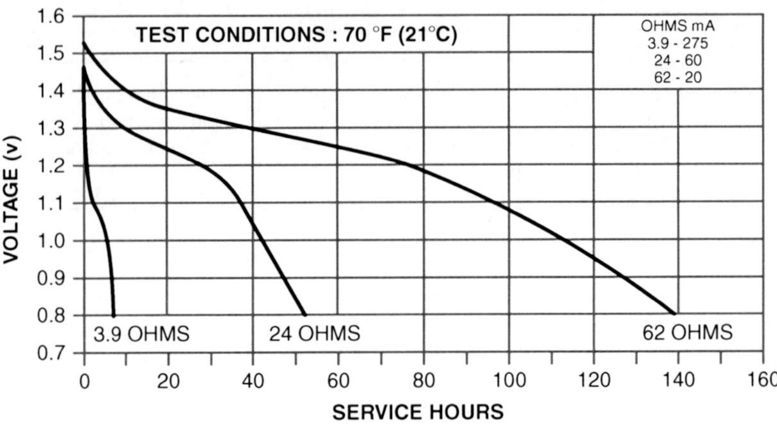

Alkaline Primary Cells, Fig. 3 Discharge of the alkaline AA-size cell (MN1500) at three discharge rates. The sloping discharge is the characteristic of the alkaline Zn–MnO_2 cell system (Courtesy of Duracell)

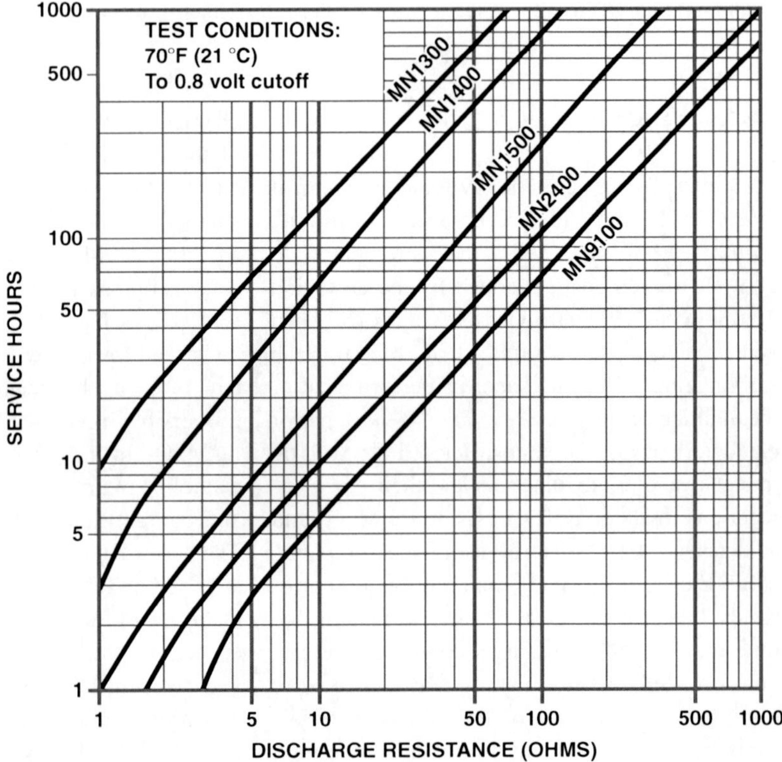

Alkaline Primary Cells, Fig. 4 Summary of the discharge performance of the D-size (MN1300), C-size (MN1400), AA-size (MN1500), AAA-size (MN2400), and N-size (MN9100) alkaline cells on resistive loads (Courtesy of Duracell). They all follow the characteristic sloping discharge depicted in Fig. 3 for the AA-size cells

The superior performance of the alkaline cell is the result of the use of electrolytic manganese dioxide, high-surface-area zinc powder, as well as the high conductivity alkaline electrolyte in the cell construction. The higher performance of the alkaline cell carries over to better low-temperature performance, as well.

In addition to competition from lower-cost zinc–carbon cells, the AA-size alkaline cell has experienced significant competition from the more expensive rechargeable nickel–metal hydride (Ni–MH) and nickel–cadmium (Ni–Cd) cells and the lithium–iron sulfide primary cell (Li–FeS_2) because of their excellent high-rate

Alkaline Primary Cells, Fig. 5 Effect of temperature on the discharge performance D-size of zinc–MnO_2 alkaline cells

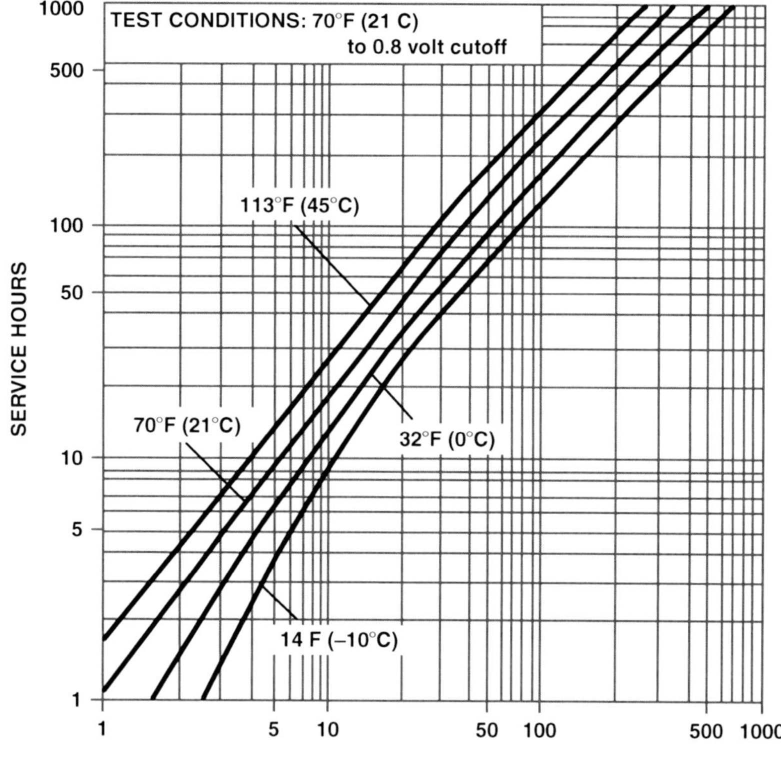

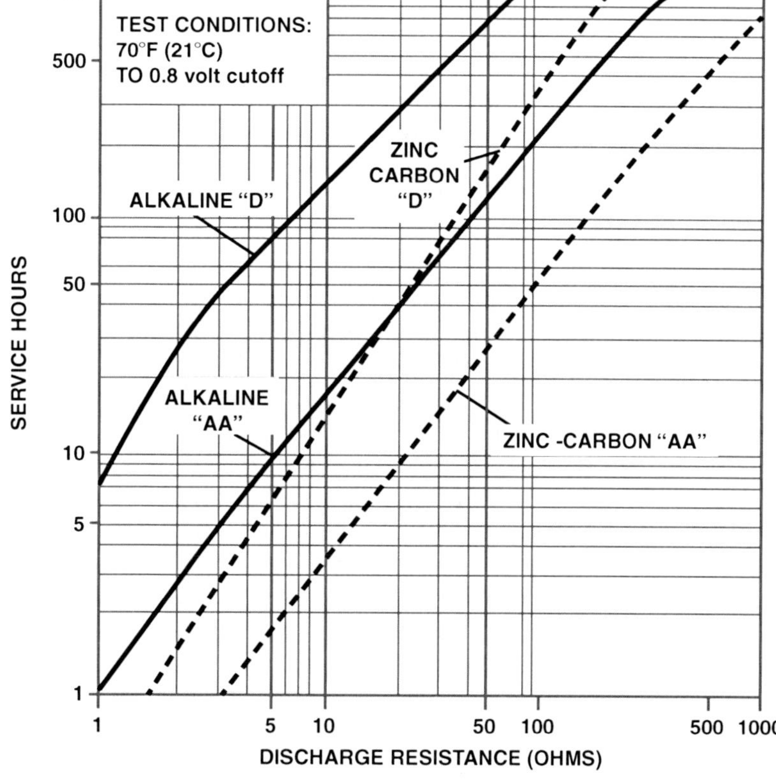

Alkaline Primary Cells, Fig. 6 Comparison of the performance of AA- and D-size alkaline and carbon–zinc cells at 0 °C and 20 °C as a function of the discharge resistance (Courtesy of Duracell). The alkaline cell clearly has superior performance at all discharge rates

capability for cameras and other electronic applications.

Alkaline Miniature Cells

The alkaline primary miniature cells were developed in response to the need for high energy storage systems to power small portable electronic devices. The energy storage capability of the three major chemistries, alkaline (Zn–MnO_2), silver (Zn–Ag_2O), and air (Zn–O_2), is illustrated in Fig. 7, as a function of temperature along with the nonaqueous lithium (Li–MnO_2) and the Leclanché cell. The Zn–air cell has the highest energy density of any alkaline cell system chemistry and considerably greater than that of the lithium–manganese dioxide coin cell system. The alkaline zinc–silver cell has essentially the same energy storage capability as that for the competing Li–MnO_2 system.

Alkaline Coin and Button Miniature Cells

Regular D-, C-, and AA-size carbon–zinc cylindrical cells were used exclusively until about 1940 when Samuel Ruben developed the small zinc–mercury oxide coin cell for military applications. The Ruben cell had high capacity and constant voltage discharge, withstood storage under tropical conditions, and had good energy density. It was used in early pacemakers, hearing aids, cameras, and small electronic equipment and as a voltage reference source. The use of mercury and cadmium in primary cells ended when they were banned worldwide in 1996 by government regulations for health reasons.

The three major small alkaline cylindrical coin/button cell systems are alkaline (Zn–MnO_2), silver–zinc (Zn–Ag_2O), and zinc–air (Zn–O_2). These cells are manufactured on an automated production line, similar to the automated cell manufacturing process outlined above in Fig. 2 for their larger cousins, alkaline zinc–manganese dioxide cell, but are modified for their smaller cell dimensions and cell chemistries. Modern production processes produce cells at a rate of several hundreds of cells per minute. Cells are available in a wide range of physical sizes and ampere-hour capacities.

Button/coin cells are smaller versions of the larger cousins the AA-, C-, and D-size cells. The construction consists of an anode subassembly, cathode subassembly, and a separator to form a layered design. The anode assembly, which becomes the top of the cell, consists of a bimetal laminate of nickel-plated steel and either copper or tin plated to protect it from the corrosive electrolyte. The cathode consists of a cup, into which the manganese dioxide–carbon active cathode material mix is pressed and consolidated under pressure. A porous barrier layer polymer separator, which allows current to flow but blocks any migration of any insoluble material or small charged carbon particles, separates the anode from the cathode. A plastic washer is placed to insulate the positive can from the negative electrode cap containing a zinc particulate anode. The anode current collector subassembly is placed into the cathode can subassembly and hermetically sealed by polymer compression seal. They follow a similar cell fabrication process flow and construction as given for larger cells but modified to produce a small cylindrical cell in the shape of a button or coin configuration.

The difference between a coin cell and button cell is subtle and related to their structure. Both are round cells with the descriptive designation, R. To a first approximation, coin cells are larger in diameter and thinner, while the button cells are smaller in diameter and thicker. For instance, a coin cell will not fit into a hearing aid but a Zn–air button cell fits very comfortably into the hearing aid unit that is placed into the ear cavity to power the hearing aid. The coin and button cells of the same chemistry have lower energy storage capability (Wh/l and Wh/cc) than the larger cylindrical cousins due to the lower percentage of active materials to container weight.

Figure 8 depicts the structure of the Zn–MnO_2 miniature cell. It uses the same basic electrode materials as its larger cousin, the D-size alkaline cell technology. The electrode reactions are found above in Eqs. 1, 2, 3, and 4.

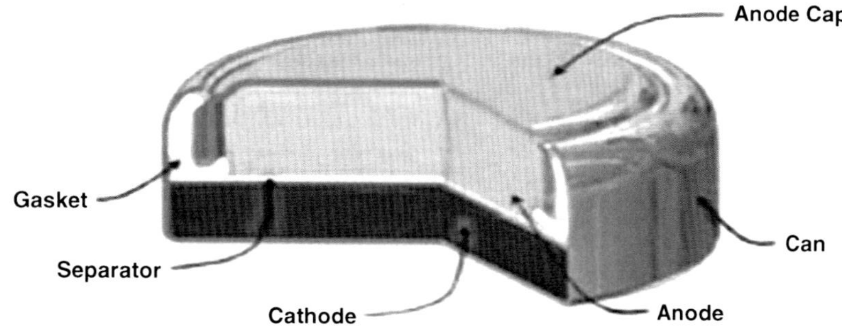

Alkaline Primary Cells, Fig. 7 Cutaway view of the alkaline $Zn–MnO_2$ coin cell construction (Courtesy of Duracell)

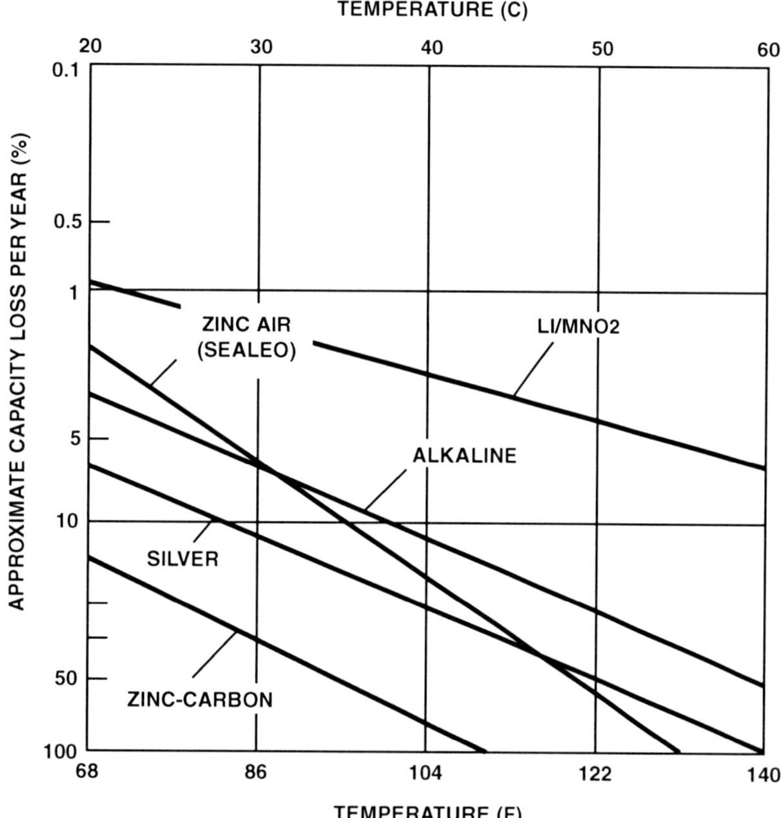

Alkaline Primary Cells, Fig. 8 Capacity loss on storage of various miniature alkaline cell systems compared to carbon–zinc and lithium–manganese dioxide cells. The lithium-based cells have significantly better storage characteristics compared to alkaline and carbon–zinc cells

Figure 9 depicts the effect of temperature on lithium–manganese dioxide compared with zinc–silver and zinc–air cell miniature cell performance. At low temperatures the electrode reactions slow down and severely limit the ability of the cell to deliver current. Cells also lose capacity slowly when stored at room temperature, due to internal self-discharge reactions which are accelerated at higher temperatures.

Zinc–Air Cells

Edison invented the first zinc–air cells in the early 1900s. The Edison Carbonaire (Zn–air) system was used to power railroad signaling devices in rural areas that were without electricity and as navigation aids. These were large box cells with a zinc plate anode and a porous carbon electrode with MnO_2 catalyst for the oxygen electrode. The present Zn–air miniature cell is an offshoot

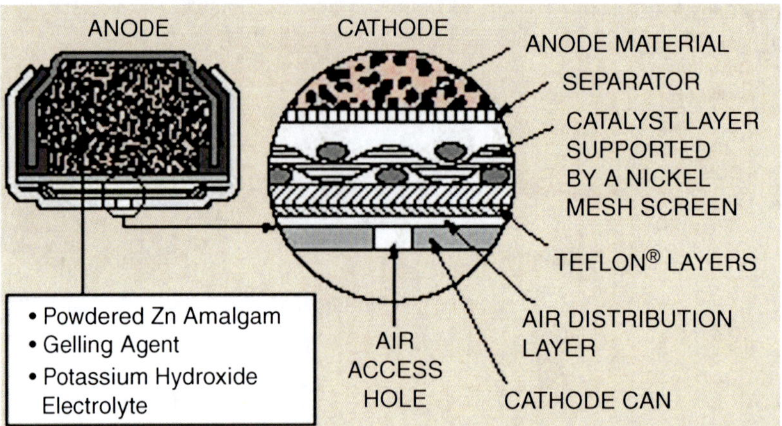

Alkaline Primary Cells, Fig. 9 Construction of a zinc–air (Zn–air) button cell with controlled air access, mainly used in hearing aids that fit into the ear hole. A tab is placed across the air access hole to prolong the shelf life of the Zn–air cell. The tab is removed to activate the cell. The cells are also made in a coin cell size with larger diameter and longer service life. The larger coin cells have a similar construction (Courtesy of Duracell)

based on the thin, high-rate, polymer-bonded, oxygen fuel cell electrodes (developed during the gasoline/energy crisis in the 1970s) and the powdered zinc materials from the alkaline Zn–MnO_2 cell.

The internal structure of the zinc–air cell is given in Fig. 10. Air is "free," so to speak, and gives the zinc–air cell a big performance advantage over the zinc–silver and the zinc–manganese alkaline cells that require the cathode reactants to be stored within the cell. Since the air electrode, itself, takes little volume, it allows for an increased amount of zinc anode material to be stored inside the cell structure, resulting in a longer service life. The zinc–air cell has the highest energy Wh/l and Wh/kg capability of any alkaline cells and is available in the coin and button cell configurations.

The zinc–air cell finds application as the energy supply for medical components, especially hearing aids that require a lightweight, maximum energy source in a small volume. Cells are produced in two main types, a coin size, like a quarter, and a button cell format with a smaller diameter and a thicker or taller format. The structure of the button cell configuration in Fig. 10 was designed specifically for hearing aid applications. The smaller electrode surface area limits the high-rate capability.

Advantages	Disadvantages
Very high energy storage	Limited life activated
Flat discharge voltage	Carbonation of the electrolyte
No environmental problems	High humidity affects performance
Long shelf life, inactivated	Activated by tape removal
Low cost	

The air electrode replaces the MnO_2 cathode, thereby creating a significantly larger zinc anode compartment and a resultant longer active service time. Since O_2 is readily available at all times from outside the cell, the cell can deliver current as long as the metallic zinc remains available for reaction. To prevent self-discharge before use, an external gas barrier layer is applied over the air (oxygen) inlets in the base of the cell, to prevent oxygen from the air to enter the cell and react directly with the zinc powder anode. Removing the barrier layer activates the cell in a matter of a few minutes that it takes for the outside air to fully permeate the air electrode.

Electrode Reactions

$$\text{Anode}: \ \mathrm{Zn} + 4\ \mathrm{OH}^- = 2\mathrm{e}^- = \mathrm{Zn(OH)_4}^{-2} \quad (5)$$

$$\mathrm{Zn(OH)_4}^{-2} = \mathrm{ZnO} + \mathrm{H_2O} + 2\ \mathrm{OH}^- \quad (6)$$

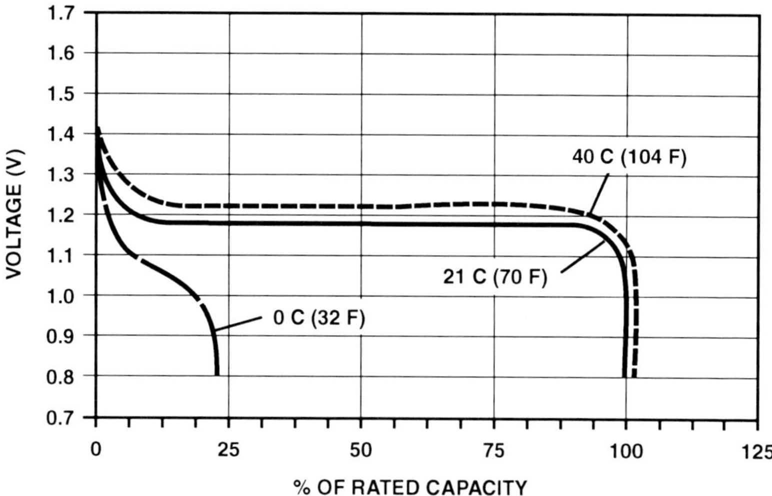

Alkaline Primary Cells, Fig. 10 Discharge characteristics of the zinc–air cell at 0 °C, 21 °C, and 40 °C at the 75-h rate. The constant voltage maintained during discharge is best for hearing aid applications (Courtesy of Duracell)

$$\text{Cathode}: \tfrac{1}{2}\, O_2 + H_2O + 2e^- = O_2H^- + OH^- \quad (7)$$

$$O_2H^- = OH^- + 1/2O_2 \quad (8)$$

$$Zn + \tfrac{1}{2}\, O_2 = ZnO \quad (9)$$

E = 1.2 V 450 Wh/l 150 Wh/kg

The Zn–air system has the highest energy storage capability, at 1,350 Wh/l and 415 Wh/kg, of any battery system. The discharge voltage is flat (unchanging) for the full discharge. The constant voltage discharge characteristic is ideal for use in powering hearing aids and other electronic equipment.

The air electrode usually contains a MnO_2 catalyst to decompose the peroxide reaction intermediate (O_2H^-) in the O_2 reduction reaction. The use of platinum catalyst makes the cell too expensive and is not needed for the relatively low-rate applications for miniature Zn–air. On active stand, the cell electrolyte will absorb water and CO_2 from a humid environment and will lose water to a dry environment, both of which will lower the cell performance.

The zinc electrode in the Zn–air cell has the same basic composition as the zinc powder gel anode, used in the assembly of the alkaline primary cell, and contains less than 0.25 mg/cell mercury as permitted by law. Organic corrosion inhibitors may be added to the electrolyte and other alloy elements added to the zinc to retard corrosion. The air electrode is a combination of Teflon-bonded carbon matrix with a nickel screen current collector adapted from fuel cell technology. The porous Teflon barrier layer separates the air electrode from the outside air supply. The hydrophobic nature of Teflon holds the aqueous KOH electrolyte inside the cell yet allows the oxygen from the air to permeate the cathode. The air (oxygen) reactant has a limitless supply, so the cell capacity is controlled by the amount of zinc contained in the cell.

The final step in cell assembly of the zinc–air cell is the application of a metalized barrier tab to cover the air holes and prolong shelf life. The cell remains inert until activated by removing the tab so oxygen (air) can enter and permeate the porous electrode structure. Once the barrier is removed, the cell will self-activate so that approximately 60 s should be allowed before use to ensure that sufficient oxygen diffuses in and is available for reaction at the positive electrode. The cells have good storage characteristic and can deliver over 80 % of their original capacity after 4-year storage at 25 °C. However, once the tab is removed, the activated cells have a short shelf life and can lose as much as 50 % of their capacity in 3 weeks on activated stand.

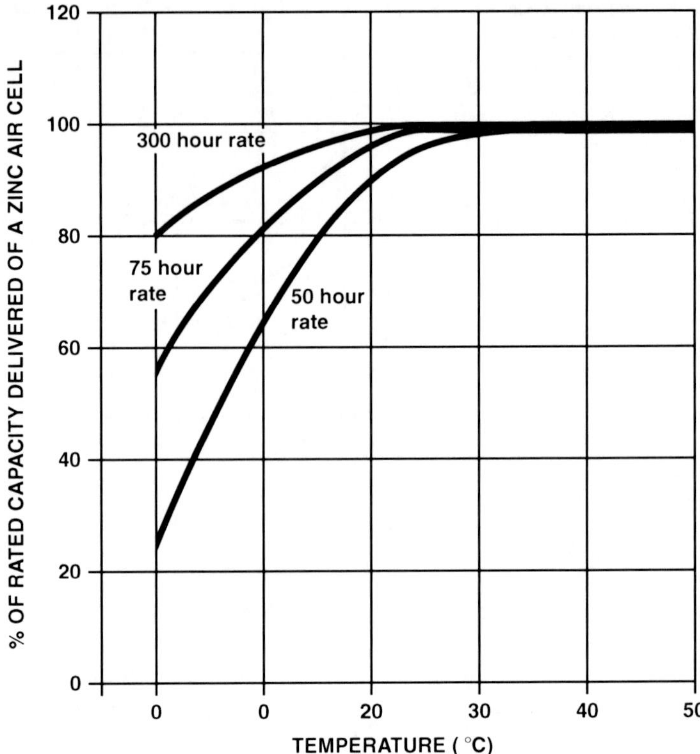

Alkaline Primary Cells, Fig. 11 Discharge characteristics of the zinc–air alkaline cell at various temperatures. The performance falls off quickly below 20 °C, due to buildup of the peroxide reaction intermediate (Courtesy of Duracell)

Figure 11 illustrates the discharge capacity of an air cell at the 300-, 75-, and 50-h rate at 40 °C, 21 °C, and 0 °C. The cell has a steady voltage throughout its useful life. The performance of the air electrode is sensitive to discharge rate, temperature, and humidity. The oxygen electrode reactions have strong temperature dependence, and performance begins to fall off below 20 °C. At lower temperatures, the peroxide intermediate in the reaction 11 becomes more stable and can interfere with normal electrode operation and shorten cell life. The cells have reasonable shelf life and retain about 90 % of the rated capacity after 2 years storage at room temperature.

Zinc–Silver Oxide

The first silver–zinc (Ag–Zn) cell, the voltaic pile, was constructed by Professor Alessandro Volta in Como, Italy, in about 1796. The Volta pile consists of alternate layers of silver and zinc sheets immersed in a salt-containing electrolyte. It was quickly adopted as a source of electricity for laboratory experiments. It was not until 1941 that Professor Henri Andre in France invented the rechargeable silver–zinc battery using a semipermeable cellophane separator to power his electric car. The separator retards the diffusion of silver ions to the negative zinc electrode, which makes the rechargeable silver–zinc feasible.

The Ag–Zn cells have very high energy density, constant voltage discharge, and excellent high-rate discharge capability. The cells are rugged and have very good low-temperature performance and good shelf life and are resistant to leakage. However, their high cost restricts their use. A comparison of the performance of silver, air, and manganese cells was given in Fig. 9.

Advantages	Disadvantages
High energy density	High cost
Flat discharge voltage	
Excellent high-rate capability	
Low, constant internal resistance	
Good shelf life	

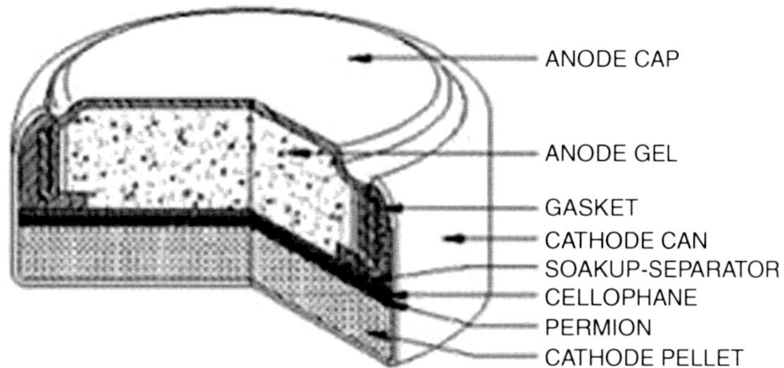

Alkaline Primary Cells, Fig. 12 Construction of a zinc–silver (Zn–Ag) (Courtesy of Duracell). The construction is very similar to that for the Li–MnO_2 coin cell but with an added cellophane layer to the regular separator material that prevents soluble silver ions from reaching the zinc anode and depositing on the zinc. The Zn–MnO_2 coin cell has an identical construction except that there is no need for the cellophane layer

The zinc–silver oxide cell (Ag–Zn) has the same basic structure as the Zn–MnO_2 cell replacing the MnO_2 with Ag_2O in a cathode pellet, as shown in Fig. 12. It has better high-rate performance but is more costly. The monovalent silver oxide (Ag_2O) is preferred as the cathode active material. A typical cell has an energy storage capability of 135 Wh/kg and 530 Wh/l. The divalent silver oxide (AgO) and trivalent silver oxides (Ag_2O_3) have higher energy density but are unstable in KOH electrolyte. Their limited shelf life is not suitable for commercial applications.

Cell Reactions

$$\text{Anode}: \ Zn + 4\ OH^- - 2e^- = Zn(OH)_4^{\ -2} \qquad (10)$$

$$Zn(OH)_4^{\ -2} = ZnO + H_2O + 2\ OH^- \qquad (11)$$

$$\text{Cathode}: \ Ag_2O + H_2O + 2e^- = 2\ Ag + 2\ OH^- \qquad (12)$$

$$\text{Overall}: \ Zn + Ag_2O = 2Ag + ZnO \qquad (13)$$

E = 1.59 V 175 Wh/kg 530 Wh/l

Cell construction of the Zn–Ag miniature cell is essentially identical to that for the Zn–MnO_2 coin/button cell with the silver oxide replacing the manganese cathode. The Zn–silver cells have a constant discharge voltage over their useful life. Silver has a small but finite solubility in KOH or NaOH electrolytes. If silver ions penetrate the separator, they will deposit on the surface of the zinc particles and decrease the amount of zinc available for discharge (self-discharge). The cellulosic membranes react with the soluble silver and, with time, lower the useful capacity of the cell. Synthetic polymer materials are preferred as barrier separators to prevent Ag^+ ion migration. A typical discharge curve is given in Fig. 14. The cells have good charge retention and deliver about 90 % of the original capacity after 5-year storage at 20 °C.

Large Silver–Zinc (Ag–Zn) Cells

Ag–Zn cells have excellent energy storage and high-rate discharge capability, good shelf life, and a flat discharge characteristic. They find many military and aerospace applications, where performance, weight, and volume are paramount over cost. The high energy density cells occupy only 25–50 % of the volume and weigh 20–30 % of Ni–Cd and lead–acid rechargeable batteries of the same capacity. The chemistry is the same as noted in Eqs. 10, 11, 12, and 13.

Ag–Zn cells have exceptional high-rate capability and are very reliable. They are rated at 175 Wh/kg and 500 Wh/l. They can deliver their full capacity at essentially constant voltage in 15 min or less. Ag–Zn cells power most torpedoes used in submarines. The largest rechargeable silver–zinc batteries were used in

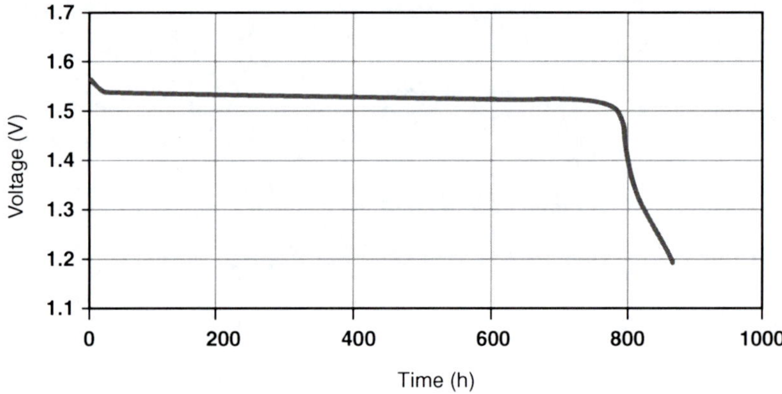

Alkaline Primary Cells, Fig. 13 Typical discharge curve for the Zn–Ag miniature cell (Courtesy of Duracell). Like the Zn–air cell, the Zn–Ag_2O cell discharges at constant voltage, a very useful characteristic for use in powering electronic devices

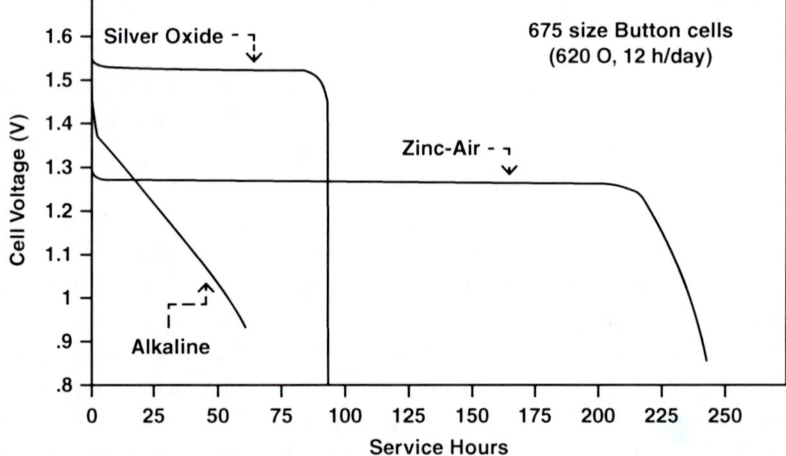

Alkaline Primary Cells, Fig. 14 Comparison of the performance of the zinc–air, silver, and alkaline miniature cells

submarines (256 t by the United States and 300 t by Russia). The main disadvantage is their limited cycle life (hundreds) compared to regular rechargeable lead acid batteries which can deliver many thousands of cycles.

Space applications for Ag–Zn cells include powering the space shuttle, space station, the lunar rover, Mars lander, and deep space probes. They were also used in the Saturn launch vehicles, the Apollo Lunar Lander, and life support jacket and to power space module operations on launch and return. In addition, they power spacecraft during launch and reentry. The lunar rover and space-walk suits are powered by the Ag–Zn battery systems. They also supply burst of power to supplement the fuel cell for operations in the command module.

Detailed technical information is available from the cell manufacturers' websites including www.energizer.com, www.duracell.com, and www.panasonic.com

Appendix

Battery Standards

The International Electrochemical Commission (IEC) is the designated organization responsible for standardization in the field of electricity, electronics, and related technologies. The American National Standards Institute (ANSI) serves as the administrator of the voluntary battery standardization system in the United States and is the US representative in the IEC. The minimum performance, the dimensions, and chemistry are also specified in the standard. The process for developing a standard requires that the cell manufacturers generate and agree on physical dimensions, cell termination, etc., as well as test methods, sizes, minimum performance levels, for each cell size and chemistry. As a result, the physical dimensions and shapes of the battery cell, e.g., AA-size cell or the 1625 coin cell, are the same in the United States as in Japan, China, Germany, etc. One can purchase the AA-size alkaline cell anywhere in the world and it will fit your device. This interchangeability has been a key element in the success and growth of the battery industry as well as the electronic devices they power.

The number in front of the cell size is the number of cells in the unit structure. The AAAA-size cells are not sold separately but packaged as a 9 V rectangular package.

Product Ban

In the past, most alkaline cells contained mercury as a corrosion inhibitor to prolong shelf life in alkaline electrolyte cells as well to act as the cathode active material in hearing aid cells. This ended on May 13, 1996. An exception was given for up to 25 mg in miniature alkaline button cells. The 1991 European commission directive 91/157, when adopted by member states, prohibited the marketing of certain types of batteries containing more than 25 mg of mercury, or, in the case of alkaline batteries, more than 0.025 % by weight of mercury. In 1998 the ban was extended to cells containing more than 0.005 % by weight of mercury.

In the United States, in 1992 the state of New Jersey prohibited sales of mercury batteries. In 1996 the United States Congress passed the Mercury-Containing and Rechargeable Battery Management Act (the Battery Act), 104–142, May 13, 1996, that prohibited further sale of mercury-containing batteries unless manufacturers provided a reclamation facility, with an exemption for alkaline zinc–air button cells.

Cross-References

► Carbon-Zinc Batteries
► Manganese Oxides
► Metal-Air Batteries

References

1. Kordesch KV (1974) Batteries, vol 1. Marcel Dekker, New York
2. Crompton TR (2000) Battery reference book, 3rd edn. Elsevier Science, Boston
3. www.duracell.com
4. www.energizer.com
5. European Community directive 91/157, 18 March 1957; 2006/66
6. U. S. Congress, Mercury-containing and rechargeable battery management act, 104–142, 13 May 1996
7. Falk SU, Salkind AJ (1969) Alkaline storage batteries. Wiley, New York
8. Himy A (1995) Silver-zinc battery: best practices, facts, and reflections. Vantage Press, New York
9. Karpinski A, Serenyi R, Salkind A, Bagotzky V (1995) The silver-zinc battery system, a 60 year prospective, from Andre, to Sputnik, to Mars. In: Proceedings of the symposium on rechargeable zinc batteries, proceedings, vol 95–14. The Electrochemical Society, Pennington
10. Reddy TB (2002) Lindens handbook of batteries, 4th edn. McGraw-Hill, New York. ISBN 0-07-135978-8
11. Root M (2011) The TAB battery book. McGraw Hill, New York

Aluminum Smelter Technology

Geir Martin Haarberg
Department of Materials Science and Engineering, Norwegian University of Science and Technology (NTNU), Trondheim, Norway

Introduction

Production of primary aluminum metal is rather unique in that the principles of the electrolysis technology that were proposed and independently patented by Hall and Heroult in 1886 are essentially unchanged. Also remarkable is the fact that the electrolytic Hall-Heroult process is the only industrial production route for aluminum. Notwithstanding, great progress has taken place over more than 100 years of development. The main improvements have been related to current efficiency, electrical energy consumption, productivity, and environmental impact.

The production of primary aluminum was ~41 million metric ton in 2010 [1]. The largest producing country was China with ~16 million ton. After a slight decrease in the annual production in 2009 due to world financial problems, there has been a slow increase due to the importance of aluminum alloys for transportation and building materials.

Alumina is dissolved and reduced to aluminum in a molten fluoride electrolyte based on cryolite (Na_3AlF_6) at ~950 °C [2]. Dissolved oxygen complexes are oxidized at consumable carbon anodes to give CO_2. Modern cells are operating at 400 kA with a current efficiency of ~95 % and an energy consumption of ~15 kWh/kg Al (Fig. 1).

Raw Materials and Electrolyte

Carbon anodes and aluminum oxide are the main consumable raw materials in the Hall-Heroult process. Small quantities of electrolyte components, mainly AlF_3, are also consumed.

There are two anode designs: the use of several prebaked anodes which are replaced regularly without disturbing the electrolysis and the use of one soderberg anode which is continuous and self-baking. Modern cells are equipped with prebaked anodes, mainly because of the size limitations and higher local pollution by using soderberg anodes. Both types of anodes are made from petroleum coke aggregate and coal tar pitch binder, the pitch content being about 13 wt% for prebaked anodes and about 25 wt% for soderberg anodes. Prebaked anodes are baked in furnaces at about 1,100 °C.

Cryolite (Na_3AlF_6), which has a melting point of 1,010 °C, is the only electrolyte known to dissolve appreciable amounts of aluminum oxide. The industrial electrolyte is modified by additions of AlF_3 and CaF_2 and in some cases LiF and MgF_2. These additions cause a reduced liquidus temperature and are known to be beneficial for obtaining a higher current efficiency.

The binary phase diagram of NaF-AlF_3 is shown in Fig. 2 [3]. The phase diagram of Na_3AlF_6-Al_2O_3 is a simple binary system where the alumina solubility is about 10 wt% at the eutectic temperature of ~960 °C [4]. The alumina solubility will decrease by adding other components such as AlF_3 and by lowering the temperature. Therefore, there is a limit to how much the temperature can be reduced.

Other important properties of the electrolyte are electrical conductivity, density, surface and interfacial tensions, vapor pressure, and viscosity.

Electrode Processes and Current Efficiency

The primary cell reaction in Hall-Heroult cells is

$$1/2\,Al_2O_3(diss) + 3/4\,C(s) = Al(l) + 3/4\,CO_2(g) \quad (1)$$

The cathode process involves the transport of the aluminum containing species from the bulk phase of the electrolyte through the diffusion boundary layer to the cathode/electrolyte interphase where the charge transfer reactions take place. It is known that sodium ions, Na^+, are the main carrier of the current in the electrolyte [2]. Al (III) is present as dissolved complexes involving aluminum, fluorine, and oxygen, the major species being $Al_2OF_6^{2-}$ and $Al_2O_2F_4^{2-}$ [5]. Concentration gradients may be established near the cathode surface. Due to the transport and electrode processes, sodium ions and fluoride ions will accumulate near the cathode surface. Therefore, the electrolyte composition is higher near the cathode/electrolyte interface than in the bulk electrolyte. This gives rise to a small but significant concentration overpotential [6]. The electron transfer for deposition of aluminum is very fast, which is a common feature for metal deposition reactions in high temperature molten salts.

Aluminum Smelter Technology, Fig. 1 Modern prebaked aluminum electrolysis cells at Sunndal, Norway

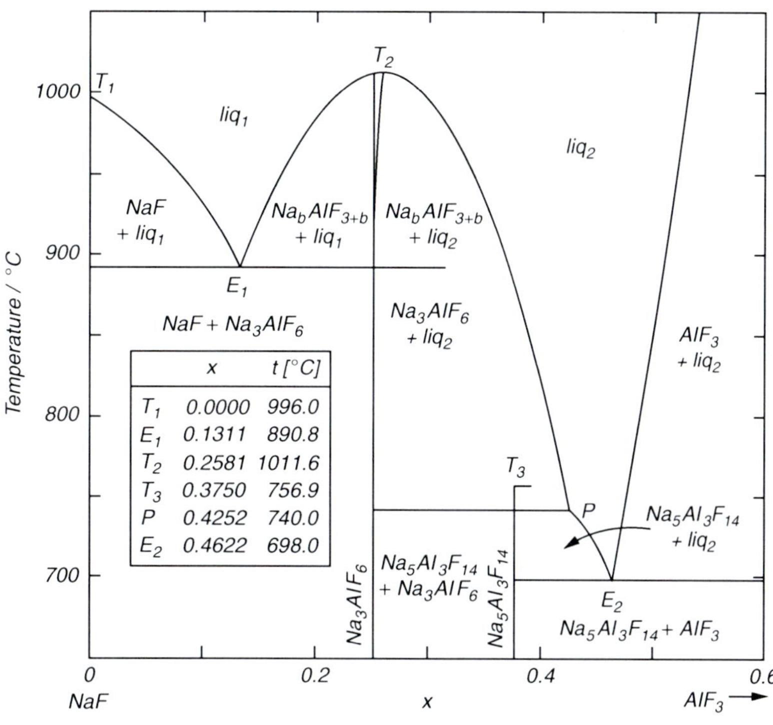

Aluminum Smelter Technology, Fig. 2 The binary phase diagram of NaF-AlF_3

The anode process involving the formation of adsorbed CO_x compounds, associated with a high overvoltage of several hundred millivolts, is reported [7] at normal current densities of about 0.7 A/cm^2. CO_2 is the primary anode product at normal current density, while CO is formed at very low anodic current densities [8].

An important feature of the process is the fact that aluminum dissolves in the molten electrolyte, which is a general phenomenon taking place when a metal is in contact with a molten salt containing the metal cation [9]. In molten cryolite-based electrolytes, dissolved Na must be considered in addition to dissolved Al. A small but significant activity of sodium is established at the metal/electrolyte interface due to the following equilibrium:

$$Al + 3\ NaF = AlF_3 + 3\ Na \qquad (2)$$

It is known that the subvalent species AlF_2^- is formed as well as dissolved Na, the latter being responsible for a small contribution to electronic conductivity [2]. Solubility studies have been carried out in laboratory experiments, and under industrial operation, the metal solubility is ~0.06 wt% Al [2]. The solubility decreases by increasing content of AlF_3 and decreasing temperature. Reliable data for the metal solubility have been published [10, 11].

The loss in current efficiency with respect to aluminum is mainly due to the so-called back reaction between dissolved Al and the anode product according to

$$Al(diss) + 3/2\ CO_2(g) = 1/2\ Al_2O_3(diss) + 3/2\ CO(g) \qquad (3)$$

The back reaction Eq. 2 takes place near the cathode surface, and the rate is controlled by the diffusion of dissolved Al through the diffusion layer [12]. Hence, the concentration of dissolved Al at the cathode/electrolyte interface and the diffusion layer thickness are important for the loss in current efficiency. The rate of the back reaction can be expressed as follows:

$$v_B = k\prime\ (dc/dx) = k\prime\ (c^o/\delta) \qquad (4)$$

where k' is a constant including the diffusion coefficient, c^o is the saturation concentration of dissolved metal at the cathode/electrolyte interface, and δ is the diffusion layer thickness. A theory for the mechanism for the loss in current efficiency for aluminum deposition in the industrial process has been presented [13]. Experimental results and model calculations for the current efficiency based on laboratory studies have been published [14, 15]. The effects of electrolyte impurities and electrolyte composition were included in the investigations. Realistic values for the current efficiency were obtained. The variation of the current efficiency with respect to current density, electrolyte composition, and temperature was essentially found to be closely linked to the metal solubility data. Recent experimental results from laboratory experiments are in agreement [16]. Current efficiencies are ranging from 85 % to 96 %, increasing by increasing the cathodic current density, increasing AlF_3 content, and lowering the temperature. Current efficiency data from industrial cells show similar trends. A revised model for calculating the current efficiency based on local variations in cathodic current density related to the three-phase flow of electrolyte at the cathode interface has also been proposed [17].

Impurities present in the Hall-Heroult process originate mainly from the raw materials, carbon and alumina. Some additional impurities come from bath components (AlF_3), tools, anode stubs, and sidelining (SiC) and refractory lining materials. The major part of these impurities is initially present in the electrolyte as dissolved fluoride or oxyfluoride complexes. The dissolution process can be expressed as follows:

$$3MO + 2AlF_3 = 3MF_2 + Al_2O_3 \qquad (5)$$

for a divalent cation. The concentration of impurities in the bath is normally well below saturation [2]. In cases where the decomposition voltage of MO is lower than that of Al_2O_3, M will be reduced at the cathode and end up in the produced aluminum. It has been shown that the transfer of such impurity elements from the bath to the cathode is mass transfer controlled [18]. According to this mechanism, an impurity element will be reduced at the cathode at its limiting current density (i_{lim}), which depends on the concentration of the element in the bath (c^o) and the mass transfer coefficient (k_m):

$$i_{lim} = nFk_mc^\circ \qquad (6)$$

The mass transfer coefficient increases with increasing convection. These impurities will mainly affect the purity of the produced aluminum, and iron and silicon are quantitatively the most important elements. Another class of impurities is exemplified by phosphorus which is less likely to be deposited at the cathode. Dissolved phosphorus complexes can exist in several different valencies and participate in cyclic oxidation and reduction reactions at the electrodes. In certain cases, up to 50 % of the loss in current efficiency due to impurities can be ascribed to phosphorus [14, 19, 20].

Cell Technologies and Operating Conditions

The different cell technologies for aluminum production depend on the nature of the carbon anodes and the current load. However, the electrolyte composition and the operation of the electrolysis is very similar for all technologies. Information about innovations and performance data related to aluminum electrolysis has traditionally been very open. Figure 3 shows a schematic drawing of a modern prebaked cell.

There are so-called soderberg cells with horizontal or vertical anode studs and prebaked cells which are arranged side by side or end to end. The preferred design is prebaked cells placed side by side to reduce the footprint and optimize the magnetic field compensation. One of the major breakthroughs in the history of developing and optimizing the Hall-Heroult process was the ability to control and compensate for the influence of the magnetic fields on the movement of the liquid aluminum metal pad. A strong magnetic field is established from the large current passing through the cell [21]. Disturbances of the liquid Al give rise to loss in current efficiency, which can be realized from Eq. 4.

The interpolar distance is quite large, varying from 3.5 to 4.0 cm in modern cells. It helps to separate the electrode products and to provide a useful heat regulating tool since no external heating is provided. The horizontal configuration combined with the very large liquid Al cathode and the magnetic fields can explain the need for operating at such a high interpolar distance. Another limitation caused by the horizontal design is the need for having a high level of liquid on top of the carbon cathode blocks, which is due to the poor wetting of liquid Al on solid carbon. An essential feature of the Hall-Heroult process is the use of a frozen layer of the electrolyte for containing the molten bath. This is due to the lack of inert container materials and is achieved and maintained by establishing a certain heat loss through the sides of the cell. The operating temperature is typically 950 °C, about 10 °C above the liquidus temperature.

Alumina is added batchwise through point feeders at a regular frequency of a few minutes. The content of dissolved alumina is estimated by measuring the electrical resistance in the bath, since there is a correlation between alumina content and conductivity.

The liquid aluminum active cathode is resting on top of carbon cathode blocks mainly consisting of semi-graphitic or graphitized carbon material. Interaction with electrolyte, solid alumina, and sodium affects the cathode blocks and may cause erosion and swelling of the carbon. The cell life is commonly limited by the lifetime of the cathode blocks, which may vary from 5 to 10 years. Disposal of spent pot lining remains an unresolved environmental challenge. Inert cathode materials such as TiB_2 are still not implemented in industrial cells but may be necessary for developing alternative cell designs such as drained cells.

Anodes are made from petroleum coke aggregate and coal tar pitch binder, prebaked anodes containing about 15 % pitch while soderberg anodes containing about 25 % pitch. The anode carbon consumption is higher than the theoretical due to some airborne and reaction with CO_2. Another adverse effect of using carbon anodes is the existence of the so-called carbon dust, which consists of unreacted coke particles fallen off the anode and may cause heating of the electrolyte. Anode effect is an undesirable situation caused by depletion of dissolved alumina. The anode potential will increase and other products such as CF_4 and C_2F_6 in addition to fluorine will

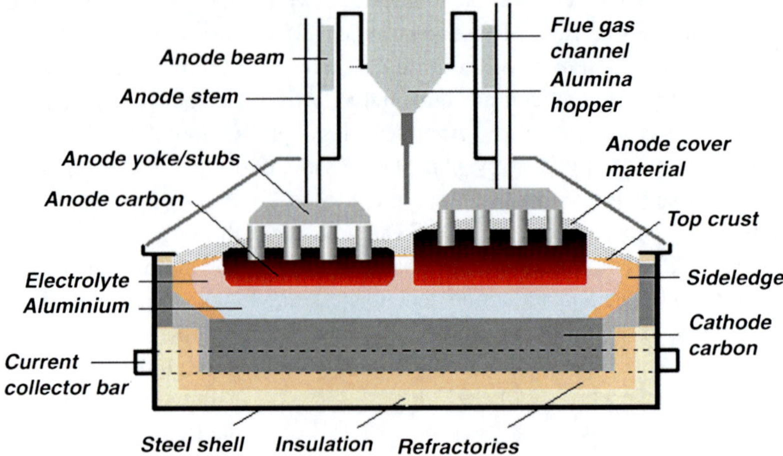

Aluminum Smelter Technology, Fig. 3 Schematic cross section of a prebaked aluminum electrolysis cell

be formed. The anode effect is undesired for several reasons: disturbance of the process, local and global environmental impact, and excess energy consumption. Some new cells are using slotted anodes to facilitate the escape of gas bubbles and minimize the resistance due to the gas bubbles underneath the anode.

The cell voltage is 4 V, which includes the reversible potential, the ohmic voltage drop of the bath, and the anodic and cathodic overvoltages. The energy efficiency is less than 50 %, mainly because of the large voltage drop of the electrolyte.

Future Directions

The trend of increasing the productivity by increasing the current load for existing cells will probably continue. Increased productivity and increased energy efficiency are key issues for developing new technologies. The development and implementation of inert anodes and capture and storage of CO_2 are likely to be postponed. The possibility to increase the cathodic current density is very attractive but very challenging, but research and development will probably be carried out. A lot of attention will be drawn to dealing with the potential risk of deteriorating quality of raw materials, especially focusing on the behavior of increasing amounts of impurities in alumina and anode carbon.

Cross-References

▶ Electrolytes for Rechargeable Batteries
▶ Electrolytes, Classification
▶ Electrolytes, History

References

1. U.S. Geological Survey. http://minerals.usgs.gov/minerals/pubs/commodity/aluminum/myb1-2010-alumi.pdf
2. Thonstad J, Fellner P, Haarberg GM, Hives J, Kvande H, Sterten Å (2001) Aluminium electrolysis. Fundamentals of the Hall-Heroult process. Aluminium-Verlag, Düsseldorf
3. Solheim A, Sterten Å (1997) Activity data for the system NaF-AlF$_3$. Proceedings of the Ninth international symposium on light metals production, Trondheim, Norway 225
4. Skybakmoen E, Solheim A, Sterten Å (1997) Met Mat Trans B 28B:81–86
5. Sterten Å (1980) Electrochim Acta 25:1673
6. Thonstad J, Rolseth S (1978) Electrochim Acta 23:223–241
7. Jarek S, Thonstad J (1987) Light Metals 1987:399–407
8. Thonstad J (1964) J Electrochem Soc 111:959
9. Bredig MA (1964) Mixtures of metals with molten salts. In: Blander M (ed) Molten salt chemistry. Interscience, New York
10. Ødegård R, Sterten Å, Thonstad J (1987) Light Metals 1987:389
11. Wang X, Peterson RD, Richards NE (1991) Light Metals 1991:323
12. Rolseth S, Thonstad J (1981) On the mechanism of the reoxidation reaction in aluminum electrolysis. Light Metals 1981:289–301
13. Sterten ÅJ (1988) Electrochem. 18:473

14. Sterten Å, Solli PA, Skybakmoen E (1998) J Appl Electrochem 28:781
15. Sterten Å, Solli PA (1995) J Appl Electrochem 25:809
16. Haarberg GM, Armoo JP, Gudbrandsen H, Skybakmoen E, Solheim A, Jentoftsen TE (2011) Current efficiency for aluminium deposition from molten cryolite-alumina electrolytes in a laboratory cell. Light Met 2011:461–463
17. Li J, Xu Y, Zhang H, Lai Y (2010) An inhomogeneous three-phase model for the flow in aluminium reduction cells. Int J Multiphase Flow 37:46–54. doi:10 1016/j.ijmultiphaseflow.2010.08009
18. Johansen HG, Thonstad J, Sterten Å (1997) Light Met 1977:253–261
19. Deininger L, Gerlach J (1979) Metall 33:131
20. Haugland E, Haarberg GM, Thisted E, Thonstad J (2001) The behaviour of phosphorus impurities in aluminium electrolysis cells. Light Met 2001:549
21. Haupin WE (1995) Principles of aluminum electrolysis. Light Met 1995:195–203

Amperometry

Jens Zosel
Kurt-Schwabe-Institut für Mess- und Sensortechnik e.V. Meinsberg, Waldheim, Germany

Introduction

Definition

The term "amperometry" describes an electroanalytical technique which is based on a constant polarization voltage ranging within the plateau of the diffusion limited current [1, 2] according to Fig. 1. According to the comparison given by Delehay et al. [3], amperometry can be distinguished easily from voltammetry by the quantity being controlled (electrode potential E and concentration c, respectively) and the quantity being measured (electrode current $I = \mathrm{f}(c)$ and $I = \mathrm{f}(E)$, respectively). The electrode potential E of an amperometric measuring electrode has to be fixed within the diffusion limited current plateau in the voltammogram $I = \mathrm{f}(E)$.

History

One of the first amperometric sensors, developed by Rideal and Evans 1913 [4], consists of a two-electrode cell with a platinum working electrode and a copper counter-electrode. Such an arrangement is still used to measure dissolved chlorine in water [5]. In the late 1950s Clark [6] and Tödt [7] developed two different approaches for measuring dissolved oxygen. The variety of possible applications of amperometry in liquid analysis was pointed out by Schwabe [8] in 1965. First applications of amperometry for gas analysis with liquid electrolyte cells were described in 1970 [9] and with solid electrolyte cells in 1966 [10]. Nowadays amperometry is a widely well introduced technique for different kind of sensors, HPLC detectors and analytical devices.

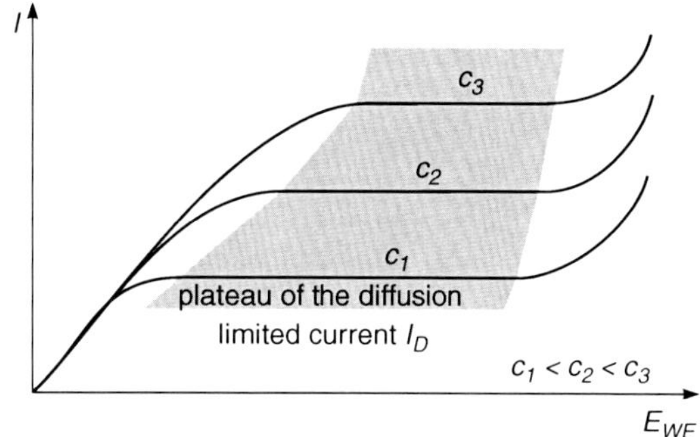

Amperometry, Fig. 1 Schematic drawing of voltammograms at different concentrations $c_1 \ldots c_3$ with diffusion limited current plateaus, symbols according to Eq. 1

Fundamentals

Amperometry in Liquids

Amperometry is performed in most cases with two electrode cells. Here the potential of the working electrode E_{WE} depends on the polarization voltage U_P, the potential of the counter electrode E_{CE} (which has to be unpolarizable as it serves also as the reference electrode in a two-electrode cell) and the Ohmic drop within the cell $I \cdot R_E$:

$$E_{WE} = U_P - E_{CE} - I \cdot R_E \qquad (1)$$

By definition E_{WE} has to be adjusted by U_P in the range where the concentration of the electroactive species (EAS) at the surface of

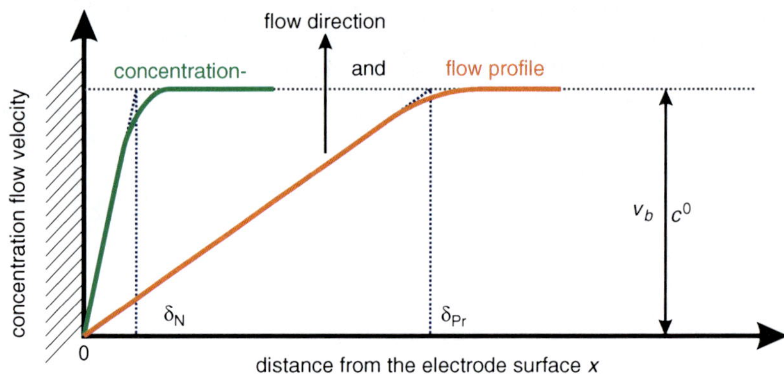

Amperometry, Fig. 2 Gradients of concentration and flow velocity near the surface of a polarized electrode in a flowing analyte: δ_N Nernst diffusion boundary layer; δ_{P_r} Prandtl flow boundary layer; v_b flow velocity in the bulk; c^0 bulk concentration of the electroactive species

the working electrode is equal to zero and the electrode current I is controlled exclusively by the diffusion of that species from the bulk to the electrode surface [11] as indicated in Fig. 2. The electrode surface A, the diffusion coefficient D of the EAS and its concentration in the bulk c^0 determine the diffusion limited current I_D:

$$I_D = z \cdot F\ A\ D\ c^0/\delta_N \tag{2}$$

with z as the number of electrons participating in the electrode reaction and F as the Faraday constant (96,485 As/mol). The thickness δ_N of the Nernst diffusion layer depends on the flow conditions near the electrode surface, the diffusion coefficient and other analyte parameters [12]. Therefore, amperometry with uncovered electrodes requires constant flow conditions at the electrode surface. On the other hand, the technique can also be used to quantify near surface flow conditions at constant and well defined concentration of EAS [13].

To avoid the flow dependency of amperometry at uncovered electrodes a defined diffusion layer thickness is often established by covering the electrode surface with a solid membrane which enables the EAS passage by solution diffusion or diffusion through perforation(s) [14]. An example of a membrane covered two-electrode amperometric cell is shown in Fig. 3.

Another fundamental advantage of that separation membrane consists in the possibility to optimize the inner electrolyte regardless of the measuring fluid.

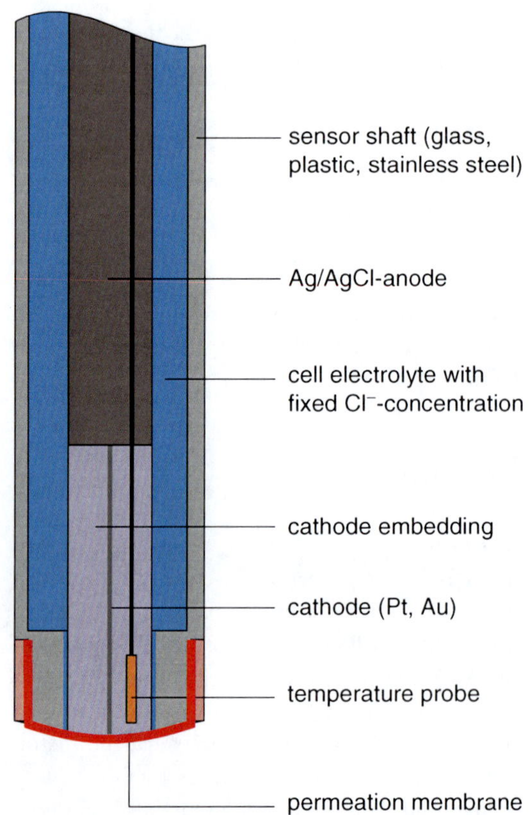

Amperometry, Fig. 3 Schematic drawing of a membrane-covered amperometric Clark sensor

The I_D in liquid analytes is determined by the membrane permeation P_M and its thickness b [15]:

$$I_D = z \cdot F \cdot c^0 \cdot A \cdot P_M/b \tag{3}$$

In case of dissolved gas measurement in liquids the concentration c has to be substituted by the partial pressure of the EAS p.

Very often polymeric membranes are utilized for amperometric two- or three electrode cells for measurements in liquids and gases. Typical materials are polyethylene (PE), polypropylene (PP) and polytetrafluoroethylene (PTFE) [16].

Amperometry in Gases

The kind of cell shown in Fig. 3 can be used in principle also for measurements in gases. Basic equations given in chapter 1.3.1 are valid also for this application. Another different form of amperometric cell that is extensively used in gases is based on solid instead of liquid electrolytes. To establish appropriate ion conductivity levels these electrolytes have to be operated at elevated temperatures in the range 500–800 °C [17]. To withstand those temperatures the diffusion barrier of these cells is often made of perforated or porous ceramics as indicated in Fig. 4.

In case of a limited current sensor with a leak aperture according to Fig. 4 the flux J of the gaseous component to be measured to the working electrode can be calculated from the following equation:

$$J = -D \cdot A \frac{dc(u)}{du} + c(u)Av \qquad (4)$$

with D is the diffusion coefficient, A is the cross section area of the leak aperture, c is the concentration of the EAS, v the mean velocity of convection through the diffusion hole and u is the distance coordinate along the leak aperture.

Integrating that equation and introducing the side conditions of the amperometric sensing principle result in the two Eqs. 5 and 6 for the diffusion limited current I_D of amperometric solid electrolyte sensors, depending on the kind of diffusion in the narrow channel(s) [18]. In case of ordinary diffusion conditions it is

$$I_D = -\frac{zFD_sAT^{0,75}}{RL} \cdot \ln[1-x] \qquad (5)$$

and under Knudsen diffusion conditions it is proportional to the molar fraction x of the EAS:

$$I_D = \frac{zFd^3\pi^{1/2}}{3L\sqrt{MRT}} \cdot x \cdot p_t \qquad (6)$$

with D_S is standard diffusion coefficient of the EAS, d, A and L are the diameter, cross section and length of the leak aperture, respectively, p_t is the total pressure, R the gas constant and M the molecular mass of the EAS. Equations 5 and 6 have been verified experimentally with good accuracy both with leak apertures and porous layers covering the measuring electrodes [19–21].

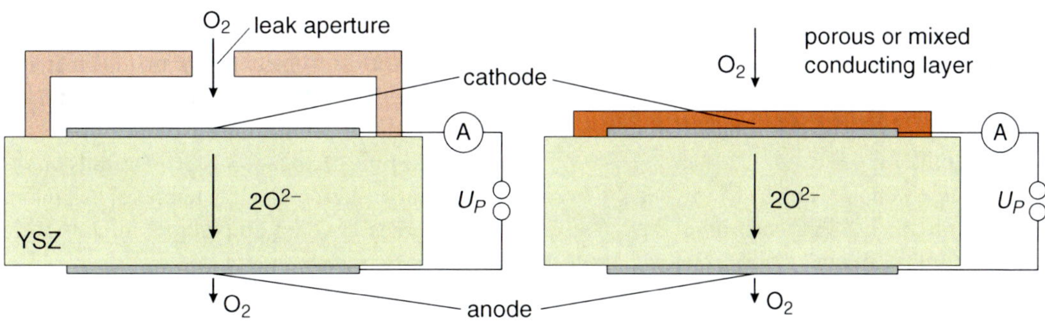

Amperometry, Fig. 4 Amperometric solid electrolyte sensors with: *left* gas-tight diffusion barrier with leak aperture, *right* porous or mixed conducting barrier, solid electrolyte = yttria stabilized zirconia (*YSZ*)

Applications

Liquid Analysis

Membrane Covered Sensors

The online measurement of dissolved oxygen is one of the main application areas of amperometry. Due to its importance, an IEC document specifies details of membrane-covered amperometric cells [22] for that purpose. As shown in Fig. 3, at the so-called Clark sensor the electrochemical sensor system is separated from the medium to be measured by a thin oxygen-permeable polymer membrane. The cathode is arranged immediately behind the membrane. If an external polarization voltage (typical −600 ... −700 mV vs. the Ag/AgCl-Ag^+ anode) within the plateau of the diffusion limited current is applied at the cathode, the oxygen permeating through the membrane is completely transformed at the cathode. In an alkaline electrolyte the following reactions take place:

$$\text{Cathode:}\ O_2 + 2H_2O + 4e^- \rightarrow 4OH^- \quad (7)$$

$$\text{Anode:}\quad 4Ag \rightarrow 4Ag^+ + 4e^- \quad (8)$$

The limited current I_D is almost independent of the applied polarization voltage and can be calculated from the Eq. 9:

$$I_D = z \cdot F \cdot p_{(O2)} \cdot A \cdot P(T)/b \quad (9)$$

where $p_{(O2)}$ is the oxygen partial pressure in the measuring solution. For PTFE the permeability amounts to $P = 3 \cdot 10^{-13}\ cm^2\ s^{-1}\ Pa^{-1}$ at 25 °C, while the value for PE ranges around $P = 1 \cdot 10^{-13}\ cm^2\ s^{-1}\ Pa^{-1}$ at the same temperature.

As an example, for a Clark sensor with a 12.5 μm thick PTFE membrane and a gold cathode with 1.5 mm diameter at the oxygen partial pressure $p_{(O2)} = 21$ kPa the resulting current would be $I_D = 1.5$ μA.

Membrane-covered amperometric sensors can be applied for measurements in gases as well as in liquids. It is worth mentioning that with this type of sensor the partial pressure of the respective species is determined and not the concentration. To get the concentration of the species in a liquid the (temperature-dependent) Henry constant k_H must be known. It is e.g., at 298 K for oxygen dissolved in water $k_H = 769.2$ L atm/mol. Dependent on the area of the cathode and the membrane parameters the oxygen consumption at the cathode the Nernst diffusion boundary layer can extend into the measuring solution. Those sensors require a certain flow of the measuring medium in front of the membrane [23].

A particular feature of membrane-covered amperometric sensors is the pronounced temperature dependence of the sensor signal that must necessarily be compensated. For this reason, generally a temperature probe (e.g., NTC or Pt 1000) is integrated in the sensor as shown in Fig. 3. According to an Arrhenius equation the permeability of the membrane material depends exponentially on temperature:

$$P_M(T) = P_{M0} \times e^{(-T_0/T)} \quad (10)$$

For the common membrane materials the value of T_0 ranges from about 3,000 K (PTFE) to 6,000 K (PP/PE) as illustrated in Fig. 5. The resulting temperature coefficients $(dI_D/I_D)/dT$ are 3.4 %/K and 6.8 %/K at 25 °C.

In fact, the temperature dependence of $P_M(T)$ is still much more complex. To achieve the required measuring accuracy, it must be individually determined for the utilized membrane material and compensated by microcomputer.

The response time of membrane-covered amperometric sensors depends on the material of the membrane and its thickness. For example, a sensor with a 20 μm thick PP membrane has a response time of about $t_{90} = 30$ s. Details on the theoretical background of the temporal behaviour of membrane covered amperometric cells were published by Mancy et al. [24].

To avoid errors caused by voltage drops in the cell, instead of the simple two-electrode cell a three-electrode cell can be used with an additional reference electrode that should be positioned as close as possible to the cathode.

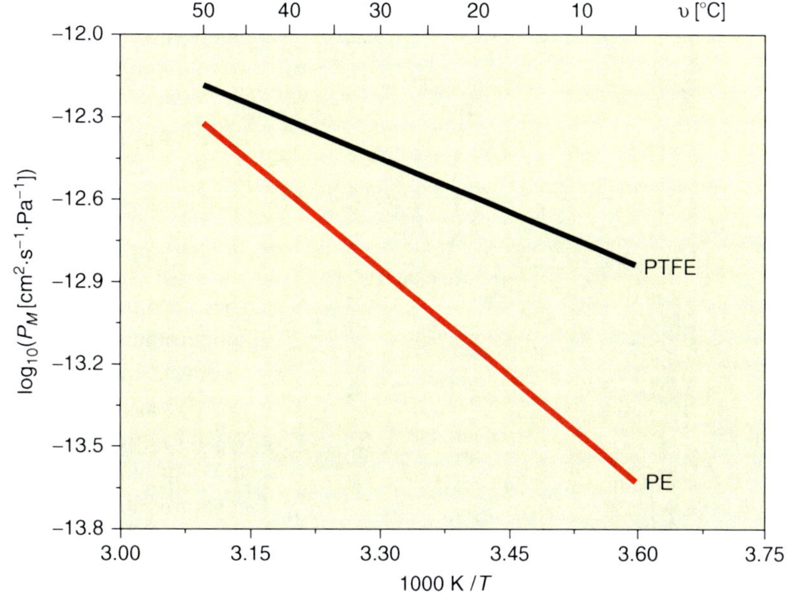

Amperometry, Fig. 5 Calculated temperature dependence of the membrane permeation P_M of PTFE and PE with P_{M0} and T_0 values given in the text

Figure 6 shows the different measuring circuits for amperometric two- and three-electrode cells. In the circuit for the three-electrode cell an electronic potentiostat applies exactly the preselected polarization voltage between the cathode and the reference electrode and compensates nearly completely the voltage drop $I_D \cdot R$ in the cell.

For special investigations in biomedical research miniaturized amperometric oxygen sensors have been developed. At the sensor shown in Fig. 7 the diameter of the Pt cathode is only 30 μm. The resulting limited current of such a sensor is in the range $I_D = 1$ nA at the oxygen partial pressure $p(O_2)\cdot = 21$ kPa.

Miniaturized amperometric sensors have been prepared also in thick-film and thin-film techniques [25]. As an example, in Fig. 8 the layout of a planar amperometric oxygen sensor in thin-film technique is shown. The KCl electrolyte and PUR membrane layers covering the electrode system are applied by dispenser technique.

In a modified membrane-covered amperometric cell a Pb anode is combined with an Ag cathode in a KOH solution [26, 27]. In this sensor, which is usually called Mackereth sensor, the following reactions take place:

a

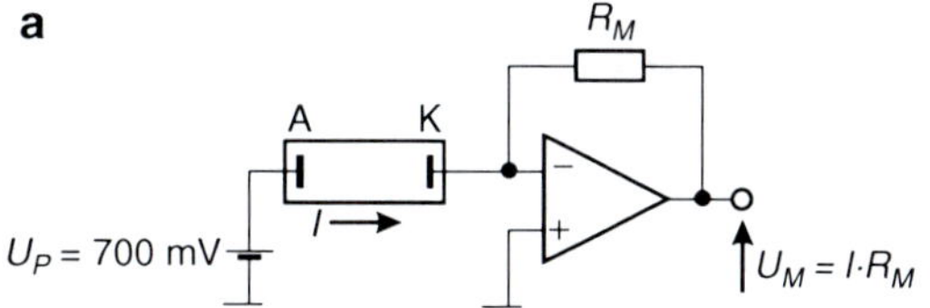

Two-electrode amperometric cell

b

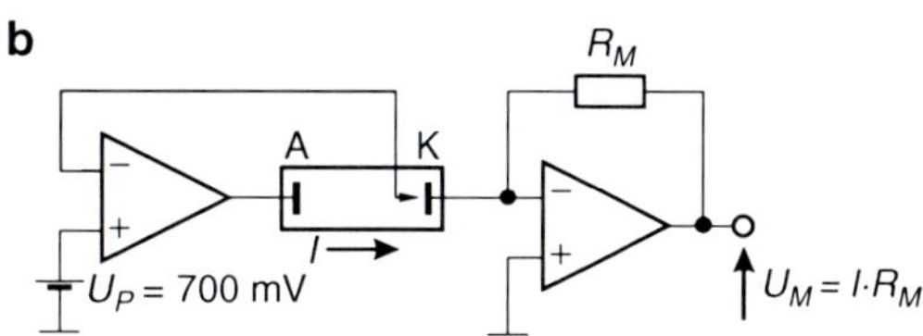

Three–electrode amperometric cell

Amperometry, Fig. 6 (**a**) Measuring circuits for amperometric sensors. Two-electrode amperometric cell. (**b**) Measuring circuits for amperometric sensors. Three-electrode amperometric cell

At the Ag cathode:

$$O_2 + 2H_2O + 4e^- \rightarrow 4OH^-$$

At the Pb anode:

$$2Pb \rightarrow 2Pb^{2+} + 4e^-$$

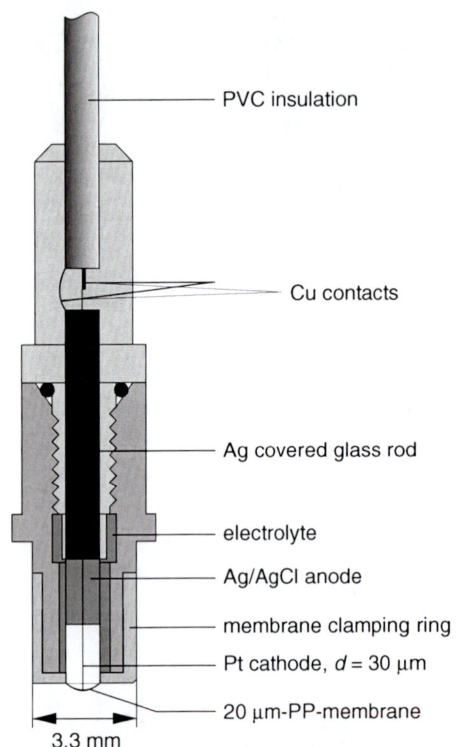

Amperometry, Fig. 7 Miniaturized membrane-covered electrochemical oxygen sensor for biomedical investigations in small measuring volumes

The advantage of this system is that its cell voltage is within the range of the plateau of the limit current of oxygen reduction. Thus, no external polarizing voltage must be applied.

Membraneless Sensors

Membraneless amperometric cells as the earliest amperometric sensors use the liquid to be measured also as the electrolyte. Therefore, the potential of the counter electrode is not fixed and (without an additional reference electrode) the potential of the working electrode as well. The electrode current is governed completely by mass transport control [28]. An example of such a cell construction is published in [7]. A similar system studied by Züllig [29] was equipped with a cleaning device for the electrodes. The other significant side condition of membraneless probes consists in the flow dependency of the signal according to Eq. 2, requiring constant analyte exchange at the electrode surface. Industrial oxygen probes have been used successfully to control the aeration of the biomass in sewage plants.

Other Applications

High-performance liquid chromatography (HPLC) as a powerful analytical tool uses detectors that provide a signal proportional to the concentration for each of the species separated by the chromatographic column. Membraneless amperometric detectors, also with glassy carbon and wax-impregnated graphite electrodes, are widely used for this purpose, especially for organic mixtures [30]. Besides high sensitivity they provide a wide dynamic range, rapid response, low signal to noise ratio and comparably small dimensions.

A variety of biochemical sensors utilize amperometric detectors as the transducer. The receptor reacts specifically with the desired component in the measuring solution. Typical products of such biochemical reactions are inorganic ions or molecules such as H^+, NH_4^+, CN^- or O_2, H_2O_2, and CO_2 [31].

One of the first biosensors using amperometry as the transducer technique was proposed by Clark [32] and has been commercialized by Yellow Springs Instruments. Here the receptor contains the enzyme glucose oxidase (GOD) providing the reaction:

$$\text{Glucose} + H_2O + O_2 \xrightarrow{GOD} \text{gluconic acid} + H_2O_2$$

The amperometric transducer oxidizes the H_2O_2 to molecular oxygen and hydronium ions providing two electrons per molecule as the signal.

Gas Analysis

Liquid Electrolyte Sensors

The construction of amperometric cells with liquid electrolytes for gas analysis does not differ principally from those for measurements in liquids. Differences can be found in the utilized membranes, electrolytes and electrode materials. Gases as carbon monoxide or nitrogen oxide can be determined only with catalytically activated working electrodes [33]. Polymer membranes can be metallized with finely distributed catalytically active metals by sputtering, chemical vapour deposition or screen printing. Another approach that enables further applications consists in the establishment of auxiliary chemical reactions in the cell electrolyte. A well described example for the measurement of acidogenic gases utilizes an iodate/iodide mixture in the electrolyte the gas to be measured can react with:

$$IO_3^- + 5I^- + 6H^+ \leftrightharpoons 3I_2 + 3H_2O$$

The formed iodine is reduced subsequently amperometrically [34].

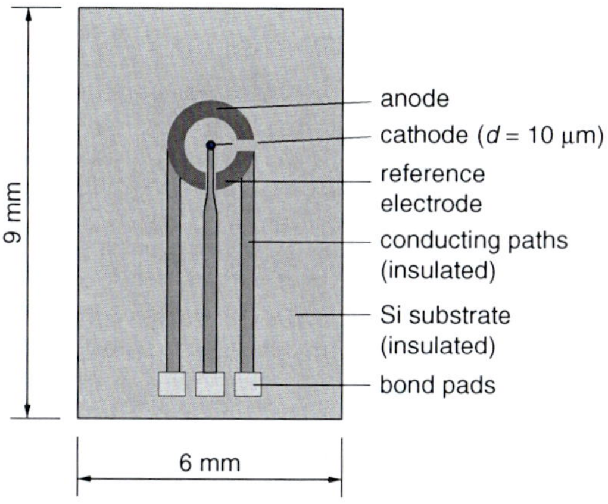

Amperometry, Fig. 8 Layout of a planar three-electrode amperometric oxygen sensor manufactured in thin-film technique

Nowadays, a broad variety of amperometric liquid electrolyte sensors is introduced successfully into the markets providing long-term stable and highly sensitive measurement of O_2, NO, NO_2, H_2S, CO, NH_3, SO_2.

Solid Electrolyte Sensors

Solid electrolyte amperometric gas sensors are particularly suited for the measurement in hot environments [35]. Due to their outstanding long-term stability, wide measuring range, low response times and potential for miniaturization they were also introduced successfully into applications, which were restricted to liquid electrolyte sensors so far. Those sensors are made from ceramic materials and utilize noble metal electrodes as described in Fig. 4. Diffusion barriers are made as gas tight layers with perforations [36], as porous layers directly applied at the electrode [37] or as gas tight layers of mixed conductors with defined permeabilities [38]. The most widely introduced solid electrolyte is stabilized zirconia [39] as an oxide ion conductor. Other examples for solid electrolytes are $SrCe_{0.95}Yb_{0.05}O_3$ [40] and NASICON, which were introduced as H^+- or Na^+-conductors, respectively [41].

To enhance selectivity and to develop multicomponent sensors the concept of amperometry with solid electrolytes was extended to multi-electrode designs [42]. This kind of sensor was introduced successfully into the market especially for the on-board

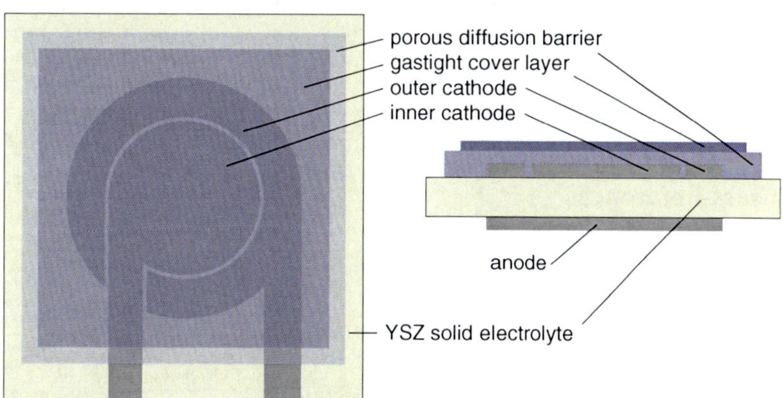

Amperometry, Fig. 9 Schematic drawing of a planar solid electrolyte amperometric sensor for NO and O_2 measurement in exhaust gases with a porous diffusion barrier

measurement of nitrogen oxide (NO) in vehicle exhaust gases [43].

As an example, a planar sensor for hydrocarbon measurement with two cathodes is illustrated in Fig. 9, which was manufactured with thick film technology [44]. The outer electrode is polarized at −400 mV to reduce the oxygen in the measuring gas, and the inner electrode reduces the NO at −600 ... −750 mV.

Another strategy to enhance selectivity of amperometric solid electrolyte sensors consists in the development of new electrode materials with tailored electrocatalytical activities for the desired components to be measured. It has been proven e.g., [45], that the introduction of Pt/Au/Rh alloys enable the selective NO measurement even at elevated oxygen concentrations.

Future Directions

Some future directions concerning amperometry will be:

- Further utilization of amperometric transducers in biosensor for medical mass diagnostics [46],
- Optimization of miniaturized amperometric cells for multicomponent lab-on-a-chip systems [47],
- Development of cheap, reliable and selective gas sensors for environmental monitoring, safety and indoor air quality control [48],
- Development of integrated systems for health monitoring [49],
- Utilization of amperometric micro probes for single cell measurements [50].

Cross-References

► Biosensors, Electrochemical
► Cell, Electrochemical
► Chronoamperometry, Chronocoulometry, and Chronopotentiometry
► Combustion Control Sensors, Electrochemical
► Defects in Solids
► Electrochemical Glucose Sensors
► Electrode
► Electrolytes, Classification
► High-Temperature Oxygen Sensor
► Oxide Ion Conductor
► Polarography
► Potentiostat
► Reference Electrodes
► Sensors
► Solid Electrolytes Cells, Electrochemical Cells with Solid Electrolytes in Equilibrium

References

1. Oehme F (1991) Liquid electrolyte sensors: potentiometry, amperometry and conductometry. In: Göpel W, Hesse J, Zemel JN (eds) Sensors-a comprehensive survey, vol 2. VCH Verlagsgesellschaft, Weinheim, p 288
2. Fabry P, Siebert E (1997) Electrochemical sensors. In: Gellings PJ, Bouwmeester HJM (eds) The CRC handbook of solid state electrochemistry. CRC Press, New York, p 354
3. Delahay P, Charlot G, Laitinen HA (1960) Classification and nomenclature of electroanalytical methods. Anal Chem 32:103A–108A
4. Rideal S, Evans UR (1913) An electrochemical indicator for oxidising agents. J Soc Public Anal 38:353–363
5. Shekhar H, Chathapuram V, Hyun SH, Hong S, Cho HJ (2003) A disposable microsensor for continuous monitoring of free chlorine in water. IEEE Sens 1:67–70
6. Clark LC (1956) Electrochemical device for chemical analysis. US Patent 2 913 386
7. Tödt F (1958) Elektrochemische Sauerstoffmessung. de Gruyter, Berlin
8. Schwabe K, Bär J, Steinhauer H (1965) Zur Systematik der elektrochemischen Analysenmethoden und amperometrische Verfahren zur Betriebskontrolle. Chem Ing Tech 37:483–492
9. Oswin H, Blurton K (1970) Electrochemical detection cell. US Patent 3 776 832
10. Bulliere C (1966) Diplome d'Etudes Superieures, Grenoble
11. Oehme F, Ertl S (1979) Industrielle Amperometrie: Messung von Diffusionsströmen an starren Elektroden als Mittel zur Konzentrationsbestimmung. Chem Tech 8:95–100
12. Heineman WR, Kissinger PT (1984) Laboratory techniques in electroanalytical chemistry. Deeker, New York
13. Deslouis C, Gil O, Tribollet B (1990) Frequency response of electrochemical sensors to hydrodynamic fluctuations. J Fluid Mech 215:85–100

14. Hitchman ML (1978) Measurement of dissolved oxygen. Wiley, New York
15. Holze R (1998) Leitfaden der Elektrochemie. Teubner, Stuttgart/Leipzig
16. Linek V, Benes P, Sinkule J, Vacek V (1988) Measurement of oxygen by membrane-covered probes: guidelines for applications in chemical and biochemical engineering. Ellis Horwood, Chichester
17. Schmalzried H (1995) Chemical kinetics of solids. Weinheim, VCH, p 368
18. Kleitz M, Siebert E, Fabry P, Fouletier J (1991) Solid-state electrochemical sensors. In: Göpel W, Hesse J, Zemel JN (eds) Sensors – a comprehensive survey, vol 2. Weinheim, VCH, p 415
19. Usui T, Asada A, Nakazawa M, Osanai H (1989) Gas polarographic oxygen sensor using an oxygen/zirconia electrolyte. J Electrochem Soc 136:534–542. doi:10.1149/1.2096676
20. Liaw BY, Weppner W (1990) Low temperature limiting-current oxygen sensors using tetragonal zirconia as solid electrolytes. Solid State Ion 40(41):428–432
21. Saji K (1987) Characteristics of limiting current-type oxygen sensor. J Electrochem Soc 134:2431–2435
22. Draft Document 660/25: Expression of performance of electrochemical analyzers. Part IV: dissolved oxygen in water utilizing membrane-covered amperometric aensors; International Electrotechnical Commission (IEC), Rue de Varembe, CH-1211 Genf 20, CH
23. Oehme F, Schuler P (1983) Gelöst-Sauerstoff-Messung. Hüthig, Heidelberg, p 130
24. Mancy KH, Okun DA, Reilley CN (1962) A galvanic cell oxygen analyzer. J Electroanal Chem 4:65–92
25. Guth U, Vonau W, Zosel J (2009) Recent developments in electrochemical sensor application and technology – a review. Meas Sci Technol 20, pp 14
26. Mackereth FJH (1964) An improved galvanic cell for determination of oxygen concentrations in fluids. J Sci Instr 41:38–41. doi:10.1088/0950-7671/41/1/311
27. Mackereth FJH (1962) Electrolytic oxygen sensor. US 3322662
28. Heitz E, Kreysa G (1986) Principles of electrochemical engineering. VCH, Weinheim/New York, p 108
29. Züllig H (1977) Gas-Wasser-Fach Wasser/Abwasser 118:227–234
30. Hughes S, Johnson DC (1981) Amperometric detection of simple carbohydrates at platinum electrodes in alkaline solutions by application of a triple-pulse potential waveform. Anal Chim Acta 132:11–22
31. Havas J (1985) Ion- and molecule-selective electrodes in biological systems. Springer, Berlin
32. Clark LC (1987) In: Turner APF, Karube J, Wilson GS (eds) Biosensors. Oxford University Press, Oxford
33. Stetter JR, Li J (2008) Amperometric gas sensors – a review. Chem Rev 108:352–366 l.c. [1]:p 306
34. Oehme F (1991) Liquid electrolyte sensors: potentiometry, amperometry and conductometry. In: Göpel W, Hesse J, Zemel JN (eds) Sensors-a comprehensive survey, vol 2. VCH Verlagsgesellschaft, Weinheim, p 306
35. Vonau W, Zosel J, Decker M, Gerlach F (2012) The impact of thick film technology on the development of electrochemical sensors. In: Panzini MI (ed) Thick films: properties, technology and applications. Nova Science, New York, p 177
36. Saji K, Takahashi H, Kondo H, Takeuchi T, Igarashi I (1984) Proceedings of 4th sensor symposium, IEE of Japan, pp 147–151
37. Schmidt-Zhang P, Sandow KP, Adolf F, Göpel W, Guth U (2000) A novel thick film sensor for simultaneous O_2 and NO monitoring in exhaust gases. Sens Actuators B 70:25–29
38. Peng Z, Liu M, Balko E (2001) A new type of amperometric oxygen sensor based on a mixed-conducting composite membrane. Sens Actuators B 72:35–40
39. Ullmann H (1993) Keramische Gassensoren: Grundlagen – Aufbau – Anwendung. Akademieverlag, Berlin
40. Katahira K, Matsumoto H, Iwahara H, Koide K, Iwamoto T (2001) A solid electrolyte hydrogen sensor with an electrochemically-supplied hydrogen standard. Sens Actuators B 73:130–134
41. Ono M, Shimanoe K, Miura N, Yamazoe N (2001) Reaction analysis on sensing electrode of amperometric NO_2 sensor based on sodium ion conductor by using chronopotentiometry. Sens Actuators B 77:78–83
42. Somov SI, Reinhardt G, Guth U, Göpel W (2000) Multi-electrode zirconia electrolyte amperometric sensors. Solid State Ion 136–137:543–547
43. Coillard V, Debéda H, Lucat C, Ménil F (2001) Nitrogen monoxide detection with a planar spinel coated amperometric sensor. Sens Actuators B 78:113–118
44. Schmidt-Zhang P, Guth U (2004) A planar thick film sensor for hydrocarbon monitoring in exhaust gases. Sens Actuators B 99:258–263
45. Schmidt-Zhang P, Zhang WF, Gerlach F, Ahlborn K, Guth U (2005) Electrochemical investigations on multi-metallic electrodes for amperometric NO gas sensors. Sens Actuators B 108:797–802
46. Dzyadevych SV, Arkhypova VN, Soldatkin AP, El'skaya AV, Martelet C, Jaffrezic-Renault N (2008) Amperometric enzyme biosensors: past, present and future. IRBM 29:171–180
47. Chin CD, Linder V, Sia SK (2007) Lab-on-a-chip devices for global health: past studies and future opportunities. Lab Chip 7:41–57
48. Tuchtenhagen D, Jung G (2006) Device for determining the characteristics of a gas. WO Patent 2006/005332 A3
49. Yang YL, Chuang MC, Lou SL, Wang J (2010) Thick-film textile-based amperometric sensors and biosensors. Analyst 135:1230–1234
50. Ewing AG, Chen TK, Chen G (1995) Voltammetric and amperometric probes for single-cell analysis. Neuromethods 27:269–304

Anion-Exchange Membrane Fuel Cells, Oxide-Based Catalysts

Youichi Shimizu
Department of Applied Chemistry, Graduate School of Engineering, Kyushu Institute of Technology, Kitakyushu, Fukuoka, Japan

Introduction

Anion-exchange membrane fuel cells have high potentials as a small-sized power generator with high efficiency. Also there are a lot of choices of materials for the fuel cell as an alkaline-based system [1]. The possibility of the use of non-platinum catalysts brings to lower the cost of fuel cells. It is also possible to produce a chemically rechargeable fuel cells when the electrochemical electrode reactions were reversible. In that case, hydrogen and oxygen evolution activities should be also required to both electrodes. As for the oxygen electrode, it means the use of "bifunctional" oxygen electrocatalyst. Here the electrode catalysts made by oxides would be introduced, and some bifunctional activities of oxide are also shown in this chapter.

Oxygen Electrocatalyst Classification

Catalysts for anion-exchange membrane fuel cells are for the use in the alkaline medium. Thus, much more materials could be applicable as catalysts than those for polymer electrolyte fuel cells, which should be stable in acidic electrolyte.

Various materials have been studied as catalysts for anion-exchange membrane fuel cells, metal-air cells, etc. These electrocatalysts which were studied and developed for alkali-based fuel cells are shown in Table 1.

Anion-Exchange Membrane Fuel Cells, Oxide-Based Catalysts, Table 1 Oxygen electrode catalyst materials

Type	Catalyst materials
Carbon	Graphite, active carbon, carbon black, carbon nanotube
Metal	Noble-metal: Pt, Ag, Au, Pd, Os, Ir
	Alloy: Pt-Pd, Pt-Ru, Pt-Au, Pt-Ag
	Base-metal: Ni, Co, Cu, Raney Ni
Oxide	Mono-oxide: NiO(+Li), RuO_2, Au_2O_3, Ag_2O, MnO_2
	Spinel type: Co_3O_4, Co_2NiO_4, $CoAl_2O_4$
	Perovskite type: $La_{1-x}Sr_xCoO_3$, $La_{1-x}Sr_xMnO_3$
	Pyrochlore type: $Pb_2Ru_2O_7$-z, $Pb_2Ir_2O_7$-z
Organometal	Porphyrine: Metal = Fe, Co, Ni, Cu, Zn, Mn
	Phthalocyanine: Metal = Fe, Co, Ru, Mn, Pd, Zn, Ag, Pt
Sulfide	Monosulfide: NiS, CoS, NiS_2, CoS_2, WS_2, MoS_2
	Thio-spinel: Co_3S_4, Ni_3S_4, $CoNi_2S_4$
Carbide	WC, Mo_2C
Nitride	Mn_4N, Ni_3N, Co_3N
Carbon alloy	From organometallic compounds (Co, Fe- Porphyrine Phthalocyanine)

Carbon (graphite) is a very good oxygen reduction catalyst, which produces hydrogen peroxide ion from oxygen, i.e., the two electrons oxygen reduction catalyst ($O_2 + 2e^- + H_2O = HO_2^- + OH^-$). There are many applications for graphite, carbon black, and acetylene black in the carbon system. In addition, new materials, such as carbon nanotube (CNT), are also examined these days [2].

Although it is most that these carbon materials are rather used as a base electrode material of an air electrode rather than an oxygen reduction catalyst, the carbons should contribute to an oxygen reduction reaction as a good two-electron oxygen reduction catalyst.

The air electrode, which consists only of carbon, should generate hydrogen peroxide ion in the discharging process. This has not only the influence on battery material in stability but also the fall in the performance in respect of electromotive force, compared with that for the high-performance oxygen reduction electrocatalysts, which shows four-electron oxygen reduction performance ($O_2 + 4e^- + 2H_2O = 4OH^-$).

The new carbon materials such as CNTs have relatively high performance for oxygen reduction, which should be coming from the oxygen diffusion characteristics of the CNTs. Performance of other new carbon-based materials should be described in other chapters in this book.

As for the four-electron oxygen reduction catalysts, (i) metal-based catalysts: noble-metals (Pt, Ag, Au), noble-metal alloys, (ii) ceramic-based catalysts: mono-metal oxides, mixed-metal oxides (spinel type, pyrochlore type, perovskite type), metal-sulfides, metal-carbides, metal-nitrides, (iii) organometallic catalysts: metal-porphyrin, metal-phthalocyanine, have been reported.

It was known that carbon catalyzed with organometallic compounds and heat treated at high temperature in an inert gas showed high electrochemical catalytic activity and high stability. These carbons, these days, are re-noticed as a new category of "carbon alloy."

As shown above, there are many choices for the catalyst materials which are non-Pt families, such as metal, inorganic materials, and organic materials, for oxide electrode anion-exchange membrane fuel cells. This is a large advantage for the reduction of the cost of the batteries [3].

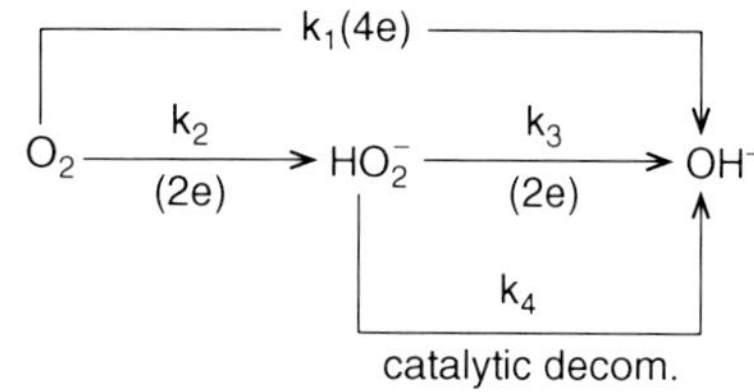

A. Direct 4-electron pathway

$$O_2 + 2H_2O + 4e^- \xrightarrow{k_1} 4OH^- \quad \text{(i)}$$

B. Peroxide pathway

$$O_2 + H_2O + 2e^- \xrightarrow{k_2} HO_2^- + OH^- \quad \text{(ii)}$$

(HO_2^- decomposition)

1. Electrochemical reduction

$$HO_2^- + H_2O + 2e^- \xrightarrow{k_3} 3OH^- \quad \text{(iii)}$$

2. Catalytic decomposition

$$2HO_2^- \xrightarrow{k_4} 2OH^- + O_2 \quad \text{(iv)}$$

Anion-Exchange Membrane Fuel Cells, Oxide-Based Catalysts, Fig. 1 Oxygen reduction mechanism in alkaline system

Oxygen Electrocatalyst Mechanism

Figure 1 shows electrochemical oxygen reduction mechanism in alkaline system. Oxygen reduction in alkaline system is considered to proceed by two overall pathways. One is the direct 4-electron pathway ($O_2 + 4e^- + 2H_2O = 4OH^-$), and the other one is the peroxide pathway ($O_2 + 2e^- + H_2O = HO_2^- + OH^-$), which produces hydrogen peroxide as an intermediate product. The HO_2^- is further reduced to OH^- by either electrochemical reaction or catalytical decomposition reaction. The reactions are dependent on the kind of electrocatalysts [4].

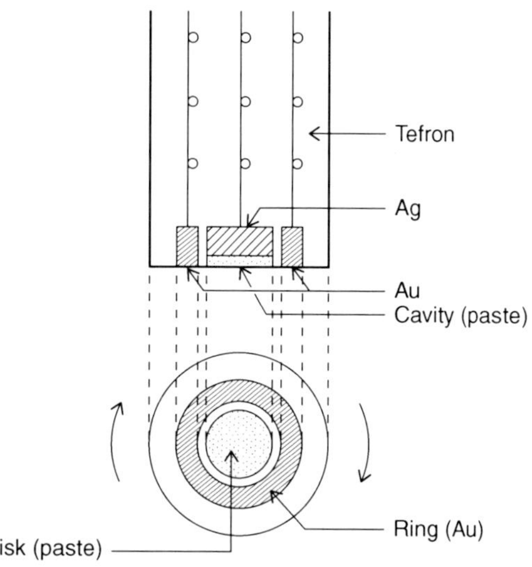

Anion-Exchange Membrane Fuel Cells, Oxide-Based Catalysts, Fig. 2 Rotating ring-disk electrode

The oxygen reduction mechanism of the electrocatalyst is usually examined by the rotating ring-disk electrode technique as shown in Fig. 2. Cathodic electrochemical oxygen reduction occurs on the disk electrode attached with electrocatalyst in the alkaline aqueous electrolyte containing dissolved oxygen. While, the ring-electrode (Au) is used to monitor the hydrogen peroxide ion which is produced on the disk electrode, by applying anodic potential which could introduce electrochemical oxidation of HO_2^- (+0.3 V vs. Hg/HgO). Almost all kinetic parameters in Fig. 1 could be determined by the ring-disk analysis.

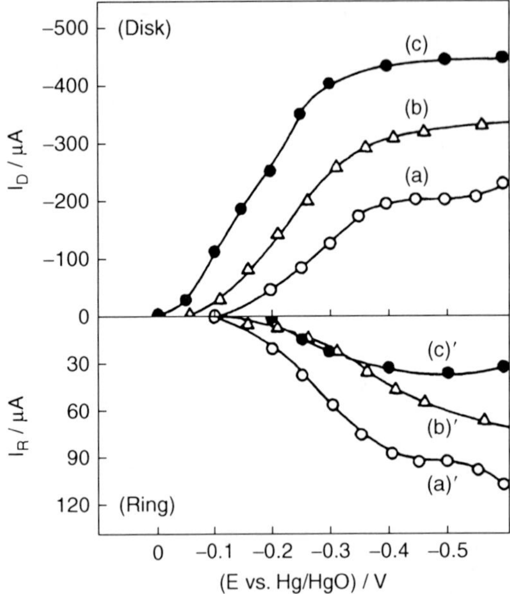

Anion-Exchange Membrane Fuel Cells, Oxide-Based Catalysts, Fig. 3 RRDE curves for (*a, a′*), carbon-only electrode; (*b, b′*), 50 wt% $LaMnO_3$-carbon electrode; (*c, c′*), 80 wt% $LaMnO_3$-carbon electrode

Figure 3 shows the analysis of ring-disk measurement for carbon-only (a, a′), 50 wt% $LaMnO_3$-carbon (b, b′), and 80 wt%$LaMnO_3$-carbon (c, c′) disks at 25 °C in 1 M KOH solution.

Disk current of oxygen reduction on carbon-only disk increased with increasing cathodic overpotential (a). Almost the same way, ring current of hydrogen peroxide oxidation (a′) for carbon-only disk increased with increasing disk current. Collection efficiency of this RRDE is 0.44, determined by geometrical calculation, i.e., 44 % of ions produced on the disk reaction could reach on the ring. Almost 44 % current of disk was observed on the ring current, and symmetrical disk-ring currents were detected; only the 2-electron oxygen reduction peroxide route occurred in the carbon-only electrode. It was found that almost all products on the carbon disk were HO_2^-. The carbon catalyst showed only the 2-electron reduction path, and the carbon could not decompose the HO_2^- to further OH^-. On the other hand, $LaMnO_3$-doped carbon showed the cathodic current of oxygen reduction at more noble potential than carbon-only electrode, and the ring current showed smaller than that of carbon-only electrode. From the other investigation, $LaMnO_3$ is the good 4-electron oxygen reduction catalyst and showed small activity to decomposition of HO_2^-. The ring current (c) comes from the carbon co-doped in the 80 wt% $LaMnO_3$-carbon disk.

For the oxygen reduction catalysts used in carbon-based electrode, it is very important that these show the direct 4-electron reduction which does not produce peroxide or these have very high catalytic activity to decompose peroxide which is produced on carbon.

It is well known that the catalysts, Pt-family, transition-metal oxides, $LaBO_3$(B: Mn, Co)-based perovskite-type oxide [5], metal-nitride, metal-sulfide, and organometallic compounds, show direct 4-electron oxygen reduction mechanism.

As for the 4-electron reduction of oxygen, it is important that the side-on adsorption on the catalyst to cut the oxygen bond (O = O) to O^-. The physical properties, such as the distance between the adsorbed positions on the catalyst, should play an important role. In the α-typed Co phthalocyanine or porphyrin, the Co-Co distances are 4A¯ as to permit formation -O-O-bridge as the side-on adsorption of oxygen, they shows good 4-electron oxygen reduction properties.

It was also found that the oxygen vacancy structure as well as nonstoichiometry in the mixed oxide systems plays important roles for the oxygen reduction performance. It is known that transition-metal oxides and mixed perovskite-type oxide system ($La_{1-x}A'_xCo_{1-y}Fe_yO_3$ A′: Sr,Ca; x, y = 0.2–0.5) have high catalytic activity for the peroxide decomposition.

Among the oxygen reduction catalysts, Pt, Pt-based catalyst, or Ag dispersed in carbon is the most powerful catalyst. However, there are some problems such as the cost for the material, cohesion of the metals, and deactivation from dropping out of the catalyst from the base electrode. Nano-sized perovskite-type oxide electrocatalyst, which could be synthesized by a wet chemical route, is one of the most promising catalysts which could substitute the conventional Pt-based catalyst.

Oxide-Catalyst for Reversible Fuel Cell

When the charging of the anion-exchange membrane fuel cells, hydrogen and oxygen are produced at anode and cathode, respectively, if these electrodes are active for the reverse reactions. The charging of the fuel cell means the chemical charge of the fuel cell, which can get hydrogen as a fuel. As for the oxygen electrode, it means that the oxygen electrocatalyst has bifunctional activities for oxygen reduction and evolution. The most active catalyst of Pt has no bifunctional activities, as its electrocatalytic activity to oxygen evolution is low. Oxide-based electrocatalysts which have bifunctional activities, such as perovskite-type oxides and pyrochlore-type oxides, were reported [5].

Figure 4 shows the oxygen reduction and oxygen evolution performances of the carbon electrode loaded with perovskite-type oxide catalyst in 7M-KOH at 25 °C. It could obtain very high current densities more than 1,000mA/cm^2 for both oxygen reduction and evolution performances. This bifunctional electrocatalyst is suitable for the rechargeable fuel cells. In the base materials, carbons are less stable for electrochemical oxidation in the anode. So, non-carbon materials, such as Ni-metal, are also investigated.

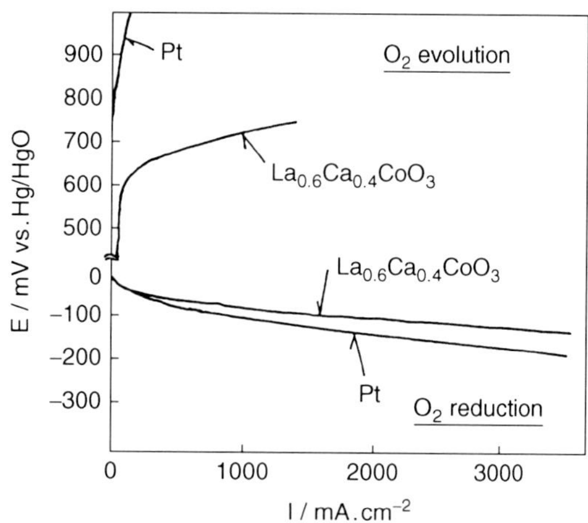

Anion-Exchange Membrane Fuel Cells, Oxide-Based Catalysts, Fig. 4 Oxygen reduction and evolution properties of the electrode loaded with $La_{0.6}Ca_{0.4}CoO_3$ catalyst

Advantages and Disadvantages

As the anion-exchange membrane fuel cell is the alkaline-based system, we can use non-platinum-based catalyst. This is a big advantage to lower the cost of fuel cells. Especially perovskite-type and pyrochlore-type oxides have high performance to oxygen-electrocatalysts which could be applicable to the cathode materials. Some oxides have also bifunctional activities as oxygen electrode catalyst to produce a reversible fuel cell; thus, future deployment is expected. While, the big problems are stability of the base electrode materials, the effect of carbon dioxide from the air.

Future Directions

Research and development of materials for oxide catalyst is going well. In the future, the preparation methods to prepare fine-oxide powders and direct deposition on electrode-based materials should be developed. Also, it should be very important to develop a three-phase interface in the electrode to increase the current density of the electrode.

Cross-References

▶ Alkaline Membrane Fuel Cells
▶ Polymer Electrolyte Fuel Cells, Membrane-Electrode Assemblies
▶ Polymer Electrolyte Fuel Cells (PEFCs), Introduction

References

1. Linden D (ed) (1984) Hand book of batteries and fuel cells. McGraw-Hill, New York

2. Barsukov IV et al (2006) New carbon based materials for electrochemical energy storage systems: batteries, supercapacitors and fuel cells, vol 229, NATO science series II. Springer, Dordrecht
3. Neburchilov V, Wang H, Martin JJ, Qu W (2010) A review on air cathodes for zinc-air fuel cells. J Power Sources 195:1271
4. Miura N, Shimizu Y, Yamazoe N, Seiyama T (1985) Kinetics of cathodic oxygen reduction on lanthanum-based perovskite-type oxides. Nippon Kagaku Kaishi 1985(4):644
5. Shimizu Y, Uemura K, Matsuda H, Miura N, Yamazoe N (1990) Bi-functional oxygen electrode using large surface area $La_{1-x}Ca_xCaO_3$ for rechargeable metal-air battery. J Electrochem Soc 137(11):3430

Anodic Decomposition of Toxic Compounds (Anodic Mineralization)

Yoshinori Nishiki
Development Department, Permelec Electrode Ltd, Kanagawa, Japan

Definition

Electrochemical oxidative treatments to degrade or to mineralize persistent organic substances by converting them to safe and low-molecular-weight substances which are positioned as an important technology to reduce the environmental impact of wastewater and to reuse the water as a resource.

Introduction

Electrochemical treatment involving anodic decomposition is a very promising method for the reduction of toxic pollutants dissolved in wastewater. It is important to select the proper anode materials to optimize this technique, because the electrolytic products strongly depend on these materials as well as the operating conditions such as the current density and temperature [1, 2].

Concerning the anodes, several types of materials have been developed, which are classified as follows: carbon (amorphous carbon, graphite), novel metal or metal oxides (Pt, IrO_2, RuO_2), and nonnovel metal oxides (PbO_2, SnO_2, TiO_x). In recent years, several researchers have been investigating the boron-doped diamond (BDD) anode in various electrochemical fields [3–12], such as electrochemical synthesis and effluent water treatment, wherein oxygen evolution should be avoided in order to obtain a high current efficiency for the target reaction. It has been reported that the BDD anode has the specific ability to destroy many kinds of organic substances.

In this section, the mechanism and examples of organic decomposition by anodic polarization are introduced.

Fundamental Consideration

It has been revealed that the decomposition of organic substances to CO_2 preferentially progresses by selecting the proper electrolytic conditions. Theoretical consideration for the electrolytic decomposition of organic substances is reviewed here, which have been summarized in previous papers [6, 7].

Limiting Current

In the sufficient novel potential region, in which the oxidation reaction of the target substance can progress, the kinetic-controlled current exponentially increases with an increase in the overvoltage wherein the Tafel relation is applicable. In a more noble potential region (higher current density), the electrolytic reaction tends to be controlled by the mass transfer of the substance, and the current density, j_{lim}(A m^{-2}), is represented by Eq. 1, where F is the Faraday constant (96,500 C mol^{-1}), k_m (m s^{-1}) is the mass transfer coefficient, and C_0 (mol m^{-3}) is the substance concentration in the bulk solution. By estimating k_m from the flow conditions in the cell, or measuring the actual diffusion current using a suitable redox ion, j_{lim} for the specific cell can be obtained.

$$j_{lim} = n\ F\ k_m C_0 \tag{1}$$

If the organic substance is represented by the general formula $C_xH_yO_z$, the mineralization reaction can be expressed by Eq. 2:

$$\begin{aligned} C_xH_yO_z &+ (2x - z)\ H_2O \\ &= xCO_2 + (4x + y - 2z)\ H^+ \\ &\quad + (4x + y - 2z)e^- \end{aligned} \quad (2)$$

The number of electrons per mole of reaction for the oxidation reaction of substances subject to CO_2, n, is expressed by Eq. 3:

$$n = 4x + y - 2z \quad (3)$$

From the relationship between C_0 and the COD (chemical oxygen demand, mol-O_2 m^{-3}),

$$j_{lim} = 4\ F\ k_m(COD) \quad (4)$$

In addition, the relationship between COD and TOC (total organic carbon, mol m^{-3}) is

$$\frac{COD}{TOC} = \frac{n}{(4x)} \quad (5)$$

However, the rates of decrease of the COD and TOC do not generally match because lower-molecular-weight intermediates are produced during the electrolysis.

Current Efficiency

When the electrolysis is carried out at constant current condition, j_{app} (A m^{-2}), the concentration of the target substance, C (mol m^{-3}), and the current efficiency, Ce (–), gradually decrease from the initial value according to the elapsed time, t (s), as expressed by Eqs. 6 and 7, where S (m^2) is the anode area and Q (m^3) is the solution volume:

$$C = C_0 \exp\left(\frac{-S\ t\ k_m}{Q}\right) \quad (6)$$

$$Ce = \frac{j_{lim}}{j_{app}} = n\ F\ k_m C_0\ \exp\frac{\left(\frac{-S\ t\ k_m}{Q}\right)}{j_{app}} \quad (7)$$

When C is high enough to suppress the water decomposition, Ce will be close to one. However, the anode surface is likely to be adsorbed by the organic substance, which leads to the induction of side reactions (for instance, polymerization) in this region. When both C and j_{lim} become lower, Ce will decrease by the increase in the oxygen gas evolution, whose current density is denoted by j_{O2}, and hence j_{app} and Ce are expressed by Eqs. 8 and 9:

$$j_{app} = j(O_2) + j_{lim} \quad (8)$$

$$Ce = \frac{j_{lim}}{(j_{O2} + j_{lim})} \quad (9)$$

Cell Voltage

The cell voltage, V (volt), is expressed by Eq. 10, where ΣIR is the sum of the ohmic drops related to the solution, separator, and electrodes. Ση is the sum of the overvoltage and V_0 is the theoretical decomposition voltage.

$$V = V_0 + \Sigma\eta + \Sigma IR \quad (10)$$

Power Consumption

Power consumption, P (Wh/mol-COD), is expressed by Eq. 11.

$$P = \frac{4\ F\ V}{(3{,}600\ Ce)} \quad (11)$$

Side Reactions

The main anodic reaction in an aqueous solution is oxygen gas evolution. Though the oxygen gas does not contribute to the decomposition of organic materials, solutes are able to be oxidized by the anodic reaction to produce active species when a potential higher than those of the theoretical decomposition values is given. It is better to understand the oxidative reaction of water and inorganic solutes before reviewing the decomposition reaction of organic substances.

Water Electrolysis

At normal industrial electrolysis current densities (100–10,000 kA m^{-2}), the overvoltages of water decomposition for novel metal catalysts are less than several hundreds of mV but that of PbO_2 is around 1 V, while that of BDD anode is surprisingly around 2 V. From the viewpoint of electric power loss, a large overvoltage is not desirable as mentioned above; however, this leads to the favorable condition that the potential is novel enough to generate reactive oxygen species, such as O_3, H_2O_2, and even OH radicals (HO·), as shown in reactions (13), (14), and (15), in addition to the occurrence of O_2 by reaction (12):

$$2H_2O = O_2 + 4H^+ + 4e^- \quad (12)$$

$$3H_2O = O_3 + 6H^+ + 6e^- \quad (13)$$

$$2H_2O = H_2O_2 + 2H^+ + 2e^- \quad (14)$$

$$H_2O = HO\cdot + H^+ + e^- \quad (15)$$

Table 1 shows the result of the current efficiencies of O_3 and H_2O_2 obtained by pure water electrolysis using an ion exchange membrane cell. It is revealed that both O_3 and H_2O_2 are produced with a high efficiency when using the BDD anode. Furthermore, the OH radical has been detected in the solution sampled from the anolyte by electron spin resonance (ESR). The advanced oxidation process (AOP) is a well-known method to destroy persistent organic substances by OH radicals that are generated by the chain reaction of O_3 with H_2O_2 (Weiss chain mechanism). It is expected that the AOP occurs in the presence of both the electrolytically generated O_3 and H_2O_2, as expressed by reaction (16):

$$O_3 + HO_2^- = HO\cdot + O_2^{\cdot -} + O_2 \quad (16)$$

Anodic Decomposition of Toxic Compounds (Anodic Mineralization), Table 1 Active species obtained from pure water electrolysis in a membrane cell at 0.1 A cm^{-2}

(Unit; %)

Species	Pt	IrO_2	PbO_2	BDD
O_3	0.2	0.05	5	2
H_2O_2	0.005	0.001	0.001	0.15
HO·	Detected	N.D.	N.D.	Detected

It has been understood that the commercially available anodes are not always suitable to use for wastewater applications because they have drawbacks, such as a limitation of current density due to high IR drops of the electrode, the dissolution of the toxic component of the electrode catalyst with a high environmental impact, and the poor oxidizing power (low overpotential). Thus the expectation for the BDD anode that can overcome these disadvantages becomes greater.

Electrolysis of Inorganic Solutes

In the presence of chloride ion (Cl^-) in a neutral effluent solution, the oxidative reaction occurs at different levels of the oxidation state depending on the anode materials and electrolytic conditions. For the IrO_2 or RuO_2 anodes, the products are limited to the hypochlorite ion (ClO^-) or chlorate ion (ClO_3^-) through the reactions (17), (18), and (19), but when using the Pt or PbO_2 anodes, the perchlorate ion (ClO_4^-) is detected, which favorably proceeds when using the BDD anode according to reaction (19):

$$^2Cl^- = Cl_2 + 2e^- \quad (17)$$

$$Cl_2 + H_2O = HClO + H^+ + Cl^- \quad (18)$$

$$Cl^- + 3H_2O = ClO_3^- + 6H^+ + 6e^- \quad (19)$$

$$ClO_3^- + H_2O = ClO_4^- + 2H^+ + 2e^- \quad (20)$$

When indirect oxidization of the organic substances using active chlorine is allowed, conventional anodes, such as the IrO_2, RuO_2, and Pt anodes, could be applied.

In the presence of sulfate ion (SO_4^{2-}), there is the possibility to produce the persulfate ion ($S_2O_8^{2-}$) according to reaction (21), which is a good oxidizing agent for organic decompositions. The BDD anode has a higher current efficiency for persulfate than the other anodes:

$$2SO_4^{2-} = S_2O_8^{2-} + 2e^- \quad (21)$$

In the presence of carbonate ion (CO_3^{2-}), the percarbonate ion ($C_2O_6^{2-}$) is produced according to reaction (22), which is widely used as a detergent or bleaching agent. The BDD anode has a several times higher efficiency than Pt:

$$2CO_3^{2-} = C_2O_6^{2-} + 2e^- \quad (22)$$

As other technologies, the mediated electrochemical oxidation (MEO) using a metallic couple as a redox reagent is known, wherein a metal ion, such as Ag^+ or Ce^{3+}, is oxidized on the anode to form a reactive oxidation state (Ag^{2+}, Ce^{4+}) and then reacts with organic substances in the bulk solution. After the reaction, the metallic ion is regenerated at the anode.

Decomposition Examples of Organic Substances

There are many reports on the decomposition of organic compounds by electrochemical oxidation applied to wastewater treatment, wherein formate and phenol are often used as examples.

Formate Decomposition

A result of the electrolytic decomposition of formic acid (HCOOH) is presented here, which is one of the final decomposition products of organic substances before mineralization. Because the theoretical decomposition potential of formic acid, expressed by reaction (23) or (24), is less noble than the oxygen generation potential, the reaction will preferably progress if the overvoltage is low.

$$HCO_2^- + 3OH^- = CO_3^{2-} + 2H_2O + 2e^- \quad (23)$$

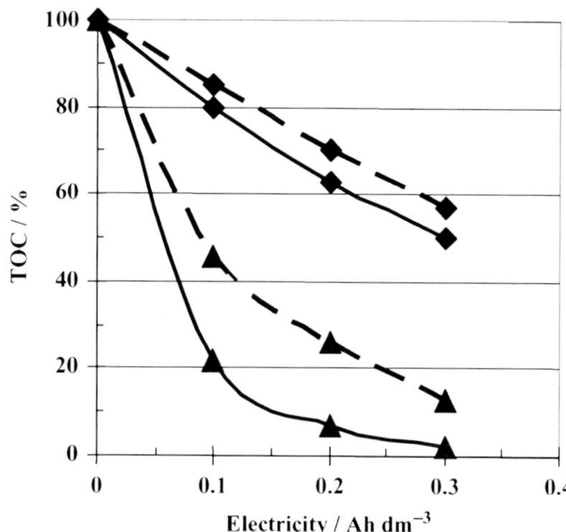

Anodic Decomposition of Toxic Compounds (Anodic Mineralization), Fig. 1 Electrolytic decomposition of formic acid (8.3mM, 100 mg dm^{-3} as TOC) at 0.05 A cm^{-2} and 50 °C in 100mM $NaHCO_3$ (*solid line*) and 100mM Na_2CO_3 (*dashed line*), using BDD (▲) and Pt (♦) nodes

$$HCO_2^- + 2OH^- = HCO_3^- + H_2O + 2e^- \quad (24)$$

Carbonate and bicarbonate were used as the electrolyte for the reason of environmental compatibility. Figure 1 shows the decay rate of TOC (corresponding to the formic acid concentration) using Pt and BDD anodes. The current efficiency of the decomposition of formic acid ranged from 10 % to 30 %. The decomposition rate obtained by the BDD was higher than that of Pt. It was confirmed that percarbonate is synthesized at the same time with a current efficiency from 10 % to 20 %. The ESR signal of the OH radical was detected in the electrolyte separated from the cell regardless of the presence of formic acid.

Phenol Decomposition

Because the decomposition potential of phenol is less noble than the potential generated O_2 phenol, the oxidation reaction (25) will easily occur on a theoretical basis:

$$C_6H_5OH + 11H_2O = 6CO_2 + 28H^+ + 28e^- \quad (25)$$

The phenol decomposition was conducted in various electrolytes as shown in Fig. 2. During the initial stage of the electrolysis, the TOC decreases in a linear mode, but the rate of the degradation of phenol at the end of the reaction showed a difference based on the electrolytes, as shown in Fig. 3. In the case of the Na_2CO_3, Na_2SO_4, and NaOH electrolytes, the rate of degradation remains constant and is higher than that of the $NaClO_4$ electrolyte, wherein related peroxides or O_3 were identified in the former three electrolytes. In addition, the current efficiencies are higher than the value estimated by j_{lim} and the signal of the OH radical has been detected in the electrolyte by ESR measurement, as shown in Fig. 4.

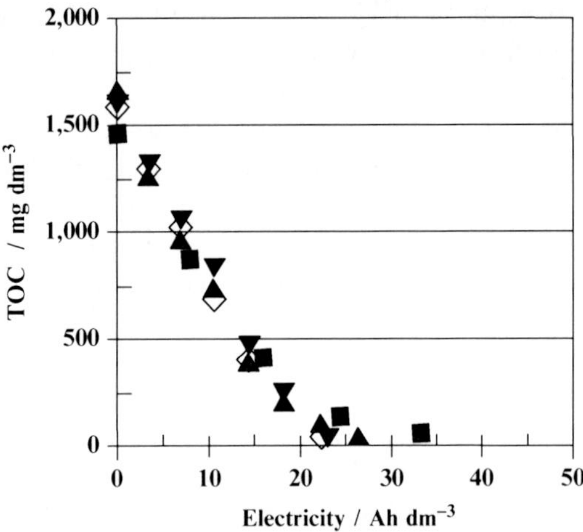

Anodic Decomposition of Toxic Compounds (Anodic Mineralization), Fig. 2 Electrolytic decomposition 100 mM phenol at 0.1 A cm^{-2} in 70mM $NaClO_4$ (■), NaOH (▲), Na_2SO_4 (▼), and Na_2CO_3 (◇)

Degradation Mechanism of Organic Substances

Taking into account the abovementioned electrolysis scheme in aqueous solutions, the electrolytic decomposition scheme of organic substances is listed below and Fig. 5:

S1. Direct electrolysis on anode

S2. Indirect electrolysis by OH radicals generated on anode

S3. Indirect electrolysis in bulk solution by active species (AS) such as ozone, hydrogen peroxide, hypochlorite, and peroxides generated from supporting electrolytes (SE) on anode

S4. Indirect electrolysis in solution by OH radicals generated from the abovementioned active species

Besides the direct electrolysis (S1), indirect decomposition by reacting with OH radicals on or near the anode (S2) and active peroxides produced on the anode will diffuse into the bulk solution to react with the organic substance (S3), or they will diffuse as a radical precursor and produce active radicals in the chain reactions (S4).

Because it is impossible for OH radicals to diffuse into the internal solution due to their short life (ca. 0.2 ms when the concentration is 1 μM), the observation of the ESR signal related to the radicals strongly suggests the existence of a radical formation mechanism even in the bulk solution, as described above.

If the current density for the degradation by indirect electrolysis in the solution is denoted by j_{SOL}, Eqs. 8 and 9 should be converted into Eqs. 26 and 27, suggesting that the Ce becomes large when the indirect decomposition contributes:

$$j_{app} = j(O_2) + j_{lim} + j_{SOL} \tag{26}$$

$$Ce = \frac{(j_{lim} + j_{SOL})}{(j_{O2} + j_{lim} + j_{SOL})} \tag{27}$$

This decomposition behavior, which continues until a low concentration occurs, as mentioned above, suggests the contribution of mechanisms (S3) and (S4). The contribution of the indirect reaction in the bulk solution is an effective means when the concentration of the substances subject to degradation or sterilization becomes low.

Other Examples of Decomposition

Examples of water treatment using the BDD anode extracted from the literature and patents are summarized in Table 2. It is revealed that

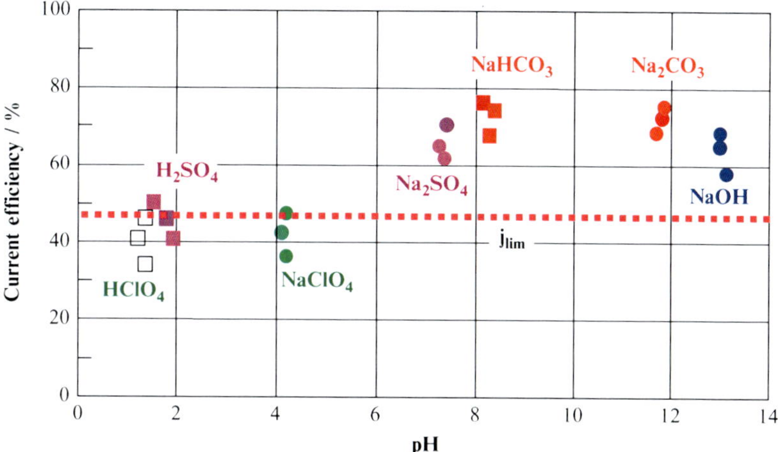

Anodic Decomposition of Toxic Compounds (Anodic Mineralization), Fig. 3 Relationship between pH and current efficiency of 20mM phenol decomposition at 0.1 A cm^{-2}

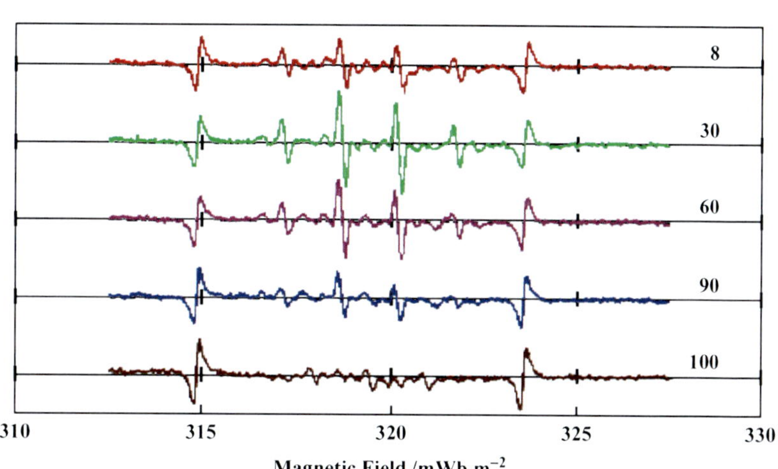

Anodic Decomposition of Toxic Compounds (Anodic Mineralization), Fig. 4 ESR signals of electrolyzed carbonate solution containing 20 mM phenol. The number is elapsed times (min) after the addition of DMPO (radical trapping reagent) to the electrolyte extracted from the cell

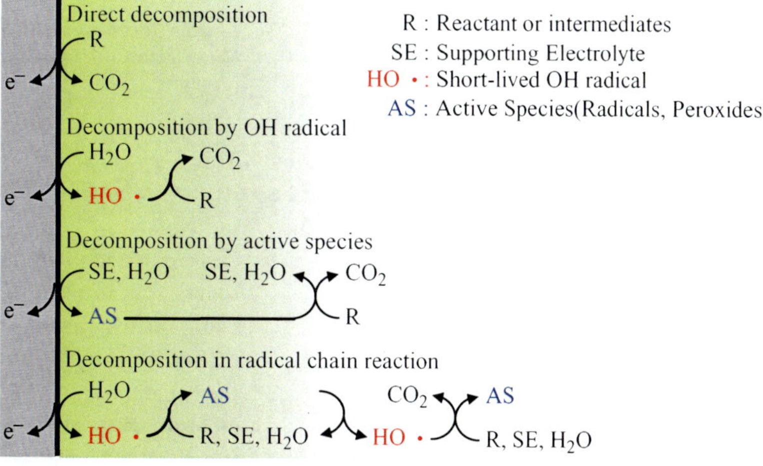

Anodic Decomposition of Toxic Compounds (Anodic Mineralization), Fig. 5 Schematic decomposition model of organic compounds

Anodic Decomposition of Toxic Compounds (Anodic Mineralization), Table 2 Examples of water treatment using BDD anode

Target fields	Target substances	Main objective
General wastewater	Aromatic compounds, aldehydes, THM's; Diphenylethane, chlorophenol, PCB, trichlorethylene, DMSO, catechol, cresol, phenol, acrylate, benzene, naphthol	TOC and COD reductions
Surfactant effluents	Anionic surfactants, nonionic surfactant, EDTA	Improvement of biodegradablity
Plating wastewater	Additives of chemical plating and electroless copper plating, Cyano-compounds	Reuse of electrolyte
Dye wastewater	Aoid blue 22, methylene blue, indigo dye, amaranth	Decolorization
Agricultural water	Herbicides, pesticides; Methamidophos, parathion, meco-prop, chloroxylenol, triazine	Detoxification
Natural water	Environment hormone; 17–estradiol	Removing harmful substances
Drinking water	*E. coli*, *Legionella*	Sterilization
Ballast water	Sea water	Prevention of aquacultural pollution
Specific effluents	Medical, water soluble machining oil, olive oil, photograph	Detoxification

Anodic Decomposition of Toxic Compounds (Anodic Mineralization), Table 3 Progress in electrolytic decomposition technologies

Technical element	Examples
Cell structure	Use of ion exchange membrane, multiple–cascade structure, fixed–bed reactor, fluidized–bed reactor
Electrolytic condition	Addition of solutes, condensation, combination of two types of electrodes
Synergistic technology	Cathodic reduction, photolysis, ultrasonic irradiation, biological treatment, electrolytic Fenton

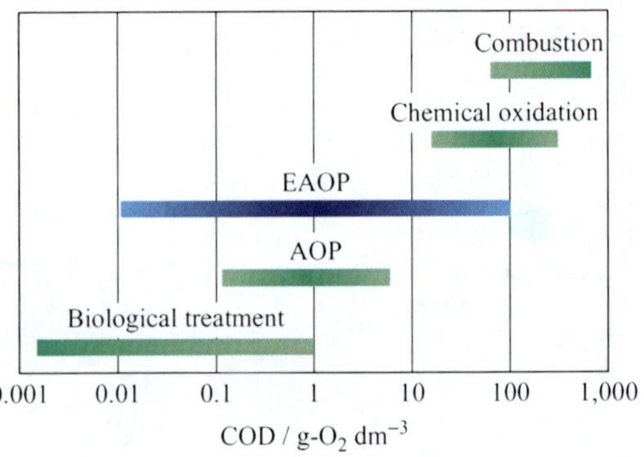

Anodic Decomposition of Toxic Compounds (Anodic Mineralization), Fig. 6 Electrolytic advanced oxidation process

there are many reports on the decomposition of organic compounds by electrochemical oxidation applied to wastewater treatment. They could be decomposed to $CO_2$with a better performance when using the BDD anode compared to the conventional anodes.

Future Directions

The progress in electrolytic technology for organic decomposition is summarized in Table 3. The concept of the electrochemical advanced oxidation process (EAOP) [13] is introduced in Fig. 6.

It was found that there is a suitable concentration region to use electrolytic processes even if using superior anodes.

The electricity cost is different depending on the location and there may be cases when the electrolytic process does not fit to the wastewater treatment, which is always a significant issue on how to supply the system with low capital and running costs, compared to the cost of conventional processes. It is essential to develop an anode which has a high oxidation ability and long durability at a low price.

It is preferable to use pretreatments, such as filtration, to eliminate suspended solid (SS) components in order to reduce the electricity and to increase the current efficiency of the target reaction. Some organic substances or metallic ions easily deposit on the anode surface and cause an activity loss. It is also effective to eliminate the hardness components because they are the source of scales on the cathode or in the separator. A rinsing process is required to maintain the cell performance.

However, in the presence of organic substances, most anodes have a tendency to rapidly lose performance due to the catalyst consumption and the deterioration of the interface between the catalyst layer and the substrate.

By overcoming these issues, the electrolytic process can be applied to more fields in conjunction with conventional processes for water treatment.

Cross-References

▶ Boron-Doped Diamond for Green Electro-Organic Synthesis

References

1. Genders DJ, Weinberg LN (1992) Electrochemistry for a cleaner environment. The Electrosynthesis Company Inc, New York, pp 271–347
2. Panizza M, Cerisola G (2009) Environmental electrochemistry. Chem Rev 109:6541–6569
3. Carey JJ, Charles SJ, Stephen NL (1995) USP 5,399,247
4. Foti G, Gandini D, Comninellis C, Perret A, Haenni W (1999) Electrochem. Solid State Lett 2:228–230
5. Fryda M, Herrmann D, Schaefer L, Klages PC, Perret A, Haenni W, Comninellis C, Gandini D (1999) New Diamond Front Carbon Technol 9:229–240
6. Boye B, Michaud AP, Marselli B, Dieng MM, Brillas E, Comninelis C (2002) New Diamond Front Carbon Technol 12:63–72
7. Canizares P, Diaz M, Dominguez AJ, Gomez GJ, Rodrigo AM (2002) Ind Eng Chem Res 41:4187–4194
8. Fujishima A, Einaga Y, Rao TN (2005) Diamond electrochemistry. Elsevier, Amsterdam, pp 534–555
9. Pedrosa VA, Miwa D, Machado SA, Avaca LA (2006) Electroanal 18:1590–1597
10. Martinez-Huitle CA, De Battisti A, Ferro S, Reyna S, Cerro-lopez M, Qiro MA (2008) Environ Sci Technol 42:6929–6935
11. Nishiki Y (2011) J Surf Finish Soc Jpn 57:157–162
12. http://condias.de/cms/

Anodic Reactions in Electrocatalysis - Methanol Oxidation

Claude Lamy
Institut Européen des Membranes, Université Montpellier 2, UMR CNRS n° 5635, Montpellier, France

Introduction

The electrocatalytic oxidation of methanol has gained much interest over a number of years, because it is the simplest alcohol which can be completely oxidized to carbon dioxide in a Direct Methanol Fuel Cell (DMFC) [1–3], thus providing the maximum energy densities (6.1 kWh kg^{-1} or 4.8 kWh dm^{-3}). The great advantage of a DMFC is that methanol is a liquid fuel, thus more easily handled and stored than hydrogen. Moreover, methanol is produced in great quantity from natural gas (NG) by methane steam reforming (MSR) at a low cost (~0.2 US$ l^{-1}) so that it is a key product in the chemical industry. Its toxicity is relatively low and its boiling point (~65 °C) makes it liquid for most utilization. The development of Proton Exchange Membrane (PEM) led to great simplification of DMFC by avoiding a fuel processor which provides a reformate gas with a low concentration of CO (<10 ppm; otherwise, it may strongly poison the platinum-based electrode catalysts used). Due to system simplicity, DMFCs are particularly efficient power sources for portable electronics (cell phones, laptop computers, cam recorders, etc.)

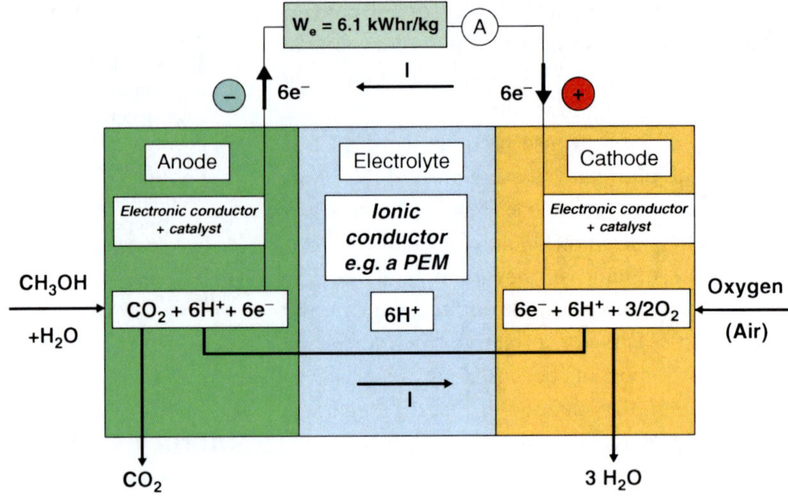

Anodic Reactions in Electrocatalysis - Methanol Oxidation, Fig. 1 Schematic diagram of a DMFC based on a proton exchange membrane

and for small size applications (micro-power sources, power sources for the soldier, propulsion of small devices, e.g., golf carts, drones, etc.).

A DMFC consists of two electrodes, a catalytic methanol anode, and a catalytic oxygen cathode, separated by an ionic conductor, preferably an acid electrolyte, for rejecting the carbon dioxide produced. Great progress was recently made by feeding methanol directly to the anodic compartment of a Proton Exchange Membrane Fuel Cell (PEMFC), in which the protonic membrane, e.g., Nafion®, plays both the role of an acidic medium and of a separator between the two electrode compartments (Fig. 1). This technology has the added advantage of thin elementary cells and hence of compact stacks.

In such a device, the electrons liberated at the anode by the electro-oxidation of methanol pass through the external electrical circuit (producing an electrical energy, $W_e = n\,F\,E_{cell}$, where E_{cell} is the cell voltage and F = 96485 C the Faraday constant) and reach the cathode, where they reduce the oxidant, usually oxygen from air. Inside the fuel cell, the electrical current is transported by migration and diffusion of the electrolyte ions (H^+, OH^-, CO_3^{2-}).

The electrochemical oxidation of methanol occurs on the anode electrocatalyst (e.g., dispersed platinum-based catalysts), which constitutes the negative pole of the cell:

in acid electrolytes

$$CH_3OH + H_2O \rightarrow CO_2 + 6H^+_{aq} + 6e^- \quad (1a)$$

in alkaline electrolytes

$$CH_3OH + 8OH^- \rightarrow CO_3^{2-} + 6H_2O + 6e^- \quad (1b)$$

whereas the electrochemical reduction of oxygen occurs at the cathode (also containing a platinum-based catalyst) which constitutes the positive pole of the cell:

in acid electrolytes

$$O_2 + 4H^+ + 4e^- \rightarrow 2H_2O \quad (2a)$$

in alkaline electrolytes

$$O_2 + 2H_2O + 4e^- \rightarrow 4OH^- \quad (2b)$$

The overall reaction corresponds thus to the catalytic combustion of methanol into oxygen, i.e.,

$$CH_3OH + 3/2O_2 \rightarrow CO_2 + 2H_2O \quad (3)$$

The cell voltage, E_{cell}, is equal to the difference between the electrode potentials of each electrode:

$$E_{cell} = E_c - E_a \quad (4)$$

where the electrode potentials E_i are defined as the difference of the internal potential at each electrode/electrolyte interface.

Thermodynamics and Energy Efficiency

One main advantage of such a power source is the direct transformation of the chemical energy of methanol combustion into electrical energy, thus avoiding a low energy efficiency given by the Carnot's theorem for thermal engines. Hence, the reversible cell voltage, E_r, can be calculated from the Gibbs energy change, ΔG, associated with the total combustion reaction of methanol (3), by equation:

$$\Delta G + n\,F\,E_r = 0 \quad \text{leading to } E_r > 0 \qquad \text{with } \Delta G < 0 \quad \text{(spontaneous reaction)} \tag{5}$$

where $n = 6$ is the number of Faradays (per mole of methanol) involved in the half-cell reaction.

Under standard conditions (25 °C), the heat of combustion, i.e., the enthalpy change ΔH° for the reaction (3), is -726 kJ mol^{-1} of methanol, and the Gibbs energy change ΔG° is -702 kJ mol^{-1} of methanol. This corresponds to a standard reversible cell voltage, E_r^o, as given by the equation:

$$E_r^o = E_c^o - E_a^o = -\frac{\Delta G^o}{nF} = \frac{702 \times 10^3}{6 \times 96485} = 1.21\ \text{V} \tag{6}$$

where E_c^o, E_a^o, are the standard electrode potentials of each electrode versus the Standard Hydrogen Electrode (SHE) used as a reference electrode.

The main features of the DMFC are its high specific energy (W_s) and high volume energy density (W_{el}), the values of which are calculated as follows:

$$W_s = \frac{(-\Delta G^o)}{3,600 \times M} = \frac{702 \times 10^3}{3,600 \times 0.032} = 6.09\ \text{kWh kg}^{-1} \tag{7}$$

and $W_{el} = W_s \times \rho = 4.82$ kWh dm^{-3}, where $M = 0.032$ kg is the molar weight of methanol and $\rho = 0.7914$ kg dm^{-3} its density.

Under standard reversible conditions (25 °C), the energy efficiency is very high:

$$\varepsilon_{rev} = \frac{W_e}{(-\Delta H^o)} = \frac{nFE_r^o}{(-\Delta H^o)} = \frac{\Delta G^o}{\Delta H^o} = \frac{702}{726} = 96.7\% \tag{8}$$

It is considerably higher than that of a H_2/O_2 fuel cell (i.e., 83 %).

Reaction Kinetics

However, under normal operating conditions, at a current density j, the electrode potentials deviate from their equilibrium values due to large overpotentials, η_i, at both electrodes (Fig. 2):

$$\eta_a = E_a(j) - E_a^o \geq 0 \quad \text{at the methanol anode} \tag{9a}$$

$$\eta_c = E_c(j) - E_c^o \leq 0 \quad \text{at the oxygen cathode} \tag{9b}$$

This results from a slow kinetics of both methanol oxidation and oxygen reduction. An additional loss is due to the cell resistance R_e (arising mainly from the proton-conducting membrane). The cell voltage under working conditions is thus

$$E(j) = E_c(j) - E_a(j) - R_e\,j = E_r - (|\eta_c| + |\eta_a| + R_e\,j) \leq E_r \tag{10}$$

so that the energy efficiency will be decreased proportionally to the so-called voltage efficiency:

$$\varepsilon_E = E(j)/E_r \tag{11}$$

For a DMFC working at 200 mA cm^{-2} and 0.5 V (e.g., with a Pt – Ru anode), this ratio will be

$$\varepsilon_E = 0.5/1.21 = 41.3\ \% \tag{12}$$

and the overall efficiency of the fuel cell will be

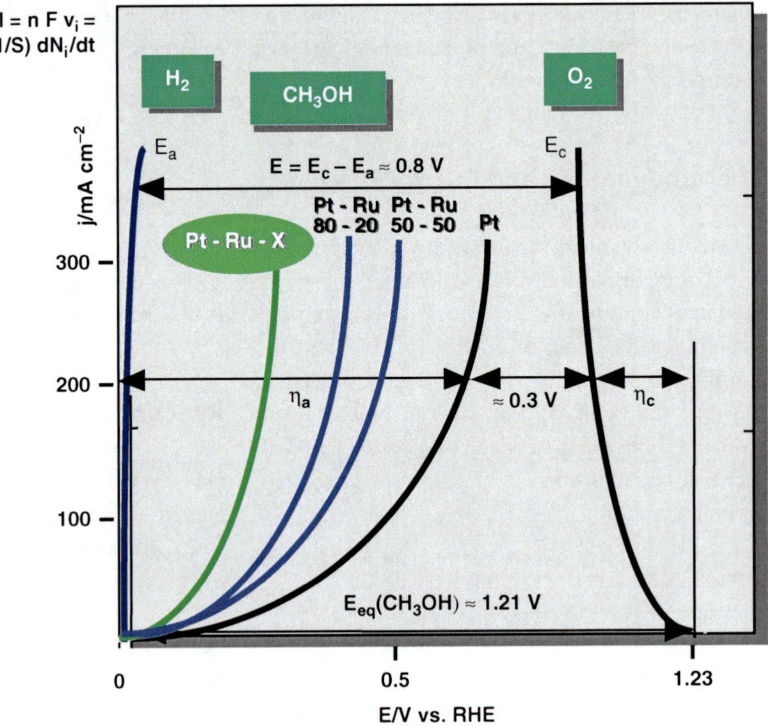

Anodic Reactions in Electrocatalysis - Methanol Oxidation, Fig. 2 Current density versus electrode potential curves for electrochemical reactions involved in a PEMFC and in a DMFC

$$\varepsilon_{cell} = \varepsilon_{rev} \times \varepsilon_E = 0.967 \times 0.413 \approx 40\ \% \quad (13)$$

assuming a Coulombic efficiency of 100 %, i.e., the total combustion of methanol. This is acceptable for an autonomous power source oxidizing methanol completely to CO_2 and giving the theoretical number of Faradays $n_{th} = 6$ F per mole of methanol – see Eq. 1. However, under some operating conditions, methanol oxidation to CO_2 is not complete, so that a Coulombic efficiency is introduced:

$$\varepsilon_F = n_{exp}/n_{th} \quad (14)$$

where n_{exp} is the number of Faraday effectively exchanged in the half-cell reaction.

The overall energy efficiency becomes:

$$\begin{aligned}\varepsilon_{cell} &= \frac{W_e}{(-\Delta H)} = \frac{n_{exp}FE(j)}{(-\Delta H)} \\ &= \frac{n_{th}FE_r}{(-\Delta H)} \times \frac{E(j)}{E_r} \times \frac{n_{exp}}{n_{th}} \\ &= \varepsilon_{rev} \times \varepsilon_E \times \varepsilon_F \end{aligned} \quad (15)$$

Therefore, the overall efficiency may be dramatically decreased, e.g., if the electro-oxidation stops at the formaldehyde stage

$$CH_3OH \rightarrow HCHO + 2H^+_{aq} + 2e^- \qquad (n_{exp} = 2),$$

or at the formic acid stage

$$CH_3OH + H_2O \rightarrow HCOOH + 4H^+_{aq} + 4e^- \qquad (n_{exp} = 4),$$

thus leading respectively to a Coulombic efficiency

$$\varepsilon_F = \frac{n_{exp}}{n_{th}} = \frac{2}{6} = 33.3\,\%$$

or

$$\varepsilon_F = \frac{n_{exp}}{n_{th}} = \frac{4}{6} = 66.6\,\%,$$

and respectively to an overall energy efficiency

$$\varepsilon_{cell} = \varepsilon_{rev} \times \varepsilon_E \times \varepsilon_F \approx 13.3\ \%$$

or

$$\varepsilon_{cell} = \varepsilon_{rev} \times \varepsilon_E \times \varepsilon_F \approx 26.6\ \%$$

Reaction Mechanisms

The mechanism of the electro-oxidation of methanol on platinum was thoroughly established, mainly after the identification of both reactive intermediates and adsorbed poisoning species [4]. In the first step, methanol is dissociatively adsorbed at Pt-based catalysts by cleavage of C – H bonds, leading to the so-called formyl-like species $-(CHO)_{ads}$. From this species, different steps can occur, but with platinum, the dissociation of $-(CHO)_{ads}$ gives rapidly adsorbed CO, which is responsible for the electrode poisoning. This is the explanation of the rather poor performance of Pt catalysts, due to the relatively high potential necessary to oxidize such CO species.

These different steps can be summarized as follows:

$$Pt + CH_3OH \rightarrow Pt - CH_3OH_{ads} \rightarrow Pt - CHO_{ads} \quad (16)$$

The vital step in the reaction mechanism appears to be the formation of the intermediate adsorbed species $-(CHO)_{ads}$, which facilitates the overall reaction. The kinetics of its further desorption and oxidation into reaction products are the key steps of the mechanism, leading to complete oxidation.

The usual path is the formation of CO_{ads}

$$Pt - CHO_{ads} \rightarrow Pt - CO_{ads} + H^+ + e^- \quad (17)$$

followed by its oxidation to CO_2 – see Eq. 23 below.

An alternative path to the spontaneous formation of the poisoning species (Eq. 17) is the oxidation of CHO_{ads} species, with OH_{ads} species coming from the dissociation of water according to the following reactions:

$$Pt + H_2O \rightarrow Pt - (OH)_{ads} + H^+ + e^- \quad (18)$$

$$Pt - CHO_{ads} + Pt - (OH)_{ads} \rightarrow 2Pt + CO_2 + 2H^+ + 2e^- \quad (19)$$

One parallel surface reaction, leading to adsorbed formate, has also been observed:

$$Pt - (CHO)_{ads} + Pt - (OH)_{ads} \rightarrow Pt + Pt - (COOH)_{ads} + H^+ + e^- \quad (20)$$

This leads to the formation of carbon dioxide by further oxidation:

$$Pt - (COOH)_{ads} \rightarrow Pt + CO_2 + H^+ + e^- \quad (21)$$

On the other hand, adsorbed CO can be oxidized through the reactions

$$Pt - (CO)_{ads} + Pt - (OH)_{ads} \leftrightarrow Pt + Pt - (COOH)_{ads} \quad (22)$$

followed by reaction (21), or through the following reaction:

$$Pt - (CO)_{ads} + Pt - (OH)_{ads} \rightarrow 2Pt + CO_2 + H^+ + e^- \quad (23)$$

This mechanism takes into account the formation of all the products, as detected by Infrared Reflectance Spectroscopy (SNIFTIRS or SPAIRS) [5] and liquid or gas chromatography [6]: formaldehyde through step (16), formic acid through steps (20) or (22), or CO_2 through steps (19), (21), or (23).

Thus, the crucial point is to determine the fate of the $-(CHO)_{ads}$ species. The different mechanisms for its oxidative removal are schematically summarized in Scheme 1.

From this scheme, it appears that desorption and oxidation of the formyl species can follow

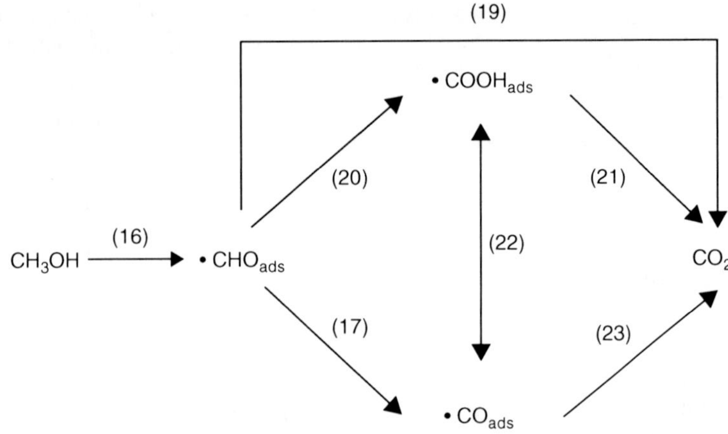

Anodic Reactions in Electrocatalysis - Methanol Oxidation, Scheme 1 Schematic representation of the mechanism of oxidation of CHO_{ads}

different pathways through competitive reactions. This scheme summarizes the main problems and challenges to improve the kinetics of the electro-oxidation of methanol.

On a pure platinum surface, step (17) is spontaneously favored, and the formation of adsorbed CO is a fast process, even at low potentials. Thus, the coverage of adsorbed CO is high and explains the poisoning phenomena encountered at a platinum electrode. This poisoning species can be removed (by oxidation through step (23) into CO_2) only at potentials at which oxygenated species are formed at the electrode surface. On pure platinum, these oxygenated species arise from dissociation of water through step (18), which occurs only at potentials more positive than 0.5 V [7]. Similarly, the direct oxidation of $-(CHO)_{ads}$ into CO_2 through step (19), or through step (20) followed by step (21) with the intermediate formation of $-(COOH)_{ads}$ species, needs again the presence of an extra-oxygen atom which can be provided only by the dissociation of water at the catalytic surface.

To lower the potential at which dissociation of water commences, a number of bimetallic Pt–M catalysts (with M = Ru, Mo, Sn, Fe, Ni, etc.) have been investigated. With Pt – Ru catalysts, which are the best bimetallic catalysts for methanol oxidation, it appears clearly from the literature, and this was fully confirmed by IR reflectance spectroscopic studies, that the presence of adsorbed $-(OH)_{ads}$ on ruthenium sites at low potentials leads to the oxidation of adsorbed CO at potentials much lower than those on pure platinum. It is also probable that $-(CHO)_{ads}$ can be oxidized directly to carbon dioxide, without the formation of adsorbed CO poisoning species [4].

At $Pt_{1-x}Ru_x (0 \leq x \leq 1)$ the following steps of methanol oxidation electrodes can be postulated (bifunctional mechanism [8]), as follows:

- Activation of the methanol molecule by Pt:

$$Pt + CH_3OH \rightarrow Pt - CH_3O_{ads} + H^+ + e^- \quad (24)$$

$$Pt - CH_3O_{ads} \rightarrow Pt - CHO_{ads} + 2H^+ + 2e^- \quad (25)$$

$$Pt - CHO_{ads} \rightarrow Pt - CO_{ads} + H^+ + e^- \quad (17)$$

- Activation of the water molecule either by Pt or by Ru:

$$Pt + H_2O \rightarrow Pt - OH_{ads} + H^+ + e^- \quad (18)$$

$$Ru + H_2O \rightarrow Ru - OH_{ads} + H^+ + e^- \quad (26)$$

- Surface reactions between the adsorbed species:

$$Pt - CHO_{ads} + Pt - OH_{ads} \rightarrow 2Pt + CO_2 + 2H^+ + 2e^- \quad (19)$$

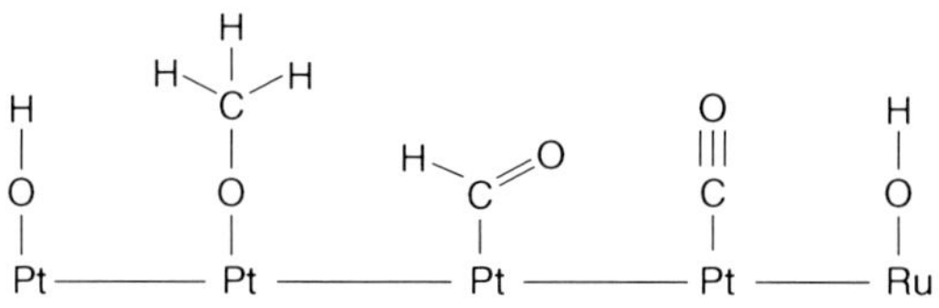

Anodic Reactions in Electrocatalysis - Methanol Oxidation, Scheme 2 Structure of the different adsorbed intermediates involved in methanol electro-oxidation

$$Pt - CHO_{ads} + Ru - OH_{ads} \rightarrow CO_2 + 2H^+ + 2e^- + Pt + Ru \quad (27)$$

$$Pt - (CO)_{ads} + Pt - (OH)_{ads} \rightarrow 2Pt + CO_2 + H^+ + e^- \quad (23)$$

$$Pt - CO_{ads} + Ru - OH_{ads} \rightarrow CO_2 + H^+ + e^- + Pt + Ru \quad (28)$$

Previously reported results provided evidence that the best activity of the Pt–Ru electrodes for methanol oxidation is obtained with about 20 % of Ru surface atoms. This can be roughly explained by the bifunctional mechanism [8], taking into account the number of surface sites necessary to accommodate all the adsorbed intermediates, as observed by IR reflectance spectroscopy [4].

A detailed analysis of the number of surface sites necessary to accommodate all the adsorbed intermediates (see Scheme 2) involved either in the oxidation of CO, or in that of methanol, should describe correctly the optimum composition found for the $Pt_{1-x}Ru_x$ alloys. In the case of CO oxidation, one Pt atom adsorbs CO, whereas one Ru atom adsorbs OH, which leads to the optimum 50/50 atomic composition observed. For methanol oxidation, the activation reactions (24, 25, 17, 18, and 26) need four Pt surface atoms (three for the CH_3OH adsorption residues and one for H_2O) and one Ru surface atom for H_2O activation leading to the formation of $Ru{-}OH_{ads}$ (Ru does not adsorb CH_3OH, nor dissociate the CH_3OH molecule), so that the optimum surface atomic composition of the Pt–Ru electrodes is four Pt for one Ru, i.e., $x = 0.2$. If we take into account the presence of other adsorbed species on Pt (e.g., $-COOH_{ads}$) or multisite adsorption (e.g., bridge bonded $> CO_B$), the number of Pt atoms could vary from 5 (corresponding to an optimum composition of 16.7 at.% Ru) to 7 (i.e., 12.5 at.% Ru). This crude model will give an optimum surface composition between 12.5 and 20 at.% Ru. This value should be compared with that proposed by Gasteiger et al. [9] which is around 10 % at room temperature.

Some Technological Developments

DMFC systems are developed by several companies all around the world: Toshiba, Sony, Panasonic, Sharp, Hitachi in Japan; Samsung in South Korea; Jet Propulsion Laboratory, MTI Microfuel Cells, Oorja Protonics, Relion in the USA; ZSW, Julich Center, Smart Fuel Cells, Baltic Fuel Cells in Germany; CMR Fuel Cells Limited in United Kingdom; Cellera Technologies in Israel. They can be used as small power sources, mainly for portable electronics (cell phone, notebook PC, music player, electric tools, soldier equipment, etc.), with liquid methanol stored either in a small cartridge or in a small tank. In most of these technologies, the DMFC provides power for recharging a Li-ion battery, thus extending greatly its autonomy and allowing a quick recharge of the system by refilling the tank with fresh methanol.

Future Directions

In order to improve the power density of the DMFC, the main challenge is to develop more efficient electrocatalysts of methanol oxidation than Pt–Ru catalysts. Ternary alloys, such as Pt–Ru–Sn, have been investigated [10], leading to some improvement by increasing the current density and decreasing the methanol oxidation overvoltage.

Moreover methanol cross-over through the polymer membrane reduces the efficiency of the Fuel Cell by decreasing the cell voltage as a result of a mixed potential at the oxygen cathode. Improved membranes with low cross-over have to be developed. Besides, in order to increase the reaction rate of methanol oxidation (and of oxygen reduction), some works propose to operate the DMFC at higher temperatures with temperature-resistant membranes [11]. Hence, new protonic membranes with higher resistance and higher conductivity above 100 °C are needed.

Finally, to favor the commercial issue of DMFC, new low cost components have to be developed, including catalysts with a lower amount of PGM or even with no noble metal, and high-temperature-resistant membranes of lower cost than sulfonated perfluorinated membrane (Nafion®) and of higher protonic conductivity and stability to operate the DMFC at higher temperatures (150–200 °C) which may increase the power density through thermal activation of the reaction kinetics.

Another approach is to consider Anion Exchange Membranes (AEMs) which may allow the use of non-noble metal catalysts relatively active in alkaline medium for methanol oxidation [12], but the AEMs available are not sufficiently stable in temperature and not enough conductive through hydroxyl anions.

Cross-References

▶ Direct Alcohol Fuel Cells (DAFCs)
▶ Polymer Electrolyte Fuel Cells (PEFCs), Introduction

References

1. Lamy C, Léger J-M, Srinivasan S (2001) Direct methanol fuel cells: from a twentieth century electrochemist's dream to a twenty-first century emerging technology. In: Bockris JOM, Conway BE, White RE (eds) Modern aspects of electrochemistry, vol 34. Kluwer/Plenum, New York, pp 53–118 (Chap 3)
2. Hamnett A (2003) Direct methanol fuel cells (DMFC). In: Vielstich W, Lamm A, Gasteiger H (eds) Handbook of fuel cells: fundamentals and survey of systems, vol 1. Wiley, Chichester, pp 305–322 (Chap 18)
3. Arico A, Baglio V, Antonucci V (2009) Direct methanol fuel cells: history, status and perspectives. In: Zhang J, Liu H (eds) Electrocatalysis of direct methanol fuel cells. Wiley-VCH, Weinheim, pp 1–78 (Chap 1)
4. Kabbabi A, Faure R, Durand R, Beden B, Hahn F, Léger J-M, Lamy C (1998) In situ FTIRS study of the electrocatalytic oxidation of carbon monoxide and methanol at platinum-ruthenium bulk alloy electrodes. J Electroanal Chem 444:41–53
5. Lamy C, Lima A, Le Rhun V, Delime F, Coutanceau C, Léger J-M (2002) Recent advances in the development of direct alcohol fuel cells (DAFC). J Power Sources 105:283–296
6. Belgsir EM, Huser H, Léger J-M, Lamy C (1987) A kinetic analysis of the oxidation of methanol at platinum-based electrodes by quantitative determination of the reaction products using liquid chromatography. J Electroanal Chem 225:281–286
7. Li NH, Sun SG, Chen SP (1997) Studies on the role of oxidation states of the platinum surface in electrocatalytic oxidation of small primary alcohols. J Electroanal Chem 430:57–67
8. Watanabe M, Motoo S (1975) Electrocatalysis by ad-atoms. 2. Enhancement of oxidation of methanol on platinum by ruthenium ad-atoms. J Electroanal Chem 60:267–283
9. Gasteiger HA, Markovic N, Ross PN, Cairns EJ (1993) Methanol electroxidation on well-characterized Pt-Ru alloys. J Phys Chern 97:12020–12029
10. Napporn WT, Laborde H, Léger J-M, Lamy C (1996) Electroxidation of C1 molecules at Pt-based catalysts highly dispersed into a polymer matrix: effect of the method of preparation. J Electroanal Chem 404:153–159
11. Aricò AS, Di Blasi A, Brunaccini G, Sergi F, Dispenza G, Andaloro L, Ferraro M, Antonucci V, Asher P, Buche S, Fongalland D, Hards GA, Sharman JDB, Bayer A, Heinz G, Zandonà N, Zuber R, Gebert M, Corasaniti M, Ghielmi A, Jones DJ (2010) High temperature operation of a solid polymer electrolyte fuel cell stack based on a new ionomer membrane. Fuel Cells 10:1013–1023
12. Varcoe JR, Slade RCT, Yee ELH, Poynton SD, Driscoll DJ (2007) Investigations into the ex situ methanol, ethanol and ethylene glycol permeabilities of alkaline polymer electrolyte membranes. J Power Sources 173:194–199

Anodic Reactions in Electrocatalysis - Oxidation of Carbon Monoxide

Elena Savinova[1], Antoine Bonnefont[2] and Frédéric Maillard[3]
[1]Institut de Chimie et Procédés pour l'Energie, l'Environnement et la Santé, UMR 7515 CNRS, Université de Strasbourg-ECPM, Strasbourg, France
[2]Institut de Chimie de Strasbourg, CNRS-Université de Strasbourg, Strasbourg, France
[3]Laboratoire d'Electrochimie et de Physico-chimie des Matériaux et des Interfaces, Saint Martin d'Héres, France

Introduction

CO electrooxidation is a reaction of great technological importance, which has been widely used as a prototype electrochemical reaction in fundamental electrocatalysis. In proton exchange membrane fuel cells (PEMFC) fed by hydrogen, presence of CO in the feed leads to significant voltage losses due to the strong CO adsorption and concomitant active site poisoning of Pt-based anode catalysts. In low-temperature fuel cells which utilize C_1- and C_2-oxygenated molecules like methanol, ethanol, formaldehyde, and formic acid as fuels, CO is formed as an intermediate which is blocking active sites and greatly impeding the anode reaction. For these applications, the use of CO-tolerant catalysts is of utmost importance. Investigation of CO electrooxidation on model single crystal and nanostructured electrodes has strongly added to the advancement of the fundamental electrocatalysis in particular in what concerns the understanding of the kinetics of electrocatalytic reactions, structural and size effects, etc. [1, 2]. CO is also widely used as a probe molecule in order to gain insight into the surface structure of electrode materials [3, 4] as well as to obtain information on their active surface areas [1, 5].

The article is organized as follows. The first part briefly summarizes the state-of-the-art understanding of CO adsorption at the metal-gas and metal-electrolyte interface. The second part is devoted to the electrooxidation of CO on Pt-group metals that are known to strongly adsorb CO, whereas its electrooxidation on weakly adsorbing metals, like gold, is discussed in the third part. Finally, the last part of this article is devoted to the discussion of CO tolerance in fuel cell electrocatalysis.

CO Adsorption

Binding of CO to Metal Surfaces

Chemisorption of CO to transition metals can be qualitatively understood within the molecular orbital model of Blyholder [6] which treats an adsorbate-metal bond as a combination of the electron donation from the filled 5σ molecular orbital of CO to d-orbitals of a metal and back-donation from the metal d-orbital to the empty $2\pi^*$ antibonding orbital of CO. State-of-the-art description of adsorption on metal surfaces is based on the Newns-Anderson model [7, 8]. Interaction of CO 5σ and $2\pi^*$ frontier orbitals with localized metal d-band results in hybridization and splitting into bonding and antibonding orbitals of a chemisorbed molecule, while interaction with delocalized metal sp-band leads to the energy level broadening. Thus, the strength of CO bonding to metal surfaces is determined by the relative position of the metal d-band and CO frontier orbitals and by their coupling. As one moves from the right to the left along a period of the periodic table, the metal d-band shifts upwards relative to the $2\pi^*$ orbital of CO leading to an increased back-donation and resulting in a stronger metal-CO bond, destabilization, and ultimately dissociation of the C-O bond [9]. The enthalpy of adsorption ΔH_{ads} is coverage dependent due to repulsive lateral interactions between adsorbed CO molecules and is thus usually tabulated in the limit of zero coverage ΔH^0_{ads}. For group IB metals (Cu, Ag, Au), d-band lies lower than the $2\pi^*$ orbital of CO, leading to a weak molecular adsorption of CO as evidenced by ΔH^0_{ads} ranging from -10 to -60 kJ mol^{-1} [10]. On the group VIIIB metals (Fe, Co, Ni,

Ru, Rh, Pd, Os, Ir, Pt), the adsorption is strong with ΔH^0_{ads} ranging from -120 to -160 kJ mol^{-1} [10]. On metals of groups VIIB (Mn, Tc, Re) and VIB (Cr, Mo, W), the interaction is so strong that depending on the conditions CO may adsorb either in a molecular or in a dissociative form the [10].

Surface Structure of CO Adlayers

Although adsorption of CO at electrified interfaces is largely similar to that at solid/gas interfaces, the strength of the adsorption and the structure of adsorbate layers are significantly affected by the interfacial potential drop, and co-adsorption of solvent molecules and anions of supporting electrolyte [11]. Thus, the strength of adsorption increases with the negative charge on a metal surface, leading to an increase of the surface coverage [12]. For example, at Pt(111)/gas interface, the maximum coverage of 0.685 monolayer (ML) can only be achieved for CO pressures above 720 Torr [13], while at Pt(111)/electrolyte interface, this coverage is stable even in the absence of CO in the electrolyte [14], and adopts a $\sqrt{19} \times \sqrt{19}R23.4° - 13CO$ ordered structure [15]. For CO pressure of 4 Torr, a coverage of 0.75 ML is achieved at Pt(111) electrode/electrolyte interface [14] with the c(2×2)–3CO adlayer structure [15]. The transition between the $\sqrt{19} \times \sqrt{19}R23.4°$ and the c(2×2) adlayer has been first observed by STM [15] and depends on the electrode potential [15], temperature [16], and presence of CO in the solution [14].

An ubiquitous method to study adsorption of CO is infrared (IR) spectroscopy which allows distinguishing different configurations of the adsorbate, linear- (a-top), bridge-, and multi-bonded CO, as well as CO adsorbed on different surface sites. For the majority of transition metals, chemisorption of CO is very sensitive to the crystallographic orientation and to the presence of low coordination surface sites, such as steps and other surface defects, that exhibit higher binding energy for CO compared to the densely packed surfaces [10]. This allows using CO as an in situ probe of the surface structure. This is particularly valuable for electrochemical systems where UHV methods are not directly applicable. One should bear in mind however that at high adsorbate coverage, the IR spectra are dominated by dipole-dipole coupling between CO molecules, yielding blue shifts of the absorption band and intensity borrowing from the low to the high frequency vibration. Decreasing the adsorbate coverage lifts vibrational coupling and delivers a wealth of information concerning the adlayer structure and the type of adsorption sites [4, 11, 17]. This is illustrated by Fig. 1 showing the evolution of the IR wavenumber (ν_{CO}) of C-O stretch upon a progressive filling of the surface sites for Pt nanoparticles (A) and stepped Pt(557) single crystal (B) with CO. The figure shows that at low surface coverages, CO preferentially occupies sites characterized by low wavenumbers (steps on Pt(557) and edges/corners on Pt nanoparticles) and, when these are largely filled, starts to populate terraces and facets. ν_{CO} is particle-size dependent decreasing for particles below 4 nm size [18]. The wavenumber of the metal-CO stretch provides a more direct probe of the surface bonding and can be detected using surface-enhanced Raman spectroscopy.

Electrooxidation of CO on Pt-Group Metals

The electrochemical oxidation of CO has been widely investigated on Pt and less on other noble metal electrodes. In this section we will mostly refer to the Pt case with some eventual digressions to other metals. Experimentally the reaction is studied either in the presence of CO (so-called "bulk CO" oxidation) or in its absence in the electrolyte. In the latter case the oxidation of adsorbed CO can be performed either by sweeping ("CO stripping") or stepping (chronoamperometry) the electrode potential positive of the initial adsorption value. Coulometry of the oxidation of adsorbed CO is widely used in electrocatalysis for measuring the surface area of metals strongly adsorbing CO. It is particularly valuable when the method of underpotential deposition of hydrogen (H_{UPD}) is inapplicable.

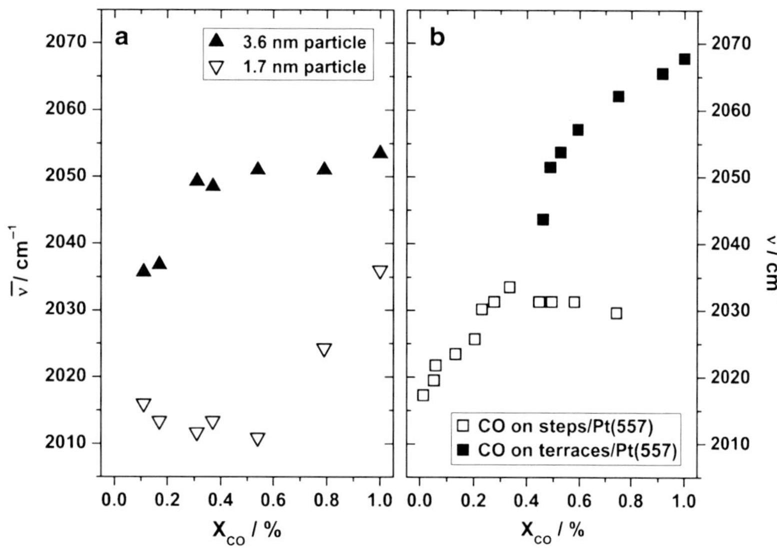

Anodic Reactions in Electrocatalysis - Oxidation of Carbon Monoxide, Fig. 1 Plot of a-top C-O band wavenumber versus the CO fractional surface coverage for (**a**) nanometer-sized Pt particles supported on pyrolytic carbon of Sibunit family and (**b**) Pt(557) = Pt [6(111)×(100)] single crystal (Reproduced from Refs. [4, 17] with permission of the American Chemical Society)

This is the case of easily oxidizable metals (such as Ni, Fe, and Ru), since H desorption is superposed with the OH adsorption; PtM alloys (M = Co [19], Ni [20], etc.), due to the electronic influence of transition metals leading to strong attenuation of the charge of H_{UPD}; and small Pt nanoparticles [1] where the H_{UPD} charge strongly decreases below 3 nm size.

Reaction Mechanism

CO oxidation on Pt occurs through a Langmuir-Hinshelwood mechanism between adsorbed CO molecules and oxygenated species [21] and can be expressed by the following reaction scheme:

$$CO_{bulk} \rightleftarrows CO_s \quad (1)$$

$$CO_s + ^* \rightleftarrows CO_{ad} \quad (2)$$

$$H_2O + ^* \rightleftarrows OH_{ad} + e^- + H^+ \quad (3)$$

$$CO_{ad} + OH_{ad} \rightarrow CO_2 + e^- + H^+ + 2^* \quad (4)$$

Here * denotes a Pt vacant site and the subscripts "ad" and "s" indicate, respectively, an adsorbed species and a species in solution close to the electrode surface. At variance with heterogeneous catalysis at solid/gas interface, where the oxidant is dissociatively adsorbed oxygen O_{ad} in electrocatalysis, oxygenated species are produced by the reversible dissociation of water molecules (Eq. 3). On densely packed surfaces, like Pt(111), reaction 3 does not occur below *ca.* 0.6 V versus RHE [1], thus rendering the CO adlayer very stable at lower electrode potentials.

CO Monolayer Oxidation

The electrooxidation of CO is one of the most structure sensitive electrochemical reactions. For extended Pt surfaces, the reaction onset is related to the formation of active oxygen-containing species (Eq. 3). There are numerous experimental evidences that reaction 3 preferentially occurs at surface defects (see [22, 23] and references therein) whose presence accelerates CO oxidation. As the density of steps increases, the onset and the peak potential of CO electrooxidation on Pt[n(111)×(111)] single crystals in acidic solutions shift towards negative potentials (Fig. 2a). This is explained by formation of OH_{ad} (Eq. 3) at steps, and fast surface diffusion of CO_{ad} to these active sites, where they react (Eq. 4) [23]. Strong structure sensitivity has also been observed for Rh single crystals [2] where contrary to the Pt case diffusion of adsorbed CO has been found to be slow leading to the tailing of stripping peaks towards positive potentials. The pH exerts strong effect on the CO_{ad} oxidation kinetics. In alkaline solutions CO_{ad} oxidation on Pt starts at lower

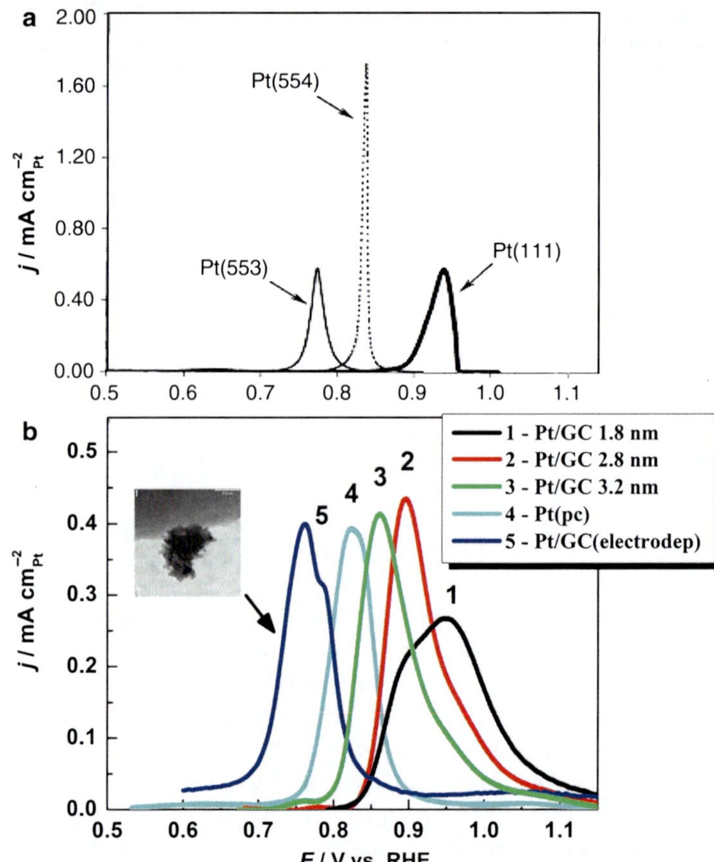

Anodic Reactions in Electrocatalysis - Oxidation of Carbon Monoxide, Fig. 2 (**a**) CO stripping voltammograms on Pt(111), Pt(553) = Pt [5(111)×(111)], and Pt (554) = Pt[10(111)×(111)] single crystals in 0.5 M H_2SO_4 at the potential sweep rate of 0.050 V s^{-1} and (**b**) background-subtracted CO stripping voltammograms on model Pt nanoparticles supported on glassy carbon (GC) in 0.1 M H_2SO_4 at the potential sweep rate of 0.100 V s^{-1} for different number average Pt particle sizes: 1.8 (1), 2.8 (2), 3.2 nm (3), polycrystalline Pt foil (4), and Pt electrodeposited on GC comprising multigrained Pt with the grain size of *ca.* 5 nm (5). T = 298 K (Reproduced from Ref. [24] and Refs. [25, 26] with permission of Elsevier and the Royal Society of Chemistry, respectively)

potentials and exhibits multiple stripping peaks on stepped electrodes [2], which is explained by surface blocking by the reaction product, carbonate species, strongly adsorbed on the surface and hindering CO_{ad} diffusion to the active sites. On nm-sized Pt electrocatalysts, the structure sensitivity of the reaction is even more striking. As the Pt particle size decreases below 3–4 nm, both the onset and the peak potentials in CO stripping voltammograms shift positive, accompanied by the broadening of the main electrooxidation peak (Fig. 2b) [1]. This is explained by the particle-size-dependent decrease of the rate constant for the $CO_{ad} + OH_{ad}$ recombination (Eq. 4) and of the surface diffusion coefficient of CO_{ad} molecules. Interestingly, it has also been shown that nanoparticle agglomeration yields formation of "multigrained" structures, composed of individual Pt nanoparticles interconnected via grain boundaries, which are extremely active for the generation of OH_{ad} species (see curve 5 in Fig. 2b) and thus accelerate the CO_{ad} and methanol electrooxidation kinetics [1]. The strong sensitivity of the CO electrooxidation reaction to the structure of carbon-supported nanoparticles renders CO stripping voltammetry useful for in situ probing the particle size and distribution, and the occurrence of particle agglomeration [1].

Bulk CO Oxidation on Polycrystalline and Single Crystalline Pt Electrodes

In the presence of CO in the electrolyte, CO adsorption (Eq. 2) and formation of adsorbed oxygenated species from water (Eq. 3) are in competition for the Pt-free sites. In combination with the CO mass transport (Eq. 1), the competitive Langmuir-Hinshelwood mechanism induces an autocatalytic loop into the electrochemical system which manifests itself by the bistable

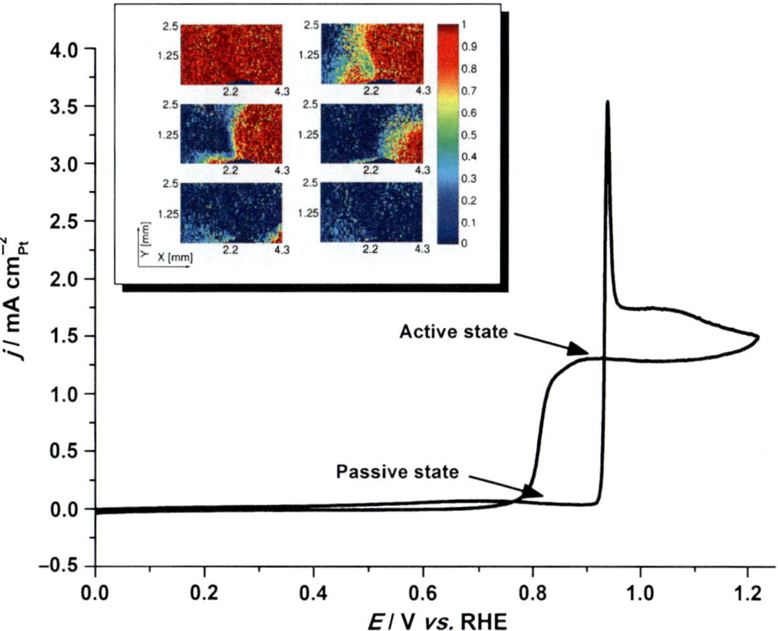

Anodic Reactions in Electrocatalysis - Oxidation of Carbon Monoxide, Fig. 3 Cyclic voltammogram of a rotating polycrystalline Pt disk electrode in CO saturated 0.1 M H_2SO_4. Potential sweep rate: 0.050 V s^{-1}; rotation rate 900 rpm. The inset shows the integrated IR peak of CO_{ad} plotted versus the position on the electrode for the applied current of 0.3, 1.4, 1.5, 1.7, 1.75, and 1.8 mA (*left to right* and *top to bottom*). The *red color* indicates a high CO_{ad} coverage and the *blue* a CO-free surface (Reproduced from Ref. [28] with permission of the American Chemical Society)

behavior in the current-potential curves [27]. In case of flat polycrystalline or single-crystal Pt rotating disc electrodes, cyclic voltammetry reveals the existence of two stable steady-state currents, for a single value of the applied potential, as shown in Fig. 3 [27]. The passive current state corresponds to the Pt electrode poisoned by CO_{ad}, whereas the active current state corresponds to the diffusion-limited oxidation current of CO from solution associated with a low CO_{ad} coverage. For an applied potential value in the bistability region, the actual state of the electrochemical system depends on its prehistory, i.e., on the potential program previously applied to the electrode. In the case of CO electrooxidation, the bistable behavior observed under strictly potentiostatic conditions is associated with an *S*-shaped negative differential resistance (S-NDR) which should be accessible under galvanostatic conditions [27]. However, the electrode potential in the case of an S-NDR electrochemical system induces a negative feedback loop with a fast characteristic time, resulting in the formation of self-organized spatial patterns, such as domain patterns, under current control, as shown by Morschl et al. using spatially resolved IR spectroscopy [28]. For intermediate values of the applied current, two stationary domains are observed: one with a high CO_{ad} coverage and the other one with a low CO_{ad} coverage (cf. inset in Fig. 3). The interaction of a fast autocatalytic reaction with a slow inhibiting process results in excitable and oscillatory behaviors. A possible inhibiting process comes from the strong adsorption of anions, such as chlorides, which are competing with CO and water for Pt-free surface sites. In particular, spontaneous current oscillations have been reported under strictly potentiostatic conditions for CO electrooxidation on polycrystalline and single crystal Pt electrodes in the presence of small amounts of chloride anions in solution [29]. Further complex phenomena may be observed when porous three-dimensional electrode layers consisting of Pt nanoparticles supported on carbon materials are utilized. To gain insight into the influence of the electrode architecture on the bistable behavior, Ruvinskiy et al. [30] studied CO bulk electrooxidation on thick 3D-ordered catalytic layers based on vertically aligned carbon nanofilaments. Spontaneous formation of a stationary CO coverage front along the nanofiber length was observed under potentiostatic conditions, the width of the front depending on the Pt coverage on the nanofilament [30].

CO Electrooxidation on Group I Metals

CO adsorption and electrooxidation on gold is strongly pH dependent. While in acidic electrolytes CO stays on the surface only in its presence in the electrolyte [31, 32], strong irreversible adsorption of CO on gold electrodes, viz., stable CO coverage in the absence of CO in solution, has been observed in alkaline electrolytes. Likewise, the electrocatalytic activity of gold towards electrochemical CO oxidation strongly increases with the pH of electrolyte, in alkaline electrolytes the reaction overpotential on Au being *ca.* 0.5 V lower than on Pt electrodes [31, 33]. Recently, it was proposed [34] that CO oxidation occurs through a self-promotion mechanism, in which adsorbed CO on the gold surface enhances the adsorption of the OH species involved in the oxidation mechanism. This mechanism offers a plausible explanation for the higher activity in alkaline solutions and is supported by the density functional theory (DFT) calculations and by the measurement of the experimental reaction orders in CO that are slightly higher in alkaline compared to acidic media. Note that copper and silver electrodes were found to be rather catalytically inactive for CO oxidation, even in alkaline media [31].

CO Tolerance in Fuel Cell Catalysis

CO, which may be present in the PEMFC feed when H_2 is obtained from hydrocarbon sources, strongly depreciates the output fuel cell voltage [35, 36]. Being present even at ppm levels, CO attains close to the saturation coverage on the surface of a Pt-based fuel cell anode due to its strong adsorption (Eq. 1) [36]. To increase the tolerance to CO, bimetallic alloys are used. PtRu remains the most widely used for improving the tolerance to CO [37], while other bimetallic compositions, such as PtMo [36] or PtSn, also show improved performance compared to pure Pt. Two effects are invoked to explain the CO tolerance of PtM surfaces, viz., the "ligand" and the "bifunctional" effect [37]. Within the ligand effect approximation, it is supposed that the addition of M to Pt leads to a downshift of the Pt *d*-band decreasing the strength of the Pt-C bond and the CO_{ad} coverage, thus increasing the number of active sites available for the dissociative adsorption of H_2. This effect is dominating, for example, for PtSn, but also for PdAu catalysts [38]. On the other hand, within the realm of a bifunctional mechanism, which is dominating for PtRu, it is assumed that M atoms provide oxygenated species by dissociating H_2O molecules at lower potentials against Pt sites (see Eq. 3), leading to an accelerated CO_2 formation through surface reaction of OH species adsorbed on M sites with CO adsorbed on Pt sites (Eq. 4), and a concomitant decrease in the CO coverage. The bifunctional effect is illustrated by Fig. 4 showing a systematic shift of the CO stripping peak upon increasing the coverage of Ru deposited on Pt particles.

Conclusions and Outlook

CO adsorption and electrooxidation are important both from the fundamental and from the applied point of view, and have been extensively studied on various mono- and bimetallic surfaces, including single crystalline, polycrystalline and nanoparticulated. Various methods have been applied, including electrochemical (e.g., cyclic voltammetry, chronoamperometry), imaging (STM), and spectroscopic (e.g., IR and Raman spectroscopy, X-ray absorption spectroscopy, sum frequency, and second harmonic generation spectroscopy). Because of the space limitations, only some of these were reviewed in this article. CO oxidation is one of a few electrocatalytic reactions whose mechanism is relatively well understood. Electrochemistry of CO also provides an interesting example of how the electrode potential and the electrolyte composition (pH, presence of strongly adsorbing anions) influence the strength of adsorption and the rate of electrooxidation. Despite vast bibliography which could not be fully cited, one may expect that CO adsorption and oxidation will continue serving the electrochemical community in refining the understanding of the role of the electrode composition, the size,

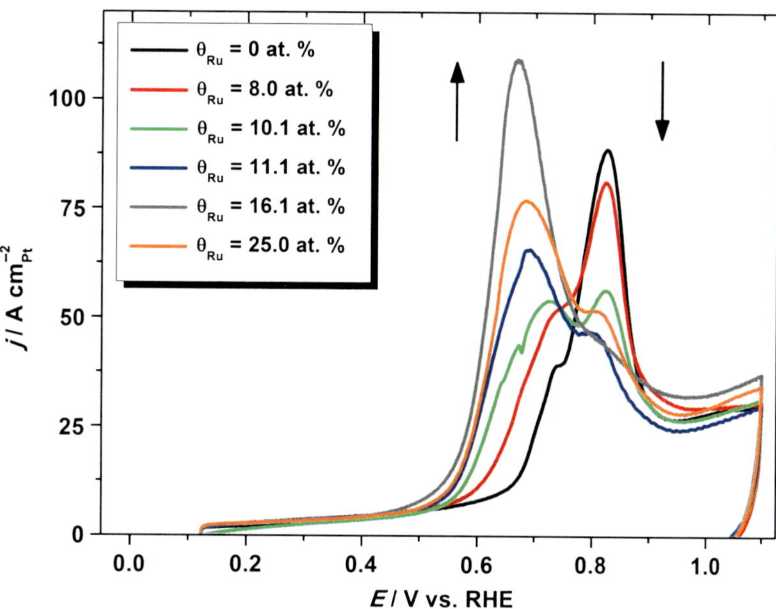

Anodic Reactions in Electrocatalysis - Oxidation of Carbon Monoxide, Fig. 4 CO stripping voltammograms in 0.1 M $HClO_4$ at a potential sweep rate of 0.020 V s^{-1} measured on a carbon-supported Pt electrode modified by different Ru coverages achieved by a spontaneous deposition (Reproduced from Ref. [37] with permission of the American Chemical Society)

and the structure in electrocatalysis, as well as in unveiling intriguing nonlinear phenomena.

Cross-References

- ▶ Electrocatalysis of Anodic Reactions
- ▶ Raman Spectroelectrochemistry

References

1. Maillard F, Pronkin S, Savinova ER (2009) Influence of size on the electrocatalytic activities of supported metal nanoparticles in fuel cells related reactions. In: Vielstich W, Gasteiger HA, Yokokawa H (eds) Handbook of fuel cells – fundamentals, technology and applications. Wiley, New York
2. Koper MTM (2011) Structure sensitivity and nanoscale effects in electrocatalysis. Nanoscale 3:2054–2073
3. Beden B, Lamy C, de Tacconi NR, Arvia AJ (1990) The electrooxidation of CO – a test reaction in electrocatalysis. Electrochim Acta 35:691–704
4. Kim CS, Korzeniewski C (1997) Vibrational coupling as a probe of adsorption at different structural sites on a stepped single-crystal electrode. Anal Chem 69:2349–2353
5. Trasatti S, Petrii OA (1992) Real surface-area measurements in electrochemistry. J Electroanal Chem 327:353–376
6. Blyholder G (1964) Molecular orbital view of chemisorbed carbon monoxide. J Phys Chem 68:2772–2778
7. Anderson PW (1961) Localized magnetic states in metals. Phys Rev 124:41–53
8. Newns DM (1969) Self-consistent model of hydrogen chemisorption. Phys Rev 178:1123–1135
9. Hoffman R (1988) Rev Mod Phys 60:601–628
10. Guczi L (1991) In: Guczi L (ed) New trends in CO activation, vol 64, Studies in surface science and catalysis. Elsevier, Amsterdam
11. Weaver MJ (1999) Binding sites and vibrational frequencies for dilute carbon monoxide and nitric oxide adlayers in electrochemical versus ultrahigh-vacuum environments: the roles of double-layer solvation. Surf Sci 437:215–230
12. Wasileski SA, Koper MTM, Weaver MJ (2001) Field-dependent chemisorption of carbon monoxide on platinum-group (111) surfaces: relationships between binding energetics, geometries, and vibrational properties as assessed by density functional theory. J Phys Chem B 105:3518–3530
13. Longwitz SR, Schnadt J, Vestergaard EK, Vang RT, Stensgaard I, Brune H, Besenbacher F (2004) High-coverage structures of carbon monoxide adsorbed on Pt(111) studied by high-pressure scanning tunneling microscopy. J Phys Chem B108:14497–14502
14. Cuesta A, del Carmen Perez M, Rincon A, Gutierrez C (2006) Adsorption isotherm of CO on Pt(111) electrodes. ChemPhysChem 7:2346–2351
15. Villegas I, Weaver MJ (1994) Carbon-monoxide adlayer structures on platinum(111) electrodes – a synergy between in-situ scanning-tunneling-microscopy and infrared-spectroscopy. J Chem Phys 101:1648–1660
16. Markovic NM, Ross PN (2002) Surface science studies of model fuel cell electrocatalysts. Surf Sci Rep 45:117–229

17. Maillard F, Savinova E, Simonov PA, Zaikovskii VI, Stimming U (2004) Infrared spectroscopic study of CO adsorption and electrooxidation on carbon-supported Pt nanoparticles: inter-particle versus intra-particle heterogeneity. J Phys Chem B 108:17893–17904
18. Park S, Wasileski SA, Weaver MJ (2001) Electrochemical infrared characterization of carbon-supported platinum nanoparticles: a benchmark structural comparison with single-crystal electrodes and high-nuclearity carbonyl clusters. J Phys Chem B105:9719–9725
19. Dubau L, Maillard F, Chatenet M, André J, Rossinot E (2010) Nanoscale compositional changes and modification of the surface reactivity of Pt_3Co/C nanoparticles during proton-exchange membrane fuel cell operation. Electrochim Acta 56:776–783
20. Stamenkovic VR, Mun BS, Mayrhofer KJJ, Ross PN, Markovic NM (2006) Effect of surface composition on electronic structure, stability, and electrocatalytic properties of Pt-transition metal alloys: Pt-skin versus Pt-skeleton surfaces. J Am Chem Soc 128:8813–8819
21. Gilman S (1964) The mechanism of electrochemical oxidation of carbon monoxide and methanol on platinum. II. The "reactant-pair" mechanism for electrochemical oxidation of carbon monoxide and methanol. J Phys Chem 68:70–80
22. Petukhov AV, Akemann W, Friedrich KA, Stimming U (1998) Kinetics of electrooxidation of a CO monolayer at the platinum/electrolyte interface. Surf Sci 404:182–186
23. García G, Koper MTM (2011) Carbon monoxide oxidation on Pt single crystal electrodes: understanding the catalysis for low temperature fuel cells. ChemPhysChem 12:2064–2072
24. Lebedeva NP, Koper MTM, Herrero E, Feliu JM, Van Santen RA (2000) CO oxidation on stepped Pt[n(111)x(111)] electrodes. J Electroanal Chem 487:37–44
25. Maillard F, Eikerling M, Cherstiouk OV, Schreier S, Savinova E, Stimming U (2004) Size effects on reactivity of Pt nanoparticles in CO monolayer oxidation: the role of surface mobility. Faraday Discuss 125:357–377
26. Maillard F, Schreier S, Hanzlik M, Savinova ER, Weinkauf S, Stimming U (2005) Influence of particle agglomeration on the catalytic activity of carbon-supported Pt nanoparticles in CO monolayer oxidation. Phys Chem Chem Phys 7:385–393
27. Koper MTM, Schmidt TJ, Markovic NM, Ross PN (2001) Potential oscillations and S-shaped polarization curve in the continuous electro-oxidation of CO on platinum single-crystal electrodes. J Phys Chem B 105:8381–8386
28. Morschl R, Bolten J, Bonnefont A, Krischer K (2008) Pattern formation during CO electrooxidation on thin Pt films studied with spatially resolved infrared absorption spectroscopy. J Phys Chem C 112:9548–9551
29. Malkhandi S, Bonnefont A, Krischer K (2009) Dynamic instabilities during the continuous electrooxidation of CO on poly- and single crystalline Pt electrodes. Surf Sci 603:1646–1651
30. Ruvinskiy PS, Bonnefont A, Bayati M, Savinova ER (2010) Mass transport effects in CO bulk electrooxidation on Pt nanoparticles supported on vertically aligned carbon nanofilaments. Phys Chem Chem Phys 12:15207–15216
31. Kita H, Nakajima H, Hayashi K (1985) Electrochemical oxidation of CO on Au in alkaline solution. J Electroanal Chem 190:141–156
32. Rodriguez P, Garcia-Araez N, Koverga A, Frank S, Koper MTM (2010) CO Electrooxidation on gold in alkaline media: a combined electrochemical, spectroscopic, and DFT study. Langmuir 26:12425–12432
33. Roberts JL, Sawyer DT (1964) Voltammetric determination of carbon monoxide at gold electrodes. J Electroanal Chem 7:315–319
34. Rodriguez P, Koverga AA, Koper MTM (2010) - Carbon monoxide as a promoter for its own oxidation on a gold electrode. Angew Chem Int Ed 49:1241–1243
35. Breiter MW (1975) Influence of chemisorbed carbon monoxide on the oxidation of molecular hydrogen at smooth platinum in sulfuric acid solution. J Electroanal Chem 65:623–634
36. Igarashi H, Fujino T, Zhu YM, Uchida H, Watanabe M (2001) CO Tolerance of Pt alloy electrocatalysts for polymer electrolyte fuel cells and the detoxification mechanism. Phys Chem Chem Phys 3:306–314
37. Maillard F, Lu GQ, Wieckowski A, Stimming U (2005) Ru-decorated Pt surfaces as model fuel cell electrocatalysts for CO electrooxidation. J Phys Chem B 109:16230–16243
38. Schmidt TJ, Jusys Z, Gasteiger HA, Behm RJ, Endruschat U, Boennemann H (2001) On the CO tolerance of novel colloidal PdAu:carbon electrocatalysts. J Electroanal Chem 501:132–140

Anodic Substitutions

R. Daniel Little
University of California, Santa Barbara, CA, USA

Introduction

Anodic substitution reactions are frequently defined by the equation shown below (Eq. 1).

The substrate, R–E, is oxidized at the anode and is subsequently intercepted by a nucleophile, Nu. A host of nucleophiles have been utilized; some are listed beneath the equation. The chemistry has been the subject of a number of excellent reviews [1–4]. The following discussion is intended, therefore, to provide an overview of the chemistry rather than an in-depth review.

$$R\text{–}E + Nu^- \xrightarrow{[O]} R\text{–}Nu + E^+ + 2e^- \quad (1)$$

nucleophiles, both charged and neutral have included:

HO^-, RO^-, RCO_2^-, NO_3^-, NO_2^-, N_3^-, ^-OCN, ^-SCN, ^-SeCN, halide, ^-CN, H_2O, ROH, RCO_2H, CH_3CN, pyridine, allyl silanes, silyl enol ethers, ketene acetals, 1,3-dicarbonyl compounds

Examples and Technological Advancements

The two examples portrayed below, each occurring via different mechanistic pathways, give one a sense of the breadth of the chemistry. In the first (Eq. 2), different nucleophiles, viz., azide and methanol, add across a double bond [5]. The more nucleophilic of the pair adds to the less substituted carbon to afford a heteroatom-stabilized cation that is then intercepted by the second nucleophile, ultimately delivering the product (Eq. 3). The second example portrays the replacement of a C–H bond on anthracene with acetonitrile to afford the acetamide adduct after hydrolysis of the nitrilium ion intermediate (Eq. 4) [6].

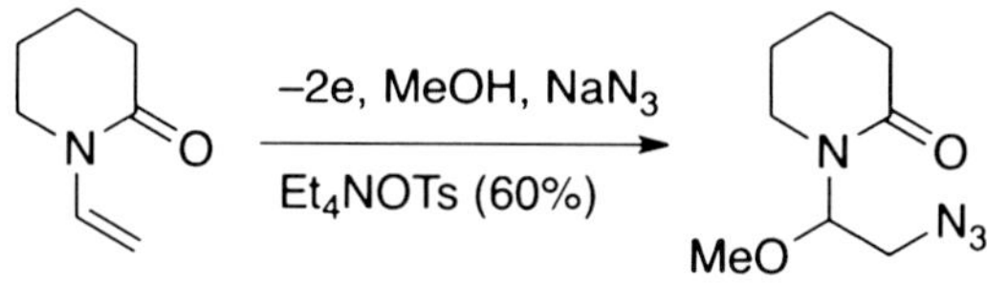

(2)

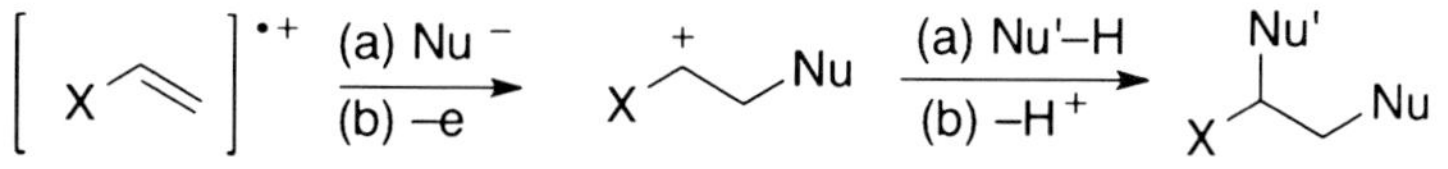

(3)

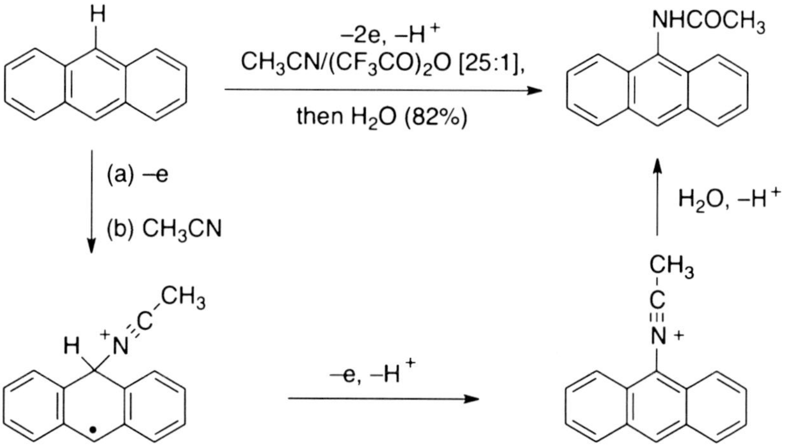

(4)

These are but two of the large number of interesting and synthetically useful transformations that can be accomplished using anodic substitution. Others range from addition to hydrocarbon frameworks [1], to cycloadditions (Eq. 5) [7], to site-specific fluorination (Eq. 6) [8], and spirocyclization [9, 10], to name a few.

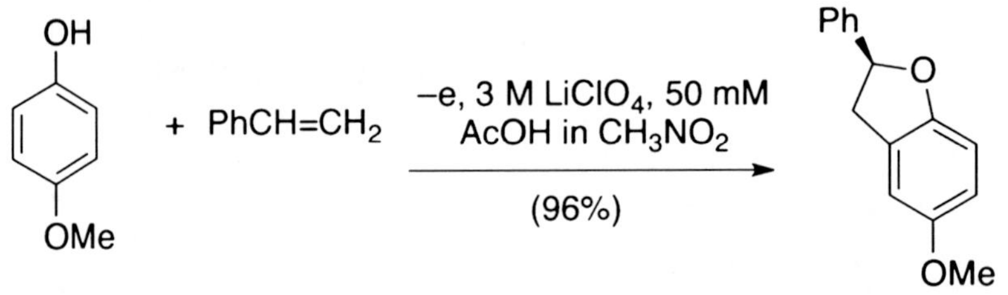

$$\text{4-MeOC}_6\text{H}_4\text{OH} + \text{PhCH=CH}_2 \xrightarrow[\text{(96\%)}]{-e,\ 3\text{ M LiClO}_4,\ 50\text{ mM AcOH in CH}_3\text{NO}_2} \text{2-Ph-5-methoxy-2,3-dihydrobenzofuran} \quad (5)$$

$$\text{N-isopropyl-benzothiazinone} \xrightarrow[\text{Et}_3\text{N·3HF, CH}_3\text{CN (88\%)}]{-2e,\ -H^+} \text{2-fluoro product} \quad (6)$$

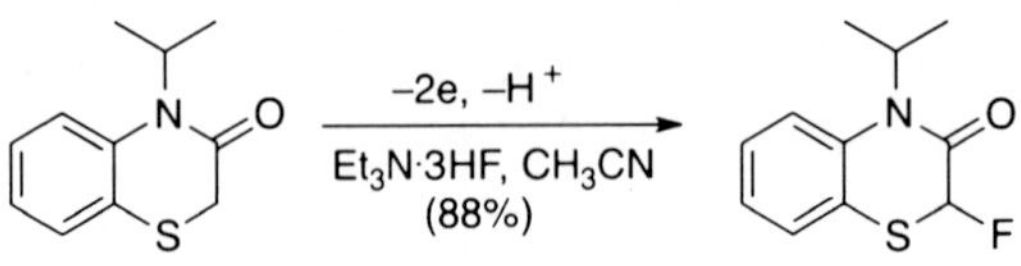

Occasionally, one encounters a situation where the nucleophile is easier to oxidize than the substrate. One way to overcome the problem is to modify the structure of the substrate, RCH_2Y, by incorporating a so-called electroauxiliary, X, at the site that is to undergo substitution [11]. In this manner, interaction of the C–X σ-orbital with the nonbonding lone pair on Y raises the energy of the HOMO, thereby making RCHXY easier to oxidize than RCH_2Y. This is illustrated by the dramatic decrease of 700 mV that accompanies introduction of the phenylthio group into the pyrrolidine framework shown below. In doing so, it proved possible to carry out a selective oxidation in the presence of allyltrimethylsilane (Eq. 7). Were it not for the electroauxiliary, the silane would be oxidized rather than the pyrrolidine [12].

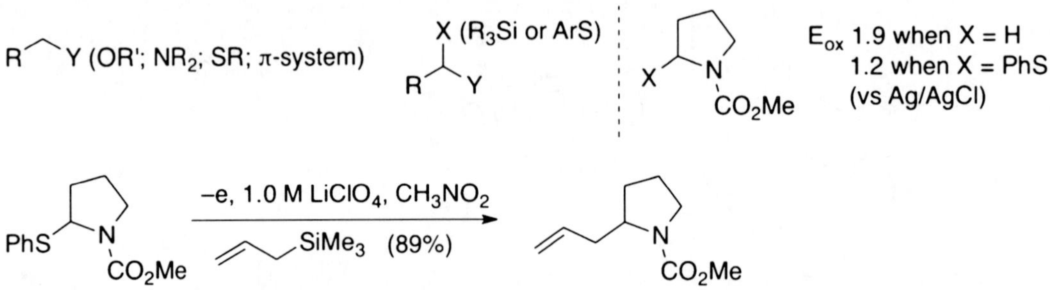

R⁀Y (OR'; NR$_2$; SR; π-system) → RCH(X)Y, X (R_3Si or ArS)

2-X-N-(CO_2Me)pyrrolidine: E_{ox} 1.9 when X = H; 1.2 when X = PhS (vs Ag/AgCl)

$$\text{2-PhS-N-(CO}_2\text{Me)pyrrolidine} \xrightarrow[\text{CH}_2\text{=CHCH}_2\text{SiMe}_3\ (89\%)]{-e,\ 1.0\text{ M LiClO}_4,\ \text{CH}_3\text{NO}_2} \text{2-allyl-N-(CO}_2\text{Me)pyrrolidine} \quad (7)$$

Other protocols have been developed to mitigate the problems encountered when the nucleophile is easier to oxidize than the substrate. Two notable approaches include the so-called cation pool method of Yoshida and coworkers, and the laminar flow method pioneered in the laboratory of Fuchigami. The cation pool method has been described in detail elsewhere [13]. Basically, the method calls for the formation and accumulation of a "pool" of the cation at low temperatures where it survives long enough to be intercepted by a nucleophile. A partial list of nucleophiles includes acetyl acetone, dialkyl malonates, cyclic and acyclic silyl enol ethers and silanes, benzyl silanes [14], and ketene acetals. Many cations have been generated and intercepted using this synthetically powerful strategy; two are shown here (Eqs. 8 and 9). The chemistry can also be carried out using a flow microreactor. This approach has the advantage of being able to monitor the chemistry by FTIR during the course of the reaction.

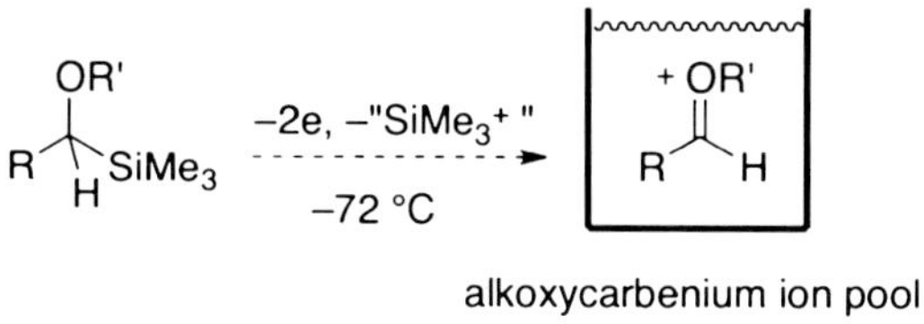

$$R\text{-}CH(OR')\text{-}SiMe_3 \xrightarrow[-72\,°C]{-2e,\ -\text{"}SiMe_3^+\text{"}} [R\text{-}CH{=}O^+R']$$

alkoxycarbenium ion pool

(8)

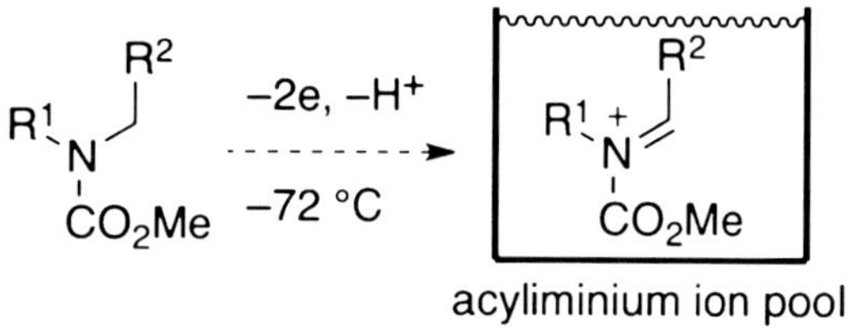

$$R^1\text{-}N(CO_2Me)\text{-}CH_2R^2 \xrightarrow[-72\,°C]{-2e,\ -H^+} [R^1\text{-}N^+(CO_2Me){=}CHR^2]$$

acyliminium ion pool

(9)

One novel feature of the chemistry stems from a realization that the initially formed cation pool intermediate can be reduced [13]. An acyliminium ion, for example, can be reduced to the corresponding radical, thereby offering a direct pathway for the formation of these intermediates. The resulting species undergo the usual sorts of reaction that characterize radicals including, for example, being intercepted by electron-deficient alkenes (Eq. 10).

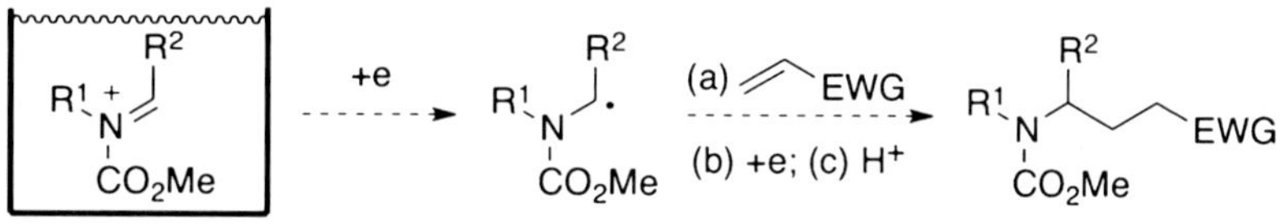

$$[R^1\text{-}N^+(CO_2Me){=}CHR^2] \xrightarrow{+e} R^1\text{-}N(CO_2Me)\text{-}\dot{C}HR^2 \xrightarrow[(b)\ +e;\ (c)\ H^+]{(a)\ CH_2{=}CH\text{-}EWG} R^1\text{-}N(CO_2Me)\text{-}CH(R^2)CH_2CH_2\text{-}EWG \quad (10)$$

The laminar flow method calls for introduction of the substrate and nucleophile into a microflow reactor through separate inlets to create a parallel stream consisting of each of the reactants. The methodology has been successfully applied to both cathodic as well as anodic substitution reactions [15, 16]. The substance to be oxidized enters through the inlet located nearest the anode. Ultimately, diffusion occurs and the streams mix with one another. Ideally, the mixing occurs after the substrate has been oxidized and before the nucleophile has an opportunity to migrate to the anode or the oxidized species is reduced at the cathode (Fig. 1).

The transformation illustrated in the accompanying equation and table highlights the fact that even though the oxidation potential of the nucleophile, in this case, allyl silane, is less than that of the carbamate (1.75 compared with 1.91 vs. Ag/AgCl), the carbamate is oxidized preferentially since it enters the microflow reactor through the input stream nearest the anode (Eq. 11) [17]. Comparison of entries 1, 2, and 3 reveals the dramatic influence of the reaction medium upon the success of the approach. Here, one sees the dramatic improvement that accompanies the use of 2,2,2-trifluoroethanol (TFE) as the solvent; its ability to solvate and stabilize carbocations is undoubtedly beneficial. Further improvement accompanies the use of an ionic liquid, in this case, 1-ethyl-3-methylimidazolium bis(trifluoromethanesulfonyl) imide ([emim][TFSI]). Presumably, it too stabilizes the cation so that it can react with the nucleophile before the cation decomposes.

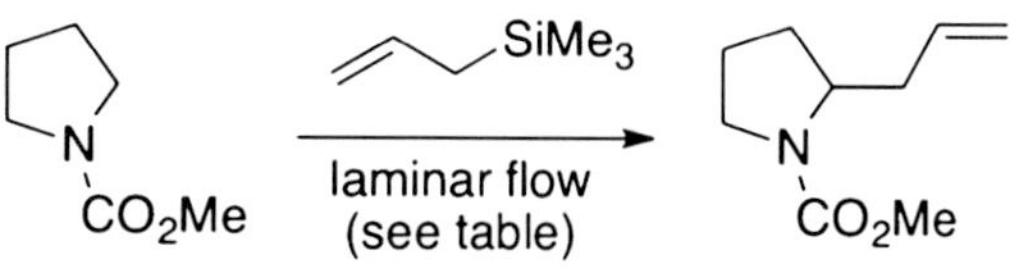

$$\text{N-}CO_2Me\text{ pyrrolidine} \xrightarrow[\text{laminar flow (see table)}]{CH_2{=}CHCH_2SiMe_3} \text{2-allyl-N-}CO_2Me\text{ pyrrolidine}$$

(11)

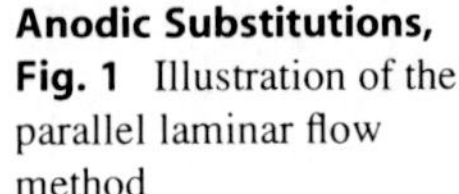

Anodic Substitutions, Fig. 1 Illustration of the parallel laminar flow method

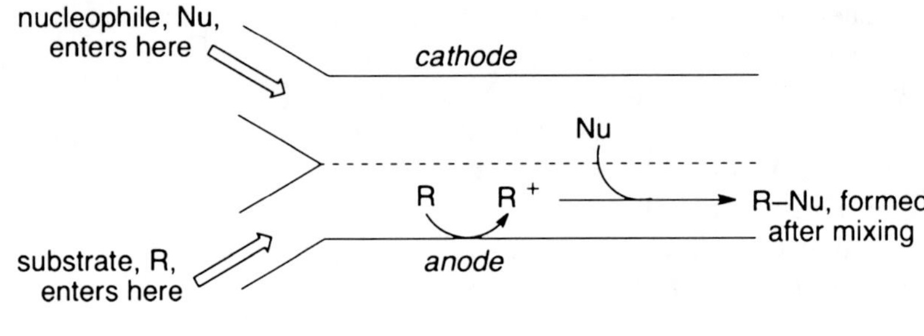

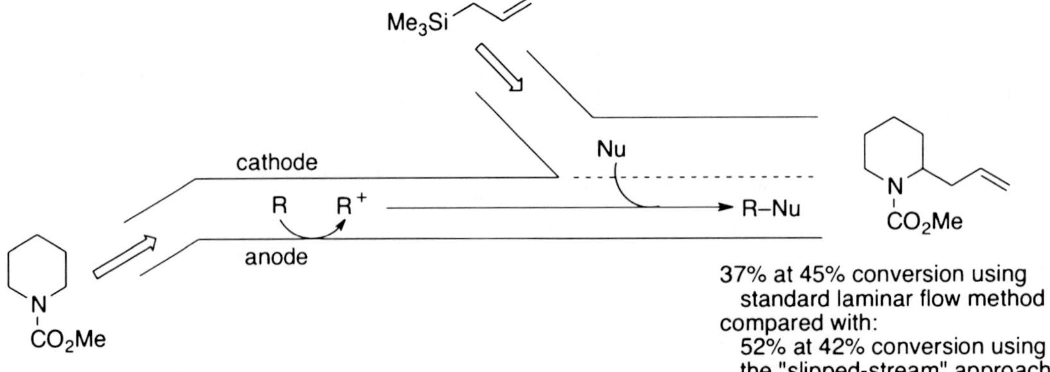

Anodic Substitutions, Fig. 2 A useful variant of the laminar flow method

Entry	Electrolytic solution	Conversion (%)	Yield (%)	Comment
1	0.1 M *n*-Bu_4NBF_4/CH_3CN	73	0.6	
2	0.1 M *n*-Bu_4NBF_4/CF_3CH_2OH	58	59	Note improvement accompanying solvent change
3	The ionic liquid, [emim][TFSI]	66	73	Use of an ionic liquid leads to improvement

[emim][TFSI] refers to the ionic liquid 1-ethyl-3-methylimidazolium bis(trifluoromethanesulfonyl)imide

The methodology has been improved through a slight modification calling for introduction of the nucleophile downstream of where the cation is first generated [17]. This is intended to provide time for the latter to be generated prior to encountering the nucleophile. The accompanying figure illustrates the concept. The substrates were introduced as solutions in an ionic liquid (Fig. 2).

Future Directions

For decades, anodic substitution reactions have received a great deal of attention. This is not surprising given the power of the chemistry to achieve a wide variety of transformations that are frequently difficult or impossible to achieve using alternative approaches. Clearly much progress has been made and a great deal of creativity has been expressed as researchers have sought to solve the difficult synthetic and mechanistic challenges that chemistry presents. It seems safe to conclude that this area of research will continue to flourish for decades to come.

Cross-References

- ► Cation-Pool Method
- ► Electroauxiliary
- ► Electrochemical Microflow Systems
- ► Electrosynthesis in Ionic Liquid

References

1. Hammerich O (2001) Anodic oxidation of hydrocarbons, Chapter 13. In: Lund H, Hammerich O (eds) Organic electrochemistry. Dekker, New York
2. Hammerich O, Utley JHP, Eberson L (2001) Anodic substitution and addition, Chapter 24. In: Lund H, Hammerich O (eds) Organic electrochemistry. Dekker, New York
3. Eberson L, Nyberg K (1976) Synthetic uses of anodic substitution reactions. Tetrahedron 32:2185–2206
4. Eberson L, Nyberg K (1973) Anodic aromatic substitution. Acc Chem Res 6:106–112
5. Fujimoto K, Tokuda Y, Matsubara Y, Maekawa H, Mizuno T, Nishiguchi I (1995) Regioselective azidomethoxylation of enol ethers by anodic oxidation. Tetrahedron Lett 36:7483–7486
6. Hammerich O, Parker VD (1974) Reaction of the anthracene cation radical with acetonitrile. A novel anodic acetamidation. J Chem Soc Chem Commun 7:275–276
7. Chiba K, Fukuda M, Kim S, Kitano Y, Tada M (1999) Dihydrobenzofuran synthesis by an anodic [3 + 2] cycloaddition of phenols and unactivated alkenes. J Org Chem 64:7654–7656
8. Konno A, Naito W, Fuchigami T (1992) Electrolytic partial fluorination of organic compounds. 6. Highly regioselective electrochemical monofluorination of aliphatic nitrogen-containing heterocycles. Tetrahedron Lett 33:7017–7020
9. Amano Y, Nishiyama S (2006) Oxidative synthesis of azacyclic derivatives through the nitrenium ion: application of a hypervalent iodine species electrochemically generated from iodobenzene. Tetrahedron Lett 47:6505–6507
10. Zhong W, Little RD (2009) Exploration and determination of the redox properties of the pseudopterosin class of marine natural products. Tetrahedron 65:10784–10790
11. Yoshida J, Kataoka K, Horcajada R, Nagaki A (2008) Modern strategies in electroorganic synthesis. Chem Rev 108:2265–2299
12. Kim S, Hayashi K, Kitano Y, Tada M, Chiba K (2002) Anodic modification of proline derivatives using a lithium perchlorate/nitromethane electrolyte solution. Org Lett 4:3735–3737
13. Yoshida J-I, Suga S (2002) Basic concepts of "cation pool" and "cation flow" methods and their applications in conventional and combinatorial organic synthesis. Chem Eur J 8:2650–2658
14. Maruyama T, Mizuno Y, Shimizu I, Suga S, Yoshida J-I (2007) Reactions of a N-acyliminium ion pool with benzylsilanes. Implication of a radical/cation/radical cation chain mechanism involving oxidative C-Si bond cleavage. J Am Chem Soc 129:1902–1903
15. Horii D, Amemiya F, Fuchigami T, Atobe M (2008) A novel electrosynthetic system for anodic substitution reactions by using parallel laminar flow in a microflow reactor. Chem Eur J 14:10382–10387
16. Amemiya F, Fuse K, Fuchigami T, Atobe M (2010) Chemoselective reaction system using a two inlet micro-flow reactor: application to carbonyl allylation. Chem Commun 46:2730–2732
17. Horii D, Fuchigami T, Atobe M (2007) A new approach to anodic substitution reaction using parallel laminar flow in a micro-reactor. J Am Chem Soc 129:11692–11693

Aqueous Rechargeable Lithium Batteries (ARLB)

Yuping Wu
Department of Chemistry, Fudan University, Shanghai, China

Definition

A rechargeable battery using lithium intercalation (insertion) compound(s) as one or two electrodes based on redox reactions and lithium-containing aqueous solution as electrolyte. During the charge and discharge process, redox reactions, not absorption and desorption, happen together with insertion/removal of lithium ions and gain/loss of electrons. When one electrode utilizes the absorption/desorption, it is called as hybrid supercapacitors instead of batteries.

Its electrolyte is based on aqueous solutions of lithium salts instead of nonaqueous solutions [1, 2], which are usually combustible [3].

Components

It is a rechargeable battery and consists of the following main materials: positive electrode, negative electrode, electrolytes, and separators. However, in some reports, the positive electrode and negative electrode are called as cathode and anode, respectively. Scientifically speaking, this calling is not pertinent.

Positive Electrode Materials

Their positive electrode materials are lithium intercalation (insertion) compounds such as

$LiCoO_2$, $LiMn_2O_4$, $LiNiO_2$, and $LiFePO_4$ [4, 5], which can also be cathode materials for lithium ion batteries. Their redox potentials are below that for oxygen evolution so that they can be stable in aqueous electrolytes. However, their behaviors in the aqueous electrolytes are a little different from that in the organic electrolytes. For example, in the case of $LiCoO_2$, its cycling is very good in the organic electrolytes. In aqueous electrolytes, their cycling behavior is poorer. In the case of $LiMn_2O_4$, due to the Jahn-Teller effect, its cycling behavior in the organic electrolytes is poor. However, in the aqueous electrolytes, its cycling performance is much better. In the case of porous $LiMn_2O_4$ without doping, its capacity retention is 93 % after 10,000 full cycles, which is shown in Fig. 1 [6].

Since the ionic conductivities of the aqueous electrolytes are about two orders of magnitudes higher than those of the organic electrolytes, they can charge much faster than in the organic electrolytes.

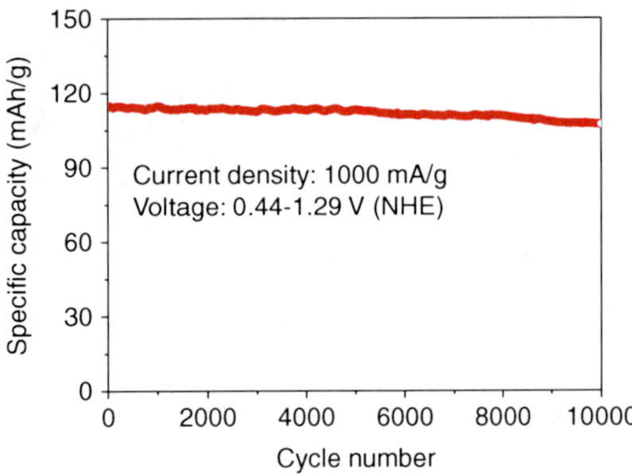

Aqueous Rechargeable Lithium Batteries (ARLB), Fig. 1 Cycling behavior of macroporous $LiMn_2O_4$ in 0.5 M Li_2SO_4 aqueous solution at 100% DOD using an activated carbon as the counter electrode

Negative Electrode Materials

Their negative electrode materials include vanadium oxides, molybdenum oxide, and some lithium intercalation compounds such as LiV_3O_8 and $LiTi_2(PO_4)_3$ [4]. Since the transitional metal oxides are not stable in the aqueous electrolyte during charge/discharge process, they will dissolve in the aqueous electrolytes leading to poor cycling performance. When they are coated with carbon or conductive polymers such as polypyrrole, their cycling performance is greatly improved [4, 7]. Of course, they can also be charged very fast, which is similar to their cathode materials.

Electrolytes

Their electrolytes include the following solutions: Li_2SO_4, $LiNO_3$, and LiOH or their mixtures. The salts are easily available, and their cost is much cheaper than that of $LiPF_6$, which is well used for lithium ion batteries. Of course, they are very friendly to the environment and will not produce environmental problems like $LiPF_6$.

Separators

There is no systematic report on this material. However, since the working condition is much less strict than those for Ni-MH and lead acid rechargeable batteries. All the separators for them can be used for ARLBs.

Electrochemical Performance

Depending on the combination of negative electrode and positive electrode, the discharge and charge curves of the ARLBs are different. For example, the average discharge voltage for $LiV_3O_8/LiCoO_2$ is 1.05 V, that for $V_2O_5/LiMn_2O_4$ is 1.20 V, and that for $MoO_3/LiMn_2O_4$ is 1.25 V. In the case of the $LiV_3O_8/LiCoO_2$, its cycling is not good. When MoO_3 is coated with polypyrrole, there is no evident capacity fading after 500 full cycles for the ARLB of $MoO_3/LiMn_2O_4$. Its rate capability is also very good. Even at 6 kW/kg, its energy density can still be above 36 Wh/kg [8].

Characteristics

ARLB has the following characteristics:

- Good availability of lithium salts.
- The rate capability of the ARLB is much better than that of traditional lithium ion batteries. It can arrive at the minute-level charge performance, which means that ARLB can be fully charged within minutes instead of hours [6, 7].

- High power density.
- Good safety, no combustibility or explosion.
- Low cost for production due to no requirements on the content of moisture.
- Low requirements on separators especially the shutdown performance.
- Easy to produce.
- Friendly to environment, completely **green**.
- Its cycling performance is excellent. For example, its positive electrode such as $LiMn_2O_4$ can retain 93 % capacity after 10,000 full cycles [6].
- Satisfactory energy density, 40–90 Wh/kg, which is dependent on insertion electrodes.

Use

It can be used for energy storage of smart grids, hybrid electric vehicles, and assistance for electric vehicles and range extenders.

Future Directions

- To improve its charge rate capability by controlling the facets or crystal orientation of their electrode materials. If the charge time can be within the range of seconds (<30 s), it will be of great attractions for a lot of high power application [5, 9].
- To broaden its working voltage by utilizing the overpotential. If it can be above 3 V, the energy density will be higher than that for lithium ion batteries [10, 11], which can be further used for mobile electronics and electric vehicles.

Cross-References

► Insertion Electrodes for Li Batteries
► Lithium-Ion Batteries

References

1. Wu L, Dahn JR, Wainwright DS (1994) Rechargeable lithium batteries with aqueous electrolytes. Science 264:1115–1117
2. Wang GJ, Fu LJ, Zhao NH, Yang LC, Wu YP, Wu HQ (2007) An aqueous rechargeable lithium battery with good cycling performance. Angew Chem Int Ed 46:295–297
3. Wu YP, Dai XB, Ma JQ, Cheng YJ (2004) Lithium ion batteries: applications and practice. Chemical Industry Press, Beijing
4. Tang W, Zhu YS, Hou YY, Liu LL, Wu YP, Loh KP, Zhu K (2013) Aqueous rechargeable lithium batteries as an energy storage system of superfast charging like filling gasoline. Energy Environ Sci 5:2093–2104
5. Wang FX, Xiao SY, Chang Z, Yang YQ, Wu YP (2013) Nanoporous $LiNi_{1/3}Co_{1/3}Mn_{1/3}O_2$ as an ultrafast charge cathode material for aqueous rechargeable lithium batteries. Chem Commun 49: 9209–9211
6. Qu QT, Fu LJ, Zhan XY, Samuelis D, Li L, Guo WL, Li ZH, Wu YP, Maier J (2011) High-rate and long-life $LiMn_2O_4$ cathode for aqueous rechargeable lithium batteries. Energy Environ Sci 4:3985–3990
7. Tang W, Liu LL, Zhu YS, Sun H, Wu YP, Zhu K (2012) An aqueous rechargeable lithium battery of excellent rate capability based on nanocomposite of MoO_3 coated with PPy and $LiMn_2O_4$. Energy Environ Sci 5:6909–6913
8. Tang W, Gao XW, Zhu YS, Yue YB, Shi Y, Wu YP, Zhu K (2012) Coated hybrid of V_2O_5 nanowires with MWCNTs by polypyrrole as anode material for aqueous rechargeable lithium battery with excellent cycling performance. J Mater Chem 22:20143–20145
9. Tang W, Hou YY, Wang FX, Liu LL, Wu YP, Zhu K (2013) $LiMn_2O_4$ nanotube as cathode material of second-level charge capability for aqueous rechargeable batteries. Nano Lett 13:2036–2040
10. Wang XJ, Hou YY, Zhu YS, Wu YP, Holze R (2013) An aqueous rechargeable lithium battery using coated Li metal as anode. Sci Rep 3:1401
11. Wang XJ, Qu QT, Hou YY, Wang FX, Wu YP (2013) An aqueous rechargeable lithium battery of high energy density based on coated Li metal and $LiCoO_2$. Chem Commun 49:6179–6181

Artificial Photosynthesis

Stenbjörn Styring, Anders Thapper and Reiner Lomoth
Department of Chemistry - Ångström Laboratory, Uppsala University, Uppsala, Sweden

Introduction: Solar Fuels and Artificial Photosynthesis

This article deals with research aimed at the development of solar fuels. The term "solar fuel" is becoming established since the beginning of the new millennium. Parts of this research are often named "artificial photosynthesis for fuel production," "solar-hydrogen research," "the artificial leaf," or something analogous. The introduction of solar fuels on a very large scale is motivated by concerns about global warming, energy security for everyone and decreased availability of oil and gas. It is also driven by recent advances in a range of scientific fields that make scientists convinced that solar fuels are possible to produce in an efficient and cheap way in a not too distant future.

A solar fuel is always made using solar energy as the only energy source. The idea is to harvest the energy that comes when the sun shines, convert it and store it as a fuel, and then use this when and for whatever we want. The raw material for the storage fuel is the second, equally important, key aspect. The raw material must be essentially inexhaustible, cheap and widely available. Most scientists target water as the raw material and this is the only really sound option. Thus, solar fuels research gathers science with the aim to provide a fuel based on solar energy and water. Both resources are essentially endless and fairly evenly spread over the entire planet. The target fuel is the third important issue. Many scientists target hydrogen as the solar fuel. When water is used as starting material this is natural from scientific reasons. An advantage here is that this hydrogen production would be CO_2 free – a huge advantage over fossil fuels. An alternative is to use CO_2 itself as a second raw material (together with water) to create a carbon-based solar fuel.

All these aspects are analogous to natural photosynthesis. In nature, solar energy is absorbed by chlorophyll containing enzymes known as reaction centers. The energy is then used to extract electrons from water which is split to molecular oxygen, protons and electrons in the key enzyme photosystem II (PSII). The electrons are then used in light-independent reactions to reduce CO_2 and other oxidized compounds to a huge variety of compounds. It is because of these fundamental analogies the term "artificial photosynthesis" is useful to describe some of the research aiming for solar fuels.

In artificial photosynthesis it is necessary to develop a system from several components and combine them such that the system can carry out the three processes depicted in Fig. 1a. The central point is the photosensitizer P. This is connected to an electron donor (D) and an electron acceptor (A). The donor and acceptor are in turn connected to two catalysts, C_D and C_A respectively. When the photon energy is absorbed by P, this is excited to an energy-rich state. This triggers electron transfer to A and the catalyst C_A that can reduce protons (or CO_2) to hydrogen (or a carbon-based fuel). Concomitantly, the electron hole is filled by an electron from D and the catalyst C_D which is able to oxidize water to oxygen, protons and electrons. The three components, P, A and D should be linked somehow such that the electron can only go in one direction (indicated by arrows in Fig. 1a). After the absorption of two photons, two electrons have been transferred to A which then is able to reduce 2 protons to molecular hydrogen. After the absorption of four photons, four electron holes have been created at D which then is able to oxidize water (a four-electron process).

Consequently, there are many electron transfer steps involved and several of them are coupled to proton release or uptake. This gives wide room for advanced studies with electrochemistry both in the oxidizing chemistry at D and in the reducing chemistry at A. Some aspects of these interesting and complex reactions are covered in this short overview article.

There are several important and complicated aspects in the process depicted in Fig. 1a that are worth discussing. We refer the interested reader to specialized articles and recent entire journal issues [1, 2]. Here we mention one important aspect. Both catalysts C_D and C_A involve transition metals as catalytic centers. This holds regardless if the envisioned system is of molecular character or if nanostructures or photocatalytic materials are involved. This is critical! For an artificial system to become a real contender for fossil fuels these catalysts must be made of abundant metals. This excludes the use of catalytically very useful metals like Ru, Rh, Ir,

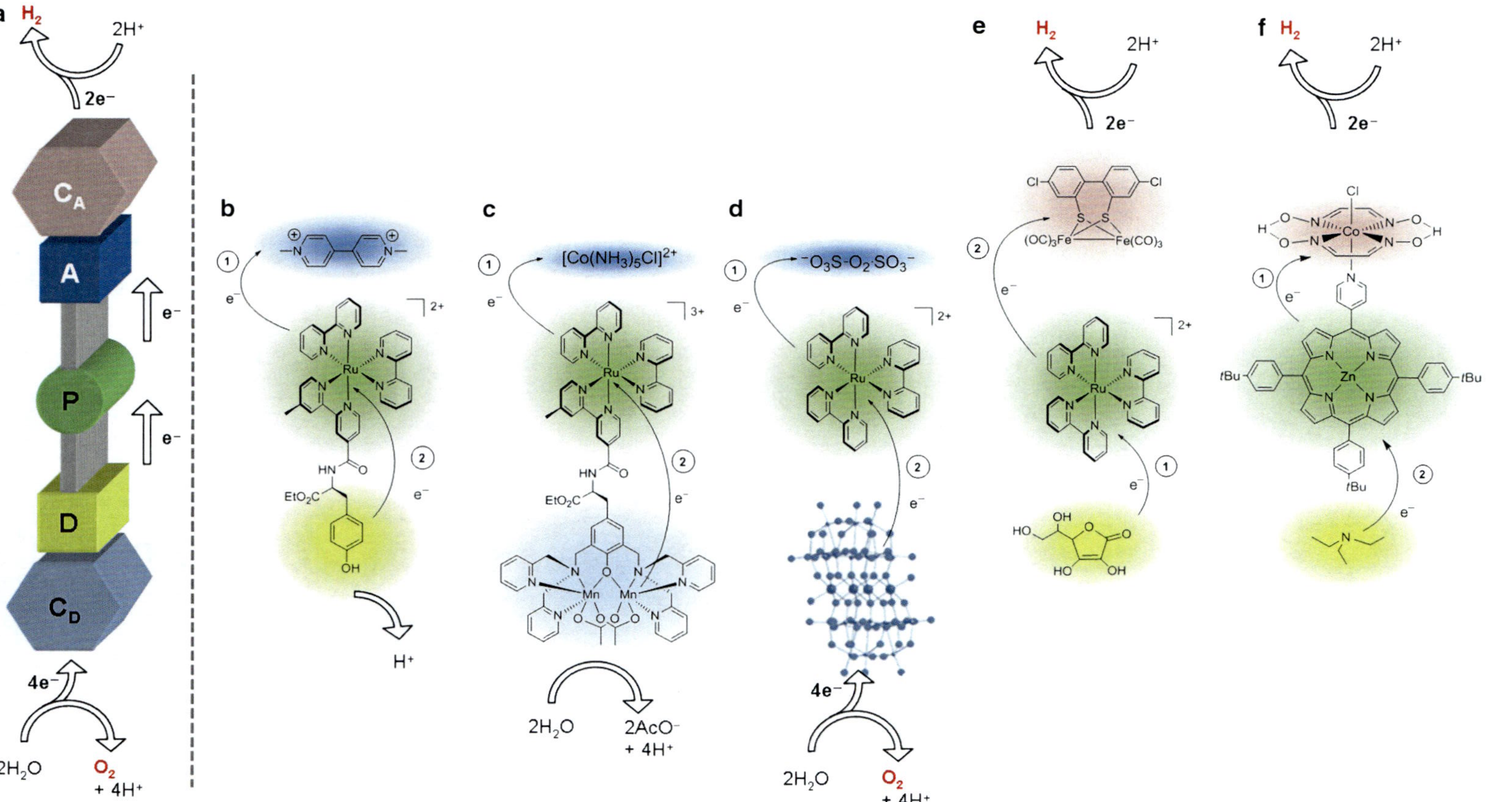

Artificial Photosynthesis, Fig. 1 Examples of molecular assemblies with donor or acceptor side functionality. Encircled numbers indicate order of electron transfer steps after photoexcitation. (**a**) Schematic representation of a water-splitting assembly for artificial photosynthesis. (**b**) Proton coupled electron transfer from a phenol donor to $[Ru(bpy)_3]^{3+}$ generated by photoinduced electron transfer to a acceptor (methyl viologen). (**c**) Accumulative ($3e^-$) oxidation of a $[Mn_2(bpmp)(\mu\text{-}OAc)_2]^+$ multi-electron donor with $[Ru(bpy)_3]^{3+}$ generated by electron transfer to a sacrificial acceptor $[Co(NH_3)_5Cl]^{2+}$. (**d**) Photocatalytic water oxidation with a tetracobalt polyoxometallate catalyst ($[Co_4(H_2O)_2\{\alpha\text{-}PW_9O_{34}\}_2]^{10-}$) and $[Ru(bpy)_3]^{3+}$ generated by photoinduced electron transfer to a sacrificial $S_2O_8^{2-}$ acceptor. (**e**) Photocatalytic H_2 formation with a [FeFe] hydrogenase mimic ($[Fe_2(Cl_2\text{-}bdt)(CO)_6]$) and $[Ru(bpy)_3]^{2+}$ sensitizer with sacrificial ascorbate donor. (**f**) Photocatalytic H_2 formation with a cobaloxime ($[Co(dmgH)_2Cl]$) catalyst by photoinduced electron transfer from a Zn porphyrin sensitizer linked via a pyridine and regenerated with a sacrificial donor (triethylamine)

In, Pt, Pd and others. In this article we have restricted ourselves to oxidizing systems (C_D) composed of Mn or Co while we will not discuss the many interesting studies using Ru- or Ir-based catalysts. The same holds for the reducing catalyst (C_A) where we only discuss systems based on Fe, Ni or Co while the efficient and well-known Pd- and Pt-based catalysts are not discussed.

Molecular Assemblies for Artificial Photosynthesis

Figures 1b–f show examples of molecular systems that feature essential functions that are required for a prospective artificial photosynthetic assembly. The examples are restricted to assemblies build from molecular units only and to abundant elements as far as catalysts are concerned.

Figure 1b exemplifies a system where excited-state electron transfer triggers proton-coupled electron transfer. The initial charge shift occurs from the photoexcited Ru^{II}–trisbipyridine unit to methyl viologen acceptor. The thermodynamics of this reaction are governed by the excited-state reduction potential of the sensitizer and the ground state reduction potential of the acceptor as illustrated in Fig. 2. The oxidized sensitizer subsequently oxidizes the attached tyrosine unit. The electron transfer reaction is coupled to deprotonation of the electron donor and affects thermodynamics and kinetics of the reaction. This resembles the situation in PSII, where oxidation of the tyrosine$_Z$ results in transfer of a proton to a nearby base, which brings the potential below that of the P_{680}^+ oxidant.

For the general concept of artificial photosynthesis it is vital that the photoinduced charge shift reactions, which are one-electron processes, can be coupled to the multi-electron chemistry that results in stable products (H_2, O_2) and avoids high-energy intermediates. This requires catalytic donor and acceptor units that can accumulate oxidation and reduction equivalents, stabilize intermediates and catalyze O–O and H–H bond formation, respectively. The accumulative oxidation or reduction of such catalysts necessitate charge-compensating reactions that avoid charge buildup and compress the potentials for accumulative electron transfer steps in a narrow range below the photooxidant's potential. In PSII, this principle is implemented in the stepwise oxidation of the Mn_4Ca complex where the electron transfer steps are coupled to charge-compensating deprotonation reactions and changes in bridging ligands [3].

A synthetic system where charge compensation allows for light-induced, accumulative electron transfer is shown in Fig. 1c [4–6]. Oxidation of the linked $Mn_2^{II,II}$ complex in three steps to $Mn_2^{III,IV}$ was driven by repeated photooxidation of the Ru^{II} – trisbipyridine unit to Ru^{III} [4]. The third oxidation step was achieved in aqueous media, where Mn oxidation was accompanied by exchange of the acetate ligands for water-derived ligands. Studies by Fourier transform infrared (FTIR) spectroelectrochemistry, electrospray ionization (ESI)–mass spectrometry online coupled to flow electrolysis [5], and extended X-ray absorption fine structure (EXAFS) measurements on electrolyzed samples [6] could show how Mn oxidation states and water content of the medium control the exchange of acetate ligands for water-derived ligands and their deprotonation (H_2O, OH^-, O^{2-}). Ultimately, these reactions result in a $Mn_2^{III,IV}$ complex with two fully deprotonated, water-derived oxo ligands and, thus, the same charge (2+) as the original bis-acetato $Mn_2^{II,III}$ complex. Owing to the charge compensation in the proton-coupled electron transfer steps, the dioxo $Mn_2^{III,IV}$ complex could be generated at somewhat more oxidizing potential than that required to generate the bis-acetato $Mn_2^{II,III}$. This effect puts all three oxidation steps in a narrow potential range of ca. 0.2 V and well below the potential of the photooxidant.

The dinuclear complex in Fig. 1c mimics key aspects of photosynthetic water oxidation, i.e., substrate binding to the Mn centers, photoinduced multi-electron oxidation of the Mn centers and deprotonation of the aquo ligands. The ultimate objective of catalytic turnover under O_2 release was however not met with this complex and molecular catalysts for light-driven water

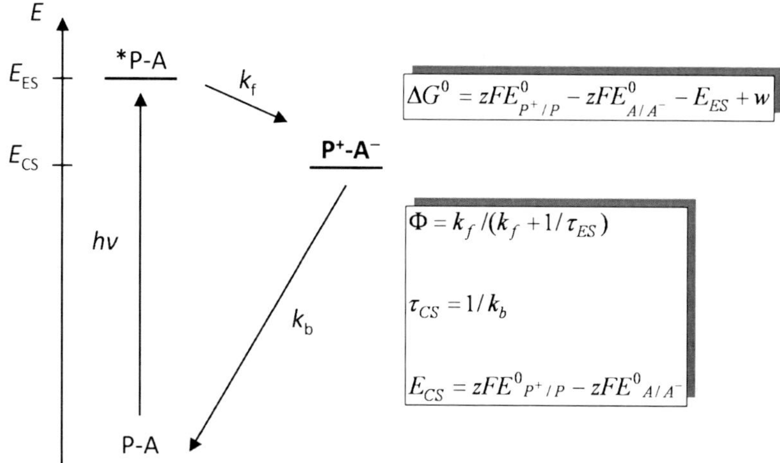

Artificial Photosynthesis, Fig. 2 Schematic state energy diagram for photoinduced charge separation in a photosensitizer–acceptor pair. Excitation to *P-A results in electron transfer with a driving force $-\Delta G^0$ that is determined by the reduction potentials for the P^+/P and A/A^- couples, the excited-state energy E_{ES}, and coulombic work w due to the charge shift. The rate of forward electron transfer k_f and the excited-state lifetime τ_{ES} determine the charge separation yield Φ. The charge-separated state stores energy E_{CS} that can be employed to drive subsequent redox reactions during its lifetime τ_{CS} that is given by the rate of recombination k_b

oxidation based on abundant metals remain generally scarce.

An example of a system for photocatalytic water oxidation with a molecular catalyst based on abundant metals is shown in Fig. 1d. The Cobalt polyoxometalate catalyst is oxidized with Ru^{III}–trisbipyridine that is generated by quenching of the photoexcited Ru^{II}–trisbipyridine sensitizer with peroxodisulfate as sacrificial electron acceptor. The system operates at pH 8 and exhibits a high (30 %) photon-to-O_2 yield while the stability of the catalyst allowed for turnover numbers >220 that were limited by depletion of electron acceptor only [7]. This performance of the abundant-metal-based catalyst is superior to that of an analogue ruthenium polyoxometalate water oxidation catalyst.

Light-driven acceptor side (H_2 formation) assemblies employ both biomimetic catalysts modeled after the hydrogenase active site (Fe, Ni complexes) as well as other catalysts based on abundant metals like cobalt but with non-biomimetic structures. The catalyst shown in Fig. 1e is an example of an iron–iron [FeFe] hydrogenase active site model that has been employed in a system for photocatalytic hydrogen formation in near-neutral H_2O/DMF (pH 5.5) with visible light [15]. Reduction of the catalyst is in this example driven by the reduced sensitizer that forms upon quenching with ascorbate as sacrificial electron donor. The biphenyldithiolate bridging ligand renders the reduced catalyst unusually stable resulting in high turnover numbers (200 equivalents of H_2 per catalyst).

Regarding non-biomimetic catalysts for H_2 formation that are based on abundant metals, cobaloxime complexes have attracted much attention. Figure 1f shows an example where a pyridine substituted Zn porphyrin sensitizer was attached to the catalyst by axial coordination to the Co center [16]. In this system electron transfer to the catalyst occurs from the excited state of the sensitizer. The oxidized sensitizer subsequently reduced with an amine donor. Turnover numbers for hydrogen evolution up to 48 were obtained with this system upon irradiation with visible light.

Another family of biomimetic complexes for proton reduction (and hydrogen oxidation) is the $[Ni(PNP)_2]^{2+}$ complexes (Fig. 3) and modifications thereof [8, 9]. These complexes contain amines in the second coordination sphere that are essential for catalysis. A pyrene modified [Ni $(PNP)_2]^{2+}$ complex has recently been physisorbed onto multiwalled carbon nanotubes and deposited on electrodes and used as electrocatalysts for hydrogen production [10]. These complexes have not so far been linked to light chemistry but are nevertheless very interesting for future artificial photosynthesis applications.

Enzymatic Assemblies for Artificial Photosynthesis

Another approach to artificial photosynthesis is to make use of natural (wild type) or mutated enzymes from organisms in a context and for a purpose that is different from their natural role. For example, photosystem II (PSII) could be incorporated in the oxidative half cell of an enzyme-based construct for artificial photosynthesis (Fig. 4). In comparison with Fig. 1a above PSII then contains both P (in the form of the entire antenna system and ultimately P680), D, and C_D, the manganese cluster responsible for water oxidation. To utilize PSII in an artificial photosynthesis context the acceptor side has to be coupled to something so that the electrons that are leaving PSII are directed to produce a chemical fuel. Hydrogenases are a good choice for reductive catalysts in this case if hydrogen is to be produced as a fuel. However, coupling hydrogenases directly to PSII is not advantageous since most hydrogenases are oxygen sensitive. Instead photosystem I (PSI) has been utilized as a "photosensitizer" with an inbuilt electron transfer pathway on the acceptor side (A in Fig. 1a) that can be coupled to for example hydrogenases (C_A in Fig. 1a). Immobilization of the enzymes on electrodes and/or surfaces has been envisaged as an attractive way to facilitate the separation of PSII-based water oxidation and PSI–hydrogenase-based proton reduction. Both PSI and PSII are naturally located in membranes and can be organized in single layers on surfaces with a built-in directionality of the electron flow through the membrane. The two compartment

Artificial Photosynthesis, Fig. 3 The general structure of the $[Ni(PNP)_2]^{2+}$ complexes

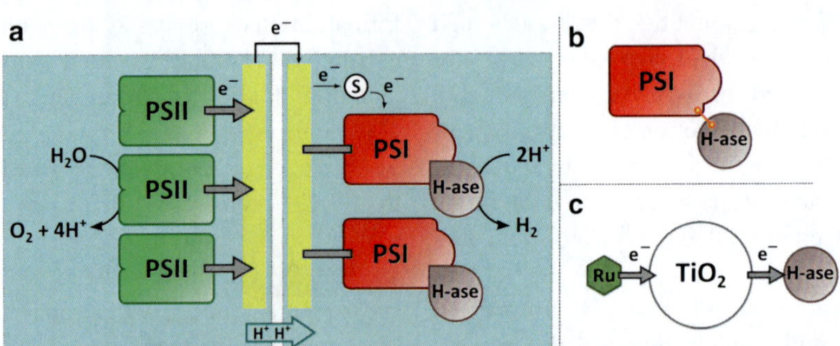

Artificial Photosynthesis, Fig. 4 (**a**) Schematic view of a possible setup for an artificial photosynthesis device based on enzymes. A photoanode made from PSII anchored on a gold electrode has been presented by Rögner and coworkers [11]. Lenz and coworkers [12] have prepared a photocathode based on a PSI-hydrogenase construct anchored to a gold electrode. H-ase = hydrogenase, S = redox shuttle. The scheme is based on figures in Refs. [11] and [12]. (**b**) Schematic representation of an alternative PSI–hydrogenase construct reported by Golbeck and coworkers [13] with an organic linker connecting the two enzymes. (**c**) Construct for light-driven proton reduction reported by Armstrong and coworkers [14] with a ruthenium-based photosensitizer and a hydrogenase, both linked to a TiO_2 particle construct would be a complete artificial photosynthetic Z-scheme with double sensitizers.

As a first step towards an artificial photosynthesis cell based on immobilized enzymes a photoanode utilizing PSII has been constructed (Fig. 4a, left side) [11]. For this a His-tagged PSII was attached to a gold electrode via a Ni(II) ion, a nitrilotriacetic acid (NTA) and an alkylthiol monolayer. This resulted in a highly photoactive electrode. Amperometric measurements showed photocurrent densities corresponding to oxygen evolution rates comparable with the highest activities measured for isolated PSII complexes (>6,000 O_2 Chl^{-1} h^{-1}).

A fully enzymatic photocathode can be envisaged by fusing an active hydrogenase to the acceptor side of PSI and attach the construct to an electrode. Two different approaches towards this have been reported in the literature. The first modified the acceptor side of PSI to create a binding site for a dithiol [13]. This dithiol was used as a linker to a similar binding site created on an [FeFe] hydrogenase that catalyzed hydrogen production (Fig. 4b). This PSI–hydrogenase construct showed moderate hydrogen evolution activity and could potentially be further attached to an electrode surface for use in artificial photosynthetic devices.

A second approach created a fusion protein from a PSI subunit (PsaE) and a nickel–iron [NiFe] hydrogenase [12]. This new protein was then assembled into a PSI mutant lacking the PsaE subunit. The fused enzymatic system was attached to a gold surface in the same way as the PSII electrode described above using a His-tag on PSI, Ni(II) and NTA functionalities on the surface (Fig. 4a, right side). A soluble electron shuttle was used to transfer electrons from the electrode to PSI. From these two approaches the fusion protein is to date the most effective "artificial" enzymatic system for photo-driven hydrogen production and the activity is comparable to the electrocatalytic activity of the hydrogenase alone immobilized directly on an electrode.

"Mixed" artificial systems can also be created. A [NiFe] hydrogenase and a ruthenium-based photosensitizer have both been attached to the same TiO_2 particles (Fig. 4c) [14]. By using a sacrificial electron donor light-driven hydrogen production could be performed. The system was stable, reaching more than 100 000 turnovers during 8 h illumination.

Future Directions

- Double sensitizer (Z-scheme)
- Immobilization and compartmentalization towards devices
- Self-assembly and self-repair
- Merging of molecular and solid-state approaches (nanoparticles and quantum dots as catalysts or sensitizer)

Cross-References

▶ Bioelectrochemical Hydrogen Production
▶ Hydrogen Evolution Reaction
▶ Oxygen Evolution Reaction
▶ Photoelectrochemical CO_2 Reduction

References

1. Hammarström L, Hammes-Schiffer S (2009) Special issue on artificial photosynthesis and solar fuels. Acc Chem Res 42:1859–2029
2. Andreiadis ES, Chavarot-Kerlidou M, Fontecave M, Artero V (2011) Artificial photosynthesis: from molecular catalysts for light-driven water splitting to photoelectrochemical cells. Photochem Photobiol 87:946–964
3. Dau H, Haumann M (2008) The manganese complex of photosystem II in its reaction cycle–Basic framework and possible realization at the atomic level. Coord Chem Rev 252:273–295
4. Huang P, Magnuson A, Lomoth R, Abrahamsson M, Tamm M, Sun L, Van Rotterdam B, Park J, Hammarström L, Åkermark B, Styring S (2002) Photo-induced oxidation of a dinuclear $Mn_2^{II,II}$ complex to the $Mn_2^{III,IV}$ state by inter- and intramolecular electron transfer to Ru^{III}*tris*-bipyridine. J Inorg Biochem 91:159–172
5. Eilers G, Zettersten C, Nyholm L, Hammarström L, Lomoth R (2005) Ligand exchange upon oxidation of a dinuclear Mn complex-detection of structural changes by FT-IR spectroscopy and ESI-MS. Dalton Trans:1033–1041
6. Magnuson A, Liebisch P, Högblom J, Anderlund MF, Lomoth R, Meyer-Klaucke W, Haumann M, Dau H (2006) Bridging-type changes facilitate successive

oxidation steps at about 1 V in two binuclear manganese complexes-implications for photosynthetic water-oxidation. J Inorg Biochem 100:1234–1243
7. Huang Z, Luo Z, Geletii YV, Vickers JW, Yin Q, Wu D, Hou Y, Ding Y, Song J, Musaev DG, Hill CL, Lian T (2011) Efficient light-driven carbon-free cobalt-based molecular catalyst for water oxidation. J Am Chem Soc 133:2068–2071
8. Kilgore UJ, Roberts JAS, Pool DH, Appel AM, Stewart MP, Dubois MR, Dougherty WG, Kassel WS, Bullock RM, Dubois DL (2011) $[Ni(PPh_2NC_6H_4X_2)_2]^{2+}$ Complexes as electrocatalysts for H_2 production: effect of substituents, acids, and water on catalytic rates. J Am Chem Soc 133:5861–5872
9. Dubois MR, Dubois DL (2009) Development of molecular electrocatalysts for CO_2 reduction and H_2 production/oxidation. Acc Chem Res 42:1974–1982
10. Tran PD, Le Goff A, Heidkamp J, Jousselme B, Guillet N, Palacin S, Dau H, Fontecave M, Artero V (2011) Noncovalent modification of carbon nanotubes with pyrene-functionalized nickel complexes: carbon monoxide tolerant catalysts for hydrogen evolution and uptake. Angew Chem Int Ed 50:1371–1374
11. Badura A, Esper B, Ataka K, Grunwald C, Wöll C, Kuhlmann J, Heberle J, Rögner M (2006) Light-driven water splitting for (bio-)hydrogen production: photosystem 2 as the central part of a bioelectrochemical device. Photochem Photobiol 82:1385–1390
12. Krassen H, Schwarze A, Friedrich B, Ataka K, Lenz O, Heberle J (2009) Photosynthetic hydrogen production by a hybrid complex of photosystem I and [NiFe]-hydrogenase. ACS Nano 3:4055–4061
13. Lubner CE, Grimme R, Bryant DA, Golbeck JH (2010) Wiring photosystem I for direct solar hydrogen production. Biochemistry 49:404–414
14. Reisner E, Fontecilla-Camps JC, Armstrong FA (2009) Catalytic electrochemistry of a [NiFeSe]-hydrogenase on TiO_2 and demonstration of its suitability for visible-light driven H_2 production. Chem Commun:550–552
15. Streich D, Astuti Y, Orlandi M, Schwartz L, Lomoth R, Hammarström L, Ott S (2010) High-turnover photochemical hydrogen production catalyzed by a model complex of the [FeFe]-hydrogenase active site. Chem Eur J 16:60–63.
16. Zhang P, Wang M, Li C, Li X, Dong J, Sun L (2010) Photochemical H2 with noblemetal-free molecular devices comprising a porphyrin photosensitizer and a cobaloximecatalyst. Chem Commun 46:8806–8808.

B

Biocorrosion

▸ Microbiologically Influenced Corrosion

Biodeterioration

▸ Microbiologically Influenced Corrosion

Bioelectrochemical Hydrogen Production

Svenja Lohner
Stanford University, Stanford, CA, USA

Introduction

Uses of hydrogen are manifold ranging from being an important industrial feedstock, a valuable energy carrier, to being a key electron donor for reduction of various oxidized water contaminants [1]. Today, most hydrogen produced is derived from energy-consuming processes based on fossil fuels such as steam reforming from natural gas, as well as from water electrolysis [2]. In contrast, biological hydrogen production from fermentation or photosynthesis provides a sustainable and carbon-neutral source for hydrogen. However, challenges such as low hydrogen yield from fermentation and the need for improved and oxygen tolerant proteins for photosynthetic or enzymatic hydrogen production prevent these approaches from being economically competitive so far.

Bioelectrochemical hydrogen production ("microbial electrolysis" or "biocatalyzed electrolysis") is a novel biological hydrogen production process, coupling microbiology and electrochemistry, which has attracted a lot of interest recently as it enables the extraction of hydrogen from the final products of dark fermentation [3]. In such an electrolysis cell electrons are generated from microbial oxidation of organic compounds and utilized to produce hydrogen at the cathode by providing additional energy from an external power supply. Already back in 1994, Kreysa et al. proposed to fuel water electrolysis with electricity generated from a microbial fuel cell (MFC). They managed to successfully generate hydrogen with a stacked MFC configuration although the energy efficiency was only 11 % [4]. In 2006, two other groups further developed this process by combining the water electrolyzer and the microbial fuel cell into one device, thus inventing the concept of microbial electrolysis cells (MECs) [5, 6]. Since then this emerging technology has been subject of extensive research and although it holds considerable promises and opportunities, it still needs to overcome significant technical and microbiological hurdles [7].

In the following, the principles of bioelectrochemical hydrogen production using microbial electrolysis cells are described, and advantages,

G. Kreysa et al. (eds.), *Encyclopedia of Applied Electrochemistry*, DOI 10.1007/978-1-4419-6996-5,

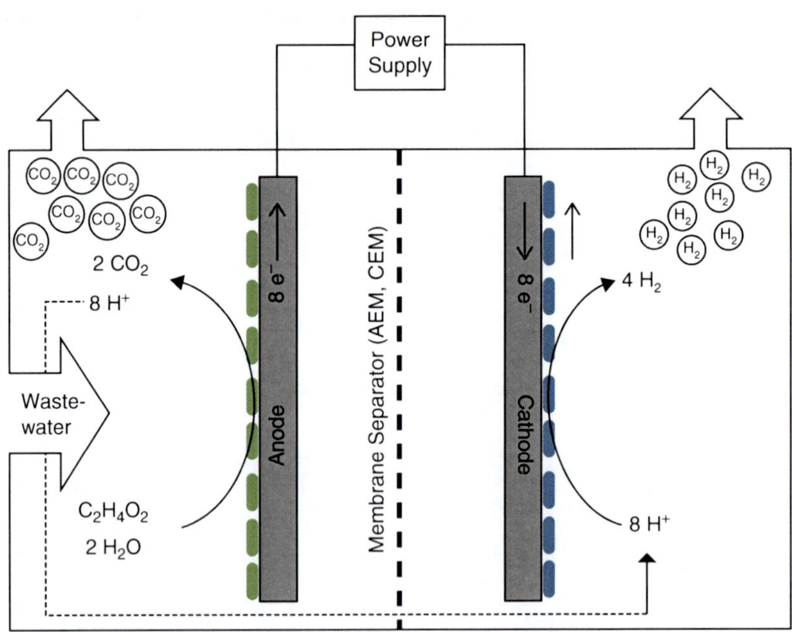

Bioelectrochemical Hydrogen Production, Fig. 1 Schematic layout of a microbial electrolysis cell (MEC) showing the anode and cathode chamber, the electrodes with attached biocatalysts, the membrane separator and the power supply, as well as the anodic and cathodic half reactions. *AEM* anion exchange membrane, *CEM* cation exchange membrane

	Microbial Fuel Cell (MFC)	$E^{0'}$ [V]	Microbial Electrolysis Cell (MEC)	$E^{0'}$ [V]	Water Electrolysis	$E^{0'}$ [V]
Anode	$C_2H_4O_2 + 2\,H_2O \Leftrightarrow 2\,CO_2 + 8\,e^- + 8\,H^+$	−0.28	$C_2H_4O_2 + 2H_2O \Leftrightarrow 2\,CO_2 + 8\,e^- + 8\,H^+$	−0.28	$2\,H_2O \Leftrightarrow 4\,H^+ + 4\,e^- + O_2$	0.81
Cathode	$2\,O_2 + 8\,e^- + 8H^+ \Leftrightarrow 4\,H_2O$	0.81	$8\,H^+ + 8\,e^- \Leftrightarrow 4\,H_2$	−0.42	$4\,H_2O + 4\,e^- \Leftrightarrow 2\,H_2 + 4\,OH^-$	−0.42
Total	$C_2H_4O_2 + 2\,O_2 \Leftrightarrow 2\,H_2O + 2\,CO_2$	1.09	$C_2H_4O_2 + 2\,H_2O \Leftrightarrow 2\,CO_2 + 4\,H_2$	−0.14	$2\,H_2O \Leftrightarrow 2\,H_2 + O_2$	−1.23

Bioelectrochemical Hydrogen Production, Fig. 2 Comparison of the electrode reactions in a microbial fuel cell and microbial electrolysis cell and for abiotic water electrolysis. $E^{0'}$ are the equilibrium potentials at microbial conditions (pH 7, acetate: 1 M, 1 bar)

disadvantages, and future perspectives of this technology are discussed.

Microbial Electrolysis Cells

Principle

A microbial electrolysis cell is based on the concept of a microbial fuel cell (MFC) and consists of an anode and cathode chamber, a membrane that electrically separates the electrodes, and an external power supply [8] (Fig. 1). The anodic reaction is the same as in a microbial fuel cell: Electrochemically active microorganisms oxidize organic compounds such as acetate, generating carbon dioxide (CO_2), protons (H^+), and electrons (e^-), using the anode as terminal electron acceptor (Fig. 2). The released electrons flow from the anode through an electrical circuit to the cathode where they can be used either to reduce oxygen in the case of a microbial fuel cell or to produce hydrogen from protons in a microbial electrolysis cell [9].

However, hydrogen production from acetate oxidation, as aimed for in an MEC, is thermodynamically not feasible. Indicated by the equilibrium potentials of the individual half reactions at microbial conditions (pH 7, acetate: 1 M, 1 bar), the electromotive force (emf) (= cathode potential − anode potential) of this reaction is −0.14 V, which means that additional electrical energy is required to support electrolytic hydrogen formation (Fig. 2). This is provided by applying a circuit voltage that is greater than 0.14 V, which is significantly less than what would be necessary in an abiotic water electrolysis cell in which electrons and protons are derived from water (1.23 V, Fig. 2).

MEC Performance: Hydrogen Yields and Coulombic Efficiencies

To compare different reactor setups and also to measure their performance, usually parameters such as the hydrogen yield Y_{H_2} (moles of H_2 produced per mol of substrate consumed), the cathodic hydrogen efficiency r_{cat} (moles of H_2 produced per mol of electrons consumed) and the coulombic efficiency CE (moles of electrons produced per mol of substrate) are reported for an MEC [10, 11]:

$$Y_{H_2} = \frac{V_{H_2} P M_S}{RT\Delta c_s}$$

V_{H_2}: volume H_2 produced [L]

P: atmospheric pressure [bar]

M_S: molecular weight of substrate [g/mol]

R: ideal gas constant 0.08314 [L bar/ K mol]

T: temperature [K]

Δc_s: substrate consumed [g]

$r_{cat} = \frac{nH_2}{n_{CE}}$ where $n_{CE} = \left(\int_{t=0}^{t} I \mathrm{d}t\right)/2F$

nH_2: moles of hydrogen actually recovered at the electrode

n_{CE}: moles of hydrogen that theoretically could have been produced from the measured current

dt: time interval over which data has been collected [s]

2: converts moles of electrons into moles of hydrogen

F: Faraday's constant 96485 [C/mol]

I: current [A]

$$CE = \frac{I\Delta t M}{F n \Delta c_s}$$

I: average current [A]

Δt: time interval during current measurements [s]

F: Faraday's constant 96485 [C/mol]

n: number of electrons transferred per mol of substrate oxidized

Δc_s: substrate consumed over a certain period of time [mol]

These values depend on many parameters such as the applied circuit voltage of the cell, the reactor design, the substrate fed to the anode, the electrode materials, and the efficiency and composition of the biocatalyst. So far, hydrogen yields up to 91 % of the theoretical maximum have been achieved for acetate (4 mol H_2 per mol of acetate) and 67–91 % for other organic substrates [12]. Reported coulombic efficiencies range from 23 % to 92 % for applied voltages that are usually in the range of 0.5 V [13]. To make bioelectrochemical hydrogen production economically attractive, at the end, the energy captured in the generated hydrogen must be greater than the energy that has been put into the system in form of electricity and organic substrates [10, 13]. Although theoretically possible, this is not the case, yet, as in practice many energy losses are observed such as potential or hydrogen losses.

Energy (Potential and Hydrogen) Losses

Theoretically, when using acetate as substrate, an applied potential of 0.14 V should be sufficient to form hydrogen in an MEC. In practice, however, at least 0.25 V has been necessary to observe hydrogen production [5]. Major causes for potential losses are the occurrence of overpotentials at the electrodes, potential drops due to the ohmic resistance of the electrochemical system (membrane, electrolyte, etc.), and the fact that the biocatalysts at the anode divert some energy to themselves for growth. One of the main contributors to increased voltage demands is a high overpotential for the cathodic hydrogen evolution reaction [6]. In addition, it is known that transport of competing cation species other than protons from the anode to the cathode chamber causes a pH gradient across the cation exchange membrane, which results in an additional potential loss of 0.06 V per pH unit difference according to the Nernst equation [14].

Hydrogen losses mainly occur via diffusion of gas through the membrane into the anode chamber, where it can be consumed by hydrogenotrophic methanogens to make methane [15, 16]. This can be especially problematic in membraneless microbial electrolysis cells, which try to reduce the potential losses due to the pH gradient across the membrane [11] or in systems with biocathodes where methanogens are present. In addition, hydrogen also is easily lost abiotically through tubings or seals of the reactor setup [10].

Application of Bioelectrochemical Hydrogen Production

An attractive application possibility of the MEC is in wastewater treatment plants, where wastewater treatment can be combined with energy recovery in the form of hydrogen [1, 17]. Such an integrated process would allow access to the energy of remaining organic compounds that otherwise have to be removed by energy-consuming aerobic treatments [18]. As microbial electrolysis operates completely anaerobically, an additional advantage would be the reduction of sludge production as well as the decrease in odor release due to the closed system operation [10]. Life-cycle assessments of an MEC relative to the “conventional” anaerobic treatment showed that such an integrated process would provide significant environmental benefits [19]. Challenges to overcome before large-scale application becomes feasible include slow or insufficient degradation of complex substrates in the wastewater at the bioanode; scale-up and optimization of reactor designs to decrease potential losses, dealing with low conductivities of domestic wastewaters which result in high ohmic losses; as well as the high capital costs for bioelectrochemical systems [10, 18].

Alternatively, the use of MECs for renewable energy production has been suggested [10] using cellulose or glucose as substrates [12]. In combination with dark fermentation, hydrogen yields from complex organic compounds can be significantly increased which might be attractive for industries producing fermentation effluents [1].

Advantages and Disadvantages

Compared to dark fermentation, bioelectrochemical hydrogen production results in much higher hydrogen yields as it is not limited by dead-end products such as acetate, butyrate, or lactate. Whereas during fermentation the maximum theoretical hydrogen yield of glucose is 4 mol hydrogen per mol glucose (provided hydrogen and acetate are the only products), the combination with an MEC has the potential to increase the yields up to ca. 8–9 mol hydrogen per mol of glucose [5]. With this, hydrogen can be produced from a range of organic substrate at very high purity without side products such as hydrogen sulfide, carbon dioxide, or carbon monoxide. Although water electrolysis also provides high purity hydrogen, the energy investment (theoretically 1.23 V) is much higher as in an MEC (theoretically 0.14 V). Hence, additional gas treatment processes are not necessary saving costs and energy [6].

Whereas the anode generally consists of a cheap carbon/graphite material to provide a good substrate for the electroactive biofilm, at the cathode the use of precious metal catalysts such as platinum (Pt) is necessary to reduce large overpotentials required for the hydrogen evolution reaction. This not only is cost intensive but also has negative environmental impacts through mining of precious metals. In addition, metal catalysts can be poisoned by wastewater components such as sulfide rendering them ineffective. Promising alternatives currently investigated are the use of low-cost, non-precious cathode catalysts [20] and the development of biocathodes [9, 21, 22]. Biocathodes make use of microorganisms or enzymes (hydrogenases) functioning as biocatalysts, therefore replacing the metal catalyst [23–26] with the additional advantage of being self-sustaining.

Another disadvantage is the occurrence of high potential losses in the MEC, as discussed above, which leads to much higher energy inputs necessary than theoretically required. At the same time, hydrogen yields need to be improved and hydrogen losses, e.g., via methanogenesis, diminished. Biocatalysts at the anode have to be able to metabolize more realistic and complex compounds to become economically competitive in larger scale. At the end, hydrogen production rates will depend on the substrate-utilization kinetics of these electroactive microorganisms as well as their physiology [22].

Future Directions

There are certainly several issues that need to be solved before bioelectrochemical hydrogen production can be considered to be a mature technology. Future research needs to focus on:

- Reducing potential losses within the system by decreasing the hydrogen evolution overpotential using biocathodes or non-precious metal catalysts. Also, optimized reactor designs such as the use of membraneless reactors (without losing the advantage of hydrogen purity) might have potential for improvements [11].
- Preventing hydrogen losses via diffusion through the membrane and subsequent consumption by methanogens. Selective inhibition of methanogens [15] might be an option as well as the development of fast hydrogen harvesting techniques.
- Scaling up the MECs as well as gaining more experience with real organic feedstocks from wastewater treatment plants or industries with fermentation effluents.
- Understanding and optimizing external electron transfer (EET) processes between biocatalysts and electrodes to improve the kinetics of electron extraction from the substrate and hydrogen evolution.

Overall the biggest challenge is the need to make bioelectrochemical hydrogen production an economically competitive technology. This means that the embodied energy in the produced hydrogen must be higher than the energy that has been provided in the form of applied voltage to drive the hydrogen evolution reaction.

Cross-References

► Biofilms, Electroactive
► Electrochemical Reactor Design and Configurations
► Hydrogen Evolution Reaction
► Overpotentials in Electrochemical Cells

References

1. Lee H-S, Vermaas WFJ, Rittmann BE (2010) Biological hydrogen production: prospects and challenges. Trends Biotechnol 28(5):262–271. doi:10.1016/j.tibtech.2010.01.07
2. Logan BE (2004) Extracting hydrogen and electricity from renewable sources. Environ Sci Technol 38(9):160A–167A. doi:10.1021/es040468s
3. Logan BE, Hamelers B, Rozendal R, Schröder U, Keller J, Freguia S, Aelterman P, Verstraete W, Rabaey K (2006) Microbial fuel cells: methodology and technology. Environ Sci Technol 40(17):5181–5192. doi:10.1021/es0605016
4. Kreysa G, Schenk K, Sell D, Vuorilehto K (1994) Bioelectrochemical hydrogen production. Int J Hydrogen Energy 19(8):673–676. doi:10.1016/0360-3199(94)90152-X
5. Liu H, Grot S, Logan BE (2005) Electrochemically assisted microbial production of hydrogen from acetate. Environ Sci Technol 39(11):4317–4320. doi:10.1021/es050244p
6. Rozendal RA, Hamelers HVM, Euverink GJW, Metz SJ, Buisman CJN (2006) Principle and perspectives of hydrogen production through biocatalyzed electrolysis. Int J Hydrogen Energy 31:1632–1640. doi:10.1016/j.ijhydene.2005.12.006
7. Rabaey K, Giguis P, Nielsen LK (2011) Metabolic and practical considerations on microbial electrosynthesis. Curr Opin Biotechnol 22:371–377. doi:10.1016/j.copbio.2011.01.010
8. Geelhoed JS, Hamelers HVM, Stams AJM (2010) Electrically-mediated biological hydrogen production. Curr Opin Microbiol 13:307–315. doi:10.1016/j.mib.2010.02.002
9. Jeremiasse AW, Hamelers HVM, Buisman CJN (2009) Microbial electrolysis cell with a microbial biocathode. Bioelectrochemistry 78(1):39–43. doi:10.1016/j.bioelechem.2009.05.005
10. Logan BE, Call D, Cheng S, Hamelers HVM, Sleutels THJA, Jeremiasse AW, Rozendal RA (2008) Microbial electrolysis cells for high yield hydrogen gas production from organic matter. Environ Sci Technol 42(23):8630–8640. doi:10.1021/es801553z
11. Tartakovsky B, Manuel M-F, Wang H, Guiot SR (2009) High rate membrane-less microbial electrolysis cell for continuous hydrogen production. Int J Hydrogen Energy 34:672–677. doi:10.1016/j.ijhydene.2008.11.003
12. Cheng S, Logan BE (2007) Sustainable and efficient biohydrogen production via electrohydrogenesis. Proc Natl Acad Sci 104(47):18871–18873. doi:10.1073/pnas.0706379104
13. Wrana N, Sparling R, Cicek N, Levon DB (2010) Hydrogen gas production in a microbial electrolysis cell by electrohydrogenesis. J Clean Prod 18: S105–S111. doi:10.1016/j.jclepro.2010.06.018
14. Rozendal RA, Hamelers HVM, Molenkamp RJ, Buisman CJN (2007) Performance of a single chamber biocatalyzed electrolysis with different types of

ion exchange membranes. Water Res 41:1984–1994. doi:10.1016/j.watres.2007.01.019
15. Chae K-J, Choi M-J, Kim K-Y, Ajayi FF, Chang I-S, Kim IS (2010) Selective inhibition of methanogens for the improvement of biohydrogen production in microbial electrolysis cells. Int J Hydrogen Energy 35:13379–13386. doi:10.1016/j.ijhydene.2009.11.114
16. Chae K-J, Choi M-J, Lee J, Ajayi FF, Kim IS (2008) Biohydrogen production via biocatalyzed electrolysis in acetate-fed bioelectrochemical cells and microbial community analysis. Int J Hydrogen Energy 33:5184–5192, doi:10.1016,j.ijhydene.2008.05.013
17. Schröder U (2008) From wastewater to hydrogen: biorefineries based on microbial fuel-cell technology. ChemSusChem 1:281–282. doi:10.1002/cssc.200800041
18. Rozendal RA, Hamelers HVM, Rabaey K, Keller J, Buisman CJN (2008) Towards practical implementation of bioelectrochemical wastewater treatment. Trends Biotechnol 26(8):450–459. doi:10.1016/j.tibtech.2008.04.008
19. Foley JM, Rozendal RA, Hertle CK, Lant PA, Rabaey K (2010) Life cycle assessment of high-rate anaerobic treatment, microbial fuel cells and microbial electrolysis cells. Environ Sci Technol 44(9):3629–3637. doi:10.1021/es100125h
20. Wang L, Chen Y, Ye Y, Lu B, Zhu S, Shen S (2011) Evaluation of low-cost cathode catalysts for high yield biohydrogen production in microbial electrolysis cell. Water Sci Technol 63(3):440–448. doi:10.2166/wst.2011.241
21. Rozendal RA, Jeremiasse AW, Hamelers HVM, Buisman CJN (2008) Hydrogen production with a microbial biocathode. Environ Sci Technol 42:629–634. doi:10.1021/es071720
22. Geelhoed JS, Hamelers HVM, Stams AJM (2010) Electricity-mediated biological hydrogen production. Curr Opin Microbiol 13:307–315. doi:19.1016/j.mib.2010.02.002
23. Morozov SV, Vignais PM, Cournac L, Zorin NA, Karyakina EE, Karyakin AA, Cosnier S (2002) Bioelectrocatalytic hydrogen production by hydrogenase electrodes. Int J Hydrogen Energy 27:1501–1505. doi:10.1016/S0360-3199(02)0091-5
24. Oh Y-K, Lee Y-J, Choi E-H, Kim M-S (2008) Bioelectrocatalytic hydrogen production using *Thiocapsa roseopersicina* hydrogenase in two-compartment fuel cells. Int J Hydrogen Energy 33:5218–5223. doi:10.1016/j.ijhydene.2008.05.015
25. Karyakin AA, Morozov SV, Karyakina EE, Zorin NA, Perelygin VV, Cosnier S ((2005) Hydrogenase electrodes for fuel cells. Biochem Soc Trans 33(L1):73–75. doi:10.1042/BST0330073
26. Lojou E, Durand MC, Dolla A, Bianco P (2002) Hydrogenase activity control of *Desulfovibrio vulgaris* cell coated carbon electrodes: biochemical and chemical factors influencing the mediated bioelectrocatalysis. Electroanalysis 14((13)):913–922. doi:1040-0397/02/1307-0913

Biofilms, Electroactive

Uwe Schroeder and Falk Harnisch
Institute of Environmental and Sustainable Chemistry, Technical University Braunschweig, Braunschweig, Germany

Definition and Occurrence

Microorganisms associated with interfaces (biofilms and bioaggregates) represent the common way of microbial life. In many biofilms the microbial cells, together with their extracellular polymeric substances (EPS), are usually associated with a solid substratum [1]. Electroactive microbial biofilms can be defined as microbial biofilms exchanging electrons with their conductive substratum via oxidation and/or reduction reactions. These biofilms can be found in different environments ranging from deep-sea oceans to man-made devices. Here, from mankind's perspective, electroactive microbial biofilms can cause severe damaging effects like ► biocorrosion. On the other hand, they can be exploited as bioelectrocatalysts in microbial fuel cells and related microbial bioelectrochemical systems (BES), a seminal technology concept for sustainable energy transformation (see section Technical Exploitation). Figure 1 shows photographic as well as scanning electron microscopic images of exemplary electroactive microbial biofilms.

Electron Transfer Mechanisms

The first description of electroactive microorganisms can be dated back to Potter in 1911 [2]. From there on and till the late 1990 it was generally assumed that artificial substances, mostly redox dyes like neutral red, are required to serve as mediators between the microbial cell and the conductive substratum – the electrode surface [3, 4]. However, after the discovery that electroactive biofilms can be formed without the help of exogenous substances [5, 6] the research

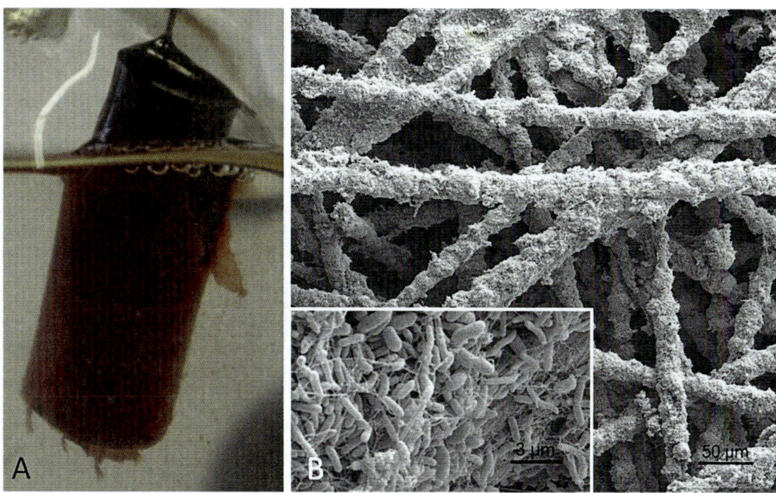

Biofilms, Electroactive, Fig. 1 (**a**) Photograph of a biofilm (dominated by *Geobacter*)-covered carbon electrode and (**b**) SEM images of an identical biofilm on carbon fiber electrodes in different magnifications

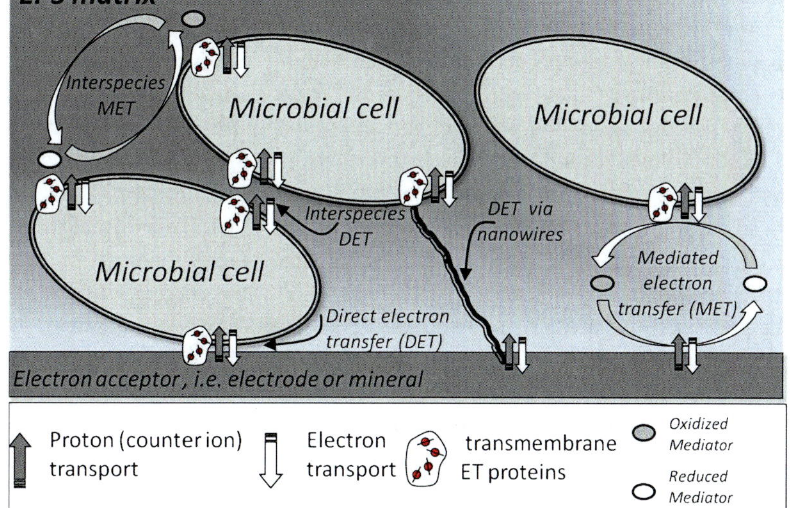

Biofilms, Electroactive, Fig. 2 Depicting the different (so far described) bacterial electron transfer mechanisms in biofilms on the example of an anodic biofilm

on the underlying mechanisms was intensified. This was mostly fostered by the prospects of microbial bioelectrochemical systems like microbial fuel cells. Thereby several mechanisms for the electron exchange between the microorganisms and electrodes have been discovered, which are summarized in Fig. 2 [7]. These mechanisms include the electron shuttling by endogenous redox mediators like phenazines [8] and the direct electron transfer by membrane-bound enzymes like cytochromes and bacterial nanowires [9–11]. Furthermore, redox-active primary metabolites like hydrogen, formate or sulfide are excreted by distinctive bacteria, which can be oxidized by other microorganism as well as at tailor-made electrocatalytic materials [12, 13]. Noteworthy, there is also growing evidence that the EPS matrix plays an important role in the electron transfer processes, e.g., [14].

Characterization of Electroactive Microbial Biofilms

The characterization of electroactive microbial biofilms is a truly transdisciplinary venue.

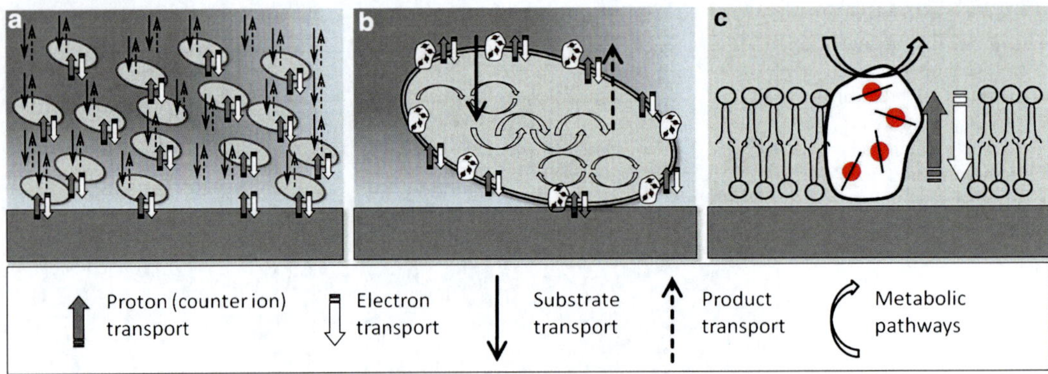

Biofilms, Electroactive, Fig. 3 Complexity and levels of biofilm characterisation on the example of an anodic biofilm, the main transport directions and DET are depicted: (**a**) biofilm level, (**b**) cellular level, (**c**) subcellular level

It comprises techniques and methods from numerous disciplines and the combined focus on different scales. As Fig. 3 illustrates, the scales of study are reaching from the macroscopic level for the complete biofilm understanding and engineering *via* the cellular scale, focusing on single microorganisms and their interaction, down to the subcellular level. In the latter, the function, interaction and regulation of single-cell molecules, e.g., bacterial transmembrane proteins or lectins of the EPS, are studied. From the microbiologist's perspective the research on electroactive microbial biofilms might be divided into two areas. The first area focuses on pure culture biofilms of model organisms like *Geobacteraceae* [15, 16] and *Shewanellaceae* [17], whereas the second part of research is devoted to biofilms in natural habitats and mixed culture derived biofilms. So far, the majority of studies are mostly focusing on the phenomenological, global level of biofilm performance in terms of current density and coulombic efficiency. Yet, for a detailed understanding and engineering it is indispensible to unreveal the fundamentals of electroactive microbial biofilms.

Electrochemical and Spectroelectrochemcial Techniques

As the electron exchange with its substratum is an inevitable prerequisite for electroactive microbial biofilms the investigation of the underlying electron transfer mechanism calls for electrochemical techniques and methods. In the recent years several techniques have been exploited, most prominently cyclic voltammetry [18, 19], but also difference pulse voltammetry, electrochemical impedance spectroscopy [20], and further more (see respective chapters in [21] for a overview). Figure 4a shows the cyclic voltammetric curve of a *Geobacter* dominated anodic biofilm for non turn-over, i.e., substrate depleted, conditions. One can clearly identify four redox couples, representing possible electron transfer sites. The respective formal potentials of these are indicated and denominated E_1^f to E_4^f. The corresponding CV for turn-over conditions, i.e., in the presence of electron donor, shows a typical (bio) electrocatalytic s-shape (Fig. 4b). Interestingly, the first derivative, shown in the inset of Fig. 4b, reveals clearly that only E_2^f to E_3^f are associated with the bacterial electron transfer. It has to be mentioned that the in-depth kinetic and thermodynamic analysis of electrochemical data needs the set-up of new models and procedures, see e.g., [22] for kinetics, as due to the high complexity of these biofilms "traditional" models from molecule or enzyme electrochemistry cannot be applied.

Purely electrochemical or microscopic techniques (see below) only allow a limited insight into the microbial electron transfer processes. Consequently, the use of spectroelectrochemical techniques, i.e., the direct coupling electrochemical and spectroscopic analysis, has been demonstrated. These techniques, including for instance the combination of voltammetry and UV/vis spectroscopy [19], attenuated total reflectance surface-enhanced infrared absorption spectroscopy

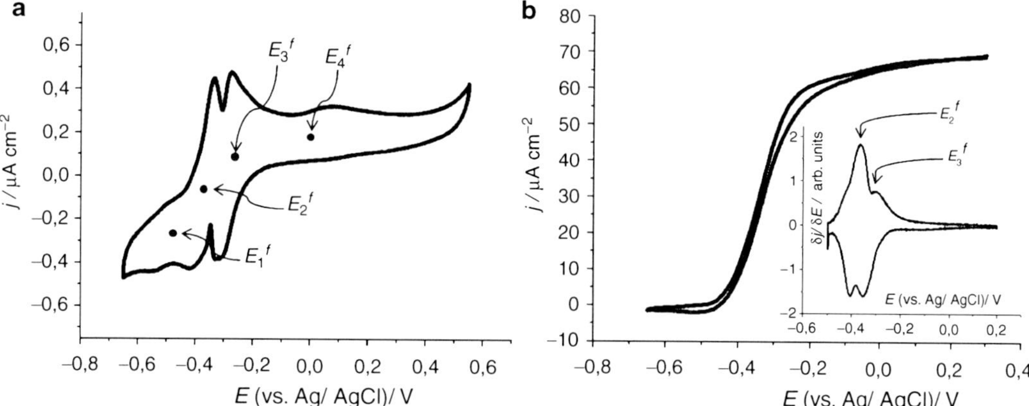

Biofilms, Electroactive, Fig. 4 (**a**) Cyclic voltammogram of a *Geobacter* biofilm grown at 0.2 V vs. Ag/AgCl on a graphite rod electrode in substrate depleted (non turn-over) conditions E_1^f to E_4^f indicate formal potentials of the four detected redox couples of the biofilm; (**b**) CV of the same biofilm in the presence of acetate (turn-over conditions) the inset shows the first derivative of the CV curves and clearly reveals that only E_2^f and E_3^f are associated with the bacterial electron transfer. (Scan rate in all cases 1 mV/s; data according to Fricke et al. [19])

(ATR-SEIRAS) [19] and surface-enhanced Raman resonance scattering (SERRS) spectroscopy [23] have been performed on suspended cells and living biofilms. They thereby unprecedented insights in the bacterial electron transfer mechanisms and the identification of the respective proteins, including the coordination of the central atom [23], was possible.

Microscopic, Microbial and Genetic Techniques

Additionally to the electrochemical investigation a whole arsenal of techniques is available for the study of electroactive biofilms. A snapshot about these techniques shall be provided in the following section.

Microscopic techniques like transmission or scanning electron microscopy (TEM and SEM) are most common and have been in use for biofilm studies for a long time (Fig. 1b shows an exemplary SEM image of an anodic biofilm). However, both techniques require an extensive sample preparation and fixation, possibly leading to nonnatural images of the biofilm. Consequently, less invasive methods are growingly popular. Here, especially confocal laser scanning microscopy (CLSM) using various probes, e.g., FISH and pH, is going to be exploited; see e.g., [24–27]. Noteworthy, further methods like AFM and immune/probe microscopy that are already used for single-cell studies [9, 28, 29] may be exploited.

For mixed culture biofilms the analysis of the bacterial composition and the identification of microbial key players are of outstanding interest. Here the respective leading edge methods of microbial ecology, like 16sRNA fingerprinting are exploited. These methods now include flow cytometry and TFRLP analysis [30] as well as transcriptomics [31] and will move on to pyrosequencing and next generation methods; see e.g., [32].

Technical Exploitation: Microbial Bioelectrochemical Systems

The development of microbial bioelectrochemical systems (BES) has recently achieved an impressive progress [33]. The elegance of microbial bioelectrochemical systems is based on the possibility to directly link the metabolic activity of living microorganisms with electrodes – for a direct conversion of chemical into electric energy

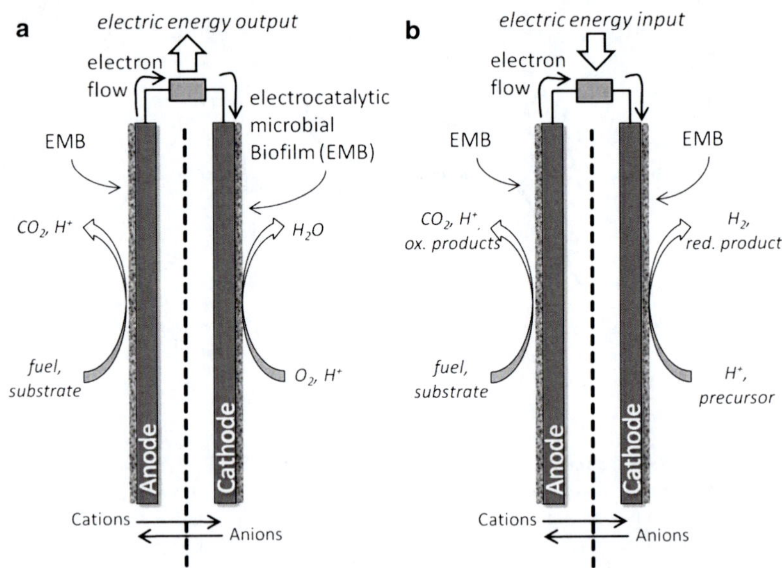

Biofilms, Electroactive, Fig. 5 Principles of the most abundant microbial bioelectrochemical systems (**a**) microbial fuel cells and (**b**) microbial electrosynthesis cells on the example of the H_2 production. (Note: Here the anodic and cathodic reactions are catalyzed by biofilms, yet as described in the text also other catalysts can be exploited)

and vice versa. The major focus of BES technology lies on exploitation of low-value biomasses, like wastewater, as a chemical energy source. For a long time, the main focus of BES research was the generation of electricity in microbial fuel cells (MFCs), e.g., [34] and Fig. 5a. Recently, a number of further fascinating application possibilities have been proposed, spanning from the production of drinking water in microbial desalination cells (MDC) [35] via ► electrochemical bioremediation [36, 37], to the production and upgrading of chemical products in microbial electrosynthesis cells (MEC). The principle of an MEC is shown in Fig. 5b. The portfolio of chemicals shown to be gained in MECs comprises, up to now, hydrogen, methane, H_2O_2 and hydroxide solution ("caustic"), which can be produced using electrocatalysts of chemical and biological origin (see, e.g., [38] and [39] for a summary).

Future Directions

The future research and development on electroactive microbial biofilms will further require combining application-driven engineering with fundamental research. This venture will include research for a deeper understanding of electrode-microbe interactions and the tailoring of the involved interfaces; the development of new, cost-effective materials; the establishment of new bioelectrocatalytic target reactions (especially involving bioelectrosynthesis); the study of environmental factors and the engineering of involved process parameters.

Cross-References

- ► Biocorrosion
- ► Bioelectrochemical Hydrogen Production
- ► Biosensors, Electrochemical
- ► Cell Membranes, Biological
- ► Cofactor Substitution, Mediated Electron Transfer to Enzymes
- ► Direct Electron Transfer to Enzymes
- ► Electrochemical Bioremediation
- ► Electrochemical Functional Transformation
- ► Microbial Corrosion
- ► Microbiologically Induced Corrosion
- ► Microbiologically Influenced Corrosion
- ► Modeling and Simulation of Biosensors
- ► Spectroelectrochemistry, Potential of Combining Electrochemistry and Spectroscopy
- ► Voltammetry of Adsorbed Proteins

References

1. Flemming H-C, Wingender J (2010) The biofilm matrix. Nat Rev Microbiol 8:623–633
2. Potter MC (1912) Electrical effects accompanying the decomposition of organic compounds. Proc R Soc Lond (B) 84:260–276
3. Aston WJ, Turner APF (1984) Biosensors and biofuel cells. In: Russell GE (ed) Biotechnology & genetic engineering reviews. Intercept, Newcastle upon Tyne, pp 89–120
4. Kano K, Ikeda T (2000) Fundamentals and practices of mediated bioelectrocatalysis. Anal Sci 16(10): 1013–1021
5. Kim BH et al (1998) Biofuel cell without using elec tron transfer medium, in Korea Advanced Institute of Science Technology
6. Kim BH et al (1999) Mediator-less biofuel cell. Patent US 5976719
7. Schröder U (2007) Anodic electron transfer mechanisms in microbial fuel cells and their energy efficiency. Phys Chem Chem Phys 9:2619–2629
8. Rabaey K et al (2005) Microbial phenazine production enhances electron transfer in biofuel cells. Environ Sci Technol 39(9):3401–3408
9. Reguera G et al (2005) Extracellular electron transfer via microbial nanowires. Nature 435:1098–1101
10. El-Naggar MY et al (2010) Electrical transport along bacterial nanowires from Shewanella oneidensis MR-1. Proc Natl Acad Sci USA 107(42):18127–18131
11. Gorby YA et al (2006) Electrically conductive bacterial nanowires produced by Shewanella oneidensis strain MR-1 and other microorganisms. Proc Natl Acad Sci USA 103:11358–11363
12. Rosenbaum M et al (2006) Interfacing electrocatalysis and biocatalysis using tungsten carbide: a high performance noble-metal-free microbial fuel cell. Angew Chem Int Ed 45:6658–6661
13. Schröder U, Nießen J, Scholz F (2003) A generation of microbial fuel cell with current outputs boosted by more than one order of magnitude. Angew Chem Int Ed 42:2880–2883
14. Magnuson TS (2011) How the xap locus put electrical "Zap" in Geobacter sulfurreducens biofilms. J Bacteriol 193(5):1021–1022
15. Lovley DR (2008) The microbe electric: conversion of organic matter to electricity. Curr Opin Biotechnol 19:1–8
16. Lovley DR (2006) Bug juice: harvesting electricity with microorganisms. Nat Rev Microbiol 4:497–508
17. Nealson KH, Scott J (2006) Ecophysiology of the genus Shewanella. In: Prokaryotes. The Prokaryotes, Springer Publishing, pp 1133–1151
18. Srikanth S, Marsili E, Flickinger MC, Bond DR (2008) Electrochemical characterization of Geobacter sulfurreducens cells immobilized on graphite paper electrodes. Biotech. Bioeng. 99(5): 1065–1073
19. Fricke K, Harnisch F, Schröder U (2008) On the use of cyclic voltammetry for the study of anodic electron transfer in microbial fuel cells. Energy Environ Sci 1(1):144–147
20. He Z, Mansfeld F (2009) Exploring the use of electrochemical impedance spectroscopy in microbial fuel cell studies. Energy Environ Sci 2: 215–219
21. Rabaey K et al (eds) (2010) Bioelectrochemical systems: from extracellular electron transfer to biotechnological application. Integrated environmental technology, ed. P. Lens. IWA Publishing, p 450
22. Torres CI et al (2010) A kinetic perspective on extracellular electron transfer by anode-respiring acteria. FEMS Microbiol Rev 34:3–17
23. Millo D et al (2011) In situ spectroelectrochemcial investigation of electrocatalytic microbial biofilms by surface-enhanced resonance raman spectroscopy. Angew Chem Int Ed 50:2625–2627
24. Read ST et al (2010) Initial development and structure of biofilms on microbial fuel cell anodes. BMC Microbiol 10:98
25. Virdis B et al (2011) Biofilm stratification during simultaneous nitrification and denitrification (SND) at a biocathode. Bioresour Technol 102(1):334–341
26. Franks AE et al (2010) Microtoming coupled to microarray analysis to evaluate the spatial metabolic status of Geobacter sulfurreducens biofilms. ISME J 4:509–519
27. Franks AE et al (2009) Novel strategy for three-dimensional real-time imaging of microbial fuel cell communities: monitoring the inhibitory effects of proton accumulation within the anode biofilm. Energy Environ Sci 2:113–119
28. Lower SK, Hochella MF Jr, Beveridge TJ (2001) Bacterial recognition of mineral surfaces: nanoscale interactions between Shewanella and ?-FeOOH. Sci 292(5520):1360–1363
29. Lower BH et al (2009) Antiobody recognition force microscopy shows the outer membrane cytochromes omcA and MtrC are expressed on the exterior surface of Shewanella oneidensis MR-1. Appl Environ Microbiol 75(9):2931–2935
30. Harnisch F, Koch C, Patil SA, Hübschmann T, Müller S, Schröder U (2011) Revealing the electrochemically driven selection in natural community derived microbial biofilms using flow-cytometry. Energy & Environmental Science 4(4):1265–1267
31. Nevin KP et al (2009) Anode biofilm transcriptomics reveals outer surface components essential for high density current production in Geobacter sulfurreducens fuel cells. PLoS One 4(5):e5628
32. Müller S, Harms H, Bley T (2010) Origin and analysis of microbial population heterogeneity in bioprocesses. Curr Opin Biotechnol 21:100–113
33. Rabaey K et al (eds) (2009) Bioelectrochemical systems: from extracellular electron transfer to biotechnological application. Integrated environmental technology, ed. P. Lens. IWA Publishing, p 450

B

34. Rabaey K et al (eds) (2010) Bioelectrochemical systems: from extracellular electron transfer to biotechnological application. Integrated environmental technology series, ed. P. Lens. IWA Publishing, London/New York
35. Cao X et al (2009) A new method for water desalination using microbial desalination cells. Environ Sci Technol 43(18):7148–7152
36. Aulenta F et al (2007) Electron transfer from a solid-state electrode assisted by methyl viologen sustains efficient microbial reductive dechlorination of TCE. Environ Sci Technol 41(7):2554–2559
37. Virdis B et al (2008) Microbial fuel cells for simultaneous carbon and nitrogen removal. Water Res 42:3013–3024
38. Harnisch F, Schröder U (2010) From MFC to MXC: chemical and biological cathodes and their potential for microbial bioelectrochemical systems. Chem Soc Rev 39:4433–4448
39. Rabaey K, Rozendal RA (2010) Microbial electrosynthesis – revisiting the electrical route for microbial production. Nat Rev Microbiol 8:706–716

Biomedical Applications of Electrochemistry, Use of Electric Fields in Cancer Therapy

Guillermo Marshall
Laboratorio de Sistemas Complejos, Departamento de Ciencias de la Computación, Facultad de Ciencias Exactas y Naturales, Universidad de Buenos Aires, Buenos Aires, Argentina

Introduction

The use of electric fields (**EF**) in tissues triggers complex electrochemical processes, thus constituting a paradigmatic example of biomedical applications of electrochemistry. In recent years, **EF**-based cancer treatments proved to be an effective therapy for certain types of tumors. Although having less undesired effects than traditional cancer therapies like radio- or chemotherapy, **EF** therapies still have some side effects that it is necessary to minimize. The first part of this entry presents a summary of **EF** solid tumor therapies that have reached clinical practice. The second part describes some examples of the underlying electrochemistry utilized to optimize them through mathematical or in silico modeling, in particular ion transport and pH fronts evolution.

Electric Field-Based Tumor Therapies

EF-based therapies – electrochemical treatment of cancer, electroporation-based therapies, nanosecond pulsed electric fields, and tumor treating fields – basically differ in the type of the **EF** being applied and in whether the **EF** is combined or not with anticancer drugs.

The electrochemical treatment of cancer (**EChT**) consists in the passage of a direct electric current through two or more electrodes inserted locally in the tumor tissue. The extreme pH changes induced are the main tumor destruction mechanism [1]. Tissue destruction by this technique has been reported in a wide range of solid tumors, with greater efficacy observed in skin cancer, oral cavity, and thyroid malignancies [2]. Some of the characteristics of **EChT** are its simplicity, effectiveness, low cost, and negligible side effects. In **EChT** typical clinical applications 6–8 V/m are used during periods from minutes to hours. At present, there are several groups working in Australia, China, Cuba, Japan, Sweden, and the USA; for a general review see [1].

In recent years, pulsed electric fields have been applied in local tumor treatment based upon electroporation (**EP**), a technique in which pulsed electric fields are employed to perturb cell membrane integrity creating pores across it. **EP**-based therapies are electrochemotherapy (**ECT**), electrogenetherapy (**EGT**), and irreversible electroporation (**IRE**). **ECT** combines a reversible **EP** (cell membrane-permeabilizing electric pulses below the irreversible threshold) with non-permeant or poorly permeant anticancer drugs, such as bleomycin, to potentiate their entry to the cell thus their intrinsic cytotoxicity [3]. Since its beginnings in the late 1980s, **ECT** has evolved into a clinically verified palliative or cytoreductive treatment for cutaneous and subcutaneous tumor nodules of different malignancies in Europe and the USA [3]. Typical **ECT** treatment in humans consists in a train of 8 square pulses of high-electric field (around 1,000 V/cm) and short duration (around 100 μs) delivered at 1Hz [4]. **EGT** (introduction of plasmids or oligonucleotides to the cell by **EP**) is a new technique intensely studied due to its potential as a nonviral gene-delivery system [5]. Typical DNA **EGT**-optimized pulse combination [6] consists in a combination of a short high-voltage pulse (HV, 1000 V/cm, 100 μs) followed by a long low-voltage pulse (LV, 100 V/cm, 400 ms). A recent derivation of **ECT** is **IRE**, introduced in [7], an irreversible **EP** (electric pulses above the irreversible threshold), without thermal effects, that leaves intact main tissue structures [8]. Typical **IRE** applications in models consist in electric fields around 500 V/cm, using around 70 ms delivered at 4Hz. An important difference between **IRE** and **ECT** is that in **IRE** cell elimination is induced by permanent membrane disruption without any drug or DNA delivery. A review of electric pulse generators for cell and tissue **EP** may be found in [9].

Recently, pulse duration attained the nanosecond range, its effects reaching the nucleus of the cell, thus allowing a direct manipulation of the nucleoplasm apparently leaving intact the membrane cell [10]. This gave rise to what is now called high-intensity nanosecond pulsed electric field (**nsPEF**) or nanosecond **EF**. Typical **nsPEF** pulses are around 100 ns and 200 kV/cm. **nsPEP** can be used as a purely electric cancer therapy eliminating tumors without hyperthermia or drugs [11]. The main characteristics of **nsPEF** are their low energy leading to very low heat production and their ability to penetrate into the cell. The effects on plasma membrane are to permeabilize it after a few nanoseconds, while the effects in subcellular membranes are to permeabilize intracellular organelles and release calcium from the endoplasmic reticulum (network of tubules and flattened sacs that serve a variety of functions in the cell), thus signaling apoptosis (programmed cell death) [10]. **nsPEF** therapy was used to treat a skin tumor in a single case of human basal cell carcinoma; a complete remission after one single treatment was obtained [12]. **nsPEP** electric pulse generator devices are complex and costly, thus not commercially available. A review of nanosecond electroporation and nanopulse generators is presented in [13].

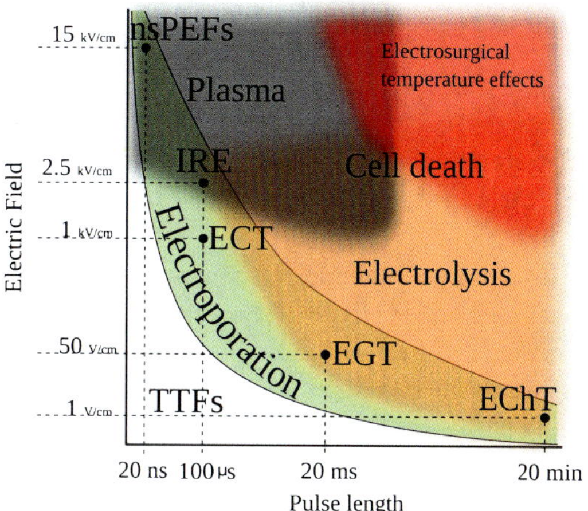

Biomedical Applications of Electrochemistry, Use of Electric Fields in Cancer Therapy, Fig. 1 Schematic relationship between solid tumor electric field intensity and pulse length (not in scale)

There is still another recent **EF**-based solid tumor therapy – tumor treating fields (**TTFs**) – worth mentioning, although its relation with electrochemistry needs to be explored further. It consists in low intensity, intermediate frequency, alternating electric fields (ac fields) applied with insulated electrodes, which slow the growth of solid tumors in vivo, and have shown promise in pilot clinical trials in patients with advanced solid tumors. Yoram Palti and collaborators have shown that 100–1,000 kHz have significant effects in cells under division. These effects during cytokinesis (the separation into two daughter cells during cell division) were shown to be the unidirectional forces induced by the inhomogeneous fields at the neck separating the daughter cells that interfere with the normal process and induce dielectrophoresis [14].

To give a global picture of the different solid tumor electric-based therapies previously described, the relationship between electric field intensities and pulse length is schematically represented in Fig. 1 from [15], modified by N. Olaiz.

We end this brief survey of **EF**-based solid tumor therapies mentioning several excellent books dealing with electroporation from different perspectives: [16–21].

Bioelectrochemistry and EChT and ECT Electric Field-Based Cancer Therapies

EF-based solid tumor therapies have some undesired side effects (loss of cell viability, uncontrolled necrosis, plasmid damage) that it is necessary to minimize. Electrochemical, specifically bioelectrochemical, methods can be used to elucidate fundamental aspects of electric field–tissue interaction and thus contribute to minimize side effects. In silico modeling, validated with in vivo and in vitro modeling, appears to be a powerful tool to study the underlying bioelectrochemical effects in tissues subject to **EF** as well as for dose-planning strategies. Therefore, in what follows, **EF** tumor-based therapies mathematical modeling and some aspects of **EChT** and ECT in terms of in vitro and in silico modeling, in particular ion transport and the role of pH, are briefly reviewed.

A note about multiscale modeling is relevant here. **EF** effects in tumors require an interdisciplinary and multiscaling approach. This implies working in vivo, in vitro, and in silico and at the tissue, cellular, and cell membrane level. Multiscale electroporation modeling is loosely defined as the use of mathematics and computational methods to describe the simultaneous dynamic interaction of electroporation phenomena at two or more scales. Multiscale electroporation modeling is being pursued near the two ends of the scale: Molecular Dynamics (MD) and continuum modeling, respectively. The bridge between these two approaches is still an open problem, though great efforts are presently devoted to this issue. Relevant information can be found in NIBIB/NIH (http://www.nibib.nih.gov/Research/ProgramAreas/MathModeling, acceded February 2013).

ECT in silico modeling in the literature includes computing electric field distributions based on 2D models that show that the applied voltage, configuration of the electrodes, and electrode position need to be chosen specifically for each individual case [22, 23]. Calculation of transmembrane potential (TMP) and electroporation density across membrane of spheroidal cells subject to ultrashort, high-intensity pulses show that the TMP induced by pulsed external voltages can be substantially higher in oblate spheroidal as compared to spherical or prolate spheroidal cells [24]. The simulation of effects of external electric fields on clusters of excitable cells shows that the stimulation of a given cell depends in part on the arrangement of cells within the field and not simply the location within the field [25]. Treatment planning for different electrode configurations and electric pulse parameters in deep-seated tumors shows that treatment by **ECT** is feasible and sets the ground for numerical treatment planning-based **ECT** [26]. **ECT** realistic simulations of pore creation and resealing at a single whole cell suggest that the model may serve to develop and test in silico protocols tailored to specific tasks such as creation of large pores for DNA uptake or small pores for drug delivery [27]. A first attempt at developing a macroscopic mathematical model for analyzing the mass transfer into cells during **ECT** by introducing a multiscale model that couples an external electric field model at tissue level, an electroporation-driven mass transfer model at a single cell level, and a macroscopic mass transfer diffusion model in tissue illustrates the value of such a model in designing optimal **ECT** protocols for a typical situation in which tumor is treated with **ECT** and bleomycin [28]. A multiscale model for describing **IRE** and **nsPEF** electroporation by means of multicellular and tissue models suggests that they may lead to a choice of preferential death: necrosis with **IRE** and apoptosis with **nsPEF** [29]. Molecular Dynamics (MD) is a deterministic computational method that simulates the Newtonian equations of motion for systems of atoms or molecules with hundreds to millions of particles (www.gromacs.org; http://ambermd.org/, acceded February 2013). MD simulations are a powerful method allowing the evaluation at an atomic level of the effects produced by electric fields, for instance, on membrane structure such as pore formation [30–34].

Experimental observations in **EChT** show that the electric field causes a flux of interstitial water from anode to cathode. Tissue surrounding the anode dehydrates and edema results at the cathode. Hydroxyl ions and hydrogen gas are liberated at the cathode while hydrogen ions, oxygen, and chlorine gas are produced at the anode. These reactions induce immediate and substantial pH changes (acidic at the anode and alkaline at the cathode) which are the main tumor destruction mechanism through necrosis and apoptosis inducement [2]. Pioneering **EChT** modeling was introduced in [1]; in a series of papers, they first modeled the tumor tissue as an ionic solution; further refinements included addition of buffering capacity, a nonspecific organic content, and considering the impact of chlorine evolution on the medium. They showed that the pH profiles obtained correlated with the necrotic area, thus suggesting that the model could be used for predicting the area of the tissue lesion induced by **EChT** [1].

More recently, extending results presented in [1], an **EChT** in silico modeling based on previous work in electrochemical deposition in thin layer cells [35] was introduced in [36]. The model consists in the one-dimensional Nernst–Planck equations for ion transport in a four-component electrolyte (Na+, OH−, Cl−, and H+) and the Poisson equation for the electrostatic potential under galvanostatic conditions; the model was solved numerically and validated with in vivo and in vitro modeling. The main finding was that, whether in vivo, in vitro, or in silico, an initial condition with almost neutral pH evolves between electrodes into extreme cathodic alkaline and anodic acidic fronts moving towards each other, leaving the possible existence of a biological pH region between them; towards the periphery, the pH falls to its neutral values.

In the classical **EChT** method, the distance between electrodes is in general in the order of centimeters, due to the general belief that an optimal distance between electrodes exists and that if the anode and cathode are placed too close to each other, chemical reactions between the different electrode reaction products are possible. At variance with this belief, the application of **EChT** to a gel model using a one-probe two-electrode device (OPTED) containing the cathode and the anode very close to each other (1 mm) reveals the emergence of two half-spherical pH fronts, one basic and the other acid (from cathode and anode, respectively), expanding towards the periphery and configuring an almost full sphere [37]. The characteristics of the OPTED are the insertion of one applicator rather than two or more (thus minimizing tissue intrusion, for instance, in the nervous system), the ability to reach tumors beyond capabilities of conventional surgery, and the minimization of electric current circulation through the treated organ.

While in **EChT** tumor necrosis is the main goal of the treatment, and in **IRE** it may contribute to tumor destruction, in ECT and **EGT** it is usually avoided because of its collateral effects. A common problem of **ECT** and **EGT** is their low cell viability and, in relation to **EGT**, its low transfection efficiency compared with other transfection methods. It has been suggested that these effects may be strongly dependent on the change of pH induced by electrolysis during the process. Strong pH variations of the medium during **EGT** may have undesired effects over the plasmids used for the delivery, as DNA denaturalization is affected by pH. In this context, ion transport in gel models under **ECT** for conditions typical to many clinical studies found in the literature was recently introduced, unveiling the presence of pH fronts emerging from both electrodes, and that these fronts were immediate and substantial [38]. Moreover, theoretical predictions of the comparison of **ECT** pH fronts with those arising in **EChT** showed a striking result: anodic acidification is larger in **ECT** than in **EChT**, suggesting that tissue necrosis could also be greater. The quantification of pH extension and evolution is relevant for optimizing **ECT** treatment, where it is desired to apply an effective dosage while minimizing pH alterations leading to necrosis and plasmid damage. It was suggested that one way to achieve this could be designing protocols minimizing voltage and pulse number while maximizing pulse lengths as far as possible [38].

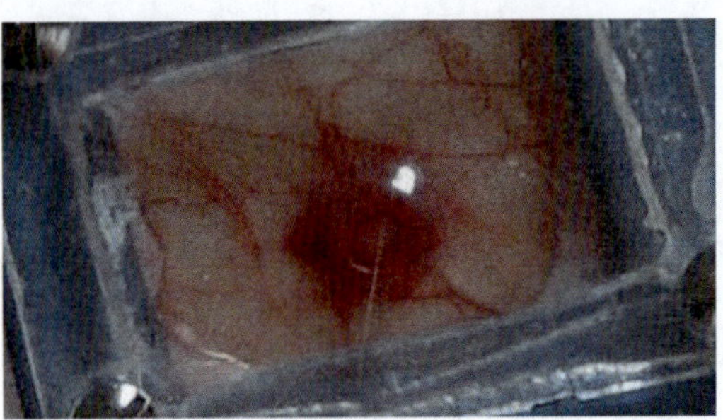

Biomedical Applications of Electrochemistry, Use of Electric Fields in Cancer Therapy, Fig. 2 Dorsal skinfold chamber (From [39])

Future Directions

The methods of electrochemisty can be further extended into in vivo modeling domains: a natural extension of a classical thin layer electrochemical cell [35] consists in a transparent thin plastic chamber implanted into the dorsal skinfold chamber (DSC) in mice subject to **EF**. Figure 2 shows an example of a DSC implanted in mice [39]. With this in vivo model that we named a bioelectrochemical cell (BECC), it is possible to measure the effects of **EF** in tumors and tissue with the help of fluorescent markers and intravital microscopy.

Finally, it is worth mentioning the possibility of analyzing the role of pH fronts in electroporation through the use of giant unilamellar vesicles (GUVs). GUVs are membrane "bubbles" formed by bending and closing up a lipid bilayer and serve as simple models for cell membranes [40]. Indeed, membrane deformation under pH fronts were recently reported in [41], and the ejection of a tubule growing towards a chemical source due to a curvature instability triggered by a pH front was reported in [42]. These findings may open a wide avenue of research into the role of pH in cell membrane electroporation.

As a general conclusion, although **EF**-based tumor therapies have experienced a sustained progress, the study of the underlying electrochemistry is still incipient. The possibility of its future development raises tantalizing prospects for optimizing **EF**-based tumor therapies.

Cross-References

- ► Biomolecules in Electric Fields
- ► Cell Membranes, Biological
- ► Conductivity of Electrolytes
- ► Electropermeabilization of the Cell Membrane
- ► Electrochemistry of Drug Release
- ► Electrogenerated Acid
- ► Electrogenerated Base
- ► Gene Electrotransfer for Clinical Use
- ► Numerical Simulations in Electrochemistry

References

1. Nilsson E, von Euler H, Berendson J, Thorne A, Wersall P, Naslund I, Lagerstedt A, Narfstrom K, Olsson J (2000) Electrochemical treatment of tumors. Bioelectrochemistry 51:1–11
2. von Euler H, Strahle K, Thorne A, Yongqing G (2004) Cell proliferation and apoptosis in rat mammary cancer after electrochemical treatment (EChT). Bioelectrochemistry 62:57–65
3. Mir L (2006) Bases and rationale of the electrochemotherapy. Eur J Canc Suppl 4:38–44
4. Sersa G, Miklavcic D, Cemazar M, Rudolf Z, Pucihar G et al (2008) Electrochemotherapy in treatment of tumors. Eur J Surg Oncol 34: 232–240
5. Mir L (2009) Nucleic acids electrotransfer-based gene therapy (electrogenetherapy): past, current, and future. Mol Biotechnol 43:167–176
6. Hojman PJ, Zibert R, Gissel H, Eriksen J, Gehl J (2007) Gene expression profiles in skeletal muscle after gene electrotransfer. BMC Mol Biol 8:56. doi:10.1186/1471-2199-8-56
7. Davalos R, Mir L, Rubinsky B (2005) Tissue ablation with irreversible electroporation. Ann Biomed Eng 33:223–231
8. Al-Sakere B, Andre F, Bernat C, Connault E, Opolon P et al (2007) Tumor ablation with irreversible electroporation. PLoS One 2:e1135
9. Rebersek M, Miklavcic D (2010) Concepts of electroporation pulse generation and overview of electric pulse generators for cell and tissue electroporation. In: Pakhomov AG, Miklavcic D, Markov MS (eds) Advanced electroporation techniques in biology and medicine, Lecture notes in physics. CRC Press/Taylor and Francis Group, London, chapter 16
10. Schonebach KH, Beebe SJ, Buescher ES (2001) Intracellular effect of ultrashort electrical pulses. J Bioelectromagn 22:440–448
11. Nuccitelli R, Pliquett U, Chen X, Ford W, Jame R (2006) Nanosecond pulsed electric fields cause melanomas to self-destruct. Biochem Biophys Res Commun 343:351–360

12. Garon EB, Sawcer D, Vernier PT, Tang T, Sung Y, Marcu L, Gundersen MA, Koeffler HP (2007) In vitro and in vivo evaluation and a case report of intense nanosecond pulsed electric field as a local therapy for human malignancies. Int J Cancer 121:675–682
13. Sundararajan R (2009) Nanosecond electroporation: another look. Mol Biotechnol 41:69–82
14. Kirson ED, Gurvich Z, Schneiderman R, Dekel E, Itzhaki A, Wasserman Y, Schatzberger R, Palti Y (2004) Disruption of cancer cell replication by alternating electric fields. Cancer Res 64:3288–3295
15. Davalos R, Rubinsky B (2007) Tissue ablation with irreversible electroporation, US Patent Appln # 2007 0043345
16. Neumann E, Sowers AE, Jordan CA (1989) Electroporation and electrofusion in cell biology. Springer
17. Pakhomov AG, Miklavcic D, Markov MS (2010) Advanced electroporation techniques in biology and medicine, Lecture notes in physics. CRC Press/Taylor and Francis Group, London
18. Rubinsky R (2010) Irreversible electroporation. Springer, New York
19. Jaroszeski MJ, Heller R, Gilbert R (2010) Electrochemotherapy, electrogenetherapy, and transdermal drug delivery, electrically mediated delivery of molecules to cells, vol 37. Humana Press, Totowa
20. Kee ST, Gehl J, Lee EW (2011) Clinical aspects of electroporation. Springer, New York
21. Spugnini E, Baldi A (2011) Electroporation in laboratory and clinical investigations. Nova Science, New York
22. Corovic S, Pavlin M, Miklavcic D (2007) Analytical and numerical quantification and comparison of the local electric field in the tissue for different electrode configurations. Biomed Eng Online 15:6–37
23. Sel D, Lebar A, Miklavcic D (2007) Feasibility of employing model-based optimization of pulse amplitude and electrode distance for effective tumor electropermeabilization. IEEE Trans Biomed Eng 54:773–781
24. Hu Q, Joshi R (2009) Transmembrane voltage analyses in spheroidal cells in response to an intense ultrashort electrical pulse. Phys Rev E Stat Nonlin Soft Matter Phys 79:011901
25. Ying W, Pourtaheri N, Henriquez C (2006) Field stimulation of cells in suspension: use of a hybrid finite element method. Conf Proc IEEE Eng Med Biol Soc 1:2276–2279
26. Miklavcic D, Snoj M, Zupanic A, Kos B, Cemazar M et al (2010) Towards treatment planning and treatment of deep-seated solid tumors by electrochemotherapy. Biomed Eng Online 23:9–10
27. Krassowska W, Filev P (2007) Modeling electroporation in a single cell. Biophys J 92:404–417
28. Granot Y, Rubinsky B (2008) Mass transfer model for drug delivery in tissue cells with reversible electroporation. Int J Heat Mass Transfer 51:5610–5616
29. Esser AT, Smith KC, Gowrishankar TR, Weaver JC (2010) Drug-free, solid tumor ablation by electroporating pulses: mechanisms that couple to necrotic and apoptotic cell death pathways. In: Pakhomov AG, Miklavcic D, Markov MS (eds) Advanced electroporation techniques in biology and medicine, Lecture notes in physics. CRC Press/Taylor and Francis Group, London, chapter 13
30. Tieleman DP, Leontiadou H, Mark AE, Marrink SJ (2003) Simulation of pore formation in lipid bilayers by mechanical stress and electric fields. J Am Chem Soc 125:6282–6383
31. Hu Q, Viswanadham S, Joshi RP, Schonebach KH, Beebe SJ, Blackmore PF (2005) Simulations of transient membrane behavior in cells subjected to a high-intensity ultrashort electric pulse. Phys Rev E Stat Nonlin Soft Matter Phys 71:031914
32. Tarek M (2005) Membrane electroporation: a molecular dynamics study. Biophys J 88:4045–4053
33. Vernier PT, Ziegler MJ, Sun Y, Chang WV, Gundersen MA, Tieleman DP (2006) Nanopore formation and phosphatidylserine externalization in a phospholipid bilayer at high transmembrane potential. J Am Chem Soc 128:6288–6289
34. Fernández ML, Marshall G, Sagués F, Reigada R (2010) Structural and kinetic molecular dynamics study of electroporation in cholesterol-containing bilayers. J Phys Chem B 114:6855–6865
35. Marshall G, Mocskos P, Swinney HL, Huth JM (1999) Buoyancy and electrical driven convection models in thin-layer electrodeposition. Phys Rev E 59:2157
36. Colombo L, Gonzalez G, Marshall G, Molina F, Soba A, Suarez C, Turjanski P (2007) Ion transport in tumors under electrochemical treatment: in vivo, in vitro and in silico modeling. Bioelectrochemistry 71:223–232
37. Olaiz N, Maglietti F, Suárez C, Molina FV, Miklavcic D, Mir L, Marshall G (2010) Electrochemical treatment of tumors using a one-probe two-electrode device. Electrochim Acta 55:6010–6014
38. Turjanski P, Olaiz N, Maglietti F, Michinski S, Suarez C, Molina FV, Marshall G (2011) The role of pH fronts in reversible electroporation. PLoS One 6(4):e17303. doi:10.1371/journal.pone.0017303
39. Olaiz N (2011) Electroquimioterapia aplicada a tumores. Dissertation, Faculty of Medicine, University of Buenos Aires
40. Dimova R, Aranda S, Bezlyepkina N, Nikolov V, Riske KA, Lipowsky R (2006) A practical guide to giant vesicles. Probing the membrane nanoregime via optical microscopy. J Phys Condens Matter 18: S1151–S1176
41. Khalifat V, Puff N, Bonneau S, Fournier J-B, Angelova MI (2008) Membrane deformation under local pH gradient: mimicking mitochondrial cristae dynamics. Biophys J 95:4924–4933
42. Fournier J-B, Khalifat N, Puff N, Angelova MI (2009) Chemically triggered ejection of membrane tubules controlled by intermonolayer friction. Phys Rev Lett 102:018102

Biomolecules in Electric Fields

Marc Tornow
Institut für Halbleitertechnik, Technische Universität München, München, Germany

Introduction

Biomolecules can be manipulated efficiently by electric fields in various ways. This article gives a brief survey of the following methodologies: DC electrokinetic transport (electrophoresis and electroosmosis), AC dielectrophoretic transport, and the electrically induced modulation of the biomolecule conformation. Selected examples for applications will be described as based on these mechanisms, including molecule transport and separation, the dielectrophoretic trapping of molecules, and two schemes in biosensing involving "smart" (switchable) bio-surfaces and nanopore-patterned membranes.

Charges of Biomolecules and Screening

Biomolecules may in general carry net electric charges at ionized groups and permanent or induced dipole moments, depending on their (frequency dependent) polarizability. As a result they may be subjected to a force, depending on the magnitude and gradient of the electric field they are placed in. As one important example, DNA features one negative charge per nucleotide at its phosphate backbone, under neutral pH conditions. Proteins, in contrast, are often amphoteric molecules: they may carry a more complex distribution of both negative and positive ionic charges at the same time. At a pH below their isoelectric point (*pI*), proteins are net positively charged while above their *pI* their net charge is negative. In electrolyte solution, the biomolecule charges are screened by a diffusive cloud of counterions or even by closely adhering counterions that have "condensed" onto the molecule. Diffusive screening is generally described by the Poisson-Boltzmann theory, and one characteristic parameter is the Debye screening length, which for a symmetrical electrolyte reads [1]

$$l_{\mathrm{D}} = \sqrt{\frac{\varepsilon\varepsilon_0 kT}{2nz^2e^2}}$$

Here, ε is the relative permittivity of the electrolyte solution, k is the Boltzmann constant, n is the concentration of each ion, z is the ion valency, and e is the elementary charge. Typical values of the Debye length are of the order of a few nanometers, e.g., $l_D = 0.96$ nm for a 1:1 electrolyte of concentration 100 mM.

Electrokinetic Transport

By interaction with DC electric fields, charged biomolecules can be directly transported using electrophoresis. Assuming a charged, spherical particle of radius R that is surrounded by a counterion shell of thickness l_D, both the inner sphere of radius R and the outer one of radius $R + l_d$ are subjected to oppositely directed forces in the external field. In equilibrium, this force is equal to the friction force in the liquid, according to Stokes law. The net velocity of the particle moving together with its double layer results from the sum of both contributions and can be calculated to [2]

$$v = c\frac{\varepsilon\varepsilon_0\zeta E}{\eta}$$

with c being a numerical factor that depends on the ratio l_D/R, η the viscosity, and ζ the zeta potential. The zeta potential is the electrical potential at a distance measured from a charged surface below which the first layers of liquid molecules or ions appear bound to this surface, when the surface is moved through the liquid (shear plane).

In electroosmosis, an electric field is applied parallel to a (e.g., negatively) charged surface making the (positive) counterions in the layers close to this surface move towards the (negative) electrode. When applied to a micro- or nanoscale, liquid-filled channel, the entire electrolyte fluid inside this channel can be forced to move as a whole, dragged by the counterions. This bulk movement of the electrolyte which in general adopts a plug flow profile is used in capillary electroosmosis applications for the transport of biomolecules dissolved in the liquid.

Dielectrophoresis

Dielectrophoretic manipulation is based on the force that an electrical dipole exhibits in a nonuniform electric field (gradient). If a biomolecule carries a permanent dipole or if it is polarized in an external field, the molecule will be attracted towards the region of higher electric field as long as the polarizability of the surrounding medium is less than the one of the particle ("positive dielectrophoresis"); otherwise it will be repelled from that region ("negative dielectrophoresis"). The dielectrophoretic force acting on a spherical particle of radius r may be expressed as

$$F = 2\pi\varepsilon_m r^3 \,\mathrm{Re}[K(\omega)]\,\mathrm{grad}\left(E^2\right)$$

with ε_m the relative permittivity of the medium, E the electric field, and $\mathrm{Re}[K(\omega)]$ the real part of the Clausius-Mossotti factor:

$$K(\omega) = \left(\varepsilon_p^* - \varepsilon_m^*\right) / \left(\varepsilon_p^* + 2\varepsilon_m^*\right)$$

with ε^*_p and ε^*_m the complex permittivities of the particle (e.g., the molecule) and the medium, respectively [3].

Examples and Applications

Electrophoretic Separation and Nanopore Translocation

The transport of biomolecules, and in particular their separation, is widely been carried out using electrophoresis. An important example is gel electrophoresis, e.g., sodium dodecyl sulfate-polyacrylamide gel electrophoresis (SDS-PAGE), where proteins are separated according to their size, smaller proteins having larger mobilities. The basic scheme has been extended to capillary electrophoresis, i.e., capillary sieving electrophoresis (CSE) or capillary gel electrophoresis (CGE) having several advantages over traditional gel electrophoresis in terms of resolution, protein quantification, and molecular weight determination [4]. If the gel features a pH gradient, proteins can be separated according to their different pI. This technique is frequently used, also based on polyacrylamide gel, and constitutes one of the two steps in 2-D gel electrophoresis, where proteins are separated according to two of their properties, here mass and pI. When performed in micro- or nanofluidic channels, electrokinetic transport and separation methods may form integral parts of micro-total analysis systems (μTAS) or Lab-on-a-Chip (LOC) platforms [5]. In these systems, sample preparation in reservoirs, injection into the channels, analyte separation, and detection/analysis are all integrated on a single, micro-, or nanofabricated chip format. They have found widespread applications in clinical diagnostics, environmental monitoring, and basic research.

A unique form of electrokinetic transport of DNA, proteins, or other biological macromolecules by virtue of their net charge is investigated in the active research field of nanopore translocation. The single nanopores under study include either biological pores [6] (membrane protein, e.g., alpha hemolysin) or nanometer-scale pores that are "drilled" into an inorganic material membrane [7] (e.g., silicon nitride) by high-resolution nanofabrication techniques. Application of an electric field across the nanopore in electrolyte solution may induce the translocation of single biomolecules through the pore. The successful transport of the bio-macromolecule is commonly observed by measuring the transient suppression or even blockade of the ionic current flow though the pore during the translocation event. Using this technique, different schemes in biosensing have been demonstrated, and a major focus of current research is devoted to a possible all-electronic sequencing of DNA [8].

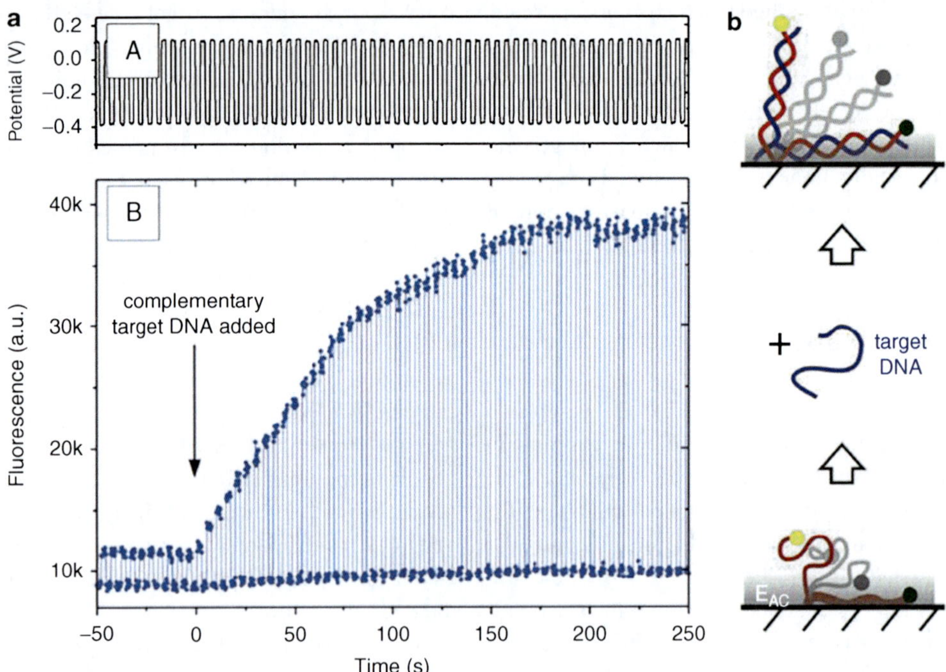

Biomolecules in Electric Fields, Fig. 1 Electrical conformation switching and hybridization of DNA layers. *Left*, (**a**): the potential applied to a gold electrode that supports a DNA layer (single stranded, 48 bases). (**b**) Fluorescence emission of the dye-labeled DNA layer. The *arrow* marks the injection of complementary targets that hybridize with the single-stranded probe layer. *Right*: schematic illustration showing how flexible single strands are only partially aligned by the short-ranged electric field (E_{AC}) at the surface. In contrast, double-stranded helices may be oriented efficiently because of their intrinsic rigidity (After [13], © by the National Academy of Sciences)

Dielectrophoretic Manipulation and Trapping

The dielectrophoretic force has been extensively used to manipulate biological objects over a large size range, from single molecules such as DNA or proteins up to viruses or whole cells. Such manipulation has found various applications in particular in AC dielectrophoretic Lab-on-a-Chip devices, for the controlled transport, separation, or concentration of the objects. The force is further increasingly used for a controlled positioning ("trapping") of single biomolecules such as DNA, peptide nanotubes, proteins, and artificial, self-assembled DNA "origami" 2D and 3D nanoscale structures on pre-patterned surfaces [3]. Here, the predominant applications are either structural (e.g., DNA stretching in AC fields) or in particular electro (−optical) studies of these biomaterials after having trapped them on appropriate nanoscale electrode pairs, by applying an AC electric voltage to these electrodes in solution.

Switchable Bio-layers

Biomolecules can be effectively immobilized on inorganic surfaces such as semiconductors, oxides, and metals, in the form of monomolecular layers. They have been widely used as receptor interface, e.g., in the field of chip-based biosensing applications. A particular functionality arises for those monolayers which may undergo a controlled and reversible change of their spatial conformation upon external trigger ("smart surfaces") [9–11]. In the case of DNA, the electrical interaction with charged (metal) surfaces in aqueous electrolytes can be employed in multiple ways, including the electrically controlled positioning of DNA, their on-surface-immobilization, hybridization, release, and in particular their alignment and induced conformation change [12]. The latter includes the controlled switching of short oligonucleotide strands bound to a solid surface, between a "lying" and a "standing" conformation, depending on whether the surface electrically attracts or repels the negatively charged strands, respectively. This technique has been demonstrated as novel scheme in label-free biomolecule detection [13, 14], e.g., for the sensitive detection of DNA hybridization (see Fig. 1). The working principle is based on oligonucleotides that are tethered to a gold surface and functionalized at their top-end with fluorescence markers. This enables the real-time optical monitoring of the oligo conformation, owing to partial quenching of the fluorescence which depends on the dye's distance to the gold electrode. The dynamics of switching single- and double-stranded DNA exhibits pronounced differences. These originate from their dissimilar flexibilities, together with the fact that sufficiently high electric fields to manipulate the DNA orientation are present only in close proximity to the electrode surface (within the Debye screening length l_D). Figure 1 illustrates how the hybridization of an electrically switched, single-stranded probe DNA with its matching, complementary target DNA results in a drastic enhancement of the recorded fluorescence switching signal.

Future Directions

Electric separation techniques for biomolecules have been continuously improved during the last decades and are expected to further develop towards even higher merit regarding speed, cost, and resolution. The emerging fields of controlled biomolecule manipulation in particular in combination with nanostructured surfaces are experiencing a strongly increasing interest both in fundamental science and in applications such as biosensing. This includes specifically both the envisaged pure electrical DNA sequencing using functional nanopores, and smart, switchable surfaces that would allow for a further integration of inorganic nanosystems with tailored bioelectronic functionality.

Cross-References

► Bioelectrochemical Hydrogen Production
► Biomedical Applications of Electrochemistry, Use of Electric Fields in Cancer Therapy

References

1. Bard AJ, Faulkner LR (2000) Electrochemical methods–fundamentals and applications, 2nd edn. Wiley, New York
2. Butt H-J, Graf K, Kappl M (2006) Physics and chemistry of interfaces. Wiley-VCH, Weinheim
3. Kuzyk A (2011) Dielectrophoresis at the nanoscale. Electrophoresis 32:2307–2313
4. Zhu Z, Lu JJ, Liu S (2012) Protein separation by capillary gel electrophoresis: a review. Anal Chim Acta 709:21–31
5. Arora A, Simone G, Salieb-Beugelaar GB, Kim JT, Manz A (2010) Latest developments in micro total analysis systems. Anal Chem 82:4830–4847
6. Bayley H, Braha O, Cheley S, Gu L-Q (2004) Engineered nanopores. In: Niemeyer CM, Mirkin CA (eds) Nanobiotechnology. Wiley-VCH, Weinheim, p 93
7. Dekker C (2007) Solid-state nanopores. Nat Nanotechnol 2:209–215
8. Branton D, Deamer DW, Marziali A, Bayley H, Benner SA, Butler T, Di Ventra M, Garaj S, Hibbs A, Huang X, Jovanovich SB, Krstic PS, Lindsay S, Ling XS, Mastrangelo CH, Meller A, Oliver JS, Pershin YV, Ramsey JM, Riehn R, Soni GV, Tabard-Cossa V, Wanunu M, Wiggin M, Schloss JA (2008) The potential and challenges of nanopore sequencing. Nat Biotechnol 26:1146–1153
9. Mendes PM (2008) Stimuli-responsive surfaces for bio-applications. Chem Soc Rev 37:2512–2529
10. Nandivada H, Ross AM, Lahann J (2010) Stimuli-responsive monolayers for biotechnology. Prog Polym Sci (Oxf) 35:141–154
11. Wong IY, Almquist BD, Melosh NA (2010) Dynamic actuation using nano-bio interfaces. Mater Today 13:14–22
12. Tornow M, Arinaga K, Rant U (2007) Electrical manipulation of DNA on metal surfaces. In: Shoseyov O, Levy I (eds) Nano biotechnology: bioinspired devices and materials of the future. Humana (Springer), Berlin, p 187
13. Rant U, Arinaga K, Scherer S, Pringsheim E, Fujita S, Yokoyama N, Tornow M, Abstreiter G (2007) Switchable DNA interfaces for the highly sensitive detection of label-free DNA targets. Proc Natl Acad Sci USA 104:17364–17369
14. Rant U, Pringsheim E, Kaiser W, Arinaga K, Knezevic J, Tornow M, Fujita S, Yokoyama N, Abstreiter G (2009) Detection and size analysis of proteins with switchable DNA layers. Nano Lett 9:1290–1295

Biosensors, Electrochemical

Marco Mascini and Ilaria Palchetti
Dipartimento di Chimica "Ugo Schiff",
Università degli Studi di Firenze,
Sesto Fiorentino, Firenze, Italy

Introduction

In early 2000, two Divisions of the International Union of Pure and Applied Chemistry (IUPAC) prepared recommendations on the definition, classification, and nomenclature related to electrochemical biosensors; these recommendations have been then extended to other types of biosensors. Following these IUPAC recommendations, "a biosensor is defined as a self-contained integrated device, which is capable of providing specific quantitative or semi-quantitative analytical information using a biological recognition element (biochemical receptor) which is retained in direct spatial contact with an electrochemical transduction element. Because of their ability to be repeatedly calibrated, a biosensor should be clearly distinguished from a bioanalytical system, which requires additional processing steps, such as reagent addition" [1].

Another important concept was introduced, some years later, by Turner and Newman in [2]; they referred to a biosensor as "*a compact analytical device incorporating a biological or biologically-derived sensing element either integrated within or intimately associated with a physicochemical transducer,*" thus including synthetic chemical compounds that mimic the biological material in the development of biosensors. Nowadays, the concept of biologically derived element is fully accepted in the scientific community.

In 2010, a technical report of IUPAC upgrades the definition of electrochemical biosensors introducing the concept of nucleic acid (NA)-based biosensor [3]. In this context, an electrochemical NA-based biosensor is a device that integrates an NA (natural and biomimetic forms of oligo- and polynucleotides) as the biological recognition element and an electrode as the physicochemical transducer.

Extending the definitions reported above, biosensors should have the following features over other analytical methods, i.e., they should be cheap, small, portable, capable of performing rapid measurements and of being used by semi-skilled operators.

Biomolecular Recognition Elements Coupled with Electrochemical Transducers

Enzymes were historically the first molecular recognition elements [2–6] included in biosensors and continue to be the basis for a significant number of publications in this field. Amperometric biosensors based on glucose oxidase (GOx) are the most famous example of biosensors applied to medical diagnostic. The three modes of oxidation reactions that occur in redox enzyme-based biosensors, like GOx, are referred to as first, second, and third generation as follows [7]:

First generation: oxygen and hydrogen peroxide sensor-based systems
Second generation: mediator-based systems
Third generation: systems based on direct electron transfer from redox enzyme

Enzyme-based biosensors may be classified according to the analytes or reactions that they monitor; for example, direct monitoring of analyte concentration of reactions producing or consuming such analytes is the direct mode. Some environmental or food contaminants selectively inhibit the activity of certain enzymes; thus, the resulting product concentration is affected. This inhibition is analytically useful and has been used advantageously in the development of many biosensing devices. The monitoring of inhibitor or activator of the enzyme is considered the indirect mode.

Biosensors based on the principle of enzyme inhibition have by now been applied for a wide range of significant analytes such as organophosphorous pesticide (OP), organochlorine pesticides, derivatives of insecticides, and heavy metals [8–13]. In general, the development of these biosensing systems relies on a quantitative measurement of the enzyme activity before and after exposure to a target analyte.

An important class of biorecognition elements is so-called whole-cell systems. Whole cells have been used since many times for environmental applications, like BOD monitoring. However, recently they have found enormous benefits from the recent improvement in recombinant DNA technology, and their use has been renewed in monitoring environmental pollution and toxicity [14, 15].

Antibodies are the most used affinity proteins for all life science applications. Antibody-based biosensors are also called immunosensors. Nowadays, antibodies have been generated which specifically bind to individual compounds or groups of structurally related compounds with a wide range of affinities. The main antibody formats currently available for use in immunosensors are polyclonal, monoclonal, and recombinant antibodies. However, the increasing experience in the field of combinatorial libraries and protein engineering has inspired researchers to develop new non-immunoglobulin affinity protein without the limitations of antibodies. Consequently, today antibodies are facing increasing competition from a large number so-called engineered protein scaffolds.

Immobilization of antibodies (or other ligands) directly on the surface of an electrode has been widely demonstrated. However, the use of the electrode surface as solid phase as well as electrochemical transducer presents some problems: a shielding of the surface by biospecifically bound antibody molecules can cause hindrance of the electron transfer, resulting in a reduced electrochemical signal. An interesting approach to increase the sensitivity involves the use of electrodes for the transduction step, whereas the affinity reaction is performed using a different support, as, for example, magnetic beads [15–20]. Obviously the use of magnetic beads is not limited to immunsensors, but other kinds of bioreceptors (like nucleic acid etc.) have been successfully coupled to magnetic beads.

As already described, NA have been incorporated into a wide range of biosensors and bioanalytical assays, due to their wide range of physical, chemical, and biological activities. As it is commonly known, NA molecules have the function of carrying and passing genetic information. This can be exploited, from an analytical point of view, for a specific identification of animal and vegetal species, genetic modified organisms, bacteria, virus, toxins etc. Genosensors, in particular, result from the integration of a sequence-specific probe (usually a short synthetic oligonucleotide) and a signal transducer. The probe, immobilized onto the transducer surface, acts as the biorecognition molecule and recognizes the target DNA or RNA via hybridization reaction. However, about 20 years ago, NA began to find a new role in the field of materials science and biotechnology. Aptamers, for instance, are single-stranded DNA or RNA ligands which can be selected for different targets starting from a huge library of molecules containing randomly created sequences. Moreover, in the recent years, NA has also been used for creating system able of catalytic activity and in particular DNA (due to the fact that, compared with RNA, DNA molecules are less susceptible to hydrolysis and thus are highly stable). The term "nucleic acid enzyme" is used to identify these nucleic acid structures that have catalytic activity. MIPs (Molecular Imprinted Polymers) have been also coupled to electrochemical transducers.

Amperometric and Voltammetric Biosensors

Amperometric and voltammetric biosensors rely on an electrochemically active analyte that can be oxidized or reduced at a working electrode. Typical electrode materials are platinum (Pt), gold (Au), and carbon. Nowadays some innovative techniques for electrode preparation, characterized by the possibility of mass production and high reproducibility, have been proposed. Among these, the equipment needed for thick-film technology is less complex and costly, and thus, this is one of the most used for sensor production. Thick-film technology consists of depositing inks on a substrate in a film of controlled pattern and thickness, mainly by screen printing. One of the main reasons to use screen-printed sensors as electrochemical transducers in biosensor technology is the possibility of making them disposable; this characteristic arises from the low cost and the mass production of these systems. In electrochemistry a disposable sensor offers the advantage of not suffering from the electrode fouling that can result in loss of sensitivity and reproducibility. The avoidance of contamination among samples is another important advantage of using disposable sensors. The micro-dimensions of these devices are important to satisfy the needs of decentralized testing. Finally, the high degree of reproducibility that is possible for these one-time use electrodes eliminates the cumbersome requirement for repeated calibration.

Amperometric or voltammetric biosensors typically rely on an enzyme system that catalytically converts electrochemically non-active analytes into products that can be oxidized or reduced at a working electrode. Although these devices are the most commonly reported class of biosensors, they tend to have a small dynamic range due to saturation kinetics of the enzyme, and a large overpotential is required for oxidation of the analyte; this may lead to oxidation of interfering compounds as well (e.g., ascorbate in the detection of hydrogen peroxide). In addition to the use in enzyme-based biosensors, amperometric transducers have also been used to measure enzyme-labelled tracers for affinity-based biosensor (mainly immunosensors and genosensors). Enzymes which are commonly used for this purpose include horseradish peroxidase (HRP) [17] and alkaline phosphatase (AP) [18, 19, 21].

Potentiometric Biosensors

Besides amperometry Guilbault and Montalvo in 1969 [22] use glass electrodes coupled with urease to measure urea concentration by potentiometric measurement. pH change or ion concentration monitoring could be possible by using potentiometric transducers. The main disadvantage is that such transducers may reach only mediocre limit of detection even if improvements have been recently reported in literature.

Impedimetric Biosensor

Electrochemical Impedance Spectroscopy (EIS) is a powerful technique for the characterization of electrochemical systems. The fundamental approach of all impedance methods is to apply a small-amplitude sinusoidal excitation signal to the system under investigation and measure the response (current or voltage or another signal of interest). An advantage of EIS compared to amperometry or potentiometry is that labels are no longer necessary, thus simplifying sensor preparation. However, the use of labels (like enzymes or nanoparticles) increases a lot the sensitivity of the method [23–26].

Electrochemical Scanning Techniques

These last years have seen increasing interest in scanning probe microscopies for biosensing. Indeed, scanning probe techniques offer the advantage of high lateral resolution and thus allow the versatile approach of hybridization detection from high-density biochip characterization to single biological event detection [27, 28]

Scanning electrochemical microscopy (SECM) sustains great interest for biomolecular recognition detection [29, 30]. This comes from the versatility of SECM methodology that offers versatile detection principles, e.g., positive or negative feedback modes together with collection mode that are compatible with unlabelled hybridization detection as well as with redox amplification strategies of DNA hybridization [31, 32].

Future Directions

Electrochemical biosensors have some advantages over other analytical transducing systems, such as the possibility to operate in turbid media, comparable instrumental sensitivity, and possibility of miniaturization. As a consequence of miniaturization, small sample volume can be required. Modern electroanalytical techniques (square-wave voltammetry, chronopotentiometry, chronoamperometry, differential pulse voltammetry) have very low detection limit. In situ or on-line measurements are both allowed. Furthermore the equipment required for electrochemical analysis are simple and cheap compared to most other analytical techniques. Nevertheless, nanotechnology will further help versatility of electrochemical biosensors. Nanomaterials are claimed to improve electrochemical transducer sensitivity. Functional DNA molecules, DNA origami, engineered proteins are examples of smart biomolecules that can greatly influence specificity and sensibility of biosensors. Thus, in the near future, biosensor technology will undoubtedly benefit from innovative areas such as nanotechnology and biotechnology.

Cross-References

► Biofilms, Electroactive
► Cell Membranes, Biological
► Electrochemical Impedance Spectroscopy (EIS) Applications to Sensors and Diagnostics
► Scanning Electrochemical Microscopy (SECM)

References

1. Thevenot DR, Toth K, Durst RA et al (2001) Electrochemical biosensors: recommended definitions and classification. Biosens Bioelectron 16:121–131
2. Newman JD, Tigwell LJ, Turner APF et al (2004) Biosensors – a clearer view. Cranfield University Publication, UK
3. Labuda J, Oliveira Brett AM, Evtugyn G, Fojta M, Mascini M, Ozsoz M, Palchetti I, Paleček E, Wang J (2010) Electrochemical nucleic acid-based biosensors: concepts, terms, and methodology (IUPAC technical report). Pure Appl Chem 82:1161–1187
4. Clark LC (1956) Monitor and control of blood and tissue oxygen tensions. Trans Am Soc Artif Intern Organs 2:41–48
5. Clark LC, Lyons C (1962) Electrode systems for continuous monitoring cardiovascular surgery. Ann N Y Acad Sci 102:29–45
6. Updike SJ, Hicks GP (1967) The enzyme electrode. Nature 214:986–988
7. Eggins BR (2002) Chemical sensors and biosensors. Wiley, Chichester
8. Cagnini A, Palchetti I, Lionti I, Mascini M, Turner APF (1995) Disposable ruthenized screen-printed biosensors for pesticides monitoring. Sens Actuators B 24–25:85–89
9. Palchetti I, Cagnini A, Del Carlo M, Coppi C, Mascini M, Turner APF (1997) Determination of anticholinesterase pesticides in real samples using a disposable biosensor. Anal Chim Acta 337:315–321
10. Hernandez S, Palchetti I, Mascini M (2000) Determination on anticholinesterase activity for pesticides monitoring using acetylthiocholine sensor. Int J Environ Anal Chem 78:3–4
11. Mascini M, Palchetti I (2001) Electrochemical biosensor for evaluation of contaminants in food for quality improvement. Arch Ind Hyg Toxicol 52:49–59
12. Cagnini A, Palchetti I, Mascini M, Turner APF (1995) Ruthenised screen-printed choline oxidase-based biosensors for measurement of anticholinesterase activity. Mikrochim Acta 121:155–166
13. Laschi S, Ogończyk D, Palchetti I, Mascini M (2007) Evaluation of pesticide-induced acetylcholinesterase inhibition by means of disposable carbon-modified electrochemical biosensors. Enzyme Microb Technol 40:485–489
14. Rogers KR (2006) Recent advances in biosensor techniques for environmental monitoring. Anal Chim Acta 568:222
15. Mascini M, Palchetti I (eds) (2011) Nucleic acid biosensors for environmental pollution monitoring. RSC Publishing, London. ISBN 978-1-84973-131-7
16. Laschi S, Miranda-Castro R, González-Fernández E, Palchetti I, Reymond F, Rossier JS, Marrazza G (2010) A new gravity-driven microfluidic-based electrochemical assay coupled to magnetic beads for nucleic acid detection. Electrophoresis 31:1–10
17. Centi S, Bonel Sanmartin L, Tombelli S, Palchetti I, Mascini M (2009) Detection of C reactive protein (CRP) in serum by an electrochemical aptamer-based sandwich assay. Electroanalysis 21:1309–1315
18. Laschi S, Palchetti I, Marrazza G, Mascini M (2009) Enzyme-amplified electrochemical hybridization assay based on PNA, LNA and DNA probe-modified micro-magnetic beads. Bioelectrochemistry 76:214–220
19. Berti F, Laschi S, Palchetti I, Rossier J, Reymond F, Mascini M, Marrazza G (2009) Microfluidic-based electrochemical genosensor coupled to magnetic beads for hybridization detection. Talanta 77:971–978
20. Centi S, Messina G, Tombelli S, Palchetti I, Mascini M (2008) Different approaches for the detection of thrombin by an electrochemical aptamer-based assay coupled to magnetic beads. Biosens Bioelectron 23:1602–1609
21. Centi S, Silva E, Laschi S, Palchetti I, Mascini M (2007) Polychlorinated biphenyls (PCBs) detection in milk samples by an electrochemical magneto-immunosensor (EMI) coupled to solid phase extraction (SPE) and disposable low density arrays. Anal Chim Acta 594:9–16

22. Bettazzi F, Lucarelli F, Palchetti I, Berti F, Marrazza G, Mascini M (2008) Disposable electrochemical DNA-array for PCR amplified detection of hazelnut allergens in foodstuff. Anal Chim Acta 614:93–102
23. Guilbault G, Montalvo J (1969) A urea specific enzyme electrode. J Am Chem Soc 9:2164–2169
24. Katz E, Willner I (2003) Probing biomolecular interactions at conductive and semiconductive surfaces by impedance spectroscopy: routes to impedimetric immunosensors, DNA-sensors, and enzyme biosensors. Electroanalysis 15:913–947
25. Daniels JS, Pourmand N (2007) Label-free impedance biosensors: opportunities and challenges. Electroanalysis 19:1239–1257
26. Katz E, Willner I, Wang J (2004) Electroanalytical and bioelectroanalytical systems based on metal and semiconductor nanoparticles. Electroanalysis 16:119–144
27. Palchetti I, Mascini M (2012) Electrochemical nanomaterial-based nucleic acid aptasensors. Anal Bioanal Chem 402:3103–3114. doi:10.1007/s00216-012-5769-1
28. Batchelor-McAuley C, Dickinson EJF, Rees NV, Toghill KE, Compton RG (2012) Anal Chem 84:669–684. doi:org/10.1021/ac2026767
29. Sinensky AK, Belcher AM (2007) Label-free and high-resolution protein/DNA nanoarray analysis using Kelvin probe force microscopy. Nat Nanotechnol 2:653–659. doi:10.1038/nnano.2007.293
30. Bard AJ, Mirkin MV (2001) Scanning electrochemical microscopy. Marcel Dekker, New York
31. Turcu F, Schulte A, Hartwich G, Schuhmann W (2004) Label-free electrochemical recognition of DNA hybridization by means of modulation of the feedback current in SECM. Angew Chem Int Ed 43:3482–3485
32. Palchetti I, Laschi S, Marrazza G, Mascini M (2007) Electrochemical imaging of localized sandwich DNA hybridization using scanning electrochemical microscopy. Anal Chem 79:7206–7213

Biphasic Electrolytic System

Kazuhiro Chiba and Yohei Okada
Tokyo University of Agriculture and Technology, Fuchu, Tokyo, Japan

Definition

An electrochemical reaction carried out in a biphasic solution is known as a "biphasic electrolytic system". Typically, a biphasic solution consists of a less polar organic solvent and a polar electrolyte solution, where the electron transfer process occurs. In this system, the polar electrolyte phase is spatially separated from the less polar organic phase, enabling a facile separation process and continuous synthesis.

Introduction

From the synthetic aspect, electrochemical approaches have been employed to regulate chemical transformations in a heterogeneous manner [1–3]. Several reactive intermediates, including ions, radicals, and radical ions,are generated byelectron transfer at the electrodes permittingnot only various functional group transformations, but also a wide variety of carbon–carbon bond formation reactions. In these approaches, the electrodes function as both oxidizing and reducing reagents to avoid consumption of additional reagents; thus, they are promising methodologies from an environmental viewpoint. On the other hand, the use of a large amount of supporting electrolyte is essential to impart electrical conductivity to polar organic solvents. However, separation and disposal of the organic solvent are necessary. Consequently, the development of more practical systems for electrochemical transformations is important to create greenerelectrochemical reactions. In this context, elegant separation techniques assisted by biphasic systems, including thermomorphic [4, 5], fluorous [6–10], and ionic [11–14] systems,have been developedas key roles in modern synthetic chemistry. On the basis of these achievements, electrochemical reactions can also be carried out in biphasic solutions, enabling a facile separation process and continuous synthesis, namely known as a "biphasic electrolytic system".

Examples and Applications

Typically, electrolyte solutions are composed of supporting electrolytes and polar organic solvents, such as methanol, acetonitrile, or

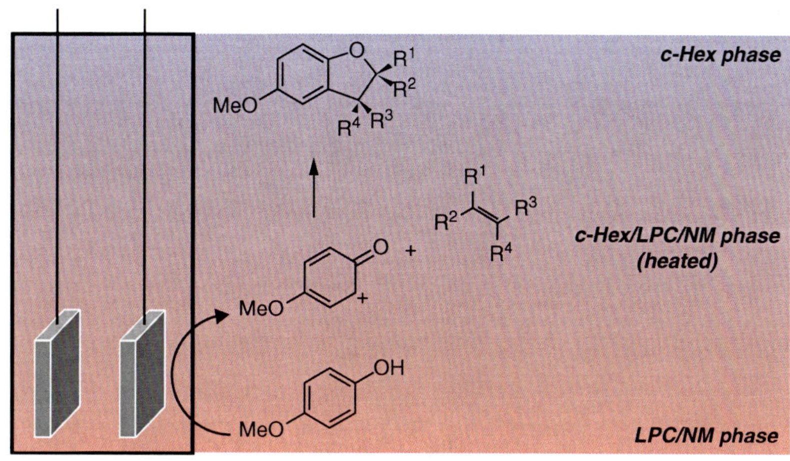

Biphasic Electrolytic System, Scheme 1 Electrochemical [3 + 2] cycloaddition reactions in thermomorphic biphasic cyclohexane (*c*-Hex)/ lithium perchlorate (LPC)/ nitromethane (NM) solution

nitromethane. Therefore, less polar organic solvents are expected to form biphasic solutions in combination with several polar electrolyte solutions. In particular, it has been demonstrated that cyclohexanescan formthermomorphic biphasic solutions with typical polar organic solvents and their reversible separation and mixing can be regulated at a moderatetemperature range [15–17]. For example, a 1:4 (v/v) mixture of cyclohexane and nitromethane shows biphasic condition at 25 °C, which form monophasic condition at *ca.* 60 °C and higher temperatures. In this system, the thermomorphic "biphasic" solution ismixed to serve as an effective homogeneous reaction field. After completion of the reaction, the thermomorphic "monophasic" solution isseparated, forminga heterogeneous mixture, and the hydrophobic products or designed hydrophobic platforms can be recovered from the cyclohexane phase.

Various electrochemical [3 + 2] cycloaddition reactions havebeen achieved between alkoxyphenols and olefin nucleophiles in lithium perchlorate/nitromethane electrolyte solution [18]. In these reactions, alkoxyphenols are anodically oxidized to generate the corresponding phenoxonium cations, which are then trapped by olefin nucleophiles to form the desired [3 + 2] cycloadducts. However, because the oxidation potentials of these [3 + 2] compounds are relatively lower than those of starting alkoxyphenols, overoxidation of the desired products can take place. To address this problem, cyclohexane is introduced into these reaction mixtures to formbiphasic electrolyte solutions [19]. There is no conductivity in the cyclohexane phase, clearly suggesting that the anodic oxidation takesplace only in the electrolyte phase. In addition, interface of the cyclohexane andlithium perchlorate/ nitromethane electrolyte solution ismixed at 60 °C to form a homogeneous area to enhance interactions between the anodically generated polar phenoxonium cations and the less polar olefin nucleophiles. The corresponding [3 + 2] cycloadducts are assembled into the cyclohexane phase because of their low polarities; thus, their overoxidation should be avoidable. After completion of the reaction, the desired [3 + 2] products can be recovered in high pure form from the cyclohexane phase through a simple liquid-liquid extraction (Scheme 1).

A wide variety of electrochemical [2 + 2] cycloaddition reactions have also been established in lithium perchlorate/nitromethane electrolyte solution [20–25]. In these reactions, enol ethers are anodically oxidized to generate the corresponding radical cations, which are then trapped by olefin nucleophiles to form the desired [2 + 2] cycloadducts. Cyclohexane is also addedto these [2 + 2] cycloaddition reactions to formbiphasic electrolyte solutions [26]. In these cases, the desired [2 + 2] cycloadducts are efficiently obtained even ina biphasic condition at 25 °C and cyclohexane plays dual roles as the substrate supply and the product assembly phase. Thus, less polar starting enol ethers and olefin

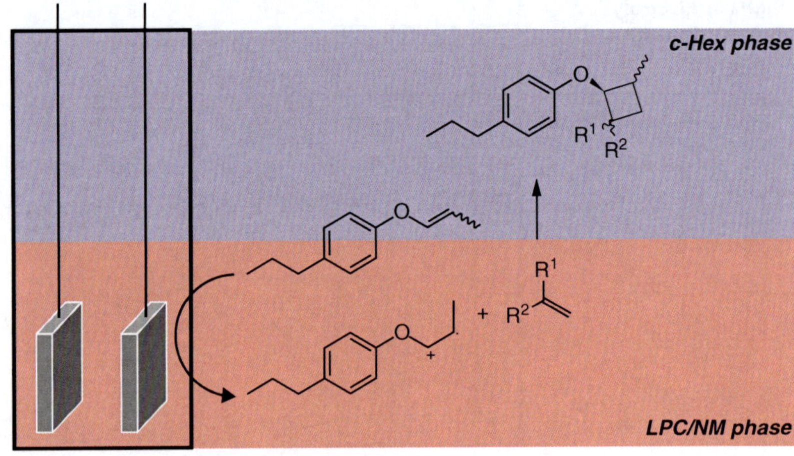

Biphasic Electrolytic System, Scheme 2 Electrochemical [2 + 2] cycloaddition reactions in biphasic cyclohexane (*c*-Hex)/lithium perchlorate (LPC)/nitormethane (NM) solution

nucleophiles are supplied from the cyclohexane phase to the electrolyte phase, where electrochemical [2 + 2] cycloaddition reactions are induced at the electrodes. The less polar resulting [2 + 2] cycloadducts are then assembled into the cyclohexane phase, which can be completely separated from the electrolyte solution, enabling the effective recycling of electrolyte solution. Based on thisbiphasic electrolytic system, afacile separation process and continuous synthesis are realized, avoiding the use of a large amount of supporting electrolytes (Scheme 2).

Future Directions

From the environmental aspect, electrochemical transformations are promising because a wide variety of chemical transformations can be regulated under mild reaction conditions without the use of additional reagents. As described in this chapter, biphasic electrolytic systems would decrease the use of electrolyte solutions and enable simple separation processes, leading to industrial applications. Further investigation oneffective thermomorphic biphasic electrolytic solutions would extend the reaction scope of these systems and combinationswith reaction devices such as microflow platforms would be great aidsfor the development of practical electrochemical synthetic systems.

Cross-References

▶ Electrochemical Microflow Systems
▶ Electrosynthesis in Ionic Liquid
▶ Electrosynthesis in Supercritical Fluids
▶ Electrosynthesis Under Ultrasound and Centrifugal Fields
▶ Electrosynthesis Using Water Suspension System

References

1. Yoshida J, Kataoka K, Horcajada R, Nagaki A (2008) Modern strategies in electroorganic synthesis. Chem Rev 108:2265–2299
2. Sperry JB, Wright DL (2006) The application of cathodic reductions and anodic oxidations in the synthesis of complex molecules. Chem Soc Rev 35:605–621
3. Moeller KD (2000) Synthetic applications of anodic electrochemistry. Tetrahedron 65:9527–9554
4. Bergbreiter DE, Tian J, Hongfa C (2009) Using soluble polymer supports to facilitate homogeneous catalysis. Chem Rev 109:530–582
5. Bergbreiter DE (2002) Using soluble polymers to recover catalysts and ligands. Chem Rev 102:3345–3384
6. Zhang W (2009) Fluorous linker-facilitated chemical synthesis. Chem Rev 109:749–795
7. Zhang W (2004) Fluorous synthesis of heterocyclic systems. Chem Rev 104:2531–2556
8. de Wolf E, van Koten G, Deelman B-J (1999) Fluorous phase separation techniques in catalysis. Chem Soc Rev 28:37–41

9. Studer A, Hadida S, Ferritto R, Kim S-Y, Jeger P, Wipf P, Curran DP (1997) Fluorous synthesis: a fluorous-phase strategy for improving separation efficiency in organic synthesis. Science 275:823–826
10. Horvath IT, Rabai J (1994) Facile catalyst separation without water: fluorous biphase hydroformylation of olefins. Science 266:72–75
11. Zakrzewska ME, Bogel-Łukasik E, Bogel-Łukasik R (2011) Ionic liquid-mediated formation of 5-hydroxymethylfurfural – a promising biomass-derived building block. Chem Rev 111:397–417
12. Isambert N, Sanchez Duque MM, Plaquevent J-C, Génisson Y, Rodriguez J, Constantieux T (2011) Multicomponent reactions and ionic liquids: a perfect synergy for eco-compatible heterocyclic synthesis. Chem Soc Rev 40:1347–1357
13. Śledź P, Mauduit M, Grela K (2008) Olefin metathesis in ionic liquids. Chem Soc Rev 37:2433–2442
14. Dupont J, de Souza RF, Paulo AZ, Suarez PAZ (2002) Ionic liquid (molten salt) phase organometallic catalysis. Chem Rev 102:3667–3692
15. Kim S, Tsuruyama A, Ohmori A, Chiba K (2008) Solution-phase oligosaccharide synthesis in a cycloalkane-based thermomorphic system. Chem Commun 15:1816–1818
16. Hayashi K, Kim S, Kono Y, Tamura M, Chiba K (2006) Microwave-promoted Suzuki–Miyaura coupling reactions in a cycloalkane-based thermomorphic biphasic system. Tetrahedron Lett 47:171–174
17. Chiba K, Kono Y, Kim S, Nishimoto K, Kitano Y, Tada M (2002) A liquid-phase peptide synthesis in cyclohexane-based biphasic thermomorphic systems. Chem Commun 16:1766–1767
18. Chiba K, Fukuda M, Kim S, Kitano Y, Tada M (1999) Dihydrobenzofuran synthesis by an anodic [3 + 2] cycloaddition of phenols and unactivated alkenes. J Org Chem 64:7654–7656
19. Kim S, Noda S, Hayashi K, Chiba K (2008) An oxidative carbon–carbon bond formation system in cycloalkane-based thermomorphic multiphase solution. Org Lett 10:1827–1830
20. Okada Y, Nishimoto A, Akaba R, Chiba K (2011) Electron-transfer-induced intermolecular [2 + 2] cycloaddition reactions based on the aromatic "redox tag" strategy. J Org Chem 76:3470–3476
21. Okada Y, Chiba K (2011) Electron transfer-induced four-membered cyclic intermediate formation: olefin cross-coupling vs.olefin cross-metathesis. Electrochim Acta 56:1037–1042
22. Okada Y, Akaba R, Chiba K (2009) EC-backward-E electrochemistry supported by an alkoxyphenyl group. Tetrahedron Lett 50:5413–5416
23. Okada Y, Akaba R, Chiba K (2009) Electrocatalytic formal [2 + 2] cycloaddition reactions between anodically activated aliphatic enol ethers and unactivated olefins possessing an alkoxyphenyl group. Org Lett 11:1033–1035
24. Arata M, Miura T, Chiba K (2007) Electrocatalytic formal [2 + 2] cycloaddition reactions between anodically activated enyloxy benzene and alkenes. Org Lett 9:4347–4350
25. Chiba K, Miura T, Kim S, Kitano Y, Tada M (2001) Electrocatalytic intermolecular olefin cross-coupling by anodically induced formal [2 + 2] cycloaddition between enol ethers and alkenes. J Am Chem Soc 123:11314–11315
26. Okada Y, Chiba K (2010) Continuous electrochemical synthetic system using a multiphase electrolyte solution. Electrochim Acta 55:4112–4119

Boron-Doped Diamond for Green Electro-Organic Synthesis

Siegfried R. Waldvogel
Johannes Gutenberg-University Mainz, Mainz, Germany

Introduction

Since only electrons serve as reagents in electrochemical transformations, almost no reagent waste is produced. The outstanding atom economy is often combined with pronounced energy efficiency, and consequently, such electrochemical processes are considered as green chemistry [1]. Therefore, electrosynthesis is of outstanding technical significance, and novel electrosynthetic transformations will be the basis for technical innovations and future applications. Boron-doped diamond (BDD) electrodes provide an exceptionally wide electrochemical window in protic media [2]. The offset potentials for the evolution of molecular hydrogen and oxygen are extraordinarily large [3, 4]. Currently, the anodic regime is almost exclusively exploited for electro-organic synthesis, which will be surveyed within this entry.

BDD anodes allow the formation of OH radicals at potentials well below the onset of oxygen evolution. Such anodic systems can be used for new oxidation reactions which are otherwise not feasible in water [5]. In electrolytes containing methanol, methoxyl radicals were discussed as reactive intermediates (Scheme 1) [6]. The formation of OH spin centers have been proved by

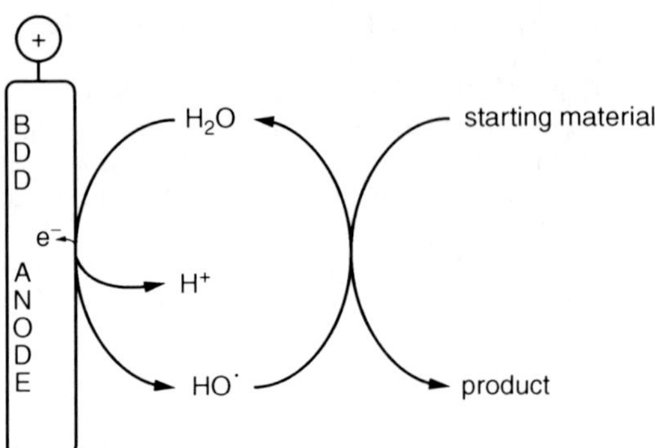

Boron-Doped Diamond for Green Electro-Organic Synthesis, Scheme 1 Working mode of boron-doped diamond anodes

spin traps [7]. The extreme reactivity of the formed radical intermediates causes the destructive performance of BDD anodes, including disinfection, detoxification, and wastewater treatment. Unfortunately, the formation of hydroxyl radicals at highly positive potentials also leads to a complete incineration of organic materials via complex degradation sequences [8, 9]. The control of the reactivity is the most important challenge in this field and can be most practically addressed by a high concentration of substrate and/or partial conversion [10, 11].

Anodic Phenol Coupling

The selective oxidative phenolic *ortho*-coupling reaction of simple methyl-substituted phenols turned out to be challenging [12]. When 2,4-dimethylphenol (1) is treated by conventional or electro-organic methods, not only the desired biphenol (2) is formed but rather a plethora of polycyclic architectures (Scheme 2) is observed. The major product is Pummerer's ketone (3) and related compounds with a wide structural diversity [13–16]. Application of a boron tether ameliorated the situation tremendously, and biphenol (2) was obtained as the major product [17, 18]. This templated anodic oxidation of 1 represents a multistep process but is suitable for the electro-organic synthesis of (2) on larger scale (see entry "▶ Electrosynthesis Using Template-Directed Methods") [19].

When (1) is anodically converted employing the so-called Comninellis conditions (dilute methanol solution of substrate), only traces of the product are found. The organic substrate is mostly incinerated. Using high concentrations of the phenol derivative provided the desired biphenol (2) in a chemoselectivity of about 90–95 %. The isolated yield of (2) can be up to 56 %. The electrolysis was carried out in almost neat phenol with supporting electrolyte and 11 % of water as an additive. With this strategy, the over-oxidation at BDD anodes could be circumvented. The chemical incineration is dramatically reduced when the electrolysis is performed to a partial conversion of about 30 %. The resulting biphenol 2 is very clean, and the starting material (1) is efficiently recovered [19, 20]. This protocol was restricted to 2,4-dimethylphenol (1) and some related compounds. Since the mass transport from the bulk to the vicinity of the BDD electrode seems to be crucial for avoiding the mineralization, the critical parameter seems to be the low viscosity of the employed phenols. Therefore, more redox-stable and protic additives such as highly fluorinated alcohols can successfully be employed. Best results were found with 1,1,1,3,3,3-hexafluoroisopropanol. A wide scope of phenols was anodically coupled, and the conversion of (1) to the biphenol (2) was achieved with a 47 % yield [21]. The beneficial role of fluorinated alcohols at BDD anodes is treated in more detail in chapter ▶ Electrosynthesis Using Diamond Electrode.

Alkoxylation Reactions

The direct alkoxylation of hydrocarbons allows the installation of functionalities without using the detour via halogenations. Consequently, this represents a very important technical field. The anodic synthesis of benzaldehyde dimethyl ketales is industrially relevant and performed on a several thousand ton scale [22, 23]. The direct anodic methoxylation of benzylic or allylic moieties can be performed at BDD anodes. The results obtained on BDD electrodes are quite similar to the ones when graphite serves

Boron-Doped Diamond for Green Electro-Organic Synthesis, Scheme 2 Direct anodic coupling of 2,4-diemthylphenol

Boron-Doped Diamond for Green Electro-Organic Synthesis, Scheme 3 Methoxylation reaction of technically relevant benzylic substrate

Boron-Doped Diamond for Green Electro-Organic Synthesis, Scheme 4 Dimethoxylation reaction of furane

Boron-Doped Diamond for Green Electro-Organic Synthesis, Scheme 5 Alternative route to trimethoxy orthoformiate

as the anode [24]. Detailed studies of the anodic methoxylation of 4-*tert*-butyltoluene (4) reveal that several intermediates are formed which are consecutively degraded to (5) (Scheme 3) [25]. At BDD electrodes, the enlarged electrochemical window seems to be beneficial [25].

The electrochemical methoxylation of furan derivatives represents another technically relevant process. Anodic alkoxylation of furan (6) in an undivided cell provides 2,5-dimethoxy-2,5-dihydrofuran (7), which serves as a building block for the synthesis of nitrogen-containing heterocycles. The current industrial electro-organic processes employ graphite electrodes and sodium bromide, which acts both as supporting electrolyte and mediator [22]. Importantly, this electrolysis of (6) can be carried out at BDD anodes, but no mediator is required (Scheme 4)! The conversion is performed with 8 % furan in MeOH and Bu_4NBF_4 as the supporting electrolyte. After application of 1.5 F, (7) is obtained in 75 % yield with excellent current efficiency [26, 27]. The absence of bromine as a mediating reagent indicates that furan is anodically oxidized in the initial step, and subsequently, methanol enters the scene.

Trimethyl orthoformate (9) represents a formic acid equivalent which is commonly applied in organic condensation reactions. The anodic methoxylation of the inexpensive formaldehyde dimethyl acetal (8)–(9) is an interesting alternative, since excessive salt waste or operation with cyanhydric acid can be avoided. The electrolysis can be conducted at BDD anodes in undivided cells. Product (9) is obtained in 75 % selectivity with a partial conversion of 27 % (8).

Boron-Doped Diamond for Green Electro-Organic Synthesis, Scheme 6 Direct incorporation of carbon dioxide into an aminoacid precursor

The electrolyte consists of 24 % MeOH, 70 % (8), and $LiN(SO_2CF_3)_2/NaOMe$ (Scheme 5) [28]. Most remarkably, the transformation cannot be performed on graphite electrodes.

Boron-Doped Diamond for Green Electro-Organic Synthesis, Scheme 7 Dominant reaction when using zinc anodes

Cathodic Carboxylation Reactions

Electric current is by far the least expensive reduction equivalent and will be abundantly available because of photovoltaic or wind power. Many anodic conversions are well studied, whereas electrochemical reduction is still less elaborated. BDD cathodes may play a key role in this aspect since they exhibit a high overpotential for the evolution of hydrogen. Furthermore, highly toxic heavy metals and eventually formed organometallic species thereof can be excluded. This paves the way for the application in greener processes. The installation of C1 building blocks into chemicals using CO_2 seems to be very attractive. Based on ecological considerations, the electrochemical fixation of CO_2 is environmentally benign. Carbon dioxide is a good electron acceptor which can be subjected to cathodic conversions [29]. 2-Hydroxy-4-methylsulfanylbutyric acid (11) represents an important technical product used on a large scale for animal feeding. Methylsulfanylpropionaldehyde (10) is a readily accessible starting material which can be cathodically carboxylated when using magnesium as a sacrificial anode [30, 31]. The conversion can also be conducted at BDD cathodes in a divided cell equipped with a Nafion™ separator (Scheme 6). Although the yield for (11) in this process is lower than on magnesium electrodes (conversion 66 %, current efficiency 22 %), the use of sacrificial anodes can be avoided [32]. Unfortunately, reduction of the aldehyde (10) to alcohol (12) is a significant side reaction.

The outstanding electron acceptor property of carbon dioxide can be used for the reductive coupling to oxalic acid ((13), Scheme 7). Electrolysis is conducted in DMF in an undivided cell using zinc as a sacrificial anode. Unfortunately, only a few details are reported about this particular process [33]. A current density of 6 mA/cm^2 with 12 mM NBu_4BF_4 as a supporting electrolyte was employed and provided a current efficiency of 60 %.

Future Directions

The use of boron-doped diamond electrodes in nondestructive and synthetic applications in preparative organic chemistry has just started. The reported examples are not yet on a technical production level. But the clear advantages indicate successful applications in the near future. Remaining challenges for the technical use of BDD electrodes are the corrosion of the support material of BDD anodes when working in nonaqueous media on one hand [2] and the competing electrochemical incineration due to the highly reactive intermediates on the other hand. In a few cases, the specific reactivity of BDD anodes might be obtained with graphite electrodes when the appropriate electrolyte system is developed [34].

Cross-References

▶ Electrochemical Functional Transformation
▶ Electrosynthesis Using Diamond Electrode
▶ Electrosynthesis Using Template-Directed Methods
▶ Organic Reactions and Synthesis

References

1. Steckhan E, Arns T, Heineman WR et al (2001) Environmental protection and economization of resources by electroorganic and electroenzymatic syntheses. Chemosphere 43(1):63–73
2. Waldvogel SR, Mentizi S, Kirste A (2012) Boron-doped diamond electrodes for electroorganic chemistry. Top Curr Chem 320:1–31
3. Brillas E, Martínez-Huitle CA (eds) (2011) Synthetic diamond films – preparation, electrochemistry, characterization and applications. Wiley-VCH, Hoboken, New Jersey
4. Francke R, Cericola D, Kötz R et al (2012) Novel electrolytes for electrochemical double layer capacitors based on 1,1,1,3,3,3-hexafluoropropan-2-ol. Electrochim Acta 62:372–380
5. Iniesta J, Michaud PA, Panizza M et al (2001) Electrochemical oxidation of 3-methylpyridine at a boron-doped diamond electrode: application to electroorganic synthesis and wastewater treatment. Electrochem Commun 3(7):346–351
6. Zollinger D, Griesbach U, Pütter H et al (2004) Methoxylation of p-tert-butyltoluene on boron-doped diamond electrodes. Electrochem Commun 6(6):600–604
7. Marselli B, Garcia-Gomez J, Michaud PA et al (2003) Electrogeneration of hydroxyl radicals on boron-doped diamond electrodes. J Electrochem Soc 150(3):D79–D83
8. Rodrigo MA, Michaud PA, Duo I et al (2001) Oxidation of 4-chlorophenol at boron-doped diamond electrode for wastewater treatment. J Electrochem Soc 148(5):D60–D64
9. Panizza M, Michaud PA, Cerisola G et al (2001) Anodic oxidation of 2-naphthol at boron-doped diamond electrodes. J Electroanaly Chem 507(1–2, Sp. Iss. SI):206–214
10. Waldvogel S, Elsler B (2012) Electrochemical synthesis on boron-doped diamond. Electrochim Acta. doi:10.1016/j.electacta.2012.03.173 (in press)
11. Waldvogel SR, Mentizi S, Kirste A (2011) Use of diamond films in organic electrosynthesis. In: Brillas E, Martínez-Huitle CA (eds) Synthetic diamond films – preparation, electrochemistry, characterization and applications. Wiley-VCH, Hoboken, New Jersey, pp 483–510
12. Waldvogel SR (2010) Novel anodic concepts for the selective phenol coupling reaction. Pure Appl Chem 82(4):1055–1063
13. Malkowsky IM, Rommel CE, Wedeking K et al (2006) Facile and highly diastereoselective formation of a novel pentacyclic scaffold by direct anodic oxidation of 2,4-dimethylphenol. Eur J Org Chem (1):241–245
14. Barjau J, Koenigs P, Kataeva O et al (2008) Reinvestigation of highly diastereoselective pentacyclic spirolactone formation by direct anodic oxidation of 2,4-dimethylphenol. Synlett 15:2309–2312
15. Barjau J, Schnakenburg G, Waldvogel SR (2011) Diversity-oriented synthesis of polycyclic scaffolds by modification of an anodic product derived from 2,4-dimethylphenol. Angew Chem Int Ed 50(6):1415–1419
16. Barjau J, Fleischhauer J, Schnakenburg G et al (2011) Installation of amine moieties into a polycyclic anodic product derived from 2,4-dimethylphenol. Chem Eur J 17(52):14785–14791
17. Malkowsky IM, Rommel CE, Fröhlich R et al (2006) Novel template-directed anodic phenol-coupling reaction. Chem Eur J 12(28):7482–7488
18. Malkowsky IM, Fröhlich R, Griesbach U et al (2006) Facile and reliable synthesis of tetraphenoxyborates and their properties. Eur J Inorg Chem (8):1690–1697
19. Griesbach U, Pütter H, Waldvogel SR, Malkowsky I (2006) Anodic electrolytic oxidative electrodimerisation of hydroxy-substituted aromatics to give dihydroxy-substituted biarylene compounds (WO 2006077204 A2), BASF AG, Ludwigshafen (Germany)
20. Malkowsky IM, Griesbach U, Pütter H et al (2006) Unexpected highly chemselective anodic ortho-coupling reaction of 2,4-dimethylphenol on boron-doped diamond electrodes. Eur J Org Chem (20):4569–4572
21. Kirste A, Nieger M, Malkowsky IM et al (2009) Ortho-selective phenol-coupling reaction by anodic treatment on boron-doped diamond electrode using fluorinated alcohols. Chem Eur J 15(10):2273–2277
22. Degner D (1988) Organic electrosynthesis in industry. Top Curr Chem 148:1–9523
23. Pütter H (2001) In: Lund H, Hammerich O (eds) Organic electrochemistry, 4th edn. Marcel Dekker, New York, p 1259
24. Pütter H, Weiper-Idelmann A, Merk C, Fryda M, Klages C, Hampel A (2000) Diamantbeschichtete Elektroden (EP 1 036 861 B1), BASF AG, Ludwigshafen (Germany)

25. Zollinger D, Griesbach U, Pütter H et al (2004) Electrochemical cleavage of 1,2-diphenylethanes at boron-doped diamond electrodes. Electrochem Commun 6(6):605–608
26. Reufer C, Möbus K, Lehmann T, Weckbecker C (2004) Method fot the anodic alkoxylation of organic compounds (WO 2004 087999 A2), DEGUSSA AG, Hanau (Germany)
27. Reufer C, Lehmann T, Sanzenbacher R, Weckbecker C (2004) Method for the anodic alkoxylation of organic substrates (WO 2004 085710 A2), DEGUSSA AG, Hanau (Germany)
28. Fardel R, Griesbach U, Pütter H et al (2006) Electrosynthesis of trimethylorthoformate on BDD electrodes. J Appl Electrochem 36(2):249–253
29. Yoshida J, Kataoka K, Horcajada R et al (2008) Modern strategies in electroorganic synthesis. Chem Rev 108(7):2265–2299
30. Lehmann T, Schneider R, Weckbecker C, Dunach E, Olivero S (WO 02/16671 A1)
31. Lehmann T, Schneider R, Reufer C et al (2001) GDCh-Monographie, Frankfurt (23):251–258
32. Bilz J, Hateley M, Lehmann T, Reufer C, Sanzenbacher R, Weckbecker C (2006) (EP 1 63 1702), DEGUSSA AG, Hanau (Germany)
33. Lehmann T, Dunach E (2009) personal communication
34. Kirste A, Hayashi S, Schnakenburg G et al (2011) Highly selective electrosynthesis of biphenols on graphite electrodes in fluorinated media. Chem Eur J 17(50):14164–14169

C

Carbon-Zinc Batteries

Brooke Schumm
Eagle Cliffs, INC

A family of cells that have a zinc anode and a manganese dioxide cathode has three variations. They are Leclanché cells, zinc chloride cells, and alkaline cells. The first two are often called carbon-zinc cells and with these we are concerned here [5]. Both have a mild acid electrolyte. All zinc-manganese dioxide cells provide from 1.58 to 1.75 V on open circuit but differ in their current supporting capability and capacity. The first commercial success in this group was invented by Georges Leclanché in 1867. It is a notable survivor, though modified, from among many cells known in the last half of the nineteenth century. Because it used cheap, easy to obtain ingredients, Leclanché cells were from the beginning a strong commercial competitor. The anode, the negative electrode in this primary cell, is a zinc alloy bar, sheet, or cup, the alloy consisting of (in this day and age) zinc and some manganese. In the beginning the alloy was zinc with a small amount of lead, cadmium, and mercury. These elements provided stiffness and corrosion resistance. The electrolyte is a saturated aqueous solution of ammonium chloride and zinc chloride, sometimes containing corrosion-inhibiting salts or organic compounds. The cathode, the positive electrode, is carbon as graphite or carbon black mixed with manganese dioxide, which is the active ingredient. The carbon components conduct electrons to the manganese dioxide hence the name carbon-zinc cell. The cell was completed with a baked carbon current collector rod in the center of the cathode mix all in a glass jar. In the 1920s a conductive carbon containing plastic sheet on the side of the cathode was used as a current collector. The cathode is typically a cake containing the carbon, manganese dioxide powder, and enough electrolyte sufficient to create a moldable mass.

The anode and cathode are spaced apart by a separator which can be porous earthenware, porous plaster, a gelled paste, or a paste-coated paper. Leclanché's cell (Fig. 1) was a commercial success because the ingredients were inexpensive in large part because they were readily available and the manganese dioxide ore was dug from the ground and bought for a few cents per pound. In Georges Leclanché's time, a European natural manganese ore was used for the manganese dioxide content [2]. Later Russian ores, African ores, and many other smaller sources have been used. Now highly pure, but more costly, electrically or chemically deposited synthetic manganese dioxides are blended with ore for premium cells. The exact formula depends on the market and the expected use. Blends can provide up to twice the service of specially chosen but low-cost natural ores.

G. Kreysa et al. (eds.), *Encyclopedia of Applied Electrochemistry*, DOI 10.1007/978-1-4419-6996-5,

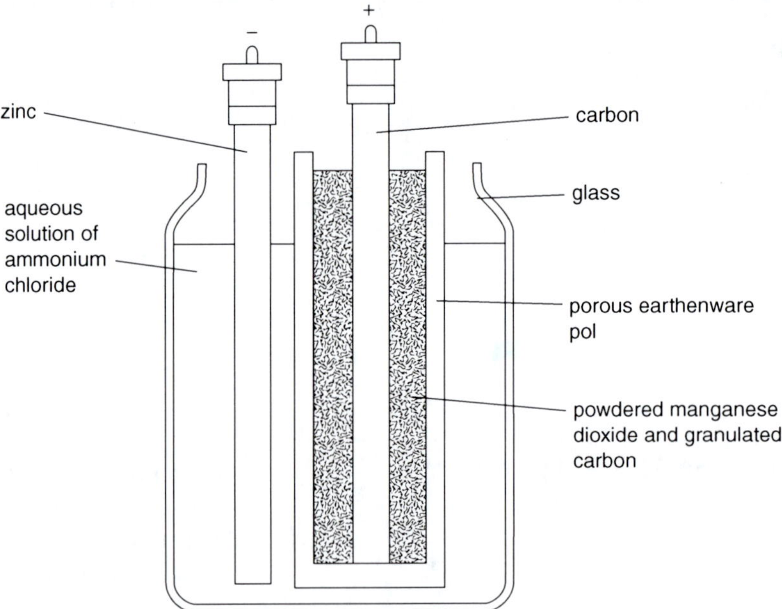

Carbon-Zinc Batteries, Fig. 1 George Leclanché's cell [1]

Despite its long history, the chemical reactions in Leclanché cells are more complex than in most common cells because two electrolyte salts are used. Ammonium chloride provides hydrogen ions to support the cathode reaction and combines with zinc ions to create a partially soluble zinc diammino-chloride which may precipitate out in needle-like crystals close to the zinc anode surface. The process proceeds as follows:

$$\text{Zinc} \rightarrow \text{Zn}^{++} + 2\,\text{e}^- \quad \text{(at the anode)} \qquad (1)$$

$$\text{Zn}^{++} + 2\text{NH}_4\text{Cl} \rightarrow \text{Zn(NH}_3)_2\text{Cl}_2 + 2\text{H}^+ \qquad (2)$$

$$2\text{H}^+ + 2\text{MnO}_2 + 2\text{e}^- \rightarrow 2\,\text{MnOOH or Mn}_2\text{O}_3.\text{H}_2\text{O} \text{ (in the cathode mixture)} \qquad (3)$$

Alternatively depending on the relative amount of each salt present, the cell reaction may proceed as follows:

$$2\text{Zn}^{++} + 2\text{H}_2\text{O} + 2\text{Cl}^- \rightarrow 2\text{Zn(OH)Cl} + 2\text{H}^+ \qquad (4)$$

$$2\text{Zn}^+ + 4\text{MnO}_2 \rightarrow 2\text{ZnO.Mn}_2\text{O}_3 - 2\text{e}^- \qquad (5)$$

The overall reaction thus can proceed through the principal path

$$\text{Zn} + 2\text{MnO}_2 + 2\text{NH}_4\text{Cl} \rightarrow \text{Zn(NH}_3)_2\text{Cl}_2 + 2\text{MnOOH} \qquad (6)$$

or other combinations of the above reactions. The diamino reaction (2) is typically dominant in the early life of the Leclanché cell and the zinc to ZnO Mn_2O_3 reaction (5) in later cell life depending on the current load and intermittency. As is always the case to produce direct current electricity from a battery, there must be electrons passing through the load, a light bulb or motor, etc. from the anode to the cathode propelled by the potential difference between the two electrodes.

Zinc Chloride Cells

The outward appearance of cells with primarily zinc chloride as the salt in the electrolyte is often the same as for Leclanché cells. The zinc chloride cells do typically contain 5 % or less ammonium chloride in the electrolyte as compared to Leclanché cells with 18–23 %. The cells were actually invented in 1899, but the cell constructions of the time were not well sealed, an important requirement for zinc chloride electrolyte cells. These cells usually contain more synthetic manganese dioxide and more electrolyte in the cathode mix than a similar Leclanché cell. Thus the typical internal reactions are more like reaction (4) at the anode and reaction (5) in the cathode mix. The cells will not discharge efficiently with a thick separator so all commercial zinc chloride cells should have a thin usually coated paper separator. (Leclanché cells often have a relatively thick separator which minimizes a tendency to leak electrolyte if the cell is over discharged.)

Carbon-Zinc Batteries, Table 1 Carbon-zinc system energy characteristics

System	Cell voltage	Energy density (Whr/kg)	Power density (W/kg)	Energy density (Whr/L)
Leclanché cells	1.5	105	20	225
Zinc chloride cells	1.5	115	25	280

The advantage of the zinc chloride design is that it performs better on heavy drain tests such as 2.25 Ω loads (for an R_2O cell) and is much less likely to leak when abused. Fortunately the rare leakage is only a mild acidic salt mixture. It is an eye irritant and care must be taken not to rub the eyes but irrigate with water. Rubbing the eyes may damage the eye surface due to the crystalline salt content of the typical exudates. Leakage generally has little or no effect on bare skin which is not the case with alkaline cells (Table 1) (Fig. 2).

Despite the many years since the original Leclanché and zinc chloride cells were invented, the availability of improved materials and blending methods coupled with more volumetrically efficient constructions has made constant progress in energy output feasible as illustrated in Fig. 3. The graph is expressed in kilojoules/liter because standard sizes make typical, relatively small, dry cell performance a function of volume rather than weight.

Another approach to cell design especially for high-voltage, low-current carbon-zinc cells is the so-called Minimax^R Construction [3].

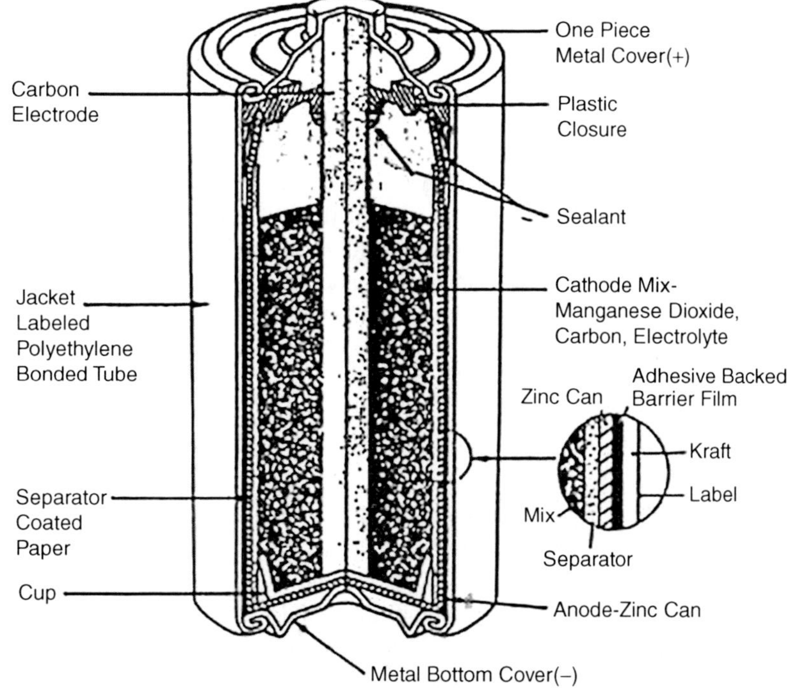

Carbon-Zinc Batteries, Fig. 2 Model heavy duty zinc chloride cell (Courtesy Eveready Battery Company) [4]

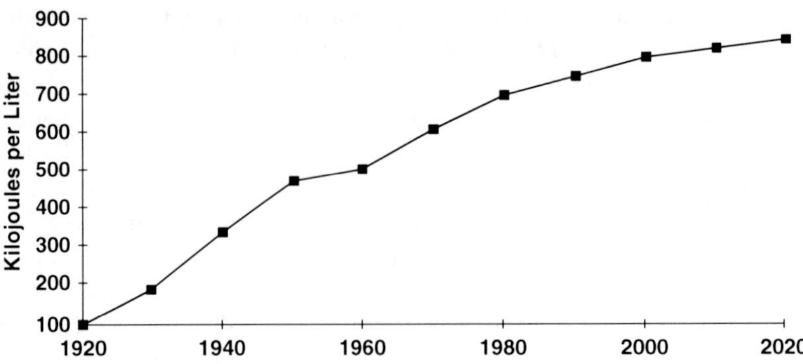

Carbon-Zinc Batteries, Fig. 3 One hundred years of increase in energy density in Leclanché and zinc chloride cells

Essentially these batteries are a bipolar stack of individual cells where the back of the internal zinc sheets is either painted with a composition including graphite or carbon black to make it conductive or a conductive carbon-containing sheet cemented to the back of the zinc anode piece with a conductive adhesive. The sides of each cell are sealed by a shrink tubing jacket so electrolyte cannot escape from one cell to another shorting out the stack. The painted side of each zinc piece contacts the cathode mix in the next cell so they can be stacked in series allowing up to a several hundred volts to be placed in a small package. The small zinc anodes can run in size from 1 × 1.5 cm up to 5 × 7.5 cm. The anode area approximately defines the width and depth of the battery without limiting its length. Thus a variety of battery heights and voltage output is possible without changing the size of each cell. In recent years the most common group is a 9 V battery with six cells. Most of the 9 V batteries available in retail stores however are alkaline batteries made from a bundle of six tall, small diameter cylindrical cells.

Future Directions

Electrolytic manganese dioxide crystal form and porosity can be further optimized, the zinc encasement (plastic coating) and jacket thickness can be made thinner there by increasing useful internal volume for active electrodes predominantly manganesedioxide cathode.

Cross-References

▶ Aerospace Applications for Primary Batteries
▶ Alkaline Primary Cells
▶ Cell, Electrochemical
▶ Data Banks of Electrolytes
▶ Electrochemical Cells, Current and Potential Distributions
▶ Electrosynthesis of Fine Chemicals
▶ Ionic Mobility and Diffusivity
▶ Ions at Solid-Liquid Interfaces
▶ Macroscopic Modeling of Porous Electrodes
▶ Manganese Oxides
▶ Overpotentials in Electrochemical Cells
▶ Oxygen Nonstoichiometry of Oxide
▶ Primary Batteries for Medical Applications
▶ Primary Batteries for Military Applications
▶ Primary Batteries, Comparative Performance Characteristics
▶ Primary Batteries, Selection and Application
▶ Primary Battery Design
▶ Thermal Effects in Electrochemical Systems

References

1. Leclanche G (1868) Les Mondes 16:5632
2. Cahoon NC (1976) Leclanche and zinc chloride cells. In: Cahoon NC, Heise GW (eds) The primary battery, vol 2. Wiley, New York
3. Schumm B Jr (1991) Zinc-carbon batteries. In: Tuck CDS (ed) Modern battery technology. Ellis Horwood, London, pp 87–111
4. Eveready Battery Product Data, www.Energizer.com
5. Schumm B Jr (2011) Zinc-Carbon batteries, Leclanche and zinc chloride systems. In: Reddy T (ed) Linden's handbook of batteries. McGraw Hill, New York

Cathodic Hydrocoupling of Acrylonitrile (Electrosynthesis of Adiponitrile)

Yoshifumi Kado
Asahi Kasei Chemicals Corporation,
Tokyo, Japan

Introduction

Cathodic hydrocoupling of acrylonitrile is one of the most well-known organic electrochemical processes because of its unique chemical reaction and industrial application. This reaction had already known since the 1940s, but the yield of product was not high in those days. In 1960 Manuel M. Baizer improved the yield and current efficiency of the coupling reaction by a discovery of using quaternary ammonium salts as supporting electrolyte [1]. The product of this coupling reaction is adiponitrile which is known as intermediate of polyamid used for engineering plastics, tire cord, and clothes. Adiponitrileis also used for synthesis of polyurethane. From the benefit of Nylon and polyurethane, cathodic hydrocoupling of acrylonitrile is absolutely essential technology for human life.

Cathodic Reaction of Acrylonitrile

The reaction mechanism for cathodic hydrocoupling of acrylonitrile has been extensively investigated by many researchers and it has been establishedas shown in Scheme 1 [2–4]. Firstly, an anion radical of acrylonitrileis generated by the first one-electron transfer to an acrylonitrile molecule from cathode. An anion radical of acrylonitrilecombines with an unreacted acrylonitrilemolecule. Finally, the resulting anion radical dimer species undergoes a second one-electron transfer leading to the formation of adiponitrile. The feature of this electrochemical reaction is that coupling is selectivity proceed by head to head dimerization.

Electrochemical Process of Adiponitrile Production

First electrochemical production of adiponitrilehad developed by Monsanto in 1963 and commercialized in 1965. Monsanto process adopted a homogeneous electrolysis system. On the other hand, Asahi Chemical Industry (presently Asahi Kasei Chemicals) started the commercial adiponitrileplant withan emulsion electrolysis system in 1971. Both of the processes were operated in the divided cell equipped with separator such as ion exchange membrane in the first stage of development and have improved into operation in undivided cell in order to reduce higher cell voltage caused by resistance of separator.

The cathodes used in this process are Hg, Pb or Cd which have higher overvoltage of hydrogen evolution, because the generation of hydrogen is the side reaction of the hydrocupling.

On the other hand, Fe, Ni, or Pb are used as an anodes which have lower overvoltage of oxygen evolution in order to prevent the anodic oxidation of organic compounds.

Supporting electrolyte is the mixture of quaternary ammonium salts and inorganic salt. Quaternary ammonium salts such as ethyltributylammoniumsalt are effective not only for increasing the selectivity of adiponitrile generation but also for protection of cathode from corrosion. Inorganic salts such as potassium phosphate and alkali metal borate increases conductivity of electrolyte and prevent corrosion of the anode.

The operation conditions and performances of this coupling process have been reported by

$$CH_2{=}CHCN \xrightarrow{e^-} \dot{C}H_2{-}\bar{C}HCN \xrightarrow{CH_2=CHCN} \begin{matrix} CH_2{-}\bar{C}HCN \\ | \\ CH_2{-}\dot{C}HCN \end{matrix} \xrightarrow{e^-,\ 2H^+} NC(CH_2)_4CN$$

Cathodic Hydrocoupling of Acrylonitrile (Electrosynthesis of Adiponitrile), Scheme 1 Mechanism for the cathodic reduction of acrylonitrile to adiponitrile

Cathodic Hydrocoupling of Acrylonitrile (Electrosynthesis of Adiponitrile), Table 1 Operation conditions and performances

Literature	Ref. [5]
Cathode	Pb
Anode	Fe (Ni 9 %)
Supporting electrolyte	$(EtBu_3N)_2HPO_4$
	K_2HPO_4
	$K_2B_4O_7$
Current density	2 kA/m^2
Temperature	55 °C
Current efficiency	90.7 %

several workers [1, 5] and an example is shown in Table 1. At present more than 300,000 t per year of adiponitrile is produced by cathodic hydrocoupling of acrylonitrile in the world.

Conclusion and Future Direction

Cathodic hydrocoupling of acrylonitrile is selectively proceed by head to head dimerization and produces adiponitrile. Manuel M. Baizer improved the yield and current efficiency of this coupling reaction by a discovery of using quaternary ammonium salts. Today, this reaction is necessary for our life & living from the point of industrial application, because adiponitrileis intermediate of polyamid used for engineering plastics, tire cord, and clothes. Therefore, Cathodic hydrocoupling of acrylonitrile might keep to be an important technology of chemical industry continuously in the future.

Cross-References

▶ Organic Electrochemistry, Industrial Aspects
▶ Organic Reactions and Synthesis

References

1. Hermann P (2001) Industrial Electroorganic Chemistry. In Lund H, Hammerich O (ed) Organic electrochemistry, 4th edn. revised and expanded. Marcel Dekker, New York, Chapter 31
2. Fry AJ (1972) Electrochemical reduction of conjugated systems. Synthetic organic electrochemistry. Harper and Row, New York, Chapter 7
3. Rifi MR (1975) In: Weinberg NL (ed) Technique of electroorganic synthesis. Wiley, New York, part 2, Chapter 8
4. Atobe M, Sasahira M, Nonaka T (2000) Ultrason Sonochem, Ultrasonic Effects on Electroorganic Processes. Part 17. Product Selectivity Control in Cathodic Reduction of Acrylonitrile 7:103–107
5. (1987) Japanese laid open Patent 62-27583

Cation-Pool Method

Seiji Suga
Okayama University, Okayama, Japan

Introduction

Carbocations are positively charged carbon-centered reactive intermediates. Although they were considered to be relatively unstable and transient species, Olah's extensive work in 1960s reveal that some carbocations can be long-lived species in superacid media [1]. Various carbocations were generated and accumulated in superacid, and they were characterized by NMR spectroscopy. However, the nature of carbocations in conventional reaction media, which are used for organic synthesis, has not been fully clarified as yet.

To generate carbocations for preparative purposes, there are two methods, that is, acid promoted reaction (Scheme 1a) and oxidative reaction (Scheme 1b). Acid-promoted reactions are the most commonly used for the generation of carbocations. In this method, a proton or a Lewis acid is used to activate a leaving group, and then the heterolysis of the bond between the carbon and the leaving group occurs to generate the carbocation. Because these steps are reversible, several species often exist in the solution as an equilibrium mixture. Then, a nucleophile, which is usually present in the solution, attacks the carbocation to give the final product. In the oxidative generation, two-electron oxidation followed by deprotonation or desilylation give

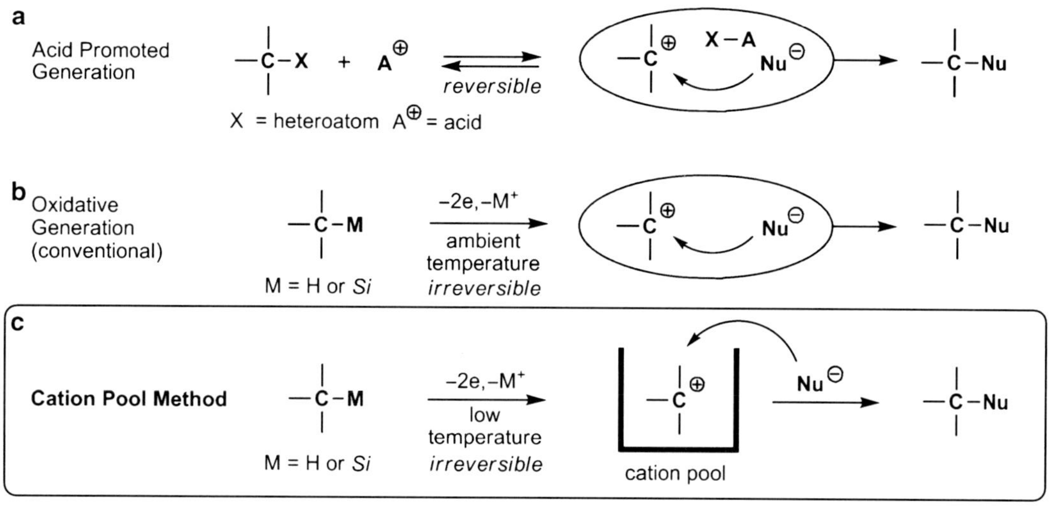

Cation-Pool Method, Scheme 1 General procedure for cation pool method

rise to the formation of the carbocation. These steps are essentially irreversible, and hence the oxidative generation could serve as a good method for a profound study on the chemistry of carbocations. However, the oxidative generation of carbocations is also usually carried out in the presence of nucleophiles because of the instability of carbocations. In this case the concentration of carbocations should be low, and therefore, it is very difficult to detect carbocation intermediates spectroscopically.

In the cation pool method [2], carbocations are generated and accumulated in relatively high concentration by the low temperature electrochemical oxidation of the corresponding precursor in the absence of nucleophiles (Scheme 1c). The electrolysis is usually carried out at low temperature such as −78 °C in order to avoid decomposition of carbocations. It seems to be difficult to carry out preparative electrolyses at such low temperature probably because of the high viscosity of the solution, which in turn disfavors the movement of ions to carry the electricity. By choosing an appropriate solvent and a supporting electrolyte, however, the electrolysis at such low temperature can be accomplished to generate and accumulate carbocations. Thus generated carbocation solutions of relatively high concentration can be utilized for the direct spectroscopic measurement such as NMR and IR spectroscopies, which exhibited similar spectra to those of the carbocations generated in super acid media.

The General Procedure for Cation Pool Method

A divided cell equipped with a sintered glass separator, a carbon fiber anode, and a platinum cathode is used in order to avoid the electrochemical reduction of anodically generated carbocations. Tetrabutylammonium tetrafluoroborate is usually used as supporting electrolyte, and dichloromethane is in most cases suitable as solvent because of less nucleophilicity and low viscosity at low temperature. Two equivalent of TfOH (trifluoromethanesulfonic acid) to a cation precursor is added in the cathodic chamber to facilitate the reduction of protons in the cathodic process. The constant current electrolysis (20 mA) was then carried out at −78 °C with magnetic stirring until 2.0–2.5 F/mol of electricity was consumed to give a cation pool. Carbamates (α-silyl carbamate) (Scheme 2a) [3], α-silyl ethers (Scheme 2b) [4], diarylmethanes (silylated diarylmethanes) (Scheme 2c) [5] can be employed as carbocation precursors. The cation pool of dications can also be generated and

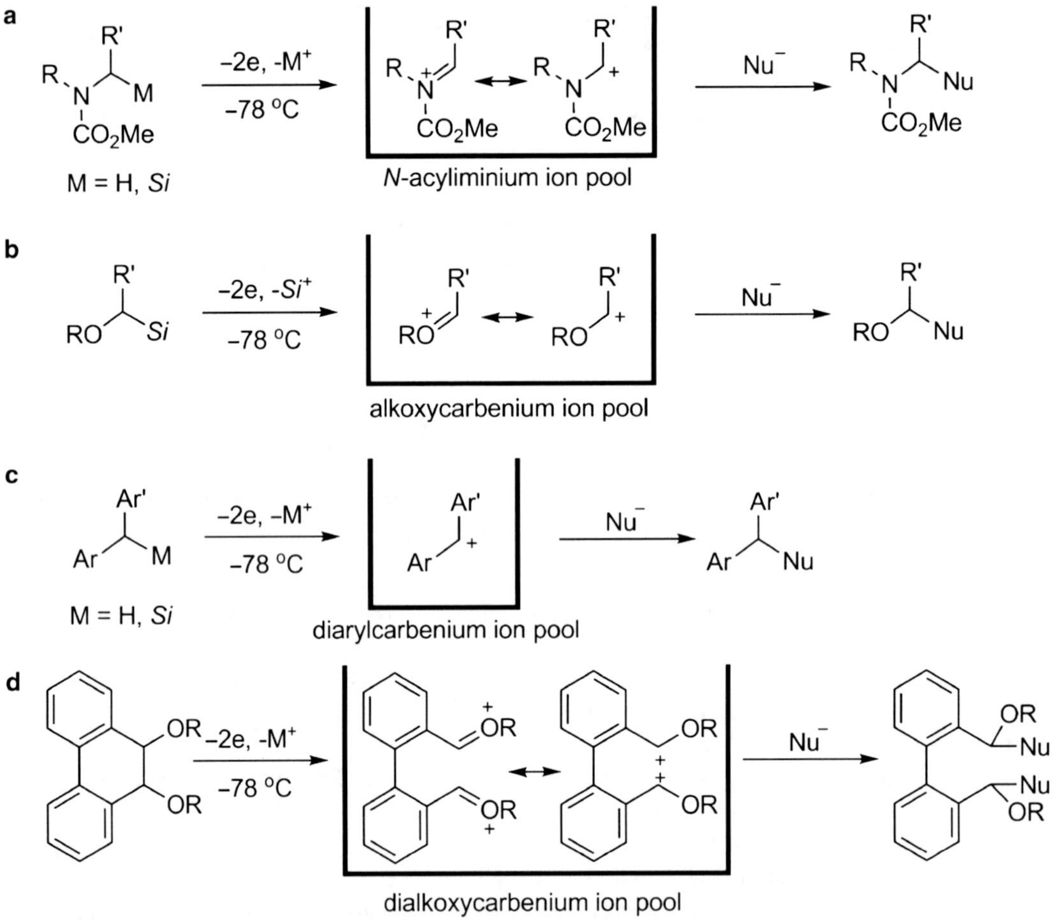

Cation-Pool Method, Scheme 2 Examples of cation pool

accumulated by the oxidative carbon-carbon bond dissociation [6] (Scheme 2d).

Synthetic Applications

The cation pool method is widely applicable to a variety of transformations otherwise difficult to perform. Especially, *N*-acyliminium ions can be utilized for the alkaloid synthesis. The examples of the *N*-acyliminium ions pools by the cation pool method are depicted in Fig. 1.

The radical mediated carbon-carbon bond formation can be performed based on the reduction of a cation pool (Scheme 3) [7].

Sequential use of the cation pool method can also be applicable to the prepearation of nitrogen-containing compounds having a quaternary carbon center, especially spiro compounds. The starategy could be utilized for the formal synthesis of cephalotaxine (Scheme 4) [8].

The reactions initiated by the addition of a carbon-carbon multiple bond to the cation pool bring in unique one-pot transformations otherwise difficult to realize, because the reactive carbocations are existing in the solution in relatively high concentration. For example, a sequential one-pot three-component coupling reactions have been developed (Scheme 5) [9, 10]. [4 + 2] cyclo addition reactions in which an *N*-acyliminium ion was used as a hetero diene [11], and cationic carbohydroxylation of alkenes and alkynes using the cation pool method [12] were also accomplished.

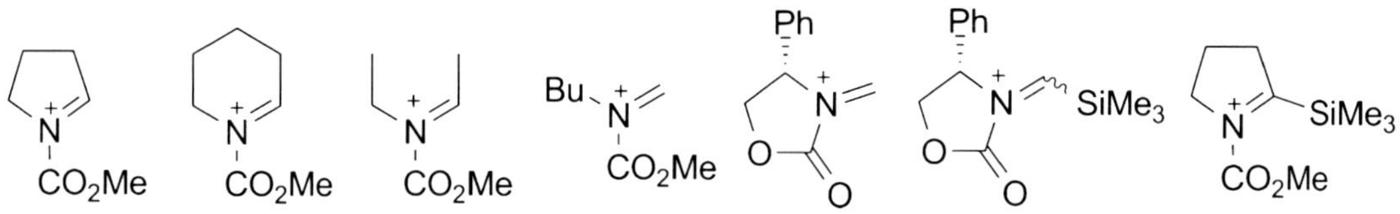

Cation-Pool Method, Fig. 1 *N*-Acyliminium ion pools

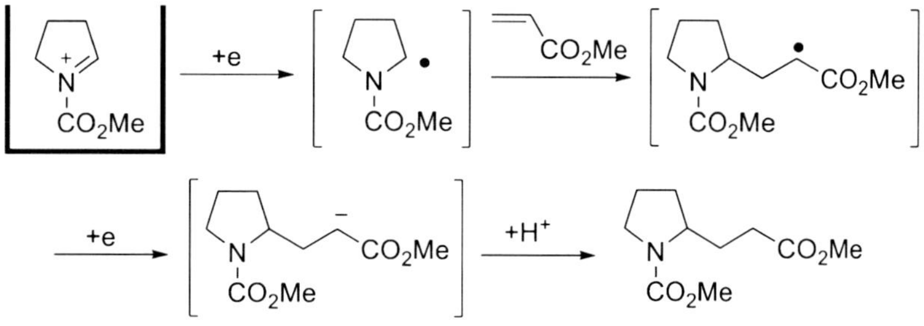

Cation-Pool Method, Scheme 3 Reduction of cation pool

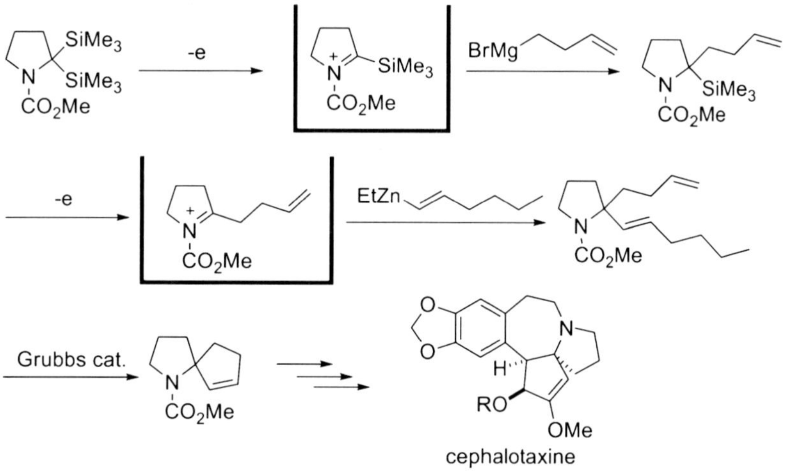

Cation-Pool Method, Scheme 4 Sequential transformation based on the cation pool method

The distannane mediated organic radical addition to *N*-acyliminium ions [13], and the benzylic radical addition to *N*-acyliminium ions which proceeds via radical/cation/radical cation chain mechanism [14] show that the cation pool can be utilized as good nucleophilic radical acceptors because of their strong electrophilic character. Iterative molecular assembly based on the cation pool method lead to the efficient formation of dendritic molecules [15]. The manipulation of the cation pool in the microflow system realized an efficient controlled/living cationic-polymerization [16] and a selective Friedel-Crafts mono-alkylation [17]. The reactions using highly reactive carbocations generated by the cation pool method have good matches to the microreaction chemistry.

By using the low temperature electrochemical oxidation, other reactive cationic species such as iodonium ions (I^+) [18] and sulfonium ion equivalents ($ArS(ArSSAr)^+$) [19] can also be generated and accumulated in a similar fashion to the cation pool method.

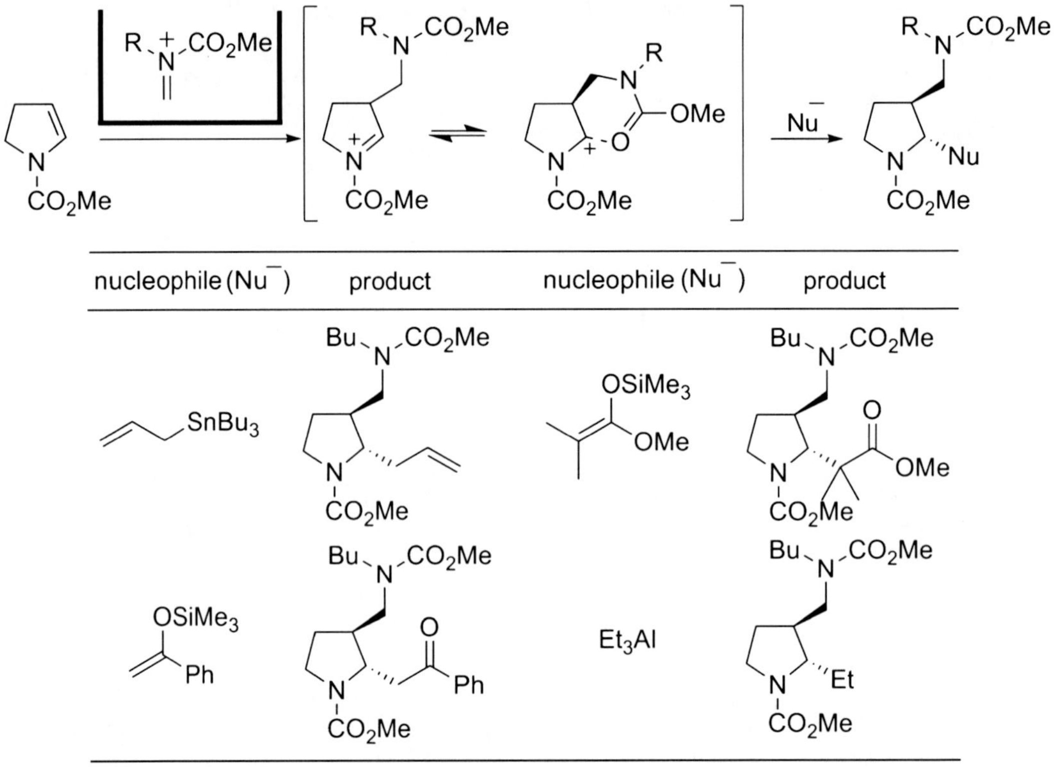

Cation-Pool Method, Scheme 5 One-pot three component coupling

Future Directions

The cation pool method opens a new aspect of the chemistry based on carbocations, which have been considered to be difficult to manipulate in normal reaction media. These methods involve the generation of carbocations in the absence of nucleophiles, spectroscopic characterization, and reactions with a variety of carbon nucleophiles to achieve direct carbon-carbon bond formation.

Although the applications to *N*-acyliminium ions, alkoxycarbeniumions, and benzylic cations were successful, it seems to be difficult to apply the method to less stabilized cations. The applicability of the cation pool method inebitably depends upon the stability of the cation that is accumulated. The cation flow method [20, 21] which involves generation of carbocations in a microflow electrochemical system should be much more favorable because of short residence times and efficient temperature control. Indirect oxidation of the cation precursors is another choice [22]. Future work aimed at expanding the scope of the cations using such a system will offer further expansion of the present means into new class of useful synthetic transformations.

The cation pool method is very powerful tool for the mechanic study of the cationic reaction because the cation intermediate can be detected by the spectroscopies. In contrast to the carbanion chemistry, it is generally difficult to get insight into the mechanism of carbocation chemistry. Further application of the cation pool method to explore the mechanistic investigation is also expected.

Cross-References

► Anodic Substitutions
► Electroauxiliary
► Electrochemical Microflow Systems

References

1. Olah GA (1995) My search for carbocations and their role in chemistry. Angew Chem Int Ed 34:1393–1405
2. Yoshida J, Suga S (2002) Basic concepts of "cation pool" and "cation flow" methods and their applications in conventional and combinatorial organic synthesis. Chem Eur J 8:2650–2658
3. Yoshida J, Suga S, Suzuki S, Kinomura N, Yamamoto A, Fujiwara K (1999) Direct oxidative carbon-carbon bond formation using "cation pool" method I. Generation of iminium cation pools and their reaction with carbon nucleophiles. J Am Chem Soc 121:9546–9549
4. Suga S, Suzuki S, Yamamoto A, Yoshida J (2000) Electrooxidative generation and accumulation of alkoxycarbenium ions and their reactions with carbon nucleophiles. J Am Chem Soc 122:10244–10245
5. Okajima M, Soga K, Nokami T, Suga S, Yoshida J (2006) Oxidative generation of diarylcarbenium ion pools. Org Lett 6:7324–7325
6. Okajima M, Suga S, Itami K, Yoshida J (2005) "Cation pool" method based on C-C bond dissociation. J Am Chem Soc 127:6930–6931
7. Suga S, Suzuki S, Yoshida J (2002) Reduction of a "cation pool". A new approach to radical mediated C-C bond formation. J Am Chem Soc 124:30–31
8. Suga S, Watanabe M, Yoshida J (2002) Electroauxiliary-assisted sequential introduction of two carbon nucleophiles on the same a-carbon of nitrogen: application to the synthesis of spiro compounds. J Am Chem Soc 124:14824–14825
9. Suga S, Nishida T, Yamada D, Nagaki A, Yoshida J (2004) Three-component coupling based on the "cation pool" method. J Am Chem Soc 126:14338–14339
10. Suga S, Yamada D, Yoshida J (2010) Cationic three-component coupling involving an optically active enamine derivative. From time integration to space integration of reactions. Chem Lett 39:404–406
11. Suga S, Nagaki A, Tsutsui Y, Yoshida J (2003) "*N*-acyliminium ion pool" as a hetero diene in [4 + 2] cycloaddition reaction. Org Lett 5:945–947
12. Suga S, Kageyama Y, Babu G, Itami K, Yoshida J (2004) Cationic carbohydroxylation of alkenes and alkynes using the cation pool method. Org Lett 6:2709–2711
13. Maruyama T, Suga S, Yoshida J (2005) Radical addition to "cation pool". Reverse process of radical cation fragmentation. J Am Chem Soc 127:14702–14703
14. Maruyama T, Mizuno Y, Shimizu I, Suga S, Yoshida J (2007) Reaction of *N*-acyliminium Ion pool with benzylsilane. Implication of a radical/cation/radical cation chain mechanism involving oxidative C-Si bond cleavage. J Am Chem Soc 129:10922–10928
15. Nokami T, Ohata K, Inoue M, Tsuyama H, Shibuya A, Soga K, Okajima M, Suga S, Yoshida J (2008) Iterative molecular assembly based on the cation-pool method. Convergent synthesis of dendritic molecules. J Am Chem Soc 130:10864–10865
16. Nagaki A, Kawamura K, Suga S, Ando T, Sawamoto M, Yoshida J (2004) "Cation pool" initiated controlled/living polymerization using microsystems. J Am Chem Soc 126:14702–14703
17. Nagaki A, Togai M, Suga S, Aoki N, Mae K, Yoshida J (2005) Control of extremely fast competitive consecutive reactions using micromixing. Selective friedel-crafts aminoalkylation. J Am Chem Soc 127:11666–11675
18. Midorikawa K, Suga S, Yoshida J (2006) Selective monoiodination of aromatic compounds with electrochemically generated I^+ using micromixing. Chem Commun (36):3794–3796
19. Matsumoto K, Suga S, Yoshida J (2011) Organic reactions mediated by electrochemically generated ArS^+. Org Biomol Chem 9:2586–2596
20. Suga S, Okajima M, Fujiwara K, Yoshida J (2001) "Cation flow" method. A New approach to conventional and combinatorial organic syntheses using electrochemical micro flow systems. J Am Chem Soc 123:7941–7942
21. Saito K, Ueoka K, Matsumoto K, Suga S, Nokami T, Yoshida J (2011) Indirect cation-flow method: flash generation of alkoxycarbenium ions and studies on the stability of glycosyl cations. Angew Chem Int Ed 50:5153–5156
22. Suga S, Matsumoto K, Ueoka K, Yoshida J (2006) Indirect cation pool method. Rapid generation of alkoxycarbenium ion pools from thioacetals. J Am Chem Soc 128:7710–7711

Cell Membranes, Biological

Ernst Niebur
Mind/Brain Institute, Johns Hopkins University, Baltimore, MD, USA

Introduction

All biological life (disregarding viruses that only show some properties of life) is composed of cells or collections of cells. At the most basic functional level, a cell consists of a collection of internal structures that are immersed in intracellular fluid and separated from the outside by a cell membrane, which is also called plasma membrane or plasmalemma. There are other membranes of great importance in biology, e.g., those surrounding the nucleus of an eukaryote cell, but this entry is focused on the electrochemistry of the membranes surrounding a cell. Geometrically, a membrane is a two-dimensional surface which is for many cells large enough to be seen with even simple light microscopes; as early as in the seventeenth century, Hooke described the cellular structure of cork and introduced the word "cell" to biology. In the third dimension, perpendicular to the surface separating a cell's inside and outside, the thickness of a membrane is below the resolution of light microscopes. Although its components were long known to be mainly phospholipids and proteins, the structure of cell membranes was not discovered until the 1970s. In a seminal paper, Singer and Nicolson [1] proposed that the matrix of a cell membrane consists of two adjacent sheets of phospholipids and that it is thus only two molecules thick. Each phospholipid molecule has a hydrophilic (phosphate) and a hydrophobic (lipid) end, and Singer and Nicolson [1] showed that the free energy of phospholipids in aqueous solution is lowest when the lipid ends of the phospholipid molecules are arranged next to each other and their phosphate groups face the aqueous solutions both inside and outside the cell, see Fig. 1a. The phospholipid bilayer is an effective barrier for ions and molecules, establishing a firm boundary between the inside and the outside of a cell. Embedded in this matrix are various protein molecules, facing either the inside or the outside of the cell, or spanning the width of the membrane. In electrochemistry, the most important proteins are those that actively move ("pump") specific ions across the membrane, or those that allow specific ions to move across the membrane following concentration gradients. Metabolically important substances, both nutrients and products, are also actively transported across the membrane.

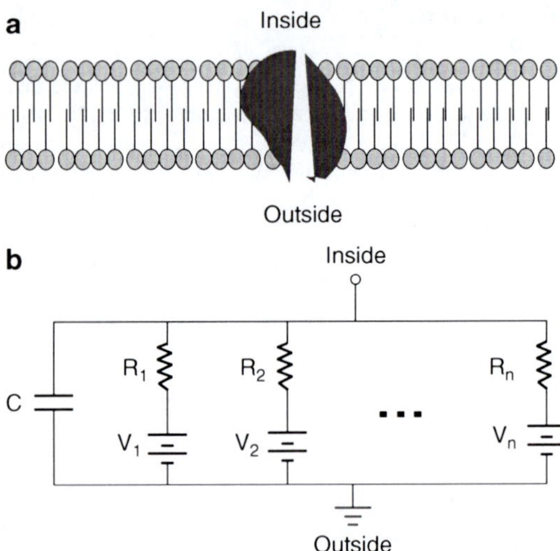

Cell Membranes, Biological, Fig. 1 (**a**) Schematic view of a cell membrane in cross section. Phospholipids are represented with hydrophilic heads (*circles*) and hydrophobic tails (*lines*). The former are exposed to the aqueous solutions inside and outside of the cell. A transmembrane protein (*large dark shape*) with a pore spans the membrane. (**b**) Equivalent circuit of membrane patch. For simplicity, all reversal potentials v_i are shown with the same polarity

Electrical Properties of Cell Membranes

Pure phospholipid bilayers are good insulators with a specific conductance per unit area of about $10^{-13}\Omega^{-1}m^{-1}$. Typically, the conductance of biological cell membranes is several orders of magnitude higher, mainly because of various ion channels provided by transmembrane proteins. The transmembrane conductance is highly variable, both in space and time, since it is by controlling the passage of ions across the plasmalemma that the cell controls the electrical potential of the cell (relative to the outside), as well as the concentration of various ion concentrations inside the cell.

Compared to the aqueous solutions both in the cytoplasm and the extracellular medium, the membrane is a relatively poor conductor. The small distance across the membrane results in a high specific capacitance, at approximately $10^{-2}Fm^{-2}$. In contradistinction to the transmembrane conductance, the specific capacitance varies little between different cell membranes and is also constant over time.

Electrical Potentials Across the Membrane

Biological cells actively generate and maintain differences in the concentrations of several ion species across the cell membrane.

Transmembrane proteins function as "ion pumps" by using metabolic energy (typically provided by the ATP → ADP process) to transport ions against their osmotic gradients across the membrane. A typical example is the $Na^+ - K^+$ pump which moves two potassium ions into the cell and, at the same time, three sodium ions out of the cell. In thermodynamic equilibrium, the probability P_{in} of finding a specific ion inside the cell, as compared to the probability P_{out} of finding it outside, depends on the energy the ion has inside (E_{in}) versus outside (E_{out}). It is given by the Boltzmann distribution,

$$\frac{P_{in}}{P_{out}} = \exp\left\{ -\frac{ze(V_{in} - V_{out})}{kT} \right\}, \qquad (1)$$

where k is the Boltzmann factor, T is the temperature, z is the valence of the ion, e is the electric unit charge, and V_{in}, V_{out} are the electrical potentials of the ion inside and outside the cell, respectively. We can use Eq. 1 to compute the electrical potential difference from ion concentrations,

$$\Delta V = V_{in} - V_{out} = \frac{kT}{ze} \ln \frac{P_{out}}{P_{in}} \qquad (2)$$

This relation is known as the Nernst Equation. Usually, it is expressed in terms of the ion concentrations but since the probability of finding an ion at some location is proportional to its concentration at this location, our derivation in terms of probabilities is completely equivalent. It is also customary to set the voltage scale such that $V_{out} = 0$.

We thus find that each ion species has its own voltage at which it is in statistical equilibrium. This voltage is commonly called the "reversal potential" of this ion because the current generated by these ions reverses its sign at this voltage. Some values typical for many neurons are listed in the following table.

Ion	z	P_{in}	P_{out}	ΔV
K^+	1	100 mM	5 mM	−80 mV
Na^+	1	15 mM	150 mM	62 mV
Ca^{2+}	2	0.2 μM	2 mM	246 mV
Cl^-	−1	13 mM	150 mM	−65 mV

Membrane Patch in Equilibrium

Consider a patch of cell membrane which may contain many ion channels but which is small enough so that the transmembrane voltage is approximately the same everywhere in the patch. Electrically, a single ion channel is equivalent to a resistance in series with a voltage source. Its resistance $r_{channel}$ is the inverse of the conductance of the ion channel pore, and the voltage v_i is the reversal potential of ion species i which can pass through the channel. If the number of channels in the patch is N, their total resistance is $R = r_{channel}/N$. Ion channels which let pass selectively different ions are electrically in parallel. For instance, if we have two kinds of ion channels in the membrane, say sodium and potassium (the more general case is treated below), we can determine the resting potential V_L of the cell from Ohm's law,

$$I_{Na} = \frac{V_{Na} - V_L}{R_{Na}} = g_{Na}(V_{Na} - V_L) \qquad (3)$$

$$I_K = \frac{V_K - V_L}{R_K} = g_K(V_K - V_L) \qquad (4)$$

where $G_{Na} = R_{Na}^{-1}$ and $G_K = R_K^{-1}$. Conservation of charge (Kirchhoff's current law) implies that all currents balance; thus, $I_{Na} + I_K = 0$. From Eqs. 3 and 4, we therefore have

$$g_K V_K - g_K V_L + g_{Na} V_{Na} - g_{Na} V_L = 0, \qquad (5)$$

resulting in

$$V_L = \frac{g_{Na} V_{Na} + g_K V_K}{g_{N_a} + g_K}. \qquad (6)$$

This can be easily generalized for more than two ion species. The resting potential is then given by the quotient of two sums over all ion species,

$$V_L = \frac{\Sigma_i g_i V_i}{\Sigma_i g_i} \qquad (7)$$

Much of the cell's behavior is governed by interactions at the plasmalemma which, in turn, are controlled to a large extent by the proteins embedded in it. For instance, action potentials are large excursions from the resting potential caused by selective exchange of ions across the membrane. The best known examples may be cardiac cells and neurons, but action potentials are found in cells of many other types, both in animals and plants. Another important interaction is communication between cells (frequently but not always neurons) through synapses. In electrical synapses, specialized protein structures in the membrane (connexins in vertebrates) establish a galvanic connection between the interiors of the two neurons, allowing current flow between them. In chemical synapses, a substance (neurotransmitter) released from the membrane of the presynaptic (sender) neuron diffuses across the space between the two neurons (synaptic cleft) and docks to receptors on the postsynaptic (receiver) neuron. The receptors are proteins in the postsynaptic membrane and can be of one of two types. In the first one, docking of a neurotransmitter molecule directly opens a selective ion channel in the protein, allowing the selected ions to flow down their potential gradients and changing the electrochemical potential of the cell. In the second synapse type, receptors modulate the production of chemical messengers in the postsynaptic cells which, among other functions, modify the permeability of other ion channels and thereby indirectly change its transmembrane potential.

Temporal Dynamics

The temporal dynamics of current flow are governed by the membrane capacitance. The capacitive current is the change of the electrical charge on the capacitor, $I_C = \frac{dQ}{dt} = \frac{d}{dt}(CV) = C\frac{dV}{dt}$, since C is constant. Kirchhoff's current law, $I_K + I_{Na} + I_C = 0$, thus gives,

$$C\frac{dV}{dt} = g_{N_a}(V_{N_a} - V) + g_K(V_K - V) \qquad (8)$$

Again, the generalization to more than two ion species is obvious. Figure 1b shows the equivalent circuit of the membrane patch. The capacitive current through the capacitance C is in parallel with a number of ionic currents with reversal potentials $V_1, V_2, \cdots V_n$ and their corresponding resistances $R_1, R_2, \cdots R_n$. Note that in equilibrium, the temporal derivative disappears, resulting in Eq. 5.

In the practically important case of time-independent conductances and reversal potentials, one can combine the ionic conductances and introduce the leakage conductance g_L and rewrite Eq. 8 as

$$C\frac{dV}{dt} = g_L(V_L - V) \qquad (9)$$

or, equivalently,

$$\tau\frac{dV}{dt} = V_L - V \qquad (10)$$

where $\tau = C/g_L$ is the *time constant* of the cell. The equation shows that the membrane will approach the resting potential V_L exponentially, with a characteristic time τ. Taking into account the spatial extent of the cell requires the solution of a partial differential equation in three dimensions. A practically important simplification is possible for long, extended cells, or for cells with long linear protuberances. In this case, variations in transmembrane voltage across the long axis of the neural process are negligible compared to differences along the axis and the three dimensional voltage equation can be approximated by a one- dimensional equation (or, for several protuberances, by a system of coupled one-dimensional equations). This situation applies to many neurons and in particular, the length of their axons can exceed their diameter by many orders of magnitude. For instance, axons running along the necks of a giraffe have a length of several meters but a diameter on the order of a few micrometers. Voltages along the neural process are governed by the same physics as those in a long electrical cable except that inductance can be neglected, and can be described as,

$$\tau \frac{\partial V}{\partial t} - \lambda^2 \frac{\partial^2 V(x)}{\partial x^2} = V_L - V \qquad (11)$$

where λ is the characteristic length along the process. This is the "Telegrapher's Equation" (with vanishing inductance) which was first described in the nineteenth century in the context of submarine cables.

Applications and Future Directions

Much of biological activity happens at the interface between cells and their surroundings, including other cells. Understanding these interactions is of utmost importance in many fields of basic and applied sciences. Although not strictly cell membranes, another application is the study of liposomes. These are lipid bilayers forming enclosed spaces which can be easily generated artificially and serve both as models for biological cells as well as for various applications, e.g., drug delivery.

Cross-References

► Artificial Photosynthesis
► Biofilms, Electroactive
► Biomolecules in Electric Fields
► Biosensors, Electrochemical
► Electric and Magnetic Fields Bioeffects
► Electropermeabilization of the Cell Membrane
► Ionic Liquids, Biocompatible
► Ionic Mobility and Diffusivity
► Ions at Biological Interfaces

References

1. Singer SJ, Nicolson GL (1972) The fluid mosaic model of the structure of cell membranes. Science 175(23):720–731

Cell, Electrochemical

Wolfram Oelßner
Kurt-Schwabe-Institut für Mess- und Sensortechnik e.V. Meinsberg, Kurt-Schwabe-Straße, Waldheim, Germany

Components of Electrochemical Cells

The General Term "Cell" in Electrochemistry

In electrochemistry the term "cell" is commonly used for a wide variety of devices with different functions, shapes, and sizes in which electrochemical reactions take place. It comprises, e. g., galvanic and electrolytic cells, standard cells, sensor cells, conductivity cells, spectroelectrochemical cells, fuel cells, electrochemical measuring cells, and two- and three-electrode cells. Generally electrochemical cells contain at least two electrodes each in contact with an electrolyte. In this sense the electrodes are, e.g., solid and liquid metals, graphite, or semiconductors that carry electric charge by electrons and exchange it with an electrolyte, whereas in the electrolyte the electric charge is carried by ions in aqueous or nonaqueous liquids, conducting polymers, molten salts, or other solid electrolytes. The electrodes can be changed in electrochemical reactions but also be inert and only serve to transfer electrons. Inert electrodes are usually made from platinum or carbon.

Unfortunately and somewhat confusing the term "electrode" has a variety of different meanings in electrochemistry. Unlike the commonly and also above-used definition, where it refers to the electronically conducting component of the cell only, it is also applied to the combination of an electronically conducting material in contact with an ionically conducting phase [1]. Thus, the half-cell described in the next chapter is often colloquially referred to as electrode. Furthermore, the term "electrode" is even used to denote complete electrochemical sensors such as glass electrode measuring chains.

Electrochemical Half-Cell

An electrode in contact with an electrolyte is called a "half-cell," often also written "half cell." Thus, a simple two-electrode electrochemical cell is composed of two half-cells that contain either the same electrolyte but different electrodes or different electrodes and electrolytes. The first type of chemical cell, where there is no phase boundary between different electrolytes, is a cell without transference. The other type, in which a liquid-liquid junction potential or diffusion potential is developed across the boundary between the two solutions, is a cell with transference. Commercially available reference electrodes can be considered "half-cells."

Salt Bridge and Separation Systems

If the electrolytes in the two half-cells are different, the two compartments can be joined by a so-called salt bridge. On the one hand the salt bridge must allow ions to pass between the two half-cells to provide electrical connection between them and to complete the cell circuit; on the other hand it should prevent mixing of the two different solutions as far as possible [2].

The classical and mostly shown embodiment for a salt bridge is an inverted U-tube, filled with a concentrated salt solution or gel, whose ends dip into the solutions of the two half-cells. When such a bridge is used and the ions in the bridge are present in large excess, they carry almost the whole of the current between the half-cells. To minimize the diffusion potential, the cation and anion of the bridge electrolyte should have similar ionic mobility and almost the same transport number [3]. Generally salts such as KCl, KNO_3, or NH_4NO_3 are suitable for this purpose. However, many other materials and shapes may also be used to separate two or more compartments of electrochemical cells mechanically and to connect them electrically at the same time. Even simply a filter paper soaked with an inert electrolyte has been applied. Commercially available reference electrodes use mostly ceramic or sleeve diaphragms.

Separators in technical electrochemical cells and processes are typically either membranes or diaphragms. A membrane has very small pores that permit only diffusional or conductive motion of the solvent or the electrolyte from one compartment to the other. A diaphragm has larger pores so that it permits the flow of the electrolyte solution from one compartment to the other but still restricts the complete intermixing of the two solutions. The separator should be as far as possible impermeable for ions and still have high electrical conductivity. To meet these contradictory demands, concerning the porosity, the size, and geometrical shape of the diaphragm, a compromise must be found, which is often quite difficult. In more detail separation systems in electrochemical cells are described and characterized in [4].

Galvanic and Electrolytic Electrochemical Cells

Main Classification and Polarities of Electrochemical Cells

Electrochemical cells are fundamentally classified into galvanic (or voltaic) cells and electrolytic cells [5].

Galvanic Cells Are Characterized by the Following Features

- In galvanic cells electrochemical reactions occur spontaneously at the electrodes when they are connected externally by a conductor.
- As a result, in a galvanic cell chemical energy is converted into electrical energy.
- According to Fig. 1 in a galvanic cell, anode is the negative and cathode is the positive electrode.
- Oxidation occurs at the anode and reduction at the cathode. Within the cell negative charge is transferred from cathode to anode. In the external circuit electrons flow from anode to cathode (also in electrolytic cells).
- Galvanic cells include primary (non-rechargeable), secondary (rechargeable), and fuel cells.
- As an example, a charged Pb-PbO_2 storage battery is a galvanic cell.

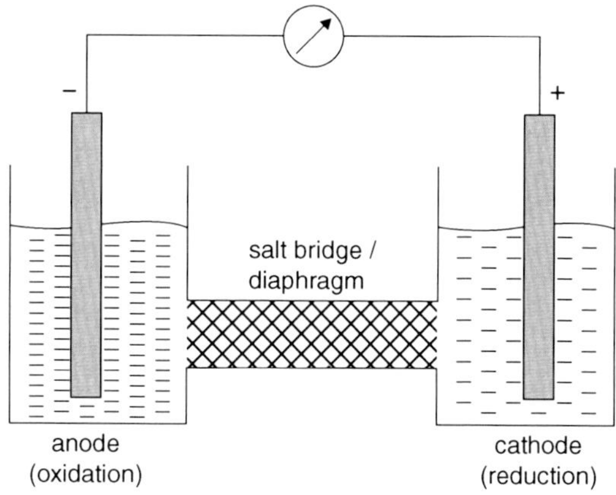

Cell, Electrochemical, Fig. 1 Schematic of a galvanic cell

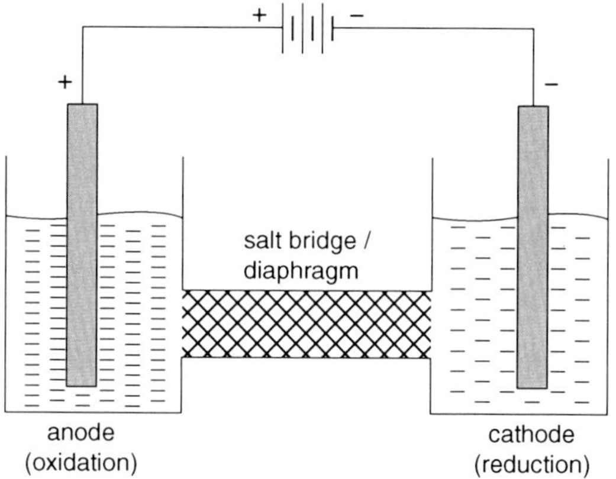

Cell, Electrochemical, Fig. 2 Schematic of an electrolytic cell

Electrolytic Cells Are Characterized by the Following Features

- In electrolytic cells electrochemical reactions do not occur spontaneously but are driven by an externally applied voltage greater than the open-circuit voltage of the cell.
- As a result, in electrolytic cells electrical energy is converted into chemical energy.
- According to Fig. 2 in an electrolytic cell, anode is the positive and cathode is the negative electrode.
- Oxidation occurs at the anode and reduction at the cathode. Within the cell negative charge is transferred from cathode to anode. In the external circuit electrons flow from anode to cathode (also in galvanic cells).
- Electrolytic cells are used for technical processes such as electrolytic syntheses, electrorefining of metals, and electroplating.
- As an example, a Pb-PbO_2 storage battery when it is being recharged is an electrolytic cell.

Individual electrochemical cells can be interconnected in series or in parallel to achieve batteries with higher voltages or current capabilities. Parallel coupling is possible only with cells having identical cell voltage.

The Daniell Cell

Despite being nowadays only of historical interest, the Daniell cell (invented in 1836) is the most often cited example of a galvanic cell [6, 7].

In the anodic half-cell a zinc electrode is immersed into a $ZnSO_4$ solution. The metallic zinc is oxidized to zinc ions, which go into the solution:

$$Zn(s) \rightarrow Zn^{2+}(l) + 2e^-.$$

The standard potential of this anodic half-reaction is −0.76 V versus SHE.

In the cathodic half-cell a copper metal electrode is immersed into $CuSO_4$ solution. Cu^{2+} ions from the solution are reduced to metallic copper, which is deposited on the Cu electrode:

$$Cu^{2+}(l) + 2e^- \rightarrow Cu(s).$$

The standard potential of this cathodic half-reaction is +0.34 V versus SHE.

Thus, the overall redox reaction of the Daniell cell is

$$Cu^{2+}(l) + Zn(s) \rightarrow Zn^{2+}(l) + Cu(s),$$

and the resulting standard cell voltage is 1.1 V.

The two half-cells of the Daniell cell are separated/joined by a diaphragm, e.g., a porous glass frit that prevents the mechanical mixing of the solution.

Notation to Describe Electrochemical Cells

The following universally accepted conventions are used to describe the structures of electrochemical cells in a shorthand notation [8].

- The anode (negative electrode) is written on the left-hand side and cathode (positive electrode) on the right-hand side.
- The anode of the cell is represented by writing the metal first and then the metal ion present in the electrolytic solution. Both are separated by a vertical line or a semicolon.
- The cathode of the cell is represented by writing the cation of the electrolyte first and then the metal. Both are separated by a vertical line or semicolon.
- The salt bridge that separates the two half-cells is indicated by two parallel vertical lines.
  - / a slash or single vertical line indicates a phase boundary.
  - // a double slash or double vertical line indicates a phase boundary whose potential is regarded as a negligible component of the overall cell potential.
  - , a comma separates two components in the same phase.

For example, the Daniell cell shown in Fig. 3 can be written compactly in a cell diagram as $Zn(s) \mid ZnSO_4(aq) \parallel CuSO_4(aq) \mid Cu(s)$.

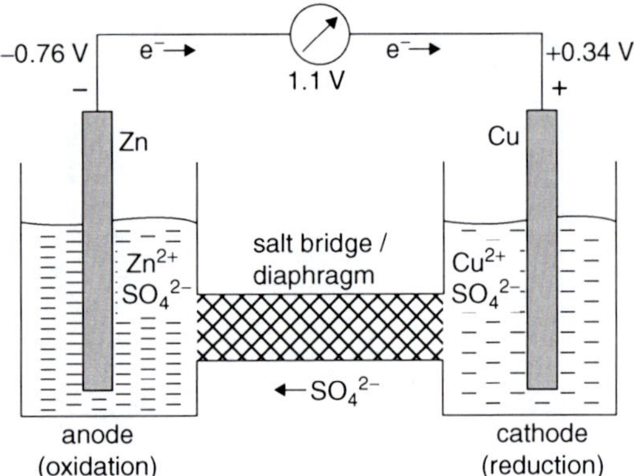

Cell, Electrochemical, Fig. 3 Schematic of the Daniell cell

Electrochemical Cell Potential

The potential or voltage of an electrochemical cell is the potential difference between the two electrodes of the cell, measured under any condition. It is composed of several parts such as [9]:

- Electrode potentials at the electrode/electrolyte interfaces
- Ohmic potential drops within the electrochemical cell
- Overvoltages from diffusion, activation, and reactions

It is worth mentioning in this context that generally single potentials of electrochemical electrodes cannot be measured. Only the difference of two electrode potentials is measurable. For this reason, the potential of the standard hydrogen electrode (SHE) was defined as origin of the electrochemical potential scale, and all electrode potentials are referred to this electrode or more customary to other reference electrodes such as the Ag/AgCl, Cl^- electrode.

The cell voltage of a galvanic cell measured when there is no current flowing through the cell is called electromotive force, abbreviated as emf. It is the maximum absolute voltage value that the cell can deliver and is expressed in the measuring unit Volt. The term "force" in this context is a historically caused misnomer as force is actually a physical quantity that is measured in Newton [10].

Fuel Cells

Fuel cells are galvanic cells that convert the energy of a chemical reaction directly into electrical energy. Unlike in primary and secondary cells, fuel and oxidizing agent are supplied continuously to the electrodes of the fuel cell. Thus, energy can be obtained from a fuel cell as long as the external supply of fuel is maintained. More basic information about fuel cells is given in Refs. [11–20].

Roughly fuel cells can be classified into low-temperature (operation temperature below 500 °C) and high-temperature (operation temperature up to 1,000 °C) cells [13]. As electrolytes molten salts or ionically conductive solids are used. Concerning the fuels, electrodes, electrolytes, and constructive details, in recent years a variety of different fuel cells of both types have been developed, e.g.,:

- Hydrogen-Oxygen Fuel Cells
- Redox Fuel Cells

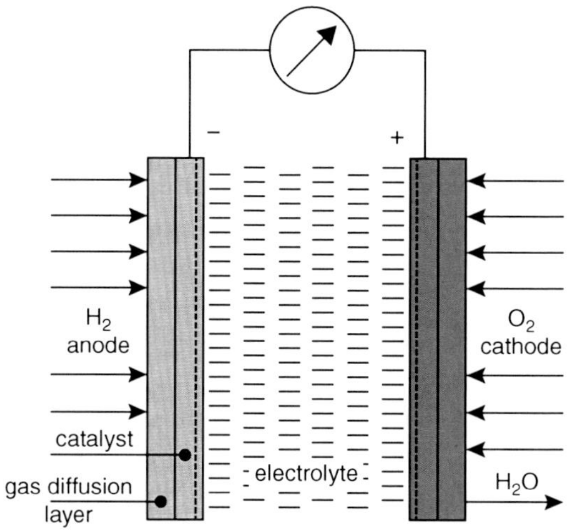

Cell, Electrochemical, Fig. 4 Schematic of a hydrogen-oxygen fuel cell according to [20]

- Phosphoric Acid Fuel Cells (PAFC)
- Direct Methanol Fuel Cells (DMFC)
- Polymer Electrolyte Membrane Fuel Cells (PEMFC)
- Solid-Oxide Fuel Cells (SOFC, HTSO)
- Molten Carbonate Fuel Cells (MCFC)

As an example Fig. 4 shows the schematic of a hydrogen-oxygen fuel cell [20]. The electrodes are made of porous carbon and the electrolyte is a resin containing concentrated aqueous sodium hydroxide solution. Hydrogen gas is fed to the anode where it is oxidized, and oxygen is fed to the cathode where it is reduced according to the following reaction equations:

$$\text{Anode}: \; 2H_2(g) + 4OH^- \rightarrow 4H_2O(l) + 4e^-$$

$$\text{Cathode}: \; O_2(g) + 2H_2O(l) + 4e^- \rightarrow 4OH(l)$$

$$\text{Overall}: \; 2H_2(g) + O_2(g) \rightarrow 2H_2O(l)$$

The overall cell reaction produces water. Thus, fuel cells are environment-friendly.

Other Electrochemical Cells

Three-Electrode Electrochemical Cells

In contrast to two-electrode cells, three-electrode electrochemical cells make it possible to measure or control the potential of an electrode exactly. This is in particular advantageous for studies of the kinetics and mechanism of electrode reactions and for the application of the manifold different voltammetric techniques and electroanalytical methods. Furthermore, three-electrode configurations are, e. g., also used in amperometric electrochemical sensors and in corrosion-measuring cells.

Generally, three-electrode cells contain a working electrode, a counter electrode, and a reference electrode. The desired potential is measured between the working and reference electrode and compared with an input or control voltage, and the current flow necessary to match both voltages is measured between the working and counter electrode. The working electrode is either a material sample to be examined or treated or provides the surface for the transfer of charge to and from the analyte. The almost non-current-carrying reference electrode is needed for measuring and controlling the potential of the working electrode. In most cases a conventional Ag/AgCl system is used for this purpose. The counter (or auxiliary) electrode serves only as second electrical connection to the electrolyte that completes the current path to the working electrode. It is usually made of an inert material, such as a noble metal or graphite.

Figure 5 shows the electrode configuration in a widely used three-electrode corrosion test cell [21]. To avoid any migration of chloride ions from the Ag/AgCl reference electrode into the measuring solution, it is inserted into an intermediate vessel IV filled with measuring solution. The potential of the working electrode is detected by means of a Luggin capillary LC that is positioned very close to the working electrode.

Four-Electrode Electrochemical Cells

Four-electrode cells are applied for electrochemical conductivity and impedance measurements, particularly in high conductivity solutions. According to Fig. 6 they contain two current electrodes CE and two voltage (sense) electrodes VE. For determining the conductivity of the solution, the alternating current I is applied to the current electrodes and the voltage U_M is measured at the voltage electrodes using a high impedance circuit. The main advantage over conventional two-electrode conductivity measuring cells is that errors due to electrode polarization and contamination as well as from cable and connector resistances are eliminated.

Spectroelectrochemical Cells

Spectroelectrochemistry (SEC) combines the techniques of electrochemistry and spectroscopy. It has become increasingly a valuable tool for studies of electron transfer processes and complex redox reactions of organic, inorganic, and organometallic compounds as well as biological systems. In a specially designed electrochemical cell, a redox-active compound is oxidized or reduced. The products of the redox reactions or electrogenerated intermediates are monitored in situ by established spectroscopic techniques such as UV–vis, IR, or Raman spectroscopy. A great advantage of spectroelectrochemistry is the possibility to obtain simultaneous information from combined electrochemical and spectroscopic experiments. Under potential control, spectroscopic information about in situ electrogenerated species can be obtained by absorption, transmission, fluorescence or reflection measurements [22–25].

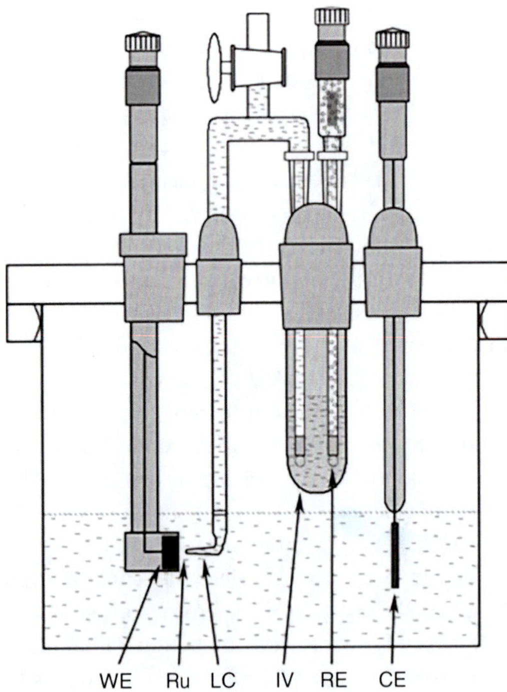

Cell, Electrochemical, Fig. 5 Electrochemical three-electrode cell. *WE* working electrode, *CE* counter electrode, *RE* reference electrode, *LC* Lugging capillary, *IV* intermediate vessel, *Ru* uncompensated resistance [21]

A variety of optically transparent thin-layer electrochemical cells have been developed for UV–vis spectroelectrochemical experiments where the beam is reflected at the highly polished surface of a carbon, gold, or platinum working electrode and led back to the detector via a system of mirrors. In another approach an optically transparent working electrode (OTE) is used that can be prepared from glass slides coated with transparent and electronically conductive indium-doped tin oxide (ITO) layer [26]. In optically transparent thin-layer electrochemical (OTTLE) cells, usually gold minigrid working electrodes are used [27]. Figures 7 and 8 show the principle of spectroelectrochemical measurements with optically transparent electrodes and a photo of an OTTLE cell with a mechanically much more stable working electrode, consisting of a partially laser-perforated 80 μm thick gold foil with an approximate 64 % optical

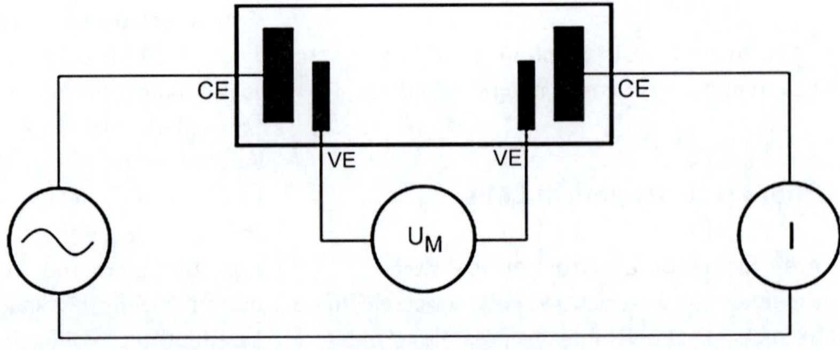

Cell, Electrochemical, Fig. 6 Schematic of a 4-electrode electrochemical conductivity measuring cell. *UM* measuring voltage, *I* measuring current, *CE* current electrodes, *VE* voltage electrodes

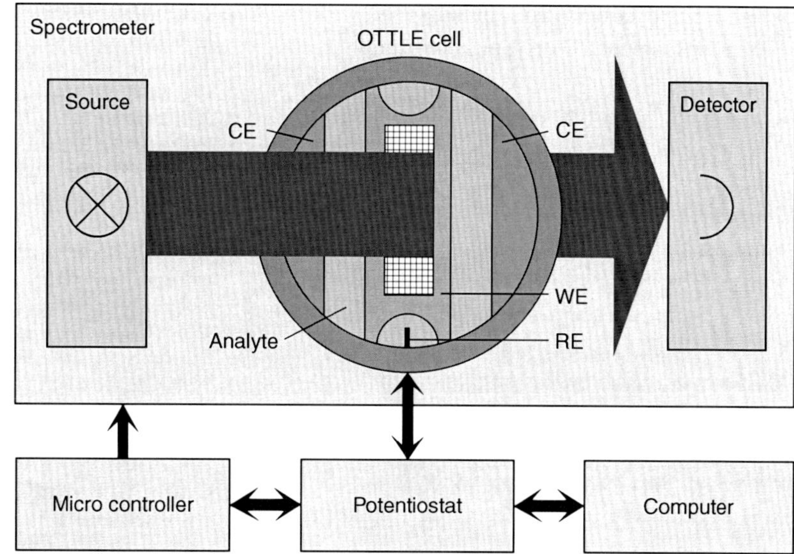

Cell, Electrochemical, Fig. 7 Computer-controlled equipment for spectroelectrochemical measurements. *WE* laser-perforated Au working electrode, *RE* Ag reference electrode, *CE* Pt counter electrodes

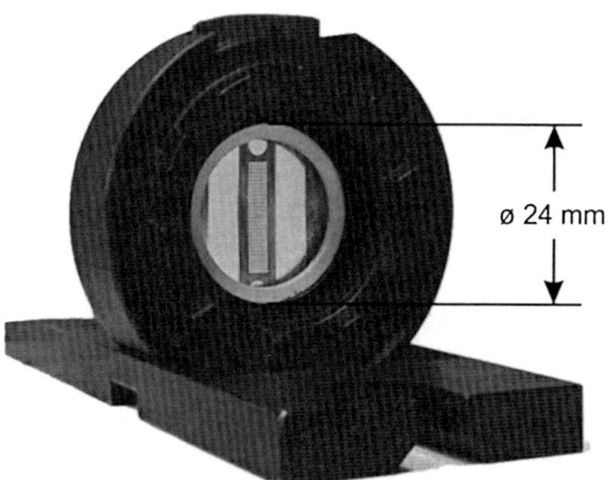

Cell, Electrochemical, Fig. 8 Photograph of the spectroelectrochemical OTTLE cell [29]

transparency. The cell has a well-defined and reproducible geometry and can be easily mounted and cleaned [28]. It has been used successfully for voltammetric and spectroelectrochemical studies on aminophenol and dithiines [29, 30].

Future Directions

Some future directions concerning electrochemical cells will be [31]:

- Optimization of electrochemical cells, particularly those with three-dimensional electrodes as well as those with membrane processes
- Design and construction of more efficient cells for electrowinning of copper to work at lower specific energy consumption and of cells for electrorefining of metals to improve their purity and to achieve economical operation
- Development of new battery systems and the improvement of existing systems
- Making fuel cells less expensive for wider both mobile and stationary applications and capable of operating and surviving in extreme conditions
- Creation of new electrochemical cells for bioelectrochemistry and bioelectrochemical engineering as subdisciplines undergoing very fast development

Cross-References

► Cell Membranes, Biological
► Conductometry
► Electrochemical Cell Design for Water Treatment
► Electrochemical Cells, Current and Potential Distributions
► Electrode
► Electrolytes, Classification
► Fuel Cells, Principles and Thermodynamics
► High-Temperature Polymer Electrolyte Fuel Cells

► Infrared Spectroelectrochemistry
► Molten Carbonate Fuel Cells, Overview
► Overpotentials in Electrochemical Cells
► Polymer Electrolyte Fuel Cells (PEFCs), Introduction
► Primary Batteries, Comparative Performance Characteristics
► Reference Electrodes
► Solid Electrolytes Cells, Electrochemical Cells with Solid Electrolytes in Equilibrium
► Solid Oxide Fuel Cells, Introduction
► Spectroelectrochemistry, Potential of Combining Electrochemistry and Spectroscopy
► UV–vis Spectroelectrochemistry

References

1. Holze R (2009) Experimental electrochemistry. A laboratory textbook. Wiley-VCH, Weinheim, p 3
2. Holze R (2009) Experimental electrochemistry. A laboratory textbook. Wiley-VCH, Weinheim, p 8
3. Atkins PW, de Paula J (2009) Physical chemistry, 8th edn. Oxford University Press, Oxford, p 216, 1019
4. Heitz E, Kreysa G (1986) Principles of electrochemical engineering. VCH, Weinheim/New York, pp 143–160
5. Bard AJ, Faulkner LR (2001) Electrochemical methods – fundamentals and applications. Wiley, New York/Chichester/Weinheim/Brisbane/Singapore/Toronto, p 18
6. Holze R (2008) Daniell, John Frederic, Daniell cell. In: Bard AJ, Inzelt G, Scholz F (eds) Electrochemical dictionary. Springer, Berlin, p 136
7. Holze R (1998) Leitfaden der Elektrochemie. Teubner, Stuttgart/Leipzig, p 95, 100
8. Inzelt G (2008) Daniell, John Frederic and Daniell cell. In: Bard AJ, Inzelt G, Scholz F (eds) Electrochemical dictionary. Springer, Berlin, p 81
9. Heitz E, Kreysa G (1986) Principles of electrochemical engineering. VCH, Weinheim/New York, pp 100–101
10. Graneau N (2006) In the grip of the distant universe. World Scientific, Hackensack, London, Singapore, p 191
11. Gileadi E (2011) Physical electrochemistry, fundamentals, techniques and applications. Wiley-VCH, Weinheim, pp 346–355
12. Holze R (2009) Experimental electrochemistry. A laboratory textbook. Wiley-VCH, Weinheim, pp 213–215
13. Schröder U (2008) Fuel cells. In: Bard AJ, Inzelt G, Scholz F (eds) Electrochemical dictionary. Springer, Berlin, p 286
14. Logan BE (2008) Microbial fuel cells. Wiley, Hoboken
15. Vielstich W, Lamm A, Gasteiger HA (eds) (2006) Handbook fuel cells, fundamentals, technology and applications. Wiley, Chichester
16. Hoogers G (2003) Fuel cell technology handbook. CRC Press, Boca Raton, Florida
17. Holze R (1998) Leitfaden der elektrochemie. Teubner, Stuttgart/Leipzig, pp 121–125
18. Kordesch K, Simander G (1996) Fuel cells and their applications. VCH, Weinheim
19. Heitz E, Kreysa G (1986) Principles of electrochemical engineering. VCH, Weinheim/New York, pp 198–201
20. Wendt H, Rohland B (1991) Design principles of fuel cells and their components. In Kreysa G (ed) Electrochemical cell design and optimization procedures. DECHEMA Monographs, vol 123. VCH, Weinheim
21. Oelßner W, Berthold F, Guth U (2006) The iR drop – well-known but often underestimated in electrochemical polarization measurements and corrosion testing. Mater Corros 57(6):455–465
22. Holze R (2009) Experimental electrochemistry. A laboratory textbook. Wiley-VCH, Weinheim, pp 201–202
23. Plieth W, Wilson GS, Gutierrez de la Fe C (1998) Spectroelectrochemistry: a survey of in situ spectroscopic techniques. Pure Appl Chem 70:1395–1414 and 2409–2412
24. Kaim W, Klein A (eds) (2008) Spectroelectrochemistry. Royal Society of Chemistry, Cambridge, UK
25. Keyes TE, Forster RJ (2007) Spectroelectrochemistry. In: Zoski CG (ed) Handbook of electrochemistry. Elsevier, Amsterdam, pp 591–633
26. Holze R (1998) Leitfaden der elektrochemie. Teubner, Stuttgart/Leipzig, pp 286–287
27. Hartl F, Luyten H, Nieuwenhuis HA, Schoemaker GC (1994) A versatile cryostated optically transparent thin-layer electrochemical (OTTLE) cell for variable-temperature UV–vis/IR spectroelectrochemical studies. Appl Spectr 48:1522
28. Oelßner W, Mitschke F, Hennig H, Kaden H (2000) Spektroelektrochemische Messzelle. Utility patent RN 200 12 374.2
29. Schwarz J, Oelßner W, Kaden H, Schumer F, Hennig H (2003) Voltammetric and spectroelectrochemical studies on 4-aminophenol at gold electrodes in aqueous and organic media. Electrochim Acta 48:2479–2486
30. Hennig H, Schumer F, Reinhold J, Kaden H, Oelssner W, Schroth W, Spitzner R, Hartl F (2006) Molecular structures and electronic transitions of 3,6-diphenyl-1,2-dithiin and its radical cation: a spectroelectrochemical and DFT study. J Phys Chem A 110(5):2039–2044
31. Stankovic´ V (2012) Electrochemical engineering – its appearance, evolution and present status. Approaching an anniversary. J Electrochem Sci Eng 2:67–75, doi:10.5599/jese.2012.0011 Open Access ISSN 1847–9286

Charged Colloids

Luc Belloni
CEA Saclay, Gif-sur-Yvette, France

Introduction

Colloidal solutions contain *mesoscopic* particles, the colloids, in the nanometer to micrometer size range, and small solvent and additional solute molecules. In presence of *polar* solvent, mainly water, the colloidal surfaces carry charges by ionization of surface groups or adsorption of bulk ions and the solution becomes a very asymmetrical mixture of *charged colloids*, *counterions* (ions of sign opposite to the colloidal charge, which must always be present in order to satisfy the whole electroneutrality of the solution), and *coions* (of same sign, which appear, for instance, by adding salt), all immersed in a structured, dielectric solvent. The structural, equilibrium, thermodynamical properties of such colloidal systems are mainly governed by excluded volume and coulombic couplings existing between the different species. Conceptually, they could be viewed as a multivalent electrolyte, the colloid playing the role of an ion of high valence, but it is clear that all concepts and approaches like the Debye-Hückel theory of electrolytes and the Bjerrum theory of ionic association, valid for simple ions, break down very soon for such big and highly charged particles. In the opposite regime, the local colloidal surface could be viewed as a piece of macroscopic planar-charged interface, monitored by the Gouy-Chapman theory of double layer, the Stern and DLVO theories, etc., described in parallel essays. This analogy remains perfectly valid as long as the size of the colloids remains much larger than all characteristic distances, mainly the Debye length and the surface-surface separation between approaching colloids. In the present essay, we will focus on the intermediate range of charged colloids, which cannot be related to the extreme cases of electrolytes or charged planes and where the size of the particles is a relevant parameter. Moreover, we will assume that the particles keep a globular, if not spherical, shape (the linear geometry of polymers is treated in other essays). Many organic/inorganic, natural, industrial, and biological systems correspond to this schematic picture: micelles of charged surfactants, soluble proteins in the eye lens, mineral oxide particles, latex in aqueous paints, etc., of size between 1 and 100 nm, belong to this category. One could also add colloids in the one-micrometer range when immersed in highly deionized conditions for which the Debye length reaches the same order of magnitude.

A complete understanding of such colloidal solutions requires a multi-scale description: how are the solvent molecules and the ions perturbed in the vicinity of the curved charged surfaces? What is the behavior of the electrostatic potential due to a colloid, at far distances? What is the *colloidal interaction*, the energy, or force felt by two approaching colloids, as a function of their separation? How are the colloids organized locally inside a solution at finite density (this question is central when one considers that light, X-ray, and neutron scattering techniques are experiments of choice for such systems)? Lastly, what is the colloidal stability, and what is the phase diagram? Since most of the experiments are sensitive to the big particles only, it is tempting to use the mesoscopic level of description which treats explicitly the colloids *only* and requires an a priori knowledge of the colloid-colloid potential of interaction that means the potential averaged over the degrees of freedom of the hidden ion and solvent species. The origin and validity of famous candidates like the screened coulombic and DLVO potentials in $\exp(-\kappa r)/r$ can be understood only by zooming in one step further and using the next-order level of description, the so-called primitive model (PM), which considers explicitly colloids as well as ions, all immersed in a continuous dielectric solvent and interacting via solvent-averaged pair potentials in $Z_iZ_j/\varepsilon r$ (Z_i is the valence of species i, ε is the dielectric constant of the solvent). In some cases, the PM is too much primitive and the precise structure of the water molecules must be explicitly introduced at the next level of description. This Born-Oppenheimer (BO) level is needed, for instance, when the hydration of the ions is perturbed in the vicinity of the highly charged colloid or when specific ionic effects take place. The complete understanding of the colloidal solution thus consists to back and forth exchange between the different levels. We will survey some key concepts in such multi-scale description.

Ionic Condensation, Charge Renormalization, Effective Charge

The *linearized* Poisson-Boltzmann (PB) theory says that the electrostatic potential around an isolated charged sphere of radius a and structural charge Z_{str} reads

$$\phi(r) = \frac{e\psi(r)}{kT} = \frac{Z_{str}L_B}{r}\frac{e^{-\kappa r}}{1+\kappa a} \quad (1)$$

where r is the distance to the center of the colloid, T the temperature, L_B the Bjerrum length, and κ the screening constant, inverse of the Debye length, due to the (salt) ions. This well-known screening behavior is based on many approximations: continuous solvent, mean-field picture of the ionic fluid (point-like ions, no ion-ion correlations inside the environment around the colloid), and, above all, linearization of the colloid-ion coupling. This last assumption, $Z_{str}L_B/a << 1$, is the most severe one and is never valid in practice. Meanwhile, even if the linearization is forbidden near the colloidal surface, it is at least valid far from it, so the asymptotic behavior should still obey the $\exp(-\kappa r)/r$ law. The notion of *effective charge* Z_{eff} is introduced by writing the coefficient in front of it in analogy with Eq. 1:

$$\phi(r) \underset{r\to\infty}{\approx} \frac{Z_{eff}L_B}{r}\frac{e^{-\kappa r}}{1+\kappa a} \quad (2)$$

The precise value of Z_{eff} is derived by numerically solving the full, nonlinearized PB equation (quite easy integration) and identifying the asymptote with Eq. 2. While Z_{eff} obviously coincides with Z_{str} at low Z_{str}, it becomes lower than Z_{str} and ultimately saturates at higher charge. What is the physical meaning of this very general *charge renormalization* and saturation effect? The counterions localized in the vicinity of the colloidal surface feel a coulombic energy so high (compared to kT) that they are *condensed* to (from the electrostatic point of view; there is no chemistry involved) and partly neutralize the bare colloid, and belong to an *effective* particle, whose charge Z_{eff} reflects the balance between the structural charge and the condensed one and characterizes the colloid seen from large distances. At the plateau, adding more surface charges is accompanied by an equivalent amount of added condensation and Z_{eff} becomes independent of Z_{str}!

In the absence of numerical PB solution, a good estimate of the saturation value is given by [1]:

$$\frac{Z_{eff}L_B}{4a} = \frac{1}{|Z_{counterion}|} \quad (3)$$

This approximation, valid at least at not too large κa (remember, this is again the intermediate size domain addressed in this essay, where the precise curvature is a key factor), somewhat generalizes the Bjerrum's concept of ionic association in multivalent electrolytes and should be advantageously compared to the equivalent, very famous Manning-Oosawa law for colloids of cylindrical geometry [2].

What happens when the colloids are no more isolated in a salt reservoir, as in colloidal solutions at *finite* concentration? The situation becomes more complex because (i) the counterion species which equilibrates the colloid one now contributes to the screening (κ becomes dependent on the colloidal density through the electroneutrality condition) and (ii) the notion of far-field environment around a given colloid has little meaning since colloidal neighbors are localized nearby. Meanwhile, a precise value of Z_{eff} can be defined and calculated using the PB approach in the *cell* geometry [3]. For more detailed information about this very general concept of ionic condensation, the interested reader could refer to the review [4].

Electrostatic Colloidal Interaction

Now, *two* spherical charged colloids are approaching. What is the pair potential of interaction $v(r)$, averaged over the solvent + ion degrees of freedom, as a function of the center-center separation r? The PB approach, although approximate, is again a method of choice and has been the subject of thousands of works since Gouy-Chapman, Debye-Hückel, DLVO, etc. Within this mean-field picture, it can be easily demonstrated that the interaction between *like*-charged colloids is always *repulsive*. It is fruitful to note that it is a *free energy* type of interaction, *subtle* balance between an *attractive* internal energy contribution (the colloids being attracted by the counterions preferentially localized in the interstitial region between the two spheres), and a *repulsive* entropy contribution (the osmotic pressure of these extra ions pushing against the colloidal walls). Beware! An inconsistent evaluation of one or both terms may break the balance and predict an incorrect sign of the interaction!

The general PB resolution around two colloids requires a numerical treatment much more difficult than around one; this explains why the systematic solution has been obtained only in the mid-1990s [5]. Fortunately, it is in general sufficient to know about the asymptotic law, valid at large separation, which can be linked to the asymptotic behavior (Eq. 2) around each colloid, taken as isolated (weak overlap approximation of the two diffuse layers). That defines the very popular DLVO-screened coulombic colloidal interaction (like-charged particles) [6]:

$$\frac{v(r)}{kT} \underset{\kappa(r-2a)>>1}{\approx} \frac{{Z_{\mathrm{eff}}}^2 L_B}{r} \frac{e^{-\kappa(r-2a)}}{(1+\kappa a)^2} \tag{4}$$

The direct $1/\varepsilon r$ repulsion is screened by the condensed ions (through the factor Z_{eff} [7]) and by the free, diffuse ones (through the $\exp(-\kappa r)$ factor), with finite size corrections in κa. This very powerful simple expression has been the starting point of any study of charged colloid interaction and stability at the mesoscopic level of description since the 1940s. It can be combined with the Van-der-Waals attraction to define the theory of charge stabilization against irreversible aggregation or flocculation [6]. Its strength and range can be monitored at will by adding or removing salt ions, especially multivalent counterions (see Eq. 3). A very wide zoology of interaction law (sign, strength, range, shape) can be obtained by adding small solute molecules inside the solution and creating depletion, steric, bridging, etc. contributions competing with the coulombic term (see, for instance, review [8]). Since the early 1980s, this pair potential expression is routinely used with great success in the fit of scattering intensity spectra, osmotic pressure equations of state, sedimentation profiles, etc. At finite concentration and/or low-added salinity, the ionic strength is dominated by the free counterion contribution and the interaction among the colloids, despite its pairwise additive decomposition, becomes of N-body nature!

What are the limits of the approximated expression Eq. 4? Mainly those due to the mean-field nature of PB. For, say, 99 % of the studied systems, the ions are monovalent, ion-ion correlations in water can be safely ignored, and the standard expression is valid. This is no more the case in presence of multivalent counterions (or monovalent ions in solvent of low ε). That opens to the fascinating concept of electrostatic *attraction* between *like*-charged colloids, subject of numerous false analyses, debates, and controversies in the literature for 30 years. Figure 1 presents Monte Carlo (MC) simulations data for the force vs. separation law within the primitive model (two latex colloids and ions in continuous solvent) in presence of counterions of increasing valence. While the PB/DLVO prediction remains everywhere repulsive, the exact MC behavior deviates at intermediate separation and develops an attractive well deeper and deeper as the valence increases above 3. This non mean-field effect is due to the repulsions and correlations among the counterions localized in the intersticial region (discreteness of the condensed layer). The same type of colloidal attraction is responsible for a liquid-gas (concentrated solution-dilute solution) phase separation, observed

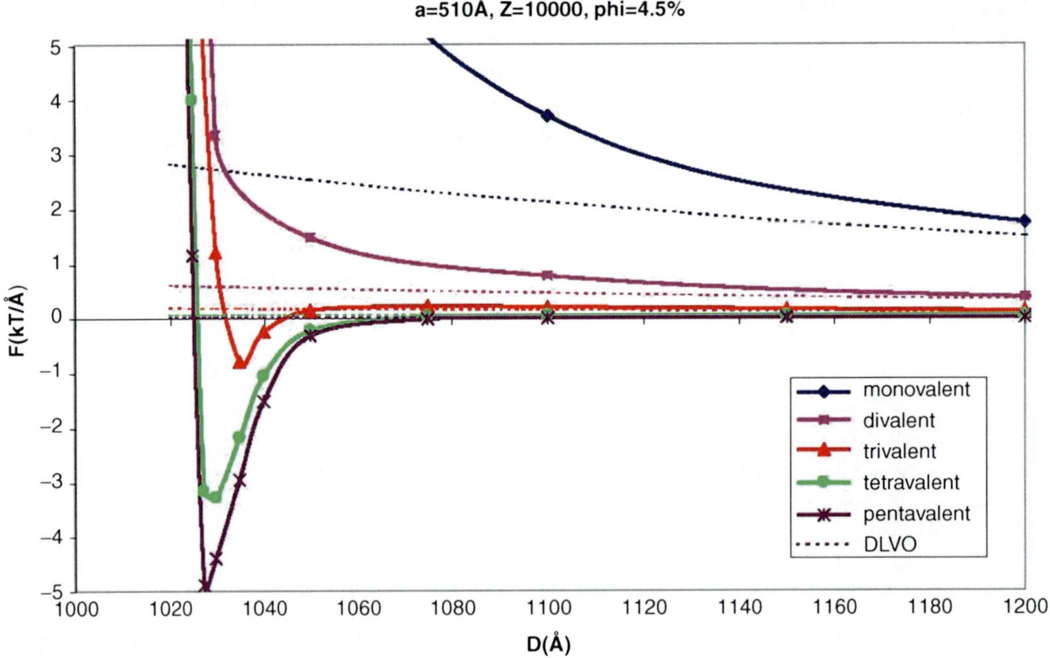

Charged Colloids, Fig. 1 Monte Carlo force versus distance F(D) between two charged colloidal spheres of radius 510 Å and structural charge 10,000 in presence of multivalent counterions. Cell geometry at 4.5 % volume fraction, water solvent at room temperature, no added salt

again by numerical simulation within the PM [9], and reminiscent of what is known for symmetrical electrolytes. It is interesting to note that, while theories and simulations have clearly demonstrated the existence of attraction and demixion, a clear, undisputed experimental evidence of such effect is still lacking, despite some claims (and controversies). This should be opposed to the case of planes (SFA measurements, clay behavior) or rods (phase separation of polyelectrolytes, formation of DNA bundles) in presence of multivalent counterions.

Future Directions

Up to now, we have not addressed the more refined BO level of description (explicit discrete solvent). The reason is technical: even now, it is still very difficult (read very CPU time consuming, with poor statistics) to perform a numerical simulation with tens of thousands of water molecules, with hundreds of ions around a structured charged macromolecule (not to mention between many ones!). There is a clear need for alternate theories, maybe less exact but with well-controlled validity and much quicker numerical resolution. One observes in the recent literature a profusion of works on such approaches, like the molecular Ornstein-Zernike integral equations, the RISM, and the dipolar-PB trying to fill the gap. Beware, *quicker* doesn't mean *easier*! These modern, up-to-date, powerful techniques require sophisticated numerical codes and will face a double challenge in the near future: improving the validity of the approximation (say the closure for integral equations) and accelerating the resolution.

Cross-References

- ► DLVO Theory
- ► Ions in Clays
- ► Polyelectrolytes, Properties
- ► Polyelectrolytes, Simulation

References

1. Belloni L, Drifford M, Turq P (1984) Counterion diffusion in polyelectrolyte solutions. Chem Phys 33:147–154
2. Manning GS (1969) Limiting laws and counterion condensation in polyelectrolyte solutions I. Colligative properties. J Chem Phys 51:924–933
3. Alexander S, Chaikin PM, Grant P, Morales GJ, Pincus P, Hone D (1984) Charge renormalization, osmotic pressure, and bulk modulus of colloidal crystals: theory. J Chem Phys 80:5776–5781
4. Belloni L (1998) Ionic condensation and charge renormalization in colloidal suspensions. Coll Surf A 140:227–243
5. Carnie SL, Chan DYC, Stankovitch J (1994) Computation of forces between spherical colloidal particles: nonlinear poisson-boltzmann theory. J Colloid Interface Sci 165:116–128
6. Verwey EJW, Overbeek JTG (1948) Theory of stability of lyophobic colloids. Elsevier, Amsterdam
7. Bell GM, Levine S, McCartney LN (1970) Approximate methods of determining the double-layer free energy of interaction between two charged colloidal spheres. J Colloid Interface Sci 33:335–359
8. Belloni L (2000) Colloidal interactions. J Phys Cond Matter 12:R549–R587
9. Rescic J, Linse P (2001) Gas-liquid phase separation in charged colloidal systems. J Chem Phys 114:10131–10136

Chlorate Cathodes and Electrode Design

Ann Cornell
School of Chemical Science and Engineering, Applied Electrochemistry, KTH Royal Institute of Technology, Stockholm, Sweden

Introduction

Sodium chlorate ($NaClO_3$) is industrially produced by electrolysis. Over 3 million tons of sodium chlorate is produced annually world wide, about 40 % of which in North America [1]. The main use of sodium chlorate is in the pulp and paper industry for generation of chlorine dioxide, which is used to bleach chemical pulp. Substitution of elemental chlorine by chlorine dioxide in the bleaching sequences can drastically reduce the formation of undesired chlorinated organic compounds in the pulp bleaching process [2]. This is the driver behind the increased demand for chlorate over the last 25 years.

Electrolytic production of sodium chlorate consumes significant amounts of electrical energy, which constitutes up to 70 % of the production costs. Close to half of the electrical energy added ends up as irreversible losses in the form of IR drops in the inter electrode gaps and overvoltages for the electrode reactions. The major part of the losses relates to the cathodes, and with increasing costs for electricity, the demand for more energy-efficient cathode materials has become increasingly important.

Here first a brief description of the rather complex electrochemistry and chemistry of the chlorate process will be given. The electrodes, in particular the cathodes, will then be discussed.

Electrode Reactions and Chlorate Chemistry

Main Reactions

Industrial chlorate electrolysis takes place in undivided cells, where sodium chlorate and hydrogen gas are formed as described by reaction 1. More detailed, reactions 2 and 3 show the main anode and cathode reactions of chloride oxidation and hydrogen evolution, respectively. Note that these electrode reactions are similar to those in a chlor-alkali cell, though while a chlor-alkali cell has a membrane or diaphragm separating an acidic anolyte from an alkaline catholyte, the chlorate cell is undivided with an electrolyte at close to neutral pH. Chlorine formed therefore dissolves as in reactions 4 and 5 and, chlorate is formed in a disproportionation reaction, number 6 below [3].

$$NaCl(s) + 3\,H_2O(l) \rightarrow NaClO_3(s) + 3\,H_2(g) \quad (1)$$

$$2Cl^- \rightarrow Cl_2 + 2e^- \quad (2)$$

$$2H_2O + 2e^- \rightarrow H_2 + 2OH^- \quad (3)$$

$$Cl_2 + H_2O \rightarrow ClOH + Cl^- + H^+ \quad (4)$$

$$ClOH \leftrightarrow ClO^- + H^+ \quad (5)$$

$$2ClOH + ClO^- \rightarrow ClO_3^- + 2H^+ + 2Cl^- \quad (6)$$

Reaction 6 proceeds at its fastest rate when the concentration ratio [ClOH]:[ClO⁻] is 2, thus at

$$pH = pK_a - \log\ 2 \quad (7)$$

where pK_a refers to reaction 5. This corresponds to a pH of 5.8–6.5 [3], and to keep reaction 6 at a high rate, the chlorate electrolyte is controlled at a pH of 6–6.5.

Chromate is added to chlorate electrolyte, where it has several functions. In addition to hindering the cathodic reduction of the hypochlorite and chlorate ions, it acts as a buffer in the pH range 5–7 [3, 4], an effect mainly related to the equilibrium in reaction 8. In acidic solution, as in the anodic diffusion layer, dichromate is formed according to reaction 9.

$$CrO_4^{2-} + H^+ \leftrightarrow HCrO_4^- \quad (8)$$

$$2HCrO_4^- \leftrightarrow Cr_2O_7^{2-} + H_2O \quad (9)$$

Although the chlorate electrolyte is close to pH neutral, the pH in the electrolyte at the electrode surfaces is far from neutral. The pH of the electrolyte varies from about 4 at the anode, to 6–7 in the electrolyte bulk and up to high values, in the range of 12–13, at the hydrogen-evolving cathode.

Side Reactions

The current efficiency in the chlorate process is commonly 93–95 % [4]. The deviation from 100 % is caused by side reactions on the electrodes and in the bulk as well as by Cl_2 escaping with the cell gas. Oxygen is the major by-product, and oxygen in the cell gas affects not only the electricity consumption but is also considered a safety risk as explosive oxygen-hydrogen gas mixtures may form.

There are many possible reactions leading to the formation of oxygen gas. In the absence of hypochlorite (in the text referred to as the sum of ClOH and ClO^-), at potentials lower than the reversible potential for chlorine evolution, the main reaction is oxygen evolution from water discharge:

$$2H_2O \rightarrow O_2 + 4H^+ + 4e^- \quad (10)$$

Hypochlorite is known as an important source for oxygen in chlorate and chlor-alkali electrolysis [5–9]. Anodic chlorate formation from oxidation of hypochlorite, under acidic conditions written as reaction 11, has been suggested and is presented in the review by Ibl and Vogt [9].

$$12ClOH + 6H_2O \rightarrow 4ClO_3^- + 8Cl^- + 24H^+ + 3O_2 + 12e^- \quad (11)$$

Other suggested electrochemical oxygen-forming reactions are reactions 12a and 12b, involving hypochlorous acid [8] and the hypochlorite ion [6, 9, 10], respectively.

$$ClOH + H_2O \rightarrow 3H^+ + Cl^- + O_2 + 2e^- \quad (12a)$$

$$ClO^- + H_2O \rightarrow 2H^+ + Cl^- + O_2 + 2e^- \quad (12b)$$

Hypochlorite may also decompose to form oxygen in a chemical reaction in the bulk according to reactions 13a and 13b, catalyzed by several transition metal ions and metal oxide particles [3, 4, 8, 11].

$$2ClOH \rightarrow O_2 + 2HCl \quad (13a)$$

$$2ClO^- \rightarrow O_2 + 2Cl^- \quad (13b)$$

Perchlorate formation by oxidation of chlorate, reaction 14, proceeds with low current efficiency, about 0.05 %, at normal chlorate operating conditions [11]. Although produced in low amounts, it remains in the electrolyte and may over time accumulate to high concentrations, which makes the chlorate process less efficient.

$$ClO_3^- + H_2O \rightarrow ClO_4^- + 2H^+ + 2e^- \quad (14)$$

Two important side reactions on the cathode are the reduction of hypochlorite and of chlorate, reactions 15 and 16, which are both suppressed by the addition of dichromate to the chlorate electrolyte [3].

$$ClO^- + H_2O + 2e^- \rightarrow Cl^- + 2OH^- \quad (15)$$

$$ClO_3^- + 3H_2O + 6e^- \rightarrow Cl^- + 6OH^- \quad (16)$$

The Chlorate Electrolyte

A typical chlorate electrolyte consists of 500–650 g/l $NaClO_3$, 100–120 g/l NaCl, 1–4 g/l NaClO, Cr(VI) corresponding to 1–6 g/l $Na_2Cr_2O_7$, at a bulk pH of 6.0–6.5 and a temperature of 70–85 °C. The electrolyte can also contain $NaClO_4$ at concentrations that should not exceed 100 g/l [11]. A high chlorate concentration is essential for the separation of $NaClO_3(s)$ by crystallization, and a high chloride concentrations is important for the anode operation. Still, the salt concentrations must have a margin to the solubility limits to avoid precipitation in, for example, the electrochemical cells. An elevated temperature is important for the chlorate formation, c.f. reaction 6 above, and in order to lower the electrode overpotentials.

The Chlorate Cathode

Industrial Cathode Materials

Cathode materials in the first years of chlorate manufacture were copper, nickel, and platinum. Today it is mainly steel, mild steel, or low-carbon steel and in some plants titanium or a Ti-0.2 % Pd alloy [3, 4]. A problem in chlorate electrolysis is the extremely corrosive electrolyte with active chlorine that attacks the steel cathodes when they are not under cathodic protection. This not only shortens the lifetime of the cathodes but also causes problems with pieces of steel in the electrolyte. They may cause short circuits and poor electrolyte circulation when getting stuck in the narrow (2–4 mm) electrode gaps, and they may also contaminate the chlorate product. However a benefit of this corrosion is that the surface is continuously renewed with removal of deposits of mainly calcium and magnesium compounds. Corrosion also increases the active surface area and thereby lowers the overvoltage. Some corrosion products may even be more catalytically active than the steel itself.

Valve metals as titanium do not corrode under open circuit conditions, but form hydrides and become brittle when working as hydrogen, evolving cathodes. Hydrogen penetrates through the lattice interstices and causes an expansion of the crystal. The lifetime for a 2 mm thick Ti cathode is about 2 years [8], whereas for the Ti-Pd alloy, the lifetime is longer. The overvoltage for hydrogen evolution is much higher on titanium than on iron and steel.

The Chromium Hydroxide Film

As mentioned above chromate is added to the chlorate electrolyte, mainly to hinder the reduction of hypochlorite and chlorate on the cathode (reactions 15 and 16). During electrolysis Cr(VI) is reduced and forms a thin film, less than 10 nm thick, of $Cr(OH)_3 \times H_2O$ on the cathode [12]. The film hinders also some other cathodic reactions as oxygen reduction, whereas hydrogen evolution can take place though with changed kinetics compared to on a bare electrode surface [13, 14]. It has been found that very low amounts of chromate in the electrolyte (in the micro-molar range) are sufficient to hinder hypochlorite reduction on polished titanium [15], whereas a corroded steel surface requires higher chromate concentrations (>3 g/l $Na_2Cr_2O_7$) for a high current efficiency for hydrogen evolution [16]. As Cr (VI) is harmful and not environmentally benign, a replacement for chromate addition is desired.

Hypochlorite reduction, reaction 15, is a mass transport controlled reaction in the absence of chromate in the electrolyte and takes place easily on most electrode materials. Chlorate reduction (reaction 16), on the other hand, is kinetically controlled and chlorate is present at a high concentration of about 6 M in chlorate electrolyte. Chlorate reduction is slow on metals as Co, Ni, Mo, and Ti [16] but rapid on certain catalytic oxides [17, 18]. In fact, when electrolyzing a solution of 550 g/l $NaClO_3$ without chromate addition with cathodes coated by RuO_2, no hydrogen gas bubbles were formed [18]. Instead of reducing water to hydrogen, the reaction of chlorate reduction was the by far dominating. When adding chromate to the electrolyte, the current efficiency raised to close to 100 %.

Cathode Development

Attempts to develop an activated cathode for chlorate cells have not yet been successful, and a material for the application faces many constraints. Some important properties for a chlorate cathode are (a) low overpotential for hydrogen evolution, (b) high stability during hydrogen evolution (resistant to the mechanical stress from gas bubbles and no detrimental hydride formation), (c) resistant during shut downs (low corrosion rate at open circuit in chlorate electrolyte), (d) low activity for hypochlorite decomposition, (e) low activity for reduction of hypochlorite and chlorate in the presence and in the absence of the chromium hydroxide film (the latter a step in the search for a chromate-free process), (f) relatively resistant to impurities in the electrolyte, (g) easy to manufacture, (h) easy to install in existing cell concepts, and (i) cost-effective.

Ruthenium dioxide is a very active catalyst for hydrogen evolution and used in cathode coatings on nickel substrates in chlor-alkali cells [19]. Extensive work has been carried out to develop an activated chlorate cathode made of a nano crystalline alloy consisting of Ti, Ru, Fe, and O, applied as a coating on an electrode substrate [20]. The alloy was prepared by high-energy ball milling and deposited, for example, by plasma spraying. In chlorate electrolyte at a current density of 2.5 kA/m^2, the overpotential was about 300 mV lower than for mild steel cathodes [20]. Ruthenium was important for the catalytic activity [21], but since it is expensive, the minimum ruthenium content to sustain appreciable activity is of interest. Experiments with varying catalyst loadings from 300 to 10 mg/cm^2 showed a constant activity down to 20 mg/cm^2 [22]. The presence of oxygen in the coating was essential for the electrode stability, probably related to a larger resistance to hydride formation in the presence of oxygen [23, 24]. Activated cathodes for chlorate production are still not available due to problems with long-term stability at industrial operating conditions.

An alternative way to lower the cell voltage of a chlorate cell is to change cathode reaction from hydrogen evolution to reduction of oxygen, reaction 17 [25]. This change reduces the cell voltage by approximately 1 V, but the technique is not yet commercial.

$$O_2 + 2H_2O + 4e^- \rightarrow 4OH^- \qquad (17)$$

The Chlorate Anode

Anode materials in the early days were platinum and magnetite, later replaced by graphite. In the late 1970s, graphite was replaced by coated metal anodes – titanium coated by a mixture of platinum and iridium (Pt-Ir anodes) or by ruthenium dioxide in combination with titanium dioxide (ruthenium-based DSA®). These metal anodes are not consumed during operation as were the graphite anodes. Today the ruthenium-based DSA® is preferred over Pt-Ir coatings due to the higher cost and shorter lifetime of the latter. DSA®s for chlorate generally consist of a catalytic coating of about 30 mol% RuO_2 and 70 mol% TiO_2 on a titanium substrate. A big era for chlorate research was in the 1960s and 1970s when graphite anodes were used. Therefore, much of the published work is done under operating conditions which differ markedly from those in use today with DSA®s; see Table 1. The anode overpotential and the rate of by-product oxygen formation vary between coatings of different morphology and chemical composition. For example, increasing the surface roughness of a DSA® lowers the anode overpotential and, at the same time, increases the selectivity for oxygen evolution.

Chlorate Cathodes and Electrode Design, Table 1 Typical operating data for chlorate cells [9]

Anodes	Graphite	Coated titanium
Cell potential, V	2.9–3.8	2.9-3.3
Current density, A/m^2	300–600	1,500–4,000
Current efficiency, %	82–87	92–95
Temperature, °C	40–45	60–80
pH	6–7	6–6.5
Concentration, g/l		
NaCl	100–310	50–310 (100–120)
$NaClO_3$	0–500	0–650 (500–650)
$Na_2Cr_2O_7$	1–6	1–6
Energy cons., kWh/t $NaClO_3$	5,000–7,000	4,600–5,400
Anode wear, kg/t $NaClO_3$	7–18	$(0.1–0.5) \times 10^{-3}$
Current concentration, A/l	2–6	20–50

In a chlorate plant, premature failure of the anode coating may occur if, for example, the electrolyte is accidentally contaminated by impurities that adsorb at active sites or that block the anode surface and thereby hinder the supply of chloride ions. In the latter case, oxygen evolution is enhanced, resulting in oxidation of ruthenium and loss of the active coating. High anodic potentials (or high cell voltages) combined with high oxygen levels indicate possible anode failure with passivation and ruthenium loss. The term passivation here relates to the formation of an irreversible, highly resistive, titanium oxide layer between the titanium metal substrate and the active oxide coating. The coating deterioration can be a fast process, in the order of days or weeks, with a costly recoating of the anodes as a consequence.

Future Directions

A forecast for the growth of sodium chlorate production until 2016 is given in Chemical Economic Handbook [1]. The main use of sodium chlorate is in the pulp and paper industry for generation of chloride dioxide, which is used to bleach chemical pulp. In the nearest years to come, an expected growth in the production of chemical pulp, and a trend towards higher-brightness paper, is anticipated to increase the demand for sodium chlorate. Asia, South America, and Russia are the regions where this growth will be strongest. The major cost in sodium chlorate production is that of electrical energy, and the price for sodium chlorate depends on the price for electricity. When electricity prices are high, it may be advantageous for pulp mills to replace some of the chlorate in the bleaching by other chemicals as hydrogen peroxide. This may affect the demand for sodium chlorate.

The high, and increasing, cost for electricity motivates the development of more energy-efficient production, in principle electrolysis with a lower cell voltage and a higher current efficiency.

Main directions to lower the cell voltage are a cathode that is stable, active, and selective for hydrogen evolution, alternatively an oxygen-depolarized cathode. Operation at higher temperature, 90–100 °C, would lower the cell voltage but is today limited by the materials used in the process.

The current efficiency could be improved by the development of an anode coating with a lower selectivity for the by-products oxygen and perchlorate. Another way is to find a catalyst for the conversion of hypochlorite to chlorate (reaction 6), which could favor the desired chlorate formation over oxygen-forming reactions and thereby increase the current efficiency. It could also allow smaller reaction volumes – more compact plant designs.

Environmental concerns motivate the development on an alternative to chromate addition to the electrolyte. The main roles of chromate are to hinder cathodic side reactions and to buffer the electrolyte, and it may be difficult to find a single method that satisfies both functions. A combination of in situ additives may be required, or a new more selective cathode material together with a suitable electrolyte buffer agent.

Cross-References

- ▶ Chlorate Synthesis Cells and Technology
- ▶ Chlorine and Caustic Technology, Overview and Traditional Processes
- ▶ Chlorine and Caustic Technology, Using Oxygen Depolarized Cathode
- ▶ Electrocatalysis of Chlorine Evolution
- ▶ Hydrochloric Acid Electrolysis
- ▶ Hydrogen Evolution Reaction

References

1. Chemical Economics Handbook (2012) http://www.ihs.com/products/chemical/planning/ceh/sodium-chlorate.aspx. Accessed 7 Jan 2013
2. Axegård P, Bergner E (2011) Environmental performance of modern ECF bleaching, international pulp bleaching conference, Portland, pp 119–126
3. Colman JE (1981) Electrolytic Production of Sodium Chlorate. In: Alkire R, Beck T (eds) Tutorial lectures in electrochemical engineering and technology, vol 77, AIChE symposium series 204, Institute of Chemical Engineers, New York, p 244
4. Viswanathan K, Tilak BV (1984) Chemical, electrochemical, and technical aspects of chlorate manufacture. J Electrochem Soc 131:1551–1559. doi:10.1149/1.2115908
5. Wanngård J (1992) Impurity Effects in Chlorate Plants. In: Wellington TC (ed) Modern chlor-alkali technology, vol 5. Elsevier Appl Sci, London/New York, p 295
6. Hardee KL, Mitchell LK (1989) The Influence of Electrolyte Parameters on the Percent Oxygen Evolved from a Chlorate Cell. J Electrochem Soc 136:3314–3318. doi:10.1149/1.2096444
7. Evdokimov SV (1999) Kinetics of oxygen evolution on dimensionally stable anodes during chlorate electrolysis. Russ J Electrochem 37:792–797. doi:10.1023/A:1016726801283
8. Kotowski S, Busse B (1986) The Oxygen Side Reaction in the Membrane Cell. Investigation of Various Oxygen and Chlorate Sources. In: Wall K (ed) Modern chlor-alkali technology, vol 3. Ellis Horwood, Chichester, p 310
9. Ibl N, Vogt H (1981) Inorganic Electrosynthesis. In: Bockris JO'M, Conway BE, Yeager E, White RE (eds) Comprehensive treatise of electrochemistry, vol 2. Plenum, New York, p 167
10. Byrne P, Fontes E, Lindbergh G, Parhammar O (2001) A simulation of the tertiary current density distribution from a chlorate cell –I. Mathematical model. J Electrochem Soc 148:D125–D132. doi:10.1149/1.1397318
11. Tilak BV, Chen C-P (1999) Electrolytic Sodium Chlorate Technology: Current Status. In: Burney HS, Furuya N, Hine F and Ota KI (eds) Chlor-alkali and chlorate technology, The electrochemical society proceedings series, PV 99–21, Pennington, p 8
12. Ahlberg Tidblad A, Lindbergh G (1991) Surface analysis with ESCA and GD-OES of the film formed by cathodic reduction of chromate. Electrochim Acta 36:1605–1610. doi:10.1016/0013-4686(91)85013-W
13. Lindbergh G, Simonsson D (1991) Inhibition of cathode reactions in sodium-hydroxide solution containing chromate. Electrochim Acta 36:1985–1994. doi:10.1016/0013-4686(91)85083-J
14. Cornell A, Lindbergh G, Simonsson D (1992) The effect of addition of chromate on the hydrogen evolution reaction and on iron oxidation in hydroxide and chlorate solutions. Electrochim Acta 37:1873–1881. doi:10.1016/0013-4686(92)85093-Z
15. Gustavsson J, Li G, Hummelgård C, Bäckström J, Cornell A (2012) On the suppression of cathodic hypochlorite reduction by electrolyte additions of molybdate and chromate ions. J Electrochem Sci Eng 2:185–198. doi:10.5599/jese.2012.021
16. Wulff J, Cornell A (2007) Cathodic current efficiency in the chlorate process. J Appl Electrochem 37:181–186. doi:10.1007/s10800-006-9263-3
17. Tilak BV, Tari K, Hoover CL (1988) Metal anodes and hydrogen cathodes – their activity towards O_2 evolution and ClO_3^- reduction reactions. J Electrochem Soc 135:1386–1392. doi:10.1149/1.2095999
18. Cornell A, Simonsson D (1993) Ruthenium dioxide as cathode material for hydrogen evolution in hydroxide and chlorate solutions. J Electrochem Soc 140:3123–3129. doi:10.1149/1.2220996
19. Wendt H, Kreysa G (1999) Electrochemical Engineering: science and technology in chemical and other industries. Springer, Berlin/Heidelberg
20. Boily S, Jin S, Schulz R, Van Neste A (1997) Alloys of Ti Ru Fe and O and use thereof for the manufacture of cathodes for the electrochemical synthesis of sodium chlorate, US 5662834A
21. Jin S, Van Neste A, Ghali E, Boily S, Schulz R (1997) New cathode materials for chlorate electrolysis. J Electrochem Soc 144:4272–4279. doi:10.1149/1.1838177
22. Gebert A, Lacroix M, Savadogo O, Schulz R (2000) Cathodes for chlorate electrolysis with nanocrystalline Ti-Ru-Fe-O catalyst. J Appl Electrochem 30:1061–1067. doi:10.1023/A:1004030706423
23. Roue L, Irissou E, Bercier A, Bouaricha S, Blouin M, Guay D, Boily S, Huot J, Schulz R (1999) Comparative study of nanocrystalline Ti2RuFe and Ti2RuFeO2 electrocatalysts for hydrogen evolution in long-term chlorate electrolysis conditions. J Appl Electrochem 29:551–560. doi:10.1023/A:1026441532352

24. Roue L, Guay D, Schulz R (2000) Hydrogen electrosorption in nanocrystalline Ti-based alloys. J Electroanal Chem 480:64–73. doi:10.1023/A:1026441532352
25. Fontes E, Håkansson B, Herlitz F, Lindstrand V (2012) Process for producing alkali metal chlorate. US 8216443 B2

Chlorate Synthesis Cells and Technology

Ann Cornell
School of Chemical Science and Engineering, Applied Electrochemistry, KTH Royal Institute of Technology, Stockholm, Sweden

Introduction

Chlorate has been industrially produced for over a century, in the beginning primarily produced as potassium chlorate ($KClO_3$) for safety matches but today mainly as sodium chlorate ($NaClO_3$) used for bleaching in the pulp and paper industry. At the pulp mills chlorate is converted to chlorine dioxide, which is used to bleach chemical pulp.

The first chlorate plant was built in Switzerland as early as in 1886 [1] and since then the technology has much improved. An important example of chlorate process development is the introduction of DSA in the late 1970s. This allowed operation at higher temperature and higher current density, see "▸ Chlorate Cathodes and Electrode Design", which triggered the development of new cell systems. Another important example is the development of closed loop plants. Earlier it was common that electrolyte from the chlorate plants was shipped as product to the pulp mills, and used in the chlorine dioxide generators. Today most often a dry product, chlorate crystals, are instead shipped and dissolved in aqueous solution at the pulp mills. Without the electrolyte shipments and other process bleeds, impurities that dissolve in the chlorate electrolyte risk building up to high levels that may damage the operation of, primarily, the electrodes. Operation of closed loop plants requires various purification steps in the process, to treat the raw materials and other process streams. Hydrogen utilization is another increasingly important issue. In the old days it was common to just vent the gas, but today the product of hydrogen is important for the production economy. Hydrogen can be used as a reagent in the production of chemicals as ammonia, hydrogen peroxide etc., or may be used as a fuel without any CO_2 emission. The possibilities for utilization depend on the plant location, on the local need for hydrogen gas.

$$NaCl(s) + 3H_2O(l) \rightarrow NaClO_3(s) + 3H_2(g) \quad (1)$$

The overall reaction for chlorate process is given in reaction 1. More description of the complex chlorate chemistry and the main electrochemical reactions can be found in the chapter "▸ Chlorate Cathodes and Electrode Design." The main steps of a typical chlorate production process will be presented in the following text, with emphasis on the electrochemical cells.

Industrial Chlorate Production: Overview of the Process

The process flow diagram in Fig. 1 illustrates a typical modern chlorate plant. Water and raw salt enter in the left hand side of the figure, hydrogen gas and a solid chlorate salt leave as red product streams.

The raw salt can be rock, solar or vacuum salt, where the latter has been purified by vacuum crystallization. Impurities as calcium and magnesium etc. in the salt may harm the electrolysis operation by precipitating on the electrodes and result in high electrode potentials. Therefore purification of the salt is needed, and the quality of the salt set requirements on necessary purification steps. In Fig. 1 the incoming salt is first dissolved and then subject to ion exchange for removal of divalent cations as Ca^{2+} and Mg^{2+}. The evaporator illustrates re-crystallization of the

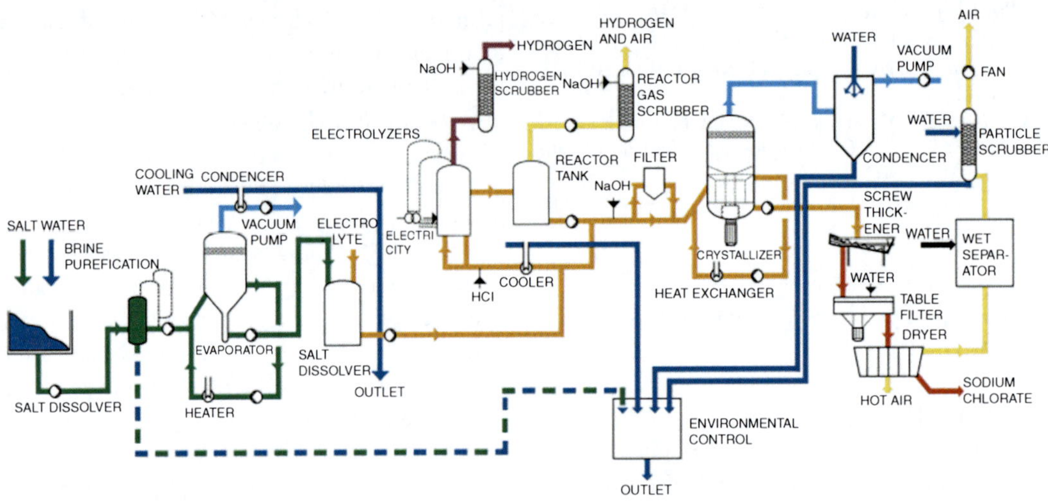

Chlorate Synthesis Cells and Technology, Fig. 1 Process flow diagram of a sodium chlorate plant (Courtesy of AkzoNobel Pulp and performance chemicals)

salt to remove sulfate ions, which can otherwise harm the anode operation and result in increased levels of the by-product oxygen. The purified salt is dissolved into chlorate electrolyte and enters the electrolyte circulation loop, orange process stream. Note that other salt purification methods than those shown in Fig. 1 can be used, for example precipitation and filtration steps are common in the brine treatment [2].

The electrolyte circulation loop runs through several steps. In Fig. 1 there is first a cooling step, necessary as excess heat is generated from irreversible losses in the electrochemical cells – about 50 % of the electrical energy added ends up as heat that can be used in the process for evaporation steps and, for example, externally for district heating. Acidification of the electrolyte by HCl addition prior to the electrolysis is necessary as active chlorine escaping with the cell gas results in an increase of the electrolyte pH. The electrolyte then enters the electrolyzers, where chloride ions are oxidized on the anodes and water reduced on the cathodes to hydrogen gas. Gas from the cells and reactor tanks needs to be purified by alkaline scrubbing to remove chlorine. Additional gas purification methods may be necessary, depending of the specific use of the hydrogen product.

An electrolyte side stream from the circulation loop is alkalized by addition of NaOH, filtered to remove precipitations that resulted from the alkalization, and fed to a crystallizer. Figure 1 shows a vacuum crystallizer with exit streams of water vapor and a slurry of sodium chlorate crystals. The chlorate stream is dewatered using a screw thickener, washed, filtered and dried. The chlorate product is then packed and ready for delivery to customers.

Careful environmental control of all outlet streams is necessary, in particular as the process contains chromium (VI) and chlorate which can be harmful to living organisms.

Cell System Configurations

The introduction of DSA® anodes in the 1970s allowed a higher operating temperature and a higher current density, see the chapter "▶ Chlorate Cathodes and Electrode Design," and a number of chlorate technologies were developed for these new conditions. By 1999 there were approximately 14 different chlorate cell technologies operating with metal electrodes (not graphite anodes) [1]. This chapter gives some principles and examples of chlorate

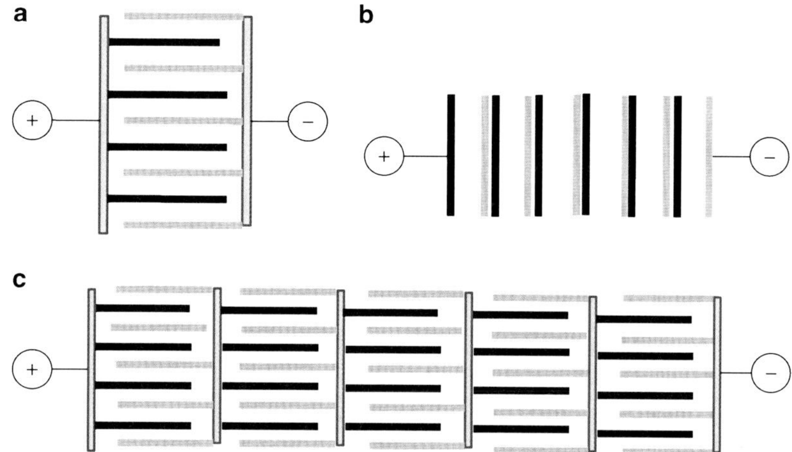

Chlorate Synthesis Cells and Technology, Fig. 2 Schematic of different electrode configurations (**a**) monopolar (**b**) bipolar (**c**) multi monopolar arrangement

systems, more details on different chlorate technologies can be found in the literature [1–5].

In principle the chlorate technologies differ with respect to electrode configuration (monopolar/bipolar), materials of construction and whether the operations of electrolysis, heat exchange and chemical conversion to chlorate are made in separate vessels or combined in single or double vessel systems [3].

There are monopolar as well as bipolar electrode configurations, see Fig. 2. The monopolar cells can be of metals as titanium or steel, whereas the bipolar cells must be made of plastic materials or well insulated metal sections [3] due to the higher voltage in the cell. Monopolar cells have a voltage of one unit cell, about 3 V, and a high current that depends on the total electrode area. Bipolar cells, on the other hand, have a lower current corresponding to one unit cell (2–3 kA/m^2 electrode area) and a total voltage depending on the number of electrode pairs. Some advantages with the monopolar arrangements are that they are relatively easy to construct and have less risk of stray currents due to the low cell voltage compared to bipolar cells. Advantages with the bipolar concepts are less current leads to the electrodes and a higher capacity per electrolyzer. A special case of monopolar cells is the multi monopolar, which combines advantages of the monopolar and bipolar technology [3].

The anodes used today are mainly ruthenium based DSA® specially designed for the chlorate process, while the cathode material is mainly steel, mild steel or low carbon steel and in some plants titanium or a Ti-0.2 % Pd alloy [3]. The Tafel slope for hydrogen evolution on steel, and the corresponding overvoltages, are much higher than the Tafel slope and overvoltages for chloride oxidation on the chlorate anodes [7]. Therefore, and as cathode steel is less expensive than coated titanium, electrode configurations with a larger cathode area than anode area (extended cathodes) have been designed and claimed to operate at lower voltage compared to cells with equal anode and cathode areas [4]. During electrolysis hydrogen may penetrate the metal and form hydrogen blisters in the steel. Steel cathodes can be perforated to avoid formation of large blisters and to enhance mass transport in the cells. The inter-electrode gap between anode and cathode blades is typically 2–4 mm.

The electrolyte circulation is induced by natural convection caused by the hydrogen gas, the so-called "gas-lift", and circulation may also be forced by a pump. The "gas-lift" can results in high electrolyte velocities, in the order of 0.2–0.8 m/s in the inter-electrode gap [8], which is advantageous as it removes the products formed in the electrolysis. Hypochlorite formed on the anodes is quickly removed from the electrodes, avoiding electrochemical side

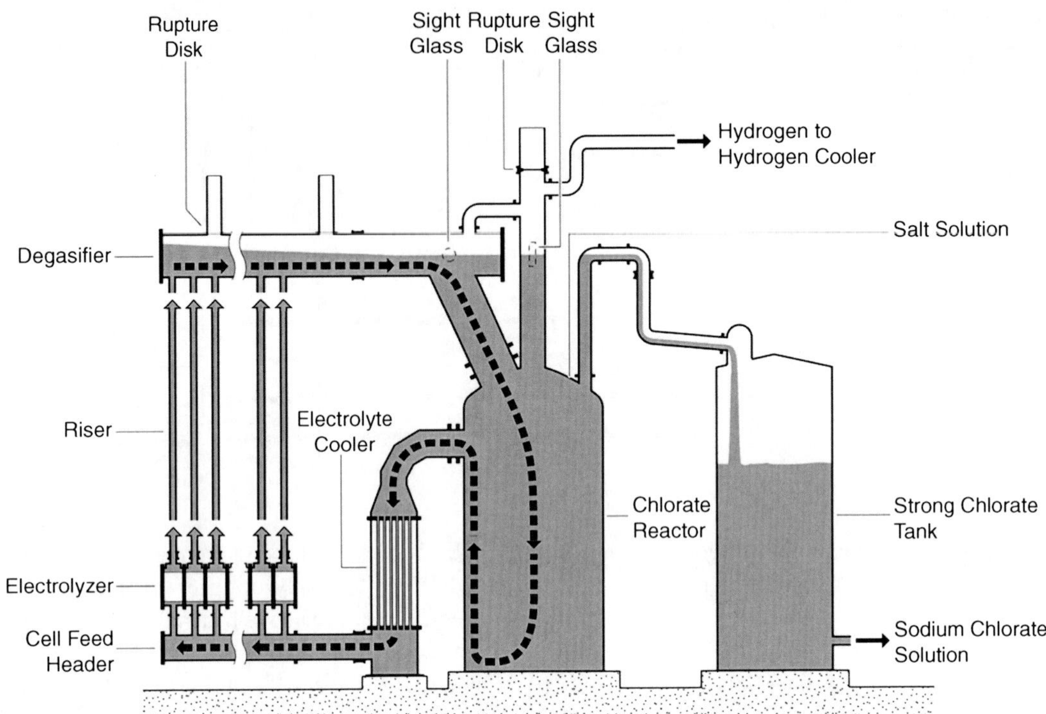

Chlorate Synthesis Cells and Technology, Fig. 3 Electrolyte circulation loop (Courtesy of Chemetics Inc.)

reactions; see details in the chapter "▶ Chlorate Cathodes and Electrode Design." Hydrogen gas bubbles generated are also removed and hindered both from building shielding gas layers on the cathode and from accumulating in the cell gap. Both the shielding and the accumulation can result in increased cell voltage. The two-phase flow in gas-lift chlorate cells of different geometries has been modeled using computational fluid dynamics (CFD) [9].

The chemical conversion of hypochlorite ($ClOH + ClO^-$) to chlorate, reaction 2, is a relatively slow reaction and is of 3rd order with respect to hypochlorite [10]. In the continuous chlorate process a reaction volume with a long enough residence time is needed for chlorate to form, and to reach a relatively low concentration of hypochlorite before the electrolyte reaches the electrochemical cells.

$$2ClOH + ClO^- \rightarrow ClO_3^- + 2H^+ + 2Cl^- \quad (2)$$

Reaction 2 is highly temperature dependent and therefore, when the extra reaction volume is in separate tanks as in Fig. 1, it is common to have a higher temperature in the reactor tank than in the electrolysis cells. Operation at very high temperatures, 90–100 °C, is limited by the choice of materials in the chlorate process.

An example of a chlorate circulation loop is given in Fig. 3, illustrating the Chemetics technology. Similar to in Fig. 1 the electrolyte circulates between electrolytic cells, a reactor tank for reaction 2 and a heat exchanger to remove excess heat. Hydrogen gas produced in the electrolyzers creates the "gas-lift" that forces the electrolyte-gas mixture through pipes (risers) to the degasifier, where the gas is separated from the liquid phase. Each electrolyzer can contain up to 18 cells and each reactor is connected to 2–6 electrolyzers [2]. A photo from a Chemetics chlorate plant showing electrolyzers with risers is given in Fig. 4.

Chlorate Synthesis Cells and Technology, Fig. 4 Chlorate cell room (Courtesy of Chemetics Inc.)

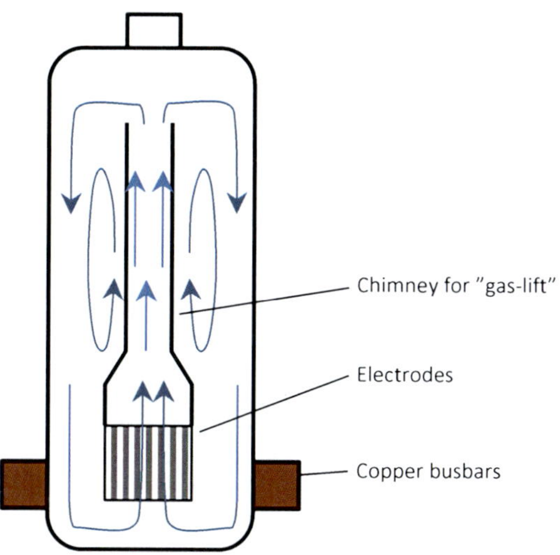

Chlorate Synthesis Cells and Technology, Fig. 5 Schematic of a cylindrical cellbox, AkzoNobel technology

Chlorate Synthesis Cells and Technology, Fig. 6 Chlorate cell room (Courtesy of AkzoNobel Pulp and performance chemicals)

Figure 5 shows a schematic of an AkzoNobel cylindrical cellbox. Electrodes are located in the lower part of the box and are fed with DC current via busbars from the rectifier. Above the electrodes is an internal pipe. This "chimney" helps in creating an excellent internal "gas-lift" circulation within the cellbox, as indicated by the blue arrows. Electrolyte enters the cell in the lower end of the box (not shown), and electrolyte together with the cell gas exit at the top. After gas separation the electrolyte is fed to a reactor tank as in Fig. 1. In Fig. 6 a photo of a line of cylindrical cellboxes from an AkzoNobel chlorate plant is shown.

These are just two examples of chlorate cell systems, recommended literature for further reading are references [1–5].

Future Directions

A forecast for the growth of sodium chlorate production until 2016 is given in Chemical Economics Handbook [11]. The main use of sodium chlorate is in the pulp and paper industry for generation of chloride dioxide, which is used to bleach chemical pulp. In the nearest years to come an expected growth in the production of chemical pulp, and a trend towards higher-brightness paper, is anticipated to increase the demand for sodium chlorate. Asia, South America and Russia are the regions where this growth will be strongest. The major cost in sodium chlorate production is that of electrical energy, and the price for sodium chlorate depends on the price for electricity. When electricity prices are high, it may be advantageous for pulp mills to replace some of the chlorate in the bleaching by other chemicals as hydrogen peroxide. This may affect the demand for sodium chlorate.

The high, and increasing, cost for electricity motivates the development of more energy efficient production, in principle electrolysis with a lower cell voltage and a higher current efficiency.

Main directions to lower the electrical energy consumption is to decrease the cell voltage by either a cathode that is stable, active and selective for hydrogen evolution, alternatively an oxygen depolarized cathode. Operation at higher temperature, 90–100 °C, would lower the cell voltage but is today limited by the materials used in the process.

The current efficiency could be improved by the development of an anode coating with a lower selectivity for the by-products oxygen and perchlorate. Another way is to find a catalyst for the conversion of hypochlorite to chlorate (Reaction 2), which could favor the desired chlorate formation over oxygen forming reactions and thereby increase the current efficiency. It could also allow smaller reaction volumes – more compact plant designs.

Environmental concerns motivate the development on an alternative to chromate addition to the electrolyte. The main roles of chromate are to hinder cathodic side reactions and to buffer the electrolyte, and it may be difficult to find a single method that satisfies both functions. A combination of in-situ additives may be required, or a new more selective cathode material together with a suitable electrolyte buffer agent.

Cross-References

▶ Chlorate Cathodes and Electrode Design
▶ Chlorine and Caustic Technology, Overview and Traditional Processes
▶ Chlorine and Caustic Technology, Using Oxygen Depolarized Cathode
▶ Electrocatalysis of Chlorine Evolution
▶ Hydrogen Evolution Reaction
▶ Hydrochloric Acid Electrolysis

References

1. Tilak BV, Chen C-P (1999) Chlor-alkali and chlorate technology. In: Burney HS, Furuya N, Hine F, Ota KI (eds) The electrochemical society proceedings series PV 99–21, Pennington, p 8
2. Colman JE, Tilak BV (1995) Sodium Chlorate In: McKetta JJ (ed) Encyclopedia of chemical processing and design. Marcel Dekker, New York, p 126
3. Colman JE (1981) In: Alkire R, Beck T (eds) Tutorial lectures in electrochemical engineering and technology, vol 77, AIChE symposium series 204, Institute of Chemical Engineers, New York, p 244
4. Viswanathan K, Tilak BV (1984) Chemical, electrochemical, and technical aspects of chlorate manufacture. J Electrochem Soc 131:1551–1559. doi:10.1149/1.2115908
5. Ibl N, Vogt H (1981) Inorganic Electrosynthesis In: Bockris JO'M, Conway BE, Yeager E, White RE (eds) Comprehensive treatise of electrochemistry, vol 2. Plenum Press, New York, p 167
6. Tilak BV, Tari K, Hoover CL (1988) Metal anodes and hydrogen cathodes – their activity towards O_2 evolution and ClO_3- reduction reactions. J Electrochem Soc 135:1386–1392. doi:10.1149/1.2095999
7. Cornell A (2002) Electrode reactions in the chlorate process. Ph.D. thesis, KTH Royal Institute of Technology
8. Wanngård J (1992) Impurity Effects in Chlorate Plants In: Wellington TC (ed) Modern chlor-alkali technology, vol 5. Elsevier Applied Science, London/New York, p 295

9. Wedin R, Dahlkild A (1999) A numerical and analytical hydrodynamic two-phase study of an industrial gas-lift chlorate reactor. Computational technologies for fluid/thermal/structural/chemical systems with industrial applications 1, PVP-vol 397–1, ASME, p 125
10. Adam LC, Fabian I, Suzuki K, Gordon G (1992) Hypochlorous acid decomposition in the pH 5–8 region. Inorg Chem 31:3534–3541. doi:10.1021/ic00043a011
11. Chemical Economics Handbook (2012) http://www.ihs.com/products/chemical/planning/ceh/sodium-chlorate.aspx. Accessed 7 Jan 2013

Chlorine and Caustic Technology, Membrane Cell Process

Jakob Jörissen
Chair of Technical Chemistry, Technical University of Dortmund, Germany

Introduction

The idea, to use an ion exchange membrane as cell separator in chlor-alkali electrolysis, is nearly as old as the invention of ion exchange membranes in the 1950s. However, membranes based on polystyrene are not stable in the presence of chlorine. A first chance for a technical realization came with Nafion® (DuPont), a perfluorosulfonic acid polymer (PFSA) with high chemical and thermal stability. It was used as an ion conductor in fuel cells of the Gemini space program in 1966. These membranes were optimized for application as ion conductors. The quality of their second property "permselectivity" (permeation selectivity), i.e., preferred transfer of Na^+ cations and rejection of OH^- anions, was far away from requirements for industrial chlor-alkali electrolysis. The worldwide research efforts to develop a membrane electrolysis process in the beginning of 1970s were dramatically enhanced by the decision of the Japanese government to restrict mercury applications including the amalgam process. Background was illness and dead of many people due to the Minamata disease which is caused by mercury poisoning. Mercury compounds of catalysts polluted waste water and the sea and were accumulated in fish, which was used as human food.

The breakthrough in membrane development and the precondition for the membrane process as an alternative for the amalgam process was in 1975 the replacement of sulfonic acid groups by carboxylic acid groups in a perfluorinated membrane polymer, initially applied by the Asahi Glass Company [1, 2].

Today, the membrane process is state of the art. Its energy demand is about 25–30 % lower; its operation is easier, safer, more efficient and flexible and needs less maintenance in comparison with the traditional processes, without environmental problems of asbestos or mercury (see entry "► Chlorine and Caustic Technology, Overview and Traditional Processes"). Investment costs and space requirements are significantly reduced. All new chlor-alkali electrolysis plants use the membrane process since about 20 years.

This entry can only give a short overview about the principle and a few characteristics of the membrane process of chlor-alkali electrolysis. Some more information is given in [3] and, especially concerning electrochemistry, in [4]. A detailed and periodically updated overview, including explicit flow sheets, pictures of cell constructions and many references, is offered in [1] and [5]. Comprehensive information of ion exchange membranes is available in [6] and [7]. All process details of chlor-alkali electrolysis are completely discussed in [8]. Actual information should be taken from the internet pages of supplying companies for membranes and electrolysis plants, e.g., (arranged in alphabetical order): Asahi Glass Company, Asahi Kasei Chemicals Corporation, Chlorine Engineers Corporation, DuPont (E. I. du Pont de Nemours and Company), Uhde (ThyssenKrupp Uhde GmbH.), and UHDENORA S.p.A.

Membrane Process

At first glance, the difference between the diaphragm process (see Fig. 1 of entry "► Chlorine and Caustic Technology, Overview and Traditional Processes") and the membrane

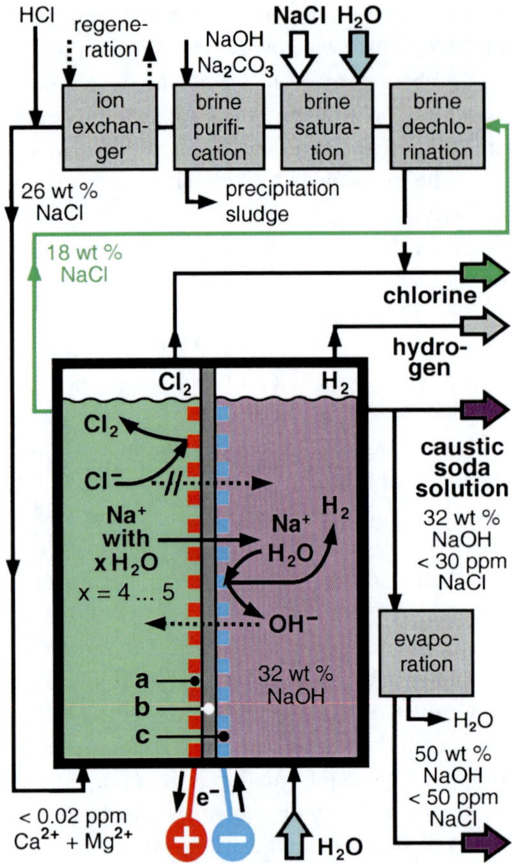

Chlorine and Caustic Technology, Membrane Cell Process, Fig. 1 Scheme of the membrane process (**a**) anode (titanium DSA), (**b**) cation exchange membrane, (**c**) cathode (nickel, activated)

process (see Fig. 1 below) is not easy to identify. The reactions are the same (see section 2 of entry "► Chlorine and Caustic Technology, Overview and Traditional Processes"). However, there are fundamental differences in the separator:

- The diaphragm is a simple porous separator of asbestos and polymer fibers. All ingredients of the electrolytes (water and ions) can be unselectively transferred:
  - By diffusion due to a concentration gradient
  - By convection due to a pressure difference
  - By migration of ions in the electrical field

  The undesired transfer of OH^- ions from catholyte into anolyte due to diffusion and migration is blocked by a countercurrent convection stream of the entire anolyte into the catholyte with sufficient flow velocity. This is adjusted by a pressure difference which results from a higher level of anolyte above catholyte. Therefore, the produced catholyte is a mixture of low concentrated caustic soda and sodium chloride solution which needs an energy-intensive and expensive treatment by evaporation.
  - The membrane has almost no usual porosity, its permeability for a pressure-driven fluid flow is small and it is nearly gastight so that intermixture of gases in the electrolysis cell is insignificant. However, operating as a cation exchange membrane, it enables migration of Na^+ cations (or K^+ cations), including a hydration shell of 4–5 water molecules per ion, and inhibits transfer of the anions OH^- and Cl^-. The permselectivity of virgin, undamaged membranes today is nearly perfect: about 98 % for Na^+ cations and only 2 % for OH^- anions (equivalent to 98 % current efficiency). The function of the applied cation exchange membranes is elucidated in section "Function of Cation Exchange Membranes Based on the Cluster-Network Model."

A closed-loop anolyte circuit with solid salt as feed for the brine saturation is necessary. The required, extremely high purity of the brine is elucidated in section "Brine Purification for the Membrane Process."

State-of-the-art membranes produce caustic soda solution of 32 wt%. This is sufficiently high for internal use in industry. Evaporation to 50 wt%, the usual concentration in traditional processes, is applied if long distance transport is necessary.

The purity of produced caustic soda solution is excellent (less than 30 ppm NaCl), equal or better than that of amalgam process. The reason is that Cl^- anions can only be transferred by diffusion into the catholyte and are rejected by the electrical field as well as due to the permselectivity of the membrane.

Function of Cation Exchange Membranes Based on the Cluster-Network Model

The upper part of Fig. 2 shows the molecular structure of the perfluorinated Nafion® cation exchange membranes with sulfonic acid as well as with carboxylic acid groups as fixed ions. These are covalently bonded at the end of side chains of the PTFE (polytetrafluoroethylene) polymer backbone. The polymer has excellent chemical and thermal stability, similar to PTFE [9].

The chemical structure in the membranes of other manufacturers is very similar (Flemion® of Asahi Glass, Aciplex® F of Asahi Kasei) [2].

The fixed ions $-SO_3^-$ or $-COO^-$ and the Na^+ counter ions, which have to be present for electroneutrality, associate with water molecules under formation of hydration shells if the membrane is immersed and swollen in water. Due to the large chain length (number x about 1,000 at top of Fig. 2) the polymer remains insoluble in water and no cross-linking is necessary between the polymer chains. Thus, they are freely movable. This is an important precondition for the membrane function. A phase separation is energetically favored: on the one hand the hydrophobic PTFE polymer backbone, on the other hand the water in the hydration shells with the included fixed and counter ions.

A scheme of the resulting cluster structure is shown, approximately true to scale, in the lower part of Fig. 2 [9]. It is comparable with small water droplets in a water-in-oil emulsion, which is stabilized by a surfactant as an emulsifier, like in a skin cream. In the membrane the flexible polymer side chains, similar to an anionic surfactant, operate as an emulsifier: their hydrophobic, perfluorinated parts are connected to the PTFE polymer backbone while their hydrophilic fixed ions are arranged in the aqueous phase near to its margin. Such a structure with dimensions in the range of 5 nm was confirmed by X-ray structure analysis. Even though the shape of this cluster structure will be nonrigid and fluctuating during

molecular structure of Nafion® cation exchange membrane:

$[-CF(-O-CF_2-CF(CF_3)-)_m-O-CF_2-CF_2-SO_3^- \; Na^+)-CF_2-(-CF_2-CF_2-)_n]_x$

n = 5–13,5
x = ca. 1000
m = 1 (≤ 3)

effective layer for chlor-alkali electrolysis:

$[-CF(-O-CF_2-CF(CF_3)-)_m-O-CF_2-COO^- \; Na^+)-CF_2-(-CF_2-CF_2-)_n]_x$

cluster structure after membrane swelling in water

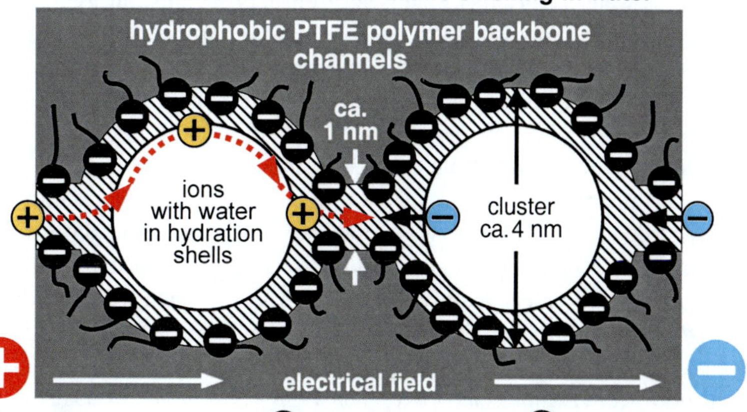

Chlorine and Caustic Technology, Membrane Cell Process, Fig. 2 Structure of the Nafion® cation exchange membrane (Based on [9])

operation it can elucidate the permselectivity of the membrane. The channels between the water clusters have a diameter in the range of 1 nm. This was calculated from measurements of water permeability as a function of an applied pressure difference.

The migration of Na^+ cations in the electrical field from anode to cathode is the desired membrane function. A Na^+ cation is attracted by the negatively charged fixed ions. But it remains movable because the distance between the fixed ions is so small that jumping from one to the next needs little energy. Thus, a Na^+ ion can nearly unhindered glide on the walls of the cluster structure, even through the small channels.

The situation is completely different in case of the migration of OH^- anions from the cathode to the anode. An anion is repelled by the negatively charged fixed ions. The shaded areas near the walls of the clusters and channels indicate locations where anions need high energy to be present. Therefore, anions have to overcome at each of the innumerable channels a considerably high energy barrier.

The different transport mechanisms for cations and anions enable the permselectivity. However, it is insufficient for sulfonic acid fixed ions. The water absorption is too high and the diameter of clusters and especially of channels is too large. An improvement was achieved by decreasing the ion exchange capacity of the membrane material, i.e., with an expanded content of inactive PTFE material (increasing of number n at top of Fig. 2). Then less water is absorbed, the size of clusters and channels is diminished and their number is enlarged. Hence, the described mechanism of permselectivity operates more effectively. But above 80 % Na^+ ion permselectivity for a catholyte with 20 wt% NaOH is not attainable using this method.

The deciding success was obtained by changing from sulfonic acid to carboxylic acid fixed ions (see upper part of Fig. 2). Then the water absorption is significantly decreased and obviously the dimensions of the cluster structure are advantageous for typically 98 % Na^+ ion permselectivity at 32 wt% caustic soda solution.

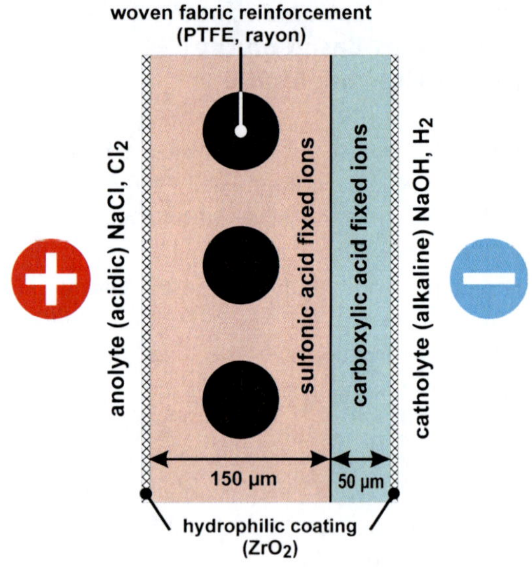

Chlorine and Caustic Technology, Membrane Cell Process, Fig. 3 Principle of technical membrane structure (Based on [10])

Technical Structure of Membranes for Chlor-Alkali Electrolysis

The small water absorption of the perfluorinated polymer with carboxylic acid fixed ions on the one hand enables the desired excellent permselectivity. On the other hand it decreases the ion conductivity and a higher voltage drop is the consequence. The voltage drop can be lowered by using a very thin layer, e.g., 50 μm in Fig. 3, without noticeable deterioration of permselectivity. However, membranes for industrial use, with up to 5 m^2 area, have to be robust. Thus, a thicker layer of the perfluorinated polymer with sulfonic acid fixed ions is laminated as a mechanical support with high ion conductivity. It has no influence on the permselectivity and the voltage drop is only increased by an acceptable value. Additionally a woven fabric is embedded for further reinforcement. PTFE fibers remain stable. Frequently rayon fibers are added which are present only while assembling and then are dissolved during operation in order to decrease the voltage drop (sacrificial fibers).

The voltage drop of the membrane at usual conditions includes an electrochemical potential difference of about 0.05–0.1 V due to the different solutions on both sides of the membrane ("Donnan potential"). The additional voltage drop shows as a first approximation ohmic behavior with a slope of up to 0.1 V per kA m^{-2} current density.

The membrane polymer contents predominantly perfluorinated, hydrophobic materials. This becomes a problem if small electrode distances are used in electrolyzers (see section "Cell Constructions and Electrolyzers for the Membrane Process"). Then gas bubbles adhere at the hydrophobic surfaces. This hinders current flow and increases voltage drop. Therefore, membrane surfaces are coated with a porous hydrophilic material, typically with zirconium oxide particles, in order to minimize adherence of gas bubbles.

iminodiacetic acid group

aminomethylphosphonic acid group

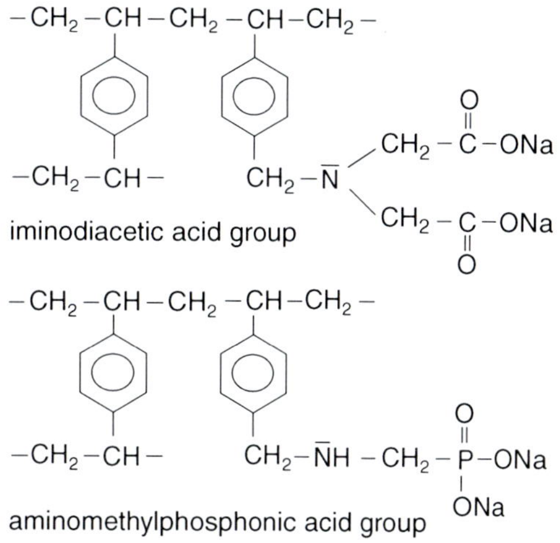

Chlorine and Caustic Technology, Membrane Cell Process, Fig. 4 Examples of tridentate chelating ion exchange resins [10]

Brine Purification for the Membrane Process

The membrane process needs brine of abnormally high purity. State of the art is less than 0.02 ppm of calcium and magnesium, i.e., 20 mg per ton of brine. The reason is that Ca^{2+} and Mg^{2+} ions (and other metal ions) are dissolved in the acidic anolyte and flow together with Na^+ ions into the membrane. The transition from the acidic medium of anolyte to the alkaline medium of catholyte occurs within the membrane, most probably near to the interface between the two layers in Fig. 3. At this position calcium carbonate (small amounts of carbonate anions are everywhere present) and magnesium hydroxide are irreversibly precipitated. Their crystals destroy the membrane structure. The usual membrane service life of 4 years is only guaranteed if the high brine purity never goes below the limit.

Such extremely high purities can be achieved using chelating ion exchange resins. Examples of tridentate fixed ions are shown in Fig. 4 with two acid functions (carboxylic acid or phosphonic acid) and one free electron pair of the nitrogen atom. Multiply charged ions like Ca^{2+}, Mg^{2+}, or heavy metal ions are removed from the brine by selective formation of very stable chelate complexes, even in presence of saturated sodium chloride solution.

The regeneration of the ion exchanger needs solutions with relatively high concentrations of hydrochloric acid (4–10 wt%) for removing the metals and of caustic soda (4 wt%) for reconstitution of the sodium form as shown in Fig. 4.

It is possible to economize the usual precipitation step for the first brine purification (see Fig. 1) and directly to use the ion exchanger for final purification of a vacuum salt solution.

Additional Special Properties of the Membrane Process

The anolyte can be depleted down to about 18 wt% NaCl (see Fig. 1). Moreover, 4–5 molecules of water are removed from the anolyte as hydration shells of the Na^+ ions during their transfer through the membrane. This so-called electroosmotic flow (EOF, also named "electroosmotic drag") is an intrinsic property of ion exchange membranes. Here, it is advantageous because it reduces the flow rate of the anolyte circuit. EOF increases with decreasing anolyte NaCl concentration which must not fall below 18 wt%, not even locally due to insufficient mixing. In this case, the increased EOF still can flow through the membrane layer with sulfonic acid fixed ions (left layer in Fig. 3), but it becomes too large for the transfer through the right layer with carboxylic acid fixed ions so that the membrane is delaminated ("blistering").

The membrane process enables addition of hydrochloric acid into anolyte for neutralization of the OH^- ions, which enter through the membrane from catholyte (see Fig. 1). Moreover, anolyte can be acidified for reduced by-product formation. Oxygen evolution then is decreased to less than 0.5 vol.% in the anode gas and generation of hypochlorite and chlorate is completely suppressed (see reactions (3) and (5)–(7) in section 2 of entry "▶ Chlorine and Caustic Technology, Overview and Traditional Processes"). However, the addition of acid has to be performed very carefully with sufficient mixing of the anolyte, usually by the mammoth pump effect of the produced chlorine gas. The pH value must nowhere fall below 2. Otherwise, the carboxylic acid fixed ions (see upper part of Fig. 2), which are the anions of a relatively weak acid, will combine with H^+ ions and lose their activity so that the membrane is damaged.

Cell Constructions and Electrolyzers for the Membrane Process

Numerous electrolyzer types were constructed in the beginning of membrane process development. One question was the electrical connection of the cells. The total voltage of an electrolysis plant has to be some 100 V for high rectifier efficiency. This needs serial connection of electrolyzers. However, individual electrolysis units can be connected:

- Parallel, so-called monopolar, where all units have the same potential but large amounts of copper are needed for high current interconnections,
- Serial, so-called bipolar, with simple electrical interconnection but with high potential differences within the electrolyzer.

Bipolar cells are state of the art today. Detrimental shunt currents, which provoke corrosion, can be caused by high potential differences within flowing electrolytes between electrolysis units. They are minimized by means of pipework with high resistance: thin and long tubes are applied for electrolyte feeding and electrolytes are exhausted as a foamy mixture with produced gases.

Today, bipolar cell elements with active areas of up to 5 m^2 are in use. Their one side is an anode compartment (titanium) the other side a cathode compartment (nickel), connected in the center by an explosive-cladded titanium nickel sheet. Up to 100 and more of such elements, separated by membranes and gasket seals, are compressed by a hydraulic system in one electrolyzer (comparable with a frame filter press). Replacement of a defective component is possible but difficult. In the "single element" technology anode and cathode compartment are carried out as two half shells which are together with a membrane and gasket seals individually tightened by a bolted flange connection. These elements are mounted in a rack and electrically connected by mechanical contact pressure. Replacement of a defective single element is relatively uncomplicated.

Membrane cells need maintenance – unless in case of unexpected disruption – typically not earlier than after 4 years. Then, membranes and gasket seals should be replaced for economic reasons due to losses in permselectivity and increased voltage drop (this is dependent on brine purity as mentioned above). Electrode coatings usually have to be regenerated after 8 years.

Today, permselectivity (current efficiency) of membranes is close to 100 %, i.e., there is no optimization potential. However, cell voltage should be minimized by further development of membranes, electrodes, and cell constructions for energy saving. Simultaneously, increased current density is desired for a high production rate per active electrode area. State-of-the-art energy consumption was in 1985 about 2,400 kWh/t NaOH at 4 kA m^{-2}, and today it is about 2,050 kWh/t NaOH at 6 kA m^{-2} [11, 12]. Current densities of up to 8 kA m^{-2} are planned.

The most part of this cell voltage reduction was achieved by optimization of electrode setup. Expanded metal electrodes with a "finite gap" of 2–3 mm were used from the beginning of membrane process development. The membrane has to be fixed at one position, otherwise it would be mechanically damaged by fluctuations. Therefore, it is slightly pressed onto the anode, using a small excess pressure in the cathode compartment. This is advantageous for low voltage drop because conductivity of catholyte is higher than that of anolyte. However, the contact between membrane and anode must not hinder brine transport which supplies the anode with Cl^- ions for chlorine evolution as well as the membrane with Na^+ ions including their hydration shells. A special structure of the anode surface can assure sufficient brine transport.

Generally, optimization of electrode layout is important for energy saving. The active electrode area should be maximized for uniform current density distribution. Bubbles of the produced gases have to disappear behind the electrode as soon as possible in order to avoid an unnecessary high voltage drop. Very fine and relatively stable bubbles of hydrogen are formed in caustic soda solution at the cathode, while chlorine bubbles from the anode are quickly coalescing in brine to increased diameters. Thus, in particular highly sophisticated anode constructions, e.g., based on louver structures, are successfully developed for decreased energy demand.

The simplest action for a lesser voltage drop seems to be reducing of the electrode distance. However, this needs additional precautions. First of all, an improved precision of the electrolyzer components is indispensable, otherwise membrane crushing is to be expected. Furthermore, the necessity of hydrophilic membrane surfaces was discussed in section "Technical Structure of Membranes for Chlor-Alkali Electrolysis." The ultimate possibility of minor electrode distance is "zero gap," i.e., both electrodes are in contact with the membrane. Actual constructions of this principle use an elastic element of fine, interwoven nickel wire for current feeding on the cathode side which presses the cathode (fine mesh) and the membrane onto the anode (e.g., [11]).

An overview of the voltage composition in a membrane cell at 6 kA m^{-2}, 90 °C, 32 wt% caustic soda solution, and 18 wt% NaCl in the anolyte is given in [12]:

Thermodynamic decomposition voltage	2.21 V
Anode overpotential	0.07 V
Membrane voltage (incl. Donnan potential)	0.51 V
Ohmic drop of cell structure	0.02 V
Ohmic drop of catholyte	0.02 V
Cathode overpotential	0.09 V
Others (gas influence etc.)	0.07 V
Sum	2.99 V

Future Directions

Within the next decade most probably nearly all chlor-alkali electrolysis plants will use the membrane process and in old plants the traditional processes will be replaced.

Higher caustic soda concentrations of up to 50 wt% are still intended. However, the membranes, which until now have been developed for this purpose, showed too high voltage drop and insufficient stability.

The optimization potential of the membrane process becomes smaller and smaller because the cell voltage approximates the theoretical value. However, the membrane voltage drop is the biggest part of the cell voltage behind the thermodynamic decomposition voltage. It seems that high permselectivity and high voltage drop are directly linked (see section "Technical Structure of Membranes for Chlor-Alkali Electrolysis"). Nevertheless, new ways to reduce the membrane voltage drop would be very interesting.

A significant saving of electrical energy (about 30 %) is possible by changing the cathode reaction: oxygen depolarized cathodes (ODC) are used if hydrogen production is not desired (see entry "▶ Chlorine and Caustic Technology, Using Oxygen Depolarized Cathode").

Cross-References

- ▶ Chlorine and Caustic Technology, Overview and Traditional Processes
- ▶ Chlorine and Caustic Technology, Using Oxygen Depolarized Cathode
- ▶ Electrocatalysis of Chlorine Evolution
- ▶ Hydrogen Evolution Reaction

References

1. Schmittinger P, Florkiewicz T, Curlin L, Lüke B, Scannell R, Navin P, Zelfel E, Bartsch R (2011) Chlorine. In: Ullmann's Encyclopedia of industrial chemistry. Wiley-VCH, Weinheim, doi:10.1002/14356007.a06_399.pub3
2. Coulter MO (ed) (1980) Modern chlor-alkali technology. Ellis Horwood, Chichester
3. Büchel KH, Moretto HH, Woditsch P (2000) Industrial inorganic chemistry, 2nd edn. Wiley-VCH, Weinheim
4. Hamann CH, Hamnett A, Vielstich W (2007) Electrochemistry, 2nd edn. Wiley-VCH, Weinheim
5. Bommaraju TV, Lüke B, O'Brien TF, Blackburn MC (2002) Chlorine. In: Kirk-othmer encyclopedia of chemical technology, Wiley, doi:10.1002/0471238961.0308121503211812.a01.pub2
6. Sata T (2004) Ion exchange membranes: preparation, characterization, modification and application. Roy Soc Chem. doi:10.1039/9781847551177
7. Tanaka Y (2007) Ion exchange membranes, fundamentals and applications, vol 12, Membrane science and technology. Elsevier, Amsterdam
8. O'Brien TF, Bommaraju TV, Hine F (2005) Handbook of chlor-alkali technology, vol 1–5. Springer, New York, doi:10.1007/b113786
9. Hsu WY, Gierke TD (1983) Ion transport and clustering in nafion perfluorinated membranes. J Membr Sci 13:307–326
10. Bergner D (1994) Current state of alkali chloride electrolysis. Part 1. Cells, membranes, electrolytes, and products. (Language: German) Chem Ing Tech 66:783–791
11. http://www.thyssenkrupp-uhde.de/publications/videos/base-chemicals/electrolysis-generation-6.html. Download 23 Apr 2013
12. http://www.emt-india.net/Presentations2011/05-Chemical_21July2011/Material/13-AsahiKasei-Chemicals-6%20Slide.pdf. Download 23 Apr 2013

Chlorine and Caustic Technology, Overview and Traditional Processes

Jakob Jörissen
Chair of Technical Chemistry, Technical University of Dortmund, Germany

Introduction

Chlorine is one of the most important base chemicals in chemical industry. The worldwide chlorine production capacity in 2008 was 63 million tons per year with significant increasing tendency [1]. About 60 % of all chemical products need chlorine during the production process. These products would be not available without chlorine or more raw materials and/or energy would be needed for production. Nevertheless, chlorine is not present in most final products. It is used due to its high reactivity for selective formation of intermediates and finally removed. Examples are polyurethanes and polycarbonates, pharmaceuticals, ultrapure silicon for electronic chips, titanium dioxide as white pigment, silicone sealing compounds or PTFE (polytetrafluoroethylene), and much more. The most important chlorine-containing product is PVC (polyvinylchloride, using round 1/3 of the produced chlorine).

About 95 % of the chlorine is manufactured by electrolysis of sodium chloride solution (chlor-alkali electrolysis; small amounts of potassium chloride and hydrochloric acid are additionally used, see essay "▶ Hydrochloric Acid Electrolysis"). Caustic soda, the coproduct of chlor-alkali electrolysis, is also an essential compound in (chemical) industry (1,13 t NaOH per ton Cl_2). Large amounts of caustic soda are necessary for neutralization of hydrochloric acid which is formed during reactions with chlorine.

This essay can only give a short overview about the principle and a few characteristics of chlor-alkali electrolysis. Some more information is given in [2] and, especially concerning electrochemistry, in [3]. A detailed and periodically updated overview, including operation conditions, explicit flow sheets, pictures of cell constructions, and many references, is offered in [1] and [4]. Comprehensive information of all process details is available in [5].

Principle and Chemistry of Chlor-Alkali Electrolysis

The overall electrolysis reaction is:

$$2\,NaCl + 2\,H_2O + 2F \rightarrow Cl_2 + 2\,NaOH + H_2 \quad (1)$$

A saturated common salt solution (brine) is delivered to the anode where chlorine gas is evolved:

$$2\,Cl^- \leftrightarrow Cl_2 + 2\,e^- \quad E_0 = 1.36\ V \qquad (2)$$

Water decomposition forming oxygen should be preferred according to thermodynamics, because its standard potential is lower:

$$H_2O \leftrightarrow 1/2\,O_2 + 2\,H^+ + 2e^- \; E_0 = 1.23\ V \quad (3)$$

However, oxygen formation is kinetically hindered at usual anode materials, i.e., it needs a high charge transfer overpotential, and therefore, oxygen is only a small by-product of chlorine evolution (in most cases less than vol 2 % in the anode gas, depending on pH value; the produced chlorine is liquefied in order to remove oxygen and then reevaporated).

Water is reduced to hydrogen at the cathode (however, in the mercury (amalgam) process, another cathode reaction is used, see section Amalgam (Mercury) Process):

$$2\,H_2O + 2e^- \leftrightarrow H_2 + 2\,OH^- \; E_0 = -0.83\ V \qquad (4)$$

The OH^- ions combine with the Na^+ ions from the anode to the coproduct NaOH.

It is essential to separate anode and cathode to avoid gas mixing (explosive mixture) as well as the side reaction under hypochlorite formation:

$$Cl_2 + 2\,NaOH \leftrightarrow NaOCl + NaCl + H_2O \quad (5)$$

Three processes, which use different ways for the separation of anolyte and catholyte, are applied in chemical industry (see sections "Diaphragm Process" "Amalgam (Mercury) Process" and essay "▸ Chlorine and Caustic Technology, Membrane Cell Process").

A more or less small amount of OH^- ions flows into the anode compartment if this separation is incomplete. In this case some chlorine is bonded by reaction (Eq. 5) and oxygen formation according to reaction (Eq. 3) is increased (up to 2 vol.% in the anode gas). Additionally, hypochlorite can disproportionate under formation of chlorate:

$$2\,HOCl + OCl^- \rightarrow 2\,H^+ + 2\,Cl^- + ClO_3^- \qquad (6)$$

It is possible to neutralize the invading OH^- ions in the membrane process by addition of hydrochloric acid into the anode compartment and to decrease the pH value down to 2. Then, hypochlorite and chlorate formation is stopped, and the oxygen content in the anode gas is reduced to 0.5 vol.%. No OH^- ions are formed within the electrolysis cell of the amalgam process so that the oxygen content of chlorine gas is less than 1 vol.%.

The anolyte is under all conditions at least slightly acidic (about pH 4–5) due to hydrolysis of chlorine gas:

$$Cl_2 + H_2O \leftrightarrow H^+ + Cl^- + HOCl \qquad (7)$$

Additionally, chlorine is physically dissolved in the anolyte. This amount and the chlorine, which is reversibly bonded in hypochlorite (reaction 5 and 7), can be removed after acidification by vacuum dechlorination. This is a precondition for brine resaturation in the anolyte circuit of amalgam and membrane process.

Anodes for Chlor-Alkali Electrolysis

The classical anode material was graphite. It is slowly oxidized to carbon dioxide, due to the formation of small amounts of oxygen (reaction 3), and the anode loses material. Thus, electrode gap and ohmic voltage losses increase, or anodes have to be mechanically readjusted.

The introduction of Dimensionally Stable Anodes (DSA®) in the beginning of the 1970s [6] resulted in about 10 % energy saving. Titanium is used as a carrier metal which is corrosion resistant due to a passivation layer of titanium dioxide, stabilized by the oxidation potential of chlorine in aqueous solution.

This isolating layer prohibits any anodic charge transfer. It is substituted by a coating of mixed oxides, typically of titanium and ruthenium (and additives). This layer retains the corrosion protection property, but additionally it becomes sufficiently conductive and provides a high electro-catalytic activity. The charge transfer overpotential for chloride ion oxidation and anodic ruthenium reoxidation is low due to the easy change of the oxidation state of ruthenium:

$$Ru^{4+} + Cl^- \leftrightarrow 1/2\, Cl_2 + Ru^{3+} \quad (8)$$

$$Ru^{3+} \leftrightarrow Ru^{4+} + e^- \quad (9)$$

The difference between the low overpotential for chlorine and the high overpotential for oxygen avoids a significant oxygen evolution. DSA today can be used for more than 8 years before the coating has to be regenerated. For details of DSA, see essay "▸ Electrocatalysis of Chlorine Evolution."

Cathodes for Chlor-Alkali Electrolysis

Iron or stainless steel is used as cathode material for reaction 4 at low caustic soda concentrations (diaphragm process), pure nickel is necessary at high concentrations (membrane process). Hydrogen evolution needs at these materials a relative high overpotential of up to 300 mV. It can be significantly decreased if coatings of, e.g., ruthenium oxides, are applied on the cathode surface (see also essay "▸ Hydrogen Evolution Reaction").

DSA only can operate with a functioning coating, while cathodes are working even with destroyed coatings, but at increased potential. Service life of cathode coatings today is more than 8 years, like DSA.

About 30 % of electrical energy can be saved in the membrane process if the coproduct hydrogen is not desired and an Oxygen Depolarized Cathode (ODC) is applied in place of a hydrogen evolving cathode (see essay "▸ Chlorine and Caustic Technology, Using Oxygen Depolarized Cathode").

Brine Treatment

Salt is available from sea water (sea salt, solar salt) or from geological salt deposits (solid rock salt from underground mining, well brine by solution mining from a bore hole). Sometimes vacuum salt is used which is pre-purified by crystallization from brine. Generally, a careful purification is necessary before the brine can be used for electrolysis. Especially calcium, magnesium and heavy metals have to be removed, typically by precipitation as hydroxides or carbonates after addition of caustic soda and soda. The precipitation sludge is separated by sedimentation and filtration, and the clear brine is neutralized with hydrochloric acid. Special requirements for the different processes are discussed below.

Diaphragm Process

The diaphragm process is the oldest version of chlor-alkali electrolysis which started up 1885 in Germany, soon after electrical energy was available due to the invention of electric generators for power stations. The diaphragm process was by far the most applied chlor-alkali process until some years ago, especially in the USA, and also today it remains important.

Figure 1 shows a scheme of the process. The electrodes a and c are made of perforated metal plate or expanded metal mesh where the gases chlorine and hydrogen are evolved. The electrodes are separated by the diaphragm b, an about 5-mm-thick layer of asbestos fibers with addition of polymer fibers for reinforcement (e.g., PTFE). It is uniformly applied to the perforated cathodes of industrial cells by a vacuum method from a slurry of asbestos and polymer fibers.

Inhalation of the very thin and short asbestos fibers is dangerous due to their carcinogenic properties. Much effort is necessary for health protection of the operating crew. Some governments restricted the application of asbestos. Starting at the end of the 1980s, noncarcinogenic fibers have been developed by different

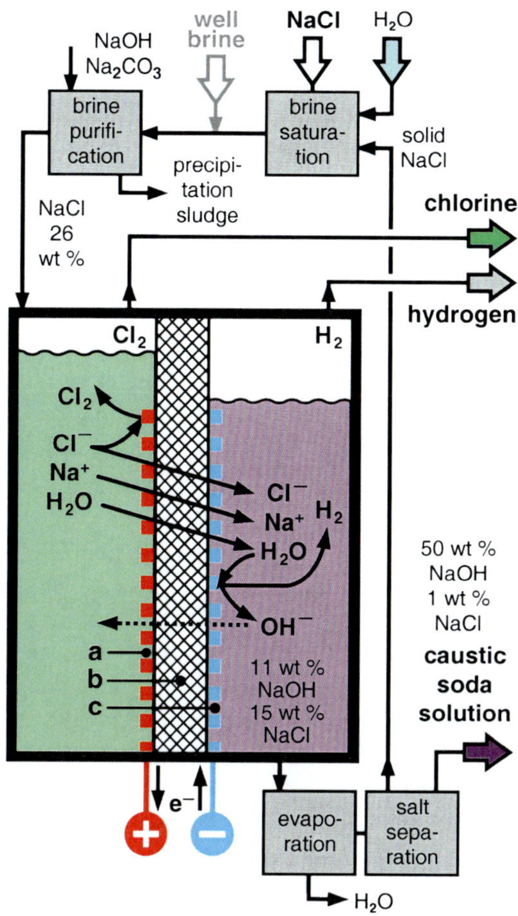

Chlorine and Caustic Technology, Overview and Traditional Processes, Fig. 1 Scheme of the diaphragm process (**a**) Anode (titanium DSA), (**b**) diaphragm (asbestos), (**c**) cathode (steel or stainless steel)

companies, e.g., of PTFE and an inorganic material like zirconium oxide particles [1, 7].

The liquid level in the anode compartment is higher than in the cathode compartment. Due to the pressure difference, a liquid flow of anolyte into the catholyte results. The level difference is continuously adjusted so that the flow velocity on the entire diaphragm area as exactly as possible rejects the migration of OH^- ions into the anolyte. Thus, side reactions 3, 5, and 6 can be minimized, but they are increased in comparison with the other processes.

The catholyte contents the produced caustic soda (about 11 wt%). It is diluted by all the brine which is delivered into the anode compartment, and there only partially is converted into chlorine. Therefore, the salt content in the catholyte (about 15 wt%) is higher than the low concentration of caustic soda. An unnecessary high flow rate through the diaphragm has to be avoided because the NaOH concentration would be further decreased, and the NaOH/NaCl ratio becomes worse. Thus, the controlling of the level difference and its adaptation to the decreasing permeability of the diaphragm during its service life of about 2 years is a challenging task for the operating crew.

Current density in diaphragm cells is relative low, usually only up to 2–3 kA m^{-2}. It is difficult to adapt it to changes in the required chlorine production. Special low-cost cell constructions allow an economic operation at decreased current density and correspondingly low electrical power consumption, in spite of the needed larger electrode area.

The catholyte has to be concentrated in multi-effect evaporators to the usual content of 50 wt% NaOH. The salt is precipitated under these conditions due to its lower solubility down to a content of 1 wt% NaCl. It is filtered and recycled for the preparation of new brine. The evaporation needs expensive, highly corrosion resistant apparatus and much thermal energy. Cost-saving well brine from solution mining can be used as alternative for solid salt as feed material because the included water is discharged by the evaporation. The brine has to be purified in all cases as described before. Purity is less important than in the other processes because the brine flows completely into the catholyte and no accumulation of impurities is possible. However, careful filtration is especially needed in the diaphragm process in order to avoid soon clogging of the diaphragm.

Caustic soda solution, produced in the diaphragm process, is not qualified for many applications, especially in chemical industry, due to its relatively high salt content of 1 wt% NaCl. A further purification by extraction with liquid ammonia is possible but expensive.

Amalgam (Mercury) Process

The amalgam process was developed in order to produce highly pure caustic soda in the beginning of industrial chlor-alkali electrolysis as well as the diaphragm process (1892). It is until now important in chemical industry, especially in Europe.

Figure 2 shows a scheme of the process. Chlorine is evolved according to reaction 2 at the anodes (DSA) and disappears upwards in the undivided horizontal cell with a small downward slope. However, the cathode reaction is completely changed compared with the other processes: at a flowing mercury film sodium metal is deposited and dissolved as sodium amalgam:

$$Na^+ + e^- + x\ Hg \leftrightarrow Na \cdot Hg_x \qquad (10)$$

It reacts outside of the electrolysis cell in the amalgam decomposer at a graphite surface with water to hydrogen and caustic soda:

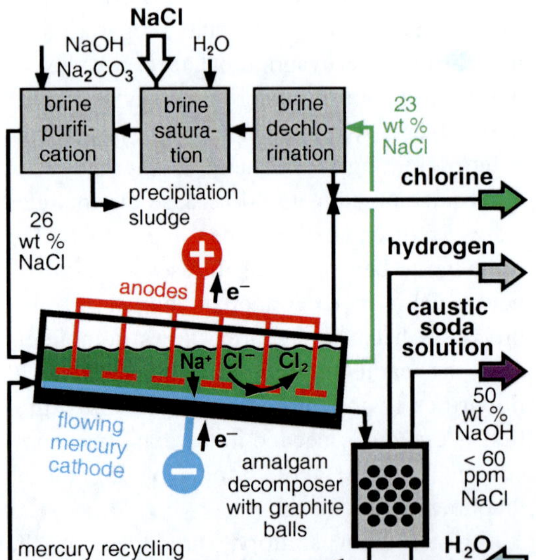

Chlorine and Caustic Technology, Overview and Traditional Processes, Fig. 2 Scheme of the amalgam (mercury) process

$$Na \cdot Hg_x + H_2O \rightarrow 1/2\ H_2 + NaOH + x\ Hg \qquad (11)$$

Cathodic deposition of sodium metal is possible in the amalgam process because the cathode potential is shifted to a higher value than hydrogen evolution under these conditions, despite of the very low standard potential of sodium $E_0 = -2.71$ V:

$$E = E_0 + (RT/F)\mathrm{In}(a_{0x}/a_{red}) \qquad (12)$$

- Sodium metal is dissolved immediately in mercury with a maximum concentration of 0.2 wt%; thus, the activity a_{red} of Na (reduced species) in the Nernst Eq. (12) is low.
- Brine is only slightly depleted from 26 to 23 wt%; thus, the activity a_{ox} of Na^+ (oxidized species) in the Nernst Eq. (12) is high.
- At a mercury surface evolution of hydrogen is strongly kinetically hindered, i.e., a very high charge transfer overpotential would be necessary (under the precondition that other heavy metals are excluded carefully).

Evolution of hydrogen is impossible under these conditions at a mercury surface in the electrolysis cell as well as in the amalgam decomposer due to the same reasons, elucidated above. It is enabled according to reaction 11 in the amalgam decomposer due to an electrical contact between sodium metal containing mercury droplets and a catalyst of graphite balls, as demonstrated in Fig. 3.

Electrons from the oxidation of sodium metal into Na^+ ions cannot reduce water into hydrogen and OH^- ions at the mercury surface. However, reduction becomes possible if the electrons flow to the surface of the graphite balls.

The caustic soda solution is directly produced in the desired concentration of 50 wt%. Its purity is excellent. The high density of mercury (13.5 g cm $^{-3}$) enables a nearly perfect separation of the amalgam at the outlet of the electrolysis cell from remaining brine droplets. This is completed by washing with water prior to entry into the amalgam decomposer.

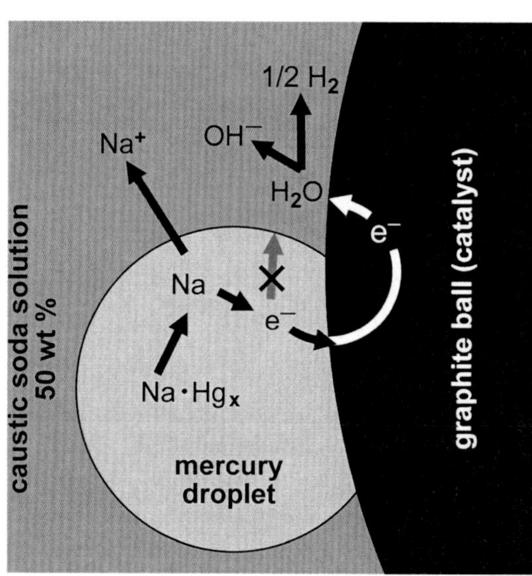

Chlorine and Caustic Technology, Overview and Traditional Processes, Fig. 3 Amalgam decomposition at a graphite surface

No OH^- ions are generated in the electrolysis cell. Thus, anodic oxygen evolution is low (about 1 vol.% in the anode gas), and no chlorate is formed.

The amalgam process needs a closed loop of anolyte in which only salt and nearly no water is converted (see Fig. 2), contrary to the diaphragm process, where all introduced brine leaves the anode compartment via the diaphragm (see Fig. 1). NaCl has to be delivered by solid salt in the brine saturation. First, the anolyte outlet must be dechlorinated by acidification and vacuum application. It is important for the brine purification to remove all heavy metals, e.g., vanadium, that can decrease the hydrogen overpotential at mercury. Otherwise hydrogen evolution becomes possible, in worst case an explosion together with chlorine can happen.

The demand of electrical energy is relatively high for the amalgam process compared with the other processes due to the cathodic deposition of sodium metal. The horizontal cells of up to 30 m^2 active area need a large floor space. Therefore, high current densities of typically 10 and up to 15 kA m^{-2} are used in order to realize a sufficient space-time yield. This is possible without an inacceptable high cell voltage. The so called k-factor is defined as increasing of the cell voltage with rising current density. Its value of amalgam cells is in the range of 0.1 V/kA m^{-2}. In consequence, the cell voltage is typically about 4.00–4.25 V [1]. An important reason for the small k-factor is that the amalgam cells have been constructed for graphite anodes and include a computer-controlled motor-driven electrode distance adjustment. It is used to regulate the electrode gap between DSA and mercury cathode to an optimal value of about 2 mm. Additionally, chlorine gas bubbles can easily be released upwards from the horizontal DSA. Hence, an unnecessary high gas content and voltage drop in the electrode gap is avoided. The current density can be changed without problems in order to fit the chlorine production to the given demand.

The usage of mercury may be a problem of environmental pollution. 70 g mercury was lost during the production of 1 t of chlorine in the beginning of 1970s. Today, this value is decreased to less than 2 g/t Cl_2 by various expensive procedures. The mercury content of the precipitation sludge is not included. It has to be deposited in the underground as hazardous waste. The mercury residue in the products today is reduced to very small traces.

Future Directions

New chlor-alkali electrolysis plants are built up since about 20 years only using the membrane process (see "▶ Chlorine and Caustic Technology, Membrane Cell Process"). However, old plants of diaphragm and amalgam process are amortized and therefore have significant financial benefit. Meanwhile, there is a clear tendency to change to the membrane process because old plants reach after many decades their maximum lifetime. This is fortified in case of amalgam process by environmental arguments.

Alcoholates are interesting niche products for the amalgam process. They can be extraordinarily economically produced with alcohols in place of water in the amalgam decomposer using special catalysts. This application most probably will be remaining in future for the amalgam process.

About 30 % of electrical energy can be saved by application of oxygen depolarized cathodes (ODC) in the membrane process, as mentioned above. The energy saving will be increased to 50 % if the conversion to this technology starts from the diaphragm or amalgam process.

Cross-References

- ▶ Chlorine and Caustic Technology, Membrane Cell Process
- ▶ Chlorine and Caustic Technology, Using Oxygen Depolarized Cathode
- ▶ Electrocatalysis of Chlorine Evolution
- ▶ Hydrochloric Acid Electrolysis
- ▶ Hydrogen Evolution Reaction

References

1. Schmittinger P, Florkiewicz T, Curlin L, Lüke B, Scannell R, Navin P, Zelfel E, Bartsch R (2011) Chlorine. In: Ullmann's encyclopedia of industrial chemistry, Wiley-VCH, Weinheim. doi:10.1002/14356007.a06_399.pub3
2. Büchel KH, Moretto HH, Woditsch P (2000) Industrial inorganic chemistry, 2nd edn. Wiley-VCH, Weinheim
3. Hamann CH, Hamnett A, Vielstich W (2007) Electrochemistry, 2nd edn. Wiley-VCH, Weinheim
4. Bommaraju TV, Lüke B, O'Brien TF, Blackburn MC (2002) Chlorine. In: Kirk-Othmer encyclopedia of chemical technology. Wiley, Headquarter USA, Hoboken, NJ. doi:10.1002/0471238961.0308121503211812.a01.pub2
5. O'Brien TF, Bommaraju TV, Hine F (2005) Handbook of chlor-alkali technology, vol 1–5, Springer, US, New York. doi:10.1007/b113786
6. Trasatti S (2000) Electrocatalysis: understanding the success of DSA®. Electrochim Acta 45:2377–2385, doi:10.1016/S0013-4686(00)00338-8
7. Dötzel O, Schneider L (2002) Non-asbestos diaphragm in chlorine-alkali electrolysis. Chem Eng Technol 25:167–171

Chlorine and Caustic Technology, Using Oxygen Depolarized Cathode

Jakob Jörissen
Chair of Technical Chemistry, Technical University of Dortmund, Germany

Introduction

Chlorine and caustic soda belong to the most important products of chemical industry. About 95 % are manufactured with high energy consumption by chlor-alkali electrolysis of common salt (see essays "▶ Chlorine and Caustic Technology, Overview and Traditional Processes" and "▶ Chlorine and Caustic Technology, Membrane Cell Process") [1, 2]. The coproduct hydrogen gas has excellent purity. Partially, it is used as a valuable compound for chemical reactions or selling in gas cylinders, but this needs expensive compression. At many locations the hydrogen is utilized as an additional fuel in a power plant in order to recover its energy content. However, sometimes the hydrogen is discharged simply into atmosphere.

Energy recuperation from undesired hydrogen using fuel cells enables higher efficiency than in conventional thermal power stations. This is successfully tested in an industrial scale; the biggest plant has 1 MW electrical power [3].

The cell voltage of the state-of-the-art membrane cells in chlor-alkali electrolysis is about 3.0 V (see Fig. 1). Theoretically, according to thermodynamics, the voltage of a hydrogen-oxygen fuel cell should be 1.23 V. However, a realistic value for operation with air today is approximately 0.7 V (see essay "▶ Fuel Cells, Principles and Thermodynamics"). Thus, if the hydrogen of electrolysis is completely consumed in fuel cells, the energy demand of this combination is equivalent to a voltage of around 2.3 V. In this case the electrolyzers can be used unchanged but high investment for fuel cells is necessary. Additionally, energy is lost two times due to hydrogen overpotentials: during the cathodic

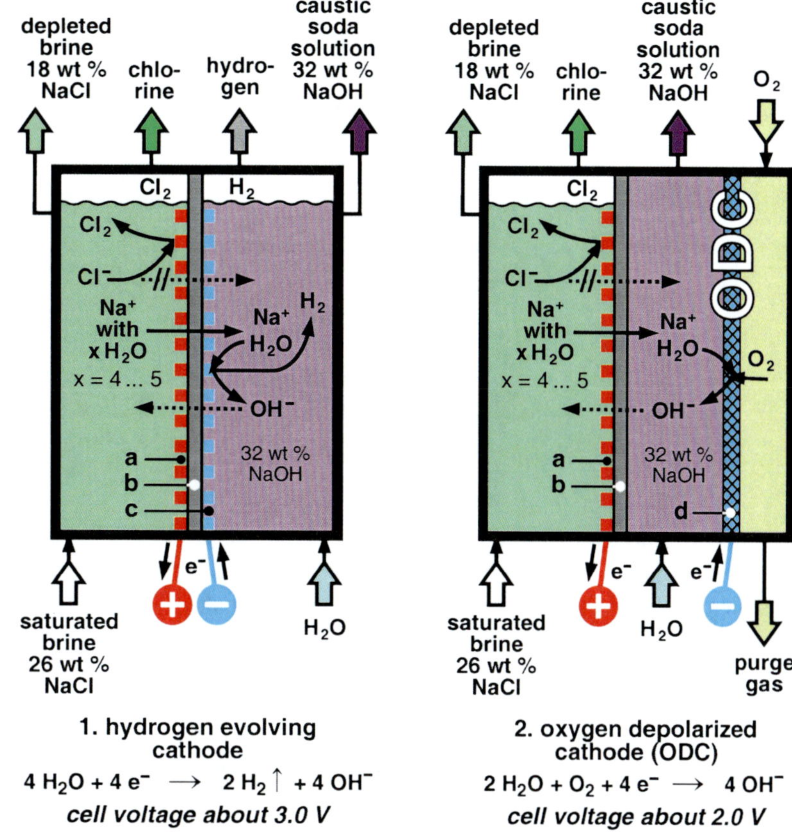

Chlorine and Caustic Technology, Using Oxygen Depolarized Cathode, Fig. 1 Scheme of the membrane process using different cathodes. *a* Anode (titanium DSA), *b* cation-exchange membrane, *c* hydrogen-evolving cathode (nickel, activated), *d* oxygen-depolarized cathode (ODC, e.g., silver-PTFE gas diffusion electrode GDE); cell voltages at 4 kA m^{-2} and 90 °C

hydrogen evolution in the electrolyzer as well as during the anodic hydrogen oxidation in the fuel cell.

Alternatively, an "oxygen-depolarized cathode" (ODC) – in principle a cathode of an "alkaline fuel cell" (AFC) – can be inserted directly into the chlor-alkali electrolysis cell instead of a hydrogen-evolving cathode as it is shown in Fig. 1. No overpotential of a hydrogen reaction occurs. Thus, the cell voltage is reduced approximately by 1.0 V, from 3.0 to 2.0 V, i.e., saving of around one-third of electrical energy is possible. This needs new highly sophisticated ODCs and electrolyzer constructions which are state of the art since 2011 in a demonstration plant of 20,000 t/a chlorine capacity [5–7]. Until now, pure or at least highly concentrated oxygen is necessary. Nevertheless, 30 % overall energy saving is realistic.

The idea of ODC application in chlor-alkali electrolysis is not new. Attempts since 1950 to establish it for industrial diaphragm cells were not successful. A patent was applied in 1959 for an ODC in a membrane cell, long time before suitable membranes were available. After the beginning of the membrane process in the 1970s, much research was started about ODC utilization, up to construction of pilot plants in an industrial scale. A comprehensive overview is given in [4]. The essay "▶ Hydrochloric Acid Electrolysis" includes the meanwhile industrially approved ODC application in this process.

Gas Diffusion Electrodes (GDE)

The electrochemical conversion of a gas is much more complicated than gas evolution, especially if a liquid electrolyte is used as in the case of chlor-alkali electrolysis. The "oxygen reduction reaction" (ORR, Fig. 1) is only possible in a "gas

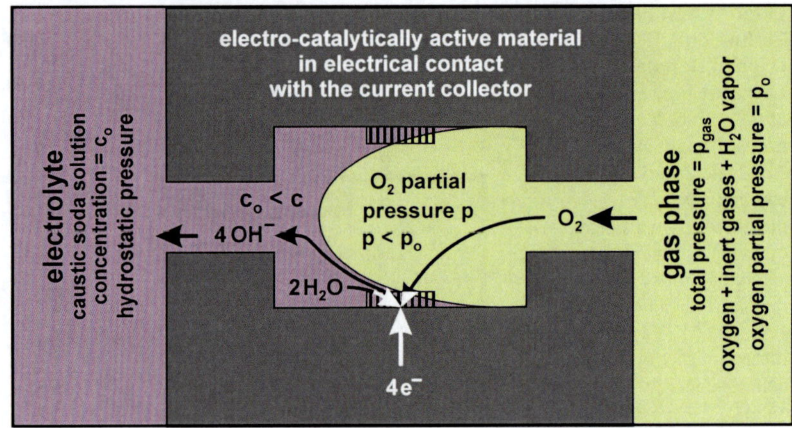

Chlorine and Caustic Technology, Using Oxygen Depolarized Cathode, Fig. 2 Scheme of one pore in a gas diffusion electrode (GDE, based on [7])

diffusion electrode" (GDE; see also essay "► Oxygen Reduction Reaction in Alkaline Solution"). The principle is shown schematically in Fig. 2.

"Three-phase zones" are indispensable for ORR where the solid electrode material needs a surface of high electro-catalytic activity and has to be simultaneously in optimal contact with:

- Current collector at low Ohmic resistance for supplying with electrons
- Gas phase for supplying with oxygen at a sufficient partial pressure
- Electrolyte for supplying with water and for discharging the produced OH^- ions at high ionic conductivity (i.e., low Ohmic resistance).

The simple model in Fig. 2 assumes a cylindrical pore where a meniscus is formed between electrolyte and gas phase in consequence of a suitable gas excess pressure. The shaded areas demonstrate the small part of the surface in the pore where all preconditions of three-phase zones are fulfilled and ORR is enabled. A highly porous structure is necessary in a GDE for sufficiently sized active areas which altogether have to be much larger than the geometrical electrode area.

The reduced diameters on both sides of the pore in Fig. 2 refer to the unavoidable hindrance of mass transport. The partial pressure p of oxygen is lower in the active three-phase zones, where oxygen is consumed, compared with p_0 in the well-mixed gas phase outside of the GDE. This difference will be increased notably in the presence of inert gases which are accumulated within the pore during the reaction of oxygen. Analogously, the concentration c of the produced caustic soda (OH^- ions from ORR and Na^+ ions which arrive for electroneutrality through the membrane b; see Fig. 1) is higher within the pores than c_0 in the bulk electrolyte. A diffusion overpotential is caused by all these effects which are strongly intensified with increasing current density (i.e., higher oxygen consumption).

Especially the supplying of ORR with oxygen is a significant problem:

- Precondition of ORR is dissolved oxygen in the electrolyte, but its solubility is very low: $9.3 \cdot 10^{-6}$ mol/l = 0.27 ml gas per liter of caustic soda solution (33 wt% NaOH, 80 °C, 1 bar) [8].
- In consequence of the small solute concentration, a sufficient oxygen transfer through the electrolyte film on the catalyst surface is only possible if this film is extremely thin (about 0.06 μm estimated in [9]).
- The disadvantageous effect of inert gases was mentioned above. At the usual elevated temperatures, water vapor is present as inert gas with high partial pressure which decreases substantially the oxygen content of the gas phase.

The three-phase zones of the simplified model in Fig. 2 are very sensitive to any change of pressure. The inevitable pressure difference in an industrial-scale electrolysis cell is elucidated

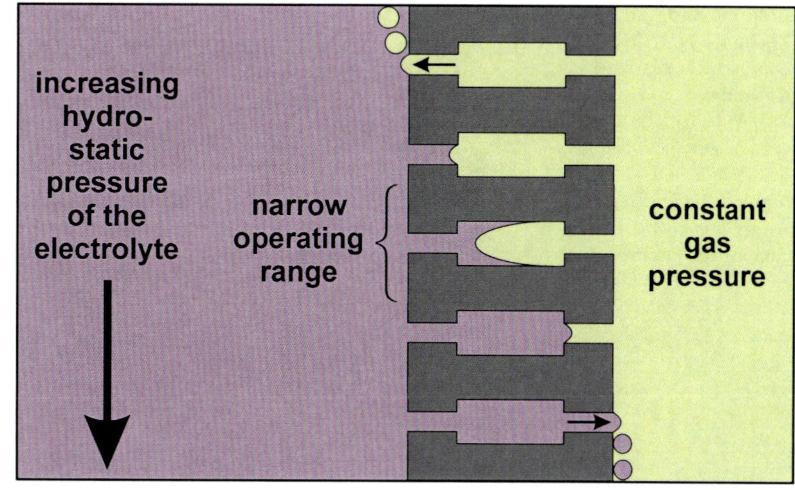

Chlorine and Caustic Technology, Using Oxygen Depolarized Cathode, Fig. 3 Pressure differences in an industrial-scale GDE (Based on [7])

in Fig. 3. While the gas pressure is constant, the hydrostatic pressure of the electrolyte increases from top to bottom of the cell due to the solution level of up to more than 1 m. Only a narrow operation range with functioning three-phase zones is available, unless additional design features for stabilization are implemented. At higher positions gas fills thoroughly the pores and finally it is blown out through the GDE. Electrolyte flows at lower positions into the complete pores and finally it trickles through the GDE. Thus, simple porous GDEs are not suitable for industrial applications.

The classical way and state of the art for stabilization of three-phase zones within a GDE is the combination of hydrophobic and hydrophilic materials; see Fig. 4.

The aqueous electrolyte cannot enter into small-sized volumes between fibers of a very hydrophobic polymer, typically PTFE (polytetrafluoroethylene), because the PTFE surface is not wettable with water and the high surface tension of aqueous solutions avoids the fragmentation into very small droplets. Thus, the way is free between the PTFE fibers for gas transport to all three-phase zones within the GDE.

Frequently, PTFE is used simultaneously as a binder for mechanical stability of the GDE which is usually some tenths of a millimeter thick.

The electrode material with electro-catalytic properties is conductive and hydrophilic so that it enables the electrical contact of all three-phase zones with the current collector mesh as well as the connection to the bulk electrolyte for ion transport by way of electrolyte films on the catalyst surface. These films are thin enough for gas transport into the reaction zone.

Part A of Fig. 4 shows an optimal structure: connected catalyst particles and PTFE fibers (gas pores) are dispersed as finely as possible for maximal surface area of three-phase zones. This enables the best GDE performance (i.e., highest current density at a given electrode potential). In reality, this is not yet achieved in state-of-the-art GDEs.

Part B of Fig. 4 shows the present state: a significant part of the catalyst particles is not separated by PTFE fibers (gas pores). Thus, catalyst particles are clustered to agglomerates and flooded by electrolyte. Exclusively the outside surfaces of these catalyst agglomerates, which are covered by a thin electrolyte film, can operate as three-phase zones. All internal catalyst surfaces within the agglomerates are nearly inactive due to insufficient gas transport through the long distance inside of electrolyte [9].

A changeover from part B to part A of Fig. 4 is an important goal of GDE optimization in order to increase performance and/or to decrease catalyst demand. Highly sophisticated techniques for high dispersion, which simultaneously guarantee the connection between catalyst particles, have to be developed.

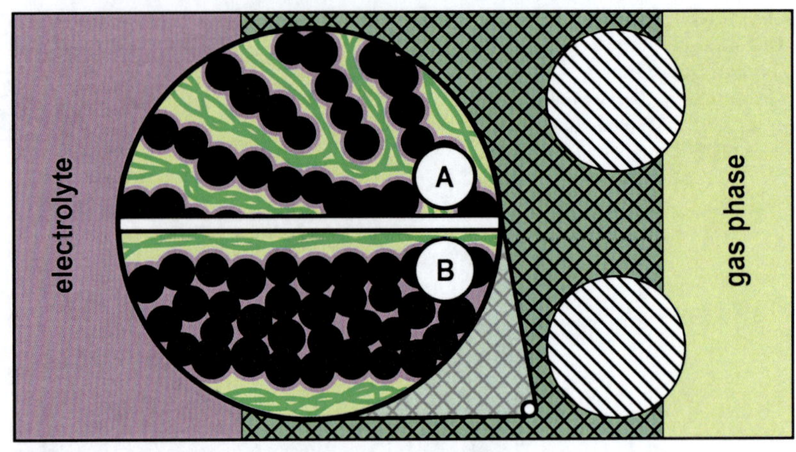

Chlorine and Caustic Technology, Using Oxygen Depolarized Cathode, Fig. 4 Schematic cross section and magnified details of a GDE structure (not to scale, based on [7]) *A* very fine distribution of catalyst and gas pores, formed by PTFE fibers, *B* clustering of catalyst particles to flooded agglomerates

Catalysts for Oxygen Reduction Reaction (ORR)

The typical catalyst in fuel cells is platinum (and possibly other platinum group metals). Usually, nanoparticles of these metals are applied on carbon black as a conductive, hydrophilic carrier. In alkaline solution also silver is suitable for ORR. The moderate price of silver permits the application of pure metal as catalyst without a carrier. Additionally, the high conductivity of silver is advantageous. No silver is lost during operation and can be recycled after the ODC service life. However, the high fixed capital may be a problem.

Much research was and is focused on ODCs for chlor-alkali electrolysis which are based on carbon carrier materials [4]. However, until now the stability of such ODCs is insufficient. The most probable reason is the formation of hydrogen peroxide species which attack the carbon and destroy the connection between catalyst metal and carbon carrier. Using pure silver as catalyst, no peroxide formation has to be expected.

The state-of-the-art ODCs in the mentioned demonstration plant use a silver catalyst.

Technical Realization of an ODC Electrolyzer

Operation of chlor-alkali electrolysis with ODC requires suitable and stable ODCs as well as adequate cell constructions. A particular problem that has to be solved is the pressure difference between electrolyte and gas phase, elucidated in Fig. 3. The state-of-the-art ODCs, discussed using Fig. 4, withstand a relatively high pressure difference, but until now the full height of an industrial electrolyzer exceeds the operating range of difference pressure in ODCs. Many constructions are discussed in literature and patents, including a horizontal cell or a cell with a "falling film" technology of catholyte and anolyte which avoids any hydrostatic pressure. Some cell constructions divide the ODC in horizontal sections and supply oxygen with different pressures, adjusted to the respective hydrostatic pressure. The simplest and most reasonable of these developments uses so-called gas pockets with self-regulating pressure because oxygen is bubbling from gas pocket to gas pocket through the catholyte. However, sealing and current supplying of these gas pockets is complicated. An overview of cell constructions is given in [4].

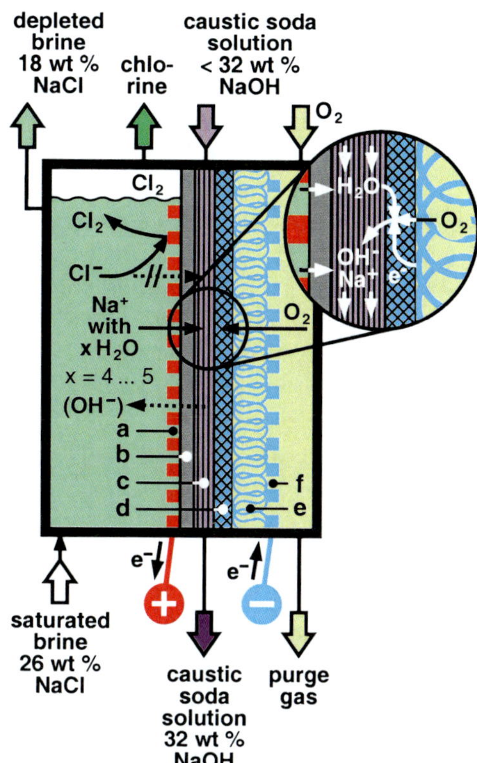

Chlorine and Caustic Technology, Using Oxygen Depolarized Cathode, Fig. 5 Scheme of the ODC percolator cell. *a* Anode (titanium DSA), *b* cation-exchange membrane, *c* percolator, *d* oxygen-depolarized cathode (ODC, silver-PTFE gas diffusion electrode), *e* elastic element (fine, interwoven nickel wire), *f* cathodic current distributor

Relatively much information is published about the mentioned demonstration electrolyzer and can be discussed here [6, 7]. Figure 5 shows a scheme of its principle.

The electrolyzer is based on the "single element" technology (ThyssenKrupp Uhde GmbH.) About 80 independent electrolysis cells of 2.7 m^2 area, individually sealed by a bolted flange connection, are mechanically combined and electrically connected in one electrolyzer. All components, including anode half shells (titanium), are used without change, and only cathode half shells (nickel) are modified for ODC operation.

The construction on the cathode side is based on the falling film technology [10]. However, the catholyte flows between the membrane b and the ODC d, made of silver and PTFE (Bayer Material Science AG, BMS), within the so-called percolator c, a woven fabric which adjusts the flow rate of caustic soda solution. The ODC d and the cathodic current distributor f are electrically connected on their entire area by the elastic element e, made of fine, interwoven nickel wire. This elastic element e presses the ODC d, the percolator c, and the membrane b with optimized pressure onto the anode a. Therefore, the usual anode half shell of the conventional membrane process, completely filled with anolyte, can be applied, and no falling film construction is necessary on the anode side as it is used in [10]. The percolator c withstands the pressure of the elastic element e and remains sufficiently permeable for the catholyte flow. Free oxygen gas transport into the ODC d is possible through the elastic element e.

Economic Aspects of ODC Application in Chlor-alkali Electrolysis

The economic situation of ODC application in chlor-alkali electrolysis is extremely dependent on the particular conditions of an industry location. Especially the value of hydrogen as a reactant or energy carrier and the price and the availability of electrical energy as well as of oxygen are the most important parameters [7].

The energy saving by ODC application decreases indirectly the carbon dioxide emission of usual coal power plants. This may be a significant argument in the environmental concept of companies.

On the one hand, if electrical energy is available and hydrogen is in demand, the conventional chlor-alkali electrolysis is the most economical electrochemical way to produce hydrogen, much better than water electrolysis which is in discussion concerning regenerative electrical energy. On the other hand, ODC application will be economically interesting if no qualified application for hydrogen is existent and/or electrical energy is limited and/or too expensive.

The best chance for ODC utilization is given if the production capacity of an electrolysis plant has to be expanded, but additional electrical energy is not available. In this case ODC application enables an increasing of production capacity by about 40 % without additional power consumption.

Future Directions

After many decades of development, ODC technology is now industrially available and still includes a large optimization potential.

It may be a dream to find really better catalyst materials for ORR in alkaline solution. However, decreasing of the local current density due to maximized active electrode surface by better distribution of catalyst and gas channels (see Fig. 4) is an important intention in case of every electrode material. An economical compromise has to be found between low energy consumption, high current density (i.e., high production rate), and low catalyst demand.

ODC operation in chlor-alkali electrolysis using air instead of pure oxygen may be another dream which could be realized possibly with optimized ODCs.

Generally, gas reactions are enhanced at elevated pressure. This was proved for ODC chlor-alkali electrolysis in laboratory scale at pressures up to 5 bar [11]. Operation with air was possible. Even though pressurized electrolysis producing the dangerous gas chlorine is risky, it could be interesting to apply new technical know-how for this challenge.

The actual cell construction (see Fig. 5) operates since 2011 without problems [5, 6]. However, cost-saving constructions may be possible by further development [7]. An ODC that can withstand the hydrostatic pressure of full electrolyzer height will be useful.

A significantly simplified construction could be realized using "zero gap" technology, i.e., membrane b and ODC d are in direct contact (for literature see [4]). Precondition would be an ODC design where the produced caustic soda solution completely can flow through the ODC without blocking oxygen supplying of ORR. In this case the water, which is needed for ORR and for dilution of caustic soda solution, only can be supplied by hydration shells of Na^+ ions during their migration through the membrane (see Fig. 5). This so-called electroosmotic flow includes 4–5 water molecules per Na^+ ion, increasing with decreasing NaCl concentration in the anolyte. It is advantageous that ORR consumes less water than hydrogen evolution (see Fig. 1). Nevertheless, at present membranes cannot operate under these conditions and are damaged (too high caustic soda concentration or blistering at too low anolyte concentrations; see essay "► Chlorine and Caustic Technology, Membrane Cell Process"). A combination of especially developed ODCs and membranes may be suitable for this interesting process variant.

Cross-References

- ► Chlorine and Caustic Technology, Membrane Cell Process
- ► Chlorine and Caustic Technology, Overview and Traditional Processes
- ► Fuel Cells, Principles and Thermodynamics
- ► Hydrochloric Acid Electrolysis
- ► Oxygen Reduction Reaction in Alkaline Solution

References

1. Schmittinger P, Florkiewicz T, Curlin L, Lüke B, Scannell R, Navin P, Zelfel E, Bartsch R (2011) Chlorine. In: Ullmann's encyclopedia of industrial chemistry, Wiley-VCH, doi:10.1002/14356007.a06_399.pub3
2. Bommaraju TV, Lüke B, O'Brien TF, Blackburn MC (2002) Chlorine. In: Othmer K (ed) Encyclopedia of chemical technology. Wiley, New York. doi:10.1002/0471238961.0308121503211812.a01.pub2
3. Press release 2011-07-26 "Nedstack makes to world's largest fuel cell to generate clean electricity for chlorine plant in Antwerp", download: http://www.nedstack.com/images/stories/news/documents/press%20release%20solvay%20powerplant_nl.pdf
4. Moussallem I, Jörissen J, Kunz U, Pinnow S, Turek T (2008) Chlor-alkali electrolysis with oxygen depolarized cathodes: history, present status and future prospects. J Appl Electrochem 38:1177–1194. doi:10.1007/s10800-008-9556-9
5. http://www.process-worldwide.com/engineering_construction/operational_excellence/energy_efficiency/articles/366423/

6. http://www.thyssenkrupp-uhde.de/publications/videos/base-chemicals/electrolysis-nacl-odc.html
7. Jörissen J, Turek T, Weber R (2011) Chlorine production with oxygen depolarized cathode. Energy saving in electrolysis (language: German). Chem Unserer Zeit 45:172–183. doi:10.1002/ciuz.201100545
8. Chatenet M, Aurousseau M, Durand R (2000) Comparative methods for gas diffusivity and solubility determination in extreme media: application to molecular oxygen in an industrial chlorine-soda electrolyte. Ind Eng Chem Res 39:3083–3089
9. Pinnow S, Chavan N, Turek T (2011) Thin-film flooded agglomerate model for silver-based oxygen depolarized cathodes. J Appl Electrochem 41:1053–1064. doi:10.1007/s10800-011-0311-2
10. Tetzlaff KH, Schmid D, Russow J (1985) Electrochemical method for treating liquid electrolytes. DE patent 3401636 A1, EP patent 0150017 B1
11. Simmrock KH, Poblotzki J (1988) Oxygen consuming cathodes of the carbon and silver type in chloralkali electrolysis. Proc Electrochem Soc 88:369–382

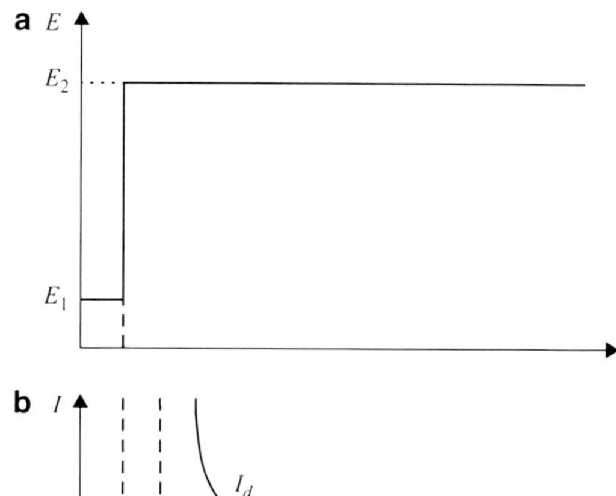

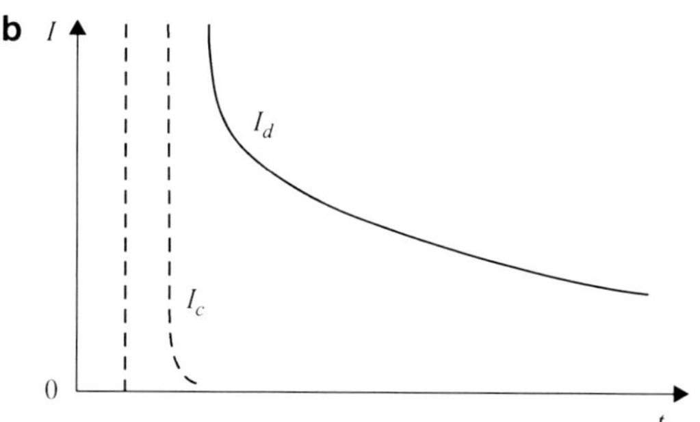

Chronoamperometry, Chronocoulometry, and Chronopotentiometry, Fig. 1 Typical waveform of the potential step (**a**) and the respective chronoamperometric response (**b**)

Chronoamperometry, Chronocoulometry, and Chronopotentiometry

György Inzelt
Department of Physical Chemistry, Eötvös Loránd University, Budapest, Hungary

Introduction

Chronoamperometry, chronocoulometry, and chronopotentiometry belong to the family of step techniques [1–4]. In chronoamperometry the current, while in chronocoulometry the charge is measured as a function of time after application of a potential step perturbation. In the case of chronopotentiometry, a current step is applied, and the change of the potential with time is detected.

Chronoamperometry

If the potential is stepped from E_1, where no current flows, i.e., the oxidation or reduction of the electrochemically active species does not take place, to E_2, where the current belongs to the electrode reaction and is limited by diffusion, the current flow at any time after application of the potential step will obey the Cottrell equation [1–4].

In contrast to steady-state conditions, the current decreases with time because the concentration gradient decreases:

$$\partial c(x,t)/\partial x = c^* (\pi D t)^{-1/2} \exp\left(-x^2/\pi D t\right) \quad (1)$$

where x is the location, t is time, D and c are the diffusion coefficient and bulk concentration of the reacting species.

The perturbation and the current response as well as the capacitive current (I_c) at short times are shown schematically in Fig. 1.

$$[\partial c(x,t)/\partial x]_{x=0} = c^* (\pi D t)^{-1/2} \quad (2)$$

For fast charge transfer, the diffusion current will change with $t^{1/2}$:

$$I(t) = nFAD_R^{1/2} c^* (\pi t)^{-1/2} \left\{1 + (D_O/D_R)^{1/2} \exp\left[(nF/RT)\left(E - E^{0'}\right)\right]\right\}^{-1} \quad (3)$$

where $E^{0'}$ is the formal electrode potential. If the potential is stepped to the diffusion limiting current region, i.e., $E >> E^{0'}$, Eq. (3) will be simplified to the Cottrell equation. From the slope of $I(t)$ versus $t^{-1/2}$ plot – knowing other quantities – D can be determined.

$$I(t) = nFDc^* \left[(\pi Dt)^{-1/2} + r_0^{-1}\right] \quad (4)$$

where r_o is the radius of the spherical electrode or microelectrode disk. Equation 4 turns into the usual Cottrell equation if $r_o \to \infty$. It can be seen that at long times, a steady-state current flows. The smaller the electrode radius, the faster the steady state is achieved.

At short times, the current's response deviates from that expected theoretically due to the charging of the double layer and possibly inadequate power of the potentiostat.

If the heterogeneous charge transfer is slow, the following expression is valid for an oxidation reaction:

$$I(t) = nFAk_{ox}c_R^* \exp\left(k_{ox}^2 t/D_R\right) \operatorname{erfc}\left(k_{ox}t^{1/2}/D_R^{1/2}\right) \quad (5)$$

If both the oxidized (O) and reduced (R) forms are present:

$$I(t) = nFA\left(k_{ox}c_R^* - k_{red}c_O^*\right) \exp\left[\left(\frac{k_{ox}}{D_R^{1/2}} + \frac{k_{red}}{D_O^{1/2}}\right)^2 t\right] \times \operatorname{erfc}\left[\left(\frac{k_{ox}}{D_R^{1/2}} + \frac{k_{red}}{D_O^{1/2}}\right) t^{1/2}\right] \quad (6)$$

This equation can be used when the step is made to any potential in the rising portion of the voltammogram, where either charge transfer control or mixed kinetic-diffusional control prevails. Figure 2 shows the I versus t responses when the potential is stepped from E_1 to $E_{22} < E_{21} < E_2$, respectively.

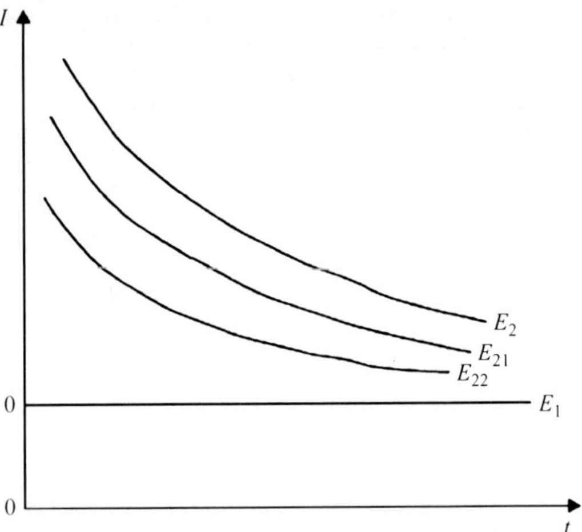

Chronoamperometry, Chronocoulometry, and Chronopotentiometry, Fig. 2 The I versus t curves when the potential is stepped from E_1 to $E_{22} < E_{21} < E_2$, respectively

Since at the foot of the voltammetric wave k_{ox} and k_{red} are small, Eqs. 5 and 6 can be linearized. For instance, Eq. 5 is simplified to:

$$I(t) = nFAk_{ox}c_R^* \left[1 - \frac{2k_{ox}t^{1/2}}{D_R^{1/2}\pi^{1/2}}\right] \quad (7)$$

The plotting $I(t)$ versus $t^{1/2}$ and extrapolating the linear plot to $t = 0$, k_{ox} can be obtained from the intercept.

In the case of a large pseudocapacitance, e.g., an electrochemically active polymer film on the surface, the current-time decay reflects the diffusion rate of the charge carriers through the surface layer, thus at shorter times the decay of the current should conform to the Cottrell equation. At long times, when $(Dt)^{1/2} \geq L$, where L is the film thickness, the concentration within surface film impacts on the film-solution boundary, the chronoamperometric current will be less than that predicted by the Cottrell equation, and a finite diffusion relationship

$$I(t) = \frac{nFAD^{1/2}c^*}{\pi^{1/2}t^{1/2}}\left[1 + 2\sum_{m=1}^{\infty}(-1)^m \exp\left(-\frac{m^2L^2}{Dt}\right)\right] \quad (8)$$

becomes appropriate [2, 6].

If the chronoamperometric response of a polymer-modified electrode is measured alone – in contact with inert supporting electrolyte – Cottrell-type response can be obtained usually for thick films only, because at short times ($t < 0.1$–1 ms), the potential is not established, while at longer times ($t >$ 10–100 ms), the finite diffusion conditions will prevail and I exponentially decreases with time. Another complication that may arise is the dependence of D on the potential in the case of conducting polymer films [7].

Specific current-time curves are obtained in the case of electrocrystallization or electrodeposition. The characteristic responses expected for different nucleation and growth processes, e.g., instantaneous nucleation, one-, two-, and three-dimensional layer or layer-by-layer growth can be found in the literature [8–10].

Chronocoulometry

Chronocoulometry gives practically the same information that is provided by chronoamperometry, since it is just based on the integration of the current-time response after the application of a potential step [1, 3, 4, 11–15]. Nevertheless, chronocoulometry offers important experimental advantages. First, unlike the current response that quickly decreases, the measured signal usually increases with time, and hence, the later parts of the transient can be detected more accurately. Second, a better signal-to-noise ratio can be achieved. Third, contributions of charging/discharging of the electrochemical double layer and any pseudocapacitance on the surface (charge consumed by the electrode reaction of adsorbed species to the overall charge passed as a function of time) can be distinguished from those due to the diffusing electroreactants. In the case of the electrochemical oxidation of a species R ($R \rightarrow O^+ + e^-$) after application of the potential step from E_1 (where no current flows) to E_2 (where the current is limited by diffusion or more precisely by the rate at which the reactant is supplied to the electrode surface), and the conditions of linear diffusion (flat electrode, unstirred solution) prevail, the current (I) flow at any time (t) will obey the Cottrell equation. The time integral of the Cottrell equation gives the cumulative charge (Q) passed in the course of oxidation of R:

$$Q_{diff}(t) = \int_{O}^{t} nFAD_R^{1/2} c_R^* \pi^{1/2} \mathrm{d}t = 2\pi FAD_R^{1/2} c_R^* \pi^{-1/2} t^{1/2} \quad (9)$$

where $Q_{\mathrm{diff}}(t)$ is the charge consumed until time t, n is the charge number of electrode reaction, F is the Faraday constant, A is the surface area, D_R and c_R* are the diffusion coefficient and the bulk concentration of the reacting species R. Equation 9 is sometimes referred to as Anson equation [13, 14] in the literature.

However, at least one additional current component has to be taken into account, because of the charging of double layer while stepping the potential from E_1 to E_2. After the application of a potential step of magnitude $E = E_2 - E_1$, the exponential decay of the current with time depends on the double-layer capacitance (C_d) and the solution resistance (R_s), i.e., on the time constant $\tau = R_s C_d$. Consequently, if we assume that C_d is constant and the capacitor is initially uncharged ($Q = 0$ at $t = 0$), for the capacitive charge (Q_c), we obtain:

$$Q_c = EC_d\left(1 - e^{-t/R_S C_d}\right) \quad (10)$$

In order to obtain a straight line in the Q versus $t^{1/2}$ plot for a relatively long period of time – which is the precondition of straightforward data evolution – the decay of the current belonging to the double layer charging should be very fast. It follows that the application of a working

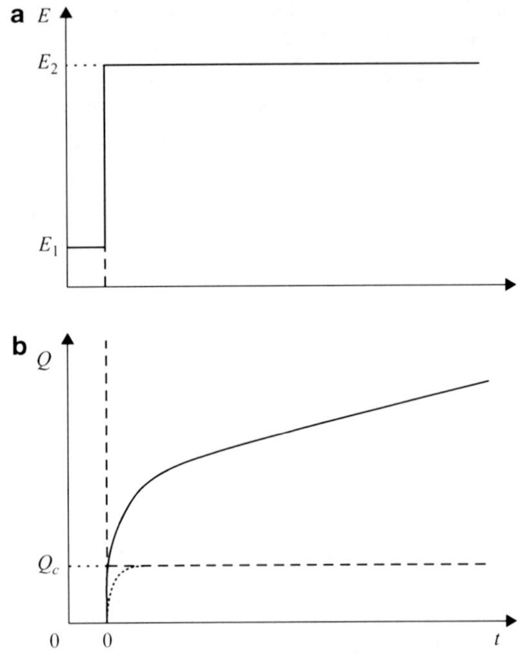

Chronoamperometry, Chronocoulometry, and Chronopotentiometry, Fig. 3 Waveform of a potential step experiment (**a**) and the respective chronocoulometric response (**b**). *Dashed* curves in (**b**) displayed for the illustration of the capacitive charge effect

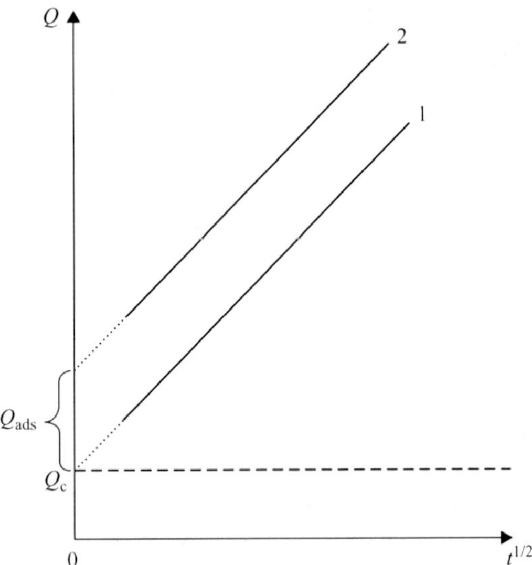

Chronoamperometry, Chronocoulometry, and Chronopotentiometry, Fig. 4 Chronocoulometric charge versus (time)$^{1/2}$ plot in the absence (1) and in the presence (2) of adsorption. The *dashed*, horizontal line represents the charge response in the absence of reactant. This representation is sometimes referred to as Anson plot in the literature

electrode of small A in order to decrease C_d and the use supporting electrolyte to lower R_s is of importance to decrease this effect and to maintain potential control. The typical waveform of a potential step experiment and the respective chronocoulometric curve are shown schematically in Fig. 3. The Q_c versus t curve, which is also displayed, can be determined by repeating the experiment in the absence of R, i.e., in pure supporting electrolyte.

If R is adsorbed at the electrode surface at E_1, the adsorbed amount will also be oxidized at E_2. This process is usually very quick compared to the slow accumulation of R by diffusion. The total charge can be given as follows:

$$Q(t) = Q_{\mathrm{diff}}(t) + Q_c(t) + Q_{\mathrm{ads}}(t) \qquad (11)$$

From Q_{ads} the adsorbed amount of R (Γ_R) can be estimated, since $Q_{ads} = n\,F\,\Gamma_R$. According to Eq. 11, the plot of Q_{diff} versus $t^{1/2}$ should be linear and the slope is proportional to the concentration of the reactant, as well as to n, A and $D^{1/2}$. It is shown in Fig. 4.

From the slope, e.g., D can be calculated provided that all other quantities are known, while the intercept gives $Q_c + Q_{ads}$. At very short times, Q versus $t^{1/2}$ curves are not linear, since neither the charging process nor the oxidation of the adsorbed amount of reactant is instantaneous, albeit under well-designed experimental conditions (see above), this period is less than 1 ms. This effect is indicated as dotted extensions of lines 1 and 2 in Fig. 4.

Chronocoulometric responses may be governed wholly or partially by the charge transfer kinetics. In some cases, the diffusion-limited situation cannot be reached, e.g., due to the insufficient power of the potentiostat and the inherent properties of the system, especially at the beginning of the potential step. If the heterogeneous rate constants, k_{ox}, or k_{red} are small, the following expression can be derived, e.g., for the oxidation:

$$Q(t) = \frac{nFAD_{R}c_{R}^{*}}{k_{ox}}\left[\exp\left(\frac{k_{ox}^{2}}{D_{R}^{1/2}}t\right)\mathrm{erfc}\left(\frac{k_{ox}}{D_{R}^{1/2}}t^{1/2}\right) + \frac{2k_{ox}t^{1/2}}{D_{R}^{1/2}\pi^{1/2}} - 1\right] \quad (12)$$

A high values of $(k_{ox}/D_R)^{1/2}$, i.e., when k_{ox} is high or D_R is small, e.g., in polymer solution of high viscosity, Eq. 9 is approached.

If the potential amplitude of the step is less than $E_2 - E_1$, i.e., the step is made to any potential in the rising portion of the voltammogram, either charge transfer control or mixed kinetic-diffusional control prevails [c_R $(x = 0)$ is not zero but smaller than $c_R{}^*$]. For the description of this curve, Eq. 12 can be applied, and k_{ox} can be calculated.

In the case of polymer-modified electrodes, due to the diffusion conditions, the chronocoulometric curve can be given as follows [7, 15]:

$$\frac{Q}{Q_T} = 1 - \frac{8}{\pi^2}\sum_{m=1}\left(\frac{1}{2m-1}\right)^2 \times \exp\left[-(2m-1)^2\pi^2\frac{Dt}{L^2}\right] \quad (13)$$

where Q_T is the total charge that can be consumed by the electroactive surface film ($Q_T = Q_{ads}$) and L is the film thickness. For 2 % accuracy, it is enough to consider the first member of the summation ($m = 1$); hence,

$$\frac{Q}{Q_T} = 1 - \frac{8}{\pi^2}\exp\left(-\pi^2\frac{Dt}{L^2}\right) \quad (14)$$

It follows that, in the presence of a thick ($L >$ 100 nm, $\Gamma > 10^{-8}$ mol cm^{-2}), electrochemically active surface layer reliable measurement can be made in a time window from ms to some seconds [3, 4, 6, 7, 15].

Double-step chronocoulometry is a powerful tool in identifying adsorption phenomena, in

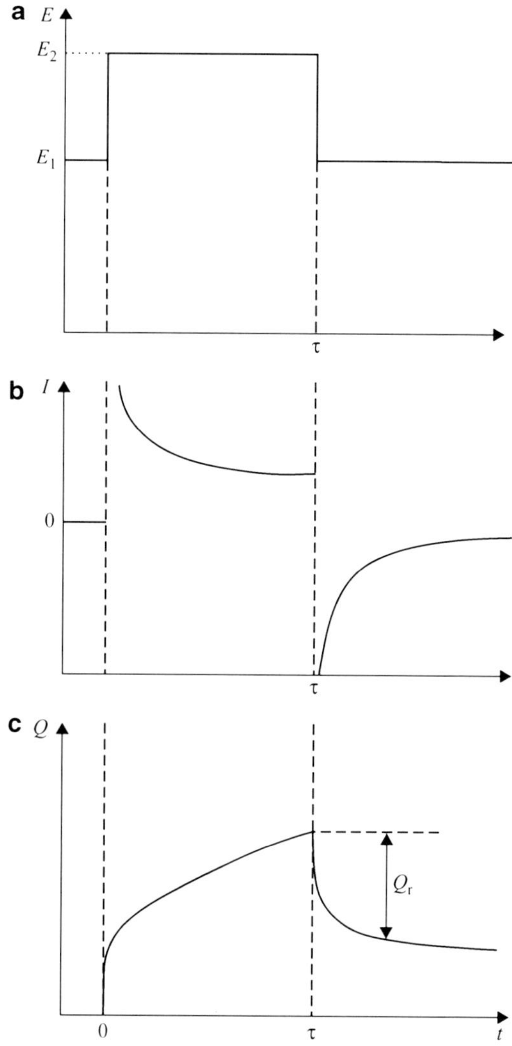

Chronoamperometry, Chronocoulometry, and Chronopotentiometry, Fig. 5 The waveform (**a**), the chronoamperometric (**b**) and chronocoulometric (**c**) responses in the case of double potential steps

obtaining information on the kinetics of coupled homogeneous reactions and for the determination of the capacitive contribution. The double potential step is executed in such a way that after the first step from E_1 to E_2, a next step is applied, i.e., the reversal of the potential to its initial value E_1 from E_2 (see Fig. 5).

It means that the product O^+ is reduced again during the second step from E_2 to E_1. If the magnitude of the reversal step is also large enough to ensure diffusion control, the chronocoulometric response for $t > \tau$ – where τ is the duration of the first step – is given by the following equation:

$$Q_{diff}(t > \tau) = 2nFAD_{\mathrm{R}}^{1/2} c_{\mathrm{R}}^{*} \pi^{-1/2} \left[t^{1/2} \quad (t \quad \tau)^{1/2} \right] \tag{15}$$

It is important to note that there is no capacitive contribution because the net potential change is zero.

The quantity of charge consumed in the reversal step is the difference $Q(\tau)$–$Q(t > \tau)$ since the second step actually withdraws the charge injected during the first step:

$$Q_{diff}(t > \tau) = 2nFAD_{\mathrm{R}}^{1/2} c_{\mathrm{R}}^{*} \pi^{-1/2} \times \left[\tau^{1/2} - (t - \tau)^{1/2} - t^{1/2} \right] \tag{16}$$

By plotting $Q(t < \tau)$ versus $t^{1/2}$ and $Q(t > \tau)$ versus $[\tau^{1/2} + (t{-}\tau)^{1/2}{-}t^{1/2}]$, two straight lines should be obtained. In the absence of adsorption, the intercept is Q_c and the two intercepts are equal (see Fig. 6).

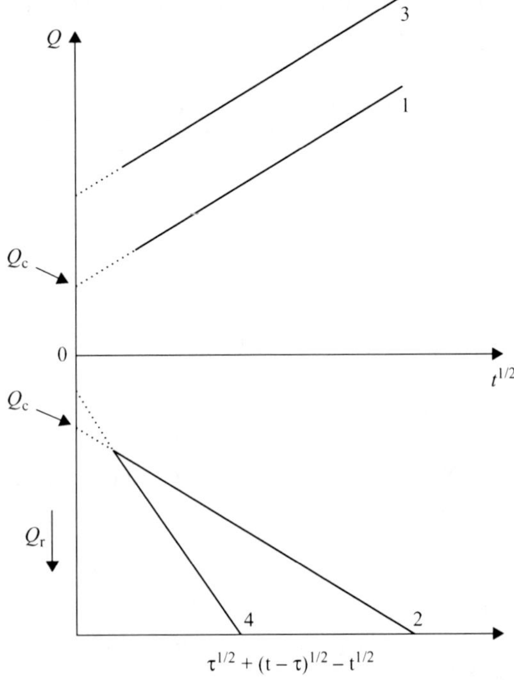

Chronoamperometry, Chronocoulometry, and Chronopotentiometry, Fig. 6 Chronocoulometric plots for double-step experiments. *Lines 1* and *2* correspond to the case when no adsorption of the reactant or product occurs. *Lines 3* and *4* depict the linear responses when the reactant is adsorbed

Chronopotentiometry

Chronopotentiometry is a controlled-current technique in which the potential variation with time (t) is measured following a current step. Other current perturbations such as linear, cyclic, or current reversals are also used [1, 3, 4, 12]. For a reversible electrode reaction following a current step, chronopotentiograms shown in Fig. 7 can be obtained.

Where no electrode reaction occurs, a small amount of charge is enough to charge the double layer. However, the electrode reaction consumes a substantial amount of charge; therefore, the potential at a planar electrode varies only to a small extent until the end of the transition time t, which corresponds to the total consumption of the electroactive species in the neighborhood of the electrode, i.e., c_{O} ($x = 0$) becomes zero. The theoretical description was given by Sand [16]. For an unstirred solution, when the reduced form (R) is initially absent, the concentration of the oxidized form will change with the time according to the following equation:

$$c_{\mathrm{o}}(0, t) = c_{\mathrm{o}}^{*} - \frac{2It^{1/2}}{nFAD_{\mathrm{o}}^{1/2} \pi^{1/2}} \tag{17}$$

where c_{O} ($x = 0$) and c_{O}* are the concentrations of the oxidized form at the electrode surface and the bulk, respectively, I is the current applied, A is the electrode area, F is the Faraday constant,

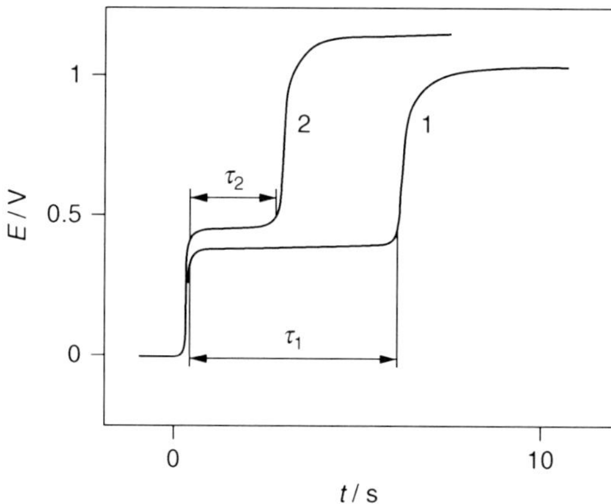

Chronoamperometry, Chronocoulometry, and Chronopotentiometry, Fig. 7 Chronopotentiometric curves. $I_2 > I_1$

D_O is the diffusion coefficient of the oxidized species, n is the number of transferred electrons. When c_O becomes zero, the following equation becomes valid

$$\frac{It^{1/2}}{c_o^*} = \frac{nFAD_o^{1/2}\pi^{1/2}}{2}, \tag{18}$$

which is called Sand-equation.

After the time τ, the flux of oxidized species does not support the applied current I anymore; the electrode potential jumps to another value where another electrode reaction starts.

For a reversible reaction

$$E = E_{\tau/4} + \frac{RT}{nF}\ln\left(\frac{\tau^{1/2} - t^{1/2}}{t^{1/2}}\right) \tag{19}$$

where

$$E_{\tau/4} = E^{o'} - \frac{RT}{2nF}\ln\left(\frac{D_o}{D_R}\right) \tag{20}$$

and $E_{\tau/4}$ is the so-called quarter-wave potential, which is identifiable with $E_{1/2}$ in a conventional voltammogram.

For an irreversible system,

$$E = E^{o'} + \frac{RT}{\alpha_c n_c F}\ln\left[\frac{2k_s}{(\pi D_o)^{1/2}}\right] + \frac{RT}{\alpha_c n_c F} \times \ln\left(\tau^{1/2} - t^{1/2}\right) \tag{21}$$

where E^o is the formal potential and α is the charge transfer coefficient.

References

1. Bard AJ, Faulkner LR (2001) Electrochemical methods, 2nd edn. Wiley, New York, pp 156–180
2. Inzelt G (2010) Kinetics of electrochemical reactions. In: Scholz F (ed) Electroanalytical methods, 2nd edn. Springer, Berlin, pp 33–56
3. Inzelt G (2010) Chronocoulometry. In: Scholz F (ed) Electroanalytical methods, 2nd edn. Springer, Berlin, pp 147–158
4. Oldham HB, Myland JC (1994) Fundamentals of electrochemical science. Akademic, San Diego
5. Rieger PH (1987) Electrochemistry. Prentice Hall, Oxford, pp 151–165
6. Murray RW (1984) Chemically modified electrodes. In: Bard AJ (ed) Electroanalytical chemistry, vol 13. Marcel Dekker, New York, pp 191–368
7. Inzelt G (2008) Conducting polymers. Springer, Berlin, pp 71–72
8. Harrison JA, Thirsk HR (1971) The fundamentals of metal deposition. In: Bard AJ (ed) Electroanalytical chemistry, vol 5. Marcel Dekker, New York, pp 67–148
9. Vargas T, Varma R (1991) Techniques for nucleation analysis in metal deposition. In: Varma R, Selman JR (eds) Techniques for characterization of electrodes and electrochemical processes. Wiley, New York, pp 707–760
10. Bockris J'OM, Khan SUM (1993) Surface electrochemistry. Plenum, New York, pp 350–376
11. Galus Z (1994) Fundamentals of electrochemical analysis, 2nd edn. Ellis Horwood/Polish Scientific Publisher PWN, New York/Warsaw
12. Brett CMA, Oliveira Brett AM (1993) Electrochemistry. Oxford University Press, Oxford, pp 206–208
13. Anson FC (1966) Innovations in the study of adsorbed reactants by chronocoulometry. Anal Chem 38:54
14. Christie JH, Anson FC, Lauer G, Osteryoung RA (1963) Determination of charge passed following application of potential step in study of electrode processes. Anal Chem 35:1979

15. Chambers JQ (1980) Chronocoulometric determination of effective diffusion-coefficients for charge-transfer through thin electroactive polymer- films. J Electroanal Chem 130:381
16. Sand HJS (1901) On the concentration at the electrodes in a solution, with special reference to the liberation of hydrogen by electrolysis of a mixture of copper sulphate and sulphuric acid. Philos Mag 1:45

Cofactor Regeneration, Electrochemical

Dirk Holtmann[1] and Svenja Kochius[2]
[1]DECHEMA Research Institute of Biochemical Engineering, Frankfurt am Main, Germany
[2]DECHEMA Research Institute, Frankfurt am Main, Germany

Introduction

Enzymes are highly specific and selective, especially for enantio- or regio-selective introduction of functional groups. Therefore, isolated enzymes such as alcohol dehydrogenases or P450 monooxygenases become more and more important for industrial production or sensor application. These enzymes can be used for the production of building blocks for synthesis of fine chemicals and important pharmaceuticals. Many of the applicable enzymes are known to be cofactor dependent. Cofactors are low-molecular compounds, which are responsible for the transfer of hydrogen, electrons, or functional groups in enzyme-catalyzed reactions [1]. Important cofactors are adenosine-5′-triphosphate (ATP), flavin adenine dinucleotide (FAD), coenzyme A (CoA), pyrroloquinoline quinone (PQQ), and nicotinamide cofactors (nicotinamide adenine dinucleotide phosphate (NADPH) and nicotinamide adenine dinucleotide (NADH)) (see Fig. 1).

In vivo the oxidized and reduced forms of the cofactors are continuously regenerated within the cellular metabolism. From an economic point of view, the stoichiometric use of cofactors is not sustainable. Therefore, several cofactor regeneration systems have been developed in the past.

Regeneration of cofactors can be accomplished through several ways: enzymatically, chemically, photochemically or electrochemically [1, 2]. In order to choose an appropriate regeneration system, several requirements have to be taken in account. First of all the method should be practical and inexpensive. Secondly the regeneration system has to be stable over a long period of time. Beyond that products need to be separated without much effort. There should be no cross-reactions between educts and products of the producing reaction or the compounds needed for cofactor regeneration. The product formation should be thermodynamically as well as kinetically preferred, and there should be no formation of by-products. Electrochemical cofactor regeneration may represent a cost-effective alternative. If the mediator, the enzyme, and perhaps the cofactor can be immobilized or used with separation in membrane reactors, an appropriate process would deliver an almost pure product by avoiding waste accumulations. The most important cofactors for technical application are NAD^+/NADH and $NADP^+$/NADPH. Therefore, the following chapter focuses on these compounds.

Principles of Electrochemical Cofactor Regeneration

In general, the electrochemical cofactor regeneration depends on a three-electrode system, consisting of a working electrode, a counter electrode, and a reference electrode. The electrochemical reaction for the cofactor regeneration occurs at the working electrode. Basically there are four principles to regenerate the cofactor [3, 4]:

- Direct regeneration at bare electrodes
- Soluble redox dye as mediator between the electrodes and the cofactor
- Immobilized mediators
- Enzyme-coupled electrochemical regeneration

An example for each of the four modes is given for the oxidation of NADH in Fig. 2. Besides the basic principles, mixed modes are also possible.

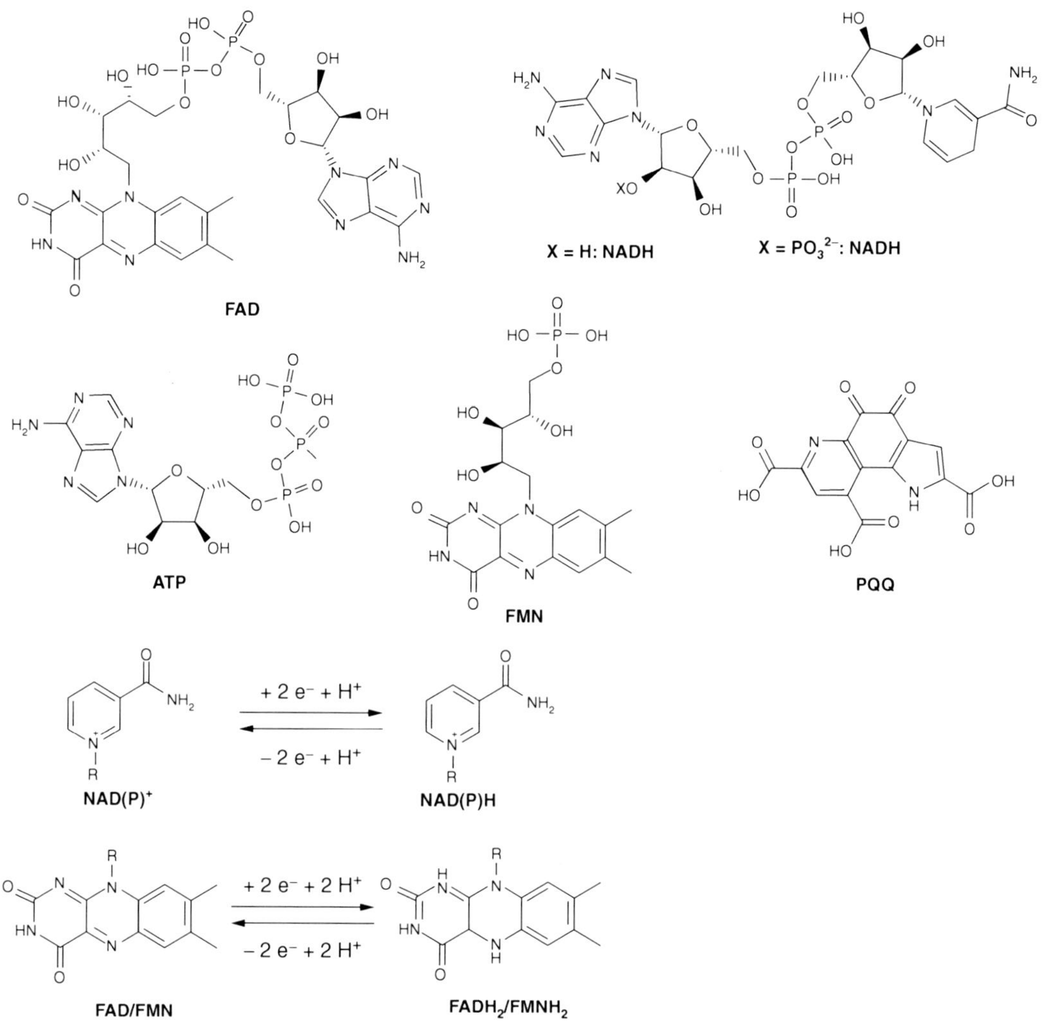

Cofactor Regeneration, Electrochemical, Fig. 1 Structure and redox reactions of important cofactors

Oxidation of NADH and NADPH

The electrochemical oxidation of NADH can be described as a two single electron and a one proton transfer (electrochemical-chemical-electrochemical mechanism – ECE) [5]. The mechanism is shown in Fig. 3.

Direct amperometric oxidation of NAD(P)H suffers from electrode fouling and large overpotentials. Therefore, this method can only be used for reactions with substrates that are oxidation stable. One successful application of the direct regeneration of NAD^+ is the synthesis of gluconolactone from glucose by the enzyme glucose dehydrogenase [3]. To avoid high overpotential, different redox mediators can be used as electron carriers. The mediator 2,2′-azino-bis(3-ethylbenzothiazoline-6-sulphonic acid) (ABTS) was used to regenerate NAD^+ to produce a chiral lactone with horse liver alcohol dehydrogenase (HLADH) [6]. The whole reaction is shown in Fig. 4. *Meso*-3,4-dihydroxy-methylcyclohex-1-ene is converted by the HLADH into (3aR,7aS)-3a, 4,7,7a-tetrahydro-3H-isobenzofuran-1-one by consuming NAD^+. The formed NADH is reoxidized by ABTS and

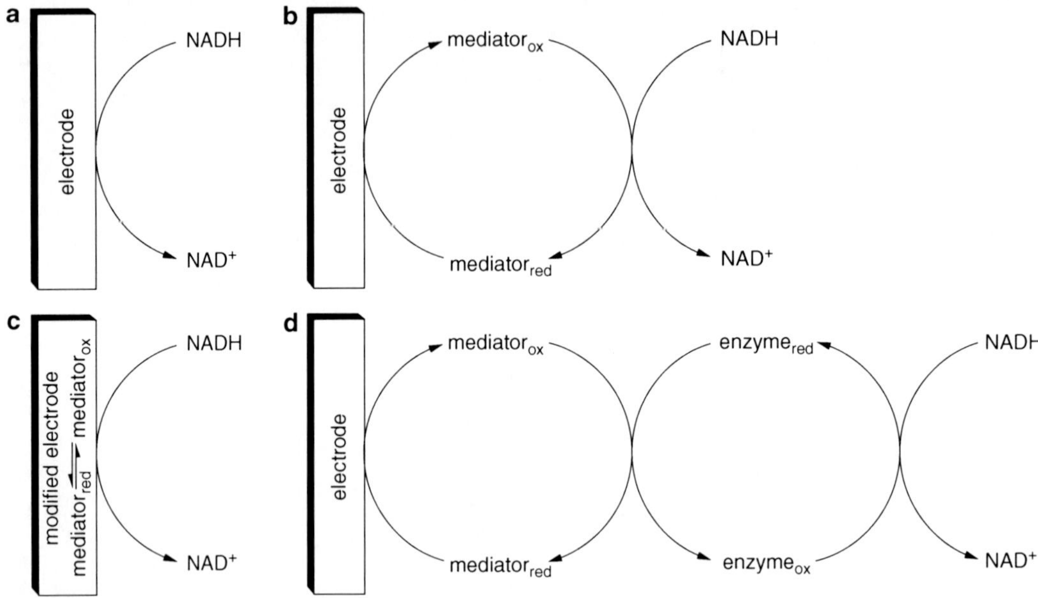

Cofactor Regeneration, Electrochemical, Fig. 2 Principles of the electrochemical cofactor regeneration ((**a**): direct oxidation of NADH at the bare electrode, (**b**): NAD^+ regeneration indirectly by a soluble mediator, (**c**): NADH is oxidized at the electrode, which was previously modified with a redox mediator, (**d**): enzyme-coupled electrochemical regeneration)

Cofactor Regeneration, Electrochemical, Fig. 3 ECE mechanism of NAD(P)H oxidation SET = single electron transfer [5]

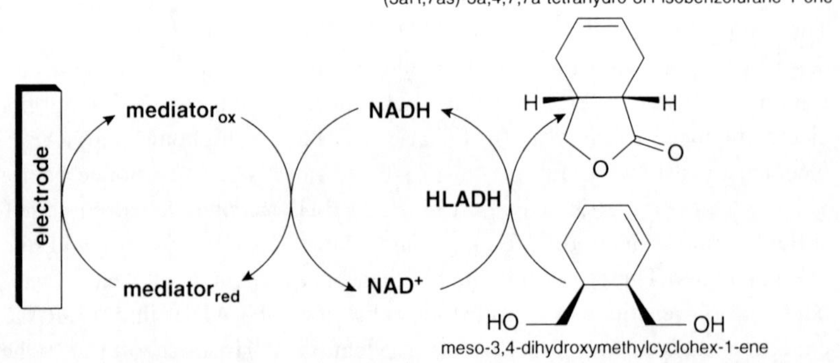

Cofactor Regeneration, Electrochemical, Fig. 4 Coupling electrochemical NAD^+ regeneration with the formation of (3aR,7aS)-3a,4,7,7a-tetrahydro-3H-isobenzofuran-1-one from *meso*-3,4-dihydroxymethylcyclohex-1-ene by the reaction of the horse liver alcohol dehydrogenase can be used in a further enzymatic reaction. To complete the regeneration cycle, the reduced form of ABTS is oxidized at the electrode.

The use of dissolved mediators for technical applications is not a very convenient strategy since the reaction system is not reagentless. The most important drawbacks of dissolved mediators in biocatalytic applications are interferences during product purification, a limited reusability of the mediators, and their cost-intensive elimination from wastewater. Therefore, the use of immobilized mediators has both economic and ecological advantages. The first publication of a mediator-modified electrode for NADH oxidation describes the immobilization of two primary amines, dopamine and 3,4-dihydroxybenzylamine, forming a monolayer onto the surface of activated glassy carbon electrodes [7]. In the meantime numerous ways for immobilization have been explored in order to provide more stable and environmental-friendly approaches [8]. Mediator immobilization has been performed by adsorption, covalent attachment, polymer modification, or inclusion of the mediator in the electrode material. While adsorption and inclusion of the mediators in the electrode material are the simplest procedures, more elaborate approaches have been developed in order to obtain immobilized mediators with enhanced operational stability. A stable electroactive film of poly(nile blue A) has been deposited on the surface of a glassy carbon electrode by cyclic voltammetry [9]. The modified electrode showed electrocatalytic activity toward NADH oxidation, with an overpotential 660 mV lower than that of the bare electrode. Electropolymerization of methylene green has been compared with adsorption of the same mediator on electrodes. The polymer-modified electrode exhibited a higher stability and better catalytic activity toward NADH oxidation [10]. Figure 5 shows the structure of suitable mediators for NAD(P)H oxidation.

The conversion of cyclohexanol into cyclohexanone by HLADH was also combined to a regeneration approach using quinone as mediator and diaphorase as regenerating enzyme [11]. The same reaction was performed by using a ferrocene/diaphorase/HLADH-immobilized electrode. The last method is a combination of the immobilization of mediators and the enzyme-coupled electrochemical regeneration [12].

Reduction of NAD^+ and $NADP^+$

The easiest way to regenerate NAD(P)H would be the direct reduction of the oxidized cofactors on electrode surfaces [13]. Unfortunately, this reaction requires a high overpotential, resulting in increasing amounts of inactive isomers or dimers. For the electrochemical reduction of nicotinamide cofactors, a single electron transfer takes place resulting in radicals. The radicals form six different dimeric species. After a second electron transfer, three different NADH isomers occur. The yield of the enzymatically active 1,4-NADH depends on the pH of the solution [14, 15]. The formation of enzymatically inactive forms can be circumvented by the use of modified electrodes. For example, by using porous tin oxide electrodes, NADH could be regenerated from NAD^+ without the formation of NAD dimers [16]. Enzymatically active NADH was also formed directly at the cholesterol-modified gold amalgam electrode, which is supposed to hinder the dimerization of the NAD radicals [17]. The direct electrochemical NAD^+ reduction process was used favorably to drive an enzymatic reduction of pyruvate to D-lactate.

Soluble mediators can be used to overcome the formation of inactive NAD(P)H as well. A suitable mediator should fulfill the following requirements [18]:

- The mediator must transfer two electrons in one step or a hydride ion.
- Its electrochemical activation has to be possible at potentials less negative than – 0.9 V versus a saturated calomel electrode in order to avoid dimer formation.
- It should only transfer the electrons or hydride ions to NAD^+ and not to the substrate.
- Solely enzymatically active NAD(P)H should be produced.

Cofactor Regeneration, Electrochemical, Fig. 5 Mediators for the oxidation of NAD(P)H. Most of the mediators can be used in soluble form as well as immobilized mediators

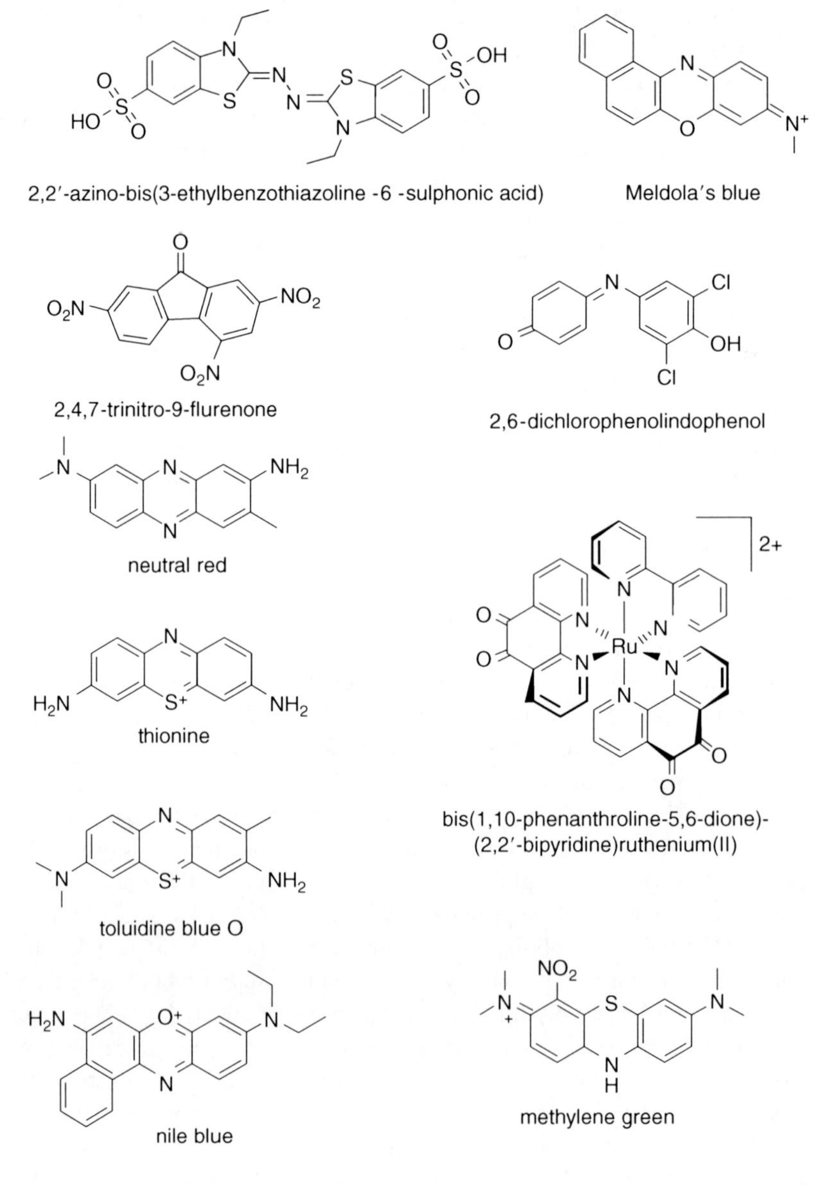

2,2′-azino-bis(3-ethylbenzothiazoline -6 -sulphonic acid)

Meldola's blue

2,4,7-trinitro-9-flurenone

2,6-dichlorophenolindophenol

neutral red

thionine

bis(1,10-phenanthroline-5,6-dione)-(2,2′-bipyridine)ruthenium(II)

toluidine blue O

nile blue

methylene green

A common class of mediators, which fulfill these requirements, are rhodium complexes (e.g., tris(2,2′-bipyridyl)- and substituted or nonsubstituted (2,2′-bipyridyl) (pentamethylcyclopentadienyl)-rhodium complexes). This regeneration system has been efficiently applied in electroenzymatic reduction of pyruvate to D-lactate and the reduction of 4-phenyl-2-butanone to (S)-4-phenyl-2-butanol [1]. In an electrochemical membrane reactor, NADH was regenerated using methyl viologen or a rhodium complex as mediator for synthesis of cyclohexanol from cyclohexanone with HLADH [19]. Figure 6 shows the mechanism for the synthesis of cyclohexanol coupled with the NADH regeneration. The HLADH converts cyclohexanone into the corresponding alcohol consuming NADH. To regenerate the cofactor, the oxidized form reacts with the reduced form of a mediator. The regenerated NADH is now

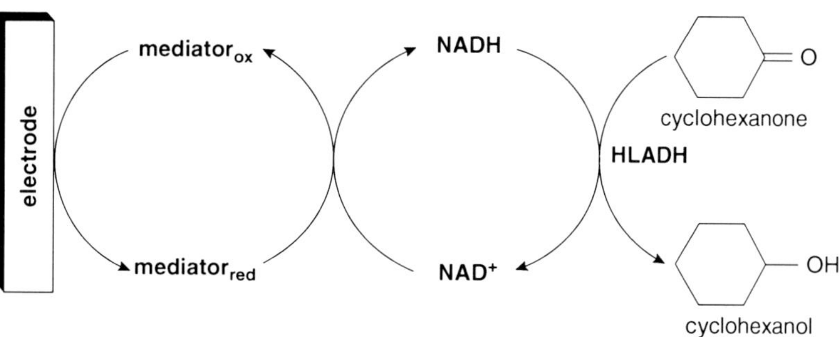

Cofactor Regeneration, Electrochemical, Fig. 6 Coupling electrochemical NADH regeneration with the enzymatic reaction of the horse liver alcohol dehydrogenase for the conversion of cyclohexanone to cyclohexanol

available for another reaction of the HLADH. The oxidized form of the mediator is reduced at the surface of the electrode closing the regeneration cycle.

The formation of inactive isomers or dimers of NAD(P)H can also be prevented by addition of regenerating enzyme. This second enzyme must be able to accept two electrons in two steps from the mediator and then transfer one electron pair to NAD(P)$^+$ [18]. Examples for such regenerating enzymes are lipoamide dehydrogenase (diaphorase), ferredoxin-NADP$^+$ reductase (FNR), enoate reductase, and viologen-accepting pyridine-nucleotide oxidoreductases (VAPOR enzymes) [1]. Comtat et al. have shown that NADH can be regenerated electrochemically from NAD$^+$ using a FAD as mediator. FAD shuttles electrons from a cathode to a regeneration enzyme (e.g., formate dehydrogenase, FDH) that regenerates NADH from NAD$^+$ [20].

Using immobilized mediators for NADH regeneration to circumvent drawbacks of dissolved mediators is also described in literature. The electropolymerization of neutral red (NR) led to formation of a redox active film on the electrode surface [21]. At negative potentials a current of NAD$^+$ reduction was observed at modified electrodes. It was shown that the main product of NAD$^+$ electroreduction was the enzymatically active NADH. Immobilized mediators can also be used in biofuel cell applications. Arechederra et al. have developed a biobattery on the basis of poly(neutral red) [22]. Poly(neutral red) was used for both oxidation of NAD(P)H and reduction of NAD(P)$^+$, which is necessary in a rechargeable biobattery environment.

Future Directions

In the literature several enzyme-coupled and substrate-coupled cofactor regenerations system are described. There are some disadvantages using an enzyme-coupled regeneration system: costs for a second enzyme and a co-substrate, interferences during product purification caused by secondary products, and reaction conditions which have to be optimized for both enzyme reactions. The substrate-coupled approach also provides disadvantages like thermodynamic limitations, costs for co-substrate, reduced catalytic efficiency caused by reaction of the enzyme with co-substrate, and interferences during product purification caused by secondary products. To overcome these drawbacks, an electrochemical mediator-based system can be used. This strategy prevents costs for second enzymes or co-substrates, thermodynamic limitations, and reduced catalytic efficiency. Furthermore reaction conditions have to be optimized for just one enzyme. Mediators used in these applications can be soluble or immobilized on the electrode material. Although dissolved mediators have benefits (simple utilization, mediator is kept in its original electrochemical properties), there are also some disadvantages within this concept: it is not reagentless, and contact of enzyme with electrode may cause denaturation processes, costs for mediator, costs for wastewater treatment, and interferences during product purification caused by secondary products. Immobilized mediators may have the potential to avoid these drawbacks. A suitable immobilized mediator for the oxidation of NAD(P)H should fulfill a number of demands [23]: a sustainable reduction of the overvoltage, high regeneration rate, long-term stability, and a high selectivity for NAD(P)H oxidation or reduction. It will be one of the most important investigations to identify an "optimal" mediator. Therefore, a systematic comparison of different mediators would be necessary. Furthermore, protein engineering can optimize the enzymes to tailor-made catalysts with increased stability in the electrochemical reaction system.

Cross-References

► Biosensors, Electrochemical
► Cofactor Substitution, Mediated Electron Transfer to Enzymes
► Direct Electron Transfer to Enzymes
► Protein Engineering for Electrochemical Applications
► Reconstituted Redox Proteins on Surfaces for Bioelectronic Applications

References

1. Weckbecker A, Gröger H, Hummel W (2010) Regeneration of nicotinamide coenzymes: principles and applications for the synthesis of chiral compounds. In: Wittmann C, Krull R (eds) Advances in biochemical engineering/biotechnology, vol 120. Springer, Berlin/Heidelberg, pp 195–242. doi:10.1007/10_2009_55
2. Wichmann R, Vasic-Racki D (2005) Cofactor regeneration at the lab scale. In: Kragl U (ed) Advances in biochemical engineering/biotechnology, vol 92. Springer, Berlin/Heidelberg, pp 225–260. doi:10.1007/b98911
3. Kohlmann C, Märkle W, Lütz S (2008) Electroenzymatic synthesis. J Mol Catal B Enzym 51(3–4):57–72. doi:10.1016/j.molcatb.2007.10.001
4. Hollmann F, Schmid A (2004) Electrochemical regeneration of oxidoreductases for cell-free biocatalytic redox reactions. Biocatal Biotrans 22(2):63–88. doi:10.1080/10242420410001692778
5. Blankespoor RL, Miller LL (1984) Electrochemical oxidation of NADH: kinetic control by product inhibition and surface coating. J Electroanal Chem Interfacial Electrochem 171(1–2):231–241. doi:10.1016/0022-0728(84)80116-3
6. Schröder I, Steckhan E, Liese A (2003) In situ NAD^+ regeneration using 2,2′-azinobis(3-ethylbenzothiazoline-6-sulfonate) as an electron transfer mediator. J Electroanal Chem 541:109–115. doi:10.1016/s0022-0728(02)01420-1
7. Tse DC-S, Kuwana T (1978) Electrocatalysis of dihydronicotinamide adenosine diphosphate with quinones and modified quinone electrodes. Anal Chem 50(9):1315–1318
8. Vasilescu A, Noguer T, Andreescu S, Calas-Blanchard C, Bala C, Marty J-L (2003) Strategies for developing NADH detectors based on Meldola blue and screen-printed electrodes: a comparative study. Talanta 59(4):751–765. doi:10.1016/s0039-9140(02)00614-8
9. Cai CX, Xue KH (1997) Electrocatalysis of NADH oxidation with electropolymerized films of nile blue A. Anal Chim Acta 343(1–2):69–77. doi:10.1016/s0003-2670(96)00592-2
10. Zhou DM, Fang HQ, Chen H-Y, Ju HX, Wang Y (1996) The electrochemical polymerization of methylene green and its electrocatalysis for the oxidation of NADH. Anal Chim Acta 329(1–2):41–48. doi:10.1016/0003-2670(96)00117-1
11. Itoh S, Fukushima H, Komatsu M, Ohshiro Y (1992) Heterocyclic o-quinones. Mediator for electrochemical oxidation of NADH. Chem Lett 21(8):1583–1586
12. Kashiwagi Y, Osa T (1993) Electrocatalytic oxidation of NADH on thin poly(acrylic acid) film coated graphite felt electrode coimmobilizing ferrocene and diaphorase. Chem Lett 22(4):677–680
13. Burnett JN, Underwood AL (1965) Electrochemical reduction of diphosphopyridine nucleotide*. Biochemistry 4(10):2060–2064. doi:10.1021/bi00886a021
14. Bresnahan WT, Elving PJ (1981) The role of adsorption in the initial one-electron electrochemical reduction of nicotinamide adenine dinucleotide (NAD^+). J Am Chem Soc 103(9):2379–2386. doi:10.1021/ja00399a039
15. Hans J (1981) 420 – A study of the products formed in the electrochemical reduction of nicotinamide-adenine-dinucleotide. Bioelectrochem Bioenerg 8(3):355–370. doi:10.1016/0302-4598(81)80018-9
16. Kim YH, Yoo YJ (2009) Regeneration of the nicotinamide cofactor using a mediator-free electrochemical method with a tin oxide electrode. Enzyme Microb Technol 44(3):129–134. doi:10.1016/j.enzmictec.2008.10.019
17. Baik SH, Kang C, Jeon IC, Yun SE (1999) Direct electrochemical regeneration of NADH from NAD^+ using cholesterol-modified gold amalgam electrod. Biotechnol Tech 13
18. Steckhan E (1994) Electroenzymatic synthesis. In: Steckhan E (ed) Topics in current chemistry, vol 170. Springer, Berlin/Heidelberg, pp 83–111
19. Délécouls-Servat K, Bergel A, Basséguy R (2004) Membrane electrochemical reactors (MER) for NADH regeneration in HLADH-catalysed synthesis: comparison of effectiveness. Bioprocess Biosyst Eng 26(4):205–215. doi:10.1007/s00449-004-0356-2

20. Durliat H, Barrau MB, Comtat M (1988) FAD used as a mediator in the electron transfer between platinum and several biomolecules. Bioelectrochem Bioenerg 19(3):413–423. doi:10.1016/0302-4598(88)80022-9
21. Karyakin AA, Bobrova OA, Karyakina EE (1995) Electroreduction of NAD^+ to enzymatically active NADH at poly(neutral red) modified electrodes. J Electroanal Chem 399(1):179–184
22. Arechederra MN, Addo PK, Minteer SD (2010) Poly (neutral red) as a NAD^+ reduction catalyst and a NADH oxidation catalyst: towards the development of a rechargeable biobattery. Electrochim Acta 56(3):1585–1590
23. Gorton L, Domínguez E (2002) Electrocatalytic oxidation of NAD(P)H at mediator-modified electrodes. Rev Mol Biotechnol 82(4):371–392. doi:10.1016/s1389-0352(01)00053-8

Cofactor Substitution, Mediated Electron Transfer to Enzymes

Anders O. Magnusson[1] and Dirk Holtmann[2]
[1]DECHEMA Research Institute, Frankfurt am Main, Germany
[2]DECHEMA Research Institute of Biochemical Engineering, Frankfurt am Main, Germany

Introduction

Enzymes have become valuable tools in asymmetric synthesis due to their high versatility and specificity as well as the mild reaction conditions required, specially benefitting synthesis with sensitive or reactive compounds. The use of oxidoreductases is steadily increasing due to their interesting reduction, oxidation, and oxyfunctionalization reactions [1, 2]. A "dilemma" utilizing oxidoreductases is their need for redox equivalents. In nature are these redox equivalents supplied through cofactors, generally NAD(P)H to deliver electrons and $NAD(P)^+$ to accept electrons, which are regenerated through the cell metabolism consuming an energy source such as glucose (Fig. 1 top).

In an efficient in vitro process with isolated enzyme, the cofactor cannot be supplied in equimolar amounts [3]. Therefore, the cofactors have to be regenerated or substituted with another system donating or accepting the redox equivalents. Examples of enzymatic and electrochemical cofactor regeneration as well as chemical and electrochemical cofactor-substitution systems are depicted in Fig. 1. Cofactor regeneration replaces the metabolic cofactor regeneration found in vivo, while cofactor substitution "shortcuts" the natural system and transfers electrons via the mediator directly to the oxidoreductase or possibly another protein in the electron-transfer chain. The natural monooxygenase system depicted here consists of three proteins, typical for several bacterial P450s, but systems with two or only one component also exist and the principles for the cofactor regeneration and substitution are the same.

All these methods have their advantages and drawbacks [4, 5]. In this chapter the cofactor substitution via mediated electron transfer (MET) is described. The natural cofactor, generally $NAD(P)^+$ or NAD(P)H, is replaced by a mediator. An electrochemical potential is applied that is high enough to regenerate the mediator at the electrode, reduce or oxidize, whereupon the mediator regenerates the redox enzyme for the next catalytic cycle. Thus, the mediator is continuously cycling between electrode and enzyme. MET benefits from a very cheap source of redox equivalents (electricity), the possibility to simplify the reaction system by omitting a redox partner(s) (e.g., a transport protein and/or a reductase), and a simplified downstream processing due to fewer components in the reaction medium. Direct electron transfer also has those benefits but in most cases is the catalytic center of the enzyme deeply buried within the protein structure. Therefore, a direct electrical communication between this site and the electrode surface is difficult to achieve. MET is, for example, utilized in biofuel cells [6, 7], biocatalysis [8, 9, 16], and biosensors [10]. For the components (mediator, enzyme, and electrode) used in a process based on MET, different requirements should be fulfilled.

Mediator

Mediators are generally small inorganic molecules that can transfer electrons to or from the

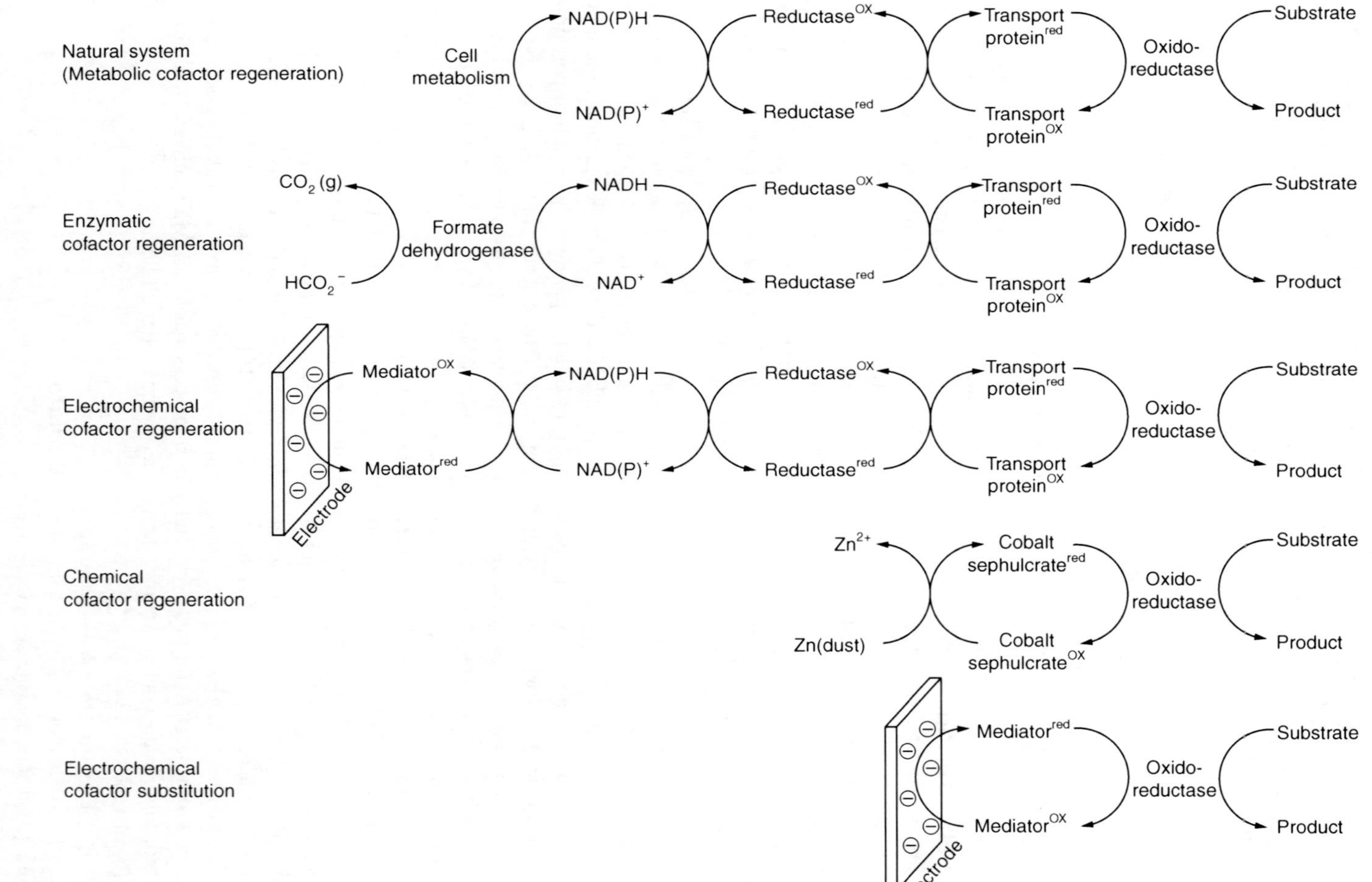

Cofactor Substitution, Mediated Electron Transfer to Enzymes, Fig. 1 (continued)

redox center of the enzyme as well as the electrode and can diffuse between the enzyme and the electrode carrying the charge. The structure of some representative mediators and their redox potentials are presented in Fig. 2. The redox potential of a mediator transferring electrons to an enzyme has to be more negative than that of the redox center of the enzyme and more positive for electron transfer in the opposite direction, but the difference in redox potential between enzyme and mediator should not be too large for an efficient process.

A mediator should have the following characteristics in order to accomplish an efficient electron transfer [10]:

- It should be able to react rapidly with the enzyme.
- It should exhibit reversible heterogeneous kinetics.
- It should be soluble in the reduced and oxidized form.
- The overpotential for the regeneration of the mediator should be low.
- It should have stable oxidized and reduced forms.
- No by-product should be formed, e.g., the reduced form should not react with oxygen to form reactive oxygen species.
- No interferences with the other reaction steps should occur.

Cyclic voltammetry can be used to investigate if the mediator has a suitable redox potential, if the redox process is reversible, and to investigate if the rate-limiting step of the electron transfer is the chemical reaction at the electrode surface or the diffusion rate of the mediator to the electrode. In the most cases freely diffusing mediators are used. Efficient electron transfer has been demonstrated with mediators immobilized on lysine residues in glucose oxidase, and the electron transfer was further improved by wiring this modified enzyme to the electrode with conducting polymers [12].

Enzyme

The natural enzymes are optimized by the nature for specific reaction conditions, but not for a bioelectrochemical application. By using an enzyme as "electro enzyme" specific requirements have to be taken in account. The enzyme performing the redox catalysis has to be stable at the required potential to regenerate the mediator. Enzyme denaturation at the electrode surface is a problem that limits both mediator regeneration and naturally the enzyme activity. Reaction conditions, e.g., buffer composition and pH, have to be compromised between optimal mediator regeneration, substrate solubility, and enzyme activity and stability. Immobilization of the enzyme on a carrier is a mean to increase its stability and prevent its fouling of the electrode. Furthermore, coating the electrode surface with a modifier such as alkane thiol can protect proteins from the electric field-driven denaturation. The used modifier should not prevent or interfere with the regeneration of the mediator at the electrode.

Directed evolution and rational design have become common techniques to improve enzyme properties such as activity, selectivity, and stability and will also gain impact in the improvement of mediated electron transfer. Mediator-enzyme interactions, electron-transfer path and rate, as well as the redox center itself are all properties of the enzyme that could be optimized. A comprehensive review covering different approaches and examples thereof to improve the electron-transfer rate between enzyme and electrode has been published by Güven et al. [13].

Cofactor Substitution, Mediated Electron Transfer to Enzymes, Fig. 1 Examples of artificial cofactor regeneration and cofactor substitution used in in vitro enzyme catalysis. Cofactor regeneration (middle) replaces the metabolic cofactor regeneration found in vivo (top), while cofactor substitution (bottom) "shortcuts" the natural system and transfers electrons via the mediator directly to the oxidoreductase or to another protein in the electron-transfer chain if available. The oxidoreductase can have two, one, or no redox partners. The examples depicted here show the transport of reduction equivalents to the oxidoreductase, but the electron transport can also be performed in the opposite direction

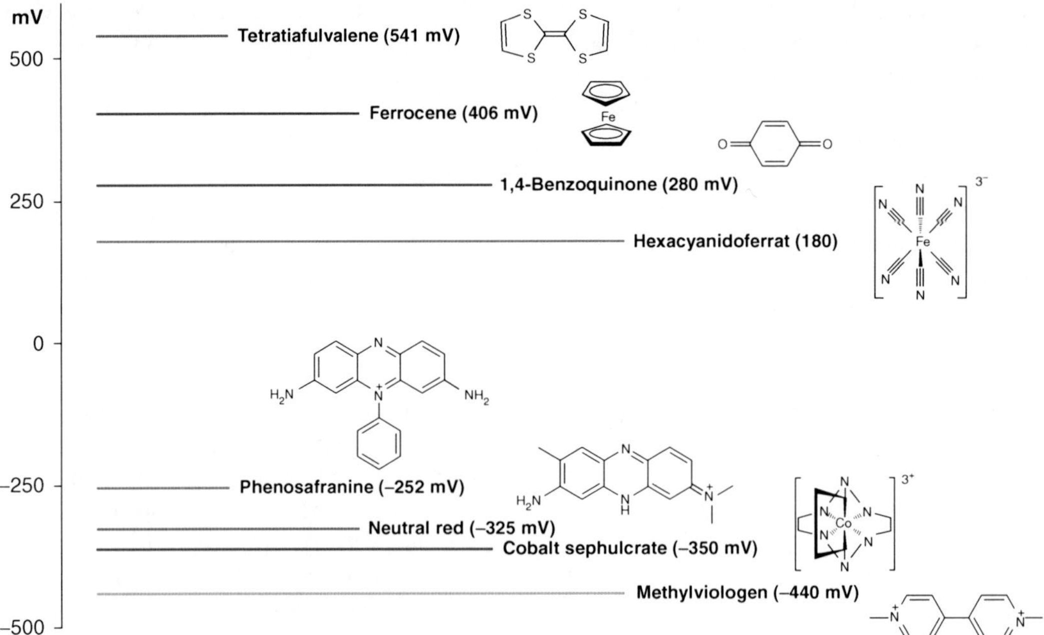

Cofactor Substitution, Mediated Electron Transfer to Enzymes, Fig. 2 Examples of the structure and redox potential (vs. SHE) of several mediators used for electron transfer between electrodes and enzymes. The mediators with positive redox potential were used in biosensors transferring electrons to the electrode [10]. The mediators with negative redox potential were used with P450 monooxygenases transferring electrons to the enzymes [11]

They specifically address the improvement of enzymes for applications in biofuel cells, but the approaches are also applicable for other processes. An example of improved MET by directed evolution is the 2.7 times increased electron-transfer rate from glucose oxidase to ferrocenemethanol [14].

Electrode and Reaction System

The choice of electrode material depends on the application. In biocatalysis and biofuel cells is a high mediator regeneration rate crucial, and a large electrode surface is thus important for an efficient process. Fixed bed reactors with glassy carbon and reticulated vitreous carbon have a high area to volume ratio and are often the material of choice. These materials are cheap and thus also good options when scaling up the process. In biosensors is the sensitivity more important than the reaction rate, and the electrode material is selected due to its specificity to react with the mediator.

Future Directions

There are still many undiscovered oxidoreductases with potentially interesting properties that can be used in enzymatic processes with mediated electron transfer. Specially plant P450 monooxygenases, the largest source of P450 enzymes, have an unused potential that can be explored once the difficulties associated with membrane proteins become less restricting. Enzymes have not been evolved for mediated electron transfer but rather to have an isolating shell preventing unspecific electron transfer. Thus, directed evolution and rational redesign of the enzymes to improve the electron-transfer rate is a promising means to improve their applicability in industrial processes. Design or modification of the mediator for better interactions with the enzyme and a more suitable redox potential is also a mean that should be used to improve mediated electron transfer. The substitution of the cofactor with mediated electron transfer can confer a work intensive testing of a huge amount of mediators in order to find one with the desired properties. Different modeling software can be used to estimate mediator-enzyme interactions, and such software will likely be improved for such applications and become a standard procedure to prescreen mediators followed by the testing of a few promising candidates [17, 18]. The rational redesign of the enzymes for improved electron transfer will also gain from specialized modeling programs that will develop with growing experience in this field.

Today many reports are missing important facts to allow for reproduction and comparison of the experiments. In descriptions of electrochemical processes one has, for example, to give concentrations, ionic strength, reference electrode, as well as electrode area and reaction volume (or less preferred the ratio). By following already existing guidelines for reporting of experimental data [15], describing experiments more carefully and reviewing manuscripts more critically, development within this field will benefit in the future.

Cross-References

- ▶ Cofactor Regeneration, Electrochemical
- ▶ Direct Electron Transfer to Enzymes
- ▶ Reconstituted Redox Proteins on Surfaces for Bioelectronic Applications

References

1. Hollmann F, Arends IWCE, Buehler K, Schallmey A, Buhler B (2011) Enzyme-mediated oxidations for the chemist. Green Chem 13:226–265
2. Hollmann F, Arends IWCE, Holtmann D (2011) Enzymatic reductions for the chemist. Green Chem 13:2285–2314
3. Holtmann D, Mangold KM, Schrader J (2009) Entrapment of cytochrome P450 BM-3 in polypyrrole for electrochemically-driven biocatalysis. Biotechnol Lett 31:765–770
4. Hollmann F, Hofstetter K, Schmid A (2006) Non-enzymatic regeneration of nicotinamide and flavin cofactors for monooxygenase catalysis. Trends Biotechnol 24:163–171
5. Wichmann R, Vasic-Racki D (2005) Cofactor regeneration at the lab scale. In: Kragl U (ed) Technology transfer in biotechnology. Springer, Berlin/Heidelberg, pp 225–260
6. Pas LM (2007) Alternative energy: enzyme-based biofuel cells. Basic Biotechnol eJournal 3:93–97
7. Hao Yu E, Scott K (2010) Enzymatic biofuel cells—fabrication of enzyme electrodes. Energies 3:23–42
8. Zengin Çekiç S, Holtmann D, Güven G, Mangold K-M, Schwaneberg U, Schrader J (2010) Mediated electron transfer with P450cin. Electrochem Commun 12:1547–1550
9. Reipa V, Mayhew MP, Vilker VL (1997) A direct electrode-driven P450 cycle for biocatalysis. Proc Natl Acad Sci USA 94:13554–13558
10. Chaubey A, Malhotra BD (2002) Mediated biosensors. Biosens Bioelectron 17:441–456
11. Holtmann D, Schrader J (2007) Approaches to recycling and substituting NAD(P)H as a CYP cofactor. In: Schmid RD, Urlacher V (eds) Modern biooxidation. Wiley, Germany, pp 269–294
12. Heller A (1990) Electrical wiring of redox enzymes. Acc Chem Res 23:128–134
13. Güven G, Prodanovic R, Schwaneberg U (2010) Protein engineering – an option for enzymatic biofuel cell design. Electroanalysis 22:765–775
14. Zhu Z, Wang M, Gautam A, Nazor J, Momeu C, Prodanovic R, Schwaneberg U (2007) Directed evolution of glucose oxidase from Aspergillus niger for ferrocenemethanol-mediated electron transfer. Biotechnol J 2:241–248
15. Gardossi L, Poulsen PB, Ballesteros A, Hult K, Svedas VK, Vasic-Racki D et al (2010) Guidelines for reporting of biocatalytic reactions. Trends Biotechnol 28:171–180
16. Ley C, Schewe H, Ströhle FW, Ruff AJ, Schwaneberg U, Schrader J, Holtmann D (2013) Coupling of electrochemical and optical measurements in a microtiter plate for the fast development of electro enzymatic processes with P450s. Journal of Molecular Catalysis B: Enzymatic 92:71–78. DOI:10.1016/j.molcatb.2013.03.019
17. Ströhle FW, Zengin Cekic S, Magnusson AO, Schwaneberg U, Roccatano D, Schrader J, Holtmann D (2013) A computational protocol to predict suitable redox mediators for substitution of NAD(P)H in P450 monooxygenases. Journal of Molecular Catalysis B: Enzymatic 88:47–51. DOI:10.1016/j.molcatb.2012.11.010
18. Ströhle SW, Zengin Cekic S, Magnusson AO, Schwaneberg U, Roccatano D, Schrader J, Holtmann D (2013) A computational protocol to predict suitable redox mediators for substitution of NAD(P)H in P450 monooxygenases. Journal of Molecular Catalysis B: Enzymatic 88:47–51. DOI:10.1016/j.molcatb.2012.11.010

Combinatorial Electrochemical Synthesis

Jun-ichi Yosida
Kyoto University, Kyoto, Japan

Introduction

Creation of new compounds having desired biological activity or functions is the key to the progress of not only life and materials sciences but also pharmaceutical and chemical industries. To meet such demand synthetic organic chemistry has been creating a variety of organic compounds through developing efficient methods and strategies. Although target-oriented synthesis has been the mainstream approach, the emergence of combinatorial chemistry [1] and diversity-oriented synthesis [2] has changed the way of planning and doing chemical synthesis as a whole. Generating molecular diversity [3] enables rapid discovery of molecules having desired biological activity or function and their optimization. In addition, there is a high probability of discovering unexpected functions or biological activity in these library-based approaches.

Definition and Principles

Combinatorial electrochemical synthesis is defined as a method in which very large numbers of compounds are synthesized as ensembles (libraries) by combining a small number of building blocks together in all combinations defined by an electrochemical reaction. Combinatorial synthesis enables diversity oriented synthesis. If two building blocks (A and B) are combined by an electrochemical reaction and three different types of A and three different types of B are used, nine (3×3) different products (A-B) are produced. 100 types of each building block will yield $100 \times 100 = 10{,}000$ products. Use of such products as substrate for the next reaction leads to formation of a large number of compounds as chemical libraries. Combinatorial electrochemical synthesis can be done in a parallel fashion using multiple electrochemical cells. Serial combinatorial synthesis can be accomplished by using a flow electrochemical cell. Mixture synthesis can also be done, in principle, by electrolyzing a mixture of several starting materials. The high-throughput screening of those libraries for compounds with desirable activity or properties serves as a rapid method for discovering new drugs or functional materials.

Advantages

To achieve combinatorial synthesis effectively, the reaction should be generally applicable to combining building blocks of a wide structural diversity. The electrochemical method enables generation of highly reactive species under mild conditions, and this feature is advantageous for allowing reactions with a wide range of building blocks. No need to use of chemical reagents are also advantageous, because separation of products from excess reagents and/or byproducts derived from the reagents are not necessary. Use of microelectrode arrays is also advantageous because only a tiny space is needed to create a chemical library, and each compound produced on the microelectrode is addressable electrochemically.

Parallel Macro Electrosynthesis

A spatially addressable electrolysis platform has been used for parallel electrosynthesis. For example, the anodic α-alkoxylation of carbamates and sulfonamides is carried out using a Teflon block with 16 wells and a set of 16 glass vials under the constant-current conditions to obtain the corresponding products in a parallel fashion (Fig. 1) [4]. The parallel electrosynthesis can also be applied to reduction. The electrochemical reductive hydrocoupling of aldimines using sacrificial Al anodes gives the corresponding 1,2-diamine derivatives [5].

The cation-pool method serves as a powerful tool for parallel combinatorial synthesis because of high reactivity of cation pools [6]. A typical example is shown in Fig. 2. A solution of a cation $A1^+$ ("cation pool"), which is generated by low-temperature electrochemical oxidation of its precursor A1, is divided into several portions. Different nucleophiles(B1–B5) are added to each portion to obtain the products of different coupling combinations(A1-B1.... A1-B5). The use of different precursor leads formation of coupling products of different combination. The procedure can be easily automated by a robotic synthesizer equipped with automated syringes and low-temperature reaction vessels. The yields of the products are essentially the same as those obtained by one-pot reactions with manual operation, which are shown in parentheses.

Combinatorial Electrochemical Synthesis, Fig. 1 A spatially addressable 16-well platform for parallel electrolysis

Parallel Micro Electrolysis

Miniaturized combinatorial electrosynthesis has been achieved by using a computer-controlled instrument equipped with a well-containing microtiter plates (Fig. 3) [7]. An electrode bundle consisting of a PTFE holder, a working electrode, a CV microdisk, a reference electrode, and a counter electrode is moved from well to well automatically. Libraries of iminoquinol ethers and triazolopyridinium ions are generated by under controlled potential conditions. Progress of the electrolyses can be monitored by microelectrode steady-state voltammetry.

A microelectrode array has been developed for molecular library construction. A 1 cm^2 chip having an array of 1,024 individually addressable Pt electrodes shown in Fig. 4 is used [8]. At first the array is coated with a porous hydroxylated polymer membrane, and 10-undecenoic acid moieties are introduced to the hydroxyl groups. The Pd-catalyzed

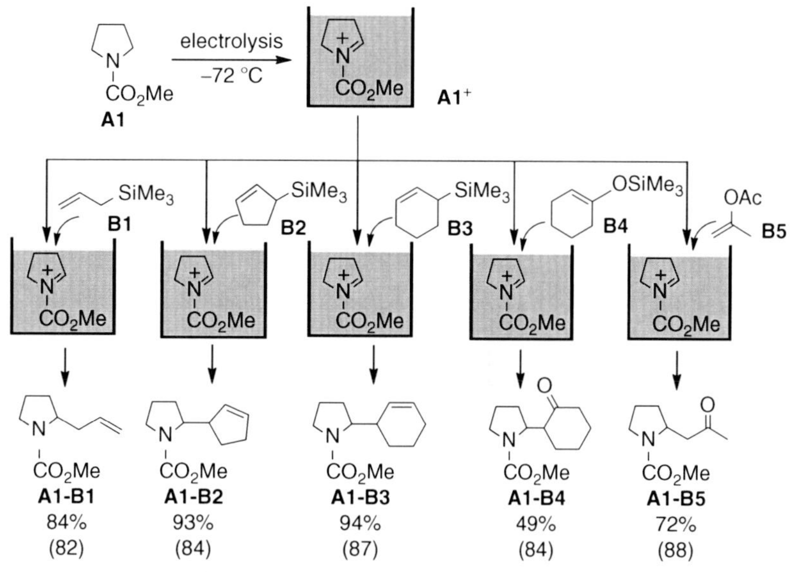

Combinatorial Electrochemical Synthesis, Fig. 2 Parallel combinatorial synthesis based on the cation-pool method

Wacker oxidation mediated by the electrochemical oxidation of a triarylamine is performed at selected electrodes to generate the ketones, which are converted to their 2,4-DNP derivatives. The 2,4-DNP recognition sites at the selected electrodes are treated with a rabbit *anti*-2,4-dinitrophenol antibody that is conjugated to a fluorescent probe. Selected electrodes can be selectively modified chemically at will, which is proved by imaging using an epifluorescence microscope. The system enables building electrochemically addressable libraries on an array of microelectrodes. This approach has been expanded to other addressable libraries, which can be monitored by mass spectrometry based on a cleavable linker [9]. Selective coumarin synthesis and real-time signaling of antibody–coumarin binding based on the microelectrode array have also been reported [10].

Combinatorial Electrochemical Synthesis, Fig. 3 Anassembly of the electrode bundle consisting of a PTFE holder

Serial Electrolysis

The serial method based on continuous flow synthesis by simple flow switching is useful. Although the parallel method needs to use many reactors in parallel, the serial method needs to use only one flow reactor. Different substrates and/or reagents are fed sequentially to the inlets of the flow reactor and different products are produced sequentially at the outlet of the reactor. Continuous operation of this system leads to synthesis of many different compounds in a combinatorial manner.

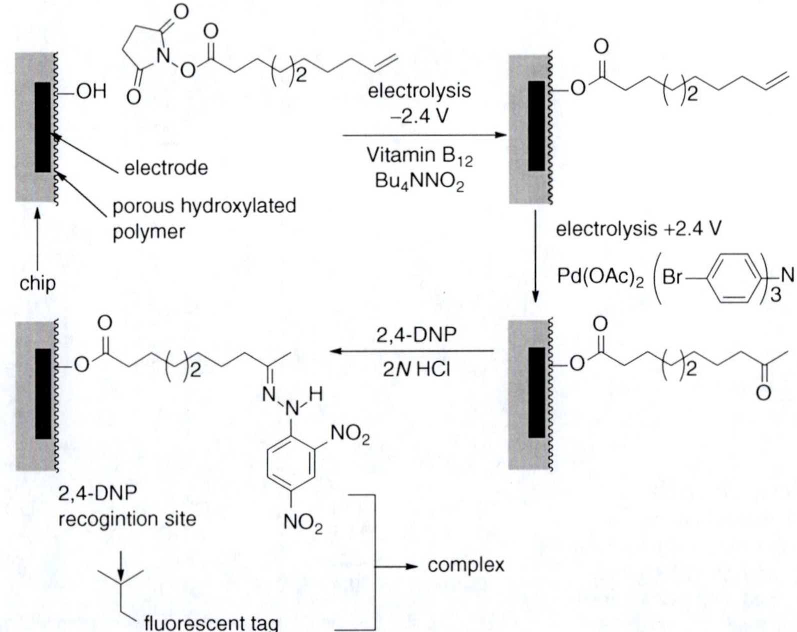

Combinatorial Electrochemical Synthesis, Fig. 4 The combination of electrochemistry and Pd chemistry for developing addressable molecular libraries on the microelectrode array

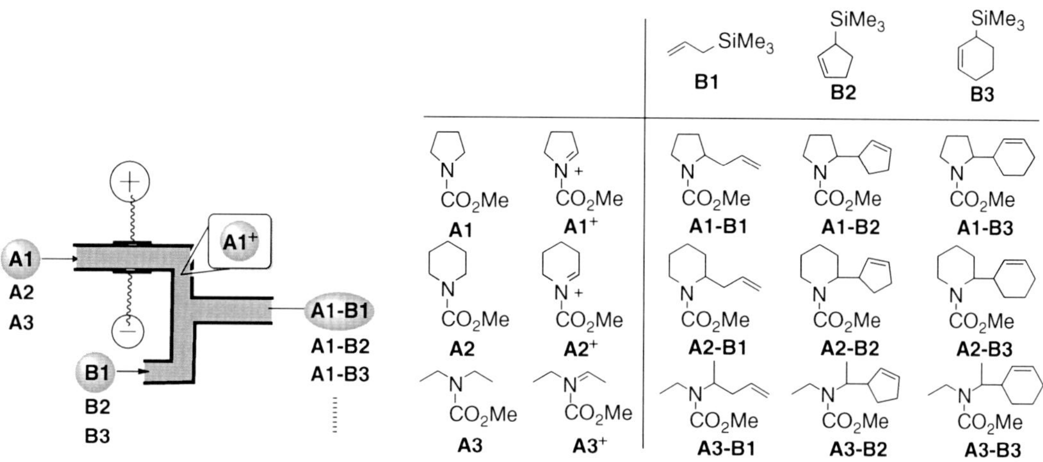

Combinatorial Electrochemical Synthesis, Fig. 5 Serial combinatorial electrochemical synthesis based on the cation-flow method

Figure 5 shows an example based on the "cation flow" method, in which three *N*-acyliminium ions are generated and reacted with three different carbon nucleophiles to achieve 3 × 3 combinatorial electrochemical synthesis in one flow[11]. In the first step, a highly reactive and unstable organic cation $A1^+$ is generated from its precursor A1 by low-temperature flow electrolysis (the "cation flow" method). In the flow system $A1^+$ is allowed to react with nucleophile B1 to obtain product A1-B1. Then, by flow switching $A1^+$ is allowed to react with nucleophile B2 to produce A1-B2. In the third step, $A1^+$ is allowed to react with nucleophile B3 to produce A1-B3. Then, the precursor of a cation is switched to A2, and $A2^+$ which is generated by flow electrolysis of A2 is allowed to react with nucleophiles B1, B2, and B3 sequentially to give A2-B1, A2-B2, and A2-B3, respectively. Then, the precursor of the cation is switched to A3, and $A3^+$ is allowed to react with nucleophiles B1, B2, and B3 sequentially to give A3-B1, A3-B2, and A3-B3, respectively.

Future Directions

Combinatorial synthesis based on organic electrochemistry serves as powerful tools for discovering and developing new biologically active compounds and functional materials. It is hoped that combinatorial electrosynthesis will enjoy a variety of applications by taking advantages of the electrochemical method such as generation of highly reactive species under mild conditions and microelectrode arrays in pharmaceutical and materials science and engineering.

Cross-References

▶ Anodic Substitutions
▶ Cation-Pool Method
▶ Electrochemical Microflow Systems

References

1. An H, Cook PD (2000) Methodologies for generating solution-phase combinatorial libraries. Chem Rev 100:3311–3340
2. Burke MD, Schreiber SL (2004) A planning strategy for diversity-oriented synthesis. Angew Chem Int Ed 43:46–58
3. Yudin AK, Siu T (2001) Combinatorial electrochemistry. Curr Opin Chem Biol 5:269–272
4. Siu T, Li W, Yudin AK (2000) Parallel electrosynthesis of α-Alkoxycarbamates, α-Alkoxyamides, and α-Alkoxysulfonamides using the spatially addressable electrolysis platform (SAEP). J Comb Chem 2:545–549
5. Siu T, Li W, Yudin AK (2001) Parallel electrosynthesis of 1,2-diamines. J Comb Chem 3:554–558

6. Yoshida J, Suga S, Suzuki S, Kinomura N, Yamamoto A, Fujiwara K (1999) Direct oxidative carbon-carbon bond formation using the "cation pool" method. 1. Generation of iminium cation pools and their reaction with carbon nucleophiles. J Am Chem Soc 121:9546–95497
7. Märkle W, Speiser B, Tittel C, Vollmer M (2005) Combinatorial micro electrochemistry part 1. Automated micro electrosynthesis of iminoquinol ether and [1,2,4]triazolo[4,3-a]pyridinium perchlorate collections in the wells of microtiter plates. Electrochimica Acta 50:2753–2762
8. Tesfu E, Maurer K, Ragsdale SR, Moeller KD (2004) Building addressable libraries: the use of electrochemistry for generating reactive Pd(II) reagents at preselected sites on a chip. J Am Chem Soc 126:6212–6213
9. Chen C, Nagy G, Walker AV, Maurer K, McShea A, Moeller KD (2006) Building addressable libraries: the use of a mass spectrometry cleavable linker for monitoring reactions on a microelectrode array. J Am Chem Soc 128:16020–16021
10. Tesfu E, Roth K, Maurer K, Moeller KD (2006) Building addressable libraries: site selective coumarin synthesis and the "real-time" signaling of antibody–coumarin binding. Org Lett 8:709–712
11. Suga S, Okajima M, Fujiwara K, Yoshida J (2001) "Cation flow" method. A new approach to conventional and combinatorial organic syntheses using electrochemical microflow systems. J Am Chem Soc 123:7941–7942

Combustion Control Sensors, Electrochemical

Ulrich Guth[1,2], Jens Zosel[1] and Pavel Shuk[3]
[1]Kurt-Schwabe-Institut für Mess- und Sensortechnik e.V. Meinsberg, Waldheim, Germany
[2]FB Chemie und Lebensmittelchemie, Technische Universität Dresden, Dresden, Germany
[3]Rosemount Analytical Inc. Emerson Process Management, Solon, OH, USA

Combustion as Chemical Process

The combustion of fossil and renewable fuels is a chemical process requiring a certain amount of oxygen or air depending on C/H ratio and stoichiometry of the reaction:

$$\begin{aligned} C_nH_m + (n + m/2)\,O_2 &\rightarrow n\,CO_2 \\ &+ m/2\,H_2O + \Delta_rH \end{aligned} \tag{1}$$

In perfect combustion process, hydrocarbon fuel (C_nH_m) would react with oxygen (O_2) producing primarily carbon dioxide (CO_2) and water (H_2O) with traces of other gases like sulfur dioxide (SO_2) and nitrogen oxides (NO_x) coming from fuel impurities and air nitrogen oxidation. The stoichiometric point with the highest efficiency and the lowest emissions will never be achieved in a real combustion process because of the imperfect fuel/air uniformity, fuel energy density, and air/fuel flow variation [1]. As the result of combustion, heat (Δ_rH) for different high temperature processes is produced. Although the chemical reaction seems to be simple and is used over the whole time of the mankind history, the combustion of fuel is a very difficult process that needs a deep knowledge in chemical thermodynamics, reaction kinetics, and fluid engineering. Even at the stoichiometric point the combustion reaction is never complete because of the oxygen/fuel mixture inhomogeneity, and a certain amount of oxygen excess is needed because of reaction rate and gas transport limitation. Oxygen deficit will cause carbon monoxide (CO) formation and oxygen excess will contribute to the formation of nitric oxides meaning that the firing process has to be conducted in a very narrow technological window. Combustion flue gas concentrations around the stoichiometric point are schematically presented in Fig. 1.

The main reason for the oxygen excess limitation is the combustion efficiency reduction. The excessive air for the combustion containing ~80 % nitrogen will be heated contributing to the efficiency loss. An important point is that the combustion process is not uniform especially in the big power stations. Depending on the fuel type, e.g., gas, oil, coal, and more or less heterogeneous mixture of fuel/air, more oxygen (air)

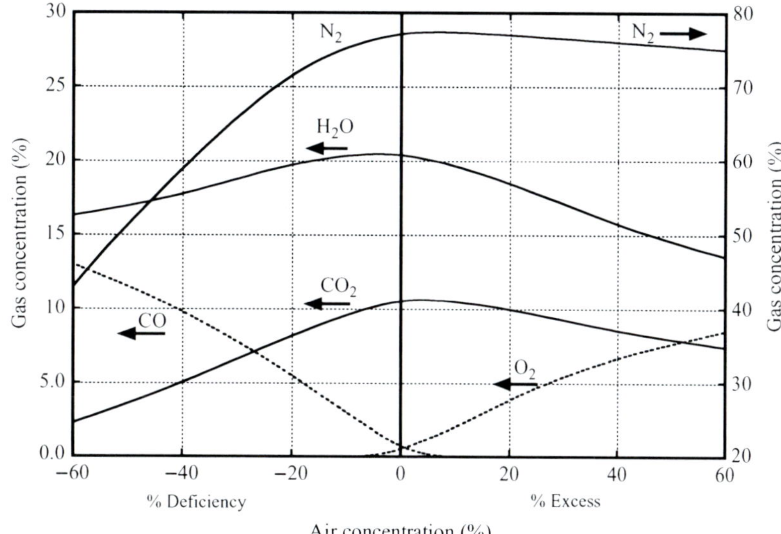

Combustion Control Sensors, Electrochemical, Fig. 1 Flue gas composition around the combustion stoichiometric point

than the stoichiometric amount is needed to complete combustion. Natural fuels are not uniform regarding their chemical composition (heat content), density, and viscosity so that the required amount of oxygen (air) during the firing process could be very abruptly changing. Additionally, the flue gas is viscose so that it does not mix in the exhaust. In such plants, the dimensions of an exhaust can be as big as $25 \times 40\ m^2$. Therefore, the composition of the exhaust gas varies depending on the site and time. Furthermore the oxygen excess in the combustion process varies on the burner load. The less power will be produced, the more oxygen excess is needed to control the combustion.

Combustion Control Using emf Signal of the Solid Electrolyte Cell

The combustion is a time-dependent reaction. Flue gas inhomogeneities across the stack with many areas of stagnation, duct bends, and expansion joints would require many in situ point oxygen measurements. After removing untruthful data related to the air leaks or one of the local burner failures, the average oxygen concentration is used for the boiler control. State of the art is the fast in situ measurement of oxygen and potentiometric oxygen sensor with zirconia solid electrolyte electrochemical cell was taking combustion market overnight fulfilling all requirements especially on sensitivity, selectivity, stability, response time, and long-term stability. Electrochemical oxygen concentration cells consist of ceramic solid electrolyte (yttria-stabilized zirconia, YSZ) in form of a tube or disc coated with platinum electrodes separated by the gastight oxygen-ion-conducting ceramics (see entry "▶ Solid State Electrochemistry, Electrochemistry Using Solid Electrolytes") (Fig. 2, left).

One electrode (measuring) is open to the flue gas, and the opposite electrode is the reference one using dried air as a reference gas. The main reason for the success of those sensors is that they can be applied directly in the combustion process up to the temperatures as high as 1,600 °C [3, 4]. Therefore, such in situ measurements allow the determination of the oxygen content in the flue gas without any gas sampling. Due to the very fast response of oxygen sensor (in ms range), real-time information can be obtained and used for the combustion process control. Quantitative determination of the reducing and oxidizing gases could be also possible using the signal of the electrochemical oxygen sensor

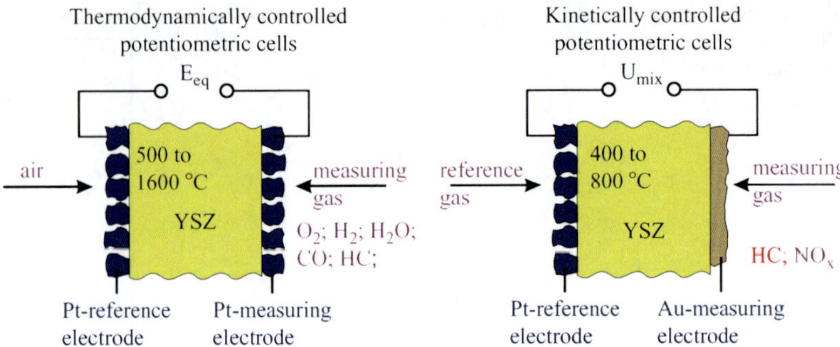

Combustion Control Sensors, Electrochemical, Fig. 2 Basic principles of solid electrolyte cells: *Left* oxygen sensor (Nernst sensor), *Right* combustible sensor (mixed potential sensor) [2]

(see entry "▶ Gas Titration with Solid Electrolytes") [5]. This oxygen titration application using electrochemical cells with solid electrolyte similar to acid–base titration was first introduced in 1966 by Möbius [5].

Other Methods Using Solid Electrolytes

Beside oxygen also CO is an important gas to be measured to control the combustion process. For this purpose also solid electrolyte cells can be used based on the mixed potential principle. Depending on the electrode material, mostly gold and gold alloys are preferred; the unburnt gas components do not equilibrate on the measuring electrode at temperatures <700 °C. These gas species are not thermodynamically stable and electrochemically active. In C_xH_y-, CO-, and O_2-containing gas mixtures, at least two electrode reactions can take place: the electrochemical reduction of oxygen and the electrochemical oxidation of carbon monoxide and hydrocarbons. The measured open-circuit voltage does not obey the Nernst's equation. Therefore, such electrode behavior is often referred to non-Nernstian electrodes (or mixed potential sensors) (Fig. 2, right). According to Miura's theory, the cell voltage mainly depends logarithmically on the concentrations [6]. The mixed potential of such solid electrolyte electrodes is in contrast to that of aqueous solution electrodes very stable and reproducible. Up to now only few companies offer such CO measurements for in situ application.

Combustion Calculation Using the Signal from the Solid Electrolyte Oxygen Sensor

For better understanding the other combustible components in the fuel like sulfur are omitted in Eq. 1. In combustion calculations the value lambda λ (commonly denoted as excess air ratio or excess air coefficient) is used as the ratio of volume between air used in the combustion and the air that is necessary for stoichiometric combustion.

$$\lambda = \frac{v_{\text{air}}}{v_{\text{air,stoich}}} \tag{2}$$

In engineering the demand of air is expressed by means of λ. In the case of stoichiometric combustion, this value amounts to $\lambda = 1$, at fuel excess $\lambda < 1$ and at air excess $\lambda > 1$.

Between λ and the oxygen, concentration relations are valid which depend on the kind of fuel and the range in which the combustion takes place [4]. In the lean range of hydrocarbon ($\lambda > 1$, excess of oxygen), combustion (C_nH_m) λ is given by

$$\lambda = \frac{1}{4n/m + 1} \frac{1 + \varphi_{O_2 \text{ in air}} \exp\left[-4U_{eq}F/(RT)\right]}{1 - \exp\left[-4U_{eq}F/(RT)\right]} \tag{3}$$

where φ is the volume concentration $\varphi_{O_2} = \frac{v_{O_2}}{v}$ and $U_{eq} = -E$ the equilibrium voltage or the negative emf of the electrochemical sensor.

According to the Nernst's equation with air (50 % r.h.) as the reference gas, U_{eq} is given by

$$\begin{aligned} -E &= U_{eq}/mV \\ &= -[0.06527 - 0.021543 \cdot \ln(\varphi_{O_2}/vol\%)] \\ &\quad \cdot T/K \end{aligned} \tag{4}$$

The oxygen concentration in the test gas φ_{O_2} is obtained for temperature T:

$$\begin{aligned} \varphi_{O_2}/vol\% &= 20.69 \\ &\quad \cdot \exp[-46.42(U_{eq}/mV)/(T/K)] \end{aligned} \tag{5}$$

In the region of an excess of hydrocarbons ($\lambda < 1$, fat range), the combustion becomes incomplete. The exhaust gas contains a mixture of CO, CO_2, H_2, and H_2O (water gas):

$$\begin{aligned} & C_nH_m + [(1-a/2)\,n + (1-b)m/4]\,O_2 \\ & \quad \rightarrow (1-a)\,n\,CO_2 + an\,CO \\ & \qquad + (1-b)\,m/2\,H_2O + b\,m/2H_2 \end{aligned} \tag{6}$$

For λ the following equation is obtained:

$$\lambda = 1 - \frac{na/2 + mb/4}{n + m/4} \tag{7}$$

With the ratio of burnt and unburnt species $Q = \frac{p_{CO_2} + p_{H_2O}}{p_{CO} + p_{H_2}}$ and V the ratio of carbon-containing species to hydrogen-containing ones $V = \frac{p_{CO} + p_{CO_2}}{p_{H_2} + p_{H_2O}}$, the following equations for Q and V are obtained:

$$Q = \frac{(1-a)n + (1-b)m/2}{an + bm/2} \tag{8}$$

$$V = 2\,n/m \tag{9}$$

In such a way, values of a and b can be calculated if the C/H ratio and the sensor signal are known. Provided that the gas composition is in equilibrium, the mass action law yields

$$K(T) = \frac{p_{CO}p_{H_2O}}{p_{CO_2}p_{H_2}} \tag{10}$$

and the constant K(T) can be expressed by a and b:

$$K = \frac{a(1-b)}{b(1-a)} \tag{11}$$

Using Eq. 12 for water gas,

$$U_{eq}/mV = 0.049606(T/K)\lg p_{O_2}* \tag{12}$$

and Eqs. 7, 8, and 9a, b, and λ can be calculated if m and n of the fuel are known where $p_{O_2}*$ is the equivalent oxygen partial pressure.

For the combustion of octane (C_8H_{18}), the relation between the λ value, the oxygen concentration, and the signal of the solid electrolyte sensor are illustrated in Fig. 3. In the "lean" region for each λ, the concentration of oxygen is given. In the "fat" region, the ratio of burnt and unburnt species can be calculated. The equilibrium oxygen concentration is very low ($<10^{-10}$ bar). Based on this behavior given in the curves of Fig. 3, the λ probe (oxygen sensor) for automotive application was developed [7].

The more homogeneous the fuel/air mixture is, the less air excess would be required to complete the reaction. The gas combustion can be carried out near to the stoichiometric point with just a slight oxygen excess. For each fuel depending on the kind of burner, an optimal λ value and therefore an optimal fuel/air ratio is recommended. Some optimized data are summarized in Table 1.

Oxygen Probes for Combustion

Electrochemical oxygen concentration cells using solid electrolyte (also called as zirconia O_2 analyzer or zirconia oxygen probe) are now well established and powerful tools. Such devices are offered in length up to 4 m by several companies, i.e., Rosemount Analytical Inc. (USA), ABB (Switzerland, Sweden), Ametek (USA), Yokogawa (Japan), Enotec (Germany),

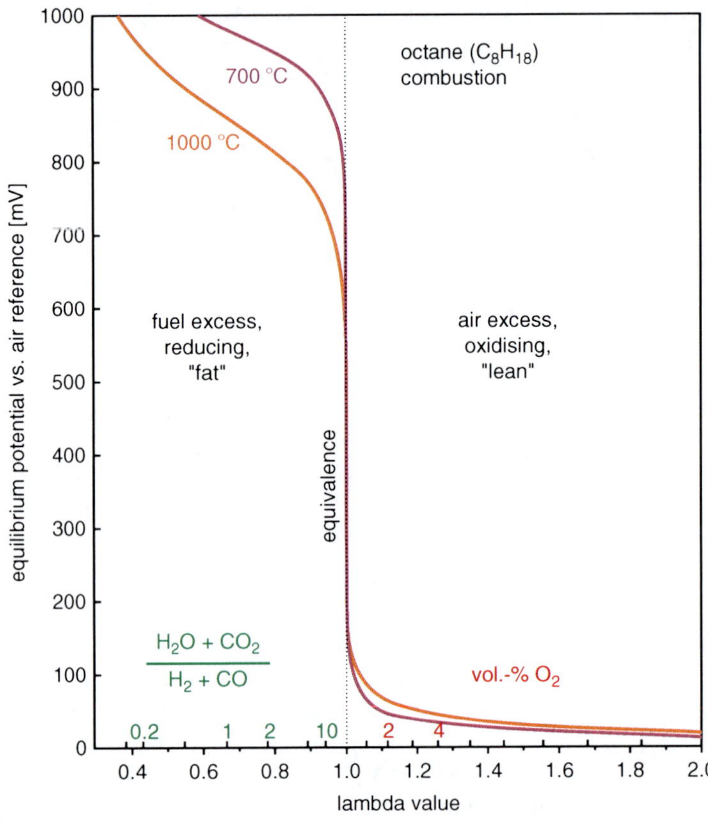

Combustion Control Sensors, Electrochemical, Fig. 3 Emf of the solid electrolyte cell versus lambda value [3]

Combustion Control Sensors, Electrochemical, Table 1 Excess air ratio for the different fuels

Fuel	Excess air ratio λ
Natural gas	1.05–1.1
Oil	1.1–1.2
Pulverized brown-coal combustion	1.1–1.3
Coal	1.4–2.0

Zirox (Germany), and LAMTEC (Germany). These probes are designed using special protective shield and diffuser at the probe end with internal calibration and reference gas lines made from special alloys like Inconel, Hastelloy, or stainless steel 316 L with excellent oxidation/corrosion resistance. Depending on temperature and composition, the lifetime of such sensors can be up to 5 years. They are the leading instrumentation in combustion process oxygen measurements because of the very high chemical stability in severe process environment and exclusive selectivity to oxygen permitting reliable in situ oxygen control very close to the fireball. Many zirconia analyzers have been built in heaters maintaining high temperature (>700 °C) for the cell operation. There are also high temperature O_2 analyzers using process heat for the cell operation and limited by the process temperatures of 550–1,450 °C. Oxygen measurement during boiler start-up or shutdown may not be possible with this type of device. Preliminary location for zirconia O_2 analyzer installation in coal-fired boiler is mostly before or shortly after the heater exchange zone also called as economizer (Fig. 4).

An array of many individual burner elements is used in industrial boilers, and various post-flame combustion control systems, such as overfire air, staging air, re-burning systems, and selective non-catalytic reduction systems, can be employed in the post-flame zone to enhance the efficiency. Emissions of carbon monoxide, unburnt carbon, nitrous oxides, or sulfur dioxide are monitored to ensure the compliance with environmental regulations. There are many events affecting combustion process, and fuel type change, boiler steam load, and ambient air humidity ultimately would change the fuel/air mixing and combustion efficiency. Variations in

Combustion Control Sensors, Electrochemical, Fig. 4 Coal-fired boiler diagram with zirconia oxygen analyzers location [10]

the burning process require minor changes also called trimming to the fuel/air mixture to maintain the correct ratio and efficient boiler operation depending on the load (Fig. 6). Oxidizing combustion will cause the heat loss and excessive nitrogen oxides pollution production, and reducing combustion will produce a sooty emission as unburnt fuel goes up the stacks leading to the greatly reduced combustion burner life span. Typical flue gas oxygen excess concentration is ~2–3 % for gas burners and ~2–6 % for the boilers and oil burners. Combustion would be the most efficient between 0.75 % and 2 % oxygen excess, and for every 1 % oxygen excess reduction in combustion process depending on the process and flue gas temperature, ~1–3 % of fuel could be saved [8–10]. Flue gas from coal-fired boiler contains a large quantity of dust, i.e., flying ash (typical 10–30 g/Nm^3), sulfur, and sulfur dioxide; SO_2 in situ measurements are a more reliable option compared to extractive systems with their handling complicated by the plug gage and condensation problem.

Combustion control is based on the monitoring of NO_x, opacity, velocity, temperature,

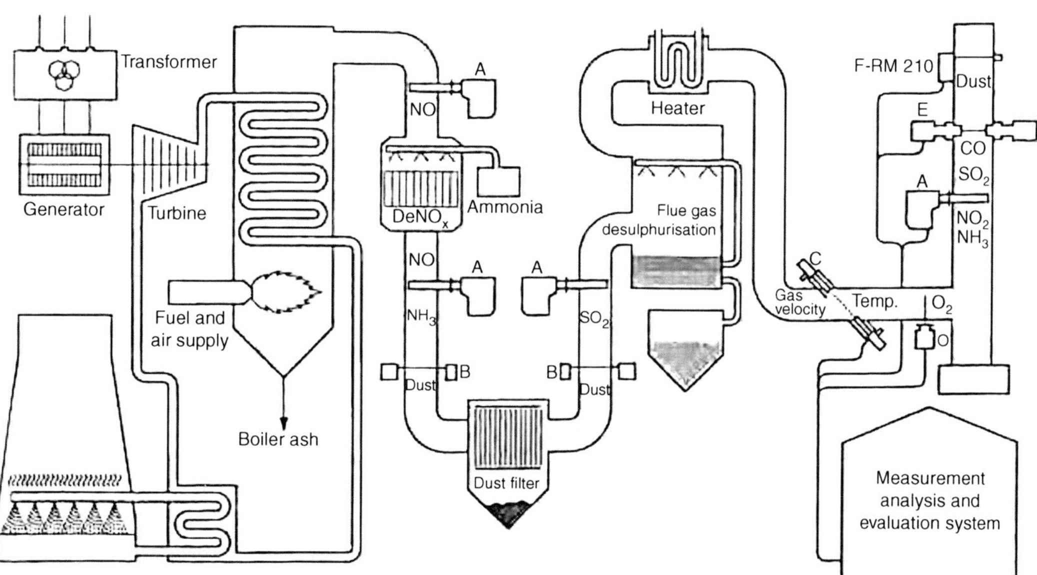

Combustion Control Sensors, Electrochemical, Fig. 5 Coal-fired boiler with different control and monitoring options [11]

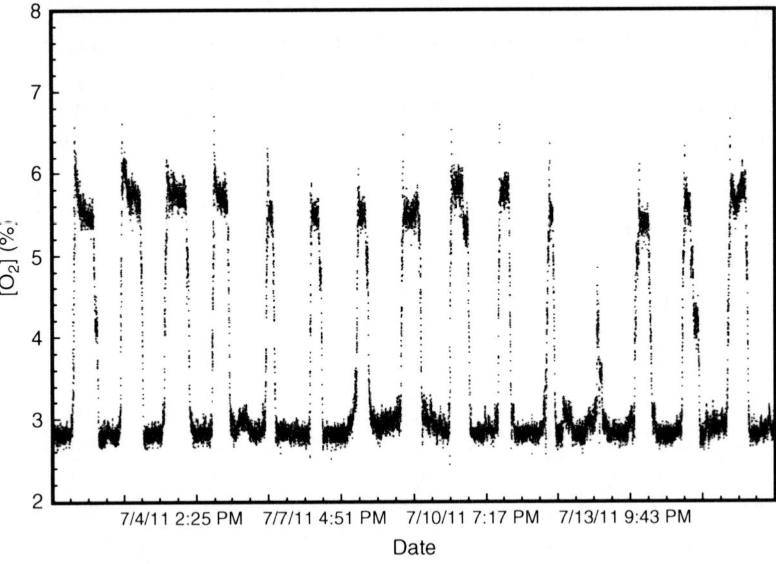

Combustion Control Sensors, Electrochemical, Fig. 6 A typical coal-fired boiler control with ~2.5 % O_2 excess during the day and ~5.5 % overnight operation depending on the load

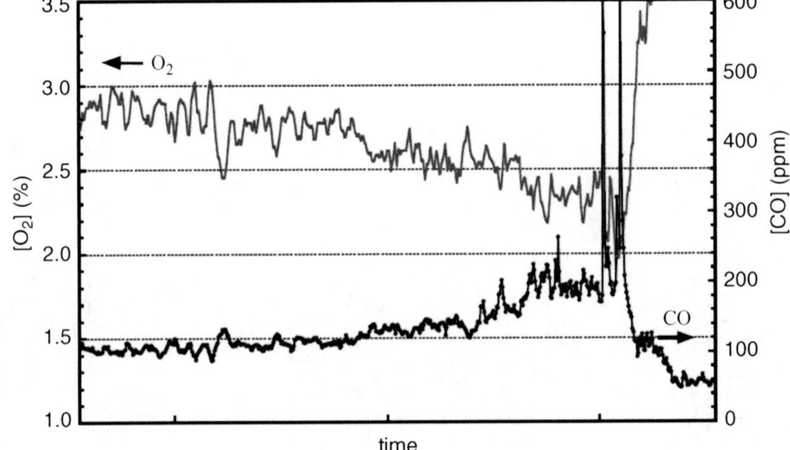

Combustion Control Sensors, Electrochemical, Fig. 7 Precise O_2 control trimming using CO measurements

oxygen, and carbon monoxide in the process environment (Fig. 5).

Boiler trimming methodology varies by boiler manufacturer, fuel type, and control scheme regulating injected air, fuel, or both. However, in any scheme, it is critical, for safety and efficiency, to know the amount of oxygen in the process. This optimization is normally based on overall excess oxygen estimation maintaining an adequate local stoichiometry for each burner, adjustment of the flame type (based on an appropriate control of the air inputs and the operation of the mills), and identification of the optimum number of active burners for each operating load.

While good combustion control can be accomplished with O_2 measurement alone, improved combustion efficiency and stability can be achieved with the concurrent measurement of carbon monoxide, CO [10]. Operation at near trace CO levels of about 100 vol.-ppm and a slight amount of excess air indicate conditions near the stoichiometric point with the highest efficiency (Figs. 6 and 7).

Besides the most reliable in situ zirconia oxygen analyzer based on potentiometric electrochemical O_2 cell, there are several other technologies to be considered like coulometric titration (Fig. 8→*coulometry*), tunable diode

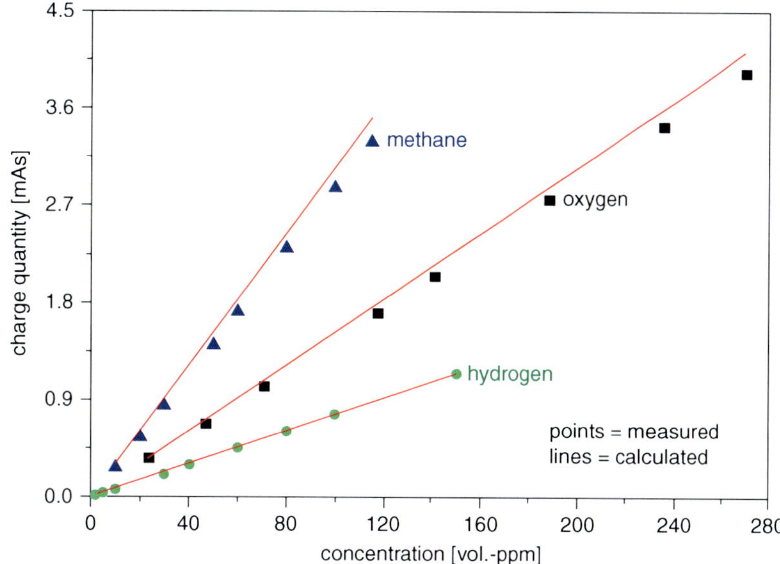

Combustion Control Sensors, Electrochemical, Fig. 8 Coulometric determination of O_2, CH_4, H_2

laser spectroscopy (TDLS) [12], and paramagnetic [13] and electrochemical cell with liquid electrolyte [14] for extractive or across the duct (TDLS) measurements. In all cases probing and cleaning of the exhaust gas are necessary.

Potentiometric solid electrolyte sensors are also used to determine the burn-off of flames and the shape of the flame [15, 16].

References

1. Docquier N, Candel S (2002) Process control and sensors: a review. Prog Energ Combust Sci 28:107–150
2. Guth U (2012) Gas sensors. In: Bard A, Inzelt G, Scholz F (eds) Electrochemical dictionary, 2nd edn. Springer, Heidelberg
3. Möbius HH (1991) Solid state electrochemical potentiometric sensors for gas analysis. In: Göpel W, Hesse J, Zemel JN (eds) Sensors a comprehensive survey. VCH, New York/Weinheim
4. Kleitz M, Siebert E, Fabry P, Fouletier J (1991) Solid state electrochemical sensors. In: Göpel W, Hesse J, Zemel JN (eds) Sensors a comprehensive survey, vol 2. VCH, New York/Weinheim
5. Möbius HH (1966) Potentiometric titration of oxydable gases with air in solid electrolyte cells, in German: Potentiometrische Titrationen oxydabler Gase mit Luft in Festelektrolytzellen. Z Phys Chem (Leipzig) 231:209–214
6. Miura N, Raisen T, Lu G, Yamazoe N (1998) Highly selective CO sensor using stabilized zirconia and a couple of oxide electrodes. Sensor Actuator B47:84–91
7. Riegel J, Neuman H, Wiedenmann HM (2002) Exhaust gas sensors for automotive emission control. Solid State Ion 152–153:783–800
8. Neumannn H, Hoetzel G, Lindermann G (1997) Advanced planar oxygen sensors for future emission control strategies. SAE Techn Pap 970459:1–9
9. Shuk P, Bailey E, Guth U (2008) Zirconia oxygen sensor for the process application: state-of-the-art. Mod Sens Technol 90:174–184
10. Shuk P (2010) Zirconia oxygen sensor for the process application: state-of-the-art. Tech Mess 77:19–23
11. Leslie L (1997) Continuous emissions monitoring for coal-fired power stations. IEA Coal Res 51 pp
12. Lackner M (2008) Gas sensing in industry by tuneable diode laser spectroscopy (TDLS). Process Eng, Wien
13. Kovacich P, Martin N, Clift M, Stocks C, Gaskin I, Hobby J (2006) Highly accurate measurement of oxygen using a paramagnetic gas sensor. Meas Sci Technol 17:1579–1585
14. Stetter J, Penrose W, Yao S (2003) Sensors, chemical sensors, electrochemical sensors, and ECS. J Electrochem Soc 150(2):S11–S16
15. Harbeck W, Guth U (1990) Determination of burn-off of gas flames by means of gas potentiometric Method (in German) Ermittlung der Ausbrandgrenze von Gasflammen mit Hilfe gaspotentiometrischer Bestimmungsmethoden. Gas Wärme Int (Essen) 39:10–24
16. Lorenz H, Tittmann K, Sitzki L, Trippler S, Rau H (1996) Gas-potentiometric method with solid electrolyte oxygen sensors for the investigation of combustion. Fresenius J Anal Chem 356:215–220

Compound Semiconductors, Electrochemical Decomposition

David Rauh
EIC Laboratories, Inc, Norwood, MA, USA

Introduction

While silicon still dominates semiconductor technology, compound semiconductors, with their nearly limitless flexibility for material design, are of ever-increasing importance. Interest in electrochemistry of compound semiconductors has followed closely their emergence in such diverse areas as high-speed electronics, *photovoltaics* and *photoelectrochemical solar energy* conversion, lasers and LEDs, quantum-confined materials, and 3D nanostructures. A focus of their electrochemistry is frequently their stability in contact with electrolytes in the dark or under illumination, at open circuit or under cathodic or anodic polarization. Issues include their environmental corrosion in microelectronic devices and photovoltaic converters, their ability to act as photocatalysts or as converters of light to chemical or electrical energy, their ability to operate continuously as electron or hole injectors in solid-state light-emitting diodes, their electrochemically controlled etching to fabricate high aspect ratio or nanoporous structures, and the anodic formation of electrically passivating or insulating surfaces for device processing.

General Principles

Compound semiconductors of current technological significance include II–VI materials such as CdTe, III–V materials such as GaAs and InP as well as their solid solutions, I–III–VI_2 materials such as $CuInSe_2$, and SiC and GaN. Despite their variety, there are certain principles which govern their electrochemical decomposition. Decomposition can occur under either anodic or cathodic polarization, and the reaction products and electrochemical potentials are strongly dependent on the nature of the electrolyte. Furthermore, n-type semiconductors will tend to form blocking contacts with electrolytes such that, within a potential window of a few volts, significant current will only flow in the dark under cathodic polarization at potentials negative of the flatband potential, while the opposite is true for p-type materials. With bandgap illumination, minority carriers are generated allowing anodic processes to occur for n-type materials and cathodic processes for p-type. The currents for these processes are dependent on the minority carrier generation rate and hence on the light intensity.

As described by the *Gerischer model* [1, 2], a first approximation of the stability of semiconductors in contact with an electrolyte may be gained by viewing a map of the interfacial band energies obtained from the semiconductor flatband potentials on an *electrochemical potential scale*, such as versus the H^+/H_2 couple (SHE). Potentials of dissolution reactions may be calculated from their free energies and added to the map, leading to the four stability conditions pictured in Fig. 1. An anodic reaction within the bandgap will "capture" a *valence band* hole, giving rise to spontaneous anodic decomposition of p-doped and degenerate semiconductors and photodecomposition of n-type. Indeed, in aqueous electrolytes, the anodic decomposition potential of most if not all known semiconductors falls negative of the interfacial potential of the valence band, rendering them thermodynamically unstable. Cathodic processes proceed by electron injection from the conduction band, and reduction reactions of the semiconductor lattice are sometimes (but not always) found in the potential region positive of the conduction band edge.

The interfacial energetics of the band edges are also affected by ion adsorption from the electrolyte, frequently as a result of pH-dependent acid-base equilibria of surface oxides that form spontaneously due to decomposition. In aqueous solutions, the pH dependence of the surface potential moves in the same direction as many of the possible dissolution reactions, so that the relative disposition of the band edges and the decomposition potentials are relatively constant under most conditions.

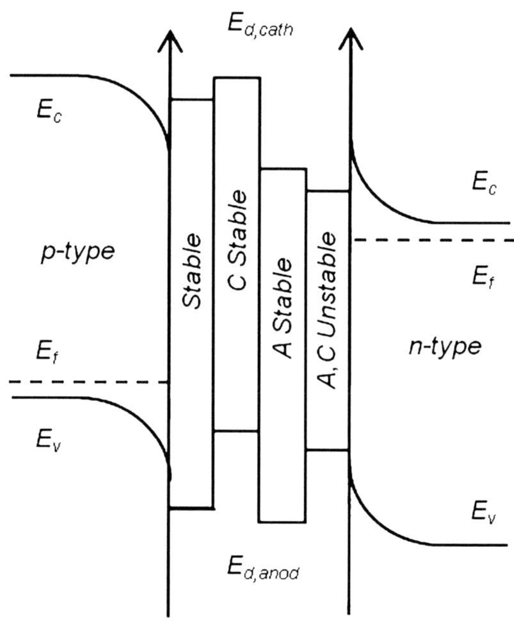

Compound Semiconductors, Electrochemical Decomposition, Fig. 1 Possible thermodynamic electrochemical stability windows for semiconductor electrodes relative to the potentials of the band edges at the electrolyte interface

The actual thermodynamic existence map of dissolution products in any given electrolyte as a function of potential and pH can be expressed as a *Pourbaix diagram*, and such diagrams are available for a broad range of common semiconductors; see, for example, [3–6]. Figure 2 provides an example for GaAs in aqueous solutions. The potentials of the band edges over the pH range are also shown. Clearly, oxidation and reduction potentials of the GaAs lattice both occur within the bandgap such that valence band holes and conduction band electrons at the solution interface can lead to decomposition. Oxidative decomposition will result in insoluble products throughout the intermediate pH range and can occur for p-GaAs in the dark and n-GaAs under illumination. Valence band holes are not sufficiently oxidizing to decompose H_2O to form O_2, but conduction band electrons are sufficiently reducing to produce H_2. However, deposition of Ga and evolution of arsine gas are favored thermodynamically. In fact, a mixed reaction occurs experimentally and kinetic factors dominate the product distribution.

Investigators have sought to assess the detailed mechanism of oxidative electrochemical decomposition of compound semiconductors, many relating to II–VI and III–V materials [1, 4, 7]. II–VI materials encompass a broad range of chalcogenides such as MX (M = Cd, Zn and X = S, Se, Te). Electrochemical decomposition stoichiometries range from 2 to 8 e^-/mol. III–V semiconductors like GaAs, GaP, InAs, and InP, along with their ternary and quaternary mixtures, have multistep anodic decomposition processes that typically consume 6 electrons/mol. Anodic oxidation reaction mechanisms have been defined by a sequence of hole capture and chemical processes, and their kinetics are frequently a strong factor in determining the product composition. It is generally accepted that anodic decomposition processes are independent of doping type.

Galvanic Corrosion

Electrochemical reactions are the basis for galvanic corrosion of semiconductors as might occur at the interface of dissimilar materials in the presence of an electrolyte. In these cases, O_2 can generally be reduced at a metal surface, to form initially O_2^-, while the semiconductor is spontaneously oxidized. Such corrosion processes can be problematic at the metallization interfaces of compound semiconductor lasers, for example. In these devices, the area ratio of metallizations to exposed semiconductors can be large, exacerbating the rate of pitting of the active elements. For example, a study of p-type InGaAs lasers with Au metallization with a 10^4 ratio of Au to InGaAs area showed that extensive pitting at the metallization interface (Fig. 3) occurred during device processing in HF and H_3PO_4 media [8]. With O_2 reduction as the cathodic process driving the reaction, dissolution of the semiconductor to In^{+3}, Ga^{+3}, and arsenates is favored. Oxidizing agents in the processing chemistries can also drive the decomposition reactions.

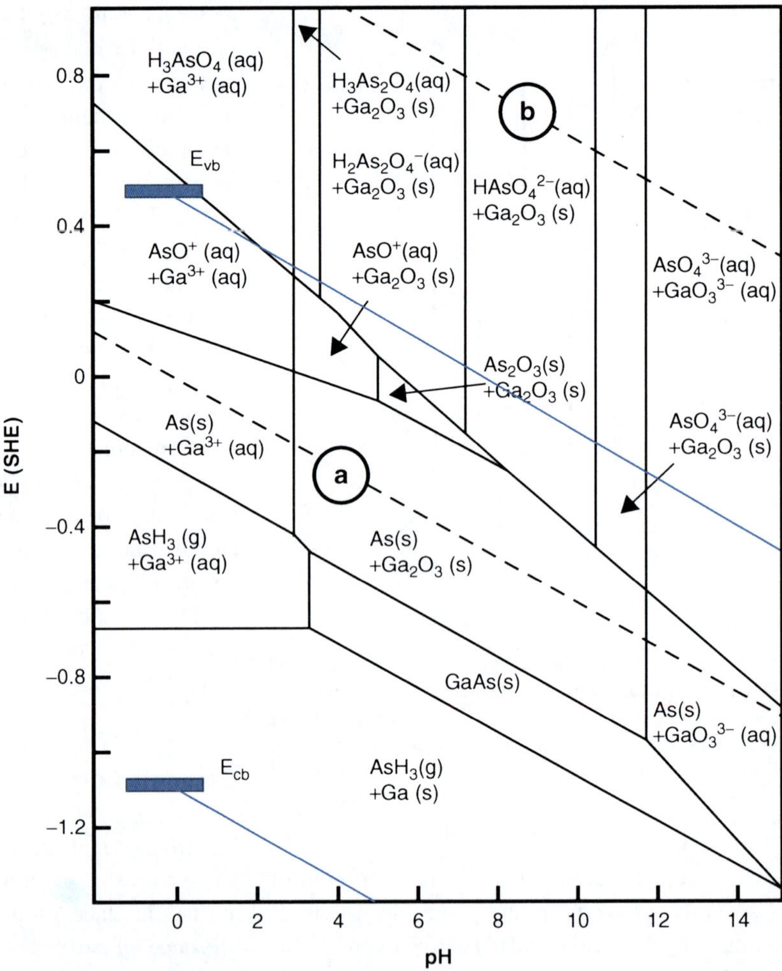

Compound Semiconductors, Electrochemical Decomposition, Fig. 2 Pourbaix diagram of equilibrium potential versus pH for GaAs. Also shown are the experimental values of the interfacial potentials of the conduction and valence bands, along with the aqueous reduction (**a**) and oxidation potentials (**b**)

Localized Avalanche Breakdown: Defect Etching and Pore Formation

Anodic polarization of an n-type electrodes and cathodic polarization of p-type electrodes in the absence of surface defects will pass current in the dark at high applied voltages exceeding the bandgap due to tunneling of electrons across the space-charge layer [9, 10]. However, the presence of microscopic surface defects can facilitate an avalanche breakdown process providing higher currents at lower voltages than would be predicted by tunneling. Avalanche breakdown, like tunneling, is dependent on the doping density. In this process, a conductive pathway becomes established through crystallographic bulk defects and terminating at surface defects. Such defects, which can be induced by scratching the surface, are electrochemically more labile than the bulk material and become sites for preferred electrochemical etching. The result is a "decoration" of defect features featuring etch pits distributed across the semiconductor surface.

Under some conditions, electrochemical (anodic) etching results can produce extensive pore formation over the surface [11]. Such pore structures are of interest for enhancing the light output and tunability of optoelectronic devices. Pores nucleate around defects and, at high current densities, often grow in the direction of the current, i.e., perpendicular to the electrode surface, following the "current line." Electrochemical etching processes are further controlled by the fact that certain crystallographic planes are more reactive

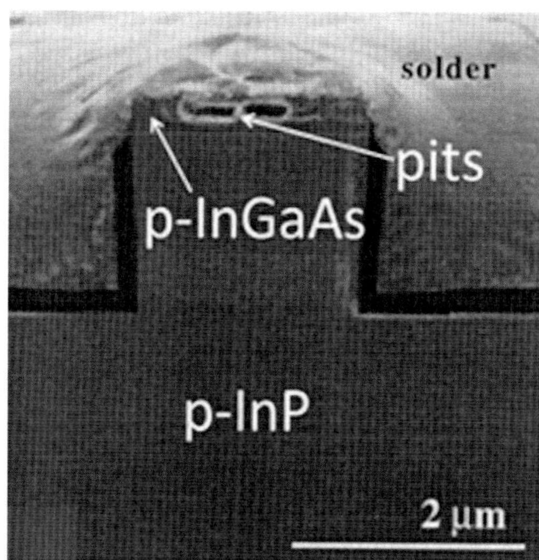

Compound Semiconductors, Electrochemical Decomposition, Fig. 3 Etch pits caused by galvanic corrosion in a p-InP-based semiconductor laser (From Ref. [8])

than others, leading to etch pits and pores that are bounded by such "stopping planes." The formation and breakdown of passivation oxide layers further modulate the pore etching process. For some materials, notably n-type InP(001), a spatially organized array of pores is observed [12] as shown in Fig. 4. The pore walls may be thinned by subsequent cathodic decomposition process:

$$InP + 3H^+ + 3e^- \rightarrow In + PH_3$$

where the metallic In dissolves in the acidic electrolyte.

Controlled Decomposition-Electrochemical Microfabrication

Electrochemical and photoelectrochemical control of semiconductor anodic decomposition reactions allow for controlled fabrication of fine structures on semiconductor surfaces, for example, diffraction gratings, lenses, and MEMs structures [13–22]. These processes are usually carried out under conditions where the oxidation products are soluble in the electrolyte. Patterns are

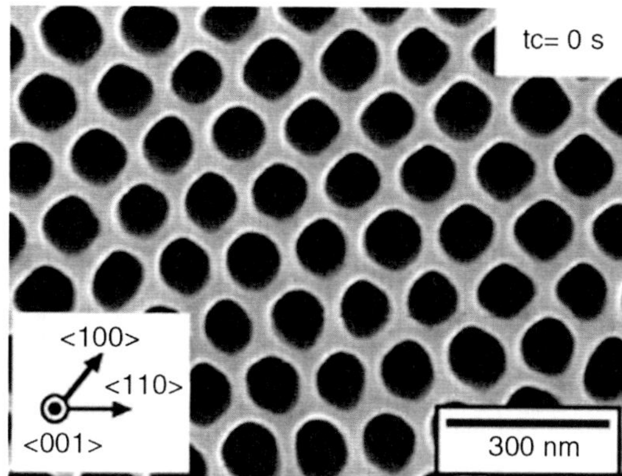

Compound Semiconductors, Electrochemical Decomposition, Fig. 4 SEM of n-InP(001) formed by anodization at 4 V in an aqueous acidic electrolyte (From Ref. [12]. Reproduced by permission of The Electrochemical Society)

produced by photoelectrochemical etching using masking or by projecting and image of the pattern on the semiconductor electrode surface. Compound semiconductors tend to show etching anisotropy under some conditions in which certain crystallographic faces are favored for dissolution, and this anisotropy extends to electrochemical dissolution as well. Anisotropy can result from differences in dissolution potentials or from favored formation of passivating products at certain crystallographic faces [14, 17, 18].

Photoelectrochemical processing is particularly useful for wide bandgap compound semiconductors such as SiC and GaN, which are notoriously difficult to etch chemically. N-SiC can be photoelectrochemically dissolved under anodic bias under mild conditions (e.g., in 0.1 M KOH [19]) according to the reaction:

$$SiC + 8OH^- + 6h^+ \rightarrow Si(OH)_2O_2^{2-} + CO + 3H_2O$$

The products are soluble/gaseous and therefore do not passivate the surface, except at high light intensities. GaN is similarly etched in dilute KOH to soluble and gaseous products:

$$2GaN + 12OH^- + 6h^+ \rightarrow 2GaO_3^{3-} + 6H_2O + N_2$$

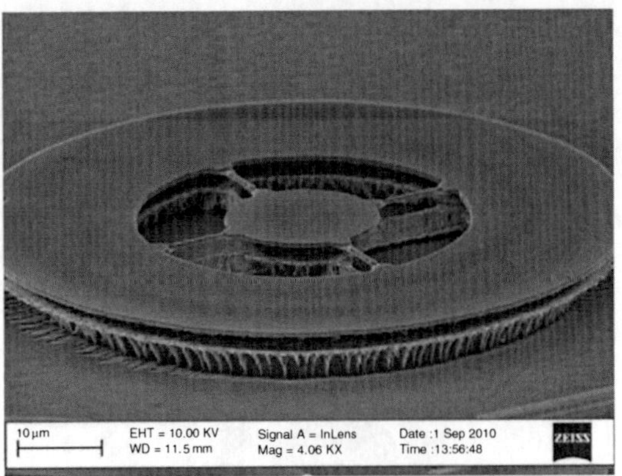

Compound Semiconductors, Electrochemical Decomposition, Fig. 5 Single crystal SiC MEMs structure fabricated by photoanodic etching of n-SiC (Ref. [22]. Reprinted with permission from Elsevier 2011)

A recent example of microfabrication of *MEMs* structures in SiC by photoelectrochemical dissolution is shown in Fig. 5 [22]. These structures were prepared by anodic photoelectrochemical undercutting of n-SiC beneath a p-SiC epitaxial layer.

Electrochemical Surface Passivation

In semiconductor device processing, it is necessary to minimize the concentration of surface defects that can act as traps for charge carriers. Anodic polarization to form a dielectric passivation layer on the semiconductor surface is one method. Here the choice of an electrolyte is of paramount importance as electrolyte composition and pH can determine the solubility of the surface reaction products. For example, several anodization protocols have been tried for passivating GaAs, a particularly important electronic material. Notably, galvanostatic polarization of GaAs in an electrolyte containing propylene glycol, H_2O, and 3 % tartaric acid results in the growth of a dense, insoluble oxide layer. The layer thickness can be monitored by the voltage increase of the electrode, with the polarization increasing by 0.05 V per Å oxide growth. The oxides formed in aqueous solutions are quite leaky electrically, however [23]. A more recent example employs liquid NH_3 as a nonaqueous electrolyte for anodic film formation on GaAs and InP. Polarization of InP leads to the formation of an extremely uniform phosphazine-type layer [24] (Fig. 6). Although the As analog on GaAs was unstable, this process is an example of non-oxide passivation by anodic decomposition.

Compound Semiconductors, Electrochemical Decomposition, Fig. 6 Passivating phosphazine film formed by anodizing n-InP in a liquid NH_3 electrolyte under bandgap illumination (From Ref. [24])

Stabilizing Compound Semiconductors Against Decomposition

Semiconductors may be stabilized against anodic or cathodic decomposition if a reactant is present that is capable of scavenging reactive electrons or holes at the surface before they have a chance to react with the crystal lattice [1–3]. Such stabilization is required for the continuous action of semiconductor photocatalysts or regenerative photoelectrochemical cells. In the special case of photoelectrolysis of water at some n-type semiconductors like TiO_2, the potential required for oxidation of water is negative of that required for decomposing the lattice, resulting in preferential capture of photogenerated holes by H_2O or OH^- and stable O_2 evolution.

In many cases, decomposition reactions compete with the redox system, and the stability towards decomposition is a matter of kinetics. Measurements of the photoelectrochemical stabilization efficiency have been made, where S = [current consumed by redox species]/[total current][25]. S can be measured directly with rotating ring-disk electrode (RRDE) experiments, where the semiconductor comprises the disk material, the ring is a stable metal electrode, and the electrolyte contains the redox species. For example, an RRDE experiment with n-GaAs as the disk and Fe^{II}EDTA in the electrolyte measured, under illumination and anodic bias, the ring current due to reduction of Fe^{III} produced by photogenerated holes as a fraction of the total anodic current from the semiconductor [23]. In general, S increases with redox concentration and with increasingly negative redox potential.

Stability of Nanocrystals and Quantum Dots

Nanocrystals (*quantum dots*, QD) of compound semiconductors are frequently used as extrinsic luminescent labels for immunoassays and other biological processes. Thus, their stability under illumination in aqueous environments is important to their operation. Because of their small volume, QDs do not form a significant charge depletion region at the surface like bulk semiconductors. The quantum confinement effect increases the effective bandgap of these materials as the nanocrystal's size decreases. The electrochemistry of semiconductor nanocrystals reflects this trend insofar as the gap between their oxidation and reduction potentials reflects approximately the HOMO-LUMO gap [26–29]. Naked QD materials are mostly unstable to oxidation due to their high surface to volume ratios and high concentration of reactive surface groups [27]. Thus, QDs such as CdSe are often capped with surface complexing molecular groups such as trioctylphosphine oxide (TOPO) to stabilize them or with a wide bandgap or insulating shell to ensure high luminescence efficiency. Under illumination, QDs become strongly oxidizing, and thus surface organic groups can become depleted photocatalytically (e.g., TOPO on CdSe [30]).

Future Directions

Optoelectronic Devices

Increasing use of LEDs for lighting and semiconductor lasers in consumer products and instrumentation will continue to require a better understanding of corrosion processes that may affect the lifetime of the component semiconductor materials.

Solar Cells

Photovoltaic devices based on compound semiconductors like $CuInSe_2$, CdTe, and quantum dots as well as liquid junction devices will require improved strategies for stabilization against environmental corrosion and dissolution.

Semiconductor Recycling

Reclamation of materials will become an issue with the widespread deployment of compound semiconductor photovoltaics [31, 32]. Electrochemical dissolution and reclamation of the elements in CIS, CdS, and CdTe have been proposed as a recycling process.

Nanoporous Semiconductors

The controlled fabrication of high-porosity compound semiconductor surfaces through electrochemical dissolution processes will provide further opportunities to develop novel and improved optoelectronic devices.

Nanofabrication

Nanostructures comprising various high aspect ratio features will be fabricated by electrochemical dissolution reactions controlled by crystallographic orientation, by masks, and/or photoelectrochemically by light.

Cross-References

- ▶ Photoelectrochemistry, Fundamentals and Applications
- ▶ Quantum Dot Sensitization
- ▶ Redox Processes at Semiconductors-Gerischer Model and Beyond
- ▶ Reference Electrodes
- ▶ Semiconductor Electrode
- ▶ Semiconductor–Liquid Junction: From Fundamentals to Solar Fuel Generating Structures

► Semiconductors Group IV, Electrochemical Decomposition
► Semiconductors, Principles
► TiO_2 Photocatalyst

References

1. Gerischer H (1979) Solar photoelectrolysis with semiconductor electrodes. In: Seraphin B (ed) Topics in applied physics. Solar energy conversion. Solid state physics aspects, vol 31. Springer, New York
2. Rajeshwar K (2001) Fundamentals of semiconductor electrochemistry and photoelectrochemistry. In: Licht S (ed) Encyclopedia of electrochemistry, chapter 1, 3–53. Wiley-VCH, Weinheim
3. Memming R (2002) Semiconductor electrochemistry. Wiley-VCH, Weinheim
4. Bouroushian M (2010) Electrochemistry of metal chalcogenides. Springer, New York
5. Park S, Barber M (1979) Thermodynamic stabilities of semiconductor electrodes. J Electroanal Chem Interfacial Electrochem 99:67–75
6. Voloshchuk A, Tsipishchuk N (2002) Equilibrium potential–pH diagram of the CdTe–H_2O system. Inorg Mater 38:1114–1116
7. Vanmaekelbergh D, Gomes W (1990) Relation between chemical and electrochemical steps in the anodic decomposition of III–V semiconductor electrodes: a comprehensive model. J Phys Chem 94:1571–1575
8. Ivey D, Luo J, Ingrey S, Moore R, Woods I (1998) Galvanic corrosion effects in InP-based laser ridge structures. J Electron Mater 27:89–95
9. Tranchart J, Hollan L, Memming R (1978) Localized avalanche breakdown on GaAs electrodes in aqueous electrolytes. J Electrochem Soc 125:1185–1187
10. Hasegawa H, Sato T (2005) Electrochemical processes for formation, processing and gate control of III–V semiconductor nanostructures. Electrochim Acta 50:3015–3027
11. Föll H, Langa S, Carstensen J, Christopherson M, Tiginyanu I (2003) Pores in III–V semiconductors. Adv Mater 15:183–198
12. Sato T, Fujino T, Hashizume T (2007) Electrochemical formation of size-controlled InP nanostructures using anodic and cathodic reactions. Electrochem Solid State Lett 10:H153–H155
13. Nowak G, Xia X, Kelly J, Weyher J, Porowski S (2001) Electrochemical etching of highly conductive GaN single crystals. J Crystal Growth 222:735–740
14. Kelly J, Philipsen G (2005) Anisotropy in the wet etching of semiconductors. Curr Opin Solid State Mater Sci 9:84–90
15. Huygens I, Strubbe K, Gomes W (2000) Electrochemistry and photoetching of n-GaN. J Electrochem Soc 147:1797–1802
16. Choi K (2010) Shape effect and shape control of polycrystalline semiconductor electrodes for use in photoelectrochemical cells. J Phys Chem Lett 1:2244–2250
17. Weyher J (2006) Characterization of wide-band-gap semiconductors (GaN, SiC) by defect-selective etching and complementary methods. Superlattices Microstruct 40:279–288
18. Kohl P, Wolowodiuk C, Ostermayer F (1983) The photoelectrochemical oxidation of (100), (111) and (111) n-InP and n-GaAs. J Electrochem Soc 130:2288–2293
19. Van Dorp D, Kelly J (2007) Photoelectrochemistry of 4H–SiC in KOH solutions. J Electroanal Chem 599:260–266
20. Su N, Tang Y, Zhang Z, Fay P (2008) Observation and control of electrochemical etching effects in the fabrication of InAs/AlSb/GaSb heterostructure devices. J Vacuum Sci Technol B 26(3):1025–1029
21. Yang B, Fay P (2006) Bias-enhanced lateral photoelectrochemical etching of GaN for the fabrication of undercut MEMS structures. J Vacuum Sci Technol B 24:1337–1340
22. Zhao F, Islam M, Huang C (2011) Photoelectrochemical etching to fabricate single-crystal SiC MEMS for harsh environments. Mater Lett 65:409–412
23. Hasegawa H, Hartnagel H (1976) Anodic oxidation of GaAs in mixed solutions of glycol and water. J Electrochem Soc 123:713–723
24. Goncalves A, Le Floch P, Mezailles N, Mathieu C, Etchberry A (2010) Fully protective yet functionalizable monolayer on InP. Chem Mater 22:3114–3120
25. Frese K, Madou M, Morrison S (1980) Investigation of photoelectrochemical corrosion of semiconductors. J Phys Chem 84:3172–3178
26. Haram S, Quinn B, Bard A (2001) Electrochemistry of CdS nanoparticles: a correlation between optical and electrochemical band gaps. J Am Chem Soc 123:8860–8861
27. Yu M, Fernando G, Li R, Papadimitrakopoulos F, Shi N, Ramprasad R (2006) First principles study of CdSe quantum dots: Stability, surface, unsaturations, and experimental validation. Appl Phys Lett 88: 231910–231910-3
28. Brus L (1983) A simple model for the ionization potential, electron affinity and aqueous redox potentials of small semiconductor crystallites. J Chem Phys 79:5566–5571
29. Lim S, Kim W, Jung S, Seo J, Shin S (2011) Anisotropic etching of semiconductor nanocrystals. Chem Mater 23:5029–5036
30. Aldana J, Wang A, Peng X (2001) Photochemical instability of CdSe nanocrystals coated by hydrophilic thiols. J Am Chem Soc 123:8844–8850

31. Menezes S (2001) Electrochemical approach for removal, separation and retrieval of CdTe and CdS films from PV module waste. Thin Solid Films 387:175–178
32. Bradwell D, Osswald S, Wei W, Barriga S, Ceder G, Sadoway D (2011) Recycling ZnTe, CdTe, and other compound semiconductors by ambipolar electrolysis. J Am Chem Soc 133:19971–19975

Compressed and Liquid Hydrogen for Fuel Cell Vehicles

Rittmar von Helmolt and Ulrich Eberle
Government Programs and Research Strategy, GM Alternative Propulsion Center, Adam Opel AG, Rüsselsheim, Germany

Introduction

Fuel cell electric vehicles (FCEV) incorporate three major technology elements or subsystems that are new to the automotive sector:

1. Electric propulsion
2. Fuel cell technology
3. Hydrogen storage

All of these technologies are part of intensive R&D and engineering programs by major international carmakers, and their principal viability could be demonstrated in demonstration fleet projects which showed that the FCEV is a real alternative to conventional power trains, with a competitive performance and reliability. The subsystems and their interactions have been refined, and so have the technical and economical requirements. During this process, several technology routes could be down-selected, and the development programs have started to converge on specific solutions to achieve the respective targets.

Basically, there have been four major options for automotive onboard hydrogen storage (see Ref. [1] and Fig. 1):

1. CGH2 compressed gaseous hydrogen, 35–70 MPa and room temperature
2. LH2 liquid hydrogen, 20–30 K, 0.5–1 MPa
3. Solid-state absorbers (such as hydrides or high-surface materials)
4. Hybrid solutions, utilizing at least two of the above mentioned technologies.

The mainstream automotive development process is currently focusing on the first option. Some activities are still ongoing with the second option or derivatives thereof. The last two technologies are described in detail in several other publications, Ref. [2] is recommended to readers

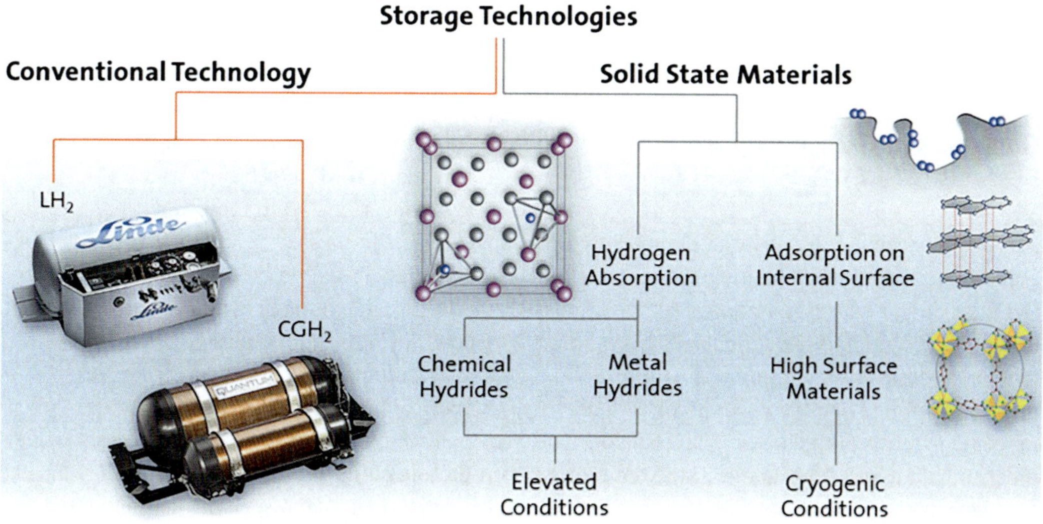

Compressed and Liquid Hydrogen for Fuel Cell Vehicles, Fig. 1 Various hydrogen storage technologies

interested in detail in hydrogen storage pathways summarized under point 3.

It also has to be stated that values for gravimetric and volumetric energy densities may correspond either to (a) just a materials approach or (b) a systems approach including all required components and mountings [1]. From an engineering perspective, the second approach is preferable; also the target values provided by the US Department of Energy are defined on this basis [3]. As mentioned above, the options (1) and (2) have been implemented by the automotive industry in recent years. As an example, the GM HydroGen3 (a multipurpose vehicle based on the Opel Zafira) was adaptable to both storage types [1]. The HydroGen3 vehicles were capable of storing either 3.1 kg H2 (70 MPa CGH2) or 4.6 kg H2 (LH2). These values correspond to ranges of 270 km or 400 km, respectively, in the New European Driving Cycle (NEDC).

The Chevrolet Equinox Fuel Cell, GM's fourth generation fuel cell vehicle, incorporates a 4.2 kg 70 MPa CGH2 system [1]. Virtually all major automotive companies that pursue hydrogen vehicle programs have meanwhile decided to adopt the 70 MPa CGH2 technology and to integrate them in existing conventional vehicle architectures. Eventually, "purpose-built" cars that are not based on any existing vehicle platform but were designed around the hydrogen propulsion and fuel storage systems [1] could be viable future solutions.

Compressed Gaseous Hydrogen Systems (CGH2)

Besides the chemical energy content of hydrogen (compared to lower heating value or LHV), the mechanical energy content of a CGH2 system is also of great interest, because this number is related to the energy required to compress the gas from ambient conditions to 35 or 70 MPa, in percentage terms of the lower heating value. The mechanical energy ΔW can be simply computed using the assumptions of an ideal gas as working gas and an isothermal compression process [2, 4]:

$$\Delta W = \int p\, dV = RT \int_{V_I}^{V_F} \frac{1}{V}\, dV = -RT \,\mathrm{In}\, \frac{p_F}{p_I}, \tag{1}$$

whereas p_I is the initial pressure and p_F the final pressure (see Fig. 2a). A mechanical energy content of about 8 MJ per kg of H2 for 70 MPa and 7 MJ per kg of H2 for 35 MPa tank systems is obtained. This rather small differential often causes surprise but is only a consequence of integrating over the ideal gas law, leading to a nonlinear behavior. When assessing a real-world hydrogen refueling procedure (taking just 3–5 min.), the compression process is obviously far from being isothermal [2, 4]. In this case, a polytropic process characterization is needed:

$$\Delta W = \int p\, dV = \frac{n}{n-1} RT \left(\left[\frac{p_F}{p_I} \right]^{\frac{n-1}{n}} - 1 \right), \tag{2}$$

whereas n is the polytropic coefficient. With $n = 1.36$, that equation leads for a 70 MPa system to an ideal value of 10.2 MJ per kg of H2. Considering again a real-world process, by contrast, additional mechanical losses occur. Hence, the engineering value to compress hydrogen is as high as circa18 MJ per kg of H2 at 70 MPa, or 14.5 MJ per kg of H2 at 35 MPa [2]. From a normalized compression-energy perspective, there is thus no significant drawback to work with substantially larger operating pressures than 35 MPa. The maximum for a reasonable operating pressure level is given by the flattening of the hydrogen mass density for pressure values greater than 70 MPa (see Fig. 2b).

When working with gases at high pressure levels, specific vessel designs are needed. Obviously, a sphere would be the preferred geometry, but due to the limited available space onboard vehicles, cylinders are typically preferred for their simpler automotive packaging properties [1, 2]. Cylindrical vessels embody the second-best geometry, in which the required wall thickness depends on the operating pressure, the tank diameter, the tank wall material's specific strength, and the winding pattern in case of fiber composite vessels. From an analysis of the stresses in thicker-walled

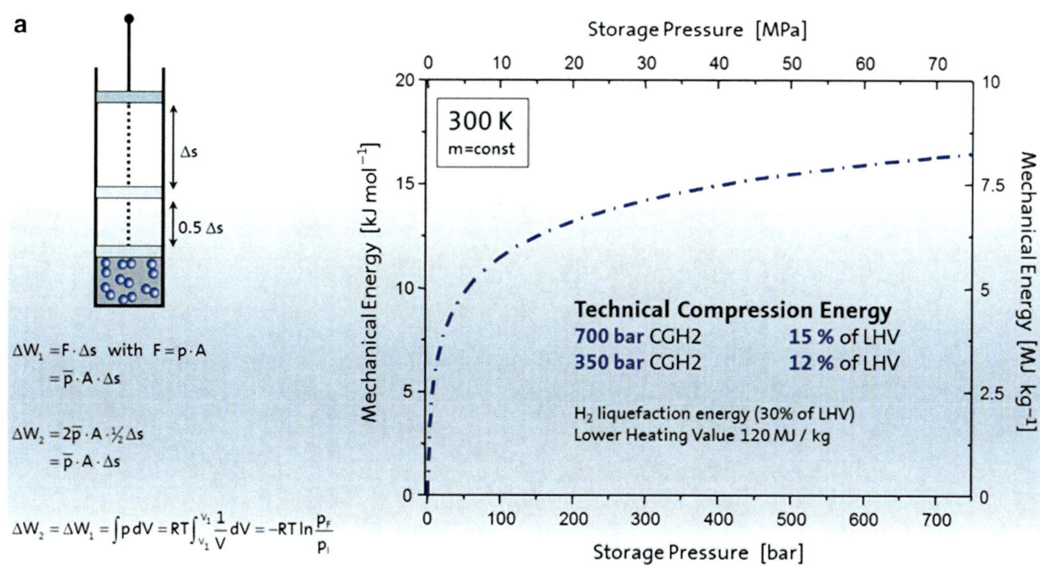

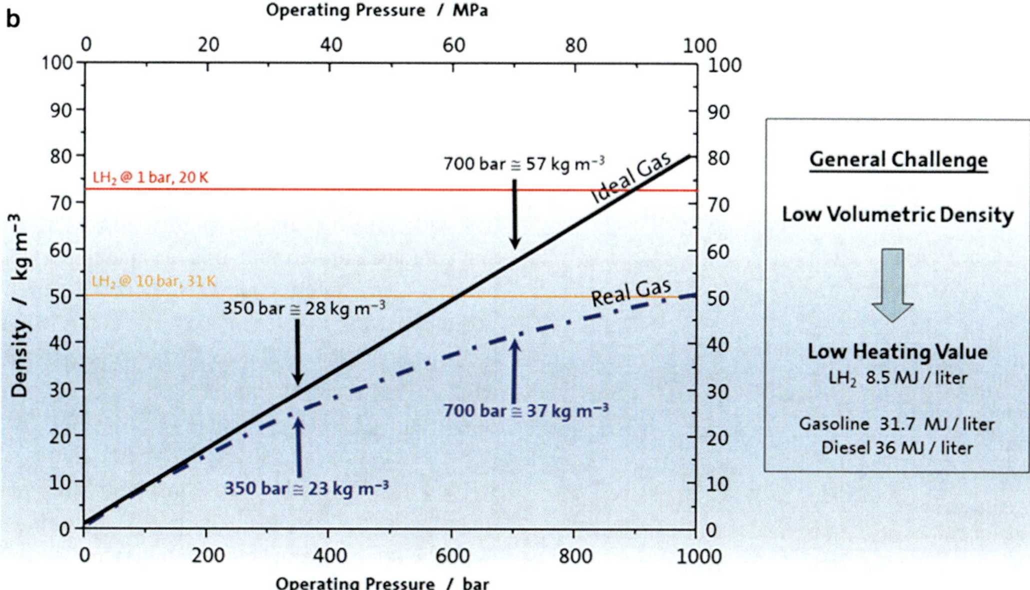

Compressed and Liquid Hydrogen for Fuel Cell Vehicles, Fig. 2 (**a**) CGH2, technical compression energy; (**b**) Gas density versus pressure level

cylinders, it follows that increasing the vessel diameter leads to an overproportional increase in wall thickness. In order to increase the volume stored inside a hydrogen system, extending the vessel length instead of diameter is the better design choice. In any case, high-strength materials, such as carbon composites, are required to reach a viable hydrogen storage density. Other design options would lead to a greater wall thickness. Therefore, small- to mid-scale pressure vessels make the most sense, for example, a three-vessel carbon-composite unit to store 4.2 kg of hydrogen at 70 MPa weighs about 135 kg (the weight of a similar steel system would be 600 kg). A schematic drawing of a typical vessel design can be found in Fig. 3.

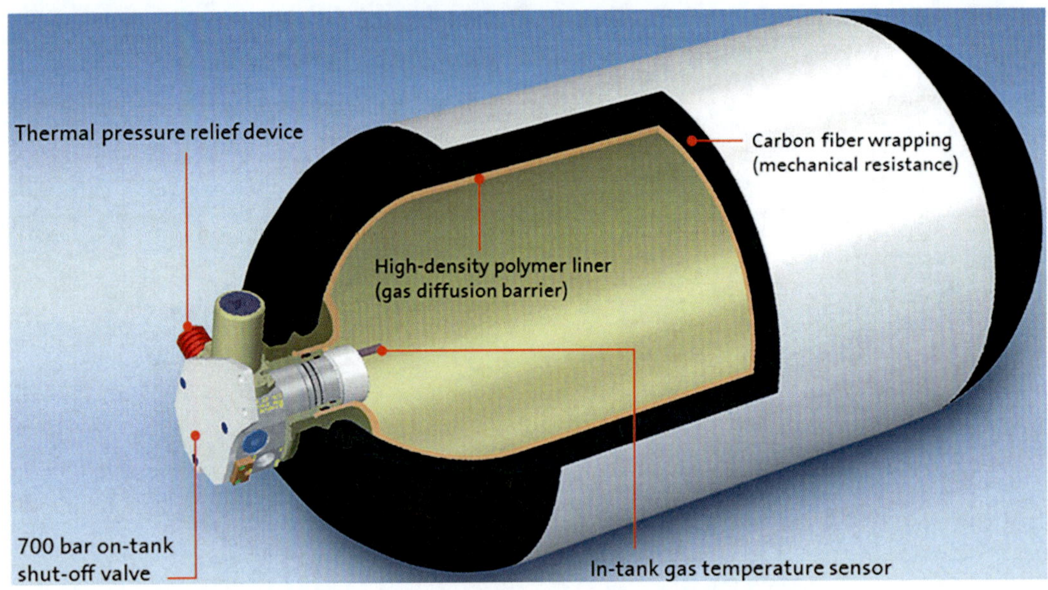

Compressed and Liquid Hydrogen for Fuel Cell Vehicles, Fig. 3 70 MPa CGH2 vessel (type IV)

Compressed and Liquid Hydrogen for Fuel Cell Vehicles, Table 1 System properties of a benchmark 70 MPa CGH2 hydrogen storage system

Capacity	6 kg H2
Volumetric energy density	260 L, 0.023 kg L^{-1}
Gravimetric energy density	125 kg, 0.048 kg kg^{-1}
Shape	Cylindrical
Production cost at large volumes	3,600 $
Boil-off losses	Not existing
Permeation rate	<1 Ncm^3/L/h
Extraction efficiency	97 %
Max. extraction rate	2 g H2 s^{-1}
Refilling time	3–5 min
Refilling efficiency	98 %
Heat exchanger capability	0 kW
Operating temperature for tank components	−60 °C and +85 °C
Operating lifetime	>10 years

The energy densities for the optimal hydrogen tank comprising a single vessel correspond to values of about 0.048 kg H2 per kg tank weight and 0.023 kg H2 per liter tank volume (Ref. [1] and Table 1). Together with the requirement of a cylindrical design (caused by the large operating pressures of about 35–70 MPa), the integration of such a tank into existing car architectures remains an important challenge. But despite all volumetric limitations of the CGH2 technology, this option yields the best overall technical performance to date and shows the highest maturity for automotive applications. Furthermore, it is feasible to refill an empty CGH2 system completely within 3–5 min. For this reason, this technology is established as the benchmark for all competing conventional and alternative storage systems [1, 2].

Currently, two types of CGH2 vessels are under discussion: the so-called type III and type IV tanks. A type III vessel consists of an inner metallic liner and sheets of carbon-fiber composites. The liner prevents hydrogen permeation and the composite is required for the necessary mechanical strength and stability. In the case of a type IV vessel, the metallic liner is replaced by a plastic one, typically high-density polyethylene (HD-PET). Most automotive companies prefer type IV vessels since they provide major advantages versus type III cylinders:

- 20 % lower weight with identical volumetric storage density
- Higher potential regarding long-term fatigue and durability (virtually no liner cracking)
- Elasticity allows for using lower-modulus and thus lower-cost carbon fibers.

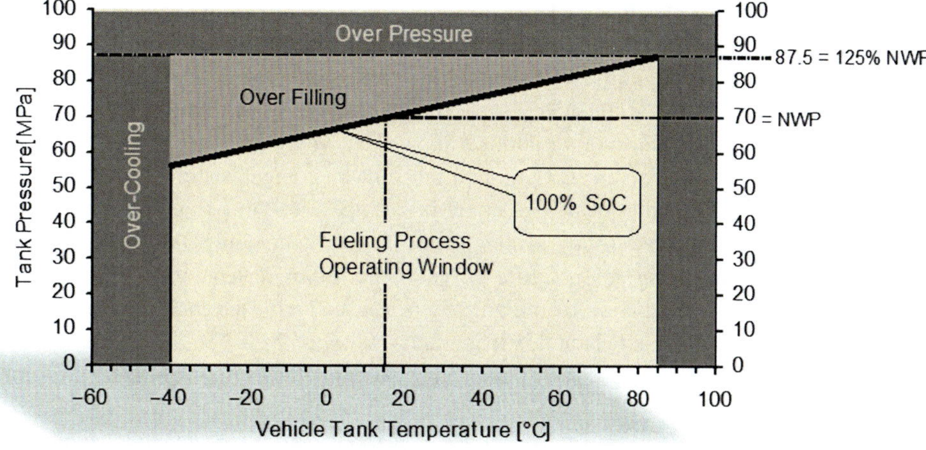

Compressed and Liquid Hydrogen for Fuel Cell Vehicles, Fig. 4 Tank pressure and state of charge (SoC) versus tank temperature for a 70 MPa CGH2 tank system

Tank cost is dominated by the vessel material, and even at a production level of only 10,000 tank systems per year, type IV tanks are by a factor of nearly 2.4 more cost-efficient, compared to type III technology. This feature is also projected to be maintained when future developments and increasing production volumes are considered. Industry estimates show that using type IV technology and considering high-volume production, a 6 kg automotive H2 system could be produced for circa US$ 3,600 [1]. A good review on carbon-composite storage vessel technology, as well as on the production and validation processes, is to be found in Ref. [5].

"Type IV" tanks combine the carbon-fiber pressure vessel and the plastic liner with metallic components mounted to the vessel via a "boss" adaptor. Many system components, such as valves, pressure relief devices, and fittings, contain metallic parts. Typical pressure–temperature combinations for a 70 MPa hydrogen tank (as specified in the "SAE J 2601" document) during fuelling [6] are shown in Fig. 4.

Figure 4 also shows "forbidden zones," to be avoided because of overpressure (at certain temperatures and pressures a maximum pressure is exceeded), or over-filling (at certain temperature and pressure levels a maximum hydrogen content is exceeded). For example, the maximum tank temperature of +85 °C at 87.5 MPa is reached after a fast fill due to the compression heat of the hydrogen gas, a value about 25 % greater than the nominal working pressure (NWP) of 70 MPa [6]. Measurements showed that this temperature decreases to ambient temperature (defined as −40 °C to + 40 °C for automotive applications) within minutes. When driving a hydrogen vehicle at full speed in cold climate at an ambient temperature of −40 °C, the tank temperature can decrease to −80 °C, which is seen as the lower limit of the tank operating conditions. Furthermore, hydrogen environment embrittlement (HEE) is possible in all subsystems where hydrogen is in direct contact with metallic materials, i.e., the hydrogen tank and the hydrogen loop of the fuel cell, the so-called anode subsystem. In general, materials used for H2 tanks are at much more risk to HEE since in this environment, the hydrogen purity is very high (better than 99.99 % as specified in ISO 14687-2), the pressure is very high (up to 87.5 MPa), and the temperature is around room temperature or lower for most of the operating times [6]. Therefore, the metallic materials need to be tested carefully under this wide operating regime in order to be qualified as viable for 70 MPa storage devices.

Liquid Hydrogen Systems (LH2)

Until circa 5–10 years ago, liquid hydrogen tanks (LH2) have also been considered to be a technology viable for automotive application. But its drawbacks concerning an efficient thermal insulation (see Fig. 5) could not be overcome in order to obtain a competitive storage system. Due to the low LH2 operating temperature of in-between 20 and 30 K compared to ambient temperature levels, an unavoidable heat flux from the outside takes place. There are three major components [1, 2]:

1. Thermal conduction
2. Convection
3. Thermal radiation.

Among those effects, points 1 and 3 are dominant, namely, the thermal conduction through mountings, pipes, and cables to the inner storage vessel and the heat radiation from the environment to the cryogenic liquid. To achieve these minimum low overall heat transport values as mentioned above, it is decisive to again work with cylindrical tank structures (compare to CGH2). This geometry is very close to the optimal surface-to-volume ratio, only beaten again by completely spherical structures. Furthermore, it is required to install a highly efficient multilayer vacuum super insulation, consisting of circa 40 layers of metal foil [1]. The wrapping procedure of the foils around the inner storage vessel in general, and around the dome areas in particular, is time-consuming and highly demanding. In addition, this process is further complicated by the in- and outlets for H2, respectively, the mountings. The remaining heat inflow of 2–3 W leads to an enhanced LH2 evaporation that eventually causes a pressure rise inside the pressure vessel. When a system pressure of about 1 MPa is reached, a valve has to be opened to vent some of the hydrogen gas to keep the system pressure below that maximum value. The typical time period between putting the vehicle into an idle or parking mode and the venting process is known as "dormancy": Values for the length of this period are typical of the order of magnitude of 1 or 2 days. After the dormancy period is exceeded, hydrogen is continuously lost to the environment via the venting valve. The lost hydrogen is known as boil-off gas [1]. The amount of the boil-off gas can be simply calculated

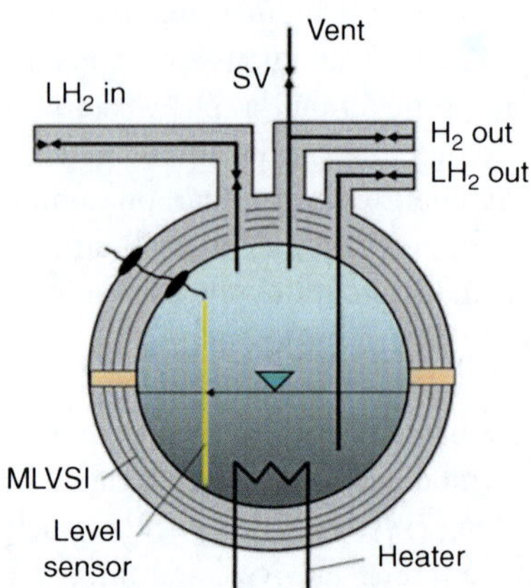

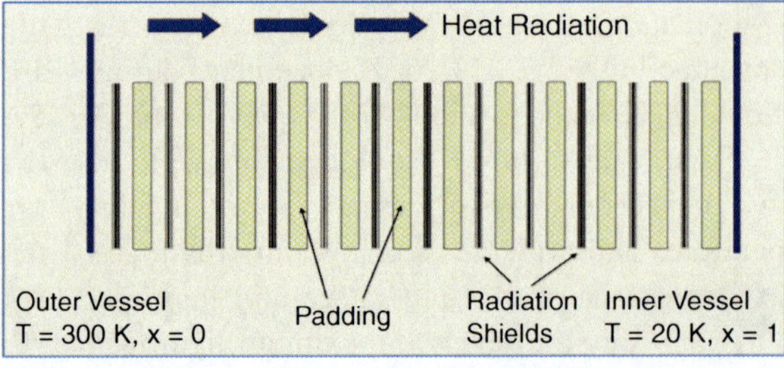

Compressed and Liquid Hydrogen for Fuel Cell Vehicles, Fig. 5 Schematic drawing of a liquid hydrogen tank system and its multilayer vacuum super insulation (MLVSI)

as a first-order-estimate by multiplying the heat flow with time period and dividing this number by the H2 heat of evaporation of circa 0.45 MJ/kg [1].

The cooling-down losses which occur during refueling (caused by hydrogen evaporation) is a related challenge: Pipes, dispensers, nozzles, and valves have to be cooled down to cryogenic temperatures before a substantial mass of LH2 can be filled into the inner storage vessel.

In combination, both these effects lead to an unacceptable amount of hydrogen losses, regardless whether the end customer or the infrastructure operator is concerned. The LH2 storage system complexity in addition to the cost- and engineering-intensive efforts to reduce the boil-off losses as much as possible leads to total system costs that are not competitive with CGH2 systems. In addition, the energy needed to liquefy hydrogen consumes 30 % of the stored chemical energy compared to just 15 % for 70 MPa CGH2 (12 % for 35 MPa CGH2), based on the H2 lower heating value of 120 MJ per kg of hydrogen [1, 2]. Furthermore, the volumetric storage density of LH2 tanks is only slightly higher compared to CGH2 systems (compare Table 1). There is not a large advantage in vehicle packaging of a LH2 tank versus a CGH2 tank, so CGH2 was evaluated by most car companies to be the superior technology.

Future Directions

All novel concepts have to beat the CGH2 figures in Table 1 in most of the categories. In principal there are three key boundary conditions for any novel storage concept [1]:

1. Limitations of available space (due to vehicle packaging needs, especially when utilizing existing architectures)
2. Technical operating parameters defined by fuel cell propulsion system requirements (e.g. hydrogen extraction rate, supply pressure, and temperature)
3. Customer demands (e.g., total cost of ownership, overall capacity, refueling time, and efficiency).

For example, solid state absorbers of hydrogen offer an impressive volumetric hydrogen density on a materials basis [2]. But unfortunately, this is only an incomplete description. Due to the required short refueling times of 3–5 min, a substantial increase in system complexity and cost emerges. Assuming a 6 kg H2 tank system and a storage material M with a heat of formation delta H of 25 MJ per kg H2, a thermal load of 150 MJ needs to be compensated during a refilling procedure:

$$M + H2 \rightarrow MH2 + Heat \tag{3}$$

In such a case, an average heat exchanger power of at least 800 kW would be required [1]. Such a huge heat exchanger is not viable to be installed onboard a vehicle since cost, volume, and weight would be prohibitive. Usually, only values substantially lower than 100 kW would be reasonable for an automotive application. In the opposite case, just to enable a H2 supply rate of $2\ g\ s^{-1}$ to the fuel cell system, under full throttle conditions, would lead to a heat management challenge of about 50 kW.

Many solid-state absorber systems require operating pressures of about or above 10 MPa, most at least during the refueling process but also during the storage phase. Therefore a pressure vessel is still needed. Furthermore, the operating temperature of a solid-state absorber has to be limited to a maximum of 70 °C since this is the waste heat level of a typical automotive PEM fuel cell. In order to obtain even higher temperatures to supply the heat of desorption and to reach the required kinetics, hydrogen would have to be either converted into electricity or be burned directly [1].

This amount of hydrogen could not be used for propulsion purposes; the effective H2 capacity and therefore also the effective range of the vehicle would be lower. Hydrogen absorbers often consist of powder materials: It is thus needed to evaluate whether the apparent density of the absorber is comparable to the crystal density of a compact block of the original compound. All these points lead to an empirical rule of thumb that roughly describes the correlation between the materials value and the respective maximum systems value: The solid-state absorber can be attributed to about 50 % of the total weight, whereas the remaining 50 % relate to the additional system components (such as pressure vessel, heat exchanger, valves, and pipes) [1]. In most realizations of proof-of-concept tanks, the mass percentage of auxiliary components even amounted to higher values than 50 %.

Consequently, future directions for materials research are:

1. Heat of formation has to be reduced as low as thermodynamically possible.
2. Operating temperature should be less than 70 °C.
3. Operating pressure should be limited to values less than 5 MPa for cryogenic or elevated temperature (up to 70 °C) operation modes.
4. Operating pressure should be less than 35 MPa for room-temperature applications using low-ΔH hydrides.

Using, e.g., the fourth path, a so called hybrid approach by combining high-pressure technology and conventional hydrides, the volumetric storage density of a 70 MPa CGH2 system could already be achieved at an operating pressure of about 35 MPa. Such an approach would simplify the vehicle packaging and infrastructure challenges substantially in the case of a future discovery of an appropriate solid state absorber [7]. Unfortunately, on the other hand, the CGH2 system gets more complex by integrating the storage compound and a heat exchanger into the pressure vessel. Higher technology complexity leads to considerably greater system cost and weight numbers compared to our defined benchmark systems (see Table 1).

A competing technology to the points above is the decomposition of hydrogen-rich, but non-reversible, compounds (such as sodium borohydride or ammonia borane) or liquid organic hydrogen carriers (e.g., cyclohexane/benzene or the closely related carbazole approach; see Ref. [2] for a general review). These systems are however not considered to be a viable alternative for the automotive industry: The inherent complex onboard system design (inter alia, concerning the treatment of waste materials), the infrastructure implications (e.g., related to the need for exchangeable fuel cartridges or the physicochemical properties of corresponding materials, resp., the required solvents), and the off-board recycling, respectively, energy issues are prohibitive for an automotive application. But there might be applications in the stationary field.

Pathways for further improvement of CGH2 technology include the development of novel fiber and resin concepts, cost-efficient tank system manufacturing processes, new liner materials with improved permeation properties or even liner-less tank concepts, improved high-pressure and low-temperature capable sealing materials, and improved metallic materials less vulnerable to hydrogen embrittlement. Additionally, work is ongoing to develop and improve the refueling technology and recycling concepts for carbon-composite vessels. Also coordinated activities between carmakers, the energy industry, technology companies, and governments to set up a 70 MPa CGH2 infrastructure are currently ongoing, e.g., in Japan or in Germany as part of the National Innovation Program and the H2Mobility initiative [8, 9].

As of October 2012, it therefore can be concluded that 70 MPa CGH2 systems are the best option available for automotive onboard hydrogen storage [1, 5, 8, 9] and offer significant room for improvement, especially on the cost level.

Cross-References

▶ Fuel Cell Vehicles
▶ Hydrogen Storage Materials (Solid) for Fuel Cell Vehicles

References

1. von Helmolt R, Eberle U (2007) Fuel cell vehicles: status 2007. J Power Sources 165:833–843
2. Eberle U, Felderhoff M, Schüth F (2009) Chemical and physical solutions for hydrogen storage. Angew Chem Int Ed 48:6608–6630
3. DoE hydrogen storage targets. http://www1.eere.energy.gov/hydrogenandfuelcells/storage/pdfs/targets_onboard_hydro_storage.pdf. Accessed Oct 2012

4. Zhang J, Fisher TS, Ramachandran PV, Gore JP, Mudawar I (2005) A review of heat transfer issues in hydrogen storage technologies. J Heat Transf 127:1391–1399
5. Sirosh N, Niedzwiecki A (2008) Development of storage tanks. In: Leon A (ed) Hydrogen technology. Springer, Heidelberg, pp 291–310. doi:10.1007/978-3-540-69925-5_10
6. Michler T, Lindner M, Eberle U, Meusinger J (2012) Assessing hydrogen embrittlement in automotive hydrogen tanks. In: Gangloff RP, Somerday BP (eds) Gaseous hydrogen embrittlement of materials in energy technologies: the problem, its characterisation and effects on particular alloy classes, vol 1. Woodhead Publishing, Cambridge, pp 94–125. ISBN 13: 978-1-84569-677-1
7. Weidenthaler C, Felderhoff M (2011) Hydrogen storage for mobile applications – quo vadis? Energy Environ Sci 4:2495–2502
8. Eberle U, von Helmolt R (2010) Energiespeicher Wasserstoff: Auf dem Weg zur Kommerzialisierung, Automobil Industrie, pp 52–55; also available in html-format. http://www.e-auto-industrie.de/energie/articles/295843/. Accessed Oct 2012
9. H2Mobility study (2012) A portfolio of power-trains for Europe: a fact-based analysis. www.zeroemission-vehicles.eu. Accessed Oct 2012

Conductive Polymers, Immobilization of Macromolecular Bio-Entities

Serge Cosnier, Michael Holzinger and Alan Le Goff
Department of Molecular Chemistry, CNRS UMR 5250, CNRS-University of Grenoble, Grenoble, France

Introduction

The immobilization of a macromolecular biomolecule on conductive surfaces is the subject of increasing research efforts for the development of electrochemical biosensors and biofuel cells. The main motivation for the functionalization of the surface of a conventional electrode is to confer molecular recognition properties or selective catalytic activity to this electrode.

An electrochemical biosensor is conventionally described as a self-contained, integrated receptor-transducer device, capable of providing selective and even quantitative analytical information, using a biological recognition macromolecule in intimate contact with the electrode. Biosensors were thus successfully applied for the detection of a wide range of environmental contaminants in water and air and have also proved their usefulness in clinical diagnosis and food analysis. Biosensors may also be an attractive tool for rapid and sensitive on-field monitoring of viruses that are among the most important causes of human disease and are of increasing concern as possible agents of bio-warfare and bioterrorism. The biomolecular recognition, ensuring the selectivity of the sensors, is obtained using five main biological compounds: whole cells, nucleic acids, immunological molecules, enzymes, and aptamers. The electrochemical transduction of the biomolecular recognition process is measured by different techniques including potentiometric, amperometric, conductometric, and impedimetric methods [1, 2]. Concerning the biofuel cells, these are mainly composed of microbial fuel cells (whole organism) and enzymatic fuel cells (enzymes) according to the biocatalyst immobilized on the cathode and anode [3, 4]. The operating mode of biofuel cells consists of combining an oxidation reaction supplying electrons at the anode with a reduction reaction consuming electrons at the cathode. This allows the conversion of chemical energy to electrical energy.

The ideal immobilization method should place the biological layer in close proximity to the electrode, but should also maintain its biological activity after its immobilization at the interface. In order to achieve a high efficiency of transduction, or the catalytic process, the immobilization procedures should lead to a high density of biomolecules on the electrode surface and even establish an electrical communication with these biomolecules.

Among the various methods employed for the biofunctionalization of electrodes, the electropolymerization of conducting polymers emerged three decades ago to be an essential and particular efficient way of making bioelectrodes. These electrogenerated polymers result from a monomer electrochemical reduction or oxidation that induces the formation of a polymer with a conjugated π-system. For instance, monomers such as pyrrole, thiophene, or aniline undergo an oxidative electropolymerization to form polypyrrole, polythiophene, and polyaniline. Apart from their important application in optoelectronics, electrogenerated polymers constitute a powerful platform for the development of chemical or biological sensors. Intrinsically conducting polymers, redox polymers, and even insulating films have aroused widespread attention considering the potential technological applications that offer these materials by combining the properties of plastics with the electric, magnetic, and optical behavior of metals, semiconductors, and organometallic complexes. The electropolymerization process is commonly used in constructing biosensors and more recently biofuel cells because it consists in a one-step procedure that is simple in preparation. Moreover, electropolymerization leads to the formation of polymer coating over electrodes whatever their size and geometry. In addition, the resulting electropolymerized films are highly robust during an operation in both aqueous and organic media.

The main strategies employed for biomolecule immobilization by electropolymerized films encompass entrapment within the polymer during its electrochemical growth, covalent binding onto polymers and noncovalent binding by specific affinity, or host-guest interactions between films and biomolecules.

Procedures of Biomolecule Immobilization Using Electrogenerated Polymers

Entrapment During the Electropolymerization Process

Twenty-five years ago, a straightforward procedure of biomolecule immobilization based on the entrapment of proteins in polymer films during their electrogeneration on electrode was, for the first time, simultaneously described by Foulds et al. [5] and Umaña et al. [6]. Entrapment in electropolymerized films remains today one of the most popular electrochemical immobilization procedures for proteins, oligonucleotides, and microorganisms. This method consists in the application of an appropriate potential to the electrode placed in aqueous media containing both macromolecular biomolecules and monomers. Biomolecules present in the immediate vicinity of the electrode surface are thus incorporated in the growing polymer. This entrapment process theoretically occurs without chemical reaction between the "in situ" formed polymers and the biomolecules, preserving thus their biological activity.

The advantage of electrochemical entrapment lies in the simplicity and speed of this one-step procedure and the commercial availability of the monomers. Furthermore, this method enables precise control of the thickness of the polymer layer via the monitoring of the electric charge used for the polymerization process. Therefore, this allows the modulation of the amount of entrapped biomolecules. It should be stressed that the entrapment procedure was obviously restricted to monomers which are soluble in aqueous solutions. Consequently, the concept of enzyme wiring is limited by the lack of solubility of monomers functionalized by hydrophobic redox mediators. The biomolecule entrapment can also be performed after the electropolymerization step by electro-induced doping process. For instance, an immunosensor for ochratoxin, a potential carcinogen produced by filamentous fungi, was fabricated by doping electrochemically an electrogenerated polyaniline/poly(vinylsulfonic acid) film with the corresponding anti-mycotoxin antibody [7]. The impedimetric response of the resulting electrode led to a detection limit of ochratoxin (10 pg/kg) which is markedly more sensitive than the limits required by the European Commission, namely, 5 mg/kg.

Another elegant two-step procedure for biomolecule entrapment is based on the specific properties of pyrrolic surfactants in their adsorbed state. The electrochemical entrapment of biomolecules consisted first, in the adsorption of an aqueous mixture of monomer and biomolecule on the electrode surface. Second, the subsequent electropolymerization of the adsorbed monomers in an aqueous electrolyte, without biomolecules and monomers, induced the physical entrapment of the adsorbed biomolecule in the "in situ" generated polypyrrole films [8].

Covalent Binding Between Polymers and Biomolecules by Chemical Reactions

Initially, this two-step strategy consisted in the electrogeneration of polymer films bearing amino or carboxylate groups, followed by their chemical activation in order to bind the biomolecule at the polymer-solution interface via the formation of an amide linkage [9]. As a consequence, electropolymerization and covalent binding can be performed in optimal conditions for each step. In particular, the electropolymerization process can be carried out in organic solvents and at high voltages that are harmful for biological compounds. Nevertheless, the presence of additional chemical reagents in aqueous solution such as water-soluble carbodiimines promoting the biomolecule attachment may partly denature the protein and lead to incomplete functionalization of the electrode surface. The activation and incubation time, indeed, requires several hours, and the maximum amount of immobilized biomolecule is limited to the formation of a biomolecule monolayer. Moreover, several linkages may be created between films and biomolecules, thus blocking their conformational mobility and hence decreasing their specific activity. Nevertheless, the chemical grafting of DNA probes functionalized by an end-terminal amino group remains a valuable approach for the design of DNA sensors. Recently, an amino-functionalized DNA was covalently grafted onto copolymer films of pyrrole and pyrrolylacrylic acid via carbodiimine activation. The resulting microelectrodes were successfully applied to the labeless impedimetric detection of ssDNA (single-stranded DNA) target with an attractive detection limit of 10 pM [10].

The second generation of polymers dedicated to the chemical grafting of biomolecules consists in films bearing activated ester groups such as *N*-hydroxysuccinimide, *N*-hydroxyphthalimide, or pentafluorophenyl esters (Figure 1). The use of such activated ester groups allows biomolecule grafting without further chemical reagent in solution. The main advantage of the covalent binding lies in the accessibility to the immobilized biomolecule. This procedure is thus currently used to produce electrochemical immunosensors or DNA or aptamer sensors.

More recently, the covalent binding of proteins onto polymers was attempted by irradiation of electrogenerated photoactivable polypyrrole films in presence of proteins. For instance, an optical fiber modified with a poly(pyrrole-benzophenone) film allowed the photochemical linking of biological receptors for the detection of many different viral infections, e.g., Ebola [11] or hepatitis C [12]. The photoreaction process for benzophenone involved a triplet-state excitation, followed by a radical recombination by hydrogen transfer forming the covalent bond with proteins at an amino acid unit with accessible C–H bonds. This photografting procedure was successfully applied to the production of biochips and optical immunosensors that may be used in the future as an important clinical tool or a field operative tool for the screening of viral infection.

Immobilization by Specific Non-Covalent Interactions Between Polymers and Biomolecules

The anchoring of biomolecules on electropolymerized films can be achieved by affinity interactions via the formation of a single attachment point between both entities. In contrast to the chemical grafting, this approach may better preserve the biological activity and provide a specific orientation of the biomolecule previously modified by genetic engineering.

Among the different non-covalent binding systems involving electropolymerized films, the main approach was based on the strong associations between biotin (a vitamin) and avidin (a protein) displaying four binding sites for biotin [13]. The biomolecule anchoring was easily achieved via the formation of avidin bridges between biotinylated biomolecules and

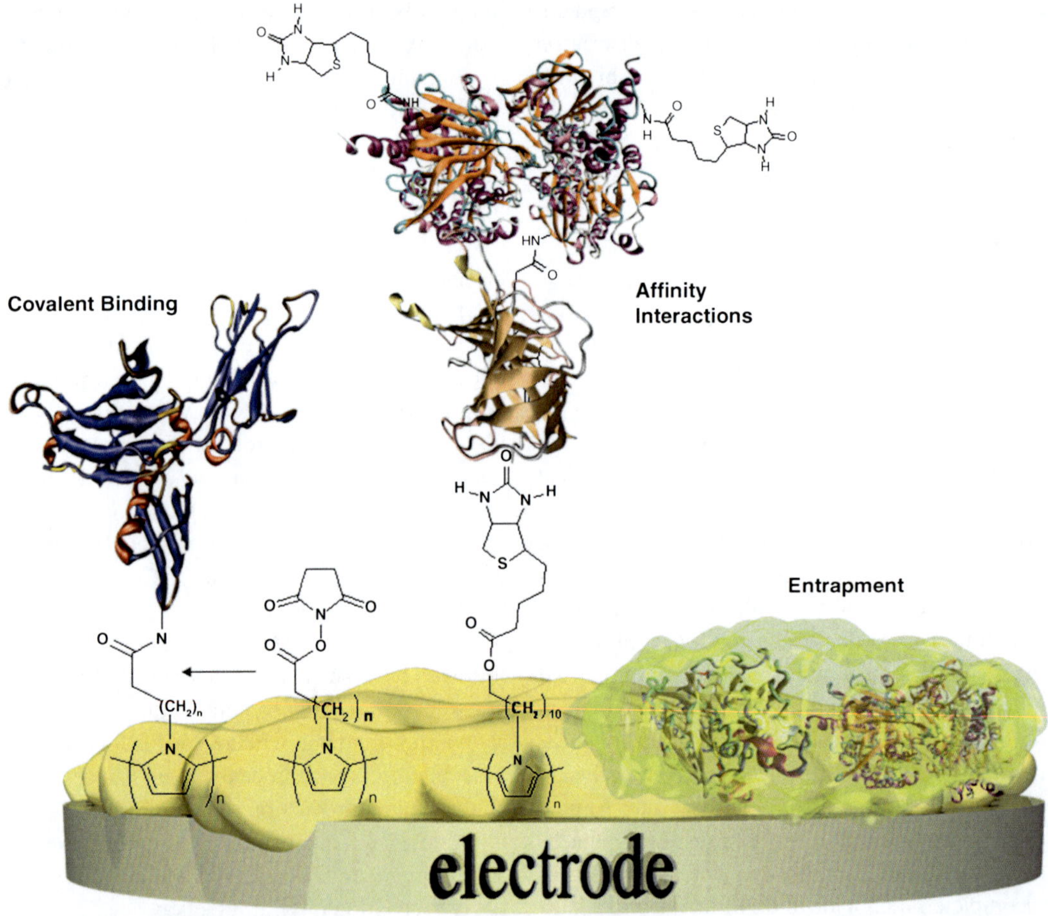

Conductive Polymers, Immobilization of Macromolecular Bio-Entities, Fig. 1 Most common immobilization methods for biomolecules using functional conductive polymers

biotinylated polymers (Figure 1). Although different electropolymerizable groups such as aniline, phenol, or carbazole were investigated, an overwhelming majority of biotinylated films are based on the polypyrrole skeleton. For instance, avidin-biotin interactions were used to modify optical fibers with biotinylated cholera toxin B molecules, for the construction of an immunosensor to detect cholera antitoxin antibodies [14]. After the immunoreaction, the target (anti-cholera toxin antibody) was transduced by a peroxidase-labeled secondary antibody conjugate. In presence of enzyme substrates (H_2O_2 and luminol), the enzymatic marker catalyzes a chemiluminescent process leading to an efficient detection limit of antibody titer of 1:1,200,000.

Owing to the geometrical repartition of the docking sites for biotin of avidin, avidin-biotin interactions can be used to build protein multilayers by successive depositions of avidin and biotinylated proteins on biotinylated polymers. This opportunity may be exploited to combine complementary biological activities.

In parallel, various affinity systems were also investigated such as the specific interactions between maltose-binding protein and positively charged polypyrrole film or a metal affinity immobilization concept based on the coordination of polyhistidine-tagged proteins or DNA to metal centers such as Ni^{2+} or Cu^{2+} chelated by polypyrrole films containing carboxylate, imidazole, or nitrilotriacetic acid groups (NTA) [15].

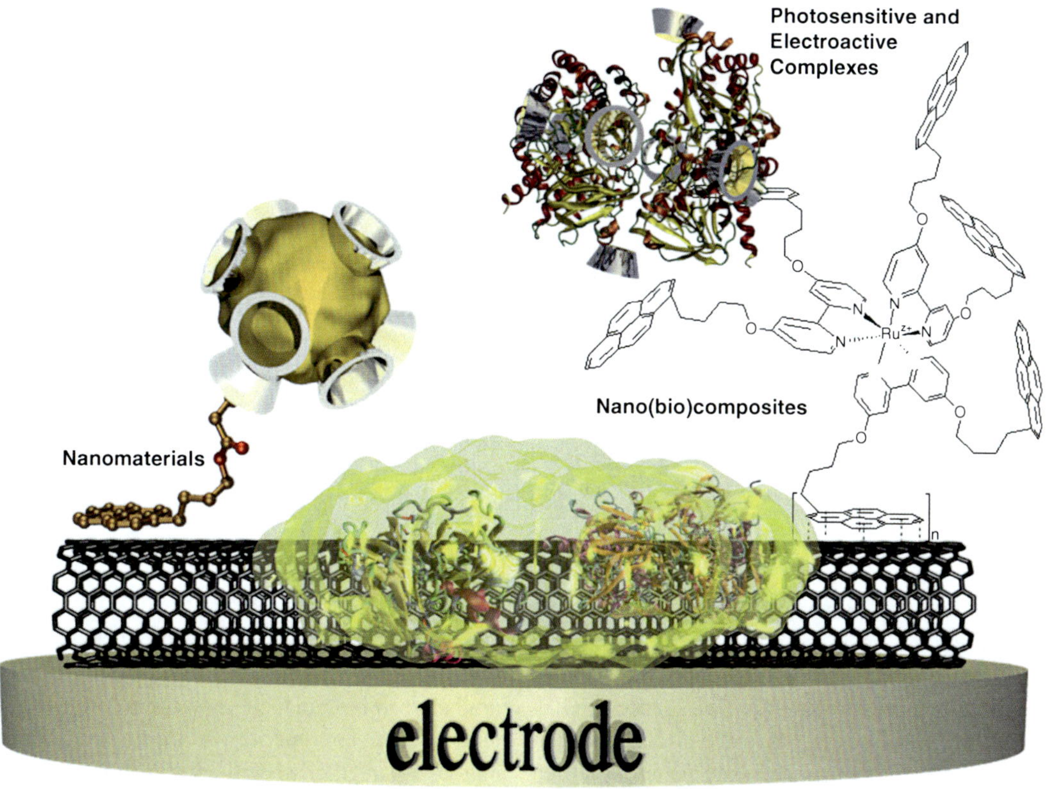

Conductive Polymers, Immobilization of Macromolecular Bio-Entities, Fig. 2 New generation of nanomaterials and photoelectrochemical compounds as interface or matrix for biomolecular entities

The fabrication of DNA sensors was thus carried out by the electropolymerization of the chelating pyrrole-NTA ligand followed by the successive coordination of Cu^{2+} and ssDNA modified by a polyhistidine tag [16]. Then, the labeless impedimetric detection of a ssDNA from the human immunodeficiency virus was performed via its hybridization onto the polypyrrole surface. The resulting detection limit (10^{-15} mol L^{-1}) was markedly better than those currently recorded for impedimetric DNA sensors.

Recently, the original possibility to coordinate biotinylated compounds on copper-poly (pyrrole-NTA) films was reported [17]. This approach combines the advantages of the affinity systems biotin/(strept)avidin and NTA/Cu^{2+}/histidine, eliminating the disadvantages of each system, and thus constitutes a real advance in the field of non-covalent immobilization. Usually, the immobilization of biotin-tagged biomolecules onto a biotinylated polymer requires a layer of avidin bridging the two entities. Combined with the poor permeability of biotinylated polymers, this additional protein layer reduces in many cases the electrochemical intensity of the molecular recognition event. Moreover, histidine-tagged compounds have to be produced locally at lab scale. This supplementary layer is not necessary with the NTA Cu^{2+}/biotin system, and a wide range of biotinylated biomolecules are commercially available.

The formation of a supramolecular inclusion complex by host-guest associations between biomolecules or nanoparticles bearing β-cyclodextrin groups and polymerized adamantane groups was used to develop new biointerfaces. The electropolymerization of an adamantane-pyrrole derivative was thus efficiently applied to the direct anchoring of β-cyclodextrin-modified glucose oxidase or to the indirect anchoring of adamantane-tagged glucose oxidase via an intermediate layer of β-cyclodextrin-modified gold nanoparticles [18]. In addition, the possibility to form an inclusion complex between β-cyclodextrin and pyrene derivative was recently used for the specific attachment of β-cyclodextrin-tagged glucose oxidase onto an electropolymerized film of a tris (bipyridine) iron(II) complex bearing six pyrene groups [19] (Fig. 2).

Future Directions

During the last two decades, biosensor and biofuel cell research progressed, thanks to the availability of numerous innovative material technologies, immobilization methods, and the development of new transducer methodologies such as electrochemiluminescence (ECL). Indeed, the introduction of a coordination complex to the polymer backbone is aimed at designing metallopolymers with multiple functionalities such as novel signal transducing abilities or electron-transfer properties. In particular, the emergence of new electrogenerated polymers functionalized by photosensitive complexes displaying ECL properties in water such as ruthenium or iridium complexes [20] may be at the origin of a new generation of electrooptical affinity sensors operating without labeling step of the target. The recent combination of electrochemical and optical biosensor fields via the development of optically transparent conductive substrates constitutes a new exciting area of research. The use of transparent conductive substrates has thus opened the way for the functionalization of optical fibers by electropolymerized films. Moreover, a combination of electrochemical and optical measurement methodologies will permit the simultaneous detection of different analytes or the detection of the same target by two different methods with a wider dynamic range and an increased control over the sensing environment.

The continuous development in nanotechnologies, in particular, the emergence of novel nanomaterials, will help in the future to obtain electrochemical biosensors with higher sensitivity and biofuel cells with increased power. The important potential of nano-objects consists in their high specific surface and conductivity [21]. Indeed, one of the main limitations of electrochemical biosensors and biofuel cells lies in the two-dimensional configuration of the biomaterials created on the electrode surface. Significant efforts are thus focalized on the development of highly porous conductive three-dimensional nanostructured bioarchitectures.

Thanks to their impressive properties (intrinsic conductivity and high specific surface of theoretically 1,000 m^2 g^{-1}), carbon nanotubes are widely used as building blocks or templates for the fabrication of biomaterials based on electropolymerized films. Moreover, carbon nanotubes may assume the direct electrical wiring between biomolecules and the bulk electrode due to the appropriate geometry of carbon nanotubes and their specific hydrophobic interactions with some regions of the proteins [22]. Among the wide range of existing functionalization methods for carbon nanotubes to attach biomolecules, a simple, soft, and fast strategy consists in the use of non-covalent π-stacking interactions between pyrene derivatives and the nanotube sidewalls. The modification of pyrenes with different affinity systems or redox groups allowed thus the non-covalent anchoring of biomolecules or their electrical wiring, respectively.

Recently, the electropolymerization of pyrene derivatives, previously adsorbed onto carbon nanotubes by dip coating, was reported as an efficient procedure to reinforce the mechanical stability of the π-stacked compounds [23]. This approach allows the formation of ultrathin polymer films over the whole surface of the nanotube structures immobilized on the electrode.

In parallel, the development of carbon-coated metal nanoparticles also appeared as a promising class of nanomaterials for promoting electron-transfer reactions in bioelectrochemical processes. For instance, carbon-coated nickel nanoparticles were adsorbed on an electrode surface and stabilized by electropolymerization of phenylenediamine. The resulting nanostructured polymer was then applied to the electrostatic immobilization of bovine serum albumin and the detection of papaverine, a natural opium alkaloid [24].

References

1. Cosnier S, Holzinger M (2011) Electrosynthesized polymers for biosensing. Chem Soc Rev 40:2146–2156. doi:10.1039/c0cs00090f
2. Nambiar S, Yeow JTW (2011) Conductive polymer-based sensors for biomedical applications. Biosens Bioelectron 26:1825–1832. doi:10.1016/j.bios.2010.09.046
3. Atanassov P, Apblett C, Banta S, Brozik S, Barton SC, Cooney M, Liaw BY, Mukerjee S, Minteer SD (2007) Enzymatic biofuel cells. Electrochem Soc Interfac 16:28–31
4. Osman MH, Shah AA, Walsh FC (2011) Recent progress and continuing challenges in bio-fuel cells. Part I Enzym cells Biosen Bioelectron 26:3087–3102. doi:10.1016/j.bios.2011.01.004
5. Foulds NC, Lowe CR (1986) Enzyme entrapment in electrically conducting polymers. Immobilisation of glucose oxidase in polypyrrole and its application in amperometric glucose sensors. J Chem Soc Faraday Trans 1 82:1259–1264. doi:10.1039/F19868201259
6. Umaña M, Waller J (1986) Protein-modified electrodes. The glucose oxidase/polypyrrole system. Anal Chem 58:2979–2983. doi:10.1021/ac00127a018
7. Muchindu M, Iwuoha E, Pool E, West N, Jahed N, Baker P, Waryo T, Williams A (2011) Electrochemical ochratoxin a immunosensor system developed on sulfonated polyaniline. Electroanalysis 23:122–128. doi:10.1002/elan.201000452
8. Cosnier S (1997) Electropolymerization of amphiphilic monomers for designing amperometric biosensors. Electroanalysis 9:894–902. doi:10.1002/elan.1140091206
9. Schuhmann W, Lammert R, Uhea B, Schmidt H-L (1990) Polypyrrole, a new possibility for covalent binding of oxidoreductases to electrode surfaces as a base for stable biosensors. Sensor Actuator B 1:537–541. doi:10.1016/0925-4005(90)80268-5
10. Kannan B, Williams DE, Booth MA, Travas-Sejdic J (2011) High-sensitivity, label-free DNA sensors using electrochemically active conducting polymers. Anal Chem 83:3415–3421. doi:10.1021/ac1033243
11. Petrosova A, Konry T, Cosnier S, Trakht I, Lutwama J, Rwaguma E, Chepurnov A, Mühlberger E, Lobel L, Marks RS (2007) Development of a highly sensitive, field operable biosensor for serological studies of Ebola virus in central Africa. Sensor Actuator B Chem 122:578–586. doi:10.1016/j.snb.2006.07.005
12. Konry T, Novoa A, Shemer-Avni Y, Hanuka N, Cosnier S, Lepellec A, Marks RS (2005) Optical fiber immunosensor based on a poly (pyrrole – benzophenone) film for the detection of antibodies to viral antigen. Anal Chem 77:1771–1779. doi:10.1021/ac048569w
13. Cosnier S, Galland B, Gondran C, Pellec AL (1998) Electrogeneration of biotinylated functionalized polypyrroles for the simple immobilization of enzymes. Electroanalysis 10:808–813. doi:1040-0397/98/1209-0808
14. Barton AC, Davis F, Higson SPJ (2008) Labeless immunosensor assay for the stroke marker protein neuron specific enolase based upon an alternating current impedance protocol. Anal Chem 80:9411–9416. doi:10.1021/ac801394d
15. Davis J, Glidle A, Cass AEG, Zhang J, Cooper JM (1999) Spectroscopic evaluation of protein affinity binding at polymeric biosensor films. J Am Chem Soc 121:4302–4303. doi:10.1021/ja984467+
16. Baur J, Gondran C, Holzinger M, Defrancq E, Perrot H, Cosnier S (2010) Label-free femtomolar detection of target DNA by impedimetric DNA sensor based on poly(pyrrole-nitrilotriacetic acid) film. Anal Chem 82:1066–1072. doi:10.1021/ac9024329
17. Baur J, Holzinger M, Gondran C, Cosnier S (2010) Immobilization of biotinylated biomolecules onto electropolymerized poly(pyrrole-nitrilotriacetic acid)–Cu^{2+} film. Electrochem Comm 12:1287–1290. doi:10.1016/j.elecom.2010.07.001
18. Holzinger M, Bouffier L, Villalonga R, Cosnier S (2009) Adamantane/β-cyclodextrin affinity biosensors based on single-walled carbon nanotubes. Biosens Bioelectron 24:1128–1134. doi:10.1016/j.bios.2008.06.029
19. Le Goff A, Gorgy K, Holzinger M, Haddad R, Zimmermann M, Cosnier S (2011) Tris(bispyrene-bipyridine)iron(II): a supramolecular bridge for the biofunctionalization of carbon nanotubes via p-stacking and pyrene/b-cyclodextrin host-guest interactions. Chem Eur J 17:10216–10221. doi:10.1002/chem.201101283
20. Le Goff A, Cosnier S (2011) Photocurrent generation by MWCNTs functionalized with bis-cyclometallated Ir(III)- and trisbipyridyl ruthenium (II)- polypyrrole films. J Mater Chem 21:3910–3915. doi:10.1039/C0JM03472J
21. Gruner G (2006) Carbon nanotube transistors for biosensing applications. Anal Bioanal Chem 384:322–335. doi:10.1007/s00216-005-3400-4
22. Zebda A, Gondran C, Le Goff A, Holzinger M, Cinquin P, Cosnier S (2011) Mediatorless high-power glucose biofuel cells based on compressed carbon nanotube-enzyme electrodes. Nat Commun 2:370. doi:10.1038/ncomms1365

23. Haddad R, Holzinger M, Villalonga R, Neumann A, Roots J, Maaref A, Cosnier S (2011) Pyrene-adamantane-β-cyclodextrin: an efficient host-guest system for the biofunctionalization of SWCNT electrodes. Carbon 49:2571–2578. doi:10.1016/j.carbon.2011.02.049
24. Feng L-J, Zhang X-H, Zhao D-M, Wang S-F (2011) Electrochemical studies of bovine serum albumin immobilization onto the poly-o-phenylenediamine and carbon-coated nickel composite film and its interaction with papaverine. Sens Actuator B 152:88–93. doi:10.1016/j.snb.2010.09.031

Conductivity of Electrolytes

Roland Neueder
Institute of Physical and Theoretical Chemistry, University of Regensburg, Regensburg, Germany

Electrolyte Solutions

Electrolyte solutions are electric conducting solutions of different compounds in mixed or pure solvents. The electric current in such solutions is carried out by the movement of ions, which are generated by more or less complete dissociation of the dissolved electrolyte. Aqueous electrolyte solutions can be found in numerous geological, biochemical, and technical processes. Nonaqueous electrolyte solutions are involved in various new technologies such as high-energy batteries, electrodeposition, nonemissive displays, solar cells, phase transfer catalysis, or electroorganic synthesis [1, 2].

Electrolytes can be classified as ionophores or as ionogenes, independent of the stoichiometry. Ionophores already exist as ionic crystals in their pure state (e.g., sodium chloride), whereas ionogenes form ions by a chemical reaction with solvent molecules (e.g., nitric acid). Equation 1 shows the dissolution process of nitric acid in water forming contact ion pairs (CIP), solvent separated ion pairs (SSIPs) and free ions [3]

$$\begin{aligned} HNO_3 \xrightarrow{H_2O} \left[H^+ \cdots NO_3^-\right](aq) &\rightarrow \left[H^+(H_2O)NO_3^-\right](aq) \\ &\rightarrow \left[H^+\right](aq) + \left[NO_3^-\right](aq) \end{aligned} \quad (1)$$

Dissolved ionophores are initially completely dissociated and the ions are solvated. However, association to ion pairs, triple ions, or even higher aggregates may occur, depending on solvent properties (e.g., permittivity) or electrolyte concentration. Equation 2 shows the different association steps and the different species for lithium tetra fluoroborate in dimethoxyethane (DME) as an example (solvated ions, ion pair, and triple ions) [4, 5]

$$\begin{aligned} LiBF_4(s) &\xrightarrow{DME} \left[Li^+\right] + \left[BF_4^-\right] \\ &\rightleftharpoons \left[LiBF_4\right]^{\circ} \begin{cases} +\left[BF_4^-\right] \rightarrow \left[BF_4^- Li^+ BF_4^-\right]^- \\ +\left[Li^+\right] \rightarrow \left[Li^+ BF_4^- Li^+\right]^+ \end{cases} \end{aligned} \quad (2)$$

These dissociation and association reactions usually are not complete but reach an equilibrium state, that is described by thermodynamic equilibrium constants, such as K_D (dissociation constant) or K_A (association constant).

It must be stated that for a series of simultaneous equilibria, see Eq. 2 for an example, the quantitative discussion is limited to simple cases. However, if sufficient precise data are available, even six equilibrium constants can be detected; see, for example, the analysis of conductance data of aqueous solutions of benzenehexacarboxylic acid (mellitic acid) [6].

Conductance Equation of Dilute Electrolyte Solutions

An external electric field acting on an electrolyte solution produces a particle flow that leads to electric conductance. For a completely dissociated binary salt, the cationic and anionic conductivities can be combined to yield the molar conductivity Λ which is related to the specific conductance κ according to

$$\Lambda = \lambda_+ + \lambda_- = \frac{\kappa}{c}, \quad (3)$$

whereas for a partially associated or incompletely dissociated binary electrolyte, the degree of dissociation α must be introduced in Eq. 3 to give

$$\Lambda = \alpha(\lambda_+ + \lambda_-). \tag{4}$$

Molar electrolyte conductivity for completely dissociated electrolytes are represented in the form

$$\Lambda = \Lambda^\infty - \Lambda^{rel} - \Lambda^{el}, \tag{5}$$

with Λ^∞ as the concentration independent limiting electrolyte conductance (no disturbance by other ions) and the concentration dependent relaxation (Λ^{rel}) and electrophoretic (Λ^{el}) terms. Relaxation and electrophoretic effects are generally expressed in the form of a truncated series expansion yielding for symmetrical electrolytes equations of type [7, 8]

$$\Lambda = \Lambda^\infty - S\sqrt{c} + Ec\ln c + J_1c - J_2c^{3/2}. \tag{6}$$

The coefficients S and E depend only on the ion charges, whereas coefficients J_1 and J_2 show additional dependence on the distance of closest ion approach (R) in the solution. For completely dissociated electrolytes, this distance parameter is the sum of the ionic radii, $R = a_+ + a_-$. The complete set of equations for the calculation of the coefficients of Eq. 6 in the expansion of Fuoss and Justice [9] applied to Barthel's chemical model can be found in part 1 of the Electrolyte Data Collection [8].

Taking into account that ion pairs of symmetrical electrolytes do not contribute to the electric conductivity, for partially associated or incompletely dissociated electrolytes, the concentration c must be replaced by the concentration of the free ion αc and Eq. 6 changes to

$$\Lambda = \alpha\left[\Lambda^\infty - S\sqrt{\alpha c} + E\alpha c\ln(\alpha c) + J_1\alpha c - J_2(\alpha c)^{3/2}\right]. \tag{7}$$

The association constant K_A (process: $C^+ + A^- \rightleftharpoons [CA]^\circ$) relates the degree of dissociation α to the electrolyte concentration c:

$$K_A = \frac{1-\alpha}{\alpha^2 c}\left(\frac{1}{y'_\pm}\right)^2 \tag{8}$$

$y'_\pm$ is the activity coefficient of the free ions which can be expressed with the help of the low concentration chemical model [7].

For unsymmetrical electrolytes, conductivity equations have been developed by Lee and Wheaton [10] and by Quint and Viallard [11]. The molar conductivity Λ of the solution is calculated from the ionic conductivities λ_i of all ions present according to

$$\Lambda = \sum_{i=1}^{n} \frac{|z_i| c_i \lambda_i}{c}. \tag{9}$$

The dependence of λ_i on ionic strength I is given by the Quint-Viallard conductance equation:

$$\lambda_i = \lambda_i^\infty - S_i\sqrt{I} + E_i I\ \ln I + J_{1i}I - J_{2i}I^{3/2} \tag{10}$$

Complete expressions for the coefficients of this equation can be found in Ref. [12] where the conductivity of phosphoric acid was analyzed taking into account the different dissociation equilibria. A general treatment of mono-, di-, and tribasic organic acids including multistep equilibria can be found in Ref. [13].

In the last two decades, new extended laws have been obtained for the concentration dependence of transport properties. It was possible [15, 16] to use the Fuoss-Onsager theory together with new, more accurate equilibrium pair distribution functions as obtained with the help of the hypernetted chain (HNC) or mean spherical approximation (MSA).

It is important for applied research that MSA permits the reproduction of conductivity data from low to moderate electrolyte concentrations by the use of an analytical expression which contains only meaningful physical parameters, c.f. Ref. [17].

An alternative theoretical approach to describe electrolyte conductivity is the use of molecular dynamic (MD) simulation [18] that more easily can take into account specific hydration of ions or the discrete nature of the solvent. For example, Calero et al. [19] studied

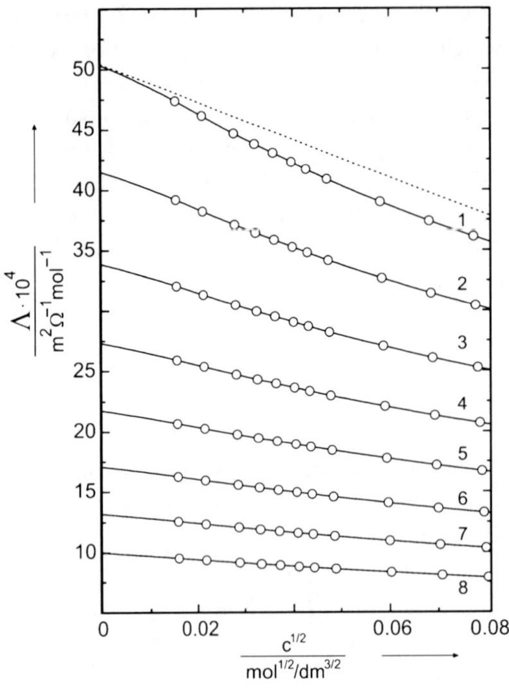

Conductivity of Electrolytes, Fig. 1 Molar conductivity of KI in ethanol temperature dependence in steps of 10 °C (1:25 °C, ..., 8:–45 °C), data from Ref. [14]. Onsager limiting law (25 °C)

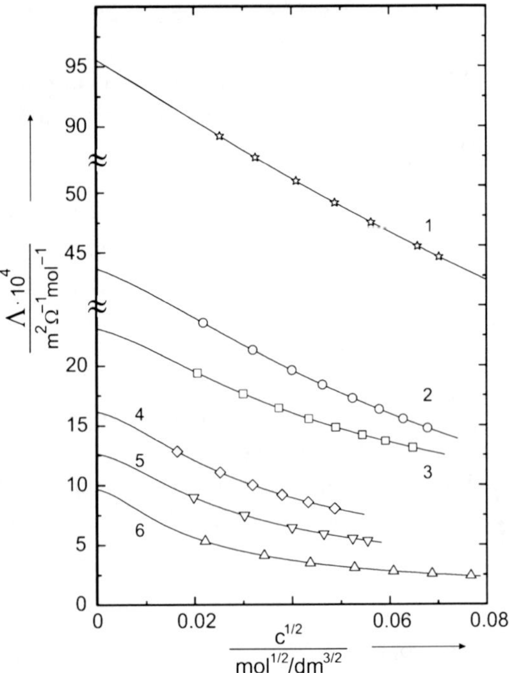

Conductivity of Electrolytes, Fig. 2 Molar conductivity of Bu_4NBr in *1* methanol, *2* ethanol, *3* 1-propanol, *4* 1-butanol, *5* i-butanol, *6* i-amyl alcohol (Data from Ref. [17])

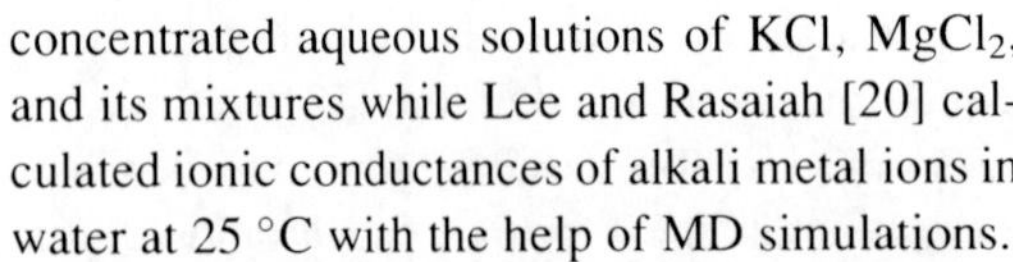

concentrated aqueous solutions of KCl, $MgCl_2$, and its mixtures while Lee and Rasaiah [20] calculated ionic conductances of alkali metal ions in water at 25 °C with the help of MD simulations.

Figure 1 shows the temperature and concentration dependence of the molar conductivity of potassium iodide in ethanol. It is typical for systems with small or moderate association constants (K_A ranges from 77 dm^3/mol at 25 °C to 29 dm^3/mol at –45 °C for KI in ethanol) and the extrapolation to zero concentration (Λ^∞) can easily be done. The Onsager limiting law (only the linear term in Eq. 6) is plotted at 25 °C for comparison.

Increasing ion pair formation (higher association constants) leads to a distinct decrease of molar conductivity with concentration yielding a pronounced increase of the graphs curvature. This pattern is clearly shown in Fig. 2 for solutions of tetrabutylammonium bromide in methanol ($K_A = 31$ dm^3/mol), ethanol ($K_A = 139$ dm^3/mol), 1-propanol ($K_A = 387$ dm^3/mol), 1-butanol ($K_A = 1034$ dm^3/mol), iso-butanol ($K_A = 1386$ dm^3/mol), and iso-amylalcohol ($K_A = 3264$ dm^3/mol).

Specific Conductance of Concentrated Electrolyte Solutions

Until now theoretically well-founded conductivity equations for concentrated solutions have not been developed. Actually, an empirical equation is used which takes into account the fact that electrolyte conductance features a maximum at moderate to high concentration.

This maximum is the result of competition between the increase of charge carrying particles (ions) and the decreasing ionic mobility at increasing electrolyte concentration and can be explained with the following treatment. Rearranging Eq. 3 and differentiation yields

$$d\kappa = \Lambda dc + c d\Lambda, \tag{11}$$

which plainly explains this behavior. With increasing concentration ($dc > 0$) the first term is always positive, whereas the second term is negative (Λ and c are always positive and Λ decreases with increasing concentration, c.f. Eq. 5), leading to a conductance maximum.

It has been proved [17, 21] that the relationship [22]

$$\kappa = \kappa_{max}\left(\frac{m}{\mu}\right)^{a} \exp\left[b(m-\mu)^2 - \frac{a}{\mu}(m-\mu)\right] \tag{12}$$

reproduces best conductivity κ as a function of molality (mol/kg) at moderate and high concentrations (up to saturation). In Eq. 12 μ is the concentration at which the conductivity maximum κ_{max} is observed. Parameters a and b are coefficients without physical meaning. The advantage of this equation is its simplicity and reliability in the reproduction of measured data.

Figure 3 features the conductance maximum at moderate to high concentrations. The dotted line specifies this maximum at different temperatures.

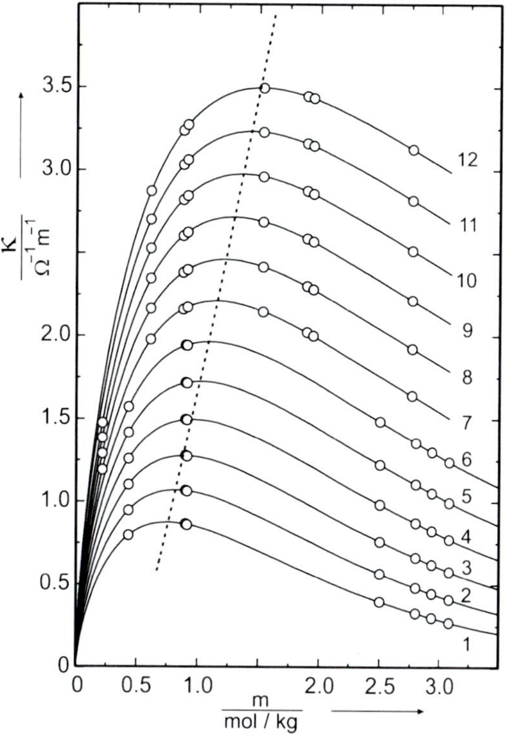

Conductivity of Electrolytes, Fig. 3 Specific conductivity of Bu_4NBr in acetonitrile (…κ_{max}) temperature steps of 10 °C (1: –35 °C, …, 7:25 °C, …, 12: 75 °C) (Data from Ref. [17])

Future Directions

Electrochemical technologies search for electrolyte solutions with particular properties adapted to the special tasks they must fulfill. For example, intensive research on lithium ion technology led to efficient batteries that guaranty the everyday life mobility that is nowadays a key issue for everyone. Electrical conductivity is a very reliable experimental method, which provides information on key properties for the performance of lithium ion cells, like voltage drop, power density as well as heat evolution during charging and discharging. Besides mixtures of nonaqueous solvents, ionic liquids (room temperature molten salts) and their blends will play an increasing role in future electrochemical technology (for example, as substitutes for traditional organic solvents). Optimization of electrolytes based on ionic liquids or blends of ionic liquids and one or more lithium salts is still under progress and needs conductivity data as one of the important key values for modeling.

Cross-References

▶ Conductometry
▶ Electrolytes, History
▶ Electrolytes, Thermodynamics
▶ Ion Mobilities
▶ Ion Properties
▶ Non-Aqueous Electrolyte Solutions

References

1. Barthel J, Gores HJ, Schmeer G, Wachter R (1983) Non-aqueous electrolyte solutions in chemistry and modern technology. Top Curr Chem 111:33–144

2. Barthel J, Buchner R (1986) Dielectric properties of nonaqueous electrolyte solutions. Pure Appl Chem 58:1077–1090
3. Wang S, Bianco R, Hynes JT (2010) Dissociation of nitric acid at an aqueous surface: large amplitude motions in the contact ion pair to solvent-separated ion pair conversion. Phys Chem Chem Phys 12:8241–8249
4. Maaser HE, Delsignore M, Newstein M, Petrucci S (1984) Thermodynamics of dimerization of Lithium salts in 1,2-dimethoxyethane. J Phys Chem 88: 5100–5107
5. Barthel J, Gerber R, Gores HJ (1984) The temperature dependence of the properties of electrolyte solutions. VI. Triple ion formation in solvents of low permittivity exemplied by lithium tetra uoroborate solutions in dimethoxyethane. Ber Bunsenges 88:616–622
6. Apelblat A, Bester-Rogac M, Barthel J, Neueder R (2006) An analysis of electrical conductances of aqueous solutions of polybasic organic acids. Benzenehexacarboxylic (mellitic) acid and its neutral and acidic salts. J Phys Chem B 110:8893–8906
7. Barthel JMG, Krienke H, Kunz W (1998) Physical chemistry of electrolyte solutions. Modern aspects, vol 5, Topics in physical chemistry. Springer, Berlin
8. Barthel J, Neueder R (1992) Electrolyte data collection. In: Eckerman R, Kreysa G (eds) DECHEMA data series. Part 1, vol XII. Dechema, Frankfurt
9. Renard E, Justice JC (1974) A comparison of the conductimetric behavior of cesium chloride in water-tetrahydrofuran, water-dioxane, and water-l,2-dimethoxyethane mixtures. J Solut Chem 3:633–647
10. Lee WH, Wheaton RJ (1979) Conductance of symmetrical, unsymmetrical and mixed electrolytes. Part 3. Examination of new model and analysis of data for symmetrical electrolytes. J Chem Soc Faraday Trans II 75:1128–1145
11. Quint J, Viallard A (1978) Electrical conductance of electrolyte mixtures of any type. J Solut Chem 7:533–548
12. Tsurko EN, Neueder R, Barthel J, Apelblat A (1999) Conductivity of phosphoric acid, sodium, potassium, and ammonium phosphates in dilute aqueous solutions from 278.15 K to 308.15 K. J Solut Chem 28:973–999
13. Apelblat A, Neueder R, Barthel J (2006) Electrolyte data collection. In: Kreysa G (ed) DECHEMA data series. Part 4c, vol XII. Dechema, Frankfurt
14. Barthel J, Neueder R, Feuerlein F, Strasser F, Iberl L (1983) Conductance of electrolytes in ethanol solutions from –45 to 25 C. J Solut Chem 12:449–471
15. Bernard O, Kunz W, Turq P, Blum L (1992) Conductance in electrolyte solutions using the mean spherical approximation. J Phys Chem 96:3833–3840
16. Barthel J, Graml H, Neueder R, Turq P, Bernard O (1994) Electrolyte conductivity from infinite dilution to saturation. Curr Topics Sol Chem 1:223–239
17. Barthel J, Neueder R (1992–2003) Electrolyte data collection. In: Eckerman R, Kreysa G (eds) DECHEMA data series. Part 1a to Part 1h, vol XII. Dechema, Frankfurt
18. Lee SH, Cummings PT, Simonson JM, Mesmer RE (1998) Molecular dynamics simulation of the limiting conductance of NaCl in supercritical water. Chem Phys Lett 293:289–294
19. Calero C, Faraudo J, Aguilella-Arzo M (2011) Molecular dynamics simulations of concentrated aqueous electrolyte solutions. Mole Simul 37:123–134
20. Lee SH, Rasaiah JC (1994) Molecular dynamics simulation of ionic mobility. I. Alkali metal cations in water at 25 °C. J Phys Chem 101:6964–6974
21. Neueder R, Gores HJ, Barthel J (2010) Electrolyte data collection. In: Sass R (ed) DECHEMA data series. Part 5a, vol XII. Dechema, Frankfurt
22. Casteel JF, Amis EA (1972) Specific conductance of concentrated solutions of magnesium salts in water-ethanol system. J Am Chem Soc 17:55–5

Conductometry

Hamidreza Sardary and Rudolf Holze
AG Elektrochemie, Institut für Chemie, Technische Universität Chemnitz, Chemnitz, Germany

The flow of electric current generally perceived as the motion of electrons in metallic, graphitic, and related materials or as the flow of negative (electrons) and positive (holes) charge carriers in semiconductors is of fundamental importance in science and technology, it is sometimes called the "lifeline of industrialized civilizations." Less popularly known is the flow of electric current proceeding via the movement of charged atomic or molecular species, i.e., ions in systems like liquid solutions or melts providing the possibility of movements of ions. Once again industrialization is hardly conceivable without this, production of many metals and basic chemicals would be impossible without solutions or melts containing mobile ions. Of equal importance is the importance of ionic conductors in energy storage and conversion systems omnipresent in daily life. The movement of an ion beyond the random Brownian motion is always effected by a gradient, in case of diffusion it is a concentration gradient, in case of convection it is a density gradient, and in case of an electric current it as an electric field gradient causing migration. The velocity of the mobile species (i.e., its movement) is related to the conductance of the matter containing this species. A higher velocity at a given field gradient translates directly into a higher conductivity. A higher number of charge carriers or a larger number of charge per moving species support a higher conductivity; further properties of the system like temperature and viscosity of the medium also affect the observed capability of sustaining an electric current.

The actually observed conductance Λ of matter is given by the current I flowing at a given applied voltage U (or electric potential E), i.e., a field (or electric potential) gradient $\partial E/\partial x$.

$$\Lambda = \frac{I}{E}$$

Values are given as mho (1 mho = 1 S), Siemens (S), or Ω^{-1}. The resistance R is the inverse property according to

$$R = \frac{U}{I} = \frac{1}{\Lambda}$$

Values are given in Ohm (Ω); this relationship is also known as Ohm's rule. When stated with respect to specific standard conditions (for a distance of 1 cm between two adjacent, parallel surfaces of exactly 1 cm^2 surface area each, as depicted below) (Fig. 1), the conductance is called specific conductance (conductivity). Values are given in $\Omega^{-1}\cdot cm^{-1}$.

The inverse values are called resistance and specific resistance ρ (resistivity), respectively; the units are $\Omega\cdot$ and $\Omega\cdot cm^{-1}$. Frequently the terms conductance, conductivity, resistance, and resistivity are liberally mixed assuming the reader is aware of the relationships between them and thus seeing no need to take care of the distinctions [1]. The relationship between resistance and resistivity taking into account the

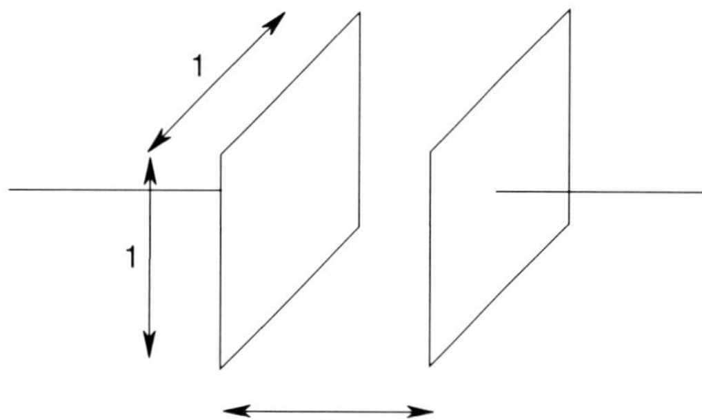

Conductometry, Fig. 1 Schematic of a conductance measurement cell

distance d between the electrodes and their geometric surface area A is given by

$$R = \rho \frac{d}{A}$$

and between conductance and conductivity

$$\Lambda = \kappa \frac{A}{d}$$

Given the preceding description the procedure for measuring conductance including specific conductance is already given: A known DC current is applied to the sample under investigation with the dimensions as depicted, the voltage drop is measured. A plot of data as schematically shown below should be obtained (Fig. 2).

The slope of the line is the conductance. The linear relationship over an extended range of applied currents (or voltages in case of resistance measurements) including the passage through the origin is called ohmic behavior. Numerous metallic conductors show this behavior. The influence of external parameters like temperature and pressure is treated in detail elsewhere. If an ionically conducting substance is inserted instead of an ohmic conductor, the situation changes dramatically. Because of conceivable chemical reactions between the two metal plates enclosing the sample under investigation, inert materials like glassy carbon or noble metals like platinum are used. Since those measurements are mostly done by

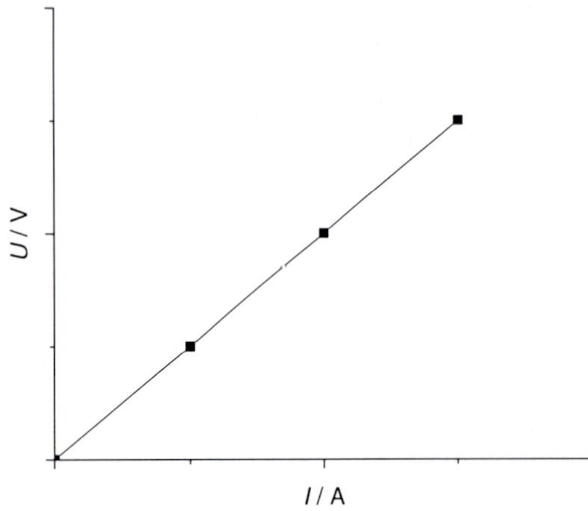

Conductometry, Fig. 2 Voltage-current relationship for an Ohmic resistor

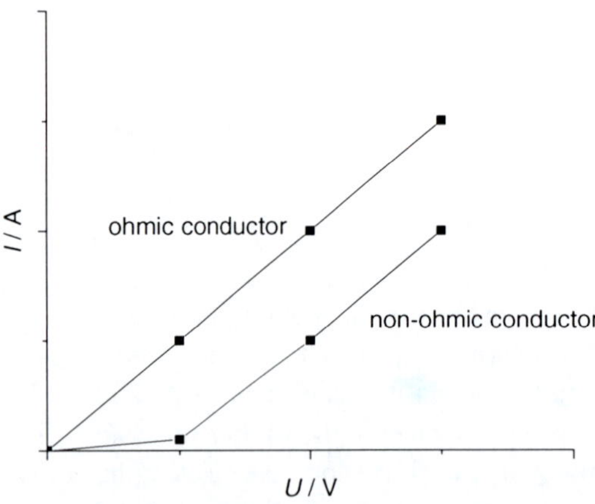

Conductometry, Fig. 3 Current-potential relationships for ohmic and non-ohmic conductors

applying a voltage instead of a current, the axes in the following picture are exchanged as compared to the preceding one (Fig. 3):

At small applied voltages almost no current flows; beyond a certain voltage (the numbers are just arbitrary examples) the current suddenly rises. Beyond this value the plot shows an almost ohmic-like behavior. The explanation has to take into account the phase boundaries at the contact interfaces. In the previous first case, two electronic conductors meet, electrons can move

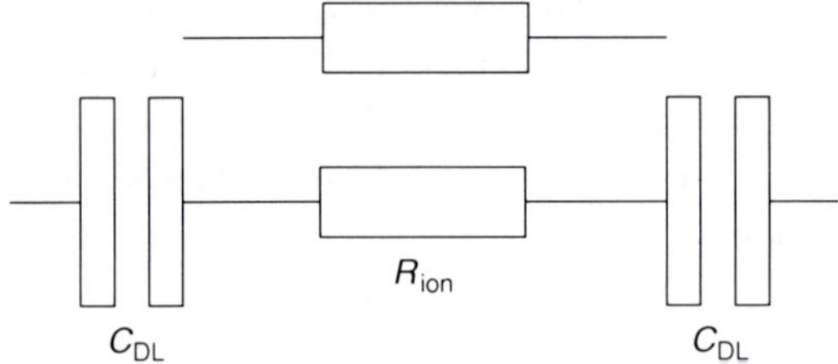

Conductometry, Fig. 4 Equivalent circuits: Ohmic resistor (top), conductance cell (bottom)

freely (taking not into account details like contact potential differences and work functions). In this second example electronic current flowing into the contact plates has to be coupled to an ionic current in the sample. This coupling proceeds via electrochemical reactions (oxidation or reduction) at the interfaces including transfer of electrons across the phase boundary. This process is not ohmic-like at all. At voltages below the threshold needed to drive this transfer, almost no current flows; at voltages above this threshold, the actual flow of current depends on numerous variables. Obviously measurement with an applied DC voltage does not yield the desired conductance data. Applying an AC voltage yields results typical of ohmic conductors. The behavior of the contact between electronic and ionic conductor – the electrochemical interface – provides the explanation. It behaves like a capacitor [2]. Thus the electrical equivalent circuit of the object of measurement changes from a simple resistor (top) to a more complicated one (bottom) as shown below (Fig. 4).

Both interfacial capacitances show a capacitive resistance depending on the actual value of the capacity and the applied frequency according to

$$R_{AC} = \frac{1}{\omega C}$$

At a sufficiently high frequency, both capacitive resistances can be neglected in comparison to the ohmic resistance of the ionic conductor between the plates. Increase of the interfacial

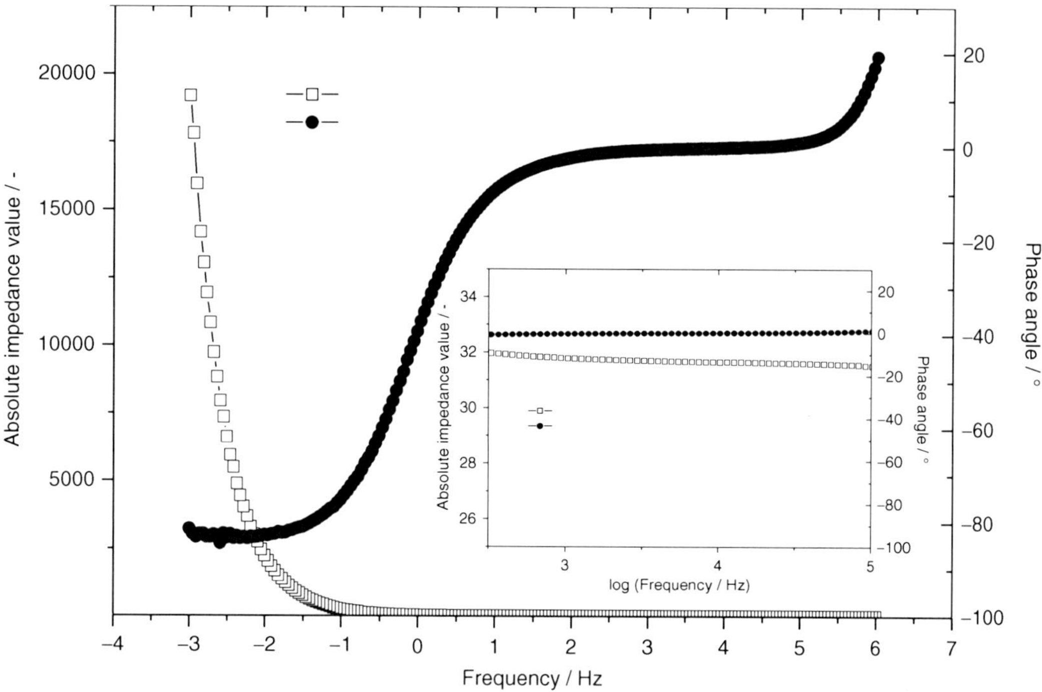

Conductometry, Fig. 5 Absolute impedance value and phase angle as a function of frequency of a conductance cell in an aqueous solution of 0.1 **M** KCl

surface area and/or reduction of the cross section of the ionic path of conduction between the contact further shifts this ratio in the desired direction. An increase of the true surface area (as compared to the geometric surface area) can be accomplished by roughening the surface or by depositing a porous layer (e.g., platinum black). The choice of the applied AC frequency and the amplitude is subject to some consideration. At very high frequencies the ions cannot follow the changing direction of charge movement enforced by the AC voltage perfectly well anymore. The observed conductance is smaller than expected; this is called Debye-Falkenhagen effect, and with aqueous solutions it is observed at frequencies higher than 5 MHz. Frequencies too small result in artificially low conductance values. These phenomena are extensively studied using dielectric spectroscopy (although strictly speaking ionically conducting systems are hardly dielectrics). A plot of various system properties as a function of frequency illustrates this situation. Although there are numerous ways to display results of such measurements (they are called electrochemical impedance measurements, frequently incorrectly impedance spectroscopy) available, a plot of the absolute value of the impedance (together with the phase angle plotted in the same figure, this is called a Bode plot) shows the significant frequency dependence (Fig. 5).

At large applied voltages (i.e., in strong electric field of about 10^4–10^5 V/cm), the velocity of ions is too large to allow perfect establishment of the cloud of solvent molecules around the ions. The actually observed conductance is larger than expected; the effect is called electrophoretic or (first) Wien effect. In solutions of weak electrolytes (i.e., incompletely dissociated ones), the large strength of the electric field may cause artificially enhanced dissociation; this also results in increased conductance. It is called dissociation

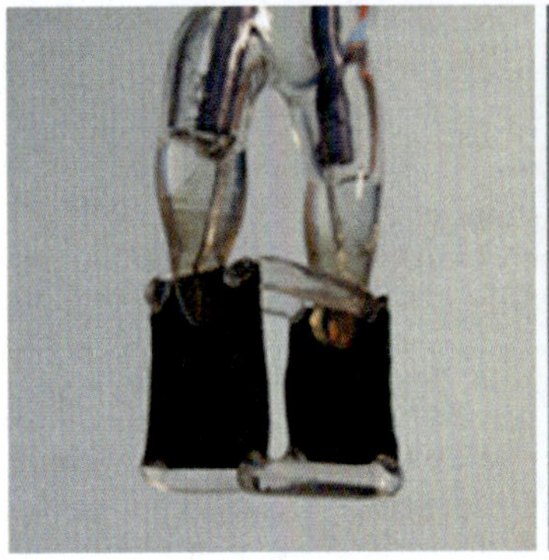
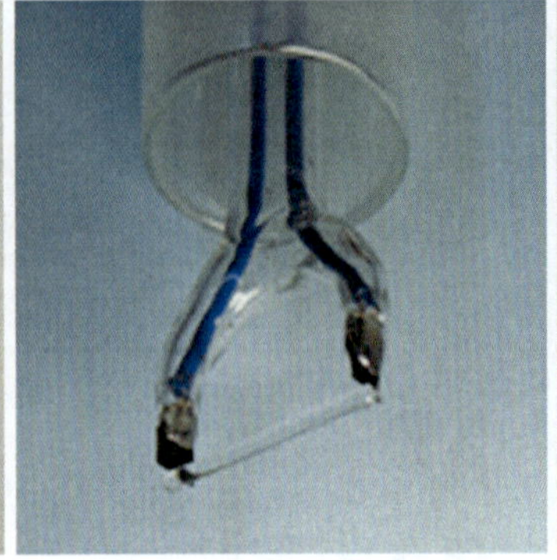
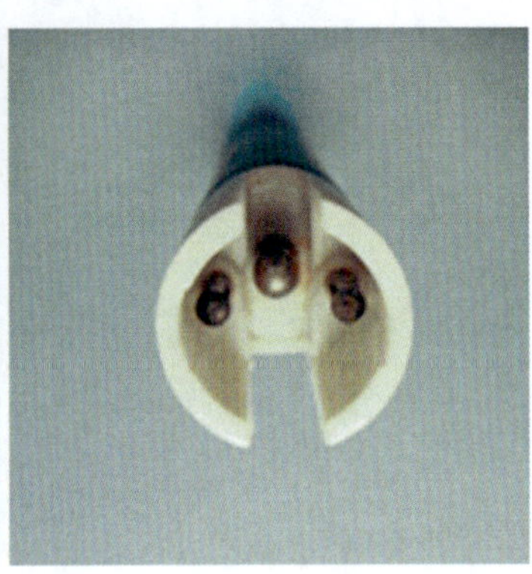

Conductometry, Fig. 6 Conductance cells: Left: Standard cell with large area electrodes for average and poorly conducting solutions; Middle: large distance cell for highly conducting solutions; Right: Glassy carbon rod electrodes in resin body with temperature sensor (middle)

field effect (or second Wien effect). Influence of this effect can be minimized easily by applying small voltages.

Ionic conduction can be observed in numerous systems: liquids showing self-dissociation (e.g., water, hydrogen sulfide), solutions containing ions formed by dissociation of salts (true or real electrolytes, e.g., NaCl) or molecules (potential electrolytes, e.g., HCl) (these systems are frequently called electrolyte, obviously this convenient simplification is misleading), molten salts, ionic liquids, ionic crystals, etc.

Measurement of conductance as an experimental tool is popular in analytical chemistry in titration methods where the equivalent point can be verified easily and in the presence of colorants in the sample solution with conductance measurements. Typical conductance cells are depicted below (Fig. 6).

Future Directions

Two aspects will merit attention in the future: Experimental methods applied in measuring conductivity and modification (improvement, enhancement) of the conductivity of electrolyte solutions as employed, e.g., in primary or secondary batteries. Combinations of liquids and electrolyte (salts) aiming at reduced viscosity in an ever wider range of temperatures and improvement of chemical stability in particular under adverse environmental conditions are just a few directions. These developments will be supported with tools of theoretical chemistry, in particular from molecular modelling.

Cross-References

- ▶ Electrochemical Cells, Current and Potential Distributions
- ▶ Electrode
- ▶ Electrolytes, Classification

References

1. Arjomandi J, Holze R (2008) Cent Eur J Chem 6:199
2. Holze R (2007) In: Martienssen W, Lechner MD (eds) Landolt-Börnstein: numerical data and functional relationships in science and technology, New series, Group IV: physical chemistry, vol 9, Electrochemistry, Subvolume A: electrochemical thermodynamics and kinetics. Springer, Berlin

Further Reading

1. Holze R. In: Martienssen W, Lechner MD (eds) Landolt-Börnstein: numerical data and functional relationships in science and technology, New series, Group IV: physical chemistry, Volume 9: electrochemistry, Subvolume B: ionic conductivities of liquid systems. Springer-Verlag, Berlin, (in preparation)

Controlled Flow Methods for Electrochemical Measurements

Martin Eichler and Rudolf Holze
AG Elektrochemie, Institut für Chemie, Technische Universität Chemnitz, Chemnitz, Germany

In every electrode reaction – which is inherently a heterogeneous reaction proceeding at a two-phase boundary – transport of reactands plays a major role. This affects both educts and products; both a lack of educts and local enrichment of products will limit and impede possible charge transfer reactions. According to the theory of overpotentials in electrochemical kinetics [1], the differences in concentration of species between the bulk value and the interfacial values cause a concentration overpotential η_{conc} which may be due to insufficient transport or slow homogeneous reactions preceding or following the charge transfer step. The former causes a diffusion overpotential η_{diff}, the latter a reaction overpotential η_{react}. Together with the charge transfer overpotential η_{ct} caused by sluggish charge transfer, the adsorption overpotential η_{ad} based on a slow adsorption or desorption step and the crystallization overpotential η_{crist} caused by slow formation of metal deposits or their dissolution, these overpotentials add up to the difference between the electrode potential at rest E_0 and the electrode potential E at flowing current:

$$\eta = E - E_0 = \eta_{conc} + \eta_{ct} + \eta_{ad} + \eta_{crist}$$
$$= \eta_{diff} + \eta_{react} + \eta_{ct} + \eta_{ad} + \eta_{crist}$$

For all technical applications, small overpotentials are desirable with corrosion reactions being the notable exception. In industrial processes (electrolysers [2, 3]), energy storage systems (e.g., redox flow batteries [4, 5]) and further systems, mass transport is generally enhanced by circulating electrolyte solutions, using three-dimensional electrodes or applying other means of artificially enhanced convection. In some cases diffusive mass transport needs to be controlled (e.g., by porous membranes in gas sensors for analytical purposes), and in analytical applications and basic studies diffusion as the sole means of mass transport should be calculable. This would enable extrapolation to the limiting case where diffusion would proceed at infinite rate. As a result, the remaining overpotential can be assigned to the other steps described above; finally, the rate of a step may be determined when all other steps are either made extremely fast or mathematically describable.

Calculating or mathematically describing (modeling) flow of species towards an interface has been possible so far for the following experimental setups:

1. Rotating electrodes
2. Wall-tube and wall-jet electrode
3. Pipe flow
4. Turbulent pipe flow
5. Channel electrodes.

Other forms of forced mass transfer, e.g., by gas evolution combined with forced circulation [6], have not been described sufficiently so far using mathematical models.

1. Mass transport towards a rotating disc electrode wherein the electrode with radius r is embedded in insulating material in the front of a rotating cylinder has been mathematically described first by Levich [7] (see also [8, 9]) (Fig. 1).

Because of the viscosity of the electrolyte solution and centrifugal action affecting the solution adhering to the rotating front surface, pumping action results with the liquid at the surface accelerated radially away (Fig. 2).

The thickness of the diffusion layer δ_D is constant across the diameter of the disc:

$$\delta_D = 0.643 \cdot D^{1/3} \cdot v^{1/6} \omega^{1/2}$$

depending on the diffusion coefficient D, the angular velocity $\omega = 2 \cdot \pi \cdot f$ in s^{-1}, and the kinematic viscosity υ given by the dynamic viscosity η divided by the density ρ according to

$$\upsilon = \frac{\eta}{\rho}$$

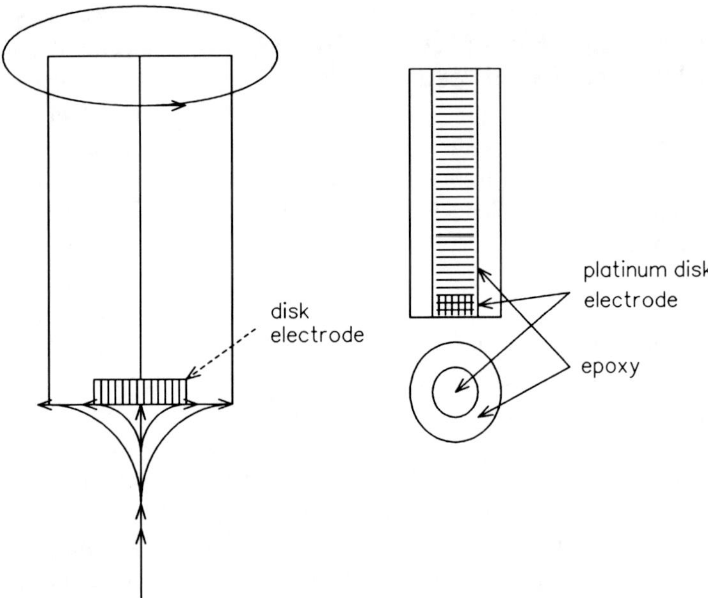

Controlled Flow Methods for Electrochemical Measurements, Fig. 1 Cross section and bottom view of a rotating disc electrode

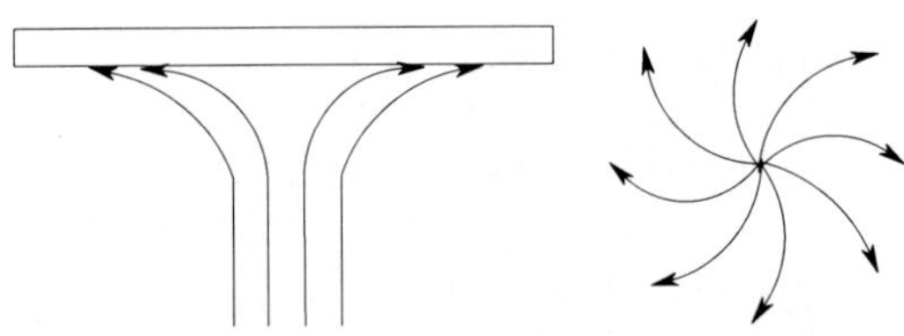

Controlled Flow Methods for Electrochemical Measurements, Fig. 2 Centrifugal transport of liquid in front of a rotating disc: *left*, vertical cross section; *right*, radial distribution

Because flux of species j towards the electrode surface is depending on transition across the diffusion layer only, it is given by

$$j = -D\frac{\partial c}{\partial z} = D\frac{(c_0 - c_s)}{\delta_\mathrm{D}}$$

with the bulk concentration c_0 and the surface concentration c_s of the reactand. When all species arrive at the electrode surface, they are consumed immediately and the surface concentration c_s will drop to zero and the diffusion-limited current $I_{\mathrm{diff.,lim.}}$ is reached. It is given by the Levich equation:

$$I_{\mathrm{diff.,lim.}} = 1.554 \cdot n \cdot F \cdot A \cdot D^{2/3} \cdot \nu^{-1/6}\omega^{1/2}c_0$$

with surface of the electrode A. A typical plot of data is shown in Fig. 3.

The respective Levich plot is shown in Fig. 4.

A mathematical description taking into account both mass transport by diffusion and charge transfer is

$$\frac{1}{I} = \frac{1}{I_{\mathrm{diff.,lim.}}} + \frac{1}{n \cdot F \cdot A \cdot k_{\mathrm{ct}} \cdot c_0}$$

with number of electrons n transferred in charge transfer reaction and with heterogeneous rate constant k_{ct}. This equation is called Koutecky-Levich equation (for details, see [10]). Beyond determination of diffusion coefficients from the Levich equation (for experimental details, see [11], for further experimental options, see [12]) extrapolation to infinite rate of rotation permits determination of k_{ct} (see also [13]). The disc can be made as an optically transparent film on a translucent rod. Light guided to the electrode may initiate photochemical reactions enabling electrochemical studies of the species thus generated. Further devices with rotating wires, semispheres, etc., have been reported; they are used mainly in electroanalytical chemistry in order to obtain enhanced mass transfer without the need for an exact mathematical description.

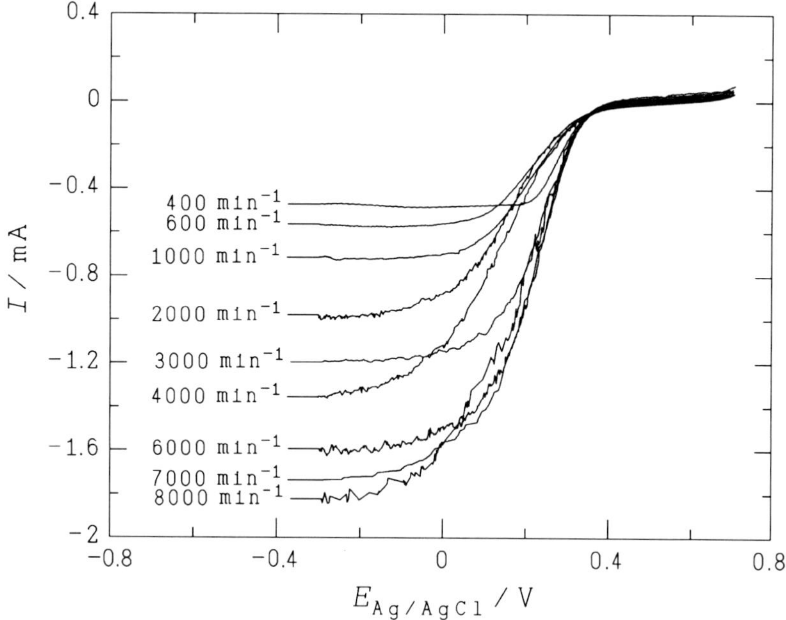

Controlled Flow Methods for Electrochemical Measurements, Fig. 3 Negative-going electrode potential scans of a platinum electrode in an aqueous solution of 5 mM $K_3Fe(CN)_6$ + 5 mM $K_4Fe(CN)_6$ + 0.5 M K_2SO_4 at $dE/dt = 0.01\ V \cdot s^{-1}$; angular velocities as indicated

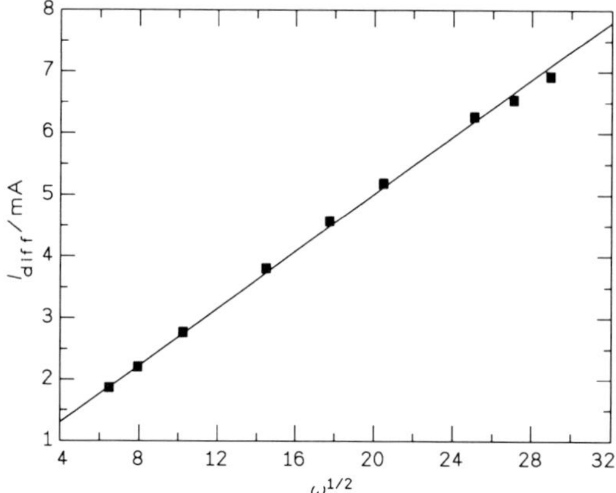

Controlled Flow Methods for Electrochemical Measurements, Fig. 4 Levich plot of I_{diff} vs $\omega^{1/2}$ from data in previous figure

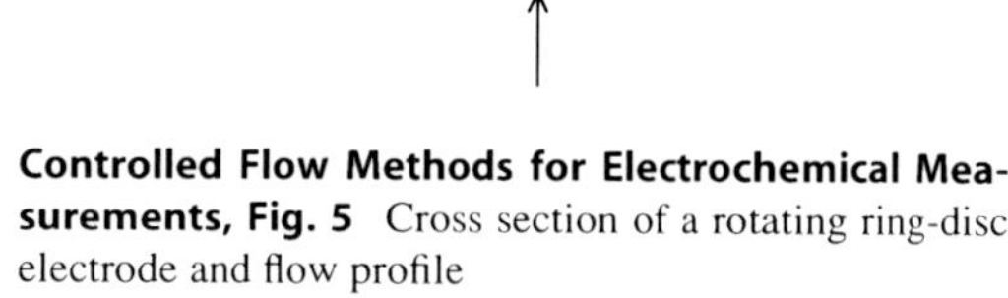

Controlled Flow Methods for Electrochemical Measurements, Fig. 5 Cross section of a rotating ring-disc electrode and flow profile

A second electrode may be placed as a ring around the disc electrode embedded in the same insulating body (Fig. 5).

Species generated at the disc electrode are transported by the flow already described above. Provided a suitably set ring electrode material and potential are applied, they can be detected there. In case of the electroreduction of Cu^{2+}-ions, the stepwise reduction via Cu^{+} depending on the disc electrode potential becomes easily visible (Fig. 6).

The amount of species actually collected as a fraction of species generated, given by

$$I_R = N \cdot I_D$$

can be calculated mathematically. Values of collection efficiency N depending on the disc radius and the inner and outer ring radius are tabulated [8, 9]; an approximation has been described [14]. In the example shown, a value of $N = 0.154$ can be calculated; the experimental result is $N = 0.19$.

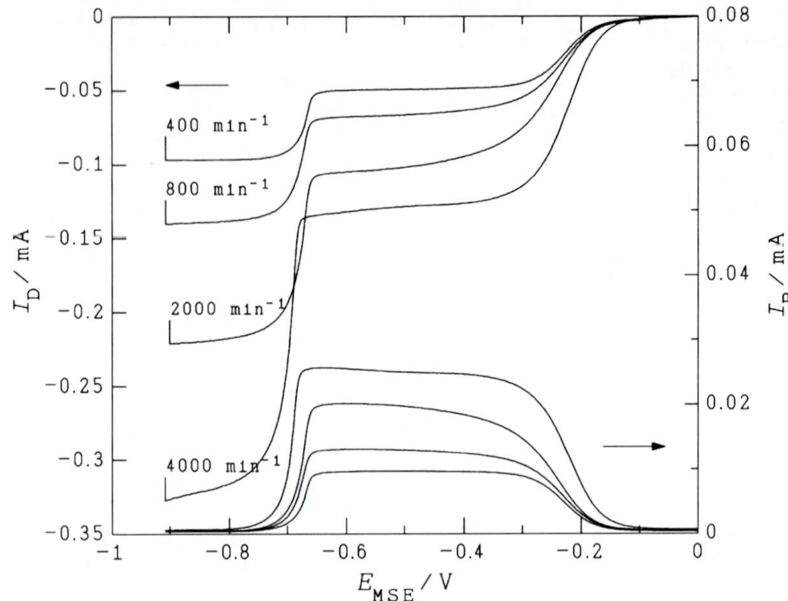

Controlled Flow Methods for Electrochemical Measurements, Fig. 6 Rotating disc and ring electrode currents (platinum electrodes) in an aqueous solution of 1 mM $CuCl_2$ + 0.5 M KCl_4 at $dE/dt = 10\ mV \cdot s^{-1}$; angular velocities as indicated; $E_{R,MSE} = 0.2$ V

When species generated at the disc electrode are consumed by a homogeneous chemical reaction, the actually collected amount at the ring will decrease; this effect will be more pronounced at lower angular velocities. This provides an additional access to kinetic data for this reaction.

2. Rotating electrodes may be inconvenient in some applications when further devices like spectrometers shall be coupled with them in hyphenated techniques or because of noise caused by the inherently necessary contact brushes. Consequently, attempts have been made to move the electrolyte solution instead in a controlled fashion. In the case of the wall-tube electrode, a jet of electrolyte solution from a nozzle with diameter d is directed towards a circular electrode with radius r (with d larger than the electrode diameter $2r$) embedded at distance h in insulating material as depicted below (Fig. 7).

The diffusion-limited current can be calculated with flow rate V_f (in $cm^3 \cdot s^{-1}$) according to

$$I_{\text{diff.,lim.}} = 12.08 \cdot n \cdot F \cdot A \cdot D^{2/3} \cdot \nu^{-1/6} c_0 \left(\frac{V_f}{\pi \cdot d^3}\right)^{1/2} \left(\frac{h}{d}\right)^{-0.054}$$

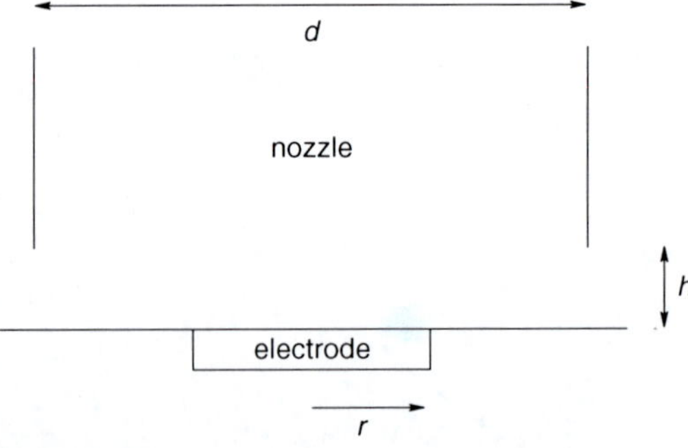

Controlled Flow Methods for Electrochemical Measurements, Fig. 7 Scheme of the wall-tube electrode configuration

In case of a nozzle with diameter d significantly smaller than the electrode diameter $2r$ the equation is slightly modified:

$$I_{\text{diff.,lim.}} = 1.38 \cdot n \cdot F \cdot A \cdot D^{2/3} \cdot \nu^{-5/12} d^{-1/2} r^{3/4} c_0 \cdot V_f^{3/4}$$

In the latter case, smaller amounts of solution are required making the device suitable for analytical applications; the arrangement is called wall-jet electrode. A second ring-shaped electrode may be added enabling studies of reaction

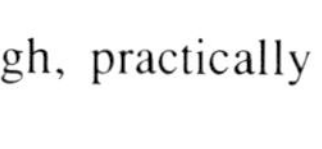

Controlled Flow Methods for Electrochemical Measurements, Fig. 8 Schematic of flow through a tube with radius r

products generated at the central electrode like the additional ring of the rotating ring-disc electrode arrangement (see above).

3. In the case of pipe flow (tubular flow, also tubular or tube electrode), the electrolyte solution is pumped through a circular tube at a rate (flow rate V_f) low enough to secure laminar flow (for the distinction from turbulent flow, see below). The working electrode is embedded as a ring (annulus) in the wall of the pipe; a double ring can be mounted also to enable mechanistic studies like with a ring-disc electrode (see above).

The setup of a ring electrode embedded in the wall of a tube with a liquid passing through is sketched below (Fig. 8).

Inside a tube under conditions of laminar flow caused by internal friction (viscosity) of the flowing medium, a velocity profile of approximately parabolic shape is established:

$$v = v_0\left(1 - \left[\frac{r}{r_0}\right]^2\right)$$

with r radius, r_0 radius of tube

with highest velocity in the center at $r = r_0$. The diffusion layer thickness δ_D in front of the electrode increases along the x-direction of flow proportional to $x^{1/3}$. Accordingly, an uneven current distribution results. As long as the diffusion layer thickness is small compared to the tube radius r, a diffusion-limited current can be calculated based on the Levich equation [7] (assuming that all other steps are proceeding at very high, practically infinite, rate):

$$I_{diff.,\text{lim.}} = 5.43 \cdot n \cdot F(D \cdot l)^{2/3} V_F^{1/3} c_0$$

The assumed flow of solution must be large enough to neglect contributions from diffusion along the direction of flow, and enforced convection is efficient enough to permit linearization of concentration gradients of species involved in the electrode reaction to be linearized. The result does not depend on the tube radius and the kinematic viscosity (as in the case of the rotating electrode). At slow flow rates and large values of l (i.e., a thick or "long" electrode), the expression is simplified to

$$I_{\text{diff.,lim.}} = n \cdot F \cdot V_F \cdot c_0$$

with complete conversion of species during passage of the electrolyte solution from edge-to-edge of the electrode.

A second ring embedded in the wall (a second annulus) will extend this device to double ring tubular electrode similar in experimental options to the ring-disc electrode. The second electrode may be employed to detect and quantify species generated at the first electrode. Selection of electrode material and operating electrode potential may be utilized as experimental variables to establish selectivity. The setup can also be used to study kinetics and mechanisms of homogeneous reactions proceeding in the solution between species generated at the first ring and solution species [15], and in analytical applications [16].

4. In the case of turbulent pipe flow, the electrolyte solution is pumped at a rate sufficiently high to establish turbulent transport [17, 18]. Turbulent flow is generally found when the Reynolds number Re passes a characteristic value:

$$Re = \frac{2 \cdot r \cdot \bar{v} \cdot \rho}{\eta}$$

with r, radius of pipe; $\bar{v}$, average streaming velocity; ρ, density of liquid; and η, viscosity of liquid. In a tube up to $Re \approx 2{,}300$ laminar flow dominates, at $2{,}300 < Re < 4{,}000$ first turbulences are observed, and at $Re > 4{,}000$ turbulent

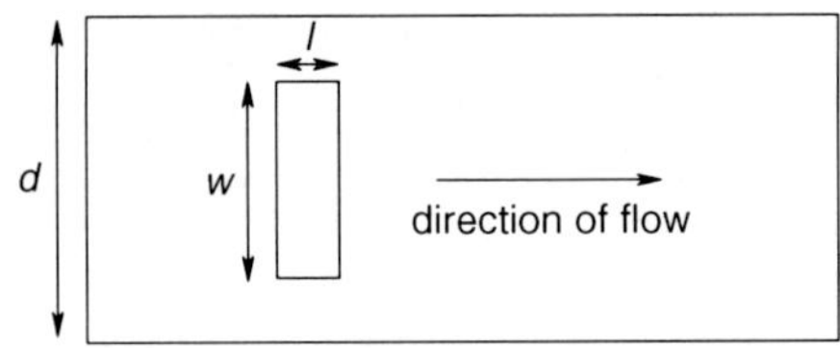

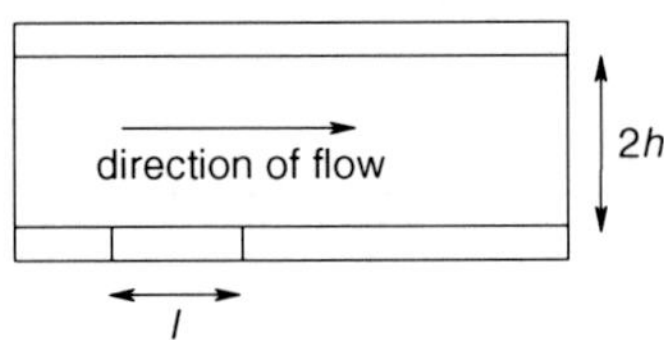

Controlled Flow Methods for Electrochemical Measurements, Fig. 9 Schematics of a channel with embedded electrode: *left*, view from *top*; *right*, cross section

pipe flow is established. Current-potential curves can be calculated (see [18]), the limiting current $I_{\text{lim.}}$ is given by

$$I_{\text{lim.}} = 0.276 \cdot A \cdot n \cdot F \cdot D^{2/3} \upsilon^{1/3} Re^{7/12} L^{-1/3} d^{-2/3} c$$

with area of the electrode A, number of electrons transferred in the reaction n, diffusion coefficient of the reacting species D, kinematic viscosity υ, Reynolds number Re, length of the electrode L, diameter of the electrode d, and reactand concentration c.

Placement of a second electrode downstream permits the same options already discussed above for the laminar flow case [19].

5. In case of a channel electrode, electrolyte solutions flow through a closed channel (i.e., a rectangular tube) of width d and height 2 h with the height variable y ranging from $-h$ at the electrode surface (and the bottom of the channel) and h at the top of the channel. In the channel bottom an electrode of width w and length l is embedded as depicted in the sketches below (Fig. 9).

Under conditions of laminar flow a parabolic velocity variation with height follows:

$$v = v_0 \left[1 - \left(\frac{y}{h} \right)^2 \right]$$

The diffusion layer thickness again varies with $x^{1/3}$ resulting in an uneven current distribution. A diffusion-limited current can be calculated based on the same assumption as with the tubular electrode (see above):

$$I_{\text{diff.,lim.}} = 0.925 \cdot n \cdot f \cdot w \cdot \left(\frac{D \cdot l}{h} \right)^{2/3} \left(\frac{V_F}{d} \right)^{1/3} c_0$$

At low flow rates and with sufficiently large (long) electrodes, complete conversion of species in the flowing solution can be accomplished suggesting use of this electrode in analytical applications.

An optically transparent window placed downstream of the electrode can be incorporated enabling spectroelectrochemical studies of species generated at the electrode and use of photochemical techniques, e.g., to study subsequently generated species electrochemically.

Like with the ring added to the disc electrode or the second electrode in a tubular electrode, a second electrode placed in the channel downstream may be used to detect and study species generated at the first electrode. The material of this second electrode (called collector electrode sometimes because of its purpose) as well as its operating potential have to be precisely selected to ensure complete conversion of arriving species. A collection efficiency N relating the current at the collector electrode to the current associated at the first electrode with the formation of species is given by

$$j_{\text{coll}} = -N j_{\text{gen}}$$

Because of the nonuniform current distribution at both electrodes and the complicated interplay of the rate of flow with the rate of charge transfer (reversible and irreversible electrode reactions) and further experimental variables equations derived for N are complicated [20].

Future Directions

The experimental methods described and reviewed above are well-established, further development beyond instrumental optimization and in particular miniaturization seems hardly conceivable. Most popular are rotating electrode devices which have been coupled to other methods (e.g., quartz crystal microbalance or differential electrochemical mass spectrometry). Thus, apparently substantial innovation can be expected from further hyphenated techniques. Because of the considerable need of experimental devices and experience, widespread application of hyphenated techniques seems to be unlikely, whereas simple rotating disc techniques will remain popular in, e.g., analytical chemistry. Turbulent pipe flow and channel electrode have not yet been applied widely, because of their rather limited application in fundamental studies of electrode kinetics this will most likely stay this way.

Cross-References

► Electrochemical Cells, Current and Potential Distributions
► Electrode
► Electrokinetics in the Removal of Chlorinated Organics from Soils
► Electrokinetics in the Removal of Hydrocarbons from Soils
► Electrokinetics in the Removal of Metal Ions from Soils
► Electrolytes, Classification

References

1. Vetter KJ, Bruckenstein S (1967) Electrochemical kinetics. Academic, New York
2. Pletcher D, Walsh FC (1993) Industrial electrochemistry. Blackie Academic & Professional, London
3. Oldham KB, Myland JC, Bond AM (2012) Electrochemical Science and Technology. Wiley, Chichester
4. Wu Y, Holze R (2013) Electrochemical energy storage and conversion. Wiley-VCH, Weinheim, for an overview see e.g.
5. Holze R (2013) In: Handbook of electrochemistry. Swider Lyons K, Breitkopf C (eds) Springer, Heidelberg
6. Hine F, Yasuda N, Ogata Y, Hara K (1984) J Electrochem Soc 131:83
7. Levič V (1962) Physicochemical hydrodynamics (фусіко-фімічеськая гідродінаміка). Prentice-Hall, Englewood Cliffs
8. Pleskov YV, Filinovskii VY (1976) The Rotating Disc Electrode. Consultants Bureau, New York
9. Albery WJ, Hitchman ML (1971) Ring-Disc Electrodes. Clarendon, Oxford
10. Bard AJ, Inzelt G, Scholz F (eds) (2012) Electrochemical dictionary, 2nd edn. Springer, Heidelberg
11. Gostisa-Mihelcic B, Vielstich W (1973) Ber Bunsenges Phys Chem 77:476
12. Bard AJ, Faulkner L (2001) Electrochemical Methods. Wiley, New York
13. Holze R (2007) In: Martienssen W, Lechner MD (eds) Landolt-Börnstein: numerical data and functional relationships in science and technology, new series, group IV: physical chemistry, volume 9: electrochemistry, subvolume A. Electrochemical thermodynamics and kinetics. Springer, Berlin
14. Filinovsky VYu, Pleskov YuV (1984) In: Yeager E, Bockris JO'M, Conway BE, Sarangapani S (eds) Comprehensive treatise of electrochemistry, Vol 9. Plenum Press, New York, p 339
15. Herrmann J, Schmidt H, Vielstich W (1984) Z Physik Chem NF 139:83
16. Schleifer GW, Blaedel WJ (1977) Anal Chem 49:49
17. Bernstein C, Heindrichs A, Vielstich W (1978) J Electroanal Chem 87:81
18. Iwasita T, Schmickler W, Herrmann J, Vogel U (1983) J Electrochem Soc 130:2026
19. Herrmann JA (1983) Entwicklung und Anwendung einer elektrochemische Methode zur Untersuchung schneller zwischengelagerter Reaktionen an Ringelektroden in turbulenter Rohrströmung, PhD-Dissertation, Rheinische Friedrich-Wilhelms-Universität zu Bonn
20. Mount AR (2003) In: Bard AJ, Stratmann M, Unwin P (eds) Encyclopedia of electrochemistry. Wiley-VCH, Weinheim, p 134

Coulometric Analysis

Chang-Jung Hsueh[1], Metini Janyasupab[1], Ying-Hui Lee[2] and Chung-Chiun Liu[1]
[1]Electronics Design Center, and Chemical Engineering Department, Case Western Reserve University, Cleveland, OH, USA
[2]Chemical Engineering Department, National Tsing Hua University, Hsinchu, Taiwan

Introduction

Coulometric analysis is an electrochemical method, in which an analyte of interest is exhaustively electrolysis adjacent to the surface of

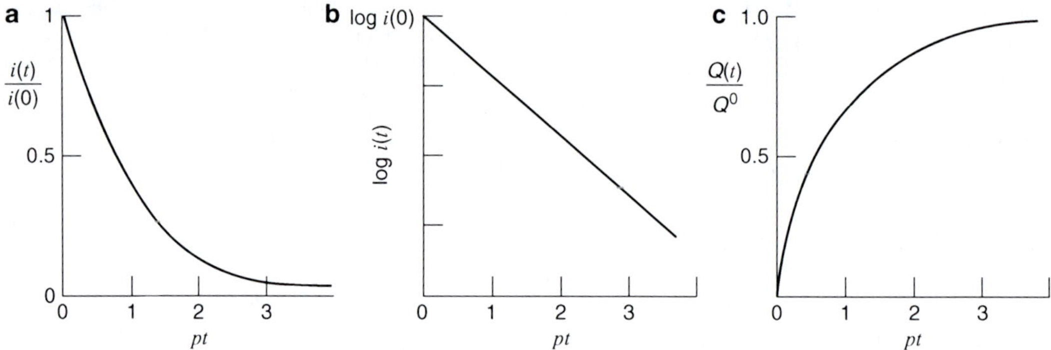

Coulometric Analysis, Fig. 1 Dimensionless current (electric charge)–time relation during controlled – potential electrolysis. (**a**) i–t (**b**) log (i)–t (**c**) Q–t [2]

electrode [1]. Due to absolute measurement of total charge via a single redox reaction of an analyte, coulometric analysis is principally governed by Faraday's law, without reference standard [1, 2]. In general, there are two types of coulometric technique based on controlled parameters: controlled-potential coulometry (CPC) and controlled-current coulometry (CCC) [2]. In this entry, the fundamentals, applications and limitations of each coulometry will be discussed, along with the recent studies performed by either type of coulometry.

Controlled-Potential Coulometry (CPC)

Controlled-potential (potentiostatic) coulometry [3] is an easy and efficient method to carry out exhaustive (complete) electrolysis, by simply applying constant potential onto the working electrode with respect to the reference electrode. Ideally, electrolysis of a single analyte-related reaction is maintained at the limiting current (maximum current value) condition with 100 % current efficiency. During electrolysis, current decays exponentially with increasing time to, ideally, zero or to the background/residual (non-faradic charging) level. Figure 1a [2] presented current–time behavior in a controlled-potential coulometry, according to Eq. 1. The relation of electrolysis charge and time scale (Fig. 1c [2]) can be interrogated by the integration of i–t curve in Fig. 1a [2] (i.e., the area under curve).

$$i(t) = i(0)\exp(-pt), \text{where } i(0) \text{ is the current at } t = 0; \quad p = AD/V\delta \tag{1}$$

p is alike 1st order of reaction rate constant and constitutes with diffusion coefficient of the analyte (D), the thickness of the diffusion layer, the surface area of the working electrode (A), and total solution volume (V).

General applicability of controlled-potential coulometry (CPC) has been utilized in the analysis of inorganic substances, individually or sequentially, due to convenience and high accuracy. Furthermore, precious metals are determined by CPC. High selectivity of relevant species is achieved based on the relative position of redox potential in order to the accompanying interferents [3]. The employment of CPC is also performed to the synthesis of organic compounds [4], in which the quantity of redox electrons (i.e., electron-transfer numbers) is accurately measured [5–7].

Conventional bulk electrolysis by CPC requires relatively long operation time to allow analyte to be completely consumed for accurate determination [2]. Based on Eq. 1, to improve the longtime electrolysis is essential to enhance the ratio of A/V, i.e., by enlarging the surface area of working electrode and reducing the volume of electrolysis solution. Furthermore, effective stirring or flowing of solution maintains the thickness of the diffusion layer (δ) [8]. In the following discussion, there are two types of

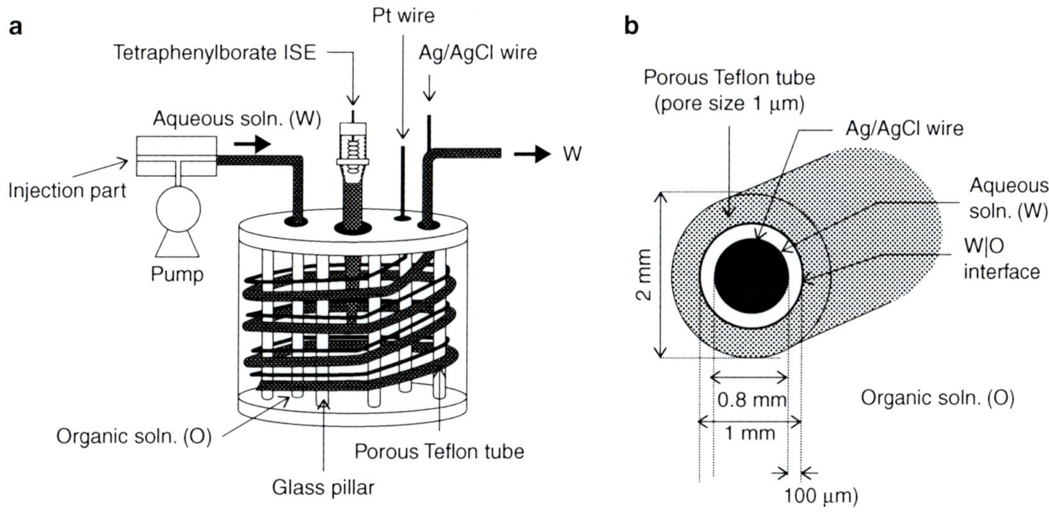

Coulometric Analysis, Fig. 2 (**a**) Scheme of the flow electrolytic cell using ISE at the water/organic (W/O) interface. (**b**) Sectional view of the ISE with the incorporation of porous Teflon tube [9]

modified techniques of CPC: thin-layer coulometry and biosensing devices, fulfilling the requirements of a small sample volume.

Thin-layer coulometry (TLC) is a desirable technique allowing small amount of testing solution electrolyzed in an electrochemical cell. With the incorporation of ion-selective membrane, the amount of introduced solution during electrolysis is confined and the species of interest is selectively assayed. Bonnick et al. [8] fabricated a thin-layer electrode configuration by screen printing technique. The testing solution of chlorine was introduced via the channel and rapidly consumed (electrolyzed) in a minute. Kihara et al. [9, 10] developed an ion selective electrode (ISE) in a flow cell, consisting of a porous Teflon tubular membrane for rapid and selective determinations of potassium (K^+), calcium (Ca^{2+}) and magnesium (Mg^{2+}). The flow electrolytic cell is schematically illustrated in Fig. 2.

The ISE comprising a Ag/AgCl inner working electrode was immersed into an organic solvent containing an ionophore, allowing selective binding of the anaytes. A small volume of aqueous solution was forced through the gap between the tabular Teflon membrane and the Ag/AgCl wire electrode. Exhaustive electrolysis was carried out under the controlled-potential mode.

Thin-layer coulometry fulfilling high ratio of A/V is an effective and rapid measurement of analyte concentration [8]. However, both calibration correction and limit of detection suffer from high background (non-faradic) signal. Bakker et al. [11–14] proposed a double pulse technique to compensate the interference effect. In Fig. 3 [14], the 1st excitation potential (E_1 = OCP + ΔE) with respect to the open circuit potential (OCP) was applied for exhaustive electrolysis of the analyte. Before the second excitation pulse, the ion selective membrane was relaxed to eliminate the buildup of polarization to the pre-electrolysis level. Assumingly, the analyte is completely electrolyzed during the 1st pulse duration, and interference is still as concentrated as the original level. After the 1st step perturbation pulse, the 2nd excitation ($E_2 = E_1$) results in the contribution of the interference.

Bakker et al. reported this method improving the limit of detection during a thin-layer coulometric analysis [12–14]. An ionophore was doped enhancing the ion-selective ability of a thin-layer membrane. Using a thin-layer coulometry coupling with double-pulse technique, the analysis of nitrate (NO_3^-), potassium (K^+), Calcium (Ca^+) were successfully undertaken in the scale of micromolar (μM). Furthermore, the advanced compensation of non-faradic charging

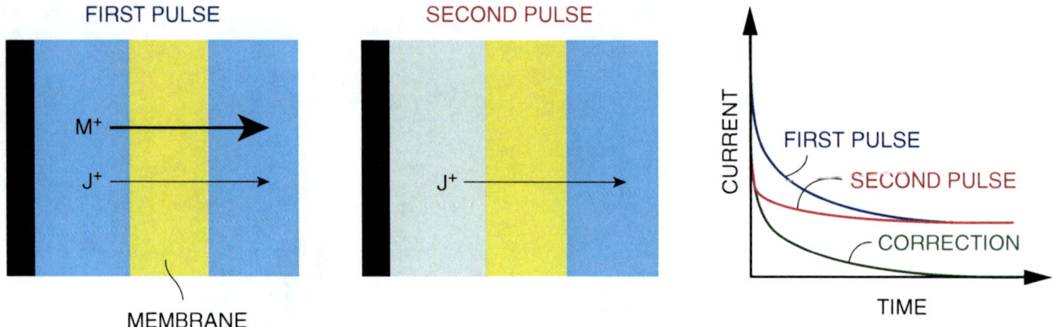

Coulometric Analysis, Fig. 3 (*left*) During the 1st potential pulse, exhaustive electrolysis of the analyte (M^+), along with partial consumption of the interference (J^+). (*middle*) Exhaustive electrolysis of the interference (J^+) in the 2nd potential pulse. (*right*) Correction of the current–time curve in the 1st electrolysis process by subtracting the non-faradic contribution from the 2nd electrolysis [14]

was achieved by switching the working electrode from Ag/AgCl to Ag/AgI, spacing inert material between the electrode and the membrane, adjusting the compositions of the outer solution [13].

Modified technique of controlled-potential coulometry is applied onto bio-sensing applications [15, 16]. Based on point-of-care and practical purpose, a small volume sample is required to deposit onto the working area of a sensor device. Therefore, exhaustive depletion of a species of interest can be rapidly undertaken in few minutes. FreeStyle Lite glucose sensor [17], one of well-known commercial blood glucose meters, is based on coulometric measurement. In a coulometric assay, lot-to-lot variation can be minimized by controlling the introduced volume of sample into the sensor, in which the channel dimensions are precisely confined. During the operation of the glucose sensor, a 300 ml of blood sample is assessed by the oxidation of an osmium mediator with glucose dehydrogenase. FreeStyle biosensor system and the coulometric response curves are schemed in Fig. 4.

Rigo et al. [15] proposed a coulometric biosensor equipping an electrochemical microflow cell. A 1 μL of H_2O_2 sample over 0.3–100 μM was introduced into the cell, in which the horseradish peroxidase (HPR)-modified electrode was installed. In the reaction, the 1, 4-benzoquinone was enzymatically generated and coulometrically electrolyzed at −0.1 V vs. Ag/AgCl. Due to the introduction of small volume sample, exhaustive depletion was conducted at 300s with good linear calibration ($r \geq 0.99$).

Zhang et al. [18] developed nano-gold (ANP)-modified screen-printed carbon electrode (SPCE) for DNA detection. Based on the coulometric measurement of enzymatic silver (Ag) deposition, the target DNA was ultrasensitively detected in the 1990s, in the dynamic range of 3.0×10^{-17} to 1.0×10^{-14} M (30 aM–100fM). The current–time curves during exhaustive electrolysis were undertaken before and after the target DNA hybridization (over 30 aM – 100fM), as shown in Fig. 5 [18].

Controlled-Current Coulometry (CCC)

Another coulometric approach is based on controlled-current technique. In this case, a constant current ($i_{app} < i_L(t = 0)$) is performed, with corresponding potential varied with time span of the electrolysis in Fig. 6a. While the i_{app} is higher than $i_L(t > 0)$, a secondary (side) reaction occurs during the electrolysis (Fig. 6b [2]). As a result, the current efficiency of the analyte falls below 100 %, due to the potential drift with increasing time.

It is difficult to prevent a secondary reaction by controlled-current coulmetry (CCC), prior to the exhaustive redox consumption of a target species. Therefore, this species may not be

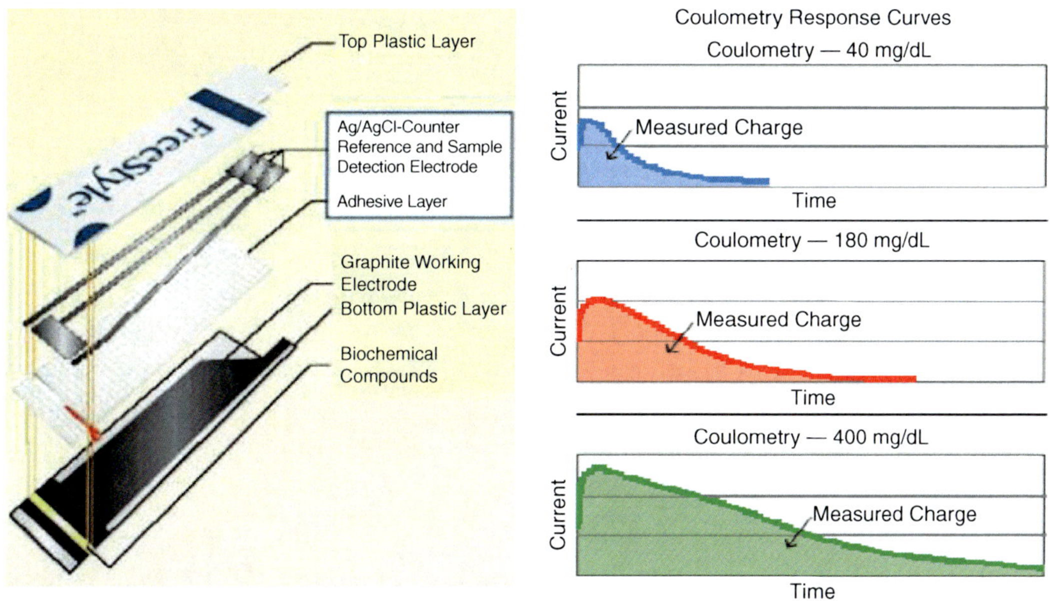

Coulometric Analysis, Fig. 4 (*left*) Schematic layer-by-layer illustration of FreeStyle glucose sensor, and (*right*) coulometric response by integration of current–time curve at a given concentration of glucose [17]

accurately determined since nearly 100 % current efficiency is not followed. However, the issue of 100 % current efficiency can be solved and maintained by simply adding a new species into the system. Given the example of Fe^{2+} oxidation, with increasing electrolysis time and potential drift, H_2O is going to being oxidized. If Ce^{3+} is added into the system, it will oxidize before the oxidation of H_2O to Ce^{4+}. The resulting Ce^{4+} will then react rapidly with the remaining Fe^{2+} in the system followed by the chemical reaction:

$$Ce^{3+}{}_{(aq)} \leftrightarrow Ce^{4+}{}_{(aq)} + e^-$$

$$Ce^{4+}{}_{(aq)} + Fe^{2+}{}_{(aq)} \leftrightarrow Ce^{3+}{}_{(aq)} + Fe^{3+}{}_{(aq)}$$

By combining the above two reactions, the net reaction in the system is the oxidation of Fe^{2+}:

$$Fe^{2+}{}_{(aq)} \leftrightarrow Fe^{3+}{}_{(aq)} + e^-$$

In this case, Ce^{3+} is a species called mediator or titrant, and this analytical method is called as coulometric titration.

End-point determination is always needed as an indicator when the exhaustive electrolysis of the analyte is completed and the titration ends up. It is convenient to end up a titrating reaction by a visual indicator [19], commonly using in a redox titration. Alternatively, potentiometric method (measurement of the potential vs the fraction of analyte's depletion) and amperometric method (measurement of the current vs. the fraction of the analyte's depletion) are both useful techniques for end-point determinations [20].

Controlled-current coulometry (coulometric titration) can be utilized to determine not-easily oxidizable (or reducible) analytes of different applications via acid–base, precipitation, complexation titrations, etc. Furthermore, it benefits short analysis time and small amount determination [2]. Dzudovic et al. [21] reviewed some studies employing acid–base titrations for the determinations of non-aqueous or water-insoluble compounds (organic and inorganic). Typically, acidimetric titrations were undertaken coulometrically based on the H^+ liberated by the oxidation of the introduced H_2O. Coulometric titrations of bases in nonaqueous solvent were performed using anodic depolarizers (titrants) to generate H^+ as a source. On the other hand, coulometrically alkalimetric

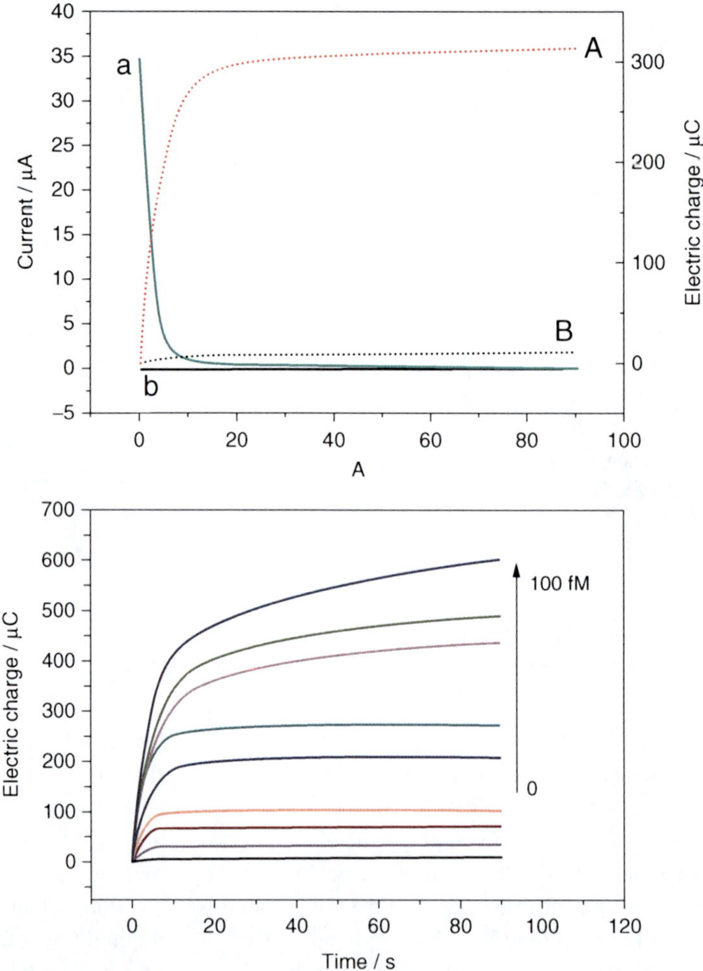

Coulometric Analysis, Fig. 5 (*top*) Current (*a* and *b*) – time and electric charge (*A* and *B*) – time response curves. *Curves a* and *A* presented after the target DNA hybridization and Ag deposition. *Curves b* and *B* before the target DNA hybridization. (*bottom*) Exhautive coulometric responses of the target DNA with varied concentrations: 0, 30 aM, 50 aM, 100 aM, 500 aM, 1 fM, 5 fM, 10 fM, and 100 fM [18]

titrations were conducted via the reduction of water using a strong base as a titrant, by generating hydroxide ions.

Asakai et al. [22] implemented coulometric titration on potassium dichromate ($K_2Cr_2O_7$), an essential reference material for volumetric analysis. Its reliability and certainty was evaluated precisely and automatically titrating with Fe^{2+}. In Fig. 7 [22], the coulometric titration cell was schematically illustrated. The counter electrode was isolated from the analytical solution preventing cross-reacting from the electrolysis products and the analyte. The oxidimetric purity of $K_2Cr_2O_7$ was obtained on the basis of Faraday's law. Furthermore, the dependences of the purity on sample mass and applied electrolysis current were investigated.

Trapp [23] and Wiegran [24] presented a CO_2 gas sensor based to coulometric titration. A thermally sputtered Ir/IrO electrode was employed to measure pH decrease, resulting from the reaction of diffused – CO_2 with the inner alkaline electrolye. As illustrated in Fig. 8 [23], the pH shift within the specific level was undertaken during the duration of measuring period. By electrochemically generating OH– ions, the initial pH value is reached and the new testing cycle resumed. According to the correlation of the partial pressure of CO_2, this sensor determines aero-CO_2 and dissolved CO_2 over the range of 200–20,000 ppm and 30–180 ppm, respectively.

Some studies have reported the applications of CCC into pharmaceutical determinations.

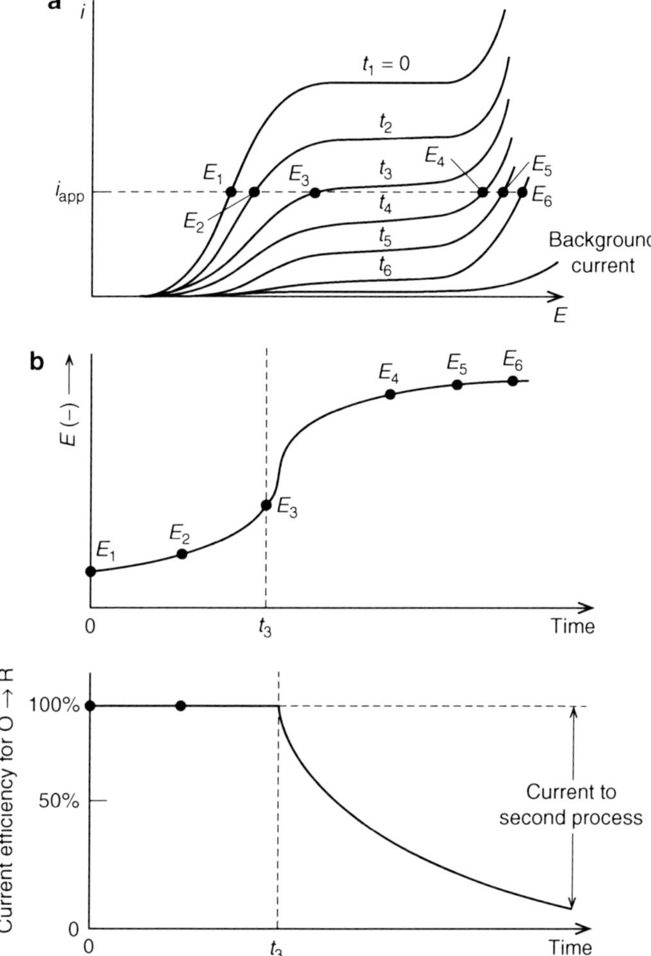

Coulometric Analysis, Fig. 6 (**a**) Current–potential curves at variable time spans (increasing from t_1 to t_6) during the electrolysis. (**b**) Corresponding illustration of (**a**) in potential–time behavior and current efficiency [2]

Tomcik et al. [25] developed an interdigitated microelectrode array (IDA) for the coulometric analysis of dosage determination of a couple of commercial available drugs: Antabus (TETD) and Celaskon. The principle was simply performed based on the quantity of TETD with electrogenerated hypobromite, AA in Celaskon with electrogenerated iodine. Ziyatdinova et al. [26] applied CCC for the detection of pharmaceutical antioxidants in surfactant media. Coulometric measurement of chain-breaking antioxidants in ionic (anionic/cationic) and neutral surfactants has been investigated titratingly via electrogenerated halogen (Hal_2) and hexacynoferrate (III) [$Fe(CN)_6 3-$].

This review session discusses the development of a couple of coulometric techniques (CPC and CCC) and provides assessment of both controlled modes for further application of some specific fields. Controlled – potential coulomtry can be modified with the incorporation of thin-layer membrane. High area/volume (A/V) ratio is achieved, and hence exhaustive electrolysis can be completed in a few minutes. Ion-selective electrode (ISE) with an ionophore doping effectively minimizes the interference during electrolysis. Low detection limit and calibration-free usage can be accomplished by eliminating background (non-faradic) charging via double-pulse potential technique. Controlled-current coulometry is an alternative for rapidly absolute measurement. Addition of an easily redox mediator (titrant) yields nearly 100 % current efficiency of an analyte's

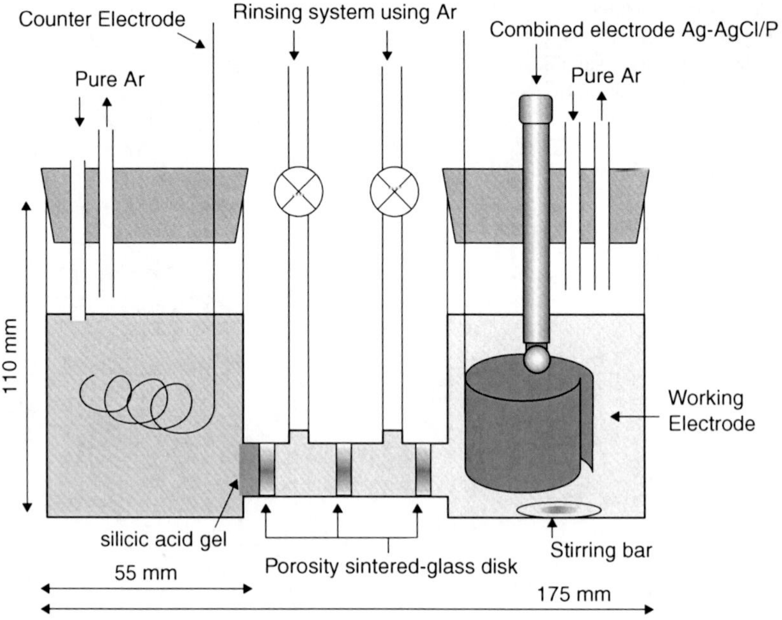

Coulometric Analysis, Fig. 7 The scheme of coulometric titration cell [22]

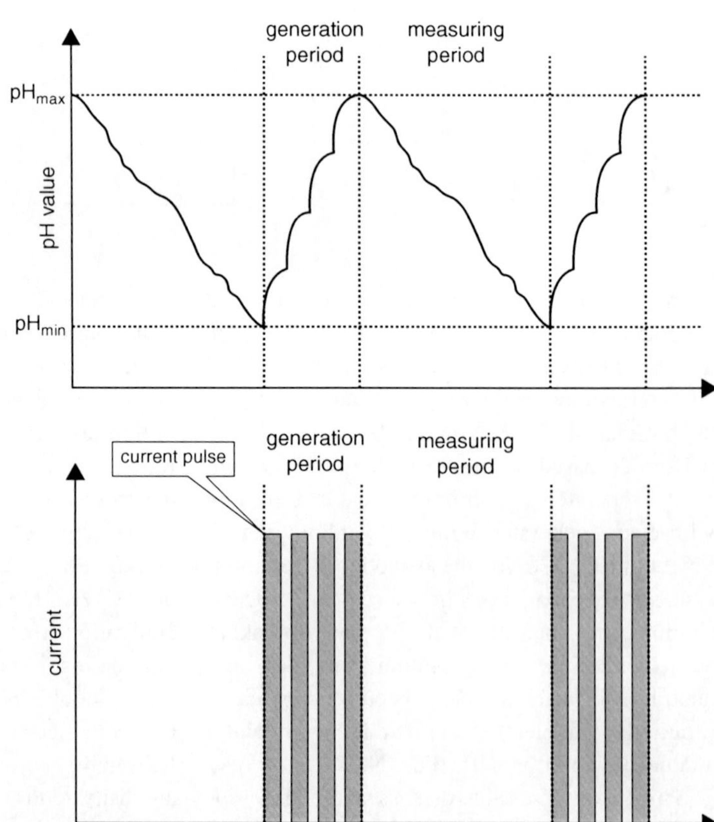

Coulometric Analysis, Fig. 8 The illustrated principle of the coulometric CO_2 sensor [23]

electrolysis. Almost non-redox compounds (organic or inorganic) can be determined via coulometric titrations, along with suitable redox titrants. CCC provides a convenient method over conventional redox titrations, due to low detection concentration and rapid measurement.

Cross-References

► Coulometry

References

1. Lewis TD (1961) Columetric methods in analysis: a review. Analyst 86:494–506
2. Bard AJ (2001) Electrochemical methods, fundamentals and applications. Wiley, New York
3. Harrar JE (1987) Analytical controlled-potential coulometry. TrAC Trend Anal Chem 6(6):152–157
4. Yang TF, Chiu KY, Cheng HC, Lee YW, Kuo MY, Su Y (2012) Studies on the structure of N-phenyl-substituted hexaaza [1_6] paracyclophane: synthesis, electrochemical properties, and theoretical calculation. J Organ Chem 77(19):8627–8633
5. Tan SLJ, Webster RD (2012) Electrochemically induced chemically reversible proton-coupled electron transfer reactions of riboflavin (Vitamin B2). J Am Chem Soc 134(13):5954–5964
6. Jain R, Dwivedi A, Mishra R (2009) Adsorptive stripping voltammetric behavior of nortriptyline hydrochloride and its determination in surfactant media. Laugmuir 25(17):10364–10369
7. Roth KM, Lindsey JS, Bocian DF, Huhr WG (2002) Characterization of charge storage in redox-active self-assembled monolayers. Laugmuir 18(10):4030–4040
8. Dennison S, Bonnick DM (1995) Stopped-flow thin-layer coulometric method for the determination of disinfectants in water. Anal Proc Anal Commun 32(1):13–15
9. Yoshizumia A, Ueharab A, Kasunoa M, Kitatsujic Y, Yoshidac Z, Kihara S (2005) Rapid and coulometric electrolysis for ion transfer at the aqueous|organic solution interface. J Electroanal Chem 581(2):275–283
10. Sohail M, De Marco R, Lamb K, Bakker E (2012) Thin layer coulometric determination of nitrate in fresh waters. Analytica Chimica Acta 744:39–44
11. Grygolowicz-Pawlak E, Bakker E (2010) Background current elimination in thin layer ion-selective membrane coulometry. Electrochem Commun 12(9):1195–1198
12. Shvarev A, Neel B, Bakker E (2012) Detection limits of thin layer coulometry with ionophore based ion-selective membranes. Anal Chem 84(18):8038–8044
13. Grygolowicz-Pawlak E, Bakker E (2011) Thin layer coulometry ion sensing protocol with potassium-selective membrane electrodes. Electrochim Acta 56(28):10359–10363
14. Grygolowicz-Pawlak E, Numnuam A, Thavarungkul P, Kanatharana P, Bakker E (2012) Interference compensation for thin layer coulometric ion-selective membrane electrodes by the double pulse technique. Anal Chem 84(3):1327–1335
15. Vianello F, Zennaro L, Rigo A (2007) A coulometric biosensor to determine hydrogen peroxide using a monomolecular layer of horseradish peroxidase immobilized on a glass surface. Biosens Bioelectron 22(11):2694–2699
16. Mizutani F, Ohta E, Mie Y, Niwa O, Yasukawa T (2008) Enzyme immunoassay of insulin at picomolar levels based on the coulometric determination of hydrogen peroxide. Sensors Actuat B Chem 135(1):304–308
17. Abbott Diabetes Care (2005) Clinical and laboratory studies FreeStyle™ Blood glucosetest strip performance. Abbott Diabetes Care
18. Liu J, Yuan X, Gao Q, Zhang C Ultrasensitive DNA detection based on coulometric measurement of enzymatic silver deposition on gold nanoparticle-modified screen-printed carbon electrode. Sensors Actuat B Chem 162(1):384–390
19. Dabke RB, Gebeyehu Z, Thor R (2011) Coulometric analysis experiment for the undergraduate chemistry laboratory. J Chem Educ 88(12):1707–1710
20. Lotz A (1998) A variety of electrochemical methods in a coulometric titration experiment. J Chem Educ 75(6):775–777
21. Mihajlovic R, Jaksic P, Lj N, Dzudovic RM (2006) Coulometric generation of acids and bases for acid-base titrations in non-aqueous solvents. Analytica Chimica Acta 557:37–44
22. Asakai T, Kakiharaa Y, Kozukaa Y, Hossakaa S, Murayamaa M, Tanakab T (2006) Evaluation of certified reference materials for oxidation-reduction titration by precise coulometric titration and volumetric analysis. Analytica Chimica Acta 567(2):269–276
23. Trapp T, Ross K, Cammann K, Schirmer E, Berthold C (1998) Development of a coulometric CO2 gas sensor. Sensors Actuat B Chem 50(2):97–103
24. Wiegran K, Trapp T, Cammann K (1999) Development of a dissolved carbon dioxide sensor based on a coulometric titration. Sensors Actuat B Chem 57(13):120–124
25. Tomcik P, Krajcikova M, Bustin D (2001) Determination of pharmaceutical dosage forms via diffusion layer titration at an interdigitated microelectrode array. Talanta 55(6):1065–1070
26. Ziyatdinova G, Ziganshina E, Budnikov H (2012) Surfactant media for constant-current coulometry. Application for the determination of antioxidants in pharmaceuticals. Analytica Chimica Acta 744:23–28

Coulometry

György Inzelt
Department of Physical Chemistry, Eötvös Loránd University, Budapest, Hungary

Introduction

In 1834 Michael Faraday described two fundamental laws of electrolysis. (In fact, Faraday's laws are based on two fundamental laws, i.e., on the conservation of matter and the conservation of charge.) According to Faraday the amount of material deposited or evolved (m) during electrolysis is directly proportional to the current (I) and the time (t), i.e., on the quantity of electricity (amount of charge) (Q) that passes through the solution (first law). The amount of the product depends on the equivalent mass of the substance electrolyzed (second law).

Theory and Methods

According to the Faraday's law,

$$m = \frac{M}{nF} Q = \frac{MIt}{nF} \tag{1}$$

where Q is the amount of charge consumed during the electrochemical transformation, n is the charge number of the electrochemical cell reaction, I is the current, and t is the duration of electrolysis.

If the current efficiency is 100 %, i.e., the total charge is consumed only by a well-defined electrode reaction; the measurement of charge provides an excellent tool for both qualitative and quantitative analysis [1, 2]. For instance, knowing m and Q, M/n can be obtained which is characteristic to a given substance and its electrode reaction. By knowing M and n, the amount of the substance in the solution can be determined. In many cases – especially in organic electrochemistry – the determination of the number of electrons (n) transferred during the electrode process is of importance regarding the elucidation of the reaction mechanism. For this purpose the total amount of charge necessary for the exhaustive electrolysis of a known amount of substrate has to be determined [3]. This method is known as coulometry. The coulometric experiment can be carried out at constant potential or at constant current.

Coulometry at Constant Potential

During coulometry at constant potential, the total amount of charge (Q) is obtained by integration of the current (I) – time (t) curve or Q can be determined directly by using a coulometer (electronic integrator). In principle, the end point $I = 0$, i.e., when the concentration of the species under study becomes zero, can be reached only at infinite time, however, in practice; the electrolysis is stopped when the current has decayed to a few percent of the initial values. The change of I and Q as a function of time at a constant potential $|E| >> |E_e|$, for a stirred solutions and for an uncomplicated electrolysis, is as follows:

$$I(t) = I(t = 0)\exp\left(-\frac{DA}{\delta V} t\right) \tag{2}$$

and

$$Q(t) = Q(t \to \infty)\left[1 - \exp\left(-\frac{DA}{\delta V} t\right)\right] \tag{3}$$

where D is the diffusion coefficient of the reacting species, A is the electrode area, δ is the diffusion layer thickness, and V is the volume of the solution. The applied potential (E) is far from the respective equilibrium potential (E_e), i.e., the current is diffusion limited. It is also possible to generate a reactant by electrolysis in a well-defined amount and then it will enter a reaction with a component of the solution. It is used in coulometric titration (invented by László Szebellédy [4]) where the end point is detected in a usual way, e.g., by using an indicator. A specific variant of coulometric titration is the gas titration used in both liquid and solid electrolytes.

Coulometry at Constant Current

Coulometry at constant current is somewhat more complicated; however, it is usually faster. Its advantage is that the charge consumed during the reaction is directly proportional to the electrolysis time. The change in concentration during electrolysis can conveniently be followed by cyclic voltammetry since the respective peak currents are gradually decreasing. Care must be taken to avoid the potential region where another electrode reaction may start. The *n* values determined by coulometry may differ from those obtained by a fast technique like cyclic (or linear) sweep voltammetry since the time scales are different and the primary reaction product may undergo a slow chemical reaction or a second electron transfer may also occur.

Application

Coulometry is, in addition to gravimetry, a primary standard analysis technique. Coulometry is used for the determination of the thickness of metallic coatings by measuring the quantity of electricity needed to dissolve the coating. Coulometric titration is applied to determine the concentration of water on the order of milligrams per liter in different samples when Karl Fischer reaction is used.

Cross-References

- ▶ Amperometry
- ▶ Coulometric Analysis
- ▶ Potentiometry

References

1. Inzelt G (2010) Chronocoulometry. In: Scholz F (ed) Electroanalytical methods, 2nd edn. Springer, Berlin, pp 147–148
2. Bard AJ, Faulkner LR (2001) Electrochemical methods, 2nd edn. Wiley, New York, pp 427–435
3. Hammerich O, Svensmark B, Parker VD (1983) Methods for the elucidation of organic electrochemical reactions. In: Baizer MM, Lund H (eds) Organic electrochemistry. Marcel Dekker, New York, pp 127–130
4. Szebellédy L, Somogyi Z (1938) Fresenius Z Anal Chem, 112:313, 323, 385, 391, 395, 400

C

Cyclic Voltammetry

Salma Bilal
National Centre of Excellence in Physical Chemistry, University of Peshawar, Peshawar, Pakistan

Introduction

Cyclic voltammetry is an elegant and simple electrochemical technique for studying redox reactions at the electrode solution interfaces and has become increasingly employed in all fields of chemistry. It is thought to be a simple, rapid, and powerful method for characterizing the electrochemical behavior of analytes that can be electrochemically oxidized or reduced. The method uses a reference electrode, working electrode, and counter electrode, which in combination are sometimes referred to as a three-electrode setup. The potential of the working electrode is measured against a reference electrode which maintains a constant potential and the resulting applied potential produces an excitation signal. Common materials for working electrodes include glassy carbon, platinum, and gold. These electrodes are generally encased in a rod of inert insulator with a disk exposed at one end. The counter electrode, also known as the auxiliary or second electrode, can be any material which conducts easily and will not react with the bulk solution. Reactions occurring at the counter electrode surface are unimportant as long as it continues to conduct current well. To maintain the observed current, the counter electrode will often oxidize or reduce the solvent or bulk electrolyte. Electrolyte is usually added to the test solution to ensure sufficient conductivity. The combination of the solvent, electrolyte, and specific working electrode material determines the range of the potential. In this technique the current that develops in an electrochemical cell by cycling the potential of a working electrode, under conditions where voltage is in excess of that predicted by the Nernst equation, is measured. Electrodes are static and sit in unstirred solutions during cyclic voltammetry. This "still" solution method results in cyclic voltammetry's characteristic diffusion controlled peaks. This method also allows a portion of the analyte to remain after reduction or oxidation where it may display further redox activity.

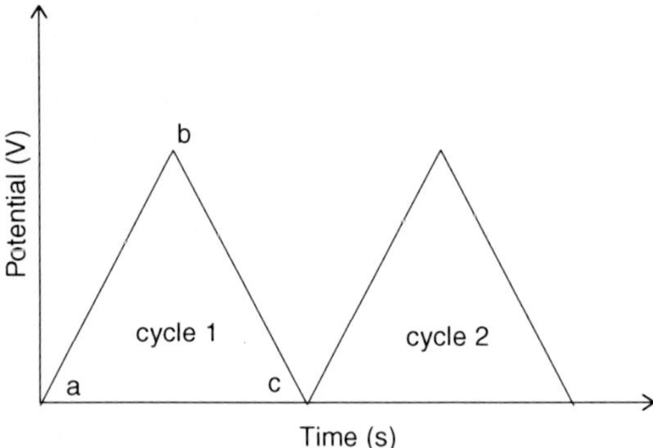

Cyclic Voltammetry, Fig. 1 Cyclic voltammetry potential waveform

Basic Principal

When a cyclic potential sweep is imposed on the working electrode, the current response is observed. A potentiostat system sets the control parameters of the experiment. Its purpose is to impose a cyclic potential sweep on the working electrode and to output the resulting current-potential curve known as cyclic voltammogram (CV). This sweep is described in general by its initial, switching, and the final potential (E_i, E_s, and E_f, respectively) and the scan rate in V/s. The scan rate may be define as the rate of change of potential as a function of time.

The potential range is scanned in one direction, starting at the initial potential (point a, Fig. 1) where no electrode reaction occurs and moving the potential where reduction or oxidation of a solute (material being studied) occurs (point b, Fig. 1). The direction of linear sweep is reversed after traversing the potential region in which one or more reactions take place and the electrode reactions of the intermediates and products, formed during the forward scan can be detected (towards point c, Fig. 1). Thus, the waveform is usually of the form of an isosceles triangle. This has the advantage that the product of the electron transfer reaction that occurred in the forward scan can be probed again in the reverse scan. The time scale of the experiment is controlled by scan rate.

Cyclic Voltammetry of Reversible Systems

The most useful aspect of this technique is its application to the qualitative diagnosis of electrode reactions, such as voltammetry of a redox couple (e.g., ferro-/ferricyanide or Fe^{2+}/Fe^{3+} in aqueous solution) or cyclic voltammetry of a redox system. A typical cyclic voltammogram recorded for Fe^{2+}/Fe^{3+} in aqueous solution presenting a reversible single electrode transfer reaction is shown below in Fig. 2. The solution contains only a single electrochemical reactant.

When the scan is reversed, we simply move back through the equilibrium positions gradually converting electrolysis product (Fe^{2+} back to reactant Fe^{3+}). The current flow is now from the solution species back to the electrode and so occurs in the opposite sense to the forward sweep seep but otherwise the behavior can be explained in an identical manner.

Important Parameter for a Reversible System

Important parameters for a cyclic voltammogram are the peak potentials E_p and peak currents i_p, which are measured with the help of peak

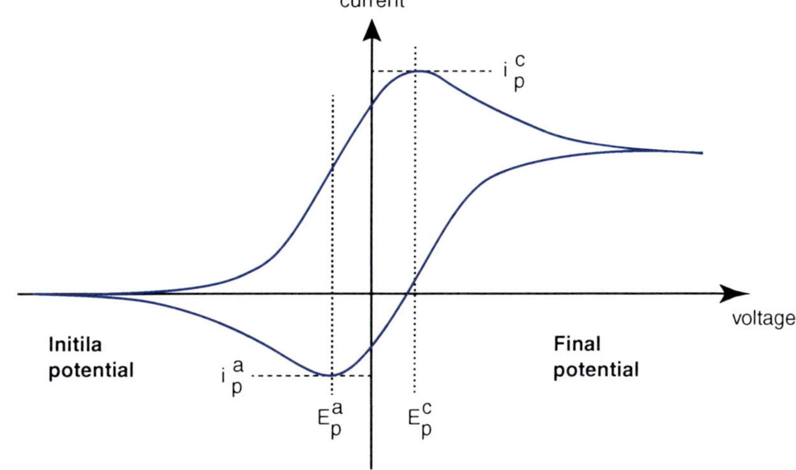

Cyclic Voltammetry, Fig. 2 Cyclic voltammogram recorded for Fe^{+2}/Fe^{+3} in aqueous solution

parameters operations. If a redox system remains in equilibrium throughout the potential scan, the redox process is said to be reversible. The main requirement of the equilibrium is that the surface concentration of the oxidized and reduced species is maintained at the values required by the Nernst equation.

The following parameters are used to characterize the cyclic voltammogram of a reversible process:

1. The separation of peak potential $E_p = E_{pc} - E_{pa} = 58/n$ mV at all scan rates at 25 °C.
2. The peak current ration $i_{pa}/i_{pc} = 1$ at all scan rates.
3. The peak current function $i_p/v^{1/2}$ (v is the scan rate) is independent of v.

The peak current is given by the equation

$$i_p = 2.69 * 10^5 n^{3/2} ACD^{1/2} V^{1/2} V^{1/2}$$

where:

n = number of electrons transferred per molecule
A = surface area of the electrode (cm^{-2})
C = concentration (mol cm^{-3})
D = diffusion coefficient ($cm^2\ s^{-1}$)

For a reversible process, E_0 is given by the mean peak potentials. Deviations from reversible behavior of a redox process can be predicted by the variations of the above parameters from the observed values for reversible processes.

Cyclic Voltammetry of Quasi-Reversible Systems

In case of quasi-reversible systems, the cyclic voltammograms show considerably different behavior from their reversible counterparts. Figure 3 shows the voltammogram for a quasi-reversible reaction for different values of the reduction and oxidation rate constants.

The first voltammogram shows the case where both the oxidation and reduction rate constants are still fast; however, as the rate constants are lowered, the redox peaks shift to more reductive potentials. Again this may be explained in terms of the equilibrium at the surface, which is no longer establishing so rapidly. In these cases the peak separation is no longer fixed but varies as a function of the scan rate. Similarly the peak current no longer varies as a function of the square root of the scan rate.

By analyzing the variation of peak position as a function of scan rate, it is possible to gain an estimate for the electron transfer rate constants.

Applications of Cyclic Voltammetry and Future Avenues

The most useful aspect of this technique is its application to the qualitative diagnosis of electrode reactions, such as voltammetry of a redox

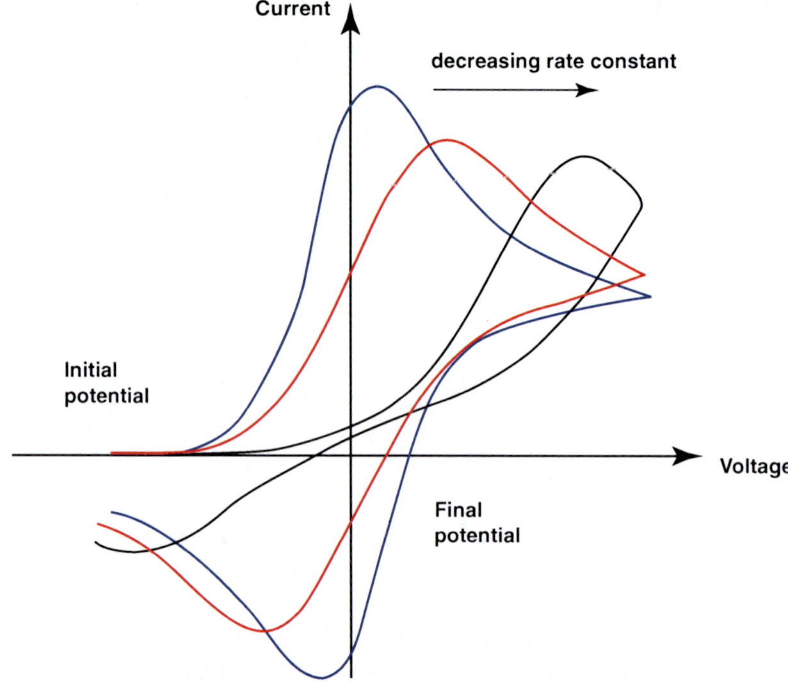

Cyclic Voltammetry, Fig. 3 Cyclic voltammograms for a quasi-reversible system for different values of reduction and oxidation rate constants

couple (e.g., ferro-/ferricyanide or Fe^{2+}/Fe^{3+} in aqueous solution) or cyclic voltammetry of a redox system. Cyclic voltammetry can be used as a technique to:

- Reveal surface contamination
- Estimate relative surface area and roughness
- Evaluate electrolyte leakage at electrode-insulator interfaces
- "Fingerprint" electrochemical reactions for benchmarking and quality control
- Estimate potentials at which reduction-oxidation reactions occur
- Make initial redox characterization such as the redox potentials, stability of different oxidation states, and for qualitative investigation of chemical reactions that accompany electron transfer
- Determine the charge storage capacity
- Evaluate electron transfer kinetics.

Disadvantages of Cyclic Voltammetry

Despite of the wide use of cyclic voltammetry, there are a number of disadvantages associated with this technique:

- The effect of slow heterogeneous electron transfer and chemical reactions cannot be separated. If both of these effects are present, then the rate constant of these processes can only be calculated using simulation methods.
- There is a background charging current throughout the experiment. This causes a restriction in the detection limit. In addition the ratio of the peak faradic current to the charging current decreases with increasing v as i_p is proportional to $v^{1/2}u^{1/2}$. This places an upper limit on the value of u that can be used.

However, indeed in some areas of research, cyclic voltammetry is one of the standard 0 techniques used for characterization.

Cross-References

- ▸ Amperometry
- ▸ Electrokinetics in the Removal of Chlorinated Organics from Soils
- ▸ Electrokinetics in the Removal of Hydrocarbons from Soils

▶ Electrokinetics in the Removal of Metal Ions from Soils
▶ Potentiometry

References

1. Bard AJ, Faulkner RL (2001) Electrochemical methods, fundamentals and applications. Wiley, New York
2. Gosser DK (1993) Cyclic voltammetry: simulation and analysis of reaction mechanisms. Wiley-VCH, New York
3. Adams RN (1968) Electrochemistry at solid electrodes. Marcel Dekker, New York
4. Feldberg SW (1969) A general methods for simulation. In: Electroanalytical chemistry series. Marcel Dekker, New York
5. Nicholson RS, Shain I (1964) Theory of stationary electrode polarography, single scan and cyclic methods applied to reversible, irreversible and kinetic systems. Anal Chem 36:706

D

Data Banks of Electrolytes

Richard Sass
DECHEMA e.V., Informationssysteme und Datenbanken, Frankfurt, Germany

Introduction

For the design and modeling of chemical plants, process calculations are carried out during the whole process life cycle. Especially in the case of the synthesis, simulation, and optimization of separation processes, the phase equilibrium of the system to be separated has to be known exactly. Chemical and process engineers are nowadays able to model or even predict a vapor-liquid equilibrium, the density or viscosity of a multicomponent mixture containing numerous different species with sufficient reliability. However, these calculations depend heavily on availability and reliability of thermophysical property data of the pure components and mixtures involved, especially when in a later status of a process design the accuracy of models (equations of state or group contribution methods) are no longer sufficient [1]. This is especially true when the mixtures contain salts. Even traces of salts have the consequence that nearly all models work with less accuracy or even tend to fail. Another area influenced greatly by electrolyte modeling is biochemical engineering. Until now, it is nearly impossible to predict quantitatively salting-out effect of proteins. Nevertheless, modeling results have a great impact on the design and construction of single chemical apparatus as well as whole plants or production lines. Another problem is that inaccurate data may lead to very expensive misjudgments, whether it is to proceed with a new process or modification of it or not to go ahead. So even today, where the usage of process simulation packages is a standard feature, many improvements can be done for the precision of the models in using experimental thermophysical properties. Therefore, chemical engineers developing new production processes are having a pressing need to access reliable thermophysical property data.

Important Properties of Electrolyte Solutions

Process as well as model development – either predicting or even only interpolating – requires multitudinous amounts of reliable thermophysical property data for electrolytes and electrolyte solutions. Among the most important property types are:

- Vapor-liquid equilibrium data
- Activity coefficients
- Osmotic coefficients
- Electrolyte and ionic conductivities
- Transference numbers
- Viscosities
- Densities
- Frequency-dependent permittivity data

How to get such data?

G. Kreysa et al. (eds.), *Encyclopedia of Applied Electrochemistry*, DOI 10.1007/978-1-4419-6996-5,

Reliable Data Sources

Print Format

Compilations of data can be found in the classical papers, e.g., by Pitzer and coworkers [2] or Barthel and coworkers [3–7], in books [8] as well as in printed data collections as the DECHEMA Chemistry Data Series [9, 10], which contains a sound collection of electrolyte data, mainly transport properties (conductivity, viscosity, dielectric properties, and phase equilibria). Part of DECHEMA's Chemistry Data is the **Electrolyte Data Collection** published by Prof. J. Barthel and his coworkers from the University of Regensburg. The printed collection and the database ELDAR (described later) have complementary functions. The data books are giving a clear arrangement of selected recommended data for each property of an electrolyte solution. The electrolyte solutions are classified according to their solvents and solvent mixtures. All solution properties have been recalculated from the original measured data with the help of compatible property equations. A typical page of the books contains for the described system:

- General solute and solvent parameters
- Fitted model parameter values
- Measured data together with deviations against the fit
- A plot
- The literature references

The Electrolyte Data Collection has meanwhile 20 volumes and consists of 7,400 printed pages. Covered properties are:

- Specific conductivities
- Transference numbers
- Limiting ionic conductivities
- Dielectric properties of water, and aqueous and no aqueous electrolyte solutions
- Viscosities of aqueous and no aqueous electrolyte solutions

The preferred way for reliable data sources today is to search within an electronic database for the components, mixtures, and properties one needs. Printed data collections or databases are typically compiled and/or maintained from individuals or groups having a well-known reputation in that field. Therefore, they have an overview of the primary literature publishing physical property data and are able to continuously add new data to their collections. In many cases, these groups are able to check the consistency of the data published in literature. Efforts are even made today by publishers to avoid the publication of data with doubts about the repeatability and uncertainty of published data during the referee process (e.g., NIST data tools [11, 12]). In most cases, these groups also use their own collections and methods for model development [13].

In the following, a survey of currently maintained databases for electrolyte properties is given [14]:

Examples of Databases with Properties of Electrolyte Solutions

DETHERM/ELDAR Database

The **El**ectrolyte **Da**tabase **R**egensburg **ELDAR** [15] is a numerical property database for electrolytes and electrolyte solutions from the Institute of Physical and Theoretical Chemistry of the University of Regensburg. It started in 1976 and contains data on pure substances and aqueous as well as organic solutions. The database was designed as a literature reference, numerical data, and also model database for fundamental electrochemical research, applied research, and also the design of production processes.

ELDAR contains data for more than 2000 electrolytes in more than 750 different solvents with a total of 56,000 chemical systems, 15,000 literature references, 45,730 data tables, and 595,000 data points. ELDAR contains data on **physical properties** such as densities, dielectric coefficients, thermal expansion, compressibility, p-V-T data, state diagrams and critical data. The **thermodynamic properties** include solvation and dilution heats, phase transition values (enthalpies, entropies and Gibbs' free energies), phase equilibrium data, solubilities, vapor pressures, solvation data, standard and reference values, activities and activity coefficients, excess values, osmotic coefficients, specific heats, partial molar values and apparent partial molar values. **Transport properties** such as electrical conductivities, transference numbers, single ion conductivities, viscosities, thermal conductivities, and diffusion coefficients are also included.

ELDAR is distributed as part of DECHEMA's numerical database for thermophysical property data DETHERM [16, 17]. To access ELDAR, one can therefore use several options.

- In-house client-server installation as part of the DETHERM database
- Internet access using DETHERM ... *on the Web*

To get an overview of the data available, the Internet access option is recommended, because existence of data for a specific problem can be checked free of charge and even without registration.

Data Bank for Electrolyte Solutions at CERE DTU Chemical Engineering

The Center for Energy Resources Engineering (CERE) of the Technical University of Denmark (DTU) is operating a data bank for electrolyte solutions [18]. It is a compilation of experimental data for (mainly) aqueous solutions of electrolytes and/or nonelectrolytes. The database is a mixture between a literature reference database and a numerical database. Currently references to more than 3,000 papers are stored in the database together with around 150,000 experimental data. The main properties are activity and osmotic coefficients, enthalpies, heat capacities, gas solubilities, and phase equilibria like VLE, LLE, and SLE. The access to the literature reference database is free of charge. The numerical values must be ordered at CERE.

The Dortmund Database DDB

The Dortmund Database published by DDBST GmbH [19] of Prof. J. Gmehling from the University of Oldenburg is well known for its data collections in the areas of vapor-liquid equilibria and related properties [20]. While the major part of the data collections is dealing with nonelectrolyte systems, two collections contain exclusively electrolyte data. They are focused on:

- Vapor-liquid equilibria
- Gas solubilities

The two collections together currently contain around 11,000 data sets. Access to these collections is possible either online using the DETHERM ... *on the Web* service or in-house using special software from DDBST or DECHEMA.

JESS

A very useful tool is provided by the Joint Expert Speciation System (JESS), a joint project of two groups in Australia and South Africa [21, 22]. JESS has a large database for physicochemical properties of electrolytes in aqueous solution. This comprises about 300,000 property values for over 100 electrolytes. Data from the literature are available for activity coefficients, osmotic coefficients, heat capacities, and densities/volumes.

Springer Materials: The Landolt-Börnstein Database

Based on the Landolt-Börnstein (New Series) book collection [23], the Springer Materials database provides also data on electrochemical systems, for example, on electrochemical processes at the boundaries of electrodes and electrolytes or electrical conductivities and equilibrium data [24].

Closed Collections

In addition to the above-described publicly available and still maintained databases, old electrolyte data collections were built up in the past. One of these is the ELYS database, which was compiled by Professor Victor M.M. Lobo, Department of Chemistry, University of Coimbra, Portugal, with thermodynamic and transport property data, such as density, viscosity and diffusion coefficients. Other noteworthy examples include the DIPPR 811 and 861 Electrolyte Database Projects [25]. But these closed collections are typically not maintained anymore and also not publicly available. Likely the references and/or data published in these collections could be also found inside the before mentioned, living collections.

A Glance at the Future of the Properties Databases

Most of the engineers in chemical companies trust in the power of their evaluation of the equations of state for the calculation of the optimal point of work. Nevertheless, the opinion that databases have less importance these days is growing up, mainly when budgetary elements come into consideration. Unfortunately, the production of many basic chemicals is today transferred into low cost countries, i.e., also not very relevant for research purposes. Many companies outsourced their measurements so that only a limited amount of experts in industry hold the knowledge for these activities.

When we look at the constraints to find new methods for the design of biologic or polymer solutions, we must be skeptical to find enough people to manage future visions for models with the knowledge what was in the past. In an age where China is replacing the United States as world's leading consumer of steel, there is the hope that the rapidly growing demand of energy increases also the interest in physical properties. Meanwhile it is a good sign that new projects are coming up in order to find a new approach to build evaluated databases.

Cross-References

- ► Activity Coefficients
- ► Conductivity of Electrolytes
- ► Electrolytes, Thermodynamics
- ► Ion Properties
- ► Thermodynamic Properties of Ionic Solutions - MSA and NRTL Models

References

1. Carlson EC (1996) Don't gamble with physical properties for simulations. Chem Eng Progr 92(10):35–46
2. Pitzer KS, Pelper JC, Busey RH (1984) Thermodynamic properties of aqueous sodium chloride solutions. J Phys Chem Ref Data 13(1):1–102
3. Barthel J, Lauermann G, Neueder R (1986) Vapor pressure measurements on non-aqueous electrolyte solutions. Part 2. Tetraalkylammonium salts in methanol. Activity coefficients of various 1-1 electrolytes at high concentrations. J Solution Chem 10:851–867
4. Barthel J, Neueder R, Poepke H, Wittmann H (1999) Osmotic coefficients and activity coefficients of nonaqueous electrolyte solutions. Part 2. Lithium perchlorate in the aprotic solvents acetone, acetonitrile, dimethoxyethane, and dimethylcarbonate. J Solution Chem 28:489–503
5. Nasirzadeh K, Neueder R, Kunz W (2005) Vapor pressures, osmotic and activity coefficients of electrolytes in protic solvents at different temperatures. 3. Lithium bromide in 2-propanol. J Solution Chem 34:9–24
6. Tsurko EN, Neueder R, Kunz W (2007) Water activity and osmotic coefficients in solutions of glycine, glutamic acid, histidine and their salts at 298.15 K and 310.15 K. J Solution Chem 36:651–672
7. Barthel J, Krienke H, Kunz W (1998) Physical chemistry of electrolyte solutions. Modern aspects. Springer, New York
8. Robinson RA, Stokes RH (2002) Electrolyte solutions, Second revised edition. Dover, Mineola
9. Barthel J et al (1992–2010) Electrolyte data collection, vol XII, Chemistry data series, Parts 1–5. DECHEMA, Frankfurt
10. Engels H (1990) Phase equilibria and phase diagrams of electrolytes, vol XI, Chemistry data series, Part 1. DECHEMA, Frankfurt
11. Frenkel M, Chirico RD, Diky V, Yan X, Dong Q, Muzny C (2005) ThermoData engine (TDE): software implementation of the dynamic data evaluation concept. J Chem Inf Model 45(4):816–838
12. Frenkel M, Chirico RD, Diky V, Muzny C, Dong Q, Marsh KN, Dymond JH, Wakeham WA, Stein SE, Königsberger E, Goodwin ARH, Magee JW, Thijssen M, Haynes WM, Watanasiri S, Satyro M, Schmidt M, Johns AI, Hardin GR (2006) New global communication process in thermodynamics: impact on quality of published experimental data. J Chem Inf Model 46(6):2487–2493
13. Barthel J, Popp H (1992) Methods of the knowledge based system ELDAR for the simulation of electrolyte solution properties. Anal Chim Acta 265: 259–266
14. Westhaus U, Sass R (2004) Reliable thermodynamic properties of electrolyte solutions – a survey of existing data sources. Z Phys Chem 218:1–8
15. http://www.uni-regensburg.de/Fakultaeten/nat_Fak_IV/Physikalische_Chemie/Kunz/eldar/eldhp.html
16. Westhaus U, Droege T, Sass R (1999) DETHERM – a thermophysical property database. Fluid Phase Equilib 158–160:429–435
17. http://www.dechema.de/en/detherm.html
18. http://www.cere.dtu.dk/Expertise/Data_Bank.aspx
19. http://www.ddbst.de
20. Onken U, Rarey-Nies J, Gmehling J (1989) The Dortmund data bank: a computerized system for retrieval, correlation, and prediction of thermodynamic properties of mixtures. J Int J Thermophys 10(3):739–747
21. http://jess.murdoch.edu.au/jess_home.htm
22. May PM, Rowland D, Königsberger E, Hefter G (2010) JESS, a joint expert speciation system – IV: a large database of aqueous solution physicochemical properties with an automatic means of achieving thermodynamic consistency. Talanta 81(1–2):142–148
23. http://www.springer.com/springermaterials
24. Holze R, Lechner MD (to be published in 2014) Electrochemistry, Subvolume B: Electrical conductivities and equilibria of electrochemical systems. Springer, Berlin, Heidelberg
25. Thomson GH, Larsen AH (1996) DIPPR: satisfying industry data needs. J Chem Eng Data 41:930–934

Defect Chemistry in Solid State Ionic Materials

H.-I. Yoo
Department of Materials Science and Engineering, Seoul National University, Seoul, Korea

Introduction

Solid state electrochemistry is concerned with all kinds of electrical phenomena associated with chemical changes, and vice versa, in solid state. These are induced by migration of charged mass particles or ions under the action of a variety of thermodynamic forces, gradients of component chemical potentials, electrical potential, temperature, stress, and the like. Naturally the solids composed of ions, viz., ionic solid compounds serve the main stages for solid state electrochemistry. In solid state, electrons may also be mobile, which makes solid state electrochemistry even more versatile and interesting.

An ionic compound, e.g., MO, consists of charged components, cations (M^{2+}), anions (O^{2-}) and electrons (e^-). These charged components are rendered mobile only via defects. It is thus a prerequisite in solid state electrochemistry to understand the defect structure of the system of interest, that is, types of defects and their concentrations against the thermodynamic variables of the system.

Defects are defined as whatsoever makes a crystalline solid deviate from its ideal crystal structure. An ideal crystal refers to an infinite crystal with component ions all sitting at their respective lattice sites in a periodic array as stipulated by its crystallographic structure, the interstitial sites all empty, electrons all at the valence band sites, and the conduction band sites all empty. For example, an "infinite" crystal $M^{2+}O^{2-}$ with M^{2+} on M^{2+} sites $\left(M^{2+}_{M^{2+}}\right)$; O^{2-} on O^{2-} sites $\left(O^{2-}_{O^{2-}}\right)$; electrons e^- on the filled valence-band e^- sites $\left(e^-_{e^-}\right)$; ionic emptiness or vacancy V^0 with actual charge 0 on interstitial sites where there should be no charge $\left(V^0_{I^0}\right)$; electronic emptiness or holes h^0 with actual charge 0 on the empty conduction band h^0 sites $\left(h^0_{h^0}\right)$.

We here denote the structure elements constructing the ideal structure MO as $S^m_{L^n}$, indicating the species S(=M, O, V, e, h), with an actual charge unit m(=⋯,−2, −1, 0, +1, +2,⋯), sitting at the site or locus L(=M, O, I, e, h) where there would have to be an actual charge unit n(=⋯,−2, −1, 0, +1, +2,⋯) if ideal. Defining the effective charge of a structure element as c = m–n and denoting it, instead of numerals, as the same number of dots (•) and primes (′) for $c > 0$ and $c < 0$, respectively and a cross (x) for $c = 0$, the structure elements constructing the ideal structure MO may be simplified in the form of S^c_L as suggested by Kröger and Vink [1] as

$$M^x_M,\ O^x_O,\ V^x_I,\ e^x_e,\ h^x_h$$

which are often called the regular structure elements.

Anything which makes the structure deviate from the ideal one or defects or irregular structure elements may then be immediately sorted out as:

Missing cation or cation vacancy V''_M; missing anion or anion vacancy $V^{\bullet\bullet}_O$; interstitial cation $M^{\bullet\bullet}_I$; interstitial anion O''_I; misplaced cation $M^{\bullet\bullet\bullet\bullet}_O$; misplaced anion O''''_M; electron in empty conduction-band e'_h; missing electron or hole in filled valance-band $h^\bullet_e$; impurity or alien ion substituting cation, A^c_M or substituting anion, A^c_O; interstitial impurity or alien ion A^c_I.

Each of these is confined to a site or point, thus, they are called point defects. There are more: dislocations disturbing the periodicity of lattice sites; grain boundaries and surfaces spatially confining the crystal which would have to be infinite if ideal; voids and inclusions that are three-dimensional aggregates of point defects of a kind. Depending on their geometries, they are often called line defects, planar defects and volume defects, respectively.

Formation of a defect always requires work exerted on the system, thus increasing the energy of the system. Therefore all these defects can only be stabilized by configurational entropy gain as temperature goes up. In this sense, only point defects are thermodynamically stable in the normal temperature range of existence of ionic solids. It means that their concentrations are uniquely determined by the thermodynamic variables of the system in equilibrium state. For higher dimensional defects, the entropy gain can hardly compensate the energy increase, if not impossible.

In this chapter, we will discuss the generation modes of thermodynamically stable point defects and the defect-chemical logic to calculate the equilibrium defect structure of a given system. As a stereotype of systems, we will consider only a binary oxide MO, but the idea and logic can be readily extended to other binary, ternary and higher systems with minor modifications [2–6].

Intrinsic or Thermal Disorders

For a given system, the ratio of the M-sites to O-sites must remain fixed in any case, as required by its crystallographic structure. For the case of e.g., MO,

$$[\text{M-sites}]/[\text{O-sites}] \equiv 1$$

where [ ] stands for the concentration of the thing therein. Whenever whatsoever defects are generated, this site ratio or structure condition must be observed by necessity. Otherwise, the structure would be no longer the structure of the system itself.

For pure MO, possible ionic defects are exhaustively V_M'', $M_I^{\bullet\bullet}$, $V_O^{\bullet\bullet}$, O_I'', $M_O^{\bullet\bullet\bullet\bullet}$, O_M''''. These defects can only be generated in pair involving either M- or O-sublattice alone or both sublattices in order to maintain the structure as

$$0 = V_M'' + V_O^{\bullet\bullet} \tag{1a}$$

$$M_M^x + O_O^x = M_I^{\bullet\bullet} + O_I'' \tag{2a}$$

$$M_M^x + V_I^x = M_I^{\bullet\bullet} + V_M'' \tag{3a}$$

$$O_O^x + V_I^x = O_I'' + V_O^{\bullet\bullet} \tag{4a}$$

$$M_M^x + O_O^x = M_O^{\bullet\bullet\bullet\bullet} + O_M'''' \tag{5a}$$

In these generation reactions, one should recognize that in addition to the structure condition or site (L) conservation, charge (c) and mass (S) are conserved, as a Kröger-Vink symbol S_L^c suggests. These three must be conserved in any case.

If the defect concentrations are small enough compared to those of the regular structure elements as is normally the case, one may apply the mass action law to write for each case in order as:

$$[V_M''][V_O^{\bullet\bullet}] = K_s \tag{1b}$$

$$[M_I^{\bullet\bullet}][O_I''] = K_{aS} \tag{2b}$$

$$[M_I^{\bullet\bullet}][V_M''] = K_F \tag{3b}$$

$$[O_I''][V_O^{\bullet\bullet}] = K_{aF} \tag{4b}$$

$$[M_O^{\bullet\bullet\bullet\bullet}][O_M''''] = K_a \tag{5b}$$

Here, the mass action law constant K_j(j = S, aS, F, aF, a) takes the form

$$K_j = K_{j,o} \exp\left(\frac{\Delta h_j}{kT}\right) \tag{6}$$

where $K_{j,0}$ and Δh_j are the pre-exponential factor and standard enthalpy change of the generation reaction of j-type and kT has the usual significance. These defect pairs are named, in order, the Schottky disorder (j = S), anti-Schottky disorder(j = aS), Frenkel disorder (j = F), anti-Frenkel disorder (j = aF), and anti-structure disorder (j = a). All these defects are only internally generated for the entropic reason and hence, called the intrinsic or thermal disorder. The defect pair which is energetically least costly in a given structure usually overwhelms the rest, thus making the majority disorder type. The anti-structure disorder, however, will be too costly in ionic solids, thus normally out of concern.

In a similar way, electronic disorders are generated internally or thermally as

$$e_e^x + h_h^x = h_e^{\bullet} + e_h' \quad (7a)$$

It is here noted that even in generating the electronic defects, the three (S_L^c) should be conserved, viz., the densities of states in the valence and conduction band, charge and mass. By defining free holes and electrons as $h^{\bullet} \equiv h_e^{\bullet} - e_e^x$ and $e' \equiv e_h' - h_h^x$, respectively, this intrinsic electronic excitation reaction is often represented more succinctly as

$$0 = e' + h^{\bullet} \quad (7b)$$

Again applying the mass action law, one may write

$$[e_h'][h_e^{\bullet}] = K_i \quad (7c)$$

where K_i is called the intrinsic electronic excitation equilibrium constant. In terms of semiconductor jargons, $[e_h'] = [e'] = n$ and $[h_e^{\bullet}] = [h^{\bullet}] = p$ in number concentration and then

$$np = K_i = N_v N_c \exp\left(-\frac{E_g}{kT}\right) \quad (7d)$$

where N_v and N_c are the effective density of states at the valence (e^- sites) and conduction band edge (h^0 sites), respectively, and E_g the band gap.

Impurity-Induced Disorders

Nothing is pure for the thermodynamic reason. An impurity can be incorporated into the host lattice either substitutionally or interstitially depending mostly on the relative size of the impurity to host ion. For simplicity sake, we here consider two types of cation impurities only: one with a lower valence and the other with a higher valence than the host ion M^{2+} in MO. It is noted that our system exchanges mass particles with the surrounding only in electrically neutral forms, because it would otherwise be electrically charged to energetically prevent further exchange. It is thus always easier to consider an impurity doping as incorporating the impurity in the form of an oxide when the host is MO.

Let us first consider the incorporation of A_2O in MO. There can be two possibilities, A_M' or $A_I^{\bullet}$, which are charge-compensated by generating oppositely charged native defects, either $V_O^{\bullet\bullet}$ or $M_I^{\bullet\bullet}$ and either V_M'' or O_I'', respectively, or by themselves if amphoteric. Incorporation reactions may thus be formulated exhaustively as:

$$A_2O = 2A_M' + V_O^{\bullet\bullet} + O_O^x \quad (8a)$$

$$A_2O = 2A'_M + M_I^{\bullet\bullet} + 2O_O^x \quad (8b)$$

$$A_2O = 2A_I^{\bullet} + V_M'' + O_O^x \quad (8c)$$

$$A_2O = 2A_I^{\bullet} + O_I'' \quad (8d)$$

$$A_2O = A_M' + A_I^{\bullet} + O_O^x \quad (8e)$$

It is informative to write down the lattice molecular formula for each case. For xA_2O added, they are in order: $M_{1-2x}A_{2x}O_{1-x}$; $M_{1-x}A_{2x}O$; $M_{1-x}A_{2x}O$; $MA_{2x}O_{1+x}$; $M_{1-x}A_{2x}O$. This means that the mechanisms, Eqs. 8b, 8c, and 8e cannot be distinguished chemically.

We will next consider the case where the impurity with higher valence E_2O_3 is incorporated. Following the same logic as before, we may write the incorporation equations exhaustively as

$$E_2O_3 = 2E_M^{\bullet} + V_M'' + 3O_O^x \quad (9a)$$

$$E_2O_3 = 2E_M^{\bullet} + O_I'' + 2O_O^x \quad (9b)$$

$$E_2O_3 = 2E_I^{\bullet\bullet\bullet} + 3V_M'' + 3O_O^x \quad (9c)$$

$$E_2O_3 = 2E_I^{\bullet\bullet\bullet} + 3O_I'' \quad (9d)$$

The lattice molecular formulae for xE_2O_3 added are in turn $M_{1-3x}E_{2x}O$; $M_{1-2x}E_{2x}O_{1+x}$; $M_{1-3x}E_{2x}O$; $ME_{2x}O_{1+3x}$. In this case, they are all chemically distinguished from each other.

Which of these multiple possibilities is really responsible in a given structure is again determined by which is the least costly. Rule of thumb is that the charge compensating native defect is determined by the majority type of thermal disorder for the given system.

Redox-Induced Disorders

No compound exists in one and only fixed composition again for the thermodynamic reason. For example, MO should exist over a range of composition or homogeneity range between the M-saturated and O-saturated composition. The compound should, thus, be represented more appropriately as $M_{1-\delta}O$ or $MO_{1+\delta}$. Its homogeneity range may then be divided, in general, into three regions: hypostoichiometric ($\delta < 0$), near-stoichiometric ($\delta \approx 0$), and hyperstoichiometric ($\delta > 0$). In this light, the stoichiometric composition MO is nothing but a special composition in the region $\delta \approx 0$ or a symbolic representation of the compound $M_{1-\delta}O$. An extreme example will be "FeO," which does not even exist within its homogeneity range ($Fe_{1-\delta}O$ with $\delta > 0$ always).

For the composition of a system to vary, component particles should be exchanged with its surrounding, and the exchange is driven by a difference in component chemical potentials across the boundary. Once component chemical potential distributions are rendered uniform, particle exchange ceases and then the system is said to be in external equilibrium. For the case of binary $MO_{1+\delta}$, the number of composition variables is only one (δ), and only one of the two component chemical potentials or activities a_M and a_{O2} can be varied independently at given temperature T and pressure P due to the Gibbs-Duhem equation:

$$\mathrm{d}\ln a_M + \frac{1}{2}(1+\delta)\mathrm{d}\ln a_{O2} = 0 \qquad (10)$$

When the external equilibrium condition is disturbed, the system MO may change its composition by exchange of both components M and O in principle, but normally by the exchange of the more volatile component which is O for the case of oxides.

When the oxygen chemical potential in the surrounding is higher than that in the solid oxide, component oxygen tends to be incorporated to raise the oxygen content by creating oxygen interstitials or metal vacancies depending on which are energetically cheaper. This process is called oxidation. In the opposite case, oxygen leaves the crystal to lower the oxygen content by leaving behind oxygen vacancies or metal interstitials depending on which are energetically cheaper. This is called reduction. What is eventually achieved is reduction or oxidation equilibrium, thus often dubbed as redox equilibrium.

It is noted that because only neutral oxygen is exchanged, what are left behind should also be electrically neutral. A neutral interstitial oxygen may be regarded as a normal oxygen ion interstitial O_I'' bearing two free holes $2h^\bullet$ within it and a neutral metal vacancy as a missing metal ion V_M'' bearing two free holes $2h^\bullet$ within it or

$$O_I^x = \left(O_I'', 2h^\bullet\right)^x;\; V_M^x = \left(V_M'', 2h^\bullet\right)^x \qquad (11)$$

Likewise,

$$M_I^x = \left(M_I^{\bullet\bullet}, 2e'\right)^x;\; V_O^x = \left(V_O^{\bullet\bullet}, 2e'\right)^x \qquad (12)$$

In this sense, the former two bearing holes (or missing electrons) are like electron acceptors and the latter two bearing extra electrons electron donors in elemental semiconductors. They are indeed so: $MO_{1+\delta}$ always tends to be of p-type when $\delta > 0$ and of n-type if $\delta < 0$. For the sake of simplicity, we here assume that once these are generated, they immediately donate all holes (or equivalently, accept electrons) or donate all electrons. This is termed "fully ionized" and actually happens at elevated temperatures.

Oxidation reactions and corresponding mass action laws may then be written as:

$$\frac{1}{2}O_{2(g)} = O_I^x = O_I'' + 2h^\bullet : \qquad (13a)$$
$$[O_I'']p^2 = K_{Ox,1}a_{O_2}^{1/2}$$

$$\frac{1}{2}O_{2(g)} = O_O^x + V_M^x = O_O^x + V_M'' + 2h^{\bullet} :$$
$$[V_M'']p^2 = K_{Ox,2}a_{O_2}^{1/2} \tag{13b}$$

Reduction reaction equilibria as:

$$O_O^x = \frac{1}{2}O_{2(g)} + V_O^x = \frac{1}{2}O_{2(g)} + V_O^{\bullet\bullet} + 2e' :$$
$$[V_O^{\bullet\bullet}]n^2 = K_{Re,1}a_{O_2}^{-1/2} \tag{13c}$$

$$M_M^x + O_O^x = \frac{1}{2}O_{2(g)} + M_I^x$$
$$= \frac{1}{2}O_{2(g)} + M_I^{\bullet\bullet} + 2e' : \tag{13d}$$
$$[M_I^{\bullet\bullet}]n^2 = K_{Re,2}a_{O_2}^{-1/2}$$

These four redox reaction equilibrium constants K_j (j = Ox,1; Ox,2; Re,1; Re,2) each take the shape as in Eq. 6. The reader, however, should note that these four K_j's are not all independent of each other owing to the internal equilibria, Eqs. 1 to 4 or $K_{Ox,1}K_{Re,1} = K_{aF}K_i^2$; $K_{Ox,1}K_{Re,2} = K_{aS}K_i^2$; $K_{Ox,2}K_{Re,1} = K_SK_i^2$; $K_{Ox,2}K_{Re,2} = K_FK_i^2$. It turns out that only one out of the four is independent and hence, only one out of the four is enough to describe the redox equilibrium of the binary system MO. If ternary, there would be two composition variables and hence, there would be two external equilibria.

Defect Structure of "Pure" Nonstoichiometric Compound, $MO_{1+\delta}$

Nothing can be absolutely pure. Here "pure" means that the majority type of intrinsic disorder overwhelms impurities in concentration. Such a compound is said to be in its "intrinsic regime."

In order to calculate the equilibrium defect structure of a compound whether intrinsic or extrinsic, one should first postulate the most likely defects on the basis of the structure of MO. Let us suppose that our system MO has the Schottky disorder as the majority type of ionic disorder. The defects of present interest may then be listed as

$$V_M'', V_O^{\bullet\bullet}; e', h^{\bullet}$$

We therefore need four equations to calculate these four concentrations as functions of the thermodynamic intensive variables of the system $MO_{1+\delta}$, T, P and a_{O_2}. These equations are formulated from the requirements that the system has to meet:

1. 2 internal equilibria; Eqs. 1b and 7c
2. 1 external equilibrium: Eq. 13a or 13c
3. 1 charge neutrality: $n + 2[V_M''] = p + 2[V_O^{\bullet\bullet}]$

One may solve, in principle, these four simultaneous equations for each defect concentration as a function of T, P (via K_j) and a_{O_2}, but analytic solution is impossible in many cases because of different algebraic character of the charge neutrality condition compared to the other constraints. If the latter, however, can be approximated in terms of only one pair of oppositely charged defects or majority disorder type depending on δ-ranges, the analytic solution will be rendered trivial in the form of

$$[S_L^c] = \prod_j K_j^{n_j} a_{O_2}^m \tag{14}$$

where n_j and m are the exponents. This trick was first proposed by Brouwer, thus called the Brouwer approximation [1–3]. It goes as follows:

The charge neutrality condition may be approximated to the 2 × 2 limiting charge neutrality conditions or majority disorder types:

$$\text{(i)}\ n \approx 2[V_O^{\bullet\bullet}](\delta < 0);$$
$$\text{(ii)}\ n \approx p(\delta \approx 0);$$
$$\text{(iii)}\ [V_M''] \approx [V_O^{\bullet\bullet}](\delta \approx 0);$$
$$\text{(iv)}\ 2[V_M''] \approx p(\delta < 0)$$

The problem is then how to allocate these majority disorder types along the axis of oxygen activity at a given temperature. It is reminded that the homogeneity range of $M_{1-\delta}O$ is generally divided into the three regions: $\delta < 0$; $\delta \approx 0$; $\delta > 0$. The nonstoichiometry as defined here as oxygen excess or metal deficit is given in the present case as

$$\delta = \left[V_M''\right] - \left[V_O^{\bullet\bullet}\right] = \frac{1}{2}(p - n) \qquad (16)$$

due to the charge neutrality condition. By using this relationship, each majority disorder type can be assigned to one of the three δ-ranges, which is already done in Eq. 15. Now one can see there are two majority disorder types (ii) and (iii) simultaneously in the near-stoichiometry region causing a logical conflict in appearance, but it is a matter of whether $K_i >> K_S$ or $K_S >> K_i$ for the system given. If the former is the case, sequence of the majority disorder types will be (i) → (ii) → (iv) with increasing oxygen activity a_{O_2}; if the latter, (i) → (iii) → (iv). The defect structure of the system across its entire range of existence can finally be constructed by combining the piecewise solutions, Eq. 14 for each disorder regime, in accord with the sequence of majority disorder types. Figure 1 shows an example for $K_i >> K_S$.

Defect Structure of "Impure" Nonstoichiometric Compound, $M_{1-2x}A_{2x}O_{1-x+\delta}$

Let us next calculate the defect structure of a more general case, A_2O-doped MO. Here we assume that A substitutes M and MO has the anti-Frenkel disorder as the majority type of ionic disorder. Then we may list the defects of the most concern as

$$A_M', V_O^{\bullet\bullet}, O_I'', e', h^{\bullet}$$

We thus need five constraints for these five unknowns, which comprise:

1. Two internal equilibria: Eqs. 4b and 7c
2. One external equilibrium, Eq. 13a or 13c
3. One mass conservation for the dopants, $[A'_M] = x$ (constant)
4. One charge neutrality condition, $n + \left[A_M'\right] + 2\left[O_I''\right] = p + 2\left[V_O^{\bullet\bullet}\right]$

These are all and exhaustive. The last charge neutrality condition may be approximated to 3 × 2 limiting conditions or majority disorder types as:

$$\begin{aligned}
&\text{(i)}\ n \approx p(\delta \approx 0);\\
&\text{(ii)}\ n \approx 2\left[V_O^{\bullet\bullet}\right](\delta < 0);\\
&\text{(iii)}\ \left[A_M'\right] \approx p(\delta > 0);\\
&\text{(iv)}\ \left[A_M'\right] \approx 2\left[V_O^{\bullet\bullet}\right](\delta \approx 0);\\
&\text{(v)}\ 2\left[O_I''\right] \approx p(\delta > 0);\\
&\text{(vi)}\ [O_I''] \approx [V_O^{\bullet\bullet}](\delta \approx 0)
\end{aligned} \qquad (17)$$

Defining oxygen nonstoichiometry of the present xA_2O-doped MO as $M_{1-2x}A_{2x}O_{1-x+\delta}$,

$$\delta = \left[O_I''\right] - \left[V_O^{\bullet\bullet}\right] + x = \frac{1}{2}(p - n) \qquad (18)$$

due to the charge neutrality condition. Each majority disorder type is now assigned to each of the three δ-regions as indicated in Eq. 17. It should first be noted that once "impure" or extrinsic, the intrinsic disorder types should not be in majority in the near-stoichiometry region (δ ≈ 0) by definition, thus, the possibilities (i) and (vi) are already ruled out. Then noting that any two contiguous disorder regimes should have one defect in common, which is termed the "continuity principle," [6] one may establish the sequence of majority disorder types with increasing a_{O_2} as (ii) → (iv) → (iii) → (v). By combining the piecewise solution, Eq. 14 in each region in this sequence, one finally ends up with the defect map for the present system as shown in Fig. 2.

Defect Structure of Impure "Stoichiometric" Compound MO

Finally, we consider a special case, a "stoichiometric" compound, which normally refers to a compound with negligible nonstoichiometry. We have seen that for a nonstoichiometry to be generated, electronic defects are always involved accompanying the component particle exchanges with the surrounding. As Eqs. 11 and 12 suggest, the range of

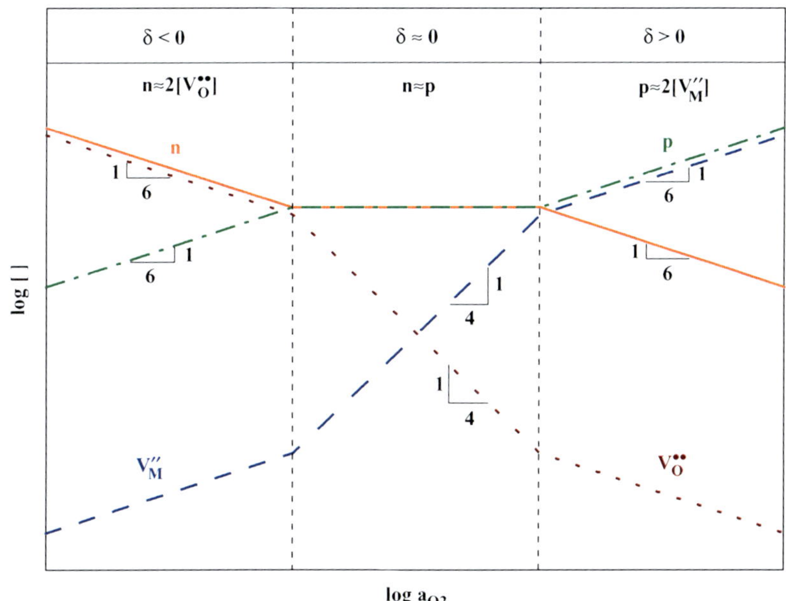

Defect Chemistry in Solid State Ionic Materials, Fig. 1 Defect structure of $M_{1-\delta}O$ versus oxygen activity at fixed temperature, as calculated assuming $K_i >> K_S$ (not to scale). The *triangles* represent the oxygen exponents "m" of the piecewise solutions

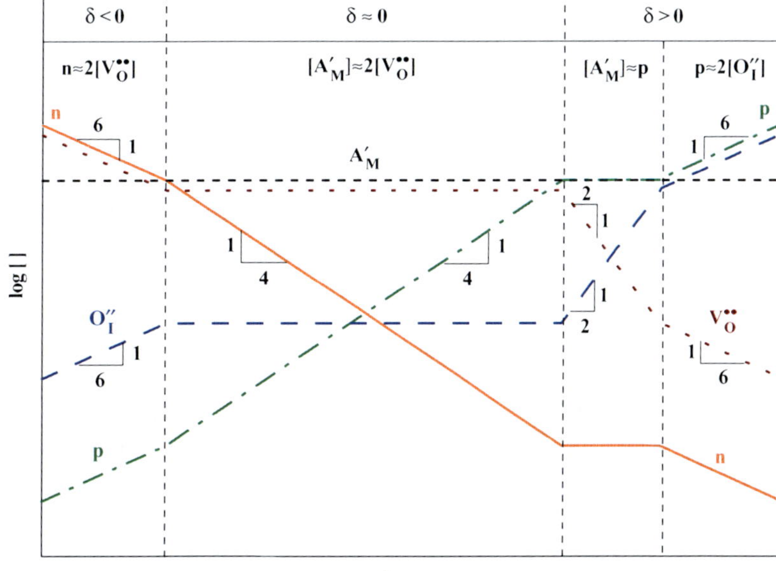

Defect Chemistry in Solid State Ionic Materials, Fig. 2 As-calculated defect structure of extrinsic $M_{1-2x}A_{2x}O_{1-x+\delta}$ versus oxygen activity at fixed temperature, not to scale. Note the continuity principle between nearest-neighboring disorder regimes. The *triangles* represent the oxygen exponents "m" of the piecewise solutions

nonstoichiometry is determined by the variability of the valences of the component ions from chemistry point of view. This is why transition metal compounds normally exhibit wider ranges of homogeneity. Therefore the compounds with almost fixed-valent ions, e.g., alkali halides, alkali earth oxides, and the like may be regarded "stoichiometric" and their electronic defects may also be neglected compared to the ionic defects. Once stoichiometric, the composition is fixed and hence, it is as if the system boundary were closed, therefore external equilibrium is no longer a concern. Then the defect structure will be determined only by the internal equilibria or

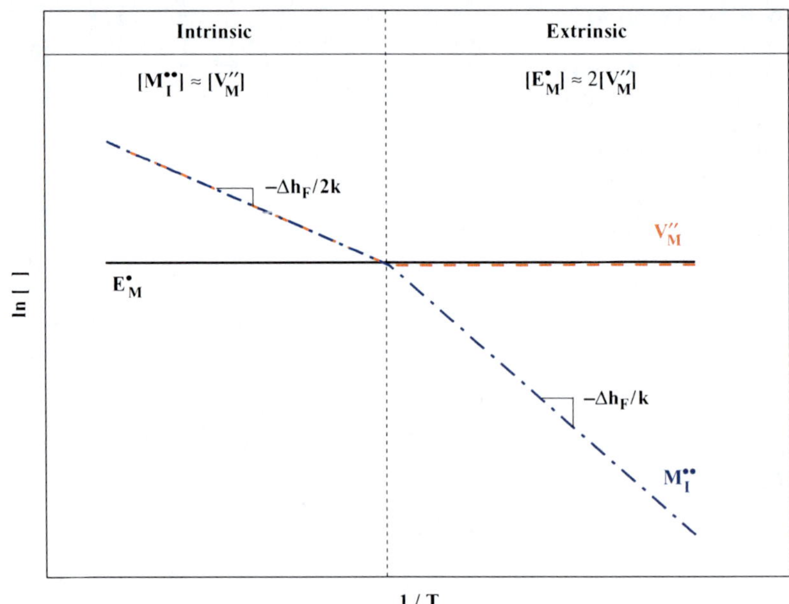

Defect Chemistry in Solid State Ionic Materials, Fig. 3 Defect structure of E_2O_3 doped, "stoichiometric" MO with the Frenkel disorder as the majority, not to scale. *Small triangles* denote the slopes. Note the intrinsic regime at the high temperatures and extrinsic regime at the low temperatures

temperature and impurity content no matter how many components there are.

Suppose that our "stoichiometric" MO has the Frenkel disorder as the majority and substitutional impurity E_2O_3 (see Eq. 9a). Then the defects of concern will be:

$$E^{\bullet}_{M},\ V''_{M},\ M^{\bullet\bullet}_{I}$$

We have now

1. One internal equilibrium: Eq. 3b
2. One mass conservation: $\left[E^{\bullet}_{M}\right] = x(\text{constant})$
3. One charge neutrality: $\left[E^{\bullet}_{M}\right] + 2\left[M^{\bullet\bullet}_{I}\right] = 2\left[V''_{M}\right]$

By solving these three equations, one can get each defect concentration as a function of T and x independently of the component activities. The present situation can be solved even analytically, but for the consistency sake, we will employ the Brouwer approximations as: $\left[E^{\bullet}_{M}\right] \approx 2\left[V''_{M}\right]$; $\left[M^{\bullet\bullet}_{I}\right] \approx \left[V''_{M}\right]$. The former prevails at low temperatures such that $x >> K_S^{1/2}$ and the latter at high temperatures such that $x << K_S^{1/2}$. By combining the piecewise solutions for each temperature region, one obtains the defect structure as in Fig. 3. Note that in the high temperature intrinsic regime, the impurity concentration is overwhelmed by the intrinsic disorder and vice versa in the low temperature extrinsic regime.

Concluding Remarks

The very basic logical framework of defect chemistry has thus far been given to a binary system. This can be readily extended to ternary or higher systems simply by adding external equilibria each corresponding to additional composition variables in addition to the exhaustive internal equilibrium conditions, mass conservation condition, and charge neutrality condition. If the nonstoichiometry is negligible, thus, the composition practically remains fixed, however, one may disregard the external equilibria.

The defect structure gets more involved as defect concentrations increase so that the mass-action-laws are no longer applicable. This basic logical framework can also be extended by taking into account defect associates or long range interactions in terms of activity coefficients of defects.

For more extended treatments, the reader is referred to the general references [1–6].

Cross-References

▶ Defects in Solids
▶ Kröger-Vinks Notation of Point Defects

References

1. Kröger FA, Vink VJ (1956) Relations between the concentrations of imperfections in crystalline solids. In: Seitz F, Turnbull D (eds) Solid state physics, vol 3. Academic, New York
2. Kingery WD, Bowen HK, Uhlmann DR (1960) Introduction to ceramics. Wiley, New York (Chap. 4)
3. Kröger FA (1974) Imperfection chemistry of crystalline solids, vol 2, 2nd edn, The chemistry of imperfect crystals. North-Holland, Amsterdam
4. Schmalzried H (1964) Point defect in ternary ionic crystals. In: Reiss H (ed) Progress in solid state chemistry, vol 2. North-Holland, Amsterdam
5. Smyth DM (2000) The defect chemistry of metal oxides. Oxford University Press, New York
6. Yoo HI (2010) Defect structure, nonstoichiometry, and nonstoichiometry relaxation of complex oxides. In: Riedel R, Chen IW (eds) Ceramic science and technology, vol 2. Wiley-VCH, Weinheim

Defects in Solids

Ulrich Guth
Kurt-Schwabe-Institut für Mess- und Sensortechnik e.V. Meinsberg, Waldheim, Germany
FB Chemie und Lebensmittelchemie, Technische Universität Dresden, Dresden, Germany

Introduction

In a perfect solid, atoms, molecules, or ions are arranged in a three-dimensional lattice. The removal of any particle from that needs the overcome of the lattice force. Therefore, the perfect lattice of a solid is an ideal state. The real solid is imperfect respectively the arrangement of the particles (disorder in arrangement), the different orientation (orientation disorder), and not uniform in vibration and rotation (movement disorder). Most important for different properties like mechanical and electrical behavior is the nonideal arrangement of the lattice building blocks. According to the extension, it can be distinguished in three-, two-, one-, and zero-dimensional disorder, respectively. Most important for the understanding of electrical conductivity as well as for the diffusion in solids is the zero-dimensional disorder (point defects). From the view of thermodynamics, the generation of defects in solids is understandable. Only at temperature 0 K, the lattice is perfect; that means all particles take their normal positions and the entropy is zero. Due to the entropy effect, the amount of defects increases with increasing temperature and therefore the Gibbs energy is diminished. At a certain concentration of defects, the free energy increases because of the forming of defects has no influence on the entropy. Those defects which are formed without changing of stoichiometry and arise from the solid system itself according to the thermodynamic conditions are called as intrinsic defects or sometimes as stoichiometric defects. According to the place in the lattice, atoms, ions, or molecules move from their normal positions in the lattice into free spaces of the lattice between particles into the so-called interstitials (Frenkel defects) or to the surface (Schottky defects) [1].

Intrinsic Disorder

Besides the Frenkel and the Schottky disorders, also the anti-Frenkel and anti-Schottky disorders exist. But more important are the Frenkel and Schottky types. In the case of sodium sulfate, sodium ions on the normal lattice position Na_{Na}^{x} (the notation of Kröger-Vink is used; see entry "▶ Kröger-Vinks Notation of Point Defects") go into free space of ions (interstitials) and sodium vacancies remain (Frenkel defects):

$$Na_{Na}^{x} \rightleftharpoons Na_{Na}^{\bullet} + V_{Na}^{\prime}$$

Additional to the particles and mass balance as well as to the electroneutrality of chemical equation, the availability of free space has to be taken into account as well so that the equation has to be written in exact manner:

$$Na^{x}_{Na} + V_i \rightleftharpoons Na^{\bullet}_{Na} + V'_{Na}$$

However, the number of interstitials is very high and does not change practically due to forming of defects of ions so that the V_i can be omitted from the equation.

According to the mass action law, the defect concentration is a function of temperature. Due to the same mole fraction x of vacancies and sodium ions on interstitials, the mass action constant can be expressed by

$$\sqrt{K_x} = x_{Na^{\bullet}_{Na}} = x_{V'_{Na}}$$

With the Gibbs free energy of formation $\Delta_f G = -RT \ln K_x$, the mole fraction (concentration) of defects $x_{Na^{\bullet}_{Na}} = x_{V'_{Na}}$ is given by

$$\begin{aligned} x_{Na^{\bullet}_{Na}} &= \sqrt{K_x} = \exp(-\Delta_f G/2RT) \\ &= \exp(-\Delta_f H/2RT)\exp(-\Delta_f S/2R) \propto \sigma \end{aligned}$$

where K_x is the mass action constant. The concentration of defects is proportional to the electric conductivity σ that can be measured in dependence on temperature where $\Delta_f H \, and \Delta_f S$ are the enthalpy and the entropy of defect formation, respectively. The slope of the plot log σ vs. 1/T is equal to $\Delta_f H/2R$

On the other hand, the electrical conductivity depends on the electrical mobility of the mobile species u_{Na} as well:

$$\sigma \propto x_{V'_{Na}} u_{Na}$$

The mobility u, in turn, depends exponentially from the temperature with the activation energy E_a which is necessary to overcome the potential barrier (diffusion by vacancy or interstitial mechanism) when the ion is moving from one site to the other one:

$$u = u_0 \exp(-E_a/RT)$$

For the temperature dependence of the electrical conductivity σ, the product of two exponential functions is obtained:

$$\sigma = const. \ \exp(-\Delta_f H/2RT)\exp(-E_a/RT)$$

The first exponential expression is thermodynamic origin and the second is the kinetic one.

In the Schottky types of disorder, a pair of ions moves from the normal sites to the surface of the solid [2–4]. For LiI the Schottky equilibrium can be expressed by

$$Null \rightleftharpoons V'_{Li} + V^{\bullet}_{I}$$

For this equation the mass action law can be applied as well. The number of Schottky defects N_V depends exponentially from the temperature. A similar expression as for Frenkel defects is obtained (N is the total number of sites):

$$N_V = const. N \exp(-\Delta_f H/2RT)$$

Defects can be also generated due to the interaction of a gas with its ion formed from it in the solid. Oxides which are in close contact with oxygen-containing gas phase are able to release or to built in oxygen as oxide ion from or into the lattice.

The release of oxygen from the normal lattice position (O^{x}_{O}) is connected with the generation of excess electrons (e'). From the chemical point of view, these electrons can be located at the ion in a lower valence state. In the case of oxides like TiO_2, SnO_2, and ZrO_2, three valent ions Ti^{3+}, Sn^{3+}, and Zr^{3+} can be formed. This is in accordance with the gray or black color which is typical for lower-valent compound and for the electronic conductivity. Physically spoken the excess electrons are in the conductivity band:

$$O^{x}_{O} \rightleftharpoons \tfrac{1}{2}O_2(gas) + V^{\bullet\bullet}_{O} + e'$$

The incorporation of oxygen can occur on interstitials with forming of electronic holes, e.g.,

$$\tfrac{1}{2}O_2(gas) + V^{\bullet\bullet}_{O} \rightleftharpoons O^{x}_{O} + 2h^{\bullet}$$

The holes can be imagined as oxygen ion with one negative charge O^{-}. In the Kröger-Vink notation, it can be expressed as $O^{\bullet}_{O} = h^{\bullet}$. In both cases the mass action law can be applied. This leads to a logarithmic expression for the oxygen partial pressure dependence on the electrical conductivity. In the double logarithmic plot of σ vs p_{O_2}, a slope of 1/4...1/6 is obtained. In the case of an excess of electrons, the sign of the slope is negative, whereas for the hole conductivity, a positive sign can be expected.

Extrinsic Disorder

Extrinsic disorder can be established by admixing of alio-valent ions to the basic substance without changing the phase (homogeneous doping). That can be done for cations as well as for anions. In such cases a built-in equation is used. For the system $(Na_2SO_4)_{0.95}(BaSO_4)_{0.05} = Na_{1.9}Ba_{0.05}SO_4$, the following equation is obtained:

$$BaSO_4 \xrightarrow{Na_2SO_4} Ba^{\bullet}_{Na} + V'_{Na}$$

The anion lattice is complete. Ba^{++} has two positive charges and can compensate double amount of Na^+ so that in the cation lattice 0.05 mol remains empty as a negatively charged sodium ion vacancy V'_{Na}. The electroneutrality is ensured.

Yttria-stabilized zirconia (YSZ), e.g., the Nernst mass $(ZrO_2)_{0.84}(Y_2O_3)_{0.08} = Zr_{0.84}Y_{0.16}O_{1.92}$, is an example for a complete cation lattice and an incomplete anion lattice:

$$Y_2O_3 \xrightarrow{ZrO_2} 2Y'_{Zr} + V^{\bullet\bullet}_O + 3O^x_O$$

That means the cation sites result 0.84 + 0.16 = 2. The cation lattice is complete because the Y^{3+} ions occupy Zr^{4+} sites ($Y^{3+}_{Zr^{4+}}$ and in the Kröger-Vink notation Y'_{Zr}). For two Y^{3+} ions, a double positively charged oxide ion vacancy $V^{\bullet\bullet}_O$ is formed, in this case 0.08. The extrinsic disorder is often called as chemical disorder, mixed phase, or nonstoichiometric disorder.

In the real solid, all kinds of defects occur and have to take into account in a defect balance so

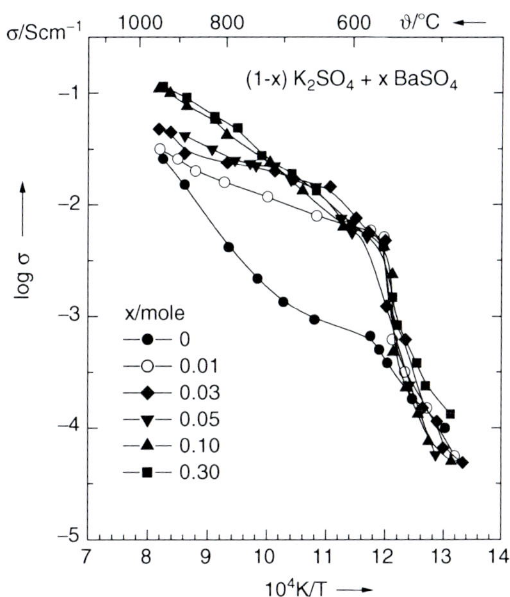

Defects in Solids, Fig. 1 Electrical conductivity in the system K_2SO_4-$BaSO_4$

that the concentrations of negatively and positively charged particles are equal. But mostly one type of defects is dominated. The role of intrinsic and extrinsic disorder can be illustrated in Fig. 1. In the range of 600–900 °C, the conductivity of pure and low-doped (up to 3 mol%) K_2SO_4 is determined by extrinsic disorder. The number of defects is nearly proportional to the concentration of dopant $BaSO_4$. At temperatures higher than 900 °C, all curves are merged by trend. In this range the conductivity is dominated by the intrinsic disorder. At higher amount of $BaSO_4$ (10–30 mol%), the system becomes heterogeneous and an increase of conductivity by heterogeneous doping (see below) can be observed.

Heterogeneous Doping

If an insulating material, e.g., γ-Al_2O_3, is mixed with an ionic conductor like LiI or Na_2SO_4, the electric conductivity increases due to the generation of defects on the surface of the ionic conductor. This effect is sometimes called as

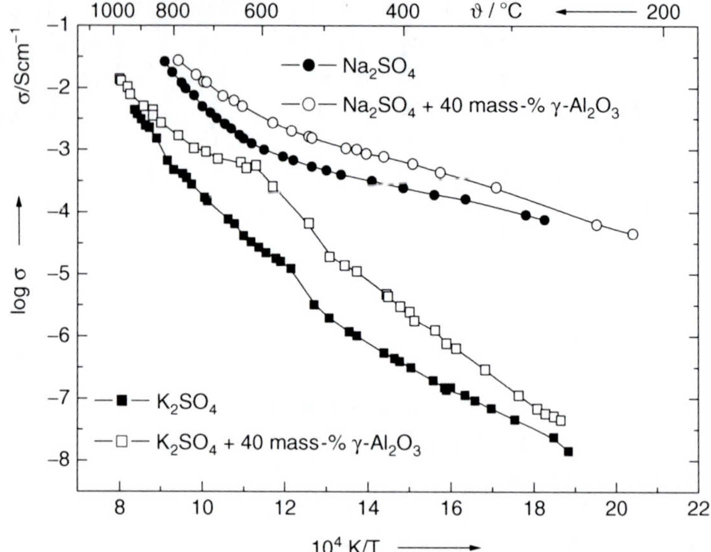

Defects in Solids, Fig. 2 Electrical conductivity of pure alkaline sulfates and composites consisting of salts with 40 mass% γ-Al_2O_3

"heterogeneous doping." That means in a two-phase system, Al^{3+} attracts counter ions in the ionic conductor so that ionic defects in the surface layer are formed [4]:

$$Na_{Na}^{x} + V_A \rightarrow Na_A^{\bullet} + V_{Na}^{'}$$
$$Na_A^{\bullet} + V' \rightarrow Na_i^{\bullet} + V_A$$

$Na_A^{\bullet}$ *and* $V_{Na}^{'}$ mean a sodium ion and vacancy at the grain boundary, respectively. This effect is caused by the interaction between the grains of electrolyte that disperse in fine grains of insulation material like γ-Al_2O_3 and depends on the volume fraction and the grain size of the dispersoid. This effect is illustrated in Fig. 2. The addition of 40 mass% γ-Al_2O_3 to sodium sulfate as well as to potassium sulfate provokes an increase in electrical conductivity.

Super Ionic Conductors

Besides the described disorder in some solids, there are more places than ions or the ions can occupy different sites. Such compounds like α-AgI or δ-Bi_2O_3 are called as super ionic conductors (see entry "▶ Solid Electrolytes"), and this disorder is described as structural disorder. Sometimes, it was distinguished between super ionic conductors of type I which show the conduction phenomenon over the whole temperature range like → β-alumina Na_2O*11 Al_2O_3, NASICON $Na_{1+x}Zr_2Si_xP_{3-x}O_{12}$ $(0 \leq x \leq 3)$ and such compounds which are super ionic conductors after the transition into the high-temperature phase like α-AgI or δ-Bi_2O_3 (type II).

Electronic Defects in Semiconductors

At 0 K the conduction band of semiconductors is completely empty (→ band theory). All electrons are in the valence band. With increasing temperature, electrons leave the valence band, overcome the band gap, and transfer into the conduction band. At given temperature an equilibrium between electrons and defect electrons is established:

$$e'\,(VB) = e'\,(CB) + V\,(VB)$$

Electron (valence band) $e'(VB)$ = electron (conduction band) $e'(CB)$ + electron vacancy (valence band) $V(VB)$

The electron vacancy is an electronic defect and can be regarded as a hole h°. This equilibrium can be expressed using the Kröger-Vink notation as

$$e' + h^{\circ} = Null = 0$$

Using the chemical potentials of electrons and holes with the concentrations of electrons and defect *[e′] [h°]* and the standard concentration *[e′]° and [h°]°*, respectively

$\mu_e = \mu_e^\circ + RT \ln \frac{[e]}{[e]^\circ}$ and $\mu_h = \mu_h^\circ + RT \ln \frac{[h]}{[h^\circ]^\circ}$,

the following expression is obtained:

$$[e'][h^\circ] = \text{const.}\ (p, T)$$

For a certain semiconductor, the product of the concentrations of electrons and holes is constant at defined partial pressure and temperature.

Experimental Methods for Defect Investigations

Defects can be discovered and determined by different experimental methods. By measurement of the electrical conductivity (see to ► Mixed Conductors, Determination of Electronic and Ionic Conductivity (Transport Numbers)) in dependence on partial pressure and temperature [5, 6] and the heat capacity in dependence on temperature [7], the defect formation could be detected. Hund investigated the defect structure in doped zirconia by measurement of specific density by means of XRD and pycnometric determination [8]. Transference measurements [9] and diffusion experiments with tracers [10–12] or colored ions [4] are suited for verifying defects.

Cross-References

► Defect Chemistry in Solid State Ionic Materials
► Kröger-Vinks Notation of Point Defects

References

1. Gellings PJ, Bouwmeester HJM (eds) (1997) Solid state chemistry. CRC Handbook, Boca Raton, New York, London, Tokyo
2. West AR (1984) Solid state chemistry and its applications. Wiley, Chichester/ New York/ Brisbane/ Toronto/Singapore
3. Rickert H (1982) Electrochemistry of solids. Springer, Berlin/Heidelberg/New York
4. Maier J (2004) Physical chemistry of ionic materials: ions and electrons in solids. Wiley, Chichester
5. Wagner C, (1936) Ionic conductivity in solid salts (in German) Die elektrische Leitfähigkeit in festen Salzen 42:635–654
6. Baumbach H-H-v V, Wagner C (1934) Electric conductivity of nickel oxide (in German) Die elektrische Leitfähigkeit von Nickeloxyd. Z Phys Chem B24:59–67
7. Nölting J (1970) Defect behavior of solids (Ionic crystals and metals) in Germ. Fehlordnungsverhalten von Festkörpern (Ionenkristallen und Metallen). Angew Chem 82:498
8. Hund F (1951) Abnormal mixed crystals in the system ZrO_2-Y_2O_3 Crystal structure of the Nernst glower (in Germany) Anomale Mischkristalle im System ZrO_2 Y_2O_3 Kristallbau der Nernst-Stifte. Z Elektrochem 55:363–366
9. Tubandt C (1932) Conductivity and transference number in solid electrolytes (in German) Leitfähigkeit und Überführungszahlen in festen Elektrolyten. In: Wien W, Harms F (eds), Handbuch der Experimantalphysik , Leipzig, pp 381–469
10. Jost W (1969) Diffusion in solids, liquids and gases. Academic, New York
11. Kingery WD, Bowen HK, Uhlmann DR (1976) An introduction to ceramic. Wiley, New York
12. Dieckmann R, Schmalzried H (1977) Defects and Cation Diffusion in Manganite (I). Ber Bunsen Phys Chem 81:344–347

Degradation of Organics, Use of Combined Electrochemical-Ultrasound

Jinren Ni
Peking University, Beijing, China

Introduction

Combined electrochemical–ultrasound systems are very efficient for organic degradation. Electrochemical oxidation processes are under mass-transport control at normal operating conditions [1, 2]. Therefore, enhancement of mass transport would be of primary importance to optimization of the processes. The ultrasound treatment, which is associated with acoustic cavitations in liquid media, is a rapid developing field in organic degradation [3, 4]. When cavitation bubbles undergo asymmetrical implosion near a solid surface, it results in a strong microjet of liquid and violent shock wave towards the solid surface. Consequently, the solid–liquid mass transfer between the electrodes and the solution is strongly enhanced and the anode fouling is reduced [5, 6]. The combination of an ultrasonic field with an electrochemical oxidation can clean electrode surface and improve mass transport, resulting in a powerful alternative for efficient degradation of organic pollutants and particularly of biorefractory organics.

Functions of Combined Systems

Ultrasound-enhanced electrochemical systems with different electrodes have been widely reported for oxidation of variety of organics such as chlorinated organics [7, 8], dyes [9–11], phenolic substances [12–15], and other recalcitrant pollutants [16]. For example, the combined electrochemical–ultrasound process developed by Yasman et al. [7] was reported successfully at nickel anode for degradation of chlorinated aromatic compounds (2,4-dichlorophenoxyacetic acid and its derivative 2,4-dichlorophenol) in environmental water. Siddique et al. [9] investigated degradation of dye pollutants with a lead oxide anode enhanced by ultrasound under varying dye concentration, pH, ultrasonic frequency, and reaction time, followed by a full discussion of reaction kinetics, organic carbon, and mechanisms. The decomposition of reactive blue 19 dye and hence decolorization by ultrasound-assisted electrochemical process proved the higher efficiency and lower energy consumption of the combined system comparing with separated electrolysis or sonochemical system. Lima Leite et al. [14] used Pt electrode associated with ultrasound activation and totally oxidized the 2,4-dihydroxybenzoic acid at low concentrations, which was hardly completed by electrochemical process solely. Zhu et al. [17] investigated the effects of low-frequency (40 kHz) ultrasound on the electrochemical oxidation of several p-substituted phenols with BDD (boron-doped diamond) and PbO_2 anodes for a comparison, and considerable enhancement of organic degradation was achieved in different systems.

Influencing Factors

In the combined electrochemical–ultrasound systems, ultrasound frequency is one of the significant factors for degradation of organics. Under low-frequency ultrasound, cavitation leads to cleaning of the electrode surface thus increasing the active electrode surface and enhancing the mass transfer rates [14]. Then, the degradation of organics can be optimized with less intermediate compounds. At higher frequencies, hydroxyl radicals are generated acceleratedly as the lifetime of bubbles is shorter and collapse of cavitation bubbles occurs rather quickly. And then the oxidation between organics and hydroxyl radicals is increased. Siddique et al. [9] tested dyes pollutants on the lead oxide anode and found the degradation rate increased with the frequency in the scope of 0–80 kHz. However, no significant improvement was detected by increasing frequency from 80 to 100 kHz. There exists an optimal acoustic power as to a certain chemical reaction and a proper transfer condition of acoustic energy through the whole liquid as to a certain reaction system [18, 19].

Other influencing factors include initial pH, applied potentials, temperature, initial organic concentrations, and supporting electrolyte. [7]. It is indicated by Bringas et al. with 20–30 mg L^{-1} of diuron at BDD anode [16] that alkaline pH favors the mineralization rate and the total organic carbon removal was 92 % after 6 h of degradation, higher than that at neutral or acidic conditions (less than 80 %). Cañizares et al. obtained similar conclusions and the influence of pH depended on both the nature of organic compounds and its concentration [20, 21]. A faster degradation of organics could be achieved with increasing applied potentials; however, much more electric energy would be consumed. Optimal state could be established for various kinds of electrochemical–ultrasound processes [7, 22]. No obvious impact of the temperature in the range between 10 °C and 40 °C was observed on the kinetics of the sonoelectrochemical mineralization of diuron [16].

Mechanism

The advantages of the combined ultrasound–electrochemical process over other techniques have been confirmed, and the mechanism of synergy has been fully investigated in the past years. Klima [23] emphasized the role of ultrasound and proposed four possible mechanisms such as electrochemistry-acoustic streaming, turbulent movement due to the cavitation, microjets resulting from the asymmetric collapse, and shock waves generated by spherical collapse of cavitating bubbles. Acoustic streaming optimizes the process of mass transfer, while other three mechanisms are helpful not only to transfer enhancement but also to electrode surface activation (depassivation). Compton et al. [24] pointed out that ultrasound can induce depassivation and erosion effects on electrode surfaces through cavitation events. They detected the huge effect of ultrasound on the mass transport by various voltammetric techniques and further described the effect using a model of an extremely thinned diffusion layer of uniform accessibility.

Some recent studies are focused on the mechanism of strong oxidation capacity dominated by hydroxyl radicals. Zhu et al. [17] made a significant contribution by revealing that anode materials determined the state of hydroxyl radicals and play an important part in electrochemical oxidation [12, 25]. In general, the hydroxyl radicals produced by water decomposition at active anodes (such as Pt, IrO_2, and RuO_2) can interact with the oxide anode and tend to form chemisorbed "active oxygen" (MO_{x+1}), leading to a weak oxidation ability [26], while at non-active anodes such as PbO_2, SnO_2, and BDD (boron-doped diamond), hydroxyl radicals do not react with anodes and thus they show a stronger degradation ability of organic compounds [27]. It is demonstrated [25, 28] that the hydroxyl radicals mainly exist as free state and react effectively with organic pollutants at BDD anodes, while adsorbed hydroxyl radicals dominate at PbO_2 anodes, and are not very effective for organics degradation. At SnO_2 anodes, the organic compounds reacted with both adsorbed hydroxyl radicals and free hydroxyl radicals. Furthermore, the effectiveness and mechanisms of combined electrochemical oxidation and ultrasound process was put forward with several p-substituted phenols as the model organics [17]. It is indicated that the enhancement of ultrasound on organic degradation was greater at the BDD anode (73–83 %) than at the PbO_2 anode (50–70 %) mainly due to the diverse effect of ultrasound on specialized existence of hydroxyl radicals at different electrodes.

Future Directions

Ultrasound plays a role in either the chemical degradation pathway or the electrochemical reaction at the electrode based on the frequency used. The increase extent is closely related with the electrode material. It seems that further investigation of the geometry of both electrodes and reactor is needed in order to optimize the ultrasound energy distribution, enhance the degradation efficiency, and reduce the energy consumption.

References

1. Polcaro AM, Mascia M, Palmas S, Vacca A (2004) Electrochemical degradation of diuron and dichloroaniline at BDD electrode. Electrochim Acta 49:649–656
2. Polcaro AM, Vacca A, Mascia M, Palmas S (2005) Oxidation at boron doped diamond electrodes: an effective method to mineralise triazines. Electrochim Acta 50:1841–1847
3. Pollet BG, Lorimer JP, Hihn JY, Phull SS, Mason TJ, Walton DJ (2002) The effect of ultrasound upon the oxidation of thiosulphate on stainless steel and platinum electrodes. Ultrason Sonochem 9:267–274
4. Del Campo FJ, Coles BA, Marken F, Compton RG, Cordemans E (1999) High-frequency sonoelectrochemical processes: mass transport, thermal and surface effects induced by cavitation in a 500 kHz reactor. Ultrason Sonochem 6:189–197

5. Kumbhat S (2000) Potentialities of power ultrasound in electrochemistry: an overview. Bull Electrochem 16:29–32
6. Bremner DH, Burgess AE, Li FB (2000) Coupling of chemical, electrochemical and ultrasonic energies for controlled generation of hydroxyl radicals: direct synthesis of phenol by benzene hydroxylation. Appl Catal A Gen 203:111–120
7. Yasman Y, Bulatov V, Gridin VV, Agur S, Galil N, Armon R, Schechter I (2004) A new sono-electrochemical method for enhanced detoxification of hydrophilic chloroorganic pollutants in water. Ultrason Sonochem 11:365–372
8. Esclapez MD, Sáez V, Milán-Yáñez D, Tudela I, Louisnard O, González-García J (2010) Sonoelectrochemical treatment of water polluted with trichloroacetic acid: from sonovoltammetry to pre-pilot plant scale. Ultrason Sonochem 17:1010–1020
9. Siddique M, Farooq R, Khan ZM, Khan Z, Shaukat SF (2011) Enhanced decomposition of reactive blue 19 dye in ultrasound assisted electrochemical reactor. Ultrason Sonochem 18:190–196
10. Rivera M, Pazos M, Sanromán MA (2009) Improvement of dye electrochemical treatment by combination with ultrasound technique. J Chem Technol Biotechnol 84:1118–1124
11. Cai M, Jin M, Weavers LK (2011) Analysis of sonolytic degradation products of azo dye Orange G using liquid chromatography-diode array detection-mass spectrometry. Ultrason Sonochem 18:1068–1076
12. Zhao G, Shen S, Li M, Wu M, Cao T, Li D (2008) The mechanism and kinetics of ultrasound-enhanced electrochemical oxidation of phenol on boron-doped diamond and Pt electrodes. Chemosphere 73:1407–1413
13. Zhao G, Gao J, Shen S, Liu M, Li D, Wu M, Lei Y (2009) Ultrasound enhanced electrochemical oxidation of phenol and phthalic acid on boron-doped diamond electrode. J Hazard Mater 172:1076–1081
14. Lima Leite RH, Cognet P, Wilhelm AM, Delmas H (2002) Anodic oxidation of 2, 4-dihydroxybenzoic acid for wastewater treatment: study of ultrasound activation. Chem Eng Sci 57:767–778
15. Trabelsi F, Aït Lyazidi H, Ratsimba B, Wilhelm AM, Delmas H, Fabre PL, Berlan J (1996) Oxidation of phenol in wastewater by sonoelectrochemistry. Chem Eng Sci 51:1857–1865
16. Bringas E, Saiz J, Ortiz I (2011) Kinetics of ultrasound-enhanced electrochemical oxidation of diuron on boron-doped diamond electrodes. Chem Eng J 172:1016–1022
17. Zhu XP, Ni JR, Li HN, Jiang Y, Xing X, Borthwick A (2010) Effects of ultrasound on electrochemical oxidation mechanisms of p-substituted phenols at BDD and PbO_2 anodes. Electrochim Acta 55:5569–5575
18. Vajnhandl S, Marechal AML (2007) Case study of the sonochemical decolouration of textile azo dye Reactive Black 5. J Hazard Mater 141:329–355
19. Price JG (1992) Current trends in sonochemistry. The Royal Society of Chemistry, Cambridge, UK
20. Cañizares P, García-Gómez J, Sáez C, Rodrigo MA (2004) Electrochemical oxidation of several chlorophenols on diamond electrodes: part II. Influence of waste characteristics and operating conditions. J Appl Electrochem 34:87–94
21. Ai Z, Li J, Zhang L, Lee S (2010) Rapid decolorization of azo dyes in aqueous solution by an ultrasound-assisted electrocatalytic oxidation process. Ultrason Sonochem 17:370–375
22. Nagata Y, Nakagawa M, Okuno H, Mizukoshi Y, Yim B, Maeda Y (2000) Sonochemical degradation of chlorophenols in water. Ultrason Sonochem 7:115–120
23. Klima J (2011) Application of ultrasound in electrochemistry. An overview of mechanisms and design of experimental arrangement. Ultrasonics 51:202–209
24. Compton RG, Eklund JC, Marken F, Rebbitt T, Akkermans RP, Waller DN (1997) Dual activation: coupling ultrasound electrochemistry-an overview. Electrochim Acta 42:2919–2927
25. Zhu XP, Tong MP, Shi SY, Zhao HZ, Ni JR (2008) Essential explanation of the strong mineralization performance of boron-doped diamond electrodes. Environ Sci Technol 42:4914–4920
26. Johnson SK, Houk LL, Feng JR, Houk RS, Johnson DC (1999) Electrochemical incineration of 4-chlorophenol and the identification of products and intermediates by mass spectrometry, Environ Sci Technol 33:2638–2644
27. Panizza M, Michaud PA, Cerisola G, Comninellis C (2001) Anodic oxidation of 2-naphthol at boron-doped diamond electrodes. J Electroanal Chem 507:206–214
28. Zhu XP, Shi SY, Wei JJ, Lv FX, Zhao HZ, Kong JT, He Q, Ni JR (2007) Electrochemical oxidation characteristics of p-substituted phenols using a boron-doped diamond electrode. Environ Sci Technol 41:6541–6546

DFT Screening and Designing of Electrocatalysts

Ping Liu
Brookhaven National Laboratory, Upton, NY, USA

Introduction

One of the focuses in electrochemistry is to develop more active, more selective, and more stable electrocatalysts with low cost. To accomplish this task requires enough knowledge of how a catalyst functions, such as the reaction mechanism and active sites. The ultimate goal is to have enough knowledge of important factors that determine the catalytic activity, which can be used to tailor catalysts atom by atom. However, it is extremely difficult to obtain all of the necessary details from current experimental techniques. Extensive theoretical and computational approaches have been employed to try to meet the goal of developing a fundamental understanding as a basis for catalyst design [1].

Development of DFT in Electrochemistry

Over the past few decades, density functional theory (DFT) starts to be recognized as an essential tool in describing electrocatalysts. Modeling the electrochemical systems using DFT presents a considerable challenge [2]. Quantum mechanical simulations are typically carried out within the canonical ensemble formalism, where the number of electrons is conserved. In order to model electrochemical systems, several factors have to be modeled simultaneously: the structure and chemistry that occurs at the anode and the cathode, the electron transfer between the two electrodes, and the local changes in the electrolytes. For the time being, this is impossible. Instead, various approximated approaches have been developed to simulate the electrochemistry, which describe the solid electrode surface, the liquid solution, the solvated ions, and the effect of changes in the chemical potential of the electrons in the solid.

Modeling the Solution

The presence of solution can dramatically alter the chemical reactions. This is clearly present in the electrochemical processes. Water has been considered critical to the performance of electrocatalysts [3–8], which may effectively stabilize both the protons and the reaction intermediates in various important electrochemical reactions (Fig. 1). Water/catalyst interfaces were described by optimizing a cluster or an ice-like solvent structure on the catalyst surfaces, partially or completely filling the volume between the upper and lower slab surfaces. In this way, the solvated proton can also be well described in the theoretical model.

Modeling the Potential

The solid–liquid interface is charge neutral as the electrolyte is conducting. The change on the solid surface will be counteracted by a countercharge built up by ions just outside the surface. The interface region, the Helmholtz layer, is approximated by a ca. 3Å thick electrical double-layer. The double-layer has a strong electrical field, which is central to the chemical activities of the interface. Although the quantum mechanical models have been developed to understand both qualitatively and quantitatively the effect of potential on the surface and intermolecular bindings [2], few ab initio efforts aimed at the potential dependence of a chemical reaction.

In order to include potential effects into considerations, Lozovio et al. developed a first-principles periodic DFT supercell approach [9], and Neurock et al. established a more advanced model [10]. In these methods, a potential across the interface is induced by tuning the charge on the electrode, where the countercharge is distributed homogeneously over the background of the cell. In this way, the water reorients according to the homogeneous background charge density and subsequently the reaction proceeds to form the well-known double-layer structure at the interface. By comparing the energies for different structures at a given potential, one is able to determine the free energy difference between these given states at a desired range of potentials. In the method of Otani and Sugino, an external dipole layer is introduced to create a charged surface, and the countercharge is therefore located approximately 10Å away from the electrode [11]. Both methods provide a large step forward towards a microscopic description of the electrochemical cell, but in the first case, the countercharge distributions are not localized, but arbitrary or broad. Anderson et al. recently have combined the DFT and a modified Poisson-Boltzmann theory, where the water molecules are not included directly but distributing the countercharge in a continuum dielectric medium.

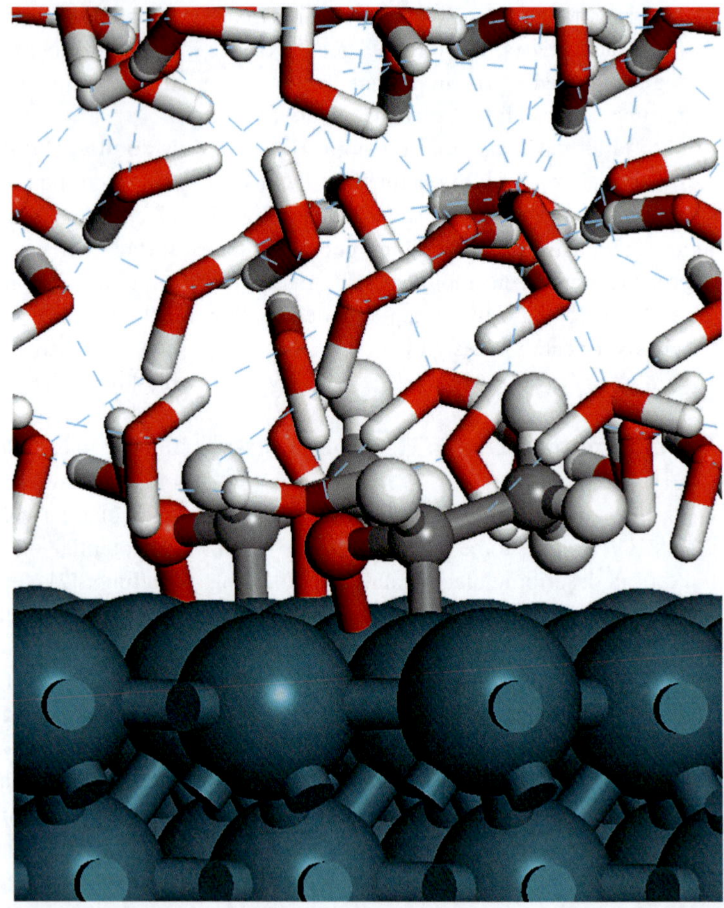

DFT Screening and Designing of Electrocatalysts, Fig. 1 Optimized CH_3CHO adsorption at the water/Rh(111) interface

It is able to predict with a useful accuracy the reversible potentials for electron transfer in acid and base aqueous solutions [12].

In order to avoid using these artificial treatments for counterchanges, Jonsson, Nørskov, and coworkers proposed to control the electrode potential by changing not only the charges of electrodes but also the number of protons to keep the system neutral [13]. However, the interactions from a long-ranged double-layer are not described. A more severe problem is that if a charge transfer reaction takes place and the unit cell is small, the potential changes significantly during the reaction. Therefore, it is very difficult to well control the electrode potential. However, these calculations are computationally expensive and limited to describe relatively simple reactions and electrocatalysts.

All the methods above cannot compare directly with experiments due to the difficulty in assigning the potential scale to the electrode. Rossmeisl et al. proposed a realistic atomic model for calculating reaction energies and activation energies for charge transfer reactions without finite-size errors [5]. It also provides a measure of the vacuum potential relative to the NHE and gives values for the interface capacitance in agreement with experiments. However, the calculations require several calculations for different unit cell sizes to extrapolate to the limit of an infinite surface unit cell. Therefore, it is computationally more expensive than the ordinary calculations for surface reactions.

Nørskov et al. developed a simple approach to describe the electrocatalytic reactions [14]. The thermochemistry of electrochemical reactions is estimated by calculating the stability of the reaction intermediate using the DFT slab model. In this method, the chemical potential for $H^+ + e^-$ in solution is sent to that of $1/2H_2$ by setting the reference potential to that of the standard hydrogen electrode (SHE). The only two ways in which the potential enters the calculations are through the binding energy of ΔE^U_{ads} and the chemical potential of the electrons. An electrical field is introduced at the surface to simulate the double-layer. The potential dependence of adsorption energy was given largely by the first-order interaction between the adsorbate dipole moment μ_A and the field $F = U/d$ (U: potential, d: width of double-layer) with

$$\Delta E^U_{ads} = \Delta E^{U-0}_{ads} - \mu_A U/d$$

For all elementary reaction steps involving an electron, the enthalpy of the state is shifted by $-eU$. The free energy of the intermediates under a certain potential is expressed as a function of enthalpy, entropy, temperature, and pH value of the liquid phase. In this way, the artificial treatments for counterchanges can be avoided. The main downside is that this methodology only describes the thermodynamics. For estimating hydrogen binding with the surface intermediates by static methods and investigating the kinetics of surface process involving a charge transfer between the slab and the electrolyte, it will not be sufficient.

Application of DFT in Electrocatalysis

Developing a fundamental understanding of how catalyst function as a basis for design of improved catalysts has become one of the grand challenges in electrocatalysis. The electrochemical processes always involve multiple reaction pathways, active sites, and products and cannot be well characterized experimentally. The development of DFT in electrochemistry, as demonstrated above, makes it possible to understand the reaction mechanism at the atomic level. Such understanding allows the theoretical screening for better catalysts.

Description of the Electrochemical Reaction

Advances in DFT make it possible to describe electrocatalytic reactions at surfaces with the detail. The method developed by Neurock et al. [10] described qualitatively various electrochemical reactions, water activation, oxygen reduction reaction (ORR), as well as methanol decomposition on metal surfaces, being able to gain insights into the reaction mechanism under potential over aqueous-metal interfaces [3, 15, 16]. The method proposed by Nørskov et al. [14] is also extensively employed to describe the electrochemistry, which gives in some cases accuracy required for computational results to compare with experiment in a meaningful way. For instance, DFT calculations and a micro-kinetic modeling are combined to describe the H_2/CO electrooxidation on Pt and Pt alloy surfaces [17]. The model is very simple and is able to express the kinetics of a promoted anode surface relative to the activity of pure Pt directly from the calculated adsorption energy differences. As shown in Fig. 2, the calculated polarization curves on different surfaces agree well with the experimental measurements.

Understanding of Trend in Activity and Reaction Mechanism

DFT-based studies also provide the understanding of variations in catalytic activity from one catalyst to another, which is also qualitatively comparable to the experimental measurement [14, 17–20]. This allows more insight into the reaction mechanism. As shown in Fig. 2, both experiment and theory show that PtRu and Pt_3Sn are better electrocatalysts than Pt, being able to oxidize H_2 and CO at lower potential; in contract, higher potential should be applied to oxide CO on Ru. Within the model, the origin of the promoting effect of alloying can be analyzed. That is, the promoting effect of alloying on H_2/CO oxidation reaction can be attributed to the fact that alloyed metals modify Pt in the surface to bond CO weaker, thus decreasing the CO coverage under working conditions of the electrode. Such detailed understanding cannot be achieved merely using experimental techniques and is very important to the rational catalyst screening.

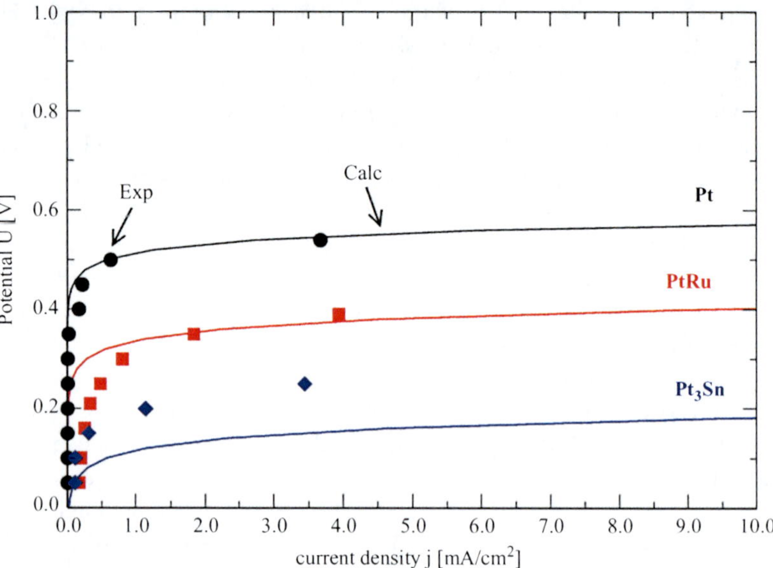

DFT Screening and Designing of Electrocatalysts, Fig. 2 Experimentally measured (*symbols*) and calculated (*curves*) overpotential as a function of current density for an anode half cell with 1 atm. of H_2 with 250 ppm of CO at 60°C (Adapted from Ref [17])

Screening of Catalysts Based on Single Descriptor

Such detailed understanding of mechanism based on DFT-based calculations leads to the theoretical screening for better catalysts using a single-scaling descriptor. A descriptor is a parameter, being able to capture the variation in overall activity or selectivity from one system to the next. One of representative examples is the study of the screening of anode catalysts for PEM fuel cell [21]. According to DFT-based kinetic studies [17], the experimentally measured activity of H_2/CO oxidation on Pt and Pt alloys is well correlated with a single descriptor including the DFT-calculated H_2 and CO binding energies. That is, the calculated descriptor is able to directly give a semiquantitative description of the experimentally measured difference in activity. The anode catalysts with better CO tolerance are the Pt alloys that bond CO more weakly than pure Pt while still dissociating H_2. Accordingly, tens of ternary catalysts, MPtRu, are screened at a theoretical level, where the activity trend predicted theoretically agrees well with the observations in the screening experiments [21]. In addition, the DFT-calculated O as well as CO and OH binding energies are considered as descriptors for ORR and methanol oxidation on metal or alloy surfaces, respectively [14, 22]. By using the single descriptor, the activity variation from one system to the next can be well explained, and more importantly new advanced catalysts are screened at a theoretical, which guides the experiment for catalyst development [23].

Future Directions

Even though some progress has been made towards understanding electrocatalytic process and screening electrocatalysts from DFT, the method has difficulty in providing quantitative numbers for detailed reaction steps. On one hand, methodological improvements are required to describe the electron transfer at solid–liquid interface, the band structure, and the excited states effectively, which is currently limitation of DFT. On another hand, the model systems in DFT studies are somewhat too simplified to model the real catalysts effectively. For instance, the real catalysts are powders, which may behave differently with size. Recently, efforts have been made to model the nanoparticles with the size of the real catalysts (<5 nm), showing indeed different behaviors from the extended surfaces even in term of trend (Fig. 3), a common model used in DFT studies [24, 25]. Thus, theoretical

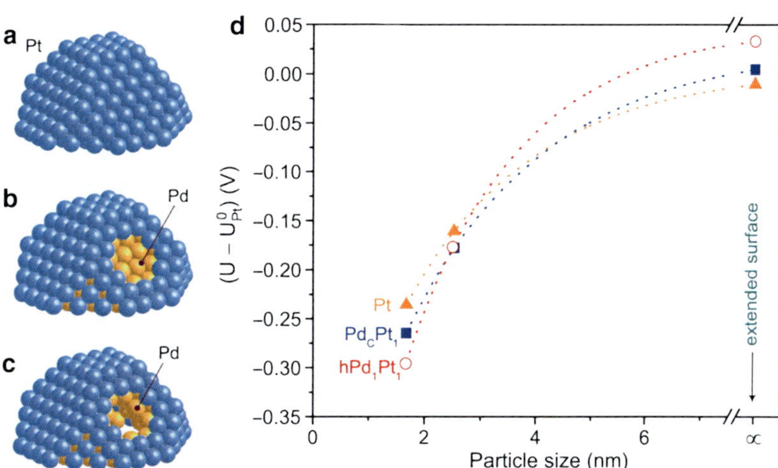

DFT Screening and Designing of Electrocatalysts, Fig. 3 Schematic illustration of surface models for the nanoparticles representing (**a**) pure Pt, (**b**) Pt shell-Pd core, (**c**) Pt shell-partially hollow Pd core, and (**d**) DFT-predicted dissolution potentials as a function of particle sizes (Adapted from Ref [24])

predictions based on the calculations of extended surfaces may not necessarily be able to describe the electrocatalysts with small size. In addition, the electrochemical processes may always be complex, including multiple reaction pathways, rate-limiting steps, and products. Given that, it will be difficult to capture the variation in overall activity or selectivity just using single descriptor. More sophisticated models are needed to capture knowledge needed to develop the rules for catalyst optimization. Finally, in contrast to the activity, the study of stability under potential and in solution is still at early stage [24, 26, 27]. Attention should be paid to ensure the stability of the screened electrocatalysts, which is of great importance for the practical application. With the further development of DFT method and models that can effectively treat more realistic catalysts and their environments, it can be envisioned that soon DFT modeling in electrochemistry will not only provide insight into the experimental measurements but also become the standard choice for designing a new catalyst for a catalytic process.

Cross-References

- ▶ Hydrogen Oxidation and Evolution on Platinum in Acids
- ▶ Oxygen Reduction Reaction in Acid Solution
- ▶ Platinum-Based Anode Catalysts for Polymer Electrolyte Fuel Cells
- ▶ Platinum-Based Cathode Catalysts for Polymer Electrolyte Fuel Cells

References

1. DOE-BES report (2007) Basic Research Needs: Catalysis for Energy: Department of Energy, http://science.energy.gov/~/media/bes/pdf/reports/files/cat_rpt.pdf.
2. Santan JA, Neurock M (2006) Molecular heterogeneous catalysis. Willey-VCH, Weinheim
3. Filhol J, Neurock M (2006) Elucidation of the electrochemical activation of water over Pd by first principles. Angew Chem Int Ed 45:402–406
4. Santan JA, Mateo JJ, Ishikawa Y (2010) Electrochemical hydrogen oxidation on Pt(110): a combined direct molecular dynamics/density functional theory study. J Phys Chem C 114:4995–5002
5. Rossmeisl J, Skulason E, Bjorketun ME, Tripkovic V, Norskov JK (2008) Modeling the electrified solid–liquid interface. Chem Phys Lett 466(1–3):68–71
6. Tian F, Jinnouchi R, Anderson AB (2009) How potential of zero charge and potentials for water oxidation to OH(ads) on Pt(111) electrodes vary with coverage. J Phys Chem C 113:17484–17492
7. Wang Y, Balbuena PB (2005) Ab initio molecular dynamics simulations of the oxygen reduction reaction on a Pt(111) surface in the presence of hydrated hydronium (H3O) + (H2O)2: direct or series pathways? J Phys Chem B 109:14896–14907
8. Hyman MP, Medlin JW (2006) Mechanistic study of the electrochemical oxygen reduction reaction on Pt (111) using density functional theory. J Phys Chem B 110:15338–15344

9. Lozovoi AY, Alavi A, Kohanoff J, Lynden-Bell RM (2001) Ab initio simulation of charged slabs at constant chemical potential. J Chem Phys 115(4): 1661–1669
10. Taylor CD, Wasileski SA, Filhol J, Neurock M (2006) First principles reaction modeling of the electrochemical interface: consideration and calculation of a tunable surface potential from atomic and electronic structure. Phys Rev B 73:165402
11. Otani M, Sugino O (2006) First-principles calculations of charged surfaces and interfaces: a plane-wave nonrepeated slab approach. Phys Rev B 73(11): 115407
12. Jinnouchi R, Anderson AB (2008) Electronic structure calculations of liquid–solid interfaces: combination of density functional theory and modified Poisson-Boltzmann theory. Phys Rev B 77: 245417
13. Skulason E, Karlberg GS, Rossmeisl J, Bligaard T, Greeley J, Jonsson H et al (2007) Density functional theory calculations for the hydrogen evolution reaction in an electrochemical double layer on the Pt(111) electrode. Phys Chem Chem Phys 9:3241–3250
14. Nørskov JK, Rossmeisl J, Logadottir A, Lindqvist L, Kitchin J, Bligaard T (2004) The origin of the overpotential for oxygen reduction at a fuel cell cathode. J Phys Chem B 108:17886–17892
15. Rossmeisl J, Nørskov JK, Taylor CD, Janik MJ, Neurock M (2006) Calculated phase diagrams for the electrochemical oxidation and reduction of water over Pt(111). J Phys Chem B 110(43):21833–21839
16. Janik MJ, Taylor CD, Neurock M (2009) First-principles analysis of the initial electroreduction steps of oxygen over Pt(111). J Electroanal Chem Soc 156(1):B126–B135
17. Liu P, Logadottir A, Nørskov JK (2003) Modeling the electro-oxidation of CO and H2/CO on Pt, Ru, PtRu and Pt3Sn. Electrochim Acta 48:3731–3742
18. Nørskov JK, Bligaard T, Logadottir A, Kitchin J, Chen JG (2005) Trends in the exchange current from hydrogen evolution. J Electrochem Soc 152(3): J23–J26
19. Kowal A, Li M, Shao M, Sasaki K, Vukmirovic MB, Zhang J et al (2009) Ternary Pt/Rh/SnO_2 electrocatalysts for oxidizing ethanol to CO_2. Nat Mater 8:325–330
20. Ingram DB, Linic S (2009) First-principles analysis of the activity of transition and noble metals in the direct utilization of hydrocarbon fuels at solid oxide fuel cell operating conditions. J Electrochem Soc 156(12): B1457–B1465
21. Strasser P, Fan Q, Devenney M, Weinberg WH, Liu P, Nørskov JK (2003) High throughput experimental and theoretical predictive screening of materials – a comparative study of search strategies for new fuel cell anode catalysts. J Phys Chem B 107(40): 11013–11021
22. Ferrin P, Nilekar AU, Greeley J, Mavrikakis M, Rossmeisl J (2008) Reactivity descriptors for direct methanol fuel cell anode catalysts. Surf Sci 602(21): 3424–3431
23. Greeley J, Stephens IEL, Bondarenko AS, Johansson TP, Hansen HA, Jaramillo TF et al (2009) Alloys of platinum and early transition metals as oxygen reduction electrocatalysts. Nat Chem 1(7):552–556
24. Sasaki K, Naohara H, Cai Y, Choi YM, Liu P, Vukmirovic MB et al (2010) Core-protected platinum monolayer shell high-stability electrocatalysts for fuel-cell cathodes. Angew Chem Int Ed 49(46): 8602–8607
25. Wang JX, Inada H, Wu L, Zhu Y, Choi Y, Liu P et al (2009) Oxygen reduction on well-defined core-shell nanocatalysts: particle size, facet, and Pt Shell thickness effects. J Am Chem Soc 131(47):17298–17302
26. Greeley J, Nøskov JK (2007) Electrochemical dissolution of surface alloys in acids: thermodynamic trends from first-principles calculations. Electrochim Acta 52(19):5829–5836
27. Tang L, Han B, Persson K, Friesen C, He T, Sieradzki K et al (2009) Electrochemical stability of nanometer-scale Pt particles in acidic environments. J Am Chem Soc 132(2):596–600

Dielectric Properties

Richard Buchner
Institute of Physical and Theoretical Chemistry, University of Regensburg, Regensburg, Germany

Static Relative Permittivity (Dielectric Constant)

Dielectric properties describe the polarization, $\boldsymbol{P}$, of a material as its response to an applied electric field $\boldsymbol{E}$ (bold symbols indicate vectors) [1–3]. In the field of solution chemistry, the discussion of dielectric behavior is often reduced to the equilibrium polarization, $\boldsymbol{P}_0 = \varepsilon_0(\varepsilon - 1)\,\boldsymbol{E}_0$ (ε_0 is the electric field constant), of the isotropic and nonconducting solvent in a static field, $\boldsymbol{E}_0$. Characteristic quantity here is the static relative permittivity (colloquially "dielectric constant"), ε, which is a measure for the efficiency of the solvent to screen Coulomb interactions between charges (i.e., ions) embedded in the medium. As such, ε enters into classical electrolyte theories, like Debye-Hückel theory or the Born model for solvation free energy [4, 5] and is used

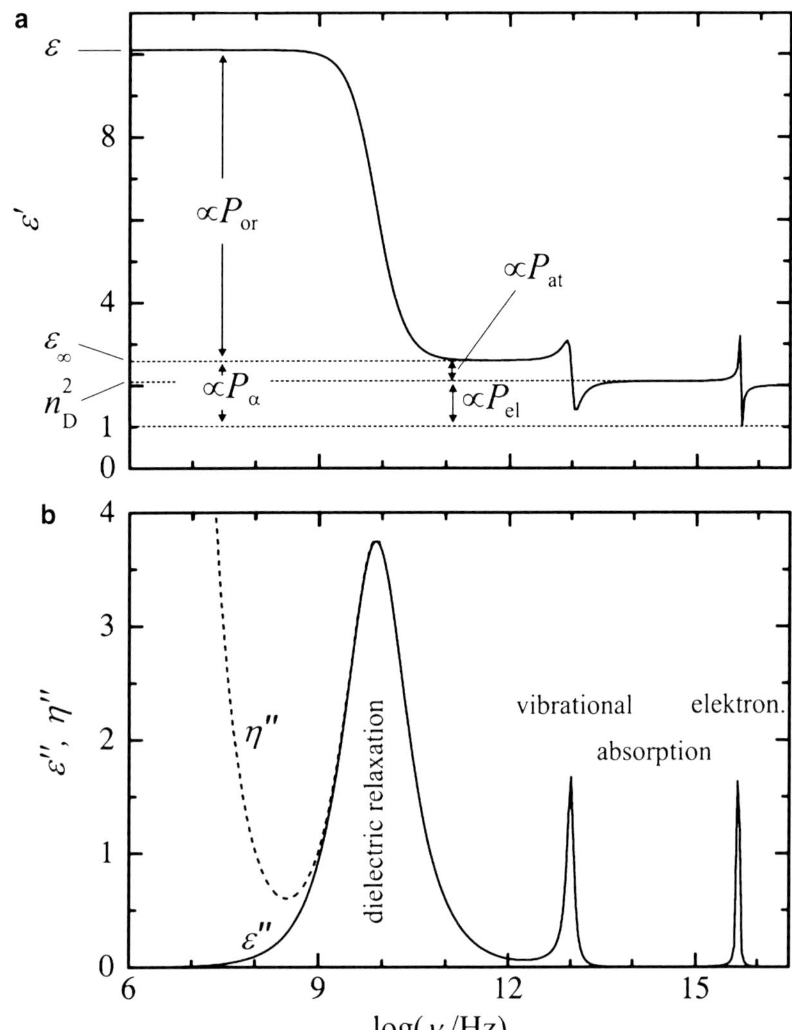

Dielectric Properties, Fig. 1 Schematic frequency dependence of (**a**) the relative permittivity, $\varepsilon'(\nu)$, and (**b**) the total loss, $\eta''(\nu)$ (*dashed line*), and the dielectric loss, $\varepsilon''(\nu)$ (*solid line*), of a sample with dc conductivity, κ_{dc}, static relative permittivity, ε, infinite-frequency permittivity, ε_∞, and optical refractive index, n_D. For simplicity, a single relaxation process and only one infrared-active vibration and one UV/vis absorption band are assumed. In (**a**), the dispersion steps associated with orientational ($\boldsymbol{P}_{or} \propto \varepsilon - \varepsilon_\infty$), atomic ($\boldsymbol{P}_{at} \propto \varepsilon_\infty - (n_D)^2$), and electronic ($\boldsymbol{P}_{el} \propto (n_D)^2 - 1$) polarization are indicated

as a scale for solvent polarity. Note that ε also influences activation-barrier heights for chemical reactions in solution due to the necessary solvent reorganization along the reaction coordinate [6].

For molecular fluids, like most solvents, $\boldsymbol{P}$ can be split into induced polarization, $\boldsymbol{P}_\alpha$, associated with intramolecular polarizability, α, and orientational polarization, $\boldsymbol{P}_{or}$, arising from the partial alignment of the permanent dipole moments, μ, of the constituting molecules against thermal motions through $\boldsymbol{E}_0$. The induced polarization can be further subdivided into a generally small contribution $\boldsymbol{P}_{at}$ from the field-induced fluctuations of the molecular geometry (atomic polarizability) and the dominating electronic polarizability, $\boldsymbol{P}_{el}$, from the deformation of the electron cloud [1, 2]. Figure 1a schematically indicates the contributions of $\boldsymbol{P}_{or}$, $\boldsymbol{P}_{at}$, and $\boldsymbol{P}_{el}$ to the static permittivity and the frequency ranges up to which they are active. For atomic fluids, like liquid argon, only $\boldsymbol{P}_{el} \neq 0$ so that $\varepsilon \approx 2$. Since the electronic polarizability can follow changes of the electrical field without delay up to frequencies, ν, corresponding to visible light, we find $\varepsilon = (n_D)^2$ where n_D is the optical refractive index. Nonpolar liquids, like tetrachloromethane, also have $\mu = 0$ but intramolecular vibrations are possible, thus $\boldsymbol{P} = \boldsymbol{P}_{at} + \boldsymbol{P}_{el}$. As a result, the static permittivity, $\varepsilon = \varepsilon_\infty \approx 2\ldots3$, is ~5–20 % higher than $(n_D)^2$. Larger values for the static permittivity, $\varepsilon \approx 3\ldots200$ are only possible for polar fluids where the molecules possess a permanent dipole moment. Orientational polarization, $\boldsymbol{P}_{or}$, is essentially proportional to μ^2 and the molar concentration, c, of the dipoles. However, in a liquid, the molecular dipoles interact with each other and their resulting orientational correlations may significantly affect $\boldsymbol{P}_{or}$ and thus ε. Especially fluids composed of molecules forming hydrogen-bonded chains, like *N*-methylformamide, exhibit very large static permittivities due to a pronounced parallel alignment of the molecular dipoles. Such dipole-dipole correlations are conveniently expressed with the help of the Kirkwood factor, but they render the theoretical prediction of ε from α and μ difficult [1, 2]. Due to cross-correlations between the dipole moments of the component molecules, theoretical expressions for the static permittivity of liquid mixtures are rather elaborate [1, 2, 7].

Dielectric Response of Electrolytes

Even for pure liquids ε is not measured through the conceptually simple experiment of applying a static field, $\boldsymbol{E}_0$, to the sample-filled capacitor. Instead a harmonic ac field, $\boldsymbol{E} = \boldsymbol{E}_0 \cos(2\pi\nu t)$, with frequency $\nu \approx 1$ MHz is generally used and it is assumed that ν is sufficiently small to allow the molecular dipoles to follow changes of the electric field without delay, i.e., the equilibrium polarization is reached at any time t [8].

In the case of electrolyte solutions, composed of the solvent and a dissolved salt, with the latter more or less completely dissociated into free solvated ions, direct determination of ε is impossible due to the notable dc conductivity, κ_{dc}, of the solution. In this case, $\boldsymbol{E}$ not only aligns the permanent dipole moments of the solvent molecules and creates induced moments in all ions and molecules but additionally also drives an electric current. This *direct current* through ion transport dominates the sample response to a static field ($\nu = 0$, corresponding to observation times $t \rightarrow \infty$) whereas the *displacement current* associated with $\boldsymbol{P}_{or} + \boldsymbol{P}_{\alpha}$ vanishes as the time required for aligning the permanent and creating induced dipole moments is finite.

From this fact, it is sometimes erroneously concluded that ε is not defined or infinite for electrolyte solutions. However, the situation is more subtle and the time, respectively, frequency dependence of the sample response has to be considered: At low ν, the molecules bearing a permanent dipole moment can follow changes of the electric field acting on them without delay, i.e., the equilibrium value of $\boldsymbol{P}_{or} + \boldsymbol{P}_{\alpha}$ and thus the static permittivity, ε, is always reached. However, with increasing ν, the dipoles will more and more lag behind $\boldsymbol{E}(\nu)$ because of intermolecular friction and inertia. Consequence is an increasing phase shift between $\boldsymbol{E}(\nu)$ and $\boldsymbol{P}(\nu)$ which can be expressed by splitting polarization into a component in phase with $\boldsymbol{E}(\nu)$ and a component shifted by 180°. The first is characterized by the frequency-dependent relative permittivity, $\varepsilon'(\nu)$, which exhibits a decrease (dispersion) from ε at $\nu \rightarrow 0$ to ε_∞ at frequencies where orientational polarization is too slow to contribute (Fig. 1a). This dispersion of $\varepsilon'(\nu)$ is accompanied by energy dissipation in the sample, which is conveniently expressed by the dielectric loss, $\varepsilon''(\nu)$ (Fig. 1b), characterizing the out-of-phase component of $\boldsymbol{P}(\nu)$. In other words, because of intermolecular frictions, the ensemble of molecular dipoles requires a certain time to relax to the new equilibrium polarization after an instantaneous change of the electric field and this process dissipates heat. For most liquids around ambient temperature, characteristic relaxation times are in the order of picoseconds to nanoseconds, corresponding to dispersion frequencies in the microwave region (ca. 10 MHz–1 THz; Fig. 2); ε_∞ is reached in the far-infrared region. The response time for $\boldsymbol{P}_{at}$ and $\boldsymbol{P}_{el}$ is much shorter and governed by the quantum mechanics of intramolecular vibrations and electronic transitions. This leads to resonance features in $\varepsilon'(\nu)$ (Fig. 1b). The translational motions of the ions also depend on time so that in general, a frequency dependence of conductivity has to be assumed.

From Maxwell's equations, it follows that only the sum of direct and displacement currents

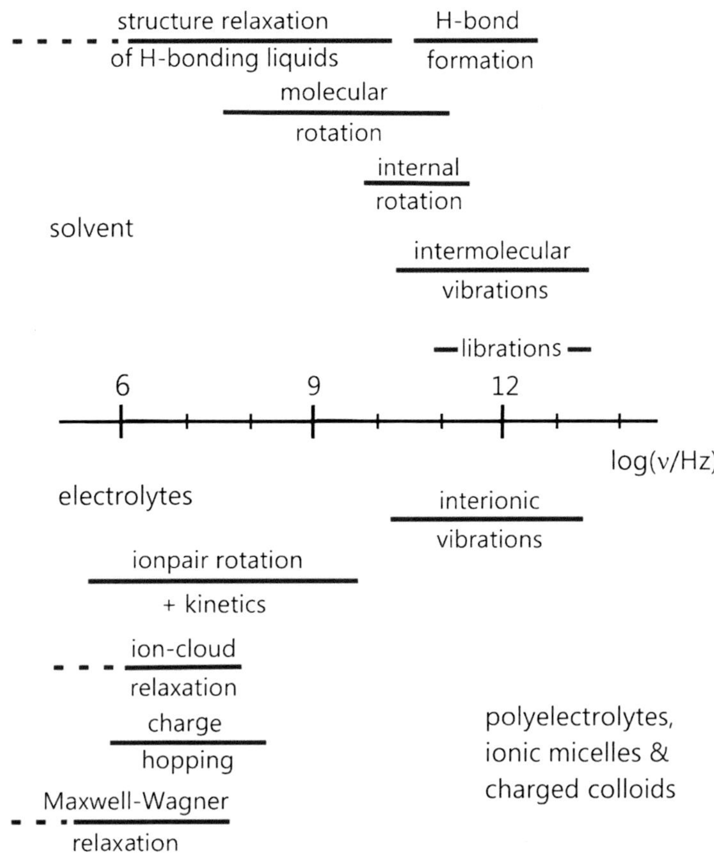

Dielectric Properties, Fig. 2 Frequency regions for typical processes contributing to the dielectric spectra of electrolyte solutions at room temperature

can be experimentally observed for a conducting sample in a time-dependent electric field [1–3]. This total dielectric response of the electrolyte solution is conveniently described by the total complex permittivity, $\eta^*(\nu) = \eta'(\nu) - i\eta''(\nu)$, where the real part, η', is in phase with $\boldsymbol{E}(\nu)$ and the total loss, η'', is shifted by 180° ($i^2 = -1$) [1, 2]. Alternatively, the total complex conductivity, $\kappa^*(\nu) = i2\pi\nu\varepsilon_0\eta^*(\nu)$, or the complex electric modulus, $M^*(\nu) = 1/\eta^*(\nu)$, can be used [3]. The causality principle requires that for $t \to \infty$ ($\nu = 0$), steady-state conditions (equilibrium values) are reached. Therefore, $\kappa^*(0) = \kappa'(0) = \kappa_{dc}$ and $\eta'(0) = \varepsilon$, so that $\eta''(0) = \lim_{\nu\to 0}\kappa_{dc}/(2\pi\nu\varepsilon_0) = \infty$ [1, 2]. Thus, despite diverging total loss for $\nu \to 0$, the static permittivity is a well-defined quantity for electrolyte solutions and its magnitude is essentially determined by orientational and induced polarization. Even more, for $\nu \to 0$, the slope of total relative permittivity vanishes, i.e., $d\eta'/d\nu = 0$, so that the required extrapolation $\varepsilon = \lim_{\nu\to 0}\eta'(\nu)$ of the experimental permittivity spectra is facilitated.

It is convenient to subtract the contribution arising from dc conductivity from the total complex permittivity, $\eta^*(\nu)$, yielding $\varepsilon^*(\nu) = \varepsilon'(\nu) - i\varepsilon''(\nu)$ where $\varepsilon'(\nu) = \eta'(\nu)$ and $\varepsilon''(\nu) = \eta''(\nu) - \kappa_{dc}/(2\pi\nu\varepsilon_0)$. In the following, we call $\varepsilon^*(\nu)$ the complex dielectric permittivity of the material (Fig. 1). Note that according to this pragmatic definition, which differs from Böttcher [1, 2], $\varepsilon^*(\nu)$ subsumes **all** processes contributing to $\boldsymbol{P}$ that explicitly depend on time/frequency, irrespective of their origin. In other words, no formal distinction is made between relaxations caused by dipole reorientation and modes originating from hindered ion translation or other possible mechanisms. This is reasonable as the origin of the features observed in the dielectric spectrum is not immediately obvious and additional information is required for their assignment to molecular-level motions. Figure 2

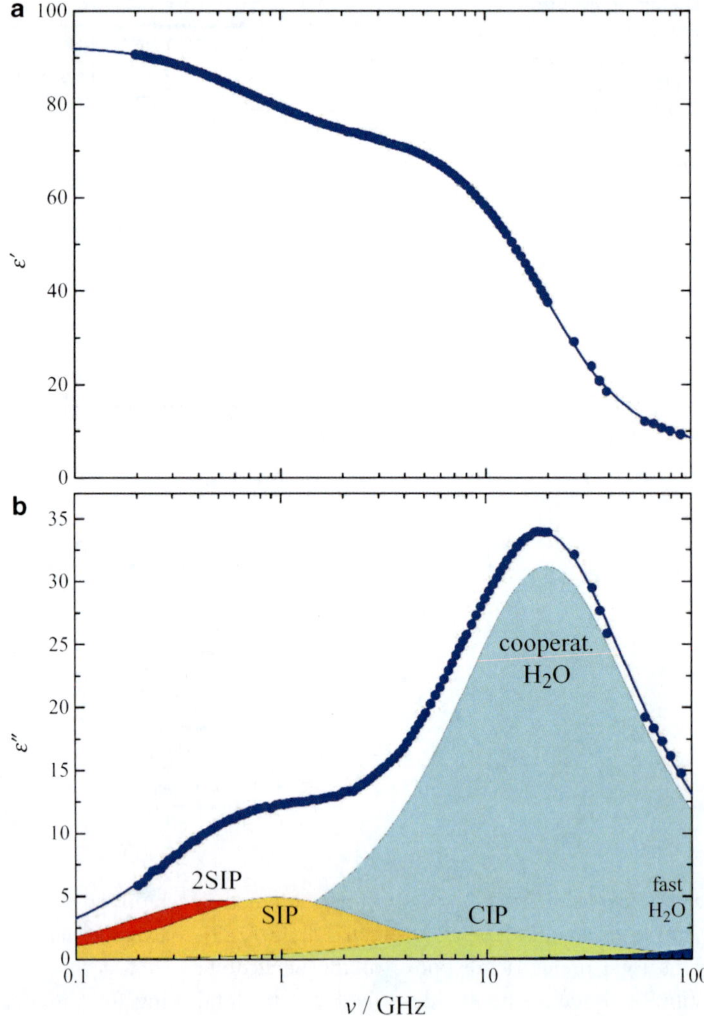

Dielectric Properties, Fig. 3 Spectra of (**a**) relative permittivity, $\varepsilon'(\nu)$, and (**b**) dielectric loss, $\varepsilon''(\nu)$, of 0.086 M $Al_2(SO_4)_3$ in water at 25 °C. Symbols represent experimental data, lines are calculated from a superposition of the five relaxation processes indicated by the *shaded areas* in (**b**)

provides an overview of dynamical processes relevant for electrolyte systems and their typical frequency ranges. Most of these modes, arising from the solvent and the solute, are relaxation processes but librations (hindered rotations), as well as intermolecular and interionic vibrations, are damped resonances.

Dielectric relaxation spectroscopy, i.e., the determination of $\varepsilon^*(\nu)$ in the appropriate frequency range, provides a wealth of information on the dynamics of liquids and solutions at a molecular to mesoscopic (micelles, colloids) level (Fig. 2) [9, 10]. However, to get this, $\varepsilon^*(\nu)$ has to be decomposed into individual modes (Fig. 3) which subsequently have to be assigned to physical processes. This is not always an easy task [10], especially as direct computer simulations of $\varepsilon^*(\nu)$ reach only qualitative agreement so far [11]. For electrolyte solutions, a typical aim of dielectric studies is ion solvation. Since the solvent contribution to $\varepsilon^*(\nu)$ (e.g., the bulk and fast water relaxations of Fig. 3) is characteristically affected, the determination of effective solvation numbers and inference on the relative strength of ion-solvent versus solvent-solvent interactions is possible [9, 10]. Ion association is a further focus as it manifests in relaxation processes specific to the ion pair species present (Fig. 3: 2SIP, SIP, CIP). In contrast to molecular spectroscopies, like NMR or

Raman, which only detect contact ion pairs (CIPs), dielectric spectroscopy is able to monitor all species with permanent dipole moment, including solvent-shared (SIP) and solvent-separated (2SIP) ion pairs [9, 10].

Note that in practice, experimental spectra – especially at low frequencies – may also contain contributions originating from the electrode/electrolyte interface. A typical example is electrode polarization arising from the formation of a diffuse double layer of ions close to a charged surface. Partly, such features depend on the dielectric properties of the electrolyte solution (our focus), but in essence, they are specific to the interface and thus are not topic of this contribution. However, such electrode processes are intensively studied in electrochemistry using, e.g., impedance spectroscopy.

Future Directions

The full potential of dielectric relaxation spectroscopy in electrolyte studies is just emerging. This is due to important progress in instrumentation during the last 5–10 years, which now allows access of the relevant frequency region of ~10 MHz–1 THz and beyond with relative ease. Also, quantitative computer simulation of $\eta^*(\nu)$ is coming into reach. This will provide a major step forward in the interpretation of dielectric spectra.

Cross-References

- ▶ Activity Coefficients
- ▶ Charged Colloids
- ▶ Electrochemical Impedance Spectroscopy (EIS) Applications to Sensors and Diagnostics

References

1. Böttcher CFJ (1973) Theory of electric polarization, vol 1. Elsevier, Amsterdam
2. Böttcher CFJ, Bordewijk P (1978) Theory of polarization, vol 2. Elsevier, Amsterdam
3. Kremer F, Schönhals A (2003) Broadband dielectric spectroscopy. Springer, Berlin
4. Bockris JO'M, Reddy AKN (1998) Modern electrochemistry 1: ionics, 2nd edn. Plenum, New York/London
5. Barthel J, Krienke H, Kunz W (1998) Physical chemistry of electrolyte solutions. Modern aspects. Springer, New York
6. Rainieri FO, Friedman HL (1999) Solvent control of electron transfer reactions. Adv Chem Phys 107: 81–189
7. Reis JCR, Iglesias TP (2011) Kirkwood correlation factors in liquid mixtures from an extended Onsager-Kirkwood-Fröhlich equation. Phys Chem Chem Phys 13:10670–10680
8. Moldover MR, Marsh KN, Barthel J, Buchner R (2003) Relative permittivity and refractive index. In: Goodwin ARH, Marsh KN, Wakeham WA (eds) Measurement of the thermodynamic properties of single phases, vol VI, Experimental thermodynamics. Elsevier, Amsterdam
9. Buchner R (2008) What can be learnt from dielectric relaxation spectroscopy about ion solvation and association? Pure Appl Chem 80:1239–1252
10. Buchner R, Hefter G (2009) Interactions and dynamics in electrolyte solutions by dielectric spectroscopy. Phys Chem Chem Phys 11:8984–8999
11. Schröder C, Steinhauser O (2010) Computational dielectric spectroscopy of charged, dipolar systems. In: Grünberg J (ed) Computational spectroscopy: methods, experiments and applications. Wiley-VCH, Weinheim

Direct Alcohol Fuel Cells (DAFCs)

Claude Lamy
Institut Européen des Membranes, Université Montpellier 2, UMR CNRS n° 5635, Montpellier, France

Introduction

In several applications, fuel cells are widely recognized as very attractive devices for producing directly electric energy from the combustion of a chemical product into oxygen, i.e., air. Low-temperature fuel cells, generally built around a proton exchange membrane, are aimed to be used for a large range of power sources. However, the final choice of the fuel is still difficult and depends greatly on the field of application. If hydrogen or hydrogen-rich gas obtained by hydrocarbon reforming (e.g. methane steam reforming) is clearly the best choice for stationary applications, the difficulty of hydrogen distribution and on board storage for mobile applications leads to the necessity to look for alternative liquid fuels suitable for a direct oxidation fuel cell (DOFC) [1–3]. Several organic fuels have been often considered despite of their rather low electrochemical reactivity in comparison to hydrogen.

In this context, hydrogen carriers like alcohols feeding a direct alcohol fuel cell (DAFC) appear advantageous for two main reasons: they are liquid (which simplifies the problems of storage and distribution) and their theoretical mass energy density is rather high, close to that of gasoline (6.1 and 8.0 kWh/kg for methanol and ethanol, respectively [4]). The most studied alcohols are methanol [5], which is the simplest mono-alcohol, and ethanol [6].

One of the main advantages of methanol is its availability, its low price and the easiness of its storage as a liquid. But methanol is a toxic compound and a pollutant, so that extensive use is inconceivable because of environmental hazards (methanol is entirely miscible with water, which can lead to water contamination). On the other hand, ethanol appears to be an interesting alternative fuel for a wide range of utilization, even if its price is actually too high. Its low toxicity and its availability (from biomass, which will not change the natural balance of carbon dioxide in the atmosphere in contrast to the use of fossil fuels [1]) are important positive points for its use as an alternative fuel to methanol even if its reactivity is slightly lower [7].

The oxidation of ethanol is more difficult than that of methanol with the necessity to break the C–C bond to obtain its complete oxidation to CO_2. It was observed by "in situ" IR reflectance spectroscopy that the dissociation of ethanol leads also to the formation of adsorbed CO [8]. This is the proof that the C–C bond can be, at least to some extent, broken at room temperature and that carbon dioxide can be obtained. However, the main oxidation products are acetaldehyde and acetic acid, as observed after long term electrolyses of ethanol solution [9]. Conversely to methanol, Pt–Ru-based electrocatalysts lead to poor performances in ethanol electroxidation. Several works have indicated that the modification of Pt by tin gives very encouraging results leading to the oxidation of ethanol at lower potentials than on pure platinum.

Thermodynamics and Kinetics of Alcohol Oxidation in a DAFC

In a DAFC the total electro-oxidation to CO_2 of an aliphatic mono-alcohol, C_xH_yO, involves the participation of water (H_2O) or of its adsorbed residue (OH_{ads}) provided by the cathodic reaction (electro-reduction of dioxygen).

The overall electro-oxidation reaction in acid medium to reject the carbon dioxide produced can thus be written as follows:

$$C_xH_yO + (2x - 1)H_2O \rightarrow xCO_2 + nH^+ + ne^- \quad (1)$$

with $n = 4x + y - 2$. Such an anodic reaction is very complicated from a kinetics point of view since it involves multi-electron transfers and the presence of different adsorbed intermediates and several reaction products and by-products. However from thermodynamic data it is easy to calculate the reversible anode potential, the cell voltage under standard conditions, the theoretical efficiency and the energy density.

Thermodynamic Data

According to reaction (1) the standard Gibbs energy change $-\Delta G_1{}^o$, allowing calculating the standard anode potential $E_1^o = \frac{-\Delta G_1^o}{nF}$ can be evaluated from the standard energy of formation ΔG_i^f of reactant (i):

$$-\Delta G_1^o = x\Delta G_{CO_2}^f - \Delta G_{C_xH_yO}^f - (2x - 1)\Delta G_{H_2O}^f \quad (2)$$

In the cathodic compartment the electro-reduction of oxygen does occur, as follows:

$$\frac{1}{2}O_2 + 2H^+ + 2e^- \rightarrow H_2O \quad (3)$$

with $\Delta G_2^o = \Delta G_{H_2O}^f = -237.1\ \text{kJ mol}^{-1}$, leading to a standard cathodic potential E_2^o:

$$E_2^o = -\frac{\Delta G_2^o}{2F} = \frac{237.1 \times 10^3}{2 \times 96485} = 1.229\ \text{V vs. SHE} \quad (4)$$

where SHE is the standard hydrogen electrode, acting as a reference electrode.

In the fuel cell the electrical balance corresponds to the complete combustion of the organic compound in the presence of oxygen, as follows:

$$C_xH_yO + \left(x + \frac{y}{4} - \frac{1}{2}\right)O_2 \rightarrow xCO_2 + \frac{y}{2}H_2O \quad (5)$$

with

$$\Delta G_r^o = \left(2x + \frac{y}{2} - 1\right)\Delta G_2^o - \Delta G_1^o = x\Delta G_{CO_2}^f + \frac{y}{2}\Delta G_{H_2O}^f - \Delta G_{C_xH_yO}^f, \quad (6)$$

leading to the equilibrium standard cell voltage:

$$E_{eq}^o = -\frac{\Delta G_r^o}{nF} = -\frac{\Delta G_2^o}{2F} + \frac{\Delta G_1^o}{nF} = E_2^o - E_1^o \quad (7)$$

Then it is possible to evaluate the specific energy W_e in kWh kg^{-1}:

$$W_e = \frac{(-\Delta G_r^o)}{3,600\,M} \quad (8)$$

with M the molecular mass of the compound, and knowing the enthalpy change ΔH_r^0 from thermodynamic data:

$$\Delta H_r^o = \left(2x + \frac{y}{2} - 1\right)\Delta H_2^o - \Delta H_1^o = x\Delta H_{CO_2}^f + \frac{y}{2}\Delta H_{H_2O}^f - \Delta H_{C_xH_yO}^f \quad (9)$$

one may calculate the reversible energy efficiency under standard conditions:

$$\varepsilon_{rev} = \frac{\Delta G_r^o}{\Delta H_r^o} \quad (10)$$

For methanol and ethanol, the electrochemical oxidation reaction and the standard anode potentials are:

$$CH_3OH + H_2O \rightarrow CO_2 + 6H^+ + 6e^- \qquad E_{MeOH}^o = 0.016\ \text{V vs. SHE} \quad (11)$$

$$CH_3CH_2OH + 3H_2O \rightarrow 2CO_2 + 12H^+ + 12e^- \qquad E_{EtOH}^o = 0.085\ \text{V vs. SHE} \quad (12)$$

This corresponds to the overall combustion reaction of these alcohols in oxygen:

$$CH_3OH + 3/2O_2 \rightarrow CO_2 + 2H_2O \quad (13)$$

$$CH_3CH_2OH + 3O_2 \rightarrow 2CO_2 + 3H_2O \quad (14)$$

with the thermodynamic data under standard conditions.

For higher alcohols, such as n-propanol, taken as an example, the following calculations can be made:

$$C_3H_7OH + 5H_2O \rightarrow 3CO_2 + 18H^+ + 18e^- \quad (15)$$

$$-\Delta G_1^o = 3\Delta G_{CO_2}^f - \Delta G_{C_3H_7OH}^f - 5\Delta G_{H_2O}^f = -3x394.4 + 168.4 + 5x237.1 = 171\ \text{kJ mol}^{-1} \quad (16)$$

so that $E_1^o = -\frac{\Delta G_1^o}{18F} = \frac{171.1\times10^3}{18\times96{,}485} = 0.098$ V vs. SHE and

$$C_3H_7OH + 9/2O_2 \rightarrow 3CO_2 + 4H_2O \quad (17)$$

with: $\Delta G_r^o = 9\Delta G_2^o - \Delta G_1^o = 3\Delta G_{CO_2}^f + 4\Delta G_{H_2O}^f - \Delta G_{C_3H_7OH}^f = -3x394.4 - 4x237.1 + 168.4 = -1{,}963\ \text{kJ mol}^{-1}$

Direct Alcohol Fuel Cells (DAFCs), Table 1 Thermodynamic data associated with the electrochemical oxidation of some alcohols (under standard conditions)

Alcohol	ΔG^o_f/kJ mol^{-1}	E^o_f/V vs. SHE	ΔG^o_r/kJ mol^{-1}	E^o_{cell}/V	W_e / kWh kg^{-1}	ΔH^o_r/kJ mol^{-1}	ε_{rev}
CH_3OH	–9.3	0.016	–702	1.213	6.09	–726	0.967
C_2H_5OH	–97.3	0.085	–1,325	1.145	8.00	–1,367	0.969
C_3H_7OH	–171	0.098	–1,963	1.131	9.09	–2,027	0.968
1-C_4H_9OH	–409	0.177	–2,436	1.052	9.14	–2,676	0.910
CH_2OH-CH_2OH	–25.5	0.026	–1,160	1.203	5.20	–1,189	0.976
CH_2OH-$CHOH$-CH_2OH	1	–0.001	–1,661	1.230	5.02	–1,650	1.01

The standard cell voltage is thus:

$$E^o_{eq} = -\frac{\Delta G^o_r}{18F} = \frac{1,963 \times 10^3}{18 \times 96,485}$$

$$= \frac{237.1 \times 10^3}{2 \times 96,485} - \frac{171 \times 10^3}{18 \times 96,485}$$

$$= 1.229 - 0.098 = 1.131 \text{ V}$$

and the specific energy is:

$$W_e = \frac{1,963 \times 10^3}{3,600 \times 0.060} = 9.09 \text{ kWh kg}^{-1} \quad (18)$$

The enthalpy change of reaction (17) is:

$$\Delta H^o_r = -3 \times 395.5 - 4 \times 285.8 + 302.6$$

$$= -2,027 \text{ kJmol}^{-1} \quad (19)$$

so that the reversible energy efficiency is:

$$\varepsilon_{rev} = \frac{\Delta G^o_r}{\Delta H^o_r} = \frac{1,963}{2,027} = 0.968 \quad (20)$$

Table 1 summarizes the results obtained with some mono-alcohols and polyols under standard conditions (25 °C, 1 bar, liquid phase).

For all the alcohols listed in Table 1, the cell voltage varies from 1.23 V to 1.05 V, which is very similar to that of a hydrogen/oxygen fuel cell (E°_{eq} = 1.23 V). The energy density varies between half to one that of gasoline (10–11 kWh kg^{-1}) so that these compounds are good alternative fuels to hydrocarbons. Furthermore the reversible energy efficiency ε_{rev} is close to 1, while that of the H_2/O_2 fuel cell is 0.83 at 25 °C (standard conditions).

However, the practical electric efficiency of a fuel cell is depending on the current which is delivered by the cell and is lower than that of the reversible efficiency. This is due to the irreversibility of the electrochemical reactions involved on the electrodes. The practical efficiency of a fuel cell can be expressed as follows:

$$\varepsilon_{cell} = \frac{n_{exp} \times F \times E(j)}{-\Delta H^o_r}$$

$$= \frac{nFE^o_{eq}}{(-\Delta H^o_r)} \times \frac{E(j)}{E^o_{eq}} \times \frac{n_{exp}}{n}$$

$$= \varepsilon_{rev} \times \varepsilon_E \times \varepsilon_F \quad (21)$$

with $E(j) = E^o_{eq} - (|\eta_a| + |\eta_c| + R_e j)$.

In most cases the anodic overpotential, η_a, is at least 0.5 V for a reasonable current density (100 mA cm^{-2}) so that the cell voltage, including an overpotential η_c = – 0.3 V for the cathodic reaction, will be of the order of 0.4 V and the voltage efficiency will be ε_E = 0.4/1.2 = 0.33, under operating conditions. Such a drawback of the direct alcohol fuel cell can only be removed by improving the kinetics of the electro-oxidation of the fuel. This needs to have a relative good knowledge of the reaction mechanisms, particularly of the rate determining step, and to search for electrode materials (Pt–X binary and Pt–X–Y ternary electrocatalysts) with improved catalytic properties.

From equation (21), it follows that the increase of the practical fuel cell efficiency can be achieved by increasing the voltage efficiency ($\varepsilon_E = E(j)/E^o_{eq}$) and the faradic efficiency ($\varepsilon_F = n_{exp}/n$), the reversible yield, ε_{rev}, being fixed by the thermodynamic data and the working conditions (temperature and pressure).

For a given electrochemical system, the increase of the voltage efficiency is directly related to the decrease of the overpotentials of the oxygen reduction reaction, η_c, and alcohol oxidation reaction, η_a, which needs to enhance the activity of the catalysts at low potentials and low temperature, whereas the increase of the faradic efficiency is related to the ability of the catalyst to oxidize completely or not the fuel into carbon dioxide, i.e. it is related to the selectivity of the catalyst. Indeed, in the case of ethanol, taken as an example, acetic acid and acetaldehyde are formed at the anode [10], which corresponds to a number of electrons involved of 4 and 2, respectively, against 12 for the complete oxidation of ethanol to carbon dioxide. The enhancement of both these efficiencies is a challenge in electrocatalysis.

Kinetics and Reaction Mechanism

The electro-oxidation of alcohols, even the simplest one, i.e. methanol, is a complex reaction, involving multi-electron transfer and several intermediate steps. The complete oxidation to CO_2, except for methanol, is never observed at room temperature. This is due to the difficulty of breaking the C–C bond, and of finding multifunctional electrocatalysts, in order to simultaneously activate one of the following main reaction steps : cleavage of C–H bonds with hydrogen adsorption and oxidation, breaking of the C–C bond, removal from the electrode active surface sites of strongly adsorbed intermediates (C_1 and/or C_n poisoning species) at low potentials, activation of the water molecule at low potentials to further oxidize the adsorbed residues, completion of the oxidation reaction to CO_2 by providing extra oxygen atoms, etc.

In order to improve the fuel utilization in a direct alcohol fuel cell (DAFC) it is important to investigate the reaction mechanism and to develop active electrocatalysts able to activate each reaction path. The elucidation of the reaction mechanism, thus, needs to combine pure electrochemical methods (cyclic voltammetry, rotating disc electrodes, etc.) with other physicochemical methods, such as "in situ" spectroscopic methods (infrared [11] and UV–VIS [12] reflectance spectroscopy), or mass spectroscopy such as DEMS [13], or radiochemical methods [14] to monitor the adsorbed intermediates and "on line" chromatographic techniques [15] to analyze quantitatively the reaction products and by-products.

The electrocatalytic oxidation of ethanol has been investigated for many years on several platinum-based electrodes, including Pt-X alloys (with X = Ru, Sn, Mo, etc.), and dispersed nanocatalysts. Pure platinum smooth electrodes are rapidly poisoned by some strongly adsorbed intermediates, such as carbon monoxide, resulting from the dissociative chemisorption of the molecule, as shown by the first experiments in infrared reflectance spectroscopy [8]. Both kinds of adsorbed CO, either linearly-bonded or bridge-bonded to the platinum surface, are observed. Besides, other adsorbed species have been identified by IR reflectance spectroscopy, including reaction intermediates, such as acetaldehyde and acetic acid, and other by-products [9].

Voltammetric results, completed by the different spectroscopic and chromatographic results, allowed us to propose a detailed reaction mechanism of ethanol oxidation, involving parallel and consecutive oxidation reactions, on Pt-based electrodes (such as Pt–Sn catalysts), where the key role of the adsorption steps was underlined.

$$Pt + CH_3 - CH_2OH \rightarrow Pt - (CHOH - CH_3)_{ads} + H^+ + e^- \tag{22}$$

$$Pt - (CHOH - CH_3)_{ads} \rightarrow Pt + CHO - CH_3 + H^+ + e^- \tag{23}$$

As soon as acetaldehyde (AAL) is formed, it can adsorb on platinum sites leading to a -$(CO–CH_3)_{ads}$ species (adsorbed acetyl):

$$Pt + CHO - CH_3 \rightarrow Pt - (CO - CH_3)_{ads} + H^+ + e^- \quad (24)$$

Then, because Sn is known to activate water at lower potentials than platinum, some OH species can be formed at low potentials on Sn sites according to reaction:

$$Sn + H_2O \rightarrow Sn - (OH)_{ads} + H^+ + e^- \quad (25)$$

and adsorbed acetaldehyde species can react with adsorbed OH species to give acetic acid (AA) according to:

$$Pt - (CO - CH_3)_{ads} + Sn - (OH)_{ads} \rightarrow Pt + Sn + CH_3 - COOH \quad (26)$$

Further oxidation to carbon dioxide is usually difficult on pure Pt electrodes at room temperature. However, carbon monoxide acting as a poisoning species, and CO_2, were clearly observed by infrared reflectance spectroscopy [8] or by DEMS [16], and CO_2 was detected by gas chromatography, whereas some traces of methane were observed at low potential ($E < 0.4$ V/RHE) by DEMS [16]. This may be explained by the following mechanism involving the dissociation of -$(CO–CH_3)_{ads}$ by breaking the C–C bond:

$$Pt + Pt - (CO - CH_3)_{ads} \rightarrow Pt - (CO)_{ads} + Pt - (CH_3)_{ads} \quad (27)$$

$$2Pt + H_2O \rightarrow Pt - H_{ads} + Pt - OH_{ads} \quad (28)$$

$$Pt - (CH_3)_{ads} + Pt - H_{ads} \rightarrow CH_4 + 2Pt \quad (29)$$

$$Pt - (CO)_{ads} + Sn - OH_{ads} \rightarrow CO_2 + H^+ + e^- + Pt + Sn \quad (30)$$

This mechanism can explain the higher efficiency of PtSn in forming AA compared to Pt at low potentials ($E < 0.35$ V vs. RHE), as was shown by electrolysis experiments. Moreover adsorbed OH species on Sn atoms can oxidize adsorbed -$(CO–CH_3)_{ads}$ species to CH_3–COOH or CO species to CO_2, according to the bifunctional mechanism [17].

This reaction mechanism of ethanol oxidation on a Pt-based electrode can be summarized by the following scheme [4] (Scheme 1).

In this mechanism the adsorbed acetyl plays a key role and its further oxidation is favored by the addition to platinum of metal atoms more easily oxidizable at low potentials, such as Ru, Sn, Mo, etc. Other electrocatalysts were considered for the electro-oxidation of ethanol, such as rhodium, iridium [18] or gold [19], leading to similar results in acid medium. The oxidation of ethanol on rhodium proceeds mainly through the formation of acetic acid and carbon monoxide, which is further oxidized to carbon dioxide when the rhodium surface begins to oxidize, at 0.5–0.7 V/RHE [18]. On gold in acid medium the oxidation reaction leads mainly to the formation of acetaldehyde [19].

Fuel Cell Results

The direct ethanol fuel cell is based on a proton exchange membrane fuel cell (PEMFC), in which the anodic compartment is fed with an ethanol-water mixture (Fig. 1).

Typical results were obtained with a membrane-electrode assembly (MEA) consisting of a Nafion®117 membrane on which are pressed the anodic electrocatalysts (Pt or Pt-based catalysts dispersed on a high surface area carbon support) and the cathodic catalyst (usually Pt/C with metal loading from 40 % to 60 %). An example of the electrical characteristics of a DEFC with Pt/C, PtSn/C or PtSnRu/C anode catalysts is given in Fig. 2.

Comparison between these three catalysts in terms of selectivity is difficult to perform for different reasons. As it can be seen in Fig. 2, the $E(j)$ polarization curves obtained with a DEFC mounted with a Pt/C anode does not achieve cell voltages high enough to be compared with the other catalysts. Thus data were recorded at

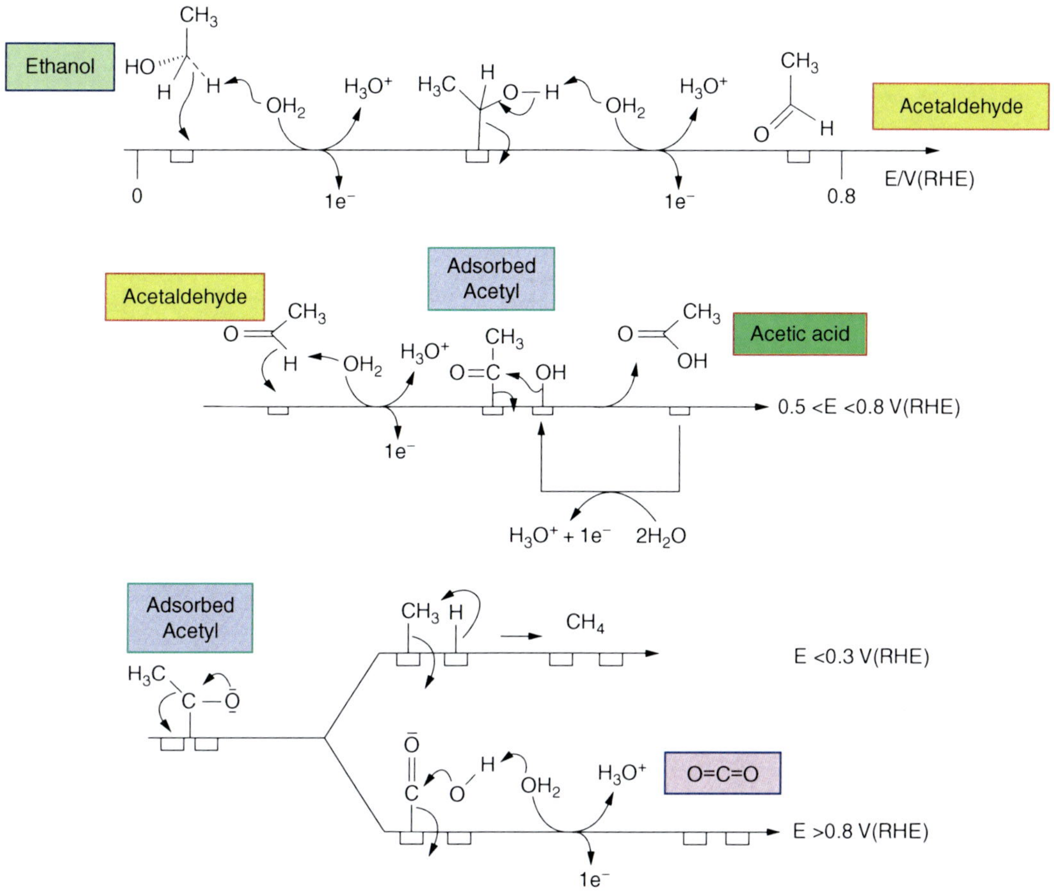

Direct Alcohol Fuel Cells (DAFCs), Scheme 1 Proposed mechanism for the electrocatalytic oxidation of ethanol on a Pt-based elecrode in acidic medium (all the species with color filling were detected either by *IR* reflectance spectroscopy or by chromatographic analysis)

cell voltages close to 0.2 V at a current density $j = 8$ mA cm^{-2} for the Pt/C anode and close to 0.48–0.58 V at $j = 32$ mA cm^{-2} for the PtSn/C and PtSnRu/C anodes.

The analysis of the reaction products at the exhaust from the anodic compartment can be made by HPLC. The analytical results are given in Table 2.

From these results it appears that the addition of tin to platinum greatly favors the formation of acetic acid comparatively to acetaldehyde. This can be explained by the bifunctional mechanism [17] where ethanol is adsorbed dissociatively at platinum sites, either via an O-adsorption or a C-adsorption process [9], followed by the oxidation of these adsorbed residues by oxygenated species formed on Sn at lower potentials giving AA.

On the other hand, the yield in CO_2 is twice higher with a Pt/C catalyst than with a Pt–Sn/C catalyst. This can be explained by the need to have several adjacent platinum sites to adsorb dissociatively the ethanol molecule and to break the C-C bond. As soon as some tin atoms are introduced between platinum atoms, this latter reaction is disadvantaged.

Future Directions

DEFC systems are claimed to be commercialized soon by some companies (e.g. Acta S.p.A. in

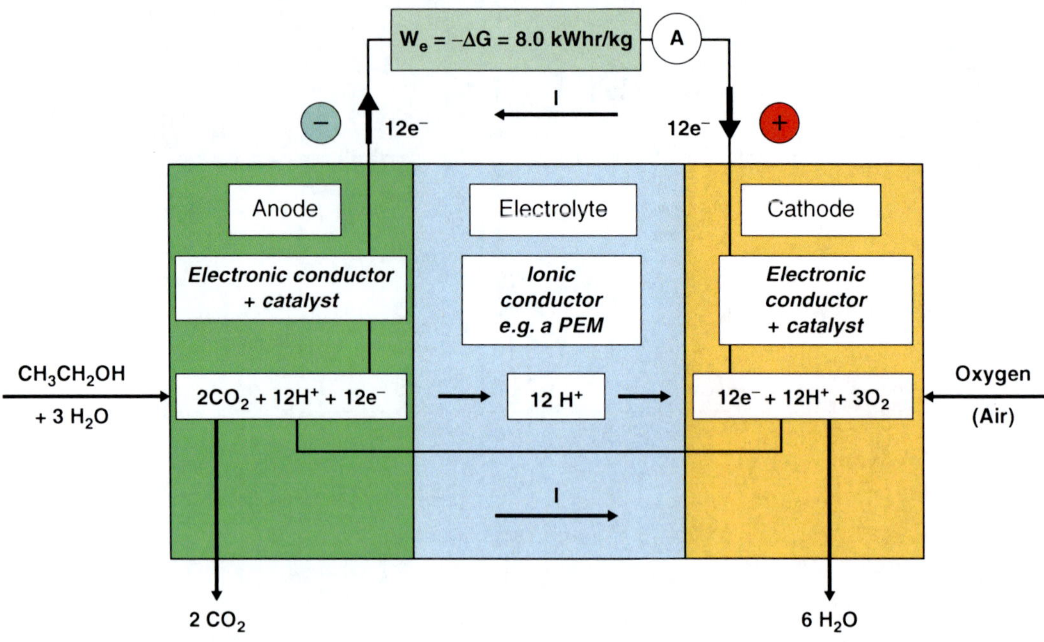

Direct Alcohol Fuel Cells (DAFCs), Fig. 1 Schematic diagram of a DEFC based on a proton exchange membrane

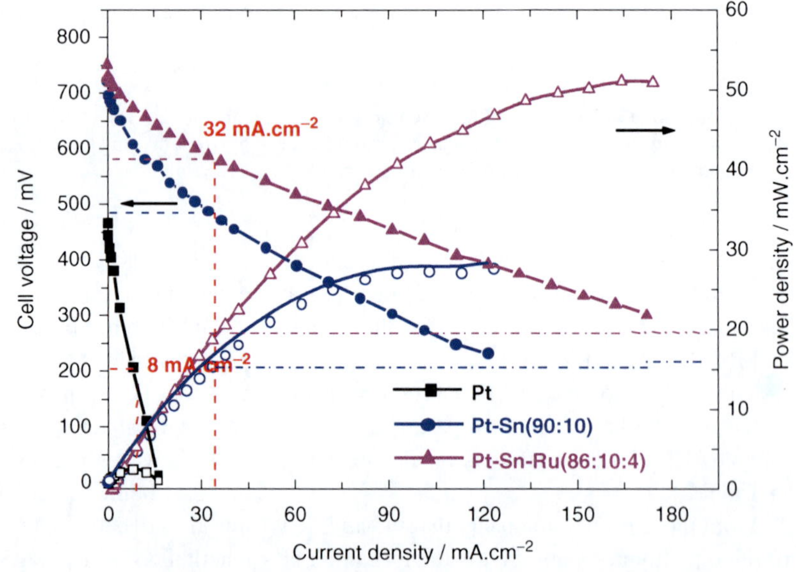

Direct Alcohol Fuel Cells (DAFCs), Fig. 2 Fuel cell characteristics of a DEFC with a 25 cm^2 surface area electrode recorded at 80 °C with different catalysts (2 M C_2H_5OH, N117)

Italy), but the technology is still immature for large-scale commercialization.

Indeed, the presence of both poisoning species (mainly CO) and intermediate reaction products (AAL, AA) decreases correspondingly the useful energy density of the fuel, and also the power density, since the oxidation current densities are lower than those obtained with the oxidation of

Direct Alcohol Fuel Cells (DAFCs), Table 2 Distribution of the reaction products resulting from the oxidation of 2 M ethanol at 80 °C after 4 h working at a current density of 32 $mA.cm^{-2}$ for PtSn and PtSnRu catalysts or of 8 $mA.cm^{-2}$ for Pt

Anode	Acetic acid	Acetaldehyde	CO_2
60 % Pt/XC72	13 mM	19 mM	8 mM
60 % Pt–Sn(90:10)/XC72	50 mM	10 mM	5 mM
60 % Pt–Sn–Ru (86:10:4)/XC72	44 mM	9 mM	6 mM

methanol and above all with that of hydrogen. To improve the kinetics of ethanol oxidation would require the development of new electrocatalysts able to break the C-C bond at low temperatures and to oxidize adsorbed CO at lower potentials, i.e. to reduce the oxidation overpotential.

In addition improvements in the sulfonated perfluorinated membranes, or development of new protonic membranes with reduced alcohol cross-over and better conductivity and stability at higher temperatures (up to 200 °C) are still challenging topics before any commercialization of DAFC systems.

Alternatively anion exchange membranes (AEM), conducting by OH^- anions, with good properties (stability, conductivity, etc.) will allow using non noble metal catalysts (transition metals) which are good electrocatalysts in alkaline medium for alcohol oxidation [20]. However the actually available AEMs are not stable above 60–70 °C and their ionic conductivity is one order of magnitude lower than that of Nafion®. Thus AEMFC needs the investigation of new OH^- conducting polymers with higher conductivity and stability at higher temperatures (above 100 °C).

Cross-References

► Alkaline Membrane Fuel Cells
► Electrocatalysis of Anodic Reactions
► Polymer Electrolyte Fuel Cells (PEFCs), Introduction

References

1. Gosselink JW (2002) Pathways to a more sustainable production of energy: sustainable hydrogen – a research objective for Shell. Int J Hydrogen Energy 27:1125
2. Takeichi N, Senoh H, Yokota T, Tsuruta H, Hamada K, Takeshita HT, Tanaka H, Kiyobayashi T, Takano T, Kuriyama N (2003) "Hybrid hydrogen storage vessel", a novel high-pressure hydrogen storage vessel combined with hydrogen storage material. Int J Hydrogen Energy 28:1121
3. Ströbel R, Oszcipok M, Fasil M, Rohland B, Jörissen L, Garche J (2002) The compression of hydrogen in an electrochemical cell based on a PE fuel cell design. J Power Sources 105:208
4. Lamy C, Belgsir EM (2003) Other direct alcohol fuel cells: fundamentals and survey of systems. In: Vielstich W, Lamm A, Gasteiger H (eds) Handbook of fuel cells, vol. 1. Wiley, Chichester, pp 323–334, Chap. 19
5. Zhang H, Liu H (eds) (2009) Electrocatalysis of direct methanol fuel cells. Wiley, Weinheim
6. Lamy C, Coutanceau C, Léger J-M (2009) The direct ethanol fuel cell: a challenge to convert bioethanol cleanly into electric energy. In: Barbaro P, Bianchini C (eds) Catalysis for sustainable energy production. Wiley, Weinheim, pp 3–46, Chap. 1
7. Lamy C, Lima A, Le Rhun V, Delime F, Coutanceau C, Léger J-M (2002) Recent advances in the development of direct alcohol fuel cells (DAFC). J Power Sources 105:283–296
8. Perez J-M, Beden B, Hahn F, Aldaz A, Lamy C (1989) "In situ" infrared reflectance spectroscopic study of the early stages of ethanol adsorption at a platinum electrode in acid medium. J Electroanal Chem 262:251–261
9. Hitmi H, Belgsir EM, Léger J-M, Lamy C, Lezna RO (1994) A kinetic analysis of the electro-oxidation of ethanol at a platinum electrode in acid medium. Electrochim Acta 39:407–415
10. Rousseau S, Coutanceau C, Lamy C, Léger J-M (2006) Direct ethanol fuel cell (DEFC): Electrical performances and reaction products distribution under operating conditions with different platinum-based anodes. J Power Sources 158: 18–24
11. Beden B, Lamy C (1988) Infrared reflectance spectroscopy. In: Gale RJ (ed) Spectroelectrochemistry – theory and practice. Plenum Press, New York, pp 189–261, Chapter 5
12. Kolb DM (1988) UV–VIS reflectance spectroscopy. In: Gale RJ (ed) Spectroelectrochemistry– theory and practice. Plenum Press, New York, pp 87–188, Chapter 4
13. Jusys Z, Massong H, Baltruschat H (1999) A new approach for simultaneous DEMS and EQCM: Electro-oxidation of adsorbed CO on Pt and Pt-Ru. J Electrochem Soc 146:1093

14. Horányi G (1999) In: Wieckowski A (ed) Interfacial electrochemistry: theory, experiment and applications. Marcel Dekker, New York, p 477
15. Belgsir EM, Bouhier E, Essis-Yei H, Kokoh KB, Beden B, Huser H, Léger J-M, Lamy C (1991) Electrosynthesis in aqueous medium: a kinetic study of the electrocatalytic oxidation of oxygenated organic molecules. Electrochim Acta 36: 1157–1164
16. Iwasita T, Pastor E (1994) A dems and FTir spectroscopic investigation of adsorbed ethanol on polycrystalline platinum. Electrochim Acta 39: 531–537
17. Watanabe M, Motoo S (1975) Electrocatalysis by adatoms: Part III. Enhancement of the oxidation of carbon monoxide on platinum by ruthenium adatoms. J Electroanal Chem 60:275–283
18. De Tacconi NR, Lezna RO, Beden B, Hahn F, Lamy C (1994) In-situ FTIR study of the electrocatalytic oxidation of ethanol at iridium and rhodium electrodes. J Electroanal Chem 379:329–337
19. Tremiliosi-Filho G, Gonzalez ER, Motheo AJ, Belgsir EM, Léger J-M, Lamy C (1998) Electrooxidation of ethanol on gold: analysis of the reaction products and mechanism. J Electroanal Chem 444:31–39
20. Varcoe JR, Slade RCT, Lam How Yee E, Poynton SD, Driscoll DJ (2007) Investigations into the ex situ methanol, ethanol and ethylene glycol permeabilities of alkaline polymer electrolyte membranes. J Power Sources 173:194–199

Direct Electrochemistry

► Direct Electron Transfer to Enzymes

Direct Electron Transfer to Enzymes

Roland Ludwig
Department of Food Science and Technology,
Vienna Institute of Biotechnology
BOKU-University of Natural Resources and
Life Sciences, Vienna, Austria

Synonyms

Direct electrochemistry

Definition

Direct electron transfer (DET) means an exchange of electrons between the cofactor of a redox-active enzyme (oxidoreductase) or a redox protein and an electrode (transducer) in the absence of redox mediators. DET is rarely observed and reported for only few redox proteins and redox enzymes. DET is of interest for fundamental studies on electron transfer in proteins and enzymes and for the development of sensitive and specific biosensors, robust biofuel cells and heterogeneous bioelectrosynthesis.

Background

Redox enzymes catalyze many biological processes, examples can be found in the bioenergetic cell metabolism which heavily depends on redox proteins and redox enzymes in the respiratory or photosynthetic chains. The importance of electron transfer reactions to sustain life is obvious. Many of these reactions occur in membranes (e.g., the mitochondrial inner membrane) where redox proteins and redox enzymes form electron transfer chains. Some of them exhibit DET when placed on electrodes. The reason is that electron transfer between proteins needs the ability of the involved proteins and enzymes to take-up or pass-on electrons from other constituents of the electron transfer chain along a potential gradient. Another group of redox-active enzymes capable of DET are extracellular enzymes involved in ligninolysis like laccase, lignin peroxidase and manganese peroxidase or cellulolysis like cellobiose dehydrogenase [1]. Obviously, an electrode surface of a suitable material or a properly modified electrode can replace redox proteins and redox enzymes, which are the natural electron transfer partners. Having redox enzymes directly connected to electrodes should allow the exploitation of their naturally high catalytic efficiency and selectivity in biosensors, biofuel cells and bioelectrosynthesis.

Research on the direct electron exchange between electrodes and proteins started in the late 1970s. In 1977 two groups independently reported the observation of DET for cytochrome *c* on the surface of a tin doped indium oxide electrode [2] and a 4,4′-bipyridyl modified gold electrode [3]. The first enzymes shown to exhibit direct electrochemistry were a laccase [4] and a peroxidase [5]. Since then a growing number of redox proteins and redox enzymes have been shown to directly exchange electrons with electrodes. Many new electrode materials and surface modifications as well as new electrochemical techniques were developed to investigate the interaction of redox proteins and redox enzymes with the electrode surfaces in detail. Simultaneously, materials and techniques were improved and used to optimize DET to achieve high current densities and to develop reliable biosensors and biofuel cells.

Scientific Fundamentals

The redox enzymes that are most promising to exhibit DET have usually natural redox partners, e.g., cytochromes, ferredoxins or quinones. The electron transfer event between two nonbonded redox proteins or redox enzymes occur at, or close to, the surface of the protein (outer-sphere electron transfer). The electron transfer kinetics can be analyzed in various ways; the most commonly applied are the Butler-Volmer model and the Marcus theory [6]. According to the Marcus theory, the electron transfer kinetics between two redox species are determined by the driving force (potential difference), the reorganization energy (reflecting the structural rigidity of the redox species) and the distance between two redox centers. A greater distance between two redox-active prosthetic groups or a prosthetic group and the electrode surface reduces the electron transfer rate. It is often stated that at a distance greater than 15 Å the electron transfer rate becomes too low for physiological reactions or useful technical applications. Therefore, the design of optimized electrodes and redox proteins and enzymes pursues a high DET rate by reducing the electron transfer distance. Examples are the deglycosylation of proteins or the use of nanoparticles to achieve a more intimate contact

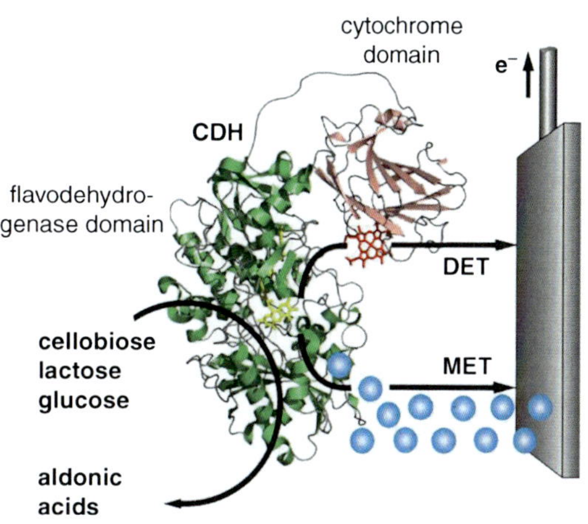

Direct Electron Transfer to Enzymes, Fig. 1 A schematic representation of a redox enzyme in vicinity to an electrode surface exemplifies two ways of electrical communication between the biorecognition element and the electrode. The shown flavocytochrome cellobiose dehydrogenase consist of two domains. The flavodehydrogenase domain (*green*) carries an FAD cofactor (*yellow*) in the catalytic site to oxidize aldoses at the anomeric carbon into aldonic acids. Connected to this catalytic domain via a short, flexible linker is the cytochrome domain (*pink*) carrying a *b*-type heme (*red*). The cytochrome domain is catalytically inactive, but can transfer electrons. Direct electron transfer (*DET*) can occur via the surface exposed heme. In this case the electrons from the reduced cofactor $FADH_2$ are transferred subsequently in one-electron steps to the electrode surface. Alternatively, electrons can be taken up directly from the $FADH_2$ by redox mediators (*blue spheres*) and transported to the electrode. Redox mediators for cellobiose dehydrogenase are one- or two-electron acceptors and can be soluble, diffusible redox mediators like redox dyes, metal complexes and quinones or Os-complex carrying redox polymers [9]

of redox center and conducting surface. If no direct electrochemistry between the enzyme's redox center and the electrode can be established, mediated electron transfer (MET) which uses diffusible redox mediators or redox polymers is the method of choice for contacting. An example of both approaches is given in Fig. 1. Two experimental approaches can be used to elucidate whether DET occurs between a redox protein or redox enzyme. Direct evidence comes from the observation of an electrochemical response of the redox-active cofactor. Indirect evidence can be obtained by observing a catalytic current in presence of an enzyme's substrate.

Active-Sites, Redox Proteins, and Redox Enzymes Exhibiting DET

DET has been observed for the following active sites of redox proteins or redox enzymes: iron-sulfur clusters (2Fe-2S, 3Fe-4S, 4Fe-4S, 4Fe-4S), flavins (FAD and FMN), hemes (*b*-type, *c*-type, etc. and subtypes) and quinones (PQQ). The best known and studied redox proteins exhibiting DET are cytochromes (cytochrome *c*, cytochrome c_3, cytochrome c_{552}, cytochrome b_5), globins (myoglobin, hemoglobin), ferredoxins, flavodoxins, plastocyanins, and cupredoxins. Redox enzymes are certainly proteins, but in contrast to redox proteins they do not store and transfer electrons, but catalyze reactions. The combination of DET and catalytic activity renders redox enzymes into valuable recognition elements for biosensors and biocatalysts for biofuel cells. Among others DET was reported for the following redox enzymes: Heme enzymes, e.g., nitrite reductase, cytochrome *c* oxidase, horseradish peroxidase, lignin peroxidase, manganese peroxidase, catalase; flavoenzymes, e.g., flavocytochrome b_2, cellobiose dehydrogenase, glucose oxidase, fumarate reductase, dimethyl sulfoxide reductase; copper-containing enzymes, e.g., laccase, tyrosinase, ascorbate oxidase, ceruloplasmin, galactose oxidase, nitrite reductase; molybdoenzymes, e.g., arsenit reductase, nitrate reductase, sulfite reductase and quinoheme alcohol dehydrogenase.

Techniques Used to Characterize Electron Exchange Processes

Bioelectrochemistry has benefited from the development of several new electroanalytic techniques like cyclic voltammetry or pulse voltammetry in the second half of the twentieth century. These technologies allowed solid macro- and microelectrodes of various materials to be used for the elucidation of electron transfer processes on various electrode materials. To overcome mass-transfer limitation rotating disc electrodes or flowcells for stationary electrodes have been developed. Cyclic voltammetry is the most widely used method but suffers sometimes from a lack of sensitivity or ill-defined redox waves. The familiar peak-like voltammograms appear for some species only at very low scan rates down to 1 mV s^{-1}. Square-wave voltammetry is a method that can be useful to overcome particular problems by being inherently more sensitive than cyclic voltammetry.

The success of protein voltammetry for DET measurements depends critically on the electrode material – how it is prepared and modified. The current response may come from (1) free diffusing molecules in solution or (2) tightly bound protein molecules. Diffusion-controlled electrochemistry (case 1) requires a transient interaction of the protein with the electrode so that the electrode is not blocked by reacted species. Nowadays the interest has turned more to electrode-bound protein molecules, allowing the study of redox proteins or redox enzymes as a stable monolayer or layer-by-layer assembly. Proteins can be immobilized by adsorption, electrostatic interaction, hydrophobic interaction or covalent bonds.

Electrode materials and surfaces that have successfully been used to study DET include: (1) metal electrodes made of gold, silver or platinum which can also be modified. (2) Metal oxide electrodes (tin(II) oxide, indium(III) oxide, zirconium(IV) oxide, ruthenium(IV) dioxide, FeO_3 nanoparticles, etc.) with or without modification. Gold and silver electrodes have been modified with alkanethiols, which form self-assembled monolayers (SAM). The head group (−SH) forms a covalent bond to the gold surface whereas the terminal group provides binding sites for the redox protein or redox enzyme. The terminal group can also be used to modulate hydrophobicity or surface charges of the SAM and therefore influence DET. The development of mixed SAM consisting of several species of head groups can be used to fine-tune the SAM properties to maximize binding and DET. Further modifications are ultrathin conductive polyion films and conductive polymer films. (3) Carbon (pyrolytic graphite, "edge" plane-oriented pyrolytic graphite, glassy carbon, or boron-doped diamond) is a suitable electrode material for many redox proteins and redox enzymes. Especially the surface roughness of pyrolytic and "edge" plane graphite provides a high specific area for protein binding. Further carbon-based materials are carbon black, single-walled carbon nanotubes (SWCNT) and multiwalled carbon nanotubes (MWCNT), carbon nanoparticles or graphene sheets. The observed enhancement of the DET current comes from an increased specific surface area available for protein binding or from the closer contact of the nanostructures (spheres, edges, tubes) with the redox enzyme's active site.

Targeted Applications

Biosensors Based on DET

A biosensor is an analytical device consisting of a biological recognition element (bioelement) and an electrode. In the particular case of an amperometric biosensor the biorecognition element is oxidized or reduced by the analyte and generates a current that is proportional to the analyte concentration. The development of amperometric biosensors or enzyme electrodes was historically started by detecting an electroactive substrate, co-substrate or reaction product of the enzyme. Later, a second approach introduced uses nonphysiological redox mediators to shuttle electrons between the enzyme's active site and the biosensor's electrode. The third approach is to directly couple redox enzymes and electrodes by DET. These three sensor designs have been classified as first-, second- and third-generation biosensors [7].

An interesting feature of DET biosensors is a simple and robust electrode architecture (e.g., no leaking of soluble redox mediators). Additionally, many interfering substances affecting the detection in first- and second-generation biosensors do not interfere with biosensors based on DET. The number of possible analytes is of course restricted to the number of available enzymes, but by use of modern protein engineering techniques the spectrum of analytes will be broadened in future. An example is the modification of the substrate specificity of cellobiose dehydrogenase, which has a natural preference for cellobiose and cello-oligosaccharides. By rational engineering the enzyme's glucose turnover rate was improved and the enzyme can now serve as biocomponent for glucose detection.

DET Based Enzymatic Fuel Cells

In an enzymatic biofuel cell (BFC) the anode, cathode, or both are modified with enzymes to achieve bioelectrocatalysis. At the bioanode the fuel is oxidized and the biocathode reduces the oxidant. The catalytic process depends on the substrate specificity of the used enzymes. By cleverly selecting from the available DET enzymes, a variety of substrates and oxidants can be combined to achieve a high power output of the BFC. Substrates for DET-based BFCs are hydrogen, carbohydrates (glucose, fructose, lactose, maltose), alcohols (ethanol), amino acids, catecholamines, or ascorbic acid, which can be oxidized. The electrons are transferred from the enzyme to the anode and after passing the resistor/load electrons are transferred from the cathode to the biocatalyst where they reduce the oxidant. Protons are transferred through the electrolyte. Suitable oxidants are molecular oxygen (O_2), hydrogen peroxide (H_2O_2) or organic peroxides (ROOR') [8]. The main target of a BFC is to achieve a high power density – the ability to generate even in a small compartment a high current at a high cell voltage. A high specific current is pursued by three-dimensional, porous or nanostructured electrodes with a high specific surface area (m^2/m^3). A high cell voltage can only be achieved by careful selection of anode and cathode biocatalysts. A high equilibrium cell voltage (open circuit potential) is desired and requires an anode biocatalyst with an as low as possible and a cathode biocatalyst with an as high as possible redox potential of the cofactor. The thermodynamic border of the anode potential is −0.4 V versus SHE at pH 7.0 when using glucose as fuel and the border of the thermodynamic cathode potential is 0.8 V versus SHE when using oxygen as an oxidant. This sums up to a theoretical cell voltage of 1.2 V. The achieved cell voltages deviate from the

Direct Electron Transfer to Enzymes, Table 1 Fuels and biocatalysts tested in DET-based enzymatic BFCs

Fuel/oxidant	Anode biocatalysts/ cathode biocatalyst
Fuels	*Anode*
Cellobiose, lactose, glucose; maltose	Cellobiose dehydrogenase (CDH, EC 1.1.99.18)
Fructose	Fructose dehydrogenase (FDH, EC 1.1.99.11)
Ethanol	Alcohol dehydrogenase (ADH, EC 1.1.1.1)
Oxidants	*Cathode*
Oxygen (O_2)	Bilirubin oxidase (BOx, EC 1.3.3.5)
Oxygen (O_2)	Laccase (EC 1.10.3.2)
Hydrogen peroxide (H_2O_2)	Peroxidase (EC 1.11.1.7)

equilibrium voltage and decrease with increasing current. Reported cell voltages for enzymatic glucose/oxygen BFCs are between 0.2 and 0.8 V at power densities up to 1 mW/cm^2.

From the great number of oxidoreductases used to modify enzymatic BFC electrodes only a minority is capable of DET, which reduces the number of fuels and oxidants (Table 1). The substrate specificity of enzymes redners half-cell separation by e.g., membranes unnecessary. DET between enzyme and electrode also stops the need for soluble redox mediators to shuttle electrons between enzyme and electrode. This results in the possibility to design membraneless, non-compartmentalized enzymatic BFCs with a simple architecture. However, so far achieved DET currents are lower than MET currents, because usually only enzyme monolayers can be contacted. Strategies to improve the current density aim at the use of high surface area electrode materials like CNTs, AuNPs etc. or the layer-by-layer approach

Bioelectrosynthesis Employing DET

The application of enzymes as biocatalysts for the synthesis of organic molecules is a rapidly growing field. The most influential features of enzymes driving their application are their high chemo-, regio- and enantioselectivity as well as their substrate specificity. The high specificity of enzymes relieves the need for protecting groups and allows the transformation of complex molecules. Additionally, the advance of genetic engineering by rational design or directed evolution has a marked impact on the availability of stable and substrate specificity optimized enzymes. Redox enzymes catalyze a very wide range of reactions of interest in organic synthesis, but especially oxidoreductases with their need for cofactor- or coenzyme regeneration are difficult to employ and control. Clearly, these cofactors and coenzymes can be regenerated (oxidized or reduced) on electrodes. This is, in principle, the reverse approach of a biofuel cell which converts chemical energy into electricity. In bioelectrosynthesis the cheapest reduction equivalents (electrons) can be used for biochemical transformations. The application of DET-based enzyme electrodes for this purpose has, however, been hampered by the often poor electron transfer kinetics between electrode and interesting catalytic enzymes.

Future Directions

Basic research will further develop techniques for studies on electron transfer phenomena in biological macromolecules. In applied research, the optimization of the direct electron transfer between bioelements and electrodes is dominant. Improvement of DET is anticipated by a better orientation and binding of redox proteins and redox enzymes and the stabilization of the bioelement. Simultaneously an increased stability is pursued. The employed genetic methods for protein engineering are based on rational design and directed evolution. With the advance of analytical methods and a better knowledge of the structure/functional relationship of DET in redox proteins and redox enzymes the engineering of electron transfer pathways within proteins will become feasible.

Cross-References

▶ Biosensors, Electrochemical
▶ Cofactor Substitution, Mediated Electron Transfer to Enzymes

References

1. Christenson A, Dimcheva N, Ferapontova EE, Gorton L, Ruzgas T, Stoica L, Shleev S, Yaropolov AI, Haltrich D, Thornley RNF, Aust SD (2004) Direct electron transfer between ligninolytic redox enzymes and electrodes. Eletroanalysis 16:1074–1092
2. Yeh P, Kuwana T (1977) Reversible electrode reaction of cytochrome c. Chem Lett 10:1145–1148
3. Eddowes MJ, Hill HAOJ (1977) Novel method for the investigation of the electrochemistry of metal proteins: cytochrome c. Chem Soc Chem Commun:771b–772
4. Berezin IV, Bogdanovskaya VA, Varfolomeev SD, Tarasevich MR, Yaropolov AI (1978) Bioelectrocatalysis. Equilibrium oxygen potential in the presence of laccase. Dokl Akad Nauk SSSR 240:615–618
5. Yaropolov AI, Malovik V, Varfolomeev SD, Berezin IV (1979) Electroreduction of hadrogen peroxide on an electrode with immobilized peroxidase. Dokl Akad Nauk SSSR 249:1399–1401
6. Leger C, Bertrand P (2008) Direct electrochemistry of redox enzymes as a tool for mechanistic studies. Chem Rev 108:2379–2438
7. Habermüller K, Mosbach M, Schuhmann W (2000) Electron-transfer mechanisms in amperometric biosensors. Fresenius J Anal Chem 366:560–568
8. Falk M, Blum Z, Shleev S (2012) Direct electron transfer based enzymatic fuel cells. Electrochim Acta 82:191–202
9. Ludwig R, Harreither W, Tasca F, Gorton L (2010) Cellobiose dehydrogenase: a versatile catalyst for electrochemical applications. Chem Phys Phys Chem 11:2774–2697

Disinfection of Water, Electrochemical

Henry Bergman
Anhalt University, Anhalt, Germany

Introduction

Disinfection is the killing of vegetative microorganisms. Accepted methods must be able to reduce microorganisms by at least five orders of magnitude. Electrochemical water disinfection (ECD) is one of these methods [1]. ECD acts by means of electrochemical disinfectant generation onsite or inline (Fig. 1). Besides the disinfection effect, benefit of ECD arises from relatively simple cell construction and easy automation.

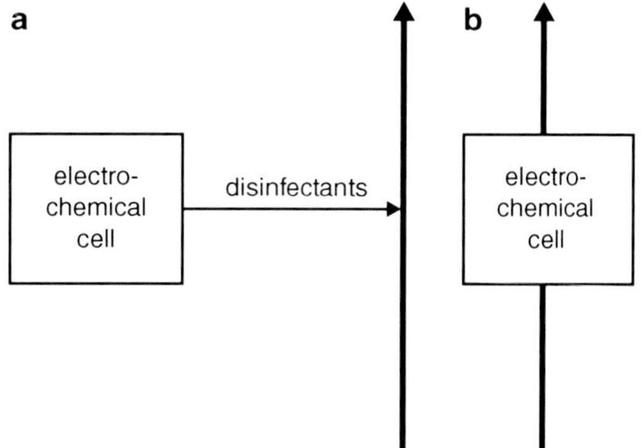

Disinfection of Water, Electrochemical, Fig. 1 Onsite (**a**) and inline addition (**b**) of disinfectants to water

The method must be distinguished from the killing or inactivation of microorganisms previously adhered to a surface (biofilm at different stages) by applying electrochemical surface polarization and from electroporation technologies working with field strength up to 10 MV m^{-1}. Whereas direct electron transfer seems to exist for adhered microorganisms [2], oxidant production is the most probable mechanism in electrochemical disinfection [3]. There were many speculations on the killing mechanisms in the past, including electrical field and pH effects. In flow-through treatment disinfection, however, immediate killing of all cells is not possible. The chemical reaction of disinfectants with microorganisms may proceed within a timescale from seconds to hours as well as outside the treatment unit.

Typical disinfectants that can be electrochemically produced are free active chlorine/free available chlorine (FAC) species as the sum of dissolved Cl_2, HOCl, and OCl^-. If these components react with selected water constituents, by-product formation occurs. In some cases, these by-products may also act as disinfectants (chloramines [4]). Other possible disinfectants obtained, depending on the electrolysis conditions, are ozone, hydrogen peroxide, chlorine dioxide, heavy metal ions, and bromine active species.

The participation of radicals, such as singlet and triplet oxygen radicals, hydroxyl radicals, or radicals resulting from consecutive reactions of radicals with water constituents, is discussed very controversially. Radical lifetimes in the range of nanoseconds in a μm-reaction zone near the anode let one conclude that direct reactions with microorganisms are not decisive for their killing. Obviously, under these conditions, the majority of microorganisms are killed in reaction with more stable products (H_2O_2, O_3, FAC). Oxidant generation is mostly related to anodic reactions. However, cathodic reactions such as the reduction of dissolved oxygen to H_2O_2 may also contribute to generation of an oxidant:

$$O_2 + 2H_2O + 4e^- \rightarrow 2H_2O_2 \qquad (1)$$

A large variety of cell constructions, often using gas diffusion electrodes, have been developed or tested [5–7]. Widespread technologies using natural salt matrices for oxidant production in a flow-through regime (Fig. 1b) are known as *inline electrolysis, tube electrolysis, anodic oxidation, low-amperic electrolysis, electrochemical water activation*, or by brand names. In addition, immersion constructions have been reported. Fixed installations and mobile systems are in use (Fig. 2).

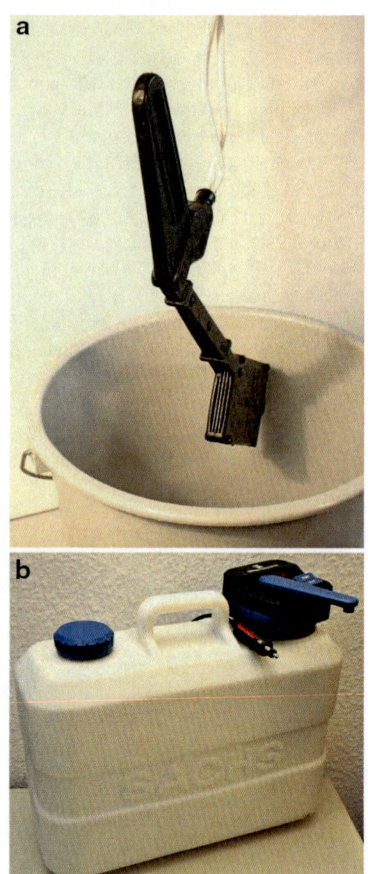

Disinfection of Water, Electrochemical, Fig. 2 (**a**) Immersion stack cell and (**b**) transportable disinfection cell with water container pump and electrolysis unit, both for discontinuous outdoor ECD; voltage sources are not shown

Inline methods are mostly related to the treatment of relatively low amounts of water per unit of time. Typical applications include the disinfection of pool and drinking water, for example, in hospitals, submarines, airplanes, senior residences, and other facilities. Potential nondrinking water applications include urban wastewater treatment, cooling water disinfection, saline wastewater disinfection, fish farming, vegetable washing, cleaning of pipe systems and containers, algae removal, tumor research, lens washing, dental implants, medical equipment, and cleaning of filters and permeates, often combined with odor abatement.

Disinfectants, if generated continuously in an electrochemical cell, may be added to a side or a mainstream of water. In special applications (i.e., drinking water), lower and upper limiting concentrations for FAC exist for the resulting concentration (often between 0.1 and 6 mg dm^{-3}). In comparison with ozone and chlorine dioxide, the main advantage of chlorine species as disinfectants is their reservoir effect in a system over increased time. The main disadvantage is the formation of halogenated by-products. Because of this, and to avoid microorganism adaptation effects, continuous use of disinfection cells without periodic interruption is not recommended.

Inline flow-through cells dominate the market. Divided and undivided cells are known. Cells have been commercialized to a remarkable extent since the 1970s, especially in Eastern Europe and Germany. Selected schematic constructions and technologies are shown in Figs. 3 and 4.

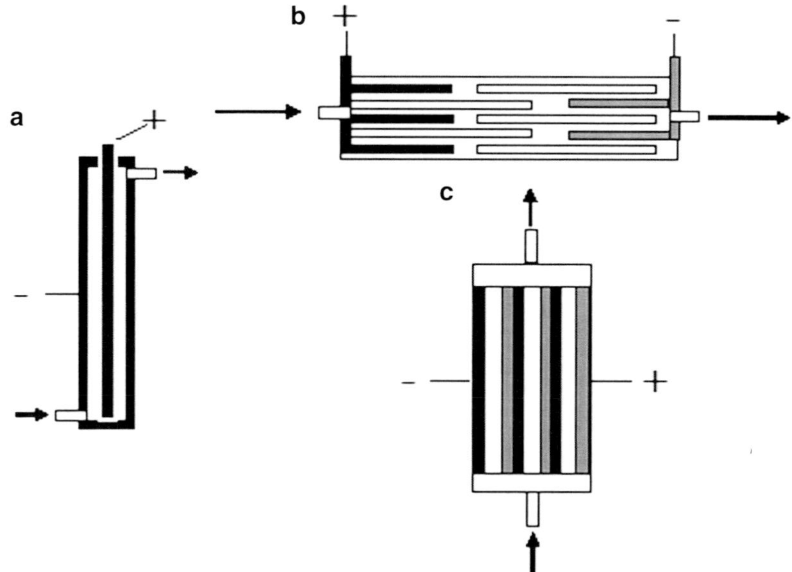

Disinfection of Water, Electrochemical, Fig. 3 Typical undivided flow-through constructions for ECD: tube electrolyzer (**a**), bipolar cell with nonoptimal current distribution (**b**), and bipolar parallel plate reactor (**c**)

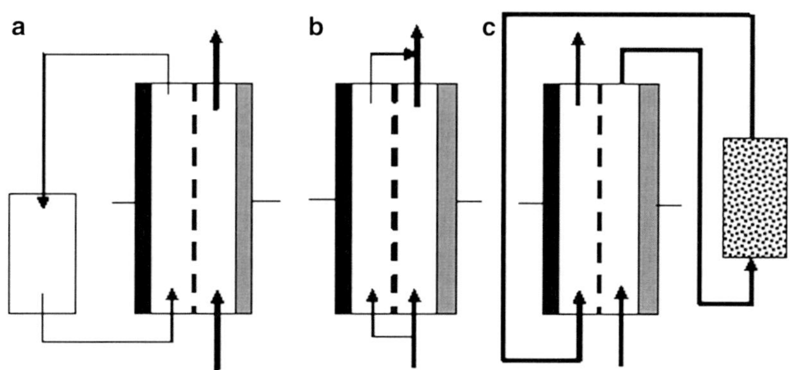

Disinfection of Water, Electrochemical, Fig. 4 Typical technologies using divided flow-through cells for ECD: with special catholyte recirculation (**a**), with addition of catholyte to the main water stream (**b**), and removal of by-products in a final fixed bed reactor (**c**)

Mostly, so-called mixed oxide (MIO) electrodes (typically TiO_2 doped with IrO_2, RuO_2, or VaO_x on Ti carrier material) are used in practical cases. TaO_x may be part of the oxide layers. Many researchers have studied other surface modifications. RuO_2-based electrodes seem to have better chlorine production compared with IrO_2 or Pt layers, but specific composition and methods of producing the activated layers are decisive for chlorine production efficiency [8, 9]. Untreated titanium, graphite, carbon cloth and glassy carbon, carbides, and nitrides were also studied. A relatively new, but sometimes problematic, material is boron-doped diamond (BDD, [10]). Applications of BDD anodes are accepted as so-called advanced oxidation methods. Summaries and disinfection-related papers can be found [8, 11–14]. Most applications and studies use direct current for disinfectant production. A few papers deal with alternating and pulsating current.

A large variety of technological schemes, electrode constructions, electrode materials, and water matrices have often resulted in confusion of health administrations in assessing applicability. There are big differences worldwide in practice and legislation with respect to the usability of several disinfectants in special water systems. For example, contrary to North America or Australia, chloramine- or bromine-based disinfection is not widely used in Europe. In Germany, H_2O_2 may be used for pipe cleaning but not for drinking water disinfection, etc. Because in most cases (ECD) chlorine species are responsible for disinfection effects, the treated water must be classified as chloride containing or as water free of chloride.

Electrochemical onsite ClO_2 generation at ppm levels is a challenging new technology. Technologies dealing with the targeted electrochemical evolution of oxygen or the cathodic generation of H_2O_2, ferrates [15], chloramines, percarbonate, peroxodisulfate [16], or bromine for disinfecting waters are not included in the scope of this paper.

Main Systems and Reactions

Electrolytes Free of Chloride

In chloride-free water, the generation of ozone is the reaction of choice. Buffer or supporting electrolyte components must be added to the anolyte. Only anode materials with high oxygen overvoltage (Pt, glassy carbon, PbO_2, BDD) are able to generate ozone in water electrolysis. An overall reaction is shown in Eq. 2:

$$3H_2O \rightarrow O_3 + 6H^+ + 6e^- \quad (2)$$

Splitting of water is the main parasitic reaction (Eq. 3):

$$2H_2O \rightarrow O_2 + 4H^+ + 4e^- \quad (3)$$

To avoid side reactions and to isolate hydrogen evolution (Eq. 4 for acidic conditions), the anolyte and catholyte must be separated each from other:

$$2H^+ + 2e^- \rightarrow H_2 \quad (4)$$

Reaction mechanisms vary with different electrode materials. Ozone formation was first shown for Pt anodes [17, 18]. Initial commercial cells were equipped with PbO_2 layers on porous Ti carriers (*Membrel* technology). BDD is a promising commercial material despite its high price. The formation of hydroxyl radicals is accepted for many electrode materials in a more or less adsorbed state and in connection with more or less density of active sites. It is also accepted that on BDD, ozone is formed as the result of stepwise radical formation in a μm reaction zone (Eqs. 5, 6, 7, and 8):

$$H_2O \rightarrow {}^{\cdot}OH + H^+ + e^- \quad (5)$$

$${}^{\cdot}OH \rightarrow O^{\cdot} + H^+ + e^- \quad (6)$$

$$2O^{\cdot} \rightarrow O_2 \quad (7)$$

$$O_2 + O^{\cdot} \rightarrow O_3 \quad (8)$$

Hydroxyl radicals may quickly form H_2O_2: [19]

$$2\,{}^{\cdot}OH \rightarrow H_2O_2 \quad (9)$$

Other radical reactions are possible. In most cases, the reaction scheme is not well understood.

Current efficiency calculated from the concentration of dissolved ozone is a function of current density [20]. Technical cells are equipped with expanded mesh anodes to increase current densities. Ozone is added onsite or inline. Immersion constructions were also reported. As compared with silent discharge, electrochemical ozone generation is characterized by the high specific energy demand of about 2 kWh per kg ozone. Much higher ozone concentrations can be adjusted, however.

In electrolytes that are extremely poor in chloride ions, formation of ozone must be considered as a side reaction. Often, the standard DPD test for chlorine analysis shows higher values than expected [21]. Further research is still necessary. Application of MIO anodes under these conditions is highly risky because the obtained by-product spectrum can be contrary to rules for drinking water. Some researchers have studied the disinfection effect in waters without chloride ions, mostly using sulfate electrolytes and Pt, stainless steel, and other electrode materials [22–25]. Discussions are sometimes questionable due to difficulties in relating the disinfection effect with species clearly responsible for it.

Electrolytes with Chloride Concentrations in the Range of Grams per Liter

MIO anodes are in widespread use for this technology. The anodic product is chlorine, with current efficiencies higher than 90 % (Eq. 10):

$$2Cl^- \rightarrow Cl_2 + 2e^- \quad (10)$$

The electrochemical reactor must be divided because hydrogen evolution is the main cathodic reaction (Eq. 4, avoiding explosive mixtures). As in industrial chlorine-alkaline electrolysis, chlorine can be directly contacted with the water to be disinfected, or it can be used for preparation of sodium hypochlorite by reacting with NaOH (Eq. 11) for subsequent addition:

$$Cl_2 + NaOH \rightarrow NaOCl + H^+ + Cl^- \quad (11)$$

Because this technology does not resolve problems with chlorine dosage from gas balloons (uncontrolled setting free of chlorine gas), it is not widely used in practice.

Electrochemical cells with MIO anodes for hypochlorite solution production as stock solutions for onsite addition of $HOCl/OCl^-$ mixtures have found increasing application worldwide (cell units producing more than 100 kg FAC per day). Divided cells and NaCl concentrations in the range of seawater (3.5 %) are typical, but undivided cells are also offered by some manufacturers. Chlorate formation inside the stock solution at longer storage times is a problem with this technology:

$$OCl^- + 2HOCl \rightarrow ClO_3^- + 2H^+ + 2Cl^- \quad (12)$$

Some plant manufacturers prefer the use of water with a chloride concentration close to that of drinking water with optional addition of NaCl salt. The production of anolyte solution for disinfection has the advantage that the resulting pH is relatively low, and HOCl may be the predominant species. Its disinfection ability is much higher than that of OCl^-.

Electrolytes with Chloride Concentrations Typical for Drinking Water

This technology for direct drinking water electrolysis has been the focus of interest for 60 years [26], but detailed kinetic studies began only at the end of 1990s. Many drinking water regulations recommend a maximum concentration of 250 mg dm^{-3} for chloride ions. This is sufficient to produce FAC with typical current efficiency amounts of 5–15 % on MIO anodes and in undivided cells. Nearly linear curve behavior for FAC production and current efficiency versus chloride concentration (Fig. 5) exist [8, 9]. Often, current efficiency maxima are observed between 100 and 200 A m^{-2} due to the competing oxygen evolution reaction. Specific electroenergy demands (for electrolysis current) of 50 kWh per kg FAC can be measured.

At very low chloride concentrations (normally lower than 10 mg[Cl^-] dm^{-3}), no FAC can be found. Two general mechanisms are responsible for this behavior:

(i) Formed FAC is converted to chloride by oxidants such as H_2O_2.
(ii) Formed FAC is oxidized to products such as chlorate and perchlorate in electrochemical reactions. Finally, only traces of chloride can be found in the treated water.

In long-term electrolysis, all prior existing Cl is finally distributed between the species FAC, ClO_3^-, and ClO_4^- [27]. Reaction rates are higher for BDD anodes as compared to MIO anodes. Intermediates are probable but not exactly known. Participation of radicals is also possible on MIO and Pt electrodes [28].

Considering electrode kinetics, the most probable first step is the oxidation of chloride to the chloride radical by electron transfer or by reaction with hydroxyl radicals:

$$Cl^- \rightarrow Cl + e^- \quad (13)$$

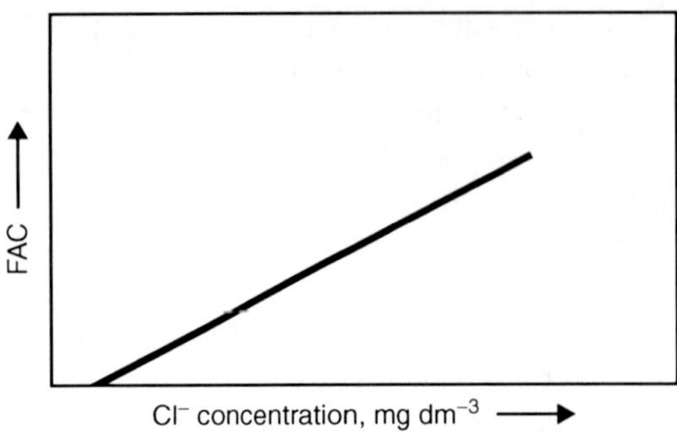

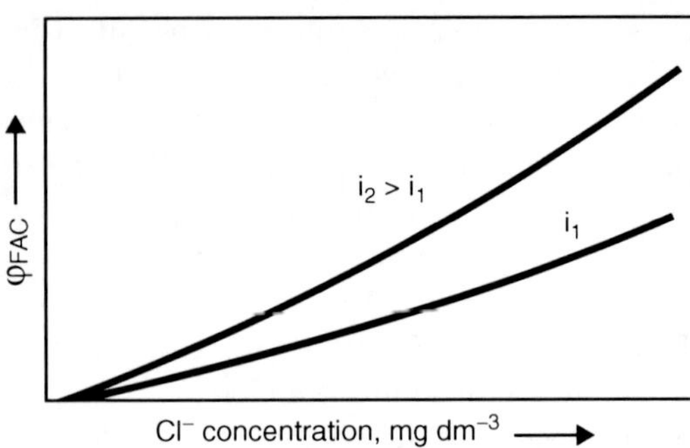

Disinfection of Water, Electrochemical, Fig. 5 Quasi-linear behavior of chlorine (FAC) production versus chloride concentration (*left*) and FAC current efficiencies versus chloride concentration at different current densities (*right*)

$$Cl^- + {}^{\cdot}OH \rightarrow Cl + OH^- \quad (14)$$

Chlorine molecules formed from the chlorine radical quickly react with water and hydroxide ions to form hypochlorous acid and hypochlorite ions with a strong pH dependence of species distribution [4].

Chlorine Dioxide Generation from Chlorite

Chlorine dioxide is an effective disinfectant that reduces the formation of organic disinfection by-products. There are different chemical and electrochemical technologies for producing and adding ClO_2 onsite, initiating the reaction from chlorite or chlorate [5]. It is noteworthy here that even traces of chlorite added to a water stream may quickly react to form chlorine dioxide at low ppm levels [29]:

$$ClO_2^- \rightarrow ClO_2 + e^- \quad (15)$$

When chloride ions are present during electrolysis, chlorine dioxide formation is accelerated [29], obviously as the result of the quick reaction with hypochlorous acid in the acidic electrode layer:

$$2HClO_2 + HClO \rightarrow 2ClO_2 + Cl^- + H_2O + H^+ \quad (16)$$

Semi-inline technologies with chlorite additions are still under research.

Future Directions

Address Problems in Electrochemical Water Disinfection

By-Product Formation

Species produced by ECD must be classified as *disinfection products* (e.g., FAC), *disinfection by-products* (named as DBPs, e.g., chloramines and THMs) resulting from disinfectant reactions with organic matter or from inorganic reaction schemes (chlorate), and *electrolysis by-products* (e.g., nitrite and ammonium ions from cathodic processes or organic intermediates on BDD anodes). Ozone may be an electrolysis product or by-product depending on the target of electrolysis. Due to health risks, some of these by-products are regulated by legislation. Many by-products are not yet anchored in national or international rules. Overall, the subject has not yet been researched enough. Concerning reactions of FAC with organic matter, by-products similar to those observed in chlorination processes can be expected [30].

Electrochemical reaction behavior of organic compounds is difficult to assess. Research works sometimes show that intermediates may be formed that are more toxic than the initial system. The formation of chlorate [31], bromate on mixed oxide electrodes [32], peroxodisulfate, and H_2O_2 [19] is well known as well as the formation of DPDs such as THMs and AOX.

In studies of Bergmann and co-workers, the following by-products of inline drinking water electrolysis were identified for the first time: chlorate, perchlorate, bromate and perbromate on BDD, and other anodes; chlorine dioxide as a by-product in disturbed chloride electrolysis; nitrite, ammonia/ammonium as cathodic products on mixed oxide electrodes; and hydrogen peroxide on BDD anodes and MIO cathodes [8, 33–36]. Many results were confirmed by other researchers [37, 38]. Risks for halogenate and perhalogenate formation increase in the order of dosage of chlorine as Cl_2 – dosage as a hypochlorite solution – inline electrolysis of FAC species.

Electrochemical methods for ammonia and ammonium reduction were suggested by Kim et al., Kapalka et al., and others [39, 40]. Modification of reaction conditions and clarification of mechanisms are the subject of current research.

The state of legislation and control concerning electrochemical disinfection devices in environmentally oriented applications is not satisfactory. Only in the USA, first attempts for perchlorate restrictions in drinking water can be observed. In Germany, a first complex project on inline electrolysis was finished in 2010. Currently, the method is still outside legislation but tolerated. The project included cell producers, water authorities, and research groups.

Formation of Deposits on Electrodes and System Surfaces

Especially in cells without a separator between anode and cathode and in processes using natural water matrices, cathodes show a tendency to be coated with deposits. Due to the increased pH in the cathodic electrode layer, mainly calcareous and hydroxidic scaling with earth alkali ions is observed ($CaCO_3$, $Mg(OH)_2$, etc.). Several antiscaling methods have been suggested, such as periodic dissolution in acids, change of polarity, rotating brushes or cleaning vanes, ultrasonic treatment, and current pulsation. Many methods for scaling quantification were suggested (potential control, quartz mass balance, etc. [41, 42]). In practical application, cell voltage and cell current monitoring are the most widespread methods used. In some cases, deposition of iron hydroxide on water pipe walls was observed when electrolysis shifted the pH to a more basic pH region.

Gas Inside the System

If the chloride concentration is in the range of milligrams per liter, the primary electrochemical reaction is the splitting mechanism of water, with oxygen and hydrogen formation. Use of separators is the best way to keep hydrogen gas out; however, oxygen still remains in the system and should be allowed to escape in a sophistic way. Increased corrosion is sometimes reported as the result of changing water decomposition and pH during electrolysis.

References

1. Martinez-Huite CA, Brillas E (2008) Electrochemical alternatives for drinking water disinfection. Angew Chem 47:1998–2005
2. Matsunaga T, Namba Y, Nakajima T (1984) Electrochemical sterilization of microbial cells. Bioelectrochem Bioenerg 13:393–400
3. Bergmann H, Iourtchouk T, Schoeps K et al (2001) What is the so-called anodic oxidation and what can it do? GWF Wasser Abwasser 142: 856–869
4. White GC (1999) Handbook of chlorination and alternative disinfectants, 4th edn. Wiley, New York
5. Oloman C (1996) Electrochemical processing for the pulp and paper industry. The Electrochemical Consultancy, Romsey
6. Drogui P, Elmaleh S, Rumeau M et al (2001) Hydrogen peroxide production by water electrolysis: application to disinfection. J Appl Electrochem 31: 877–882
7. Qiang Z, Chang JH, Huang CP (2002) Electrochemical generation of hydrogen peroxide from dissolved oxygen in acidic solutions. Water Res 36:85–94
8. Bergmann MH (2009) Drinking water disinfection by inline electrolysis-product and inorganic by-product formation. In: Comninellis C, Chen G (eds) Electrochemistry for the environment. Springer, New York, pp 163–205
9. Kraft A, Stadelmann M, Blaschke M, Kreysig D et al (1999) Electrochemical water disinfection, part I: hypochlorite production from very dilute chloride solutions. J Appl Electrochem 29:861–868

10. Alkire RC, Kolb DM (eds) (2003) The electrochemistry of diamond, vol 8. Wiley-VCH, Weilheim
11. Kraft A, Wuensche M, Stadelmann M et al (2003) Electrochemical water disinfection. Recent Res Dev Electrochem (India) 6:27–55
12. Fujishima A, Einaga Y, Rao TN, Tryk DA (eds) (2005) Diamond electrochemistry. Elsevier, Amsterdam/Tokyo
13. Kraft A (2008) Electrochemical water disinfection: a short review. Platinum Met Rev 52: 177–185
14. Brillas E, Martinez-Huitle CA (eds) (2011) Synthetic diamond films: preparation, electrochemistry, characterization, and applications. Wiley, Hoboken
15. Mácová Z, Bouzek K, Híveš J et al (2009) Research progress in the electrochemical synthesis of ferrate (VI). Electrochim Acta 54:2673–2683
16. Comninellis C, Michaud S, Savall A et al (2002) Electrochemical preparation of peroxodisulfuric acid using boron doped diamond thin film electrodes. Electrochim Acta 48:431–436
17. Foller PC, Tobias CW (1982) The anodic evolution of ozone. J Electrochem Soc 129:1982
18. Da Silva LM, Santana MHP, Boodts JFC (2003) Electrochemistry and green chemical processes: electrochemical ozone production. Quim Nova 26:880–888
19. Marselli B, Garcia-Gomez J, Michaud PA et al (2003) Electrogeneration of hydroxyl radicals on boron doped diamond electrodes. J Electrochem Soc 150:D79–D83
20. Kraft A, Stadelmann M, Wünsche M et al (2006) Electrochemical ozone generation using diamond anodes and a solid polymer electrolyte. Electrochem Comm 8:883–886
21. Bergmann MEH (2006) On the applicability of DPD method for drinking water disinfection analysis. GWF Wasser Abwasser 147:780–786
22. Rosenberg B, van Camp L, Krigas T (1965) Inhibition of cell division in *Escherichia coli* by electrolysis products from a platinum electrode. Nature 205:698–699
23. Kerwick MI, Reedy SM, Chamberlain AHL et al (2005) Electrochemical disinfection, an environmentally acceptable method of drinking water disinfection? Electrochim Acta 50:5270–5277
24. Barashkov NN, Eiseinberg D, Eiseinberg S (2010) Electrochemical chlorine-free AC disinfection of water contaminated with *Salmonella typhimurium* bacteria. Russ J Electrochem 46:306–311
25. Li H, Zhu X, Ni J (2010) Inactivation of *Escherichia coli* in Na_2SO_4 electrolyte using boron doped diamond anode. Electrochim Acta 56:448–453
26. Reis A (1951) The anodic oxidation as an inactivator of pathogenic substances and processes. Klin Wschr 29:484–485
27. Bergmann MEH, Rollin J, Iourtchouk T (2009) The occurrence of perchlorate during drinking water electrolysis using BDD electrodes. Electrochim Acta 54:2102–2107
28. Jeong J, Kim C, Yoon J (2009) The effect of electrode material on the generation of oxidants and microbial inactivation in the electrochemical disinfection processes. Water Res 43:895–901
29. Bergmann H, Koparal AS (2005) Problems of chlorine dioxide formation during electrochemical disinfection. Electrochim Acta 50:5218–5228
30. Oh BS, Oh SG, Hwang YY et al (2010) Formation of hazardous inorganic by-products during electrolysis of seawater as disinfection process for desalination. STOTEN 408:5958–5965
31. Czarnetzki LR, Janssen LJJ (1992) Formation of hypochlorite, chlorate and oxygen during NaCl electrolysis from alkaline solutions at a RuO_2/TiO_2 anode. J Appl Electrochem 22:315–324
32. Cettou P, Robertson PM, Ibl N (1984) On the electrolysis of aqueous bromide solutions to bromate. Electrochim Acta 29:875–885
33. Bergmann MEH, Rollin J (2007) Product and by-product formation in disinfection electrolysis of drinking water using boron-doped diamond anodes. Catal Today 124:198–203
34. Bergmann MEH, Koparal AT, Koparal AS et al (2008) The influence of products and by-products obtained by drinking water electrolysis on microorganisms. Microchem J 89:98–107
35. Bergmann MEH, Iourtchouk T, Schmidt W et al (2011) Perchlorate formation in electrochemical water disinfection. Paperback Imprint Nova Science, New York
36. Bergmann MEH, Iourtchouk T, Rollin J (2011) The occurrence of bromate and perbromate on BDD anodes during electrolysis of aqueous systems containing bromide-first systematic studies. J Appl Electrochem 41:1109–1123
37. Polcaro AM, Vacca A, Mascia M et al (2008) Product and by-product formation in electrolysis of dilute chloride solutions. J Appl Electrochem 38:979–984
38. Sáez C, Cañizares P, Sánchez-Carretero A, Rodrigo MA (2010) Electrochemical synthesis of perbromate using conductive-diamond anodes. J Appl Electrochem 40:1715–1719
39. Kim KW, Kim YJ, Kim IT et al (2006) Electrochemical conversion characteristics of ammonia to nitrogen. Water Res 40:1431–1441
40. Kapałka A, Joss L, Anglada Á (2010) Direct and mediated electrochemical oxidation of ammonia on boron-doped diamond electrode. Electrochem Comm 12:1714–1717
41. Gabrielli C, Maurin G, Perrot H (2002) Investigation of electrochemical calcareous scaling – potentiostatic current- and mass-time transients. J Electroanal Chem 538(539):133–143
42. Kadyk T (2005) Comparative analysis of measuring techniques to investigate the scaling of the cathode in direct water disinfection electrolysis. Diploma thesis, Anhalt University, Köthen/Anh

DLVO Theory

Dominik Horinek
Institute of Physical and Theoretical Chemistry, University of Regensburg, Regensburg, Germany

DLVO Theory

DLVO theory [1–3] describes the stabilization of colloidal dispersions by an interplay of van der Waals and electrostatic forces (as opposed to steric repulsions of colloids by polymeric solubilizers). The theory was developed in the 1940s by Derjaguin and Landau [4] and by Verwey and Overbeek [5]. In DLVO theory, the two determining interactions for the stability of a colloidal system are the attractive van der Waals interactions between the colloidal particles and the repulsive electrostatic Coulomb interactions. When salt is added, the alteration of the electrostatic interactions affects the stability.

The strength of the van der Waals interactions is determined by the size and the shape of the colloidal particles and by the chemical composition of the system, which is described by the Hamaker constant A. Between two similar particles, the van der Waals forces are always attractive and A is a positive constant. For spherical particles of radius R at separation d, the van der Waals energy is given as

$$V(d) = -AR/12d.$$

This relation is valid until the particles are in contact, where a steep repulsion prevents steric overlap. For other shapes, the dependence on the separation d is different, but for all cases, a $1/d^n$ dependence is observed. The attractive van der Waals forces, which depend only weakly on the salt concentration, are strong at small separations and give the dominant energy contribution.

The second interaction that is relevant for the stability is of electrostatic nature: In the presence of inert salts (salts that are composed of ions that do not adsorb on the colloidal surface), an electrostatic double layer is formed around a charged colloidal particle. The simplest description, which holds for monovalent salts at low concentrations, is the Gouy-Chapman solution of the Poisson-Boltzmann equation, which yields that the resulting electrostatic potential decays exponentially with the distance from the colloid. How quick this decay is depends on the salt environment and is characterized by the screening length k^{-1}. According to the Gouy-Chapman model, the extent of the repulsion becomes shorter with increasing salt concentration: $\mathrm{k}^{-1} \sim \chi^{-1/2}$. During the coagulation process, two identical particles approach each other, and the electrostatic interaction between the two spherical particles, including their double layers, is

$$V(d) = RZ/2\ \mathrm{Exp}[-\mathrm{k}d]$$

In the following, we will consider particles with similar surface chemistry, for which Z is a positive constant that depends on the surface potential of the particles. For different shapes, rather similar expressions that depend on $\mathrm{Exp}[-\mathrm{k}d]$ are obtained. The electrostatic forces are therefore repulsive and act against coagulation. The sign and magnitude of the surface potential can be influenced by so-called potential-determining ions, which react with or bind to the surface. In many cases, colloidal particles have acidic or basic surface groups, and the surface potential control is achieved by changing the pH.

Two different electrostatic boundary conditions are frequently combined with DLVO theory: In the constant charge boundary condition, the surface charge does not depend on the particle separation, in contrast to the constant potential boundary condition, which accounts for changes in the surface charge of one particle that are induced by the potential of the other one. Under constant charge, the predicted repulsive force is an upper limit for the experimental force, while a lower estimate is given under constant potential conditions. Charge-regulation models are closer to the real situation, but they require additional input.

The DLVO approach to colloidal stability identifies the interactions between particles as the sum of the van der Waals energy and the screened Coulomb energy. Depending on their

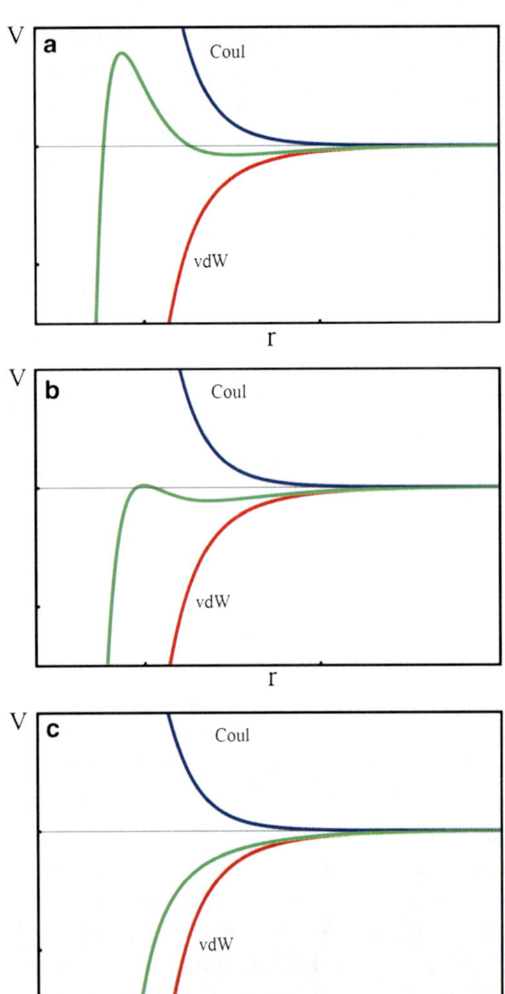

DLVO Theory, Fig. 1 Coulomb and vdW Energy between Colloidal Particles

relative strength, different behavior arises: For high surface potentials, there are two minima: a "primary" minimum at contact of the particles in the coagulated state, and a "secondary" minimum that describes flocculation, where a loose agglomerate of colloids is reversibly formed. Both minima are separated by an electrostatic barrier that prevents coagulation. This case is shown in Fig. 1a. For weaker surface potentials, this barrier gets smaller and eventually becomes zero. In this case, there is still a weak secondary minimum, but the colloidal dispersion is metastable, and rapid coagulation sets in. This case is shown in Fig. 1b. When the surface potential gets even smaller, the secondary minimum totally disappears. Particles always attract each other in this case, and coagulation from this unstable system will occur instantly. This case is shown in Fig. 1c.

In summary: High surface potentials stabilize colloidal systems. The addition of inert salt leads to stronger screening and destabilizes the system. The point at which rapid coagulation (case b) sets in is defined as the critical coagulation concentration (ccc). One key result of DLVO theory is the explanation of the Schultze-Hardy rule, which states that the ccc depends on the counterion valency z like $1/z^6$.

For the quantitative application of DLVO theory, it is important to notice that the origin, $d = 0$, of the van der Waals and Coulomb potential do not need to coincide. The screening layer starts at the outer Helmholtz plane (OHZ), which occurs at distances that are larger by a constant d than the origin of the van der Waals potential. This can drastically change the total potential: The primary minimum can be eliminated with values of d in the range of a few Ang. In this case, the system will be stable regardless of the extension of the screening layer.

At the times when DLVO theory was developed, the direct measurement of forces between colloidal particles and surfaces in solution was not possible, and the macroscopic observation of colloidal stability was the only experimental reference data. With increasing technological advancement, setups have been developed for the direct observation of such forces: The surface force apparatus (SFA) allows for the measurements of forces between surfaces in solution [6], and with an atomic force microscope (AFM), forces on a colloidal particle can be detected [7]. It is a major success that DLVO theory predicts forces that agree nicely with the measured forces for large particle separations (more than 3–10 nm), but at the same time, it is obvious that in the regime of short particle separations, not all effects are captured by DLVO. When the barrier for coagulation occurs at such low separations, the DLVO prediction for colloidal stability is not accurate (Fig. 2).

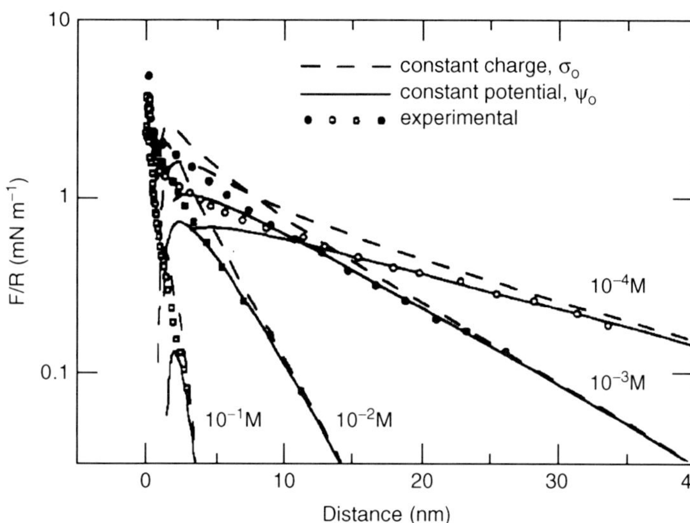

DLVO Theory, Fig. 2 Experimental Results

Despite its success, DLVO theory has various other limitations beside the failure at low particle separations. Intrinsically highly charged surfaces, where the ion's electrostatic interactions with the surface dominate, are often well described by DLVO theory. However, in the case of surfaces with weak intrinsic charge, non-electrostatic forces between ions and the surface become important. DLVO theory does not account for these ion-specific effects [8], which have their origin in ion-type-dependant interactions of the ions with the colloidal particles and with the surrounding solvent. Consequently, DLVO is not able to predict the stability of proteins [9] or of simple air bubbles in salt solutions [10], where ion-specific effects cannot be neglected. Especially ions that are attracted to the particle surfaces are enriched in the particle solvation shells and induce deviations from DLVO theory: For surfaces with hydrophobic character, large ions like I^- or SCN^- show such deviations, at the surface of hydrophilic colloidal particles binding between specific ions and specific functional surface groups is important, an important example is ion pairing at ionized carboxyl groups on a protein surface.

Finally, another limitation of DLVO theory is that ion-ion correlations, which are important for multivalent ions or at high salt concentrations, are also not included.

Future Directions

The main shortcoming of DLVO theory is that it treats the interactions between the ions and the colloidal particle as purely electrostatic. Ions close to the particle are also subject to ion-type specific, often solvent-mediated attractive forces. Their inclusion on the level of an extended Poisson-Boltzmann equation leads to a more complex scenario, in which the salt concentration also changes the effective surface charge of the colloid [9]. This can induce resolubilization at high salt concentrations when a potential barrier reappears due to overcharging. The inclusion of ion-colloid dispersion interactions into the Poisson-Boltzmann equation can also induce similar effects.

Cross-References

- ▶ Electrolytes, History
- ▶ Specific Ion Effects, Evidences
- ▶ Specific Ion Effects, Theory

References

1. Evans DF, Wennerström H (1999) The colloidal domain, 2nd edn. Wiley-VCH, New York
2. Hunter RJ (2000) Foundations of colloid science. Oxford University Press, New York

3. Israelachvili JN (2010) Intermolecular and surface forces, 3rd edn. Academic, Burlington
4. Derjaguin BV, Landau LD (1941) Theory of the stability of strongly charged lyophobic sols and of the adhesion of strongly charged particles in solution of electrolytes. Acta Physicochim URSS 14:633
5. Vervey EJW, Overbeek JTG (1948) Theory of stability of lyophobic colloids. Elsevier, Amsterdam
6. Israelachvili JN, Adams GE (1978) Measurements of forces between 2 mica surfaces in aqueous electrolyte solutions in the range 0–100 nm. J Chem Soc Faraday Trans I 74:975
7. Ducker WA, Senden TJ, Pashley RM (1992) Measurement of forces in liquids using a force microscope. Langmuir 8:1831
8. Boström M, Williams DRM, Ninham BW (2001) Specific ion effects: why DLVO theory fails for biological and colloidal systems. Phys Rev Lett 87:168103
9. Kunz W, Henle J, Ninham BW (2004) 'Zur Lehre von der Wirkung der Salze' (about the science of the effect of salts): Franz Hofmeister's historical papers. Curr Opin Coll Inter Sci 9:19
10. Craig VSJ, Ninham BW, Pashley RM (1993) Effects of electrolytes on bubble coalescence. Science 364:317
11. Schwierz N, Horinek D, Netz RR (2010) Reversed anionic Hofmeister series: the interplay of surface charge and surface polarity. Langmuir 26:7370

DNA/Electrode Interface, Detection of Damage to DNA Using DNA-Modified Electrodes

Jan Labuda[1] and Vlastimil Vyskocil[2]
[1]Institute of Analytical Chemistry, Faculty of Chemical and Food Technology, Slovak University of Technology, Bratislava, Slovakia
[2]UNESCO Laboratory of Environmental Electrochemistry, Faculty of Science, Department of Analytical Chemistry, Charles University in Prague, Prague, Czech Republic

Introduction

Deoxyribonucleic acid (DNA) itself and rather large scale of nucleic acids are utilized as biorecognition elements at DNA biosensors which represent often a type of DNA-modified electrodes with typical advantages of electrochemical sensors [1]. While electrodes covered by single-stranded DNA (ssDNA) are utilized as hybridization sensors (genosensors), electrodes with a layer of ssDNA or double-stranded DNA (dsDNA) can serve at the detection of damage to DNA. DNA damage is feature of normal body cells existence. However, as serious DNA damage can be induced by variety of physical or chemical agents occurring in the environment, generated in the organisms as by-products of metabolism or used as therapeutics, its detection is of great importance for human health and its protection. Damage to DNA generally represents an alteration in DNA chemical structure which include interruptions of the sugar–phosphate backbone (leading to strand breaks), release of the DNA bases due to hydrolysis of *N*-glycosidic bonds (resulting in abasic sites) and a variety of nucleobase lesions (resulting from reactions of DNA with genotoxic substances). Moreover, specific binding of a guest molecule to the double helix by intercalation, i.e., an insertion between the stacked base pairs, can cause a subtle damage like lengthening of the DNA helix, perturbation of the phosphate backbone and even untwisting of the helix [2].

Together with chromatographic, electrophoretic, and mass spectrometric methods of DNA damage detection [3, 4], electroanalytical techniques utilizing electrodes with immobilized DNA represent highly sensitive tools due to sensitivity of DNA response to its relatively small structural changes [5]. The electrochemical DNA-based biosensors have been used not only to detect but also to induce and control DNA damage at the electrode surface via electrochemical generation of the damaging (usually radical) species [2, 6].

Construction of DNA Biosensors

An electrode (transducer) material should be chosen with respect to the electrochemical process of interest and technique of DNA immobilization. Typically, mercury-based (mercury film, solid amalgams), carbon-based (glassy carbon (GCE), carbon paste (CPE), graphite and pyrolytic graphite (PGE), graphite–epoxy composite) and some other (gold, platinum, indium tin oxide (ITO)) electrodes together with various thin- and thick-film electrodes like screen-printed carbon (SPCE) and gold electrodes are used in single or interdigitated (array) arrangement [7]. Typically, after a chemical or electrochemical pretreatment of the bare electrode, either thin DNA layer or thicker layers of DNA gel are formed as the electrode surface coverage. Techniques of the DNA immobilization vary depending on the kind of transducer and the biosensor application and play major role in the overall biosensor performance [8, 9]. Nanostructured interface between the bare electrode and DNA formed by various nanomaterials like gold nanoparticles and carbon nanomaterials represent an approach to enhancement of the biosensor response due to inherent electroactivity, effective electrode surface area, etc. [10, 11]. Nanometer-scale complex films of DNA, enzymes, polyions, and redox mediators were suggested for tests of genotoxic activity of various chemicals [12].

Methods of DNA Damage Detection

Voltammetric and chronopotentiometric detection modes are mostly used [20]. Together with them, electrochemical impedance spectroscopy (EIS) becomes to be popular at DNA-based biosensors [13]. According to electrochemically active species which responses are evaluated at the detection of damage to DNA, the experimental techniques can be classified as follows [1]:

(a) Label-free and often reagentless techniques which represent the work with no additional chemical reagents (indicators, redox mediators, enzyme substrates) needed to generate measured response
(b) Techniques which employ redox indicators either non-covalently bound to the DNA (groove binders, intercalators, anionic or cationic species interacting with DNA electrostatically) or present in the solution phase (e.g., hexacyanoferrate anions)
(c) Techniques which employ electrochemically active labels (enzymes, nanomaterials) covalently bound to DNA

Combination of these principles allows obtaining of more complex information on DNA changes [14].

The first group of techniques utilizes surface activity or redox activity of DNA itself [15]. The electrochemical activity is based on the presence of redox active sites at nucleobases and sugar residues. Adenine, cytosine and guanine possess reduction responses at mercury-based electrodes where the guanine residue yielded also an anodic signal due to electrooxidation of its reduction product. On the other hand, the common nucleobases undergo electrochemical oxidation at carbon and some other solid electrodes. As both the electrochemical reduction and oxidation of DNA bases are irreversible, measurements cannot be performed repeatedly. Initial increase in the anodic guanine response after short time incubation of the biosensor in prooxidative agents can indicate opening of the original dsDNA structure, while decrease in this response is an evidence for the deep DNA degradation [14]. Decrease of the anodic guanine peak height or area relative to that yielded by intact DNA was suggested as a measure representing damage to this nucleobase and proposed as a screening test for environmental toxicants present in water or wastewater samples [16]. Redox mediators such as rhodium or ruthenium complexes are sometimes used to shuttle electrons from guanine residues in distant parts of DNA chains to the electrode [17, 18]. Some of products of the DNA damage exhibit characteristic electrochemical signals (e.g., anodic oxidation of 8-oxoguanine (8-OG) [19] and 2,8-dihydroxyadenine [20]) which can be evaluated with better sensitivity than the change in original guanine response. An Os(III/II)-mediated electrochemical oxidation of 8-OG was reported [21]. The damaged DNA modified by "bulky" adducts with the nucleobases (e.g., mitomycin C and other drugs whose pharmacological effects involve DNA modification) is also able to give specific electrochemical response [22].

The second group of techniques employs electroactive compounds added to the measured system and interacting with DNA non-covalently as its indicators (cationic indicators, intercalators and groove binders). They typically possess electrochemical response at a "safe" electrode potential and the response is high due to their accumulation at the immobilized DNA layer. Using the scheme of indicator accumulation – voltammetric measurement – chemical regeneration of the DNA layer (by removal/desorption of the accumulated indicator particles), the response of the indicator can be measured repeatedly [8]. Decrease in the intercalator or groove binder response indicates strand breaks and helix destruction. The redox indicators may be also used as diffusionally free species present in the solution phase. For instance, the hexacyanoferrate(III/II) anions, $[Fe(CN)_6]^{3-/4-}$, indicate the presence of DNA layer on the electrode surface on the basis of electrostatic repulsion between the indicator anion and negatively charged DNA backbone [23, 24].

Electroactive labels introduced into DNA also possess electrochemical signals at less extreme potentials than intrinsic DNA responses. An example is electroactive osmium tetraoxide with 2,2′-bipyridine bound to free 3′-ends of the ss regions created by a DNA repair enzyme exonuclease III, which responds to the extent of DNA damage [25]. The technique is capable of detection of one lesion per $\approx 10^5$ nucleotides in supercoiled plasmid DNA. DNA-hybridization biosensors were proposed for studies of DNA damage by common toxicants and pollutants where voltammetric transduction was achieved by coupling ferrocene moiety to streptavidin linked to biotinylated target DNA [26].

Detection of DNA Damage by Bioactivated Xenobiotics

DNA damage induced by environmental pollutants is a major endogenous toxicity pathway in biological system [27]. Most of organic pollutants may not directly cause DNA damage, but their metabolized products by enzyme reactions are genotoxic and may cause the DNA lesion [28]. Electrochemical DNA biosensors enabling detection of such DNA damage could serve as a basis for in vitro genotoxicity screening for new organic chemicals at an early stage in their commercial development. For example, styrene is one of the most widely used industrial chemicals and itself shows little genotoxicity [29]. However, after being metabolized by liver cytochrome P450 enzymes, its oxidized product styrene oxide can induce DNA damage by formation of DNA adducts [27, 29]; styrene oxide is classified as a probable human carcinogen (group 2A) [30].

Electrochemical genotoxicity biosensors based on enzyme–DNA films that detect relative DNA damage rates resulting from enzyme bioactivation of chemicals (e.g., styrene [31–33], epoxide [34], or arylamines [35]) and direct damage of DNA have been described [36]. A layer-by-layer technique was employed to immobilize DNA and enzymes on the sensor surfaces and film construction was verified by measuring frequency decrease using a quartz crystal microbalance. DNA damage was detected using catalytic square wave voltammetry with soluble $[Ru(bpy)_3]^{2+}$ (bpy = 2,2′-bipyridine), which catalyzes oxidation of guanine moieties [37]. Sensor response versus enzyme reaction time provided relative DNA damage rates that were related to indices of animal toxicity and were validated by directly measured rates of nucleobase adduct formation in DNA films using LC–MS/MS [38].

A bright future is expected for applications of enzyme–DNA films to toxicity biosensors and for the development of catalytic sensors for oxidative stress. The approach to sensor construction is achievable by room temperature solution processing and is amenable to automation, perhaps by using robotic solutions spotters. Electrode arrays could be developed to provide many tests simultaneously. For example, future sensor arrays could be configured to detect toxicity of metabolites generated by a range of human cytochromes P450 [36].

Future Directions

The DNA-modified electrodes already represent very effective and at the same time simple, fast, cheap, miniaturized and mass-producible analytical devices for evaluation and maybe also classification of modes of genotoxic effects of individual compounds and complex food and environmental samples as well as for prescreening of new drugs and newly synthesized chemicals. Moreover, the evaluation of DNA protection (antioxidative) capacity of various natural and synthetic chemical substances together with tee and plant extracts is also possible using the detection of DNA damage by prooxidants [11, 39, 40]. Electrochemical signal formation and transduction remain to play a key role in both sensitivity and selectivity of these investigations with DNA-modified electrodes.

It can be anticipated that, in a near future, complex biorecognition layers will be suggested to detect potentially risk compounds and to improve further abilities of the biosensors to detect damage to DNA. The advanced level of medical and clinical diagnosis will be largely dependent on the successful development and implementation of new materials and technologies envisaging the fabrication of state-of-the-art biosensors [41]. The attractive properties of electrochemical devices are extremely promising for improving the efficiency of environmental screening, diagnostic testing, and therapy monitoring even more today with the construction of very large multiplexed arrays. Future DNA biosensors will require the development of new reliable devices or the improvement of the existing ones for use by nonspecialized personnel without compromising accuracy and reliability. Compact and portable devices will constitute another future area of multidisciplinary research on sensors [42].

Acknowledgements This work was supported by the Scientific Grant Agency VEGA of the Slovak Republic (Project 1/0182/11), by the Ministry of Education, Youth and Sports of the Czech Republic (Projects MSM 0021620857 and RP 14/63), and by Charles University in Prague (Project UNCE 2012/44).

Cross-References

▶ Biosensors, Electrochemical

References

1. Labuda J, Oliveira-Brett AM, Evtugyn G, Fojta M, Mascini M, Ozsoz M, Palchetti I, Palecek E, Wang J (2010) Electrochemical nucleic acid-based biosensors: concepts, terms, and methodology (IUPAC technical report). Pure Appl Chem 82:1161–1187
2. Fojta M (2005) Detecting DNA damage with electrodes. In: Palecek E, Scheller F, Wang J (eds) Electrochemistry of nucleic acids and proteins – towards electrochemical sensors for genomics and proteomics. Elsevier, Amsterdam
3. Pouget JP, Douki T, Richard MJ, Cadet J (2000) DNA damage induced in cells by gamma and UVA radiation as measured by HPLC/GC-MS and HPLC-EC and comet assay. Chem Res Toxicol 13:541–549
4. European Standards Committee on Oxidative DNA Damage (ESCODD) (2003) Measurement of DNA oxidation in human cells by chromatographic and enzymic methods. Free Radical Biol Med 34:1089–1099
5. Palecek E, Fojta M (2001) Detecting DNA hybridization and damage. Anal Chem 73:74A–83A
6. Vyskocil V, Labuda J, Barek J (2010) Voltammetric detection of damage to DNA caused by nitro derivatives of fluorene using an electrochemical DNA biosensor. Anal Bioanal Chem 397:233–241
7. Sassolas A, Leca-Bouvier BD, Blum LJ (2008) DNA biosensors and microarrays. Chem Rev 108: 109–139
8. Labuda J, Fojta M, Jelen F, Palecek E (2006) Electrochemical sensors with DNA recognition layer. In: Grimes CA, Dickey EC, Pishko MV (eds) Encyclopedia of sensors. American Scientific, Stevenson Ranch
9. Wang J (2005) Electrochemical nucleic acid biosensors. In: Palecek E, Scheller F, Wang J (eds) Electrochemistry of nucleic acids and proteins – towards electrochemical sensors for genomics and proteomics. Elsevier, Amsterdam
10. Ferancova A, Labuda J (2008) DNA biosensors based on nanostructured materials. In: Eftekhari A (ed) Nanostructured materials in electrochemistry. Wiley-VCH, Weinheim
11. Galandova J, Ziyatdinova G, Labuda J (2008) Disposable electrochemical biosensor with multiwalled carbon nanotubes–chitosan composite layer for the detection of deep DNA damage. Anal Sci 24: 711–716
12. Rusling JF (2005) Sensors for genotoxicity and oxidized DNA. In: Palecek E, Scheller F, Wang J (eds) Electrochemistry of nucleic acids and proteins – towards electrochemical sensors for genomics and proteomics. Elsevier, Amsterdam
13. Park JY, Park SM (2009) DNA hybridization sensors based on electrochemical impedance spectroscopy as a detection tool. Sensors 9:9513–9532
14. Ziyatdinova G, Labuda J (2011) Complex electrochemical and impedimetric evaluation of DNA

D

damage by using DNA biosensor based on a carbon screen-printed electrode. Anal Methods 3: 2777–2782

15. Palecek E, Jelen F (2005) Electrochemistry of nucleic acids. In: Palecek E, Scheller F, Wang J (eds) Electrochemistry of nucleic acids and proteins – towards electrochemical sensors for genomics and proteomics. Elsevier, Amsterdam
16. Mascini M, Palchetti I, Marrazza G (2001) DNA electrochemical biosensors. Fresenius J Anal Chem 369:15–22
17. Mugweru A, Rusling JF (2002) Square wave voltammetric detection of chemical DNA damage with catalytic poly(4-vinylpyridine)–$Ru(bpy)_2^{2+}$ films. Anal Chem 74:4044–4049
18. Popovich N, Thorp H (2002) New strategies for electrochemical nucleic acid detection. Interface 11:30–34
19. Oliveira-Brett AM, Piedade JAP, Serrano SHP (2000) Electrochemical oxidation of 8-oxoguanine. Electroanalysis 12:969–973
20. Oliveira SCB, Corduneanu O, Oliveira-Brett AM (2008) In situ evaluation of heavy metal-DNA interactions using an electrochemical DNA biosensor. Bioelectrochemistry 72:53–58
21. Holmberg RC, Tierney MT, Ropp PA, Berg EE, Grinstaff MW, Thorp HH (2003) Intramolecular electrocatalysis of 8-oxo-guanine oxidation: secondary structure control of electron transfer in osmium-labeled oligonucleotides. Inorg Chem 42:6379–6387
22. Tian L, Wei WZ, Mao Y (2004) Kinetic studies of the interaction between antitumor antibiotics and DNA using quartz crystal microbalance. Clin Biochem 37:120–127
23. Galandova J, Ovadekova R, Ferancova A, Labuda J (2009) Disposable DNA biosensor with the carbon nanotubes–polyethyleneimine interface at a screen-printed carbon electrode for tests of DNA layer damage by quinazolines. Anal Bioanal Chem 394:855–861
24. Labuda J, Ovadekova R, Galandova J (2009) DNA-based biosensor for the detection of strong damage to DNA by the quinazoline derivative as a potential anticancer agent. Microchim Acta 164:371–377
25. Havran L, Vacek J, Cahova K, Fojta M (2008) Sensitive voltammetric detection of DNA damage at carbon electrodes using DNA repair enzymes and an electroactive osmium marker. Anal Bioanal Chem 391:1751–1758
26. Nowicka AM, Kowalczyk A, Stojek Z, Hepel M (2010) Nanogravimetric and voltammetric DNA-hybridization biosensors for studies of DNA damage by common toxicants and pollutants. Biophys Chem 146:42–53
27. Scharer OD (2003) Chemistry and biology of DNA repair. Angew Chem Int Ed 42:2946–2974
28. Turesky RJ (2002) Heterocyclic aromatic amine metabolism, DNA adduct formation, mutagenesis, and carcinogenesis. Drug Metab Rev 34:625–650
29. Speit G, Henderson L (2005) Review of the in vivo genotoxicity tests performed with styrene. Mutat Res-Rev Mutat Res 589:67–79
30. Vodicka P, Koskinen M, Arand M, Oesch F, Hemminki K (2002) Spectrum of styrene-induced DNA adducts: the relationship to other biomarkers and prospects in human biomonitoring. Mutat Res-Rev Mutat Res 511:239–254
31. Zhou LP, Yang J, Estavillo C, Stuart JD, Schenkman JB, Rusling JF (2003) Toxicity screening by electrochemical detection of DNA damage by metabolites generated in situ in ultrathin DNA–enzyme films. J Am Chem Soc 125:1431–1436
32. Zhang Y, Hu NF (2007) Cyclic voltammetric detection of chemical DNA damage induced by styrene oxide in natural dsDNA layer-by-layer films using methylene blue as electroactive probe. Electrochem Commun 9:35–41
33. Zu Y, Hu NF (2009) Electrochemical detection of DNA damage induced by in situ generated styrene oxide through enzyme reactions. Electrochem Commun 11:2068–2070
34. Wang BQ, Rusling JF (2003) Voltammetric sensor for chemical toxicity using $Ru(bpy)_2$poly(4-vinylpyridine)$_{10}Cl^+$ as catalyst in ultrathin films. DNA damage from methylating agents and an enzyme-generated epoxide. Anal Chem 75:4229–4235
35. So M, Hvastkovs EG, Bajrami B, Schenkman JB, Rusling JF (2008) Electrochemical genotoxicity screening for arylamines bioactivated by N-acetyltransferase. Anal Chem 80:1192–1200
36. Rusling JF (2004) Sensors for toxicity of chemicals and oxidative stress based on electrochemical catalytic DNA oxidation. Biosens Bioelectron 20: 1022–1028
37. Thorp HH (1998) Cutting out the middleman: DNA biosensors based on electrochemical oxidation. Trends Biotechnol 16:117–121
38. So MJ, Hvastkovs EG, Schenkman JB, Rusling JF (2007) Electrochemiluminescent/voltammetric toxicity screening sensor using enzyme-generated DNA damage. Biosens Bioelectron 23:492–498
39. Ferancova A, Heilerova L, Korgova E, Silhar S, Stepanek I, Labuda J (2004) Anti/pro-oxidative properties of selected standard chemicals and tea extracts investigated by DNA-based electrochemical biosensor. Eur Food Res Technol 219:416–420
40. Ziyatdinova G, Galandova J, Labuda J (2008) Impedimetric nanostructured disposable DNA-based biosensors for the detection of deep DNA damage and effect of antioxidants. Int J Electrochem Sci 3: 223–235
41. Teles FRR, Fonseca LR (2008) Trends in DNA biosensors. Talanta 77:606–623
42. Cagnin S, Caraballo M, Guiducci C, Martini P, Ross M, SantaAna M, Danley D, West T, Lanfranchi G (2009) Overview of electrochemical DNA biosensors: new approaches to detect the expression of life. Sensors 9:3122–3148

DSP of Biomolecules

Matthias Franzreb[1] and Dirk Holtmann[2]
[1]Institute of Functional Interfaces, Karlsruhe Institute of Technology, Eggenstein-Leopoldshafen, Germany
[2]DECHEMA Research Institute of Biochemical Engineering, Frankfurt am Main, Germany

Introduction

Downstream processing of biomolecules means recovery and purification of the substances from natural sources such as animal or plant tissue or fermentation broth. It is an essential step in the manufacture of pharmaceuticals (e.g., of antibiotics, hormones, antibodies, and vaccines), natural fragrance and flavor compounds, amino and organic acids or enzymes. Often, 50–70 % of the total cost of a biotechnological process comes from downstream processing. Hence, it is important to develop the latter as an integral part of the overall process. Different downstream techniques based on electrochemical methods have been developed. The key advantages of electrochemical steps in downstream processes are inexpensive cell constructions and peripheral equipment, often no influence onto product stability, easy scale-up, high energy efficiency, and easy automation.

Electrokinetic Chromatography

Electrokinetic chromatography (EKC) is a separation technique based on the combination of electrophoresis and interactions of the analytes with additives which form a dispersed phase moving at a different velocity [1]. In order to achieve separation, either the analytes or the secondary phase should be charged. A special case of EKC is the micellar electrokinetic chromatography (MEKC), in which the secondary phase is a micellar dispersed phase in the capillary. As an example, the separation capillary is filled with an SDS micellar solution [2]. Under neutral or basic conditions, the solution migrates toward the cathode by the electroosmotic flow (EOF) when a high voltage is applied, while the micelle is forced toward the anode by electrophoresis. Normally, the EOF is stronger than the electrophoretic migration of the SDS micelle, and hence, the micelle migrates toward the cathode at a slower velocity than the aqueous phase. When a neutral analyte is injected into the micellar solution at the anodic end, it will be distributed between the micelle and the aqueous phase. An analyte that is not incorporated into the micelle at all migrates at the same velocity as the EOF toward the cathode, whereas an analyte that is totally incorporated into the micelle migrates at the lowest velocity or the same velocity as the micelle toward the cathode. As long as the analyte is electrically neutral, it migrates at a velocity between the two extremes or between the velocity of the EOF and that of the micelle. The analytes are detected in an increasing order of the distribution coefficients at the cathodic end. MEKC can be performed by adding an ionic micelle to the running solution without modifying the instrument. MEKC is a useful technique particularly for the separation of small molecules, both neutral and charged, and yields high-efficiency separation in a short time with minimum amounts of samples and reagents [3, 4]. Figure 1 shows a schematic representation of the separation principle of MEKC.

A further technique based on electrokinetic chromatography is the microemulsion electrokinetic chromatography (MEEKC). MEEKC is an electro-driven separation technique that offers the possibility of highly efficient separation of both charged and neutral analytes covering a wide range of water-soluble substances [5]. Microemulsions are solutions containing nanometer-size droplets of an immiscible liquid dispersed in an aqueous buffer. The droplets are coated with a surfactant to reduce the surface tension between the two liquid layers allowing an emulsion to form. The surface tension of the droplet is further lowered by addition of a short-chain alcohol, such as butanol, that stabilizes the system. A microemulsion containing ionic surfactant allows chromatographic

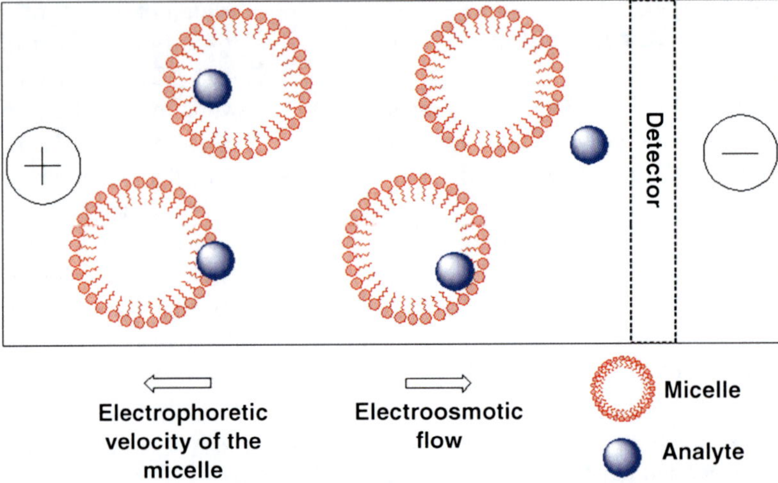

DSP of Biomolecules, Fig. 1 Schematic illustration of the micellar electrokinetic chromatography

separation to be obtained as analytes can partition between the charged oil droplets and the aqueous buffer phase. Water-insoluble compounds will favor inclusion into the oil droplets rather than into the buffer phase.

Capillary Electrophoresis and Capillary Electrochromatography

Capillary electrophoresis (CE) is an electro-driven separation technique. The main advantages are low reagent consumption, high efficiency, and selectivity with reasonably short analysis time [6]. In CE, the capillary is filled with a suitable buffer and after injecting analytes from the anode side, a relatively high voltage up to 30 kV is applied at its both ends. The positively or negatively charged analytes will move with different velocity and can be separated based on their electrophoretic mobility. Figure 2 shows a scheme of a CE experiment.

Capillary electrochromatography (CEC) is a hybrid technique utilizing the principles of electromigration techniques and chromatography [7]. The separation process is based on interactions of the analyte between the stationary and the mobile phases. In CEC, the flow of mobile phase is driven through the column by an electric field. The electroosmotic flow is generated by applying a high voltage across the column. Positive ions of the added electrolyte accumulate in the electrical double layer of particles of column packing, move toward the cathode, and drag the liquid mobile phase with them. As in capillary electrophoresis and micellar electrokinetic chromatography, small diameter columns with favorable surface area-to-volume ratio are employed to minimize thermal gradients from ohmic heating, which can have an adverse effect on band widths [8]. Table 1 shows a comparison of different electrochemically driven separation techniques.

Electrofiltration

Electrofiltration is a hybrid technology, combining pressure filtration and electrophoresis within a single device. In most cases, this combination is achieved by using a dead-end filtration system in which an electric field is applied parallel to the flow direction of the filtrate [9]. However, there also exist reports according to which a cross-flow filtration system is used. In the first case, higher concentration factors of the target molecules can be achieved [10], while the second case results in a continuous operation but also higher shear forces [11]. The idea behind electrofiltration is to impose an electrophoretic mobility onto charged biomolecules which points in the opposite direction of the filtration flow (see Fig. 3).

By this, the formation of a filtration cake can be retarded or even prevented and the filtration process can be greatly accelerated. For

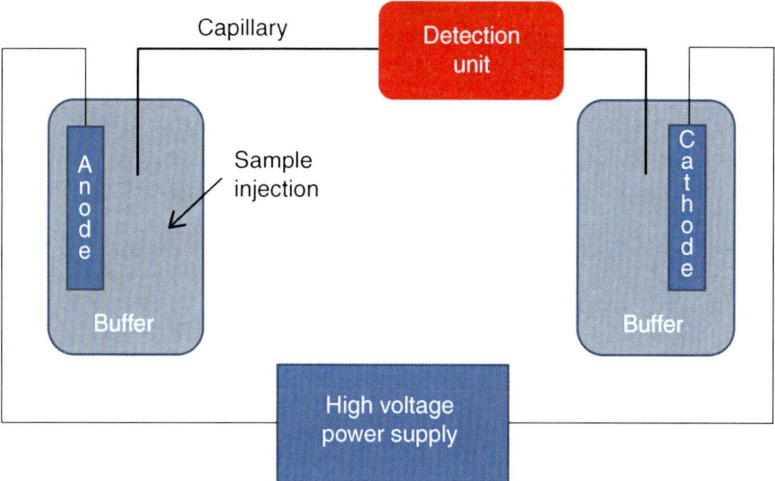

DSP of Biomolecules, Fig. 2 Scheme of a capillary electrophoresis. A CE instrument consists of a high-voltage power supply, a silica capillary with an internal diameter ranging from 20 to 200 μm, two buffer reservoirs, two electrodes connected to the power supply, and a detector. The capillary is filled with a buffer. The sample is injected from the anode side and both ends of the capillary and the electrodes are placed into buffer reservoirs; finally the voltage is applied across the capillary to start electrophoresis [6]

a theoretical description of electrofiltration of biomolecules, theories originally developed for colloidal inorganic particle suspensions can be adapted. In order to describe experimental data, the electroosmotic flow within the pores of the filtration membrane and, if present, in the filter cake has to be considered in addition in these theories [12]. Electrofiltration can be applied to charged biomolecules, especially to highly viscous, colloidal suspensions of biopolymers. The benefits of electrofiltration have been demonstrated for different polysaccharides, like poly(3-hydroxybutyrate), hyaluronic acid, chitosan and xanthan [13]. In the latter case, the filtration time is reduced from hours to minutes, making the filtration process economically feasible as compared to the usual precipitation step.

Electromembrane Extraction

The electromembrane extraction (EME) method extracts charged substances from a small sample volume through a thin membrane of organic solvent immobilized in the wall of a hollow fiber and into a solution inside the lumen of the hollow fiber. This extraction process is forced by an applied potential difference across the membrane, and this combination of well-known liquid–liquid extraction processes with electrokinetic migration yields a rapid and selective sample preparation method for ionic substances. The potential difference is utilized to extract charged analytes of interest from the sample across the organic membrane [14]. In EME using porous hollow fibers, target analytes are extracted as charged species from an aqueous sample, typically with a volume of 100–300 μl [15]. Figure 4 shows the scheme of an EME device.

Potential-Controlled Liquid Chromatography

Potential-controlled liquid chromatography (PC-LC) or electrochemically modulated liquid chromatography (EM-LC) is a further combination of electrochemical methods with liquid chromatography. The principle of PC-LC is related to ion exchange chromatography: The separation of molecules depends on their own electric charge and on the potential of the stationary chromatographic phase. The electrical

DSP of Biomolecules, Table 1 Comparison of capillary electrophoresis (*CE*), micellar electrokinetic chromatography (MEKC), capillary electrochromatography (*CEC*), potential-controlled liquid chromatography (*PC-LC*), and electrodialysis (*ED*) modified from [8]

	CE	MEKC	CEC	PC-LC	ED
(Main) separation principle	Different mobilities of ions in electric field	Partition between bulk solution and micelle moving in opposite direction to analyte	Partition between solid stationary phase and mobile phase	Different electrosorption of analytes	Selective transfer of ions through semi-permeable membranes
Stationary phase	None	None	Particles with bonded groups	Un-modified and modified conductive matrices	None
Sample type	Charged species	Neutrals	Neutral and charged species	Charged	Charged species

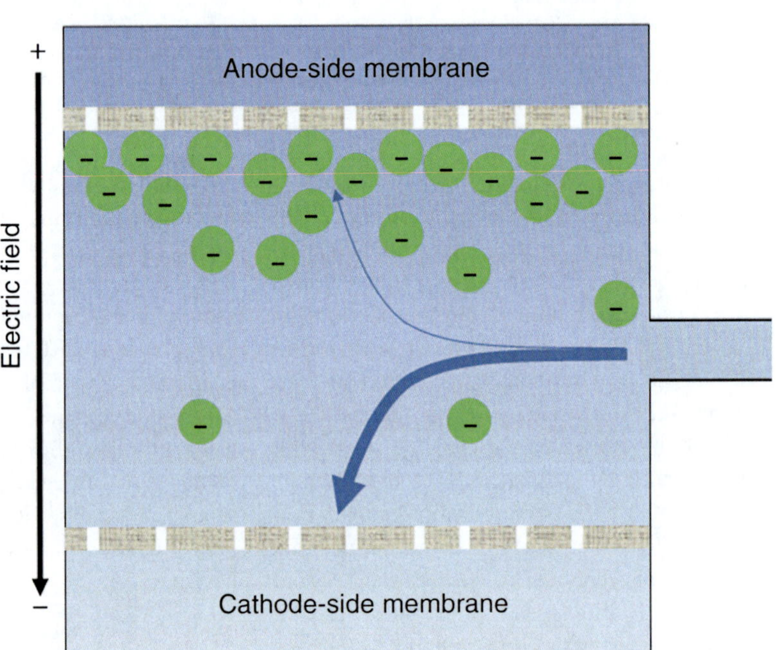

DSP of Biomolecules, Fig. 3 Electrofiltration of a negatively charged biopolymer. If the electric field is strong enough, the cathode-side membrane will be kept free of the biopolymer which will accumulate at the anode-side membrane, while the fluid flux will mainly pass through the cathode-side membrane. The figure shows a simplified principle without the ions of the electrolyte keeping electroneutrality and without flushing chambers of the electrodes

potential of this stationary phase can be easily controlled. In principle, a stationary phase can be used for either cation or anion exchange applications, depending on the selected electrode potential. In contrast to classical ion exchange chromatography, the electrical charge of the stationary chromatographic phase does not depend on immobilized charge carriers but on an adjustable abundance or a deficiency of electrons within the stationary phase itself. The principle of potential-controlled chromatography is illustrated in Fig. 5. Using an applied potential to a conductive packing (e.g., porous graphitic carbon), chromatographic resolution and retention are manipulated by altering the donor–acceptor interactions between the

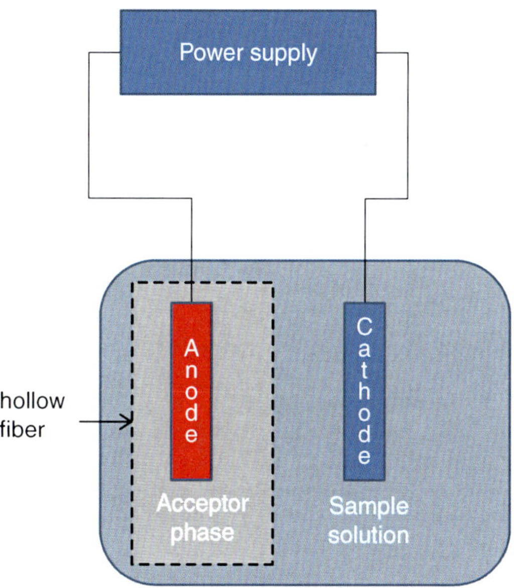

DSP of Biomolecules, Fig. 4 Scheme of electromembrane extraction (EME, adapted from [16]). Two electrodes are placed in the sample and the acceptor solution, respectively, and connected to a power supply. EME was found to be compatible with a wide range of biological matrices, e.g., whole blood and urine, preparing clean extracts in a short period of time with simple and inexpensive equipment

packing and analytes. In potential-controlled liquid chromatography, the conductive stationary phases like glassy carbon and porous graphitic carbon are packed into an HPLC column that is also configured to function as a three-electrode electrochemical cell. As such, the packing acts both as a chromatographic stationary phase and a high surface area working electrode. This dual function results in the unique ability to manipulate the surface charge density of the conductive packing through changes in applied potential, which, in turn, alters analyte retention. Stationary phases in PC-LC must possess several properties: High surface area, compositional and microstructural stability at the applied potentials, and high electrical conductivity [17–20].

Electrodialysis

An electrodialysis (ED) stack is composed of several flow chambers separated by ion exchange membranes and superimposed by an electric field. In most cases, an alternating arrangement

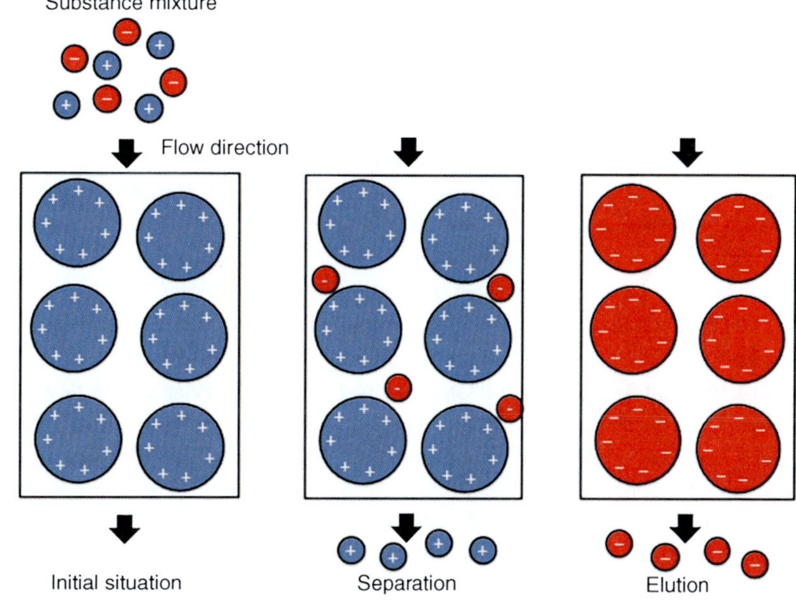

DSP of Biomolecules, Fig. 5 Simplified scheme of the separation of charged substances in potential-controlled liquid chromatography

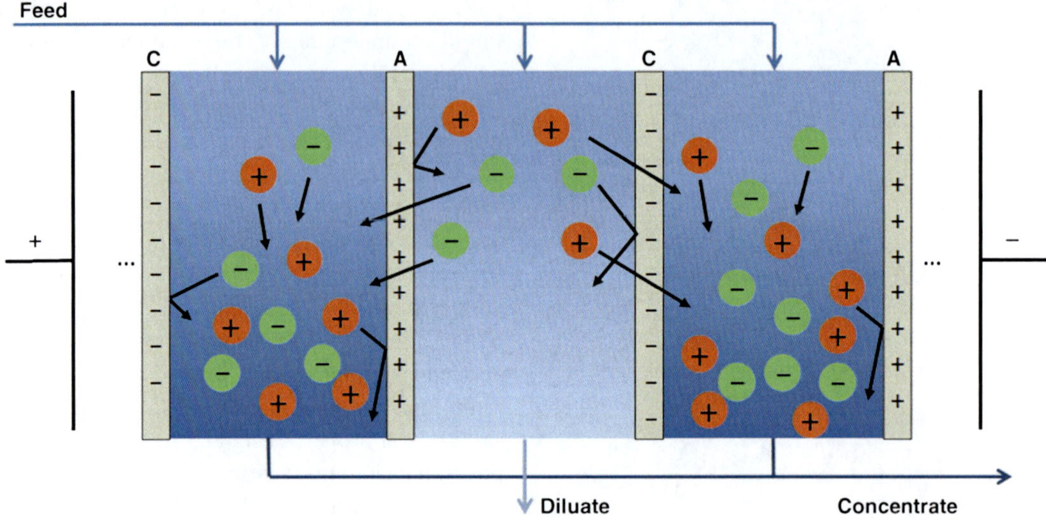

DSP of Biomolecules, Fig. 6 Working principle of electrodialysis. The illustration shows three cells of an electrodialysis stack and the respective trajectories of charged molecular species

of cation and anion exchange membranes is used. While a solution flows through the chamber, charged molecules are influenced by the electric field, resulting in a movement of anions toward the anode and cations toward the cathode. On their way, anions can penetrate anion exchange membranes but not cation exchange membranes and vice versa (see Fig. 6).

Given this arrangement, the charged species of the feed will deplete in one group of flow chambers and will concentrate in the adjacent chambers. Therefore, depending on which effluent is used as product stream, ED can be used for at least a partial desalting or a concentration of the charged species. The principle of ED is well known from the field of water technology. Here, the process is used for desalination of seawater as well as e.g., for recycling of rinsing water in wastewater treatment. However, for about two decades, the interest in ED as an efficient tool for downstream processing in biotechnology has constantly been increasing. Thinking about the application of ED, one has to be aware that conventional ion exchange membranes are practically non-porous and therefore molecules above approx. 500 Da are not able to penetrate the membrane irrespective of their charge [21]. Consequently, classical ED can be used for the separation of small biomolecules, like lactic acid from fermentation broth [22], but not for the fractionation of peptide or protein solutions. In order to extend the field of application, the ion exchange membranes in ED stacks can be replaced by nano- or ultrafiltration membranes [23, 24]resulting in a process often called EDUF (electrodialysis ultrafiltration) or electrophoretic separation.

Future Directions

The flexibility of separation processes based on electrochemical techniques, where the driving force can be easily controlled by an external power supply, makes these techniques more and more promising for application in biotechnology processes. Further improvements of the stationary phases, membranes and capillary systems are necessary to develop tailor-made processes for highly selective downstream processes. Furthermore, the understanding of protein adsorption at charged surfaces is important for a wide range of scientific disciplines including surface engineering, separation sciences, and pharmaceutical sciences [25]. Therefore, further investigations of the mechanism of the potential-dependent protein adsorption are essential for knowledge-based processes. Besides further investigations in the laboratory, the development of reliable mathematical models to predict the interactions of the biomolecules with the charged electrochemical systems is important. In general, the development of electrochemical methods in downstream processing of biomolecules is an interdisciplinary approach to be used by (electro)-chemists, physicists, biotechnologists, and engineers.

Cross-References

- ► Electrophoresis
- ► Membrane Processes, Electrodialysis

References

1. Riekkola M-L, Jönsson JÅ, Smith RM (2004) Terminology for analytical capillary electromigration techniques. Pure Appl Chem 76(2):443–451
2. Otsuka K, Terabe S (2004) Chiral micellar electrokinetic chromatography. Methods in Molecular Biology 243:355-363
3. Terabe S (2009) Capillary separation: micellar electrokinetic chromatography. Ann Rev Anal Chem 2(1):99–120
4. Foley JP (1990) Optimization of micellar electrokinetic chromatography. Anal Chem 62(13):1302–1308
5. Mahuzier P-E et al (2003) An introduction to the theory and application of microemulsion electrokinetic chromatography. LC-GC Eur 16:22–29
6. Rizvi SAA, Do DP, Saleh AM (2011) Fundamentals of micellar electrokinetic chromatography (MEKC). Eur J Chem 2(2):276–281
7. Mikšík I, Sedláková P (2007) Capillary electrochromatography of proteins and peptides. J Sep Sci 30(11):1686–1703
8. Cikalo MG et al (1998) Capillary electrochromatography. Analyst 123:87R–102R
9. Iritani E, Mukai Y, Kiyotomo Y (2000) Effects of electric field on dynamic behaviors of dead-end inclined and downward ultrafiltration of protein solutions. J Membr Sci 164(1–2):51–57
10. Hofmann R, Posten C (2003) Improvement of dead-end filtration of biopolymers with pressure electrofiltration. Chem Eng Sci 58(17):3847–3858
11. Enevoldsen AD, Hansen EB, Jonsson G (2007) Electro-ultrafiltration of industrial enzyme solutions. J Membr Sci 299(1–2):28–37
12. Yukawa H et al (1976) Analysis of batch electrokinetic filtration. J Chem Eng Japan 9:396–401
13. Gözke G, Posten C (2010) Electrofiltration of biopolymers. Food Eng Rev 2(2):131–146
14. Gjelstad A, Pedersen-Bjergaard S (2011) Electromembrane extraction: a new technique for accelerating bioanalytical sample preparation. Bioanalysis 3(7):787–797
15. Petersen N et al (2010) On-chip electro membrane extraction. Microfluid Nanofluid 9(4):881–888
16. Petersen NJ et al (2011) Electromembrane extraction from biological fluids. Anal Sci 27(10):965
17. Muna GW et al (2008) Electrochemically modulated liquid chromatography using a boron-doped diamond particle stationary phase. J Chromatogr A 1210(2):154–159
18. Keller DW, Ponton LM, Porter MD (2005) Assessment of supporting electrolyte contributions in electrochemically modulated liquid chromatography. J Chromatogr A 1089(1–2):72–81
19. Knizia M et al (2003) Potential-controlled chromatography of short-chain carboxylic acids. Electroanalysis 15(1):49–54
20. Kocak F et al (2005) Potential-controlled chromatography for the separation of amino acids and peptides. J Appl Electrochemistry 35(12):1231–1237
21. Galier S, Roux-de Balmann H (2001) Study of the mass transfer phenomena involved in an electrophoretic membrane contactor. J Membr Sci 194(1):117–133
22. Boniardi N et al (1997) Lactic acid production by electrodialysis Part I: experimental tests. J Appl Electrochemistry 27(2):125–133
23. Horvath ZS et al (1994) Multifunctional apparatus for electrokinetic processing of proteins. Electrophoresis 15(1):968–971
24. Firdaous L et al (2009) Concentration and selective separation of bioactive peptides from an alfalfa white protein hydrolysate by electrodialysis with ultrafiltration membranes. J Membr Sci 329(1–2):60–67
25. Hartvig RA et al (2011) Protein adsorption at charged surfaces: the role of electrostatic interactions and interfacial charge regulation. Langmuir 27(6):2634–2643

Dye-Sensitization

Laurence (Laurie) Peter
Department of Chemistry, University of Bath, Bath, UK

Definition

Photosensitization can be defined as a process in which light absorption by a photosensitizer molecule leads to a photophysical or photochemical change in a second molecule or system.

History

The scientific term *sensitization* referred originally to the process by which a photographic film or plate was made more sensitive to particular wavelengths of light. The history of dye-sensitization began in 1873 with the discovery by Hermann Wilhelm Vogel (1834–1898) that the sensitivity of silver halide photographic plates to green and red light was greatly enhanced by the presence of dyes in the photographic emulsion. Using a "cocktail" of different colored dyes, Vogel was able to achieve tone balance in black and white photographs [1]. It is now generally accepted that the photoexcited state of the dye injects electrons into the silver halide, leading to the formation of silver atoms. The oxidized dye can be regenerated by electron transfer from a "supersensitizer" such as hydroquinone or iodide.

At the end of the same century Oscar Raab reported the first observation of the phototoxic action of dyes, a discovery that led to the development of photodynamic therapy for treatment of tumors [2]. This therapy also relies on dye-sensitization. One of the mechanisms involved in the phototoxic effect is the creation of singlet oxygen by energy transfer to the (triplet) ground state of dioxygen from the triplet state of the photoexcited dye. Free radical formation by electron or hydrogen transfer reactions involving the triplet state can also occur. Dye-sensitization is also used more generally in photochemistry to generate reactive triplet states by energy transfer from a photoexcited dye to reactant molecules.

Dye-sensitization is also employed in electrostatic printing. This process, which is the basis of dry photocopying, involves electrostatic charging of small particles of a photoconductor such as zinc oxide dispersed in a resin binder. Illumination of the semiconductor discharges the photosensitive film so that it does not pick up the toner, and as a consequence only dark areas are printed. In 1962 RCA patented an invention by Harold Greig, who showed that the spectral sensitivity of the ZnO photoconductor could be enhanced in the visible region by addition of a wide range of organic dyes.

Photosensitization of (mostly single crystal) semiconductor electrodes was studied in a number of laboratories the 1960s [3]. Widespread interest in the process was sparked by the invention of the dye-sensitized solar cell, which was developed in the research group of Michel Grätzel at the Ecole Federal Polytechnique, Lausanne, Switzerland [4], and patented in 1992 [5]. The following 20 years have seen an exponential expansion of publications and patents dealing with dye-sensitized solar cells, and attempts are being made to commercialize the technology.

Dye-Sensitization of Semiconductors and Insulators

Dye-sensitization of semiconductors and insulators (including molecular crystals) generally involves photoexcitation of dye molecules adsorbed at the interface between the solid and an electrolyte solution. The photoexcited state of the dye may inject electrons or holes into the electrode or, if the emission spectrum overlaps with the absorption energy of optical transitions (e.g., from bulk or surface defect states) in the electrode, it may transfer its energy to the electrode, exciting electrons to higher energies [6]. The quantum efficiency of the first route can approach 100 %, whereas the second route generally leads to quenching of the excited state without significant photocurrent generation.

Depending on the positions of the highest occupied molecular orbital (HOMO) and lowest occupied molecular orbital (LUMO) of the dye, the photoexcited state may inject either electrons or holes into the conduction or valence band, respectively, of the solid [7] as shown in Fig. 1. (Note the broken arrows in Fig. 1 which show the direction in which *electrons* move).

Dye-sensitization of ZnO single-crystal electrodes was studied extensively in the late 1960s by Gerischer and Tributsch [8], Hauffe [9], and others. Illumination of ZnO electrodes coated with adsorbed monolayers of dyes such as rhodamine B, eosin, and methylene blue generates anodic current due to electron injection from the photoexcited dye. Around the same time,

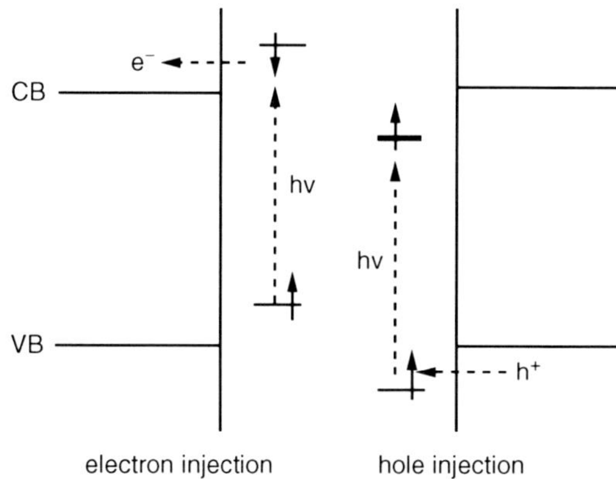

Dye-Sensitization, Fig. 1 Electron and hole injection from excited states of adsorbed dye molecules. Note that for electron injection, the LUMO level of the dye must lie above the conduction band. For hole injection, the HOMO level must lie below the valence band edge

injection of holes (i.e., extraction of valence band electrons) was observed for aromatic molecular crystals such as anthracene and perylene sensitized by similar dyes [10]. In both cases the external quantum efficiency is very low due to the very weak light absorption by the dye monolayer (typically <1 %). Figure 2 shows an example of the photocurrent excitation spectrum and the solution absorption spectrum for a mixture of three different cyanine dyes adsorbed on the (101) surface of a single crystal of the anatase from of TiO_2 [11].

Injection of electrons or holes from the illuminated layer of sensitizer dye leaves the dye in its oxidized or reduced state, respectively. As a consequence the dye layer usually degrades under illumination. In order to prevent this, a "supersensitizer" is added to the electrolyte. In the case of injection of electrons, the supersensitizer is an electron donor species such as iodide which regenerates the dye from its oxidized state. In the case of hole injection by the dye, an electron acceptor can be used as a supersensitizer. Provided that dye regeneration is faster than degradation reactions, the photosensitized system is stable.

This opens up the possibility of designing regenerative photoelectrochemical cells based on dye-sensitization. The sequence of reactions taking place in a regenerative cell is as follows (for the case of electron injection):

$D + h\nu \rightarrow D^*$ Photoexcitation of dye

$D^* \rightarrow D^+ + e^-$ (CB) Electron injection into conduction band

$D^+ + R \rightarrow D + O$ Regeneration of dye

$O + e^-$ (CE) $\rightarrow$ R Regeneration of R at the counter electrode (CE)

Dye-Sensitization of Mesoporous Wide-Band-Gap Oxides

As Fig. 2 shows, the external quantum efficiency or IPCE (incident photon to current conversion efficiency) for dye-sensitization of flat single-crystal surfaces is much less than 1 %. Attempts to enhance the IPCE by using thicker layers of dyes have not been successful because quenching occurs. In order to design an efficient solar cell based on the regenerative dye-sensitization scheme shown above, it is necessary to enhance the IPCE by at least two orders of magnitude. An obvious way around this problem is to increase the surface area of the semiconductor. In 1976, Tsubomura et al. [12] showed that a sintered ZnO pellet could be sensitized with rose Bengal to give much higher photocurrents. Their cell, which used an aqueous iodide/triiodide electrolyte to regenerate the dye, achieved a monochromatic power efficiency of 1.5 %. However no further progress was made until 1991, when O'Regan and Grätzel published a paper in Nature that demonstrated that sensitization of a mesoporous film of anatase by a ruthenium polypyridyl dye could be used as the basis for fabrication of dye-sensitized solar cells with solar conversion efficiencies of up to 7 % [4]. This remarkable breakthrough paved the way for the development of dye-sensitized solar cells with AM 1.5 efficiencies that have now exceeded 12 % [13]. One of the keys to the success of the DSC is the high internal surface area of the mesoporous anatase films used in the device. A 10-μm-thick mesoporous anatase layer formed by sintering particles with a size of 10–30 nm can have an internal surface that is more than 500 times greater than its geometric area.

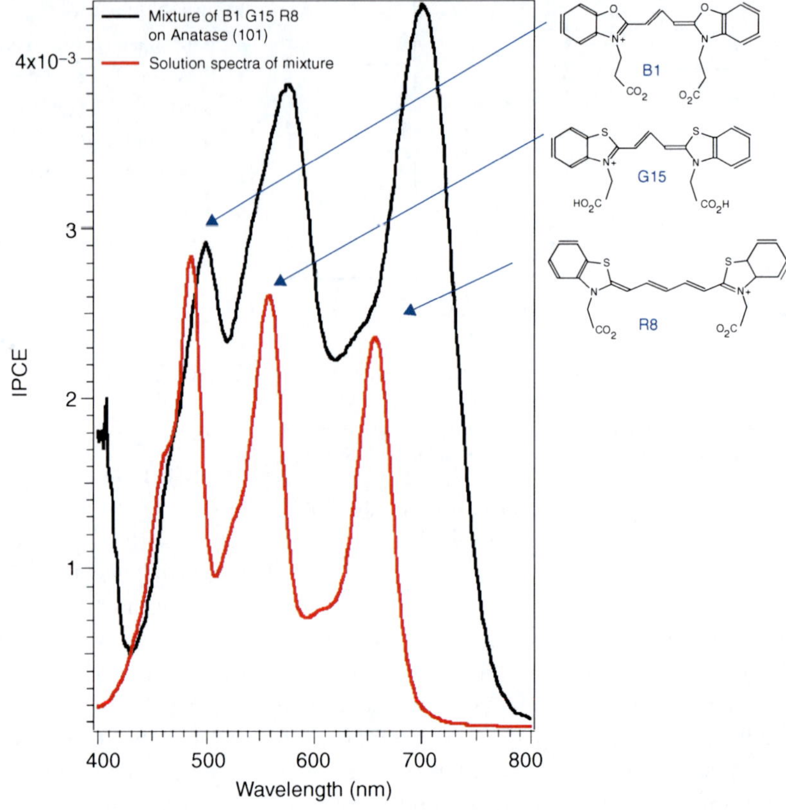

Dye-Sensitization, Fig. 2 Photocurrent spectrum for TiO_2 (101) anatase sensitized with a mixture of the three cyanine dyes shown on the right of the figure. The solution spectrum is shown for comparison. IPCE stands for incident photon to current efficiency, i.e., it is the external quantum efficiency (Reproduced with permission from Ref. [11])

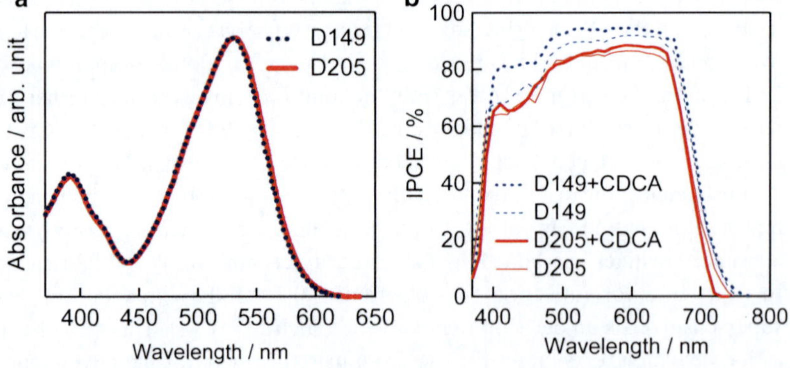

Dye-Sensitization, Fig. 3 IPCE spectra for two metal-free organic indoline dyes (Reproduced with permission from Ref. [15])

Considerable efforts have gone into the engineering of efficient panchromatic sensitizer dyes [14]. An enormous number of metal-ligand and metal-free dyes have been synthesized with the objective of improving the spectral response, which ideally should extend to around 900 nm for optimum performance under AM1.5 solar illumination. Figure 3 is an example of the very high IPCEs that are obtained with mesoporous anatase films and, in this case, the organic metal-free dye shown in Fig. 4 [15].

Dye-Sensitization, Fig. 4 Molecular structure of the indoline dyes D149 and D205 (Reproduced with permission from Ref. [15])

Codename	R=
D149	$-C_2H_5$
D205	$-C_8H_{17}$-n

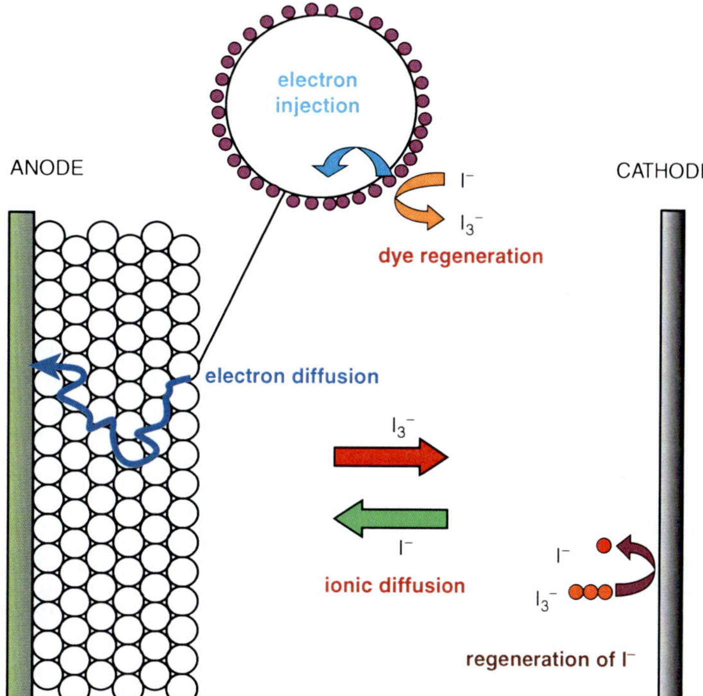

Dye-Sensitization, Fig. 5 Summary of the processes taking place in a DSC under illumination

The Dye-Sensitized Solar Cell or Grätzel Cell

The dye-sensitized solar cell (DSC), often called the Grätzel cell, is constructed as follows. A layer of titania paste consisting of anatase nanoparticles is spread onto a glass plate that is coated with a thin conducting layer of fluorine-doped SnO_2. The paste contains organic binders that are removed when the film is calcined in air (typically at 450–500 °C), leaving a mesoporous titania film (ca. 10 μm thick) with a porosity of around 50 %. The titania film is immersed in a dilute dye solution to impregnate it with dye. The counter electrode (cathode) consists of another sheet of conducting glass covered with a thin layer of platinum. The cell is assembled as a sandwich structure using a thin thermoplastic spacer, and the electrolyte is vacuum filled into the gap. The most widely used electrolyte is the I^-/I_3^- redox couple in organic solvents or ionic liquids. Several other redox systems have been used [16]; the most successful of these is a cobalt complex which outperforms the iodide electrolyte [13]. The processes taking place in the illuminated DSC are illustrated in Fig. 5.

The generic concept of the DSC has been adapted by replacing the dye with inorganic sensitizers costing of thin semiconductor layers or layers of quantum dots. The liquid electrolyte has been replaced by organic and inorganic hole conductors and the titania by other materials such as ZnO and SnO_2. The morphology of the oxide films has also been changed, with DSCs fabricated from a range of nanostructures including nanorods and nanotubes.

Kinetics of Electron Injection and Dye Regeneration

Transient optical and microwave absorbance measurements have established that electron injection can occur on a sub-picosecond time scale [17], competing effectively with other routes for deactivation of the excited state of the dye. Provided that geminate recombination does not occur (i.e., the injected electron transfers to the HOMO level of the dye), the injection efficiency can be high, approaching 100 % for the best ruthenium polypyridyl complexes. Regeneration of the dye from its oxidized state is much slower, taking place on a time scale of microseconds [18]. Nevertheless, regeneration is generally faster than back electron transfer from electrons present in the TiO_2, leading to high IPCE values. Competition between the rates of regeneration and decomposition of the oxidized state determines the turnover number for the dye. Since each dye molecule in the DSC absorbs around 1 photon per second, turnover numbers in excess of 10^8 are required for a 10-year lifetime.

Future Perspectives

Dye-sensitization is an example of a broader class of photophysical processes that involve charge transfer from an excited state. Other examples include sensitization of wide-band-gap semiconductors by quantum-confined nanoparticles (quantum dots) and by thin layers of semiconductors (extremely thin absorber layer – ETA cells). Charge transfer at the interface between donor and acceptor species in organic solar cells involves an analogous process in which mobile "excitons" dissociate at the interfaces to generate electrons and holes in the adjacent phases. These similarities are stimulating research into new types of solar cell that exploit the fundamental process of light-driven electron transfer. The challenge now is to move from the laboratory to commercial exploitation [19].

Cross-References

- ► Dye-sensitized Electrode, Photoanode
- ► Photoelectrochemistry, Fundamentals and Applications
- ► Photography, Silver Halides
- ► Quantum Dot Sensitization
- ► Semiconductor Electrode
- ► Semiconductor–Liquid Junction: From Fundamentals to Solar Fuel Generating Structures
- ► Solid State Dye-Sensitized Solar Cell

References

1. Vogel WH (2009) Die Photographie Farbiger Gegenstände in den Richtigen Tonverhältnissen. BiblioBazaar, Charleston
2. Szeimies R-M, Dräger J, Abels C, Landthaler M (2001) History of photodynamic therapy. In: Clazavera-Pinton P-G, Szeimies R-M, Ortel B (eds) Dermatology in photodynamic therapy and fluorescence diagnosis in dermatology. Elsevier, Amsterdam
3. Gerischer H, Tributsch H (1968) Electrochemistry of ZnO monocrystal spectral sensitivity. Ber Bunsenges Phys Chem 72:437–45
4. O'Regan B, Grätzel M (1991) A low-cost high-efficiency solar cell based on dye-sensitized colloidal TiO_2 films. Nature 353:737–740
5. Grätzel M, Liska P (1992) Photo-electrochemical cell and process of making same. US Patent 5084365
6. Memming R, Tributsch H (1971) Electrochemical investigations of spectral sensitization of gallium phosphide electrodes. J Phys Chem 75:562–570
7. Gerischer H, Michel-Beyerle ME, Rebentrost F, Tributsch H (1968) Sensitization of charge injection into semiconductors with large band gap. Electrochim Acta 13:1509–1515
8. Tributsch H, Gerischer H (1969) Electrochemical investigations on mechanism of sensitization and supersensitization of ZnO monocrystals. Ber Bunsenges Phys Chem 73:251

9. Hauffe K, Danzmann HJ, Pusch H, Range J, Volz H (1970) New experiments on the sensitization of zinc oxide by means of the electrochemical cell technique. J Electrochem Soc 117:993–999
10. Gerischer H, Willig F (1976) Reaction of excited dye molecules at electrodes. Top Curr Chem 61:31–84
11. Spitler MT, Parkinson BA (2009) Dye sensitization of single crystal semiconductor electrodes. Acc Chem Res 42:2017–2029
12. Tsubomura H, Matsamura M, Nomura Y, Amamiya T (1976) Dye-sensitized zinc oxide: aqueous electrolyte: platinum photocell. Nature 26:402–403
13. Yell A, Lee HW, Tsao HN (2012) Porphyrin-sensitized solar cells with cobalt(III/II)-based redox electrolyte exceed 12 percent efficiency. Science 334:629–634
14. Nazeeruddin K, Pechy P, Renouard T et al (2001) Engineering of efficient panchromatic sensitizers for nanocrystalline TiO_2-based solar cells. J Am Chem Soc 123:1613–1624
15. Ito S, Miur H, Uchida, et al. (2008) High-conversion efficiency organic dye-sensitized solar cells with a novel indoline dye. Chem Commun 5194–5196
16. Hamann TW, Ondersma JW (2011) Dye-sensitized solar cell redox shuttles. Energy Environ Sci 4:370–381
17. Huber R, Moser JE, Grätzel M et al (2002) Real-time observation of photoinduced adiabatic electron transfer in strongly coupled dye/semiconductor colloidal systems with a 6 fs time constant. J Phys Chem B 106:6494–6499
18. Pelet S, Moser JE, Grätzel M (2000) Cooperative effect of adsorbed cations and iodide on the interception of back electron transfer in the dye sensitization of nanocrystalline TiO_2. J Phys Chem B 104:1791–1795
19. Peter LM (2011) The Gratzel cell: where next? J Phys Chem Lett 2:1861–1867

Dye-Sensitized Electrode, Photoanode

Tsutomu Miyasaka
Graduate School of Engineering, Toin University of Yokohama, Yokohama, Kanagawa, Japan

Introduction

Study of dye-sensitized semiconductor electrodes has started in the late 1960s as an extension of photographic science where silver halide grains are photosensitive materials to be spectrally sensitized. Dye molecules adsorbed on the surface of silver halide crystals and photoexcited by absorption of visible light act as electron donors to silver halide and spectrally sensitize the formation of silver image. Gerischer and co-workers explored various semiconductors of metal oxides and inorganic compounds which can be sensitized with organic dyes in the structure of electrochemical cell [1, 2]. As the result of sensitization, n-type semiconductors such as ZnO, TiO_2, and CdS generate anodic photocurrents, and p-type semiconductors such as GaP generate cathodic photocurrents with their action spectra following the absorption spectra of the sensitizing dyes. The dye-sensitized photocurrent is generally larger in density and efficiency in the n-type sensitization than the p-type. In 1971, Tributsch and Calvin demonstrated photocurrent generation by a thin film of natural chlorophyll deposited on n-type semiconductor ZnO [3] to mimic spectral sensitivity of photosynthesis. This study was synchronized with the discovery by Fujishima and Honda (1972) of water photolysis on n-type semiconductor TiO_2 [4]. These studies established the field of photoelectrochemistry as a simple model of photosynthetic energy conversion. In the following years, the Langmuir-Blodgett method [5–7] has revealed that a single monolayer of adsorbed dye (sensitizer) only can contribute to photoexcited electron injection to n-type semiconductor electrode. This principle agreed with the theory of dye sensitization then established for silver halide photographic material [8]. Light absorption and energy conversion attained by a single dye layer, however, was too small to be applied to photovoltaic power generation. This application was later realized when the Grätzel group invented in 1991 a mesoporous semiconductor electrode with high surface area which enhances light absorption by adsorbed dye molecules [9]. Dye-sensitized cells constructed with mesoporous semiconductor photoelectrodes have achieved high efficiency in energy conversion and joined a class of utility-type solar cells.

Sensitization Mechanism

Semiconductor electrode in which electron is the dominant carrier for conductivity is classified as n-type, which has a Felmi level located close to

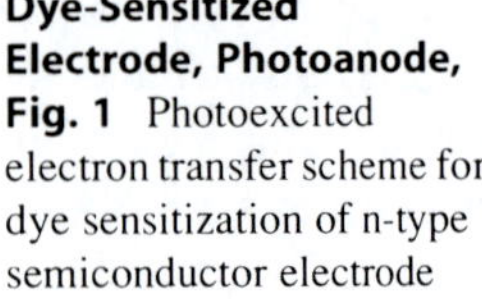

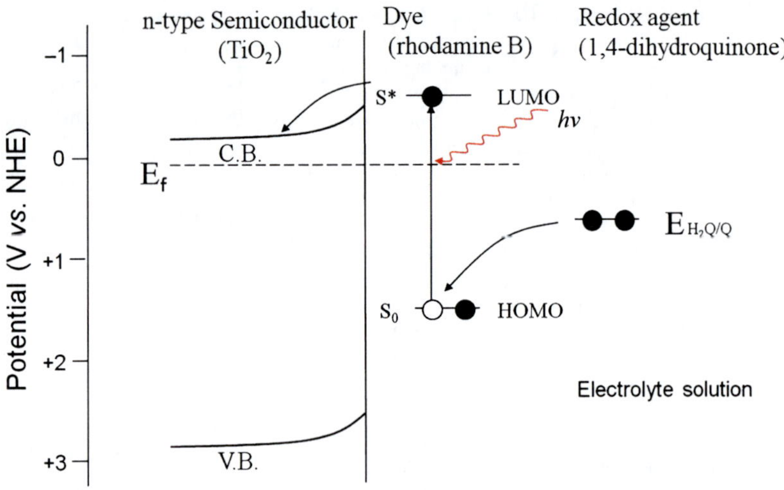

Dye-Sensitized Electrode, Photoanode, Fig. 1 Photoexcited electron transfer scheme for dye sensitization of n-type semiconductor electrode

the conduction band. Study of semiconductor electrodes in the early days has started by using single crystals of semiconductors such as TiO_2, ZnO, and CdS. In contact with an ionic electrolyte, surface potentials develop at the semiconductor surface through chemical equilibrium between metal oxide and ionic redox spices in electrolyte. With aqueous electrolytes, surface electrochemical potentials are pinned at oxide semiconductors as determined by the work function of the semiconductor and the electrolyte pH, the latter causing a potential shift according to the Nernst equation with a slope, −59 mV/pH. On polarization of the semiconductor electrode positively or negatively by means of a potentiostat, there occurs a potential bending (gradient) from the surface to the bulk of semiconductor.

Metal oxide semiconductors have their intrinsic sensitivity to light only in the ultraviolet and/or short wavelength regions due to their band gaps (e.g., 3.5 eV for TiO_2 and eV for ZnO). This limits utilization of incident photons to a low level for photovoltaic conversion. When the excited states of a dye molecule possess sufficiently long lifetimes and electrochemically more negative potentials than the conduction band edge potential, photoexcited electrons are injected to the conduction band leading to charge separation at the semiconductor-dye interface. Separated charges can however be recombined being influenced by the state of potential gradient. In the conditions of no potential gradient (flat-band condition) and cathodic polarization with a negative slope of potential, charge recombination takes place by back transfer of electrons to the dye. Anodic polarization that causes a positive potential slope favors unidirectional electron transfer from dye to the bulk of semiconductor. In a suitable condition of anodic polarization, i.e., external bias voltage, rectified photocurrent occurs as a result of dye sensitization. Action spectrum of the dye-sensitized photocurrent thus follows optical absorption spectrum of the dye adsorbed on the semiconductor surface, which covers some area of the visible light wavelength region, 400–800 nm.

The efficiency and density of the dye-sensitized photocurrent is a function of electrode potential applied by external bias (voltage). More positively regulated electrode potential enhances photocurrent by suppression of back electron transfers. More positive potential in effect enhances the potential barrier, i.e., a Schottky barrier, at the semiconductor surface in aspects of energy gap (eV) as well as in the thickness of space charge layer. This potential barrier is crucial to control the rectification of dye-sensitized photocurrent. Figure 1 shows the schematic diagram for dye-sensitized photocurrent generation on a TiO_2 single crystal electrode. Here, rhodamine B is presented as a sensitizer which strongly absorbs light at around 550 nm. It is generally known that the sensitized photocurrent tends to saturate in the existence of a potential barrier (gradient) of more than 0.2 eV. Based on the carrier (donor) density of TiO_2 (10^{18} cm^{-3}), this polarization corresponds to formation of a space charge layer of a thickness of 50 nm. The maximum photovoltage of dye-sensitized electrode is obtained in the open-circuit condition of the cell in which the conduction band (CB) is filled with injected electrons. For a cell composed of a working photoelectrode and a counter electrode, open-circuit photovoltage is defined as the potential difference between the Felmi level of TiO_2 and the redox potential of electrolyte species that corresponds to the potential of counter electrode.

Dye-sensitized n-type semiconductors act as photoanodes which generate anodic photocurrents driven by generation of negative photovoltages. The quantum efficiency of sensitized photocurrent is influenced by several processes of energy dissipation which reduce the rectified photocurrent. Back electron transfers from electrode to the dye-oxidized state or to the electrolyte redox species are the main causes of photocurrent decrease. In addition, deactivation of the dye-excited state can occur before electron injection in case where the dye-excited state is not sufficiently high enough with respect to the CB level of semiconductor and/or its lifetime is too short. It is normally required that the excited state is positioned more than 0.1 eV higher than CB. Short lifetime of an excited dye can cause internal conversion to the ground state. This type of deactivation is caused by molecular aggregation of dyes as well as by energy transfer between dye molecules accompanied by photochemical quenching. The latter has been demonstrated by use of a monomolecular film of dye (chlorophyll) on SnO_2 electrode in which intermolecular distance is controlled by insertion of spacer molecules (lipids). Suppression of dissipative energy transfer and quenching largely improves the quantum efficiency of photocurrent [6]. This practice has led to the method of enhancing the performance of dye-sensitized solar cells by use of co-absorbents in dye adsorption.

Dye-Sensitized Mesoporous Photoanodes and Solar Cells

The theory of electron exchange reaction as well as experimental demonstration by means of Langmuir-Blodgett films teaches that the dye molecule in direct contact with the semiconductor surface only is capable of electron injection to electrode. Optical absorption by a single dye monolayer (1–3 % of incident light), however, is not sufficient to collect photons. Harvesting photons can be realized by using a mesoporous surface that gives 500 times or more large surface areas for dye loading per semiconductor thickness of 10 mm. Mesoporous semiconductor electrodes of TiO_2 and ZnO are prepared, e.g., by means of sintering (450–550 °C) of a TiO_2 particle-dispersed paste coated on a glass substrate or electrochemical deposition of ZnO from Zn^{2+}-containing aqueous electrolytes. A family of Ru bipyridyl complexes [10–14] has been widely employed as sensitizers on TiO_2 by taking the advantage of their broad absorption bands due to metal-ligand charge transfer as well as of their high stability to light. As a result of enhanced light scattering, action spectrum of the dye-sensitized photocurrent covers the visible region (400–900 nm) peaking at the green region (around 530 nm), giving a spectral sensitivity similar to those of amorphous silicon solar cells. With use of iodine/iodide redox couples in organic electrolytes (typically, acetonitrile), photocurrent density of 10–20 mA cm^{-2} in the short-circuit condition and open-circuit photovoltage of 0.6–0.8 V can be normally obtained under exposure to 1,000 W m^{-2} sun light; the output performance depends on the thickness and porosity of semiconductor and the composition of electrolyte. Figure 2 depicts the power generation scheme of the mesoporous solar cell. A unique characteristic of the mesoporous electrode is that the dyed (colored) electrode is optically semitransparent. This nature enables to construct a tandem photocell in which incident light is used twice by two superposed cells. After optimization for the cell structure, the power conversion efficiency of dye-sensitized solar cell has reached 11 % and more [15].

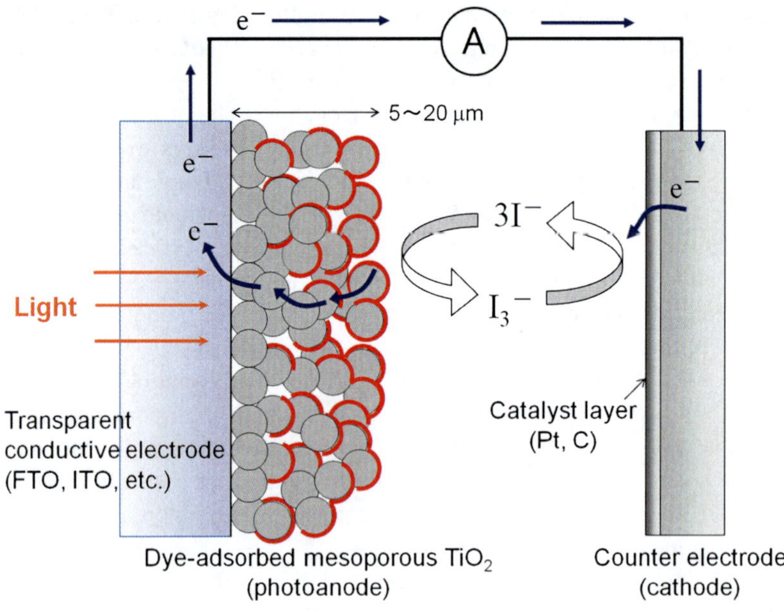

Dye-Sensitized Electrode, Photoanode, Fig. 2 Dye-sensitized solar cell based on mesoporous TiO_2 photoanode

Sensitizer plays a key role in expanding the spectral sensitivity, enhancing photon absorption, and determining the range of photovoltage. On TiO_2, Lowest unoccupied molecular orbital (LUMO) level of the excited state sensitizer should be normally higher than the conduction band by 0.15 eV or more. Sensitizers of lower LUMO levels tend to decrease not only the photocurrent density but also photovoltage. Among various kinds of dye sensitizers, indoline-type dyes [16–18], oxazole-type dyes [19], and porphyrin families [20, 21] possess high extinction coefficients and work as efficient sensitizers on TiO_2. Successful combination of sensitizer and redox agent allows further improvement of conversion efficiency. Use of metal porphyrins has achieved highest conversion efficiency, 12 %, when coupled with cobalt bipyridyl complexes as redox compounds which are capable of yielding high voltage around 1.0 V in combination with TiO_2 photoanode [21]. Various derivatives of Ru complex dyes have also been synthesized for improvement of extinction coefficient and chemical stability against degradation caused by photocatalytic reaction of TiO_2. Durability of dye-sensitized TiO_2 electrode has been examined under high temperature and continuous soaking to light. It is however generally found that exposure of the cell to ultraviolet (UV) light deteriorates the electrode performance. Lifetime of the cell under accelerated conditions can reach more than thousands of hours by applying UV filters [15, 22]. Inorganic nanocrystalline particle and quantum dots also work as spectral sensitizers. Their main advantage is addressed to the tunability of absorption wavelength (band gap) by changing the particle size as the quantum confinement effect. For examples, CdS [23], CdSe [24], and PbS [25] have been studied as quantum dot sensitizers to TiO_2. These inorganic sensitizers are intrinsically more stable against light soaking while their chemical stability against liquid electrolytes is subject of improvement. As an organic inorganic hybrid compound, an organo metal halide, $CH_3NH_3PbX_3$(X=Br, I) was found to act as high-efficiency photovoltaic material and light absorber on mesoporous metal oxide electrodes [26]. It forms a perovskite crystalline film on the surface of Al_2O_3 and TiO_2. A solid-state photovoltaic cell based on $CH_3NH_3PbI_2Cl$-coated Al_2O_3 yields 10.9% conversion efficiency [27]. Further high efficiencies, 12-15%, of the perovskite-based solar cell have been currently achieved with $CH_3NH_3PbI_3$-coated TiO_2 electrodes [28, 29]. This type of solid-state

Dye-Sensitized Electrode, Photoanode, Fig. 3 Dye-sensitized solar cell of bifacial photovoltaic activity fabricated with thin plastic substrates

photovoltaic cell, however, is accepted as a new type, hybrid semiconductor solar cell rather than a family of dye-sensitized cell.

Future Perspectives

For industrial applications, a strong merit of dye-sensitized solar cell over the existing solid-state solar cells has been emphasized on account of its low-cost manufacturing processes without the high vacuum and nanoscale manipulation as required by the solid-state pn junction cells. Dye-sensitized photoelectrodes are fabricated not only on glass or metal substrates but also on plastic film substrates by applying low-temperature coating technologies for mesoporous TiO_2 and ZnO. Fabrication of lightweight plastic photoelectrodes has been subject of intense study and achieved conversion efficiencies of 5-6% for TiO_2 [30, 31]. Flexible solar cells and large-area modules have been designed with plastic electrodes to demonstrate them as low-cost printable solar cell for versatile applications including consumer's electronics. Figure 3 displays an example of the solar cell, which has a semitransparent body due to mesoporous electrodes and is capable of utilizing light on the both sides of the cell [32]. For future applications to energy industry, high sensitivity of dye-sensitized power generation to weak indoor light is especially advantageous for versatile applications to small power sources, which include power supply to computers and other IT equipment.

Cross-References

- ▶ Dye-Sensitization
- ▶ Dye-Sensitized Electrode, Photoanode
- ▶ Photoelectrochemistry, Fundamentals and Applications
- ▶ Solid State Dye-Sensitized Solar Cell

References

1. Tributsch H, Gerischer H (1969) The use of semiconductor electrodes in the study of photochemical reactions. Ber Bunsenges Phys Chem 73:850–854 and references therein
2. Memming R (2001) Semiconductor electrochemistry. Wiley-VCH, Weinheim
3. Tributsch H, Calvin M (1971) Electrochemistry of excited molecules: photoelectrochemical reaction of chlorophylls. Photochem Photobiol 14:95–112
4. Fujishima A, Honda K (1972) Electrochemical photolysis of water at a semiconductor electrode. Nature 238:37–38

5. Kuhn H, Möbius D, Bücher H (1972) Spectroscopy of monolayer assemblies. Physical Methods in Chemistry. In: A. Weissberger and B. W. Rossiter (eds.), Part IIIB Optical, Spectroscopic, and Radioactivity Methods. Wiley-Interscience, New York, 577–578
6. Miyasaka T, Watanabe T, Fujishima A, Honda K (1979) Highly efficient quantum conversion at chlorophyll a-lecithin mixed monolayer coated electrode. Nature 277:638–640
7. Miyasaka T, Watanabe T, Fujishima A, Honda K (1980) Photoelectrochemical study of chlorophyll-a multilayers on SnO_2 electrode. Photochem Photobiol 32:217–222
8. Tani T (1995) Photographic sensitivity: theory and mechanisms. Oxford University Press, New York
9. O'Regan B, Grätzel M (1991) A low-cost, high-efficiency solar cell based on dye-sensitized colloidal TiO_2 films. Nature 335:737–740
10. Nazeeruddin MK, Pechy P, Grätzel M (1997) Efficient panchromatic sensitization of nanocrystalline TiO_2 films by a black dye based on a trithiocyanato-ruthenium complex. J Chem Commun 18:1705–1706
11. Wang P, Zakeerruddin SM, Comte P, Charvet R, Humphry-Baker R, Grätzel M (2003) Enhance the performance of dye-sensitized solar cells by co-grafting amphiphilic sensitizer and hexadecylmalonic acid on TiO_2 nanocrystals. J Phys Chem B 107:14336–14341
12. Chen CY, Wu SJ, Wu CG, Chen JG, Ho KC (2006) A ruthenium complex with superhigh light-harvesting capacity for dye-sensitized solar cells. Angew Chem Int Ed 45:5822–5825
13. Jiang KJ, Masaki N, Xia JB, Noda S, Yanagida S (2006) A novel ruthenium sensitizer with a hydrophobic 2-thiophen-2-yl-vinyl-conjugated bipyridyl ligand for effective dye sensitized TiO_2 solar cells. Chem Comm 23:2460–2462
14. Wang P, Klein C, Humphry-Baker R, Zakeerruddin SM, Grätzel M (2005) A high molar extinction coefficient sensitizer for stable dye-sensitized solar cells. J Am Chem Soc 127:808–809
15. Chen CY, Wang M, Li JY, Pootrakulchote N, Alibabaei L, Ngoc-le C, Decoppet JD, Tsai JH, Grätzel C, Wu CG, Zakeeruddin SM, Grätzel M (2009) Highly efficient light-harvesting ruthenium sensitizer for thin-film dye-sensitized solar cells. ACS Nano 3:3103–3109
16. Horiuchi T, Miura H, Sumioka K, Uchida S (2004) High efficiency of dye-sensitized solar cells based on metal-free indoline dyes. J Am Chem Soc 126:12218–12219
17. Ito S, Zakeeruddin SM, Humphry-Baker R, Liska P, Charvet R, Comte P, Nazeeruddin MK, Péchy P, Takata M, Miura H, Uchida S, Grätzel M (2006) High-efficiency organic-dye- sensitized solar cells controlled by nanocrystalline-TiO_2 electrode thickness. Adv Mater 18:1202–1205
18. Kuang D, Uchida S, Humphry-Baker R, Zakeeruddin SM, Grätzel M (2008) Organic dye-sensitized ionic liquid based solar cells: remarkable enhancement in performance through molecular design of indoline sensitizers. Angew Chem Int Ed 47:1923–1927
19. Zhang XH, Wang ZS, Cui Y, Koumura N, Furube A, Hara K (2009) Organic sensitizers based on hexylthiophene-functionalized indolo[3,2-b]carbazole for efficient dye-sensitized solar cells. J Phys Chem C 113:13409–13415
20. Ikegami M, Ozeki M, Kijitori Y, Miyasaka T (2008) Chlorin-sensitized high-efficiency photovoltaic cells that mimic spectral response of photosynthesis. Electrochem 76:140–143
21. Yella A, Lee HW, Tsao HN, Yi C, Chandiran AK, Nazeeruddin MK, Diau EWG, Yeh CY, Zakeeruddin SM, Grätzel M (2011) Porphyrin-sensitized solar cells with cobalt (II/III)–based redox electrolyte exceed 12 percent efficiency. Science 334:629–634
22. Yu Q, Zhou D, Shi Y, Si X, Wang Y, Wang P (2010) Stable and efficient dye-sensitized solar cells: photophysical and electrical characterizations. Energy Environ Sci 3:1722–1725
23. Chang CH, Lee YL (2007) Chemical bath deposition of CdS quantum dots onto mesoscopic TiO_2 films for application in quantum-dot-sensitized solar cells. Appl Phys Lett 91:053503–1–053503–3
24. Diguna LJ, Shen Q, Kobayashi J, Toyoda T (2007) High efficiency of CdSe quantum-dot-sensitized TiO_2 inverse opal solar cells. Appl Phys Lett 91:023116–1–023116–3
25. Plass R, Pelet S, Krueger J, Grätzel M, Bach U (2002) Quantum dot sensitization of organic–inorganic hybrid solar cells. J Phys Chem B 106:7578–7580
26. Kojima A, Teshima K, Shirai Y, Miyasaka T (2009) Organometal halide perovskites as visible-light sensitizers for photovoltaic cells. J Am Chem Soc 131:6050–6051
27. Lee MM, Teuscher J, Miyasaka T, Murakami TN, Snaith HJ (2012) Efficient hybrid solar cells based on meso-superstructured organometal halide perovskites. Science 338:643–647
28. Noh JH, Im SH, Heo JH, Mandal TN, Seok SI (2013) Chemical management for colorful, efficient, and stable inorganic−organic hybrid nanostructured solar cells. Nano Lett 13:1764–1769
29. Grätzel M (2013) Perovskite nano-pigments and new molecularly engineered porphyrin light harvesters for mesoscopic solar cells. Hybrid and Organic Photovoltaics Conference (HOPV13), Sevilla, Spain, 7 May 2013
30. Miyasaka T, Ikegami M, Kijitori Y (2007) Photovoltaic performance of plastic dyesensitized electrodes prepared by low-temperature binder-free coating of mesoscopic titania. J Electrochem Soc 154:A455–A461
31. Lee KM, Wu SJ, Chen CY, Wu CG, Ikegami M, Miyoshi K, Miyasaka T, Ho KC (2009) Efficient and stable plastic dye-sensitized solar cells based on a high light-harvesting ruthenium sensitizer. J Mater Chem 19:5009–5015
32. Miyasaka T (2011) Toward printable sensitized mesoscopic solar cells: Light-harvesting management with Thin TiO_2 films. J Phys Chem Lett 2:262–269

Dynamic Methods in Solid-State Electrochemistry

Ulrich Guth
Kurt-Schwabe-Institut für Mess- und Sensortechnik e.V. Meinsberg, Waldheim, Germany
FB Chemie und Lebensmittelchemie, Technische Universität Dresden, Dresden, Germany

Definition

Although the most of the methods using solid electrolytes cells are those in which the measured values are time independent (stationary), there are also techniques in which the excitation signal (sensor response) is time dependent (in-stationary). According to the values that are changed, it can be distinguished between controlled potential techniques (voltammetric techniques) and controlled current techniques (coulostatic techniques) [1]. The excitation function can be changed with time periodically (impedance) or according to a special time regime (voltammetry). In solid-state electrochemistry linear sweep mainly voltammetric and impedimetric techniques are applied. Current as a function of a controlled electrode potential and time is recorded as a so-called voltammogram.

Voltammetric Techniques

Mostly the voltage is changed linearly with constant sweep rate (dU/dt) or as cyclic voltammetry where the current response of the electrode potential is measured in increasing direction up to the turning point and decreasing to the starting potential [2]. When an electrochemical active component is present, an anodic current peak can be detected. In the reverse scan a corresponding cathodic peak may be observed. The peak potential is typical for the electrode reaction and the peak current depends on the scan rate, the number of electrons, and the diffusion coefficient of the active species. Symmetric bell-shaped oxidation–reduction peaks are obtained for reversible reactions so for the oxygen reduction and the oxidation of O^{2-} ions are approximately reversible reactions [3]. Also differential techniques like differential pulse voltammetry (DPV) and square wave voltammetry (SQV) are in use to enlarge the response signal.

Linear sweep voltammograms are suited to investigate the electrode activity [4]. As an example the polarization behavior of metal electrodes is shown in Fig. 1 [5]. For the catodic O_2 reduction, the following row is obtained using the charge transfer resistance calculated by means of the current density (i) vs. voltage (U) curves: $R_{ct}(Pt) < R_{ct}(PtAu) < R_{ct}(PtAuRh)$. Regarding to the O_2-reduction, the PtAuRh/YSZ-electrode is ideally polarizable because the current density j is nearly zero in the range $0\ V > U > -0.22\ V$. In contrast to that a remarkable cathodic currents due to the NO reduction can be detected in this potential range shown in the differential change of j-U-curves.

Alternating Current Measurements (Impedance)

Up to now mostly the electrochemical impedance spectroscopy (► *EIS*) was used to investigate the electrochemical cell regarding the electrolyte and the electrochemical transfer reaction. If the relaxation times ($RC = \left[\frac{UIt}{IU}\right] = [t]$ has the dimension time) for single processes are different, then their separation is possible. Otherwise an overlap of semicircles can be observed. The separation may be achieved by changing of temperature because the temperature dependence for single processes is mostly different.

The Ohm's Law for alternating current can be expressed as a vector equation:

$$\vec{U} = \vec{Z} \cdot \vec{I} \tag{1}$$

For sinusoidal voltage and current are valid:

$$U(t) = U_{\max} e^{j\omega t} \tag{2}$$

$$I(t) = I_{\max} e^{j(\omega t - \varphi)} \tag{3}$$

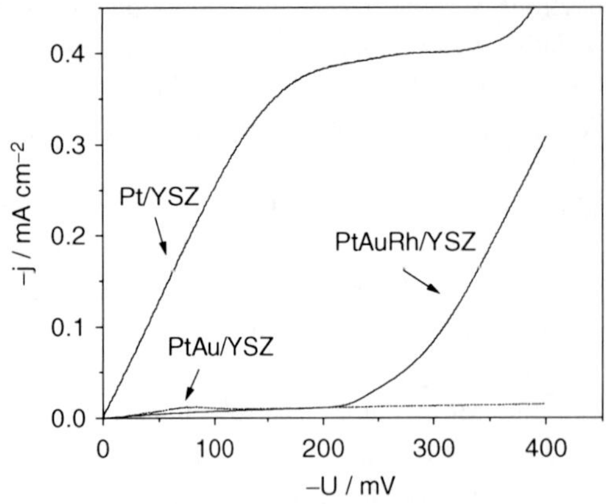

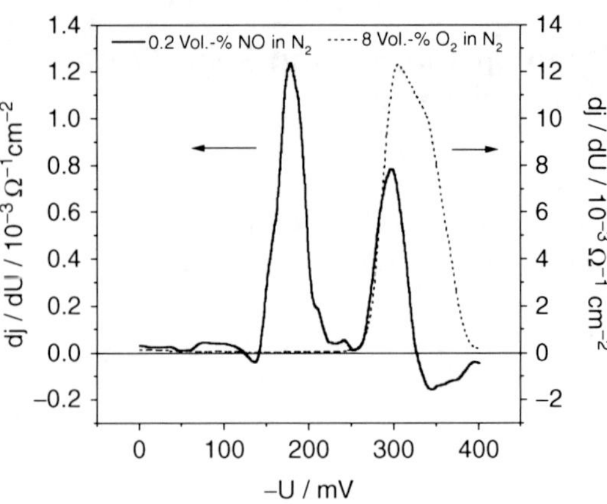

Dynamic Methods in Solid-State Electrochemistry, Fig. 1 Voltammetric investigation on different electrode materials in N_2, 1 vol.-% O_2 (*left*), dj/dU-U-curves of a PtAuRh/YSZ-electrode at 550 °C in 8 vol.-% O_2 in N_2 und 0.2 Vol.-% NO in N_2 (*right*)

where ω is angular frequency, and φ is the phase shift between voltage and current due to the capacitive effect. With the Euler's formula

$$e^{j\omega t} = \cos\omega t + j\sin\omega t \text{ and } j^2 = -1 \quad (4)$$

the impedance can be written as a complex number $\vec{Z} = a + bj$

$$\vec{Z} = |Z|\cos\varphi + j|Z|\sin\varphi \quad (5)$$

$|Z|$, $|Z|\cos\varphi$, and $|Z|\sin\varphi$ are the modulus (norm), real part, and imaginary part of the impedance $\vec{Z}$, respectively.

In Nyquist plots (imaginary part vs. real part of the impedance) for each separated physical process, a semicircle is obtained. Mostly, the imaginary part is plotted as an ordinate in positive direction because the inductivities do not play an important role in solid electrolyte cells. Ideally, the middle point of the semicircle is located on the real axis. This equates an *RC* combination. In reality, due to the inhomogeneity of the electrode, depressed semicircles are obtained with middle points below the real axis. The intercepts with the real axis and maxima are related to a resistance and capacitor. The complex impedance diagram is not unambiguous. It is useful to interpret this by means of an equivalent circuit in terms of values of physical processes [6]. In Fig. 2 the Nyquist diagram for a simple R(RC) combination is shown. For a solid electrolyte cell, this circuit gives an easy interpretation: R_1 is attributed to the solid electrolyte resistance and the circle with the maximum of the imaginary part $\omega_{(extr.\ Im(Z))} = 1/(R_2C)$ corresponds to the electrode impedance. The total resistance (second intercept) equates the dc cell resistance. Variations of the cell geometry (diameter and thickness), the cell temperature, and the gas partial pressure are proper means to verify physically the equivalent circuit. The resistance of solid electrolytes is proportional to the thickness (l) and reverse proportional to the electrode area (A), whereas the polarization resistance depends only on the electrode area (A). Furthermore results of dc techniques like steady-state current voltage measurements may help to support the model.

In Bode diagrams phase angle (phase shift between current and voltage) and the modulus of the impedance or its logarithm are plotted vs. log frequency (really spectroscopic plots). For the circuit given in Fig. 2, R_1 (high frequency) and $R_1 + R_2$ (low frequencies) are plateaus parallel to the frequency axis in modulus versus log frequency plot. In between the curve is bent in an s-shape manner. The inflection point of the curve corresponds to the minimum in the phase angle versus log frequency curve at the angular frequency $\omega_{(\text{extr.}\ \varphi)}$ ($\omega = 2\pi\upsilon$) where υ is the

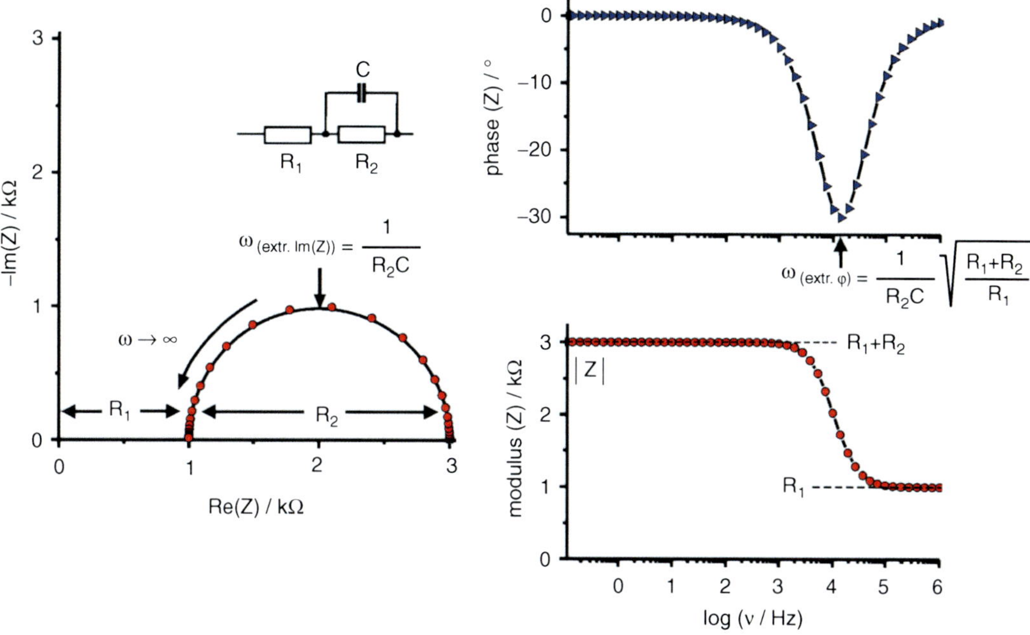

Dynamic Methods in Solid-State Electrochemistry, Fig. 2 Impedance plot for a simple $R_1(R_2C)$ combination, Nyquist diagram (*left*), Bode diagram (*right*)

frequency of the alternating voltage. At the phase angle minimum

$$\omega_{(\text{extr.}\varphi)} = \frac{1}{R_2C}\sqrt{\frac{R_1 + R_2}{R_1}} \tag{6}$$

is obtained.

For detailed investigations of electrodes, more complex equivalent circuits are necessary which are described in the literature [7, 8]. In solid-state electrochemistry mostly Nyquist impedance plots are used, which are suited to:

- Separate the bulk resistance and the grain boundary resistance in solid electrolytes
- Separate the solid electrolyte resistance from the electrode impedance
- Separate charge transfer process from transport phenomena
- Determine the rate limiting step of the electrochemical process

First, who introduced the alternating current measurements and used Nyquist plots in the solid-state electrochemistry, was Bauerle [9]. As earlier as 1967, Hartung and Möbius described

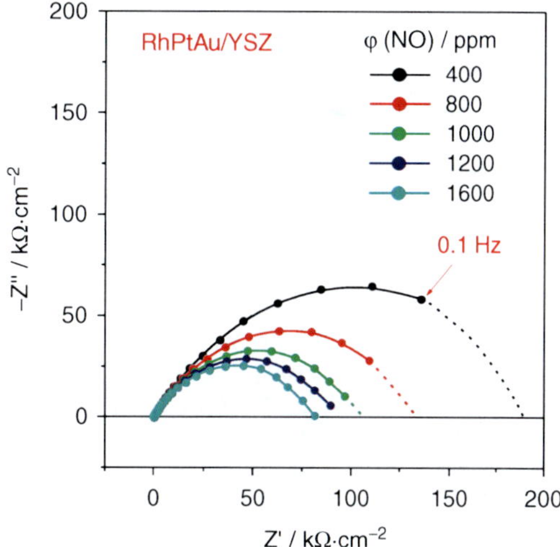

Dynamic Methods in Solid-State Electrochemistry, Fig. 3 Nyquist plots of a solid electrolyte electrode, NO, O_2, RhPtAu/YSZ at various NO concentrations [12]

polarization effect on Pt/zirconia electrodes by means of impedance measurements [10]. Now this technique is used for both the development of sensors [11] and as a sensor principle demonstrated in Fig. 3 [12]. The impedance of the solid electrolyte gas electrode, RhPtAu/YSZ, depends on both the concentrations of oxygen and nitric oxide. At (nearly fixed) O_2 only the NO sensitivity has an influence on the diameter of the semicircle. It is not necessary to measure the whole frequency range. Measurements at only a few selected frequencies are sufficient to extrapolate mathematically a semicircle and to determine the intercepts for $\omega \to 0$. Such sensors are also denoted as impedimetric (impedance) sensors or impedancemetric sensors [13].

Cross-References

► Controlled Flow Methods for Electrochemical Measurements
► Cyclic Voltammetry
► Electrochemical Impedance Spectroscopy (EIS) Applications to Sensors and Diagnostics
► Voltammetry of Adsorbed Proteins

References

1. Scholz F (2012) Dynamic techniques. In: Bard A, Inzelt G, Scholz F (eds) Electrochemical dictionary, 2nd edn. Springer, Heidelberg, pp 238–239
2. Marken F (2012) Cyclic voltammetry. In: Bard A, Inzelt G, Scholz F (eds) Electrochemical dictionary, 2nd edn. Springer, Heidelberg, pp 183–185
3. Shoemaker EL, Vogt MC, Dudek FJ (1996) Cyclic voltammetry applied to an oxygen-ion-conducting solid electrolyte as an active electro-catalytic gas sensor. Solid State Ion 92:285–292
4. Kenjo T, Yamakoshi Y, Wada K (1993) An estimation of the electrode-electrolyte contact area by linear sweep voltammetry in Pt/ZrO_2 oxygen electrodes. J Electrochem Soc 140:2151–2157
5. Schmidt-Zhang P, Zhang W, Gerlach F, Ahlborn K, Guth U (2006) Electrochemical investigations on Pt-alloy/YSZ electrodes for amperometric gas sensors (in German). Abh Sächs Akad Wiss Leipzig Math-naturw Klasse 63:121–126
6. Rickert H (1982) Electrochemistry of solids. Springer, Berlin/New York/Heidelberg
7. Barsoukov E, Macdonald JR (2005) Impedance spectroscopy theory, experiment, and application, 2nd edn. Wiley, Hoboken
8. Orazem WE, Tribollet B (2008) Electrochemical impedance spectroscopy. Wiley, Hoboken
9. Bauerle JE (1969) Study of solid electrolyte polarization by a complex admittance method. J Phys Chem 30:2657–2670
10. Hartung R, Möbius HH (1967) About alternating current polarization in platinum electrodes on oxygen ion conducting solid electrolyte (in German). Z Chem (Leipzig) 7:325
11. Matsui N (1981) Complex-impedance analysis for the development of zirconia oxygen sensors. Solid State Ion 3–4:525–529
12. Guth U, Zosel J (2004) Electrochemical solid electrolyte gas sensors – hydrocarbon and NO_x analysis in exhaust gases. Ionics 10:366–377
13. Nakatou M, Miura N (2006) Detection of propene by using new-type impedancemetric zirconia-based sensor attached with oxide sensing-electrode. Sens Actuator 120:57–62

Dynamics of Mobile Ions in Materials with Disordered Structures - the Case of Silver Iodide and the Two Universalities

Klaus Funke
Institute of Physical Chemistry, University of Muenster, Muenster, Germany

Introduction

In solid electrolytes that exhibit the key property of structural disorder, three types of nonvibrational motion of the mobile ions may be discerned. These are:

1. A "liquid-like" motion, as for instance in alpha silver iodide
2. A correlated hopping motion, which leads to macroscopic transport
3. A correlated localized motion, which creates the Nearly Constant Loss effect

By comparison, the situation is much simpler in ionic crystals with comparatively low degrees of disorder, in which the mobile point defects may be regarded as "random walkers." An example of such a material is crystalline silver bromide at 200 °C [1].

The Case of Alpha Silver Iodide

In marked contrast to silver bromide, silver iodide in its high-temperature alpha phase,

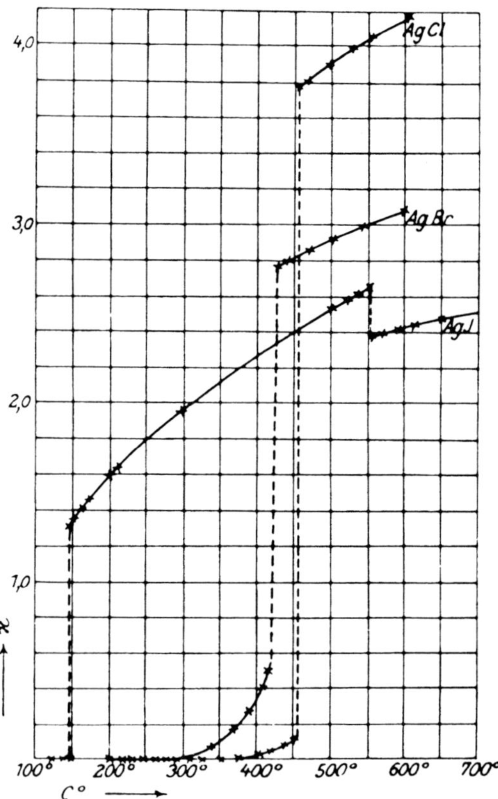

Dynamics of Mobile Ions in Materials with Disordered Structures - the Case of Silver Iodide and the Two Universalities, Fig. 1 Ionic conductivity of the silver halides, original plot of Tubandt and Lorenz [2]

α-AgI, is *structurally* disordered, see below. It is indeed the archetypal fast ion conductor. The unexpected properties of α-AgI were discovered by C. Tubandt and E. Lorenz in 1914 [2], on the occasion of their measurements of the electric conductivities of the silver halides, AgCl, AgBr, and AgI. Their original plot is reproduced in Fig. 1.

From the figure it is seen that the highly conducting α-phase of AgI is stable between 147 °C and 555 °C. At the β to α phase transition, the conductivity increases by more than three orders of magnitude up to 1.3 Ω^{-1} cm^{-1}. Within the α-phase, it increases only by a factor of two and then drops upon melting. From their measurements of transference numbers and from interdiffusion experiments, Tubandt and his coworkers concluded that the charge was carried by the cations [3]. Note that the extraordinarily high value of the ionic conductivity in α-AgI is comparable to the best conducting liquid electrolytes.

Since Tubandt's times, the silver ions in α-AgI have, therefore, been regarded as moving in a "liquid-like" fashion within the crystallographic framework provided by the anions. A first structural analysis of α-AgI was presented by L.W. Strock in 1934 and 1936 [4], who assigned a body-centered cubic (bcc) structure to the iodide sublattice, while considering as many as 42 possible positions for the two silver ions in the bcc unit cell.

In 1977, a contour map of the probability density of the silver ions in α-AgI, $\rho(\underline{r})$, was constructed on the basis of single-crystal neutron-diffraction data [5], showing flat maxima of $\rho(\underline{r})$ at the tetrahedral voids and flat minima at the octahedral positions. It may thus be concluded that the periodic potential barriers provided by the anions for the translational diffusion of the cations are only of the order of the thermal energy, which corroborates the view of a liquid-like motion.

Three years later, quasielastic neutron-scattering spectra taken on single-crystalline α-AgI were reproduced (including their anisotropy) by a model that approximated the actual liquid-like motion of the silver ions by a spatial convolution of two processes [6]. One of them was a fast diffusive motion in a local cage of about 0.1 nm radius, while the other was a random hopping via tetrahedral positions.

Remarkably, the ionic conductivity of α-AgI displays no frequency dependence up to at least 40 GHz [7]. This implies that the mobile silver ions move so fast that any memory of individual movements is erased after a time which is the inverse of $2\pi \cdot 40$ GHz, i.e., after 4 ps.

The Two Universalities

In contrast to α-AgI, structurally disordered ionic materials usually exhibit strongly frequency-dependent ionic conductivities. Examples

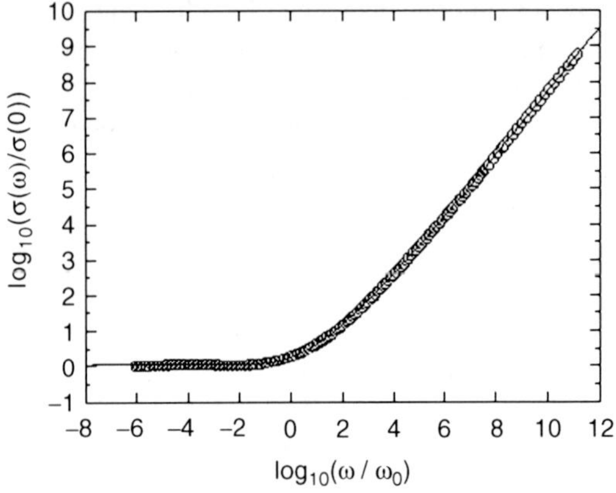

Dynamics of Mobile Ions in Materials with Disordered Structures - the Case of Silver Iodide and the Two Universalities, Fig. 2 First universality: This scaled representation [8] of experimental and model conductivities (*circles* and *solid line*, respectively, with data from 0.45 LiBr · 0.56 Li_2O · B_2O_3 glass) is characteristic of many disordered ion conductors which largely differ in their structures and compositions. Its slope increases continuously, slowly tending towards unity

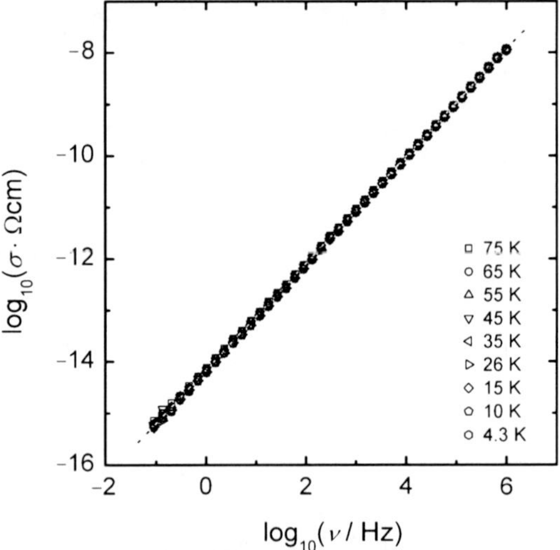

Dynamics of Mobile Ions in Materials with Disordered Structures - the Case of Silver Iodide and the Two Universalities, Fig. 3 Second universality (Nearly Constant Loss): low-temperature conductivity isotherms displaying a linear frequency dependence and essentially no temperature dependence (Data from 0.3 Na_2O · 0.7 B_2O_3 glass [11])

include crystalline, glassy, and polymeric electrolytes [8, 9] and even molten salts [9] and ionic liquids [10]. While these materials greatly differ in phase, structure, and composition, their broadband conductivity spectra are characterized by an unexpected degree of similarity. In particular, two surprising "universalities" have been detected, see Figs. 2 and 3.

The "first universality" is a fingerprint of activated hopping along interconnected sites, while the "second universality" reflects non-activated, localized movements of interacting ions. The former is observed at sufficiently high temperatures, while the other is seen at low ones, e.g., in the cryogenic temperature regime.

In disordered solid electrolytes, conductivity spectra taken at different temperatures typically show a gradual transition from one universality to the other, as seen in the example of Fig. 4.

Notably, the transition is much easier to locate on the temperature scale, when iso-frequency conductivity data are plotted versus inverse temperature, see Fig. 5 and the text further below. The crossover points marked in the two figures are identical.

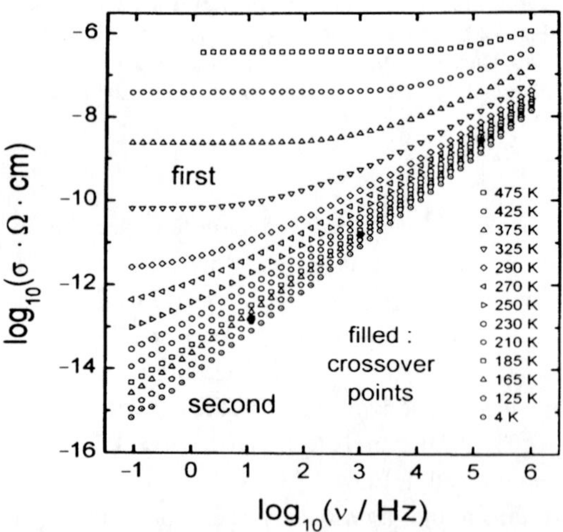

Dynamics of Mobile Ions in Materials with Disordered Structures - the Case of Silver Iodide and the Two Universalities, Fig. 4 The two universalities, here demonstrated on the basis of experimental frequency-dependent conductivities of 0.3 Na_2O · 0.7 B_2O_3 glass taken at different temperatures [12]. The three *darkened points* show the crossover between "ordinary" correlated hopping (first universality) and strictly localized ionic motion (second universality), cf. Fig. 5.

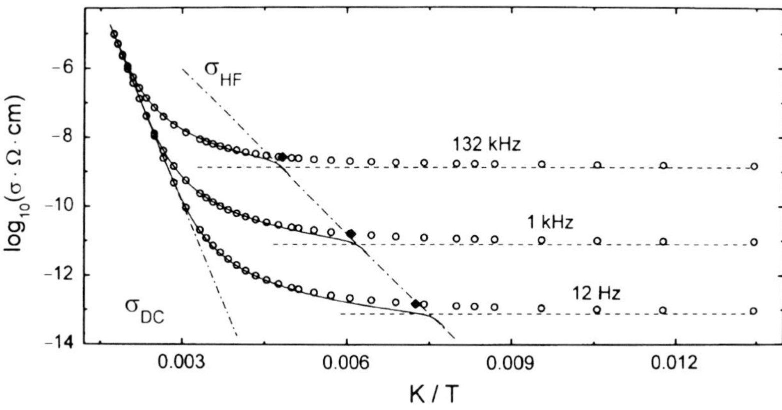

Dynamics of Mobile Ions in Materials with Disordered Structures - the Case of Silver Iodide and the Two Universalities, Fig. 5 Iso-frequency representation of log σ versus $1/T$ for 0.3 $Na_2O \cdot 0.7B_2O_3$ glass. The dash-dotted lines indicate the Arrhenius temperature dependence of the DC conductivity, σ_{DC}, and the high-frequency conductivity, σ_{HF}. The filled diamonds represent the crossover from the first to the second universality. For details, see main text and Ref. [12]

First Universality

For any fast ion conductor featuring an activated hopping of ions, a scaled master curve like the one of Fig. 2 can be constructed by suitably shifting its conductivity isotherms in a log-log plot of conductivity times temperature, $\sigma \cdot T$, versus frequency, ν. If the conduction mechanism does not change with temperature, the isotherms are found to collapse when shifted along a straight line with a slope of one. This (most frequent) case of "time-temperature superposition" [13–15] is called "Summerfield scaling" [16]. The underlying reason is obvious. The macroscopic coefficient of self-diffusion, which is proportional to DC conductivity times temperature, $\sigma_{DC} \cdot T$, varies at the same rate as the frequencies that characterize the motion of the mobile ions. The first universality now states that virtually identical master curves are obtained for a multitude of ionic materials including crystals, glasses, polymers, and ionic liquids, cf. Fig. 2.

This was first noted by A.K. Jonscher in 1975 [17, 18]. Based on his observations, he introduced his famous power-law description, $\sigma(\nu) - \sigma_{DC} \propto \nu^p$, with a constant exponent p. He also coined the terms *Universal Dielectric Response* and *Universal Dynamic Response*, often abbreviated as UDR. In the meantime, however, it has become apparent that the simple power-law approach is at variance with both experiment and theory, see, e.g., Ref. [19].

Shortly after Jonscher's findings, K.L. Ngai formulated his *Coupling Concept* [20, 21], which emphasized the importance of ion-ion interactions without violating linear response theory [22]. However, the *Coupling Concept* also features the unrealistic power-law increase of $\sigma(\nu)$, before a high-frequency plateau, σ_{HF}, is eventually attained.

In their Monte Carlo studies, A. Bunde et al. considered the effects of structural disorder and long-range Coulomb interactions systematically [23, 24]. In particular, they reproduced the frequent occurrence of correlated forward-backward hopping sequences, which are the main cause of the first-universality phenomenon. Disorder and interactions were also taken into account in the *Counterion Model* developed by W. Dieterich et al., who derived realistic spectra, $\sigma(\nu)$, from their numerical simulations [25].

In the *Random Barrier/Random Energy Models*, the non-random ion dynamics have been modeled in a formal fashion by considering individual mobile ions in random potential landscapes that are static in time [26, 27], again yielding realistic shapes of $\sigma(\nu)$.

The alternative viewpoint was to assume that the site potentials of the ions were not static, but varied in time, thus reflecting their changing momentary arrangements and interactions. This view led to the so-called Jump Relaxation Models, which are based on simple rate equations for the ion dynamics [8, 9, 28]. The solid lines included in Fig. 2 and Fig. 5 have been derived from the most recent model version, called the *MIGRATION concept* [29].

Second Universality

In 1991, A.S. Nowick and his coworkers discovered a new, "second" universality, which is ubiquitous in disordered ionic materials ("present in every plastic bag") but becomes visible only at sufficiently low temperatures and/or high frequencies [30]. The phenomenon is also called *Nearly Constant Loss* (NCL) effect, since the dielectric loss function, $\varepsilon'' \propto \sigma/\nu$, appears to be virtually independent of both frequency and temperature, cf. Fig. 3.

There is a broad consensus that the second universality reflects cooperative localized movements of a large number of ions. In ionic crystals, the effect is seen to develop from ordinary Debye relaxation as the number density of locally mobile ionic defects is drastically increased [31].

In early interpretations, static distributions of asymmetric double-well potentials (ADWP) were assumed to exist for the locally mobile ions, providing a wide range of relaxation times. This model, which was based on ideas of Pollak and Pike [32], has often been used to fit experimental data [33, 34].

The decisive step forward was made when W. Dieterich and his coworkers treated the locally mobile ions as reorienting electric dipoles, with Coulomb interactions between them [35, 36]. In their Monte Carlo simulations, they considered collections of such interacting dipoles, which were randomly distributed in space, and studied their localized reorientational movements. The resulting conductivity spectra show the following features. As frequency is increased, the slope of $\log\sigma$ versus $\log\nu$ first changes from two to one and later from one to zero. The linear NCL regime thus lies between two crossover points. The frequency range spanned by the NCL is found to increase as the ratio of Coulomb energy by thermal energy increases, that is, with increasing number density and with decreasing temperature.

Very similar results have also been obtained from a suitably modified version of the *MIGRATION concept* [9, 37], see also Ref. [8]. According to this treatment, the Coulomb interactions seem to provide rapid see-saw-type variations of the local potentials that are experienced by the individual ions, resulting in their collective localized movements [11, 37].

Cross-References

- ▶ Conductivity of Electrolytes
- ▶ Defect Chemistry in Solid State Ionic Materials
- ▶ Defects in Solids
- ▶ Dynamic Methods in Solid-State Electrochemistry
- ▶ Electrolytes, Classification
- ▶ Electrolytes, History
- ▶ Ion Mobilities
- ▶ Ionic Mobility and Diffusivity

References

1. Funke K, Lauxtermann T, Wilmer D, Bennington SM (1995) Creation and recombination of Frenkel defects in AgBr. Z Naturforsch A 50:509
2. Tubandt C, Lorenz E (1914) Molekularzustand und elektrisches Leitvermögen kristallisierter Salze. Z Physik Chem 87(513):543
3. Tubandt C (1932) Leitfähigkeit und Überführungszahlen in festen Elektrolyten. In: Wien W, Harms F (eds) Handbuch der Experimentalphysik XII, part 1. Akadem Verlagsges, Leipzig
4. Strock LW (1934) Kristallstruktur des Hochtemperatur-Jodsilbers α-AgJ. Z physik Chem B 25 (1934) 411 and B 31 (1936) 132
5. Cava RJ, Reidinger F, Wuensch BJ (1977) Single-crystal neutron-diffraction study of AgI between 23° and 300° C. Solid State Comm 24:411

6. Funke K, Höch A, Lechner RE (1980) Quasielastic neutron scattering from a single crystal of alpha silver iodide. J de Physique 41:C6–17
7. Funke K, Roemer H, Schwarz D, Unruh H-G, Luther G (1983) On the microwave conductivity of alpha silver iodide, part II: complex conductivity by measurement of the complex transmission factor. Solid State Ion 11:254
8. Funke K, Banhatti RD, Brückner S, Cramer C, Krieger C, Mandanici A, Martiny C, Ross I (2002) Ionic motion in materials with disordered structures – conductivity spectra and the concept of mismatch and relaxation. Phys Chem Chem Phys 4:3155
9. Funke K, Banhatti RD (2006) Ionic motion in materials with disordered structures. Solid State Ion 177:1551
10. Šantić A, Wrobel W, Mutke M, Banhatti RD, Funke K (2009) Frequency-dependent fluidity and conductivity of an ionic liquid. Phys Chem Chem Phys 11:5930
11. Laughman DM, Banhatti RD, Funke K (2010) New nearly constant loss feature detected in glass at low temperature. Phys Chem Chem Phys 12: 14102
12. Banhatti RD, Laughman D, Badr L, Funke K (2011) Nearly constant loss effect in sodium borate and silver meta-borate glasses: new insights. Solid State Ion 192:70
13. Taylor HE (1956) The dielectric relaxation spectrum of glass. Trans Faraday Soc 52:873
14. Isard JO (1970) Dielectric dispersion in amorphous conductors. J Non-Cryst Solids 4:357
15. Kahnt H (1991) Ionic transport in oxide glasses and frequency dependence of conductivity. Ber Bunsenges Phys Chem 95:1021
16. Summerfield S (1985) Universal low-frequency behaviour in the a.c. hopping conductivity of disordered systems. Philos Mag B 52:9
17. Jonscher AK (1975) The interpretation of non-ideal dielectric admittance and impedance diagrams. Phys Status Solidi A 32:665
18. Jonscher AK (1977) The 'universal' dielectric response. Nature 267:673
19. Funke K, Banhatti RD (2008) Translational and localized ionic motion in materials with disordered structures. Solid State Sci 10:790
20. Ngai KL (1979) Universality of low-frequency fluctuation, dissipation and relaxation properties of condensed matter, parts I and II. Comments Solid State Phys 9 (1979) 127 and 9 (1980) 141
21. Ngai KL (2003) The dynamics of ions in glasses: importance of ion-ion interactions. J Non-Cryst Solids 323:120
22. Kubo R (1957) Linear response theory of irreversible processes. J Phys Soc Jpn 12:570
23. Meyer M, Maass P, Bunde A (1993) Spin-lattice relaxation: non-Bloembergen-Purcell-Pound behavior by structural disorder and Coulomb interactions. Phys Rev Lett 71:573
24. Maass P, Meyer M, Bunde A (1995) Nonstandard relaxation behavior in ionically conducting materials. Phys Rev B 51:8164
25. Knödler D, Pendzig P, Dieterich W (1996) Ion dynamics in structurally disordered materials: effects of random Coulombic traps. Solid State Ionics 86–88:29
26. Dyre JC (1988) The random free-energy barrier model for ac conduction in disordered solids. J Appl Phys 64:2456
27. Schrøder TB, Dyre JC (2002) Computer simulations of the random barrier model. Phys Chem Chem Phys 4:3173
28. Funke K (1993) Review: Jump relaxation in solid electrolytes. Prog Solid State Chem 22:111
29. Banhatti RD, Funke K (2004) Dielectric function and localized diffusion in fast-ion conducting glasses. Solid State Ion 175:661
30. Lee W-K, Liu JF, Nowick AS (1991) Limiting behavior of ac conductivity in ionically conducting crystals and glasses: A new universality. Phys Rev Lett 67:1559
31. Nowick AS, Lim BS (2001) Electrical relaxations: simple versus complex ionic systems. Phys Rev B 63:184115
32. Pollak M, Pike GE (1972) ac conductivity of glasses. Phys Rev Lett 28:1449
33. Jain H, Krishnaswami S (1998) Composition dependence of frequency power law of ionic conductivity of glasses. Solid State Ion 105:129
34. Jain H (1999) 'Jellyfish' atom movement in inorganic glasses. Met Mater Process 11:317
35. Rinn B, Dieterich W, Maass M (1998) Stochastic modeling of ion dynamics in complex systems: dipolar effects. Philos Mag B 77:1283
36. Höhr T, Pendzig P, Dieterich W, Maass P (2002) Dynamics of disordered dipolar systems. Phys Chem Chem Phys 4:3168
37. Laughman DM, Banhatti RD, Funke K (2009) Nearly constant loss effects in borate glasses. Phys Chem Chem Phys 11:3158

D

E

Electric and Magnetic Fields Bioeffects

Marko S. Markov
Research International, Williamsville, NY, USA

Electromagnetic Fields and Life

It is well accepted now that the first primitive cell originated and further the life on the planet develops and exists in the presence of a number of physical factors including magnetic and electromagnetic fields. As a part of the Universe, the Earth has been exposed to the influence of radiation with space origin that includes ionizing and nonionizing radiation. However, the discussions of the origin and evolution of life have failed to provide documented or reachable data for the values of the magnetic field during the life-span of the planet Earth. The science now has evidence that the geomagnetic field serves as a protector against ionizing radiation and magnetic fields reaching the atmosphere, thereby protecting terrestrial life.

It should also be taken into account that during the twentieth and the twenty-first centuries, the biosphere has been exposed to increasing number and variety of electromagnetic fields related to innovations in technology, communication, transportation, home equipment, and education.

The Earth magnetosphere is determined by the Earth's magnetic field, as well as by the solar and interplanetary magnetic fields. In the magnetosphere, a mix of ozone molecules, free ions, and electrons from the Earth's ionosphere is confined by electromagnetic forces. Life on Earth developed and is sustained under the protection of this spatially and time-variable magnetosphere. Fast forward, the magnetobiology has evidence that when living creature is placed in an environment that is shielded from ambient magnetic field, some changes in the organisms are observed [1].

Space Electromagnetic Fields and Their Influence on Biosphere

To completely understand the role of space factors and especially the role of the solar magnetic field, attention should be paid to little known research of the Russian scientist Leonid Chizhevsky. Being an expert in heliophysics, biophysics, space biology, cosmobiology, and geobiology, through study of the impact of cosmic physical factors on processes in living nature, Chizhevsky found a relationship between solar activity cycles and many phenomena in the biosphere. He demonstrated that the physical fields of the Earth and its surroundings should be taken into account as being among the main factors influencing the state of the biosphere. He claimed variations of solar activity and dependent geomagnetic oscillations have impact on any type of life. Chizhevsky proposed that human history is shaped by the 11-year cycles in the Sun's activity that triggered solar magnetism and further geomagnetic storms manifesting in power

G. Kreysa et al. (eds.), *Encyclopedia of Applied Electrochemistry*, DOI 10.1007/978-1-4419-6996-5,

shortages, plane crashes, epidemics, grasshopper infestations, upheavals, and revolts [2].

Endogenous and Exogenous Magnetic Fields

The centuries of development of natural sciences provide enough evidence to claim: "Life is an electromagnetic event." Contemporary biology knows that all physiological processes are performed with movement of electrical charges, ions, and dipoles within the cell interior, through the plasma membranes and in communication between different cells. The movement of charges generates electric currents, and as result magnetic fields could originate within living tissues. These magnetic fields are commonly classified as endogenous fields.

On the other hand, every magnetic field generated outside the biological system is exogenous field. Usually these are fields connected with physical means of generation. It includes solar and terrestrial magnetic fields, as well as electromagnetic fields generated by industrial and communication systems. In this category, however, may fall some fields with biological origin. For example, the magnetic field created by one organ in human body would be endogenous to another organ.

The necessity to distinguish endogenous and exogenous magnetic fields is also related to the use of two terms: biomagnetism and magnetobiology, which mistakenly are used as synonyms. The semantics of the words, however, should suggest the difference. Biomagnetism is an area of science which deals with magnetic fields generated by biological systems, while magnetobiology studies the effects of exogenous magnetic fields when applied to biological objects.

Hazard from EMF

Very often the news media discuss how dangerous electromagnetic/magnetic field might be for human and environmental health, especially in relation to cancer initiation. The hazard should be considered in respect to the continuous exposure to electromagnetic fields in workplace and/or occupational conditions, while at the same time short, controlled exposure to specific electromagnetic fields makes possible therapeutic benefit.

The hazard issue in the western scientific community has been discussed during the last half a century, beginning with the power-line electromagnetic fields and continuing with wireless communications. Three very important features must be pointed out:

Every evaluation of the "hazard" as well as every standard for the permissible level of exposure should be done following the precautionary principle: If we do not know that a given food, drink, medication, physical, or chemical factor is safe, we should treat it as potentially hazardous.

Having this in mind, the evaluation and prediction of the potential adverse effects from using wireless communications (any mobile device, as well as laptops), especially for children, becomes a question of crucial importance.

So far, the potential hazard from electromagnetic fields is discussed in regard to the human population. However, especially with satellite and wireless communication, all living creatures in the biosphere are exposed to electromagnetic fields.

The hazard issue is frequently represented as "controversial." It is not controversial; it is conflict of interest of industry on one side and mankind and environment on the other. It is remarkable that IARC (International Agency of Research on Cancer) in the summer of 2011 classified radiofrequency EMF as possible cancerogene.

Benefits of Clinical Application of EMF

The human history provides numbers of evidence for benefit obtained by using various magnetic and electromagnetic fields. Despite the fact that magnetic field use for treatment of various health problems has long and widespread history, the western medicine is still skeptical and reluctant to accept the magnetotherapy as an effective, even complimentary, method for helping patients in cases when pharmaceutical or other therapies failed.

The application of magnetic fields for treating specific medical problems such as arthritis, fracture unification, chronic pain, wound healing, insomnia, headache, and others has steadily increased during the last decades. In contrary to pharmacotherapy, magnetotherapy provides noninvasive, safe, and easily applied methods to **directly** treat the site of injury, the source of pain, and inflammation.

There is a large body of basic science and clinical evidence that time-varying magnetic fields can modulate molecular, cellular, and tissue function in a physiologically and clinically significant manner, most recently summarized in several books and review articles [3–7].

Mechanisms of Detection and Response to EMF

The fundamental question for engineers, scientists, and clinicians is to identify the biochemical and biophysical conditions under which applied magnetic fields could be recognized by cells in order to further modulate cell and tissue functioning. It is also important for the scientific and medical communities to comprehend that different magnetic fields applied to different tissues could cause different effects. Nevertheless, hundreds of studies had been performed in search of one unique mechanism of action of magnetic field on living systems. It would be fair to say that these efforts are determined to fail. Why? For the same reason that during the millions of years of evolution of life, enormous number of different living systems originated. It is difficult to believe that the same response will be seen at bone and soft tissue, at elephant and butterfly, and at microorganism and buffalo. Biology knows that the geographical and climate conditions created genetic and physiological differences in the organisms from the same species.

The problem of mechanisms of interactions might be discussed from different points of view, engineering and physics, biology, and medicine. More plausible is to follow the signal-transduction cascade that postulates that in any biological system the modifications that may occur as a result of the influence of the applied magnetic field on structures such as cellular membrane or specific proteins, conformational changes, and/or charge redistribution could be initiated and by signal-transduction mechanism can be spread over the cell or tissue. Discussing the theoretical feasibility of radical-pair mechanism, Eichwald and Walleczek [8] affirmed that this model is capable of accounting for bioelectromagnetic phenomena which depend on the frequency in a nonlinear, resonance-like fashion (frequency window), field amplitude (amplitude window), the combination of appropriate AC and DC fields, and the biodynamic state of the biological system exposed to EMF.

Does Threshold Exist?

The research of effects of electromagnetic field is going in parallel with studying the effects of ionizing radiation. The common denominator of ionizing and nonionizing radiation is exactly the word "radiation." As basic physics teaches, radiation constitutes energy, and for that reason the energy interactions with any physical or biological body are connected with damage or heating of the body when the intensity of the radiation is above certain threshold level. For decades the thermal effects in bioelectromagnetics have been the subject of intensive discussions, related to specific absorption rate (SAR) as useful criteria. It is clear that SAR requires a threshold value determination.

However, hundreds of studies and publications reported biological and clinical effects at low-intensity and low-frequency electromagnetic fields, as well as at static magnetic fields. At these interactions it is very unlikely, or even impossible, to expect thermal effects, and the threshold level approach is not reasonable. Several other mechanisms of interactions as ion-pair and free radical formation, heat shock proteins, and calcium-calmodulin interactions have been proposed. One of the most reasonable hypotheses is "window hypothesis" to be discussed next.

Window Hypothesis

The concept of "biological windows" was introduced in attempt to demonstrate that during evolution Mother Nature created preferable levels of recognition of the signals from exogenous magnetic fields. The "biological windows" could be identified by amplitude, frequency, and their combinations [9–11]. The research in this direction requires assessment of the response in a range of amplitudes and frequencies. While the publication of Ross Adey group was on frequency resonance and the other two groups are reporting amplitude resonance, the world window became slowly accepted to identify that certain biological systems or tissues require that applied magnetic field should be with defined values in order to achieve optimal response. It is important – the biological response may be detected outside the window, but it will be smaller than at resonance/window level. If we look in another way – the amplitude or frequency window is the naturally selected optimization of electromagnetic field stimulation.

The pioneering work of Ross Adey's group identified the frequency window for calcium ion, and further ion cyclotron resonance method proposed by Abe Liboff [12] identified resonance frequencies for a number of biologically important ions.

It has been also shown that at least 3 amplitude windows exist: at 50–10,000 μT (0.5–10 G), 15–20 mT (150–200 G), and 45–50 mT (450–500 G). The best review could be found elsewhere [13].

The suggested existence of specific "permitted" states which biosystems could attain under the action of selected magnetic fields, having define physical parameters, more likely are related to the informational status of the system, manifested as defined conformational state of important proteins.

When the frequency/amplitude/information is adequate to that necessary for the transition/conformational change, the system may achieve a new "stationary" state at which it can remain for a certain period of time. Any other frequency/amplitude/information (lower or larger) would bring the system to a state different than the stationary one which is unstable and therefore the effects would be smaller and would quickly disappear.

One very recent conformation of existence of biological windows has been published recently: Yuan et al. [14] reported window effects in studying magnetic field-induced angiogenesis and improvement in cardiac function in rats.

In evaluating the hazard and/or benefit of EMF interactions, it is very important to avoid generalizing statements based upon studying one or even several signals. This often happens in publications that report specific experiment/clinical trial or even in review papers. By not saying that some or selected PEMF could initiate hazardous or plausible therapeutic effects, we simply lead everyone to think that "**all**" magnetic fields could achieve the goals.

The art here is that the identified "windows of opportunities" that exist naturally could be used for treatment of specific medical problems. Once again, the magnetic field with specific physical parameters might be most successful for treatment of specific medical problems. The suggested existence of specific "permitted" levels which biosystems could attain under the magnetic field stimulation is critical for selecting the appropriate device and modes of therapy.

Necessity of Accurate Dosimetry at the Target Site

To accurately evaluate the potential of EMF, one should correctly identify the physical parameters of the applied fields. There are general advantages for the evaluation of the effects of static and low-frequency magnetic field, related to fact that magnetic properties of the living tissues at low frequency are similar to the properties of the air. This allows the development of physical and biophysical dosimetry. The physical dosimetry relates to characterization of the magnetic device in respect to engineering and physics parameters, not the actual field distribution inside the target tissue. Biophysical dosimetry is accounting for the "dose" received by the target tissue. From biological and clinical viewpoint the only valuable information is that which describes the field at the target site.

It has been shown [15] that any study of MF action on a particular biological system has to consider a number of parameters:

- Type of magnetic field
- Magnetic field intensity or induction
- Component (electric or magnetic)
- Spatial vector (dB/dx)
- Temporary gradient (dB/dt)
- Pulse shape
- Frequency and repetition time
- Localization
- Time of exposure
- Depth of penetration

Electroporation

It is already 40 years since the first, theoretical papers have been published on the possibility of high-frequency EMF to initiate reversible pores in plasma membranes. The recent advances in theoretical and experimental work, as well as application with therapeutic purposes, were summarized by Pakhomov et al. [16]. This method allows transport through the membrane of small ions and large molecules, which is otherwise impossible. The electric field required to achieve electroporation depends on duration of the pulse and the amplitude of the applied electric field.

In general, biological membrane is not only a separator of the cell interior from surrounding media but a "transporter" of material, energy, and information. Therefore, biological membrane represents a powerful amplifier in the signal transduction.

Electroporation has been studied in different systems as artificial lipid bilayers, lipid vesicles, and animal and plant cells, and the effect has been considered at the level of plasma membranes. During the last two decades nanoelectroporation has been developed. The extremely short pulses applied in this approach allow the "poration" to occur at the level of subcellular structures.

Contemporary "classical" and "nano" electroporation offers new avenues in clinical medicine by providing conditions for electrochemotherapy, gene transfer, and other modalities to treat problems within cell interior. As one may expect, the largest interest and application is in the cancer therapy.

Future Perspectives

The electric, magnetic, and electromagnetic fields became a serious factor of everyday life of all living systems, starting from occupational and everyday life of human population, as well as for the entire biosphere. While considering environmental impact of these factors, science must dedicate serious attention to potential benefit of using EMF for human health.

Obviously, the science and technology of coming decades will be developing in two distinct directions. On one hand, the assessment and prediction of the potential hazard of the power-line low-frequency electromagnetic fields and high-frequency wireless communication fields should be developed in coordination between public health and industry interests. No doubt that the future development of the diagnostic and therapeutic modalities would provide medicine and public health with new devices and protocols for their use.

Cross-References

► Biomolecules in Electric Fields

References

1. Pavlovich SA (1975) Shielding magnetic fields may cause mutations in microorganisms. Nauka, Moscow, 98 p. (in Russian)
2. Chizhevsky L (1976) The terrestrial echo of solar storms. Nauka, Moscow, 366 pp
3. Markov MS (2002) Can magnetic and electromagnetic fields be used for pain relief? Bull Am Pain Soc 12(1):3–7
4. Shupak N (2003) Therapeutic uses of pulsed magnetic-field exposure: a review. Radio Sci Bull 307:9–32
5. Rosch PJ, Markov MS (eds) (2004) Bioelectromagnetic medicine. Marcel Dekker, New York, 850 pp
6. Barnes F, Greenebaum B (eds) (2007) Handbook of biological effects of electromagnetic fields, 3rd edn. CRC Press, Boca Raton

7. Lin J (ed) (2011) Electromagnetic fields in biological systems. CRC Press, Boca Raton, p 381
8. Eichwald C, Walleczek J (2000) Model for magnetic field effects on radical pair recombination in enzyme kinetics. Science 287(5451):273–278
9. Bawin SM, Adey WR (1976) Sensitivity of calcium binding in cerebral tissue to weak environmental electric fields oscillating at low frequency. Proc Natl Acad Sci USA 73:1999–2003
10. Ukolova MA, Kvakina EB, Garkavi LH (1975) Stages of magnetic field action. In: Problems of action of magnetic fields on biological systems, vol 1. Nauka, Moscow, pp 57–71
11. Markov MS, Todorov SI, Ratcheva MR (1975) Biomagnetic effects of the constant magnetic field action on water and physiological activity. In: Jensen K, Vassileva Y (eds) Physical bases of biological information transfer. Plenum Press, New York, pp 441–445
12. Liboff AR (1985) Cyclotron resonance in membrane transport. In: Chiabrera A, Nicolini C, Schwan HP (eds) Interactions between electromagnetic fields and cells. Plenum Press, New York, pp 281–296
13. Markov MS (2005) "Biological Windows": a tribute to W. Ross Adey. The Environmentalist 25:67–74
14. Yuan Y, Wei L, Li F, Guo W, Li W, Lv A, Wang H (2010) Pulsed magnetic field induces angiogenesis and improves cardiac function of surgically induced infarcted myocardium in Sprague–Dawley rats. Cardiology 117:57–63
15. Markov MS (1994) Biological effects of extremely low frequency magnetic fields. In: Ueno S (ed) Biomagnetic stimulation. Plenum Press, New York, pp 91–102
16. Pakhomov A, Miklavcic D, Markov M (eds) (2010) Advanced electroporation techniques in biology and medicine. CRC Press, Boca Raton, 507 pp

Electrical Double-Layer Capacitors (EDLC)

Masanobu Chiku
Department of Applied Chemistry, Graduate School of Engineering, Osaka Prefecture University, Osaka, Japan

Introduction

At the interface between an electrode as an electronic conductor and electrolyte as an ionic conductor, a layer of ions confronts a layer of opposite and equal electronic charges. These two layers of charge, which are called an electrical double layer, constitute a capacitor. Because the separation of the layers is atomically small, the capacitance of an electrical double layer is huge. Electrical double-layer capacitors (EDLCs) are energy storage devices which utilize the electric charge of the electrical double layer.

EDLC consists of a pair of electrodes which are called the positive and negative electrodes. The positive charges are stored on the positive electrode, and anions in the electrolyte adsorb on the electrode surface. On the other hand, the opposite phenomenon takes place at the negative electrode.

Basic Principle

Figure 1a shows a basic construction of an EDLC which have two high surface area electrodes such as activated carbon and acidic aqueous electrolytic solution. EDLCs are usually operated with cell voltages below 1.2 V to prevent water decomposition. Immediately after a voltage is applied, an EDLC generates an electric field between positive and negative electrodes (Fig. 1b) and then cations and anions move to opposite directions and large current flow. The cations and anions are concentrated near the negative and positive electrode surfaces, respectively, and accumulated on the electrode surface due to the lack of any faradaic reaction. Large voltage develops at the electrode/electrolyte interface and also large electric fields are generated only at the interface (Fig. 1c). This process finish within $10 \sim 10^2$ ms with the electrolyte solution containing enough amount of electrolyte. Finally, the voltage applied to EDLC cells is approximately equivalent to that at the electrode/electrolyte interface (Fig. 1d).

EDLCs Configuration

Electrode

The highest capacitance of an electrical double layer is generated on the metal surface which is

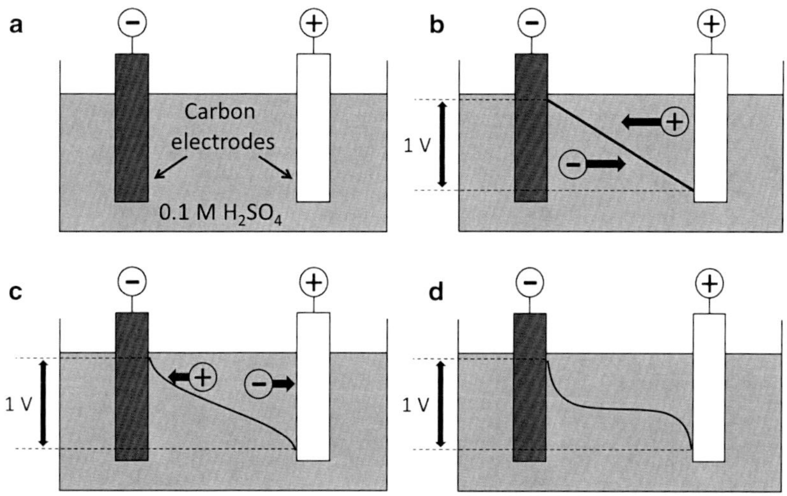

Electrical Double-Layer Capacitors (EDLC), Fig. 1 A model of charging EDLC

clean and flat at the atomic level. The mercury surface is proximate to realize the highest capacitance. The capacitance of mercury surface is about 200 ~ 350 mF m^{-2}. Naturally, the actual electrode surface has less electrical double-layer capacitance than the mercury one. The electrode material which is used as practical devices has large electrode/electrolyte interface to increase the capacitance of EDLCs. From this point of view, most EDLCs use activated carbon as the electrode material. Activated carbon has an extremely high surface area of about 800 ~ 2,500 $m^2\ g^{-1}$. EDLCs using organic electrolyte with quaternary alkylammonium salt have a capacitance about 60 ~ 80 F g^{-1}, and aqueous electrolyte like sulfuric acid solution realizes the capacitance of about 120 F g^{-1}.

Except the part-like activated carbon which directly generates electrical double layer, some ingredients are included in the electrode. For example, the binder material is used to maintain the shape of electrodes. In general, polytetrafluoroethylene (PTFE) and polyvinylidene fluoride (PVdF) are used as binder materials. PTFE has a fiber-like form and disperses to cover the activated carbon surface. PVdF disperses to form domain structure and hold the activated carbon like glue. Generally these binder materials are used about 5 ~ 20 wt% of electrode. The conductive auxiliary agents like acetylene black, Ketjenblack, carbon nanofiber, and carbon nanotube are also used about 1 ~ 10 wt% of electrode.

Electrolyte

Several electrolytes are used for organic solvents. Ammonium, phosphonium, and imidazolium salts are used for cation, while tetrafluoroborate, hexafluorophosphate, and bis (fluoroalkanesulfonyl) imide are used for anion. Aqueous solutions are also used for EDLCs. Electrolytes for aqueous solution are sulfonic acid as strong acid solution, potassium hydroxide as strong base solution, and several strong electrolytes as neutral pH solution.

Solvents

Solvents for EDLCs are divided into two categories: organic solvents and aqueous solvents. Organic solvents are required to have high dielectric constant to accelerate dissociation of electrolyte and low viscosity to improve electrical conductivity. Mixed solvents are also used for EDLCs. In general, the organic solvents used for rechargeable lithium ion batteries like propylene carbonate, ethylene carbonate, diethyl carbonate, and γ-butyrolactone. For example, adding ethylene carbonate to propylene carbonate increases dielectric constant. About 30 ~ 50 wt% sulfuric acid and about 20 ~ 30 wt%

potassium hydroxide solutions are typical electrolytes with high electric conductivity. Aqueous EDLCs can operate with cell voltages up to ~1.5 V because of the limitation of water decomposition potential, but they have several advantages like no flammability, low cost, and high electrical conductivity.

Separators

Separators are inserted between the positive and negative electrodes for preventing a short circuit. In addition, separators are required to permeate electrolyte solutions. Cellulosic paper or grass fibers are often used as a separator for EDLCs.

Future Perspectives

There already exist the commercial EDLCs manufactured by several companies and almost all of them use activated carbon electrodes. Several carbon materials are investigated by researchers. For example, the activated carbon material, Alonso et al. suggested activated carbon derived from a highly functionalized pitch showed 400 F g^{-1} capacity with KOH aqueous solution [1]. Carbon nanotubes, carbon aerogels, and graphenes are prospective candidates for EDLC active materials. Bordjiba et al. introduced their carbon nanotube-carbon aerogel composite electrode as high capacitance electrode for EDLC and it showed 524 F g^{-1} capacity in KOH aqueous electrolyte [2]. Wang et al. suggested the capability of graphene materials for elecfhightrode with 205 F g^{-1} capacity in KOH aqueous electrolyte [3]. It would need some more innovation for the practical use of these high-performance electrode materials.

Cross-References

► Electrolytes for Electrochemical Double Layer Capacitors

References

1. Alonso A, Ruiz V, Blanco C, Santamaria R, Granda M, Menedez R, de Jager SGE (2006) Activated carbon produced from sasol-lurgi gasifier pitch and its application as electrodes in supercapacitors. Carbon 44:441–446
2. Bordjiba T, Mohamedi M, Dao LH (2008) New class of carbon-nanotube aerogel electrodes for electrochemical power sources. Adv Mater 20:815–819
3. Wang Y, Shi Z, Huang Y, Ma Y, Wang C, Chen M, Chen Y (2009) Supercapacitor devices based on graphene materials. J Phys Chem C 113:13103–13107

Electroauxiliary

Jun-ichi Yosida
Kyoto University, Kyoto, Japan

Introduction

Electrochemical processes serve as powerful methods for making and breaking chemical bonds in organic synthesis. In order to achieve electrochemical reactions in a selective manner, the following two points should be critical:

1. Electron transfer should occur selectively at the position in a substrate molecule that is needed for the subsequent chemical process.
2. The subsequent chemical process should occur selectively to cleave the specific bond or make a bond in the specific position.

Methods using functional groups that control the reactivity of substrate molecules and reaction pathways are often used in organic synthesis. They are quite effective for driving otherwise difficult reactions and controlling reaction pathways to obtain the desired products selectively. A method for such control has also been developed in electrochemical organic synthesis, and such functional groups are called electroauxiliaries [1].

Definition of Electroauxiliary

Electroauxiliaries are defined as functional groups that promote the electron transfer and control the reaction pathways of electrochemically generated reactive species to give desired products selectively. Because synthetic applications of electroauxiliaries in cathodic reduction are rather rare, only the electroauxiliaries for anodic oxidation are discussed here.

Principles of Electroauxiliaries

From a viewpoint of molecular orbitals, the oxidation process is explained by electron transfer from the highest occupied molecular orbital (HOMO) of a substrate molecule to the anode. Therefore, an increase in the HOMO level is the most straightforward method for activating the substrate toward anodic oxidation. When the substrate that we wish to oxidize has a similar or lower HOMO level than that of the other species, it is, in principle, very difficult or impossible to accomplish selective oxidation of the substrate. In this case, selective increase of the HOMO level of the substrate that we wish to oxidize by the introduction of an electroauxiliary group serves as a powerful method for accomplishing the desired selective oxidation. In the case where several sites are susceptible to oxidation in a single molecule, a similar argument can be applied, and the use of an electroauxiliary to activate a particular site that we wish to oxidize is effective for selective oxidation.

The orbital interaction is quite effective for producing activation by increasing the HOMO level. In principle, the interaction of the HOMO with a high-energy filled orbital increases its energy level according to the theory of orbital interaction. For example, the energy level of a C-Si σ-orbital is usually much higher than that of C-H and C-C σ orbitals. Therefore, the C-Si σ orbital interacts effectively with a nonbonding *p*-orbital of a heteroatom such as N, O, and S, which is often the HOMO of a heteroatom compound (σ-n interaction), if two orbitals align in the same plane (Fig. 1) [2, 3]. Therefore, a silyl group at the α-position activates a heteroatom compound toward anodic oxidation. One electron oxidation gives the radical cation, in which the C-Si σ orbital interacts with a half-vacant *p*-orbital of the heteroatom to stabilize the system. Such interaction also weakens the C-Si bond, and therefore, the C-Si bond is cleaved selectively without affecting the C-H or C-C bond. The resulting carbon radical undergoes further oxidation to give the carbocation, which is trapped by a nucleophile to give the desired product Eq. 1. Therefore, the silyl group not only activates substrates toward oxidation but also controls the reaction pathway. The interaction of the C-Si σ orbital is also effective for raising the energy level of the adjacent π-systems (σ-π interaction), and the C-Si bond is cleaved selectively without affecting allylic or benzylic C-H and C-C bonds. Therefore, the silyl group serves as an electroauxiliary for oxidation of the π-systems. A stannyl group also serves as an electroauxiliary for the electrochemical oxidation of heteroatom compounds and π-systems.

R_3Si, R, Y $\xrightarrow{-e}$ [R_3Si, R, Y]$^{\bullet +}$ $\xrightarrow{-\,{}^{+}SiR_3}$ R, •, Y $\xrightarrow{-e}$ R, +, Y $\xrightarrow{NuH}$ Nu, R, Y (1)

Y = N, O, S, π-system

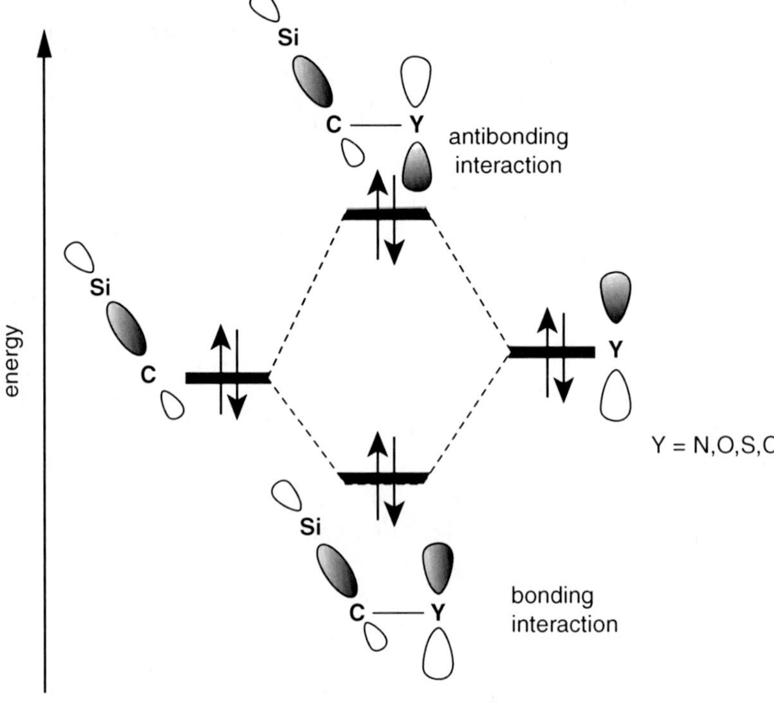

Electroauxiliary, Fig. 1 Interaction between the C–Si σ-orbital and the orbital of Y to increase the HOMO level

The introduction of an arylthio (ArS) group at an α-position of a heteroatom compound containing N or O causes a significant decrease in the oxidation potential, because the energy level of the S *p*-orbital is usually much higher than that of N and O *p*-orbital. In other words, the S *p*-orbital becomes the HOMO of the compound. One electron oxidation of such compounds leads to the formation of the radical cation, in which the C-S bond is cleaved selectively to generate a carbocation adjacent to the heteroatom Eq. 2. Therefore, the ArS group serves as an electroauxiliary for the oxidation of heteroatom compounds containing N and O, although the principle is completely different from the case of the silyl electroauxiliary [4]. One of the advantages of the ArS electroauxiliary is the easy tuning of the activity (or the oxidation potential) by changing the substituent on the aryl group. For example, the introduction of an electron donating group on the aryl group such as a methoxy group causes a significant decrease in the oxidation potential.

$$\mathrm{ArS(R)CH{-}Y} \xrightarrow{-e} \left[\mathrm{ArS(R)CH{-}Y}\right]^{\bullet+} \xrightarrow{-\mathrm{ArS}\bullet} \mathrm{R\overset{+}{C}H{-}Y} \xrightarrow{\mathrm{NuH}} \mathrm{Nu(R)CH{-}Y} \quad (2)$$

Y = N, O, π-system

Oxidative electron transfer in solution is generally assisted by the stabilization of the resulting radical cation by the coordination of solvent molecules or ions that are present in the solution. However, if a substrate molecule has a specific coordinating site to stabilize the developing charge, the electron transfer should be assisted by intramolecular coordination. Such coordination also facilitates subsequent chemical processes such as fragmentation. The pyridyl group serves as an effective coordinating group for the oxidation of compounds containing heteroatoms and, therefore, functions as an electroauxiliary [5].

Applications of Electroauxiliaries

Some typical examples of the use of electroauxiliaries are shown below, although a rich variety of examples are reported in the literature.

The electrochemical methoxylation of unsymmetrically substituted carbamates usually leads to formation of a mixture of two regioisomeric products Eq 3. However, the introduction of a silyl group as an electroauxiliary controls the reaction pathway to bias the formation of the desired product Eq. 4 [6]. It is also noteworthy that the oxidation potential of a silyl-substituted carbamate is much less positive than that of the parent carbamate.

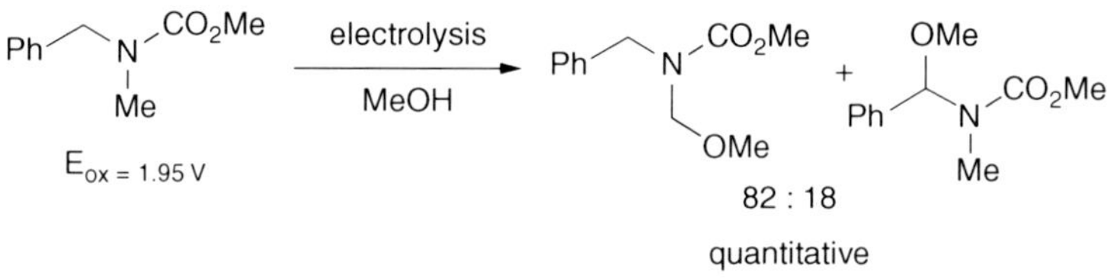

Ph N CO_2Me Me — electrolysis, MeOH → Ph N CO_2Me OMe + OMe Ph N CO_2Me Me (3)

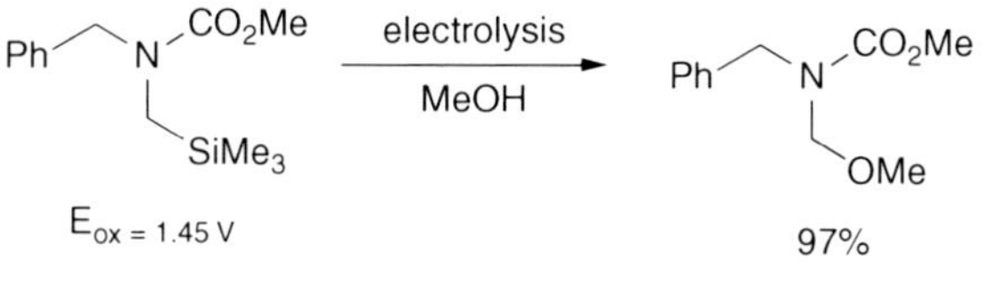

E_{ox} = 1.95 V; 82 : 18, quantitative

Ph N CO_2Me $SiMe_3$ — electrolysis, MeOH → Ph N CO_2Me OMe (4)

E_{ox} = 1.45 V; 97%

The concept of electroauxiliary has been effectively used to construct functionalized peptidomimetics Eq. 5. A dipeptide analogue containing the $SiMe_2Ph$ group is electrochemically oxidized to form the desired methoxylated product, which is allowed to react with $BF_3{\cdot}OEt_2$ to obtain the bicyclic compound [7]. The silyl group is also effective for regioselective functionalization of β-lactams by anodic oxidation Eq. 6 [8, 9]. An alkoxyl group is selectively introduced on the carbon to which the silyl group has been attached, whereas the benzylic carbon is oxidized in the absence of the silyl group. Cyanide anion is also effective as a nucleophile in electroauxiliary-assisted electrochemical oxidation of nitrogen-containing compounds. For example, the anodic oxidation of cyclic α-silyl amines in the presence of sodium cyanide gives rise to effective cyanation under mild conditions Eq. 7 [10].

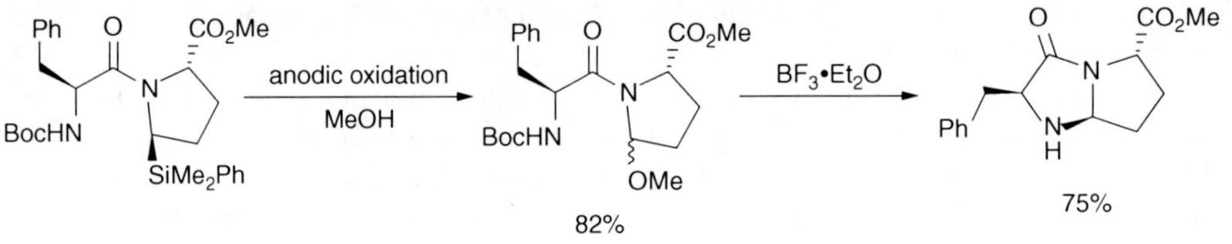

(5)

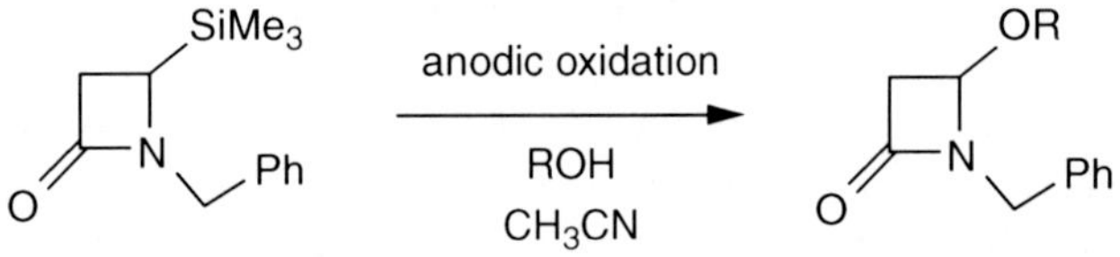

(6)

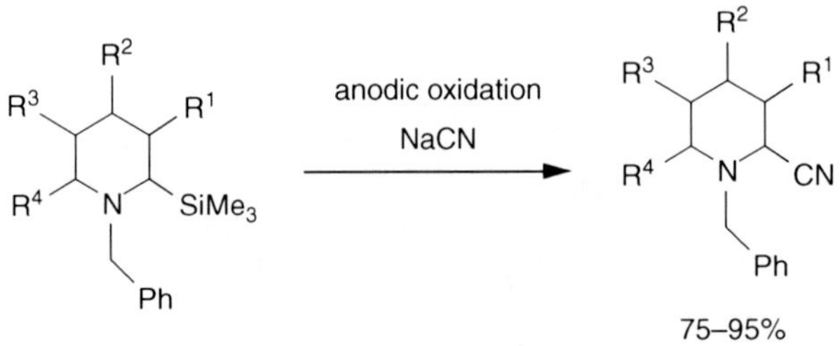

(7)

The use of an ArS group as electroauxiliaries expands the scope of nucleophiles, the in situ use of carbon nucleophiles. Thus, ArS-substituted carbamates can be anodically oxidized in the presence of allyltrimethylsilane to give the corresponding allylated products directly Eq. 8 [11].

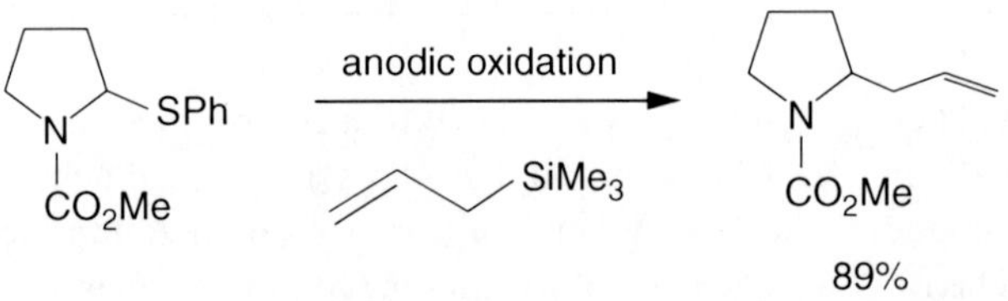

(8)

Aliphatic ethers are usually difficult to oxidize because their oxidation potentials are very high. The use of an electroauxiliary enables the anodic oxidation of aliphatic ethers under mild conditions. For example, the anodic oxidation of α-silyl ethers in an alcohol as solvent leads to the formation of acetals via selective C-Si bond cleavage. Similar to the carbamate cases, α-stannyl- and ArS-substituted ethers have lower oxidation potentials than those of α-silyl ethers. Therefore, carbon-carbon bond formation using carbon nucleophiles such as allylsilanes can be achieved.

Intramolecular carbon-carbon bond formation has also been achieved by the anodic oxidation of molecules containing an electroauxiliary and a tethered olefin Eq. 9 [12]. Fluoride ion derived from BF_4 is introduced to the cyclized cation to give cyclic fluorides.

EA anodic oxidation Bu_4NBF_4 (9)

EA=			
$SiMe_3$	68%	(55 :	45)
$SnBu_3$	98%	(74 :	26)
SMe	64%	(87 :	13)

A silyl group serves as an effective electroauxiliary for allylic oxidation of alkenes Eq. 10. The electron transfer takes place only for the allylsilane moiety without affecting the other carbon-carbon double bond, because the silyl group activates the neighboring carbon-carbon double bond. In the radical cation thus generated, the C-Si bond is cleaved selectively without affecting other allylic C-H bonds to generate the corresponding allylic radical, which is further oxidized to give the allylic cation in a selective manner. The attack of a nucleophile such as an alcohol, water, or acetic acid gives the final product as a mixture of two regioisomers [13, 14].

anodic oxidation NuH (10)

NuH =	
MeOH	69% (68:32)
EtOH	56% (63:37)
H_2O	62% (60:40)
AcOH	26% (67:37)

The silyl group and ArS groups are effective for the oxidation of the benzylic position. The anodic oxidation of benzylic sulfides in the presence of allylsilanes takes place smoothly, giving rise to selective C-S bond cleavage and introduction of an allyl group on the benzylic carbon Eq. 11 [15].

anodic oxidation, $SiMe_3$; 75% (11)

Future Directions

The concept of electroauxiliary enables selective electron transfer and precise control of pathways of subsequent reactions. It is hoped that the use of electroauxiliaries provides a solution to the problems of conventional electrochemical processes and serves as a powerful method for electrochemical organic synthesis. Hopefully, a wide range of electroauxiliaries based on different principles including electroauxiliaries for cathodic reduction will be exploited and will work together to meet the great demands for electrochemical organic synthesis in the near future.

Cross-References

- ► Anodic Substitutions
- ► Cation-Pool Method
- ► Electrosynthesis Using Template-Directed Methods

References

1. Yoshida J, Kataoka K, Horcajada R, Nagaki A (2008) Modern strategies in electroorganic synthesis. Chem Rev 108:2265–2299
2. Yoshida J, Nishiwaki K (1998) Redox selective reactions of organosilicon and -tin compounds. J Chem Soc Dalton Trans 2589–2596
3. Yoshida J, Maekawa T, Murata T, Matsunaga S, Isoe S (1990) The origin of β-silicon effect in electron-transfer reactions of silicon-substituted heteroatom compounds. Electrochemical and theoretical studies. J Am Chem Soc 112: 1962–1970
4. Yoshida J, Sugawara M, Tatsumi M, Kise N (1998) Electrooxidative inter- and intramolecular carbon-carbon bond formation using organothio groups as electroauxiliaries. J Org Chem 63:5950–5961
5. Yoshida J, Izawa M (1997) Intramolecular assistance of electron transfer. Oxidative cleavage of the carbon-tin bond of tetraalylstannanes. J Am Chem Soc 119:9361–9365
6. Yoshida J, Isoe S (1987) Electrochemical oxidation of α-silylcarbamates. Tetrahedron Lett 28: 6621–6624
7. Sun H, Moeller KD (2002) Silyl-substituted amino acids: new routes to the construction of selectively functionalized peptidomimetics. Org Lett 4:1547–1550
8. Suda K, Hotoda K, Watanabe J, Shiozawa K, Takanami T (1992) An efficient and regioselective preparation of 4-oxyazetidin-2-ones from 4-trimethylsilylazetidin-2-ones by use of anodic oxidation. J Chem Soc Perkin Trans 1:1283–1284
9. Fuchigami T, Tetsu M, Tajima T, Ishii H (2001) Indirect anodic monofluorodesulfurization of β-phenylsulfenyl β-lactams using a triarylamine mediator. Synlett 1269–1271
10. Le Gall E, Hurvois JP, Sinbandhit S (1999) Regio- and diastereoselective synthesis of α-cyanoamines by anodic oxidation of 6-membered α-silylamines. Eur J Org Chem 1999:2645–2653
11. Kim S, Hayashi K, Kitano Y, Chiba K (2002) Anodic modification of proline derivatives using a lithium perchlorate/nitromethane electrolyte solution. Org Lett 4:3735–3737
12. Yoshida J, Ishichi Y, Isoe S (1992) Intramolecular carbon-carbon bond formation by the anodic oxidation of unsaturated α-stannyl heteroatom compounds. Synthesis of fluorine-containing heterocyclic compounds. J Am Chem Soc 114:7594–7595
13. Yoshida J, Murata T, Isoe S (1986) Electrochemical oxidation of organosilicon compounds 1. Oxidative cleavage of carbon-silicon bond in allylsilanes and benzylsilanes. Tetrahedron Lett 27:3373–3376
14. Koizumi T, Fuchigami T, Nonaka T (1989) Anodic oxidation of (trimethylsilyl)methanes with π-electron substituents in the presence of nucleophiles. Bull Chem Soc Jpn 62:219–225
15. Chiba K, Uchiyama T, Kim S, Kitano Y, Tada M (2001) Benzylic intermolecular carbon-carbon bond formation by selective anodic oxidation of dithioacetals. Org Lett 3:1245–1248

Electrobioremediation of Organic Contaminants

Lukas Y. Wick and Jinyi Qin
Department of Environmental Microbiology, Helmholtz Centre for Environmental Research - UFZ, Leipzig, Germany

Introduction

Electrobioremediation is a general name for a hybrid technology coupling bioremediation to electrokinetics. It comprises a large group of engineered cleanup methods that apply electrokinetic phenomena for the directed transport of contaminants, nutrients, electron acceptors, and contaminant-transforming microorganisms in the subsurface.

Engineered bioremediation thereby is a tightrope walk that needs to assure the well-being of the transforming microbial communities and the appropriate flux of contaminants to their microorganisms. Apart from the substrate, optimal microbial activity requires a proper supply of nutrients and terminal electron acceptors (TEA), ideal pH, temperature, and water conditions as well as the absence of cell-toxic contaminants and metabolites that may affect the welfare of the microbial communities. In most contaminated soils, a low contact probability of contaminants and microorganisms reduces the success of bioremediation approaches. This is due to the heterogeneity of the geo-matrix, a patchy distribution of contaminants (i.e. separate phase contaminant spot sources and/or preferential contaminant sorption to soil constituents), the association of microcolonies of microbes with the solid phase, and the immobility of microbes, particularly when the soil pore space is partially filled with air. Distances to be bridged to render contaminant molecules, bioavailable are in the submillimeter to centimeter range. Higher degrees of heterogeneity exist at locations where contaminants are present as nonaqueous phase liquids (NAPL) or solids and where the distribution of nutrients and electron acceptors is fragmented. In order to effectively overcome the bioavailability constraints (i.e., the availability of a contaminant for uptake and transformation by a microorganism), any biotechnological method has to ensure rapid dispersal of microorganisms and/or chemicals at least over the distances typically separating hotspots of pollution and the enhanced release of entrapped organic contaminants, respectively.

Principles of Electrobioremediation

The fundamental idea of applying electrokinetics to bioremediation is to stimulate contaminant biodegradation by effective homogenization of microorganisms and contaminants without further extensive mechanical treatment of the subsurface matrix. Electrokinetic dispersal by electromigration, electrophoresis, and electroosmosis thereby will need to transport physiologically relevant chemicals (contaminants, nutrients, TEA) to catabolically active microorganisms [1] or vice versa (Fig. 1). Contrary to electrokinetic extraction methods, a microscale (i.e., millimeter to centimeter scale) rather than meter-scale transport of chemicals is required to bridge the distances needed due to the ubiquity of bacteria in the subsurface [2].

Over the last two decades, many reports on the electrokinetic stimulation of bioremediation at the laboratory and field scale have been published (for a review: [1, 3–7]). Electrokinetic approaches have been found to be of particular relevance in fine-grained geo-matrices, where the conductivity for pressure-driven water flow is low; i.e. where quasi-stagnant soil pore water prevails, and poor nutrient and carbon flows limit contaminant biodegradation rates. Next to enhanced transport of chemicals, electric fields may further induce other biologically relevant physicochemical phenomena in soil, such as desiccation, electrolysis of water and subsequent H_2–and O_2–generation, formation of pH-gradients and reactive oxygen species (ROS), precipitation of salts, or the decomposition of minerals. Electrochemical processes have also been used to heat soil and boost bioremediation of contaminated soil in cold climate zones or to stimulate sequential reductive/oxidative microbial degradation in contaminated environments.

The "electrobioremediation tetrahedron" in Fig. 2 conceptualizes the impact of an electric field on the ecology of subsurface bioremediation: Its base triangle represents the biogeochemical interactions governing the ecology of contaminant biodegradation, whereas the side triangles reflect the electrokinetic effects on bacterial dispersal, contaminant bioavailability, and mobilization. Please refer to the literature for further reading on physical [8], physicochemical [9], engineering [3], bioavailability [1], and remediation aspects [10, 11] of electrokinetics in soil remediation.

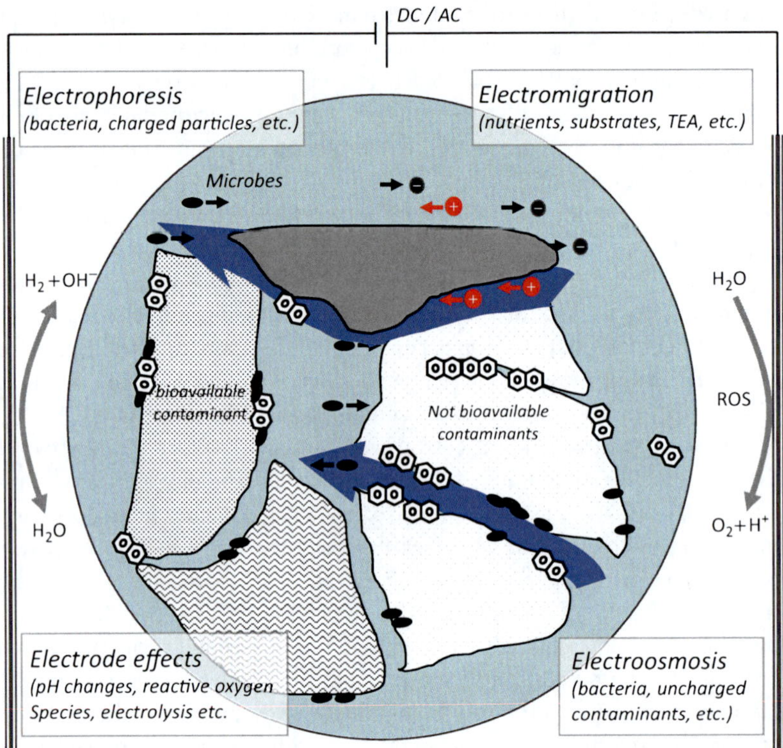

Electrobioremediation of Organic Contaminants, Fig. 1 Schematic representation of electrokinetic processes in the geo-subsurface. When an electric field is applied to a geo-matrix, it invokes electromigration, electrophoresis, and electroosmosis. Electromigration is the transport of ionic substances to the electrode of opposite charge. Electrophoresis is the transport of charged bacteria and particles under the influence of an electric field, and electroosmosis is the movement of pore water to the cathode. Electrolysis of pore fluids at both electrodes will generate pH changes, oxygen, hydrogen, and ROS formation

Electrokinetic Influence on Contaminant Biotransformation

Electrokinetically enhanced biotransformation requires that the use of electric fields has no negative effect on the biocatalytic activity of the microbial communities involved. From a microorganism's viewpoint, this means that neither physical nor (physico-)chemical phenomena of the electric field are harmful to its activity. Although electrobioremediation depends on the stimulation of indigenous contaminant degrading microorganisms, poor information on the effect of weak electric fields on microorganisms and their ecology during electrobioremediation measures is available to date [1]. Current data suggest that electric field gradients as typically used in electrobioremediation approaches ($X = 1\text{–}2\ \mathrm{V\ cm^{-1}}$) have no harmful consequence on contaminant-degrading bacteria given that secondary effects such as pH changes or the formation of toxic reactive oxygen or chlorine species are avoided. Relevant for bioremediation measures is the electrokinetically enhanced mass transfer of nutrients, substrates, or metabolites to and from the microbial cells. Growth stimulation of immobilized cells by 140 % because of the electrokinetic removal of inhibitory products and electroosmotically enhanced carbon substrate supply has been reported. Such phenomena may also explain the increased removal of contaminants in soil.

Electrokinetic stimulation may also be achieved by transport of immobilized indigenous or bioaugmented bacteria to biogeochemical niches of suitable chemical and environmental conditions (Fig. 2). Several studies have demonstrated centimeter- to meter-scale electrokinetic transport of bacteria and yeast cells through sand, soil, and aquifer sediments by either electrophoresis or electroosmosis. Transport direction and rates depend on the type of the subsurface matrix, the environmental conditions, and the size and

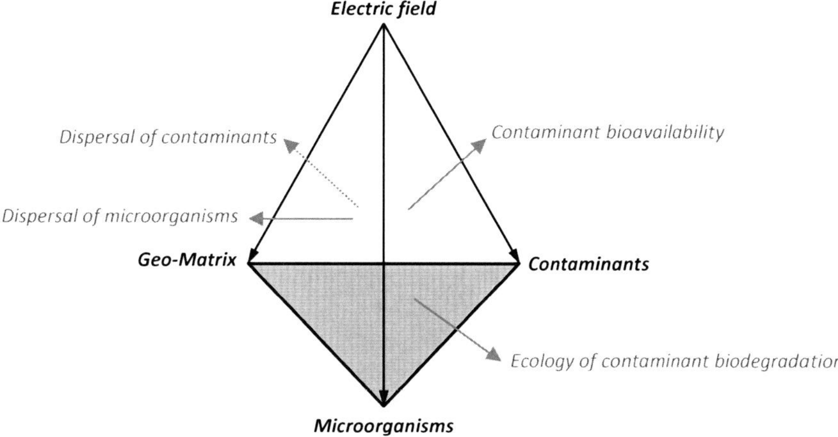

Electrobioremediation of Organic Contaminants, Fig. 2 Conceptualization of the impact of an electric field on the ecology of subsurface bioremediation ("electrobioremediation tetrahedron"): the base triangle of the tetrahedron represents the interactions governing the ecology of contaminant to degrading microorganisms, whereas the side triangles reflect the electrokinetic effects on bacterial dispersal, contaminant bioavailability, and mobilization

physicochemical cell surface properties of the microorganisms. Highly negatively charged cells are predominantly transported by electrophoresis at rates of 0.02 $cm^2\ h^{-1}\ V^{-1}$ for yeast and up to 4 $cm^2\ h^{-1}\ V^{-1}$ for bacterial cells. By contrast, poorly charged organisms are prone to electroosmotic water flow to the cathode. Limited information is available on the mechanisms governing microbial retardation and detachment in geo-matrices exposed to electric fields with the current studies suggesting no influence on the deposition efficiency and consequent retardation of bacteria tested.

Electrokinetic Influence on Contaminant and Nutrient Transport

Electrokinetics stimulates microbial activity by electromigration of charged species, the electrophoretic translocation of particle-bound chemicals, and/or electroosmotic transport of uncharged molecules to active cells. Successful electroosmotic flushing has been applied to subsurfaces contaminated with, e.g., phenolic compounds, hexachlorobenzene, chlorinated solvents, or low molecular weight fuel components. Electroosmotic transport of more hydrophobic organic contaminants (HOC) over long distances, however, was generally feasible only when the HOC were present as droplets, colloidal particles or solubilized by surfactants, and cyclodextrins or chelating agents. As outlined above, long-distance contaminant transport may not be needed to enhance the bioremediation efficiency. For HOC intra- and interparticle diffusion mass, transfer resistances at the microscale often impose serious limitations on the rate biotransformation in geo-matrices. Sorption to geo-sorbents makes contaminants temporarily immobile and limits their release to microorganisms [12]. Such interactions are matrix and contaminant specific, thus increasing the variability, complexity, and difficulty of remediation technologies. Nonetheless, stimulated release by electroosmosis may increase the local HOC-bioavailability and the HOC-bioremediation efficiency, especially in stagnant water zones often found under hydraulic flow regimes restricted by low permeability in fine-grained geo-matrices. A lack of adequate provision of nutrients (e.g., nitrogen phosphorus, potassium) and terminal electron acceptors (e.g., oxygen, nitrate, sulfate, or Mn^{2+} and Fe^{3+}) to catabolically active communities has often been encountered as additional bottleneck for efficient in situ bioremediation. Other studies showed successful injection and electrokinetic transport of charged nutrients, co-substrates, water, and/or terminal electron acceptors into soil. Finally, tailor-made electrokinetic mobilization of contaminants into defined treatment areas such as reactive permeable barrier zones hence can lead to optimal biodegradation efficiency.

Future Directions

The microbial ecology of electrokinetically stimulated contaminant biotransformation in geomatrices is influenced by at least three interrelating levels of complexity: (i) the bioavailability (driven by sorption, desorption, transport, and transformation) of organic chemicals at the geo-sorbent microsites; (ii) the abiotic environmental, physical, and chemical conditions of the physical geo-environment; and (iii) the functional redundancy, cooperation, and process reliability of the degrading microbial communities. Thus, a process-based (i.e., quantifiable, transferable, and scalable) understanding of the interplay of subsurface structure and contaminant-transforming function of microbial associations and their catabolic relationship with various contaminant bioavailability scenarios is important for controlling the extent of engineered subsurface natural attenuation of contaminants. Particular attention should be paid to a process-driven mechanistic understanding and control of dispersal processes at the microbe scale, i.e., at the matrix-contaminant and microorganism-contaminant interface for improved bioremediation results. This comprises (i) advanced understanding of the electroosmosis-driven contaminant release and mobilization in nano- and micropores of geomatrices, (ii) improved knowledge of the impact of low electric fields/currents on the physiology and deposition behavior of (prokaryotic and eukaryotic) microorganisms at a single-cell level, (iii) enhanced information of the (aut)ecological adaptations of contaminant-degrading microbial consortia to electrokinetically enhanced fluxes of mixtures of compounds containing both toxic (e.g., heavy metals) and beneficial chemicals (substrates and nutrients), and (iv) improved electrokinetic management of the subsurface heterogeneities and anomalies often found in situ (e.g., large quantities of iron oxides, large rocks, or gravel). Such knowledge will help to overcome the existing limitations of the electrobioremediation technology for cost-effective utility applications at varying scales.

Cross-References

► Electrochemical Bioremediation

References

1. Wick LY (2009) Coupling electrokinetics to the bioremediation of organic contaminants: principles and fundamental interactions. In: Reddy KR, Cameselle C (eds) Electrochemical remediation technologies for polluted soils, sediments and groundwater. Wiley, Hoboken, pp 369–387
2. Whitman WB, Coleman DC, Wiebe WJ (1998) Prokaryotes: the unseen majority. Proc Nat Acad Sci USA 95(12):6578–6583
3. Lageman R, Clarke RL, Pool W (2005) Electro-reclamation, a versatile soil remediation solution. Eng Geol 77(3–4):191–201
4. Kim BK, Baek K, Ko SH, Yang JW (2011) Research and field experiences on electrokinetic remediation in South Korea. Sep Purif Technol 79(2):116–123
5. Pazos M, Rosales E, Alcantara T, Gomez J, Sanroman MA (2010) Decontamination of soils containing PAHs by electroremediation: a review. J Hazard Mater 177(1–3):1–11
6. Megharaj M, Ramakrishnan B, Venkateswarlu K, Sethunathan N, Naidu R (2011) Bioremediation approaches for organic pollutants: a critical perspective. Environ Int 37(8):1362–1375 (in English)
7. Yeung AT, Gu YY (2011) A review on techniques to enhance electrochemical remediation of contaminated soils. J Hazard Mater 195:11–29
8. Yeung AT (1994) Electrokinetic flow processes in porous media and their applications. In: Advances in porous media, vol 2. Elsevier, Amsterdam, pp 309–395
9. Saichek RE, Reddy KR (2005) Electrokinetically enhanced remediation of hydrophobic organic

compounds in soils: a review. Crit Rev Environ Sci Technol 35:115–192
10. Virkutyte J, Sillanpää M, Latostenmaa P (2002) Electrokinetic soil remediation – critical overview. Sci Tot Environ 289:97–121
11. Page MM, Page CL (2002) Electroremediation of contaminated soils. Journal of. Environmental Engineering-Asce 128(3):208–219
12. Luthy RG et al (1997) Sequestration of hydrophobic organic contaminants by geosorbents. Environ Sci Technol 31(12):3341–3347

Electrocatalysis - Basic Concepts, Theoretical Treatments in Electrocatalysis via DFT-Based Simulations

YongMan Choi
Chemistry Department, Brookhaven National Laboratory, Upton, NY, USA

Introduction

Computational modeling in electrocatalysis affords detailed information at the atomic level that cannot be obtained easily from experiments, so greatly helping to rationalize the design of catalytic materials with high activity, low loading, and enhanced durability. In particular, density functional theory (DFT) [1] is used extensively in electrocatalysis because this approach entails far less computational time than do conventional wave-function methods, while offering reliable results. In particular, owing to the recent enormous decrease in the costs of hardware and to the continuing development of algorithms, expectedly there will be a rapid rise in the popularity of employing DFT methods for designing electrocatalysts. Furthermore, such calculations provide substantial insights into electrode reaction mechanisms (e.g., the oxygen reduction reaction (ORR) [2–9], the hydrogen evolution reaction (HER) [10], and the oxidation of alcohols [11, 12]). Reaction mechanisms are critical for designing novel high-activity electrocatalysts. As Bockris and Khan wrote [13], quantum chemical analyses can be a powerful means of formulating new and better electrocatalysts, and proper computational modeling approaches can efficiently guide us in designing superior electrode materials. For example, as Fig. 1 shows, the DFT simulation model very satisfactorily the practical size of core-shell nano-electrocatalysts. In the next sections, a brief discussion of the DFT methodology is given and then several significant advances in electrocatalysis using DFT-based calculations are described.

DFT Calculations in Electrocatalysis

DFT Modeling

In this section, we briefly summarize the DFT method and then review efforts to examine the ORR. A detailed derivation of the DFT method is available elsewhere [14]. Essentially, the DFT approach computes electron density and then predicts other properties, while conventional wave-function methods solve the Schrödinger equation to obtain the wave function and then the electron density [15]. As is well known, one significant issue in the DFT method is obtaining reliable exchange-correlation energies that are expressed by the electron density. Among numerous approximations, such as the local density approximation (LDA) and the generalized gradient approximation (GGA) [16], the standard GGA functionals are used widely in electrocatalysis because of their higher accuracy. Another major step in applying DFT calculations lies in constructing a surface model to properly interpret the experimental findings or to formulate a rational design. In general, cluster and periodic slab models are applicable in resolving these problems (Fig. 2). We note that because the cluster model must be validated systematically to avoid unreasonable boundary effects due to convergence [17], the slab model with the periodic boundary condition (PBC) is more widely used in electrocatalysis, for example, for simulating the electric effects of the interaction of methanol in aqueous solution on Pt(111) [11] and the interaction of water on Pt(111). Furthermore, to elucidate the mechanisms underlying the electrochemical reaction, researchers much favor

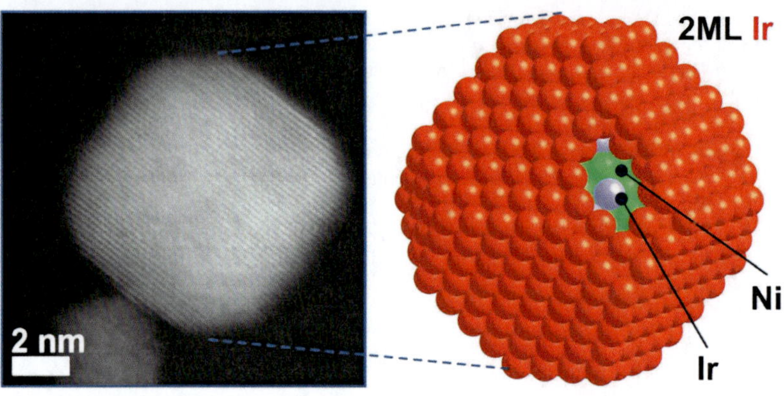

Electrocatalysis - Basic Concepts, Theoretical Treatments in Electrocatalysis via DFT-Based Simulations, Fig. 1 A core-shell model for a DFT simulation of an Ir shell and a Ni core (Courtesy of the scanning transmission electron microscopy (*STEM*) image from Drs. Kurian A. Kuttiyiel and Dong Su)

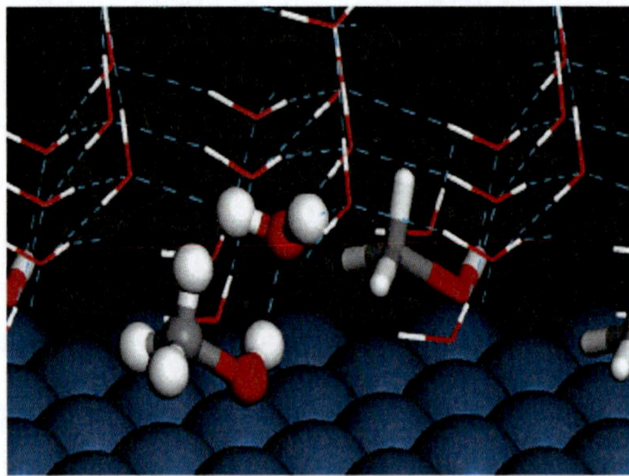

Electrocatalysis - Basic Concepts, Theoretical Treatments in Electrocatalysis via DFT-Based Simulations, Fig. 2 Illustration of representative slab models used for CH_3OH optimization in an aqueous-solvated system [11] (Reprinted with permission from Cao D, Lu GQ, Wieckowski A, Wasileski SA, Neurock M (2005) Mechanisms of methanol decomposition on platinum: a combined experimental and ab Initio approach. J Phys Chem B 109:11622–11633. Copyright 2005 American Chemical Society)

the nudged elastic band (NEB) method [18] that locates a transition state between a reactant and a product. The transition state is identified by the number of imaginary frequencies (NIMG = 1). For other studies, such as kinetic modeling using the transition state theory (TST) [19], zero-point energy (ZPE) corrections can be applied. To examine charge transfer between an adsorbate and an electrocatalyst's surface, the Bader analysis software [20, 21], based on Bader's theory of atoms in molecules (AIM) [22], is readily used. Based on optimized geometries, the density of states (DOS) frequently is employed to analyze the electronic structure of the interactions between an adsorbate and an electrocatalyst and to estimate band gaps. In particular, the DOS findings are very suitable for calculating the d-band centers [9, 23]. Various software packages are available for DFT modeling, including CPMD (Car-Parrinello Molecular Dynamics) [24], QUANTUM ESPRESSO (opEn Source Package for Research in Electronic Structure, Simulation, and Optimization) [25], CASTEP (Cambridge Serial Total Energy Package) [26], and VASP (Vienna ab initio simulation package) [27, 28]. However, the invariable recommendation is to choose a proper computational method and surface model so to generate reasonable computational results through a validation process.

Elucidation of Electrochemical Reactions Using DFT Calculations

Over the years, there have been numerous reports of using DFT computations to study electrochemical reactions [2–7, 9–12, 17, 29–31]. In this section, we introduce our approach to employing the DFT method to gain an understanding of the ORR at the atomic level. Here, the introduction of the d-band model may be among the most crucial progresses made to date in the computational modeling of electrocatalysis [9, 23]. For example, Nørskov and coworkers reported the trend of the ORR activity [4] using DFT calculations (Fig. 3a);

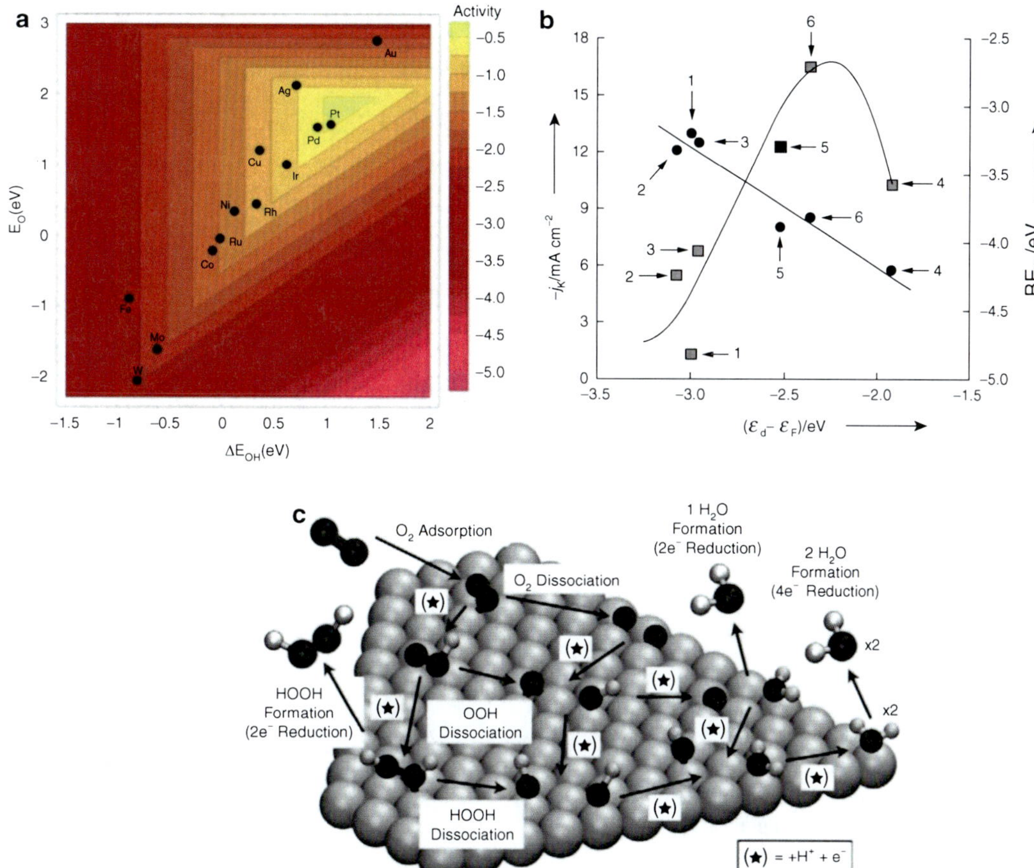

Electrocatalysis - Basic Concepts, Theoretical Treatments in Electrocatalysis via DFT-Based Simulations, Fig. 3 (**a**) Correlation of the ORR activity as a function of O and the OH binding energies [4] (Reprinted with permission from Nørskov JK, Rossmeisl J, Logadottir A, Lindqvist L, Kitchin JR, Bligaard T, Jónsson H (2004) Origin of the overpotential for oxygen reduction at a fuel-cell cathode. J Phys Chem B 108:17886–17892. Copyright 2004 American Chemical Society). (**b**) Kinetic currents measured at 0.8 V for the ORR on the platinum monolayer on single crystal surfaces in 0.1 M $HClO_4$ versus the binding energies of atomic O (BEO) as a function of the d-band center [5] (Zhang J, Vukmirovic MB, Xu Y, Mavrikakis M, Adzic RR (2005) Controlling the catalytic activity of platinum-monolayer electrocatalysts for oxygen reduction with different substrates. Angew Chem Int Ed 44:2132–2135. Copyright Wiley-VCH Verlag GmbH & Co. KGaA. Reproduced with permission). (**c**) Schematic of ORR mechanisms on Pt(111) [29] (Keith JA, Jacob T (2010) Theoretical studies of potential-dependent and competing mechanisms of the electrocatalytic oxygen reduction reaction on Pt(111). Angew Chem Int Ed 49:9521–9525. Copyright Wiley-VCH Verlag GmbH & Co. KGaA. Reproduced with permission)

they successfully applied this approach to explain the finding from experiments with single crystals [5] (Fig. 3b). Based on the correlation curve with the d-band center and adsorption of oxygen (or the binding energy of oxygen), we can now rationally design more active electrode materials before carrying out experiments. As we noted, gaining an understanding of the ORR is difficult owing to the high complexity of the electrochemical system (Fig. 3c). Indeed, in considering multicomponent electrocatalysts, like a platinum monolayer core-shell electrocatalyst, the system's tortuousness is even more increased and modeling is difficult. However, thanks to the systematic study by

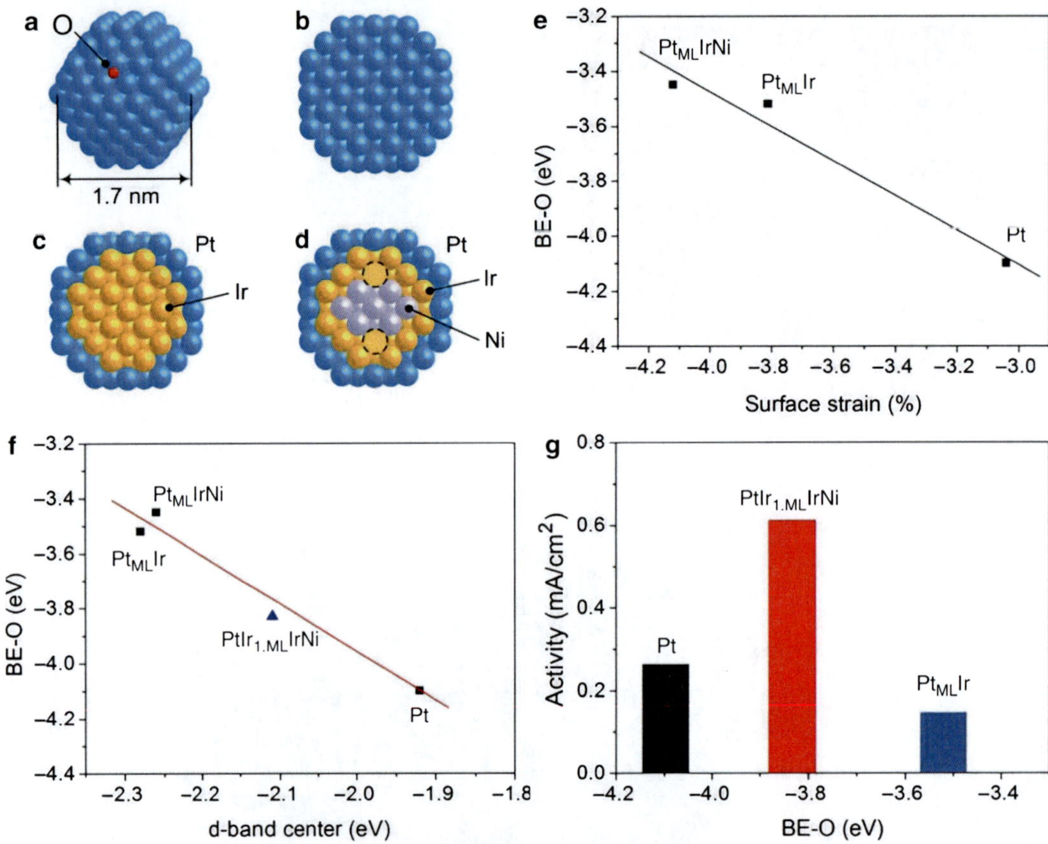

Electrocatalysis - Basic Concepts, Theoretical Treatments in Electrocatalysis via DFT-Based Simulations, Fig. 4 Geometrical illustration of (**a**) a sphere-like nanoparticle, (**b**) Pt, (**c**) $Pt_{ML}Ir$, and (**d**) $Pt_{ML}IrNi$. (**e**) The predicted binding energy of oxygen (BEO) against strain. (**f**) The correlation of BEO and the d-band center of metals. (**g**) The specific activity versus BEO of $Pt_{Ir1.ML}IrNi$, $Pt_{ML}Ir$, and Pt [33] (Kuttiyiel KA, Sasaki K, Choi Y, Su D, Liu P, Adzic RR (2012) Bimetallic IrNi core platinum monolayer shell electrocatalysts for the oxygen reduction reaction. Energy Environ Sci 5:5297–5304. Reproduced by permission of The Royal Society of Chemistry)

Ruban and coworkers [32], experimentalists can rationally design practical core-shell electrocatalysts.

Based on the previous pioneering studies using extended surfaces, nano-cluster core-shell models [2, 3, 31] were advantageously applied to interpreting the enhanced activity and durability of platinum monolayer core-shell electrocatalysts (i.e., $Pt_{ML}/M/C$; M is a core metal and C is carbon). For example, spin-polarized DFT calculations were adopted to examine the experimental finding that $Pt_{ML}/IrNi/C$ shows higher activity and more enhanced durability than $Pt_{ML}/Ir/C$ and Pt/C [33]. For these calculations, only the Γ-point for **k** sampling was considered after putting the sphere-like nano-cluster (~1.7 nm) in a sufficiently large box (i.e., 28 × 28 × 28 Å). The geometric effect was examined by assessing surface strains (Fig. 4e), while the estimated binding energy of atomic O was used as a descriptor for the ORR activity [3]. As shown in Fig. 4f, the correlation between the BEO and the d-band center (Pt and $Pt_{ML}Ir$) is consistent with that for Pt(111) and $Pt_{ML}Ir(111)$ [5]. Furthermore, the partial replacement of Ir with Ni caused a slightly weaker BEO ($Pt_{ML}Ir$ vs. $Pt_{ML}IrNi$). However, the calculated geometric and electronic effects cannot fully interpret the experimental finding. Thus, the anti-segregation under the ORR condition was taken into account by interchanging one Ir atom in the core with a Pt atom in the shell ($PtIr_{1,ML}IrNi$), resulting in a reasonable correlation between the BEOs and the experimental-specific activities (Fig. 4g). It can be concluded that while Pt interacts with O too strongly and $Pt_{ML}Ir$ does so too weakly, $Pt_{ML}IrNi$ supports a moderate BEO, leading to the efficient O-O dissociation and the ease of O removal.

Concluding Remarks

Although electrochemical reactions have been widely examined theoretically to support experimental findings, their high complexity leaves numerous questions unanswered. Therefore, researchers have attempted to combine the DFT method with other theoretical ones to assure the rational design of novel electrocatalysts in fuel cells and to bridge the gap between experiment and theory. It is crucial in this endeavor that we gain an understanding of the electrochemical processes at the interfaces (i.e., the adsorption of chemical species at the liquid–solid interface) with the practical size of nanoparticles (~5.0–10 nm). However, at the same time, the costs of molecular modeling are steep. Hence, it is anticipated that quantum mechanics/molecular mechanics (QM/MM) methodology may help model electrochemical reactions at the interfacial region with a high accuracy and in a reasonable computational time. Also, simulating a cyclic voltammogram [34] using ab initio calculations can assist in quantitatively understanding electrochemical reactions, while the coupling the DFT approach and a genetic algorithm [35] can enable us to create better core-shell electrocatalysts.

Cross-References

▶ Electrocatalysis of Anodic Reactions
▶ Electrocatalysis of Chlorine Evolution
▶ Electrocatalysis, Fundamentals - Electron Transfer Process; Current-Potential Relationship; Volcano Plots

References

1. Kohn W, Sham LJ (1965) Self-consistent equations including exchange and correlation effects. Phys Rev B 140:A1133
2. Wang JX et al (2011) Kirkendall effect and lattice contraction in nanocatalysts: a new strategy to enhance sustainable activity. J Am Ceram Soc 133:13551–13557
3. Wang JX, Inada H, Wu L, Zhu Y, Choi Y, Liu P, Zhou W-P, Adzic RR (2009) Oxygen reduction on well-defined core-shell nanocatalysts: particle size, facet, and Pt shell thickness effects. J Am Chem Soc 131:17298–17302
4. Nørskov JK, Rossmeisl J, Logadottir A, Lindqvist L, Kitchin JR, Bligaard T, Jónsson H (2004) Origin of the overpotential for oxygen reduction at a fuel-cell cathode. J Phys Chem B 108:17886–17892
5. Zhang J, Vukmirovic MB, Xu Y, Mavrikakis M, Adzic RR (2005) Controlling the catalytic activity of platinum-monolayer electrocatalysts for oxygen reduction with different substrates. Angew Chem Int Ed 44:2132–2135
6. P S et al (2010) Lattice-strain control of the activity in dealloyed core–shell fuel cell catalysts. Nat Chem 2:454–460
7. Ramírez-Caballero GE, Balbuena PB (2010) Dissolution-resistant core-shell materials for acid medium oxygen reduction electrocatalysts. J Phys Chem Lett 1:724–728
8. Tang W, Henkelman G (2009) Charge redistribution in core-shell nanoparticles to promote oxygen reduction. J Chem Phys 130:194504
9. Nørskov JK, Bligaard T, Rossmeisl J, Christensen CH (2009) Towards the computational design of solid catalysts. Nat Chem 1:37–46
10. Hinnemann B, Moses PG, Bonde J, Jørgensen KP, Nielsen JH, Horch S, Chorkendorff I, Nørskov JK (2005) Biomimetic hydrogen evolution: MoS_2 nanoparticles as catalyst for hydrogen evolution. J Am Chem Soc 127:5308–5309
11. Cao D, Lu GQ, Wieckowski A, Wasileski SA, Neurock M (2005) Mechanisms of methanol decomposition on platinum: a combined experimental and ab initio approach. J Phys Chem B 109:11622–11633
12. A K et al (2009) Ternary $Pt/Rh/SnO_2$ electrocatalysts for oxidizing ethanol to CO_2. Nat Mater 8:325–330
13. Bockris JOM, Khan SUM (1993) Surface electrochemistry: a molecular level approach. Plenum Press, New York
14. Levine IN (1991) Quantum chemistry, 4th edn. Prentice-Hall, New Jersey
15. Huang P, Carter EA (2008) Advances in correlated electronic structure methods for solids, surfaces, and nanostructures. Annu Rev Phys Chem 59:261–290
16. Jensen F (1999) Introduction to computational chemistry. Wiley, New York
17. Jacob T, Goddard WA III (2004) Adsorption of atomic H and O on the (111) surface of Pt_3Ni alloys. J Phys Chem B 108:8311–8323

18. Henkelman G, Uberuaga BP, Jönsson H (2000) A climbing image nudged elastic band method for finding saddle points and minimum energy paths. J Chem Phys 113:9901
19. Laidler KJ (1987) Chemical kinetics, 3rd edn. Harper and Row, New York
20. Henkelman G, Arnaldsson A, Jonsson H (2006) A fast and robust algorithm for Bader decomposition of charge density. Comput Mater Sci 36:354–360
21. http://theory.cm.utexas.edu/bader/
22. Bader RFW, Beddall PM (1972) Virial field relationship for molecular charge distributions and spatial partitioning of molecular properties. J Chem Phys 56:3320–3329
23. Hammer B, Nørskov JK (1995) Electronic factors determining the reactivity of metal surfaces. Surf Sci 343:211–220
24. (Copyright MPI für Festkörperforschung Stuttgart 1997–2001)
25. Giannozzi P et al (2009) QUANTUM ESPRESSO: a modular and open-source software project for quantum simulations of materials. J Phys Condens Matter 21(39):395502
26. Clark SJ, Segall MD, Pickard CJ, Hasnip PJ, Probert MJ, Refson K, Payne MC (2005) First principles methods using CASTEP. Z Kristallogr 220:567–570
27. Kresse G, Hafner J (1993) Ab initio molecular dynamics for liquid metals. Phys Rev B 47:558–561
28. Kresse G, Furthmuller J (1996) Efficient iterative schemes for ab initio total-energy calculations using a plane-wave basis set. Phys Rev B 54:11169–11186
29. Keith JA, Jacob T (2010) Theoretical studies of potential-dependent and competing mechanisms of the electrocatalytic oxygen reduction reaction on Pt (111). Angew Chemie Int Ed 49:9521–9525
30. Rossmeisl J, Skúlason E, Björketun ME, Tripkovic V, Nørskov JK (2008) Modeling the electrified solid–liquid interface. Chem Phys Lett 466:68–71
31. Sasaki K, Naohara H, Cai Y, Choi YM, Liu P, Vukmirovic MB, Wang JX, Adzic RR (2010) Core-protected platinum monolayer shell high-stability electrocatalysts for fuel-cell cathodes. Angew Chem Int Ed 49:8602–8607
32. Ruban AV, Skriver HL, Nørskov JK (1999) Surface segregation energies in transition-metal alloys. Phys Rev B 59:15990–16000
33. Kuttiyiel KA, Sasaki K, Choi Y, Su D, Liu P, Adzic RR (2012) Bimetallic IrNi core platinum monolayer shell electrocatalysts for the oxygen reduction reaction. Energy Environ Sci 5:5297–5304
34. Karlberg GS, Jaramillo TF, Skúlason E, Rossmeisl J, Bligaard T, Nørskov JK (2007) Cyclic voltammograms for H on Pt(111) and Pt(100) from first principles. Phys Rev Lett 99:126101
35. Froemming NS, Henkelman G (2009) Optimizing core-shell nanoparticle catalysts with a genetic algorithm. J Chem Phys 131:234103

Electrocatalysis of Anodic Reactions

Kotaro Sasaki and Meng Li
Chemistry Department, Brookhaven National Laboratory, Upton, NY, USA

Introduction

The present section describes recent advances in electrocatalysis for anodic reactions in low-temperature fuel cells that use hydrogen/carbon monoxide, methanol, and ethanol as fuels in acidic media. Electrocatalysis in alkaline fuel cells are not discussed.

H_2/CO Oxidation

Pure hydrogen (H_2) is an ideal fuel for proton exchange membrane fuel cells (PEMFCs). Its chemical energy can be electrochemically converted to electrical energy with zero emissions and high efficiency. The reaction rate of hydrogen oxidation reaction (HOR, see Eq. 1), measured by the exchange current density i_0 (the current density at the equilibrium potential), greatly depends on the electrode materials [1, 2]. Platinum undoubtedly is the best catalyst for HOR, and its overpotential is negligibly small. However, H_2 gas for PEMFCs is commonly produced by re-forming methanol or hydrocarbons, and inevitably contains small amounts of carbon monoxide (CO). CO level as low as 5 ppm in H_2 can cause significant degradation of the HOR activity because CO molecules adsorb strongly on the active Pt sites, thus blocking them for HOR; this is commonly referred to as CO poisoning. Therefore, CO tolerance is of importance for anode catalysts in H_2-feed fuel cells.

$$H_2 \rightarrow 2H^+ + 2e^- \qquad (1)$$

Methanol Oxidation

Liquid fuels, especially methanol and ethanol, are considered as potential alternatives to hydrogen fuel in PEMFCs due to their high energy density, likely production from renewable sources, and most importantly, the ease of their storage and transportation. Methanol, containing only one carbon atom, is the simplest alcohol and its electrocatalysis is also the simplest; therefore, there is a rising interest in direct methanol fuel cells (DMFCs) as potential power sources for portable electronic devices and for transportation usages. The total methanol oxidation reaction (MOR) can be written as

$$CH_3OH + H_2O \rightarrow CO_2 + 6H^+ + 6e^- \quad (2)$$

Methanol has a low theoretical oxidation potential (0.03 V) comparable to that of hydrogen (0 V), and thus, in principle, it can be an efficient fuel at low temperatures. However, CO poisoning also takes place during MOR. This reaction on a pure Pt electrode is best described as a parallel pathway mechanism forming a number of reaction intermediates and products (CO_2, HCOOH, HCHO); among them adsorbed CO can block Pt sites [3–5].

$$Pt - CH_3OH_{ad} \rightarrow Pt - CHO_{ad} + 3H^+ + 3e^- \quad (3)$$

$$Pt - CHO_{ad} \rightarrow Pt - CO_{ad} + H^+ + e^- \quad (4)$$

The tardiness of the MOR at the anode is more prominent than that of the oxygen reduction reaction (ORR) at the cathode, thus adversely affecting the performance of DMFCs; consequently, a large amount of Pt is required to enhance the anodic performance of DMFCs.

Improving electrocatalysts for CO tolerance and methanol oxidation is a challenging task given the need to ensure the oxidation of CO to CO_2 to allow turnover of the electrocatalytic sites at relatively low potentials. Oxidizing CO requires oxygen atoms, usually supplied by the dissociation of water. However, Pt does not chemisorb H_2O at potentials lower than 0.7 V; thus pure Pt is a poor electrocatalyst for CO and alcohol oxidation at low potentials. Alloying Pt with other oxophilic metals, including Ru, Os, Ir, Rh, and Sn, was shown to reduce CO poisoning and increase the catalytic activity [6–9]. This enhanced activity of the Pt alloys can be explained either by the bifunctional model and/or the electronic effect (also called ligand effect) [4, 10]. The former was proposed by Watanabe and Motoo based on their electrochemical studies of the Pt–Ru system. Ru provides active OH species by dissociating water at the Ru sites at lower potentials compared with the Pt sites; this accelerates CO_2 formation and lowers the CO poisoning, thereby improving CO tolerance [4].

$$Ru + H_2O \rightarrow Ru - OH_{ad} + H^+ + e^- \quad (5)$$

$$CO_{ad} + OH_{ad} \rightarrow CO_2 + H^+ + e^- \quad (6)$$

Also operative could be the electronic effect on Pt atoms caused by interference from Ru atoms. X-ray absorption spectroscopy (XAS) measurements demonstrated a considerable increase in *d*-band vacancies on Pt in the PtRu/C electrocatalyst compared with that for pure Pt [11]. This is likely caused by an effective transfer of *d* electrons from the Pt to the neighboring Ru. The increase in *d*-band vacancies broadens the width of the *d* band and lowers its center [12–15]. As a consequence, the back donation of Pt *d* electrons to the CO $2\pi^*$ orbitals is hampered and CO bonding to Pt is weakened.

Ethanol Oxidation

Ethanol is a nontoxic renewable energy source that can be produced from agricultural products; its theoretical energy density (8.0 kWh/kg) is higher than that of methanol (6.1 kWh/kg). However, the commercialization of direct ethanol fuel cells (DEFCs) is considered more difficult than that of DMFCs, because the kinetics of ethanol oxidation reaction is slower than that of methanol oxidation even on the best available catalysts.

The ethanol electrooxidation mechanism on platinum electrodes in acidic solution has been studied by various techniques and a number of adsorbed intermediates have been identified [16–25]. Carbon dioxide (CO_2), acetaldehyde (CH_3CHO), and acetic acid (CH_3COOH) are the main products of the reaction. The global oxidation mechanism can be summarized in the following parallel reactions based on the foregoing studies [26]:

$$CH_3CH_2OH \rightarrow [C_2H_5OH]_{ad} \rightarrow C1_{ad}, C2_{ad} \rightarrow CO_2 \text{ (total oxidation)} \quad (7)$$

$$CH_3CH_2OH \rightarrow [C_2H_5OH]_{ad} \rightarrow CH_3CHO \rightarrow CH_3COOH \text{ (partial oxidation)} \quad (8)$$

The parallel reactions considerably lower the efficiency in fuel utilization to generate electricity. In total oxidation (Eq. 7), the formation of CO_2 goes through adsorbed intermediates $C1_{ad}$ and $C2_{ad}$, representing fragments with one and two carbon atoms. The activation of the C–C bonds is more difficult than that of the C–H bonds, and cleavage of the former bond is a major challenge in ethanol electrocatalysis. Furthermore, Pt itself is rapidly poisoned by the strongly adsorbed intermediates generated from the dissociative adsorption of ethanol [27]. In a similar manner to that described above, Pt is thus alloyed with other elements, such as Ru and Sn, that can supply the oxygen-containing species to remove the blocking intermediates [27]. However, Behm et al. [28] reported that the addition of Sn or Ru to Pt, though beneficial for the overall activity for ethanol oxidation, does not enhance the activity for C–C bond breaking.

Pt–Ru-Based System

Among various anode catalysts developed, Pt–Ru alloys are generally considered as the best candidates for H_2/CO and alcohol oxidation; these alloy catalysts show high CO tolerance and acceptable durability under FC operating conditions. Several commercial Pt–Ru alloy nanoparticles supported on carbon black have been available for applications in PEMFCs, DMFCs, and DEFCs. Efforts to improve the activity and stability of Pt–Ru alloy catalysts continuously are being made. Recently, the nanocapsule method has been successfully employed to synthesize Pt–Ru nanoparticles with a uniform chemical composition and a narrow size distribution [29]. The Pt_2Ru_3/C nanoparticles developed by this method exhibited higher CO tolerance and enhanced activity compared with commercial Pt_2Ru_3 catalysts.

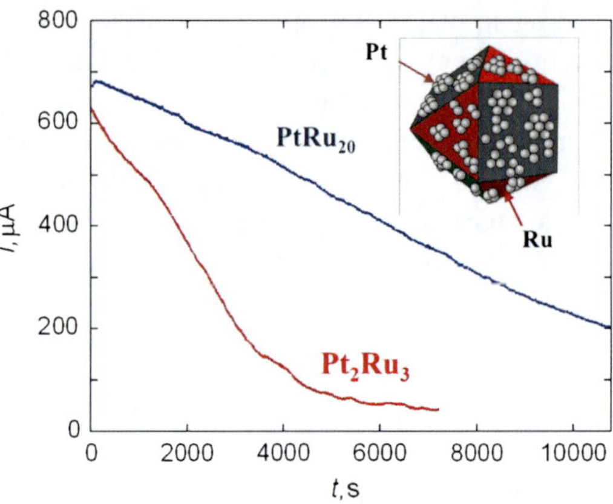

Electrocatalysis of Anodic Reactions, Fig. 1 Comparison of CO tolerance of two catalysts based on the current as a function of time for the oxidation of H_2 with 1,000 ppm of CO at 60 °C for $PtRu_{20}$ (1 $\mu g_{Pt}/cm^2$) and commercial Pt_2Ru_3 (4 $\mu g_{Pt}/cm^2$) electrocatalysts at 0.05 V. Electrolyte: 0.5 M H_2SO_4. Rotation rate: 2,500 rpm

Adzic et al. developed a new approach for designing and synthesizing anode electrocatalysts that dramatically reduces the Pt content while enhancing their catalytic activity and stability, by spontaneous deposition of a submonolayer of Pt on carbon-supported Ru nanoparticles [30–35]. Spontaneous deposition of a noble metal on another noble metal surface was first demonstrated by Wieckowski and co-workers [36, 37] for the deposition of Ru adlayers on a Pt(*hkl*) single crystal surface. Figure 1 shows a comparison of the CO tolerance of the new catalyst with atomic ratio of Pt:Ru ($PtRu_{20}$) with that of a commercial Pt_2Ru_3 alloy electrocatalyst for the oxidation of H_2 containing 1,000 ppm CO. Substantially enhanced stability is seen under these conditions for the $PtRu_{20}$ electrocatalyst, despite the lower Pt loading (1 $\mu g_{Pt}/cm^2$) than in the commercial Pt_2Ru_3 alloy electrocatalyst (4 $\mu g_{Pt}/cm^2$). The inset of Fig. 1 displays the model of the $PtRu_{20}$ electrocatalyst, a cubo-octahedron Ru nanoparticle of 2.5 nm in diameter, with a Pt submonolayer on its surface. The model is derived from data obtained by electrochemical methods, in situ XAS and ex situ transmission electron microscopy (TEM). In long-term stability tests in a fuel cell setup, there was no detectable loss in performance over 870 h, even though the Pt loading was approximately 1/10 of the standard loading [33].

The approach to decorate Ru nanoparticle surfaces by Pt adatoms is beneficial in reducing the Pt loadings and increasing the activity of catalysts for methanol oxidation. Hwang et al. [38] prepared Pt-decorated Ru nanoparticle catalysts by a redox-transmetalation process, and the catalysts showed higher mass activity for MOR than that of commercial Pt–Ru catalysts. Sasaki et al. have synthesized electrocatalysts comprising a monolayer amount of Pt on Ru nanoparticles by galvanic displacement of Cu under potential deposition (UPD) adlayer [39]; the catalysts showed three times higher Pt mass activity and improved durability for methanol oxidation compared with a commercial Pt–Ru alloy catalyst.

Pt–Ru catalysts also received much attention in ethanol oxidation [40–43]. Camara et al. [42] investigated the EOR activity of Pt–Ru electrodeposits as a function of their atomic composition and demonstrated that the catalytic activity of Pt–Ru is strongly dependent on the Ru content. The optimum Pt:Ru composition was 3:2. At low Ru concentration, there are insufficient Ru sites to effectively assist the oxidation of adsorbed residues, and the oxidation current remains almost at the levels obtained for pure Pt. This finding is in line with the observation from Lamy et al. [27] showing a poor activity with the Pt:Ru atomic ratio of 4:1. Ru concentrations higher than *ca* 40 % caused a decrease in current, and this effect can be rationalized in terms of inhibition of ethanol adsorption, presumably due to the diminution of Pt sites.

It was found that addition of a third metal (W, Ni, Mo, Pb, etc.) [44–50] or a metal oxide (RuO_2, IrO_2, etc.) [51] to Pt–Ru catalysts can further improve catalytic activity of this system.

Pt–Sn-Based System

Pt–Sn catalysts have long attracted significant attention due to their high CO tolerance. Pt_3Sn(111) is known as one of the most active systems for CO oxidation [52]. A theoretical study based on density functional theory (DFT) calculations also showed that the activation barrier for CO oxidation on the Pt_3Sn(111) surface is lower than that of Pt(111), thus leading to 2–4 orders of magnitude higher reaction rate [53]. Eichhorn et al. have developed Pt–Sn intermetallic and random alloy nanoparticles by chemical reaction and/or potential cycling [54]. PtSn core-shell nanoparticles with Pt shells were also made through potential cycling of the PtSn intermetallic nanoparticles in CO-saturated H_2SO_4 solutions. Their results demonstrate that the order of CO tolerance decreases in the order of Pt–Sn core-shell > intermetallic > alloy. Therefore, the nanoscale architectures have a significant effect on the reactivity. They discussed that the bifunctional mechanism plays an important role in promoting CO-tolerance for PtSn intermetallics and alloys, while the electronic effect may be more effective because the CO-tolerance for the PtSn core-shell nanoparticles with Pt shells is the higher.

Pt–Sn catalysts only show modest improvement in catalyzing MOR compared to pure Pt catalysts, despite the superior performance of Sn as a co-catalyst to enhance CO oxidation [55–58]. Generally, comparisons between Pt–Ru and Pt–Sn catalysts indicate that the former are more active for the MOR, so that the performance of DMFCs with Pt–Ru/C anode catalysts is substantially better compared to that with PtSn/C catalysts under similar operating conditions [58–62].

On the other hand, Pt–Sn catalysts have been found to be the most active binary systems for ethanol oxidation, established by electrochemical techniques and fuel cell measurements [62–64]. Polyol method [65, 66] and "Bönneman" method [27, 67] were employed to synthesize Pt–Sn/C alloy and Pt–SnO_x/C catalysts, and Jiang et al. [68] claimed that the greater activity was from Pt–SnO_x/C due to the presence of both sufficiently large Pt ensembles for ethanol dehydrogenation and C–C bond splitting and of SnO_x for OH generation. Considerable research has been devoted to optimizing the atomic ratio of Pt-Sn electrocatalysts. Lamy et al. [27] found the alloy

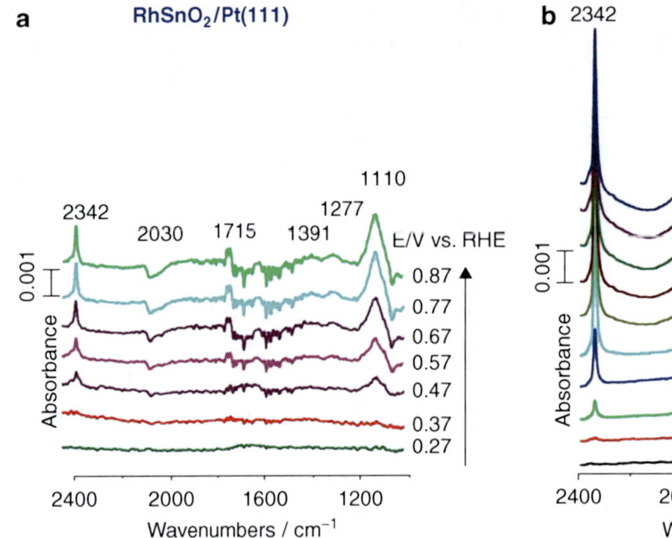

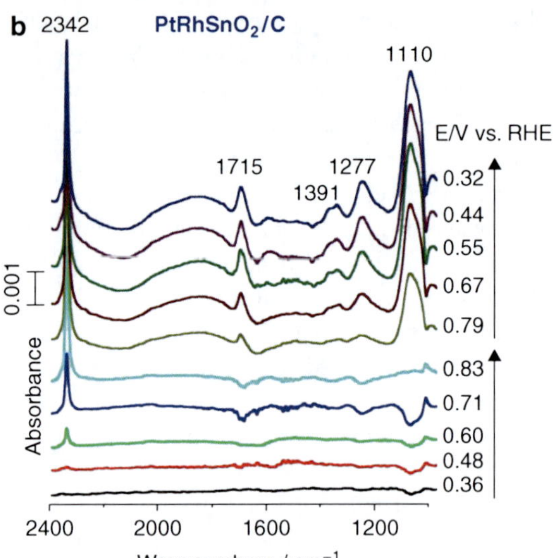

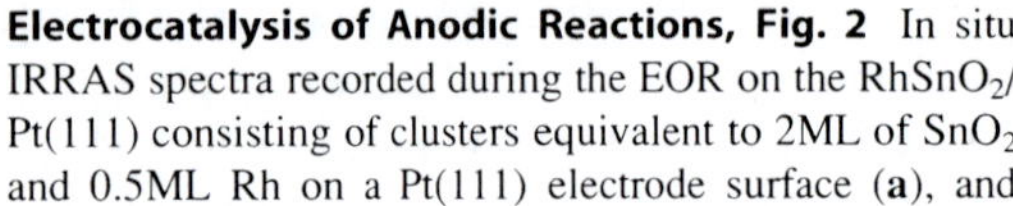

Electrocatalysis of Anodic Reactions, Fig. 2 In situ IRRAS spectra recorded during the EOR on the $RhSnO_2$/Pt(111) consisting of clusters equivalent to 2ML of SnO_2 and 0.5ML Rh on a Pt(111) electrode surface (**a**), and $PtRhSnO_2$/C nanoparticle catalyst (**b**) in a 0.2 M ethanol in 0.1 M $HClO_4$ solution. 128 interferograms (resolution 8 cm^{-1}) were collected and co-added into each spectrum [70]

Pt–Sn/C catalysts with 10–20 at. % of Sn exhibited best activity at low potentials. Results from Jiang et al. [68] showed that the Pt–SnO_x/C catalyst with 30 at.% of Sn was the most active among four different Pt/Sn ratios. An integrated surface science and electrochemistry study of the SnO_x/Pt(111) model catalysts indicate a "volcano" dependence of the EOR activity on the surface composition, with the maximum at the SnO_x coverage of 37 % [69]. Despite the improved overall EOR activity of optimized Pt–Sn system, on-line differential electrochemical mass spectroscopy (DEMS) studies have shown that acetic acid and acetaldehyde represent the dominant products with CO_2 formation contributing only 1–3 % [68].

The major challenge for the electrocatalysis of ethanol is to facilitate its total oxidation to CO_2 at low overpotentials, which cannot be achieved by existing Pt-based binary catalysts. Ternary Pt–Rh–SnO_2 catalyst developed by Adzic et al. [70, 71] showed an unprecedented activity for the EOR with the onset of reaction occurring at low overpotentials explained in terms of synergistic effect between its three constituents. In situ infrared reflection–absorption spectra (IRRAS) indicated that CO_2 is the major product in both model and nanoparticle systems (Fig. 2), while DFT calculation suggested that the optimal reaction pathway goes through an oxametallacyclic conformation (CH_2CH_2O). More recently, $PtRhSnO_2$/C electrocatalysts with different compositions were characterized by in situ XAS studies [72]. The results suggest that the metal–metal oxide interaction increases oxidation state of both Pt and Rh so that a too high content of oxide (i.e. SnO_2) lowers the EOR activity.

Other ternary PtSn-based systems including Pt–Sn–Ni, Pt–Sn–Ru, Pt–Sn–Ir and Pt–Sn–Ce [73–79] were also studied to improve the EOR activity.

Pt-M System (M=Ni, Co, Fe)

Pt alloys with first-row transition metals, such as Ni, Co, and Fe, have also been explored for CO tolerance and MOR electrocatalysts [80–82]; their activities are generally considered comparable to those of Pt–Ru alloy catalysts. Various experiments demonstrated that the non-noble metals in the binary Pt–M alloys can be leached out in acidic electrolyte, thus leaving a core-shell structure comprising a Pt "skin." Such preferential dissolution of a less noble component, commonly referred as "dealloying," has also been employed to develop active bimetallic catalysts (e.g., Pt–Cu) for oxygen reduction reaction [83]. We note too that the formation of a Pt shell also can be induced by thermal treatments; Markovic et al. [84] have improved the catalytic properties of Pt and reduced its contents in Pt–M alloys essentially based on the core-shell approach, where a layer of Pt is formed by the surface segregation of Pt, for example, in the Pt_3Ni alloy by thermal annealing. The high activity of the Pt–skin catalysts for CO tolerance and MOR has been ascribed to the electronic effects of the transition metals in the cores that may decrease the CO adsorption energy due to the shift in the *d*-band center of Pt on surfaces [80, 81, 84].

Non-Pt System

In spite of the similar electronic properties to Pt, the electrocatalytic activity of Pd for the HOR is lower than that of Pt [85]; this is presumably due to the stronger interaction of Pd–H_{ad} than that of Pt–H_{ad} [86]. However, by changing the H_{ad} binding energy through alloying Pd with Pt, the HOR rate on the catalytic surfaces can be increased, which may be comparable to that of Pt [87]. On the other hand, some Pd alloys (e.g., Pd–Fe, Pd–Co) were found to have little activity to MOR; these Pd-M alloys are however, considered to be the excellent methanol-toletant materials for solving the methanol crossover in DMFCs [88].

Ir–Ni core-shell nanoparticles developed by chemical reduction and subsequent thermal annealings show formation of two atomic layers of Ir on Ir–Ni solid-solution alloy cores (*ca* 4 nm diameter), as verified by various experimental methods including synchrotron radiation techniques [89]. Although the HOR activity of Ir alone is much lower than that of Pt [1, 2], the activity of the Ir–Ni nanoparticles was found to be slightly higher than that of a commercial Pt/C catalyst. This advantage predominantly is due to the Ni core-induced Ir shell contraction that makes the surface less reactive for Ir–OH formation, and the resulting more metallic Ir surface becomes more active for H_2 oxidation.

Ir- and Rh-based catalysts [90–93] have been investigated for ethanol oxidation reaction, and there have been some interesting and reasonably promising results. de Tacconi et al. carried out in situ FTIR study on polycrystalline Ir and Rh electrodes and showed ethanol oxidation on Ir leading selectively to either acetic acid or acetaldehyde, while Rh is a better catalyst in the total oxidation of ethanol to CO_2 [90]. Cao et al. studied carbon-supported Ir_3Sn nanoparticle electrocatalyst, and the fuel cell test results showed that the overall performance of Ir_3Sn/C was comparable to that of the Pt_3Sn/C catalyst [92].

Future Directions

The present chapter summarized the fundamental aspects and recent advances in electrocatalysts for the oxidation reactions of H_2/CO, methanol, and ethanol occurring at fuel cell anodes; emphases were placed on the state-of-the-art Pt–Ru- and Pt–Sn-based catalytic systems. Pt-based catalysts are still considered to be the most viable for the anodic reactions in acidic media. The major drawback however, is the price and limited reserves of Pt. To lower the Pt loading, the core-shell structure comprising Pt shells is more beneficial than the alloy structure, since all the Pt atoms on the nanoparticle surfaces can participate in the reactions (and those in cores do not); particularly, the Pt submonolayer/monolayer approach would be an ultimate measure to minimize the Pt content [30–35]. The architectures in nanoscale also have a significant effect on the reactivity and durability [54, 94] and thus should be explored continuously in the future. As for the ethanol oxidation, Rh addition is shown to enhance the selectivity towards C–C bond splitting [70, 71]; however, Rh is even more expensive than Pt, and thus less expensive constituents replacing Rh are necessary to be found.

Cross-References

- ► Direct Alcohol Fuel Cells (DAFCs)
- ► Electrocatalysis - Basic Concepts, Theoretical Treatments in Electrocatalysis via DFT-Based Simulations
- ► Electrocatalysis, Fundamentals - Electron Transfer Process; Current-Potential Relationship; Volcano Plots

References

1. Conway BE, Jerkiewicz G (2000) Relation of energies and coverages of underpotential and overpotential deposited H at Pt and other metals to the 'volcano curve' for cathodic H_2 evolution kinetics. Electrochim Acta 45:4075–4083
2. Trasatti S (1972) Work function, electronegativity, and electrochemical behavior of metals. J Electroanal Chem 39:163–184
3. Lamy C, Leger JM, Srinivasan S (2001) Direct methanol fuel cells: from a twentieth century electrochemist's dream to a twenty-first century emerging technology. In: Bockris JOM, Conway BE, White RE (eds) Modern aspects of electrochemistry, vol 34. Kluwer Academic/Plenum, New York, pp 53–118
4. Watanabe M, Motoo S (1975) Electrocatalysis by ad-atoms: part II. Enhancement of the oxidation of methanol on platinum by ruthenium ad-atoms. J Electroanal Chem 60:267–283
5. Dubau L, Coutanceau C, Garnier E, Leger JM, Lamy C (2003) Electrooxidation of methanol at platinum–ruthenium catalysts prepared from colloidal precursors: atomic composition and temperature effects. J Appl Electrochem 33:419–429
6. Watanabe M, Uchida M, Motoo S (1987) Preparation of highly dispersed Pt + Ru alloy clusters and the activity for the electrooxidation of methanol. J Electroanal Chem 229:395–406
7. Gurau B, Viswanathan R, Liu R, Lafrenz TJ, Ley KL, Smotkin ES, Reddington E, Sapenza A, Chan BC, Mallouk TE, Sarangapani S (1998) Structural and electrochemical characterization of binary, ternary, and quaternary platinum alloy catalysts for methanol electro-oxidation. J Phys Chem B 102:9997–10003
8. Choi JH, Park KW, Park IS, Nam WH, Sung YE (2004) Methanol electro-oxidation and direct methanol fuel cell using Pt/Rh and Pt/Ru/Rh alloy catalysts. Electrochim Acta 50:787–790
9. Traprailis H, Birss VI (2004) Sol–gel derived Pt-Ir mixed catalysts for DMFC applications. Electrochem Solid State Lett 7:A348–A352
10. Hammer B, Nørskov JK (2000) Theoretical surface science and catalysis – calculations and concepts. Adv Catal 45:71–129
11. Sasaki K, Zhang J, Wang J, Uribe F, Adzic R, Sasaki K, Zhang J, Wang J, Uribe F, Adzic RR (2006) Platinum submonolayer–monolayer electrocatalysts: an electrochemical and X-ray absorption spectroscopy study. Res Chem Intermediate 32:543–559
12. Bae IT, Scherson DA (1996) *Insitu* core-electron spectroscopy of carbon monoxide adsorbed on high-area platinum in an acid electrolyte. J Phys Chem 100:19215–19217
13. BaeI T, Scherson DA (1998) *In situ* real-time Pt L-III-edge x-ray absorption spectra of carbon monoxide adsorbed on platinum particles dispersed in high-area carbon in aqueous electrolytes. J Electrochem Soc 145:80–83
14. Blyholder GJ (1964) Molecular orbital view of chemisorbed carbon monoxide. J Phys Chem 68:2772–2777
15. Wong YT, Hoffmann R (1991) Chemisorption of carbon monoxide on three metal surfaces: nickel (111), palladium(111), and platinum(111): a comparative study. J Phys Chem 95:859–867
16. de Souza JPI, Queiroz SL, Bergamaski K, Gonzalez ER, Nart FC (2002) Electro-oxidation of ethanol on Pt, Rh, and PtRh electrodes. A study using DEMS and in-situ FTIR techniques. J Phys Chem B 106:9825–9830
17. Iwasita T, Pastor E (1994) DEM Sand FTIR spectroscopic investigation of adsorbed ethanol on polycrystalline platinum. Electrochim Acta 39:531–537
18. Xia XH, Liess HD, Iwasita T (1997) Early stages in the oxidation of ethanol at low index single crystal platinum electrodes. J Electroanal Chem 437: 233–240
19. Bittins-Cattaneo B, Wilhelm S, Cattaneo E, Buschmann HW, Vielstich W (1998) Intermediates and products of ethanol oxidation on platinum in acid-solution. Ber Bunsen-Ges Phys Chem 92:1210–1218
20. Gootzen JFE, Visscher W, Van Veen JAR (1996) Characterization of ethanol and 1,2-ethanediol adsorbates on platinized platinum with Fourier transform infrared spectroscopy and differential electrochemical mass spectrometry. Langmuir 12:5076–5082
21. Bittins-Cattaneo B, Cattaneo E, Königshoven P, Vielstich W (1991) In: Bard AJ (ed) Electroanalytical chemistry: a series of advances, vol 17. Marcel Dekker, New York
22. Wolter O, Heitbaum J (1984) The adsorption of CO on a porous Pt-electrode in sulfuric-acid studied by DEMS. Ber Bunsen-Ges Phys Chem 88:6–10
23. Fujiwara N, Friedrich KA, Stimming U (1999) Ethanol oxidation on Pt–Ru electrodes studied by differential electrochemical mass spectrometry. J Electroanal Chem 472:120–125
24. Iwasita T, Rasch B, Cattaneo E, Vielstich W (1989) A sniftirs study of ethanol oxidation on platinum. Electrochim Acta 34:1073–1079
25. Shao MH, Adzic RR (2005) Electrooxidation of ethanol on a Pt electrode in acid solutions: in situ ATR-SEIRAS study. Electrochim Acta 50:2415–2422

26. Iwasita T, Dalbeck R, Pastor E, Xia X (1994) Progress in the study of electrocatalytic reactions of organic species. Electrochim Acta 39:1817–1823
27. Lamy C, Rousseau S, Belgsir EM, Coutanceau C, Léger JM (2004) Recent progress in the direct ethanol fuel cell: development of new platinum–tin electrocatalysts. Electrochim Acta 49:3901–3908
28. Wang Q, Sun GQ, Jiang LH, Xin Q, Sun SG, Jiang YX, Chen SP, Jusys Z, Behm RJ (2007) Adsorption and oxidation of ethanol on colloid-based Pt/C, PtRu/C and Pt_3Sn/C catalysts: in situ FTIR spectroscopy and on-line DEMS studies. Phys Chem Chem Phys 9:2686–2696
29. Sato T, Kunimatsu K, Okaya K, Yano H, Watanabe M, Uchida H (2011) *In situ* ATR-FTIR analysis of the CO-tolerance mechanism on Pt_2Ru_3/C catalysts prepared by the nanocapsule method. Energy Environ Sci 4:433–438
30. Wang JX, Brankovic SR, Adzic RR (2001) Pt submonolayers on Ru nanoparticles – A novel low Pt loading, high CO tolerance fuel cell electrocatalyst. Electrochem Solid-State Lett 4:A217–A220
31. Brankovic SR, Wang JX, Zhu Y, Sabatini R, McBreen J, Adzic RR (2002) Electrosorption and catalytic properties of bare and Pt modified single crystal and nanostructured Ru surfaces. J Electroanal Chem 524:231–241
32. Wang JX, Brankovic SR, Zhu Y, Hanson JC, Adzic RR (2003) Kinetic characterization of PtRu fuel cell anode catalysts made by spontaneous Pt deposition on Ru nanoparticles. J Electrochem Soc 150:A1108–A1117
33. Sasaki K, Mo Y, Wang JX, Balasubramanian M, Uribe F, McBreen J, Adzic RR (2003) Pt submonolayers on metal nanoparticles – novel electrocatalysts for H_2 oxidation and O_2 reduction. Electrochim Acta 48:3841–3849
34. Sasaki K, Wang JX, Balasubramanian M, McBreen J, Uribe F, Adzic RR (2004) Ultra-low platinum content fuel cell anode electrocatalyst with a long-term performance stability. Electrochim Acta 49:3873–3877
35. Brankovic SR, Wang JX, Adzic RR (2001) Metal monolayer deposition by replacement of metal adlayers on electrode surfaces. Surf Sci 477:L173–L179
36. Chrzanowski W, Wieckowski A (1997) Ultrathin films of ruthenium on low index platinum single crystal surfaces: an electrochemical study. Langmuir 13:5974–5978
37. Chrzanowski W, Kim H, Wieckowski A (1998) Enhancement in methanol oxidation by spontaneously deposited ruthenium on low-index platinum electrodes. Catal Lett 50:69–75
38. Chen CH, Sarma LS, Wang DY, Lai FJ, Al Andra CC, Chang SH, Liu DG, Chen CC, Lee JF, Hwang BJ (2010) Platinum-decorated ruthenium nanoparticles for enhanced methanol electrooxidation. Chem Cat Chem 2:159–166
39. Sasaki K, Adzic RR (2008) Monolayer-level Ru- and NbO_2-supported platinum electrocatalysts for methanol oxidation. J Electrochem Soc 105:B180–B186
40. Antolini E (2007) Catalysts for direct ethanol fuel cells. J Power Sources 170:1–12
41. Zhou Z, Wang S, Zhou W, Wang G, Jiang L, Li W, Song S, Liu J, Sun G, Xin Q (2003) Novel synthesis of highly active Pt/C cathode electrocatalyst for direct methanol fuel cell. Chem Commun 3:394–395
42. Camara GA, de Lima RB, Iwasita T (2004) Catalysis of ethanol electro oxidation by PtRu: the influence of catalyst composition. Electrochem Commun 6:812–815
43. Markovic NM, Gasteiger HA, Ross PN, Jiang X, Villegas I, Weaver MJ (1995) Electro-oxidation mechanisms of methanol and formic acid on Pt-Ru alloy surfaces. Electrochim Acta 40:91–98
44. Tanaka S, Umeda M, Ojima H, Usui Y, Kimura O, Uchida I (2005) Preparation and evaluation of a multi-component catalyst by using a co-sputtering system for anodic oxidation of ethanol. J Power Sources 152:34–39
45. Wang Z, Yin G, Zhang J, Sun Y, Shi P (2006) Co-catalytic effect of Ni in the methanol electro-oxidation on Pt-Ru/C catalyst for direct methanol fuel cell. Electrochim Acta 51:5691–5697
46. Paulus UA, Wokaun A, Scherer GG, Schmidt TJ, Stamenkovic V, Radmilovic V, Markovic NM, Ross PN (2002) Oxygen reduction on carbon-supported Pt-Ni and Pt-Co alloy catalysts. J Phys Chem B 106:4181–4191
47. Park KW, Choi JH, Kwon BK, Lee SA, Sung YE, Ha HY, Hong SA, Kim HS, Wieckowski A (2002) Chemical and electronic effects of Ni in Pt/Ni and Pt/Ru/Ni alloy nanoparticles in methanol electrooxidation. J Phys Chem B 106:1869–1877
48. Oliveira Neto A, Franco EG, Arico E, Linardi M, Gonzalez ER (2003) Electro-oxidation of methanol and ethanol on Pt–Ru/C and Pt–Ru–Mo/C electrocatalysts prepared by Bönnemann's method. J Eur Ceram Soc 23:2987–2992
49. Kelaidopoulou A, Abelidou E, Kokkinidis G (1999) Electrocatalytic oxidation of methanol and formic acid on dispersed electrodes: Pt, Pt-Sn and Pt/M (upd) in poly(2-hydroxy-3-aminophenazine). J Appl Electrochem 29:1255–1261
50. Li G, Pickup PG (2006) The promoting effect of Pb on carbon supported Pt and Pt/Ru catalysts for electro-oxidation of ethanol. Electrochim Acta 52:1033–1037
51. Calegaro ML, Suffredini HB, Machado SAS, Avaca LA (2006) Preparation, characterization and utilization of a new electrocatalyst for ethanol oxidation obtained by the sol–gel method. J Power Sources 156:300–305
52. Stamenkovic VR, Arenz M, Lucas CA, Gallagher ME, Ross PN, Markovic NM (2003) Surface chemistry on bimetallic alloy surfaces: adsorption of anions and oxidation of CO on $Pt_3Sn(111)$. J Am Chem Soc 125:2735–2745
53. Dupont C, Jugnet Y, Loffreda D (2006) Theoretical evidence of PtSn alloy efficiency for CO oxidation. J Am Chem Soc 128:9129–9136

54. Liu Z, Jackson GS, Eichhorn BW (2010) PtSn Intermetallic, core-shell, and alloy nanoparticles as CO-tolerant electrocatalysts for H_2 oxidation. Angew Chem Int Ed 49:3173–3176
55. Colmati F, Antolini E, Gonzalez ER (2005) Pt-Sn/C electrocatalysts for methanol oxidation synthesized by reduction with formic acid. Electrochim Acta 50:5496–5503
56. Bonneman H, Brijoux W, Brinkmann R, Dinjus E, Jouen T, Korall B (1991) Formation of colloidal transition metals in organic phases and their application in catalysis. Angew Chem Int Ed Engl 30:1312–1314
57. Liu Z, Guo B, Hong L, Lim TH (2006) Microwave heated polyol synthesis of carbon supported PtSn nanoparticles for methanol electrooxidation. Electrochem Commun 8:83–90
58. Wang K, Gasteiger HA, Markovic NM, Ross PN (1996) On the reaction pathway for methanol and carbon monoxide electrooxidation on Pt-Sn alloy versus Pt-Ru alloy surfaces. Electrochim Acta 41:2587–2593
59. Götz M, Wendt H (1998) Binary and ternary anode catalyst formulations including the elements W, Sn and Mo for PEMFCs operated on methanol or reformate gas. Electrochim Acta 43:3637–3644
60. Morimoto Y, Yeager EB (1998) Comparison of methanol oxidations on Pt, Pt|Ru and Pt|Sn electrodes. J Electroanal Chem 444:95–100
61. MacDonald JP, Gualtieri B, Runga N, Teliz E, Zinola CF (2008) Modification of platinum surfaces by spontaneous deposition: methanol oxidation electrocatalysis. Int J Hydrogen Energ 33:7048–7061
62. Zhou WJ, Zhou B, Li WZ, Zhou ZH, Song SQ, Sun GQ, Xin Q, Douvartzides S, Goula M, Tsiakaras P (2004) Performance comparison of low-temperature direct alcohol fuel cells with different anode catalysts. J Power Sources 126:16–22
63. Lamy C, Belgsir EM, Léger JM (2001) Electrocatalytic oxidation of aliphatic alcohols: application to the direct alcohol fuel cell (DAFC). J Appl Electrochem 31:799–809
64. Colmati F, Antolini E, Gonzalez ER (2006) Effect of temperature on the mechanism of ethanol oxidation on carbon supported Pt, PtRu and Pt_3Sn electrocatalysts. J Power Sources 157:98–103
65. Song SQ, Zhou WJ, Zhou ZH, Jiang LH, Sun GQ, Tsiakaras P, Xin Q, Leonditis V, Kontou S, Tsiakaras P (2005) Direct ethanol PEM fuel cells: the case of platinum based anodes. Int J Hydrogen Energ 30:995–1001
66. Jiang L, Sun G, Sun S, Liu J, Tang S, Li H, Zhou B, Xin Q (2005) Structure and chemical composition of supported Pt–Sn electrocatalysts for ethanol oxidation. Electrochim Acta 50:5384–5389
67. Bonneman H, Britz P, Vogel W (1998) Structure and chemical composition of a surfactant stabilized Pt3Sn alloy colloid. Langmuir 14:6654–6657
68. Jiang L, Colmenaresa L, Jusysa Z, Sunb GQ, Behma RJ (2007) Ethanol electrooxidation on novel carbon supported Pt/SnOx/C catalysts with varied Pt:Sn ratio. Electrochim Acta 53:377–389
69. Zhou WP, Axnanda S, White MG, Adzic RR, Hrbek J (2011) Enhancement in ethanol electrooxidation by SnOx nanoislands grown on Pt(111): effect of metal oxide-metal interface sites. J Phys Chem C 115:16467–16473
70. Kowal A, Li M, Shao M, Sasaki K, Vukmirovic MB, Zhang J, Marinkovic NS, Liu P, Frenkel AI, Adzic RR (2009) Ternary Pt/Rh/SnO_2 electrocatalysts for oxidizing ethanol to CO_2. Nat Mater 8:325–330
71. Li M, Kowal A, Sasaki K, Marinkovic NS, Su D, Korach E, Liu P, Adzic RR (2010) Ethanol oxidation on the ternary Pt-Rh-SnO_2/C electrocatalysts with varied Pt:Rh:Sn ratios. Electrochim Acta 55: 4331–4338
72. Li M, Marinkovic NS, Sasaki K (2012) In situ characterization of ternary Pt-Rh-SnO_2/C catalysts for ethanol electrooxidation. Electrocatalysis 3(3–4): 76–385
73. Spinace EV, Linardi M, Oliveira Neto A (2005) Co-catalytic effect of nickel in the electro-oxidation of ethanol on binary Pt-Sn electrocatalysts. Electrochem Commun 7:365–369
74. Rousseau S, Coutanceau C, Lamy C, Leger JM (2006) Direct ethanol fuel cell (DEFC): electrical performances and reaction products distribution under operating conditions with different platinum-based anodes. J Power Sources 158:18–24
75. Antolini E, Colmatia F, Gonzalez ER (2007) Effect of Ru addition on the structural characteristics and the electrochemical activity for ethanol oxidation of carbon supported Pt–Sn alloy catalysts. Electrochem Commun 9:398–404
76. Sine G, Smida D, Limat M, Foti G, Comninellis C (2007) Microemulsion synthesized Pt/Ru/Sn nanoparticles on BDD for alcohol electro-oxidation. J Electrochem Soc 154:B170–B174
77. Neto AO, Dias RR, Tusi MM, Linardi M, Spinace EV (2007) Electro-oxidation of methanol and ethanol using PtRu/C, PtSn/C and PtSnRu/C electrocatalysts prepared by an alcohol-reduction process. J Power Sources 166:87–91
78. Du W, Wang Q, LaScala CA, Zhang L, Su D, Frenkel AI, Mathura VK, Teng X (2011) Ternary PtSnRh–SnO_2 nanoclusters: synthesis and electroactivity for ethanol oxidation fuel cell reaction. J Mater Chem 21:8887–8892
79. Colmati F, Antolini E, Gonzalez ER (2008) Preparation, structural characterization and activity for ethanol oxidation of carbon supported ternary Pt-Sn-Rh catalysts. J Alloy Compd 456:264–270
80. Watanabe M, Zhu Y, Igarashi H, Uchida H (2000) Mechanism of CO tolerance at Pt-alloy anode catalysts for polymer electrolyte fuel cells. Electrochemistry 3:244–251
81. Uchida H, Izumi K, Aoki K, Watanabe M (2009) Temperature-dependence of hydrogen oxidation reaction rates and CO-tolerance at carbon-supported

Pt, Pt-Co, and Pt-Ru catalysts. Phys Chem Chem Phys 11:1771–1779
82. Antolini E, Salgado JRC, Gonzalez ER (2006) The methanol oxidation reaction on platinum alloys with the first row transition metals – The case of Pt-Co and -Ni alloy electrocatalysts for DMFCs: a short review. Appl Catal B 63:137–149
83. Strasser P, Koh S, Anniyev T, Greeley J, More K, Yu C, Liu Z, Kaya S, Norlund D, Ogasawara H, Toney MF, Nilsson A (2010) Lattice-strain control of the activity in dealloyed core–shell fuel cell catalysts. Nat Chem 2:454–460
84. Stamenkovic VR, Fowler B, Mun BS, Wang GF, Ross PN, Lucas CA, Markovic NM (2007) Improved oxygen reduction activity on Pt3Ni(111) via increased surface site availability. Science 315: 493–497
85. Shao MH (2011) Palladium-based electrocatalysts for hydrogen oxidation and oxygen reduction reactions. J Power Sources 196:2433–2444
86. Norskov JK, Bligaard T, Logadottir A, Kitchin JR, Chen JG, Pandelov S (2006) Response to "comment on 'Trends in the exchange current for hydrogen evolution' [J Electrochem Soc 152:J23 (2005)]". J Electrochem Soc 153:L33–L33
87. Cho YH, Choi B, Cho YH, Park HS, Sung YE (2007) Pd-based PdPt(19: 1)/C electrocatalyst as an electrode in PEM fuel cell. Electrochem Commun 9:378–381
88. Shao MH, Huang T, Liu P, Zhang J, Sasaki K, Vukmirovic MB, Adzic RR (2006) Palladium monolayer and palladium alloy electrocatalysts for oxygen reduction. Langmuir 22:10409–10415
89. Sasaki K, Kuttiyiel KA, Barrio L, Su D, Frenkel AI, Marinkovic N, Mahajan D, Adzic RR (2011) Carbon-supported IrNi core-shell nanoparticles: synthesis, characterization, and catalytic activity. J Phys Chem C 115:9894–9902
90. de Tacconi NR, Lezna RO, Beden B, Hahn F, Lamy C (1994) In-situ FTIR study of the electrocatalytic oxidation of ethanol at iridium and rhodium electrodes. J Electroanal Chem 379:329–337
91. Fujiwara N, Siroma Z, Ioroi T, Yasuda K (2007) Rapid evaluation of the electrooxidation of fuel compounds with a multiple-electrode setup for direct polymer electrolyte fuel cells. J Power Sources 164:457–463
92. Cao L, Sun G, Li H, Xin Q (2007) Carbon-supported IrSn catalysts for a direct ethanol fuel cell. Electrochem Commun 9:2541–2546
93. Choi Y, Liu P (2011) Understanding of ethanol decomposition on Rh(111) from density functional theory and kinetic Monte Carlo simulations. Catal Today 165:64–70
94. Sasaki K, Naohara H, Cai Y, Choi YM, Liu P, Vukmirovic MB, Wang JX, Adzic RR (2010) Core-protected platinum monolayer shell high-stability electrocatalysts for fuel-cell cathodes. Angew Chem Int Ed 49:8602–8607

Electrocatalysis of Chlorine Evolution

Branislav Ž. Nikolić[1] and Vladimir Panić[2]
[1]Department of Physical Chemistry and Electrochemistry, University of Belgrade, Faculty of Technology and Metallurgy, Belgrade, Serbia
[2]University of Belgrade, Institute of Chemistry, Technology and Metallurgy, Belgrade, Serbia

Electrochemical Production of Chlorine

The production of chlorine by electrochemical oxidation of chlorides (chlor–alkali technology, CAT) is nowadays one of the largest processes in industrial electrochemistry. The process, which spends two moles of electricity (2 F), is based on the following stoichiometric equation:

$$2NaCl + 2H_2O + 2F \rightarrow Cl_2 + 2NaOH + H_2 \quad (1)$$

The main products – chlorine and the corresponding hydroxide – are, besides sulfuric acid and ammonia, the most important industrial inorganic chemicals. The third product – hydrogen (as a by-product) – is very valuable from the standpoints of contemporary power sources ("hydrogen economy;" fuel cells) and different industrial synthesis (mostly organic).

Historical Overview

Although the possibility of electrolytic decomposition of sodium chloride was indicated by W. Cruikshank back in 1800, the first electrolysis plant (diaphragm-type cells) was introduced in 1888 in the company Griesheim (Germany). A few years later (1892), H. Y. Castner and C. Kellner patented independently CAT cells with flowing mercury cathodes. During last decades of twentieth century, membrane-type cells were developed and applied by Hooker

Chemicals and Plastics (USA) and Diamond Shamrock Corp. (USA) [1]. This type of cells is becoming dominant in the CAT plants of the future.

Anode Materials

Without going into details of the three types of CAT cells, it should be noted that the anodic process – the chlorine evolution reaction (CER) from chlorides – is the same for all of them:

$$2Cl^- \rightarrow Cl_2 + 2e^- \quad E^0 = 1.359 \text{ V} \qquad (2)$$

However, the electrochemical oxygen evolution reaction (OER) from the water in acid solutions is thermodynamically slightly more favorable than the CER given by Eq. 2 (the E^0 value for the OER is 1.229 V). Bearing in mind that electrocatalysis essentially implies the activity and selectivity of various electrode materials toward a certain electrode reaction [2], the correct choice of electrode material that is of high electrocatalytic activity for the CER and of lower activity for the OER is of extreme importance. Anyhow, CAT is always subjected to current loss due to the OER as a side reaction. This fact sets additional demands toward the anode material for the CER, i.e., its stability toward the OER and consequently its durability in CAT.

The first anode materials in CAT cells were magnetite and carbon, but the methods for manufacture of synthetic graphite in electric furnaces were the most important discoveries in CATs since the first age of industrial chlorine production. Graphite, usually impregnated with linseed oil, has been used almost exclusively in CAT cells since the end of the nineteenth century.

Dimensionally Stable Anodes

In late 70s of the twentieth century, dimensionally stable anodes, DSA®, were introduced in CAT, bringing forth one of the most important breakthroughs in industrial electrolytic processes of the century. DSA were invented by H. B. Beer (Brit. Patent 1-147-442 (1965)), but O. De Nora was the first to report their industrial application [2]. The contributing paper by V. De Nora and A. Nidola, at the Electrochemical Society Meeting, Los Angeles, CA, 1970, opened this subject to the scientific community.

The beginning of industrial chlorine production has been constantly followed by intensive investigations toward the most suitable electrocatalytic material for CER. CER kinetics has been studied on a variety of anode materials, starting from different types of graphite, different metals (Pt, Ir, Rh, and their alloys), and metal oxides (Co_3O_4, WO_3, Fe_3O_4, PbO_2, MnO_2, PtO_x, IrO_2, and RuO_2) [2–4]. Figure 1, which shows typical polarization curves for the CER at different anode materials, illustrates the supreme electrocatalytic properties of TiO_2–RuO_2 DSA [5]. The overpotential in the CER, η, is considerably lower than at other materials, especially at current densities, j, typical for CAT (0.2–1 A cm^{-2}). In comparison to the formerly applied CAT anodes, the energy savings with DSA can reach 1 kW h per cm^2 of anode surface. Thus, extreme activity, in addition to good selectivity and stability, led toward the exclusive application of DSA in CAT. Essentially, RuO_2 is combined in DSA with TiO_2. The similarity of the crystal lattices parameters of the two oxides, the high corrosion resistivity of TiO_2 in aggressive media and the excellent electrocatalytic properties and durability of RuO_2 in the CER, led to the introduction of TiO_2–RuO_2 DSA as almost ideal anodes for CAT [3].

It is believed that such DSA owe their high activity not only to the overlapping of the d-orbitals of Ru in higher oxidation states with oxygen p-orbitals but also to the hydrous nature of its oxide. RuO_2 maintains its high degree of hydration even at temperatures as high as 500 °C in DSA fabrication [3].

Chlorine Evolution Kinetics at DSA

It was the high activity of DSA for the CER that enabled an accurate evaluation of the kinetic

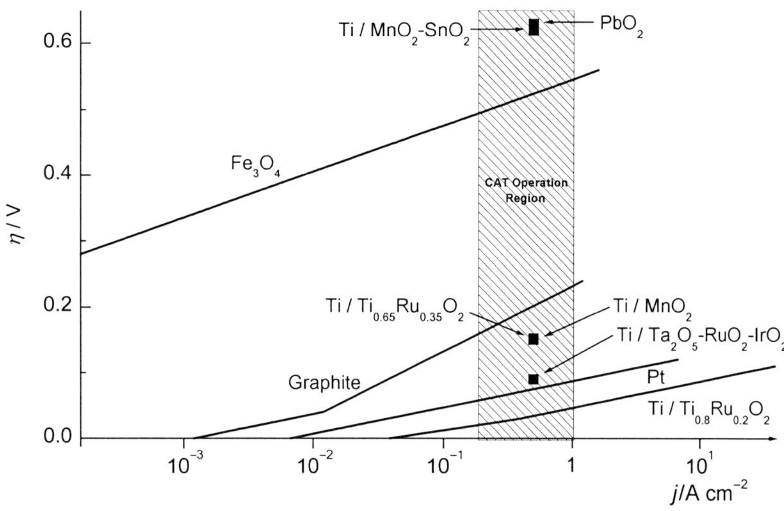

Electrocatalysis of Chlorine Evolution, Fig. 1 Polarization curves for the chlorine evolution from chlorides at different anode materials (constructed from the data in Refs. [4] and [5])

parameters and mechanism of the CER according to the following experimental findings [6, 7]:

- The Tafel slope changes from 30 mV at lower overpotentials to 40 mV at the higher,
- The limiting current density of Cl_2 reduction (which is not assignable to mass transport limitations) equals anodic exchange current density related to the 30 mV slope,
- The anodic reaction orders are 1 per Cl^- and 0 per Cl_2, whereas the cathodic ones are 1 per Cl_2 and −1 per Cl^-.

Concerning the CER kinetics, it was accordingly concluded by the pioneer in this field, L. I. Krishtalik that the CER proceeds on a DSA as a barrier-less reaction at low over potentials. Taking into account all the possibilities, the most probable mechanism follows the so-called Volmer–Krishtalik–Tafel mechanism:

$$Cl^- \rightarrow Cl_{ads} + e^- \quad (3)$$

$$Cl_{ads} \rightarrow Cl^+ + e^- \quad (4)$$

$$Cl^+ + Cl^- \rightarrow Cl_2 \quad (5)$$

with step (4) as the rate-determining one. Cl^+ could be adsorbed as a cation, but most likely Cl^+ is present as hypochlorite, with oxygen originating from crystal lattice of the oxide, which could explain the high CER activity of RuO_2. In this sense, it is believed that step (4) proceeds simultaneously with the reversible oxidation of Ru.

Another possibility is a two-step mechanism involving a very fast step (3) and recombination (30 mV slope) or electrochemical desorption as the rate-determining step (40 mV slope) [9, 10]. In the case of graphite, step (3) could be rate determining, which explains the 120 mV slope.

More detailed experiments enabled slight modifications of the Volmer–Krishtalik–Tafel mechanism to involve changes in the electrocatalytic material itself [8]:

$$(-S^Z) + H^+ \rightarrow (-H^{Z+1}) \quad (6)$$

$$(-S^Z) \rightarrow (-S^{Z+1}) + e^- \quad (7)$$

$$(-S^{Z+1}) + Cl^- \rightarrow (-SCl^{Z+1}) + e^- \quad (8)$$

$$(-SCl^{Z+1}) + Cl^- \rightarrow (-S^Z) + Cl_2 \quad (9)$$

$$2(-SCl^{Z+1}) \rightarrow 2(-S^{Z+1}) + Cl_2 \quad (10)$$

where $(-S^Z)$ denotes a RuO_2 active site. Step (6) is intrinsic to the electrocatalyst itself, irrespective of the nature of the reaction occurring on the catalyst surface. This extended mechanism is able to explain the dependence of the CER on pH, which was a shortcoming of the basic Volmer–Krishtalik–Tafel mechanism.

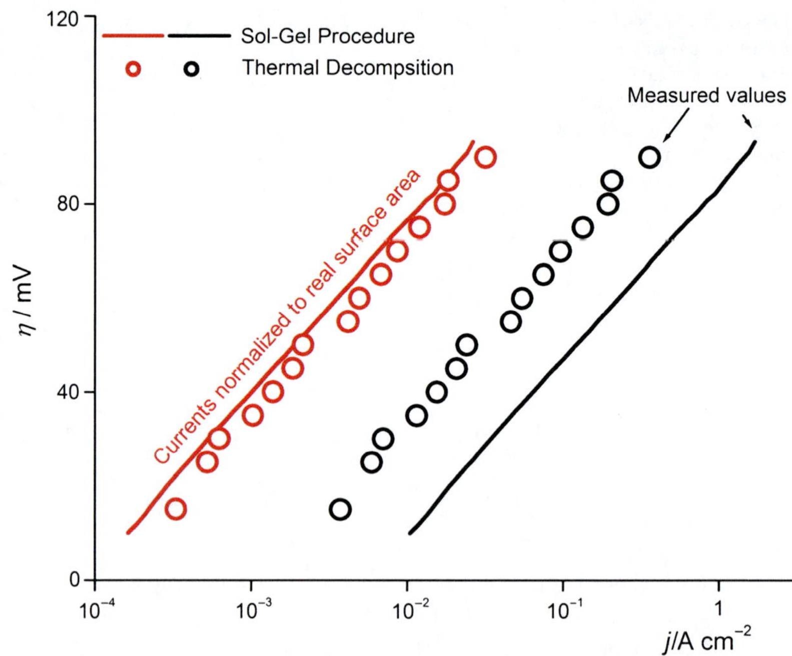

Electrocatalysis of Chlorine Evolution, Fig. 2 The measured and normalized Tafel plots of thermally and sol–gel prepared DSA for the chlorine evolution reaction. The normalized data correspond to the "electronic" electrocatalytic factor

DSA and Nanotechnology

Novel types of synthesis of modern electrocatalysts revealed that the properties of electrode materials can be affected by the controlled formation of nano-sized, finely dispersed, electrocatalyst particles. In the case of DSA, already the traditional preparation procedure involves the thermal decomposition of the corresponding chlorides after dissolution in an appropriate solvent (usually a solvent of low viscosity, e.g., 2-propanol) [3]. Recently, sol–gel synthesis was introduced for DSA preparation, with the main effect being related to the increase in the real surface area of the anode [9, 10]. The effect is recognized as the "geometric" factor of increased electrocatalytic ability in addition to an "electronic" factor related to the chemical structure of electrocatalyst [2, 4], which is essential for step (6). The "geometric" factor is important since the measure for the reaction rate is the current density, i.e., the current per surface area of the electrode available for the reaction. Thus, the reaction rate can be considerably increased by the application of nano-3D electrodes, which are porous systems with an extended real surface area. The polarization curves for the CER on DSA prepared by the traditional thermal decomposition and the sol–gel procedure, as typical for the synthesis of uniform nano-sized oxide particles, are compared in Fig. 2 [9].

In order to resolve the "geometric" factor, the registered currents have to be divided by the real surface area. The real surface area can be determined by physical adsorption methods (e.g., the Brunauer–Emmett–Teller, BET, method), but the obtained value is difficult to apply to electrochemistry due to restricted accessibility of reacting species to a 3D surface. As a rule, the adsorbing species in the BET method can more easily reach the whole surface, which is hardly accessible to electrochemically reacting species [2]. Hence, electrocatalysis recognizes the so-called electrochemically active real surface area (EARSA), the measure of which is the double layer capacitance of a 3D surface (electrochemical capacitor). This capacitance can be easily quantified by standard electrochemical measuring methods, e.g., cyclic voltammetry [2], and used to normalize the measured currents in order to check "electronic" and "geometric" factors of electrocatalytic ability of an anode. As it is seen in Fig. 2, the sol–gel prepared DSA shows higher currents for the CER, which is of practical importance. However, if currents are

Electrocatalysis of Chlorine Evolution, Table 1 The durability of different DSA coatings as obtained by the accelerated stability test (AST)

DSA coating [references]	Coating amount (mg cm^{-2})	Preparation procedure	AST conditions	Durability (h)
Industrial DSA® [15]	≈2	Thermal decomposition of chlorides	2 M H_2SO_4 25 °C 1.5 A cm^{-2}	≈10
RuO_2 [16, 17]	2–4	Thermal decomposition of chloride; Pechini method	1 M $HClO_4$ 50 °C 0.8 A cm^{-2}	8
	1.0	Inorganic sol–gel procedure	0.5 M NaCl pH 2; 33 °C 2 A cm^{-2}	3–4
$Ti_{0.6}Ru_{0.4}O_2$ [18, 19]	1.2	Thermal decomposition of chlorides	0.5 M NaCl 25 °C 3 A cm^{-2}	9
	1.0	Thermal decomposition of chlorides	0.5 M NaCl pH 2; 33 °C 2 A cm^{-2}	4–5
		Inorganic sol–gel procedure		7–17
$Ti_{0.7}Ru_{0.3}O_2$ [20, 21]	Below 1	Thermal decomposition of chlorides	0.5 M NaCl 25 °C 0.6 A cm^{-2}	18
	1.2	Pechini method	1 M $HClO_4$ 25 °C 0.4 A cm^{-2}	45
$Ti_{0.4}Ru_{0.3}Ir_{0.3}O_2$ [20]	Below 1	Thermal decomposition of chlorides	0.5 M NaCl 25 °C 0.6 A cm^{-2}	30
$Ti_{0.6}Ru_{0.3}Ir_{0.1}O_2$ [22]	1.0	Inorganic sol–gel procedure	Sea water 16 °C 0.6 A cm^{-2}	20
$Sn_{0.7}Ru_{0.3}O_2$ [21]	1.2	Pechini method	1 M $HClO_4$ 25 °C 0.4 A cm^{-2}	80

divided by EARSA, practically the same CER activity of sol–gel and thermally prepared anode is obtained (i.e., these anodes are of similar "electronic" activity).

The Stability and Durability of DSA

The term "dimensionally stable" relates to the whole anode assembly including the titanium substrate (of different shapes) and the electrocatalytic coating, which, however, suffers from wear processes. The stability implies the ability of the DSA to function under a constant overpotential at a constant imposed current. The anode loses electrocatalytic activity (the end of its service life) with a sudden increase in overpotential under a constant current regime. The literature [11–14] recognizes two causes for the loss of electrocatalytic activity: (a) the electrochemical dissolution of the active component in the coating – RuO_2 – as a side reaction to the CER and (b) the growth of an insulating TiO_2 layer in the coating/Ti substrate interphase due to oxidation of the substrate. The latter cause becomes dominant at higher operational current densities [11].

The DSA durability depends not only on the coating thickness and composition but also on the preparation conditions of the coating. The most stable state of a TiO_2–RuO_2 DSA is reached for coatings thermally treated at 450–550 °C that contain 30–60 mol % of RuO_2. The stability considerably decreases for coatings with a RuO_2 content above 60 mol %. This finding indicates the important role of TiO_2 as the stabilizing component of a coating. Table 1 summarizes some of the data related to the durability of different types of DSA prepared by different procedures and tested for stability under different conditions (involving the CER and/or the OER) in the so-called accelerated stability test (AST) [12]. It follows that the characteristics of the electrolyte are essential for faster coating degradation.

DSAs can operate for years in hot saturated brine solution, whereas the data from the Table 1 show that service life is considerably shorter in cold dilute chloride or chloride-free solutions. This introduces the OER as the crucial factor for the loss of coating activity. Another important finding is that RuO_2 durability is negligibly affected by increasing the amount of coating above 2 mg cm^{-2}. This is because cause (b) is dominant for the loss with respect to cause (a) for thick coatings [14]. Bearing in mind that noble metal oxides are rather expensive, this finding is important for the projection of CAT costs.

Besides the good effects on the CER activity (Fig. 2), the controlled synthesis of rather uniform oxide phases by application of the sol–gel procedure is also beneficial for coating stability (Table 1). Sol–gel processed TiO_2–RuO_2 coatings last 30–40 % longer than those prepared by the traditional thermal decomposition of chlorides. Table 1 also evidences that insertion of IrO_2 and/or SnO_2 improves the durability of DSA. Hence, these oxides are unavoidable components of industrial DSA.

Future Directions

Noble metal oxides of dimensionally stable anodes are today best-known electrocatalysts with fairly long service lives in the chlorine evolution reaction. Similar to platinum, which is the best catalyst for the majority of electrochemical reactions of practical importance, e.g., those applied in energy storage devices, their main disadvantages are low abundance and consequently high price. On the other hand, it is to believe that some chip replacement for these electrode materials could be discovered in the near future; hence, the relationship between activity–stability and the cost should be considered as judicial in the search for the solution for sustainable electrochemical processes in the future. However, the tailoring of the synthesis of electrode materials at the nanolevel, in order to increase the catalyst efficiency and/or to reduce the catalyst consumption, is seen as a valuable future direction in the development of advanced materials in chlorine electrocatalysis.

Cross-References

► Chlorate Synthesis Cells and Technology
► Chlorine and Caustic Technology, Overview and Traditional Processes
► Cyclic Voltammetry
► Electrocatalysis - Basic Concepts, Theoretical Treatments in Electrocatalysis via DFT-Based Simulations
► Electrocatalysis of Anodic Reactions
► Hypochlorite Synthesis Cells and Technology, Sea Water

References

1. Coulter MO (ed) (1980) Modern chlor–alkali technology. Ellis Horwood, Chichester
2. Trasatti S (2000) Electrocatalysis: understanding the success of DSA®. Electrochim Acta 45:2377–2385. doi:10.1016/S0013-4686(00)00338-8
3. Trasatti S, O'Grady WE (1982) Properties and applications of RuO_2-based electrodes. In: Gerisher H, Tobias CW (eds) Advances in electrochemistry and electrochemical engineering, vol 12. Wiley, New York
4. Trasatti S (1999) Interfacial electrochemistry of conductive oxides for electrocatalysis. In: Wieckowski A (ed) Interfacial electrochemistry – theory, experiment and applications. Marcel Dekker, New York
5. Cardarelli F (2008) Materials handbook: a concise desktop reference, 2nd edn. Springer, London

6. Krishtalik LI (1981) Kinetics and mechanism of anodic chlorine and oxygen evolution reactions on transition metal oxide electrodes. Electrochim Acta 26:329–337. doi:10.1016/0013-4686(81)85019-0
7. Krishtalik LI (1983) Kinetics of electrochemical reactions at metal–solution interfaces. In: Conway B, Bockris J, Yeager E, Khan S, White R (eds) Kinetics and mechanisms of electrode processes, vol 7, Comprehensive treatise of electrochemistry. Plenum Press, New York
8. Fernández JL, de Chialvo MRG, Chialvo AC (2002) Kinetic study of the chlorine electrode reaction on Ti/RuO_2 through the polarisation resistance. Electrochim Acta 47:1129–1152. doi:10.1016/S0013-4686(01)00837-4; 10.1016/S0013-4686(01)00838-6; 10.1016/S0013-4686(01)00839-8
9. Panić V, Dekanski A, Mišković-Stanković VB, Nikolić B, Milonjić S (2005) The role of sol–gel procedure conditions in electrochemical behavior and corrosion stability of Ti/[RuO_2–TiO_2] anodes. Mater Manuf Process 20:89–103. doi:10.1081/AMP-200041645
10. Panić VV, Nikolić BŽ (2008) Electrocatalytic properties and stability of titanium anodes activated by the inorganic sol–gel procedure. J Serb Chem Soc 73:1083–1112. doi:10.2298/JSC0811083P
11. Beck F (1992) Wear mechanisms of anodes. Electrochim Acta 34:811–822. doi:10.1016/0013-4686(89)87114-2
12. Gajić-Krstajić LM, Trišović TL, Krstajić NV (2004) Spectrophotometric study of the anodic corrosion of Ti/RuO_2 electrode in acid sulfate solution. Corros Sci 46:65–74. doi:10.1016/S0010-938X(03)00111-2
13. Jovanović VM, Dekanski A, Despotov P, Nikolić BŽ, Atanasoski RT (1992) The roles of the ruthenium concentration profile, the stabilizing component and the substrate on the stability of oxide coatings. J Electroanal Chem 339:147–165. doi:10.1016/0022-0728(92)80449-E
14. Panić V, Dekanski A, Mišković-Stanković VB, Milonjić S, Nikolić B (2005) On the deactivation mechanism of RuO_2–TiO_2/Ti anodes prepared by the sol–gel procedure. J Electroanal Chem 579:67–76. doi:10.1016/j.jelechem.2005.01.026
15. Pilla AS, Cobo EO, Duarte MME, Salinas DR (1997) Evaluation of anode deactivation in chlor–alkali cells. J Appl Electrochem 27:1283–1289. doi:10.1023/A:1018444206334
16. Terezo AJ, Pereira EC (2002) Preparation and characterisation of Ti/RuO_2 anodes obtained by sol–gel and conventional routes. Mater Lett 53:339–345. doi:10.1016/S0167-577X(01)00504-3
17. Panić V, Dekanski A, Milonjić SK, Atanasoski R, Nikolić B (2000) The influence of the aging time of RuO_2 sol on the electrochemical properties of the activated titanium anodes obtained by sol–gel procedure. In: Uskoković DP, Battiston GA, Nedeljković JM, Milonjić SK, Raković DI (eds) Trends in advanced materials and processes, vol 352, Materials science forum. Trans Tech, Zurich
18. Panić VV, Dekanski A, Milonjić SK, Atanasoski RT, Nikolić BŽ (1999) RuO_2–TiO_2 coated titanium anodes obtained by the sol–gel procedure and their electrochemical behaviour in the chlorine evolution reaction. Colloids Surf A 157:269–274. doi:10.1016/S0927-7757(99)00094-1
19. Panić V, Dekanski A, Milonjić S, Atanasoski R, Nikolić B (2000) The influence of the aging time of RuO_2 and TiO_2 sols on the electrochemical properties and behavior for the chlorine evolution reaction of activated titanium anodes obtained by the sol–gel procedure. Electrochim Acta 46:415–421. doi:10.1016/S0013-4686(00)00600-9
20. Fathollahi F, Javanbakht M, Norouzi P, Ganjali MR (2011) Comparison of morphology, stability and electrocatalytic properties of $Ru_{0.3}Ti_{0.7}O_2$ and $Ru_{0.3}Ti_{0.4}Ir_{0.3}O_2$ coated titanium anodes. Russ J Electrochem 47:1281–1286. doi:10.1134/S1023193511110061
21. Forti JC, Olivi P, de Andrade AR (2001) Characterisation of DSA®-type coatings with nominal composition Ti/$Ru_{0.3}Ti_{(0.7-x)}Sn_xO_2$ prepared via a polymeric precursor. Electrochim Acta 47:913–920. doi:10.1016/S0013-4686(01)00791-5
22. Panić VV, Nikolić BŽ (2007) Sol–gel prepared active ternary oxide coating on titanium in cathodic protection. J Serb Chem Soc 72:1393–1402. doi:10.2298/JSC0712393P

Electrocatalysis, Fundamentals - Electron Transfer Process; Current-Potential Relationship; Volcano Plots

Svetlana B. Strbac[1] and Radoslav R. Adzic[2]
[1]ICTM-Institute of Electrochemistry, University of Belgrade, Belgrade, Serbia
[2]Chemistry Department, Brookhaven National Laboratory, Upton, NY, USA

Introduction

Electrocatalysis is the science exploring the rates of electrochemical reactions as a function of the electrode surface properties. In these heterogeneous reactions, the electrode does not only accepts or supplies electrons (electron transfer), as in simple redox reactions, but affects the reaction rates interacting with reactants, intermediates, and reaction products, i.e., acts as a catalyst remaining unchanged upon its completion.

The term *electrocatalysis*, an extension to electrochemistry of the term catalysis (Greek *kata* (down) and *lyein* (to let)), was apparently first used in 1934[1]. The beginning of intensive research in this area can be traced back to early 1960s in connection with the broadening fuel cell research. Many electrocatalytic reactions have great importance. These include hydrogen, oxygen, and chlorine evolution; oxygen reduction oxidation of small organic molecules suitable for energy conversion (methanol, ethanol, formic acid); and reactions of organic syntheses. The limitations of the operating temperature for aqueous solutions in electrocatalysis, compared those in catalysis, are compensated by the possibility to increase the reaction rates by applied potential. The rates of some reactions can be increased several orders of magnitude by small change of potential. Such an increase in the rate of chemical reaction would require very high temperatures. Important features of electrocatalytic reactions, facilitated by the application of the electrode potential, include (i) high reaction rates that can be achieved, (ii) high selectivity at defined potentials, and (iii) the unique direct energy conversion in fuel cells that are likely to become one of the major sources of clean energy. The main events in an electrocatalytic reaction are adsorption/desorption, electron transfer, and bond breaking/formation.

Electron Transfer Process

Electrochemical reactions are heterogeneous chemical reactions in which electrons are exchanged between the electrode and the molecules or ions in the electrolyte. The electrode is metal or other electronic conductive material, while the electrolyte is purely ionic conductor which includes water and nonaqueous solvents and melt or solid electrolytes. In the course of an electrochemical reaction, the electron transfer occurs through the electrode/electrolyte interface. Electrons can be transferred through the interface in both directions. Particle in the electrolyte becomes either reduced when it accepts an electron from the electrode or oxidized when it gives an electron to the electrode. Thus, the electrochemical reaction involves the passage of electrical current. Currents corresponding to the oxidation and reduction reactions are called a partial anodic and partial cathodic current, respectively. When the electrode potential is equal to the equilibrium potential, partial anodic and partial cathodic currents are equal, so that the total current is zero. However, when the imposed electrode potential is more positive or more negative than the equilibrium potential, the total current that passes through the electrode is the anodic or cathodic current, respectively. The corresponding electrodes are called anodes or cathodes.

The simplest electrochemical reactions are those in which the electron transfer causes only the change of the oxidation state of a reactant, no bond formation or splitting takes place. Much more common are cases in which the electron transfer is followed by or occurs simultaneously with the adsorption and/or chemical changes of a reactant, reaction intermediates, or products. Thus, the electrochemical reactions are divided into two classes: (i) outer-sphere one-electron transfer with the solution-phase electron donors or acceptors in the outer Helmholtz plane (OHP) of the electrical double layer where the electron transfer occurs and (ii) more complex processes where more than one electron may be transferred. Class 1 of electrochemical reactions involves a simple ionic redox process in which only the change of oxidation state of reactants positioned in the OHP is involved. Class 2 reactions often involve multiple steps, some can be chemical. When a reaction occurs in a series of consecutive steps, the overall reaction rate is determined by the rate of the slowest step, called the rate-determining step. All other preceding and following steps can be considered to be in equilibrium. If the slowest step in the reaction mechanism is the exchange of electron, then electrochemical reaction takes place under electrochemical or activation control. Many electrochemical reactions of organic molecules and reactions accompanied by gas evolution or dissolution are in a class 2.

Just as for chemical reactions in general, the charge transfer is controlled by the existence of the energy barrier between oxidized and reduced

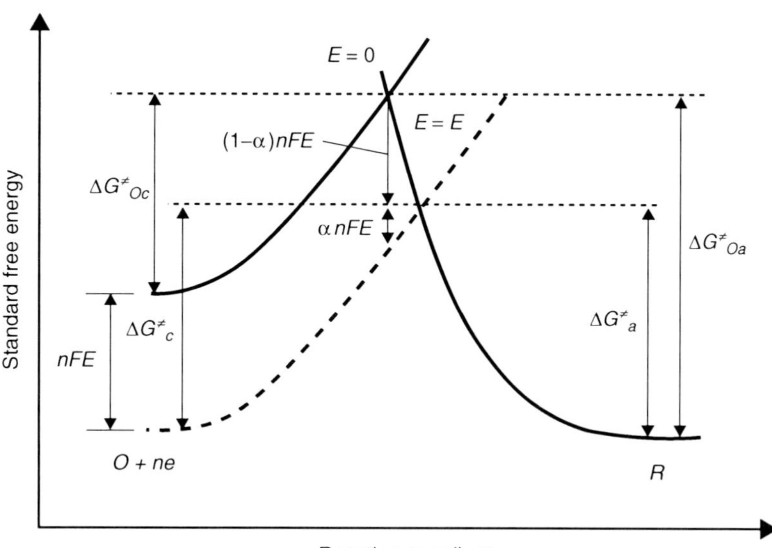

Electrocatalysis, Fundamentals - Electron Transfer Process; Current-Potential Relationship; Volcano Plots, Fig. 1 Effect of electrode potential on the free energy versus coordinate curves for an electron reactant at two electrode potentials: $E = Ee$ and $E < Ee$ (*broken line parabola*)

states. A unique feature of the electrode reactions is that the height of this barrier can be decreased or increased by changing the potential across the interface.

The Rate of Electron Transfer

Because of the heterogeneous nature of electrode reactions, the rate of charge transfer across the electrochemical interface depends not only on the potential but also on the double layer structure and the adsorption of reactants, intermediates and products, and other eventual solution phase species. Mass-transport limitations are not considered here. The expression relating current to the electrode potential can be obtained from the absolute rate theory applied to the electrochemical interface. For that electrochemical reaction case, the heights of the free energy barriers are functions of the potentials drop across the interface in accordance with the absolute rate theory.

In the simplest case of one-electron transfer reaction:

$$O + e \Leftrightarrow R \qquad (1)$$

A shift in the electrode potential from 0 to a value E causes the changes depicted in Fig. 1.

The barrier for the oxidation $\Delta G^{\neq}$ is decreased by a fraction α of the energy change nFE, while the barrier for reduction is increased by $(1 - \alpha)$ nFE. Rate constants for the reduction and oxidation are k_{red} and k_{ox}, respectively. Assuming that there is an arbitrary amount of oksidant (O) and reductant (R) species in the solution, the total current flowing j is the sum of the partial cathodic j_c, and partial anodic j_a, currents:

$$j = j_c + j_a = nFAk_{red}[O]_o - nFAk_{ox}[R]_o \qquad (2)$$

where A is the electrode area, F is the Faraday constant, n is the number of electrons transferred, and $[O]_O$ and $[R]_O$ are the surface concentrations of (O) and (R), respectively. According to the transition state theory from chemical kinetics, rate constants are related to the free energies of activation, which are related to the potential. From these relations the basic equation is derived, which describes how the overall current on an electrode depends on the applied potential.

Current–Potential Relationship

The equation which describes the fundamental relationship between the electrical current on an electrode and the electrode potential, assuming that both a cathodic and an anodic reaction occur on the same electrode, is called the Butler–Volmer equation:

$$j = j_0\left\{\exp\left(\frac{(1-\alpha)nF\eta}{RT}\right) - \exp\left(\frac{-\alpha nF\eta}{RT}\right)\right\} \tag{3}$$

where j_0 is the exchange current density, T is absolute temperature, R is universal gas constant, α is so-called symmetry factor or charge transfer coefficient, and η is the overpotential. Overpotential is the extent to which the reaction is driven beyond the equilibrium potential, Eeq:

$$\eta = E - E_{eq} \tag{4}$$

At high anodic overpotential, partial cathodic current can be excluded compared to the anodic, meaning that the Butler–Volmer Eq. (3) simplifies to

$$j = j_a = j_0 \exp\left[\frac{(1-\alpha)nF\eta}{RT}\right] \tag{5}$$

Partial anodic currents can be excluded from Butler–Volmer equation at high cathodic overpotential:

$$j = j_c = -j_0 \exp\left[-\frac{\alpha nF\eta}{RT}\right] \tag{6}$$

From the Eqs. (5) and (6) for the high anodic or cathodic overpotential, the corresponding Tafel equations can be derived:

$$\eta = -2.303\frac{RT}{(1-\alpha)nF}\log j_0 + 2.303 \times \frac{RT}{(1-\alpha)nF}\log j_a \tag{7}$$

$$\eta = 2.303\frac{RT}{\alpha nF}\log j_0 - 2.303\frac{RT}{\alpha nF}\log j_c \tag{8}$$

Since the partial cathodic current is negative under the conventions and a logarithm of a negative number is undefined, it is assumed in Eq. (8) that value j_c is in fact absolute value $|j_c|$.

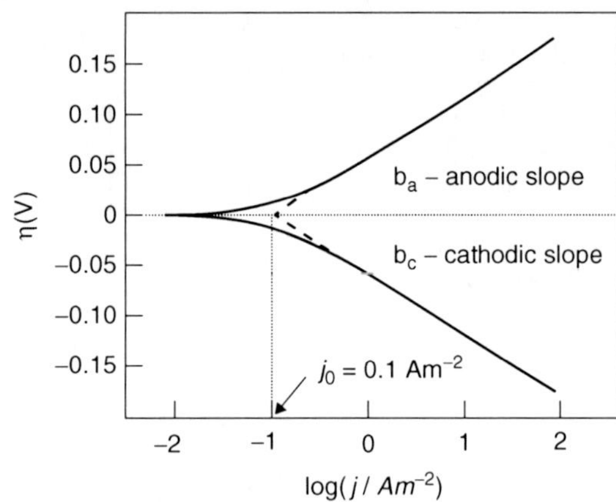

Electrocatalysis, Fundamentals - Electron Transfer Process; Current-Potential Relationship; Volcano Plots, Fig. 2 Tafel plot

Tafel Slope

Logarithmic dependence of η on j given by Eqs. (7) and (8) can be simplified to

$$\eta = a + b\log j \tag{9}$$

where a and b are constants. In its graphical presentation (Fig. 2), the value of a is the intercept and b is a slope. From the intercept, one can calculate the exchange current density, j_0, by the extrapolation of the potential to the equilibrium potential. Slope b is called the Tafel slope. For anodic reaction Tafel slope is positive, while for the cathodic reaction it is negative. This value is important in electrochemical kinetics, because it allows the calculation of the symmetry factors for elementary reactions and makes it possible to predict the reaction mechanism of complex reactions. The linear dependence takes place at $\eta \approx 50$–100 mV.

Tafel slope value also indicates the number of electrons exchanged in the electrochemical reaction. For a reaction, where the number of electrons exchanged equals 1 and the charge transfer coefficient is 0.5, the theoretical Tafel slope at 25 °C is $\pm$ 118 mV per decade. Thus, Tafel slopes provide valuable information regarding the mechanism of a reaction and indicate the identity of a rate-determining step of the overall reaction.

Theories of Charge Transfers

Theoretical reactions at electrodes as well as between redox couples in solutions can be interpreted in terms of electron tunneling. Microscopic aspects of the charge transfer in the phenomenological treatment are contained in the rate constants k and the transfer coefficient a.

Descriptions of charge transfer of charge transfer (electrons or protons) in electrochemical reactions, have been a subject of much interest since the work of Gurney [2] who suggested that electron transfer has two main approaches to the problem. In the first, which can be termed "molecular," attention is focused on the behavior of one chemical bond that is modified in the interfacial reaction. The theory assumes that the energy of activation is determined by the distribution of thermal energy in the various internal modes of the reacting species but ignores the dynamic behavior and dielectric relaxation properties of the solvent [3, 4].

The second approach, referred to as a "continuum theory," has been pursuit by theorists using classical or quantum statistical mechanics with the common focus on solvent dipole fluctuation as a major factor controlling the charge transfer [5, 6].

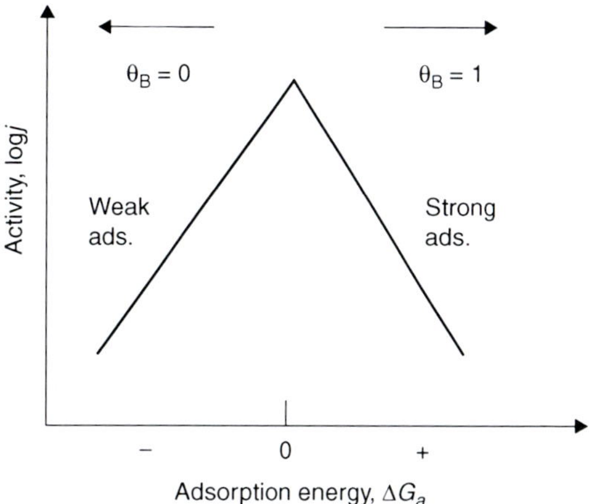

Electrocatalysis, Fundamentals - Electron Transfer Process; Current-Potential Relationship; Volcano Plots, Fig. 3 Schematics of a *volcano* curve in electrocatalysis with adsorption of B being the rate-determining step

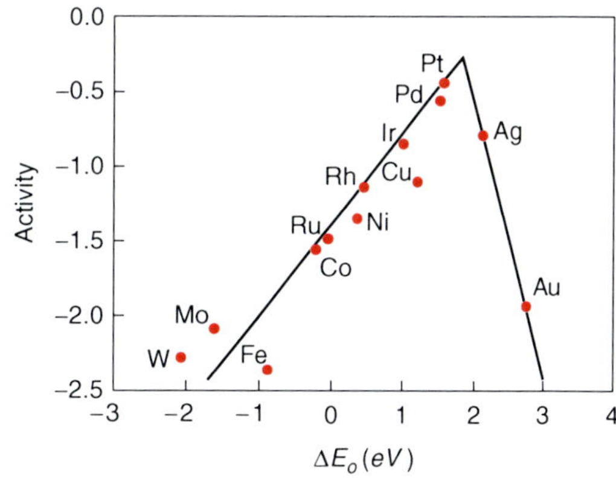

Electrocatalysis, Fundamentals - Electron Transfer Process; Current-Potential Relationship; Volcano Plots, Fig. 4 Oxygen reduction activity plotted as a function of the oxygen binding energy [14]

Volcano Plots

Electrocatalytic reactions involve the strong interactions of reactants and/or intermediates with the electrode surface. As a consequence, the rate of these reactions shows pronounced dependence on the nature of the electrode material. A good example is hydrogen evolution reaction, for which the rate of various metals varies by 11 orders of magnitude. The plots of the catalyst activity (reaction rate) against a descriptor of the adsorption properties such as the adsorption energies or adsorption bond strength of the reactant or reaction intermediates pass through a maximum. These are volcano plots that are generally based on the Sabatier principle [7], which states that the interactions between the catalyst and the adsorbate should be neither too strong nor too weak. If the interaction is too weak, the adsorbate will fail to bind to the catalyst and no reaction will take place. On the other hand, if the interaction is too strong, the catalyst gets blocked by adsorbate or product that fails to desorb. Volcano correlations are important for describing trends in reactivity that are helpful in designing new more active catalysts. Early treatments of volcano plots in catalysis are due to Balandin [8], in

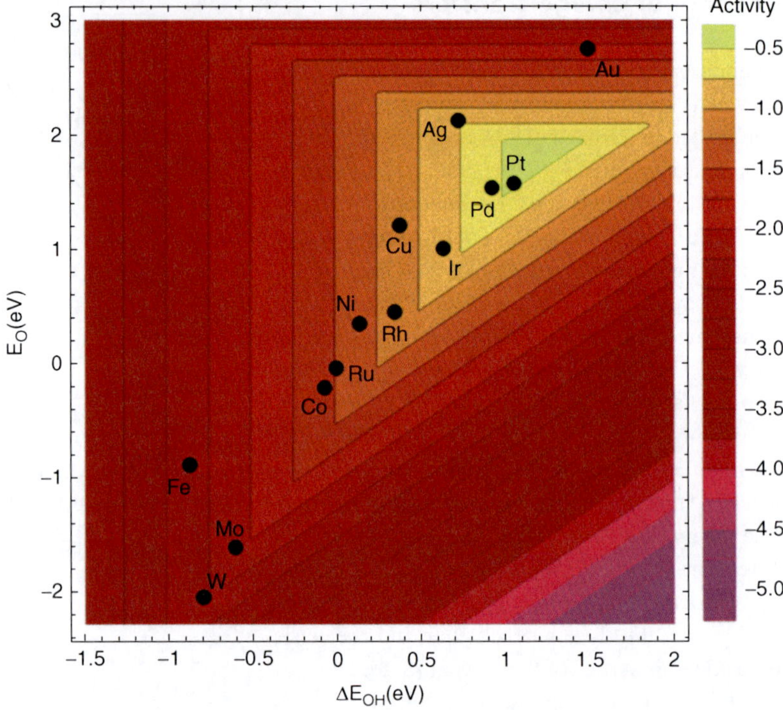

Electrocatalysis, Fundamentals - Electron Transfer Process; Current-Potential Relationship; Volcano Plots, Fig. 5 Oxygen reduction activity plotted as a function of both the O and the OH binding energies [14]

electrocatalysis Parsons [9], Gerischer[10], Krishtalik[11], Trasatti[12], and Appleby[13].

From the basic relationship between kinetic current, *i*, and potential, *E*, consistent with the Butler–Volmer approach, the rate expression for electrocatalytic reaction:

$$A + e \rightarrow B^{-} \rightarrow C \tag{10}$$

can be written in the following form: $\Delta G^{\neq}$

$$I = constC_A(1 - \theta_B)\exp\big(1 - \Delta G^{\neq}/ RT - \alpha \Delta G_B^O \alpha FE/RT\big) \tag{11}$$

Assuming formation of B to be the rate-determining step, if B is adsorbed on the electrode surface, it will form with a lower activation Gibbs energy than in the absence of adsorption. For the limiting cases when $\theta_B \sim 0$ and $\theta_B \sim 1$, Eq. (11) at constant potential, E, becomes $ln\, i \sim \Delta G_B$ or $ln\, i \sim \Delta G_B$, respectively.

As illustrated schematically in Fig. 3, the logarithm of the reaction rate varies linearly with ΔG, increasing from weak adsorption (positive) to very strong adsorption (negative), which predicts a linear decrease of the reaction rate as result of the blocking effect.

As an example, volcano plot for the activity of different metal catalysts for oxygen reduction reaction (ORR) versus the respective metal–oxygen bond strength is shown in Fig. 4.

Analogous two-dimensional plots can also be built against two different descriptors, such as the adsorption bond strength of the two intermediates. In that case the plot is generally shown as a contour plot and is called a volcano surface [14]. Figure 5 shows volcano surface for the activity of different metal catalysts for ORR versus both metal–O and metal–OH binding energies.

Cross-References

- ▶ Electrocatalysts for the Oxygen Reaction, Core-Shell Electrocatalysts
- ▶ Electrode
- ▶ Hydrogen Evolution Reaction

References

1. Kobozev NI, Monblanova V (1934) The mechanism of the electrodiffusion of hydrogen through palladium, Acta Physicochim, URSS 1:611–650
2. Gurney RW (1937) The quantum mechanics of electrolysis, Proc Roy Soc A134:137–154
3. Horiuti J, Polanyi M (1935) The basis of a theory of proton transfer. Electrolytic dissociation; prototropy; spontaneous ionization; electrolytic evolution of hydrogen; hydrogen-ion catalysis, Acta Physicochim URSS 2:505–532
4. Bockris JOM, Khan SUM (1993) Surface electrochemistry, a molecular level approach. Plenum, New York
5. Dogonadze RR, Kuznetsov AM, Levich VG (1968) Theory of hydrogen-ion discharge on metals: Case of high overvoltages, Electrochim Acta 13:1025–1044
6. Marcus RA (1956) On the theory of oxidation-reduction reactions involving electron transfer I, J Chem Phys 24:966–978
7. Sabatier P (1911) Hydrogenation and dehydrogenation by catalysis, Ber Dtsch Chem Ges 44:1984–2001.
8. Balandin AA (1969) Modern state of the multiplet theory of heterogeneous catalysis. In: Eley DD, Pines H, Haag WO (eds) Advances in catalysis, vol 19. Academic, New York, pp 1–210
9. Parsons R (1958) Rate of electrolytic hydrogen evolution and the heat of adsorption of hydrogen, Trans Faraday Soc 54:1053–1063
10. Gerischer H (1956) Relation between the mechanism of electrolytic deposition of hydrogen and the energy of adsorption of atomic hydrogen on different metals, Z Phys Chem N. F. 8: 137–153; (1958) Mechanism of electrolytic discharge of hydrogen and adsorption energy of atomic hydrogen, Bull Soc Chim Belg 67: 506–527
11. Krishtalik LI (1957) A contribution to the slow discharge theory, Zh Fiz Khim 31:2403–2413; (1959) Velocities of the elementary stages of the hydrogen evolution mechanism on the cathode I, Zh Fiz Khim 33:1715–1725
12. Trasati S (2000) Electrocatalysis: understanding the success of DSA®, Electrochim Acta 45: 2377–2385
13. Appleby AJ (1974) Electrocatalysis. In: Bockris BE, Conway BE (eds) Modern aspects of electrochemistry, vol 3. Butterworth, London, pp 369–478
14. Nørskov JK, Rossmeisl J, Logadottir A, Lindqvist L, Kitchin JR, Bligaard T, Jónsson H (2004) Origin of the Overpotential for Oxygen Reduction at a Fuel-Cell Cathode, Phys J Chem B 108(46): 17886–17892

Electrocatalysis, Novel Synthetic Methods

Stanko Brankovic
Electrical and Computer Engineering Department, Chemical and Bimolecular Engineering Department, and Chemistry Department, University of Houston, Houston, TX, USA

Introduction

In the last several decades, the electrochemical synthesis and electrodeposition became the enabling fabrication methods behind the train of hi-tech enterprise [1, 2]. There are many examples where electrochemical synthesis provides convenient if not *the only* approach to deliver the desired structures, materials, or catalytic surfaces. In recent years, the scientific community has witnessed the numerous examples where electrochemical synthesis is used to grow multilayered metallic thin films and nanostructures [3–5], nanoscale metallic architectures [6–12], and high-quality single-crystal overlayers [13–17]. The most recent developments suggest that the electrochemical methods become an attractive fabrication route for catalyst synthesis for fuel cells and metal-air batteries [18, 19]. These new applications make the future of the research in electrochemical material science a seemingly interesting and quite exciting endeavor.

The Metal Deposition via Surface Limited Redox Replacement of Underpotentially Deposited Metal Monolayer

Recently, Brankovic et al. [20] have developed the deposition protocol which has been extensively used for synthesis of noble metal monolayers and catalyst materials [21–26]. The basic steps describing the concept behind this new deposition protocol are illustrated in Fig. 1, [20]. The method is based on *surface limited*

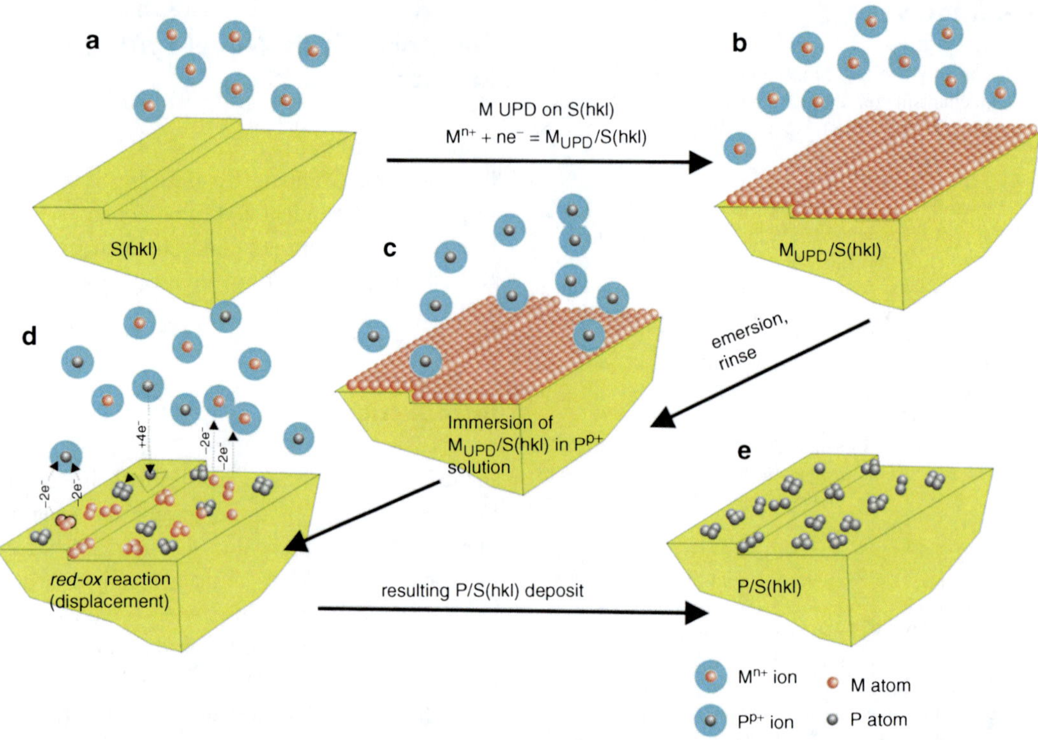

Electrocatalysis, Novel Synthetic Methods, Fig. 1 The basic steps of the deposition method described in Ref. [20]. (**a–b**) formation of *M* UPD ML on *S(h,k,l)*, (**b**) $M_{UPD}/S(h,k,l)$ surface emersion from solution at certain potential ensuring the desired *M* UPD ML coverage of S(*h,k,l*), (**c**) transfer to P^{p+}-containing solution at open-circuit potential (*OCP*), (**d**) *redox* or displacement reaction at OCP, (**e**) final morphology of *P/S(h,k,l)* deposit

redox replacement (SLRR) of underpotentially deposited (UPD) monolayer (ML) of metal *M* on substrate *S(h,k,l)* by more noble metal *P*. The amount of deposited metal *P* is controlled by reaction stoichiometry, structure, and coverage of the UPD ML (Eq. 1) [27].

$$\left(\theta_M \cdot \rho_M^{UPD}\right) \cdot M_{UPD}^0/S(h,k,l) + \left(\theta_M \cdot \rho_M^{UPD}\right) \cdot \left(\frac{m}{p}\right) P_{solv}^{p+} \Rightarrow \left(\theta_M \cdot \rho_M^{UPD}\right) \cdot M_{solv}^{m+} + \left(\theta_M \cdot \rho_M^{UPD}\right) \cdot \left(\frac{m}{p}\right) P_s^0/S(h,k,l) \quad (1)$$

Here, m and p are the oxidation states of UPD metal M and more noble metal P. The factors, q_M and ρ_M^{UPD}, are introduced to accurately express the amount of deposited metal P in ML units with respect to the areal density of the *S(h,k,l)* atoms. They are respectively the UPD ML coverage and the packing density of M atoms in complete UPD ML with respect to the substrate *S(h,k,l)*. The subscripts s and solv. indicate the physical state of the metal (solv. = solution phase and s = deposited). The stoichiometry of SLRR is also dependent on specific experimental conditions which can favor one or other oxidation state of the metal constituting the UPD ML. For example, in the case of Cu UPD ML, the Cu(I) ions are thermodynamically more favorable than Cu(II) if the presence of Cl^- is the only complexing anion in the near region of the electrode/solution interface [27].

If sequence A–E (Fig. 1) is repeated an arbitrary number of times, a multilayer homo- or heteroepitaxial films are obtained [16]. The thin-film growth using this SLRR synthetic protocol can be completely automated with experimental apparatus for *electrochemical*

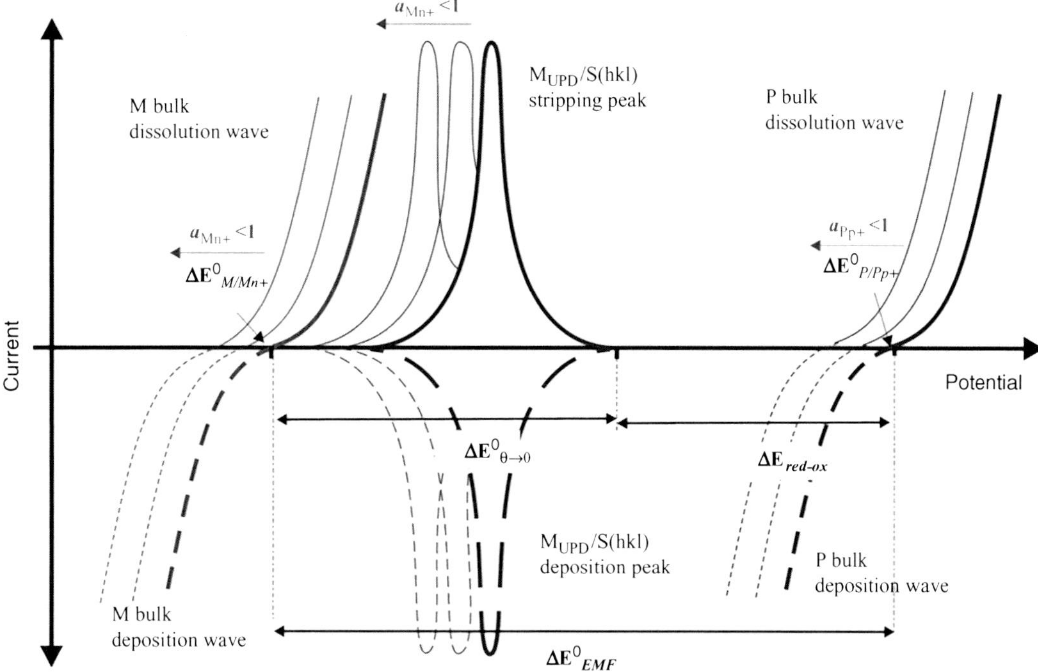

Electrocatalysis, Novel Synthetic Methods, Fig. 2 The schematics of the thermodynamic entities and their relations determining the ΔE_{redox}, Eq. 2, for displacement reaction. The *gray lines* are current–potential dependence for bulk M metal electrode and UPD ML of M on substrate S. The *dark line* is the current–potential dependence for bulk noble metal electrode. The terms from Eq. 2 are identified in the figure

atomic layer epitaxy (ECALE) developed by *Stickney* et al. [28]. The additional variations of the SLRR protocol were recently introduced by Dimitrov et al. for the thin-film growth application [29]. In this approach the SLRR step in Fig. 1c was combined with a short potential pulse representing a co-deposition of P and UPD ML of metal M on the substrate surface so that the *one cell, one solution* concept can be implemented.

The UPD represents a potential-dependent adsorption process, and the UPD ML coverage is controlled effectively down to a fraction of a monolayer [30]. According to stoichiometry of SLRR Eq. 1, the same accuracy for deposition of metal *P* can be achieved as well. This important advantage gives the whole array of new opportunities for application of catalyst monolayer design and synthesis via SLRR of UPD ML.

The electrochemical driving force for redox replacement reaction (galvanic displacement) between the metal P and the UPD ML of metal M is the positive difference between the equilibrium potential of *P* metal and the equilibrium potential of $M_{UPD}/S(h,k,l)$ at its coverage approaching a zero limit, $\theta_{UPD} \to 0$. This condition is defined as [31] (Fig. 2):

$$\Delta E_{red-ox} = \Delta E^0_{EMF} - \Delta E^0_{\theta\to 0} - \frac{RT}{F}\ln\frac{[a_{M^{n+}}]^n}{[a_{P^{p+}}]^p} > 0. \tag{2}$$

Here, $\Delta E^0_{EMS}\left(\Delta E^0_{EMF} = E^0_{Pp+/P} - E^0_{Mn+/M}\right)$ represents the electromotive force for the bulk *M* and *P* galvanic couple at standard conditions (Fig. 2). The $\Delta E^0_{\theta\to 0}$ represents the equilibrium potential of $M_{UPD}/S(h,k,l)$ at $\theta_{UPD} \to 0$ limit at standard conditions ($a_M{}^{n+} = 1$, $a_P{}^{p+} = 1$, Fig. 2), while the logarithmic term provides the correction for the departure from standard conditions. Inspection of Eq. 2 offers the evidence that the driving force for the

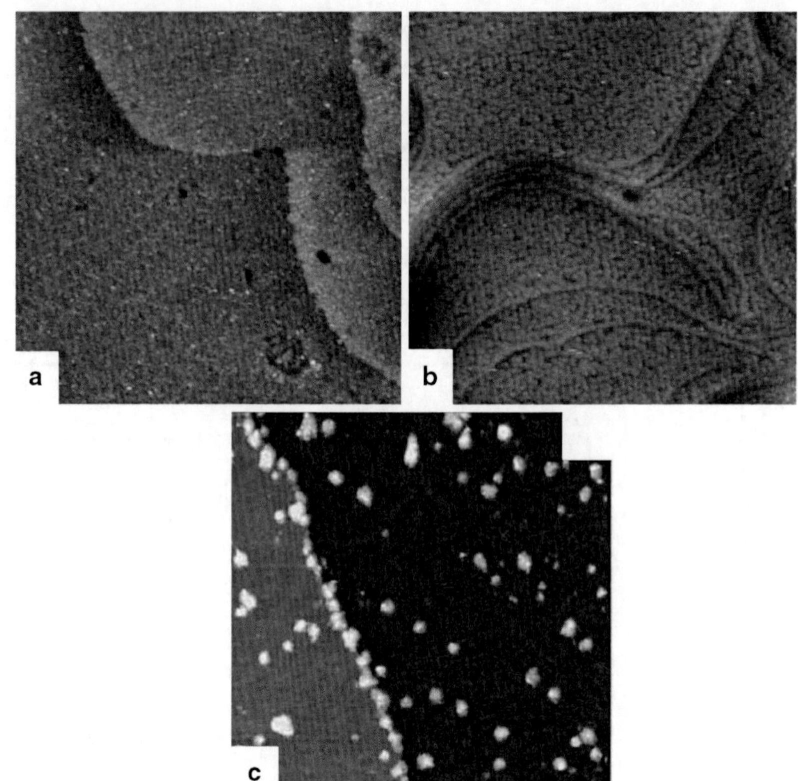

Electrocatalysis, Novel Synthetic Methods, Fig. 3 The STM images of deposit morphology of (**a**) ~1 ML of Pd on Au (111) via SLRR of Cu_{UPD}/Au(111), reaction: $Cu_{UPD}/Au(111) + Pd^{2+} = Cu^{2+} + Pd/Au(111)$, $q - 1$, $\theta = 1$, $m/p = 1$, image size 200 nm × 200 nm; (**b**) ~0.5 ML of Pt on Au(111) *via* SLRR of Cu_{UPD}/Au (111), reaction: $Cu_{UPD}/Au(111) + Pt^{4+} = Cu^{2+} + 0.5Pt/Au(111)$, $\theta = 1$, $q = 1$, $m/p = 0.5$, image size 320 nm × 320 nm; (**c**) Pt submonolayer on Au (111), image size 150 nm × 150 nm (Images **a** and **b** are from Ref. [20]; image **c** is from Ref. [32]). Size of Pt clusters is ~3 nm. The achieved nucleation density is ~10^{13} cm^{-1}

redox reaction can be modified by adjusting the activities of M^{n+} and P^{p+} ions in the immersing (displacement) solution. If there are no M^{n+} ions in the immersing solution, the logarithmic term in Eq. 2 dominates the value of ΔE_{redox}(very large and positive), which means that extremely high overpotentials for nucleation could be achieved resulting in high nucleation rates and large nucleation density of the metal P on the substrate S.

Depending on the combination of the $M_{UPD}/S(h,k,l)$ and P^{p+} ions involved in the *redox* reaction, a different amount and coverage of *P* monolayer deposit is formed. This is illustrated in Fig. 3 using the example of Pd and Pt deposit obtained by displacement of Cu_{UPD}/Au(111). The deposit morphology varies from an atomic monolayer thin-film to monolayer high nanoclusters with very narrow size distribution (Fig. 3a, b).

Particularly interesting result in Fig. 3 is the spatially uniform coverage of Pt nanoclusters on Au(111) shown in image B. This suggests that nucleation of the depositing metal (Pt) is independent on thermodynamically favorable nucleation sites on substrate surface such as steps, for example. This is evident if one compares the morphology of Pt SML deposit on Au (111) obtained by electrodeposition at low overpotentials [32] (Fig. 3c) and the morphology of Pt sub-ML and Pd ML deposits shown in Fig. 3a, b. In the latter case, one can notice that Pt and Pd nanoclusters are equally distributed across the surface regardless the presence of surface steps where preferential nucleation and growth is typically found during electrodeposition process (Fig. 3c). This fact represents one of the major advantages of this deposition protocol over the traditionally used ones for catalyst synthesis application. By manipulation of the experimental conditions for SLRR, it was demonstrated that design and synthesis of different catalyst ML can be achieved [22, 23], as well as sub-monolayers (SML) with different coverage and morphology [23, 24, 33],

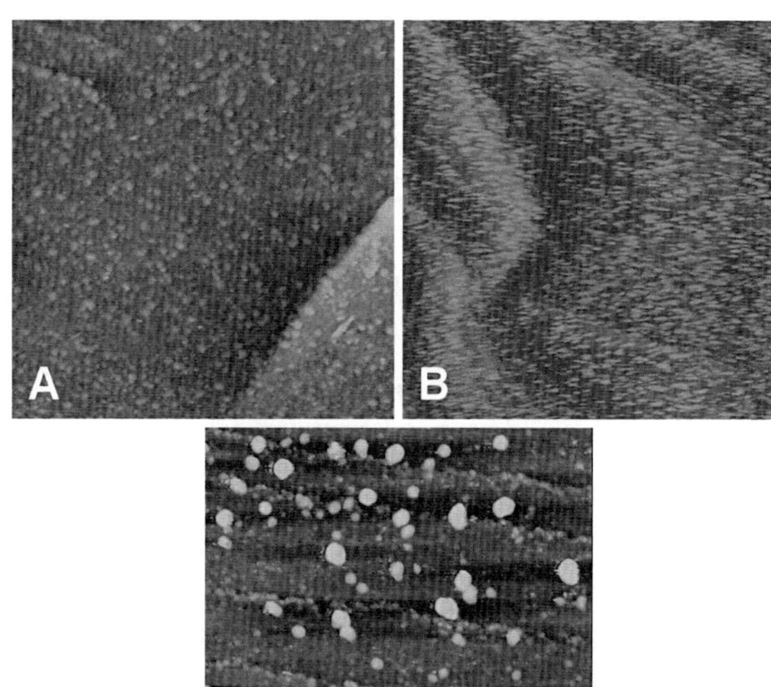

Electrocatalysis, Novel Synthetic Methods, Fig. 4 Morphology of the Pt deposit on Ru(1010) (**a**) and Ru(0001) (**b**, **c**). The spontaneous *NMonNM* deposition is used in image (**a**) and (**b**) [43], and the potentiostatic deposition is used in image (**c**), $\eta = 30$ mV. The depositing solution was 10^{-2} M $PtCl_6^{2-} + 10^{-2}$M $HClO_4$. The nucleation density in (**a**) and (**b**) is approximately ~500 times higher than in (**c**). (**a**) Image size; 280 × 280 nm, 2D, ~3 nm Pt clusters; (**b**) image size; 130 × 130 nm, 2D, ~5 nm Pt clusters; and (**c**) image size; 115 × 115 nm, 3D, ~10–15 nm Pt clusters

heteroepitaxial ultra-thin films, and ultra-thin film alloys [16, 29, 34].

Spontaneous Noble Metal on Noble Metal Deposition (NMonNM)

The spontaneous noble metal deposition on noble metal surface occurs when freshly prepared and clean noble metal electrode is immersed in the solution containing different noble metal ions. This phenomenon has been reported in various systems like Ru^{2+}/Pt(*h,k,l*) [35, 36], Pt^{4+}/Ru(*h,k,l,m*) [37, 38], Pd^{2+}/Ru(0001) [39], and Pd^{2+}/Pt(*h,k,l*) [40, 41]. The morphology of the deposit varies from monolayer high nanoclusters to larger 3D structures. The spontaneous *noble metal on noble metal* (*NMonNM*) deposition occurs as a result of an *irreversible surface controlled redox reaction* among depositing noble metal ions and noble metal substrate. The substrate surface in this reaction becomes oxidized by noble metal ions to form the certain type of surface oxide/hydroxide. The entire process is represented by the following reaction [37]:

$$\frac{m}{n}p^{n} + S^{0}(hkl) + mH_2O \rightarrow \frac{m}{n}p^{0}_{dep} + [S^{m+}][OH^{-}]_{m} + [mH^{+}] \quad (3)$$

where P^{n+}, P^{0}, S^{m},$^{+}$ and $S^{0}(h,k,l)$ are, respectively, the noble metal ions, the noble metal deposit, the substrate in the surface hydroxide, and the noble metal substrate. The conductive nature of Ru-oxide/hydroxide and ability to form more than one oxide/hydroxide monolayer without hindering the electron transport across the interface has been discussed as the reason for spontaneous formation of multilayer Pt and Pd deposits on Ru (0001) surface [37, 39]. The thermodynamic driving force for *NMonNM* deposition can be expressed in terms of difference between the equilibrium potential of the noble metal electrode in contact with its ions in

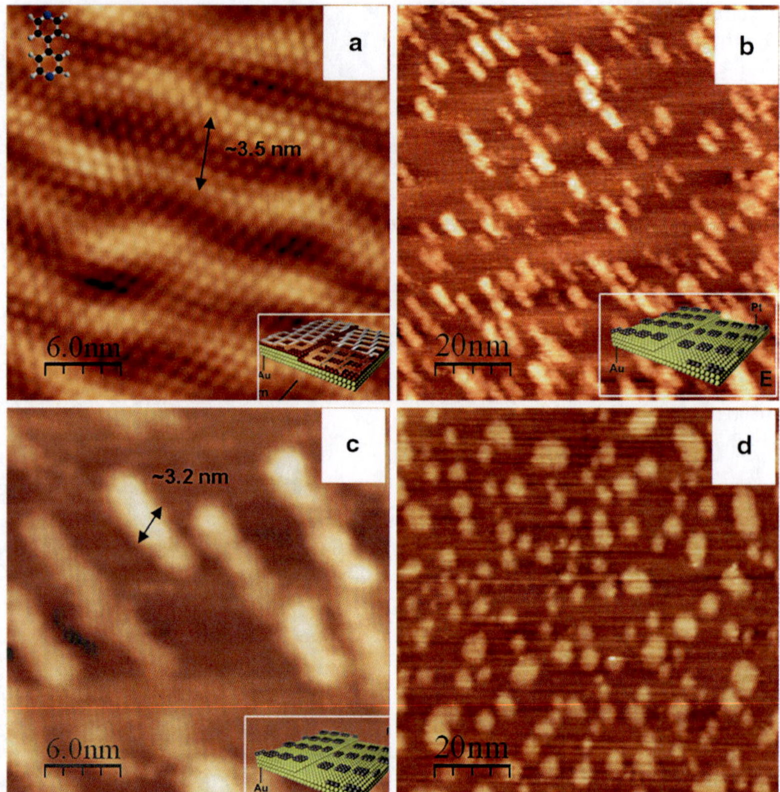

Electrocatalysis, Novel Synthetic Methods, Fig. 5 The STM images of (**a**) 4,4′-bipyridine layer adsorbed on Cu_{UPD}/Au (111), E = −0.V versus SCE in 0.1 M $HClO_4$, image size: 30 × 30 nm; (**b**) Pt on Au(111) after SLRR of Cu UPD ML by Pt^{4+} through the 4,4′-bipyridine adlayer, image size: 100 × 100 nm; (**c**) same as in **b**, image size 30 × 30 nm; (**d**) Pt deposit on Au(111) after SLRR of Cu UPD ML by Pt^{4+}, no organic template is present during SLRR. In image **a**, model of 4,4′-bipyridine is shown in *upper left corner*. In images (**a**, **b**, **c**), *lower right angle*, the cartoons of the corresponding structures are shown

the solution and the equilibrium potential for oxidation of the noble metal substrate [37, 39]:

$$\Delta E_{red-ox} = \Delta E_{P^{n+}/P^0} - \Delta E_{[S^{m+}]_x[OH]_y/S^0} > 0. \qquad (4)$$

According to Eq. 4, the change in concentration of the ions of depositing noble metal or pH of the solution alters the effective driving force for spontaneous *NMonNM* deposition. This fact is conveniently used to achieve different noble metal nucleation densities, regime of nucleation, and deposit morphology (Fig. 4). The *NMonNM* deposition offers advantage over the existing deposition methods due to the fact that 2D morphology of depositing noble metal is easily achieved without significant control of the deposition process (simple timed immersion). The example of this advantage is illustrated in Fig. 4 by comparison of ~1 ML Pt/Ru(0001) deposit obtained by potentiostatic deposition and ~0.7 ML Pt/Ru(0001) and 0.4 ML Pt/Ru (1010) obtained by spontaneous *NMonNM* deposition: Fig. 4a–c.

The introduction of *NMonNM* deposition method has initiated applications as possible route to produce catalysts for fuel cells with improved performance [37, 38, 42]. Different experimental methods were used to characterize electrosorption characteristics and activity of these modified bimetallic noble metal surfaces for different reactions [38, 43, 44]. These efforts have led to the design and characterization of one of the most efficient catalysts known today for polymer electrolyte membrane fuel cell anodes [4, 45–47].

Future Directions

The future prospect of the catalyst monolayer synthesis using metal deposition via SLRR reaction or NMonNM deposition techniques is quite exciting. Different variations of these methods could be applied in order to better control morphology of the catalyst monolayers and their activity [48]. For example, the SLRR deposition guided by the presence of organic phase adsorbed on the surface with UPD monolayer is one approach that deserves attention [49]. The concept is based on the idea of spatially controlling the nucleation sites for catalyst monolayer clusters and their shape evolution by introducing an organic phase on the electrode surface. The example of Pt ML deposits obtained using this approach is shown in Fig. 5. The 4,4′-bipyridine, adsorbed on CuUPD/Au(111) surface and serving as a template, is shown in Fig. 5a. As one can see, the 4,4′-bipyridine forms a rippled layer with hexagonal lattice on CuUPD/Au(111) surface. The nearest neighbor distance of the hexagonal lattice of the adsorbed 4,4′-bipyridine phase is 1.5 nm, while the average width and periodicity of the ripples is 3.5 nm. After redox replacement of CuUPD ML by Pt^{4+} through the 4,4′-bipyridine phase, the Pt deposit formed on Au(111) is shown in Fig. 5b, c [50]. The presence of adsorbed bipyridine phase has an obvious effect on the shape and orientation of deposited Pt nanoclusters (compare Fig. 5b, c with Fig. 5d). The average width of the Pt clusters is approximately the same ($\sim$3.2 nm) as the periodicity of bipyridine ripples shown in Fig. 5a, ($\sim$3.5 nm). The elongated shape of Pt clusters and their propagating direction replicate the arrangement and symmetry of bipyridine ripples. These observations do indicate that adsorbed organic phase serving as template has decisive influence on nucleation and growth of Pt deposit during SLRR reaction.

Cross-References

- ▶ Electrocatalysis - Basic Concepts, Theoretical Treatments in Electrocatalysis via DFT-Based Simulations
- ▶ Electrocatalysis, Fundamentals - Electron Transfer Process; Current-Potential Relationship; Volcano Plots

References

1. Edelstein A, Cammarata R (1996) Nanomaterials: synthesis, properties and applications. IOP, Bristol
2. Cao G (2004) Nanostructures and nanomaterials: synthesis, properties and applications. Imperial College Press, London
3. Nicewarner-Peña S et al (2001) Submicrometer metallic barcodes. Science 294:137–141
4. Schwarzacher W (1999) Metal nanostructures, a new class of electronic devices. Electrochem Soc Interface 8:20–24
5. Pauling H, Juttner K (1992) Top-on-top monolayer formation of foreign metals on gold single crystal surfaces. Electrochim Acta 37:2237–2244
6. Whitney T et al (1993) Fabrication and magnetic properties of arrays of metallic nanowires. Science 261:1316–1319
7. Sung M et al (2001) Electrodeposition of magnetic nanoparticle arrays with ultra-uniform length in ordered alumite. Appl Phys Lett 78:2964–2966
8. Bartlett P (2004) Electrodeposition of nanostructured films using self-organizing templates. Electrochem Soc Interface 13:28–33
9. Whitaker J, Nelson J, Schwartz D (2005) Electrochemical printing: software reconfigurable electrochemical microfabrication. J Micromech Microeng 15:1498–1503
10. Kolb D, Ullmann R, Will T (1997) Nanofabrication of small copper clusters on gold(111) electrodes by a scanning tunneling microscope. Science 275:1097–1099
11. Zach M, Ng K, Penner R (2000) Molybdenum nanowires by electrodeposition. Science 290: 2120–2123
12. Li C et al (1999) Fabrication of stable metallic nanowires with quantized conductance. Nanotechnology 10:221–223
13. Yong F et al (1999) Large magnetoresistance of electrodeposited single-crystal bismuth thin films. Science 284:1335–1337
14. Sieradzki K, Brankovic S, Dimitrov N (1999) Electrochemical defect-mediated thin-film growth. Science 284:138–141
15. Brankovic S, Dimitrov N, Sieradzki K (1999) Surfactant mediated electrochemical deposition of Ag on Au (111). Electrochem Solid State Lett 2:443–445
16. Vasilic R, Dimitrov N (2005) Epitaxial growth by monolayer-restricted galvanic displacement. Electrochem Solid State Lett 8:C173–C176
17. Hwang S, Oh H, Kwak J (2001) Electrodeposition of epitaxial Cu (111) thin films on Au (111) using defect-mediated growth. J Am Chem Soc 123:7176–7177
18. Adzic RR et al (2007) Platinum monolayer fuel cell electrocatalysts. Top Catal 46:249
19. Sasaki K et al (2010) Recent advances in platinum monolayer electrocatalysts for oxygen reduction reaction: scale-up synthesis, structure and activity of Pt shells on Pd cores. Electrochim Acta 55:2645

20. Brankovic S, Wang J, Adzic R (2001) Metal monolayer deposition by replacement of metal adlayers on electrode surfaces. Surf Sci 474:L173–L179
21. Brussel V et al (2003) Oxygen reduction at platinum modified gold electrodes. Electrochim Acta 48:3909–3919
22. Zhang J et al (2005) Controlling the catalytic activity of platinum-monolayer electrocatalysts for oxygen reduction with different substrates. Angew Chem Int Ed 44:2132–2135
23. Zhang J et al (2005) Mixed-metal Pt monolayer electrocatalysts for enhanced oxygen reduction kinetics. J Am Chem Soc 127:12480–12481
24. Zhang J et al (2007) Stabilization of platinum oxygen-reduction electrocatalysts using gold clusters. Science 315:220–222
25. Park S et al (2002) Transition metal-coated nanoparticle films: vibrational characterization with surface-enhanced Raman scattering. J Am Chem Soc 124:2428–2429
26. Kowal A et al (2009) Ternary Pt/Rh/SnO2 electrocatalysts for oxidizing ethanol to CO2. Nature materials advanced Online Publication http://www.nature.com/nmat/index.html
27. Gokcen D, Bae SE, Brankovic SR (2010) Stoichiometry of Pt submonolayer deposition via galvanic displacement of underpotentially deposited Cu monolayer. J Electrochem Soc 157:D582
28. Huang B et al (1995) Preliminary studies of the use of an automated flow-cell electrodeposition system for the formation of CdTe thin films by electrochemical atomic layer epitaxy. J Electrochem Soc 142: 3007–3016
29. Fayette M, Liu Y, Bertrand D, Nutariya J, Vasiljevic N, Dimitrov N (2011) From Au To Pt via surface limited redox replacement of Pb UPD in one-cell configuration. Langmuir 27:5650
30. Swathirajan S, Burckenstein S (1983) Thermodynamics and kinetics of underpotential deposition of metal monolayers on polycrystalline substrates. Electrochim Acta 28:865–877
31. Gokcen D, Bae S, Brankovic S (2007) Nucleation and growth of low-dimensional noble metal structures using galvanic displacement of UPD monolayers. Abstract #1381, 212th ECS Meeting, Washington, DC, 7–12 Oct 2007
32. Waibel H et al (2002) Initial stages of Pt deposition on Au (111) and Au (100). Electrochim Acta 47:1461–1467
33. Bae S-E, Gokcen D, Liu P, Mohammadi P, Brankovic SR (2012) Size effects in monolayer catalysis. Electrocatal. doi:10.1007/s12678-012-0082-5
34. Kim YG, Kim JY, Vairavapandian D, Stickney JL (2006) Platinum nanofilm formation by EC-ALE via redox replacement of UPD copper: studies using in-situ scanning tunneling microscopy. J Phys Chem B 110:17998
35. Chrzanowski W, Wieckowski A (1997) Ultra-thin films of ruthenium on low index platinum single crystal surfaces: an electrochemical study. Langmuir 13:5974–5978
36. Chrzanowski W, Kim H, Wieckowski A (1998) Enhancement in methanol oxidation by spontaneously deposited ruthenium on low index platinum electrodes. Catal Lett 50:69–75
37. Brankovic S, McBreen J, Adzic R (2001) Spontaneous deposition of Pt on Ru (0001) surface. J Electroanal Chem 503:99–104
38. Brankovic S et al (2002) Electrosorption and catalytic properties of bare and Pt modified single crystal and nanostructured Ru surfaces. J Electroanal Chem 524–525:231–241
39. Brankovic S, McBreen J, Adzic R (2001) Spontaneous deposition of Pd on Ru (0001). Surf Sci 479: L363–L368
40. Attard G, Bannister A (1991) The electrochemical behaviour of irreversibly adsorbed palladium on Pt (111) in acid media. J Electroanal Chem 300:467–485
41. Llorka M et al (1993) Electrochemical structure-sensitive behavior of irreversibly adsorbed palladium on Pt (100), Pt (111) and Pt (110) in an acidic medium. J Electroanal Chem 351:299–319
42. Strbac S, Johnston CM, Lu GQ, Crown A, Wieckowski A (2004) In situ STM study of nanosized Ru and Os islands spontaneously deposited on PT (111) and Au(111) electrodes. Surf Sci 573:80
43. Brankovic S et al (2002) Carbon monoxide oxidation on bare and Pt-modified Ru(1010) and Ru(0001) single crystal electrodes. J Electroanal Chem 532:57–66
44. Inoue H, Brankovic SR, Wang JX, Adzic RR (2002) Oxygen reduction on bare and Pt monolayer-modified Ru(0001), Ru(1010) and Ru nanostructured surfaces. Electrochim Acta 47:3777
45. Brankovic SR, Wang JX, Adzic RR (2001) Pt submonolayer on Ru nanoparticles – a novel low Pt loading, high CO tolerance fuel cell electrocatalyst. Electrochem Solid State Lett 4:A217
46. Wang JX, Brankovic SR, Zhu Y, Adzic RR (2003) Kinetic characterization of PtRu fuel cell anode catalysts made by spontaneous Pt deposition on Ru nanoparticles. J Electrochem Soc 150:1108
47. Brankovic SR, Wang JX, Adzic RR (2002) The CO tolerant electrocatalyst with low platinum loading and a process for its application. US Patent 132154
48. Gokcen D, Bae S-E, Gokcen D, Liu P, Mohammadi P, Brankovic SR (2012) Size effects in monolayer catalysis – model study: Pt submonolayers on Au (111). Electrocatal. doi:10.1007/s12678-012-0082-5
49. Gokcen D, Miljanic O, Brankovic SR (2010) - Morphology control of Pt sub-monolayers, 218th ECS Meeting. Abstract #2015, Las Vegas, NV. 10–15 Oct 2010
50. Gokcen D, Miljanić OŠ, Brankovic SR (2009) Nano-organization and morphology control of metal monolayers deposited by galvanic displacement of UPD monolayers. 216th The Electrochemical Society Meeting, Abstract *#2443*, Vienna, 4–9 Oct 2009

Electrocatalysts for Carbon Dioxide Reduction

Dmitry E. Polyansky
Chemistry Department, Brookhaven National Laboratory, Upton, NY, USA

Introduction

The major *scientific challenges* identified with the chemical reduction of CO_2 are the incredible thermodynamic stability of this molecule and the requirement to transfer multiple electrons and protons in the course of the chemical reaction. As a result, reaction pathways involving transfer of a single electron proceed through the formation of highly energetic intermediates, and these processes are thermodynamically highly unfavorable (e.g., reaction 2 below). This in turn significantly diminishes the efficiency of overall CO_2 reduction and can manifest itself in, e.g., high overpotential for the electrochemical process. On the other hand, proton-coupled multielectron reactions (e.g., reactions 3–6) are considerably more favorable; however, in order to implement them catalysis has to be involved. In addition, the use of water as the reaction medium does not only allow facile proton delivery but also enables coupling of the CO_2 reduction half-reaction to the water oxidation half-reaction in the complete electrochemical cell architecture.

At pH 7 vs. NHE [1]:

$$2H^+ + 2e^- \rightarrow H_2 \quad E^{\circ\prime} = -0.41\ V \tag{1}$$

$$CO_2 + 1e^- \rightarrow CO_2^{\bullet-} \quad E^{\circ\prime} = -2\ V \tag{2}$$

$$CO_2 + 2H^+ + 2e^- \rightarrow CO + H_2O \quad E^{\circ\prime} = -0.52\ V \tag{3}$$

$$CO_2 + 2H^+ + 2e^- \rightarrow HCOOH \quad E^{\circ\prime} = -0.61\ V \tag{4}$$

$$CO_2 + 6H^+ + 6e^- \rightarrow CH_3OH + H_2O \quad E^{\circ\prime} = -0.38\ V \tag{5}$$

$$CO_2 + 8H^+ + 8e^- \rightarrow CH_4 + 2H_2O \quad E^{\circ\prime} = -0.24\ V \tag{6}$$

The use of water as a solvent, however, requires very careful identification of each component's acid–base equilibrium. In particular, equilibrium of CO_2 with carbonates in aqueous solution is important (reactions 7–9), since CO_2 was identified as the primary electroactive species [2, 3].

$$CO_2 + H_2O \rightarrow HCO_3^- + H^+ \quad pK_a = 6.4 \tag{7}$$

$$H_2CO_3 \rightarrow HCO_3^- + H^+ \quad pK_a = 3.6 \tag{8}$$

$$HCO_3^- \rightarrow CO_3^{2-} + H^+ \quad pK_a = 10.3 \tag{9}$$

In addition to thermodynamic requirements for CO_2 reduction reactions, a significant kinetic barrier is associated with large reorganization energy (0.35 eV), which is mainly due to the geometry change from the linear CO_2 molecule to the bent formate radical anion [4]. This significant structural change is one of the factors contributing to the large overpotential required for electrochemical CO_2 reduction.

Reduction of CO_2 on Metal Electrodes

The electrochemical reduction of CO_2 on metal electrodes has been extensively studied in the past [3, 5–8]. Its mechanism varies substantially depending on the type of metal used as the working electrode. An important correlation between the metal's overpotential in the hydrogen evolution reaction (HER) and its ability to reduce CO_2 has been observed. According to their ability to facilitate HER, metals can be divided into three major groups [6–8]. Metals in the first group demonstrate high overpotential for the HER (e.g., Hg, Cd, Pb, Tl, In, Sn), and the main product of electrochemical CO_2 reduction at these metals is the formate anion. In the second group, metals with moderate HER overpotential (e.g., Cu, Au, Zn, Ag) mainly yield mixtures of formate and CO. And, finally, the third group of metals has low overpotential for the HER (Pt, Ni, Pd), and CO_2 reduction on these metals usually results in the formation of surface-bound CO. This correlation is not simply phenomenological but is defined by the mechanism of the CO_2 reduction reaction at the metal surface, and this mechanism depends on the electronic structure of the metal. The key steps of CO_2 reduction for each group of metals are outlined below.

For metals with high and medium HER overpotential, the formation of formate upon CO_2 reduction was proposed to proceed through two consecutive one-electron reductions (Eqs. 10–11) [8, 9]. Reaction 11 or 11' (where BH is a proton donor, e.g., a buffer) was proposed to be rate limiting [2]. The intermediate species adsorbed on the surface of metal electrodes were identified by surface spectroscopy [10–12] in support of the proposed mechanism. Alternatively, the mechanism of formate formation on Cu was proposed to proceed through transfer of surface hydrogen (H_{ads}) to the formate radical anion ($CO_2{}^{\bullet -}{}_{ads}$) adsorbed on the metal surface [5].

$$CO_2 + e^- \rightarrow CO_2{}^{\bullet -}{}_{ads} \qquad (10)$$

$$CO_2{}^{\bullet -}{}_{ads} + BH + e^- \rightarrow HCO_2^- + B^- \quad pH > 4 \qquad (11)$$

$$CO_2{}^{\bullet -}{}_{ads} + H^+ + e^- \rightarrow HCO_2^- \quad pH < 4 \qquad (11')$$

In addition to formate, comparable yields of CO were detected if medium HER overpotential electrodes were used [13]. The mechanism for CO formation was proposed to involve discrete protonation and electron transfer steps (reactions 12–13) or surface hydrogen transfer to $CO_2{}^{\bullet -}{}_{ads}$ (reaction 14) [6].

$$CO_2{}^{\bullet -}{}_{ads} + H^+ \rightarrow \left({}^{\bullet}COOH\right)_{ads} \qquad (12)$$

$$\left({}^{\bullet}COOH\right)_{ads} + e^- \rightarrow CO_{ads} + OH^- \qquad (13)$$

$$CO_2{}^{\bullet -}{}_{ads} + H_{ads} \rightarrow CO_{ads} + OH^- \qquad (14)$$

Nevertheless, all proposed mechanisms share the common intermediate species $CO_2{}^{\bullet -}{}_{ads}$, which is the product of a highly energetic one-electron process (reaction 10). This explains the high overpotentials required for successful reduction of CO_2 on metal electrodes with high and medium HER overpotentials.

On the other hand, the electrochemical reduction of CO_2 on metal electrodes with low HER overpotential is completely different. It has been shown that CO_2 reacts with H_{ads}, leading to tightly bound CO_{ads} (reactions 15–16) [14–16]. Since the formation of H_{ads} on the Pt surface requires little overpotential, the overall CO_2 reduction reaction is achieved very efficiently, though only stoichiometrically. The formed CO_{ads} binds very tightly to the Pt and poisons the surface of the metal, a problem well known in catalysis on Pt surfaces.

$$Pt(CO_2)_{ads} + Pt{-}H_{ads} \rightarrow Pt{-}COOH \qquad (15)$$

$$Pt{-}COOH + Pt{-}H_{ads} \rightarrow Pt{-}CO + H_2O \qquad (16)$$

Another problem associated with the use of Pt-group metals for CO_2 reduction is the very efficient HER which diminishes yields of CO_2 reduction products [17]. Such an efficient HER is attributed to the fast recombination of two H_{ads} located on neighboring atoms on the surface [18].

$$H_{ads} + H_{ads} \rightarrow H_2(\text{recombination})$$

However, the HER can be easily suppressed by blocking adjacent surface Pt atoms with foreign molecules [18]. This seems to be a useful strategy to increase the efficiency of CO_2 reduction in protic environments.

Reduction of CO_2 Promoted by Molecular Catalysts

Chemical (or related electrochemical and photochemical) reduction of carbon dioxide promoted by molecular catalysts is another major field of CO_2 chemistry. The vast majority of molecular catalysts used for CO_2 reduction are coordination complexes of transition metals. Properties of these complexes are determined by the unique combination of the metal center and the set of ligands and can be conveniently tuned in a predictable way to affect their reactivity towards CO_2. The common feature which unites most efficient CO_2 reduction catalysts is the ability of the transition metal center to adopt multiple oxidation states within a single complex. This allows the catalyst to act as a donor of two or more electrons, permitting the reduction reactions to proceed along low energy pathways (reactions 3–6) and avoid high energy intermediates. Several classes of transition metals complexes most commonly used in catalytic CO_2 reduction are discussed below. The catalysts are organized by the ligand type rather than the nature of the metal center (as opposed to bulk metal electrodes), since common reactivity trends can be observed within these groups [19, 20]. More detailed reviews of CO_2 reduction catalyzed by transition metal complexes can be found elsewhere [19–23].

Macrocyclic Complexes of Transition Metals

One of the representative and probably most studied classes of molecular catalysts for CO_2 reduction includes coordination complexes of transition metals with tetraaza macrocyclic ligands. The nature of the ligand as pointed out by Collin and Sauvage [20] plays an important role in the complex redox chemistry. Most aromatic macrocyclic ligands (e.g., porphyrins and phthalocyanines) bear significant negative charge (−2 for porphyrin-free base), compared to neutral nonaromatic tetraaza macrocyclic ligands (e.g., cyclam). In addition, the reduction of complexes with aromatic porphyrin or phthalocyanine ligands can result in ligand-localized redox states, thus diminishing the reactivity of the metal center.

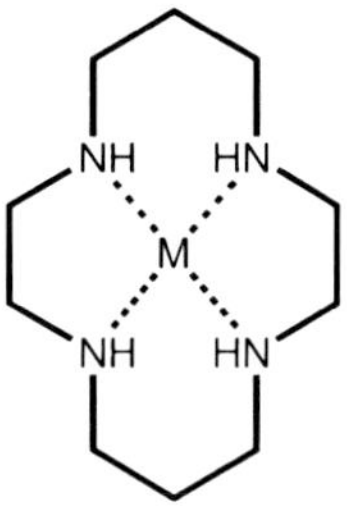

M(cyclam)

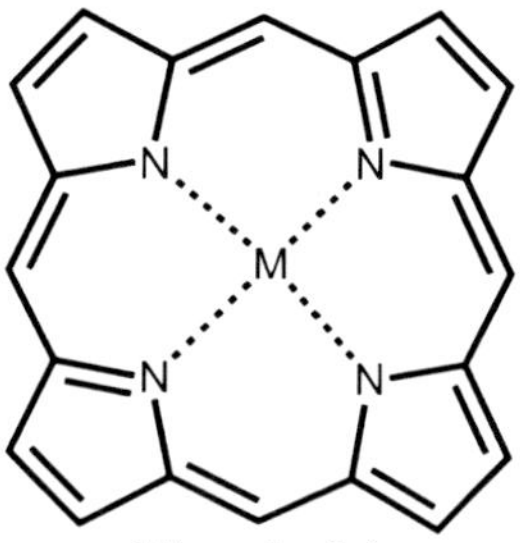

M(porphyrin)

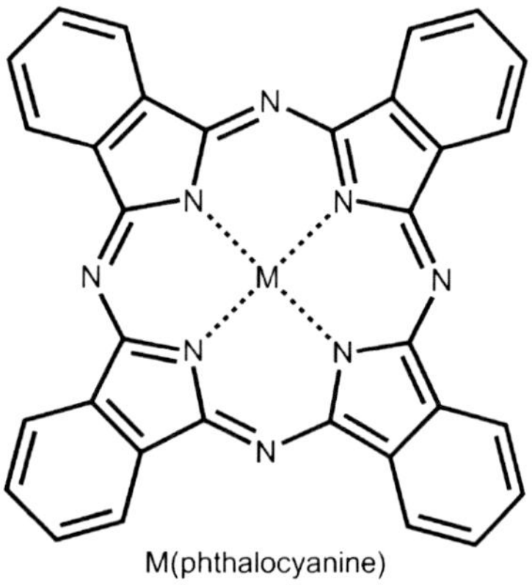

M(phthalocyanine)

The CO_2 reduction catalyzed by non-porphyrinic tetraaza macrocyclic complexes of Co(II) and Ni(II) yields CO and formic acid as main reduction products together with H_2 as a by-product. The commonly accepted mechanism involves the reduction of the metal center M(II)L to form the one-electron-reduced species M(I)L followed by the direct reaction with CO_2 to form an intermediate M(III)CO_2L, which is further protonated and reduced to form CO and water and recover the starting form of the catalyst (Fig. 1).

Alternatively M(I)L can react with a proton to form a metal hydride species M(III)HL followed by the insertion of CO_2 and formation of metal-bound formate intermediate. Additional reduction of the formate intermediate liberates formate and recovers the starting complex. The reduction and protonation of M(III)HL yields H_2 as a by-product [21]. The mechanism described above clearly demonstrates how the metal center can facilitate multielectron reactions by adopting multiple oxidation states (e.g., M(I), M(II), and M(III)). As a result, reaction (3), e.g., can be driven at fairly low overpotentials (−0.9 V for CO production catalyzed by the Ni(cyclam) complex at pH 4 was reported by Sauvage's group [24]).

Catalytic reduction of CO_2 by porphyrinic tetraaza macrocyclic complexes of Fe and Co was reported to yield CO, while Ag- and Pd-catalyzed reduction mainly resulted in oxalate formation [19, 20]. The mechanism of CO_2 reduction by these complexes is quite similar to that described above; however, deeper reduction of a metal center might be required (oxidation state M(I) or lower) [25].

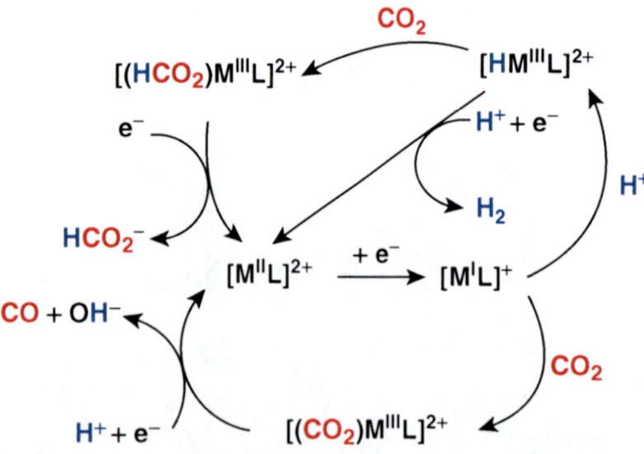

Electrocatalysts for Carbon Dioxide Reduction, Fig. 1 CO_2 reduction catalyzed by macrocyclic metal complexes

Transition Metal Complexes Coordinated to 2, 2′–Bipyridine (bpy)

Re and Ru 2,2′–bipyridine complexes can catalyze electrochemically (e.g., $Ru(bpy)_2(CO)_2$ at ca. −1.3 V, water/DMF 10:1, pH 6 or at −1.3 V for $Re(I)(bpy)(CO)_3Cl$ in acetonitrile) [26] or photochemically driven CO_2 reduction to yield primarily CO, formate, and H_2. The catalytic mechanism varies depending on the nature of the metal center and the ligand set, but can be generally represented by the reactivity of Re(I) $(bpy)(CO)_3Cl$ as proposed by Sullivan and Meyer based on their electrochemical studies in organic solvents [27] (Fig. 2).

The electrochemical reduction of CO_2 catalyzed by rhodium and iridium bpy complexes at ca. –1.35 V primarily yields formate [19].

Transition Metal Complexes Coordinated to Phosphine Ligands

Despite the fact that phosphine complexes are widely used in homogeneous catalysis, there are

$Re^I(bpy)(CO)_3Cl$
$+e^-$
$[Re^I(bpy^{\bullet-})(CO)_3Cl]^-$
Cl^-
e^-
Cl^-
A = an oxide ion acceptor
$CO + CO_3^{2-}$
$Re^0(bpy)(CO)_3$
CO_2
$[Re^0(bpy^{\bullet-})(CO)_3]^-$
$A + e^-$
$CO_2 + 2e^-$
$Re(CO_2)(bpy)(CO)_3$
$[Re(CO_2)(bpy)(CO)_3]^-$
$CO + [AO]^-$

Electrocatalysts for Carbon Dioxide Reduction, Fig. 2 Mechanism of CO_2 reduction reaction catalyzed by Re(bpy) $(CO)_3Cl$ complex

only a few CO_2 reduction catalysts reported in this series [20]. $Rh(dppe)_2Cl$ (dppe = 1,2-bis(diphenylphosphino)ethane) complex was one of the first reported to catalyze CO_2 reduction at −1.35 V in DMF yielding formate as the final product [19, 20]. Later DuBois' group examined a series of tridentate triphosphine complexes and found no activity for Pt and Ni complexes, but Pd complexes were found to be active electrocatalysts [23]. Electrochemical reduction of CO_2 in acidic acetonitrile solutions catalyzed by Pd(triphos) complexes was reported at potentials as positive as −1.1 V versus ferrocene [28] (a conversion factor of +0.5 V versus NHE was suggested [29]), resulting mainly in CO and H_2. The mechanism by which Pd(triphos) catalysts operate was proposed to be somewhat similar to Co or Ni tetraaza macrocycles. The initial reduction of the complex is followed by the addition of CO_2 molecule to the metal center, which liberates CO and a water molecule after additional reduction and two protonations [23]. In spite of the high catalytic rates (up to 300 M^{-1} s^{-1}) and low overpotentials for CO production, triphosphine complexes were found not to be particularly stable, resulting in low turnover numbers (up to 100). Contrary to the inactivity of Ni(triphos) complexes, Kubiak's group has reported CO_2 reduction catalyzed by poly-nuclear Ni complexes (e.g., $[Ni_3(\mu_3\text{-I})(\mu_3\text{-CNMe})(\mu_2\text{-dppm})_3]^+$, dppm = bis(diphenylphosphino)methane) in acetonitrile solutions [19]. The only products observed after electrolysis at ca. −0.85 V were CO and CO_3^{2-}. The same group has also reported that binuclear copper complexes, such as $[Cu_2(\mu\text{-PPh}_2\text{bpy})_2\text{-}(MeCN)_2]^{2+}$, ($PPh_2bpy$ = 6-diphenylphosphino-2,2′-bipyridyl) can catalyze CO_2 reduction to CO at potentials below −1.3 V in acetonitrile.

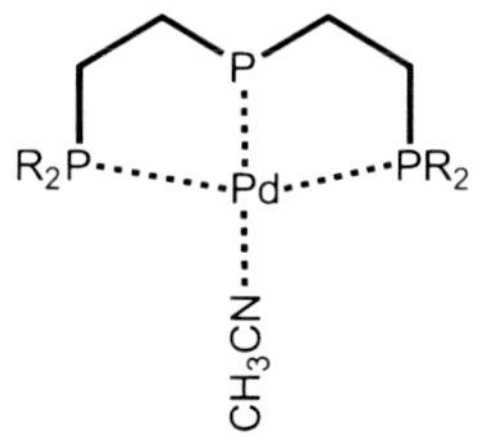

Pd(triphos)(CH_3CN)

Future Directions

In summary, despite decades of extensive research, the electrochemical reduction of CO_2 on metal electrodes still remains far beyond practical applicability. Significant improvements in Faradic efficiencies (over 95 %) have been achieved by optimization of the reaction conditions (e.g., operating under high pressure or optimizing electrolyte composition). However, these advances still did not overcome the major drawback of known electrochemical systems – the low energy efficiency of CO_2 reduction – which manifests itself in high overpotentials to obtain substantial currents. This drawback is inherently linked to the fundamental problem of CO_2 reduction at metal surfaces and is associated with pathways leading to high energy intermediates. In order to avoid highly energetic intermediates, multielectron proton-coupled reactions have to be implemented. This requires reaction control on the molecular level. Some early studies by several research groups have demonstrated a significant enhancement of electrocatalytic activity towards CO_2 reduction in the presence of pyridine or polyaniline [30–32]. The electrochemical reduction of CO_2 to methanol at nearly thermodynamic potential with high Faradic efficiencies was reported [33]. While these results demonstrate that the use of Lewis bases represents a promising new approach for efficient electrochemical CO_2 reduction, the exact mechanism (including identification of the key intermediates) remains unclear.

Molecular catalysts, such as transition metal coordination complexes, are capable of mediating multielectron and proton-coupled reactions because of the ability of transition metals to accommodate multiple redox states. In addition, electronic and acid/base properties of these catalysts can be conveniently tuned through the ligand design. Very low overpotentials for transition metal-catalyzed CO_2 reduction have been demonstrated. However, the main drawback of molecular catalysts is their relatively low stability resulting in low turnover numbers (TON). The benchmark for industrial application is in the range 10^6–10^7 TON [34], which is far beyond currently reported TON for molecular catalysts for CO_2 reduction. Immobilization on solid substrates (e.g., electrodes) was found to increase the stability of molecular catalysts. A better understanding of linker properties, linker stability under reaction conditions and the rates of electron and mass transport at the surface of immobilized catalysts is still needed to develop more stable and efficient catalytic systems for the reduction of carbon dioxide.

Cross-References

- ▶ Electrocatalysis - Basic Concepts, Theoretical Treatments in Electrocatalysis via DFT-Based Simulations
- ▶ Electrocatalysis, Fundamentals - Electron Transfer Process; Current-Potential Relationship; Volcano Plots
- ▶ Electrochemical Fixation of Carbon Dioxide (Cathodic Reduction in the Presence of Carbon Dioxide)

References

1. Willner I, Maidan R, Mandler D, Durr H, Dorr G, Zengerle K (1987) Photosensitized reduction of CO_2 to CH_4 and H_2 evolution in the presence of ruthenium and osmium colloids – strategies to design selectivity of products distribution. J Am Chem Soc 109(20):6080–6086
2. Vassiliev YB, Bagotzky VS, Osetrova NV, Khazova OA, Mayorova NA (1985) Electroreduction of carbon-dioxide 1. The mechanism and kinetics of electroreduction Of CO_2 in aqueous-solutions on metals with high and moderate hydrogen overvoltages. J Electroanal Chem 189(2):271–294
3. Hori Y (2008) Electrochemical CO_2 reduction on metal electrodes. Mod Aspects Electrochem 42: 89–189
4. Schwarz HA, Creutz C, Sutin N (1985) Homogeneous catalysis of the photoreduction of water by visible-light 4. Cobalt(I) polypyridine complexes – redox and substitution kinetics and thermodynamics in the aqueous 2,2′-bipyridine and 4,4′-dimethyl-2,2′-bipyridine series studied by the pulse-radiolysis technique. Inorg Chem 24(3):433–439
5. Cook RL, Macduff RC, Sammells AF (1988) On the electrochemical reduction of carbon-dioxide at in situ electrodeposited copper. J Electrochem Soc 135(6):1320–1326
6. Gattrell M, Gupta N, Co A (2006) A review of the aqueous electrochemical reduction of CO_2 to hydrocarbons at copper. J Electroanal Chem 594(1):1–19
7. Hara K, Kudo A, Sakata T (1995) Electrochemical reduction of carbon-dioxide under high-pressure on various electrodes in an aqueous-electrolyte. J Electroanal Chem 391(1–2):141–147
8. Jitaru M, Lowy DA, Toma M, Toma BC, Oniciu L (1997) Electrochemical reduction of carbon dioxide on flat metallic cathodes. J Appl Electrochem 27(8):875–889
9. Sammels AF, Cook RL (1993) Electrocatalysis and Novel Electrodes for High Rate CO_2 reduction under Ambient Conditions. In: Sullivan BP, Krist K, Guard HE (eds) Electrochemical and electrocatalytic reduction of carbon dioxide. Elsevier, Amsterdam, p 247
10. Batista EA, Temperini MLA (2009) Spectroscopic evidences of the presence of hydrogenated species on the surface of copper during CO_2 electroreduction at low cathodic potentials. J Electroanal Chem 629(1–2):158–163
11. Koga O, Matsuo T, Yamazaki H, Hori Y (1998) Infrared spectroscopic observation of intermediate species on Ni and Fe electrodes in the electrochemical reduction of CO_2 and CO to hydrocarbons. Bull Chem Soc Jpn 71(2):315–320
12. Koga O, Matsuo T, Yamazaki H, Hori Y (1998) Infrared spectroscopic study of CO_2 and CO reduction at metal electrodes. Advances in chemical conversions for mitigating carbon dioxide, Elsevier, Amsterdam, vol 114, pp 569–572
13. Hori Y, Murata A, Takahashi R (1989) Formation of hydrocarbons in the electrochemical reduction of carbon-dioxide at a copper electrode in aqueous-solution. J Chem Soc Faraday Trans I 85:2309–2326
14. Bagotzky VS, Vassiliev YB, Khazova OA (1977) Generalized scheme of chemisorption, electrooxidation and electroreduction of simple organic-compounds on platinum group metals. J Electroanal Chem 81(2):229–238
15. Taguchi S, Aramata A (1994) Surface-structure sensitive reduced CO_2 formation on Pt single-crystal electrodes in sulfuric-acid-solution. Electrochim Acta 39(17):2533–2537
16. Vassiliev YB, Bagotzky VS, Osetrova NV, Mikhailova AA (1985) Electroreduction of carbon-dioxide 3. Adsorption and reduction of CO_2 on platinum metals. J Electroanal Chem 189(2):311–324
17. Tomita Y, Hori Y (1998) Electrochemical reduction of carbon dioxide at a platinum electrode in acetonitrile-water mixtures. Advances in chemical conversions for mitigating carbon dioxide, Elsevier, Amsterdam, vol 114, pp 581–584
18. Adzic RR, Spasojevic MD, Despic AR (1979) Hydrogen evolution on platinum in the presence of lead, cadmium and thallium adatoms. Electrochim Acta 24(5):569–576
19. Benson EE, Kubiak CP, Sathrum AJ, Smieja JM (2009) Electrocatalytic and homogeneous approaches

to conversion of CO_2 to liquid fuels. Chem Soc Rev 38:89–99
20. Collin JP, Sauvage JP (1989) Electrochemical reduction of carbon-dioxide mediated by molecular catalysts. Coord Chem Rev 93(2):245–268
21. Morris AJ, Meyer GJ, Fujita E (2009) Molecular approaches to the photocatalytic reduction of carbon dioxide for solar fuels. Acc Chem Res 42:1983–1994
22. Doherty MD, Grills DC, Muckerman JT, Polyansky DE, Fujita E (2010) Toward more efficient photochemical CO_2 reduction: use of $scCO_2$ or photogenerated hydrides. Coord Chem Rev 254:2472–2482
23. DuBois MR, DuBois DL (2009) Development of molecular electrocatalysts for CO_2 reduction and H-2 production/oxidation. Acc Chem Res 42(12): 1974–1982
24. Beley M, Collin JP, Ruppert R, Sauvage JP (1986) Electrocatalytic reduction of carbon dioxide by nickel cyclam2+ in water: study of the factors affecting the efficiency and the selectivity of the process. J Am Chem Soc 108(24):7461–7467
25. Grodkowski J, Neta P, Fujita E, Mahammed A, Simkhovich L, Gross Z (2002) Reduction of cobalt and iron corroles and catalyzed reduction of CO_2. J Phys Chem A 106(18):4772–4778
26. Ishida H, Tanaka K, Tanaka T (1987) Electrochemical CO_2 reduction catalyzed by ruthenium complexes $[Ru(bpy)2(CO)_2]^{2+}$ and [Ru(bpy)2(CO)Cl]+. Effect of pH on the formation of CO and HCOO. Organometallics 6(1):181–186
27. Sullivan BP, Bolinger CM, Conrad D, Vining WJ, Meyer TJ (1985) One- and two-electron pathways in the electrocatalytic reduction of CO_2 by fac-Re(bpy) $(CO)_3Cl$ (bpy = 2,2[prime or minute]-bipyridine). J Chem Soc Chem Commun 20:1414–1416
28. DuBois DL, Miedaner A, Haltiwanger RC (1991) Electrochemical reduction of carbon dioxide catalyzed by [Pd(triphosphine)(solvent)](BF4)2 complexes: synthetic and mechanistic studies. J Am Chem Soc 113(23):8753–8764
29. DuBois DL, Miedaner A (1987) Mediated electrochemical reduction of CO_2. Preparation and comparison of an isoelectronic series of complexes. J Am Chem Soc 109(1):113–117
30. Barton EE, Rampulla DM, Bocarsly AB (2008) Selective solar-driven reduction of CO_2 to methanol using a catalyzed p-GaP based photoelectrochemical cell. J Am Chem Soc 130(20):6342–6344
31. Ogura K, Nakayama M, Kusumoto C (1996) In situ Fourier transform infrared spectroscopic studies on a metal complex-immobilized polyaniline Prussian blue modified electrode and the application to the electroreduction of CO_2. J Electrochem Soc 143(11):3606–3615
32. Seshadri G, Lin C, Bocarsly AB (1994) A new homogeneous electrocatalyst for the reduction of carbon-dioxide to methanol at low overpotential. J Electroanal Chem 372(1–2):145–150
33. Cole EB, Lakkaraju PS, Rampulla DM, Morris AJ, Abelev E, Bocarsly AB (2010) Using a one-electron shuttle for the multielectron reduction of CO_2 to methanol: kinetic, mechanistic, and structural insights. J Am Chem Soc 132(33):11539–11551
34. Hagen J (2005) Industrial catalysis. Weinheim, Wiley-VCH, p 507

Electrocatalysts for the Oxygen Reaction, Core-Shell Electrocatalysts

Miomir B. Vukmirovic
Chemistry Department, Brookhaven National Laboratory, Upton, NY, USA

E

Introduction

The power generation and transportation heavily depend on fossil fuels. Therefore, fossil fuel availability, efficiency of conversion of their energy, and environmental effects are of great concern. A promising alternative to using fossil fuels that can solve problems associated with that can be achieved using fuel cells, because they are inherently clean and efficient energy conversion devices compatible with renewable energy sources.

Despite recent advances [1–7], existing fuel cell technology still suffers from inadequately efficient energy conversion due to sluggish oxygen reduction reaction (ORR) kinetics. In acid solution, Pt is the best electrode material for ORR due to its activity and stability (dissolution). But, to surmount problems with low levels of both, high Pt loadings are required, thereby impeding the widespread usage of fuel cells.

A way to reduce the amount of Pt is by alloying Pt with less noble metals [8]. However, the best alloys have high Pt content [8]. Since the catalytic activity of the electrode is determined by the properties of surface layers, the better way to reduce the amount of Pt is to remove the unused Pt that lies below the surface layers of the metal nanoparticles and replace it with some other (less expensive) metal. In this way, Pt will

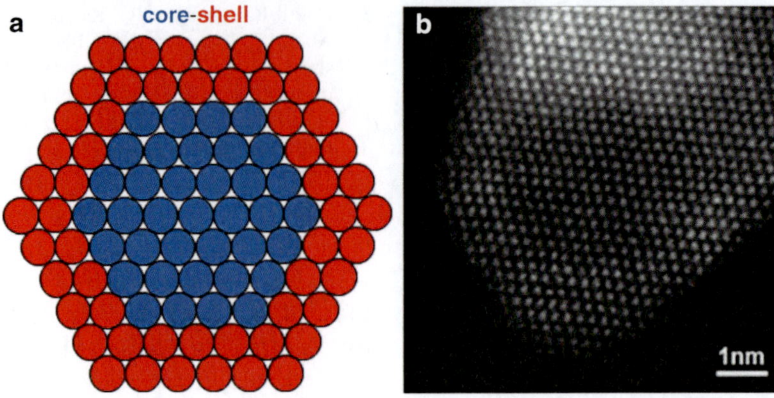

Electrocatalysts for the Oxygen Reaction, Core-Shell Electrocatalysts, Fig. 1 (**a**) Schematic representation of a core-shell nanoparticle. (**b**) High-angle annular dark field scanning transmission electron microscopy picture of Pt shell on a Pd core nanoparticle (Reprinted with permission from Ref. [9] (Copyright 2009 American Chemical Society))

be restricted to the surface in the shell of a core-shell catalyst, Fig. 1. In addition, such nanoparticles enable tuning the properties of Pt shell by changing its thickness or by choosing appropriate metal for the core. One of the most attractive types of core-shell catalysts is a Pt monolayer (Pt_{ML}) electrocatalysts for ORR [2, 10] which shell is only a monolayer thick. Developments and properties of this catalyst stimulated numerous studies of core-shell catalysts [2, 9–14].

The concept of Pt_{ML} electrocatalysts offers a possible solution to the impasse caused by a slow ORR kinetics and inadequate stability causing large Pt content of conventional electrocatalysts. A Pt_{ML} shell on a nanoparticle substrate core ensures that every Pt atom is available for catalytic activity; in other words, a Pt_{ML} achieves the ultimate reduction in Pt loading and complete Pt utilization. Also, through geometric and electronic interaction with the substrate, a Pt_{ML} can change its electronic properties and be more active and durable than pure Pt electrocatalysts.

The synthesis of Pt_{ML} electrocatalysts was facilitated by our new synthesis method which allowed us to deposit a monolayer of Pt on various metals or alloy nanoparticles [10, 15, 16] for the cathode electrocatalyst. In this synthesis approach, Pt is laid down by galvanically displacing a Cu monolayer, which was deposited at underpotentials (UPD) in a monolayer-limited reaction on appropriate metal substrate, with Pt after immersing the electrode in a K_2PtCl_4 solution.

The Role of Substrate in Determining the Activity of a Pt_{ML}

The ORR is a complex multistep reaction involving the exchange of four electrons, whose detailed mechanism still defies formulation [17]. The overall four electron reduction of O_2 in acid aqueous solutions is

$$O_2 + 4H^+ + 4e \rightarrow 2H_2O, \ (E_{NHE})_{298K} = 1.229 \ V$$

Irrespective of the microscopic mechanism, a four-electron process must involve the breaking of an O–O bond and the formation of O–H bonds [17]. Surfaces that strongly bind an adsorbate tend to enhance the kinetics of bond-breaking steps. On the other hand, surfaces that bind species weakly tend to facilitate the kinetics of bond-making steps. Hence, according to principle of Sabatier [18], the catalyst which strikes the best balance between O_2 adsorption and ORR intermediates removal will be the most active for ORR.

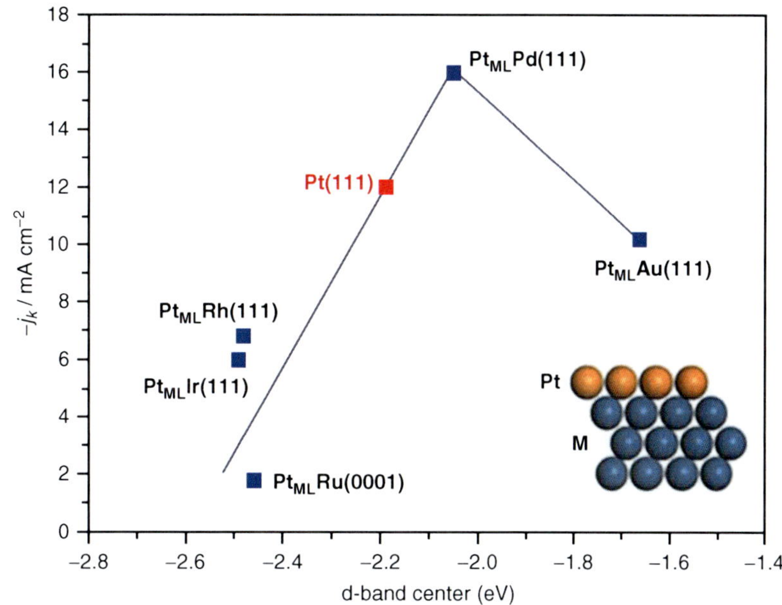

Electrocatalysts for the Oxygen Reaction, Core-Shell Electrocatalysts, Fig. 2 Kinetic currents (j_K) at 0.8 V versus RHE for ORR on the Pt_{ML} supported on different single-crystal surfaces in a 0.1 M $HClO_4$ solution as functions of calculated d-band center (ε_d-ε_F; relative to the Fermi level) of the respective clean Pt_{ML}. The current data for Pt(111) is taken from Ref. [22] and included for comparison

The role of substrate in determining the activity of a Pt_{ML}, as a way to fine tune this balance, was first demonstrated in the studies of the ORR on Pt_{ML} deposited on five different single-crystal surfaces and confirmed with nanoparticle supports [10, 19]. The ORR kinetic data and the density functional theory calculations (DFT) of the binding energy of atomic oxygen on the six surfaces result in the volcano dependence of kinetic currents as a function of the d-band center energy (ε_d) having the Pt_{ML}/Pd (111) surface at the top of the curve [19], Fig. 2. Its activity was higher than that of the Pt(111) surface. The positions of the ε_d of the Pt_{ML} do not correlate strictly with the amount of strain because the position of the ε_d for these monolayers depends both on the strain (geometric effects) and on the electronic interaction between the Pt_{ML} and its substrate (ligand effect) [20, 21].

Electrocatalysts for the Oxygen Reaction, Core-Shell Electrocatalysts, Table 1 ORR activities at 0.9 V versus RHE measured at 10 mV/s in oxygen-saturated 0.1 M $HClO_4$ solutions as a function of Pt shell thickness on Pd core and Pt catalysts [9].

Number of Pt layers	Specific activity (mA/cm^2)	Pt mass (mA/μg)
	4.0 nm Pd/C (10 wt%)	
1	0.50	0.96
2	0.52	0.59
3	0.51	0.43
	3 nm Pt/C (20 wt%)	
0	0.25	0.17

Effects of Pt Shell Thickness

Important issue to explore is the effects of Pt-shell thickness on ORR activity. To generate Pt shells with controlled thickness on a Pd core [9], viz, Pt_nPd nanoparticles (with n being the number of Pt layers in the shell on Pd cores), a Cu-UPD mediated electrodeposition method was utilized [23]. Table 1 lists Pt mass and specific activities for Pt_nPd and Pt catalysts [9]. Specific activity does not change much with the Pt-shell thickness, and is twice as high as one for pure Pt catalyst. Adding additional Pt layers relieves the strain between Pt top layer and Pd core, caused by the difference in lattice constant, causing geometrical effect less significant in Pt (top layer)- Pd(core) interaction. Therefore, almost constant value of specific activity means that Pd core predominately through ligand effect

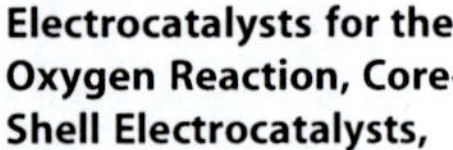

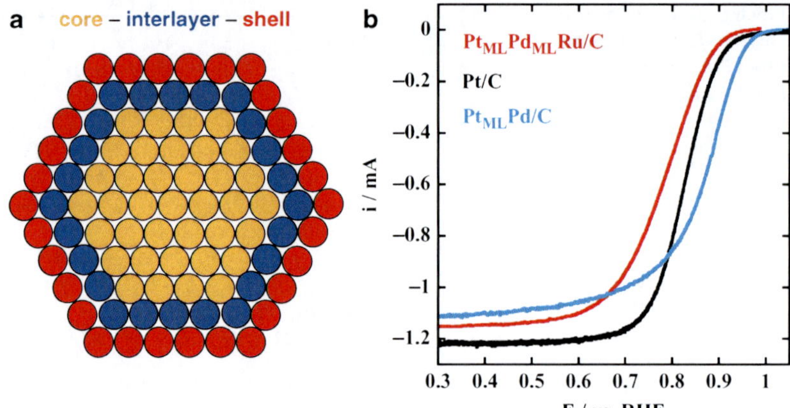

Electrocatalysts for the Oxygen Reaction, Core-Shell Electrocatalysts, Fig. 3 (**a**) Schematic representation of a core-interlayer-shell nanoparticle. (**b**) The voltammetry curves for $Pt_{ML}Pd_{ML}Ru/C$, Pt/C, and $Pt_{ML}/Pd/C$ recorded in oxygenated 0.1 M $HClO_4$ solutions with 10 mV/s scan rate. Rotation rate 1,600 rpm

influence catalytic properties of Pt(top layer). Since there is no much change in Pt(top layer)-Pd(core) interaction with shell thickness increase, only one Pt layer is necessary. This is observed in decrease of Pt mass activity with shell thickness increase due to increase of Pt mass.

Tuning Pt_{ML}-core Interaction Via Pd Interlayer

A Pt_{ML}-core interaction can be modified by placing Pd layer between a Pt_{ML} and a core or by Pd surface segregation in Pd alloy cores, Fig. 3a. A Pd interlayer approach was demonstrated by depositing Pd layer on IrCo (Ir_3Co) [24], Re [12] and Ru [25] cores and using Pd_3Ir [25], Pd-Au [13, 26], Pd_3Co [9], $Pd_2Co(111)$ [27], and $Pd_3Fe(111)$ [28] alloys, but only Ru and $Pd_3Fe(111)$ case will be discussed.

The activity of Pt_{ML} on Ru with a Pd interlayer is much larger then without it [19], but is smaller than on Pt_{ML} on Pd (Fig. 3b). The influence of the Pd interlayer on activity for ORR could be explained by the position of the Pt_{ML} d-band center [20, 21]. The Pt_{ML} on Ru(0001) is compressed by 2.5 % compared to Pt(111) causing down-shift in energy of ε_d. As a consequence oxygen bind less strongly on that Pt_{ML} then on Pt(111) leading to slow step of breaking the O–O bond which results in low activity for ORR compared to Pt(111) [19]. The Pd_{ML} on Ru (0001) is compressed by 1.65 % compared to Pd (111), while Pt_{ML} on Pd(111) is compressed by 0.85 % compared to Pt(111). Thus, introducing Pd layer as a buffer between Pt and Ru will alleviate some of the compressive strain causing upshift in energy of ε_d, thus boosting activity for ORR. In addition, Pd interlayer decreases the number of low coordination sites and decreases the curvature-induced strain; all of these effects increase the electrocatalyst's activity.

An annealed $Pd_3Fe(111)$ single-crystal alloy support has a segregated Pd layer, verified using low-energy ion scattering, that has a structure as Pd(111) shown by low-energy electron diffraction techniques [28]. It is considerably more active than Pd(111), and its ORR kinetic is comparable to that of a Pt(111) surface [28]. The enhanced catalytic activity of the segregated Pd layer compared to bulk Pd apparently reflects the modification of Pd surface's electronic properties by underlying Fe. This behavior was explained by DFT calculations which showed that Pd–OH interaction is weaker on Pd/Pd_3Fe compared to bulk Pd(111) [28]. A Pt_{ML} supported on Pd/annealed-$Pd_3Fe(111)$ shows the highest ORR kinetics compared to Pt(111) and $Pt_{ML}/Pd(111)$ surfaces, as demonstrated by their corresponding specific activities at 0.9 V versus RHE (Fig. 4) [28]. Our DFT studies suggest that the observed enhancement of ORR activity originates mainly from weaker Pt–OH interaction on Pt_{ML}/Pd_3Fe (111) compared to $Pt_{ML}/Pd(111)$ [28].

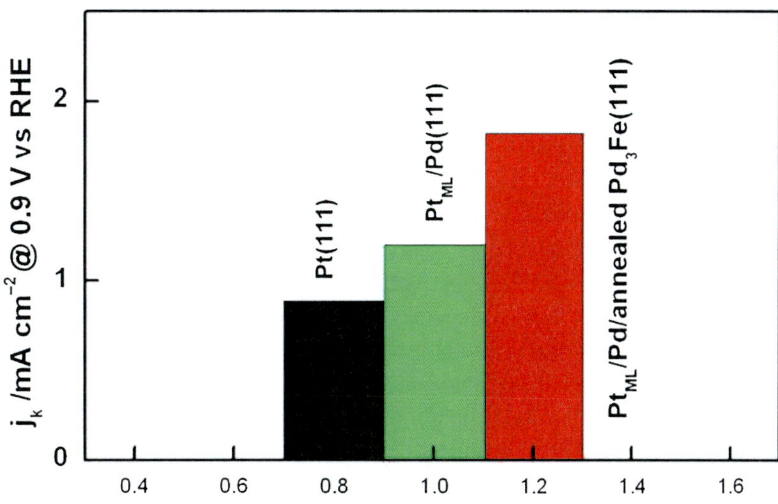

Electrocatalysts for the Oxygen Reaction, Core-Shell Electrocatalysts, Fig. 4 Pt specific activities of Pt(111), Pt_{ML}/Pd(111) and Pt_{ML}/Pd/annealed/Pd_3Fe(111) at 0.9 V versus RHE (Reprinted with permission from Ref. [28] (Copyright 2009 American Chemical Society))

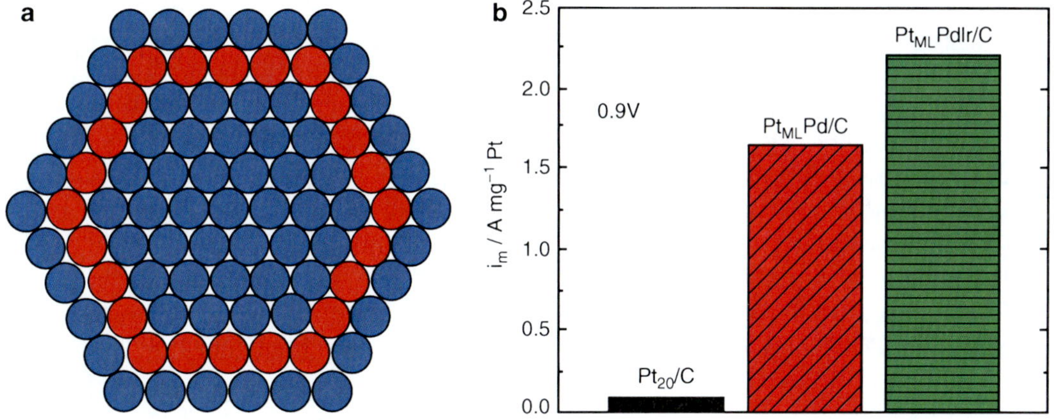

Electrocatalysts for the Oxygen Reaction, Core-Shell Electrocatalysts, Fig. 5 (**a**) Schematic representation of a subsurface modification of core. (**b**) Pt mass activities of Pt/C, Pt_{ML}Pd/C, and Pt_{ML}PdIr/C at 0.9 V versus RHE [29]

Subsurface Modification of Pd Core

The core-shell interaction can be modified either by changing shell, mixed-metal (Pt + M) [14], or by introducing interlayer between shell and core. A last possibility to influence core-shell interaction is to change a core. One way to do this is by subsurface modification of core, Fig. 5a. As an example of subsurface modification of Pd, a Pd–Ir sample will be discussed.

The subsurface layer of Ir in Pd support, obtained by galvanic displacement of a Cu monolayer by Ir and subsequent annealing at 350 °C to induce segregation of Pd to the surface and Ir into the particle, enhances the ORR kinetics at Pt_{ML} on Pd–Ir nanoparticles (Fig. 5b) [29]. The subsurface Ir causes a Pd skin to contract, so Pt_{ML} on Pd–Ir is compressed more than on Pd. As a consequence, the Pt–OH bond will weaken causing an increase in ORR activity. These results are supported by DFT calculations which showed that OH binding energy on Pt_{ML} on Pd–Ir nanoparticles is weaker than on Pt_{ML}Pd/C and Pt/C [29]. Weaker binding of OH leads to lower OH coverage, less poisoning for ORR, and thereby higher ORR activity.

Future Directions

The electrocatalysts for various catalytic processes contain noble metals. Due to high prices and limited resources of noble metals, reduction of their content in catalysts is needed. Core-shell catalysts offer an attractive choice since noble metals (active component) are restricted to the surface in the shell of a catalyst, while core can be made of less expensive metals. In this way ultimate reduction in noble metal loading and its complete utilization can be achieved, while core-shell interaction can change electronic properties of noble metal shell making it more active and durable. One of the most attractive types of core-shell catalyst is a Pt_{ML} electrocatalysts for ORR.

Pt_{ML} electrocatalysts for ORR show to be promising in the search for viable candidates of core-shell catalysts in fuel cell technology by simultaneously reducing Pt loading as well as enhancing overall performance. The results present ample illustration of the broad possibilities for designing the ORR electrocatalysts by tuning core-shell interaction. Therefore, these findings indicate the broad applicability of the Pt_{ML} catalysts and the possibility of extending this concept to the catalysts based on other noble metals which could prolong their availability for future use.

Acknowledgments This work is supported by US Department of Energy, Division of Chemical Sciences, Geosciences and Biosciences Division, under the Contract No. DE-AC02-98CH10886.

Cross-References

- ► Electrocatalysis - Basic Concepts, Theoretical Treatments in Electrocatalysis via DFT-Based Simulations
- ► Electrocatalysis, Fundamentals - Electron Transfer Process; Current-Potential Relationship; Volcano Plots
- ► Fuel Cell Vehicles
- ► Fuel Cells, Principles and Thermodynamics
- ► Oxygen Reduction Reaction in Acid Solution
- ► Oxygen Reduction Reaction in Alkaline Solution
- ► Polymer Electrolyte Fuel Cells (PEFCs), Introduction

References

1. Gasteiger HA, Kocha SS, Sompalli B, Wagner FT (2005) Activity benchmarks and requirements for Pt, Pt-alloy, and non-Pt oxygen reduction catalysts for PEMFCs. Appl Catal Environ 56:9–35
2. Adzic RR, Zhang J, Sasaki K, Vukmirovic MB, Shao M, Wang JX, Nilekar AU, Mavrikakis M, Valerio JA, Uribe F (2007) Platinum monolayer fuel cell electrocatalysts. Top Catal 46:249–262
3. Stamenkovic VR, Mun BS, Arenz M, Mayrhofer KJJ, Lucas CA, Wang G, Ross PN, Markovic NM (2007) Trends in electrocatalysis on extended and nanoscale Pt-bimetallic alloy surfaces. Nat Mater 6:241–247
4. Gottesfeld S, Zawodzinski TA (1997) Polymer electrolyte fuel cells. In: Alkire RC, Gerischer H, Kolb DM, Tobias CW (eds) Advances in electrochemical science and engineering, vol 5. Wiley-VCH, New York
5. Zhang J (2011) Recent advances in cathode electrocatalysts for PEM fuel cells. Front Energy 5:137–148
6. Gewirth AA, Thorum MS (2010) Electroreduction of dioxygen for fuel-cell applications: materials and challenges. Inorg Chem 49:3557–3566
7. Debe MK (2012) Electrocatalyst approaches and challenges for automotive fuel cells. Nature 486:43–51
8. Stephens IEL, Bondarenko AS, Grønbjerg U, Rossmeisl J, Chorkendorff I (2012) Understanding the electrocatalysis of oxygen reduction on platinum and its alloys. Energy Environ Sci 5:6744–6762
9. Wang JX, Inada H, Wu L, Zhu Y, Choi YM, Liu P, Zhou WP, Adzic RR (2009) Oxygen reduction on well-defined core – shell nanocatalysts: particle size, facet, and Pt shell thickness effects. J Am Chem Soc 131:17298–17302
10. Zhang J, Mo Y, Vukmirovic MB, Klie R, Sasaki K, Adzic RR (2004) Platinum monolayer electrocatalysts for O_2 reduction: Pt monolayer on Pd(111) and on carbon-supported Pd nanoparticles. J Phys Chem B 108:10955–10964
11. Lima FHB, Zhang J, Shao MH, Sasaki K, Vukmirovic MB, Ticianelli EA, Adzic RR (2007) Catalytic activity – d-band center correlation for the O_2 reduction reaction on Pt in alkaline solutions. J Phys Chem C 111:404–410
12. Sasaki K, Vukmirovic MB, Wang JX, Adzic RR (2010) Platinum monolayer electrocatalysts: improving structure and activity. In: Wieckowski A, Nørskov JK (eds) Fuel cell science: theory, fundamentals, and biocatalysis. Wiley, Hoboken

13. Sasaki K, Naohara H, Cai Y, Choi YM, Liu P, Vukmirovic MB, Wang JX, Adzic RR (2010) Core-protected platinum monolayer shell high-stability electrocatalysts for fuel-cell cathodes. Angew Chem Int Ed 49:8602–8607
14. Zhang J, Vukmirovic MB, Sasaki K, Nilekar AU, Mavrikakis M, Adzic RR (2005) Mixed-metal Pt monolayer electrocatalysts for enhanced oxygen reduction kinetics. J Am Chem Soc 127:12480–12481
15. Zhang J, Lima FHB, Shao MH, Sasaki K, Wang JX, Hanson J, Adzic RR (2005) Platinum monolayer on nonnoble metal – noble metal core – shell nanoparticle electrocatalysts for O_2 reduction. J Phys Chem B 109:22701–22704
16. Vukmirovic MB, Bliznakov ST, Sasaki K, Wang JX, Adzic RR (2011) Electrodeposition of metals in catalyst synthesis: the case of platinum monolayer electrocatalysts. Electrochem Soc Interface 20(2): 33–40
17. Adzic RR (1998) Recent advances in the kinetics of oxygen reduction. In: Lipkowski J, Ross PN (eds) Electrocatalysis. Wiley–VCH, New York
18. Masel RI (1995) Principles of adsorption and reaction on solid surfaces. Wiley, New York
19. Zhang J, Vukmirovic MB, Xu Y, Mavrikakis M, Adzic RR (2005) Controlling the catalytic activity of platinum-monolayer electrocatalysts for oxygen reduction with different substrates. Angew Chem Int Ed 44:2132–2135
20. Hammer B, Nørskov JK (2000) Theoretical surface science and catalysis-calculations and concepts. Adv Catal 45:71–129
21. Greeley J, Nørskov JK, Mavrikakis M (2002) Electronic structure and catalysis on metal surfaces. Annu Rev Phys Chem 53:319–348
22. Markovic NM, Gasteiger HA, Grgur BN, Ross PN (1999) Oxygen reduction reaction on Pt(111): effects of bromide. J Electroanal Chem 467:157–163
23. Sieradzki K, Brankovic SR, Dimitrov N (1999) Electrochemical defect-mediated thin-film growth. Science 284:138–141
24. Gong K, Chen WF, Sasaki K, Su D, Vukmirovic MB, Zhou WP, Izzo EL, Perez-Acosta C, Hirunsit P, Balbuena PB, Adzic RR (2010) Platinum-monolayer electrocatalysts: palladium interlayer on IrCo alloy core improves activity in oxygen-reduction reaction. J Electroanal Chem 649:232–237
25. Vukmirovic MB, Zhou WP, Chen WF, Sasaki K, Jiao J, Mavrikakis M, Adzic RR (2009) Platinum monolayer electrocatalysts: tuning the Pt monolayer-supporting core interaction by a Pd interlayer. In: 216th ECS Meeting in Vienna, Austria
26. Xing Y, Cai Y, Vukmirovic MB, Zhou WP, Karan H, Wang JX, Adzic RR (2010) Enhancing oxygen reduction reaction activity via Pd-Au alloy sublayer mediation of Pt monolayer electrocatalysts. J Phys Chem Lett 1:3238–3242
27. Zhou WP, Vukmirovic MB, Sasaki K, Adzic RR (2008) Oxygen reduction reaction on a Pt monolayer on a $Pd_2Co(111)$ single crystal surface. In: ECS transactions – fundamentals of energy storage and conversion, 213th ECS meeting, vol 13. Phoenix, AZ
28. Zhou WP, Yang X, Vukmirovic MB, Koel BE, Jiao J, Peng G, Mavrikakis M, Adzic RR (2009) Improving electrocatalysts for O_2 reduction by fine-tuning the Pt – support interaction: Pt monolayer on the surfaces of a $Pd_3Fe(111)$ single-crystal alloy. J Am Chem Soc 131:12755–12762
29. Knupp SL, Vukmirovic MB, Haldar P, Herron JA, Mavrikakis M, Adzic RR (2010) Platinum monolayer electrocatalysts for O_2 reduction: Pt monolayer on carbon-supported PdIr nanoparticles. Electrocatalysis 1:213–223

Electrocatalytic Hydrogenation

Jean Lessard
Universite de Sherbrooke, Quebec, Canada

Introduction

Electrocatalytic hydrogenation (*ECH*) involves the generation of chemisorbed hydrogen, M $(H)_{ads}$ (M is an adsorption site), by reduction of water or hydronium ions Eq. 1, then the reaction of $M(H)_{ads}$ with an adsorbed organic substrate, M $(Y = Z)_{ads}$ Eq. 3, followed by desorption of the hydrogenated product, $M(YH–ZH)_{ads}$ Eq. 4. If the YH–ZH sigma bond is sufficiently weak, its hydrogenolysis may also take place Eq. 5 as it is the case with hydroxylamines (YH–ZH = RNHOH). *ECH*s have been carried out since the beginning of the last century [1, 2]. It has been suggested quite early that electrochemical reductions of unsaturated organic compounds in aqueous media, on cathodes of spongy transition metals, might involve the reaction of the unsaturated substrate with $M(H)_{ads}$ [3]. Most of the organic functional groups which have been hydrogenated by heterogeneous catalytic hydrogenation (CH_{het}) have also been hydrogenated by *ECH* [4–12]. A large variety of catalytically active materials have been used as cathode [4–12]. A number of reviews on the *ECH* of organic compounds have been published [4–12].

$$H_2O(H_3O^+) + 2e^- + 2M \Leftrightarrow 2M(H)_{ads} + 2HO^-(H_2O) \text{ (Volmer)} \quad (1)$$

$$Y = Z + M \Leftrightarrow M(Y = Z)_{ads} \quad (2)$$

$$M(Y = Z)_{ads} + 2M(H)_{ads} \Leftrightarrow M(YH - ZH)_{ads} + 2M \quad (3)$$

$$M(YH - ZH)_{ads} \Leftrightarrow YH - ZH + M \quad (4)$$

$$M(YH - ZH)_{ads} + 2M(H)_{ads} \Leftrightarrow YH_2 + ZH_2 + 3M \quad (5)$$

Mechanistic Aspects

General Considerations

The mechanism of *ECH* of an unsaturated substrate is described by the Eqs. 1 (the Volmer step of the hydrogen evolution reaction (*HER*)) 2, 3, and 4 above. The steps (2) to (4) are exactly the same as in classical heterogeneous catalytic hydrogenation (CH_{het}). There are two main differences between *ECH* and CH_{het}. The first main difference comes from the fact that, in *ECH*, the active hydrogen species, $M(H)_{ads}$, is generated directly on the surface of the electrode using electrical energy Eq. 1 while in CH_{het}, $M(H)_{ads}$ is generated by thermal dissociation of dihydrogen, H_2 (the reverse of Eq. 7, the Tafel step of *HER*). The second main difference is due to the competition between hydrogenation of the organic substrate Eqs. 1, 2, 3 and 4 and the electrochemical desorption of $M(H)_{ads}$ (the Heyrovský step of *HER*) and/or the thermal desorption of $M(H)_{ads}$ Eq. 7. On Raney Ni cathodes for example, the *HER* proceeds via the Volmer-Heyrovský mechanism Eqs. 1 and 6 [13] so, in *ECH*, the competition between hydrogenation and desorption of $M(H)_{ads}$ involves mainly the electrochemical desorption Eq. 6 [2, 14].

$$M(H)_{ads} + H_2O(H_3O^+) + e^- \Leftrightarrow H_2 + HO^-(H_2O) + M \text{ (Heyrovský)} \quad (6)$$

$$2M(H)_{ads} \Leftrightarrow H_2 + 2M \text{(Tafel)} \quad (7)$$

Factors Influencing the Efficiency (the Current Efficiency) of *ECH*

The current efficieny (c.e.) of *ECH* is directly related to the V_{hydr}/V_{desorp} ratio, that is the ratio of the rate of hydrogenation of the organic substrate $Y = Z$, V_{hydr} (overall rate for steps (2) to (4)), to the rate of desorption of $M(H)_{ads}$, V_{desorp} (overall rate of steps (6) and/or (7)) [2, 14]. The V_{hydr}/V_{desorp} ratio is affected by the organic substrate $Y = Z$ (by the strength of the bond to be hydrogenated) and by the factors which influence the coverage of the electrode by $M(H)_{ads}$ and the activity of $M(H)_{ads}$, and, *for a given* $Y = Z$, the factors which influence the adsorption of $Y = Z$ (amount of or coverage by $M(Y = Z)_{ads}$): (1) the electrode (the electrode material and the electrode surface); and, *for a given electrode*: (2) the current density (polarization of the cathode), (3) the periodic current control, (4) the pH of the solution, (5) the organic co-solvent and the presence of any organic molecule which can be adsorbed on the electrode (which may compete with $Y = Z$ for the adsorption sites), (6) surfactants, (7) the supporting electrolyte (both the cation and the anion), (8) the temperature, and (9) the concentration of $Y = Z$ [2, 15, 16].

Generally, the least organic the **co-solvent**, the better as would be expected from the competition between the adsorption of $Y = Z$ and the adsorption of the organic co-solvent. For a given organic substrate and given conditions (substrate concentration, cathode, pH, co-solvent, supporting electrolyte), there is an optimum temperature and an optimum current density. The **optimum temperature** corresponds to a high coverage of the cathode by both $M(Y = Z)_{ads}$ and $M(H)_{ads}$, and to a high mobility of the adsorbed species on the cathode surface (high probability of reaction (encounter) of $M(Y = Z)_{ads}$ and $M(H)_{ads}$), so the V_{hydr}/V_{desorp} ratio (the c.e.) is also optimum. At lower temperatures, the adsorbed species are less mobile and the probability of reaction between $M(Y = Z)_{ads}$ and $M(H)_{ads}$ are lower, hence lower is the V_{hydr}/V_{desorp} ratio (lower c.e.). The rate of the electrochemical desorption Eq. 6 is not significantly affected by the temperature. At temperatures higher than the optimum temperature, the amount of $M(Y = Z)_{ads}$ and $M(H)_{ads}$

becomes lower and so is the V_{hydr}/V_{desorp} ratio (lower c.e.). At the **optimum current density**, the coverage of the cathode by $M(H)_{ads}$ is also optimum and the rate of electrochemical desorption of $M(H)_{ads}$ Eq. 6 is low so the V_{hydr}/V_{desorp} ratio (the c.e.) is optimum. If the current density is too low, the amount of $M(H)_{ads}$ is not sufficient to ensure a frequent encounter between $M(Y = Z)_{ads}$ and $M(H)_{ads}$ sees mainly the electrons, the electrochemical desorption of $M(H)_{ads}$ Eq. 6 becomes faster and the V_{hydr}/V_{desorp} ratio (c.e.) is lower. At current densities higher than the optimum, the potential of the cathode being more negative and the rate constant of electron transfer being related exponentially to the potential, the electrochemical desoprtion of $M(H)_{ads}$ Eq. 6 is faster and the V_{hydr}/V_{desorp} ratio (c.e.) is lower. Note that the V_{hydr}/V_{desorp} ratio is equal to the maximum current efficiency ($c.e._{max}$) when the cathode surface is completely covered by $M(Y = Z)_{ads}$. As the organic substrate is consumed, its bulk concentration decreases and a point is reached when the catalyst surface is no more saturated with $M(Y = Z)_{ads}$. From that point on, V_{hydr}/V_{desorp} decreases to reach zero at complete conversion of the organic substrate. So the c.e. for any *ECH* process might be substantially lower than the $c.e._{max}$.

Electrocatalytic Hydrogenation (*ECH*) *Versus* Heterogeneous Catalytic Hydrogenation (*CH_{het}*)

ECH presents advantages but also disadvantages with respect to CH_{het}. Both will be discussed briefly. However, for a truly worthwhile comparison between the two methods, the comparisons should be made by the same experimenters, on the same catalysts, in the same solvent, etc. Such comparisons have been very rarely made if ever made at all.

Advantages of *ECH* Over CH_{het}

The two main advantages of *ECH* over CH_{het} comes from the generation of $M(H)_{ads}$ *in situ*, directly on the catalytic cathode surface Eq. 1. Firstly, the kinetic barrier due to the thermal splitting of H_2 is completely bypassed so *ECH* can be performed under mild reaction conditions, at room temperature and at atmospheric pressure, even on catalysts which have a low activity in CH_{het}. Secondly, the mass transport of insoluble dihydrogen gas through the solution and its manipulation are also completely bypassed in *ECH* [1, 8]. Here are two examples of the first advantage. *Example 1*: Under the best conditions (T = 50 °C; J = 6.25 mA/cm^2; periodic current control (T_1 = 1 s (potential applied), T_2 = 8 s (open circuit)); H_3BO_3 0.1 M and NaCl 0.1 M in ethylene glycol-water 95:4 (v/v); pH = 3.5–6.8; pressure of an inert gas ($P_{(N2)}$ = 3.5 atm)), the hydrogenation of phenanthrene at a Raney Ni (RNi) cathode (in a two-compartment H-cell with a Nafion-324 membrane) gives a 1:1 mixture of the two possible octahydrophenanthrenes in 92 % yield and 77 % current efficiency after complete conversion (Q = 10 F) [15]. These conditions are much milder than those required for the CH_{het} of phenanthrene to octahydrophenathrenes at a RNi catalyst: T = 110–120 °C, $P_{(H2)}$ = 175–260 atm [17]. *Example 2*: The commercial CH_{het} of nitrobenzene to aniline is carried out on Cu catalysts in the gas phase (T > 150 °C) and under a pressure of dihydrogen ($P_{(H2)}$ = 200 atm) [17]. The use of a moderately active Cu catalyst ensures that the phenyl ring is not hydrogenated. On the other hand, the electrohydrogenation of nitrobenzene on a Raney Cu (RCu) cathode at room temperature and atmospheric pressure, in basic or neutral aqueous ethanol, gives aniline in 99–100 % yield with high current efficiencies (c.e. = 94–95 %) [18].

Disadvantages of *ECH* Over CH_{het}

The main disadvantage comes from the competition between hydrogenation Eqs. 2, 3, and 4 and H desorption Eqs. 6 and/or 7. For an organic substrate too difficult to hydrogenate on a given cathode, for which the V_{hydr}/V_{desorp} ratio is too low (e.g. < 0.05), no hydrogenation would take place and only hydrogen evolution Eqs. 1 and 6 and/or 7 would occur. Here are two examples of such cases. *Example 1*.- On a RNi cathode, it is not possible to hydrogenate a single and non-conjugated aromatic ring presumably because of the strong aromatic character of the benzene ring. For example, the *ECH* of phenanthrene stops at the octahydrophenanthrenes stage [1, 14]. On the other hand, perhydrophenanthrenes have been obtained in good yields (68–85 %) by CH_{het} of phenanthrene at RNi at T = 175–200 °C and $P_{(H2)}$ = 175–250 atm [17]. Interestingly, the ECH of a 20:60:20 mixture of octahydro-, decahydro-, and dodecahydrophenanthrenes at RNi under the best conditions [15] gave, after Q = 8 F, a 58:33:9 mixture of octahydro-, dodecahydro-, and perhydrophenanthrenes. About 38 % of decahydrophenanthrenes were dehydrogenated to the octahydrophenanthrenes, which sit in a well since they cannot be electrohydrogenated on RNi. The dehydrogenation generated about 12 % (0.1 F) of the total dihydrogen evolved! *Example 2.-* Although the *ECH* of lignin models (hydrogenolysis of β-O-4 bonds) in aqueous alkali at 50 °C were moderately efficient (100 % conversion, 88 % yield of products, 25 % c.e.) at low current densities [19], the ECH of lignin itself under the same conditions was totally inefficient and only H_2 evolution occurred [20]. This is because most of the hydrogenolysable β-O-4 bonds of a lignin adsorbed on the cathode are not accessible to react with $M(H)_{ads}$. Indeed, lignins are macromolecules which resemble a ball of wool so most of the β-O-4 bonds are embedded within the ball.

Conclusions on the Comparison Between *ECH* and CH_{het}

In general, despite the extensive amount of studies on *ECH* of organic compounds on a large array of catalytically active cathodes, there are probably very rare cases for which *ECH* proves to be a better and more selective method of hydrogenation than classical CH_{het}. For instance, there was hope that, because of the milder conditions, the *ECH* of ketones on a cathode modified by a chiral adsorbent would give a higher enantiomeric excess than the CH_{het} on the same catalyst. This is not the case for the hydrogenation of ketones on RNi modified with optically pure tartaric acid (see ref. [21] for *ECH* and ref. [22] for CH_{het}) as well as for the hydrogenation of ethylpyruvate on Pd/C modified with the alkaloid cinchonin (see ref. [23] for *ECH* and ref. [24] for CH_{het}). However, the production of $M(H)_{ads}$ directly from water makes *ECH* a greener method than CH_{het} even if the use of a supporting electrolyte is required in *ECH* but not for for CH_{het}.

Electrohydrogenation of Nitrocompounds

The electrohydrogenation of nitro compounds to the corresponding amines on an RCu electrode in neutral or basic aqueous alcohol is a very selective method. Nitroaryl and nitroalkyl groups in molecules containing a triple bond, a double bond, a nitrile or an easily cleavable sigma bond have been electrohydrogenated to the corresponding amino group with very high yields and very high selectivities (without hydrogenating the other functional group), for example: 4-nitro-4-methyl-2-pentynal dimethylacetal to 4-amino-4-methyl-2-pentynal dimethylacetal (70 % yield) [11]; 3-nitro-3-methylbutene to 3-amino-3-methylbutene (100 % yield) [11]; 4-cyano-nitrocumene to 4-cyano-aminocumene (91 % yield) [11]; 2-iodonitrobenzene to 2-iodoaniline (90–97 % yield) [11, 25]. One reason for this high selectivity might be that the reduction of the nitro compound to the dihydroxylamine Eqs. 8, 9, and 10 [and/or to the hydroxylamine via the nitroso intermediate Eqs. 11a and 11b] would involve an electronation-protonation (*EP*) mechanism on ***adsorbed species***, the nitro group being one of the most easily reducible organic functional group (the nitroso group is still easier to reduce Eq. 11b). Alternatively, the reason might also be that the reaction of the nitro group of the adsorbed molecule with $M(H)_{ads}$ Eq. 12 would be faster than the reaction of any other functional group present in the molecule (double bond, triple bond, carbonyl) with $M(H)_{ads}$. In neutral and basic media, the dihydroxylamine and the hydroxylamine are not protonated and are not reducible through an *EP* mechanism but their electrocatalytic hydrogenolysis (ECH_{sis}) (Eqs. 11 and 13 respectively) is rapid on RCu [18]. It is noteworthy that the protonation of the adsorbed 4-cyano-nitrocumyl radical anion (Eq. 9, R = 4-cyanocumyl) is faster than its homolytic cleavage Eq. 14, the rate constant of which is $k = 5 \times 10^6$/s [26].

$$M(RNO_2)_{ads} + e^- \Leftrightarrow M(RNO_2^{\bullet -})_{ads} \quad (8)$$

$$M(RNO_2^{\bullet -})_{ads} + H_2O \Leftrightarrow M(RNO_2H^{\bullet})_{ads} + HO^- \quad (9)$$

$$M(RNO_2H^{\bullet})_{ads} + e^- + H_2O \Leftrightarrow M(RN(OH)_2)_{ads} + HO^- \quad (10)$$

$$M(RN(OH)_2)_{ads} + 2M(H)_{ads} \Leftrightarrow M(RNHOH)_{ads} + 2M + H_2O \text{ and/or} \quad (11)$$

$$M(RN(OH)_2)_{ads} \Leftrightarrow M(RN=O)_{ads} + H_2O \quad (11a)$$

$$M(RN=O)_{ads} + 2e^- + 2H_2O \Leftrightarrow M(RNHOH)_{ads} + 2HO^- \quad (11b)$$

$$M(RNO_2)_{ads} + 2M(H)_{ads} \Leftrightarrow M(RN(OH)_2)_{ads} \quad (12)$$

$$M(RNHOH)_{ads} + 2M(H)_{ads} \Leftrightarrow RNH_2 + 3M + H_2O \quad (13)$$

$$4-CNC_6H_4C(CH_3)_2NO_2^{\bullet -} \Leftrightarrow 4-CNC_6H_4C(CH_3)_2^{\bullet} + NO_2^- \quad (14)$$

The electrohydrogenation of nitro compounds on RCu electrodes, in aqueous in neutral or basic alcohol, is the best method for the highly selective and highly efficient conversion of $R_1R_2R_3C{-}NO_2$ to $R_1R_2R_3C{-}NH_2$ (R_1, R_2, R_3 being alkyl groups) [27].

Future Directions

Despite many years of research on *ECH* [4–12], the process is still not fully understood and a better molecular-level understanding of how the electronic nature and surface geometry of the electrode influence the reaction mechanism and direction is needed. From the work of Balthruschat and coworkers [28], Morallon and coworkers [29], and Jerkiewicz and coworkers [30, 31] on the adsorption and *ECH* of organic compounds on polycrystalline and single-crystal electrodes, it seems that the active species which hydrogenates an unsaturated organic molecule would be the over-potential deposited H (H_{OPD}) and not the under-potential deposited H (H_{UPD}). However, experimental research that comprises well-defined electrode materials (geometry, electronic structure, adsorption behavior) and simple unsaturated organic compounds is still needed in order to shed more light on the mechanistic and kinetic aspects of the process.

Cross-References

- ▶ Electrocatalysis - Basic Concepts, Theoretical Treatments in Electrocatalysis via DFT-Based Simulations
- ▶ Electrocatalysis, Fundamentals - Electron Transfer Process; Current-Potential Relationship; Volcano Plots
- ▶ Electrocatalytic Synthesis
- ▶ Electrochemical Asymmetric Synthesis

References

1. Fokin S (1906) The role of the metal hydrides with reduction reactions and new data for the explanation of the question over the composition of some fats and fish-oils. Z Elektrochem 12:749–762
2. Robin D et al (1990) The electrocatalytic hydrogenation of fused polycyclic aromatic compounds at Raney nickel electrodes: the influence of catalyst activation and electrolysis conditions. Can J Chem 68:1218–1227 (and references in the Introduction)
3. Leslie WM, Butler JAV (1936) Mechanism of electrolytic processes. III. Irreversible reductions. Trans Faraday Soc 32:989–998
4. Birkett MD, Kuhn AT, Bond GC (1983) The catalytic hydrogenation of organic compounds - a comparison between the gas-phase, liquid-phase, and electrochemical routes. In: Bond GC, Webb G. (eds) Royal Society of Chemistry, Catalysis 6:61–69
5. Kokkinidis G (1986) Underpotential deposition and electrocatalysis. J Electroanal Chem 201:217–236
6. Motram CA, Pletcher D, Walsh FC (1990) Electrode materials for electrosynthesis. Chem Rev 90:837–865
7. Moutet JC (1992) Electrocatalytic hydrogenation on hydrogen-active electrodes. A review. Org Prep Proced Int 24:309–325
8. Chapuzet JM, Lasia A, Lessard J (1998) Electrocatalytic hydrogenation of organic compounds. In: Lipkowski J, Ross PN (eds) Electrocatalysis. Wiley-VCH, Pennignton, pp155–196

9. Horányi G (2003) In Horiuti's footsteps: links between catalysis and electrocatalysis. J Mol Catal A Chem 199:7–17
10. Navarro DMAF, Navarro M (2004) Hydrogenation of organic compounds by an electrochemical method for in situ hydrogen generation: electrocatalytic hydrogenation. Quim Nova 27:301–307
11. Lessard J (2005) 2004 Murray Raney award: electrocatalytic hydrogenation of organic compounds at Raney metal electrodes: scope and limitations. In: Sowa JR (ed) Chemical industries (catalysis of organic reactions), vol 104. CRC Press, Taylor and Francis Group, Boca Raton, Fl, pp 3–17
12. Korotaeva LM, Rubinskaya TY (2005) Electrocatalytic hydrogenation of unsaturated organic molecules: some features and preparative opportunities. In: Nuñez M (ed) Focus on electrochemistry research. Nova, New York, pp 145–166
13. Chen L, Lasia A (1992) Influence of the adsorption of organic compounds on the kinetics of the hydrogen evolution reaction on Ni and Ni-Zn alloy electrodes. J Electrochem Soc 139:1058–1064
14. Mahdavi B, Chapuzet JM, Lessard J (1993) The electrocatalytic hydrogenation of phenanthrene at Raney nickel electrodes: the effect of periodic current control. Electrochim Acta 38:1377–1380 (and references therein)
15. Menini R et al (1998) The electrocatalytic hydrogenation of phenanthrene at Raney nickel electrodes: the influence of an inert gas pressure. Electrochim Acta 43:1697–1703
16. Chambrion P et al (1995) The influence of surfactants on the electrocatalytic hydrogenation of organic compounds in micellar, emulsified and hydroorganic solutions at Raney nickel electrodes. Can J Chem 73:804–815
17. Hudlicky M (1984) Reduction in organic chemistry. Wiley, New York
18. Cyr A, Huot P, Belot G, Lessard J (1990) The efficient electrochemical reduction of nitrobenzene and azoxybenzene to aniline in neutral and basic aqueous methanolic solutions at Devarda copper and Raney nickel electrodes: electrocatalytic hydrogenolysis of N-O and N-N bonds. Electrochim Acta 35:147–152
19. Cyr A et al (2000) Electrocatalytic hydrogenation of lignin models at Raney nickel and palladium-based electrodes. Can J Chem 78:307–315
20. Jeanson P (2001) Études de l'hydrogénation électrocatalytique de modèles de lignine et de la lignine. M.Sc. Dissertation, Université de Sherbrooke
21. Fujihira M, Yokosawa A, Kinoshita H, Osa T (1982) Asymmetric synthesis by modified Raney nickel powder electrodes. Chem Lett 11:1089–1092
22. Izumi Y (1971) Methods of asymmetric synthesis. Enantioselective catalytic hydrogenation. Angew Chem Int Ed Engl 10:871–881
23. Vago M, Williams FJ, Calvo EJ (2007) Enantioselective electrocatalytic hydrogenation of ethyl pyruvate on carbon supported Pd electrodes. Electrochem Commun 9:2725–2728
24. Collier PJ et al (1998) Solvent and substituent effects on the sense of the enantioselective hydrogenation of pyruvate esters catalyzed by Pd and Pt in colloidal and supported forms. Chem Comm 14:1451–1452
25. Coté B et al (1993) The electrohydrogenation of *o*-iodonitrobenzene at mercury and Raney metal electrodes in aqueous methanolic solutions. J Electroanal Chem 355:219–233
26. Zheng Z-R, Evans D, Chan-Shing ES, Lessard J (1999) Cleavage reactions of radical anions that range from homolytic to heterolytic within the same family of compounds. J Am Chem Soc 121:9429–9434
27. Chapuzet JM et al (1996) The chemoselective reduction of nitro compounds: scope of the electrochemical method. J Chim Phys 93:601–610
28. Schmiemann U, Jusys Z, Baltruschat H (1994) The electrochemical stability of model inhibitors: a dems study on adsorbed benzene, aniline and pyridine on mono- and polycrystalline Pt, Rh and Pd electrodes. Electrochim Acta 39:561–576
29. Montilla F, Morallon E, Vazquez JL (2002) Electrochemical study of benzene on Pt of various surface structures in alkaline and acidic solutions. Electrochim Acta 47:4399–4406
30. DeBlois M, Lessard J, Jerkiewicz G (2005) Influence of benzene on the H_{UPD} and anion adsorption on Pt (110), Pt(100) and Pt(111) electrodes in aqueous H_2SO_4. Electrochim Acta 50:3517–3523
31. Obradović MD, Lessard J, Jerkiewicz G (2010) Cyclic-voltammetry behavior of Pt(111) in aqueous $HClO_4$ + C_6H_6: influence of C_6H_6 concentration, scan rate and temperature. J Electroanal Chem 649:248–256

Electrocatalytic Synthesis

Ichiro Yamanaka
Department of Applied Chemistry, Ookayama, Graduate School of Science and Enginnering, Tokyo Institute of Technology, Ookayama, Meguro-ku, Tokyo, Japan

Introduction

Active control of a formation rate and selectivity of product are an attractive subject for chemical synthesis from a viewpoint of green and sustainable chemistry. In a conventional catalytic reaction, it is hard to change a formation rate and selectivity of product in operation. When

Electrocatalytic Synthesis, Fig. 1 Schematic diagram of fuel cell for Wacker oxidation of ethylene to acetaldehyde

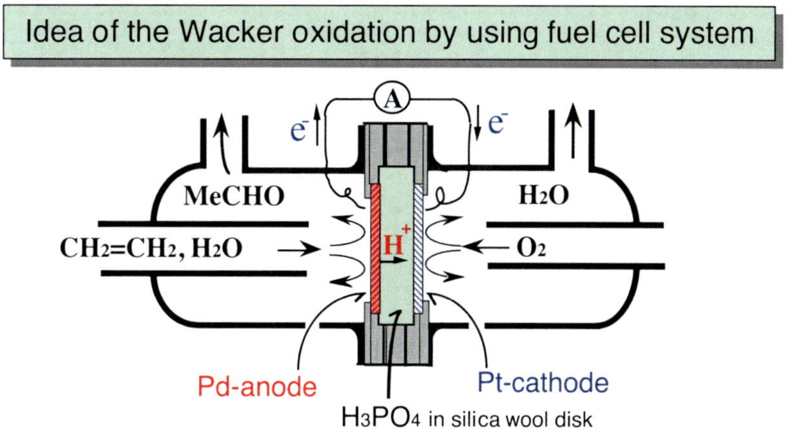

a catalyst and reaction conditions are chosen, a formation rate and selectivity of product are usually fixed. In a conventional electrochemical reaction, a formation rate of product can be easily changed in operation by controlling electrode potentials. If a suitable electrocatalyst is chosen, a desirable formation rate and selectivity of product are able to obtain by tuning electrode potential [1–3]. Electrocatalysis is essential for electrochemical active control of a formation rate and selectivity of product.

Catalytic Reaction and Fuel Cell Reaction

Fuel cell is a device to convert Gibbs free energy in chemical reaction into electricity through electrochemical cell reactions. In an H_2-O_2 fuel cell, electricity is obtained through formation of water from O_2 and H_2. When an acidic electrolyte is used, electrochemical oxidation of H_2 to e^- and H^+ occurs at an anode and reduction of O_2 with e^- and H^+ to H_2O occurs at a cathode. The net reaction is formation of water from H_2 and O_2. In other words, catalytic reaction of water formation can be decomposed to two electrochemical reactions at an anode and cathode. This principle indicates that catalytic oxidation and reduction in chemical synthesis can convert fuel cell reactions at an anode and cathode. For example, the Wacker oxidation of ethylene to acetaldehyde with O_2 would be able to perform using fuel cell reactions.

Wacker Oxidation Using Fuel Cell Reactions

Oxidation of ethylene to acetaldehyde, the Wacker oxidation process, is one of the most important processes in the current chemical industry. The Wacker oxidation is catalyzed by redox couples of Pd^{2+}/Pd^0 and Cu^{2+}/Cu^+ in HCl solutions. This catalytic oxidation Eq. 1 can be decomposed into oxidation of ethylene to acetaldehyde with water at an anode Eq. 2 and reduction of O_2 to water at a cathode Eq. 3.

$$\text{Overall}: \ C_2H_4 + H_2O + 1/2\ O_2 \rightarrow CH_3CHO + H_2O \quad (1)$$

$$\text{Anode}: \ C_2H_4 + H_2O \rightarrow CH_3CHO + 2H^+ + 2e^- \quad (2)$$

$$\text{Cathode}: \ 1/2\ O_2 + 2\ H^+ + 2\ e^- \rightarrow H_2O \quad (3)$$

A fuel cell reactor was designed for oxidation of ethylene with O_2, as shown in Fig. 1. The cell was assembled using a membrane anode prepared from Pd-black supported carbon, a membrane cathode prepared from Pt-black supported carbon, and an electrolyte membrane of H_3PO_4/SiO_2–wool. Selective oxidation of ethylene to acetaldehyde with a 95 % selectivity was performed using ethylene–O_2 fuel cell reaction at 373 K. Electrocatalysis of the Pd/C anode for the partial oxidation of ethylene is essential [1, 4].

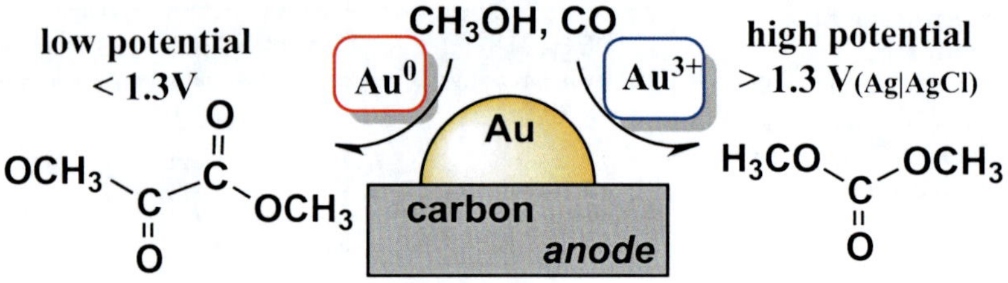

Electrocatalytic Synthesis, Fig. 2 Electrocarbonylation of methanol to dimethyloxalate (DMO) and dimethylcarbonate (DMC) at Au/activated carbon anode

When propylene was used for a fuel instead of ethylene, partial oxidation of propylene to acrolein was performed with a 96 % selectivity under short circuit conditions. Acrolein is a π-allyl oxidation product at a Pd^0 catalyst and is not a Wacker oxidation product at a Pd^{2+} catalyst. When the oxidation rate of propylene was accelerated by an applied voltage, acetone was produced with 90 % selectivity [5]. The anode potentials in operation were lower than a redox potential of Pd (+0.74 V (Ag/AgCl)) under short circuit conditions. On the other hand, the potentials were higher than the redox potential under applying voltage conditions [6, 7]. The oxidation state of Pd at the anode was Pd^0 under the former conditions and was Pd^{2+} under the latter conditions. The product selectivities to acrolein and acetone were able to control in the propylene oxidation by tuning anode potentials in operation.

Electrocarbonylation of Alcohol and Selectivity Control

Carbonylation of alcohol group, methanol to dimethyl carbonate (DMC) and dimethyl oxalate (DMO), phenol to diphenyl carbonate (DPC), is very important chemical process in the current chemical industry. Dialkoxyl carbonate is key material for phosgene free process. The electrocarbonylation has great advantages to compare with a conventional catalytic carbonylation with O_2. A particular advantage of electrocarbonylation is to be able to suppress CO_2 formation by oxidation of CO because oxidizing power can be controlled as finely as one millivolt and there is no oxygen.

Selective electrochemical carbonylation of methanol to DMC was reported at a Pd supported carbon-fiber anode [8–10]. Except Pd anodes, Au supported activated carbon (AC) anode showed electrocarbonylation of methanol to DMO and DMC [11]. The Au/AC anode has a particular electrocatalysis for the carbonylation that selectivities to DMO and DMC can be controlled by anode potentials, as shown in Fig. 2.

DMO was selectively produced at low potentials $< +1.3$ V, in contrast, DMC was selectively produced at high potentials $> +1.3$ V. This drastic change in the carbonylation selectivities was due to changes of oxidation state of Au. Au^0 was active phase for the DMO formation. On the other hand, Au^{3+} was active phase for the DMC formation. The carbonylation selectivities to DMO and DMC at the Au/AC anode were able to control by electrochemical potentials as well as the carbonylation activity [12].

Direct carbonylation of phenol to DPC is very difficult synthesis because phenol shows a low electrophilicity and a high reactivity for oxidation. Electrocarbonylation of phenol to DPC has recently succeeded by using a $PdCl_2$/AC anode and a promoter of sodium phenoxide under mild conditions (P(CO) = 1 atm, T = 25 °C) [13, 14]. Finely control of oxidation potential is essential for the electrocarbonylation of phenol to DPC.

Hydrogen Peroxide Synthesis Using Fuel Cell Reactions

H_2O_2 is one of essential chemicals for pulp breaching and waste treatment. H_2O_2 is expected to use as an oxidant for selective oxidation of hydrocarbons and a disinfectant for living environments. The industrial chemical process of H_2O_2 synthesis is limited to use of the anthraquinone method; however, the production cost and transport

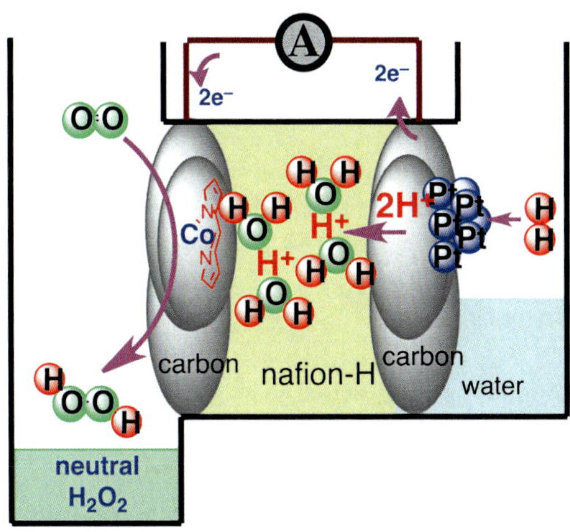

Electrocatalytic Synthesis, Fig. 3 H_2-O_2 fuel cell for synthesis of netral hydrogen peroxide solution using heat-treated Co-porphyrin/carbon electrocatalyst

limitations of H_2O_2 are serious for its use in new applications. A method for direct production of H_2O_2 from O_2 and H_2 has been desired.

H_2/O_2 fuel cell system for direct formation of H_2O_2 has been reported [15]. The fuel cell system can be operated safely for the H_2O_2 production because H_2 and O_2 are separated by a NaOH or H_2SO_4 electrolyte membrane [16–18]. Both alkaline and acid H_2O_2 solutions are useful; however, a neutral H_2O_2 aqueous solution is the most useful and flexible form. Formation of a neutral H_2O_2 aqueous solution has recently succeeded using a new conceptual solid-polymer-electrolyte membrane reactor and a new electrocatalyst [19, 20], as shown in Fig. 3.

A particular electrocatalyst of Co-TPP/VGCF (TPP: 5, 10, 15, 20-tetrakis(phenyl)-21*H*, 23*H*-porphyrin, VGCF: vapor growing carbon fiber) heat-treated in He at 1,073 K enhanced the formation of neutral H_2O_2 remarkably. The maximum concentration of 4.0 M was obtained with 90 mA cm^{-2} and 42 % current efficiency at 278 K. Electrocatalysis of the Co-TPP/VGCF was excellent; turnover frequency of Co as 14 s^{-1} and total turnover number as 4×10^5 in 8 h. Various characterization data propose that electrochemical active site for reduction of O_2 to H_2O_2 was CoN_2Cx structure on carbon surface [21]. Direct production of the neutral and halide-free H_2O_2 solution from O_2 and H_2 can be achieved.

Future Directions

Simple and conventional catalytic synthesis method and stoichiometric organic or inorganic synthesis methods have an economic advantage as industrial process to compare with conventional electrochemical synthesis methods. A serious disadvantage of electrochemical method is complicated cell structure and components. Thus, excellent performances of product yield and selectivity by using unique electrocatalysis is essential to achieve a new green and sustainable electrochemical process in future.

Cross-References

- ▶ Electrocatalysis - Basic Concepts, Theoretical Treatments in Electrocatalysis via DFT-Based Simulations
- ▶ Electrocatalysis, Fundamentals - Electron Transfer Process; Current-Potential Relationship; Volcano Plots
- ▶ Electrocatalytic Hydrogenation
- ▶ Electrochemical Asymmetric Synthesis

References

1. Otsuka K, Yamanaka I (1998) Electrochemical cellsas reactors for selective oxygenation of hydrocarbons at low temperature. Catal Today 41:311–352
2. Otsuka K, Yamanaka I (2000) Oxygenation of alkanes and aromatics by reductively activated oxygen during H_2-O_2 cell reaction. Catal Today 57:71–86
3. Yamanaka I (2008) Direct synthesis of H_2O_2 by a H_2/O_2 fuel cell. Catal Surv Asia 12:78–87
4. Otsuka K, Shimizu Y, Yamanaka I (1998) Selective synthesis of acetaldehyde using a fuel cell system in the gas phase. J Chem Soc Chem Comm 137:2076–2081
5. Otsuka K, Shimizu Y, Yamanaka I, Komatsu T (1989) Wacker type and π-allyl type oxidations of propylene controlled by fuel cell system in the gas phase. Catal Lett 3:365–370
6. Yamanaka I, Nishi A, Otsuka K, (1998) Selective synthesis of MeCHO by C2H4-(O2+NO) cell system, Chem Commun 19:2105–2106
7. Yamanaka I, Naba Y, Otsuka K (2003) Electrochemical studies of the alkene-NOx fuel cell

for organic synthesis. J Electrochem Soc 150: D129–D133
8. Otsuka K, Yagi T, Yamanaka I (1994) Dimethyl carbonate synthesis by electrolytic carbonylation of methanol in the gas phase. Electrochim Acta 39:2109–2115
9. Yamanaka I, Funakawa A, Otsuka K (2004) Electro-catalytic synthesis of DMC over the Pd/VGCF membrane anode by gas–liquid–solid phase-boundary electrolysis. J Catal 221:110–118
10. Funakawa A, Yamanaka I, Otsuka K (2006) High efficient electrochemical carbonylation of methanol to dimethyl carbonate by Br_2/Br^- mediator system over Pd/C anode. J Electrochem Soc 153:D68–D73
11. Funakawa A, Yamanaka I, Takenaka S, Otsuka K (2004) Selective control of carbonylation of methanol to dimethyl oxalate and dimethyl carbonate over gold anode by electrochemical potential. J Am Chem Soc 126:5346–5347
12. Funakawa A, Yamanaka I, Otsuka K (2005) Active control of methanol carbonylation selectivity over Au/carbon anode by electrochemical potential. J Phys Chem B 109:9140–9147
13. Murayama T, Arai Y, Hayashi T, Yamanaka I (2010) Direct synthesis of diphenyl carbonate by electrocarbonylation of phenol at a Pd^{2+} supported anode. Chem Lett 39:418–419
14. Murayama T, Arai Y, Hayashi T, Yamanaka I (2011) Direct synthesis of diphenyl carbonate by mediated electrocarbonylation of phenol at Pd^{2+}-supported activated carbon anode. Electrochim Acta 56: 2926–2933
15. Otsuka K, Yamanaka I (1990) One step synthesis of hydrogen peroxide through fuel cell reaction. Electrochim Acta 35:319–322
16. Yamanaka I, Hashimoto T, Otsuka K (2002) Direct synthesis of hydrogen peroxide (>1wt%) over the cathode prepared from active carbon and vapor-grown-carbon-fiber by a new H_2-O_2 fuel cell system. Chem Lett 8:852–853
17. Yamanaka I, Onizawa T, Takenaka S, Otsuka K (2003) Direct and continuous production of hydrogen peroxide (7 wt%) with 93 % selectivity using the fuel cell system. Angew Chem Int Ed 42:3653–3655
18. Yamanaka I, Hashimoto T, Ichihashi R, Otsuka K (2008) Direct synthesis of H_2O_2 acid solutions over carbon cathode prepared from activated carbon and vapor-growing-carbon-fiber by a H_2/O_2 fuel cell. Electrochim Acta 53:4824–4832
19. Yamanaka I, Murayama T (2008) Neutral H_2O_2 synthesis by electrolysis of water and air. Angew Chem Int Ed 10:1900–1902
20. Yamanaka I, Tazawa S, Murayama T, Ichihashi R, Hanaizumi N (2008) Catalytic neutral H_2O_2 synthesis from O_2 and H_2 by a fuel cell reaction. Chem Sus Chem 1:988–990
21. Yamanaka I, Tazawa S, Murayama T, Takenaka S (2010) Catalytic synthesis of neutral hydrogen peroxide at CoN_2C_x cathode of PEMFC. Chem Sus Chem 3:59–62

Electrochemical Asymmetric Synthesis

Osamu Onomura
Nagasaki University, Nagasaki, Japan

Introduction

A lot of optically active compounds having one or more chiral carbon atom(s) and/or axial chilarity (ies) have been used as functional molecules containing pharmaceuticals, agrochemicals, and perfumes. In many cases, their different enatiomers or diastereomers often have different biological activities and fragrances. Accordingly, in pharmaceutical's market having 500 billion $/year, the ratio of optically active compounds is over 30 % and continuously increasing. Methodologies which can supply the compounds in economically and environmentally advantaged manner are very useful. At present asymmetric synthesis is one of the powerful methods and most attractive topics in organic synthesis. Electrochemical synthesis is a promising method to oxidize and reduce organic molecules since it has an essential advantage of accomplishing the oxidation or reduction without any oxidants or reductants. Accordingly, many trials of electrochemical methods have often been applied to asymmetric synthesis.

Electrochemical Asymmetric Synthesis

Electrochemical asymmetric synthesis is electro-organic synthesis that introduces one or more new and desired elements of chirality. There are three main approaches to electrochemical asymmetric synthesis starting from chiral substrates in enantiospecific or diastereoselective manner, or non-chiral substarates in enantioselective manner as shown in Scheme 1.

Electrochemical Enantiospecific Substituition of Chiral Substrates (Memory of Chirality)

"Memory of chirality" can be defined as a phenomenon in which the chirality of a substrate having

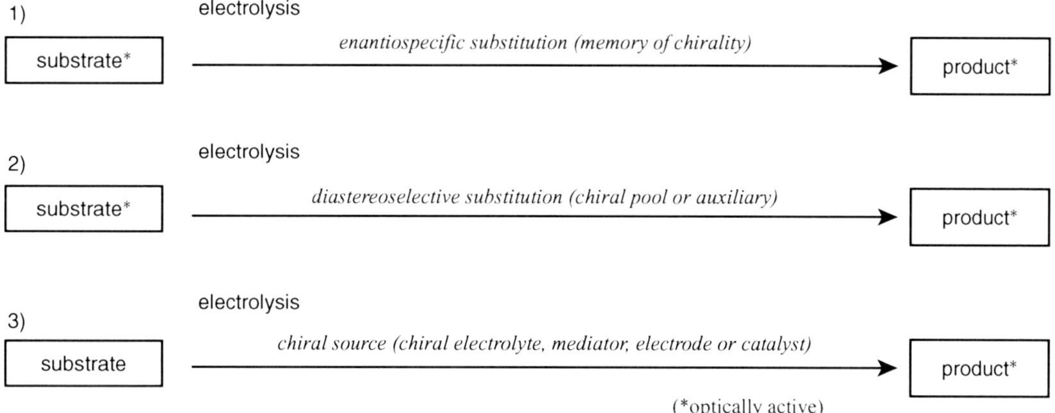

Electrochemical Asymmetric Synthesis, Scheme 1 Electrochemical asymmetric synthesis

Electrochemical Asymmetric Synthesis, Scheme 2 Loss of optical purity via carbenium ion

chiral —2e→ [sp² carbon intermediate] → racemic

a chiral sp^3-carbon is preserved in the reaction product even though the reaction proceeds at the chiral carbon as a reaction center through reactive intermediates such as carbanion, singlet monoradicals, biradicals, or carbenium ions [1].

Electrochemical Diastereoselective Substituition of Chiral Substrates

In chiral pool or chiral auxiliary method, chiral substrates could be diastereoselectively transformed by using electrochemical reaction that retain its chirality into the desired target molecule. This is especially attractive for target molecules having similar chirality to a relatively inexpensive naturally occurring building-block such as α-amino acid [2].

Electrolysis of Racemic, Meso, or Prochiral Substrates

Electrochemical synthesis of non-chiral substrates using suitable chiral source might occur by asymmetric induction to afford optically active product. Chiral electrolytes, chiral mediators, chiral modified electrodes, and/or chiral catalysts as chiral sources were examined [3].

Examples of Electrochemical Asymmetric Synthesis

Recent representative results concerning electrochemical asymmetric synthesis are shown below.

Eantiospecific Substitution (Memory of Chirality)

Although transformation of optically active α-amino acid into active intermediates without any loss of optical purity is useful for synthesis of optically active nitrogen-containing compounds, intermediary iminium ion which is a typical sp^2cation, might lose the original chirality to afford racemic product (Scheme 2).

In the memory of chirality via carbenium ion chemistry, we reported that when *N-o*-phenylbenzoylated oxazoline and thiazoline derivatives were electrochemically oxidized, optically active products (83 % and 91 % enantiomeric excess (ee), respectively) were obtained (Scheme 3) [4, 5].

Similarly, the memory of chirality was observed in the electrochemical substitution of

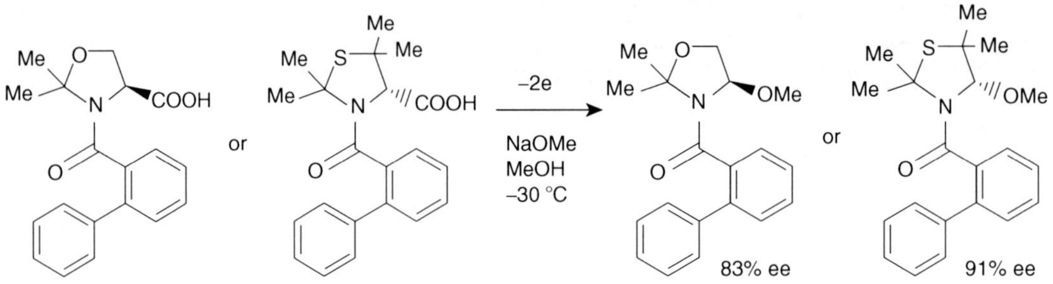

Electrochemical Asymmetric Synthesis, Scheme 3 Memory of chirality for α-amino acid

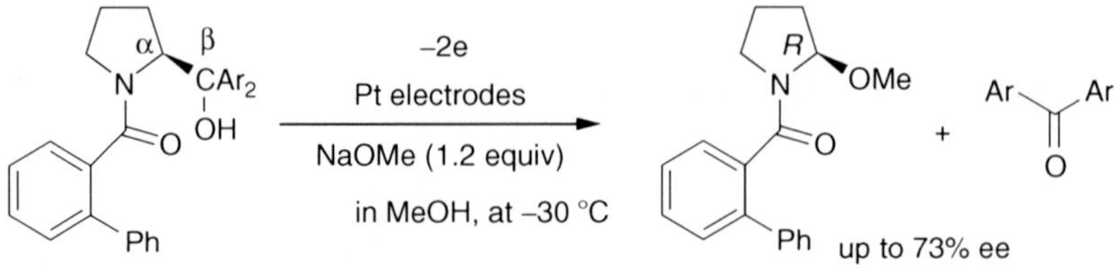

Electrochemical Asymmetric Synthesis, Scheme 4 Memory of chirality for amino alcohol

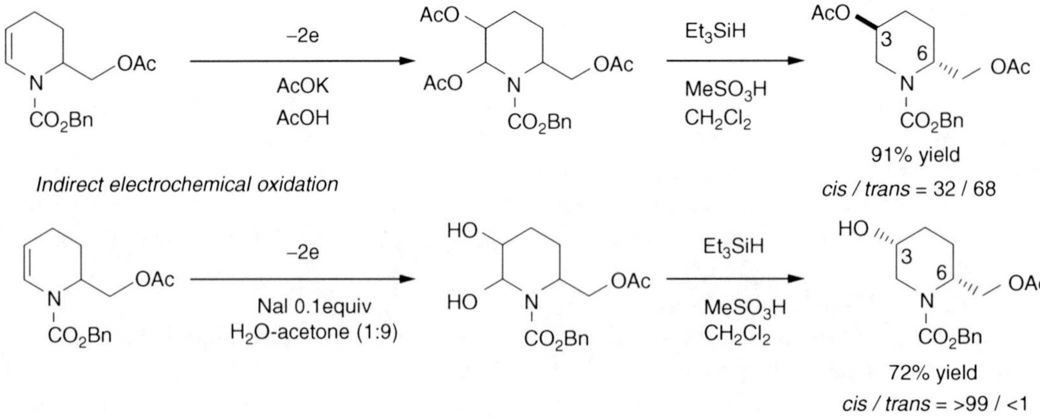

Indirect electrochemical oxidation

Electrochemical Asymmetric Synthesis, Scheme 5 Diastereoselective hydroxylation

optically active amino alcohol derivatives (Scheme 4) [6].

Diastereoselective Substitution

Diastereoselective Hydroxylation

Direct electrochemical oxidation of 6-acetoxymethyl-2,3-didehydropiperidine derivative afforded 3,6-*trans* isomer, while indirect one gave 3,6-*cis* isomer in high diastereoselectivity (Scheme 5) [7]. Indirect method using I^- as a mediator proceeded via inversion of the stereochemistry. Although iodohydroxylation afforded 3-*trans*-iodinated intermediate, succesive epoxidation by electrogenerated base (EGB) occured with the inversion of the stereochemistry at the 3-position (Scheme 6).

Diastereoselective Cyanation

Electrochemical cyanation of L-proline derivative proceeded to afford 5-*cis* substituted product in excellent diastereoselectivity (Scheme 7) [8].

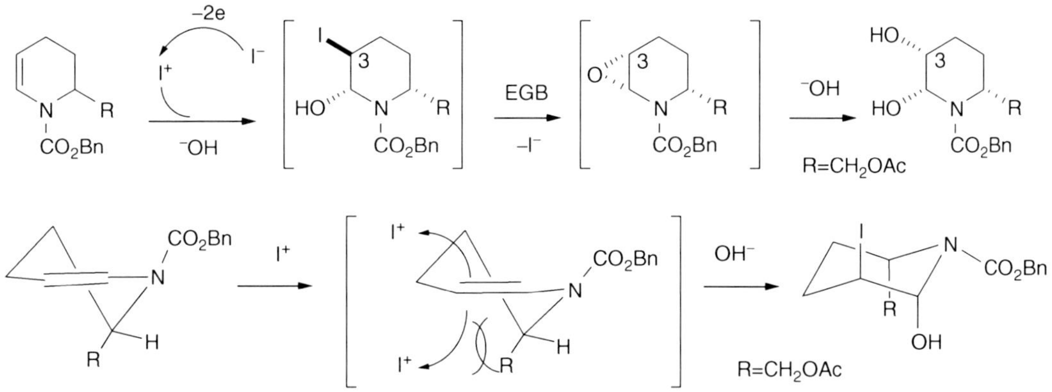

Electrochemical Asymmetric Synthesis, Scheme 6 Stereochemical course for *cis*-selective hydroxylation

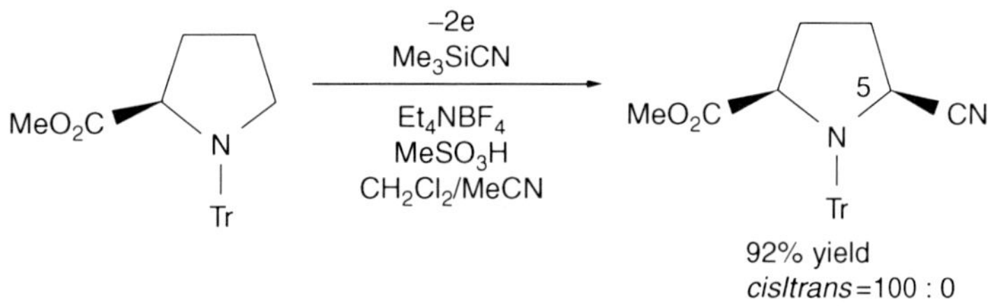

Electrochemical Asymmetric Synthesis, Scheme 7 *cis*-Selective cyanation

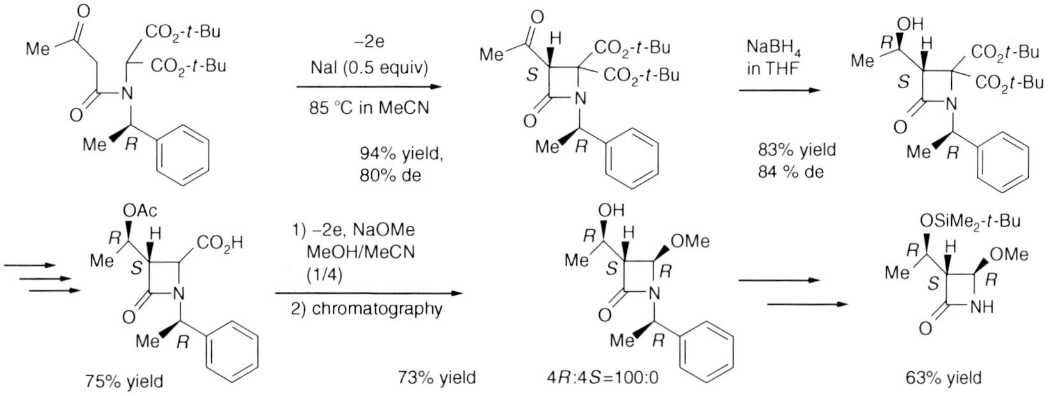

Electrochemical Asymmetric Synthesis, Scheme 8 Diastereoselective cyclization

Diastereoselective Cyclization

Important intermediate for preparation of carbapenam antibiotics was synthesized by electrochemical intramolecular carbon-carbon bond forming reaction (Scheme 8) [9]. In this cyclization, (*R*)-phenylethyl group works as a good chiral auxiliary.

Diastereoselective Substitution in Regiospecific Manner

Electrochemical oxidation of bicyclic amine prepared from (*S*)-prolinol and trifluoroacetaldehyde proceeded to afford enantiomerically pure methoxylated compound in excellent regioselectivity. This product was easily

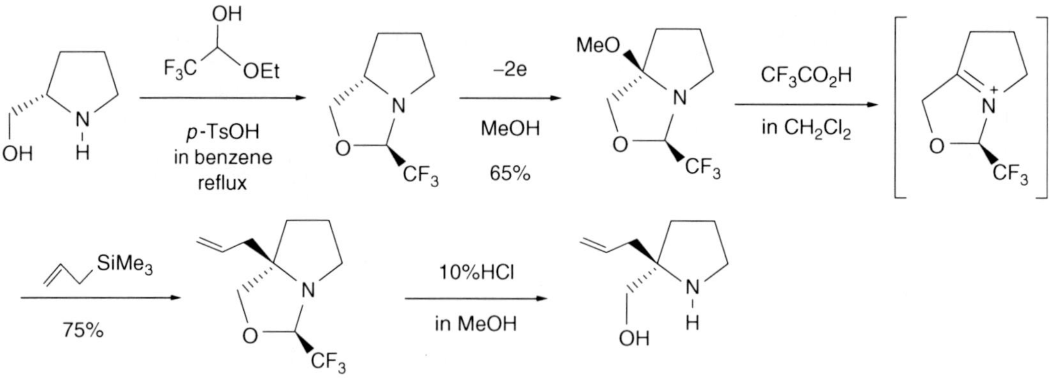

Electrochemical Asymmetric Synthesis, Scheme 9 Preparation of (*S*)-α-allylprolinol

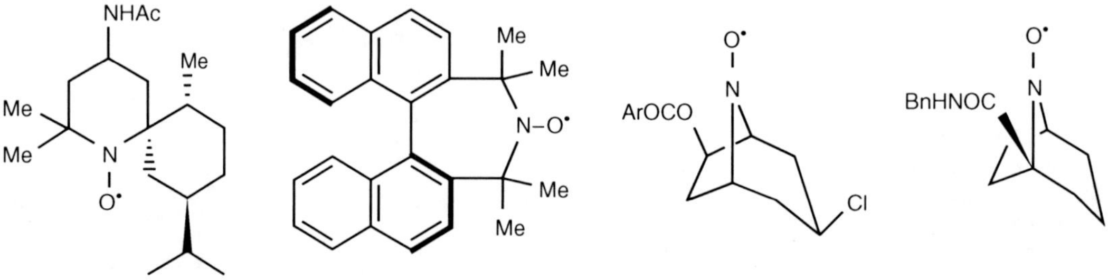

Electrochemical Asymmetric Synthesis, Fig. 1 Recent developed chiral *N*-oxyls

transformed into (*S*)-α-allylprolinol (Scheme 9) [10].

Enantioselective Oxidation

Chiral Electrolyte

Chiral electrolyte derived from quinine is effective for the enantioselective reduction of ethynyl ketone to afford propargyl alcohol in 70 % ee [11].

Chiral Solvent

Any effective chiral solvent for electrochemical enantioselective reaction have not been found until now.

Chiral Modified Electrode

Poly-L-valine coated anode promoted enantioselective oxidation of sulfide to sufoxide in 93 % ee [12].

Chiral Mediator

Recently, some chiral *N*-oxyls were effective for asymmetric desymmetrization or kinetic resolution to afford optically active lactone or alcohols (Fig. 1) [13–16].

Chiral Catalyst

Chiral Lewis acid coordinated with diol or its analogue to activate the O-H bond by chelate formation. Since the complex was easily deprotonated to generate the corresponding alkoxide ion, the complex was more oxidizable than the original alcohol as shown in Table 1. Succesive oxidation afforded the corresponding optically active carbonyl compounds (Scheme 10).

Based on this concept, electrochemical oxidative kinetic resolution of cyclic diol, aminoalcohol, or aminoaldehyde efficiently proceeded to afford optically active compounds (Scheme 11) [17].

Electrochemical Asymmetric Synthesis, Table 1 Oxidation potentials (V vs. $Ag/AgNO_3$)

Entry	Material	Oxidation potential (V)
1	$Cu(OTf)_2$	>3.0
2	(*R*,*R*)-Ph-BOX	2.07
3	$Cu(OTf)_2$-(*R*,*R*)-Ph-BOX complex	>3.0
4	*cis*-1,2-cyclohexanediol	2.10
5	*cis*-1,2-cyclohexanediol-$Cu(OTf)_2$-(*R*,*R*)- Ph-BOX complex	1.80

Electrochemical Asymmetric Synthesis, Scheme 10 Asymmetric oxidation of diol

Electrochemical Asymmetric Synthesis, Scheme 11 Kinetic resolution of diol and its analogue

Future Directions

Before the year 2000, enantioselective conversion methods in organic electrochemistry were much less developed than those in chemical synthesis. Recently, the potential of electrochemical asymmetric synthesis increases to show developments as outlined above. As a result, optically active compounds have been obtained in moderate to excellent selectivities. Since electrochemical reaction usually occurs on surface of electrode, in future, the synthesis in heterogeneous medium might afford different progress from chemical synthesis which is usually in homogeneous medium.

Cross-References

- ▶ Anodic Substitutions
- ▶ Electrocatalytic Synthesis
- ▶ Electrochemical Functional Transformation
- ▶ Electrosynthesis Using Mediator

References

1. Zhao H, Hsu DC, Carlier PR (2005) Memory of chirality. An emerging strategy for asymmetric synthesis. Synthesis 1–16. doi:10.1055/s-2004-834931
2. Fuchigami T, Nonaka T, Schäfer HJ (2001) Selectivity in electyrochemical reactions. In: Schäfer HJ (ed) Organic electrosynthesis. Wiley-VCH, Weinheim
3. Matsumura Y (2004) Asymmetric synthesis by organo-electrolytic method. In: Fuchigami T (ed) New developments in organic electrosynthesis. CMC Publishing, Tokyo, in Japanese
4. Wanyoike GN, Onomura O, Maki T, Matsumura Y (2002) Highly enhanced enantioselectivity in the memory of chirality via acyliminium ions. Org Lett 4:1875–1877
5. Wanyoike GN, Matsumura Y, Kuriyama M, Onomura O (2010) Memory of chirality in the electrochemical oxidation of thiazolidine-4-carboxylic acid derivatives. Heterocycles 80:1177–1185. doi:10.3987/COM-09-S(S)101
6. Wanyoike GN, Matsumura Y, Onomura O (2009) Memory of chirality in the electrochemical oxidation of *n-o*-phenylbenzoylated prolinols. Heterocycles 79:339–345. doi:10.3987/COM-08-S(D)10
7. Libendi SS, Ogino T, Onomura O, Matsumura Y (2007) Stereoselective introduction of hydroxyl group to piperidine ring using electrochemical method. J Electrochem Soc 154:E31–E35. doi:10.1149/1.2401038
8. Libendi SS, Demizu Y, Onomura O (2009) Direct electrochemical a-Cyanation of N-Protected cyclic amines. Org Biomol Chem 7:351–356. doi:10.1039/b816598
9. Minato D, Mizuta S, Kuriyama M, Matsumura Y, Onomura O (2009) Diastereoselective construction of azetidin-2-ones by electrochemical intramolecular C-C bond forming reaction. Tetrahedron 65: 9742–9748
10. Onomura O, Ishida Y, Maki T, Minato D, Demizu Y, Matsumura Y (2006) Electrochemical oxidation of L-Prolinol derivative protected with 1-Alkoxy-2,2,2-trifluoroethyl group. Electrochemistry 74:645–648
11. Yadav AK, Manju M, Chhinpa PR (2003) Enantioselective cathodic reduction of some prochiral ketones inthe presence of (−)-N, N-dimethylquininium tetrafluoroborate atmercury cathode. Tetrahedron Asymmetry 14:1079–1081
12. Komori T, Nonaka T (1984) Stereochemical studies of the electrolytic reactions of organic compounds. 25. Electroorganic reactions on organic electrodes. 6. Electrochemical asymmetric oxidation of unsymmetric sulfides to the corresponding chiral sulfoxides on poly(amino acid)-coated electrodes. J Am Chem Soc 106:2656–2659. doi:10.1021/ja00321a028
13. Kashiwagi Y, Kurashima F, Chiba S,Anzai J, Osa T, Bobbitt TM (2003) Asymmetric electrochemical lactonization of diols on a chiral 1-azaspiro[5.5] undecane *N*-oxyl radical mediator-modified graphite felt electrode. Chem Commun114–115. doi:10.1039/B209871G
14. Tanaka H, Kawakami Y, Goto K, Kuroboshi M (2001) An aqueous silica gel disperse electrolysis system. N-Oxyl-mediated electrooxidation of alcohols. Tetrahedron Lett 42:445–448
15. Shiigi H, Mori H, Tanaka T, Demizu Y, Onomura O (2008) Chiral azabicyclo-*N*-oxyls mediated enantioselective electrooxidation of sec-alcohols. Tetrahedron Lett 49:5247–5251. doi:10.1016/j.tetlet.2008.06.112
16. Demizu Y, Shiigi H, Mori H, Matsumoto K, Onomura O (2008) Convenient synthesis of enantiomerically pure bicyclic proline and its *N*-oxyl derivatives. Tetrahedron Asymmetry 19:2659–2665. doi:10.1016/j.tetay.2008.12.011
17. Minato D, Arimoto H, Nagasue Y, Demizu Y, Onomura O (2008) Asymmetric electrochemical oxidation of 1,2-diols, aminoalcohols, and aminoaldehydes in the presence of chiral copper catalyst. Tetrahedron 64:6675–6683

Electrochemical Bioremediation

Klaus-Michael Mangold
DECHEMA-Forschungsinstitut, Frankfurt am Main, Germany

Introduction

Contamination of soil and groundwater is an important problem facing industrialized countries and emerging economies. The main sources are spills during industrial processes, waste treatment, oil extraction, and inadequate storage of goods. Contaminants are heavy metals, polycyclic aromatic hydrocarbons (PAHs), aromatic hydrocarbons (BTEX), phenols, chlorinated hydrocarbons (CHCs), and mineral oil. The European Environmental Agency estimated for Europe about 250,000 contaminated sites which needed to be remediated. This number may increase by more than 50 % in 2025. In the last 30 years, only about 80,000 sites have been cleaned up [1]. Well-established methods are excavation and treatment of soil or pump and treat, i.e., groundwater extraction by wells or drains, treatment of the water, e.g., by activated carbon or air stripping, and pumping back underground or into a nearby stream. However, the costs of ex situ processes are very high. A cost-efficient way to clean up the soil is natural attenuation, i.e., the reduction in toxicity, mass and/or mobility of a contaminant without human intervention owing to degradation, or immobilization of contaminants by microbiological, chemical, and physical processes, e.g., dilution, sorption, volatilization [2]. Bioremediation technologies use microorganisms to degrade contaminants. A comprehensive review is found in [1]. The aerobic or anaerobic biodegradation by microorganisms is limited by several parameters: biocompatible pH, redox potential, and suitable temperature. The contaminant needs to be available for microorganisms. Problems arise if contaminant concentration is too high, i.e., toxic, or too low. The concentration of contaminants available for microorganism is determined by mass transfer, sorption, and desorption at the soil. Some contaminants cannot be metabolized by microorganisms, e.g., quaternary carbon compounds and some pharmaceuticals (diclofenac, diatrizoate). Microorganisms also need electron acceptors, electron donors, and nutrients, e.g., N- or P-containing compounds, in adequate amounts. A continuous water film in soil is important for the transport of contaminants and nutrients and enables microorganisms to move to more favorable locations. In dry soil, activity of microorganisms is decreased or even inhibited.

Natural attenuation is a cheap but slow process. Economic utilization of contaminated areas needs an enhancement of natural remediation processes. Techniques of enhanced natural attenuation (ENA) may involve the addition of electron acceptors, electron donors, or nutrients to stimulate naturally occurring microbial populations (biostimulation) or could introduce specific microorganisms aimed at enhancing the biodegradation of the target compound (bioaugmentation) [3]. Bioaugmentation can involve the ex situ stimulation of naturally occurring microbial populations that are reinjected into the contaminated site, the addition of wild-type strains or mixed cultures not native to the site that are capable of biodegrading or co-metabolizing the target compound, or the addition of genetically modified organisms.

How Can Electrochemistry Enhance Natural Attenuation?

There are two approaches for electrochemical bioremediation: electrokinetic remediation and electrochemical biostimulation.

Electrokinetic Remediation

Electrokinetic soil remediation is based on the principles of electrokinetic effects. Contaminants are moved by electrokinetic effects to defined zones, e.g., a well, and removed. An extensive review is found in [4]. Applying a DC electric field to a solution of ions results in movement of ions to the electrode of opposite charge. Cations are attracted by the negatively charged cathode and anions are attracted by the positively charged anode. The transport of ions induced by an electric field is called migration. Similar to ions, charged particles or colloids are also attracted by electrodes of opposite charge. This is called electrophoresis. The transport of ions, charged particles, and colloids will induce convection of the solution due to the net viscous drag on the solvent enclosing the charged species. This is called electro-osmosis. Nonionic species are transported by electro-osmotic flow or by adsorption on charged particles. The dominant electrochemical process occurring at the electrodes is electrolysis of groundwater, which results in a decrease of pH at the anode and an increase of pH at the cathode. In order to prevent precipitation of metal hydroxides and carbonates, there is the necessity to introduce acid in the soil. At the electrodes, contaminants may be removed by electroplating, precipitation, pumping, or complexing with ion exchanger resign. In most cases, heavy metals or radio nuclides are removed by electrokinetic remediation. However, it has also been demonstrated that simultaneous removal of heavy metals and organic contaminants [5], e.g., polycyclic aromatic hydrocarbons (PAH), as well as removal of pure PAH [6] is possible. Neutral organic contaminants are moved to the cathode by electro-osmotic flow. The process might be enhanced by the use of surfactants or reagents to improve mobility of contaminants. The efficiency of electrokinetic remediation is dependent on type, structure, water content, and saturation of the soil [4].

Furthermore electrokinetic effects also can be used to enhance bioremediation. At many contaminated sites for contaminants, nutrients, electron acceptors, and microorganisms, a nonuniform distribution is observed. Transport of compounds predominately occurs in groundwater flow direction and only limited transversal mixing takes place. Hence, the availability of nutrients and electron acceptors, e.g., sulfate or nitrate, is often limiting the bioremediation. The application of electrokinetic effects by arranging electrodes parallel to groundwater flow direction is an approach to overcome transversal mass transfer limitations [7].

Electrochemical Biostimulation

For soil and groundwater remediation, several bioremediation techniques can be applied [8]. Unassisted metabolism of contaminants by indigenous microorganisms is called intrinsic bioremediation. In this case, only regular monitoring is necessary. If indigenous microorganisms are unable to degrade contaminants, suitable bacteria cultures can be added to the contaminated medium. This technique is called bioaugmentation. The stimulation of microbial activity of indigenous microbial populations can be achieved by drawing oxygen through the soil (bioventing) or by providing nutrients (biostimulation). Electrochemistry is a promising method for biostimulation. A descriptive example is the sequential reductive and oxidative electrochemical biostimulation of microbial degradation of chlorinated hydrocarbons [9]. Electrolysis of groundwater provides hydrogen at the cathode and oxygen at the anode. Both compounds

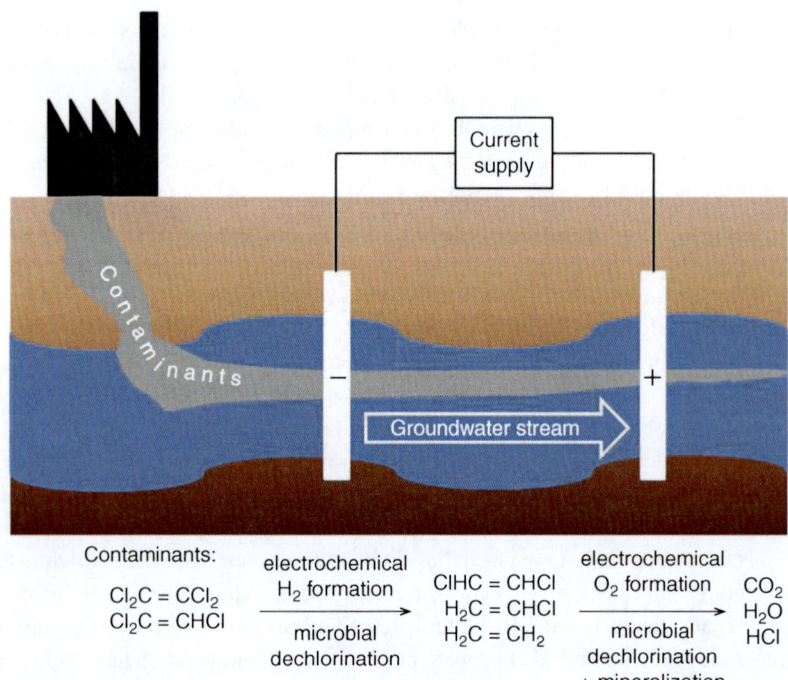

Electrochemical Bioremediation, Fig. 1 Scheme of the sequential degradation of chlorinated ethenes by anaerobic and aerobic metabolism of microorganisms. Biostimulation is achieved by providing electron donor and acceptor by electrolysis of groundwater. At the cathode (−) hydrogen as an electron donor and at the anode (+) oxygen as an electron acceptor are produced

are necessary for microbial degradation of chloroethenes. *Dehalococcoides* species can convert perchloroethene via trichloroethene and *cis*-dichloroethene to vinyl chloride and finally to ethene by reductive dechlorination. Hydrogen as an electron donor is able to enhance the microbial dechlorination. However, specific stimulation of dechlorination is challenging due to competing hydrogen consumption by methanogenesis or sulfate reduction. A constant rate of hydrogen formation at low concentrations is beneficial, because of the lower hydrogen threshold of dechlorination compared to the other reactions. Electrolysis is able to provide a defined hydrogen concentration by applying an appropriate current. Subsequent *cis*-dichloroethene and vinyl chloride can be mineralized by microbial oxidative degradation. To accomplish the mineralization, microorganisms need oxygen which is provided by groundwater electrolysis at the anode. Placing the cathode and anode in a row in the groundwater stream enables a sequential electrochemical bioremediation, as shown in Fig. 1. The feasibility of this anaerobic/aerobic bioelectrochemical process is demonstrated in laboratory columns [9].

Electrochemically synthesized hydrogen and oxygen are electron mediators, i.e., compounds that shuttle electrons between electrode and microorganism. An alternative is the direct electron transfer between microorganism and electrode without electron mediator. Some bacteria, e.g., *Shewanella* and *Geobacter* species, can exchange electrons directly with an electrode by different mechanisms [10]. Concerning dechlorination, *Geobacter lovleyi* is able to transform tetrachloroethene to *cis*-dichloroethene by accepting electrons from a cathode instead of utilizing hydrogen [11].

Future Directions

Electrochemical bioremediation is a new and promising method to enhance natural attenuation. Several electrochemical techniques for bioremediation are in different stages of development from laboratory studies, e.g., electrochemical biostimulation, to field application, e.g., electrokinetic remediation. All electrochemical methods require the placement of electrodes belowground. Concerning costs, it is important how electrodes are inserted in the ground. Excavation of wells is an expensive technique. Direct push technologies which are already used for soil and groundwater sampling are a cost-efficient option if appropriate electrodes are developed.

Cross-References

▶ Electrochemical Cell Design for Water Treatment

References

1. Guimaraes BCM, Arends JBA, van der Ha D, Van de Wiele T, Boon N, Verstraete W (2010) Microbial services and their management: recent progress in soil bioremediation technology. Applied Soil Ecol 46:157–167
2. Röling WFM, van Verseveld HW (2002) Natural attenuation: what does the subsurface have in store? Biodegradation 13:53–64
3. Scow KM, Hicks KA (2005) Natural attenuation and enhanced bioremediation of organic contaminants in groundwater. Curr Opin Biotechnol 16:246–253
4. Virkutyte J, Sillanpää M, Latostenmaa P (2002) Electrokinetic soil remediation – critical overview. Sci Total Environ 289:97–121
5. Maturi K, Reddy KR (2008) Cosolvent-enhanced desorption and transport of heavy metals and organic contaminants in soil during electrokinetic remediation. Water Air Soil Pollut 189:199–211
6. Gan S, Lau EV, Ng HK (2009) Remediation of soils contaminated with polycyclic aromatic hydrocarbons (PAHs). J Hazard Mater 172:532–549
7. Tiehm A, Augenstein T, Ilieva D, Schell H, Weidlich C, Mangold K-M (2010) Bio-electro-remediation: electrokinetic transport of nitrate in a flow-through system for enhanced toluene biodegradation. J Appl Electrochem 40:1263–1268
8. Boopathy R (2000) Factors limiting bioremediation technologies. Bioresour Technol 74:63–67
9. Lohner ST, Becker D, Mangold K-M, Tiehm A (2011) Sequential reductive and oxidative biodegradation of chloroethenes stimulated in a coupled bioelectro-process. Environ Sci Technol 45:6491–6497

10. Yang Y, Xu M, Guo J, Sun G (2012) Bacterial extracellular electron transfer in bioelectrochemical systems. Process Biochem 47:1707–1714
11. Strycharz SM, Woodard TL, Johnson JP, Nevin KP, Sanford RA, Löffler FE, Lovley DR (2008) Graphite electrode as a sole electron donor for reductive dechlorination of tetrachloethene by geobacter lovelyi. Appl Environ Microbiol 74:5943–5947

Electrochemical Cell Design for Water Treatment

Frank C. Walsh and Carlos A. Ponce de Leon
Electrochemical Engineering Laboratory, University of Southampton, Faculty of Engineering and the Environment, Southampton, Hampshire, UK

Introduction

In the current demand for green technologies, the electrochemical methods for wastewater treatment offer a great advantage, since no additional chemicals are required to remove the pollutants [1], and the electron may be considered as a "green," controllable reagent. The use of electrochemical techniques presents several challenges, due to the nature of many wastewaters. For example, such solutions:

(i) May have a variable composition with time
(ii) May be diluted in electroactive species
(iii) Can be poorly conductive electrolytes
(iv) Can have a complex, multicomponent composition
(v) May be corrosive to electrodes, membranes, and constructional materials

Electrochemical methods are considered by many to be a relatively new technology that is developing rapidly, thanks to the continuing discovery of new electrode materials and applications to ever more diverse industries. The capacity of an electrochemical reactor to treat a certain volume of water by oxidizing an organic compound or recovering metal ions depends on the electrode area. Therefore, the design of such reactors should be aiming to provide large surface active area per unit volume of reactor, defined as $A_S = A/V_R$, where A is the electrode area and V_R the volume of the reactor. In addition to the high area/volume ratio, other important performance parameters that the design of the reactor should be aiming for are the mass transport coefficient $k_L\,(= I_L/nFAc)$, where I_L is the limiting current, n the number of electrons interchanged, c is the concentration of electroactive species, and F the Faraday constant. Space-time and the space-velocity are defined as Q/V_R and $1/\tau_{ST}$, respectively. The space-time yield, ρ_{ST}, is also an important figure of merit that defines the mass of species converted per unit time, t in the reactor and is generally expressed as $(\Phi i)/nFV_R$, where ϕ is the current efficiency. The reactor design and the selection of materials must provide uniform potential and current distribution through the electrodes to achieve high reaction rates. In order to achieve this, the electrical resistance, R, of the cell components, such as electrodes, electrolyte, and external electric circuits, should be minimized. Also, the losses caused by the electrochemical process itself, including the additional potential needed to start the reaction, i.e., the overpotential η_{act}, and the concentration polarization overpotential, η_{conc}, should be as low as possible. If the reactor design includes the division of the electrodes, the separator must have high ionic conductivity. The terms mentioned above have great influence in the overall cell potential value, E_{cell}, across the anode and cathode, which is related to the energy required to drive the electrochemical reaction [2]:

$$E_{cell} = E^0{}_{cell} - |\eta^a{}_{act}| - |\eta^c{}_{act}| - \Sigma_k(IR)_k - |\eta^a{}_{conc}| - |\eta^c{}_{conc}| \quad (1)$$

where the super indices a and b indicate anode and cathode, respectively. In electrochemical reactors designed for water treatment containing organic material, the oxidation will occur at the anode, while in solutions containing metal ions, the recovery occurs at the cathode. In the case of mediated oxidation processes, high oxidant species, for example, hydrogen peroxide, can be generated in the cathode electrode, which reacts with the organic matter in solution. In these systems, the mass transport of species also determines the rate of oxidation or recovery, and the electrochemical reactors must be designed to increase the supply and removal of the species to and from the electrodes. Typical reactors currently used for water treatment will be indicated below.

Cell Design

Fundamental electrochemical studies for water treatment processes are generally carried out in a three-electrode compartment cell with small volumes of electrolyte, typically 50–100 cm^3. When the process is scaled up, some general guidelines should be followed [3]: the cell design should be simple to minimize cost and avoid complicated routine maintenance procedures, use small cell voltage and minimization of peripheral equipment, such as motors and stirrers, and have low pressure drop to minimize pumping cost.

A typical example of an electrochemical cell designed to minimize the cell potential is the Chemelec cell [4]. This cell consists of flat or expanded metal mesh with alternated cathode and anode electrodes separated by ≈ 4 mm. Typical reactor dimensions are $0.5 \times 0.6 \times 0.7$ cm, which can contain up to 6 anodes and 7 cathodes providing an area of up to 3.3 m^2. The cell contains nonconductive glass beads, which are fluidized when the electrolyte is pumped from the bottom of the cell through flow distributors, and these help to increase the mass transport of the electroactive species. The cell has been recently used to remove 8.89×10^{-4} mol dm^{-3} of Cd^{+2} and 4.82×10^{-4} mol dm^{-3} of Pb^{+2} in 1 mol dm^{-3} NaCl reaching 99.0 % and 94 % lead and cadmium recovery, respectively. Other cell designs for metal recovery include a cell containing a rotating cylinder electrode [5–7] and cells containing three-dimensional electrodes, such as reticulated vitreous carbon (RVC) or stainless steel [8–10].

Cells design with two- and three-dimensional electrodes have been also used for the oxidation of organic compounds. Among the two-dimensional electrodes, the cells containing dimensional stable anodes (DSA) and boron-doped diamond (BDD) electrodes have been the most successful contenders for the oxidation of organic molecules contained in wastewater.

Electrocoagulation with Fe and Al sacrificial electrodes has also been used to achieve more than 90 % decolorization of wastewater and helps to avoid the excess chemicals used in chemical coagulation. The fate of the organic pollutants in the samples treated chemically and electrochemically deserves much attention because the possibility of enhancing the toxicity of wastewaters during chemical and electrochemical treatment is high. Hazardous chemicals, such as carcinogenic lipophilic aromatic amines (by splitting of the $N = N$ group or desulfonation of hydrophilic amines), can be generated. Sulfonated aromatic amines are considered to be nontoxic. Anodic oxidation AOX (due to the high halogen loads) is an area which has often received very little attention. The adsorption of materials or polymers on the electrode surface can result in poisoning or clogging, and divided cells are often used to avoid the formation of halogenated compounds due to evolution of chlorine at the anode. The experience has shown that AOX may not necessarily be formed in an undivided cell, as halogenated compounds can be dehalogenated at the cathode. This demonstrates that each wastewater system must be examined by running experiments in parallel rigs with divided and undivided cells.

Future Directions

The ever increasing chemical complexity of waste solutions and effluents together with more stringent discharge limits are placing increasing demands on existing technology. Further developments in electrochemical cell design will require research and development activities in the following areas: 1) low-cost, efficient, and durable electrodes; current electrodes include precious metal based coatings and expensive boron-doped diamond; carbon polymer composite materials and lower cost catalyst coatings require development 2) high surface area, catalytic nanostructured surfaces 3) highly permselective yet lower cost and durable membranes; In many cells the membrane is based on costly perflourinated polymer material. In some applications, even lower cost micro-porous separator materials might work as well 4) electrode structure and cell design that minimize transport losses, improve efficiency, and reduce cost; cell fabrication cost reduction will be useful here 5) designs with minimal pumping and electrode movement to minimise costs and simplify maintenance 6) hybrid reactors which combine electrochemical unit processes with complementary techniques, e.g., ion exchange, UV photolysis and dialysis in a single package are attractive 7) published case studies showing the deployment of electrochemical techniques to real environmental waste and recycling problems.

Cross-References

► Electrochemical Bioremediation

References

1. Comninellis C, Chen G (eds) (2010) Electrochemistry for the environment. Springer, New York
2. Walsh FC (1993) A first course in electrochemical engineering. RSC Publishing, Cambridge
3. Pletcher D, Walsh FC (1990) Industrial electrochemistry, 2nd edn. Blackie A & P, Cambridge
4. Segundo JEDV, Salazar-Banda GR, Feitoza ACO, Vilar EO, Cavalcanti EB (2012) Cadmium and lead removal from aqueous synthetic wastes utilizing Chemelec electrochemical reactor: study of the operating conditions. Separ Purif Tech 88:107–115
5. Low CTJ, Ponce de Leon C, Walsh FC (2005) The Rotating Cylinder Electrode (RCE) and its application to the electrodeposition of metals. Aust J Chem 59:246–262
6. Reade GW, Ponce-de-Leon C, Walsh FC (2006) Enhanced mass transport to a reticulated vitreous carbon rotating cylinder electrode using jet flow. Electrochim Acta 51:2728–2736
7. Terrazas-Rodriguez JE, Gutierrez-Granados S, Alatorre-Ordaz MA, Ponce de Leon C, Walsh FC (2011) A comparison of the electrochemical recovery of palladium using a parallel flat plate flow-by reactor and a rotating cylinder electrode reactor. Electrochim Acta 56:9357–9363
8. Maja P, Damijana K, Lestan D (2011) Electrochemical EDTA recycling after soil washing of Pb, Zn and Cd contaminated soil. J Hazard Mater 192:714–721
9. Reade GW, Bond P, Ponce de Leon C, Walsh FC (2004) The application of reticulated vitreous carbon rotating cylinder electrodes to the removal of cadmium and copper ions from solution. J Chem Tech Biotechnol 79:946–953
10. Szpyrkowicz L, Zilio-Grandi F, Kaul SN, Rigoni-Stern S (1998) Electrochemical treatment of copper cyanide wastewaters using stainless steel electrodes. Water Sci Tech 38:261–268

Electrochemical Cells, Current and Potential Distributions

Alan West
Department of Chemical Engineering Columbia University, New York, NY, USA

Introduction

The spatial variation on the electrode of current density i is often referred to as the "current distribution." Since the current density is related to reaction rate through Faraday's law, the current distribution is thus a manner of expressing the variation of reaction rate within an electrochemical cell. As for traditional chemical reactors, nonuniformities in reaction rate may be anticipated if the fluid flow is inadequate to prevent concentration gradients. However, electrical field effects also influence the current distribution in an electrochemical cell, and thus reaction rates can be nonuniform even if perfect mixing is achieved in the reactor. Electrochemical cells of course have two electrodes, and sometimes optimizing a current distribution of one electrode is more important than the other. Depending on the proximity of the two electrodes, the current distributions of the electrodes may or may not influence each other.

When taking an electrochemical process from the laboratory into manufacturing, the current distribution is a key design consideration. For example, in an electrodeposition system, film thickness (proportional to local current density through Faraday's law) may be uniform on a small electrode but highly nonuniform on a large electrode. In a perhaps less obvious example, capital costs in an electrolyzer may be greatly influenced by nonuniformities if significant portions of an electrode surface do not carry out a reaction at its optimal rate.

While many engineers are experienced with the challenges of scaling up a process, "scaling down" may also be a challenge, requiring an understanding of the relative magnitude of transport phenomena on different scales. For electrochemical microfabrication processes, the current distribution may be greatly influenced by both the reactor design and the features on the substrate. In fact, it is common to consider multi-scale current distributions for applications in microelectronics.

For the scale up of a chemical reactor, inadequate mixing may result in spatial variations in, for example, reactant composition or temperature. An electrochemical reactor (cell) is a chemical reactor where the reduction and oxidation reactions are spatially separated on cathodes and anodes. The flow of ionic current through the electrolyte results in an electric field through the electrolyte. Since charged species move in response to an electrical field [1–3] and since the potential difference across the double layer impacts reaction rate, electrical field effects can significantly impact current distribution. Thus, in contrast to a chemical reactor, perfect mixing to eliminate all concentration fields does not necessarily result in uniform reaction rates.

Electrical field effects are an example of a transport phenomenon that does not arise in most chemical reactors, and these field effects often dictate the current distribution. Usually, electrical field effects are more important in the (ionically conducting) electrolyte than in the (electronically conducting) electrodes. However, as is the case of porous electrodes for fuel cells and batteries, significant potential variations in the electrodes may result if the electrodes are very thin, very large, or have high specific resistivity. Current distributions where the potential drop in the electrode is important were first studied in 1953 [4]; the phenomenon is called the "terminal effect" or "resistive substrate effect."

Fundamentally, the electrical field in the electrolyte is coupled to concentration fields of all of the ionic species and ensures that the electrolyte is electrically neutral at all locations. Rigorous calculations of these coupled fields are possible, although it typically requires numerical simulations. In the literature, the resulting current distribution is often coined the "tertiary current distribution." The resulting distribution is often dependent on multiple dimensionless groups, and it may be difficult to draw general conclusions from such studies.

Order of magnitude estimates of dimensionless groups may provide a sound basis to determine whether nonuniformities are anticipated and whether concentration or electrical field effects are the primary consideration. These dimensionless groups have well-understood physical meaning and are discussed below. Similar concepts emerge in the analysis and design of porous electrodes that may be used in a battery or fuel cell.

Concentration Fields

Concentration Fields – Concentration fields may influence the current distribution. Specifically, when a reaction rate is high relative to the rate at which the reactant can be transported to the surface, near-surface concentration gradients are established. In the extreme, the local current density is limited by the rate of mass transfer of a reactant to the surface. This current density is known as the limiting current density i_{lim}, and the dimensionless ratio of the spatial-average applied current density i_{avg} (total current divided by electrode area) to the limiting current density gives an estimate of the importance of concentration gradients.

When the ratio

$$\frac{i_{avg}}{i_{lim}} \to 1$$

the current distribution is normally dictated by the concentration field, which is strongly influenced by fluid flow. For example, the impingement flow of a rotating disk electrode leads a uniform current distribution, while the limiting current distribution on an electrode in the presence of a shear flow may be highly nonuniform with reaction rates much higher at the leading edge of the electrode.

When $\frac{i_{avg}}{i_{lim}} << 1$, concentration variations can be neglected, and electrical field effects dictate the current distribution. In practice, this approximation of assuming no concentration variations may be excellent for small but significant fractions of the limiting current.

Electrical Field Effects in the Electrolyte

When concentration variations are unimportant, the current distribution is governed by a ratio of the resistance to charge transfer to the ohmic resistance to current flow. When the ratio is large, the current distribution is uniform. When it is small, it can be very nonuniform depending on the geometry of the electrochemical cell. This ratio is known as the Wagner number and is given by

$$\mathrm{Wa} = \frac{\text{charge transfer resistance}}{\text{ohmic resistance}}$$

In the limit that $\mathrm{Wa} \to 0$, the current distribution is the most nonuniform. This is known as the primary current distribution in the literature.

In all cases, the ohmic resistance scales with ℓ/κ, where ℓ is a characteristic length, often the electrode size, and κ is the electrolyte conductivity. Thus, the Wagner number decreases as the system size increases. Current distributions thus tend to be less uniform with increasing scale. Likewise, the addition of supporting electrolyte may increase κ, reducing the ohmic resistance and thus causing a more uniform current distribution.

The charge transfer resistance is estimated from

$$R_{ct} = \left.\frac{\partial \eta_s}{\partial i}\right|_{i_{avg}}$$

where η_s is the surface overpotential (roughly a deviation from an equilibrium potential). In general, we can anticipate that a Butler-Volmer equation provides a means of estimating the relationship between overpotential and the current density. The limiting cases of low and high overpotentials of the Butler-Volmer equation leads to the approximations of linear and Tafel kinetics, respectively. These two cases allow for facile evaluation of the Wagner number for linear kinetics:

$$Wa_L = \frac{RT\kappa}{i_o(\alpha_a + \alpha_c)F\ell}$$

where i_o is the exchange current density, α_a and α_c are anodic and cathodic transfer coefficients, F is Faraday's constant, R is the gas constant, and T is the absolute temperature. For (cathodic) Tafel kinetics

$$Wa_T = \frac{RT\kappa}{|i_{avg}|\alpha_c F\ell}$$

In both cases, the current distribution becomes uniform as Wagner number becomes large; the current distribution may be nonuniform, depending on geometry, when the Wagner number goes to zero. The Tafel-kinetics Wagner number may be particularly interesting because it may be more typically a better approximation and because all of the parameters in it are readily estimated or measured.

Summary

Current distributions refer to the spatial variations of reaction rate on an electrode. Typically, current distribution uniformity increases with decreasing size of the electrode. In general concentration and electrical fields govern current distributions. While detailed simulations are feasible, back of the envelope calculations can often allow for rapid estimates of whether a nonuniform current distribution is anticipated.

References

1. Newman JS, Thomas-Alyea KE (2004) Electrochemical systems. Wiley-Interscience, Hoboken
2. Prentice G (1991) Electrochemical engineering principles. Prentice Hall, Englewood Cliffs

3. West AC (2012) Electrochemistry and electrochemical engineering. An introduction. CreateSpace
4. Tobias CW, Wijsman R (1953) J Electrochem Soc 100:459

Electrochemical Chain Reaction

Kazuhiro Chiba and Yohei Okada
Tokyo University of Agriculture
and Technology, Fuchu, Tokyo, Japan

Definition

"Electrochemical chain reactions" are autoactivating chemical transformations induced by electron transfer. Initially, the starting material is activated at the surface of electrodes, while the resulting intermediates are responsible for the following activation to drive the chain cycle. Typically, anodic oxidation of the starting material triggers electrochemical reactions and the resulting intermediates function as subsequent oxidants. In this reaction mechanism, only a catalytic amount of electricity is required to complete the reaction.

Introduction

Chain reactions have been widely utilized in both academic and industrial synthetic chemistry to efficiently produce various organic molecules. Generally, there are three stages in these reactions: initiation, propagation, and termination. For example, in a radical chain reaction, a small amount of chemical radical initiator is required in the initiation stage to generate the reactive radical substrates, which are responsible for the generation of subsequent reactive radicals through chemical transformations in the propagation stage. In the termination stage, the resulting radicals are consumed by several chemical reactions such as radical coupling. In organic synthesis, electrochemical processes have been employed to regulate chemical transformations in a heterogeneous manner [1–3]. Various functional group transformations and carbon–carbon bond formations have been achieved through electron transfer at the electrodes, generating several reactive intermediates, including ions, radicals, and radical ions. Since the electrodes act as both oxidizing and reducing reagents, additional chemical oxidants or reductants are not required, enabling straightforward reaction mechanisms. Electrochemical processes are also used to initiate chain reactions, also known as electrochemical chain reactions.

Examples and Applications

Typically, anodic oxidation is employed as an initiator to trigger electrochemical chain reactions. Starting materials are anodically oxidized to generate reactive radicals, cations, or radical cations. Through chemical transformations, these reactive species can oxidize another starting material without electron transfer at the electrodes, enabling the chain process. For example, electrochemical [2 + 2] cycloadditionsin lithium perchlorate/nitromethane electrolyte solution are shown to involve such chain mechanism [4–9]. In these reactions, enol ethers are oxidized at the anode to generate the radical cations in the initiation stage. These radical cations are then trapped by olefin nucleophiles possessing a substituted aromatic ring through intermolecular reaction to give the corresponding aromatic radical cations. Because the oxidation potentials of these aromatic radical cations are higher than those of the starting enol ethers, they can oxidize the starting enol ethers to generate radical cations, producing the desired [2 + 2] cycloadducts. Thus, the reaction needs only a catalytic amount of electricity to efficiently give the [2 + 2] cycloadducts (Scheme 1).

Electrochemically induced cation chain reactions are also demonstrated. For example, ArS $(ArSSAr)^+$, an equivalent of ArS^+, can be generated through anodic oxidation of ArSSAr in dichloromethane electrolyte solution at low temperature, which reacts with various thioacetals to give the corresponding alkoxycarbenium ions and ArSSAr [10–16]. These alkoxycarbenium ions are then trapped by olefin nucleophiles through intramolecular reaction to form the

Electrochemical Chain Reaction, Scheme 1 The chain mechanism of the intermolecular carbon–carbon bond formation by electrochemical reaction

A: Initiation, B: Propagation, C: Termination

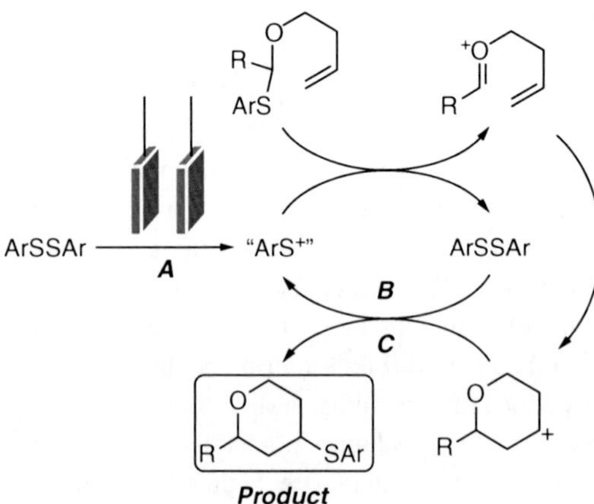

A: Initiation, B: Propagation, C: Termination

Electrochemical Chain Reaction, Scheme 2 The chain mechanism of intramolecular carbon–carbon bond formation by electrochemical reaction

corresponding reactive carbocation intermediates, regeneratingArS(ArSSAr)$^+$ with the formation of the final products, enabling the chain process (Scheme 2).

Future Directions

Electrochemical chain reactions have been proven to be environmentally friendly in organic synthesis since they require only a small amount of electricity for completion. As described in this chapter, both intra- and intermolecular carbon–carbon bond formations are achieved under mild electrolytic conditions. Moreover, electroanalytical methods, such as cyclic voltammetry, are combined with these chemical transformations to enable rational designs of novel reactions based on their mechanisms.

Cross-References

- ▶ Electrocatalytic Synthesis
- ▶ Electrogenerated Acid
- ▶ Electrogenerated Base
- ▶ Electrogenerated Reactive Species
- ▶ Electrosynthesis Using Mediator

References

1. Yoshida J, Kataoka K, Horcajada R, Nagaki A (2008) Modern strategies in electroorganic synthesis. Chem Rev 108:2265–2299
2. Sperry JB, Wright DL (2006) The application of cathodic reductions and anodic oxidations in the synthesis of complex molecules. Chem Soc Rev 35:605–621

3. Moeller KD (2000) Synthetic applications of anodic electrochemistry. Tetrahedron 65:9527–9554
4. Okada Y, Nishimoto A, Akaba R, Chiba K (2011) Electron-transfer-induced intermolecular [2 + 2] cycloaddition reactions based on the aromatic "Redox Tag" strategy. J Org Chem 76:3470–3476
5. Okada Y, Chiba K (2011) Electron transfer-induced four-membered cyclic intermediate formation: olefin cross-coupling versus. Olefin cross-metathesis. Electrochim Acta 56:1037–1042
6. Okada Y, Akaba R, Chiba K (2009) EC-backward-E electrochemistry supported by an alkoxyphenyl group. Tetrahedron Lett 50:5413–5416
7. Okada Y, Akaba R, Chiba K (2009) Electrocatalytic formal [2 + 2] cycloaddition reactions between anodically activated aliphatic enol ethers and unactivated olefins possessing an alkoxyphenyl group. Org Lett 11:1033–1035
8. Arata M, Miura T, Chiba K (2007) Electrocatalytic formal [2 + 2] cycloaddition reactions between anodically activated enyloxy benzene and alkenes. Org Lett 7:4347–4350
9. Chiba K, Miura T, Kim S, Kitano Y, Tada M (2001) Electrocatalytic intermolecular olefin cross-coupling by anodically induced formal [2 + 2] cycloaddition between enol ethers and alkenes. J Am Chem Soc 123:11314–11315
10. Matsumoto K, Suga S, Yoshida J (2011) Organic reactions mediated by electrochemically generated ArS^+. Org Biomol Chem 9: 2586–2596
11. Fujie S, Matsumoto K, Suga S, Nokami T, Yoshida J (2010) Addition of ArSSAr to carbon–carbon multiple bonds using electrochemistry. Tetrahedron 66:2823–2829
12. Matsumoto K, Ueoka K, Suzuki S, Suga S, Yoshida J (2009) Direct and indirect electrochemical generation of alkoxycarbenium ion pools from thioacetals. Tetrahedron 65:10901–10907
13. Matsumoto K, Fujie S, Suga S, Nokami T, Yoshida J (2009) Addition of ArSSAr to dienes *via* intramolecular C–C bond formation initiated by a catalytic amount of ArS^+. Chem Commun (36):5448–5450
14. Fujie S, Matsumoto K, Suga S, Yoshida J (2009) Thiofluorination of carbon–carbon multiple bonds using electrochemically generated ArS $(ArSSAr)^+BF_4^-$. Chem Lett 38:1186–1187
15. Matsumoto K, Fujie S, Ueoka K, Suga S, Yoshida J (2008) An electroinitiated cation chain reaction: intramolecular carbon–carbon bond formation between thioacetal and olefin groups. Angew Chem Int Ed 47: 2506–2508
16. Suga S, Matsumoto K, Ueoka K, Yoshida J (2006) Indirect cation pool method. Rapid generation of alkoxycarbenium ion pools from thioacetals. J Am Chem Soc 128:7710–7711

Electrochemical Fixation of Carbon Dioxide (Cathodic Reduction in the Presence of Carbon Dioxide)

Hisanori Senboku
Hokkaido University, Sapporo, Hokkaido, Japan

Introduction

Electrochemical fixation of carbon dioxide means an incorporation of carbon dioxide in organic molecules as functional groups by the use of electrochemical method. Fixation of carbon dioxide can be principally classified into two categories; one is the fixation with a C–C bond formation between carbon dioxide and organic molecules to yield carboxylic acid. Most of the reports on electrochemical fixation of carbon dioxide are involved in this category, and consequently, electrochemical fixation of carbon dioxide yielding carboxylic acid is often called as "electrochemical carboxylation" or "electrocarboxylation." The other one is the fixation of carbon dioxide with a C-heteroatom bond formation between a carbon of CO_2 and a heteroatom, such as oxygen or nitrogen atom, in organic molecules. As heteroatom sources, alcohol and amine are used for this type of the fixation to form carbonate and carbamate ions ($-X\text{-}CO_2^-$), respectively. Since the resulting carbonic acid and carbamic acid cannot generally be isolated due to their rapid decarboxylation, they are obtained and isolated as their esters, carbonate and carbamate, respectively, by the reaction of the forming carbonate or carbamate ions with appropriate electrophiles, such as alkyl halides, in situ after the fixation of CO_2 (Scheme 1).

Among the two electrochemical fixation of carbon dioxide in organic molecules, this entry focuses on electrochemical fixation of carbon dioxide with C-C bond formation giving carboxylic acid.

C-C bond formation between carbon dioxide and organic molecules is considered to proceed via two different pathways. When the reduction

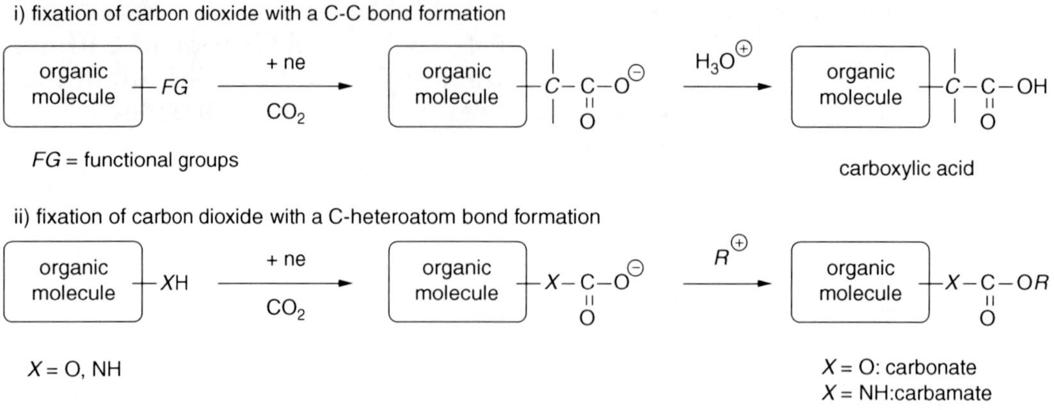

i) fixation of carbon dioxide with a C-C bond formation

organic molecule–FG → (+ ne, CO_2) organic molecule–C–C(=O)–$O^{\ominus}$ → ($H_3O^{\oplus}$) organic molecule–C–C(=O)–OH

FG = functional groups

carboxylic acid

ii) fixation of carbon dioxide with a C-heteroatom bond formation

organic molecule–XH → (+ ne, CO_2) organic molecule–X–C(=O)–$O^{\ominus}$ → ($R^{\oplus}$) organic molecule–X–C(=O)–OR

X = O, NH

X = O: carbonate
X = NH:carbamate

Electrochemical Fixation of Carbon Dioxide (Cathodic Reduction in the Presence of Carbon Dioxide), Scheme 1 Divergent Electrochemical Fixation of Carbon Dioxide

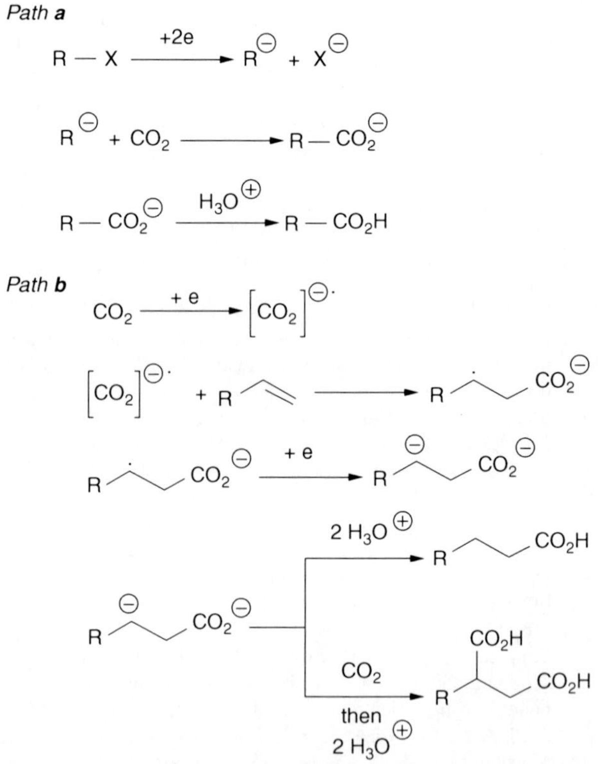

Path ***a***

$$R-X \xrightarrow{+2e} R^{\ominus} + X^{\ominus}$$

$$R^{\ominus} + CO_2 \longrightarrow R-CO_2^{\ominus}$$

$$R-CO_2^{\ominus} \xrightarrow{H_3O^{\oplus}} R-CO_2H$$

Path ***b***

$$CO_2 \xrightarrow{+e} [CO_2]^{\ominus\cdot}$$

$[CO_2]^{\ominus\cdot}$ + R–CH=CH$_2$ → R–ĊH–CH$_2$–$CO_2^{\ominus}$

R–ĊH–CH$_2$–$CO_2^{\ominus}$ → (+ e) R–CH$^{\ominus}$–CH$_2$–$CO_2^{\ominus}$

R–CH$^{\ominus}$–CH$_2$–$CO_2^{\ominus}$ → (2 $H_3O^{\oplus}$) R–CH$_2$–CH$_2$–CO_2H

R–CH$^{\ominus}$–CH$_2$–$CO_2^{\ominus}$ → (CO_2 then 2 $H_3O^{\oplus}$) R–CH(CO_2H)–CH$_2$–CO_2H

Electrochemical Fixation of Carbon Dioxide (Cathodic Reduction in the Presence of Carbon Dioxide), Scheme 2 Two Different Pathways in Electrochemical Fixation of Carbon Dioxide with C-C Bond Formation

potential of an organic substrate is more positive than that of carbon dioxide, electrochemical reduction of the organic substrate predominantly occurs to generate anionic species. Nucleophilic attack of the resulting anionic species on carbon dioxide yields carboxylic acid (*path* ***a*** in Scheme 2). On the other hand, when the reduction potential of the organic substrate is more negative than that of carbon dioxide, one-electron reduction of carbon dioxide predominantly occurs to generate the radical anion of carbon dioxide, which reacts with organic substrates, typically

Br, O, + 2e, CO_2, in the presence of Ni catalyst or electron transfer mediator, CO_2H, O

Electrochemical Fixation of Carbon Dioxide (Cathodic Reduction in the Presence of Carbon Dioxide), Scheme 3 Electrochemical Carboxylation subsequent to Cyclization

such as alkenes, to yield the corresponding mono- and/or dicarboxylic acids (*path **b*** in Scheme 2).

Advantage

Fixation of carbon dioxide in organic molecules with C-C bond formation gives carboxylic acid. Carbon dioxide used as a source of a carboxyl group is not only abundant and economical but also nontoxic and attractive as an environmentally benign C1 chemical reagent for organic synthesis. By the use of electrochemical method, efficient fixation of carbon dioxide in appropriate organic molecules can be achieved even under an atmospheric pressure of carbon dioxide under mild and almost neutral conditions. Especially, electrochemical reduction in the presence of carbon dioxide using magnesium or aluminum as a sacrificial anode in one-compartment electrochemical cell is found to be the most convenient, useful, and effective method for efficient fixation of carbon dioxide to yield carboxylic acid in high yields [1–4]. Carbon dioxide is also used for synthesizing carboxylic acid in conventional methods. For example, a reaction of Grignard reagent or organolithium reagent with carbon dioxide in ether or tetrahydrofuran, flammable organic solvents, gives carboxylic acid. However, the reaction proceeds under highly basic conditions, and the use of these highly reactive reagents makes both limitation of usable functional groups in the reagents themselves and necessity of special cautions in their handling. As well as carbon dioxide, cyanide ion, such as KCN and NaCN, and carbon monoxide (CO) are also effective C1 carbon sources for synthesis of nitrile and ester in conventional methods, which are the precursors of carboxylic acid and are readily converted to carboxylic acid by hydrolysis. However, these C1 reagents are unfortunately toxic, and from the viewpoint of green and sustainable chemistry, the use of these reagents is unfavorable. On the other hand, electrochemical method is an environmentally benign method, and there is no need to use such toxic reagents and flammable solvents with special cautions in electrochemical fixation of carbon dioxide to yield carboxylic acid in high yields. Although transition-metal-catalyzed fixation of carbon dioxide in organic and organometallic compounds has recently been developed, electrochemical method can achieve efficient fixation of carbon dioxide without such expensive and air-sensitive metal catalysts by using electron as an essential reagent.

Cathode Materials

Platinum, grassy carbon, graphite, stainless steel, carbon fiber, silver, lead, mercury pool, and some other metals are reported to be usable as cathode materials in electrochemical carboxylation. Among them, platinum, stainless steel, carbon fiber, and graphite are frequently used for an efficient formation of carboxylic acid by electrochemical fixation of carbon dioxide with a sacrificial anode, such as magnesium and aluminum [1–4], as a couple in an undivided cell (one-compartment cell).

Synthetic Applications

Efficient fixation of carbon dioxide in various kinds of organic molecules has been successfully carried out by electrochemical method with C-C

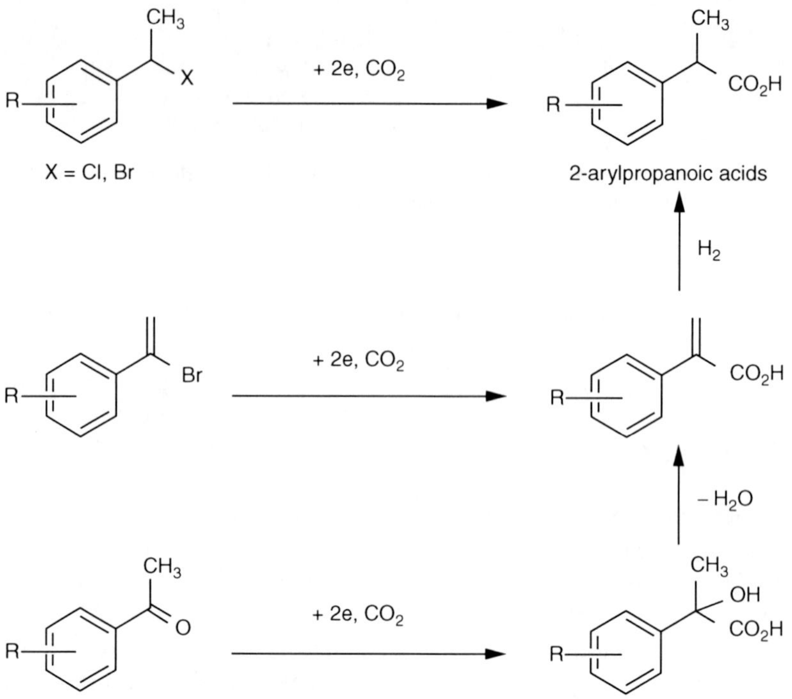

Electrochemical Fixation of Carbon Dioxide (Cathodic Reduction in the Presence of Carbon Dioxide), Scheme 4 Synthetic Routes to NSAIDs having 2-Arylpropanoic Acid Skeletons by Electrochemical Carboxylation

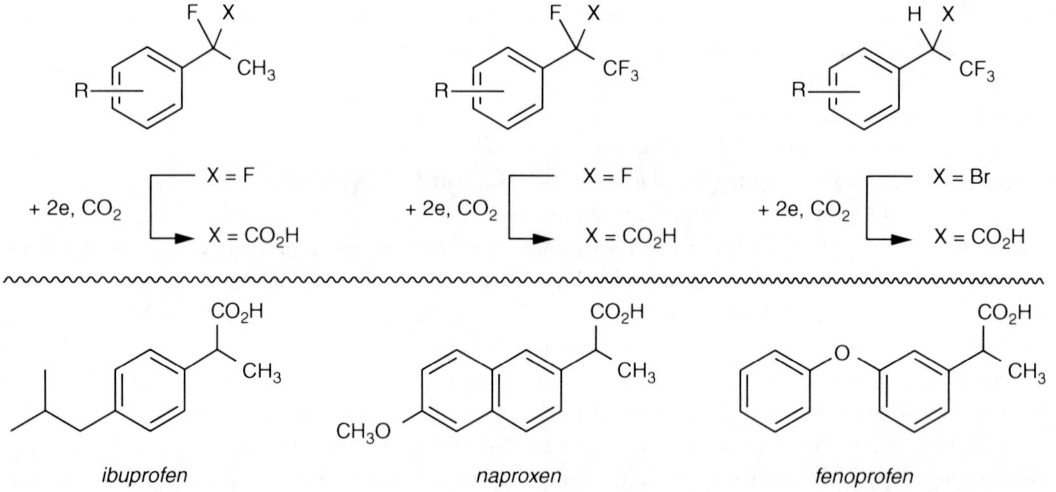

X = Cl, Br

+ 2e, CO_2

2-arylpropanoic acids

H_2

+ 2e, CO_2

$- H_2O$

+ 2e, CO_2

+ 2e, CO_2 : X = F → X = CO_2H

+ 2e, CO_2 : X = F → X = CO_2H

+ 2e, CO_2 : X = Br → X = CO_2H

ibuprofen *naproxen* *fenoprofen*

Electrochemical Fixation of Carbon Dioxide (Cathodic Reduction in the Presence of Carbon Dioxide), Scheme 5 Syntheses of Fluorinated 2-Arylpropanoic Acids, Fluorinated Analogues of NSAIDs by Electrochemical Carboxylation

bond formation; alkenes and activated alkenes, alkynes, enynes, aromatic and aliphatic halides, allyl, propargyl and benzyl halides, benzyl carbonates, vinyl bromides, aldehydes and ketones, imines, epoxides and aziridines, aryl and vinyl triflates, and other organic substrates are efficiently carboxylated by electrochemical method to afford various kinds of useful carboxylic acids. Electrochemical carboxylation is sometimes carried out in the presence of transition metal

Electrochemical Fixation of Carbon Dioxide (Cathodic Reduction in the Presence of Carbon Dioxide), Scheme 6 Synthesis of α-Fluorinated Ibuprofen by Electrochemical Carboxylation of Difluoroethylarene

catalyst such as Pd and Ni complexes, and the use of which often results in drastic enhancement of chemo- and regioselectivity and efficiency of CO_2 fixation. Carboxylation also occurs with electrochemically induced other chemical reactions such as cyclization (Scheme 3) [5, 6].

One of great synthetic applications of electrochemical fixation of carbon dioxide is synthesis of 2-arylpropanoic acids, nonsteroidal anti-inflammatory drugs (NSAIDs), and their derivatives. Electrochemical carboxylations of benzyl halides [1, 3, 8–13], aryl methyl ketones [14, 15], and α-bromostyrenes [16] are reported to be successfully applied to the synthesis of several NSAIDs, such as ibuprofen and naproxen, and their precursors and derivatives (Scheme 4).

Especially, syntheses of fluorinated analogues of NSAIDs (Schemes 5 and 6) by electrochemical fixation of carbon dioxide [11–13] indicate usefulness, significance, advantage, and convenience of electrochemical method because the syntheses by using fixation of carbon dioxide seem to be difficult by the conventional chemical methods.

Future Directions

Although there are so many reports on electrochemical carboxylation, only a few studies on diastereoselective [17, 18] and asymmetric electrochemical carboxylation [19, 20] have been reported with unsatisfied results. Further developments of these subjects will be desirable.

Use of supercritical carbon dioxide [21] as both a reagent and a solvent and ionic liquid as a solvent [22, 23] for the electrochemical fixation of carbon dioxide is still under development and would also be one of future directions in this area.

Cross-References

► Electrocatalysts for Carbon Dioxide Reduction
► Electrosynthesis in Supercritical Fluids
► Reactive Metal Electrode

References

1. Silvestri G, Gambino S, Filardo G, Gulotta A (1984) Sacrificial anodes in the electrocarboxylation of organic chlorides. Angew Chem Int Ed Engl 23:979–980. doi:10.1002/anie.198409791
2. Silvestri G, Gambino S, Filardo G (1991) Use of sacrificial anodes in synthetic electrochemistry. Processes involving carbon dioxide. Acta Chem Scand 45:987–992. doi:10.3891/acta.chem.scand.45-0987
3. Sock O, Troupel M, Périchon J (1985) Electrosynthesis of carboxylic acids from organic halides and carbon dioxide. Tetrahedron Lett 26:1509–1512. doi:10.1016/S0040-4039(00)98538-1
4. Chaussard J, Folest JC, Nédélec JY, Périchon J, Sibille S, Troupel M (1990) Use of sacrificial anodes in electrochemical functionalization of organic halides. Synthesis 369–381. doi:10.1055/s-1990-26880
5. Senboku H, Michinishi J, Hara S (2011) Facile synthesis of 2,3-dihydrobenzofuran-3-ylacetic acids by novel electrochemical sequential aryl radical cyclization-carboxylation of 2-allyloxybromobenzenes using methyl 4-tert-butylbenzoate as an electron-transfer mediator. Synlett 1567–1572. doi: 10.1055/s-0030-1260794
6. Olivero S, Dunach E (1999) Electrochemical intramolecular reductive cyclisation catalysed by electrogenerated $Ni(cyclam)^{2+}$. Tetrahedron Lett 36: 4429–4432
7. Olivero S, Dunach E (1999) Selectivity in the tandem cyclization – carboxylation reaction of unsaturated haroaryl ethers catalyzed by electrogenerated nickel complexes. Eur J Org Chem 1885–1891
8. Isse AA, Ferlin MG, Gennaro A (2005) Electrocatalytic reduction of arylethyl chlorides at silver cathodes in the presence of carbon dioxide: synthesis of 2-srylpropanoic acids. J Electroanal Chem 581:38–45

9. Isse AA, Ferlin MG, Gennaro A (2003) Homogeneous electron transfer catalysis in the electrochemical carboxylation of arylethyl chlorides. J Electroanal Chem 541:93–101
10. Fauvarque JF, Jutand A, Francois M (1988) Nickel catalysed electrosynthesis of anti-inflammatory agents part I – synthesis of 2-arylpropanoic acids, under galvanostatic conditions. J Appl Electrochem 18:109–115
11. Yamauchi Y, Fukuhara T, Hara S, Senboku H (2008) Electrochemical carboxylation of α,α-difluorotoluene derivatives and its application to the synthesis of α-fluorinated nonsteroidal anti-inflammatory drugs. Synlett 438–442
12. Yamauchi Y, Sakai K, Fukuhara T, Hara S, Senboku H (2009) Synthesis of 2-aryl-2,3,3,3-tetrafluoropropanoic acids, tetrafluorinated fenoprofen and ketoprofen by electrochemical carboxylation of pentafluoroethylarenes. Synthesis 3375–3377
13. Yamauchi Y, Hara S, Senboku H (2010) Synthesis of 2-aryl-3,3,3-trifluoropropanoic acids using electrochemical carboxylation of (1-bromo-2,2,2-trifluoroethyl) arenes and its application to the synthesis of β, β, β-trifluorinated non-steroidal anti-inflammatory drugs. Tetrahedron 66:473–479
14. Chan ASC, Huang TT, Wagenknecht JH, Miller RE (1995) A novel synthesis of 2-arylacetic acids via electrocarboxylation of methyl aryl ketones. J Org Chem 60:742–744
15. Silvestri G, Gambino S, Filardo G (1986) Electrochemical carboxylation of aldehydes and ketones with sacrificial aluminum anodes. Tetrahedron Lett 27:3429–3430
16. Kamekawa H, Senboku H, Tokuda M (1997) Facile synthesis of aryl-substituted 2-alkenoic acids by electroreductive carboxylation of vinylic bromides using a magnesium anode. Electrochim Acta 42: 2117–2123
17. Feroci M, Orsini M, Palombi L, Sotgiu G, Colapietro M, Ineci A (2004) Diastereoselective electrochemical carboxylation of chiral a-bromocarboxylic acid derivatives: an easy access to unsymmetrical alkylmalonic ester derivatives. J Org Chem 69:487–494
18. Orsini M, Feroci M, Sotgiu G, Inesi A (2005) Stereoselective electrochemical carboxylation: 2-phenylsuccinates from chiral cinnamic acid derivatives. Org Biomol Chem 3:1203–1208
19. Zhang K, Wang H, Zhao S, Niu D, Lu J (2009) Asymmetric electrochemical carboxylation of prochiral acetophenone: an efficient route to optically active atrolactic acid via selective fixation of carbon dioxide. J Elctroanal Chem 630:35–41
20. Zhao S, Zhu M, Zhang K, Wang H, Lu J (2011) Alkaloid induced asymmetric electrochemical carboxylation of 4-methylpropiophenone. Tetrahedreon Lett 52:2702–2705
21. Sasaki A, Kudoh H, Senboku H, Tokuda M (1998) Electrochemical carboxylation of several organic halides in supercritical carbon dioxide. In: Torii S (ed) Novel trends in electroorganic synthesis. Springer, Tokyo, pp 245–246
22. Hiejima Y, Hayashi M, Uda A, Oya S, Kondo H, Senboku H, Takahashi K (2010) Electrochemical carboxylation of α-chloroethylbenzene in ionic liquids compressed with carbon dioxide. Phys Chem Chem Phys 12:1953–1957. doi:10.1039/b920413j
23. Wang H, Zhang G, Liu Y, Luo Y, Lu J (2007) Electrocarboxylation of activated olefins in ionic liquid $BMIMBF_4$. Electrochem Commun 9:2235–2239

Electrochemical Functional Transformation

Shigeru Nishiyama
Department of Chemistry, Keio University, Hiyoshi, Yokohama, Japan

Introduction

There has been a great deal of progress in organic synthesis using electrolysis since the development of the Kolbe protocol for coupling of electrochemically generated carbon radicals. The successful results in this field include the synthesis of adiponitrile by electrochemical dimerization of acrylonitrile to act as an important synthetic intermediate of Nirone 6,6. Both syntheses involve carbon-carbon bond formation reactions, with the former being an oxidative procedure, while the latter is reductive. In principle, organic redox reactions involving electron transfer processes may be candidates for electrochemical reaction to provide functional group transformation, although detailed regulation of electrolytic reaction conditions is required to achieve procedures that are superior to standard chemical reactions. In this entry, representative anodic and cathodic functional transformations, leading to several natural products with diverse functionalities, will be discussed.

Anodic Functional Group Transformation

Generally, anodic oxidation of functional groups in organic molecules provides corresponding

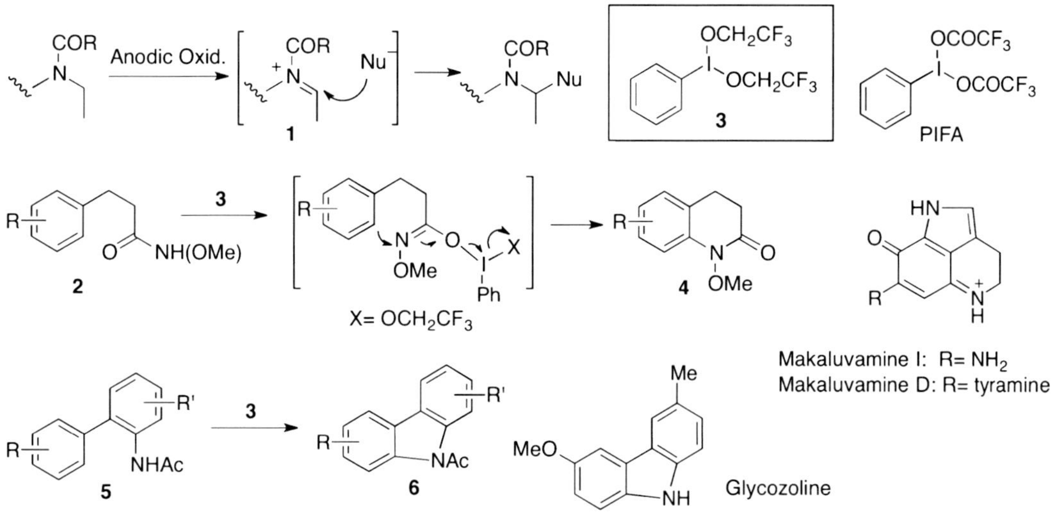

Makaluvamine I: R= NH_2
Makaluvamine D: R= tyramine

Glycozoline

Electrochemical Functional Transformation, Scheme 1 Amide oxidation

products similar to those of chemical oxidation, e.g., oxidation of alcohols gives ketonic compounds and/or acetals [1], depending on the reaction conditions, and the vicinal diols provide C–C bond cleavage leading to diketonic products [2] or acyloins [3] by tandem abstraction of electrons and protons. While electrochemical protocols can be applied to a range of oxidation reactions, there are characteristic reactions in the field of oxidation of heteroatoms.

Oxidation of Amide Nitrogen

Introduction of nucleophiles at the adjacent carbon. When amides are oxidized under anodic oxidation conditions [4], electron-deficient nitrogen functions produce N-acyliminium ion **1**, which shows potent reactivity with nucleophiles to yield diverse carbon frameworks attached with nitrogen functions, such as alkaloids.

Oxidative arylation of amide functions using electrochemically generated hypervalent iodine. The alkaloid family is widespread in nature and includes quinoline derivatives, which show a variety of biological activities. Among the enormous number of synthetic routes to obtain these molecules, oxidative cyclization of the phenylalkylamide derivative **2**–**4** by the electrochemically generated hypervalent iodine oxidant (**3**, PIFE) [5, 6]. The structure of the oxidant **3** was elucidated by mass spectroscopic similarity to that of phenyliodine bis(trifluoroacetate) (PIFA). Trifluoroethanol, which is stable under electrolytic and oxidative conditions, was bound to the oxidized iodine moiety as a ligand. The iodine oxidant can be produced without use of toxic and explosive oxidants. Total synthesis of makaluvamines, the tetrahydropyrroloiminoquinone alkaloids of marine origin, was reported by employing the quinolinone derivative (type **4**) synthesized by this oxidative cyclization strategy [7]. Synthesis of carbazoles (**5** → **6**) was accomplished by essentially the same procedure as the case described above (**2** → **4**). In construction of the relatively constrained tricyclic structure, PIFE exhibited better results than the case of PIFA due to the moderate electron-withdrawing property of the 2,2,2-trifluoromethoxy ligand. The synthetic application to glycozoline, isolated from the roots of *Glycosmis pentaphylla* (Retz.) DC, was carried out [8] (Scheme 1).

Phenolic Oxidation

Phenolic oxidation can be seen in a wide range of biosyntheses. In plants, propylbenzene derivatives, which are produced from shikimic acid, are enzymatically oxidized to yield lignans, neolignans, and lignins. In contrast to the dimeric structures of the former, lignins possess

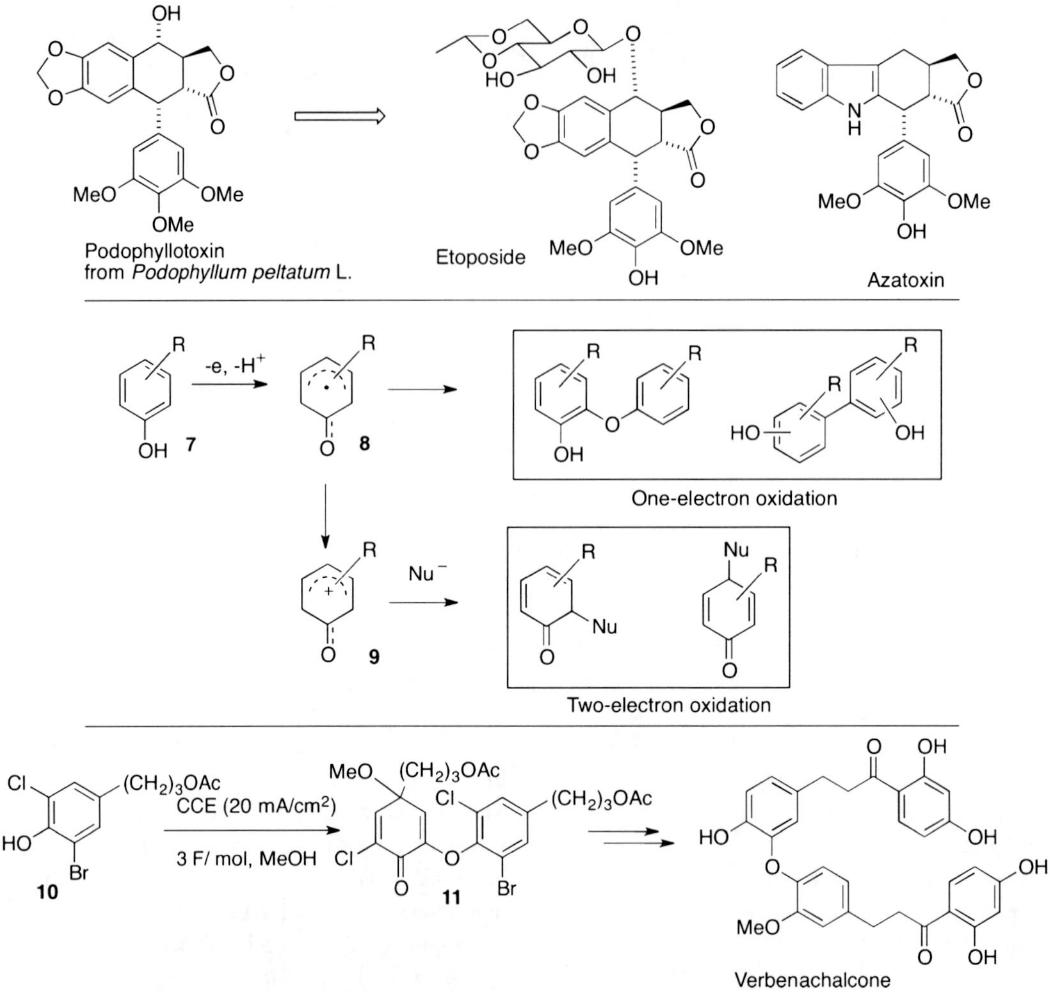

Electrochemical Functional Transformation, Scheme 2 Phenolic oxidation and products

polymeric structures to support the plant body. Lignans and neolignans have been investigated as new leads for chemotherapeutic agents: Etoposide (inhibitor of topoisomerase II) and azatoxin (inhibitor of tubulin polymerization and topoisomerase II) were developed from podophyllotoxin (Scheme 2) [9].

Synthesis of diaryl ethers. Dimerization or polymerization reactions may proceed by phenolic oxidation, which is initiated by one-electron oxidation of the phenol **7** to give the radical **8**. The radicals react successively with each other to yield dimeric structures. When the radicals are further oxidized, the cation **9** is produced, and reacts with nucleophiles, leading to nonphenolic dienones. Synthesis of verbenachalcone, isolated from *Verbena littoralis* H.B.K., was accomplished using the diaryl ether **11**, a one-electron oxidation product of the corresponding phenol **10** [10].

Anodic Oxidation of Bromide Ion and Its Use for Dethioacetalization

When oxidized under anodic oxidation conditions, the bromide ions of salts such as LiBr, NaBr, and nBu_4NBr effected in situ generation of bromonium ions and/or bromine molecules, which reacted with appropriate substrates as oxidants or brominating reagents. Scheme 3 shows examples of spiroacetalization of the dithioacetal derivative (**12**–**13**) under anodic oxidation conditions in the

Electrochemical Functional Transformation, Scheme 3 Reactions using bromonium ion in situ generated

Electrochemical Functional Transformation, Scheme 4 Pinacol type reactions

presence of LiBr [11]. Bromonium ions generated in situ effected the transannular reaction of germacrene D **14**–**15** [12].

Cathodic Functional Group Transformation

Cathodic reduction consists of electron supply from the cathode, followed by acquisition of protons by anionically charged substrates. Thus, ready accessibility of the reductive electron transfer processes leads to a variety of reactions, including reductive abstraction of halogens [13, 14], reduction of alcohols to alkanes [15], reduction of imines, oximes, and hydrazones to the corresponding amines [16], and conversion of nitro groups to the corresponding nitroso, hydroxyamino, and amino groups [17]. In addition to the typical unit process, the reductive process was utilized for production of complicated molecules.

Pinacol Type Coupling Reaction: Reduction of Ketones

Pinacol coupling is an electron transfer process using Mg or SmI_2 to produce the corresponding vicinal diols in an intra- and intermolecular manner. The coupling of aliphatic and aromatic ketones under cathodic reduction conditions has been reported [18]. As part of this series, intramolecular reaction of phthalimide **16** with aromatic aldehyde in the presence of chlorotrimethylsilane and triethylamine was investigated to produce the corresponding cyclic product **17**. Total synthesis of lennoxamine was demonstrated using this reductive procedure (Scheme 4) [19].

Electrochemical Functional Transformation, Scheme 5 Reduction of aromatic rings

Cathodic Reduction of Aromatic Compounds: Conversion of Aromatic Compounds to Cyclohexadienes

Birch reduction is a standard reductive procedure of aromatic molecules to non-aromatic derivatives. Similar to the two-electron oxidation of phenol derivatives to the corresponding dienones, the Birch protocol is considered to involve reductive conversion of aromatic compounds to cyclohexadienes. When aromatic compounds were subjected to cathodic reduction in the presence of a proton source, Birch-type reduction products were produced (Scheme 5) [20 – 22].

Future Direction

Manipulation and accumulation of experimental data will demonstrate the applicability of electrochemical functional transformation to organic synthesis. In particular, electrochemical redox reactions may be incorporated into catalytic cycles in the field of organometallics. Active species produced by electrolytic conditions may lead to unprecedented and effective reaction pathways, which could not be realized by chemical reactions. Electrochemical protocol will open up new perspective of molecular conversion.

Cross-References

- ► Anodic Substitutions
- ► Electrochemical Asymmetric Synthesis
- ► Electrosynthesis Using Mediator

References

1. Sundholm G (1971) The anodic oxidation of absolute methanol and ethanol. J Electroanal Chem Interfacial Chem 31:265–267
2. Shono T, Matsumura Y, Hashimoto T, Hibino K, Hamaguchi H, Aoki T (1975) Electroorganic chemistry. XXII. Novel anodic cleavage of glycols to carbonyl compounds. J Am Chem Soc 97: 2546–2548
3. Maki T, Fukae K, Hasegawa H, Ohishi T, Matsumura Y (1998) Selective oxidation of 1,2-diols by electrochemical method using organotin compound and bromide ion as mediators. Tetrahedron Lett 39:651–654
4. Moeller KD (2000) Synthetic applications of anodic electrochemistry. Tetrahedron 56:9527–9554
5. Amano Y, Nishiyama S (2006) Oxidative synthesis of azacyclic derivatives through the nitrenium ion: application of a hypervalent iodine species electrochemically generated from iodobenzene. Tetrahedron Lett 47:6505–6507
6. Amano Y, Inoue K, Nishiyama S (2008) Oxidative access to quinolinone derivatives with simultaneous rearrangement of functional groups. Synlett 19:134–136
7. Inoue K, Ishikawa Y, Nishiyama S (2010) Synthesis of tetrahydropyrroloiminoquinone alkaloids

based on electrochemically generated hypervalent iodine oxidative cyclization. Org Lett 12:436–439
8. Kajiyama D, Ishikawa Y, Nishiyama S (2010) A synthetic approach to carbazoles using electrochemically generated hypervalent iodine oxidant. Tetrahedron 66:9779–9784
9. Apers S, Vlietinck A, Pieters L (2003) Lignans and neolignans as lead compounds. Phytochem Rev 2:201–207
10. Tanabe T, Doi F, Ogamino T, Nishiyama S (2004) A total synthesis of verbenachalcone, a bioactive diaryl ether from *Verbena littoralis*. Tetrahedron Lett 45:3477–3480
11. Honjo E, Kutsumura N, Ishikawa Y, Nishiyama S (2008) Synthesis of a spiroacetal moiety of antitumor antibiotic ossamycin by anodic oxidation. Tetrahedron 64:9495–9506
12. Ogamino T, Mori K, Yamamura S, Nishiyama S (2004) Electrochemical bromination of germacrene D. Electrochim Acta 49:4865–4869
13. Triebe FM, Borhani KJ, Hawley MD (1979) Electrochemistry of the carbon-halogen bond. 9-Fluorenyl and benzhydryl halides in dimethylformamide at platinum and vitreous carbon electrodes. J Am Chem Soc 101:4637–4645
14. Nonaka T, Ota T, Odo K (1977) Stereochemical studies of the electrolytic reactions of organic compounds. II. Electrolytic reduction of optically active 6-chloro-2,6-dimethyloctane to the corresponding alkane. Bull Chem Soc Jpn 50: 419–421
15. Nazar-ul-Islam, Sopher DW, Utley JHP (1987) Electro-organic reactions. Part 28. Preparative applications of the oxalate cathodic cleavage reaction including one-pot conversions of aldehydes and ketones. Tetrahedron 43: 2741–2748
16. Pienemann T, Schäfer H-J (1987) Reductive amination of ketones and aldehydes at the mercury cathode. Synthesis 1987:1005–1007
17. Gomez JRO (1991) Electrosynthesis of N-methylhydroxylamine. J Appl Electrochem 21: 331–334
18. Torii S (2006) Electroorganic reduction synthesis, vol 1. Kodansha Tokyo, Wiley-VCH Weinheim, pp 60–68
19. Kise N, Isemoto S, Sakurai T (2011) Electroreductive intramolecular coupling of phthalimides with aromatic aldehydes: application to the synthesis of lennoxamine. J Org Chem 76:9856–9860
20. Torii S (2006) Electroorganic reduction synthesis, Vol 1. Kodansha Tokyo, Wiley-VCH Weinheim, pp 161–167
21. Swenson KE, Nanjundiah DZC, Keriv-Miller E (1983) Birch reductions of methoxyaromatics in aqueous solution. J Org Chem 48:1777–1779
22. K.-Miller E, Swenson KE, Lehman GK, Andruzzi R (1985) Selective cathodic Birch reductions. J Org Chem 50:556–560

Electrochemical Glucose Sensors

Chang-Jung Hsueh[1], Metini Janyasupab[1], Ying-Hui Lee[2] and Chung-Chiun Liu[1]
[1]Electronics Design Center, and Chemical Engineering Department, Case Western Reserve University, Cleveland, OH, USA
[2]Chemical Engineering Department, National Tsing Hua University, Hsinchu, Taiwan

Diabetes mellitus is a leading disease around the world, resulting from metabolic blood-sugar dysfunction (insulin deficiency or hyperglycemia), relevant to blood glucose level [1, 2]. Either high or low glucose concentration (normally 80–120 mg/dL of blood glucose) suffers from such a disorder [3]. Biosensor provides an easily used and convenient avenue for regularly monitoring and strictly managing blood glucose of a human body. Clark et al. [4, 5] initially presented the glucose oxidase (GOD)-attached electrode for continuous monitoring in cardiovascular surgery. The enzymatically consumed oxygen was assessed via electroreduction, as shown in the following reaction:

$$\text{Glucose} + O_2 \xrightarrow{GOD} \text{gluconic acid} + H_2O_2$$
$$O_2 + 4H^+ + 4e^- \longrightarrow 2H_2O$$

Since the Clark-type biosensor is commercially released, electrochemical-based glucose sensor has been tremendously performed toward point of care of diabetes control. In addition to diabetes management, such a device exhibits great promise in a variety of applications of bioprocess and food industries [6, 7]. Variable approaches on exploration of electrochemical glucose electrodes have been reported. This includes the blood glucose used for patient care in diabetic patient management. This review section discusses the fundamental operations of electrochemical glucose sensors, particularly based on amperometric technique. Given the developing history of different generation of glucose biosensor, the pros and cons of each sensor design

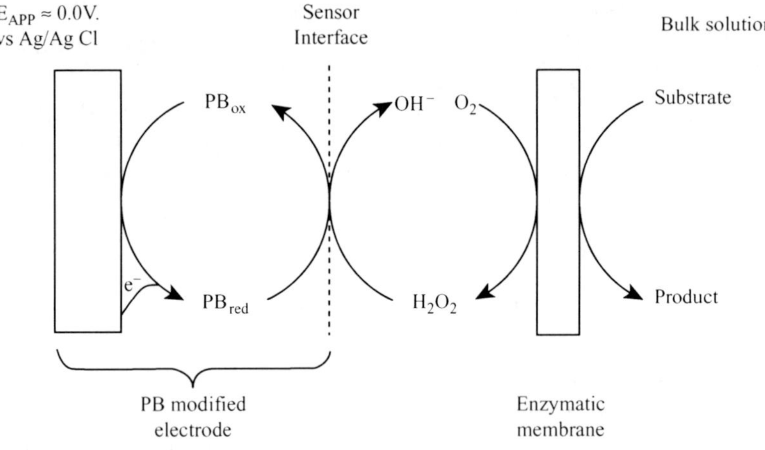

Electrochemical Glucose Sensors, Fig. 1 Schematic illustration of Prussian blue (PB)-modified electrode [16]

and performance are presented and examined. The further advancement of development nonenzymatic glucose sensor is also discussed.

First-Generation (H_2O_2-Based) Glucose Sensor

The first-generation glucose sensor relies on the amperometric measurement of enzymatically liberated hydrogen peroxide (H_2O_2) [8]. Based on the stoichiometry of scheme 1, blood glucose over the interested testing ranges can be quantified. Typically, the oxidation potential for H_2O_2 on platinum (Pt) (typically 0.6 vs. SCE) is high relative to reference electrode [9]. Consequently, other coexisting electroactive interference, like ascorbic acid (AA), uric acid (UA), and acetaminophen, is unavoidable during the anodic (oxidizing) measurement. This may result in producing undesirable interfering currents and lowering sensor selectivity.

There are various strategies applied to minimize the effect of electroactive interference. The first one is to employ anti-interference membranes. These membranes possess the perm-selective property characterized by the charge and/or pore size to block interferents out toward the electrode surface [3]. Nafion, a cationic exchanger membrane, is widely performed to overcome interfering problem on glucose sensor [10]. Electropolymerized film is also applied to screen out interferents based on pore size exclusion [11, 12]. Some report [12, 13] utilized a composite polymer consisting of electropolymerized layer and Nafion. This technique has combinational benefits of eliminating the influence of electroactive interferents and stabilizing the enzyme electrode.

Another method diminishing electroactive interference is based on preferential electrocatalytic detection of the generated H_2O_2. Horseradish peroxidase (HRP)-modified electrodes coupled with glucose oxidases (GOD) have been proposed for the development of glucose biosensor [14]. In these systems, enzymatically generated H_2O_2 is electroreduced by the HRP [15]. HRP is then electro-catalyzed (i.e., reoxidation) at the electrode surface. Therefore, the incorporation of HRP with GOD promises the detection of glucose at a low potential, and thus, interference of coexisting electroactive constituents is suppressed. Similar operating rationale shown in Fig. 1, modifying an electrode with Prussian blue [16] (PB, ferric ferrocyanide) facilitates and electroreduces H_2O_2 at an applied potential around/below 0.0 V versus Ag/AgCl, exhibiting better sensor performance (high sensitivity and selectivity) than others such as Pt or horseradish peroxidase (HRP) [1].

Metallized carbon electrode permits a membrane-free approach toward selectively screening glucose concentration, by simply incorporating nano-metallic catalysts into the electrode development [3, 17]. These dispersed metal particles exhibit favorable electrocatalytic oxidation of H_2O_2 at relatively low potentials, at which undesirable co-oxidation of the interferents is

Electrochemical Glucose Sensors, Fig. 2 The schematic principle of mediator-based electrode [21]

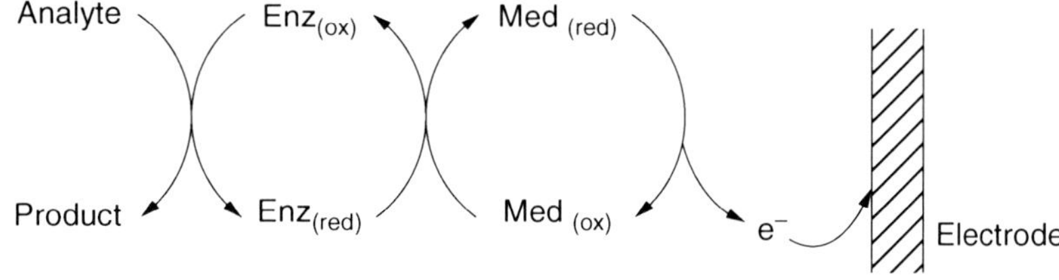

negligible during biosensing operation. Alternatively, the reduction of H_2O_2 (cathodic current) on the metallized carbon transducer is undertaken for sensing of glucose, without significantly reducing O_2 [18, 19]. Without overlapping additional diffusion barrier resulting from anti-interference membrane, metallized carbon electrode allows rapid detection of glucose.

Second-Generation (Mediator-Based) Glucose Sensors

Given the abovementioned electroactive interference, the first-generation glucose biosensor suffers from oxygen deficiency. Oxygen is a co-substrate and physiological electron acceptor in an oxidase-based enzymatic reaction. Therefore, resulting errors may derive from the fluctuations of oxygen tensions and stoichiometric limits [2, 20]. In order to diminish these issues, incorporating artificial (synthetic) electron transfer, in place of O_2, is capable of electrically connecting (electron shuttle) between the redox centers of enzyme (i.e., FAD active site) to the surface of electrode. In Fig. 2 [21], the general operating principle of this mediator-based glucose sensor is depicted.

Based on Fig. 2, without H_2O_2 involved in the reaction, the reduced form of mediator gives a current output via the transducer, proportional to the glucose concentration. A variety of mediators [21] have been employed in the development of glucose sensor, like ferrocene derivatives, ferricyanide, quinone compounds, and conducting organic salts. Most of them possess characteristics being a suitable mediator: facilitating redox reaction with the reduced enzyme, being electrochemically reversible and chemically stable (either reduced or oxidized form), and being oxygen independent. At present, the commercial glucose sensors are developed based on this strategy. Ferricyanide and ferrocene derivatives are primarily performed into the electrode fabrication [22]. Some utilizes glucose dehydrogenase (GDH) instead of GOD due to its oxygen independence.

Redox hydrogel polymer provides a nondiffusional and non-leachable mediating medium. It can be tightly bound with proteins onto the electrode surface via a long flexible hydrophilic polymer backbone [2, 23]. In this approach, the enzyme (GOD) is entrapped into a water-swollen and cross-linked polymer network. In Fig. 3 [23, 24], Heller introduced a $Os(bpy)_2(PVP)Cl_2^+$ redox hydrogel covalently linking osmium-complex electron relays (depicted as R in Fig. 3 [23, 24]) with a polymeric matrix (PVP). Polycationic treatment of the polymer permitted electrostatic attraction with the polyanonic enzyme. When the redox polymer folded along the enzyme and electrically connected the inner FADH active center with its relay site, due to shorter distance accessing the FADH site of the enzyme, electron transfer was shuttling via the redox hydrogel to the electrode transducer.

Third-Generation (Direct Electron-Transfer) Glucose Sensors

This generation of the glucose biosensor diminishes the application of mediators, simplifying the reaction complexation of both the first- and second-generation sensors. The mediator-free

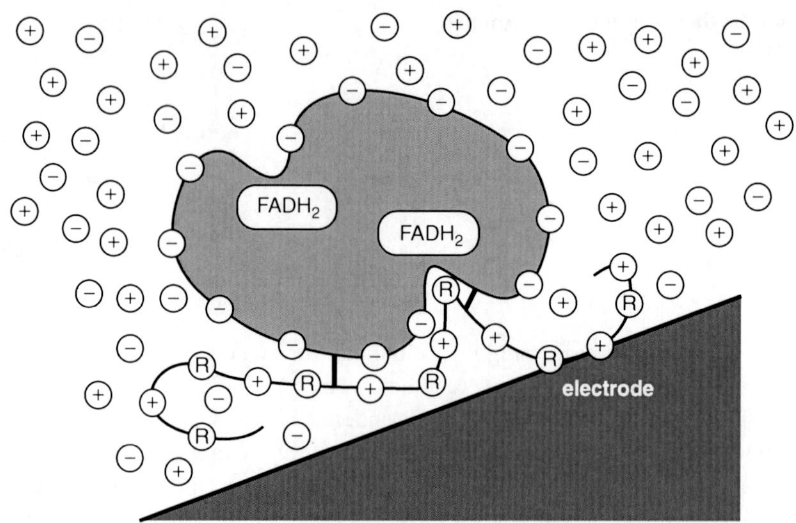

Electrochemical Glucose Sensors, Fig. 3 Schematic representation of a GOD electrode containing $FADH_2$ centers, modified with a redox polycationic polymer possessing redox groups, *R*

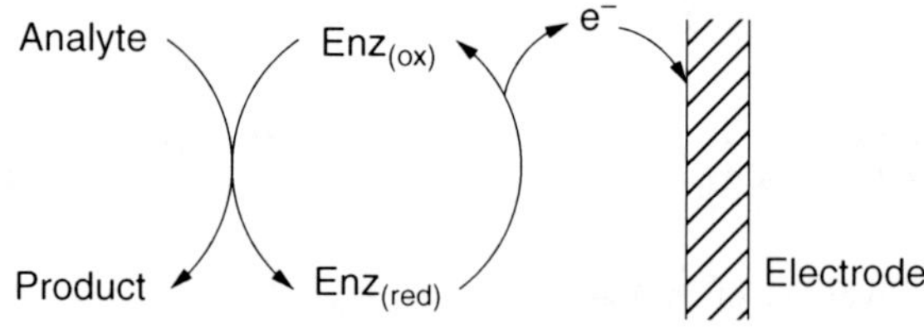

Electrochemical Glucose Sensors, Fig. 4 The schematic principle of mediator-free electrode [21]

(reagentless) glucose biosensor is developed at a low operating potential (high selectivity). The electrons are transferred directly from glucose to the electrode via the active sites of the enzyme (GOD). The redox center within biomolecules is usually embedded deeply into the large three-dimensional structure of enzyme molecule [25]. The primary challenge is to overcome any difficulty of this direct electron-transfer (DET) route. There are reports of such a mediator-free electron transfer between GOD and the electrode. The optimally designed system is to ensure that the electron-transfer distance between the immobilized enzyme and the sensing electrode surface is as short as possible (Fig. 4).

There has been an exploration of interest applying carbon nanomaterials [25–28] (particularly carbon nanotube (CNT) and graphene) into the development of electrochemical biosensor. CNT- and graphene-modified electrodes both have recently been performed for studying the direct electrochemistry of proteins/enzymes; it is due to their unique characteristics: electronic and structure properties, high surface–volume ratio, and high conductivity. These promise carbon nanomaterials a novel avenue for rapid communication with the redox active centers of proteins. The operating principle of direct electrochemistry of glucose is based on the following reaction:

$$\text{Glucose} + \text{GOD(FAD)} \rightarrow \text{gluconolactone} + \text{GOD(FADH}_2\text{)}$$

$$\text{GOD(FAD)} + 2e^- + 2H^+ \leftrightarrow \text{GOx-FADH}_2$$

Ionic liquid (IL) possesses tunable behaviors based on asymmetric ion-pair combinations. Along with its unique characteristics, such as high conductivity and wide potential window, it will be desirable to applying IL into the fields of bio-electrochemistry [29]. Incorporating IL with nanomaterials (metallic nanoparticles and carbon materials) into the development of modified

Electrochemical Glucose Sensors, Fig. 5 The aqueous equilibrium of glucose with ratio of α-, β-, and γ-glucose is 37:53:0.003 in the physiological condition and the schematic pathways of electrooxidation electron transfer [32]

electrode provides a new DET approach for biosensing application of glucose [30].

Nonenzymatic (Enzymeless) Glucose Sensors

Nonenzymatic glucose sensor is a new trend preventing sensor performance from suffering of the enzyme (GOD) instability. Due to the intrinsic properties of enzymes, the catalytic activity of GOD is susceptible to both environmental and operation conditions. Therefore, complicated immobilization procedures may further hinder and destabilize enzyme activity [31]. Enzyme-free glucose detection has been potentially undertaken based on the oxidation of glucose directly at the electrode surface. As shown in Fig. 5 [32], two-electron oxidation of three-typed glucose is attained with stable gluconic acid produced, regardless of whether the gluconolactone involved is an intermediate or not.

A number of nonenzymatic glucose study have been reported by amperometrically measuring the current response of direct glucose oxidation [33, 34]. The electrocatalytic activity of the electrode material is the key factor affecting the sensitivity and selectivity of glucose detection for nonenzymatic glucose sensors.

This review session discusses the developing history of different generation of electrochemical glucose sensors. The first-generation glucose sensor is based on the measurement of the enzymatically generated H_2O_2. Various techniques have been performed to minimize the electroactive interference. To overcome oxygen dependency, the second-generation glucose sensor is developed to incorporate an artificial redox mediator, in place to O_2. Furthermore, the employment of redox hydrogels can resolve the leachable problem of redox mediators. Shortening and facilitating the electron-transfer route between the redox active sites of enzyme and the electrode permits the third-generation glucose sensor operating via direct electron-transfer technique. The mediators and reaction complexation can be avoided. Nonenzymatic glucose sensor is operated without enzyme incorporation. Glucose oxidation is performed directly at the surface of electrode which possesses excellent electrocatalytic activity.

Cross-References

- ▶ Biosensors, Electrochemical
- ▶ Electrochemical Sensor of Gaseous Contaminants
- ▶ Electrochemical Sensors for Aerospace Applications
- ▶ Electrochemical Sensors for Environmental Analysis
- ▶ Electrochemical Sensors for Monitoring Conditions of Lubricants
- ▶ Electrochemical Sensors for Water Pollution and Quality Monitoring

References

1. Yoo EH, Lee SY (2010) Glucose biosensors: an overview of use in clinical practice. Sensor 10(5): 4558–4576
2. Wnag J (2008) Electrochemical glucose biosensors. Chem Rev 108(2):814–825
3. Shen J, Dudik L, Liu C-C (2007) An iridium nanoparticles dispersed carbon based thick film electrochemical biosensor and its application for a single use, disposable glucose biosensor. Sensor Actuat B-Chem 125(1):106–113
4. Clark LC Jr (1956) Monitor and control of blood and tissue oxygen tensions. T Am Soc Art Int Org 2:41–48
5. Clark LC Jr, Lyons C (1962) Electrode systems for continuous monitoring in cardiovascular surgery. Ann N Y Acad Sci 102:29–45
6. Mello LD, Kubota LT (2002) Review of the use of biosensors as analytical tools in the food and drink industries. Food Chem 77(2):237–256
7. Prodromidis MI, Karayannis MI (2002) Enzyme based amperometric biosensors for food analysis. Electroanalysis 14(4):241–261
8. Wang J (2001) Glucose biosensors: 40 years of advances and challenges. Electroanalysis 13(12):983–988
9. Garjonyte R, Malinauskas A (2000) Glucose biosensor based on glucose oxidase immobilized in electropolymerized polypyrrole and poly(o-phenylenediamine) films on a Prussian Blue-modified electrode. Sensor Actuat B-Chem 63:122–128
10. Yuan CJ, Hsu CL, Wang SC, Chang KS (2005) Eliminating the interference of ascorbic acid and uric acid to the amperometric glucose biosensor by cation exchangers membrane and size exclusion membrane. Electroanalysis 17(24):2239–2245
11. Jia WZ, Wang W, Xia XH (2010) Elimination of electrochemical interferences in glucose biosensors. Trac Trend Anal Chem 29(4):306–318
12. Yuqing M, Jianrong C, Xiaohua W (2004) Using electropolymerized non-conducting polymers to develop enzyme amperometric biosensors. Trends Biotechnol 22(5):227–231
13. Xu JJ, Yu ZH, Chen HY (2002) Glucose biosensors prepared by electropolymerization of p-chlorophenylamine with and without Nafion. Anal Chim Acta 463(2):239–247
14. Lei CX, Wang H, Shen GL, Yu RQ (2004) Immobilization of enzymes on the Nano-Au film modified glassy carbon electrode for the determination of hydrogen peroxide and glucose. Electroanalysis 16(9):736–740
15. De Benedetto GE, Palmisano F, Zambonin PG (1996) One-step fabrication of a bienzyme glucose sensor based on glucose oxidase and peroxidase immobilized onto a poly(pyrrole) modified glassy carbon electrode. Biosens Bioelectron 11(10): 1001–1008
16. Ricci F, Palleschi G (2005) Sensor and biosensor preparation, optimisation and applications of Prussian Blue modified electrodes. Biosens Bioelectron 21(3):389–407
17. Wang J, Chen Q, Pedrero M, Pingarron JM (1995) Screen-printed amperometric biosensors for glucose and alcohols based on ruthenium-dispersed carbon inks. Anal Chim Acta 300(1–3):111–116
18. Wang J, Rivas G, Chicharro M (1996) Iridium-dispersed carbon paste enzyme electrodes. Electroanalysis 8(5):434–437
19. Arjsiriwat S, Tanticharoenb M, Kirtikaraa K, Aokic K, Somasundrumet M (2000) Metal-dispersed conducting polymer-coated electrode used for oxidase-based biosensors. Electrochem Commun 2(6):441–444
20. Ghindilis AL, Atanasov P, Wilkins E (1997) Enzyme-catalyzed direct electron transfer: fundamentals and analytical applications. Electroanalysis 9(9):661–674
21. Chaubey A, Malhotra BD (2002) Mediated biosensors. Biosens Bioelectron 17(6–7):441–456
22. Heller A, Feldman B (2008) Electrochemical glucose sensors and their applications in diabetes management. Chem Rev 108(7):2482–2505
23. Heller A (1990) Electrical wiring of redox enzymes. Acc Chem Res 23(5):128–134
24. Bard AJ (2001) Electrochemical methods fundamentals and applications. Wiley, New York
25. Kang X, Wang J, Wu H, Aksay IA, Liu J, Lin Y (2009) Glucose oxidase–graphene chitosan–modified electrode for direct electrochemistry and glucose sensing. Biosens Bioelectron 25(4):901–905

26. Zhang J, Feng M, Tachikawa H (2007) Layer-by-layer fabrication and direct electrochemistry of glucose oxidase on single wall carbon nanotubes. Biosens Bioelectron 22(12):3036–3041
27. Jacobs CB, Peairs MJ, Venton BJ (2010) Review: carbon nanotube based electrochemical sensors for biomolecules. Anal Chim Acta 662(2):105–127
28. Shao Y, Wang J, Wu H, Liu J, Aksay IA, Lin Y (2010) Graphene based electrochemical sensors and biosensors: a review. Electroanalysis 22(10):1027–1036
29. Shiddiky MJ, Torriero AA (2011) Application of ionic liquids in electrochemical sensing systems. Biosens Bioelectron 26(5):1775–1787
30. Opallo M, Lesniewski A (2011) A review on electrodes modified with ionic liquids. J Electroanal Chem 656(1–2):2–16
31. Sassolas A, Blum LJ, Leca-Bouvier BD (2012) Immobilization strategies to develop enzymatic biosensors. Biotechnol Adv 30(3):489–511
32. Park S, Boo H, Chung TD (2006) Electrochemical non-enzymatic glucose sensors. Anal Chim Acta 556(1):46–57
33. Rahman MM, Ahammad AJS, Jin JH, Ahn SJ, Lee JJ (2010) A comprehensive review of glucose biosensors based on nanostructured metal-oxides. Sensor 10(5):4855–4886
34. Lu LM, Zhang L, Qu FL, Lu HX, Zhang XB, Wu ZS, Huan SY, Wang QA, Shen GL, Yu RQ (2009) A nano-Ni based ultrasensitive nonenzymatic electrochemical sensor for glucose: enhancing sensitivity through a nanowire array strategy. Biosens Bioelectron 25(1):218–223

Electrochemical Impedance Spectroscopy (EIS) Applications to Sensors and Diagnostics

Vadim F. Lvovich
NASA Glenn Research Center, Electrochemistry Branch, Power and In-Space Propulsion Division, Cleveland, OH, USA

Concept of Complex Impedance

The concept of electrical impedance was first introduced by Oliver Heaviside in the 1880s and was soon after developed in terms of vector diagrams and complex number representation by A. E. Kennelly and C. P. Steinmetz [1]. Since then the technique gained in exposure and popularity, propelled by a series of scientific advancements in the field of electrochemistry, improvements in instrumentation performance and availability, more elaborate mathematical methods for the data analysis, and increased exposure to ever widening range of practical applications. Transformational advancements that have occurred over the last 30 years in electrochemical equipment and computer technology allowed for digital automated impedance measurements to be performed with significantly higher quality, better control, and versatility than what was available during the early years of EIS. One can argue that these advancements completely revolutionized the field of impedance spectroscopy (and in broader sense the field of electrochemistry), allowing the technique to be applicable to exploding universe of practical applications [2].

In spite of ever expanding use of EIS in analysis of practical and experimental systems, the impedance (or complex electrical resistance for a lack of a better term) fundamentally remains a simple concept. Electrical resistance R is related to the ability of a circuit element to resist the flow of electrical current. Ohm's law (Eq. 1) defines resistance in terms of the ratio between input voltage V and output current I:

$$R = \frac{V}{I} \tag{1}$$

While this is a well-known relationship, its use is limited to only one circuit element – the ideal resistor. An ideal resistor follows Ohm's law at all current, voltage, and AC frequency levels. The resistor's characteristic resistance value R [ohm] is independent of AC frequency, and AC current and voltage signals through the ideal resistor are "in phase" with each other. Let us assume that the analyzed sample material is ideally homogeneous and completely fills the volume bounded by two external current conductors ("electrodes") with a visible area A that is placed apart at uniform distance d is formed, as shown in Fig. 1. When external voltage V is applied, a uniform current I passes through the sample, and the resistance is defined as

$$R = \rho \frac{d}{A} \tag{2}$$

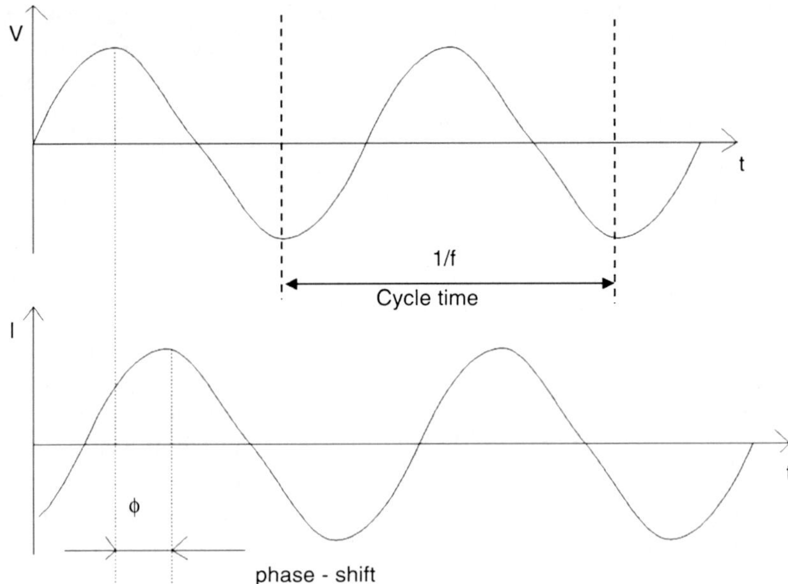

Electrochemical Impedance Spectroscopy (EIS) Applications to Sensors and Diagnostics, Fig. 1 Impedance experiment – sinusoidal voltage input *V* at a single frequency *f* and current response *I*

where ρ [ohm cm] is characteristic electrical resistivity of a material, representing its ability to resist the passage of the current. The inverse of resistivity is conductivity σ [1/(ohm cm)] or [Sm/cm], reflecting the material's ability to conduct electrical current between two bounding electrodes.

An ideal resistor can be replaced in the circuit by another ideal element that completely rejects any flow of current. This element is referred as "ideal" capacitor (or "inductor" which stores magnetic energy created by applied electric field), formed when two bounding electrodes are separated by a nonconducting (or "dielectric") media. The AC current and voltage signals through the ideal capacitor are completely "out of phase" with each other, with current following voltage. The value of the capacitance presented in Farads [F] depends on the area of the electrodes *A*, the distance between the electrodes *d*, and the properties of the dielectric reflected in a "relative permittivity" parameter ε as

$$C = \frac{\varepsilon_o \varepsilon A}{d} \quad (3)$$

where ε_0 is the constant electrical permittivity of vacuum ($8.85 \cdot 10^{-14}$ F/cm). The relative permittivity value represents a characteristic ability of the analyzed material to store electrical energy. This parameter (often referred as simply permittivity or "dielectric") is essentially a convenient multiplier of the vacuum permittivity constant ε_0 that is equal to a ratio of the material's permittivity to that of the vacuum. The permittivity values are different for various media: 80.1 (at 20°C) for water, between 2 and 8 for many polymers, and 1 for ideal vacuum.

Impedance is a more general concept than either pure resistance or capacitance, as it takes the phase differences between the input voltage and output current into account. Like resistance, impedance is the ratio between voltage and current, demonstrating the ability of a circuit to resist the flow of electrical current, represented by the "real impedance" term, but it also reflects the ability of a circuit to store electrical energy, reflected in the "imaginary impedance" term. Impedance can be defined as a complex resistance encountered when current flows through a circuit composed of various resistors, capacitors, and inductors. This definition is applied to both direct current (DC) and alternating current (AC).

In experimental situations the electrochemical impedance is normally measured using excitation AC voltage signal *V* with small amplitude V_A (expressed in Volts) applied at frequency

f (expressed in Hz or 1/s). The voltage signal $V(t)$, expressed as a function of time t, has the form:

$$V(t) = V_A \sin(2\pi f t) = V_A \sin(\omega t) \quad (4)$$

In this notation a "radial frequency" ω of the applied voltage signal (expressed in radians/s) parameter is introduced, which is related to the applied AC frequency f as $\omega = 2\pi f$.

In linear or pseudo-linear systems, the current response to a sinusoidal voltage input will be a sinusoid at the same frequency but "shifted in phase" (either forward or backward depending on the system's characteristics) that is determined by the ratio of capacitive and resistive components of the output current (Fig. 1). In a linear system, the response current signal $I(t)$ is shifted in phase (ϑ) and has a different amplitude, I_A :

$$I(t) = I_A \sin(\omega t + \varphi) \quad (5)$$

An expression analogous to Ohm's law allows to calculate the complex impedance of the system as the ratio of input voltage $V(t)$ and output measured current $I(t)$:

$$Z* = \frac{V(t)}{I(t)} = \frac{V_A \sin(\omega t)}{I_A \sin(\omega t + \varphi)} = Z_A \frac{\sin(\omega t)}{\sin(\omega t + \varphi)} \quad (6)$$

Using Euler's relationship where j is the imaginary number:

$$\exp(j\phi) = \cos\vartheta + j\sin\vartheta \quad (7)$$

it is possible to express the impedance as a complex function. The potential $V(t)$ and the current $I(t)$ responses are described as:

$$V(t) = V_A e^{j\omega t} \quad (8)$$

$$I(t) = I_A e^{j\omega t - j\varphi} \quad (9)$$

The impedance is then represented as a complex number that can also be expressed in complex math format as a combination of "real," or in-phase (Z_{REAL}), and "imaginary," or out-of-phase (Z_{IM}) parts,

$$Z* = \frac{V(t)}{I(t)} = Z_A e^{j\varphi} = Z_A(\cos\varphi + j\sin\varphi) = Z_{REAL} + jZ_{IM} \quad (10)$$

and the phase angle ϕ at a chosen radial frequency w is a ratio of the imaginary and real impedance components,

$$\tan\varphi = \frac{Z_{IM}}{Z_{REAL}} \text{ or } \varphi = \arctan\left(\frac{Z_{IM}}{Z_{REAL}}\right) \quad (11)$$

while the magnitude of the impedance signal $/Z/$ at a particular frequency becomes

$$Z_A = \sqrt{Z_{REAL}^2 + Z_{IM}^2} \quad (12)$$

In the case of a frequency sweep, the phase angle and total impedance values are calculated at each frequency.

The impedance is therefore expressed in terms of a magnitude (absolute value), $Z_A = /Z/$, and a phase shift, ϑ. Impedance is an alternating current (AC) phenomenon that is usually specified at a particular frequency. By measuring impedance across a number of frequencies, a valuable data about an element can be extracted. This is the basis of impedance spectroscopy, and it is the fundamental concept underlying many industrial, instrumentation, and automotive sensors.

If we plot the applied sinusoidal voltage signal on the X-axis of a graph and the sinusoidal response signal $I(t)$ on the Y-axis, an oval known as a "Lissajous figure" will appear (Fig. 2a). Analysis of Lissajous figures on oscilloscope screens was the accepted method of impedance measurement prior to the availability of lock-in amplifiers and frequency response analyzers. Modern equipment allows automation in applying the voltage input with variable frequency and collecting the output impedance (and current) responses as the frequency is scanned from very high (MHz–GHz) values

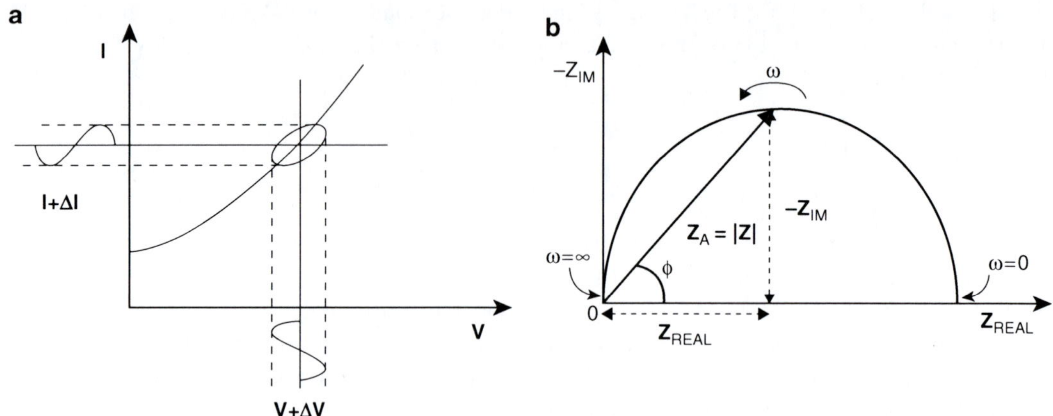

Electrochemical Impedance Spectroscopy (EIS) Applications to Sensors and Diagnostics, Fig. 2 Impedance data representations: (**a**) Lissajous figure; (**b**) complex impedance or "Nyquist" plot

where time scale of the signal is in micro- and nanoseconds to very low frequencies (μHz) with time scale of order of hours. The resulting data of imaginary versus real portion of impedance is presented as the complex impedance or "Nyquist" plot (Fig. 2b).

"The Path of the Least Impedance to Current" in Impedance Measurements

In a basic electrochemical impedance experiment, there are at least two electrodes bounding a sample (such as electrolyte solution), with external potential (voltage) difference applied between the electrodes. For the overall analyzed system, the voltage difference between two electrodes and the resulting current can be measured with an ordinary voltmeter and ampere meter (Fig. 3).

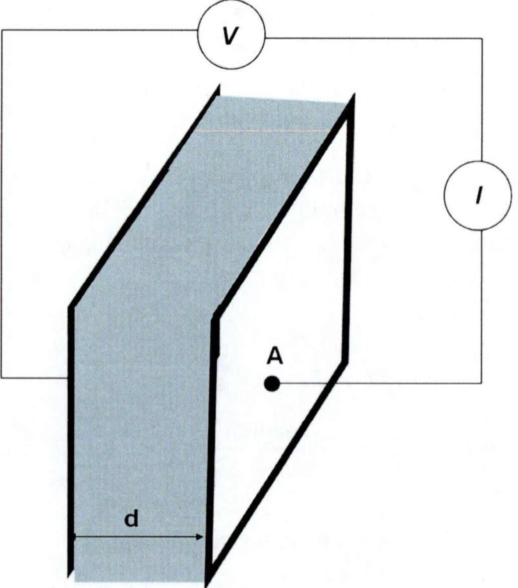

Electrochemical Impedance Spectroscopy (EIS) Applications to Sensors and Diagnostics, Fig. 3 Fundamental electrochemical experiment

The impedance output is fundamentally determined by characteristic of an electrical current conducted through the system and is based on a concept that the obtained data represents the sample's "least impedance to the current." In any given experimental system, there are always several competing paths for the current to travel through a sample. The current, however, chooses one or several closely matched predominant "paths of least resistance" between two electrical conductors (electrodes) under applied conditions such as AC frequency and voltage amplitude, DC voltage, electrode geometry and configuration, sample composition and concentration of main conducting species, temperature, pressure, convection, and external magnetic fields. Only the conduction through this predominant path of least resistance, or, to be more exact,

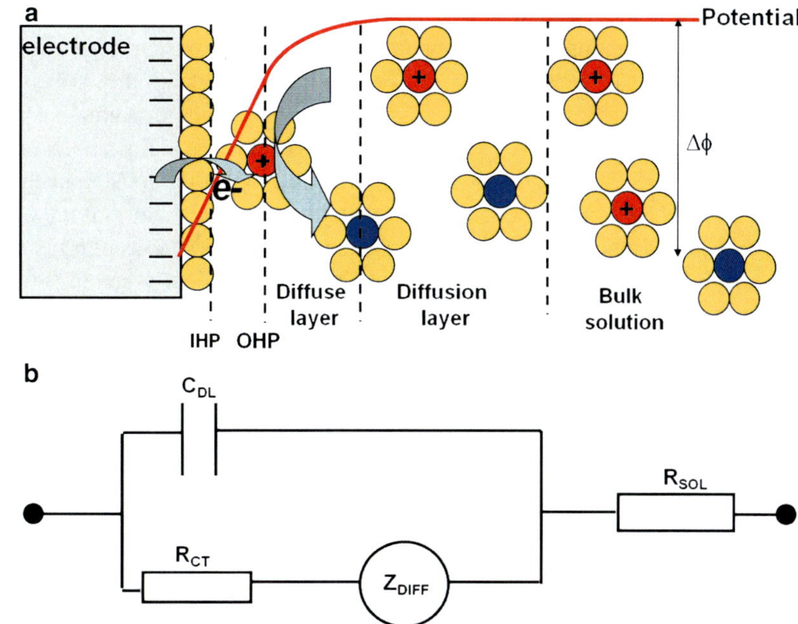

Electrochemical Impedance Spectroscopy (EIS) Applications to Sensors and Diagnostics, Fig. 4 (**a**) Interfacial electrochemical reaction with diffusion and double-layer components; (**b**) representative electrical circuit

the path of the least impedance, is being always measured by the EIS.

For a sample represented by a parallel combination of a capacitor C and a resistor R, at high applied frequencies ω, the impedance to current is the lowest through the capacitive component where impedance is inversely proportional to the frequency as $Z_i^* \sim (\omega C)^{-1}$ and therefore is smaller than the impedance of the finite resistor R. At the lower frequencies, the opposite becomes true – the capacitive impedance component becomes large, and the current predominantly flows through the resistor with the total measured impedance reflecting the resistance value as $Z_i^* \sim R$. The detected impedance output is determined by measuring the current passing through the least impeding segment of the circuit. The characteristic parameters of this ideal system can therefore be determined from the total impedance response $Z^* = f(\omega)$ as pure capacitance C at the higher frequencies and pure resistance R at the low frequencies. Characteristic relaxation frequency f_C corresponds to a value where switch between the two conduction mechanisms occurs. The inverse of this frequency is characteristic "relaxation time" $\tau = 1/\omega_C$ for this circuit.

Under the influence of the electric field, the current passes through a complete circuit, transported by electrons in the metal conductors and the electrodes and by ionic migration in the electrolyte. The resistance to ionic migration current in the aqueous bulk solution within the frequency range of a typical impedance measurement can be simplified by a small "solution resistance" component R_{SOL} (Fig. 4). Current in the bulk solution is transported predominantly by migration which is dependent on the applied external voltage V (often referred in the electrochemical literature as "electrochemical potential difference Δϕ"). For mobility process involving electroactive species of type i, the Kohlrausch equation applies as familiar Ohm's law equation, and for flux of charged species J_i [mol/s] under the influence of electrochemical potential gradient $\Delta\phi/\Delta x$,

$$J_i = \frac{\Delta V(x)}{\Delta x} AF \sum z_i u_i C_i^* = \frac{\Delta \varphi(x)}{\Delta x} A \sum \sigma_i \tag{13}$$

where A is the surface area of electrode; z, full charge of electroactive species of type i; F, Faradaic constant (96,500 C/mol) representing free charge of 1 mol of elementary charges; C^*_i, concentration (mol/cm^3) of ions in solution participating in migration; u_i, the mobility constant; and $\sigma_i = F\sum z_i u_i C^*_i$, bulk solution conductivity of this type of ions [Sm/cm].

Other types of samples can be subjected to electrochemical analysis, such as aqueous and nonaqueous solutions of $\sim 1\ \mu m$ sized colloidal particles, whereas charged double layer of "counterions" surround each particle and the particles can be regarded as "macro-ions." The colloidal particles will migrate and contribute to the solution conductivity. Solid ionic electrolytes and composite materials are also a possibility. There are also mixed conductors with both ionic and electronic conductance, such as many polymers. Lastly, semiconductors possess "forbidden gaps" that prevent electrons from entering conduction bands resulting in low conductivity. The conduction can be significantly and selectively enhanced by adding impurities and creating local energy centers.

At the electrode–electrolyte interface separating the electrodes from the sample, there is a charge carrier shift – a transition between electronic and ionic conduction. The transfer of electric charge across the solution–electrode interface is accompanied by electrochemical reaction at each electrode, a phenomenon known as electrolysis. Electrochemical discharge of the species at the interface is determined by their electrochemical properties, such as their ability to release or accept an electron at a voltage (or potential) of this electrode. The electrode potential is essentially a measure of an excess of electrons at the electrode which is charged negatively or lack of electrons on a positively charged electrode. In electrical terms the impedance of the system to combined current generated by the discharge processes occurring in a very thin "Helmholtz" layer separating the electrode and the sample can be represented by the so-called "charge transfer" resistance R_{CT} (Fig. 4).

As a result of chemical and electrochemical reactions, depletion or accumulation of matter and charge may occur next to the electrodes. This reaction consumes or releases additional ions or neutral species from or into the bulk solution, resulting in a concentration gradient when concentration of the species in the bulk solution is different from that in the vicinity of the electrodes. However, all species always attempt to maintain equal concentration distribution inside of any sample volume. For instance, when the local concentration gradient is created in the vicinity of the electrodes as a result of the species' consumption over the course of electrochemical reaction, the more abundant solution species always attempt to replenish continuously depleting species at the interface by moving or "diffusing" to the electrode surface. Alternatively the release of new species as a result of completion of the electrochemical reaction leads to their diffusion away from the electrode into the bulk solution where their concentration is lower than that at the electrode–solution interface. This concentration gradient results in a diffusion mass-transport process occurring in a thick "diffusion layer" that can be measured by electrochemical impedance and represented by a complex diffusion impedance element Z_{DIFF} (Fig. 4).

Deposition of ions at the interface with no or minimal electrochemical discharge countered by electronic charges of the opposite sign on the electrodes interface produce a "double-layer capacitance" with a value C_{DL}. In conductive solution an electric double layer is formed at the electrode as soon as the electrode is wetted. It may be useful to remember that at all interfaces (such as transition areas between the electrode and solution, tissue, gel, solution and surface of colloidal particle, or inside the sample), there will be a nonuniform distribution of charges and resulting electrochemical potential gradient. Maxwell–Wagner, Helmholtz, Gouy–Chapman, Stern, and Grahame theories have been used to describe the interfacial and double-layer dynamics [3].

In addition to the double layer, electrochemical reactions, and diffusion effects, specific adsorption (or chemisorption) can be present in some systems. Adsorption is a process where species are chemically bound to the metal surface of an electrode due to their chemical affinity, and not due to coulombic forces based on charges difference or polarity. Adsorption and electrochemical reactions take place on the electrode surface as a function of applied DC electrochemical potential and are determined by specific electrochemical properties of participating species. Unlike heterogeneous electrochemical reactions that typically occur at the interface, the overall mass transport in the bulk of a sample is a homogeneous phase phenomenon that has to be carefully controlled. The Ohm's law is completely valid under the assumption of homogeneous and isotropic bulk sample medium, when the current direction and the field are coinciding, and other effects, primarily related to electrochemical interfacial reactions (such as diffusion, electrodes polarization, est.), are absent. In the bulk solution with free ions, the electroneutrality condition $\sum z_i C*_i = 0$ applies and the sum of charges is zero. Electroneutrality does not prevail, however, at the interfacial boundaries with space charge regions. The hypothesis often made is that of a dilute solution for which the flux of a species i can be separated into a flux due to diffusion and a flux due to migration in an electric field. At the electrochemical double-layer region, the current is being transported by both migration and diffusion; the latter is driven by the concentration gradient at applied DC potential. In some situations an external mixing, or "convection," is added, moving the sample with a flow rate of v. Overall current equation which includes diffusion (first-term)–migration (second-term)–convection (third-term) can be expressed as the Nernst–Planck equation [4, 5]:

$$\begin{aligned} I &= -R_G TF \sum \left(z_i \Delta C_i^* u_i\right) - F^2 \sum \left(z_i^2 C_i^* u_i \Delta\phi\right) \\ &\quad + Fv \sum \left(z_i C_i^*\right) \\ &= -F \sum \left(z_i \Delta C_i^* D_i\right) - F \sum \left(z_i \sigma_i \Delta\varphi\right) \\ &\quad + Fv \sum \left(z_i C_i^*\right) \end{aligned} \tag{14}$$

where the mobility constant u_i is related to the diffusion coefficient D_i and gas constant R_G through the Nernst–Einstein equation $u_i = D_i / R_G T$ and conductivity is $\sigma_i = F\sum z_i C*_i u_i$.

In principle the mobility can be viewed as a balance between a drag force $F_{DRAG} = 6\pi\eta v_i a_P$ on a particle of size a_P in a sample of viscosity η, which is moving under the influence of an electric field $F_{EL} = z_i e_0 \Delta\varphi$. Assuming that the transport properties (D_i, u_i) are uniform in the solution bulk, and hence are independent of C_i, the concentration change of species type i in the absence of a chemical reaction becomes

$$\frac{\partial C_i^*}{\partial t} = D_i \nabla^2 C_i^* + z_i F u_i \nabla (C_i^* \Delta\phi) - v \nabla C_i^* \tag{15}$$

Practical electrochemistry always attempts to analyze a complicated system by minimizing the effects of some of its components until limited number of unknowns that can be solved. The aim of the electrochemist is to be able to study each elementary phenomenon in isolation from the others. Hence, a technique should be employed capable to extract the data which allows these phenomena to be separated. Mobility dominates the conduction when electrolyte has mobile ionic species on reversible electrodes and at high AC frequencies where only the bulk solution resistance is visible in the impedance spectrum and polarization at the electrodes does not develop. If the electroneutrality of the solution is assured in the presence of major ionic species (often referred to as "supporting electrolyte") which are not taking part and consumed in the electrochemical reaction and therefore are not participating in concentration driven diffusion, a small migration impedance term essentially does not change with applied electrochemical potential. Hence the constant migration impedance term can be relatively easily identified and subtracted. Concentration calculation in the presence of convection is difficult unless very particular hydrodynamic conditions are fulfilled. The well-known example of the rotating disc electrode introduces a steady convection and a constant concentration gradient at the electrode–solution interface. However, in the majority of situations, analyzed media is stagnant or moves at a constant speed, and the contribution of convection becomes negligible or easily identifiable. If the effects of constant mobility and convection can be identified and subtracted from the overall impedance response, the remaining diffusion process results in a concentration gradient located in so-called "Nernst" or "diffusion" layer of thickness L_{DIFF}, within which the liquid is nearly motionless ($v = 0$). For such conditions the transport Eq. 15 can then be reduced with good accuracy to the Fick's diffusion equation:

$$\frac{\partial C_i^*}{\partial t} = D_i \nabla^2 C_i^* \tag{16}$$

When the above conditions of effects of convection and migration are realized, the resulting current is limited only by the diffusion-driven transport of the electroactive species to the electrode–solution interface. At the interface charge transfer reactions take place that can be studied as a function of the electrochemical potential (Fig. 4). That experimental situation contains the fundamental premise of the traditional impedance analysis – make mass-transport effect on interfacial impedance exclusively diffusion limited (Z_{DIFF}) and investigate the interfacial kinetic phenomena composed of the diffusion and electrochemical reaction (discharge or electrolysis R_{CT}) impedances that can be combined in the so-called Faradaic impedance.

The combined data is often being presented as a complete "equivalent circuit" (such as the one shown in Fig. 4b). Equivalent circuit is composed of combination of series and parallel resistances, capacitances, and inductors, representing combined impedance to the current passage through the bulk solution (at high frequencies) and through the interfacial region (at low frequencies). Equivalent circuit is idealized and often imperfect, but extremely visually appealing representation of analyzed experimental system where a combination of ideal electrical elements such as ideal resistors, capacitors, or inductors are used to approximate real physical and chemical processes. Inductors have an impedance of $j\omega L$ and capacitors of $1/j\omega C$. The impedances of these components combine using complex number arithmetic. When exposed to a signal of increasing frequency, capacitive impedance decreases, and inductive impedance increases, leading to changes in overall impedance as a function of frequency. The impedance of a pure resistor R does not change with frequency.

The path of the least impedance through a real-life sample placed between two conducting metal electrodes can be represented by a combination of chemical and mechanistic elements that can only to some degree be approximated as ideal electrical elements. These conducting venues through the sample may be the most plentiful and mobile electrons, ions, and particles; double layer capacitive charging effects; specific adsorption; charge transfer "resistance" to tunneling electrons crossing the interface between the sample and electrodes; and transport of discharging species to the surface of the electrode through diffusion layer concentration gradient at the electrode/sample interfaces. To interpret the EIS results, scientists and engineers have to intelligently devise the experiment that extracts the needed analytical information from the experimentally obtained impedance. This interpretation includes initial development of the relationships between the electrical parameters representing the path of the least impedance to the current and the "real life" investigated by chemical and mechanistic components that result in the measured-impedance response. Through these relationships it may be possible to analyze the actual chemical, physical, and mechanistic processes inside the analyzed system.

To a great benefit of analytical chemist concerned with the accurate analysis of concentrations of analytes, and material scientist preoccupied with the sample properties, the mechanisms of transporting current are largely determined by limited types of species that are present in the analyzed sample at significant levels. With that in mind, one cannot always determine "all" species present in the experimental system as not "all" species may participate in the conduction through the path of the least impedance, even if their response is dependent on the controlled experimental parameters, such as AC frequency. As compared to many other spectroscopy methods, EIS has an advantage of often being able to separate and identify several types of species inside an experimental sample, as various species may create a path of least impedance through the sample at different (and variable) experimental conditions such as temperature, pressure, applied electrochemical potential, applied AC frequency and amplitude range, geometry of the electrodes, and the sample. However, many potential analytes may not be detectable as they do not conduct current or are present in such low concentrations that their contribution to the overall conduction process may be negligible (however, they may have a disproportionably strong activity at the electrode sample interface). The same general limitations are present in optical and other "spectroscopies" based on "detection of the most prominent contributors" principle. It is a task of good experimental scientist to develop experimental design that would provide the necessary information about the task at hand. In that respect AC impedance is indeed a "spectroscopy"-type technique, where a combination of carefully conducted experiments, some degree of initial preconceived knowledge of the analyzed sample, selection of experimental conditions, and careful results interpretation are required to allow determining the sample composition and mechanisms responsible for the impedance response, quantify them, and develop a mechanistic model of the entire system's physical and chemical response to external electric field.

Electrochemistry in general and the EIS in particular are often used to analyze both bulk sample conduction mechanisms and interfacial processes, where electron transfer, mass transport, and adsorption are often present. EIS analysis has often treated the bulk and interfacial processes separately [4]. The analysis is achieved on the basis of selective responses of bulk and interfacial processes to sampling AC frequencies. The features appearing in the impedance AC frequency spectrum can be described according to the theory of impedance relaxations. Again, as in the case of any other spectroscopy method, the subject of the EIS analysis is the detection and interpretation of these spectrum features.

Impedance Data Analysis

To examine the impedance of an element when swept at different frequencies, one usually has to examine the response signal in either the time or the frequency domain. Analog signal analysis techniques, such as AC-coupled bridges, were commonly used to examine the signal in the frequency domain, but the advent of high-performance A/D converters has led to data collection in the time domain with subsequent conversion to the frequency domain.

A number of integral transforms to convert data into the frequency domain can be utilized, with fast Fourier transform (FFT) analysis being a common approach. This technique takes a time-series representation of a signal and applies an integral transform to map the representation into its frequency spectrum. One can use the technique to provide a mathematical description of the relationship between any two signals. In impedance analysis, the relationship between the excitation voltage $V(t)$ and the current output $I(t)$ signal is of interest. If a system is linear, the ratio of the Fourier transforms (Φ) of the measured time-domain voltage and current is equal to the complex impedance $Z^*(j\omega)$, and it is expressed as a complex number. The real and imaginary components of the resulting complex number form a key piece of the subsequent data analysis:

$$\frac{\Phi[V(t)]}{\Phi[I(t)]} = Z^*(j\omega) \tag{17}$$

Impedance is only properly defined as a transfer function when the system under investigation fulfils the conditions of causality, linearity, stability, and consistency during the measurement [1–4]. Early investigators question the validity of EIS measurements on the basis of system stability and linearity requirements over the time of the measurement. FFT method applies resulting voltage signal from mixed AC waves of several dozen selected frequencies to a system on top of DC-voltage bias. The current signal obtained in the time domain is converted back into AC signals in the frequency domain by the Fourier transformation. The recent results showed that the FFT impedance technique indeed records instantaneous impedance data, whereas traditional single frequency impedance techniques present a time-averaged data. Fast impedance methods are indeed fast enough for the

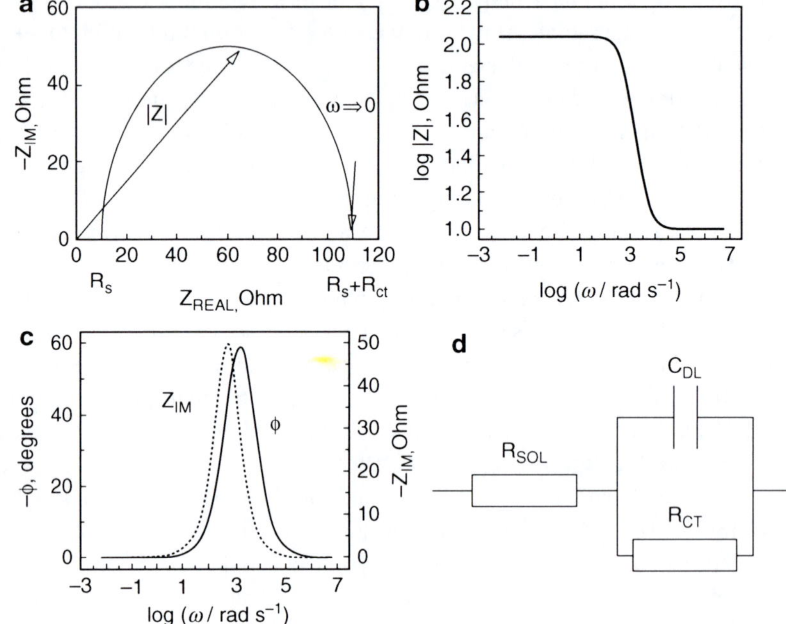

Electrochemical Impedance Spectroscopy (EIS) Applications to Sensors and Diagnostics, Fig. 5 (**a**) Nyquist and Bode (**b**) total impedance and (**c**) Z_{IM} and phase angle plots. (**d**) R_{SOL}- R_{CT}/C_{DL} circuit diagram

electrochemical reaction timescale [6, 7], making it available for impedance-based sensors and in situ measurements.

Commonly plots of impedance against frequency are generated as part of impedance data analysis. A Nyquist plot (Fig. 2b) is a parametric plot of the real and imaginary parts of the transfer function in the complex plane as the frequency is swept over a given range. Real impedance Z_{REAL} is plotted on the X-axis and the imaginary impedance Z_{IM} on the Y-axis of a chart (Z_{IM} is typically expressed as a negative value). A representation of the impedance at each frequency is obtained. In other words, each point on the plot is the impedance at a particular frequency. The total impedance */Z/* is calculated as a length of the vector Z_A, and the angle between this vector and the X-axis is phase angle ϕ. Figure 2b shows a typical Nyquist plot for a resistor and capacitor in parallel.

Although often used, Nyquist plots don't provide information on frequency; it is therefore impossible to tell what frequency was used for particular impedance. For this reason Nyquist plot (Fig. 5a) is usually supplemented with the Bode plots (Fig. 5b, c). The Bode plot presents impedance and phase shift as a function of frequency. In this case, the log of frequency is plotted on the X-axis, and both the absolute value of the impedance |Z| and the phase shift are plotted on the Y-axis. Usually, the Nyquist and Bode plots are used together to understand a sensor element's transfer function.

The impedance data is often being visualized as an "equivalent circuit." For example, Randles circuit is one of the simplest and most common cell models used for many aqueous, conductive, and ionic solutions. It includes only solution resistance R_{SOL} and a parallel combination of a double-layer capacitor C_{DL} and a charge transfer or polarization resistance R_{CT} (Fig. 5d). The Randles circuit is characterized by high-frequency purely resistive solution component R_{SOL} where the phase angle is 0° and the current is $I = V/R_{SOL}$. At lower frequencies the current flow becomes distributed between the double-layer capacitor C_{DL} at medium frequencies with corresponding increase in the absolute value of the phase angle and the finite charge transfer resistor R_{CT} at the lower frequencies where the phase angle approaches 0° and the current approaches $I = V/(R_{SOL} + R_{CT})$. The time constant for the circuit equals $\tau = R_{CT}C_{DL}$, with imaginary impedance reaching maximum value at the critical frequency $f_C = 1/(2\pi R_{CT} C_{DL})$. The simplified Randles cell model is often the starting point for other more complex models, mainly for the charge transfer kinetic analysis in highly conductive solution systems not impeded by migration and diffusion mass-transport effects.

The Nyquist plot for a Randles cell is always a semicircle (Fig. 5a). The solution resistance R_{SOL} can found by reading the real axis value at the high-frequency intercept while $Z_{REAL} \rightarrow R_{SOL}$ and $Z_{IM} \sim 1/\omega C_{DL} \rightarrow 0$. The real axis value at the low-frequency intercept is $Z_{REAL} \rightarrow R_{SOL} + R_{CT}$ equaling the sum of the charge transfer resistance and the solution resistance $R_{SOL} + R_{CT}$, while $Z_{IM} \rightarrow 0$. The diameter of the semicircle is therefore equal to the charge transfer resistance R_{CT}. The Bode impedance magnitude plot (Fig. 5b) shows the solution resistance R_{SOL} at high and the sum of the solution resistance R_{SOL} and the charge transfer resistance R_{CT} and low frequencies. The phase angle is resistive (0°) at high frequencies, changes to significant negative values when the impedance becomes partially capacitive at medium frequencies, and becomes again completely resistive at low frequencies (Fig. 5c).

Bode and Nyquist plots are useful in examining the frequency response of the sensor. Examining impedance over a number of frequencies provides a more accurate result than a single-point measurement because it helps average out noise. It also allows identifying the optimal operating point by examining the frequency response of the capacitive and inductive components under particular conditions. To visualize the data generated by the impedance sweep, an equivalent circuit model of the system composed of resistors, capacitors, and inductors connected in such a way as to mimic the electrical behavior of the system can be developed. An empirical model should be based on the best match between the empirical model and the measured data. Empirical models can be developed by successively adding and subtracting component impedances until you obtain the best fit. This is usually conducted on the basis of a nonlinear least-squares (NLLS) fitting algorithm. With the aid of a computer, the NLLS algorithm takes initial estimates of the model parameters, successively changes each of the model parameters, and evaluates the resulting fit. The software progresses iteratively until the fit is considered acceptable [1].

Often one can construct several different empirical models and have all of them provide the same impedance profile. The equivalent circuit model with the best match between experimental and mode data may not always correspond to physical characteristics of the process and can be constructed solely to give a best fit. A good least-squares fit can often be obtained for a completely incorrect model that is not representative of the physical system. And the NLLS fit algorithm can ignore part of, or fail to converge on, the measured profile. This is because many algorithms attempt to optimize the fit over the entire spectrum but overlook poor fits at various points of the spectrum. It is usually possible to build a good-fitting model by adding components, but this does not suggest it is representative of the system's electrochemical processes. In general, empirical models should have minimum number of components, and, where possible, physical models based on theoretical underpinnings of system processes should be identified before the fit is attempted. An important point to make is that the choice of particular ideal electrical components and interconnections should always be based on knowledge of the physical characteristics of the process. One can use an existing model from the literature, or you can empirically develop a new model [2].

Even larger problem of the impedance analysis difficulties is related to the fact that in many real-life applications the narrowly defined experimental conditions, such as limiting the measured-impedance response by diffusion mass transport or rejection of the interfacial effects, cannot be imposed or controlled. Very frequently an unknown system or phenomenon has to be characterized, and the insuring complexity of analysis with many potential unknowns has to be faced. Therefore, often a preliminary concept for the investigated system has to be developed even before the system is analyzed.

While EIS is a powerful and versatile method applicable to many areas of science and

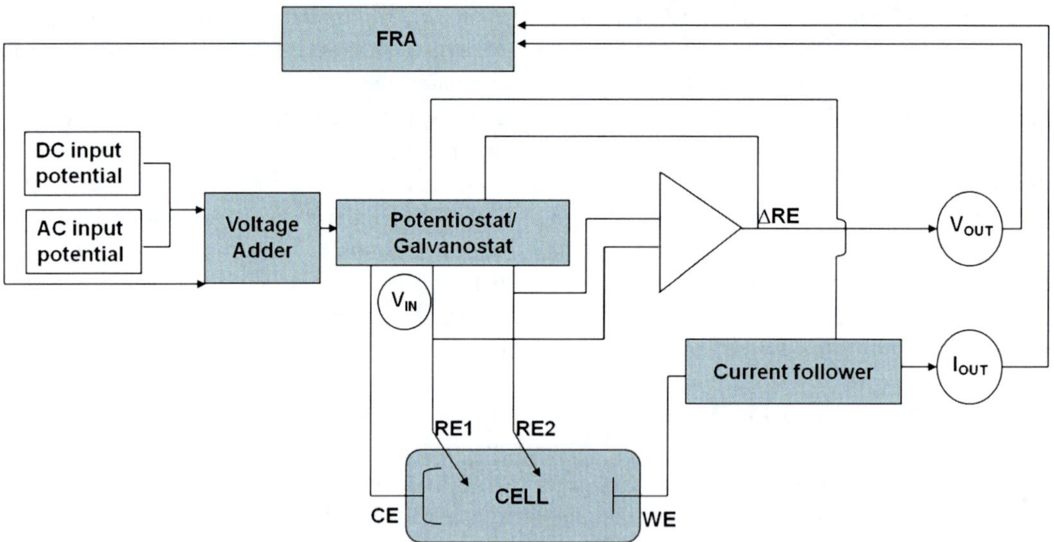

Electrochemical Impedance Spectroscopy (EIS) Applications to Sensors and Diagnostics, Fig. 6 Example for impedance measurement setup with a frequency response analyzer and a 4-electrode cell

technology, it is a completely inapplicable method to blind analysis of systems that the investigator has no preliminary knowledge about. At least in very general terms, one almost has to have a significant amount of advanced knowledge about the analyzed subject. A series of basic questions about the system has to be answered. Is the system a solution, solid, or gas? Is it aqueous or nonaqueous solution? Is the analyzed media composed of ions, conductive or polarizable particles, or is it a semiconductor? Are there any electrochemically active components in the system that may be able to discharge and/or adsorb on the metal electrodes? Interpretation of the data depends greatly on initial understanding of basic components of the system and their projected response to the applied AC and DC electric fields. In response to these questions, it is highly recommended to develop a series of preliminary expectations and assumptions of the chemical, physical, and mechanical characteristics of the analyzed system and the anticipated general type of the experimental setup, expected-impedance data, and possible interpretation strategy. Intelligent characterization of a studied system is typically based on iterative comparison of the obtained experimental impedance data with these expectations. The expectations are typically based on previously published examples of electrochemical, physical, mechanical, and chemical analysis for similar types of systems, both in application-driven and laboratory experimentation environments.

Electronic Circuitry

Once the equivalent circuit model is validated, the electronic data acquisition system has to be designed to perform the frequency sweep and capture the data. This is often a complex and time-consuming activity, requiring significant electronic skills to optimize the electronic-circuit designs.

Impedance measurement system usually integrates an AC measurement unit such as frequency response analyzer (FRA), a potentiostat or galvanostat of suitably high bandwidth, and the electrochemical cell composed of 2, 3, or 4 electrodes in contact with an investigated sample (Fig. 6). The analyzed electrochemical interface is located between the sample and the working electrode (WE). A counter electrode (CE) is used to supply a current through the cell. Where there is a need to control the potential difference across the interface, one or two reference electrodes

(RE1 and RE2) with a constant and reproducible potential is used.

In the case of EIS measurements, the potentiostat is not only responsible for maintaining a defined DC potential level but also for applying predetermined AC voltage to the analyzed system. The DC potential and the AC perturbation are added together (V_{IN}) and applied to the electrochemical cell at the counter electrode. The voltage difference between the reference and the working electrodes is measured and fed back to the control loop, which corrects the voltage V_{IN} applied to the counter electrode and the current flowing through the working electrode until the required potential difference between the working and the reference electrodes is achieved. The voltage measured between the working and the reference electrodes V_{OUT} and the current measured at the working electrode I_{OUT} are amplified by the potentiostat and fed into the FRA as voltage signals. The voltage and current amplifiers should have a high common mode rejection and a wide bandwidth. In addition, they need to have an offset capability in order to eliminate the DC component of the analyzed signal. The potentiostat can provide DC rejection as well as compensate for the solution resistance (IR compensation) if required. The summing amplifier is used with a gain of 0.01 for the analyzing signal (this gain is necessary because digital generators have a low signal-to-noise ratio at very low level and it is better to use a 1 V signal reduced by a factor of 100 than a 0.01 V signal) [1].

For many dielectric study applications such as analysis of polymers, composite materials, and biomedical tissues, where it is not essential to maintain a DC-voltage control during the impedance measurement, a potentiostat is not required. In this case the AC measurement (dielectric) unit can be used more efficiently by itself since it usually allows more accurate measurements and the use of higher frequencies. Potentiostats, however, are useful additions to the impedance test equipment when the system under investigation has low and often also high impedance. Potentiostats have high input impedance rejecting any current flow through the reference electrode and good current sensitivity.

In the present state-of-the-art equipment, it is possible to measure and plot the electrochemical impedance automatically. Electronic circuitry is designed to generate the frequency sweep of a desired resolution over the range of interest. The generator can be programmed to sweep from a maximum to a minimum frequency in a number of required frequency steps. The commonly used modern equipment AC impedance measurement techniques can be subdivided into two main groups – single sine (lock-in amplification and frequency response analysis) and multiple sine techniques such as fast Fourier transforms (FFT).

The A/D converter must then acquire the system response to the excitation frequency. Some designs require two A/D converters to capture the excitation and response signals. This can be complex because simultaneous sampling of the A/D converters is required to allow the detection of phase changes between the signals. In addition, the overall system must remain linear. The overall bandwidth of the system must be adequate, and signal sizes must be large enough to get good measurements – but not so large that they over-range the A/D converter or any other components and cause distortion. Because the element under test often has an unknown impedance range, some trial-and-error experimentation is often initially needed to optimize the system and ensure that it remains linear.

Once converted to the digital format, the response signal is usually presented to a computer for subsequent analysis. Automatic plotting of the experimental data can be performed by computer software in different graphical representations – either as the Nyquist or Bode plots for impedance, modulus, or permittivity notations, which also allow subsequent digital processing and circuit and mechanistic parameter identification. Newer approaches perform much of the analysis on the chip (such as AD5933) by extracting the real and imaginary components of the response signal before subsequent processing on a computer. This relieves the computer of performing significant arithmetic functions and improves data collection because the analog signal processing circuitry is optimized to operate with the other functional blocks. Although computers can easily provide results extending to four or more digits, unless the measurement of the analog signals is valid and attention is given to ensuring that the overall system remains linear, the result will contain errors. Careful system design and validation to obtain valid measurement are critical to the accuracy of the result.

Impedance-Based Sensors and Diagnostics

Under normal conditions, a sensor's impedance signature is based on a combination of capacitive, inductive, and resistive component-simulated properties of the analyzed system. If a change in the system's environment causes any of these values to change, the sensor's impedance will change. By examining the sensor element over frequency, a new impedance profile will result as a consequence of this change. A relatively simple technique for doing this is to compare a measured-impedance profile with a predetermined profile.

For example, a metal-detection sensor uses Eddy currents as its operating principle. A high-frequency AC signal source is connected to a coil embedded in the sensor housing. The resulting electromagnetic field generated by the coil induces Eddy currents in a conductive target. This in turn interacts with the sensor coil and changes its impedance. Examining the impedance of the coil over frequency provides a number of benefits. For example, because the permeability of the material will affect the impedance of the coil, one can use empirical impedance signatures to draw conclusions about the type of metal the sensor is detecting. This process makes it possible for the sensor to detect metals of different permeability. Permeability change to measure stress in metals also can be employed because changing stress will change permeability, which in turn will change impedance.

Comparing measured-impedance with expected-impedance profiles can be applied to many different impedance-based sensor technologies, where resistive, capacitive, or inductive changes are provoked. Common applications include biological and pathological cells detection in clinical analysis, predicting performance aspects of chemical and biological sensors, development of gas chemical sensors, moisture detection using capacitive-based sensors, monitoring polymeric coatings degradation, corrosion analysis of metal structures, optimization of fuel-cell performance, prediction of battery health, and characterization of electrochemical performance of a material.

Impedance analysis can involve more than simply comparing an impedance response to an expected profile. EIS is also commonly used to characterize complex physical, chemical, and mechanical systems and to extract valuable information about them. This includes solid–solid (in the case of many chemical sensors) or solid–liquid (when examining concentration of a species in a liquid) interfaces. EIS takes advantage of the fact that a small voltage potential applied to the interface will polarize the interface. The manner in which the interface polarizes, combined with the rate at which it changes when the applied potential is reversed, characterizes the interface. This allows extracting information about the system interface, such as adsorption/reaction rate constants, diffusion coefficients, and capacitance. And you can estimate information on the dielectric constant, conductivity, mobility of charge equilibrium, constituent concentrations, and bulk generation/recombination rates of the element (i.e., the sensor or substance under investigation). Some of these impedance applications, such as analysis of electrical conduction mechanisms in bulk polymers and biological cells suspensions, have been actively practiced since the 1950s [8, 9]. Others, such as localized studies of surface corrosion kinetics and analysis of the state of biomedical implants, have come into prominence only relatively recently [10–12].

Impedance Applications: Biosensors and Clinical Diagnostic Devices

Very often electrochemical-based methods are being developed as higher quality substitutes for fluorescent staining, magnetic counting, microdialysis, plate-culture techniques, and other clinical laboratory methods. For example, electrochemical biosensors offer a number of advantages over optical, ultrasonic, magnetic, and other diagnostic principles employed in clinical and biomedical settings. The increased interest in application of electrochemical technology as a basis for point-of-care biomedical diagnostic devices and sensors arises from high sensitivity, selectivity, picomolar detection limit, temporal and spatial resolution, rapid response, simplicity of rapid screening procedures, label-free noninvasive sensing, cost effectiveness, versatility, flexibility of design, easiness of integration, compatibility with microfabrication technology, high throughput screening, and ultimately an ability (either real of potential depending on biocompatibility) to perform in vivo and respond adequately to the dynamic nature of living systems [13]. Small size of electrochemical devices allows them to be used in microfluidic products and sensor arrays when simultaneous detection of several analytes present in low-volume sample is required. One of the examples of in vivo application of electrochemical technology is the use of fast scan voltammetry with implanted microelectrodes, which produced a unique fingerprint for dopamine with excellent selectivity and sensitivity [14]. Microelectrode arrays can often reliably record neural activity for several months after the implantation.

Since the pioneering works of Schwan in 1950s [15], the foundation was laid for the impedance analysis and interpretation of biological cells dispersions. Since then many electrochemical researchers have characterized biological colloidal suspensions [16–19], developed "Coulter" counters and capacitive cytometers for bioparticles detection [20–22], designed enzyme-based biosensors for glucose monitoring [23, 24], and practiced electrophoretic and dielectrophoretic separations of drugs, proteins, cells, DNA, and pathogenic bacteria [25–28]. Impedance spectroscopy has been used to study biomedical and pathogenic cell cultures, which is extremely useful for both medical diagnostics of many major clinical complications and early detection and prevention of infectious diseases.

The field of electrochemical biosensing has also become one of the most important methods in the detection of bacterial and blood cells. For instance, the need for pathogen detection nowadays arises in areas as diverse as food industry, water or sludge treatment, or even national security. Prevention and early detection leading to prompt treatment are essential since minor infections can rapidly turn life threatening. Diagnostic systems therefore play an integral role in facilitating an effective response to be provided against these infections. In most of these cases, the electrochemical technique was used to record changes in impedance or capacitance near the electrode surface. Overall the contribution of electrochemical techniques to the detection of clinical and pathogenic cells is mainly present in the detection step. However, that contribution not only involves the generation of an electrochemical signal but is also utilized in steps such as immobilization of bacteria on the electrode surface and cell lyses. In an attempt to deliver higher selectivity and specificity, and speed up the detection process, either the sensor surface has been immunomodified to help the capturing of cells or dielectrophoretic trapping or manipulation steps have been used to preconcentrate the target particles [26–30].

In an aqueous solution, charge is carried between any two electrodes. The presence of particulate matter (such as blood or bacterial cells) physically obstructs the movement of these charge-carrying ions and thereby leads to higher impedance between the electrodes. Typical bacterial cells ($\sim$1,000 cells/ml) have a volume fraction of $\sim 10^{-12}$, and consequently the corresponding impedance change is not significant. It may, however, be discernable if the low-volume suspension is made to pass between the two electrodes separated by a narrow slit, only slightly larger than the size of a cell, or if the target cells can be made to adhere to or congregate at a surface of the electrodes. Concentrations of 10^5 cells/ml can be detected using the impedance technique with a detection time of 2–3 h [29, 30]. However, the abovementioned electrical detection techniques, such as the "Coulter" counter, suffer from low throughput and slit clogging.

The impedance measurements on biosensor can provide information about type, spreading, attachment, and morphology of the cultured cells. Typically as membranes are insulative, the "adsorption-type" capacitive response is developing at the interfaces of the electrodes in the absence of significant externally applied electrochemical overpotential ΔV. The main effect of cells on the sensor signal is due to the insulating property of the cell membrane. The presence of intact cell membranes on the electrodes and their distance to the electrodes determine the current flow and thus the sensor signal. If cells grow directly on an electrode, this effectively reduces the electrode area reached by the solution ionic current and the interfacial impedance increases. The aqueous gap between the cell membrane and the substrate prevents direct influence of the cell membrane capacity on the interface impedance of the electrodes. Nevertheless, the increase in the interfacial impedance should be expected. Effects of unspecific protein adsorption from the medium should be small compared to a closely packed lipid bilayer. The difference between dead and alive cells was observed using the interfacial impedance changes [31, 32].

Long-term stability, biocompatibility, and selectivity in complex biochemical environments are often an issue with implantable electrochemical devices. Biomedical industry is continuously searching for materials and devices that can maintain their functionality for prolonged periods of time (days to years) after being implanted. Optimization and performance improvement of electrochemical sensors are usually based on development of modified surfaces, often utilizing enzyme membranes combined with electropolymerization [33]. Biosensor typically contains a biologically sensitive element, such as protein or peptide, and is being placed in contact with a tissue. Biofouling, adsorption of biomolecules, and temperature instability of enzyme-based membranes contribute to decrease in biomedical device performance [34]. In order for biosensors and other implantable medical devices to function properly, the mutual interactions of the device and the surrounding tissue must not influence the performance of the device.

For all in vivo measurements, the implanted device perturbs the environment and initiates inflammatory response of tissue resulting in encapsulation of the implant [35]. The acute inflammatory response starts immediately after the implantation, with fluid carrying plasma proteins, and inflammatory cells migrate to the site of the implant and adsorb at the implant interface. These cellular events result in the development of a compact sheath of cells and accumulations of extracellular protein matrix material surrounding the implant. The tissue damage following the implantation may result in alteration of functionalities of the damaged cells in plasma and changes in local concentration of the analytes (such as glucose or oxygen) in the vicinity of the implant resulting from a wound-healing process. The extracellular environment around implanted electrodes changes due to insertion-related damage and sustained response promoted by the presence of the device. This tissue encapsulation can cause changes in the electrical properties of the tissue adjacent to an implant. This encapsulation tissue was found to typically have a higher resistance than normal tissue. In case of biosensors, this tissue response often leads to a significant modification of the sensor functionality making its response difficult or impossible to interpret and in the case on medical implants can lead to implant performance degradation due to corrosion, surface modification, and tissue modification/damage around the implant. These issues are particularly felt in biosensor industry where reliable implantable glucose monitoring sensor with a lifetime even on the order of days is yet to be developed [24]. The attempts to reduce biofouling have been primarily concentrated on development of specialized coatings (such as hydrogels, Nafion^{TM}, polyethylene oxide/polyethylene glycol) on the outermost membrane surface that inhibits protein adhesion to the surface.

Considering these mechanisms of interaction between an electrochemical device and a tissue, it is necessary to investigate overall tissue and tissue-implant interfacial impedance responses for several reasons. Firstly, studying interfacial impedance response allows determining the dynamics of biofouling which is usually related to the proteins and cell adsorption kinetics at foreign body surfaces of sensors and implants and the tissue response change. This is important in studies of the implantable in vivo sensors and device performance, determination of their reliability, and studies of their interactions with the tissue. Considering that, for instance, glucose and other enzyme-based sensors' performance is based on multilayered membrane including an additional biocompatible protection layer, the resulting structure presents itself to an impedance analysis as a complicated electrochemical surface kinetic system. This system may combine electron transfer through several mediating enzyme layers and external protection layer, porosity-modified electron and mass transport, and the effects of protein adsorption on bare metallic surfaces of the medical implants or over the sensing surfaces of the sensors. Secondly, investigations of bulk solution and interfacial impedance responses open a window into an interesting study of cellular colloidal suspensions containing components that are quite useful as biomarkers for clinical diagnostics of such diseases as cancer, diabetes, and thrombosis. These studies are also essential for establishing technologies for preconcentrating of pathogenic bacteria and cellular biomarkers and development of portable field devices capable of their rapid detection at low concentrations.

Electrochemical impedance analysis, due to its AC frequency-dependent nature, allows combining in a single measurement the interfacial studies of membrane-mediated electron and mass-transport kinetics, protein and cell surface adsorption, processes at the implant surface (such as corrosion and surface deactivation), and bulk solution "dielectric" studies of biological and bacterial cells colloidal suspensions. EIS capability to study temporary and spatial details of bulk media and interfacial kinetics in real time – either as a method of solution species characterization, biofouling monitoring, or membranes degradation – is an important advantage in in vivo clinical applications. Application of EIS allows characterization of different types of biological cells, microparticles, proteins, and bacteria by estimating their concentrations, sizes, diffusion coefficients, and chemical changes. The 2-electrode, 3-electrode, and 4-electrode [25, 36–38] methods with reference electrodes to measure and excluded nonspecific changes in the test module have been routinely used for impedance characterization of biological membranes, adsorption, and colloidal phenomena. Impedance sensors have an advantage of being able to be effectively packaged with the dielectrophoretic and electrophoretic electrodes in a single device combining biomarker preconcentration and detection stages. The reference module can serve as a control for temperature changes, evaporation, changes in amounts of dissolved gases, and degradation of culture medium during incubation. Construction of realistic model of the analyzed system is important for extraction of relevant cellular characteristics such as concentration, diffusion coefficients, sizes, and membrane properties. Impedance monitoring of protein adsorption is essential for qualification of embedded sensors and other implantable devices.

Traditionally electrode impedance has been utilized for testing of potency of the electrode insulation after implantation. This was done by measuring the magnitude and phase at a single frequency of $\sim$1 kHz, mainly because this is the fundamental frequency of a neuronal action potential, corresponding to a time period of 1 ms. It was believed that this would be the frequency component of the electrode impedance that would primarily affect the recorded signal from surrounding neurons. Due to the fact that most traditional neurophysiology electrode impedance monitoring equipment is set up to measure the impedance magnitude at 1 kHz only, it would be useful to be able to correlate 1 kHz impedance to the state of tissue around an implanted electrode. Typically these types of implant monitoring demonstrate a general increase in the measured-impedance magnitude at 1 kHz that reaches a peak several days post-implant. Histological analyses suggest that tissue reactions are confined to localized regions of the implants around the monitoring electrodes [34].

The broadband impedance spectroscopy data was used to develop a more comprehensive model for a typical impedance response of an implanted biosensor (Fig. 7). The media around an implant consists of biological cells and adsorbing proteins. A three-component model consisting of an electrode–tissue interface, an encapsulation (protein adsorption) region, and a neural cellular tissue component is sufficient to represent an implant–tissue system. The impedance profiling of the system reveals several semicircular arcs in the Nyquist space, typical of a distributed network of lumped parallel resistive and capacitive pathways. Due to the commutative property of electrical components in series, models based upon impedance measurements cannot be used to explicitly determine spatial distribution, although when paired with histological observations, spatial distributions of model components may be inferred.

The electrode-tissue interface of this model simplified as $Z_{INTERFACE}$ has been well established. It consists of a capacitive element C_{DL} representing double-layer interfacial charging, in parallel with Faradaic component representing a poorly defined charge transfer resistance (R_{CT}) and diffusion impedance Z_{DIFF} [39]. The $Z_{INTERFACE}$ component reflects current distribution due to variability in length of conducting passes between the electrode surface and electroactive components in the tissue, electrode surface roughness, and porosity effects. Diffusion impedance to mass transport of discharging species to the interface appears at the lowest frequencies. Additionally, a shunt capacitance or inductance (C_{SH}), apparent only at very high frequencies, may arise from the finite capacitive coupling between the electrode lead wires and the surrounding electrolyte solution. In experiments using microwire arrays, the uncompensated impedance of the cables is only apparent at frequencies exceeding 1 MHz. C_{SH} will have a more substantial effect when using thin-film electrodes, which have relatively thin dielectric layers and close lead spacing. These interfacial impedance kinetics and cell geometry-related values can be experimentally determined based upon pre-implant impedance tests in physiological buffer solution.

As the current propagates toward ground in vivo, it would next encounter the protein-rich encapsulation (adsorption) layer. This medium frequency impedance region is composed primarily of extracellular matrix proteins and is considered to be largely resistive [40–43]. This zone can be modeled as a parallel combination of proteins adsorption resistance (R_{ADS}) and capacitance (C_{ADS}) but also can be simplified as encapsulation resistance $R_{EN} = R_{ADS}$.

Subsequently, at higher frequencies the current would propagate outward and come across the bulk media that can be represented by a parallel combination of a resistive extracellular fluid and cell membranes. The "bulk" solution current at high frequencies can be conducted along two pathways, either around or between cells in extracellular media or through and across cells (intracellular). The extracellular media is assumed to be principally resistive and is represented by a purely resistive element (R_{EX}). At low cell densities, there are sufficient numbers of low-resistance pathways for current flow to follow, such that the capacitance contribution is negligible. As the cell density increases, the extracellular space decreases, thus diminishing the low-resistance pathway. The current flows through capacitive (membrane) and resistive (cytoplasm) cellular compartments, presenting an opportunity for characterization of cellular biomarkers. An impedance of this pathway is simplified by impedance Z_{CELL}, which is characterized by a membrane conductance (G_{MBR}) and capacitance (C_{MBR}) scaled by the total cell membrane area (A_{MBR}). The conductance and capacitance are often constant parameters drawn from the literature [39]. Qualitatively, the impedance spectral signatures match those reported for chronic implantation of various types of electrodes in vitro [44–47].

The impact of the surrounding tissue on the impedance implant measurement is heavily influenced by the volume immediately surrounding the electrode (primarily proteins' adsorption), as the voltage drops off in all directions as it spreads through the tissue volume to the ground. Therefore, the influence of reactive cells and proteins will have a more significant effect as the cell density immediately adjacent to the electrode surface increases and clear impedance spectral

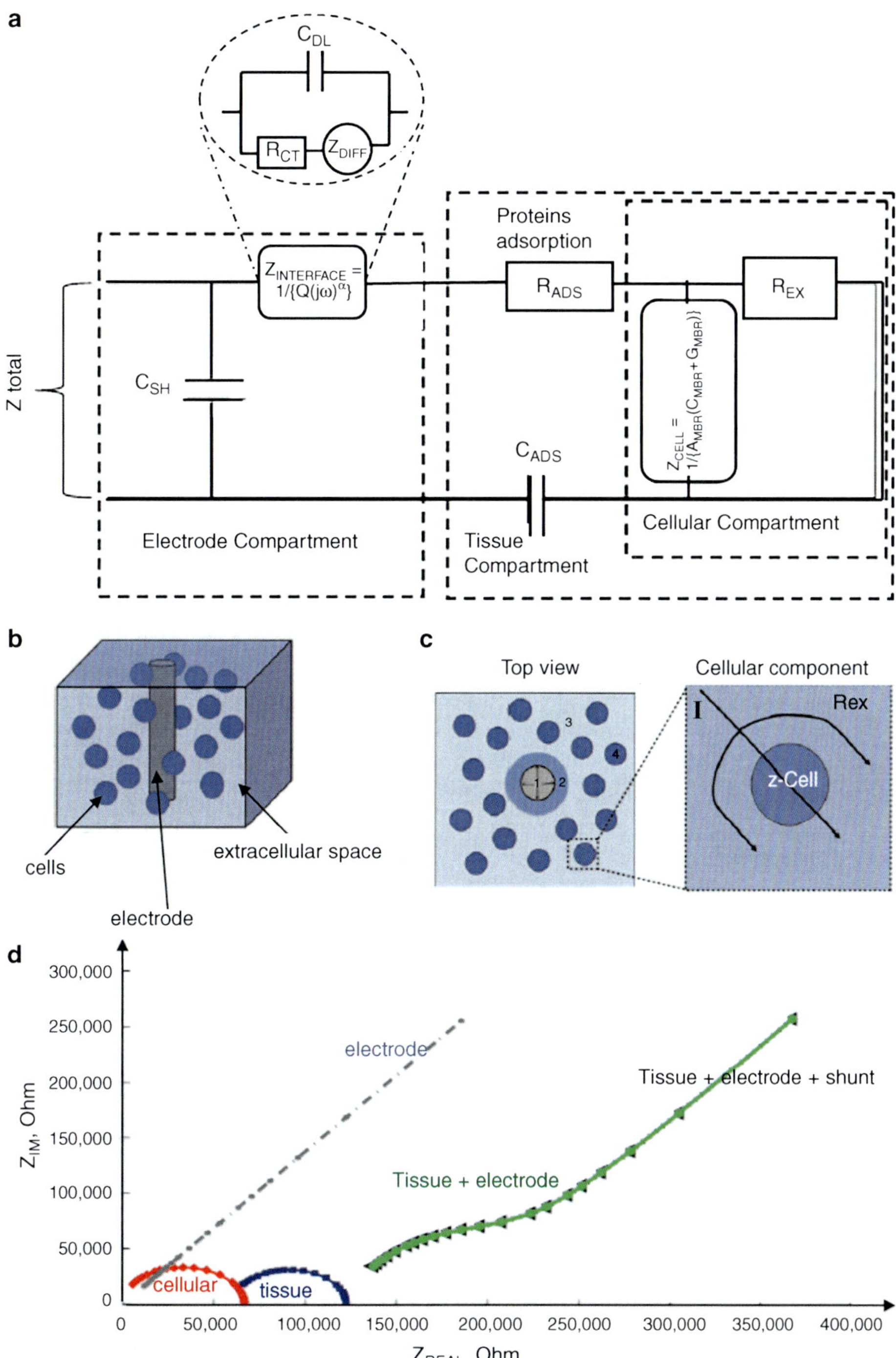

Electrochemical Impedance Spectroscopy (EIS) Applications to Sensors and Diagnostics, Fig. 7 Model circuit diagram and physical interpretation for device implanted in living tissue. (**a**) Circuit model representing the impedance variation commonly found during in vivo impedance spectroscopy.; (**b**) 3-dimensional and (**c**) Two-dimensional rendering of theorized physical components of the model. (**d**) Nyquist representation of expected model results from a typical in vivo impedance spectrum (*Partially reproduced from* [34] *with permission from IOPP*)

signatures arise in cases where there is significant accumulation of dense reactive cells and proteins in the immediate volume adjacent to the electrode. The model also implied the infiltration of reactive cells and increased cellular density in close proximity to the electrode site. These results suggest that changes in impedance spectra are directly influenced by cellular distributions around implanted electrodes over time and that impedance measurements may provide an online assessment of cellular reactions to implanted devices [34].

In this model, one of the fundamental assumptions made is that the impedance of the electrodes does not change significantly over time. Alternatively, it could be proposed that the interaction of cells and tissue with the electrode surface fundamentally alters the properties of the electrode–surface interactions. In the literature, models have been proposed which suggest that the interaction between electrodes and cells could possibly affect the properties of both the electrode and the cell membrane [48–50]. There are observations that support these assumptions, namely, that the change in the electrode impedance in saline from pre-implantation measurements does not change significantly compared to the changes in the impedance spectra observed over the implant duration. This is also supported by experimental studies in culture that also suggest that intimate cell–electrode interactions did not significantly alter electrode parameters [38–40]. While this does not necessarily imply that interactions between the electrode tissue and the adjacent tissue do not alter the fundamental properties of the electrode characteristics, it appears that the changes in the electrical properties of the surrounding volume and frequent formation of scar tissue contribute more significantly to the observed changes in impedance over time. What is not often understood are the long-term effects of implantation on the electrode state. Further experiments and a better understanding of the mechanisms involved in electrode–tissue interactions will be useful for future electrode–tissue model development and refinement.

Future Directions of Impedance Technology

In a broad sense, EIS is an extraordinarily versatile, sensitive, and informative technique broadly applicable to studies of electrochemical kinetics at the electrode-media interfaces and determination of conduction mechanisms in various materials through bound or mobile electronic, ionic, semiconducting, and mixed charges. Impedance analysis is fundamentally based on a relatively simple electrical measurement that can be automated and remotely controlled. In contrast to many other analytical techniques, EIS is essentially a noninvasive technique that can be used for investigating and monitoring of online processes. The method offers the most powerful online and off-line analysis of the status of investigated media, electrodes, and probes in many different complex time – and space-resolved processes that occur in electrochemical laboratory experiments or over a lifetime of monitored samples, devices, or materials. For instance, EIS technique has been broadly practiced in the development of sensors for monitoring rates of materials' degradation, such as metals' corrosion and biofouling of implantable medical devices.

EIS is useful as an empirical quality control procedure that can also be employed to interpret fundamental electrochemical and electronic processes. Experimental impedance results can be correlated with many practically useful chemical, physical, mechanical, and electrical variables. The main strength of the method lies in its ability to interrogate a variety of relaxation phenomena with time constants ranging over several orders of magnitude from minutes to microseconds within a time frame of a single measurement.

During the most recent decades, the EIS technique advances have been closely linked to those of computer science and electrical engineering, enabling the great progress achieved in the development of commercial EIS instrumentation. That enabled further advancements for the method – both in expanding to an even broader universe of emerging fields of use and in improvements in understanding of the method's underlying principles, best experimental practices, and data interpretation in multitude of practical applications.

EIS changed the ways electrochemists interpret the electrode–solution interface. With impedance analysis, a complete description of an electrochemical system can be achieved using equivalent circuits as the data contains all necessary electrochemical information. The technique offers the most powerful analysis on the status of electrodes, monitors, and probes in many different processes that occur during electrochemical experiments, such as adsorption, charge and mass transport, and homogeneous reactions. EIS offers huge experimental efficiency, and the results that can be interpreted in terms of Linear Systems Theory, modeled as equivalent circuits, and checked for discrepancies by the Kramers–Kronig transformations [1].

EIS is playing an absolutely critical role in solving the great challenges of the twenty-first century. Among others, health-care advancements through therapies and biomedical devices, developing new renewable sustainable sources of energy such as batteries and fuel cells, and monitoring of mechanical stability of structures from as large as bridges and refineries to as small as interdigitated implantable sensors, all rely on the impedance-based devices. The future of the impedance method lays in its applicability to studies of almost unlimited number of experimental processes and applied systems. These EIS applications are directly addressing many industries of paramount worldwide importance such as corrosion and anodic behavior of metals and composite materials, states of electrodes during charging/discharging cycles of batteries and fuel cells, surface characterization of polymer-modified structures, biomedical implants, and sensors. EIS applicability to fast real time in situ measurements, and utilization of multiple hyphenated techniques such as combined spectroelectrochemical/gravimetric/impedance measurements during electrochemical experiments, allows for continuous development of impedance-based technologies and products. Incredible flexibility of the method, its ability to be combined with other techniques for in situ and ex situ analysis, fueled by the continuous scientific instrumentation advancements present a significant opportunity to resolve the remaining interpretation challenges of electrochemical impedance spectroscopy analysis. Developing a better understanding and limiting the effects of the most significant EIS limitations, such as the data interpretation ambiguities and pattern recognition problems, will result in better interpretation of the EIS response related to studies of physical, chemical, mechanical, and electrical properties of the studied experimental systems [2].

Cross-References

- ▶ Biomedical Applications of Electrochemistry, Use of Electric Fields in Cancer Therapy
- ▶ Biomolecules in Electric Fields
- ▶ Biosensors, Electrochemical
- ▶ Electrocatalytic Synthesis
- ▶ Environmental Energy Technologies
- ▶ Fuel Cells, Principles and Thermodynamics
- ▶ Fuel Cell Vehicles
- ▶ Green Electrochemistry
- ▶ Primary Batteries, Selection and Application
- ▶ Wastewater Treatment, Electrochemical Design Concepts

References

1. Orazem ME, Tribollet B (2008) Electrochemical impedance spectroscopy. Wiley, Hoboken
2. Lvovich VF (2012) Impedance spectroscopy: applications to electrochemical and dielectric phenomena. Wiley, Hoboken
3. Bard AJ, Faulkner LR (2001) Electrochemical methods, fundamentals and applications. Wiley, New York
4. Barsukov E, MacDonald JR (2005) Impedance spectroscopy. Wiley, Hoboken
5. Gabrielli C (1988) Identification of electrochemical processes by frequency responseanalysis, Solartron analytical technical report 004/83, pp 1–119
6. Yoo J-S, Park S-M (2000) An electrochemical impedance measurement technique employing Fourier transform. Anal Chem 72:2035–2041
7. Park S-M, Yoo J-S (2003) Electrochemical impedance spectroscopy for better electrochemical measurements. Anal Chem 21:455A–461A
8. Grimnes AS, Martinsen OG (2000) Bioimpedance and bioelectricity basics. Academic, London
9. Asami K (1995) *Evaluation of colloids by* dielectric spectroscopy, HP Application Note 380–3:1–20

10. Krause S (2001) Impedance method. In: Bard AJ (ed) Encyclopedia of electrochemistry, vol 3. Wiley-VCH, Weinheim
11. Lasia A (1999) Electrochemical impedance spectroscopy and its applications. In: Conway BE, Bockris J, White R (eds) Modern aspects of electrochemistry, vol 32. Kluwer Academic/Plenum, New York, pp 143–248
12. Conway BE (1999) Electrochemical supercapacitors. Kluwer Academic, New York
13. Wilson GS, Gifford R (2005) Biosensors for real-time in vivo measurements. Biosens Bioelectron 20:2388–2403
14. Wightman RM, Runnels P, Troyer K (1999) Analysis of chemical dynamics in microenvironments. Anal Chim Acta 400(1):5–12
15. Bothwell TP, Schwan HP (1956) Studies of deionization and impedance spectroscopy for blood analyzer. Nature 178(4527):265–266
16. Jones T (2003) Basic theory of dielectrophoresis and electrorotation. IEEE Eng Med Biol 6:33–42
17. Gimsa J *Characterization of particles and biological cells by* AC electrokinetics, Interface. Electrokinetics and Electrophoresis 13:369–400
18. Asami K, Yonezawa T, Wakamatsu H, Koyanagi N (1996) Dielectric spectroscopy of biological cells. Bioelectroch Bioelectr 40:141–145
19. Hanai T (1968) Electrical properties of emulsions, chapter 5. In: Sherman P (ed) Emulsion science. Academic, London, pp 467–470
20. Sohn LL, Saleh OA, Facer GR, Beavis AJ, Allan RS, Notterman DA (2007) Capacitance cytometry: measuring biological cells one by one. Proc Nat Acad Sci 97(20):10687–10690
21. Gagnon Z, Gordon J, Sengupta S, Chang H-C (2008) Bovine red blood cell starvation age discrimination through a glutaraldehydeamplified dielectrophoretic approach with buffer selection and membrane cross-linking. Electrophoresis 29: 2272–2279
22. Qiu Y, Liao R, Zhang X (2009) Impedance-based monitoring of ongoing cardiomyocyte death induced by tumor necrosis factor-A. Biophys J 96:1985–1991
23. Caduff A, Dewarrat F, Talary M, Stalder G, Heinemann L, Yu F (2006) Non-invasive glucose monitoring in patients with diabetes: a novel system based on impedance spectroscopy. Biosens Bioelectron 22(5):598–604
24. Chia CW, Saudek CD (2004) Glucose sensors: toward closed loop insulin delivery. Endocrinol Metab Clin North Am 33:175–195
25. Brett MAC, Brett AMO (1993) Electrochemistry: principles, methods and applications. Oxford University Press, New York
26. Varshney M, Yang LJ, Su XL, Li YB (2005) Magnetic nanoparticle-antibody conjugates for the separation of *Escherichia Coli* O157: H7in ground beef. J Food Prot 68:1804–1812
27. Varshney M, Li YB, Srinivasan B, Tung S (2007) A label-free, microfluidics and interdigitated array microelectrode-based impedance biosensor in combination with nanoparticles immunoseparation for detection of *Escherichia coli* O157:H7 in food samples. Sensor Actuat B Chem 128:99–107
28. Boyaci IH, Aguilar ZP, Hossain M, Halsall HB, Seliskar CJ, Heineman WR (2005) Amperometric determination of live *Escherichia coli* using antibody-coated paramagnetic beads. Anal Bioanal Chem 382:1234–1241
29. Yang L, Bashir R (2008) Electrical/electrochemical impedance for rapid detection of foodborne pathogenic bacteria. Biotechnol Adv 26:135–150
30. Ivnitski D, Abdel-Hamid I, Atanasov P, Wilkins E (1999) Review – biosensors for detection of pathogenic bacteria. Biosens Bioelectron 14: 599–624
31. Schwan HP, Morowitz HJ (1962) Electrical properties of the membranes of the pleuropneumonialike organism A 5969. Biophys J 2:395–407
32. Schwan HP, Takashima S, Miyamoto VK, Stoeckenius W (1970) Electrical properties of phospholipid vesicles. Biophys J 10:1102–1119
33. Lvovich VF, Scheeline A (1997) Amperometric sensors for simultaneous superoxide and hydrogen peroxide detection. Anal Chem 69:454–462
34. Williams JC, Hippensteel JA, Dilgen J, Shain W, Kipke DR (2007) Complex impedance spectroscopy for monitoring tissue responses to inserted neural implants. J Neural Eng 4:410–423
35. Pethig R (1979) Dielectric and electronic properties of biological materials. Wiley, New York
36. Yang M, Zhang X (2007) A novel impedance assay for cardiac myocyte hypertrophy sensing. Sens Actuat A 136:504–509
37. Ehret R, Baumann W, Brischwein M, Schwinde A, Stegbauer K, Wolf B (1997) Monitoring of cellular behaviour impedance measurements on byinterdigitated electrode structures. Biosens Bioelectron 12(1):29–41
38. DeSilva M, Zhang Y, Hesketh PJ, Maclay GJ, Gendel SM, Stetter JR (1995) Impedance based sensing of the specific binding reaction between Staphylococcus enterotoxin B and its antibody on an ultra-thin platinum film. Biosens Bioelectron 10:675–682
39. Houssin T, Follet J, Follet A, Dei-Cas E, Senez V (2010) Label-free analysis of water-polluting parasite by electrochemical impedance spectroscopy. Biosens Bioelectron 25(5):1122–1129
40. McAdams ET, Josinette J (1995) Tissue impedance: a historical overview. Physiol Meas 16:A1–A13
41. Buitenweg JR, Rutten WLC, Willems WPA, Van Nieuwkasteele JW (1998) Measurement of sealing resistance of cell-electrode interfaces in neuronal cultures using impedance spectroscopy. Med Biol Eng Comput 36:630–637

42. Stensaas SS, Stensaas LJ (1978) Histopathological evaluation of materials implanted in the cerebral cortex. Acta Neuropathol 41:145–155
43. Edell DJ, Toi VV, McNeil VM, Clark LD (1992) Factors influencing the biocompatibility of insertable silicon microshafts in cerebral cortex. IEEE Trans Biomed Eng 39:635–643
44. Turner JN, Shain W, Szarowski DH, Andersen M, Martins S, Isaacson M, Craighead H (1999) Cerebral astrocyte response to micromachined silicon implants. Exp Neurol 156:33–49
45. Grill WM, Mortimer JT (1994) Electrical properties of implant encapsulation tissue. Ann Biomed Eng 22:23–33
46. Johnson MD, Otto KJ, Kipke DR (2005) Repeated voltage biasing improves unit recordings by reducing resistive tissue impedances. IEEE Trans Neural Syst Rehabil Eng 13:160–165
47. Johnson MD, Otto KJ, Williams JC, Kipke DR (2004) Bias voltage at microelectrodes change neural interface properties in vivo. In: Proceedings of 26th annual international conference IEEE EMBS(September 2004) San Francisco, CA
48. Kyle AH, Chan CT, Minchinton AL (1999) Characterization of three-dimensional tissue cultures using electrical impedance spectroscopy. Biophys J 76: 2640–2648
49. Liu X, McCreery DB, Carter RR, Bullara LA, Yuen TG, Agnew WF (1999) Stability of the interface between neural tissue and chronically implanted intracortical microelectrodes. IEEE Trans Rehabil Eng 7:315–326
50. Grattarola M, Martinoia S (1993) Modeling the neuron-microtransducer junction: from extracellular to patch recording. IEEE Trans Biomed Eng 40–41:35–41

Electrochemical Mass Spectrometry

Helmut Baltruschat and Abd El Aziz Abd-El-Latif
Institute of Physical and Theoretical Chemistry, University of Bonn, Bonn, Germany

For electrochemical studies, differential electrochemical mass spectrometry (DEMS) has become an indispensable tool not only for the qualitative and quantitative detection of volatile products or intermediates of continuous faradaic reactions, but also for determination of the amount of adsorbates (sub- or monolayer) at different electrode surfaces by means of their desorption.

In 1971, Bruckenstein and Gadde [1] were the first to qualitatively detect the electrochemically generated gaseous products using a hydrophobic porous electrode by in situ electrochemical mass spectrometry (EMS) with a time constant of about 20 s. In order to use EMS for quantitative studies (current efficiency and kinetic information), Wolter and Heitbaum [2, 3] improved the vacuum system of the EMS and reduced the delay time of detection. Thus, the mass spectrometric signal of a species became proportional to its entrance rate into the vacuum system.

In a typical DEMS experiment, the ion current corresponding to a given species of interest is recorded in parallel to the faradaic electrode current during the potential sweep (cyclic voltammogram), yielding the so-called mass spectrometric voltammograms (MSCV). Other electrochemical techniques, such as potentiostatic [4, 5] or galvanostatic ones [6] and even pulsed voltammetry for short time (1 s), have also been combined with DEMS.

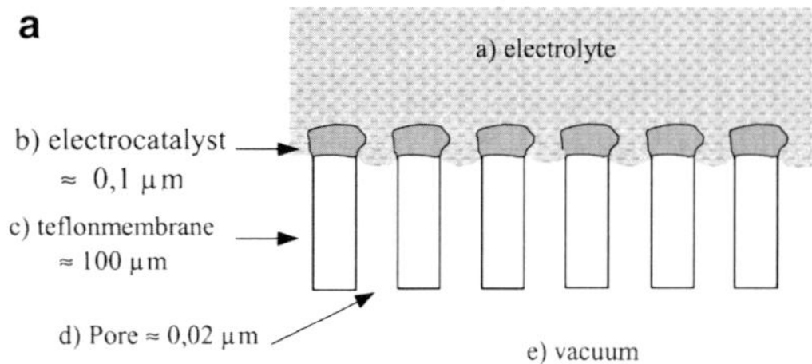

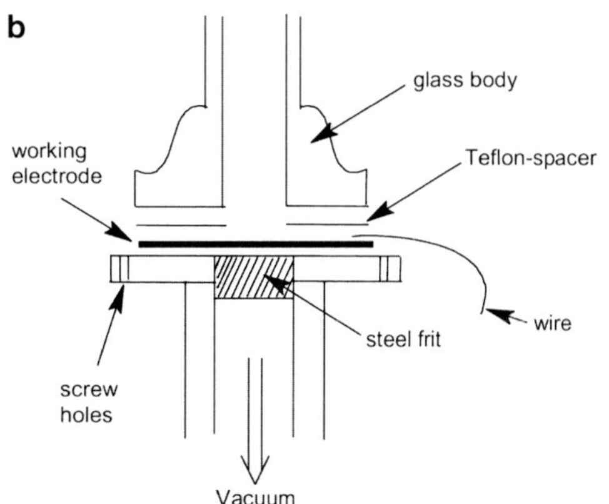

Electrochemical Mass Spectrometry, Fig. 1 Schematic representation of a (*top*: **a**) sputtered membrane electrode and (*bottom*: **b**) conventional cell for DEMS

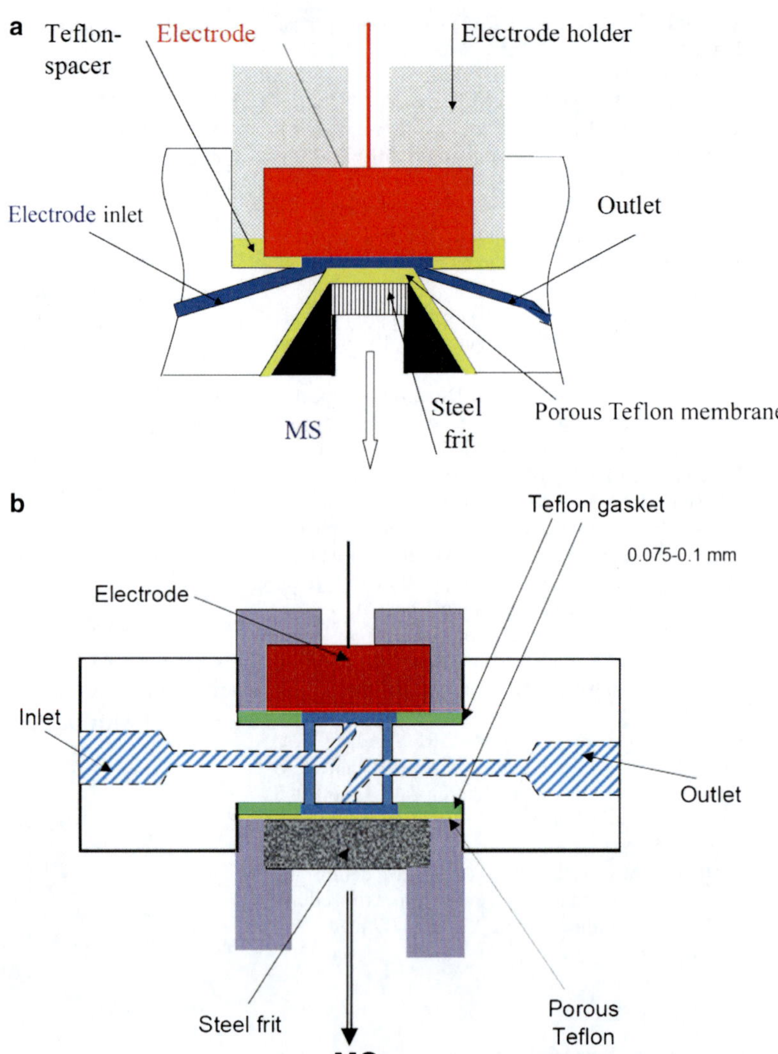

Electrochemical Mass Spectrometry, Fig. 2 (*Top*: **a**) Thin-layer flow cell. (*bottom*: **b**) Dual thin-layer flow through cell

Types of DEMS Cells

In order to detect species which are produced at an electrode surface by mass spectrometry, they have to be transferred from the electrolyte phase to vacuum. Fortunately, for aqueous systems [2] and also for some organic electrolytes with a high surface tension, e.g., propylene carbonate [7], the separation of the electrolyte from the vacuum can be achieved by using porous Teflon® membranes. Due to their hydrophobicity, the liquid does not penetrate into the pores, whereas dissolved gaseous and other volatile species readily evaporate in them. The critical pore size depends on the surface tension of the liquid and the contact angle between the liquid and Teflon. For water it is $r < 0.8$ μm [2]. A typical Teflon membrane (e.g., Gore-Tex) is 75 μm thick and has a nominal pore width of 20 nm with a porosity of 50 %. The Teflon membrane is usually supported by a glass or steel frit. Different types of working electrodes such as lacquer, sputter-deposited, supported nanoparticles, smooth and massive ones were used for DEMS. Several cell types have been described as reviewed in detail in Ref. [8, 9] and references therein.

In the conventional cell, the electrocatalyst layer, e.g., Pt, is deposited onto the Teflon

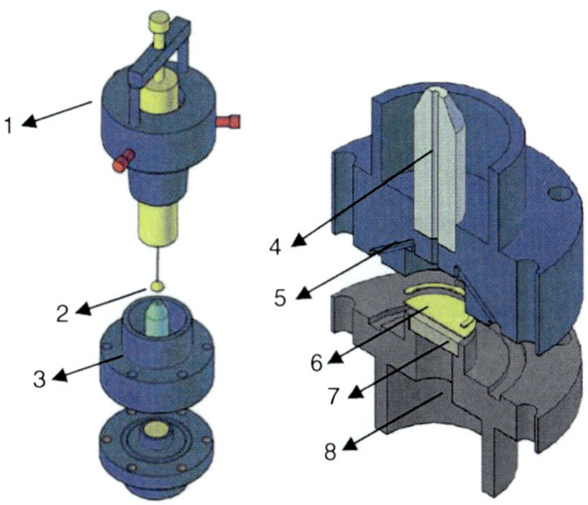

Electrochemical Mass Spectrometry, Fig. 3 Sketch for bead crystal flow through cell. *1* 3D crystal holder, *2* bead crystal, *3* Kel-F support, *4* glass capillary, *5* six-outlet capillaries, *6* Teflon membrane, *7* steel frit, *8* stainless steel connection to MS

membrane with a typical thickness of 50–80 nm (Fig. 1a). A typical cell for these electrodes is shown in Fig. 1b. The response time for an electrochemical experiment was determined to be about 0.1 s [10]. When using volatile reactants (such as the often studied CO), it must be kept in mind that the electrochemical reaction and evaporation are competing processes. A rotating inlet system (similar to the rotating disc electrode) is described in [11].

For the use of massive electrodes for DEMS, e.g., single-crystal electrodes, the thin-layer cells of Fig. 2 were developed [12, 13]. In Fig. 2a, volatile species produced at the massive electrode (with a diameter of 1 cm) diffuse through a 50–100 μm thick electrolyte layer within 2 s to reach the Teflon membrane.

Two capillaries serve as electrolyte inlet and outlet and as a connection to the reference and counter electrodes. This cell is well suited for desorption experiments under stagnant conditions. During a continuous flow of electrolyte, however, the considerable part of the product formed close to the outlet of the cell is transported out of the thin-layer volume before it can reach the Teflon membrane. A modification of a thin-layer cell is described in Ref. [14].

In the dual thin-layer flow cell (Fig. 2b), the electrochemical compartment is separated from a mass spectrometric compartment [8, 15]. The product species are transferred from the upper compartment to the lower compartment through six capillaries (d = 0.5 mm) by constant flow of the electrolyte. With its small electrolyte volume (≈3.5 μL), this cell is well suited for continuous faradaic reactions and also for the detection of adsorbates after their desorption, as the product concentration of desorbing species and thus the sensitivity for their detection are high. This flow cell can also be combined with a quartz crystal microbalance [15] or IR spectroscopy. To do so, the working electrode is replaced by either a quartz or by a prism with a thin layer of the electrode material for ATR (attenuated total reflection) [5, 16, 17]. Furthermore, the Teflon membrane in the bottom compartment may be covered by Au or Pt, thus serving simultaneously as a detection electrode like the ring in a rotating ring-disc assembly.

In another approach, Kita and coworkers [18] used a hanging meniscus configuration for massive electrodes, using a pinhole as the gas inlet, located at the hemispherical end of a glass tube, which is covered by a Teflon film. Similarly, Koper et al. [19] placed a small Teflon tip as inlet close to a bead single-crystal electrode in a hanging meniscus arrangement. This setup has a long delay time of 10–15 s and does not allow to work under convection.

A new approach combines the advantage of defined convection of a dual thin-layer cell with the possibility to use small bead crystals as shown in Fig. 3 [20]. A cone-shaped capillary is placed in the usual hanging meniscus very close to the bead electrode with a gap of about 200 μm. The electrolyte is sucked continuously through the gap and the capillary to the mass spectrometric detection compartment, and fresh one is injected to the cell at the same rate, in order to ensure a stable hanging meniscus at the electrode.

Recently, Abruna and coworkers described a double-band-electrode channel flow cell with interface to the mass spectrometer between the working electrode at the electrolyte inlet and a detection electrode at the outlet [21]. In the wall-jet geometry described by Scherson and coworkers, the working electrode is surrounded

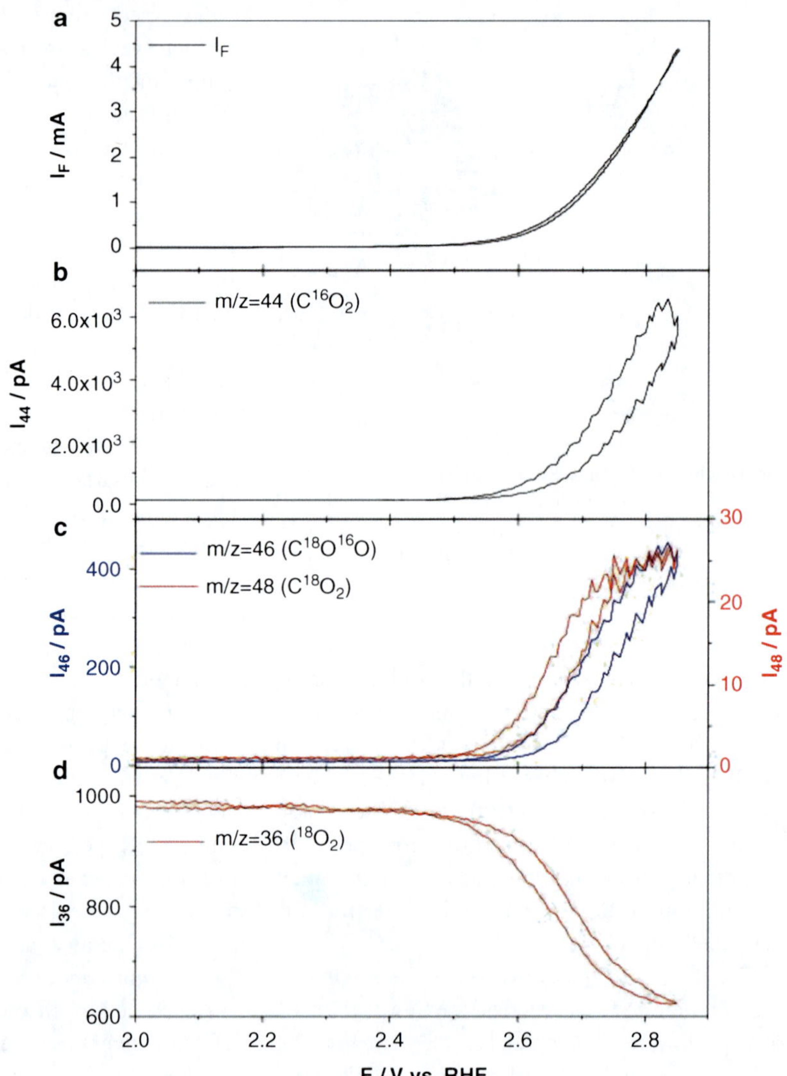

Electrochemical Mass Spectrometry, Fig. 4 Simultaneously recorded CV (**a**) and MSCV for (**b**) $C^{16}O_2$, (**d**) $^{18}O_2$, (**c**) $C^{18}O_2$, and $C^{16}O^{18}O$; $^{18}O_2$-saturated solution of 50 mM acetic acid in 1 M $HClO_4$; scan rate 10 mV s^{-1}, flow rate 5 μL s^{-1}

by a thin, porous Teflon cylinder, through which gaseous products are transferred to the vacuum of the MS [22].

Calibration of DEMS

Product formation rates are monitored by recording the corresponding mass spectrometrical ion current I_i. It is directly proportional to the incoming flow J_i = dn/dt in mol s^{-1} of that species i, and therefore,

$$I_i = K^\circ J_i \tag{1}$$

K° contains all settings of the mass spectrometer and the ionization and fragmentation probability of the corresponding species. When the species is produced by a known electrochemical reaction (such as H_2 evolution or CO oxidation), J_i is given by the faradaic current I_F corresponding to that process:

$$J_i = NI_F/(zF) \tag{2}$$

where z is the number of electrons, F is the Faraday constant, and N is the transfer efficiency, i.e., the ratio of the amount of the species entering

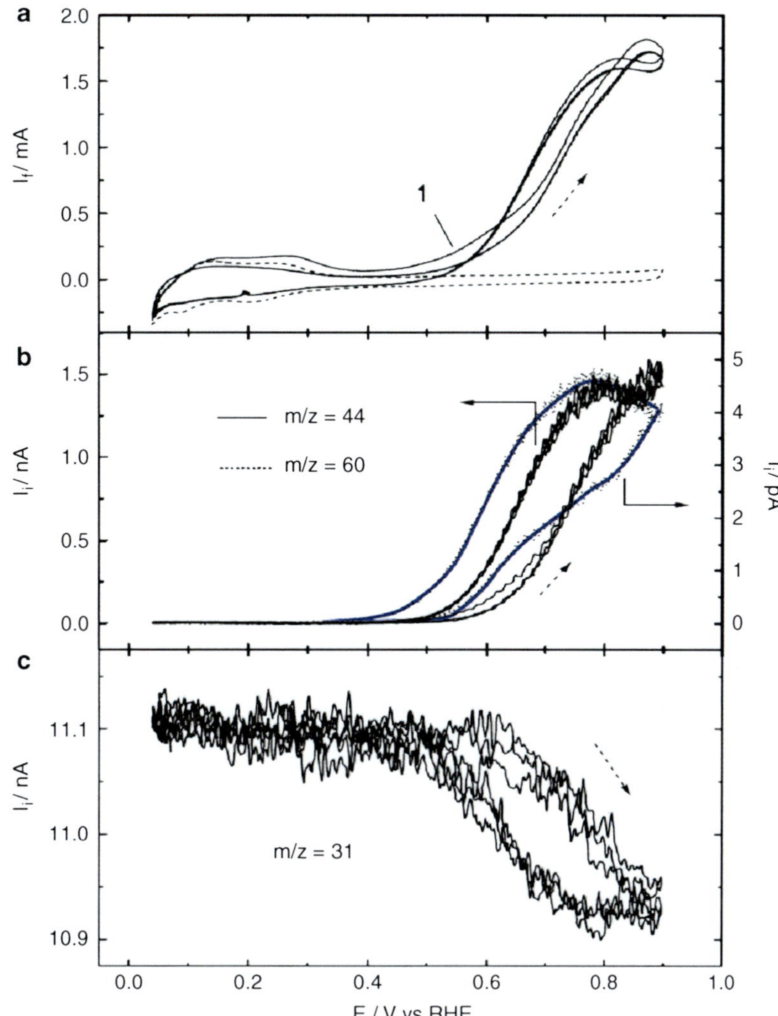

Electrochemical Mass Spectrometry, Fig. 5 Simultaneously recorded (**a**) CV; (**b**) MSCVs $m/z = 44$ and 60; and (**c**) MSCV $m/z = 31$ during methanol oxidation on Vulcan XC-72/40 % Pt supported on GC in 0.1 M methanol + 0.5 M H_2SO_4 solution. $v = 10\ mV\ s^{-1}$; $u = 5\ \mu L\ s^{-1}$. (- - - -) the CV in 0.5 M H_2SO_4 solution

the vacuum system to the total amount of species produced.

When oxidizing adsorbed CO, the double-layer charging effects amount to 20 % of the oxidation charge even after background subtraction. The origin is mainly the different double-layer charge at a given potential with and without adsorbed CO [3, 23, 24]. N may be less than 1 because a part of the produced species diffuses away from the electrode into the electrolyte and is not recovered at the membrane to the vacuum. Therefore,

$$I_i = (K^*/z)I_F,\ with\ K^* = K^\circ N/F \qquad (3)$$

When the current efficiency is less than 100 %, I_F has to be replaced by its product with the current efficiency.

Some Examples

Boron-doped diamond (BDD) is an interesting electrode material because of the wide potential range even in aqueous electrolyte. One of the possible applications is the use for wastewater treatment. At the positive potential limit, OH-radicals are formed which oxidize nearly every organic compound. DEMS is used here because the oxygen evolution can be distinguished from the parallel oxidation of the organic species; current efficiencies can be separately evaluated for O_2 and CO_2. Very helpful is the use of isotopically labelled species: In the example of Fig. 4, acetic acid is oxidized in the presence of $^{18}O_2$ [25]. Interestingly, not only CO_2 with m/z = 44 is found, but also $C^{16}O^{18}O$ and $C^{18}O_2$ (m/z = 46 or 48, resp.). Obviously, after an electrochemical initiation, O_2 is involved in the oxidation reaction. In the first step, a radical is formed from the organic molecule either by direct electron transfer and proton abstraction or by a reaction with an OH-radical, which then reacts with O_2 in a further step.

$$R - H \overset{\bullet OH}{\rightarrow} R\bullet \overset{O_2}{\rightarrow} CO_2$$

When studying oxygen evolution at oxide catalysts such as RuO_2 or IrO_2, an important question is whether the oxide-O atom participates in the reaction. Using isotopic labelling, it could be shown that lattice oxygen is exchanged and that corresponding labelled atoms are found in the evolved O_2 [26–29].

The most frequent use of DEMS is for studies of possible fuels in fuel cells. Figure 5 shows the faradaic and ion currents for CO_2 and methylformate during methanol oxidation at carbon-supported Pt nanoparticles. Note that the formation of methylformate starts at a slightly lower potential than that of CO_2. The ratio of the CO_2 formation rate to the faradaic current yields a current efficiency of 90 % in this case. Under flow and at smooth Pt electrodes, the current efficiency for CO_2 remains at 30 % for all flow rates [4]. This proves the parallel reaction mechanism suggested by Bagotsky [30]. One path leads to formaldehyde and formic acid. Under flow, these molecules diffuse away from the electrode, while under stagnant conditions as in the pores of a porous electrode, they are further oxidized to CO_2. The other path leads to CO_2 via adsorbed CO and is independent of flow rate.

Often, the formation of methylformate was taken as an indication of the formation of formic acid, which would react with methanol.

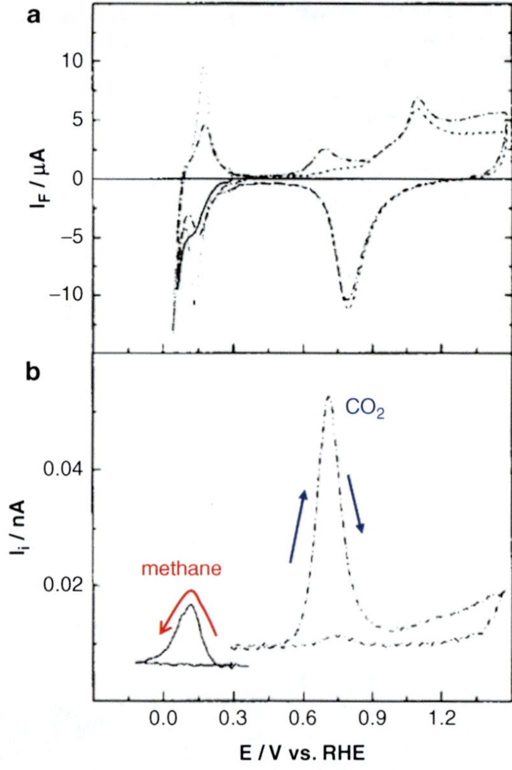

Electrochemical Mass Spectrometry, Fig. 6 Cathodic desorption of preadsorbed ethanol from Pt(110) and subsequent oxidation of the remaining adsorbate $E_{ad} = 0.3$ V, [EtOH] = 0.1 M in 0.1 M H_2SO_4. (**a**) *Solid*, 1st sweep in cathodic direction; *dashed dot*, subsequent oxidation; and *dotted*, Pt(1 10) in supporting electrolyte. (**b**) MSCV

However, this reaction is very slow [31], and therefore, methylformate is formed in another parallel reaction at the surface.

Another important application of DEMS is the elucidation of reaction products during ethanol oxidation. The main product detectable by DEMS is acetaldehyde. In cyclic voltammetric experiments, also CO_2 is detected, besides very small amounts of ethyl acetate. Under defined bulk oxidation conditions, hardly any CO_2 is formed at constant potential, but only from adsorbates formed during cyclic voltammetry [32].

Adsorbates at electrode surfaces usually are detected using a spectroscopic technique such as FTIR. Because of the high sensitivity of DEMS, it can also be used for the detection of adsorbates. One possibility to do so is to oxidize completely the organic adsorbate to CO_2, which is quantitatively

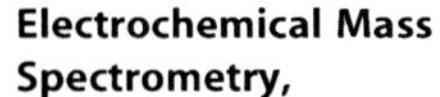

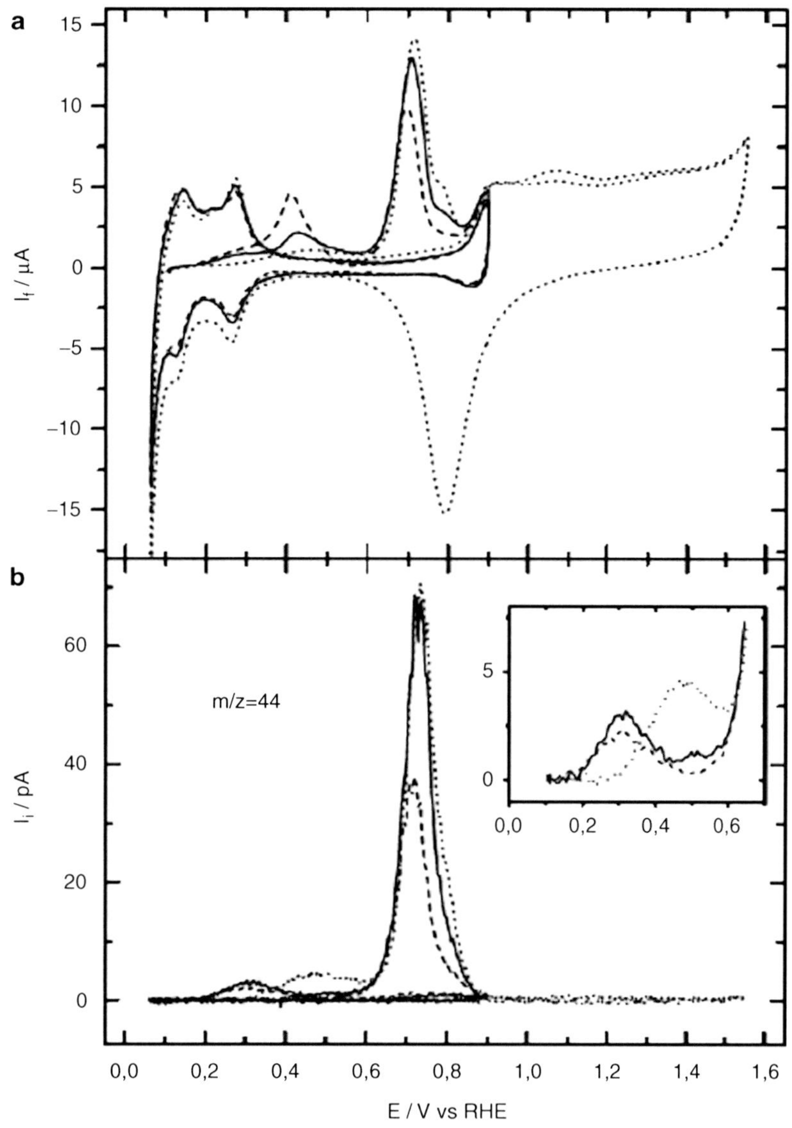

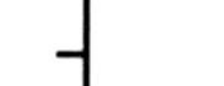

Electrochemical Mass Spectrometry, Fig. 7 Simultaneously recorded CV (**a**) and MSCV (**b**) m/z = /44 for the oxidation of preadsorbed CO on smooth Pt(pc) (*dashed*) and Mo-modified smooth Pt(pc) ($\Theta_{Mo} = 0.07$, *solid*; $\Theta_{Mo} = 0.41$, *dotted*) in 0.5 M H_2SO_4 solution in the dual thin-layer cell. $v = 10\ mV\ s^{-1}$, $u = 5\ \mu L s^{-1}$. (*inset*) Expanded current scale for low potential region

monitored by the change of the ionic current in the mass spectrometer. Many adsorbates can also be desorbed at certain potentials as such or as the hydrogenated product, allowing a more direct characterization of the adsorbate. The latter has been demonstrated for several unsaturated hydrocarbons such as benzene and ethene not only at Pt, but also for model inhibitors, at less noble metals [33, 34] and at monoatomic rows of Pd decorating the steps of Au single crystals [35, 36].

Already in early experiments, it was shown that the abovementioned adsorption of ethanol and the breaking of a C-C bond, ultimately leading to CO_2, are slow [37]. By isotopic marking it was shown that adsorbed CO is formed from the α-C of ethanol, whereas an adsorbed CH_x species is formed from the methyl group. The latter desorbs as methane at negative potentials, but can also be slowly oxidized to CO_{ad} (and ultimately CO_2) above 0.6 V [37] (Fig. 6).

The above results demonstrate the importance of a fast oxidation of adsorbed CO. This oxidation rate is increased in the presence of cocatalysts such as Ru, Sn, or Mo. Figure 7 shows the

CV and the MSCV for CO_2 during the stripping of CO_{ad} from Pt(pc)- and Mo-modified Pt surface [38]. The presence of Mo shifts the onset of CO_2 formation by about 200 mV to lower potentials (0.2 V), although the oxidation of the largest part of CO remains unaffected. This "early" oxidation of the so-called weakly adsorbed CO (which is masked in the CV by the surface oxidation of Mo itself) results in a sufficient number of free surface sites for the oxidation of H_2, which in case of the PEM fuel cell frequently contains some residual CO. No positive effect for direct methanol oxidation is observed. However, when Ru-modified surfaces are used, the main oxidation peak for CO oxidation is shifted downwards, which, together with the faster adsorption of methanol to form CO_{ad}, results in a preferential oxidation of methanol in the reaction path via CO_{ad}, yielding higher current efficiencies even at smooth electrodes [39–41].

Future Directions

At present, there are two main limitations of DEMS: the detection of nonvolatile species and the use of nonaqueous electrolytes. In principle, the first can be circumvented by using electrospray ionization [42–45]. This technique is used by many authors now for organic systems, but for aqueous electrolytes such as sulfuric acid, a derivatization step has to be used [46]. For the use of nonaqueous electrolyte, a headspace approach has recently been designed [47, 48].

These examples demonstrate that the development of the method continuous and that its use becomes more and more widespread.

Cross-References

- ▶ Boron-Doped Diamond for Green Electro-Organic Synthesis
- ▶ Electrocatalysis of Anodic Reactions
- ▶ Electrochemical Reactor Design for the Oxidation of Organic Pollutants
- ▶ Ethanol Oxidation, Electrocatalysis of Fuel Cell Reactions
- ▶ Infrared Spectroelectrochemistry

References

1. Bruckenstein RR, Gadde J (1971) Use of a porous electrode for in situ mass spectrometric determination of volatile electrode reaction products. J Am Chem Soc 93:793
2. Wolter O, Heitbaum J (1984) Differential electrochemical mass spectroscopy (DEMS) - a new method for the study of electrode processes. Ber Bunsenges Phys Chem 88:2–6
3. Wolter O, Heitbaum J (1984) The adsorption of CO on a porous Pt-electrode in sulfuric acid studied by DEMS. Ber Bunsenges Phys Chem 88:6–10
4. Wang H, Wingender C, Baltruschat H, Lopez M, Reetz MT (2001) Methanol oxidation on Pt, PtRu, and colloidal Pt electrocatalysts: a DEMS study of product formation. J Electroanal Chem 509:163–169
5. Heinen M, Chen YX, Jusys Z, Behm RJ (2007) CO adsorption kinetics and adlayer build-up studied by combined ATR-FTIR spectroscopy and on-line DEMS under continuous flow conditions. Electrochim Acta 53(3):1279–1290
6. Lanova B (2009) Oxidation of methanol and carbon monoxide on platinum surfaces. The influence of foreign metals. Ph.D., Rheinische Friedrich-Wilhelms Universität Bonn, Germany
7. Eggert G, Heitbaum J (1986) Electrochemical reactions of propylenecarbonate and electrolyes solved therein - a DEMS study. Electrochim Acta 31(11):1443–1448
8. Baltruschat H (2004) Differential electrochemical mass spectrometry. J Amer Soc Mass Spectrometry 15:1693–1706
9. Baltruschat H (1999) Differential electrochemical mass spectrometry as a tool for interfacial studies. In Wieckowski A (ed) Interfacial electrochemistry. Marcel Dekker, New York/Basel, pp 577–597
10. Dülberg A (1994) Die Adsorption und Zersetzung halogenierter Kohlenwasserstoffe an Elektrokatalysatoren, Untersuchungen mit elektrochemischer Massenspektrometrie. Dissertation, Universität Witten-Herdecke, Witten-Herdecke
11. Tegtmeyer D, Heindrichs A, Heitbaum J (1989) Electrochemical on line mass spectrometry on a rotating electrode inlet system. Ber Bunsenges Phys Chem 93:201–206
12. Baltruschat H, Schmiemann U (1993) The adsorption of unsaturated organic species at single crystal electrodes studied by Differential Electrochemical Mass Spectroscopy. Ber Bunsenges Phys Chem 97(3):452–460
13. Hartung T, Baltruschat H (1990) Differential electrochemical mass spectrometry using smooth electrodes: adsorption and H/D-exchange reactions of benzene on Pt. Langmuir 6(5):953–957
14. Smith SPE, Casado-Rivera E, Abruna HD (2003) Application of differential electrochemical mass spectrometry to the electrocatalytic oxidation of formic acid at a modified Bi/Pt electrode surface. J Solid State Electrochem 7(9):582–587

15. Jusys Z, Massong H, Baltruschat H (1999) A new approach for simultaneous DEMS and EQCM: electro-oxidation of adsorbed CO on Pt and Pt-Ru. J Electrochem Soc 146:1093
16. Heinen M, Chen Y-X, Jusys Z, Behm RJ (2007) Room temperature COad desorption/exchange kinetics on Pt electrodes - a combined in situ IR and mass spectrometry study. ChemPhysChem 8(17): 2484–2489
17. Heinen M, Chen YX, Jusys Z, Behm RJ (2007) In situ ATR-FTIRS coupled with online DEMS under controlled mass transport conditions—A novel tool for electrocatalytic reaction studies. Electrochim Acta 52:5634–5643
18. Gao Y, Tsuji H, Hattori H, Kita H (1994) New on-line mass spectrometer system designed for platinum single crystal electrode and electroreduction of acetylene. J Electroanal Chem 372:195–200
19. Wonders AH, Housmans THM, Rosca V, Koper MTM (2006) On-line mass spectrometry system for measurements at single-crystal electrodes in hanging meniscus configuration. J Appl Electrochem 36(11):1215–1221
20. Abd-El-Latif A-E-A, Xu J, Bogolowski N, Königshoven P, Baltruschat H (2012) New cell for DEMS applicable to different electrode sizes. Electrocatalysis 3(1):9
21. Wang H, Rus E, Abruna HD (2010) New double-band-electrode channel flow differential electrochemical mass spectrometry cell: application for detecting product formation during methanol electrooxidation. Anal Chem 82(11):4319–4324
22. Treufeld I, Jebaraj AJJ, Xu J, Martins de Godoi D, Scherson D (2012) Porous teflon ring-solid disk electrode arrangement for differential mass spectrometry measurements in the presence of convective flow generated by a jet impinging electrode in the wall-jet configuration. Anal Chem 84(12): 5175–5179
23. Willsau J, Heitbaum J (1986) Analysis of adsorbed intermediates and determination of surface potential shifts by DEMS. Electrochim Acta 31(8):943–948
24. Clavilier J, Albalat R, Gómez R, Orts JM, Feliu JM, Aldaz A (1992) Study of the charge displacement at constant potential during CO adsorption on Pt(110) and Pt(111) electrodes in contact with a perchloric acid solution. J Electroanal Chem 330:489–497
25. Kapalka A, Lanova B, Baltruschat H, Fóti G, Comninellis C (2008) Electrochemically induced mineralization of organics by molecular oxygen on boron-doped diamond electrode. Electrochem Commun 10(9):1215–1218
26. Wohlfahrt-Mehrens M, Heitbaum J (1987) Oxygen evolution on Ru and RuO_2 electrodes studied using isotope labelling and on-line mass spectrometry. J Electroanal Chem 237(2):251–260
27. Fierro S, Nagel T, Baltruschat H, Comninellis C (2007) Investigation of the oxygen evolution reaction on Ti/IrO_2 electrodes using isotope labelling and on-line mass spectrometry. Electrochem Commun 9(8):1969–1974
28. Petrykin V, Macounova K, Shlyakhtin OA, Krtil P (2010) Tailoring the selectivity for electrocatalytic oxygen evolution on ruthenium oxides by zinc substitution. Angew Chem Int Ed 49(28):4813–4815
29. Macounova K, Makarova M, Krtil P (2009) Oxygen evolution on nanocrystalline RuO_2 and $Ru0.9Ni0.1O_2$-delta electrodes - DEMS approach to reaction mechanism determination. Electrochem Commun 11(10):1865–1868
30. Bagotzky VS, Vassiliev YB, Khazova OA (1977) Generalized scheme of chemisorption, electrooxidation and electroreduction of simple organic compounds on platinum group metals. J Electroanal Chem 81:229
31. Abd-El-Latif AA, Baltruschat H (2011) Formation of methylformate during methanol oxidation revisited: the mechanism. J Electroanal Chem 662(1):204–212
32. Abd-El-Latif AA, Mostafa E, Huxter S, Attard G, Baltruschat H (2010) Electrooxidation of ethanol at polycrystalline and platinum stepped single crystals: a study by differential electrochemical mass spectrometry. Electrochim Acta 55(27): 7951–7960
33. Schmiemann U, Jusys Z, Baltruschat H (1994) The electrochemical stability of model inhibitors: a DEMS study on adsorbed benzene, aniline and pyridine on mono- and polycristalline Pt, Rh and Pd electrodes. Electrochim Acta 39(4):561–576
34. Zinola CF, Vasini EJ, Müller U, Baltruschat H, Arvia AJ (1996) Detection of CO desorbing from the Ni-electrode surface by DEMS. J Electroanal Chem 415(1–2):165–167
35. Steidtner J, Hernandez F, Baltruschat H (2007) The electrocatalytic reactivity of Pd monolayers and monoatomic chains on Au. J Phys Chem C 111:12320–12327
36. Sanabria-Chinchilla J, Baricuatro JH, Soriaga MP, Hernandez F, Baltruschat H (2007) Electrocatalytic hydrogenation and oxidation of aromatic compounds studied by DEMS: benzene and p-dihydroxybenzene at ultrathin Pd films electrodeposited on Au(hkl) surfaces. J Colloid Interface Sci 314:152–159
37. Schmiemann U, Müller U, Baltruschat H (1995) The influence of the surface structure on the adsorption of ethene, ethanol and cyclohexene as studied by DEMS. Electrochim Acta 40(1):99–107
38. Samjeské G, Wang H, Löffler T, Baltruschat H (2002) CO and methanol oxidation at Pt-electrodes modified by Mo. Electrochim Acta 47(22–23):3681–3692
39. Wang H (2001) In Baltruschat H (ed) Co-Ad coverage and current efficiency of methanol oxidation studied by DEMS and potential step. DMFC Symposium, Meeting of the Electrochemical Society 2001, Washington, DC; Narayanan SR (ed) The Electrochemical Society, Washington, DC
40. Wang H, Baltruschat H (2007) DEMS study on methanol oxidation at poly- and single crystalline platinum electrodes: the effect of anion, temperature, surface

structure, Ru adatom and potential. J Phys Chem C 111(19):7038–7048
41. Samjeské G, Xiao X-Y, Baltruschat H (2002) Ru decoration of stepped Pt single crystals and the role of the terrace width on the electrocatalytic CO oxidation. Langmuir 18(12):4659–4666
42. Zettersten C, Lomoth R, Hammarstrom L, Sjoberg PJR, Nyholm L (2006) The influence of the thin-layer flow cell design on the mass spectra when coupling electrochemistry to electrospray ionisation mass spectrometry. J Electroanal Chem 590(1):90–99
43. Zettersten C, Sjöberg PJR, Nyholm L (2009) Oxidation of 4-chloroaniline studied by on-line electrochemistry electrospray ionization mass spectrometry. Anal Chem 81(13):5180–5187
44. Lu WZ, Xu XM, Cole RB (1997) On-line linear sweep voltammetry electrospray mass spectrometry. Anal Chem 69(13):2478–2484
45. Deng HT, Van Berkel GJ (1999) A thin-layer electrochemical flow cell coupled on-line with electrospray-mass spectrometry for the study of biological redox reactions. Electroanalysis 11(12):857–865
46. Zhao W, Jusys Z, Behm RJ (2010) Quantitative online analysis of liquid-phase products of methanol oxidation in aqueous sulfuric acid solutions using electrospray ionization mass spectrometry. Anal Chem 82(6):2472–2479
47. Hahn M, Wursig A, Gallay R, Novak P, Kotz R (2005) Gas evolution in activated carbon/propylene carbonate based double-layer capacitors. Electrochem Commun 7(9):925–930
48. McCloskey BD, Bethune DS, Shelby RM, Girishkumar G, Luntz AC (2011) Solvents' critical role in nonaqueous lithium-oxygen battery electrochemistry. J Phys Chem Lett 2(10):1161–1166

Electrochemical Microflow Systems

Frank Marken
University of Bath, Bath, UK

Introduction to Flow Electrochemical Processes

Flow (convection) plays a major role in electrochemistry and electrochemical engineering in conjunction with migration and diffusion in providing transport in a fundamentally heterogeneous process [1]. The topic of flow and convection has dominated in the design of electrochemical systems, in particular after seminal contributions from Levich [2]. A large body of literature is available on topics like electrochemical processes under high speed flow [3, 4], gravity/hydrostatic pressure driven [5], pump-driven, or electro-osmotic flow-driven systems, and in general micro-fluidic flow systems [6]. Micro-fluidic system control is of particular importance in fuel cell technology [7], in flow electrosynthesis, and in flow electroanalysis. With new materials and fabrication technologies becoming available (e.g., nano-lithography and 3D printing [8]), this field of research and development is rapidly expanding, for example, into smaller dimensions and more effective designs [9].

When considering thin layer flow in a rectangular duct (see Fig. 1a), a fully developed parabolic flow (laminar) can be assumed for conditions of the Reynolds number $Re < 10^5$ (with $Re = \frac{average\ flow\ velocity \times characteristic\ length}{kinematic\ viscosity}$). For cases of a higher Re, the onset of turbulence can be delayed and a "useful" laminar zone can be exploited [10]. The current density under mass transport limiting conditions is high at the initial edge of the electrode, and it then decays with $x^{-1/3}$ toward the trailing edge to give a nonuniformly accessible electrode and a limiting current expressed in Eq. 1 [11].

$$I_{\lim} = 0.925nFcD^{2/3}\left(\frac{wx}{h}\right)^{2/3}V_f^{1/3} \quad (1)$$

In this equation, the mass transport limited current $I_{\lim}$ is given by n, the number of electrons transferred per molecule diffusing to the electrode surface, F, the Faraday constant, c, the bulk concentration, D, the diffusion coefficient, w, the electrode width, x, the electrode length, h, the electrode half height, and V_f, the volume flow rate. By comparison with the total flux of redox-active material in the channel, $V_f \times c \times n \times F$, the degree of conversion can be expressed (Eq. 2).

$$conversion(\%) = 92.5D^{2/3}\left(\frac{wx}{h}\right)^{2/3}V_f^{-2/3} \quad (2)$$

For example, a cell of 1-cm width, 50-micrometer half height, a typical diffusion coefficient of 10^{-9} m^2s^{-1}, and a volume flow

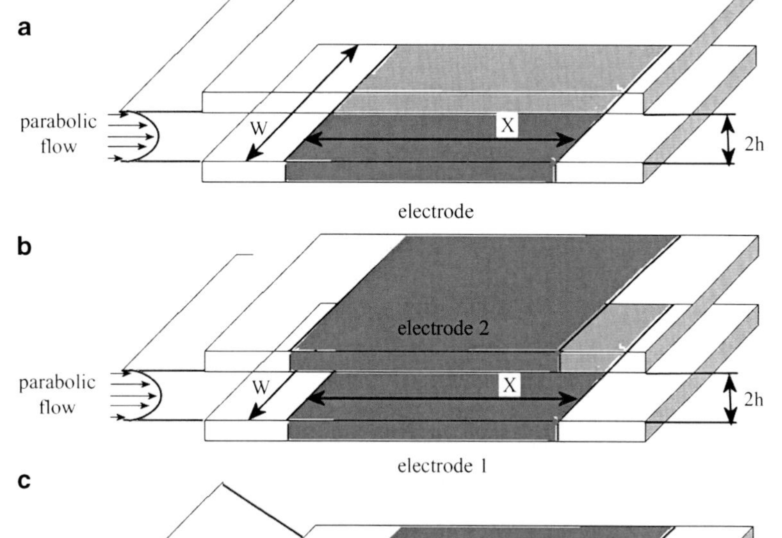

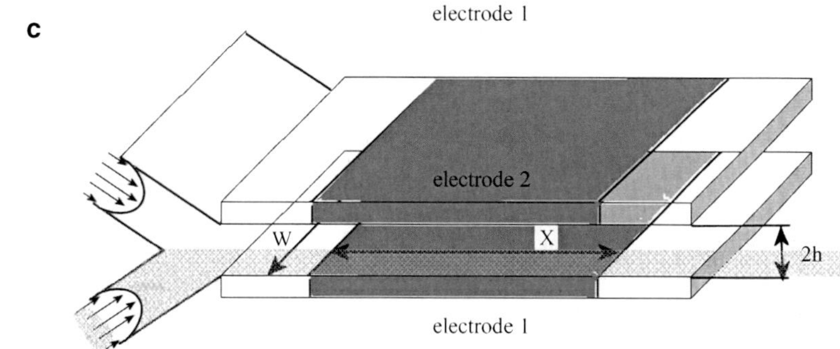

c

electrode 2

W

X

2h

electrode 1

Electrochemical Microflow Systems, Fig. 1 Schematic drawing of a thin layer flow cell with (**a**) one working electrode or (**b**) two opposite working electrodes, and (**c**) a confluence cell

Electrochemical Microflow Systems, Table 1 Common flow geometries and their corresponding limiting current expressions (in SI units)

Type	Electrode and geometric parameters	Flow parameter	Limiting current I_{lim}
Tube	Length x	V_f	$I_{\text{lim}} = 5.43nFc(Dx)^{2/3}V_f^{1/3}$
Channel	Length x, width w, half height h	V_f	$I_{\text{lim}} = 0.925nFcD^{2/3}\left(\frac{wx}{h}\right)^{2/3}V_f^{1/3}$
Wall-tube	Radius r, tube radius R, kinematic viscosity v	V_f	$I_{\text{lim}} = 1.92nFcD^{2/3}r^2v^{-1/6}\left(\frac{V_f}{R}\right)^{1/2}$
Wall-jet	Radius r, jet radius R, kinematic viscosity v	V_f	$I_{\text{lim}} = 1.59nFcD^{2/3}r^{3/4}v^{-5/12}R^{-1/2}(V_f)^{3/4}$

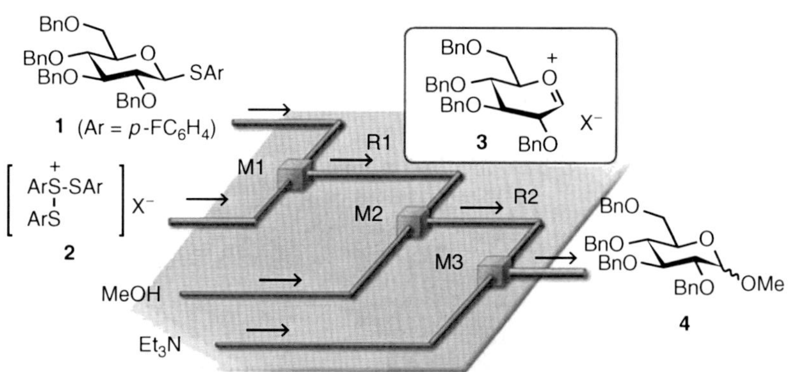

Electrochemical Microflow Systems, Fig. 2 Microflow reactor glycosylation process by Yoshida (Redrawn from [38])

rate of 20 $microLmin^{-1}$, the length of the electrode has to be $x = 1.6$ mm to give 90 % conversion and 1.9 mm to give 99 % conversion (assuming mass transport limiting conditions). In practice, kinetic factors and migration are likely to play an impeding role. Expressions for limiting currents for other types of flow reactor cell geometries s are summarized in Table 1 [12].

A key consideration in the design of microflow cell configuration is often the positioning of counter and reference electrodes. For systems under potentiostatic control, a reference and a counter electrode are usually placed upstream and downstream, respectively, of the working electrode, although improved more symmetric designs are beneficial. Electrical instability in the potentiostatic control circuit can arise and sometimes additional measures, like a capacitance link from reference to counter, can help in stabilizing the system. Placing the counter electrode opposite to the working electrode (see Fig. 1b) has some considerable advantages in providing a more uniform potential distribution. This arrangement has been exploited in particular in "paired electrolysis" [13] where both anode and cathode (or electrode 1 and electrode 2, see Fig. 1b) are active and an overlap of the diffusion layer is exploited to generate new products. The average extent of the diffusion layer $d_{diffusion}$ can be obtained based on the Nernst model [14], with Eq. 1 for the limiting current substituted into $I_{\lim} = \frac{nFDAc}{\delta_{diffusion}}$ (see Eq. 3).

$$\delta_{diffusion} = 1.081\left(\frac{wxh^2D}{V_f}\right)^{1/3} \quad (3)$$

For example, a cell of 1-cm width, 50-micrometer half height, a diffusion coefficient of 10^{-9} m^2s^{-1}, a volume flow of 20 $microLmin^{-1}$, and the length of the electrode $x = 1.6$ mm, the *average* diffusion layer thickness is approximately $\delta_{diffusion} = 50$ micrometer (note that here the nonuniformly accessible diffusion layer increases toward the trailing edge).

Confluence microflow reactors (see Fig. 1c), biphasic [15, 16] and single-phase [17], have been proposed for various applications. Atobe and coworkers [18] further developed this concept to the parallel laminar flow reactor with application such as anodic substitution. An excellent review describing the literature and various types of electrochemical micro-reactor systems and application has been published by Ziogas et al. [19].

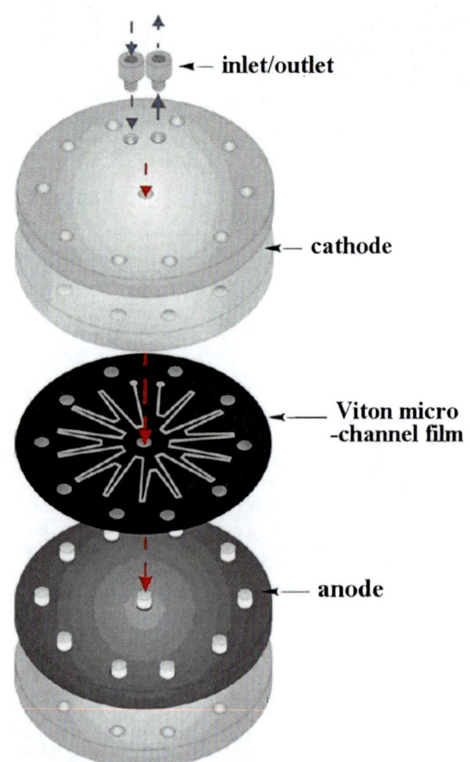

Electrochemical Microflow Systems, Fig. 3 Viton microchannel flow electrolysis cell (Redrawn from [42]) for two-electrode configuration electrolysis

The driving force for microflow system often is provided by syringe pumps or peristaltic pumps with the experimental problem of maintaining uniform pulse-free flow. Elegant electro-osmotic methods have been proposed [20] and novel reactor designs reviewed recently by Zimmerman [21]. Effective flow electrolysis in micro-reactors is also contributing to analytical technologies, and effective microflow cell designs are reviewed by Pletcher and Walsh [22], by Iwaska [23], by Vickers [24], by Trojanowicz [25], and by Gun et al. [26].

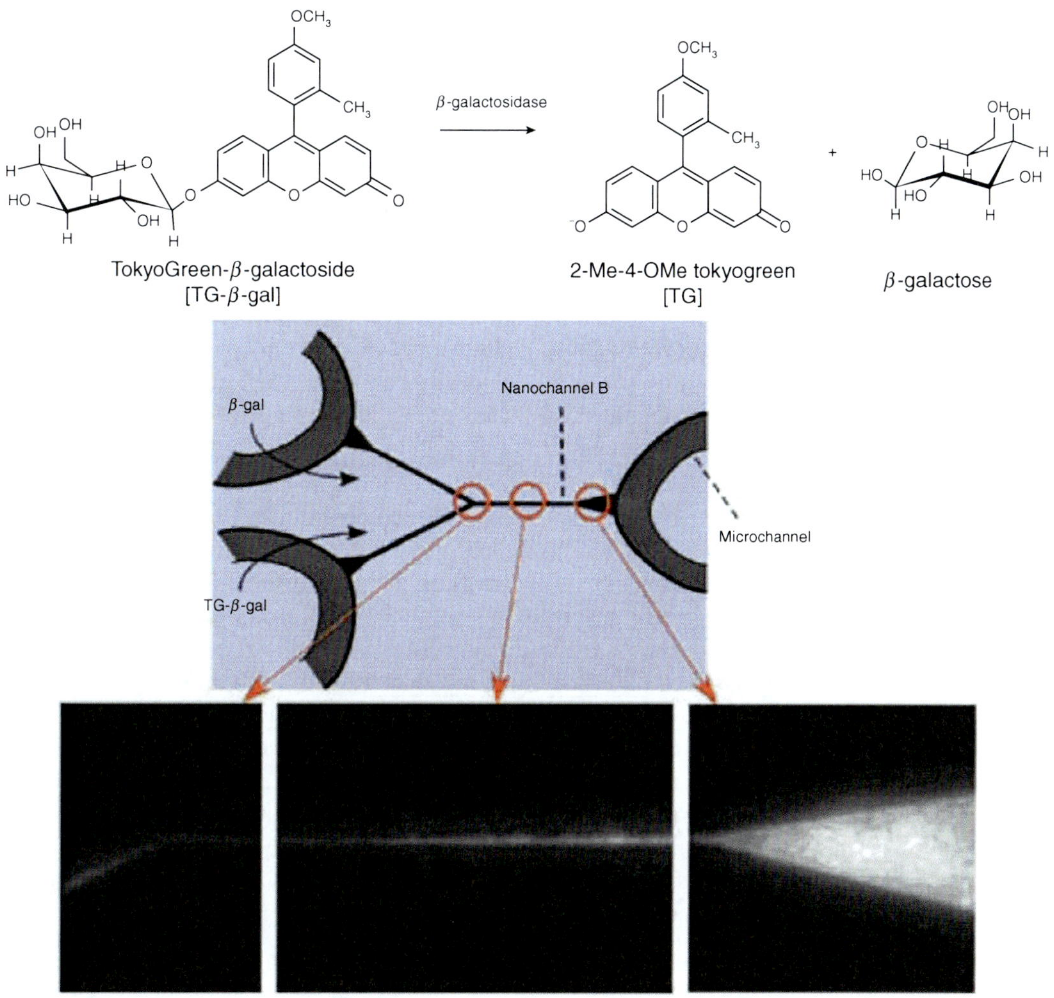

Electrochemical Microflow Systems, Fig. 4 Galactosidase reaction monitored by fluorescence in a nano-channel reactor (Redrawn from [58])

Microflow Electrochemical Systems

Electro-organic processes offer versatile and powerful tools which are often classed as "green" [27, 28] or sustainable chemistry [29]. Millimeter-spaced electrodes in flow reactors have been employed for many years with major applications in industry [30], but recently, more interest in sub-millimeter or microflow systems has arisen due to further potential benefits such as (i) lower flow volume, (ii) a faster conversion with better control over flow and diffusion layer, (iii) better heat exchange, (iv) the use of co-flow and rapid mixing, (v) better control over multiphase reactions, and (vi) the opportunity to work under conditions of "paired electrode processes" [13] and low supporting electrolyte concentration or "self-supported" [31]. In particular, biphasic micro-reactor systems allow operation in the absence of intentionally added electrolyte, but in contrast to micro-batch biphasic test reactors [32, 33], biphasic flow systems [34] also allow continuous operation and they have better potential for scale-up.

Wiles and Watts [35] have reviewed progress in micro-reaction technologies including microflow electrolysis, highlighting, for example, the work by Yoshida and coworkers [36, 37]. Figure 2 shows an elegant Yoshida glycosylation process [38] where an electrochemically generated reactive cation **2** is rapidly mixed with a thioacetal to give a cyclic alkoxycarbenium intermediate for glycosylation of appropriate nucleophiles.

Simple flow through micro-reactor cells for operation in two-electrode mode have been proposed based on low pressure [39, 40] and high pressure [41] designs. Recently Birkin and coworkers proposed a versatile design based on a Viton microchannel foil [42] (see Fig. 3) for acetoxylation and TEMPO-mediated oxidations [43].

In order to minimize the use of chemical reagents and supporting electrolyte, processes have been suggested where the electrolysis is producing ionic intermediates to provide some conductivity [44–47]. A disadvantage in this approach can be seen in the additional energy requirements due to a higher cell voltage and the additional instability encountered in scale-out [48].

Biphasic microflow systems allow the product flow to be separated from the electrolyte flow in order to eliminate work-up steps and to allow recycling of electrolyte [49]. Finally, a "salt cell" design was suggested with supporting electrolyte present as a separate phase to allow redox processes even in humidified hexane [50].

Nano-Flow Electrochemical Systems

Flow pipes of 10–100-nm diameter attract attention [51] in part due to the development of new fabrication methods. Sensing based on nanofluidic devices is an emerging field [52, 53], and electrokinetic phenomena play a significant role [54, 55]. Nanofluidic slippage phenomena have been observed [56], which in future may lead to nanoflow devices that are more effective when compared to theoretical prediction based on simply scaling microflow characteristics.

On-chip nanofluidic systems [57] and extended nanofluidic systems [58] have been developed by Kitamori and coworkers. For example, enzyme reactions studied in 100-nm scale flow channels (see Fig. 4) were monitored by fluorescence microscopy, and enzyme activity in the nano-channel was shown to be higher compared to bulk solution processes. Initial electrochemical experiments in extended nanochannels have been reported by the same group [59].

Future Directions

Microflow devices provide powerful new tools to synthetic and analytical chemists. In conjunction with in situ analysis (e.g., employing micro-spectro-electrochemistry [60]), processes can be controlled and optimized. Coupled cooling and heating are possible with rapid temperature transients applied in a small space. New 3D printing technologies will pave the way to more complex microflow reactor designs where multistep reactions are performed fast and efficiently. In the parallelization and scale-up of microflow electrochemical reactions, this will help in eliminating reaction steps and providing diversity in product ranges. Biphasic microflow systems naturally eliminate purification and work-up steps, and therefore, in future, these will be of major importance. Most exciting at this point of the development is probably the emerging nanofluidic technology which may dramatically change the way chemical processes are performed.

Cross-References

► Anodic Substitutions
► Combinatorial Electrochemical Synthesis

References

1. Oldham KB, Bond A, Myland J (2011) Electrochemical science and technology: fundamentals and applications. Wiley, New York
2. Levich VG (1963) Physicochemical hydrodynamics. Longman Higher Education, New York
3. Giovanelli D, Lawrence NS, Compton RG (2004) Electrochemistry at high pressures: a review. Electroanalysis 16:789–810
4. Prieto F, Aixill WJ, Alden JA, Coles BA, Compton RG (1997) Voltammetry under high mass transport conditions. The high-speed channel electrode and transient measurements. J Phys Chem B 101:5540–5544

5. Goodridge F, King CJH (1974) In: Weinberg NL (ed) Technique of electroorganic synthesis, vol V. Wiley, New York, p 7, Part 1
6. Ehrfeld W, Hessel V, Löwe H (2000) Microreactors. Wiley-VCH, Weinheim
7. Kjeang E, Djilali N, Sinton D (2009) Microfluidic fuel cells: a review. J Power Sources 186:353–369
8. Bonyar A, Santha H, Ring B, Varga M, Kovacs JG, Harsanyi G (2010) In: Jakoby B, Vellekoop MJ (eds) 3D Rapid Prototyping Technology (RPT) as a powerful tool in microfluidic development. Eurosensors XXIV Conference, Procedia Engineering, vol 5. Linz, Austria, pp 291–294
9. Waldbaur A, Rapp H, Lange K, Rapp BE (2011) Let there be chip-towards rapid prototyping of microfluidic devices: one-step manufacturing processes. Anal Methods 3:2681–2716
10. Rees NV, Dryfe RAW, Cooper JA, Coles BA, Compton RG, Davies SG, Mccarthy TD (1995) Voltammetry under high-mass transport conditions -a high-speed channel electrode for the study of ultrafast kinetics. J Phys Chem 99:7096–7101
11. Brett CMA, Brett AMO (1993) Electrochemistry. Oxford University Press, Oxford, p 152
12. Pletcher D, Walsh FC (1993) Industrial electrochemistry. Chapman & Hall, London, p 629
13. Paddon CA, Atobe M, Fuchigami T, He P, Watts P, Haswell SJ, Pritchard GJ, Bull SD, Marken F (2006) Towards paired and coupled electrode reactions for clean organic microreactor electrosyntheses. J Appl Electrochem 36:617–634
14. Compton RG, Banks CE (2007) Understanding voltammetry. World Scientific, London, p 94
15. Yunus K, Marks CB, Fisher AC, Allsopp DWE, Ryan TJ, Dryfe RAW, Hill SS, Roberts EPL, Brennan CM (2002) Hydrodynamic voltammetry in microreactors: multiphase flow. Electrochem Commun 4:579–583
16. MacDonald SM, Watkins JD, Bull SD, Davies IR, Gu Y, Yunus K, Fisher AC, Page PCB, Chan Y, Elliott C, Marken F (2009) Two-phase flow electrosynthesis: comparing N-octyl-2-pyrrolidone-aqueous and acetonitrile-aqueous three-phase boundary reactions. J Phys Org Chem 22:52–58
17. Fulian Q, Stevens NPC, Fisher AC (1998) Computer-aided design and experimental application of a novel electrochemical cell: the confluence reactor. J Phys Chem B 102:3779–3783
18. Horii D, Amemiya F, Fuchigami T, Atobe M (2008) A novel electrosynthetic system for anodic substitution reactions by using parallel laminar flow in a microflow reactor. Chem Eur J 14:10382–10387
19. Ziogas A, Kolb G, O'Connell M, Attour A, Lapicque F, Matlosz M, Rode S (2009) Electrochemical microstructured reactors: design and application in organic synthesis. J Appl Electrochem 39:2297–2313
20. Watts P, Haswell SJ, Pombo-Villar E (2004) Electrochemical effects related to synthesis in micro reactors operating under electrokinetic flow. Chem Eng J 101:237–240
21. Zimmerman WB (2011) Electrochemical microfluidics. Chem Eng Sci 66:1412–1425
22. Pletcher D, Walsh FC (1993) Industrial electrochemistry Chapman & Hall, London, p 630
23. Ivaska A, Kolev SD, Mckelvie ID (2008) Electrochemical detection. Compr Anal Chem 54:441–459
24. Vickers JA, Henry CS (2008) In: Gomez FA (ed) Biological applications of microfluidics. Wiley, New York, pp 435–450
25. Trojanowicz M (2011) Recent developments in electrochemical flow detections-a review part II. Liquid chromatography. Anal Chim Acta 688:8–35
26. Gun J, Bharathi S, Gutkin V, Rizkov D, Voloshenko A, Shelkov R, Sladkevich S, Kyi N, Rona M, Wolanov Y, Rizkov D, Koch M, Mizrahi S, Pridkhochenko PV, Modestov A, Lev O (2010) Highlights in coupled electrochemical flow cell- mass spectrometry, EC/MS. Israel J Chem 50:360–373
27. Schäfer HJ (2011) Contributions of organic electrosynthesis to green chemistry. C R Chim 14:745–765
28. Frontana-Uribe BA, Little RD, Ibanez JG, Palma A, Vasquez-Medrano R (2010) Organic electrosynthesis: a promising green methodology in organic chemistry. Green Chem 12:2099–2119
29. Yoshida JI, Kim H, Nagaki A (2011) Green and sustainable chemical synthesis using flow microreactors. ChemSusChem 4:331–340
30. Pletcher D, Walsh FC (1993) Industrial electrochemistry. Chapman & Hall, London, p 152
31. He P, Watts P, Marken F, Haswell SJ (2006) Self-supported and clean one-step cathodic coupling of activated olefins with benzyl bromide derivatives in a micro flow reactor. Angew Chem Int Ed 45:4146–4149
32. Watkins JD, Ahn SD, Taylor JE, Bull SD, Bulman-Page PC, Marken F (2011) Liquid-liquid electro-organo-synthetic processes in a carbon nanofibre membrane microreactor: triple phase boundary effects in the absence of intentionally added electrolyte. Electrochim Acta 56:6764–6770
33. Watkins JD, Taylor JE, Bull SD, Marken F (2012) Mechanistic aspects of aldehyde and imine electro-reduction in a liquid–liquid carbon nanofiber membrane microreactor. Tetrahedron Lett 53:3357–3360
34. Watkins JD, MacDonald SM, Fordred PS, Bull SD, Gu YF, Yunus K, Fisher AC, Bulman-Page PC, Marken F (2009) High-yield acetonitrile | water triple phase boundary electrolysis at platinised Teflon electrodes. Electrochim Acta 54:6908–6912
35. Wiles C, Watts P (2011) Recent advances in micro reaction technology. Chem Commun 47:6512–6535
36. Yoshida J, Kataoka K, Horcajada R, Nagaki A (2008) Modern strategies in electroorganic synthesis. Chem Rev 108:2265–2299
37. Yoshida J, Nagaki A (2009) In: Hessel V (ed) Micro process engineering, vol 2. Wiley-VCH, Weinheim, pp 109–130

38. Saito K, Ueoka K, Matsumoto K, Suga S, Nokami T, Yoshida J (2011) Indirect cation-flow method: flash generation of alkoxycarbenium ions and studies on the stability of glycosyl cations. Angew Chem Int Ed 50:5153–5156
39. Paddon CA, Pritchard GJ, Thiemann T, Marken F (2002) Paired electrosynthesis: micro-flow cell processes with and without added electrolyte. Electrochem Commun 4:825–831
40. Bouzek K, Jiricny V, Kodym R, Kristal J, Bystron T (2010) Microstructured reactor for electroorganic synthesis. Electrochim Acta 55:8172–8181
41. Attour A, Dirrenberger P, Rode S, Ziogas A, Matlosz M, Lapicque F (2011) A high pressure single-pass high-conversion electrochemical cell for intensification of organic electrosynthesis processes. Chem Eng Sci 66:480–489
42. Kuleshova J, Hill-Cousins JT, Birkin PR, Brown RCD, Pletcher D, Underwood TJ (2011) A simple and inexpensive microfluidic electrolysis cell. Electrochim Acta 56:4322–4326
43. Hill-Cousins JT, Kuleshova J, Green RA, Birkin PR, Pletcher D, Underwood TJ, Leach SG, Brown RCD (2012) TEMPO-mediated electrooxidation of primary and secondary alcohols in a microfluidic electrolytic cell. ChemSusChem 5:326–331
44. He P, Watts P, Marken F, Haswell SJ (2005) Electrolyte free electro-organicsynthesis: the cathodic dimerisation of 4-nitrobenzylbromide in a micro-gap flow cell. Electrochem Commun 7:918–924
45. Horii D, Atobe M, Fuchigami T, Marken F (2005) Self-supported paired electrosynthesis of 2,5-dimethoxy-2,5-dihydrofuran using a thin layer flow cell without intentionally added supporting electrolyte. Electrochem Commun 7:35–39
46. He P, Watts P, Marken F, Haswell SJ (2007) Electrosynthesis of phenyl-2- propanone derivatives from benzyl bromides and acetic anhydride in an unsupported micro-flow cell electrolysis process. Green Chem 9:20–22
47. Horii D, Atobe M, Fuchigami T, Marken F (2006) Self-supported methoxylation and acetoxylation electrosynthesis using a simple thin-layer flow cell. J Electrochem Soc 153:D143–D147
48. He P, Watts P, Marken F, Haswell SJ (2007) Scaling out of electrolyte free electrosynthesis in a micro-gap flow cell. Lab Chip 7:141–143
49. MacDonald SM, Watkins JD, Gu Y, Yunus K, Fisher AC, Shul G, Opallo M, Marken F (2007) Electrochemical processes at a flowing organic solvent aqueous electrolyte phase boundary. Electrochem Commun 9:2105–2110
50. Watkins JD, Hotchen CE, Mitchels JM, Marken F (2012) Decamethylferrocene redox chemistry and gold nanowire electrodeposition at salt crystal electrode nonpolar organic solvent contacts. Organometallics 31:2616–2620
51. Mattia D, Gogotsi Y (2008) Review: static and dynamic behavior of liquids inside carbon nanotubes. Microfluid Nanofluid 5:289–305
52. Lemay SG (2009) Nanopore-based biosensors: the interface between ionics and electronics. ACS Nano 3:775–779
53. Rassaei L, Mathwig K, Goluch ED, Lemay SG (2012) Hydrodynamic voltammetry with nanogap electrodes. J Phys Chem C 116:10913–10916
54. Siwy ZS, Howorka S (2010) Engineered voltage-responsive nanopores. Chem Soc Rev 39:1115–1132
55. Pumera M (2011) Nanomaterials meet microfluidics. Chem Commun 47:5671–5680
56. Lee KP, Leese H, Mattia D (2012) Water flow enhancement in hydrophilic nanochannels. Nanoscale 4:2621–2627
57. Tsukahara T, Mawatari K, Kitamori T (2010) Integrated extended-nano chemical systems on a chip. Chem Soc Rev 39:1000–1013
58. Mawatari K, Tsukahara T, Sugii Y, Kitamori T (2010) Extended-nano fluidic systems for analytical and chemical technologies. Nanoscale 2:1588–1595
59. Tsukahara T, Kuwahata T, Hibara A, Kim HB, Mawatari K, Kitamori T (2009) Electrochemical studies on liquid properties in extended nanospaces using mercury microelectrodes. Electrophoresis 30:3212–3218
60. Flowers PA, Strickland JC (2010) Easily constructed microscale spectroelectrochemical cell. Spectroscopy Lett 43:528–533

Electrochemical Monitoring of Cellular Metabolism

Jennifer R. McKenzie[1], David E. Cliffel[1] and John P. Wikswo[2]

[1]Department of Chemistry, Vanderbilt University, Nashville, TN, USA

[2]Department of Physics and Astronomy, Vanderbilt University, Nashville, TN, USA

Introduction

The application of electrochemical techniques for providing insight into biological processes has become common practice over the last several decades. The study of cellular metabolism, largely ignored by the fields of molecular biology and toxicology until now [1, 2], is a vast field exploring how cellular energetics respond to internal or external influences, as well as how the metabolic state of the cell influences cellular regulation [3]. Traditionally, metabolism has been studied by following the uptake of radioactive metabolites [4, 5] or by quantifying consumption or production of analytes in flasks containing millions of cells [6–8]. Designing electrochemical experiments to study cellular metabolism requires extensive considerations of the cellular environment and sensor design. Many biosensors designed today are intended for point-of-care measures, where a sample of blood, urine, or cerebral spinal fluid is obtained and diluted in a buffer that allows for optimal electrode performance and a single measurement performed. When studying the real-time metabolism of living cells, no alteration of the cellular environment can be performed; thus the sensor must be capable of operating in these environments. A number of electrochemical methods have been developed and applied to advance the study of metabolism [7, 9–13]. One of these methods, multi-analyte microphysiometry, will be presented as an example of a successful method for investigations of cellular metabolism. This review will discuss current techniques employed in microphysiometry, including the application of the instrument to the study of ischemic neurons [14], the contributions of insulin and glucose toward cellular energetic states [15] and the investigation of the metabolic effects of protein toxins [16].

Multi-Analyte Microphysiometry

Microphysiometry is a technique where a small population of cells or tissues is sealed in a micro or nanofluidic environment capable of sustaining cellular activity while a sensing element is employed to provide real-time and continuous measurement of extracellular analytes. The multi-analyte microphysiometer (MAMP) was developed to allow for real-time detection of changes in cellular metabolism, specifically monitoring four analytes central to aerobic and anaerobic respiration. The instrument, which is comprised of up to eight discrete chambers, measures extracellular levels of glucose, lactate, and oxygen, and acid. Cell inserts containing $\sim 10^5$ cells are sealed between a light-addressable potentiometric sensor (LAPS) for acid detection and a sensor head containing platinum electrodes modified for amperometric detection of glucose, lactate and oxygen (Fig. 1a). Each sensor head also contains inlet and outlet tubing and once sealed, creates a 3 μL chamber allowing for continuous perfusion of media over the cells for hours or days and allowing for introduction of agents under study [10, 17]. Depletion of vital nutrients, such as glucose and oxygen, and accumulation of cellular waste can lead to an alteration in the basal metabolism of the cells under study [8]. Microfluidic chambers with integrated electrodes offer many advantages, including complete customization, precise manipulation of fluid and cellular movement, small volumes, parallelization, and reduction in electrode size and instrument footprint [9, 13, 18–20].

Amperometric glucose, lactate and oxygen detection is performed with a multi-chamber bipotentiostat and related LabVIEW software designed by the *Vanderbilt Institute for Integrative Biosystems Research and Education* (VIIBRE) which enables monitoring of multiple analytes in four to eight chambers simultaneously, a feat not currently possible with commercial potentiostats. In the MAMP, oxygen is detected at −0.45 V versus Ag/AgCl (2 M KCl) through direct reduction at a 127 μm diameter platinum electrode modified with Nafion. Two 0.5 mm diameter Pt electrodes are coated with glucose oxidase (GOx) and lactate oxidase (LOx) films for indirect detection through the oxidation of hydrogen peroxide produced by their respective immobilized oxidase enzymes. A low-buffered (1 mM PO_4^{3-}) and low-glucose (5 mM) RPMI 1640 is used in most experiments, where low buffering allows for detectable changes in extracellular pH, and low glucose reduces the background against which consumption must be measured, as well as conforms to physiologically relevant glucose conditions in vivo.

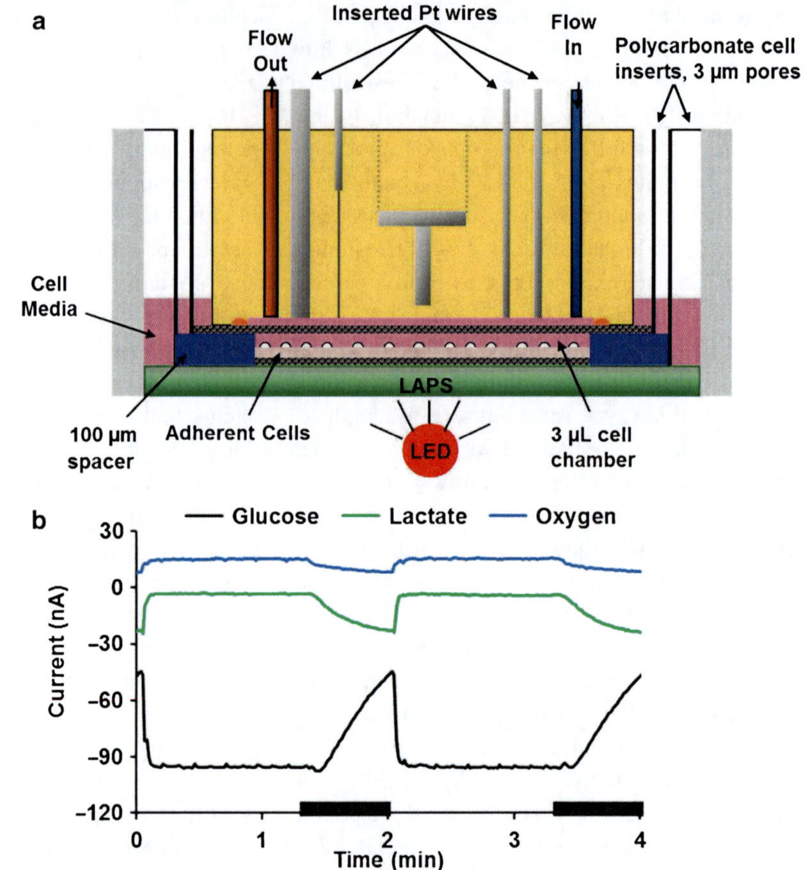

Electrochemical Monitoring of Cellular Metabolism, Fig. 1 (**a**) Cross-view of a MAMP cell chamber. Cells are cultured on a polycarbonate insert that is placed in the sensor cup. A spacer and additional insert are placed on top of the cells, and the sensor head is lowered over the assembly. The sensor head features inlet and outlet tubes to allow for 100 μL/min flow through the 3 μL chamber, as well as the Pt wires for amperometric detection and a stainless steel counter electrode for the LAPS circuit. A Ag/AgCl (2 M KCl) reference electrodes is placed downstream of the chamber. (**b**) Raw amperometric signals of cells during 80 s 100 μL flow periods and 40 s stop-flow periods (*black bar*)

To further enhance the sensitivity of the MAMP, as well as enable calculation of metabolic rates, a flow/stop-flow pattern (80 s on, 40 s off) is employed to measure real-time consumption and accumulation of the cellular metabolites. During each flow period, the current or potential at each electrode reaches a steady state. When flow is halted, depletion of glucose and oxygen in the chamber is observed as a decrease in current magnitude and accumulation of lactate is measured as an increase in current magnitude (Fig. 1b). Accumulation of extracellular acid is measured as a decrease in open circuit potential over time, and the extracellular acidification rate (ECAR) is calculated as -μV/s. Rates are calculated for each stop-flow period, allowing for monitoring of changes in metabolic rate in response to stimuli over the course of several hours. By accounting for the microfluidic volume, the number of cells present and the length of the stop-flow period, metabolic rates for each stop-flow period can be calculated in terms of $\text{mol}\bullet\text{cell}^{-1}\bullet\text{s}^{-1}$ and can be compared to rates calculated using traditional methods [6, 16]. Further detail of this process [16], as well as mathematical modelling [21], has been previously discussed.

Discrimination of Select List Agents

One of the early investigations of cellular metabolism in the MAMP concerned the discrimination of select list agents based on resulting changes in cellular metabolism upon exposure. The metabolic profiles of neuroblastoma, macrophage, and kidney cell lines exposed to ricin, botulinum neurotoxin, cholera toxin (CTx), muscarine, alamethicin and anthrax protective antigen (PA) were performed [22], and each presented a different metabolic profile. For example, macrophages treated with alamethicin and PA, both pore-forming agents, exhibited distinguishable changes in metabolism. Treatment with alamethicin leads to rapid loss of lacate and acid release related to cellular death and PA treatment reflects uncoupling of lactate and acid production, possibly from the intracellular sequestration of protons in lysosomes. The individual metabolic profiles of each agent suggest that the MAMP could be a useful tool in discrimination of select list agents in unknown samples.

Metabolic Response of Neurons to Cholera Toxin Exposure

The metabolic effects of CTx were further explored in the MAMP to determine the metabolic response of several cell types to CTx [16]. Although CTx exposure is typically studied with intestinal cell lines, the largest metabolic response was found using a neuronal cell line, PC-12. Previous electrophysiology studies of intestinal cell lines exposed to CTx revealed a secretion of chloride ions from the cell beginning roughly 30 min to an hour after exposure, with the lag-time attributed to the time required for the toxin to transverse the cell and activate continuous cAMP production [23]. Using the MAMP to investigate CTx exposure in PC-12 neurons, significant increases of extracellular acidification and lactate production, as well as decreases in oxygen consumption, were seen in as little as 10 min. Additional use of inhibitors and other effectors which target specific cellular functions relavant to CTx exposure demonstrated that the observed metabolic response was due to toxin binding and cAMP activation instead of some downsteam effect of toxin entry or transport within the cell. Additionally, these experiments show that the MAMP is a useful tool for determining the acute metabolic effects of biological toxins in real-time and for exploring the signaling pathways triggered by exposure to a toxin or other agents to determine the root cause for metabolic disruption.

Environmental Toxins

In one study, CHO cells were exposed to chromate, a potent environmental toxin which enters the cytoplasm and interferes with redox enzymes [24]. Ultimately, CHO cells exhibited a rapid and drastic decrease in lactate production, followed by a slower decrease in glucose consumption and acid production. This result could be due to an immediate inhibition of lactate dehydrogenase, halting anaerobic respiration and lowering lactate production, followed by irreversible cell damage which reduced the metabolic rate of glucose consumption and acid production. Preliminary work has also be performed to investigate the metabolic response of cells to pesticides, such as parathion. The MAMP can be used to probe the metabolic disruption caused by exposure to environmental toxins.

Metabolic Compensation of Ischemic Neurons

Stroke is the second leading cause of death and the largest cause of long-term adult disability, with a majority of strokes being ischemic in nature, where a loss of oxygenated blood flow results in neuronal death [25]. The MAMP was utilized in the development of a new model of transient ischemic attack (TIA), also described as "mini-strokes," where primary neurons given short 5 min deprivations of oxygen and glucose in the days leading up to an otherwise lethal 90 min deprivation achieved increased survival rates over those receiving only the lethal deprivation [14]. Traditional toxicology performed in this study revealed an increase in cellular ATP 24 h after each nonlethal 5 min deprivation and a total loss of ATP production in neurons deprived for 90 min, indicating that cellular metabolism may play a role in the protective mechanisms occurring in TIA. This was confirmed in the MAMP where neuronal oxygen consumption was significantly increased less than 1 h after 5 min glucose deprivation but permanently decreased in neurons receiving the lethal deprivation. Lactate production in nonlethally deprived neurons recovered to control levels within 30 min, while those receiving the lethal deprivation failed to fully recover. The increase in aerobic respiration in neurons receiving nonlethal stress, as indicated by increased oxygen consumption and ATP production, suggests that the protective pathways upregulated in this ischemic model include immediate increased production of energy stores through aerobic respiration, which may help to prevent cellular starvation upon subsequent stress.

Mamp Modification for Insulin Detection for Islet Analysis

Detection of insulin secretion is crucial to the study of diabetes mellitus, particularly for pancreatic islet metabolism. The MAMP sensor head was modified to allow for direct and sensitive detection of insulin released from islets [15]. As discussed in Snider et al., several electrochemical sensors for direct detection of insulin had been created, but due to film thickness and adherence of these films under flow conditions, a new insulin-sensitive electrode was devised and used to obtain a dynamic profile of insulin release from pancreatic islets undergoing glucose stimulation. The typical MAMP sensor head was modified with a 1 mm glassy carbon rod, which was then coated in multi-walled carbon nanotubes dispersed in 3,4,-dihydro-2H-pyran (DHP), resulting in a thin-film insulin sensor capable of measuring changes in insulin secretion through electrochemical oxidation of the analyte with a response time on the order of 1 s. Numerous pancreatic islets (75–125) were immobilized within the cellular insert and insulin was measured in the MAMP as islets were perfused with 2.8 mM glucose in Hank's balanced salt solution (HBSS) and then stimulated with 16.7 mM glucose. The stimulation resulted in a rapid increase of insulin secretion of 228 $\pm$ 1 % compared to control islets and demonstrated the feasibility of performing direct, real-time metabolic analysis of islets in the MAMP. Modification of the MAMP for specialized purposes, such as insulin detection, is achievable due the versatility of the sensor head and instrumentation, providing a wealth of opportunities for future work with this instrument.

New MAMP Platform for Detection of Immunoresponse

A novel screen-printed platinum electrode (SPE) containing four modifiable working electrodes was designed and paired with a PDMS microfluidic chamber to create a new MAMP platform. The SPE was initially used in the development of a novel sensor for superoxide (SO), a molecule released by macrophages during an immune response known as oxidative burst [26]. SO has a very short lifetime and is typically detected as its downstream products, such as hydrogen peroxide. By coupling MAMP methods with a new platform and sensor, it was possible to detect real-time superoxide release from RAW macrophages treated with phorbol myristate acetate (PMA), a known promoter of oxidative burst. Upon treatment, a 46 $\pm$ 19 % increase in current was measured over a 30 min time period, or an increase of 8.8 $\pm$ 3.2 attomoles O_2/cell/s over control levels, demonstrating successful detection of sustained macrophage oxidative burst. The SPE MAMP platform can further be modified for the detection of glucose, lactate, oxygen, and acid to allow for detection of oxidative burst in conjunction with cellular immune responses.

Future Directions

The potential applications of multianalyte microphysiometry are numerous, including topics not discussed herein, including further investigations of the metabolic compensation of nutrient-challenged neurons [27], the investigation of macrophage response to bacterial infections [28–29], as well as the metabolic action of drug targets [30]. The VIIBRE potentiostat has also been modified to add potentiometric detection simultaneously with amperometric detection. This advancement will allow for expansion beyond the LAPS currently used for acid detection, including new sensing platforms capable of microphysiometry techniques such as the SPE. This improvement will also allow for addition of other potentiometric sensors, including ions key to cellular metabolism, such as H^+, K^+, Ca^{2+}, and Na^+, which are in constant flux in cellular physiology [31]. As the sensitivity and durability of electrochemical sensors continues to improve and evolve, these sensors can be applied to increasingly complex studies of cellular metabolism, advancing our understanding of metabolic contributions to cellular physiology.

Cross-References

- ▶ Amperometry
- ▶ Biosensors, Electrochemical
- ▶ Electrochemical Glucose Sensors
- ▶ Electrochemical Oxygen Sensors for Operation at Ambient Temperature
- ▶ Enzymatic Electrochemical Biosensors
- ▶ pH Electrodes - Industrial, Medical, and Other Applications
- ▶ Potentiometric pH Sensors at Ambient Temperature

References

1. Ray LB (2010) Metabolism is not boring. Science 330:1337
2. Wong W (2010) Focus issue: metabolic signals. Sci Signal 3:eg12
3. McKnight SL (2010) On getting there from here. Science 330:1338–1339
4. Bittner CX, Loaiza A, Ruminot I, Larenas V, Sotelo-Hitschfeld T, Gutierrez R, Cordova A, Valdebenito R, Frommer WB, Barros LF (2010) High resolution measurement of the glycolytic rate. Front Neuroenerg 2:32
5. Barros LF (2010) Towards single-cell real-time imaging of energy metabolism in the brain. Front Neuroenerg 2:4
6. Casciari JJ, Sotirchos SV, Sutherland RM (1992) Variations in tumor cell growth rates and metabolism with oxygen concentration, glucose concentration, and extracellular pH. J Cell Phys 151:386–394
7. Sanfeliu A, Paredes C, Cairo JJ, Godia F (1997) Identification of key patterns in the metabolism of hybridoma cells in culture. Enzyme Microb Technol 21:421–428
8. Goudar C, Biener R, Zhang C, Michaels J, Piret J, Konstantinov K (2006) Towards industrial application of quasi real-time metabolic flux analysis for mammalian cell culture. Adv Biochem Eng Biotechnol 101:99–118
9. Satoh W, Takahashi S, Sassa F, Fukuda J, Suzuki H (2009) On-chip culturing of hepatocytes and monitoring their ammonia metabolism. Lab Chip 9:35–37
10. Eklund SE, Taylor D, Kozlov E, Prokop A, Cliffel DE (2004) A microphysiometer for simultaneous measurement of changes in extracellular glucose, lactate, oxygen, and acidification rate. Anal Chem 76:519–527
11. Dransfeld C-L, Alborzinia H, Woelfl S, Mahlknecht U (2010) Continuous multiparametric monitoring of cell metabolism in response to transient overexpression of the sirtuin deacetylase SIRT3. Clin Epigenet 1:55–60
12. Burdallo I, Jimenez-Jorquera C (2009) Microelectrodes for the measurement of cellular metabolism. Procedia Chem 1:289–292
13. Ges IA, Baudenbacher F (2010) Enzyme electrodes to monitor glucose consumption of single cardiac myocytes in sub-nanoliter volumes. Biosens Bioelectron 25:1019–1024
14. Zeiger SLH, McKenzie JR, Stankowski JN, Martin JA, Cliffel DE, McLaughlin B (2010) Neuron specific metabolic adaptations following multi-day exposures to oxygen glucose deprivation. Biochimica et Biophysica Acta Mol Basis Dis 1802:1095–1104
15. Snider RM, Ciobanu M, Rue AE, Cliffel DE (2008) A multiwalled carbon nanotube/dihydropyran composite film electrode for insulin detection in a microphysiometer chamber. Analytica Chimica Acta 609:44–52
16. Snider RM, McKenzie JR, Kraft L, Kozlov E, Wikswo JP, Cliffel DE (2010) The effects of cholera toxin on cellular energy metabolism. Toxins 2:632–648
17. Eklund SE, Cliffel DE, Kozlov E, Prokop A, Wikswo J, Baudenbacher F (2003) Modification of the cytosensor(TM) microphysiometer to simultaneously measure extracellular acidification and oxygen consumption rates. Analytica Chimica Acta 496:93–101
18. Ges IA, Ivanov BL, Werdich AA, Baudenbacher FJ (2007) Differential pH measurements of metabolic cellular activity in nl culture volumes using

microfabricated iridium oxide electrodes. Biosens Bioelectron 22:1303–1310
19. Thomas PC, Halter M, Tona A, Raghavan SR, Plant AL, Forry SP (2009) A noninvasive thin film sensor for monitoring oxygen tension during in vitro cell culture. Anal Chem (Washington, DC, US) 81:9239–9246
20. Gordito MP, Kotsis DH, Minteer SD, Spence DM (2003) Flow-based amperometric detection of dopamine in an immobilized cell reactor. J Neurosci Methods 124:129–134
21. Velkovsky M, Snider R, Cliffel DE, Wikswo JP (2011) Modeling the measurements of cellular fluxes in microbioreactor devices using thin enzyme electrodes. J Math Chem 49:251–275
22. Eklund S, Thompson R, Snider R, Carney C, Wright D, Wikswo J, Cliffel D (2009) Metabolic discrimination of select list agents by monitoring cellular responses in a multianalyte microphysiometer. Sensors 9:2117–2133
23. Fishman PH (1980) Mechanism of action of cholera toxin: studies on the lag period. J Membr Biol 54:61–72
24. Eklund SE, Snider RM, Wikswo J, Baudenbacher F, Prokop A, Cliffel DE (2006) Multianalyte microphysiometry as a tool in metabolomics and systems biology. J Electroanal Chem 587:333–339
25. Flynn RWV, MacWalter RSM, Doney ASF (2008) The cost of cerebral ischaemia. Neuropharmacology 55:250–256
26. Hiatt LA, McKenzie JR, Deravi LF, Harry RS, Wright DW, Cliffel DE (2012) A printed superoxide dismutase coated electrode for the study of macrophage oxidative burst. Biosens Bioelectron. 33:128–133
27. McKenzie JR, Palubinsky AM, Brown JE, McLaughlin B, Cliffel DE (2012) Metabolic Multianalyte Microphysiometry Reveals Extracellular Acidosis is an Essential Mediator of Neuronal Preconditioning. ACS Chem Neurosci 3:510–518
28. Harry RS, Hiatt LA, Kimmel DW, Carney CK, Halfpenny KC, Cliffel DE, Wright DW (2012) Metabolic Impact of 4-Hydroxynonenal on Macrophage-Like RAW 264.7 Function and Activation. Chem Res Toxicol 25:1643–1651
29. Kimmel DW, Meschievitz ME, Hiatt LA, Cliffel DE (2013) Multianalyte Microphysiometry of Macrophage Responses to Phorbol Myristate Acetate, Lipopolysaccharide, and Lipoarabinomannan. Electroanalysis 25:1706–1712
30. Kimmel DW, Dole WP, Cliffel DE (2013) Application of multianalyte microphysiometry to characterize macrophage metabolic responses to oxidized LDL and effects of an apoA-1 mimetic. Biochem Biophys Res Commun 431:181–185
31. Wu Y, Wang P, Ye X, Zhang Q, Li R, Yan W, Zheng X (2001) A novel microphysiometer based on MLAPS for drugs screening. Biosens Bioelectron 16:277–286

Electrochemical Oxygen Sensors for Operation at Ambient Temperature

Chen-Wei Liu[1], Metini Janyasupab[2], Ying-Hui Lee[3] and Chung-Chiun Liu[2]
[1]Institute for Material Science and Engineering, National Central University, Jhongli City, Taoyuan County, Taiwan
[2]Electronics Design Center, and Chemical Engineering Department, Case Western Reserve University, Cleveland, OH, USA
[3]Chemical Engineering Department, National Tsing Hua University, Hsinchu, Taiwan

Electrochemical Sensors

Electrochemical sensors, well established in analytical and clinical chemistry, have been used extensively as an integral part of a chemical and biomedical sensing element, including specially for PO_2 sensing. In this chapter, we focus on the electrochemical oxygen sensor workable at ambient temperature. Sensing of dissolved oxygen at ambient condition becomes increasingly important in various applications for bioprocess control and medical diagnosis.

Electrochemical sensor is basically an electrochemical cell which employs a two- or three-electrode arrangement, containing working and counter (reference) electrodes or working, counter, and reference electrodes, respectively. Electrochemical sensors combine the characteristic of an electrochemical recognition process and an electrochemical transducer, which converts chemical energy into electrical energy, when a chemical reaction occurs on the working electrode. Typically, the electrochemical sensor is composed of two metal electrodes connected by an ionically conducting phase such as electrolytes. The most commonly used electrolytes of oxygen sensors are in aqueous or solid state. Panametric sensors, typically, use aqueous electrolyte. The electrochemical cell requires a specific design for the different principles of measurements. In the case of galvanic cells with solid and liquid electrolytes, the desirable electrodes are of primary importance.

Based on electrochemical principles, the Clark type (Clark 1956) of polarographic oxygen sensor is a well-known practical biomedical sensor for measuring dissolved oxygen. The Clark-type oxygen sensor consists of a sensing platinum electrode (cathode) and a reference silver electrode (anode), and it is enclosed within an oxygen permeable membrane. When an electrochemical potential of 800 mV is applied, oxygen can be reduced into hydroxide ions, resulting in a current that is proportional to the oxygen concentration. The electrode reactions are given below [1].

Based on the working signal produced by the cell, electrochemical sensors for oxygen detection can be divided in two main categories of sensors: amperometric and potentiometric. The amperometric and potentiometric sensors are characterized by their current-potential relationship within the electrochemical system. The applied current or potential for electrochemical sensors depends on the mode of operation.

Potentiometric Sensors

The potential established in an electrode-electrolyte interface where a redox reaction takes place in an electrochemical cell is the basis for a potentiometric sensor. For the fundamental of potentiometric oxygen sensors, the potential may be used to quantify the concentration of the oxygen species involved in the reaction. The potentiometric sensor is based on the physical properties of a galvanic cell, in which electrical current is produced by a spontaneous reaction occurring inside the sensor [2, 3]. The classic potentiometric oxygen gas sensor consists of a reference partial pressure (PO_2^{Ref}) and unknown partial pressure (PO_2) on the two side ends of the electrolyte. This electrolyte separates two compartments with different oxygen partial pressures. An open-circuit potential, governed by the "Nernst equation," measures the variation of oxygen concentrations between these two electrodes. The measured and the reference oxygen partial pressures are described as

$$E = (RT/nF)\ln\left(PO_2/PO_2^{Ref}\right) \quad (1)$$

where E is the measured potential, R is the universal gas constant, T is the operating temperature in absolute temperature scale, F is the Faraday's constant, and n is the number of electrons transferred. If PO_2^{Ref} at the reference electrode is maintained constant, the voltage at constant temperature is then directly related to PO_2. For sensing purposes, only one of half-cell reactions (those occurring at the working electrode) should involve the species of interest, in this case, the oxygen reaction. The other half-cell reaction is preferably reversible and noninterfering. According to the Eq. 1, a linear relation can be observed by the measured potential E and the natural logarithm of the ratio of the activities of the reactant and product. This slope value governing the sensitivity of the potentiometric sensor can also be inferred by the Eq. 1. However, anything will lead to an ionic interference in a potentiometric sensor, contributing to the measured potential in Eq. 1. Consequently, the potential measured in this galvanic cell cannot be used to quantify the ions of interest. Therefore, the surface of the active electrode often incorporates a specific functional membrane which exhibits ion-selective, ion-permeable, or ion-exchange properties in order to minimize the ionic interference [4].

Therefore, potentiometric sensors typically rely on the use of ion-selective electrodes, which are generally membrane-based devices, consisting of a permselective ion-conducting material. Ion-selective electrodes are used to separate the sample solution from the inside of the electrode providing an analytical signal. This membrane is designed to produce a potential difference due to the analyte of interest, in this case, the oxygen. The membranes with reagent selectively bind the target analyte, while the interaction of the reagent and analyte produces a charge separation at the surface of the membrane by measuring the potential versus the reference electrode. The activity potential of the analyte of interest in the sample electrolyte solution arises due to a potential difference. Ion-selective electrodes are based on the Nernstian behavior, and the potential can be related to the activity of the analyte, a_i, by the equation:

$$E = C + (2.303RT/nF) \log a_i \quad (2)$$

where C is a constant that contains the interface potential in the cell, R is the universal gas constant, T is the operating temperature in absolute temperature scale, F is the Faraday's constant, and n is the number of electrons transferred on the analyte of interest.

At low analyte concentrations, the activity of interfering ions must be considered. Ion-selective electrodes can be divided into categories based on the nature of the membrane material. These categories include glass membrane electrodes, liquid membrane electrodes, and solid-state membrane electrodes.

The reaction for potentiometric sensors requires that thermodynamic equilibrium conditions be maintained. Therefore, potentiometric measurements are performed under conditions of near-zero current. Consequently, a high-input impedance electrometer is needed for the measurements. In addition, a long response time may be required for a potentiometric sensor to reach equilibrium conditions in order to obtain a meaningful measurement.

The development of novel potentiometric sensors is becoming increasingly important for biomedical applications.

Amperometric Sensors

Amperometric sensors with the linear relationship of the current response have received great attention because of their high sensitivity, wide detection range, and short response time [5–9]. Amperometric sensors are based on the current-potential relationship of the electrochemical cell, in which a non-spontaneous reaction is driven by an external source of current. In amperometric sensors, the transduction mode works by operating the potential of the working electrode at a fixed value, relative to the reference electrode, and observing the current as a function of time. The applied potential assists to drive the electron transfer reaction of the electroactive species, and the resulting current is a direct measure of the rate of the electron transfer reaction. The measured rate can be used to quantify the concentrations of species involved in the reaction.

The current-potential characteristics of electroactive species can be directly affected by the mass transfer of detecting oxygen species in this case in the kinetics of the faradaic or charge transfer reaction at the electrode surface [4]. This mass transfer can be accomplished through (a) an ionic migration as a result of an electric potential gradient, (b) a diffusion under a chemical potential difference or concentration gradient, and (c) a bulk transfer by natural or forced convection [4]. The rate of the faradaic process in an electrochemical cell can be influenced by the electrode reaction kinetics and the mass transfer processes.

Fick's law of diffusion can be used to describe when a preferred mass transfer condition is total diffusion control. When measuring the rate of faradaic process at an electrode surface, the cell current usually increases with increases in the electrode potential. When it reaches maximum mass transfer rate, the current approaches a limiting value. Under this condition, the zero concentration, which is diffusional mass transfer, of the detecting species at the electrode surface can be obtained. Consequently, the limiting current and the bulk concentration of the detecting species can be related by [4]

$$i = ZFkmC^* \quad (3)$$

where km is the mass transfer coefficient and C^* is the bulk concentration of the detecting species. At the extreme condition, when compared with the mass transfer rate, the slow kinetics of electrochemical system, usually corresponding to a small overpotential, is observed in the reaction kinetic control regime. The limiting current and the bulk concentration of the detecting species can be related as [4]

$$i = ZFkcC^* \quad (4)$$

where kc is the kinetic rate constant for the electrode process. The limiting current or relatively small overpotential conditions are usually operated in an amperometric sensor. When operating at fixed electrode potential of amperometric sensors, the cell current can be correlated with the bulk concentration of the detecting species. The external cell voltage of amperometric sensors is sufficiently high to maintain a zero oxygen concentration at the cathodic surface. Therefore, the sensor current response is diffusion controlled. Furthermore, the amperometric sensors show some selectivity since the reduction or oxidation potential is characteristic of the species being analyzed.

According to the peak current, the concentrations of the reactant can be quantified when the diffusivity is negligible. The potential at which the peak current occurs can be used to identify the reaction. Based on the half-cell potential of the electrochemical reactions, the identification for reactions or reactants is listed extensively in handbooks and references. Amperometric sensors can be used very effectively to carry out qualitative and quantitative analyses of chemical and biochemical species.

Ambient Temperature Oxygen Sensors

Ambient temperature oxygen sensors have generated much interest for chemical and biomedical sensing applications. Solid-state oxygen sensors have been developed using oxide ion conductors in order to control the oil-gas ratio in an engine. However, the original design required elevated temperatures, well above 400 °C, which was not suitable for bioprocess control and medical application. Electrochemical oxygen sensors using liquid electrolytes, the so-called Clark-type oxygen electrodes as mentioned above, were used for the detection of oxygen at ambient temperature. In contrast to the solid electrolyte gas sensor, gas sensors use liquid electrolytes, providing a reliable response at ambient temperature. The main disadvantage of Clark-type gas electrode is the consumption of oxygen in the form of OH^-resulting in an electrolyte with an alkaline level which can cause the solid polymer electrolyte (SPE) film dry. The AgCl coating on the silver anode reduced the reaction area over time, and this fabrication process is not suitable for microfabrication techniques [1, 10]. Therefore, developing oxygen sensors capable of functioning at an ambient temperature and microfabrication technique is important for biomedical application. Attempts have been made to lower the operating temperature of solid-state electrolyte oxygen sensors. For example, a field-effect transistor (FET) sensor using YSZ thin film was reported to operate at substantially lower temperature, but operation at an ambient temperature remained impossible or unstable [11]. In contrast to oxide ion conductors, halide ion conductors and proton conductors were considered to be used for solid-state oxygen sensors at room temperature. LaRoy et al. reported that they manufactured an oxygen sensor based on a non-oxidic material using an evaporated LaF_3 film [12]. The amperometric response in the sensor cell changed with oxygen partial pressure at room temperature. Table 1 summarized the different types of oxygen sensors operated in ambient temperature [10]. Similar amperometric sensors based on LaF_3 [13] or beta-PbF_2 [14] were subsequently reported. These types of amperometric sensors with limited current may have an intrinsic difficulty in obtaining a stable output signal due to a fluoride ion, the electric carrier of the conductors. Furthermore, metal halides, such as $PbSnF_4$, LaF_3, and SrO_2-doped $SrCl_2$, exhibited high ionic conductivity and therefore could be utilized as the solid-state electrolyte for the gaseous oxygen sensor and operate in temperature lower than 250 °C. The oxygen sensor utilizing $PbSnF_4$ as the solid-state electrolyte worked at 100 °C, but the response time was long, i.e., 340 min [15]. The study adopted the construction of Sn, SnF_2/$PbSnF_4$/FePc (reference electrode)/(solid electrolyte)/(sensing electrode), which reduced the response time to 10 min, lowering the operation temperature to 20 °C [16]. LaF_3 represented a more promising material to be used as the solid-state electrolyte in oxygen sensors. LaF_3 pellet operated at the temperatures as low as 50 °C. However, the response time was relatively slow (30 min) [17]. In order to improve the sensing performance, single crystal LaF_3 was

Electrochemical Oxygen Sensors for Operation at Ambient Temperature, Table 1 Oxygen sensors operates at ambient temperature [10]

Sensor configuration	Sensing mode	PO_2 range	90 % response time	Reference
Bi/LaF_3 (evap film)/Au	Amperometric (applied voltage)	$N_2 \leftarrow\rightarrow O_2$ (linear)	4 min (drift)	[12]
Au/LaF_3 (evap film)/La	Amperometric (applied voltage)	ca 1–100 % (logarithmic)	3 min	[13]
Au/PbF_2 (evap film)/Au	Amperometric (applied voltage)	–	ca 2–3 min	[14]
FePc/$PbSnF_4$ (+BaO_2)/Sn,SnF_2	Potentiometric (emf)	ca 3–100 % (logarithmic)	10 min	[16]
Pt/LaF_3 (single cryst)/Sn,SnF_2	Potentiometric (emf)	ca 2–100 % (logarithmic)	ca 2–3 min	[18–20]
Pt/LaF_3 (sputtered film)/Sn,SnF_2	Potentiometric (emf)	ca 10–100 % (logarithmic)	20 s	[22, 23]
Pt/Nafion/Ag	Potentiometric (emf)	ca 2–100 % (logarithmic)	30 min	[30]
Pt/Nafion/Pt	Amperometric (applied voltage)	ca 0–80 % (linear)	3 min	[31]
(EMIBF4) porous polyethylene membrane-coated	Amperometric (applied voltage)	$N_2 \leftarrow\rightarrow O_2$ (linear)	2.5 min	[37]
Pt/THTDP-AQS/BMIM–NTF_2	Amperometric (applied voltage)	$N_2 \leftarrow\rightarrow O_2$ (linear)	10–15 s	[38]

utilized in place of LaF_3 pellet. The sensor configuration was Sn, SnF_2/single crystal LaF_3/Pt black (ref. electrode)/(solid electrolyte)/(sensing electrode) [18]. The oxygen sensor could operate at 25 °C with the 90 % response time as short as 3 min, whereas no response was observed for LaF_3 pellet under the same conditions. The improvement was attributed to the high conductivity of single crystal LaF_3 even though the resistance for single crystal LaF_3 and LaF_3 pellet was found to be 107 Ω and 109 Ω, respectively. Similar performance was observed when changing the sensing electrode to Pb phthalocyanine. The response time was only 30 s when operating at 30 °C [19]. In addition, further treatment of water vapor exposure at 150 °C for 2 h formed a partially hydroxylated surface on the single crystal LaF_3 (formation of $La(OH)_3$) and therefore shortened the response time from 3 to 2 min. However, the hydroxylation had an adverse effect on the sensing performance [20]. Therefore, sputtered LaF_3 film is employed as an alternative, considering the fabrication cost, the complexity of the microfabrication processes, and the sensor integration. Sputtered LaF_3 film showed a faster response compared to single crystal LaF_3. Similar performance was observed when utilizing the CuPc black as the sensing electrode [21]. With the optimal water vapor treatment at 90 °C for 1 h, the response time was reduced to 30 s [22]. In addition to the detection of gaseous oxygen, the sputtered LaF_3 film also showed great performance for the detection of dissolved oxygen. Short response time of 30 s was obtained even without the surface pretreatment of water vapor. [23] Nevertheless, the continuous use of the sensor would result in an increased response time; reactivation would be required to recover the sensitivity. Heat treatment of merely 300 ns might reduce the response time back to ca. 35 s [24]. Previous studies indicated that the sensor showed stable EMF change to the relative humidity (RH), yet the decrease of EMF was observed when the RH increases to 97 % [25].

In addition to halide ion conductors, proton conductors such as Nafion membrane enabled the development of a unique amperometric oxygen sensor operating at ambient temperature.

Researchers had attempted to develop solid-state gaseous sensors, which could be operated at room temperature [26–31]. The advantage of a solid polymer electrolyte (SPE), such as Nafion membrane, was that it did not involve liquid electrolyte minimizing the leakage problem. The disadvantages of solid polymer electrolyte film included that the wetness of solid polymer electrolyte film dramatically affected by its conductivity [32]. Also, the drying out problem of solid polymer electrolyte films remained to be solved. Recently, room temperature ionic liquids (RTILs) have been used, including organic and inorganic synthesis, separation processes, and electrochemistry [33–36]. Room temperature ionic liquids with low melting point, a wide range of detection windows, and high solubility for a numerous materials were desirable for preparing gas sensors. Furthermore, a relatively low vapor pressure of ambient temperature ionic liquids would minimize the drying out problem of the liquid electrolyte solution, a major problem for the sensors using liquid electrolyte. Therefore, through the use of room temperature ionic liquids, a high stability and a long lifetime can be observed. Rong Wang et al. [37] reported that supported 1-ethyl-3-methylimidazolium tetrafluoroborate (EMIBF4) porous polyethylene membrane-coated electrodes using as a solid-state ionic conductor at room temperature. The transient reduction current ($I_{c,\ trans}$), steady-state reduction current ($I_{c,\ s}$), and transient oxidation currents ($I_{a,\ trans}$) obtained by potential-step chronoamperometry were used for determining the O_2 concentration in the gas phase. This sensor showed a wide detection range, a high sensitivity, and a good reproducibility. Furthermore, R. Toniolo et al. [38] found that the addition of quinone moieties dissolved in these low-melting salts was able to promote oxygen reduction at relatively lower potentials with respect to its standard potential, conceivably through an inner sphere electron transfer proceeding via a covalent linkage between the two redox partners [39, 40]. The effective amperometric sensors with suitable functionalized (RTILs) (trihexyltetradecylphosphonium- 9,10-anthraquinone- 1-sulphonate; THTDP-AQS), and 1-butyl-3-methylimidazolium bis(trifluoromethylsulfonyl) imide (BMIM–NTF2) were developed by adding small amounts of a further low-melting salt bearing a quinone moiety. This allowed the reduction of O_2 to occur through an electrocatalytic pathway at quite lower potentials than those required by its direct reduction [38].

The advantages of using room temperature ionic liquid (RTIL) electrolyte in membrane-coated electrode as an O_2 gas sensor compared to solid electrolyte gas sensors and classic Clark-type gas sensors included easy construction and the ability to operate at ambient temperature. This would be the direction for future ambient temperature oxygen sensor development.

Cross-References

▶ Electrochemical Sensor of Gaseous Contaminants
▶ Electrochemical Sensors for Environmental Analysis

References

1. Ramamoorthy R, Dutta PK, Akbar SA (2003) J Mater Sci 38:4271
2. Yang P, Zhao D, Margolese D, Chmelka B, Stucky G (1998) Nature 396:152
3. Yang P, Zhao D, Margolese D, Chmelka B, Stucky G (1999) Chem Mater 11:2813
4. Liu CC Electrochemical sensors. The biomedical engineering handbook, 2nd edn
5. Bakker E, Telting-Diaz M (2002) Anal Chem 74:2781
6. Eggins BR (2002) Chemical sensors and biosensors. Wiley, Chichester
7. Chang SC, Stetter JR, Cha CS (1993) Talanta 40:461
8. Docquier N, Candel S (2002) Prog Energy Combust Sci 28:107
9. Yamazoe N, Miura N (1999) MRS Bull 24:37
10. Noboru Yamazoe, Norio Miura (1990) Trends Analyt Chem 9:170
11. Miyahara Y, Tsukuda K, Miyagi H (1987) In: Proceedings transducers '87, IEEJ, Tokyo, p 648
12. LaRoy BC, Lilly AC, Tiller CO (1973) J Electrochem Soc 120:1668
13. Yamaguchi A, Matsuo T (1981) In: Proceedings 1st sensor Symposium, Tsukuba, IEEJ, Tokyo, p 99
14. Coutuner G, Danto Y, Grbaud R, Salardenne J (1981) Solid State Ionics 5:621
15. Siebert E, Fouletier J (1983) Solid State Ionics 9:1291
16. Siebert E, Fouletier J, Kleitz M (1987) J Electrochem Soc 134:1573
17. Kuwata S, Miura N, Yamazoe N, Seiyama T (1983) Denki Kagaku 51:947

18. Kuwata S, Miura N, Yamazoe N, Seiyama T (1984) Chem Lett 981
19. Lukaszewicz JP (1992) Sens Actuators B 9:55
20. Yamazoe Y, Hisamoto J, Miura N (1987) Sens Actuators 12:415
21. Tan GL, Wu XJ, Wang LR, Chen YQ (1996) Sens Actuators B 34:417
22. Miura N, Hisamoto J, Kuwata S, Yamazoe N (1987) Chem Lett 1477
23. Miura N, Hisamoto J, Yamazoe N (1989) Sens Actuators 16:301
24. Moritz W, Krause S, Roth U, Klimm D, Lippitz A (2001) Anal Chim Acta 437:183
25. Sun G, Wang H, Jiang Z (2011) Rev Sci Instrum 82:083901
26. McRipley MA, Linsenmeier RA (1996) J Electroanal Chem 414:235
27. Limoges B, Degrand C, Brossier P (1996) J Electroanal Chem 402:175
28. Do JS, Shieh RY (1996) Sens Actuators B 37:19
29. Katakura K, Onoma A, Ogumi Z, Takemara Z (1990) Chem Lett 1291
30. Miura N, Kato H, Yamazoe N, Seiyama T (1984) Denki Kagaku 52:376
31. Kuwata S, Miura N, Yamazoe N (1988) Chem Lett 1197
32. Wallgren K, Sotiropoulos S (2001) Electrochim Acta 46:1523
33. Li R, Wang J (2002) Huagong Jinzhan 21:43
34. Tobishima S (2002) Electrochem 70:198
35. Ue M, Takeda M (2002) Electrochem 70:194
36. Matsumoto H, Matsuda T (2002) Electrochem 70:190
37. Wang R, Okajima T, Kitamura F, Ohsaka T (2004) Electroanalysis 16:66
38. Toniolo R, Dossi N, Pizzariello A, Pizzariello AP, Susmel S, Bontempelli G (2012) J Electroanal Chem 670:23
39. Hubig SM, Rathore R, Kochi JK (1999) J Am Chem Soc 121:617
40. Houmam A (2008) Chem Rev 108:2180

Electrochemical Perfluorination

Akimasa Tasaka
Doshisha University, Kyotanabe, Kyoto, Japan

Introduction

The presence of fluorine atoms in inorganic and organic compounds has quite unique and specific effects on various properties because fluorine has the highest electronegativity and the sterically second smallest van der Waal's radius [1, 2]. Therefore, fluorinated inorganic and organic compounds have wide application in various fields such as dry etchant, cleaning gas, refrigerant, fire extinguisher gases, surface-active agents, agricultural chemicals, and medicines.

Electrochemical fluorination is a useful process for production of perfluorinated organic compounds with functional groups in starting materials as well as perfluorinated inorganic compounds [3]. In this method, elementary fluorine is neither produced nor employed. Characteristics of electrochemical fluorination are as follows [4]:

1. Liquid hydrogen fluoride (anhydrous hydrofluoric acid), which is industrially produced on a large scale, can be used as a fluorine source.
2. Perfluorinated compounds as objective can be obtained in one step process.
3. Although the cleavage of carbon-carbon bond in a fluorinating organic compound takes place in a certain degree during electrochemical fluorination, perfluorinated organic compounds obtained by this process have function groups such as –COF, $-SO_2F$ and so on in starting materials.
4. System for manufacturing perfluorinated compounds is relatively simple.
5. Both batch and continuous processes can be utilized for manufacturing perfluorinated compounds.
6. Not only perfluorinated compound but also many kinds of by-products can be produced in this process.
7. In general, the yield of perfluorinated organic compounds using electrochemical fluorination is low. However, the yield of perfluorinated organic compounds using chemical reaction is much lower than that using electrochemical fluorination.
8. Generally, any partially fluorinated compound can be hardly obtained by this process.
9. A nickel anode is available in liquid hydrogen fluoride (anhydrous hydrofluoric acid), but anodic dissolution may take place vigorously by several percentage of passing electricity. A carbon anode is also available in the melt having hydrogen fluoride such as KF•2HF and NH_4F•2HF, but anode effect may occur often.

Definition

Electrochemical fluorination is a process for substitution of all hydrogen atoms in organic and inorganic compounds as a starting material to fluorine atoms without elementary fluorine generation during electrolysis and employing it in the reaction.

Principles

When electrolysis is conducted at potentials in the range of 5–8 V, fluorination can take place at an anode, that is, on the anode and/or in an electrolyte near the anode. Theoretical standard potential for fluorine gas generation in every electrolyte is 2.87 V. This value is independent of temperature and is almost constant at every temperature. A fluoride anion in an electrolyte can be discharged on a carbon anode at potentials higher than ca. 3.5 V to form fluorine radical (atomic fluorine). Then it fluorinates a starting material solved in a HF solvent or a melt having HF stepwise. Fluorine radical reacts with a fluorinating organic or inorganic compound to extract a hydrogen atom in a starting material, resulting in HF formation. Substrate radical formed from a starting material couples with another atomic fluorine to form a partially fluorinated compound (a monofluorinated compound) together with HF generation. That is, substitution of one hydrogen atom in a fluorinating substance with one fluorine atom needs two fluorine atoms. A monofluorinated compound reacts successively with atomic fluorine to form more fluorinated compound with HF generation. The similar reaction is repeated in many steps, and finally, the perfluorinated compound is formed. Because a partially fluorinated compound has polarity, it stays in a solution such as liquid hydrogen fluoride and a melt, and so, only a perfluorinated compound can be separated from an electrolyte, that is, a mixed solution composed of hydrogen fluoride, a few partially fluorinated compounds, a starting material, and so on. On the other hand, a fluoride anion can be also discharged on a nickel anode at potentials higher than ca. 3.5 V to form fluorine radical (atomic fluorine). Fluorine radical oxidizes nickel fluoride formed on the nickel anode as well as a starting material to a highly oxidized nickel fluoride and a monofluorinated compound with HF formation, respectively. A highly oxidized nickel fluoride acts as a mediator at a nickel anode during electrochemical fluorination. After that, both fluorine radical and a highly oxidized nickel fluoride work as a fluorinating agent and fluorinate a partially fluorinated compound to form more fluorinated compound and nickel fluoride. The similar reaction takes place successively, and finally, a perfluorinated compound is formed. In this case, a partially fluorinated organic or inorganic compound has also polarity and it stays in an electrolyte. In contrast, the perfluorinated organic and inorganic compounds become nonpolar and then, they precipitate from liquid hydrogen fluoride on the cell bottom as a liquid compound and volatilize from a solution composed of hydrogen fluoride, a partially fluorinated compound, a starting material, and so on as a gaseous compound.

Applications

Some typical examples of the use of Electrochemical Perfluorination are shown below, although a rich variety of examples are reported in the literature.

Electrolysis of Liquid Hydrogen Fluoride Containing an Organic Starting Material

This method is called as Simons' process [5, 6]. The electrochemical process consists of passing current through a simple one-compartment cell containing organic chemical compound or compounds, and an electrically conducting solution containing hydrogen fluoride. Many organic compounds, particularly those containing oxygen or nitrogen, have relatively high solubility in liquid hydrogen fluoride and form an electrically conducting solution. Others, such as hydrocarbons, are not very soluble and do not form conducting solutions, but fortunately, the addition of some materials such as KF and NaF which imparts conductivity enables the process to be used even with hydrocarbon starting materials. When current is passed, electrochemical reaction is mainly the formation of fluorine radical through a discharge of a fluoride anion. Then the formed fluorine radical fluorinates a starting material stepwise, and finally, a perfluorinated compound is separated from a medium. At first, electrochemical reaction takes place at an anode and then chemical reaction for fluorination and electrochemical reaction take place alternately.

The cells can be constructed of any material, such as iron or copper, not readily attacked by hydrogen fluoride. The electrodes can be in a compact pack with alternating cathodes and anodes. The cell body can serve as a part of the cathodes. The anodes are usually nickel. Cathodes can be nickel, iron, copper, and many other conductors. Hydrogen is generated, and gaseous or readily volatilized products escape with hydrogen. Liquid products from a layer in the bottom of the cell and are drained off. Voltages between 5 and 8 are useful but between 5 and 6 are preferred. Sometimes even lower voltages are possible. Current density in the order of 0.02 Ampere per square centimeter (0.02 A/cm^2) can be employed. Temperature of an electrolyte should be kept below 0 °C, because of the higher vapor pressure of HF (ca. 50.6 kPa even at 0 °C).

Under conditions of interest, HF is in equilibrium, as shown in Eqs. 1, 2 and 3 [7–9].

$$HF \rightleftharpoons H^+ + F^- \tag{1}$$

$$H^+ + mHF \rightleftharpoons H(FH)_m{}^+ \tag{2}$$

$$F^- + nHF \rightleftharpoons F(HF)_n{}^- \tag{3}$$

where $F(HF)_n{}^-$ and $H(FH)_m{}^+$ are the solvated ions of F^- and H^+, respectively, and are now written as F^- and H^+ for simplicity. Because highly pure HF has no electric conductivity, KF and NaF are added to liquid hydrogen fluoride and dissociated to form K^+ and Na^+ cations and F^- anion in the solution. Fluoride anion, F^-, comes of KF and NaF is also coordinated with HF in the solution to form $F(HF)_n{}^-$ according to Reaction (3).

For example, mechanism on fluorination of $CH_3(CH_2)_nSO_2F$ and $CH_3(CH_2)_nCOF$ to $CF_3(CF_2)_nSO_2F$ and $CF_3(CF_2)_nCOF$, which are surface-active agents produced by 3 M Company in USA, is as follows:

$$F^- \rightarrow \bullet F + e^- \tag{4}$$

In the case of $C_mH_{2m+1}SO_2F$, fluorination takes place according to Reactions (5) and (6).

$$C_mH_{2m+1}SO_2F + \bullet F \rightarrow \bullet C_mH_{2m}SO_2F + HF \tag{5}$$

$$\bullet C_mH_{2m}SO_2F + \bullet F \rightarrow C_mH_{2m}FSO_2F \tag{6}$$

Total reaction for fluorination of $C_mH_{2m+1}SO_2F$ is written as Reaction (7).

$$C_mH_{2m+1}SO_2F + 2\bullet F \rightarrow C_mH_{2m}FSO_2F + HF \tag{7}$$

In the case of fluorinating $C_mH_{2m+1}COF$, similar reactions take place.

$$C_mH_{2m+1}COF + \bullet F \rightarrow \bullet C_mH_{2m}COF + HF \tag{8}$$

$$\bullet C_mH_{2m}COF + \bullet F \rightarrow C_mH_{2m}FCOF \tag{9}$$

Total reaction for fluorination of $C_mH_{2m+1}COF$ is also written as Reaction (10).

$$C_mH_{2m+1}COF + 2\bullet F \rightarrow C_mH_{2m}FCOF + HF \tag{10}$$

The ECF reaction similar to Reactions (4), (5), (6) and (7) and Reactions (4), (8), (9) and 10) is repeated successively and total reactions for electrolytic production of these surfactants are written as Reactions (11) and (12).

$$CH_3(CH_2)_nSO_2F + (6+4n)HF - (6+4n)e^- \rightarrow CF_3(CF_2)_nSO_2F + (3+2n)HF + (3+2n)H_2 \tag{11}$$

$$CH_3(CH_2)_nCOF + (6 + 4n)HF - (6 + 4n)e^- \rightarrow CF_3(CF_2)_nCOF + (3 + 2n)HF + (3 + 2n)H_2 \quad (12)$$

Another fluorination with respect to a highly oxidized nickel fluoride takes place in liquid hydrogen fluoride using a nickel anode, similar to electrolytic production of nitrogen trifluoride using a nickel anode in an $NH_4F{\bullet}2HF$ melt, as mentioned below [7–14]. Anode consumption of nickel is caused according to Reaction (13). Nickel fluoride, NiF_2, on a nickel anode is a protective film against corrosion, and it is formed according to Reactions (14) and (14′). When K^+ cation is present in a solution, $KNiF_3$ is formed on a nickel anode instead of NiF_2 [15, 16]. $KNiF_3$ is a brittle compound and is not resistant to corrosion of a nickel anode in a solution. Hence, K^+ is very detrimental to corrosion of a nickel anode. Highly oxidized nickel fluorides seem to form during electrolysis using a nickel anode in liquid hydrogen fluoride, according to Reactions (15), (16), (17), and (18).

$$Ni \rightarrow Ni^{2+} + 2e^- \quad (13)$$

$$Ni^{2+} + 2F^- \rightleftharpoons NiF_2 \quad (14)$$

$$Ni + 2F^- \rightarrow NiF_2 + 2e^- \quad (14')$$

$$2NiF_2 + F^- \rightarrow Ni_2F_5 + e^- \quad (15)$$

$$NiF_2 + F^- \rightarrow NiF_3 + e^- \quad (16)$$

$$NiF_2 + 2F^- \rightarrow NiF_4 + 2e^- \quad (17)$$

$$2NiF_3 + 2F^- \rightarrow NiF_2 + NiF_6{}^{2-} \quad (18)$$

Highly oxidized nickel fluorides such as Ni_2F_5, NiF_3, and NiF_4 on a nickel anode and fluorocomplex ion such as $NiF_6{}^{2-}$ in a solution may act as a fluorinating agent similar to fluorine radical, so that highly oxidized nickel fluorides on a nickel anode and/or a fluorocomplex ion in a solution also fluorinate an organic starting material to form a perfluorinated organic compound.

Electrolysis of a Molten Salt With and Without a Fluorinating Compound

Phillips' Process

Phillips Petroleum Company in Germany developed this process, and so it is called as Phillips' process [17]. Substrates of choice should be at least moderately volatile and not particularly soluble in $KF{\bullet}2HF$ electrolyte. Electrolysis is conducted in a $KF{\bullet}2HF$ melt similar to fluorine. It is thought that elemental fluorine is generated as the anode reaction. The anode is porous carbon (not graphite), and the process is thought to involve the electrolytic generation of elementary fluorine and reaction of that fluorine with substrate within the porous carbon anode. Suitable substrates include alkanes, cycloalkanes, chloroalkanes, acyl fluorides, and esters. Substrate is introduced continuously into the network of pores through the gas cap cavity at the bottom of the anode; from there, it passes up through the network of pore and reaches with fluorine as it moves up. The products appear to come from the statistical replacement of substrate hydride with fluorine and range from monofluoro- to perfluoro-compounds. The mild steel cell case is the cathode. Hydrogen from the cathode lifts the electrolyte upward in the anode/cathode gap; it escapes into the vapor space; and the electrolyte returns through the downcomer. The electrolysis is run at 90–100 °C.

The process is efficient for many fluorinations. While passing a vapor of insoluble hydrocarbons through network of pores in the porous carbon anode, it is fluorinated and a few kinds of partially fluorinated organic compounds together with a perfluorinated organic compound are formed. For example, ethane is frequently run as a model substrate. It runs well, and the product fluoroethanes can be analyzed completely. All the fluoroethanes are produced. Conversion of perfluorinated ethane is smaller than total conversion of a few of partially fluorinated ethanes. The ratio of perfluorinated ethane to a few of partially fluorinated ethanes is dependent upon the ratio of ethane flow rate to passing current and conversion of perfluorinated ethane increased with the increase in passing electricity.

Mechanism of fluorination in this process is "In situ" reaction with fluorine generation, that is, organic substrate of hydrocarbon reacts with free radical of fluorine and so, this process is classified in direct fluorination.

Mitsui Chemicals' Process [7–14, 18]

Nitrogen trifluoride (NF_3) is a stable gas at room temperature and has a strong oxidation power at higher temperatures. Therefore, it has been already used as an oxidizing agent for rocket fuels and as a stable fluorinating agent. Recently, much effort has been devoted to use NF_3 as a dry etching gas (dry etchant) and a cleaning gas for apparatus used in the CVD technique. At present, a large amount of NF_3 is used as an etchant and a cleaning gas in the world.

This process is employed at present for electrolytic production of NF_3 in an industrial scale in Japan. On the bases of papers reported by O. Glemser et al. [19], J. Massonne [20], and the author [21–23], the author studied electrolysis of $NH_4F{\cdot}2HF$ melt with a nickel anode in order to develop the electrolytic process as one for manufacture of NF_3. Mitsui Chemicals Inc. Company developed this process in the facility site at Shimonoseki on the advice of the author and with full supports of the author and this process is called as Mitsui Chemicals' process.

The cells can be constructed of any material, such as nickel or iron coated with PTFE layer, not readily attacked by hydrogen fluoride. The cell body made of nickel can serve as a cathode. Nickel and carbon are available as the anode for electrochemical fluorination of molten fluorides containing HF, because the oxidized layers formed on these materials have an electronic conductivity and a high resistance to corrosion. The anode gas is composed of gaseous products generated at a nickel anode. Cathodes can be nickel and Monel (Ni–Cu alloy). Hydrogen is generated on a cathode. A metallic skirt is provided between an anode and a cathode to separate the anode gas from hydrogen generated at a cathode, so that explosion and loss of NF_3 are prevented. When the cell wall is used as a cathode, the cell bottom is covered with the PTFE sheet to avoid hydrogen evolution. The electrolyte is charged in the cell placed in air, and hence, the water content seems to be more than 0.02 wt% at start of electrolysis.

$NH_4F{\cdot}2HF$ is prepared with extra pure HF and NH_4F (99.9 %, at laboratory) or highly pure NH_3 (at industry) in a dry box. Electrolysis is conducted with the electrolytic cells at 100 or 120 °C. Although the water content is high before start-up, it may be decreased by electrolysis to less than 0.02 wt% within 80 h [15]. Voltages between 5 and 8 are useful but between 5 and 6 are preferred. Current density in the order of 0.10–0.20 A per sq. cm. (0.10–0.20 A/cm^2) can be employed. Temperature of an electrolyte is kept at 100–120 °C.

Electrochemical fluorination is conducted on a nickel anode. Water in the melt is electrolyzed in preference to fluorination of NH_4^+ for some 30 h after start-up. NF_3 is the main product, followed by N_2 with a small amount of O_2, N_2O, N_2F_2, and N_2F_4. The current efficiency for NF_3 formation increases with the increase in the current density and reaches constant values at 0.120 A/cm^2 [24]. The maximum current efficiency on a nickel anode is 68 % [24].

Under the conditions of interest, NH_4^+ and NH_3 are in equilibrium, as shown in Eqs. 19 and 20,

$$NH_4F + nHF \rightleftharpoons NH_4^+ + F(HF)_n^- \qquad (19)$$

$$NH_4^+ + mHF \rightleftharpoons NH_3 + H(FH)_m^+ \qquad (20)$$

where $F(HF)_n^-$ and $H(FH)_m^+$ are also the solvated ions of F^- and H^+, respectively, and are written as F^- and H^+ for simplicity. The amount of NH_3 in the melt is considered negligible. Raman and Infrared spectra reveal that the n value in fluorohydrogenate anions of $F(HF)_n^-$ is 1–3, and $F(HF)_2^-$ is mainly present in $NH_4F{\cdot}mHF$ melts [25].

Mechanism on fluorination of NH_4^+ and/or NH_3 to NF_3 is as follows:

At first, a fluoride anion is discharged at potentials higher than 3.5 V to form fluorine radical (atomic fluorine), according to Reaction (4).

$$F^- \rightarrow {\cdot}F + e^- \qquad (4)$$

Fluorine radical reacts with NH_4^+ and/or NH_3 to form NF_3, according to Reactions (21), (21′), (22), (23), (24), (25), and (26).

$$NH_4^+ + \bullet F \rightarrow \bullet NH_2 + H^+ + HF \quad (21)$$

$$NH_3 + \bullet F \rightarrow \bullet NH_2 + HF \quad (21')$$

$$\bullet NH_2 + \bullet F \rightarrow NH_2F \quad (22)$$

$$NH_2F + \bullet F \rightarrow \bullet NHF + HF \quad (23)$$

$$\bullet NHF + \bullet F \rightarrow NHF_2 \quad (24)$$

$$NHF_2 + \bullet F \rightarrow \bullet NF_2 + HF \quad (25)$$

$$\bullet NF_2 + \bullet F \rightarrow NF_3 \quad (26)$$

Total reaction for electrochemical formation of NF_3 is written as Eqs. 27 and 27′.

$$NH_4^+ + 7F^- \rightarrow NF_3 + 4HF + 6e^- \quad (27)$$

$$NH_3 + 6F^- \rightarrow NF_3 + 3HF + 6e^- \quad (27')$$

By-products such as N_2F_2 and N_2F_4 are formed, according to Reactions (28) and (29).

$$\bullet NHF + \bullet NF_2 \rightarrow N_2HF_3 \rightarrow N_2F_2 + HF \quad (28)$$

$$2\bullet NF_2 \rightleftharpoons N_2F_4 \quad (29)$$

Another fluorination with respect to a highly oxidized nickel fluoride also takes place in a $NH_4F\bullet 2HF$ using a nickel anode similar to electrolytic production of organic perfluorinated compound using a nickel anode in a liquid hydrogen fluoride as mentioned above [5, 6]. In this process, anodic dissolution takes place according to Reaction (13). Nickel fluoride, NiF_2, on a nickel anode is a protective film against corrosion and it is formed according to Reactions (14) and (14′).

Highly oxidized nickel fluorides in Mitsui Chemicals' process also seem to form during electrolysis using a nickel anode according to Reactions (15), (16), (17) and (18) and Reactions (30), (31), (31′) and (32).

$$Ni \rightarrow Ni^{2+} + 2e^- \quad (13)$$

$$Ni^{2+} + 2F^- \rightleftharpoons NiF_2 \quad (14)$$

$$Ni + 2F^- \rightarrow NiF_2 + 2e^- \quad (14')$$

$$2NiF_2 + F^- \rightarrow Ni_2F_5 + e^- \quad (15)$$

$$NiF_2 + F^- \rightarrow NiF_3 + e^- \quad (16)$$

$$NiF_2 + 2F^- \rightarrow NiF_4 + 2e^- \quad (17)$$

$$2NiF_3 + 2F^- \rightarrow NiF_2 + NiF_6^{2-} \quad (18)$$

$$Ni^{2+} + 3F^- \rightarrow NiF_3^- \quad (30)$$

$$NiF_3^- + 3F^- \rightarrow NiF_6^{3-} + e^- \quad (31)$$

$$NiF_3 + 3F^- \rightleftharpoons NiF_6^{3-} \quad (31')$$

$$NiF_6^{3-} \rightarrow NiF_6^{2-} + e^- \quad (32)$$

Highly oxidized nickel fluorides such as Ni_2F_5, NiF_3, and NiF_4 on a nickel anode and fluorocomplex ions such as NiF_6^{3-} and NiF_6^{2-} in a $NH_4F\bullet 2HF$ may act as a fluorinating agent similar to fluorine radical so that highly oxidized nickel fluorides on a nickel anode and/or fluorocomplex ions in the melt fluorinate NH_4^+ and/or NH_3 to form a perfluorinated inorganic compound of NF_3.

The characteristics of the process with a nickel anode are as follows:

1. NF_3 with high purity free from CF_4 can be obtained.
2. The yield in this process is higher than that in the chemical process.
3. No anode effect takes place during electrolysis.
4. The weight loss caused by Ni anodic dissolution is lower than 3 % of passing electricity during electrolysis.
5. The sludge of NH_4NiF_3 is stored in the electrolytic cell.

6. Vapor pressures of NH_3 and HF on a molten NH_4F-HF system are relatively high, so that the melt losses and the outlet line of gas is packed with NH_4F condensed from NH_3 and HF.
7. Removal of the condensed NH_4F from the inside of the gas line is needed.

Tasaka and Watanabe's Process [21–23, 26]

The molten NH_4F-KF-HF system has a higher surface energy and a higher electrolytic conductivity. Also, it has a relatively low vapor pressure of HF and a higher viscosity so as to suppress the penetration of HF into the pore of carbon anode and the breakdown of carbon anode. Hence, a carbon anode can be used as an anode for electrolytic production of NF_3.

The composition gas of anode gas is dependent upon the current density, the concentration of starting material such as NH_4F (or NH_3) and amine, and the anode potential [7, 21–23]. NH_4F (or NH_3) is the best starting material for production of NF_3 with a lower concentration of CF_4. When a melt contains a certain concentration of water, N_2, N_2O, O_3, and O_2 are formed at the carbon anode and NF_3 is produced after electrolysis for more than week. Hence, the NH_4F-KF-HF system containing a trace of water should be used as the electrolyte in order to obtain NF_3 by electrolysis for shorter duration.

The electrolytic conditions for electrolytic production of NF_3 from the molten NH_4F-KF-HF system using a carbon anode (FE-5) are the NH_4F concentration of 15–25 mol% in the melt containing LiF, the desirable current density range of 0.010–0.020 A/cm^2, temperature range of 120–150 °C, and the cell voltage of ca. 6–7 V. The maximum current efficiency for NF_3 formation is ca. 60 % under the conditions as mentioned above. The current efficiency for NF_3 formation increases with current density and reaches constant values at current density of 0.050 A/cm^2 [24]. When electrolysis is conducted at current densities higher than 0.030 A/cm^2, the current efficiency for NF_3 formation on a carbon anode is lower than that on a nickel anode [24]. This is because fluorinating agents are highly oxidized nickel fluorides as well as fluorine radical in electrolysis using a nickel anode, though it is only fluorine radical in electrolysis using a carbon anode.

Mechanism of formation of NF_3 on a carbon anode is similar to that due to fluorine radical in Mitsui Chemicals' process. Electrochemical reaction is only formation of fluorine radical according to Reaction (4), and others such as Reactions (21), (21′), (22), (23), (24), (25), and (26) are chemical reactions.

$$F^- \rightarrow \bullet F + e^- \tag{4}$$

$$NH_4^+ + \bullet F \rightarrow \bullet NH_2 + H^+ + HF \tag{21}$$

$$NH_3 + \bullet F \rightarrow \bullet NH_2 + HF \tag{21′}$$

$$\bullet NH_2 + \bullet F \rightarrow NH_2F \tag{22}$$

$$NH_2F + \bullet F \rightarrow \bullet NHF + HF \tag{23}$$

$$\bullet NHF + \bullet F \rightarrow NHF_2 \tag{24}$$

$$NHF_2 + \bullet F \rightarrow \bullet NF_2 + HF \tag{25}$$

$$\bullet NF_2 + \bullet F \rightarrow NF_3 \tag{26}$$

Total reaction for electrochemical formation of NF_3 is written as Reactions (27) and (27′).

$$NH_4^+ + 7F^- \rightarrow NF_3 + 4HF + 6e^- \tag{27}$$

$$NH_3 + 6F^- \rightarrow NF_3 + 3HF + 6e^- \tag{27′}$$

Characteristics of this process with a carbon (FE-5) anode are as follows:

1. The molten NH_4F-KF-HF system has a higher electrolytic conductivity.
2. The molten NH_4F-KF-HF system has a relatively low vapor pressure of HF on the melt.
3. Existence of KF in the melt suppresses penetration of HF into the pore of a carbon anode.
4. Current efficiency for NF_3 formation in this process is higher than that in a chemical reaction.
5. Carbon steel can be used as the structural material for electrolytic cell.

6. Anode effect takes place during electrolysis.
7. Carbon anode sometimes breaks down during electrolysis.
8. NF_3 is contaminated with CF_4, whose concentration is lower than 1 %.

Recently, boron-doped diamond (BDD) anodes have been developed to suppress the anode effect and a gradual increase in electrolytic voltage in the production of F_2. A carbon anode cannot be operated as an anode at current densities higher than 0.032 A/cm^2, because of occurrence of anode effect. In contrast, anode potential on a BDD electrode is much lower and no anode effect takes place up to 1 A/cm^2. This means that a BDD can be used as an anode even at 1 A/cm^2 without dissolution of anode and occurrence of anode effect. The maximum current efficiency for NF_3 formation on a BDD anode is 72.4 %, and its value is the highest among those on the anode materials studied by authors [9]. BDDs are promising candidates for anode materials in electrochemical synthesis of NF_3.

Process for Perfluorotrimethylamine [($(CF_3)_3N$)] Production [27–29]

Perfluorotrimethylamine, $(CF_3)_3N$, easily decomposes to release perfluoromethyl radicals, $\bullet CF_3$, and/or perfluoromethyl groups, $-CF_3$, which react with organic compounds and promote lipophilicity of the resulting products. It is therefore considered that $(CF_3)_3N$ is an important fluorine source for synthesis of many useful organic fluorocompounds such as medicines and agricultural chemicals. $(CF_3)_3N$ is a potential fire extinguish gas [30, 31]. In addition, $(CF_3)_3N$ is expected as an etching gas for SiO_2 film on Si wafer instead of hexafluoroethane, C_2F_6, in semiconductor industry.

Recently, a new process has been developed for electrolytic synthesis of $(CF_3)_3N$ at room temperature using nickel as an anode and a mixed melt of $(CH_3)_4NF{\bullet}m HF$ and $CsF{\bullet}2.0HF$ as an electrolyte [27, 28]. In this process, electrochemical fluorination of $(CH_3)_4N^+$ cation takes place according to Reaction (33).

$$(CH_3)_4N^+ + 25F^- \rightarrow (CF_3)_3N + CF_4 + 12HF + 24e^- \tag{33}$$

Because the amount of a by-product, CF_4, produced in reaction (33) is equal to that of $(CF_3)_3N$, the current efficiency for $(CF_3)_3N$ formation is theoretically low. In order to improve the current efficiency for $(CF_3)_3N$ formation, another room temperature molten salt system based on $(CH_3)_3N{\bullet}m HF$ is preferable theoretically and used as an electrolyte. Fluorohydrogenate anions such as $F(HF)_2^-$, $F(HF)_3^-$ and $(FH)(FHF)(HF)_2^-$ are present in a room temperature molten salt of $(CH_3)_3N{\bullet}m HF$ (m = 3–5) [25]. In this process, nickel and BDD are available as an anode. Voltages between 4 and 8 are useful but between 4 and 7 are preferred. Current density in the order of 0.005–0.020 A per sq. cm. (0.005–0.020 A/cm^2) can be employed. Temperature of an electrolyte is kept at room temperature. Hydrogen is generated on a nickel cathode.

Under the conditions of interest, $(CH_3)_3N$ and $(CH_3)_3NH^+$ are in equilibrium, as shown in Eq. 34. $(CH_3)_3NH^+$ coming from $(CH_3)_3N$ seems to be fluorinated electrochemically according to Reaction (35).

$$(CH_3)_3N + HF \rightleftharpoons (CH_3)_3NH^+ + F^- \tag{34}$$

$$(CH_3)_3NH^+ + 19F^- \rightarrow (CF_3)_3N + 10HF + 18e^- \tag{35}$$

In addition, cesium fluorohydrogenate, $CsF{\bullet}2.3HF$, in which fluorohydrogenate anions such as $F(HF)^-$ and $F(HF)_2^-$ are present, is added to $(CH_3)_3N{\bullet}m HF$ because it suppresses the passivation of Ni anode [29]. The maximum current efficiency for $(CF_3)_3N$ formation on a nickel anode is obtained in electrolysis at 0.020 A/cm^2 in the mixed melt of $(CH_3)_3N{\bullet}3HF$ + 70 % $CsF{\bullet}2.3HF$ and its value is 66.2 %.

Mechanism on fluorination of $(CH_3)_4N^+$ and $(CH_3)_3N$ is similar to that in Simons' process. In the mixed melt of $(CH_3)_3N{\bullet}m HF$ and $CsF{\bullet}2.3HF$, the highly oxidized nickel fluoride of $CsNi_2F_6$ is formed on a nickel anode and it also fluorinates $(CH_3)_3N$ to form $(CF_3)_3N$.

Future Directions

The concept of electrochemical perfluorination can substitute all hydrogen atoms in a starting material to all fluorine atoms and can increase the current efficiency for formation of perfluorinated organic and inorganic compounds. It is hoped that the use of BDD anode for electrochemical fluorination provides a solution to the problems in conventional electrochemical processes for fluorination and serves as a powerful method for electrolytic synthesis of organic and inorganic fluorine-containing substances. Hopefully, a wide range of electrochemical perfluorination based on different principles for anodic oxidation will be exploited, and will work together to meet the great demands for electrochemical synthesis of organic and inorganic fluorine-containing substances in the near future.

Cross-References

► Anodic Substitutions
► Selective Electrochemical Fluorination

References

1. Chambers RD (2004) General discussion of organic fluorine chemistry. In: Fluorine in organic chemistry. Wiley-Blackwell, pp 1–19
2. Braendin HP, Mabee ET (1960) Effects of adjacent perfluoroalkyl groups on carbonyl reactivity. In: Stacey M, Tatlow JC, Sharp AG (eds) Advances in fluorine chemistry, vol 3. Butterworths, London, pp 1–18
3. Simons TC, Hoffman FW, Beck RB, Holler HV, Katz T, Koshar RJ, Larson ER, Wulvaney JE, Paulson KE, Rogers FE, Singleton B, Sparks RE (1957) Fluorocarbon Derivatives. II. Cyclic Nitrides. J Am Chem Soc 70:3429
4. Nagase S, Inugai K (1980) Denkai Fussoka Han-nou (Electrochemical fluorination). Kagaku Sosetsu (27):17–36 [Japanese]
5. Simons JH (1949) The electrochemical process for the production of fluorocarbons. J Electrochem Soc 95:47–67
6. Rudge AJ (1971) Electrochemical fluorination. In: Kuhn AT (ed) Industrial electrochemical processes. Elsevirer, London, pp 71–88
7. Tasaka A (2004) Electrochemical fluorination of molten fluorides containing HF with nickel and carbon anodes. Curr Top Electrochem 10:1–36
8. Tasaka A (2007) Electrochemical synthesis and application of NF_3. J Fluorine Chem 128:296–310
9. Tasaka A (2007) Anodic behavior and anode performance of nickel, nickel-based composite and carbon electrodes for electrochemical fluorination in a few molten fluorides. Electrochemistry 75(12):934–944
10. Zemva B, Lutar K, Chacon L, Fele-Beuermann M, Allman J, Shen C, Bartlett N (1995) Synthses and some properties of new nickel fluorides. J Am Chem Soc 117:10025
11. Bartlett N, Chambers RD, Roche AJ, Spink RCH, Chacon L, Whalen JM (1996) New fluorination of organic compounds using thermodynamically unstable nickel fluorides. Chem Commun:1049
12. Tramsek M, Zemva B (2002) Higher fluorides of nickel: syntheses and some properties of Ni_2F_5. Acta Chim Slov 49:209
13. Sartori P, Ignat'ev N (1980) The actual state of our knowledge about mechanism of electrochemical fluorination in anhydrous hydrogen fluoride (Simons process). J Fluorine Chem 15:231
14. Ignat'ev N, Welz-Biermann U, Heider U, Kucheryna A, Von Ahsen S, Habel W, Sartori P, Willner H (2003) Carbon-chain isomerization during the electrochemical fluorination in anhydrous hydrogen fluoride – a mechanistic study. J Fluorine Chem 124:21–37
15. Tasaka A, Kobayashi H, Negami M, Hori M, Osada T, Nagasaki K, Ozaki T, Nakayama H, Katamura K (1997) Effect of trace elements on the electrolytic production of NF_3. J Electrochem Soc 144:192
16. Tasaka A, Osada T, Kawagoe T, Kobayashi M, Takamuku A, Ozasa K, Yachi T, Ichitani T, Morikawa K (1998) Effect of metal fluorides in the electrolyte on the electrolytic production of NF_3. J Fluorine Chem 87:163
17. Fox HM, Ruehlen FN, Childs WV (1971) Electrochemical fluorination. J Electrochem Soc 118:1246
18. Colburn CE (1963) Nitrogen fluorides and their inorganic derivatives. In: Stacey M, Tatlow JC, Sharp AG (eds) Advances in fluorine chemistry, vol 3. Butterworth, London, pp 92–116
19. Glemser O, Schroeder J, Knaak J (1966) Notiz zur Darstellung von Stickstofftrifluorid durch Elektrolyse von geschmolzene Ammoniumhydrogenfluorid. Chem Ber 99:371
20. Massonne J (1969) Herstellung und Reinigung von Stickstofftrifluorid in einer Versuchanlage. Chem Ing Tech 41:695
21. Tasaka A (1970) Denkaihou ni yoru Sanfukkachisso no Gousei ni kansuru Kenkyu (Study on synthesis of nitrogen trifluoride by electrolytic method). Doctor thesis, Kyoto University [Japanese]
22. Tasaka A, Watanabe N (1970) Denkaihou ni yoru NF_3 no Gousei (Synthesis of NF_3 by electrolytic method). Yoyuen (presently Molten Salts) 13:152–173 [Japanese]
23. Tasaka A (1981) Sanfukkachisso no Gousei to Ouyou (Synthesis and application of nitrogen trifluoride). Yoyuen (presently Molten Salts) 24:195–226 [Japanese]

24. Tasaka A, Kawagoe T, Takuwa A, Yamanaka M, Tojo T, Aritsuka M (1998) Effect of anode materials on NF_3 formation. J Electrochem Soc 145:1160
25. Isogai T, Nakai T, Nakanishi K, Inaba M, Tasaka A (2009) Ionic conductivity and viscosity of low temperature molten fluorides containing HF. Electrochemistry 77(8):713
26. Tasaka A, Yamanaka (Miki) M, Morimoto E, Nagamine S, Mimoto A, Fujikawa T, Abe M, Kobayashi A, Takebayashi H, Tojo T, Inaba M (2006) Anodic behavior of LiF-impregnated carbon and surface analysis of pristine carbon (FE-5) electrode polarized at various potentials in dehydrated melts of NH_4F•KF•mHF (M = 3 and 4). J New Mat Electrochem Syst 9:297
27. Tasaka A, Yachi T, Makino T, Hamano K, Kimura T, Momota K (1999) Anodic behaviors of nickel and platinum in a mixed molten salt of $(CH_3)_4NF$•4HF and CsF•2HF at room temperature. J Fluorine Chem 97:253
28. Shodai Y, Inaba M, Momota K, Kimura T, Tasaka A (2004) Electrolysis of mixed melt of $(CH_3)_4NF$•*m*HF + *x* wt.% CsF•2.0HF with nickel anode. Electrochim Acta 49:2131
29. Tasaka A, Nakanishi K, Masui N, Nakai T, Ikeda K, Momota K, Saito M, Inaba M (2011) Effect of CsF-concentration on electrolytic conductivity, viscosity and anodic reaction of nickel electrode in $(CH_3)_3N$-CsF-HF system at room temperature. Electrochim Acta 56:4335
30. Fukaya H, Ono T, Abe T (1998) Theoretical study of reaction of trifluoromethyl radical with hydroxyl and hydrogen radicals. J Comput Chem 19:277
31. Fukaya H, Ono T, Abe T (1995) New fire suppression mechanism of perfluoroalkylamines. J Chem Soc Chem Commun:1207

Electrochemical Processes for Gaseous Sulfur Oxides (SO and SO_x) Removal

Ulker Bakir Ogutveren
Anadolu University, Eskişehir, Turkey

Introduction

Oxides of sulfur occur in the forms of gas, liquid, aerosols, and the constituents of fine atmospheric particles in the atmosphere. They are all active components in atmospheric processes influencing air quality. Most of the sulfur dioxide results from human activities, particularly the burning of fossil fuels in power plants, heating, and combustion units. Smelting processes of nonferrous metals, petroleum refining, and sulfuric acid production are also important sources of sulfur dioxide in the air. Sulfur dioxide in the atmosphere is known to be oxidized to sulfur trioxide at a rate of 0.5–10 % per hour. The sulfur trioxide then reacts with water to form sulfric acid which is one of the main causes of smog over industrialized urban areas and acid rain which damages ecosystems.

Sulfur dioxide emissions have been controlled through the use of the so-called flue gas desulfurization (FGD) processes. Current FGD processes commercially used and the developments of conventional FGD processes can be found in the literature.

The environmental importance of the removal of gaseous pollutants from flue and waste gases has led to development of more efficient processes. Therefore a number of electrochemical processes which do not necessitate the continuous use of chemical reagents, which is the main problem besides the generation of huge amount of waste in some traditional FGD processes, have been proposed and most of them have been included in this text [1].

Electrochemical sulfur dioxide removal processes are divided mainly into two groups: direct processes and indirect processes. Direct processes can be applied by (i) adsorption and regeneration of the adsorbant by electrochemical SO_2 oxidation, (ii) absorption in a separate vessel and transferring the solution to an electrochemical reactor for SO_2 conversion, (iii) absorption and SO_2 conversion within an electrochemical reactor, and (iv) electrochemical reaction at a gas diffusion electrode. Indirect processes can be applied by (i) using homogeneous redox mediators usually in an outer cell, (ii) using heterogeneous redox mediators, (iii) catalytic oxidation with oxygen and electrochemical regeneration of the catalyst, and (iv) chemical absorption with electrochemically produced acid/alkali [2].

In direct processes at industrial scale, the pollutant usually needs to be transferred from the gas phase into a liquid phase by absorption. The subsequent step is the electrochemical conversion of dissolved species to harmless and/or potentially valuable products in an inner cell or outer cell processes.

Homogeneous and heterogeneous catalysts can be applied as redox mediators for indirect processes. Use of heterogeneous mediators such as oxides has the advantage of being the reaction products in a separate phase. Electrolytic cell arrangements similar to that in the fuel cell technology have also been used for gaseous pollutant removal.

Electrochemical Processes Used for Gaseous Sulfur Oxide Removal

Electrochemical removal of sulfur oxides from the waste gases has been extensively investigated and reported in the literature [1–7].

Direct Electrochemical Conversion of SO_2

Direct electrochemical conversion of SO_2 by an inner cell process has been carried out in electrochemical absorption columns consisting of packed bed of conducting materials as anode divided by a porous diaphragm or ion exchange membrane from the cathode. The dissolved pollutant usually in sulfuric acid is directly converted at the surface of the packed bed anode [8–12]. Similar arrangement consisting of bipolar electrodes has also been successfully tested for electrochemical absorption of sulfur dioxide using graphite and duriron (a ferrosilicon alloy) as anode materials [13]. Direct oxidation typically requires the use of membrane separators to prevent the side reactions deactivating the system. An undivided cell operated without cathodic sulfur formation was proposed as an attractive alternative to divided cells in terms of lower cell cost, energy consumption, and lower reactor installation cost [14]. A small pilot scale sieve plate electrochemical reactor, which is an undivided cell consisting of the monopolar connected electrodes, was used for direct oxidation of SO_2 absorbed in the sulfuric acid solution. The anode material was platinized titanium or lead dioxide and cathode material was zirconium metal or Ebonex [15]. The product of all these processes was sulfuric acid.

Electrochemical oxidation of aqueous SO_2 on the electrode surface has also aroused interest in the large-scale hydrogen production processes via a hybrid sulfur cycle which was patented as Westinghouse Process. The electrochemical oxidation of aqueous sulfur dioxide with respect to the hybrid sulfur cycle has recently been reviewed and shown the importance of the mechanism of the oxidation of sulfur dioxide on the electrode surface. Platinum, gold, and carbon materials as electrocatalysts have been reviewed to compare the catalytic activity for SO_2 oxidation [16]. The first step in the Westinghouse Process is the dissociation of H_2SO_4 into SO_2 and O_2 by thermal cracking at about 1,000 °C:

$$H_2SO_4 \rightarrow H_2O + SO_2 + {}^1\!/_2\, O_2$$

Second step is the oxygen recovery in an evaporator and third step is the hydrogen production from the electrochemical reaction between SO_2 and water. Overall reaction:

$$SO_2 + H_2O \leftrightarrow H_2SO_3 \text{ then } SO_2 + 2H_2O \rightarrow H_2SO_4 + H_2$$

SO_2/H_2SO_4 couple has a standard potential of 0.17 V which is much smaller than that of 1.23 V for water (H_2O/H_2). Although this is an advantage for this process, some technical constraints should be overcome for the design. Thus, the use of an electrolyzer with proton exchange membrane, in which the transport rate of reactants to the electrode surface is expected to enhance, was proposed. This method differs from the Westinghouse Process from the aspect of performing the anode reactions in the gas phase [17].

Indirect Electrochemical Conversion of SO_2

Using the electrochemically regenerated redox mediators as the oxidizing agents for indirect oxidation of gaseous pollutants eliminates the consumption of expensive and unstable absorbents or oxidizing agents used in conventional FGD techniques.

Indirect electrochemical process with homogeneous redox mediator called as "Peracidox" process (Saraberg-Holter-Lurgi-SHL) has been developed by Lurgi [18] using peroxodisulfate as the redox mediator, and the modified Mark 13A (Ispra Mark III) process has been developed at the Joint Research Center for the European Community in Ispra (Italy). A pilot installation at the 10 MWe level has been in operation in Saras refinery in Sardinia, Italy, since 1989, using bromine mediator for the indirect oxidation of SO_2 [19].

Electrochemical generation of CeIII/CeIV as redox mediators has been carried out in a batch reactor by testing different anode materials (TSIA, particle group anode with activated carbon-supported CeO_2, polymer-modified graphite electrodes) and followed by scrubbing of SO_2 by the redox mediators produced [20]. Oxidation of SO_2 produced the significant amount of dithionate and sulfuric acid in dilute sulfuric acid solutions, whereas SO_2 was observed to be oxidized into sulfuric acid in 5 M media. In the presence of air, sulfur dioxide was observed to be oxidized by air oxygen under the catalytic action of both CeIII and CeIV: these processes were shown to be of significant efficiency for the gas treatment. Experiments have been carried out in industrial scale by Socrematic S.A. in France [21].

Since flue gases not only contain SO_2 but to a certain extent also NO_x, the processes for the simultaneous removal of both components have been developed in many studies. The lead dioxide-dithionite process has combined direct and indirect conversion of SO_2 and NO_x, respectively. In the first step dithionite was used as homogeneous redox mediator for the indirect reduction of NO_x. SO_2 has been led to pass the NO absorption column without reaction and entered an electrochemical cell where it was oxidized to sulfric acid at the lead dioxide anode. A pilot plant for the treatment of 100 Nm^3h^{-1} of flue gas having the NO concentration of 600 ppm has been tested on an industrial site [22].

CeIV as homogeneous redox mediator for the simultaneous oxidation of SO_2 and NO_x to sulfuric acid and nitric acid, respectively, has been used in an indirect outer cell. The electrochemical regeneration of CeIV was also studied at a platinized titanium surface in a batch filter press cell, and the oxidation was found to be linked to other oxidative processes such as O^{-2} formation [23].

Use of AgI/AgII redox couple for simultaneous removal of NO_x and SO_2 has recently been investigated, and the continuous removal of NO_x and SO_2 has been observed without producing by-products. Concentration of AgII was maintained constant by continuous electroregeneration, therefore AgI/AgII system could have been reused [24, 25].

Indirect electrochemical treatment of SO_2 in which aqueous absorbed SO_2 species were reduced to elemental sulfur by electrogenerated aqueous reductants such as Ti^{3+}, V^{2+}, or Cr^{2+} ions has been conducted, and the reductants could be regenerated electrochemically in a divided electrochemical reactor after separation and recovery of the sulfur product [26].

The electrocatalytic oxidation behaviors by using iodine were found different at the platinum and graphite electrodes in a few studies performed. The reaction overpotential for SO_2 oxidation was observed to decrease significantly on both platinum and graphite materials by the addition of small amounts of HI [27, 28]. The iodide ions regenerated from the chemical reaction were used again for oxidation. The use of iodide especially considering the activity on graphite seemed to be advantageous by eliminating the need for a noble metal catalyst [29].

Although conventional heterogeneous catalysis of oxidation of SO_2 is a promising process, it has some disadvantages such as the low oxidation rate, the high production cost since most of them are metal-supported catalysts, the short lifetime due to catalyst deactivation, and the difficulties to control their activity during the process. However the use of electrocatalysis to eliminate these disadvantages is attracting an increasing interest for electrochemical treatment of sulfur dioxide [3, 26, 30–33]. It was observed in the electrochemical oxidation of sulfur dioxide on carbon gas diffusion electrodes modified with cobalt phthalocyanine (CoPc) that when operating with mixtures of air containing up to 20 % by volume SO_2, permeation of SO_2 into the electrolyte which is a basic problem of the electrooxidation of SO_2, could be resolved [32]. The catalytic activity of CoPc deposited onto activated carbon "Norit-NK" for oxidation of sulfur dioxide in sulfuric acid media was also investigated and found that gas diffusion electrodes (GDE), catalyzed with Co-Pc, pyrolyzed at 700 °C ensure 500 h operation at current density of 60 $mAcm^2$ [33].

Chemisorptive and catalytic properties of metal catalysts have been discovered to be effected significantly by using electrochemical promotion of catalysis and metal-support interactions with ionically conducting (Y_2O_3-stabilized-ZrO_2, YSZ) or mixed ionic-electronic conductors (ZrO_2, CeO_2, TiO_2, W^{+6} doped-TiO_2) in the early 1980s. This effect is called as electropromotion of catalysis "EPOC" or non-Faradaic electrochemical modification of catalytic activity "NEMCA effect." The non-Faradaic activation of heterogeneous catalytic reactions is proposed as a promising application of electrochemistry particularly in industrial production and in exhaust gas treatment.

When using YSZ as the solid electrolyte, the promoting ionic species ($O^{-\delta}$) are generated in an electrochemical step at the catalyst-gas-solid electrolyte interface (three-phase boundaries, tpb) [34]

$O^{-2}(YSZ) \rightarrow [O^{-\delta} + \delta]Pt + 2e^-$ at a rate I/2 F where I is current and F is the Faraday's constant.

In the case of SO_2 oxidation, a second electrochemical reaction simultaneously occurs at the anode tpb is:

$$SO_2 + O^{-2} \rightarrow SO_3 + 2e^-$$

One at the cathode is:

$$O_{ads} + 2e^- \rightarrow O^{-2}$$

It was found that positive potential application, i.e., O^{-2} supply to catalyst surface could lead to increase up to 200 % in the catalytic oxidation rate of SO_2 with Faradaic efficiency values up to 30 at flow rates as high as 30 L/min over a thin (≈ 40 nm) Pt catalyst electrodes interfaced with YSZ in a monolithic electropromoted reactor [35].

Another aspect of electrochemical promotion is the use of a liquid phase catalytic reaction. In the case of liquid phase catalysis, it is not the catalyst which is polarized but an otherwise catalytically inactive electrode. A cell construction similar to that of a fuel cell with molten Na_2S-Li_2S electrolyte at high temperature for the removal of SO_2 and H_2S from the gas effluents has been used [36–39]. A membrane separation process has been developed for simultaneous SO_x/NO_x abatement, in which the cell consisted of an immobilized $K_2S_2O_7$-based molten salt electrolyte and porous gas diffusion electrodes formed from a ceramic $La_{0,8}Sr_{0,8}CoO_3$ [40]. Similar system with lithiated nickel oxide porous gas diffusion electrodes was used to remove SO_x from flue gas streams at high temperatures. The technology was shown to be economically competitive in a 500 MW plant [41]. The catalytic activity of the V_2O_5-$K_2S_2O_7$ molten catalyst with and without polarization has been determined at 440 and 460 °C to investigate the effect of electrochemical promotion on SO_2 oxidation [42]. Since more than 70 different catalytic reactions have been electrochemically promoted on several catalysts deposited on O^{-2}(YSZ), Na^+ (β"-Al_2O_3), H^+($CaZr_{0,9}In_{0,1},O_{3-\alpha}$), F^-(CaF2), aqueous or molten salt, and mixed ionic-electronic conductors, EPOC is said not to be limited to any particular class of conductive catalyst, catalytic reaction, or ionic support. If the support is both electronically and ionically conductive, NEMCA is induced even without external wire connection of the catalyst and counter electrode [43].

Future Directions

Of all these methods indirect oxidation of SO_2 using redox mediators and electrochemically promoted oxidation seem to be the nearest alternatives to the commercial applications. The removal of SO_2 by using redox mediators has already been tested in commercial scale in the processes named Peracidox and Ispra Mark III. Westinghouse Process is another commercial application in terms of hydrogen production via hybrid sulfur cycle.

Direct utilization of electrochemical promotion in commercial reactors has some constraints as discussed below and future researches have to be focused on:

1. Material cost: Utilization of usually noble metals as efficient catalyst materials is the main cost item in electrocatalysis. The use of electrocatalyst as thin as possible or dispersed catalysts will be the solution of this problem. EPOC has been studied very recently with thicknesses as low as 30 nm prepared by sputtering pulsed laser deposition or impregnation [34]. The production of active, inexpensive, robust, durable, electrochemically promotable nanoscale structures will make EPOC applicable for commercial utilization.
2. Electrical connection: Efficient electrical current collection should be provided with electrical connections entering to the reactor and interconnects as low as possible. Recent discovery of "bipolar" or "wireless" NEMCA provides electrochemical promotion induced on catalyst films deposited on a solid electrolyte but not directly connected to an electronic conductor, i.e., wire [34].
3. Reactor design: Since electrochemically promoted reactor is a chemical reactor, the design of reactors will be the main factor effecting the use of electrochemically promoted reactors in commercial scale [43]. Recently designed bipolar monolithic electrochemically promoted reactors seem to provide an attractive solution for practical applications [35, 44].

Cross-References

▶ Electrochemical Promotion for the Abatement of Gaseous Pollutants
▶ Electrochemical Removal of H_2S

References

1. Kreysa G, Jüttner K (1994) Fundamental-studies on a new concept of flue-gas desulfurization. In: Lapique F, Storck A, Wragg AA (eds) Electrochem eng and energy. Plenum, New York
2. Scott K (1995) Electrochemical processes for clean technology. The Royal Society of Chemistry, London
3. Kreysa G, Bisang JM, Kochanek W, Linzbach G (1985) Fundamental-studies on a new concept of flue-gas desulfurization. J Appl Electrochem 15: 639–647
4. Bockris JOM, Bhardwaj RC, Tennakoon CLK (1994) Electrochemistry of waste removal – a review. Analyst 119:781–789
5. Juttner K, Galla U, Schmieder H (2000) Electrochemical approaches to environmental problems in the process industry. Electrochim Acta 45(15–16):2575–2594. doi:10.1016/S0013-4686(00)00339-X
6. Robin DR, Kenneth RS, Sergeĭ V (2003) Green industrial applications of ionic liquids. Kluwer Academic Publisher, New York
7. Dennis JMH, Winnick J (1994) Electrochemical membrane process for flue gas desulfurization. AIChE 40(1):143–151. doi:10.1002/aic.690400116
8. Kreysa G, Kulps HJ (1983) Anew electrochemical gas purification process. Ger Chem Eng 6:325–336
9. Kreysa G, Kulps HJ (1983) An electrochemical absorption process for gas purification. Chem Ing Tech 55:58–59
10. Kreysa G, Kulps HJ (1990) Electrochemical processes in environmental protection. Chem Ing Tech 62:357–365
11. Tezcan Ün Ü, Koparal AS, Bakır Öğütveren Ü (2007) Electrochemical desulfurization of waste gases in a batch reactor. J Environ Eng 133:13–19. doi:10.1061/(ASCE)0733-9372(2007)133:1(13)
12. Tezcan Ün Ü, Koparal AS, Bakır Öğütveren Ü (2007) Sulfur dioxide removal from fue gases by electrochemical absorption. Sep Purif Technol 53(1):57–63. doi:10.1016/j.seppur.2006.06.016
13. Stauffer JE (1994) Bipolar process for removal of sulfur dioxide from waste gases. US Patent 5344529
14. Scott K, Taama W (1997) Electrolysis of simulated flue gas solutions in an undivided cell. J Chem Technol Biotechnol 70(1):51–56. doi:10.1002/(SICI)1097-4660(199709)70:1<51
15. Scott K, Taama W, Cheng H (1999) Towards an electrochemical process for recovering sulphur dioxide. Chem Eng J 73(2):101–111. doi:10.1016/S1385-8947(99)00023-6
16. O'Brien JA, Hinkley JT, Donne SW, Lindquist SE (2010) The electrochemical oxidation of aqueous sulfur dioxide: a critical review of work with respect to the hybrid sulfur cycle. Electrochim Acta 55(3):573–591. doi:10.1016/j.electacta.2009.09.067
17. Sivasubramanian PK, Ramasamy RP, Freire FJ, Holland CE, Weidner JW (2011) Electrochemical hydrogen production from thermochemical cycles using a proton exchange membrane electrolyzer. Int J Hydrogen Energy. doi:10.1016/j.ijhydene.2006.06.056
18. Deutsches Patent und Markenamt, Lurgi Peracidox (1973) Register Number: 906171
19. van Velzen D, Langenkamp H, Moryoussef A (1990) HBr electrolysis in the Ispra Mark-13A flue-gas desulfurization process – electrolysis in a dem cell. J Appl Electrochem 20:60–68

20. Devadoss V, Kotteswaran P, Kottaisamy M, Vasudevan T (2010) Desulfurisation of flue gas using Ce3+/Ce4+ redox mediators, Kalasalingam University, India, Research Project. http://www.kalasalingam.ac.in/nano/Environmental.pdf. Accessed14 Oct 2011
21. Aurousseau M, Roizard C, Storck A, Lapicque F (1996) Scrubbing of sulfur dioxide using a cerium IV-containing acidic solution: a kinetic investigation. Ind Eng Chem Res 35(4):1243–1250. doi:10.1021/ie950553w
22. Juttner K, Kreysa G, Kleifges KH, Rottmann R (1994) Electrochemical waste cleanup process for simultaneous removal of SO_2 and NO_x. Chem Ing Tech 66:82–85. doi:10.1002/cite.330660115
23. Nzikou JM, Aurousseau M, Lapicque F (1995) Electrochemical investigations of the CeIII/CeIV couple related to a CeIV assisted process for SO_2/NO_x abatement. J Appl Electrochem 25:967–972
24. Raju T, Chung SJ, Moon IS (2008) Novel process for simultaneous removal of NOx and SO2 from simulated flue gas by using a sustainable AgI/AgII redox mediator. Environ Sci Technol 42(19):7464–7469. doi:10.1021/es801174k
25. Chung SJ, Raju T, Kim SM, Moon IS (2008) Removal of NO_x and SO_x in mediated electrochemical oxidation using AgI/AgII redox system. Theories Appl Chem Eng 14(1):299–302
26. Kelsall GH, Robbins DJ (1991) Sulfur-dioxide removal from flue-gases by electrogenerated aqueous reductants 1. Outline of the components of the proposed process. Process Saf Environ 69:43–49, B1
27. Struck BD, Junginger R, Boltersdorf D, Gehrmann J (1980) The anodic oxidation of sulfur dioxide in the sulfuric acid hybrid cycle. Int J Hydrogen Energy 5:487–497. doi:10.1016/0360-3199(80)90055-5
28. Yen SC, Chapman TW (1985) Indirect electrochemical processes at a rotating-disk electrode-oxidation of sulfite catalyzed by iodide. J Electrochem Soc 132(9):2149–2156. doi:10.1149/1.2114307
29. Cho BW, Yun KS, Chung IJ (1987) A study on the anodic oxidation of iodide mediated sulfur dioxide solution. J Electrochem Soc:Electrochem Sci Technol 134(7):1664–1667. doi:10.1149/1.2100732
30. Vitanov T, Budevski E, Nikolov I, Petrov K, Nidener V, Christov I (1990) Electrocatalytic oxidation of sulfur-dioxide (the Elcox process). Chem Eng Symp Ser 116:251–260
31. Bart H, Morr R, Burtscher K (1991) Absorption of NO_x and SO_2 without waste products Abstract 35-13, 4th World Congress in Chem Eng, Karlsruhe
32. Nikolov I, Petrov K, Vitanov T (1996) Low temperature electrochemical oxidation of sulfur dioxide. J Appl Electrochem 26(7):703–709
33. Petrov K, Nikolov I, Vitanov T, Uzun D, Ognjanov V (2010) Pyrolyzed Co-phthalocyanine as a catalyst for the oxidation of sulphur dioxide. Bulgarian Chem Commun 42(3):189–193
34. Vayenas CG, Bebelis S, Pliangos C, Brosda S, Tsiplakides D, Koutsodontis CG (2008) Non-faradaic electrochemical activation of catalysis. J Chem Phys 128(18):182506
35. Hammad A, Souentie S, Papaioannou EI et al (2011) Electrochemical promotion of the SO_2 oxidation over thin Pt films interfaced with YSZ in a monolithic electropromoted reactor. Appl Catal B Environ 103:336–342. doi:10.1016/j-apcatb.2011.01.040
36. Townley D, Winnick J (1983) Electrochemical sulfur-dioxide concentrator for flue-gas desulfurization. Electrochim Acta 28(3):389–393. doi:10.1016/0013-4686(83)85139-1
37. Lim HS, Winnick J (1984) Electrochemical removal and concentration of hydrogen-sulfide from coal-gas. J Electrochem Soc 131(3):562–568. doi:10.1149/1.2115627
38. Weaver D, Winnick J (1987) Electrochemical removal of H_2S from hot gas streams – nickel nickel-sulfide cathode performance. J Electrochem Soc 134(10):2451–2458. doi:10.1149/1.2100220
39. Winnick J (1990) Electrochemical separation of gases. In: Gerischer H, Tobias CW (eds) Advances in electrochemical science and engineering 1. VCH-Verlagsgesellschaft, NewYork
40. Franke M, Winnick J (1989) Membrane separation of sulfur-oxides from hot gas. Ind Eng Chem Res 28(9):1352–1357. doi:10.1021/ie00093a012
41. Schmidt DS, Winnick J (1998) Electrochemical membrane flue-gas desulfurization:K_2SO_4/V_2O_5 electrolyte. AIChE J 44(2):323–331. doi:10.1002/aic.690440210
42. Petrushina IM, Bandur VA, Cappel F, Bjerrum NJ (2000) Electrochemical promotion of sulfur dioxide catalytic oxidation. J Electrochem Soc 147(8):3010–3013. doi:10.1149/1.1393640
43. Vayenas CG, Bebelis S, Pliangos C, Brosda S, Tsiplakides D (2001) Electrochemical activation of catalysis: promotion and metal support interactions, Kluwer Academic/Plenum Publishers, New York
44. Katsaounis A (2010) Recent developments and trends in the electrochemical promotion of catalysis(EPOC). J Appl Electrochem 40:885–902. doi:10.1007/s10800-009-9938-7

Electrochemical Promotion for the Abatement of Gaseous Pollutants

Constantinos G. Vayenas
Department of Chemical Engineering, University of Patras, Patras, Achaia, Greece

Introduction

The phenomenon of electrochemical promotion of catalytic activity (EPOC) or non-Faradaic electrochemical modification of catalytic activity (NEMCA effect) refers to the pronounced and usually reversible change in catalytic activity and selectivity observed upon electrical current or potential application between a catalyst interfaced with an electrolyte and a second electrode interfaced with the same electrolyte. The effect was first reported in solid electrolyte systems [1, 2, 4–11], but several NEMCA studies also exist using aqueous electrolyte systems [1, 12–14] or Nafion membranes [1]. The EPOC phenomenon leads to apparent Faradaic efficiencies, Λ, well in excess of 100 % (values up to 10^5 have been measured in solid state electrochemistry and up to 10^2 in aqueous electrochemistry). This is due to the fact that, as shown by a variety of surface science and electrochemical techniques [1, 2, 4, 5, 15–19], the NEMCA effect is due to electrocatalytic (Faradaic) introduction of promoting species onto catalyst-electrode surfaces [1, 4], each of these promoting species being able to catalyze numerous (Λ) catalytic turnovers. The main experimental features and theory of the electrochemical promotion of catalysis have been reviewed several times, e.g., [1, 4, 6, 10, 17], and summarized lucidly by Sanchez and Leiva [18, 19].

Basic Phenomenology

The basic phenomenology of EPOC when using O^{2-}-conducting supports is given in Fig. 1. The (usually porous) metal (Pt) catalyst-electrode, typically 40 nm to 4 mm thick, is deposited on an 8 mol % Y_2O_3-stabilized-ZrO_2 (YSZ) solid electrolyte. Under open-circuit operation ($I = 0$, no electrochemical rate), there is a catalytic rate, r_0, of ethylene consumption for oxidation to CO_2 (Fig. 1).

Application of an electrical current, I, or potential (U_{WR}) between the catalyst and a counter electrode, thus changing the catalyst potential, U_{WR}, with respect to a reference electrode, causes very pronounced and strongly non-Faradaic (i.e., $\Delta r \gg I/2F$) alterations to the catalytic rate (Fig. 1).

Two parameters are commonly used to describe the magnitude of electrochemical promotion: first, the apparent Faradaic efficiency, Λ, defined from:

$$\Lambda = \Delta r_{catalytic}/(I/2F) \tag{1}$$

where $\Delta r_{catalytic}$ is the current-or potential-induced change in catalytic rate, I is the applied current, and F is Faraday's constant.

And second, the rate enhancement, ρ, which is defined from:

$$\rho = r/r_o \tag{2}$$

where r is the electropromoted catalytic rate, and r_o is the unpromoted (open-circuit) catalytic rate.

A reaction exhibits electrochemical promotion when $|\Lambda| > 1$, while electrocatalysis is limited to $|\Lambda| \leq 1$. A reaction is termed electrophobic when $\Lambda > 1$ and electrophilic when $\Lambda < -1$. In the former case, the rate increases with catalyst potential, U, while in the latter case, the rate decreases with catalyst potential. Λ values up to 3×10^5 [1, 20] and ρ values up to 150 [1] have been found for several systems. More recently, ρ values between 300 and 1,200 [21, 22] have been measured for C_2H_4 oxidation on Pt.

In the experiment of Fig. 1, $\Lambda = 74{,}000$ and $\rho = 26$, i.e., the rate of C_2H_4 oxidation increases by a factor of 25, and the increase in the rate of O consumption is 74,000 times larger than the rate, I/2F, of O^{2-} supply to the catalyst.

So far, more than 100 different catalytic reactions (oxidations, hydrogenations, dehydrogenations, isomerizations, decompositions) have been electrochemically promoted on Pt, Pd, Rh, Ag, Au, Ni, IrO_2, and RuO_2 catalysts deposited on O^{2-} (YSZ), Na^+ (β''-Al_2O_3), H^+ ($CaZr_{0.9}In_{0.1}O_{3-\alpha}$, Nafion), F^- (CaF_2), aqueous, molten salt, and mixed ionic-electronic (TiO_2, CeO_2) conductors [1].

These studies have shown that, quite often, over relatively wide (e.g., 0.3–0.5 V) ranges of potential, the catalytic rates depend on catalyst-electrode potential in an exponential manner, similar to the high field approximation of the Butler-Volmer equation, i.e.,

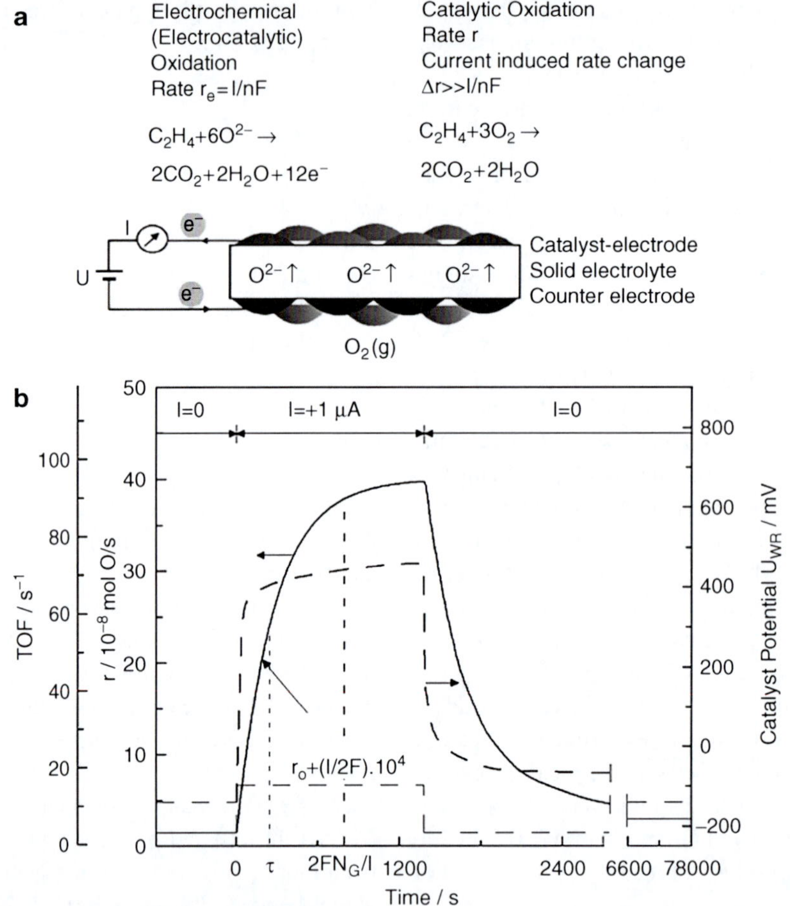

Electrochemical Promotion for the Abatement of Gaseous Pollutants, Fig. 1 (a) Basic experimental setup and operating principle of electrochemical promotion with O^{2-}-conducting supports. (**b**) Catalytic rate, r, and turnover frequency, TOF, response of C_2H_4 oxidation on Pt deposited on YSZ, an O^{2-} conductor, upon step changes in applied current. T = 370 °C, p_{O_2} = 4.6 kPa, $p_{C_2H_4}$ = 0.36 kPa. Also shown (*dashed line*) is the catalyst-electrode potential, U_{WR}, response with respect to the reference, R, electrode. The catalytic rate increase, Δr, is 25 times larger than the rate, r_0, before current application and 74,000 times larger than the rate, I/2F, of O^{2-} supply to the catalyst-electrode. N_G is the Pt/gas interface surface area in mol Pt, and TOF is the catalytic turnover frequency (mol O reacting per surface Pt mol per s). Reprinted with permission from Ref. [1]. Copyright © 2001 Kluwer/Plenum Publishers

$$r/r_o = \exp\left(\frac{\alpha e \Delta U}{k_b T}\right) = \exp\left(\frac{\alpha \Delta \Phi}{k_b T}\right) \tag{3}$$

where r_o is the unpromoted (i.e., open-circuit) catalytic rate, ΔU is the applied overpotential, ΔΦ is the overpotential-induced change in the catalyst-electrode work function and α (typically |α| ≈ 0.2 to 1) is a parameter which is positive for electrophobic reactions (∂r/∂Φ > 0, Λ > 1) and negative for electrophilic reactions (∂r/∂Φ < 0, Λ < − 1).

The second equality (3) holds because, as shown by Kelvin probe [1, 2] and UPS [1] work function measurements, the equality,

$$e\Delta U_{WR} = \Delta\Phi, \tag{4}$$

is valid over wide (e.g., 0.5–1 V) catalyst-electrode potential ranges in solid state electrochemistry. This equation is identical with that reported in aqueous electrochemistry for emersed electrodes [23, 24].

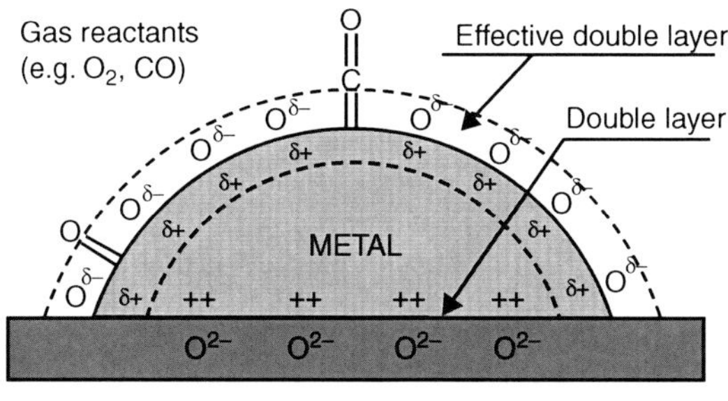

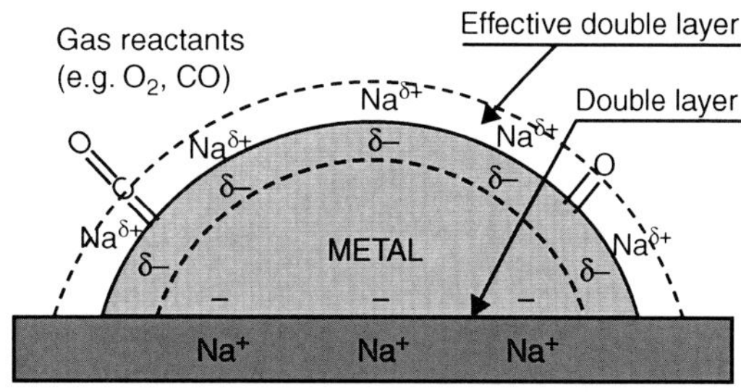

Electrochemical Promotion for the Abatement of Gaseous Pollutants, Fig. 2 Schematic representation of a metal catalyst-electrode deposited on a O^{2-}-conducting and a Na^+-conducting solid electrolyte, showing the location of the metal-solid electrolyte double layer and of the effective double layer created at the metal/gas interface due to potential-controlled ion migration (back-spillover) (Reprinted with permission from Ref. [1]. Copyright © 2001 Kluwer/Plenum Publishers)

Equation 4 is a limiting case of the general equation:

$$e\Delta U_{WR} = -\Delta\overline{\mu}(= -\Delta E_F) = \Delta\Phi + e\Delta\Psi \quad (5)$$

It is valid for any electrochemical cell [1], where $\overline{\mu}$ is the electrochemical potential of electrons in the catalyst electrode, E_F (= $\overline{\mu}$) is the Fermi level of the catalyst-electrode and Ψ is the outer (Volta) potential of the metal catalyst-electrode in the gas outside the metal/gas interface. The latter vanishes ($\Psi = 0$, $\Delta\Phi = 0$) when no net charge resides at the metal/gas interface [1, 25]. Thus, the experimental Eq. 4 manifests the formation of a neutral double layer, termed "effective" double layer, at the metal/gas interface (Fig. 2). At the molecular level, the stability of the effective double layer, and thus the validity of Eq. 4, requires that the migration (back-spillover) of the promoting ion ($O^{\delta-}$, $Na^{\delta+}$) is fast relative to its desorption or catalytic consumption. When this condition is not met (e.g., high T or non-porous electrodes), or also when the limits of zero or saturation coverage of the promoting ion are reached (at very positive or negative ΔU_{WR}), then deviations from Eq. 1 are observed [1, 25].

In view of Eq. 4, it follows that in electrochemical promotion, the work function of the catalyst surface can be in situ controlled via the applied potential $U_{WR,}$ and thus, electrochemical promotion is, simply, catalysis in the presence of an electrochemically controllable double layer at the catalyst/gas interface [1]. The effective double layer affects the binding strength of chemisorbed reactants and reaction intermediates and thus affects the catalytic rate in a very pronounced and reversible manner.

The molecular origin of electrochemical promotion is currently understood on the basis of the sacrificial promoter mechanism [1]. NEMCA results from the Faradaic (i.e., at a rate I/nF) introduction of promoting species ($O^{\delta-}$ in the case of O^{2-} conductors, H^+ in the case of H^+ conductors) on the catalyst surface. This electrochemically introduced O^{2-} species acts as a promoter for the catalytic reaction (by changing the catalyst work function and affecting the chemisorptive bond strengths of co-adsorbed reactants and intermediates) and is eventually consumed at a rate equal, at steady state, to its rate of supply (I/2F), which is Λ times smaller than the rate of consumption of the catalytic reactant, e.g., atomic O originating from the gas phase [1].

Figure 3 shows the validity of the sacrificial promoter concept for the galvanostatic transient of Fig. 1 by presenting O_2 TPD (Fig. 3b) and cyclic voltammetric (Fig. 3c) spectra obtained at times corresponding to those of the NEMCA galvanostatic transient of C_2H_4 oxidation (Fig. 1) under high vacuum conditions. One clearly observes, both with TPD and with cyclic voltammetry, the Faradaic introduction over

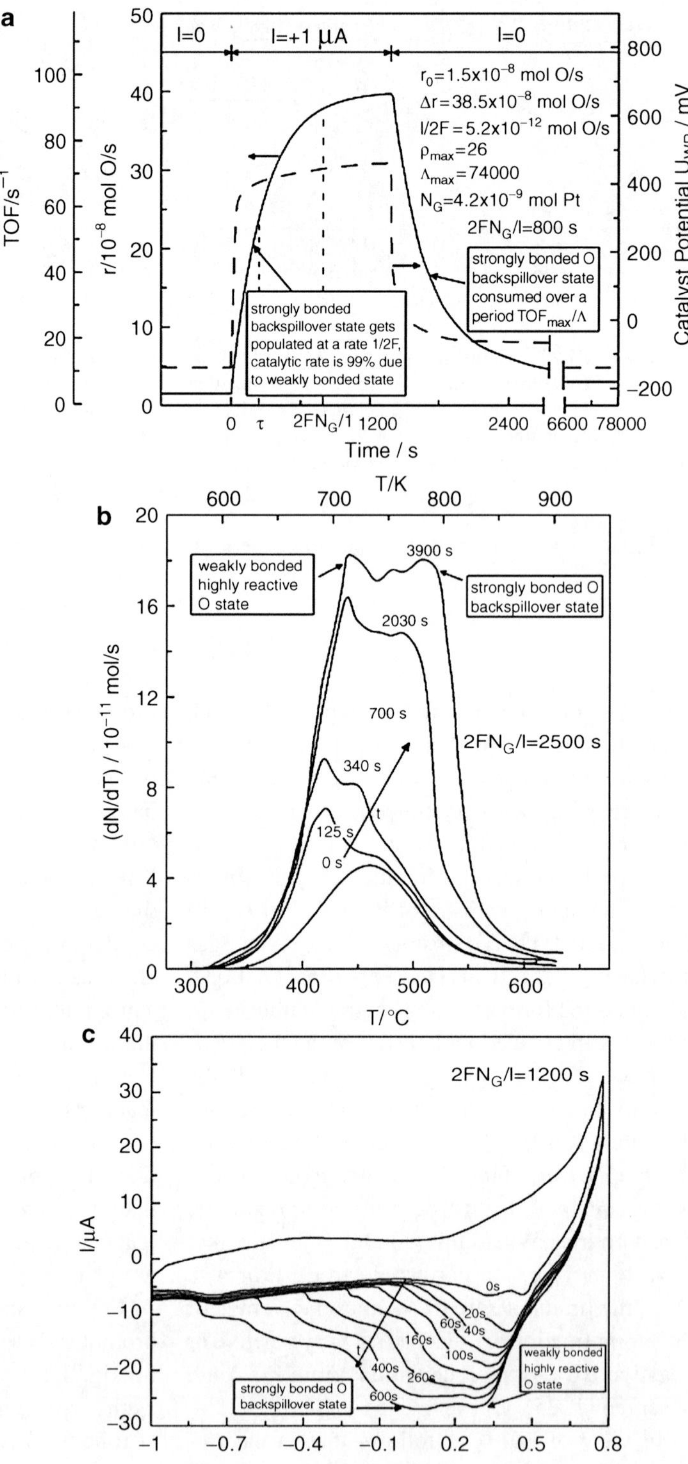

Electrochemical Promotion for the Abatement of Gaseous Pollutants, Fig. 3 NEMCA and its origin on Pt/YSZ catalyst electrodes [1]. Transient effect of the application of a constant current (**a**, **b**) or constant potential U_{WR} (**c**) on (**a**) the rate, r, of C_2H_4 oxidation on Pt/YSZ (also showing the corresponding U_{WR} transient) (**b**) the O_2 TPD spectrum on Pt/YSZ after current (I = 15 μA) application for various times t. (**c**) the cyclic voltammogram of Pt/YSZ after holding the potential at U_{WR} = 0.8 V for various times t. Reprinted with permission from Ref. [1]. Copyright © 2001 Kluwer/Plenum Publishers

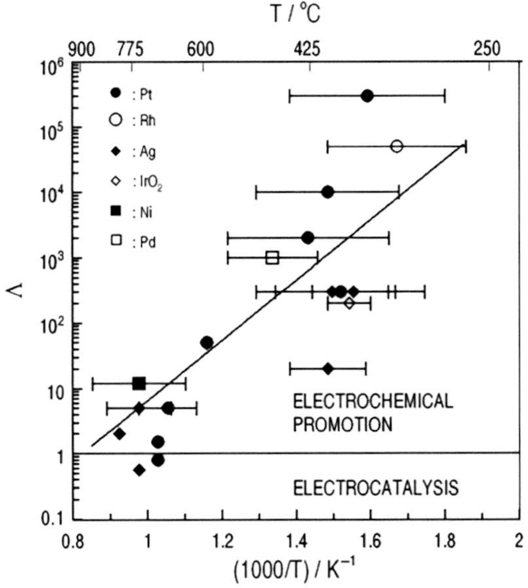

Electrochemical Promotion for the Abatement of Gaseous Pollutants, Fig. 4 Effect of temperature on the Faradaic efficiency, Λ, values measured in electrochemical promotion (NEMCA) studies of C_2H_4 oxidation on various metals [1]. Reprinted with permission from Ref. [1]. Copyright © 2001 Kluwer/Plenum Publishers

a time period $2FN_G/I$ (where N_G is the catalyst surface area expressed in mol, and thus $2FN_G/I$ is the time required to form a monolayer of O^{2-} on the catalyst surface) of a second (back-spillover) strongly bonded oxygen species on the Pt catalyst surface that displaces the normally chemisorbed oxygen state to lower desorption temperatures. This displacement, which results from strong repulsive lateral interactions between O^{2-} and more covalently bonded atomic oxygen [1], causes the observed pronounced enhancement in the catalytic rate. The back-spillover O^{2-} state acts as a sacrificial promoter. This molecular picture has been confirmed by the use of $^{18}O_2$ TPD [26].

The limits of electrocatalysis ($|\Lambda| \leq 1$) and electrochemical promotion ($|\Lambda| > 1$) are defined by the parameter $2Fr_o/I_0$ (Eq. 1; [1, 18], which provides an estimate of $|\Lambda|$). In general, $|\Lambda|$ decreases with temperature until the limit of electrocatalysis and electrochemical promotion are reached, as shown in Fig. 4.

Future Directions

The NEMCA effect has allowed for the extraction of general rules for the selection of promoters in heterogeneous catalysis [1, 17, 27]. The practical application of the EPOC phenomenon is currently sought via the design of novel ceramic monolithic reactors [16] that allow for compact arrangement of highly porous impregnated catalyst electrodes on thin ceramic components with a minimum, down to two, electrical connections. Application is sought both in automotive exhaust treatment and in chemicals synthesis [28–30].

Cross-References

- ▶ Electrochemical Processes for Gaseous Sulfur Oxides (SO and SO_x) Removal
- ▶ Electrochemical Removal of H_2S

References

1. Vayenas CG, Bebelis S, Pliangos C, Brosda S, Tsiplakides D (2001) Electrochemical activation of catalysis: promotion, electrochemical promotion and metal-support interactions. Kluwer/Plenum, New York
2. Vayenas CG, Bebelis S, Ladas S (1990) The dependence of catalytic activity on catalyst work function. Nature 343:625–627
3. Pritchard J (1990) Electrochemical Promotion Nature 343:502
4. Vayenas CG, Jaksic MM, Bebelis S, Neophytides SG (1996) The electrochemical activation of catalysis. In: Bockris JOM, Conway BE, White RE (eds) Modern aspects of electrochemistry, vol 29. Kluwer/Plenum, New York
5. Lambert RM, Williams F, Palermo A, Tikhov MS (2000) Modeling alkali promotion in heterogeneous catalysis: in situ electrochemical control of catalytic reactions. Top Catal 13:91–98
6. Foti G, Bolzonella I, Comninellis C (2003) Electrochemical promotion of catalysis. In: Vayenas CG, Conway BE, White RE (eds) Modern aspects of electrochemistry, vol 36. Kluwer/Plenum, New York
7. Cavalca CA, Haller GL (1998) Solid electrolytes as active catalyst supports: electrochemical modification of benzene hydrogenation activity on Pt/β″(Na) Al_2O_3. J Catal 177:389–395
8. de Lucas-Consuegra A, Princivalle A, Caravaca A, Dorado F, Marouf A, Guizard C, Valverde JL,

Vernoux P (2009) Preparation and characterization of a low particle size Pt/C catalyst electrode for the simultaneous electrochemical promotion of CO and C_3H_6 oxidation. Appl Catal A Gen 365:274–280
9. Vernoux P, Gaillard F, Bultel L, Siebert E, Primet M (2002) Electrochemical promotion of propane and propene oxidation on Pt/YSZ. J Catal 208:412–421
10. Metcalfe I (2001) Electrochemical promotion of catalysis: I: thermodynamic considerations. J Catal 199:247–258
11. Dorado F, de Lucas-Consuegra A, Vernoux P, Valverde JL (2007) Electrochemical promotion of platinum impregnated catalyst for the selective catalytic reduction of NO by propene in presence of oxygen. Appl Catal B: Environ 73:42–50
12. Neophytides S, Tsiplakides D, Stonehart P, Jaksic M, Vayenas CG (1994) Electrochemical enhancement of a catalytic reaction in aqueous solution. Nature 370:45–47
13. Ploense L, Salazar M, Gurau B, Smotkin ES (1997) Proton spillover promoted isomerization of *n*-butylenes on Pt-black cathodes/nafion. J Am Chem Soc 19:11550
14. Baltruschat H, Anastasijevic NA, Beltowska-Brzezinska M, Hambitzer G, Heitbaum J (1990) Electrochemical detection of organic gases: the development of a formaldehyde sensor. Ber Bunsenges Phys Chem 94:996–1000
15. Neophytides SG, Vayenas CG (1995) TPD and cyclic voltammetric investigation of the origin of electrochemical promotion in catalysis. J Phys Chem 99:17063–17067
16. Tsiplakides D, Balomenou S, Katsaounis A, Archonta D, Koutsodontis C, Vayenas CG (2005) Electrochemical promotion of catalysis: mechanistic investigations and monolithic electropromoted reactors. Catal Tod 100:133–14
17. Vayenas CG, Brosda S, Pliangos C (2003) The double-layer approach to promotion, electrocatalysis, electrochemical promotion, and metal–support interactions. J Catal 216:487–504
18. Sánchez C, Leiva E (2003) The NEMCA effect. In: Vielstich W, Lamm A, Gasteiger HA (eds) Handbook of fuel cells. J. Wiley & Sons Ltd, Chichester, UK
19. Sánchez C, Leiva E (2003) Theory of the NEMCA effect. In: Vielstich W, Lamm A, Gasteiger HA (eds) Handbook of fuel cells. J. Wiley & Sons Ltd, Chichester, UK
20. Bebelis S, Vayenas CG (1989) Non-faradaic electrochemical modification of catalytic activity: 1. The case of ethylene oxidation on Pt. J Catal 118:125–146
21. Kokkofitis C, Karagiannakis G, Stoukides M (2007) Electrochemical promotion in O_2 cells during propane oxidation. Top Catal 44:361–368
22. Kotsionopoulos N, Bebelis S (2007) In situ electrochemical modification of catalytic activity for propane combustion of Pt/β″-Al_2O_3 catalyst-electrodes. Top Catal 44:379–389
23. Rath DL, Kolb DM (1981) Continuous work function monitoring for electrode emersion. Surf Sci 109:641–647
24. Kolb DM (1987) UHV techniques in the study of electrode surfaces. Z Phys Chem Neue Folge 154:179–199
25. Riess I, Vayenas CG (2003) Fermi level and potential distribution in solid electrolyte cells with and without ion spillover. Solid State Ion 159:313–329
26. Katsaounis A, Nikopoulou Z, Verykios XE, Vayenas CG (2004) Comparative isotope-aided investigation of electrochemical promotion and metal–support interactions 1. $^{18}O_2$ TPD of electropromoted Pt films deposited on YSZ and of dispersed Pt/YSZ catalysts. J Catal 222:192–206
27. Brosda S, Vayenas CG, Wei J (2006) Rules of chemical promotion. Appl Catal B Environ 68:109–124
28. Vayenas CG, Koutsodontis CG (2008) Non-Faradaic electrochemical activation of catalysis. J Chem Phys 128:182506
29. Vernoux P, Vayenas CG (2011) Note on electrochemical promotion of catalytic reactions. Prog Surf Sci 86:83–93
30. Anastasijevic NA (2009) NEMCA – from discovery to technology. Catal Tod 146:308–311

Electrochemical Quartz Crystal Microbalance

Adriana Ispas and Andreas Bund
FG Elektrochemie und Galvanotechnik, Institut für Werkstofftechnik, Technische Universitüt Ilmenau, Ilmenau, Germany

Introduction

The electrochemical quartz microbalance (EQCM) is a powerful measuring technique commonly used by electrochemists as well by scientists coming from other fields, like physicists and biologists. One can find descriptions of the EQCM technique and of its applications in textbooks dealing with fundamentals of electrochemistry [1–4]. This fact proves the importance of this technique within the electrochemical scientific community.

The EQCM can determine very accurately small amounts of electrodeposited or adsorbed species [5–7] (on the order of magnitude of ng/cm^2) as well as viscoelastic properties of the adjacent medium [8–19]. Also one can get information in situ about the roughness of the deposited layer [20–24].

From all applications of the EQCM, in situ microgravimetry is the most popular. Thus, the EQCM technique allows in situ monitoring of the mass changes on the surface of the working electrode, which usually is an Au electrode deposited on a quartz crystal. The sensitivity with which the mass is detected depends on the resonance frequency of the quartz crystal. Thus, if the resonance frequency of the quartz is high, lower masses changes of order ng/cm^2 can be detected. Typical fundamental resonance frequencies are 5, 9, and 10 MHz, but resonance frequencies from kilohertz until gigahertz were already used in different sensors [36].

Biochemists, physicists, materials scientists, and electrochemists published articles and reviews about the electrochemical quartz crystal microbalance. We mention here just a limited selection of seminal papers that appeared recently on the (E)QCM [25–35]. Due to the impressive number of reports dealing with application of the EQCM technique for different in situ observations, we decided to focus in this entry mainly on some works that have been made so far using the EQCM technique combined with other in situ measuring techniques.

History

The principle of functionality of the EQCM is based on the fact that quartz, like some other materials (tourmaline, topaz, Rochelle salt, langasite, niobate, gallium-orthophosphate), is a piezoelectric material. Piezoelectricity was discovered in 1880 by the Curie brothers, Pierre and Jacques. It did not have the immediately impressive spread and use in the scientific community as other physical phenomena based on high frequency or on electromagnetic fields. However, in the USA, quartz crystal resonators were used in World War I in improving the ultrasonic submarine detectors [36].

Piezoelectricity means "pressure electricity" and it is derived from Greek [36]. It denotes the appearance of an electric charge on the opposite surfaces of a crystal when a mechanical stress is applied. The converse piezoelectric effect also exists, and it defines the appearance of a mechanical stress/movement of the two parallel planes, when an electric current is applied on the opposite surfaces of the material.

The converse piezoelectric effect is used in EQCM devices. Almost all commercial EQCM devices use monocrystalline quartz crystals that oscillate in a thickness-shear mode. Usually the quartz crystals are cut from a big block along specific crystallographic directions. The most used crystals for mass measurements are the so-called AT- and BT-cut quartz crystals. The BT-cut crystals are 50 % thicker than the AT-cut quartzes for a given resonance frequency. The BT crystals make an angle of $-49°$ with the z-axis, and the AT-cut present an angle of $49°44'$ with the z-axis. The AT-cut crystals are preferred due to their lower changes in the resonance frequency when changing the temperature, while the resonance frequency of the BT-cut quartzes changes parabolic with the temperature.

Because quartz crystals can vibrate with minimal energy dissipation, they can be used to build very stable oscillator circuits [26]. The quartz oscillator has a strong preference to vibrate at a characteristic resonant frequency (f_0), which depends on the shear modulus (μ_q), density (ρ_q), and the thickness of the quartz (t_q), $f_0 = (\mu_q/\rho_q)^{1/2}/(2t_q)$.

More details on the physical properties of quartz crystals or how one can build oscillating circuits using quartz crystals can be found in [36–40].

EQCM Fundamentals

Sauerbrey [41] described the dependence of the variation of the resonance frequency of quartz crystals when a rigid mass is deposited. His equation is known since then as the Sauerbrey equation (1). It is valid for thin, rigid layers and cannot be applied without corrections for viscoelastic or extremely rough layers. Furthermore, the Sauerbrey equation assumes that the particle displacement and the shear stress are continuous across the interface of the deposited layer:

$$\Delta f = -2f_0^2 \frac{\Delta m}{A\sqrt{\mu_q \rho_q}} \quad (1)$$

In Eq. 1 f_0 is the frequency of the unloaded quartz crystal, Δm the mass change, Δf the frequency shift, A the mass sensitive area, μ_q the shear modulus, and ρ_q the density of the quartz.

However, not only changes of the mass on the quartz surface will induce a change of the resonance frequency. Immersing the quartz crystal into a fluid will also produce a shift of the resonance frequency, due to the damping of the mechanical wave in the fluid. As an example the resonance frequency of a 10 MHz quartz crystal decreases by ca. 2 kHz when it is immersed in water.

Kanazava and Gordon [8, 9, 33] showed in this case that the shift of the resonance frequency is proportional to the square root of the density–viscosity product of the liquid Eq. 2:

$$\Delta f = -f_0^{3/2} \sqrt{\frac{n\rho_l \eta_l}{\pi \rho_q \mu_q}} \quad (2)$$

In Eq. 2 n is the number of the overtone ($n = 1,3,5,7,\ldots$), ρ_l is the density of the liquid, and η_l is the viscosity of the liquid.

The shift of the resonance frequency due to the viscous adjacent medium can be used in determining the viscosity–density product of the adjacent layer. Thus, the viscosity of solutions can be determined by using usually small volumes of liquids when compared to typical rotational viscosimeters.

In fact, it could be shown that Eq. 2 (and more general relations) can be successfully applied to determine the viscosity or viscoelastic properties of not only aqueous or organic systems [10–13, 15, 17], or of conducting polymers [11, 42–46], but also of ionic liquids [14, 18, 19]. Ionic liquids are very viscous electrolytes, having a viscosity 50–100 times larger than that of aqueous systems.

Besides the adsorbed or deposited mass, roughness, and the viscoelastic properties of the adjacent medium, the shift of the resonance frequency of an EQCM system depends also on temperature, pressure, or slippage effects [31]. Moreover, the effective viscosity of electrolytes is increased in the electric field of the double layer, which indicates that the resonance frequency of a quartz crystal will also be influenced by the thickness of the double layer, which, at its turn, depends on the electrolyte concentration [31]. All these factors can make difficult the interpretation of all the nuances of the EQCM signal. However, if one records simultaneously the dissipated energy of the quartz crystal and its resonance frequency, for example, with a network analyzer, one can separate the effects due to deposited mass from those caused by density–viscosity changes of the electrolyte [11, 15, 47]. With a network analyzer, the admittance spectrum of the quartz resonator near its resonance frequency can be recorded. The dissipated energy of the quartz crystal can be obtained from monitoring the full width at half maximum (FWHM) of the admittance spectra.

Generally, electroacoustical resonators can be described by mechanical and electrical equivalent circuits. For the quartz, two electrical models are often used: the transmission line model and the Butterworth–van Dyke circuit (BVD circuit). These models were made in order to describe the propagation of the acoustic wave in analogy with the electrical waves. More detailed descriptions of electrical equivalent circuits can be found, for example, in [4, 11, 26, 48, 49].

Another aspect, which is quite important for EQCM designers and users, is the simulations and the experimental measurements that have been made in order to determine the mass sensitivity of the EQCM and how the mass of the electrodes influence the sensitivity of the quartz crystal [50–57].

EQCM alone is already a powerful technique. By combining it with other in situ techniques one can get more useful information of electrochemical phase formation, corrosion, surface roughness effects during electrochemical deposition/dissolution, etc. Some combinations of EQCM with other in situ techniques are described in [4, 35]. In the following, we will focus on some specific ones.

EQCM and AFM/STM

Atomic force microscopy (AFM) is a nondestructive technique that is used for imaging the topography of materials as well as for studying the nanotribological effects [58]. A flexible cantilever with a very low spring constant is brought close to the surface of the sample. Due to the interaction forces between the cantilever and the surface, the cantilever will be deflected. By (e.g., optically) monitoring the deflection of the cantilever, one can reproduce the morphology of the layers with accuracy that can scale from atomic resolution up to several tens micrometers. One can get also information on the origin of atomic friction, wear, or adhesion by analyzing the lateral forces between the tip of the cantilever and the surfaces [59]. Therefore, it is obvious that by combining AFM with EQCM techniques, important information on tribology of the layers or crystal nucleation and growth can be followed in situ, during the electrodeposition or dissolution processes, besides other information characteristic to the layers, such as the roughness or topography of the layers. However, to combine the two techniques and to record simultaneously the data coming from a potentiostat, EQCM and AFM is not very trivial. This is the main reason in our opinion for the scarce literature reports on this topic.

The first works reported in the following have more of a historical importance in developing and understanding the combination between the EQCM and AFM, as no electrochemistry was used in these works. From the end of 1980, Krim and coworkers studied the sliding friction coefficients and the slip times for different physisorbed liquid films (such as krypton, xenon, nitrogen, water, cyclohexane) on Au and Ag substrates, with in situ QCM [60]. This group showed that the QCM can be used for determining the interfacial viscosity of adsorbed films. Later, the same research team developed a pioneering STM–QCM system [61], with which they were able to accurately detect the amplitude of an oscillating quartz crystal by measuring the surface oscillatory motion with the STM (scanning tunneling microscopy). Different lubricants were analyzed by the STM–QCM system, from which some were immobile (iodobenzene on copper) and some were mobile (*t*-butyl phenyl phosphate on platinum, tricresyl phosphate, and benzene on copper) [61]. Better resolution STM images have been obtained when the quartz of the QCM was oscillating as when the QCM was "off" for the mobile lubricants, while the quality of the STM image did not change when the QCM was "on" in the case of the immobile lubricant. From this type of experiments, one can learn more about the frictional transport coefficients of lubricants and thus identify when the sliding rates are comparable with the diffusion rates of the lubricants.

Kautek et al. studied the adsorption of silver halides using a special AFM–EQCM system [62]. From the AFM data, the signals due to topography could be separated from those due to the lateral forces (signals gave by lateral force microscopy), and thus the authors could identify the nanotribological changes within the double layer due to formation of adsorbed submonolayers. With the EQCM, they could quantify when the formation of a monolayer occurred. Thus, by combining the AFM with the EQCM information, the authors could gain more knowledge about the phase formation of Ag and Ag halogenides in different electrolytes, and the anodic behavior and the electropolishing process of Ag layers.

Friedt and coworkers have coupled the AFM with the EQCM [63]. They showed that the EQCM does not affect the AFM imaging (as the vibration amplitude of the EQCM is below the resolution of the AFM), but the AFM cantilever motion can affect the EQCM stability (as the cantilever holder can reflect the longitudinal acoustic waves generated by EQCM and thus induce standing waves which can change the boundary conditions of the longitudinal acoustic waves) [63]. Friedt and coworkers investigated the electrodeposition of Ag and Cu on Au from an aqueous electrolyte with an AFM–EQCM system [64], as well as adsorption of proteins, such as human plasma fibrinogen or antihuman immunoglobulin [65]. They monitored simultaneously the shift in the frequency for overtones as well as the damping of the quartz crystal (EQCM), but also the topography of the substrate (AFM). From recording the changes of all these parameters, one can identify accurately the conditions when the shift in resonance frequency can be transformed in deposited/adsorbed mass. In the case of biomolecules, one can identify furthermore the adsorption schemes or the binding mechanisms of the bioproteins to metal surfaces, both on cm^2 scale (QCM) and on μm^2 (AFM). The scientists mentioned above decided to measure mainly on the third and fifth overtone of the resonance frequency, as the fundamental resonance frequency of the quartz crystal proved to be highly unstable under environmental changes, such as flow and viscosity of the liquid.

Deposition and corrosion of Cu thin layers from an aqueous sulfate-based electrolyte were studied in situ with an AFM–EQCM system by Bund et al. [66]. The shift of the resonance frequency and of the damping of the quartz crystal (EQCM), on one side, and the evolution of surface roughness (AFM), on the other side, could be thus monitored simultaneously. No disturbance of the AFM response by the oscillating quartz or vice versa was observed in this study. The authors proved that a quantitative separation of internal (information given by EQCM) and external friction (which is the roughness contribution, obtained by AFM) of the solid materials is possible by combining AFM and EQCM techniques.

Smith et al. [67] studied the Zn electrodeposition from deep eutectic solvents (DES) by an AFM–EQCM system. The DES chosen in this study were the solvents based on a eutectic mixture of choline chloride (ChCl) and ethylene glycol (EG). The peculiarity of the DES is their very high viscosity at room temperature when compared to that of water. The authors proved by combining chronoamperometry, AFM, and EQCM techniques that the initial phases of Zn nucleation from DES resemble the 3D progressive model, as described mathematically by Sharifker and Hills [68].

Recently, Inoue et al. described how one can determine with a combined AFM–EQCM system the energy dissipated by different oscillating substrates (mica, HOPG) due to a sliding motion [69]. With the AFM, one can control and measure the loading force applied to the substrate, and with the EQCM, one can measure simultaneously the changes in the quality factor, Q, of the quartz crystal. The amplitude of the oscillations, as well as the force applied on the surface with the AFM cantilever, has been varied. The Q-values can be used in determining the dissipated energy as a function of applied force and of amplitude of oscillation.

The simultaneous in situ observation of electrodeposition processes with AFM and EQCM is a very important characterization method. In one single experiment, one can monitor in situ and in real time simultaneously the topography of the working electrode (AFM) and the mass changes, roughness changes of the working electrode, and the viscosity–density product of the electrolyte. Thus, correlations can be made between the nucleation and growth processes and the surface morphology of the deposits. Furthermore, one can relatively easily characterize the recrystallization of the metastable structures which can be prepared by electrochemical methods.

The combination of the simple QCM, without its electrochemical part, with the AFM can help in identifying and understanding the adsorption schemes and the binding mechanisms for different bioproteins on metal surfaces [70].

EQCM and Rotating Disk Electrode

The rotating disk electrode (RDE) and the rotating ring-disk electrode (RRDE) are two devices quite popular among electrochemists. They offer the advantage of controlled hydrodynamic conditions (in terms of reproducible mass-transfer conditions) at the working electrode [1, 3]. Rotating the electrode in the solution is more efficient than stirring of the solution with a magnetic stirrer or by gas bubbling. One other advantage of using the RDE is that the flow of the solution at the rotating disk is laminar. All these facts induced the possibility to solve the hydrodynamics equations and the convective-diffusion equations for the steady state case of the RDE. The RRDE can be successfully used to analyze intermediary species that can occur during electrochemical experiments and that have a short life time [1, 3]. Thus, by combining the (E)QCM with the RDE or RRDE, one can characterize in situ the mass deposited/dissolved on the working electrode under well-controlled convection conditions. Often in the literature the term "rotating (electrochemical) quartz crystal microbalance" [r(E)QCM] is used to define the combination of the RDE/RRDE with the (E)QCM. The rEQCM is in fact a useful balance, which offers simultaneous information on the mass deposited/dissolved and charge that passed.

A first attempt to use a rEQCM comes from 1987, when Grzegorzewski and Heusler characterized the kinetics of manganese dioxide with a RRDE/EQCM system at 40 °C. In these experiments, the disk was a platinum or gold disk covered with manganese dioxide layer, which was grown electrochemically. The AT-cut quartz crystal was fixed to the disk on the rotating cylinder with silicone glue, and the electric alternative potential needed for its oscillation was applied via Ag wires. At the disk, the mass changes were observed, while, at the ring electrode, one could detect the changes of the flux of hydrogen ions induced by the anodization reaction at the disk electrode of the pre-deposited manganese oxide layer [71]. The stability of the resonance frequency decreased from 0.1 to ca. 10 Hz when the electrode was rotated, due to the vibrations in the rotating shaft. The authors showed that the exchange current of the manganese ions can be evaluated and that it is bigger than that of oxygen ions. However, this first attempt was criticized by later literature reports because of the way the connections between the EQCM and RRDE were done, which in fact did not justify calling this system a "real" rEQCM.

Another attempt to combine EQCM and hydrodynamic techniques was done by Gabrielli and his colleagues and Itagaki and his colleagues [72]. In their case, the quartz crystal did not rotate, and the electrolyte was pushed against the electrode by impinging jet technique. From all the flow techniques, the impinging jet gives the closest hydrodynamic conditions to the RRDE, and the expression for the electrical current is the same. Gabrielli et al. studied the dissolution of Cu in NaCl+ $NaHCO_3$ solution with such a rEQCM. The electrodes on the quartz crystals ($f_0 \sim 6$ MHz) were sputtered in a special designed mask, so that a ring and a disk electrode were on one side of a quartz and only a disk on the opposite side. Electrochemical impedance at the disk, the emission efficiency at the disk, and the electrogravimetric transfer function could be measured simultaneously in one experiment. Thus, the authors could identify the chemical reactions that are associated with the dissolution of Cu and the species adsorbed/desorbed during the dissolution process. Also Cu dissolution in HCl solutions was studied by Itagaki and colleagues. The average valence of the dissolved ions was estimated from the changes in the resonance frequency of the quartz crystal, and this value was compared to the one determined from the currents at the ring electrode. The authors found a good agreement between these two average valences. Both Cu(I) and Cu(II) released from the disk electrode could be detected on the ring electrode, and the mass changes produced by dissolution could be separated from those due to adsorption/desorption processes.

Landolt and coworkers developed also a rotating ring-disk electrochemical quartz crystal microbalance for studying oxide film formations and growth on titanium substrates or the adsorption of organic corrosion inhibitors on iron substrates [73]. They designed a special electrode holder that allowed fast and easy exchanges of the working electrodes and of the quartz crystals. The collection efficiency of the system was calibrated in a ferricyanide solution with Au rings and Au disks, and it was found to be close to the theoretical value. The collection efficiency was also measured for oxygen produced during anodic polarization of the disk electrodes in a mixed sulfuric acid (0.1 M) and sodium sulfate (0.4 M) solution. In this case, it was found to be less than the theoretical value, and also, the authors found out that it varied with the rotation rate, a fact probably due to the formation of bubbles, but also due to the anodic reactions at the disk electrode and cathodic reactions at the ring electrode. Oxygen formation and its reduction were also studied during titanium dissolution, with Pt rings and Ti disks, or for detecting the adsorption of organic inhibitors on iron substrates [73]. The authors showed that during anodization of titanium, the film growth dominates below 3 V, while above 3 V, they observed both Ti dissolution and oxygen formation. In the case that adsorption of corrosion inhibitors on iron substrates was studied, two cases were taken into account: when iron substrate was in the active or passive potential region. The authors explained the positive frequency shift of the quartz crystal which results from the adsorption of a monolayer of inhibitor with a water replacement model. Thus, the large organic molecules that are adsorbed on the electrodes will displace the smaller and ordered water molecules, leading to a decrease of the viscosity near the resonator surface. The EQCM will sense both the adsorption of a monolayer of inhibitor and the decrease in viscosity. The data obtained during the corrosion inhibition measurements were fitted with a Langmuir–Freundlich isotherm model. The authors found higher adsorption constants in the active potential region than in the passive potential region for the corrosion inhibitors chosen, but also that the amine base had the highest adsorption constant.

Some reports on using rotating QCM were made in the literature by Zheng et al. and by Marlot and Vedel [74, 75]. Thus, kinetic studies of gold dissolution in oxygenated solutions that contained cyanide, ammonia, and copper, as well as copper electrodeposition from cyanide solutions and the leaching of copper in cyanide solutions, were reported in [74]. The copper cyanide species could be easily identified, and the authors showed that these complexes take active part in the dissolution of Au in cyanide electrolytes, but also the rates for Au leaching in the cyanide solutions were determined. The authors showed that by using a rotating QCM, the electrochemical window is extended. Another advantage of this system is that reaction kinetics can be easily studied, in order to find out if a reaction is diffusion controlled. In ref. [75] Marlot and Vedel reported on using a rotating QCM system during the electrodeposition of Cu–Se alloys from an aqueous solution of $CuSO_4$ and H_2SeO_3. The authors observed that the composition of this alloy depends very much on mass transport. Therefore, the authors designed a rotating QCM in order to study the deposition of Cu–Se alloys under determined hydrodynamic conditions, and implicit, under controlled diffusion [75]. Information of the morphology of deposits has been obtained from monitoring the mass increase (with the EQCM) and the current that flow [75].

EQCM and Ultrasound

Sonoelectrochemistry is the science that couples ultrasound to electrochemical processes. Ultrasound baths and ultrasound transducer have been used by chemists, electrochemists, and physicists since the 1930s [76]. Many reviews on sonoelectrochemistry have been published, a fact that reflects the large spread of this technique in electrochemistry [76]. The useful effect of the ultrasound waves comes from the existence and propagation of acoustic cavitation events. The ultrasound induces pressure fluctuations that will produce partial or total collapse – the so-called cavitation – of small bubbles in the liquid. Cavitation is a source of local high thermal energy that can cleave chemical bonds of species present in the bubbles or in the vicinity of the bubble. Cavitation that is generated close to a solid/liquid interface can produce the erosion at the solid surface, a fact that is useful in cleaning the surfaces from impurities or adherent films. Ultrasound can also influence the mass transport at phase boundaries, thus promoting the transport processes from or towards the working electrodes. From the beneficial effects of ultrasound in electrochemistry, one can mention the formation of denser, brighter, and harder deposits, synthesis of nanoparticles by electrochemical methods, electropolymerization, sonoelectroanalysis, and others (see [76] and references therein).

Schneider et al. combined EQCM and sonoelectrochemistry techniques, and they performed basic studies for Cu electrodeposition from sulfate and chloride baths as well as for codeposition of Ni and Co matrixes with CeO_2 particles [77]. They determined the optimum conditions in terms of electrode–horn separation and ultrasound intensity so that no strong cavitation takes place on the Au electrode of the quartz crystal, and thus, the resonance spectra acquired with the EQCM can still be evaluated, despite the small noise that occurs in the EQCM signal in the presence of ultrasound. The cyclic voltammograms recorded in the Cu^{2+}-containing electrolyte presented a cathodic peak (associated to Cu deposition) and an anodic one (due to Cu dissolution) under silent ultrasound condition. No cathodic peak was observed in the presence of ultrasound, and the current density values recorded during the potentiodynamic experiments were higher than in the absence of ultrasound, a fact that proves the enhanced mass transport towards the electrode in the presence of ultrasound [77]. Application of ultrasound increased the amount of Cu deposited under identical electrochemical condition from a sulfate electrolyte but without ultrasound, by a factor up to 7, and made the process of Co deposition completely reversible and of Co/CeO_2 codeposition more reversible (in the absence of ultrasound, the processes for pure Co layers and Co/CeO_2 layers were completely irreversible: less material was dissolved than was deposited). The ultrasound influenced also the roughness of the electrode: thus, the gold surface exposed to ultrasound was much rougher than the original one. However, ultrasound has very little effect on deposition and dissolution potential of Cu from sulfate electrolyte, a fact that indicates that ultrasound does not affect the charge transfer kinetics or the nucleation overpotential of Cu. Schneider et al. showed that EQCM performed in parallel to sonoelectrochemistry is very useful for quantifying the streaming rate and the thickness of the diffusion layer boundary in electrochemical reactions. Furthermore, an accurate determination of the streaming rates was obtained when comparing the experiments performed in sonoelectrochemical conditions with analogue ones performed with a rotating disk electrode (RDE). Ultrasound slightly affected the current efficiencies during Cu deposition, but it did not affect the dissolution of Cu layer.

EQCM and Surface Plasmon Resonance

Surface plasmon resonance spectroscopy (SPR) is a technique used to determine the evolution of optical thickness of thin films, such as DNA, proteins, or conducting polymers on a noble metal. SPR measurements are based on the reflection of optical waves at the interface substrate/deposited film. The principle of this technique is based on measuring the reflectivity, respectively, of the absorbance of an incident light onto a thin film. The surface plasmons that are excited by the incident light propagate along the surface of the substrate. For a given SPR angle, also called the coupling angle, the obtained reflectivity curve presents a sharp minimum. If a film is present on the surface, the refractive index of the medium will be different from that of the substrate, and thus, changes of the reflectivity will be measured in terms of shifts of the coupling angle. The changes of the coupling angle can be transformed into changes of film thickness.

The EQCM also gives the thickness of films, but in terms of an acoustic thickness. Some reports showed that it is very useful to combine EQCM and SPR, as the acoustic thickness can be different from the optical thickness [78–80]. Thus, by knowing both the acoustic and the optical thickness, one can elucidate the mechanisms that govern changing in the structure of adsorbed species in time. This fact is quite important in biological samples, but also for conducting polymers.

Johannsmann et al. showed that the acoustic thickness measured for the adsorption of streptavidin on self-assembled monolayers of biotinylated alkyl thiols is smaller than the optical thickness [78]. Moreover, the acoustic thickness leveled off after 30 min, but the optical thickness increased continuously within the timescale of the experiment. The authors attributed this effect to the roughness of the substrate and to the densification of the streptavidin film in time. Johannsmann et al. studied also by combining SPR and EQCM the adsorption of poly(N-isopropyl acrylamide), pNIPAM, on gold substrates and the electropolymerization of ethylene dioxythiophene, EDOT [78]. The adsorption of pNIPAM produced a parallel increase in the optical, but also in the acoustical, thickness, even if the absolute value of the acoustic thickness exceeded slightly the optical one. From the data obtained with a combined EQCM–SPR system, the authors could suggest the growth mechanism of the polymer film. Thus, it seems that the pNIPAM grows in thickness rather than to become denser in time. The obtained results have been attributed to the dense phase of the polymer, which wets the hydrophobic Au substrate at a temperature of 31 °C. This temperature is just below the critical temperature of the pNIPAM, where it is still soluble in the electrolyte. For this system, the authors could identify the adsorption/desorption process separate from the swelling/de-swelling phenomena [78]. The electropolymerization of EDOT was studied in aqueous solution, without any surfactant or containing anionic (SDS) or cationic surfactant (Triton-X100), by the potentiostatic pulse method and by cyclic voltammetry. In the case of SDS, the authors investigated two concentrations: one below and one above the cmc. The authors showed from the EQCM data that the film obtained by potentiostatic pulse method keeps its mechanical properties during the measurements independent if a surfactant was used, while the films obtained by cyclic voltammetry became softer with increasing cycling number for the solutions containing micelles. At the same time, the potential, at which the electropolymerization set in, shifts towards more cathodic potentials with adding and increasing the amount of surfactants in the electrolyte. From the SPR data, the authors could also calculate the optical thickness of the polymer layer in its oxidized and in its reduced state, and this was compared with the thickness given by the EQCM data. Fair agreement between the two values (optical and acoustical thickness) was obtained for electropolymerization of EDOT.

Bund et al. used an EQCM–SPR system for studying in situ the electropolymerization and the doping/dedoping of poly(pyrrole), ppy [79]. In this case, in the EQCM data, no changes of the dissipated energy of the quartz crystal were noticed during the polymer relaxation associated with switching off the current. However, the SPR signal presented a small drift in the reflectivity when the current was turned off. This indicated that the mechanical impedance and the complex shear modulus of the quartz crystal were constant. The first reduction step always indicated a mass decrease, which suggested that the ppy acts as an exchange membrane for a combination of cation and anions. A continuous increase in the reflectivity was observed in the doping steps, which was attributed to the growth of the ppy film. The imaginary part of the dielectric constant of the polymer film increased with oxidation. This fact was attributed to be a consequence of the formation of polaronic band. However, the optical properties of the film varied strongly with the state of the film (oxidation or reduction), a fact that made difficult a quantitative analyze for the ppy films, in terms of optical thickness and absolute values of the dielectric constants of the layers.

Damos et al. studied the polyaniline films, PANI, doped with copper tetrasulfonated phthalocyanide, CuTsPc, with an SPR–EQCM system [80]. The authors took into account that both the cations and the anions participate in the charge compensation, and they observed that the presence of CuTsPc suppresses the anion transport, which was noticed in the pure PANI films. The deposited PANI–CuTsPc behaved as a rigid layer: no significant contributions of its viscoelastic properties or of its roughness could be noticed, nor in the EQCM signal, or in the SPR angle. Moreover, the SPR was sensitive to the transition of the polymer films from the insulator to the conductive state, a transition that occurred during the potentiodynamic measurements in the cathodic potential region.

In summary, combining EQCM and SPR can give important information on the mechanisms of growth for conducting polymers, on the adsorption mechanism for proteins onto metal surfaces, or on the doping/dedoping mechanism of conducting polymers. Moreover, an optical and an acoustical thickness can be obtained, which can elucidate the growth of thin isolating films on metal surfaces and their properties.

Other In Situ Combinations of EQCM

Calvo and Etchenique summarized in their review some further in situ combinations of EQCM with non-electrochemical techniques (see [35] and references therein). For example, EQCM was also combined with ellipsometry in order to study the nucleation and growth of polyaniline films (reference 24 in [35]) or the viscoelastic behavior of poly(γ-methyl-L-co γ–n-octadecyl-L-glutamate) [17]. EQCM was combined with UV-visible absorption spectroscopy, in order to investigate the redox reactions of viologens. A combination of EQCM and probe beam deflection, PBD, was also reported in the literature (references 29, and 30 in [35], and [81]). PBD can discriminate between anion, cation, and solvent fluxes that might be generated on the electrode surface.

Spectro-electrochemistry at grazing incidence angle was also combined with EQCM for studying the Cu deposition from ammonium solutions, and also, there were some reports on photochemically induced changes in CuO_2 layers which were measured in parallel to EQCM [35]. Bard et al. and Ward et al. combined the EQCM with scanning electrochemical microscopy in order to get information on specific ions that induce the change in the resonance frequency of an EQCM [82]. Gabrielli and Keddam et al. developed a so-called ac-quartz electrogravimetry, which allowed measuring of electrochemical impedance spectroscopy, EIS, in parallel to mass–voltage transfer functions. Thus, information on the intermediate species that are adsorbed on the electrodes was obtained in parallel to EIS data [35].

Future Perspective

Most of the reports that one can find now in the literature deal with using the EQCM technique for the determinations of deposited or absorbed masses from the shift of the resonance frequency, based solely on the Sauerbrey equation. Many authors do not take into account that Sauerbrey equation cannot be applied for every experimental situation. For example, changes in the viscosity of adjacent layers or in the roughness of the electrodes will also induce a change in the resonance frequency. One solution to this problem and which can improve significantly the accuracy of the reported experimental data is to monitor the energy dissipated by the quartz crystal, besides the resonance frequency. This will also help in knowing exactly if one can apply the Sauerbrey equation to calculate the masses involved in the experiments. The dissipated energy can be obtained from recording the full bandwidth at half maximum in an impedance spectrum with a network analyzer or from the motional resistance of the quartz crystal from the data obtained with sophisticated oscillator circuits.

EQCM can be successfully applied for studying the nucleation and growth of metals. Usually, for the electrodeposition of most of metals from aqueous solution, side reactions, such as hydrogen evolution, occur in parallel. However, most of the theories describing different models of nucleation and growth take into account just the partial currents due to metal depositions, and not the total applied current. Unfortunately, one can find many reports in the literature where the total current is considered when calculating the type of nucleation and growth, a fact that can mislead when discussing electrocrystallization processes. One elegant solution is to use the EQCM, as already reported in some works [83, 84]. From the shift in frequency, one can get information on the mass deposited, and by derivation of this mass in time, one can get the partial current of the metal deposition. This current can be further used in comparing the obtained experimental data with different theoretical models that describe the nucleation and growth of metals.

The quartz crystals can be electrically driven on their fundamental resonance frequency, but also on their overtones. The overtones are odd multiples of the resonance frequency. Higher resonance frequency implies better sensitivity of the EQCM signal, but most of the time also higher noise. Very few reports can be found in the literature on using the overtones of the EQCM [17, 65], even if these few reports proved that using the overtones is quite useful sometimes. Therefore, if one uses electrochemical cells where a good signal/noise ratio can be obtained even for the overtones, this will help in obtaining clearer and more complete information of the system to be analyzed.

Most of the literature reports use systems with only one resonator. However, sometimes, it can be useful using two or more resonators in parallel [85]. One of the resonators can be used as a reference to study the effects of various factors (such as temperature, viscosity, hydrostatic pressure) on the resonance frequency. Those can then be subtracted from the signal recorded on the second resonator which is under electrochemical control (e.g., metal deposition or dissolution). In this way, clear information on the deposited mass can be obtained, without the influence of the adjacent medium or on the environmental factors.

Even if quartz crystals are now the most popular in microgravimetric sensors, maybe in future resonators made from other materials will also be used. For example, it was shown that langasite ($La_3Ga_5SiO_{14}$) and gallium-orthophosphate ($GaPO_4$) are promising candidates for thickness-shear mode resonators, as their useable temperature range is much larger compared to quartz [86].

In conclusion the EQCM is a powerful technique that can help us in understanding better electrodeposition processes and to characterize both the deposits and the properties of the adjacent media. Combining EQCM with other electrochemical and non-electrochemical techniques is not all the time a trivial thing, but it can help in gaining more information on the investigated systems. The authors of this entry encourage the scientists to further develop combinations of EQCM and other techniques but also to use the EQCM technique alone. However, one should be aware that the simple linear relation between frequency decrease and deposited mass (Sauerbrey equation) cannot be applied in all cases. More sophisticated evaluation schemes are available and they should be used if needed.

References

1. Gileadi E (2011) Physical electrochemistry, fundamentals, techniques and applications. Wiley-VCH Verlag GmbH, Weinheim, pp 253–264; pp 44–52
2. Plieth W (2008) Electrochemistry for material science. Elsevier, Niederland, pp 121–122
3. Bard AJ, Faulkner LR (2001) Electrochemical methods: fundamentals and applications, 2nd edn. Wiley, New York, pp 725–728; pp 331–364; pp 516–528
4. Hillman RA (2003) The electrochemical quartz crystal microbalance. In: Bard AJ, Stratmann M (eds) Instrumentation and electroanalytical chemistry, Encyclopedia of electrochemistry. Weinheim, Wiley-VCH
5. Tsionsky V, Gileadi E (1994) Use of the quartz crystal microbalance for the study of adsorption from the gas phase. Langmuir 10(8):2830–2835
6. Goubaidoulline I, Vidrich G, Johannsmann D (2005) Organic vapor sensing with ionic liquids entrapped in alumina nanopores on quartz crystal resonators. Anal Chem 77:615
7. Łukaszewski M, Siwek H, Czerwiński A (2010) Analysis of the electrochemical quartz crystal microbalance response during oxidation of carbon oxides adsorption products on platinum group metals and alloys. J Solid State Electrochem 14(7):1279–1292
8. Kanazawa KK, Gordon JG (1985) Frequency of a quartz microbalance in contact with liquid. Anal Chem 57:1770–1771
9. Kanazawa KK, Gordon JG (1985) The oscillation frequency of a quartz resonator in contact with a liquid. Anal Chim Acta 175:99–105
10. Johannsmann D (2008) Viscoelastic, mechanical, and dielectrical measurements on complex samples with the quartz crystal microbalance. Phys Chem Chem Phys 10:4516–4534; König AM, Düwel M, Du B, Kunze M, Johannsmann D (2006) Measurements of interfacial viscoelasticity with a quartz crystal microbalance: influence of acoustic scattering from a small crystal-sample contact. Langmuir 22:229–233
11. Johannsmann D (1999) Viscoelastic analysis of organic thin films on quartz resonators. Macromol Chem Phys 200:501–516
12. Martin SJ, Frye GC, Wessendorf KO (1994) Sensing liquid properties with thickness-shear mode resonators. Sens Actuat A Phys 44(3):209–218
13. Goubaidoulline I, Reuber J, Merz F, Johannsmann D (2005) Simultaneous determination of density and viscosity of liquids based on quartz-crystal resonators covered with nanoporous alumina. J Appl Phys 98:014305–1–014305–4
14. McHale G, Hardacre C, Ge R, Doy N, Allen RWK, MacInnes JM, Brown MR, Newton MI (2008) Density – viscosity product of small-volume ionic liquid samples using quartz crystal impedance analysis. Anal Chem 80:5806–5811
15. Bund A, Schwitzgebel G (1998) Viscoelastic properties of low-viscosity liquids studied with thickness-shear mode resonators. Anal Chem 70(13):2584–2588
16. Bund A, Schneider M (2002) Characterization of the viscoelasticity and surface roughness of electrochemically prepared conducting polymer films by impedance measurements at quartz crystals. J Electrochem Soc 149(9):E331–E339

17. Johannsmann D, Mathauer K, Wegner G, Knoll W (1992) Viscoelastic properties of thin films probed with a quartz crystal resonator. Phys Rev B 46:7808–7815
18. Bund A, Zschippang E (2007) Nickel electrodeposition from a room temperature eutectic melt. ECS Trans 3(35):253–261
19. Ispas A, Pölleth M, Hoa Tran Ba K, Bund A, Janek J (2011) Electrochemical deposition of silver from 1-ethyl-3-methylimidazolium trifluoromethanesulfonate. Electrochim Acta 56(28):10332–10339
20. Urbakh M, Daikhin L (1994) Roughness effect on the frequency of a quartz crystal resonator in contact with a liquid. Phys Rev B 49:4866–4870
21. Urbakh M, Daikhin L (1994) Influence of the surface morphology on the quartz crystal microbalance response in a fluid. Langmuir 10:2836–2841
22. Urbakh M, Daikhin L (1997) Influence of surface roughness on the quartz crystal microbalance response in a solution. A new configuration for QCM studies. Faraday Discuss 107:27–38
23. Daikhin LI, Gileadi E, Katz G, Tsionsky VM, Urbakh M (2002) Influence of roughness on the admittance of the quartz crystal microbalance immersed in liquids. Anal Chem 74:554–561
24. Bund A (2004) Application of the quartz crystal microbalance for the investigation of nanotribological process. J Solid State Electrochem 8:182–186
25. Schumacher R (1999) Die Schwingquarzmethode. Ein sensibles Meßprinzip mit breitem Anwendungsspektrum. Chem Unserer Zeit 33:268–278
26. Buttry DA, Ward MD (1992) Measurement of interfacial processes at electrode surfaces with the electrochemical quartz crystal microbalance. Chem Rev 92:1355–1379
27. Deakin MR, Buttry DA (1989) Electrochemical applications of the quartz crystal microbalance. Anal Chem 61(20):1147A–1154A
28. Henry C (1996) Measuring the masses: quartz crystal microbalance. Anal Chem News & Features 68:625A–628A
29. Ward MD, Buttry DA (1990) In situ interfacial mass detection with piezoelectric transducers. Science 249:1000–1007
30. Buttry DA (1991) Applications of the quartz crystal microbalance to electrochemistry. In: Bard AJ (ed) Electroanalytical chemistry: a series of advances, vol 1. Marcel Dekker, New York, pp 1–85
31. Tsionsky V, Daikhin L, Urbach M, Gileadi E (2003) Ch 22: Looking at the Metal/Solution Interface with the Electrochemical Quartz-Crystal Microbalance: Theory and Experiment. In: Bard AJ, Rubinstein I (eds) Electroanalytical chemistry, a series of advances, vol 22. CRC Press, Marcel Dekker, New York, pp 1–99
32. Wudy F, Stock C, Gores HJ (2009) Measurement methods electrochemical: quartz microbalance. In: Encyclopedia of electrochemical power sources, vol 3. Elsevier, Oxford, UK, pp 660–672
33. Kanazawa K, Cho N-J (2009) Quartz crystal microbalance as a sensor to characterize macromolecular assembly dynamics. Hindawi Publishing Corporation. J Sens 2009. Article ID 824947, pp 1–17
34. Marx KA (2003) Quartz crystal microbalance: a useful tool for studying thin polymer films and complex biomolecular systems at the solution-surface interface. Biomacromolecules 4(5):1099–1120
35. Calvo EJ, Etchenique RA (1999) Chapter 12: kinetic applications of the electrochemical quartz crystal microbalance (EQCM). In: Compton RG, Hancock G (eds) Comprehensive chemical kinetics, vol 37. Elsevier B.V., Amsterdam, The Netherlands, pp 461–486
36. Bottom VE (ed) (1982) Introduction to quartz crystal unit design. Van Nostrand Reinhold, New York
37. Mason WP (ed) (1965) Physical acoustics- principles and methods, vol. II- part A, properties of gases, liquids, and solutions. Academic, New York/London
38. Salt D (ed) (1987) Hy-Q handbook of quartz crystal devices. Van Nostrand Reinhold, Wokingham
39. Bond WL (1943) The mathematics of the physical properties of crystals. Bell Syst Tech J XXII:1–72
40. Mecea VM, Carlsson JO, Bucur RV (1996) 'Extensions of the quartz-crystal-microbalance technique. Sens Actuat A 53:371–378
41. Sauerbrey G (1956) Verwendung von Schwingquarzen zur Wägung dünner Schichten und zur Mikrowägung. Z Phys 155:206–222
42. Lucklum R, Hauptmann P (1997) Determination of polymer shear modulus with quartz crystal resonators. Faraday Discuss 107:123–140; Bandey L, Hillman AR, Brown MJ, Martin S (1997) Viscoelastic characterization of electroactive polymer films at the electrode/solution interface. Faraday Discuss 107:105–121; Calvo EJ, Etchenique R, Bartlett PN, Singhal K, Santamaria C (1997) Quartz crystal impedance studies at 10 MHz of viscoelastic liquids and films. Faraday Discuss 107:141–157; Lucklum R, Behling C, Cernosek RW, Martin SJ (1997) Determination of complex shear modulus with thickness shear mode resonators. J Phys D Appl Phys 30:346; McHale G, Lücklum R, Newton MI, Cowen JA (2000) Influence of viscoelasticity and interfacial slip on acoustic wave sensors. J Appl Phys 88(12):7304
43. Hillman AR, Efimov I, Ryder KS (2005) Timescale- and temperature-dependent properties of viscoelastic PEDOT films. J Am Chem Soc 127:16611; Mohamoud MA, Hillman AR, Efimov I (2008) Film mechanical resonance phenomenon during electrochemical deposition of polyaniline. Electrochim Acta 53(21):6235–6243; Hillman AR, Dong Q, Mohamoud MA, Efimov I (2010) Characterization of viscoelastic properties of composite films involving polyaniline and carbon nanotubes. Electrochim Acta 55(27):8142–8153; Efimov I, Hillman AR, Schultze JW (2006) Sensitivity variation of the electrochemical quartz crystal microbalance in response to energy trapping. Electrochim Acta

51(12):2572–2577; Efimov I, Koehler S, Bund A (2007) Temperature dependence of the complex shear modulus of cation and anion exchanging poly (pyrrole) films. J Electroanal Chem 605(1):61–67; Ispas A, Peipmann R, Adolphi B, Efimov I, Bund A (2011) Electrodeposition of pristine and composite poly(3,4-ethylenedioxythiophene) layers studied by electro-acoustic impedance measurements. Electrochim Acta 56(10):3500–3506; Koehler S, Ueda M, Efimov I, Bund A (2007) An EQCM study of the deposition and doping/dedoping behavior of polypyrrole from phosphoric acid solutions. Electrochim Acta 52(9):3040–3046; Ispas A, Peipmann R, Bund A, Efimov I (2009) On the p-doping of PEDOT layers in various ionic liquids studied by EQCM and acoustic impedance. Electrochim Acta 54(20):4668–4675; Koehler S, Bund A, Efimov I (2006) Shear moduli of anion and cation exchanging polypyrrole films. J Electroanal Chem 589(1):82–86

44. Arnau A, Jimenez Y, Fernández R, Torres R, Otero M, Calvo EJ (2006) Viscoelastic characterization of electrochemically prepared conducting polymer films by impedance analysis at quartz crystal study of the surface roughness effect on the effective values of the viscoelastic properties of the coating. J Electrochem Soc 153:C455–C466
45. Schweiss R, Lübben JF, Johannsmann D, Knoll W (2005) Electropolymerization of ethylene dioxythiophene (EDOT) in micellar aqueous solutions studied by electrochemical quartz crystal microbalance and surface plasmon resonance. Electrochim Acta 50(14):2849–2856
46. Hillman AR, Jackson A, Martin SJ (2001) The problem of uniqueness of fit for viscoelastic films on thickness-shear mode resonator surfaces. Anal Chem 73:540–549
47. Bund A, Schwitzgebel G (2000) Investigations on metal depositions and dissolutions with an improved EQCMB based on quartz crystal impedance measurements. Electrochim Acta 45:3703–3710
48. Behling C (1999) The non-gravimetric response of thickness shear mode resonators for sensor applications. Dissertation, Magdeburg University, Germany, p 20
49. Bund A (1999) Die Quarzmikrowaage in Rheologie und Elektrochemie: Fortschritte in der Signalauswertung durch Netzwerkanalyse. Dissertation, Universität des Saarlandes, Saarbrücken, Germany
50. Piefort V (2001) Finite element modelling of piezoelectric active structures. Dissertation, UniversitéLibre de Bruxelles, Brussels, Belgium
51. Johannsmann D, Heim LO (2006) A simple equation predicting the amplitude of motion of quartz crystal resonators. J Appl Phys 100:094505; Johannsmann D (2001) Deviation of the shear compliance of thin films on quartz resonators from comparison of the frequency shifts on different harmonics: a perturbation analysis. J Appl Phys 89(11):6356–6364
52. Mecea VM (2005) From quartz crystal microbalance to fundamental principles of mass measurements. Anal Lett 38:753–767; Bucur RV, Carlsson J-O, Mecea VM (1996) Quartz-crystal mass sensors with glued foil electrodes. Sens Actuators B 37:91–95; Bucur RV, Mecea VM, Carlsson J-O (2003) EQCM with air-gap excitation electrode. Calibration tests with copper and oxygen coatings. Electrochim Acta 48:3431–3438; Mecea V, Bucur RV (1979) The mechanism of the interaction of thin films with resonating quartz crystal substrates: the energy transfer model. Thin Solid Films 60:73–84; Mecea V, Bucur RV, Indrea E (1989) On the possibility of thin film structure study with a quartz crystal microbalance. Thin Solid Films 171:367–375
53. Weihnacht M, Bruenig R, Schmidt H (2007) 5D-3 more accurate simulation of quartz crystal microbalance (QCM) response to viscoelastic loading. In: Ultrasonics symposium, IEEE, New York, NY (USA), pp 377–380
54. Lee PCY, Huang R (2002) Mechanical effects of electrodes on the vibrations of quartz crystal plates. IEEE T Ultrason Ferr 49:604–611
55. Martin BA, Hager HE (1989) Flow profile above a quartz crystal vibrating in liquid. J Appl Phys 65(7):2627–2629
56. Couturier G, Boisgard R, Jai C, Aimé JP (2007) Compressional wave generation in droplets of water deposited on a quartz crystal: experimental results and numerical calculations. J Appl Phys 101:093510
57. Josse F, Lee Y, Martin SJ, Cernosek RW (1998) Analysis of the radial dependence of mass sensitivity for modified-electrode quartz crystal resonators. Anal Chem 70:237–247
58. AFM User's Manual from Digital Instruments Inc., Santa Barbara, USA
59. Carpick RW, Salmeron M (1997) Scratching the surface: fundamental investigations of tribology with atomic force microscopy. Chem Rev 97:1163–1194
60. Krim J, Widom W (1988) Damping of a crystal oscillator by an adsorbed monolayer and its relation to interfacial viscosity. Phys Rev B 38(17):12184–12189; Widom A, Krim J (1994) Spreading diffusion and its relation to sliding friction in molecularly thin adsorbed films. Phys Rev E 49(5):4154–4156
61. Borovsky B, Mason BL, Krim J (2000) Scanning tunneling microscope measurements of the amplitude of vibration of a quartz crystal oscillator. J Appl Phys 88(7):4017–4021; Abdelmaksoud M, Lee SM, Padgett CW, Irving DL, Brenner DW, Krim J (2006) STM, QCM, and the windshield wiper effect: A joint theoretical–experimental study of adsorbate mobility and lubrication at high sliding rates. Langmuir 22:9606–9609
62. Kautek W, Dieluweit S, Sahre M (1997) Combined scanning force microscopy and electrochemical quartz microbalance in-situ investigation of specific adsorption and phase change processes at the silver/halogenide interface. J Phys Chem B 101(14):2709–2715

63. Friedt JM, Choi KH, Francis L, Campitelli A (2002) Simultaneous atomic force microscope and quartz crystal microbalance measurements: interactions and displacement field of a quartz crystal microbalance. Jpn J Appl Phys 1 41(6A):3974–3977
64. Friedt J-M, Choi KH, Frederix F, Campitelli A (2003) Simultaneous AFM and QCM measurements methodology validation using electrodeposition. J Electrochem Soc 150(10):H229–H234
65. Choi K-H, Friedt J-M, Frederix F, Campitelli A, Borghs G (2002) Simultaneous atomic force microscope and quartz crystal microbalance measurements: Investigation of human plasma fibrinogen adsorption. Appl Phys Lett 81(7):1335–1337; Choi K-H, Friedt J-M, Laureyn W, Frederix F, Campitelli A, Borghs G (2003) Investigation of protein adsorption with simultaneous measurements of atomic force microscope and quartz crystal microbalance. J Vac Sci Technol B 21(4):1433–1436
66. Bund A, Schneider O, Dehnke V (2002) Combining AFM and EQCM for the in situ investigation of surface roughness effects during electrochemical metal depositions. Phys Chem Chem Phys 4:3552–3554
67. Smith EL, Barron JC, Abbott AP, Ryder KS (2009) Time resolved in situ liquid atomic force microscopy and simultaneous acoustic impedance electrochemical quartz crystal microbalance measurements: a study of Zn deposition. Anal Chem 81(20):8466–8471
68. Scharifker B, Hills G (1983) Theoretical and experimental studies of multiple nucleation. Electrochem Acta 28:879–889
69. Inoue D, Hosomi N, Taniguchi J, Suzuki M, Ishikawa M, Miura K (2010) Development of a combined atomic force microscope with an A T-cut quartz resonator. J Phys: Conference Series, International Conference on Science of Friction 258:012019
70. Westwood M, Kirby AR, Parker R, Morri VJ (2012) Combined QCMD and AFM studies of lysozyme and poly-l-lysine–poly-galacturonic acid multilayers. Carbohydrate Polymers 89(4):1222–1231; Richter RP, Brisson A (2004) QCM-D on mica for parallel QCM-DsAFM studies. Langmuir 20:4609–4613; Gurdak E, Dupont-Gillain CC, Booth J, Roberts CJ, Rouxhet PG (2005) Resolution of the vertical and horizontal heterogeneity of adsorbed collagen layers by combination of QCM-D and AFM. Langmuir 21:10684–10692
71. Grzegorzewski A, Heusler KE (1987) A kinetic investigation of the manganese dioxide electrode with a rotating quartz frequency balance. J Electroanal Chem 228:455–470
72. Gabrielli C, Keddam M, Minouflet-Laurent F, Perrot H (2000) Simultaneous EQCM and ring-disk measurements in AC regime. Application to copper dissolution. Electrochem Solid State Lett 3(9):418–421; Itagaki M, Kadowaki J, Watanabe K (2000) Analysis of active dissolution of copper in acidic solution by EQCM/Wall jet split ring disk electrode. Electrochemistry 68(9):684–688
73. Kern P, Landolt D (2000) Design and characterization of a rotating electrochemical quartz-crystal-microbalance electrode J Electrochem Soc 147:318–325; Vergé M-G, Mettraux P, Olsson C-O A, Landolt D (2004) Rotating ring-disk electrochemical quartz crystal microbalance: a new tool for in situ studies of oxide film formation. J Electroanal Chem 566(2):361–370; Kern P, Landolt D (2001) Adsorption of organic corrosion inhibitors on iron in the active and passive state. A replacement reaction between inhibitor and water studied with the rotating quartz crystal microbalance. Electrochim Acta 47(4):589–598
74. Zheng J, Ritchie IM, La Brooy SR, Singh P (1995) Study of gold leaching in oxygenated solutions containing cyanide-copper-ammonia using a rotating quartz crystal microbalance. Hydrometallurgy 39:277–292; Jeffrey MI, Zheng J, Ritchie IM (2000) The development of a rotating electrochemical quartz crystal microbalance for the study of leaching and deposition of metals. Meas Sci Technol 11:560–567
75. Marlot A, Vedel J (1999) Electrodeposition of copper-selenium compounds onto gold using a rotating electrochemical quartz crystal microbalance. J Electrochem Soc 146:177–183
76. Compton RG, Eklund JC, Marken F, Rebbitt TO, Akkermans RP, Waller DN (1997) Dual activation: coupling ultrasound to electrochemistry- an overview. Electrochim Acta 42(19):2919–2927; Compton RG, Eklund JC, Marken F (1997) Sonoelectrochemical processes: a review. Electroanalysis 9(7):509–522; González-García J, Esclapez MD, Bonete P, Hernández YV, Garretón LG, Sáez V (2010) Current topics on sonoelectrochemistry. Ultrasonics 50:318–322; Cobley AJ, Mason TJ, Saez V (2011) Review of effect of ultrasound on electroless plating processes. Trans Inst Met Finish 89(6):303–309
77. Schneider O, Matić S, Argirusis Chr (2008) Application of the electrochemical quartz crystal microbalance technique to copper sonoelectrochemistry – Part 1. Sulfate-based electrolytes. Electrochim Acta 53(17):5485–5495; Argirusis Chr, Matić S, Schneider O (2008) An EQCM study of ultrasonically assisted electrodeposition of Co/CeO2 and Ni/CeO2 composites for fuel cell applications. Phys Stat Sol (a) 205(10):2400–2404; Schneider O, MartensS, Argirusis C (2010) Electrochemical Quartz Crystal Microbalance Technique in Sonoelectrochemistry. ECS Transactions 25(28):69–80
78. Schweiss R, Lübben JF, Johannsmann D, Knoll W (2005) Electropolymerization of ethylene dioxythiophene (EDOT) in micellar aqueous solutions studied by electrochemical quartz crystal microbalance and surface plasmon resonance. Electrochim Acta 50(14):2849–2856; Plunkett MA, Wang Z, Rutland MW, Johannsmann D (2003) Adsorption of

pNIPAM Layers on Hydrophobic Gold Surfaces, Measured in Situ by QCM and SPR. Langmuir 19:6837–6844; Laschitsch A, Menges B, Johannsmann D (2000) Simultaneous determination of optical and acoustic thicknesses of protein layers using surface plasmon resonance spectroscopy and quartz crystal microweighing. Appl Phys Lett 77(14):2252–2255
79. Bund A, Baba A, Berg S, Johannsmann D, Lübben J, Wang Z, Knoll W (2003) Combining surface plasmon resonance and quartz crystal microbalance for the in situ investigation of the electropolymerization and doping/dedoping of poly(pyrrole). J Phys Chem B 107:6743–6747
80. Damos FS, de Cássia Silva Luz R, Tanaka AA, Kubota LT (2006) Investigations of nanometric films of doped polyaniline by using electrochemical surface plasmon resonance and electrochemical quartz crystal microbalance. J Electroanal Chem 589(1):70–81
81. Henderson MJ, Hillman AR, Vieil E, Lopez C (1998) Combined electrochemical quartz microbalance (EQCM) and probe beam defection (PBD): validation of the technique by a study of silver ion mass transport. J Electroanal Chem 458:241–248; Henderson MJ, Hillman AR, Vieil E (1999) Ion and solvent transfer discrimination at a poly(o-toluidine) film exposed to HClO4 by combined electrochemical quartz crystal microbalance (EQCM) and probe beam deflection (PBD). J Phys Chem B 103:8899–8907; Henderson MJ, French H, Hillman AR, Vieil E (1999) A combined EQCM and probe beam defection study of salicylate ion transfer at a polypyrrole modified electrode. Electrochem Solid State Lett 2(12):631–633
82. Cliffel DE, Bard AJ (1998) Scanning electrochemical microscopy. 36. A combined scanning electrochemical microscope-quartz crystal microbalance instrument for studying thin films. Anal Chem 70(9):1993–1998; Hillier AC, Ward MD (1992) Scanning electrochemical mass sensitivity mapping of the quartz crystal microbalance in liquid media. Anal Chem 64(21):2539–2554; Cliffel DE, Bard AJ, Shinkai S (1998) Electrochemistry of tert-Butylcalix[8]arene-C(60) films using a scanning electrochemical microscope-quartz crystal microbalance. Anal Chem 70(19): 4146–4151
83. Ispas A, Matsushima H, Bund A, Bozzini B (2009) Nucleation and growth of thin nickel layers under the influence of a magnetic field. J Electroanal Chem 626(1–2):174–182; Bund A, Ispas A, Mutschke G (2008) Magnetic field effects on electrochemical metal depositions. Sci Technol Adv Mater 9:024208
84. Koza JA, Uhlemann M, Gebert A, Schultz L (2008) Nucleation and growth of the electrodeposited iron layers in the presence of an external magnetic field. Electrochim Acta 53(27):7972–7980
85. Berg S, Johannsmann D (2001) Laterally coupled quartz resonators. Anal Chem 73(6):1140–1145; Stafford GR, Bertocci U (2009) In situ stress and nanogravimetric measurements during hydrogen adsorption/absorption on Pd overlayers deposited onto (111)-Textured Au. J Phys Chem C 113:13249–13256; Way AS (1993) Quartz resonator techniques for simultaneous measurement of areal mass density, lateral stress, and temperature in thin films. Vacuum 44(3/4):385–388
86. Fritze H, Tuller HL (2001) Langasite for high-temperature bulk acoustic wave applications. Appl Phys Lett 78:976; Elam JW, Pellin MJ (2005) $GaPO_4$ sensors for gravimetric monitoring during atomic layer deposition at high temperatures. Anal Chem 77:3531–3535

Electrochemical Reactor Design and Configurations

Eric M. Stuve
Department of Chemical Engineering, University of Washington, Seattle, WA, USA

Introduction

Electrochemical processes combine chemistry and electricity to meet the needs of today's society. Chlorine [303748, 303736, 303737], lye [304624], aluminum [304624], hydrogen [303854, 303855], oxygen [303858, 305263, 303855], copper, and other chemicals are vital for consumers and industry. Batteries and fuel cells supply demands for efficient and portable power. At the heart of all electrochemical processes is the electrochemical reactor.

Table 1 lists several electrochemical processes and their reactions, feeds, and products. The chlor-alkali process consumes approximately 2 % of the electricity generated in the USA. The process involves electrolysis of a brine solution to produce Cl_2 at the anode and NaOH at the cathode. In the Hall process for aluminum refining, Al_2O_2 reacts with a carbon electrode to form Al and CO_2. Water electrolysis is a widespread technology for generating pure H_2 and O_2. While reforming of methane or other

Electrochemical Reactor Design and Configurations, Table 1 Selected electrochemical processes, reactions, feeds, and products

Process	Reaction	Feed	Products
Chlor-alkali		Brine	Cl_2, NaOH
Anode	$2Cl^- \rightarrow Cl_2 + 2e^-$		
Cathode	$2H_2O + 2e^- \rightarrow H_2 + 2OH^-$		
Overall	$2NaCl + 2H_2O \rightarrow Cl_2 + 2NaOH + H_2$		
Aluminum		Al_2O_3, C	Al
Anode	$2O^{2-} + C \rightarrow CO_2 + 4e^-$		
Cathode	$Al_2O_3 + 6e^- \rightarrow 2Al + 3O^{2-}$		
Overall	$2Al_2O_3 + 3C \rightarrow 4Al + 3CO_2$		
Water electrolysis		H_2O	H_2, O_2
Anode	$2H_2O \rightarrow O_2 + 4H^+ + 4e^-$		
Cathode	$2H^+ + 2e^- \rightarrow H_2$		
Overall	$2H_2O \rightarrow 2H_2 + O_2$		
Copper refining		Cu	Cu (ref.)
Anode	$Cu \rightarrow Cu^{2+} + 2e^-$		
Cathode	$Cu^{2+} + 2e^- \rightarrow Cu$ (refined)		
Overall	$Cu \rightarrow Cu$ (refined)		
Li-ion battery		Elec.	Elec.
Anode (dis.)	$LiC \rightarrow Li^+ + C + e^-$		
Cathode (dis.)	$Li^+ + e^- + MO_x \rightarrow LiMO_x$		
Overall	$LiC + MO_x \rightleftharpoons C + LiMO_x$		
H_2/O_2 Fuel cell		H_2, O_2	Elec.
Anode	$H_2 \rightarrow 2H^+ + 2e^-$		
Cathode	$O_2 + 4H^+ + 4e^- \rightarrow 2H_2O$ $2H_2 + O_2 \rightarrow 2H_2O$		
Overall	$2H_2 + O_2 \rightarrow 2H_2O$		

hydrocarbon gas is a cheaper alternative for H_2 production, electrolysis produces H_2 in much higher purity with the additional benefit of high purity O_2. Copper refining involves dissolution of an anode of 98 wt% Cu in a sulfuric acid electrolyte and cathodic plating as pure (99.95 %) Cu. This refined copper can be further purified by a melting/casting procedure to produce oxygen-free high thermal conductivity (OFHC) copper.

Batteries and fuel cells produce electricity from energy that was either stored (battery) or converted (fuel cell) from the chemical energy of a fuel. During discharge, lithium in a carbon intercalation anode is oxidized to lithium ions. The lithium ions then migrate out of the carbon electrode structure into the electrolyte and across the separator to react with a transition metal oxide (MO_x, where M is a transition metal) cathode, forming $LiMO_x$. In so doing, M is reduced by one electron to account for the positively charged Li being incorporated into the cathode. During charge, the process is reversed, with $LiMO_2$ now serving as the anode and the carbon intercalation electrode as the cathode.

Configurations of Electrochemical Reactors

Electrochemical reactors share many of the same features as the basic chemical engineering reactors: batch, continuous stirred tank (CSTR), and plug flow (PFR). These reactor designs are invoked so far as possible for electrochemical reactors. Electrochemical processes encompass a wide range of process chemicals, temperature, phases, and, of course, potentials. Accordingly, electrochemical reactors are not easily categorized into a general operational unit in the chemical engineering sense. Instead, reactor engineering is based on the electrochemical fundamentals of reaction kinetics, ohmic resistances of electrolyte and electrodes, and mass transfer limitations. Design of the reactor hardware rests upon judicious choice of materials to handle corrosion and other reactions in oxidizing and reducing conditions, flow fields that establish desired flow conditions, and experience in reactor fabrication.

Several basic flow configurations in electrochemical reactors are depicted in Fig. 1. Flow through a porous layer, as would occur in a fuel cell, is shown in (a). Flow along a single plate and through two parallel plates is shown in (b) and (c). A rotating disk electrode is shown in (d). This configuration reduces mass transfer

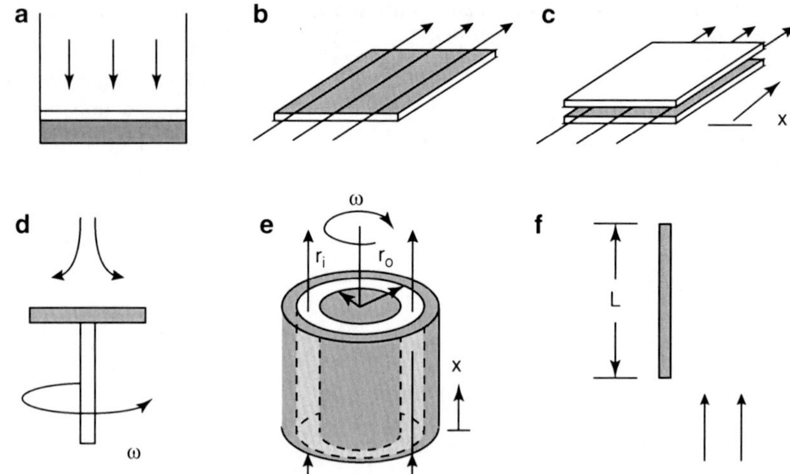

Electrochemical Reactor Design and Configurations, Fig. 1 Flow configurations in electrochemical reactors: (**a**) through a porous layer (*white*) upstream of an electrode (*gray*), (**b**) along a free plate, (**c**) through two parallel plates, (**d**) impinging on a rotating disk electrode, (**e**) through a rotating annular electrode, and (**f**) free convection along a vertical plate

limitations. A rotating annular electrode is shown in (e). Electrolyte flows in the annular gap and a rotating inner electrode improves mass transfer in the cell. Finally, free convection along a vertical electrode is shown in (f).

Figure 2 shows several designs of electrochemical reactors. The bipolar plate reactor (a) is common in fuel cells. Each electrode (except for the end electrodes) serves as the positive terminal of the cell to its left and the negative terminal of the cell to its right, hence the term bipolar. Electrical connections are easier with a bipolar cell; one needs to connect only to the end electrodes. In the monopolar cell (b), each electrode has one polarity and is common to two cells. The monopolar reactor has more complicated electrical connections, but simpler electrolyte flow pathways. In some cases, a monopolar design helps reduce shunt currents relative to a bipolar design.

A bipolar rotating disk electrode is shown in (c). Electrolyte impinges on both sides of a rotating electrode (shaded). Each side is a separate electrochemical cell. The rotating disk is bipolar in that it serves as the positive electrode for the upper cell (the electrolyte between the upper stator (+) and the rotor) and the negative electrode of the lower cell. A fluidized-bed reactor based on a shell and tube heat exchanger is shown in (d). Catholyte fluid enters the bottom of the reactor (tube side) and flows upward to fluidize the bed around the cathode electrode. Anolyte fluid flows through the shell side.

Fundamentals of Electrochemical Reactor Design

Definitions

One of the main goals of electrochemical reactor design is to minimize the overpotential for the reactor. The cell overpotential, which expresses the deviation from ideality, is defined according to the thermodynamic convention as

$$\eta = E - E_{rev} \tag{1}$$

where η is the cell overpotential, E the cell potential, and E_{rev} the reversible cell potential. In the thermodynamic convention, cell potential is defined as

$$E = E_c - E_a \tag{2}$$

where E_c is the cathode potential and E_a the anode potential. From Eq. 1 it follows that cell overpotential is always negative.

Cell overpotential can be divided into components associated with the cathode and anode,

$$\eta = \eta_c - \eta_a \tag{3}$$

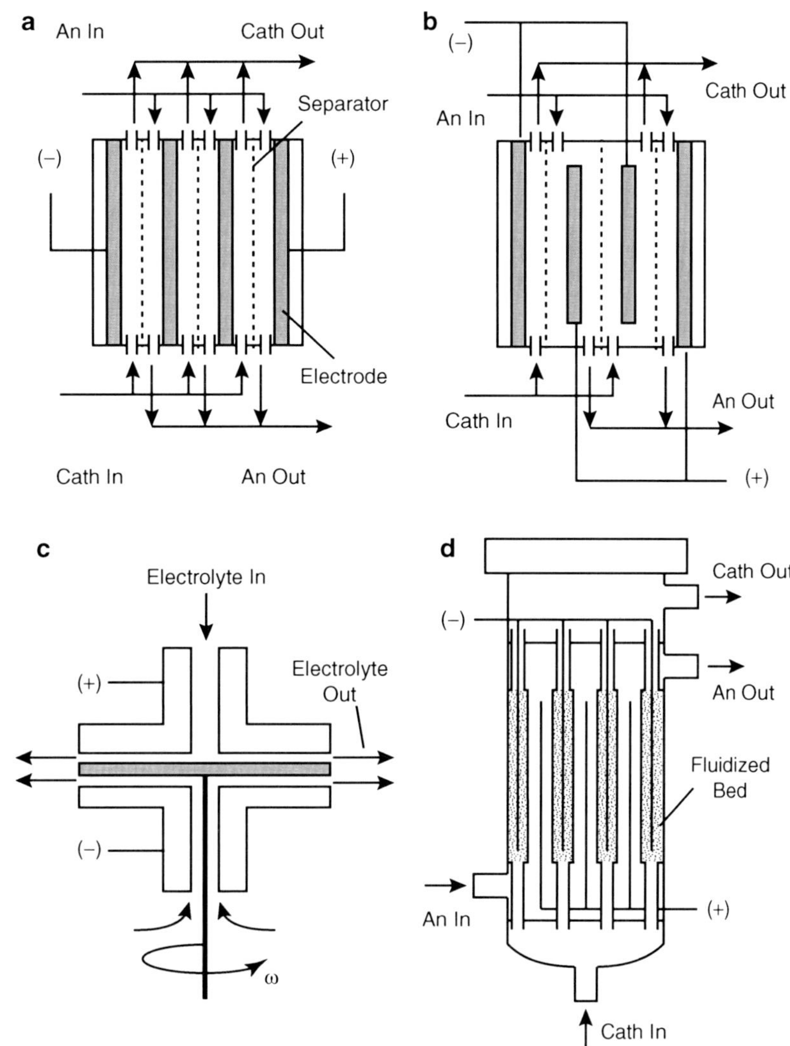

Electrochemical Reactor Design and Configurations, Fig. 2 Types of electrochemical reactors: (**a**) bipolar parallel plate reactor, (**b**) monopolar parallel plate reactor, (**c**) bipolar rotating disk reactor, and (**d**) fluidized-bed reactor. Anolyte and catholyte are abbreviated as An and Cath, respectively (Drawings adapted from Pletcher and Walsh [1])

where η_c and η_a are the cathode and anode overpotentials, respectively. Cathode overpotentials are always negative, since a more negative potential is more reducing. Conversely, anode overpotentials are positive, since a more positive potential is more oxidizing. Because anode overpotential enters into Eq. 3 by subtraction, one can see that both cathode and anode potentials act in the same direction for their influence on cell overpotential.

Component Overpotentials

The overpotential can be divided into four functional components representing η_y, electrolyte resistance; η_k, activation or reaction kinetics limitations; η_e, electrode resistance; and η_t, transport limitation or reactant depletion. The cell overpotential is the sum of these components

$$\eta = \eta_y + \eta_k + \eta_e + \eta_t. \tag{4}$$

The electrolyte and electrode resistances are ohmic and can be combined into a single overpotential for ohmic resistance η_r as

$$\eta_r = \eta_y + \eta_e. \tag{5}$$

Activation Overpotential

The activation overpotential is a function of current and determined by the Butler-Volmer equation. For a generic oxidation reaction of

$$R \rightleftharpoons O^+ + e^- \quad (6)$$

where R is the reduced species and O^+ the oxidized species, the Butler-Volmer equation is

$$j = j_o\left\{\exp\left[\frac{(1-\beta)nF\eta_k}{RT}\right] - \exp\left(\frac{-nF\eta_k}{RT}\right)\right\} \quad (7)$$

where j is the current, j_o is the exchange current density for the reaction of interest, β is the symmetry factor (usually assumed to be 1/2), n is the number of electrons transferred in the reaction, F is Faraday's constant, R is the gas constant, and T is temperature. The sign of the activation overpotential is important in Eq. 7. For oxidation reactions, $\eta_k > 0$, and the first term is dominant. For reduction, $\eta_k < 0$, and the second term is dominant. The activation overpotential for each electrode is given by writing the corresponding Butler-Volmer equation for that electrode,

Anode:

$$j = j_{o,a}\left|\exp\left[\frac{(1-\beta_a)\, n_a F\eta_{k,a}}{RT}\right] - \exp\left(\frac{-\beta_a n_a F\eta_{k,a}}{RT}\right)\right| \quad (8)$$

Cathode:

$$j = j_{o,c}\left|\exp\left[\frac{(1-\beta_c)n_c\eta_{k,c}}{RT}\right] - \exp\left(\frac{-\beta_c n_c F\eta_{k,c}}{RT}\right)\right| \quad (9)$$

In Eqs. 8 and 9, the anode and cathode have their own current densities, $j_{o,a}$ and $j_{o,c}$; symmetry factors, β_a and β_c; and numbers of electrons transferred, n_a and n_c. The same current flows through the electrodes, however, albeit with different signs, formally. The absolute values in Eqs. 8 and 9 remove the sign dependence of j. For sufficiently high current,

$$\text{Anode: } \ln\left|\frac{j}{j_{o,a}}\right| > 1.5 \quad (10)$$

$$\text{Cathode: } \ln\left|\frac{j}{j_{o,c}}\right| > 1.5 \quad (11)$$

a single-term approximation of the Butler-Volmer equation can be used,

$$\text{Anode: } \eta_{k,a} \approx \frac{RT}{(1-\beta_a)n_a F}\ln\left|\frac{j}{j_{o,a}}\right| \quad (12)$$

$$\text{Cathode: } \eta_{k,c} \approx \frac{-RT}{\beta_c n_c F}\ln\left|\frac{j}{j_{o,c}}\right| \quad (13)$$

Equations 12 and 13 are explicit expressions for the activation overpotentials. If the current limits in Eqs. 10 or 11 do not hold, then the full Butler-Volmer equations, Eqs. 8 and 9, must be solved iteratively or through an appropriate curve fit. If only one of Eqs. 10 or 11 holds, then the single-term expression, Eqs. 12 or 13, for that electrode can be used.

Overpotential for Mass Transfer Limitations

Electrochemical reactor performance is generally most affected by mass transfer limitations. Most design effort is therefore focussed on minimizing mass transfer problems. There are two main types of limitations: (1) the transport limit, which is the maximum rate of reactant transfer to the electrode, and (2) reactant depletion, in which reactant conversion reduces the bulk fluid concentration relative to the inlet concentration.

The transport limit is characterized by zero, or very small, concentration of reactant at the electrode surface, although sufficient reactant concentration exists in the bulk fluid. This condition arises from diffusional limitations near and up to the electrode surface, whether due to diffusion in the reacting or electrolyte fluid, diffusion through a porous layer or porous electrode, formation of passivating layers, liquid flooding of gas phase electrodes, or other causes. Some transport problems can be mitigated by changes in operating conditions, but the best approach is to minimize them in the design stage.

Electrochemical Reactor Design and Configurations, Table 2 Constants and exponents used in the limiting current equation (Eq. 15). All flow is laminar except for the porous medium and free convection, which are turbulent. The flow geometries are illustrated in Fig. 1

Flow	α	γ	θ	ε	ξ	λ	σ
Porous medium[a]	1	1	0	1	0	0	0
Flat plate	0.3387	2/3	1/2	0	1/2	0	1/6
Parallel plates[b]	0.78009	2/3	1/3	1/3	1/3	0	0
Annulus[c]	0.8546	2/3	1/3	1/3	1/3	0	0
Rotating disk[d]	0.62045	2/3	0	0	0	1/2	1/6
Rotating cylinders[e]	0.0487	0.644	0	1	0	0.70	0.344
Vertical plate[f, g]	0.19	−1/3	0	−1/3	0	0	1/3

[a] $h = \delta/\phi^{1.5}$; δ, thickness of porous medium; ϕ, porosity of porous medium

[b] h: distance between plates

[c] $h = (1-\kappa)R/\phi$; $\kappa = R_{in}/R$; R, R_{in}, radii of outer and inner electrodes;
outer electrode, $\phi^{1/3} = 1.145 - 0.09e^{-\kappa/0.4} - 0.05e^{-\kappa/0.05}$;
inner electrode, $\phi^{1/3} = 1.583\kappa^4 - 4.453\kappa^3 + 4.718\kappa^2 - 2.362\kappa + 1.660$

[d] High Schmidt number limit, $Sc = \nu/D_i$

[e] $h = d_R^{1.1}/d_L^{0.70}$;
d_R, diameter of inner (rotating) cylinder;
d_L, diameter of cylinder with limiting current

[f] Free convection; turbulent flow

[g] $h = g(\rho_\infty - \rho_o)/\rho_\infty$; g, gravitational constant;
ρ_∞, fluid density far from surface; ρ_o, fluid density at surface

Reactant depletion is a function of reactant conversion and exercises rate control when the reduction of reactant concentration becomes significant. A significant reduction could occur for conversions of 20–100 %, depending on the process. Reactant starvation occurs for conversions near 100 %, causing loss of reaction current and possibly the onset of undesirable reactions. In fuel cells, for example, fuel or oxygen starvation can lead to cell reversal and onset of corrosion reactions that may be irreversible.

Both mechanisms of mass transfer limitation lead to a reduced concentration of reactant at the electrode surface and give rise to an overpotential for transport limitations. This overpotential is also known as the "concentration overpotential," as it arises from the loss of reactant concentration at the surface. The transport-limited overpotential is given by [2]

$$\eta_{tl} = \frac{RT}{nF} \ln\left(1 - \frac{j}{j_L}\right) \tag{14}$$

where j_L is the transport-limited current at the electrode. Limiting currents have been studied for a variety of flow geometries [3]. The general equation for transport-limiting current is

$$j_L = \alpha \cdot \frac{nFC_i}{s_i} \cdot \frac{D_i^\gamma < v >^\theta \omega^\lambda}{h^\varepsilon x^\xi v^\sigma} \tag{15}$$

where α is a constant, C_i the concentration of species i in bulk solution, s_i the stoichiometric coefficient of species i, D_i the diffusivity of species i, $< v >$ the average fluid velocity, h the geometry descriptor, x the distance from the leading edge of flow over the electrode, ω the rotation speed, and v the kinematic viscosity. Table 2 lists constants and exponents for Eq. 15 for several flow geometries. All of the flows are laminar, except for the porous medium and for free convection, which is turbulent. More flow geometries and situations are given in Newman and Thomas-Alyea [3].

Note that the term nFC_i/s_i is common to all flow geometries. That is, the limiting current is proportional to bulk reactant concentration. The limiting current is increased upon increases in diffusivity, flow velocity, and rotational speed of rotating electrodes. Likewise, it is decreased upon increases in h, distance along the electrode (x), and kinematic viscosity. With the exception of free convection, the term h is a characteristic length of mass transfer for each flow geometry. For free convection, h is the density ratio (Table 2).

The concentration term in Eq. 15 is of the bulk reactant at the point (x) where mass transfer is considered. The bulk concentration will change, however, along the length of the electrode due to reactant conversion. Thus, the bulk concentration at a given conversion X is given by

$$C_i = (1 - X)C_{i,in} \tag{16}$$

where $C_{i,in}$ is the concentration of species i at the reactor inlet. The limiting current can be written in terms of the limiting current based on inlet conditions $j_{L,in}$ as

$$j_L = (1 - X)j_{L,in}. \tag{17}$$

To compute $j_{L,in}$, C_i in Eq. 15 is replaced by $C_{i,in}$. Substitution of Eq. 17 into Eq. 14 gives

$$\eta_{tl} = \frac{RT}{nF} \ln\left[1 - \frac{j}{(1 - X)j_{L,in}}\right]. \tag{18}$$

For the cell transport-limited overpotential, Eq. 18 is written for both cathode (c) and anode (a)

$$\eta_{tl,cell} = \frac{RT}{nF} \ln\left[1 - \frac{j}{(1 - X_c)j_{L,c,in}}\right] + \frac{RT}{nF} \times \ln\left[1 - \frac{j}{(1 - X_a)j_{L,a,in}}\right]. \tag{19}$$

In these equations, the conversion terms $(1 - X_c)$ and $(1 - X_a)$ account for reaction depletion, and the limiting current terms $j_{L,c,in}$ and $j_{L,a,in}$ account for diffusional transport limitations.

The Polarization Curve

The influence of the component overpotentials on cell potential is shown in Fig. 3a for a simulated polarization curve of a fuel cell. Model parameters for this curve are listed in Table 3. The solid curve is the cell potential as a function of current density. The dashed line shows the reversible potential at 1.160 V, and the component overpotential curves are shown for η_k, η_e, and η_{tl}. These three component overpotentials sum to the cell overpotential η. The activation overpotential dominates over the others at low current densities. At higher current density, the effects of ohmic resistance (of the electrolyte and electrodes) and transport limitations become of similar magnitude as the kinetics resistance. Reducing activation overpotential is largely the responsibility of laboratory and computational research. Ohmic overpotentials require both laboratory research and proper design of the reactor. Transport limitations are mostly in the purview of reactor design. All three aspects must be minimized in order to achieve the highest efficiency of reaction.

Polarization curves of proton exchange membrane fuel cells deviate from the simulated curve in Fig. 3a. Most significant is the low open-circuit potential of 0.9–1.0 V, as opposed to the reversible potential of approximately 1.2 V. The low open-circuit potential has a number of contributing factors, which include multiple reactions that set up a mixed potential, crossover of H_2 or O_2 through the membrane, and finite resistance effects of voltage measurement devices. These effects cannot be modeled with the overpotential method discussed here. Nonetheless, it is interesting to see how well the overpotential model can approximate a real curve.

Figure 3b shows a comparison with an experimental curve obtained by Debe [4]. The low open-circuit potential of 0.9 V is far removed from the reversible potential, making it difficult to obtain a reasonable fit. (Table 3 lists the parameters of the model fit). The exchange current densities are 0.1 and 0.01 A/cm^2 for the anode and cathode, respectively. There is little slope to the experimental curve, which was modeled by a low ohmic resistance of 0.05 Ω cm^2. The transport limitations were well modeled with reasonable limiting current densities of 1.6 and 1.65 A/cm^2 for the anode and cathode, respectively. The model polarization curve matches the experimental curve in the central region, which was the intent, but does not agree well with the low and high current density regions.

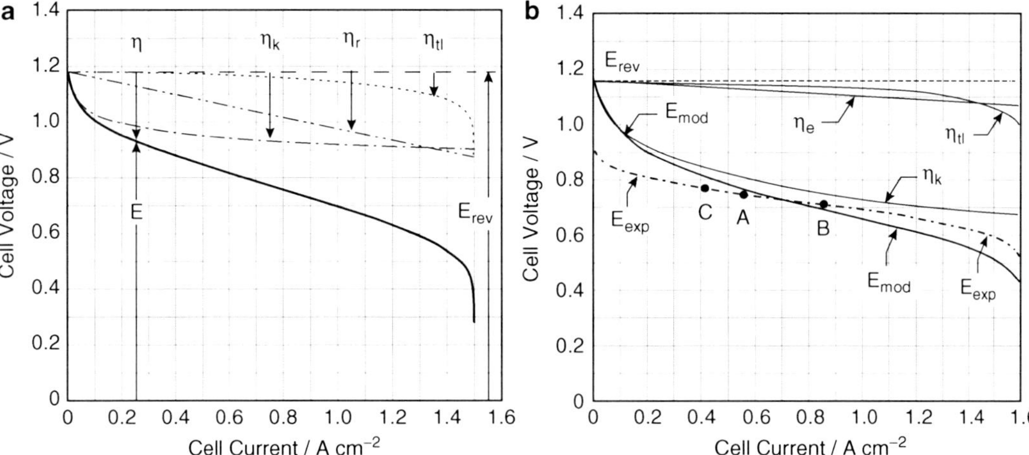

Electrochemical Reactor Design and Configurations, Fig. 3 Contributions of component overpotentials to a fuel cell polarization curve: (**a**) simulated polarization curve for water-saturated H_2 and air at 80 °C and 1 atm. The curves are (–) cell potential E, (– –) reversible potential E_{rev}, (···) transport-limited overpotential η_{tl}, (– ·· –) ohmic resistive overpotential η_r, and (– · –) activation overpotential η_k. The cell overpotential is given by the relation $\eta = \eta_r + \eta_k + \eta_{tl} = E - E_{rev}$. (**b**) Comparison of the polarization curve model with an experimental polarization curve (*dash-dot*) [4]. Parameters for the model polarization curves are listed in Table 3. Points A, B, and C are discussed in the example

Electrochemical Reactor Design and Configurations, Table 3 Parameters for the model polarization curves of Fig. 3. The same value of β was used for the anode and cathode. Effects of reactant conversion have been neglected in j_L

Parameter Units	R_y Ω cm^2	$j_{o,a}$ A/cm^2	$j_{o,c}$ A/cm^2	β	$j_{L,a}$ A/cm^2	$j_{L,c}$ A/cm^2
Figure 3a	0.2	10^{-2}	10^{-3}	0.5	1.495	1.495
Figure 3b	0.05	0.1	0.01	0.5	1.6	1.65

Estimating Reactor Costs

Although electrochemical reactors exhibit much variety in design, the basic tenets of chemical engineering reactor design [5] serve well to develop a framework for cost estimation. Profitability of a project is given by the sum of cash flows over the project lifetime and can be measured before or after taxes and with or without discounting. The basic, nondiscounted, before-tax cash flow equation is

$$\sum_{j=1}^{n} C_{f,j} = -C_{TCI} + \sum_{j=1}^{n} (R_j - E_j) + V_S \quad (20)$$

where n is the project lifetime in years, $C_{f,j}$ the cash flow in year j, C_{TCI} the total capital investment, R_j and E_j the revenue and expenses in year j, and V_S the salvage value at the end project. Here, revenue would be the value of any products or electricity that can be sold. Expenses are the cost of manufacture, which includes direct costs such as reactants, utilities, and labor; fixed costs such as taxes and overhead; and general expenses such as research and development.

Equation 20 applies to an entire plant or project, in which revenue is determined by the value of products sold. In designing specific units within the plant, revenue is not considered, and the focus is on equipment and utilities costs for that specific unit. In a typical chemical plant, the separation system has the highest cost for both equipment and utilities. The reactor is a minor cost in comparison, but it has a large effect on separation system costs. Good reactor design is crucial in order to produce a valuable product with the highest selectivity and therefore minimize separation costs.

A fuel cell system is a good example of a small electrochemical plant. It uses a chemical fuel to produce electricity that can be either sold or used. The entire fuel cell plant contains a variety of fuel and air handling processes including preheating, vaporization, pressurization, filtration, and purification. The electricity from the fuel cell must be combined with a power controller to produce an electricity output consistent with the needs of the application: direct versus alternating current, voltage, phase, allowable variation in load, and others. The fuel cell plant will be treated as a chemical plant so that Eq. 20 applies. Alternatively, it could be one unit of a larger plant that would have to be optimized in the context of the revenue of that larger plant.

Example: Design of a Fuel Cell Plant

A proton exchange membrane (PEM) fuel cell plant of 100 kWe capacity is to be designed in conjunction with the experimental polarization curve of Fig. 3b. The stack voltage requirement is 200 V, and cell area is limited to 30×30 cm^2 set by the method of assembly. The stack runs on humidified H_2 and air at 1 atm and 80 °C. The cost of H_2 is $5/kg. The stack operates 8,150 h/year and has a lifetime of 5 years with no salvage value. (In reality, the Pt in the stack would be of value). The objective of this example is to determine the number of cells, cell area, and power density of the stack and costs of the stack, fuel cell plant, fuel, and electricity (per kWhe).

For the fuel cell plant, Eq. 20 simplifies to

$$\sum_{j=1}^{5} C_{f,j} = -C_{TCI} - \sum_{j=1}^{5} E_j \tag{21}$$

where it is assumed that the electricity will be used, rather than sold, so that revenue is neglected. Equation 21 therefore represents only costs, which are negative cash flows. The fuel cell is assumed to be purchased by the user, so that its cost represents the cost of manufacture (equipment, labor, overhead, fees, etc.) and profit of the manufacturer. The TCI of the fuel cell, which is the price the purchaser pays, is given by

$$C_{TCI} = L_f(C_{fcs} + C_{bop}) \tag{22}$$

where L_f is the Lang factor, typically used for rough estimates of TCI based on equipment costs; C_{fcs} is the cost of the fuel cell stack; and C_{bop} is the cost of the balance of plant. A typical value of L_f of 4 will be assumed. In general, costing of a fuel cell and its balance of plant is a detailed task. For simplicity, the cost correlations assumed for this example are

$$C_{fcs} = (\$50/\text{l})V_{fcs} = (\$50/\text{l})n_c t_c A_c \tag{23}$$

$$C_{bop} = C_{fcs} \tag{24}$$

where V_{fcs} is the volume of the stack, n_c the number of cells in the stack, t_c the cell thickness, and A_c the cell area. The value of $50/l gives costs ($C_{fcs} + C_{bop}$) similar to the value of $49/kWe reported by the US Department of Energy in 2011 [6].

Scenario A: Design of Optimum Fuel Cell

Specific details of the design are listed in Table 4. The most electrically efficient (highest cell potential) will be that of the highest cell area (lowest current density). For a 200 V stack voltage, a current of 500 A is needed to produce 100 kWe. At the maximum cell area of 900 cm^2, the current density is 0.556 A/cm^2, which then locates point A on the polarization curve of Fig. 3. The potential at this point is 0.75 V, and therefore 267 cells are required to achieve 200 V. With an assumed cell thickness of 0.3 cm, the stack volume is 72.03 l, thus giving a power density of 1.39 kWe/l and a fuel cell plant cost of $7,203 according to Eq. 23 and TCI of $28,813 by Eqs. 22 and 24.

With an allowance of 95 % fuel utilization, fuel flow rate $\dot{n}_f$ is given by

$$\dot{n}_f = \frac{P_e}{u_f nFE} \tag{25}$$

where P_e is the electrical power output and u_f is the fuel utilization. For scenario A, the fuel flow rate is 0.719 mol H_2/s, which translates to fuel cost of $191,732 per year. The total cash flow is the sum of the yearly fuel cost for 5 years and the TCI, which is –$987,474. This amounts to an electricity cost of $0.242/kWh.

Electrochemical Reactor Design and Configurations, Table 4 Design of a 100 kWe, 200 V, H_2/air fuel cell based on the polarization curve of Fig. 3b. Design details are given in the text

	Units	A	B	C
Cell potential	V	0.750	0.710	0.772
Current density	A/cm^2	0.556	0.850	0.410
Cells per stack		267	282	267
Area per cell	cm^2	900.0	590.5	900.0
Stack volume	l	72.03	49.76	72.03
Power density	kWe/l	1.39	2.01	1.06
Plant cost	$	7,203	4,976	7,203
TCI	$	28,813	19,906	28,813
Fuel (H_2) flow	mol/s	0.719	0.759	0.530
Fuel cost	$/year	191,732	202,504	141,885
Total annual cost	$/year	197,495	206,485	147,148
Cum. cash flow	$	−987,474	−1,032,424	−735,739
Electricity cost	$/kWh	0.242	0.253	0.181

Scenario B: Repeat Scenario A for a Minimum Power Density of 2.0 kWe/l

This design requires an iteration of operating points on the polarization curve until the number of cells and cell area gives a volume of 50 l. The corresponding operating point (B) is 0.71 V at 0.85 A/cm^2 for which 282 cells of 590.5 cm^2 are required. The resulting cost of electricity is $0.253/kWh.

Scenario C: Repeat Scenario A for 75 kWe

The number of cells and their area are fixed as in Scenario A. The difference is that the flow rate of H_2 is reduced, and hence the fuel cost is reduced. The operating point (C) is found by iteration about the polarization curve or by calculating the power curve as jE versus j. Point C is located at 0.772 V and 0.41 A/cm^2, which gives a cost of electricity of $0.181/kWh. The stack voltage is 206 V (267 cells of 0.772 V each), which is close enough to 200 V to not cause problems with a power controller.

Discussion

The fuel cell in this example does not optimize in the usual fashion of equipment versus operating costs. The low cost of the fuel cell, about 3 % of the total fuel cost, all but eliminates equipment cost from the optimization. As the cell voltage increases, electrical efficiency increases and fuel costs decrease accordingly. Cell area, however, increases with increasing cell voltage to the extent that cell area controls the design, either by a limitation in the maximum area that can be fabricated ("Scenario A: Design of Optimum Fuel Cell") or by volume constraints of the given application ("Scenario B: Repeat Scenario A for a Minimum Power Density of 2.0 kWe/l"). Scenario C illustrates the cost advantage of running a larger stack at lower power or, conversely, the cost increase of running a smaller stack at a larger power.

The polarization curve is assumed to be fixed for the various scenarios, but this will not be the case in practice. The polarization curve is a function of operating conditions, which themselves must be optimized. A more complete design involves optimization of the fuel cell with detailed cost data, variations in operating conditions, and variations in polarization curves.

Cross-References

- ▶ Aluminum Smelter Technology
- ▶ Chlorine and Caustic Technology, Overview and Traditional Processes
- ▶ Electrochemical Reactor Design for the Oxidation of Organic Pollutants

References

1. Pletcher D, Walsh FC (1990) Industrial electrochemistry, 2nd edn. Chapman and Hall, London
2. O'Hayre R, Cha SW, Colella W, Prinz FB (2006) Fuel cell fundamentals. Wiley, Hoboken
3. Newman J, Thomas-Alyea KE (2004) Electrochemical systems, vol Electrochemical society series, 3rd edn. Wiley, Hoboken
4. Debe MK (2013) Advanced cathode catalysts and supports for PEM fuel cells, 2011 annual merit review DOE hydrogen and fuel cells and vehicle technologies programs. http://www.hydrogen.energy.gov/pdfs/review11/fc001_debe_2011_o.pdf. Accessed 15 March 2013
5. Turton R, Bailie RC, Whiting WB, Shaewitz JA (2009) Analysis, synthesis, and design of chemical processes, 3rd edn. Prentice Hall, Upper Saddle River
6. The Department of Energy Hydrogen and Fuel Cells Program Plan, Fuel Cell Technologies Program, U.S. Department of Energy (2011) http://www1.eere.energy.gov/hydrogenandfuelcells/pdfs/program_plan2011.pdf. Accessed 14 March 2013

Electrochemical Reactor Design for the Oxidation of Organic Pollutants

André Savall
Laboratoire de Génie Chimique, CNRS,
Université Paul Sabatier, Toulouse, France

Introduction

In recent years, there has been an increasing interest in the use of electrochemical processes for the treatment of wastewaters containing toxic and biorefractory organic pollutants. The aim of this entry is to present the principle of the electrochemical destruction of organics by electro-oxidation (EO) and to give some examples of reactor design conceived for this kind of treatment. However, it is widely believed that the deployment of this technique in combination with a biological posttreatment could be a suitable solution to this issue. Indeed, when the effluent is first subjected to electro-oxidation, biodegradability of the organic pollutants can be enhanced, improving the performance of the subsequent biologic process [1]. During the last two decades, research works have been focused on the efficiency in oxidizing various pollutants on different electrode materials. The main objective was to improve the electrocatalytic activity and electrochemical stability of electrode materials [2–7]. On the other hand, experimental investigations were conducted in the field of kinetics of pollutant degradation, on factors affecting the process performance [8], and on the energy consumption [9].

Pollutants treated in synthetic solutions include phenols, aromatic amines, quinones, glucose, carboxylic acids, tannic acid, herbicides, pesticides, surfactants, dyes, etc. [2–5]. Some papers have considered the treatment of real effluents including human wastes, landfill leachates, tannery wastes, dye plant effluents, and herbicide manufacture effluents [4–7].

Definitions, Terminologies

The instantaneous current efficiency (ICE) is an important global parameter commonly used to estimate the progress and the efficiency of electrochemical treatments. ICE is calculated from the experimental variation of chemical oxygen demand (COD) according to Eq. (1) [4, 6]:

$$ICE = \frac{\Delta COD_{exp}}{\Delta COD_{tk}} = 4\,F\,V\,\frac{[COD(t) - COD(t + \Delta t)]}{I\,\Delta t\,M_{o_2}} \quad (1)$$

[COD(t)] and [COD(t + Δt)] are respectively the concentrations of COD expressed in g L^{-1} at times t and t + Δt, M_{02} is the dioxygen molar weight (g mol^{-1}), V the volume of solution (L), C is the Faraday's constant (96,485 C mol^{-1}), and 4 is the number of faradays per mol of O_2.

For an electrochemical reactor working in galvanostatic conditions, the electrolysis energy is given by Eq. 2:

$$E = I \int_t U(t)\, dt \quad (2)$$

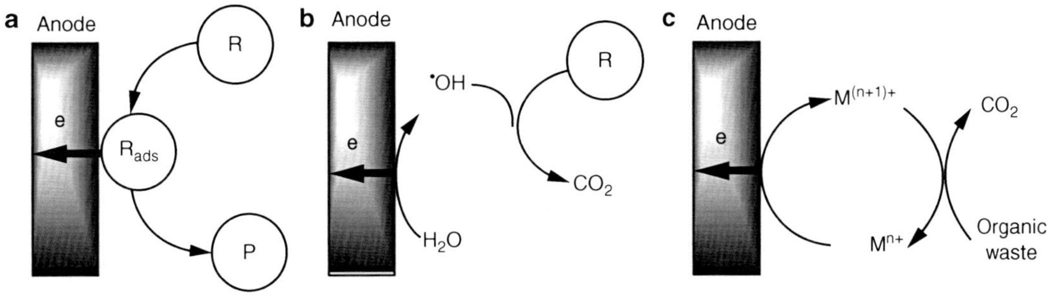

Electrochemical Reactor Design for the Oxidation of Organic Pollutants, Fig. 1 (**a**) Direct electro-oxidation. (**b**) Electrochemical mineralization. (**c**) Indirect electrochemical oxidation

The voltage of the cell U(t) is given by the following sum of terms:

$$U(t) = U^0 + \eta_A + \eta_C + \sum R(t)I \qquad (3)$$

where U^0 is the potential at nil current, η_A and η_C are respectively the anodic and cathodic overvoltages, and R(t) I is the ohmic drop through a solution section of resistance R. The variation of the cell voltage results mainly from the change of the ohmic drop of the solution due to its change in composition. As this variation is generally weak (around 10 %), an average value of the cell voltage can be used to evaluate energy consumption.

Direct Electrochemical Oxidation

In direct electro-oxidation, pollutants are oxidized after adsorption on the anode surface by mechanisms involving only electrons (Fig. 1a):

$$R \rightarrow R_{ads} \rightarrow P + \ z\,e \qquad (4)$$

Direct electro-oxidation is theoretically possible at potential before water oxidation; for example, the standard thermodynamic potential of the reversible cell based on the complete mineralization of phenol is 0.108 V [8], a value much lower than that of water oxidation (1.23 V/SHE), but the reaction rate usually has low kinetics. Although the oxidation rate of certain organic pollutants is consistent, when materials such as Pt or metal oxide anodes (Ti/IrO_2, Ti/RuO_2-TiO_2, etc.) are used, numerous reactions driven under potentiostatic mode are strongly slowed down by the poisoning of the electrode; this phenomenon often results from the forming of polymers adsorbed on the anode surface, especially when aromatics are present in the waste water [4].

Electrochemical Mineralization

In electrochemical mineralization, oxygen atoms are transferred from water to the organic pollutant R via the strong oxidizing hydroxyl radicals ($E° = 2.74$ V/SHE) produced by water discharge (Fig. 1b):

$$H_2O \ + S[\,] \ \rightarrow S\left[{}^{\bullet}OH\right] \ + H^+ \ + e \qquad (5)$$

A concomitant reaction is oxygen evolution through water oxidation:

$$S\left[{}^{\bullet}OH\right] \ + H_2O \ \rightarrow S[\,] \ + O_2 \ + 3\ H^+ \ + 3\,e \qquad (6)$$

The transfer rate of the oxygen atom depends on the nature of the anode material. Anodes characterized by low oxygen overvoltage (Table 1, case of RuO_2, IrO_2, Pt) favor oxygen evolution and have low chemical reactivity with respect to mineralization of organics. On these materials, the active oxygen (${}^{\bullet}OH$) is strongly adsorbed (chemisorbed) on the metal oxide (MO_x) and

Electrochemical Reactor Design for the Oxidation of Organic Pollutants, Table 1 Potential for oxygen evolution reaction on different electrode materials in H_2SO_4. Standard potential for oxygen evolution is 1.23 V versus standard hydrogen electrode (*SHE*) [4]

Anode	Value versus SHE (V)	Conditions
RuO_z	1.47	0.5 M H_zSO_4
IrO_z	1.52	0.5 M H_zSO_4
Pt	1.6	0.5 M H_zSO_4
Oriented pyrolytic graphite	1.7	0.5 M H_zSO_4
SnO_z	1.9	0.05 M H_zSO_4
PbO_z	1.9	1 M H_zSO_4
Boron doped diamond (BDD)	2.3	0.5 M H_zSO_4

can penetrate in the oxide lattice to form MO_{x+1}, giving rise to selective oxidation products. Inversely, on anodes with high oxygen overvoltage that are poor catalysts for oxygen evolution, hydroxyl radicals are weakly adsorbed (physical adsorption) and are available to participate in the transfer of oxygen [10]. It is the case of SnO_2, PbO_2, and boron doped diamond (BDD) which favor complete combustion to CO_2 (Table 1). For example, on a PbO_2 anode, a phenol molecule can undergo several oxidation steps, leading successively to hydroquinone, benzoquinone, maleic acid (Fig. 2b), and even CO_2, without desorption of the intermediate products [11], which proves a behavior situated between cases **a** and **b** in Fig. 1. Attempts to use dimensionally stable anodes (DSAs) of the type Ti/SnO_2 were successful for the COD abatement, but these electrodes suffer from very low service life [6]. Titanium covered with PbO_2 forms electrodes which are a little less effective than Ti/SnO_2 during mineralization of organics, but their long service life favors their use, especially since these materials are commercially available [12].

Anodic oxidation is the most effective in the case of the electrode of BDD [4] because it ends in the complete mineralization of the substrate (see Figs. 1b and 2a) [4]:

$$R + S\left[{}^{\bullet}OH\right] \rightarrow S + CO_2 + zH^+ + ze \qquad (7)$$

Kinetic Model of Organics Mineralization on a BDD Anode

Figure 2 presents typical results obtained during mineralization of phenol in an undivided cylindrical flow cell using successive disks made respectively of Si covered with BDD and of Ti covered with PbO_2 under the same hydrodynamic conditions. Other parameters being equal, phenol species disappear at the same rate on these two materials, but the BDD anode was more efficient to eliminate COD; less intermediates were formed with BDD compared to PbO_2 irrespective of the current density [9].

Considering that the simplified reaction scheme (Eqs. 5 and 7) can be identified as a direct oxidation for the BDD anode, a kinetic model was developed assuming that the global rate of the electrochemical mineralization of organics is a fast reaction controlled by mass transport [13]. As a result, the limiting current density for the electrochemical mineralization of an organic compound (or a mixture of organics) under given hydrodynamic conditions can be written as:

$$I_{lim} = 4\ F\ k_m[COD] \qquad (8)$$

where k_m is the average mass transfer coefficient of the reactor (m s^{-1}). Results presented in Fig. 2a permitted to adapt, for the BDD anode, engineering models developed for systems recycling the electrolyte between a reservoir and the electrochemical reactor operating under galvanostatic conditions [13]. Table 2 presents equations of the modeling that describe COD and ICE as a function of time during electro-oxidation of organics at BDD anode.

Electrochemical Reactor with Diamond Anode for Efficient Mineralization of Organics

A large variety of electrochemical systems have been tested for the treatment of wastewaters by EO. Electrochemical oxidation over BDD electrodes has received special attention, because it exhibits high efficiencies. Flow cells with parallel

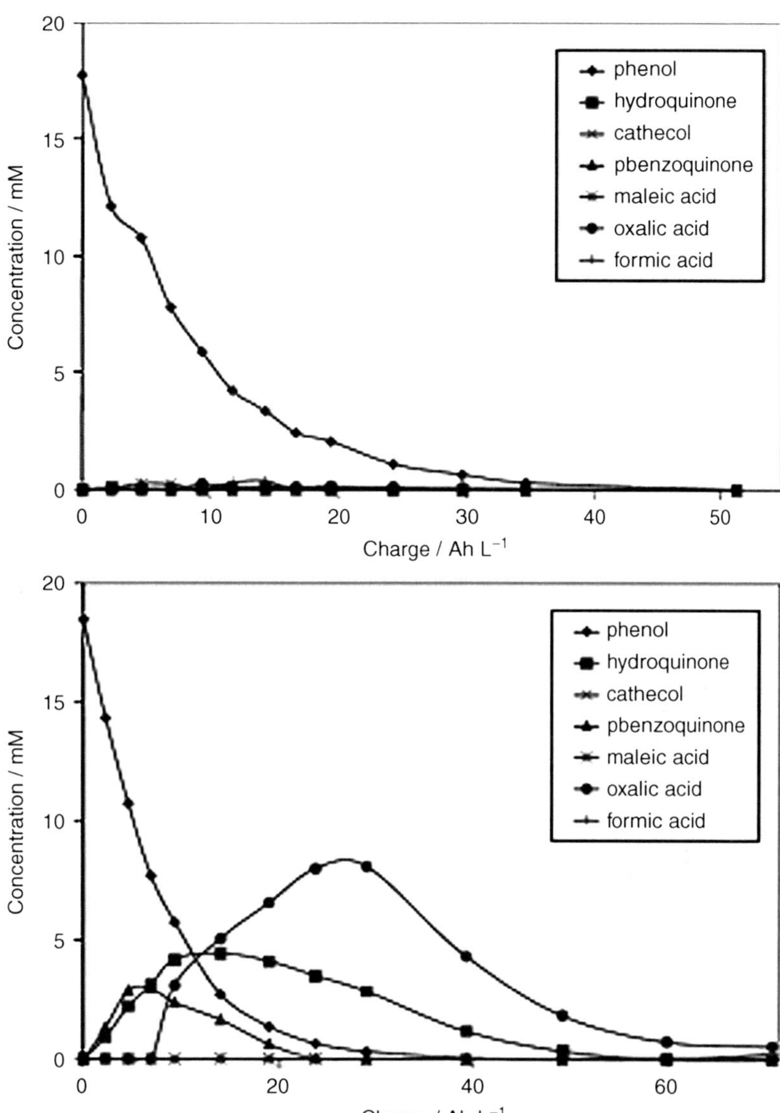

Electrochemical Reactor Design for the Oxidation of Organic Pollutants, Fig. 2 Concentration variation of organic compounds during phenol oxidation. Electrolysis of 1 L of 0.02 M phenol solution in 0.1 M H_2SO_4. Flow rate = 200 L h^{-1}. Reactor without separator. Anode (63.6 cm^2): (**a**) BDD; (**b**) PbO_2. Cathode: Zr. Current density: i = 142 mA cm^{-2}; i^0_{lim} (or i^0_{lim}) = 108 mA cm^{-2} (Eq. 8) (From Ref. [9])

Electrochemical Reactor Design for the Oxidation of Organic Pollutants, Table 2 Equations for COD and ICE variation during mineralization of organics on a BDD anode [13]. V_R is the reservoir volume ($V_R > V_{cell}$); A the electrode area (m^2), COD^0 the initial chemical oxygen demand, $\alpha = i/i^0_{lim}$ and i^0_{lim} is the initial limiting current density (A m^{-2})

	Instantaneous current efficiency (ICE)	Chemical oxygen demand COD (mol $O_2 m^{-3}$)
$i_{appl} < i_{lim}$ (current limited control)	ICE = 1	$COD(t) = COD^0\left(1 - \frac{\alpha A k_m}{V_R}t\right)$ (9)
$i_{appl} > i_{lim}$ (mass-transport control)	$ICE = \exp\left(-\frac{Ak_m}{V_R}t + \frac{1-\alpha}{\alpha}\right)$ (10)	$COD(t) = \alpha COD^0 \exp\left(-\frac{Ak_m}{V_R}t + \frac{1-\alpha}{\alpha}\right)$ (11)

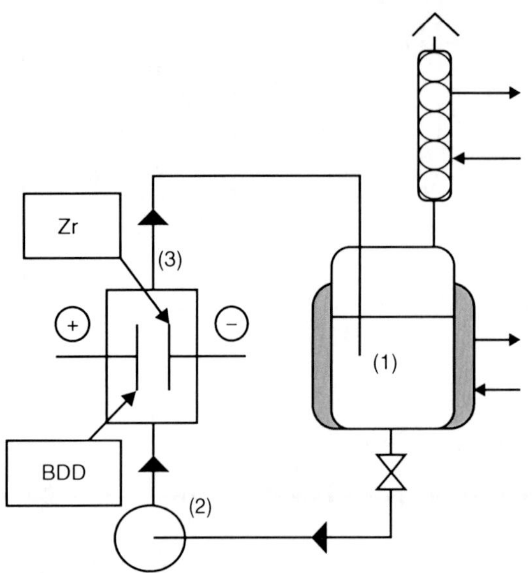

Electrochemical Reactor Design for the Oxidation of Organic Pollutants, Fig. 3 Discontinuous process with a single compartment electrochemical reactor for oxidation of organics on a diamond anode. (*1*) thermoregulated reservoir (volume V_R); (*2*) pump; (*3*) electrochemical cell with BDD anode and Zr cathode

electrodes with or without a reservoir have been employed [4]. Figure 3 illustrates a typical electrochemical cell with planar monopolar electrodes operating in batch mode with recycling. This simple design allows hydrodynamic conditions to reach a suitable mass-transport coefficient and makes easy the scale-up either by size increase or by the duplication of elementary cells. The use of electrodes formed with grids of Ti or Nb covered with BDD [14] improves the mass transfer coefficient k_m by increasing the buoyancy.

By using the model presented in Table 2, the required ratio A/V_R to obtain a desired COD conversion can be calculated for 1 h of electrolysis of 1 m^3 of a polluted solution under a constant current density equal to the limiting value calculated at $t = 0$. Figure 4a shows that very high surface areas of anode must be used when the desired COD conversion tends toward 1. Figure 4b shows how the electrolysis time varies for reaching a given conversion value, while Fig. 4c shows the specific energy as a function of conversion. Figure 4d shows the variation of ICE as a function of time for three initial galvanostatic conditions. This model, which does not include any adjustable parameter, was validated for many pollutants [4]. On account of the encouraging results obtained at laboratory scale, electro-oxidation of wastewaters by means of Si/BDD electrodes has been implemented at pilot scale including 150 DiaCell® modules for a total anode surface of 1.05 m^2 [15].

Possibilities of new developments are hoped with BDD anodes on substrates with great specific area. Thus, materials under shapes of grids, tubes, or rods can be elaborated from valve metals such as titanium or niobium [14]. For example, a bipolar trickle bed tower reactor filled with Raschig ring shaped BDD anode was proposed for the treatment of textile wastewater [16].

Indirect Electrochemical Oxidation

The principle of indirect electrolysis using a redox couple is shown in Fig. 1c. Metal ions in acidic solutions are oxidized on the anode from their stable oxidation state (M^{n+}) to the higher reactive oxidation state ($M^{(n+1)+}$) capable to attack the organic feed, breaking it down into carbon dioxide, insoluble inorganic salts, and water. This mediated electrochemical oxidation (MEO) is preferably used to treat solid wastes (like obsolete pesticides, radioactive contaminated solid organic wastes, etc.) or concentrated solutions, in order to avoid or limit subsequent separation. For total organics oxidation, a redox pair with a high oxidation potential, such as Ag(II)/Ag(I) ($E^\circ = 1.98$ V), Co(III)/Co(II) ($E^\circ = 1.82$ V), or Ce(IV)/Ce(III) ($E^\circ = 1.44$ V), must be chosen [4]. A MEO diagram representing the Ag(II) regeneration process is presented in Fig. 5. In the divided electrochemical cell, Ag (II) is generated by oxidation of Ag (I), forming the complex $Ag(NO_3)^+$ in 6 M HNO_3. Platinum is the anode generally used; however, Ti/Pt and Nb/BDD showed good performance [17].

DSA-type anodes coated with a layer of RuO_2 or IrO_2 can be used efficiently for organic disposal by indirect electrolysis generating in situ active chlorine (Cl_2, HClO, or ClO, depending on pH) by the oxidation of chloride ions present in the solution according to the following reaction:

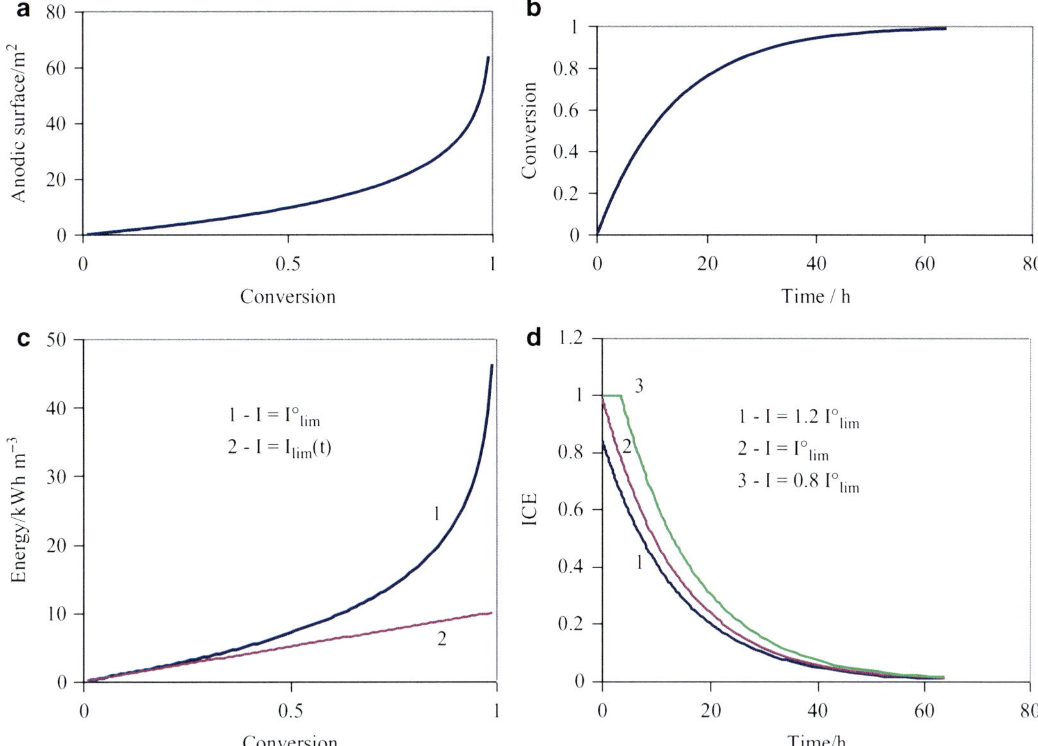

Electrochemical Reactor Design for the Oxidation of Organic Pollutants, Fig. 4 (**a**) Surface needed as a function of conversion $X = \{[COD](t = 0) - [COD](t)\}/[COD](t = 0)$; $A = 1\ m^2$; $V = 1\ m^3$; $I = I°_{lim.} = 241\ A\ m^{-2}$. (**b**) Conversion X as a function of time; same conditions as (**a**). (**c**) Energy as a function of X for two galvanostatic conditions. (**d**) Influence of I on ICE. $U = 3$ V; $k_m = 2 \times 10^{-5} ms^{-1}$; $A = 1\ m^2$; $V = 1\ m^3$; $[COD](t = 0) = 1{,}000$ ppm

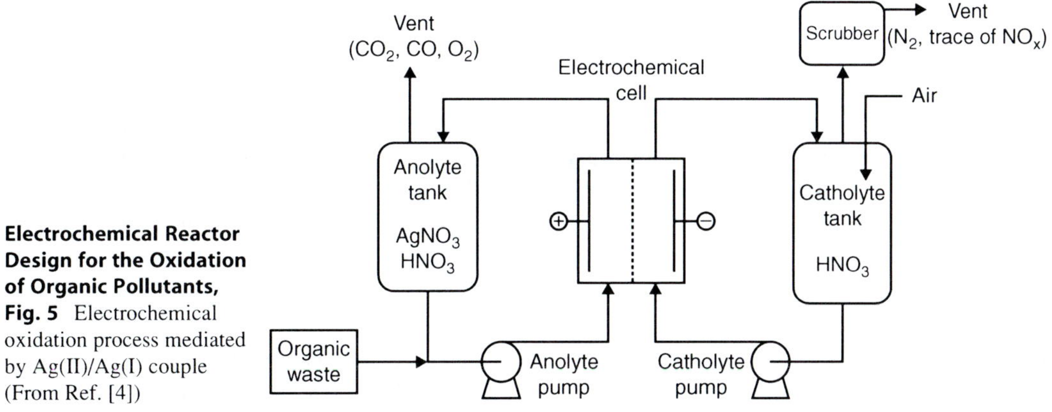

Electrochemical Reactor Design for the Oxidation of Organic Pollutants, Fig. 5 Electrochemical oxidation process mediated by Ag(II)/Ag(I) couple (From Ref. [4])

$$2Cl^- \rightarrow Cl_2 + 2e \tag{9}$$

This process can effectively oxidize many pollutants at chloride concentrations typically larger than 3 gL^{-1} [4]; however, it has the drawback of permitting the formation of chlorinated organic compounds during the electrolysis [3, 4, 6].

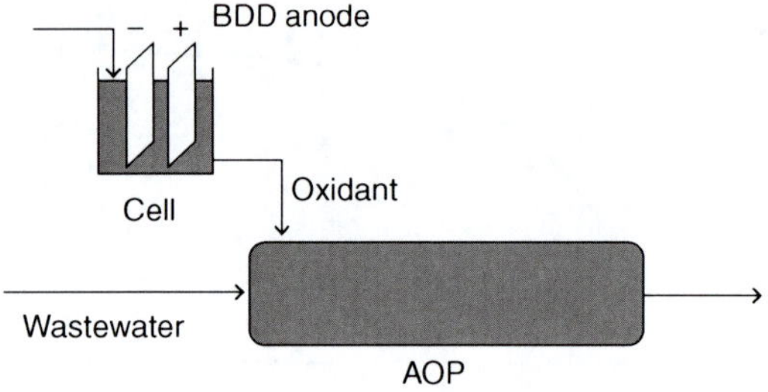

Electrochemical Reactor Design for the Oxidation of Organic Pollutants, Fig. 6 Combination of oxidant production on a BDD anode and an advanced oxidation process (*AOP*) (From Ref. [20])

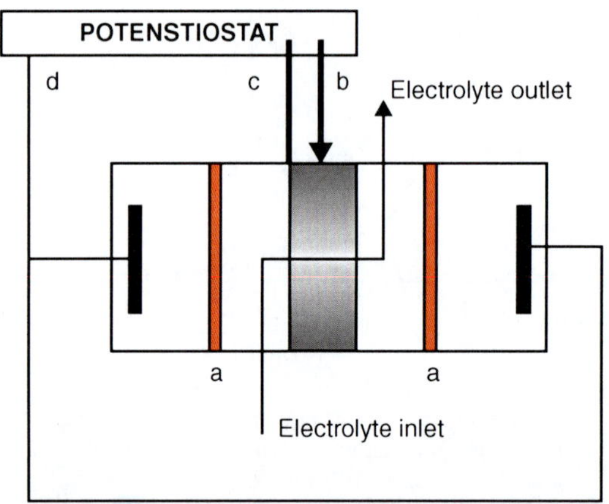

Electrochemical Reactor Design for the Oxidation of Organic Pollutants, Fig. 7 Diagram of the laboratory percolation cell: (*a*) cationic membranes; (*b*) saturated calomel electrode (*SCE*); (*c*) working electrode (disk of graphite felt:10-mm diameter, 10-mm thickness); (*d*) auxiliary counterelectrode (From Ref. [22])

Other oxidizers generated electrochemically (O_3, $S_2O_8^{2-}$, etc.) were successfully used for disinfection or treatment of effluents containing low concentrations of biorefractory contaminants [18–20]. Electrolytic production of ozone may become more attractive using a new simple electrode assembly of a "sandwich" configuration: diamond anode/solid polymer electrolyte (SPE)/ cathode, with 47 % current efficiency [18]:

$$3\ H_2O \rightarrow O_3 + 6\ e + 6\ H^+ \qquad (10)$$

Peroxodisulfuric acid ($H_2S_2O_8$) can be produced at high concentration without any problem of mass transfer limitations by oxidation of concentrated sulfuric acid [19]; it can be used for wastewater treatment in a separate chemical reactor as shown in Fig. 6.

Future Directions

1. The lack of ideal anodes, cheap, with high oxygen overvoltage, good activity, and mechanical stability, is a critical problem in electrochemical oxidation for organic wastewater treatment. Attempts to develop new materials endowed with these properties have not yet succeeded. The BDD films deposited on Si, Ta, and Nb by CVD have shown excellent electrochemical stability [14]; however, Si/BDD electrodes are very brittle and have a weak conductivity. For the moment, large-scale usage of Nb/BDD and Ta/BDD is not favored in reason of the high cost of Nb and Ta substrates [3]. One can, however, think that a thin layer of Nb or Ta electrochemically deposited in molten salt on a substrate of steel could reduce manufacturing costs [21].
2. Complete mineralization of a pollutant can be achievable by combining biological and electrochemical methods. For example, tetracycline can be completely biologically mineralized after a single pass through commercially available graphite felt into a percolated electrochemical reactor (Fig. 7) [22].
3. As has been shown previously, the low concentrations of the organic species in the wastewater limit the efficiency of the anode made of a high oxygen overvoltage material by reason of mass-transport limitation. An alternative method was proposed according to which a pre-concentration is performed by adsorption on granular active carbon (GAC). However, GAC suffers from its relatively high cost and its difficult regeneration. Electrochemical regeneration of GAC has been studied as an alternative to thermal regeneration [23]. The electrochemical regeneration technique possesses several potential advantages over thermal regeneration, such as in situ regeneration, minimal GAC losses, and destruction of the contaminants desorbed via oxidation at the anode.

Cross-References

► Boron-Doped Diamond for Green Electro-Organic Synthesis
► Electrochemical Reactor Design and Configurations
► Overpotentials in Electrochemical Cells

References

1. Anglada A, Urtiaga A, Ortiz I (2009) Contributions of electrochemical oxidation to waste-water treatment: fundamentals and review of applications. J Chem Technol Biotechnol 84:1747–1755
2. Jüttner K, Galla U, Schmieder H (2000) Electrochemical approaches to environmental problems in the process industry. Electrochim Acta 45:2575–2594
3. Chen G (2004) Electrochemical technologies in wastewater treatment. Sep Purif Technol 38:11–41
4. Panizza M, Cerisola G (2009) Direct and mediated anodic oxidation of organic pollutants. Chem Rev 109:6541–6569
5. Martínez-Huitle CA, Brillas E (2009) Decontamination of wastewaters containing synthetic organic dyes by electrochemical methods: a general review. Appl Catal B Environ 87:105–145
6. Panizza M (2010) Importance of electrode material in the electrochemical treatment of wastewater containing organic pollutants. In: Comninellis C, Chen G (eds) Electrochemistry for the environment. Springer, New York
7. Li X, Pletcher D, Walsh FC (2011) Electrodeposited lead dioxide coatings. Chem Soc Rev 40:3879–3894
8. Kapałka K, Foti G, Comninellis Ch (2010) Basic principles of the electrochemical mineralization of organic pollutants for wastewater treatment. In: Comninellis Ch, Chen G (eds) Electrochemistry for the environment. Springer, New York
9. Weiss E, Groenen-Serrano K, Savall A (2008) A comparison of electrochemical degradation of phenol on boron doped diamond and lead dioxide anodes. J Appl Electrochem 38:329–337
10. Comninellis C (1994) Electrocatalysis in the electrochemical conversion/combustion of organic pollutants for waste water treatment. Electrochim Acta 39:1857–1862
11. Belhadj Tahar N, Savall A (1998) Mechanistic aspects of phenol electrochemical degradation by oxidation on a Ta/PbO_2 anode. J Electrochem Soc 145:3427–3434
12. Ueda M, Watanabe A, Kameyama T, Matsumoto Y, Sekimoto M, Shimamune T (1995) Performance characteristics of a new type of lead dioxide-coated titanium anode. J Appl Electrochem 25:817–822
13. Panizza M, Michaud PA, Cerisola G, Comninellis C (2001) Anodic oxidation of 2-naphthol at boron-doped diamond electrodes. J Electroanal Chem 507:206–214
14. Haenni W, Rychen P, Fryda M, Comninellis Ch (2004) Industrial applications of diamond electrodes. In: Semiconductors and semimetals, thin-film diamond II, vol 77. Elsevier, Amsterdam, pp 149–196, (Ch 5)
15. Anglada A, Urtiaga A, Ortiz I (2010) Laboratory and pilot plant scale study on the electrochemical oxidation of landfill leachate. J Hazard Mater 181:729–735
16. Koparal AS, Yavuz Y, Gürel C, Ögütveren ÜB (2007) Electrochemical degradation and toxicity reduction of CI Basic Red 29 solution and textile wastewater by using diamond anode. J Hazard Mater 145:100–108
17. Racaud C, Savall A, Rondet P, Bertrand N, Groenen Serrano K (2012) New electrodes for silver(II) electrogeneration: comparison between Ti/Pt, Nb/Pt, and Nb/BDD. Chem Eng J 211–212:53–59
18. Kraft A (2008) Electrochemical water disinfection: a short review. Platinum Metals Rev 52:177–185
19. Serrano K, Michaud PA, Comninellis C, Savall A (2002) Electrochemical preparation of peroxodisulfuric acid using boron doped diamond thin film electrodes. Electrochim Acta 48:431–436
20. Vatistas N, Comninellis C, Serikawa RM, Prosperi G (2005) Oxidant production on BDD anodes and advanced oxidation processes. In: Fujishima A et al (eds) Diamond electrochemistry. Elsevier, Amsterdam
21. Cardarelli F, Taxil P, Savall A, Comninellis C, Manoli G, Leclerc O (1998) Preparation of oxygen evolving electrodes with long service life under extreme conditions. J Appl Electrochem 28:245–250
22. Belkheiri D, Fourcade F, Geneste F, Floner D, Aït-Amar H, Amrane A (2011) Feasibility of an electrochemical pre-treatment prior to a biological treatment for tetracycline removal. Sep Purif Technol 83:151–156
23. Wang L, Balasubramanian N (2009) Electrochemical regeneration of granular activated carbon saturated with organic compounds. Chem Eng J 155:763–768

Electrochemical Reduction of Nitrate

David Reyter
INRS Energie, Matériaux et Télécommunications, Varennes, Quebec, Canada

Introduction

Due to the increasing use of synthetic nitrogen fertilizers, livestock manure in intensive agriculture, and industrial and municipal effluent

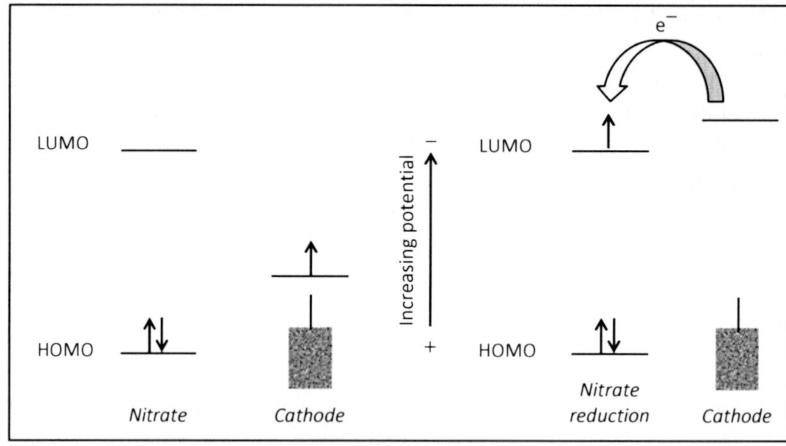

Electrochemical Reduction of Nitrate, Fig. 1 Molecular orbitals during nitrate electroreduction process

discharge, nitrate contamination in ground and surface water is a common problem in numerous worldwide countries. This pollution represents a risk not only to aquatic ecosystems (e.g. eutrophication) but also for human health (e.g. digestive system cancer and methemoglobinemia) [1, 2]. For this reason, the World Health Organization recommends a maximum limit of 50 $mg.l^{-1}$ (NO_3^-) for nitrate concentration in drinking water.

To date, several methods have been used to remove nitrate. Biological denitrification is attractive for converting nitrate to nitrogen but cannot treat toxic or nitrate-rich sources (e.g. more than 2000 $mg.l^{-1}$) [3]. Extractive methods (e.g. electrodialysis, reverse osmosis, ion exchange resins) are very effective but produce nitrate concentrates [4], which must be treated later. As electrochemical reduction of nitrate has high treatment efficiency, no sludge production, small area and relatively low investment costs, it has been the focus of a large number of researchers recently [5]. These techno-economic benefits are offset by a difficult scientific challenge, which is to reduce efficiently and selectively the nitrate to a harmless compound like nitrogen. This chapter will give an overview on 7nitrate electroreduction, including generalities on nitrate electrochemistry, electrochemical methods used to evaluate the electrocatalytic activity, and state of the art and future trends in this field.

The Electrochemistry of Nitrate

During nitrate electroreduction, charge transfer occurs from the cathode to the nitrate ions when the cathode potential is sufficient to promote electrons into the lowest unoccupied molecular orbital (LUMO) of the anion (Fig. 1).

To induce this reaction, the kinetic inhibition of the reaction must be overcome by applying an overpotential, which must be minimized. This reaction, in which electrons are transferred across the metal-solution interface with a resulting nitrate reduction, is called a faradaic process. Also, the complexity of the interfacial system is such that other phenomena do occur that can affect the electrode behavior. These processes include adsorption, desorption, and charging of the interface as a result of changing electrode potential; these are called non-faradaic processes. Both the efficiency and the selectivity of nitrate electroreduction strongly depend on several parameters such as the electrode composition, physicochemical properties of the electrolyte (pH, coexisting species, temperature, etc.) and the applied potential.

The nitrate electroreduction is very complex owing to the coexistence of several more or less stable intermediate and final products. A summary of some redox reactions occurring in alkaline and neutral media during nitrate reduction is given below [6].

$$NO_3^- + H_2O_{(l)} + 2e^- \leftrightarrow NO_2^- + 2OH^-$$
$$E^0 = 0.01V$$

$$NO_3^- + 3H_2O_{(l)} + 5e^- \leftrightarrow \frac{1}{2}N_{2(g)} + 6OH^-$$
$$E^0 = 0.26\ V$$

$$NO_2^- + 5H_2O_{(l)} + 6e^- \leftrightarrow NH_{3(g)} + 7OH^-$$
$$E^0 = -0.16\ V$$

$$NO_2^- + 4H_2O_{(l)} + 4e^- \leftrightarrow NH_2OH + 5OH^-$$
$$E^0 = -0.45\ V$$

$$2NO_2^- + 4H_2O_{(l)} + 6e^- \leftrightarrow N_{2(g)} + 8OH^-$$
$$E^0 = 0.41\ V$$

$$2NO_2^- + 3H_2O_{(l)} + 4e^- \leftrightarrow N_2O_{(g)} + 6OH^-$$
$$E^0 = 0.15V$$

$$NO_2^- + H_2O_{(l)} + 2e^- \leftrightarrow NO_{(g)} + 2OH^-$$
$$E^0 = 0.46V$$

$$N_2O + 5H_2O_{(l)} + 4e^- \leftrightarrow 2NH_2OH + 4OH^-$$
$$E^0 = -1.05V$$

Potential cathodic side reactions are hydrogen peroxide evolution in oxygen-enriched water and hydrogen evolution.

A number of basic electrochemical research works related to nitrate ion reduction have been reported in the last two decades. In most studies, the authors attempt to reduce the overpotential and increase the current for nitrate reduction by synthesizing materials with functional compositions (e.g. alloys, monocrystals, etc.) and highly reactive surfaces (nanoparticles, nanowires, porous materials, etc.).

How Can the Nitrate Electroreduction Be Investigated?

Numerous electrochemical methods and analytical tools can be employed to study the electroreduction of nitrate.

Although one of the more complex electrochemical techniques, cyclic voltammetry is very frequently used because it offers a wealth of experimental information and insights into both the kinetic and thermodynamic details of nitrate electroreduction. In a typical voltammogram, there can be several peaks related to the formation of different intermediate or final nitrate reduction products. From the sweep-rate dependence of the peak amplitudes, widths and potentials of the peaks observed in the voltammogram, it is possible to investigate the role of adsorption, diffusion, and coupled homogeneous chemical reaction mechanisms [7, 8]. The Rotating Disk Electrode (RDE) and the Rotating-Ring Disk Electrode (RRDE) are commonly used to describe multi-electron processes occurring during the electrocatalytic reduction of nitrate ions. Detection of non-gaseous oxidizable (e.g. NH_3, NO_x, etc.) products can be performed in the RRDE setup [9, 10] by applying a constant potential to the disk and scanning the platinum ring, for example. From the Levich and Koutecky plot in RDE, a rate constant and the number of electrons transferred during the different nitrate reduction steps can be determined [7, 11]. In situ IR spectroelectrochemistry (FTIR) has been employed for the identification of possible reaction intermediates [12]. On line electrochemical mass spectroscopy (OLEMS) is a powerful technique for following the volatile products (NO, N_2O, N_2) in real time during the electrochemical experiments [13, 14]. An electrochemical quartz crystal microbalance (EQCM) can be used to detect the adsorption of N-species on the electrode surface and even determine the surface coverage of these compounds [7].

Also, nitrate electroreduction is estimated by prolonged electrolysis by controlling either the potential or the current. The electrode performance for nitrate electrolysis is evaluated by measuring the nitrate destruction yield, the current efficiency, the selectivity and the specific energy consumption [15, 16]. During electrolysis, products can be accurately detected and quantified by ionic/gaseous chromatography, and UV–Vis spectroscopy.

Electrochemical Reduction of Nitrate, Fig. 2 Nitrate electroreduction pathways on transition metals

$NO_3^-(aq) \xleftrightarrow{(1)} NO_3^-(ads) \xrightarrow{(2)} NO_2^-(ads) \xleftrightarrow{(3b)} NO_{(ads)}$

$NO_2^-(ads) \xrightarrow{(3a)} NO_2^-(aq)$

$NO_{(ads)} \xrightarrow{(4a)} No_{(aq)}$

$NO_{(ads)} \rightarrow$ (4d) $N_2O \xrightarrow{(5)} N_2$; (4c) NH_2OH; (4b) NH_3

State of the Art for the Electrocatalytic Reduction of Nitrate

Nitrate electroreduction has been extensively studied over the last few decades. This reaction is a multi-electron transfer process showing different mechanisms as a function of pH, nitrate and supporting electrolyte concentration, chemical composition and structure of the catalyst. In recent years, nitrate electroreduction has been widely studied over diamond and many monometallic electrodes such as Pb, Ni, Zn or Rh, Ru, Ir, Pd, Cu, Ag and Au. Because none of the common pure metals is able to provide high selectivities for nitrogen, bimetallic alloys or monometals modified with foreign metal adatoms were prepared and evaluated for the reduction of nitrate. More recently, an electrochemical process in which nitrate ions are reduced to ammonia at the cathode, and where the produced ammonia is oxidized at the anode to nitrogen with the contribution of hypochlorite ions, has been evaluated.

Diamond Electrodes

Several studies have shown that boron doped diamond (BDD) electrodes are able to reduce nitrate to ammonia in alkaline, neutral or acidic media, at potentials close to the hydrogen evolution reaction. In 2003, an interesting result was reported by Lévy-Clément et al. who succeeded in converting nitrate to nitrogen with a selectivity of 50 % after an electrolysis at −2 V vs SCE (saturated calomel electrode). The authors suggested that a chemical nitrate reduction by hydrogen could occur in addition to the nitrate electroreduction [17]. Electrochemical nitrate reduction on BDD cathodes was investigated in a series of experiments by Georgeaud et al. Their results showed a good reduction rate of nitrate into almost exclusively N_2, with a final nitrate concentration lower than 50 mg/L with a low energy consumption of under 25 kWh/kg NO_3^- [18].

Monometallic Catalysts

In one of the most relevant papers in this field, Dima et al. [13] studied the electrocatalytic behavior of different polycrystalline metals such as Ru, Rh, Ir, Pd, Pt, Cu, Ag and Au for nitrate (100 mM) reduction in 0.5 M H_2SO_4. On the basis of the peak current density related to nitrate reduction on cyclic voltammograms, the activities of each electrode were compared. It was determined that rhodium is the most active catalyst among the noble metals for the reduction of nitrate, with the activity decreasing in the order Rh, Ru, Ir, Pt, Pd and Cu, Ag, Au for transition metals. The high electrocatalytic performance of Rh for nitrate reduction was also observed by Brylev et al. [19]. By using Differential Electrochemical Mass Spectrometry (DEMS), a reduction mechanism for nitrate reduction has been determined for transition metals (Fig. 2).

Tafel plots, effect of anions (SO_4^- and ClO_4^-) and reaction order allowed determination that nitrate adsorption (Fig. 2, reaction 1) and nitrate to nitrite reduction (Fig. 2, reaction 2) were the rate determining steps, whereas the conversion of NO_2^- to NO is fast. Subsequent steps will determine the selectivity. For the different metals investigated, it was noted that the kinetics for nitrate reduction increases toward the upper left of the transition metals in the periodic table. This trend is consistent with the capacity of these metals to adsorb species such as nitrates.

Petri and Safonova [20] reported the poor efficiency of palladium and platinum electrodes for nitrate reduction, explaining that adsorbed hydrogen covers the electrode surface at potentials where nitrate electroreduction should occur. This blockage happens because the adsorption enthalpy of hydrogen is higher than that of nitrate anions. On platinum in acidic media, Groot and Koper [21] showed by using the DEMS that N_2 and N_2O where produced at a slow rate before the hydrogen evolution reaction region, between 0.2 and 0.4 V vs RHE (Reversible Hydrogen Electrode). Gold is shown to be a poor electrocatalyst for nitrate reduction since the reaction is practically not detectable by cyclic voltammetry. This is because NO_3^- ions are not adsorbed on Au, which is negatively charged at the reduction potentials of nitrate, its potential for zero charge being 0.23 V vs RHE. As a result, a hydrogen evolution reaction is predominant, and only a small quantity of nitrite can be detected [13, 22]. Ohmori et al. [23] found that nitrate ions are effectively reduced to nitrite and ammonia at a Au electrode in acidic and basic $NaNO_3$ and $CsNO_3$ solutions of pH higher than 1.6. Under these conditions, the Na and Cs adatoms display residual positive charges of 0.22 and 0.44, respectively, allowing the NO_3^- to be adsorbed on the electrode. Several studies revealed that Rh presents a great activity for nitrate reduction due to a strong N-Rh bond [24, 25]. Tucker et al. [25] described an activation of the rhodium surface for nitrate reduction by cycling the potential in a KCl alkaline electrolyte. This potential cycling was applied to increase the surface area of the rhodium electrodes through oxidative dissolution and reduction of rhodium ions near the surface. This resulting nanostructured sponge-like structure hindered the hydrogen evolution reaction and favored the nitrate electroreduction. The nitrate reduction products were not analyzed. Bockris and Kim [26] compared Ni, Pb and Fe cathodes and found that all of them lead to the production of nitrite with ammonia (only 6 % of N_2) as the final product in 1.33 M NaOH. The potential was controlled between −0.6 and −1.2 V vs RHE.

Among the coinage metals, copper presents the most active surface for nitrate reduction both in neutral and alkaline media but produces nitrite, hydroxylamine and ammonia. On copper, nitrite and nitrate reduction display similar voltammetric curves, but nitrite reduction starts at a potential about 0.2–0.3 V more negative than nitrate electroreduction. It was found that ammonia is the main nitrate-reduction product at high negative overpotential whereas nitrite is preferentially formed at lower negative overpotential. From polarization and EQCM measurements, the deactivation of copper was attributed to the adsorption of nitrate-reduction products, blocking the electrode surface and slowing down the nitrate electroreduction rate. Cu electrodes can be reactivated by the periodic application of a square wave potential at higher potentials close to the open circuit potential of copper, which causes the desorption of poisoning species [7]. Also, Reyter et al. have reported methods for activating copper for nitrate reduction, such as high-energy ball milling [27] and electrochemical pretreatment in an alkaline solution [28].

Katsoumaros et al. [29] mentioned that tin at high potentials (−2.9 V vs Ag/AgCl) is the most effective cathode to have been reported in the literature so far for the electrochemical conversion of nitrate to nitrogen, since it combines both high selectivity for nitrogen (more than 90 %) and high rate of reduction.

Bimetallic Catalysts

Several platinum-based bimetallic alloys have been investigated as cathodes for nitrate electroreduction. Pt-Ir [30], Pt-Sn [31] and Pt-Pd [32] alloy catalysts are more active toward nitrate reduction compared to pure metals. This enhancement has been explained by a synergistic effect between the two metals, providing a bifunctional property to the cathode. In acidic media, nitrate was mainly converted to ammonia with these alloys. Polatides and Kyriacou [33] studied $Cu_{15}Sn_{85}$ and $Cu_{60}Zn_{40}$ cathodes in neutral media. Electrolysis results clearly show that $Sn_{85}Cu_{15}$ is an efficient electrocatalyst for the reduction of nitrate, since both the reduction rate and the selectivity for N_2 are high (43 %). For an electrodeposited $Cu_{45}Tl_{55}$ composite cathode, Cassela and Gatta [34] reported that the addition of thallium to copper increases the electrocatalytic activity for nitrate electroreduction, while pure thallium has no activity for nitrate electroreduction. This enhancement was attributed to the presence of thallium, which could promote the adsorption of nitrate.

The electrocatalytic properties of copper modified with platinum [35], palladium [9, 36, 37] or gold [38], either as alloys or as pure Pt, Pd or Au surfaces modified with small amounts of Cu, were investigated. Among these materials, copper-palladium materials appeared to be the most efficient for the reduction of nitrate to nitrogen. De Vooys et al. [9] demonstrated that a palladium electrode partially covered with copper allowed the reduction of nitrate to nitrogen with a selectivity of 60 %. At this bimetallic electrode, nitrate ions are first adsorbed and reduced to nitrite on copper sites, which is the rate determining step. The rapid conversion of NO_2^- to NO also occurs on copper. While reduction of NO to N_2O takes place either on copper or palladium, palladium remains the most active catalyst for the reduction of N_2O to N_2, copper having no activity for this reaction. More recently, Ghodbane et al. [36] obtained a maximum N_2 selectivity of 70 % for a Pd-Cu modified graphite electrode with a Cu-Pd composition of 95 at.% Pd – 5 at.% Cu. A biphasic Cu-Pd electrode composed of 77 % $Pd_{80}Cu_{20}$ + 23 % Cu successfully reduced nitrate to nitrogen with a current efficiency approaching 76 % [37]. Bimetallic Cu–Pd electrodes showed interesting electrocatalytic properties with a good selectivity toward nitrate electroreduction to nitrogen, however, the Pd/Cu surface ratio and the electrode potential have to be accurately controlled, otherwise nitrite or ammonia is produced. Moreover, under these conditions, the nitrate destruction rate remained very slow.

Future Directions

Recently, the simultaneous reduction of nitrate and the oxidation of the produced ammonia have been investigated in various cell configurations [39–44]. The paired electrolysis approach seems to be the most efficient electrochemical method for converting nitrate to nitrogen. In this process, nitrate is reduced to ammonia at the cathode while chlorine is generated at the anode and immediately transformed to hypochlorite. Hypochlorite reacts with ammonia to produce nitrogen at pH higher than 9 according to the reaction:

$$2ClO^- + 2NH_3 + 2OH^- \leftrightarrow N_2 + 2Cl^- + 4H_2O$$

The electrooxidation of chloride to hypochlorite, and hence of ammonia to nitrogen, is achieved at dimensionally stable anode (DSA) type electrodes such as Ti/RuO_2, Ti/IrO_2 or other mixed oxides on titanium [45, 46]. Li et al. [43] investigated the electrochemical reduction of nitrate on Fe, Cu and Ti as cathodes and Ti/IrO_2-Pt as anode in an undivided cell. In neutral solution, the nitrate removal was 87 % and selectivity to nitrogen was 100 % in 3 h with a Fe cathode in the presence of NaCl. Yu and Kupferle [44, 47] have studied this two-stage sequential electrochemical treatment in a divided cell containing a copper cathode and a Ti/Pt anode. The influence of the applied current was examined for the electrolytic oxidation of ammonium ions in the presence of chloride. In the anodic compartment, they observed that nitrate and chlorate formation (undesirable byproducts) increased with increasing applied current. In this context, the main difficulty is to find the proper conditions to perform both the reduction of the nitrate to ammonia and the oxidation of ammonia to nitrogen. Ideally, it should be performed by using an undivided cell reactor (i.e., without membrane) in order to avoid the problems associated with membrane deterioration (e.g., blocking by carbonates or organic compounds). Corbisier et al. [42] cited an energy consumption of 45–71 kWh/kg NO_3^- by paired electrolysis in a two-compartment electrolyzer with copper and Ti/RuO_2–TiO_2 as cathode and anode materials, respectively. Up to now, copper is mainly selected as cathode material because of its good activity for nitrate electroreduction. However, at a pure copper cathode, nitrite is generated in addition to ammonia. In an undivided cell, nitrite ions are oxidized to nitrate at the anode, drastically decreasing the efficiency of the paired electrolysis. Moreover, copper displays a poor corrosion resistance in the presence of chloride, nitrate and ammonia. Paired electrolysis performed by using $Cu_{70}Ni_{30}$ cathodes and Ti/IrO_2 anodes, without a membrane, permits the conversion of nitrate to nitrogen from more than 600 to less than 50 ppm NO_3^- with a power consumption as low as 20 kWh/kg NO_3^-. This performance was attributed to the use ofa $Cu_{70}Ni_{30}$ cathode displaying good corrosion resistance and a high efficiency along with selectivity for the reduction of nitrate to ammonia [48].

In water treatment, the target range for nitrate is 1–50 ppm, and the volume of effluent is usually large. To be efficient, the nitrate electrochemical removal technology will have to run in well-designed electrolyzers, containing three-dimensional electrodes with high surface areas, or will include a nitrate preconcentration step (e.g. ion exchange). In the future, efforts should be made toward the synthesis of tridimensional electrodes (e.g. rough and porous materials), highly efficient for nitrate electroreduction, and on the design of the electrolyzer. Also, the influence of co-existing species like heavy metals, phosphate, carbonates or organic compounds on the efficiency of the nitrate reduction process and the electrode life (scaling, metal deposition) will have to be evaluated.

Cross-References

- ► Chronoamperometry, Chronocoulometry, and Chronopotentiometry
- ► Cyclic Voltammetry
- ► Electrocatalysis of Chlorine Evolution
- ► Electrochemical Cell Design for Water Treatment
- ► Electrochemical Quartz Crystal Microbalance
- ► Electrochemical Reactor Design and Configurations

References

1. Carpenter SR, Caraco NF, Correll DL, Howarth RW, Sharpley AN, Smith VH (1998) Nonpoint pollution of surface waters with phosphorus and nitrogen. Ecol Appl 8:559–568
2. Wolfe AH, Patz JA (2002) Reactive nitrogen and human health: acute and long-term implications. Ambio 31:120–125
3. Tchobanoglous G, Burton FL, Stensel HD (2003) Wastewater engineering, treatment, and reuse. McGraw-Hill, New York
4. Kapoor A, Viraraghavan T (1997) Nitrate removal from drinking water – Review. J Environ Eng 123:371–380
5. Rajeshwar K, Ibanez JG (1997) Environmental electrochemistry: fundamentals and applications in pollution abatement. Academic, Waltham
6. Bard AJ (1973) Encyclopedia of electrochemistry of the elements. M. Dekker, New York
7. Reyter D, Bélanger D, Roué L (2008) Study of the electroreduction of nitrate on copper in alkaline solution. Electrochim Acta 53:5977–5984
8. Gabriela Elena B (2009) Electrocatalytic reduction of nitrate on copper electrode in alkaline solution. Electrochim Acta 54:996–1001
9. de Vooys ACA, van Santen RA, van Veen JAR (2000) Electrocatalytic reduction of NO3− on palladium/copper electrodes. J Mol Catal A: Chem 154:203–215
10. Chen Y, Zhu H, Rasmussen M, Scherson D (2010) Rational design of electrocatalytic interfaces: the multielectron reduction of nitrate in aqueous electrolytes. J Phys Chem Lett 1:1907–1911
11. Aouina N, Cachet H, Debiemme-chouvy C, Tran TTM (2010) Insight into the electroreduction of nitrate ions at a copper electrode, in neutral solution, after determination of their diffusion coefficient by electrochemical impedance spectroscopy. Electrochim Acta 55: 7341–7345
12. Figueiredo MC, Souza-Garcia J, Climent V, Feliu JM (2009) Nitrate reduction on Pt(111) surfaces modified by Bi adatoms. Electrochem Commun 11:1760–1763
13. Dima GE, de Vooys ACA, Koper MTM (2003) Electrocatalytic reduction of nitrate at low concentration on coinage and transition-metal electrodes in acid solutions. J Electroanal Chem 554–555:15–23
14. Yang J, Duca M, Schouten KJP, Koper MTM (2011) Formation of volatile products during nitrate reduction on a Sn-modified Pt electrode in acid solution. J Electroanal Chem 662:87–92
15. Li HL, Chambers JQ, Hobbs DT (1988) Electroreduction of nitrate ions in concentrated sodium hydroxide solutions at lead, zinc, nickel and phthalocyanine-modified electrodes. J Appl Electrochem 18:454–458
16. Peel JW, Reddy KJ, Sullivan BP, Bowen JM (2003) Electrocatalytic reduction of nitrate in water. Water Res 37:2512–2519

17. Lévy-Clément C, Ndao NA, Katty A, Bernard M, Deneuville A, Comninellis C, Fujishima A (2003) Boron doped diamond electrodes for nitrate elimination in concentrated wastewater. Diamond Relat Mater 12:606–612
18. Georgeaud V, Diamand A, Borrut D, Grange D, Coste M (2011) Electrochemical treatment of wastewater polluted by nitrate: selective reduction to N 2on Boron-Doped Diamond cathode. Water Sci Technol 63:206–212
19. Brylev O, Sarrazin M, Bélanger D, Roué L (2006) Rhodium deposits on pyrolytic graphite substrate: physico-chemical properties and electrocatalytic activity towards nitrate reduction in neutral medium. Appl Catal Environ 64:243–253
20. Petrii OA, Safonova TY (1992) Electroreduction of nitrate and nitrite anions on platinum metals: a model process for elucidating the nature of the passivation by hydrogen adsorption. J Electroanal Chem 331:897–912
21. De Groot MT, Koper MTM (2004) The influence of nitrate concentration and acidity on the electrocatalytic reduction of nitrate on platinum. J Electroanal Chem 562:81–94
22. Da Cunha MCPM, Weber M, Nart FC (1996) On the adsorption and reduction of NO3 – ions at Au and Pt electrodes studied by in situ FTIR spectroscopy. J Electroanal Chem 414:163–170
23. Ohmori T, El-Deab MS, Osawa M (1999) Electroreduction of nitrate ion to nitrite and ammonia on a gold electrode in acidic and basic sodium and cesium nitrate solutions. J Electroanal Chem 470:46–52
24. Wasberg M, Horányi G (1995) Electrocatalytic reduction of nitric acid at rhodized electrodes and its inhibition by chloride ions. Electrochim Acta 40:615–623
25. Tucker PM, Waite MJ, Hayden BE (2004) Electrocatalytic reduction of nitrate on activated rhodium electrode surfaces. J Appl Electrochem 34:781–796
26. Bockris JOM, Kim J (1997) Electrochemical treatment of low-level nuclear wastes. J Appl Electrochem 27:623–634
27. Reyter D, Chamoulaud G, Bélanger D, Roué L (2006) Electrocatalytic reduction of nitrate on copper electrodes prepared by high-energy ball milling. J Electroanal Chem 596:13–24
28. Reyter D, Odziemkowski M, Bélanger D, Roú L (2007) Electrochemically activated copper electrodes: surface characterization, electrochemical behavior, and properties for the electroreduction of nitrate. J Electrochem Soc 154:K36–K44
29. Katsounaros I, Ipsakis D, Polatides C, Kyriacou G (2006) Efficient electrochemical reduction of nitrate to nitrogen on tin cathode at very high cathodic potentials. Electrochim Acta 52:1329–1338
30. Ureta-Zañartu S, Yáñez C (1997) Electroreduction of nitrate ion on Pt, Ir and on 70:30 Pt: Ir alloy. Electrochim Acta 42:1725–1731
31. Shimazu K, Goto R, Tada K (2002) Electrochemical reduction of nitrate ions on tin-modified platinum and palladium electrodes. Chem Lett 2:204–205
32. Gootzen JFE, Peeters PGJM, Dukers JMB, Lefferts L, Visscher W, Van Veen JAR (1997) The electrocatalytic reduction of NO- 3 on Pt, Pd and Pt + Pd electrodes activated with Ge. J Electroanal Chem 434:171–183
33. Polatides C, Kyriacou G (2005) Electrochemical reduction of nitrate ion on various cathodes – Reaction kinetics on bronze cathode. J Appl Electrochem 35:421–427
34. Casella IG, Gatta M (2004) Electrochemical reduction of NO3 – and NO 2 – on a composite copper thallium electrode in alkaline solutions. J Electroanal Chem 568:183–188
35. Vilà N, Brussel MV, D'Amours M, Marwan J, Buess-Herman C, Bélanger D (2007) Metallic and bimetallic Cu/Pt species supported on carbon surfaces by means of substituted phenyl groups. J Electroanal Chem 609:85–93
36. Ghodbane O, Roué L, Bélanger D (2008) Study of the electroless deposition of Pd on Cu-modified graphite electrodes by metal exchange reaction. Chem Mater 20:3495–3504
37. Reyter D, Bélanger D, Roué L (2009) Elaboration of Cu-Pd films by coelectrodeposition: application to nitrate electroreduction. J Phys Chem C 113:290–297
38. Xing X, Scherson D (1989) Electrocatalytic properties of metal adatoms in a potential range negative to Nernstian bulk deposition. J Electroanal Chem 270:273–284
39. Vanlangendonck Y, Corbisier D, Van Lierde A (2005) Influence of operating conditions on the ammonia electro-oxidation rate in wastewaters from power plants (ELONITA™ technique). Water Res 39:3028–3034
40. Szpyrkowicz L, Daniele S, Radaelli M, Specchia S (2006) Removal of NO3 – from water by electrochemical reduction in different reactor configurations. Appl Catal Environ 66:40–50
41. Cheng H, Scott K, Christensen PA (2005) Paired electrolysis in a solid polymer electrolyte reactor – simultaneously reduction of nitrate and oxidation of ammonia. Chem Eng J 108:257–268
42. Corbisier D, Vereist L, Vanlangendonck Y, Van Lierde A (2006) Electro-degradation of nitrate ions and ammonia from power station effluents the ELONITA technique. VGB PowerTech 86: 98-101-106
43. Li M, Feng C, Zhang Z, Sugiura N (2009) Efficient electrochemical reduction of nitrate to nitrogen using Ti/IrO2-Pt anode and different cathodes. Electrochim Acta 54:4600–4606
44. Yu J, Kupferle MJ (2008) Two-stage sequential electrochemical treatment of nitrate brine wastes. Water Air Soil Pollut: Focus 8:379–385
45. Chen J, Shi H, Lu J (2007) Electrochemical treatment of ammonia in wastewater by RuO 2-IrO2-TiO2/Ti electrodes. J Appl Electrochem 37:1137–1144
46. Xu LL, Shi HC, Chen JL (2007) Electrochemical oxidation of ammonia nitrogen wastewater using Ti/RuO2-TiO2-IrO2-SnO2 electrode. Huanjing Kexue/Environ Sci 28:2009–2013

47. Yu J, Kupferle MJ (2009) Impact of cathode conditions on coupled electrochemical treatment of nitrate brine concentrates. Water Air Soil Pollut: Focus 9:245–251
48. Reyter D, Bélanger D, Roué L (2011) Optimization of the cathode material for nitrate removal by a paired electrolysis process. J Hazard Mater 192:507–513

Electrochemical Removal of H_2S

Geoffrey H. Kelsall
Department of Chemical Engineering, Imperial College London, London, UK

Process Options for HS Treatment

Fossil fuels normally contain sulfur, as sulfur compounds in oil and in the case of raw natural gas, as H_2S, COS, and mercaptans from traces to 80 vol %, associated with CO_2. H_2S is also present in off-gases of some processes, such as from hydrodesulfurization of those heavy hydrocarbons in which hydrogen is reacted catalytically to desorb sulfur as H_2S, which:

- Is highly toxic, 800 ppm being the lethal concentration for 50 % of humans after 5 min exposure; fortunately the recognition odor threshold is 0.47 ppb.
- Is corrosive to metals such as steel, copper, and silver.
- Combusts to produce SO_2, which itself must be treated, otherwise leading to acid rain.
- Has a lower flammability/explosive limit of ca. 4.3 vol% in air.

Hence, H_2S has to be removed from raw natural gas and a range of process gases, usually by chemical oxidation processes that convert it to elemental sulfur, e.g., for subsequent production of sulfuric acid.

Claus Processes

Claus processes [1, 2] are used for large-scale (>5 t sulfur day^{-1}) industrial treatment of H_2S-containing gases, with extension processes and tail gas treatment for conversions > 97 %, the limit otherwise imposed by equilibria in the oxidative conversion of H_2S to elemental sulfur with atmospheric oxygen:

$$H_2S + 0.5O_2 \rightleftarrows H_2O + (1/x)S_x \quad (1)$$

Chemical solvents, e.g., alkanolamines, or physical solvents may be used as concentrating processes from which desorbed H_2S can be fed to a Claus process and may also be used to deplete CO_2 selectively when a raw gas has too high a CO_2:H_2S ratio for a Claus plant to handle efficiently. The initial thermal stage (950–1,250 °C) of Claus processes oxidizes H_2S to SO_2, the latter then oxidizing more of the former and producing predominantly S_2 species, as predicted by Fig. 1:

$$H_2S + 3/2O_2 \rightarrow SO_2 + H_2O \quad (2)$$

$$2SO_2 + 4H_2S \rightarrow 3S_2 + 4H_2O \quad (3)$$

The product stream is fed to catalytic stages with activated Al_2O_3 or TiO_2 catalysts at 170–350 °C:

$$SO_2 + 2H_2S \rightleftarrows 3/xS_x + 2H_2O \quad (4)$$

The sulfur product is then condensed from the vapor phase.

Several "liquid redox" processes have been developed for smaller-scale (ca. 0.25–20 t sulfur day^{-1}) applications, achieving 99.9+ % recovery. For example, SulFerox® and ARI-LO-CAT® use complexed Fe^{III} species to oxidize absorbed H_2S in acidic solutions, and Stretford, Unisulf, and Sulfolin use vanadium (V) in slightly alkaline solutions as the oxidant for absorbed HS^- ions. Atmospheric oxygen is the ultimate oxidant in all such processes, as it reoxidizes the reduced form of the dissolved metal species, the concentrations of which are not so limited as that of dissolved oxygen:

$$S_8 + 16H^+ + 16e^- \leftarrow 8H_2S \quad (5)$$

$$16(Fe^{III}EDTA^- + e^- \rightleftarrows Fe^{II}EDTA^{2-}) \quad (6)$$

$$4(O_2 + 2H_2O + 4e^- \rightarrow 4OH^-) \quad (7)$$

At scales < ca. 0.25 t sulfur day^{-1}, adsorbents such as ZnO could be used for H_2S depletion.

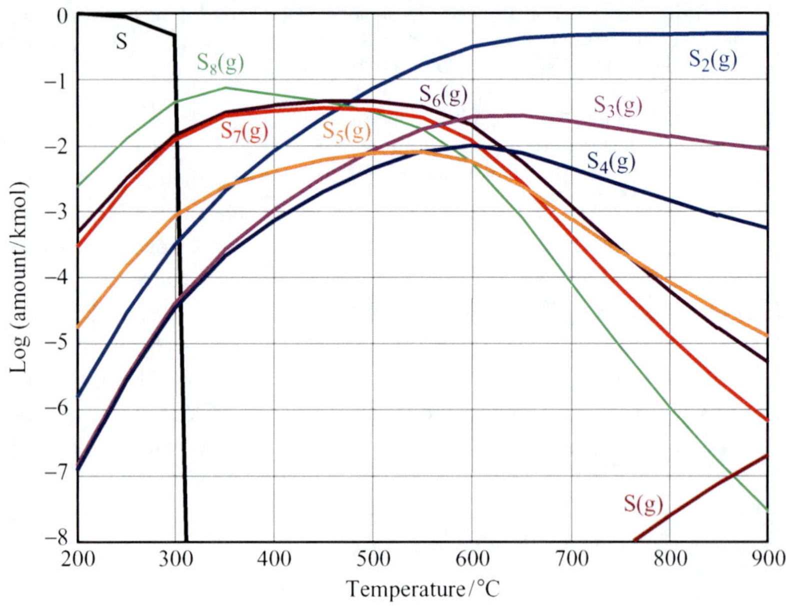

Electrochemical Removal of H_2S, Fig. 1 Effect of temperature on equilibrium composition of sulfur species

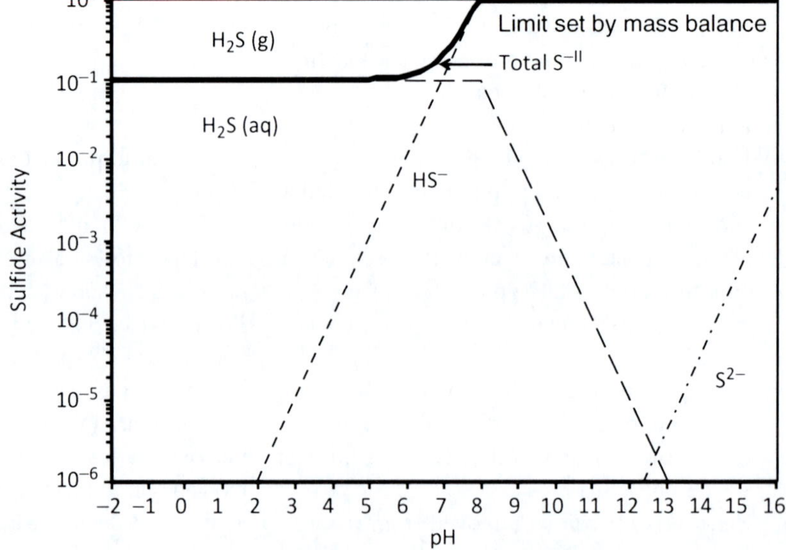

Electrochemical Removal of H_2S, Fig. 2 Predicted effect of pH on speciation and solubility of sulfur(−II) species at 298 K, unit total dissolved activity and $P(H_2S) = 1$

Thermodynamics of Sulfur–Water Systems

As there appears to be convincing evidence that *p*Ka (HS^-) is in the range 17–19, rather than the widely quoted value of 13, a value of $pK_a(HS^-) = 19$ (corresponding to $\Delta_f G^\theta$ (S^{2-}) = 120.5 kJ mol^{-1}) was used in the calculations [3] of the data plotted in Figs. 2, 3, and 4. Figure 1 shows the effect of pH on the speciation and solubility of H_2S in aqueous solutions at 298 K and $P(H_2S) = 1$; hydrolysis at pH > 6 increases total dissolved sulfide activities due to HS^- formation, for which a mass balance limit was set arbitrarily to unity. Hence, H_2S absorption processes have larger driving forces in alkaline solutions:

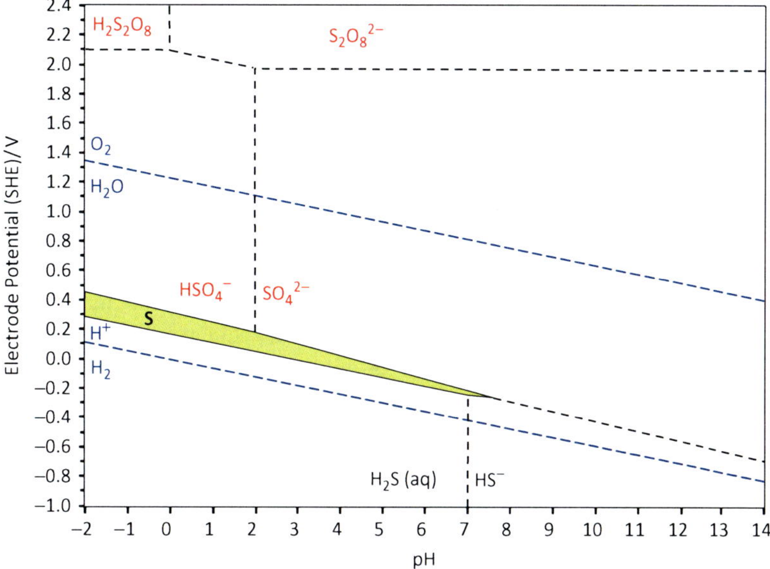

Electrochemical Removal of H_2S, Fig. 3 Potential–pH diagram for sulfur–water system at 298 K: dissolved sulfur activity = 0.1; $P = 1$

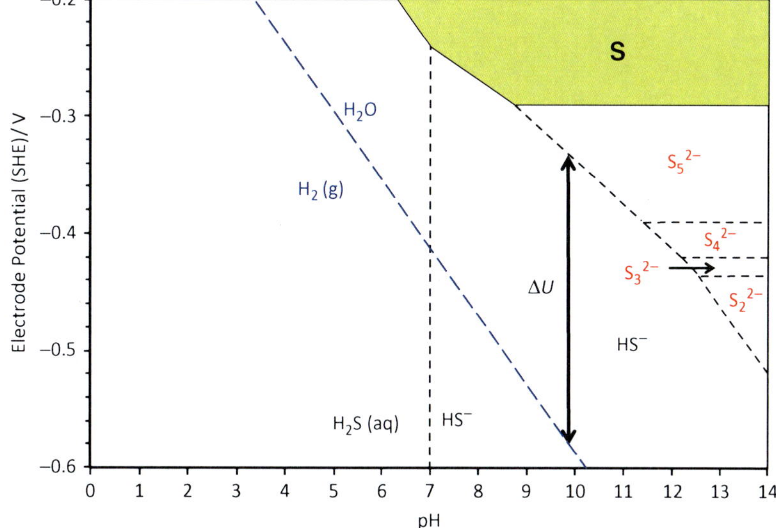

Electrochemical Removal of H_2S, Fig. 4 Potential–pH diagram for metastable sulfur–water system (sulfoxy species excluded) at 298 K and P = 1

$$H_2S(g) + OH^- \rightleftarrows HS^- + H_2O \tag{8}$$

Figure 3 shows a potential–pH diagram for the S-H_2O system at 298.15 K, predicting that only the −2, 0, and +6 oxidation states are truly stable in water; peroxodisulfate species $H_2S_2O_8/S_2O_8^{2-}$ are predicted to be capable of oxidizing water to oxygen.

Oxidation of hydrosulfide ions at pH > 8 is predicted to form sulfate ions. However, the large activation energy for that reaction is known to result in metastable products such as thiosulfate and sulfite ions [3]. In addition, yellow-colored polysulfide (S_n^{2-}) solutions can result from the atmospheric oxidation of hydrosulfide ions in mildly alkaline solutions. The conditions under which polysulfides form may be predicted by excluding all sulfoxy species from the calculations, resulting in the potential–pH diagram shown in Fig. 4.

Electrochemical Oxidation Kinetics and Mechanisms

Results have been reported of using stationary gold [4] and carbon [5] electrodes, as well as rotating ring-disc electrodes with Au [6, 7], and Pt [8, 9] discs to investigate the kinetics of sulfide oxidation in alkaline aqueous solutions. X-ray photoelectron spectroscopy of Au surfaces detected sulfur deposited by the anodic oxidation of sulfur(−II) species in solution; the initial layer behaved as gold sulfide, but multilayers of sulfur had a lower volatility and a smaller electron binding energy than bulk elemental sulfur, indicating an interaction with the underlying gold or gold sulfide [6]. Polysulfide ions (Fig. 4) were formed as intermediates in both sulfide oxidation and polysulfide/sulfur reduction processes by reaction (9), with intermediates being detected at the ring electrode:

$$nHS^- + nOH^- \underset{\text{ring}}{\overset{\text{disc}}{\rightleftarrows}} S_n^{2-} + nH_2O + (2n-2)e^- \tag{9}$$

Polysulfides were also produced by chemical reaction of deposited sulfur with sulfur(−II) species in solution, providing a method of de-passivating anodes on which elemental sulfur is formed unintentionally. The kinetics of the coupled chemistry–electrochemistry of polysulfide reactions has been modelled and studied experimentally [10–12].

Electrochemical Oxidation Processes

Aqueous Processes

Fuel Cell Mode

Figures 3 and 4 predict that coupling an oxygen gas diffusion cathode, at which the reaction occurs, in principle could drive reaction at a suitable anode, forming the basis of a fuel cell with the overall reaction:

$$5HS^- + 2O_2 \rightarrow S_5^{2-} + H_2O + 3OH^- \tag{10}$$

However, the extent of oxidation would have to be controlled to avoid deposition of elemental sulfur on the anode, and diffusion of sulfide species through the membrane to the cathode would poison most cathode materials. Similar reactions are utilized in "liquid redox" processes [1, 2] for absorbed H_2S oxidation, but with additional mediators such as (complexed) Fe^{III}/Fe^{II} in acidic aqueous solutions and V^V/V^{IV} [13] in slightly alkaline solutions, to catalyze reaction (7) and to enhance the concentrations of oxidant over that achievable with dissolved oxygen.

Electrolytic Reactor Mode

In electrolyser mode, a reactor such as that shown schematically in Fig. 5 can be used to oxidize hydrogen sulfide ions [14, 15, 16] at suitable anodes coupled to hydrogen evolution at the cathode, with a reactor equilibrium potential difference (ΔU) far lower (Fig. 4) than that required when oxygen is the anodic product (Fig. 3); high purity hydrogen and sulfur were produced at a current density of 3 kA m^{-2} and reactor potential difference U of ca. 1.0 V. Operating temperatures were typically 80–90 °C, further decreasing operating potential differences and decreasing the tendency for oxidative deposition of insulating elemental sulfur that would passivate an anode. That reaction is avoided by judicious control of the anode potential and solution composition (Fig. 4), but if that fails, sulfur can be stripped reductively to reactivate at least carbon fiber electrodes [17].

An electrochemical reactor with anolyte and catholyte divided by a cation-permeable membrane or semipermeable separator is required to preclude the diffusion of sulfide/polysulfide anions from anolyte to catholyte and to avoid reduction of polysulfide formed at the anode by reaction (9), being reduced at the cathode by the reverse reaction. Similar processes have been reported [18] for producing polysulfide (and hydroxide) solutions at current densities of 5 kA m^{-2} for Kraft paper pulping processes.

As reaction (9) consumes hydroxide ions and forms water, the anolyte pH decreases with time, exacerbated by H_2S absorption, while the catholyte pH increases. The $[OH^-]$:$[HS^-]$

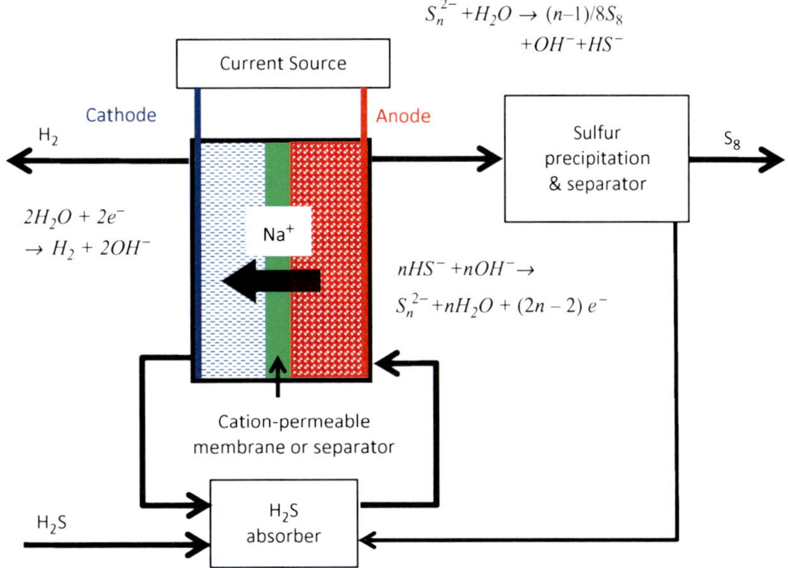

Electrochemical Removal of H_2S, Fig. 5 Electrochemical process for sulfide oxidation coupled to hydrogen production

concentration ratio needs to be maintained at about unity, to avoid the pH decreasing beyond the range at which S_n^{2-} ions are stable, causing their disproportionation in the reactor in which the resulting sulfur would then cause anode passivation:

$$S_n^{2-} + H_2O \rightarrow (n-1)/8\ S_8 + OH^- + HS^- \tag{11}$$

Results of the performance of various anode materials for reaction (9) have been reported, including graphite [19], lanthanum strontium manganite ($La_{0.8}Sr_{0.2}MnO_{3\text{-}d}$- LSM) [16], and metal sulfides, such asMoS_2,Co_xS_y, and Ni_xS_y [20].
Gas absorption:

$$n(H_2S + OH^- \rightarrow HS^- + H_2O) \tag{12}$$

Oxidation:

$$nHS^- + nOH^- \rightarrow S_n^{2-} + nH_2O + (2n-2)e^- \tag{13}$$

Reduction:

$$(2n-2)H_2O + (2n-2)^{e^-}c \rightarrow (n-1)H_2 + (2n-2)OH^- \tag{14}$$

Disproportionation:

$$S_n^{2-} + H^+ \rightarrow HS^- + (n-1)/8S_8 \tag{15}$$

Overall:

$$nH_2S + OH^- \rightarrow H_2O + HS^- + (n-1)/8S_8 + (n-1)H_2 \tag{16}$$

High-Temperature Membrane Processes

High-temperature electrolysis (Fig. 6) with a molten salt electrolyte/membrane, through which sulfide ions are transported, has been advocated [21–28] to achieve better thermal integration with high-temperature processes producing H_2S and to produce elemental sulfur and hydrogen:
Membrane:

$$(Li_{0.62}K_{0.38})_2CO_3 + H_2S \rightarrow (Li_{0.62}K_{0.38})_2S + CO_2 + H_2O \tag{17}$$

Cathode:

$$2H_2S + 4e^- \rightarrow 2H_2 + 2S^{2-} \tag{18}$$

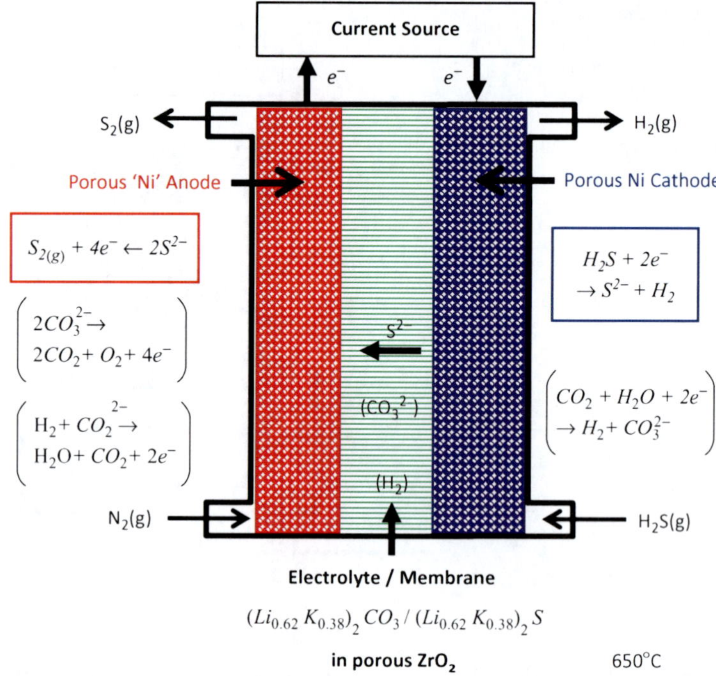

Electrochemical Removal of H_2S, Fig. 6 Electrochemical reactor using immobilized molten electrolyte for H_2S splitting by its reduction to H_2 and oxidation to S_2

Anode:

$$2S^{2-} \rightarrow S_2 + 4e^- \tag{19}$$

Reactor potential difference:

$$U = -\left(E_{S_2/S^{2-}} + \eta_{S_2} - E_{H_2S/H_2} - \eta_{H_2} + \sum_i \frac{Id_i}{\bar{\kappa}_i A_i} \right) \tag{20}$$

Specific electrical energy consumption:

$$w_{e,H_2}/\mathrm{kWh\,Nm^{-3}}\,H_2 = \frac{2FU}{3.6\times10^6 \Phi^e_{H_2} V_{M,H_2}} \tag{21}$$

where $V_{M,H2}$ is the molar volume (Nm^3) of H_2 at STP. Current efficiencies ($\Phi_{H2}{}^e$) were predicted to be ca. 1.0 for 1,000 ppm H_2S levels at 90 % conversion with reactor potential differences of ca. 0.5 V; however, with 100 or 10 ppm H_2S at the inlet, predicted current efficiencies decreased to 0.93 and 0.4, respectively.

Photo-Electrochemical Oxidation Processes

Sulfide oxidation can also be driven photo-electrochemically, using photon absorption at suitable semiconductors to generate electron–hole pairs [29, 30], the latter oxidizing sulfide ions by reaction (24):

Gas absorption:

$$n(H_2S + OH^- \rightarrow HS^- + H_2O) \tag{22}$$

Photoactivation:

$$(2n-2)\left(\mathrm{Semiconductor} + h\nu \underset{\mathrm{recombination}}{\overset{\mathrm{absorption}}{\rightleftarrows}} \mathrm{Semiconductor}(h^+_{VB}, e^-_{CB}) \right) \tag{23}$$

Oxidation:

$$nHS^- + nOH^- + (2n-2)h^+_{VB} \rightarrow S_n^{2-} + nH_2O \tag{24}$$

Reduction:

$$(2n-2)H_2O + (2n-2)e^-_{CB} \rightarrow (n-1)H_2 + (2n-2)OH^- \quad (25)$$

Future Directions

If there is a significant change in the economics of hydrogen production by steam reforming of natural gas:

$$CH_4 + 2H_2O \rightarrow CO_2 + 4H_2 \quad (26)$$

e.g., due to a carbon tax being introduced, then the processes described above for simultaneous oxidation of H_2S to sulfur and production of hydrogen would become more attractive.

References

1. Ullmann's Encyclopedia of industrial chemistry (2005). Wiley-VCH Verlag, Weinheim, Germany
2. Eow JS (2002) Recovery of sulfur from sour acid gas: a review of the technology. Environ Prog 21:143–162. doi:10.1002/ep.670210312
3. Kelsall GH, Thompson I (1993) The redox chemistry of H_2S oxidation by the british gas stretford process. I. Thermodynamics of sulfur-water systems. J Appl Electrochem 23:279–286
4. Hamilton IC, Woods R (1983) An investigation of the deposition and reactions of sulfur on gold electrodes. J Appl Electrochem 13:783–794
5. Nygren K, Atanasoski R, Smyrl WH, Fletcher EA (1989) Hydrogen and sulfur from hydrogen sulfide-V. Anodic oxidation of sulfur on activated glassy carbon. Energy 14:323–332
6. Buckley AN, Hamilton IC, Woods R (1987) An investigation of the sulfur(−II)/sulfur(0) system on gold electrodes. J Electroanal Chem 216:213–227
7. Kelsall GH, Thompson I (1993) The redox chemistry of H_2S oxidation by the british gas stretford process. II. Electrochemical kinetics of HS-/S redox systems. J Appl Electrochem 23:287–295
8. Szynkarczuk J, Komorowski G, Donini JC (1994) Redox reactions of hydrosulfide ions on the platinum electrode – I. The presence of intermediate polysulfide ions and sulfur layers. Electrochim Acta 39:2285–2289
9. Behm M, Simonsson D (1997) Electrochemical production of polysulfides and sodium hydroxide from white liquor Part I: experiments with rotating disc and ring-disc electrodes. J Appl Electrochem 27:507–518
10. Lessner P, Winnick J, McLarnon FR, Cairns EJ (1986) Kinetics of aqueous polysulfide solutions I. Theory of coupled electrochemical and chemical reactions, response to a potential step. J Electrochem Soc 133:2510–2516
11. Lessner P, Winnick J, McLarnon FR, Cairns EJ (1986) Kinetics of aqueous polysulfide solutions II. Electrochemical measurement of the rates of coupled electrochemical and chemical reactions by the potential step method. J Electrochem Soc 133:2517–2522
12. Lessner P, Winnick J, McLarnon FR, Cairns EJ (1987) Kinetics of aqueous polysulfide solutions part III. Investigation of homogeneous and electrode kinetics by the rotating disk method. J Electrochem Soc 134:2669–2677
13. Kelsall GH, Thompson I (1993) The redox chemistry of H_2S oxidation by the british gas stretford process. IV. Thermodynamics of V-S-H_2O systems and the electrochemical behaviour of V^V/V^{IV} couples in alkaline solutions. J Appl Electrochem 23:417–426
14. Anani AA, Mao Z, White RE, Srinivasan S, Appleby AJ (1990) Electrochemical production of hydrogen and sulfur by low-temperature decomposition of hydrogen sulfide in an aqueous alkaline solution. J Electrochem Soc 137:2703–2709
15. Mao Z, Anani A, White RE, Srinivasan S, Appleby AJ (1991) A modified electrochemical process for the decomposition of hydrogen sulfide in an aqueous alkaline solution. J Electrochem Soc 138:1299–1303
16. Petrov K, Srinivasan S (1996) Low temperature removal of hydrogen sulfide from sour gas and its utilization for hydrogen and sulfur production. Int J Hydrogen Energy 21:163–169. doi:10.1016/0360-3199(95)00003-8
17. Dutta PK, Rozendal RA, Yuan Z, Rabaey K, Keller J (2009) Electrochemical regeneration of sulfur loaded electrodes. Electrochem Comm 11:1437–1440. doi:10.1016/j.elecom.2009.05.024
18. Behm M, Simonsson D (1997) Electrochemical production of polysulfides and sodium hydroxide from white liquor Part II: electrolysis in a laboratory scale flow cell. J Appl Electrochem 27:519–528
19. Behm M, Simonsson D (1999) Graphite as anode material for the electrochemical production of polysulfide ions in white liquor. J Appl Electrochem 29:521–524
20. Lessner PM, McLarnon FR, Winnick J, Cairns EJ (1992) Aqueous polysulfide flow-through electrodes: effects of electrocatalyst and electrolyte composition on performance. J Appl Electrochem 22:927–935
21. Lim HS, Winnick J (1984) Electrochemical removal and concentration of H_2S from coal gas. J Electrochem Soc 131:562–568. doi:10.1149/1.2115627
22. Waver D, Winnick J (1987) Electrical removal of H_2S from hot gas streams. J Electrochem Soc 134: 2451–2458

23. Burke A, Li SW, Winnick J et al (2004) Sulfur-tolerant cathode materials in electrochemical membrane system for H_2S removal from hot fuel gas. J Electrochem Soc 151:D55–D60. doi:10.1149/1.1758815
24. Burke A, Winnick J, Xia CR et al (2002) Removal of hydrogen sulfide from a fuel gas stream by electrochemical membrane separation. J Electrochem Soc 149:D160–D166
25. Alexander SR, Winnick J (1994) Electrochemical polishing of hydrogen-sulfide from coal synthesis gas. J Appl Electrochem 24:1092–1101
26. Alexander SR, Winnick J (1994) Removal of hydrogen sulfide from natural gas through an electrochemical membrane separator. AIChE J 40:613–620. doi:10.1002/aic.690400406
27. Robinson JS, Winnick J (1998) Theoretical limiting prediction of H_2S removal efficiency from coal gasification streams using an intermediate temperature electrochemical separation process. J Appl Electrochem 28:1343–1349. doi:10.1023/A:1003416500001
28. Robinson JS, Smith DS, Winnick J (1998) Electrochemical membrane separation of H_2S from reducing gas streams. AIChE J 44:2168–2174. doi:10.1002/aic.690441006
29. Lessner P, McLarnon FR, Winnick J, Cairns EJ (1988) Solution processes in photoelectrochemical cells. Appl Phys Lett 53:1185–1187
30. Ardoin N, Winnick J (1988) Polysulfide solution chemistry at a CdSe photoanode. J Electrochem Soc 135:1719–1722

Electrochemical Sensor of Gaseous Contaminants

Dimitrios Tsiplakides
Department of Chemistry, Aristotle University of Thessaloniki, Thessaloniki, Greece

Introduction

The detection of gaseous contaminants in ambient air or in gas streams has found widespread applications in the sensitive and selective real-time detection of trace gases in diverse fields, including urban (e.g., automotive and air quality control), industrial (e.g., petrochemical industry, combustion sites and waste incinerators), and rural (e.g., greenhouses and agroecosystems) emissions, manufacturing processes (e.g., chemical analysis and control of semiconductor fabrications) and atmospheric chemistry studies (e.g., global studies for monitoring levels of greenhouse gases in the environment) [1]. Gaseous contaminants can be divided into two main categories: primary and secondary pollutants. Primary pollutants are compounds that are emitted into the atmosphere directly from the source of the pollutant and retain the same chemical form. Typical examples of primary pollutants produced by human activity are sulfur dioxide and sulfuric acid vapor, nitrogen oxides, carbon monoxide and partially oxidized organic compounds, volatile organic compounds, hydrogen chloride and hydrogen fluoride, hydrogen sulfide and other reduced sulfuric compounds, and ammonia. Secondary pollutants are gaseous and vapor phase compounds that form when primary pollutants react with each other or interact with naturally occurring compounds in the atmosphere. Examples of a secondary pollutant include ozone, which is formed when hydrocarbons and nitrogen oxides combine in the presence of sunlight; nitrogen dioxide, which is formed as NO combines with oxygen in the air; and sulfuric and/or nitric acid, which is formed when sulfur dioxide or nitrogen oxides react with water. The growing trend to lower the levels of concentration at which potential gaseous pollutants have to be monitored and controlled according to environmental regulations and industrial standards has led to the need for reliable, sensitive, selective and stable sensors.

Among the various types of chemical sensors, defined as *devices that transform chemical information ranging from concentration of a specific sample component (analyte) to total compositional analysis into an analytically useful signal* [2], electrochemical sensors constitute the largest group in terms of both sensor literature volume and technological applications. They represent approximately 58 % of the total; other types include optical (24 %), mass (12 %) and thermal (6 %). As the name implies, electrochemical sensors utilize the effect of the electrochemical interaction between an analyte and an electrode in order to provide continuous information about analyte concentration. Electrochemical gas sensors can be categorized into three main classes according to their operating principle (measured electric parameter): amperometric, potentiometric and conductometric sensors. Each class contains many sensor types. In practical applications, a variety of different types of potentiometric and amperometric sensors (also known as electrochemical gauges) cover the detection of almost all gaseous contaminants; the operating principles, the range of their application and typical examples of these two types are outlined in the following sections.

Potentiometric Sensors

Potentiometric sensors use the voltage at zero current (open-circuit potential) of an indicator electrode against a reference one as the output electrical signal corresponding to the concentration of a particular chemical species. Open-circuit potential is typically representative of an equilibrium electrochemical process and arises because an electrochemical reaction can occur at surfaces or at membranes in solid, liquid, or condensed phases. Because the sensor signal is taken for a process at equilibrium, the ultimate signal is less influenced by mass transport characteristics or sensor dimension and provides a reading reflecting the local equilibrium conditions. The generated signal is an electromotive force (EMF) that is dependent on the activity of the analyte and is described by Nernst's equation. Response time depends mostly on how fast equilibrium can be established at the sensor interface.

The key component of a potentiometric sensor is the ion-selective electrode (ISE), an electrode or electrode assembly with a potential that is dependent on the concentration of an ionic species in the test solution and is used for electroanalysis. This interfacial potential at the electrode surface is caused by the selective ion exchange reaction. Ion-selective electrodes are often membrane type electrodes. The well-known glass pH electrode, which is selectively sensitive to hydrogen ions, is a typical ISE and has been used for years for the measurement of acidity or basicity of aqueous solutions in

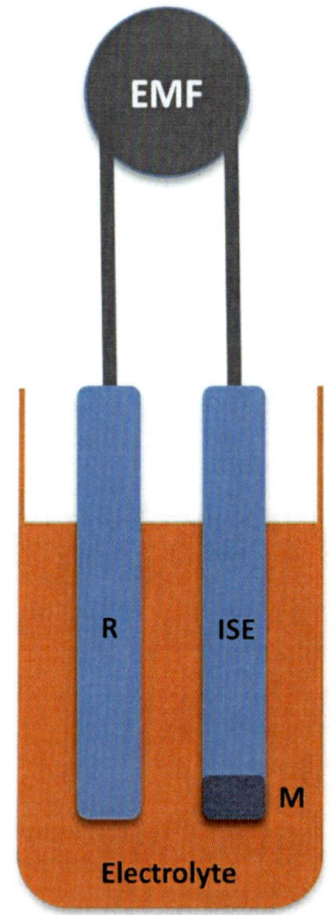

Electrochemical Sensor of Gaseous Contaminants, Fig. 1 Typical potentiometric sensor composed of an ion-selective electrode (ISE) and a reference electrode

a dynamic range of ~36 pH units [3]. The importance of ISE in the development of potentiometric sensors is thus evident; however, one should also take into account that the voltage signal from the sensor is obtained from the measurement of the voltage, E_{cell}, of the cell, which is composed of an ISE and a reference electrode (Fig. 1):

$$E_{cell} = E_{ISE} - \left(E_{ref} \pm E_j\right)$$

where E_{ISE} is the potential of the ISE, E_{ref} is the potential of the reference electrode, and E_j is the liquid junction potential (usually incorporated in the reference potential). Therefore, half of the information comes from the indicator electrode (the ion-selective electrode) and the other half from the reference electrode, including the liquid junction. Various kinds of stable, reversible and reproducible reference electrodes exist, including the standard hydrogen electrode (SHE), the

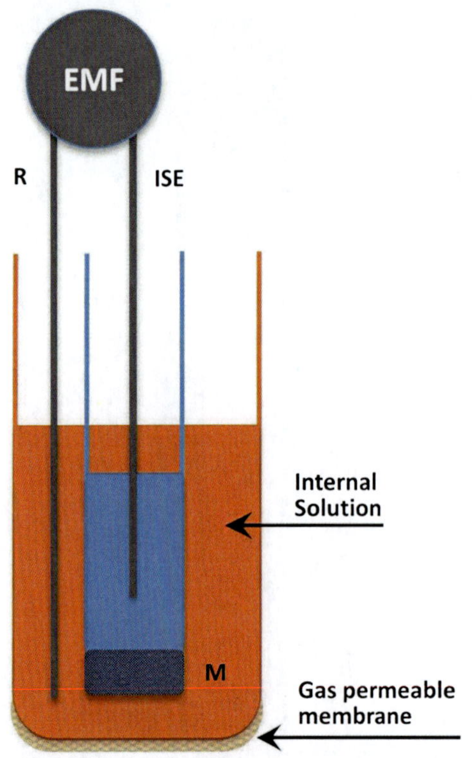

Electrochemical Sensor of Gaseous Contaminants, Fig. 2 Schematic of a Severinghaus-type sensor

silver/silver chloride electrode (Ag/AgCl) and the saturated calomel electrode (SCE).

Ion-selective electrodes are used mainly as ion sensors [3], that is for the detection of ionic species in liquid solutions, since the potential is always generated by separation of charge, for example, by the partitioning of ions at the selective membrane/solution interface. However, the detection of neutral species, such as gases, can be realized with ISEs, assuming that the gas is in equilibrium with an ion in an intermediate solution that is in contact with the external medium through a gas-permeable membrane. The original design of such a device was developed by Severinghaus and Bradley [4] in 1958 for sensing carbon dioxide (Fig. 2). The concept is that the gas can penetrate though a permeable membrane into the internal solution compartment of the sensor, where it undergoes a chemical transformation (e.g., hydrolysis) in which some detectable ion or ions are produced. These ions are detected by the ISE of the sensor. The Severinghaus-type electrodes present a simple construction and operating principle and are used for the detection of various gases, including O_2, Cl_2, I_2, H_2, CO_2, SO_2–SO_3, NO_2, NO_x and H_2O. A considerable drawback in many applications is their relatively low response time constant (typically 10–20 s), because partitioning between the sample and the relatively large internal volume, as well as dissolution equilibria, is involved [3].

An emerging technology which may confront many of the drawbacks of liquid electrolytes is the development of solid electrolyte potentiometric sensors [5–8]. The core of these devices is a dense solid-state membrane, which is an ionic conductor and serves as the electrolyte. Most of them are crystalline materials (e.g. YSZ, NASICON or β''-Al_2O_3) with an operating temperature as high as 500–1,000 °C, a temperature range where they exhibit sufficiently high conductivity. The existence of low temperature ceramic materials (250–400 °C) or even solid polymer ion-conductive structures (e.g. Nafion®) operating at room temperature should be mentioned. The advantages of solid-state sensors include robustness, good thermal stability, operation in harsh environments, fast response and miniaturization potential. Therefore, they are especially suitable for measuring environmental gaseous pollutants such as CO_2, CO, NO_x, SO_x, H_2, Cl_2, and NH_3 [9].

Equilibrium Potentiometric Sensors

A conventional solid-state potentiometric gas sensor consists of two electrodes (usually comprised of porous metal layers) attached to both sides of a solid electrolyte (Fig. 3a). One electrode is exposed to the gas to be detected, while the other electrode (reference) is facing a reference gas with a constant concentration. The measured gas is converted to the predominant mobile ion of the solid electrolyte, i.e., the electrolyte should have a common species with the electrode or gas phase (gas electrode). The electrode (interface) potential of such type of electrodes, referred to as electrodes of the *first*

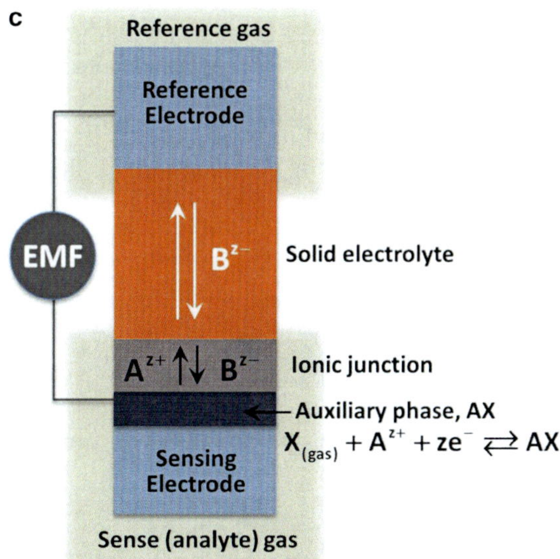

Electrochemical Sensor of Gaseous Contaminants, Fig. 3 Typical structure of (**a**) Type I, (**b**) Type II and (**c**) Type III potentiometric gas sensors

kind, is derived from the condition of local equilibrium at the interface. The potential difference established between the two sides of the solid electrolyte is dependent on the difference in activity of the species across the solid electrolyte. This potential difference, also called cell electromotive force (EMF), is determined by the Nernst equation and corresponds to the signal of the sensor. A typical example of such sensors is the oxygen sensor, where an oxygen ion conductor, such as yttria-stabilized zirconia (YSZ), is used. In this case, the electrode reaction

$$O_{2(gas)} + 4e^- \rightleftharpoons 2O^{2-}$$

takes place at the three-phase boundaries (metal/electrolyte/gas), giving rise to an EMF:

$$E = \frac{RT}{4F} \ln \frac{p}{p_o}$$

where p and p_o are the oxygen partial pressures at the measuring and reference compartments, respectively. When p_o maintains a known, constant value, either by exposure in a reference gas or by using an "oxygen buffer" [8], the above equation may be used for the determination of oxygen partial pressure. A variety of solid electrolytes, including H^+, $F^-$and Cl^- conductors, can be used for the detection of the corresponding gas (H_2, F_2 and Cl_2, respectively). The potentiometric sensors described above, which are essentially composed of two electrodes of the first kind, constitute the *Type I* sensors [10].

In order to gain more flexibility in the detection of gaseous chemical species other than the mobile species of the electrolyte, electrodes of the *second kind* are used. The second kind of electrodes occurs when the analyte reacts reversibly with the mobile ions in the electrolyte, forming an intermediate phase, separate or dissolved phase in the electrolyte. An equilibrium is thus established between the gas and the intermediate phase. For the reference electrode, an electrode of the first kind is utilized. Typical examples of the so-called *type II* sensors (Fig. 3b) are CO_2 and SO_3 sensors with K_2CO_3 [11] and Ag_2SO_4 [7] solid electrolytes, respectively.

Type II sensors are still restricted by the limited combinations of available solid ionic conductors containing an immobile species with a known equilibrium reaction with the gaseous species to be detected. To overcome this problem, *type III* sensors (Fig. 3c) were developed with the application of an additional auxiliary phase. The sensor is thus composed of an electrode of the second kind and an ionic junction between the electrolyte and the auxiliary phase. The ionic junction allows the measurement of the concentration of the chemical species that are not present in the electrolyte. The concentration of NO_2 or CO_2 can be measured with a type III sensor, utilizing NASICON or β''-Al_2O_3 as a Na^+ solid electrolyte, $NaNO_3$ or Me_2CO_3 ($Me = Li$, Na, K, Cs, Ca) as the auxiliary phase and the mixture of Na_2ZrO_3/ZrO_2 or Pt for the reference electrode [8].

Nonequilibrium Potentiometric Sensors (or Mixed Potential Sensors)

According to the above analysis, it may concluded that potentiometric sensors follow a Nernstian behavior, meaning that the dependence of sensor signal (EMF) on the partial pressure of the detecting gas is governed by the Nernst equation, i.e. the sensor signal is a logarithmic function of gas partial pressure. Discrepancies from Nernstian behavior have been reported and were attributed to (a) local non-equilibria of the reactions occurring at the electrodes and auxiliary phase due to low operating temperature, (b) mixed electronic-ionic conductivity of the solid electrolyte and (c) coupling of two or more competitive anodic and cathodic electrochemical reactions at the sensing electrode. The later reflects the *mixed potential* phenomenon and was first observed by Flemming [12] in zirconia-based oxygen sensors when CO was also present in the gas mixture. In this case, two electrochemical reactions are taking place simultaneously at the sensing electrode:

$$O_2 + 4e^- \rightleftharpoons 2O^{2-}$$

$$CO + O^{2-} \rightleftharpoons CO_2 + 2e^-$$

Additionally, the following chemical (catalytic) reaction occurs, leading to a decrease in the local oxygen concentration at the sensing electrode:

$$CO + \frac{1}{2}O_2 \rightleftharpoons CO_2$$

The steady-state potential of this sensing electrode (mixed potential) and the corresponding EMF of the sensor are established when the rates of the two electrochemical reactions are equal [13]. In order to estimate the mixed potential, one should consider the absolute values of the cathodic and anodic currents, expressed by the Butler–Volmer equation, taking also into account the rate from the catalytic reaction which determines the amounts of adsorbed species at the three-phase boundaries. A detailed analysis [8] concluded either a logarithmic or linear dependence for the concentration of reacting gas on mixed potential. The former behavior is expected when kinetics are determined by electrochemical reactions, while the later appears when both slow mass transport of chemical species toward the electrode for one electrochemical reaction with a large overpotential and Tafel-type kinetics for another reaction with negligible overpotential are involved.

Amperometric Sensors

Amperometric sensors can be considered as a subclass of voltammetric sensors. Whereas amperometric sensors operate at a fixed potential, voltammetric sensors can operate in other modes, such as linear and cyclic voltammetric modes; the respective current–potential response for each mode will be different. Amperometric sensors are based on electrochemical reactions that are governed by the diffusion of the electroactive species through a barrier consisting of a small orifice (hole) or a porous layer (membrane). The applied voltage is fixed on the diffusional plateau of the I(U) curve (Fig. 4), which is defined by the limited supply of gas through the barrier. The corresponding limiting current is proportional to the concentration of the gas and thus can be used as the sensor signal. While potentiometric sensors commonly allow the measuring of chemical activities over a very wide range of many orders of magnitude, amperometric sensors generally cover a quite limited range but have a much higher resolution and do not depend on chemical equilibria at the electrode/electrolyte interface [7].

The basic configuration of an amperometric electrochemical sensor [14], sometimes called a limiting current-type sensor, is illustrated in Fig. 5. It is composed of a working electrode, a counter electrode and, optionally, a reference electrode in electrolytic contact (through an ionic conductive electrolyte) embedded within a cell enclosure containing a small aperture (or diffusion hole) or a porous, gas-permeable membrane. The main role of the aperture or membrane is to control the rate of gas flow into the sensor and thus control the amount of gas molecules reaching the electrode surface. Additionally, gas

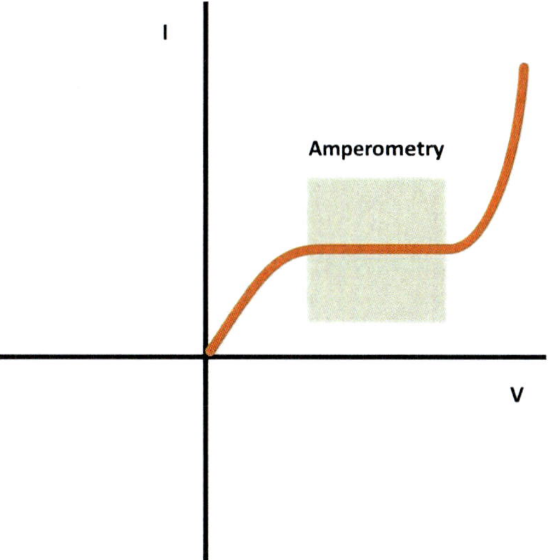

Electrochemical Sensor of Gaseous Contaminants, Fig. 4 Current–voltage curve indicating the domain of amperometric sensors

membranes can aid selectivity, allowing only the analyte gas to pass, as well as provide a barrier to prevent leakage of the electrolyte, when liquid, from the interior of the sensor. Several types of membranes exist, made of polymeric or inorganic materials; most are commercially available thin solid Teflon® films, microporous Teflon® films or silicon membranes. An ideal gas membrane will have a constant permeability to the target gaseous contaminant during sensor operational life over a wide temperature range and process mechanical, chemical and environmental stability. The earliest successful amperometric gas sensor was developed by Clark et al. [15] in 1953. The original Clark sensor, introduced for oxygen determination in blood samples, utilized a cellophane membrane-covered platinum electrode to regulate oxygen diffusion on the electrode, while at the same time, the cellophane membrane prevented interference from blood cells and other gases. The same concept is applied in many commercial sensors, not only for the detection of oxygen.

The electrode reaction at the working–sensing electrode refers usually, but not exclusively, to the anodic oxidation of the analyzed gas. The selection of the working electrode material is

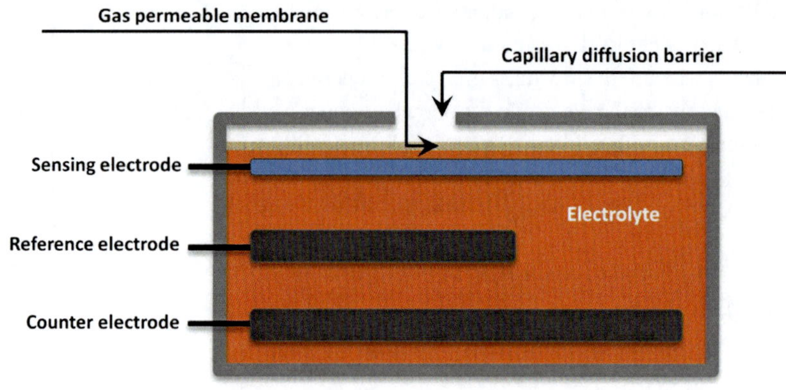

Electrochemical Sensor of Gaseous Contaminants, Fig. 5 Typical setup of an amperometric gas sensor

very important. It has to be a catalyzed material which performs the aforementioned half-cell reaction over a long period of time. Typically, the electrode is made of a noble metal, such as platinum or gold, which is capable of making a defined interface with the electrolyte in the cell and is porous to allow efficient diffusion of the gas phase to a large and reactive electrode/ electrolyte interface. The counter electrode must also be stable in the electrolyte and efficiently perform the complementary half-cell reaction that is opposite of the analyte reaction. In some cases, the counter electrode may be a sacrificial one, meaning that during operation it is transformed to a non-conductive phase. This is the case for an oxygen sensor where the lead counter electrode is oxidized to the corresponding oxide (PdO) through the electrode reaction:

$$2Pb + 4OH^- \rightarrow 2PbO + 2H_2O + 4e^-$$

In many practical applications, the control of gas inflow is accomplished by a diffusion hole [5]. The general current–voltage relationship for a one component conduction system is the following:

$$U = iR + (RT/nF)\ln\left(P_i^o/P_i^{el}\right)$$

where P_i^o and P_i^{el} are the pressure of the component i in the analyzed gas and at the electrode, respectively. For a reaction limited by the mass transport process, the general flux equation in a one-dimensional model is:

$$J_i = -D_i\frac{\partial P_i}{\partial x} + \alpha_{conv}P_i$$

where α_{conv} is the convection coefficient of the gas forming the ambient atmosphere. Depending on the hole diameter, two types of diffusion can be distinguished:

1. If the hole is very small ($<\sim 1$ μm), a Knudsen mechanism is prevalent, and the flux is equal to $J_i = K'_d(P_i^o - P_i^{el})$, where K'_d is the Knudsen coefficient defined by $K'_d = \frac{d}{3L\sqrt{\pi M_i RT}}$, where d and L are the diameter and length of the hole, respectively, and M_i is the molar mass of gas i. In the diffusional limited regime, all the electroactive species are consumed at the electrode faradaically and thus $P_i^{el} = 0$. Then the limiting current is given by:

$$I_L = 2zK'_dF\pi d^2P_i^o$$

where z is the charge number of ions involved in the electrode reaction. The limiting current is thus proportional to analyte partial pressure.

2. If the hole diameter is $> \sim 10$ μm, the limiting current is given by [5]:

$$I_L = \frac{2zFD^\circ T^{3/4}\pi d^2}{RL}\ln(1 - x_i)$$

where D^o is the standard diffusion coefficient of species i in the analyzed gas, and x_i is the molar fraction of species i. Hence, the limiting current becomes proportional to $\ln(1-x_i)$.

Two major achievements in the evolution of modern amperometric gas sensors involve the development of gas diffusion electrodes and the use of solid-state electrolytes. Gas diffusion electrodes (GDEs), introduced by Niedrach [16], make use of high-surface area supports for metal catalysts, as well as solid inert materials with high gas permeability. They are usually composed of a porous PTFE membrane and a porous electrode, consisting of highly interlocked matrices of gas pores, electrolyte channels, electronically conducting paths, and electrocatalytic surfaces [14]. The fact that the real surface of the gas diffusion electrodes can be several orders of magnitude higher than the geometric area allows species with relatively poor electroactivity to produce measurable currents (apparent current amplification ranging up to around 100 times that obtained on an equivalent simple geometric area), thus leading to higher sensor sensitivity at extremely low levels (ppb). Because GDE designs have been derived from fuel cell technology, these devices have been incorrectly called fuel cell sensors.

The use of solid polymer or ceramic electrolytes may confront inefficiencies and maintenance requirements of liquid electrolytes while offering versatility for the detection of different gases under harsh conditions (e.g. monitoring of CO, NO_x, SO_x and hydrocarbons in automotive exhausts), robustness, and mechanical stability. Nafion® solid polymer electrolyte has also been used for the detection of various gases [7], while yttria-stabilized-zirconia (YSZ), β''-Al_2O_3 and NASICON are common ceramic solid electrolytes utilized for the construction of sensors [9].

Summary

A gas sensor has to meet a number of diverse requirements, such as high sensitivity and selectivity, large dynamic range, compactness, high reliability, insensitivity to external (ambient) conditions, low power consumption, relative ease of use and cost-effectiveness in terms of both initial and maintenance cost. The aforementioned requirements are a challenging research and development task. Besides fundamental research in materials development, sensors are adopting many engineering aspects from the current trend in integrated electronic circuit design and miniaturization technology in order to fabricate advanced, multiple-component devices. The advanced characteristics of electrochemical sensors make them very popular for online, continuous monitoring of gaseous contaminants, and thus they are widely used for portable and bench instruments.

Cross-References

▶ Electrochemical Sensors for Environmental Analysis

References

1. Tittel FK (1999) Detection of trace gas contaminants using diode laser based methods. In: Lasers and electro-optics. CLEO/Pacific Rim '99. The Pacific Rim conference on lasers and electo-optics, IEEE, Seoul
2. Hulanicki A, Glab S, Ingman F (1991) Chemical sensors definitions and classification. Pure Appl Chem 63(9):1247–1250
3. Janata J (2009) Principles of chemical sensors. Springer, New York
4. Severinghaus JW, Bradley AF (1958) Electrodes for blood P_{O2} and P_{CO2} determination. J Appl Physiol 13:515–520
5. Fabry P, Siebert E (1997) Electrochemical sensors. In: Gellings PJ, Bouwmeester HJM (ed) The CRC handbook of solid state electrochemistry. CRC Press, Boca Raton
6. Fergus JW (2009) Electrochemical sensors: fundamentals, key materials, and applications. In: Kharton VV (ed) Solid state electrochemistry. Wiley-VCH, Weinheim
7. Park CO, Fergus JW, Miura N, Park J, Choi A (2009) Solid-state electrochemical gas sensors. Ionics 15(3):261–284
8. Pasierb P, Rekas M (2009) Solid-state potentiometric gas sensors-current status and future trends. J Solid State Electrochem 13(1):3–25
9. Stetter JR, Penrose WR, Yao S (2003) Sensors, chemical sensors, electrochemical sensors, and ECS. J Electrochem Soc 150(2):S11–S16
10. Weppner W (1992) Advanced principles of sensors based on solid-state ionics. Mat Sci Eng B-Solid State Mater Adv Tech 15(1):48–55

11. Gauthier M, Chamberland A (1977) Solid-state detectors for potentiometric determination of gaseous oxides. J Electrochem Soc 124(10):1579–1583
12. Fleming WJ (1977) Physical principles governing non-ideal behavior of zirconia oxygen sensor. J Electrochem Soc 124(1):21–28
13. Garzon FH, Mukundan R, Brosha EL (2000) Solid-stale mixed potential gas sensors: theory. experiments and challenges. Solid State Ionics 136:633–638
14. Stetter JR, Li J (2008) Amperometric gas sensors – a review. Chem Rev 108(2):352–366
15. Clark L, Wolf R, Granger D (1953) Continuous recording of blood oxygen tensions by polarography. J Appl Physiol 6:189–193
16. Niedrach LW, Alford HR (1965) Novel fuel cell structure. US Patent 3905832, Chem Abstr 62: 11416c

Electrochemical Sensors for Aerospace Applications

Gary W. Hunter
NASA Glenn Research Center,
Cleveland, OH, USA

Introduction

Sensors are devices that produce a measurable change in output to an input stimulus. This stimulus can be a physical stimulus like temperature and pressure or a chemical stimulus provided by the concentration of a specific chemical or biochemical species. The output signal is typically proportional to the input stimulus, and thus a correlation can be established between the stimuli and the output signal. Categories of solid state sensors are broad and include (1) physical sensors for measurement of properties such as temperature, pressure, or acceleration; (2) chemical sensors for the measurement of a chemical species such as ethanol, carbon monoxide, or other molecules; (3) biosensors for measurement of biologically active substances such as bacteria or viruses; and (4) radiative sensors for detection of emitted particles or waves such as alpha and beta particles, gamma and X-rays, as well as other forms of electromagnetic radiation [1]. A single sensor may be influenced by more than one stimulus and provide a more complex response beyond straightforward measurement of the targeted parameter. For example, a chemical sensor can be affected by temperature and the measured response needs to be corrected for that influence. Within these measurements categories, the method by which the sensor detects the parameter of interest can vary. For example, the input stimulus that changes the sensor output can be based on optical, mechanical, thermal, magnetic, electronic, or electrochemical transduction mechanism [1].

This chapter discusses gas sensors associated with aerospace applications that have electronic and electrochemical transduction mechanisms. That is, sensors that respond to chemical reactions between the sensor element and the surrounding gas with a resulting change in the electronic or electrochemical properties of the sensor. There are a range of sensor types that operate in this manner including sensors constructed as resistors, electrochemical cells, calorimetric devices, and Schottky diodes [2]. This chapter will first discuss the measurement needs associated with aerospace applications, where aerospace refers to both aeronautic and space applications. Examples are given of an electronic sensor and an electrochemical sensor that have relevance to aerospace applications. Challenges associated with implementation of sensors in aerospace applications will be discussed, as well as future technology directions.

Measurement Needs

Aerospace applications require a range of sensor technology to enable improved performance, increased safety, better awareness of the vehicle health state, and to enable increased automation and autonomy [3]. The following are three examples of aerospace applications involving the detection of gas species: one associated with space propulsion and fuel safety, one related to aircraft emissions monitoring, and the third involves both space and aircraft fire detection. Briefly described is each application as well as the shortcomings of traditional technology.

Leak Detection

The detection of low concentrations of fuel leaks such as hydrogen (H_2) at potentially low temperatures is important, for example, in operation of space launch vehicles such as the Space Shuttle [4, 5]. In 1990, hydrogen leaks on the Space Shuttle while on the launch pad temporarily grounded the fleet until the leak source could be identified. Other events since then have also impacted launch availability and schedule. Concern continues that fuels leaks in propulsion and fuel storage systems are a potential risk for space flight systems, fuel storage, and a range of space exploration applications [6]. The requirements of launch vehicle operation meant that standard hydrogen sensor technology was not viable for this application since these technologies often need oxygen or depend upon moisture to operate [7, 8]. Thus, the development of a leak detection system able to operate in a range of environments, able to determine the presence of hazardous conditions, and the isolate the location of a leak is necessary for space applications.

Emissions Monitoring

The control of emissions from aircraft engines is an important component of the development of the next generation of these engines [4, 5, 9]. The ability to monitor the type and quantity of emissions generated by an engine is important for qualifying engine technology, as well as potentially determining the health status of the engine, and eventually for emission control. Ideally, an array of sensors placed in the emissions stream close to the engine could provide information on the gases being emitted by the engine. While commercially available oxygen (O_2) sensors based on potentiometric electrochemical cells using a zirconium dioxide (ZrO_2) electrolyte have been used for years in automobile engines [10] to decrease emissions, using this sensor technology in aeronautic applications is challenging. Operation of these ZrO_2 sensors in this potentiometric mode limits the oxygen detection range, and the sensor power consumption on the order of several watts. Further, the availability of other high-temperature emission sensors beyond oxygen to more fully characterize engine emissions has historically been problematic motivating the development of high-temperature engine emission monitoring systems [9].

Fire Safety and Environmental Monitoring

Fire detection equipment used in the cargo holds of many commercial aircraft rely on the detection of smoke [11]. Although highly developed, these sensors can be subject to false alarms. These false alarms may be caused by a number of sources including changes in humidity, condensation on the fire detector surface, and contamination from animals, plants, or other contents of the cargo bay [4, 5]. A second method of fire detection to complement existing techniques, such as the measurement of chemical species indicative of a fire, can help reduce false alarms and improve aircraft safety. Further, the detection of the combustion products associated with fires addresses a critical risk for space exploration vehicles and habitats due to the fact that a fire is a significant risk to crew safety and health both during and after the fire event [12, 13]. The hazard from a fire can be significantly reduced if fire detection is rapid and occurs in the early stages of fire development and an understanding of the environment after a fire is important for astronaut health as well as for further monitoring of hazardous conditions.

Sensor Examples

Two examples of sensor technology designed for aerospace applications are described below: a hydrogen sensor and an oxygen sensor. Each one is designed for minimal size, weight, and power consumption and fabricated by using microelectromechanical systems (MEMS) or microfabrication technology. The overall approach is to enable miniature, mobile sensor systems that can be applied throughout ground, flight, and in-space operations with minimal impact to the vehicle.

Hydrogen Sensor Technology

In response to the hydrogen leak problems, development has been ongoing to improve propellant

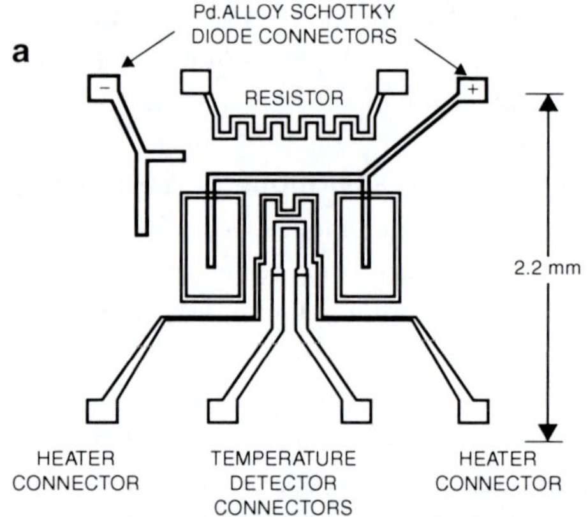

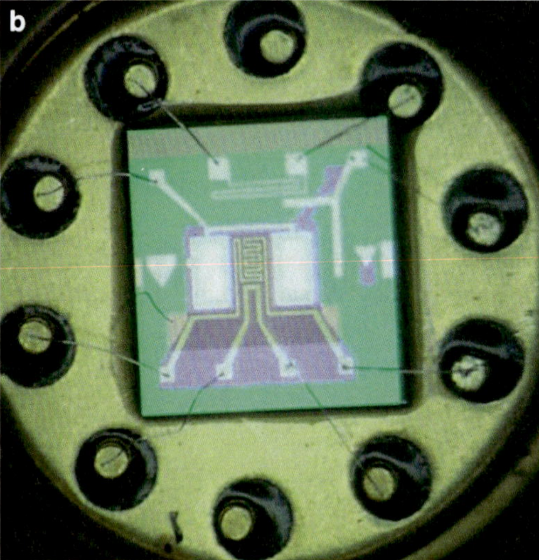

Electrochemical Sensors for Aerospace Applications, Fig. 1 (**a**) Schematic diagram of the silicon-based hydrogen sensor. The Pd alloy Schottky diode (*rectangular regions*) resides symmetrically on either side of a heater and temperature detector. (**b**) Picture of the packaged sensor [6]

leak detection capabilities. In particular, efforts to develop an automated hydrogen leak detection system using micro hydrogen sensors. One approach is to meet this application need is a palladium (Pd) alloy hydrogen sensor shown in Fig. 1a. The structure includes a Pd alloy Schottky diode and resistor, a temperature detector, and a heater all incorporated in the same chip. Shown in Fig. 1b is a picture of the packaged sensor. This figure shows features present in the sensor platforms developed for multiple aerospace applications: small size, multiple structures on the same chip, and integration with temperature control capabilities [4–6, 9, 11–13]. This hydrogen sensor has been demonstrated on the Space Shuttle and NASA Helios vehicle, matured for launch vehicle applications, and has been qualified for use in a criticality one function on the International Space Station (ISS) [14]. In each case, the sensor and supporting hardware were tailored for the application. This is an example of an electronic sensor developed for aerospace applications.

Oxygen Sensor Technology

The detection of oxygen is an important measurement for range of aerospace application including leak detection, fire detection, environmental monitoring, and emissions monitoring. A microfabricated version for aerospace applications of the ZrO_2-based oxygen sensor design is shown in Fig. 2. The operation of ZrO_2 as an electrolyte occurs at higher temperatures, e.g., 600 °C, so heating of the sensor is necessary. However, the design is intended to minimize the thermal mass and decease the power consumption, e.g., to a level on the order of 200 mW, rather than watts for commercial sensors. The sensor is operated in an amperometric mode and has a nearly linear response to various concentrations of oxygen. Oxygen sensors using this basic miniaturized design have been demonstrated for use in International Space Station (ISS) environmental monitoring applications, in jet engine emission applications, and as part of a fire detection system [15] and operated for extended times on the outside of the ISS as part of a Material International Space Station Experiment (MISSE). This is an example of an electrochemical sensor developed for aerospace applications.

Challenges in Application of Aerospace Sensors

As noted in this chapter, there are a range of challenges in the use of gas sensor technology in aerospace applications. The type of challenge varies from application to application. For example, in measurement cryogenic fluid leaks, sensor exposure to low temperatures and operation in vacuum environments is necessary. In some harsh environment engine applications, the sensor must be able to operate reliably in conditions of high temperatures, high vibration, and significant thermal shock. These are vastly different operational environments and, in each case, the sensor system must be tailored for the application. Standard off-the-shelf commercial technology is often not designed for the operational conditions necessary for aerospace applications and often new technology must be developed.

In particular, across the range of aerospace applications, there is a continuing drive for smaller size, reduced power, ease of integration, multiparameter detection, and an ability to easily interface to a range of communication standards. The maturation and implementation of small, smart, multifunctional, standalone sensor systems that can handle a range of aerospace applications is a significant challenge. An example of such a sensor system designed for aerospace applications is the leak sensor system shown in Fig. 3a. This leak sensor system contains the hydrogen and oxygen sensors described above, as well as a hydrocarbon sensor, to measure both fuel and oxygen concentrations within the same sensor array. A microprocessor makes this sensor system a "smart sensor," and the overall design allows sensor temperature control, data processing, signal conditioning, communication, and even the addition of battery power.

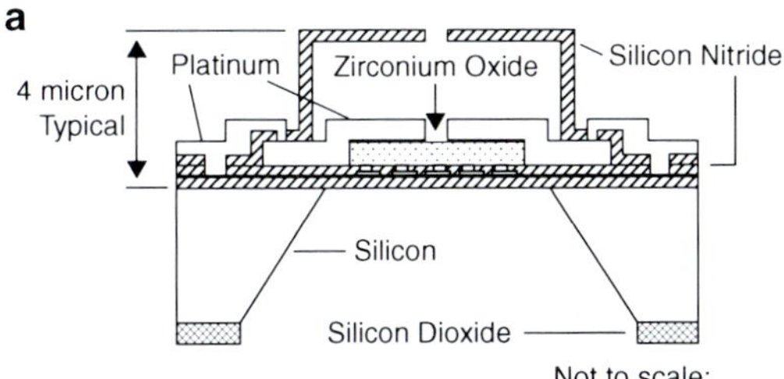

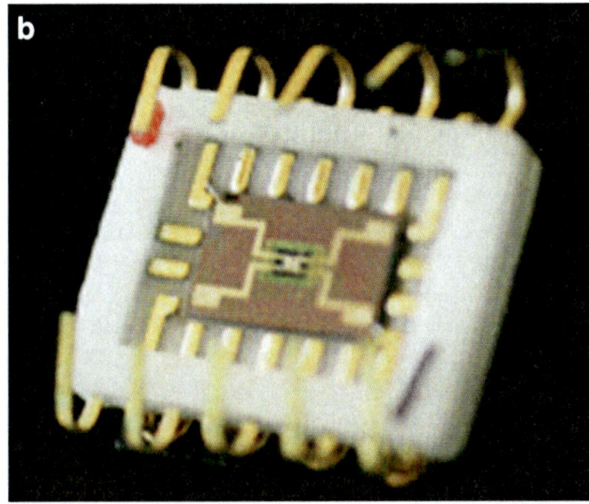

Electrochemical Sensors for Aerospace Applications, Fig. 2 (**a**) Schematic diagram of the zirconia-based oxygen microsensor. The sensor design is aimed towards limiting the flow of oxygen to the electrodes, as well as minimal size, weight, and power consumptions. (**b**) Picture of packaged zirconia-based O_2 sensor [4]

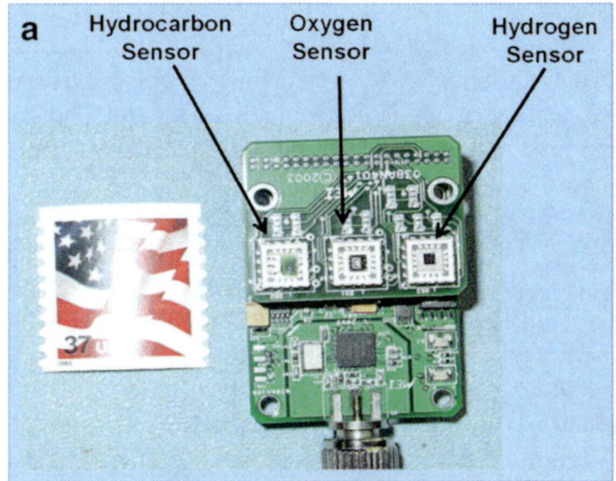

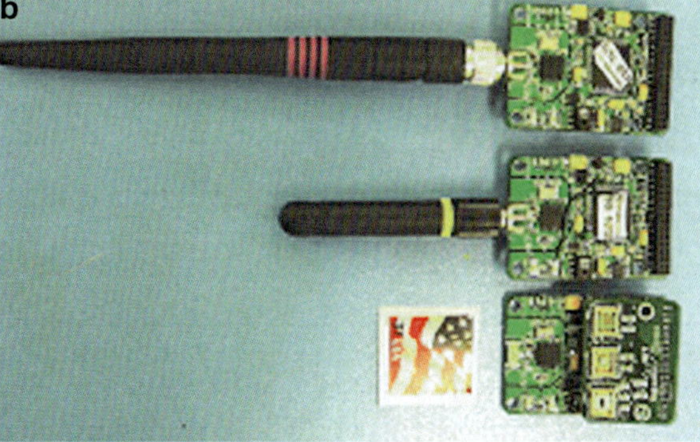

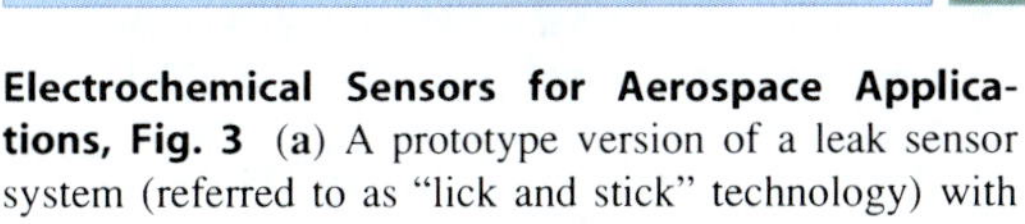

Electrochemical Sensors for Aerospace Applications, Fig. 3 (**a**) A prototype version of a leak sensor system (referred to as "lick and stick" technology) with supporting electronics including signal conditioning and telemetry. (**b**) "Lick and stick" system configured with different wireless communication antennae [6]

Figure 3b shows the adaptability of this sensor system: multiple communication antennas can be implemented starting from the same core design.

Overall, for any sensor to be implemented in an aerospace application, reliability is a core and fundamental feature. Aerospace applications tend to include operation in extreme environments with an inability to simply cease operation if the sensor system is producing confusing data. Whether the sensor is on an airplane flying at high altitudes, or on a space station above the earth, it is crucial that the data being provided by the sensor is reliable. In order for sensor to qualify for such applications, significant application testing is necessary. Reliability must be proven for the sensor system to be implemented in vehicle systems.

Future Directions

In order to reach the promise of electronic and electrochemical sensor systems, further advancements in both micro and nanotechnology, as well as associated Smart Sensor hardware and software, are necessary. The future objective is that new sensor technology may truly enable the ability to "Lick and Stick" sensor systems wherever and whenever needed. The drive is to make sensor systems that are small, smart, highly miniaturized, integrated, and self-contained units whose presence is unnoticed and whose implementation is as simple as placing the sensor where you want it. This is a challenging vision for even conventional technology, but significant technology advancements are needed in order to realize such a vision for the range of often extreme operating environments present in aerospace applications.

Acknowledgements The authors would like to acknowledge the contributions of Dr. J. Xu, Dr. A. Biaggi-Labiosa, L. Evans, Dr. L. Matus, and Dr. Mary Zeller of NASA Glenn Research Center (GRC); Dr. C. Chang of ASRC Aerospace/NASA GRC; Dr. L. Chen of OAI; Dr. B. Ward and Dr. D. Makel of Makel Engineering, Inc.; and Prof. P. Dutta of the Ohio State University.

References

1. Stetter J, Hesketh P, Hunter GW (2006) Sensors: engineering structure and materials from micro to nano. ECS interface, (Vol. 15), pp 66–69
2. Liu CC, Hesketh P, Hunter GW (2004) Chemical microsensors. ECS interface, pp 22–29
3. Hunter GW, Oberle LG, Baakalini G, Perotti J, Hong T (2011) System health management: with aerospace applications. In: Johnson SB, Gormley T, Kessler S, Mott C, Patterson-Hine A, Reichard K, Scandura P Jr (eds) Integrated system health management in exploration applications. Wiley, Chichester
4. Hunter GW, Xu JC, Liu CC, Makel DB (2005) Microfabricated chemical sensors for aerospace applications. In: Gad-El-Hak M (ed) The MEMS handbook, vol 2, 2nd edn, Design and fabrication. CRC Press, Baton Rouge, Chapter 11
5. Hunter GW, Xu JC, Makel DB (2008) Case studies in chemical sensor development. In: Hesketh PJ (ed) BioNanoFluidic MEMS. Springer, New York, pp 197–231
6. Hunter GW, Xu JC, Neudeck PG, Makel DB, Ward B, Liu CC (2006) Intelligent chemical sensor systems for in-space safety applications. In: 42nd AIAA/ASME/SAE/ASEE joint propulsion conference and exhibit, Sacramento, 9–12 July 2006, Paper AIAA-2006-4356
7. Hunter GW (1992) A survey and analysis of commercially available hydrogen sensors, NASA technical memorandum 105878
8. Hunter GW (1992) A survey and analysis of experimental hydrogen sensors, NASA technical memorandum 106300
9. Ward BJ, Wilcher K, Hunter G (2010) Gas microsensor array development targeting enhanced engine emissions testing. In: AIAA Infotech@Aerospace 2010, AIAA, Reston, 20–22 Apr 2010, AIAA 2010–3327
10. Logothetis EM (1991) Automotive oxygen sensors. In: Yamazoe N (ed) Chemical sensor technology, vol 3. Kodansha, Tokyo, pp 89–104
11. Hunter GW, Xu JC, Greenberg P, Ward B, Carranza S, Makel D, Liu CC, Dutta P, Lee C, Akbar S, Blake D (2004) Miniaturized sensor systems for aerospace fire detection applications. In: The 4th triennial international aircraft fire and cabin safety research conference, Lisbon, Nov 2004
12. Hunter GW, Greenberg PS, Xu JC, Ward B, Makel D, Dutta P, Liu CC (2009) Miniaturized sensor systems for early fire detection in spacecraft. In: 39th international conference on environmental systems, SAE, Warrendale, 12–16 July 2009, 09ICES-0335
13. Hunter GW, Xu JC, Dungan L, Ward B, Dutta P, Adeyemo AD, Liu CC, Gianettino DP (2010) Smart chemical sensor systems for fire detection and environmental monitoring in spacecraft. In: 40th international conference on environmental systems, AIAA, Reston, 11–15 July 2010, AIAA766637

14. Msadoques G, Makel D (2005) Flight hydrogen sensor for use in the ISS oxygen generation assembly. In: International conference on evolable systems, 05 ICES-350, Barcelona, Sept 2005
15. Hunter GW, Xu JC, Evans L, Biaggi-Labiosa A, Ward BJ, Rowe S, Makel DB, Liu CC, Dutta P, Berger GM, Vander Wal RL (2010) The development of micro/nano chemical sensor systems for aerospace applications, SPIE Newroom, June 2010. doi:10.1117/2.1201006.002984

Electrochemical Sensors for Environmental Analysis

Margarita Stoytcheva and Roumen Zlatev
Instituto de Ingenieria, Universidad Autonoma de Baja California, Mexicali, Baja California, Mexico

Introduction

The electrochemical sensors, according to IUPAC definitions and classification, are a category of chemical sensors [1], designed by coupling the receptor part of the device to an electrochemical transducer. The transducer transforms the analytical information originating from the electrochemical interaction analyte-electrode into a measurable electrical signal. A large number of electrochemical sensors, including biosensors, are based on chemically modified electrodes [2, 3].

The electrochemical sensors are compact, portable, and simple to handle instruments, able to provide analytical information in a real time, without or with a minimum sample preparation. These performances, in concert with their sensitivity, selectivity, and low cost, make them suitable for *infield* and *online* environmental analysis [4–7] and an excellent complement to the expensive and time-consuming *off-site* chromatographic and adsorption or emission spectrometric methods, currently applied in environmental monitoring [8].

Electrochemical Sensors Classification

The electrochemical sensors for environmental analysis, according to the applied electrochemical transduction mode, are mainly [5, 6, 9, 10]:

- Potentiometric sensors, in which the potential of an indicator electrode is measured against a reference electrode at zero current conditions. The potential of the indicator electrode varies as a function of the analyte concentration, according to the Nernst equation. The logarithmic character of the response of the indicator electrode determines the extended concentration range of the calibration plot (3–4 decades) but also the unsatisfactory accuracy of the analysis.
- Voltammetric, including amperometric sensors, in which current is measured controlling the potential of the indicator electrode. The current response of the indicator electrode is proportional to the concentration of the present electroactive species. The advantages of the technique consist in the high sensitivity and precision of the determinations, coupled with the large linear concentration range, the ability of simultaneous determination of several analytes by controlling the process through the applied electrode potential, and the rapid analysis time.
- Conductometric sensors, measuring the variation of the solution conductance due to electrical charge concentrations changes. The method is simple but not selective, since the conductance depends on the ionic concentration of all of the present species. In view of the fact that no electrochemical processes take place, conductometric sensors are not strictly electrochemical ones. They are considered, according to IUPAC chemical sensors definitions and classification [1], as a subclass of electrical devices in which the signal results from the change of electrical properties caused by the interaction of the analyte. Nevertheless, electrical devices are frequently put into one category with the electrochemical devices [1].

Environmental Applications of the Electrochemical Sensors

Inorganic Pollutants Determination

Heavy and Toxic Metals

Potentiometric sensors are attractive analytical tools for environmental monitoring, because of the low cost and operation convenience of the equipment. Thus, potentiometric sensors, namely, ion-selective electrodes, were applied for trace level quantification of a number of metals (Ag, Cd, Cs, Cu, Pb, etc.). Nonetheless, potentiometric sensors for metal determination present some limitations in terms of LOD, dynamic range, response time, and *online* environmental analysis application [11].

Voltammetric stripping-based sensors are considered as more appropriate for metals contamination monitoring. They are well suited for the determination of about 30 metal ions with a detection limit close to that of AAS and ICP analysis. In addition, in contrast to the currently applied methods, voltammetry allows the selective speciation and quantification of the metals in their various oxidation states, which exhibit differing toxicities and mobilities in the environment. Stripping voltammetric techniques involve two steps: metal preconcentration, in which trace metals are accumulated onto/into the working electrode, and measurement, in which metals are "stripped" from the electrode by scanning the applied potential. The unique sensitivity of the method is due to the metal preconcentration. Stripping-based sensors [5, 6, 12–18] were applied for the determination of all of the metals (As, Sb, Be, Cd, Cr, Cu, Pb, Hg, Ni, Se, Ag, Tl, and Zn), included in the EPA list of the 129 priority pollutants of the environment [19]. Stripping voltammetric methods for Pb, Cd, and Zn determination are approved by EPA [20].

Promising voltammetric technique, which allows metal species distinction in complex multicomponent matrices without any preliminary separation, is the differential alternative pulses voltammetry (DAPV) [21].

Biosensors application for heavy and toxic metals determination is extensively reviewed in some recent works [7, 22–24]. Relevant examples include inhibition-based enzyme sensors, DNA-, antibody-, and whole-cell-based sensors.

Other Inorganic Pollutants

Electrochemical sensors are commonly applied to water analysis. The "Standard Methods for the Examination of Water and Wastewater" [20] include the following approved by EPA voltammetric sensors-based techniques for inorganic nonmetallic constituents monitoring: dissolved oxygen evaluation by membrane electrode method, voltammetric iodide determination, polarographic iodate analysis, and cyanides quantification by flow injection, ligand exchange, and amperometric detection methodology.

The accepted by EPA potentiometric methods involve selective electrodes for fluorides, cyanides, nitrates, ammonia, and sulfides detection [20]. The potentiometric characteristics of the anion-selective sensors are strongly dependent on the anion receptor design and properties [25]. At this time, molecular recognition of anions by synthetic receptors is an expanding research area.

Volatile inorganic contaminants are monitored using selective potentiometric sensors for gases (NH_3, SO_2, H_2S, NO_2, CO_2, HF, etc.). Most of them are commercialized.

Inorganic nitrogen is also quantified using conductometric sensors [20].

Advances in electrochemical biosensors application for inorganic nonmetallic pollutants analysis (inorganic phosphate and nitrate) are considered in the review [7].

Organic Pollutants Determination

Pesticides Determination

The electrochemical sensors with chemical recognition applied to pesticides analyses take advantage of the analytical features of the differential pulse voltammetry, square-wave voltammetry, and stripping voltammetry [26–30], which allow pesticides detection at residues levels. Current efforts are directed toward sensitivity and selectivity improvement by chemical modification of the electrode surface by molecularly imprinted polymers and micro- and nanostructured materials [31].

The electrochemical biosensors for pesticides monitoring involve enzyme, antibodies, and cells-based sensors [7, 24, 27, 30, 32]. The application of various enzymes – acetylcholinesterase, acid phosphatase, alkaline phosphatase, organophosphorus hydrolase, and tyrosinase – for the quantification of organochlorine, organophosphate, and carbamate pesticides in the environment is extensively revised by Van Dyk et al. [33]. Exhaustive reviews on enzyme inhibition-based sensors, including inhibition determination in organic phase, are provided by Amine et al. [34], Lopez et al. [35], and Palchetti et al. [36]. Immunosensors design, fabrication, analytical capabilities, and drawbacks, limiting their application in environmental analysis, are commented by Jiang et al. [32]. Microbial sensors development and applications are outlined in some relevant recent reports [37–40].

Other Organic Pollutants

Pulse polarography and stripping voltammetry are considered as methods of choice for the quantification of environmentally significant organic molecules (carbonyls, nitroaromatics, phenols, carboxylic acids, sulfonates, azomethines, thiols, amines, etc.) using electrochemical sensors with chemical recognition [28]. The current capabilities of biosensors for organic pollutants analysis are discussed by [7].

Biological Pollutants Determination

Pathogenic viruses, bacteria, protozoa, and helminthes are typically quantified using amperometric biosensors. Besides their use in enzyme-based sensors, amperometric transducers were also applied to measure enzyme-labeled tracers for affinity-based biosensors (immunosensors and genosensors) [41]. Other biosensors involve the monitoring of microbial metabolism by oxygen consumption measurement using the amperometric Clark-type oxygen electrode or by quantifying electroactive metabolites. Achievements and problems facing biosensors for biological pollutants determination are summarized in the reviews [7, 42, 43].

Future Directions

Future directions in environmental electrochemical sensors development could be summarized as follows:

- Validation of the electrochemical sensors approaches to environmental pollutants monitoring, applying approved standard protocols
- Rational (bio)molecular electrode design, involving nanotechnology and bioengineering, leading to sensors sensitivity, selectivity, and stability improvement
- Development of sensors arrays for simultaneous multicomponent analysis
- Miniaturization of the electrochemical sensors and analyzers with regard of their *infield* application
- Implementation of automated systems with remote control for continuous contaminant detection and early pollution prevention
- Mass production and commercialization of the electrochemical biosensors.

Cross-References

► Electrochemical Sensor of Gaseous Contaminants
► Electrochemical Sensors for Water Pollution and Quality Monitoring

References

1. Hulanicki A, Glab S, Ingman F (1991) Chemical sensors definition and classification. Pure Appl Chem 63:1247–1250
2. Thévenot D, Toth K, Durst R, Wilson G (1999) Electrochemical biosensors. Recommended definitions and classification. Pure Appl Chem 17:2333–2348
3. Durst R, Bäumner A, Murray R, Buck R, Andrieux P (1997) Chemically modified electrodes: recommended terminology and definitions. Pure Appl Chem 69:1317–1323
4. Brett C (2001) Electrochemical sensors for environmental monitoring. Strategies and examples. Pure Appl Chem 73:1969–1977
5. Hanrahan G, Wang J (2004) Electrochemical sensors for environmental monitoring: design, development and applications. J Environ Monit 6:657–664
6. Wang J (2004) Electrochemical sensors for environmental monitoring: a review of recent technology. NERL U.S. EPA, Las Vegas
7. Rodriguez-Mozaz S, Marco M-P, Lopez de Alda MJ, Barceló D (2004) Biosensors for environmental applications: future development trends. Pure Appl Chem 76:723–752
8. Official methods of analysis of AOAC International. http://www.eoma.aoac.org/. Accessed 27 Sept 2011
9. Bakker E (2004) Electrochemical sensors. Anal Chem 76:3285–3298

10. Bakker E, Quin Y (2006) Electrochemical sensors. Anal Chem 78:3965–3984
11. Gupta V (2005) Potentiometric sensors for heavy metals-an overview. Chimia Int J Chem 59: 209–217
12. Cavicchioli A, La-Scalea MA, Gutz IGR (2004) Analysis and speciation of traces of arsenic in environmental, food and industrial samples by voltammetry: a review. Electroanalysis 16:698–711
13. Toghill K, Min L, Compton R (2011) Electroanalytical determination of antimony. Int J Electrochem Sci 6:3057–3076
14. Niedzielski P, Siepak M (2003) Analytical methods for determining arsenic, antimony and selenium in environmental samples. Polish J Environ Stud 12:653–667
15. Locatelli C (1997) Anodic and cathodic stripping voltammetry in the simultaneous determination of toxic metals in environmental samples. Electroanalysis 9:1014–1017
16. Nriagu J (1998) Thallium in the environment. Advances in environmental science and technology, volume 30. Wiley, N.Y., USA
17. Brisson M, Ekechkwu A (2009) Beryllium: environmental analysis and monitoring. RSC Publishing, Cambridge
18. Morita M, Yoshinaga J, Edmondst J (1998) The determination of mercury species in environmental and biological samples. Pure Appl Chem 70: 1585–1615
19. USEPA (1979) Water related fate of the 129 priority pollutants, vol. 1 EP-440/4-79-029A. Washington
20. Standard methods for the examination of water and wastewater. http://www.standardmethods.org/. Accessed 27 Sept 2011
21. Zlatev R, Stoytcheva M, Valdez B, Magnin J-P, Ozil P (2006) Simultaneous determination of species by differential alternative pulses voltammetry. Electrochem Commun 8:1699–1706
22. Verma N, Singh M (2005) Sensors for heavy metals. Biometals 18:121–129
23. Turdean G (2011) Design and development of biosensors for the detection of heavy metal toxicity. Int J Electrochem 2011:15 doi:10.4061/2011/343125
24. Jaffrezic-Renault N, Dzyadevych S (2008) Conductometric microbiosensors for environmental monitoring. Sensors 8:2569–2588
25. Gupta V (2010) Potentiometric sensors for inorganic anions based on neutral carriers-an invited review article. Arab J Sci Eng 35(2A):7–25, 10
26. Manisankar P, Viswanathan S, Vedhi G (2010) Analysis of pesticide residue using electroanalytical techniques. In: Rathore H, Nollet L (eds) Handbook of pesticides. Methods of pesticides residues analysis. CRC Press/Taylor & Francis, Boca Raton
27. Garrido E, Delerue-Matos C, Lima J, Brett A (2004) Electrochemical methods in pesticides control. Anal Lett 3:1755–1791
28. Smyth F, Smyth M (1987) Electrochemical analysis of organic pollutants. Pure Appl Chem 59: 245–256
29. Tonle I, Ngameni E (2011) Voltammetric analysis of pesticides. In: Stoytcheva M (ed) Pesticides in the modern world-Trends in pesticides analysis. InTech, Croatia
30. Stoytcheva M (2011) Organophosphorus pesticides analyses. In: Stoytcheva M (ed) Pesticides in the modern world-trends in pesticides analysis. InTech, Croatia
31. Goicolea M, Gómez-Caballero A, Barrio R (2011) New materials in electrochemical sensors for pesticides monitoring. In: Stoytcheva M (ed) Pesticides in the modern world-trends in pesticides analysis. InTech, Croatia
32. Jiang X, Li D, Xu X, Ying Y, Li Y, Ye Z, Wang J (2008) Immunosensors for detection of pesticides residues. Biosens Bioelectron 23: 1577–1587
33. Van Dyk JS, Pletschke B (2011) Review on the use of enzymes for the detection of organochlorine, organophosphate and carbamate pesticides in the environment. Chemosphere 82:291–307
34. Amine A, Mohammadi H, Bourais I, Palleschi G (2006) Enzyme inhibition-based biosensors for food safety and environmental monitoring. Biosens Bioelectron 21:1405–1423
35. López M, López-Cabarcos E, López-Ruiz B (2006) Organic phase enzyme electrodes. Biomol Eng 23:135–147
36. Palchetti I, Laschi S, Mascini M (2009) Electrochemical biosensor technology: application to pesticide detection. Methods Mol Biol 504:115–126
37. Xu X, Ying Y (2011) Microbial biosensors for environmental monitoring and food analysis. Food Rev Int 27:300–329
38. Su L, Jia W, Hou C, Lei Y (2011) Microbial biosensors: a review. Biosens Bioelectron 26: 1788–1799
39. Mulchandani A (2011) Microbial biosensors for organophosphate pesticides. Appl Biochem Biotechnol 165:687–699
40. Stoytcheva M (2010) Enzyme vs. bacterial electrochemical sensors for organophosphorus pesticides quantification. In: Somerset V (ed) Intelligent and biosensor. InTech, Croatia
41. Palchetti B, Mascini M (2008) Electroanalytical biosensors and their potential for food pathogen and toxin detection. Anal Bioanal Chem 391: 455–471
42. Ivnitski D, Abdel-Hamid I, Atanasov P, Wilkins E (1999) Biosensors for detection of pathogenic bacteria. Biosens Bioelectron 14:599–624
43. Leonard P, Hearty S, Brennan J, Dunne L, Quinn J, Chakraborty T, O'Kennedy R (2003) Advances in biosensors for detection of pathogens in food and water. Enzyme Microb Technol 32:3–13

Electrochemical Sensors for Monitoring Conditions of Lubricants

Chung-Chiun Liu[1] and Laurie Dudik[2]
[1]Electronics Design Center, and Chemical Engineering Department, Case Western Reserve University, Cleveland, OH, USA
[2]Case Western Reserve University, Cleveland, OH, USA

Introduction

Lubrication has improved from simple oils and greases to multiple viscosity lubricants with new, machinery and application-targeted additives. Rather than make the task of machine reliability easier, lubricant selection options have become more complex. Coinciding with the growing complexity of lubricants the ability to detect and predict equipment failure has been an area of growing interest. The commercial and industrial sector is now embracing condition-based maintenance and prognostic technologies as the key to enhanced competitiveness.

Lubricants provide vital functions in mechanical systems such as separating moving parts, suspending contaminate, neutralizing corrosive acids, protecting wear surfaces, dissipating heat, and other performance-enhancing features. Ultimately, lubricants reach the end of their useful life due to a variety of degradation mechanisms leading to the increase in oxidation and nitration, base depletion, acid build up, soot contamination, water, fuel and air containment, and viscosity changes [1–3]. The proper maintenance of a lubricant is a critical piece of equipment operation. A breakdown in the lubricant can lead to system failure and loss of the equipment.

Smolenski and Schwartz made the first attempt to classify lubricant analysis techniques and interpret changes in lubricant condition with respect to the source of contamination. Table 1 lists the engine oil problems and the analyses recommended to determine analytically the degradation of the oil in each category. In addition to analysis of viscosity, TAN, TBN, and contaminants as lubricant-condition properties, major directions for lubricant analysis methodologies (e.g., specific gravity, density, Brookfield viscosity, pour point, flash point, distillation characteristics, sulfated ash, chlorine, sulfur, nitrogen, nickel, silver, and ferrographic analysis) and sensors (conductivity, impedance, or dielectric properties and to measure optical properties such as refractive index and visible light absorption) were identified [4].

Electrochemical Sensors for Monitoring Conditions of Lubricants, Table 1 Engine oil problems and recommended analyses [4]

Problem	Analyses
Oil thickening	Viscosity, TAN, TBN, insolubles, glycol, metal, oxidation
Oil thinning	Fuel contamination, molecular weight distribution
Wear	Metals, viscosity, insolubles, TAN, TBN, water, glycol, oxidation
Deposits	Metals, TAN, TBN, viscosity
Corrosion	TAN, TBN, metals, water, glycol
Low-temperature sludge	Water, coagulated pentane insolubles, glycol

The useful life of an industrial lubricant depends on several factors such as the base oil formulation, the type and amount of lubricant additives, the equipment size and operating conditions [1]. Although the bulk of the lubricant is the base oil, simply changing the base oil cannot satisfy the lubrication needs of modern equipment. The presence of additives is required to improve the chemical and the physical properties, the performance and the long-term stability. During its life, a lubricant will undergo substantial chemical changes due to oxidative degradation and contamination by water, ethylene glycol, fuel, soot, wear metals, and viscosity changes. Common degradation of a lubricant results from its exposure to high temperature and the presence of nitrogen oxides, moisture, and air [2]. Chemically active additives, such as dispersants, detergents, oxidation inhibitors, and antiwear agents, interact with contaminants and oxidative by-products of lubricant degradation (high molecular weight aldehydes, ketones, and carboxylic acids) inactivating them.

The complex nature of lubricants and variability in types of industrial equipment make it difficult to foresee every possible pathway leading to failure. In everyday practice, lubricants are changed after a given service time or engine mileage without prior testing; however, assumptions based on miles driven or hours of operation do not provide complete protection. Current methods used to determine lubricant quality, routinely performed by major engine and lubricants manufacturers, are frequent, repetitive, and time-consuming physical and chemical tests. These tests include determination of oil viscosity, total acid number (TAN), total base number (TBN), insolubles (such as soot) content, fuel and water dilution, glycol contamination, and metals content. Validity and interpretation of the results of these tests is often ambiguous [5].

Broadly there are four main classes of lubricant properties, namely, physical-mechanical, electromagnetic, chemical/contamination, and optical that have been identified as being used as sensing parameters for lubrication condition monitoring. Viscosity and specific density are important physical-mechanical properties of lubricant. The electrical properties of lubricating fluids include conductivity, dielectric constant, and magnetic properties. The chemical properties providing the key index to monitoring the lubricant condition is the total acid number (TAN) or total base number (TBN), other important chemical properties to monitor are corrosiveness and the presence of contaminants such as water, coolant, metal particles, and soot. Optical properties include color, transparency, reflection, refractive index, and absorbance.

Electrochemical methods are in general free from the difficulties associated with the current industry testing methods and present an opportunity for a relatively quick, simple, and inexpensive approach, free of temperature limitations and sample preparation issues. Electrochemistry has been previously employed to inspect lubricant condition over the life of engine oils [6, 7]. For example, electrochemical impedance spectroscopy (EIS) has been used for characterization of both engine oils [8] and nonaqueous colloidal dispersions [9].

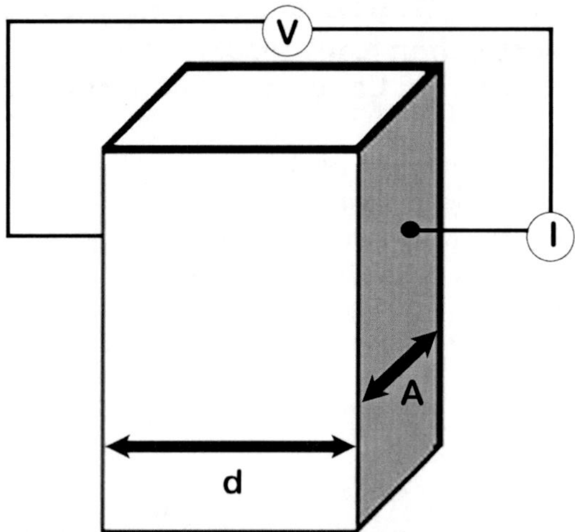

Electrochemical Sensors for Monitoring Conditions of Lubricants, Fig. 1 Fundamental impedance experiment

Electrochemical Impedance Spectroscopy Method of Determining Lubricant Properties

The Electrochemical Impedance Spectroscopy (EIS) technique has gained in exposure and popularity in recent years propelled by a series of scientific advancements in the field of electrochemistry, improvements in instrumentation performance and availability, and increased exposure to an ever-widening range of practical applications. Electrical resistance, R, is related to the ability of a circuit element to resist the flow of electrical current. Ohm's Law defines resistance in terms of the ratio between input voltage V and output current I, $R = V/I$. This well-known relationship is limited to only one circuit element – the ideal resistor which follows Ohm's Law at all current, voltage, and AC frequency levels. Assume that the analyzed sample material is ideally homogenous and completely fills the volume bounded by two external current conductors (electrodes) with area A that are placed apart at uniform distance d as shown in Fig. 1. When external voltage V is applied, a uniform current passes through the sample and the resistance is defined as $R = \rho\, d/A$ where

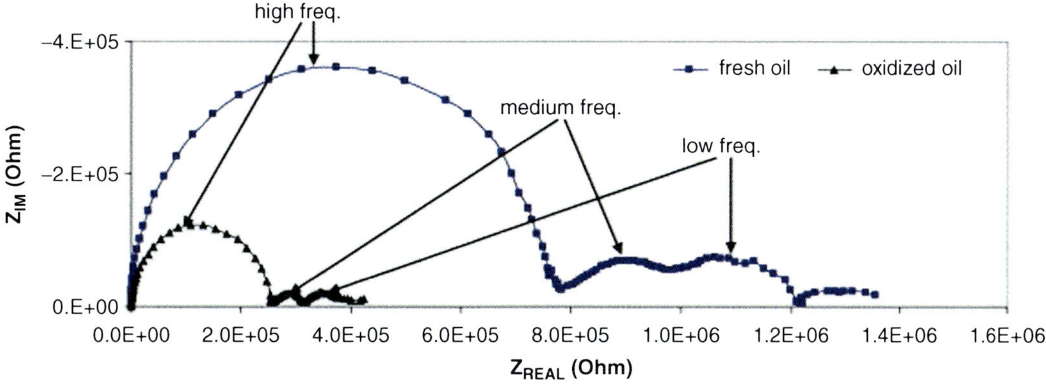

Electrochemical Sensors for Monitoring Conditions of Lubricants, Fig. 2 Impedance spectra for fresh and oxidized oils tested at 120 °C [5]

ρ (ohm cm) is the characteristic electrical resistivity of a material. The ideal resistor can be replaced by an ideal capacitor (or inductor); the AC current and voltage signals through the ideal capacitor are completely "out of phase" with each other, with current following voltage. The value of the capacitance presented in farads (F) depends on the area of the electrodes A, the distance between the electrodes d, and the properties of the dielectric reflected in a "relative permittivity" parameter ε as $C = (\varepsilon_0 \varepsilon A)/d$ where $\varepsilon_0 =$ constant electrical permittivity of a vacuum. The relative permittivity value represents a characteristic ability of the analyzed material to store electrical energy [10].

Typical fully formulated lubricating oil is composed of a combination of mineral or synthetic base oils and specialized dispersed additives designed to improve long-term stability and enhance performance in an aggressive environments. Any commercial lubricant can be viewed as a nonaqueous highly resistive polymeric colloid with low, primarily ionic electrical conductivity. This system is composed of nonpolar base oil and suspended polar molecules and structures. The suspended phase includes both specialized oil additives blended into base oils to enhance the lubricant performance and polar contaminants such as soot, water, and oxidant products that ingress into the lubricant with age and use. The dipolar nature of nonaqueous lubricants and other industrial colloids permits investigation of their properties using high-frequency impedance/dielectric analysis [11].

EIS has been used for general characterization of engine oils [2], studies of lubricants' oxidation [5], and monitoring oil degradation due to its contamination by glycol [3], water [11], and soot [12]. The combination of EIS and multivariate data analysis can be used to simultaneously determine the amounts of soot and diesel in engine oil [13] as well as the concentration and pH of an industrial cutting fluid [14].

EIS allows one to resolve a complicated nonaqueous colloidal system both spatially and chemically and to analyze specific parts of that system based on relaxation frequencies. The impedance diagram of a typical industrial lubricant can be separated into at least three regions representing several independent relaxation phenomena: a very prominent high-frequency region (10 MHz–10 Hz), medium-frequency region (10 Hz–100 MHz), and low-frequency region (100–1 MHz) (see Fig. 2). Identification and characterization of these features is based on initial knowledge of chemical composition and physical properties of industrial lubricants [5].

Electrochemical Humidity Sensors for Detection of Water in Lubrication Fluid

Water contamination in hydraulic fluids can pose a severe threat to the hydraulic system. However,

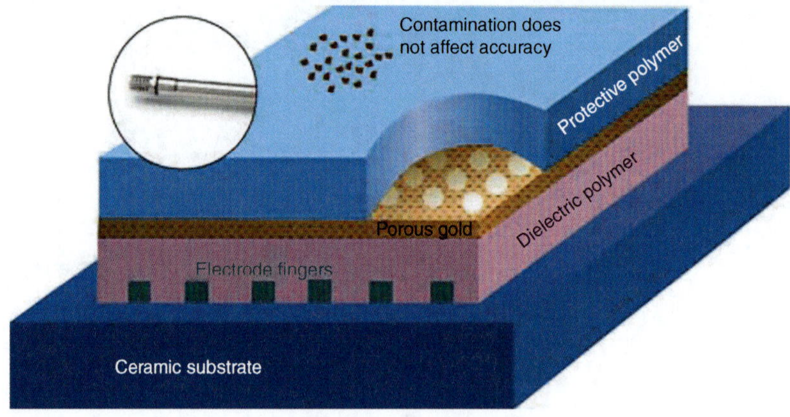

Electrochemical Sensors for Monitoring Conditions of Lubricants, Fig. 3 Capacitive water sensor technology [17]

one has to distinguish between dissolved water, emulsified water, and free water. Dissolved water often does not harm machine components. Emulsified water may, as well as free water, cause severe machine problems. Water can cause hydrolytic splitting of additive or ester molecules and accelerate oil ageing [15].

A capacitive water sensor offers real-time monitoring and can be used as a control device. The most common relative humidity sensors are capacitors with a polymer dielectric, sandwiched between platinum, nickel, or gold metallic electrodes [16]. The lower electrode is deposited upon a ceramic substrate, which is impervious to moisture while the upper capacitor plate – a dielectric polymer – allows the transmission of water molecules. Water molecules migrate into or out of this layer depending on the humidity of the polymer relative to the fluid. This alters its dielectric constant and thus the capacitance of the capacitor. This change in capacitance is then converted into a signal proportional to the saturation level of the fluid, in a percent range of 0–100, where 100 % saturation corresponds to the solubility limit of water in the fluid at the given temperature. The water saturation levels can be related to the absolute water concentration (ppm) via an algorithm that uses fluid parameters determined for the specific fluid brand in conjunction with the fluid temperature via a calibration curve [17]. An RTD (resistance temperature detector) sensor in the capacitive water detection probe compensates for temperature changes. Measuring free-water content as compared to dissolved water can require modifications to the sensing capacitor. In particular, a tube-shape capacitor can replace the polymer-based sensor. Fluid flows axially through the sensing zone and oil acts as the dielectric medium. Capacitance changes are correlated to free-water content in bulk oil [18] (Fig. 3).

In 2009, the ASTM International Committee D02 on Petroleum Products and Lubricants has approved a new ASTM method which covers the determination of moisture content in new and in-service lubricants and additives by relative humidity sensor. The method, D7546 – 09 Standard Test Method for Determination of Moisture in New and In-Service Lubricating Oils and Additives by Relative Humidity Sensor, offers users an alternative to moisture testing by Karl Fischer titration.

Electrochemical Measurement of TAN/TBN

The determination of the total acid and total base number (TAN and TBN, respectively) are important markers of oil degradation. In general, the TAN will increase and the TBN will decrease during oil degradation. Wang et al. [2, 6, 19] developed in situ, macro, and micro lubricant-condition sensors. The macro sensor was composed of two circular-shaped, gold-plated iron electrodes, while the micro sensor was made of aluminum and silicon layers deposited on top of a silicon wafer.

Smiechowski and Lvovich monitored the levels of acidity and basicity in industrial lubricant. The sensor was based on electrochemical impedance methodology. An iridium oxide potentiometric sensor was developed in both a conventional and MEMS configuration. Tests of the sensors in diesel lubricant showed good correlation between TAN, TBN, and the voltage output of each sensor [7]. Widera et al. used a potentiometric iridium oxide electrode as an indicating electrode with a silver/silver chloride reference electrode for the off-line monitoring of fuel acidity. The data showed that the iridium oxide sensor responds to compounds present in fuel that have acid–base character and it is possible to determine the acidity of different fuels and discriminate between unstressed and thermally stressed fuels. Experimental results indicated the ability to correlate the response of the iridium oxide sensor with the total acid numbers of different fuels [20].

Electrochemical Measurement of Viscosity

Saloka and Meitzler [21] have developed an experimental design of an engine mounted, capacitive sensor to monitor changes in the dielectric constant of the engine lubricant. The sensing element was a small, air-gap capacitor, mounted in a spacer ring that fit between the lubricant filter and the engine block. The sensor could monitor a change in dielectric constant, which increased with accumulated mileage, of the lubricant as it aged.

Turner and Austin reported a study in which changes to the dielectric properties of a lubricant was assessed as a method of measuring the degradation of engine lubricant. The relationship between lubricant use, measured by accumulated mileage, and lubricant viscosity was also measured. An interleaved, parallel plate capacitor (see Fig. 4) was constructed as the sensor. The dielectric properties (Dielectric constant $= C_{oil}/C_{air}$) of the lubricant correlated reasonably well with viscosity and this was proposed as the basis of a sensing technique [22].

Electrochemical Detection of Wear Particles

Failure of a lubricant often results in accelerated metal wear and the release of wear debris in the lubricant. Early detection of abnormal metal wear is important for fault detection and failure prevention. An electrochemical cell can be operated in a lubricating fluid in such a way that the operating characteristics of the cell can provide an indication of the chemistry of the fluid. For example, certain ions in the fluid, such as wear metal ions, will react to particular potential values applied to electrodes in the electrochemical cell. By applying a changing potential across the electrodes in an electrochemical cell and observing the resulting current, it is possible to detect and identify the ionic species present in the lubricating fluid [23].

Electrochemical Detection of Corrosion

A crucial property of lubrication oil is its corrosiveness as it directly relates to the corrosive wear of engine parts. With the increasing market share of biofuels, the corrosiveness of used engine oils increases in particular. Agoston et al. investigated the utilization of devices consisting of multiple copper thin films with different thicknesses on glass substrates. Upon immersion in the lubricating oil, corrosive material loss occurs on the surfaces of the thin films, which represented the progress of the corrosion. This loss was monitored electrically in terms of the film resistance. The sensor indicates both the acid and corrosive sulfur compound contents in oil. The surface treatment of thin films is crucial and was seen a potential source for measurement errors [24, 25]. Peratec also reported that an electrical resistive sensor that was used to determine the level of corrosivity of the fluid [26].

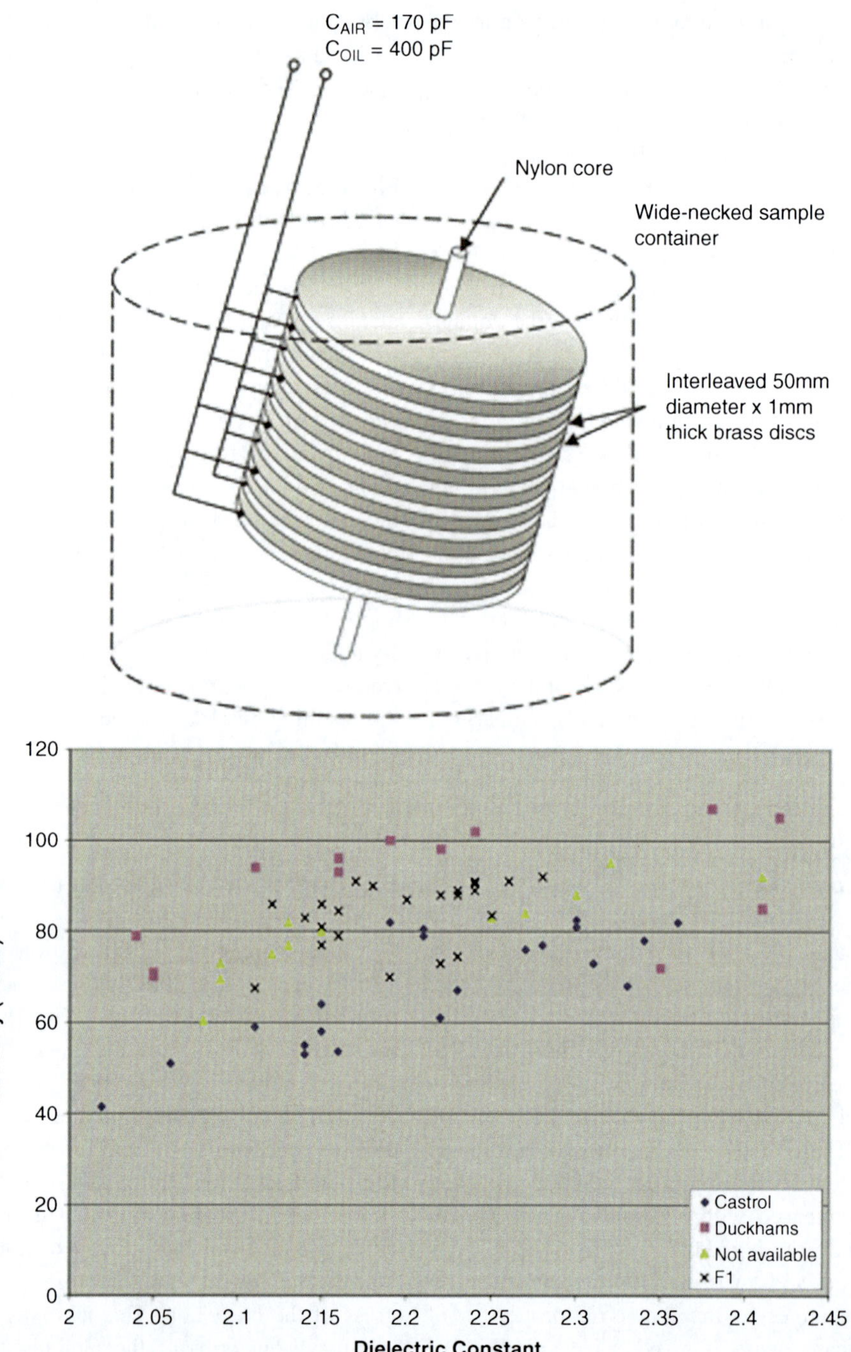

Electrochemical Sensors for Monitoring Conditions of Lubricants, Fig. 4 Interleaved parallel plate capacitor for measurement of dielectric constant and lubricating oil viscosity versus dielectric constant [22]

Future Directions

Corrosion is a multifaceted process in lubricants. In situ corrosion sensors are most practicably to be contained within a multi element sensor array. Different sensors in the array could monitor the oxidation of various metals the lubricant comes into contact with, the conductivity of the fluid, while other sensors could monitor the TAN/TBN of the fluid.

Cross-References

► Electrochemical Impedance Spectroscopy (EIS) Applications to Sensors and Diagnostics

References

1. Cipris D, Walsh A, Palanisamy T (1990) In: Turner DR (ed) Sensors for motor oil quality. The electrochemical society proceedings series, Pennington, PV 87–9:401
2. Wang SS, Lee HS, Smolenski DJ (1994) The development of in situ electrochemical oil-condition sensors. Sens Actuators B 17:179–185. doi:10.1016/0925-4005(93)00867-X
3. Wang SS, Lee HS (1997) The application of a.c. impedance technique for detecting glycol contamination in engine oil. Sens Actuators B 40:193–197. doi:10.1016/S0925-4005(97)80261-4
4. Smolenski DJ, Schwartz SE (1994) Automotive engine-lubricant condition monitoring. In: Booser ER (ed) CRC handbook of lubrication and tribology, vol 3., pp 17–32
5. Lvovich VF, Smiechowski MF (2006) Impedance characterization of industrial lubricants. Electrochim Acta 51:1487–1496. doi:10.1016/j.electacta.2005.02.135
6. Wang S (2001) Road tests of oil condition sensor and sensing technique. Sens Actuators B 73:106–111. doi:10.1016/S0925-4005(00)00660-2
7. Smiechowski M, Lvovich V (2003) Iridium oxide sensors for acidity and basicity detection in industrial lubricants. Sens Actuators B 96:261–267. doi:10.1016/S0925-4005(03)00542-2
8. Smiechowski M, Lvovich V (2002) Electrochemical monitoring of water-surfactant interactions in industrial lubricants. J Electroanal Chem 534:171–180. doi:10.1016/S0022-0728(02)01106-3
9. Smiechowski MF, Lvovich VF (2005) Characterization of non-aqueous dispersions of carbon black nanoparticles by electrochemical impedance spectroscopy. J Electroanal Chem 577:67–78. doi:10.1016/j.jelechem.2004.11.015
10. Lvovich VF (2012) Impedance spectroscopy: applications to electrocehmical and dielectric phenomena. Wiley, New Jersey
11. Jakoby B, Vellekoop MJ (2004) Physical sensors for water-in-oil emulsions. Sens Actuators A 110:28–32. doi:10.1016/j.sna.2003.08.005
12. Boyle FP, Lvovich VF, Humphrey BK, Goodlive SA (2004) On-board sensor systems to diagnose condition of diesel engine lubricants-focus on soot. SAE technical paper 2004-01-3010. doi: 10.4271/2004-01-3010
13. Ulrich C, Petersson H, Sundgren H, Björefors F, Krantz-Rülker C (2007) Simultaneous estimation of soot and diesel contamination in engine oil using electrochemical impedance spectroscopy. Sens Actuators B 127:613–618. doi:10.1016/j.snb.2007.05.014
14. Ulrich C et al (2012) Evaluation of industrial cutting fluids using electrochemical impedance spectroscopy and multivariate data analysis. Talanta 97:478–472. doi:10.1016/j.talanta.2012.05.001
15. Seyfert C, Kesseler A, Luther R, Meindorf T (2002) Sensors for on-line monitoring of bio-degradable hydraulic oils. Olhydraul Pneum 47:1–15
16. Duchowski JK, Mannebach H (2006) A novel approach to predictive maintenance: a portable, multi-component MEMS sensor for on-line monitoring of fluid condition in hydraulic and lubricating systems. Tribol Trans 49:545–553. doi:10.1080/10402000600885183
17. Day M, Bauer C (2007) Water contamination in hydraulic and lube systems. Practicing oil analysis. http://www.machinerylubrication.com/Read/1084/water-contamination-lube. Accessed 11 Dec 2012
18. Duchowski J (2006) Sensors eye fluids and lubricants. Mach Des 78:88–96
19. Lee HS, Wang S, Smolenski D, Viola M, Klusendorf E (1994) In situ monitoring of high-temperature degraded engine oil condition with microsensors. Sens Actuators B 20:49–54. doi:10.1016/0925-4005(93)01168-4
20. Wideraa J, Riehlb BL, Johnson JM, Hansenb DC (2008) State-of-the-art monitoring of fuel acidity. Sens Actuators B 130:871–881. doi:10.1016/j.snb.2007.10.056
21. Saloka GS, Meitzler AH (1991) A capacitive lubricant deterioration sensor. SAE paper #910497, proceedings of SAE international congress, Detroit, 25 Feb – March 1, pp 137–146
22. Turner JD, Austin L (2003) Electrical techniques for monitoring the condition of lubrication oil. Meas Sci Technol 14:1794–1800. doi:10.1088/0957-0233/14/10/308
23. Discenzo F et al (2007) Dissolved wear metal monitoring in lubricating fluids. In: ASME/STLE 2007 International joint tribology conference, San Diego, pp 989–991. doi:10.1115/IJTC2007-44102
24. Agoston A, Dörr N, Jakoby B (2007) Corrosion sensors for engine oils – laboratory evaluation and field tests. Sens Actuators B 127:15–21. doi:10.1016/j.snb.2007.07.041

25. Agoston A, Svasek E, Jakoby B (2005) Novel sensor monitoring corrosion effects of lubrication oil in an integrating manner. In: Proceedings of the IEEE sensors 2005, 0-7803-9057-1, pp 1120–1123
26. Peratec (1994) Automatic on-line fluid monitoring. Ind Lubr Tribol 46:8–9. doi:10.1108/00368799410781124

Electrochemical Sensors for Water Pollution and Quality Monitoring

Ying-Hui Lee and Chi-Chang Hu
Chemical Engineering Department, National Tsing Hua University, Hsinchu, Taiwan

Introduction

Several decades of industrialization have changed the environment drastically, leading to all sorts of pollution. Water pollution, being one of most important issues related to daily life, has always been addressed and monitored by various means of analytical tools. Different electrochemical sensors for the detection of pollutants in water have been well established, which can be categorized into the following: (i) potentiometric sensors, (ii) amperometric sensors, (iii) voltammetric sensors, and (iv) conductometric sensors. In this chapter, we will introduce the fundamentals, applications, advantages, limitations, and recent trends for the development of each type of sensors.

Potentiometric Sensors

The most representative potentiometric sensors are ion-selective electrodes (ISEs), as shown in Fig. 1. They are common in environmental analysis today and some typical ISEs are summarized in Table 1. The measurement of the ion of interest depends on the potential difference across the membrane between the sample solution and the inner reference electrolyte (Fig. 1 left, inner reference electrode). The membrane potential obtained from the difference between the ISE and the reference electrode (Fig. 1 right, external reference electrode) under no current flow where the equilibrium is reached can be used to determine the amount of ion of interest by Nernst equation.

The membrane electrodes should reach equilibrium rapidly, response only to the ion of interest, and change linearly with different concentrations of the ion. Different types of membranes have been developed to fabricate ISEs. The major ones are glass membranes, solid-state membranes, liquid membranes, and polymeric membranes [2].

Glass membranes are the most conventional membranes and well known for the determination of the activity of proton (i.e., pH value) [3]. They are chosen due to the high selectivity, low detection limit, long-term stability, and independence of redox interferences. The selectivity towards different cations of interest depends on the composition of the glass membranes. Hence, other cations such as alkali metal ions (e.g., Li^+, Na^+, K^+) and heavy metal ions (e.g., Cu^{2+}, Pb^{2+}, Cd^{2+}, Ag^+) can be measured besides proton. The representative glass membranes for heavy metal detection in industry are chalcogenide glass membranes. Different compositions of chalcogenide glass membranes have been developed and the miniaturization of the sensors is underway to combine with other microelectronics and microactuators for the waste water monitoring [4].

Solid-state membranes are mostly composed of insoluble inorganic salts. The most well-known membrane is the single crystal LaF_3 membrane for the detection of fluoride ion. The LaF_3 membrane, first introduced by Frant and Ross [5], is usually doped with europium to improve the ionic conductivity. The doping of other divalent cations is also reported to enhance the ion mobility and selectivity [6]. Slightly soluble salts, such as AgCl, AgBr, AgI, and Ag_2S, when dispersed in inert matrices (e.g., PVC, methacrylate, or epoxy), polycrystalline membranes which exhibit ionic conductivity are formed and suitable for the detection of Br^-, I^-, Cu^{2+}, Pd^{2+}, etc. [7–9]. Recently, solid contact has been utilized in the sensors incorporating the solid-state membrane. The absence of conventional internal electrolyte introduces the

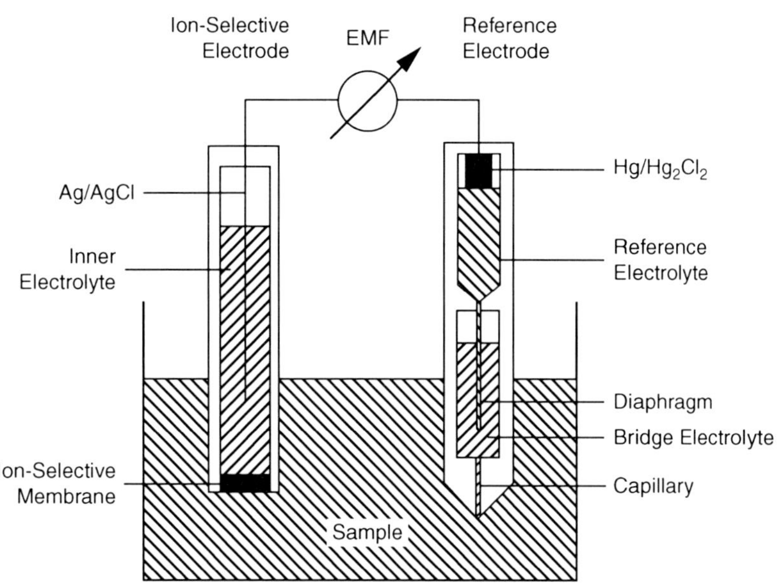

Electrochemical Sensors for Water Pollution and Quality Monitoring, Fig. 1 Typical potentiometric sensor setup incorporating an ion-selective membrane electrode [1] (Reprinted from Electroanalysis, 1999. **11**(13): p. 915–933 with permission from John Wiley & Sons)

Electrochemical Sensors for Water Pollution and Quality Monitoring, Table 1 Examples of applications of ISEs for environmental analysis [17] (Reprinted from Electroanalysis, 2007. **19**(19–20), p. 1987–2001 with permission from John Wiley & Sons)

ISE	Environmental application
Cu ISE	Analysis of Cu^{2+} in natural waters
Cu ISE	Copper complexation capacity in natural waters and soil peats
Fe ISE	Analysis of Fe^{3+} in natural waters
Cd ISE	Analysis of Cd^{2+} in natural waters
Hg ISE	Analysis of Hg^{2+} in natural waters
F ISE	Analysis of F^- in nature waters
Pb ISE	Analysis of Pb^{2+} in natural waters and culture media
Cr ISE	Analysis of $HCrO_4^-$ in natural waters
CN ISE	Analysis of CN^- in natural waters
UO_2^{2+} ISE	Analysis of UO_2^{2+} in tap water and seawater
Anion, cation, and gas ISEs	Analysis of CN^-, NO_2^-, F^-, NH_4^+, and NH_3 in natural waters
$H_2PO_4^-$ and Sm^{3+}ISE	Analysis of PO_4^{3-} directly or indirectly by titration in natural waters
Mg ISE	Analysis of Mg^{2+} in seawater
Cu ISE	Determination of water hardness
TFP ISE	Back titration of excess 2-aminoperimindium ion for SO_4^{2-} in seawater
Anion ISEs	Analysis of corrosive anions (e.g., CI^-, NO_3^-, and SO_4^{2-}) on electronic equipment or in reinforced concrete structures

robustness (e.g., sustainability under high temperature and pressure) and the possibility for miniaturization of the sensors [9, 10].

Liquid membranes are formed by dissolving an ionophore in a viscous organic liquid membrane. The ionophores, either synthetically or naturally obtained, are neutral or charged macrocyclic ion carriers which help transport the ions of interest into the membranes [11]. The most prominent example is the Ca^{2+}-selective electrodes which combine different ionophores into lipophilic organophosphoric acid as the membranes [12, 13]. Similar to liquid membranes, the ionophores can also be incorporated into polymers such as PVC to make polymeric membranes [1]. The polymeric membrane has replaced many liquid membranes due to their high selectivity. To further improve the ion-to-electron transduction between the membrane and the electronic conductor, conducting polymers have been integrated to the membrane as the solid contact (ion-to-electron transducer) to replace the conventional liquid internal electrolyte [14, 15]. Another approach is to directly utilize the conducting polymers as the membranes by introducing ion-recognition sites into the polymer matrix, allowing the ion-recognition sites and ion-to-electron transducer within the same material [16].

The ISEs are simple and inexpensive instruments which are highly desirable for environmental analysis. However, challenges appear when employing ISEs for the analysis of complex environmental samples (e.g., lakes, rivers, seawater), which are electrode carryover, electrode drift, electrode dissolution, electrode fouling, and electrode passivation [17]. Electrode carryover can be easily overcome by cleansing the electrode with a low concentration of the analyte solution. Electrode drift and dissolution can be ameliorated by preconditioning the ISE in the analyte in advance to produce the stable, reproducible, and fast response of the ISE. Frequent calibration and rejuvenation of the ISE is required for electrode fouling and passivation. The problems can be alleviated by utilizing the ISE in flow analysis (FIA or CFA) where the hydrodynamic flow helps remove the adsorbed foulants.

Amperometric/Coulometric Sensors

The working principle for amperometric sensors and coulometric sensors is the same. A fixed potential (the value on the plateau region from the voltammogram) is applied and the current passing through the cell is recorded as a function of time. The current measured can be related to the concentration of the analyte. The only difference between amperometric and coulometric sensors is that the fraction of electroactive species electrolyzed is usually much less than unity for the former, whereas it is unity for the latter.

A prototype amperometric sensor is the Clark-type electrode (composed by Pt electrodes) in which a gas permeable membrane or porous barrier separates the measuring cathode from the O_2-containing gas phase. This type of sensor is commercially available and has been designed for the detection of CO [18], NO [19], H_2S [20], etc. Similar sensors for the detection of dissolved gas in aqueous solution have also been developed. For example, H_2 is often added to the high-temperature water to prevent corrosion of metals during the operation of power plants and the monitoring of the concentration is very important. Thus, amperometric sensor which utilizing a proton-conducting solid electrolyte was fabricated for the first time to monitor the content of H_2 in situ in the effluent [21]. Clark-type microelectrodes have also been constructed to detect dissolved O_2 in sedimentary pore water in situ [22] and more recently for the oceanographic analysis with very low detection limit of 0.9 μM and low concentration relative error [23]. The detection of analytes other than gas in the solution was also widely reported. For example, the determination of H_2O_2 in a swimming pool has been achieved by an ultramicroelectrode array. With the coating of Nafion®, the interference from other species can be reduced and further enhances the sensitivity of the sensor [24]. Even in the electrolyte-free liquid sample, the measurement of H_2O_2, hydroquinone (H_2Q), Cl^-, and Br^-, are accomplished by Pt electrode encased in Pyrex tube and with polymer sealing at the bottom [25]. Since the detection of biological and organic pollutants in drinking and industrial water is highly concerned, sensors for continuous monitoring of biocides such as hypochlorous acid/hypochlorite were developed [26]. Pt oxide formed during the measurement can be removed by cathodic cycling and therefore extends the life time of Pt electrode.

In addition, the mixtures of noble metals, e.g., Ru and Pt as the electrodes, enable broader sensitivity for various organic pollutants such as formaldehyde [27].

The metal electrodes further modified by polymers have also shown good sensitivity for the pollutants, e.g., Cl^- and choline, in drinking water [28, 29]. The monitoring for hazardous chemicals like bisphenol A showed promising results [30]. The measurement of pharmaceutical contaminants in water such as acetaminophen has been achieved by poly(L-serine) film-modified electrode; more importantly, the chemical can be degraded during the detection by increasing the current and pH value of the solution [31]. Similar detection and degradation of acetaminophen can be accomplished by utilizing a biomimetic catalyst to modify the electrode [32]. The coupling of flow injection analysis system (FIA) further enabled real-time measurement of the chemical. Besides polymers, other smaller organic molecules were employed to modify the electrode, such as Prussian Blue (PB) for the detection of H_2O_2 [33] and the mixture of chitosan and ionic liquid for the measurement of phosphate in the fertilizer [34]. The highly toxic chemicals, such as phenylhydrazine and hydrazine, can also be simultaneously detected by the electrode modified by the paste of ferrocene and CNTs [35].

Due to the difficulty of detecting increasing number of pollutants in the environment, biosensors, which immobilized with enzymes, nonenzymes, antibodies, and even whole cells, have been fabricated to monitor heavy metal ions and various organic pollutants in water [36–38]. Due to the characteristics of proteins, the sensitivity for this type of sensor is strongly dependent on temperature, pH, and ionic strength of the solution. Successful heavy metal ion monitoring utilizing either enzymes or microorganisms was reported [36]. For example, an enzyme-based sensor was developed for the detection of highly toxic CrO_4^- in ground water with little interference from other organic molecules and heavy metal ions [39]. Interestingly, biosensors without the immobilization of enzymes on the electrode, i.e., free enzymes dispersed in the solution, have also been demonstrated. Compared to conventional analytical methods, the enzyme immobilization-free sensors exhibit very low detection limit for Hg^{2+} (0.1 ppb), Cd^{2+} (0.6 ppb), and As^{3+} (1.5 ppb) [40]. Organic pollutants, such as sulfide, phenols, herbicides, and pesticides, are noxious to human health, leading to the fabrication of numerous enzyme-based biosensors. Biosensors employing different types of enzymes for the detection of phenols (common waste from industrial processes) have shown promising sensitivity [41–44]. The sensors are possible to perform field monitoring [44] and the stability may last as long as a year [43]. The measurements of organophosphate, chlorsulfuron, atrazine, carbamate, etc. in herbicides and pesticides have been widely reported by different enzyme-based biosensors [45–51]. The biosensors can be employed for on-site remote field monitoring [45, 52, 53] which cannot be achieved by conventional gas or liquid chromatography. The detection limit for organophosphorus pesticide can reach as low as 0.145 ppb [45]. Examples of biosensors for water monitoring are summarized in Table 2.

Biosensors demonstrate the specificity, reliability, portability, simplicity, and the advantage of real-time analysis [36]. However, the adhesion, mechanical stability, and reproducibility of the protein or whole cell immobilization on electrodes should be considered. One of the solutions is to confine protein layer in an anisotropically etched silicon groove [54]. Also for the sensors with lower detection concentration range, a diffusion-membrane such as polyurethane and silicone can be utilized to extend the detectable range. Interference is another problem for biosensors. To minimize interference, two approaches have been proposed. The first one would be the utilization of perm-selective membrane, e.g., Nafion® and cellulose acetate. The second choice is to fabricate a pair of working electrodes. One immobilized with active protein and the other immobilized with deactivated protein. The signal of the working electrode is obtained by subtracting the signal from the deactivated electrode to ensure the high sensitivity of the sensor [55].

Electrochemical Sensors for Water Pollution and Quality Monitoring, Table 2 Examples of biosensors for water monitoring

Detection type	Biomaterials immobilized on electrode	Detection target	Detection limit	Ref
Heavy metal	Pyruvate oxidase	Hg^{2+}	50.0 nM	[36]
	Escherichia coli	Cu^{2+}	1.00 μM	[36]
	Cytochrome c3	Cr^{6+}	3.85 μM	[39]
	Urease (immobilization-free)	Hg^{2+}	0.10 ppb	[40]
		Cd^{2+}	0.60 ppb	
		As^{3+}	1.50 ppb	
Organics	Poly(allylamine hydrochloride)-wrapped MWNTs/ horseradish peroxide	Catechol	0.06 μM	[41]
	Pyrrole amphiphilic monomer-tyrosinase	Phenol	10.0 nM	[43]
	Acetylcholinesterase functionalized MWNTs	Paraoxon	0.50 nM	[45]
	Horseradish peroxidase on anti-chlorsulfuron antibodies	Chlorsulfuron	0.03 nM	[52]
	Chlorella vulgaris (microalga)	Atrazine	12.0 μM	[51]
		3-(3,4-dichlorophenyl)-1,1-diethylurea	1.00 μM	

Voltammetric Sensors

A voltammetric sensor measures the current response as a function of the potential applied to the electrochemical cell. Two steps are involved for the measurement, the preconcentration and the stripping of the ions of interest, which lead to the enhancement of sensitivity and selectivity of the sensor. The ions of interest are first collected onto or in a working electrode for preconcentration; during the stripping the collected ions will be oxidized or reduced back to solution. This type of sensor demonstrates extremely low detection limits (10^{-10}–10^{-12} M), which is suitable for monitoring ultralow levels of trace metals in natural water. Other advantages include multielements measurement, trace element speciation and distribution, and continuous in situ detection. The sensors for two voltammetric stripping techniques are discussed below: anodic stripping voltammetry (ASC) and adsorptive cathodic stripping voltammetry (AdCSV).

Anodic stripping voltammetry is a well-established and widely used technique. Common electrodes employed for the technique are hanging mercury drop electrode (HMDE) and mercury film electrode (MFE). HMDE demonstrates high reliability for each test due to the formation of each new drop (i.e., new electrode surface), whereas MFE exhibits robustness and shows high selectivity owing to the high surface area to volume ratio. During the preconcentration step, the deposition potential of 0.3–0.4 V more negative than the reduction potential of the metal ion of interest is applied. Then the reverse voltammetric scan towards more positive potential results in the oxidation of the metal in mercury. The recorded oxidation peak potential indicates the metal analyzed and the peak intensity is proportional to the concentration of the metal ion. The mechanism of ASV for mercury electrodes is shown in Eq. 1 [56]:

$$M^{n+} + ne^{-} + Hg \leftrightarrow M(Hg) \quad (1)$$

Mercury electrodes have been reported for the detection of various metal ions. The sensitivity of MFE is better than HMDE because smaller volume of MFE leads to greater amount of the metal collected into mercury. With the microelectrode design covered with an antifouling agarose membrane, MFE is able to achieve excellent detection limit for Pb^{2+}, Cd^{2+}, and Cu^{2+} in seawater [57]. Due to the environmental issues faced by mercury electrodes, other material-based electrodes were also demonstrated. Carbon paste electrode is able to detect Ag^+ with extremely low detection limit (2.5 pM) and almost free from the interference of 20 metal species [58]. Gold electrodes which exhibit comparable sensing ability to that of mercury electrodes were fabricated for the detection of heavy metals such as Pb^{2+}, As^{3+}, and Hg^{2+} [59–62]. With the ultramicroelectrode design, the on-site field analysis and long-term (1 month) stability for the detection of As^{3+} in ground water was accomplished [60]. Gold nanoparticles can also be homogeneously dispersed in 3D matrix (e.g., organic fibers) and 2D sheets (e.g., graphene) to enhance the sensing performance for Hg^{2+} in tap and river water. These sensors showed the detection limit much below the guideline value of World Health Organization (WHO), i.e., 1 ppb for Hg^{2+}, and effectively inhibited the interference from Cu^{2+}, Cr^{3+}, Co^{2+}, Fe^{2+}, and Zn^{2+} [61, 62]. Solid amalgam electrodes, different from toxic liquid mercury electrodes, were also utilized as the electrodes for the measurement of several heavy metals such as Fe^{2+}, Zn^{2+}, Pb^{2+}, and Cu^{2+} in seawater or costal water [63–65]. Heavy metals such as Fe have high reduction overpotential on liquid mercury due to the insolubility in mercury, which results in the overlap of reduction signals from both the metals and hydrogen. However, the solid amalgam electrodes show higher overpotential for hydrogen evolution and further separate the reduction signals of heavy metals from hydrogen, leading to the enhancement of sensitivity [63].

Adsorptive cathodic stripping voltammetry is very sensitive for numerous trace metal analyses. Specific added ligand (AL) is needed to form an adsorptive complex (adsorbed on the working electrode) with the metal of interest, i.e., preconcentration step. Effective preconcentration with short adsorptive times results in fast and extremely sensitive measurement. Since the formation of the complex is pH-dependent, the pH of the solution is usually controlled by adding buffer solution. The adsorption potential applied is slightly more positive (ca. 0.1 V or more) than the reduction potential of the metal-AL complex. The reactions are shown in Eqs. 2 and 3 [56]. Then reverse voltammetric scan towards more negative potential leads to the reduction of the metal from the complex (Eq. 4 [56]). Accordingly, the type and the amount of the metal can be determined from the reduction peak potential and the peak intensity, respectively:

$$yM^{n+} + zAL^{m-} \leftrightarrow M_y(AL)_z^{(yn-zm)} \quad (2)$$

$$M_y(AL)_z^{(yn-zm)} \leftrightarrow M_y(AL)_z^{(yn-zm)}(\text{adsorbed}) \quad (3)$$

$$M_y(AL)_z^{(yn-zm)}(\text{adsorbed}) + e^- \leftrightarrow yM^{(n-1)} + zAL^{m-} \quad (4)$$

Both HMDE and MFE are commonly applied for AdCSV [57, 66–68], especially HMDE. With the adsorptive time of only 30 s using suitable added ligand, extremely low detection limit for heavy metals can be obtained, e.g., 5.4 pM for Co^{2+} [67] and 2.5 pM for Cr^{6+} [68]. The interference from organics, ionic, and nonionic surfactants in water can be further minimized by the addition of Amberlite XAD-7 resin [68]. The measurement of Co^{2+}, Ni^{2+}, and Cr^{6+} has also been achieved by bismuth film electrodes [69, 70]. The bismuth film electrodes are promising for on-site monitoring with comparable sensing performance (~ nM level) to that of mercury electrodes. The interference-free detection of U^{6+} was accomplished using bismuth film electrode [71]. Four times lower detection limit for U^{6+} can be obtained by utilizing bismuth-coated carbon fiber electrode with larger surface area [72]. In addition to the detection for heavy metals, bismuth electrodes demonstrate sensing ability to neonicotinoid insecticide such as imidacloprid and acetamiprid. But the pH of the solution should be optimized for the best sensitivity, i.e., pH 8.0 for imidacloprid and pH 3.0 for acetamiprid [73]. The examples of both ASV and AdCSV for water monitoring are summarized in Table 3.

Conductometric Sensors

A conductometric sensor determines the amount of ions or molecules of interest through monitoring

Electrochemical Sensors for Water Pollution and Quality Monitoring, Table 3 Examples of voltammetric sensors for water monitoring

Technique	Electrode	Detection target	Detection limit	Ref
ASV	MFE microelectrode	Pb^{2+}	0.05 nM	[57]
		Cd^{2+}	0.05 nM	
		Cu^{2+}	0.02 nM	
	Carbon paste electrode	Ag^{+}	2.50 pM	[58]
	Au electrode	Pb^{2+}	0.39 nM	[59]
	Au electrode	As^{3+}	0.10 ppb	[60]
	Au-Pt electrode	Hg^{2+}	8.00 ppt	[61]
	Au on graphene electrode	Hg^{2+}	6.00 ppt	[62]
	Amalgam electrode	Fe^{2+}	0.90 nM	[63]
	Amalgam electrode	Fe^{2+}	0.90 nM	[64]
		Zn^{2+}	1.53 nM	
		Pb^{2+}	0.48 nM	
		Cu^{2+}	1.57 nM	
AdCSV	HMDE	Co^{2+}	5.40 pM	[67]
	HMDE	Cr^{6+}	0.13 nM	[68]
	Bismuth film electrode	Co^{2+}	1.36 nM	[69]
		Ni^{2+}	4.43 nM	
	Bismuth film electrode	Cr^{6+}	0.30 nM	[70]
	Bismuth-coated carbon fiber electrode	U^{6+}	1.26 nM	[72]
	Bismuth film electrode	Imidacloprid	2.86 μM	[73]
		Acetamiprid	3.95 μM	

the electrical conductivity change of the selective layer on the electrode. The fabrication for conductometric sensor is simple and inexpensive, suitable for miniaturization and mass production. However, this type of sensor is least selective among different electrochemical techniques since several resistances need to be considered during the measurement (electrical resistance, the inverse of conductivity, is what actually being measured), e.g., contact resistance between the electrodes and the selective layer, the bulk resistance, the surface resistance of the selective layer. Any one or any combination of the above can be influenced by the sensing interaction between the selective layer and the analyte, which results in the difficulty of distinguishing different ions or molecules of interest. Hence, highly selective recognition elements should be incorporated into the selective layer to enhance the performance of the sensor. For example, similar to potentiometric sensors, polymeric membranes containing ionophores have been employed as the selective membrane in conductometric sensors to detect K^+, Ca^{2+}, NH_4^+, and Li^+. The sensors showed the sensitivity around μM level with fast response time (<10 s) and long-term stability (several weeks) [74]. Conductometric biosensors with the immobilization of enzymes or whole cells targeted for toxic organics and heavy metals have also been demonstrated. Through the inhibition of enzyme activity, a series of organophosphorous pesticides can be measured with the detection limit ranging from μM to nM level [75]. And the employment of microalgae immobilization was able to achieve the detection limit of 1 ppb for Cd^{2+} since the self-assembled monolayer (SAMs) linked between microalgae and the electrode provided sufficient bonding of microalgae and eliminated diffusion barrier for the analyte compared to conventional physical trapping of microalgae with gelling agents [76]. Considering the instability and sensitivity of biomolecules to physical and chemical influences, synthetic polymers mimicking biological receptors were developed. The molecular imprinted polymers (MIPs) were fabricated based on a template molecule (i.e., the molecule of interest). After removing the template, the functional groups capable of binding to the template molecule are retained, which constitutes the high selectivity of MIPs for the target molecule. Organic pollutants such as atrazine and haloacetic acids have been successfully determined in drinking water using MIPs [77, 78].

Typical electrochemical sensors for water pollution and quality monitoring are reviewed in this chapter. The well-established potentiometric sensors (ISEs) are simple and inexpensive. The employment of solid contact introduces the robustness of the sensors and the feasibility for miniaturization. However, the stability and the reproducibility of the electrodes should be monitored constantly after long-term measurement in complex environmental samples. Amperometric biosensors show high sensitivity and selectivity for various heavy metal ions and toxic organic pollutants in water. But the binding and the stability of the biomaterials are easily influenced by the environment; optimal operation condition is required to ensure the long-term sensing performance. Voltammetric sensors are promising for obtaining extremely low detection limit of pollutants, especially heavy metal ions, due to the preconcentration step. Several materials have been utilized to replace the common yet toxic mercury electrodes with comparable sensing abilities. Although the conductometric sensors demonstrate the advantages for miniaturization and mass production, the sensibility needs to be improved by the incorporation of more selective elements. Studies of these electrochemical sensors for more precise on-site and real-time field monitoring are still in progress to ensure the water quality both in the industry and the environment.

Cross-References

▶ Electrochemical Sensors for Environmental Analysis

References

1. Bakker E, Bühlmann P, Pretsch E (1999) Polymer membrane ion-selective electrodes – What are the limits. Electroanalysis 11:915
2. Rajeshwar K, Ibanez JG, Swain GM (1994) Electrochemistry and the environment. J Appl Electrochem 24:1077
3. Spitzer P, Pratt KW (2010) The history and development of a rigorous metrological basis for pH measurements. J Solid State Electrochem 15:69
4. Schöning MJ, Kloock JP (2007) About 20 years of silicon-based thin-film sensors with chalcogenide glass materials for heavy metal analysis: Technological aspects of fabrication and miniaturization. Electroanalysis 19:2029
5. Frant MS, Ross JW (1966) Electrode for sensing fluoride ion activity in solution. Science 154:1553
6. Wang XD, Shen W, Cattrall RW, Nyberg GL, Liesegang J (1996) The effects of doping on the selectivity and response time of fluoride ion-selective electrodes. Aust J Chem 49:897
7. Mascini M, Liberti A (1971) Preparation, analytical evaluation and applications of a new heterogeneous membrane electrode for copper(II). Anal Chim Acta 53:202
8. Mascini M, Liberti A (1972) Preparation and analytical evaluation of a new lead(II) heterogeneous membrane electrode. Anal Chim Acta 60:405
9. Tan J, Bergantin JH, Merkoçi A, Alegret S, Sevilla F (2004) Oil dispersion of AgI/Ag2S salts as a new electroactive material for potentiometric sensing of iodide and cyanide. Sens Actuators B Chem 101:57
10. Bralic M, Radic N, Brinic S, Generalic E (2001) Fluoride electrode with LaF3-membrane and simple disjoining solid-state internal contact. Talanta 55:581
11. Bakker E, Buhlmann P, Pretsch E (1997) Carrier-based ion-selective electrodes and bulk optodes. 1. General characteristics. Chem Rev 97:3083
12. Nair SG, Hwang ST (1991) Selective transport of calcium-ion from a mixed cation solution through an hdehp N-dodecane supported liquid membrane. J Membr Sci 64:69
13. Ross JW (1967) Calcium-selective electrode with liquid ion exchanger. Science 156:1378
14. Bobacka J (2006) Conducting polymer-based solid-state ion-selective electrodes. Electroanalysis 18:7
15. Rahman MA, Kumar P, Park DS, Shim YB (2008) Electrochemical sensors based on organic conjugated polymers. Sensors 8:118
16. Bobacka J, Ivaska A, Lewenstam A (2003) Potentiometric ion sensors based on conducting polymers. Electroanalysis 15:366
17. De Marco R, Clarke G, Pejcic B (2007) Ion-selective electrode potentiometry in environmental analysis. Electroanalysis 19:1987

18. Kuchnicki TC, Campbell NER (1983) Amperometric response to carbon-monoxide by a clark-type oxygen-electrode. Anal Biochem 131:34
19. Liu X, Liu Q, Gupta E, Zorko N, Brownlee E, Zweier JL (2005) Quantitative measurements of NO reaction kinetics with a Clark-type electrode. Nitric Oxide 13:68
20. Horn JJ, McCreedy T, Wadhawan J (2010) Amperometric measurement of gaseous hydrogen sulfide via a Clark-type approach. Anal Methods 2:1346
21. Kriksunov LB, Macdonald DD (1996) Amperometric hydrogen sensor for high-temperature water. Sens Actuator B Chem 32:57
22. Luther GW, Reimers CE, Nuzzio DB, Lovalvo D (1999) In situ deployment of voltammetric, potentiometric. and amperometric microelectrodes from a ROV to determine dissolved O-2, Mn, Fe, S(-2), and pH in porewaters. Environ Sci Technol 33:4352
23. Sosna M, Denuault G, Pascal RW, Prien RD, Mowlem M (2007) Development of a reliable microelectrode dissolved oxygen sensor. Sens Actuators B Chem 123:344
24. Schwake A, Ross B, Cammann K (1998) Chrono amperometric determination of hydrogen peroxide in swimming pool water using an ultramicroelectrode array. Sens Actuators B Chem 46:242
25. Toniolo R, Comisso N, Bontempelli G, Schiavon G, Sitran S (1998) A novel assembly for perfluorinated ion-exchange membrane-based sensors designed for electroanalytical measurements in nonconducting media. Electroanalysis 10:942
26. Ordeig O, Mas R, Gonzalo J, Del Campo FJ, Muñoz FJ, de Haro C (2005) Continuous detection of hypochlorous acid/hypochlorite for water quality monitoring and control. Electroanalysis 17:1641
27. Sun W, Sun G, Qin B, Xin Q (2007) A fuel-cell-type sensor for detection of formaldehyde in aqueous solution. Sens Actuators B Chem 128:193
28. Lee HJ, Beattie PD, Seddon BJ, Osborne MD, Girault HH (1997) Amperometric ion sensors based on laser-patterned composite polymer membranes. J Electroanal Chem 440:73
29. Mehta A, Shekhar H, Hyun SH, Hong S, Cho HJ (2006) A micromachined electrochemical sensor for free chlorine monitoring in drinking water. Water Sci Technol 53:403
30. Huang JD, Zhang XM, Liu S, Lin Q, He XR, Xing XR, Lian WJ (2011) Electrochemical sensor for bisphenol A detection based on molecularly imprinted polymers and gold nanoparticles. J Appl Electrochem 41:1323
31. Chen TS, Huang KL (2012) Electrochemical detection and degradation of acetaminophen in aqueous solutions. Int J Electrochem Sc 7:6877
32. de Oliveira MCQ, Tanaka AA, los Lanza MRD, Sotomayor MDT (2011) Studies of the electrochemical degradation of acetaminophen using a real-time biomimetic sensor. Electroanalysis 23:2616
33. de Mattos IL, Gorton L, Ruzgas T (2003) Sensor and biosensor based on Prussian Blue modified gold and platinum screen printed electrodes. Biosens Bioelectron 18:193
34. Berchmans S, Karthikeyan R, Gupta S, Poinern GEJ, Issa TB, Singh P (2011) Glassy carbon electrode modified with hybrid films containing inorganic molybdate anions trapped in organic matrices of chitosan and ionic liquid for the amperometric sensing of phosphate at neutral pH. Sens Actuators B Chem 160:1224
35. Afzali D, Karimi-Maleh H, Khalilzadeh MA (2011) Sensitive and selective determination of phenylhydrazine in the presence of hydrazine at a ferrocene-modified carbon nanotube paste electrode. Environ Chem Lett 9:375
36. Verma N, Singh M (2005) Biosensors for heavy metals. Biometals 18:121
37. Farre M, Kantiani L, Perez S, Barcelo D (2009) Sensors and biosensors in support of EU Directives. Trac-Trend Anal Chem 28:170
38. Kroger S, Piletsky S, Turner APF (2002) Biosensors for marine pollution research, monitoring and control. Mar Pollut Bull 45:24
39. Michel C, Ouerd A, Battaglia-Brunet F, Guigues N, Grasa JP, Bruschi M, Ignatiadis I (2006) Cr(VI) quantification using an amperometric enzyme-based sensor: Interference and physical and chemical factors controlling the biosensor response in ground waters. Biosens Bioelectron 22:285
40. Pal P, Bhattacharyay D, Mukhopadhyay A, Sarkar P (2009) The detection of Mercury, Cadium, and Arsenic by the deactivation of Urease on Rhodinized Carbon. Environ Eng Sci 26:25
41. Liu LJ, Zhang F, Xi FN, Lin XF (2008) Highly sensitive biosensor based on bionanomultilayer with water-soluble multiwall carbon nanotubes for determination of phenolics. Biosens Bioelectron 24:306
42. Solna R, Skladal P (2005) Amperometric flow-injection determination of phenolic compounds using a biosensor with immobilized laccase, peroxidase and tyrosinase. Electroanalysis 17:2137
43. Cosnier S, Fombon JJ, Labbe P, Limosin D (1999) Development of a PPO-poly(amphiphilic pyrrole) electrode for on site monitoring of phenol in aqueous effluents. Sens Actuators B Chem 59:134
44. Svitel J, Miertus S (1998) Development of tyrosinase based biosensor and its application for monitoring of bioremediation of phenol and phenolic compounds. Environ Sci Technol 32:828
45. Joshi KA, Tang J, Haddon R, Wang J, Chen W, Mulchandani A (2005) A disposable biosensor for organophosphorus nerve agents based on carbon nanotubes modified thick film strip electrode. Electroanalysis 17:54
46. Marty JL, Mionetto N, Noguer T, Ortega F, Roux C (1993) Enzyme sensors for the detection of pesticides. Biosens Bioelectron 8:273

47. Bernabei M, Chiavarini S, Cremisini C, Palleschi G (1993) Anticholinesterase activity measurement by a choline biosensor – application in water analysis. Biosens Bioelectron 8:265
48. Neufeld T, Eshkenazi I, Cohen E, Rishpon J (2000) A micro flow injection electrochemical biosensor for organophosphorus pesticides. Biosens Bioelectron 15:323
49. Campanella L, Cubadda F, Sammartino MP, Saoncella A (2001) An algal biosensor for the monitoring of water toxicity in estuarine environments. Water Res 35:69
50. Ghosh D, Dutta K, Bhattacharyay D, Sarkar P (2006) Amperometric detection of pesticides using polymer electrodes. Environ Monit Assess 119:481
51. Shitanda I, Takamatsu S, Watanabe K, Itagaki M (2009) Amperometric screen-printed algal biosensor with flow injection analysis system for detection of environmental toxic compounds. Electrochim Acta 54:4933
52. Dzantiev BB, Yazynina EV, Zherdev AV, Plekhanova YV, Reshetilov AN, Chang SC, McNeil CJ (2004) Determination of the herbicide chlorsulfuron by amperometric sensor based on separation-free bienzyme immunoassay. Sens Actuators B Chem 98:254
53. Wang J, Chen L, Mulchandani A, Mulchandani P, Chen W (1999) Remote biosensor for in-situ monitoring of organophosphate nerve agents. Electroanalysis 11:866
54. Rohm I, Kunnecke W, Bilitewski U (1995) UV-Polymerizable screen-printed enzyme pastes. Anal Chem 67:2304
55. Suzuki H (2000) Advances in the microfabrication of electrochemical sensors and systems. Electroanalysis 12:703
56. Achterberg EP, Braungardt C (1999) Stripping voltammetry for the determination of trace metal speciation and in-situ measurements of trace metal distributions in marine waters. Anal Chim Acta 400:381
57. Tercier-Waeber ML, Belmont-Hebert C, Buffle J, Graziottin F, Fiaccabrino GC, Koudelka-Hep M (1998) Real-time continuous Mn(II) monitoring in lakes using a novel voltammetric in situ profiling system. Oceans'98 – Conf Proc 1–3:956
58. Svancara I, Kalcher K, Diewald W, Vytras K (1996) Voltammetric determination of silver at ultratrace levels using a carbon paste electrode with improved surface characteristics. Electroanalysis 8:336
59. Richter EM, Pedrotti JJ, Angnes L (2003) Square-wave quantification of lead in rainwater with disposable gold electrodes without removal of dissolved oxygen. Electroanalysis 15:1871
60. Feeney R, Kounaves SP (2002) Voltammetric measurement of arsenic in natural waters. Talanta 58:23
61. Gong J, Zhou T, Song D, Zhang L, Hu X (2010) Stripping Voltammetric Detection of Mercury(II) Based on a Bimetallic Au-Pt Inorganic-Organic Hybrid Nanocomposite Modified Glassy Carbon Electrode. Anal Chem 82:567
62. Gong JM, Zhou T, Song DD, Zhang LZ (2010) Monodispersed Au nanoparticles decorated graphene as an enhanced sensing platform for ultrasensitive stripping voltammetric detection of mercury(II). Sens Actuators B Chem 150:491
63. Mikkelsen O, Schroder KH (2004) Voltammetric monitoring of bivalent iron in waters and effluents, using a dental amalgam sensor electrode. Some preliminary results. Electroanalysis 16:386
64. Mikkelsen O, Skogvold SM, Schroder KH (2005) Continouos heavy metal monitoring system for application in river and seawater. Electroanalysis 17:431
65. Mikkelsen O, Skogvold SM, Schroder KH, Gjerde MI, Aarhaug TA (2003) Evaluations of solid electrodes for use in voltammetric monitoring of heavy metals in samples from metallurgical nickel industry. Anal Bioanal Chem 377:322
66. Wang J, Setiadji R (1992) Selective determination of trace uranium by stripping voltammetry following adsorptive accumulation of the uranium cupferron complex. Anal Chim Acta 264:205
67. Colombo C, van den Berg CMG, Daniel A (1997) A flow cell for on-line monitoring of metals in natural waters by voltammetry with a mercury drop electrode. Anal Chim Acta 346:101
68. Grabarczyk M (2008) A catalytic adsorptive stripping voltammetric procedure for trace determination of Cr (VI) in natural samples containing high concentrations of humic substances. Anal Bioanal Chem 390:979
69. Hutton EA, Hocevar SB, Ogorevc B, Smyth MR (2003) Bismuth film electrode for simultaneous adsorptive stripping analysis of trace cobalt and nickel using constant current chronopotentiometric and voltammetric protocol. Electrochem Commun 5:765
70. Lin L, Lawrence NS, Thongngamdee S, Wang J, Lin YH (2005) Catalytic adsorptive stripping determination of trace chromium(VI) at the bismuth film electrode. Talanta 65:144
71. Kefala G, Economou A, Voulgaropoulos A (2006) Adsorptive stripping voltammetric determination of trace uranium with a bismuth-film electrode based on the U(VI) -> U(V) reduction step of the uranium-cupferron complex. Electroanalysis 18:223
72. Lin L, Thongngamdee S, Wang J, Lin YH, Sadik OA, Ly SY (2005) Adsorptive stripping voltammetric measurements of trace uranium at the bismuth film electrode. Anal Chim Acta 535:9
73. Guzsvany V, Kadar M, Papp Z, Bjelica L, Gaal F, Toth K (2008) Monitoring of photocatalytic degradation of selected neonicotinoid insecticides by cathodic voltammetry with a bismuth film electrode. Electroanalysis 20:291
74. Cammann K, Ahlers B, Henn D, Dumschat C, Shulga AA (1996) New sensing principles for ion detection. Sens Actuators B Chem 35:26

75. Dzyadevych SV, Soldatkin AP, Arkhypova VN, El'skaya AV, Chovelon JM, Georgiou CA, Martelet C, Jaffrezic-Renault N (2005) Early-warning electrochemical biosensor system for environmental monitoring based on enzyme inhibition. Sens Actuators B Chem 105:81
76. Guedri H, Durrieu C (2008) A self-assembled monolayers based conductometric algal whole cell biosensor for water monitoring. Microchim Acta 163:179
77. Sergeyeva TA, Piletsky SA, Brovko AA, Slinchenko EA, Sergeeva LM, El'skaya AV (1999) Selective recognition of atrazine by molecularly imprinted polymer membranes. Development of conductometric sensor for herbicides detection. Anal Chim Acta 392:105
78. Suedee R, Intakong W, Dickert FL (2006) Molecularly imprinted polymer-modified electrode for online conductometric monitoring of haloacetic acids in chlorinated water. Anal Chim Acta 569:66

Electrochemical Systems - Scaling, Dimensionless Groups

Shriram Santhanagopalan
National Renewable Energy Laboratory, Golden, CO, USA

Dimensionless Groups

Physical laws governing transport of fluids are independent of the characteristic dimensions, as long as the criteria for continuity are satisfied. For example, Fick's laws of diffusion do not depend on the reactor size. Consequently, the conservation equations for mass, momentum, charge, or energy can be reduced to dimensionless forms that are independent of channel lengths or time constants for the reaction. The flux balances at the boundaries introduce additional constraints that incorporate information about the influence of the design parameters on the transport processes under investigation. For each variable in the governing equation representing a transport law, there exists one well-defined constraint. If fewer constraints are specified, mathematical combinations of some of the variables in the governing equation, resulting in fewer variables, are adequate to characterize the problem. On the other hand, if constraints are specified in excess, the dimensionless formulation of the governing equation results in introduction of dimensionless parameters. Such dimensionless groups identified in commonly encountered engineering problems are sometimes named after engineers who first popularized their use. A list of these groups can be found in standard references. The use of dimensionless groups is a technique well established in several disciplines. Dimensional analysis offers significant benefits to both theoretical analyses of complex engineering problems and serves as a quick design tool to arrive at order of magnitude estimates before performing rigorous calculations. Standard dimensionless groups, such as the Reynolds number, used in characterizing transport phenomena find similar utility in designing electrochemical reactors:

1. The most prominent use of dimensionless parameters is to understand the behavior of the system under limiting conditions. For example, if the Peclet number is small, then advection effects within the reactor can be neglected compared to limitations imposed by diffusion.
2. The equations involving limiting cases are simpler to solve mathematically – there often exist analytical or asymptotic solutions that can be used to make quick calculations, thus avoiding the tedium of developing sophisticated computational tools.
3. The process of scaling up of reactors from pilot or lab-size equipment to production-size is streamlined with the use of dimensionless groups to characterize the operating regime. For example, in order to ensure a safe operating regime similar to that observed in the laboratory, the design engineer has to determine the right characteristic length for the reactor that matches the range of Nusselt numbers determined in the lab-scale experiments, so that heat loss due to convection is comparable for the two cases.
4. In some instances, some of the parameters used in calculations are not intuitive to an experimentalist; hence, it is difficult to determine the numerical values for such parameters from experiments. One example is the value for reaction rate coefficients under different conditions. In such instances, it is easier to determine the values for such parameters relative to

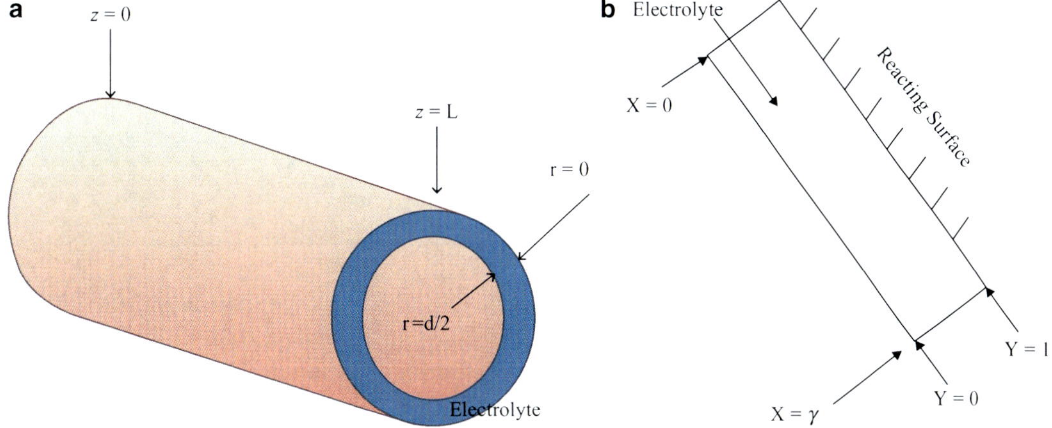

Electrochemical Systems - Scaling, Dimensionless Groups, Fig. 1 Concentration and potential distribution within the electrolyte undergoing an electrochemical reaction on the surface of a cylinder: definition of the geometric coordinates

a reference condition: for example, determining the local current densities as a fraction of the average current density is often used in electroplating calculations – and compares well with the throwing power observed experimentally.

Dimensional Analysis

In general, the number of parameters used in the analysis is reduced because of the combination of two or more parameters into dimensionless groups.

To illustrate this idea, let us consider the potential and concentration distribution across a cylindrical flow channel, on the inner walls of which an electrochemical reaction takes place.

For this case, the geometry is shown in Fig. 1 and the mass balance is given by

$$\frac{\partial c}{\partial t} = D\left(\frac{\partial^2 c}{\partial z^2} + \frac{1\partial}{r\partial r}\left(r\frac{\partial c}{\partial r}\right)\right) \tag{1}$$

and the charge conservation is given by Laplace's equation:

$$\left(\frac{\partial^2 \phi}{\partial z^2} + \frac{1\partial}{r\partial r}\left(r\frac{\partial \phi}{\partial r}\right)\right) = 0 \tag{2}$$

The initial concentration is set to c^{ref} and the flux at boundaries far from the reacting surface (i.e., at $r = 0$), and those at the exit (i.e., at $z = L$) are set to zero. The potential at the entrance ($z = 0$) is set to the equilibrium value less the product of the potential sweep rate time. The boundary conditions for this case are shown in Table 1.

Electrochemical Systems - Scaling, Dimensionless Groups, Table 1 Boundary conditions corresponding to Eqs. 1 and 2

$\left(\frac{\partial c}{\partial r}\right)_{r=0} = 0$	$\left(\frac{\partial \phi}{\partial r}\right)_{r=0} = 0$
$-D\left(\frac{\partial c}{\partial r}\right)_{r=d/2} = \frac{i_n}{nF}$	$\left(k\frac{\partial \phi}{\partial z}\right)_{r=d/2} = i_n$
$\left(\frac{\partial c}{\partial z}\right)_{z=0} = -k_M\left(c^{ref} - c\right)$	$\phi_{z=0} = \phi^{ref} - vt$
$\left(\frac{\partial c}{\partial z}\right)_{z=L} = 0$	$\left(\frac{\partial \phi}{\partial z}\right)_{z=L} = 0$

The flux at the inlet is related to the bulk concentration (c^{ref}) through a mass transfer coefficient (k_M), and the reaction current is related to the potential at the surface (ϕ_0) by

$$i_n = nFk^0\left\{c_O \exp\left[-\alpha\frac{nF}{RT}\left(\phi_0 - \phi^{ref}\right)\right] - c_R \times \exp\left[(1-\alpha)\frac{nF}{RT}\left(\phi_0 - \phi^{ref}\right)\right]\right\} \tag{3}$$

Introducing dimensionless variables as shown in Eq. 4

$$C = \frac{c}{c^*} \quad E = \frac{\phi - \phi^{ref}}{\phi^*} \quad \tau = \frac{t}{t^*}$$
$$X = \frac{z}{z^*} \quad Y = \frac{r}{r^*} \tag{4}$$

Equations 1, 2, and 3 can be rewritten as

$$\boxed{\frac{(r^*)^2}{Dt^*}}\frac{\partial C}{\partial \tau} = \boxed{\left(\frac{r^*}{z^*}\right)^2}\frac{\partial^2 C}{\partial X^2} + \frac{1}{Y}\frac{\partial}{\partial Y}\left(Y\frac{\partial C}{\partial Y}\right) \tag{5}$$

$$\left(\frac{r^*}{z^*}\right)^2\frac{\partial^2 E}{\partial X^2} + \frac{1\partial}{Y\partial Y}\left(Y\frac{\partial E}{\partial Y}\right) = 0 \tag{6}$$

$$\boxed{\frac{i_n}{nFk^0c^*}} = \left\{C\exp\left[-\alpha\boxed{\frac{nF}{RT\phi^*}}E\right] - \left(\frac{1+\xi-\xi C}{\xi}\right) \times \exp\left[(1-\alpha)\frac{nF}{RT\phi^*}E\right]\right\} \tag{7}$$

The dimensionless boundary conditions are listed in Table 2. The dimensionless mass transfer coefficient K is defined as $k_M d/2$, and the parameter ξ(see Eq. 7) relating the concentration of the oxidized and reduced species, c_O and c_R, respectively, is determined using the criterion for equal diffusivity using the expression $\xi=(c_o^{ref}/c_R^{ref})$. There are ten dimensionless groups, each shown within a box in Eqs. 5, 6, and 7 and in Table 2; since we have defined only five dimensionless variables in Eq. 4, we need five dimensionless parameters that characterize the problem. To determine these parameters, we define the characteristic quantities as shown in Eq. 8:

$$c^* = c^{ref} \quad r^* = \frac{d}{2} \quad z^* = \frac{d}{2} \quad t^* = \frac{r^{*2}}{D}$$
$$\phi^* = \frac{nF}{RT} \tag{8}$$

The dimensionless form of the governing equations is as follows:

$$\frac{\partial C}{\partial \tau} = \frac{\partial^2 C}{\partial X^2 +} + \frac{1\partial}{Y\partial Y}\left(Y\frac{\partial C}{\partial Y}\right) \tag{9}$$

Electrochemical Systems - Scaling, Dimensionless Groups, Table 2 Dimensionless form of the boundary conditions shown in Table 1

$\left(\frac{\partial C}{\partial Y}\right)_{Y=0} = 0$	$\left(\frac{\partial E}{\partial Y}\right)_{Y=0} = 0$
$\left(\frac{\partial C}{\partial Y}\right)_{Y=\boxed{\frac{d}{2r^*}}} = -\boxed{\frac{r^*}{Dc^*}\frac{i_n}{nF}}$	$\left(\frac{\partial E}{\partial Y}\right)_{Y=\frac{d}{2r^*}} = \boxed{\frac{r^*}{\phi^*}\frac{i_n}{k}}$
$\left(\frac{\partial C}{\partial X}\right)_{X=0} = -K\left(\boxed{\frac{c^{ref}}{c^*}} - C\right)$	$E_{X=0} = -\boxed{\frac{vt^*}{\phi^*}}\tau$
$\left(\frac{\partial C}{\partial X}\right)_{X=\boxed{\frac{L}{z^*}}} = 0$	$\left(\frac{\partial E}{\partial X}\right)_{X=\frac{L}{z^*}} = 0$

$$\frac{\partial^2 E}{\partial X^2} + \frac{1\partial}{Y\partial Y}\left(Y\frac{\partial E}{\partial Y}\right) = 0 \tag{10}$$

The boundary conditions shown in Table 2 become

$$\text{a. } \left(\frac{\partial C}{\partial Y}\right)_{Y=0} = 0 \quad \text{b. } \left(\frac{\partial C}{\partial Y}\right)_{Y=1} = -\sqrt{\sigma}I_n$$
$$\text{c. } \left(\frac{\partial C}{\partial C}\right)_{X=0} = -K\,(1-C) \quad \text{d. } \left(\frac{\partial C}{\partial X}\right)_{z=\gamma} = 0$$
$$\text{e. } \left(\frac{\partial E}{\partial Y}\right)_{Y=0} = 0 \quad \text{f. } \left(\frac{\partial E}{\partial Y}\right)_{Y=1} = \left(\frac{I_n}{J\Lambda}\right)$$
$$\text{g. } E_{X=0} = -\sigma\tau \quad \text{h. } \left(\frac{\partial E}{\partial X}\right)_{z=\gamma} = 0 \tag{11}$$

where the five dimensionless parameters are defined as shown in Table 3.

The parameter J is sometimes referred to as Wagner number and Λ corresponds to the Biot number for mass transfer. It is possible to combine these two parameters to define a single dimensionless group – which can be used to represent the ratio between the ohmic and mass transfer resistances. These groups are typically useful in determining whether or not tertiary current distribution is significant for a given set of operating conditions.

Scaling

Computational analysts often prefer the use of dimensionless variables for numerical efficiency

Electrochemical Systems - Scaling, Dimensionless Groups, Table 3 Dimensionless parameters required to characterize an electrochemical reaction taking place on the surface of a cylinder as described in Eqs. 1, 2, 3, and 4

Dimensionless parameter	Expression	Physical significance
J	$\frac{2\kappa/d}{n^2F^2k^0c^{ref}/RT}$	$\frac{\text{kineticresistance}}{\text{ohmicresistance}}$
Λ	$\frac{k^0}{\left(n\left(\frac{F}{RT}\right)vD\right)^{\frac{1}{2}}}$	$\frac{\text{mass} - \text{transfer resistance}}{\text{kineticresistance}}$
σ	$\frac{d^2nFv}{4RTD}$	$\frac{\text{timeconstant fordi fusion}}{\text{timeconstant for voltagesweep}}$
γ	$\frac{2L}{d}$	Form factor
I_n	$\frac{i_n}{nFc^{ref}\left(n\left(\frac{F}{RT}\right)vD\right)^{\left(\frac{1}{2}\right)}}$	Dimensionlesscurrentdensity

and stability – particularly in the case of problems that pose numerical stiffness owing to a the large span of time constants for the various physical phenomena taking place within the individual phases, or as in the case of heterogeneous reactions, sharp gradients in the species flux at the interface. Electrochemical reactions involve charge transfer at the interface, which induces a nonlinear relationship between the flux of the species and the potential difference across the interface. A kinetic rate equation, usually the Butler-Volmer expression, is used to capture this nonlinearity. As shown in Eq. 3, there exists a logarithmic relation between the surface overpotential and the derivative of the potential at the electrode surface. For systems limited by kinetics, there is a steep change in the reaction current when a potential sweep experiment is simulated. In such numerical simulations, it is imperative to find an initial guess value for the distribution of the potential and concentration values solved for that are sufficiently close to the actual values to ensure convergence of the iteration routine. A typical instance where the scaling of individual species concentrations and potentials is important is the solution of tertiary current-distribution problems. The potential drop in these instances depends on the spatial distribution of the individual species taking part in the reactions at the surface. Depending on the number of reactions involved and the absolute concentrations of the individual species, it is often difficult to arrive at the initial values for all the variables of interest. In such cases, it is advisable to solve the generic problems of primary and secondary distributions to obtain the scaled potential distribution in the absence of concentration variations in the vicinity of the electrode surface. Subsequently, the dimensionless potential values can be adopted as initial guess values for the tertiary distribution problem of interest. In addition to this, the electroneutrality condition, which introduces an algebraic constraint, must also be satisfied. Oftentimes, among the set of species of interest are supporting electrolytes, whose concentrations are an order or two higher than other species in the system, as are additives and leveling agents, whose concentrations are lower by orders of magnitude compared to the reference species. In such instances, one cannot use a global tolerance limit for all the variables solved for, but the use of a vector array of absolute tolerances for the individual species flux introduces fluctuations in the stiffness matrix. The convergence rate of such iterations is extremely slow and sometimes outside the numerical limits of accuracy specified on the machine. Thus, a highly nonlinear set of differential algebraic equations used to represent electrochemical systems necessitates the use of dimensionless variables for rapid numerical convergence as illustrated in the following example. The generic convective diffusion equation, scaled by the properties corresponding to the reference species R, is shown below:

$$\frac{D_i d^2 C_i}{D_R d\zeta^2} + 3\zeta^2 \frac{dC_i}{d\zeta} + \frac{z_i D_i F}{RTD_R}\left(C_i \frac{d^2E}{d\zeta^2} + \frac{dC_i dE}{d\zeta d\zeta}\right) + \frac{\delta_D^2}{D_R} R_i = 0 \tag{12}$$

The two step oxidation of tetrathiafulvalene (TTF) in aqueous cetyl-trimethyl ammonium chloride (CTAC) solution involves the following reactions:

$$\begin{aligned} TTF &\overset{i_{o,1}}{\rightarrow} TTF^{+} + e^{-} \\ TTF^{+} &\overset{i_{o,2}}{\rightarrow} TTF^{2+} + e^{-} \\ TTF + TTF^{2+} &\overset{k_f}{\rightleftarrows} 2TTF^{+} \end{aligned} \tag{13}$$

Electrochemical Systems - Scaling, Dimensionless Groups, Table 4 Summary of kinetic parameters and bulk concentrations for the reactions (13): note the several orders of magnitude differences in the reaction rate constants and the concentration values

Kinetic parameters	Bulk properties
$i_{o,1}$ = 7.87e-6 A/cm^2	c_{CTAC} = 0.02 M
$i_{o,1}$ = 9.48e-12 A/cm^2	c_{TTF2+} = 1e-13 M
K_{eq} = 1e5	c_{TTF+} = 9.1e-8 M
k_f = 1e8 cm^6/mol-s	c_{TTF} = 1e-3 M

Equation 12 to represent the individual species balances, together with the properties shown in Table 4, results in reaction currents that may be as much as ten orders of magnitude apart – depending on the value of the equilibrium constant for the homogeneous reaction shown in Eq.13, as shown in Fig. 2. This is a typical example of a situation wherein, the scaling of the individual concentrations is absolutely critical for convergence of the numerical simulations.

The use of dimensionless groups and the need for numerical manipulation of the equations solved for has been considerably reduced over the recent years, due to the use of sophisticated software capable of integrating computational

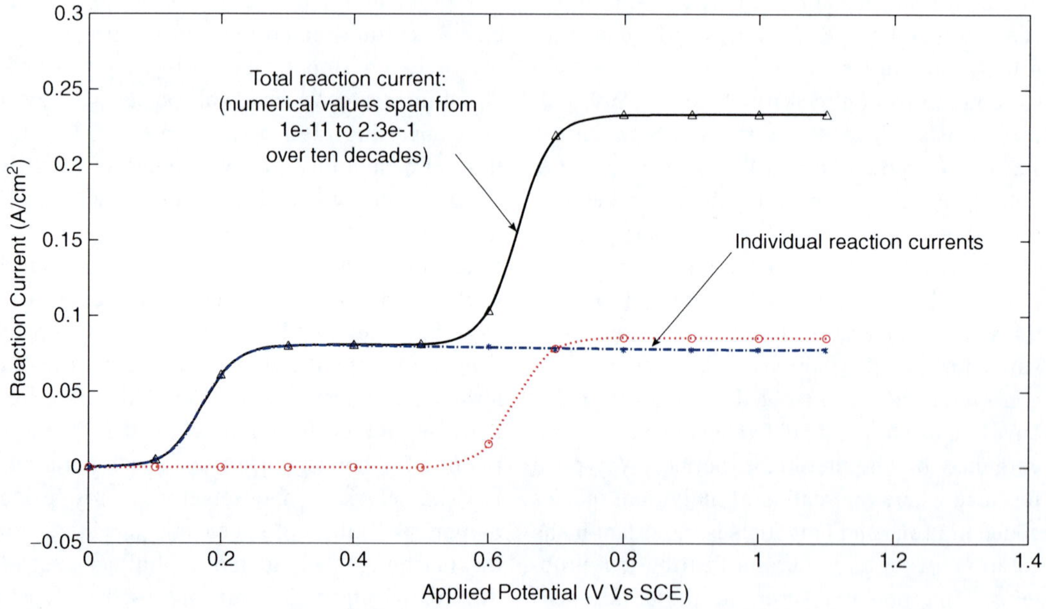

Electrochemical Systems - Scaling, Dimensionless Groups, Fig. 2 Reaction currents corresponding to reactions (13): the value of current at the electrode surface rises sharply as the potential is swept from 0 to 1.2 V versus SCE and can change by as much as 14 orders of magnitude: the use of scaled equations helps overcome numerical stiffness

fluid dynamics simulations with electrochemical reaction models. These software tools often include inherent capability to scale variables as required for numerical efficacy and have been quite successful at that. The attractiveness of closed-form solutions, especially for control applications, still has motivations for one to coin limiting-case analyses or specific solutions valid under a limited range of operating conditions (such as reduced order models or proper orthogonal decomposition-based solutions). The validity of the solution under a given set of constraints is often reiterated by the evaluation of a dimensionless group of interest (e.g., the Reynolds number for laminar flow), as part of the design process, before the subsequent integration with the control algorithm is carried out. The scaling of the resulting variables is still the best practice in place, to interpret the changes to the response with respect to the baseline operating conditions. Recent numerical methods, such as domain decomposition, often segregate the mass-matrices used the iterative solution scheme, based on the time constants associated with the constitutive elements for numerical efficiency, thus introducing newer means to exploit benefits from the scaling schemes used in the past.

List of Symbols

c	Concentration (mol/m^3)
C	Dimensionless concentration
d	Radial dimension of the cylindrical flow channel (m)
D	Diffusion coefficient (m^2/s)
E	Dimensionless potential
F	Faraday's constant (96487 C)
i_n	Current density in the normal direction (A/m^2)
i_0	Exchange current density for a charge transfer reaction (A/m^2)
I_n	Dimensionless current density in the normal direction
J	Wagner number (see Table 3)
k^0	Rate constant for the charge transfer reaction (m/s)
k_f	Rate constant for the forward reaction shown in (13c) (m^6/mol-s)
k_M	Mass transfer coefficient (m^{-1})
K	Dimensionless mass transfer coefficient
K_{eq}	Equilibrium constant for reaction (13c)
	(continued)
L	Axial dimension of the cylindrical flow channel (m)
M	Molarity (1 M = 1e-3 mol/m^3)
n	Number of electrons transferred
r	Radial coordinate (m)
R	Universal gas constant (8.314 J/mol-K)
t	Time (s)
T	Temperature (K)
X	Dimensionless axial coordinate
Y	Dimensionless radial coordinate
Z	Axial coordinate (m)
α	Transfer coefficient
δ_D	Dimensionless length for a rotating disc electrode
ζ	Dimensionless position coordinate for a rotating disc electrode
γ	Dimensionless axial coordinate
Λ	Biot number
κ	Conductivity (S/m)
v	Voltage sweep rate (V/s)
ϕ	Potential (V)
ξ	Dimensionless ratio of surface concentrations (see Eq. 7)
σ	Dimensionless time constant
τ	Dimensionless time
Superscripts	
*	Characteristic value
ref	Reference value
Subscript	
0	Surface property

Cross-References

▶ Electrochemical Cells, Current and Potential Distributions

Further Reading

1. Landau U (1994) Novel dimensionless parameters for the characterization of electrochemical cells. In: Newman JS, White RE (eds) Proceedings of the Douglas N. Bennion memorial symposium: topics in electrochemical engineering, vol 94–22, Proceedings. The Electrochemical Society, Pennington
2. Fedkiw PS (1990) Characterization of reaction kinetics in a porous electrode. NASA Report NASA-CR-186504 available from http://openlibrary.org/books/OL15278624M/Characterization_of_reaction_kinetics_in_a_porous_electrode
3. Fahidy TZ (1985) Principles of electrochemical reactor analysis, Chemical engineering monographs. Elsevier Science Ltd, New York. ISBN 0444424512

4. Monroe C, Newman J (2005) A method for determining self-similarity. Chem Eng Educ 39(1):42–47, Winter 2005
5. Speiser B Numerical simulations in electrochemistry. Available from http://www.springerreference.com/index/chapterdbid/303485. Accessed March 2013
6. Prentice GA, Tobias CW (1982) A survey of numerical methods and solutions for current distribution problems. J Electrochem Soc 129(1):72–78
7. Newman JS, Thomas Alyea KE (2004) Electrochemical systems, 3rd edn. Wiley Interscience, Hoboken

Electrochemical Treatment of Landfill Leachates

Ane Urtiaga and Immaculada Ortiz
Department of Chemical Engineering and Inorganic Chemistry, University of Cantabria, Santander, Cantabria, Spain

Introduction

Leachates generated from municipal landfills are complex effluents that contain high concentrations of organic pollutants, ammonium, chloride, and many other soluble compounds. Table 1 lists the typical concentrations found in the leachate of a mature landfill site [1, 2]. With regard to the value of BOD_5/COD, this leachate may be classified as a poorly biodegradable wastewater.

As a result, the use of different integrated processes combining bio-physico-chemical [3] and bio-advanced-oxidation [4] steps has been investigated. Among the processes studied, electrochemical oxidation stands out for its robustness, versatility, and amenability to automation and for its little or no need for addition of chemicals [5]. In Table 1, it is worth noticing the high conductivity value, permitting the application of electrochemical oxidation without the addition of extra electrolytes. Electrochemical oxidation has proved to be capable of eliminating both the organic and ammonia contaminant load [6, 7]. The choice of anodic material, together with the effect of operating variables and characteristics of the effluent on the electrochemical oxidation of landfill leachates, will be discussed in the following sections.

Electrochemical Treatment of Landfill Leachates, Table 1 Physical-chemical properties of landfill leachates

Parameter	Value
pH	8.4
Conductivity (mS/cm)	22.6–12.8
TOC (mg/L)	2,780
COD (mgO_2/L)	3,385–4,430
BOD_5 (mg/L)	500–1,195
BOD_5/COD	0.15–0.27
[$N\text{-}NH_4^+$] (mgN/L)	1,235–1,939
Chloride (mg/L)	2,587–3,230

Anode Materials and Oxidation Mechanisms

In relation with the anodic material, oxide-coated (PbO_2/Ti, SnO_2/Ti, TiO_2-RuO_2/Ti, SPR, DSA) anodes were initially employed. Cossu et al. [8] compared the efficiency of PbO_2 /Ti with SnO_2/Ti anode to oxidize landfill leachate that had been previously treated on-site by aerobic lagooning, denitrification, and activated sludge process and did not observe a significant influence. Chiang et al. [9] treated landfill leachate by electrochemical oxidation assisted by different anodic materials: PbO_2/Ti, DSA, SPR, and graphite anodes. Among the investigated anodes, the SPR anode gave the best efficiency for landfill leachate treatment. This was assumed to be due to the higher electrocatalytic activity of SPR for chlorine/hypochlorite production. Zhang et al. [10] investigated the removal of ammoniacal nitrogen and COD from landfill leachate in a three-dimensional electrochemical reactor comprised of a RuO_2-IrO_2/Ti anode and packed with activated carbon. COD and ammonia nitrogen removal reached values as high as 26 % and 81 %, respectively.

More recently, boron-doped diamond (BDD) electrodes have been extensively studied for the oxidation of both organic compounds and ammonia contained in landfill leachates [11, 12]. Several technological properties of BDD electrodes make this material an outstanding candidate for the treatment of landfill leachates: (i) an inert surface with low adsorption properties and a strong tendency to resist deactivation, (ii) corrosion stability in aggressive media in comparison to conventional electrodes such as graphite or glassy carbon electrodes, and (iii) the large electrochemical window of BDD electrodes enables the generation of strong oxidants such as ozone, peroxydisulfate, and hydroxyl radicals. The latter property leads to excellent chemical oxygen demand (COD) removal efficiency. Moreover, BDD is also very efficient in the generation of free chlorine in the presence of chlorides, and hence, high ammonium removal efficiencies are obtained at the same time [13, 14].

It is well documented [15] that oxidation of organic compounds on BDD anodes, in the absence of other oxidants, takes place primarily by means of electrogenerated hydroxyl radicals. The fast electrochemical mineralization of organics by hydroxyl radical-mediated oxidation and the instable nature of this species confine the reaction zone to the thin aqueous layer adjacent to the anode surface [16], and the overall oxidation kinetics are controlled by the mass transport of the organics to the electrode surface. Generation of other strong oxidizing agents is also possible. Chloride ions present in the leachate may react on BDD to give dissolved chlorine. The latter species hydrolyses rapidly in water to form hypochlorous acid and hypochlorite. Thus, in the presence of chloride, oxidation of organics contained in landfill leachates occurs through two different pathways: hydroxyl radicals and chlorine-mediated mechanisms [17]. Finally, hypochlorous acid rapidly reacts with ammonia [13].

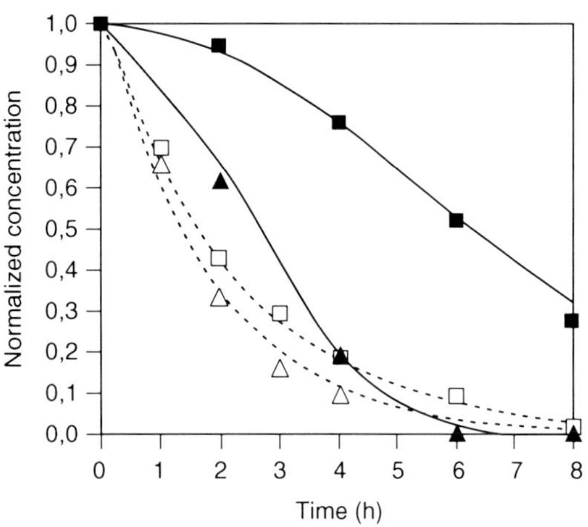

Electrochemical Treatment of Landfill Leachates, Fig. 1 Normalized COD (*empty symbols*) and NH_4^+ (*filled symbols*) concentration profiles during the electrochemical treatment of landfill leachates using BDD anodes. Rectangles represent data of raw leachate and triangles of ½ diluted leachate. Raw leachate: $[COD]_0 = 3{,}275\ mgL^{-1}$, $[NH_4^+]_0 = 2{,}453\ mgL^{-1}$, and $[Cl^-]_{0,av} = 2{,}417\ mgL^{-1}$. $J = 600\ Am^{-2}$. Adapted from reference [1]

Influence of Process Variables and Scale-Up

Electrochemical oxidation of landfill leachates is significantly dependant on two process variables: (i) the applied current density and (ii) the composition of the aqueous waste. Figure 1 shows the effect of the initial concentrations of COD and ammonia on the oxidation kinetics of both compounds, using BDD anodes. The raw leachate and a second sample obtained by dilution are compared. NaCl was added to the diluted batch, to obtain the same Cl^- concentration as in the raw leachate. It is observed that ammonium oxidation occurs at a slower rate than COD oxidation. Faster removals are obtained at lower initial concentrations of both pollutants. Interestingly, the effect of the initial contaminant load is much higher on ammonium than on COD removal. Keeping in mind the competitive nature of the chlorine evolution reaction and the oxidation of COD at the anode surface, it happens that by lowering down the initial COD concentration while maintaining the initial concentrations of chloride ions constant, chloride evolution is enhanced which in turn results in higher ammonium oxidation rates.

Regarding the effect of current density, the rate of COD and ammonium oxidation increases with current density, as it is exemplified in Fig. 2 [18]. This suggests that, at higher current densities, indirect oxidation by means of chlorine-mediated reactions plays an important role in the overall electrochemical oxidation of organics. With this in mind, several authors [4, 19] have increased the original concentration of chloride

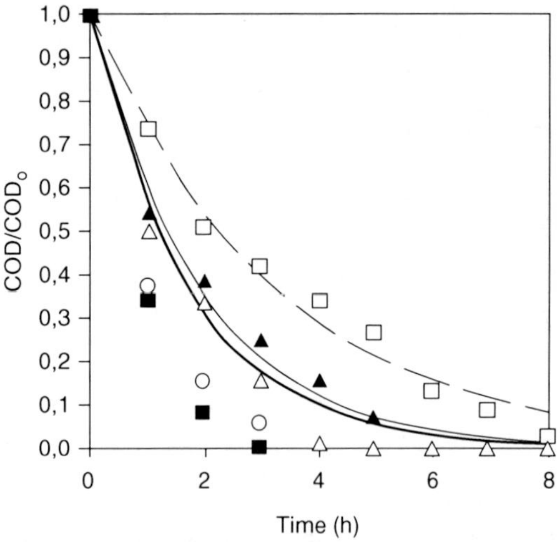

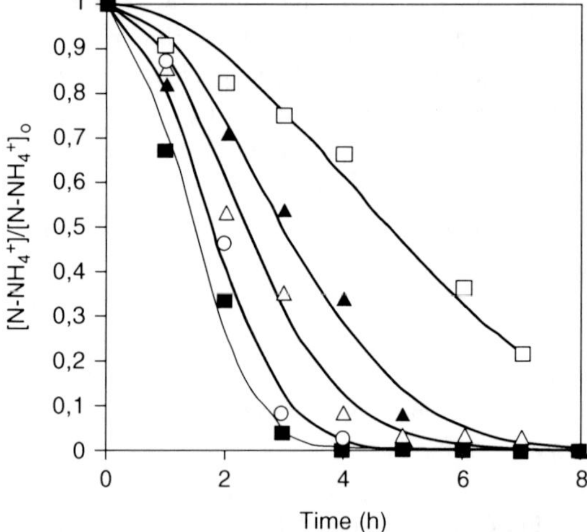

Electrochemical Treatment of Landfill Leachates, Fig. 2 Normalized COD and NH_4^+ concentration profiles during the electrochemical treatment of biologically pretreated landfill leachate, at different current densities: (□)300 Am^{-2}; (▲)450 Am^{-2}; (Δ)600 Am^{-2}; (o)900 Am^{-2}; and (■)1,200 Am^{-2}. Operation conditions: $[COD]_0 = 1{,}350\ mgL^{-1}$, $[NH_4^+]_0 = 2{,}453\ mg\ L^{-1}$, and $[Cl\text{-}]_0 = 2{,}070\ mg\ L^{-1}$. $J = 600\ Am^{-2}$. Adapted from Ref. [18]

ions present in the leachate with the hope of enhancing the oxidation efficiency. Overall an increase in the rate of COD and ammonia oxidation has been observed upon addition of chloride ions, although significant differences are observed when using different anodic materials [1].

Recent progress in the development of consistent mathematical models that describe the complex scenarios and reactions involved in the electrooxidation of multicomponent mixtures of various pollutants such as those found in landfill leachates and its integration with knowledge about reactor hydrodynamics will undoubtedly contribute to the practical implementation of this technology [20, 21]. The on-site operation at the landfill site of pilot-scale (1.05 m^2 of anodic area) electrochemical treatment has been reported showing excellent removal efficiencies of organic and ammonia pollutants. The energy consumption needed to reach the disposal limits to the sewer ($[NH_4^+] < 100\ mg\ L^{-1}$) was estimated at 86 kWh m^{-3} [22]. Nonetheless, integration of electrooxidation with biological treatment and the potential use of solar radiation to power the process encourages the sustainability of this kind of treatment [23].

Formation of Oxidation By-Products

The formation of oxidation by-products such as nitrate ions as well as chlorinated compounds is a matter of concern when considering the electrochemical treatment of landfill leachates [24]. With regard to the use of BDD anodes, nitrogen gas and nitrate ions appear as the main oxidation products of ammonium oxidation [25, 26]. Nitrite has been found to appear as an intermediate compound although its presence may not be detected due to its fast oxidation rate. The role of chloride seems to be dependant on its concentration since it has been observed that higher chloride concentrations have a positive effect in the formation of nitrogen as the main oxidation product. On the contrary, higher chloride concentration favors the formation of chlorinated organics, with chloroform as the main compound generated during the electrooxidation of leachates using BDD anodes, followed by 1,2-dichloroethane [2]. With regard to the formation of chlorate and perchlorate, once again the concentration of chloride exerts a strong influence on the formation kinetics of the oxidation by-products, and whereas at low chloride concentrations, chlorate appears like an intermediate compound leading to the formation of perchlorate; at high chloride concentrations, chlorate formation can be delayed significantly [25]. Therefore, although during the treatment of landfill leachates total mineralization of organic matter and partial oxidation of ammonia to nitrate ions occurs, the formation of oxidation by-products will require of a subsequent separation step in order to obtain an effluent with suitable characteristics for its reuse or for its discharge to natural water sources. In this context, deployment of electrochemical oxidation in combination with ion exchange or reverse osmosis [27] has been proposed.

Future Directions

Future development will require research and development activities in the following areas:

1. The superior technical properties of BDD electrodes for the simultaneous removal of organic and ammonia pollutants are hindered by their high cost. Alternative methods of preparation of diamond electrodes that lead to lower production costs and a substrate that is not brittle like silicon or too expensive like tantalum, niobium and tungsten should be found.
2. Determination of the optimal operational conditions to achieve high elimination of ammonia with minimum conversion to nitrate and perchlorate.
3. Integration of electrooxidation with biological processes and/or other AOPs in order to cut down treatment costs of landfill leachates.

Cross-References

▶ Electrochemical Cell Design for Water Treatment
▶ Organic Pollutants in Water Using BDD, Direct and Indirect Electrochemical Oxidation

References

1. Cabeza A, Urtiaga A, Ortiz I (2007) Electrochemical treatment of landfill leachates using a boron-doped diamond anode. Ind Eng Chem Res 46:1439–1446
2. Anglada A, Urtiaga A, Ortiz I, Mantzavinos D, Diamadopoulos E (2011) Boron-doped diamond anodic treatment of landfill leachate: evaluation of operating variables and formation of oxidation by-products. Wat Res 45:828–838
3. Rivas F, Beltran F (2005) Study of different integrated physical-chemical-adsorption processes for landfill leachate remediation. Ind Eng Chem Res 44:2871–2878
4. Wang P, Lau IWC, Fang HHP (2001) Electrochemical oxidation of leachate pretreated in an upflow anaerobic sludge blanket reactor. Environ Technol 22:373–381
5. Deng Y, Englehardt JD (2007) Electrochemical oxidation for landfill leachate treatment. Waste Manage 27:380–388
6. Vlyssides AG, Karlis PK (2001) Treatment of leachate from a domestic solid waste sanitary landfill by an electrolysis system. Environ Technol 22:1467–1476
7. Chiang LC, Chang JE (2001) Electrochemical oxidation combined with physical-chemical pretreatment processes for the treatment of refractory landfill leachate. Environ Eng Sci 18:369–379
8. Cossu R, Polcaro AM (1998) Electrochemical treatment of landfill leachate: oxidation at Ti/PbO_2 and Ti/SnO_2 anodes. Environ Sci Technol 32:3570–3573
9. Chiang LC, Chang JE, Wen TC (1995) Indirect oxidation effect in electrochemical oxidation treatment of landfill leachate. Wat Res 29:671–678
10. Zhang H, Wu X, Zhang Y, Zhang D (2010) Application of response surface methodology to the treatment of landfill leachate in a three dimensional electrochemical reactor. Waste manage 30:2096–2102
11. Anglada A, Urtiaga A, Ortiz I (2009) Pilot scale performance of the electrooxidation of landfill leachate at boron-doped diamond anodes. Environ Sci Technol 43:2035–2040
12. Urtiaga A, Rueda A, Anglada A, Ortiz I (2009) Integrated treatment of landfill leachates including electrooxidation at pilot plant scale. J Hazard Mat 166:1530–1534
13. Cabeza A, Urtiaga A, Rivero MJ, Ortiz I (2007) Ammonium removal from landfill leachate by anodic oxidation. J Hazard Mat 144:715–719
14. Kapalka A, Joss L, Anglada A, Comninellis C, Udert KM (2010) Direct and mediated electrochemical oxidation of ammonia on boron-doped diamond electrode. Electrochem Comm 12:1714–1717
15. Mascia M, Vacca A, Palmas S, Polcaro AM (2007) Kinetics of the electrochemical oxidation of organic compounds at BDD anodes: modeling of surface reactions. J App Electrochem 37:71–76
16. Cañizares P, García-Gómez J, Lobato J, Rodrigo MA (2004) Modeling of wastewater electro-oxidation processes part I. General description and application to inactive electrodes. Ind Eng Chem Res 43:1915–1922
17. Polcaro AM, Vacca A, Mascia M, Palmas S, Rodriguez Ruiz J (2009) Electrochemical treatment of waters with BDD anodes: kinetics of the reactions involving chlorides. J Appl Electrochem 39:2083–2092

18. Anglada A (2011) Electro-oxidation on boron-doped diamond anodes of ammonia and organic pollutants in landfill leachate. PhD dissertation thesis, University of Cantabria, Spain
19. Chiang LC, Chang JE, We TC (1995) Electrochemical treatability of refractory pollutants in landfill leachate Haz. Waste Hazard Mat 12:71–82
20. Anglada A, Urtiaga AM, Ortiz I (2010) Laboratory and pilot plant scale study on the electrochemical oxidation of landfill leachate. J Hazard Mat 181: 729–735
21. Moraes PB, Bertazzoli R (2005) Electrodegradation of landfill leachate in a flow electrochemical reactor. Chemosphere 58:41–46
22. Anglada A, Urtiaga AM, Ortiz I (2010) Electrochemical oxidation of landfill leachates at pilot scale. Evaluation of energy needs. Wat Sci Technol 61:2211–2217
23. Dominguez RA, Aldaco R, Irabien A (2010) Photovoltaic solar electrochemical oxidation (PSEO) for treatment of lignosulfonate wastewater. J Chem Technol Biotechnol 85:821–830
24. Comninellis C, Chen G (2010) Electrochemistry for the environment. Springer, New York
25. Pérez G, Sáiz J, Ibañez R, Urtiaga A, Ortiz I (2012) Assessment of the formation of oxidation by-products during the electrocatalytic treatment of ammonium from landfill leachates. Wat Res 46:2579–2590
26. Díaz V, Ibañez R, Gómez P, Urtiaga AM, Ortiz I (2011) Kinetics of electro-oxidation of ammonia-N, nitrites and COD from a recirculating aquaculture saline water system using BDD anodes. Wat Res 45:125–134
27. Cabeza A, Urtiaga AM, Ortiz I (2007) Definition of a clean process for the treatment of landfill leachates integration of electrooxidation and ion exchange technologies. Sep Sci Technol 42:1585–1596

Electrochemical Treatment of Reverse Osmosis Concentrates

Jelena Radjenovic, Arseto Bagastyo, Damien Batstone and Jurg Keller
Advanced Water Management Centre (AWMC), The University of Queensland, Brisbane, QLD, Australia

Introduction

Reverse osmosis (RO) membranes are increasingly widely used for municipal wastewater reclamation [1]. However, RO treatment generates a waste stream, reverse osmosis concentrate (ROC), which contains 15–25 % of the feed flow and almost all the original salts and dissolved organics. Most wastewater reclamation plants currently rely on direct or indirect (i.e., via municipal sewers) discharge of ROC into water bodies [2]. Yet, the ROC increases salt levels in the downstream sewage treatment plant [3] and contains significant amounts of dissolved organics. As a concentrated point source, ROC should be considered as an opportunity for reducing the environmental risks by implementing brine treatment, or further treatment to recovery. For this purpose, different options are being actively investigated, such as coagulation, adsorption, ozonation, sonolysis, and advanced oxidation processes (AOPs) [4–7]. Among these treatments, AOPs have been identified as the most efficient in removing the organic matter. They are characterized by the generation of highly reactive hydroxyl radicals ($OH^•$) through a range of processes that generally consume substantial amounts of energy and/or chemicals.

As an alternative, $OH^•$ as well as a variety of other oxidants (e.g., O_3, H_2O_2) can be formed during the electrolysis of water at an anode. The capability of electrochemical oxidation to directly generate oxidants in situ without the need for reagents is a compelling advantage over competing technologies that are at least partly relying on chemical additions. Therefore, electrooxidation can overcome the limitations of AOPs, not only in terms of chemical usage but also in terms of the oxidizing abilities, because a range of oxidant species can be generated at the anode poising different anode potentials. In addition, operation is at ambient temperature and pressure [8].

Electrode Material

The choice of electrode material affects efficiency and selectivity of electrochemical treatment [9]. One of the most commonly investigated electrodes for environmental applications of electrochemical oxidation is mixed metal oxide (MMO) electrodes, also known under the trade name Dimensionally Stable Anode, DSA®. MMOs have been investigated for the treatment of pesticide-contaminated water, landfill leachate, organic petroleum wastewater, and other difficult to treat waste streams [9]. They consist of corrosion-resistant base materials such as titanium or tantalum, coated with a layer of transient metal oxides (e.g., IrO_2, SnO_2). The particular type of coating determines the electrocatalytic activity of the electrodes. According to their interaction with the electro-generated $OH^•$, they are classified as "active" and "non-active" anodes [10]. For example, IrO_2– and RuO_2–based anodes are considered to be "active" due to their strong interaction with the electro-generated $OH^•$, which results in a low overpotential for O_2 evolution [9]. High oxidation power anodes such as Ti/PbO_2 and Ti/SnO_2 are "non-active," characterized by a weak electrode –$OH^•$ interaction and lower electrochemical activity for O_2 evolution, thus providing a higher current efficiency for organics oxidation. As postulated by Comninellis [11], non-active electrodes are theoretically capable of mineralizing the organic compounds via $OH^•$, in contrast to active anodes which only partially oxidize the organics by the adsorbed active oxygen species.

The latest technological advancement in the area of technical electrode materials is the invention of boron-doped diamond (BDD). BDD electrodes are characterized by a wider potential range in aqueous electrolytes compared to other electrode materials (ca. – 1.25 to + 2.3 V versus the normal hydrogen electrode – NHE) [12]. The interaction between the surface of BDD electrodes and electro-generated $OH^•$ is weak, allowing them to mineralize organic pollutants in the vicinity of the electrode surface. These quasi-free $OH^•$ represent the main advantage of BDD electrodes over conventional materials, as they enhance the degradation rates of organic pollutants. The high cost of the substrate onto which the BDD film is deposited (e.g., Nb, W, Ta) represents a major impediment to their large-scale application. While the comparatively cheaper Ti possesses all the necessary features of a good substrate material (i.e., good electrical conductivity, mechanical strength and electrochemical inertness), the deposition of BDD films on Ti with satisfactory stability was found to be difficult so far [13].

Effect of Chloride Ions on Electrochemical Treatment Efficiency

It is generally considered that direct anodic oxidation has a minor contribution to the overall oxidation of organics at potentials higher than the threshold for water electrolysis, i.e., 1.23 V versus NHE (pH = 0) [9]. Above this potential, direct electron transfer between the adsorbed organic compound and the electrode is less likely due to evolution of O_2, as well as mass transfer limitations. Thus, electrochemical oxidation will mainly rely on oxidation via $OH^•$ in the vicinity of the electrode surface and on the indirect mechanisms via other reactive oxygen species (ROS). In the presence of Cl^- ions, indirect oxidation of organic contaminants via $Cl_2/HClO/ClO^-$ will be promoted. $HClO/ClO^-$ is a long-lived oxidant able to diffuse away into the bulk liquid and react chemically with the organics [14]. In the case of MMO electrodes, particularly active anodes with low capability for $OH^•$ generation, addition of Cl^- has been reported to enhance the removal of a range of organic compounds, including dyes, pesticides, endocrine-disrupting compounds (EDCs), and other persistent contaminants [15–21]. Several studies have indicated that the superiority of non-active MMO and BDD electrodes over active MMOs such as RuO_2 practically vanishes in the presence of active chlorine species [22–24]. However, most of these studies were performed with supporting electrolyte solutions, and concerns have been raised about the increase in toxicity in chlorine-mediated electrolysis of real waste streams [25–28]. The insertion of chlorine substituent in an organic compound will increase its lipophilicity and thus toxicity to living organisms [29]. In order to minimize the formation of chlorinated organic compounds, different operational strategies were proposed in literature, such as application of long residence times [30, 31], activated carbon polishing treatment [30], and oversaturation of the solution by chlorine [21]. Nevertheless, in the electrochemical oxidation of the herbicide glyphosate at MMO electrodes, the absorbable organic halogen (AOX) concentrations increase directly proportional with the concentration of Cl^- ions [32]. The formed organic halides were not further degraded even at long electrolysis times and high current densities. Furthermore, the detrimental effect of chloride ions is not exclusive to MMO electrodes. A recent study [33] has demonstrated that during oxidation of sulfamethoxazole using BDD electrodes, electrochemical hypochlorination and oxidation by $OH^{\bullet}$ occurs simultaneously. Chlorinated by-products of sulfamethoxazole were more resistant to oxidation, as reflected in the significant AOX fraction remaining in the solution.

Electrochemical Treatment of ROC

Table 1 summarizes the existing studies on the electrochemical treatment of ROC. Van Hege et al. [34, 35] proposed electrochemical oxidation as a promising method for the treatment of problematic waste streams such as ROC. Electrochemical remediation of ROC offers an obvious advantage of low energy consumption due to its high electric conductivity. In their first study, Van Hege et al. [34] observed complete removal of chemical oxygen demand (COD) and total ammonium nitrogen (TAN) after supplying 1.5 and 2 Ah L^{-1} of specific electrical charge (Q) galvanostatically at 167 A m^{-2}, using an undivided cell. Nevertheless, although absorbance at visible wavelength was completely removed, the decrease in absorbance at 254 nm was less than 50 %, implying incomplete degradation of the aromatic fraction of organic matter. When comparing the performance of MMO and BDD electrodes for electrooxidation of ROC [35], the highest COD removal was noted for BDD and Ti/SnO_2 and maximal TAN removal for BDD and Ti/RuO_2 anodes. This was explained by the predominant mechanism of electrochlorination for TAN removal, as the highest concentration of active chlorine was measured using the BDD electrode, followed by Ti/RuO_2. Also, removal of color was complete, and, as noted by Panniza et al [36], likely achieved through chlorine-mediated indirect oxidation mechanisms. On the other hand, in the case of Ti/PbO_2 and Ti/SnO_2 electrodes, cathodic increase in pH was higher than the anodic pH decrease in a nondivided electrochemical cell, causing the precipitation of Ca^{2+} and Mg^{2+} ions present in the ROC, and electrode scaling.

Dialynas et al. [4] used a DiaCell® single-compartment electrolytic flow cell equipped with BDD electrodes to electrochemically oxidize 8 L of ROC in batch recirculation mode. At a current density (J) of 514 A m^{-2}, only 30 % of dissolved organic carbon (DOC) was removed after 30 min of electrolysis, while a further increase to 2,543 A m^{-2} did not result in any significant improvement. It should be noted that the electrolysis was able to oxidize the organic matter with absorbance below 250 nm, but it did not affect significantly the absorbance peak at 310 nm, often related to naphthalene-like compounds [37].

Chaplin et al. [38] investigated electrochemical destruction of N-nitrosodimethylamine (NDMA) in ROC using a rotating disk electrode (RDE) BDD and a flow-through Mini DiaCell® reactor. In the RDE experiments with decreased mass transfer limitations, NDMA oxidation was not affected by the DOC, Cl^- and HCO_3^- concentrations. However, galvanostatic experiments performed at 20 A m^{-2} in a flow-through reactor revealed five times lower NDMA removal rate constants. In the latter case, the estimated energy requirement for removing 75 % of DOC was 6.9 kWh m^{-3}. It should be noted that the anode surface area (25 cm^2) was quite large in comparison to the active volume of the reactor (15 mL), thus creating strongly oxidizing conditions inside the Mini DiaCell® reactor.

When comparing $Ti/IrO_2{-}Ta_2O_5$, BDD, and $Ti/IrO_2{-}RuO_2$ electrodes in treating ROC, Zhou et al. [24] observed a complete COD removal at 250 A m^{-2} after 1.5 h and 2 h for BDD and $Ti/IrO_2{-}RuO_2$, respectively, while partial removal was noted for $Ti/IrO_2{-}Ta_2O_5$. The $Ti/IrO_2{-}RuO_2$ electrode was suggested as the most suitable for full-scale application due to its

Electrochemical Treatment of Reverse Osmosis Concentrates, Table 1 Summary of studies of electrochemical treatment of reverse osmosis concentrate (ROC)

Experiment	A_{AN} (cm^2)	V_{ACT} (L)	V_{TOT} (L)	d (mm)	J ($A\ m^{-2}$)	Q ($Ah\ L^{-1}$)	q ($L\ h^{-1}$)	Key results	Ref.
^{1}BDD, $^{2}Ti/RuO_2$:	60	0.2	1	10	167	2	15	^{1}R(COD, TAN) = 100 % for 2 Ah L^{-1}.	[34]
Undivided cell, batch mode	^{1}R = 40–50 % (UV_{254}, UV_{280})							CE(COD) = ^{1}25 %, ^{2}35 % CE(TAN) = ^{1}10 %, ^{2}14 % ^{1}R = 7090 % (UV_{360}, UV_{455}, UV_{405}) ^{1}188 kWh kg_{COD}^{1}, ^{2}780 kWh kg_{TAN}^{1} $^{1}ClO_3$ formation.	
^{1}BDD, $^{2}Ti/RuO_2$, $^{3}Ti/PbO_2$, $^{4}Ti/SnO_2$: Undivided cell, batch mode.	50	0.2	1	10	100–300	1–2	15	^{3,4}Ca and Mg precipitates (rise in pH) R(UV_{455}, 1 Ah L^{-1}) = ^{1,2}74 % R(COD, 1 Ah L^{-1}) = ^{1}56 %, ^{2}12 %, ^{3}19 %, ^{4}22 % R(TAN, 1 Ah L^{-1}) = ^{1}48 %, ^{2}42 %, ^{3}14 %, ^{4}6 % ^{1}R(COD, TAN) = 100 % for 2 Ah L^{-1}. Better performance of BDD over Ti/RuO_2 was assigned to higher formation of FAC	[35]
BDD (Diacell): Undivided, batch mode.	70	0.0078	8	10	^{a}514, ^{b}2543	^{a}0.23, ^{b}1.11	1200	^{a}R = 30 % (DOC) ^{b}R = 36 % (DOC) Oxidation of $UV_{\leq 250}$, but not UV_{310}.	[4]
BDD (Diacell): Undivided, batch mode.	25	0.015	0.25	3	20	Up to 2	6	CE(DOC) = 68–75 %, 6.9 kWh m^{-3}	[36]
BDD (Diacell): Undivided, batch mode.	70	0.0078 (min)	2	5	^{1}50, ^{2}100, ^{3}200	Upto 4.9	600	R(COD) = 100 % (^{1}0.5, ^{2}0.75, ^{3}1 Ah L^{-1}) ^{1}59 kWh kg_{COD}^{-1} ClO_3, NO_3 and THMs formation R(TrOCs) ≥ 90 %, not affected by J	[37]
Ti/ $Ru_{0.7}Ir_{0.3}O_2$:	24	0.114	10	4	^{1}250,	^{1}1.46	^{1}9.72	R(DOC) = ^{1}33 %, ^{2}9–26 %	[25]
Divided, ^{1}batch and ^{2}continuous mode					^{2}1250	^{2}max 0.77	^{2}0.78	^{1}Initial increase in $SUVA_{254}$, ^{2}R ($SUVA_{254}$) = 29–42 %. Higher R(DOC) for ROC with higher initial	

(*continued*)

Electrochemical Treatment of Reverse Osmosis Concentrates, Table 1 (continued)

Experiment	A_{AN} (cm^2)	V_{ACT} (L)	V_{TOT} (L)	d (mm)	J ($A\ m^{-2}$)	Q ($Ah\ L^{-1}$)	q ($L\ h^{-1}$)	Key results	Ref.
								$SUVA_{254}$. Complete removal of persistent TrOCs only in batch m. Increase in toxicity by *V. fischeri* in both modes, higher in continuous	
[1]Ti/$Ru_{0.7}Ir_{0.3}O_2$, [2]$TiSnO_2$-Sb: Divided, batch mode	24	0.114	1	4	100, 250	Up to 1.8	9.72	Identical oxidation by-products detected for both anodes and J. Faster degradation rates for [2] and higher J. Abundant formation of Cl^- and Br by-products of model compound metoprolol. Increase in toxicity by *V.fischeri* and *Pseudokirchneriella subcapitata* (V_{TOT} = 10 L)	[26]
[1]Ti/$Ru_{0.7}Ir_{0.3}O_2$, [2]$TiSnO_2$-Sb, [3]Ti/PbO_2, Ti/Pt-IrO_2, [5]Ti/IrO_2-Ta_2O_5: Divided, batch mode	24	0.114	10	4	100	0.550	9.72	CE(COD) = [1]16 %, [2]32 %, [3]25 %, [4]45 %, [5]12 %. The highest formation of FAC, fastest removal of TAN, colour at [2] and [4]. R (DOC) = [2,4]16 %, [1,3,5]9 %. Formation of HAAs and THMs at all anodes. ClO_3^- formation at all anodes	[27]
[1]BDD, [2]Ti/$IrRuO_2$, [3]Ti/IrO_2-Ta_2O_5: Undivided, batch mode.	3	0.05	0.5	10	83, 167, 250	1, 2, 3	–	R(COD, 250 $A\ m^{-2}$) = [1]100 % (2.25 $Ah\ L^{-1}$), [2]100 % (30 $Ah\ L^{-1}$), [3]33 % (3 $Ah\ L^{-1}$). Ti/$IrRuO_2$ had the lowest energy consumption e. g. for J = 167 $A\ m^{-2}$: [1]203, [2]66 and [3]130 kWh $kgCOD^{-1}$	[24]

A_{AN}-anode surface (cm^2), V_{ACT} active volume, i.e., electrolytic cell volume (L), V_{TOT}-total volume (L), d-inter-electrode distance (mm), J-current density ($A\ m^{-2}$), Q-specific electrical charge ($Ah\ L^{-1}$), q-volumetric recirculation flow rate ($L\ h^{-1}$), R-removal (%), *CE* current efficiency (%), *FAC* free available chlorine, *COD* chemical oxygen demand, *TAN* total ammonia nitrogen, *DOC* dissolved organic carbon, *UV*-ultraviolet absorbance, $SUVA_{254}$ specific UV absorbance at 254 nm, *THMs* trihalomethanes, *HAAs* haloacetic acids

lower cost and lower energy consumption compared to BDD (e.g., 66 versus 203 kWh $kgCOD^{-1}$ for Ti/IrO_2–$RuO_2$and BDD at 250 A m^{-2}, respectively).

ROC streams have very high concentrations of Cl^-, typically from 1.4 up to even 8 g L^{-1} [2], and besides their positive effect in terms of lowered ohmic resistance and thus decreased energy consumption, they can lead to prevalent electrochlorination in the bulk liquid and generation of toxic chlorinated by-products. Research conducted within our center on electrochemical oxidation of ROC [25–27] revealed discouraging results in terms of application of MMO electrodes for electrochemical remediation of ROC. When oxidizing ROC at a $Ti/Ru_{0.7}Ir_{0.3}O_2$ electrode, we observed incomplete DOC removal in both batch and continuous modes of operation. Namely, at a current density of 250 A m^{-2}, only 9 % of DOC was removed in a single-pass continuous mode, while operating the reactor in batch mode resulted in 30 % of DOC removal. Furthermore, the efficiency of oxidation was found to depend strongly on the characteristics of the organic fraction in the ROC, because under the same conditions (i.e., continuous operation, $J = 250$ A m^{-2}), 26 % of DOC removal was achieved using a different ROC with a higher specific UV absorbance at 254 nm ($SUVA_{254}$) (i.e., 2.3 L mg^{-1} m^{-1} versus 1.6 L mg^{-1} m^{-1}). Oxidation of most of the target pharmaceuticals and pesticides in continuous mode required $J > 100$ A m^{-2}, with the exception of compounds having electrophilic substituents (e.g., triclopyr). However, bioassays performed with *Vibrio fischeri* showed an up to 50-fold increase in toxicity in both operational modes, indicating the formation of toxic, likely chlorinated by-products. A study of oxidation pathways of a model compound (β-blocker metoprolol) in ROC [26] revealed the formation of identical oxidation by-products at $Ti/Ru_{0.7}Ir_{0.3}O_2$ and $Ti/SnO_2{-}Sb$ anodes. Among 25 identified compounds, 16 were formed either by direct attacks of reactive halogen species (RHS) (e.g., $Cl^{\bullet}$, $Br^{\bullet}$, HOBr, Br_2, $Cl_2^{\bullet -}$) at metoprolol or its oxidation by-products. ROS such as $OH^{\bullet}$, O_2, H_2O_2, and/or O_3 were mainly responsible for the formation of primary oxidation by-products. Consecutive attacks of active chlorine and bromine converted these primary by-products into their halogenated derivatives at either active (i.e., $Ti/Ru_{0.7}Ir_{0.3}O_2$) or non-active electrodes (i.e., $Ti/SnO_2{-}Sb$). The Ti/SnO_2 electrode has higher oxidizing power for the same applied specific electrical charge, thus the generated derivatives were degraded faster than in the case of the $Ti/Ru_{0.7}Ir_{0.3}O_2$. However, an increase in nonspecific toxicity was observed even at higher current applied at Ti/SnO_2.

Besides electrochlorination of higher molecular weight organic matter, formation of smaller by-products such as trihalomethanes (THMs) and haloacetic acids (HAAs) was reported for electrooxidation of ROC, even when using a BDD anode. Perez et al. [39] have reported generation of THMs, NO_3^- and ClO_3^- using a DiaCell® setup for the oxidation of ROC. While the authors determined 100 A m^{-2} as the optimum current density for minimization of THM concentrations, generation of NO_3^- and ClO_3^- was increased at the higher currents. Other studies have previously reported a risk of formation of toxic ClO_3^-, ClO_4^- [40], BrO_3^- and BrO_4^- [41] at BDD electrodes. Additionally, at 50 A m^{-2}, the energy consumption for the total removal of COD was estimated at 59 kWh $kgCOD^{-1}$ (i.e., 6.4 kWh m^{-3}), similar to the values previously reported by Van Hege et al. [34, 35]. For a complete TAN removal, higher current was required (i.e., 100 A m^{-2}). The authors [39] have also reported nearly complete disappearance of eight pharmaceuticals and two stimulant drugs already at the lowest current density tested.

Another study by Bagastyo et al. [27] compared the performances of $Ti/Ru_{0.7}Ir_{0.3}O_2$, $Ti/IrO_2{-}Ta_2O_5$, Ti/PbO_2, $Ti/Pt\text{-}IrO_2$, and $Ti/SnO_2{-}Sb$ anodes in oxidizing ROC in a divided electrolytic cell. The best oxidation performance in terms of COD, DOC, $SUVA_{254}$, and TAN was observed for $Ti/Pt\text{-}IrO_2$ and $Ti/SnO_2{-}Sb$ anodes. However, this performance was correlated to the production of active chlorine, and inevitably the highest formation of THMs and HAAs was observed for $Ti/Pt{-}IrO_2$ and $Ti/SnO_2{-}Sb$. On the other hand, the lowest concentration of THMs and HAAs was recorded for the worst performing anodes, $Ti/IrO_2{-}Ta_2O_5$.

Future Directions

The large variety and complexity of brine streams studied so far makes overall evaluation of electrochemical treatment rather difficult. From the available studies, it seems that complete COD, DON, and color removal in electrochemical oxidation of ROC can be achieved in an economical and technically feasible way due to the presence of Cl^- ions. Although chlorine-mediated electrolysis enhances the rates of disappearance and possibly mineralization of contaminants, electro-generated chlorine and bromine may outweigh the contribution of $OH^•$ and other ROS in the bulk and direct the oxidation pathways towards the formation of more toxic and persistent halogenated by-products. Several studies reviewed in this manuscript reported the formation of AOX and halogenated by-products such as THMs and HAAs in electrooxidation using active and non-active MMOs but also novel BDD anodes. The increased toxicity of electrochemically treated effluent may diminish the beneficial effects of chloride ions (e.g., decreases in ohmic resistance), irrespective of electrode material used. Further research is needed to explore the performance of different reactor configurations (e.g., undivided versus divided electrolytic cells) and operational parameters (e.g., flow rate and mixing conditions, residence time, pH) to ensure a safe application of electrochemical processes for the treatment of ROC.

References

1. Bellona C et al (2004) Factors affecting the rejection of organic solutes during NF/RO treatment – a literature review. Water Res 38(12):2795–2809
2. Khan SJ et al (2009) Management of concentrated waste streams from high-pressure membrane water treatment systems. Crit Rev Env Sci Tec 39(5):367–415
3. Voutchkov N (2005) Alternatives for ocean discharge of seawater desalination plant concentrate. In: 20th Annual WateReuse Symposium, WateReuse Association, Denver, 18–21 Sept 2005
4. Dialynas E et al (2008) Advanced treatment of the reverse osmosis concentrate produced during reclamation of municipal wastewater. Water Res 42(18):4603–4608
5. Bagastyo AY et al (2011) Characterisation and removal of recalcitrants in reverse osmosis concentrates from water reclamation plants. Water Res 45(7):2415–2427
6. Westerhoff P et al (2009) Oxidation of organics in retentates from reverse osmosis wastewater reuse facilities. Water Res 43(16):3992–3998
7. Benner J et al (2008) Ozonation of reverse osmosis concentrate: kinetics and efficiency of beta blocker oxidation. Water Res 42(12):3003–3012
8. Ángela A et al (2009) Contributions of electrochemical oxidation to waste-water treatment: fundamentals and review of applications. J Chem Technol Biot 84(12):1747–1755
9. Comninellis C, Chen G (2010) Electrochemistry for the environment. Springer, New York/London
10. Marselli B et al (2003) Electrogeneration of hydroxyl radicals on boron-doped diamond electrodes. J Electrochem Soc 150(3):D79–D83
11. Comninellis C (1994) Electrocatalysis in the electrochemical conversion/combustion of organic pollutants for waste water treatment. Electrochim Acta 39(11–12):1857–1862
12. Panizza M, Cerisola G (2005) Application of diamond electrodes to electrochemical processes. Electrochim Acta 51(2):191–199
13. Lawrence KW et al (2007) Advanced physicochemical treatment technologies, in The Handbook of Environmental Engineering. Humana Press, Springer, New York
14. Pillai KC et al (2009) Studies on process parameters for chlorine dioxide production using IrO_2 anode in an un-divided electrochemical cell. J Hazard Mater 164(2–3):812–819
15. Panizza M et al (2007) Electrochemical degradation of methylene blue. Sep Purif Technol 54(3):382–387
16. Malpass GRP et al (2006) Oxidation of the pesticide atrazine at DSA® electrodes. J Hazard Mater 137(1): 565–572
17. Bonfatti F et al (2000) Electrochemical incineration of glucose as a model organic substrate II. Role of active chlorine mediation. J Electrochem Soc 147(2):592–596
18. Panizza M, Cerisola G (2003) Electrochemical oxidation of 2-naphthol with in situ electrogenerated active chlorine. Electrochim Acta 48(11):1515–1519
19. Martínez-Huitle CA, Brillas E (2009) Decontamination of wastewaters containing synthetic organic dyes by electrochemical methods: a general review. Appl Catal B-Environ 87(3–4):105–145
20. Comninellis C, Nerini A (1995) Anodic oxidation of phenol in the presence of NaCl for wastewater treatment. J Appl Electrochem 25(1):23–28
21. Gallard H et al (2004) Chlorination of bisphenol A: kinetics and by-products formation. Chemosphere 56(5):465–473
22. Kim S et al (2010) Effects of electrolyte on the electrocatalytic activities of RuO_2/Ti and Sb-SnO_2/Ti anodes for water treatment. Appl Catal B-Environ 97(1–2):135–141
23. Wu M et al (2009) Applicability of boron-doped diamond electrode to the degradation of chloride-mediated and chloride-free wastewaters. J Hazard Mater 163(1):26–31

24. Zhou M et al (2011) Treatment of high-salinity reverse osmosis concentrate by electrochemical oxidation on BDD and DSA electrodes. Desalination 277(1–3):201–206
25. Radjenovic J et al (2011) Electrochemical oxidation of trace organic contaminants in reverse osmosis concentrate using RuO_2/IrO_2−coated titanium anodes. Water Res 45(4):1579–1586
26. Radjenovic J et al (2011) Electrochemical degradation of the β-blocker metoprolol by $Ti/Ru_{0.7}Ir_{0.3}O_2$ and Ti/SnO_2-Sb electrodes. Water Res 45(10):3205–3214
27. Bagastyo AY et al (2011) Electrochemical oxidation of reverse osmosis concentrate on mixed metal oxide (MMO) titanium coated electrodes. Water Res 45(16):4951–4959
28. Anglada Á et al (2011) Boron-doped diamond anodic treatment of landfill leachate: evaluation of operating variables and formation of oxidation by-products. Water Res 45(2):828–838
29. Escher BI, Fenner K (2011) Recent advances in environmental risk assessment of transformation products. Environ Sci Technol 45(9):3835–3847
30. Rajkumar D et al (2005) Indirect electrochemical oxidation of phenol in the presence of chloride for wastewater treatment. Chem Eng Technol 28(1):98–105
31. Shao L et al (2006) Electrolytic degradation of biorefractory organics and ammonia in leachate from bioreactor landfill. Water Sci Technol 53(11):143–150
32. Aquino Neto S, De Andrade AR (2009) Electrochemical degradation of glyphosate formulations at DSA® anodes in chloride medium: an AOX formation study. J Appl Electrochem 39(10):1863–1870
33. Boudreau J et al (2010) Competition between electrochemical advanced oxidation and electrochemical hypochlorination of sulfamethoxazole at a boron-doped diamond anode. Ind Eng Chem Res 49(6):2537–2542
34. Van Hege K et al (2002) Indirect electrochemical oxidation of reverse osmosis membrane concentrates at boron-doped diamond electrodes. Electrochem Commun 4(4):296–300
35. Van Hege K et al (2004) Electro-oxidative abatement of low-salinity reverse osmosis membrane concentrates. Water Res 38(6):1550–1558
36. Chaplin BP et al (2010) Electrochemical destruction of N -Nitrosodimethylamine in reverse osmosis concentrates using boron-doped diamond film electrodes. Environ Sci Technol 44(11):4264–4269
37. Perez G et al (2010) Electro-oxidation of reverse osmosis concentrates generated in tertiary water treatment. Water Res 44(9):2763–2772
38. Panizza M et al (2000) Electrochemical treatment of wastewater containing polyaromatic organic pollutants. Water Res 34(9):2601–2605
39. Silverstein RMC et al (1991) Spectrophotometric identification of organic compounds. Wiley, New York
40. Bergmann MEH et al (2009) The occurrence of perchlorate during drinking water electrolysis using BDD anodes. Electrochim Acta 54(7): 2102–2107
41. Bergmann MEH, Iourtchouk T, Rollin J (2011) The occurrence of bromate and perbromate on BDD anodes during electrolysis of aqueous systems containing bromide: first systematic experimental studies. J Appl Electrochem 41(9):1109–1123

Electrochemical Treatment of Swimming Pools

Jean Gobet
Adamant-Technologies, La Chaux-de-Fonds, Switzerland

Introduction

Microbiological contamination, mainly due to fecal contamination, is the main cause of illness or infection associated with swimming pool water. Acceptable microbiological water quality is maintained, in particular, by good hygiene practices, efficient filtration, and disinfection. An exhaustive analysis of risks and preventive measures in recreational water environments is provided in Ref. [1].

Chlorine is the most commonly used disinfecting agent in swimming pools. It is a very efficient disinfectant (a free chlorine residual of 0.5 ppm kills 99.9 % of Escherichia coli in less than 1 min) and has a long-lasting effect (hours), unless destroyed by direct exposure to sunlight or by organic load addition. Large public pools frequently use chlorine gas, but its toxicity imposes strict and complex security measures. In private pools, on-site generation of chlorine by electrolyzing common salt solutions (saltwater chlorination) is an increasingly popular alternative to sodium hypochlorite (household bleach) or chloroisocyanurate salt (stabilized chlorine) addition. Economical saltwater chlorination of pools was made possible by the advent of dimensionally stable anodes, DSA™, developed for the industrial chlor-alkali process. Silver and copper ions production by anodic dissolution is an alternative electrochemical disinfection method.

Saltwater Chlorination

Salt chlorinators, commercialized since the 1970s, generate free chlorine through oxidation of chloride ions added to the pool water:

$$2\ Cl^- \rightarrow Cl_2 + 2\ e^-$$

A sodium chloride concentration within 3–6 g/L is the typical recommended addition.

At usual pool pH, the chlorine molecule reacts very rapidly with water to produce a hypochlorous acid:

$$Cl_2 + H_2O \leftrightarrow HOCl + HCl$$

HOCl is a weak acid with a pKa of 7.4 that dissociates to form the hypochlorite ion:

$$HOCl <-> H^+ + OCl^-$$

The total concentration of HOCl + OCl^-, expressed as the equivalent Cl_2 concentration in ppm or mg/L, is called "free chlorine." The free chlorine disinfection level in residential swimming pools is maintained between 0.5 and 3 ppm and is usually controlled by simple colorimetric analysis. The biocidal activity of the acid form, HOCl or "active chlorine," is more than 20 times stronger than the activity of the ionized OCl– form [2]. It is therefore critical to maintain a pH lower than 7.6–7.8 where the HOCl form is significant (see Fig. 1). The minimum recommended pH is ~7 for bather comfort and pool material compatibility reasons.

Water oxidation is the competing anodic reaction:

$$2H_2O \rightarrow O_2 + 4e^- + 4H^+$$

Water reduction producing hydrogen gas takes place on the cathode:

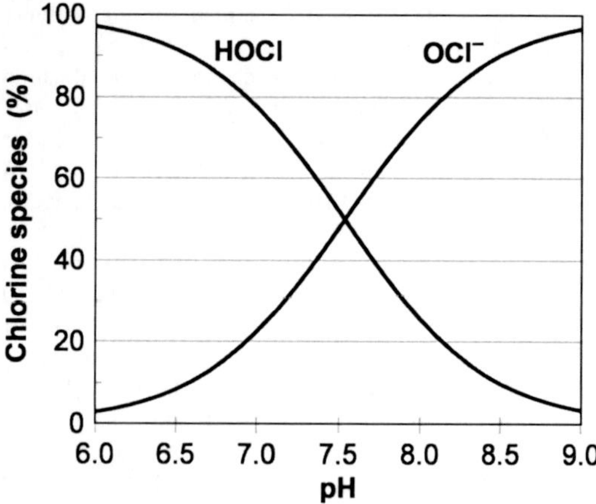

Electrochemical Treatment of Swimming Pools, Fig. 1 HOCl/OCl^- equilibrium at 25 °C, equilibrium constant from White [3]

$$2H_2O + 2e^- \rightarrow 2OH^- + H_2$$

The high pH at the negative electrode surface promotes calcium carbonate or magnesium hydroxide deposition, and regular cleaning is necessary. Polarity reversal, every 2–8 h depending on the water hardness, will remove the deposits, but this operation decreases the electrode lifetime (see below). If the electrode material does not allow polarity reversal (titanium or stainless steel cathodes), frequent acid cleaning is necessary.

High chlorine production yield and suitable lifetime are achieved with mixed metal oxide electrodes (MMO), initially developed for the chlor-alkali process [4]. They consist of a base metal, a titanium plate, or grid for saltwater chlorination coated with a combination of at least two metal oxides including titanium, ruthenium, iridium, or tantalum [5]. Elaborated proprietary production processes are used to achieve reliable electrodes. The chlorine faradaic yield under typical pool working conditions (3 g/L NaCl, current density 50 mA/cm^2) is around 60 % [6]. Good quality electrodes achieve several years of service time (several thousands of hours of continuous operation) with a 3 h reversal time and at 60 % of the nominal maximum power. Lifetime is reduced by high current densities and by frequent polarity reversal. Operation at

Electrochemical Treatment of Swimming Pools, Fig. 2 Example of a three electrode stack: 50 × 90 mm grids (Sterilor HC100)

temperatures below 15 °C is not allowed, as the chlorine faradic yield decreases with a concomitant detrimental increase of oxygen production.

Electrode plates or grids are assembled in stack of cells (see Fig. 2). There is no membrane separation between the electrodes. The electrodes are inserted "inline" directly into the water flow after the filtering unit. Monopolar, bipolar, and combined bipolar-monopolar electrode configurations are used. The required maximum chlorine production rate, for residential pools 2–3 g of Cl_2 per hour per 10 m^3 of water, defines the size and number of cells.

Chloramines resulting from the reaction of chlorine with nitrogen-containing compounds (ammonia, urea) are the main chemicals responsible for "chlorine smell" and eye/skin irritation in swimming pools. Saltwater chlorination lowers the chloramine level by a mechanism similar to "breakpoint chlorination." The excess concentration of chlorine relative to ammonia close to the electrode surface results in chloramine decomposition prior to injection of treated water into the pool [7]. In the presence of organics, saltwater chlorination, as all chlorine-based disinfections, can generate unhealthy halogenated by-products, in particular trihalomethanes and haloacetic acids.

On-site hypochlorite or chlorine gas electrochemical generators using brine solutions have larger production capacities and are not commonly used in residential pools.

Silver/Copper Ionization

The efficiency of silver and copper ions to disinfect water has been known since antiquity. Controlled anodic dissolution of silver and copper anodes or of silver/copper alloy anodes allows the release of the required amount of silver ($Ag+$, 5–50 ppb) and copper ($Cu+$, Cu^{2+}, 0.4–0.8 ppm) in the pool water.

The advantage of disinfection based only on metal ions is the absence of chlorine disinfection by-products [8]. The inactivation rate of viruses and bacteria by silver/copper is one or two orders of magnitude slower than with chlorine, and these ions have no oxidation effect on organic contamination. For this reason, small chlorine residuals (0.5 ppm) are still recommended [9].

Future Directions

Conductive boron-doped diamond (BDD) thin films, deposited on doped silicon or on self-passivating metals such as niobium or tantalum, form a new class of electrodes with outstanding properties. BDD electrodes have very high electrochemical corrosion stability and exhibit large overpotentials for oxygen evolution enabling efficient generation of highly oxidizing hydroxyl radicals. Because of the extremely powerful oxidation properties of hydroxyl radicals, however, operating conditions of disinfection systems based on boron-doped electrodes must be adjusted to minimize the chloride oxidation to unwanted chlorate or perchlorate higher oxides. In addition to by-product formation, the tendency exists for cathodes to be coated with scaling deposits. Future efforts should aim at solving these problems.

Cross-References

- ▶ Electrochemical Sensors for Water Pollution and Quality Monitoring
- ▶ Electrochemical Treatment of Urine

References

1. World Health Organization (2006) Guidelines for safe recreational waters – volume 2: swimming pools and similar environments. World Health Organization, Geneva
2. Kott Y, Nupen EM, Ross WR (1975) Effect of pH on efficiency of chlorine disinfection and virus enumeration. Water Res 9(10):869–872
3. White GC (1972) Handbook of chlorination. Van Nostrand Reinhold Company, New York
4. Beer H (1980) The invention and industrial development of metal anode. J Electrochem Soc 127:303C–307C
5. Xinyong Li LG, Chen G (2010) Techniques for electrode fabrication. In: Comninellis C, Chen G (eds) Electrochemistry for the environment. Springer, New York, pp 55–62, Chapter 3
6. Carlson R, Hardee K, Lau S (2011) (A1) – electrochlorination method for above-ground swimming pools. WO2011107561
7. Kapałka A, Katsaounis A, Michels NL, Leonidova A, Souentie S, Comninellis C, Udert K (2010) Ammonia oxidation to nitrogen mediated by electrogenerated active chlorine on Ti/PtOx-IrO2. Electrochem Commun 12:1203–1205
8. Beer W, Guilmartin L, Mcloughlin T, White T (1999) Swimming pool disinfection: efficacy of copper/silver ions with reduced chlorine levels. J Environ Health 61:9–12
9. Yahya MT, Landeen LK, Messina MC, Kutz SM, Schulze R, Gerba CP (1990) Disinfection of bacteria in water systems by using electrolytically generated copper:silver and reduced levels of free chlorine. Can J Microbiol 36(2):109–116

Electrochemical Treatment of Urine

Kai M. Udert
Eawag, Dübendorf, Switzerland

Aim of Electrochemical Urine Treatment

Separation at the source and decentralized treatment of urine, feces, and greywater is a novel concept for wastewater management [1]. This approach is particularly suitable for regions where the availability of water resources is low or the infrastructure for centralized wastewater treatment is missing. Urine is the main contributor of nutrients to municipal wastewater (80 % of nitrogen and 50 % of phosphorus, [2]). Since the urine volume is small (1.3 $L \cdot s^{-1} \cdot d^{-1}$, [2]), nutrient removal (or recovery) could be achieved in small decentralized reactors.

Some properties of electrochemical processes make them particularly suitable for urine treatment in a small decentralized reactor: no chemicals have to be added, the reaction can be easily started and terminated, a wide range of compounds can be oxidized or reduced, and the electric parameters current and voltage can be used for process automation and remote process control. However, electrochemical processes also have some drawbacks: the energy consumption is higher than for comparable biological processes, side reactions can produce unwanted by-products, and efficient electrodes are often expensive.

The main focus of this article is on the electrochemical removal of nitrogen and phosphorus from source-separated urine. Bioelectrochemical systems for urine treatment will not be discussed, because the research is still at an early stage. Some possible applications of bioelectrochemical systems for urine treatment have been presented in Udert et al. [3].

Composition of Urine

High conductivity (up to 40 $mS \cdot cm^{-1}$, [4]) and high chloride concentrations (up to 3.8 $g \cdot L^{-1}$, [5]) make urine a suitable solution for electrochemical processes, especially for indirect oxidation via chlorine. However, researchers and engineers also have to consider that biological degradation processes change the composition of urine considerably. Electrochemical processes, which can be used to treat fresh urine, might be unsuitable for stored urine and vice versa. During the collection of urine in toilet systems, the main nitrogen compound, urea, is quickly hydrolyzed by the enzyme urease [6]. Due to urea degradation, the pH increases from

6.2 to 9.0, nearly all nitrogen is present as ammonia (about 8.1 gN · L^{-1}), the total carbonate concentration reaches values of around 3.2 gC L^{-1}, and the alkalinity is close to 0.5 mol · L^{-1} [5]. Due to the pH increase, phosphate minerals are oversaturated, and most of the magnesium and calcium precipitates with phosphate as struvite ($MgNH_4PO_4 \cdot 6H_2O$) and hydroxylapatite ($Ca_{10}(PO_4)_6(OH)_2$) [6]. In undiluted urine, around 30 % of the phosphate (initial concentration around 740 mgP · L^{-1}) is removed. More phosphate can be easily precipitated as struvite by dosing a magnesium source [7]. It should be noted that the concentrations given in Udert et al. [5] are based on medical data. In real urine-collecting systems, the urine concentrations can be substantially lower due to ammonia volatilization and dilution [8].

Some organic compounds are fermented in the collection tank. Fresh urine contains a wide variety of different organic compounds, with creatinine being the most abundant (25 % of the chemical oxygen demand, COD, calculated with data from [9]). After storage, more than half of the organic compounds are volatile fatty acids (VFAs). Udert et al. [3] reported 47 % acetic acid, 4 % propionic acid, and 6 % butyric acid. For comparison, fresh urine contains only about 0.7 % acetic acid, 0.1 % propionic acid, and 1.7 % butyric acid (calculated with data from [9]).

Nitrogen Removal

Indirect Oxidation

Indirect oxidation via chlorine is the most common electrochemical process for ammonia removal. At high anode potentials, chloride is oxidized to chlorine (Cl_2) which reacts with water to form hypochlorous acid (HOCl). The latter is a strong oxidant that reacts with ammonia to form chloramines and, at sufficiently high chlorine dosage, to elemental nitrogen (N_2) and chloride [10]. Chlorine-mediated oxidation of ammonia has been described for various electrodes such as boron-doped diamond electrodes (BDD, [11, 12]) and dimensionally stable anodes (DSA, e.g., Ti/PtOx-IrO_2, [10]).

Electrochemical chlorine formation can also be used to oxidize urea. The products are N_2 and carbon dioxide and minor amounts of nitrate [13]. Kim et al. [14] used an electrochemical cell consisting of a BiOx-TiO_2 anode and a steel cathode to produce hydrogen from urea in fresh urine. The energy was provided by photovoltaic arrays. Ikematsu et al. [15] used a laboratory reactor consisting of two platinum-iridium DSA as anodes and an iron electrode as cathode. The experiments were conducted with diluted samples of fresh and synthetic urine. Since the current density was high (40 mA.cm^{-2}), indirect oxidation was the most probable oxidation mechanism. About 95 % of the total nitrogen could be removed. The residual dissolved nitrogen was nitrate.

Amstutz et al. [16] compared indirect nitrogen oxidation in synthetic solutions representing fresh and stored urine. They used thermally prepared IrO_2 anodes. During electrolysis of fresh urine, 72 % of the nitrogen was removed, and 24 % was converted to nitrate. The current efficiency for urea oxidation was 70 %. However, in stored urine, only about 3 % of ammonia was removed, and 9 % was oxidized to nitrate. The researchers showed that the high carbonate concentrations were responsible for the low ammonia oxidation in stored urine. They argued that carbonate oxidation prevented the formation of chlorine.

Direct Oxidation

Ammonia

Direct ammonia oxidation is well understood for platinum and some other noble metals [17]. The actual reactant is free ammonia (NH_3); therefore more ammonia is oxidized at higher pH values. At low overpotentials, nearly all ammonia is oxidized to N_2; as the potential increases, some nitrate and nitrite are produced, and at even higher potentials, the electrodes are poisoned by nitrogen adsorption. Of all tested noble metal electrodes, those with platinum and iridium deposits exhibit a particularly high performance (see, e.g., [18]). Direct ammonia oxidation has also been reported for electrodes without noble metals. Two examples are BDD [12] or

$Ni/Ni(OH)_2$ [19]. To our knowledge, no study on direct electrochemical ammonia oxidation in urine has been reported so far. Nicolau et al. [20] proposed an electrolyzer for urine treatment in space missions, which consists of a first reactor with immobilized urease to produce ammonia, which is later oxidized on a platinized BDD to hydrogen and nitrogen. However, all experiments were conducted with urea solutions and not urine.

Much of the recent research on direct ammonia electrooxidation aimed at using ammonia as an energy source. Ammonia could be suitable for hydrogen storage because the energy demand for hydrogen production from ammonia is low: under standard conditions, the minimum voltage for hydrogen production from ammonia is only 0.06 V compared to 1.23 V for water splitting [18]. Boggs and Botte [21] used a hybrid system consisting of an ammonia electrolytic cell and a breathable proton exchange membrane fuel cell to directly recover electric energy from ammonia. The anode in the ammonia electrolytic cell consisted of carbon fiber paper electrodes supported by titanium foil, covered with platinum and iridium deposits.

Lan and Tao [22] successfully applied a novel fuel cell type with an alkaline membrane to oxidize ammonia at room temperature. Compared to solid oxide fuel cells, the alkaline membrane fuel cell is less brittle and can be operated at low temperatures. As an advantage of alkaline membrane fuel cells over conventional alkaline fuel cells, no KOH-based electrolyte is needed. The researchers used two types of anodes: first platinum and ruthenium deposited on carbon and second chromium-decorated nickel. The ammonia sources were either ammonia gas or a 35 wt% aqueous ammonia solution.

Urea

Boggs et al. [23] investigated electrolytic hydrogen production from urea. At standard conditions, hydrogen production from urea would require only 0.37 V compared to 1.23 V for water splitting. They conducted experiments with urea in an alkaline KOH solution and found that inexpensive nickel was more efficient than Pt, Pt-Ir, or Rh anodes. Electrodes with even higher performances could be produced by depositing noble metals on a nickel substrate [24] by using nickel hydroxides [25] or nickel-cobalt hydroxides [26]. It is unclear whether direct urea electrooxidation on nickel-type electrodes would also work with urine. Boggs et al. [23] used cyclic voltammetry to prove urea degradation in a 1 mol $\cdot$ L^{-1} KOH solution spiked with urine, but long-term urea electrooxidation has not been described in the scientific literature so far.

Lan et al. [27] used alkaline membrane fuel cells for electricity production from urea. Power densities of up to 0.10 mW $\cdot$ cm^{-2} were achieved with a urine feed and a Ni/C anode. The performance was substantially increased by using nano-sized nickel particles instead of nickel powder [28]. Again, no long-term experiments were conducted to prove the resilience and steadiness of the process.

The studies show that direct electrochemical oxidation of urea (or ammonia) could be used to recover energy from urine. However, the energy recovery from one person's urine is rather low (less than 1 $\cdot$ W p^{-1} for a urea fuel cell, [3]). This process might only be economically interesting at places with high urine production [3].

Phosphate Removal

Electrochemical dissolution of sacrificial anodes, for example, iron, aluminum, or magnesium, has been proposed for phosphate removal from urine. Ikematsu et al. [15] used an electrochemical reactor consisting of two DSA and one iron electrode for combined nitrogen oxidation and phosphate precipitation. First, urea was oxidized at the DSA, then the current direction was changed, and phosphate was precipitated by dissolving the iron electrode. Zheng et al. [29, 30] used synthetic and real fresh urine for their experiments with iron and aluminum electrodes. With both types of electrodes, complete phosphate removal was achieved. At 40 mA $\cdot$ cm^{-2} and a gap width of 5 mm, 1.3 mol Fe . mol P^{-1} had to be dosed to remove 98 % of the phosphate (calculated by assuming a current efficiency of 100 %). The energy demand was 8.7 Wh · gP^{-1}. At lower anode potentials the energy demand was substantially lower: 2.4 Wh · gP^{-1} at 10 mA · cm^{-2}. The energy demand for aluminum electrodes was generally higher, e.g., 4.6 Wh · gP^{-1} at 10 mA · cm^{-2}.

When using magnesium as a sacrificial electrode, phosphate can be precipitated as struvite [31]. Struvite is a preferred product over iron or aluminum phosphates, because it is a better phosphate fertilizer [32]. Without base addition, the process can only be applied to stored urine, because a high pH value is required for struvite formation.

Hug and Udert [31] investigated the use of electrochemical struvite precipitation with magnesium electrodes in stored urine with a magnesium plate as anode, a steel plate as cathode and a gap width of 55 mm. They found that a minimum anode potential of −0.9 V vs. NHE was necessary to overcome passivation by hydroxide films. In a long-term experiment with 13 subsequent cycles and an anode potential of −0.6 V vs. NHE, they measured a current efficiency of 118 ± 10 % and a current density of 5.5 ± 0.7 mA · cm^2. In a batch experiment with the same anode potential, the energy demand was 1.7 Wh · gP^{-1}, which is similar to the energy demand for iron dissolution. 90 % of the phosphate could be removed immediately when 1 mol magnesium was dosed per 1 mol of initial phosphate. Nearly all phosphate could be removed (about 96 %) when the precipitation process continued for some hours during storage after the current had been switched off.

Future Directions

Optimizing Nitrogen Removal

Direct electrooxidation of urea and ammonia is more attractive than indirect electrooxidation, since the energy demand and the risk of producing harmful by-products is lower. However, most electrodes for direct oxidation are expensive. Therefore, research is needed to identify and test cheaper electrodes for direct oxidation. The experiments have to be done with fresh and stored urine, because both solutions have very different compositions. Finally, long-term electrolysis experiments are required to assess the long-term suitability of the electrodes.

Removal of Pathogens and Micropollutants

Electrochemical treatment could also be used to disinfect urine before it is used as a fertilizer. The most likely process is electrochemical chlorine formation. However, indirect electrooxidation via chlorine can lead to unwanted by-products such as chlorate, perchlorate, and halogenated organic compounds [12]. More research is needed to determine whether and how the formation of unwanted by-products can be prevented.

Another possible application of electrochemical reactors for urine treatment is the removal of micropollutants such as pharmaceutical residues. Depending on the characteristics of the micropollutants, direct oxidation at the anode [33] or reduction at the cathode (e.g., for the removal of halogenated organic compounds such as trihalomethanes, [34]) can be a suitable process.

References

1. Larsen TA, Udert KM, Lienert J (2013) Source separation and decentralization for wastewater management. IWA Publishing, London
2. Friedler E, Butler D, Alfiya Y (2013) Wastewater composition. In: Larsen TA, Udert KM, Lienert J (eds) Source separation and decentralization for wastewater management. IWA Publishing, London
3. Udert KM, Brown-Malker S, Keller J (2013) Electrochemical systems. In: Larsen TA, Udert KM, Lienert J (eds) Source separation and decentralization for wastewater management. IWA Publishing, London
4. Ronteltap M, Maurer M, Hausherr R, Gujer W (2010) Struvite precipitation from urine – influencing factors on particle size. Water Res 44(6):2038–2046
5. Udert KM, Larsen TA, Gujer W (2006) Fate of major compounds in source-separated urine. Water Sci Technol 54(11–12):413–420
6. Udert KM, Larsen TA, Biebow M, Gujer W (2003) Urea hydrolysis and precipitation dynamics in a urine-collecting system. Water Res 37(11):2571–2582
7. Etter B, Tilley E, Khadka R, Udert KM (2011) Low-cost struvite production using source-separated urine in Nepal. Water Res 45(2):852–862
8. Siegrist H, Laureni M, Udert KM (2013) Transfer into the gas phase: ammonia stripping. In: Larsen TA, Udert KM, Lienert J (eds) Source separation and

decentralization for wastewater management. IWA publishing, London
9. Ciba-Geigy (1977)WissenschaftlicheTabellen Geigy, Teilband (Körperflüssigkeiten (Scientific tables Geigy. Volume body fluids), 8th edn. Basel, Switzerland. (in German)
10. Kapałka A, Katsaounis A, Michels NL, Leonidova A, Souentie S, Comninellis C, Udert KM (2010) Ammonia oxidation to nitrogen mediated by electrogenerated active chlorine on Ti/PtOx-IrO_2. Electrochem Commun 12(9):1203–1205
11. Anglada A, Ortiz D, Urtiaga AM, Ortiz I (2010) Electrochemical oxidation of landfill leachates at pilot scale: evaluation of energy needs. Water Sci Technol 61(9):2211–2217
12. Kapałka A, Joss L, Anglada A, Comninellis C, Udert KM (2010) Direct and mediated electrochemical oxidation of ammonia on boron-doped diamond electrode. Electrochem Commun 12(12):1714–1717
13. Simka W, Piotrowski J, Nawrat G (2007) Influence of anode material on electrochemical decomposition of urea. Electrochim Acta 52(18):5696–5703
14. Kim J, Choi WJK, Choi J, Hoffmann MR, Park H (2012) Electrolysis of urea and urine for solar hydrogen. Catal Today 199:2–7. doi:10.1016/j.cattod.2012.02.009
15. Ikematsu M, Kaneda K, Iseki M, Matsuura H, Yasuda M (2006) Electrolytic treatment of human urine to remove nitrogen and phosphorus. Chem Lett 35(6):576–577
16. Amstutz V, Katsaounis A, Kapalka A, Comninellis C, Udert KM (2012) Effects of carbonate on the electrolytic removal of ammonia and urea from urine with thermally prepared IrO_2 electrodes. J Appl Electrochem 42(9):787–795
17. Bunce NJ, Bejan D (2011) Mechanism of electrochemical oxidation of ammonia. Electrochim Acta 56(24):8085–8093
18. Vitse F, Cooper M, Botte GG (2005) On the use of ammonia electrolysis for hydrogen production. J Power Sources 142(1–2):18–26
19. Kapałka A, Cally A, Neodo S, Comninellis C, Wächter M, Udert KM (2010) Electrochemical behavior of ammonia at Ni/$Ni(OH)_2$ electrode. Electrochem Commun 12(1):18–21
20. Nicolau E, González-González I, Flynn M, Griebenow K, Cabrera CR (2009) Bioelectrochemical degradation of urea at platinized boron doped diamond electrodes for bioregenerative systems. Adv Space Res 44(8):965–970
21. Boggs BK, Botte GG (2009) On-board hydrogen storage and production: an application of ammonia electrolysis. J Power Sources 192(2):573–581
22. Lan R, Tao S (2010) Direct ammonia alkaline anion-exchange membrane fuel cells. Electrochem Solid-State Lett 13(8):B83–B86
23. Boggs BK, King RL, Botte GG (2009) Urea electrolysis: direct hydrogen production from urine. Chem Commun 32:4859–4861
24. King RL, Botte GG (2011) Investigation of multi-metal catalysts for stable hydrogen production via urea electrolysis. J Power Sources 196(22):9579–9584
25. Wang D, Yan W, Botte GG (2011) Exfoliated nickel hydroxide nanosheets for urea electrolysis. Electrochem Commun 13(10):1135–1138
26. Yan W, Wang D, Botte GG (2012) Nickel and cobalt bimetallic hydroxide catalysts for urea electro-oxidation. Electrochim Acta 61(1):25–30
27. Lan R, Tao S, Irvine TS (2010) A direct urea fuel cell – power from fertiliser and waste. Energ Environ Sci 3:438–441
28. Lan R, Tao S (2011) Preparation of nano-sized nickel as anode catalyst for direct urea and urine fuel cells. J Power Sources 196(11):5021–5026
29. Zheng XY, Kong HN, Wu DY, Wang C, Li Y, Ye HR (2009) Phosphate removal from source separated urine by electrocoagulation using iron plate electrodes. Water Sci Technol 60(11):2929–2938
30. Zheng XY, Shen YH, Ye HR, Yan L, Zhang YJ, Wang C, Kong HN (2010) Phosphate removal from source-separated urine by electrocoagulation using aluminium plate electrodes. Fresen Environ Bull 19(4 A):635–640
31. Hug A, Udert KM (2013) Struvite precipitation from urine with electrochemical magnesium dosage. Water Res 47(1):289–299
32. Römer W (2006) Vergleichende Untersuchungen zur Pflanzenverfügbarkeit von Phosphat aus verschiedenen P-Recycling-Produkten im Keimpflanzenversuch (Plant availability of P from recycling products and phosphate fertilizers in a growth-chamber trial with rye seedlings). J Plant Nutr Soil Sci 169(6):826–832 (in German with English summary)
33. Radjenovic J, Bagastyo A, Rozendal R, Mu Y, Keller J, Rabaey K (2011) Electrochemical oxidation of trace organic contaminants in reverse osmosis concentrate using RuO_2/IrO_2-coated titanium anodes. Water Res 45(4):1579–1586
34. Radjenovic J, Farré MJ, Mu Y, Gernjak W, Keller J (2012) Reductive electrochemical remediation of emerging and regulated disinfection byproducts. Water Res 46(6):1705–1714

Electrochemiluminescence

Norihisa Kobayashi
Chiba University, Chiba, Japan

Introduction

Electrochemiluminescence (also called electrogenerated chemiluminescence and abbreviated ECL) is a light-emitting phenomenon whereby species electrochemically generated on the anode and the cathode undergo electron-transfer reactions to form excited states that emit light [1, 2]. The light emission can be found by the application of a bias voltage to an electrolyte solution containing ECL materials (e.g., $Ru(bpy)_3^{2+}$, where bpy is 2,2′-bipyridine) [3]. This light-emitting process is available to detect the molecules related emitting with the concentration lower than of nanomolar levels. Therefore, ECL is useful for analytical applications to detect some important biological molecules, such as protein, antibodies, and DNA [4]. The first detailed ECL study was reported in the mid-1960s [5, 6], and today, ECL is used as highly sensitive and selective analytical method for many analytes.

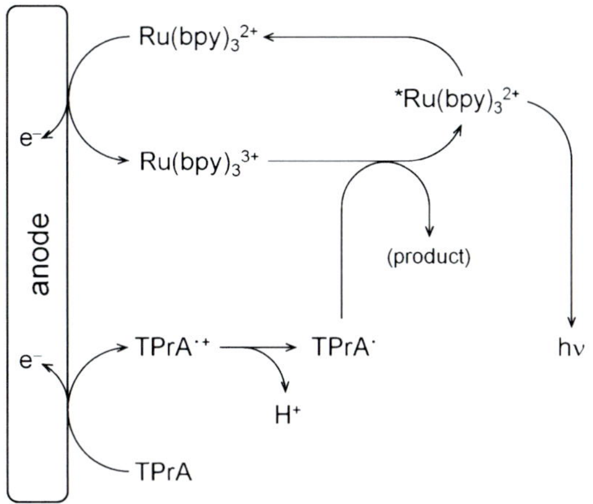

Electrochemiluminescence, Fig. 1 Proposed mechanism for the excited state formation and the light emission of $Ru(bpy)_3^{2+}$/TPrA ECL system

Principle

Both ECL and chemiluminescence (CL) are light-emitting phenomena from excited luminescent species generated by electron-transfer reactions. However, the trigger to obtain luminescence in ECL and CL is different. The trigger in CL is just a mixing of appropriate reagents with different potential energies. In contrast, the trigger in ECL to apply potential to the material is to induce the electrochemical reaction.

There are two reaction processes to generate ECL. The first detailed study about ECL was focused on "annihilation" mechanism induced by electron-transfer reactions between an oxidized and a reduced species. 9,10-Diphenylanthracene (DPA) was used as the emitting molecule for the annihilation mechanism in the early stage [5, 6]. A proposed mechanism is shown below.

$$DPA - e^- \rightarrow DPA^{\cdot+}$$

$$DPA + e^- \rightarrow DPA^{\cdot-}$$

$$DPA^{\cdot+} + DPA^{\cdot-} \rightarrow {}^*DPA + DPA$$

$${}^*DPA \rightarrow DPA + h\nu(\sim 430nm)$$

Oxidized and reduced states of DPA electrochemically generated on the anode and the cathode, respectively, collide with each other to generate the excited state of DPA. Consequently, luminescence is obtained from the excited state of DPA.

The other ECL mechanism using a "coreactant" with ECL materials is also widely accepted. The intermediate generated by the oxidation or reduction of coreactant works as the strong reducing and oxidizing agent for ECL emitting molecules to generate its excited state. The well-known example of this coreactant system is $Ru(bpy)_3^{2+}$/tri-*n*-propylamine(TPrA) system (as shown Fig. 1) [7, 8]. In this system, TPrA works as a coreactant.

$$Ru(bpy)_3^{2+} - e^- \rightarrow Ru(bpy)_3^{3+}$$

$$TPrA - e^- \rightarrow \left[TPrA^{\cdot}\right]^+ \rightarrow TPrA^{\cdot} + H^+$$

$$Ru(bpy)_3^{3+} + TPrA^{\cdot} \rightarrow {}^*Ru(bpy)_3^{2+} + Products$$

$${}^*Ru(bpy)_3^{2+} \rightarrow Ru(bpy)_3^{2+} + h\nu(\sim 620nm)$$

Oxidation of both $Ru(bpy)_3^{2+}$ and TPrAon the anode is the trigger to generate ECL in this system. TPrA radical working as a strong reducing agent is formed via the generation of

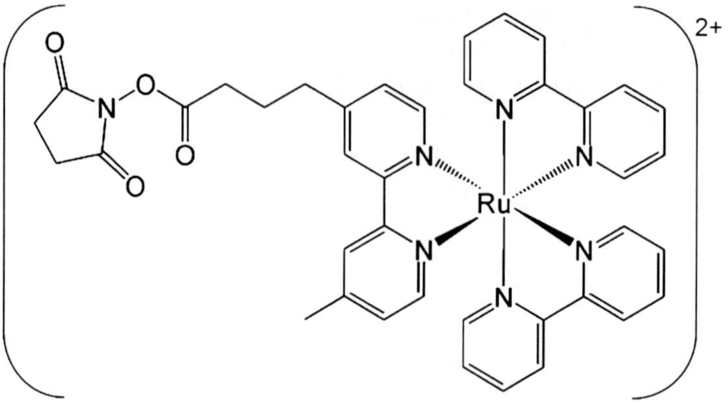

Electrochemiluminescence, Fig. 2 An example of $Ru(bpy)_3^{2+}$ derivative which has NHS ester (where NHS is *N*-Hydroxysuccinimide) for labeling to biomolecules such as proteins and peptides

short-lived radical cation ($[TPrA^{\cdot}]^{+}$) after oxidation. This radical reduces generated $Ru(bpy)_3^{3+}$ (oxidized state of the emitting molecules) to form its excited state by the electron transfer from TPrA radical to the π*-orbital of a Ru complex ligand. Consequently, $^{*}Ru(bpy)_3^{2+}$ formed gives luminescence.

The $Ru(bpy)_3^{2+}$ is the most well-known and valuable ECL material. This is because of its strong luminescence and good solubility in a variety of aqueous and nonaqueous solvents and so on. Moreover, Ru complexes which have effective ligands to link to other molecules (an example is shown in Fig. 2) can act as a label for analysis by attaching to analytes. TPrA is known to provide more efficient ECL in $Ru(bpy)_3^{2+}$/coreactant system compared with other coreactants. Therefore, this combination plays a key role in many analytical applications for specific biomolecules.

Applications

Diagnostic assays are the most well-known and important ECL application [4]. Compared with fluorescence (PL) methods that are also commonly used as diagnostic assays, light sources are not necessary in ECL methods, resulting in no negative effect of optical noises on assay sensitivity. Furthermore, ECL methods are more convenient as an analytical technique than CL methods that also do not need light sources. Since the redox reactions and the generations of excited states in ECL can be controlled by electrode potentials, ECL is highly selective method for some analytes over CL. Other advantage of ECL methods is that ECL materials can be regenerated after the luminescence in many cases. Therefore, one emitter can emit light repeatedly and gives enough emission intensity to detect even if ECL materials present at low concentration.

ECL assays are effective for determining the concentration of an emitter or a coreactant. The concentration of an emitter (or a molecule attached to ECL label such as Ru complex) is determined by detecting the emission intensity under the constant concentration of coreactant. The concentration of the coreactant is also measured under the constant emitter concentration. ECL emission intensity is generally proportional to the analyte concentration in both cases.

ECL is often combined with other assay techniques to detect the concentration of a coreactant or an emitter more precisely. For example, in order to determine the precise coreactant concentration, ECL assays are used with the combination of some analytical methods for the separation of molecules such as high-performance liquid chromatography (HPLC) and capillary electrophoresis (CE) [9, 10]. On the other hand, the detection of the ECL label as the emitter is often combined with magnetic bead-based assays (Fig. 3) that enables to

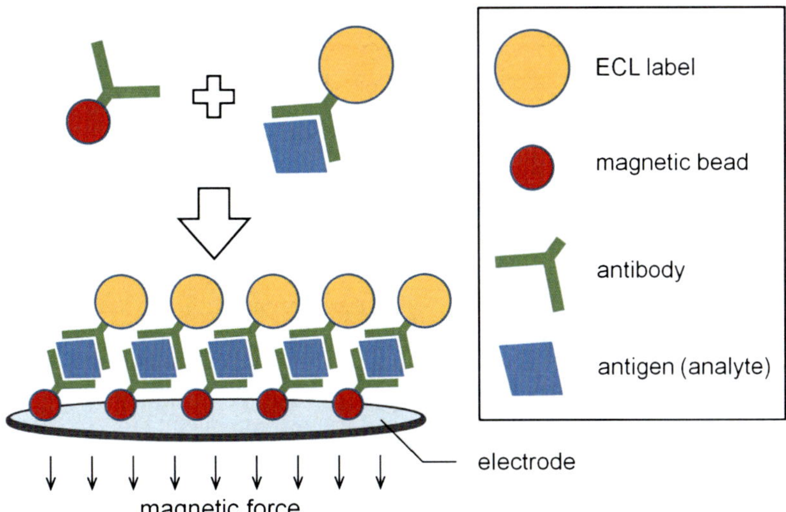

Electrochemiluminescence, Fig. 3 Schematic representation of sandwich type immunoassay using ECL label and magnetic bead for the detection of the analyte. Emission is observed from ECL label in the presence of a coreactant when bias voltage is applied to the electrode located the magnetic field

detect the limit of an analyte concentration lower than nanomolar levels [11]. This combination is widely used for commercial clinical tests.

Future Directions

Since mid-1960s, ECL had attracted much attention because of its scientific interests and practical applications. Although ECL has been developed with the progress in analytical methods for biomolecules to date, nowadays, ECL assays combined with other techniques are used as a powerful tool for the understanding of the detailed molecular structures and intercellular components.

In contrast, there is an interest in novel light-emitting devices based on the annihilation ECL. These devices are called ECL cell or light-emitting electrochemical cells (LECs). The first report about solid-state LECs using electrochemically active polymers was in 1995 [12], and since then, LECs have taken interests as possible alternatives for organic light-emitting diodes (OLEDs) because of specific advantages of LECs. These advantages attribute to the difference in the operation mechanism between LECs and OLEDs. The mechanism of LECs is based on the diffusion of ions generated electrochemically, while that of OLEDs is based on the injection and the migration of electrons and holes. Particularly, LECs do not merely depend on the work function of the electrodes because the electric double layer contributes to inject charges effectively into the active layer consisting of ECL molecules and electrolytes. Therefore, the electrode material in LECs is not limited in its work function. Typical LECs also have a very simple structure where only a single active layer is sandwiched between a pair of electrodes. The thickness of the active layer ranging from nanometer to submillimeter can be applicable in these LECs. Additionally, LECs can be generally turned on at lower voltage with the help of the electric double layer. LECs have some specific advantages over OLEDs.

Solid-state LECs are broadly divided into two categories such as polymer-based LECs and ionic transition metal complex (iTMC)-based LECs. In particular, iTMC-based LECs have been stimulating much attention for device application [13]. On the other hand, solution state and gel state LECs have been also studied toward a practical use for light-emitting devices. The application of alternating current (AC) to LECs is one of the techniques to improve the device properties

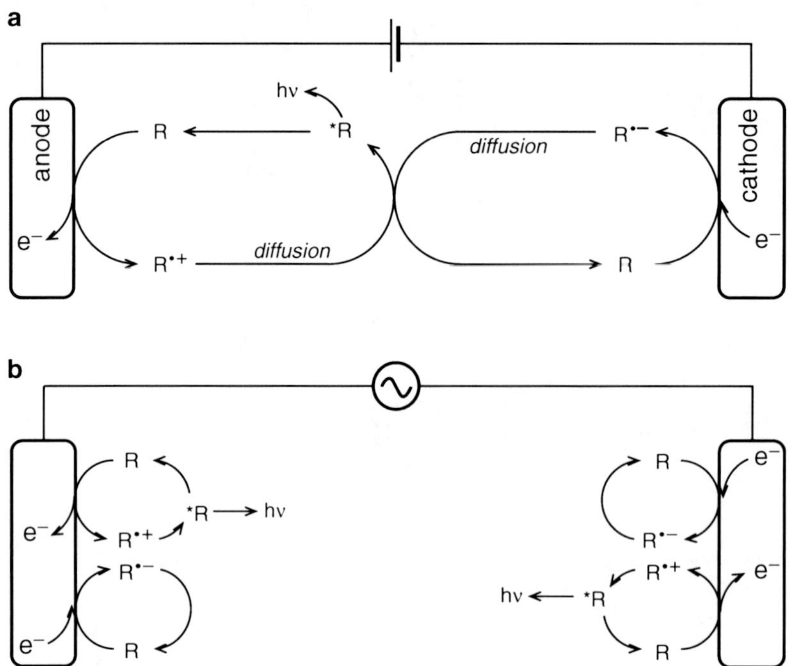

Electrochemiluminescence, Fig. 4 Proposed mechanism for the excited state formation and the light emission of annihilation ECL system in the LEC under application of (**a**) DC or (**b**) AC bias voltage

(shown in Fig. 4) [14]. Light-emitting devices, not only LECs but also OLEDs or other devices, are generally driven by application of direct current (DC). Especially, DC operation in LECs depended on the diffusion of ions, resulting in seriously slow turn-on response. In contrast, AC operation in LECs can achieve quick response ($\sim$ a few ms). This is because both of oxidized species and reduced species are generated on the same electrode surface that is switched reversibly between the anode and the cathode in an AC cycle, and it results in the generation of annihilation ECL without long distance diffusion of species.

Further, LECs driven by AC are expected to avoid the accumulation of ionic impurities near the electrodes leading to quenching process. In this way, LECs are expected to have enough potential as practical light-emitting device.

Cross-References

▶ Biomedical Applications of Electrochemistry, Use of Electric Fields in Cancer Therapy

References

1. Bard AJ (2004) Electrogenerated chemiluminescence. Marcel Dekker, New York
2. Richter MM (2004) Electrochemiluminescence (ECL). Chem Rev 104:3003–3036. doi:10.1021/cr020373d
3. Tokel NE, Bard AJ (1972) Electrogenerated chemiluminescence. IX. Electrochemistry and emission from systems containing Tris(2,2-bipyridine)ruthenium(II) Dichloride. J Am Chem Soc 94:2862–2863. doi:10.1021/ja00763a056
4. Hu L, Xu G (2010) Applications and trends in electrochemiluminescence. Chem Soc Rev 39:3275–3304. doi:10.1039/B923679C
5. Visco RE, Chandross EA (1964) Electroluminescence in solutions of aromatic hydrocarbons. J Am Chem Soc 86:5350–5354. doi:10.1021/ja01077a073
6. Santhanam KSV, Bard AJ (1965) Chemiluminescence of electrogenerated 9,10-diphenylanthracene anion radical. J Am Chem Soc 87:139–140. doi:10.1021/ja01079a039
7. Leland JK, Michael PJ (1990) Electrogenerated chemiluminescence: an oxidative-reduction type ECL reaction sequence using Tripropyl Amine. J Electrochem Soc 137:3127–3131. doi:10.1149/1.2086171
8. Miao W, Choi J-P, Bard AJ (2002) Electrogenerated chemiluminescence 69: the tris(2,2′-bipyridine)

ruthenium(II), ($Ru(bpy)_3^{2+}$)/tri-n-propylamine (TPrA) system revisited – a new route involving $TPrA^{\bullet+}$ cation radicals. J Am Chem Soc 124:14478–14485. doi:10.1021/ja027532v
9. Yin X-B, Wang E (2005) Capillary electrophoresis coupling with electrochemiluminescence detection: a review. Anal Chim Acta 533:113–120. doi:10.1016/j.aca.2004.11.015
10. Du Y, Wang E (2007) Capillary electrophoresis and microchip capillary electrophoresis with electrochemical and electrochemiluminescence detection. J Sep Sci 30:875–890. doi:10.1002/jssc.200600472
11. Miao W, Bard AJ (2004) Electrogenerated chemiluminescence. 80. C-reactive protein determination at high amplification with $[Ru(bpy)_3]^{2+}$-containing microspheres. Anal Chem 76:7109–7113. doi:10.1021/ac048782s
12. Pei Q, Yu G, Zhang C, Yang Y, Heeger AJ (1995) Polymer light-emitting electrochemical cells. Science 269:1086–1088
13. Bolink HJ, Coronado E, Costa RD, Ortí E, Sessolo M, Graber S, Doyle K, Neuburger M, Housecroft CE, Constable EC (2008) Long-living light-emitting electrochemical cells – control through supramolecular interactions. Adv Mater 20:3910–3913. doi:10.1002/adma.200801322
14. Nobeshima T, Morimoto T, Nakamura K, Kobayashi N (2010) Advantage of an AC-driven electrochemiluminescent cell containing a $Ru(bpy)_3^{2+}$ complex for quick response and high efficiency. J Mater Chem 20:10630–10633. doi:10.1039/C0JM02123G

Electrochemistry of Drug Release

Ellis Meng and Tuan Hoang
University of Southern California, Los Angeles, CA, USA

Introduction

Rate-controlled drug delivery pumps, particularly those at the microscale, can provide exquisite control of released drug profiles and thus closely approach the goal of maintaining the therapeutic drug concentration over the entire duration of treatment. This is in contrast to oral or topical routes that, while both convenient and noninvasive, are not suitable routes for many novel pharmaceutical agents including biologics, biosimilars, and other small molecules. Pumps have been prescribed for acute and chronic conditions including cancer, chronic pain, spasticity, and diabetes. Uses include the administration of antibiotics, chemotherapy, analgesics and opioids, nutrition formulas, insulin, lipids, vasopressors, blood products, and other drugs for which controlled rate of delivery is required [1]. The ability to modulate delivery rate is significant and allows for the design of new treatments that better synchronize with the dynamic biological processes occurring within the body. In turn, improved drug efficacy can be achieved with a customized dosing regimen. Drug pumps release pharmaceutical agents directly into the body either to the systemic circulation or locally at the site of treatment. Drug pumps can be classified based on their placement: external or internal.

Commercially available implantable pumps have employed different strategies to address these design considerations. While small in comparison to most external high-flow drug infusion pumps, the size of current implantable pumps still limits placement to within the abdomen. Catheters are required to route the drug to the desired site of action; not all sites are accessible using this method. These implantable pumps are predominantly driven by vapor pressure (fluorocarbon propellant) or electromechanical motors. Vapor pressure-based pumps were developed for implantable applications and provide limited control of drug delivery; drugs are infused at a fixed rate. Motorized pumps are used both internally and externally and can be controlled by microprocessors. Advanced systems have the ability to store specific programmed regimens. For implantable applications, the lifetime of pumps is currently limited by the battery. New classes of pumps have been enabled by developments in the field of microelectromechanical systems (MEMS). MEMS is a method for creating electromechanical devices with micro- and nanoscale features using microfabrication processes borrowed from the integrated circuit industry [2–4]. Miniaturized MEMS pumps are now possible that have small form factor so as to be less invasive and new accurate actuation methods to drive drug delivery [5–13]. Widely used mechanisms with electrical control of pumping include shape memory allow (SMA) [14–18],

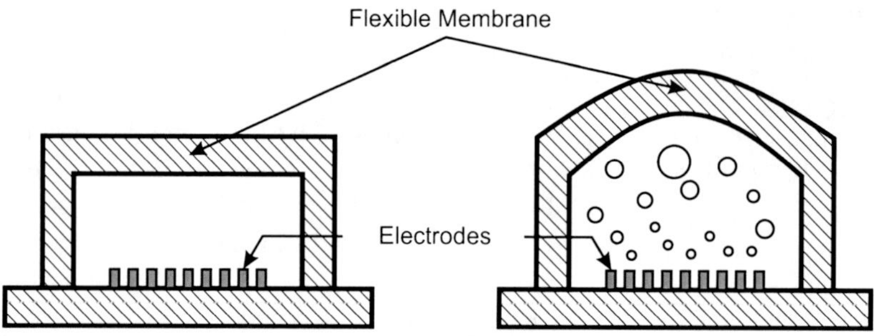

Electrochemistry of Drug Release, Fig. 1 Cross-sectional illustration of an electrolysis actuator consisting of a housing with a flexible membrane containing interdigitated electrodes in contact with an electrolyte solution. Application of current to the electrodes results in the evolution of gases, an increase in pressure, and deflection of the flexible membrane

piezoelectric [19–24], electrostatic [25–29], and thermopneumatic [30, 31].

Recent advances in MEMS electrochemical actuators enable a new class of drug pumps having a wide range of performance, miniature form factor, and low power consumption. In particular, this combination of features is well suited for implantable applications. Overall, electrochemical pumps can be harnessed to advance the management of chronic conditions as well as for research, preclinical, and clinical studies. Here, the principles behind electrochemically driven pumping and their application to novel implantable drug pumps are introduced and discussed.

Electrochemically Driven Drug Pumps

Electrolysis has long been recognized as an efficient process for the direct conversion of electrical energy to pressure-volume changes [32, 33]. Its advantages over other forms of actuation for implantable drug delivery applications include low power consumption, low heat dissipation, large volume change, continuous flow, reversible reaction, mechanical latching, and ON/OFF control [34–38]. The combination of these features allows precise pumping which is a requirement for drug delivery applications [39–42]. Implantable electrolysis-based micropumps have been explored for applications in ocular drug delivery [43–45], gene delivery to combat cancer [45], and laboratory animal research [46, 47].

Large-scale electrolysis pumps typically use a pair of wire or rod electrodes (typically gold, palladium, or platinum) to form the actuator [48–53]. Using microfabrication techniques, it is possible to achieve greater control over the placement and geometry of the wires as well as achieve accurately defined thin film interdigitated electrodes. When using water as the electrolyte in contact with the electrodes, current application induces the formation of oxygen at the anode and hydrogen at the cathode (Eqs. 1–3). This occurs at a rate that is proportional to the magnitude of the current applied (Eq. 4). While is it is possible to use voltage control, current control is preferred for drug delivery applications for its direct linear relationship to flow rate and to avoid variations in flow rate with changes in the system impedance. When the current application ceases, the reverse reactions take place and the hydrogen and oxygen gases recombine to form water. This reverse reaction can be catalyzed using platinum. If the electrolysis reaction is conducted in an appropriate pressure vessel containing an outlet, the liquid displaced is equivalent in quantity to the gas generated. Although direct electrolysis of the drug can be used for pumping, this may result in undesirable reaction products or alter the drug. A more useful electrolysis actuator design entails separation of the electrolysis reaction from the drug to be pumped (Fig. 1). Instead, a flexible membrane can be displaced by the electrolysis reaction and act on drug stored in an adjacent chamber. A further modification of the actuator may be useful in specific instances in which it is desired that the displaced membrane position be preserved when the actuator is powered off. By separating the gases generated by anode and cathode, recombination can be prevented.

A key advantage of electrolysis-based pumping is the wide range of flow rates that can be achieved and the ability to maintain constant flow rate or variable flow rates on demand. The scheme works well on fluids having a range of viscosities.

In the electrolysis reaction, three moles of gas is generated for every four moles of electrons according to the reactions shown below:

At the anode(oxidation):

$$2H_2O(l) \underset{\text{recombination}}{\overset{\text{electrolysis}}{\rightleftarrows}} O_2(g) + 4H^+(aq) + 4e^- \quad (1)$$

At the cathode(reduction):

$$2H_2O(l) + 2e^- \underset{\text{recombination}}{\overset{\text{electrolysis}}{\rightleftarrows}} 2OH^-(aq) + H_2(g) \quad (2)$$

$$\text{Net reaction}: 2H_2O(l) \underset{\text{recombination}}{\overset{\text{electrolysis}}{\rightleftarrows}} O_2(g) + 2H_2(g) \quad (3)$$

Therefore, hydrogen is produced at twice the volume of oxygen given that temperature and pressure are equal for both gases. If we assume that all gases are evolved as bubbles, the fluid pumping rate at atmospheric pressure is (in m^3/s):

$$Q_{\text{electrolysis}} = \frac{3}{4}\frac{i}{F}V_m \quad (4)$$

where i is the constant current (in A), F is Faraday's constant (96.49×10^3 C/mol), and V_m is the molar gas volume at 25 °C and atmospheric pressure (24.7×10^{-3} m^3/mol). In the ideal case, the total pumped drug volume is a function of the duration of applied current and is calculated as:

$$V_{\text{theoretical}} = Q_{\text{electrolysis}} t \quad (5)$$

However, efficiency of the reaction is usually impacted by a number of factors including electrode properties, electrochemical cell design, friction, heat dissipation, and other operating parameters [34]. The reaction efficiency is calculated using the following equation [54]:

$$\eta = \frac{V_{\text{experimental}}}{V_{\text{theoretical}}} \quad (6)$$

In addition to water, other electrolytes have also been explored for electrolysis-based pumping [34–36, 48]. When using an electrolyte solution, continuous contact between the electrodes and fluid maybe impeded by the presence of gas or postural variations of the pump, for example, if used in ambulatory settings. The first issue can be addressed by coating the electrodes with a solid electrolyte such as the polymer Nafion. The Nafion coating possesses high gas solubility so that gases rapidly diffuse away from the electrode surface [46, 47]. Separation of the electrode from the liquid electrolyte prevents direct contact of the generated gases on the electrode. Alternatively, use of an electrolytic hydrogel has been explored [53]. The hydrogel also alleviates issues related to postural variations of the actuator.

For the purpose of drug delivery, the most practical pump construction consists of a drug reservoir, electrolysis electrodes, electrolyte, flexible membrane, and catheter (Fig. 2). In some applications, it may also be desirable to include a one-way check valve or other forms of flow regulation in line with the catheter. Pumping commences when the generation of gas induces the deflection of the flexible membrane. The membrane then forces drug out of the catheter to the site of delivery. After the pump is turned off, the hydrogen and oxygen gases recombine to allow the membrane to return to its original resting position. In such a manner, the pump can be turned ON and OFF to achieve the desired dosing regimen. The large pneumatic pressures produced allow pumping even the presence of physiologically relevant backpressures and of viscous drugs [44].

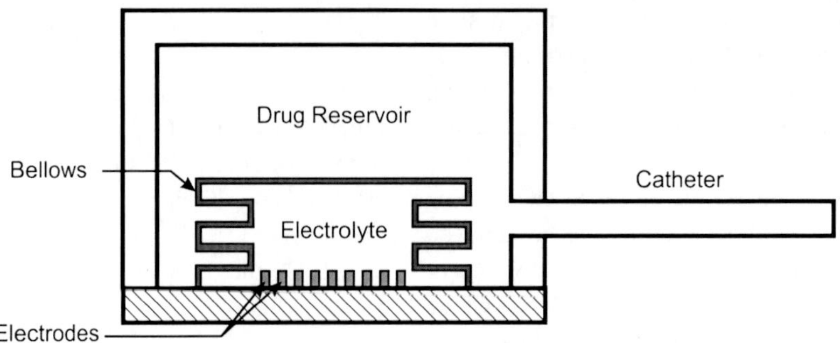

Electrochemistry of Drug Release, Fig. 2 Cross-sectional illustration of an electrolysis-based pump with an actuator having a flexible bellows diaphragm. The bellows mechanically couples the electrolysis actuation to the adjacent drug reservoir. Inflation of the bellows resulting from the electrolysis reaction pushes drug out of the rigid drug reservoir through an attached catheter directly to the site of treatment

Future Directions

Electrochemical pumps powered by electrolysis reactions enable rate-controlled drug infusion over a wide range of flow rates and operating conditions. In particular, these pumps offer many advantages over existing mechanisms used in implantable pumps. The ability to administer drug on demand based on biological feedback or according to circadian scheduling (chronotherapy) are significant in the context of improved management of chronic conditions. In the immediate future, the dynamic processes that occur during the electrolysis reactions may give rise to slight dosing inaccuracies that can be addressed by closed-loop control [39]. Electrolysis-based dosing systems and drug pumps are on track to move beyond simple open-loop systems through the incorporation of novel sensors that track pressure, flow rate, or other parameters that impact delivery [39, 55]. Further advances in biological sensors are required to realize the "holy grail" of true closed-loop feedback based on physiological cues such that the appropriate drug invention can be administered and the associated therapeutic effect can be confirmed. Such systems may play a role in emerging telemedicine and wireless health.

References

1. Graham A, Holohan T (1994) External and implantable infusion pumps. Agency for Health Care Policy and Research, Rockville
2. Kovacs GTA (1998) Micromachined transducers sourcebook. McGraw-Hill, Boston
3. Madou M (1997) Fundamentals of microfabrication. CRC Press, Boca Raton
4. Meng E (2010) Biomedical microsystems. CRC Press, Boca Raton
5. Ziaie B, Baldi A, Lei M, Gu YD, Siegel RA (2004) Hard and soft micromachining for biomems: review of techniques and examples of applications in microfluidics and drug delivery. Adv Drug Deliv Rev 56:145–172
6. Grayson ACR, Shawgo RS, Li YW, Cima MJ (2004) Electronic MEMS for triggered delivery. Adv Drug Deliv Rev 56:173–184
7. Razzacki SZ, Thwar PK, Yang M, Ugaz VM, Burns MA (2004) Integrated microsystems for controlled drug delivery. Adv Drug Deliv Rev 56:185–198
8. Tao SL, Desai TA (2003) Microfabricated drug delivery systems: from particles to pores. Adv Drug Deliv Rev 55:315–328
9. Nuxoll EE, Siegel RA (2009) Biomems devices for drug delivery improved therapy by design. Ieee Eng Med Biol Mag 28:31–39
10. Tsai NC, Sue CY (2007) Review of MEMS-based drug delivery and dosing systems. Sens Actuators a-Phys 134:555–564
11. Deo S, Moschou E, Peteu S, Eisenhardt P, Bachas L, Madou M, Daunert S (2003) Responsive drug delivery systems. Anal Chem 75:207A–213A
12. Prausnitz MR (2004) Microneedles for transdermal drug delivery. Adv Drug Deliv Rev 56:581–587
13. Prausnitz MR, Mikszta JA, Cormier M, Andrianov AK (2009) Microneedle-based vaccines. Curr Top Microbiol Immunol 333:369–393
14. Döring C, Grauer T, Marek J, Mettner MS, Trah H-P, Willmann M (1992) Micromachined thermoelectrically driven cantilever structures for fluid jet deflection. Paper presented at MEMS '92, Travemünde, 4-7 February 1992
15. Jerman H (1990) Electrically-activated, micromachined diaphragm valves. Paper presented at the 1990 solid

state sensor and actuator workshop, Hilton Head Island, 4-7 June 1990
16. Kohl M, Skrobanek KD (1998) Linear microactuators based on the shape memory effect. Sens Actuators A 70:104–111
17. Reynaerts D, Peirs J, VanBrussel H (1997) An implantable drug-delivery system based on shape memory alloy micro-actuation. Sens Actuators a-Phys 61:455–462
18. Benard WL, Kahn H, Heuer AH, Huff MA (1998) Thin-film shape-memory alloy actuated micropumps. J Microelectromechan Syst 7:245–251
19. Esashi M, Shoji S, Nakano A (1989) Normally closed microvalve and micropump fabricated on a silicon wafer. Sens Actuators A 20:163–169
20. Mescher M, Abe T, Brunett B, Metla H, Schlesinger TE, Reed M (1995) Piezoelectric lead-zirconate-titanate actuator films for microelectromechanical system applications. Paper presented at the MEMS '95, Amsterdam, 29 January - 2 February 1995
21. Maillefer D, van Lintel H, Rey-Mermet G, Hirschi R (1999) A high-performance silicon micropump for an implantable drug delivery system. Paper presented at the MEMS '99, Orlando, 17-21 January 1999
22. Cao L, Mantell S, Polla D (2001) Design and simulation of an implantable medical drug delivery system using microelectromechanical systems technology. Sens Actuators a-Phys 94:117–125
23. Su GG, Pidaparti RM (2010) Drug particle delivery investigation through a valveless micropump. J Microelectromech Syst 19:1390–1399
24. Su GG, Pidaparti RM (2010) Transport of drug particles in micropumps through novel actuation. Microsyst Technol-Micro- Nanosyst-Inform Storage Process Syst 16:595–606
25. Branebjerg J, Gravesen P (1992) A new electrostatic actuator providing improved stroke length and force. Paper presented at the MEMS '92, Travemünde, 4-7 February 1992
26. Sato K, Shikida M (1992) Electrostatic film actuator with a large vertical displacement. Paper presented at the MEMS '92, Travemünde, 4-7 February 1992
27. Bourouina T, Bosseboeuf A, Grandchamp JP (1997) Design and simulation of an electrostatic micropump for drug-delivery applications. J Micromechan Microengin 7:186–188
28. Yih TC, Wei C, Hammad B (2005) Modeling and characterization of a nanoliter drug-delivery mems micropump with circular bossed membrane. Nanomedicine 1:164–175
29. Teymoori MM, Abbaspour-Sani E (2005) Design and simulation of a novel electrostatic peristaltic micromachined pump for drug delivery applications. Sens Actuators a-Phys 117:222–229
30. Grosjean C, Yang X, Tai Y-C (1999) A practical thermopneumatic valve. In: (ed) MEMS '99, Orlando, 17-21 January 1999
31. Jeong OC, Tang SS (2000) Fabrication of a thermopneumatic microactuator with a corrugated p + silicon diaphragm. Sens Actuators A 80:62–67
32. Faraday M (1834) On electrical decomposition. Philos Trans 124:77–122
33. Nicholson W (1800) Account of the new electrical or galvanic apparatus of sig. Alex. Volta, and experiments performed with the same. J Nat Philos Chem Arts 4:179–187
34. Cameron CG, Freund MS (2002) Electrolytic actuators: alternative, high-performance, material-based devices. Proc Nat Acad Sci USA 99:7827–7831
35. Neagu C, Jansen H, Gardeniers H, Elwenspoek M (2000) The electrolysis of water: an actuation principle for mems with a big opportunity. Mechatronics 10:571–581
36. Neagu CR, Gardeniers JGE, Elwenspoek M, Kelly JJ (1996) An electrochemical microactuator: principle and first results. J Microelectromechan Syst 5:2–9
37. Stanczyk T, Ilic B, Hesketh PJ, Boyd JG (2000) A microfabricated electrochemical actuator for large displacements. J Microelectromechan Syst 9:314–320
38. Pang C, Tai YC, Burdick JW, Andersen RA (2006) Electrolysis-based diaphragm actuators. Nanotechnology 17:S64–S68
39. Bohm S, Timmer B, Olthuis W, Bergveld P (2000) A closed-loop controlled electrochemically actuated micro-dosing system. J Micromechan Microeng 10:498–504
40. Suzuki H, Yoneyama R (2002) A reversible electrochemical nanosyringe pump and some considerations to realize low-power consumption. Sens Actuators B-Chem 86:242–250
41. Suzuki H, Yoneyama R (2003) Integrated microfluidic system with electrochemically actuated on-chip pumps and valves. Sens Actuators B-Chem 96:38–45
42. Bohm S, Olthuis W, Bergveld P (1999) An integrated micromachined electrochemical pump and dosing system. Biomed Microdevices 1:121–130
43. Meng E, Shih J, Li P-Y, Lo R, Humayun M, Tai Y-C (2006) Electrolysis-driven drug delivery for treatment of ocular disease. Paper presented at the Micro total analysis systems 2006, Tokyo, 5-9 November 2006
44. Li PY, Shih J, Lo R, Saati S, Agrawal R, Humayun MS, Tai YC, Meng E (2008) An electrochemical intraocular drug delivery device. Sens Actuators A-Phys 143:41–48
45. Gensler H, Sheybani R, Li PY, Lo R, Zhu S, Yong K-T, Roy I, Prasad PN, Masood R, Sinha UK, Meng E (2010) Implantable mems drug delivery devices for cancer radiation reduction. Paper presented at the MEMS 2010, Hong Kong, 24-28 January 2010
46. Sheybani R, Meng E (2011) High efficiency wireless electrochemical actuators: design, fabrication and characterization by electrochemical impedance spectroscopy. Paper presented at the MEMS 2011, Cancun, 23-27 January 2011

E

47. Sheybani R, Gensler H, Meng E (2011) Rapid and repeatable bolus drug delivery enabled by high efficiency electrochemical bellows actuators. Paper presented at the Transducers 2011, Beijing, 5-9 June 2011
48. Young DB, Jackson TE, Pearce DH, Guyton AC (1977) A portable infusion pump for use on large laboratory animals. IEEE Trans Biomed Eng 24:543–545
49. Nalecz M, Lewandowski J, Werynski A, Zawicki I (1978) Bioengineering aspects of the artificial pancreas. Artif Organs 2:305–309
50. Janocha H (1988) Neue aktoren. Proc Actuator 88:389
51. O'Keefe D, Oherlihy C, Gross Y, Kelly JG (1994) Patient-controlled analgesia using a miniature electrochemically driven infusion-pump. Br J Anaesth 73:843–846
52. Groning R (1997) Computer-controlled drug release from small-sized dosage forms. J Control Release 48:185–193
53. Kim HC, Bae YH, Kim SW (1999) Innovative ambulatory drug delivery system using an electrolytic hydrogel infusion pump. IEEE Trans Biomed Eng 46:663–669
54. Xie J, Miao YN, Shih J, He Q, Liu J, Tai YC, Lee TD (2004) An electrochemical pumping system for on-chip gradient generation. Anal Chem 76:3756–3763
55. Gutierrez C, Sheybani R, Meng E (2011) Electrochemically-based dose measurement for closed-loop drug delivery applications. Paper presented at the Transducers 2011, Beijing, 5-9 June 2011

Electrode

Ping Gao and Rudolf Holze
AG Elektrochemie, Institut für Chemie, Technische Universität Chemnitz, Chemnitz, Germany

An electron-conducting material brought into contact with an ionically conducting phase establishes an electrode. In case of semiconductors in addition to electrons, holes may act as means of charge propagation. The ionically conducting phase may be an electrolyte solution composed of a dissociated electrolyte and a solvent, an ionic liquid, a molten salt, or a solid electrolyte. At the established interphase an equilibrium is established. Assuming at the instance of contact a non-equilibrium between both phases and the chemical potentials of the involved species, two $\overline{\mu}_i$ possibilities may be considered; one is depicted below (Fig. 1):

Considering as an electrode silver metal and silver ions (of, e.g., $AgNO_3$ dissolved in water), the electrode reaction at the interface may be

$$\mathrm{Ag} \rightleftarrows \mathrm{Ag}^+ + \mathrm{e}^-$$

At equilibrium the condition

$$\sum_i \nu_i \overline{\mu}_i = 0$$

has to be matched. For every species involved this implies

$$\overline{\mu}_i^1 = \overline{\mu}_i^2$$

i.e.,

$$\overline{\mu}_{\mathrm{Ag}} = \overline{\mu}_{\mathrm{Ag}^+} + z \cdot \overline{\mu}_{\mathrm{e}^-}$$

with the added line in the symbol indicating that electric charges have to be taken into account for all participating chemical species. The electrochemical potential of the silver ions is

$$\overline{\mu}_{\mathrm{Ag}^+} = \mu_{\mathrm{Ag}^+} + z \cdot F \cdot E_{sol}.$$

the others are

$$\overline{\mu}_{\mathrm{e}^-} = \mu_{\mathrm{e}^-} + z \cdot F \cdot E_{\mathrm{Ag}} \text{ and } \overline{\mu}_{\mathrm{Ag}} = \mu_{\mathrm{Ag}}$$

The electrical potential difference (although a quantity not accessible to direct measurement) is

$$\Delta E = \frac{1}{z \cdot F}\left(\mu_{\mathrm{Ag}^+} + z \cdot \mu_{\mathrm{e}^-} - \mu_{\mathrm{Ag}}\right)$$

With standard values

$$\mu_{\mathrm{Ag}^+} = \mu^0_{\mathrm{Ag}^+} + R \cdot T \cdot ln\, a_{\mathrm{Ag}^+}$$

this can be rearranged into

$$\Delta E = \frac{1}{z \cdot F}\left(\mu^0_{\mathrm{Ag}^+} + z \cdot \mu_{\mathrm{e}^-} - \mu_{\mathrm{Ag}}\right) + \frac{R \cdot T}{z \cdot F} \cdot ln\, a_{\mathrm{Ag}^+}$$

Electrode, Fig. 1 Metal and solution (*left*) before brought into contact (*right*) after being brought into contact

With index 00 for standard conditions and 0 for equilibrium conditions, one obtains

$$\Delta E_{00} = \frac{1}{z \cdot F}\left(\mu^0_{\mathrm{Ag^+}} + z \cdot \mu_{\mathrm{e^-}} - \mu_{\mathrm{Ag}}\right)$$

and

$$\Delta E_0 = \Delta E_{00} + \frac{R \cdot T}{z \cdot F} \cdot \ln a_{\mathrm{Ag^+}}$$

This potential difference is generally called the electrode potential E [as stipulated by IUPAC this symbol is written – like all symbols – in italics (Größen, Einheiten und Symbole in der Physikalischen Chemie (IUPAC Ed.), VCH, Weinheim 1996; see also Pure Appl. Chem. 37 (1974) 499.); the equation is called the Nernst equation. Quite obviously considering only the electronically conducting phase in describing an electrode is insufficient (although frequently done); this has already been pointed out by W. Nernst.

Numerous classifications of electrodes have been established. Depending on the purpose reference electrodes, working electrodes and auxiliary or counter electrodes as used in electrochemical three-electrode arrangements (see entries potentiostat, reference electrode) are distinguished. Various reactants are used in naming: A gas electrode employs at least one gaseous reactant in the electrode potential-determining redox couple:

$$\mathrm{Cl_2 + 2e^- \rightleftarrows 2Cl^-}$$

In a redox electrode both reactants are chemically closely related with different states of oxidation being the discerning property:

$$\mathrm{Fe^{3+} + e^- \rightleftarrows Fe^{2+}}$$

A metal electrode is composed of a metal in its elementary form and a corresponding ion in the ionically conducting phase:

$$\mathrm{Cu \rightleftarrows Cu^{2+} + 2e^-}$$

Sometimes electrodes are just named for the chemical identity of the material constituting the electronically conducting phase: carbon electrode, gold electrode, and silver electrode. As already pointed out above, this is apparently incomplete, even misleading. The lead electrode behaves entirely different when lead metal is brought into contact with sulfuric acid:

$$\mathrm{Pb + SO_4^{2-} \rightleftarrows PbSO_4 + 2e^-}$$

and with perchloric acid:

$$\mathrm{Pb \rightleftarrows Pb^{2+} + 2e^-}$$

with a solid lead sulfate formed directly on the lead electrode upon lead oxidation in the former case and the formation of soluble lead ions in the perchloric acid solution in the latter case.

Because of the relationship between the electrode potential-determining species and the electrode potential, electrodes may be employed as sensors (ion-sensitive and -selective detectors) in analytical chemistry, medicine, process control, and environmental monitoring.

According to their size, in particular to their typical dimensions, electrodes are called macroelectrodes with typical dimension (e.g., diameter of a disc-shaped electrode, length of the edge of a sheet electrode) in the range of mm or cm, microelectrodes (with μm), and nanoelectrodes (with nm). Electrodes for particular methods are called rotating disc electrodes (see entry "► Controlled Flow Methods for Electrochemical Measurements"), optically transparent electrodes (OTL, see entry "► UV–Vis Spectroelectrochemistry"), and thin-layer electrodes (TLE for electrolysis with a limited solution volume present as a thin layer of liquid).

Future Directions

Depending on the specific application and the constituting materials, new names may come up. The use of the term in the broader Nernstian sense is highly recommended in order to avoid confusion caused by lack of precision.

Cross-References

- ► Controlled Flow Methods for Electrochemical Measurements
- ► Micro- and Nanoelectrodes
- ► pH Electrodes - Industrial, Medical, and Other Applications
- ► Potentiostat
- ► Reference Electrodes
- ► Sensors
- ► Three-Dimensional Electrode
- ► UV–Vis Spectroelectrochemistry

References

1. Bard AJ, Inzelt G, Scholz F (2012) Electrochemical dictionary. Springer, Heidelberg

Electrode Catalysts for Direct Methanol Fuel Cells

Hideo Daimon
Advanced Research and Education, Doshisha University, Kyotanabe, Kyoto, Japan

Introduction

On January 2011, Japanese major automobile makers of Toyota, Honda, and Nissan have made a joint statement that they will start to manufacture fuel cell vehicle (FCV) from 2015 together with construction of hydrogen fueling stations [1]. The fuel cell used in the FCV is polymer electrolyte fuel cells (PEFCs) in which anode and cathode fuels are hydrogen and oxygen, respectively. Although the PEFCs can work under ambient condition, very expensive Pt system catalysts are used to promote anode reaction (hydrogen oxidation reaction; HOR) and cathode one (oxygen reduction reaction; ORR). Since kinetics of the ORR is slower than that of the HOR, much more Pt catalysts are needed in the cathode, increasing cost of the PEFCs as well as the FCV. Therefore, current research topics are focused on development of highly active cathode catalysts toward the ORR, such as core/shell-structured catalysts [2–6] and PtM-alloyed catalysts (M: 3d transition metals such as Fe, Co, Ni, and Cu) [7–14].

There is another fuel cell working under the ambient condition, that is, direct methanol fuel cells (DMFCs). Difference in the PEFCs and DMFCs is their anode fuels (the cathode fuel is oxygen in both cases). In the DMFCs, methanol (CH_3OH) is supplied to the anode instead of the hydrogen for the PEFCs and this difference is crucial for their cell performances. Although the Pt is known to be an active catalyst for both HOR and methanol oxidation reaction (MOR), kinetics of the MOR is much slower than that of the HOR and ORR on the Pt catalyst, which increases anode overpotential and gives an inferior cell performance in the DMFCs as demonstrated in Fig. 1. Therefore, an important research topic is

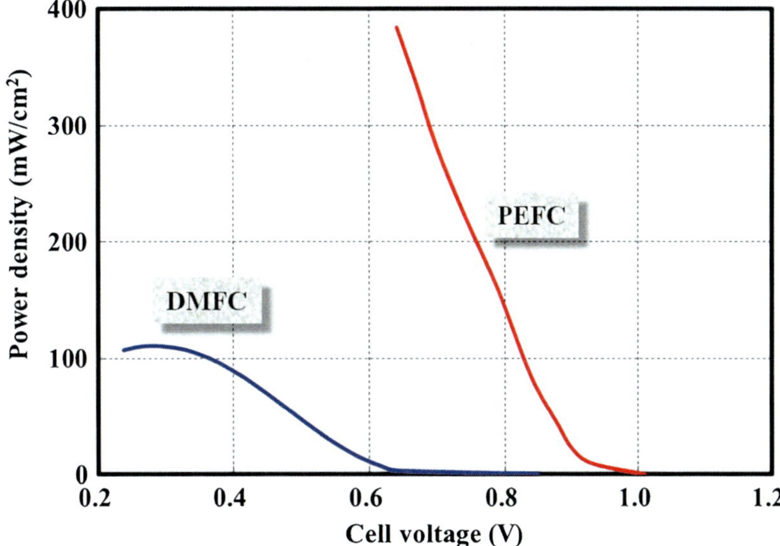

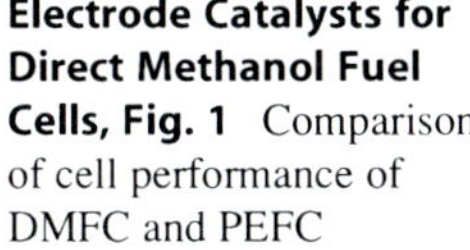
Electrode Catalysts for Direct Methanol Fuel Cells, Fig. 1 Comparison of cell performance of DMFC and PEFC

development of highly active anode catalysts toward the MOR in the DMFCs.

The slow kinetics of the MOR in the DMFCs is caused from carbon monoxide (CO) poisoning of the Pt catalyst. The methanol is not directly oxidized to carbon dioxide (CO_2) in the MOR, and elementary reactions on the Pt catalyst are as follows.

The CO, an intermediate in the MOR, is a strong poison for the Pt catalyst, and it strongly chemisorbs on the surface of the Pt ($Pt–CO_{ads.}$ in formula (1)).

$$Pt + CH_3OH \rightarrow Pt - CO_{ads.} + 4H^+ + 4e^- \quad (1)$$

In the PEFCs using reformed hydrogen as the anode fuel, trace amount of CO exists even after reforming and CO preferential oxidation (PROX) reactions, which also poisons the Pt catalyst. The chemisorbed CO is oxidized to CO_2 by a hydroxyl functional group formed on other active sites of the Pt by chemical reaction with a water molecule (formulas (2) and (3)).

$$Pt + H_2O \rightarrow Pt - OH + H^+ + e^- \quad (2)$$

$$Pt - CO_{ads.} + Pt - OH \rightarrow CO_2 + H^+ + e^- + 2Pt \quad (3)$$

The kinetics of the formulas (2) and (3) is much slower than that of the HOR and ORR, which is a main reason for the inferior cell performance of the DMFCs. Although the cell performance of the DMFCs doesn't surpass that of the PEFCs due to the slow kinetics of the MOR, the direct electrooxidation of the methanol in the DMFCs is attractive from a perspective of generating electric power directly from the fuel without the reforming and PROX reactions. Furthermore, the methanol is much safer and easier to carry and store in comparison with the hydrogen.

Owing to these features of the DMFCs, extensive studies have been conducted on highly CO-tolerant anode catalysts in the MOR. The highly CO-tolerant catalysts were strongly desirable for the MOR; however, it was difficult to realize the character with a monometallic Pt catalyst under an ambient condition. Fortunately, it has been found that the MOR activity of the Pt can be enhanced with a second component such as Ru [15–19] and Sn [20, 21]. In the 1960s, Adlhart and Heuer have shown that the Pt–Ru alloy catalyst shows a lower overpotential toward the MOR in 1 mol/l H_2SO_4 aqueous solution and that the MOR polarizations of the Pt–Ru alloy and the Pt catalysts are 0.24 V and 0.44 V versus NHE@20 mA/cm^2 at 373 K, respectively [15].

At present, the Pt–Ru bimetallic system is recognized as the most promising CO-tolerant anode catalyst for the DMFCs. A large body of literature exist demonstrating improvement of the CO oxidation on the Pt–Ru alloy and Ru-modified Pt catalysts. The superior CO tolerance of the Pt–Ru bimetallic catalysts compared with the monometallic Pt catalyst is frequently explained with concepts of bifunctional mechanism [17] and ligand effect [22, 23]. The former mechanism proposed by Watanabe and Motoo is widely accepted. They claimed that the Ru has higher reactivity with water than Pt and that formation of Ru–OH at a lower potential promotes the electrooxidation of the chemisorbed CO on the Pt (formulas (4) and (5)).

$$Ru + H_2O \rightarrow Ru - OH + H^+ + e^- \quad (4)$$

$$Pt - CO_{ads.} + Ru - OH \rightarrow CO_2 + H^+ + e^- + Pt + Ru \quad (5)$$

However, the bifunctional mechanism presumes that the Pt and the Ru behave similarly as they are individual components. It has been well discussed that catalytic properties of a material are influenced with its electronic structure, the so-called ligand effect. For example, a coupling between adsorbate (CO) valence states and transition metal d-states defines the adsorption energy. A lower *d*-band center in the catalyst is expected to weaken the catalyst-CO bonding. Theoretical studies on the overlayer systems based on density functional theory (DFT) have provided useful data about changes in the d-states for a number of transition elements [24–29]. According to Nørskov and his coworkers, the *d*-band center for Pt overlayer on Ru substrate was calculated to be 0.61 eV lower relative to that of bulk Pt surface [30]. This theoretical study of electronic modification and reduction of the CO adsorption energy in the Pt overlayer configuration is supported by several electrochemical experiments [31–33]. Therefore, it is possible to tune activity of the bimetallic Pt–Ru catalysts with controlling their electronic structure by changing their composition and atomic arrangement.

Although the ligand effect was not taken into consideration in the original bifunctional mechanism, it has been reported that the effect accounts for approximately 25 % of the change in the catalytic activity of the Pt–Ru alloy catalysts [34, 35]. Either way, there is no doubt that the bifunctional mechanism and the ligand effect are basis for understanding the enhanced MOR activity in the Pt–Ru bimetallic catalysts. For these bases, microstructure of the Pt–Ru bimetallic catalysts is a key determining degree of actions of the two concepts. This article describes control of the microstructures in the Pt–Ru bimetallic catalysts for the MOR. The MOR activity and durability of the catalysts are discussed in terms of their microstructures.

Core/Shell Microstructured Pt–Ru Bimetallic Catalysts

It is natural that Pt^{2+} cation is easier to be reduced than Ru^{3+} one according to their standard reduction potentials E^0 (formulas (6) and (7)).

$$Pt^{2+} + 2e^- = Pt, E^0: 1.19 \text{ V } vs. \text{ NHE} \quad (6)$$

$$Ru^{3+} + 3e^- = Ru, E^0: 0.68 \text{ V } vs. \text{ NHE} \quad (7)$$

When these cations are simultaneously reduced with a reducing agent (co-reduction process), the Pt^{2+} cation is preferentially reduced and Pt-rich core/Ru-rich shell microstructure is likely to be formed. Since the methanol is oxidized on the Pt site, it is expected that the MOR activity in the Pt–Ru bimetallic catalysts with the Ru-enriched shell is not sufficiently improved. Consequently, it is general to increase feeding ratio of the Pt^{2+} cation for enrichment of the surface Pt composition. Therefore, we need a technique to measure the surface composition of the Pt–Ru bimetallic catalysts to clarify an influence of the surface composition on the MOR activity. Unfortunately, there is no way to directly measure the surface Pt and Ru composition for nanosized Pt–Ru bimetallic catalysts (Pt–Ru catalyst 2 nm in size is shown in Fig. 2 [36]).

Green and Kucernak have demonstrated that oxidation potentials of Cu monolayer from Pt and Ru sites are different each other [37, 38].

They formed the Cu monolayer on the Pt–Ru bimetallic nanoparticles using Cu under potential deposition (Cu-UPD) technique and the Cu monolayer was stripped. Each coulombic amount of the Cu oxidation from the Pt and Ru sites can be divided as shown in Fig. 3.

Therefore, the surface Pt and Ru composition (θ_{Pt}, θ_{Ru}) in the Pt–Ru bimetallic nanoparticle catalysts are explained with the following equations:

$$\theta_{Pt} = \zeta_{Pt}/(\zeta_{Pt} + \zeta_{Ru}) \times 100\% \quad (8)$$

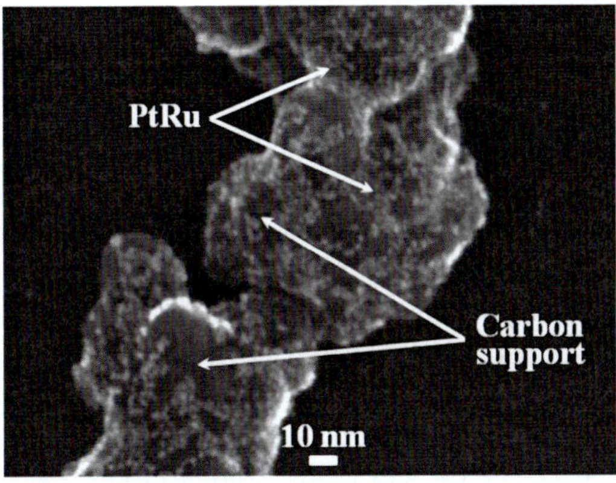

Electrode Catalysts for Direct Methanol Fuel Cells, Fig. 2 HR-SEM image of Pt–Ru catalyst

$$\theta_{Ru} = \zeta_{Ru}/(\zeta_{Pt} + \zeta_{Ru}) \times 100\% \quad (9)$$

where ζ_{Pt} is the coulombic amount of the Cu oxidation from the Pt site and ζ_{Ru} is that from the Ru site. Figure 4 shows a relationship between the MOR activity and the Pt composition in the Pt–Ru bimetallic nanoparticle catalysts. The bulk and the surface Pt composition were determined with XRF and the Cu-UPD/Cu-stripping technique, respectively. The bulk Pt composition increased with increasing the feeding ratio of the Pt precursor ($Pt(acac)_2$ in this study). The feeding ratio of $[Pt^{2+}]/[Ru^{3+}] = 50/50$ (sample (a)) gave the bulk composition of $Pt_{49}Ru_{51}$ (atomic %). Whereas, the surface composition of the sample (a) is $Pt_{28}Ru_{72}$, indicating that the Pt^{2+} cation was preferentially reduced and consumed in forming core part of the catalyst. The MOR activity was improved with increase of the Pt^{2+} feeding ratio, and the maximum MOR activity was obtained with the bulk and surface composition of $Pt_{73}Ru_{27}$ and $Pt_{53}Ru_{47}$, respectively [39].

According to the bifunctional mechanism, the surface Pt and Ru atomic ratio of 50/50 is favorable toward the CO oxidation reaction as shown in the formula (5). Suzuki and his colleagues have also reported that the maximum MOR activity was achieved with the surface composition around $Pt_{50}Ru_{50}$ in the Pt–Ru bimetallic

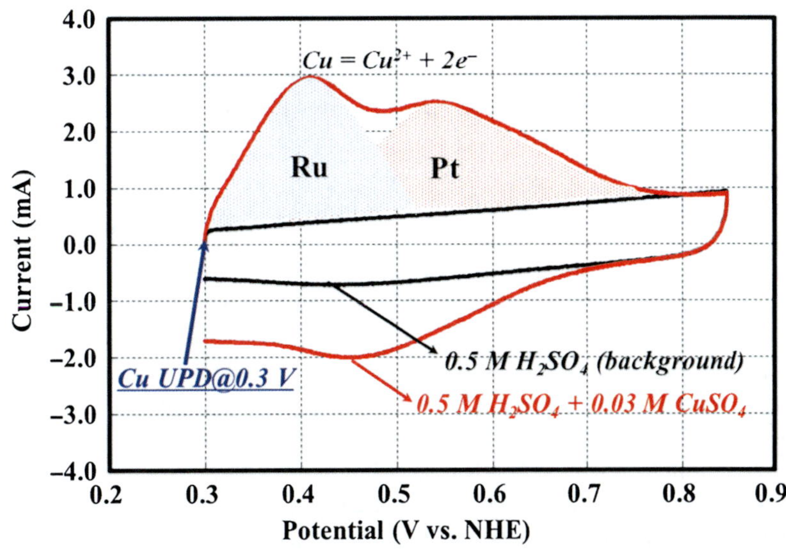

Electrode Catalysts for Direct Methanol Fuel Cells, Fig. 3 Cu stripping voltammogram from Pt–Ru catalyst

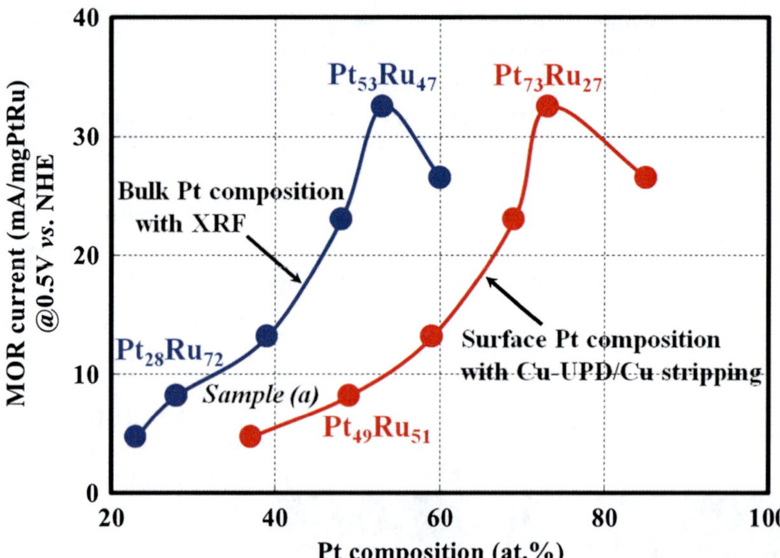

Electrode Catalysts for Direct Methanol Fuel Cells, Fig. 4 MOR activity versus bulk and surface Pt composition of Pt–Ru catalysts

nanoparticle catalysts [40]. Gasteiger and his coworkers, however, have demonstrated that the optimum surface composition was around $Pt_{90}Ru_{10}$ using sputtered Pt–Ru polycrystalline bulk catalysts [19]. In the bifunctional mechanism, the surface Ru composition is essential for the CO oxidation, and the CO molecule chemisorbed on the Pt sites diffuses to the hydroxide (Ru–OH) formed on the Ru sites. However, it seems that the CO bonding energies on the Pt nanoparticle and on the Pt plate/film are different due to the electronic and/or geometric (edge and corner) effects. The stronger bonding energy results in lower CO mobility on the Pt sites.

Lebedeva and his coworkers have estimated that coefficient of CO diffusion (D_{CO}) on the Pt plate is larger than 1×10^{-11} cm^2/s [41]. On the contrary, Maillard and Babu have reported that the D_{CO} on the Pt nanoparticles (3–7 nm in size) is 1×10^{-13} to 1×10^{-14} cm^2/s [42, 43]. These reports imply that higher surface Ru composition is required in the Pt–Ru nanoparticles since the D_{CO} on the nanoparticles may have the lower values compared with those on the Pt plate/film. Therefore, an optimum surface Ru composition in the Pt–Ru nanoparticle catalysts becomes higher compared with the Pt–Ru plate/film ones. The higher surface Ru composition in the Pt–Ru nanoparticle catalysts results in a lower specific MOR activity than the plate/film catalysts (Fig. 5 [40]), because the active center for the MOR is the surface Pt sites.

Here, a simple calculation is performed on the Pt–Ru nanoparticles 2 nm in size (Fig. 2). Beginning with a Pt sphere 2 nm in size, the sphere has 276 Pt atoms, and the surface Pt monolayer and the Pt core part are composed of 168 and 108 Pt atoms, respectively. Supposed that the surface Pt monolayer is replaced with a $Pt_{50}Ru_{50}$ monolayer, the bulk composition of this core/shell microstructured sphere is calculated to be $Pt_{70}Ru_{30}$, which is close to the bulk composition of $Pt_{73}Ru_{27}$ showing the highest MOR activity in Fig. 4. This simple calculation supports that a microstructure close to Pt-rich core/$Pt_{50}Ru_{50}$ shell was formed in the Pt–Ru bimetallic nanoparticles.

Well-Mixed Pt–Ru Bimetallic Catalysts

The co-reduction of the Pt and Ru cations resulted in the formation of the core/shell microstructured catalysts due to the preferential reduction of the Pt cation. In this section, a strategy for synthesizing well-mixed Pt–Ru bimetallic nanoparticle catalysts is presented. A key for the synthesis is decreasing difference of effective reduction potentials between the Pt

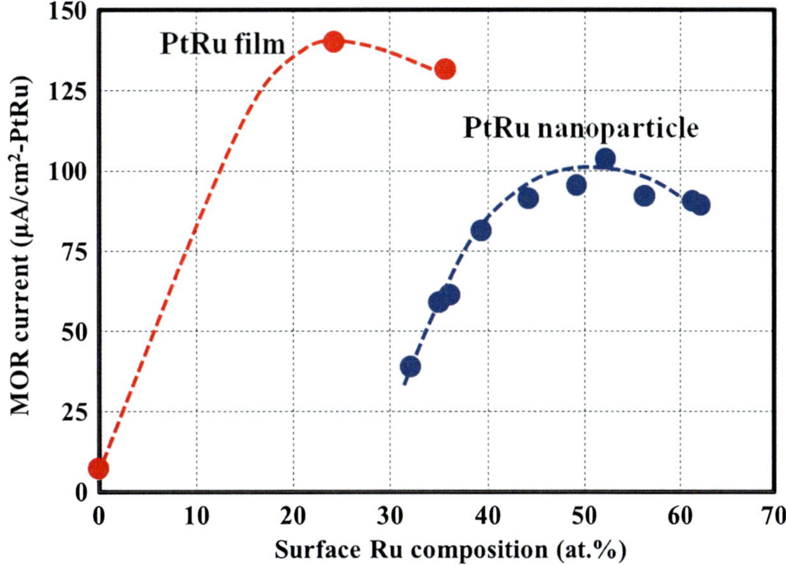

Electrode Catalysts for Direct Methanol Fuel Cells, Fig. 5 MOR activity of film and nanoparticle Pt–Ru catalysts versus surface Ru composition

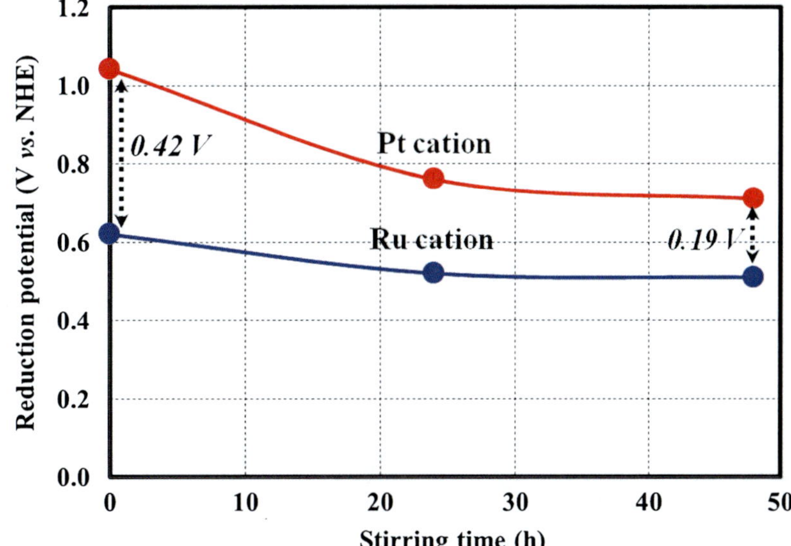

Electrode Catalysts for Direct Methanol Fuel Cells, Fig. 6 Changes in effective reduction potentials of Pt and Ru cations with stirring time by addition of D,L-tartaric acid (TA)

and Ru cations via an addition of chelate ligand [44, 45].

The synthesis was carried out in water using H_2PtCl_6 and $RuCl_3$ precursors and HPH_2O_2 as a reducing agent at 363 K under atmosphere. The chelate ligand of D,L-tartaric acid (TA) was added to the synthetic solution for controlling the effective reduction potentials of the Pt^{4+} and Ru^{3+} cations. Figure 6 shows changes in the effective reduction potentials of the cations with stirring time by the addition of the TA at room temperature [45]. Two important things are seen in this figure. First, the addition of the chelate ligand decreases each effective reduction potential. Second, it takes time for decreasing the effective reduction potentials, indicating that replacement reactions of Cl^- ligand with TA one slowly proceed in the Pt and Ru precursors. Finally, the difference in their effective reduction potentials decreased from 0.42 V to 0.19 V

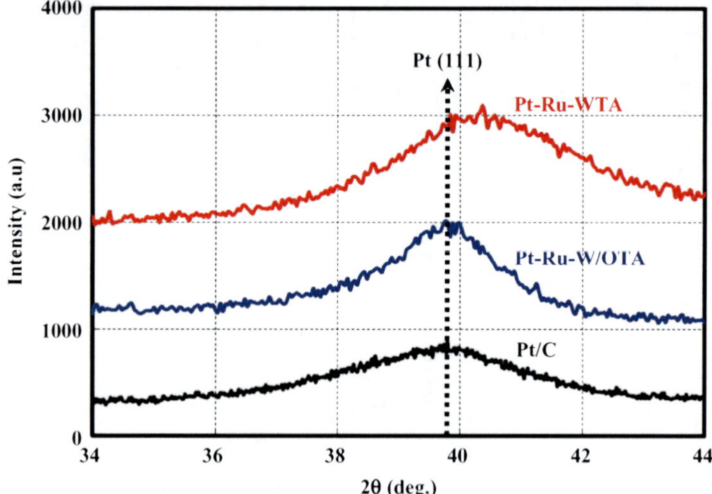

Electrode Catalysts for Direct Methanol Fuel Cells, Fig. 7 XRD patterns of Pt–Ru catalysts synthesized with and without addition of D,L-tartaric acid together with Pt/C catalyst

after 48 h's stirring. The UV–vis spectroscopy measurements revealed that little changes were observed in the spectra after 48 h's stirring [45].

The carbon-supported Pt–Ru bimetallic nanoparticle catalysts were synthesized at the feeding ratio of [Pt]/[Ru] = 50/50 with and without addition of TA. The XRD patterns of the catalysts are shown in Fig. 7 together with a carbon-supported Pt catalyst (Pt/C). The Pt–Ru bimetallic nanoparticle catalysts synthesized with TA (Pt–Ru-WTA) and without TA (Pt–Ru-W/OTA) showed diffraction patterns of fcc solid solution phase. It should be noted that (111) diffraction of the Pt–Ru-WTA clearly shifts toward higher angle relative to the Pt/C catalyst, indicating that the Pt and Ru atoms are mixed in the Pt–Ru-WTA catalyst. On the contrary, the shift in the Pt–Ru-W/OTA is negligible. The surface Pt and Ru compositions of the catalysts were analyzed with the Cu-UPD/Cu-stripping technique. Although the bulk compositions of the two catalysts measured with XRF were $Pt_{51}Ru_{49}$, the surface compositions of the Pt–Ru-WTA and Pt–Ru-W/OTA were $Pt_{46}Ru_{54}$ and $Pt_{38}Ru_{62}$, respectively. These XRD and surface composition data strongly indicate that the Pt and Ru cations were simultaneously reduced and well-mixed Pt–Ru bimetallic nanoparticle catalyst was synthesized with the addition of the chelate ligand TA [45].

The mixing states of the Pt and Ru atoms in the catalysts were analyzed using EXAFS technique. In the analysis, paring factors of P_{Pt} and P_{Ru} were introduced to evaluate the mixing states [46]. The factors are defined as the following equations:

$$P_{Pt} = N_{Pt-Ru}/(N_{Pt-Ru} + N_{Pt-Pt}) \quad (10)$$

$$P_{Ru} = N_{Ru-Pt}/(N_{Ru-Pt} + N_{Ru-Ru}) \quad (11)$$

where the N_{Pt-Ru} is coordination number of Ru atom viewing from Pt atom, the N_{Pt-Pt} is coordination number of Pt atom viewing from Pt atom, the N_{Ru-Pt} is coordination number of Pt atom viewing from Ru atom, and the N_{Ru-Ru} is coordination number of Ru atom viewing from Ru atom. According to the equations, values of the pairing factors increase when the mixing of the Pt and Ru atoms is advanced in the catalyst. Table 1 summarizes the pairing factors of the Pt–Ru-WTA and Pt–Ru-W/OTA catalysts together with their bulk and surface composition. It is clear that both of the pairing factors in the Pt–Ru-WTA catalyst are larger than those in the Pt–Ru-W/OTA one. Therefore, it was clarified that the mixing state of the Pt and Ru atoms in the Pt–Ru bimetallic nanoparticle catalyst was promoted by the synthesis with the addition of the chelate ligand TA. The MOR activities

Electrode Catalysts for Direct Methanol Fuel Cells, Table 1 Pairing factors, bulk, and surface composition of Pt–Ru catalysts

Catalyst	P_{Pt}	P_{Ru}	Bulk composition	Surface composition
Pt–Ru without TA[a]	0.072	0.171	$Pt_{51}Ru_{49}$	$Pt_{38}Ru_{62}$
Pt–Ru with TA[a]	0.308	0.424	$Pt_{51}Ru_{49}$	$Pt_{46}Ru_{54}$

[a]TA: D,L-tartaric acid

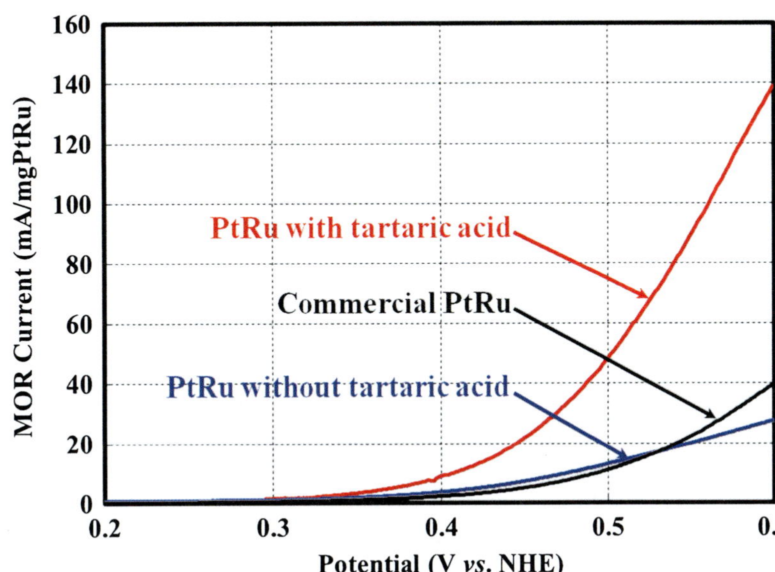

Electrode Catalysts for Direct Methanol Fuel Cells, Fig. 8 MOR activities of Pt–Ru catalysts synthesized with and without addition of D,L-tartaric acid together with a commercial Pt–Ru catalyst

of the Pt–Ru bimetallic nanoparticle catalysts are demonstrated in Fig. 8 together with a commercial Pt–Ru catalyst. Since the surface composition of the Pt–Ru-WTA and Pt–Ru-W/OTA catalysts are $Pt_{46}Ru_{54}$ and $Pt_{38}Ru_{62}$, respectively, the Pt–Ru-WTA catalyst showed much higher MOR activity than that of the Pt–Ru-W/OTA one.

Durability of Pt-Ru Bimetallic Catalysts

As explained above, it is possible to enhance the MOR activity of the Pt–Ru bimetallic catalysts with both core/shell and well-mixed microstructures by controlling their surface compositions around $Pt_{50}Ru_{50}$. In addition to the enhancement of the MOR activity, durability is an important issue for practical catalysts. In this section, an importance of the well-mixed Pt–Ru bimetallic catalyst is demonstrated in order to improve the durability.

The durability test was conducted with a potential cycling (0.2–1.1 V vs. NHE) in 1.5 mol/l H_2SO_4 aqueous solution at 298 K under nitrogen atmosphere, and the MOR activity was periodically measured by a linear sweep voltammetry [44]. Figure 9 shows changes in the MOR activity of the Pt–Ru bimetallic nanoparticle catalysts synthesized with and without D, L-tartaric acid. The MOR activity of the Pt–Ru-W/OTA catalyst declined within a 100 potential cycling (pink line), and the MOR activity became the same level as the Pt/C catalyst (black line). In the Pt–Ru-W/OTA catalyst, Pt-rich core/Pt–Ru shell microstructure was formed because the Pt cation was preferentially reduced with a large difference in the effective reduction potentials of the Pt and Ru cations as illustrated in Fig. 6. Therefore, it is considered that the Pt–Ru shell was dissolved out during the early stage of the potential cycling, resulting in the equivalent MOR activity to the Pt/C catalyst. On the

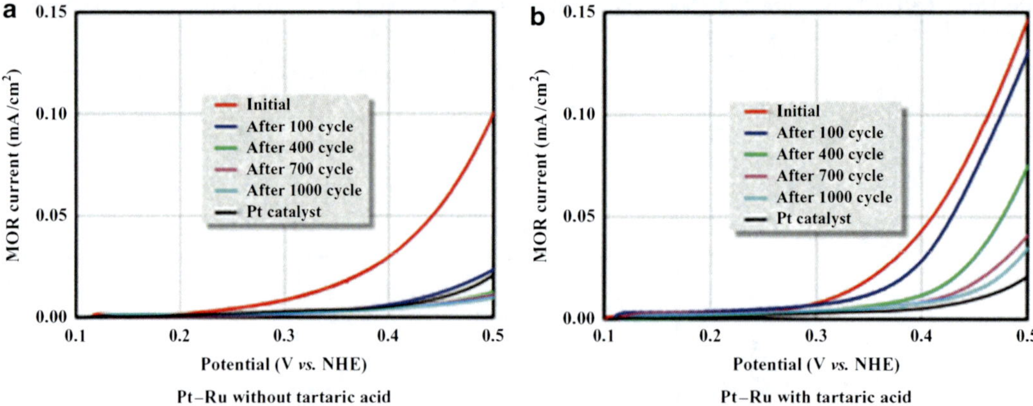

Electrode Catalysts for Direct Methanol Fuel Cells, Fig. 9 Changes in MOR activities of Pt–Ru catalysts synthesized with and without addition of D,L-tartaric acid by potential cycling durability test

contrary, the decay in the MOR activity of the Pt–Ru-WTA catalyst was much suppressed. In the Pt–Ru-WTA catalyst, the Ru atoms exist not only at the surface but also in the bulk due to the well-mixed microstructure, which improved the durability of the catalyst. It is of great importance that the well-mixed microstructure is important for long-term durability in the Pt–Ru bimetallic catalysts [44].

Future Directions

The correlation of the MOR activity of the Pt–Ru bimetallic catalysts with their microstructures was described. On the basis of the bifunctional mechanism and ligand effect, the control of the microstructures in the catalysts is crucial for the enhancement of the CO tolerance, the MOR activity, and the durability. The co-reduction of the Pt and Ru precursors is a realistic process in the industrial production of the Pt–Ru bimetallic catalysts. In the synthesis, a key is decreasing the difference in the effective reduction potentials of the Pt and Ru cations. The decrease promotes the simultaneous reduction of the cations and realizes the well-mixed microstructure, which enhances the MOR activity and the durability of the Pt–Ru bimetallic catalysts.

Even though size reduction is effective for the enhancement of the catalytic activity, it should be noted that interface of the catalysts contacting with an electrolyte increases with the size reduction, resulting in accelerated dissolution of the catalysts. Consequently, the size of the Pt–Ru bimetallic catalysts should be designed with consideration on the dissolution issue. Especially, the easier dissolution nature of the Ru should be concerned.

Cross-References

▸ Direct Alcohol Fuel Cells (DAFCs)

References

1. Ministry of Economy, Trade and Industry of Japan (2010) Joint statement on introduction of fuel cell vehicle to domestic market and servicing of infrastructure of hydrogen station http://www.meti.go.jp/press/20110113003/20110113003-2.pdf. Accessed 13 Jan 2010
2. Zhang J, Mo Y, Vukmirovic MB, Klie R, Sasaki K, Adzic RR (2004) Platinum monolayer electrocatalysts for O_2 reduction: Pt monolayer on Pd(111) and on carbon-supported Pd nanoparticles. J Phys Chem 108:10955–10964
3. Zhang J, Vukmirovic MB, Xu Y, Mavrikakis M, Adzic RR (2005) Controlling the catalytic activity of platinum-monolayer electrocatalysts for oxygen reduction with different substrates. Angew Chem Int Ed 44:2132–2135
4. Vukmirovic MB, Zhang J, Sasaki K, Nilekar AU, Uribe F, Mavrikakis M, Adzic RR (2007) Platinum

monolayer electrocatalysts for oxygen reduction. Electrochim Acta 52:2257–2263
5. Sasaki K, Naohara H, Cai Y, Choi YM, Liu P, Vukmirovic MB, Wang JX, Adzic RR (2010) Core-protected platinum monolayer shell high-stability electrocatalysts for fuel-cell cathodes. Angew Chem Int Ed 49:8602–8607
6. Wang JX, Ma C, Choi YM, Su D, Zhu Y, Liu P, Si R, Vukmirovic MB, Zhang Y, Adzic RR (2011) Kirkendall effect and lattice contraction in nanocatalysts: a new strategy to enhance sustainable activity. J Am Chem Soc 133:13551–13557
7. Stamenkovic V, Schmidt TJ, Ross PN, Markovic NM (2002) Surface composition effects in electrocatalysis: kinetics of oxygen reduction on well-defined Pt_3Ni and Pt_3Co alloy surfaces. J Phys Chem B 106:11970–11979
8. Stamenkovic V, Mun BS, Mayrhofer KJ, Ross PN, Markovic NM, Rossmeisl J, Greeley J, Nørskov JK (2006) Changing the activity of electrocatalysts for oxygen reduction by tuning the surface electronic structure. Angew Chem Int Ed 45:2897–2901
9. Stamenkovic VR, Fowler B, Mun BS, Wang G, Ross PN, Lucas CA, Markovic NM (2007) Improved oxygen reduction activity on $Pt_3Ni(111)$ via increased surface site availability. Science 315:493–497
10. Stamenkovic VR, Mun BS, Arenz M, Matrhofer KJJ, Lucas CA, Wang G, Ross PN, Markovic NM (2007) Trends in electrocatalysis on extended and nanoscale Pt-bimetallic alloy surfaces. Nat Mat 6:241–247
11. Wang C, Vliet D, More KL, Zaluzec NJ, Peng S, Sun S, Daimon H, Wang G, Greeley J, Pearson J, Paulikas AP, Karapetrov G, Strmcnik D, Markovic NM, Stamenkovic VR (2011) Multimetallic Au/$FePt_3$ nanoparticles as highly durable electrocatalyst. Nano Lett 11:919–926
12. Wang C, Chi M, Li D, Strmcnik D, Vliet D, Wang D, Komanicky V, Chang KC, Paulikas AP, Tripkovic D, Pearson J, More KL, Markovic NM, Stamenkovic VR (2011) Design and synthesis of bimetallic electrocatalyst with multilayered Pt-skin surfaces. J Am Chem Soc 133:14396–14403
13. Wang C, Li D, Chi M, Pearson J, Rankin RB, Greeley J, Duan Z, Wang G, Vliet D, More KL, Markovic NM, Stamenkovic VR (2012) Rational development of ternary alloy electrocatalysts. J Phys Chem Lett 3:1668–1673
14. Strasser P, Koh S, Anniyev T, Greeley J, More K, Yu C, Liu Z, Kaya S, Nordlund D, Ogasawara H, Toney MF, Nilsson A (2010) Lattice-strain control of the activity in dealloye core–shell fuel cell catalysts. Nature Chem 2:454–460
15. Adlhart OJ, Heuer KO (1962) Fuel cell catalysis. Contact No. DA 36-39 SC-90691 U.S. Army Electronics Research and Development Laboratories
16. Bockris JOM, Wrobloma H (1964) Electrocatalysis. J Electroanal Chem 7:428–451
17. Watanabe M, Motoo S (1975) Electrocatalysis by ad-atoms: part II enhancement of the oxidation of methanol on platinum by ruthenium ad-atoms. Electroanal Chem 60:267–273
18. Watanabe M, Uchida M, Motoo S (1987) Preparation of highly dispersed Pt + Ru alloy cluster and the activity for the electrooxidation of methanol. J Electroanal Chem 229:395–406
19. Gasteiger HA, Markovic NM, Ross PN, Cairns EJ (1993) Methanol electrooxidation on well-characterized Pt–Ru alloys. J Phys Chem 97:12020–12029
20. Tillmann S, Samjesk G, Friedrich KA, Baltruschat H (2003) The adsorption of Sn on Pt(111) and its influence on CO adsorption as studied by XPS and FTIR. Electrochim Acta 49:73–83
21. Arenz M, Stamenkovic V, Blizanac BB, Mayrhofer KJ, Markovic NM, Ross PN (2005) Carbon-supported Pt-Sn electrocatalysts for the anodic oxidation of H_2, CO, and H_2/CO mixtures part II The structure-activity relationship. J Catal 232:402–410
22. Honji A, Gron LU, Chang JR, Gates BC (1992) Ligand effects in supported metal carbonyls: X-ray absorption spectroscopy of ruthenium subcarbonyls on magnesium oxide. Langmuir 8:2716–2719
23. Rodriguez JA (1996) Physical and chemical properties of bimetallic surfaces. Surf Sci Rep 24:223–287
24. Hammer B, Nørskov J (2000) Theoretical surface science and catalysis-calculations and concepts. Adv Catal 45:71–129
25. Hammer B (2006) Special sites at noble and late transition metal catalysts. Top Catal 37:3–16
26. Koper MTM, Shubina TE (2002) Periodic density functional study of CO and OH adsorption on Pt–Ru alloy surfaces: implications for CO tolerant fuel cell catalysts. J Phys Chem B 106:686–692
27. Liao MS, Cabrera CR, Ishikawa Y (2000) A theoretical study of CO adsorption on Pt, Ru and Pt-M (M = Ru, Sn, Ge) clusters. Surf Sci 445:267–282
28. Christoffersen E, Ruban PLA, Skriver HL, Nørskov JK (2001) Anode materials for low temperature fuel cells: A density functional theory study. J Catal 199:123–131
29. Tsuda M, Kasai H (2006) Ab initio study of alloying and straining effects on CO interaction with Pt. Phys Rev B Condens Matter Mater Phys 73:155405–155413
30. Ruban A, Hammer B, Stoltze P, Skriver HL, Nørskov JK (1997) Surface electronic structure and reactivity of transition and noble metals. J Mol Catal A Chem 115:421–429
31. Brankovic SR, Marinkovic NS, Wang JX, Adzic RR (2002) Carbon monoxide oxidation on bare and Pt-modified Ru(1010) and Ru(0001) single crystal electrodes. J Electroanal Chem 532:57–66
32. Davies JC, Hayden BE, Pegg DJ (2000) The modification of Pt(110) by ruthenium: CO adsorption and electro-oxidation. Surf Sci 467:118–130
33. Inoue M, Nishimura T, Akamaru S, Taguchi A, Umeda M (2009) CO oxidation on non-alloyed Pt and Ru electrocatalysts prepared by the polygonal barrel-sputtering method. Electrochim Acta 21:4764–4771

34. Lu C, Rice C, Masel RI, Babu PK, Waszczuk P, Kim HS, Oldfield E, Wieckowski A (2002) UHV, electrochemical NMR and electrochemical studies of platinum/ruthenium fuel cells catalysts. J Phys Chem 106:9581–9589
35. Lu C, Masel RI (2001) The effect of ruthenium on the binding of CO, H_2 and H_2O on Pt(110). J Phys Chem B 105:9793–9797
36. Daimon H, Korobe Y (2006) Size reduction of PtRu catalyst particle deposited on carbon support by addition of non-metallic elements. Catal Today 111:182–187
37. Green CL, Kucernak A (2002) Determination of the platinum and ruthenium surface areas in platinum-ruthenium alloy electrocatalysts by underpotential deposition of copper. 1. Unsupported catalysts. J Phys Chem B 106:1036–1047
38. Green CL, Kucernak A (2002) Determination of the platinum and ruthenium surface areas in platinum-ruthenium alloy electrocatalysts by underpotential deposition of copper. 2. Effect of surface composition on activity. J Phys Chem B 106:11446–11456
39. Daimon H, Onodera T, Nakagawa T, Nitani H, Yayamoto TA (2010) Methanol oxidation activity of nanosized PtRu catalysts and their microstructures. J Nanoelectron Optoelectron 5:1–5
40. Suzuki S, Onodera T, Kawaji J, Mizukami T, Takamori Y, Daimon H, Morishima M (2011) Optimum surface composition of platinum-ruthenium nanoparticles and sputter-deposited films for methanol oxidation reaction. Electrochem 79:602–608
41. Lebedeva NP, Koper MTM, Feliu JM, Santen RA (2002) Mechanism and kinetics of CO adlayer oxidation on stepped platinum electrodes. J Phys Chem B 106:12938–12947
42. Maillard F, Eikerling M, Cherstiouk OV, Schreier S, Savinova E, Stimming U (2004) Size effects on reactivity of Pt nanoparticles in CO monolayer oxidation: the role of surface mobility. Faraday Discuss 125:357–377
43. Babu PK, Chung JH, Oldfield E, Wieckowski A (2008) CO surface diffusion on platinum fuel cell catalysts by electrochemical NMR. Electrochim Acta 53:6672–6679
44. Daimon H, Onodera T, Honda Y, Nitani H, Seino S, Nakagawa T, Yamamoto TA (2008) Activity and durability of PtRuP catalysts and their atomic structures. Electrochem Soc Trans 11:93–100
45. Onodera T, Suzuki S, Takamori Y, Daimon H (2010) Improved methanol oxidation activity and stability of well-mixed PtRu catalysts synthesized by electroless plating method with addition of chelate ligands. Appl Catal A Gen 379:69–76
46. Nitani H, Nakagawa T, Daimon H, Kurobe Y, Ono T, Honda Y, Koizumi A, Seino S, Yamamoto TA (2007) Methanol oxidation catalysis and substructure of PtRu bimetallic nanoparticles. Appl Catal A Gen 326:194–201

Electrodeposition of Electronic Materials for Applications in Macroelectronic- and Nanotechnology-Based Devices

I. M. Dharmadasa and Obi Kingsley Echendu
Electronic Materials and Sensors Group, Materials and Engineering Research Institute, Sheffield Hallam University, Sheffield, UK

Introduction

Electrodeposition, otherwise known as electroplating, is a well-known industrial process for extraction, purification, and coating of metals for centuries. It was not until the late 1970s that the application of electrodeposition as a semiconductor growth technique was known for the first time [1–5]. The first family of semiconductors grown by this method at the time was the II–VI semiconductor family. This eventually led to the fabrication of one of the first high-efficiency CdTe-based solar cells in the early 1980s with cell efficiency greater than 10 % [6]. These initial results of electrodeposition of CdS/CdTe solar cell triggered serious research and development activities in electrodeposition of semiconductors in general. In 2002, Dharmadasa et al. published 18 % efficiency [7] (unconfirmed) for laboratory-scale CdS/CdTe-based solar cells using electrodeposited CdTe.

Attempts have also been made to electrodeposit elemental semiconductors like silicon [8], as well as compound semiconductors like III–V nitrides [9]. Other II–VI semiconductors like CdS [10], ZnSe [11], ZnTe [12–14], and ZnO [15, 16] as well as SnS [17], which is a IV–VI compound semiconductor, have also been electrodeposited in addition to ternary compounds like $CuInSe_2$ [18–22] and quaternary compounds like $CuInGaSe_2$ [23–26] and Cu_2ZnSnS_4 [27].

This article reviews the capabilities and advantages of electrodeposition as a reliable and simple technique for the growth of semiconductor nanomaterials and fabrication of large-area macroelectronic devices such as photovoltaic solar panels and large-area display devices as well as for emerging nanotechnology applications.

Electrodeposition of Semiconductors

The electrodeposition of semiconductors basically requires an electrolyte which contains the appropriate ions of the semiconductor elements to be deposited. The sources of these ions are usually high-purity chemical compounds. These compounds are made into aqueous or nonaqueous solutions in a beaker or tank as needed. A low direct current (DC) power, usually in the milliwatt range, is applied to the laboratory-scale electrolyte through appropriate electrodes by means of a potentiostat. One of the electrodes is the working electrode (usually the cathode) and the second electrode is the counter electrode (the anode). In some cases as is the common practice in electrochemistry, a third electrode (the reference electrode) is involved. The reference electrode helps to stabilize and monitor the applied current or voltage during deposition. The most commonly used reference electrodes are the Ag/AgCl and the Hg/$HgCl_2$ (saturated calomel) reference electrodes. In the electrodeposition of semiconductor materials, the working electrode is usually glass substrate with a transparent conducting oxide (TCO) coating on one side. The TCO usually serves as the front electrical contact (ohmic contact) for a fully fabricated superstrate-type thin-film solar cell. Commonly used TCOs include fluorine-doped tin oxide (FTO) and indium-doped tin oxide (ITO). However, in principle, any conducting surface can be used as the working electrode provided the electrode material does not dissolve in the acidic or alkaline electrolyte.

Once the electrolyte is prepared, the pH is adjusted to the desired value (usually in the acidic range, although pH in the alkaline range can be used for certain semiconductors). The pH adjustment is usually done using appropriate dilute acids and bases as the pH of the freshly made electrolyte may be higher or lower than the desired value. It is important to note that the control of the pH of a deposition electrolyte is very crucial as this has a profound influence on the applied deposition potential as well as on the quality of the material deposited. Next in the process is estimation of the right deposition potential range. This can be a real challenge unlike in the electroplating of single metals since most semiconductors that can be deposited today are compound semiconductors consisting of two or more elements. A way of getting over this problem is the use of cyclic voltammetry, which is a very important tool in electrochemistry. In this process, a range of potentials is applied across the electrolyte through the electrodes, using a computerized potentiostat. The computer records the data of deposition condition which can be used to produce a plot of current versus cathodic potential for the particular semiconductor. This is then used to study the deposition mechanism. From this study, the approximate range of the deposition potential is obtained for the particular semiconductor. In this deposition range, a number of samples can be electrodeposited and then characterized to obtain the best (optimum) deposition potential.

Effect of Impurities on Electrodeposited Semiconductors

Prior to the deposition of good quality semiconductors, an important step known as the pre-deposition purification step is carried out. This involves depositing at a potential slightly lower than the actual deposition potential for several hours or even few days depending on the purity grade of the starting chemicals. This is done to remove unwanted ions (impurities) which may be present in the chemicals used in preparing the electrolyte. This process is essential because of the fact that impurities in part per million (ppm) levels can adversely affect the electrical properties of semiconductors. It is important to note a chemical or perhaps a bulk semiconductor target may have purity in the range of 99.999 % but this does not make it impurity-free for semiconductor devices application. The work done by Emziane et al. [28] shows that even with the so-called high-purity chemicals, detrimental impurities can still be incorporated into deposited semiconductors from these chemicals. It should be noted that the pre-deposition purification is carried out at the pH, temperature, and stirring rate at which the desired semiconductor is intended to be deposited for optimum result. Also in the case of aqueous deposition electrolytes, deionized water is used to minimize impurities coming from the solvent.

For the same reason of impurities as stated above, the 3-electrode system can sometimes pose a problem, as leakage through the porous glass frit or ceramic fiber junction during deposition process can "poison" the electrolyte and thus deteriorate the electrical properties of the deposited semiconductor. It can be recalled that different ions (elements) can act as impurities in different semiconductors. For example, Cu^{2+}, Ag^{+}, K^{+}, Na^{+}, etc. can act as detrimental impurities in electroplated CdTe solar cells [29]. Some of these ions such as K^{+} are contained in both Ag/AgCl and $Hg/HgCl_2$ reference electrodes [30, 31], while Ag^{+} is present in Ag/AgCl reference electrode [30]. Again Na^{+} can leach into the electrolyte from glassware such as glass beakers in which case plasticware can be used in place of glassware.

After the pre-deposition purification process, the desired semiconductor can then be electrodeposited from the purified electrolyte. As electroplating progresses, further self-purification of the electrolyte continues to take place, making it possible to deposit very pure semiconductors using this technique. As part of the strengths of electrodeposition, the parameters such as pH, temperature, stirring rate, and ion concentration can be varied as desired in the same electrolyte without making a new solution. This helps to minimize waste and the consequent environmental pollution and cost of waste management. This continuous process therefore makes electrodeposition an efficient and low-cost technique in a production line.

Effect of Temperature on Electrodeposited Semiconductors

The effect of temperature on the quality of electrodeposited semiconductors is very crucial. With aqueous electrolytes, deposition temperatures cannot exceed about 90 °C to prevent the water in the electrolyte from boiling away. For this reason, semiconductors grown by electrodeposition method are usually polycrystalline in nature with small crystallites and grain sizes unlike those grown at high temperatures which have larger crystallites and grain sizes. To increase the deposition temperature in some cases, nonaqueous solvents, such as ethylene glycol and dimethyl sulfoxide (DMSO), which are organic solvents, are used in preparing the electrolytes. This has the advantage of allowing the use of higher deposition temperatures up to 170 °C for improved crystalline qualities of the deposited semiconductors. This is possible due to the high boiling point of these solvents which is higher than that of water. In any case, the crystallites are still small compared to those obtained by high-temperature growth techniques with temperatures reaching 500–600 °C such as in close space sublimation. For this reason therefore, electrodeposited materials usually consist of crystallites or particles with sizes in the nanometer range making them suitable for applications in nanotechnology [32]. Figure 1 shows the 3-D atomic force microscopy (AFM) image of electrodeposited CdS showing nanocrystallites in the form of closely packed and unidirectional nanorods. Sima et al. [33] have reported the production of standing CdTe nanowires in track-etch membranes using both acidic and alkaline solutions. This capability could trigger and open up doors for many device applications.

Owing to the low deposition temperature, annealing of electrodeposited semiconductor materials is inevitable for the ultimate improvement of the structural, electronic, optical, and other properties of these layers. The annealing process can be done at temperatures up to 450 °C, taking into consideration the transition temperature of the substrate (usually glass) used. Interestingly, electrodeposition is usually carried out in normal laboratory conditions requiring no vacuum systems. However, highest discipline is required as is customary in semiconductor growth in general.

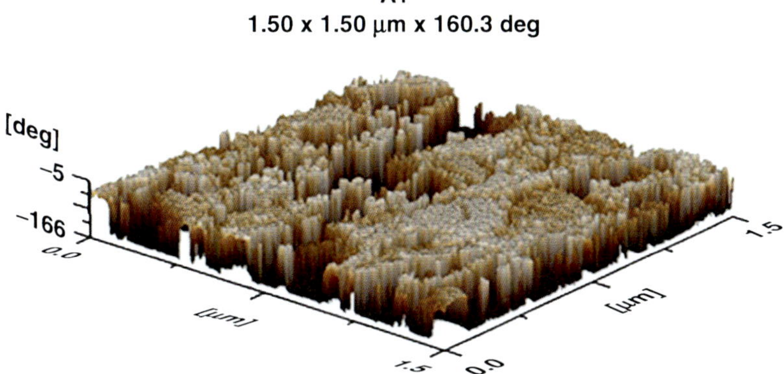

Electrodeposition of Electronic Materials for Applications in Macroelectronic- and Nanotechnology-Based Devices, Fig. 1 3-D AFM image of electrodeposited CdS showing highly oriented and closely packed nanorods standing perpendicular to the substrate

Advantages of Electrodeposition

There are many capabilities deliverable by electrodeposition as well as advantages over some of the conventional semiconductor deposition techniques. Some of these advantages are highlighted below.

Large-Area Deposition, Easy Process Control, Low Cost, and Versatility

One major advantage of electrodeposition as a macroelectronics deposition technique lies in its ability to be scaled up easily to large area such as in thin-film photovoltaic solar panels and large display devices. This capability has been demonstrated by BP Solar in the manufacturing of 0.9 m^2 solar panels using electrodeposited CdTe with conversion efficiencies over 10 % [34]. What is required in the scaling-up process is a large tank to contain the electrolyte and a multi-plate cathode consisting of a large number of substrates. Depending on the size of the tank, a number of panels can be electrodeposited in one tank at the same time for increased throughput.

The electrodeposition process is comparatively easy to control. Virtually any size and shape of substrate can be plated once it is conductive. This method can therefore be used to deposit semiconductors on patterned substrates. By adjusting the pH of the deposition bath, undesired effects such as sulphur precipitation can be controlled in certain electrolytes such as ones containing elements like sulphur and selenium. By simply varying the stirring rate, the deposition current density and hence the deposition rate can be controlled.

Electrodeposition of semiconductors is a low-cost process. The most expensive equipment used in the process is a computerized potentiostat which may cost up to £5,000 compared to techniques like molecular beam epitaxy (MBE) or metal organic vapor phase epitaxy (MOVPE) in which the cost of the machines is in the order of £1 million. In addition, electrodeposition is versatile in application in the sense that many semiconductor materials can be electrodeposited using the same equipment. The only change required is basically the replacement of the deposition electrolyte by the desired one at any time. Evidence of a variety of electrodeposited semiconductors includes CdTe [32–42], CdS [10, 32], ZnSe [11], ZnTe [12–14], SnS [17], $CuInSe_2$ [18–22], $CuInGaSe_2$ [23–27], and nitrides [9, 43].

Stability, Long Bath Lifetime, Minimum Waste Generation, and Self-Purification of Electrolytes

In some cases due to the composition of the deposition electrolyte, precipitation of metal hydroxides, sulphides, or sulphur is inevitable. This is a major issue in wet chemical methods such as electrodeposition and chemical bath deposition (CBD). In CBD in particular, the deposition electrolyte becomes useless after the first round of deposition, and a large amount of waste is consequently generated. This batch process makes this technique more expensive considering the amount of resources required for frequent waste management and disposal. On the other hand, electrodeposition provides a way of prolonging bath lifetime (a continuous process) and thus reducing this kind of waste. One way of ensuring this is by controlling the pH of the bath quickly before large amounts of precipitates are formed. It can be recalled that in electrodeposition, the reaction is driven by an applied electric field, so the reaction can easily be controlled by varying the applied voltage. The bath can also be filtered from time to time when necessary if precipitation takes place, and the pH and perhaps temperature readjusted. This is not the case with techniques like CBD where the reaction rapidly proceeds to the end once initiated, thus rendering the solution impossible to be used again and therefore generating large volumes of waste containing toxic chemicals in some cases.

When the concentration of ions in the electrolyte goes down as is usually signalled by low deposition current density, the ions can be replenished or topped up by adding calculated amounts of the chemicals containing the ions from a high-purity stock or from a pre-purified solution. This process can be controlled more precisely by the use of certain ion-sensitive devices which can give an indication of the amount of ions required or depleted in the electrolyte. With this, the deposition electrolyte is reinvigorated and therefore can last for a long time before replacement. This serves to minimize the amount of waste generated especially when toxic and environmentally hazardous chemicals are used. It also saves the cost of huge and frequent waste management and disposal.

Another built-in advantage of the electrodeposition process is the fact that as the plating process proceeds over time, the electrolyte itself gets purified more by a gradual and constant removal of detrimental impurities. As a result, subsequently deposited semiconductors become purer with enhanced properties.

Superior Material Qualities; Deposition of p-, i-, and n-Type Semiconductors; and Bandgap Engineering Capabilities

Another strength of electrodeposition also lies in its ability to produce materials with superior qualities as well as its capability in allowing for the deposition of p-, i-, and n-type semiconductors from the same bath and simultaneous tailoring of their bandgaps. As an evidence of this enhanced properties, a comprehensive study of the impurities in CdTe carried out by Lyon et al. [44] shows that electrodeposited CdTe has higher level of purity than those obtained from other techniques using even ultra-high-purity crystals. This conclusion was drawn based on the experimental evidence obtained from secondary ion mass spectroscopy (SIMS) studies carried out on various samples under similar conditions. The work carried out by our group in the past on CdS/CdTe solar cells has also proven the high quality of our electrodeposited CdTe [7, 45]. In this work (especially in Ref. [7]), a conversion efficiency of 18 % (unconfirmed) was published with high short-circuit current density reported for CdS/CdTe solar cells. This unusually high short-circuit current density has also continued to be observed from time to time in our electrodeposited CdS/CdTe solar cells research. The reproducibility of these devices with high short-circuit current densities has not been established yet, but the frequency of observation of these high currents has increased in recent research.

Detailed electrodeposition and characterization of ZnSe thin films have also been carried out in our group [11] as well as in other groups [46–48]. A comparison of the electrodeposited ZnSe (ED-ZnSe) grown in our laboratory and MBE-grown ZnSe (MBE-ZnSe) by means of X-ray diffraction (XRD) and photoluminescence (PL) studies reveals that the ED-ZnSe is of higher quality than the MBE-ZnSe in terms of crystallinity and presence of defect levels. Figures 2 and 3 show these results.

It is worthy of note that the MBE-ZnSe was grown on well-ordered GaAs(100) surface whereas the ED-ZnSe was grown on polycrystalline ITO substrate. A close observation of Fig. 2 shows that the widths of the XRD peaks point out the fact that the ED-ZnSe has better crystalline qualities than the MBE-ZnSe with the ED-ZnSe peaks having narrower widths. From Fig. 3, it is clear again that there are at least five defect levels

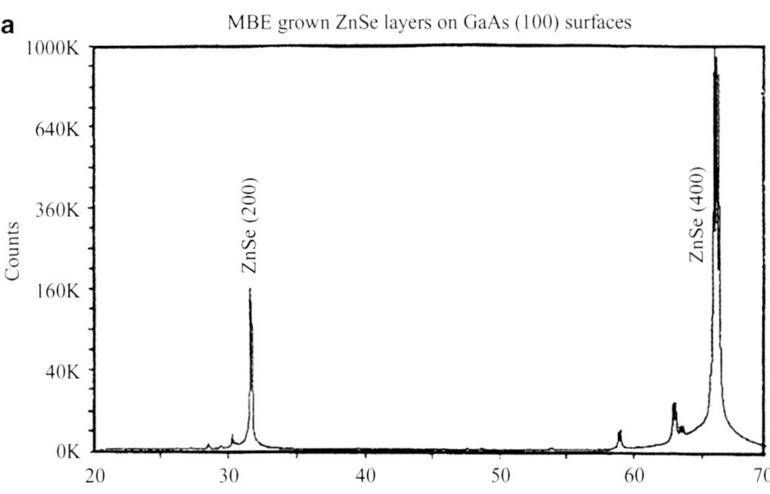

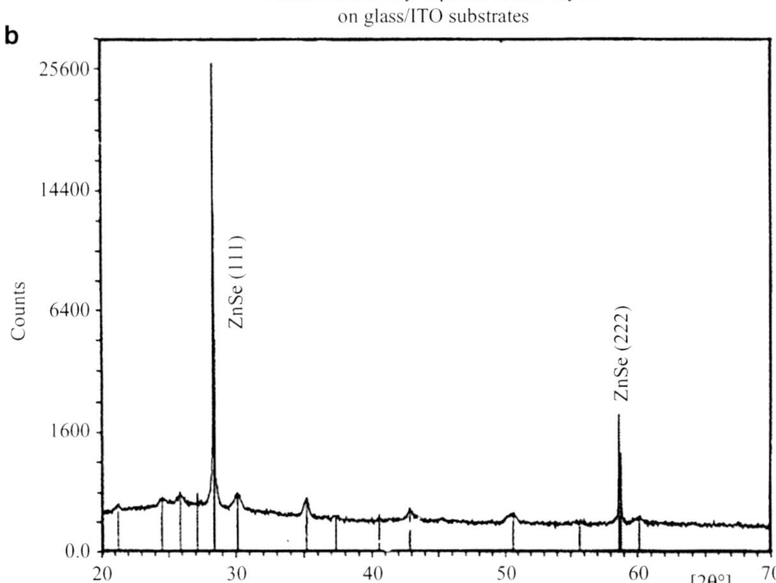

Electrodeposition of Electronic Materials for Applications in Macroelectronic- and Nanotechnology-Based Devices, Fig. 2 X-ray diffractograms of (**a**) MBE-ZnSe and (**b**) ED-ZnSe showing narrower and sharp XRD peaks for ED-ZnSe layers grown on glass/ITO substrate (Ref. [11])

in the MBE-ZnSe as against only one defect level in the ED-ZnSe, according to the number of PL peaks observed. However, there is a broadening of the peak at 0.75 eV for the ED-ZnSe perhaps due to a distribution of defect levels in that region, although these defect levels are not distinctly visible. A number of properties inherent in the electrodeposition process must be responsible for these superior qualities of electrodeposited semiconductor materials. One of these properties is the fact that electrodeposition is a liquid-to-solid transition process.

It is no doubt that nature prefers the liquid-to-solid transition in wet chemical methods compared to transitions from gas-to-solid in vapor deposition techniques. Most vacuum deposition techniques such as MBE and MOVPE involve the second type of transition. It is therefore not a surprise that electrodeposited materials like ZnSe and CdTe can have superior qualities over layers grown by dry methods. Another built-in property of electrodeposition is the hydrogen passivation mechanism [49]. This is because H^+ ions mainly from the aqueous solution are attracted to

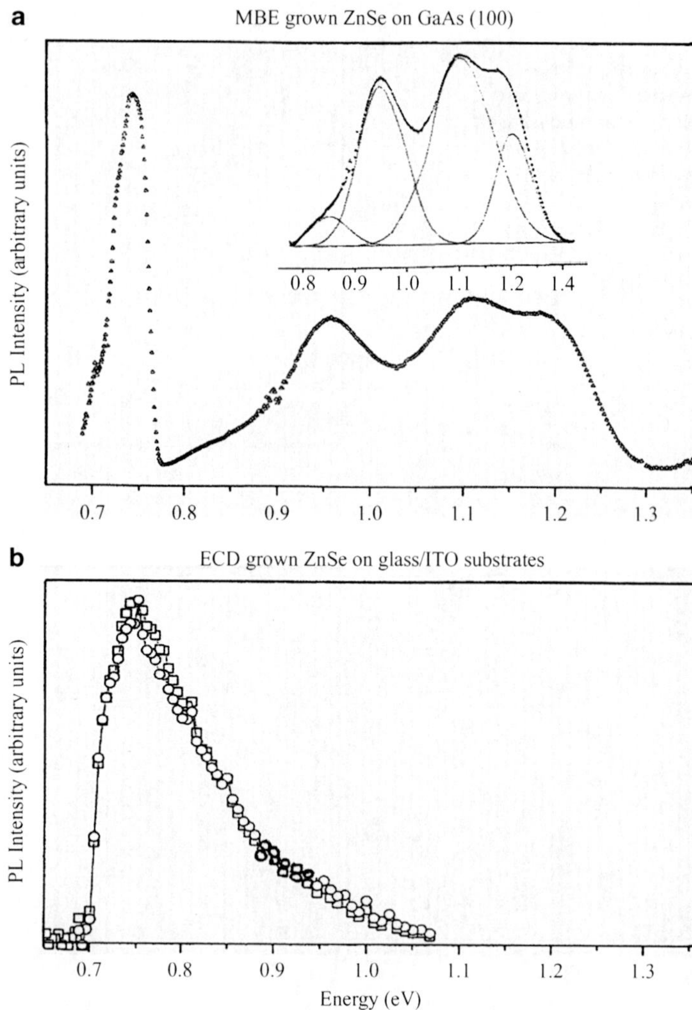

Electrodeposition of Electronic Materials for Applications in Macroelectronic- and Nanotechnology-Based Devices, Fig. 3 Photoluminescence spectra of (**a**) MBE-ZnSe and (**b**) ED-ZnSe showing reduced defect levels in the bandgap of ED-ZnSe layers (Ref. [11])

and discharged at the cathode alongside the semiconductor element ions. As a result, any dangling bonds or defects formed during the semiconductor formation at the cathode are immediately passivated by the more active hydrogen atoms produced during the discharge of H^+ ions. The mechanism of hydrogen and sulphur passivation is well known in the growth of silicon as well as in the surface preparation of GaAs [50, 51].

Electrodeposition provides a convenient way of making p-, i-, and n-type semiconductors from the same bath by changing the deposition voltage slightly. For some semiconductors, such as CdTe and $CuInGaSe_2$, the conductivity type is to a great extent dependent on the stoichiometry (composition) of the compound. In the mechanism of electrodeposition of compound semiconductors, the element with more positive electrochemical potential is first deposited at the lower cathodic potential, while the element with more negative electrochemical potential is deposited at a higher cathodic potential. Thus by changing the deposition potential slightly, around the optimum potential, the stoichiometry of the semiconductor can be varied. If the conductivity of the material is stoichiometry dependent, this can lead to the deposition of p-, i-, and n-type materials as desired from just one deposition electrolyte.

The above concept has been employed in the electrodeposition of p^+-, p-, i-, n-, and n^+-type $CuInSe_2$ [18, 19], $CuInGaSe_2$ [23], and CdTe [52] in our laboratory. For example, in the case of $CuInSe_2$ which is a I-III-VI_2 semiconductor, the stoichiometric material is obtained by having 25 % of the group I element (i.e., Cu), 25 % of the group III element (i.e., In), and 50 % of the group VI element (i.e., Se). Increasing the Cu content slightly therefore makes the material p-type, while slightly increasing the In content makes the material n-type. The p can go to p^+ and the n can go to n^+ depending on the amount of the group I and group III elements incorporated, respectively. The incorporation of these elements is simply done by slightly changing the deposition voltage to the desired level in each case after identifying the voltage at which the stoichiometric material is deposited. Similar results have also been observed for CIGS alloy compound. The result of the electrodeposited p^+-, p-, i-, n-, and n^+- type CIGS material is shown in Fig. 4 below. The figure shows that at lower cathodic voltages (0.40–0.65) V, more Cu is deposited (p-type) while at higher cathodic voltages (0.90–1.30) V more In and Ga are deposited (n-type). At intermediate voltages, a stoichiometric material is obtained with intrinsic or insulating properties.

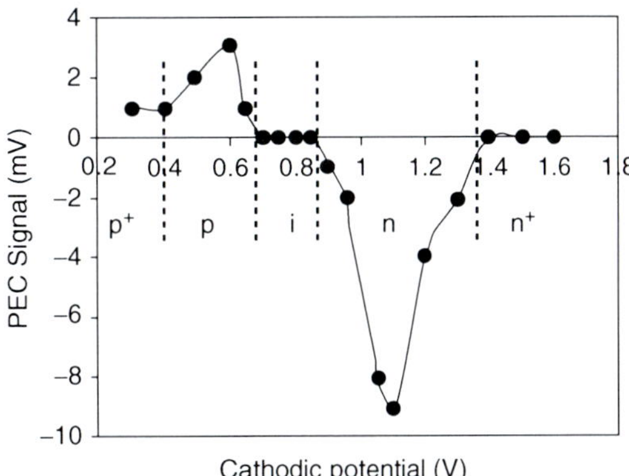

Electrodeposition of Electronic Materials for Applications in Macroelectronic- and Nanotechnology-Based Devices, Fig. 4 PEC signal of electrodeposited CIGS material showing the deposition of p^+, p, i, n, and n^+ materials at different deposition potentials (Ref. [23])

The conductivity type has been determined using photoelectrochemical (PEC) cell measurement. This involves making a solid/liquid junction using the semiconductor material as the solid and a suitable electrolyte as the liquid. The semiconductor material serves as one electrode, while another conductor (e.g., carbon rod) serves as the counter electrode both in contact with the electrolyte. The setup is then placed in an illumination chamber with the electrodes connected to a voltmeter. When the semiconductor is illuminated, an electrical potential V_L is developed between the electrodes as a result of band bending at the semiconductor/liquid junction. The sign of the resulting voltage depends on the direction of this band bending and reflects the conductivity type of the semiconductor involved. Under dark condition, voltage V_D is measured between the electrodes. The difference in the measured voltages ($V_L - V_D$) gives the PEC signal. A negative PEC signal indicates an n-type semiconductor, while a positive PEC signal indicates a p-type semiconductor. Zero PEC signal indicates a material layer with insulating or metallic properties. The PEC cell setup should however be calibrated before use with semiconductors with well-known conductivity types.

From the foregoing, it becomes clear that a complete device can be fabricated from just one electrolyte. For example, a p-i-n diode structure using CIGS material can be fabricated by applying voltages at 0.60 V, 0.75 V, and 1.10 V, respectively, for given periods of time [53]. A similar trend in conductivity type has also been observed for electrodeposited CdTe thin films [52] as shown in Fig. 5. P-, i-, and n-type CdTe can also be electrodeposited from just one electrolyte by simply changing the deposition voltage.

Figure 5 shows that at lower cathodic potentials, Te-rich (p-type) CdTe is produced, while at higher cathodic potential, Cd-rich (n-type) CdTe is produced. At intermediate potentials, stoichiometric (intrinsic) CdTe is obtained. This type of doping of semiconductors to obtain p-, i-, and n-type conductivity without introducing external dopants is known as intrinsic doping. The potential of i-CdTe however could vary slightly due to the concentration of the two elements Cd and Te. Electrodeposition can as well be used to do extrinsic doping of semiconductors simply by adding into the deposition electrolyte,

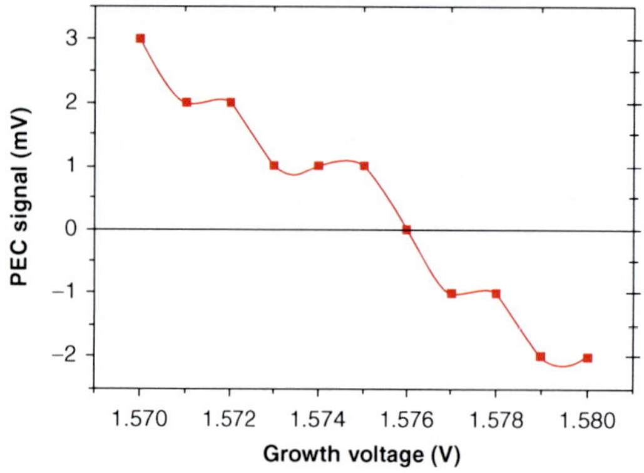

Electrodeposition of Electronic Materials for Applications in Macroelectronic- and Nanotechnology-Based Devices, Fig. 5 PEC signal of electrodeposited CdTe thin films showing the possibility of producing p-, i-, and n-type CdTe from one electrolytic bath by simply varying the deposition potential. Deposition was done using a 2-electrode system (Ref. [52])

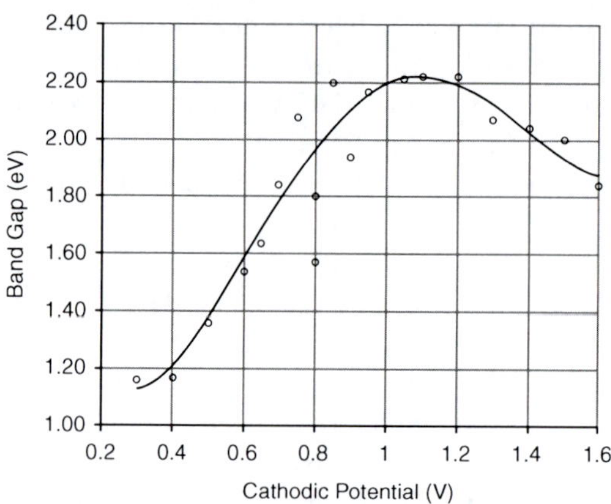

Electrodeposition of Electronic Materials for Applications in Macroelectronic- and Nanotechnology-Based Devices, Fig. 6 Variation of bandgap with deposition potential for electrodeposited $CuInGaSe_2$ (Ref. [23])

appropriate chemicals containing ions of the desired doping elements.

Another interesting strength of electrodeposition is its ability to be applied in bandgap engineering. For certain semiconductors, the bandgap can be varied by incorporating appropriate atoms into their crystal lattice or even by increasing the concentration of one of the constituent atoms. A good example of this type of semiconductors is $CuInSe_2$ whose energy bandgap is $\sim 1.00 \pm 0.1$ eV. By incorporating indium and gallium atoms into this material, its bandgap can be increased to ~2.22 eV with the formation of $CuInGaSe_2$. Electrodeposition provides an easy means of adding these gallium atoms simply by increasing the deposition voltage as described earlier. Figure 6 shows the variation of bandgap with cathodic deposition potential for electrodeposited $CuInGaSe_2$. This figure clearly shows that the bandgap of this material can be tuned from ~1.10 to 2.22 eV by simply changing the deposition voltage [23].

Another material with similar property is GaAs in which the addition of aluminum atoms in the crystal lattice results in the increase in the bandgap with the formation of $Al_xGa_{(1-x)}As$. Electrodeposition therefore provides a convenient platform for bandgap tailoring in some semiconductor materials and devices.

Shortcomings of Electrodeposition of Semiconductors

It will be incomplete if all the merits of electrodeposition are presented without mentioning the difficulties and shortcomings encountered in this technique. Although its advantages outnumber its disadvantages, there are a number of issues in the electrodeposition process.

One of the issues in the electrodeposition process is the control of the concentration of ions in the deposition electrolyte. Although prolonged bath lifetime is an advantage of this technique as mentioned earlier, a convenient in situ method is needed to accurately measure and monitor how ions are depleted in the bath as deposition proceeds. Replenishing of the ions by adding appropriate chemicals is therefore not as precise as it ought to be. It is in most cases a guess work which can subsequently result to the production of materials of slightly different stoichiometry from batch to batch.

Another issue is that of accurate determination of the thickness of deposited layers. Unlike some techniques such as MBE and MOVPE, there is no existing way of measuring the thickness of electrodeposited layers in situ. The use of deposition time as a way of ensuring equal thickness for subsequent depositions is not accurate enough.

This is because the variation of ion concentration in the bath coupled with variation in the resistivity of the particular substrate being used affects the deposition current densities (or deposition rates) of consecutive depositions. As a result, average deposition current densities and therefore deposition rates are not exactly the same, and so thickness of deposited layers differs even for equal deposition times.

Again as stated earlier, electrodeposition can only be carried out on a conducting or at least semiconducting substrate. The starting substrate usually used in electrodeposition (especially in the superstrate configuration in solar cells) is a transparent conducting oxide (TCO) coated on nonconducting normal glass for mechanical strength and support. This glass/TCO starting substrate cannot be made by electrodeposition as normal glass is nonconducting. Other techniques such as sputtering or spray pyrolysis are used to produce this glass/TCO starting substrate. In addition, the metal back contact such as in solar cells is usually made by vacuum evaporation.

In most cases, electrodeposition is carried out with acidic solutions in the pH range ~ (1.4–6.0), although alkaline solutions are also used [33, 54]. In fabricating multilayer semiconductor devices, such as multilayer solar cells, each deposited semiconductor layer except the last layer will have to go through the acidic solution at least one more time for certain periods of time. As a result of this, as well as for pH difference, some of the earlier deposited layers may stand a risk of dissolving or becoming loose in the acidic solution. This becomes an issue in electrodeposition and in wet chemical methods in general, in fabrication of electronic device structures. A way of minimizing this problem is to have the pH of the various deposition solutions equal or similar. Despite the above highlighted issues, electrodeposition still has huge advantages and potentials as a promising and low-cost future semiconductor growth technique for macroelectronic and nanotechnology applications.

Conclusion

A general procedure for the electrodeposition of semiconductor materials and devices was presented. The potentials and benefits of this simple and low-cost process as a macroelectronic and nanotechnology material deposition technique were also highlighted. These capabilities were supported by experimental evidence available for a number of electrodeposited semiconductor materials. These materials can easily find applications in large-area devices such as photovoltaic solar panels and large-area display panels. Such evidences are seen in the ability to obtain semiconductors with superior structural properties, different conductivity types, and varying bandgaps from one deposition bath. The ability to produce semiconductor nanorods that are unidirectional, tightly packed, and normal to the substrate could trigger many new applications in the nanotechnology area.

Acknowledgements The authors wish to acknowledge the contribution of Ajith Weerasinghe, Fijay Fauzi, Dahiru Diso, and Hussein Salim to the research program resulting to the preparation of this manuscript. Our appreciation also goes to our collaborators in Kazakstan, led by Prof. M. B. Dergacheva, for their invaluable contribution in materials characterization. O. K. Echendu wishes to thank the Federal University of Technology, Owerri, Nigeria, for financial support.

References

1. Kröger FA (1978) Cathodic deposition and characterization of metallic or semiconducting binary alloys or compounds. J Electrochem Soc 125: 2082
2. Panicker MPR, Knaster M, Kröger FA (1978) Cathodic deposition of CdTe from aqueous electrolytes. J Electrochem Soc 125:566
3. Danaher WJ, Lyons LE (1978) Photoelectrochemical cell with cadmium telluride film. Nature (London) 271:139
4. Fulop GF, Taylor RM (1985) Electrodeposition of Semiconductors. Ann Rev Mater Sci 15:197
5. Ortega J, Herrero J (1989) Preparation of In X (X = P, As, Sb) thin films by electrochemical methods. J Electrochem Soc 136:3388
6. Basol BM (1984) High-efficiency electroplated heterojunction solar cell. J Appl Phys 55:601
7. Dharmadasa IM, Samantileke AP, Chaure NB, Young J (2002) Semicond Sci Technol 17:1238 http://shura.shu.ac.uk/1270/ New ways of developing glass/conductingglass/CdS/CdTe/metal thin-film solar cells based on a new model
8. Nohira T, Yasuda K, Ito Y (2003) Pinpoint and bulk electrochemical reduction of insulating silicon

dioxide to silicon. Nat Mater 2:397. www.nature.com/Naturematerials
9. Griffiths LE, Lee MR, Mount AR, Kondoh H, Ohta T, Pulham CR (2001) Chem Commun (Cambridge). Low temperature electrochemical synthesis of titanium nitride (6):579
10. Diso DG, Muftah GEA, Patel V, Dharmadasa IM (2010) Growth of CdS layers to develop all-electrodeposited CdS/CdTe thin-film solar cells. J Electrochem Soc 157(6):H647–H651
11. Dharmadasa IM, Samantilleke AP, Young J, Boyle MH, Bacewicz R, Wolska A (1999) Electrodeposited p-type and n-type ZnSe layers for light emitting devices and multi-layer tandem solar cells. J Mater Sci: Mater Electron 10:441
12. Neumann-Spallart M, Konigstein C (1995) Electrodeposition of zinc telluride. Thin Solid Films 265:33
13. Chaure NB, Jayakrishnan R, Nair JP, Pandey RK (1997) Electrodeposition of ZnTe films from a nonaqueous bath. Semicond Sci Technol 12:1171
14. Arico AS (1997) Electrodeposition of thin film ZnTe semiconductors for photovoltaic applications. Adv Perform Mater 4:115
15. Pauporte T, Lincot D (2000) Electrodeposition of semiconductors for optoelectronic devices: results on Zinc oxide. Electrochim Acta 45:3345
16. Yoshida T, Komatsu D, Shimokawa N, Minoura H (2004) Mechanism of cathodic electrodeposition of zinc oxide thin films from aqueous zinc nitrate baths. Thin Solid Films 451–452:166
17. Mathews NR (2010) Charge transport in a pulse-electrodeposited SnS/Al Schottky device. Semicond Sci Technol 25:105010 (6 pp)
18. Chaure NB, Young J, Samantilleke AP, Dharmadasa IM (2004) Electrodeposition of p-i-n type CuInSe2 multilayers for photovoltaic applications. Sol Energy Mater Sol Cells 81:125
19. Dharmadasa IM, Burton RP, Simmond M (2006) Electrodeposition of of CuInSe2 layers using a two-electrode system for applications in multi-layer graded bandgap solar cells. Sol Energy Mater Sol Cells 90:2191–2200
20. Bhattacharya RN (1983) Solution growth and electrodeposited $CuInSe_2$ thin films. J Electrochem Soc 130(10):2040
21. Araujo J, Ortiz R, Lopez-Rivera A, Ortega JM, Montilla M, Alarcon D (2007) Electrochemical growth of CuInSe2 thin film on different substrates from alkaline medium. J Solid State Elechtrochem 11:407
22. Bouraiou A, Aida MS, Tomasella E, Attaf N (2009) ITO substrate resistivity effect on the properties of CuInSe2 deposited using two-electrode system. J Mater Sci 44:1241
23. Dharmadasa IM, Chaure NB, Tolan GJ, Samantilleke AP (2007) Development of p(+), p, I, n and (n+)-type CuInGaSe2 layers for applications in graded bandgap multilayer thin-film solar cells. J Electrochem Soc 154(6):H466–H471
24. Taunier S, Sicx-Kurdi J, Grand PP, Chomont A, Ramdani O, Parissi L, Panheleux P, Naghavi N, Hubert C, Ben-Farah M, Fauvarque JP, Connolloy J, Rousel O, Mogensen P, Mahe E, Guillemoles JF, Lincot D, Kerrec O (2005) Cu(In,Ga)(S,Se)2 solar cells and modules by Electrodeposition. Thin Solid Films 480–481:526
25. Donglin X, Man X, Jianzhuang L, Jiann ZX (2006) Co-electrodeposition and Characterization of Cu (In, Ga)Se2 thin films. J Mater Sci 41:1875
26. Ihlal A, Bouabid K, Soubane D, Nya M, Ait-Taleb-Ali O, Amira Y, Outzourhit A, Nouet G (2007) Comparative study of sputtered and electrodeposited CI(S,Se) and CIGSe thin films. Thin Solid Films 515:5852
27. Pawar SM, Pawar BS, Moholkar AV, Choi DS, Yung JH, Moon JH (2010) Single step electrosynthesis of Cu2ZnSnS4 (CZTS) thin films for solar cell application. Electrochim Acta 55:4057
28. Emziane M, Ottley CJ, Durose K, Halliday DP (2004) Impurity analysis of CdCl2 used for thermal activation of CdTe-based solar cells. J Phys D: Appl Phys 37:2962
29. Dennison S (1994) Dopant and impurity effects in electrodeposited CdS/CdTe thin films for photovoltaic applications. J Mater Chem 4(1):41
30. http//:www.en.Wikipedia.org/wiki/silver_chloride_e/ectrode (Silver chloride electrode)
31. http//:www.en.wikipedia.org/wiki/saturated_calomel_electrode (saturated calomel electrode)
32. Chen F, Qiu W, Chen X, Yang L, Jiang X, Wang M, Chen H (2011) Large-scale fabrication of CdS nanorod arrays on transparent conductive substrates from aqueous solutions. Sol Energy 85:2122
33. Sima M, Enuculescu I, Trautmann C, Neumann R (2004) Electrodeposition of CdTe nanorods in ion track membranes. J Optoelectron Adv Mater 6(1):121
34. Cunningham D, Rubcich M, Skinner D (2002) Cadmium telluride PV module manufacturing at BP Solar. Prog Photovolt: Res Appl 10:159. doi:10.1002/pip.417
35. Morris GC, Das SK(1993) 0-7803-1220-1/93$3.00© (1993) Influence of CdCl2 treatment of CdS on the properties of electrodeposited CdS/CdTe thin film solar cells. *IEEE*
36. Fulop G, Doty M, Meyers P, Betz J, Liu CH (1982) High-efficiency electrodeposited cadmium telluride solar cells. Appl Phys Lett 40:327
37. Lincot D, Kampmann A, Mokili B, Vedel J, Cortes R, Froment M (1995) Epitaxial electrodeposition of CdTe films on InP from aqueous solutions: Role of a chemically deposited CdS intermediate layer. Appl Phys Lett 67:2355
38. Sugimoto Y, Peter LM (1995) Photoeffects during cathodic electrodeposition of CdTe. J Electroanal Chem 386:183
39. Meulenkamp EA, Peter LM (1996) Mechanistic aspects of the electrodeposition of stoichiometric

CdTe on semiconductor substrates. J Chem Soc Faraday Trans 92(20):4077
40. Peter LM, Wang RL (1999) Channel flow cell electrodeposition of CdTe for solar cells. Electrochem Commun 1:554
41. Bhattacharya RN, Rajeshwar K (1985) Heterojunction CdS/CdTe solar cells based on electrodeposited p-CdTe thin films: Fabrication and characterization J Appl Phys 58:3590
42. Basol B (1988) Solar cells electrodeposited CdTe and HgCdTe solar cells 23:69
43. Lin WY, Wuu DS, Pan KF, Huang SH, Lee CE, Wang WK, Hsu SC, Su YY, Huang SY, Horng RH (2005) High-power GaN-mirror-Cu light-emitting diodes for vertical current injection using laser liftoff and electroplating tchniques. IEEE Photonics Technol Lett 17(9)
44. Lyons LE, Morris GC, Horton DH, Kayes JG (1984) Cathodically electrodeposited films of cadmium telluride. J Electroanal Chem Interfacial Electrochem 168:101
45. Chaure NB, Samantilleke AP, Dharmadasa IM (2003) The effects of inclusion of iodine in CdTe thin films on material properties and solar cell performance. Sol Energy Mater Sol Cells 77:303
46. Natarajam C, Sharon M, Levy-Clement C, Neumann-Spallart M (1994) Electrodeposition of zinc selenide. Thin Solid Films 237:118
47. Riveros G, Gomez H, Henriquez R, Marotti RE, Dalchiele EA (2001) Electrodeposition and characterization of ZnSe semiconductor thin films. Sol Energy Mater Sol Cells 70:255
48. Kathalingam A, Mahalingam T, Sanjeeviraja C (2007) Optical and structural study of electrodeposited zinc selenide thin films. Mater Chem Phys 106:215
49. Abdurakhmanov BM, Bilyalov RR (1995) Hydrogen passivation of defects in polycrystalline silicon solar cells. Renew Energy 6(3):303
50. Chen YF, Chen WS, Huang SU, Juang FY (1991) Photoluminescence studies of hydrogen passivation of GaAs grown on InP substrates by molecular-beam epitaxy. J Appl Phys 69(5)
51. Ohno T (1991) Sulfur passivation of GaAs surfaces. Phys Rev B 44(12):6306–6311
52. Diso DG Ph D thesis (2011) Research and development of CdTe-based thin film PV solar cells, Sheffield Hallam University, United Kingdom. http://shura.shu.ac.uk/4941/
53. Dharmadasa IM, Haigh J (2006) Strengths and advantages of electrodeposition as a semiconductor growth technique for applications in macroelectronic devices. J Electrochem Soc 153(1):G47–G52
54. Hsiu SI, Sun IW (2004) Electrodeposition behaviour of cadmium telluride from 1-ethyl-3-methylimidazolium chloride tetrafluoroborate ionic liquid. J Appl Electrochem 34:1057

Electrodisinfection of Urban Wastewater for Reuse

Manuel A. Rodrigo
Department of Chemical Engineering, Faculty of Chemical Sciences and Technology, Universided de Castille la Mandne, Ciudad Real, Spain

Introduction

Development of the society is strongly related to water. This is the most valuable chemical compound for humans, even more than fuels, in spite of its lower price. It is not only required for life but also for agriculture and industry. This strengthens the importance of this compound and shows the significance of having sources with good quality.

Climate change is suspected to modify the rain patterns around the world, and the lack of water is becoming a serious issue in many countries because of the increasing draught periods. As a consequence, the search for new sources of water is a topic of main interest, and much attention has been paid during the recent years in order to look for alternatives.

Reuse of urban wastewater is a very promising option, because it could provide large amounts of water just at the point in which people require it. This source is robust because it is not affected by the climate, but mainly by the population and their degree of development. Thus, unitary flows ranging from 150 to 350 l inhabitant^{-1} day^{-1} are typically expected.

Wastewater produced in a municipality is usually collected in the municipal sewer system from where it flows to a wastewater treatment facility (WWTF). This plant consists of a series of treatments in which the quality of the wastewater is progressively improved. These treatments are classified into four categories: pretreatments (removal of large objects, usually by grids or sieves), primary treatments (removal of particulate pollutants and suspended solids, habitually by settlers), secondary treatments (degradation of the organic content of the sewage, typically by biological reactors), and tertiary treatments (additional improvement of the quality of the treated wastewater either for discharge or for reuse). With those treatments, quality of the effluents meets the standards required for the discharge into a receiving reservoir (rivers, lakes, sea, etc.). These standards are typically proposed in the national regulations of every country, and they could depend on many factors, including the degree of development, the quality of the receiving ecosystem (i.e., susceptible to be eutrophicated), or simply the implementation degree of the technologies at a given time.

If an effluent is going to be reused in any application, further treatments should be applied in order to raise its quality up to the standards required, which would depend on the particular use which it is looked for. In addition, a very careful assessment should be done in order to determine the maximum amount of effluent which could be reused in a particular application, without affecting the ecologic quality of the receiving reservoir (i.e., biodiversity). These additional treatments could be done in the WWTF, and they will be included in the tertiary treatment (in this case they are not looking for discharge quality but for reuse quality). However, they can also be grouped in a new type of facility, which in several countries has started to be called as water regeneration facility or wastewater reuse or recycling facility (WRF). This last option is a very interesting choice when a high demand of reused wastewater is required at a given place, and the effluents of many WWTF have to be merged. An example of this can be the irrigation of large farms and large agricultural areas in dry landscapes.

The flow scheme of these facilities is very similar to the one of a conventional water supply facility which treats surface water (in spite of the very different use). It usually consists of a physicochemical treatment (to reduce pollution associated with the colloids that escape from the secondary clarifiers of the WWTF) and a disinfection unit (to remove pathogens and prevent health issues related to the wastewater reuse). The first treatment follows a four-stage scheme: coagulation, flocculation, clarification, and filtration. Membrane technology (i.e., reverse osmosis or electrodialysis) is sometimes proposed as an additional treatment. It looks to raise water quality to very high standards following a microfiltration stage or to reduce salt content in the produced water. For the second treatment, controversy arises. Chlorination, the most widely used disinfection technology in water supply, shows many significant drawbacks in water reuse which cannot be easily overcome (i.e., production of hazardous by-products such as organochlorinated compounds). Nonpersistent technologies (i.e., UV disinfection) can only assure disinfection in the treatment unit and not later.

Electrochemical Technologies for Wastewater Reuse

Environmental applications of electrochemical technology have been growing during the recent years, because of the good characteristics and prospects of this technology. It is efficient, clean, robust, and easy to be automated. This makes this technology particularly attractive for the reuse of wastewaters. In addition to the well-known electrokinetic technologies for soil remediation, electrochemical technologies for the treatment of industrial wastes have been a hot topic of research for many years, and currently, many applications are in use. These good prospects have pushed researchers in the search for new applications.

Three electrochemical technologies are being proposed as alternatives for wastewater reuse treatments: electrocoagulation, electrodialysis, and electrolysis. All these technologies have shown good efficiencies in other applications (such as industrial waste treatment or desalination), and they have promising features for the reuse of wastewaters [1]. Consequently, they are now being assessed for this new application. Electrocoagulation is currently assessed as an alternative to the physicochemical treatment. Electrodialysis seems to be a competitive choice to substitute reserve osmosis in many cases, and electrolysis shows good prospects for treated wastewater disinfection. In particular, electrocoagulation has shown very good features for this application because besides its colloid removal capability, it buffers pH, reduces salt content of the effluent, and exhibits some sort of disinfection capabilities. These disinfection capabilities are related to the production of small amounts of oxidizing reagents and to the enmeshment of microorganisms into flocs (although in this case it is not an actual disinfection because microorganisms are not killed but transferred to a floc and part of microorganisms can be released back to the water).

Electrolyses as a Disinfection Technology

Disinfection can be defined as the removal of pathogens (i.e., disease-related organisms) from water, either by using chemical or physical methods.

Chemical methods mean the dose of specific reagents, such as halogen derivatives, highly oxidizing compounds, metal ions, quaternary ammonium compounds, or other more specific reagents, while physical methods imply the use of electromagnetic radiations, particle radiation, or electric current [2]. Not all these disinfection technologies have the same maturity. Among these methods, dosing of halogen derivatives (or highly oxidizing compounds) and UV radiation are the most widely used for water and wastewater disinfection.

In every case, disinfection means the killing of pathogens. This can be accomplished either by the destruction or the modification of some of their cell components (i.e., destruction of the cellular membranes, modification of nucleic acids, etc.).

Two different types of technologies are found, depending on the duration of the effects of disinfection. Persistent disinfection means that disinfection is assured in the point in which the treatment is applied and downstream. Nonpersistent disinfection only assures disinfection during the treatment, but not later. Persistent disinfection is usually related to the dosing of oxidizing reagents such as chlorine, hypochlorite, chlorine dioxide, or chloramines which not only reacts but also remains in water for a long time. On the other hand, nonpersistent disinfection is related to very strong oxidizing agents such as ozone which react and disappear rapidly after the dosing or to UV disinfection.

For water supply, it is always required persistent disinfection in order to prevent any health issue. Nonpersistent disinfection only should be applied in the discharge of wastewaters to reservoirs trying to avoid the negative effects of the oxidizing agents dosed during the treatment in the environment, in particular of disinfection by-products. For this reason, UV disinfection has been widely proposed for this application. For reuse, persistent disinfection is normally required, especially in those applications in which reused water is going to be in contact with humans and health issues could arise. At this point, it should be taken into account that regulations in many countries prevent against the direct use of reuse water as drinking water, but not about the use in the irrigation of public gardens and in other applications in which it is an easy direct interaction between humans and water.

Oxidants production is the key point in the relationship between electrolysis and disinfection [3]. As it is known, electrolysis is related to the production of strong oxidant species, and this makes this technology a good choice for disinfection [4]. Electrolysis is the reference technology for the production of many chemicals, including oxidants. Thus, commodities such as chlorine or hypochlorite can be produced by electrochemical processes using three types of very well-known and mature technologies (mercury cell, diaphragm, or membrane cells). Other strong oxidizing agents such as hydrogen peroxide, ozone, peroxocarbonates, peroxosulfates, peroxophosphates, and ferrates are also produced efficiently by electrochemical processes. High reactivity of some of them and the harmless characteristics of their reduction pairs make them good candidates for disinfection.

Many electrode materials have been proposed for electrochemical disinfection, including graphite, platinum, and carbon cloth. At this point, the appearance of dimensionally stable anodes for the production of chlorine was a milestone in this technology. Its robustness, lifetime, and good efficiency have made the number of applications grow, and so, electrolysis with such electrodes has been proposed for the disinfection of water supply in small municipalities, and its use is also widely proposed for disinfection of recreational water, particularly in swimming pools [5]. In the recent years, new electrodes such as conductive diamond have also improved the efficiency in the production of these oxidants [6, 7], due to the large overpotentials that they exhibit for water oxidation. This feature makes the production of oxidants in aqueous solutions particularly interesting in terms of efficiency. In addition, these electrodes achieve the production of hydroxyl radicals: this means that electrolysis is being considered as an Advanced Oxidation Technology [8]. These radicals are known to be very powerful oxidizing reagents, and for sure, they have to exhibit good features for the destruction of pathogens. In addition they help to produce many other oxidants in the treated water such as persulfates, percarbonates, and perphosphates [9]. In this context, the term "mixed oxidants" is widely used to describe the large amount of different oxidant that can be formed during an in-line electrochemical disinfection (not only with conductive diamond but also with any electrode material), due to the oxidation of species contained in the water and the formation of different oxidants [10]. However, this concept is not always seen as an advantage, but sometimes it may be seen as a disadvantage because this mixture may contain some harmful by-products such as organochlorinated species or perchlorates [11].

Use of Electrochemical Disinfection for Reuse

One interesting point to assess the results of disinfection with different technologies is to take into account the typical composition of the effluents discharged by WWTPs. With existing technology, the effluent of a municipal WWTP typically contains small concentrations of organic matter (quantified as COD), nitrates (even in the case of a nutrient removal process), different salts (those coming from the raw wastewater and others added during the treatments), and particles (including suspended solids and colloids) escaping from the secondary settlers. The first stage of the reuse treatment usually consists of the removal of these solids. Results will depend on the filtering technology. This technology can be as exigent as reverse osmosis, but normally it consists of microfiltration, because of the cost (which increases significantly as the pore size decreases). This is particularly accurate when the objective of the treatment is not very exigent. This means that in the disinfection stage, some organic matter and nitrates may be present in the water. Their reactivity should be accounted in the technology assessment, in particular in terms of the occurrence of by-products generated during the disinfection treatment.

Two types of technologies can be considered for electrochemical disinfection:

- In-Line Technologies. They consist of the direct electrolysis of the water to be reused in an electrochemical cell, without dosing any reagents except for electricity. Typically, the production of chlorine and hypochlorite from the chlorides contained in the water, and also the reduction of some of the nitrates to ammonium [12], is expected. Hypochlorite reacts preferentially with ammonium, producing chloramines (breakpoint chlorination), and they become the main disinfection reagents in this technology. Special attention should be paid to the production of large amounts of chlorine in order to prevent oxidation of organic matter and subsequent formation of organochlorinated by-products. In addition, this overproduction of chlorine could also yield large amounts of chlorates (the production of which is related to the ageing of the hypochlorite) and even in some cases perchlorates. Both species are also harmful, and their production should be prevented. To avoid this generation, low current charges and low current densities should be applied. In addition, depending on the type of electrode, other oxidants such as peroxosalts or hydroxyl radicals can be expected. However, this occurrence would only be promoted at large current densities and applied electric current charges [13].
- Off-Line Technologies. Electrochemical technologies for the production of oxidant solutions are mature and well known, and they can be easily designed at any size.

This opens the possibility of production of these oxidants in the treatment facility, not in the waterline but in a separate setup. Once produced, these reagents can be dosed directly in the waterline for disinfection. This avoids the transportation of hazardous reagents and their storage, two operations typically found in chemical disinfection processes. Both facts make this technological approach particularly attractive. Hence, an off-line technology integrates the electrochemical production of oxidants from an optimum raw material and under optimum conditions (a significant difference with respect to in-line technologies in which the raw matter is the water which is being treated) and then in the dosing of these oxidants to the water which is going to be disinfected. At this point, the production of chlorine or hypochlorite from a brine solution (and its later dosing) initially seems to be a prospectively good alternative, taking into account the good disinfectant characteristics of chlorine reagents. The main advantage of this option is the high efficiency and the maturity of the electrochemical technology. The main problem related to this option is the production of organochlorinated by-products, coming from the combination of organic matter and chlorine, once chlorine is dosed to the wastewater. It is important to take into account that opposite to the in-line technology, no ammonium is expected to be present in the water (but only nitrates), and hence, in spite of an accurate dosing, reaction of chlorinated reagents with organics may occur. Thus, just in the cases in which organic matter is not going to be present, this drawback can be avoided, and this possibility can be a very interesting choice. In other cases, a detailed assessment should be done. Other alternatives are the production of ozone (although it is nonpersistent) and the production of peroxocompounds and/or hydrogen peroxide. However, this possibility is still a hot research topic, because in spite these oxidants avoid the drawbacks of the formation of hazardous by-products, it is suspected that they are not efficient enough in the disinfection. Ferrate is also a very promising option, because it is a very energetic oxidant and its reduction product is ferric ion, which it is known to be a good coagulant reagent. This may lead to an integrated technology for the reuse of water with very good scenarios of application (simultaneous disinfection and coagulation). However, a significant price of this reagent is presently a major drawback for this technology, and much research work has to be done in the near future to improve economic viability of this technology.

Future Directions

Reuse of urban wastewater is a very important issue, especially in countries in which the lack of water is (or is going to be) a serious problem. Reuse water is a robust source of water for many applications, and for this reason, in the next years it is expected a growing market for this technology. Disinfection is a very important treatment for reuse, and electrochemical technology shows good prospect for this application, either using in-line or off-line technologies. Main weak point (and hence the point in which the research stress should be pointed) is the formation of by-products during the electrolysis, in particular those derived from the oxidation of chlorides. Main advantage can be the use of alternate oxidants or even in the particular case of the in-line technology, the formation of chloramines. This is a topic of the major interest because with low applied current charges and current densities, in-line disinfection can obtain good disinfection efficiencies with no hazardous by-products production.

Cross-References

- ▶ Disinfection of Water, Electrochemical
- ▶ Electrochemical Treatment of Swimming Pools
- ▶ Wastewater Treatment by Electrocoagulation
- ▶ Wastewater Treatment by Electrogeneration of Strong Oxidants Using Borondoped Diamond (BDD)
- ▶ Water Treatment with Electrogenerated Fe(VI)

References

1. Rodrigo MA, Cañizares P, Buitrón C, Saez C (2010) Electrochemical technologies for the regeneration of urban wastewaters. Electrochim Acta 55:8160–8164
2. Rajeshwar K, Ibanez J (1997) Environmental electrochemistry, fundamentals and applications in pollution abatement. Academic, San Diego
3. Bergmann H, Iourtchouk T, Schoeps K, Bouzek K (2002) New UV irradiation and direct electrolysis-promising methods for water disinfection. J Chem Eng 85:111–117
4. Polcaro AM, Vacca A, Mascia M, Palmas S, Pompej R, Laconi S (2007) Characterization of a stirred tank electrochemical cell for water disinfection processes. Electrochim Acta 52: 2595–2602
5. Rychen P, Provent C, Pupunat L, Hermant N (2010) Domestic and industrial water disinfection using boron-doped diamond electrodes. In: Comninellis C, Chen G (eds) Electrochemistry for the environment. Springer, New York
6. Cañizares P, Saez C, Sanchez-Carretero A, Rodrigo MA (2009) Synthesis of novel oxidants by electrochemical technology. J Appl Electrochem 39:2143–2149
7. Bezerra-Rocha JH, Martínez-Huitle CA (2011) Application of diamond films to water disinfection. In: Brillas E, Martínez-Huitle CA (eds) Synthetic diamond films. Preparation, electrochemistry and applications. Wiley, Hoboken
8. Marselli B, Garcia-Gomez J, Michaud PA, Rodrigo MA, Comninellis C (2003) Electrogeneration of hydroxyl radicals on boron-doped diamond electrodes. J Electrochem Soc 150:D79–D83
9. Serrano K, Michaud PA, Comninellis C, Savall A (2002) Electrochemical preparation of peroxodisulfuric acid using boron doped diamond thin film electrodes. Electrochim Acta 48:431–436
10. Kerwick MI, Reddy SM, Chamberlain AHL, Holt DM (2005) Electrochemical disinfection, an environmentally acceptable method of drinking water disinfection? Electrochim Acta 50: 5270–5277
11. Bergmann MEH, Rollin J, Iourtchouk T (2009) The occurrence of perchlorate during drinking water electrolysis using BDD electrodes. Electrochim Acta 54:2102–2107
12. Cano A, Cañizares P, Barrera C, Sáez C, Rodrigo MA (2011) Use of low current densities in electrolyses with conductive-diamond electrochemical – Oxidation to disinfect treated wastewaters for reuse. Electrochem Commun 13:1268–1270
13. Bergmann MEH (2010) Drinking water disinfection by in-line electrolysis: product and inorganic by-product formation. In: Comninellis C, Chen G (eds) Electrochemistry for the environment. Springer, New York

Electro-Fenton Process for the Degradation of Organic Pollutants in Water

Enric Brillas and Ignasi Sirés
Laboratory of Electrochemistry of Materials and Environment, Department of Physical Chemistry, Faculty of Chemistry, University of Barcelona, Barcelona, Spain

Fundamentals of the Electro-Fenton Process

In 1876, Henry J.H. Fenton publicly announced that the use of a mixture of H_2O_2 and Fe^{2+} (thereafter so-called Fenton's reagent) allowed the destruction of an organic compound, namely, tartaric acid [1]. Such discovery triggered an intense research to elucidate the mechanistic fundamentals and propose different variants and applications of the Fenton process. The possible formation of Fe(IV) as an active Fenton intermediate, as well as the modeling of the real structure of the iron aqua complexes, is still the subject of discussion [2, 3]. However, at present, it is quite well established that the classical Fenton's reaction (1) involves the production of highly oxidative hydroxyl radicals ($^{\bullet}OH$) in the bulk as the main reactive species, and its optimum pH value is 2.8–3.0 [1]:

$$Fe^{2+} + H_2O_2 + H^+ \rightarrow Fe^{3+} + H_2O + {}^{\bullet}OH \tag{1}$$

Only a catalytic amount of Fe^{2+} is required, because this ion can be slowly regenerated from the Fenton-like reaction (2), giving rise to the weaker oxidant hydroperoxyl radical ($HO_2^{\bullet}$), as well as from reduction by $HO_2^{\bullet}$, $R^{\bullet}$, and/or the superoxide radical ($O_2^{\bullet-}$):

$$Fe^{3+} + H_2O_2 \rightarrow Fe^{2+} + HO_2^{\bullet} + H^+ \tag{2}$$

Since the degradation ability of the system is preeminently based on the nonselective action of

$^{\bullet}OH$ onto the inorganic and organic molecules, the Fenton-based processes are considered as advanced oxidation processes (AOPs). $^{\bullet}OH$ is the second strongest oxidizing agent known, with a standard reduction potential of $E^0(^{\bullet}OH/H_2O) = 2.8$ V/SHE. Nowadays, the classical Fenton process that involves the addition of H_2O_2 as a chemical reagent is mostly employed for two different purposes: (i) synthesis of organic molecules and (ii) remediation of waters, solids, and soils polluted by inorganic and organic compounds. Focusing on the degradation of organic pollutants, it has been proven that the attack of $^{\bullet}OH$ to saturated (RH) or aromatic (ArH) molecules gives dehydrogenated or hydroxylated derivatives according to reactions (3) and (4), respectively:

$$RH + {}^{\bullet}OH \rightarrow R^{\bullet} + H_2O \quad (3)$$

$$ArH + {}^{\bullet}OH \rightarrow ArHOH^{\bullet} \quad (4)$$

The existence of parasitic reactions, such as those involving $^{\bullet}OH$ with Fenton's reactants (Fe^{2+}, H_2O_2) and natural radical scavengers (Cl^-, SO_4^{2-}, CO_3^{2-}, etc.), is detrimental because they cause a decrease in the oxidation power of the Fenton system. The efficiency is also a function of temperature, pH, and H_2O_2 and catalyst concentration.

From a conceptual, operative/technical, and environmental standpoint, the electrochemical alternative to the chemical Fenton process seems more appealing because it allows (i) the on-site electrogeneration of H_2O_2, thus avoiding problems and costs related to externalized production, transportation, handling, and storage, and (ii) a much more efficient regeneration of Fe^{2+} by cathodic reduction of Fe^{3+} by reaction (5):

$$Fe^{3+} + e^- \rightarrow Fe^{2+} \quad (5)$$

The electro-Fenton (EF) process was the first proposed electrochemical AOP (EAOP) based on the continuous supply of H_2O_2 to an acidic aqueous solution from the two-electron reduction of injected oxygen gas at a carbonaceous cathode:

$$O_{2(g)} + 2H^+ + 2e^- \rightarrow H_2O_2 \quad (6)$$

The addition of an iron catalyst to the treated solution allows the formation of $^{\bullet}OH$ via Fenton's reaction (1). In 1986, M. Sudoh et al. were the first to apply the method to wastewater treatment. Since then, graphite, carbon-PTFE O_2 diffusion, carbon felt, activated carbon fiber (ACF), reticulated vitreous carbon (RVC), carbon sponge, and carbon nanotubes (CNTs) have been used as cathode materials [1].

The nature of the anode material can also play a significant role whenever an undivided cell/reactor is used to carry out the water treatment. In EF, a stable anode (M) is always employed. Therefore, the presence of a low oxidation power anode such as Pt, RuO_2, or IrO_x or a high oxidation power anode such as boron-doped diamond (BDD), PbO_2, or SnO_2 leads to the concomitant action of reactive oxygen species (ROS) such as (primarily, but not exclusively) adsorbed hydroxyl radicals ($M(^{\bullet}OH)$) formed from water discharge at the anode surface according to reaction (7). Their action on the pollutants and by-products competes with the simultaneous evolution of oxygen from reaction (8) and their self-destruction via reaction (9):

$$M + H_2O \rightarrow M(^{\bullet}OH) + H^+ + e^- \quad (7)$$

$$2\,M(^{\bullet}OH) \rightarrow 2\,M + O_2 + 2\,H^+ + 2e^- \quad (8)$$

$$2\,M(^{\bullet}OH) \rightarrow 2\,M + H_2O_2 \quad (9)$$

Several configurations have been used in order to enhance the oxidation ability of the EF technology, with promising results for two- and three-electrode divided and undivided electrolytic cells. The efficiency of the process is always a function of temperature, pH, O_2 feeding, stirring or liquid flow rate, electrolyte composition, applied potential or current, and catalysts and pollutant concentrations.

Classification of the Fenton-Based Electrochemical Processes

Since the mid-1990s, the EF process has become the starting point for implementing several

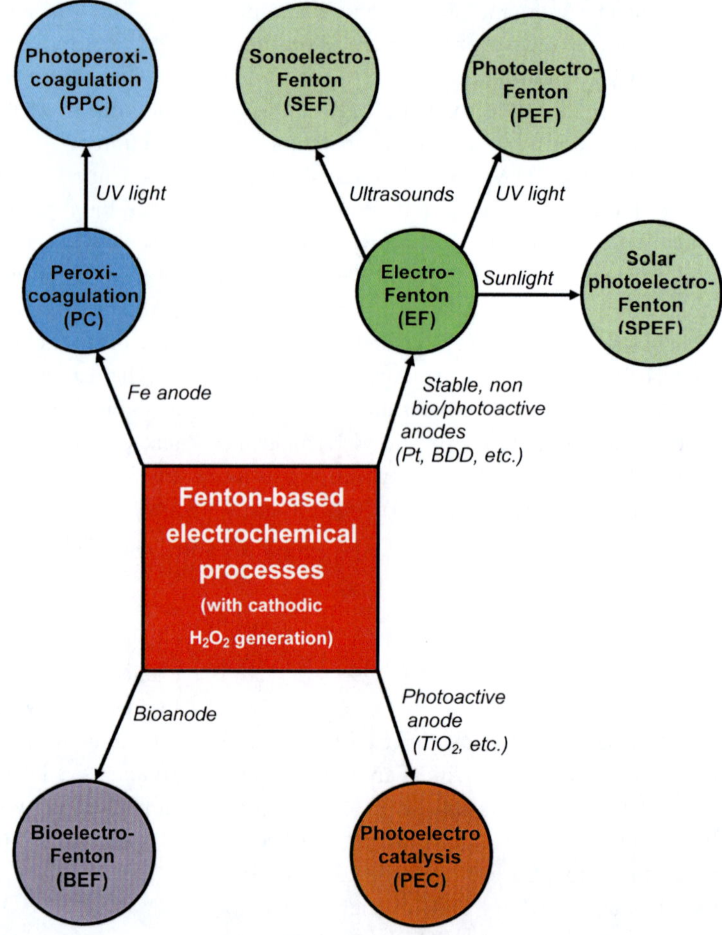

Electro-Fenton Process for the Degradation of Organic Pollutants in Water, Fig. 1 Main Fenton-based electrochemical processes

conceptual and/or technical modifications that have led to a faster and more quantitative degradation of the water pollutants. Figure 1 summarizes all the combined electrochemical processes that rely on the in situ electrogeneration of H_2O_2 to allow the production of $^{\bullet}OH$ by the Fenton's reaction (1). The synergy between various catalysts and oxidants is responsible for the increased efficiency of the treatments. Note that there exists another set of Fenton-based electrochemical processes where H_2O_2 is directly introduced as a chemical reagent [1].

In the photoelectro-Fenton (PEF) process, the solution treated under EF conditions is simultaneously irradiated with UVA light. The production of $^{\bullet}OH$ by reaction (1) and the aforementioned reactions for the regeneration of Fe^{2+} are still valid, but two key additional events take place under the action of photons [1]: (i) the reductive photolysis of $[Fe(OH)]^{2+}$, which is the predominant Fe(III) species at pH 2.5–4.0, according to reaction (10), and (ii) the photodecarboxylation, via reaction (11), of refractory complexes formed between Fe(III) and some carboxylates such as oxalate. Both reactions lead to the quicker regeneration of Fe^{2+}, thus accelerating the production of $^{\bullet}OH$ by Fenton's reaction (1):

$$[Fe(OH)]^{2+} + h\nu \rightarrow Fe^{2+} + {}^{\bullet}OH \quad (10)$$

$$[Fe(OOCR)]^{2+} + h\nu \rightarrow Fe^{2+} + CO_2 + R^{\bullet} \quad (11)$$

A clear practical limitation of the PEF technology is the prohibitive economical cost arising from the use of commercial lamps. Since the minimization of the economical requirements is mandatory to create a niche market for the newly developed water remediation technologies in large-scale systems, a much more appealing PEF process has been lately developed. It is the so-called solar photoelectro-Fenton (SPEF), since it uses sunlight as a free, limitless, renewable UV/Vis source, combined with Pt/gas diffusion (GDE) or BDD/GDE electrochemical reactors. The great performance of SPEF has already been demonstrated for solutions of cresols, pharmaceuticals, and azo dyes [1].

Sonochemical hybrid AOPs such as sonoelectro-Fenton (SEF) also possess a high efficacy for water remediation [1]. The main effect of ultrasounds applied to aqueous solutions is cavitation, which consists of the formation, growth, and collapse of microbubbles that concentrate the acoustic energy into microreactors. The consequent extreme conditions of temperature and pressure, along with the action of $^{\bullet}OH$ formed from water sonolysis via reaction (12), cause the pyrolysis of organic matter:

$$H_2O +))) \rightarrow {}^{\bullet}OH + H^{\bullet} \quad (12)$$

Furthermore, the ultrasounds contribute very significantly to the enhancement of the mass transport regime by convection, i.e., the transport of Fenton's reactants and products towards/from the cathode.

The peroxi-coagulation (PC) process, firstly proposed by Brillas' group, uses an undivided cell similar to that employed for EF but containing a sacrificial iron anode instead of a stable one [1]. Under such conditions, soluble Fe^{2+} is released to the solution from the electrically induced dissolution of Fe by reaction (13), which follows a faradaic behavior:

$$Fe \rightarrow Fe^{2+} + 2\,e^{-} \quad (13)$$

A large $Fe(OH)_3$ precipitate then appears due to the massive production of Fe^{2+}. The organic molecules can then be mineralized by $^{\bullet}OH$ formed from Fenton's reaction (1), although a significant percentage of them also coagulates with the $Fe(OH)_3$ precipitate, depending on pH and applied current. The PC treatment is thus much more efficient than the classical electrocoagulation performed with an Fe anode but in the absence of H_2O_2. The same experimental system can be completed with UVA lamps to operate under photoperoxi-coagulation (PPC) conditions. Unfortunately, the PPC is rather ineffective in practice because the light is mainly absorbed (or dispersed) by the large precipitate in suspension, as shown for various chlorophenoxy herbicides [1].

The development of microbial fuel cells (MFCs) with a bioactive anode, i.e., an anodic electrode material modified with a microorganism, has given rise to the bioelectro-Fenton (BEF) process [4]. The MFCs consist of two-compartment cells separated by an ion-exchange membrane. Biodegradable organic matter contained in the anodic chamber is oxidized at the bioanode, therefore originating an electric current that is conveyed through an external circuit to the cathode immersed in the cathodic chamber, which contains an aerated solution contaminated with iron ions. H_2O_2 is then produced in the cathodic compartment from reaction (6), and organic pollutants are destroyed with $^{\bullet}OH$ formed from Fenton's reaction (1). Large-area cathode materials such as CNTs have been used in BEF, whose performance has been well proven for the degradation of *p*-nitrophenol, amaranth, Orange II, and swine wastewater, among others.

Finally, photoelectrocatalytic (PEC) technologies based on the use of a photoactive anode illuminated with UVA or solar light have also been recently developed for the destruction of pollutants [1]. In classical TiO_2 photocatalysis, which has been the most explored variant, these energy sources promote one electron from the valence to the conduction band (e^{-}_{CB}) producing a hole (h^{+}) by reaction (14). The adsorbed organic matter can then be directly oxidized under the action of h^{+}, by reaction (15), or $^{\bullet}OH$ produced by reaction (16):

$$TiO_2 + h\nu \rightarrow h^{+} + e^{-}_{CB} \quad (14)$$

$$h^{+} + RH_{(ads)} \rightarrow R^{\bullet} + H^{+} \quad (15)$$

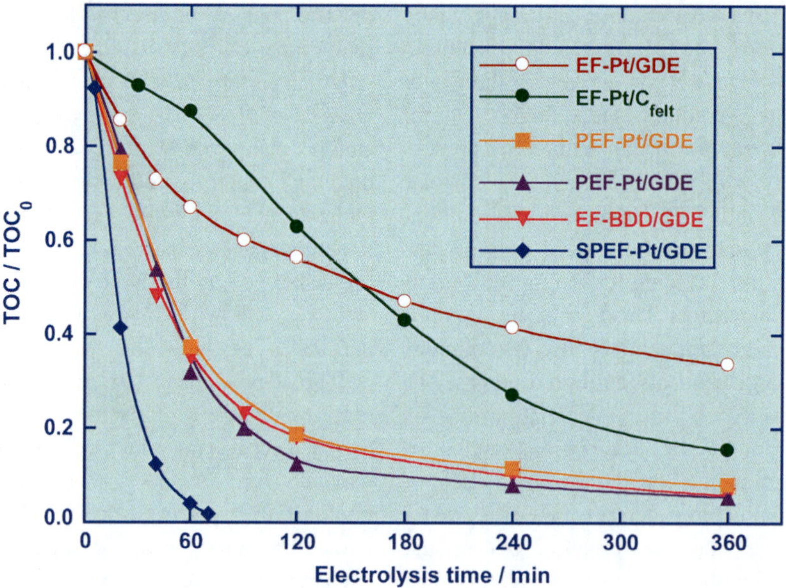

Electro-Fenton Process for the Degradation of Organic Pollutants in Water, Fig. 2 Normalized total organic carbon (*TOC*) abatement for the treatment of organic pollutants in 0.05 M Na_2SO_4 at pH 3.0 by Fenton-based electrochemical processes at 300 mA. (○) 100 mL of 1.0 mM paracetamol and 1.0 mM Fe^{2+} [5]. (●) 220 mL of 0.62 mM sulfamethoxazole and 0.2 mM Fe^{2+} [6]. (■) 100 mL of 0.83 mM clofibric acid and 1.0 mM Fe^{2+} [7]. (▲) 100 mL of 0.47 mM indigo carmine and 1.0 mM Fe^{2+} + 0.25 mM Cu^{2+} [8]. (▼) 100 mL of 1.0 mM 2,4-dichlorophenoxyacetic acid and 1.0 mM Fe^{2+} [9]. (◆) 100 mL of 1.2 mM salicylic acid and 0.5 mM Fe^{2+} [10]

$$h^+ + H_2O \rightarrow {}^{\bullet}OH + H^+ \quad (16)$$

The major efficiency loss arises from the recombination of (e^-_{CB}) with either unreacted h^+ or adsorbed $^{\bullet}OH$. This is minimized in PEC by applying a constant current or anodic potential to a semiconductor-based thin film anode subjected to UV/Vis illumination. Since the photoinduced electrons are continuously extracted from the anode through the external electrical circuit, parasitic reactions are inhibited, and the production of h^+ from reaction (14) and $^{\bullet}OH$ from reaction (16) is accelerated. The coupling of PEC with either EF or PEF is much more effective, due to the synergistic oxidative action of the photoanode and $^{\bullet}OH$ formed in the bulk via Fenton's reaction (1). PEC has been performed with planar or annular semiconductor photoanodes and three-dimensional nanotube arrays, being mainly applied to decolorization of dyes.

Degradation of Organic Pollutants in Waters

A vast number of synthetic organic pollutants, including pesticides, dyestuffs, pharmaceutical and personal care products (PPCPs), and commonly used industrial chemicals, are routinely discharged into water streams. In agreement with the precautionary principle that aims to protect the natural resources from potential contamination, a broad set of water treatment technologies has been devised and tested in recent years. The compared mineralization ability of several EAOPs discussed above is depicted in Fig. 2.

EF with a GDE cathode and a Pt anode is unable to achieve the complete TOC removal of polluted solutions. In Fig. 2, this is exemplified by the treatment of the pharmaceutical paracetamol in the presence of 1.0 mM Fe^{2+}, which only allows 66 % mineralization after 360 min.

The organic matter can be only partially transformed into CO_2 + H_2O + inorganic ions under the action of $^{\bullet}OH$ produced by reaction (1) because of the formation of refractory by-products such as complexes of carboxylic acid intermediates with Fe^{3+}. Further modifications of this EF process then focused on the role of the cathode, anode, metal ion catalyst, and light irradiation in order to reach a faster, almost complete TOC removal.

The use of a carbon-felt cathode instead of a GDE has been shown to allow the almost total mineralization of solutions containing all kinds of organic pollutants [1]. Figure 2 shows the trend of the normalized TOC for the treatment of the antibiotic sulfamethoxazole. The addition of only 0.2 mM Fe^{2+} is enough to yield > 80 % mineralization in 360 min. This significant amelioration compared to the system with GDE can be explained by the much more effective cathodic reduction of Fe^{3+} by reaction (5). This allows (i) the acceleration of the $^{\bullet}OH$ production from reaction (1) and (ii) the formation of Fe (II)-carboxylate complexes, which are much less refractory than their Fe(III) counterparts.

The use of the Pt/GDE system with UVA irradiation, i.e., the PEF process, is illustrated in Fig. 2 for the treatment of clofibric acid with 1.0 mM Fe^{2+}. The much quicker TOC abatement with > 90 % mineralization at 360 min is due to the contribution of reactions (10) and (11). The photolysis of Fe(III)-oxalate by-products that are quite refractory to the $^{\bullet}OH$-mediated oxidation is the crucial, distinctive event. For some pollutants, the efficiency of the PEF process can be further enhanced by using a metal ion cocatalyst. The action of Co^{2+}, Mn^{2+}, and Ag^{+}, among others, has been surveyed, but the most interesting effect has been demonstrated for the Cu^{2+}/Cu^{+} couple. In the case of indigo carmine dye shown in Fig. 2, for example, the positive synergistic effect of Fe^{2+} and Cu^{2+} that promotes the quicker mineralization is accounted for by the easier oxidation of some nitrogenated complexes of Cu(II) (e.g., with oxamate) compared with the competitively formed Fe(III) complexes, which are more slowly removed by $^{\bullet}OH$ and can only be photodegraded. In this system, Cu^{+} can be formed from the reduction of Cu^{2+} with $HO_2^{\bullet}$ and/or with organic radicals $R^{\bullet}$ [1]. Then, the Cu^{2+}/Cu^{+} couple can contribute to (i) the production of $^{\bullet}OH$ in the bulk from the Fenton-like reaction (17) and (ii) the regeneration of Fe^{2+} from reaction (18), which prolongs the Fe^{3+}/Fe^{2+} catalytic cycle:

$$Cu^{+} + H_2O_2 \rightarrow Cu^{2+} + {}^{\bullet}OH + OH^{-} \quad (17)$$

$$Cu^{+} + Fe^{3+} \rightarrow Cu^{2+} + Fe^{2+} \quad (18)$$

The viability of using sunlight instead of UVA lamps has been confirmed for the SPEF degradation of various pollutants such as salicylic acid. A very fast mineralization, with > 95 % TOC removal in ca. 60 min, can be achieved due to the action of natural UV/Vis photons.

Finally, the substitution of Pt by BDD leads to a dramatic enhancement of the oxidation ability of the Fenton-based EAOPs, as observed for the pesticide 2,4-dichlorophenoxyacetic acid in Fig. 2. At present, BDD is the best anode material to oxidize organic pollutants, since it yields a high concentration of physisorbed hydroxyl radicals (BDD($^{\bullet}$OH)) at a very positive anode potential from reaction (7). Therefore, in the EF with BDD, the refractory organic molecules and their complexes with metal ions can be oxidized by the combined action of BDD($^{\bullet}$OH) formed at the anode and $^{\bullet}OH$ produced in the bulk.

Advantages, Disadvantages, and Future Directions

The EF and related processes allow the oxidation of organic pollutants at ambient conditions, thus constituting a safe and simple technology that can be appealing for the industry. Large volumes of wastewaters can be treated if large electrodes and/or cell stacks are employed, which is advantageous for their implementation in water treatment facilities. As explained, quick and very high percentages of organic matter degradation are achieved, avoiding biological posttreatment that usually requires large areas. The main drawbacks are those inherent to the Fenton technology: need of pH adjustment and regulation and final neutralization with sludge production prior to discharge. Progress in self-regulating systems and near-neutral pH Fenton alternatives can minimize these steps, however.

Two main challenges can be highlighted: (i) the cut in electrode prices, particularly for BDD, and (ii) the development of energy-saving systems. Concerning the latter, the BEF process is an interesting choice, but the use of renewable energy sources to power processes must be encouraged in order to improve the sustainability of these EAOPs.

Cross-References

► Boron-Doped Diamond for Green Electro-Organic Synthesis
► Wastewater Treatment by Electrocoagulation

References

1. Brillas E, Sirés I, Oturan MA (2009) Electro-Fenton process and related electrochemical technologies based on Fenton's reaction chemistry. Chem Rev 109:6570–6631
2. Pang SY, Jiang J, Ma J (2011) Oxidation of sulfoxides and arsenic(III) in corrosion of nanoscale zero valent iron by oxygen: evidence against ferryl ions (Fe(IV)) as active intermediates in Fenton reaction. Environ Sci Technol 45:307–312
3. Zakharov II, Kudjukov KY, Bondar VV et al (2011) DFT-based thermodynamics of fenton reactions rejects the 'pure' aqua complex models. Comput Theor Chem 964:94–99
4. Zhu X, Ni J (2009) Simultaneous processes of electricity generation and *p*-nitrophenol degradation in a microbial fuel cell. Electrochem Commun 11:274–277
5. Sirés I, Garrido JA, Rodríguez RM et al (2006) Electrochemical degradation of paracetamol from water by catalytic action of Fe^{2+}, Cu^{2+}, and UVA light on electrogenerated hydrogen peroxide. J Electrochem Soc 153:D1–D9
6. Dirany A, Sirés I, Oturan N et al (2010) Electrochemical abatement of the antibiotic sulfamethoxazole from water. Chemosphere 81:594–602
7. Sirés I, Arias C, Cabot PL et al (2007) Degradation of clofibric acid in acidic aqueous medium by electro-Fenton and photoelectro-Fenton. Chemosphere 66:1660–1669
8. Flox C, Ammar S, Arias C et al (2006) Electro-Fenton and photoelectro-Fenton degradation of indigo carmine in acidic aqueous medium. Appl Catal B-Environ 67:93–104
9. Brillas E, Boye B, Sirés I et al (2004) Electrochemical destruction of chlorophenoxy herbicides by anodic oxidation and electro-Fenton using a boron-doped diamond electrode. Electrochim Acta 49:4487–4496
10. Guinea E, Arias C, Cabot PL et al (2008) Mineralization of salicylic acid in acidic aqueous medium by electrochemical advanced oxidation processes using platinum and boron-doped diamond as anode and cathodically generated hydrogen peroxide. Water Res 42:499–511

Electrogenerated Acid

Naoki Kise
Department of Chemistry and Biotechnology, Graduate School of Engineering, Tottori University, Tottori, Japan

Introduction

In electrolysis, it has been well known that the vicinity of an anode becomes acidic and that of a cathode becomes basic. The acid and base generated in the vicinity of the electrodes are called electrogenerated acid (EGA) and base (EGB), respectively. Since EGA is generated in the specific reaction field, it has the potential to posses the specific reactivity compared with the conventional chemical acids (HCl, H_2SO_4, AcOH, etc.).

Applications of EGA to Organic Reactions

In 1969, Eberson and Olafsson reported that the electrooxidation of hexamethylbenzene **1** in MeCN-H_2O-Et_4NBF_4 gave benzyl alcohol **2** and that in MeCN-H_2O-$NaClO_4$ gave benzylacetamides **3** as shown in Scheme 1 [1]. In 1972, Mayeda and Miller disclosed that the transformation of **2**–**3** was catalyzed by EGA generated in the MeCN-H_2O-$NaClO_4$ system [2].

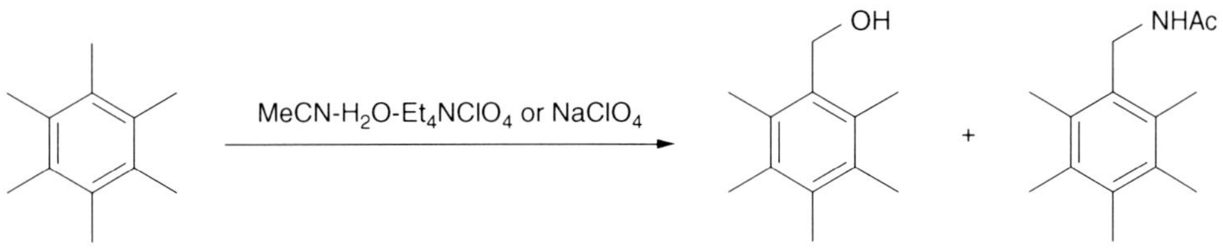

MeCN-H_2O-Et_4NClO_4 or $NaClO_4$

1 2 3

Electrogenerated Acid, Scheme 1 Electrooxidative transformation of **1** to **2** and **3**

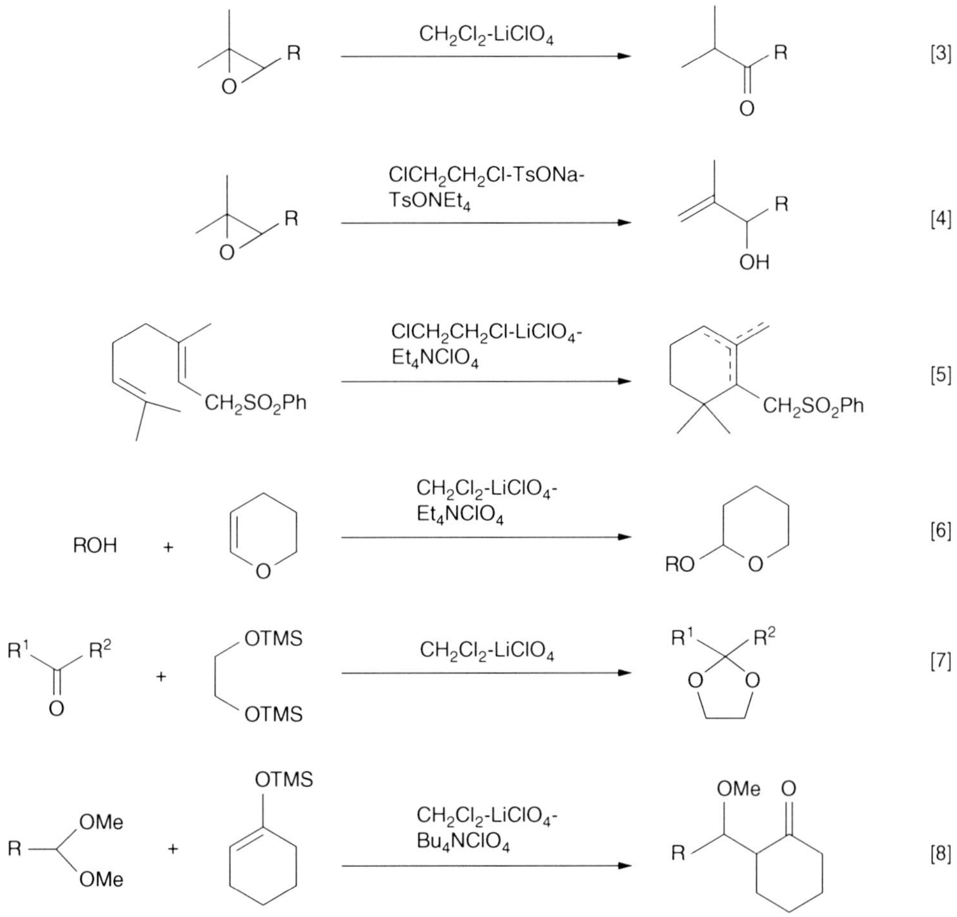

CH_2Cl_2-$LiClO_4$ [3]

$ClCH_2CH_2Cl$-TsONa-TsONEt$_4$ [4]

$ClCH_2CH_2Cl$-$LiClO_4$-Et_4NClO_4 CH_2SO_2Ph [5]

ROH + CH_2Cl_2-$LiClO_4$-Et_4NClO_4 RO [6]

R^1 R^2 + OTMS OTMS CH_2Cl_2-$LiClO_4$ [7]

R OMe OMe + OTMS CH_2Cl_2-$LiClO_4$-Bu_4NClO_4 OMe O R [8]

Electrogenerated Acid, Scheme 2 Pioneering works for EGB-catalyzed organic reactions by Torii's group

In 1980s, Torii's group energetically investigated acid-catalyzed organic reactions utilizing EGA, such as selective ring opening of epoxides to ketones [3] and allylic alcohols [4], cyclization of isoprenoids [5], protection of alcohols with dihydropyran [6], acetalization of ketones [7], and aldol-type reaction of acetals [8] (Scheme 2). These reactions proceeded by conducting a catalytic amount of electricity. In most cases, EGA provided better results than the conventional chemical acids, after the optimization of the electrolysis conditions. The nature and acidity

Electrogenerated Acid, Scheme 3 EGB-catalyzed organic reactions

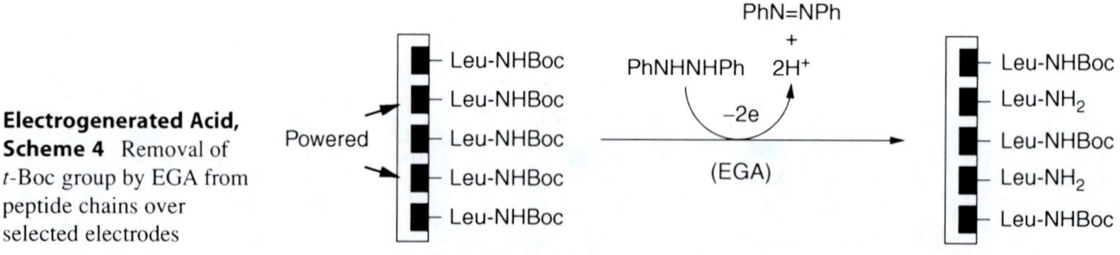

Electrogenerated Acid, Scheme 4 Removal of *t*-Boc group by EGA from peptide chains over selected electrodes

of EGA particularly depends on the combination of electrolytes and solvents. For example, a combination of metal perchlorates and aprotic solvents such as CH_2Cl_2 generates a strongly acidic EGA. The generated EGA is assumed to be anhydrous perchloric acid and, therefore, to be more acidic than commercially available aqueous perchloric acid. After that, EGA was also effectively used for endo-selective ionic Diels-Alder reaction [9], acylation of electron-rich aromatics [10], allylation of β-trifluoromethylated *O,S*-acetals [11], and preparation of 5-hydroxymethylfurfural from sucrose or fructose [12] (Scheme 3). In some cases, the anode solution after the electrolysis with a separated cell was used as an EGA solution to promote acid-catalyzed reactions.

New Applications of EGA

EGA is generated in the vicinity of the anode surface. Taking advantage of the feature, several works have recently been reported. Shinagawa and co-workers reported the electrodeposition of Ag_2O semiconductor films on the anode by EGA [13]. Maurer and co-workers used EGA generated by the electrooxidation of 1,2-diphenyhydradine for the removal of the *t*-Boc group from immobilized peptides [14] (Scheme 4). Using the same EGA, Moeller and co-workers generated reactive *N*-acyliminium ion intermediates on a microelectrode array [15] (Scheme 5). Furthermore, Daasbjerg and co-workers created well-defined aryl-tethered films on carbon surfaces by the EGA [16] (Scheme 6).

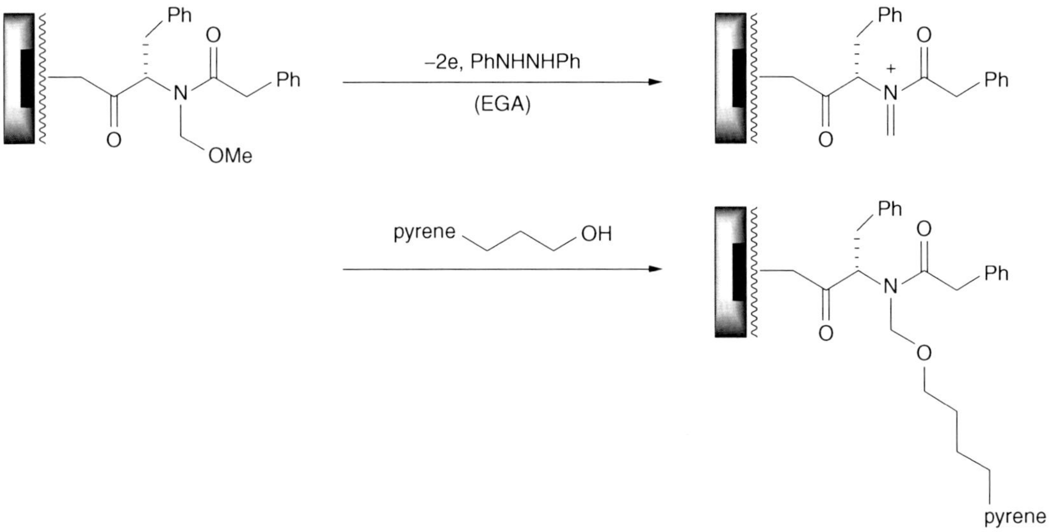

Electrogenerated Acid, Scheme 5 Generation of *N*-acyliminium ion intermediates on a microelectrode array with EGA

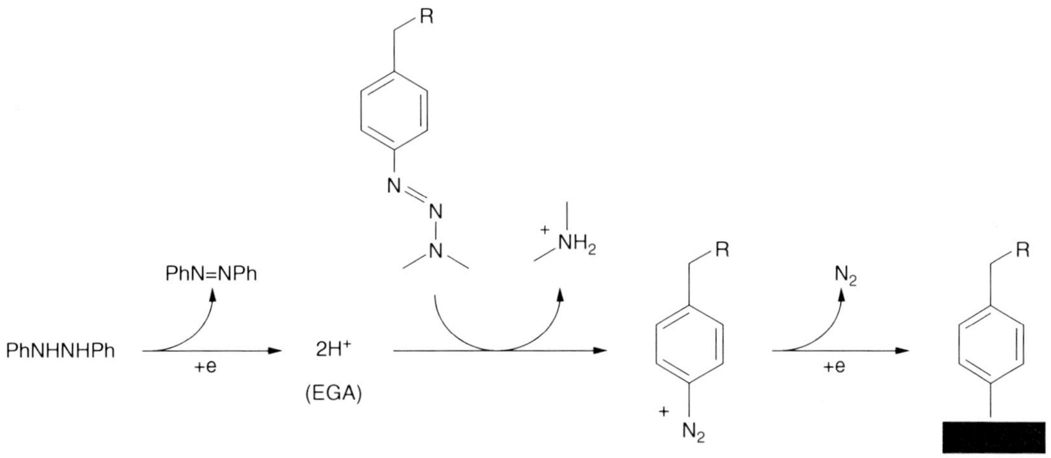

Electrogenerated Acid, Scheme 6 Electrografting of aryl groups on carbon surfaces by exploiting the reaction of aryltriazenes with EGA

Future Directions

Since the nature and acidity of EGA are different from that of the conventional chemical acids, it is expected that EGA promotes the already known acid-catalyzed reactions more effectively with high product selectivity and, moreover, unprecedented reactions can be realized by EGA. In addition to the application of EGA to organic synthesis, the modification of electrode surface with EGA seems to grow up to be a promising area.

Cross-References

▶ Electrogenerated Base

References

1. Eberson L, Olofsson B (1969) Studies on electrolytic substitution reactions. VI. On the mechanism of anodic acetamidation. Acta Chem Scand 23:2355–2366. doi:10.3891/acta.chem.scand.23-2355

2. Mayeda EA, Miller LL (1972) The origin of electrolyte effects in electrooxidative substitution reactions. Tetrahedron 28:3375–3380. doi:10.1016/0040-4020(72)88098-0
3. Uneyama K, Isimura A, Fujii K, Torii S (1983) Electrogenerated acid as a powerful catalyst for transformation of epoxides to ketones and acetonides. Tetrahedron Lett 24:2857–2860. doi:10.1016/S0040-4039(00)88043-0
4. Uneyama K, Nisiyama N, Torii S (1984) Electrochemical procedure directed to the selective ring opening of epoxides to allylic alcohols. Tetrahedron Lett 25:4137–4138. doi:10.1016/S0040-4039(01)90202-3
5. Uneyama K, Isimura A, Torii S (1985) Electrogenerated acid-catalyzed cyclization of isoprenoids. Bull Chem Soc Jpn 58:1859–1860. doi:10.1246/bcsj.58.1859
6. Torii S, Inokuchi T, Kondo K, Ito H (1985) Electrogenerated acid as an efficient catalyst for the protection and deprotection of alcohols with dihydropyran and transesterification of glyceride. Bull Chem Soc Jpn 58:1347–1348. doi:10.1246/bcsj.58.1347
7. Torii S, Inokuchi T (1983) Electrogenerated acid as an efficient catalyst for acetalization of carbonyl group with 1,2-bis (trimethylsiloxy) ethane. Chem Lett 12:1349–1350. doi:10.1246/cl.1983.1349
8. Torii S, Inokuchi T, Takagishi S, Horike H, Kuroda H, Uneyama K (1987) Electrogenerated acid-catalyzed reactions of acetals, aldehydes, and ketones with organosilicon compounds, leading to aldol reactions, allylations, cyanations, and hydride additions. Bull Chem Soc Jpn 60:2173–2188. doi:10.1246/bcsj.60.2173
9. Inokuchi T, Tanigawa S, Torii S (1990) An endo-selective ionic Diels-Alder reaction of a, b-enone and a, b-enal acetals catalyzed by electrogenerated acid. J Org Chem 55:3958–3961. doi:10.1021/jo00299a050
10. Gatti N (1990) Electrogenerated acid-catalyzed acylation of electron-rich aromatics. Tetrahedron Lett 31:3933–3936. doi:10.1016/S0040-4039(00)97510-5
11. Fuchigami T, Yamamoto K, Yano H (1992) Electrolytic reactions of fluoro organic compounds. 11. Anodic preparation and synthetic applications of β-trifluoromethylated O, S-acetals. J Org Chem 57:2946–2950. doi:10.1021/jo00036a037
12. Caruso T, Vasca E (2010) Electrogenerated acid as an efficient catalyst for the preparation of 5-hydroxymethylfurfural. Electrochem Commun 12:1149–1153. doi:10.1016/j.elecom.2010.05.040
13. Maurer K, McShea A, Strathmann M, Dill K (2005) The removal of the *t*-BOC group by electrochemically generated acid and use of an addressable electrode array for peptide synthesis. J Comb Chem 7:637–640. doi:10.1021/cc0498175
14. Ida Y, Watase S, Shinagawa T, Watanabe M, Chigane M, Inaba M, Tasaka A, Izaki M (2008) Direct electrodeposition of 1.46 eV bandgap silver (I) oxide semiconductor films by electrogenerated acid. Chem Mater 20:1254–1256. doi:10.1021/cm702865r
15. Kesselring D, Maurer K, Moeller KD (2008) Building addressable libraries: site-selective formation of an *N*-acyliminium ion intermediate. Org Lett 10:2501–2504. doi:10.1021/ol8007827
16. Kongsfelt M, Vinther J, Malmos K, Ceccato M, Torbensen K, Knudsen CS, Gothelf KV, Pedersen SU, Daasbjerg K (2011) Combining aryltriazenes and electrogenerated acids to create well-defined aryl-tethered films and patterns on surfaces. J Am Chem Soc 133:3788–3791. doi:10.1021/ja111731d

Electrogenerated Base

Shigenori Kashimura[1] and Kouichi Matsumoto[2]
[1]Kinki University, Higashi-Osaka, Japan
[2]Faculty of Science and Engineering, Kinki University, Higashi-osaka, Japan

Introduction

Electroreduction of organic compounds usually generates anionic species and most of them act as nucleophiles, but in some cases, anionic species act as not nucleophiles but bases, which have interesting reactivities in organic synthesis. They are called electrogenerated bases (EGB). Although some structures of EGB are clear, most of other structures are not clear. The organic compounds which generate EGB under the conditions of electroreduction are generally called probases. The use of these EGB often promotes useful synthetic transformations such as condensations, rearrangements, and alkylations of compounds having active hydrogen atoms.

Classification of EGB

There are some types of EGB and they are classified as shown below (type (I)–(IV)):

I. The structures of EGB are unknown.

II. The electroreduction of probase (AB) generates the corresponding radical anions or dianions, which serve as EGB (Eq. 1).

$$AB \xrightarrow{+e \text{ or } +2e} [AB]^{\cdot-} \text{ or } [AB]^{2-} \quad (1)$$

III. The electroreduction of probase (AB) generates anionic species (B^-) involving the dissociation of A–B covalent bond, which serves as EGB (Eq. 2).

$$AB \xrightarrow{+e} A\cdot + B^- \quad (2)$$

IV. When A equals H in Eq. 2 of type (III), electroreductive deprotonation generates anionic species (B^-), which serve as EGB (Eq. 3).

$$HB \xrightarrow{+e} B^- + 1/2\ H_2 \quad (3)$$

Some Synthetic Reactions Using EGB

A variety of EGB mediated synthetic reactions have been reported so far. The typical examples based on classification above mentioned are as follows.

For example, electroreduction of the mixture of dimethyl succinate and ketone promotes Stobbe condensation (Eq. 4) [1]. This type of reaction is classified into type (I), because the structure of EGB is not always clear. One of the plausible reaction mechanisms might be the formation of EGB by the reduction of residual water in the solvent according to type (IV).

Cyclohexanone + dimethyl succinate (CO_2Me, CO_2Me) $\xrightarrow[\text{DMF}]{+e}$ product (HO_2C, CO_2Me) 78% (4)

These types of EGBs promote many base-catalyzed reactions such as Dieckmann condensation [1],Wittig reaction[1], aldol condensation [2], alkylation of compounds having active hydrogen atoms[1], and Stevens rearrangement [3] etc. Equations 5, 6, 7, 8 and 9.

MeO(O)C–(CH₂)₄–C(O)OMe $\xrightarrow[\text{DMF}]{+e}$ 2-(CO_2Me)cyclopentanone 66% (5)

$^{n}Bu\text{–}\overset{+}{P}Ph_3$ + (CH₃CH₂)₂CH–CHO $\xrightarrow[CH_3CN]{+e}$ alkene 82% (cis only) (6)

CH₃CH₂CH₂CHO $\xrightarrow[\text{DMF}]{+e}$ enal (CHO) 76% (7)

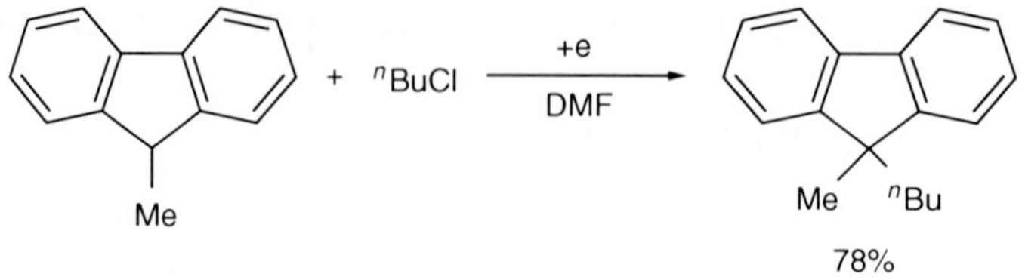

$$\text{9-methylfluorene} + {}^{n}\text{BuCl} \xrightarrow[\text{DMF}]{+e} \text{9-Me-9-}{}^{n}\text{Bu-fluorene} \; (78\%) \tag{8}$$

$$\text{PhCOCH}_2\overset{+}{\text{N}}(\text{Me})_2\text{CH}_2\text{Ph} \xrightarrow[\text{DMF}]{+e} \text{PhCO}\overset{-}{\text{C}}\text{H}\overset{+}{\text{N}}(\text{Me})_2\text{CH}_2\text{Ph} \longrightarrow \text{PhCOCH(CH}_2\text{Ph)N(Me)}_2 \; (36\%) \tag{9}$$

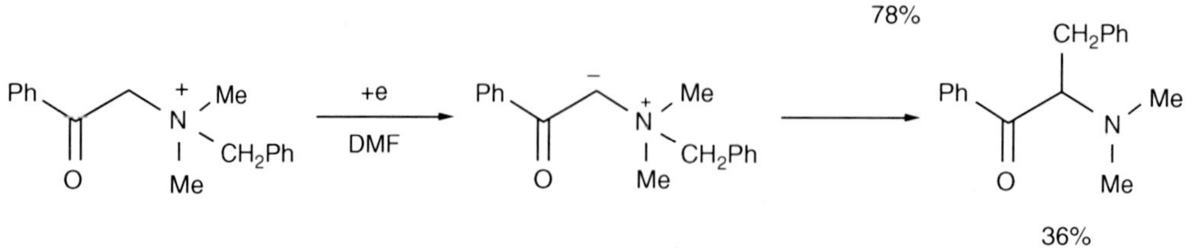

Typical examples which are classified into type (II) are the formation of EGB by the electroreduction of azobenzene (AZ). The electroreduction of AZ generates the corresponding radical anions which serve as EGB (Eq. 10).

$$C_6H_5N{=}NC_6H_5 \xrightarrow{+e} [\,C_6H_5N\text{-}NC_6H_5\,]^{\cdot -} \tag{10}$$

For example, this EGB promotes Wittig reaction. Because azobenzene is electrochemically reduced at −0.9 V (Ag/AgCl), Wittig reaction of aldehyde and phosphonium salt in the presence of azobenzene under the constant voltage reduction at −0.9 V smoothly proceeds to give the corresponding product (Eq. 11) [4].

$$C_6H_5N{=}NC_6H_5 \xrightarrow{+e} [\,C_6H_5N\text{-}NC_6H_5\,]^{\cdot -}$$
$$Ph_3\overset{+}{P}CH_2C_6H_5 \xrightarrow{[\,C_6H_5N\text{-}NC_6H_5\,]^{\cdot -}} Ph_3\overset{+}{P}\overset{-}{C}HC_6H_5 \xrightarrow{PhCHO} PhCH{=}CHC_6H_5 \; (98\%) \tag{11}$$

Other synthetic reactions such as alkylations of compounds having active hydrogen atoms [5] and α-carboxylations with 2,2-di-*tert*-buthylazobenzene as the probase [6] also give the corresponding products in good yields.

Electroreduction of oxygen forms superoxide anion ($O_2.^-$) (Eq. 12).

$$O_2 \xrightarrow{+e} O_2^{\cdot -} \tag{12}$$

This species acts as EGB and promotes useful synthetic reactions. The typical example is shown below (Eq. 13) [7]. The superoxide anion generated by the electroreduction of O_2 serves as base to deprotonate a hydrogen in the α-position in the ester. The anionic species thus generated is allowed to react with O_2 to give the substituted product. The high current efficiency in this reaction indicates the existence of a chain reaction, although the detail of mechanism is not clear yet.

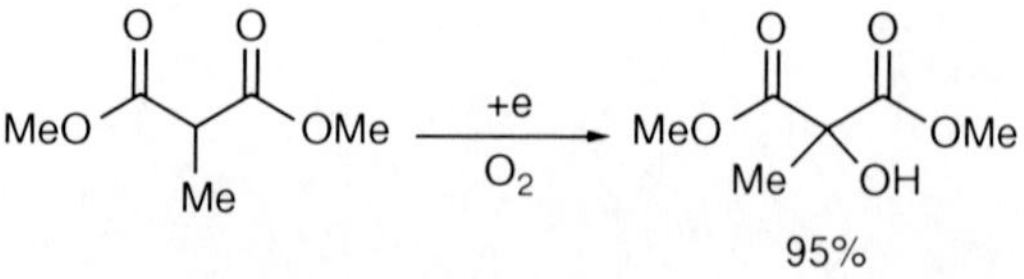

$$\text{MeOOC-CH(Me)-COOMe} \xrightarrow[O_2]{+e} \text{MeOOC-C(Me)(OH)-COOMe} \; (95\%) \tag{13}$$

Electroreduction of unsaturated compounds such as 9-(dicyanomethylene)-fluorenes [8] and fullerenes (C_{60}) [9] also generates radical anionic species and dianion, respectively, which also serve as EGB (Eqs. 14 and 15).

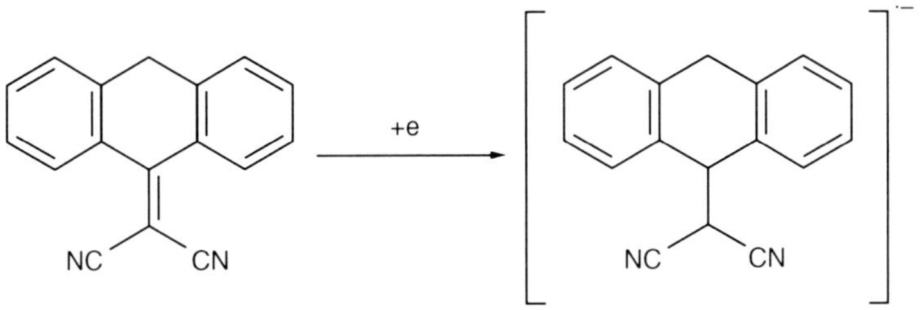

$$\text{(9-dicyanomethylene-9,10-dihydroanthracene)} \xrightarrow{+e} \left[\text{(9-dicyanomethyl-9,10-dihydroanthracene radical)}\right]^{\cdot-} \quad (14)$$

$$C_{60} \xrightarrow{+2e} C_{60}^{2-} \quad (15)$$

Typical example which is classified into type (III) is the electroreduction of carbon tetrachloride (CCl_4) to form trichloromethyl anion (CCl_3^-), which acts as EGB (Eq. 16) [10–12]. This EGB promotes anionic chain reaction with high current efficiency (Eq. 17).The catalytic amount of CCl_4 is electrochemically reduced to generate CCl_3^-, which reacts with aldehyde to give the intermediate. After the protonation of this intermediate from $CHCl_3$, CCl_3^- is regenerated, which causes the next cycle to take place.

$$CCl_4 \xrightarrow{+2e} CCl_3^- + Cl^- \quad (16)$$

$$n\text{-}C_3H_7CHO + CCl_4 + CHCl_3 \xrightarrow{+e} n\text{-}C_3H_7CH(OH)CCl_3 \quad (17)$$

88%

current efficiency 732%

Electroreduction of aryl halides generates aryl anion, which also acts as EGB. For example, the electroreduction of iodobenzene (Ph-I) gives phenyl anion, which deprotonates fluoroform. The trifluoromethyl anion derived from the fluoroform reacts with aldehydes to give the trifluoromethylated alcohols as coupling products (Eqs. 18 and 19) [13].

$$Ph\text{-}I \xrightarrow{+2e} Ph^- + I^- \quad (18)$$

$$PhCHO + CF_3H \xrightarrow[Ph\text{-}I]{+2e} PhCH(OH)\text{-}CF_3 \quad (19)$$

It is also known that the electrogenerated triphenylmethyl anions generated by the electroreduction of triphenylmethane also serve as electrogenerated base for the alkylation of arylacetic esters (Eq. 20) [14].

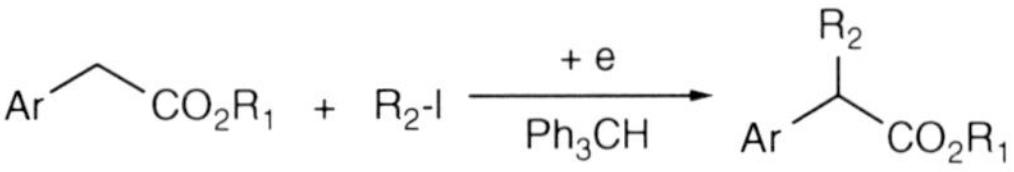

$$ArCH_2CO_2R_1 + R_2\text{-}I \xrightarrow[Ph_3CH]{+e} ArCH(R_2)CO_2R_1 \quad (20)$$

One of the typical examples which are classified into type (IV) is the generation of cyanomethyl anion by the electroreduction of acetonitrile. The electroreduction of acetonitrile in the presence of Et_4NPF_6 as a supporting electrolyte gives the corresponding $Et_4N^+ \ ^-CH_2CN$, which can be utilized in the synthesis of β-lactams involving the carbon-carbon bond formation (Eqs. 21 and 22) [15].

$$CH_3CN\text{-}Et_4NPF_6 \xrightarrow{+e} Et_4N^+ \text{-} CH_2CN + 1/2\ H_2 \quad (21)$$

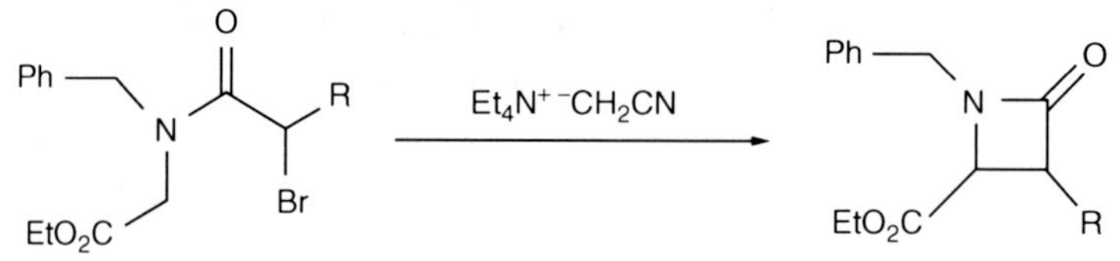

(22)

Another examples of type (IV) are the electrogenerated bases, which are generated by the electroreduction of 2-pyrrolidone (Eq. 23) [16].

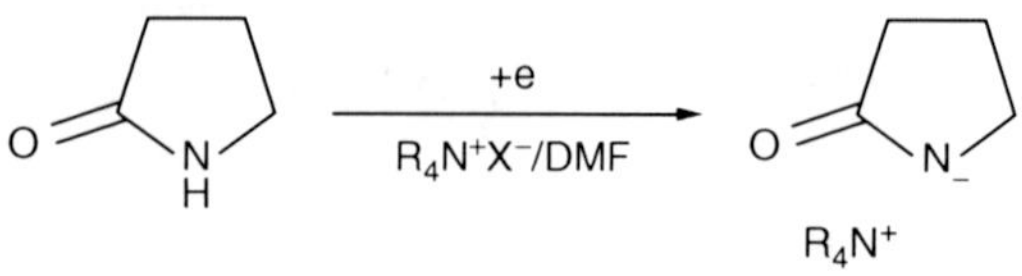

(23)

This EGB promotes a variety of useful synthetic reactions shown below. The exclusive mono-alkylation of methyl-2-phenylacetate is achieved using this EGB (Eq. 24), although the selective mono-alkylation is generally not easy with usual bases such as NaH, and often results in the contamination of the di-alkylated by-products (Eq. 25). The products of mono-alkylation are useful compounds, because they show anti-inflammatory and analgesic activities [16].

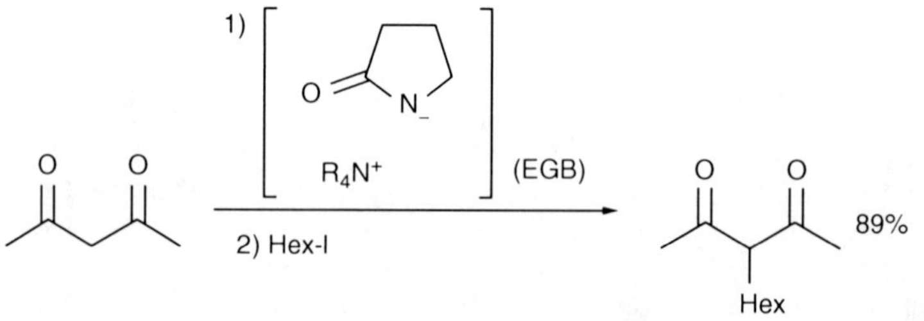

(24)

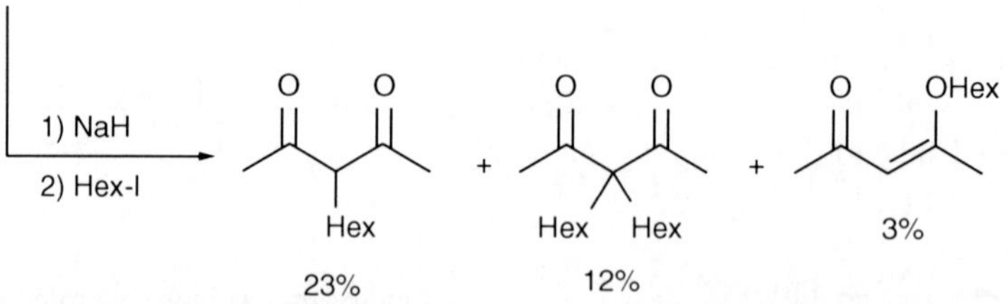

(25)

2-Pyrrolidone anion as EGB generates (trifluoromethyl) malonic ester enolate from malonic ester bearing a CF_3 moiety at the α-position (Eq. 26). Generally, it is difficult to generate the enolate anion with α-CF_3 group, because the enolate anion undergoes the decomposition such as β-elimination of F^-. This EGB promotes a generation of (trifluoromethyl) malonic ester enolate as a stable intermediate, which is allowed to react with electrophile to smoothly give the product in good yield [17].

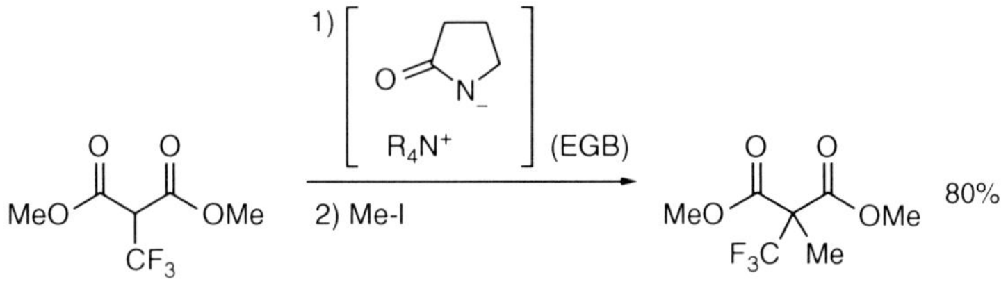

(26)

The EGB derived from 2-pyrrolidone promotes the formation of trihalomethyl anions from trihalomethane and their addition to carbonyl compounds (Eq. 27) [18, 19]. The trifluoromethylation of carbonyl compounds is remarkably useful for the synthesis of fluorine containing compounds.

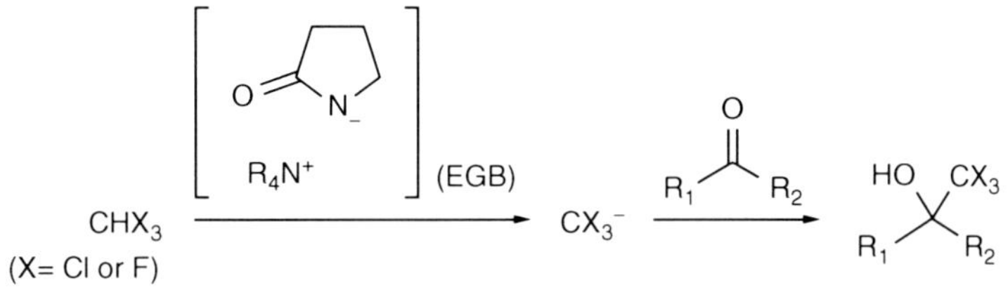

(27)

The EGB (R = Et) promotes the esterification of carboxylic acid having complex structures (Eq. 28) [20]. The high yield and selectivity of the ester might be explained by the reason that the reaction of carboxylic acid with EGB generates the carboxylate anion having ammonium counter-cation and this moiety bears high nucleophilicity. This method possesses high advantage compared to the esterification using diazo compounds.

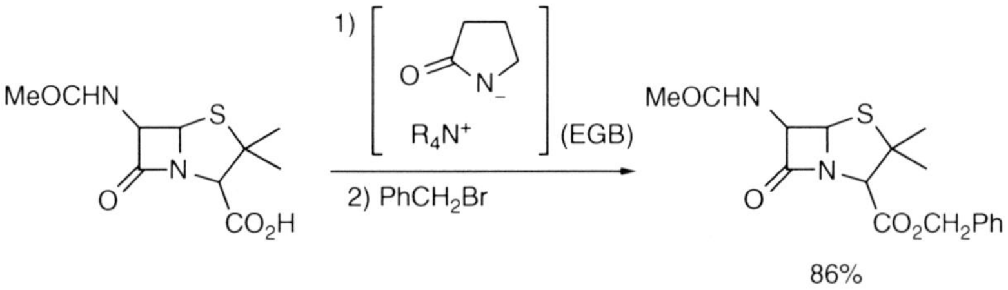

(28)

The use of this EGB (R = Bu) for the synthesis of macrolides is rather effective and gives the corresponding macrolide as the main product. Moreover, the use of EGB (R = Oct) having bulky counter-cation to 2-pyrrolidone anion results in the complete formation of macrolide without dimer (Eq. 29) [20]. As shown below, the increase of the size of the counter-cation results in the increase of the selectivity for the formation of the macrolide. The reason of unique reactivity of this EGB is found to stem from the ammonium counter-cation (R_4N^+), which is also used as a supporting electrolyte ($R_4N^+X^-$) in the electrolysis. The bulky alkyl group (R) in R_4N^+ greatly affected the reactivity of the EGB and subsequent anionic reactions. The increase of the size of the counter-cation increases nucleophilicity and basicity of anionic species. This might be one of the significant merits of EGB, because the control of the reactivity of the anionic species can be easily achieved by the change of the counter-cation, which can be chosen when the electrolysis is set up.

$$BrCH_2(CH_2)_9CH_2CO_2H \xrightarrow{[\text{2-pyrrolidinone anion } R_4N^+]\ (EGB)} \text{macrolactone } (CH_2)_{11} + \text{dilactone } (CH_2)_{11}, (CH_2)_{11}$$

	ratio		total yield
R = Et	75	25	64%
R = Bu	93	7	72%
R = Oct	100	0	66%

(29)

Future Directions

The number of the reaction mediated by the electrogenerated bases is introduced in this entry. Although the method using EGB always competes with the chemical method using bases such as NaH, BuLi, and so on, the advantage of EGB is a variety of probases, and they can be suitably chosen in each case to achieve the desirable reactions. It is also noteworthy that it is quite easy to adjust the reactivity of the electrogenerated bases with change in the electrochemical conditions such as supporting electrolytes and solvents. As future's directions, EGB might become a more useful and powerful method if the new type of EGB and reaction systems is developed, which accomplishes regio- and stereo-selective reactions including asymmetric reactions.

Cross-References

- ▶ Electroreduction of Carbon Dioxide
- ▶ Electrogenerated Acid

References

1. Kashimura S (1985) Synthetic reactions using bases generated under the conditions of electroreduction. J Synth Org Chem Jpn 43:549–556
2. Shono T, Kashimura S, Ishizaki K (1984) Electroinduced aldol condensation. Electrochim Acta 29:603–605
3. Iversen PE (1971) Electrolytic generation of strong bases. II. Stevens rearrangement1. Tetrahedron Lett 12:55–56
4. Iversen PE, Lund H (1969) Electrolytic generation of strong bases I. Wittig reaction. Tetrahedron Lett 10:3523–3524
5. Troll T, Baizer MM (1975) Synthetic utilization of electrogenerated bases. II. Reduced azobenzene as base and nucleophile. Electrochim Acta 20:33–36
6. Hallcher RC, Baizer MM (1977) Synthetic utilization of electrogenerated bases. III. Electrocarboxylation. III. Carboxylation of weak CH-acids. Liebigs Ann Chem 5:737–746
7. Allen PM, Hess U, FooteC S, Baizer MM (1982) Electrogenerated bases. IV. Reaction of electrogenerated superoxide with some carbon acids. Synthetic Commun 12:123–129
8. Nugent ST, Baizer MM, Little RD (1982) Electroreductive cyclization. A comparison of the electrochemical and analogous chemical (MIRC) reaction. Tetrahedron Lett 23:1339–1342
9. Niyazymbetov ME, Evans DH (1995) Use of anions of C_{60} as electrogenerated bases. J Electrochem Soc 142:2655–2658
10. Shono T, Ohmizu H, Kawakami S, Nakano S, Kise N (1981) Electroorganic chemistry. 47. A novel chain reaction induced by cathodic reduction. Addition of trichloromethyl anion to aldehydes or vinyl acetate. Tetrahedron Lett 22:871–874
11. Shono T, Kise N, Yamazaki A, Ohmizu H (1982) Electroorganic chemistry. 61. Novel selective synthesis of α-chloromethyl, α, α-dichloromethyl, and α, α, α-trichloromethyl ketones from aldehyde utilizing electroreduction as key reaction. Tetrahedron Lett 23:1609–1612

12. Shono T, Ohmizu H, Kise N (1982) Electroorganic chemistry. 64. Novel synthesis of carbohydrates using electroreduction as key reactions. Tetrahedron Lett 23:4801–4804
13. Barhdadi R, Troupel M, Perichon J (1998) Coupling of fluoroform with aldehydes using an electrogenerated base. Chem Commun 1251–1252
14. Suzuki S, Kato M, Nakajima S (1994) Application of electrogenerated triphenylmethyl anion as a base for alkylation of arylacetic ester and arylacetonitriles and isomerization of allylbenzenes. Can J Chem 72:357–361
15. Feroci M (2007) Synthesis of *b*-lactams by 4-exo-tert cyclization process induced by electrogenerated cyanomethyl anion, part 2: stereochemical implications. Adv Synth Catal 349:2177–2181
16. Shono T, Kashimura S, Sawamura M, Soejima T (1988) Selective C-alkylation of β-diketones. J Org Chem 53:907–910
17. FuchigamiT NY (1987) Electrolytic transformation of fluoroorganic compounds. 2. Generation and alkylation of a stable (trifluoromethyl)malonic ester enolate using an electrogenerated base. J Org Chem 52:5276–5277
18. Shono T, Kashimura S, Ishizaki IO (1983) A new electrogenerated base. Condensation of chloroform with aliphatic aldehydes. Chem Lett 1311–1312
19. Shono T, Ishifune M, Okada T, Kashimura S (1991) Electrogenerated chemistry. 130. A novel trifluoromethylation of aldehydes and ketones promoted by an electrogenerated base. J Org Chem 56:2–4
20. Shono T, Ishige O, Uyama H, Kashimura S (1986) A novel base useful for synthesis of esters and macrolides. J Org Chem 51:546–549

Electrogenerated Reactive Species

Albert J. Fry
Weslayan University, Middletown, CT, USA

Introduction

Much of modern synthetic organic chemistry involves trapping short-lived reaction intermediates in known chemical reactions in order to divert reactions into other paths producing different products. Electrochemistry is ideally suited for this purpose because single electrons are generally transferred to or from the working electrode. The initial intermediate in most organic electrode reactions is an ion radical; the initial reduction step affords an anion radical and the initial oxidation step, a cation radical. Both types of species are quite reactive and often undergo rapid follow-up chemical reactions to afford neutral radicals, as shown in Fig. 1 [1, 2].

$$X + e^- \longrightarrow X^{\cdot -} \xrightarrow{+H^+} X^{\cdot}$$
$$Y - e^- \longrightarrow Y^{\cdot +} \longrightarrow Y^{\cdot}$$

Electrogenerated Reactive Species, Fig. 1 Typical decomposition routes of electrogenerated ion radicals

$$R\text{-}Hal + e^- \longrightarrow R^{\cdot} + Hal^-$$
$$R^{\cdot} + e^- \longrightarrow R^-$$
$$R_1R_2C{=}O \xrightarrow{H^+} R_1R_2C{=}OH^+$$

Electrogenerated Reactive Species, Fig. 2 Typical electrochemical routes to reactive electrochemical intermediates

The radicals are often themselves electroactive and can therefore undergo further reduction or oxidation to form anions or cations, respectively. Radicals can also be produced in other ways. One well-known example of this is alkyl halides. They are reduced to radicals at a cathode by a process in which electron transfer and bond breaking are concerted. The electrogenerated radical is often then reduced to a carbanion (Fig. 2). Reduction of carbonyl compounds in acidic media proceeds via the conjugate acid of the carbonyl compound (Fig. 2), which can then undergo reduction to a ketyl radical. This entry will address the variety of ion radicals and radicals that can be readily generated electrochemically, as well as typical chemical reactions.

Anodically Generated Species

Alkene Cation Radicals

The simplest type of cation radical is that from an alkene. The alkene gives up an electron from its highest occupied molecular orbital to produce a cation radical in which the charge is shared between the two carbons of the double bond. Alkenes in which the four groups attached to the

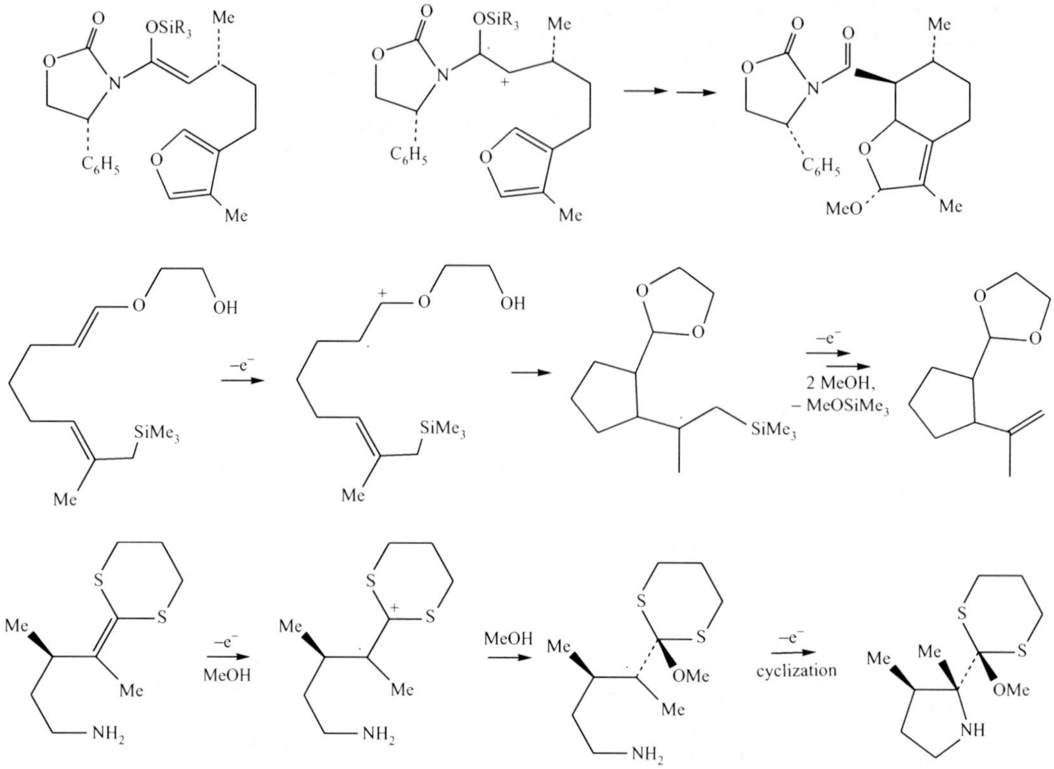

Electrogenerated Reactive Species, Fig. 3 Intramolecular anodic cyclization of *bis*-alkenes

double bond are either alkyl or hydrogen generally have unacceptably high oxidation potentials. For this reason, alkenes are used bearing one or two electron-supplying groups (alkoxy, silyl, etc.) at one end of the double bond. Not only do such groups greatly lower the oxidation potential, they also polarize the cation radical such that the charge is localized primarily at one end of the double bond, ensuring selectivity. A wide variety of applications have been reported in which an alkene cation radical reacts with another nucleophilic component of the medium to form a new carbon-carbon bond. Selective oxidations are possible of substances in which two unsaturated groups of differing oxidation potential are present in a molecule. The electroactive group of lower oxidation potential is converted selectively to its cation radical, which then reacts with the other group intramolecularly. Remarkably complex structures can thus be assembled in a single step (Fig. 3) [3–5].

Carbocations

As illustrated in Fig. 1, one prominent path followed by electrogenerated cation radicals involves loss of a proton to form a neutral radical ($R^{\cdot}$). Such radicals are easier to oxidize than the initial alkene. This second one-electron oxidation produces a carbocation ($R^{\cdot} - e^{-}\ R^{+}$). If the electrochemical oxidation is carried out in an inert solvent such as dichloromethane, the carbocation is highly reactive. Many useful reactions have been carried out using carbocations generated in this manner (Fig. 4) [1, 6, 7]. Sometimes the carbocation may be unstable at temperatures near ambient, decomposing before it can react with an added reactant. Concern over this problem has led to the very useful concept known as the "cation pool" method in which the electrochemical reaction is carried out at a low temperature (typically −78 °C) at which the carbocation is stable. The resulting solution is known as the *cation pool.*

Electrogenerated Reactive Species, Fig. 4 Ring formation via an electrogenerated carbocation

Electrogenerated Reactive Species, Fig. 5 Use of electrogenerated *N*-acyliminium ions for heterocyclic synthesis

At this point, one may add a second substance to the pool at low temperature to react with the carbocation [8]. The more versatile *cation flow* strategy involves pumping the cold solution of the cation into a mixing chamber of a microreactor into which the other reactant is simultaneously being pumped. The flow rates of the two species can be controlled such that reaction is complete within the residence time of the microreactor chamber. This mode of operation can be elaborated, further mixing the product with a third reactant downstream of the microreactor chamber. The technology can be readily adapted to combinatorial synthesis of families of related substances [8].

Alpha-Amino Carbocations (*N*-Acyliminium Ions)

One of the most widely applied applications involving anodic conversion of a neutral substance to a carbocation involves oxidation of amides, of which the most widely studied are cyclic amides (lactams). This relies on the fact that the species being oxidized need not be an alkene; it can be a heteroatom bearing an unshared pair of electrons. Thus, oxidation of amides affords *N*-acyliminium species, e.g., **1** [9]. These are good electrophiles, permitting anodic syntheses of substances containing nitrogen, a fact of great interest with respect to

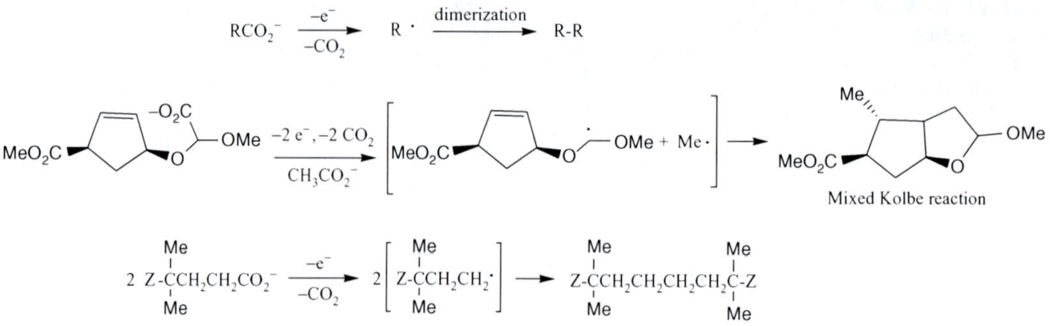

Electrogenerated Reactive Species, Fig. 6 Representative Kolbe reactions

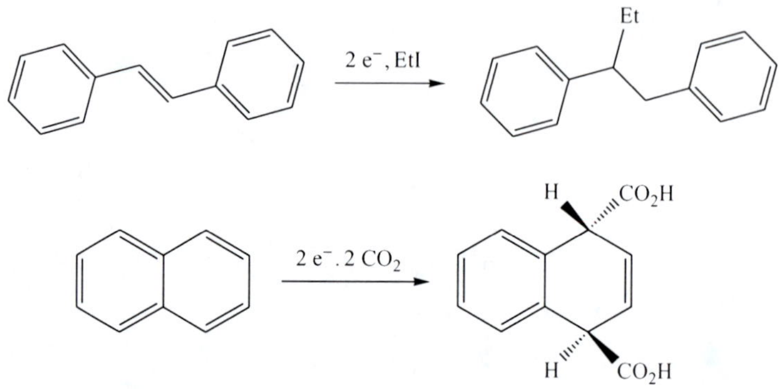

Electrogenerated Reactive Species, Fig. 7 Reactions of the electrochemically generated naphthalene anion radical

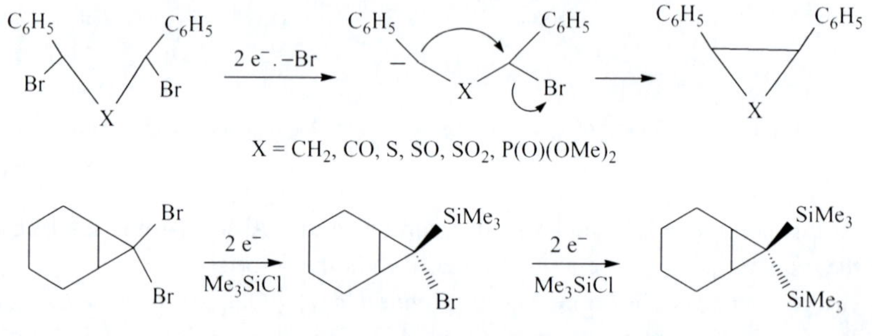

Electrogenerated Reactive Species, Fig. 8 Trapping of electrogenerated carbanions from alkyl halides

synthesis of nitrogenous natural products (Fig. 5) [9, 10].

Alkyl Radicals: The Kolbe Reaction

An excellent method for producing free radicals for synthetic applications, known for many years but still one of the best, is to oxidize a carboxylate ion (RCO_2^-) at an anode. The initial carboxyl radical rapidly loses CO_2 to afford a radical R˙. It is generally accepted that the radical is adsorbed on the electrode, extending its lifetime and favoring dimerization compared to situations in which the same radical is generated in solution. Many synthetic applications of this so-called *Kolbe* reaction have been reported (Fig. 6) [1, 10].

Cathodically Generated Species

Arene Anion Radicals and Dianions

Although the reduction potentials of alkenes are too high to permit reduction, it is possible to reduce polycyclic aromatic hydrocarbons and alkenes bearing groups that can stabilize anions in aprotic media to afford the corresponding anion radicals and dianions. These species are good nucleophiles and can be captured by added electrophiles such as protons, alkyl halides, silyl halides, and carbon dioxide (Fig. 7) [11].

Alkyl Carbanions

Carbon-heteroatom bonds can readily be cleaved by electron transfer from a cathode. Because of the relative weakness of carbon-halogen bonds and also because organohalides are abundantly available in a variety of structural types, the most well-studied example of this process is the cathodic reduction of alkyl and aryl halides, as shown in Fig. 1 [1]. The first step involves transfer of a single electron to form a free radical. A second electron transfer can take place to convert the radical to a carbanion. In all but a few cases, the reduction potential of the radical (E_2) is positive of that of the starting halide (E_1) so that the reduction to all intents and purposes is an overall two-electron process. In a few exceptional cases such as benzylic and tertiary butyl bromides and iodides, E_2 is negative of E_1, the process occurs in two discrete steps. It is not generally possible to trap the intermediate radical because of its short lifetime. However, electrogenerated carbanions are readily trapped by added electrophiles. In fact, since electrogeneration is carried out under neutral conditions, it can be used to prepare carbanions from base-sensitive substrates that would not survive other procedures for producing carbanions. Representative examples of use of electrogenerated carbanions are shown in Fig. 8.

Reduction of Carbonyl Compounds

Figure 2 illustrates the formation of an *alpha*-hydroxy radical by reduction of a carbonyl compound in the presence of a proton donor. Two such radicals generally couple to form pinacols (vicinal diols). In the absence of proton sources, carbonyl compounds are reduced to anion radicals and, at a more negative potential, dianions [1] (Fig. 9).

Electrogenerated Reactive Species, Fig. 9 Intermediates produced by reduction of carbonyl compounds in aprotic media

Electrogenerated Reactive Species, Fig. 10 Mechanism and applications of hydrodimerization of activated alkenes

Hydrodimerization of α,β-Unsaturated Carbonyl Compounds

Cathodic reduction of α,β-unsaturated carbonyl compounds and related alkenes bearing an electron-withdrawing group leads to anion radicals that exhibit a high degree of radical character at the *beta*-carbon. This in turn favors dimerization. The overall conversion involves addition of the elements of hydrogen (two electrons and two protons); hence, the process is termed *hydrodimerization*. A wide variety of substances have been shown to undergo hydrodimerization in good yield (Fig. 10) [1]. As the figure shows, intramolecular versions are known, often proceeding in excellent stereoselectivity.

Future Directions

Increasingly, organic electrosynthetic studies rely on mechanistic analysis to identify the short-lived intermediates (often more than one) that are involved in the complex reaction cascades that constitute the overall electrode process. Realization of the power of mechanism-based synthetic design, coupled with new electrode materials such as boron-doped diamond [12], techniques such as low temperature electrolysis [8] and flow microreactors [8, 13], with the increased control of reaction conditions that these permit, have led to rapid recent advances in the art of organic electrosynthesis. This may be expected to continue at a faster pace in the future. Another area that will prove increasingly important will involve applications that reduce the amount of waste products from electrochemical reactions as well as the energy needed to drive them. The recent demonstration that many common electrode reactions can be carried out using inexpensive photovoltaic devices [14] will reduce the energy costs of synthetic electrochemistry.

Cross-References

► Cation-Pool Method
► Combinatorial Electrochemical Synthesis
► Electrochemical Fixation of Carbon Dioxide (Cathodic Reduction in the Presence of Carbon Dioxide)
► Electrochemical Microflow Systems
► Kolbe and Related Reactions

References

1. Fry AJ (1989) Synthetic organic electrochemistry, vol 2. Wiley, New York
2. Lund H, Baizer MM (eds) (1991) Organic electrochemistry, 3rd edn. Dekker, New York
3. Reddy SHK, Chiba Sun YK, Moeller KD (2001) Anodic oxidation of electron-rich olefins: radical cation approaches to the synthesis of bridged bicyclic systems. Tetrahedron 57: 5183–5197
4. New DG, Tesfai Z, Moeller KD (1996) Intramolecular anodic olefin coupling and the use of electron-rich aryl rings. J Org Chem 61:1578–1598
5. Xu H-C, Moeller KD (2010) Intramolecular hydroamination of dithioketene acetals: an easy route to cyclic amino acid derivatives. Org Lett 12:5174–5177
6. Sainsbury M, Wyatt J (1976) Intramolecular coupling of diaryl amides by anodic oxidation. J Chem Soc Perkin Trans 1:661–664
7. Kotani E, Takeuchi N, Tobinaga S (1973) Total synthesis of the alkaloids (+)-oxocrinine and (+)-maritidine by anodic oxidation. Chem Commun 550–551
8. Yoshida J-i, Suga S (2002) Basic concepts of "cation pool" and "cation flow" methods and their applicatons in conventional and combinatorial organic synthesis. Chem Eur J 8: 2650–2658
9. Shono T, Matsumura Y, Kanazawa T (1983) Electroorganic chemistry. 69. A general method for the synthesis of indoles bearing a variety of substituents at the b-position, and its appication to the synthesis of L-tryptophan. Tetrahedron Lett 24: 1259–1262
10. Shono T, Matsumura Y, Inoue K (1983) Electroorganic chemistry. 71. Anodic a-methoxylation of *N*-carbomethoxylated or *N*-acylated a-aminoacid esters and aamino-b-lactams. J Org Chem 48: 1388–1389
11. Wawzonek S, Blaha EW, Berkey R, Runner ME (1955) Polarographic studies in acetonitrile and dimethylformamide. II. Behavior of aromatic olefins and hydrocarbons. J Electrochem Soc 102: 235–242
12. Waldvogel SR, Mentizi S, Kirste A (2012) Boron-doped diamond electrodes for electroorganic synthesis. Topics Curr Chem 320:1–32

13. Fumihiro A, Hidyuki M, Keishi F, Tsuneo K, Chiaki K, Fuchigami T, Atobe M (2011) Product selectivity control induced by using liquid-liquid parallel laminar flow in a microreactor. Org Biomol Chem 9:4256–4265
14. Anderson LA, Redden A, Moeller KD (2011) Connecting the dots: using sunlight to drive electrochemical oxidations. Green Chem 13:1652–1654

Electrokinetic Barriers for Preventing Groundwater Pollution

Sergio Ferro
Department of Chemical and Pharmaceutical Sciences, University of Ferrara, Ferrara, Italy

Introduction

Groundwater is water located below the ground surface, in soil pore spaces, and in the fractures of rock formations. The water moves down into the ground because of gravity, passing between particles of soil, sand, gravel, or rock until it reaches a depth where the ground is filled, or saturated, with water. The area that is filled with water is called the saturated zone, and the top of this zone is called the water table. A porous or an unconsolidated deposit is called an aquifer when it can yield a usable quantity of water.

A groundwater pollutant is any substance that, when it reaches an aquifer, makes the water unclean or otherwise unsuitable for a particular purpose. Sometimes the substance is a manufactured chemical, but just as often it might be microbial contamination. Contamination also can occur from naturally occurring mineral and metallic deposits in rock and soil.

Groundwater pollution caused by human activities usually falls into one of two categories: point source pollution and nonpoint source pollution. Chemicals used in agriculture, such as fertilizers, pesticides, and herbicides, are examples of nonpoint source pollution because they are spread out across wide areas. They may eventually reach underlying aquifers, particularly if the aquifer is shallow and not "protected" by an overlying layer of low permeability material, such as clay. Conversely, point source pollution refers to contamination originating from a single tank, disposal site, or facility. Industrial waste disposal sites, accidental spills, leaking gasoline storage tanks, and dumps or landfills are examples of point sources.

One of the best-known classes of groundwater contaminants includes chemicals known as chlorinated solvents; one example of this class of compounds is dry cleaning fluid, also known as perchloroethylene (PCE). As a general rule, the chlorine present in chlorinated solvents makes them more toxic than the respective non-chlorinated organic molecules. Unlike petroleum-based fuels, solvents are usually heavier than water and thus tend to sink to the bottoms of aquifers, forming the so-called dense non-aqueous phase liquid (DNAPL); DNAPLs usually represent a long-term secondary source of pollution. This makes solvent-contaminated aquifers much more difficult to clean up.

Typically, groundwater pollution takes place when rainfall soaks into the ground, comes in contact with buried waste or other sources of contamination, picks up chemicals, and carries them into the groundwater. Sometimes, the volume of a spill or leak is large enough that the chemical itself can reach the groundwater without the help of infiltrating water.

In consideration of the presence of a soil matrix, groundwater tends to move very slowly and with little turbulence, dilution, or mixing. Therefore, once contaminants reach groundwater, they tend to form a concentrated plume that flows along with the groundwater. Despite the slow movement of contamination through an aquifer, groundwater pollution often goes undetected for years and, as a result, can spread over a large area.

Different approaches can be followed when dealing with a contaminated site:

- To contain the contaminants, preventing them from migrating from their source
- To remove the contaminants from the aquifer

- To remediate the aquifer by either immobilizing or detoxifying the contaminants while they are still in the aquifer
- To treat the groundwater at its point of use
- To abandon the use of the aquifer and find an alternative source of water.

In order to allow for removal, contaminants must be mobile. When dispersed or dissolved in the groundwater, the pollutants can be extracted by means of extraction wells and then treated above ground (*pump-and-treat* technology). *Air sparging* can be applied when dealing with volatile chemicals: air is pumped into the aquifer, allowing for the evaporation of the volatile chemicals. Then, the contaminated air rising to the top of the aquifer is collected using vapor extraction wells.

To deal with organic, nonvolatile contaminants, a stimulation of natural attenuation (*bioremediation*) can be advisable, as it is comparatively inexpensive and involves minimal construction or disturbance. By adding nutrients or oxygen, the ability of naturally occurring or properly selected microorganisms to break down some forms of contamination into less toxic or nontoxic substances can be exploited.

Depending on the complexity of the aquifer and the types of contamination, some groundwater cannot be restored to a safe drinking quality. Under these circumstances, the only way to regain use of the aquifer is to treat the water at its point of use by means of quite costly treatment units consisting of special filters or based on reverse osmosis devices [1].

Electrokinetic Barriers

To prevent the migration of contaminants from their source, different approaches can be followed: the physical one makes use of underground barriers (made of clay, concrete, or steel); the hydraulic solution implies the utilization of pumping wells (to prevent contaminants from moving beyond the wells); while the chemical approach exploits the use of reactive chemicals for destroying or immobilizing the contaminants.

Another possibility to take into consideration is the recourse to electrokinetics. Electrokinetic phenomena are a family of several different effects requiring a porous solid in contact with a quite dilute liquid solution, in such a way that an interfacial "double layer" of charges is formed. At the surface of a solid particle (e.g., a soil particle), negative charges are commonly present because of the abrupt interruption of the ordered distribution of atoms (the crystalline lattice). These surface charges are balanced at the solution side by an accumulation of mainly positive ions, which extend from the solid surface towards the bulk of the solution, forming the so-called diffuse layer. Electrokinetic phenomena originate when an external force acting on the diffuse layer generates tangential motion of the fluid with respect to the adjacent charged surface. The force may be due to pressure or concentration gradients or may result from the application of an electric field.

In the latter case, the effect has basically two main components: one is electrophoresis and the other is the component due to the electroosmotic drag of the spatial ionic charge facing the charged surface of soil particles (the so-called electroosmotic flow, EOF). When the soil solution has a high ionic strength, electroosmosis becomes negligible, and electrophoresis may be somewhat hindered as a result of the presence of different ions. In that case, electromigration turns out to be the main mechanism of transport, and the electric field has essentially the role of "driving force" for the movement of ionic species through the water impregnating the dispersed solids, the soil in particular. The effect remains even if the solid matrix is finely divided. Both the induced movement of species (ions or charged particles in the case of electromigration and electrophoresis, respectively) and the electroosmotic flow can be used in opposition to the spread of pollutants moving in groundwater under a natural hydraulic gradient and represent the theoretical background in the application of electrokinetic barriers to prevent groundwater pollution.

An electrokinetic barrier can be created by the continuous or periodic application of an electrical gradient. In soils with low permeability, the

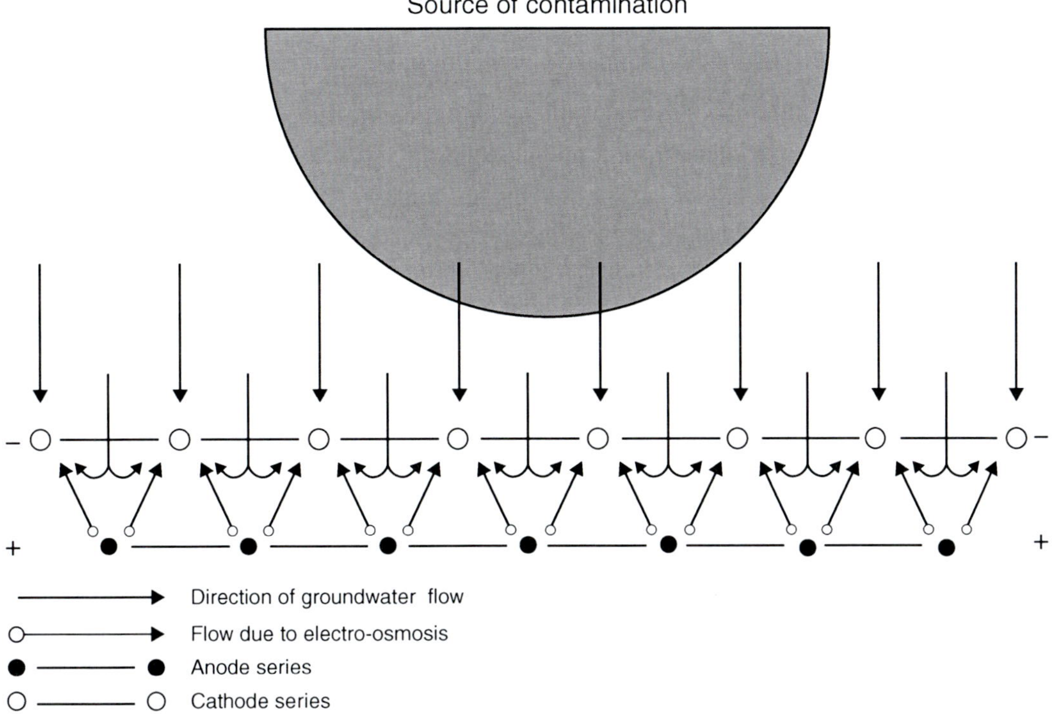

Electrokinetic Barriers for Preventing Groundwater Pollution, Fig. 1 Schematic field arrangement of an electrokinetic barrier (Reproduced from Ref. [2], with permission from Elsevier)

electroosmotic flow consequent to the electrical gradient may be able to move inorganic species much more effectively than a hydraulic gradient, thus allowing for an effective counterflow that prevents the migration of polluting species beyond the barrier. Since the barrier is used to keep the pollutants from entering an unpolluted region, its installation is particularly suitable when dealing with waste disposal facilities or contaminated sites where no adequate control measures have been implemented (for newly designed waste disposal sites, the prevention of contaminant migration is typically accomplished by using engineered containment systems based on compacted clay liners and/or geomembrane liners, sometimes integrated with leachate collection systems).

A schematized representation of an electrokinetic barrier installation is shown in Fig. 1 [2]. A DC electric generator is used to establish a low-intensity electrical potential gradient in the soil (continuously or periodically) by placing two parallel rows of electrodes on the interior and exterior surfaces of the barrier. Electrodes are arranged as shown in Fig. 1, with cathodes (negative electrodes) on the barrier side exposed to the source of contamination (internal side of the barrier) and the anodes (positive electrodes) on the external side of the barrier. The electric potential gradient induces a net water flow towards the cathode due to electroosmosis. On the other hand, electromigration can take place, which counteracts against the advection of cations (positive ions) while accelerating the advection of anions (negative ions) across the barrier. As a result, the application of an electric field may effectively retard the motion of the former but accelerate the motion of the latter when the effect of ionic migration prevails on electroosmotic advection.

In order to be effective, the activity of an electrokinetic barrier must comply with the simple concept schematically expressed in Fig. 2, i.e., the contaminant ion movement by

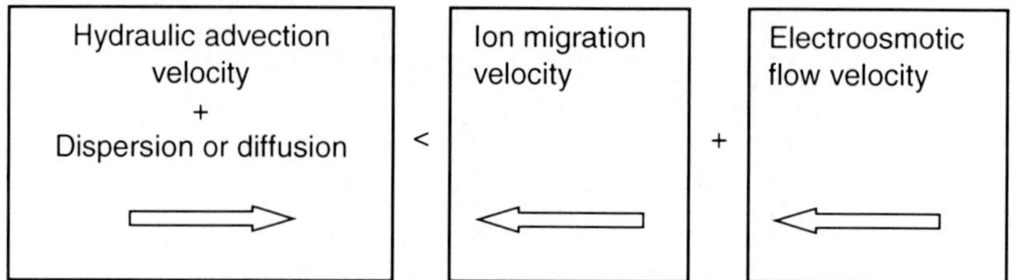

Electrokinetic Barriers for Preventing Groundwater Pollution, Fig. 2 Schematic representation of a solid (metal)-liquid interface (*left image*) and pertaining potential profile across the double-layer region (*right image*) (Reproduced from http://web.nmsu.edu/~snsm/classes/chem435/Lab14/double_layer.html with permission from the author, Sergei Smirnov, Dept. Chemistry and Biochemistry, New Mexico State University)

conventional hydraulic processes should not be greater than the sum of the electromigration velocity plus the EOF velocity [3].

From an analytical point of view, the above relationship may be expressed as follows [4]:

$$v_{hyd} + v_{dif} < v_{ion} + v_{EOF}$$

where

$$v_{dif} = -D_j \cdot A \cdot \Delta c_j / \Delta \mathrm{L}$$

$$v_{hyd} = -k_h \cdot i$$

$$v_{ion} = u^* \cdot (\Delta V / \Delta \mathrm{L}) = u \cdot (\Delta V / \Delta \mathrm{L}) \cdot n \cdot \tau$$

$$u = D_j \cdot z \cdot F / RT$$

$$v_{EOF} = -k_e \cdot (\Delta V / \Delta \mathrm{L})$$

v_{hyd} is the advection velocity due to hydraulic processes, v_{dif} is the diffusion velocity (expressed in terms of the first Fick's law; in the above relation, A represents the aquifer section), v_{ion} is the electromigration velocity, and v_{EOF} is the electroosmotic flow velocity. The last two terms depend on the strength of the applied electric field, $\Delta V/\Delta \mathrm{L}$.

k_h is the hydraulic permeability; k_e is the electroosmotic permeability; i is the hydraulic gradient, which is the hydraulic load difference (Δh) divided by the distance between two points of interest (aquifer length, L); u^* is the effective ion mobility, defined as the ion mobility (u) corrected for the soil porosity (n) and the soil tortuosity (τ); D_j is the diffusion coefficient of species j at infinite dilution; z is the ionic charge of the species j; T is the temperature; and F and R are the Faraday and the gas constants, respectively.

While the hydraulic permeability may vary between about six orders of magnitude (from 10^{-8} to 10^{-2} m/s), the electroosmotic permeability is relatively constant and usually comprised in the range of 10^{-5}–10^{-4} cm^2/Vs.

As previously anticipated, electroosmosis turns out to be negligible as the ionic strength of the soil solution increases, being the result of a minimization of the thickness of the interfacial double layer (see Fig. 3: the concentration of positive ions at the so-called outer Helmholtz plane is sufficient to balance the negative charge present at the soil particle surface).

Such a situation may arise as a result of a continuous application of the electric current over a long period of time. As a matter of fact, electrochemical reactions take place at the electrodes, and in the absence of depolarizing species, water molecules are oxidized at the anodes and reduced at the cathodes (electrolytic reactions), with formation of protons and hydroxyl anions, respectively. Once formed, these species tend to migrate under the effect of both potential and concentration gradients, allowing the development of an acidic front from the anode towards the cathode and of an alkaline front in the opposite direction (since the ionic mobility of H^+ ions is 1.75 times that of OH^- ions, the movement of protons will dominate the system chemistry).

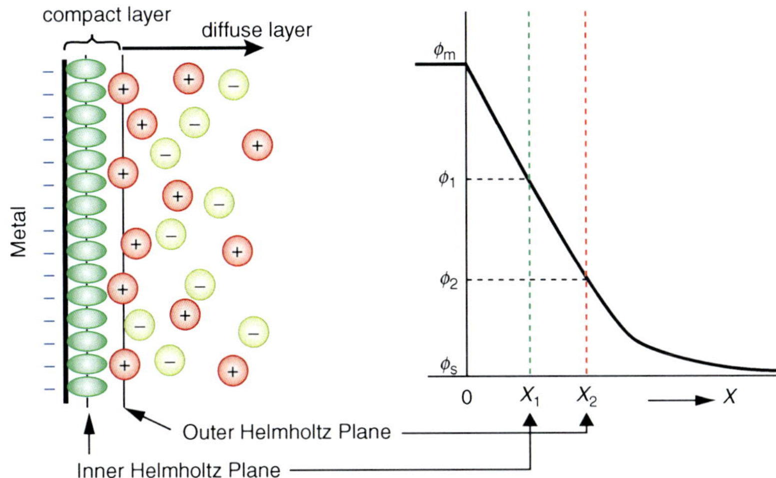

Electrokinetic Barriers for Preventing Groundwater Pollution, Fig. 3 Conceptual conditions for the beneficial exploitation of an electrokinetic barrier (Reproduced from ref. [3], with permission from Wiley)

Advantages and Disadvantages

Containment technologies are relatively expensive to construct and may require the use of specialized equipment, suitably exploitable when new waste disposal facilities are projected. Conversely, the installation of electrodes surrounding existing contaminated zones can be easily accomplished with portable equipment and may represent a quite promising and effective approach. However, the quantitative prediction of the different phenomena occurring in the soil upon the activation of an electric field and, in particular, of the evolution in space and time of the concentrations of contaminant species across an electrokinetic barrier is of paramount importance for the evaluation of effectiveness and proper design from both a technical and an economical point of view. Being a containment approach, it may be useful where natural attenuation is particularly effective in reducing the risk of long-term contamination. Wishing to solve the contamination problems, other practical solutions should be considered (e.g., recourse to permeable reactive barriers).

Cross-References

- ▶ Electrochemical Treatment of Landfill Leachates

References

1. Adapted from http://www.waterencyclopedia.com/Oc-Po/Pollution-of-Groundwater.html
2. Narasimhan B, Sri Ranjan R (2000) Electrokinetic barrier to prevent subsurface contaminant migration: theoretical model development and validation. J Contam Hydrol 42:1–17
3. Lynch R (2009) Electrokinetic barriers for preventing groundwater pollution. In: Reddy KR, Cameselle C (eds) Electrochemical remediation technologies for polluted soils, sediments and groundwater. Wiley, Hoboken, pp 335–356
4. Acar YB, Alshawabkeh AN (1993) Principles of electrokinetic remediation. Environ Sci Technol 27:2638–2647

Electrokinetic Remediation, Cost Estimation

Claudio Cameselle
Department of Chemical Engineering, University of Vigo, Vigo, Spain

Electrokinetics has been developed as an environmental technique since the late 1980s. From the very beginning, this technology attracted the interest of scientists, technicians, and governmental agencies. Their interest resulted in a fast development of the technology, and multiple applications have been reported from numerous research groups in the USA, Europe, and Korea–Japan. Most of the studies focus on lab or pilot tests, and only a few of the published works include information about the application of electrokinetic remediation at field scale. Thus, cost information for electrokinetic remediation is rather limited [1].

Cost information from lab or pilot tests are not appropriate for the estimation of the possible cost at field scale, since there is close supervision in small systems that results in large costs that can be clearly reduced as the size of the system increases. On the other hand, demonstration projects at field scale also report higher costs than a standardized commercial technology for similar reasons. The problem to define clearly the cost associated with electrokinetic remediation relies on the limited number of companies that applied the technology and also the limited number of studies with information available in the literature about large-scale tests using electrokinetics. In the USA, field projects were carried out or funded by the USEPA, DOE, ITRC, and US Army Environmental Center [2–4], as well as companies like Electropetroleum Inc. [5], Terran Corporation, Monsanto, DuPont, and General Electric, which developed the Lasagna™ technology [6–8]. In Europe, more field projects with electrokinetic remediation have been carried out, specially associated with the commercial activity of the Hak Milieutechniek Company [9–10]. Recently, some field experiences were reported in Japan and Korea [11]. Considering the information available in literature, the cost of field application of electrokinetic remediation is about 200 $\$/m^3$ for both organic and inorganic contaminants. However, it must be kept in mind that electrokinetic remediation, like any other remediation technology, is site specific, and the costs can vary from less than 100 to more than 400 $\$/m^3$ (Table 1) [12].

The application of electrokinetics at field scale implies several activities, such as preparation of the site, equipment installation, permits, and bureaucracy. During the operation time, the costs are associated with the electric energy expenditure, other utilities, control of operation and supervision, and waste management. These factors, along with their contribution to the final cost for electrokinetic remediation, are in Table 2.

Electrokinetic Remediation, Cost Estimation, Table 1 Cost estimates for electrokinetic remediation

Contaminant class	Range $\$/m^3$ ($\$/yd^3$)	Average $\$/m^3$ ($\$/yd^3$)
Inorganic	115–400 (90–300)	200 (150)
Organic	90–275 (75–200)	200 (150)

Electrokinetic Remediation, Cost Estimation, Table 2 Cost breakdown for typical electrokinetic remediation systems

Item	Typical range	Average
Electricity	7–25 %	15 %
Site preparations	5–25 %	10 %
Installation (labor, equipment, materials)	10–60 %	40 %
Operation, less electricity (labor, expendables)	15–50 %	25 %
Waste management, permits, oversight	5–20 %	10 %

References

1. Oonnittan A, Sillanpaa M, Cameselle C, Reddy KR (2009) Field applications of electrokinetic remediation of soils contaminated with heavy metals. In: Reddy KR, Cameselle C (eds) Electrochemical remediation technologies for polluted soils, sediments and groundwater. Wiley, Hoboken, pp 609–624
2. USEPA (1998) Guide to documenting and managing cost and performance information for remediation projects. EPA-542-B-98-007
3. USEPA (2000) Cost and performance report: electrokinetic extraction at the unlined chromic acid pit. Sandia National Laboratories, New Mexico
4. USEPA (2000) Innovative remediation technologies: field-scale demonstration projects in North America. EPA-542-B-00-004. Office of Solid Waste and Emergency Response
5. Wittle JK, Pamukcu S, Bowman D, Zanko LM, Doering F (2009) Field studies on sediment remediation. In: Reddy KR, Cameselle C (eds) Electrochemical remediation technologies for polluted soils, sediments and groundwater. Wiley, Hoboken, pp 661–696

6. Athmer CJ, Ho SV (2009) Field studies: organic-contaminated soil remediation with Lasagna technology. In: Reddy KR, Cameselle C (eds) Electrochemical remediation technologies for polluted soils, sediments and groundwater. Wiley, Hoboken, pp 625–646
7. Ho SV, Athmer CJ, Sheridan PW, Shapiro AP (1997) Scale-up aspects of the lasagna® process for in situ soil decontamination. J Hazard Mater 55(1–3):39–60
8. Ho SV, Athmer C, Sheridan PW, Hughes BM, Orth R, Mckenzie D, Shoemaker S (1999) The lasagna technology for in situ soil remediation. 2. Large field test. Environ Sci Technol 33(7):1092–1099
9. Lageman R, Clarke RL, Pool W (2005) Electro-reclamation, a versatile soil remediation solution. Eng Geol 77(3–4 Spec Iss):191–201
10. Lageman R, Pool W (2009) Electrokinetic biofences. In: Reddy KR, Cameselle C (eds) Electrochemical remediation technologies for polluted soils, sediments and groundwater. Wiley, Hoboken, pp 357–366
11. Chung HI, Lee MH (2009) Coupled electrokinetic PRB for remediation of metals in groundwater. In: Reddy KR, Cameselle C (eds) Electrochemical remediation technologies for polluted soils, sediments and groundwater. Wiley, Hoboken, pp 647–659
12. Athmer CJ (2009) Cost estimates for electrokinetic remediation. In: Reddy KR, Cameselle C (eds) Electrochemical remediation technologies for polluted soils, sediments and groundwater. Wiley, Hoboken, pp 583–587

Electrokinetic Transport in Soil Remediation

Claudio Cameselle
Department of Chemical Engineering, University of Vigo, Vigo, Spain

Transport Mechanisms in Electrokinetic Remediation

Electrokinetic remediation is an environmental technique especially developed for the removal of contaminants in soil, sediments, and sludge, although it can be applied to any solid porous material. Electrokinetic remediation is based on the application of a direct electric current of low intensity to the porous matrix to be decontaminated [1]. The effect of the electric field induces the mobilization and transportation of contaminants through the porous matrix towards the electrodes, where they are collected, pumped out, and treated. Main electrodes, anode and cathode, are inserted into the soil matrix, normally inside a chamber which is filled with water or the appropriate solution to enhance the removal of contaminants (Fig. 1). Typically, a voltage drop of 1 VDC/cm is applied.

The application of an electric field to a porous matrix induces the following transport mechanisms [2, 3]:

- Electromigration is defined as the transportation of ions in solution in the interstitial fluid towards the electrode of the opposite charge. Cations move towards the cathode (negative electrode), and anions move towards the anode (positive electrode). The ionic migration or electromigration depends on the size and charge of the ion and the strength of the electric field.
- Electroosmosis is the net flux of water or interstitial fluid induced by the electric field. Electroosmosis is a complex transport mechanism that depends on the electric characteristics of the solid surface and the properties of the interstitial fluid. The electroosmotic flow transports out of the porous matrix any chemical species in solution.
- Electrophoresis is the transport of charged particles of colloidal size and bound contaminants due to the application of a low direct current or voltage gradient relative to the stationary pore fluid. Compared to ionic migration and electroosmosis, mass transport by electrophoresis is negligible in low-permeability soil systems. However, mass transport by electrophoresis may become significant in soil suspension systems, and it is the mechanism for the transportation of biocolloids (i.e., bacteria) and micelles.
- Diffusion refers to the ionic and molecular constituent forms of the contaminants moving from areas of higher to areas of lower concentration because of the concentration gradient or chemical kinetic activity. Estimates of the ionic mobilities from the diffusion coefficients using the Nernst–Einstein–Townsend relation indicate that ionic mobility of a charged species is much higher than the diffusion coefficient (about 40 times the product of its charge and the electrical potential gradient). Therefore, diffusive transport is often neglected.

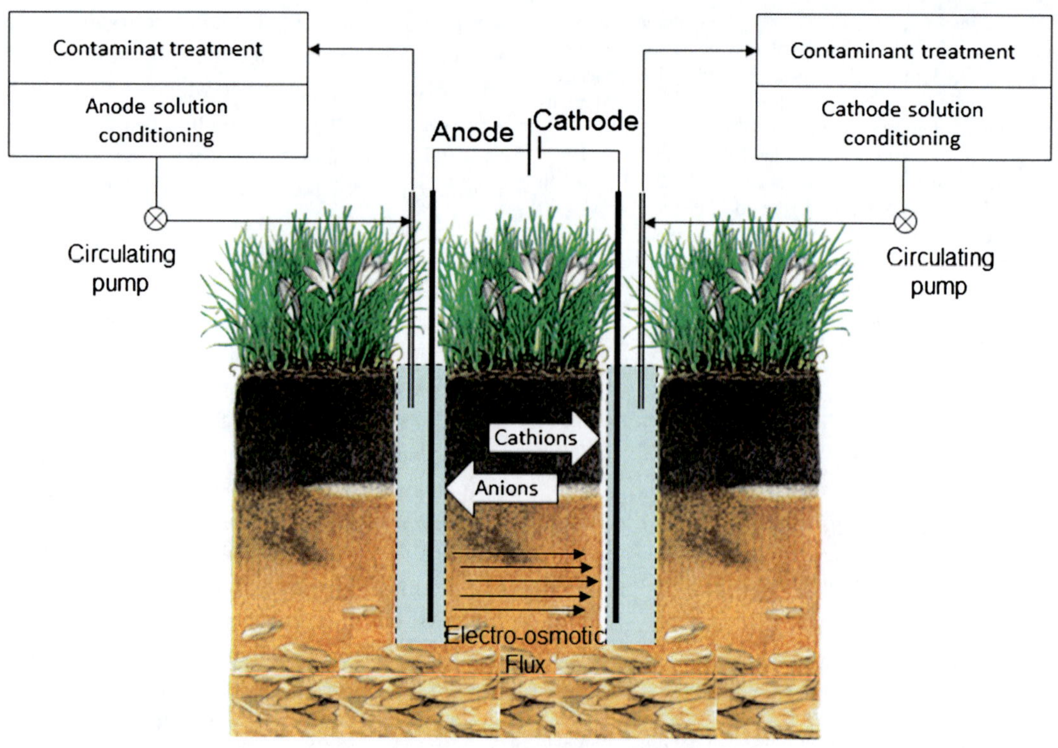

Electrokinetic Transport in Soil Remediation, Fig. 1 Application of the electrokinetic remediation in a contaminated site

The application of an electric flied to moisten a porous matrix also induces chemical reactions in the soil and upon the electrodes. Chemical reactions include acid–alkaline reactions, redox reactions, adsorption–desorption reactions, and dissolution–precipitation reactions. Such reactions dramatically affect the speciation of the contaminants and therefore affect the transportation and contaminant removal efficiency [4].

Electromigration

Electromigration is the movement of the dissolved ionic species which are present in the pore fluid towards the opposite electrode (Fig. 2). Anions move towards the anode, and cations move towards the cathode. Electromigration depends on the mobility of ionic species and the strength of the electric field [1, 5]. The electromigration flow of an ionic species can be expressed with Eqs. 1 and 2.

$$J_j^m = u_j^* c_j \nabla(-E) \tag{1}$$

$$u_j^* = u_j \tau n = \frac{D_j^* Z_j F}{RT} \tag{2}$$

Where:

Electromigration flux: J_j^m
Effective ionic mobility: u_j^*
Concentration: c_j
Electric potential: E
Ionic mobility: u_j
Porosity: n
Tortuosity: τ
Effective diffusivity: D_j^*
Ionic charge: Z_j
Faraday constant: F
Ideal gas constant: R
Temperature: T

The extent of electromigration of a given ion depends on the conductivity of the soil, soil porosity, pH gradient, applied electric potential,

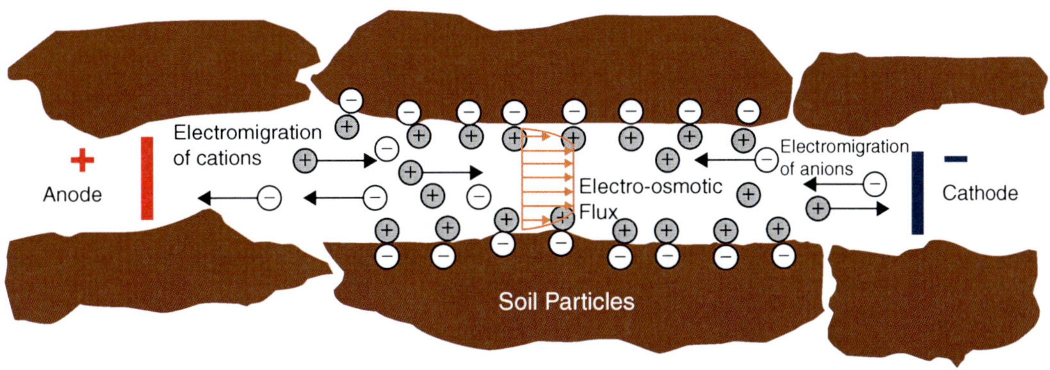

Electrokinetic Transport in Soil Remediation, Fig. 2 Transport mechanisms in electrokinetic remediation

Electrokinetic Transport in Soil Remediation, Table 1 Diffusion, ionic mobility, and effective ionic mobility for selected cationic and anionic species. Effective mobility was calculated for porosity n = 0.6 and tortuosity t = 0.35 [1]

Species	$D_j \times 10^6$ cm^2/s	$u_j \times 10^6$ cm^2/Vs	$u_j^* \times 10^6$ cm^2/Vs
Cations			
H^+	93	3,625	760
Na^+	13	519	109
Ca^{2+}	8	617	130
Cd^{2+}	9	736	155
Pb^{2+}	7	560	118
Cr^{3+}	6	694	146
Anions			
OH^-	53	2,058	432
NO_3^-	19	740	155
CO_3^{2-}	10	746	156
SO_4^{2-}	11	413	87
PO_4^{3-}	6	715	150

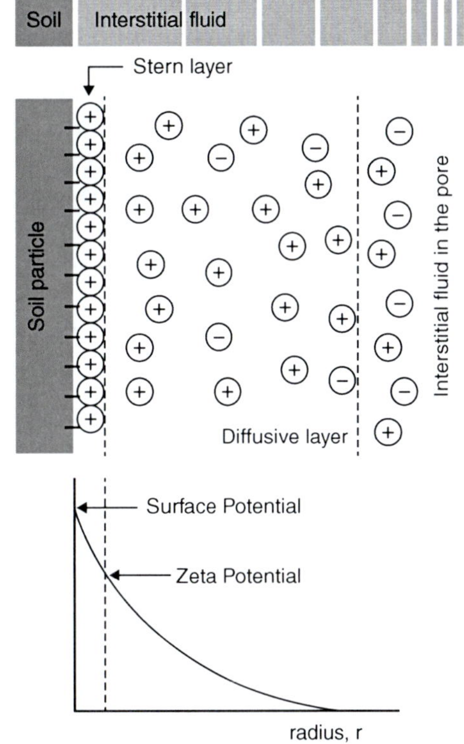

Electrokinetic Transport in Soil Remediation, Fig. 3 Electric double layer in the interphase soil-interstitial fluid

initial concentration of the specific ion, and the presence of competitive ions. Electromigration is the major transport process for ionic metals, polar organic molecules, ionic micelles, and colloidal electrolytes (Table 1).

Electroosmosis

Electroosmosis is the net flux of pore fluid in the soil induced by the electric field [1, 6]. Electroosmosis is able to remove all dissolved ionic and nonionic species in the pore fluid. Generally, soil particle surfaces are charged, and counterions (positive ions or cations) concentrate within a diffuse double-layer region adjacent to the particle surface (Fig. 3). Under an electric potential, the locally excess ions migrate in a plane parallel to the particle surface towards the

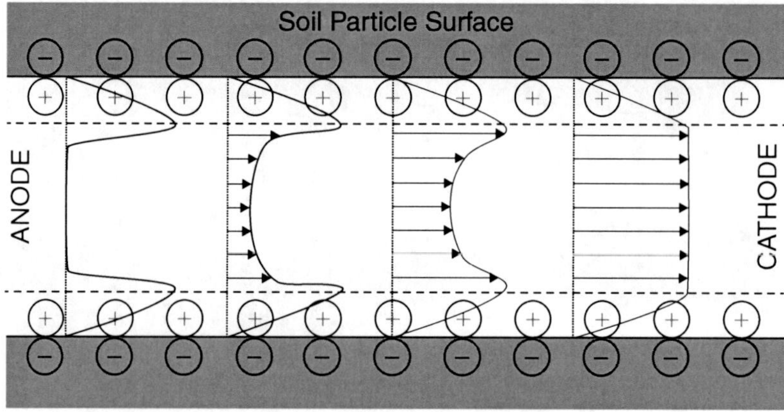

Electrokinetic Transport in Soil Remediation, Fig. 4 Development of the electroosmotic flow

oppositely charged electrode. As they migrate, they transfer momentum to the surrounding fluid molecules via viscous forces producing electroosmotic flow (Fig. 4). Since soil particle surfaces are generally negatively charged, the counterions that neutralize the surface charge are cations. Under the effect of the electric field, those cations move towards the cathode, inducing an electroosmotic flow towards the cathode [7].

The Helmholtz–Smoluchowski theory equation describes the electroosmotic flow (q_e) as it is shown in Eqs. 3 and 4.

$$q_e = k_e \nabla(-E) \tag{3}$$

$$k_e = \frac{\varepsilon\zeta}{\eta} n \tag{4}$$

Where:
Electro-osmotic permeability: k_e
Permitivitiy: ε
Zeta potential: ζ
Viscosity: η

The electroosmotic flow depends on the dielectric constant and viscosity of pore fluid, as well as the surface charge of the solid matrix represented by zeta potential (the electric potential at the junction between the fixed and mobile parts in the double layer). The zeta potential is a function of many parameters, including the types of clay minerals and ionic species that are present, as well as the pH, ionic strength, and temperature. If the cations and anions are evenly distributed, equal and opposite flow occurs, causing the net flow of zero. However, when the momentum transferred to the fluid in one direction exceeds the momentum of the fluid traveling in the other direction, electroosmotic flow is produced.

The pH changes induced in the soil by the electrolysis reactions affect the zeta potential of the soil particles, thereby affecting the electroosmotic flow. The low pH near the anode may be less than the PZC (point of zero charge) of the soil, and the soil surfaces are positively charged, while high pH near the cathode may be higher than PZC of the soil, making the soil more negative. Electroosmotic flow may be reduced and even ceased as the soil is acidified near the anode. If the majority of the soil is acidified, the electroosmotic flow direction may even be reversed, from typical anode to cathode to cathode to anode. This phenomenon is known as electroendosmosis. Understanding of such electroosmotic flow variations is critical when remediating organic pollutants [2, 6].

Electroosmosis is considered the dominant transport process for both organic and inorganic contaminants that are in dissolved, suspended, emulsified, or such similar forms. Besides, electroosmotic flow through low-permeability regions is significantly greater than the flow achieved by an ordinary hydraulic gradient, so the electroosmotic flow is much more efficient in low-permeability soils [8].

Electrochemical Reactions

The application of an electric field to a contaminated soil induces the movement of ions and water (the pore fluid) towards the electrodes. At the same time, the electric field also induces electrochemical reactions upon the electrodes and into the soil [2, 4].

The main reaction in the electrochemical/electrokinetic systems is the decomposition of water that occurs at the electrodes. The electrolytic decomposition of water reactions generates oxygen gas and hydrogen ions (H^+) due to oxidation at the anode and hydrogen gas and hydroxyl (OH^-) ions due to reduction at the cathode, as shown in Eqs. 5 and 6.

At anode (oxidation):

$$2H_2O \longrightarrow O_{2(gas)} + 4H+_{(aq)} + 4e^- E^0 = -1.229V \quad (5)$$

At cathode (reduction):

$$4H_2O + 4e^- \longrightarrow 2H_{2(gas)} + 4OH^-_{(aq)} \quad E^0 = -0.828V \quad (6)$$

Essentially, acid is produced at the anode, and alkaline solution is produced at the cathode; therefore, pH in the cathode is increased, while pH at the anode is decreased. The migration of H^+ from the anode and OH^- from the cathode into the soil leads to dynamic changes in soil pH. H^+ is about twice as mobile as OH^-, so the protons dominate the system, and an acid front moves across the soil until it meets the hydroxyl front in a zone near the cathode where the ions may recombine to generate water. Thus, the soil is divided in two zones with a sharp pH jump in between: a high-pH zone close to the cathode and a low-pH zone on the anode side. The actual soil pH values will depend on the extent of transport of H^+ and OH^- ions and the geochemical characteristics of the soil.

The implications of these electrolysis reactions are enormous in the electrokinetic treatment since they impact the contaminant migration, the evolution of the electroosmotic flow, and the transformation and degradation of contaminants into the soil. The most important geochemical reactions that must be considered include:

- Sorption–desorption reactions
- Precipitation–dissolution reactions
- Oxidation–reduction reactions

Contaminants are often present in the soil adsorbed to solid particle surfaces and/or attached to the organic matter in the soil. The transportation and removal of contaminants out of soil by electroosmosis and electromigration implies the desorption and solubilization of the contaminant in the pore fluid. The equilibrium between the solid phase and the liquid phase is dependent on the chemical nature of the contaminant, soil, and pore fluid composition. pH largely affects the sorption equilibrium. For instance, heavy metals can be removed at low pH because metal ions can be exchanged from the solid surface when the H^+ concentration is increased. The pH-dependent sorption–desorption behavior is generally determined by performing batch experiments using the soil and contaminant of particular interest [9].

The precipitation and dissolution of the contaminant species during the electrokinetic process can significantly influence the removal efficiency of the process. The solubilization of precipitates is affected by the hydrogen ions generated at the anode migrating across the contaminated soil, favoring the acidification of soil, and the dissolution of metal hydroxides and carbonates, among others. However, in some types of soils, the migration of the hydrogen ions will be hindered due to the relatively high buffering capacity of the soil. In a high-pH environment, heavy metals will precipitate, and the movement of the contaminants will be impeded. During the electrokinetic treatment, heavy metals migrate towards the cathode until reaching the high-pH zone where heavy metals accumulate and eventually precipitate, clogging soil pores and hindering the remediation process. For efficient contaminant removal, it is essential to prevent precipitation and to have the contaminants in dissolved form during the electrokinetic process.

Oxidation and reduction reactions are important when dealing with some contaminants that can be transformed in the conditions developed upon the electrodes or in the soil during the electrokinetic treatment. Thus, organic contaminants can be degraded in the soil or in the electrodes by oxidation or reductive degradation. The degradation of organic contaminants results in simpler organic compounds, usually less toxic that the former contaminants, improving the effectiveness of the electrokinetic treatment [10].

Metallic contaminants such as chromium are also affected by redox reactions. Chromium exists most commonly in two valence states: trivalent chromium (Cr(III)) and hexavalent chromium (Cr(VI)). Cr(III) exists in the form of cationic hydroxides such as $Cr(OH)_2^{-1}$ and will migrate towards the cathode during electrokinetic remediation. However, Cr(VI) exists in the form of oxyanions such as CrO_4^{2-}, which migrate towards the anode. The valence state depends on the soil composition, especially the presence of reducing agents such as organic matter and Fe(II) and/or oxidizing agents such as Mn(IV), so it is important to know the valence state of metals and their possible redox chemistry to know the chemical speciation of the contaminants and their movement across the soil. Finally, arsenic is a very toxic element difficult to remove due to its complex chemistry. Arsenic forms a variety of anionic and cationic compounds that migrate towards the cathode and anode, respectively. The electric field and the pH value induce redox reactions on arsenic compounds, changing the direction of migration and making difficult their total removal from soil.

Future Direction

The effectiveness of the electrokinetic treatment for the remediation of contaminated soils relies in the effective transportation by electromigration or electroosmosis of the contaminants. Both transportation mechanisms require the contaminants to be in solution in the interstitial fluid. It is important to use enhancing chemicals that solubilize the contaminants and keep them in solution. For heavy metals and other inorganic contaminants, the pH in the soil will be a key parameter to achieve a complete contaminant solubilization, followed by the electromigration of the ionic species in solution. The use of chelating agents also plays an important role in heavy metal removal since the metal-complex ions are more stable in solution than the metal alone.

Most of organic contaminants are not soluble in water and their removal by electrokinetic remediation requires their solubilization in the interstitial fluid with the use of surfactants, co-solvents, or enhancing chemicals. Then, the removal of organic contaminants can be achieved by electroosmosis. In this case, it is critical to maintain a high electroosmotic flow. Therefore, pH, ionic strength, and the nature of electrolytes in the interstitial fluid will be critical to maintain a high zeta potential (z) and as a result, a high electroosmotic flow. Alternatively, the organic contaminants can be degraded in situ by transporting active oxidants into the soil. In this case, the operating conditions of the electrokinetic treatment have to be adjusted to improve the transportation of the oxidants into the soil from the electrode chambers.

Thus, the operating conditions in electrokinetic remediation have to be selected to enhance the solubilization of the contaminants and/or maintain a high electroosmotic flow. Research is necessary to determine the optimum operating conditions in each contaminated soil, since the behavior largely depends on the chemical nature of the contaminants and the geochemical interactions of soil and interstitial fluid.

Cross-References

- ▸ Electrokinetic Barriers for Preventing Groundwater Pollution
- ▸ Electrokinetic Remediation, Cost Estimation
- ▸ Electrokinetics in the Removal of Chlorinated Organics from Soils
- ▸ Electrokinetics in the Removal of Hydrocarbons from Soils
- ▸ Electrokinetics in the Removal of Metal Ions from Soils

References

1. Acar YB, Alshawabkeh AN (1993) Principles of electrokinetic remediation. Environ Sci Technol 27(13): 2638–2647
2. Acar YB, Gale RJ, Putnam GA, Hamed J, Wong RL (1990) Electrochemical processing of soils: theory of pH gradient development by diffusion, migration, and linear convection. J Environ Sci Health Part A Environ Sci Eng 25(6):687–714
3. Acar YB, Gale RJ, Alshawabkeh AN, Marks RE, Puppala S, Bricka M, Parker R (1995) Electrokinetic remediation: basics and technology status. J Hazard Mater 40(2):117–137
4. Jacobs RA, Sengun MZ, Hicks RE, Probstein RF (1994) Model and experiments on soil remediation by electric fields. J Environ Sci Health Part A Environ Sci Eng 29(9):1933–1955
5. Alshawabkeh AN, Acar YB (1992) Removal of contaminants from soils by electrokinetics: a theoretical treatise. J Environ Sci Health Part A Environ Sci Eng 27(7):1835–1861
6. Probstein RF, Hicks RE (1993) Removal of contaminants from soils by electric fields. Science 260(5107):498–503
7. Maturi K, Reddy KR (2008) Cosolvent-enhanced desorption and transport of heavy metals and organic contaminants in soils during electrokinetic remediation. Water Air Soil Pollut 189(1–4):199–211
8. Saichek RE, Reddy KR (2005) Electrokinetically enhanced remediation of hydrophobic organic compounds in soils: a review. Crit Rev Environ Sci Technol 35(2):115–192
9. Reddy KR, Chaparro C, Saichek RE (2003) Iodide-enhanced electrokinetic remediation of mercury-contaminated soils. J Environ Eng 129(12):1137–1148
10. Reddy KR, Darko-Kagya K, Cameselle C (2011) Electrokinetic-enhanced transport of lactate-modified nanoscale iron particles for degradation of dinitrotoluene in clayey soils. Sep Purif Technol 79(2):230–237

Electrokinetics in the Removal of Chlorinated Organics from Soils

Claudio Cameselle
Department of Chemical Engineering,
University of Vigo, Vigo, Spain

Chlorinated Organic Compounds

The term chlorinated organic compounds (COCs) refers to the substitution of one or more hydrogen atoms in an organic molecule by one or more chlorine atoms. Chlorinated organic compounds include aliphatic, cyclic, and aromatic structures. Typical chlorinated organics are the following: trichloroethylene (TCE), vinyl chloride (VC), and carbon tetrachloride (CT) in the aliphatic group; lindane (hexachlorocyclohexane) and DDT with a cyclic structure; and pentachlorophenol (PCP) and polychlorinated biphenyls (PCBs) in the aromatic group. Chlorinated organics have been used extensively in many industrial processes including the chemical industry, medicine, electronics, and pesticides. The main uses include solvents, pesticides, preservative agents, and intermediates in the synthesis of pharmaceuticals and dyes. This kind of compounds can be found in many industrial fields and in many products in the market. Such a wide family of chemical compounds shows different physical, chemical, and biological properties, but in general, it can be stated that chlorinated compounds are toxic or poisonous for living organisms. It is known that the substitution of hydrogen atoms by chlorine atoms in hydrocarbons clearly increases the toxicity of the compound. The solubility in water also decreases with the presence of chlorine atoms in the organic molecule. So, chlorinated organics are nonsoluble in water, but they are soluble in the lipid phase. The combination of toxicity and low solubility in water makes chlorinated organics difficult to degrade and persistent in the environment. They tend to accumulate in the lipid fraction of the living organisms, creating serious health problems once chlorinated organics enter the trophic chain and accumulate in living organisms. The main harmful effects for humans and other living organisms are endocrine disturbances, behavioral effects, immunity system alterations, and cancer.

Remediation of Contaminated Sites with COCs

Contamination by chlorinated compounds in the environment is associated with the use of pesticides and herbicides, but it is also associated with the lack of management of industrial wastes in past and accidental spills. Due to low solubility in water, polychlorinated organics can be found in soils attached to the soil particles but mainly to the organic matter in the soil. The removal of chlorinated organics is not easy considering their low solubility in water and high toxicity that prevent the use of remediation techniques such as soil washing, soil flushing, pump and treat, and bioremediation.

Electrokinetic remediation has been proposed and tested for the removal of organic compounds in soil, sediments, and groundwater. The electrokinetic treatment relies on the application of a low intensity electric field that mobilizes the contaminants and induces their transportation towards the electrodes (anode and cathode). The main transportation mechanisms in electrokinetic remediation are electromigration and electroosmosis. Electromigration is the movement of ions towards the electrode of opposite charge. In general, COCs are not soluble in water (which is the interstitial fluid in natural soils) and are neither ionic nor ionizable molecules. Therefore, electromigration cannot be considered as the transport mechanisms for COCs. Electroosmosis is the net flux of water in the soil matrix that flows through the soil from one electrode to the other due to the effect of the electric field. Electroosmotic flow moves towards the cathode in electronegatively charged soils, which is the most common case. Again, COCs are not soluble in water, and therefore their elimination from soils cannot be achieved in an unenhanced electrokinetic treatment. In order to achieve an effective removal or elimination of COCs from soils, their solubility has to be enhanced with the use of cosolvents, surfactants, or any other chemical agent. Alternatively, the removal or elimination of COCs can be achieved by the combination of electrokinetics and other remediation techniques such as chemical oxidation/reduction, permeable reactive barriers, electrolytic reactive barriers, or thermal treatment.

Electrokinetic Remediation of TCE

Some COCs commonly used in the industry can be found in many contaminated sites. Trichloroethylene (TCE) is an organic solvent commonly found as a contaminant in soils and groundwater in many industrial areas. Several studies have dealt with the removal of TCE using electrokinetics. TCE, like other organics with relatively high solubility in water (benzene, toluene, m-xylene), can be achieved electrokinetics if the electroosmotic flow is enhanced. Bruell et al. [5] demonstrated the removal of TCE in laboratory spiked kaolinite samples. Weng et al. [23] confirmed the removal of TCE from clay soil artificially contaminated using synthetic groundwater as the processing fluid. TCE removal increases with the voltage applied from 86 % at 1 DCV/cm to 91 % at 2 DCV/cm in 5 days of operation. Chang et al. [3] used acetate buffer to control the pH on the cathode. pH 6 on the cathode resulted in a stable electroosmotic flow and removal efficiencies of about 85–98 % for TCE and other organic contaminants (chloroform, carbon tetrachloride, and perchloroethylene) in 2 weeks of treatment. Removal efficiency directly depended on the solubility of the organic contaminant.

Enhancement of TCE removal and degradation in low permeability soils by electrical fields can be achieved by the combination of electrokinetics and chemical oxidation. Electrokinetic phenomena can be used to transport into the soil oxidants such as Fenton reagent. Yang and Liu [24] demonstrated that Fenton reagent can be introduced into the soil by electroosmosis, and the TCE degradation results were closely related to the electroosmotic permeability. Yang and Yeh [25] used sodium persulfate as an oxidant in the electrokinetic treatment of TCE contaminated soil, meeting the local regulatory standards. On the contrary, the electric field can be used not only for transportation but also to induce redox reactions that transform the contaminant in less harmful compounds. Thus, Chen et al. [4] demonstrated the reduction of TCE to chloromethane at a granular graphite cathode. More recently, nanoscale zero valent iron (NZVI) was used for the reductive dechlorination of TCE. NZVI was supplied to the soil as a reactive barrier close to the polluted area. A surfactant, Triton X, was used to improve the solubility and mobility of the TCE that was effectively transported towards the reactive barrier, where it was completely dechlorinated.

Field scale studies and projects were carried out by a consortium of industrial partners (Monsanto, DuPont, and General Electric), the USEPA and the DOE [1]. The electrokinetic process designed and operated for the removal of TCE at Paducah (KY, USA) was called the Lasagna™ process. It utilizes a DC electric field to move pore water and contaminants uniformly through the soil mass to treatment zones emplaced within the contaminated area. The treatment materials emplaced are typically iron, coke, and kaolin. The process results in little or no wastes. The system was operated over a 2-year period when the cleanup target concentration of TCE (5.6 mg/kg) was met. The final results show an average TCE concentration of 0.38 mg/kg.

Enhanced Electrokinetic Remediation of COCs

Despite the relatively good results found in literature for soil contaminated with TCE, the removal and elimination of COCs requires enhanced electrokinetic technologies. They comprise the use of solubilizing agents such as cosolvents, surfactants, or cyclodextrins. The other possible alternative for the removal of COCs from soil implies the combination of electrokinetic with other remediation techniques such as chemical and electrochemical oxidation/reduction, permeable reactive barriers, electrolytic barriers, and electric heating.

Cosolvents

Most of the COCs are practically insoluble in water, but they are highly soluble in organic solvents; therefore, the extraction and solubility of COCs from contaminated soils can be carried out with the combination of water and an organic solvent as the processing fluid. Of course, the cosolvent has to meet some important criteria: it has to be miscible with water, safe for the environment, and easy to recover after the treatment. These limitations narrow the possible cosolvents to be selected. The use of cosolvents also presents some other drawbacks. They affect the solubility of salts and decrease the conductivity of the interstitial fluid, therefore, decreasing the electric current intensity. The cosolvents also modify the viscosity of the processing and alter the interaction soil-interstitial fluid. All these aspects affect negatively the electroosmotic flow, which is the main transport mechanism for COC removal. Anyway, enhancement of COC solubility combined with the electroosmotic flow results in the effective removal of COCs from polluted soils. Thus, Wan et al. [21] assayed different conditions for hexachlorobenzene, and the best removal results were achieved with 50 % ethanol as a cosolvent, despite the important decrease in the electroosmotic flow compared to deionized water as a processing fluid. However, the electroosmotic flow can be improved by controlling the pH, the ionic strength, and the voltage gradient (Cameselle and Reddy 2012). Other cosolvents used for the removal of COCs are n-propanol, n-butylamine, heptane (1 %), and others.

Surfactants

Surfactants are wide group of substances that lower the surface and interfacial tension of water, improving the solubility of hydrophobic organics through a process called micellar solubilization. The name surfactant is the short version of "surface-active agent," which is the main property of this group of chemicals. Basically, a surfactant is a chemical compound whose molecule includes a hydrophilic group on one side and on the opposite side, a hydrophobic group or chain. The interaction of the hydrophilic group with water assures its solubility, whereas the interaction of the hydrophobic group with the organic compound permits the solubilization of hydrophobic organics. The hydrophobic group or chain in the surfactant molecule is repelled by water, so the surfactant molecules tend to form spherical structures with the hydrophilic group outside and the hydrophobic chains inside. These spherical structures are called micelles. Thus, the surfactant creates a hydrophobic environment very appropriate for the solubilization of organic compounds. The formation of micelles depends on the concentration of surfactant, and it is defined as a critical micelle concentration that corresponds with the maximum formation of micelles. Therefore, the solubilization of hydrophobic organics will depend on the surfactant dosage and micelle formation.

Surfactants can be classified by the charge of the hydrophilic group as cationic, anionic, neutral, and zwitterionic (includes positive and negative charges in the same molecule). The use of surfactants for the removal of COC by electrokinetics has been demonstrated in literature [17]. Several surfactants have been used such as Brij 35, Triton X-100, Sodium dodecyl sulfate, Tween 80, and Igepal CA-720, but COC removal results largely depend on surfactant nature and dosage, characteristics of the soil and contaminant, and operation conditions. Therefore, it is necessary to test the surfactant for each application [8].

The addition of surfactants to the processing fluid usually increases its viscosity and modifies the interaction with the soil particle surface. It results in a reduction of the electroosmotic flow, which is the main transportation mechanism. The use of neutral surfactants has been preferred for low toxicity, which is a very important property to consider in the selection of the surfactant. Anionic surfactants have a great solubilizing potential and do not interact with soil, so the retention of the surfactant in the soil is very low. However, anionic surfactants migrate in the opposite direction of electroosmotic flow. Besides, they are much more toxic, especially for aquatic organisms. Cationic surfactants have not been used in soil electrokinetics.

Cyclodextrins

Cyclodextrins are a group of oligosaccharides formed by glucose monomers with a cyclic structure of 6, 7, or 8 glucose molecules linked by α-1,4-glicosidic bonds. The structure of cyclodextrins is like a truncated cone (bottomless bowl), and the internal cavity has different diameters depending on the number of glucose units. The inner diameter of the molecule ranged from 0.45 to 0.53 nm for α-cyclodextrin (ring of 6 glucose molecules), 0.60–0.65 nm for β-cyclodextrin (ring of 7 glucose molecules), and 0.75–0.85 nm for γ-cyclodextrin (ring of 8 glucose molecules). Cyclodextrin shows an amphiphilic behavior due to the rings of –OH groups present at the both ends of the molecule. The hydroxyl groups are polar and confer to the cyclodextrin the solubility in water. However, the inner surface of the molecule is hydrophobic, and cyclodextrins can accommodate different nonpolar, hydrophobic molecules such as aliphatic, aromatic, or lipophilic compounds.

Cyclodextrins have been used for the removal of COC from soils by electrokinetics alone or in combination with other remediation techniques [12, 14, 15, 26]. In general, cyclodextrins are facilitating agents that improve the removal of COC from soil compared to other experiments with unenhanced electrokinetics, but results from cyclodextrin tests are usually less effective than tests with surfactants, iron nanoparticles, or with chemical oxidants.

Chemical Oxidation/Reduction

The elimination of COC from soil can be achieved by a totally different approach. Instead of removing the contaminants from the soil and accumulating them in the anode or cathode reservoirs, COC can be chemically oxidized or reduced in the soil, transforming the harmful COC into less toxic compounds. An additional advantage of this technology is that no wastes are generated.

Powerful oxidants such as ozone, hydrogen peroxide, or persulfate can be introduced into the soil by electromigration or electroosmosis. In this case, the electric field is not used for the transportation of contaminants but for the transportation of chemical reagents that eliminate the contaminants. Alternatively, chemical oxidation can be performed with the generation of oxidants upon the electrodes; the electric field is used simultaneously for oxidant generation and transportation. Chemical oxidation is the preferred option for most of the organic contaminants with the objective to finally transform them into CO_2 and water. However, chlorinated organics can be treated by reductive dechlorination. The selective removal of Cl atoms from the COC molecules largely reduces their toxicity, and the reactions products can be easily degraded by bioremediation or other methods. Chemical reduction of COC can be performed upon the cathode electrode or with the introduction into the soil the appropriate chemicals, such as Fe^0.

The combination of electrokinetics and chemical oxidation was tested in a contaminated soil with hexachlorobenzene [12, 13]. Hydrogen peroxide was supplied to the soil from the anode in a Fenton-like process where the iron content in the soil was sufficient to activate the decomposition of H_2O_2 for the generation of hydroxyl radicals (·OH). 60 % of HCB was eliminated from the soil in 10 days treatment avoiding the deactivation of the Fenton reagent at high pH values. Higher removal can be achieved at longer treatment time, controlling the pH in the optimum range for Fenton reagent which is slightly acid environments. At alkaline pH, H_2O_2 decomposes into water and oxygen and does not form · OH radicals.

Reductive dechlorination of HCB can be achieved with the combination of electrokinetics with the appropriate catalysts such as nanoscale zero valent iron (NZVI). The electric field can be used as a mechanism for the delivery of NZVI into the soil [10]. Reductive dechlorination can be achieved in the soil or even upon the cathode if the contaminant reaches the cathode compartment. Other metallic catalysts such as Cu/Fe or Pd/Fe bimetal microscale particles were satisfactorily used with the same purpose. Dechlorination of hexachlorobenzene up to 98 % was achieved with Cu/Fe [27] and only 60 % with Pd/Fe [20].

Permeable Reactive Barriers

PRBs consist of digging a trench in the path of flowing groundwater and then filling it with a selected permeable reactive material. As the contaminated groundwater passes through the PRB, contaminants may be degraded, and clean groundwater exits the PRB. For the remediation of COC, the most common material used for dechlorination is zero valent iron nanoparticles [22], but also Pd/Fe and Cu/Fe microscale particles were used and reported in literature.

As it was stated before, one of the main difficulties for the remediation of COC is its low solubility in water. In order to enhance the mobility and transportation of COC with the groundwater towards the PRB, some solubilizing agents can be used. Huang and Cheng [9] combined the surfactant Triton X-100 with a PRB filled with NZVI. Using microscale Cu/Fe particles, the removal of hexachlorobenzene achieved 98 % [27], whereas Pd/Fe microscale particles led to a 60 % HCB removal (all these tests combining PRB-EK and Triton X-100 as a facilitating agent).

Semkiw and Barcelona [19] demonstrated the feasibility of the PRB for the in situ biodegradation of chlorinated ethenes. In order to improve the microbiological activity, the PRB was supplied with diary whey. The efficiency of the treatment depends on the chlorinated compound, being 100 % removal for trichloroethene and 60 % for vinyl chloride.

Electrolytic Barriers

Electrolytic reactive barriers consist of permeable electrodes installed in a trench perpendicular to the direction of groundwater flow. The electrolytic barrier intercepts the groundwater contaminant plume, degrading the contaminants and inducing oxidation reactions at the anode and reduction reactions at the cathode. A wide range of redox-sensitive contaminants, such as chlorinated ethenes (TCE, TCA), other organics, and even inorganic contaminants, may be treated using the electrolytic barriers. The main advantages of this method are no need of chemicals, low operation and maintenance costs, and possibility of inversing the polarity to improve redox degradation or avoid precipitate accumulation [18].

Electrokinetic Biofence

Electrokinetic biofence (EBF) consists of a row of alternating cathodes and anodes with a mutual distance of 5 m. Upstream of the line of electrodes, a series of infiltration wells were installed, which have been periodically filled with nutrients. The aim of the EBF is to enhance biodegradation of the VOCs in the groundwater at the zone of the fence by electrokinetic dispersion of the dissolved nutrients in the groundwater. After running the EBF for nearly 2 years, clear results have been observed. The concentration of nutrients in the zone has increased, the chloride index is decreasing, and VOCs are being dechlorinated by bioactivity.

The electrical energy for the EBF is being supplied by solar panels [7].

Electrical Heating

Soil is not a good electrical conductor. Soil conductivity directly depends on moisture content and ionic concentration in the interstitial fluid. Considering the low conductivity of soils, the application of an electric field may cause electrical heating. In order to improve the electrical heating, alternate current is used instead of direct current to move the contaminants. The heating of soil favors desorption and vaporization of volatile and semi-volatile organic compounds. Those organic vapors can be recovered from the soil by suction, trapping the vapors in activated carbon, for example. Electrical heating was used in the remediation of a contaminated site in Zeist, the Netherlands [11]. The site was severely polluted with chlorinated solvents such as perchloroethylene (PCE) and trichloroethylene (TCE), and their degradation products are cis-1,2-dichloroethene (C-DCE) and vinyl chloride (VC). Satisfactory results were obtained in the application of electrical heating of soil and groundwater in the source areas, combined with soil vapor extraction and low-yield groundwater pumping and enhancing biodegradation in the groundwater plume area. Two years of heating and 2.5 years of biodegradation has resulted in near-complete removal of the contaminants.

A full-scale implementation of six-phase electrical heating technology was used in Sheffield (UK). Terra Vac Ltd. demonstrated how remediation timescales can be reduced from months/years to weeks, with an electrical heating capable of remediation of soil in difficult geological conditions and in densely populated urban areas. TCE and VC were remediated by electrical heating up to 99.99 % [6].

Enhancing the Electroosmotic Flow

The electroosmotic flow is the dominant transport mechanism for the removal of organic contaminants. For the removal of organic contaminants, both solubilization of the contaminants and adequate electroosmotic flow are required, which appear to be quite challenging to accomplish simultaneously. The electroosmotic flow is found to be dependent on the magnitude and mode of electric potential application. The electroosmotic flow is higher initially under higher electric potential, but it reduces rapidly in a short period of time. Interestingly, the use of an effective solubilizing agent (surfactant) and periodic voltage application was found to achieve the dual objectives of generating high and sustained electroosmotic flow and, at the same time, induce adequate mass transfer into the aqueous phase and subsequent removal. Periodic voltage application consists of a cycle of continuous voltage application followed by a period of "downtime" where the voltage is not applied. This was found to allow time for the mass transfer, or the diffusion of the contaminant from the soil matrix, to occur and also to polarize the soil particles. Several laboratory studies have demonstrated such desirable electroosmotic flow behavior in a consistent manner, but field demonstration projects are needed to validate these results under scale-up field conditions [2, 16].

Future Perspectives

The scientific knowledge accumulated in the last 20 years led to several lessons learned that must be kept in mind in the design of projects for the remediation of contaminated sites. Thus, the remediation of soils contaminated with chlorinated organics is site specific. The results obtained in the remediation of a site cannot be assumed for other contaminated sites. This is due to the large influence of the physicochemical properties of the soil and its possible interactions with the organic contaminants in the results of the electrokinetic remediation treatment. Besides, the chemicals used for enhancing the electrokinetic treatment may complicate the behavior of the system, and the removal results may largely vary from one site to another. Recently, it has been considered that the combination of several remediation techniques may improve the remediation results, especially in sites with complex contamination, including recalcitrant organic compounds and inorganic contaminants [28]. The combination of electrokinetics with bioremediation, phytoremediation, chemical oxidation, or electrical heating presents very interesting perspectives for the remediation of difficult sites. It is expected that the combination of remediation technologies will improve the remediation results, saving energy and time.

Cross-References

► Electrokinetic Barriers for Preventing Groundwater Pollution
► Electrokinetic Remediation, Cost Estimation
► Electrokinetic Transport in Soil Remediation
► Electrokinetics in the Removal of Hydrocarbons from Soils

References

1. Athmer C (2004) In-situ remediation of TCE in clayey soils. Soil and Sediment Contam 13(5):479–488
2. Cameselle C, Reddy KR (2012) Development and enhancement of electro-osmotic flow for the removal of contaminants from soils. Electrochimica Acta 86:10–22
3. Chang J-H, Qiang Z, Huang C-P (2006) Remediation and stimulation of selected chlorinated organic solvents in unsaturated soil by a specific enhanced electrokinetics. Colloids Surf A Physicochem Engin Aspects 287(1–3):86–93
4. Chen J-L, Al-Abed SR, Ryan JA, Li Z (2002) Groundwater and soil remediation using electrical fields. ACS Symp Ser 806:434–448
5. Bruell CJ, Segall BA, Walsh MT (1992) Electroosmotic removal of gasoline hydrocarbons and TCE from clay. J Environ Eng 118:68–83
6. Fraser A (2009) Remediation of contaminated site using electric resistive heating. First use in the UK. EREM 2009. Lisbon (Portugal) 47–48
7. Godschalk MS, Lageman R (2005) Electrokinetic biofence, remediation of VOCs with solar energy and bacteria engineering. Geology 77(3–4 SPEC. ISS):225–231
8. Gomes HI, Dias-Ferreira C, Ribeiro AB (2012) Electrokinetic remediation of organochlorines in soil: enhancement techniques and integration with other remediation technologies. Chemosphere 87(10): 1077–1090
9. Huang Y-C, Cheng Y-W (2012) Electrokinetic-enhanced nanoscale iron reactive barrier of trichloroethylene solubilized by Triton X-100 from groundwater. Electrochimica Acta 86:177–184
10. Reddy KR, Darko-Kagya K, Cameselle C (2011) Electrokinetic-enhanced transport of lactate-modified nanoscale iron particles for degradation of dinitrotoluene in clayey soils. Sep Purif Technol 79(2):230–237
11. Lageman R, Godschalk MS (2007) Electro-bioreclamation. A combination of in situ remediation techniques proves successful at a site in Zeist, the Netherlands. Electrochim Acta 52(10 SPEC. ISS):3449–3453
12. Oonnittan A, Shrestha RA, Sillanpää M (2008) Remediation of hexachlorobenzene in soil by enhanced electrokinetic fenton process. J Environ Sci Health A Tox Hazard Subst Environ Eng 43(8):894–900
13. Oonnittan A, Shrestha RA, Sillanpää M (2009) Removal of hexachlorobenzene from soil by electrokinetically enhanced chemical oxidation. J Hazard Mater 162(2–3):989–993
14. Pham TD, Shrestha RA, Sillanpää M (2010) Removal of hexachlorobenzene and phenanthrene from clayey soil by surfactant- and ultrasound-assisted electrokinetics. J Environ Eng 136(7):739–742
15. Reddy KR, Ala PR, Sharma S, Kumar SN (2006) Enhanced electrokinetic remediation of contaminated manufactured gas plant soil Engineering. Geology 85(1–2):132–146
16. Reddy KR, Darko-Kagya K, Al-Hamdan AZ (2011) Electrokinetic remediation of pentachlorophenol contaminated clay soil. Water Air Soil Pollut 221(1–4):35–44
17. Saichek RE, Reddy KR (2005) Electrokinetically enhanced remediation of hydrophobic organic compounds in soils: a review. Crit Rev Env Sci Tec 35(2):115–192
18. Sale T C, Gilbert D M, Petersen M A (2005) "Cost and performance report: electrically induced redox barriers for treatment of groundwater." ESTCP Project CU0112
19. Semkiw ES, Barcelona MJ (2011) Field study of enhanced TCE reductive dechlorination by a full-scale whey PRB. Ground Water Monit Remed 31(1):68–78
20. Wan J, Li Z, Lu X, Yuan S (2010) Remediation of a hexachlorobenzene-contaminated soil by surfactant-enhanced electrokinetics coupled with microscale Pd/Fe PRB. J Hazard Mater 184(1–3):184–190
21. Wan J, Yuan S, Chen J, Li T, Lin L, Lu X (2009) Solubility-enhanced electrokinetic movement of hexachlorobenzene in sediments: a comparison of cosolvent and cyclodextrin. J Hazard Mater 166(1):221–226
22. Warner SD, Bablitch D, Frappa RH (2012) PRB treatment for contaminated groundwater. Mil Eng 104(675):53–54
23. Weng C-H, Yuan C, Tu H-H (2003) Removal of trichloroethylene from clay soil by series-electrokinetic process. Pract Periodical Hazard Tox Radioactive Waste Manag 7(1):25–30

E

24. Yang GCC, Liu C-Y (2001) Remediation of TCE contaminated soils by in situ EK-fenton process. J Hazard Mater 85(3):317–331
25. Yang GCC, Yeh C-F (2011) Enhanced nano-Fe 3O 4/S 2O 8 2- oxidation of trichloroethylene in a clayey soil by electrokinetics. Sep Purif Technol 79(2):264–271
26. Yuan S, Tian M, Lu X (2006) Electrokinetic movement of hexachlorobenzene in clayed soils enhanced by Tween 80 and β-cyclodextrin. J Hazard Mater 137(2):1218–1225
27. Zheng Z, Yuan S, Liu Y, Lu X, Wan J, Wu X, Chen J (2009) Reductive dechlorination of hexachlorobenzene by Cu/Fe bimetal in the presence of nonionic surfactant. J Hazard Mater 170(2–3):895–901
28. Reddy KR, Cameselle C (2009) Electrochemical remediation technologies for polluted soils, sediments and groundwater. Wiley, New York

Electrokinetics in the Removal of Hydrocarbons from Soils

Erika Bustos
Centro de Investigacióny Desarrollo Tecnológico en Electroquímica, S. C., Sanfandila, Pedro Escobedo, Querétaro, México

Introduction

The development of an industrial society is accompanied by an increase in environmental pollution – the introduction of elements or compounds into the environment that have an unacceptable risk to humans or the environment [1] – caused by diverse toxicants widespread in all parts of the environment: air, water, soil, and sediments. Consequently, many researchers have investigated new technologies over the past two decades to address this environmental problem, particularly in soils since polluted soil can reduce the usability of land [2].

Soils are a complex part of an ecosystem, capable of absorbing toxic chemicals, such as heavy metals, organic compounds, and other hazardous materials from nature or from human activities. Examples of materials found at polluted sites have been metals like cadmium, copper, mercury, chromium, nickel, zinc, strontium, uranium, etc., and hydrocarbons such as petroleum residues, solvents, pesticides, wood preservatives, volatile organic compounds (VOCs) such as benzene, toluene, and trichloroethylene, semi-volatile organics, polycyclic aromatic hydrocarbons (PAHs), and polychlorinated biphenyls (PCBs) [3].

Interest in obtaining non-polluted soils has progressed, not only by regulating soil pollution, but also in developing methodologies for soil assessment and remediation. Several technologies have been proven for remediation of polluted sites, but very few have been proven successful. These can be grouped as either ex situ or in situ technologies. Ex situ remediation technologies involve removing the polluted soil from the subsurface and treating it *on-site* or *off-site*. Conversely, in situ remediation technologies involve treating the polluted soil in place without removing it from the subsurface. This is often preferred because of minimal site disturbance and increased safety, simplicity, and cost-effectiveness by avoiding dredging and transporting processes [4].

Existing soil treatments offering solutions for most pollutants include physical, chemical, thermal, and biological techniques. Most physical treatment processes remove pollutants from the soil–water complex for further treatment or disposal in a more concentrated form. There are, however, some pollutants that are difficult to remove using conventional remediation technologies. Some of them are not only persistent or toxic but also have low solubility and strong adsorption to soil surfaces and organic matter in low-permeability clayey soils [5].

One of the most attractive, innovative, promising, and cost-effective methods of treating soil, sediment, mud, sludge, and marine dredge spoils for inorganic and organic compounds is electrokinetic remediation (EKR) because of its benefits to public health, the environment, and economic concerns [3, 6]. Although the fundamental concept of electrokinetic extraction can be easily understood, the selection of EKR for soil requires detailed information about how pollutants can be removed and an understanding of soil–pollutant relationships during electrokinetic remediation before this technology can be fully utilized.

Electrokinetic Treatment

Electrokinetic treatment is based on passing a direct current flow through a contaminated soil. The resulting processes (electrolysis, electro-osmosis, electrophoresis, and electromigration) remove toxic components from the soil. Although this method has been known for quite some time, its application in soil remediation began only in the second half of the twentieth century. It has recently come to the attention of environmental professionals because of its potential to remove pollutants from low-permeability soils, since processes that govern electrokinetic contaminant transport are not as easily hindered by low hydraulic conductivity.

Electrokinetic remediation involves the installation of electrodes into multiple wells within a contaminated zone, followed by the application of a low electric potential. Ideally, the contaminants migrate toward the electrodes due to different transport mechanisms, and, upon reaching the wells, contaminant-laden liquids are extracted and treated. Although implementation is simple, the geochemical processes that occur within soils during electrokinetic remediation are complex and dependent on system variables such as soil type, pollutant type, treatment time, electrolyte solution, and applied voltage.

Electromigration involves the movement or transport of ions under electrostatic attraction to the electrode of opposite charge, where the direction of travel depends on the charge of the ion. Electrophoresis is the transport of charged particles or colloids under the influence of an electric field. Electro-osmosis is simply the flow of water through a charged soil medium under the influence of an electric field. In water-saturated soil, the movement of water relative to the soil is under the influence of an imposed electric gradient, i.e., the net movement of pore water from the anode to the cathode. Electro-osmosis is attributed to the excess charges on the soil surface. Under the electrical field, the hydrated ions in the double layer of the soil are driven from one end to the other, which simultaneously moves the pore-liquid in the soils. In low-permeability soils with a surface charge, electric conduction involves the one-way movement of counter-ions, which are ions with a charge that is opposite to the mineral surface charge. These ions carry waters of hydration and bulk water with them, generating electro-osmotic flow. The pollutant can then desorb from soils into the pore-liquid and flush along with it [7].

Besides electrokinetic transport, chemical reactions also occur at the electrode surfaces (i.e., water electrolysis reactions with production of H^+ at the anode and ^-OH at the cathode). Common mass-transport mechanisms like diffusion or convection and physical and chemical interactions of the species with the medium also occur. In a low-permeable porous medium under an electrical field, the major transport mechanism through the soil matrix during treatment for nonionic chemical species consists mainly of: electro-osmosis, electrophoresis, molecular diffusion, hydrodynamic dispersion (molecular diffusion and dispersion varying with the heterogeneity of soils and fluid velocity [8]), sorption/desorption, and chemical or biochemical reactions. Since related experiments are conducted in a relatively short period of time, the chemical and biochemical reactions that occur in the soil water are neglected [9].

Many experiments on electrokinetic remediation are carried out on a laboratory or pilot plant scale, with artificially polluted clay media like kaolin. This is because clay soils usually contain a variety of other substances that are present in smaller or trace amounts, such as organic matter, iron oxides, quartz, feldspars, aluminum and manganese hydroxides, titanium oxides, carbonates, and calcite, which could affect electrokinetic response by decreasing the resistance to the flow of water through the sediment [10].

Such experiments, however, could represent a disadvantage, because species reactivities are not considered for reactions in the liquid phase. Different industrial tests could fail as a result, having also other limitations, such as pollutant solubility, its desorption from the soil matrix, the use of different water miscible solvents (especially surfactants or co-solvents [11]), and application time, which can vary from several days to years.

It should also be mentioned that EKR is mainly used to remove and concentrate pollutants in a small portion of soil. The pollutants that are adsorbed on the soil, or are present as precipitates or immiscible liquids, cannot be effectively removed by an electrokinetic remediation technique unless they are in an aqueous phase in the pore fluid or are weakly sorbed onto the soil surface. This is an important aspect, because electroremediation applied at a laboratory level represents a useful tool that allows knowledge of different balances that occur inside the soil particles [12, 13].

In order to have promising laboratory and pilot scale studies and experiments, it is necessary to investigate the important role that several aspects play during soil electrochemical remediation [14–16]: (1) physical and chemical characteristics of pollutants – pollutant concentrations, extension and depth of the contaminant plume, and time period of contamination; (2) environmental conditions of the polluted site – mineral content and type, soil texture, moisture content, and surface charge of the soil; (3) electrochemical conditions of the electrokinetic treatment – physical and chemical characteristics of electrodes, geometry of the electrochemical reactor, type and concentration of electrolyte, electric current and cell potential distribution in the electrochemical reactor, pH and temperature changes, assistance of chemical agents during electrochemical treatment; and (4) technical and economic evaluation of electrokinetic treatment, including power consumption and annual cost.

Hydrocarbon Removal from Soil with Electrokinetic Treatment

Electrokinetic remediation is highly dependent *on site*-specific geochemical conditions, such as soil composition, native electrolytes, pollutant aging, and pollutant mixtures. Several studies have integrated electrokinetic approaches for effective remediation of challenging polluted sites. Many of these studies are limited to bench-scale studies, and more field studies are needed to determine cost and effectiveness in field applications. Although several technologies have been developed to remediate polluted sites, many of them are not applicable for sites containing low-permeability soils, heterogeneous soils, or mixed pollutants.

Electrokinetic remediation technology has great potential for in situ remediation of low-permeability and heterogeneous soils that have been polluted by organics, heavy metals, or a combination of these. It is necessary, however, to understand soil–pollutant–electrolyte interactions in order to take advantage of what the technique can provide. In this way, many in situ EKR technologies have been studied for the extraction and/or destruction of contaminant organic compounds in soil. Extractive methods may be preferable when the extraction results in sufficient recovery of valuable liquid hydrocarbons. When applied at a constant low electric gradient, EKR may prove to have advantages over other methods for transporting hydrophobic liquid hydrocarbons in low-permeable clayey soils, which depend on a number of physico-chemical properties of the soil, the liquid hydrocarbon, and the aqueous solution in the soil pores [17].

The effects of electrokinetic treatment have been studied in the presence of other components, such as surfactants like Triton X–114. This has yielded a hydrocarbon removal percentage of about 49 % after 6 h and 66 % after 24 h. This compares with individual chemical and electrokinetic treatments after 6 h of treatment at 1.5 mL min^{-1}, showing 12 % and 35 % removal, respectively. The coupled system was very successful, although inconvenient in the permanency of Triton X-114 in the soil following remediation [18].

Similarly, there are other chemical compounds used to remove organic compounds [17]: electrokinetically enhanced persulfate oxidation of PCBs in low-permeability clayey soils, yielding 77.9 % degradation of PCBs with thermal activation after 7 days; remediation of 1,2-dichloroethane contaminated soils using electrokinetic-assisted nano $Fe_3O_4/S_2O_8^{2-}$ processes, with a removal efficiency of more than 96 % after 14 days; and degradation of sorbed hexachlorobenzene using an electrokinetic Fenton process.

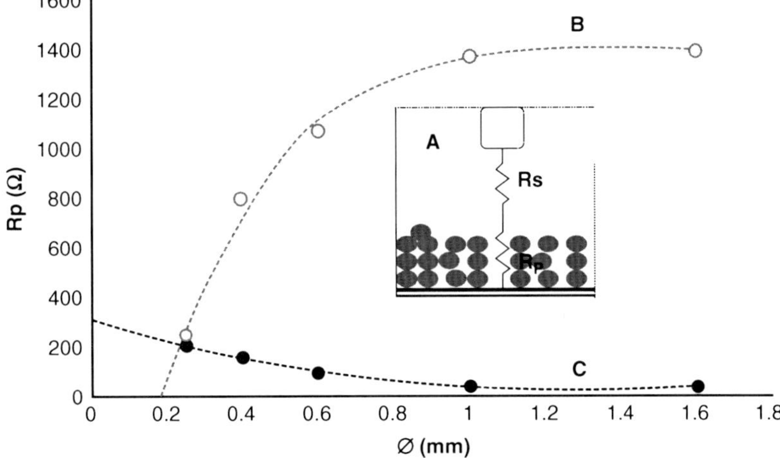

Electrokinetics in the Removal of Hydrocarbons from Soils, Fig. 1 Representation of the variation of resistance to the transference of ions in the solution (**a**), $R_t = R_S + R_P$, where R_s is similar to the resistance obtained with the target (impedance measured with the cell without soil), R_t is the obtained resistance of the measurements, and R_p is the resistance to the transfer of ions through the soil pores with respect to particle size in an organic (**b**) and inorganic (**c**) medium

Future Directions

EKR has been applied in different types of polluted soils, where direct current has been applied using different configurations of electrodes, such as 1D and 2D, between others of graphite, titanium, reticulated vitreous carbon (RVC), RVC–TiO_2, stainless steel, and dimensional stable anodes like PbO_2 $|Ti$, $PbO_2 - Co|$ Ti, $SnO_2|$ Ti, $SnO_2 - Sb_2O_3|$ Ti and $|IrO_2 - Ta_2O_5|$ Ti. The presence and action of electro-osmotic flow under the influence of an electrical field has been seen, evidenced in a change of soil pH in electrode regions. Such actions may affect the electromigration of complex organic pollutant agents [17–21].

Different analytical techniques have been developed to evaluate quantities of organic pollutants using electrochemical techniques based on solvent polarities. Changes in the resistance of ionic transference originating from pollutants in the form of hydrocarbons present in soil are verified with the use of electrochemical impedance spectroscopy (EIS). Good response of EIS for evaluating hydrocarbon-type pollutants in industrial waste soils was verified across the hydrocarbon–soil interface (Fig. 1a), where the clean soil showed increased ionic transference in a logarithmic manner with increasing particle size (Fig. 1b). In contrast, polluted soil showed decreased ionic transference in an exponential manner with increasing particle size (Fig. 1c). These results were verified with Soxhlet extraction and quantification of hydrocarbons in polluted soil [19].

Cross-References

- ▶ Electrokinetic Barriers for Preventing Groundwater Pollution
- ▶ Electrokinetic Remediation, Cost Estimation
- ▶ Electrokinetic Transport in Soil Remediation

References

1. Doménech X (1995) Química del suelo. El impacto de los contaminantes, 3rd edn. Miraguano, Madrid
2. Lal R, Shukla M-K (2004) Principles of soil physics. Marcel Dekker, New York
3. Kim HT (1990) Principles of soil chemistry, 3rd edn. Marcel dekker, New York
4. Russell BJ (1994) Description and sampling of contaminated soils. A field guide, 2nd edn. Lewis Publishers, Boca Raton

5. Huang PM, Iskandar IK (2000) Soil and groundwater pollution and remediation. Lewis publisher, Boca Raton
6. Maturi K, Reddy KR (2006) Simultaneous removal of organic compounds and heavymetals from soils by electrokinetic remediation with a modified cyclodextrin. Chemosphere 63:1022–1031
7. Lemaire T, Moyne C, Stemmelen D (2007) Modelling of electro-osmosis in clayey materials including pH effects. Phys Chem Earth 32:441–452
8. Changa J H, Qiang Z, Huang C- P (2006) Remediation and stimulation of selected chlorinated organic solvents in unsaturated soil by a specific enhanced electrokinetics. Colloids Surf A Physicochem Eng Aspects 287:86–93
9. Yu JW, Neretnieks I (1996) Modelling of transport and reaction processes in a porous medium in an electrical field. Chem Eng Sci 51(19):4355–4368
10. Grundl T, Reese C (1997) Laboratory study of electrokinetic effects in complex natural sediments. J Hazard Mater 55:187–201
11. Sogorka DB, Gabert H, Sogorka BJ (1998) Emerging technologies for soils polluted with metals: electrokinetic remediation. Hazard Ind Waste 30:673–685
12. Ottosen LM, Hansen HK, Hansen CB (2000) Water splitting at ion-exchange membranes and potential differences in soil during electrodialytic soil remediation. J Appl Electrochem 30(11):1199
13. Page MM, Page CL (2002) Electroremediation of contaminated soils. J Environ Eng 128(3):206
14. Murillo-Rivera B, Labastida I, Barrón J, Oropeza-Guzman MT, González I, Teutli-Leon MMM (2009) Influence of anolyte and catholyte composition on TPHs removal from low permeability soil by electrokinetic reclamation. Electrochim Acta 54:2119–2124
15. Alcántara MT, Gómez J, Pazos M, Sanromán MA (2008) Combined treatment of PAHs polluted soils using the sequence extraction with surfactant–electrochemical degradation. Chemosphere 70:1438–1444
16. Virkutytea J, Sillanpää M, Latostenmaa P (2002) Electrokinetic soil remediation: critical overview. Sci Total Environ 289:97–121
17. (2011) Developments in electrokinetic remediation of soils, sediments and construction materials. In: The 10th symposium on electrokinetic remediation
18. Méndez E, Castellanos D, Alba GI, Hernández G, Solís S, Levresse G, Vega M, Rodríguez F, Urbina E, Cuevas MC, García MG, Bustos E (2011) Effect in the physical and chemical properties of gleysol soil after an electro – kinetic treatment in presence of surfactant triton X – 114 to remove hydrocarbon. Int J Electrochem Sci 6:1250–1268
19. Anical L, Ana C, Dumitrache R, Staniloaie D (2009) Some aspects regarding oil polluted soils decontamination involving in-situ bioremediation and electrokinetic remediation procedures. In: Proceedings of the 11th international conference on environmental science and technology. Charia, Crete, 3–5 Sept B-33 – B-40
20. Ramírez V, Sánchez JA, Hernández G, Solís S, Antaño R, Manríquez J, Bustos E (2011) A promising electrochemical test for evaluating the hydrocarbon – type pollutants contained in industrial waste soils. Int J Electrochem Sci 6:1415–1437
21. Méndez E, Bustos E, Feria R, García G, Teutli M (2011) Electrode materials a key factor to improve soil electroremediation. In: Electrochemical cells. InTech – Open Access Publisher, Rijeka, Croasia. ISBN 978-953-308-12-5

Electrokinetics in the Removal of Metal Ions from Soils

Lisbeth M. Ottosen
Department of Civil Engineering, Technical University of Denmark, Lyngby, Denmark

Principle in Remediating Heavy Metal-Polluted Soil by Application of an Electric DC Field

Removal of heavy metals from soils by means of an applied electric DC field is termed "electrokinetic remediation." When the electric potential is applied to a moist soil, the electric current is carried by ions in the pore solution (electromigration). The direction of the ionic electromigration is towards the corresponding electrode. Anions will move towards the anode, and cations will move towards the cathode. Good control of the flow direction for the electromigrating ions can generally be achieved as the ions move along the field lines mainly defined by the electrode placement. When the polluting heavy metals are present as ions or ionic species in the soil pore solution or in the electric double layer, these will electromigrate towards the electrodes where they concentrate. In aged industrially polluted soils, however, heavy metals are unlikely to be present in ionic state because such mobile heavy metals are washed out to groundwater or surface water. Aging of the pollution implies that the heavy metals adsorb to the soil particles or precipitate in the form of various minerals, and it is crucial

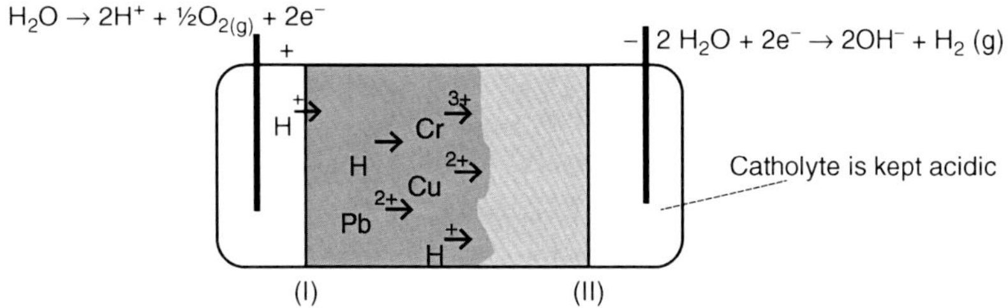

Electrokinetics in the Removal of Metal Ions from Soils, Fig. 1 Principle of electrokinetic soil remediation. The soil is acidified from the anode end (illustrated with the *darker grey area*), and in the acidified soil, the heavy metals are mobilized and transported towards the cathode

that the heavy metals are desorbed from the soil particles and the precipitates are dissolved in order for the heavy metals to be transported out from the soil by electromigration.

Most often, inert electrodes are chosen, and the major electrode reactions are electrolyses, resulting in acidification around the anode and an alkaline environment around the cathode. Subsequently, an acidic front develops in the soil from the anode towards the cathode as the remediation progresses. The analogous alkaline front from the cathode is, however, hindered in developing in the opposite direction by maintaining the catholyte as acidic. This means that over time the whole soil volume between the electrodes becomes acidified, and in the acidic conditions, most heavy metals are desorbed and precipitates dissolved. The hereby mobilized heavy metals are transported into the processing solutions at the electrode by electromigration.

A typical laboratory setup for electrokinetic remediation is shown in Fig. 1. The soil is placed in the central compartment of the cell. The metallic electrodes are placed in electrode compartments, and in each a solution is circulated. The configuration of the electrode compartments can include physical (and passive) separators towards the soil compartment, allowing the transport of liquid and ions in both directions (e.g., [1, 14]), or the separators could have direct influence on the transport, such as with ion exchange membranes [8, 17]. When using ion exchange membranes, separator I is an anion exchange membrane, and separator II is a cation exchange membrane, and the setup is called "electrodialytic remediation." In electrodialytic remediation, the soil is acidified from H^+ ions from water splitting at the anion exchange membrane.

As the electrokinetic remediation process proceeds, the heavy metals are depleted from the soil and concentrated around the electrodes, where they are collected in the processing liquid, and the heavy metals can subsequently be removed from the site.

The First Research Conducted on Electrokinetic Remediation

Transport of heavy metals in soil in an applied electric field was first reported in 1980 [24]. The researchers attempted to dewater a dredged material disposal site and its embankment foundations in order to increase the capacity of the disposal site. In the liquid samples from the process, they found various heavy metals, and even though this was only noted briefly in the actual work, the finding may have inspired other researchers to use an applied electric field for soil remediation because in the end of the 1980s, different teams started independently to develop electrokinetic soil remediation methods. The majority of the published research conducted in the late 1980s and early 1990s originated from research teams at Geokinetics [13], Louisiana State University [2], Lehigh University [20], and Massachusetts Institute of Technology [21].

Major Experimental Findings in Relation to Soil Types and Heavy Metals

Electrokinetic removal of heavy metals from spiked model soils (mainly kaolinite) has been studied intensively (e.g., [26]). In general, very high removal efficiency can be obtained for heavy metals in spiked soils, and the duration of the treatment is short. These works provide the basic insight into the electrochemical processes responsible for remediation; however, the remediation results are generally not comparable to aged polluted soils [7, 10, 18]. In spiked/model soils, the heavy metals are adsorbed less strongly than in aged industrially polluted soils, and the remediation results obtained with spiked soils may be misleading. Desorption and dissolution processes are often the rate-limiting step for the remediation when dealing with industrially polluted soils.

Electromigration is independent of the pore size, so the process can be used in both coarse and fine-grained soils. If the heavy metal species present in the pore solution are uncharged, they cannot be removed by electromigration. In such case, electroosmotic flushing can be utilized in fine-grained soils. Electromigration is, however, demonstrated to be the major transport mechanism for charged species under electrical fields [1], and most often, electrokinetic remediation of heavy metal-polluted soil is based on electromigration as the transport mechanism for the pollutants out from the soil. An important strength of electrokinetic remediation is that the method can be used for remediation of fine-grained soils, where other methods, such as soil washing or pump and treat, tend to fail.

Ottosen et al. [16] compiled literature with results obtained on laboratory and pilot scale with electrokinetic removal of heavy metals from industrially polluted soil. The general trend was that for Cd, Cu, and Zn; good removal was obtained in most investigated soil types by applying the electric potential to the soil, utilizing the acidic front developing from the anode for heavy metal mobilization, as described in Fig. 1. In many cases, 80–98 % of the heavy metals are removed (both on lab scale and field scale). For soils with a high-buffering capacity, the acidification, however, progresses slowly, and the remediation of Cd, Cu, and Zn becomes very time-consuming. In such case, enhancement is necessary so the heavy metals are mobilized and form charged complexes in the nonacidic soil. The acidification of the soil does not cause the pollutants Cr(VI), As(III), and Hg^0 to be ions or in ionic form and following mobilization for electromigration. For these pollutants, an enhancement is also necessary to obtain transport in ionic form in the applied potential. Enhancement is also necessary when the heavy metals are precipitated in compounds which are as insoluble or little soluble in acid. Further, for mixed contaminations with, e.g., heavy metals and organic compounds, enhancement can also be necessary. Enhancement can involve addition of additives to the soil for chemical manipulation of redox condition or to aid desorption by formation of complexes with the heavy metals. Enhancement can also be focused on shortening the duration of the action, either by placement of the electrodes or by combining the electrokinetic soil remediation method with other methods.

Remediation of heavy metal-polluted soil is highly complicated since a huge variety of adsorption types and pollution origins exist. However, research has overcome some major difficulties, and it is possible to remediate most heavy metal-polluted soils by means of electrokinetic remediation.

Future Directions

When intelligently used, electrokinetic soil remediation methods can be used for remediation of even highly complicated cases of soil pollution, and this is where research is developing. Yeung and Gu [25] compiled the research literature related to enhancing electrokinetic remediation. Use of enhancement solutions during electrokinetic remediation in order to solubilize the contaminants and keep them in the mobile state is one major research line, and the other is to combine electrokinetic remediation synergistically with other soil remediation techniques.

Enhancement solutions are either supplied in the electrode compartments and transported into the soil by electromigration or electroosmosis or mixed into the soil. The focuses for using enhancement solutions are (I) to mobilize different heavy metals at the same time, even though they are not mobile under same pH conditions as, e.g., Cu, Pb, and As [23], (II) to overcome a high acid-buffering capacity by mobilizing the heavy metals at neutral or alkaline conditions, e.g., the use of EDTA for removal of Pb from a calcareous soil [4], and (III) simultaneous removal of organic pollution and heavy metals, e.g., a combination of Tween 80 and EDTA for removal of lead and phenanthrene [3].

An intelligent combination of electrokinetic remediation and other heavy metal remediation techniques can be a good solution to different situations. (I) For low-level contamination over large areas, electrokinetic remediation in combination with phytoremediation can be a solution. Bi et al. [5] showed that the application of electrical fields (AC and switched polarity DC) influenced positively both the plant biomass production and the metal uptake by the plants for some species. (II) Soil washing is a soil remediation method which separates the clean coarser fraction and the fine fraction with the adsorbed pollutants, which subsequently must be deposited. The fine fraction can instead be decontaminated by electrodialytic remediation in a stirred suspension [9]. (III) Electrokinetics can be used to supply nano-Fe into the soil [19] for reduction of the highly toxic and mobile Cr(VI) to the less toxic and less mobile Cr (III). (IV) In the case of groundwater and groundwater aquifer pollution, a combination of reactive barriers and electrokinetics is also a possibility, and here the electric field is used for directing the contaminants into the reactive barrier [6], where the target reactions are to transform Cr(VI) to Cr(III) or As(III) to As(V), i.e., from the more to the less toxic and mobile form.

Development of electrokinetic methods similar to those for heavy metal removal from soil is also carried on for other particulate materials as, e.g., municipal solid waste incineration fly ash [11], bio ash [15], mine tailings [22], and harbor sediment [12].

Cross-References

- ▶ Electrokinetic Barriers for Preventing Groundwater Pollution
- ▶ Electrokinetic Remediation, Cost Estimation
- ▶ Electrokinetic Transport in Soil Remediation

References

1. Acar YB, Alshawabheh AN (1993) Principles of electrokinetic remediation. Environ Sci Technol 27:2638–2647. doi:10.1021/es00049a002
2. Acar YB, Gale RJ, Hamed J, Putnam GA (1990) Acid/base distributions in electrokinetic soil processing. Transport Res Rec 1228:23–34
3. Alcántara MT, Gómez J, Pazon M, Sanromán MA (2012) Electrokinetic remediation of lead and penanthrene polluted soils. Geoderma 173–174: 128–133. doi:10.1016/j.geoderma.2011.12.009
4. Amrate S, Akretche DE, Innocent C, Seta P (2005) Removal of Pb from a calcareous soil during EDTA-enhanced electrokinetic extraction. Sci Total Environ 349:56–66. doi:10.1016/j.scitotenv.2005.01.018
5. Bi R, Schlaak M, Siefert E, Lord R, Connolly H (2011) Influence of electrical fields (AC and DC) on phytoremediation of metal polluted soils with rapeseed (*Brassica napus*) and tobacco (*Nicotiana tabacum*). Chemosphere 83:318–326. doi:10.1016/j.chemosphere.2010.12.052
6. Chung HI, Lee MH (2008) A new method for remedial treatment of contaminated clayey soils by electrokinetics coupled with permeable reactive barriers. Electrochim Acta 52:3427–3431. doi:10.1016/j.electacta.2006.08.074
7. Cox CD, Shoesmith MA, Ghosh MM (1996) Electrokinetic remediation of mercury-contaminated soils using iodine/iodide lixiviant. Environ Sci Technol 30:1933–1938. doi:10.1021/es950633r
8. Hansen HK, Ottosen LM, Kliem BK, Villumsen A (1997) Electrodialytic remediation of soils polluted with Cu, Cr, Hg, Pb and Zn. J Chem Technol Biotechnol 70:67–73
9. Jensen PE, Ottosen LM, Ferreira C (2006) Kinetics of electrodialytic extraction of Pb and soil cations from contaminated soil fines in suspension. J Hazard Mater 138:493–499. doi:10.1016/j.jhazmat.2011.12.006
10. Jensen PE, Ottosen LM, Harmon TC (2007) The effect of soil type on the electrodialytic remediation of lead-contaminated soil. Environ Eng Sci 24:234–244. doi:10.1089/ees.2005-0122
11. Jensen PE, Ferreira CMD, Hansen HK, Rype JU, Ottosen LM, Villumsen A (2010) Electroremediation of air pollution control residues in a continous reactor. J Appl Electrochem 40:1173–1181. doi:10.1007/s10800-010-0090-1

12. Kirkelund GM, Ottosen LM, Villumsen A (2010) Investigations of Cu, Pb and Zn partitioning by sequential extraction in harbour sediments after electrodialytic remediation. Chemosphere 79:997–1002. doi:10.1016/j.chemosphere.2010.03.015
13. Lageman R (1989) Theory and Praxis og Electroreclamation. NATO/CCMS Study. Demonstration of remedial technologies for contaminated land and groundwater, ATV Denmark, Copenhagen, 8–9 May 1989
14. Lageman R (1993) Electroreclamation. Environ Sci Technol 27:2648–2650. doi:10.1021/es00049a003
15. Lima AT, Ottosen LM, Ribeiro AB, Hansen HK (2008) Electrodialytic removal of Cd from straw ash in a pilot plant. J Environ Sci Health Part A 43:844–851. doi:10.1080/10934520801974327
16. Ottosen LM, Hansen HK, Jensen PE (2009) Electrokinetic removal of heavy metals. In: Reddy KR, Cameselle C (eds) Electrochemical remediation technologies for polluted soils, sediments and groundwater, 1st edn. Wiley, New York
17. Ottosen LM, Hansen HK, Laursen S, Villumsen A (1997) Electrodialytic remediation of soil polluted with copper from wood preservation industry. Environ Sci Technol 31:1711–1715. doi:10.1021/es9605883
18. Ottosen LM, Lepkova K, Kubal M (2006) Comparison of electrodialytic removal of Cu from spiked kaolinite, spiked soil and industrially polluted soil. J Hazard Mater 137:113–120. doi:10.1016/j.jhazmat.2005.04.044
19. Pamucku S, Hannum L, Wittle JK (2008) Delivery and activation of nano-iron by DC electric field. J Environ Sci Health A 43:934–944. doi:10.1080/10934520801974483
20. Pamukcu S, Khan LI, Fang H-Y (1990) Zinc detoxification of soils by electro-osmosis. Transport Res Rec 1288:41–46
21. Probstein RF, Renaud PC (1987) Electroosmotic control of hazardous wastes. J Physicochem Hydrol 9:345–360
22. Rojo A, Gu YY, Guerra P (2009) Electrodialytic remediation of copper mine tailing pulps. Sep Sci Technol 44:2234–2244. doi:10.1080/01496390902979578
23. Ryu B-G, Park G-Y, Yang J-W, Baek K (2011) Electrolyte conditioning for electrokinetic remediation of As, Cu and Pb-contaminated soil. Sep Purif Technol 79:170–176. doi:10.1016/j.seppur.2011.0.025
24. Segall BA, O'Bannon CE, Matthias JA (1980) Electro-osmosis chemistry and water quality. J Geotech Eng Div 106(GT10):1148–1152
25. Yeung AT, Gu YY (2011) A review on techniques to enhance electrochemical remediation of contaminated soils. J Hazard Mater 195:11–29. doi:10.1016/j.jhazmat.2011.08.047
26. Yeung AT, Hsu C, Menon RM (1996) EDTA-enhanced electrokinetic extraction of lead. ASCE J Geotech Eng 122:666–673

Electrolytes at the Air-Water Interface

Hubert Motschmann
Institute of Physical and Theoretical Chemistry, University of Regensburg, Regensburg, Germany

The classical textbook picture of electrolytes at the air-water interface is shaped by surface tension measurements. Most electrolytes with the exception of some acids increase the surface tension beyond the level of the air-water interface as shown in Fig. 1 [1, 2]. The equilibrium surface tension isotherm $\gamma(c)$ is related by a fundamental thermodynamic law to the surface excess Γ. Usually, Γ is defined within the framework of Gibbs dividing plane which projects the prevailing three-dimensional concentration profiles of the species onto a fictitious plane whose location is defined by the solvent. A conceptional more elegant way is the volume model which defines surface excess in the following experiment [3]: Take a defined volume from the bulk and count the number of ions. Then take the same volume from the interfacial region and the adjacent bulk phase and count again the number of ions. The difference between both quantities specifies the surface excess or depletion Γ. Gibbs adsorption equation states that Γ is directly proportional to the slope of the $\sigma_e - \ln(a)$ isotherm.

$$\Gamma = -1/(mRT)\frac{\mathrm{d}\sigma_e}{\mathrm{d}\ln a} \tag{1}$$

where Γ denotes the surface excess, and σ_e is the equilibrium surface tension, m is the number of independent components, and a is the activity of the solute.

Since the slope of most surface tension isotherm is positive, it has been concluded that the interfacial region is depleted of ions in agreement with the intuitive reasoning that ions prefer the bulk where they are completely hydrated.

Early theories attempt to explain this phenomenon as a consequence of the electrostatic

Electrolytes at the Air-Water Interface, Fig. 1 Surface tension of aqueous monovalent electrolyte solutions [2]

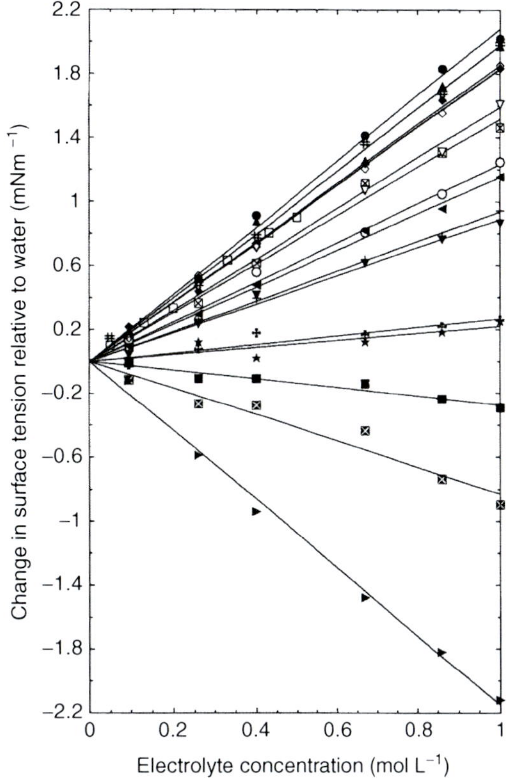

Effect of electrolyte concentration on the change in surface tension relative to water for 1:1 electrolytes. Bubble interval, 1.5 s. Experimental error in data points is ±0.1 mN m^{-1}. HCl (■), LiCl (▲), NaCl (●), KCl (◆), CsCl (⊠), NaF (□), NaI (○), NH_4Cl (▽), NaBr (◇), HNO_3 (⊠), $(CH_3)_4NCl$ (+), NH_4NO_3 (◀), $HClO_4$ (▶), $NaClO_3$ (▼), $LiClO_4$ (+), $NaClO_4$ (★), KOH (#),

interaction [4]. At the air-water interface, there is an abrupt change in the dielectric constant ε_r. The electrostatic interaction arising from the discontinuity can be accounted for by image charges that repel the ions from the air-water interface. Unfortunately, the long-range nature of the Coulomb interaction leads to a diverging solution and it was Wagner [4] who introduced the concept of screened image forces as an explanation for the surface tension behavior of salt solutions.

Onsager and Samaras inspired by success of the Debye-Hückel theory refined this argument and developed a limiting law which holds for diluted electrolyte solutions [5]. The Onsager-Samaras theory derives, subject to certain simplifying assumptions on the nature of the screening length, an analytical expression for the surface tension of the electrolytes. It predicts identical concentration profiles for the anion and cation, and commonly, it is assumed that the profile varies in a monotonic fashion from one bulk phase to the adjacent one. The theory holds up to electrolyte concentrations of about $c = 0.1$ mol/l and then breaks down as a consequence of the assumption that the screening length is independent of the distance to the interface.

Onsager-Samaras theory was believed to capture the relevant physical interactions even though some shortcomings were obvious from the very beginning. The surface tension isotherms reveal an ion specificity that cannot be accounted for within the framework of pure electrostatics. The measured surface tension depends on the nature of the electrolyte: For instance, sodium fluoride shows a stronger increase in the surface tension than an equimolar solution of

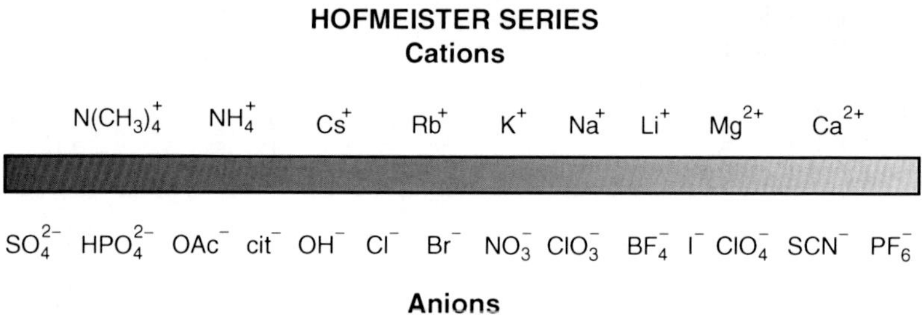

Electrolytes at the Air-Water Interface, Fig. 2 Typical ordering of ions in a Hofmeister series [14]

HOFMEISTER SERIES

Cations

$N(CH_3)_4^+$ NH_4^+ Cs^+ Rb^+ K^+ Na^+ Li^+ Mg^{2+} Ca^{2+}

SO_4^{2-} HPO_4^{2-} OAc^- cit^- OH^- Cl^- Br^- NO_3^- ClO_3^- BF_4^- I^- ClO_4^- SCN^- PF_6^-

Anions

sodium iodide. Specific ion effects are omnipresent in chemistry and biology, and there are many reports of pronounced differences in the properties of charged monolayers [6], micelles [7], vesicles [8], protein solubility [9], enzyme reaction rate [11], or polyelectrolyte multilayers [10] using different but identically charged counter-ions. The effect of ions on a particular set of experiments can be put in a characteristic ordering, the so-called Hofmeister series and its existence suggests a common underlying principle. Ion specificity is the result of a subtle interplay of several competing interactions such as electrostatics, dispersion forces, thermal motion, fluctuations, hydration, ion size effects, as well as the impact of interfacial water. The underlying mechanism is not yet completely resolved, but there are indications that the ion-water interaction plays a crucial role. A major achievement within the last decade is the identification of the so-called law of matching water affinities that serves as a simple rule of thumb with predictive power [12]. Ions are classified as hard or soft, depending on their size and polarizability. Ion-specific effects are attributed to the formation of contact ion pairs as a consequence of the ion-ion interaction mediated by the water, hard anion, and hard cation, and soft anion-soft cation do form contact ion pairs [13]. Even though convincing experimental evidence is still missing, this rule of thumb has been proven to classify the outcome of a wide range of experiments [14] (Fig. 2).

The traditional picture of the liquid-air interface of aqueous electrolyte solutions as a zone that is depleted from ions in a monotonic fashion has been challenged in the last decade. The puzzle started with research on atmospheric chemistry. Hu et al. investigated the uptake of gaseous chlorine Cl_2 and Bromine Br_2 by aqueous solutions of sodium bromide and sodium iodide [15]. The interpretation of the kinetics required the introduction of a reaction between the halogen gas and the halide ions. For this to occur, ions must reside in considerable amount at the air-water interface in contrast to the Onsager-Samaras theory. Knipping et al. investigated the reaction of hydroxyl radicals generated by ozone with chloride ions in solution. The formation of chlorine could only be explained if there is a significant amount of chloride ions at the interface in contrast to the Onsager-Samaras picture [16]. Jungwirth and Tobias provided corroborative evidence for the adsorption of ions at the liquid-air interface by molecular dynamic (MD) simulations [17, 18]. They introduced a polarizable force field and were able to demonstrate that the propensity of halides to the surface increases with size and polarizability of the ions. The non-polarizable Fluoride favors the bulk, whereas the highly polarized Iodide shows even an enrichment at the surface beyond the bulk concentration, contradicting the classical picture. A strong drawback of MD simulations is the fact that the results depend strongly on the force fields which are not exactly known. Horinek and Netz could demonstrate that a properly optimized force field without the inclusion of polarization may also lead to a non-monotonous ion profiles with an enrichment at the surface [19]. This does not mean that polarizability can be neglected but demonstrates the sensitivity of the results to the force field (Fig. 3).

Hence, scattering techniques and nonlinear optical techniques have been pushed to the limits to elucidate details on the concentration profile of

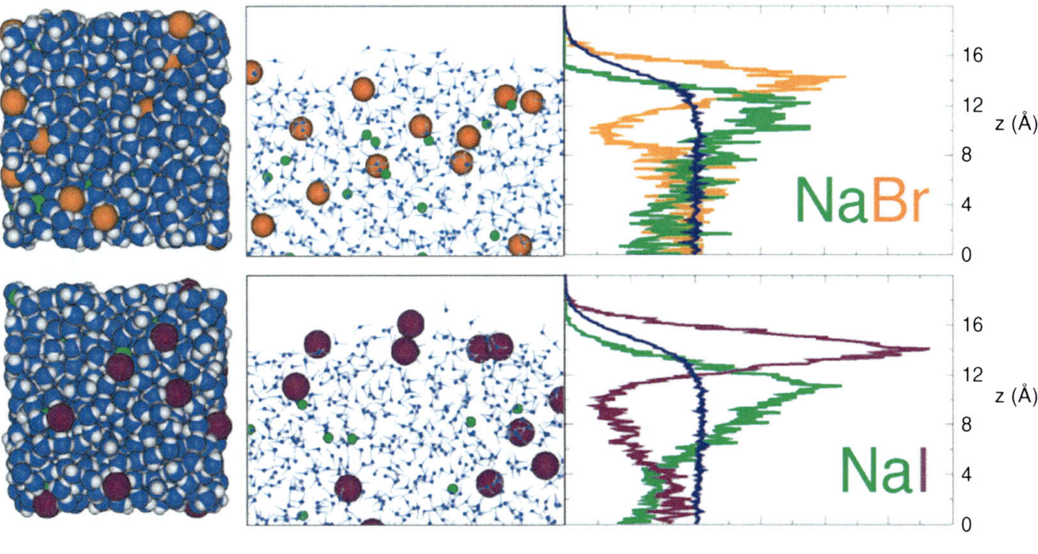

Electrolytes at the Air-Water Interface, Fig. 3 Snapshot (*top* and *side views*) of the solution/air of 1.2 M aqueous halides from the molecular dynamics simulations and density profiles (number densities) of water oxygen atoms and ions plotted vs distance from the center of the slabs in the direction normal to the interface, normalized by the bulk water density [18]

the ions at the interface [20]. Nonlinear optical techniques based on a $\chi^{(2)}$ effect possess an inherent surface specificity as molecules in the bulk phase do not contribute toward the signal [21]. Infrared-Visible sum frequency spectroscopy has been used to study the surface of aqueous sodium halides solutions. These are indirect measurements that probe the impact of the ion on the water network. The spectra revealed profound changes: In particular, the band at 3,400 cm^{-1} is enhanced while the band at 3,200 cm^{-1} is reduced in the order of NaCl, NaCl, and NaJ. Allen and coworkers attributed the observed changes to Iodide disrupting the interfacial hydrogen bonding network and the creation of more asymmetrically bonded H-molecules [22]. This interpretation is subject to some controversy, and Richmond suggested an alternative interpretation of the spectral features by a blueshift and a narrowing of the 3,400 cm^{-1} band [23].

Direct experimental evidence for a non-monotonous ion profile has been provided by the IR-VIS SFG investigations of sodium thiocyanate solution by Viswanath and Motschmann [24]. The stretching vibration of the thiocyanate ion was monitored by polarization-dependent measurements and orientation of the ion, an estimate of the surface coverage and the impact on the water structure have been analyzed in relation to the bulk concentration. The combined data are a convincing piece of evidence for the propensity of the ion to adsorb at the interface [25] (Figs. 4 and 5).

Not only the anions but also the cations show an ion specificity at the air-water interface. Wang et al. could demonstrate that the SFG spectra of the water molecules of aqueous electrolyte solution of NaF and KF exhibit different spectral features and different concentration dependences [26]. These data reveal clear cation effects.

Further corroborative evidence has been provided by X-ray gracing incidence fluorescence at the air-water interface. This is a direct, element specific surface analytical technique [27] that can be used to determine quantitatively the interfacial composition as done in the study of Viswanath et al. for the alkali halides [28]. The data provide clear evidence for the propensity of the halides toward the interface.

An interesting effect has been reported at very dilute electrolyte concentration. Ray and Jones observed a non-monotonous behavior of the

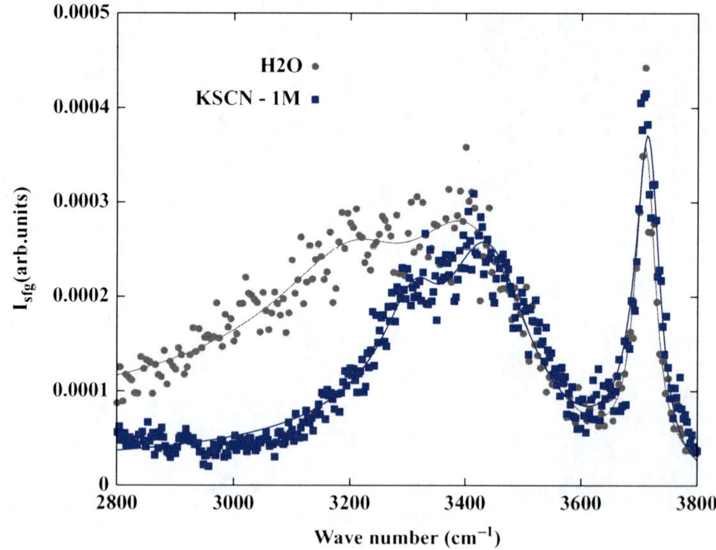

Electrolytes at the Air-Water Interface, Fig. 4 Vibrational sum frequency spectra of water and 1 M potassium thiocyanate solution. The *points* and *continuous lines* represent the experimental datas and fits, respectively

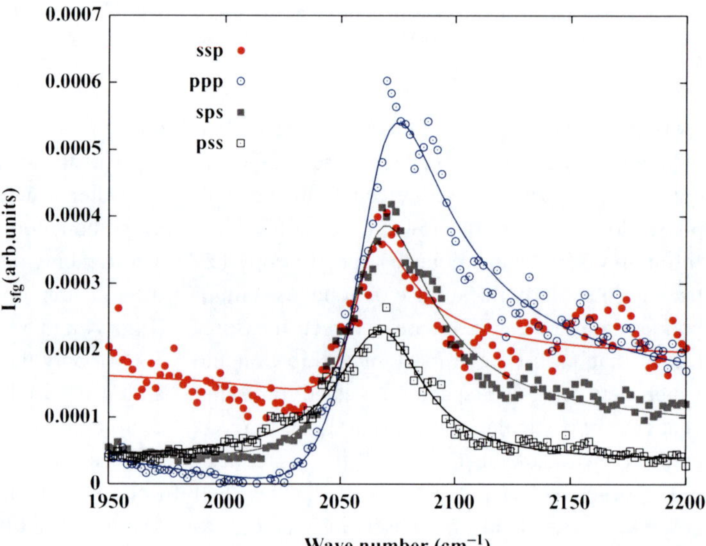

Electrolytes at the Air-Water Interface, Fig. 5 Vibrational sum frequency spectra showing CN stretch of thiocyanate anion for 1 M potassium thiocyanate solution. The *points* and *continuous lines* represent the experimental datas and fits, respectively

surface tension with an initial decrease [29]. The data obtained with a capillary rise technique were reproduced by Dole with a specially designed twin-ring surface tensiometer [30]. The decrease in the surface tension implies an overall enrichment of the ion within the interfacial region in striking difference to the behavior at high concentration adsorption, where the total surface excess is negative and enhancement is restricted to the outermost surface layer. The Ray Jones effect is controversial, discussed in the literature [31, 32], and it is not yet clear if it is an artifact caused by impurities.

In short, one of the most significant findings in the fields of electrolytes is the view that ions may adopt a non-monotonous concentration profile at the air-water interface. Please note that this is not in contradiction to thermodynamic analysis of the surface tension isotherms. Thermodynamics does not predict a distinct concentration profile; instead, it measures an integral quantity from the fluid to the gas phase. There are many concentration profiles that yield a surface excess in agreement with Gibbs adsorption equation. The introduction of an enrichment in a topmost layer, however, imposes a non-monotonous concentration profile with a depletion zone in an adjacent subsurface layer to match the overall depletion. This new point of view of aqueous electrolyte solutions is meanwhile widely accepted and represents a milestone in this field within the last decade.

References

1. Jarvis N, Scheiman M (1968) Surface potentials of aqueous electrolyte solutions. J Phys Chem 72(1):74
2. Weissenborn P, Pugh R (1996) Surface tension of aqueous solutions of electrolytes: relationship with ion hydration, oxygen solubility, and bubble coalescence. J Colloid Interface Sci 184(2):550
3. Guggenheim (1949) Thermodynamics: an advanced treatment for chemists and physicists. North-Holland, Amsterdam
4. Zeit WC (1924) Die Oberflächenspannung verdünnter Elektrolytlösungen (The surface tension of dilute solutions of electrolytes). Physik 25:474
5. Onsager L, Samaras NNT (1934) The surface tension of Debye-Hckel electrolytes. J Chem Phys 2:528–536
6. Viswanath P, Aroti A, Motschmann H, Leontidis E (2009) Vibrational sum frequency generation spectroscopic investigation of the interaction of thiocyanate ions with zwitterionic phospholipid monolayers at the air-water interface. J Phys Chem B 113:14816
7. Abezgauza L, Kuperkarb K, Puthusserickal A, Hassanc OR, Bahadurb P, Daninoa D (2010) Effect of Hofmeister anions on micellization and micellar growth of the surfactant cetylpyridinium chloride. J Colloid Interface Sci 342(1):83
8. Clarke RJ, Lupfert C (1999) Influence of anions and cations on the dipole potential of phosphatidylcholine vesicles: a basis for the Hofmeister effect. Biophys J 76(5):2614
9. Lo Nostro P, Ninham BW (2012) Hofmeister phenomena: an update on ion specificity in biology. Chem Rev 112(4):2286
10. Wong JE, Zastrow H, Jaeger W, Klitzing R (2012) Specific ion versus electrostatic effects on the construction of polyelectrolyte multilayers. Langmuir 25(24):14061
11. Pinna MC, Salis A, Monduzzi M, Ninham BW (2005) Hofmeister series: The hydrolytic activity of *Aspergillus niger* lipase depends on specific anion effects. J Phys Chem B 109:5406
12. Kunz W (2006) Specific ion effects in liquids, in biological systems, and at interfaces. Pure Appl Chem 78(8):1611–1617
13. Collins K, Neilson G, Enderby JE (2007) Ions in water: characterizing the forces that control chemical processes and biological structure. Biophys Chem 128(2–3):95–104
14. Kunz W (2010) Specific ion effects in colloidal and biological systems. Curr Opin Colloid Interface Sci 15(1–2):34–39
15. Hu JH, Shi Q, Davidovitis P (1995) Reactive uptake of cl2(g) and br2(g) by aqueous surfaces as a function of br- and i- ion concentration – the effect of chemical-reaction at the interface. J Phys Chem 99(21):8768
16. Knipping EM, Lakin MJ, Foster KL, Jungwirth P, Tobias DJ, Gerber RB, Dabdub D, Finlayson-Pitts BJ (2000) Experiments and simulations of ion-enhanced interfacial chemistry on aqueous NaCl aerosols. Science 288(5464):301
17. Jungwirth P, Tobias DJ (2000) Surface effects on aqueous ionic solvation: a molecular dynamics simulation study of NaCl at the air/water interface from infinite dilution to saturation. J Phys Chem B 104(32):7702
18. Jungwirth P, Tobias DJ (2006) Specific ion effects at the air/water interface. Chem Rev 106(4):1259–1281
19. Netz R, Horinek D (2012) Progress in modeling of ion effects at the vapor/water interface. Ann Rev Phys Chem 63:401
20. Koelsch P, Viswanath P, Motschmann H, Shapovalov VL, Brezesinski G, Moehwald H, Horinek D, Netz RR, Daillant J, Guenoun P (2007) Specific ion effects in physicochemical and biological systems: simulations, theory and experiments. Colloids Surf A-Physicochem Eng Asp 303(1–2):110–136
21. Lambert AG, Davies PB, Neivandt DJ (2005) Implementing the theory of sum frequency generation vibrational spectroscopy: a tutorial review. Appl Spectros Rev 40:103
22. Gopalakrishnan S, Jungwirth P, Tobias DJ, Allen HC (2005) Air-liquid interfaces of aqueous solutions containing ammonium and sulfate: spectroscopic and molecular dynamics studies. J Phys Chem 109:8861–8872
23. Walker DS, Richmond GL (2008) Interfacial depth profiling of the orientation and bonding of water molecules across liquid-liquid interfaces. J Phys Chem C 112:201–209
24. Viswanath P, Motschmann H (2007) Oriented thiocyanate anions at the air-electrolyte interface and its implications on interfacial water – a vibrational sum frequency spectroscopy study. J Phys Chem C 111(12):4484
25. Viswanath P, Motschmann H (2007) Effect of interfacial presence of oriented thiocyanate on water. J Phys Chem C 112(6):2099
26. Ganghua D, Wang H, Guo Y (2010) Nonlinear optical spectroscopy studies on water and electrolyte aqueous solution interfaces-specific ion effect at air/electrolyte aqueous solution interfaces. Prog Chem 24(10): 1865–1879

27. Wojciechowski K, Gutberlet T, Konovalov O (2013) Anion-specificity at water-air interface probed by total reflection X-ray fluorescence. Coll Surf A-Physicochem Eng Asp 413:184–190
28. Padmanabhan V, Daillant J, Belloni L (2007) Specific ion adsorption and short-range interactions at the air aqueous solution interface. Phys Rev Lett 99(8):086105
29. Jones G, Ray J (1939) A theoretical and experimental analysis of the capillary rise method for measuring the surface tension of solutions of electrolytes. J Am Chem Soc 59:187
30. Dole M (1940) Surface tension of dilute solutions of electrolytes. J Am Chem Soc 6:904
31. Mamatkulov S, Fyta M, Netz RR (2013) Force fields for divalent cations based on single-ion and ion-pair properties. J Chem Phys 138(2):024505
32. Petersen PB, Saykally RJ (2005) Saykally adsorption of ions to the surface of dilute electrolyte solutions: the Jones – ray effect revisited. J Am Chem Soc 127(44):15446–15452

Electrolytes for Electrochemical Double Layer Capacitors

Heiner Jakob Gores[1] and
Hans-Georg Schweiger[2]
[1]Institute of Physical Chemistry, Münster Electrochemical Energy Technology (MEET), Westfälische Wilhelms-Universität Münster (WWU), Münster, Germany
[2]Faculty of Electrical Engineering and Computer Science, Ingolstadt University of Applied Sciences, Ingolstadt, Germany

Definition and Main Properties

Electrochemical double-layer capacitors (EDLCs) (also called supercapacitors) [1–8] are high specific power density electrical energy storage devices that can be charged and discharged up to $>10^6$ times without remarkable decrease of performance. In contrast to rechargeable batteries, they offer only a small energy density. For theoretical and practical limits of such devices, see Ref. [6]. As an auxiliary energy device, they are able to handle peak currents instead of the high-energy-density source (battery or fuel cell) that reaches in these hybrid configurations longer lifetimes. The working principle of an ideal electrochemical double-layer capacitor is based on charging and discharging the electrochemical double layer at carbon electrodes of very high surface (typically up to about 1,000–2,000 m^2/g). In contrast to ideal electrochemical double-layer capacitors, both real and especially pseudocapacitors with redox-active electrode materials include chemical redox reactions of the electrode materials near to the surface in addition to physical charge–discharge processes at the surface of the electrode. Pseudocapacitors offer much higher energy density but a lower power density [1].

The energy content of EDLCs which may be related to mass and volume to receive the corresponding energy density is given by

$$E = \frac{1}{2} C U_0^2 \tag{1}$$

where C measured in F or (A s/V) is the capacity of the capacitor and U_0 is the maximal voltage that can be applied without deterioration of the electrolyte. The maximal voltage is directly related to the voltage window (also called electrochemical stability limits or electro-inactivity window) of a given electrolyte. Whereas aqueous electrolytes are limited to <1.2 V due to the possible decomposition of water, nonaqueous electrolyte solutions are promising voltage windows of up to >4 V. However, real commercially available capacitors are rated with about <3 V, only. The reason for this voltage limitation is caused by minor reactions at the electrodes that sum up. Nevertheless, some recent experiments show that higher voltages may be attainable. For example, Krause and Balducci [9] report on N-butyl-N-methyl pyrrolidinium bis(trifluoromethanesulfonyl)imide solutions in propylene carbonate that were cycled 100,000 times at up to 3.5 V with a capacitance loss of 5 %, only.

The maximal power of a capacitor is also related to U_0 that can be interpreted as the initial potential of the fully charged capacitor but includes also a term related to the velocity of transport of the ions to and away from the mesopores of the electrode material. This effect is represented by the equivalent series resistance (ESR) R_s of the capacitor. At the initial voltage U_0, the current I is zero; upon increasing the current, the voltage of the capacitor U decreases:

$$U = U_0 - IR_s \tag{2}$$

The corresponding power P is

$$P = IU_0 - I^2R_s \tag{3}$$

yielding for $dP/dI = 0$ the maximum of P, P_{max} at $U_0/2R_s$

$$P_{max} = U_0^2/(4R_s) \tag{4}$$

Equation 4 shows the importance of R_s that is mainly determined by the resistance of the electrolyte. Whereas U_0 may approximately be doubled or perhaps tripled by choosing a nonaqueous electrolyte instead of an aqueous one, the corresponding resistance is decreased by about two orders of magnitude [10]; see also table. Therefore, EDLCs based on aqueous solutions show a superior power density but an inferior energy density when compared with those based on nonaqueous solutions or even ionic liquids. For many possible uses including regenerative braking for electric vehicles, uninterruptable power supplies, storage for renewable energy sources, and power sources for starter motors, a broad usable temperature range (typically −30 °C to 60 °C) is requested as well. This restriction limits the electrolyte often to nonaqueous systems. Finally, the solubility of the salt [11] has to be taken into account as during charging the concentration and thereby the conductivity of the electrolyte decreases as the anions and cations are driven to the anode and cathode of the EDLC, respectively. It should be stressed that a solubility smaller than about 0.5 M is already critical [1]. In addition, it is stressed that the solubility often decreases drastically with temperature.

The fundamental equation describing a capacitor is

$$Q = CU \tag{5}$$

Its derivative is

$$dQ/dt = I = CdU/dt = Cv_{scan} \tag{6}$$

where Q is the charge, C the capacity of the capacitor, and U is the voltage. Equation 6 describes a method to directly obtain C by cyclic voltammetry at a given scan rate v_{scan}; I is the current.

To evaluate the performance of electrolytes in EDLCs, various electrochemical methods are available, including:

- Cyclic voltammetry [1]. Recently Wang and Pilon presented a theoretical approach for this method [12]. Stepwise cyclic voltammetry studies at cells with carbon electrodes are also used in EDLCs to evaluate the useful voltage window and the portions of unwanted electrochemical faradaic and wanted capacitive currents [13–15]. The experiments are performed by stepwise increase of the voltage range of CVs in cathodic and independently in anodic direction. These experiments yield reliable voltage windows and reveal that the useful range in anodic and cathodic direction – starting from open-circuit voltage – is not equal.
- Open-circuit voltage decay (self-discharge) is an important quality benchmark [1, 16] for capacitors measured after charging the EDLC to a preselected voltage limit. Obtained $U(t)$ data pairs can be interpreted in terms of three models proposed by Conway [1], including faradaic self-discharge (linear relations U vs. $\log(t)$), discharge by an Ohmic leak ($\log U$ vs. t), and diffusion controlled leakage currents U vs. ($t^{1/2}$).
- Electrochemical impedance spectroscopy at several preselected temperatures covering the temperature range of interest is performed before and after cycling experiments and frequency-dependent yield capacitances C and series resistances R_s [17]. For a discussion of the temperature dependence of EDLCs, see also Ref. [18]. Useful models for the interpretation of the results of impedance measurements are those by de Levie (transmission line model) [19, 20] and by Song et al.

Electrolytes for Electrochemical Double Layer Capacitors, Table 1 Specific conductivities of capacitor electrolytes

Salts	Specific conductivity (mS cm^{-1}) in (solvent) at (°C), concentration (m, M. . .)	Reference
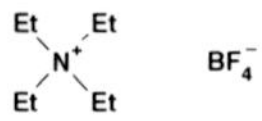 Et Et N^+ Et Et BF_4^-	63.65 mS cm^{-1}in AN at 25 °C, 2.5 mol/kg	*
	12.94 mS cm^{-1}in PC at 25 °C, 1.0 M	28
	4.3 mS cm^{-1}in ADN, 0.7 M	38
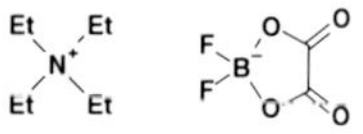 Et Et N^+ Et Et F F B^- O O O O	14.46 mS cm^{-1}in PC at 25°C,1.6 M	28
	57 mS cm^{-1}in AN at 25°C,2.8 mol/kg	
	17.8mS cm^{-1}in AN at -35°C,3.6 mol/kg	26
	22.3mS cm^{-1}in AN at -35°C,1.89 mol/kg	26
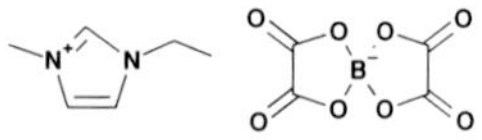 N^+ N O O O O B^- O O O O	41.68 mS cm^{-1}in AN at 25°C,2.05 mol/kg	17
	13.74 mS cm^{-1}in AN at -35°C,1.16mol/kg	
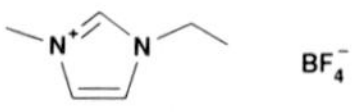 N^+ N BF_4^-	13.74 mS cm^{-1}25°C (pure IL)	2

PC propylene carbonate, *AN* acetonitrile, *ADN* adiponitrile, * = unpublished data

(including also the pore size) [21, 22]. Recently, also a model directly deduced from the Maxwell's theory was proposed for describing the dynamic behavior of EDLCs in the frequency range 1 mHz to 19.5 Hz [23]. Wang and Pilon [24] have shown that the EIS measurements have intrinsic limitations and are inadequate for accurately determining EDL capacitances for practical applications. In accordance with previous studies from other authors (cited in Ref. [24]), they suggest to use more reliable techniques such as galvanostatic charge/discharge and cyclic voltammetry measurements.

- Galvanostatic cycling (charge/discharge) parameters (current, temperature, electrolyte concentration).
- Electrochemical quartz crystal microbalance (EQCM) studies showing that also solvation of ions is of importance for EDLC behavior [25].
- Measurement of ionic conductivity of electrolytes (solvent(s) + salt(s) or ILs) [10, 17, 26].

Nonaqueous Electrolytes

Whereas aqueous electrolytes were mainly studied in the past, investigations of nonaqueous electrolytes are continuing to our days. The most often applied salts are based on electrochemically stable onium ions such as tetraalkylammonium ions and anions that show a good charge delocalization. In contrast to lithium-ion battery electrolytes where not only the conductivity is the most important variable but the transference number has to be taken into account, for EDLCs electrolytes it is the conductivity of the electrolyte, only, that decides on the performance of the EDLC as it determines mainly the equivalent series resistance R_s. Therefore, large anions (of the weakly coordinating type) are not useful due to their too large radius. Those anions could not reach the carbon surface in the mesopores. For investigations [1] of the influence of ionic radii, see also Ref. [27] where the authors have shown that also steric effects are involved when ions are adsorbed into the pores of the electrodes. According to Li et al. [28], currently the most favorable electrolyte solutions are based on tetraethylammonium tetrafluoroborate (TEABF4) and acetonitrile (AN) where an excellent conductivity is reached; see Table 1. A drawback is the high vapor pressure of the solvent. However, substituting the low viscous and low-boiling solvent AN by other electrochemically stable solvents known from lithium battery research such as propylene carbonate or other carbonates and their blends entails two other drawbacks: a much higher viscosity of the solvent causing a lower conductivity (see Table 1) and a lower solubility of the salt [28]. The same problem was recently addressed by a joint research effort trying to improve the efficiency of dye-sensitized solar cells [29]. The problems have triggered numerous investigations [26, 30–34] in a chase for better electrolytes. The following concepts and approaches were checked:

I. Salts with anions of reduced symmetry dissolved in solvents as electrolytes
II. Salts with cations of reduced symmetry dissolved in solvents as electrolytes
III. Pure ionic liquids as electrolytes
IV. Blends of ionic liquids as electrolytes
V. Ionic liquids dissolved in solvents as electrolytes
VI. 2:1 salts dissolved in solvents
VII. New solvents
VIII. New salts

Concepts I and II entail lower melting points of salts and therefore also a better solubility. For example, exchanging the symmetrical anion BF_4^- of tetraethylammonium tetrafluoroborate by the less symmetrical difluoromono(oxalate)borate reduces the melting point from 382 °C to 33 °C for tetraethylammonium difluoromono(oxalate) borate (TEADFB); further reduction of the symmetry of the cation, e.g., resulting in ethylmethylimidazoliumdifluoromono(oxalate) borate, scarcely reduces the melting point further to 22 °C [35, 36]. The solubility of TEADFB exceeds 3.5 mol/kg in AN at −35 °C, at even higher concentration than the concentration where the maximum of the conductivity (22.3 mS/cm) at about 1.9 mol/kg is obtained at this temperature [26]. III and IV address the problems of high vapor pressure of several solutions. V is a compromise, as ILs (III and IV) have rather high viscosity and therefore a low conductivity [2].

Approach VI is used to circumvent the problem of asymmetric voltage windows and of nonmatching ion radii of anions and cations [39]. However, the 2:1 salts behave mainly as an ion pair of the doubly charged cation and the anion (charge +1) in addition to the free anion entailing generally a lower conductivity, when also the large ion diameter of the cation is taken into account. VII and VIII are only sometimes contributing to remarkable advances. For recent examples of a new solvent, see Ref. [37] (reduced flammability) and [38] (large voltage window of up to 3.75 V with $TEABF_4$).

Future Directions

Despite many advances and an outstanding book [1], we are convinced that we have just begun to evaluate this field. Major advances can be expected from new electrolytes, better purification procedures, new electrode materials, and also from capacitors that include very fast electrochemical reactions (pseudocapacitors). Finally the problem of unsymmetrical voltage window deserves more attention.

Cross-References

▶ Conductivity of Electrolytes
▶ Cyclic Voltammetry
▶ Electrochemical Impedance Spectroscopy (EIS) Applications to Sensors and Diagnostics
▶ Ionic Liquids
▶ Non-Aqueous Electrolyte Solutions

References

1. Conway BE (1999) Electrochemical supercapacitors, scientific fundamentals and technological applications. Kluwer/Plenum, New York
2. Ue M (2005) Applications of ionic liquids to double layer capacitors. In: Ohno H (ed) Electrochemical aspects of ionic liquids. Wiley, Hoboken, pp 205–223
3. Koetz R, Carlen M (2000) Principles and applications of electrochemical capacitors. Electrochim Acta 45(15–16):2483–2498
4. Sharma P, Bhatti TS (2010) A review on electrochemical double-layer capacitors. Energy Convers Manag 51:2901–2912
5. Burke A (2000) Ultracapacitors: why, how, and where is the technology. J Power Sources 91:37–50
6. Lewandowski A, Galinski M (2007) Practical and theoretical limits for electrochemical double-layer capacitors. J Power Sources 173:822–828
7. Burke A (2007) R&D considerations for the performance and application of electrochemical capacitors. Electrochim Acta 53:1083–1091
8. Miller JR, Burke A (2008) Electrochemical capacitors: challenges and opportunities for real-world applications. Electrochem Soc Interface 17:53–57

9. Krause A, Balducci A (2011) High voltage electrochemical double layer capacitor containing mixtures of ionic liquids and organic carbonate as electrolytes. Electrochem Commun 13:814–817
10. Gores HJ, Barthel J, Zugmann S, Moosbauer D, Amereller M, Hartl R, Maurer A (2011) Liquid nonaqueous electrolytes. In: Daniel C (ed) Handbook of battery materials, 2nd edn. Wiley-VCH, Weinheim, pp 525–626
11. Zheng JP, Jow TR (1997) The effect of salt concentration in electrolytes on the maximum energy storage for double layer capacitors. J Electrochem Soc 144(7):2417–2420
12. Wang H, Pilon L (2012) Physical interpretation of cyclic voltammetry for measuring electric double layer capacitances. Electrochim Acta 64:130–139
13. Xu K, Ding SP, Jow TR (1999) Toward reliable values of electrochemical stability limits for electrolytes. J Electrochem Soc 146(11):4172–4178
14. Xu K, Ding SP, Jow TR (2001) A better quantification of electrochemical stability limits for electrolytes in double layer capacitors. Electrochim Acta 46(12): 1823–1827
15. Moosbauer D, Jordan S, Wudy F, Zhang SS, Schmidt M, Gores HJ (2009) Determination of electrochemical windows of novel electrolytes for double layer capacitors by stepwise cyclic voltammetry experiments. Acta Chim Slov 56(1):218–224
16. Conway BE, Pell WG (2002) Power limitations of supercapacitor operation associated with resistance and capacitance distribution in porous electrode devices. J Power Sources 105:169
17. Bruglachner H, Jordan S, Schmidt M, Geissler W, Schwake A, Barthel J, Conway BE, Gores HJ (2006) New electrolytes for electrochemical double layer capacitors I. Synthesis and electrochemical properties of 1-ethyl-3-methylimidazolium bis [1,2-oxalato(2-)-O, O'] borate. J New Mater Electrochem Syst 9:209–220
18. Fletcher SI, Sillars FB, Carter RC, Cruden AJ, Mirzaeian M, Hudson NE, Parkinson JA, Hall PJ (2010) The effects of temperature on the performance of electrochemical double layer capacitors. J Power Sources 195:7484–7488
19. de Levie R (1963) On porous electrodes in electrolyte solutions: I. Capacitance effects. Electrochim Acta 8(10):751–780
20. de Levie R (1964) On porous electrodes in electrolyte solutions. IV. Electrochim Acta 9(9): 1231–1245
21. Song HK, Sung JH, Jung YH, Lee KH, Dao LH, Kim MH, Kim HN (2004) Electrochemical porosimetry. J Electrochem Soc 151(3):E102–E109
22. Song HK, Jung YH, Lee KH, Dao LH (1999) Electrochemical impedance spectroscopy of porous electrodes: the effect of pore size distribution. Electrochim Acta 44(20):3513–3519
23. Barsali S, Ceraolo M, Marracci M, Tellini B (2010) Frequency dependent parameter model of supercapacitor. Measurement 43:1683–1689
24. Wang H, Pilon L (2012) Intrinsic limitations of impedance measurements in determining electric doublelayer capacitances. Electrochim Acta 63:55–63
25. Kim I-T, Egashira M, Yoshimoto N, Morita M (2011) On the electric double-layer structure at carbon electrode/organic electrolyte solution interface analyzed by ac impedance and electrochemical quartz-crystal microbalance responses. Electrochim Acta 56:7319–7326
26. Herzig T, Schreiner C, Bruglachner H, Jordan S, Schmidt M, Gores HJ (2008) Temperature and concentration dependence of conductivities of some new semichelatoborates in acetonitrile and comparison with other borates. J Chem Eng Data 53(2):434–438
27. Segalini J, Iwama E, Taberna P-L, Gogotsi Y, Simon P (2012) Steric effects in adsorption of ions from mixed electrolytes into microporous carbon. Electrochem Commun 15:63–65
28. Lai Y, Chen X, Zhang Z, Li J, Liu Y (2011) Tetraethylammonium difluoro(oxalato)borate as electrolyte salt for electrochemical double-layer capacitors. Electrochim Acta 56:6426–6430
29. Hinsch A, Behrens S, Berginc M, Boennemann H, Brandt H, Drewitz A, Einsele F, Fassler D, Gerhard D, Gores H, Haag R, Herzig T, Himmler S, Khelashvili G, Koch D, Nazmutdinova G, Opara-Krasovec U, Putyra P, Rau U, Sastrawan R, Schauer T, Schreiner C, Sensfuss S, Siegers C, Skupien K, Wachter P, Walter J, Wasserscheid P, Wuerfel U, Zistler M (2008) Material development for dye solar modules: results from an integrated approach. Prog Photovolt 16(6):489–501
30. McEwen AB, McDevitt SF, Koch VR (1997) Nonaqueous electrolytes for electrochemical capacitors: imidazolium cations and inorganic fluorides with organic carbonates. J Electrochem Soc 144:L84–L86
31. McEwen AB, Ngo HL, LeCompte K, Goldman JL (1999) Electrochemical properties of imidazolium salt electrolytes for electrochemical capacitor applications. J Electrochem Soc 146:1687–1695
32. Ue M, Takeda M, Toriumi A, Kominato A, Hagiwara R, Ito Y (2003) Application of low-viscosity ionic liquid to the electrolyte of double-layer capacitors. J Electrochem Soc 150:A499–A502
33. Sato T, Masuda G, Takagi K (2004) Electrochemical properties of novel ionic liquids for electric double layer capacitor applications. Electrochim Acta 49: 3603–3611
34. Kim YJ, Matsuzawa Y, Ozaki S, Park KC, Kim C, Endo M, Yoshida H, Masuda G, Sato T, Dresselhaus MS (2005) High energy-density capacitor based on ammonium salt type ionic liquids and their mixing effect by propylene carbonate. J Electrochem Soc 152:A710–A715
35. Herzig T, Schreiner C, Gerhard D, Wasserscheid P, Gores HJ (2007) Characterisation and properties of new ionic liquids with the difluoromono [1,2-oxalato (2-)-O,O′] borate anion. J Fluor Chem 128(6): 612–618

36. Schreiner C, Amereller M, Gores HJ (2009) Chloride-free method to synthesize new ionic liquids with mixed borate anions. Chem – Eur J 15(10):2270–2272
37. Francke R, Cericola D, Kötz R, Weingarth D, Waldvogel SR (2012) Novel electrolytes for electrochemical double layer capacitors based on 1,1,1,3,3,3-hexafluoropropan-2-ol. Electrochim Acta 62:372–380
38. Brandt A, Isken P, Lex-Balducci A, Balducci A (2012) Adiponitrile-based electrochemical double layer capacitor. J Power Sources 204:213–219
39. Jänes A, Kurig H, Romann T, Lust E (2010) Novel doubly charged cation based electrolytes for non-aqueous supercapacitors. Electrochem Commun 12:535–539

Electrolytes for Rechargeable Batteries

Heiner Jakob Gores[1], Hans-Georg Schweiger[2] and Woong-Ki Kim[2]
[1]Institute of Physical Chemistry, Münster Electrochemical Energy Technology (MEET), Westfälische Wilhelms-Universität Münster (WWU), Münster, Germany
[2]Faculty of Electrical Engineering and Computer Science, Ingolstadt University of Applied Sciences, Ingolstadt, Germany

Introduction

The choice of the electrolyte is one of the most important tasks in designing a cell for a battery. The electrolyte electronically separates the electrodes from reacting directly in a chemical reaction, it transports electrochemically active species to/from the electrodes, and it is responsible for the Ohmic resistance of the cell that determines Joule's heating and the loss in power and usable electrical energy. In several cell types, the electrolyte takes even its own part in the main electrochemical reactions of the cell. Then, the electrolyte is defined by the specific cell reaction. In other cases, only concentrations of the electrolyte components can be varied within a limited range. Even if the electrolyte does not take part in the main electrochemical reactions, it still has a strong impact on the performance of the cell. Its chemical and electrochemical properties including its conductivity, its liquid range limited by its freezing and boiling temperature, and the concentration of the salts affect the performance and lifetime of a battery. The behavior of the electrolyte can be fine-tuned by additives. In this entry electrolytes are described that are used in commercially available secondary (rechargeable) batteries. There are two major types of electrolytes aqueous and nonaqueous electrolytes.

Aqueous Electrolytes

Aqueous Solutions of Sulfuric Acid

The sulfuric acid is an oxoacid of sulfur, molecular formula H_2SO_4. At standard conditions for temperature and pressure, the density of pure H_2SO_4 is 1. 84 g/cm^3; it freezes at 10.35 °C and boils at 340 °C [1, 2]. The concentration of the sulfuric acid solution in lead acid batteries is usually in the range of 30–38 wt % H_2SO_4 in the full-charged condition [3, 4].

During the discharge, the sulfuric acid is consumed [4]:

$$\text{(Cell reaction : Pb (anode)} + \text{PbO}_2\text{(cathode))} + 2\text{H}_2\text{SO}_4 \underset{Charge}{\overset{Discharge}{\rightleftarrows}} 2\text{PbSO}_4 + 2\text{H}_2\text{O)}$$

and water is produced so that the solution becomes more diluted (down to about 16 wt %; relative density, 1.11 g/cm^3; freezing point, −10 °C; boiling point, 102.9 °C) [1, 2, 4]. During the charging of this cell, H_2SO_4 is regenerated resulting in a loss of water (concentration, 38 wt %; relative density, 1.28 g/cm^3; freezing point, −59.55 °C; boiling point, 111.8 °C) [1, 2, 4]. The cell voltage of 2.04 V is larger than voltage window of the water that normally limits aqueous systems to 1.23 V [5]. The resulting voltage difference results in some oxygen formation at the positive electrode and hydrogen at the negative electrode. But as hydrogen formation as well as the oxygen formation at lead electrodes have high overpotentials resulting in very small reaction rates,

Electrolytes for Rechargeable Batteries, Table 1 The properties of KOH solution in some primary and secondary cells and batteries

	Concentration of KOH solutions	Additive (typical example)
Nickel-metal hydride battery	About 30 wt % aqueous potassium hydroxide solution [21]	Lithium hydroxide for improved charging efficiency [21]
Nickel-cadmium battery	Usually between 20 wt % (1.19 g/cm^3) and 32 wt % (1.30 g/cm^3) [22]	Lithium hydroxide [22]
Primary nickel-zinc battery	20 wt % [9]	1 wt % of lithium hydroxide to suppress the solubility of zinc in the electrolyte [9]
Nickel-hydrogen battery	21 – 35 wt % [8, 22]	Generally lithium hydroxide [23]
Primary alkaline battery	30 –45 wt % [24]	–

it is still possible to have 2 V of cell voltage in a lead battery in an aqueous electrolyte. The small loss of water can be reduced via recombination, alternatively using gelled electrolytes and absorbed glass mat (AGM) in a sealed lead acid battery. The specific conductivity of the sulfuric acid is about 800 mS/cm in the 35 wt % concentration range and reduced to 600 mS/cm in 16 wt % (discharged) at room temperature. At 40 °C, the specific conductance increases to about 1,000 mS/cm. A lower conductivity results for aged electrolytes [3, 5].

Electrolytes for Rechargeable Batteries, Table 2 Boiling point, melting point, viscosity, and dielectric permittivity of some solvents [11]

	EC	PC	DMC	EMC	DEC	GBL
Boiling temperature (°C)	248	242	90	109	126	206
Melting temperature (°C)	39	−48	4	−55	−43	−43
Viscosity (cP)	1.86 (40 °C)	2.5	0.59	0.65	0.75	1.75
Dielectric permittivity	89.6 (40 °C)	64.4	3.12	2.9	2.82	39

Potassium Hydroxide and Sodium Hydroxide Solutions

The acidic sulfuric acid-based electrolyte is corrosive to many metals; in the alkaline or neutral electrolytes, metals are more stable. Therefore, those materials are used in many primary and secondary batteries. The aqueous potassium hydroxide (KOH) solution is preferred rather than sodium hydroxide (NaOH) because KOH offers not only a higher electrical conductance (at least 40 % larger conductivity in the usual concentration range when compared to NaOH solutions) but also lower freezing temperatures [3, 5]. Frequently the 20–40 wt % concentration of KOH solution is used in batteries showing a conductance of about 500–600 mS/cm at 25 °C [5]. At lower temperatures the conductance decreases (concentration: 20–40 wt %, conductivity: 400–500 mS/cm at 15 °C) [5]. Table 1 gives some properties of KOH solutions in the various batteries. Batteries with KOH-based electrolytes can operate in a high-temperature range [6] due to the low vapor pressure of the solution, e.g., a 37.5 wt % solution shows a vapor pressure of about 0.7 kPa mmHg and a 23.1 wt % solution about 2.0 kPa at ambient temperature [7].

The open-circuit voltage of the nickel-cadmium cell is changed insignificantly in the concentration range 20–30 wt% [8]. The cycle life of nickel hydrogen battery strongly depends on the concentration of KOH. The high concentration of the hydroxide increases the stability of the oxides of nickel and the capacity of the electrode. While the low concentration reduces the corrosion of the sintered nickel substrate, it decreases however the capacity of the nickel hydrogen battery [8].

Nonaqueous Electrolytes

Tarascon and Armand [31] stressed about 10 years ago:“Although the role of the electrolyte is often

Electrolytes for Rechargeable Batteries, Table 3 Conductivity of mixture of salts and solvents

	Lithium perchlorate $LiClO_4$	Lithium hexafluoroarsenate $LiAsF_6$	Lithium tetrafluoroborate $LiBF_4$	Lithium fluoroalkyphosphates LiFAP	Lithium bis(oxalato) borate LiBOB
Conductivity mS/cm, solute conc.	About 9 EC/ DMC (1:2) at 1 mol/L [26]	11.1 in 1 M EC/ DME (1:1)	4. 4(EC/DEC 1:2) 2.6 (PC/EC/EMC1:1:3) [27, 28]	8.6 in 0.8 M EC/DMC	8–9 1 M PC/EC
		7.6 in 0.9 M PC/ DEC [29]		[25]	14.9 DME [30]

considered trivial, its choice is actually crucial . . ..” The role of the electrolyte that cannot be neglected for aqueous electrolytes is even more important and demanding for nonaqueous electrolytes [10].

The Electrolyte in the Lithium Ion Battery

Electrolytes of lithium ion batteries typically consist of lithium salts and solvent(s), generally a blend of several solvents (typically ternary up to five solvents in current market) [11]. The most often used salt is lithium hexafluorophosphate ($LiPF_6$) showing the following advantages: relative high ionic conductivity of its solutions, good protective SEI (solid electrolyte interphase) formation at the anode, and good prevention of aluminum-current collector corrosion. However, its low thermal stability and the production of HF upon hydrolysis have inspired the search for other lithium salts [25]. The nonaqueous organic solvents that are frequently used include dimethyl carbonate (DMC), ethylene carbonate (EC), and ethyl methyl carbonate (EMC). Propylene carbonate (PC) and γ-butyrolactone (GBL) show some drawbacks when compared to other carbonates but may be used with appropriate additives [11]. In Table 2 the boiling point, melting point, viscosity, and dielectric permittivity of these solvents are given.

The conductivity of a $LiPF_6$ solution with EC/DEC (1:2) is 7.0 mS/cm and in EC/DMC (1:2) 10.0 mS/cm at 1 mol/L at ambient temperature [12] (Table 3).

The electrolyte additives that are commonly in the range of 1–5 mass percent in electrolyte solution offer not only a reduction of capacity loss but also the better thermal stability and lifetime (cycling stability) of electrolyte [13] and are fulfilling several tasks [10].

Future Directions

The trend to new nonaqueous battery systems with higher voltages for improving the energy density will make the choice of the electrolyte even more demanding. Reducing the amount of electrolyte in the battery will be also one of the major trends in the future to improve the energy density of batteries. Another route to higher energy densities is the use of other materials instead of lithiated carbon for the anode and the use of lithium metal electrodes.

This approach requires other host materials such as silicon or tin [14] along with measures to prevent adverse effects of high-volume change on lithiation or lithium salts with an increased cycle life of lithium by an effective SEI. Lithium difluoro(monooxalato)borate could be such a salt for lithium metal [15]. The use of ionic liquids [25] instead of solvents or in mixtures with solvents [16] will reduce the vapor pressure and thus adds to the security of the nonaqueous battery. Optimizing cells requires knowledge of many properties; concerning the electrolytes, conductivities for many systems at various temperatures are well known and even optimization procedures have been investigated [17, 18], but other properties such as the lithium ion transference number and the diffusion coefficient of lithium salts are scarcely studied, and often published results are strongly depending on the measurement method used [19, 20]. In these fields, much more work (and understanding) is requested.

Cross-References

- ► Conductivity of Electrolytes
- ► Electrolytes for Electrochemical Double Layer Capacitors
- ► Ionic Liquids
- ► Lithium-Ion Batteries
- ► Non-Aqueous Electrolyte Solutions
- ► Transference Numbers of Ions in Electrolytes

References

1. Gabel CM, Betz HF, Maron SH (1950) Phase equilibria of the system sulfur trioxide-water. J Am Chem Soc 72:1445–1448
2. Perry RH, Green DW (2008) Perry's chemical engineers' handbook, 8th edn. McGraw-Hill, New York
3. Ronald D, Rand DAJ (2001) Understanding batteries. Royal Society of Chemistry, Cambridge
4. Salkind A, Zguris G (2010) Lead-acid batteries. In: Reddy TB (ed) Linden's handbook of batteries, 4th edn. McGraw-Hill, New York
5. Blomgren GE (2010) Battery electrolytes. In: Reddy TB (ed) Linden's handbook of batteries, 4th edn. McGraw-Hill, New York
6. David LR (2003) CRC handbook of chemistry and physics, 84th edn. CRC Press, Boca Raton
7. Lawrence HT, Albert HZ (1996) Electrolyte management considerations in modem nickel/hydrogen and nickel/cadmium cell and battery designs. J Power Sources 63:53–61
8. Shukla AK, Venugopalan S, Hariprakash B (2001) Nickel-based rechargeable batteries. J Power Sources 100:125–148
9. Coates D, Ferreira E, Charkey A (1997) An improved nickel/zinc battery for ventricular assist systems. J Power Sources 65:109–l 15
10. Ue M (2009) Role assigned electrolytes: additives. In: Yoshio M, Brodd RJ, Kozawa A (eds) Lithium ion batteries: science and technologies. Springer Science/ BusinessMedia LLC, New York
11. Dahn J, Ehrlich GM (2010) Lithium-ion batteries. In: Reddy TB (ed) Linden's handbook of batteries, 4th edn. McGraw-Hill, New York/London
12. Schmidt M, Heider U, Kuehner A, Oesten R, Jungnitz M (2001) Lithium fluoroalkylphosphates: a new class of conducting salts for electrolytes for high energy lithium-ion batteries. J Power Sources 97–98:557
13. Schweiger HG, Multerer M, Schweizer-Berberich M, Gores HJ (2008) Optimization of cycling behaviour of lithium ion cells at 60 °C by additives for electrolytes based on lithium bis[1,2-oxalato(2-)-O, O'] borate. Int Electrochem Sci 3:427–443
14. Huggins RA (2011) Lithium alloy anodes. In: Besenhard JO, Daniel C (ed) Handbook of battery materials, 2nd edn. Wiley-VCH, Weinheim
15. Schedlbauer T, Krüger S, Schmitz R, Schmitz RW, Schreiner C, Gores HJ, Passerini S, Winter M (2013) Lithium difluoro(oxalato)borate: a promising salt for lithium metal based secondary batteries? Electrochim Acta 92:102–107
16. Moosbauer D, Zugmann S, Amereller M, Gores HJ (2010) Effect of ionic liquids as additives on lithium electrolytes: conductivity, electrochemical stability, and aluminum corrosion. J Chem Eng Data 55:1794–1798
17. Gores HJ, Schweiger HG, Multerer M Optimizing the conductivity of electrolytes for lithium ion cells, in advanced materials and methods for lithium Ion batteries(Ed. S. S. Zhang): advanced materials and methods for lithium ion batteries, Transworld Research Network, Kerala, India, 2007, published 2008, chapter 11, pp. 257–277
18. Schweiger HG, Multerer M, Schweizer-Berberich M, Gores HJ (2005) Finding conductivity optima of battery electrolytes by conductivity measurements guided by a simplex algorithm. J Electrochem Soc 152:A577–A582
19. Zugmann S, Fleischmann M, Amereller M, Gschwind RM, Winter M, Gores HJ (2011) Salt diffusion coefficients, concentration dependence of cell potentials, and transference numbers of lithium difluoromono (oxalato)borate-based solutions. J Chem Eng Data 56:4786–4789
20. Zugmann S, Fleischmann M, Amereller M, Gschwind RM, Wiemhöfer HD, Gores HJ (2011) Measurement of transference numbers for lithium ion electrolytes via four different methods, a comparative study. Electrochim Acta 56:3926–3933
21. Fetcenko M, Koch J (2010) Nickel-metal hydride batteries. In: Reddy TB (ed) Linden's handbook of batteries, 4th edn. McGraw-Hill, New York
22. Berndt D (2003) Electrochemical energy storage. In: Kiehne HA (ed) Battery technology handbook, 2nd edn. Marcel Dekker, New York
23. Hosung K, Ikhyun O (2008) Electrochemical behavior of the surface-treated nickel hydroxide powder and electrolyte additive LiOH for Ni-MH batteries. J Korean Electrochem Soc 11:115–119
24. Nishio K, Furukawa N (2011) Practical batteries. In: Besenhard JO, Daniel C (ed) Handbook of battery materials. 2nd edn, Wiley-VCH, Weinheim
25. Gores HJ, Barthel J, Zugmann S, Moosbauer D, Amereller M, Hartl R, Maurer A (2011). In: Daniel C (Ed) Handbook of battery materials, 2nd edn. Wiley-VCH, Weinheim, Ch. 17, pp. 525–626
26. Tarascon JM, Guyomard D (1994) New electrolyte compositions stable over the 0 to 5 V voltage range and compatible with the Li1+xMn2O4/ carbon Li-ion cells. Solid State Ion 69:293

27. Reddy TB, Hossain S (2002) Rechargeable lithium batteries (ambient temperature), Chapter 34. In: Reddy TB (ed) Linden's handbook of batteries, 4th edn. McGraw-Hill, New York.
28. Zhang SS (2007) Lithium oxalyldifluoroborate as a salt for the Improved electrolytes of Li-ion batteries. ECS Trans 3:59
29. Moumouzias G, Ritzoulis G, Siapkas D, Terzidis D (2003) Comparative study of LiBF4, LiAsF6, LiPF6, and LiClO4 as electrolytes in propylene carbonate–diethyl carbonate solutions for Li/LiMn2O4 cells. J Power Sources 122:57
30. Xu K, Zhang SS, Jow T, Xu W, Angell C (2002) LiBOB as salt for lithium-ion batteries: a possible solution for high temperature operation. Electrochem Solid-state Lett 5:A26
31. Tarascon JM, Armand M (2001) Issues and challenges facing rechargeable lithium batteries. Nature 414:359–367

Electrolytes, Classification

Werner Kunz
Institut für Biophysik, Fachbereich Physik, Johann Wolfgang Goethe-Universität Frankfurt am Main, Frankfurt am Main, Germany

Definition

Since the pioneering studies by Faraday, Berzelius, Hittorf, and others in the first half of the nineteenth century, electrolytes can be classified as a substance class that shows electrical conductivity mainly through the transport of ions and not (or less) by the transport of electrons (see entry Electrolytes: History). *Electrolytes* can be liquid or solid. Although unusual, liquid electrolytes can be divided into *lyotropic* ones (ions dissolved in a solvent) or *thermotropic* ones (*molten salts* and *Ionic Liquids*) in analogy to surfactants. They can consist of small ions or they can be charged polymers, i.e., *polyelectrolytes*. *Surfactants* can also be positively or negatively charged or both (zwitterions) but may be neutral, nonionic compounds. Whereas at concentrations below their critical micellar concentration, they behave like classical organic electrolytes, surfactants above this threshold form more or less defined structures that are considered in colloidal chemistry and will not be further discussed here. However, the counterion behavior is of significant importance and is discussed in various other entries in this volume.

According to the definition, electrostatic interactions must play an important role. From the Debye-Hückel theory, it can be concluded that in dilute systems these interactions should be long ranged, whereas in concentrated or neat systems, they are screened and therefore short ranged. However, it should be stressed that very often other interactions (e.g., due to the ions' polarizability and geometry) are of the same order of magnitude and must not be neglected.

About classical electrolytes several recommendable monographs exist [1–5]. The reader is referred to this literature for further information.

Classifications

Electrolyte Solutions versus Pure Electrolytes

Electrolytes in a solvent can further be classified either according to the solvents in which they are dissolved or according to their characteristics in the neat state:

- *Aqueous electrolyte solutions*. Electrolytes in water are still the most important class. Examples are seawater, hard water, and biological solutions such as intra- and extracellular water. Ions are ubiquitous in aqueous systems, although often they are only unwanted impurities present only in very small amounts. But even then they can influence phase equilibria because of their strong interactions with water molecules and other dipoles and ions.
- *Nonaqueous electrolyte solutions* [5, 6]. The development of modern energy-storing devices (batteries and capacitors) is closely related to the utilization of nonaqueous electrolyte solutions. Lithium tetrafluoroborate in propylene carbonate is a prominent example. Nonaqueous solutions play also a major role in other fields of electrochemistry such as electroplating, electrodeposition, or electrochemistry.
- *Ionophores* [5]. Ionophores are substances that consist of ions in the pure state. When they are dissolved in an appropriate solvent, the ions

are first solvated and possibly form solvated ion pairs in a second step. NaCl is a typical example; $LiBF_4$ in dimethoxyethane is an example of an ionophore that shows significant ion pair formation.

- *Ionogenes* [5]. Ionogenes do not contain ions in their pure state. They only form ions through chemical reactions, either with the solvent or with suitable added species in the solution. In the pure state at the temperature and pressure of investigation, they are neutral molecules. Usually, the ion-forming reaction is not complete so that the dilution leads to an equilibrium between neutral molecules and ions. An example is the dissolution of acetic acid or ammonia in water.
- *Weak and strong electrolytes*. Often electrolytes are classified according to their dissociation or reassociation behavior, independently of their ionophoric or ionogenic nature. Weak electrolytes are only partly dissociated (such as weak acids); strong electrolytes are fully dissociated. Of course, the dissociation behavior of an electrolyte depends on the solvent. In water, NaCl is a strong electrolyte, whereas in methanol it is rather a weak one.
- *Thermotropic electrolytes*. Ionophores can form electrolytes above their melting points without the addition of a solvent. For example, NaCl is an insulator in its solid state, but above its melting point of 801 °C, it conducts electrical current. Depending on their melting points, classical molten salts can be distinguished from Ionic Liquids, which per definition, have melting points below 100 °C.
- *High-temperature molten salts (see corresponding entry)*. Because of the high temperatures and the corresponding high amount of energy, today only few processes are based on such systems, the most prominent being the production of aluminum in the Hall-Héroult process. Other examples of minor importance are molten salt nuclear reactors, molten salt heat transfer for energy storage, and the use of molten salts in molten carbonate fuel cells with a mixture of liquid lithium carbonate and sodium carbonate as the electrolyte.
- *Low-temperature molten salts or Ionic Liquids* (see corresponding entries). That some salts have very low melting points is known since more than a century. The most prominent example is ethyl ammonium nitrate (EAN), where an unsymmetrical cation is combined with the nitrate, whose salts are known to have relatively low melting points. As a result the melting point of EAN is only 11 °C in its neat state. In the last years, an enormous amount of literature has been published about Ionic Liquids and their possible use as "designer solvents" or "green solvents" (because of their very low vapor pressure). However, only very few industrial applications of this class of electrolytes are known so far. On an industrial level essentially BASF's BASIL (biphasic acid scavenging utilizing ionic liquids) process is known and very recently the dissolution of cellulose in Ionic Liquids, performed by the same manufacturer.
- *Deep Eutectics*. Some years ago it was discovered that by mixing salts, e.g., choline chloride (melting point 302 °C), with urea (melting point 133 °C) in a molar ratio of 1:2, the melting point of the mixture drops down to 12 °C [7]. These novel systems are somewhere in between Ionic Liquids and electrolyte solutions and have promising properties.

Cosmotropes Versus Chaotropes

Although these expressions are usually related to ions, they are also used sometimes for entire salts. For a long time the words *cosmotropes* and *chaotropes* were misunderstood. A cosmotrope is not an ion (salt) that necessarily structures the solvent beyond the first solvation shell. It is merely meant that the solvation shell is energetically tightly bound. Correspondingly, a chaotrope is an ion with an energetically only loosely bound solvation shell. Not the special extension of the solvation or the amount of solvation molecules makes the difference, it is the energy (see specific ion effects: evidences).

In water this concept is related to Pearson's attempt of classifying organic and inorganic molecules and ions as *hard* and *soft* and the postulate that hard Lewis acids will preferably interact with hard Lewis bases and soft Lewis acids with soft Lewis bases [8, 9]. However, there are some significant differences, as will be further discussed in the entry about specific ion effects. What is comparable is the fact that "like seeks like" meaning that in water cosmotropes preferably form ion pairs with cosmotropic counterions and chaotropes and chaotropic counterions. Further, the concept of cosmotropes and chaotropes can be extended to charged head groups and their interactions with ions. The concept of chaotropes and cosmotropes applies also in the context of the *Salting-in* and *Salting-Out* behavior of salts as discussed also in a separate entry.

Inorganic Versus Organic Electrolytes

NaCl is a classical inorganic electrolyte. As such it is highly soluble in water, but less soluble (and dissociated) in many organic solvents. Surprisingly, inorganic electrolytes are not very soluble in many Ionic Liquids [10] or they lead to precipitation of mixed salts. Organic electrolytes of course are often very soluble in organic solvents. They offer the additional possibility of smearing the charge over aromatic rings which can lead either to the formation of Ionic Liquids (case of substituted imidazolium cations with convenient polarizable or organic cations like 1-butyl-3-methylimidazolium octylsulfate) or to advantageous electrolytes for energy storage devices (e.g., 1-ethyl-3-methylimidazolium bis [1,2-oxalato(2)-O,O']borate) [11].

Monovalent Versus Multivalent Ions

Electrostatic interactions should dominate in salts with multivalent ions. As a consequence they are nearly insoluble in nonaqueous solvents, except in solvents such as dimethyl sulfoxide that solvate di- or trivalent cations strongly. Uranyl nitrate dissolves even in diethyl ether. However, recent simulation results suggest that even sulfate ions are quite polarizable and that the polarization of water by the sulfate molecules is most important. The properties of sulfate salts such as their water solubility cannot be described properly without taking into account the very significant water polarizability [12].

Solid Electrolytes

Many solid electrolytes are known today and it can be expected that their importance will further increase, especially for electrochemical devices. For example, beta-alumina solid electrolyte (BASE) is a fast ion conductor, which is used as a membrane in electrochemical cells. It can contain small ions like sodium, which show a high mobility. More classical examples are electrolytes based on lithium or silver iodide where the small cations are very mobile [13]. Note that solid polymer electrolytes are also a rapidly growing field [14].

Polyelectrolytes

Polymers with charged groups in their repeating units are called polyelectrolytes. They have properties that are both defined by their polymer structure and their charged groups. Proteins and DNA are most prominent examples, but synthetic ones are also of utmost importance, for example, in detergency products, for flocculation or as coating thin films. Their properties depend subtly on the degree of dissociation of the charged groups and hence on pH and ionic strength. More details are given the corresponding entry is Polyelectrolytes: Properties.

Future Directions

Today, the behavior of ions in bulk systems and near surfaces is widely understood. This is not so for confined media, such as ion channels, where much work remains to be done, especially as far as transport properties of ions in confinement is concerned. Solid electrolytes will continue to play an important role for electrochemical devices and polyelectrolytes are already ubiquitous in industrial applications. By contrast, the future of Ionic Liquids is not easy to predict. These systems have still to prove that their undisputed advantages weigh more heavily than their disadvantages.

Cross-References

► Conductivity of Electrolytes
► Electrolytes, History

► High-Temperature Molten Salts
► Ionic Liquids
► Non-Aqueous Electrolyte Solutions
► Polyelectrolytes, Properties
► Solid Electrolytes
► Specific Ion Effects, Evidences

References

1. Robinson RA, Stokes RH (2003) Electrolyte solutions, 2nd rev edn. Dover Publications, New York
2. Lee LL (2008) Molecular thermodynamics of electrolyte solutions. World Scientific Publishing, Singapore
3. Bockris JO'M, Reddy AKN (1998) Modern electrochemistry 1: ionics, 2nd edn. Plenum Press, New York/London
4. Kunz W (ed) (2010) Specific ion effects. World Scientific Publishing, Singapore
5. Barthel J, Krienke H, Kunz W (1998) Physical chemistry of electrolyte solutions. Modern aspects. Springer, New York
6. Gores HJ, Barthel J (1995) Nonaqueous electrolyte solutions: new materials for devices and processes based on recent applied research. Pure Appl Chem 67:919–930
7. Abbott AP, Capper G, Davies DL, Rasheed RK, Tambyrajah V (2003) Novel solvent properties of choline chloride/urea mixtures. Chem Comm 70–71
8. Pearson RG (1963) Hard and soft acids and bases. J Am Chem Soc 85:3533–3543
9. Pearson RG (1997) Chemical hardness: applications from molecules to solids. Wiley-VCH, Weinheim, pp 1–195
10. Yang JZ, Wang B, Zhang QG, Tong J (2007) Study on solid–liquid phase equilibria in ionic liquid: 1. The solubility of alkali chloride (MCl) in ionic liquid EMISE. Fluid Phase Equilib 251:68–70
11. Bruglachner H, Jordan S, Schmidt M, Geissler W, Schwake A, Barthel J, Conway BW, Gores HJ (2006) New electrolytes for electrochemical double layer capacitors I. Synthesis and electrochemical properties of 1-ethyl-3-methylimidazolium bis[1,2-oxalato(2-)-O, O']borate. J New Mat Electr Syst 9:209–220
12. Wernersson E, Jungwirth P (2010) Effect of water polarizability on the properties of solutions of polyvalent ions: simulations of aqueous sodium sulfate with different force fields. J Chem Theory Comput 6:3233–3240
13. Andersson DA, Simak SI, Skorodumova NV, Abrikosov IA, Johansson B (2006) Optimization of ionic conductivity in doped ceria. PNAS 103:3518–3521
14. Barbosa PC, Rodrigues LC, Silva MM, Smith MJ (2011) Characterization of $pTMC_nLiPF_6$ solid polymer electrolytes. Solid State Ionics 193:39–42

Electrolytes, History

Werner Kunz
Institut für Biophysik, Fachbereich Physik, Johann Wolfgang Goethe-Universität Frankfurt am Main, Frankfurt am Main, Germany

Introduction

Salts play an important role in the development of mankind. Probably since more than 10,000 years, man uses sodium chloride to conserve food. It was often called "the white gold," an expression that highlights the importance and value of this chemical. Workers were in medieval times paid with salt, hence the word "salary" for their income. Many other processes like leather tanning with mineral salts depended and still depend on salts. Acids (vinegar) and bases (soaps) also have played important roles in human culture. And in general, life on earth is not imaginable without electrolytes.

Still, the molecular composition of electrolytes in general and their behavior in solvents is known only since several decades, and still its properties are not fully understood. Man is capable of flying to the moon, but the prediction of activity coefficients of NaCl in water is still a challenge.

History

The history of scientific comprehension of electrolytes begins in the nineteenth century. Even at the end of the eighteenth century, after Galvani's first electrochemical experiments, Volta speculated that molecules can separate in charged species. In 1802, Jöns Jacob Berzelius found that an electric current can split molecules and especially salts into a positive and a negative part. He concluded that these parts are inherent in any salt, for example, that potassium sulfate consists of positive KO and negative SO_3 (originally written as SO^3). Berzelius wrote in 1819: "…every chemical combination is wholly and solely dependent on two opposing forces, positive and negative electricity, and every chemical compound must be composed of two parts combined by the agency of their electrochemical reaction, since there is no third force. Hence it follows that every compound body, whatever the number of its constituents, can be divided into two parts, one of which is positively and the other negatively electrical" [1].

Some years later, Faraday termed a solvent or liquid showing electrical conductivity an electrolyte. It was also known that there must be positive and negative charges in it. Faraday called these species ions and distinguished between *cations* and *anions* [2]. In Faraday's laws, also the valency of charged species was recognized. It was evident from conductivity measurements that salts put into water yield an electrolyte solution; however, why and how was unknown in the first part of the nineteenth century.

Nevertheless, in the 1850s, Johann Wilhelm Hittorf published a series of papers, of which the first [3] was translated to English ("On the migration of ions during hydrolysis") and reprinted in 1899 [4]. This collection of papers [4] written by Faraday, Hittorf, and Kohlrausch (all in English) is a very inspiring source of information about the early investigations on electrolytes so closely related to electrochemistry.

Nevertheless, the breakthrough came not only from electrochemistry, but also from thermodynamics, in particular from Wilhelm Pfeffer's measurements of osmotic pressures of salt and sugar solutions, first described in 1877 [5]. Some years later, Franz Hofmeister summarized these experiments (in the translation from German by the author of this entry): "For these experiments he [Pfeffer] used an unglazed cell of clay. Its wall was modified by saturating it with $CuSO_4$ and with $K_4[Fe(CN)_6]$. This resulted in the building of a layer of $Cu_2[Fe(CN)_6]$. The modified wall allowed the passage of water, but not the passage of other solute substances. Such a vessel can be filled with a sugar or salt solution, and after closing with a cork, which is tagged with a manometer, it can be immersed in water. Gradually water will intrude into the solution, and this will cause an increase of pressure, which finally will remain constant. This increase of pressure can be read off on the manometer. It marks the 'osmotic pressure' for the dissolved substance at the given concentration and temperature" [6].

Based on these fundamental experiments, Jacobus van't Hoff postulated the relation between the osmotic pressure and the number of solute species in solution. This reasoning van't Hoff deduced from the analogy of gases and solutions [7]. In Hofmeister's words: "As van't Hoff calculated from Pfeffer's experiments, for some single cases, e.g., sucrose, this osmotic pressure is equal to the vapour pressure of a gas volume containing the same number of molecules as the sugar solution in the same volume. According to van't Hoff's idea, the mechanism, which causes the vapour pressure for gases and the osmotic pressure for solutions, is basically the same. In the first case it is caused by collisions of the gas molecules against the vessel's wall, in the latter case only by collisions of the molecules of the dissolved substance against the membrane of the cell, which is permeable for water. The collisions of the solvent molecules are not considered in this case, they can pass the wall" [6].

Then it was only straightforward to postulate that salts in solution must fall apart into two or more species that, as follows from conductivity measurements, must carry charges. This was recognized and summarized in the theories of electrolytes by Svante Arrhenius [8] and by Wilhelm Ostwald [9] who generalized Arrhenius' ideas. Arrhenius based his theory both on van't Hoff's interpretations of Pfeffer's experiments and on his own experiments performed with Friedrich Wilhelm Georg Kohlrausch's equipment [10] to measure the electrical conductivity of a liquid with alternating current. It was not before the late 1880s that Arrhenius's idea of dissociation of salts into ions entered into the scientific literature. Ostwald's experiments on dissociation constants of "weak" electrolytes finally made Arrhenius' ideas generally accepted by the scientific community. It can be stated that the clever combination of thermodynamic (osmotic pressure) and electrochemical (electrical conductivity) quantities and their interpretations were the basis of modern electrolyte chemistry. Fortunately, both quantities change tremendously upon addition of salts so that even very dilute solutions were accessible to experimental observation in the nineteenth century and the beginning of the twentieth century.

By the way, the very important contributions by Walther Nernst [11] to the relation between thermodynamics and electrochemistry and by Paul Walden [12] for his studies of nonaqueous electrolyte solutions should also be emphasized.

In the 1920s, Petrus Debye and Erich Hückel [13] developed their landmark theory of electrolyte solutions. Their merit is the discovery of universal laws. As they could correctly predict, at a given temperature, salt activity coefficients vary with the square root of salt concentration. Any modern theory must be in agreement with these limiting laws. Another major achievement of the Debye-Hückel (DH) theory consists of their successful link of molecular properties to macroscopically observable quantities. Of course, the DH theory has severe weaknesses. Further to some inconsistencies in the theory, it is especially the underlying model that treats the solvent as a continuum and the ions as charged hard spheres. These restrictions limit the applicability of the DH theory to very dilute solutions. Although this is common knowledge today, engineering models and approaches used in colloidal science (DLVO theory) and biology still contain DH terms even when concentrated electrolyte systems are to be described. It is only since a few years that it is recognized that the contribution of electrostatic interactions was overestimated as compared to the granularity of the solvent, specific ion-solvent interactions, the detailed geometry of the species, and other most relevant interactions, e.g., those based on the polarizability of the ions and of the solvents. We can say that the immense success of the DH theory misguided generations of scientists especially in biology and colloidal science.

Several extensions and modifications of the electrolyte theory in the first half of the twentieth century should be mentioned: Bjerrum [14] introduced the concept of limited electrostatic dissociation (ion pair formation), Onsager and Fuoss extended the DH approach and the ideas of Debye about the electrophoretic and the relaxation effect on transport properties such as electrical conductivity and diffusion coefficients [15]. As already mentioned, the DH description is also the basis of one of the two constituting parts of the DLVO theory in colloidal chemistry.

In the 1950s and with the development of modern statistical mechanics, it was possible to propose alternatives based not on differential equations (like the DH theory), but on integral equations. They all depended on the Ornstein-Zernike equation, as published in 1914 [16]. In this context landmark papers were published by Harald Friedman and coworkers in the late 1960s and in the 1970s [17]. They considerably extended the validity of the description of electrolyte activity coefficients up to moderately concentrated solutions (about 0.1 M). For higher concentrations, adjustable parameters had to be introduced. The extension of the theory of electrolytes (osmotic and activity coefficients) by Kenneth S. Pitzer [18] to high concentrations, of the order of 6 mol-kg^{-1}, and to electrolyte mixtures has been widely employed. A unification of advanced models (but still considering the solvent as a continuum) with the theory of Onsager and Fuoss appeared 20 years later [19].

In the last 40 years, tremendous progress in the description of electrolyte properties was achieved. In part, this is a consequence of modern computer science. Molecular Dynamics [20] and Monte Carlo [21] simulations allow one to consider explicitly the molecular structure of the species in electrolytes. The integral equations as more or less rigorous theories (in contrast to the "computer experiments") also include now the granularity of the solvent. However, roughly speaking, all these approaches merely sum up the direct interactions between the species that were put into the program from the beginning. Therefore, a common criticism of modern electrolyte modeling is that "you get out what you put in." Indeed the careful choice of force fields is of utmost importance and some force fields even for simple ions are still questionable [22]. Nevertheless, computer simulations made a most relevant contribution to our modern picture of electrolytes.

Concerning experiments, of course, also major progress has been made. From X-ray [23] and Neutron scattering [24], we learned a lot about ion-ion and ion-solvent interactions, the same is true for advanced spectroscopic techniques, such as dielectric relaxation spectroscopy [25], terahertz and femtosecond infrared spectroscopy [26]. Further to the bulk behavior the behavior of ions at interfaces must be taken into account to correctly describe most of the systems, for example, ions near phospholipid membranes, the air-water interface, or near electrodes. This evidence was rapidly clear, and consequently, the first models about electrolyte interface layers came up in the second half of the nineteenth century by Hermann von Helmholtz in 1853 [27] and later on by Louis Georges Gouy in 1910 [28], David Leonard Chapman in 1913 [29], and finally Otto Stern in 1924 [30], who unified the different approaches. With new experimental techniques [31, 32], our understanding of ions near interfaces could be significantly refined leading to a much more detailed picture of such interfaces [33].

With the combination of more and more results both from simulation and experiments, a profound knowledge is gained now about the behavior of electrolytes, and especially about how to describe specific ion effects beyond Coulomb interactions. This is of utmost importance in engineering, electrochemistry, and biology [34].

Whereas most of the research is focused on aqueous electrolyte solutions, two other important fields should be mentioned that gained significant importance over the last decades. One is concerned with ions in special organic solvents, for example, lithium salts in batteries like $LiPF_6$ in propylene carbonate. The other one is the area of ionic liquids, i.e., solvent-free salts with a melting point lower than 100 °C. Both topics are described in other entries of this volume.

At the end, a hint at a very valuable web site: http://electrochem.cwru.edu/estir/history.htm, with a compilation of several of the most relevant historical papers in electrochemistry.

Future Directions

What remains to be done? It is clear that the behavior of electrolyte systems is dominated by a very subtle balancing of different effects. Even when the effects are big in energy, the resulting balance is often only of the order of kT and therefore difficult to predict. As a consequence, it is very difficult to describe mixtures of ions and especially their temperature dependence and their behavior at interfaces. Second, we are far from a proper ab initio description of ion pairing and ion association, especially for weak bases and acids. The reason is that a proper modeling requires huge quantum-mechanical calculations. This is true whenever protons are involved. Finally, transport properties of electrolytes are still not really well understood, especially in confined media, although their importance, e.g., in the understanding of ion transport in channels is evident [35].

Cross-References

▶ Conductivity of Electrolytes
▶ DLVO Theory
▶ Electrode
▶ Electrolytes, Thermodynamics
▶ Ions at Solid-Liquid Interfaces
▶ Specific Ion Effects, Evidences
▶ Specific Ion Effects, Theory

References

1. Berzelius JJ (1819) Essai sur la théorie des proportions chimiques et sur l'influence chimique de l'électricité. Paris, p 98
2. Faraday M (1834) On electrochemical decomposition (Reprinted in [4], pp 11–44, Original paper: Faraday M (1834) Experimental researches in electricity. Seventh series. Phil Trans R Soc Lond 124:77–122)
3. Hittorf W (1853) Über die Wanderung der Ionen während der Elektrolyse. Pogg Ann 89:177–211
4. Goodwin HM (Ed) (1899) The fundamental laws of electrolytic conduction. Memoirs by Faraday, Hittorf and F Kohlrausch. Harper & brothers, New York/ London
5. Pfeffer W (1877) Osmotische Untersuchungen. Studien zur Zellmechanik, Leipzig, Wilhem Engelmann (Reprint 1921)
6. Hofmeister F (1888) Ueber die wasserentziehende Wirkung der Salze. Naunyn-Schmiedebergs Archiv für experimentelle Pathologie und Pharmakologie 25:1–30; Citation taken from: Kunz W, Henle J, Ninham BW (2004) Zur Lehre von der Wirkung der Salze (About the science of the effect of salts): Franz Hofmeister's historical papers. Curr Op Coll Interf Sci 9:37

7. Hoff JH (1884) Etudes de dynamique chimique, Frederik Muller, Oxford University; Die Rolle des osmotischen Druckes in der Analogie zwischen Lösungen und Gasen (1887) Z phys Chem 1:481–508; see also: Planck M (1887) Ueber die molare Konstitution verdünnter Lösungen. Z phys Chem 1:577–582
8. Arrhenius S (1887) Über die Dissociation der in Wasser gelösten Stoffe. Z Phys Chem 1:631–648
9. Ostwald W (1988) Über die Dissociationstheorie der Elektrolyte. Z Phys Chem 2:270–283
10. Kohlrausch F (1876) On the conductivity of electrolytes dissolved in water in relation to the migration of their compounds in [4], pp 83–92; Kohlrausch FWG, Holborn LFC (1898) Das Leitvermögen der Elektrolyte, insbesondere der Lösungen. Methoden, Resultate und chemische Anwendungen. Teubner BG, Leipzig
11. Nernst WH (1889) Die elektromotorische Wirksamkeit der Ionen. Z Phys Chem Stöchiom Verwandtschaftslehre 4:129–181
12. Walden P (1924) Elektrochemie nichtwässeriger Lösungen. Bredigs Handb d angew physikal Chemie, 13. Bd. Barth, Leipzig
13. Debye P, Hückel E (1923) Zur Theorie der Elektrolyte. Physik Z 24:185–206
14. Bjerrum N (1926) Untersuchungen über Ionenassoziation. Kgl Danske Vidensk Math-Fysiske Medd VIII 9:1–47
15. Onsager L, Fuoss RM (1932) Irreversible processes in electrolytes. Diffusion, conductance and viscous flow in arbitrary mixtures of strong electrolytes. J Phys Chem 36:2689–2778
16. Ornstein LS, Zernike F (1914) Accidental deviations of density and opalescence at the critical point of a single substance. Proc Acad Sci (Amsterdam) 17:793–804
17. Ramanathan PS, Friedman HL (1971) Study of a refined model for aqueous 1-1- electrolytes. J Chem Phys 54:1086–1099
18. Pitzer KS (1979) Theory: ion interaction approach. In: Pytkowicz RM (ed) Activity coefficients in electrolyte solutions, vol 1. CRC Press, Boca Raton, pp 158–208
19. Bernard O, Kunz W, Turq P, Blum L (1992) Self-diffusion in electrolyte solutions using the mean spherical approximation. J Phys Chem 96:398–403; Conductance in electrolyte solutions using the mean spherical approximation. ibid 96:3833–3840
20. Heinzinger K (1985) Computer simulations of aqueous electrolyte solutions. Physica 131B:196–216
21. Levesque D, Weis JJ, Hansen JP (1986) In: Binder K (ed) Monte Carlo methods in statistical physics (topics in current physics), 2nd edn. Springer, Berlin
22. Fyta M, Kalcher I, Dzubiella J, Vrbka L, Netz RR (2010) Ionic force field optimization based on single-ion and ion-pair solvation properties. J Chem Phys 132:024911/1–024911/10
23. Palinkas G, Kalman E (1981) X-ray diffraction on electrolyte solutions in the low angle range. Z Naturforsch Teil A 36A:1367–1370
24. Neilson GW, Enderby JE (1979) Neutron and x-ray diffraction studies of concentrated aqueous electrolyte solutions. Annu Rep Prog Chem Sect C Phys Chem 76:185–220
25. Buchner R, Barthel J (2001) Dielectric relaxation in solutions. Annu Rep Prog Chem Sect C Phys Chem 97:349–382
26. Tielrooij KJ, Garcia-Araez N, Bonn M, Bakker HJ (2010) Cooperativity in ion hydration. Science 328:1006–1009
27. Helmholtz H (1853) Ueber einige Gesetze der Vertheilung elektrischer Ströme in körperlichen Leitern mit Anwendung auf die thierisch-elektrischen Versuche. Ann Phys Chem (Leipzig) 165 (ser 2, vol 89, ser 3, vol 29):211–233
28. Gouy LG (1910) Gouy. Sur la constitution de la charge électrique à la surface d'un électrolyte. J Phys Théor Appl Ser 4 9:457–467
29. Chapman DL (1913) A contribution to the theory of electrocapillarity. Lond Edinb Dublin Philos Mag J Science Ser 6 25:475–481
30. Stern O (1924) Zur Theorie der Elektrolytischen Doppelschicht. Z Elektrochem Angew Phys Chem 30:508–516
31. Motschmann H, Koelsch P (2010) Linear and nonlinear optical techniques to probe ion profiles at the air-water interface. In [33], pp 119–147
32. Padmanabhan V, Girard L, Daillant J, Belloni L (2010) X-ray studies of ion specific effects. In [33], pp 149–169
33. Jungwirth P, Winter B (2008) Ions at aqueous interfaces: from water surface to hydrated proteins. Annu Rev Phys Chem 59:343–366
34. Kunz W (ed) (2010) Specific ion effects. World Scientific, Singapore
35. Krauss D, Eisenberg B, Gillespie D (2011) Selectivity sequences in a model calcium channel: role of electrostatic field strength. Eur Biophys J 40:775–782

Electrolytes, Thermodynamics

Christoph Held and Gabriele Sadowski
Department of Biochemical and Chemical Engineering, Technische Universität Dortmund, Dortmund, Germany

Introduction

Electrolytes are part of our daily life: They are present in natural systems as well as in many biological and chemical processes. Inorganic salts of low molecular weight belong to the "simplest" class of electrolytes. On the one hand, they are systematically used in technical processes for (1) the recovery of biomolecules (e.g., protein precipitation) or (2) as auxiliary material in separation units (e.g., reactive distillation with an electrolyte serving as entrainer). In (3) electrolysis they serve as educts for the production of certain compounds (e.g., aluminum). On the other hand, there are technical applications where salts have to be separated from a product stream (e.g., waste-/drinking-water treatment, reverse osmosis, wet-flue gas scrubbing). Moreover, ions/salts are invariably present in natural processes (e.g., causing osmotic pressure in cells or during biochemical reactions). Another class of electrolytes which is important in the engineering domain is the one of acids. The total world production of hydrochloric acid is estimated at 20 Miot per year. Acids like aqueous solutions of HCl, HBr, or HI are used for the production of organic and inorganic compounds or for pH control and neutralization. Generally, electrolytes can either fully dissociate into their ions ("strong electrolytes," e.g., NaCl) or only partially dissociate ("weak electrolytes," e.g., sodium acetate).

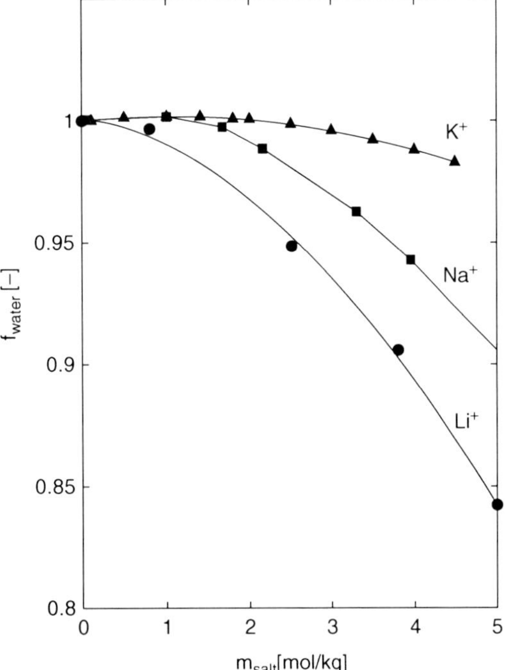

Electrolytes, Thermodynamics, Fig. 1 Influence of alkali iodides on the activity coefficient of water at 25 °C

Meaning and Significance of Solvent Activity Coefficients

Thermodynamic properties are determined by the interactions among molecules. By introducing a salt into a solution, ions will influence the solvent/solvent interactions, thereby strongly influencing the activity coefficient of the solvent. This quantity is thus directly related to the interactions in electrolyte solutions. Adding salt causes an initial slight increase of the solvent activity coefficient which – however – is followed by a continuous decrease at higher salt concentrations (Fig. 1). Obviously, two effects are present in electrolyte solutions: (1) At low salt concentrations, salts completely dissociate into their ions, i.e., small species of high charge density are introduced in the system. This causes a rearrangement of the solvent molecules. As a result, solvent/solvent interactions in the solution decrease which causes an increase in solvent activity coefficients. (2) At higher salt concentrations, solvation of the ions becomes more important. This means the strong ion/solvent interactions retain the solvent in the solution, which causes decreasing solvent activity coefficients. Figure 1 shows that solvent activity coefficients are influenced by the concentration of the salt (and therewith its ionic strength). Moreover, they also differ remarkably for adding various salts/ions. Both, concentration as well as ion-specific influences has to be considered for thermodynamic modeling.

As it can be observed from Fig. 1, electrolyte solutions are highly nonideal from the thermodynamic point-of-view, i.e., the solvent activity coefficient as well as the salt activity coefficient strongly deviates from unity (ideal solution). Activity coefficients f_i depend on temperature and composition of the considered electrolyte solution. They describe the nonideal behavior of solutions and are crucial for phase-equilibrium calculations, e.g.,:

- Calculation of vapor pressures p of electrolyte solutions:

$$p = x_{solvent} \cdot f_{solvent} \cdot p^{LV}_{0solvent}(T)$$

where $p^{LV}_{0solvent}$ is the vapor pressure of pure solvent and $x_{solvent}$ is the mole fraction of the solvent in the mixture.

- Calculation of the salt distribution between two phases I and II:

$$x^{I}_{salt} \cdot f^{I}_{salt} = x^{II}_{salt} \cdot f^{II}_{salt}$$

- Calculation of the solubility m_{salt} of a salt in a solvent:

$$m_{salt} = \left[\exp\left(-\frac{\Delta g^{+}}{RT}\right) \cdot \left(x_{solvent} f_{solvent}\left(f^{*}_{salt}\right)^{\nu\pm}\right)^{-1}\right]^{1/\nu\pm}$$

where Δg^{+} is the standard Gibbs energy of the ion-pair formation $\nu_{+}Cat^{+} + \nu_{-}An^{-} \leftrightarrow Cat_{\nu+}An_{\nu-}$ [1] and $\nu_{\pm}$ is the sum of the stoichiometric coefficients of cations and anions in the salt ($\nu_{\pm} = \nu_{+} + \nu_{-}$).

For electrolytes, the mean ionic activity coefficient f^{*}_{salt} is defined as the geometrical mean of the rational activity coefficients of the ions in solution:

$$f^{*}_{salt} = \left((f_{+}{}^{*})^{\nu+} \cdot (f_{-}{}^{*})^{\nu-}\right)^{\frac{1}{(\nu_{+}+\nu_{-})}} \tag{1}$$

The decisive difference between solvent ($f_{solvent}$) and salt activity coefficients (f^{*}_{salt}) is that $f_{solvent}$ becomes unity for the pure-component state whereas f^{*}_{salt} becomes unity for the infinite-diluted state:

$$\lim_{x_{i=solvent} \to 1} f_i = 1 \quad \text{for solvents}$$
$$\lim_{x_{i=salt} \to 0} f^{*}_{i} = 1 \quad \text{for salts} \tag{2}$$

Obtaining Activity Coefficients from Thermodynamic Models

The activity coefficients of both, solvent and salt, can be obtained by various models. The latter can be divided into two groups: excess Gibbs energy (G^E) models and Helmholtz-energy models (equations of state; EOS). The main difference between G^E models and equations of state is that G^E models are in general unable to account for volume or pressure effects, which, in contrast is possible via the application of an EOS.

Theories that account for the Coulomb forces among the ions are, e.g., provided by the work of Debye and Hückel (DH) in 1923 [2] or by the Mean Spherical Approximation (MSA) introduced by Waisman and Lebowitz [3] in 1970. However, both, DH and MSA, only consider the Coulomb long-range interactions due to the ion charges.

In addition to that, state-of-the-art models also account for short-range (SR) interactions between the ions and the solvent as well as for those among the solvent molecules. Both the Coulomb long-range (LR) and the SR contribution are usually treated as independent of each other.

$$G^E(T, p) = G - G^{id\,solution} = G^{SR} + G^{LR} \tag{3}$$

$$A^{res}(T, V) = A - A^{idgas} = A^{SR} + A^{LR} \tag{4}$$

From G^E models, the activity coefficients are directly obtained by (Eq. 5) via derivation of G^E with respect to the mole number n of the considered component i:

$$\ln f_i = \frac{1}{RT}\left(\frac{\partial G^E}{\partial n_i}\right)_{T, p, n_{j \neq i}} \tag{5}$$

Based on an EOS, the activity coefficients are obtained from fugacity coefficients:

$$f_{solvent} = \frac{\varphi_{solvent}(T, p, x_{solvent})}{\varphi_{0solvent}(T, p, x_{solvent} = 1)} \text{ for solvents and}$$
$$f^{*}_{salt} = \frac{\varphi_{salt}(T, p, x_{salt})}{\varphi^{\infty}_{salt}(T, p, x_{salt} \to 0)} \text{ for salts} \tag{6}$$

where "0solvent" denotes the pure solvent and "∞" means that the salt is infinitely diluted in the solvent. The system pressure p (or the real gas factor Z which is defined as pV/RT for V being the volume of the system) as well as the fugacity coefficients φ_i of the components in a mixture can be calculated as derivatives of the residual Helmholtz energy A^{res} with respect to volume and mole number n_i, respectively:

$$\ln \varphi_i = \left(\frac{\partial A^{res}/RT}{\partial n_i}\right)_{T,V,n_{j\neq i}} - \ln Z$$
$$\text{with } Z = 1 + p^{res}\frac{V}{RT} \text{ and } p^{res} = -\left(\frac{\partial A^{res}}{\partial V}\right)_{T,n_i} \quad (7)$$

G^E Models for Electrolyte Solutions

Nonelectrolyte G^E models only account for the short-range interaction among non-charged molecules (→G^{SR}). One widely used G^E model is the Non-Random-Two-Liquid (NRTL) theory developed in 1968. To extend this to electrolyte solutions, it was combined with either the DH or the MSA theory to explicitly account for the Coulomb forces among the ions. Examples for electrolyte models are the electrolyte NRTL (eNRTL) [4] or the Pitzer model [5] which both include the Debye-Hückel theory. Nasirzadeh et al. [6] used a MSA-NRTL model [7] (combination of NRTL with MSA) as well as an extended Pitzer model of Archer [8] which are excellent models for the description of activity coefficients in electrolyte solutions. Examples for electrolyte G^E models which were applied to solutions with more than one solvent or more than one solute are a modified Pitzer approach by Ye et al. [9] or the MSA-NRTL by Papaiconomou et al. [7]. However, both groups applied ternary mixture parameters to correlate activity coefficients. Salimi et al. [10] defined concentration-dependent and salt-dependent ion parameters which allows for correlations only but not for predictions or extrapolations.

Equations of State for Electrolyte Solutions

Despite the simplicity and accuracy of G^E models also, equations of state have been established for modeling phase equilibria in industry and research. This is due to a central shortcoming of G^E models: the inability of calculating densities. The latter are important system properties for designing of reservoirs or calculating volumetric fluxes and viscosities. Furthermore, they are typically used for validation of model consistencies. Well-known nonelectrolyte EOS are the Soave-Redlich-Kwong (SRK, 1972) and the Peng-Robinson (PR, 1976) approaches, and more recently also the Perturbed-Chain Statistical Associating Fluid Theory (PC-SAFT, 2001 [11]). Such models were combined with approaches describing the Coulomb interactions (DH, MSA), yielding electrolyte equations of state. Myers et al. [12] developed an electrolyte model based on the PR EOS. Fürst and Renon [13] proposed the combination of a modified SRK EOS with the MSA. The shortcoming of models like SRK and PR is the fact that they assume the species to show no or only small deviations from spherical shape which is certainly valid for small inorganic ions. However, this assumption is not justified for solvents (e.g., water, alcohols) or other components which may be present in electrolyte solutions (e.g., amino acids, sugars). Here segment-based models appear to be more appropriate. One of the most famous models of this kind is the SAFT approach [14] which accounts for repulsive as well as attractive interactions (via van der Waals or hydrogen bonds) as well as for the nonspherical shape of the species. There exist a whole series of modifications and extension of SAFT such as SAFT-VR (variable range) [15], PC-SAFT [11], and SAFT 1 [16]. Galindo et al. [17] successfully extended the SAFT-VR to electrolyte solutions by combining it with MSA. Radosz et al. recently published the electrolyte models SAFT1 and SAFT2 [18, 19] EOS obtained by combining the nonelectrolyte SAFT models and PC-SAFT was combined with a DH contribution, yielding the

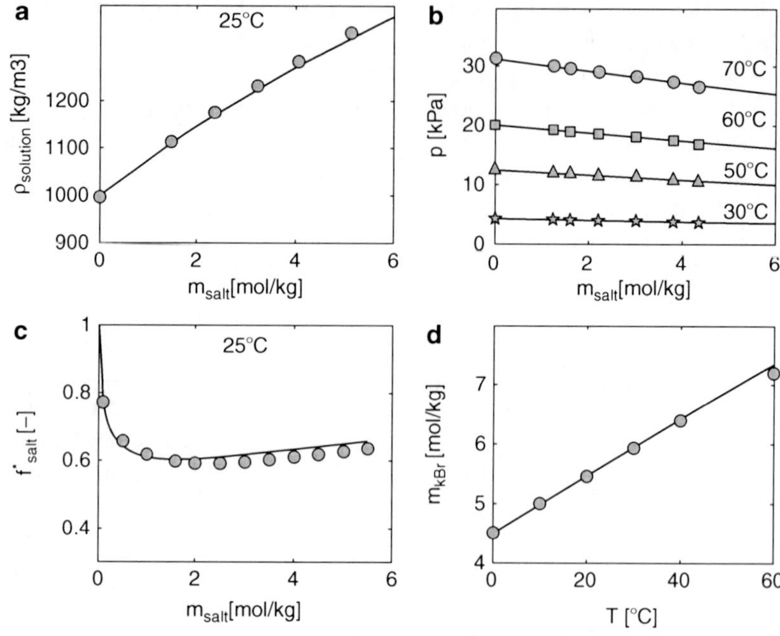

Electrolytes, Thermodynamics, Fig. 2 Solution densities, vapour pressures, salt activity coefficients, and solubility behaviour of KBr/water solutions. Symbols represent experimental data, lines are ePC-SAFT modelling

so-called electrolyte PC-SAFT (ePC-SAFT, 2005 [20]). It describes components that form hydrogen bonds (e.g., water, alcohols) using five pure-component parameters whereas strong electrolytes are described by two ion-specific parameters only. The latter are universally valid for all salts (i.e., the Na^+ parameters are the same in NaCl, $NaNO_3$, etc.). It allows for a quantitative description of thermodynamic properties of electrolyte solutions [21, 22]. Besides fully dissociated electrolytes, also weak electrolytes were modeled by considering association/dissociation equilibria [23].

An example for calculating thermodynamic properties is given in Fig. 2 for aqueous KBr solutions for a wide temperature and concentration range.

It can be observed from Fig. 2 that ePC-SAFT allows for quantitative descriptions of solution densities (a), vapor pressures (b), salt activity coefficients (c), and solubilities (d).

Future Directions

Based on the knowledge obtained from binary inorganic electrolyte solutions, the future goals are (1) the ability for modeling also organic electrolyte solutions of high molecular weight (polyelectrolytes), (2) measuring and modeling protein solutions containing salts, (3) describing the salt influence on the phase behavior of solvent mixtures.

Cross-References

► Activity Coefficients
► Data Banks of Electrolytes
► Gas Solubility of Electrolytes
► Ion Properties
► Thermodynamic Properties of Ionic Solutions - MSA and NRTL Models

References

1. Li JD, Lin YZ, Gmehling J (2005) g(E) model for single- and mixed-solvent electrolyte systems. 3. Prediction of salt solubilities in aqueous electrolyte systems. Ind Eng Chem Res 44:1602–1609
2. Debye P, Hückel E (1923) The theory of electrolytes I. The lowering of the freezing point and related occurrences. Phys Z 9:185–206
3. Waisman E, Lebowitz JL (1970) Exact solution of an integral equation for structure of a primitive model of electrolytes. J Chem Phys 52:4307–4311

4. Chen CC, Britt HI, Boston JF, Evans LB (1982) Local composition models for excess Gibbs energy of electrolyte systems. AIChE J 28:588–596
5. Pitzer KS (1973) Thermodynamics of electrolytes: I. Theoretical basis and general equations. J Phys Chem 77:268–277
6. Nasirzadeh K, Neueder R, Kunz W (2005) Vapor pressures and osmotic coefficients of aqueous LiOH solutions at temperatures ranging from 298.15 to 363.15 K. Ind Eng Chem Res 44:3807–3814
7. Papaiconomou N, Simonin JP, Bernard O, Kunz W (2002) MSA-NRTL model for the description of the thermodynamic properties of electrolyte solutions. Phys Chem Chem Phys 4:4435–4443
8. Archer DG, Phys J (1991) Modification of the Pitzer Model to Calculate the Mean Activity-Coefficients of Electrolytes in a Water-Alcohol Mixed-Solvent Solution. Thermodynamic properties of the NaBr + H2O system. Chem Ref Data 20:509–555
9. Ye S, Xans P, Lagourette B (1994) Modification of the Pitzer Model to Calculate the Mean Activity-Coefficients of Electrolytes in a Water-Alcohol Mixed-Solvent Solution. J Solution Chem 23:1301–1315
10. Salimi HR, Taghikhani V, Ghotbi C (2005) Application of the GV-MSA model to the electrolyte solutions containing mixed salts and mixed solvents. Fluid Phase Equilib 231:67–76
11. Gross J, Sadowski G (2001) Perturbed chain SAFT: an equation of state based on a perturbation theory for chain molecules. Ind Eng Chem Res 40:1244–1260
12. Myers JA, Sandler SI, Wood RH (2002) Ind Eng Chem Res 41:3282–3297
13. Fürst W, Renon H (1993) Representation of excess properties of electrolyte solutions using a new equation of state. AIChE J 39:335–343
14. Chapman WG, Gubbins KE, Jackson G, Radosz M (1990) New reference equation of state for associating liquids. Ind Eng Chem Res 29:1709–1721
15. Gil-Villegas A, Galindo A, Whitehead PJ, Mills SJ, Jackson G, Burgess AN (1997) Statistical associating fluid theory for chain molecules with attractive potentials of variable range. J Chem Phys 106:4168–4186
16. Adidharma H, Radosz M (1998) Prototype of an engineering equation of state for heterosegmented polymers. Ind Eng Chem Res 37:4453–4462
17. Galindo A, Gil-Villegas A, Jackson G, Burgess AN (1999) SAFT-VRE: Phase behavior of electrolyte solutions with the statistical associating fluid theory for potentials of variable range. J Phys Chem B 103:10272–10281
18. Tan SP, Ji XY, Adidharma H, Radosz M (2006) Statistical associating fluid theory coupled with restrictive primitive model extended to bivalent ions. SAFT2: 1. Single salt plus water solutions. J Phys Chem B 110:16694–16699
19. Ji XY, Adidharma H (2006) Ion-based SAFT2 to represent aqueous single- and multiple-salt solutions at 298.15 K. Ind Eng Chem Res 45:7719–7728
20. Cameretti LF, Sadowski G, Mollerup JM (2005) Modeling of aqueous electrolyte solutions with perturbed-chain statistical associated fluid theory. Ind Eng Chem Res 44:3355–3362; ibid., 8944
21. Held C, Cameretti LF, Sadowski G (2008) Modeling aqueous electrolyte solutions. Part1: Fully dissociated electrolytes. Fluid Phase Equlilib 270:87–96
22. Held C, Prinz A, Wallmeyer V, Sadowski G (2012) Measuring and Modeling Alcohol/Salt Systems. Chem Eng Science submitted 68:328–339
23. Held C, Sadowski G (2009) Modeling aqueous electrolyte solutions. Part 2. Weak electrolytes. Fluid Phase Equilib 279:141–148

Electropermeabilization of the Cell Membrane

Justin Teissie and Muriel Golzio
CNRS; IPBS (Institut de Pharmacologie et de Biologie Structurale), Toulouse, France

Introduction

The permeability of a cell membrane can be transiently increased when a micro-millisecond external electric field pulse is applied on a cell suspension [1–4]. Under suitable conditions depending mainly on the pulse parameters (field strength, pulse duration, number of pulses), the viability of the cell can be preserved. The resulting electropermeabilization is a powerful electrochemical tool to gain access to the cytoplasm and to introduce chosen foreign molecules or to extract metabolites [5–10].

If this approach is routinely used in cell and molecular biology, one should nevertheless know that very few is known about what is really occurring in the cell and its membranes at the molecular levels [11–14]. Electropermeabilization is now proposed as a very efficient way for drug, oligonucleotides, antibodies, and plasmids delivery in vivo for clinical biotechnological applications [15–17]. New developments for the food and environmental industries have been proposed [18]. A safe use of this approach requires a better knowledge of the molecular processes affecting the membrane organization. Most investigations during the last 30 years were mostly focused on pure lipid bilayered models [19, 20]. Clearly there are many limits in the conclusions that were obtained in order to describe what occurred in cells and tissues. Conductance and optical methods can easily and accurately assay electropermeabilization. The processes can be followed from microseconds up to days. Kinetic studies of electropermeabilization led to a description in a multistep process:

1. "Charging step" – a cell was considered as a spherical shell with a dielectric membrane and with external and internal (cytoplasmic) conducting buffers. As a spherical dielectric, a position-dependent transmembrane potential was induced when the cell is submitted to an external field. This was a fast process.
2. "Induction step" – the field induced membrane potential difference increase reached a critical value at the polar position facing the electrodes, and this gave local defects (may be due to kinks in the lipid chains). A mechanical stress was present with a magnitude that depends on the buffer composition. These defects could be associated with water wires. Due to the potential charging time, the structural transition of the membrane affected a defined cap size on the cell surface.
3. "Expansion step" – the density of the defects increased within the affected cap as long as the field was present at a strength larger than a critical value. Again an electromechanical stress remained present.
4. "Stabilization step" – as soon as the field intensity was lower than the critical value that is mentioned in step 2, a stabilization process was taking place within a few milliseconds, which brought the membrane to the permeabilized state for small molecules.
5. "Resealing step" – a slow resealing was then occurring on a scale of seconds and minutes. Molecular transport is present.
6. "Memory effect" – some changes in the membrane properties remained present on a longer time scale (hours), but the cell behavior was finally back to normal.

Membrane Potential Difference Modulation

The source of the membrane structural modification is a modulation of the membrane potential difference. This is due to the dielectric character of the membrane. From a soft matter point of view, a cell can be described as an insulating shell containing a conducting solution (the cytoplasm with a conductivity λi) and in suspension in a conducting buffer (the external solution with a conductivity λo). A cell in a field behaves as a charging spherical capacitor. The induced potential difference $\Delta\Psi_E$ can be written as (when steady states are reached):

$$\Delta\Psi_E = 1.5 g(\lambda) r\, E \cos\theta \qquad (1)$$

where the vesicle shape is assumed to be a sphere; g a complex functions to the conductivities, λ, of the membrane and of the buffers; r is the radius of the sphere; E the field strength; and Θ the angle between the normal to the membrane and the direction of the field. Being dependent on an angular parameter, the field effect is position dependent on the surface. Therefore, one side of the vesicle is going to be hyperpolarized, while the other side is depolarized. These physical predictions were checked experimentally by videomicroscopy on lipid vesicles by using potential sensitive fluorescent probes [21, 22]. Very large transmembrane fields resulted from low external applied fields due to the reduced thickness of the biological membrane (average 5 nm).

This membrane potential difference alteration was reached after a very short charging time (in the microsecond time range):

$$\Delta\Psi_E(t) = 1.5 g(\lambda) r\, E \cos\theta (1 - \exp.(-t/\tau)) \qquad (2)$$

The charging time τ depended on the dielectric properties of the membrane, which was more complex than a lipid bilayer and the conductances (i.e., ionic content) of the cytoplasm and external buffer. Under the simplifying assumptions that the membrane was a pure dielectric, then

$$\tau = rC_{memb.}(\lambda_{int.} + 2\lambda_{out})/(2\lambda_{int.}\lambda_{out}) \quad (3)$$

It was calculated and checked to be of the order of microseconds or less for cells under physiological conditions. The consequence is that very large fields must be applied to get a significant transmembrane potential when using very short pulses (ns). In "classical" electropulsation, as the rise time of voltage pulses in most electropulsators was of the order of microseconds [23], it is the limiting step in the transmembrane modulation.

Electropulsation and its results on a cell are under the control of cellular parameters and experimental settings (field strength, pulse duration). It is a noninvasive way to alter in a selective manner the membrane potential difference on the microsecond time range.

The external field induces a position-dependent modulation of the membrane potential difference linearly related to the intensity of the applied field. Theoretical predictions from Eq. 1 are valid only under the hypothesis that:

1. The cell shape is spherical.
2. The membrane is a dielectric.

Hypothesis 2 must then be corrected to take into account the leaks that are present in a membrane.

$g(\lambda)$ is associated to the electrical conductivities of the cell membrane, of the internal and of the external buffer. Its expression is:

$$g(\lambda) = [\lambda o \lambda i(2d/r)]/[(2\lambda o + \lambda i)\lambda m + (2d/r)(\lambda o - \lambda m)(\lambda i - \lambda m)] \quad (4)$$

where d is the membrane thickness.

From Eq. 4, as λo and λi are always rather high in experiments (salts are always present when working with cells), it is clear that $g(\lambda)$ is under the control of λm. Another consequence of the membrane leakiness is that it affects the loading time of the membrane when the field is applied. Its physical definition is given by:

$$\tau = rCm(\lambda i + 2\lambda o)/(2\lambda i \lambda o + r\lambda m(\lambda i + 2\lambda o)/d) \quad (5)$$

As λm is dependent on the membrane leakiness, the loading time of the membrane will decrease with an increase in the membrane leakiness.

Another problem must be taken into account in the description of the induced potential modulation. The vesicle shape is not a sphere for cells. A spheroid is a more accurate description. The effect of the field is therefore dependent on the ratio of the relative axis of the spheroid and on the orientation of the field relative to the main axis. Recent simulations predicted complex cell responses that were fairly assayed experimentally [24].

Due to the physiological resting potential, $\Delta\Psi 0$, being about −40 to −60 mV, the electric field modulation of the potential difference across the cell membrane brings a resulting complex asymmetrical distribution (Fig. 1). A surface lateral gradient in potential difference (and associated transmembrane field) is present during the field pulse. Another cell deformation associated to electropermeabilization was a post-pulse effect: swelling due to the osmotic unbalance resulting from the inflow of water. The physical result was either a loss of surface ruffling or (and) an increase in surface tension [4, 25].

Membrane Electropermeabilization: The Facts

From experiments on planar bilayer membranes (BLM), it was known that lipid bilayers were not able to withstand an increase in the applied voltage above a threshold value. A conductive state followed by a rupture was observed for values of the order of 200 mV. Electropulsation induces a transmembrane potential modulation, bringing a similar membrane instability. Indeed experiments on pure lipid vesicles showed that upon the field pulse the lipid bilayer could become leaky. This was observed on line by the associated increase in conductance of a salt-filled vesicle suspension [26]. But larger molecules could leak out and be directly detected outside the vesicles as observed with radiolabelled sucrose [27] or fluorescent dyes [28]. A very fast detection of the induction of membrane leakage is obtained by electrical conductance and light scattering

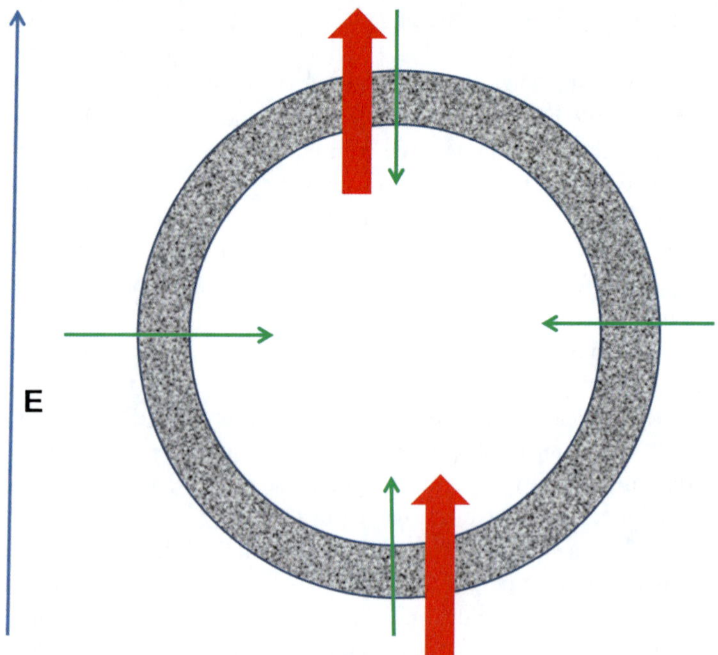

Electropermeabilization of the Cell Membrane, Fig. 1 Field induced transmembrane potential modulation. The resting transmembrane potential is negative inside (green arrow) all over the cell plasma membrane (pictured in gray). The field induced modulation keeps the orientation of the external trigger (red arrow). Therefore an additive effect on the resulting transmembrane potential is present on the side facing the cathode and a subtractive effect on the opposite cap. No modulation is present at the equator

experiments. The process is present in less than a microsecond after (during) the onset of the pulse [29, 30].

But molecular transport of charged molecules such as propidium iodide is detected only several tens of microseconds after the voltage onset [31]. This transport of charged molecules during the pulse is clearly mostly driven by the electric field but delayed by the membrane organization (a transmembrane electrophoresis). But the amazing feature is that in cells the transport remains present when the field is switched off. Most of the transport for polar molecules up to several kDa occurs in this post-pulse "resealing" phase where the field is not present anymore.

Permeabilization is controlled by the field strength (i.e., the induced potential). Permeabilization occurred only on the part of the cell surface where the membrane potential difference has been brought at its critical rupturing value. A cap in the part of the cell facing the electrodes is affected. The size of the cap is under the control of the applied field. Then a strong modulation is brought by the pulse duration. Indeed ns pulses are able to trigger a leaky state when a very high voltage (field) is delivered. This was indicative that the transition of the membrane to the leaky state was very fast (a few ns) but a contribution of the electromechanical stress might be present [32–34].

Molecular Transport

The field strength controls the extent of permeabilized membrane, whereas the efficacy of the permeabilization is under the control of the number and the duration of electric pulses [35–37]. This electro-induced permeabilization of cell membrane can be quantified in terms of the flow Fs of molecule S diffusing through the plasma membrane during the post-pulse resealing. In the case of inflow, the molecules can then diffuse freely in the cytoplasm. Fick's law and experimental data allowed to establish that:

$$\mathrm{Fs}(t) = \mathrm{Psx}(\mathrm{N},\mathrm{T})\mathrm{A}/2(1-\mathrm{Ep}/\mathrm{E})\Delta\mathrm{S} \times \exp(-\mathrm{k}(\mathrm{N},\mathrm{T})) \tag{6}$$

where Ps is the permeation coefficient of the molecule S across the membrane, x depends on the number (N) and duration (T) of electric pulses. It represents the probability of permeabilization ($0 < x < 1$) (the density of induced defects),

A is the cell surface, E is the applied electric field intensity, Ep the field threshold for membrane permeabilization ((1−Ep/E) reflects the cap where permeabilization is present), and ΔS is the concentration difference of S between the cell cytoplasm and external medium.

The final term reflects the membrane slow resealing where k is the time constant of the membrane resealing process and t is the time after the electric pulse. The total accumulation is under the control of the pulse duration that acts on the density of defects (x) and the lifetime of resealing (k). The transport is dependent on the nature of the target molecule (through Ps). Therefore, as Ps is dependent on its size, a larger transport is obtained for small molecules. This size effect is more complex with macromolecules (oligonucleotides, proteins). In the case of siRNA (small interfering RNA, MW about 20 kDa), no post-pulse transport is detected [38]. Transport occurs only during the pulse supported by the induced electrophoretic drift (accumulation in the cytoplasm is therefore present only on one side of the pulsed cells) telling that a dramatic membrane structural alteration must be present to allow the transmembrane transport of a 2 nm wide cylinder. The transport of plasmid is even more complex. The field-associated electric drift is just bringing an interfacial local accumulation of plasmids [39]. These aggregates remain stuck for minutes at the membrane level before being transported to the cytoplasm.

The conclusion is that electropermeabilization-associated transmembrane transports are complex and depend on the nature of the transferred molecules [40] (Fig. 2).

Membrane Electropermeabilization: Structural Aspects

It is the post-pulse state that is relevant of most effects dealing with delivery. Its structural organization remains unclear. Electropermeabilization alters cell plasma membrane structure. Very few direct experimental investigations have been performed on the structural or dynamical organization of the electropermeabilized membrane. Digitized video-microscopy proved experimentally that the induced potential difference is indeed position dependent and is controlled by the membrane properties [41]. The increase in membrane conductance associated to electropermeabilization is affecting only a limited part of the cell membrane. Most video observations at the single cell level under the microscope show that permeabilization is homogeneous on the part of the cell membrane (cap) which is altered [37, 42].

Electron microscopy studies showed that very short-lived cracks were present on the red blood cell surface, but they disappeared before resealing started. Villi and blebs were observed on the electropermeabilized cell surface in a post-pulse process [43]. Large pores were observed only on erythrocytes pulsed under a hypoosmotic stress and again in a post-pulse process [44]. They were relevant to the osmotic swelling, a back effect of electropermeabilization [4].

Getting structural information on the molecular supports of electropermeabilization was not easy. A key property of biological membranes is their dynamics. At a substructural level, 31P NMR spectroscopy showed that a tilt of the orientation of the phospholipid polar head region was present in the electropermeabilized state of the membrane [45, 46]. The consequence of the interfacial water organization was proposed to be associated with a decrease of the hydration forces and the observed fusogenic state of electropermeabilized surfaces. At a more collective level, phospholipid flip-flop between the two faces of the plasma membrane was observed in the case of electropermeabilized erythrocytes [47].

Membrane Electropermeabilization: A Thermodynamic Description

An electropermeabilized membrane was tentatively described based on the stability of a planar lipid bilayer. It was first suggested [20, 48] that the compression of the entire

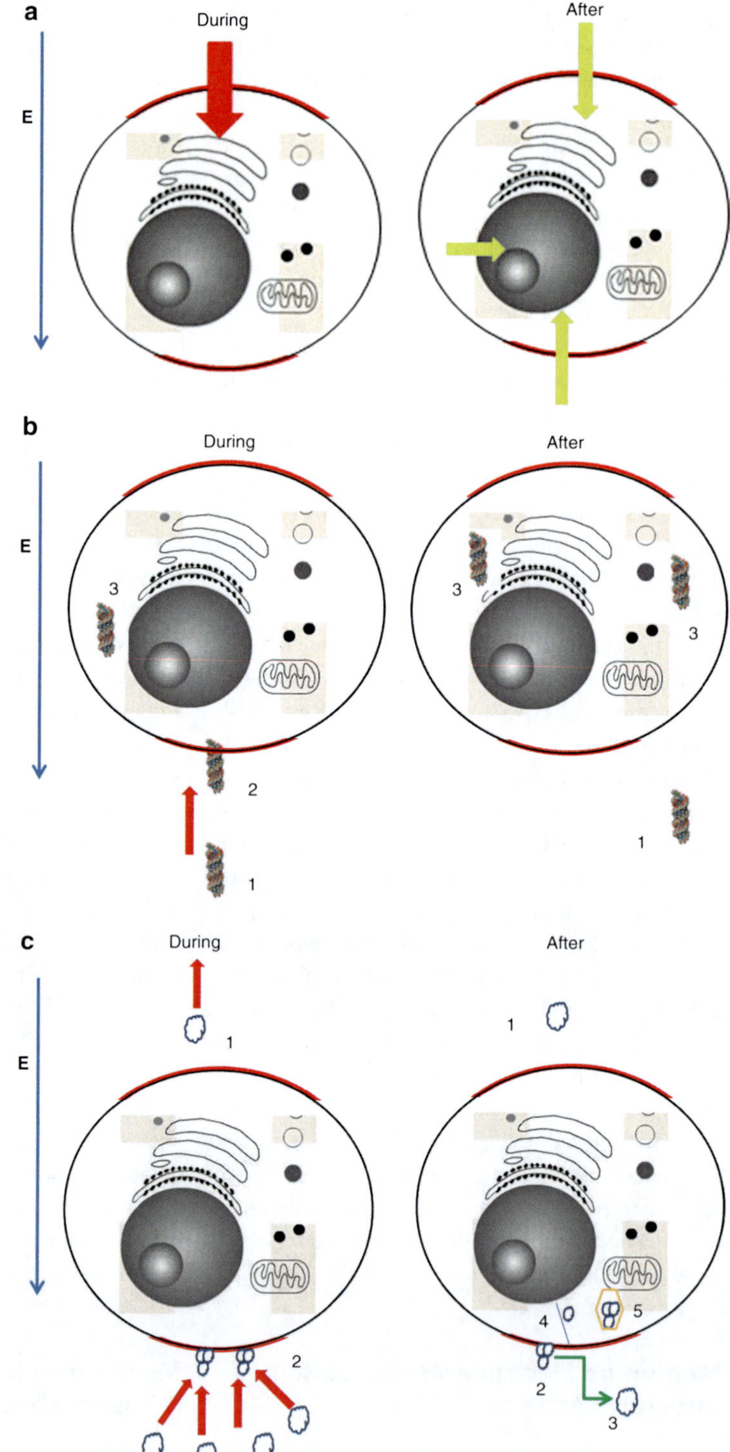

Electropermeabilization of the Cell Membrane, Fig. 2 Pathways for electrotransfer (**a**) for small molecules such as Propidium Iodide, an electrophoretic drift (red arrow) supports the transport during the pulse across the permeabilized cap facing the cathode (in red). After the pulse, transport is supported by diffusion across the two permeabilized caps and a free diffusion occurs in the cytoplasm and the nuclei (yellow arrows). (**b**) For siRNA, negatively charged, the electrophoretic forces (red arrow) push the oligonucleotides (1) across the permeabilized cap of the plasma membrane in the opposite orientation of the field (2) into the cytoplasm (3). After the pulse, no further loading is present but the internalized siRNA (3) can freely diffuse in the cytoplasm with no penetration in the nuclei. (*c*) For pDNA, the electrophoretic forces (red arrow) push the plasmids (1) in the opposite orientation of the field. A limited number of pDNA aggregates are formed in contact with the permeabilized cap of the plasma membrane (2). After the pulse, the pDNA can be released from those aggregates (3), can cross the membrane by a slow process and/or can be carried by molecular motors along the microtubules to the nuclear surface (4) or can be trapped in actin vesicles (5).

membrane by the electric field might cause mechanical collapse leading to membrane rupture in a destructive way. This was called the dielectric breakdown but could not explain why the permeabilization was reversible [48]. One model considered the membrane as a viscoelastic fluid which might rupture due to the electric stress [49]. Another description took into account that lipids were not in fixed positions but were able to move giving rise to the occurrence of so-called hydrophobic pores [50]. When a voltage was applied, it increased the energy of the membrane where an increase in size of these pores took place up to a transition to "hydrophilic pores" where the free diffusion took place [51–54]. Reversibility should occur after the pulse due to the thermodynamic instability of such pores. But such pores had never been experimentally observed, and they could not explain the stabilizing properties of the cytoskeleton as they were proposed to be purely lipidic. Several authors pointed out the fact that a membrane was a fluctuating assembly where mismatches were present. Theoretical descriptions of the phase transition processes showed that such defects were prone to support the exchange of species across the membrane. It was then proposed that electropermeabilization was associated to the induction of local reorganizations which interfaces with native structural membrane organization would support the transmembrane flow [55]. Thermal fluctuations played a role in defects formation [56]. If the effect of a transmembrane electric field was included under the assumption that lipids in the original membrane were replaced by water in the defect, i.e., changing the membrane capacitance, the energy term became dependent on the square of the defect radii and of the transmembrane voltage [57]. The model confirmed that, under appropriate conditions, pore growth was controlled and that, when the external field was removed, the membrane returned to a stable equilibrium.

Another model for electroporation based on the Smoluchowski equation predicted a rapid increase in pore density within the first microseconds of the pulse, a saturation of transmembrane voltage owing to increased conduction through the pores and a recovery time of 20 s (much longer than the experimental observations) [58]. In the theories outlined above, the field was simply a source of energy. A quite different approach considered the total stress F generated in the membrane by the field [59]. The application of the external electric field reduced the steady state energy of the membrane and increased the area per molecule. This model offered a quite different description of the role of the transmembrane voltage, as it linked the transverse field to a reduction of the lateral tension in the membrane and a consequent decrease in the interaction between the phospholipid molecules at the membrane/water interface. Defect formation resulted from a competition between the hydrophobic and hydrophilic forces at the interface.

Electropermeabilization In Silico

Atomic-scale molecular dynamics simulations were used to address ion transport through transient water pores in phospholipid membranes (but only during a few ns). The formation of a water defect was observed induced by the transmembrane ionic charge imbalance and a resulting transmembrane potential difference. The resulting transport of ions through the pore discharges the imbalance and makes the water file metastable, leading eventually to its sealing [60].

Molecular dynamics simulations of lipid bilayers under a high transverse electrical field were reported. The applied transmembrane electric fields (resulting in potential differences of 2.5–5.0 V) induced also the formation of water channels across the membrane. Switching of the external transmembrane potential for few nanoseconds brought a complete reconstitution of the bilayer [61]. The likelihood of water defects formation appeared to be increased by local membrane defects involving lipid head groups [62]. Field-directed rotation of the head group dipoles in the plane of the incipient water channel, in combination with water dipole and solvation interactions at the aqueous–lipid interface, was part of the coordinated ensemble of electropermeabilization events [63, 64]. Coarse-grained (CG) simulations using a polarizable coarse-grained water model were compared to atomistically detailed models but on a longer time. The lipid bilayer reorganization could be observed with smaller transmembrane potentials (1.5–3.5 V). Again water-filled pathways are formed. The defect formation time, however, increases with decreasing field strength and may about a microsecond in the case of a system at 1.5 V. Defects disappear as soon as the potential drop is less than 0.2 V. No water defect formation is observed on a microsecond time scale at 1 V [65].

Conclusion

Electropermeabilization of cell membranes appears not as punching holes in a lipid layer (the so-called electroporation hypothesis). Theoretical modelling and experimental data on black lipid membranes suggest the creation of aqueous pathways, a bioelectrochemical process. Other descriptions suggest a more complex process where mostly the membrane–solution interface is altered. The recovery (resealing) is dependent on the cell metabolism. This biological aspect must be taken into account in its use in in vivo drug delivery.

Future Directions

There is still a need of basic research for the investigations of the structural membrane alterations supporting the transport of charged molecules across the biological membranes. The development of this methodology for clinical applications is very promising (electrochemotherapy, gene therapy, DNA vaccines, hybridoma production), but one cannot neglect the use for biotechnology (food processing) and environment (pathogen eradication). One open question is the difference that may be present between classical electropermeabilization (as described in this chapter) and the new nanosecond electropulsation where very high fields are applied on a very short (ns or even ps) duration.

References

1. Smith J, Jones M Jr, Houghton L et al (1999) Future of health insurance. N Engl J Med 965:325–329
2. Zimmermann U, Pilwat G, Riemann F (1974) Dielectric breakdown of cell membranes. Biophys J 14:881–899
3. Neumann E, Rosenheck K (1972) Permeability changes induced by electric impulses in vesicular membranes. J Membr Biol 10:279–290
4. Kinosita K Jr, Tsong TY (1977) Formation and resealing of pores of controlled sizes in human erythrocyte membrane. Nature 268:438–441
5. South J, Blass B (2001) The future of modern genomics. Blackwell, London
6. Neumann E, Sowers AE, Jordan CA (1989) Electroporation and electrofusion in cell biology. Plenum, New York
7. Allen MJ, Cleary SF, Sowers AE, Shillady DD (1992) Charge and field effects in biosystems – 3. Birkhaüser, Boston
8. Chang DC, Chassy BM, Saunders JA, Sowers AE (1992) Guide to electroporation and electrofusion. Academic, San Diego
9. Pakhomov AG, Miklavcic D, Markov MS (2010) Advanced electroporation techniques in biology and medicine. CRC Press, Boca Raton
10. Zimmermann U (1982) Electric field mediated fusion and related electrical phenomena. Biochim Biophys Acta 694:227–277
11. Belehradek M, Domenge C, Luboinski B, Orlowski S, Belehradek J, Mir LM (1993) Electrochemotherapy, a new antitumor treatment; First clinical phase I-II trial. Cancer 72:3694–3700
12. Sixou S, Teissié J (1990) Specific electropermeabilization of leucocytes in a blood sample and application to large volumes of cells. Biochim Biophys Acta 1028:154–160
13. Wolf H, Rols MP, Neumann E, Teissié J (1994) Control by pulse parameters of electric field mediated gene transfer in mammalian cells. Biophys J 66:524–531
14. Zeira M, Tozi PF, Moumeine Y, Lazarte J, Sneed L, Volsky DJ, Nicolau C (1991) Full length CD4 electroinserted in the red blood cell membrane as a long-lived inhibitor of HIV infection. Proc Natl Acad Sci U S A 88:4409–4413
15. Gehl J (2003) Electroporation: theory and methods, perspectives for drug delivery, gene therapy and research. Acta Physiol Scand 177:437–447
16. Gothelf A, Mir LM, Gehl J (2003) Electrochemotherapy: results of cancer treatment using enhanced delivery of bleomycin by electroporation. Cancer Treat Rev 29:371–387
17. Orlowski S, Mir LM (1993) Cell electropermeabilization: a new tool for biochemical and pharmacological studies. Biochim Biophys Acta 1154:51–63
18. Teissie J, Eynard N, Vernhes MC, Benichou A, Ganeva V, Galutzov B, Cabanes PA (2002) Recent biotechnological developments of electropulsation. A prospective review. Bioelectrochemistry 55:107–112

19. Neumann E, Kakorin S, Toensing K (1999) Fundamentals of electroporative delivery of drugs and genes. Bioelectrochem Bioenerg 48:3–16
20. Weaver J, Chizmadzhev Y (1996) Theory of electroporation: a review. Bioelectrochem Bioenerg 41: 135–160
21. Lojewska Z, Farkas D, Ehrenberg B, Loew LM (1989) Analysis of the effect and membrane conductance on the amplitude and kinetics of membrane potentials induced by externally applied electric fields. Biophys J 56:121–128
22. Gross D, Loew LM, Webb WW (1986) Optical imaging of cell membrane potential changes induced by applied electric fields. Biophys J 51:339–348
23. Puc M, Corovic S, Flisar K, Petkovsek M, Nastran J, Miklavcic D (2004) Techniques of signal generation required for electropermeabilization. Survey of electropermeabilization devices. Bioelectrochemistry 64:113–124
24. Valic B, Golzio M, Pavlin M, Schatz A, Faurie C, Gabriel B, Teissié J, Rols MP, Miklavcic D (2003) Effect of electric field induced transmembrane potential on spheroidal cells : theory and experiments. Eur Biophys J 32:519–528
25. Golzio M, Mora MP, Raynaud C, Delteil C, Teissie J, Rols MP (1998) Control by osmotic pressure of voltage-induced permeabilization and gene transfer in mammalian cells. Biophys J 74:3015–3022
26. Kakorin S, Redeker E, Neumann E (1998) Electroporative deformation of salt filled lipid vesicles. Eur Biophys J 27:43–53
27. Teissié J, Tsong TY (1981) Electric field induced transient pores in phospholipid bilayer vesicles. Biochemistry 20:1548–1554
28. Raffy S, Teissié J (1995) Insertion of Glycophorin A, a transmembraneous protein, in lipid bilayers can be mediated by electropermeabilization. Eur J Biochem 230:722–732
29. Neumann E, Kakorin S (1996) Electroptics of membrane electroporation and vesicle shape deformation. Curr Opin Colloid Interface 1:790–799
30. Kakorin S, Neumann E (1998) Kinetics of the electroporative deformation of lipid vesicles and biological cells in an electric field. Ber Bunsenges Phys Chem 102:670–675
31. Pucihar G, Kotnik T, Miklavcic D, Teissié J (2008) Kinetics of transmembrane transport of small molecules into electropermeabilized cells. Biophys J 95:2837–2848
32. Frey W, White JA, Price RO, Blackmore PF, Joshi RP, Nuccitelli R, Beebe SJ, Schoenbach KH, Kolb JF (2006) Plasma membrane voltage changes during nanosecond pulsed electric field exposure. Biophys J 90:3608–3615
33. Sukhoruko VL, Mussauer H, Zimmermann U (1998) The effect of electrical deformation forces on the electropermeabilization of erythrocyte membranes in low- and high-conductivity media. J Membr Biol 163:235–245
34. Muller KJ, Sukhorukov VL, Zimmermann U (2001) Reversible electropermeabilization of mammalian cells by high-intensity, ultra-short pulses of submicrosecond duration. J Membr Biol 184:161–170
35. Gabriel B, Teissié J (1997) Direct observation in the millisecond time range of fluorescent molecule asymmetrical interaction with the electropermeabilized cell membrane. Biophys J 73:2630–2637
36. Rols MP, Teissie J (1990) Electropermeabilization of mammalian cells. Quantitative analysis of the phenomenon. Biophys J 58:1089–1098
37. Gabriel B, Teissie J (1999) Time courses of mammalian cell electropermeabilization observed by millisecond imaging of membrane property changes during the pulse. Biophys J 76:2158–2165
38. Paganin-Gioanni A, Bellard E, Escoffre JM, Rols MP, Teissié J, Golzio M (2011) Direct visualization at the single-cell level of siRNA electrotransfer into cancer cells. Proc Natl Acad Sci U S A 108:10443–10447
39. Golzio M, Teissie J, Rols MP (2002) Direct visualization at the single-cell level of electrically mediated gene delivery. Proc Natl Acad Sci U S A 99:1292–1297
40. Teissie J, Golzio M, Rols MP (2005) Mechanisms of cell membrane electropermeabilization: a minireview of our present (lack of) knowledge? Biochim Biophys Acta 1724:270–280
41. Hibino M, Shigemori M, Itoh H, Nagayama K, Kinosita K (1991) Membrane conductance of an electroporated cell analyzed by submicrosecond imaging of transmembrane potential. Biophys J 59:209–220
42. Hibino M, Itoh H, Kinosita K (1993) Time courses of cell electroporation as revealed by submicrosecond imaging of transmembrane potential. Biophys J 64:1789–1800
43. Escande-Geraud ML, Rols MP, Dupont MA, Gas N, Teissié J (1988) Reversible plasma membrane ultrastructural changes correlated with electropermeabilization in CHO cells. Biochim Biophys Acta 939:247–259
44. Chang DC, Reese TS (1990) Changes in membrane structure induced by electroporation as revealed by quick freezing electron microscopy. Biophys J 58:1–12
45. Lopez A, Rols MP, Teissié J (1988) 31P NMR analysis of membrane phospholipid organization in viable, reversibly electropermeabilized Chinese hamster ovary cells. Biochemistry 27:1222–1228
46. Stulen G (1981) Electric field effects on lipid membrane structure. Biochim Biophys Acta 640: 621–627
47. Dressler V, Schwister K, Haest CWM, Deuticke B (1983) Dielectric breakdown of the erythrocyte membrane enhances transbilayer mobility of phospholipids. Biochim Biophys Acta 732:304–307
48. Crowley JM (1973) Electrical breakdown of bimolecular lipid membranes as an electromechanical instability. Biophys J 13:711–724
49. Dimitrov S, Jain RK (1984) Membrane stability. Biochim Biophys Acta 779:437–468

50. Sugar IP, Neumann E (1984) Stochastic model for electric field-induced membrane pores—electroporation. Biophys Chem 19:211–225
51. Abidor IG, Arakelyan VB, Chernomordik LV, Chizmadzhev Y, Pastushenko VF, Tarasevich MR (1979) Electric breakdown of bilayer lipid membranes. I: the main experimental facts and their qualitative discussion. Bioelectrochem Bioenerg 6:37–52
52. Chernomordik LV, Sukharev SI, Abidor IG, Chizmadzhev Y (1983) Breakdown of lipid bilayer membranes in an electric field. Biochim Biophys Acta 736:203–213
53. Chernomordik LV, Sukharev SI, Popov SV, Pastushenko VF, Sokirko AV, Abidor IG, Chizmadzhev Y (1987) The electric breakdown of cell and lipid membranes; the similarity of phenomenologies. Biochim Biophys Acta 902:360–373
54. Weaver JC, Powell KT, Mintzer RA, Ling H, Sloan SR (1984) The electrical capacitance of bilayer membranes: the contribution of transient aqueous pores. Bioelectrochem Bioenerg 12:393–412
55. Cruzeiro-Hanson L, Mouritsen OG (1988) Passive ion permeability of lipid membrane modelled via lipid domain interfacial area. Biochim Biophys Acta 944:63–72
56. Taupin C, Dvolaitzky SC (1975) Osmotic pressure induced pores in phospholipid vesicles. Biochemistry 14:4771–4775
57. Joshi RP, Hu Q, Schoenbach KH, Hjalmarson HP (2002) Improved energy model for membrane electroporation in biological cells subjected to electrical pulses. Phys Rev E65:041920-1–041920-8
58. Neu JC, Krassowski W (1999) Asymptotic model of electroporation. Phys Rev E 59:3471–3482
59. Lewis TJ (2003) A model for bilayer membrane electroporation based on resultant electromechanical stress. IEEE Trans Dielectr Electr Insul 10:754–768
60. Gurtovenko AA, Vattulainen I (2005) Pore formation coupled to ion transport through lipid membranes as induced by transmembrane ionic charge imbalance: atomistic molecular dynamics study. J Am Chem Soc 127:17570–17571
61. Tarek M (2005) Membrane electroporation: a molecular dynamics simulation. Biophys J 88:4045–4053
62. Tieleman DP (2004) The molecular basis of electroporation. BMC Biochem 5:10
63. Vernier PT, Ziegler MJ (2007) Nanosecond field alignment of head group and water dipoles in electroporating phospholipid bilayers. J Phys Chem B 111:12993–12996
64. Levine ZA, Vernier PT (2010) Life cycle of an electropore: field-dependent and field-independent steps in pore creation and annihilation. J Membr Biol 236:27–36
65. Yesylevskyy SO, Schäfer LV, Sengupta D, Marrink SJ (2010) Polarizable water model for the coarse-grained MARTINI force field. PLoS Comput Biol 6: e1000810

Electrophoresis

Reiner Westermeier
SERVA Electrophoresis GmbH, Heidelberg, Germany

Introduction

The term electrophoresis is defined by the migration of charged particles under the influence of an electric field. This physical phenomenon can be applied as electrophoretic deposition technology in industry on coating or staining of metal components or as a method for separating chemical compounds, biomolecules, subcellular particles or even intact cells. In the following article, which is positioned under bioelectrochemistry, solely electrophoretic separation methods are described.

Principle

Electrophoretic separation methods are based on the differences of migration velocities of dispersed, differently charged compounds in the electric field with direct current (DC). These compounds are mostly biological macromolecules like nucleic acids, proteins, and glycans; the separations are mainly performed for analytical purposes. Small and highly charged particles migrate faster than large and low charged sample components. There is a very simple equation to describe the velocity of an individual compound:

$$v = m \cdot E$$

v = migration velocity [cm/s]
m = electrophoretic mobility [cm^2/V·s]
E = electric field strength [V/cm]

The driving force in electrophoresis is the electric field strength, which is defined in Volt/cm. As shown in Fig. 1, the electrophoretic mobility of a molecule is dependent on many factors: its intrinsic properties and composition of the medium.

In order to stay within the scope of this entry, the physicochemical details and equations are reduced to a minimum; more details can be found in textbooks [1–3].

The charged compound homologues migrate in distinct zones, thus forming a band pattern. The separation can take place either in free solution or in a stabilizing medium like a membrane or a gel. Depending on the medium there are several ways to detect the separated zones. Some compounds can be directly detected with UV light. Substrates like nucleic acids, glycans, proteins, and peptides can be pre-labeled with a fluorescent tag for direct detection or scanning of the stabilizing matrix. Material which has been separated in gels or membranes can be stained subsequently to electrophoresis with fluorescent or visible dyes. It is also possible to visualize compound zones specifically with functional enzyme-substrate reactions or by probing with a ligand, like an antibody or a glycoprotein.

Electrophoresis, Fig. 1 The dependence of the electrophoretic mobility of an ionized compound (here a protein molecule) on its own properties and the properties of the separation medium

Generally, the resolution achieved with electrophoresis techniques is markedly higher than with other separation methods like chromatography.

There are basically four different electrophoretic separation methods:

1. *Zone electrophoresis* is carried out in a homogeneous buffer to ensure a constant pH value. Under the influence of the electric field the ionized sample compounds migrate at different velocities towards the anodal or the cathodal electrode, depending on their charge sign (see Fig. 2). The pH of the buffer system determines the charges on the molecules. The migration distances of the zones are relative to

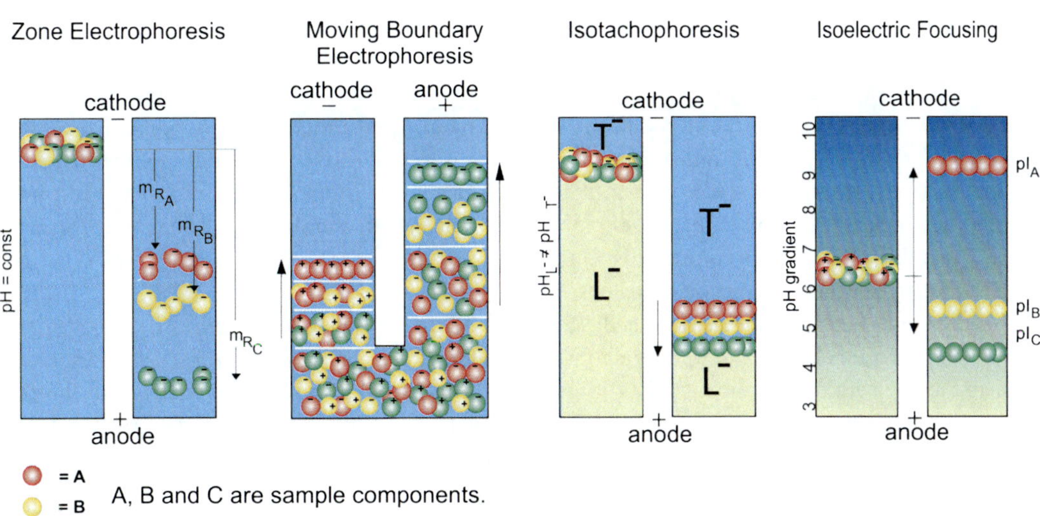

Electrophoresis, Fig. 2 The four different electrophoretic separation methods. For explanations see text

the electrophoretic mobilities of the different compounds. The separations are mostly carried out in a stabilizing medium in order to reduce diffusion effects. In order to maintain a stable pH value and buffering power, electrophoresis systems are built with buffer reservoirs. It can be applied on acidic (e.g. nucleic acids) or basic compounds, as well as amphoteric molecules like peptides and proteins.

2. *Moving boundary electrophoresis*: The sample, e.g. a mixture of proteins, is applied into the horizontal lower part of a U-shaped glass tube. The two vertical parts are filled with buffer solution; the electrodes are positioned at their upper ends. Under the influence of the applied voltage the ionized compounds migrate at different velocities towards the anode in one limb and the cathode in the second, depending on their charges (see Fig. 2). The upper zone in either limb contains solely one compound, the fastest migrating cations or anions, respectively. The second zones in either limb contain mixtures of the cations or anions of the highest mobilities and those of the second highest mobilities. The next zones are composed of these ions plus ions with the next lower mobilities and so on. The refractive indices at the boundaries are usually detected at both ends using Schlieren optics [4]. Moving boundary electrophoresis is very rarely employed in basic research to determine exact electrophoretic mobilities.
3. *Isotachophoresis* takes place in a discontinuous buffer system [5]. The ionized sample is applied between a leading ion with a higher mobility and a trailing ion with a lower mobility than all ionized sample compounds. Under the influence of the electric field all ions are forced to migrate with the same speed: If an ion with a high mobility would migrate faster than the ion with the next lower mobility and would leave it behind, an ion gap would form between them; this would lead to an interruption of the electric circle. Consequently, because $v = m \cdot E$ (see above), the electric field strength in the area of the trailing ions with very low mobility will be high, and in the area of the leading ions with very high mobility it will be low. The sample ions are migrating within a gradient of electric field strength, and while they migrate they will be separated into distinct zones of increasing mobilities, each zone following directly the next one (see Fig. 2). This is causing a concentration-regulating and a zone-sharpening effect which works against diffusion. The bands do not have the shape of Gaussian curves, but rectangular spikes. The isotachophoresis effect is mostly applied for stacking of the samples during the first phase of disc electrophoresis (from "discontinuous"); in capillary electrophoresis it is used as a separation method for small molecules.
4. Isoelectric focusing works only for the separation of amphoteric substances such as peptides and proteins, which have an "*isoelectric point*." The separation takes place in a pH gradient. Under the influence of the electric field the ionized molecules move towards the anode respectively the cathode until they arrive at the pH value of their isoelectric point. At this position in the pH gradient the net charge of the molecule is zero; therefore it has no electrophoretic mobility anymore (see Fig. 2). Also this method has a band-sharpening – a "focusing" – effect: When a molecule moves away from this position, it will become ionized again; thus it will migrate back to its isoelectric point under the influence of the electric field. Isoelectric focusing is mostly performed in polyacrylamide and agarose gels; however – due to the focusing effect – it is also a preferred method in free-solution technologies. Isoelectric focusing systems do not need buffer reservoirs.

There are two ways to generate a pH gradient: *Carrier ampholyte pH gradients* are formed by a heterogeneous mixture of amphoteric buffers under the influence of the electric field [6]. Immobilized pH gradients are generated by copolymerization of acrylamide derivatives containing buffering groups and acrylamide monomers during the preparation of a polyacrylamide gel matrix [7].

The current flow in an electrophoresis system causes the development of *Joule heat*. And, the pK values of the buffering groups of the buffers components as well as of the sample compounds are temperature dependent. Because of these two reasons electrophoretic techniques need in many cases cooling and exact temperature control. The heat development is the major reason why electrophoresis separations cannot be easily upscaled for preparative purposes; therefore they are mostly applied for analytical applications.

Another, mostly disturbing, phenomenon is electroendosmosis. Almost every material used for the static support: the surface of the separation equipment such as glass plates, tubes or capillaries and the separation medium carry charged groups: silicium oxide on glass surfaces, carboxylic groups in polyacrylamide gels, sulfonic groups in agarose gels. In presence of basic and neutral buffers these groups become deprotonated. Thus they are negatively charged. Under the influence of the electric field these charges are attracted by the anode. However, they cannot migrate, because they are a part of the solid matrix. To compensate this effect there is a transport of H_3O^+ ions towards the cathode, which results in a liquid transport into this direction, which carries solubilized substances along. This electroosmotic effect results in blurred zones and drying of the gel in the anodal area of flatbed gels. In acidic buffers, when fixed groups are positively charged, the electroosmotic flow is directed towards the anode.

Technical Variations

Electrophoresis in Free Solution

Fre- flow electrophoresis is the only continuous electrophoretic separation method. The electrical field is applied perpendicular to a continuous stream of buffer film, which flows through a 0.5–1.0 mm wide cuvette. At one end the sample is injected at a defined location, and at the other end, the fractions are collected by an array of tubes. The different electrophoretic mobilities of the sample components lead to differently strong but constant deviations inside the stream

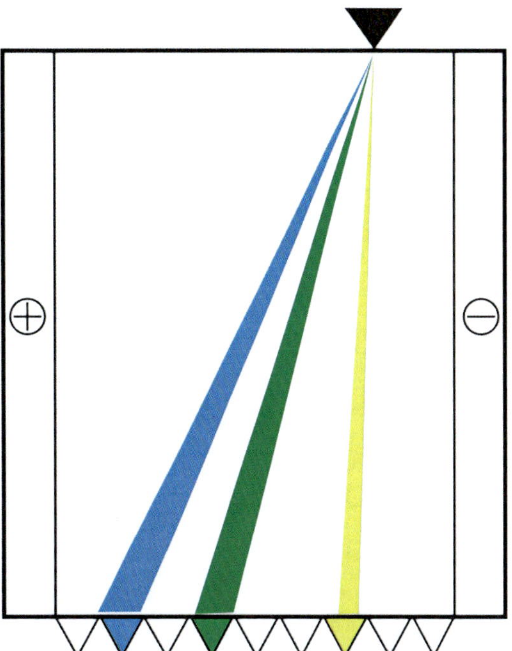

Electrophoresis, Fig. 3 Simplified diagram of free flow electrophoresis. For explanation see text

(see Fig. 3). At the end of the separation chamber they arrive at different but stable positions. The method is particularly useful for the separation of subcellular components, because there are no particle size restrictions. For dispersed molecules the best results in free-flow electrophoresis are obtained with isoelectric focusing, mainly because of the focusing effect.

Preparative isoelectric focusing systems in free solutions like recycling isoelectric focusing in rotating tubes [8] and multicompartment electrolysers employing isoelectric membranes [9] have been developed for prefractionation and purification of proteins from highly heterogeneous mixtures.

Capillary electrophoresisis usually carried out in fused silica capillaries with internal diameters of 50–100 μm [10]. Both ends are immersed in buffer reservoirs into which the electrodes are built (see Fig. 4). The amount of reagents and sample volume needed is very low, usually not more than 2–4 nL of injected material, which means nanograms of sample material. Very high field strengths of up to 1 kV/cm are applied,

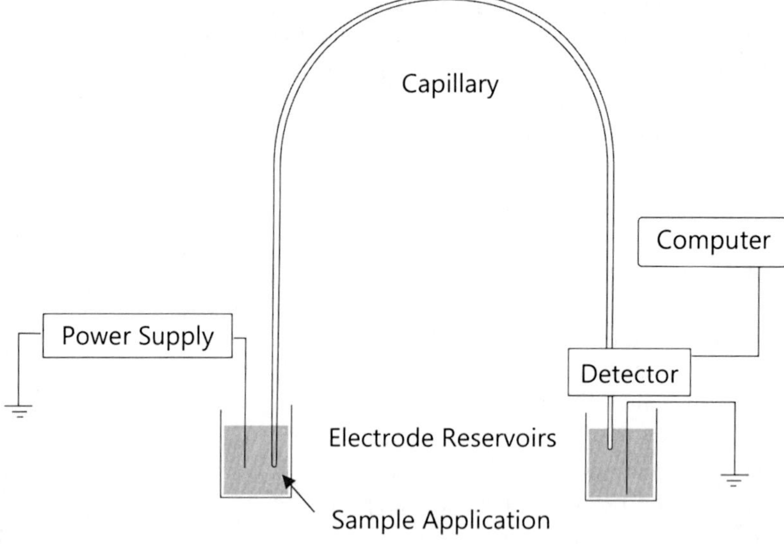

Electrophoresis, Fig. 4 Simplified diagram of capillary electrophoresis. For explanation see text

because Joule heat can be dissipated very efficiently from the capillaries with a fan. There is a choice of many different detection methods such as UV/VIS, fluorescence, conductivity, and electrochemistry. Limits of detection as low as attomoles are reported. The inside of the capillary is often coated with linear polyacrylamide or methyl cellulose in order to reduce electroendosmosis effects and to prevent adsorption of components on the surface. Capillary electrophoresis can easily be automated and equipped with an autosampler. Capillary electrophoresis instruments are used for zone electrophoresis, isotachophoresis and isoelectric focusing.

Electrophoresis in Cellulose Acetate Membranes

Cellulose acetate membranes have large pore sizes, thus the matrix exerts little effect on diffusion during the separation of proteins [11]. The separation is mainly based on the charge. This technique is used for routine clinical analysis for the analysis of human serum. The cellulose acetate membranes are easy to handle and can be quickly stained for the detection of the proteins.

Electrophoresis in Gels

Because of the sieving effect of gel pores highest resolution is achieved, when an electrophoretic separation is carried out in a gel matrix. The pore sizes can be varied via the content of solid material in the gel. In the beginning phase of electrophoresis techniques starch gels and granulated dextran gels have been used, which have nowadays been replaced by polyacrylamide and agarose gels.

Agarose Gels

Agarose is a natural product, a polysaccharide originating from red seaweed by removal of agaropectin. Gels of varying electroendosmosis and degrees of purity can be obtained. Agarose gels are prepared by dissolving the agarose powder in boiling water and subsequent chilling. The quality is determined by the melting point and the degree of electroendosmosis.

For the separation of nucleic acids 5–10 mm-thick gels are prepared, which are run in simple horizontal buffer trays under a thin layer of buffer solution. The samples are applied into little pockets, which have been preformed with a comb during casting of the gel. The separated zones are detected via fluorescent staining. However, large DNA molecules with molecular weights above 20 kb migrate with the same mobility in an agarose gel resulting in one non-resolved zone. But there is a unique variation of this *submarine* technique for separating complete chromosomes: pulsed-field gel electrophoresis [9, 12]. When an electric field changes the direction, the helical structure of DNA molecules is first stretched and then compressed again. The higher the molecular weight, the longer time is needed by the molecule for completion of this effect. Additionally, small molecules need less time to change the migration direction than large ones. Thus larger molecules have less time left for electrophoretic migration. In this way also very large DNA molecules up to the magnitude of 10 megabases are separated.

Agarose gels for zone electrophoresis or isoelectric focusing of protein samples are mostly applied in the area of clinical diagnostics. For this purpose the gels are cast on horizontal glass plates or plastic films as thin layers of about 1 mm.

Polyacrylamide Gels

Polyacrylamide gels are chemically inert and mechanically stable and show very low electroendosmosis. They are prepared by co-polymerization of acrylamide monomers with a cross-linker – mostly N,N'-methylenebisacrylamide [13]. The pore size is smaller than in agarose gels and is exactly determined by the total acrylamide concentration and the degree of cross-linking. For sample application wells are formed during polymerization using a comb.

In molecular biology these gels are used for DNA fragment analysis. However the main application is the separation of proteins and peptides; a number of methodical varieties exist:

Discontinuous electrophoresis has been developed in order to overcome the formation of protein aggregates and precipitates due to a sudden concentration of the protein mixture during sample entry [14, 15]. For this technique the gel is divided into two areas: a stacking and a resolving gel. The resolving gel has small pores sizes and a Tris-chloride buffer pH 8.8; the stacking gel is made with large pore sizes and has a pH value 6.8. In the stacking gel the sample is run under the conditions for *isotachophoresis*; see above. At the edge of the resolving gel, at pH 8.8 the glycine will suddenly have a much higher mobility, and the proteins will from now on become separated under the conditions of zone electrophoresis.

Porosity gradient gels are prepared by continuously changing the acrylamide concentration in the polymerization solution. Gradient gels have a zone-sharpening effect and can be used to determine the molecular diameter of proteins in their native state. The most important application of pore gradient gels is blue native electrophoresis of intact protein complexes [16].

Sodium dodecyl sulfate (SDS) electrophoresis affords a separation only according to the molecular sizes. The anionic detergent SDS forms anionic micelles with the proteins and causes unfolding and stretching of the polypeptide chains [17]. All these micelles are negatively charged; the mobilities are only dependent on the molecule sizes. The treatment with SDS causes denaturation of proteins. In practice SDS electrophoresis is mostly combined with disc electrophoresis [18].

Isoelectric focusing in polyacrylamide gels protein mixtures can be performed under native or denaturing conditions. For denaturing conditions the gels need to contain a high concentration of non-charged chaotropes, like urea and thiourea. Denaturing conditions are mainly chosen, when very complex protein mixtures and hydrophobic proteins have to be analyzed.

Two-dimensional electrophoresis reaches the highest resolution for protein mixtures, when denaturing isoelectric focusing is combined with SDS electrophoresis [19, 20].

Future Directions

Electrophoresis systems are further developed for easier handling and more automation.

References

1. Andrews AT (1986) Electrophoresis, theory techniques and biochemical and clinical applications. Clarendon, Oxford
2. Mosher RA, Saville DA, Thormann W (1992) The dynamics of electrophoresis. VCH, Weinheim
3. Westermeier R (2004) Electrophoresis in practice, 4th edn. WILEY-VCH, Weinheim

4. Tiselius A (1937) A new apparatus for electrophoretic analysis of colloidal mixtures. Trans Faraday Soc 33:524–531
5. Everaerts FM, Becker JM, Verheggen TPEM (1976) Isotachophoresis, theory, instrumentation and applications, vol 6, J chromatogr library. Elsevier, Amsterdam
6. Svensson H (1961) Isoelectric fractionation, analysis and characterization of ampholytes in natural pH gradients. The differential equation of solute concentrations as a steady state and its solution for simple cases. Acta Chem Scand 15:325–341
7. Bjellqvist B, Ek K, Righetti PG, Gianazza E, Görg A, Westermeier R, Postel W (1982) Isoelectric focusing in immobilized pH gradients: principle, methodology and some applications. J Biochem Biophys Methods 6:317–339
8. Bier M, Long T (1992) Recycling isoelectric focusing: use of simple buffers. J Chromat 604:73–83
9. Wenger P, de Zuanni M, Javet P, Righetti PG (1987) Amphoteric, isoelectric Immobiline membranes for preparative isoelectric focusing. J Biochem Biophys Methods 14:29–43
10. Jorgenson JW, Lukacs KD (1981) Zone electrophoresis in open-tubular glass capillaries. Anal Chem 53:1298–1302
11. Kohn J (1957) A cellulose acetate supporting medium for zone electrophoresis. Clin Chim Acta 2:297
12. Schwartz DC, Cantor CR (1984) Separation of yeast chromosome-sized DNA by pulsed field gradient gel electrophoresis. Cell 37:67–75
13. Raymond S, Weintraub L (1959) Acrylamide gel as a supporting medium for zone electrophoresis. Science 130:711
14. Ornstein L (1964) Disc electrophoresis. I. Background and theory. Ann N Y Acad Sci 121:321–349
15. Davis BJ (1964) Disc electrophoresis. 2, Method and application to human serum proteins. Ann N Y Acad Sci 121:404–427
16. Schägger H, von Jagow G (1991) Blue native electrophoresis for isolation of membrane protein complexes in enzymatically active form. Anal Biochem 199:223–231
17. Shapiro AL, Viñuela E, Maizel JV (1967) Molecular weight estimation of polypeptide chains by electrophoresis in SDS-polyacrylamide gels. Biochem Biophys Res Commun 28:815–822
18. Lämmli UK (1970) Cleavage of structural proteins during the assembly of the head of bacteriophage T4. Nature 227:680–685
19. O'Farrell PH (1975) High-resolution two-dimensional electrophoresis of proteins. J Biol Chem 250: 4007–4021
20. Görg A, Postel W, Günther S (1988) Review. The current state of two-dimensional electrophoresis with immobilized pH gradients. Electrophoresis 9:531–546

Electroreduction of Carbon Dioxide

Maria Jitaru[1] and Daniel Lowy[2]
[1]Research Institute for Organic Auxiliary Products (ICPAO), Medias, Romania
[2]FlexEl, LLC, College Park, MD, USA

Fixation of CO_2: Avenues Toward Artificial Photosynthesis

Chemical fixation of carbon dioxide is attractive both for more effectively exploiting carbon-based energy sources and reducing the CO_2 concentration in the atmosphere [1]. Photosynthesis is a common example of CO_2 fixation; this process builds organic compounds from carbon dioxide and water using solar energy with chlorophyll acting as a catalyst [2]. Over the past several decades, chemists have studied this naturally occurring carbon fixation process as a model for manufacturing synthetic fuels [3]. Efforts towards obtaining synthetic hydrocarbon fuels from CO_2 have been motivated by the unparalleled energy density of hydrocarbons, which constitute the backbone of our present-day energy infrastructure [4], and by the need to cope with increasing atmospheric CO_2 released by the burning of fossil fuels.

As anthropogenic carbon dioxide production exceeds the planet's carbon dioxide recycling capability, it causes significant environmental harm [2]. Therefore, the natural carbon cycle should be supplemented by an "artificial photosynthesis" process, which recycles carbon dioxide into hydrocarbons using renewable energy sources, reducing reliance on fossil fuels. Widely available water and CO_2 are used as the renewable starting materials, while the required energy for the synthetic carbon cycle can be supplied by any nonfossil-fuel energy source such as solar, wind, geothermal, or nuclear energy. The "artificial photosynthesis" process begins by capturing CO_2 from natural sources, including the ambient atmosphere, or industrial sources such as smokestacks. This CO_2 can then be converted by chemical or electrochemical reactions into fuels, such as methanol, dimethyl ether, synthetic hydrocarbons, or proteins for animal feed. This concept is the basis of the so-called methanol economy [2]. In a broader context, the capture of CO_2 from the atmosphere would enable a closed-loop carbon-neutral fuel cycle. Using renewable or nuclear energy, carbon dioxide and water can be recycled into liquid hydrocarbon fuels in nonbiological processes, which constitute the reverse of fuel combustion, as they remove oxygen from CO_2 and H_2O [5].

Several processes have been developed for CO_2 fixation; however, many of them are expensive, since they either require ultrahigh purity CO_2 or are energy intensive. Furthermore, many of the chemical methods show low product selectivity. Such limitations can be overcome by applying electrochemical procedures, which increase reaction pathway selectivity and reduce cost, since they allow direct control of the surface free energy by setting the electrode potential [6]. Feasible routes toward synthetic fuels are either from direct electroreduction of CO_2 [4] or indirect procedures, where H_2O and CO_2 are electrolyzed in solid oxide electrolysis cells to yield syngas, and then the syngas is converted to gasoline or diesel fuel by Fischer-Tropsch synthesis [5]. While liquid fuel preparation is in its early development stage, more mature are the procedures of electrochemical CO_2 reduction to a wide variety of useful products, including carboxylic acids, aldehydes, alcohols, and carbon monoxide. Such compounds can be obtained under mild conditions but at an energy conversion efficiency of only 30–40 % [1].

Thermodynamics and Kinetics of CO_2 Electroreduction

While the standard reduction potential for the conversion of CO_2 to CO is −0.52 V vs. SHE (at pH 7 and 298 K), practical CO_2 reduction proceeds at much greater cathodic polarization; CO formation only begins at −0.80 V vs. SHE (at pH 7) and becomes efficient at −1.10 V vs. SHE (at pH 7); the experiment was conducted on Au electrode, in $KHCO_3$ solution, at room temperature [7]. This means that a significant overpotential needs to be applied to drive CO_2 electroreduction, which often causes hydrogen reduction as well. To suppress H_2 production, one can use cathodes made of metals with high hydrogen overpotentials (e.g., Hg and Pb) or one can perform CO_2 electroreduction in organic media, as discussed in a later section.

While a wide variety of reduction mechanisms have been proposed, it is unanimously accepted that, whatever the pathway of the process is, the first step consists of a single electronation of the CO_2 molecule, forming an anion radical (Eq. 1) [8, 9]:

$$CO_2 + e^- \rightarrow CO_2\cdot^- \quad (1)$$

This reaction is the rate-determining step, and the $CO_2\cdot^-$ subsequently accepts protons (from water, which acts as a proton donor) and one or more electrons (from the electrode); a number of reaction products are possible, such as formate ions [8, 9], as shown in (Eqs. 2) and (3):

$$CO_2\cdot^-(ads) + H_2O \rightarrow HCOO\cdot(ads) + OH^- \quad (2)$$

$$HCOO\cdot(ads) + e^- \rightarrow HCOO^- \quad (3)$$

In nonaqueous media, the radical anion generated as shown in (Eq. 1) may undergo dimerization with a neutral CO_2 molecule (Eq. 4) and then undergo a second electronation to oxalate ion (Eq. 5) [1, 8, 9]:

$$CO_2\cdot^-(ads) + CO_2 \rightarrow {}^-OOC - COO\cdot(ads) \quad (4)$$

$${}^-OOC - COO\cdot(ads) + e^- \rightarrow {}^-OOC - COO^- \quad (5)$$

Experimental Aspects of CO_2 Electroreduction

The electrochemical conversion of CO_2 into fuels requires one to supply CO_2 gas to the electrolysis cell, by bubbling Hhe gas continuously in the electrolyte, at a constant flow rate. This can be done at atmospheric pressure (1 atm) or at elevated pressures (30–60 atm). Operating the electrolysis at high pressure and/or low temperature increases the otherwise low solubility of CO_2 in the electrolyte solution, and organic solvents which have higher CO_2 solubility than water can also be used [10]. Following electrolyte sorption, CO_2 typically undergoes competitive reactions that proceed simultaneously on the electrode surface, yielding different products, with a product distribution primarily determined by (i) the composition of the electrode and (ii) the composition of the electrolyte (aqueous or organic with various salts and additives). The rate of the electrochemical process can be enhanced by using high-throughput gas-diffusion electrodes [11–13], which usually consist of Teflon®-bound catalyst particles. Other effective technological means for improving the electroreduction process include the use of polymer electrolyte membrane cells, which enable the gas-phase electrolysis of CO_2 by enhancing the CO_2 transport [14, 15].

While an earlier electrode classification was based on whether the cathode metal belonged to the *sp*- or the *d*-metal group [9], Hori considered that the performance of various metals is loosely related to the periodic table. For aqueous electrolytes, Hori suggested regrouping the electrode metals into two categories: (1) CO formation metals (Cu, Au, Ag, Zn, Pd, Ga, Ni, and Pt) and (2) metals that yield formate (Pb, Hg, In, Sn, Cd, and Tl). In nonaqueous electrolytes (such as propylene carbonate), Hg, Tl, and Pb yield oxalate, while Cd, Sn, and In generate CO [1]. In aqueous solution, the electrode potentials of CO_2 reduction correlate with the heats of fusion (HoF) of the electrode metals: low HoF metals (Hg, Tl, Pb, In, Cd, and Zn) yield formate, while high HoF metals (Pt, Pd, Ni, Au, Cu, Ag, Zn, Sn, and Ga) form CO [16].

The purity of the electrolyte is essential, as even trace amounts of impurities (e.g., 5 ppm of lead or iron) interfere with either the surface processes that take place at the electrodes or the ones that occur in the heterogeneous interfacial phase. Therefore, purification of the aqueous solutions by pre-electrolysis is recommended. Often, CO_2 reduction is conducted in organic electrolytes, as they offer several advantages over aqueous media: (i) the unwanted hydrogen discharge can be suppressed, (ii) the concentration of water as a reagent can be accurately controlled, (iii) the solubility of CO_2 in organic solvents is much greater than in water, (iv) the electrolysis can be conducted at much more negative polarization, and (v) at low water concentration, organic solutions allow for dimerization reactions, yielding oxalic acid, glyoxylic acid, and glycolic acid [1].

Synthesis of Formate Salts/Formic Acid

This process appears to be the most practical application of CO_2 electroreduction, the closest to commercialization. Laboratory work performed on the 100 A scale demonstrated that reduction of CO_2 to formate can be accomplished in a trickle-bed continuous electrochemical reactor, under industrially viable conditions; conceptual flow sheets were proposed for two process options, each converting CO_2 at a rate of 100 t per day [14]. Large-scale electrochemical reduction of CO_2 to formate salts and formic acid on tin or proprietary catalysts can be performed in a flow-through reactor. This process has been shown feasible from both engineering and economic standpoints [17]. Continuous electrochemical conversion of CO_2 to formate was successfully conducted in a polymer electrolyte membrane cell, where an alkaline ion-exchange membrane was sandwiched between two catalytic electrodes that contained lead and indium for suppressing hydrogen evolution. Given the 80 % efficiency of CO_2 conversion to formate, this cell is particularly attractive for large-scale implementation. Performing the electrolysis in a pulsed mode mitigated mass transport limitations and provided high efficiency [18].

Liquid Fuels from CO_2

Copper electrodes enable the electroreduction of CO_2 to hydrocarbons and alcohols. Initially, CO_2 is reduced to CO, which interacts with the Cu

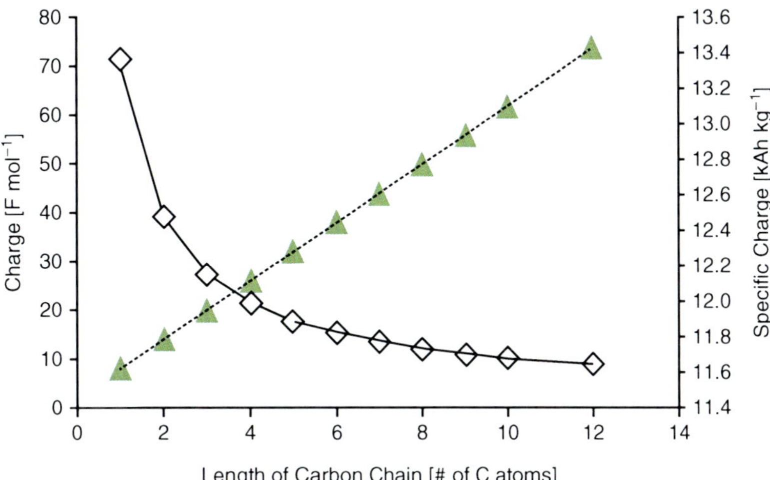

Electroreduction of Carbon Dioxide, Fig. 1 Charge consumed in the electrochemical synthesis of alkanes of various chain length: (main axis, dotted line) – charge in Faradays per mole of hydrocarbon, (secondary axis, solid line) – specific charge in kAh per kg of hydrocarbon; the vertical dotted line shows the separation between gas-phase hydrocarbons (chain length $\leq C_4$) and the liquid-phase hydrocarbons (chain length $> C_4$)

surface in a unique manner, such that further reduction to alcohols and hydrocarbons can take place, so that methane and even long-chain hydrocarbons may form. The product distribution is affected by the cations and anions present in the electrolyte solution, and product selectivity can be improved by the modification of the Cu electrode surface with halides, oxygen, or sulfur atoms. Alkanes with chain lengths up to C_{12} can be formed; the charge (in F mol^{-1}) and specific charge (in kAh kg^{-1}) needed for the electrochemical syntheses are shown in Fig. 1. As revealed by the graph, while the amount of charge needed per molecule of obtained alkane (expressed in Faraday/mol) increases linearly with the chain length, the specific gravimetric charge (kAh/kg) decreases, when the electrochemical reduction of CO_2 yields longer chain length alkanes (the vertical dotted line represents the separation between gas-phase and liquid-phase alkanes).

Using Pt nanoparticles deposited on carbon black (Vulcan XC-72) and/or carbon cloth as the electrodes, it is possible to convert CO_2 to long carbon-chain hydrocarbons ($>C_5$), at room temperature and atmospheric pressure, in a continuous flow cell, where the working electrode is in direct contact with gas-phase CO_2. This system integrates in a photoelectrochemical device, which uses solar energy and water to convert CO_2 into fuels [19].

An analysis of the energy balance and economics of CO_2 recycling to hydrocarbon fuels estimated that the full system can feasibly operate at 70 % electricity-to-liquid fuel efficiency, and the price of electricity needed to produce synthetic gasoline is of only 1.0–1.5 % of the present sales price of gasoline. In regions where inexpensive renewable electricity is currently available, such as Iceland, fuel production may already be economical. The dominant costs of the process are the electricity cost and the capital cost of the electrolyzer, where this capital cost is significantly increased when operating intermittently (on renewable power sources such as solar and wind) [5].

Future Directions

Conversion of CO_2 to fuels or chemical feedstock is attractive for combating the increasing concentration levels of atmospheric CO_2 and storing energy [18]. Effort is made toward developing large "artificial trees" that can absorb CO_2 directly from air. Once CO_2 is "harvested" from the artificial trees, it can be recycled into synthetic alcohols or synthetic fuels, balancing the hydrocarbon burning of cars and trucks and thus making these into essentially zero-emissions vehicles [20]. The *Green Freedom* concept proposed by Los Alamos National Laboratory scientists aims to remove CO_2 from the air and convert it into gasoline. Air would be blown over a liquid solution of $KHCO_3$, which absorbs the CO_2, and then it would be extracted and subjected to processes that turn CO_2 into fuel: methanol, gasoline, or jet fuel [21]. From various available procedures, the "one-pot" electrochemical synthesis of hydrocarbons from CO_2 represents an attractive and potentially useful process [22]. Ideally, the conversion should be carried out with electrical energy from a renewable generation system such that "carbon-neutral" fuels are obtained [18]. Fuel generation from CO_2 on Pt nanoparticle catalysts may be used to produce fuels electrochemically from ambient air in a device for Mars missions [19].

Cross-References

▶ Electrochemical Fixation of Carbon Dioxide (Cathodic Reduction in the Presence of Carbon Dioxide)

References

1. Hori Y (2008) Electrochemical CO2 reduction on metal electrodes. In: Vayenas C et al (eds) Modern aspects of electrochemistry, 42nd edn. Springer, New York, pp 89–189
2. Olah GA, Prakash GKS, Goeppert A (2011) Anthropogenic chemical carbon cycle for a sustainable future. J Am Chem Soc 133:12881–12898
3. Bockris JO'M, Khan SUM (1993) Surface electrochemistry. A molecular level approach. Plenum Press, New York/London, pp 534–541
4. Peterson AA, Abild-Pedersen F, Studt F, Rossmeisl J, Nørskov JK (2010) How copper catalyzes the electroreduction of carbon dioxide into hydrocarbon fuels. Energy Environ Sci 3:1311–1315
5. Graves C, Ebbesen SD, Mogensen M, Lackner KS (2011) Sustainable hydrocarbon fuels by recycling CO_2 and H_2O with renewable or nuclear energy. Renew Sustain Energy Rev 15:1–23
6. Spinner NS, Vega JA, Mustain WE (2012) Recent progress in the electrochemical conversion and utilization of CO_2. Catal Sci Technol 2:19–28
7. Hori Y, Murata A, Takahashi R, Suzuki S (1987) Electrochemical reduction of carbon dioxide to carbon monoxide at a gold electrode in aqueous potassium hydrogen carbonate. Chem Commun 10:728–729
8. Chaplin RPS, Wragg AA (2003) Effects of process conditions and electrode material on reaction pathway for carbon dioxide electroreduction with particular reference to formate formation. J Appl Electrochem 33:1107–1023
9. Jitaru M, Lowy DA, Toma BC, Toma M, Oniciu L (1997) The electrochemical reduction of carbon dioxide on flat metallic electrodes. J Appl Electrochem 27:875–889
10. Pérez-Rodriquez S, García G, Calvillo L, Celorrio V, Pastor E, Lázaro MJ (2011) Carbon-supported Fe catalyst for CO2 electroreduction to high-added value products: a DEMS study: effect of the functionalization of the support. Int J Electrochem. Article ID 249804, 1–13. doi:10.4061/2011/249804
11. Machunda RL, Ju HKK, Lee J (2011) Electrocatalytic reduction of CO2 gas at Sn based gas diffusion electrode. Curr Appl Phys 11:986–988
12. Machunda RL, Lee JG, Lee J (2010) Microstructural surface changes of electrodeposited Pb on gas diffusion electrode during electroreduction of gas-phase CO_2. Surf Interface Anal 42:564–567
13. Whipple DT, Finke EC, Kenis PJA (2010) Microfluidic reactor for the electrochemical reduction of carbon dioxide: the effect of pH. Electrochem Solid-State Lett 13:B109–B111
14. Oloman C, Li H (2008) Electrochemical processing of carbon dioxide. Chem Sus Chem 1:385–391
15. Yano H, Shirai FM, Nakayama M, Ogura K (2002) Electrochemical reduction of CO2 at three-phase (gas/liquid/solid) and two-phase (liquid/solid) interfaces on Ag electrodes. J Electroanal Chem 533:113–118
16. Hori Y, Wakebe H, Tsukamoto T, Koga O (1994) Electrocatalytic process of CO selectivity in electrochemical reduction of CO_2 at metal electrodes in aqueous media. Electrochim Acta 39:1833–1839
17. Agarwal AS, Zhai YM, Hill D, Sridhar N (2011) The electrochemical reduction of carbon dioxide to formate/formic acid: engineering and economic feasibility. Chem Sus Chem 4:1301–1310
18. Narayanan SR, Haines B, Soler J, Valdez TI (2011) Electrochemical conversion of carbon dioxide to formate in alkaline polymer electrolyte membrane cells. J Electrochem Soc 158:A167–A173
19. Centi G, Perathoner S, Winè G, Gangeri M (2007) Electrocatalytic conversion of CO_2 to long carbon-chain hydrocarbons. Green Chem 9:671–678
20. Yegulalp TM, Lackner KS, Ziock HJ (2001) A review of emerging technologies for sustainable use of coal for power generation. Int J Surf Mining Reclamation Environ 15:52–68. doi:10.1076/ijsm.15.1.52.3423
21. Martin FJ, Kubic WL (2007) Green freedom, a concept for producing carbon-neutral synthetic fuels and chemicals. Los Alamos National Laboratory, LA-UR-07-7897 (Nov. 2007)
22. Gattrell M, Gupta N, Co A (2006) A review of the electrochemical reduction of CO_2 to hydrocarbons on copper. J Electroanal Chem 594:1–19

Electrosynthesis in Ionic Liquid

Toshio Fuchigami
Department of Electrochemistry, Tokyo Institute of Technology, Midori-ku, Yokohama, Japan

Introduction

Room-temperature molten salts, namely ionic liquids, have proved to be a new class of promising solvents because of their good electroconductivity, nonflammability, thermal stability, nonvolatility, and reusability [1–3]. They consist of cations and anions without any solvent, and they are in a liquid state around room temperature. Typical examples of ionic liquids and their abbreviation are shown in Fig. 1. Ionic liquids are classified into hydrophilic and hydrophobic ones. Ionic liquids with BF_4^- or $CF_3SO_3^-$ (TfO^-) are hydrophilic. On the other hand, those with PF_6^- or $(CF_3SO_2)N^-$ (Tf_2N^-) are hydrophobic. Hydrophobic ionic liquids do not dissolve into either water or organic solvents like ether and hexane. Therefore, they form three liquid phases. Hence, products can be separated by liquid–liquid extraction. This is one of big advantages of ionic liquids. Furthermore, if a combination of cation and anion is appropriately made, aprotic media having a wide electrochemical window can be obtained. Therefore, when ionic liquids are used as electrolytic media, organic electrolytic reactions, particularly electroorganic synthesis, should be possible without any organic solvents. However, there have still been a limited number of papers dealing with organic electrolytic reactions and electrosynthesis [4–8].

Cyclic Voltammetry in Ionic Liquids

Cyclic voltammetry of nickel(II) salen was carried out at a glassy carbon electrode in [BMIM][BF_4]. Nickel(II) salen exhibited one-electron, quasi-reversible reduction to nickel(I) salen, and the latter species served as a catalyst for cleavage of carbon–halogen bonds in iodoethane and 1,1,2-trichlorotrifluoroethane (Freon^R 113) [9]. It is notable that the diffusion coefficient for nickel(II) salen in the ionic liquid at room temperature is more than 500 times smaller than that ($1.0\ 10^{-5}\ cm^2\ s^{-1}$) in a typical organic solvent–electrolyte system such as DMF containing 0.1 M $Et_4N \times BF_4$. This is due to the high viscosity of the ionic liquid.

Interestingly, 1,4- and 1,2-dinitrobenzenes exhibit two one-electron waves in acetonitrile but a single two-electron wave in the ionic liquid [BMIM][BF_4] [10]. The latter effect is ascribed to strong ion pairing between the imidazolium cation and the dinitrobenzene dianion.

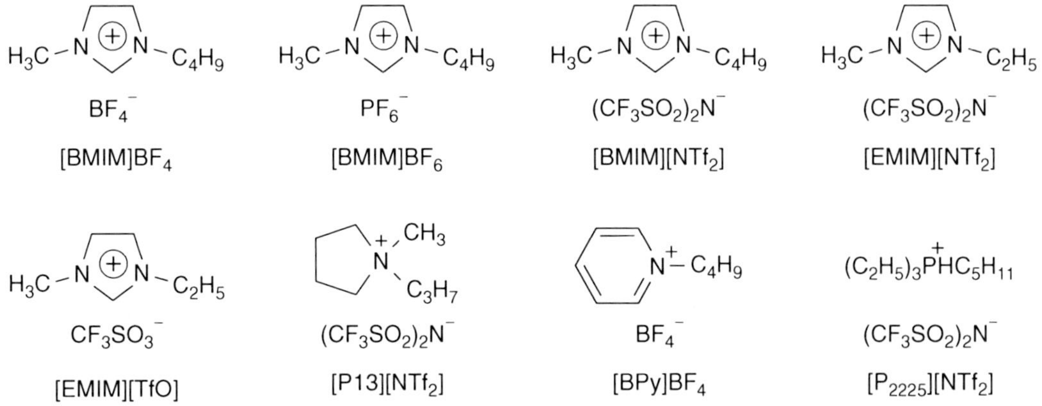

BF_4^- [BMIM]BF_4 | PF_6^- [BMIM]BF_6 | $(CF_3SO_2)_2N^-$ [BMIM][NTf_2] | $(CF_3SO_2)_2N^-$ [EMIM][NTf_2]

$CF_3SO_3^-$ [EMIM][TfO] | $(CF_3SO_2)_2N^-$ [P13][NTf_2] | BF_4^- [BPy]BF_4 | $(C_2H_5)_3\overset{+}{P}HC_5H_{11}$ $(CF_3SO_2)_2N^-$ [P_{2225}][NTf_2]

Electrosynthesis in Ionic Liquid, Fig. 1 Examples of ionic liquids and their abbreviation

Ph–C(=O)–Me → 2e (1.0 F/mol), divided cell → Ph(HO)(Me)C–C(Me)(OH)Ph

[BMIM][NTf_2] 92%, 6% d.e.
[Et_3BuN][NTf_2] 82%, 58% d.e.
[Me_3BuN][NTf_2] 80%, 54% d.e.

Electrosynthesis in Ionic Liquid, Scheme 1 Diastereoselective cathodic dimerization of acetophenone in various ionic liquids

Electrosynthesis in Ionic Liquid, Scheme 2 Cathodic reduction of phthalimide in ionic liquid with and without sonication

$2e^-$, $2H^+$
[EMIM][NTf_2]
0.11 F/mol

Sonication: 89% (C. Eff.)
Non sonication: 66% (C. Eff.)

Application of Ionic Liquids to Electrosynthesis

Cathodic reduction of carbonyl compounds like benzaldehyde and acetophenone was investigated in ionic liquids, and notably their dimerization proceeded predominantly [11, 12]. For acetophenone, the corresponding pinacol was formed as a diastereomeric mixture, and the diastereoselectivity is greatly affected by ionic liquids used as shown in Scheme 1 [11].

Since ionic liquids have generally much higher viscosity, mass transport is quite slow as described before. This is a disadvantage for electrosynthesis in ionic liquids. However, it was found that electroreduction of N-methylphthalimide was promoted under ultrasonication resulting higher conversion and current efficiency as shown in Scheme 2 [13]. This is as due to facilitated mass transport of the substrate under ultrasonication. Similar but more pronounced ultrasonication effect was observed in electrochemical difluorination of ethyl α-(phenylthio)acetate in ionic liquid Et_3N–3HF to provide the corresponding α,α-difluoro products [14].

Cyclic carbonates are prepared by the reduction of CO_2 at −2.4 V versus Ag/AgCl in the presence of epoxides in various ionic liquids like [EMIM]

Bn(Me)N–H → electrolysis; [BMIM][BF_4], divided cell, CO_2 (0.1 MPa), 55 °C, −2.4 V vs. Ag/AgCl, then EtI → Bn(Me)N–C(=O)OEt

Electrosynthesis in Ionic Liquid, Scheme 3 Electrochemical synthesis of carbamate derivative in ionic liquid

[BF_4], [BMIM][PF_6], and *N*-butylpyridinium tetrafluoroborate [BPy][BF_4] using a Cu cathode and Mg or Al anode [15].

Electrochemical synthesis of carbamates was also achieved by the electrolysis of a solution of CO_2 and amine in [BMIM][BF_4] followed by the addition of alkylating agent as shown in Scheme 3 [16].

Electrocatalytic homo-coupling of PhBr and $PhCH_2Br$ proceeds well in the presence of NiCl (bpy) complex in [BMIM][NTf_2] as shown in Scheme 4 [17].

Interestingly, Pd nanoparticles generated cathodically in ionic liquid were shown to be a highly effective ligand-free catalyst for coupling of aryl halides [18].

Electroreductive dehalogenation of *vic*-dihalides using a Co(II)salen complex in [BMIM][BF_4] was also achieved as shown in Scheme 5 [19]. The product isolation is much easier compared to the similar dehalogenation in ordinary molecular solvents since the Co(II)salen complex remains in the ionic liquid phase during product extraction with nonpolar organic solvents like diethyl ether. Furthermore, the recyclability of the catalyst/ionic liquid system was demonstrated.

N-heterocyclic carbenes can be generated by cathodic reduction of imidazolinium-based ionic liquids [20–22]. Feroci and her co-workers reported that the resulting carbenes are stable bases able to catalyze the Henry reaction as shown in Scheme 6 [22].

Tempo-mediated anodic oxidation of alcohols to aldehydes and ketones was successfully carried out in ionic liquids [23, 24]. High viscosity of the ionic liquids limiting the mass transport of the catalyst was readily overcome by addition of base and alcohols, and high yields and current efficiencies were achieved [23].

Anodic oxidation of various aromatic compounds like anisole, mesitylene, and fused aromatic compounds like naphthalene and anthracene was carried out in various ionic liquids such as $[BMIM][PF_6]$ and $[BMIM][NTf_2]$ to provide the corresponding dimmers in moderate to good yields [25]. Under similar conditions, anodic oxidation of 1,2-dimethoxybenzene leads to the corresponding trimer as shown in Scheme 7 [25].

Electrooxidative polymerization of pyrrole, thiophene, and aniline was achieved in various imidazolium ionic liquids [26–28]. 1-Butyl-3-methylimidazolium tetrafluoroborate and hexafluorophosphate ($[BMIM])[BF_4]$ and $[BMIM][PF_6]$) for electropolymerization and notably π-conjugated polymers thus obtained are highly stable, and they can undergo electrochemical doping and dedoping in the ionic liquids up to million cycles. In addition, the polymers have cycle-switching speeds as fast as 100 m.

Another ionic liquid [EMIM][OTf] was also employed for the electropolymerization. The polymerization of pyrrole in the ionic liquid

$$2\,Ph(CH_2)_xBr \xrightarrow[\substack{\text{cat. } NiCl_2\text{ (bpy)}\\ [BMIM][Tf_2N]\\ -1.4\text{ V vs Ag/Ag}^+\\ 2\text{ F/mol}}]{2e} Ph(CH_2CH_2)_xPh \quad \begin{matrix} x = 0 : 35\,\% \\ 1 : 75\,\% \end{matrix}$$

Electrosynthesis in Ionic Liquid, Scheme 4 Electrocatalytic homo-coupling using Ni-complex in ionic liquid

Electrosynthesis in Ionic Liquid, Scheme 5 Electrocatalytic dehalogenation using Co(II)salen complex in ionic liquid

Electrosynthesis in Ionic Liquid, Scheme 6 Cathodic generation of N-heterocyclic carbene in ionic liquid and its catalytic use for Henry reaction

$$3\ \text{1,2-(MeO)}_2\text{C}_6\text{H}_4 \xrightarrow{-6e,\ -6H^+} [\text{(MeO)}_2\text{C}_6\text{H}_2]_3$$

Electrosynthesis in Ionic Liquid, Scheme 7 Anodic trimerization of 1,2-dimethoxybenzene in ionic liquid

proceeds much faster than that in conventional media like aqueous and acetonitrile solutions containing 0.1 M [EMIM][OTf] as a supporting electrolyte, [27, 28]. It is notable that the morphology and some physical properties of the resulting polypyrrole are quite characteristic. The polymer film prepared in the ionic liquid has a higher electrochemical density and highly regulated morphological structures.

The utility of the ionic liquid as a recyclable medium for the polymerization was also demonstrated. More than 90 % of [EMIM][OTf] after the polymerization was easily recovered simply by extracting the remaining pyrrole monomer with chloroform. The recovered [EMIM][OTf] could be reused five times without significant loss of reactivity for the polymerization [27].

Furthermore, electrosynthesis of poly (3,4-ethylenedioxythiophene) (PEDOT) and polyphenylene in ionic liquids was reported [29, 30].

Future Directions

Electroorganic synthesis has been recognized as an environmentally friendly process since the 1970s. However, new methodologies must be developed to achieve modern electrosynthesis as a green and sustainable chemistry. In this aspect, ionic liquids are promising as green electrolytic media.

Cross-References

► Electrocatalytic Synthesis
► Electrochemical Fixation of Carbon Dioxide (Cathodic Reduction in the Presence of Carbon Dioxide)
► Electrosynthesis of Conducting Polymer
► Electrosynthesis Using Mediator
► Selective Electrochemical Fluorination

References

1. Welton T (1999) Room-temperature ionic liquids. Solvents for synthesis and catalysis. Chem Rev 99:2071–2084
2. Wassersheid P, Keim W (2000) Ionic liquids—new "solutions" for transition metal catalysis. Angew Chem Int Ed 39:3772–3789
3. Rogers RD, Seddon KR, Volkov S (eds) (2002) Green industrial applications of ionic liquids. Kluwer Academic, Dordrecht
4. Fuchigami T (2007) Unique solvent effects on selective electrochemical fluorination of organic compounds. J Fluorine Chem 128:311–316
5. Fuchigami T, Inagi S (2011) Selective electrochemical fluorination of organic molecules and macromolecules in ionic liquids. Chem Commun 47: 10211–10223
6. Fuchigami T, Inagi S (2011) Electrolytic reactions. In: Ohno H (ed) Electrochemical aspects of ionic liquids, 2nd edn. Wiley, Hoboken
7. Hapiot P, Lagrost C (2008) Electrochemical reactivity in room-temperature ionic liquids. Chem Rev 108:2238–2264
8. Yoshida J, Kataoka K, Horcajada R, Nagaki A (2008) Modern strategies in electroorganic synthesis. Chem Rev 108:2265–2299
9. Sweeny BK, Peters DG (2001) Cyclic voltammetric study of the catalytic behavior of nickel(I) salen electrogenerated at a glassy carbon electrode in an ionic liquid (1-butyl-3-methylimidazolium tetrafluoroborate, $BMIM^+BF_4^-$). Electrochem Commun 3:712–715
10. Fry AJ (2003) Strong ion-pairing effects in a room-temperature ionic liquid. J Electroanal Chem 546:35–39
11. Doherty AP, Brooks CA (2004) Electrosynthesis in room-temperature ionic liquids: benzaldehyde reduction. Electrochim Acta 49:3821–3826
12. Lagrost C, Hapiot P, Vaultier M (2005) The influence of room-temperature ionic liquids on the stereoselectivity and kinetics of the electrochemical pinacol coupling of acetophenone. Green Chem 7:468–474
13. Villagrán C, Banks CE, Pitner WR, Hardacre C, Compton RG (2005) Electroreduction of *N*-methylphthalimide in room temperature ionic liquids under insonated and silent conditions. Ultrason Sonochem 12:423–428
14. Sunaga S, Atobe M, Inagi S, Fuchigami T (2009) Highly efficient and selective electrochemical fluorination of organosulfur compounds in $Et_3N \cdot 3HF$ ionic liquid under ultrasonication. Chem Commun 956–958

15. Yang H, Gu Y, Deng Y, Shi F (2002) Electrochemical activation of carbon dioxide in ionic liquid: synthesis of cyclic carbonates at mild reaction conditions. Chem Commun 274–275
16. Feroci M, Orsini M, Rossi L, Sotgiu G, Inesi A (2007) Electrochemically promoted C-N bond formation from amines and CO_2 in ionic liquid BMIm-BF_4: synthesis of carbamates. J Org Chem 72:200–203
17. Mellah M, Gmouh S, Vaultier M, Jouikov V (2003) Electrocatalytic dimerisation of PhBr and $PhCH_2Br$ in [BMIM] + NTf_2 – ionic liquid. Electrochem Commun 5:591–593
18. Duran Pachon L, Elsevier CJ, Rothenberg G (2006) Electroreductive palladium-catalysed ullmann reactions in ionic liquids: scope and mechanism. Adv Synth Catal 348:1705–1710
19. Shen Y, Atobe M, Tajima T, Fuchigami T (2004) Electrocatalytic debromination of organic bromides using a cobalt(II)salen complex in ionic liquids. Electrochemistry 72:849–851
20. Chowdhury S, Mohan RS, Scott JL (2007) Reactivity of ionic liquids. Tetrahedron 63:2363–2389
21. Feroci M, Chiarotto I, Orsini M, Sotgiu G, Inesi A (2008) Reactivity of electrogenerated N-heterocyclic carbenes in room-temperature ionic liquids. Cyclization to 2-azetidinone ring via C-3/C-4 bond formation. Adv Synth Catal 350:1355–1359
22. Feroci M, Elinson MN, Rossi L, Inesi A (2009) The double role of ionic liquids in organic electrosynthesis: precursors of N-heterocyclic carbenes and green solvents. Henry reaction. Electrochem Commun 11:1523–1526
23. Barhdadi R, Comminges C, Doherty AP, Nédélec J-Y, O'Toole S, Troupel M (2007) The electrochemistry of TEMPO-mediated oxidation of alcohols in ionic liquid. J Appl Electrochem 37:723–728
24. Kuroboshi M, Fujisawa J, Tanaka H (2004) *N*-Oxyl-mediated electrooxidation in ionic liquid. A prominent approach to totally closed system. Electrochemistry 72:846–848
25. Mellah M, Zeitouny J, Gmouh S, Vaultier M, Jouikov V (2005) Anodic self-coupling of aromatic compounds in ionic liquids. Electrochem Commun 7:869–874
26. Lu W, Fadeev AG, Qi B, Smela E, Mattes BR, Ding J, Spinks GM, Mazurkiewicz J, Zhou D, Wallace GG, MacFarlane DR, Forsyth SA, Forsyth M (2002) Use of ionic liquids for π-conjugated polymer electrochemical devices. Science 297:983–987
27. Sekiguchi K, Atobe M, Fuchigami T (2002) Electropolymerization of pyrrole in 1-ethyl-3-methylimidazolium trifluoromethanesulfonate room temperature ionic liquid. Electrochem Commun 4:881–885
28. Sekiguchi K, Atobe M, Fuchigami T (2003) Electrooxidative polymerization of aromatic compounds in 1-ethyl-3-methylimidazolium trifluoromethanesulfonate room-temperature ionic liquid. J Electroanal Chem 557:1–7
29. Randriamahazaka H, Plesse C, Chevrot D (2004) Ions transfer mechanisms during the electrochemical oxidation of poly(3,4-ethylenedioxythiophene) in 1-ethyl-3-methylimidazolium bis((trifluoromethyl) sulfonyl)amide ionic liquid. Electrochem Commun 6:299–305
30. Abedin SZE, Borissenko N, Endres F (2004) Electropolymerization of benzene in a room temperature ionic liquid. Electrochem Commun 6:422–426

Electrosynthesis in Supercritical Fluids

Mahito Atobe
Graduate School of Environment and Information Sciences, Yokohama National University, Yokohama, Japan

Introduction

The potential advantages associated with an electrosynthesis include high material utilization and significantly less energy requirement, ease of control of the reaction, less hazardous process, and the ability to perform wide range of oxidation and reduction reactions. Therefore, many electrosynthetic reactions have been reported so far [1]. However, only a few have been employed industrially. The commercialization of electrosynthetic processes has been restricted by the limited solubility of substrates and products in conventional electrolytic solutions, the poor interphase mass transport characteristics associated with two phase system in which the reaction occurs at solid (electrode)–liquid (electrolyte) interfaces, the low selectivity for desired reaction products, and the complex processing schemes often used to recover products.

In these circumstances, supercritical fluid solvents would overcome many of the limitations associated with the conventional solvents such as water and organic solvents. Supercritical fluids have traditionally been applied to chromatography, extraction, cleaning, and so on [2]. Moreover, they are becoming widely recognized

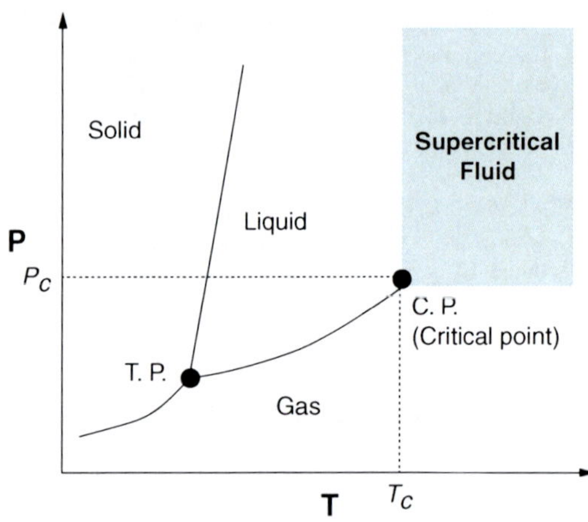

Electrosynthesis in Supercritical Fluids, Fig. 1 Pressure-temperature phase diagram

as useful media for organic and polymer syntheses in a range of laboratory and industrial processes because of low toxicity, ease of solvent removal, potential for recycling, and variation of reaction rates [3, 4].

Supercritical fluid can be defined as any substance that is above its critical temperature (*Tc*) and critical pressure (*Pc*) and exists as a single phase (Fig. 1). The physicochemical properties of a supercritical fluid are in between those of liquids and gases, and they can vary between those exhibited by gases to liquid-like values by small changes in pressure and temperature of supercritical fluid [2]. For example, supercritical fluid exhibits larger diffusibility and lower viscosity compared to those of conventional liquids. Consequently, interphase mass transfer resistance is also lower relative to the liquid solvent. On the other hand, substrates and products that are sparingly soluble in the liquid phase can become significantly more soluble in supercritical fluids. Such unique and useful properties of the fluids make them viable solvent media for electrosynthesis.

Electrosynthesis in Supercritical Organic Mixtures

Conventional supercritical solvents, such as carbon dioxide and ethane, have mild critical conditions, but they lack the ability to solvate polar electrolytes. Therefore, a reaction medium should generally be designed by mixing with cosolvent such as methanol [5], ethanol [6], acetone [6], acetonitrile [7], or DMF [8] to produce a suitable electrolytic solution.

$$2\,CH_3OH + CO \xrightarrow[\substack{Bu_4NBF_4 \\ scCO_2\text{-}CH_3OH}]{\text{electrolysis, 90 °C}} (CH_3O)_2CO$$

Electrosynthesis in Supercritical Fluids, Scheme 1 Electrochemical synthesis of dimethyl carbonate in a supercritical carbon dioxide-methanol medium

Dombro et al. reported that dimethyl carbonate has been electrosynthesized from carbon monoxide and methanol in a supercritical carbon dioxide-methanol medium (Scheme 1) [9]. They added 0.35 mol fraction of methanol to supercritical carbon dioxide to produce a mixture capable of dissolving an ammonium salt. Such a mixture is capable of ionic conductivity, and actually the electrolysis proceeded smoothly to afford dimethyl carbonate in an excellent current efficiency (near 100 %).

Tokuda et al. also reported that electrochemical carboxylation of benzyl chloride, cinnamyl chloride, and 2-chloronaphthalene was smoothly proceeded in supercritical carbon dioxide-DMF media (Scheme 2) [8]. A mixture used in this work also exhibited the ability to solvate polar electrolytes like ammonium salts. In these reactions, since carbon dioxide plays the role of a reactant, CO_2-rich conditions such as supercritical conditions are quite suitable for improving the product yields.

Electrosynthesis in Homogeneous Supercritical Fluids

As mentioned above, conventional supercritical fluids like supercritical carbon dioxide are the low dielectric constant of the fluid which limits the solubility and conductivity for many electrolytes. Therefore, a polar organic solvent should be added to the fluids to produce suitable electrolytic solutions. However, the inclusion of

Cl + $scCO_2$ (40 °C, 80 kg cm^{-2}) → [Pt (−) / Mg (+); 3 F mol^{-1}; 37.5 mA cm^{-2}; Bu_4NBr / $scCO_2$ + DMF] → COOH, Yield: 81 %

Cl + $scCO_2$ (46 °C, 80 kg cm^{-2}) → [Pt (−) / Mg (+); 3 F mol^{-1}; 37.5 mA cm^{-2}; Bu_4NBr / $scCO_2$ + DMF] → COOH, Yield: 64 %

Cl + $scCO_2$ (43 °C, 80 kg cm^{-2}) → [Pt (−) / Mg (+); 3 F mol^{-1}; 37.5 mA cm^{-2}; Bu_4NBr / $scCO_2$ + DMF] → Cl, Yield: 61 %

Electrosynthesis in Supercritical Fluids, Scheme 2 Electrochemical carboxylation of several organic chlorides in supercritical carbon dioxide-DMF media

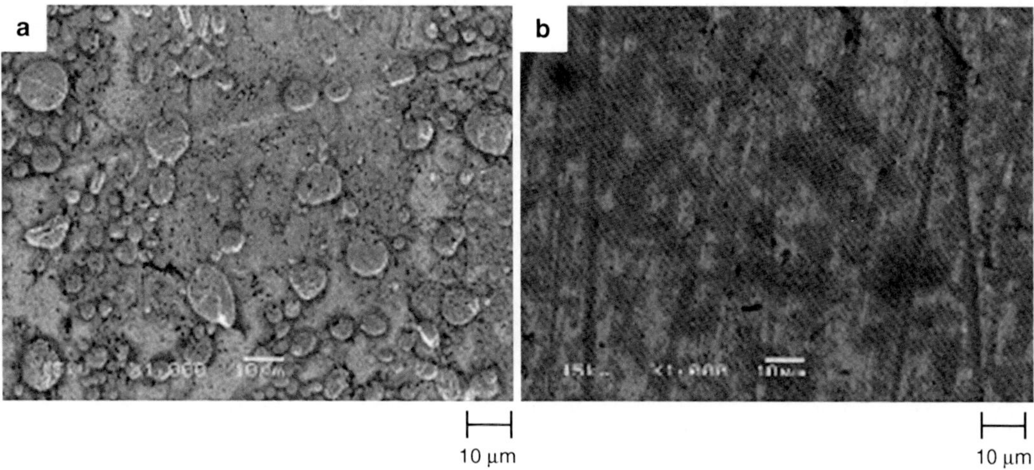

Electrosynthesis in Supercritical Fluids, Fig. 2 SEM photographs of polypyrrole films polymerized in (**a**) acetonitrile and (**b**) supercritical fluoroform solutions

a cosolvent limits the accessible potential range and introduces complications. The use of supercritical fluorinated hydrocarbons is an attractive alternative which avoids these problems. For example, supercritical fluoroform exhibits relatively high solubility, and its dielectric constant can be controlled from 1 to 7 by manipulating either the temperature or the pressure of $scCHF_3$ without any additives like polar solvents [10]. Therefore, supercritical fluoroform has been used for a reaction medium of electrochemical syntheses.

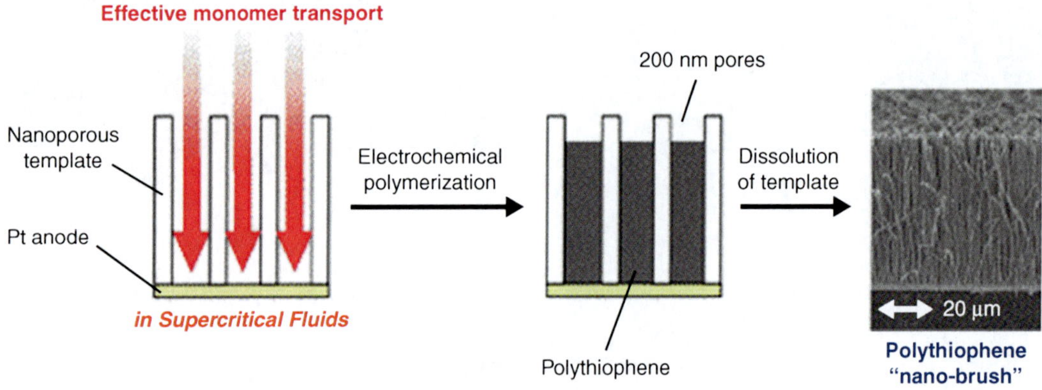

Electrosynthesis in Supercritical Fluids, Fig. 3 Preparation of polythiophene nano-brush using template electrochemical polymerization in supercritical fluoroform

Atobe et al. reported that the electrochemical synthesis of polypyrrole and polythiophene was successfully proceeded in supercritical fluoroform [11]. Both the corresponding monomers (pyrrole and thiophene) can be electropolymerized apparently faster in supercritical fluoroform than in the conventional organic media like an acetonitrile solution, and the obtained films have highly uniform structure (Fig. 2).

Moreover, Atobe et al. successfully prepared polythiophene nano-brush using template electrochemical polymerization in supercritical fluoroform [12]. In this work, nanoporous alumina membranes (60 µm thick, 200 nm pore size) coated on one side with evaporated Pt (ca. 500 nm thick) were employed as a template electrode for polythiophene electrodeposition into pores. The use of the special properties of the fluids such as higher diffusivity and lower viscosity enabled the effective monomer transport into the porous template and the nanoprecise filling with the polymers. Consequently, the solid polythiophene "nano-brush" could be obtained after removal of the alumina membrane (Fig. 3).

Future Directions

Despite many potential advantages of supercritical fluid solvent, there are relatively few examples of electrosynthesis in supercritical fluids. Consequently, the technologies are relatively new and unfamiliar. Further development will require research and development actities in the following areas: (1) easy-to-handle system for electrosynthesis in supercritical fluids, (2) highly polar supercritical fluids capable of dissolving polar electrolytes, and (3) unique and useful electrosynthetic reactions in supercritical fluids.

The technologies serve as one of the solutions for the limitations of commercialization of electrosynthetic processes and would offer a new aspect in the industrial electrochemical synthesis.

Cross-References

- ▶ Electrochemical Fixation of Carbon Dioxide (Cathodic Reduction in the Presence of Carbon Dioxide)
- ▶ Electrosynthesis of Conducting Polymer
- ▶ Electrosynthesis under Ultrasound and Centrifugal Fields

References

1. Lund H, Hammerich O (eds) (2001) Organic electrochemistry. Macel Dekker, New York
2. Vögtle F, Stoddart JF (2000) Supercritical fluids for organic synthesis. Wiley-VCH, Weinheim
3. Jessop PG, Leitner W (1999) Chemical synthesis using supercritical fluids. Wiley-VCH, Weinheim

4. Kendall JL, Canelas DA, Young JL, DeSimone JM (1999) Polymerizations in supercritical carbon dioxide. Chem Rev 99:543–564
5. Jun J, Fedkiw PS (2001) Ionic conductivity of alkali-metal salts in sub- and supercritical carbon dioxide + methanol mixtures. J Electroanal Chem 515:113–122
6. Wu W, Zhang J, Han B, Chen J, Liu Z, Jiang T, He J, Li W (2003) Solubility of room-temperature ionic liquid in supercritical CO_2 with and without organic compounds. Chem Commun 39:1412–1413
7. Yan H, Sato T, Komago D, Yamaguchi A, Oyaizu K, Yuasa M, Otake K (2005) Electrochemical synthesis of a polypyrrole thin film with supercritical carbon dioxide as a solvent. Langmuir 21:12303–12308
8. Sasaki A, Kudoh H, Senboku H, Tokuda M (1998) Electrochemical carboxylation of several organic halides in supercritical carbon dioxide. In: Torii S (ed) Novel trends in electroorganic synthesis. Springer, Tokyo
9. Dombro RA Jr, Prentice GA, McHugh MA (1988) Electro-organic synthesis in supercritical organic mixtures. J Electrochem Soc 135:2219–2223
10. Mori T, Li M, Kobayashi A, Okahata Y (2002) Reversible control of enzymatic transglycosylations in supercritical fluoroform using a lipid-coated β-D-galactosidase. J Am Chem Soc 124:1188–1189
11. Atobe M, Ohsuka H, Fuchigami T (2004) Electrochemical synthesis of polypyrrole and polythiophene in supercritical trifluoromethane. Chem Lett 33:618–619
12. Atobe M, Iizuka S, Fuchigami T, Yamamoto H (2007) Preparation of nanostructured conjugated polymers using template electrochemical polymerization in supercritical fluids. Chem Lett 36:1448–1449

Electrosynthesis of Conducting Polymer

Shinsuke Inagi
Tokyo Institute of Technology, Midori-ku, Yokohama, Japan

Introduction

Conjugated polymers in which electron can delocalize through the polymer main chain are defined as conducting polymers. Polyacetylene, polyphenylene, polypyrrole, etc. have π-conjugated structure, whereas polysilanes have σ-conjugation in the polymer chain. They are readily available by chemical polymerization such as polyaddition and polycondensation of appropriate monomers. On the other hand, the electrochemical polymerization (electropolymerization) of aromatic or heteroaromatic compounds is quite easy but powerful method to produce conducting polyarenes film on electrode [1].

Definition

Conducting polymers described here represent intrinsically conducting polymers, which can be prepared by anodic oxidation or cathodic reduction of electroactive aromatic monomers. Polyarenes such as polyphenylene, polypyrrole, and polythiophene and poly(aromatic amine)s such as polyaniline are representative intrinsically conducting polymers.

Principles

Anodic one-electron oxidation of an aromatic monomer generates its radical cation. The radical coupling and the following deprotonation lead to form a neutral dimer. Further sequential oxidation of the dimer and subsequent radical coupling results in the formation of oligomer and polymer (Fig. 1). They usually deposit on the anode due to their low solubility in the electrolytic media.

Potential sweep technique with cyclic voltammetry is usually the most efficient for electropolymerization of monomers. Potentiostatic (constant potential) electrolysis and galvanostatic (constant current) electrolysis are also used.

The polymerization of less reactive monomer requires higher concentration of the monomer solution. Ionic liquids as electrolytic media give positive effect on the properties of polymer films [2]. Boron trifluoride-ether complex (BF_3•OEt_2) is known to reduce oxidation potential of aromatic monomers when used as an electrolytic medium [3].

Cathodic reduction of aromatic dihalides also undergoes to form conducting polymers on electrode. Moreover, α,α,α',α'-tetrabromo-*p*-xylene can be electropolymerized to afford poly(*p*-phenylene-vinylene) by cathodic reaction [4].

$X = O, S, NR$

Electrosynthesis of Conducting Polymer, Fig. 1 Oxidative polymerization of heteroarene monomer

Doping and Dedoping

When a conducting polymer on electrode is positively (anodically) charged, polaron (radical anion) or bipolaron (dication) species are generated in the polymer main chain [5]. To compensate the positive charges, counter anions derived from supporting electrolytes are introduced into polymer chain as dopants. This p-doped state can be recovered to be neutral state by cathodic reduction. Conversely, application of cathodic potential to conducting polymers gives n-doped polymers.

These doping and dedoping behaviors can be easily confirmed by cyclic voltammetric analysis showing pair of broad current responses, which are usually stable and show good cycle properties. However, the application of much high potential to conducting polymers often results in over-oxidation or over-reduction to form a nonconducting state (nonconjugated state) and/or to cause degradation of polymer chain by undesirable attack of nucleophiles. The stable color change of conducting polymers can be monitored spectroscopically by spectroelectrochemical measurement, i.e., the UV-visible absorption measurement, in which a constant potential is applied to the polymer thin film on a transparent electrode.

Advantage

Polyarenes generally have low solubility in common organic solvents. Thus the film formation of chemically prepared conducting polymers with solution process is difficult. On the other hand, electropolymerization of arene monomers provides a thin film on the surface of working electrode. The deposited film is relatively strong, and film thickness can be tuned by the amount of charge passed. This facile film formation on electrode is preferable for use as a polymer film electrode, thin-layer sensor, in microtechnology.

Conductivity

Among the various interesting and useful properties of conducting polymers, their switchable electrical conductivity has proven the most attractive. The use of the conventional dc four-prove method or the ac impedance technique in a metal-polymer-metal sandwich structure for measurements of the conductivity of dry polymer samples is straightforward.

Conjugated polymers in neutral state are normally semiconducting materials. However, a partial doping of conducting polymers improves electric conductivity. The conductivity of a doped conducting polymer is of the order of $\sim 10^3$ S/cm depending on polymer structures and doping levels.

Application

Redox processes of conducting polymers combined with the doping of counterions can affect their features directly to switch the chemical, optical, electrical, magnetic, mechanical, and ionic properties [6]. The following applications are representative ones.

Electrochromism

The drastic color change upon doping-dedoping cycles is one of the promising characteristics applicable to electrochromic materials [7].

The multicolor property and stable chromism are advantageous compared to the other electrochromic systems. The simple device is composed of a sandwiched structure of a conducting polymer film and an electrolyte with a transparent electrode and a metal electrode.

Sensors

Modification of electrode with conducting polymers can improve sensitivity, impart selectivity, and provide a support matrix for sensor molecules. These approaches are intensively studied for gas sensors, electroanalysis, and biosensors. The composite film of conducting polymers and metal nanoparticles on electrode often catalyzes electrode reactions (electrocatalysis).

Energy Device

The reversible redox properties of conducting polymers are suitable for application to rechargeable batteries. For lithium ion batteries, conducting polymers can be used as cathode materials. Conducting polymers are also effective for fuel cell application as protective layers on anodes. The excellent redox and electric properties of conducting polymers are promising candidates for capacitor application.

Actuator

The intercalation of dopants to conducting polymer chains leads to an increase in volume of up to 30 % [8]. This property is used in actuators (polymer-based artificial muscles). Bilayer structure of polypyrrole-based anode and cathode is a simple model. At anode, p-doping of polymer occurs to swell, while the other side shrinks because of the expulsion of counterions. This volume changes promote a bend of the layers. The change of poles cancels the volume changes and gives rise to the movement in the opposite direction.

Future Directions

In addition to conventional applications such as electrocatalysis, electronic devices, solar cells, and electrochromic windows, conducting polymers have proven to be fascinating in nanoscience field that touches the limits of macroscopic laws. The combination of conducting polymer and bipolar electrochemistry is also interesting for development of the related fields [9, 10].

Cross-References

- ► Electrosynthesis in Ionic Liquid
- ► Electrosynthesis in Supercritical Fluids
- ► Electrosynthesis of Polysilane
- ► Electrosynthesis Under Ultrasound and Centrifugal Fields

References

1. Heinze J, Frontana-Uribe BA, Ludwigs S (2010) Electrochemistry of conducting polymers-persistent models and new concepts. Chem Rev 110:4724–4771
2. Sekiguchi K, Atobe M, Fuchigami T (2002) Electropolymerization of pyrrole in 1-ethyl-3-methylimidazolium trifluoromethanesulfonate room temperature ionic liquid. Electrochem Commun 4:881–885
3. Chen W, Xue G (2005) Low potential electrochemical syntheses of heteroaromatic conducting polymers in a novel solvent system based on trifluroborate–ethyl ether. Prog Polym Sci 30:783–811
4. Utley JHP, Gruber J (2002) Electrochemical synthesis of poly(p-xylylenes) (PPXs) and poly(p-phenylene-vinylenes) (PPVs) and the study of xylylene (quinodimethane) intermediates; an underrated approach. J Mater Chem 12:1613–1624
5. Brédas JL, Street GB (1985) Polarons, bipolarons, and solitons in conducting polymers. Acc Chem Soc 18:309–315
6. Inzelt G (2012) Conducting polymers 2nd edition. Springer-Verlag, Berlin Heidelberg
7. Beaujuge PM, Reynolds JR (2010) Color control in π-conjugated organic polymers for use in electrochromic devices. Chem Rev 110:268–320
8. Kaneto K, Kaneko M, Min Y, MacDiarmid AG (1995) "Artificial muscle": electromechanical actuators using polyaniline films. Synth Met 71:2211
9. Inagi S, Ishiguro Y, Atobe M, Fuchigami T (2010) Bipolar patterning of conducting polymers by electrochemical doping and reaction. Angew Chem Int Ed 49:10136–10139
10. Ishiguro Y, Inagi S, Fuchigami T (2012) Site-controlled application of electric potential on a conducting polymer "canvas". J Am Chem Soc 134:4034–4036

Electrosynthesis of Fine Chemicals

Hideo Tanaka
Okayama University, Okayama, Japan

Electrosynthesis

Electrosynthesis is a quite unique tool for synthesis of fine chemicals, wherein highly selective, efficient, and/or environmental benign reactions are performed without use of any oxidizing and reducing reagents under atmospheric conditions. Many successful applications have been accumulated so far [1–3], and some electrosyntheses of useful chemicals in an industrial scale, such as electroreductive coupling acrylonitrile leading to adiponitrile, have been realized [4].

Kolbe reaction is frequently utilized for synthesis of useful fine chemicals. The Kolbe electrolysis approach does not always afford high yields of the products but presents some advantages over conventional chemical methods, since it furnishes the desired compounds in one single step and many different alkanoic acids can be used for the coupling reaction. Moreover, electrolysis conditions are very mild so that many functional groups, e.g., halogen, ketone, aldehyde, alcohol, ether, ester, lactone, and alkenyl groups, are preserved during the electrolysis. Kolbe reaction of half esters of dicarboxylic acids gives another dicarboxylic acid diester. For instance, the electro-decarboxylation of adipic acid half esters to the corresponding sebatic acid diesters has been studied intensively and realized in an industrial scale [4]. Kolbe cross-coupling of two different carboxylic acids may permit a variety of applications, e.g., electrosyntheses of long-chain carboxylic acids and dicarboxylic acid [5] (Fig. 1).

Electrolysis of carboxylates in the presence of olefins may afford the radical addition products [5]. Intramolecular radical addition may furnish a simple and straightforward access to cyclic compounds. Mixed decarboxylative coupling of 6-alkenoic acids and various carboxylic acids giving cyclic compounds is utilized for synthesis of useful chemicals. For instance, a stereoselective synthesis of prostaglandin precursors is achieved successfully by mixed Kolbe-type decarboxylation [6] (Fig. 2).

Epoxies, halohydrin, 1,2-dihalides, and 1,2-dihydroxides are useful fine chemicals in organic

$$2\,RO_2C\text{–}(CH_2)_n\text{–}CO_2^- \xrightarrow[-2e^-]{-2\,CO_2} RO_2C\text{–}(CH_2)_{2n}\text{–}CO_2R$$

$$R^1\text{–}CO_2^- + R^2O_2C\text{–}(CH_2)_n\text{–}CO_2^- \xrightarrow[-2e^-]{-2\,CO_2} R^1\text{–}(CH_2)_n\text{–}CO_2R^2$$

$$R^1O_2C\text{–}(CH_2)m\text{–}CO_2^- + R^2O_2C\text{–}(CH_2)_n\text{–}CO_2^- \xrightarrow[-2e^-]{-2\,CO_2} R^1O_2C\text{–}(CH_2)_{m+n}\text{–}CO_2R^2$$

Electrosynthesis of Fine Chemicals, Fig. 1 Kolbe Reaction and Hofer-Moest Reaction

Electrosynthesis of Fine Chemicals, Fig. 2 Decarboxylative Radical Addition

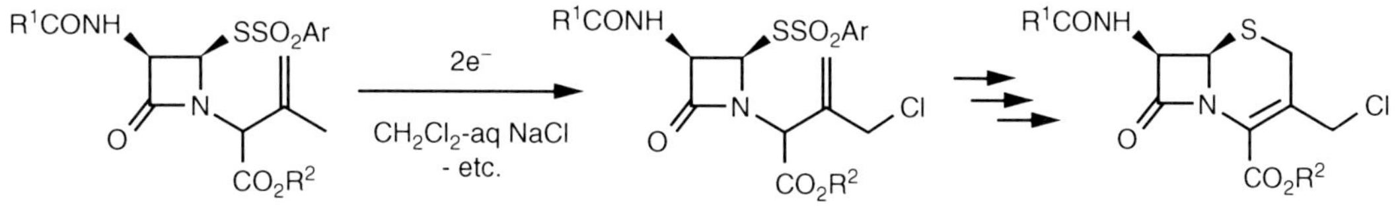

Electrosynthesis of Fine Chemicals, Fig. 3 Ene-type Chlorination

synthesis, and electrooxidation of olefins is a promising access to this class of chemicals [4, 7]. For instance, a regioselective *omega*-epoxydation of isoprenoids has been realized by the electrooxidation in a mixed solution of MeCN/THF/H_2O containing sodium bromide in an undivided cell. Passage of 2–4 F/mol of electricity affords good to excellent yields of the corresponding *omega*-epoxyisoprenoids, a prominent class of synthetic intermediates for useful alicyclic terpenoids [8]. In general, epoxydation of olefins proceeds in neutral or basic media, while bromohydrins are obtained preferentially achieved in acidic media. Electrooxidative ene-type chlorination of isoprenoids is also successful; thus, regio- and chemoselective ene-type chlorination has been realized by electrolysis in a CH_2Cl_2/ aqueous NaCl two-phase system [8]. The electrolytic ene-type chlorination is used as a key step in the synthesis of a versatile intermediate for cephalosporin antibiotics [9, 10] (Fig. 3).

Electrooxidation of aromatic compounds has been intensively investigated, and many useful fine chemicals have been prepared by both side-chain and aromatic nucleus oxidation. Side-chain oxidation of alkylbenzenes may furnish benzyl alcohols, benzyl acetates, benzyl methyl ethers, *N*-benzyl acetamides, benzaldehydes, benzoic acids, and so on. For instance, electrooxidation of *p*-methoxytoluene affords *p*-methoxybenzyl methyl ether, *p*-methoxybenzaldehyde, and/or its dimethylacetal depending on the choice of electrolysis media [3]. Many examples of electrooxidation of aromatic nucleus have been also reported. *p*-Quinones and their methyl acetals and semiquinones are prepared by electrooxidation of phenol derivatives and hydroquinones [3]. Nucleus-nucleus coupling of methoxybenzene derivatives proceeds regioselectively to afford the para-para coupling products (4,4′-methoxybiphenyl derivatives) [11].

Many electrochemical approaches to biaryls, e.g., electroreductive dimerization of aryl halides [12, 13] and electrooxidative dimerization of arylboronic acids [14], have been investigated. Substituted or non-substituted benzoic acids are electrochemically reduced to the corresponding alcohols or aldehydes. For instance, electroreduction of *m*-phenoxybenzoic acid proceeds smoothly to afford *m*-phenoxybenzaldehyde, an intermediate for insecticide [2, 15].

Electrooxidation of hetero-aromatic compounds furnishes an efficient access to useful chemicals; for instance, electrolysis of 2-(1-hydroxyethyl)furan in methanol affords the corresponding 2,5-dimethoxy-2,5-dihydrofuran, and subsequent few steps of the reactions lead to maltol, an important flavor [16].

Electrochemical oxidation and reduction of hetero atom compounds, such as N, S, and P compounds, has been intensively studied and utilized for synthesis of many fine chemicals [1–4]. Electrooxidative S-S, S-N, S-P, and N-P bond formation is performed successfully by electrolysis of thiols, disulfide/amine, disulfide/phosphate, amine/ phosphate, and so on, affording useful chemicals, e.g., thiuram disulfide [17], sulfenamide [18], sulfenimides [19], phosphorothiolates [20], phosphoramidate [21], and so on. For instance, cross-coupling of phthalimide and dicyclohexy disulfide is performed by electrolysis in acetonitrile containing a catalytic amount of sodium bromide under a constant applied voltage (3 V, 0.7–0.9 V vs. SCE) to afford *N*-(cyclohexylthio)phthalimide, an important prevulcanization inhibitor in the rubber industry, in quantitative yield [19] (Fig. 4).

Electrosynthesis of Fine Chemicals, Fig. 4 Electrooxidative N-S Bond Formation

Future Direction

Fine chemicals involving unsaturated linkages, e.g., C = C, C = O, C = N, and N = N, and/or hetero atoms, e.g., N, O, P, S, Cl, Br, and I, are target molecules for electroorganic synthesis, wherein the unsaturated linkages and/or hetero-atoms undergo electron release or electron uptake to give electron-deficient or electron-excessive active species and subsequent reactions may achieve highly selective and effective functionalization and/or carbon-carbon bond formation without use of any oxidizing and reducing agents. The electrosynthesis of a wide variety of fine chemicals is currently investigated, and many successful examples will be accumulated, thereby offering one of the most prominent and reliable tools for obtaining useful chemicals.

Cross-References

► Kolbe and Related Reactions

References

1. Lund H, Hammerich O (1991) Organic electrochemistry, 4h edn, Revised and Expanded. Marcel Dekker, New York
2. Torii S (2006) Electroorganic reduction synthesis. Kodansha/Wiley-VCH, Tokyo/Weinheim
3. Torii S (1985) Electroorganic syntheses, methods and applications, part 1: oxidations. Kodansha/VCH, Tokyo/Weinheim
4. Degner D (1988) Organic electrosyntheses in industry. In: Stckhan E (ed) Electrochemistry III, Topics in current chemistry 148. Springer, Berlin/Heidelberg/New York/London/Paris/Tokyo
5. Torii S, Tanaka H (1991) 14 carboxylic acid. In: Lund H, Hammerich O (eds) Organic electrochemistry, 4th edn, Revised and Expanded. Marcel Dekker, New York
6. Becking L, Schafer HJ (1988) Synthesis of a prostaglandin precursor by mixed Kolbe electrolysis of 3-(cyclopent-2-enyloxy)propionate. Tetrahedron Lett 29:2801–2802
7. Torii S, Uneyama K, Tanaka H, Yamanaka Y, Yasuda T, Ono M, Kohmoto Y (1981) Efficient conversion of olefins into epoxides, bromohydrins, and dibromides with sodium bromide in water-organic solvent electrolysis. J Org Chem 48:3312–3315
8. Torii S, Uneyama K, Ono M, Tazawa H, Matsunami S (1979) A regioselective omega-epoxydation of polyisoprenoids by the sodium bromide promoted electrochemical oxidation. Tetrahedron Lett 20:4661–4662
9. Torii S, Tanaka H, Inokuchi T (1988) Role of the electrochemical method in the transformation of beta-lactam antibiotics and terpenoids. In: Stckhan E (ed) Electrochemistry II, Topics in current chemistry 148. Springer, Berlin/Heidelberg/New York/London/Paris/Tokyo
10. Torii S, Tanaka H, Saitoh N, Siroi T, Sasaoka M, Nokami J (1982) Penicillin-cephalosporin conversion III. A novel route to 3-chloromethyl-d3-cephems. Tetrahedron Lett 23:2187–2188
11. Ronlan A, Bechgaard K, Parker VD (1973) Electrochemistry in media of intermediate acidity. Part VI. Coupling reactions of simple aryl ethers. Acta Chem Scand 27:2375–2382
12. Rollin Y, Troupel M, Perichon J, Fauvarque JF (1981) Electroreduction of nickel (II) salts in tetrahydrofuran-hexamethylphosphoramide mixtures. Application to electrochemical activation of carbon-halogen bonds. J Chem Res (S): 322–323
13. Amatore C, Carre E, Jutand A, Tanaka H, Ren S, Torii S (1996) Oxidative addition of aryl halides to transient anionic sigma-aryl-palladium(0) intermediates-application to palladium-catalyzed reductive coupling of aryl halides. Chem Eur J 2:957–966
14. Mitsudo K, Kaide T, Nakamoto E, Yoshida K, Tanaka H (2007) Electrochemical generation of cationic Pd catalysts and application to Pd/TEMPO double-mediatory electrooxidative wacker-type reaction. J Am Chem Soc 129:2246–2247
15. Chaintreau A, Adrian G, Couturier D (1981) Benzyl derivatives of 3-phenoxytoluene. Synth Commun 11:439–442
16. Torii S, Tanaka H, Anoda T, Simizu Y (1976) A convenient preparation of maltol, ethylmaltol and pyromeconic acid from 2-alkyl-6-methoxy-2H-pyran-3(6H)-ones. Chem Lett 5:495–498
17. Torii S, Tanaka H, Mishima K (1978) Electrosynthesis of hetero-hetero atom bonds 1. A direct preparation of bis(dialkylthiocarbamoyl) disulfides from dialkylamines and carbon disulfide. Bull Chem Soc Jpn 51:1575–1576

18. Torii S, Tanaka H, Ukida M (1978) Electrosynthesis of hetero-hetero atom bonds 2. An efficient preparation of (2-benzothiazolyl)- and thiocarbamoylsulfenamides by electrolytic cross-coupling reaction of 2-mercaptobenzothiazole, bis(2-benzothiazolyl) disulfide, and/or bis(dialkylthiocarbamoyl) disulfides with various amines. J Org Chem 43:3223–3227
19. Torii S, Tanaka H, Ukida M (1979) Electrosynthesis of hetero-hetero atom bonds 3. Sodium bromide promoted electrolytic cross-coupling reaction of imides with disulfide. N-(cyclohexylthio)phthalimide, an important prevulcanization inhibitor. J Org Chem 44:1554–1557
20. Torii S, Tanaka H, Sayo N (1979) Electrosynthesis of hetero-hetero atom bonds 4. Direct cross-coupling of dialkyl(or diaryl)phosphites with disulfides by a sodium bromide promoted electrolytic procedure. J Org Chem 44:2938–2941
21. Torii S, Sayo N, Tanaka H (1979) Electrosynthesis of hetero-hetero atom bonds 5. Direct cross-coupling of dialkylthosphites with amines by an iodonium ion-promoted electrolytic procedure. Tetrahedron Lett 20:4471–4474

Electrosynthesis of Polysilane

Manabu Ishifune
Kinki University, Higashi-Osaka, Osaka, Japan

Introduction

Polysilanes (Fig. 1) [1] have attracted considerable attention due to their usefulness as precursors for thermally stable ceramics [2, 3] or a material for microlithography [4, 5] and also due to their potentiality in preparation of new types of material showing semiconducting, photoconducting, or nonlinear optical property [6–8].

In contrast to growing interest with polysilanes, their preparation method is still limited. They have been traditionally prepared by the Wurtz-type condensation of dichlorosilanes with alkali metal (Kipping method). This method, however, requires drastic reaction conditions and, hence, is often limited in the type of substituents that are allowed to be located on the dichlorosilane monomer and also has a disadvantage in controlling the unit structure in the copolymerization. Several modified or alternative methods have been reported such as Wurtz-type reductive coupling of dichlorosilanes under well-controlled reaction conditions [9–11], transition metal-catalyzed dehydrogenative polymerization of hydrosilanes [12, 13], anionic polymerization of masked disilanes [14, 15], and ring-opening polymerization of silacycloalkanes [16, 17].

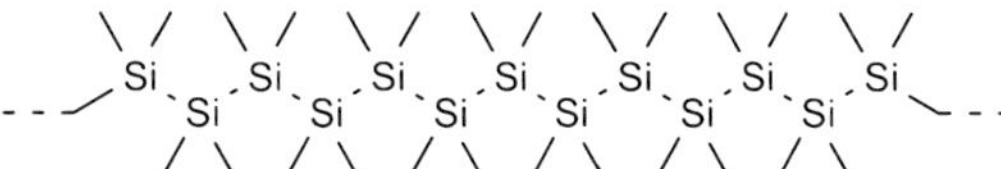

Electrosynthesis of Polysilane, Fig. 1 Polysilanes

Electrochemical method is another promising alternative for the synthesis of polysilanes. *Electrosynthesis of polysilane is the repeated Si-Si bonds formation under electrochemical conditions and can be categorized as cathodic and anodic reactions.* Two types of electrosyntheses of polysilanes, that is, *(1) electroreductive condensation polymerization of dichlorosilanes* and *(2) electrochemical condensation of dihydrosilanes*, have been developed. Hydrosilanes are active on the anode, and silyl radical cations are probably formed, then attack other hydrosilanes to form Si-Si bonds. The Si-Si bond is, however, also reactive on the anode. Therefore, the use of the reaction on the anode is limited for the electrosynthesis of high molecular weight polysilanes. For instance, electrolysis of methylphenylsilane under constant current conditions (cathode and anode; Pt, supporting electrolyte; Bu_4NBF_4, solvent; DME) affords the corresponding oligosilanes up to pentamer [18, 19]. The electroreductive polymerization of dichlorosilanes by using sacrificial electrodes is a preferable method to obtain high molecular weight polysilanes.

Electroreductive Condensation Polymerization of Dichlorosilanes

Electrode Materials

The electroreductive coupling of chlorosilanes with mercury electrode has been reported by Hengge in 1976 as a method to form disilanes [20]; however, this method is not effective in the preparation of polysilanes [21, 22]. The material of electrode is one of the most important factors to control the formation of Si-Si bond. When a solution of dichloromethylphenylsilane (**1a**) in dry THF containing $LiClO_4$ as a supporting electrolyte is electrochemically reduced with Mg cathode and anode under constant current conditions (current density = 30 mA/cm^2, supplied electricity = 4.0 F/mol), polymethylphenylsilane (**2a**) is obtained in 43 % yield after reprecipitation (Scheme 1) [23, 24]. Mg is a remarkably effective material of electrode, whereas Al gives low yield (15 %) and other materials such as Cu and Ni are rather ineffective. The electroreduction systems using Al or Cu anode in other electrolytes (Al anode/Bu_4NCl/DME [25], Al anode/LiCl/THF-HMPT [26], or Cu/Bu_4NClO_4/DME [27]) have been also reported; however, the molecular weight of the resulting polysilanes is relatively low. Other than Mg electrodes, the use of Ag anode and Pt cathode in DME containing Bu_4NCLO_4 is also reported to be effective to obtain high molecular weight polysilane [28, 29].

$$\underset{\mathbf{1a}}{Cl-\underset{Me}{\overset{Ph}{Si}}-Cl} \xrightarrow[\text{LiClO}_4\text{ / THF, sacrificial electrodes}]{+e} \underset{\mathbf{2}}{\left(\underset{Me}{\overset{Ph}{Si}}\right)_n}$$

Electrosynthesis of Polysilane, Scheme 1 Electroreductive polymerization of dichloromethyphenylsilane (**1a**)

Typical Experimental Procedure

The electrolysis of dichlorosilanes can be carried out in a 30 mL three-necked flask equipped with Mg electrodes (rods, 1 cm × 1 cm × 2.5 cm) (Fig. 2) [24]. Into this cell is placed 1.0 g of $LiClO_4$ and the content is dried in vacuo for 3 h. Chlorotrimethylsilane (0.5 mmol) and 15 mL of dry THF are then added under an argon atmosphere. After the solution is magnetically stirred for 3 h to remove residual water as hexamethyldisiloxane, 140 C of electricity is passed through the cell (pre-electrolysis) under a constant current

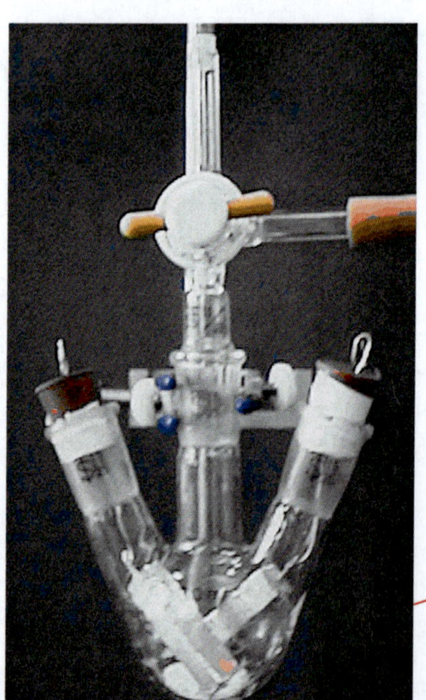

Electrosynthesis of Polysilane, Fig. 2 Electrolysis cell and typical polymerization conditions

condition (50 mA), and the polarity of the electrodes is alternated at an interval of 15 s using a commutator. During the electrolysis, the flask is externally sonicated (47 kHz) with cooling by running water. The sonication of ultrasound and the alternation of anode and cathode are effective for continuous supply of electric current. The dichlorosilane monomer (10 mmol) is then syringed into the cell, and the solution is further electrolyzed. After 4 F/mol of electricity is passed, the reaction is quenched with 5 mL of ethanol, and the resulting mixture is poured into ice cold 1 N HCl (50 mL). The aqueous solution is extracted with diethyl ether (50 mL × 3). The combined ethereal layers are washed twice with 50 mL of saturated aqueous NaCl, dried over anhydrous $MgSO_4$, and concentrated. The resulting crude polymer is dissolved in 3–5 mL of benzene and reprecipitated from ethanol (200 mL).

Effect of Monomer Concentration

Monomer concentration is another important factor to obtain high molecular weight polysilane [24]. The molecular weight of **2** becomes higher with an increase in the concentration of dichlorosilane monomer **1a**. The molecular weight (Mn) of **2a** is, for instance, 31,000 when the electroreduction of **1a** is carried out under high-concentration condition (1.2 mol/L) at 0.5 F/mol of supplied electricity though the material yield of **2a** decreased. The most satisfactory result, in which material yield is 79 % and molecular weight (Mn) was 9,900, is obtained when the concentration of **1a** is 0.67 mol/L. The polysilane **2a** obtained here shows relatively sharp monomodal distribution of molecular weight in the elution profile of gel permeation chromatograph, whereas the polysilanes prepared by the alkali metal condensation method usually show broad bimodal distribution.

Electroreductive Polymerization of Dichlorooligosilanes (Control of Repeat Units Sequence)

The electroreductive polymerization of the dichlorooligosilanes is useful for the synthesis of sequence-ordered polysilanes [30]. The electroreduction of dichlorodisilane **3** affords the corresponding polysilane **4** consisting of disilane units (Scheme 2); however, the yield of the resulting polymer is low (1–4 %). It is probably due to high reactivity of the disilene intermediate formed by the electroreduction of **3**. In fact, the addition of naphthalene, which could make a masked disilene intermediate, into the reaction system slightly increases the yield of the polysilane **4** (13 %).

Cl–Si(Ph)(Me)–Si(Me)(Me)–Cl (**3**) —[+e; $LiClO_4$ / THF; Mg electrodes; *sonication*]→ –(Si(Ph)(Me)–Si(Me)(Me))$_n$– (**4**)

Electrosynthesis of Polysilane, Scheme 2 Electroreductive polymerization of 1,2-dichloro-1,1,2-trimethyl-2-phenyldisilane (**3**)

Cl–Si(Me)(Me)–Si(Ph)(Ph)–Si(Me)(Me)–Cl (**5**) —[+e; $LiClO_4$ / THF; Mg electrodes; Ultrasound]→ –(Si(Me)(Me)–Si(Ph)(Ph)–Si(Me)(Me))$_n$– (**6**)

Electrosynthesis of Polysilane, Scheme 3 Electroreductive polymerization of 1,3-dichloro-1,1,3,3-tetramethyl-2,2-diphenyltrisilane (**5**)

Dichlorooligosilanes, such as dichlorotrisilane **5**, are good monomers for the electroreductive synthesis of the polysilanes having longer sequence units (Scheme 3). The temperature control is very important in the electroreductive polymerization of **5**. The reaction at higher temperature, the backbiting reaction of the propagating polymer, proceeds forming cyclohexasilane as a by-product at room temperature. This side reaction can be suppressed when the reaction is carried out below 0 °C, and polysilanes **5** having relatively high molecular weight are obtained (Mn = 4,400–5,500). Under the optimized reaction conditions (polymerization temperature; −10 °C), the electroreduction of dichlorotetrasilane **7** gives the corresponding polysilane **8**, units of which are ordered in four sequences in satisfactory yield (Scheme 4). The polymerizability of dichlorooligosilanes under the

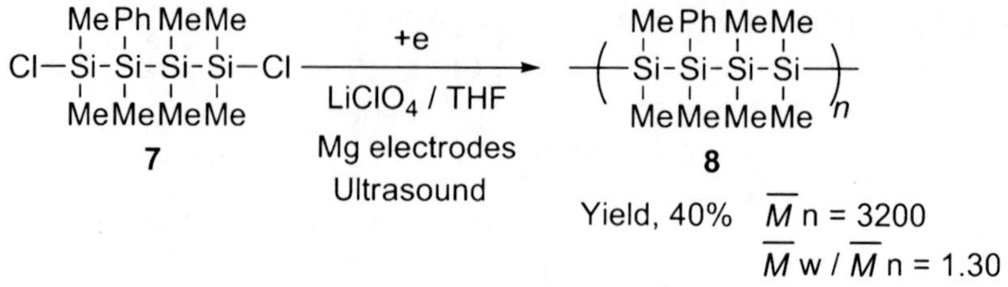

Electrosynthesis of Polysilane, Scheme 4 Electroreductive polymerization of 1,4-dichloro-1,1,2,3,3,4,4-heptamethyl-2-phenyltetrasilane (**7**)

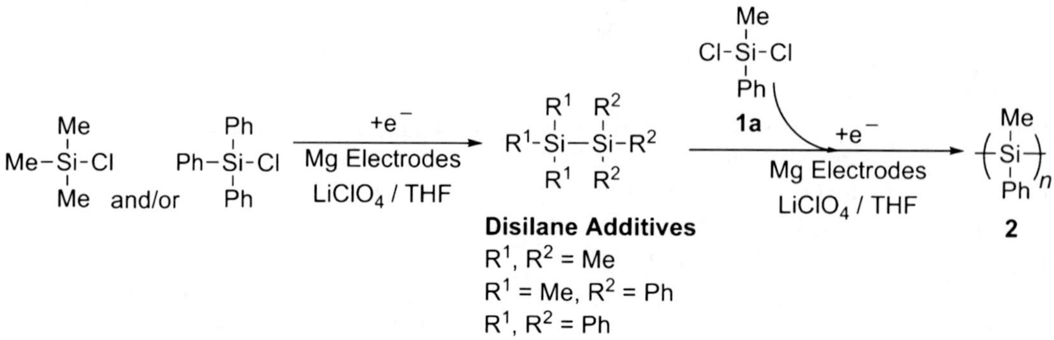

Electrosynthesis of Polysilane, Scheme 5 Electroreductive polymerization of dichloromethyphenylsilane (**1a**) in the presence of disilane additives

electroreduction conditions seems to be mainly affected by the substituents on the chlorinated terminal silicon atom, and this fact provides a wide possibility to design the oligosilane sequences of the inner silicon atoms.

Electroreductive Polymerization of Dichlorosilanes in the Presence of Disilane Additives (Control of Molecular Weight Distribution)

The disilane additives, which are generated in situ in electroreductive coupling of the corresponding chlorosilanes, are effective to the promotion of the electroreductive polymerization of dichloromethylphenylsilane (**1**) and the control of the molecular weight distribution of the resulting polysilanes (Scheme 5) [31]. The electroreduction of dichlorosilane **1** in the presence of 1,1,1-trimethyl-2,2,2-triphenyldisilane gives the corresponding polysilane **2** in 56 % yield, and the number average molecular weight and the molecular weight distribution are determined by GPC to be 3,000 and 1.10, respectively. The reduction of dichloromethylphenylsilane (**1**) by Wurtz-type condensation using metal lithium in the presence of catalytic amount of 1,1,1-trimethyl-2,2,2-triphenyldisilane affords five- and six-member ring products [32]. On the other hand, the cyclosilanes are not detected under the electroreductive conditions. The use of 1,1,1,2,2,2-hexaphenyldisilane affords the polysilane **2** in 59 % yield, and the Mw/Mn is 1.08. Thus, the polysilanes prepared in the presence of the disilane additive containing triphenylsilyl group show narrower molecular weight distributions than the polysilane prepared without the disilane additive.

Electroreductive Block Copolymerization Using Triphenylsilyl Group-Terminated Polysilane

The triphenylsilyl group-terminated polysilanes can be synthesized by the electroreductive

1a (Cl–Si(Ph)(Me)–Cl) → [+e⁻; $LiClO_4$, THF; Mg electrodes; $Ph_3SiSiPh_3$; 4 F/mol] → [+e⁻, Ph_3SiCl; *electroreductive termination*; 2.5 F/mol] → 2a $(-Si(Ph)(Me)-)_m$ → [1b (Cl–Si(Bu)(Bu)–Cl), +e⁻; 3 F/mol; *in the presence of or without* disilane additives] → 9 $(-Si(Ph)(Me)-)_m(-Si(Bu)(Bu)-)_n$

2a: *Mn* = 3350, *Mw* / *Mn* = 1.4, yield 16%

9: *Mn* = 4730, *Mw* / *Mn* = 1.2, *m* : *n* = 75 : 25, yield 38%

Electrosynthesis of Polysilane, Scheme 6 Electroreductive block copolymerization with dibutyldichlorosilane (**1b**) using triphenylsilyl group-terminated polymethylphenylsilane (**2a**) as a macroinitiator

1b (Cl–Si(Bu)(Bu)–Cl) → [+e⁻; $LiClO_4$, THF; Mg electrodes; $Ph_3SiSiPh_3$; 4 F/mol] → [+e⁻, Ph_3SiCl; *electroreductive termination*; 2.5 F/mol] → 2b $(-Si(Bu)(Bu)-)_m$ → [1a (Cl–Si(Ph)(Me)–Cl), +e⁻; 3 F/mol; *in the presence of or without* disilane additives] → 9' $(-Si(Bu)(Bu)-)_m(-Si(Ph)(Me)-)_n$

2b: *Mn* = 3950, *Mw* / *Mn* = 1.7, yield 19%

9': *Mn* = 4390, *Mw* / *Mn* = 1.3, *m* : *n* = 61 : 39, yield 25%

Electrosynthesis of Polysilane, Scheme 7 Electroreductive block copolymerization with dichloromethylphenylsilane (**1a**) using triphenylsilyl group-terminated polydibutylsilane (**2b**) as a macroinitiator

polymerization of dichloromethylphenylsilane (**1a**) in the presence of 1,1,1,2,2,2-hexaphenyldisilane, and the electroreductive termination with chlorotriphenylsilane is effective to ensure the terminus of the resulting polysilane for triphenylsilyl group [31]. By using the isolated triphenylsilyl group-terminated poly(methylphenylsilane) **2a** as a macroinitiator, the electroreductive copolymerization with dibutyldichlorosilane (**1b**) proceeds affording the corresponding block copolymer, polymethylphenylsilane-*block*-polydibutylsilane (**9**), in 38 % yield (Scheme 6) [33]. Under these conditions dichlorosilane **1b** is first electroreduced to form the corresponding oligomeric silyl anion, which attacks triphenylsilyl group-terminated polysilane **1**, and the electroreductive condensation with dichlorosilane **1b** continues. The molecular weight of the copolymer obtained from the triphenylsilyl group-terminated polysilane **2a** (Mn = 3,350, Mw/Mn = 1.4) increases to 4,730, while the molecular weight of the polymer obtained from the polysilane terminated with ethanol is not much increased. The GPC profile of the resulting copolymer **9** is monomodal, and the polydispersity index value (Mw/Mn) is 1.2. The repeat unit ratio (–Si(Me)Ph-:-$SiBu_2$-)

of the copolymer **9** prepared is determined by ^{1}H NMR to be 75:25, which shows a good agreement with the ratio (74:26) calculated on the assumption that the molecular weight of the macroinitiator is not changed.

Polydibutylsilane-*block*-polymethylphenylsilane (**9'**) can be also obtained by using triphenylsilyl group-terminated polydibutylsilane (**2b**) as a macroinitiator (Scheme 7). The electroreductive polymerization of dibutyldichlorosilane (**1b**) in the presence of 1,1,1,2,2,2-hexaphenyldisilane followed by electroreductive termination with chlorotriphenylsilane affords the macroinitiator **2b** (Mn = 3,950, Mw/Mn = 1.7). The electroreductive polymerization of dichloromethylphenylsilane (**1a**) initiates from **2b** producing the corresponding copolymer **9'** in 25 % yield, and the molecular weight of **9'** is 4,390. The polydispersity index value (Mw/Mn) is 1.3, and the repeat unit ratio ($-SiBu_2$- : -Si(Me)Ph-) is 61:39.

Multiblock polysilane copolymers can be made by modified Wurtz-type condensation reactions of dichlorosilanes [34]; however, it is difficult to obtain simple di-block polysilane copolymers. The electroreductive method in this study provides a new procedure to synthesize well-controlled di-block polysilane copolymers. It is possible to control the sequence length of each block by tuning supplied electricity and monomer concentrations.

Future Directions

Electroreductive polymerization of dichlorosilanes has several advantages as an alternative method of Wurtz-type reductive coupling reaction. The molecular weights of polysilanes are tunable from thousands to millions by controlling the supplied electricity. The molecular weight distribution is monomodal, and the polydispersity index value (Mw/Mn) is usually less than 2.0, which can be controlled by catalytic amount of disilane additives. Several kinds of end groups can be introduced by selection of quenching conditions such as electroreductive terminations, and the triphenylsilyl group-terminated polysilanes react as macroinitiators under electroreductive conditions affording the block copolymers. Regarding structural control of linear polysilanes, stereochemistry (tacticity) of the polysilane main chains in the electroreductive polymerization must be one of the most important research topics. The electroreductive polymerization is adaptable for the synthesis of polysilanes having various functional groups including chiral substituents [35]. Formation of chiral helical structure of the polysilanes having the chiral side chains has been studied [36]; however, there are few synthesized examples. The electroreductive polymerization of the dichlorosilane monomers having chiral and coordinating substituents will provide valuable examples of the tacticity control of the polysilanes. Electrosynthesis of network polysilanes [37, 38] or polysilane dendrimers are also interesting topics from the viewpoints of the optical properties of polysilanes.

Cross-References

- ► Electrogenerated Reactive Species
- ► Electrosynthesis of Conducting Polymer
- ► Electrosynthesis Under Ultrasound and Centrifugal Fields
- ► Reactive Metal Electrode

References

1. West R (1986) The polysilane high polymers. J Organomet Chem 300(1–2):327–346
2. Yajima S, Hasegawa Y, Hayashi J, Iimura M (1978) Synthesis of continuous silicon carbide fiber with high tensile strength and high Young's modulus. Part 1. Synthesis of polycarbosilane as precursor. J Mater Sci 13(12):2569–2576
3. Hasegawa Y, Okamura K (1985) Silicon carbide-carbon composite materials synthesized by pyrolysis of polycarbosilane. J Mater Sci Lett 4(3):356–358
4. Miller RD, Willson CG, Wallroff GM, Clecak N, Sooriyakumaran R, Michl J, Karatsu T, McKinley AJ, Klingensmith KA, Downing J (1989) Polysilanes: photochemistry and deep-UV lithography. Polym Eng Sci 29(13):882–886
5. Miller RD, Michl J (1989) Polysilane high polymers. Chem Rev 89(6):1359–1410
6. West R, David LD, Djurovich PI, Stearley KL, Srinivasan KSV, Yu H (1981) Phenylmethylpolysilanes:

formable silane copolymers with potential semiconducting properties. J Am Chem Soc 103(24):7352–7354
7. Kepler RG, Zeigler JM, Harrah LA, Kurtz SR (1987) Photocarrier generation and transport in s-bonded polysilanes. Phys Rev B 35(6):2818–2822
8. Baumert JC, Bjorklund GC, Jundt DH, Jurich MC, Looser H, Miller RD, Rabolt J, Soorijakumaran R, Swalen JD, Twing RJ (1988) Temperature dependence of the third-order nonlinear optical susceptibilities in polysilanes and polygermanes. Appl Phys Lett 53(13):1147–1149
9. Matyjaszewski K, Greszta D, Hrkach JS, Kim HK (1995) Sonochemical synthesis of polysilylenes by reductive coupling of disubstituted dichlorosilanes with alkali metals. Macromolecules 28(1):59–72
10. Jones RD, Holder SJ (2006) High-yield controlled syntheses of polysilanes by the Wurtz-type reductive coupling reaction. Polym Int 55(7):711–718
11. Koe J (2008) Contemporary polysilane synthesis and functionalisation. Polym Int 58(3):255–260
12. Tilley TD (1993) The coordination polymerization of silanes to polysilanes by a "σ-bond metathesis" mechanism. Implications for linear chain growth. Acc Chem Res 26(1):22–29
13. Minato M, Matsumoto T, Ichikawa M, Ito T (2003) Dehydropolymerization of arylsilanes catalyzed by a novel silylmolybdenum complex. Chem Commun 24:2968–2969
14. Sanji T, Kawabata K, Sakurai H (2000) Alkoxide initiation of anionic polymerization of masked disilenes to polysilanes. J Organomet Chem 611(1–2):32–35
15. Sanji T, Isozaki S, Yoshida M, Sakamoto K, Sakurai H (2003) Functional transformation of poly (dialkylaminotrimethyldisilene) prepared by anionic polymerization of the masked disilenes. The preparation of a true polysilastyrene. J Organomet Chem 685(1–2):65–69
16. Cypryk M, Gupta Y, Matyjaszewski K (1991) Anionic ring-opening polymerization of 1,2,3,4-tetramethyl-1,2,3,4-tetraphenylcyclotetrasilane. J Am Chem Soc 113(3):1046–1047
17. Suzuki M, Kotani J, Gyobu S, Kaneko T, Saegusa T (1994) Synthesis of sequence-ordered polysilane by anionic ring-opening polymerization of phenylnonamethylcyclopentasilane. Macromolecules 27(8): 2360–2363
18. Kimata Y, Suzuki H, Satoh S, Kuriyama A (1994) Synthesis of oligosilanes by electrolysis of hydrosilanes. Chem Lett 7:1163–1164
19. Kimata Y, Suzuki H, Satoh S, Kuriyama A (1995) Electrochemical polymerization of hydrosilane compounds. Organomet 14(5):2506–2511
20. Hengge E, Litscher G (1976) A new electrochemical method for the formation of silicon-silicon bonds. Angew Chem 88(12):414
21. Hengge E, Litscher G (1978) Electrochemical formation of di-, oligo- and polysilanes. Monatsh Chem 109(5):1217–1225
22. Hengge E, Firgo H (1981) An electrochemical method for the synthesis of silicon-silicon bonds. J Organomet Chem 212(2):155–161
23. Shono T, Kashimura S, Ishifune M, Nishida R (1990) Electroreductive formation of polysilanes. J Chem Soc Chem Commun 17:1160–1161
24. Kashimura S, Ishifune M, Yamashita N, Bu HB, Takebayashi M, Kitajima S, Yoshihara D, Kataoka Y, Nishida R, Kawasaki S, Murase H, Shono T (1999) Electroreductive synthesis of polysilanes, polygermanes, and related polymers with magnesium electrodes. J Org Chem 64(18):6615–6621
25. Umezawa M, Takeda M, Ichikawa H, Ishikawa T, Koizumi T, Nonaka T (1991) Electroreductive polymerization of mixtures of chloromonosilanes. Electrochim Acta 36(3–4):621–624
26. Biran C, Bordeau M, Pons P, Leger MP, Dunogues J (1990) Electrosynthesis, a convenient route to di- and polysilanes. J Organomet Chem 382(3):C17–C20
27. Kunai A, Kawakami T, Toyoda E, Ishikawa M (1991) Electrochemistry of organosilicon compounds. 2. Synthesis of polysilane oligomers by a copper electrode system. Organometallics 10(6):2001–2003
28. Okano M, Takeda K, Toriumi T, Hamano H (1998) Electrochemical synthesis of polygermanes. Electrochim Acta 44(4):659–666
29. Yamada K, Okano M (2006) Electrochemical synthesis of poly(cyclotetramethylenesilylene). Electrochemistry 74(8):668–671
30. Ishifune M, Kashimura S, Kogai Y, Fukuhara Y, Kato T, Bu HB, Yamashita N, Murai Y, Murase H, Nishida R (2000) Electroreductive synthesis of oligosilanes and polysilanes with ordered sequences. J Organomet Chem 611(1–2):26–31
31. Ishifune M, Kogai Y, Uchida K (2005) Effect of disilane additives on the electroreductive polymerization of organodichlorosilanes. J Macromol Sci Part A Pure and Appl Chem 42(7):921–929
32. Chen SM, David LD, Haller KJ, Wadsworth CL, West R (1983) Isomers of $(PhMeSi)_6$ and $(PhMeSi)_5$. Organometallics 2(3):409–414
33. Ishifune M, Sana C, Ando M, Tsuyama Y (2011) Electroreductive block copolymerization of dichlorosilanes in the presence of disilane additives. Polym Int 60(8):1208–1214
34. Kawabe T, Naito M, Fujiki M (2008) Multiblock polysilane copolymers: One-pot Wurtz synthesis, fluoride anion-induced block-selective scission experiments, and spectroscopic characterization. Macromolecules 41(6):1952–1960
35. Kashimura S, Ishifune M, Bu HB, Takebayashi M, Kitajima S, Yoshihara D, NishidaR KS, Murase H, Shono T (1997) Electroorganic chemistry. 153. Electroreductive synthesis of some functionalized polysilanes and related polymers. Tetrahedron Lett 38(26):4607–4610
36. Fujiki M (2003) Switching handedness in optically active polysilane. J Organomet Chem 685(1–2): 15–34

37. Huang K, Vermeulen LA (1998) First electrochemical synthesis of network silane and silane-germane copolymers: $(C_6H_{11}Si)_x(PhSi)_y$ and $(C_6H_{11}Si)_x(PhGe)_y$. Chem Commun 2:247–248
38. Okano M, Nakamura K, Yamada K, Hosoda N, Wakasa M (2006) An improvement electrochemical synthesis of network polysilanes. Electrochemistry 74(12):956–958

Electrosynthesis Under Photo-Irradiation

Hisashi Shimakoshi and Yoshio Hisaeda
Department of Chemistry and Biochemistry, Kyushu University, Graduate School of Engineering, Fukuoka, Japan

Introduction to Light-Assisted Electrosynthesis

Electroorganic chemistry is significantly expanding into various interdisciplinary fields because of its economical as well as environmental advantages [1]. One of the great advantages of the electrochemistry associated with transition metal catalysis is a convenient alternative to the usual methods to generate in situ low-valent species which are not easily prepared and/or handled. For example, the electrochemically generated Co(I) species is known as a supernucleophile that forms an alkylated complex by reaction with electrophiles such as an organic halide (RX). The alkylated complex is a useful reagent for forming radical species as the cobalt–carbon bond is readily cleaved homolytically by electrolysis, thermolysis, and photolysis [2, 3]. Therefore, the application of the alkylated complex to organic synthesis is quite interesting from the viewpoint of a radical-forming reagent in place of the conventional chemical reagent such as tin hydride. In the electrochemical system, the formation of radical species is facilitated during photoirradiation as the cobalt–carbon bond of the alkylated complex is cleaved by photolysis (see Scheme 1). Thus, electrolysis of an electrophilic reagent in the presence of a cobalt complex as a mediator will form a radical intermediate under a lower applied potential with the support of light energy (see Fig. 1). A variety of radical-mediated electroorganic syntheses have been developed.

Electrochemical Acylation of Activated Olefins

Light-assisted electrosynthesis was generally carried out using a vitamin B_{12} derivative as the mediator. Vitamin B_{12} is a term found in health and nourishment studies and has the chemical name of cyanocobalamin [4]. However, this name is typically used for a cobalamin derivative. The vitamin B_{12} derivative denotes the cobalamin family here. Vitamin B_{12} consists of a cobalt atom coordinated to the tetrapyrrole ring system (corrin ring). One of the unique properties of vitamin B_{12} is possessing a stable cobalt–carbon bond in its biological function, and vitamin B_{12} is associated with

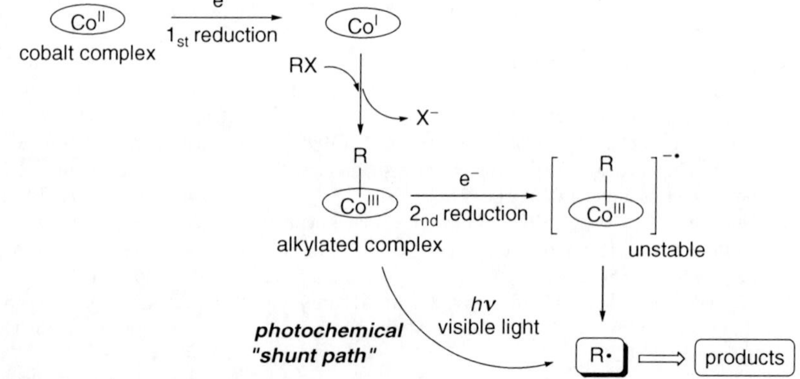

Electrosynthesis Under Photo-Irradiation, Scheme 1 Scheme for cobalt catalyzed radical reaction

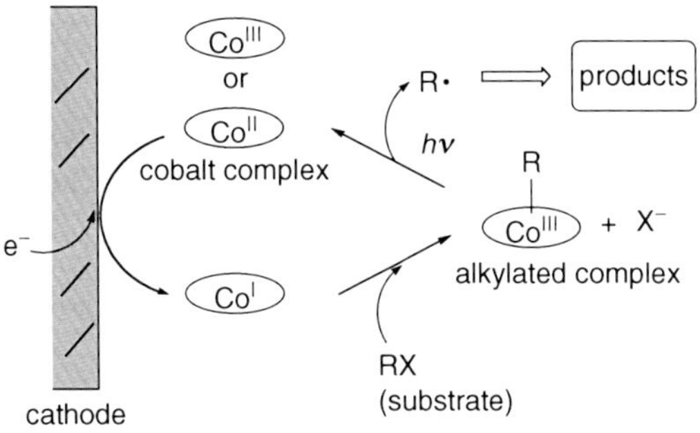

Electrosynthesis Under Photo-Irradiation, Fig. 1 Electrochemical reactions mediated by cobalt complex under photoirradation

radical-mediated enzymatic reactions [5, 6]. The application of vitamin B_{12} in catalytic chemistry was developed in electroorganic synthesis [7, 8]. Hydroxocobalamin hydrochloride (vitamin B_{12a}) as shown in Fig. 2 was used for the electrosynthesis of 1,4-dioxo compounds under photoirradiation Eq. 1 [9, 10]. 1,4-Dioxo compounds are valuable precursors for the synthesis of cyclopentanoids and furanoids. The reaction of the catalyst involves the formation and cleavage of a cobalt–carbon bond. The electrochemically formed Co(I) complex at −0.8 to −1.0 V versus SCE in DMF reacts with carboxylic anhydrides

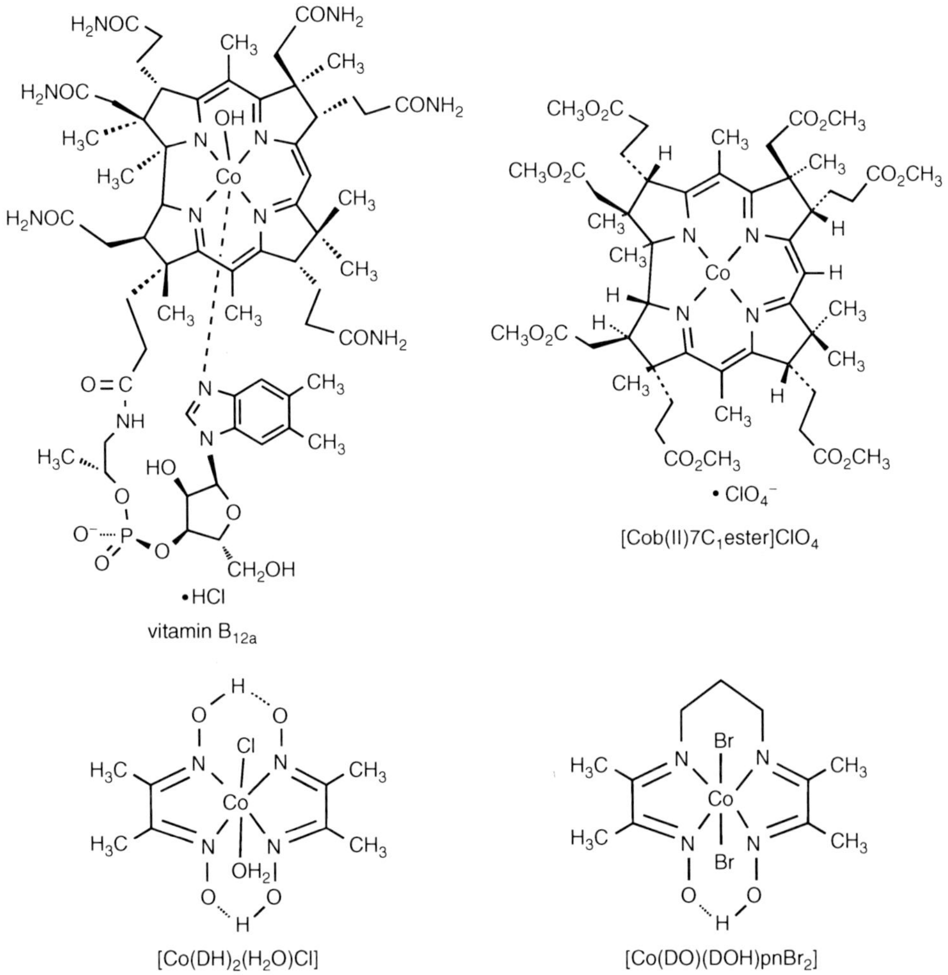

Electrosynthesis Under Photo-Irradiation, Fig. 2 Structure of vitamin B_{12} derivative and model complexes

Electrosynthesis Under Photo-Irradiation, Fig. 3 Electrochemical acylation of activated olefins mediated by vitamin B_{12a} under photoirradiation

to form the Co(III) acyl compound. Cleavage of the cobalt–carbon bond of the Co(III) acyl compound by visible light irradiation ($\lambda = 400$–550 nm) forms an acyl fragment, and it adds to the activated olefin to produce the desired 1,4-dioxo compound (see Fig. 3). The mechanism of such a light-assisted nucleophilic acylation of activated olefins was kinetically investigated [11].

$$(RCO)_2O + CH_2{=}CH{-}CHO \xrightarrow[B12_a]{h\nu,\ -0.95\ V\ vs.\ SCE} RCO{-}CH_2CH_2{-}CHO \quad (1)$$

1,4-dioxo compound

Yield: 47 % (R = CH_3)
71 % (n-C_7H_{15})

Electrochemical Reduction of Organic Bromide

The simple reduction of an organic bromide was also catalyzed by the vitamin B_{12} derivative, heptamethyl cobrinate ($Cob(II)7C_1ester$) (see Fig. 2) [12, 13]. Electrolysis of 2,2-bis (ethoxycarbonyl)-1-bromopropane was carried out at −1.0 V versus SCE in DMF under visible light irradiation (tungsten lamp, 300 W) in the presence of a catalytic amount of heptamethyl cobrinate. The simply reduced product, 2,2-bis(-ethoxycarbonyl)propane, was formed in 9–12 % yields Eq. 2. The reaction did not proceed under dark conditions since the corresponding alkylated cobalt complex is stable under this condition. The same reaction was also carried out using the simple vitamin B_{12} model complex, $[Co(DO)(DOH)pnBr_2]$ (see Fig. 2), but the yield of 2,2-bis(ethoxycarbonyl)propane was low [14, 15].

$$BrH_2C{-}C(R)(CO_2C_2H_5){-}CH_3 \xrightarrow[Cob(II)7C_1ester]{h\nu,\ -1.0\ V\ vs.\ SCE} H_3C{-}C(R)(CO_2C_2H_5){-}CH_3 \quad (2)$$

Yield: 10 % (R = $CO_2C_2H_5$)
5 % (R = CN)

Electrochemical 1,2-Migration of Functional Groups

The 1,2-migration of a functional group is a significantly important process in synthetic organic chemistry since the carbon-skeleton rearrangement creates various synthetic intermediates for fine chemicals synthesis. This carbon-skeleton rearrangement reaction is mediated by adenosylcobalamin in methylmalonyl-CoA mutase, glutamate mutase, and methyleneglutarate mutase in an enzymatic process [4, 5]. To mimic this enzymatic reaction, a vitamin B_{12} derivative-mediated electrolysis under photoirradiation was effectively developed. For example, the electrolysis of 2-acetyl-1-bromo-2-ethoxycarbonylpropane at −1.0 V versus SCE in DMF under visible light irradiation in the presence of a catalytic amount of heptamethyl cobrinate (Cob(II) $7C_I$ester) affords the 10 % yield of acetyl-migrated product as main product Eq. 3 [16].

$$H_2C(Br)-C(COCH_3)(CO_2C_2H_5)-CH_3 \xrightarrow[\text{Cob(II)7C}_1\text{ester}]{h\nu,\ -1.0\text{ V vs. SCE}} H_2C(COCH_3)-CH(CO_2C_2H_5)-CH_3 \ (\text{Yield: } 10\ \%) + H_3C-C(COCH_3)(CO_2C_2H_5)-CH_3 \ (3\ \%) \quad (3)$$

The acyl migration reaction was applied to a ring-expansion reaction. The electrolysis of alicyclic ketones (5-,6-,7-, and 8-membered rings) with a carboxylic ester and a bromomethyl group was carried out in DMF in the presence of a catalytic amount of heptamethyl cobyrinate (Cob(II)$7C_I$ester) at −1.0 V versus SCE under photoirradiation Eq. 4 [17]. The ring-expanded products were obtained in 5–11 % yields. A similar ring-expansion reaction via a 1,2-acyl migration occurred by the electrolysis of 2-alkyl-2-(bromomethyl)cycloalkanones with [Co $(DH)_2(H_2O)Cl$] (see Fig. 2) as the mediator. The corresponding ring-expanded 3-alkyl-2-alkenones were obtained by constant current electrolysis at 20 mA/cm^2 under photoirradiation (tungsten lamp, 750 W) with moderate yields Eq. 5 [18].

2-(bromomethyl)-2-($CO_2C_2H_5$)cycloalkanone $\xrightarrow[\text{Cob(II)7C}_1\text{ester}]{h\nu,\ -1.0\text{ V vs. SCE}}$ ring-expanded 3-($CO_2C_2H_5$)cycloalkanone + 2-CH_3-2-($CO_2C_2H_5$)cycloalkanone (4)

	Yield	
n = 1 (5-membered ring)	6 %	1 %
n = 2 (6-membered ring)	5 %	3 %
n = 3 (7-membered ring)	11 %	1 %
n = 4 (8-membered ring)	7 %	1 %

2-(bromomethyl)-2-R-cycloalkanone $\xrightarrow[[Co(DH)_2(H_2O)Cl]]{h\nu,\ 20\text{ mA/cm}^2}$ 3-R-2-cycloalkenone + 3-R-cycloalkanone + 2-CH_3-2-R-cycloalkanone (5)

R = n-C_6H_{13}, n = 1 (5-membered ring)	74 %	4 %	17 %
R = n-C_4H_9, n = 1 (5-membered ring)	52 %	3 %	15 %
R = n-$C_{11}H_{23}$, n = 1 (5-membered ring)	69 %	2 %	18 %
R = n-C_6H_{13}, n = 2 (6-membered ring)	51 %	trace	23 %

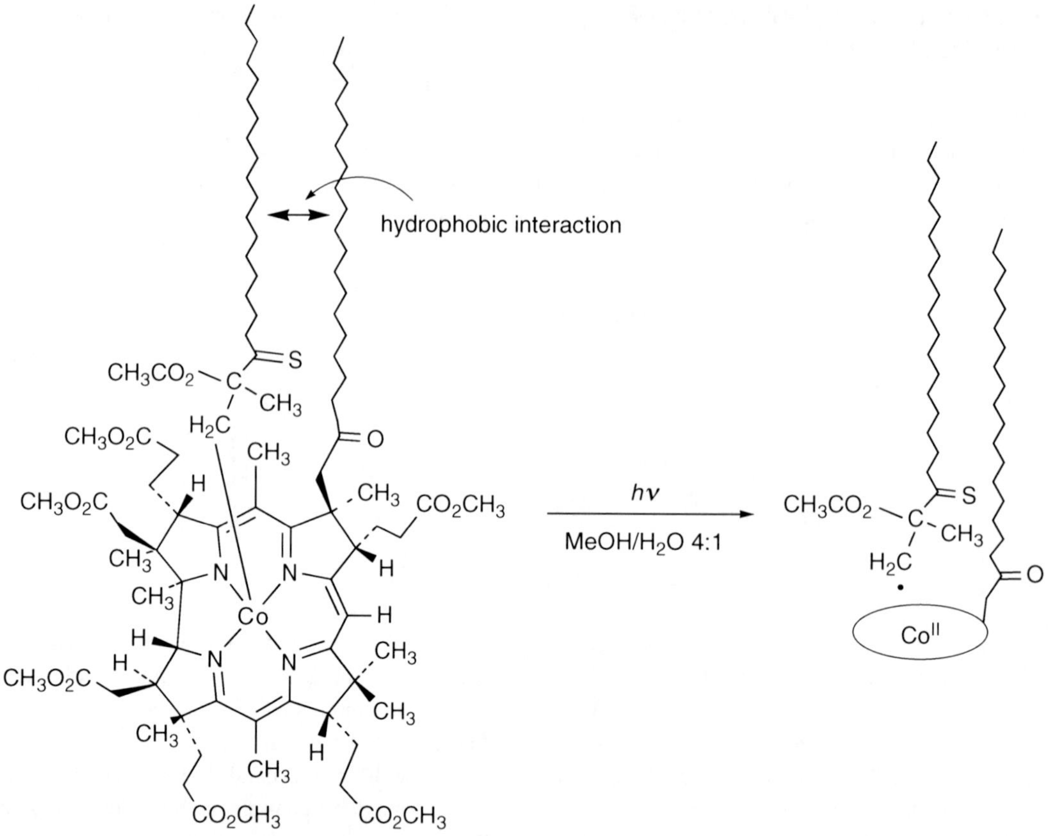

Electrosynthesis Under Photo-Irradiation, Fig. 4 Schematic illustration of hydrophobic peripheral association in alkylated complex

As for the model studies of the coenzyme B_{12} catalyzed methylmalonyl to succinyl rearrangement, the interaction between a vitamin B_{12} derivative containing a peripheral C_{18} alkyl chain and a (methyl)thiomalonate substrate bearing long alkyl chains at the thioester group was effectively used for electrolysis under photoirradiation (see Fig. 4). The electrolysis of thioesters at −0.85 V versus SCE in MeOH/H_2O (*v*/*v*, 4:1) in the presence of the 5 % vitamin B_{12} derivative, hexamethyl c-octadecyl cobyrinate, under photoirradiation afforded a 22 % of a rearranged product Eq. 6 [19, 20]. The absence of a long alkyl chain in the catalyst, heptamethyl cobyrinate (Cob(II)7C_1ester), results in a sluggish reaction with a low yield and poor reproducibility. Noncovalent associations between the substrate and the catalyst are essential for this reaction.

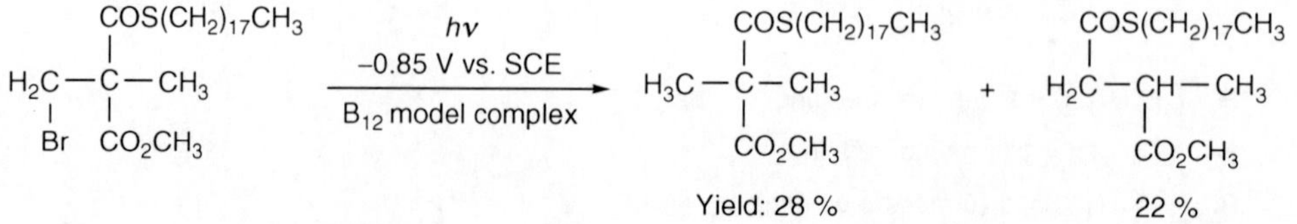

$$\mathrm{BrH_2C{-}C(CH_3)(CO_2CH_3){-}COS(CH_2)_{17}CH_3} \xrightarrow[\text{B}_{12}\text{ model complex}]{h\nu,\ -0.85\text{ V vs. SCE}} \underset{\text{Yield: 28 \%}}{\mathrm{H_3C{-}C(CH_3)(CO_2CH_3){-}COS(CH_2)_{17}CH_3}} + \underset{\text{22 \%}}{\mathrm{COS(CH_2)_{17}CH_3{-}H_2C{-}CH(CO_2CH_3){-}CH_3}} \qquad (6)$$

Electrochemical Intramolecular Cyclization

The reductive radical cyclizations of bromo acetals and (bromomethyl)silyl ethers of terpenoid alcohols were reported (Eqs. 7, 8 and 9) [21]. Electrolysis at −1.2 V versus SCE in DMF in the presence of catalytic amounts of vitamin B_{12a} (3–5 mol %) under photoirradiation (halogen lamp, 250 W) produced cyclic products.

hν, −1.2 V vs. SCE, vitamin B_{12a}

Yield: 78 % (*cis:trans* = 15:1) + 15 % (7)

hν, −1.2 V vs. SCE, vitamin B_{12a} (CO_2CH_3)

Yield: 52 % (*cis:trans* = 11:1) + 4 % + 34 % (8)

hν, −1.2 V vs. SCE, vitamin B_{12a}

Yield: 6 % (*trans:cis* = 1:2) (9)

Such an intramolecular cyclization was also catalyzed by a vitamin B_{12} derivative. Electrolysis of 2-(4-bromobutyl)-2-cyclohexen-1-one under photoirradiation (xenon lamp, 200 W with 350 nm cutoff filter) formed the *cis* and *trans* 1-decalone Eq. 10 [22]. In this reaction, the vitamin B_{12} derivative having six carboxylic groups was immobilized on a TiO_2 electrode and was used as the working electrode.

hν, −0.9 V vs. SCE, $TiO_2/B_{12}(COO^-)_6$

Yield: 65 % (*trans:cis* = 14:1) (10)

The electrolysis of bromoalkyl acrylates was carried out in DMF in the presence of a catalytic amount of heptamethyl cobyrinate (Cob(II) $7C_1$ester) Eq. 11 [23]. A series of 6-, 10-, and 16-membered cyclic lactones was obtained under photoirradiation (tungsten lamp, 500 W). These macrocyclic lactones are used in the pharmaceutical industry. A photo-labile alkylated complex was detected by UV–vis and ESI-mass analyses during electrolysis, and a radical intermediate was trapped by an ESR spin-trapping technique.

$$Br(CH_2)_nOC(=O)CH{=}CH_2 \xrightarrow[\text{Cob(II)7C}_1\text{ester}]{h\nu,\ -0.9\ \text{V vs. SCE}} \text{lactone } (CH_2)_n \tag{11}$$

Yield: 42 % (n = 12)
11 % (n = 6)
5 % (n = 2)

The intramolecular cyclization was also applied to a sequential radical reaction. The electrolysis of 3-(2'-bromo-1'-ethoxy)cyclopenten with 1-cyanovinyl-acetate in the presence of vitamin B_{12a} (0.75 mol %) at -1.1 V versus SCE in DMF under photoirradiation (halogen lamp, 150 W) produced a bicyclic compound Eq. 12 [24]. This product is the precursor of methyl jasmonate or epituberolide.

$$\xrightarrow[\text{vitamin } B_{12a}]{CH_2{=}C(CN)OAc,\ h\nu,\ -1.1\ \text{V vs. SCE}} \tag{12}$$

Substrate: OC_2H_5, Br, O, cyclopentene. Product: OC_2H_5, O, CN, OAc.

Yield: 63 %

Future Directions

The combined use of electrolysis and photoirradiation provided unique radical-mediated organic reactions with a cobalt complex as the mediator. The carbon–carbon bond formation accompanied with the carbon-skeleton rearrangement and intramolecular cyclization produced valuable organic compounds. Furthermore, the best combination of electrochemistry and photochemistry will achieve energy-saving electroorganic syntheses. The future direction of electrolysis under photoirradiation is the smart design of an illuminated cell for engineering processes. Efficient photoirradiation of an electrolyte solution is important for increasing the reaction yield. For example, utilization of a microflow system will facilitate photoirradiation during electrolysis. Electrosynthesis under photoirradiation would be expanded to versatile synthetic organic chemistry.

Cross-References

► Electrochemical Microflow Systems
► Electrosynthesis Using Mediator

References

1. Grimshaw J (2007) Electrochemical reactions and mechanisms in organic chemistry. Elsevier, Amsterdam
2. Toscano PJ, Marzilli LG (1984) B_{12} and related organocobalt chemistry: formation and cleavage of cobalt carbon bonds. Prog Inorg Chem 31:105–204
3. Giese B, Hartung J, He Jianing H, Hüter O, Koch A (1989) On the formation of "free radicals" from alkylcobalt complexes. Angew Chem Int Ed Engl 28:325–327
4. Banerjee R, Ragsdale SW (2003) The many faces of vitamin B_{12}: catalysis by cobalamin-dependent enzymes. Ann Rev Biochem 72:209–247
5. Brown KL (2005) Chemistry and enzymology of vitamin B_{12}. Chem Rev 105:2075–2149
6. Hisaeda Y, Shimakoshi H (2010) Bioinspired catalysts with B_{12} enzyme functions. Handbook of

porphyrin science, vol 10. World Scientific, Singapore, pp 313–370
7. Kräutler B (1999) B_{12} electrochemistry and organometallic electrochemical synthesis. Chemistry and biochemistry of B_{12}. Wiley, New York, pp 315–339
8. Hisaeda Y, Nishioka T, Inoue Y, Asada K, Hayashi T (2000) Electrochemical reactions mediated by vitamin B_{12} derivatives in organic solvents. Coord Chem Rev 198:21–37
9. Scheffold R (1983) Modern synthetic methods, vol 3. Wiley, Frankfurt
10. Scheffold R, Orlinski R (1983) Carbon-carbon bond formation by light-assistedB12 catalysis. Nucleophilic acylation of Michael olefins. J Am Chem Soc 105:7200–7202
11. Walder L, Orlinski R (1987) Mechanism of the light-assisted nucleophilic acylation of activated olefins catalyzed by vitamin B_{12}. Organomet 6:1606–1613
12. Murakami Y, Hisaeda Y, Tashiro T, Matsuda Y (1985) Electrochemical carbon-skeleton rearrangement as catalyzed by hydrophobic vitamin B_{12} in nonaqueous media. Chem Lett 14:1813–1816
13. Murakami Y, Hisaeda Y, Tashiro T, Matsuda Y (1986) Reaction mechanisms for electrochemical carbon-skeleton rearrangement as catalyzed by hydrophobic vitamin B_{12} in nonaqueous media. Chem Lett 15:555–558
14. Murakami Y, Hisaeda Y, Fan SD (1987) Characterization of a simple vitamin B_{12} model complex and its catalysis in electrochemical carbon-skeleton rearrangement. Chem Lett 16:655–658
15. Murakami Y, Hisaeda Y, Fan SD, Matsuda Y (1989) Redox behavior of simple vitamin B_{12} model complexes and electrochemical catalysis of carbon-skeleton rearrangements. Bull Chem Soc Jpn 62:2219–2228
16. Murakami Y, Hisaeda Y, Ozaki T, Tashiro T, Ohno T, Tani Y, Matsuda Y (1987) Hydrophobic vitamin B_{12}. V. Electrochemical carbon-skeleton rearrangement as catalyzed by hydrophobic vitamin B_{12}: reaction mechanisms and migratory aptitude of functional groups. Bull Chem Soc Jpn 60:311–324
17. Hisaeda Y, Takenaka J, Murakami Y (1997) Hydrophobic vitamin B_{12}. Part 14. Ring-expansion reactions catalyzed by hydrophobic vitamin B_{12} under electrochemical conditions in nonaqueous medium. Electrochimca Acta 42:2165–2172
18. Inokuchi T, Tsuji N, Kawafuchi H, Torii S (1991) Indirect electroreduction of 2-alkyl-2-(bromomethyl) cycloalkanones via 1,2-acyl migration. J Org Chem 56:5945–5948
19. Wolleb-Gygi A, Darbre T, Siljegovic V, Keese R (1994) The importance of peripheral association for vitamin B_{12} catalysed methylmalonyl-succinyl-rearrangement. Chem Commun 7:835–836
20. Darbre T, Keese R, Siljegovic V, Wolleb-Gygi A (1996) Model studies for the coenzyme-B12-catalyzed methylmalonyl-succinyl rearrangement. The importance of hydrophobic peripheral associations. Helv Chim Acta 79:2100–2113
21. Lee ER, Lakomy I, Bigler P, Scheffold R (1991) Reductive radical cyclizations of bromo acetals and (bromomethyl)silyl ethers of terpenoid alcohols. Helv Chim Acta 74:146–162
22. Mbindyo JKN, Rusling JF (1998) Catalytic electrochemical synthesis using nanocrystalline titanium dioxide cathodes in microemulsions. Langmuir 14:7027–7033
23. Shimakoshi H, Nakazato A, Hayashi T, Tachi Y, Naruta Y, Hisaeda Y (2001) Electroorganic syntheses of macrocyclic lactones mediated by vitamin B_{12} model complexes. Part 17. Hydrophobic vitamin B_{12}. J Electroanal Chem 507:170–176
24. Busato S, Scheffold R (1994) Vitamin B_{12} catalyzed C-C bond formation: synthesis of jasmonates via sequential radical reaction. Helv Chim Acta 77:92–99

Electrosynthesis Under Ultrasound and Centrifugal Fields

Mahito Atobe
Graduate School of Environment and Information Sciences, Yokohama National University, Yokohama, Japan

Introduction

Reaction control is very important in electrosynthetic chemistry. Because electron transfer takes place on electrode surface and/or vicinity, that is, electrode interface, the chemical functionality-modification of electrode surface has been intensively studied for this purpose so far [1]. On the other hand, mechanical energies such as ultrasound and centrifugal force cannot drive chemical reactions but control them. From this point of view, the mechanical energy modification of electrode interface has been also developed for the reaction control of electrosynthetic processes especially in the last two decades [2, 3].

Electrosynthesis Under Ultrasound Fields

The many benefits of ultrasound in chemical processes are well known and have been

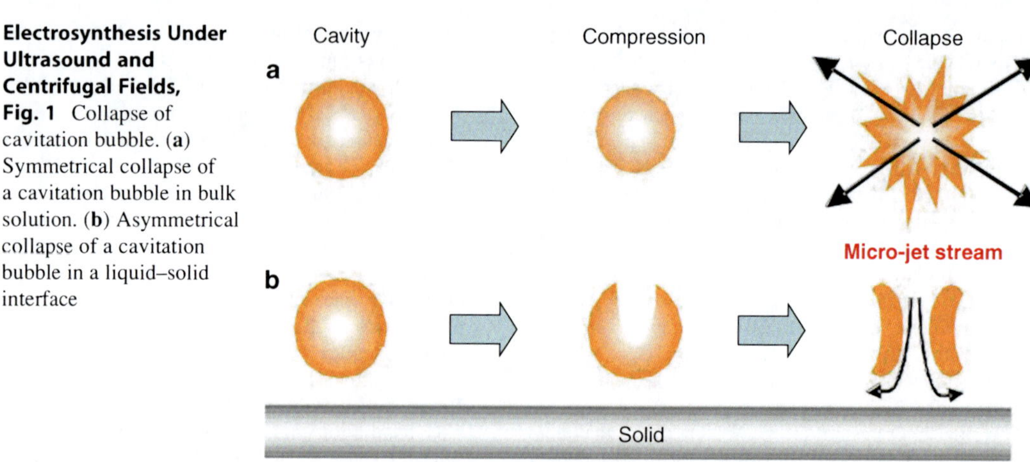

Electrosynthesis Under Ultrasound and Centrifugal Fields, Fig. 1 Collapse of cavitation bubble. (**a**) Symmetrical collapse of a cavitation bubble in bulk solution. (**b**) Asymmetrical collapse of a cavitation bubble in a liquid–solid interface

$$\mathrm{ArCHO} \xrightarrow[\substack{20\ \mathrm{mA\ cm^{-2}} \\ 0.5\ \mathrm{F\ mol^{-1}}}]{e^- + H^+ / 2e^- + 2H^+} \frac{1}{2}\,\underset{[HD]}{\mathrm{ArCH(OH)\text{-}CH(OH)Ar}} \;/\; \underset{[HM]}{\mathrm{ArCH_2OH}}$$

Ar: p-MeC_6H_4

	Current efficiency / %	Selectivity for HD / %
Still standing	35	0
Mechanical stirring	74	67
Ultrasonication	89	98

Electrosynthesis Under Ultrasound and Centrifugal Fields, Scheme 1 Electrochemical reduction of *p*-methylbenzaldehyde to the corresponding hydrodimeric (*HD*) and hydromonomeric (*HM*) products

investigated in a variety of chemical fields, but perhaps the most striking influence of ultrasound concerns heterogeneous reaction systems, particularly those with a solid–liquid interfaces where particle size modification, the modification of particle dispersion, the enhancement of mass transport, the cleaning of surfaces, or the formation of fresh surfaces are among the beneficial processes [4].

Since an electrochemical synthetic process is a typical heterogeneous one in a solid (electrode)–liquid(electrolytic solution) interface, various effects of ultrasound, particularly promotion of mass transport, would be induced by ultrasonication [4]. Major contribution to the mass transport exists micro-jet stream resulting from asymmetric collapse of a cavitation bubble (Fig. 1) [5]. Suslick et al. reported that velocity of the stream reaches an excess of 100 m s^{-1} in a water–solid interface [6].

Such a mass transfer promotion by ultrasonication provides an increase in the current efficiency for a variety of electrosyntheses. For example, Atobe et al. reported that a significant ultrasonic effect on the current efficiency was found in the electroreduction of *p*-methylbenzaldehyde (Scheme 1) [7]. The current efficiency was dramatically increased under ultrasonication. Furthermore, the product selectivity for the hydrodimeric product was also increased by ultrasonication, and the effects could be rationalized experimentally and theoretically as due to the promotion of mass transport of the substrate molecule to the electrode surface from the electrolytic solution by ultrasonic cavitation [8, 9].

Electropolymerization Under Ultrasonication

Conducting polymers exhibit not only electroconductivity but also unique optical and chemical properties [10]. The diversity of properties exhibited by conducting polymers offers these materials to be used in numerous technological applications. Generally, the properties of polymers originate from their chemical (molecular) and physical (morphological) structures. Therefore, it follows that the structures of

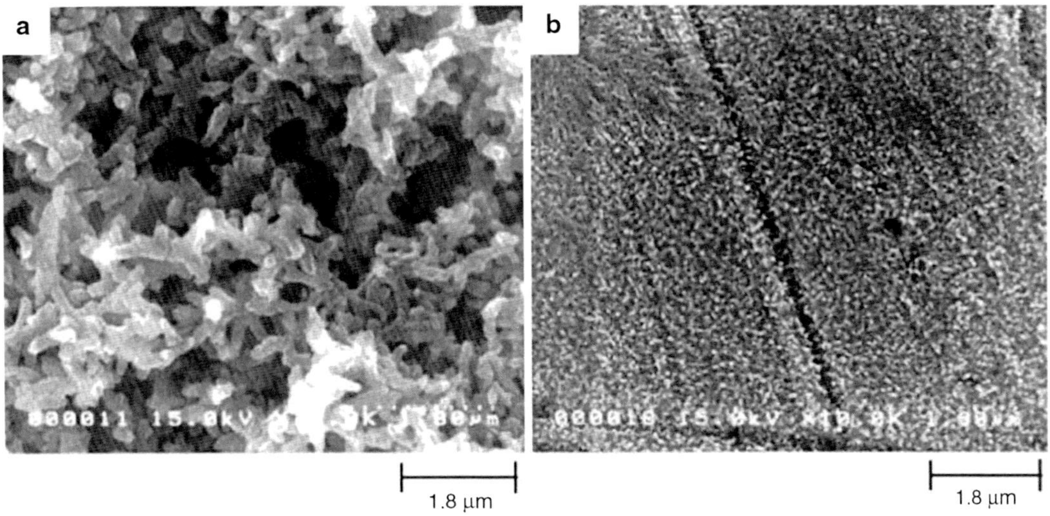

Electrosynthesis Under Ultrasound and Centrifugal Fields, Fig. 2 SEM images of polyaniline films prepared (**a**) without and (**b**) with ultrasonication

conducting polymers should be able to be controlled in order to tailor them to the purposes of their utilization. Their chemical structures can be controlled by changing the molecular structures of the corresponding monomers and by selecting conditions and procedures for polymerization [10]. On the other hand, methods for controlling their physical structures have been relatively limited, but, recently, many studies were focused on applying ultrasound to polymerization processes, particularly electropolymerization, for this purpose.

Osawa et al. have found that the quality of polythiophene films electropolymerized on an anode can be enhanced by ultrasound. By conventional methodology, the films become brittle, but by using ultrasound from a 45 kHz cleaning bath, flexible and tough films (tensile modulus 3.2 GPa and strength 90 MPa) can be obtained [11].

On the other hand, the work of Atobe and co-workers was probably the first "modern" example investigating electropolymerization under sonication in a complete series of papers at low frequencies. Starting from electro-organic reactions under ultrasonic fields [12], polymerization of aniline was studied both in electrochemical [13] and chemical route [14, 15] as well as synthesis of nanoparticle synthesis [16, 17].

The behavior of polypyrrole films electropolymerized under ultrasonication was also investigated, and unique properties in the doping–undoping processes were highlighted. The authors attributed these results to the elaboration of highly dense film under sonication and also deplored the degradation of the film due to high cavitation at 20 kHz (Fig. 2) [18]. The Besançon group also studied the use of high-frequency ultrasound (500 kHz, 25 W) for electropolymerization of 3,4-ethylenedioxythiophene (EDOT) or polypyrrole in aqueous medium in order to investigate its effects on conducting polymer properties. They showed that (i) mass transfer enhancement induced by sonication improves electropolymerization and that (ii) mass transfer effect is not the only phenomenon induced by ultrasound during electrodeposition [19, 20].

Electrosynthesis Under Centrifugal Fields

In recent years, a variety of experiments in high gravity fields have received much attention from a broad area in science and technology. The forces of 2–10 g can be generated for short time of 10^{-2}–10^{2} s in beginning and ending periods of

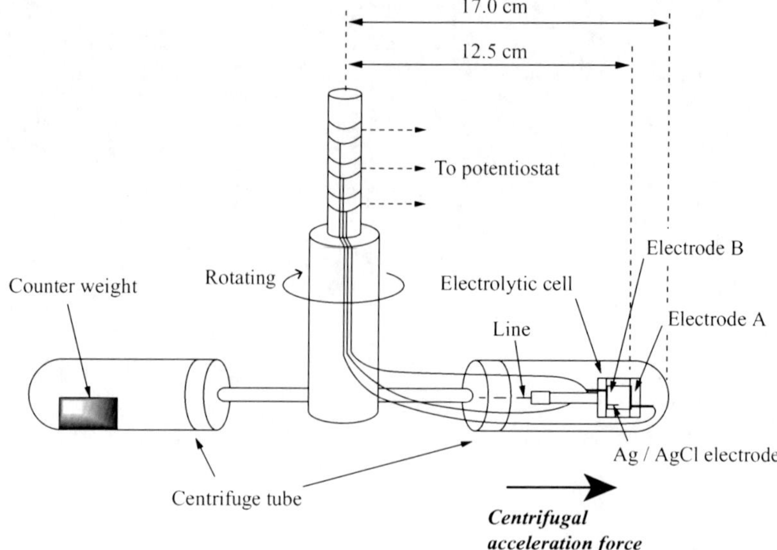

Electrosynthesis Under Ultrasound and Centrifugal Fields, Fig. 3 Centrifuge facilities equipped with an electrolytic cell

motion of falling bodies, aircrafts, and rockets [21]. This kind of forces is not constant in the course of time. Hence, it has rarely been used for experiments in high gravity fields. Contrary, the experiments can be much more easily and cheaply performed by using centrifuges which generate centrifugal acceleration force as extremely high as several hundred gravities (g's). The intensity of the force is very uniform and stable and also can be continuously and precisely controlled in arbitrarily small increments by adjusting length and rotation speed of the arm.

In a centrifugal field, the local flow of fluids occurs due to convection, acceleration gradient, the Coriolis force, etc. Therefore, centrifugal effects on physical and chemical phenomena, particularly caused by mass transport, should be interesting from basic and practical aspects. In fact, considerably big centrifuges with 3 and 18 m of arms were constructed for experiments of crystal growth in Erlangen-Nürnberg and Moscow, respectively, and consequently some interesting and purposive results could be obtained around 30 g of acceleration force [22–24]. However, it is noted that centrifugal effects on electrochemical processes, particularly electrosynthesis, have rarely reported so far, although the effects seem to be powerful and useful for controlling the processes.

Atobe et al. report successful fabrication of a centrifuge by improving commercially available centrifuge equipment with a 17-cm arm to generate a stable acceleration of roughly 300 g during electrochemical deposition of polyaniline, polypyrrole, and polythiophene [3, 25]. The experimental setup consists of a cylindrical electrolytic cell made of polytetrafluoroethylene resin, 7-mm long and 14 mm in diameter. Contacts between classical three electrode cell and potentiostat/galvanostat were created through silver rotating rings and carbon brushes. The entire system is suspended from the lid of a centrifuge tube with a polyethylene line as shown in Fig. 3. Surfaces of the platinum electrodes A and B face inward and outward, respectively, to the centrifugal acceleration force, as shown in Fig. 3. The forces on the electrodes A and B are calculated to be 315 and 290 g at 1,500 rpm, respectively.

Typical cyclic voltammograms were recorded during electropolymerization of aniline at various centrifugal accelerations (Fig. 4). Although their shape seems to be similar, oxidation and reduction peak currents are increased at 315 g (electrode A) and decreased at 290 g (electrode B – with a reversed direction force), compared to electropolymerization performed under 1 g. Using the anodic peak current around 0.5 V as an indicator, polymerization rate appears to be

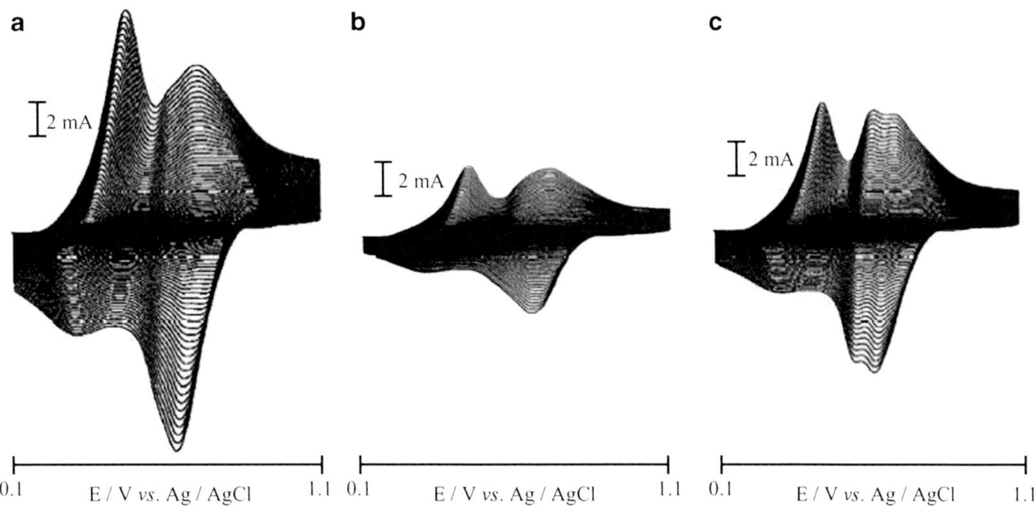

Electrosynthesis Under Ultrasound and Centrifugal Fields, Fig. 4 Cyclic voltammograms in the course of electrooxidative polymerization of aniline at 50 cycles of potential scanning at various centrifugal acceleration forces. (**a**) At 315 g on electrode A. (**b**) At 290 g on electrode B. (**c**) At 1 g on electrode A

greatly affected. Higher rates of polymer deposition at electrodes were acknowledged for each monomer. This enhancement of the formation rate also leads to a higher electrochemical capacity on the electrode situated opposite the centrifugal force. Scanning electron microscopy pictures reveal a high degree of sensitivity of the deposit morphology, which appears to be aggregated into large bundles. Chemical bonding and, finally, conductivity of the obtained polymers were considerably affected by the gravity fields. The clear anisotropy of the effects leads the authors to propose mechanisms for electropolymerization under centrifugal forces based on the basic differences between the solvent and macromolecules, shown concretely by different inertial behaviors. The accumulation of the oligomers produced on the electrode opposite the centrifugal force results in an increase in nucleation and precipitation at the surface. As a final result, the high density of oligomers closer to the surface leads to a denser polymer film.

Future Directions

Despite many potential advantages of centrifugal fields, there are relatively few examples of electrosynthesis in centrifugal fields. Consequently, the technologies are relatively new and unfamiliar. Therefore, further development will require research and development activities not only in the organic electrochemistry but also in the various electrochemical areas.

On the other hand, electrosynthesis under ultrasonication is an expanding field of study which continues to thrive on outstanding laboratory results. Compared with the past, there is now far greater contact and cooperation between the scientific disciplines interested in the effects of cavitation, and so the future of electrosynthesis under the influence of ultrasound is rosy, both from the point of view of a greater interest in the fundamental principles and industrial applications of this technology. Nowadays there is a greater availability of scale-up systems for use in processing, and this removes one of the real barriers to the large-scale use of ultrasonically enhanced electrosynthesis in industry.

Cross-References

► Electrosynthesis in Supercritical Fluids
► Electrosynthesis of Conducting Polymer
► Electrosynthesis Using Modified Electrodes

References

1. Murray RW (1984) Chemically modified electrodes. In: Bard AJ (ed) Electroanalytical chemistry, vol 13. Marcel Dekker, New York
2. Walton DJ, Mason TJ (1998) Organic sonoelectrochemistry. In: Luche J-L (ed) Synthetic organic sonochemistry. Plenum, New York
3. Atobe M, Hitose S, Nonaka T (1999) Chemistry in centrifugal fields. Part 1. Electrooxidative polymerization of aniline. Electrochem Commun 1:278–281
4. Walton DJ, Phull SS (1996) Sonoelectrochemistry. In: Mason TJ (ed) Advances in sonochemistry, vol 4. JAI Press, London
5. Lorimer P, Mason TJ (1999) The application of ultrasound in electroplating. Electrochem 67:924–930
6. Suslick KS, Gawienowski JJ, Schubert PF, Wang HH (1983) Alkane sonochemistry. J Phys Chem 87: 2299–2301
7. Matsuda K, Atobe M, Nonaka T (1994) Ultrasonic effects on electroorganic processes. Part 1. Product-selectivity in electroreduction of benzaldehydes. Chem Lett 23:1619–1622
8. Atobe M, Matsuda K, Nonaka T (1996) Ultrasonic effects on electroorganic processes. Part 4. Theoretical and experimental studies on product-selectivity in electroreduction of benzaldehyde and benzoic acid. Electroanal 8:784–788
9. Atobe M, Nonaka T (1997) Ultrasonic effects on electroorganic processes. Part 8. Cavitation threshold values of ultrasound-oscillating power. Chem Lett 26:323–324
10. Heinze J (2001) Electrochemistry of conducting polymers. In: Lund H, Hammerich O (eds) Organic electrochemistry. Macel Dekker, New York
11. Osawa S, Ito M, Tanaka K, Kuwano J (1987) Electrochemical polymerization of thiophene under ultrasonic field. Synth Met 18:145–150
12. Atobe M, Nonaka T (1998) New developments of sonoelectrochemistry - electroorganic reactions under ultrasonic fields. Nippon Kagaku Kaishi 1998(4): 219–230
13. Atobe M, Kaburagi T, Nonaka T (1999) Ultrasonic effects on electroorganic processes. XIII. Role of ultrasonic cavitation in electrooxidative polymerization of aniline. Electrochem 67:1114–1116
14. Atobe M, Chowdhury AN, Fuchigami T, Nonaka T (2003) Preparation of conducting polyaniline colloids under ultrasonication. Ultrason Sonochem 10:77–80
15. Chowdhury AN, Atobe M, Nonaka T (2004) Studies on solution and solution-cast film of polyaniline colloids prepared in the absence and presence of ultrasonic irradiation. Ultrason Sonochem 11:77–82
16. Park JE, Atobe M, Fuchigami T (2005) Sonochemical synthesis of conducting polymer-metal nanoparticles nanocomposite. Electrochim Acta 51:849–854
17. Park JE, Atobe M, Fuchigami T (2005) Sonochemical synthesis of inorganic–organic hybrid nanocomposite based on gold nanoparticles and polypyrrole. Chem Lett 34:96–97
18. Atobe M, Tsuji H, Asami R, Fuchigami T (2006) A study on doping-undoping properties of polypyrrole films electropolymerized under ultrasonication. J Electrochem Soc 153:D10–D13
19. Taouil AE, Lallemand F, Hihn JY, Blondeau-Patissier V (2011) Electrosynthesis and characterization of conducting polypyrrole elaborated under high frequency ultrasound irradiation. Ultrason Sonochem 18:907–910
20. Taouil AE, Lallemand F, Hihn JY, Melot JM, Blondeau-Patissier V, Lakard B (2011) Doping properties of PEDOT films electrosynthesized under high frequency ultrasound irradiation. Ultrason Sonochem 18:140–148
21. Regel LL, Wilcox WR (eds) (1994) Material processing in high gravity. Plenum, New York
22. Weder W, Neumann G, Müller G (1990) Stabilizing influence of the coriolis force during melt growth on a centrifuge. J Cryst Growth 100:145–158
23. Amato I (1991) The high side of gravity. Science 253:30–32
24. Burdin BV, Regel LL, Turchaninov AM, Shumaev OV (1992) The peculairities of material crystallization experiments in the CF-18 centrifuge under high gravity. J Cryst Growth 119:61–65
25. Atobe M, Murotani A, Hitose S, Suda Y, Sekido M, Fuchigami T, Chowdhury AN, Nonaka T (2004) Anodic polymerization of aromatic compounds in centrifugal fields. Electrochim Acta 50:977–984

Electrosynthesis Using Diamond Electrode

Siegfried R. Waldvogel and Bernd Elsler
Johannes Gutenberg-University Mainz,
Mainz, Germany

Introduction

The most intriguing feature of boron-doped diamond electrodes is represented by the unusually wide electrochemical window [1, 2]. In aqueous media, the overpotential for the evolution of molecular oxygen is higher than on every other common anode material and in the cathodic regime comparable to mercury. Consequently, these unique properties might lead to the replacement of either very costly noble metals or highly toxic heavy metals. In an aqueous electrolyte, the electrochemical window accounts to about 3.2 V, whereas electrolytes based on highly fluorinated alcohols such as 1,1,1,3,3,3-hexafluoroisopropanol show the largest electrochemical window (Fig. 1) observed in protic media [4, 5]. The latter was also evaluated in capacitor studies [6]. Boron-doped diamond electrodes are new tools for electroorganic synthesis. Crucial processes in these transformations, like electron transfer, have been recently studied in detail.

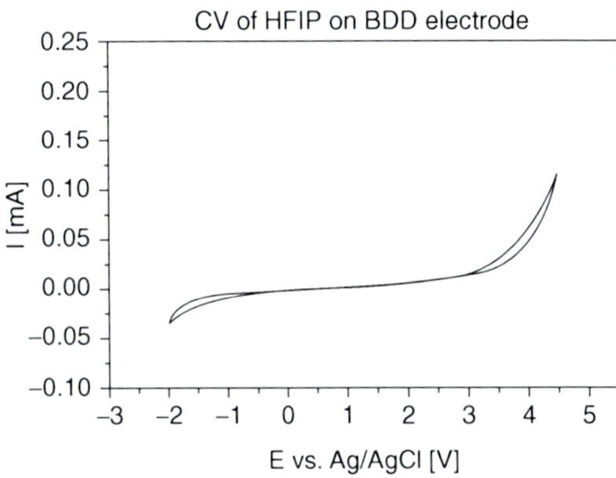

Electrosynthesis Using Diamond Electrode, Fig. 1 Cyclovoltammogram of 1,1,1,3,3,3-hexafluoroisopropanol on BDD electrode with 0.1 M Et_3NCH_3 O_3SOCH_3 (From [3] with permission from Springer)

Anodic Transformations at BDD

At boron-doped diamond (BDD) anodes, alcohols and water are efficiently converted to oxyl radicals in high concentration (Scheme 1) [7]. The outstanding high oxidizing power of these oxyl spin intermediates is successfully employed in waste water treatment for the degradation of organic materials [2, 8–11]. Therefore, an application of these intermediates seems to be self-contradictory in terms of synthetic use. The generated hydroxyl or alkoxyl species may be involved either as radicals and implemented into the product or serve as strong oxidants which transform the substrate. Due to the outstanding reactivity, the electrochemical mineralization reaction is the dominant side reaction. In most cases, the electrolyses are conducted at low current densities to enable mass transport from bulk to the surface of the anode and vice versa [3]. For an efficient use of BDD anodes and control of the reactivity of the formed intermediates, three successful strategies have been elaborated:

First, the use of a large excess of substrate and only partially performed electroorganic conversion exploits statistics. Essentially, the electrolysis is conducted in almost neat substrate. This was realized with 2,4-dimethylphenol and is discussed in the entry "► Boron-Doped Diamond for Green Electro-Organic Synthesis."

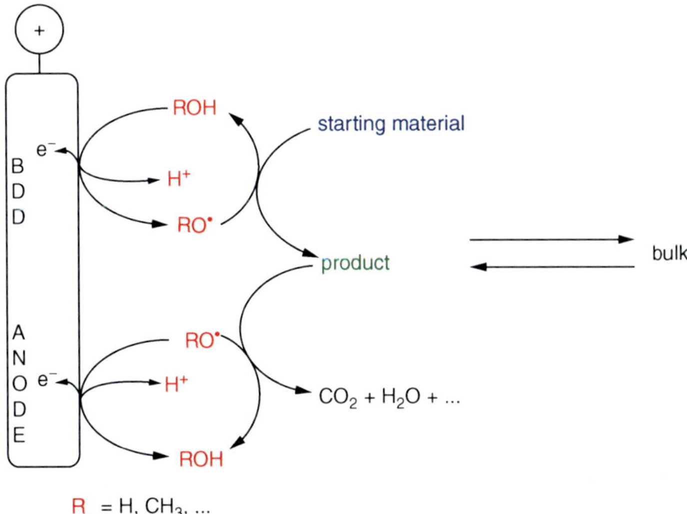

Electrosynthesis Using Diamond Electrode, Scheme 1 Generation of oxyl radicals at BDD and their possible use for electrosynthesis

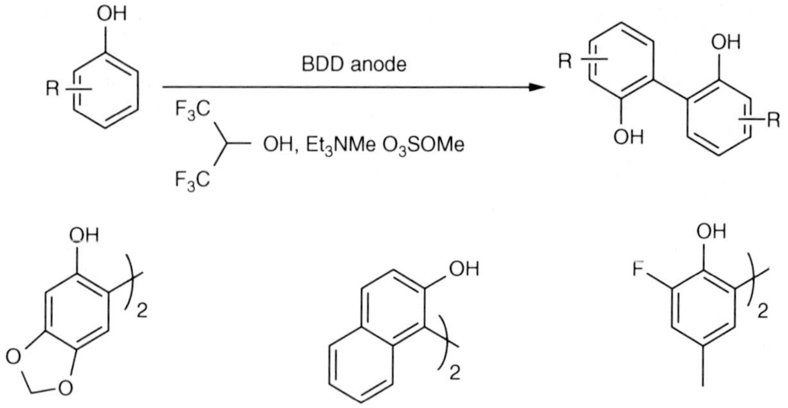

Electrosynthesis Using Diamond Electrode, Scheme 2 Phenol homo-coupling reactions in fluorinated media

BDD anode

F_3C, F_3C OH, Et_3NMe O_3SOMe

74 % **1**

41 % **2**

14 % **3**

Secondly, additives based on fluorinated alcohols turned out to be beneficial for the lifetime of the intermediate radical species [12, 13]. It is anticipated that the known stabilization of intermediates causes a strongly enhanced lifetime, reducing the electrochemical mineralization by over-oxidation.

Thirdly, the mass transport can be efficiently supported by microfluidics. Electrolysis at BDD combined with micro-flow conditions opens up completely novel reactions pathways [25].

Using Fluorinated Alcohols as Additives

Several fluorinated alcohols turned out to be suitable for the anodic coupling of phenols to the corresponding 2,2′-biphenols [14]. 1,1,1,3,3,3-Hexafluoroisopropanol together with supporting electrolytes creates a powerful electrolyte system for the anodic homo-coupling of phenols (Scheme 2). When only 1 F electric current is applied, a clean reaction mixture is obtained consisting of desired product, starting material, and electrolyte. The protocol is particularly useful for very electron-rich phenols as well as halogenated substrates. Sesamole is smoothly coupled to 1 in an unusually high yield. For BINOL (2), yield is lower because of limited solubility in the electrolyte. Very surprisingly, even fluorinated phenols were anodically coupled. These unique compounds found application as borate salts in energy storage devices [15, 16].

When subjecting guaiacol (4) to this protocol, the anticipated 2,2′-biphenol was not found, but rather a 2,3′-coupling reaction occurred (Scheme 3). Biaryl 5 was the only product which was isolated. The unprecedented connectivity indicated a reaction pathway which involves the attack of a phenoxyl radical onto a closed shell species. The electrolysis of syringol (6) gave the biaryl 7 with exclusive selectivity. Since no leaving functionalities are involved, the selectivity and yield are remarkable [17].

The selective electrochemical oxidation of one reaction partner gave rise to the first anodic phenol-arene cross-coupling (Scheme 4) [4]. The presence of additional water or methanol in the electrolyte turned out to be beneficial for the yield as well as selectivity [18]. In many cases, the ratio for the mixed biaryl (AB) vs. biphenyl (BB) exceeded 100:1. Since no leaving functionalities are required, simple starting materials can be employed, and the 1,1,1,3,3,3-hexafluoroisopropanol is almost quantitatively recovered. Biaryls 8–10 with different substitution patterns are feasible in good isolated yields. The transformation is compatible with a variety of functional groups and tolerates sulfide and tertiary alkyl moieties in the substrates. Compound 11 and 12 were the only observed mixed biaryls in the reaction mixture. Products originating

Electrosynthesis Using Diamond Electrode, Scheme 3 Formation of non-symmetrical biphenols in HFIP

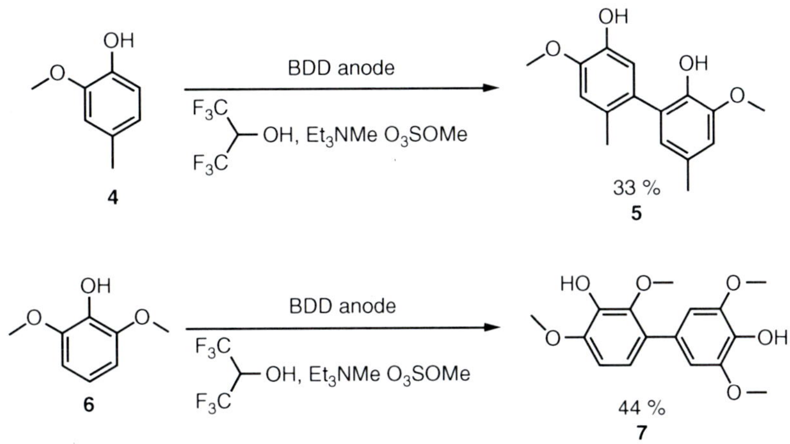

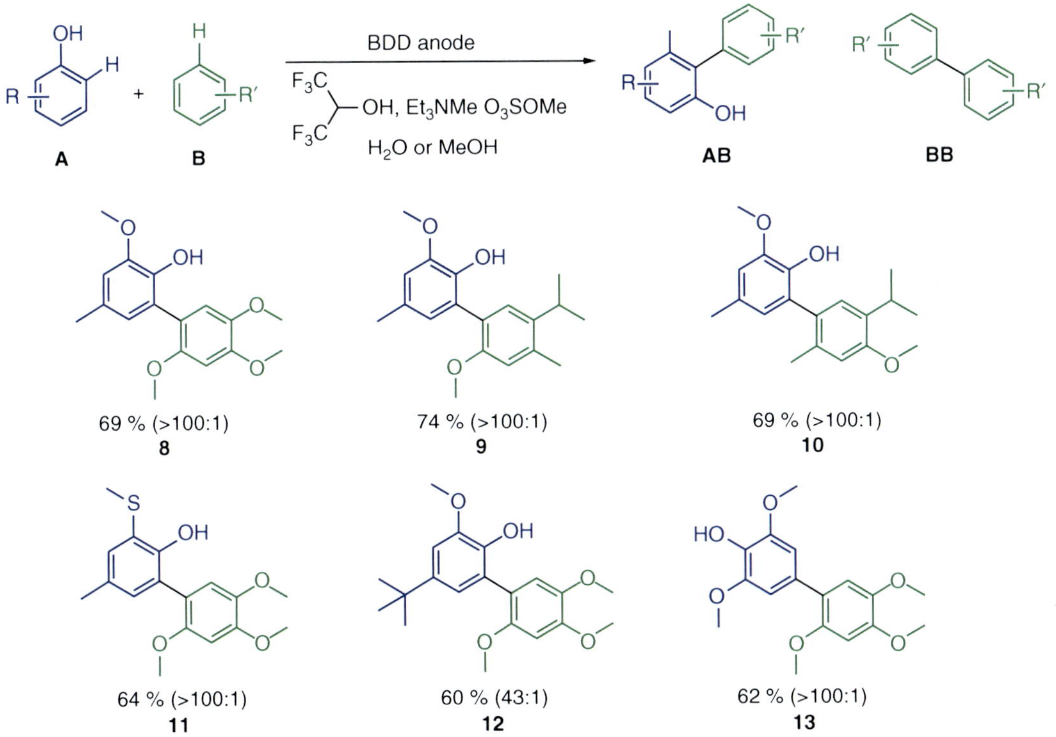

BDD anode

F_3C, F_3C >—OH, Et_3NMe O_3SOMe

4 → 33 % 5

6 → 44 % 7

BDD anode; HFIP, Et_3NMe O_3SOMe, H_2O or MeOH

A + B → AB, BB

69 % (>100:1) **8**

74 % (>100:1) **9**

69 % (>100:1) **10**

64 % (>100:1) **11**

60 % (43:1) **12**

62 % (>100:1) **13**

Electrosynthesis Using Diamond Electrode, Scheme 4 Efficient cross-coupling reactions of phenols by adding protic additives to HFIP

from over-oxidation or alkyl transfer were not observed. Cross-coupling via a position *para* to the phenol moiety is also possible and provides 13 in 62 % yield. This transformation gained a wide scope and represents a valuable tool for the synthetic repertoire in biaryl formation [5].

Application of Micro-flow Conditions

The appealing oxidative power of oxyl species generated boron-doped diamond is attractive for the synthesis of naturally occurring products. Even when the substrate 14 is easily oxidized,

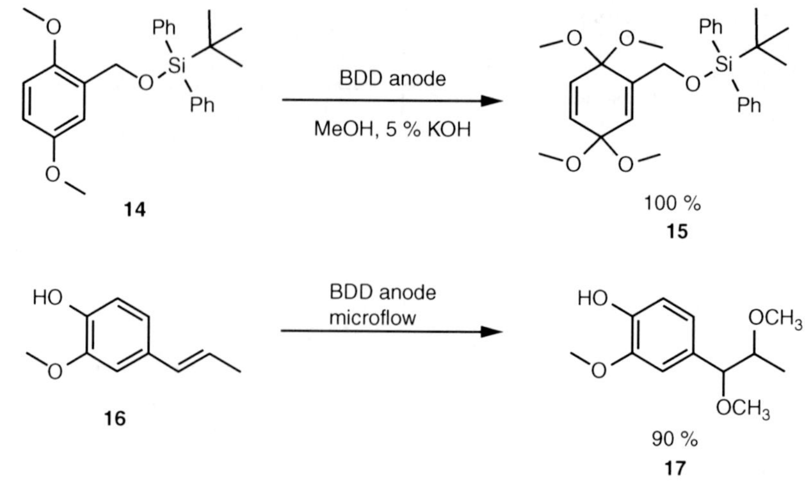

Electrosynthesis Using Diamond Electrode, Scheme 5 Anodic methoxylation of aromatic compounds and its application in micro-flow reactors

the methoxyl radicals are involved in the mechanism and evidence was indirectly found by ESR methods [19]. Here the initial radical cation recombines with a methoxyl radical which undergoes the anticipated reaction pathway to 15 (Scheme 5) [19]. BDD turned out to be the superior anode material and provides the intermediates for the synthesis of parasitenone. If isoeugenol (16) is treated in a similar way, a complex product mixture is obtained with a dehydrodimer (licarin A) as major component. When conducting the same conversion under micro-flow conditions, an anodic 1,2-dimethoxylation occurred with outstanding selectivity and yield to compound 17.

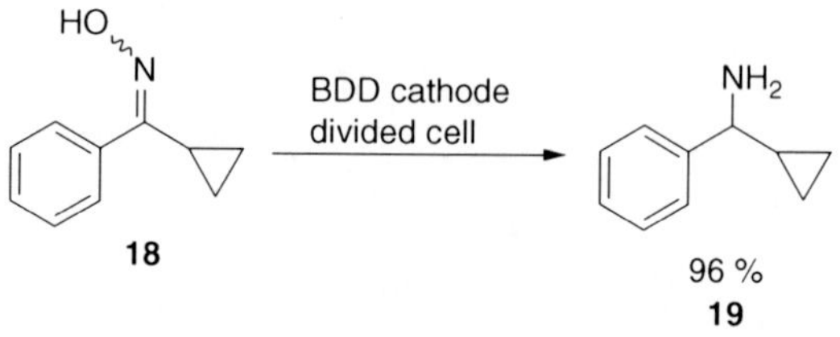

Cathodic Conversions on BDD

Since electric current is by far the least expensive reduction equivalent and no reagent waste is formed, the cathodic reduction of organic compounds is highly attractive. For an electroorganic reduction, a source of protons is required. In general, protic media allow only cathode materials with high overpotentials for the evolution of hydrogen. Consequently, late transition metals such as lead, mercury, and cadmium most often show the best results. BDD offers the possibility for the replacement of these highly toxic cathode materials. The preparation of amines starting from the corresponding oximes is a very common and general method providing a product with the full carbon skeleton [20]. Mainly reductions at a mercury pool were successful [21]. The high overpotential of BDD for the H_2 evolution by electrolysis of protic electrolytes should be beneficial for this cathodic transformation [22]. The electrochemical and cathodic reduction of cyclopropylphenylketone oxime (18) to amine 19 was performed in 96 % yield at BDD (Scheme 6). The electrolysis was conducted in a divided cell using MeOH/NaOMe as electrolyte. Performing this transformation by hydrogenation on noble metal catalyst provides significant amounts of ring-opened compounds causing problems in the purification of 19. Application of this protocol to sterically more congested oximes failed [23]. Most likely, a successful conversion of substrates on BDD cathodes requires good access to the diamond surface without spatial restrictions [24].

Electrosynthesis Using Diamond Electrode, Scheme 6 Cathodic reduction of cyclopropylphenylketone oxime in high yields

Future Directions

The application of boron-doped diamond electrodes in electroorganic synthesis is a strongly evolving field. With practical strategies in hand for controlling the extreme reactivity of the intermediate oxyl radicals, several unique transformations have been established. The competing electrochemical incineration of the organic substrates on the BDD anode can be efficiently suppressed. Consequently, many novel electrochemical transformations are expected in near future. Since many of these anodic conversions heavily depend on the employed electrolyte, developments in this respective field seem to be crucial. The use of BDD cathodes is currently underrepresented. Because of the unique electrochemical window in the reductive region, this will be an active field of future research.

Cross-References

► Boron-Doped Diamond for Green Electro-Organic Synthesis
► Electrochemical Functional Transformation
► Organic Reactions and Synthesis

References

1. Brillas E, Martínez-Huitle CA (eds) (2011) Synthetic diamond films – preparation, electrochemistry, characterization and applications. Wiley-VCH, New York
2. Fujishima A, Einaga Y (eds) (2005) Diamond electrochemistry. BKC/Elsevier, Amsterdam
3. Waldvogel SR, Mentizi S, Kirste A (2012) Boron-doped diamond electrodes for electroorganic chemistry. Top Curr Chem 320:1–31
4. Kirste A, Schnakenburg G, Stecker F et al (2010) Anodic phenol-arene cross-coupling reaction on boron-doped diamond electrodes. Angew Chem Int Ed 49(5):971–975
5. Waldvogel SR, Elsler B (2012) Electrochemical synthesis on boron-doped diamond. Electrochim Acta (in press) doi: 10.1016/j.electacta.2012.03.173
6. Francke R, Cericola D, Kötz R et al (2012) Novel electrolytes for electrochemical double layer capacitors based on 1,1,1,3,3,3-hexafluoropropan-2-ol. Electrochim Acta 62:372–380
7. Wilson NR, Clewes SL, Newton ME, Unwin PR, Macpherson JV (2006) Impact of grain-dependent boron uptake on the electrochemical and electrical properties of polycrystalline boron doped diamond electrodes. J Phys Chem B 110:5639–5646
8. Comninellis C, Chen G (eds) (2010) Electrochemistry for the environment. Springer, New York
9. Angus JC (2011) Electrochemistry on diamond: history and current status. In: Brillas E, Martínez-Huitle CA (eds) Synthetic diamond films – preparation, electrochemistry, characterization and applications. Wiley-VCH, New York, pp 3–19
10. Panizza M (2011) Application of synthetic diamond films to electro-oxidation processes. In: Brillas E, Martínez-Huitle CA (eds) Synthetic diamond films – preparation, electrochemistry, characterization and applications. Wiley-VCH, New York, pp 333–351
11. Scialdone O, Galia A (2011) Modeling of electrochemical process for water treatment using diamond films. In: Brillas E, Martínez-Huitle CA (eds) Synthetic diamond films – preparation, electrochemistry, characterization and applications. Wiley-VCH, New York, pp 261–280
12. Eberson L, Hartshorn MP, Persson O (1995) Detection and reaction of radical cations generated by photolysis of aromatic compounds with tetranitromethane in 1,1,1,3,3,3-hexafluoro-2-propanol at room temperature. Angew Chem Int Ed 34(20):2268–2269
13. Eberson L, Hartshorn MP, Persson O (1995) 1,1,1,3,3,3-Hexafluoropropan-2-ol as a solvent for the generation of highly persistent radical cations. J Chem Soc Perkin Trans II(9):1735–1744
14. Kirste A, Nieger M, Malkowsky IM et al (2009) ortho-Selective phenol-coupling reaction by anodic treatment on boron-doped diamond electrode using fluorinated alcohols. Chem-Eur J 15(10):2273–2277
15. Francke R, Cericola D, Kötz R et al (2011) Bis(2,2′-biphenoxy)borates for electrochemical double-layer capacitor electrolytes. Chem-Eur J 17(11):3082–3085
16. Waldvogel SR, Malkowsky IM, Griesbach U et al (2009) Novel fluorine-free electrolyte system for supercapacitors. Electrochem Commun 11(6): 1237–1241
17. Kirste A, Schnakenburg G, Waldvogel SR (2011) Anodic coupling of guaiacol derivatives on boron-doped diamond electrodes. Org Lett 13(12): 3126–3129
18. Kirste A, Elsler B, Waldvogel SR (2012) Efficient anodic and direct phenol-arene C, C cross-coupling. J Am Chem Soc 134(7):3571–3576
19. Sumi T, Saitoh T, Natsui K et al (2012) Anodic oxidation on a boron-doped diamond electrode mediated by methoxy radicals. Angew Chem Int Ed 51:5443–5446
20. Looft J, Vössing T, Ley J, Backes M, Blings M (2008) EP 1989944 A1
21. Tafel J, Pfeffermann E (1902) Elektrolytische Reduction von Oximen und Phenylhydrazonen in

schwefelsaurer Lösung. Ber Dtsch Chem Ges 35(2):1510–1518
22. Martin HB, Argoitia A, Landau U et al (1996) Hydrogen and oxygen evolution on boron-doped diamond electrodes. J Electrochem Soc 143(6): L133–L136
23. Kulisch J, Nieger M, Stecker F et al (2011) Efficient and stereodivergent electrochemical synthesis of optically pure menthylamines. Angew Chem Int Ed 50(24):5564–5567
24. Patten HV, Meadows KE, Hutton LA James et al (2012) Electrochemical mapping reveals direct correlation between heterogeneous electron-transfer kinetics and local density of states in diamond electrodes. Angew Chem Int Ed 51(28):7002–7006
25. Kashiwagi T, Elsler B, Waldvogel SR, Fuchigami T, Atobe M (2013) Reaction condition screening by using electrochemical microreactor: application to anodic phenol-arene C, C cross-coupling reaction in high acceptor number media. J Electrochem Soc 160(7):G3058–G3061

Electrosynthesis Using Mediator

Hideo Tanaka
Okayama University, Okayama, Japan

Introduction

Electroorganic synthesis (EOS) is one of the most promising tools for environmentally benign reactions without any oxidizing or reducing reagents, thereby saving resources and reducing the amount of the undesired wastes. EOS is initiated by electron transfer (*E process*, *electron release or uptake*) on the vicinity of the electrodes, generating electron-deficient or electron-excessive species (Sub*) which may subsequently undergo chemical reactions (*C process*) to afford the final products [1–3]. Electrosynthesis Using Mediator (indirect electrooxidation and electroreduction) has been intensively investigated, in which the mediator releases or takes up electron(s) to form electron-deficient or electron-excessive species (M*). Subsequent reaction with substrates would drive a wide variety of reactions depending on the choice of the mediatory system (M/M*). Accordingly, *design of the redox mediatory system* may expand the scope of EOS significantly and enhance reactivity and selectivity.

Various kinds of mediators have been developed for the redox mediatory systems of electroorganic synthesis. For instance, halogen redox, organic redox, and metal redox have been utilized frequently for the mediatory system, and combinations of different redoxes have been also used efficiently. Significant efforts toward the development of new redox mediatory systems have been devoted to promote highly selective and efficient molecular transformation reactions (Fig. 1).

Halogen Mediatory System

Electrosynthesis using *halogen mediator* has been intensively investigated. Electrooxidation of halide ions offers various kinds of active cationic species (M*), e. g., Cl_2, HOCl, Cl_2O, Br_2, NaOBr, I_2, I_3^-, and I_3^+, which can be utilized successfully for various synthetic purposes. The potentialities of such electrogenerated active halogen species as a mediator have been demonstrated by highly selective functionalization of olefins, e.g., epoxidation [4], halohydroxylation [4],

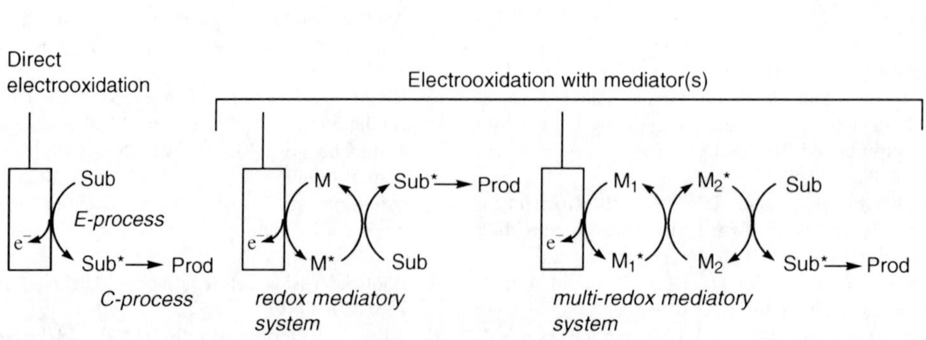

Electrosynthesis Using Mediator, Fig. 1 Electrosynthesis using mediator(s)

1,2-dihalogenation [4], and ene-type chlorination [5–7], heteroatom–heteroatom bond formations, and carbon–heteroatom bond cleavage. For instance, electrochemical epoxidation of olefin has been developed for the production of ethylene and propylene oxides in aqueous sodium chloride or bromide solution. Regioselective epoxidation of the terminal olefin of polyisoprenoids is achieved successfully by electrooxidation in an aqueous THF/MeCN solution of NaBr [8]. Electrooxidative formation of various heteroatom–heteroatom bonds has been performed by proper choice of halide mediator; thus, S–N bond formation, e.g., cross-coupling of disulfide and imide, and S–P bond formation, e.g., cross-coupling of disulfide and diarylphosphite, proceed smoothly in an electrolysis media containing bromide salt as a mediator to afford the corresponding sulfenamide and phosphorothiolate, respectively [9, 10]. P–N bond formation, e.g., cross-coupling of dialkylphosphite with amine, is efficiently achieved by electrolysis with iodide ion as a mediator to afford dialkylphosphoramidate in good to excellent yields [11]. Halide ion-mediated electrooxidative C–S bond cleavage is successfully applied to synthesis of versatile intermediates for carbapenems [12], stereoselective *N*-glycosylation of 2,3-deoxyglycosides [13], and so on.

Metal-Redox Mediatory System

Electrosynthesis with metal redox as a mediatory system has been intensively investigated, and in situ generated active metal species can be utilized for a number of synthetic purposes. Metal complex redoxes, e.g., Ni(II)/Ni(0), Co(III)/Co(I), Fe(III)/Fe(II), Pd(II)/Pd(0), Rh(III)/Rh(II), Ru(II)/Ru(0), Cr(III)/Cr(II), Mn(II)/Mn(0), Ti(iV)/Ti(III), Pb(II)/Pb(0), and Zn(II)/Zn(0), have been utilized as mediators for various electroreductive C–C bond formation reactions [14]. For instance, reductive coupling of aryl halides is achieved successfully by electroreduction with either Ni(II)/Ni(0) or Pb(II)/Pb(0) redox as a mediatory system [15–17], and Michael-type addition of alkyl halides to alpha,beta-enones is realized by Co(III)/Co(I) redox-mediated electroreduction [18]. Oxidative transformations, e.g., oxidation of side chain of aromatic compounds, 1,2-dihydroxylation of olefins, and oxidation of alcohols and amines, are attained by electrooxidation with CAN, Cr_2O_3, $Mn(SO_4)_2$, $TiClO_4$, OsO_4, RuO_2, and so on as a mediator [19].

Organic-Redox Mediatory System

Many organic redox-mediated electrosyntheses have been developed so far; for instance, 2,2,6,6-tetramethylpiperidine nitroxyl (TMPO) is used efficiently as a mediator of electrooxidation of alcohols to aldehydes and ketones [20]. Electrooxidation of divalent sulfur compounds is performed with triarylamine as a mediator. The successive electrooxidation of two sulfur moieties of 1,3-dithiane is performed successfully by use of a catalytic amount of tris(p-tolyl)amine, affording the corresponding carbonyl compound together with 1,2-dithiacylopentane [21]. Various extended pi-conjugated compounds, e.g., anthraquinone, anthracene, pyrene, phenanthrene, benzonitrile, and phthalonitrile, have been utilized as a mediator for various electroreductive transformations; for instance, electroreductive removal of *cpo*-toluenesulfonyl group of sulfonamides, leading to the corresponding amines, has been well studied, and pyrene, anthracene, and stilbene have been shown to be efficient as mediators for such a purpose [19]. Phenanthrene works as a good mediator for electroreductive cyclization of aryl halides having an olefin moiety [22].

Multi-Redox Mediatory System

Multi-redox-mediated electrosynthesis may significantly expand the scope of the electrosynthesis. Various combinations of different kinds of mediators have been studied intensively so far. Nicotinamide adenine dinucleotide (NAD)/methyl viologen double mediatory systems are utilized successfully for various enzymatic reductions [14]. The combinations of halide ions and 2,2,6,6-tetramethylpiperidine nitroxyl (TMPO)

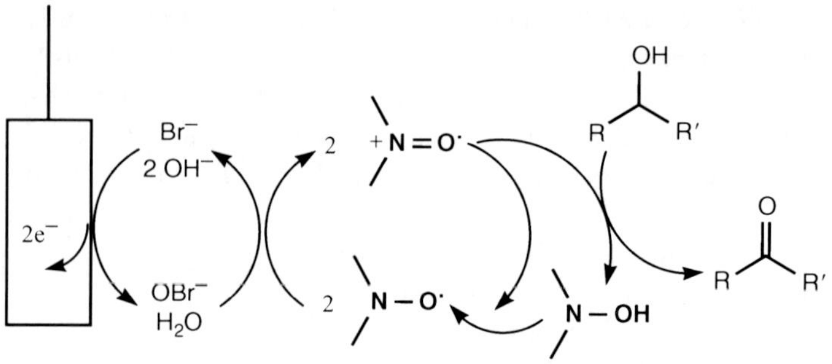

Electrosynthesis Using Mediator, Fig. 2 Electrooxidation of alcohols in Br−/TEMPO double mediatory system

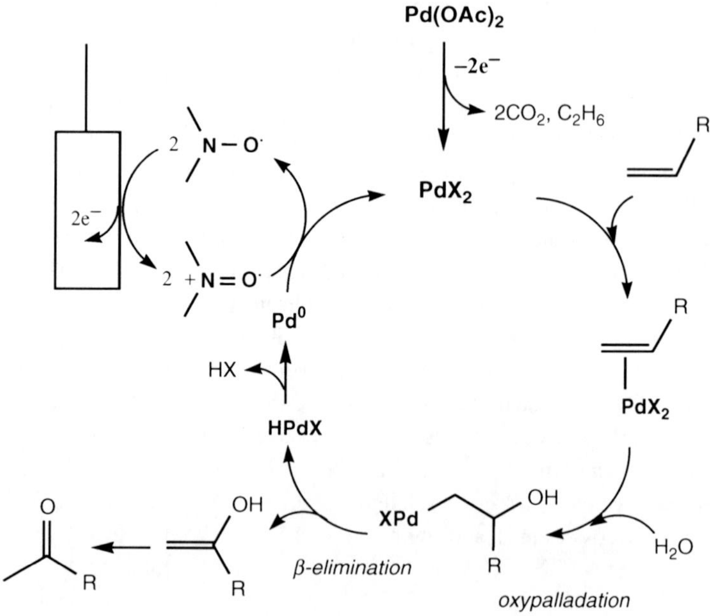

Electrosynthesis Using Mediator, Fig. 3 Electrooxidative Wacker-type reaction in a double mediatory system

offer a simple and highly efficient procedure for electrooxidation of alcohols; thus, a mixture of alcohols and a catalytic amount of TEMPO in aqueous NaBr-$NaHCO_3$/dichloromethane two-phase solution are electrolyzed in a beaker-type undivided cell under a constant current to afford the corresponding carbonyl compounds in good to excellent yields [23] (Fig. 2). Similar double mediatory electrooxidation of alcohols can be performed efficiently in disperse systems with TEMPO-immobilized silica gel or polymer particles as a disperse phase and aqueous NaBr-$NaHCO_3$as a disperse media [24]. A combination of palladium(II) acetate and TEMPO is utilized efficiently as a double mediatory system for electrooxidative Wacker-type reaction of higher molecular weight terminal alkenes [25], which are hardly oxidized in the conventional Wacker reaction, and the homo-coupling of arylboronic acids, leading to the corresponding biaryls [26].

Future Directions

Design of new mediatory systems for electrooxidative and electroreductive transformation of complex organic molecules is currently investigated intensively, and many successful examples will be accumulated, thereby offering highly selective and efficient tools for synthesis of various useful chemicals through electrosynthesis using mediators.

Cross-References

▶ Electrochemical Functional Transformation
▶ Electrosynthesis Using Water Suspension System

References

1. Lund H, Hammerich (1991) Organic electrochemistry, 4th edn, Revised and Expanded. Marcel Dekker, New York
2. Torii S (2006) Electroorganic reduction synthesis. Kodansha &Wiley-VCH, Tokyo/Weinheim
3. Torii S (1985) Electroorganic syntheses, methods and applications, part 1: oxidations. Kodansha & VCH, Tokyo/Weinheim
4. Torii Uneyama K, Tanaka H, Yamanaka Y, Yasuda T, Ono M, Kohmoto Y (1981) Efficient conversion of olefins into epoxides, bromohydrins, and dibromides with sodium bromide in water-organic solvent electrolysis. J Org Chem 48:3312–3315
5. Torii S, Tanaka H, Saitoh N, Siroi T, Sasaoka M, Nokami J (1981) Chemoselective electrolytic chlorination of methyl group of 3-methyl-3-butenoate moiety of thiazoline-azetodinone homologues. Tetrahedron Lett 22:3193–3196
6. Torii S, Tanaka H, Saitoh N, Siroi T, Sasaoka M, Nokami J (1982) Penicillin-chephalosporin conversion III, a novel route to 3-choloromethyl-delta3-cephems. Tetrahedron Lett 23:2187–2188
7. Torii S, Uneyama K, Nakai T, Yasuda T (1981) An electrochemical chlorinative ene-type reaction of isoprenoids. Tetrahedron Lett 22:2291–2294
8. Torii S, Uneyama K, Ono M, Tazawa H, Matsunami S (1979) A regioselective w-epoxidation of polyisoprenoids by the sodium bromide promoted electrochemical oxidation. Tetrahedron Lett 20:4661–4662
9. Torii S, Tanaka H, Ukida M (1979) Electrosynthesis of hetero-hetero atom bonds 3. Sodium bromide promoted electrolytic cross-coupling reaction of imides with disulfides. Convenient synthesis of N-(Cyclohexylthio) phthalimide, an important prevulcanization inhibitor. J Org Chem 44:1554–1557
10. Torii S, Tanaka H, Sayo N (1979) Electrosynthesis of hetero-hetero atom bonds 4. Direct cross-coupling of dialkyl(or diaryl)phosphites with disulfides by a sodium bromide promoted electrolytic procesure. J Org Chem 44:2938–2941
11. Torii S, Sayo N, Tanaka H (1979) Electrosynthesis of hetero-hetero atom bonds 5, Direct cross-coupling of dialkylphospites with amines by an iodonium ion-promoted electrolytic procesure. Tetrahedron Lett 20:4471–4474
12. Kuroboshi M, Miyada M, Tateyama S, Tanaka H (2008) Electrooxidative desulfurization/chlorination. A facile synthesis of 4-chloro-2-azetidinones, a potent intermediate for carbapenems. Heterocycles 76:1471–1484
13. Mitsudo K, Kawaguchi T, Miyahara S, Matsuda W, Kuroboshi M, Tanaka H (2005) Electrooxidative Glycosylation through C-S bond cleavage of 1-arylthio-2,3-dideoxyglycosides. Synthesis of 2',3'-dideoxynucleosides. Org Lett 7:4649–4652
14. Torii S (2006) Electroorganic reduction synthesis, Vol 2. Kodansha & Wiley-VCH, Tokyo/Weinheim, pp 553–634
15. Rollin Y, Troupel M, Perichon J, Fauvarque JF (1981) Electroreduction of nickel(II) salts in tetrahydrofuran-hexamethylphosphoramide mixtures. Application to electrochemical activation of carbon-halogen bonds. J Chem Res (S) 322–323
16. Torii S, Tanaka H, Morisaki K (1985) Pd(0)-catalyzed electro-reductive coupling of aryl halides. Tetrahedron Lett 26:1655–1658
17. Amatore C, Carre E, Jutand A, Tanaka H, Ren S, Torii S (1996) Oxidative addition of aryl halides to transient anionic sigma-aryl-palladium(0) intermediates-application to palladium-catalyzed reductive coupling of aryl halides. Chem Eur J 2:957–966
18. Gao J, Rusling JF, Zhou D (1996) Carbon-carbon bond formation by electrochemical catalysis in conductive microemulsions. J Org Chem 61:5972–5877
19. Torii S (1985) Indirect electrooxidation and the use of electron carriers (Mediators). In: Electroorganic syntheses, methods and applications, part 1: oxidations. Kodansha & VCH, Tokyo/Weinheim, pp 279–306
20. Smmelhack MF, Chou CS, Cores DA (1983) Nitroxyl-mediated electrooxidation of alcohols to aldehydes and ketones. J Am Chem Soc 105:4492–4494
21. Platen M, Steckhan E (1980) Indirect electrochemical processes 10. An effective and mildelectrocatalytic procedure for the removal of 1,3-dithiane protecting groups. Tetrahedron Lett 21:511–514
22. Kurono N, Honda E, Komatsu F, Orito K, Tokuda M (2004) Regioselective synthesis of substituted 1-indanols, 2,3-dihydrobenzofurans and 2,3-dihydroindoles by electrochemical radical cyclization using an arene mediator. Tetrahedron 60:1791–1801
23. Inokuchi T, Matsumoto T, Nishiyama T, Torii S (1990) Indirect electrooxidation of alcohols by a double mediatory systems with two redox couples of $[R_2N^+ = O]/R_2NO^·$ and $[Br^·$ or $Br^+]/Br^-$ in an organic-aqueous two-phase solution. J Org Chem 56:2416–2421
24. Tanaka H, Kuruboshi M, Mitsudo K (2009) Design of redox-mediatory systems for electro-organic synthesis. Electrochemistry 77:1002–1009
25. Mitsudo K, Kaide T, Nakamoto E, Yoshida K, Tanaka H (2007) Electrochemical generation of cationic Pd catalysts and application to Pd/TEMPO double-mediatory electrooxidative wacker-type reaction. J Am Chem Soc 129:2246–2247
26. Mitsudo K, Shiraga T, Tanaka H (2008) Electrooxidative homo coupling of arylboronic acids catalyzed by electrogenerated cationic palladium catalysts. Tetrahedron Lett 49:6593–6595

Electrosynthesis Using Modified Electrodes

Albert J. Fry
Weslayan University, Middletown, CT, USA

Introduction

It has long been known that the chemical composition of an electrode used for an electrochemical synthesis can have a great influence upon the nature and efficiency of the electrode reaction. Although this clearly suggests that the nature of the electrode surface can influence electrochemical behavior, historically little attention was paid to ways to modify the electrode surface. It is now recognized that the selectivity and efficiency can be affected to a major degree by proper pretreatment of the electrode surface. In fact, a well-characterized electrochemical process can sometimes even be diverted into an entirely different direction simply by electrode modification.

Modification Methods

Electrode modification can be carried out by methods that vary greatly. A reaction can be affected simply by addition to the electrolysis solution of a substance that is readily adsorbed onto the electrode surface. Thus, addition of a thiocyanate salt to the medium diverts the anodic oxidation of carboxylates from decarboxylative dimerization (Kolbe reaction) to peracid formation [1]. Often, a polymer solution containing an electrocatalyst is placed on a surface, and the solvent evaporated or a monomer is electrochemically polymerized in situ from solution onto the surface. Electrocatalysts deposited in this manner include organometallic electrocatalyst complexes such as vitamin B_{12} [2], oxidizable heterocycles such as pyrrole or thiophene, or metal ions [3]. Successive layers of complementary materials may be laid down on an electrode to achieve the desired immobilization effect. Thus, a polymer (PDAA; polydimethyldiallyl ammonium chloride) bearing cationic sites can be evaporated onto an electrode, after which it can be immersed in a solution of OL-1(a polyanionic manganese(II) oxide), which is held in place by strong electrostatic forces. The process may be continued, laying down alternate layers of PDAA and OL-1 [4].

Aryl radicals generated by cathodic reduction of aryldiazonium salts in aprotic media bond with high efficiency to the electrode surface [5]. Although this technology was initially applied to modification of glassy carbon, aryldiazonium ions can also modify the surface spontaneously or by photochemical initiation [6]. These methods have since been applied not only to a variety of forms of carbon, such as graphite, graphene, carbon nanotubes, and diamond, but also to metals, including Pd, Si, Fe, Cu, Au, Co, Ni, Cu, and Zn, and to composites such as GaAs and $Li_{1.1}V_3O_8$. Such aryl-modified electrodes can be used to effect electrocatalytic transformations [7], but the surface modification can be taken advantage of in other ways. For example, the layer of phenyl groups laid down by electrochemical reduction of phenyldiazonium ion at iron has been shown to protect the surface against corrosion [8].

Inorganic Syntheses

Photocathodic stripping of a pre-deposited tellurium film on a gold electrode in 0.1 M Na_2SO_4 electrolyte containing Cd^{2+} ions and poly(vinyl pyrrolidone) (PVP) was used as a route to the photoelectrosynthesis of CdTe nanoparticles entrained in PVP [9]. A similar approach has been used to generate a variety of other metal chalcogenides with potentially useful electronic properties, including CdSe, CdTe, In_2S_3, FeS_2, CdZnSe, and ZnO nanodots [10, 11].

Organic Syntheses

PbO particles electrogenerated on a porous Ni/Fe electrode (to afford high surface area) can be used for the cathodic conversion of 2-chloronitrobenzene to 2,2′-dichlorohydrazobenzene, an important chemical intermediate [12]. A similar system can oxidize 2,5-*bis*[hydroxymethyl]furan to the corresponding dialdehyde in good yield [13]. A zeolite prepared by anodic oxidation of Ni(II) dispersed in a carbon paste electrode modified with Ni(II) was found to be effective for electrocatalytic oxidation of methanol in NaOH solution [3]. The active oxidant in this process is a Ni(III) species, often formulated as NiOOH; this potent oxidizing agent has been electrogenerated for a variety of transformations requiring high anodic potential. It has an organic counterpart, 2,2,6,6-tetramethylpiperidinoxy (TEMPO), a stable free radical, which can be anodically oxidized to produce a strong oxidizing species. TEMPO can be attached to an electrode surface by a variety of methods such as incorporation into a polymer electrogenerated at the electrode surface. Oxidation of alcohols to carbonyl compounds can be carried out by carrying out enantioselective oxidation of racemic secondary alcohols at a TEMPO-modified electrode containing (−)-sparteine as a second constituent [14]. (S)-enantiomers are oxidized at this electrode, but (R)-enantiomers are not; hence, the (S)-ketones can be produced in >99 % enantiopurity. An electrode bearing a Nafion film containing TEMPO was found to oxidize carbohydrates to the corresponding uronic acids [15]. Similar results are found when the TEMPO is immobilized in a sol–gel on the electrode [16].

Substantial efforts have been made to achieve epoxidation of alkenes at modified electrodes. Although hydrogen peroxide is typically the source of oxygen, a variety of catalysts have been examined. The manganese in a PDAA/OL-1-layered electrode (see Modification Methods) can both promote the formation of H_2O_2 from oxygen passed through the medium and catalyze the epoxidation of alkenes by H_2O_2 [17]. Substitution of the redox enzyme myoglobin for OL-1 also results in epoxidation [18]. Highly efficient asymmetric epoxidation can be effected using a chiral Mn(III) salen derivative immobilized in a layered ammonium/zinc polystyrenephosphonate film [19]. Enantiomeric efficiencies up to >99 % with 99 % conversion could be achieved with alpha-methylstyrene. A cobalt corrin polycarboxylate can be incorporated electrostatically into a poly-L-lysine film, permitting access to the varied organic chemistry of the vitamin B_{12} system. Such electrodes can catalyze a variety of useful carbon-carbon bond formation reactions [20–22].

Electroenzymatic Syntheses

Attachment of an enzyme to an electrode surface can be used to catalyze organic redox processes. For example, inclusion of glucose oxidase in a polypyrrole film permits conversion of glucose to gluconic acid on preparative scale [23]. The use of electrode films containing myoglobin to catalyze epoxidation of alkenes has already been mentioned [17, 18]. Many reductive biotransformations involve cycling the cofactor between its active reduced form and the oxidized form, which must then be reduced to continue the cycle. Several research groups have co-immobilized a reductase enzyme and the cofactor on an electrode. This enzyme catalyzes the selective reduction of NAD^+ to NADH, and a second enzyme is chosen so as to carry out the desired transformation. Passage of a cathodic current is then used to drive the biochemical process, e.g., pyruvic acid can be converted to lactic acid [24]. Similarly, alanine and phenylalanine are produced from the corresponding keto acids by inclusion of an amine oxidase and electron mediator in the surface layer [25]. An additional benefit of this approach is that immobilization often substantially extends the lifetime of the enzyme.

Future Directions

The advantages of modified electrodes have been repeatedly and increasingly demonstrated in recent years, and this trend will undoubtedly continue. The high degree of selectivity inherent in enzymes will be increasingly explored for electrochemical applications in both synthesis and development of new sensors. The study of small clusters of atoms, i.e., nanoparticles, and the discovery that chemistry of substances in this size regime is often greatly different from that of the bulk materials will be exploited to develop new electrode processes. New forms of carbon such as carbon nanotubes and graphene have potential of entraining and/or adsorbing small molecules and offer another possible avenue for selective transformations. One disadvantage of modified electrodes is that the rate of electrolysis is governed by the area of the electrode. Flow systems are being used increasingly in synthetic electrochemistry and offer an attractive avenue for overcoming the electrode area problem.

Cross-References

- ▶ Electrocatalysis, Novel Synthetic Methods
- ▶ Electrocatalytic Synthesis
- ▶ Electrosynthesis Using Mediator
- ▶ Kolbe and Related Reactions

References

1. Khidirov SS, Khibiev KS (2005) Kolbe synthesis on a platinum anode modified with thiocyanate ions. Russ J Electrochem 41:1176–1179
2. Ruhe A, Walder L, Scheffold R (1987) Modification of carbon electrodes by vitamin B_{12} polymers. Makromol Chem 8:225–233
3. Raoof JB, Azizi N, Ojani R, Ghodrati S, Abrishamkar M, Chekin F (2011) Synthesis of ZSM-5 zeolite: electrochemical behavior of carbon paste electrode modified with Ni(II)-zeolite and its application for electrocatalytic oxidation of methanol. Int J Hydrogen Energy 36:13295–13300
4. Gao Q, Suib SL, Rusling JF (2002) Colloids, helices, and patterned films from heme proteins and manganese oxide. Chem Commun 19:2254–2255
5. Allongue P, Delamar M, Desbat B, Fagebaume O, Hitmi R, Pinson J, Saveant J-M (1997) Covalent modification of carbon surfaces by aryl radicals generated from the electrochemical reduction of diazonium salts. J Am Chem Soc 119:201–207
6. Pinson J, Podvorica F (2005) Attachment of organic layers to conductive or semiconductive surfaces by reduction of diazonium salts. Chem Soc Rev 34: 429–439
7. Mayers BT, Fry AJ (2006) Construction of catalytic electrodes bearing the triphenylamine nucleus covalently bound to carbon. A halogen dance in protonated aminotriphenylamines. Org Lett 8:411–414
8. Combellas C, Delamar M, Kanoufi F, Pinson J, Podvorica FI (2005) Spontaneous grafting of iron surfaces by reduction of aryldiazonium salts in acidic or neutral aqueous solution. Application to the protection of iron against corrosion. Chem Mater 17:3968–3975
9. Ham S, Paeng K-J, Park J, Myung N, Kim S-K, Rajeshwar K (2008) Photoinduced synthesis of CdTe nanoparticles using Te-modified gold electrode in poly(vinyl pyrrolidone)-containing electrolyte. J Appl Electrochem 38:203–206
10. Choi B, Myung N, Rajeshwar K (2007) Double template electrosynthesis of ZnO nanodot array. Electrochem Commun 9:1592–1595
11. Ham S, Jeon S, Lee U, Paeng K-J, Myung N (2008) Photoelectrochemical deposition of CdZnSe thin films on the Se-modified Au electrode. Bull Korean Chem Soc 29:939–942
12. Zhao Z, Meng Q, Li P, Cao B (2010) Electrochemical synthesis of 2,2'-dichlorohydrazobenzene from o-chloronitrobenzene on a porous Ni/Fe electrode. Electrochim Acta 56:1094–1098
13. Kokoh KB, Belgsir EM (2002) Electrosynthesis of furan-2,5-dicarboxaldehyde by programmed potential electrolysis. Tetrahedron Lett 43:229–231
14. Kashiwagi Y, Yanagisawa Y, Kurashima F, Anzai J-i, Osa T, Bobbitt JM (1996) Enantioselective electrocatalytic oxidation of racemic alcohols on a TEMPO-modified graphite felt electrode by use of chiral base. Chem Commun 24:2745–2746
15. Belgsir EM, Schafer HJ (2001) Selective oxidation of carbohydrates on Nafion-TEMPO-modified graphite felt electrodes. Electrochem Commun 3:32–35
16. Palmisano G, Mandler D, Ciriminna R, Pagliaro M (2007) Structural insight on organosilica electrodes for waste-free alcohol oxidations. Catal Lett 114: 55–58
17. Espinal L, Suib SL, Rusling JF (2004) Electrochemical catalysis of styrene epoxidation with films of MnO_2 nanoparticles and H_2O_2. J Am Chem Soc 126:7676–7682
18. Vaze A, Rusling JF (2005) Interfacial and mass transport enhancement effects on rates of styrene epoxidation catalyzed by myoglobin films in microemulsions. Faraday Discuss 129:265–274
19. Huang J, Fu X, Wang G, Ge Y, Miao Q (2012) A high efficient large-scale asymmetric epoxidation of unfunctionalized olefins employing a novel type of chiral salen Mn(III) immobilized onto layered crystalline aryldiamine modified zinc poly(styrene-phenylvinylphosphonate)-phosphate. J Mol Catal A Chem 357:162–173
20. Njue CK, Rusling JF (2002) Organic cyclizations in microemulsions catalyzed by a cobalt corrin-polyion-scaffold on electrodes. Electrochem Commun 4: 340–343
21. Njue CK, Nuthakki B, Vaze A, Bobbitt JM, Rusling JF (2001) Vitamin B_{12}-mediated electrochemical cyclopropanation of styrene. Electrochem Commun 3:733–736
22. Scheffold R, Abrecht S, Orlinski R, Ruf HR, Stamouli P, Tinembart O, Walder L, Weymuth C (1987) Vitamin B12-mediated electrochemical

reactions in the synthesis of natural products. Pure Appl Chem 59:363–372
23. Gros P, Bergel A (2005) Electrochemically enhanced biosynthesis of gluconic acid. AIChE J 51:989–997
24. Sobolov SB, Leonida MD, Bartoszko-Malik A, McKinney F, Kim J, Voivodov KI, Fry AJ (1996) Cross-linked LDH crystals for lactate synthesis coupled to electroenzymatic regeneration of NADH. J Org Chem 61:2125–2128
25. Kawabata S, Iwata N, Yoneyama H (2000) Asymmetric synthesis of amino acid using an electrode modified with amino acid oxidase and electron mediator. Chem Lett 110–111

Electrosynthesis Using Solid Polymer Electrolytes (SPE)

Jakob Jörissen
Chair of Technical Chemistry, Technical University of Dortmund, Germany

Introduction, History, and Definition of Solid Polymer Electrolytes

The first patent application for using a membrane made of an ion exchange polymer as a "solid polymer electrolyte" (SPE) rather than a conventional conductive liquid electrolyte was in 1955 [1]. The goal was to overcome the problems of liquid electrolytes in fuel cells with gaseous reactants. Cation exchange materials – working as proton conductors like sulfonated polystyrene – as well as anion exchange materials, working as hydroxyl ion conductors, were proposed, and the principle was verified experimentally. Later on, an important step for its technical realization was Nafion® (Dupont), a perfluorosulfonic acid (PFSA) polymer with high chemical and thermal stability, which was used in fuel cells of the Gemini space program 1966. Today, research dedicated to the "proton exchange membrane fuel cell" (PEMFC, also called "polymer electrolyte membrane fuel cell" or simpler "polymer electrolyte fuel cell" (PEFC)) has increased significantly. It has the potential to become a robust power source for use in daily life, e.g., in electric cars. An interesting alternative to this could be the "direct methanol fuel cell" (DMFC) using methanol as a fuel, and other alcohols may also be possible (these reactions are, in principle, organic electrosyntheses). Additional information concerning these fuel cells is provided in the related entries in this encyclopedia (see entry "► Fuel Cells, Principles and Thermodynamics").

Ion exchange membranes work in the temperature range of conventional fluid electrolytes, e.g., in electric cars from –40 °C to +80 °C and perhaps in the future up to +130 °C. This must not be confused with "solid electrolytes," which are used in "solid oxide fuel cells" (SOFC) as oxygen ion conductors at up to 1,000 °C. Lithium ion-conducting polymers are important components of high-power lithium ion secondary batteries, but that is not object of this entry.

The reverse process of PEMFC, the SPE electrolysis of pure, nonconductive water, was developed in the 1970s. Today it is once again being considered as an alternative for alkaline water electrolysis for hydrogen production in "power to gas" processes using unsteady regenerative electrical energy due to its high energy efficiency and stability.

SPE Technology for Organic Synthesis

In the beginning of the 1980s, Ogumi first used SPE technology, which was well proven in inorganic water electrolysis, for organic synthesis without a conductive fluid that contained a supporting electrolyte [2]. Initially, he studied cathodic reductions and especially the cathodic hydrogenation of olefins. He demonstrated the possibilities of the SPE technology with different anodic and cathodic reactions in subsequent publications. Several other groups also published research in this area in the 1980s (e.g., [3]).

General advantages of the SPE technology for organic synthesis can be expected:

- To economize the separation and recycling of a supporting electrolyte
- To avoid any contamination and side reaction with a supporting electrolyte

An adequate conductivity of the ion exchange membrane is a precondition for using SPE technology in organic electrosynthesis. This polymer contains covalently bonded "fixed ions," which are electrically neutralized by detached "counter ions." If water – or probably another sufficiently polar solvent – is absorbed and solvates the fixed and counter ions to enable their separation to an adequate degree, the counter ions become mobile in liquid-filled pores. Only then can an ion exchange membrane function as an ion conductor, i.e., as an SPE.

It is not possible to provide a comprehensive review of all the research done in this area in the context of this entry. Its purpose is to demonstrate the characteristics and possibilities of SPE technology for organic synthesis using examples from published research. This is intended to motivate scientists to apply this method to solve particular problems.

Principle of SPE Technology for Organic Synthesis

The principle and typical conditions of SPE technology will be discussed by means of the alkoxylation of *N*-alkylamides in a nonaqueous solution, which is as an example that works nearly perfectly using simple graphite felt electrodes pressed onto the surface of a Nafion® 117 membrane [4].

Figure 1 shows the scheme of an SPE cell for the above-mentioned example reaction. The electrode reactions take place in electrochemically active layers at the interfaces between the ion exchange membrane and permeable electrodes. The electrical current is delivered to the electrodes by suitable current collectors. A special electrocatalytic layer may be not necessary if the electrochemical activity is an inherent property of the electrode material, e.g., of graphite felt for both electrodes. Protons are formed as counter ions during the anodic oxidation reaction and migrate together with a solvation shell through the cation exchange membrane. It works like immobilized sulfuric acid due to sulfonic acid groups, which serving as fixed ions after absorbing a solvent. The protons are reduced to gaseous hydrogen at the cathode.

In this example, the mixture of methanol and *N,N*-dimethylformamide (DMF) obviously serves as a polar solvent in the Nafion® 117 membrane and is comparable with water. The membrane was pretreated in DMF for 10 min at 110 °C. A nearly ideal reaction selectivity and current efficiency of 98 % was achieved with a methanol to DMF ratio of 10:1 at 15 % DMF conversion. The cell voltage was about 5 V at 50 mA cm^{-2} current density and 60 °C. During an 8-month test, no significant decrease in current efficiency and only a slight increase of cell voltage was observed. A scale-up to a cell of 250 cm^2 active membrane area was successful. Other *N*-alkyl-amides and other alcohols could have been used [4].

Electroosmotic Flow: An Intrinsic Property of SPE Technology

A typical and perhaps unexpected feature of SPE technology is the mass transport in ion exchange membranes due to "electroosmotic flow" (EOF, also called "electroosmotic stream" or "electroosmotic drag") illustrated by the arrow in Fig. 1. EOF generally occurs in an electrochemical double layer where a charged liquid, adhering to an oppositely charged surface, is moved by a parallel electrical field. These effects are limited to layers of several nanometers thickness and are hence irrelevant in conventional electrochemical cells. In ion exchange membranes, the polymer is charged due to the fixed ions. All liquid-filled pores, including the inversely charged counter ions, have nanometer dimensions. Thus, an EOF is always present in all ion exchange membrane applications. Therefore, an ion exchange membrane is never a simple cell separator (this is sometimes not considered in literature). Due to the direct connection between membrane and electrodes in SPE technology, EOF has additional significant influence and is the reason for certain special properties.

The difference between conventional and solid polymer electrolytes is illustrated by Fig. 2. It demonstrates the transport of solvated ions at the surface of one particle of a porous anode, e.g., at a single fiber of graphite felt

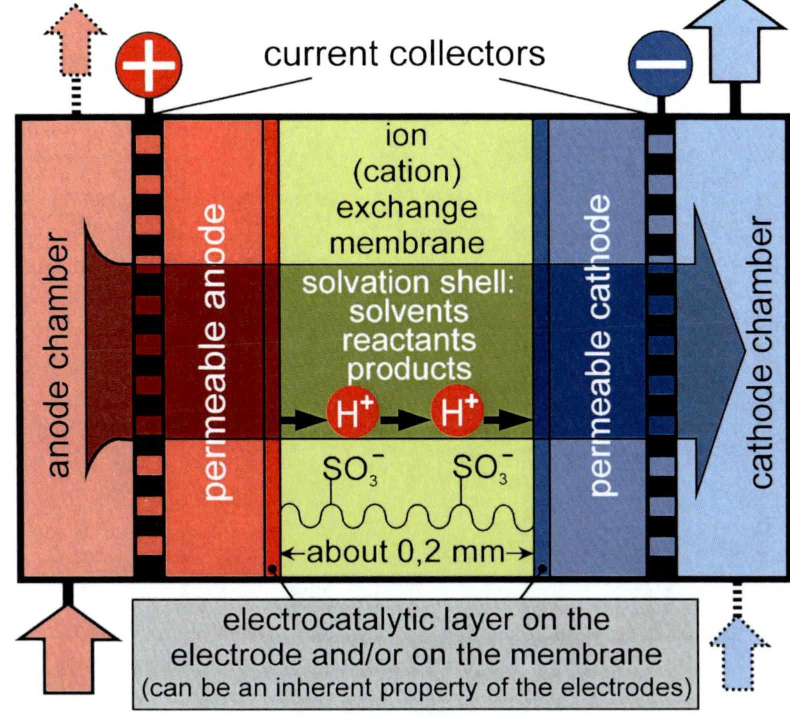

Electrosynthesis Using Solid Polymer Electrolytes (SPE), Fig. 1 Principle of solid polymer electrolyte technology [5]

example: methoxylation of N,N-dimethylformamide

reactions at the membrane / electrode interface

anode: $HC(=O)-N(CH_3)_2 + CH_3OH \longrightarrow HC(=O)-N(CH_3)(CH_2-O-CH_3) + 2H^+ + 2e^-$

H^+ ions migrate through the membrane

cathode: $2H^+ + 2e^- \longrightarrow H_2\uparrow$

(not to scale). The particle is surrounded by a stagnant diffusion layer in which convection, e.g., by stirring, in the bulk phase of the cell liquid is not present.

In a conventional electrolyte (upper part of Fig. 2), cations and anions migrate in opposite directions, and, for the most part, no effective mass transfer is generated by their solvation shells (if transference numbers of cations und anions are not too different). Uncharged reactants and products are carried by diffusion alone within the diffusion layer. Limited diffusion rates for reactants and/or products of the desired electrode reaction usually generate diffusion over potentials and can enhance side reactions.

In the case of SPE technology, the cell liquid is free of ions, which are only present in the membrane (lower part of Fig. 2). The fixed ions (negatively charged in case of a cation exchange membrane) are bonded to the polymer and are thus immobile, including their solvation shells. Charge transfer is performed, at least

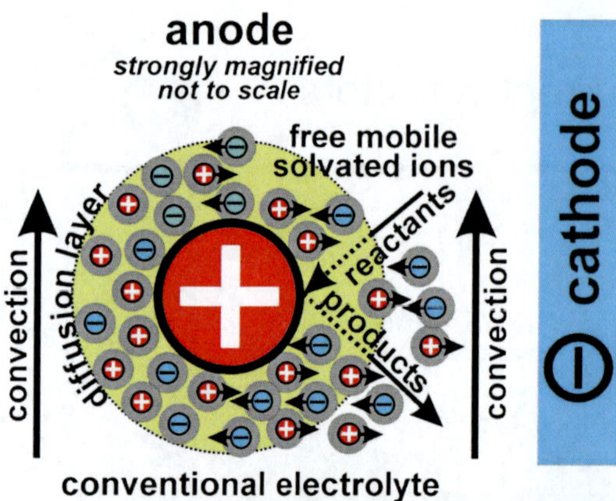

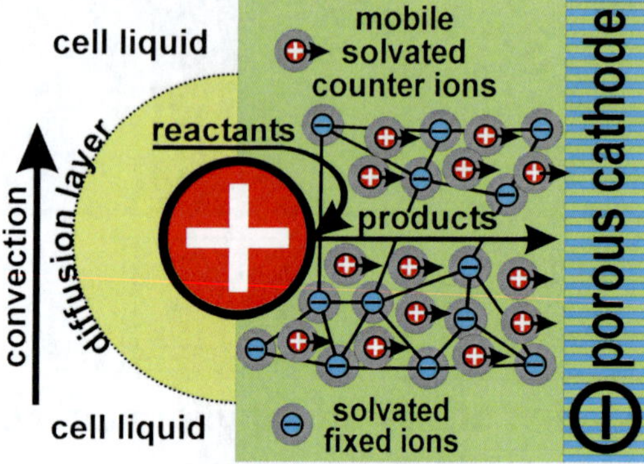

Electrosynthesis Using Solid Polymer Electrolytes (SPE), Fig. 2 Illustration of electroosmotic flow (EOF) [5]

predominantly, by counter ions, e.g., protons. These ions, which are generated during the reaction at the electrode surface, need molecules to form a solvation shell for further transfer through the membrane, which causes EOF. All compounds, such as solvents, reactants, and products, are included with the actual concentrations at the working electrode.

Thus, in contrast to a conventional electrolyte with a diffusion layer at the electrodes, in the case of SPE technology, a convective mass transport occurs directly at the electrode surface due to EOF. It delivers reactants and removes products as indicated by the arrows in the lower part of Fig. 2. This is presumably the cause of the frequent observation of higher selectivities of electrode reactions using SPE technology compared with conventional electrolytes where diffusion is the only transport mechanism. EOF, and therefore the enhancement of mass transport, is intensified proportionally with increased current density. As a result, relatively high current densities may be obtained using SPE technology without increasing mass transport problems, which would otherwise usually be observed. Of course, mass transport by diffusion takes place in addition to EOF, but it may be of lower significance.

Mass transport due to EOF may be advantageous, of no consequence or detrimental, depending on the particular application. The direction of EOF can be determined by choosing a cation or an anion exchange membrane. The amount of EOF is very dependent on the composition of the solution in the cell and on the type, preparation, and ion exchange capacity of the membrane. EOF can be enhanced by swelling the membrane in amides like *N,N*-dimethylformamide (DMF) at an elevated temperature, especially using perfluorosulfonic acid membranes like Nafion®, where the polymer molecules are not cross-linked. In this case, the EOF increases proportionally to the enlarged area of the membrane caused by swelling (this is advantageous for the reaction in Fig. 1). In experiments, the amount of EOF was observed from 2 up to 20 and more molecules per carrier ion, e.g., per proton [4, 5].

Operation Alternatives of SPE Technology

EOF works in SPE technology like an implemented dosing pump, and the function of the ion exchange membrane as a cell separator is limited. Therefore, as borderline cases, the different simplified flow schemes of operation alternatives in Fig. 3 are possible [5].

In configuration (a), the anode chamber is filled with water, which is the usual medium for ion exchange membranes (another sufficiently polar liquid may also be possible). Water is oxidized at the anode to oxygen gas and H^+ ions, which migrate together with a hydration shell through the cation exchange membrane (CEM).

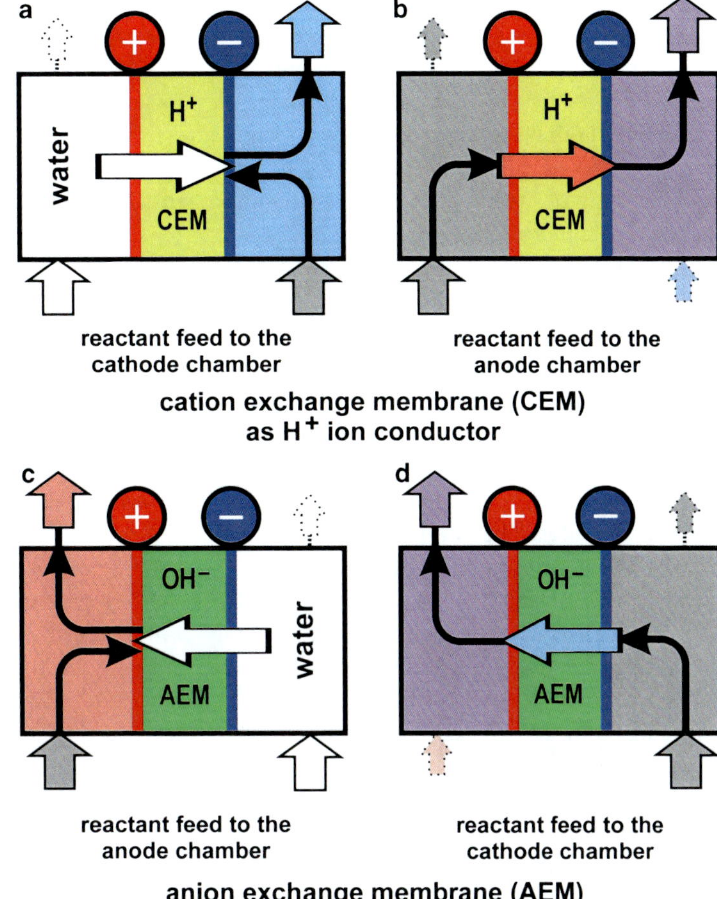

Electrosynthesis Using Solid Polymer Electrolytes (SPE), Fig. 3 Operation alternatives of SPE technology

Reactants are delivered into the cathode chamber and reduced at the cathode along with the arriving H^+ ions. Due to EOF, the entire membrane works with a suitable liquid, in this case water. Thus, it may be possible to use a reaction medium in the cathode chamber that is not able to solvate ions within the membrane, e.g., a nonpolar reactant and/or solvent. The first experiments of Ogumi are examples for this configuration [2].

EOF (in this example consisting only of water) is enhanced by diffusion in the same direction due to the existing concentration gradient and enters the cathode chamber at the cathode surface. This may generate problems with contamination and may hinder the desired cathode reaction. Moreover, this may enable or enhance undesired side reactions (a well-known example is the crossover of methanol from anode to cathode in direct methanol fuel cells (DMFC), which is detrimental for the cathodic oxygen reduction).

But EOF can be of additionally significant influence. On the one hand, it constricts the access of reactants to the cathode and therefore downgrades the performance of cathodic reactions. On the other hand, EOF can reject the opposite diffusion of reactants to the anode to a large extent and thus avoid undesired anodic reactions. That means, in this case the function of the membrane as a cell separator is intensified so that disadvantageous diffusion is blocked. The actual behavior is not easy to estimate because it is a complex result of all combined conditions, especially membrane properties, concentrations, current density, and diffusion rates (experiments are needed).

In configuration (b), the EOF is directly used as a dosing pump. The flow of reactants and solvents is exposed first to the anode and then, including products, to the cathode. The mass transport at both electrodes is enhanced. Configuration (b) is suitable if the reactions at both electrodes are desired or if the one reaction has no negative influence on the other. In the case of Fig. 1, the cathodic hydrogen evolution does not attack the anodic products. If one of the reactions is undesired, it may be changed by choosing another electrode material with an electrocatalytic activity for a more suitable reaction. For example, with platinum as a cathode material, hydrogenation is favored; with graphite, it is usually avoided.

The feed into the anode chamber in configurations (a) and (b) may be free flowing, or an excess may be used for rinsing. In configuration (b), the transport of reactants is directly determined by the EOF and in consequence also their conversion due to electrode reactions. A high flow rate of the EOF results in a low conversion. A partial recycling from the cathode compartment outlet into the anode chamber feed enables an increased conversion.

In configurations (c) and (d), an anion exchange membrane (AEM) is used as an alternative, e.g., as an OH^- ion conductor. The ion migration, the EOF, and the succession of reactions change their directions compared with a CEM as an H^+ ion conductor. However, the discussed effects of configurations (a) and (b) remain valid if the inversion is considered. These inverted directions may influence the performance of the cell significantly.

For example, the oxidation of γ-butyrolactone to succinic acid in an aqueous solution at a lead dioxide anode on titanium as a carrier metal is possible using a CEM in configuration (b). However, it is more effective using configuration (d): initially, the ring of γ-butyrolactone is opened at a graphite felt cathode to γ-hydroxybutyrate anions, which migrate through the AEM and are then transferred directly onto the anode surface, where they are oxidized to succinic acid [4].

According to the discussion of configuration (a), it is possible using configuration (c) to decrease the methanol crossover from the anodic methanol oxidation to the cathode due to the countercurrent EOF within the AEM together with OH^- ions [5]. This may be interesting for direct methanol (alcohol) fuel cells using an AEM.

Ion Exchange Membrane Materials

Ion exchange membranes are primarily produced in large amounts for electrodialysis in aqueous media, e.g., for seawater desalination. They are typically made of polystyrene, cross-linked with divinylbenzene, as a base polymer. Sulfonic acid groups are introduced as fixed ions for CEM and quaternary ammonium groups for AEM. These materials may be of limited stability in combination with organic compounds due to increased swelling or even chemical attack, particularly at elevated temperatures. Simple AEMs of this type will be destroyed in alkaline media, i.e., they cannot work for longer time as OH^- conductor. However, many membranes made from newly developed materials are also available. Different manufacturers should be contacted to identify suitable membranes for a particular task.

As mentioned above, perfluorosulfonic acid (PFSA) membranes like Nafion® (Dupont) are CEMs of very high chemical and thermal stability. In a highly sophisticated composition, such membranes are used as separators in industrial chlor-alkali electrolysis. Most research in the area of SPE electrosynthesis has also been done using Nafion® membranes.

CEMs such as Nafion® remain state-of-the-art H^+ ion conductors in fuel cells to this day. However, much research is being done to improve CEMs in PEMFC and DMFC. AEMs as OH^- ion conductors are also being investigated for alkaline hydrogen and alcohol fuel cells (see entry "► Fuel Cells, Principles and Thermodynamics"). Such new membranes will likely be suitable for SPE electrosynthesis in the future.

The fixed ions in the above-mentioned ion exchange membranes are covalently bonded and cannot be lost as long as the membrane is stable. There is only an ionic linkage in "solid-supported electrolytes," e.g., in a film of a basic polymer like polybenzimidazole impregnated with concentrated phosphoric acid. Such membranes are promising for PEMFC with gaseous reactants. Acid leaching is a problem, especially in cells using liquids. An interesting solid-supported electrolyte was realized by Fuchigami: the combination of a substrate such as silica gel with a base such as piperidine becomes in situ conductive in the presence of the reactant, e.g., methanol. This means that this material works as a fixed supporting electrolyte comparable with an ion exchange membrane. In the laboratory, e.g., anodic methoxylation reactions have been performed successfully (with discontinuous separation and reconditioning) [6].

Electrodes for SPE Technology

It is essential to minimize any voltage loss in fuel cells and water electrolysis for energy conversion. As a result, highly sophisticated "membrane electrode assemblies" (MEA) have been developed to optimize contact between the membrane and the electrocatalyst in the electrodes. Similarly, for the first organic SPE electrosyntheses [2], electrically conductive porous platinum or gold layers inside of the surfaces of Nafion® membranes were used, prepared by a chemical method similar to [7].

In the case of SPE technology for organic electrosynthesis, a low cell voltage is desired, but high reaction selectivity and product yield as well as high stability of the SPE cell in the presence of organic reactants are much more important. Thus, porous electrodes that are simply pressed on the membrane surfaces, such as graphite felt in Fig. 1, are suitable, and a somewhat higher cell voltage may be approved. This increases the variety of applicable electrode materials significantly. Even some swelling of the membrane, which is unavoidable with organic compounds, is acceptable, whereas it would destroy a MEA of a fuel cell. The Nafion® membrane becomes very soft and vulnerable after pretreatment by enhanced swelling for the reaction in Fig. 1, but this does not result in problems during mounting using soft graphite felt electrodes [4].

Metal wire mesh, e.g., of platinum, or sintered metal may also be used as an electrode material. A membrane surface that is soft due to a pretreatment may be suitable. Combining it with graphite felt assures an optimal contact by pressing them together. For example, titanium wire mesh or also porous sintered titanium metal coated with lead dioxide was used as an anode and graphite felt as a cathode for the previously mentioned reaction of γ-butyrolactone [4].

An interesting effect of SPE technology is the surprising increase in the stability of metal oxide electrodes, e.g., of lead dioxide or manganese dioxide electrochemically applied on a titanium carrier. Reasons for this may be:

- No aggressive electrolyte is present, such as sulfuric acid, which could dissolve the metal oxide.
- No oxide material can break free because it is compressed by the membrane.
- There is no mechanical stress from a turbulent flow, which usually exists in front of a conventional electrode, but only EOF generates a slow, continuous flow.

It is also possible to combine one gas diffusion electrode as in a fuel cell – a hydrogen-consuming anode or an oxygen-consuming cathode – with an electrode, which is described here for organic electrosynthesis. Under suitable conditions, even a complete membrane electrode assembly (MEA) of a PEMFC or DMFC can be used for organic electrosynthesis, usually with one side working as a hydrogen-consuming anode or an oxygen-consuming cathode, respectively.

Selected Example Reactions from Literature

Various anodic methoxylation reactions are applied in the chemical industry using a supporting electrolyte. Accordingly, several investigations using SPE technology also deal with methoxylations. The paper [6] and the alkoxylation of *N*-alkylamides were mentioned previously. Another example is the anodic dimethoxylation of furan where yields of up to 80 % of 2,5-dimethoxy-2,5-dihydrofuran have been reached [3]. A comparison in [4] of data from literature shows that very intensive mixing is necessary in order to avoid the decomposition of the product in the conventional ammonium bromide solution. This is no problem in the inert solution of SPE technology.

The di-methoxylation of *p*-methoxytoluene is also interesting for industrial applications. To do so, using SPE technology alone requires an extremely high cell voltage. It was possible to decrease this to approximately 5 V by using *N*, *N*-dimethylformamide (DMF) as a cosolvent and an optimized membrane pretreatment as discussed for the alkoxylation of *N*-alkylamides in paragraph 4. It is astonishing that no reaction of DMF, which can be easily methoxylated according to Fig. 1, was detectable. However, cell voltage increased more and more under these conditions, most probably due to thin porous layers of polymer precipitation in the graphite felt anode. Cell voltage was stabilized by the addition of a very small concentration of sulfuric acid and by increasing temperatures so that the solution within the anode started to boil slightly [8, 9].

For hydrogenation, SPE technology is applied in several research activities, in principle using a PEMFC with a complete MEA as gas diffusion electrodes. Hydrogen is used at the anode to decrease cell voltage. Examples are cathodic reduction using a lead catalyst, of *N,N*-diacetyl-L-cysteine to *N*-acetyl-L-cysteine [10] or electrochemical hydrogenation of soybean oil where the formation of unhealthy trans isomers in the product is decreased compared with the usual industrial hydrogenation process [11]. Even the generation of electricity in such fuel cells is possible, e.g., during hydrogenation of unsaturated alcohols and organic acids [12].

The final examples refer to the detoxification of waste water or of oil residues. The cathodic hydrodehalogenation of 2,4-dichlorophenol was investigated in [13], including the comparison of paraffin oil and water as reaction media. The anodic and/or cathodic decomposition of 2-chlorophenol was studied with platinum/iridium gauze as an anode and different types of graphite or carbon felt as a cathode. All variants of SPE technology illustrated in Fig. 3 were compared. Variant (b) with a Nafion® 117 CEM was most effective, i.e., first the anodic and then the cathodic reaction. Perforating the membrane enabled increased flow rates and less energy consumption independent of EOS. Nearly complete detoxification and extensive mineralization of organic bonded chlorine into chloride ions was achieved [14].

Boron-doped diamond electrodes attain very high potentials as anodes as well as cathodes in water. They are therefore very suitable for use in the electrochemical decomposition of organic toxic compounds. SPE technology also enables such applications in less conductive media like waste water. Benzyl alcohol was used as a model substance in [15].

Future Directions

The research in the area of SPE electrosynthesis shows interesting features and a large optimization potential of this technology. However, until now a commercial application is not known. The parallel development of fuel cells, which are based on the SPE technology, is a matter of very big commercial interest. New materials, e.g., membranes, and increased know-how of fuel cell research can stimulate a progress of SPE electrosynthesis.

Cross-References

- ► Alkaline Membrane Fuel Cells, Membranes
- ► Boron-Doped Diamond for Green Electro-Organic Synthesis
- ► Electrosynthesis Using Diamond Electrode
- ► High-Temperature Polymer Electrolyte Fuel Cells
- ► Hydrocarbon Membranes for Polymer Electrolyte Fuel Cells
- ► Polymer Electrolyte Fuel Cells (PEFCs), Introduction

► Polymer Electrolyte Fuel Cells, Membrane-Electrode Assemblies
► Polymer Electrolyte Fuel Cells, Perfluorinated Membranes

References

1. Grubb WT (1955) Fuel cell. US 2913511 General Electric
2. Ogumi Z, Nishio K, Yoshizawa S (1981) Application of the spe method to organic electrochemistry. II. Electrochemical hydrogenation of olefinic double bonds. Electrochim Acta 26:1779–1782
3. Raoult E, Sarrazin J, Tallec A (1985) Use of ion exchange membranes in preparative organic electrochemistry. II. Anodic dimethoxylation of furan. J Appl Electrochem 15:85–91
4. Jörissen J (1996) Ion exchange membranes as solid polymer electrolytes (spe) in electroorganic synthesis without supporting electrolytes. Electrochim Acta 41:553–562
5. Jörissen J (2003) Electro-organic synthesis without supporting electrolyte: possibilities of solid polymer electrolyte technology. J Appl Electrochem 33: 969–977
6. Tajima T, Fuchigami T (2005) Development of an electrolytic system using solid-supported bases for in situ generation of a supporting electrolyte from methanol as a solvent. J Am Chem Soc 127: 2848–2849
7. Takenaka H, Torikai E, Kawami Y, Wakabayashi N (1982) Solid polymer electrolyte water electrolysis. Int J Hydrogen Energy 7:397–403
8. Steckhan E, Arns T, Heineman WR, Hilt G, Hoormann D, Jörissen J, Kröner L, Lewall B, Pütter H (2001) Environmental protection and economization of resources by electroorganic and electroenzymatic syntheses. Chemosphere 43:63–73
9. Hoormann D, Kubon C, Jörissen J, Kröner L, Pütter H (2001) Analysis and minimization of cell voltage in electro-organic syntheses using the solid polymer electrolyte technology. J Electroanal Chem 507:215–225
10. Montiel V, Saez A, Exposito E, Garcia-Garcia V, Aldaz A (2010) Use of MEA technology in the synthesis of pharmaceutical compounds: the electrosynthesis of *N*-acetyl-L-cysteine. Electrochem Commun 12:118–121
11. Pintauro PN, Gil MP, Warner K, List G, Neff W (2005) Electrochemical hydrogenation of soybean oil with hydrogen gas. Ind Eng Chem Res 44:6188–6195
12. Yuan X, Ma Z, Bueb H, Drillet JF, Hagen J, Schmidt VM (2005) Cogeneration of electricity and organic chemicals using a polymer electrolyte fuel cell. Electrochim Acta 50:5172–5180
13. Cheng H, Scott K, Christensen PA (2004) Electrochemical hydrodehalogenation of 2,4-dichlorophenol in paraffin oil and comparison with aqueous systems. J Electroanal Chem 566:131–138
14. Heyl A, Jörissen J (2006) Electrochemical detoxification of waste water without additives using solid polymer electrolyte (SPE) technology. J Appl Electrochem 36:1281–1290
15. Kraft A, Stadelmann M, Wünsche M, Blaschke M (2006) Electrochemical destruction of organic substances in deionized water using diamond anodes and a solid polymer electrolyte. Electrochem Commun 8:155–158

Electrosynthesis Using Template-Directed Methods

Siegfried R. Waldvogel and Nina Welschoff
Johannes Gutenberg-University Mainz, Mainz, Germany

Introduction

The definition of templates originates from coordination chemistry and supramolecular science. The more modern use of the term template describes a molecular entity that brings structures together in a manner which would not happen in a non-templated fashion. This will result in a specific interaction or selective chemical transformation. In most examples, the template will be incorporated, but this is not necessarily the case [1, 2]. Electrodes and their surfaces may also be considered as two-dimensional templates [3]. Since only the electrosynthetic aspects will be treated here, they are not part of this survey.

The templates can be covalently bound to the electrophores/substrates and serve as tethers or are more weakly attached to the reaction partners by coordinative bonds. In electroorganic synthesis, this concept was successfully applied and demonstrated in several cases. The templates work if in the course of the redox transformation either a compensation of charge is achieved or a radical center is stabilized in the neighborhood

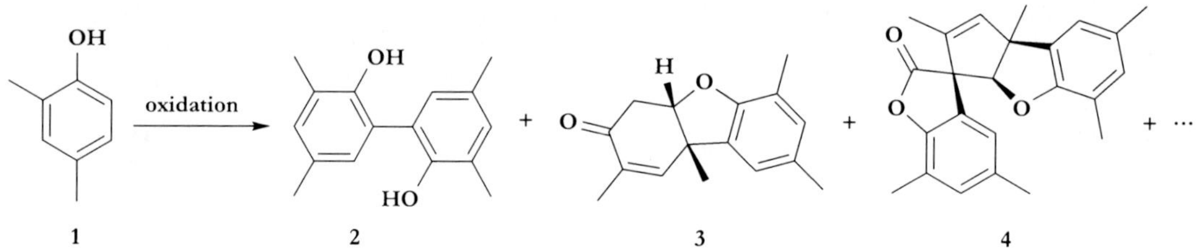

Electrosynthesis Using Template-Directed Methods, Scheme 1 Non-templated anodic oxidation of 2,4-dimethylphenol

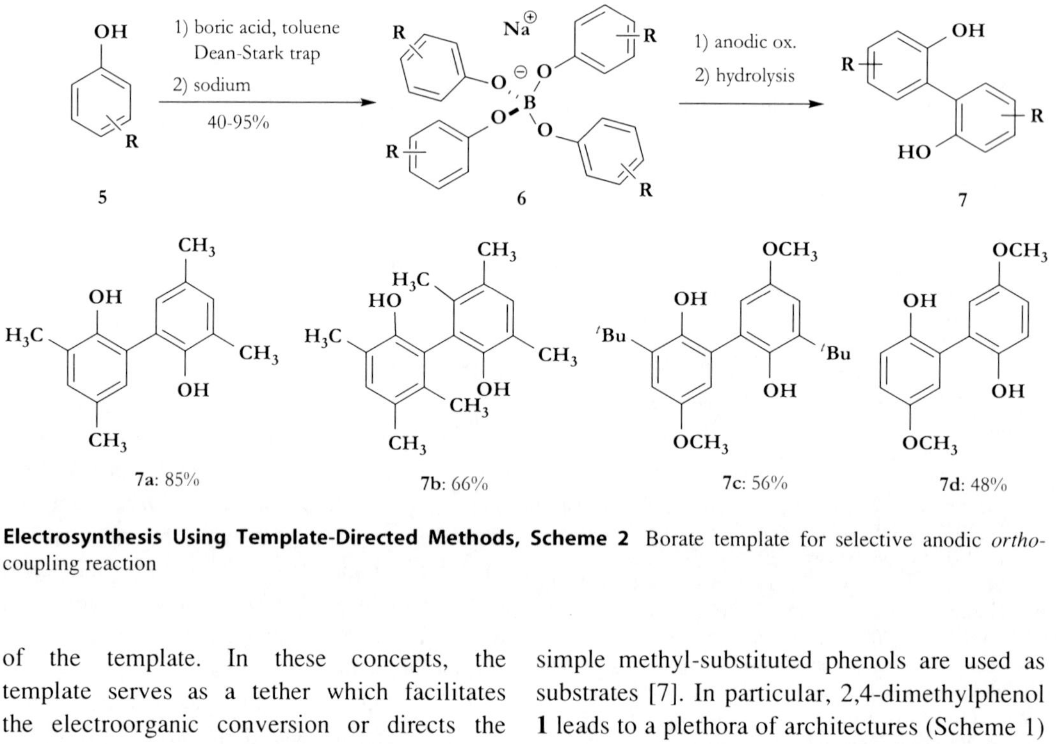

Electrosynthesis Using Template-Directed Methods, Scheme 2 Borate template for selective anodic *ortho*-coupling reaction

of the template. In these concepts, the template serves as a tether which facilitates the electroorganic conversion or directs the subsequent reaction pathway.

Within the course of reaction or upon work-up, the template is mostly cleaved off. Since the template is usually not recognized as such in the product, these systems should be named traceless templates. This naming is in analogy to the corresponding linker systems in solid-phase synthesis [4–6].

Boron Template

The selective oxidative phenolic *ortho*-coupling reaction turned out to be challenging when simple methyl-substituted phenols are used as substrates [7]. In particular, 2,4-dimethylphenol **1** leads to a plethora of architectures (Scheme 1) which are structurally related to Pummerer's ketone **3** [8, 9]. This non-templated anodic oxidation of **1** gives rise to a wide structural diversity, and the biphenol **2** is only obtained in traces [9–11]. Introduction of boron as a tetragonal template changed the situation tremendously [12]. From conventional synthesis, tethers based on silicon have been explored [13–15]. However, the tetraphenoxy borate strategy consists of two key features: First, the template phenols can be obtained in a one-pot procedure (Scheme 2) [16]. Secondly, the sodium tetraphenoxyborates **6** represent not only the substrates but also the supporting electrolyte for the electrolysis. Upon

Electrosynthesis Using Template-Directed Methods, Scheme 3 Traceless silyl template for electrochemical cyclization reaction

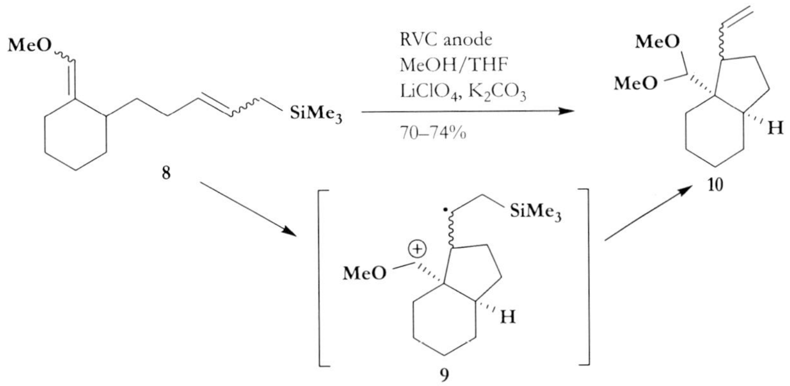

RVC anode
MeOH/THF
$LiClO_4$, K_2CO_3
70–74%

Electrosynthesis Using Template-Directed Methods, Scheme 4 Silyl tether as template

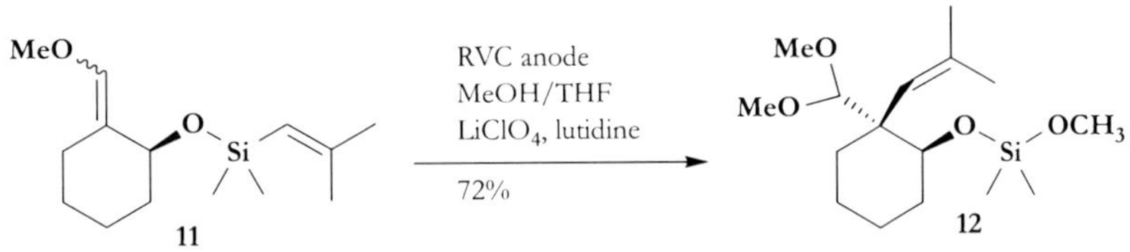

RVC anode
MeOH/THF
$LiClO_4$, lutidine
72%

anodic treatment in acetonitrile, the desired 2,2'-biphenol **7** is liberated by hydrolysis at elevated temperature. The protocol is viable for performing the electroorganic transformation on a larger scale [17].

Detailed investigations reveal that only the oxidative coupling occurs on **6**. The slow ligand exchange causes disproportionation and provides the starting material and the bis (2,2'-biphenoxy)borate. The latter exhibits sufficient stability and was used in supercapacitor studies [18, 19]. The ligand exchange unfortunately impedes the use in an anodic cross-coupling reaction.

Silicon Template

The direct anodic coupling reaction using simple alkyl-substituted olefins as substrates usually generates a complex product mixture. The high reactivity of the intermediate alkyl radicals causes a lack of selectivity [20, 21]. An elegant way to achieve regiochemical control is the use of allylsilanes as reaction partners [21–25]. In the course of the cyclization reaction, a distonic radical cation **9** is formed (Scheme 3), whereby the positive charge is located in the vicinity of the methoxy moiety and the spin center in position β to the silyl group [24]. Upon the second oxidation step, the silyl fragment is cleaved off, furnishing the product **10**. The observed selectivity cannot be found without a temporary silyl template.

Using the silyl tether in a more classical template fashion **11** brings both reaction partners together (Scheme 4). The transformation yields a silyl-protected hydroxyl derivative **12** [23]. The excellent stereocontrol is attributed to the bicyclic distonic radical cation which is formed as intermediate.

The anodic oxidation is performed at reticulated vitreous carbon (RVC) which is essentially macroporous glassy carbon. The electrolysis is operated in an undivided cell under constant current electrolysis conditions, with a platinum cathode, in an electrolyte consisting of lithium perchlorate in methanol/THF (1:1). 2,6-Lutidine serves as a proton scavenger [25]. In related oxidation-cyclization sequences, the desired bicyclic product **14** was only accomplished when the silyl moiety was present (Scheme 5). In contrast to the other examples, the silyl fragment serves as electro-auxiliary and facilitates the formation of the intermediate acyliminium species which undergoes the cyclization reaction [26].

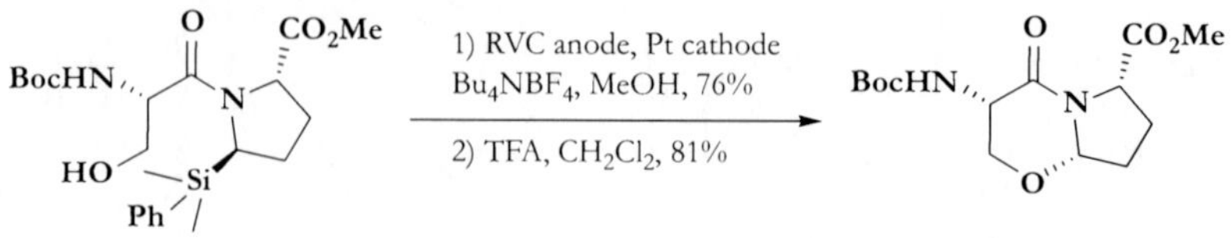

Electrosynthesis Using Template-Directed Methods, Scheme 5 Traceless silyl template

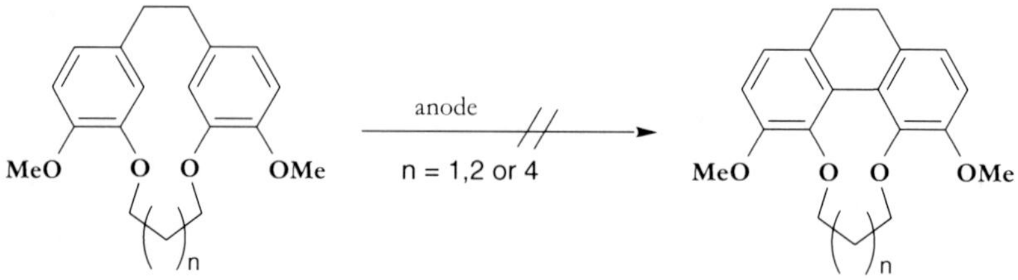

Electrosynthesis Using Template-Directed Methods, Scheme 6 Non successful carbon template in anodic biaryl coupling

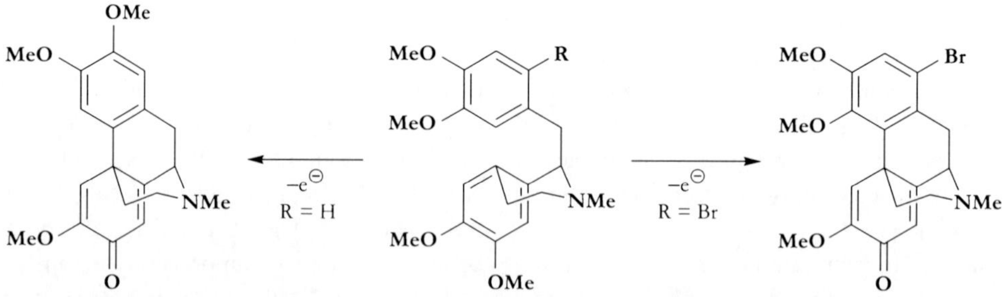

Electrosynthesis Using Template-Directed Methods, Scheme 7 Alternative directing the reaction pathway by steric influence

Carbon Templates

Tethers based on carbon chains are neutral and have electronically almost no influence. A forced alignment of the electron-rich aryl moieties in **15** does not result upon anodic treatment in the desired *ortho,ortho*-coupled product. Short tethers in **15** promote intermolecular coupling reactions, whereas longer spacers allow the formation of *ortho,para* products [27]. The electronic activation in position *para* is dominant and a simple alkylidene tether is not exploitable for synthesis (Scheme 6).

The intramolecular oxidative coupling reaction of benzyltetrahydroisoquinolines, like laudanosine **17**, provides an attractive synthetic access to the morphine skeleton [28]. Unfortunately, the position *para* to the methoxy group on the electron-rich aryl moiety yields **18** [29–33]. For the selective generation of the isomeric and desired product **19**, the activated position has to be blocked. Best results were obtained with a bromo substituent [34, 35]. Since the bulky group only provides steric demand and does not bring the reaction partners together, this cannot be considered as a template system (Scheme 7).

In conclusion, carbon-based templates could not yet be exploited in electrosynthesis. In these cases the course of reaction is efficiently

Electrosynthesis Using Template-Directed Methods, Scheme 8 Metal template for anodic cyclization reaction

Pt anode
Hg cathode
94%

20

21

influenced by protective groups, which mask reactive positions in the substrate.

Metal Templates

For the ring closure of macrocycles, the use of templates is a common tool to enhance the yield and selectivity. Substrates using metal ions as templates were successfully converted to the macrocyclic product. The templating metal center remains in the product. The nickel complex **20** is oxidized on platinum, undergoes hydrogen shift, and is trapped by water providing nickel secocorrinate **21** [36] (Scheme 8).

Future Directions

The electrochemical transformations employing templates are solely focused on anodic conversions. Extension of this concept to electroorganic reductions appears promising but requires further development. In this aspect, non-covalently attached tethers might be most practical.

Cross-References

- ▶ Boron-Doped Diamond for Green Electro-Organic Synthesis
- ▶ Electroauxiliary
- ▶ Electrochemical Functional Transformation
- ▶ Electrophoresis
- ▶ Organic Reactions and Synthesis

References

1. Busch DH (2005) First considerations; principles, classification, and history. Top Curr Chem 249:1–65
2. Laughrey ZR, Gibb BC (2005) Macrocycle synthesis through templation. Top Curr Chem 249:67–125
3. Becker C, Wandelt K (2009) Surfaces: two-dimensional templates. Top Curr Chem 287:45–86
4. Bräse S, Enders D, Köbberling J, Avemaria F (1998) A surprising solid-phase effect: development of a recyclable "Traceless" linker system for reactions on solid support. Angew Chem Int Ed 37:3413–3415
5. Bräse S, Dahmen S (2000) Traceless linkers – only disappearing links in solid-phase organic synthesis? Chem Eur J 6:1899–1905
6. Wiehn MS, Jung N, Bräse S (2008) Safety-catch and traceless linkers in solid phase organic synthesis. In: Tulla-Puche J, Albericio F (eds) The power of functional resins in organic synthesis. Wiley, Weinheim
7. Waldvogel SR (2010) Novel anodic concepts for the selective phenol coupling reaction. Pure Appl Chem 82:1055–1063
8. Malkowsky IM, Rommel CE, Wedeking K, Fröhlich R, Bergander K, Nieger M, Quaiser C, Griesbach U, Pütter H, Waldvogel SR (2006) Facile and highly diastereoselective formation of a novel pentacyclic scaffold by direct anodic oxidation of 2,4-dimethylphenol. Eur J Org Chem 241–245
9. Barjau J, Königs P, Kataeva O, Waldvogel SR (2008) Reinvestigation of highly diastereoselective pentacyclic spirolactone formation by direct anodic oxidation of 2,4-dimethylphenol. Synlett 2309–2311
10. Barjau J, Schnakenburg G, Waldvogel SR (2011) Diversity-oriented synthesis of polycyclic scaffolds by post-modification of an anodic product derived from 2,4-dimethylphenol. Angew Chem Int Ed 50:1415–1419
11. Barjau J, Fleischhauer J, Schnakenburg G, Waldvogel SR (2011) Installation of amine moieties into polycyclic anodic product derived from 2,4-dimethylphenol. Chem Eur J 17:14785–14791
12. Malkowsky IM, Rommel CE, Fröhlich R, Griesbach U, Pütter H, Waldvogel SR (2006) Novel

template-directed anodic phenol-coupling reaction. Chem Eur J 12:7482–7488
13. Schmittel M, Burghart A, Malisch W, Reising J, Söllner R (1998) Diastereoselective enolate coupling through redox umpolung in silicon and titanium bisenolates: a novel concept based on intramolecularization of carbon-carbon bond formation. J Org Chem 63:396–400
14. Schmittel M, Haeuseler A (2002) One-electron oxidation of metal enolates and metal phenolates. J Organomet Chem 661:169–179
15. Schmittel M, Haeuseler A (2003) Access to unsymmetric binaphthols through oxidative coupling of silicon bisnaphtholates. Z Naturforsch B 58:211–216
16. Malkowsky IM, Fröhlich R, Griesbach U, Pütter H, Waldvogel SR (2006) Facile and reliable synthesis of tetraphenoxyborates and their properties. Eur J Inorg Chem 1690–1697
17. Griesbach U, Pütter H, Waldvogel SR, Malkowsky IM (2006) Anodic electrolytic oxidative electrodimerisation of hydroxy-substituted aromatics to give dihydroxy-substituted biarylene compounds. PCT Int Appl WO 2006077204 A2
18. Waldvogel SR, Malkowsky IM, Griesbach U, Pütter H, Fischer A, Hahn M, Kötz R (2009) Novel fluorine-free electrolyte system for supercapacitors. Electrochem Commun 11:1237–1242
19. Francke R, Schnakenburg G, Cericola D, Kötz R, Waldvogel SR (2011) Novel Bis(2,2'-biphenoxy) borates for electrochemical double layer capacitor electrolytes. Chem Eur J 17:3082–3085
20. Moeller KD, Marzabadi MR, Chiang MY, New DG, Keith S (1990) Oxidative organic electrochemistry: a novel intramolecular coupling of electron-rich olefins. J Am Chem Soc 112:6123–6124
21. Hudson CM, Marzabadi MR, Moeller KD, New DG (1991) Intramolecular anodic olefin coupling reactions: a useful method for carbon-carbon bond formation. J Am Chem Soc 113:7372–7385
22. Moeller KD, Hudson CM (1991) Intramolecular anodic olefin coupling reactions: the use of allylsilanes. Tetrahedron Lett 32:2307–2310
23. Moeller KD, Hudson CM, Tinao-Wooldridge LV (1993) Intramolecular anodic olefin coupling reactions: the use of allyl- and vinylsilanes in the construction of quaternary carbons. J Org Chem 58:3478–3479
24. Hudson CM, Moeller KD (1994) Intramolecular anodic olefin coupling reactions and the use of vinylsilanes. J Am Chem Soc 116:3347–3356
25. Frey DA, Reddy SHK, Moeller KD (1999) Intramolecular anodic olefin coupling reactions: the use of allylsilane coupling partners with allylic alkoxy groups. J Org Chem 64:2805–2813
26. Sun H, Moeller KD (2002) Silyl-substituted amino acids: new routes to the construction of selectively functionalized peptidomimetics. Org Lett 4: 1547–1550
27. Schäfer HJ (2001) Electrolytic oxidative coupling. In: Lund H, Hammerich O (eds) Organic electrochemistry. Marcel Dekker, New York, p 926
28. Schäfer HJ (1981) Anodic and cathodic CC-bond formation. Angew Chem Int Ed 20:911–934
29. Miller LL, Sternitz FR, Falck JR (1971) Electrooxidative cyclization of laudanosine. A novel nonphenolic coupling reaction. J Am Chem Soc 93: 5941–5942
30. Miller LL, Sternitz FR, Falck JR (1973) Electrooxidative cyclization of 1-benzyltetrahydro-isoquinolines. A novel nonphenol coupling reaction. J Am Chem Soc 95:2651–2656
31. Bentley TW, Morris SJ (1986) Electrosynthesis in a beaker: an efficient route to morphinandienones avoiding potentiostats for control of electrode potentials. J Org Chem 51:5007–5010
32. Kotani E, Tobinaga S (1973) Total synthesis of the morphinandienone alkaloids, amurine, flavinantine, and pallidine by anodic oxidation. Tetrahedron Lett 48:4759–4762
33. Klünenberg H, Schäffer C, Schäfer HJ (1982) A remarkable influence of the electrolyte in anodic cyclization of 1-benyltetrahydroisoquinolines to neospirodienones or morphinandienones. Tetrahedron Lett 23:4581–4584
34. Falck JR, Miller LL, Sternitz FR (1974) Electrooxidative synthesis of morphinandienones from 1-benzyltetrahydroisoquinolines. Tetrahedron 30:931–934
35. Miller LL, Stewart RF (1978) Synthesis of morphinandienones, a dihydrophenanthrone, and pummerer's ketones by anodic coupling. J Org Chem 43:1580–1586
36. Kräutler B, Pfaltz A, Nordmann R, Hodgson KO, Dunitz JD, Eschenmoser A (1976) Experiments on a simulation of the photochemical A/D-secocorrin to corrin cycloisomerization by redox processes. Electrochemical oxidation of nickel(II)-1-methylidene-2,2,7,7,12,12-hexamethyl-15-cyano-1,19-secocorrinate perchlorate. Helv Chim Acta 59:924–937

Electrosynthesis Using Water Suspension System

Manabu Kuroboshi
Okayama University, Okayama, Japan

Introduction

Water is one of the most ideal solvents for electrolysis because water has a high dielectric

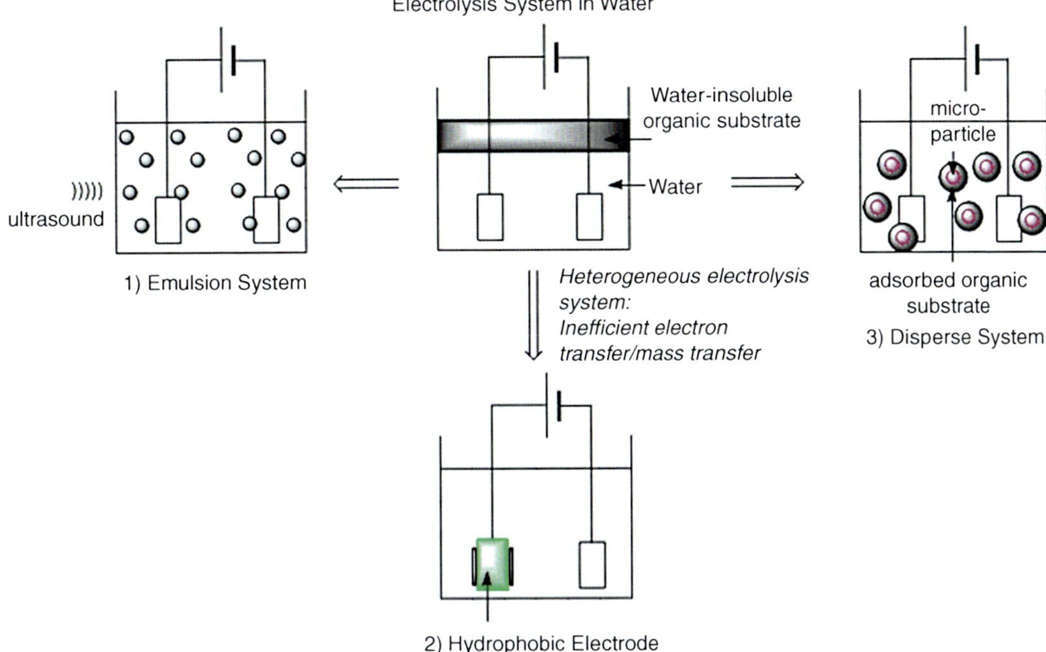

Electrosynthesis Using Water Suspension System, Fig. 1 Electrolysis system in water

constant and can dissolve many salts (supporting electrolyte). Water is readily available, safe (incombustible, nontoxic, etc.), and environmentally benign. However, water has some disadvantages as a medium for electrolysis. Potential window of water is narrower than organic solvents, and electrolysis of water occurs as a side reaction. Moreover, many organic substrates and organic mediators (electrochemical catalysts) are insoluble in water. Therefore, highly polar organic solvents, such as DMF and DMSO, and/or two-phase systems consisting of water-organic solvents, e.g., water-CH_2Cl_2 and water-AcOEt, have been used for electrolysis.

Examples of Electrosynthesis Using Water Suspension System

To overcome these disadvantages and obtain high current efficiency for the desired products in water, several methods have been developed. For instance, electrolysis has been performed

Electrosynthesis Using Water Suspension System, Fig. 2 A plausible mechanism of electro-oxidation of alcohols mediated by *N*-oxy/Br^- double mediatory system

(1) in the emulsion formed by sonication of the electrolyte [1–3], (2) on hydrophobic electrodes prepared by the composite-plating of a metal with hydrophobic particles [4, 5], and (3) by microparticle-disperse system in which the organic substrate is adsorbed on the microparticles (Fig. 1).

In the microparticle-disperse system, microparticles having large surface area, e.g., silica gel for column chromatography (*ca.* 50-μm diameter, surface area: 500 m^2/g), polystyrene particle (*ca.* 500 m^2/g), and activated carbon (*ca.* 1,000 m^2/g), are frequently used. The microparticles and the electrode, however, seldom contact, and electron transfer between the electrode and the substrate on the microparticle is retarded: Therefore, indirect electrolysis using water-soluble mediators is used in this system.

In bromide ion/*N*-oxyl-double mediated electro-oxidation of alcohols (Eq. 1) [6, 7], bromide ion (the primary mediator) is oxidized at the anode to form active bromine species [Br^+], which oxidizes *N*-oxyl **3** (secondary mediator) to *N*-oxoammonium **4**; **4** oxidizes alcohol **1** to the corresponding carbonyl compound **2** (Fig. 2). This electro-oxidation can be performed by using *N*-oxyl-immobilized microparticles/water-disperse system [8].

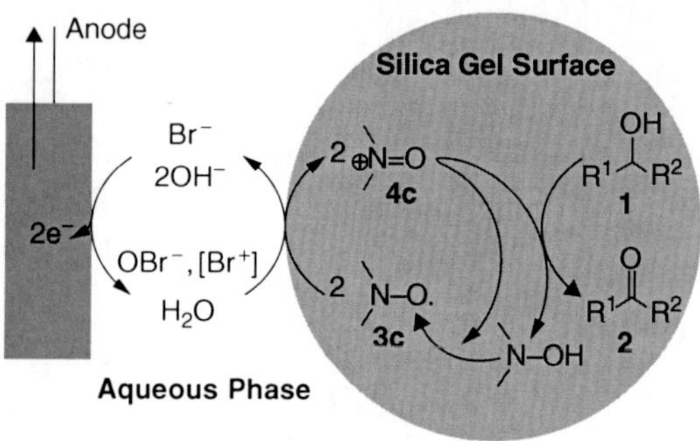

Electrosynthesis Using Water Suspension System, Fig. 3 A plausible mechanism of *N*-oxyl modified silica gel

cat. N–O· **3**

OH R^1 R^2 **1** → aq. NaBr-$NaHCO_3$/CH_2Cl_2 (Pt)-(Pt), Undiv. Cell → O R^1 R^2 **2** (1)

N-Oxyl-immobilized silica gel **3c** is prepared from **3a** with silylating reagent **5** (**3c**: 0.6 mmol *N*-oxyl/g silica gel) (Eq. 2). Active bromine species [Br^+], formed by electro-oxidation of bromide ion, readily oxidizes **3c** to give *N*-oxoammonium-modified silica gel **4c**. As an advantage of the microparticle/water-disperse system, organic mediator and alcohol **1** are gathered on these microparticles to form the highly concentrated reaction sphere, and **1** is smoothly oxidized with **4c** to give the corresponding carbonyl compound **2** (Fig. 3). After the electrolysis, **2** is still adsorbed on the silica gel. As another advantage of this system, the aqueous solution is recovered easily by filtration, and **2** and the microparticle are separated by a simple rinse. The recovered aqueous solution and *N*-oxyl-modified silica gel **3c** are reused repeatedly. Therefore, this method offers a "completely closed system" for electro-organic synthesis (Fig. 4).

$(EtO)_3Si(CH_2)_3NCO$ **5** → H_2N–(N–O•) **3a**, Benzene, Room Temp., 9h → $(EtO)_3Si$ … **3b** → Silica Gel, Benzene, reflux, 61 h → Silica Gel … **3c** (2)

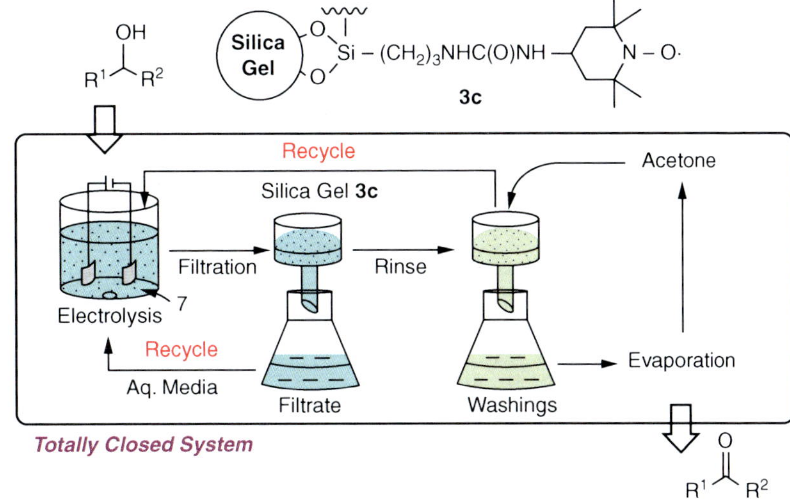

Electrosynthesis Using Water Suspension System, Fig. 4 Closed system for electro-oxidation of alcohol

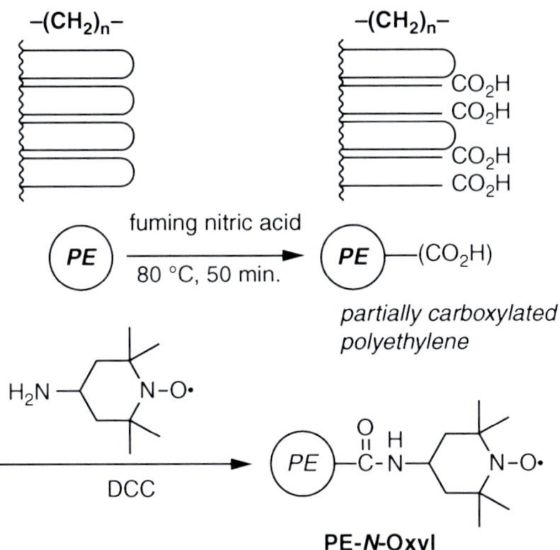

Electrosynthesis Using Water Suspension System, Fig. 5 PE-*N*-oxyl

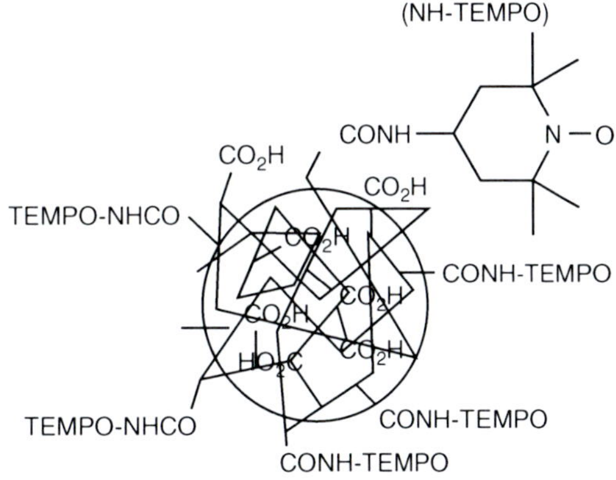

Electrosynthesis Using Water Suspension System, Fig. 6 Poly(ethylene-co-acrylic acid)-TEMPO (*5 wt% acrylic acid*)

Though *N*-oxyl-immobilized silica gel **3c** is easily prepared and used repeatedly [8], mechanical strength of the silica gel is not enough for practical use [9]. This method can be, however, easily expanded to polymer chemistry. *N*-Oxyl-modified polymers (partially carboxylated polyethylene (Fig. 5), polyethylene-polyacrylic acid copolymer (Fig. 6), or poly(*p*-phenylene-co-benzobisthiazole) (PBZT, Fig. 7), etc.) can be used as an absorbent [10, 11]. These polymer particles are tough enough and *N*-oxyl moiety is not lost by hydrolysis.

Future Directions

Many types of microparticles are supplied, and reagents/mediators/catalysts are easily introduced on the microparticles. The substrate is concentrated on the microparticles to accelerate the reaction. The products are obtained by simple filtration/rinse, and the recovered microparticles and the aqueous phase are reused repeatedly to offer a complete closed system. Therefore, this "Electrosynthesis Using Water Suspension System" can be a powerful candidate for waste-free green-sustainable synthetic method.

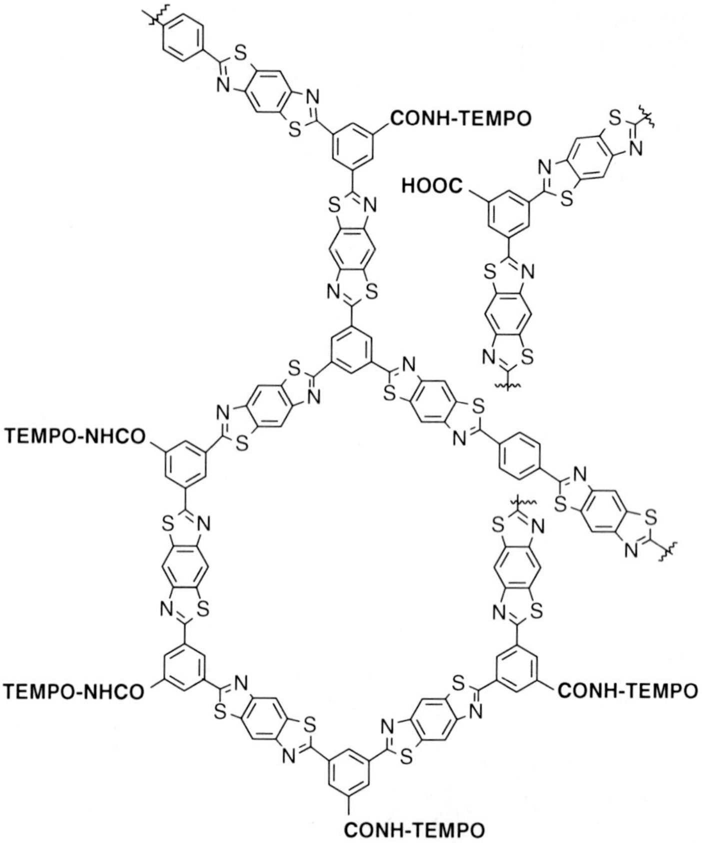

Electrosynthesis Using Water Suspension System, Fig. 7 PBZT-TEMPO

Cross-References

- ► Electrosynthesis Under Ultrasound and Centrifugal Fields
- ► Electrosynthesis Using Mediator

References

1. Wadhawan JD, Del Campo FJ, Compton RG, Foord JS, Marken F, Bull SD, Davies SG, Walton DJ, Ryley S (2001) Emulsion electrosynthesis in the presence of power ultrasound. Biphasic Kolbe coupling processes at platinum and boron-doped diamond electrodes. J Electroanal Chem 507:135–143
2. Wadhawan JD, Marken F, Compton RG, Bull SD, Davies SG (2001) Sono-emulsion electrosynthesis: electrode-insensitive Kolbe reactions. Chem Commun 87–88
3. Atobe M, Ikari S, Nakabayashi K, Amemiya F, Fuchigami T (2010) Electrochemical reaction of water-insoluble organic droplets in aqueous electrolytes using acoustic emulsification. Langmuir 26:9111–9115
4. Kunugi Y, Chen P-C, Nonaka T, Chong Y-B, Watanabe N (1993) Electrolysis of emulsions of organic compounds on hydrophobic electrodes. J Electrochem Soc 140:2833–2836
5. Ono Y, Kim S-H, Yasuda M, Nonaka T (1999) Kolbe electrolysis of carboxylates on a hydrophobic platinum electrode composite-plated with PTFE particles. Electrochemistry 67:1042–1045 (Tokyo)
6. Semmelhack MF, Chou CS, Cortes DA (1983) Nitroxyl-mediated electrooxidation of alcohols to aldehydes and ketones. J Am Chem Soc 105:4492–4494
7. Inokuchi T, Matsumoto S, Torii S (1991) Indirect electrooxidation of alcohols by a double mediatory system with two redox couples of $[R_2N^+ = O]/R_2NO\cdot$ and $[Br\cdot$ or $Br^+]/Br^-$ in an organic-aqueous two-phase solution. J Org Chem 56:2416–2421
8. Kuroboshi M, Goto K, Tanaka H (2009) Electrooxidation of alcohols in *N*-oxyl-immobilized silica gel/water disperse system: approach to totally closed system. Synthesis: 903–908
9. Palmisano G, Ciriminna R, Pagliaro M (2006) Waste-free electrochemical oxidation of alcohols in water. Adv Synth Catal 348:2033–2037, *N*-Oxyl-immobilized ITO anode was also developed

10. Tanaka H, Kubota J, Miyahara S, Kuroboshi M (2005) Electrooxidation of alcohols in an *N*-oxyl-immobilized poly(ethylene-*co*-acrylic acid)/ water disperse system. Bull Chem Soc Jpn 78:1677–1684
11. Kubota J, Ido T, Kuroboshi M, Tanaka H, Uchida T, Shimamura K (2006) Electrooxidationof alcohols in an *N*-oxyl-immobilized rigid network polymer particles/water disperse system. Tetrahedron 62: 4769–4773

Elements of Electrocatalysts for Oxygen Reduction Reaction

Junliang Zhang, Fengjuan Zhu and Fengjing Jiang
Institute of Fuel Cells, School of Mechanical Engineering, Shanghai Jiao Tong University, Shanghai, People's Republic of China

Introduction

In a PEMFC, if pure hydrogen is used as the fuel, the anode reaction is then the hydrogen oxidation reaction (HOR) at the surface of the anode platinum electrocatalyst. The hydrogen oxidation reaction (HOR) and hydrogen evolution reaction (HER) are by far the most thoroughly investigated electrochemical reaction system [1]. Due to the fast electrode kinetics of hydrogen oxidation at platinum surface [1–3], the anode platinum loading can be reduced down to $0.05mg_{Pt}/cm^2$ without significant performance loss [4]. The cathode reaction in a PEM fuel cell is the oxygen reduction reaction (ORR) at platinum surface in an acidic electrolyte. In contrast to HER at the anode, the cathode ORR is a highly irreversible reaction even at temperatures above 100 °C at the best existing catalyst – the platinum surface [5]. Gasteiger et al. [6] found that $0.4mg_{Pt}/cm^2$ was close to the optimal platinum loading for the air electrode using the state-of-the-art Pt/C catalyst and an optimized electrode structure. Further reduction of the cathode platinum loading will result in cell voltage loss at low current densities that follows the ORR kinetic loss. The high platinum loading at the cathode originates from the slow kinetics of (ORR) at platinum surface. To make the fuel cell vehicles commercially viable on the market, the platinum loading on the cathode has to be reduced significantly. Recent increases in Pt prices suggest that one should be striving for at least an eight-times improvement in the Pt-mass-specific activity (denoted as Pt-mass activity hereafter), in order to meet the Pt loading target set for the mass production of fuel cell vehicles.

ORR on Platinum Surfaces

While the detailed mechanism of ORR still remains elusive [7], it is widely accepted that the ORR on platinum surfaces is dominantly a multistep four-electron reduction process with H_2O being the final product. The overall four-electron reduction of O_2 in acid aqueous solutions is

$$O_2 + 4H^+ + 4e^- \rightarrow H_2O;$$
$$E_0 = 1.229V \text{ versus NHE at } 298K$$

where E_0 denotes the standard electrode potential for the reaction; NHE is the normal hydrogen electrode potential. Since the four-electron reduction of oxygen is highly irreversible, the experimental verification of the thermodynamic reversible potential of this reaction is very difficult. The irreversibility of ORR imposes serious voltage loss in fuel cells. In most instances the current densities practical for kinetic studies are much larger than the exchange current density of ORR; therefore, the information obtained from current-potential data are confined only to the rate-determining step. On the other hand, in the ORR kinetic potential region, the electrode surface structure and properties strongly depend on the applied potential and the time held at that potential, which makes the reaction more complicated. While the relationship between the overall kinetics and the surface electronic properties is not well understood, it is widely accepted that in the multistep reaction, the first electron transfer is the rate-determining step, which is accompanied by or followed by a fast proton transfer [7]. Two Tafel slopes are usually observed for ORR on Pt in RDE tests in perchloric acid, from −60 mV/decade at low current density transitioning to −120 mV/decade at high current density. The lower Tafel slope of the ORR in perchloric acid at low current density has been attributed to the potential dependent Pt oxide/hydroxide coverage at high potentials [8].

ORR on Pt-Alloy Catalysts

Great progress has been made in past decades in developing more active and durable Pt-alloy catalysts and in understanding the factors attributing to their activity enhancements. Two- to threefold-specific activity enhancements versus pure Pt were typically reported in literature [9]. As far as which alloy and what alloy compositions confer the highest ORR activity, there seems to be lack of a general agreement. This is probably because the measured activity depends highly on the catalyst surface and near-surface atomic composition and structure, on impurities on the surface and on particle size and shape, all of which could be affected by the preparation method, heat treatment protocol, and testing conditions. For example, to achieve the optimal alloy structure for maximum activity, different Pt-alloy particles may require different annealing temperature protocols to accommodate the distinctions between metal melting points and particle sizes [10]. Several representative mechanisms have been proposed in the literature to explain the enhanced activities observed on Pt-alloy catalysts: (1) a surface roughening effect due to leaching of the alloy base metal [11, 12]; (2) decreased lattice spacing of Pt atoms due to alloying [13]; (3) electronic effects of the neighboring atoms on Pt, such as increased Pt d-band vacancy [14] or depressed d-band center energy upon alloying[15]; and/or (4) decreased Pt oxide/hydroxide formation at high potential [16]. The increased Pt surface roughness alone may help increase Pt-mass activity, but will not increase the Pt-specific activity. Other mechanisms are correlated with each other, for example, the decreased lattice spacing may affect the electronic structure of Pt atoms, which in turn may inhibit the Pt oxide/hydroxide formation.

ORR on Pt Monolayer Electrocatalysts

The Pt monolayer electrocatalyst has been one of the key concepts in reducing the Pt loading of PEM fuel cells in recent years. Pt submonolayers deposited on Ru nanoparticles had been earlier demonstrated to give superior performance with ultra-low Pt loading compared to commercial Pt/C or Pt-Ru alloy catalysts for the anode CO-tolerant hydrogen oxidation reaction[17]. More recently, Adzic and co-workers applied this concept in making novel Pt monolayer catalysts for the cathode ORR. In general, the new method of synthesizing Pt monolayer catalysts involves underpotential deposition (UPD) of a monolayer of a sacrificial less noble metal on a more noble metal substrate, such as Cu UPD on Au or Pd, followed by a spontaneous and galvanic replacement of the less noble metal with a noble metal in a solution containing the noble metal cation, such as the Pt cation [18]. The whole procedure can be repeated in order to deposit multilayers of Pt (or another noble metal) on the substrate metal.

The advantages of Pt monolayer catalysts include (1) full utilization of the Pt atoms which are all on the surface and (2) that the Pt activity and stability can be tailored by the selection of the substrate metals. When a Pt monolayer is deposited onto different substrate metals [19], due to the lattice mismatch between the metals, it can experience either compressive stress or tensile stress, which is known to affect the Pt activity by adjusting its d-band center energy [20] and consequently its ORR activity.

Zhang et al. investigated Pt monolayer deposits on Pd(111) single crystals (Pt/Pd(111)) and on Pd/C nanoparticles (Pt/Pd/C) for ORR [18]. The ORR reaction mechanism of the monolayer catalysts was found to be the same as that on pure Pt surface. Pt/Pd(111) was found to have a 20 mV improvement in half-wave potential versus Pt(111), and the Pt/Pd/C had a Pt-mass activity 5–8 times higher than that of Pt/C catalyst. The enhanced ORR activity is attributed to the inhibited OH formation at high potential, as evidenced from XAS measurements. In a real fuel cell test, $0.47g_{Pt}/kW$ was demonstrated at 0.602 V [21]. In a related study, the ORR on platinum monolayers supported on Au(111),

Ir(111), Pd(111), Rh(111), and Ru(0001) single-crystal surfaces was investigated [22]. The trend of the ORR activities increases in the sequence Pt/Ru(0001) < Pt/Ir(111) < Pt/Rh(111) < Pt/Au(111) < Pt(111) < Pt/Pd(111).

Pt monolayer catalysts show a promising pathway toward solving one of the major problems facing PEM fuel cells by enhancing the Pt-specific activity and the utilization of Pt atoms and therefore reducing the cost of the cathode catalyst, although more fuel cell tests of durability are needed before the monolayer catalysts can be put in fuel cell vehicles. There is still a need for a reduction in the total noble metals in these catalysts.

ORR on Facet- and Shape-Controlled Pt-Alloy Nanocrystal Electrocatalysts

Wu et al. [23] recently reported an approach to the preparation of truncated-octahedral Pt_3Ni (t,o-Pt_3Ni) catalysts that have dominant exposure of {111} facets. Three sets of Pt_3Ni nanocrystals were generated with various truncated-octahedral crystal populations.

The particle size is on the order of 5–7 nm. Only two types of facets are exposed of all the nanocrystals, i.e., the {111} and {100}. The fractions of the {111} surface area over the total surface area could be calculated based on the geometries of the shapes and the population statistics. Almost linear correlations were obtained for both mass activities and specific activities versus the fraction of the (111) surface area over the total surface area. While the {111} facets of the nanocrystals showed much higher specific activity than the {100} facets, in agreement with trend found on bulk Pt_3Ni single-crystal disks, the absolute values of the specific activities of the nanocrystals are still far below those observed on bulk single crystal surfaces [24].

Future Directions

Low platinum loading, high activity, and more durable catalysts still remain as critical challenges for PEMFCs for automotive applications. Further fundamental understanding of the correlations between activity, stability, and structural properties at the atomic level are most desired from both theoretical and experimental perspectives. Studies of the connections between the activities of controlled-facet-orientation nanoparticles and extended single-crystal surfaces would be helpful. Structure- and surface-controlled syntheses of catalysts (Pt monolayer catalysts, nanostructured catalysts and electrodes, size- and facet-controlled Pt-alloy nanocrystals, combined with core-shell structure) should provide a practical viable path to achieving fuel cell catalyst loadings required for large-scale commercialization.

Cross-References

► Oxygen Reduction Reaction in Acid Solution
► Oxygen Reduction Reaction in Alkaline Solution
► Polymer Electrolyte Fuel Cells (PEFCs), Introduction

References

1. Conway BE, Tilak BV (2002) Electrochim Acta 47(22–23):3571–3594
2. Gasteiger HA, Markovic NM, Ross PN (1995) J Phys Chem 99(45):16757–16767
3. Mukerjee S, McBreen J (1996) J Electrochem Soc 143(7):2285–2294
4. Neyerlin KC et al (2007) J Electrochem Soc 154(7): B631–B635
5. Tarasevich MR, Sadkowski A, Yeager E (1983) Oxygen electrochemistry. In: Conway BE et al (eds) Comprehensive treatise in electrochemistry. Plenum, New York, p 301
6. Gasteiger HA, Panels JE, Yan SG (2004) J Power Sources 127(1–2):162–171
7. Adzic RR (1998) Recent advances in the kinetics of oxygen reduction. In: Lipkowski J, Ross PN (eds) Electrocatalysis. Wiley-VCH, New York, pp 197–241
8. Markovic NM et al (1999) J Electroanal Chem 467(1):157–163
9. Mukerjee S et al (1995) J Electrochem Soc 142(5):1409–1422
10. Koh S et al (2008) J Electrochem Soc 155(12): B1281–B1288
11. Gottesfeld S (1986) J Electroanal Chem 205(1–2):163–184
12. Paffett MT, Beery JG, Gottesfeld S (1988) J Electrochem Soc 135(6):1431–1436

13. Jalan V, Taylor EJ (1983) J Electrochem Soc 130(11):2299–2302
14. Toda T, Igarashi H, Watanabe M (1999) J Electroanal Chem 460(1–2):258–262
15. Stamenkovic VR et al (2007) Nat Mater 6(3):241–247
16. Uribe FA, Zawodzinski TA (2002) Electrochim Acta 47(22–23):3799–3806
17. Brankovic SR, Wang JX, Adzic RR (2001) Electrochem Solid State Lett 4(12):A217–A220
18. Zhang J et al (2004) J Phys Chem B 108(30):10955–10964
19. Adzic RR et al (2007) Top Catal 46(3–4):249–262
20. Hammer B, Norskov JK (2000) Theoretical surface science and catalysis – calculations and concepts. In: Gates BC, Knozinger H (eds) Advances in catalysis, vol 45. Academic, San Diego, pp 71–129
21. Zhang J et al (2005) J Serb Chem Soc 70(3):513–525
22. Zhang JL et al (2005) Angew Chem Int Ed 44(14):2132–2135
23. Wu J et al (2010) J Am Chem Soc 132(14):4984–4985
24. Stamenkovic VR et al (2007) Science 315(5811): 493–497

Ellipsometry

Somayeh Moradi and Rudolf Holze
AG Elektrochemie, Institut für Chemie, Technische Universität Chemnitz, Chemnitz, Germany

Light emitted from a glowing wire (an incandescent or tungsten halogen lamp) or from most other natural as well as artificial sources shows no preferred polarization, i.e., the electric as well as the magnetic vector (both are coupled, their planes of oscillation enclose an angle of 90°) of emitted electromagnetic waves are randomly oriented. When such a beam of light is reflected at a surface, the actual orientation of the electric vector with respect to the plane of reflection (defined by the incoming and the outgoing beam) has considerable influence on the properties of the outgoing wave. This influence can be studied more precisely with polarized light. Polarized electromagnetic radiation is generated per se, e.g., by certain types of lasers (gas ion lasers) employing windows mounted at the Brewster angle (so-called Brewster windows) at the ends of the plasma tube. From other sources, polarized light can be obtained by means of a polarizer or a polarization filter. Upon reflection of polarized light, both the amplitude (i.e., the magnitude of the electric field vector) and the phase might undergo changes. This depends on the complex refractive index N_l of the material designated 1 according to

$$N_1 = n_1 - ik_1$$

with the refractive index n_1 and the extinction coefficient k_1. The refractive index n_1 can be measured using an Abbé refractometer; the extinction coefficient k_1 is related to the absorption coefficient α_1 according to

$$a_1 = 4\pi k_1/\lambda$$

Further understanding of the interaction is most straightforward when only incident light with its electromagnetic vector parallel to the plane of reflection (E_p) and with its vector perpendicular (E_s) is considered. After superposition of the electric field vectors of these two waves, the resulting vector describes a circle before reflection, provided they were in phase and of the same amplitude. After reflection the

$$r = (\tan Y)e^{iD}$$

mentioned changes result in a superposition wherein the resulting electric field vector describes an ellipse, thus the name of the method. The changes caused by the reflection can be expressed in various ways. A very convenient and popular description using the physical parameters gives the amplitude ratio as

$$\tan\Psi = \frac{|E_p|}{|E_s|}$$

The difference Δ of the time-independent phases ε of the two components is

$$\Delta = \varepsilon_p - \varepsilon_s$$

Both parameters are combined in the basic equation of ellipsometry

$$\rho = (\tan \Psi)\, e^{i\Delta}$$

These parameters are related to the optical properties of the investigated films etc. Textbooks on ellipsometry are available. Thorough reviews of fundamentals including selected applications related to electrochemistry have been published.

An ellipsometer is used to determine the change of the polarization of light effected by reflection at a surface as expressed with ρ. The principal optical elements are polarizer and compensators. The former device has been introduced above. The latter devices (also known as retarders) introduce a defined phase difference (shift) between two orthogonal components of a passing wave. The difference may be fixed (e.g., 90 ° or quarter-wave) or variable. Compensators are manufactured from crystalline birefringent materials like mica, calcite, or quartz. Measurements can be performed at fixed angle of incidence or at variable angles; the used wavelength can be a fixed one or can be variable. In the latter case, the instrument is called a spectroscopic ellipsometer. Instruments can be grouped into two types: compensating and photometric (noncompensating) types. The basic components of a spectroscopic ellipsometer of commercial design of the former type are depicted below (Fig. 1).

In the compensating instrument, the phase change caused by the reflection is counteracted by a compensator yielding in effect linearly polarized light. From the position of the fixed polarizer and the compensator in the incoming beam and the polarizer in the reflected beam, the required information is extracted. This setup is also called nulling ellipsometer because the correct adjustment of the three optical components resulting in an intensity minimum at the detector is searched. The sensitivity of the photomultipliers frequently used as detectors has to be taken into account. Obviously this approach is too slow for practical application. By adding Faraday modulators in incoming and reflected beam, rapid oscillation of the plane of polarization is possible. Phase-sensitive detection of the signal at the detector with respect to the modulation allows faster determination of the minimum and the relevant position of the corresponding settings of the three optical devices. Further compensation of mechanical adjustment errors is possible by adding two further Faraday modulators operated with a constant current.

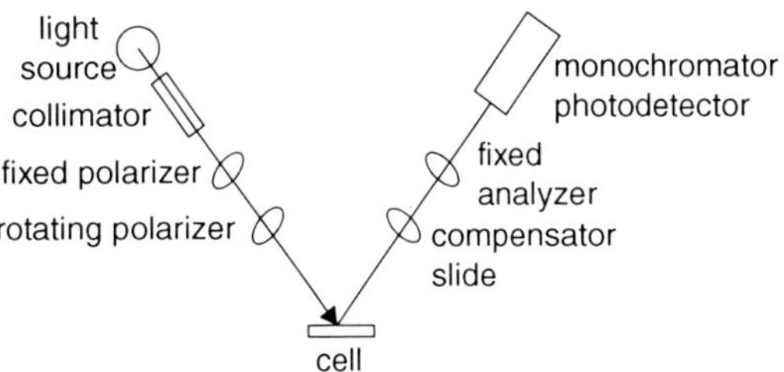

Ellipsometry, Fig. 1 Components of a spectrometric rotating polarizer ellipsometer

A further type of fast automatic ellipsometer for electrochemical investigations has been described [1]; an experimental approach to observe fast transients with ellipsometry was reported [2]. In the photometric mode, the intensity of the reflected light is measured as a function of the position of polarizer and sometimes compensator in the incoming beam; for further details and an overview, see general references. A general overview of instrumental developments has been provided [3].

The process of an ellipsometric experiment can be separated into five steps [3]:

1. Providing an incident beam of known state of polarization
2. Allowing this beam to interact with the interface/interphase under investigation
3. Measuring the state of the emerging light
4. Determination of parameters like Ψ and Δ or equivalents
5. Inferring properties and changes from these parameters

The final step almost always involves formulation of a model which describes the system under investigation. The parameters of this model are compared with those measured. By successive approximation, the parameters of the model are optimized until agreement is reached. In the most simple case of, e.g., a clean, homogeneous, and isotropic surface with N_2 that is covered with a medium of known value of N_1, a two-layer model can be applied. Already from a single ellipsometric measurement, the value of N_2 can be determined. When the surface is covered with a layer of a foreign material, a three-layer model has to be used. Although the values of N_1 and N_3 can be determined in many cases in a separate experiment where the intermediate layer is absent, the number of parameters is considerably larger requesting multiple experiments. Numerous models combining information about optical parameters of ideal films with their real morphology and composition have been used to evaluate ellipsometric data; nevertheless, in many studies, ellipsometry is used mostly to determine film thickness. A modification of an ellipsometer enabling microscopically resolved measurements (ellipsomicroscopy) has been described [4] and applied to studies of the spreading of corrosion on stainless steel.

Ellipsometry at noble metal electrode/solution interfaces has been used to test theoretically predicted microscopic parameters of the interface. Investigated systems include numerous oxide layer systems [5–10], metal deposition processes, adsorption processes [11], and polymer films on electrodes [12–14]. Submonolayer sensitivity has been claimed. Expansion and contraction of polyaniline films was monitored with ellipsometry by Kim et al. [15]. Film thickness as a function of state of oxidation of redox active polyelectrolyte layers has been measured with ellipsometry [16]. The deposition and electroreduction of MnO_2 films has been studied [17]; below a thickness of 150 nm, the anodically formed film behaved like an isotropic single layer with optical constants independent of thickness. Beyond this limit, anisotropic film properties had to be assumed. Reduction was accompanied by a thickness increase it started at the oxide/solution interface.

Future Directions

Instrumental developments enabling measurements at several wavelengths (spectroscopic ellipsometry) and with thin layers and submonolayer coverages combined with enhanced software making measurements and evaluation possible even for users not too familiar with electrooptical theory will help spreading the use of this method.

Cross-References

► Spectroelectrochemistry, Potential of Combining Electrochemistry and Spectroscopy

References

1. Cahan BD, Spanier RF (1969) Surf Sci 16:166
2. Droog JMM, Bootsma GA (1979) J Electroanal Chem 105:261
3. Hauge PS (1980) Surf Sci 96:108
4. Dornhege M, Punckt C, Hudson JL, Rotermund HH (2007) J Electrochem Soc 154:C24
5. Gottesfeld S, Maio G, Floriano JB, Tremiliosi-Filho G, Ticianelli EA, Gonzalez ER (1991) J Electrochem Soc 138:3219
6. Ohtsuka T, Heusler KE (1979) J Electroanal Chem 100:319
7. Parsons R, Visscher WHM (1972) J Electroanal Chem 36:329
8. Droog JMM, Alderliesten PT, Bootsma GA (1979) J Electroanal Chem 99:173
9. Gottesfeld S, Srinivasan S (1978) J Electroanal Chem 86:89
10. Ord JL (1982) J Electrochem Soc 129:335
11. Kunimatsu K, Parsons RH (1979) J Electroanal Chem 100:335
12. Arwin H, Aspnes DE, Bjorklund R, Ljundström I (1983) Synth Met 6:309
13. Pettersson LAA, Johansson T, Carlsson F, Arwin H, Inganäs O (1999) Synth Met 101:198
14. Christensen P, Hamnett A (2000) Electrochim Acta 45:2443
15. Kim DR, Cha W, Paik WK (1997) Synth Met 84:759
16. Tagliazucchi M, Grumelliz D, Calvo EJ (2006) Phys Chem Chem Phys 8:5086
17. Hernández Ubeda M, Pérez MA, Mishima HT, Villullas HM, Zerbino JO, López de Mishima BA, López Teijelo M (2005) J Electrochem Soc 152:A37

Further Reading

Archer RJ (1968) Manual on ellipsometry. Gaertner, Chicago

Azzam MA, Bashara NM (1977) Ellipsometry and polarized light. North-Holland, Amsterdam

Gottesfeld S (1989) In: Bard AJ, Bard AJ (eds) Electroanalytical chemistry, vol 15. Marcel Dekker, New York, p 143

Greef R (1984) In: White RE, Bockris JOM, Conway BE, Yeager E (eds) Comprehensive treatise of electrochemistry, vol 8. Plenum Pres, New York, p 339

Holze R (2007) Surface and interface analysis: an electrochemists toolbox. Springer, Heidelberg
Muller R (1976) Surf Sci 56:19
Muller R (1991) In: Varma R, Selman JR (eds) Techniques for characterization of electrodes and electrochemical processes. Wiley, New York, p 31
Neal WEJ (1982) Ellipsometry. Theory and applications. Plenum, New York
Plieth W, Kozlowski W, Twomey T (1992) In: Lipkowski J, Ross PN (eds) Adsorption of molecules at metal electrodes. VCH Publishers, New York, p 239

Environmental Energy Technologies

Francisco Alcaide
Energy Department, Fundación CIDETEC,
San Sebastián, Spain

Introduction

Energy is determinant in the economic growth of countries, but the world is still largely dependent on fossil fuels to satisfy the increasing global energy demand. Energy supply security, climate change, and economic competitiveness are issues still unresolved, and they are also the main drivers for energy research.

Energy production by conventional technologies and its consumption exert considerable pressures on the environment. The use of fossil fuels, which seems set to continue in the near and medium term, has a significant impact on the environment; it is the main cause of the greenhouse effect and global warming. Even more, life cycle assessments recently applied to energy systems have revealed an environmental impact, unpredictable until now. Within this perspective, international environmental policy focuses on the deployment of environmental energy technologies as a solution to overcome these problems. In this context, environmental electrochemistry plays a fundamental role in avoiding and monitoring pollution, monitoring process efficiencies, cleaner processing, and developing modern techniques for electrical energy storage and conversion.

Environmental Energy Technologies (EETs)

The term environmental energy technologies (EETs), in its broader meaning, involves those technologies which are focused on efficiency improvement and environmental impact reduction related to energy systems. These goals can be achieved by:

- Developing better energy technologies
- Reducing the adverse environmental impacts of energy production
- Performing energy analyses that influence energy policies in various countries.

EETs include, but are not limited to, the following technologies: biofuels, biomass power, carbon capture and storage, solar power concentration, electric vehicles, geothermal power, hydrogen use, hydro power, solar photovoltaics, solar thermal heating, and wind power. The deployment of EETs around the world is progressing rapidly; renewable energy technologies have shown an evident growth rate in recent years, the implementation of energy efficiency policies is improving, and governments are beginning to set goals to support the development of advanced vehicle markets.

Environmental energy technologies could be tentatively classified into the following main areas:

- Energy supply, transmission, and storage (advanced energy technologies)
- Pollution management
- Energy analysis
- Monitoring.

Research activities in each of the above areas are helping to overcome techno-economic barriers for their implementation, and several electrochemical processes, developed in the context of environmental electrochemistry, contribute to this research. The interconversion of chemicals and electricity via reactions at electrode-electrolyte interfaces offers the following distinctive advantages over corresponding chemical technologies:

- Environmental compatibility (pollution-free operation since electrons are clean reagents)
- Energy efficiency (Electrochemical processes take place under lower temperatures and

pressures than those in chemical processes; electrochemical cells can be designed to minimize power losses)
- Safety (electrochemical processes occur under mild conditions)
- Cost-effectiveness (use of less expensive starting materials)
- Selectivity (by manipulating the electrode potential)
- Versatility (regarding the nature of chemicals treated, their quantity, or the type of electrochemical reactor used)
- Easy of automation (the main electrochemical parameters, cell voltage, current, and design, are suited for process automation and control).

All of these advantages are used in the development of electrochemical technologies directly related to the improvement of EETs:
- Alternative sources of energy, that is, the electrochemical production of hydrogen using renewable energy sources as well as its direct conversion to electricity in fuel cells and photovoltaic devices
- The cleaning of gases, wastewaters, and soils by electrolysis (electroremediation)
- The removal of environmental contaminants like metal ions by selective electrodeposition
- The monitoring of pollution in air, solid, and liquids (electrochemical sensors).

Application of Environmental Electrochemistry to Environmental Energy Technologies

Environmental electrochemistry has great potential to contribute to the development of EETs, improving their deployment, energy efficiency, and policy progress in various countries. This section provides an overview of the areas in which EETs are classified, paying special attention to the electrochemical technologies that contribute to each one.

Energy Supply, Transmission, and Storage (Advanced Energy Technologies)

The goal of this area of research is to develop more environmentally friendly technologies for generating and storing energy, including use of better batteries and fuel cells. Aqueous-based advanced batteries for electric vehicles will cause a lower environmental impact than the current battery technology [1]. Fuel cells are energy conversion devices. Hydrogen fuel cells use hydrogen (or hydrogen-rich fuel) and oxygen to generate electricity. Hydrogen has a great potential to be a clean and nearly limitless fuel; it can be produced from various resources, including fossil fuels, biological sources, and water. Fuel cell research is focused on technologies for the successful commercialization of polymer-electrolyte and solid oxide fuel cells (SOFCs) for automotive and stationary applications [2]. In addition, photovoltaics provide renewable solar electricity that may be delivered at the point of utilization and may be applied almost everywhere. However, one of the major constraints to their commercial use is the high cost and low efficiency of generation.

Pollution Management

The electrochemical technologies involved in this research area refer to the end-of-pipe technologies for treatment of pollution after it has been generated in air, water, and soil. Greenhouse gases (GHG) like carbon dioxide (CO_2), methane (CH_4), and nitrous oxide (N_2O), as well as fluorinated gases, contribute to air pollution. The major driving force behind the increase in GHG emissions is the burning of fossil fuels, and CO_2 is the largest contributor to the greenhouse effect. Other sources of greenhouse gases include methane emissions from cattle, nitrous oxides from agricultural soils, and methane emissions from landfill wastes, as well as emission of fluorinated gases from manufacturing processes. It is possible to reduce the concentration of GHG by electrochemical methods. Thus, the electrochemical reduction of CO_2 leads to production of hydrocarbon fuels or useful chemicals like formic and oxalic acid and/or methanol [3]. The electrochemical oxidation of CH_4 in a SOFC reactor leads to the chemical cogeneration of hydrogen cyanide, ethylene, and C_2 hydrocarbons [4]. N_2O can be reduced to N_2 in a fuel cell type electrochemical reactor [5], as well as, fluorinated gases like CFC11 and CFC113, which can be electrochemically reduced to dechlorinated derivatives, methane and difluroethane, by using an electrochemical flow cell [6]. Furthermore, with regard to water recycling and processing, there are some advanced technologies in water and wastewater treatment like the electro-Fenton process and related electrochemical technologies based on Fenton's reaction chemistry [7].

Electrochemical methods for the treatment of a wide variety of polluted soils are based on electrokinetic phenomena (electrophoresis, electroosmosis, and electromigration) where an electrical current is passed through electrodes buried underground. This originates the movement of charged (and some uncharged) species. Electrokinetic remediation has been used to treat organic substances and inorganic species, as well as radioactive substances.

Energy Analysis

This area of research involves aspects of energy use in various countries, including use of energy in appliances, commercial and residential buildings, industry, the public sector, electricity markets, and transmission and distribution. The effect of proposed policy measures on energy use is also analyzed. Furthermore, methods for assessing the potential for mitigating greenhouse gas emissions and for global GHG modeling are studied.

Monitoring

Monitoring refers to measurement devices and analytical equipment that detect pollutants and changes in the environment. In this context, the development of novel arrays of electrochemical sensors will increase the ability to monitor the state of health of the environment [8].

Future Directions

EETs strengthen the economy, increase the efficiency of energy use, protect the environment, and reduce the dependence on fossil fuels. However, additional efforts are needed to ensure their implementation. The challenges of EETs for the next 15 years are as follows: (i) to increase energy efficiency; (ii) to reduce greenhouse gas emissions; (iii) to increase the electricity and the overall energy obtained from renewable energy sources; (iv) to establish other storage technologies for portable, stationary, and transport (including hydrogen fuel cell storage technology); and (v) to make power grids smart and independent so that users can produce and share energy in an open-access way. Environmental electrochemistry will play a capital role in achieving these goals through advancements in electrochemical technologies related to energy conversion and storage, pollution abatement, and improvement of electrochemical sensors.

Cross-References

▶ Fuel Cells, Principles and Thermodynamics
▶ Sensors

References

1. Huggins RA (2010) Energy storage. Springer, New York
2. Sammes N (ed) (2006) Fuel cell technology: reaching towards commercialization. Springer, London
3. Gattrell M, Gupta N, Co A (2006) A review of the aqueous electrochemical reduction of CO_2 to hydrocarbons at copper. J Electroanal Chem 594:1–19
4. Alcaide F, Cabot PL, Brillas E (2006) Fuel cells for chemicals and energy cogeneration. J Power Sources 153:47–60
5. Aziznia A, Bonakdarpour A, Gyenge EL, Oloman CW (2011) Electroreduction of nitrous oxide on platinum and palladium: towards selective catalysts for methanol–nitrous oxide mixed-reactant fuel cells. Electrochim Acta 56:5238–5244
6. Cabot PL, Segarra L, Casado J (2004) Electrodegradation of chlorofluorocarbons in a laboratory-scale flow cell with a hydrogen diffusion anode. J Electrochem Soc 151:B98–B104
7. Brillas E, Sirés I, Cabot PL (2010) Use of both anode and cathode reactions in wastewater. In: Comninellis C, Chen G (eds) Electrochemistry for the environment. Springer, New York
8. Janata J (2009) Principles of chemical sensors, 2nd edn. Springer, New York

Environmentally Accepted Processes for Substitution and Reduction of Cr(VI)

Armando Gennaro and Christian Durante
Department of Chemical Sciences, University of Padova, Padova, Italy

Introduction

Chromium is a largely diffused element in the earth's crust, mainly present as chromite ore ($FeCr_2O_4$). The trivalent oxidation state Cr(III) is the most stable form of chromium and is essential to mammals in trace concentrations for sugar, lipid, and protein metabolism [1]. Cr(III) is relatively immobile in aquatic systems, due to its low water solubility. On the contrary, hexavalent Cr(VI) is mobile because of its good solubility in water, and, unfortunately, it is highly toxic and carcinogenic; in particular, it is a strong oxidizing agent that causes severe damage to cell membranes [2]. Even if Cr(III) is the most thermodynamically stable species, the presence of oxidizing agents in the soil, such as, for instance, MnO_2, can promote the oxidation of Cr(III) to Cr(VI), which at high pH is bioavailable and highly mobile and, for this reason, a potential danger for groundwater contamination [3].

Industrial activities like electroplating, metal cleaning, dye processing, and leather tanning are the major sources of chromium release into the environment [4]. The use of chromates and other chromium compounds has been limited since 1982 due to their carcinogenic effects. In particular, the US Environmental Protection Agency (EPA), which is the main regulator of chromate emissions, has adopted several acts [5, 6]. On the other hand, the latest provision of the European Community in this regard is to decrease Cr discharge in any chemical form until complete abatement by 2020 [7].

To decrease the amount of chromium release, two different approaches have been adopted: the substitution of chromium-based processes with alternative processes employing nontoxic or less toxic compounds and the decrease of chromium amounts by revamping or optimizing the processes. We will describe some examples of these different approaches in the two main processes employing chromium, namely, coating and tanning processes.

In any case, to fulfill the provisions of different regulators, the development of abatement processes is a subject of increasing interest.

Alternative Electroplating Chromium Processes

Chromium coatings have been widely used for decorative and protection purposes in engineering industries, because they provide high hardness, excellent wear, corrosion resistance, and strong adhesive ability with substrates [8, 9]. The first example of electrodeposition of chromium dates back to a patent of 1848; however, it was not before 1928 that chromium electroplating became a commercially sustainable process. Cr electrodeposition is traditionally based on Cr(VI), the plating bath consisting of a solution of chromic acid and sulfuric acid, where SO_4^{2-} anions are necessary to sustain the cathodic deposition of chromium. Chromium plating may be divided into two general categories. In the first category, the chromium is deposited as a thin coating to serve for decorative purposes on metal articles. In the second category, or "hard" chromium plating (EHC), heavy coatings are plated so that the electrodeposit of this type is particularly resistant to heat and has a low coefficient of friction and high resistance to wear and resistance to corrosion and erosion. In these industrial applications, chromium is usually deposited directly on the metal, without intermediate coatings of other metals.

One of the major drawbacks of EHC is the quite low current efficiency (10–20 %), so therefore high amounts of the highly toxic Cr(VI) must be employed. As a consequence, alternative and environmentally friendly technologies are highly desirable due to the toxicity and carcinogenicity of Cr(VI). Corrosion resistance, hardness, and wear or abrasion resistance are important properties to ensure a useful service life to parts, components, and products; therefore, alternative coatings to EHC have to fulfill these requirements. However, several alternative technologies involve substances which are themselves on the list of carcinogenic or toxic chemicals. Furthermore, physicochemical properties of alternative coatings are frequently not comparable to those of EHC. A number of approaches are possible for reducing chromium employment in corrosion or wear resistant applications. These approaches may be (i) reduced chromium content in coatings, (ii) chromium-free coatings, and (iii) substrate surface modification. While the search for acceptable alternatives is underway, there are studies that propose the use of chromium coatings electrodeposited from a bath formulated with trivalent chromium Cr(III) salts instead of hexavalent salts. The regulations governing the use and disposal of Cr(III) salts are less restrictive than those for Cr(VI) salts, so some surface finishers have chosen to use "chromium-based" alternatives. However, to date the results of corrosion testing of plated products based on Cr(III) have not been too promising [10], and there are still many problems that need to be solved. First, it is difficult to improve the thickness of chrome deposits in Cr(III) plating [11]. Second, the Cr(III) plating process is unstable, and the deposited coating quality tends to deteriorate significantly in the presence of contaminant metal ions [12]. Furthermore, Cr(III) is easily oxidizable to Cr(VI) at anode electrodes, and this causes a serious contamination of the plating bath. These problems restrict the application and expansion of the trivalent chromium plating process. Pulsed current plating has been tried to obtain better, thicker Cr(III)-based coatings for engineering applications with some success. Furthermore, a laser-alloying technique [13] and a more traditional co-diffusion technique [14] were used for the coating of stainless steels with Cr–Ni, Cr–Si, or Cr–P [15] to improve their corrosion resistance and stability, in particular at high temperature oxidation and in the presence of halides.

Alternatives to EHC, up to certain levels, consider metal alloys especially of nickel, such as Ni–Mo, Ni–W, and Ni–P plating. Nickel and nickel alloys are typically deposited by electroplating or electroless techniques. Electroplated nickel coatings exhibit relatively good corrosion resistance, especially in alkaline environments, but they are not high in performance in sulfur-containing atmospheres and in marine environments. Electroless nickel (EN) coatings containing phosphorus (Ni–P) or small amounts of boron (Ni–B) form a barrier to the environment and often are used under physical vapor deposition (PVD) or chemical vapor deposition (CVD) coatings to increase corrosion resistance. Alloy coatings containing tin and cobalt have received considerable attention as substitutes for EHC coatings. Various tin alloys have been suggested, but the only candidate proposed as a possible alternative for chromium in some applications is the alloy containing nickel. Sn–Ni alloy coatings, especially those containing 33–35 % nickel (corresponding to the intermetallic compound in the system), exhibit good corrosion resistance in strong acids [16]. Cobalt-based alternatives [17] have shown, from potentiodynamic polarization and electrochemical impedance spectroscopy (EIS) tests, corrosion resistance 1.6–2 times higher than EHC. This improvement in corrosion resistance was attributed to the surface oxide layer formed during annealing.

Another alternative to EHC, which can work well in several cases, is thermal spray deposition. Thermal spraying is a dry deposition technology that deposits several coatings such as WC [18] or WC–Co [19, 20] and coatings containing mixtures of Ni, Cr, Fe, Si, and sometimes other elements. Thermal spray processes can deposit thick films with very good characteristics (hardness, wear, and corrosion resistance) and good flexibility in the choice of coating materials, allowing the optimization of applications to different materials. The drivers for their development at the present time are the commercial and military aircraft sectors. Another option is plasma-enhanced chemical vapor deposition, which can be used to deposit a modified quartz film that resembles an inorganic polymer. Although the films are clear and hard, they exhibit some flexibility [21]. These protective SiO_2-based coatings have been applied to brass and aluminum substrates. Their corrosion-related performance is excellent, in particular the resistance to salt fog exposure. Low temperature cationic plasma deposition produces an ultrathin hydrophobic barrier through plasma-polymerized films with thicknesses of several hundreds of nanometers and with practically defect-free matrices [22]. Also, sol–gel deposition allows the deposition of thin oxide films by using different metal alkoxides or nitrates as precursors [23].

Organic coatings such as nylon are not suitable candidates as alternatives to chromium, since, although they provide excellent resistance to salt fog exposure, water, and other chemicals, they are usually not very hard, and so they perform poorly in wear and erosion tests. Diamond-like coatings (DLC) also have been proposed as an alternative to chromium, since they show to be "inert" with respect to chemical attack [24].

Chromium-Free Tanning Processes

The tanning of leather traces its roots to ancient times, and it consists of the treatment of animal skins to obtain valuable material that can resist putrefaction and water or heat treatment. Animal hide has the tendency to shrink when it is put into water at temperature higher than 60 °C, so the tanning process allows one to confer hydrothermal stability to leather. The shrinking is due to the collapse of the collagen fibers because of the breakdown of the hydrogen bonding between fibers. The discovery of chrome tanning dates back to 1858, when Knapp succeeded in tanning leather by means of chromium salts and with excellent results. From then on, the chrome tanning process gained in importance until becoming at present the most important tanning process, covering more than 90 % of the global leather market. The success of chrome tanning relies on the fact that it is simple, fast, and cheap; presents minor drawbacks; and allows the obtaining of high quality leather with high shrinking temperature. The high hydrothermal stability of Cr(III) tanned leather is due to covalent bonds and hydrogen interaction between Cr species and carboxylic and amino functional groups of collagen protein. Therefore, it is assumed that mineral tanning agents alternative to Cr(III) should be able to interact covalently with collagen and to confer high shrinking temperature to leather. If we glance over the periodic table, only transition metals, along with some rare earths and *p* block elements, have a complex chemistry suitable for tanning [25]. However, many of the *d* and *p* block elements give poor interaction with collagen functional groups, and the result is a reversible tanning process which produces unstable materials. Some improvements have been gained by introducing a tannin treatment which allows introduction of multifunctional groups onto collagen before metal tanning, and this allows firmer grafting of the minerals. However, notwithstanding the tannage capability, the choice of a suitable alternative relies on the low toxicity and high availability of the tanning agents. This is a constraint that reduces drastically the number of candidates. As matter of fact, only titanium, zirconium, iron, and aluminum have practical possibilities, since they are a good compromise between tanning capability, toxicity, and availability. Generally, aluminum requires a pretreatment to give sufficient interaction with collagen, but since the introduction of a further procedure yields complicated and sometimes impractical results, aluminum tanning is restricted to niche materials such as lamb or buck skin. The problem with iron is that it is subjected to oxidation, and the precipitation of iron hydroxide gives an unaesthetic effect of maroon stains. Titanium and zirconium have an important complex chemistry and give covalent bonds with collagen, but the shrinking temperature is definitely lower with respect to chrome tanned leather. In conclusion, alternative metal or semimetal elements are far from being a real and practical alternative to chromium; therefore, the search for new viable routes must head to non-mineral tanning such as vegetable, oil, or aldehyde tanning. However, in the case of aldehyde tanning, there are health and safety implications with formaldehyde, which is banned as a tanning agent, whereas oil tanning is limited only to specific applications, such as cleaning cloth. Therefore, the only practical way for a chromium-free leather remains with vegetable tanning. The vegetable tanning agents are generally polyphenols that can interact with collagen functional groups via hydrogen, electrostatic, and covalent interactions that confer a shrinkage temperature of collagen between 75 °C and 80 °C. The research in this direction seems the most fruitful and practicable; actually, the study and the developing of synthetic and artificial polyphenols is an active field, but the complete giving up of chrome tanning for vegetable tanning is long to come.

An alternative approach to a radical banishing of chromium is the improvement of methods for chromium fixation or its recovery and reutilization from tannery baths and sludge. This would decrease both the amount of chromium discharged to the environment and the amount of raw chromium extracted from minerals. With the current methods of tanning, 30–40 % of Cr salt does not react with the hide, and it is thus discharged. This is primarily due to the limited access to reactive sites and/or to the limited presence of active functional groups in the tanning materials. The residual chromium in the bath cannot be reused directly, because the leather obtained would present a poor appearance and would not meet the required physical properties. Therefore, the improvement of chromium fixation is highly desirable, and several studies report an increased chromium uptake, that in some cases exceeds 90 %, when a strategic pretreatment aimed at increasing active sites of collagen or the introduction of co-tanning agents are considered [26, 27]. Alternatively, chromium reuse can be accomplished either by recovery and reuse of chromium salts [28] or through direct chrome liquor recycling [29]. The recovery of chromium from tanning baths can be accomplished by chemical precipitation [30], solvent extraction [31, 32], ion exchange [33, 34], adsorption [35, 36], and membrane technology [37–39]. However, in those cases in which the recovery is unfeasible for economic or process reasons, abatement becomes the only way to follow.

Abatement of Chromium in Wastewaters

Process waste streams containing chromium originate not only from electroplating and tanning but also from mining operations and from chemicals employed as fungicides in wood preservation or as mordents in the textile industry. Besides the technique mentioned above and employed for the recovery of chromium, other processes suitable for the abatement of Cr(III) or Cr(VI) [40] from industrial effluents are precipitation, coagulation, and adsorption on carbon [41] or vegetal adsorbents [42, 43]. The treatment of waste solutions of chromium (VI) is based on a two-stage process. In the first stage, the Cr(VI) is reduced to Cr(III), and in the second stage, the Cr(III) is precipitated as hydroxide and disposed as sludge. The traditional method for reduction of Cr(VI) involves the addition of a reducing agent such as sodium bisulfate, sodium thiosulfate, sulfur dioxide, iron sulfate, and aluminum powder in acidic solution. The reduction of Cr(VI) can also be accomplished by employing other methodologies; the most promising are enzymatic and biological reduction and photochemical reduction on TiO_2. After completed reduction, the waste solution containing Cr(III) species is collected for precipitation. Precipitation is carried out with hydrated lime or sodium carbonate. Settlement of the sludge is improved by adding aluminum sulfate, ferric chloride, etc. After settlement, the sludge is separated (e.g., by decanting) and disposed.

The adsorption process consists of the concentration of chromium ions on the surface of the sorbent. In comparison with conventional methods, such as membrane filtration or ion exchange, it has significant advantages like low cost, availability, and ease of operation. A variety of natural and synthetic materials has been used as Cr(VI) sorbents, including activated carbon, biological materials, zeolites, chitosan, and agricultural or industrial wastes. Biosorption of chromium from aqueous solutions is a relatively new process that has proven very promising in the removal of contaminants from aqueous effluents.

Electrochemical methods represent a good alternative to the conventional techniques employed in wastewater treatment. In fact, electrocoagulation and electroprecipitation have shown to be promising technologies for the removal of Cr(III) from a vast range of effluents, being more efficient than other conventional techniques, such as chemical coagulation or absorption [44]. The main handicap associated with Cr(III) abatement is the formation of stable complexes with the organic substrates, which enhances Cr(III) solubility and hence reduces the efficacy of the abatement processes [45]. It is therefore very difficult to achieve complete Cr (III) abatement as long as organic substrates are present in the wastewater. Organic compounds are often present simultaneously with chromium in the wastewater undertreatment. These have mainly three possible origins: (i) organic compounds, like leather residuals or tanning mask agents, coexisting with chromium in the effluent; (ii) excess of nutrients added in the bioreactor to support microbial growth during treatment; and (iii) organic compounds metabolically produced by the microbial biomass in the bioreactor environment. In this case, complete removal of recalcitrant fractions of chromium can be successfully achieved by an oxidative pretreatment, for example, at a boron-doped diamond electrode followed by electrocoagulation performed at either Fe–Fe or Fe–Al electrodes. In the oxidation process, Cr(VI) is formed as chromate or dichromate, which means that the metal cannot be complexed by either residual products. In the second stage, Fe^{2+} ions are added into the solution by anodic dissolution of Fe. The experimental results point out that Cr(VI) formed during the pretreatment is mainly reduced to Cr(III) by Fe(II) according to the following reaction:

$$CrO_4^{2-} + 3Fe^{2+} + 8H^+ \rightarrow Cr^{3+} + 3Fe^{3+} + 4H_2O \qquad \Delta E^o = +0.76$$

Once Cr is reduced back to the trivalent form, it can either precipitate as $(CrOH)_3$ or coprecipitate with Fe(III) ions via hydroxide or polyhydroxide formation. Another option is the formation of an insoluble Cr(III)–Fe(II) oxide, such as $FeCr_2O_4$, or Cr(VI) salt, such as $NH_4Fe(CrO_4)_2$.

Future Directions

As mentioned, the fulfillment of the provisions of different regulators requires a radical revision of different industrial processes based on chromium. Of course, the main goal is the development of alternative processes, which could be carried out with less or nontoxic agents. However, it seems that the quality of the obtained products is, very often, not comparable and not as good as that of the chromium-based processes. Therefore, it would be of great utility to lower the amount of chromium employed by revamping or optimization of the process and also by the recycling of the chromium bath.

Since the regulators' trend is towards the complete elimination of chromium release, the development of efficient abatement processes is a very important challenge. In this regard, some combined oxidation–electrocoagulation processes appear very promising.

In any case, the abatement of chromium does not represent a real solution to the pollution. In fact, the problem is only postponed, since the disposal of the sludge is still an unresolved question, and the oxidation to Cr(VI) is always possible. Thus, the recovery of Cr from the sludge needs to be considered with great attention.

Cross-References

► Electrokinetics in the Removal of Metal Ions from Soils

References

1. Chauhan D, Sankararamakrishnan N (2011) Modeling and evaluation on removal of hexavalent chromium from aqueous systems using fixed bed column. J Hazard Mater 185:55–62. doi:10.1016/j.jhazmat.2010.08.120
2. Marchese M, Gagneten AM, Parma MJ, Pavé PJ (2008) Accumulation and elimination of chromium

by freshwater species exposed to spiked sediments. Arch Environ Contam Toxicol 55:603–609
3. Avudainayagam S, Megharaj M, Owens G, Kookana RS, Chittleborough D, Naidu R (2003) Chemistry of chromium in soils with emphasis on tannery waste sites. Rev Environ Contam Toxicol 178:53–91
4. Belay AA (2010) Impacts of chromium from tannery effluent and evaluation of alternative treatment options. J Environ Prot 1:53–58. doi:10.4236/jep.2010.11007
5. EPA Federal Register (1995) National Emission Standards for Chromium emissions from hard and decorative chromium electroplating and chromium anodizing tanks, vol 60. Research Triangle Park, pp 4547–4993
6. Environmental Pollution Control Alternatives (EPA) (1990) EPA/625/5-90/025. EPA/625/4-89/023, Cincinnati
7. European Commission (2000) EC water framework directive (2000/60/EC). Off J Eur Commun L327(43):1–72
8. Saiddington JC (1978) Effect of plating interruptions on the surface appearance of electrodeposited chromium. Plat Surf Finish 65:45–49
9. Lausmann GA (1996) Electrolytically deposited hardchrome. Surf Coat Technol 87:814–820. doi:10.1016/S0257-8972(96)02973
10. Renz RP, Zhou CD, Taylor EJ, Marshall RG, Stortz EC, Grant B (1996) Functional chromium plating from a trivalent chromium Bath. In: proceedings of the AESF annual technical conference, vol 96, pp 1–10
11. Song YB, Chin DT (2000) Pulse plating of hard chromium from trivalent baths. Plat Surf Finish 87:80–82
12. Mcdougall J, EL-Shrif M, Ma S (1998) Chromium electrodeposition using a chromium(III) glycine complex. J Appl Electrochem 28:929–934
13. Lindsey N, Vasanth KL (1999) The study of corrosion behavior of Laser Induced Surface Improvement (LISI) on steel and aluminum substrates. Corrosion 99:30–45
14. Bayer GT (1998) Chromium-silicon codiffusion coating. Adv Mater Process 153:25–28
15. Li B, Lin A, Gan F (2006) Preparation and characterization of Cr-P coatings by electrodeposition from trivalent chromium electrolytes using malonic acid as complex. Surf Coat Technol 201:2578–2586. doi:10.1016/j.surfcoat.2006.05.001
16. Shahin GE (1998) Alloys are promising as chromium or cadmium substitutes. Plat Surf Finish 85:8–14
17. Eskin S, Berkh O, Rogalsky G, Zahavi J (1998) Co-W alloys for replacement of conventional hard chromium. Plat Surf Finish 85:79–84
18. Ko PL, Robertson MF (2002) Wear characteristics of electrolytic hard chrome and thermal sprayed WC-10Co-4Cr coatings sliding against Al-Ni-bronze in air at 21 degrees C and at −40 degrees C. Wear 252:880–893. doi:10.1016/S0043-1648(02)00052-2
19. Picas JA, Forn A, Matthaus G (2006) HVOF coatings as an alternative to hard chrome for pistons and valves. Wear 261:477–484. doi:10.1016/j.wear.2005.12.005
20. Nascimento MP, Souza RC, Miguel IM, Pigatin WL, Voorwald HJC (2001) Effects of tungsten carbide thermal spray coating by HP/HVOF and hard chromium electroplating on AISI 4340 high strength steel. Surf Coat Technol 138:113–124. doi:10.1016/S0257-8972(00)01148-8
21. Liu WJ, Wang RC (2011) Novel low temperature atmospheric pressure plasma jet systems for silicon dioxide and poly-ethylene thin film deposition. Surf Coat Technol 206:925–928. doi:10.1016/j.surfcoat.2011.04.043
22. Coclite AM, Milella A, Palumbo F, Le Pen C, D'Agostino R (2010) Plasma deposited organosilicon multistacks for high-performance low-carbon steel protection. Plasma Process Polym 7:802–812. doi:10.1002/ppap.201000017
23. Guglielmi M (1997) Sol–gel coatings on metals. J Sol–Gel Sci Technol 8:443–449. doi:10.1007/BF02436880
24. Sundaram VS (2006) Diamond like carbon film as a protective coating for high strength steel and titanium alloy. Surf Coat Technol 201:2707–2711. doi:10.1016/j.surfcoat.2006.05.046
25. Covington AD (2009) Tanning chemistry: the science of leather. Royal Society of Chemistry, London
26. Karthikeyan R, Balaji S, Chandrababu NK, Sehgal PK (2008) Horn meal hydrolysate-chromium complex as a high exhaust chrome tanning agent-pilot scale studies. Clean Techn Environ Policy 10:295–301. doi:10.1007/s10098-007-0119-2
27. Sundarapandiyan S, Brutto PE, Siddhartha G, Ramesh R, Ramanaiah B, Saravanan P, Mandala AB (2011) Enhancement of chromium uptake in tanning using oxazolidine. J Hazard Mater 190:802–809. doi:10.1016/j.jhazmat.2011.03.117
28. Kanagaraj J, Chandra Babu NK, Mandal AB (2008) Recovery and reuse of chromium from chrome tanning waste water aiming towards zero discharge of pollution. J Clean Prod 16:1807–1813. doi:10.1016/j.jclepro.2007.12.005
29. Money CA (2008) Tannery waste minimization. J Am Leather Chem Assoc 86:229–244
30. Panswad T, Chavalparit O, Sucharittham Y, Charoenwisedsin S (1995) A bench-scale study on chromium recovery from tanning wastewater. Water Sci Technol 31:73 81. doi:10.1016/0273-1223(95)00408-F
31. Senthilnathan J, Mohan S, Palanivelu K (2005) Recovery of chromium from electroplating wastewater using DI 2-(ethylhexyl) phosphoric acid. Sep Sci Technol 40:2125–2137. doi:10.1081/SS-200068492
32. Sze YKP, Xue LZ (2003) Extraction of zinc and chromium(III) and its application to treatment of alloy electroplating wastewater. Sep Sci Technol 38:405–425. doi:10.1081/SS-120016582

33. Yalcin S, Apak R, Hizal J, Afsar H (2001) Recovery of copper (II) and chromium (III, VI) from electroplating-industry wastewater by ion exchange. Sep Sci Technol 36:2181–2196. doi:10.1081/SS-100105912
34. Kocaoba S, Akcin G (2005) Removal of chromium (III) and cadmium (II) from aqueous solutions. Desalination 180:151–156. doi:10.1016/j.desal.2004.12.034
35. Cerjan-Stefanovic S, Siljeg M, Bokic L, Stefanovic B, Koprivanac N (2004) Removal of metal-complex dyestuffs by Croatian clinoptilolite. Stud Surf Sci Technol 154:1900–1906
36. Covarrubias C, Assiagada R, Yanez J, Garcia R, Angelica M, Barros SD, Arroya P, Sousa-Aguilar EF (2005) Removal of chromium(III) from tannery effluents, using a system of packed columns of zeolite and activated carbon. J Chem Technol Biotechnol 80:899–908. doi:10.1002/jctb.1259
37. Ortega LM, Lebrun R, Noël IM, Hauslera R (2005) Application of nanofiltration in the recovery of chromium(III) from tannery effluents. Sep Sci Technol 44:45–52. doi:10.1016/j.seppur.2004.12.002
38. Cassano A, Adzet J, Molinari R, Buonomenna MG, Roig J, Drioli E (2003) Membrane treatment by nanofiltration of exhausted vegetable tannin liquors from the leather Industry. Water Res 37:2426–2434. doi:10.1016/S0043-1354(03)00016-2
39. Shaalan HF, Sorour MH, Tewfik SR (2001) - Simulation and optimization of a membrane system for chromium recovery from tanning wastes. Desalination 141:315–324. doi:10.1016/S0011-9164(01)85008-6
40. Hawley EL, Deeb RA, Kavanaugh MC, Jacobs JA (2005) Treatments technologies for chromium (VI). In: Guertin J, Jacobs JA, Avakian CP (eds) Handbook chromium(VI). CRC Press, London
41. Park SJ, Jung WY (2001) Adsorption behaviors of chromium(III) and (VI) on electroless Cu-plated activated carbon fibers. J Colloid Interface Sci 243:316–320. doi:10.1006/jcis.2001.7910
42. Miretzky P, Fernandez Cirelli A (2010) Cr(VI) and Cr(III) removal from aqueous solution by raw and modified lignocellulosic materials: a review. J Hazard Mater 180:1–19. doi:10.1016/j.jhazmat.2010.04.060
43. Park D, Yun YS, Jo JH, Park JM (2006) Biosorption process for treatment of electroplating wastewater containing Cr(VI): laboratory-scale feasibility test. Ind Eng Chem Res 45:5059–5065. doi:10.1021/ie060002d
44. Durante C, Isse AA, Sandonà G, Gennaro A (2010) Exhaustive depletion of recalcitrant chromium fractions in a real wastewater. Chemosphere 78:620–625. doi:10.1016/j.chemosphere.2009.10.046
45. Durante C, Cuscov M, Isse AA, Sandonà G, Gennaro A (2011) Advanced oxidation processes coupled with electrocoagulation for the exhaustive abatement of Cr-EDTA. Water Res 45:2122–2130

Enzymatic Electrochemical Biosensors

Metini Janyasupab and Chung-Chiun Liu
Electronics Design Center, and Chemical Engineering Department, Case Western Reserve University, Cleveland, OH, USA

Introduction

Biosensors are commonly referred to as a device that quantifies a biological recognition into an analytical measurement. According to the standard definition provided by the International Union of Pure and Applied Chemistry (IUPAC), biosensor is an integrated self-contained receptor-transducer device, able to provide selectively quantitative analytical information using a biologic recognition element, which is in direct spatial contact with a transducer element [1]. The device is widely used in food industry [2–6], environment monitoring [7–9], and medical diagnosis [10–12]. It can be categorized by different types of biological recognitions. For example, antibody, DNA, and enzyme can be used to quantify antigen, DNA segment, and enzyme substrate, respectively. Among these different types of proteins, the enzymatic biosensor has been well developed in the commercial markets, owning an excellent specificity and efficiency as a *key-lock* combination to quantify an analyte from a drop of sample. Furthermore, there are many analytical transduction approaches, broadly described such as (1) optical-based biosensors, e.g., chemiluminescence [13–15], fluorescence [16–19] quenching [20, 21], and spectrophotometry [22–24]; (2) mass variation-based biosensors, e.g., piezoelectricity [25–28], quartz crystal microbalance (QCM) [29–32], and magnetisms [33, 34]; and (3) electrochemical-based biosensors [35–37], e.g., voltammetry, coulometry, and amperometry. While the former two approaches are commonly used in clinical analysis, they require a separation of plasma and serum in blood in order to minimize the

Enzymatic Electrochemical Biosensors, Fig. 1 Scheme of enzymatic detection by amperometric-based biosensors

interference. In addition, laborious laboratory tasks and equipment maintenance are necessary in operations. On the other hand, the latter approach provides more simplicity, convenience, robustness, and low cost maintenance, allowing patients to operate as a point of care or at home management. Therefore, the electrochemical-based biosensor becomes the most widely used in practical applications [10, 38]. One of the best examples is the stripe glucose sensor, first successfully developed by Clarks in 1962 [39] and remains the most well-known applications in today's biosensor market worldwide. The device is incorporated with the oxidoreductase enzyme, coupled with redox reduction, and suitable for electrochemical characterizations. Because of the market domination in this class of device, this section aims to focus on more insightful information of electrochemical techniques in this biosensor fabrication and collection of recent oxidoreductase enzymatic biosensors using amperometry.

Considering general enzymatic detection, the electrochemical-based biosensors are involved into biological recognition (e.g., enzymatic reaction) and electron transfer process (e.g., surface reaction on working electrode) as illustrated in Fig. 1. The scheme demonstrates an overview concept connecting between biological samples, mostly in liquid phase (e.g., blood, plasma, urine) and electrical signals occur on the surface of biosensors. Different architecture as well as different material membrane influence how enzymatic reaction occurs and can improve the efficiency of the adsorption and dissociation of enzyme interaction. Oftentime, the by-product of the enzymatic reaction is electrochemically active such as hydrogen peroxide (H_2O_2), nicotinamide adenine dinucleotide (NADH), and others, easily interacting with the metal or conductive polymeric surface. As a result, the electrons are capable to transfer from biological molecule to the surface of electrode, sensing in terms of either potential, charge, or current. This is also known as an enzyme-catalyst reaction that generally can be can describe in two processes below:

$$Substrate + Oxygen \overset{Enzyme}{\longrightarrow} Product + H_2O_2 \quad (1)$$

The reaction (1) is an enzymatic reaction whose analyte is the target species to quantify. After the enzyme interacts with the substrate, generated H_2O_2 is either oxidized or reduced onto the metal surface of the working electrode demonstrated by reaction (2, 3) and (4), respectively.

$$H_2O_2 \rightarrow O_2 + 2H^+ + 2e^- \quad (2)$$

$$Metal^{ox} + 2e^- \rightarrow Metal^{red} \quad (3)$$

or

$$Metal^{red} + H_2O_2 \rightarrow Metal^{\alpha x} + 2e^- + O_2 + 2H^+ \quad (4)$$

Similar to many electrochemical sensors, a three-electrode system is usually used and assessed by applying different electric waves and measuring the current as an output. If an input wave is a constant selected at one potential throughout the test, it is called amperometry. However, if a pattern or linear potentiometric wave is applied, the measurement is called voltammetry. Although these different characterizations aim to different aspects in electrochemistry, they share a similar principle of analytical quantification. Especially, the determination of H_2O_2 is one of the most employed strategies, serving as an important key in this class of device to translate this change analyte into electrical communication process. This concept does not only used in most commercial glucose sensors but also many other important analytes of enzymatic biosensors. Before discussing about fabrication aspect of the electrochemical biosensor, the following common criteria are necessary to define and describe the performance variables in each study.

1. *Sensitivity* is a quantitative value expressed the relationship of current and analyte concentration that the biosensor can detect.
2. *Selectivity* is a study of an ability where the biosensor detects the target analyte in the presence of many electroactive species or interferents.
3. *Limit of detection (LOD) and limit of quantification (LOQ)* are theoretically estimated as the minimum target analyte concentration that the biosensor can detect. Conventionally, LOD and LOQ are referred to the smallest concentration at the current above signal-to-noise ratio of 3 and 10, respectively.
4. *Response time* is often reported when reaches 95 % of steady-state signal (current).
5. *Long-term stability* implies an ability of storage life to maintain the performance of biosensors. It is usually reported in days, weeks, and months.
6. *Michaelis-Menten constant (Km)* corresponds to the enzyme affinity subjected to biocatalysts, temperature, pH, and ionic strength of the biosensors. The higher value of Km implies the lower affinity, describing a slower process of substrate-enzyme binding. The Km is estimated in many different models; however, Lineweaver-Burk (double reciprocal) plot is commonly applied in research studies.

Considering these performance quantifications, the following two important analytes species for medical diagnosis are demonstrated and discussed in detail of enzymatic sensor configurations, immobilization techniques, and other technology integrations in recent developments.

Cholesterol Detection

A rapid and simple determination of cholesterol is clinically important and economically viable. Due to a high correlation to the risk factor of cardiovascular disorders (e.g., atherosclerosis, coronary heart disease, hypertension), obesity, and molecular trafficking across the lipid plasma membrane, estimation of cholesterol allows patients to closely monitor their health conditions and adjust proper nutrition in their daily basis [40]. In human serum, normal level of total cholesterol is in the range of 1.3–2.0 $mg.mL^{-1}$. The borderline and high-risk levels are 2.0–2.39 $mg.mL^{-1}$ and above 2.40 $mg.mL^{-1}$, respectively [41, 42]. In addition, approximately 30 % of the total cholesterol can be classified as free-form cholesterol and 70 % are in the esterified form, containing a series of lipoproteins in term of very low-density lipoprotein (VLDL), low-density lipoprotein (LDL), and high-density lipoprotein (HDL) [41, 43, 44]. Therefore, the major enzymatic mechanisms are involved in the two following steps:

$$Esterified\ Cholesterol + H_2O \overset{Cholesterol\ Esterase}{\longrightarrow} Cholesterol + Fattyacid \tag{5}$$

$$Cholesterol + O_2 \overset{Cholesterol\ Oxidase}{\longrightarrow} Cholest - 4 - en - 3 - One + H_2O_2 \tag{6}$$

The process of conversion involved two enzymes: cholesterol esterase (ChE) and cholesterol oxidase (ChOx). The former enzyme converts the majority of total cholesterol from esterified cholesterol into free form. The latter will then convert free cholesterol into cholest-4-en-3-one, an important steroid form found in intracellular communications. The by-product at the end of the reaction (6) is the generated H_2O_2 that can be electrochemically oxidized or reduced on the surface of cholesterol biosensors.

Several developments focus on the efficacy of these mechanism conversions by enhancing enzyme immobilization on the working electrode

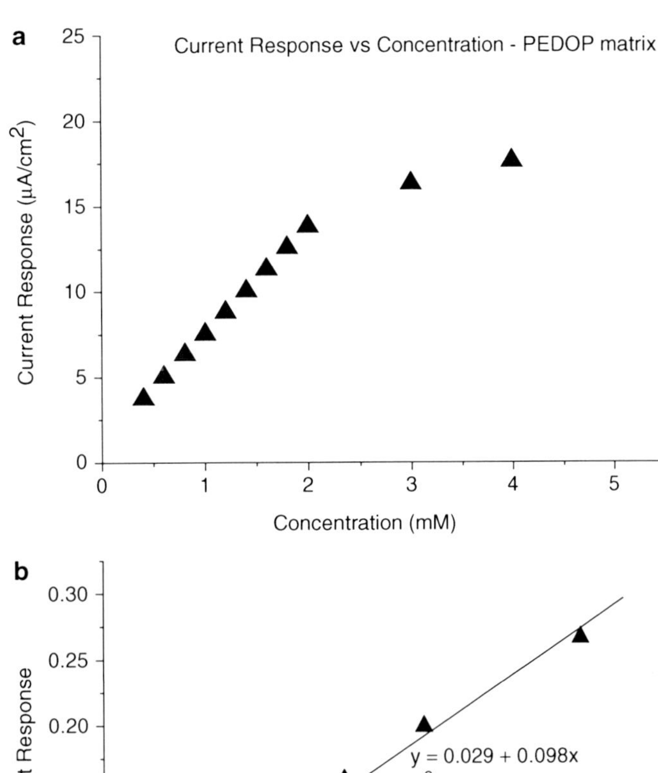

Enzymatic Electrochemical Biosensors, Fig. 2 Current versus concentration of cholesterol (**a**) and double reciprocal (Lineweaver-Burk) plot (**b**) [46] (Reprinted from The Lancet, Sensors and Actuators B: Chemical, 2009. 136(2): p. 484–488., Copyright (2009), with permission from Elsevier)

surface or improving electron transfer process in order to translate the level of H_2O_2 accurately. For instance, polymeric-based cholesterol biosensors such as polyaniline [45], pyrrole [46, 47], or glutaraldehyde [48–50] modified membranes have been used in enzyme immobilization. One of recent reports by Turkanslan and coworkers utilized poly(3,4-ethylenedioxypyrrole) (PEDOP) to construct an amperometric cholesterol biosensor. As shown in Fig. 2, the calibration plot of amperometry at +0.70 V versus Ag/AgCl shows a linear relationship of current-cholesterol (left) and a double reciprocal plot, Lineweaver-Burk, (right), with the sensitivity of 10 $\mu A.mM^{-1}.cm^{-2}$, and $R^2 = 0.99$. The Km and response time values are estimated to 3.4 mM and 150 s, respectively, with the relative maximum activity of 20 days. While the sensor shows a good linearity detecting normal level of cholesterol up to 5 mM (<2 $mg.mL^{-1}$), it can be seen that the enzyme kinetics primarily affect the performance of sensor. Influences by high overpotential, pH, and temperature change certainly become the major factors to specify. In this case (or elsewhere), pH of 7.0 at 37 °C at +0.70 V shows the optimal condition for the pyrrole base cholesterol sensor.

Alternatively, self-assembly monolayer approach has also been incorporated to link the protein amine group with the gold surface of electrode. Many thiol-compound-based cholesterol biosensors [51–54] constantly gained

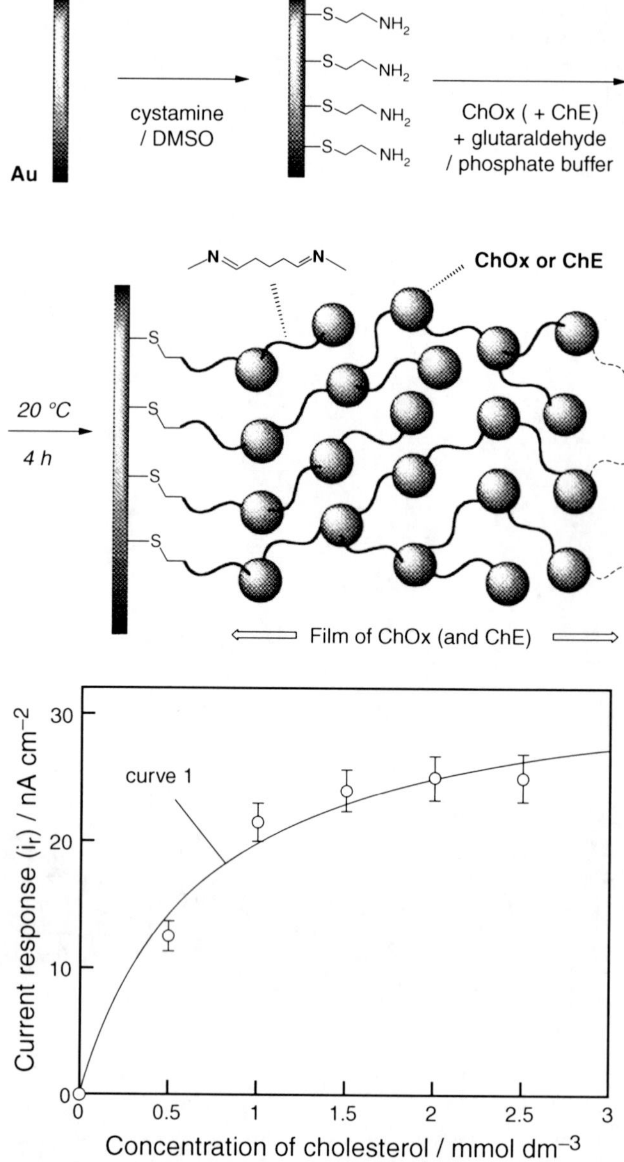

Enzymatic Electrochemical Biosensors, Fig. 3 Configuration of sensor fabrication using 2-aminoethanethiolate covalently cross-linked with ChOx and ChE (*left*) and current response of cholesterol concentrations in PBS pH 7.0 on gold electrode in the presence of thionin as mediator (*right*) [55] (Reprinted with permission from Analytical Chemistry, 1999. 71(5). Copyright (1999) American Chemical Society)

scientific interest and advantages of molecularly binding for enzyme immobilization. For example, Nakaminami et al. fabricated self-assembly monolayer (2-aminoethanethiolate) covalently immobilized with ChOx and ChE on a gold electrode as shown in Fig. 3. With the aid of glutaraldehyde, as a surface stabilizer, the amperometric response in the presence of thionin acting as a redox mediator can be found in the range of 0.5 and 2.5 mM cholesterol. The Km value of 0.59 mM found in this sensor shows a high affinity of enzyme-substrate binding which implies a fast process of the reaction. However, the sensor severely suffered a long-term stability, only remained the performance within 40 h. Similarly, Shen and Liu developed a mediator-free gold cholesterol screen-printed sensor as shown in Fig. 4. The modified surface of gold is coupled with alkyl thiol group, connected from carboxylic group (−COOH) terminal to the amine ($-NH_2$) group of the enzyme. The result shows a relatively low applied

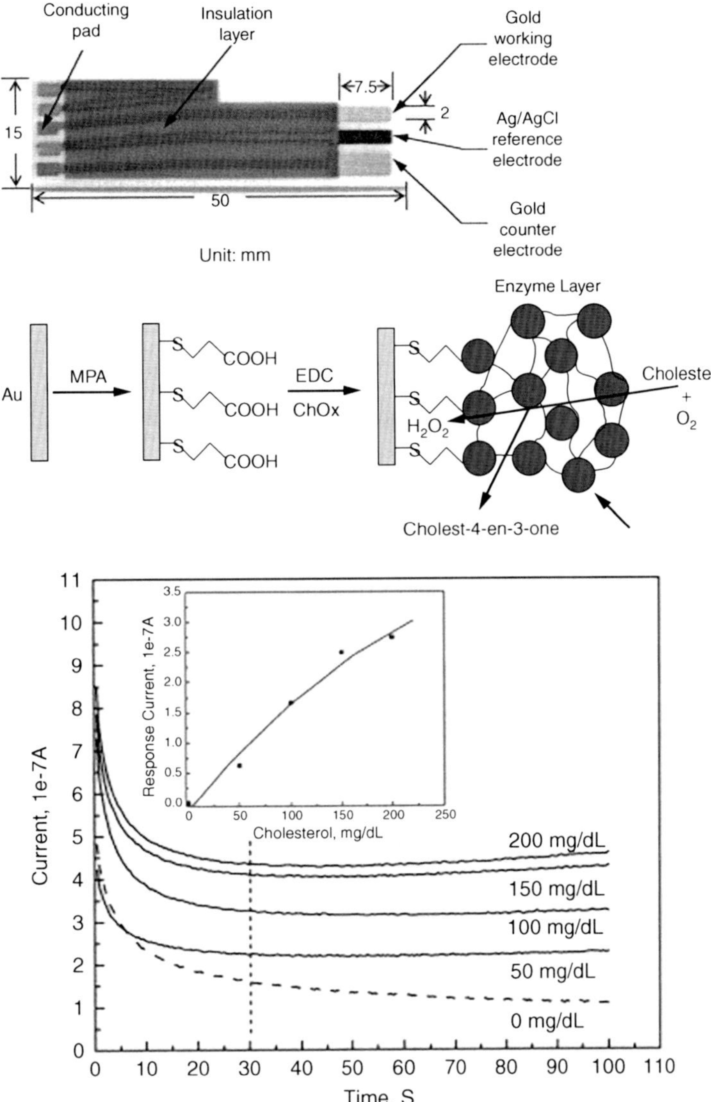

Enzymatic Electrochemical Biosensors, Fig. 4 Amperometric response of printed screen gold sensor immobilized with ChOx [41] (Reprinted from The Lancet, Sensors and Actuators B: Chemical, 2007. 120(2): p. 417–425. Vol. number, Copyright (2007), with permission from Elsevier)

potential of +0.35 V versus Ag/AgCl could be used to establish the relationship of steady-state current (evaluated at 30 s) and the concentration of cholesterol between 0 and 2 mg.mL^{-1}. This method directly utilizes the generated H_2O_2 to quantify the level of cholesterol in PBS. In spite of an excellent linear correlation of the sensor's performance, the stability substantially decreases after 4 days.

Another integration contributed to a major improvement of the electron transfer process is an incorporation of advanced carbon materials such as carbon nanotube or graphene. Tsai and coworkers reported an enzymatic ChOx with chitosan on Pt/multiwalled carbon nanotube (MWCNT). As shown in Fig. 5 (left), the modified sensor was deposited by reduced Pt nanoparticles decorated on MWCNT. An electrochemically amperometric titration was also performed at +0.4 V (versus Ag/AgCl) applied potential, with the sensitivity of 0.044 A.M^{-1}.cm^{-2} as shown in Fig. 5 (right). The integration of MWCNT allowed an enhancement of electron transfer process shown by the sensitivity values higher than other types of material. In addition, the sensor also exhibited a good selectivity

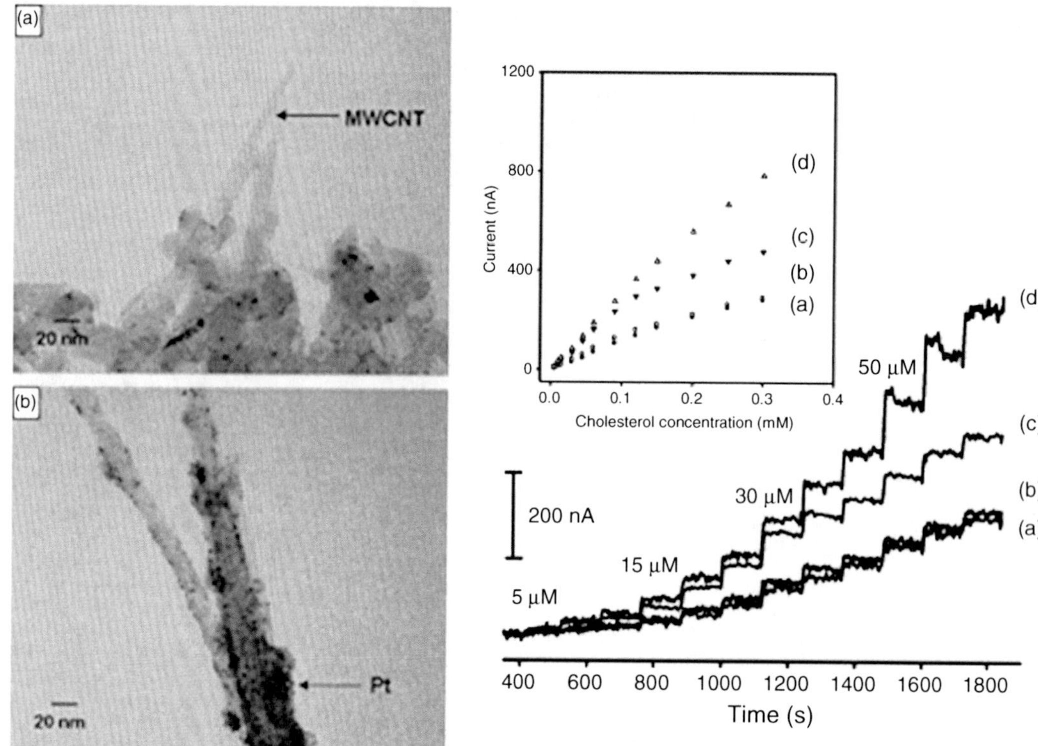

Enzymatic Electrochemical Biosensors, Fig. 5 *(left)* TEM images of multiwalled carbon nanotube with chitosan composite film before (**a**) and after (**b**) Pt complex reduction; *(right)* amperometric responses for increasing cholesterol concentrations at various compositions in 0.1 M phosphate buffer solution (pH 7). ChOx: (**a**) 6 $mgml^{-1}$, (**b**) 8 $mgml^{-1}$, (**c**) 10 $mgml^{-1}$, and (**d**) 12 $mgml^{-1}$. The inset shows the corresponding current-concentration calibration curves for cholesterol at various compositions [56] (Reprinted from The Lancet, Sensors and Actuators B: Chemical, 2008. 135(1): p. 96–101. Copyright (2008), with permission from Elsevier)

among the coexisting electroactive species in blood. By injecting 100 μM glucose, 1 μM ascorbic acid, and 2 μM uric acid, no catalytic activity change was observed in the presence of 100 μM cholesterol. Furthermore, the sensor had a good long-term stability, more than 50 % of performance could be maintained after 7 days.

With a different architecture of advanced carbon materials, graphene or graphene sheet is very attractive material to accelerate fast electron transfer process on the cholesterol sensor. As shown in Fig. 6, the sensor was composed of Pt catalyst supported by chitosan-graphene sheet. Both enzyme ChE and ChOx were physically entrapped layer by layer with Nafion polymer. A close look of morphology on the film could be observed by field emission SEM images, showing a thin layer of graphene. In terms of amperometric performance, both free-form cholesterol and total cholesterol were able to detect on this modified electrode. As shown in Fig. 6 (A), free-form cholesterol would undergo the oxidation of cholesterol by ChOx (reaction (6)) and generated H_2O_2 can be quantified at the maximum 35 μM free-form cholesterol with the sensitivity of 1.43 $\mu A.\mu M^{-1}.cm^{-2}$. On the other hand, if injected the esterified form (insoluble cholesterol), two-step processes (reaction 5–6) will be undergone on the surface of the electrode, thereby producing more H_2O_2. Thus, the total cholesterol can be detected in the range of 0.2 to 35 μM with the sensitivity of 2.07 $\mu A.\mu M^{-1}.cm^{-2}$ and Km of 5 mM. Although this Km value does not shown a strong affinity of enzyme, future

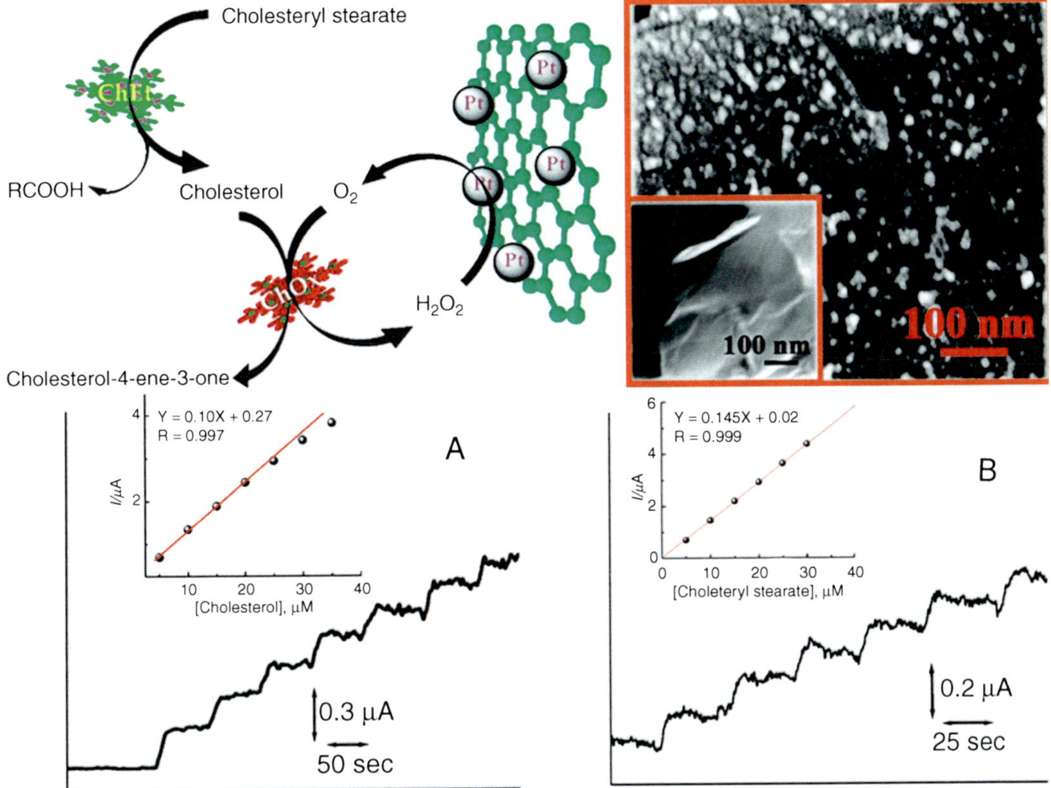

Enzymatic Electrochemical Biosensors, Fig. 6 *(top)* Scheme of configuration and FESEM image of cholesterol ester with Pt decorated graphene nanosheet; *(bottom)* amperometric response of free cholesterol (**a**) and cholesterol ester (**b**) injected by 5 μM for each step response [57] (Reprinted with permission from The Journal of Physical Chemistry C, 2010. 114(49). Copyright (2010) American Chemical Society)

improvement can be accomplished on more efficient method of incorporate polymeric layer (chitosan) into graphene sheet.

Without doubt, cholesterol detection is one of the most important medical diagnoses. However, the challenges of practicality issues for applying this sensor into real human sample still remain. For example, free-form cholesterol requires a pretreatment such as dissolving into a Triton-X or other surfactants. The insoluble cholesterol of human serum has low affinity to interact with enzyme and the needs of more practical experiments must be assessed. The use of (1) physical entrapment by conductive polymers, (2) covalent bond with self-assembly monolayer, or (3) enlargement of surface area by advanced carbon materials provides several advantages of enzyme immobilization, facilitating electron transfers [58]. While the combinations of these enhancements certainly yield a better performance, a proper integration of these surface modifications to overcome denaturation of enzymes is necessary. Although the stability of these oxidoreductase enzymes has constantly been improving, extension shelf life of the device is still considered as a real challenge for commercialization. Overall, the cholesterol biosensors certainly hold a great promise, potentially be successful similar to glucose sensor if these limitations can be overcome.

We have demonstrated a concept of enzymatic biosensor by using electrochemical detection. The process generally involved two mechanisms of (1) chemical reaction and (2) electrochemical reaction. The initial chemical reaction, resulted from enzyme and substrate, is often found by interacting with multiple enzymes as shown by the example of cholesterol esterase and cholesterol oxidase. Therefore, different property of each enzyme and efficiency of each step conversion can cause error propagation when translating into electrical signals. The immobilization of enzyme to the surface of electrodes by either physical entrapment, covalent bodingly, or electrochemical deposition plays a crucial role to complete this process. It is necessary that one of the products resulted by the enzymatic reaction is electrochemically active such as H_2O_2, NADH, or ferri/ferrocyanide ($Fe(CN)^{(3-/4-)}{}_6$). Afterward, these species will be oxidized or reduced in the electrochemical reaction, coupled with the initial chemical reaction, providing an electron transfer via catalytic property of membranes or metals. In the case of conductive membrane, a good arrangement of polymeric molecule and its functional group can both be housing the enzyme and provide electrical path to the surface of electrode. Otherwise, another redox mediator must be assisted in this transfer process. On the other hand, the metallic catalysts with a highly catalytic property such as platinum or palladium need to address other possible oxide formation that competitively occurs during the electron transfer of enzymatically generated H_2O_2. The control of their crystalline surface may be addressed to improve this property, and advanced functional carbon materials can enhance the electrochemical specific active surface ratio, increasing possibility of binding interaction to overall reactions. Furthermore, the performance of enzymatic biosensor, which is conventionally quantified in term of detection sensitivity, selectivity, long-term stability, Michaelis-Menten constant, response time, and limit of detection, is subject to the electrolyte, temperature, pH, and other experimental conditions. Therefore, the fabrication or choice of design must be considered based on the practicality and validity of application. Overall, the future of electrochemical enzymatic biosensor can provide much more opportunities to explore its feasibility to the other fields of studies, especially for a new discovery of biomarkers. Due to its simplicity and high reliability, the device remains a preferred choice in commercial design as the most efficient method to employ and still dominates the current market.

Cross-References

► Biosensors, Electrochemical
► Sensors

References

1. Thévenot DR, Toth K, Durst RA, Wilson GS (2001) Electrochemical biosensors: recommended definitions and classification. Biosens Bioelectron 16(1–2):121–131
2. Keow CM et al (2007) An amperometric biosensor for the rapid assessment of histamine level in tiger prawn (Penaeus monodon) spoilage. Food Chem 105(4): 1636–1641
3. Rawal R, Chawla S, Devender CS, Pundir CS (2012) An amperometric biosensor based on laccase immobilized onto Fe3O4NPs/cMWCNT/PANI/Au electrode for determination of phenolic content in tea leaves extract. Enzyme Microb Technol 51(4):179–185
4. Barroso MF, Delerue-Matos C, Oliveira MBPP (2012) Electrochemical evaluation of total antioxidant capacity of beverages using a purine-biosensor. Food Chem 132(2):1055–1062
5. Capannesi C, Palchetti I, Mascini M, Parenti A (2000) Electrochemical sensor and biosensor for polyphenols detection in olive oils. Food Chem 71(4):553–562
6. McGrath TF, Andersson K, Campbell K, Fodey TL, Elliott CT Development of a rapid low cost fluorescent biosensor for the detection of food contaminants. Biosens Bioelectron (0)
7. Li Y-F et al (2006) A mediator-free phenol biosensor based on immobilizing tyrosinase to ZnO nanoparticles. Anal Biochem 349(1):33–40
8. Fan Q, Shan D, Xue H, He Y, Cosnier S (2007) Amperometric phenol biosensor based on laponite clay–chitosan nanocomposite matrix. Biosens Bioelectron 22(6):816–821
9. Tan Y, Guo X, Zhang J, Kan J (2010) Amperometric catechol biosensor based on polyaniline–polyphenol oxidase. Biosens Bioelectron 25(7):1681–1687
10. D'Orazio P (2003) Biosensors in clinical chemistry. Clin Chim Acta 334(1–2):41–69
11. Sergeyeva TA et al (1999) Selective recognition of atrazine by molecularly imprinted polymer membranes. Development of conductometric sensor for herbicides detection. Anal Chim Acta 392(2–3):105–111
12. Crumbliss AL, Stonehuerner JG, Henkens RW, Zhao J, O'Daly JP (1993) A carrageenan hydrogel stabilized colloidal gold multi-enzyme biosensor electrode utilizing immobilized horseradish peroxidase and cholesterol oxidase/cholesterol esterase to detect cholesterol in serum and whole blood. Biosens Bioelectron 8(6):331–337
13. Martin AF, Nieman TA (1997) Chemiluminescence biosensors using tris (2, 2′-bipyridyl) ruthenium (II)

and dehydrogenases immobilized in cation exchange polymers. Biosens Bioelectron 12(6):479–489
14. Zhou GJ, Wang G, Xu JJ, Chen HY (2002) Reagentless chemiluminescence biosensor for determination of hydrogen peroxide based on the immobilization of horseradish peroxidase on biocompatible chitosan membrane. Sens Actuators B Chem 81(2):334–339
15. Xu Z, Guo Z, Dong S (2005) Electrogenerated chemiluminescence biosensor with alcohol dehydrogenase and tris (2, 2′-bipyridyl) ruthenium (II) immobilized in sol–gel hybrid material. Biosens Bioelectron 21(3):455–461
16. Bozym RA, Thompson RB, Stoddard AK, Fierke CA (2006) Measuring picomolar intracellular exchangeable zinc in PC-12 cells using a ratiometric fluorescence biosensor. ACS Chem Biol 1(2):103–111
17. Taitt CR, Anderson GP, Ligler FS (2005) Evanescent wave fluorescence biosensors. Biosens Bioelectron 20(12):2470–2487
18. Cotton GJ, Muir TW (2000) Generation of a dual-labeled fluorescence biosensor for Crk-II phosphorylation using solid-phase expressed protein ligation. Chem Biol 7(4):253–261
19. Ligler FS (2002) Optical biosensors: present and future. Elsevier Science
20. Fan C, Plaxco KW, Heeger AJ (2002) High-efficiency fluorescence quenching of conjugated polymers by proteins. J Am Chem Soc 124(20):5642–5643
21. Chen L et al (1999) Highly sensitive biological and chemical sensors based on reversible fluorescence quenching in a conjugated polymer. Proc Natl Acad Sci 96(22):12287–12292
22. Hatch WR, Ott WL (1968) Determination of submicrogram quantities of mercury by atomic absorption spectrophotometry. Anal Chem 40(14): 2085–2087
23. Williams C, David D, Iismaa O (1962) The determination of chromic oxide in faeces samples by atomic absorption spectrophotometry. J Agric Sci 59(03):381–385
24. Oke J, Gunn J (1983) Secondary standard stars for absolute spectrophotometry. Astrophys J 266:713–717
25. Babacan S, Pivarnik P, Letcher S, Rand A (2000) Evaluation of antibody immobilization methods for piezoelectric biosensor application. Biosens Bioelectron 15(11):615–621
26. Davis KA, Leary TR (1989) Continuous liquid-phase piezoelectric biosensor for kinetic immunoassays. Anal Chem 61(11):1227–1230
27. Tombelli S, Mascini M, Sacco C, Turner APF (2000) A DNA piezoelectric biosensor assay coupled with a polymerase chain reaction for bacterial toxicity determination in environmental samples. Anal Chim Acta 418(1):1–9
28. Abad J, Pariente F, Hernandez L, Abruna H, Lorenzo E (1998) Determination of organophosphorus and carbamate pesticides using a piezoelectric biosensor. Anal Chem 70(14):2848–2855
29. Martin SP, Lamb DJ, Lynch JM, Reddy SM (2003) Enzyme-based determination of cholesterol using the quartz crystal acoustic wave sensor. Anal Chim Acta 487(1):91–100
30. Shen Z et al (2007) Nonlabeled quartz crystal microbalance biosensor for bacterial detection using carbohydrate and lectin recognitions. Anal Chem 79(6):2312–2319
31. Cooper MA, Singleton VT (2007) A survey of the 2001 to 2005 quartz crystal microbalance biosensor literature: applications of acoustic physics to the analysis of biomolecular interactions. J Mol Recognit 20(3):154–184
32. O'sullivan C, Guilbault G (1999) Commercial quartz crystal microbalances–theory and applications. Biosens Bioelectron 14(8):663–670
33. Meyer MHF et al (2007) CRP determination based on a novel magnetic biosensor. Biosens Bioelectron 22(6):973–979
34. Chemla Y et al (2000) Ultrasensitive magnetic biosensor for homogeneous immunoassay. Proc Natl Acad Sci 97(26):14268–14272
35. Wang J (2004) Carbon-nanotube based electrochemical biosensors: a review. Electroanalysis 17(1):7–14
36. Wang J (2006) Electrochemical biosensors: towards point-of-care cancer diagnostics. Biosens Bioelectron 21(10):1887–1892
37. Sasso SV, Pierce RJ, Walla R, Yacynych AM (1990) Electropolymerized 1, 2-diaminobenzene as a means to prevent interferences and fouling and to stabilize immobilized enzyme in electrochemical biosensors. Anal Chem 62(11):1111–1117
38. Tothill IE (2001) Biosensors developments and potential applications in the agricultural diagnosis sector. Comput Electron Agric 30(1–3):205–218
39. Clark LC Jr, Lyons C (1962) Electrode systems for continuous monitoring in cardiovascular surgery. Ann N Y Acad Sci 102:29–45
40. Ansari AA, Alhoshan M, Alsalhi M, Aldwayyan A (2010) Nanostructured metal oxides based enzymatic electrochemical biosensors
41. Shen J, Liu C-C (2007) Development of a screen-printed cholesterol biosensor: comparing the performance of gold and platinum as the working electrode material and fabrication using a self-assembly approach. Sens Actuators B Chem 120(2):417–425
42. Ansari AA, Kaushik A, Solanki PR, Malhotra BD (2009) Electrochemical cholesterol sensor based on tin oxide-chitosan nanobiocomposite film. Electroanalysis 21(8):965–972
43. MacLachlan J, Wotherspoon ATL, Ansell RO, Brooks CJW (2000) Cholesterol oxidase: sources, physical properties and analytical applications. J Steroid Biochem Mol Biol 72(5):169–195
44. Allain CC, Poon LS, Chan CSG (1974) Enzymatic determination of total serum cholesterol. Clin Chem 20(4):470–475
45. Khan R, Kaushik A, Mishra AP (2009) Immobilization of cholesterol oxidase onto electrochemically

polymerized film of biocompatible polyaniline-Triton X-100. Mater Sci Eng C 29(4):1399–1403
46. Türkarslan Ö, Kayahan SK, Toppare L (2009) A new amperometric cholesterol biosensor based on poly (3,4-ethylenedioxypyrrole). Sens Actuators B Chem 136(2):484–488
47. Solanki PR, Arya SK, Singh SP, Pandey MK, Malhotra BD (2007) Application of electrochemically prepared poly-N-methylpyrrole-p-toluene sulphonate films to cholesterol biosensor. Sens Actuators B Chem 123(2):829–839
48. Shumyantseva V et al (2004) Cholesterol amperometric biosensor based on cytochrome P450scc. Biosens Bioelectron 19(9):971–976
49. Lin C-C, Yang M-C (2003) Cholesterol oxidation using hollow fiber dialyzer immobilized with cholesterol oxidase: preparation and properties. Biotechnol Prog 19(2):361–364
50. Torabi S-F, Khajeh K, Ghasempur S, Ghaemi N, Siadat S-OR (2007) Covalent attachment of cholesterol oxidase and horseradish peroxidase on perlite through silanization: activity, stability and co-immobilization. J Biotechnol 131(2):111–120
51. Singh S, Chaubey A, Malhotra BD (2004) Preparation and characterization of an enzyme electrode based on cholesterol esterase and cholesterol oxidase immobilized onto conducting polypyrrole films. J Appl Polym Sci 91(6):3769–3773
52. Arya SK et al (2006) Application of octadecanethiol self-assembled monolayer to cholesterol biosensor based on surface plasmon resonance technique. Talanta 69(4):918–926
53. Arya SK et al (2007) Poly-(3-hexylthiophene) self-assembled monolayer based cholesterol biosensor using surface plasmon resonance technique. Biosens Bioelectron 22(11):2516–2524
54. Arya SK et al (2007) Cholesterol biosensor based on *N*-(2-aminoethyl)-3-aminopropyl-trimethoxysilane self-assembled monolayer. Anal Biochem 363(2):210–218
55. Nakaminami T, Ito S-i, Kuwabata S, Yoneyama H (1999) Amperometric determination of total cholesterol at gold electrodes covalently modified with cholesterol oxidase and cholesterol esterase with use of thionin as an electron mediator. Anal Chem 71(5):1068–1076
56. Tsai Y-C, Chen S-Y, Lee C-A (2008) Amperometric cholesterol biosensors based on carbon nanotube–chitosan–platinum–cholesterol oxidase nanobiocomposite. Sens Actuators B Chem 135(1):96–101
57. Dey RS, Raj CR (2010) Development of an amperometric cholesterol biosensor based on graphene – Pt nanoparticle hybrid material. J Phys Chem C 114(49):21427–21433
58. Srivastava RC, Sahney R, Upadhyay S, Gupta RL (2000) Membrane permeability based cholesterol sensor – a new possibility. J Membr Sci 164(1–2):45–49

Ethanol Oxidation, Electrocatalysis of Fuel Cell Reactions

Nebojsa Marinkovic
Synchrotron Catalysis Consortium,
University of Delaware, Newark, DE, USA

Introduction

Easy storage and handling, high energy density, and wide availability are features that make alcohols attractive fuel cell liquid combustible and the most promising alternative power sources for transportation, portable electronics, and stationary applications. However, major obstacles have restrained the more rapid development of direct alcohol fuel cells, e.g., alcohol crossover from the anode to the cathode, relatively low activity and complex reaction mechanism of most alcohols, high costs of precious metal catalysts (Pt and Pt/Ru based catalysts), and CO poisoning of Pt catalysts at lower temperature in acidic media. In addition, it is particularly difficult to break the C–C bond in an alcohol (apart from methanol) during electrochemical oxidation.

The complete oxidation of any aliphatic mono-alcohol can be written as:

$$C_nH_{2n+1}OH + (2n-1)H_2O \rightarrow n\,CO_2 + 6nH^+ + 6ne^- \quad (1)$$

For direct ethanol fuel cell (DEFC), the alcohol oxidation reaction proceeds on the anode:

$$C_2H_5OH + 3H_2O \rightarrow 2CO_2 + 12H^+ + 12e^- \quad E'_o = 0.085\,V \quad (2a)$$

Reaction on the cathode is usually oxygen reduction:

$$3O_2 + 12H^+ + 12e^- \rightarrow 6H_2O \quad E''_o = 1.229\,V \quad (2b)$$

where E'_o and E''_o are standard electrode potentials for the anodic and cathodic reactions, respectively.

The mechanism of the simplest alcohol (methanol) oxidation on Pt has been studied for several decades as it is the only alcohol that does not need the C–C bond breaking for a full oxidation to CO_2. However, it can be seen that the number of electrons per molecule of alcohol increases from 6 to12 when exchanging methanol with ethanol as the fuel. Under standard conditions, $\Delta G^o = -1{,}325$ kJ/mol and standard emf at equilibrium is $E_{eq}{}^o = E''_o - E'_o = -\Delta G^o/nF = 1.144$ V, assuming that the process leads to total oxidation to carbon dioxide with exchange of $n = 12$ electrons. Ethanol possesses intrinsic advantages over methanol in applications of direct ethanol fuel cells (DEFCs) such as low toxicity, comparable electrochemical activity, high theoretical mass/energy density, and easy production by fermentation from sugar-containing materials. Thus, unlike methanol, ethanol is a renewable fuel.

Under reversible conditions, the theoretical efficiency ε_{theor} of a direct ethanol fuel cell, defined as the ratio of the maximum energy produced, $W_e = \Delta G^o = -nFE_o$, and the heat of combustion of the alcohol ($\Delta H^o = -1.366$ kJ), equals $\varepsilon_{theor} = W_e/\Delta H = 0.97$. However, the practical energy efficiency of a DEFC is much lower than the theoretical energy efficiency. It can be expressed as $\varepsilon_{cell} = \varepsilon_F\ \varepsilon_E\ \varepsilon_{theor}$, where ε_F is the ratio of effective number of electrons transferred in the reaction to that of the theoretical number of electrons for a complete reaction (12 for ethanol oxidation), and ε_E is the practical cell voltage needed to obtain a certain current density. Under working conditions, with a current density j, the cell voltage $E(j)$ is lower than E_o, because of a number of factors, including the overvoltage required for both electrode reactions, ohmic drop, ethanol crossover from anodic to cathodic side, etc. Thus, the practical efficiency of a DEFC working at 0.5 V with the current density of 100 mA/cm^2 is lowered by a factor of $\varepsilon_E = 0.5$ V/1.144 V than the theoretical efficiency, yielding the practical efficiency $\varepsilon_{cell} = 0.424$, assuming complete oxidation of ethanol to CO_2 with transfer of 12 electrons. This efficiency is similar to that of the best thermal engines (diesel engine). However, due to the complex mechanism of the ethanol oxidation, the reaction may stop at an intermediate stage with a transfer of less than 12 electrons, and the practical efficiency will be lowered by a factor of $\varepsilon_F = n_{eff}/12$, where n_{eff} is the effective number of electrons transferred. For example, if the ethanol oxidation stops at acetaldehyde stage with the transfer of four electrons, the practical efficiency ε_{cell} would be only 1/3 of the above practical efficiency, or about 0.15.

Schematic Diagram of DEFC

Schematic diagram of the ethanol/oxygen fuel cell in acidic medium is presented in Fig. 1. The electrochemical cell consists of two electrodes containing electron-conductive catalysts, separated by electrolyte or proton conductor. Ethanol is supplied at the anode side, and the oxygen (either pure gas or from air) is supplied at the cathode side. Electrons liberated at the anode by the oxidation of the fuel pass through the external electrical circuit and arrive at the cathode where they are used in the reduction of oxygen. The circuit is closed by the transport of protons from the anode to the cathode.

Reaction Mechanisms in Acidic Medium

The most common anode electrocatalyst for the ethanol oxidation is Pt. It is generally accepted that the first step that occurs at potentials lower than 0.6 V (vs. RHE) at platinum surface is the dissociative adsorption of ethanol; however, there is no consensus on the origin of the hydrogen atom that leaves the alcohol molecule first. Thus, two reactions can be written:

$$Pt + CH_3CH_2OH \rightarrow Pt\text{-}(OCH_2CH_3)_{ads} + H^+ + e^- \quad (3a)$$

and

$$Pt + CH_3CH_2OH \rightarrow Pt\text{-}(CHOHCH_3)_{ads} + H^+ + e^- \quad (3b)$$

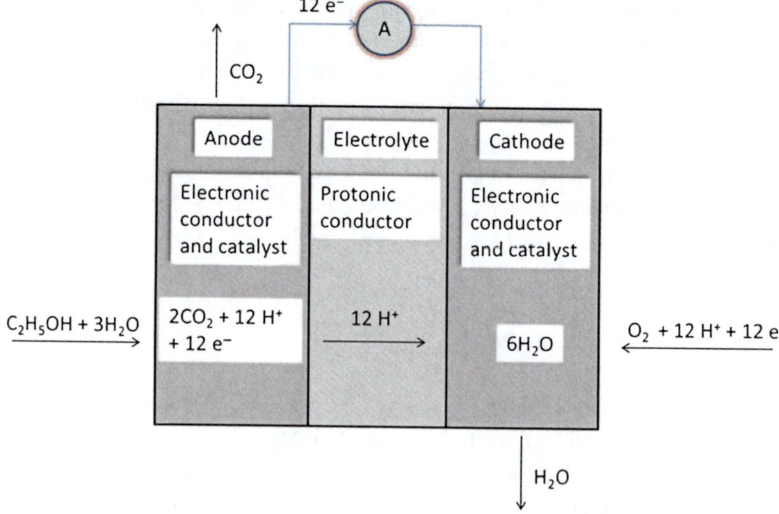

Ethanol Oxidation, Electrocatalysis of Fuel Cell Reactions, Fig. 1 Schematic diagram of direct ethanol fuel cell

Nevertheless, both above reactions lead to acetaldehyde ($CHOCH_3$) with dissociation of another hydrogen atom from C_1.

The reaction may proceed further if the molecule is re-adsorbed on the Pt surface:

$$Pt + CHOCH_3 \rightarrow Pt\text{-}(CHOCH_3)_{ads} \quad (4)$$

Vibrational techniques provided evidence that the C–C bond of acetaldehyde can be broken at potentials lower than 0.4 V, yielding adsorbed CH_x species and CO. Using SNFITIRS experiments, Iwasita and Pastor proposed the following mechanism [1]:

$$\begin{aligned} Pt + Pt\text{-}(CHOCH_3)_{ads} &\rightarrow Pt\text{-}(COCH_3)_{ads} \\ + Pt\text{-}(H)_{ads} &\rightarrow Pt\text{-}(CO)_{ads} + Pt + CH_4 \end{aligned} \quad (5)$$

Based on their SEIRS experiments, Lai et al. recently concluded that the C–C bond can be broken at potentials as low as 0.1 V, by the following mechanism [2]:

$$\begin{aligned} Pt\text{-}(CHOCH_3)_{ads} &\rightarrow Pt\text{-}(COCH)\text{-}Pt_{ads} \\ + 3H^+ + 3e^- &\rightarrow Pt\text{-}(CH)_{ads} + Pt\text{-}(CO)_{ads} \end{aligned} \quad (6)$$

where $Pt\text{-}(CH)_{ads}$ can be slowly oxidized to $Pt\text{-}(CO)_{ads}$ at potentials lower than 0.45 V.

Strongly adsorbed CO species can be oxidized further at potentials higher than 0.6 V:

$$Pt + H_2O \rightarrow Pt\text{-}(OH)_{ads} + H^+ + e^- \quad (7)$$

$$Pt\text{-}(CO)_{ads} + Pt\text{-}(OH)_{ads} \rightarrow 2Pt + CO_2 + H^+ + e^- \quad (8)$$

Simultaneously, at $E > 0.6$ V acetaldehyde can be directly oxidized by the adsorbed hydroxyl groups, yielding acetic acid without breaking the C–C bond:

$$Pt\text{-}(COCH_3)_{ads} + Pt\text{-}(OH)_{ads} \rightarrow 2Pt + CH_3COOH \quad (9)$$

Reactions (5) and (6) are slow and their products are only confirmed with highly selective vibrational techniques. Main products of ethanol oxidation on Pt electrode at potentials up to 0.6 V are acetaldehyde, acetic acid, and carbon dioxide, with electron conversions of 2, 4, or 12 respectively. Since practical potentials for DEFC are lower than 0.6 V at which the total oxidation to CO_2 proceeds slowly, the buildup of unreacted CO quickly poisons the Pt electrode surface. A study by Behm et al. showed that the ethanol oxidation on Pt/C yields acetic acid, acetaldehyde and carbon dioxide in the range of 20–65 %, 27–79 % and 0.7–7.5 %, respectively, depending on the initial concentration of ethanol [3]. Such a low selectivity of the catalyst toward CO_2 is not only a practical problem to be addressed in designing DEFCs, but also a theoretical problem as it is in contrast to the general consensus in chemistry, stating that kinetics of any catalytic reaction is expected to be faster if it leads to a more stable product. Since CO_2 is significantly more stable than the major products of the ethanol oxidation, acetic acid, and acetaldehyde, the low selectivity of the EOR to CO_2 is also a fundamental question. A recent first-principle calculations have shown that in the oxidation of ethanol on pure Pt, the formation of CO readily occurs at low potentials, but it cannot be oxidized further to CO_2 because of unavailability of oxidants (OH and/or O), whereas the presence of these oxidants at higher potentials inhibits the cleavage of the C–C bond [4]. Thus, platinum alone is not a good choice of catalyst for the EOR.

Numerous studies have been conducted to alter the rate of ethanol oxidation and the relative amount of products. $Pt_{73}Rh_{10}$ catalyst shows a high CO_2 production activity although a similar oxidation rate to pure Pt [5]. Addition of Ru, Rh, or Sn increased the rates of reaction leading to acetaldehyde and acetic acid, but decreased the rate of reaction leading to CO_2 [6], while addition of Pd had no effect [7]. Of binary alloys, Pt-Sn shows the best oxidation rates that are about three times higher than on Pt, as Sn and its oxides supply OH from water for the oxidative removal of CO-like species on the Pt surface [8]. However, the oxidation of ethanol is incomplete and C_2 products are still formed. The catalyst composed of Pt and SnOx showed slightly higher selectivity compared to Pt-Sn alloys, but the product is still acetaldehyde and selectivity toward CO_2 is even lower than on Pt [9, 10]. On the basis of the overview of bimetallic alloy catalysts for oxidation of ethanol and their correlation to the shift of d-band centers theoretically predicted by Nørskov et al. [11], Demirci [12] proposed Pd-Ni as a potential candidate, but concluded that further investigation should be focused on ternary catalysts. Addition of a third element indeed enhances the oxidation currents with respect to the Pt-Sn, as shown for Pt-Ru-W, Pt-Sn-Ni, and Pt-Sn-Rh [13–15]. A review of platinum-based ternary catalysts for ethanol oxidation summarizes the work up to 2007 [16].

A breakthrough in the ethanol oxidation in acidic solution appeared in 2009 when Adzic et al. found that the addition of Rh to Pt-SnO_2/C enhanced the catalyst capacity to break C–C bonds while enhancing the electric current for ethanol oxidation [17, 18]. The oxidative current at 0.3 V versus reversible hydrogen electrode (RHE) at 60 °C is 7.5 $mAcm^{-2}$ for the Pt/Rh/SnO_2 catalyst, in comparison to Pt/Ru catalyst that has negligible current density, Fig. 2a. Optimized geometry of the PtRh/Sn(110) surface and the proposed mechanism are depicted in Fig. 2b, c. The optimal pathway (in blue color) leads through an oxametallacyclic conformation (CH_2CH_2O) that entails direct breakage of the C–C bond with a reasonable barrier of 1.29 V. Red-colored pathway leading through acetaldehyde (CH_3CHO) is unfavorable as its barrier is by 0.66 V higher than that for CH_2CH_2O intermediate, and furthermore, the C–C bond splitting from acetaldehyde requires extremely high energy of 3.82 V. Thus, the optimal reaction pathway is:

$$\begin{aligned} {}^*CH_3CH_2OH &\rightarrow {}^*CH_3CHO + H^* \\ &\rightarrow {}^*CH_2CH_2O + 2H^* \\ &\rightarrow {}^*CH_2 + {}^*CH_2O + 2H^*. \end{aligned}$$

In situ IR study (Fig. 2d) of the ethanol oxidation confirmed these conclusions, showing the production of CO_2 (at ~2,340 cm^{-1}) at potentials as low as 0.2 V with a negligible production of CO (~2,050 cm^{-1}), and comparably small acetaldehyde and acetic acid (1,715 and 1,391 cm^{-1}).

Reaction Mechanism in Alkaline Medium

Much less attention has been given to ethanol oxidation in DEFC in alkaline medium. A problem with alkaline fuel cells is the carbonation of the solution due to CO_2 production of the fuel oxidation and from air, which can cause solid

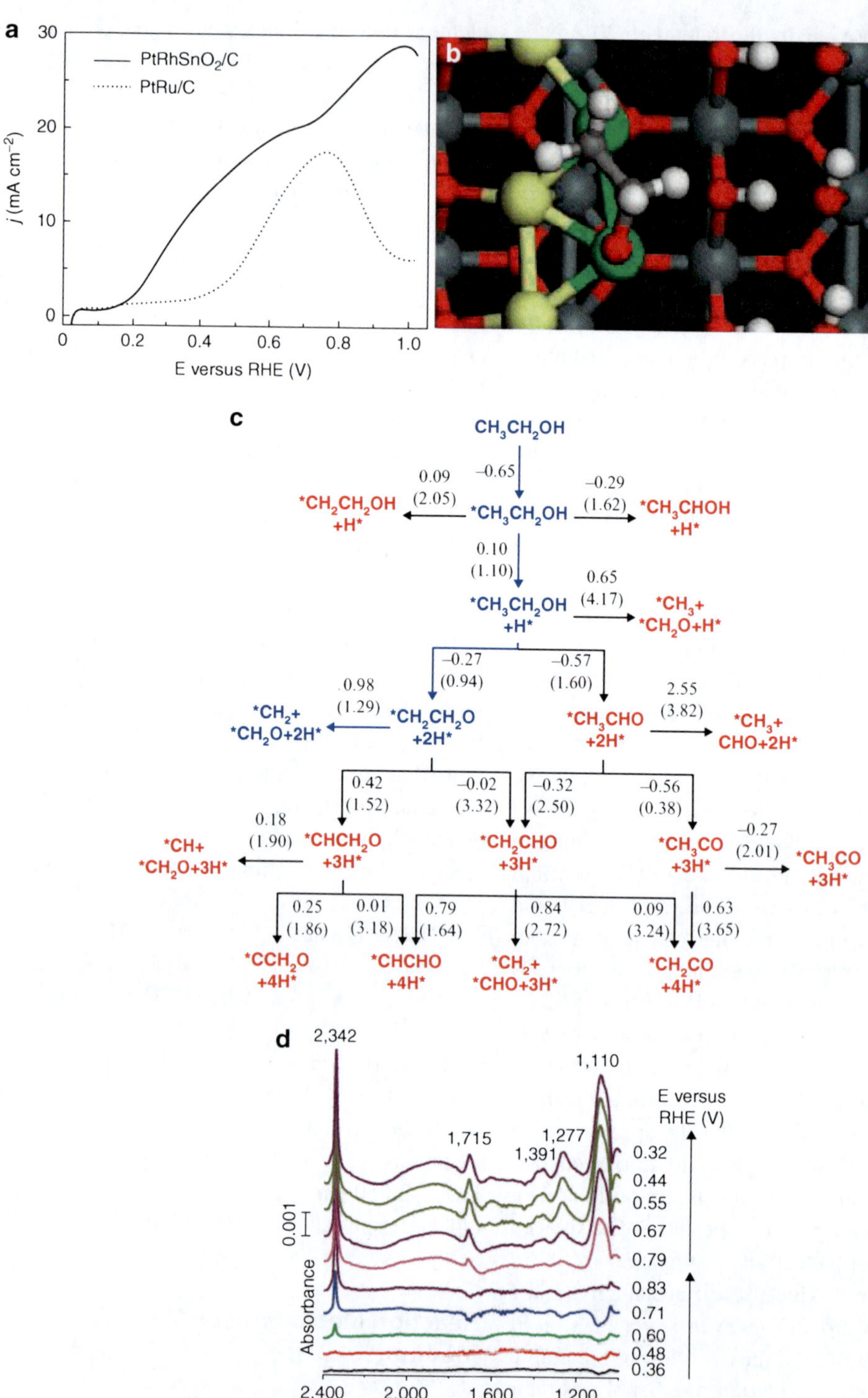

Ethanol Oxidation, Electrocatalysis of Fuel Cell Reactions, Fig. 2 (continued)

precipitation of carbonate salts on the porous electrode and a pH decrease in the alkaline electrolyte solution, leading to a reduction in reactivity for fuel oxidation in the system. On the other hand, non-precious metals that are unstable in acidic environment, e.g., Pd, Ag, Ni, and perovskite-type oxides, may be used, reducing the cost of the catalyst. In addition, unlike in acidic medium, the electrode does not suffer severely from poisoning in alkali because bonding of the chemisorbed intermediates on platinum is weak and the amount of the suggested poisoning species CO_{ads} is smaller than in an acidic medium, which leads to higher activity for the oxidation of organic fuels [19].

Schematic diagram of the ethanol/oxygen fuel cell in alkaline medium is similar to the one shown in Fig. 1. The total oxidation pathway can be summarized as follows:

$$\text{Anode}: \quad C_2H_5OH + 12OH^- \rightarrow 2CO_2 + 9H_2O + 12e^- \tag{10a}$$

$$\text{Cathode}: \quad 3O_2 + 6H_2O \rightarrow 12OH^- \tag{10b}$$

The oxidation of alcohol with hydroxyl ions produces electrons that are transported through the external circuit to the cathode. The electrons are used at the cathode to reduce oxygen and produce hydroxyl groups that migrate to the anode. Carbon dioxide and water are extracted from the anode side.

A DEMS study reported that the contribution of CO_2 to the total current on Pt nanoparticle catalyst is about 55 % in alkaline medium, compared to less than 10 % in acidic medium [20]. The reaction mechanism goes through acetaldehyde and acetic acid, similar to the acidic medium, with the rate determining step being the removal of adsorbed ethoxy-intermediate by adsorbed hydroxyl ions to form acetate.

Generally, for alcohol oxidation under alkaline solutions, the reaction current densities obtained during alcohol oxidation at certain potentials are at least an order of magnitude greater than in acidic electrolytes. The combination of alloying elements such as Ru, Mo, Sn, Re, Os, Rh, Pb, Bi as well as metal oxides such as ZrO_2 and MnO_2 with platinum gives tolerance to the poisoning species compared to platinum alone. In the ternary $Pt_{30}Pd_{38}Au_{32}/C$ electrocatalyst in which metals that barely show a catalytic efficiency towards ethanol oxidation are alloyed to Pt, a pronounced increase in the quantity of oxidation products, such as acetate and carbonate, was observed. The proposed mechanism leads through adsorption of acetate ion through C_1 atom onto OH^--covered surface, yielding $(CH_2OH)_{ads}$ and CO_2 [21–25].

Future Directions

Despite the renewed interest in ethanol oxidation, there is still a long way to go before an active, cost-effective catalyst is found for the large-scale adaptation in fuel cells. First, the understanding of the low selectivity of the reaction towards CO_2 on Pt is still lacking. Second, challenges in designing a good, low-cost catalyst with high activity, prolonged stability, negligible poisoning and total oxidation of ethanol to CO_2 need to be addressed. At present, the best catalysts in both acidic and alkaline solutions contain a certain amount of noble metals. Particularly, the catalyst in acidic solution that leads to preferential production of CO_2 as the main product consists of nanoparticle Pt/Rh solid solution [17], where the

Ethanol Oxidation, Electrocatalysis of Fuel Cell Reactions, Fig. 2 (**a**) Polarization curves for the oxidation of 0.2 M ethanol in 0.1 M $HClO_4$ on $PtRhSnO_2/C$ and PtRu/C (1:1 atomic ratio on carbon) at 60 °C and with sweep rate of $50 mVs^{-1}$. (**b**) Optimized geometry of CH_2CH_2O adsorption on a $PtRh/SnO_2(110)$ surface; (*Sn* large grey, *Pt* large yellow, *Rh* large green, *C* small gray, *O* small red, *H* small white). (**c**) Calculated possible pathways for the C–C bond breaking of ethanol on the $PtRh/SnO_2(110)$ surface; the reaction energies and parenthesized barriers in the figure are expressed in electronvolts. (**d**) IRRAS spectra recorded during ethanol electro-oxidation on the $PtRhSnO_2/C$ in the solution as in (**a**); 128 spectra at 8 cm^{-1} resolution are co-added for each spectrum (Reproduced from [17] with permission of Nature Publishing Group)

latter metal is several times more expensive than Pt. Third, a number of factors like overvoltage, ohmic drop, ethanol crossover, and others are yet to be resolved in order to increase the overall activity of DEFC in various applications.

Cross-References

► Direct Alcohol Fuel Cells (DAFCs)

References

1. Iwasita T, Pastor E (1994) A Dems and FTir spectrosopic investigation of adsorbed ethanol on polycrystalline platinum. Electrochim Acta 39:531. Ibid., D/H exchange of ethanol at platinum electrodes. Electrochim Acta 39:547
2. Lai SCS, Kleyn SEF, Rosca V, Koper MTM (2008) Mechanism of the dissociation and electrooxidation of ethanol and acetaldehyde on platinum as studied by SERS. J Phys Chem C 112:19080
3. Wang H, Yusus Z, Behm RJ (2004) Ethanol electrooxidation on a carbon-supported Pt catalyst? Reaction kinetics and product yields. J Phys Chem B 108:19413
4. Kavanagh R, Cao X-M, Lin W-F, Hardacre C, Hu P (2012) Origin of low CO_2 selectivity on platinum in the direct ethanol fuel cell. Angew Chem 124:1604
5. de Souza JPI, Queiroz SL, Bergamaske K, Gonzalez ER, Nart FC (2002) Electro-oxidation of ethanol on Pt, Rh, and PtRh electrodes. A study using DEMS and in-situ FTIR techniques. J Phys Chem B 106:9825
6. Nakagawa N, Kaneda Y, Wagatsuma M, Tsujiguchi T (2012) Product distribution and the reaction kinetics at the anode of direct ethanol fuel cell with Pt/C, PtRu/C and PtRuRh/C. J Power Sources 199:103
7. Liu J, Ye J, Xu C, Jiang SP, Tong Y (2007) Kinetics of ethanol electrooxidation at Pd electrodeposited on Ti. Electrochem Commun 9:2334
8. Lamy C, Rousseau S, Belgsir EM, Coutanceau C, Léger JM (2004) Recent progress in the direct ethanol fuel cell: development of new platinum–tin electrocatalysts. Electrochim Acta 49:3901
9. Rousseau S, Coutanceau C, Lamy C, Léger JM (2006) Direct ethanol fuel cell (DEFC): Electrical performances and reaction products distribution under operating conditions with different platinum-based anodes. J Power Sources 158:18
10. Wang Q, Sun GQ, Jiang LH, Xin Q, Sun SG, Jiang YX, Chen SP, Jusys Z, Behm R (2007) Adsorption and oxidation of ethanol on colloid-based Pt/C, PtRu/C and Pt3Sn/C catalysts: in situ FTIR spectroscopy and on-line DEMS studies. J Phys Chem Chem Phys 9:2686
11. Greely J, Norskov JK, Maurikakis M (2002) Electronic Structure and Catalysis on Metal Surfaces Annu Rev Phys Chem 53:319
12. Demirci UB (2007) Theoretical means for searching bimetallic alloys as anode electrocatalysts for direct liquid-feed fuel cells. J Power Sources 173:11
13. Tanaka S, Umeda M, Ojima H, Usui Y, Kimura O, Uchida I (2005) Preparation and evaluation of a multicomponent catalyst by using a co-sputtering system for anodic oxidation of ethanol. J Power Sources 152:34
14. Spinacé EV, Linardi M, Oliveira Neto A (2005) Co-catalytic effect of nickel in the electro-oxidation of ethanol on binary Pt–Sn electrocatalysts. Electrochem Commun 7:365
15. Colmati F, Antolini E, Gonzalez ER (2008) Preparation, structural characterization and activity for ethanol oxidation of carbon supported ternary Pt–Sn–Rh catalysts. J Alloys Compd 456:264
16. Antolini E (2007) Platinum-based ternary catalysts for low temperature fuel cells: Part II. Electrochemical properties. Appl Catal B 74:337
17. Kowal A, Li M, Shao M, Sasaki K, Vukmirovic MB, Zhang J, Marinkovic NS, Liu P, Frenkel AI, Adzic RR (2009) Ternary Pt/Rh/SnO_2 electrocatalysts for oxidizing ethanol to CO_2. Nat Mater 8:325
18. Li M, Kowal A, Sasaki K, Marinkovic N, Su D, Korach E, Adzic RR (2010) Ethanol oxidation on the ternary Pt–Rh–SnO_2/C electrocatalysts with varied Pt:Rh:Sn ratios. Electrochim Acta 55:4331
19. Beden B, Leger JM, Lamy C (1992) Electrocatalytic oxidation of oxygenated alphatic organic compounds at noble metal electrodes. In: Bockris JO'M, Conway BE, White RE (eds) Modern aspects of electrochemistry, vol 22. Plenum, New York, p 97
20. Rao V, Hariyanto H, Cremers C, Stimming U (2007) Investigation of the ethanol electro-oxidation in alkaline membrane electrode assembly by differential electrochemical mass spectrometry. Fuel Cells 7:417
21. Choban ER, Markoski LJ, Wieckowski A, Kenis PJA (2004) Microfluidic fuel cell based on laminar flow. J Power Sources 128:54
22. Lang CM, Kim K, Kohl PA (2006) High-energy density, room-temperature carbonate fuel cell batteries, fuel cells, and energy conversion. Electrochem Solid State Lett 9:A545
23. Bai YX, Wu JJ, Xi JY, Wang JS, Zhu WT, Chen LQ, Qiu XP (2005) Electrochemical oxidation of ethanol on Pt–ZrO_2/C catalyst. Electrochem Commun 7:1087
24. Verma A, Basu S (2007) Direct alkaline fuel cell for multiple liquid fuels: Anode electrode studies. J Power Sources 174:180
25. Datta J, Dutta A, Mukherjee S (2011) The beneficial role of the cometals Pd and Au in the carbon-supported PtPdAu catalyst toward promoting ethanol oxidation kinetics in alkaline fuel cells: temperature effect and reaction mechanism. J Phys Chem C 115:15324

Ex-Cell-Mediated Oxidation (via Persulfate) of Organic Pollutants

Nicolaos Vatistas
DICI, Università di Pisa, Pisa, Italy

Introduction

The efficient use of current supplied during the oxidation of bio-refractory organic pollutants by electrochemical treatment imposes to adopt an electrocatalytic material such as boron-doped diamond (BDD) that produces hydroxyl radicals. At the end of electrochemical treatment, a low transfer of pollutants occurs toward the hydroxyl radicals due to the reduced concentration of pollutants in the bulk, and this condition reduces the degree of utilization of BDD anodes. Ex-cell-mediated oxidation (via persulfate) of organic pollutants can be applied after electrochemical treatment, and its effects are a higher utilization of BDD anodes and an efficient removal of pollutants at the end of treatment, when their concentrations achieve low values. This oxidation consists of various steps: after the electrochemical production of a precursor of active oxidants, it is initially mixed with the wastewater and successively activated by thermal or ultraviolet energy in order to generate active oxidants that efficiently oxidize the pollutants.

Kinetic and Mass Transfer Limits of Electrochemical Treatment

The high electrocatalytic activity of boron-doped diamond (BDD) is due to the generated highly active hydroxyl radicals that are present in a thin layer adjacent to the anodic surface [1, 2]. Such radicals efficiently oxidize the organic pollutants that achieve this layer. A BDD anode, operating at an overpotential lower than a critical value η_{cr}, is covered after some time by a polymeric film that inhibits the occurrence of electrochemical oxidation [3]. A critical value of current density i_{cr} corresponds to this critical overpotential.

The oxidation of organic pollutants in a thin layer adjacent to the anodic surface imposes that these pollutants should be transferred to this layer. If a relatively high rate of radicals is generated, the oxidation rate will be controlled by the rate of pollutant transfer toward this thin layer of hydroxyl radicals. Such transfer rate can be expressed as the limiting current, i_{lim}, by assuming that pollutants that reach the thin layer are completely oxidized to inorganic species [4]:

$$i_{lim} = 4FkCOD \quad (1)$$

where F is Faraday's constant, k is the mass transfer coefficient, and COD is the concentration of organic species expressed as chemical oxygen demand. The limiting current imposes a limit on the efficient utilization of the applied current. The applied current density is completely utilized to oxidize the pollutants if its value is less than or equal to the limiting current. The limiting current density achieves the critical current density ($i_{lim} = i_{cr}$) when the concentration becomes COD_{cr}. This value represents the minimum concentration reached, when it is imposed, where the supplied current is used for the oxidation of pollutants.

Operating Regions of Oxidation with BDD Anodes

Technologically advanced anodes, such as BDD anodes, have a chance to be successfully used only when they efficiently utilize the applied electrical current, while at the same time operating at high current density values. In the case of organic pollutant oxidation, such efficiency is evaluated when the mass transfer coefficient is given ($k = 3.0 \cdot 10^{-5}$ m $\cdot$ s^{-1}); then the limiting current density, i_{lim}, during elimination of the organic species can be calculated by Eq. 1. Current density vs. concentration creates a linear diagram, as shown in Fig. 1, and this diagram separates the entire region of operating conditions into upper and lower regions. In the upper region, the supplied electric current is partially used to oxidize the pollutants, while in the lower

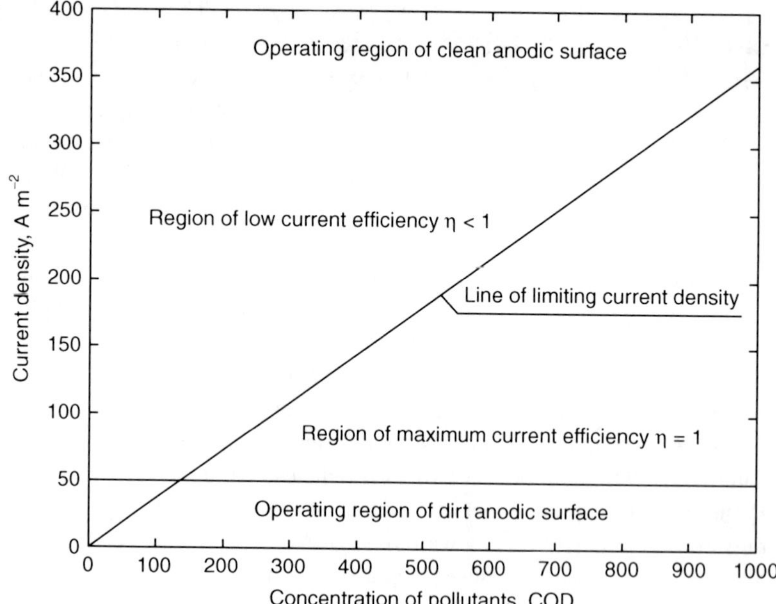

Ex-Cell-Mediated Oxidation (via Persulfate) of Organic Pollutants, Fig. 1 Operating regions of electrochemical treatment

region, the supplied current is totally utilized by this oxidation of pollutants.

Also shown in the figure is the horizontal line that represents the critical value of current density, i_{cr}. This line also separates the entire region of operating conditions into a lower region and an upper region. In the lower region, a polymeric film is formed over the anodic surface after an interval of time that isolates the anode from the electrolyte, and any electrochemical reaction on this anode is blocked. In the upper region, the anodic surface remains clean, and the occurrence of electrochemical treatment is assured.

The value of the limiting current density is defined as the maximum value of applied current density at which all supplied current is used to oxidize the organic species. Its value also indicates another condition that concerns the degree of its utilization. In fact, a low value of limiting current imposes to apply a low current density if its full use for the oxidation of pollutants is imposed. In other words, a low value of limiting current density indicates a low degree of utilization of the adopted anodic surface. As Fig. 1 shows, this condition occurs at the end of treatment, when the wastewater contains low concentrations of organic pollutants. The low degree of utilization is an important factor that inhibits the wide use of electrochemical treatment with BDD anodes in this field.

Electrochemical and Ex-Cell-Mediated Pollutant Oxidation

Ex-cell-mediated oxidation operates efficiently when the concentration of organic pollutants is low and when, because of its properties, this treatment is complementary with respect to electrochemical treatment, which operates with low efficiency when the pollutant concentration low. With the ex-cell oxidation method, a strong oxidant precursor is initially produced, which is mixed with the wastewater to assure the proximity of precursor organic species; afterwards, the action of thermal, ultraviolet, or another form of energy transforms the precursor to active oxidants. The previously achieved proximity between precursor and organic pollutants becomes an active oxidant-pollutant proximity that assures efficient oxidation, even when the concentration of organic species is low.

Ex-cell-mediated oxidation has similarities and differences with respect to the indirect electrochemical oxidation known in this field. The principal similarity is that both methods use an active oxidant that oxidizes the organic species, i.e., both methods assure the proximity of active oxidant-organic species; then, oxidation occurs efficiently even when the concentration of the organic species is low. The principal difference concerns the control of the operating conditions of the steps that compose the mediated oxidation occurring in these cases. In indirect oxidation, the steps of precursor production, mixing, activation, and oxidation occur in the electrochemical cell, where even direct oxidation can occur. In ex-cell-mediated oxidation, a specific piece of equipment is dedicated at every step of the process, and the control of each step allows its efficient realization.

Persulfate as a Precursor in Ex-Cell-Mediated Oxidation

The persulfate ion, thanks to its properties, is a suitable precursor of ex-cell-mediated oxidation treatment. It is produced by electrochemical methods and is easily and efficiently produced as $H_2S_2O_8$ using a two-compartment cell and BDD anodes. It is stable at ambient conditions and therefore can be mixed with wastewater without degradation. Its successive activation to generate sulfate and hydroxyl radicals occurs easily using thermal and ultraviolet energy, and the generated sulfate and hydroxyl radicals efficiently oxidize the organic species in a short interval of time [5–8].

Electrochemical treatment with BBD anodes has shown that when the wastewater contains sulfate ions, a small increase of current efficiency is observed at the end of the process that has been attributed to indirect oxidation of the persulfate ions generated. This small contribution of indirect oxidation is due to the operating conditions of the electrochemical cell used. A small quantity of persulfate can be generated by the low concentration of sulfate ions, and the temperature of the cell is not sufficient to transform persulfate to sulfate and generate hydroxyl radicals.

As previously reported, when BDD anodes are used, $H_2S_2O_8$ is generated, and the byproducts of the oxidation of pollutants that remain in the solution are protons and sulfate ions. Both ions are easily eliminated by the introduction of an equivalent quantity of calcium oxide (CaO):

$$H_2SO_4 + CaO \rightarrow CaSO_4 + H_2O \qquad (2)$$

The low solubility of calcium sulfate assures that only a small fraction of calcium sulfate remains in the wastewater (about 2 $gr \cdot dm^{-3}$), while a high fraction of the produced salt is eliminated by precipitation.

Combined Electrochemical and Ex-Cell-Mediated Treatment

The electrochemical and ex-cell-mediated oxidation method with the use of persulfate as a precursor is described in Fig. 2, which shows the steps of this combined treatment. Initially, direct electrochemical treatment is applied efficiently in the cell (1) by using BDD anodes and a not low concentration in the cell. After this treatment, ex-cell-mediated oxidation is applied by using as an oxidant the persulfate generated locally in a two-compartment cell with BDD anodes (2). The mixing of wastewater with the generated persulfate occurs in the mixer (3). The oxidation of the pollutants occurs in the oxidation reactor (4), with the use of thermal or ultraviolet energy to generate active sulfate and hydroxyl radicals that oxidize the organic pollutants. At the end of the oxidation step, the wastewater contains sulfuric acid as a byproduct of the oxidation step. The equivalent quantity of CaO is introduced in the final step of this treatment (5) that neutralizes the treated wastewater and eliminates the sulfate ions by precipitation of the generated $CaSO_4$.

The positive effect of ex-cell-mediated oxidation on the treatment can be quantified by considering a specific treatment of organic pollutants. A flow, Q, of wastewater is assumed

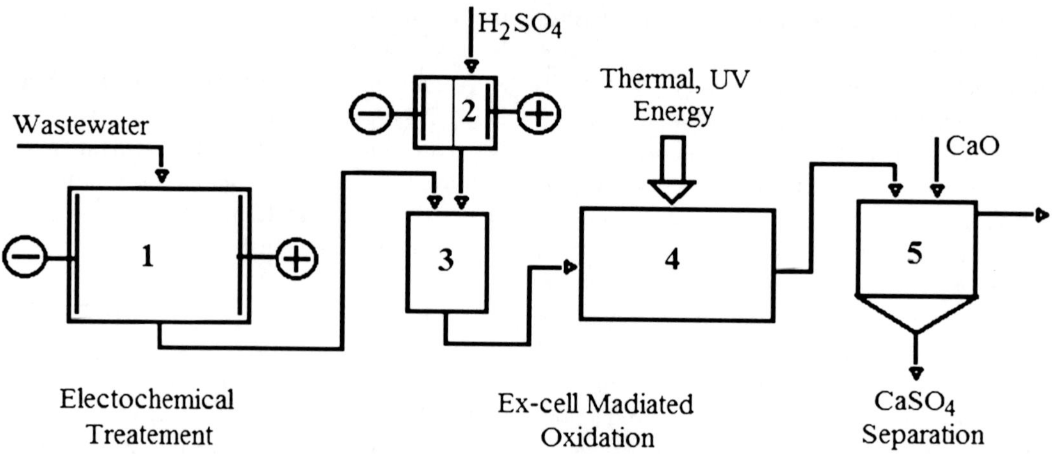

Ex-Cell-Mediated Oxidation (via Persulfate) of Organic Pollutants, Fig. 2 Combined treatment: (*1*) electrochemical treatment cell, (*2*) ex-cell, (*3*) mixer, (*4*) oxidation reactor, and (*5*) separator

to contain a given value of pollutant concentration ($COD_{in} = 1{,}000$), while a low value of this concentration is required after treatment ($COD_{fin} = 10$). Such a low final value cannot be reached efficiently by using electrochemical treatment, and so the combined treatment is applied.

The flow enters the cell for electrochemical treatment, and a concentration of pollutants in this cell exit is adopted (COD_{cell}). The concentration in the cell strongly influences the anodic surface required for electrochemical treatment, as well as the quantity of chemical that is concerned with the ex-cell-mediated oxidation. If the assumed COD_{cell} is close to COD_{in}, the electrochemical treatment eliminates a small fraction of pollutants. A small anodic surface is required, and a high quantity of chemicals is consumed if COD_{cell} is close to CODcr. Electrochemical treatment eliminates a high fraction of pollutants by using a high anodic BDD surface and consuming small quantities of chemicals.

The entire anodic surface S_t, concerning the electrochemical treatment cell and the ex-cell, is estimated by the following equation:

$$S_t = \frac{Q}{k}\left(\frac{COD_{in}}{COD_{cell}} - 1\right) + \frac{4FQ}{\eta_{ex} i_{ex}}(COD_{cell} - COD_{fin}) \qquad (3)$$

where η_{ex} is the current efficiency and i_{ex} is the current density of the ex-cell.

The value G, the calcium sulfate produced, indicates the quantity of sludge, as well as the quantities of the consumed chemicals (sulfuric acid and calcium oxide). As the following equation shows, this quantity increases with increasing COD_{cell}:

$$G = Q\frac{4PM_{CaSO4}}{PM_{O_2}}(COD_{cell} - COD_{fin}) \qquad (4)$$

These equations are tools to determine the value of the concentration of COD_{cell} that minimizes the cost of treatment.

In conclusion, the advantages of ex-cell-mediated oxidation are that (1) a low value of pollutant concentration can be reached, and (2) both cells of BDD anodes operate efficiently; the disadvantages are that (1) thermal or ultraviolet energy is required to activate the persulfate, (2) the generated radicals in the oxidation reactor impose to use a reactor with a highly resistive lining, and (3) when thermal energy is supplied, the pressure of the oxidation reactor must be controlled, and a separator is required to eliminate the gaseous products.

Future Directions

The combined treatment of wastewaters containing bio - refractory organic compounds is proposed, respect to the place single electrochemical treatment, in order to avoid the loss of a high fraction of the furnished electric charge when the final concentration of these compounds achieves low concentrations.

This kind of wastewaters require a low concentration after the treatment and the combined treatment thanks to its effectiveness respect to the single electrochemical treatment has a higher possibility to be used, moreover alternative combined treatments with other oxidants can be proposed in this field.

Cross-References

- ▶ Boron-Doped Diamond for Green Electro-Organic Synthesis
- ▶ Electrokinetic Barriers for Preventing Groundwater Pollution
- ▶ Electrokinetics in the Removal of Hydrocarbons from Soils

References

1. Kapalka A, Foti G, Comninellis C (2009) The importance of electrode material in environmental electrochemistry: formation and reactivity of free hydroxyl radicals on boron-doped diamond electrodes. Electrochim Acta 54:2018–2023
2. Vatistas N (2010) Adsorption layer and its characteristic to modulate the electro-oxidation runway of organic species. J Appl Electrochem 40: 1743–1750
3. Iniesta J, Michaud PA, Panizza M et al (2001) Electrochemical oxidation of phenol at boron-doped diamond electrode. Electrochim Acta 46:3573–3578
4. Gherardini L, Michaud PA, Panizza M et al (2001) Electrochemical oxidation of 4-chlorophenol for wastewater treatment: definition of normalized current efficiency. J Electrochem Soc 148:D78–D82
5. House DA (1962) Kinetics and mechanism of oxidation by peroxydisulfate. Chem Rev 62:185–203
6. Berlin AA (1986) Kinetics of radical-chain decomposition of persulfate in aqueous solutions of organic compounds. Kinet Catal 27:34–39
7. Kronholm J, Metsala H, Hartonen K et al (2001) Oxidation of 4-chloro-3-methylphenolin pressurized hot water/supercritical with potassium persulfate as oxidant. Environ Sci Technol 35: 3247–3251
8. Huang KC, Zhao Z, Hoag GE et al (2005) Degradation of volatile organic compounds with thermally activated persulfate oxidation. Chemosphere 61: 551–560